BIOLOGY

SECOND EDITION

CLAUDE A. VILLEE

HARVARD UNIVERSITY

ELDRA PEARL SOLOMON

CENTER FOR RESEARCH IN BEHAVIORAL MEDICINE
AND HEALTH PSYCHOLOGY
UNIVERSITY OF SOUTH FLORIDA

CHARLES E. MARTIN

RUTGERS UNIVERSITY

DIANA W. MARTIN

RUTGERS UNIVERSITY

LINDA R. BERG

UNIVERSITY OF MARYLAND

P. WILLIAM DAVIS

HILLSBOROUGH COMMUNITY COLLEGE

SAUNDERS COLLEGE PUBLISHING

PHILADELPHIA   FT. WORTH   CHICAGO   SAN FRANCISCO
MONTREAL   TORONTO   LONDON   SYDNEY   TOKYO

Text Typeface: Palatino
Compositor: York Graphic Services
Acquisitions Editor: Edward Murphy
Developmental Editors: Lynne Gery and Gabrielle Goodman
Project Editor: Carol Field
Copy Editors: Martha Hicks-Courant, Joseph Freedman, Becca
            Gruliow, and Donna Walker
Art Director: Carol C. Bleistine
Art Assistant: Doris Bruey
Text Design and Layout: Emily Harste
Cover Designer: Lawrence C. Didona
Text Artwork: J & R Technical Services
Photo Editor: Robin Bonner
Production Manager: Merry Post
Cover Credit: O'Keefe, *Untitled*, Private Collection. © Art
            Resource

Printed in the United States of America

ISBN 0-03-023417-4

Library of Congress Catalog Card Number: 88-043139

012 063 9876543

To our families, friends, and colleagues who gave freely of their love, support, knowledge, and time as we labored over this revision of BIOLOGY . . .

Especially, to . . .

. . . Dorothy Villee, M.D.

. . . Rabbi Theodore and Freda Brod
Kathleen M. Heide, Ph.D.

. . . Margaret Martin
and in memory of Margaret Y. Menzel, Ph.D

. . . Alan Berg
Jennifer Berg

. . . Karen Davis

PREFACE

One of our principal goals in writing BIOLOGY is to share with beginning biology students our sense of excitement about modern biological science. We seek to impart an understanding and appreciation of the vast diversity of living things, their special adaptations to their environment, and their evolutionary and ecological relationships. We emphasize the basic unity of life and the fundamental similarities of the problems that have been faced and solved by all living organisms. We have been very aware of our responsibility to impress upon our students that we share planet Earth with many thousands of varieties of living things. Indeed, we are dependent upon countless organisms for our very survival, and they in turn depend upon us. This interdependence is emphasized throughout the book.

PHILOSOPHY AND APPROACH

The principles of biology can be learned using as a model the frog, shark, daisy, or even the colon bacillus. We have chosen a comparative approach. Most students have a special interest in human biology—especially the structure, function, and development of the human body—generated perhaps by their plans for a career in medicine, dentistry, or one of the allied health sciences, or simply by an interest in how their own bodies are put together and how they work. For this reason, we make frequent use of the human as a biological model, and give attention to the human aspects of biology. These same students can also benefit from our comparative, concept-oriented approach; for as they continue their professional education, they may have little additional exposure to such subjects as plant biology, invertebrate biology, ecology, or evolution.

THE AUTHOR TEAM

We are in the midst of an information explosion in biology. Revolutionary advances are being made in genetics, cell biology, immunology, botany, and many other subdisciplines. Researchers are generating thousands of journal articles each year reporting findings of their studies. Biology has become such a broad, complex science that it is very difficult for a biologist to stay abreast of all of its subdisciplines. Yet we wanted to give our students—especially those majoring in biology or planning a career in the sciences—accurate information and concepts on the cutting edge of each field. For this reason, the authors and editors of BIOLOGY elected to expand the author team for the second edition. The multi-author strategy contributes greatly to the strength of the second edition of BIOLOGY.

We selected two geneticists to write the genetics unit and a botanist to write the chapters that focus on plant diversity and life processes in plants. Experts were drafted in the fields of immunology, animal behavior, and several other areas to help revise chapters in their fields of expertise. Particular thanks to Kent Rylander for his work on the behavior chapter and to Susan Pross for assistance with our immunology chapter. We are very fortunate in having a team of authors who work well together and whose writing styles are so similar that reviewers were not able to detect differences in authorship.

ORGANIZATION

This book attempts neither to be encyclopedic nor cursory, but presents the concepts of biology and their relevance to human beings in an interesting and understandable fashion. There is no general agreement among biologists as to the sequence in which the several major topics in a general biology course should be taught. This lack of consensus is understandable, for reasonable arguments can be advanced for each of the many possible combinations and permutations. The various aspects of biology are intimately related, and each could be grasped much more readily if all the other aspects had been learned previously. Because this feat cannot be accomplished (except perhaps by a student repeating the course!), each instructor must choose the sequence that seems optimal to him or her. For this reason, we have taken special pains to write each chapter and each part so that they do not depend heavily upon preceding chapters and parts. The eight parts and their chapters can be taken up in any of a number of sequences with pedagogic success.

Chapter 1   A View of Life

Chapter 1 introduces the student to several unifying key concepts. The unity of life is emphasized by describing the basic characteristics and organization of living things. An overview of the diversity of life and an introduction to evolution are presented. An appreciation of science requires not only a grasp of the product of science but also an insight into the processes by which scientific knowledge is acquired. An introduction to the methods of science is given in Chapter 1, and throughout later chapters examples of experimental work are presented to illustrate modern methods of biological research.

Part I   The Organization of Life

Chapters 2 and 3 discuss the molecular organization of living things and lay the foundations in chemistry needed for an understanding of biological processes. Chapters 4 and 5 describe the cellular organization of living organisms, with special emphasis on recent advances in cellular biology.

Part II   Energy in Living Systems

The flow of energy through the world of life is traced in Chapter 6. Then, Chapters 7 and 8 delve into the grand metabolic adaptations by which living systems obtain and utilize energy by cellular respiration and photosynthesis.

Part III   Genetics

Part III deals with the continuity of life beginning with a discussion of mitosis and meiosis in Chapter 9. Basic principles of heredity and human genetics are presented in Chapters 10 and 11. Chapter 12 describes the genetic code, and Chapter 13 describes RNA and protein synthesis. In Chapter 14, the focus is on new advances in this rapidly expanding field such as recombinant DNA and its applications. What is known about how genes are regulated is presented in Chapter 15. Chapter 16 is a unique presentation for a general biology text, an introduction to molecular development, including the very latest findings in this exciting area of research.

Part IV Evolution

Part IV, a clear presentation of the concept of organic evolution, provides the student with a fundamental framework for understanding the origin and diversity of life. Chapter 17 presents Charles Darwin's contributions, along with some evidence for evolution. Chapter 18 summarizes the concepts of population genetics. Chapter 19 focuses on speciation. Chapter 20 discusses the origin and history of life. The evolution of the primates is summarized in Chapter 21.

Part V The Diversity of Life

In this unit, we introduce basic concepts of taxonomy and survey the kingdoms of living organisms. In Chapter 22 we discuss how living things are classified, and why. Chapter 23 is devoted to the viruses, and separate chapters are devoted to Kingdom Monera, Kingdom Protista, and Kingdom Fungi. Chapters 27 and 28 present the members of the Plant kingdom and their relationships, and Chapters 29 through 31 focus on the diversity of animals on our planet. The discussion of each group of organisms focuses on their interrelationships and on their structural and functional adaptations.

Part VI Structures and Life Processes in Plants

This unit, which integrates plant structure and function, begins with a discussion of growth and differentiation in plants. Then, Chapter 33 focuses on leaves and photosynthesis; Chapter 34 on stems and plant transport; and Chapter 35 on roots and plant nutrition. Chapter 36 discusses reproduction in flowering plants and Chapter 37, plant hormones and responses.

Part VII Structures and Life Processes in Animals

Part VII emphasizes the structural, functional, and behavioral adaptations that animals have evolved to meet a multitude of environmental challenges. As each system of the animal body is discussed, a comparative approach is used to examine how various animal groups have solved similar and diverse problems. The unit begins with a chapter devoted to the architecture of the animal body, emphasizing types of tissues and organ systems.

Chapter 39 examines three systems—integumentary, skeletal, and muscle—with emphasis on how these systems typically function together. Chapter 40 is devoted to how animals process food, emphasizing adaptations for various nutritional life styles. Chapter 41 focuses on basic nutrition and on the fate of nutrients in the body. In Chapter 42, we discuss blood and different types of circulatory systems. Chapter 43 presents the basics of immunology and includes a focus box on AIDS. Chapter 44 examines the various strategies that have evolved for gas exchange, and Chapter 45 compares adaptations for solving problems of fluid balance and disposal of metabolic wastes. Neural control is discussed in two chapters: Chapter 46 is devoted to the structure and function of the neuron, and Chapter 47 compares different types of nervous systems with emphasis on the function of the vertebrate nervous system. Sense organs are described in Chapter 48, and endocrine regulation is discussed in Chapter 49. Chapter 50 is devoted to reproduction and Chapter 51 to development. The unit concludes with a chapter that introduces the student to the study of animal behavior.

Part VIII   Ecology

The final section of this book presents the principles of ecology from the standpoint of populations, communities, and ecosystems. Chapters 53 and 54 discuss the characteristics of earth, water, and air and the life zones of our planet. Population ecology is introduced in Chapter 55, and communities and ecosystems are discussed in Chapter 56. In the final chapter on human ecology, we consider some of our present predicaments involving overpopulation, pollution, and the depletion of our natural resources.

LEARNING AIDS

Many pedagogic aids have been included to help the student in the challenging task of mastering the principles of biology. A **chapter outline** at the beginning of each chapter shows the student how the material is organized and divides the subject matter into manageable chunks. **Learning objectives** at the beginning of each chapter indicate exactly what the student must be able to do to demonstrate mastery of the material in the chapter.

Important new terms are set in **boldface** for emphasis throughout the text. Numerous **tables,** many of them illustrated, organize and summarize information presented in the text. **Focus boxes** present enrichment material or introduce applications of material presented in the text.

Illustrations have been carefully designed and selected to support and clarify the concepts presented. Conceptual diagrams, photographs, photomicrographs, electron micrographs, and medical illustrations are included to help the student accurately visualize concepts presented in the text. Full color is used throughout the book to increase teaching value of the illustrations and add visual appeal to the text.

At the end of each chapter, a **summary** in outline form is provided to help the student review the main concepts presented in the chapter. There is also an objective **Post-Test** (with answers at the back of the book) so that the student can evaluate his or her mastery of the material in the chapter. **Review questions** give the student the opportunity to check understanding of important concepts, apply them, and synthesize some of the material presented. At the end of each chapter there is also a list of supplementary readings.

A **glossary of terms** giving the definitions of many important biological terms used in the text is included in the back of the book. An appendix of common prefixes, suffixes, and word roots helps the student dissect and understand biologic terms.

SUPPLEMENTS

The **Study Guide** that accompanies BIOLOGY has been designed around the learning objectives. Each chapter includes a list of key concepts, a scientific vocabulary matching test, questions testing mastery of each learning objective, a test evaluating overall mastery of the chapter, and a set of comprehensive questions. Diagrams for the student to label are included. Answers are provided for questions and practice tests.

The **Instructor's Resource Manual** includes suggestions for course organization, lists sources for audiovisual materials, and includes references. For each chapter, there is an overview as well as suggestions for enrichment, including readings, films, topics for class discussion, and essay questions usable in tests or homework assignments.

The **Test Bank** includes both chapter and unit tests. Answers are provided for all questions. Tests are presented so that the instructor can duplicate them directly from the printed page. A **Computerized Test Bank** is also available for the Apple II series and the IBM PC series.

A set of 105 **Overhead Transparencies** includes two- and four-color illustrations that were chosen and designed for optimum utility in the classroom and laboratory. There is also a set of 100 **35-mm Slides,** all in full color. A **Laboratory Manual** and an accompanying **Laboratory Instructor's Manual** are available. Also, a set of 40 **Overhead Transparencies of Electron Micrographs** is available.

ACKNOWLEDGMENTS

We are grateful to the editorial staff of Saunders College Publishing for assembling the author team and for their help and support throughout the project. Writing, illustrating, and producing a majors' level biology textbook is a Herculean task. There were times when we were not at all certain that we could complete the project in time for the 1989 publication date.

Our Acquisitions Editor Ed Murphy rescued the project several times, exhibiting great talent and tenacity in seeking out coauthors, reviewers, editors, and artists needed to develop and produce this second edition of BIOLOGY. His perseverance was rewarded with just the right people— individuals who were willing to share their abilities and invest long hours and great effort in completing this project.

We are indebted to Carol Field, our Project Editor, who meets every new challenge with flexibility, expertise, and optimism. Our Developmental Editor Lynne Gery worked side by side with us on the beginning phases of the project, and her experience and abilities were greatly missed when she left Saunders. Gabe Goodman, who took over Lynne's duties, quickly became an integral part of the project. We appreciate her valuable input. We are also grateful to Photo Editor Robin Bonner for helping us find just the right photographs and to Art Director Carol Bleistine for her expertise in putting the art program and design together.

We thank all of the Saunders editorial team—Ed Murphy, Carol Field, Lynne Gery, Gabe Goodman, Carol Bleistine, Merry Post, and other members of the staff who contributed to the project—for their talent, their support, and their willingness to help us make order out of the chaos resulting from a multiple-author team, 57 chapters, and hundreds of illustrations. This book is as much a product of their work as ours.

We also acknowledge the patience and support of our colleagues, families, and friends. We have often enlisted the help of those closest to us—to read a section and offer input, to help with word processing, to locate needed journal articles, or to help keep the word processing equipment functional.

Our colleagues and students have provided us with valuable input and have played an important role in shaping BIOLOGY. We thank them and ask for comments and suggestions from instructors and students who use this new edition. You can reach us through our editors at Saunders College Publishing.

REVIEWERS

We gratefully acknowledge the many professors and researchers who have read the manuscript during its preparation and provided valuable suggestions for improving it. Their input contributed significantly to this final product.

James Moore, University of California/San Diego
Andrew Hill, Yale University
Keith Clay, Indiana University
Gary Ogden, Moorpark College
Manuel Molles, University of New Mexico
Peter Dixon, University of California/Irvine
Andrew Dobson, University of Rochester
Kathleen Scott, Rutgers University/New Brunswick
Millicent Ficken, University of Wisconsin/Milwaukee
Steven Heidemann, Michigan State University/East Lansing
Ted Sargent, University of Massachusetts/Amherst
Charles Henry, University of Connecticut
Darrell Galloway, Ohio State University
Peggy Redshaw, Austin College
David Fromson, California State University/Fullerton
Gerald Bergtrom, University of Wisconsin/Milwaukee
Kenneth Shull, Appalachian State University
Robert Kitchin, University of Wyoming
Robert Stockhouse, Pacific University
Robert Evans, Rutgers University/Camden
Ronald Phillips, Seattle Pacific University
Robert Patterson, San Francisco State University
David Whetstone, Jacksonville State University (Alabama)
Trevor Price, University of California/San Diego
William Zimmerman, Amherst College
Stephen Dina, St. Louis University
Dwayne Wise, Mississippi State University
John Romeo, University of South Florida
John Olsen, Rhodes College
James French, Rutgers University/Busch Campus
Patricia Gensel, University of North Carolina/Chapel Hill
Carl Schlicting, Pennsylvania State University/University Park
James Hayward, Andrews University
Varley Wiedeman, University of Louisville
Paul Lago, University of Mississippi
Kenneth Kardong, Washington State University
Randal Johnston, University of Calgary
Louis Held, Texas Tech University
Karl Mattox, Miami University of Ohio
Arthur Repak, Quinnipiac College
Penny Bauer, Colorado State University
Judith Goodenough, University of Massachusetts/Amherst
Steve Vessey, Bowling Green State University
John Tudor, St. Joseph's University
Donald Keefer, Loyola College/Maryland
Robert Hurst, Purdue University

Bruce Smith, Brigham Young University
Rob Dorit, Harvard University
Lee Meserve, Bowling Green State University
Robert Cordero, St. Joseph's University
Anne Penney Newton and Edward Morgan, Temple Junior College
    (Texas)

CLAUDE A. VILLEE
ELDRA PEARL SOLOMON
CHARLES E. MARTIN
DIANA W. MARTIN
LINDA R. BERG
P. WILLIAM DAVIS

November 1988

CONTENTS OVERVIEW

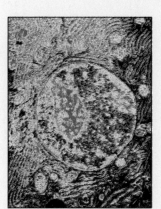

CONTENTS

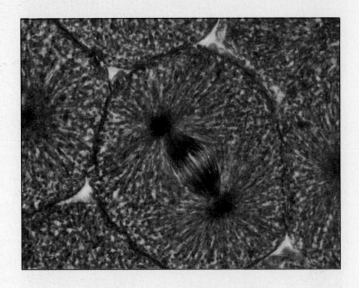

PART I

THE ORGANIZATION OF LIFE

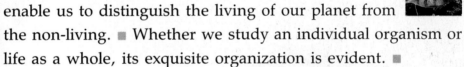

What is life? ■ In the world of biology we find
that most living things share a number of characteristics that
enable us to distinguish the living of our planet from
the non-living. ■ Whether we study an individual organism or
life as a whole, its exquisite organization is evident. ■

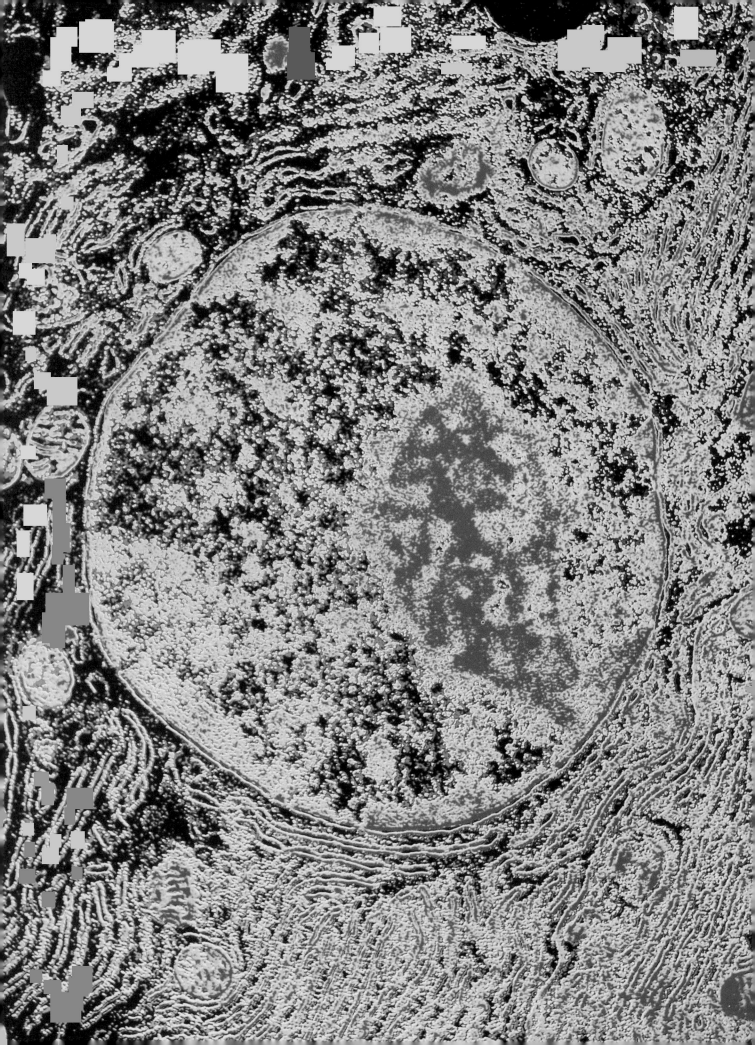

Planet Earth seen from Apollo II, about 98,000 nautical miles away

1

A View of Life

■ **LEARNING OBJECTIVES**

After you have read this chapter you should be able to:

1. Distinguish between living and nonliving things, describing the features that characterize living things.
2. Define *metabolism* and *homeostasis,* and give examples of these processes.
3. Define *adaptation*, and describe its function in promoting perpetuation of a species.
4. List in sequence and briefly describe each of the levels of biological organization.
5. Describe the roles and interdependence of producers, consumers, and decomposers.
6. Identify the five kingdoms of living organisms, and give examples for each group.
7. Give a brief overview of the theory of evolution and explain why it is a key concept in biology.
8. Design an experiment to test a given hypothesis using the procedure and terminology of the scientific method.

During the past century **biology,** the study of life, has undergone rapid change and has had a significant impact on the way we live. We are now able to produce vaccines and antibiotics, transplant hearts, and manipulate genes. Biologists today are working on such vital projects as increasing world food supply, improving environmental quality, identifying factors that contribute to health and longevity, and conquering killers like cancer, heart disease, and AIDS. Biologists also continue their interest in the study of interrelationships of the diverse forms of living things that inhabit our planet. These scientists have enhanced our awareness of the exquisite complexity that characterizes all living things, and have helped us better appreciate our own impact on other living things and on the environment (Figure 1–1).

This book is a starting point for your exploration of biology. It will provide you with the tools that will enable you to become a part of this fascinating science. Perhaps you will decide to become a research biologist and help unravel the complexities of the human brain, breed disease-resistant strains of wheat or rice, identify new species of animals or bacteria, or

3

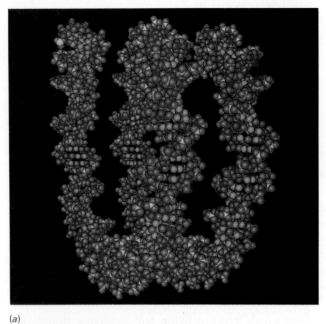

(a)

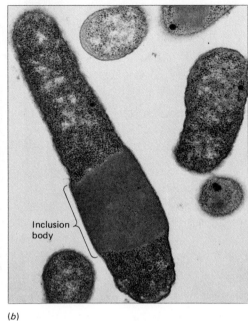

Inclusion body

(b)

(c)

FIGURE 1–1 Modern biology examines the world of life in all its details and interactions. An organism's ability to reproduce itself is essential to the continuity of life. DNA, the hereditary material of life, is shown in (a) a computer-drawn simulation of the colored plastic space-filling molecular models used by biologists. The colored balls represent the different atoms that make up DNA: *dark blue,* carbon; *red,* oxygen; *white,* hydrogen; *blue,* nitrogen; *yellow,* phosphorus. (b) A genetically engineered *Escherichia coli* bacterium (magnified 68,000 times). Its hereditary material has been modified so that the bacterium produces large quantities of the hormone human insulin. Individuals with the disease diabetes mellitus have insufficient amounts of this hormone, which is necessary for the normal metabolism of sugar. Since the bacterium itself has no use for the insulin, and has no way of excreting it, instead the hormone accumulates in an "inclusion body" within the bacterial cell. The insulin can be harvested by destroying the bacterium. (c) An angelfish guarding a section of coral reef. Biologists are concerned with the physical characteristics of the fish, how its body functions, its behavior, and its interaction with other living things in its environment, (d) Biologists also study the effects of human activities on the environment. Because human populations are expanding rapidly in the world's tropical areas, rainforests are being cleared to provide more land for agriculture. Unfortunately, tropical soils are infertile and can produce crops for only a few years. The cleared areas of rain forests may never recover from this destruction, and many species of plants and animals may become extinct as a result.

(d)

FIGURE 1–2 This sea cucumber is composed of millions of cells that are organized to perform specialized functions.

discover a cure for cancer. Or perhaps you will choose to enter an applied field of biology such as dentistry, medicine, or veterinary medicine. Even if you are not planning a career in one of the biological sciences, learning about this exciting science will enable you to better understand yourself, your environment, and the organisms with which you share your planet. As you become biologically literate you will increase your understanding of the impact biology continues to have on life and society.

In this first chapter we examine some of our everyday assumptions about living things, and start to formulate a structure for organizing our knowledge. First, as we begin our study of living things, we need to develop a deeper understanding of what life is.

WHAT IS LIFE?

It is relatively easy to determine that a human being, an oak tree, and a grasshopper are living whereas rocks are not. Yet it remains difficult to define life. At one time it was believed that a living system could be distinguished from a nonliving system by its possession of a special "vital force." Now, after centuries of searching, we understand that there is no single substance or force that is unique to living things. Perhaps the best we can do toward defining life is to list the features that living things have in common. When we do this, we find that the characteristics that distinguish most living things from nonliving things include a precise kind of organization, a variety of chemical reactions we term *metabolism*, the ability to maintain an appropriate internal environment even when the external environment changes (a process referred to as *homeostasis*), movement, responsiveness, growth, reproduction, and adaptation to environmental change. We consider each of these characteristics in the following sections.

Specific Organization

The cell theory, one of the fundamental concepts of biology, states that all living things are composed of basic units called **cells** and of cell products. Although organisms (living things) vary greatly in size and appearance, all (except the viruses[1]) are composed of these small building blocks. Some of the simplest organisms, such as bacteria, are unicellular; that is, they consist of a single cell. In contrast, the body of a human or an oak tree is made of billions of cells. In such complex multicellular organisms the processes of the entire organism depend on the coordinated functions of the constituent cells (Figure 1–2).

---

[1] As we will see in Chapter 23, viruses can carry on metabolism and reproduce only by using the metabolic machinery of *the cells they parasitize,* and so are said to be on the borderline between living and nonliving things.

Metabolism

In all living organisms chemical reactions take place that are essential to nutrition, growth and repair of cells, and conversion of energy into usable forms. The sum of all the chemical activities of the organism is called **metabolism.** Metabolic reactions occur continuously in every living organism (Figure 1–3); when they cease, the organism dies.

Each individual cell of an organism constantly takes in new substances, alters them chemically in a variety of ways, and builds new cellular components. Some nutrients are used as "fuel" for cellular respiration, a process during which some of their stored energy is captured for use by the cells. Each chemical reaction is regulated by a specific **enzyme,** a chemical catalyst; enzymes are discussed in Chapter 6. Life on earth involves a never-ending flow of energy within cells, from one cell to another, and from one organism to another.

FIGURE 1–3 Metabolic reactions occur continuously in every living organism. (*a*) Relationships of some metabolic activities. Some of the nutrients provided by proper nutrition are used to synthesize needed materials and cell parts; other nutrients are used as fuel for cellular respiration, a process that captures energy stored in food. This energy is needed for synthesis and for other forms of cellular work. Cellular respiration also requires oxygen, which is provided by the process of gas exchange. Wastes from the cells such as carbon dioxide and water must be excreted from the body. (*b*) Like most organisms, this one-celled amoeba must take in nutrients and oxygen in order to stay alive. The dark spots within the cell are food.

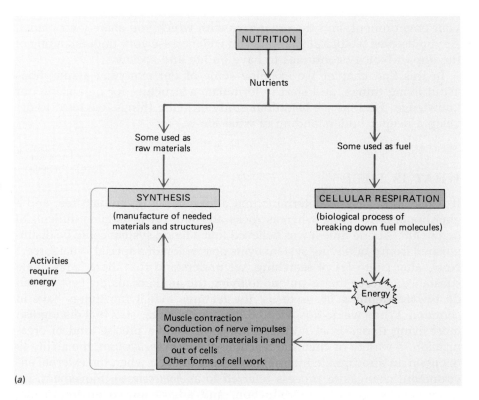

(a)

(b)

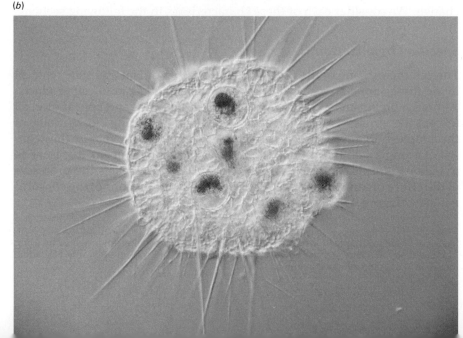

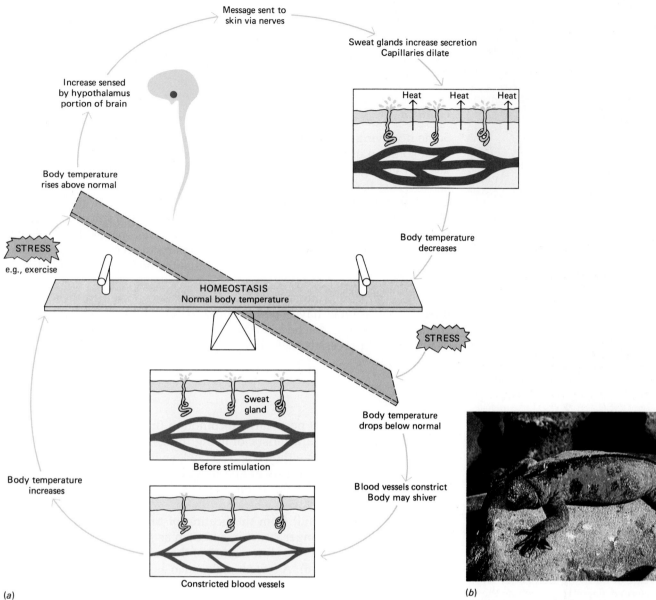

Message sent to
skin via nerves

Sweat glands increase secretion
Capillaries dilate

Heat   Heat   Heat

Increase sensed
by hypothalamus
portion of brain

Body temperature
rises above normal

Body temperature
decreases

STRESS

e.g., exercise

HOMEOSTASIS
Normal body temperature

STRESS

Sweat
gland

Body temperature
drops below normal

Before stimulation

Body temperature
increases

Blood vessels constrict
Body may shiver

Constricted blood vessels

(a)

(b)

Homeostasis

In all organisms the various metabolic processes must be carefully and
constantly regulated to maintain a balanced internal state. When enough of
some cellular component has been made, its production must be decreased
or turned off. When the supply of energy in a cell declines, appropriate
processes for making more energy available must be turned on. These
self-regulating control systems are remarkably sensitive and efficient. The
tendency of organisms to maintain a constant internal environment is
termed **homeostasis,** and the mechanisms that accomplish the task are
known as **homeostatic mechanisms.**

The regulation of body temperature in the human being is an example of
the operation of homeostatic mechanisms. When body temperature rises
above the normal 37°C, the temperature of the blood is sensed by special
cells in the brain that function like a thermostat. These cells send nerve
impulses to the sweat glands to increase the secretion of sweat (Figure
1–4). The evaporation of the sweat from the body surface lowers body
temperature. Other nerve impulses cause the dilation of small blood vessels (capillaries) in the skin, making it appear flushed. The increased blood
flow brings more heat to the body surface to be radiated away.

FIGURE 1–4 (*a*) Regulation of body
temperature in the human by homeostatic mechanisms. An increase
in body temperature above the normal range stimulates special cells in
the brain to send messages to
sweat glands and capillaries in the
skin. Increased circulation of blood
in the skin and increased sweating
are mechanisms that help the body
get rid of excess heat. When body
temperature falls below the normal
range, blood vessels in the skin
constrict so that less heat is carried
to the body surface. Shivering, in
which muscle contractions generate
heat, may also occur. (*b*) The sun-
ning behavior of this marine iguana, *Amblyrhynchus cristatus,* a native of the Galapagos Islands, is
homeostatic. The animal positions
itself to maximize the heat it receives from the sun, thus increasing its body temperature.

FIGURE 1–5 Biological growth involves the refashioning of raw materials to construct the organism as determined by its DNA. This female spectacled langur (*Presbytis obscuris*) is holding a young langur, which will eat and grow until it reaches adult size.

FIGURE 1–6 Movement is characteristic of all living things. (*a*) The euglenoid flagellates, such as the *Strombomonas conspersa* shown here, move about by beating their long whip-like flagella. (*b*) The amoeba moves about by extending its cell contents, a process known as *amoeboid motion.*

When body temperature falls below normal, the sensor in the brain initiates nerve impulses to constrict the blood vessels in the skin, reducing heat loss. If body temperature falls lower, nerve impulses may be sent to the muscles of the body, stimulating the rapid muscular contractions we call *shivering,* a process that generates heat.

Growth

Some nonliving things appear to grow. Crystals may form in a supersaturated solution of a salt; as more of the salt comes out of solution, the crystals may enlarge. However, this is not growth in the biological sense. Biologists restrict the term **growth** to those processes that increase the amount of living substance in the organism. Growth, therefore, is an increase in cellular mass that is brought about by an increase in the *size* of the individual cells, by an increase in the *number* of cells, or by both (Figure 1–5). Growth may be uniform in the several parts of an organism, or it may be greater in some parts than in others so that the body proportions change as growth occurs.

Some organisms—most trees, for example—continue to grow indefinitely. Many animals have a defined growth period that terminates when a characteristic size is reached in adulthood. One of the remarkable aspects of the growth process is that each part of the organism continues to function as it grows.

Movement

Movement, though not necessarily locomotion (moving from one place to another), is another characteristic of living things. The movement of most animals is quite obvious—they wiggle, crawl, swim, run, or fly. The movements of plants are much slower and less obvious but occur nonetheless. The streaming motion of the living material in the cells of the leaves of plants is known as **cyclosis.**

Locomotion may result from the beating of tiny hairlike extensions of the cell called *cilia* or longer structures known as *flagella,* from the contraction of muscles, or from the slow oozing of the cell, a process called **amoeboid motion** (Figure 1–6). A few animals, such as sponges, corals, oysters, and certain parasites, do not move from place to place as adults. Most of these have free-swimming larval stages. Even in the sessile (firmly attached, not free to move about) adults, however, cilia or flagella may beat rhythmically, moving the surrounding water past the organism; in this way food and other necessities of life are brought to the organism.

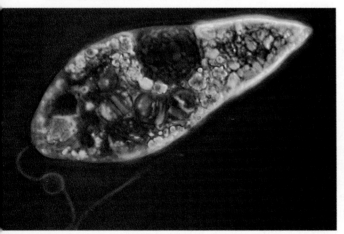

(*a*)

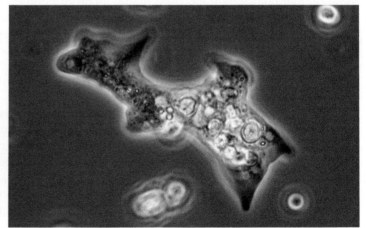

(*b*)

FIGURE 1–7 A few plants, such as the Venus's flytrap, can respond to the touch of an insect by trapping it. Here a leaf of the Venus's flytrap is shown attracting and capturing a lacewing. The leaves of this plant have a scent that attracts insects. When trigger hairs on the leaf surface detect the presence of an insect, the leaf, hinged along its midrib, folds. The edges come together and hairs interlock, preventing the escape of the prey. The leaf then secretes enzymes that kill and digest the insect.

Responsiveness

Living things respond to **stimuli,** physical or chemical changes in their internal or external environment. Stimuli that evoke a response in most organisms are changes in the color, intensity, or direction of light; changes in temperature, pressure, or sound; and changes in the chemical composition of the surrounding soil, air, or water. In complex animals such as humans, certain cells of the body are highly specialized to respond to certain types of stimuli; for example, cells in the retina of the eye respond to light. In simpler organisms, such specialized cells may be absent, but the whole organism may respond to stimuli. Certain single-celled organisms respond to bright light by retreating.

The responsiveness of plants may not be as obvious as that of animals, but plants do respond to light, gravity, water, and other stimuli principally by the growth of different parts of their bodies. The streaming motion of the cytoplasm in plant cells may be speeded up or stopped by changes in the amount of light. A few plants, such as the Venus's flytrap of the Carolina swamps (Figure 1–7), are remarkably sensitive to touch and can catch insects. Their leaves are hinged along the midrib and possess a scent that attracts insects. The presence of an insect on the leaf, detected by trigger hairs on the leaf surface, stimulates the leaf to fold. In true "Little Shop of Horrors" fashion, the edges come together and the hairs interlock to prevent the escape of the prey. The leaf then secretes enzymes that kill and digest the insect. These plants are usually found in soil that is deficient in nitrogen. Flytrapping enables these plants to obtain part of the nitrogen they require for growth from the prey they "eat."

Reproduction

Although at one time worms were believed to arise from horsehairs in a water trough, maggots from decaying meat, and frogs from the mud of the Nile, we now know that each can come only from previously existing organisms. One of the fundamental tenets of biology is that "all life comes only from living things." If any one characteristic can be said to be the very essence of life, it is the ability of an organism to reproduce its kind.

In simple organisms such as the amoeba, reproduction may be **asexual** (Figure 1–8)—that is, without sex. When an amoeba has grown to a certain size it reproduces by splitting into two to form two new amoebas. Before it divides, an amoeba makes a duplicate copy of its hereditary material (genes), and distributes one complete set to each new cell. Except for size, each new amoeba is identical to the parent cell. Unless eaten by another organism or destroyed by adverse environmental conditions such as pollution, an amoeba does not die.

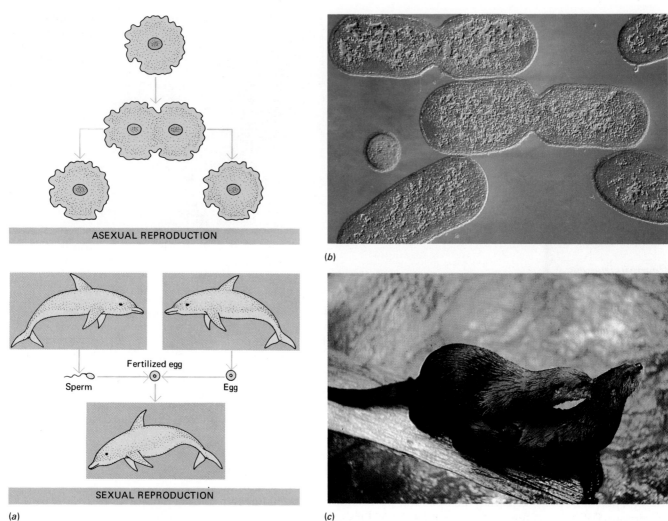

FIGURE 1–8 Approaches to reproduction. (*a*) In asexual reproduction, one individual gives rise to two or more offspring—all identical to the parent. In sexual reproduction, two parents each contribute a sex cell; these join to give rise to the offspring, which is a combination of the traits of both parents. (*b*) Asexual reproduction in the *Shigella* species of bacteria. (The bacteria in this electron micrograph have been magnified approximately 13,300 times.) (*c*) Mating in river otters.

In most plants and animals, **sexual reproduction** is carried out by the production of specialized egg and sperm cells that unite to form the fertilized egg, from which the new organism develops. With sexual reproduction, each offspring is not a duplicate of a single parent but is the product of the interaction of various genes contributed by both the mother and the father. Genetic variation is the raw material for the vital processes of evolution and adaptation.

Adaptation

The ability of a species[1] to adapt to its environment is the characteristic that enables it to survive in a changing world. **Adaptations** are traits that enhance an organism's ability to survive in a particular environment (Figure 1–9). They may be structural, physiological, or behavioral, or a combination of all of these. The long, flexible tongue of the frog is an adaptation for catching insects, and the thick fur coat of the polar bear is an adaptation for surviving frigid temperatures. Every biologically successful organism is a complex collection of coordinated adaptations.

Adaptation involves changes in species rather than in individual organisms. If every organism of a species were exactly like every other, any change in the environment might be disastrous to all, and the species would become extinct. Most adaptations occur over long periods of time

---

[1]A species is a group of similar organisms that freely interbreed and produce fertile offspring.

FIGURE 1–9 The scorpion fish blends with its background so well that it looks like a rock on the ocean floor. It is well adapted to make dinner of any small organism that unwarily swims by.

and involve many generations. Adaptations are the result of evolutionary processes (introduced later in this chapter).

THE ORGANIZATION OF LIFE

One of the striking features of life is its organization. We have already noted the cellular level of organization, but within each individual organism we can identify several other levels. Even when considering the interactions within and between groups of organisms, we can recognize a hierarchy of increasing complexity, as illustrated in Figure 1–10.

Organization of the Organism

The **chemical level** is the simplest level of organization. It includes the basic particles of all matter, **atoms,** and combinations of atoms called **molecules.** An atom is the smallest unit of a chemical element (fundamental substance) that retains the characteristic properties of that element. For example, an atom of iron is the smallest possible amount of iron. Atoms combine chemically to form molecules. For example, two atoms of hydrogen combine with one atom of oxygen to form one molecule of water.

At the **cellular level** we find that many diverse molecules may associate with one another to form complex and highly specialized structures within cells called **organelles.** The cell membrane that surrounds the cell and the nucleus that contains the hereditary material are examples of organelles. The cell itself is the basic structural and functional unit of life. The cell is the simplest part of living matter that can carry on all of the activities necessary for life. Each cell consists of a discrete body of jelly-like cytoplasm surrounded by a cell membrane. The organelles are suspended within the cytoplasm (Figure 1–11).

In most multicellular organisms cells associate to form **tissues,** such as muscle tissue or nervous tissue. Tissues, in turn, are arranged into functional structures called **organs,** such as the heart or the stomach. Each major group of biological functions is performed by a coordinated group of tissues and organs, called an **organ system.** The circulatory and digestive systems are examples of organ systems. Functioning together with great precision, the organ systems make up the complex multicellular organism.

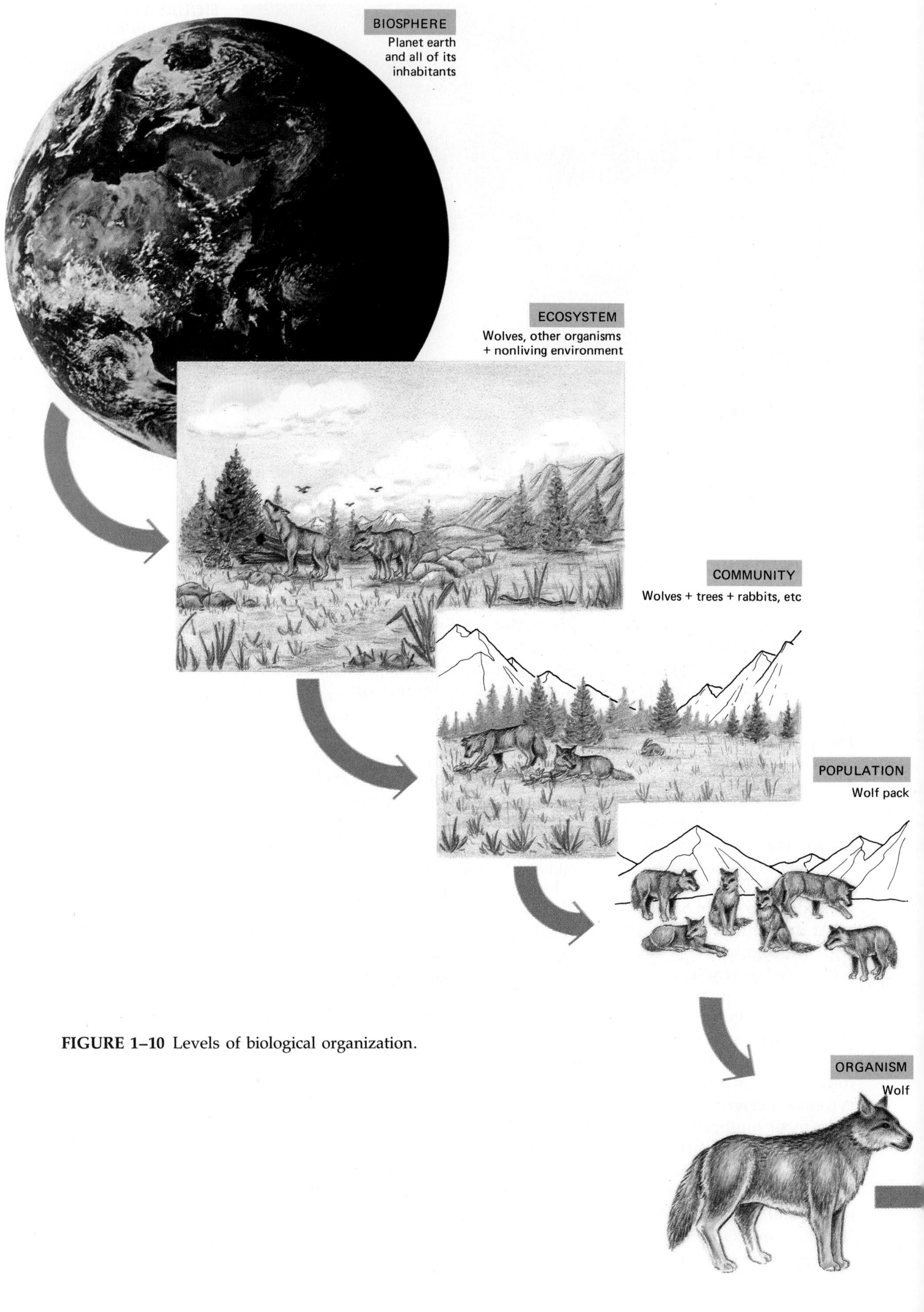

BIOSPHERE
Planet earth and all of its inhabitants

ECOSYSTEM
Wolves, other organisms + nonliving environment

COMMUNITY
Wolves + trees + rabbits, etc

POPULATION
Wolf pack

ORGANISM
Wolf

FIGURE 1–10 Levels of biological organization.

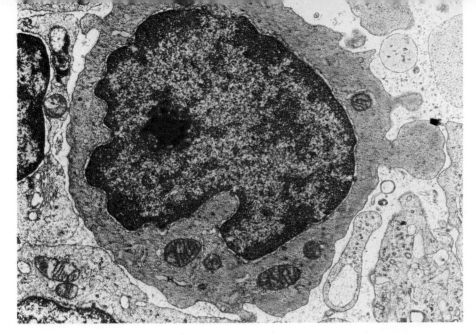

FIGURE 1–11 A human lymphocyte, a type of white blood cell (approximately ×3000). This false-color transmission electron micrograph shows the large, central nucleus (orange-brown) and darker colored chromatin (genetic material) within the nucleus toward the nuclear wall. The cytoplasm (colored green) contains a few organelles called *mitochondria* (small, brown, cigar-shaped bodies) which are the site of energy production within the cell.

Ecological Organization

Organisms interact to form still more complex levels of biological organization. All of the members of one species that live in the same area make up a **population.** The environment in which an organism or population lives is known as its **habitat.** The populations of organisms that inhabit a particular area and interact with one another form a **community.** Thus, a community can be composed of hundreds of different types of life forms. The study of how organisms of a community relate to one another and with their nonliving environment is called **ecology.** A community together with its nonliving environment is referred to as an **ecosystem.**

A self-sufficient ecosystem contains three types of organisms—producers, consumers, and decomposers—and has a physical environment appropriate for their survival (Figure 1–12). **Producers,** or **autotrophs,** are algae, plants, and certain bacteria that can produce their own food from simple raw materials. Most of these organisms use sunlight as an energy source and carry out photosynthesis. During photosynthesis the energy from sunlight is used to synthesize complex molecules from carbon dioxide and water. The light energy is transformed into chemical energy, which is stored within the chemical bonds of the food molecules produced. Oxygen, which is required not only by plant cells but also by the cells of most other organisms, is produced as a byproduct of photosynthesis.

Carbon dioxide + Water + Energy $\longrightarrow$ Food + Oxygen

Animals, including human beings, are **consumers.** The consumers, as well as the decomposers, are **heterotrophs,** organisms that are dependent on producers for food, energy, and oxygen. However, these organisms

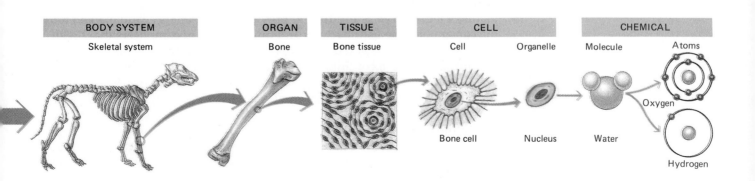

BODY SYSTEM	ORGAN	TISSUE	CELL		CHEMICAL	
Skeletal system	Bone	Bone tissue	Cell	Organelle	Molecule	Atoms
			Bone cell	Nucleus	Water	Oxygen
						Hydrogen

FIGURE 1–12 Interdependence of producers, consumers, and decomposers. The producers provide oxygen and food containing energy and nutrients for the consumers. In turn, the consumers provide the carbon dioxide needed for photosynthesis by the producers. The decomposers break down wastes and dead bodies so that minerals are recycled.

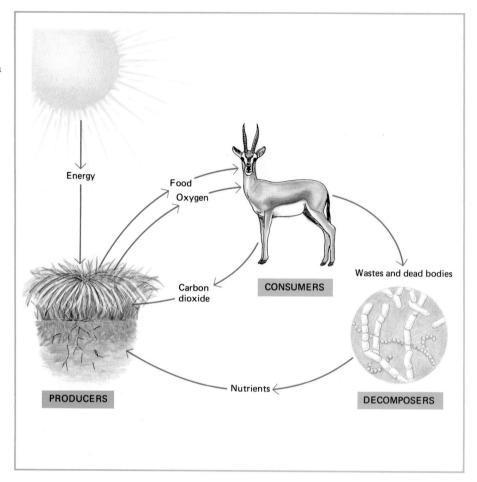

also contribute to the balance of the ecosystem. Like all living things (including producers), they obtain energy by breaking down food molecules originally produced during photosynthesis. The biological process of breaking down these fuel molecules is known as **cellular respiration.** When chemical bonds are broken during cellular respiration, their stored energy is made available for life processes (Figure 1–13).

Food + Oxygen $\longrightarrow$ Carbon dioxide + Water + Energy

Gas exchange between producers and heterotrophs by way of the nonliving environment helps maintain the life-sustaining mixture of gases in the atmosphere.

Decomposers—the bacteria and fungi—are an important component of an ecosystem because they break down the wastes and the bodies of dead organisms, making their components available for reuse. If decomposers (and scavengers, such as vultures) did not exist, nutrients would become locked up in the dead bodies of plants and animals, and the supply of elements required by living systems would soon be exhausted.

A familiar example of an ecosystem is a balanced aquarium. Green plants or algae serve as producers, providing food, energy, and oxygen for the consumers, the fish. Bacteria and fungi are the decomposers that break down waste products and dead bodies, permitting nutrients to be recycled. As long as there is a continuing input of light energy for the plants, such a system may continue to thrive for months or even years. Eventually, though, it will collapse. A larger, natural ecosystem such as a pond is far more stable and is likely to last much longer. Its stability is greater, for unlike the balanced aquarium, the natural ecosystem consists of a great diversity of organisms. Chemicals and energy have a multitude of alternative pathways within such an ecosystem. Should one type of organism die

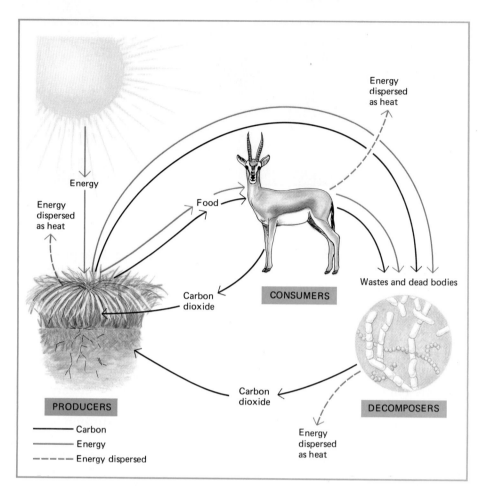

Energy
dispersed
as heat

Energy

Energy
dispersed
as heat

Food

Carbon
dioxide

CONSUMERS

Wastes and dead bodies

PRODUCERS

Carbon
dioxide

DECOMPOSERS

Energy
dispersed
as heat

——— Carbon
——— Energy
- - - - - Energy dispersed

FIGURE 1–13 Flow of energy and carbon through the biosphere. Carbon and many other chemical elements are continuously recycled. Carbon is converted from carbon dioxide gas to food by producers. Consumers obtain carbon by eating producers. Carbon leaves the consumers and producers in the form of wastes and dead material, which is broken down by decomposers. Consumers break down carbon-containing compounds during cellular respiration, converting them in part back to carbon dioxide, and the cycle begins anew. Energy, on the other hand, cannot be recycled. Some of it is dispersed as heat during every energy transaction. For this reason a constant energy input from the sun is required to keep the biosphere in operation.

out, blocking one pathway, other organisms take its place, and the system as a whole is little disturbed.

An ecosystem can be as small as a pond (or even a puddle) or as vast as the Great Plains of North America or the Arctic Tundra. The largest ecosystem is the planet Earth with all its inhabitants—the **biosphere.** (The term *ecosphere* is sometimes used instead of *biosphere.*)

THE VARIETY OF ORGANISMS

In order to study life we need a system for naming and classifying its myriad forms. The basic unit biologists have agreed upon in their classification of organisms is the **species.** It is difficult to give a definition of this term that can be applied uniformly throughout the living world, but we will define a species as a population of similar individuals, alike in their structural and functional characteristics, that in nature freely interbreed and produce fertile offspring.

Closely related species are grouped together in the next higher unit of classification, the **genus** (plural, *genera*). Each organism is given a scientific name consisting of two words, the genus and the species, both in Latin. The scientific name of the American white oak is *Quercus alba*, whereas the name of the European white oak is *Quercus robur.* Another tree, the white willow, *Salix alba,* belongs to a different genus. Our own scientific name is *Homo sapiens.*

Organisms are assigned to increasingly broad categories in which they have fewer characteristics in common (see Chapter 22). The broadest category commonly used is the **kingdom.** In this book five kingdoms are recognized: Monera, Protista, Fungi, Plantae, and Animalia. A review of the kingdoms can be found in Chapter 22 (Table 22–2), and a more detailed discussion of them in Chapters 23 through 31. We will refer to these groups

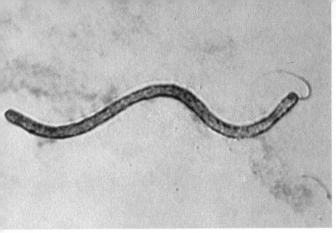

(a)

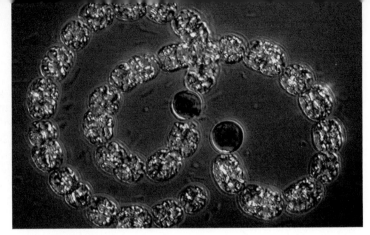

(b)

FIGURE 1–14 Members of the kingdom Monera. (*a*) The bacterium *Spirillum volutans,* which has been magnified several hundred times in this photomicrograph, propels itself by means of the whiplike flagellum at each of its ends. (*b*) These blue-green algae, *Anabaena spiroides,* form a colony that shows some specialization of its cells. The cell at the end fixes nitrogen (converts nitrogen gas to forms usable by plants and other organisms). The greatly enlarged cell is a spore, which can survive long periods of unfavorable environmental conditions such as dryness.

repeatedly throughout the text as we consider the many kinds of problems faced by living things and the various adaptations that have evolved in response to these problems.

Kingdom Monera

The single-celled **bacteria** and **cyanobacteria** (blue-green algae) of the kingdom **Monera** differ from all other organisms in that they lack a nuclear membrane as well as other membrane-bounded organelles (Figure 1–14). They are referred to as **prokaryotes.** All other organisms are **eukaryotes,** organisms with cells that have distinct nuclei surrounded by nuclear membranes, as well as a variety of other organelles that are surrounded by membranes.

Bacteria are microscopic organisms that serve as decomposers in ecosystems. Some bacteria cause diseases in human beings or in other organisms. The cyanobacteria are structurally similar to bacteria but contain the green pigment chlorophyll (as well as other pigments), which traps the energy of sunlight and enables these organisms to carry on photosynthesis. (Some true bacteria are also photosynthetic but employ alternative versions of the process.)

Kingdom Protista

Members of the kingdom **Protista** are primarily single-celled eukaryotes, but some species form loose aggregations of cells called **colonies** and some are multicellular. The animal-like protists, the **protozoa,** are generally larger than bacteria and have various means of locomotion. The plantlike protists are the algae; like plants, algae contain chlorophyll and carry on photosynthesis (Figure 1–15). However, they lack other plant characteristics such as multicellular reproductive organs, and they do not produce embryos like plants. The fungus-like protists resemble fungi in some ways but also have distinct features; some groups, for example, have flagella.

FIGURE 1–15 Members of the kingdom Protista. (*a*) Living radiolaria. These one-celled organisms (amoeboid protozoans) secrete elaborate and beautiful skeletons made of silica. These skeletons become part of the mud on the ocean floor and eventually are compressed and converted into siliceous rock. (*b*) Red algae.

(a)

(b)

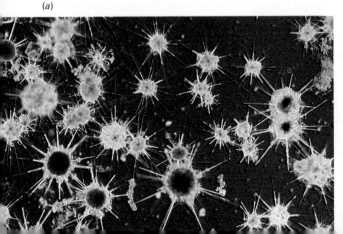

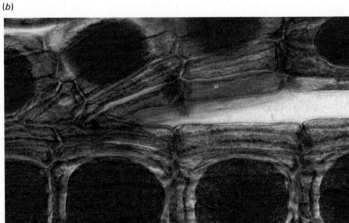

Kingdom Fungi

The **fungi** are a diverse group of eukaryotes that obtain their food by absorbing it through their surface rather than ingesting it as animals do (Figure 1–16). Some are ecologically important as decomposers, absorbing nutrients from decaying organic matter; others are parasitic. Fungi may produce both sexual and asexual spores during reproduction. The fungi include the unicellular yeasts and the multicellular molds, mushrooms, and bracket fungi.

(a)

Kingdom Plantae

Plants are multicellular organisms adapted to carry out photosynthesis. Their photosynthetic pigments, such as chlorophyll, are located within membranous organelles called **chloroplasts.** Plant cells are surrounded by rigid cell walls containing cellulose and typically have large, fluid-filled sacs called *vacuoles.* The kingdom Plantae includes the bryophytes and vascular plants (Figure 1–17).

Mosses, liverworts, and their relatives are **bryophytes.** These terrestrial plants require a moist environment to complete their reproductive cycle. Because they lack an efficient system of internal transport, bryophytes are small organisms.

The vascular plants include the ferns, the conifers **(gymnosperms),** and the flowering plants **(angiosperms).** Their efficient internal transport system moves water and nutrients from one part of the plant to another, enabling them to grow to very large sizes.

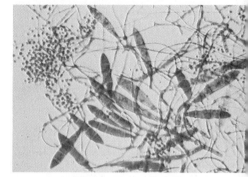

(b)

FIGURE 1–16 Members of the Kingdom Fungi. (*a*) Mushrooms. (*b*) The fungus *Trichophyton mentagrophytes,* which can cause athlete's foot.

(a)

(b)

(c)

(d)

FIGURE 1–17 The plant kingdom claims many beautiful and diverse forms. (*a*) Liverwort, a bryophyte. Like other bryophytes, most liverworts are small plants that live in moist areas. (*b*) A cycad, palmlike gymnosperm. The fossil record shows that cycads were very abundant during the Mesozoic era, more than 100 million years ago. (*c*) Flowers of *Gazania* sp (an angiosperm). This is a composite flower; its flowers are very small and are grouped into compact structures that may be mistaken for a single flower. (*d*) Water lilies. The water lily, *Nymphaea,* is an angiosperm with a large flower consisting of six petal-like parts.

Kingdom Animalia

All animals are multicellular heterotrophs. Lacking photosynthetic pigments, animals must obtain nutrients by eating other organisms. Complex animals have a high degree of tissue specialization and body organization, which have evolved along with motility, complex sense organs, nervous systems, and muscular systems. Some ten major groups, or **phyla** (singular, *phylum*), of animals are recognized (Figure 1–18). Among these are the following:

Sponges. The sponges are the simplest animals. They are aquatic and sessile. The sponge body is perforated with many pores, and food particles are strained from the water passing through.

Cnidarians. The cnidarians include jellyfish, sea anemones, and corals. These aquatic, mostly marine, animals typically have stinging cells. The body is basically a simple sac with only one opening to the digestive cavity. The mouth (which must also serve as an anus) is usually surrounded by a circle of tentacles bearing stinging cells.

Flatworms. Like the cnidarians, flatworms have a digestive cavity connected to the outside by a single opening. They live in fresh or salt water, or may be terrestrial. Flatworms are **bilaterally symmetrical,** which means the body may be divided into roughly similar right and left halves. There is a concentration of nervous tissue and sense organs in the anterior (front) end of the animal, a definite advantage to an organism that moves in a forward direction. This phylum includes planarians, flatworms, and flukes.

Mollusks. Mollusks include oysters, clams, scallops, octopods, snails, slugs, and squid. These animals have a complex body plan that is quite different from that of other animals. Most have a hard, calcareous (calcium-containing) shell, which serves as protection but makes locomotion difficult. They typically have a broad, muscular foot used in locomotion.

Annelids. The segmented worms, or annelids, are found in the oceans, in fresh water, or in moist, shady habitats. This group includes earthworms, leeches, and a variety of marine worms. The annelid body is composed of a series of rings or segments; both the body wall and the internal organs are segmented.

Arthropods. Spiders, lobsters, insects, centipedes, and millipedes are among the most familiar arthropods. There are more arthropods in terms of both number and species—there are about a million species, mainly insects—than organisms in any other phylum. Arthropods live in a greater variety of habitats and can eat a greater variety of foods than the members of any other phylum. The term *arthropod* ("jointed foot") refers to their paired jointed appendages.

Echinoderms. The spiny-skinned echinoderms include the sea stars, sea urchins, and sea cucumbers. These marine animals are radically different from all other animals but appear to be related to the chordates. The skin of echinoderms contains calcareous spine-bearing plates.

Chordates. The chordates have a skeletal stiffening rod (notochord), a tubular nerve cord, and paired gill slits. These structures (or their rudiments) are present in all chordate embryos but may be lost or transformed during development. The major chordate subphylum, the vertebrates, are characterized by a cartilaginous or bony vertebral column that surrounds and usually replaces the notochord. Vertebrates include sharks, bony fish, amphibians (frogs and salamanders), reptiles (snakes, lizards, turtles, alligators), birds, and mammals. Chordates are less diverse and much less numerous than insects but rival them in their adaptations to many lifestyles.

(a)

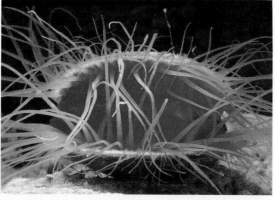
(b)

(c)

(d)

(e)

(f)

FIGURE 1–18 Some diverse life forms in the animal kingdom. (a) This tube coral, *Dendrophyllia gracilis,* a cnidarian, is a marine animal that looks like a flower. (b) The file clam is a mollusk, an animal whose soft body is covered by a protective shell. (c) The painted lady butterfly, an arthropod, migrates great distances; millions cross the Mediterranean Sea between Europe and Africa each year. (d) The lion fish, *Pterois volitans,* a chordate, is a native of coral reefs; its unusual fins are armed with glands that secrete a powerful toxin poisonous to humans as well as to other animals. (e) Another chordate, the roseate spoonbill *Ajaia ajaja,* a wading bird, uses its spoonlike beak to gather shellfish and aquatic insects from tidal areas. (f) African lions *Panthera leo,* among the fiercest of animals, are also among the most sociable; they live peaceably in prides (groups) of as many as 35.

EVOLUTION: A KEY CONCEPT

Biology is more than the science of describing and naming organisms and life processes. Biologists are concerned not only with the existence of structural similarities but with what these similarities (and differences) may tell us about how organisms are related to one another, and how things have come to be as they are.

Every organism is the product of complex interactions between its genes and environmental conditions. How organisms have changed, or **evolved,** over time has been a basic focus of investigation and debate. The theory of evolution has become one of the great unifying concepts of biology.

The theory of evolution is discussed in depth in Chapters 17 through 21. However, in this first chapter of a book that is designed to introduce you to modern biology, it is important to present a brief overview of evolution. Although evolution is itself a subdiscipline of biology, some element of an evolutionary perspective is present in almost every specialized field within biology. Biologists in almost every subdiscipline try to understand the features and functions of organisms and their constituent cells and parts by considering them in light of the long, continuing process of evolution. Additionally, biologists are constantly checking for verification of the evolutionary relationships among different organisms.

Although the concept of evolution had been discussed by philosophers and naturalists through the ages, **Charles Darwin** first brought the theory of evolution to general attention and suggested a plausible mechanism to explain it. In his book *On The Origin of Species by Means of Natural Selection,* published in 1859, Darwin synthesized many new findings in geology and biology and delineated a comprehensive theory of evolution that has helped shape the nature of biological science to the present day. Darwin presented a wealth of evidence that the present forms of life on earth descended with modifications from previously existing forms. His book raised a storm of controversy in both religion and science, some of which still lingers. It also generated a great wave of scientific research and observation that has provided much additional evidence that evolution is responsible for the great diversity of organisms present on our planet.

Darwin based his theory of **natural selection** on the following observations: (1) Individual members of a species show some variation from one another. (2) Many more organisms are produced than can possibly find food and survive into adulthood (Figure 1–19). (3) There is then a struggle for survival among the many individuals produced. Those individuals that possess characteristics that give them some advantage in the struggle for existence will be more likely to survive than those that lack these characteristics. (4) The survivors will pass these advantageous characteristics on to their offspring and to future generations.

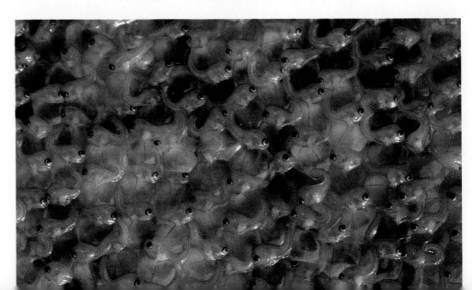

FIGURE 1–19 Eggs of the midshipman fish. Random events might be largely responsible for determining which of these developing organisms will reach adulthood and reproduce. However, certain desirable or undesirable traits that each organism might have will also contribute to its probability for success in its environment. Although not all organisms are as prolific as the midshipman fish, the generalization that more organisms are born than survive is true throughout the living world.

We now understand that differences among individual organisms are a result of different genes. Random mutations (permanent changes in genes) provide the raw material for evolution. Natural selection favors organisms with traits that best enable them to cope with pressures exerted by the environment. These organisms are most likely to survive and produce offspring. As these successful organisms pass on their genetic recipe for survival, their traits become more widely distributed in the population. Over long periods of time, as organisms continue to change (and as the environment itself changes, bringing different selective pressures), the members of the population become increasingly unlike their ancestors.

A successful organism is adapted to its environment; adaptations, and thus well-adapted organisms, are the products of evolution. The long neck of the giraffe, for example, is an adaptation for reaching leaves on trees (Figure 1–20). Ancestors of modern-day giraffes did not have elongated necks. Through the ages the giraffes with the shortest necks were at a disadvantage for obtaining food, and did not survive. Those with longer than average necks were best able to reach the leaves on trees. These giraffes survived and reproduced, passing on their genes for long necks. Through countless generations, the giraffe neck became longer and longer.

FIGURE 1–20 A successful organism is adapted to its environment. The long neck of the giraffe is an adaptation for reaching leaves high on trees.

HOW BIOLOGY IS STUDIED

This book is about the systematic study of living things—the science of biology. What distinguishes science is its insistence on rigorous methods to examine a problem and its attempts to devise experiments to validate its findings. The essence of the scientific method is the posing of questions and the search for answers to those questions. But the questions must arise from observations and experiments, and the answers must be potentially testable by further observation and experiment.

What makes science systematic is the attention it gives to organizing knowledge so it is readily accessible to all those who wish to build on its foundation. In this way science is both a personal and a social endeavor. Science is not mysterious; by its sets of rules and procedures it makes itself open to all who wish to take on its challenges. Science seeks to give us precise knowledge about those aspects of the world that are accessible to its methods of inquiry. It is not a replacement for philosophy, religion, or art, and being a scientist does not prevent one from participating in these other fields of human endeavor.

Systematic Thought Processes

The systematic thought processes on which science is based generally fall into two categories: deduction and induction. With **deductive reasoning,** we begin with supplied information, called *premises,* and draw conclusions on the basis of that information. Deduction proceeds from general principles to specific conclusions. For example, if we accept the premise that all birds have wings, and the second premise that sparrows are birds, we can conclude deductively that sparrows have wings.

Induction is almost the opposite of deduction. With **inductive reasoning,** we begin with specific observations from which we seek to draw a conclusion or discover a unifying rule or general principle. The inductive method can be used to organize raw data into manageable categories by answering the questions, "What do all these facts have in common?" A weakness of this method of reasoning is that conclusions contain *more* information than the reported facts on which they are based. We go from many observed examples to all possible examples when we formulate the general principle. This is known as the **inductive leap.** Without it, we could not arrive at generalizations. However, we must be sensitive to the possibility that the conclusion is not valid. The extra information that inductive conclusions contain can come only from the creative insight of a human mind, and creativity, however admirable, is not infallible. Here is an example of inductive reasoning: When released from support, apples, oranges, rocks, and trees fall to the ground; therefore, a force acting on these objects attracts them to the ground (i.e., the force of gravity).

Even if a conclusion is based on thousands of observations, it is still possible that new observations can challenge the conclusion. However, the greater the number of cases that are employed, the more likely we are to draw accurate scientific conclusions. The scientist seeks to state with confidence that any specific conclusion has a certain statistical probability of being correct.

Designing an Experiment

The ultimate sources of all the facts of science are careful close observations and experiments, made free of bias and with suitable controls, and carried out in as quantitative a fashion as possible. The data collected may then be analyzed so that the observed phenomena may be brought into some sort of order. The data can be synthesized or reassembled so that whatever relationships may exist can be discovered. On the basis of these initial observations the scientist makes a generalization, or constructs a hypothesis. A **hypothesis** is a trial idea about the nature of the collected data, or possibly about connections in a chain of events, or even about cause-and-effect relationships between events. Predictions made on the basis of a hypothesis can be further tested by controlled experiments.

It is in the construction of hypotheses and the ability to discern relationships among seemingly disparate events that scientists differ most. The ability to look at a mass of data and suggest a reason for their interrelations is rare. Science does not advance by the mere accumulation of facts, nor by the mere postulation of hypotheses. The two must go hand in hand in scientific investigations: observations, hypothesis, more observations, revised hypothesis, further observations, refining of the hypothesis, and so on, in a continuous process of intellectual feedback.

Let us follow the process of setting up an experiment. Suppose a pharmaceutical company wants to test a new drug to determine whether it will improve memory in elderly patients with memory problems. To test the drug, the company solicits the cooperation of physicians who work with such patients. The physicians administer a memory test and then prescribe the drug to 500 patients for a period of 2 months. They then administer another memory test, and find that the patients demonstrate a 20% increase in their ability to remember things. Can the drug company legitimately conclude that its hypothesis is correct, that the drug does indeed improve memory in elderly patients? Alternative explanations might be possible. The attention paid to the patients might in itself have stimulated them to be more attentive, for instance. Therefore, the conclusion cannot be considered valid.

To avoid such objections, a properly designed experiment must have a **control;** that is, a second experiment must be performed under the same conditions as the first, except that the one factor being tested should be varied. Thus, another similar group of patients must be given a **placebo,** a harmless starch pill similar in size, shape, color, and taste to the pill being tested. Neither group of patients should be told which pill—the drug or the placebo—has been given. In fact, to prevent bias, most medical experiments today are carried out in "double-blind" fashion: Neither the patient nor the physician knows who is getting the experimental compound and who is getting the placebo. The pills or treatments are coded in some way

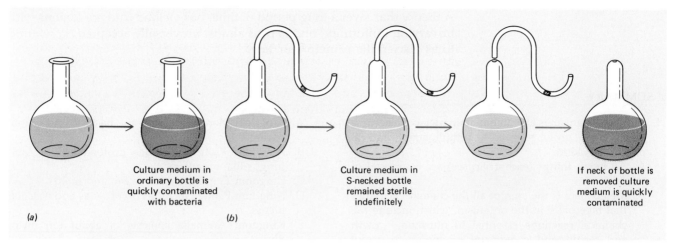

Culture medium in
ordinary bottle is
quickly contaminated
with bacteria

Culture medium in
S-necked bottle
remained sterile
indefinitely

If neck of bottle is
removed culture
medium is quickly
contaminated

(a)

(b)

FIGURE 1–21 Pasteur's experiments disproving the spontaneous generation of microorganisms. Nutrient broth (sugar and yeast) was placed in two types of flasks and boiled to kill any bacteria present. (*a*) As his control, Pasteur used flasks with straight necks that permitted bacteria to settle into the broth. In these flasks, the broth was soon teeming with bacteria. Pasteur's experimental flasks, shown in (*b*), had long, S-shaped necks that did not permit bacteria to enter, even though the flasks were open to the air. Bacteria did not grow in such flasks unless the necks were removed.

unknown to physician or patient. Only after the experiment is over and the results are in is the code broken to identify the control and experimental patients. Another example of a controlled experiment is shown in Figure 1–21.

Not all experiments can be so neatly designed; for one thing, it is often difficult to establish appropriate controls. For example, we know that the carbon dioxide content of the Earth's atmosphere is increasing because of the combustion of fossil fuels and because of widespread clearing and burning of forests. This increased carbon dioxide in the atmosphere produces a "greenhouse effect," trapping heat from solar radiation. Some scientists have warned that this thermal blanket around the globe may increase the average temperature of the Earth and ultimately alter its climate. Yet in the recent past scientists have also argued that the accumulation of particulates in the atmosphere (soot) would moderate or cancel the effect of the carbon dioxide increase. Even if the temperature of the Earth does increase, however, given other variables in the Earth's atmosphere, how can we be certain that the temperature change has resulted from human activities?

This raises an important practical question. Obviously we do not have a second unindustrialized Earth whose climate could be compared with our own. Without such a control, scientists have had to base their predictions of the future climate on mathematical modeling techniques that fall short of perfection. Should we postpone action pending the development of a perfectly predictive model? Clearly, that would involve a long wait, and by then it might be impossible to act effectively.

How a Hypothesis Becomes a Theory

A hypothesis supported by a large body of observations and experiments becomes a **theory,** defined in the dictionary as "a scientifically acceptable general principle offered to explain phenomena; the analysis of a set of facts in their ideal relations to one another." A good theory relates facts that previously appeared to be unrelated and that could not be explained on common ground. A good theory grows; it relates additional facts as they become known; it may even suggest practical applications. It predicts new facts and suggests new relationships among phenomena.

A good theory, by showing the relationships among classes of facts, simplifies and clarifies our understanding of natural phenomena. Einstein wrote, "In the whole history of science from Greek philosophy to modern physics, there have been constant attempts to reduce the apparent complexity of natural phenomena to simple, fundamental ideas and relations."

A theory that, over a long period of time, has yielded true predictions with unvarying uniformity, and is thus almost universally accepted, is referred to as a scientific **principle** or **law.**

■ SUMMARY

I. A living organism is able to maintain metabolic homeostasis, move, grow, respond to stimuli, adapt, and reproduce its kind.
   A. All living things (except viruses) are composed of cells.
   B. Metabolism is the sum of all the chemical activities that take place in the organism, which include the chemical reactions essential to nutrition, growth and repair, and conversion of energy to useful forms.
   C. Homeostasis is the tendency of organisms to maintain a constant internal environment.
   D. Living things grow by increasing the size and number of their cells.
   E. Movement, though not necessarily locomotion, is characteristic of living things.
   F. Living things respond to stimuli and adapt to their environment.
   G. Reproduction may be asexual, in which the offspring are identical with the parent, or sexual, in which the offspring reflect the characteristics of two parents.

II. There is a hierarchy of biological organization.
   A. A complex organism is organized at the chemical, cellular, tissue, organ, and organ system levels.
   B. The basic unit of ecological organization is the population. Various populations form communities; a community and its physical environment is referred to as an ecosystem. The planet Earth and all of its inhabitants may be regarded as a giant ecosystem, the biosphere.

III. Living organisms may be classified into five kingdoms. Each kind of organism is assigned to a genus and a species.

A. Kingdom Monera includes the bacteria and cyanobacteria.
B. Kingdom Protista includes protozoa, algae, and fungus-like organisms.
C. Kingdom Fungi includes yeasts and molds.
D. Kingdom Plantae includes bryophytes and vascular plants.
E. Kingdom Animalia consists of about ten major phyla including sponges, cnidarians, flatworms, mollusks, annelids, arthropods, echinoderms, and chordates.

IV. Populations of organisms evolve over time in response to changes in the environment.
   A. Natural selection favors organisms with traits that enable them to cope with environmental changes; these organisms are most likely to survive and produce offspring.
   B. As successful organisms pass on their genes for survival, their traits become more widely distributed in the population.
   C. Well-adapted organisms are the products of evolution.

V. The scientific method is a system of observation, hypothesis, experiment, more observation, and revised hypothesis.
   A. Deductive reasoning and inductive reasoning are two categories of systematic thought process used in the scientific method.
   B. A hypothesis is a trial idea about the nature of an observation or relationship.
   C. A properly designed scientific experiment must have a control and must be as free as possible from bias.

■ POST-TEST

1. The sum of all the chemical activities of the organism is termed _____ .
2. The tendency of organisms to maintain a constant internal environment is termed _____ .
3. A _____ is a physical or chemical change in the internal or external environment that evokes a response in an organism.
4. Cilia and flagella are used by some organisms for _____ .
5. The splitting of an amoeba into two is an example of _____ _____ .
6. An organism must be able to _____ to stimuli in the environment in order to survive.

Match the following terms in Column A with their descriptions in Column B.

Column A	Column B
7. Atom	a. Group of tissues arranged into a functional structure
8. Cell	b. Combination of two or more atoms
9. Molecule	c. Smallest particle of an element that retains the characteristic properties of that element
10. Organ	d. Specialized structures within cells
11. Organ system	e. Association of similar cells to carry out a specific function
12. Organism	f. Structural and functional unit of life
13. Organelles	g. Groups of organs that function together to carry out one or more of the major life functions
14. Tissues	h. Group of coordinated organ systems

15. In an ecosystem we can distinguish producers, consumers, and decomposers. The principal producers are _____ and _____. Consumers include _____. Decomposers include _____ and _____.

16. Organisms that lack a nuclear membrane and membrane-bounded organelles are termed _____.

17. The kingdom Monera includes the _____ and the _____.

18. Single-celled or colonial eukaryotic organisms are classified in the kingdom _____.

19. A trial idea about the nature of or connections within a chain of events is termed a _____.

20. A _____ is a scientifically accepted, well-tested hypothesis or group of related hypotheses offered to explain phenomena.

■ REVIEW QUESTIONS

1. Contrast a living organism with a nonliving object.
2. In what ways might the metabolism of an earthworm and a tiger be similar? What would be the metabolic consequences if an organism's homeostatic mechanisms failed?
3. Identify the components of a balanced forest ecosystem. In what ways are consumers dependent on producers? On decomposers?
4. To which kingdom would each of the following belong?
   a. an *Escherichia coli* bacterium
   b. a mold
   c. an amoeba
   d. a rose bush
5. How might you explain the thick coats of polar bears in terms of natural selection?

6. Contrast a hypothesis and a law.
7. How would you go about testing the hypothesis that beri-beri is caused by a deficiency of the vitamin thiamine? What would you consider to be proof that beri-beri is caused by thiamine deficiency?
8. How would you describe the mode of operation of the scientific method?
9. What is meant by a "controlled" experiment?
10. Devise a suitably controlled experiment to show:
    a. whether a strain of mold found in your garden produces an effective antibiotic.
    b. whether the rate of growth of a bean seedling is affected by temperature.

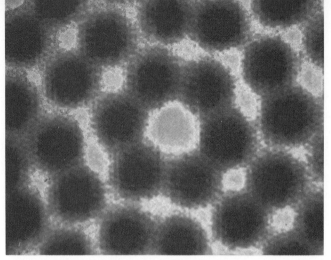
Silicon atoms viewed through scanning tunneling microscope

2

Atoms and Molecules: The Chemical Basis of Life

■ LEARNING OBJECTIVES

After you have read this chapter you should be able to:

1. Identify the chemical elements important in living things.
2. Describe the properties and roles of electrons, protons, and neutrons in determining atomic structure.
3. Distinguish among the terms *atomic number, mass number, atomic mass,* and *molecular mass.*
4. Define the term *electron orbital,* and relate orbitals to energy levels; relate the number of valence electrons to the chemical properties of the elements.
5. Distinguish between the types of chemical bonds that join atoms to form ionic and covalent compounds, and give the characteristics of each type.
6. Define and use the terms *cation* and *anion.*
7. Distinguish between and apply the terms *oxidation* and *reduction.*
8. Discuss the properties of water molecules and their importance in living things.
9. Define the terms *acid* and *base,* and discuss their properties.
10. Use the pH scale in describing the hydrogen ion concentration in living systems, and describe how buffers help minimize changes in pH.
11. Describe the composition of a salt, and explain why salts are important in living organisms.

Like everything else on our planet, living things are made of atoms and molecules. In living things, these basic components are *organized* in a very specific way. Atoms and molecules also *interact* with one another very precisely to maintain the energy flow essential to life. Much of modern biology is concerned with **molecular biology**—that is, the chemistry and physics of the molecules that constitute living things. In order to understand life processes, we must know the basic principles of chemistry. For this reason, we begin our study of life by learning about its simplest components, electrically charged atoms and molecules, and continue with a study of how atoms and molecules are assembled to form cells and tissues.

As molecular biologists have learned more about biologically important molecules, metabolic reactions, and the genetic code, our understanding of

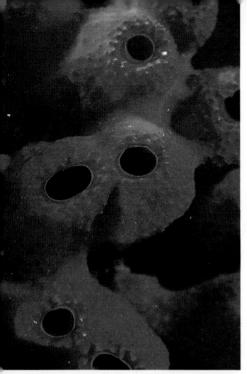

(a)

(b)

FIGURE 2–1 The chemical composition and metabolic processes of all living things are remarkably similar. (*a*) The metabolic reactions that take place within the cells of these strikingly beautiful sponges are similar to those carried on within the cells of the more complex asiatic lizard (*b*).

living organisms has increased dramatically. Two important generalizations have emerged:

1. Even though living things are strikingly diverse, their chemical composition and metabolic processes are remarkably similar (Figure 2–1). This explains why much of what biologists learn from studying bacteria or mice in their laboratories can be applied to other organisms, including humans.
2. The physical and chemical principles governing living systems are the same as those governing nonliving systems.

CHEMICAL ELEMENTS

The term **element** is applied to substances that cannot be decomposed into simpler substances by chemical reactions. The matter of the universe is composed of 92 naturally occurring elements, ranging from hydrogen, the lightest, to uranium, the heaviest. In addition to the naturally occurring

Name	*Chemical Symbol*	*Approximate Composition of Human Body by Mass (%)*	*Importance or Function*
Table 2–1 ELEMENTS THAT MAKE UP THE HUMAN BODY			
Oxygen	O	65	Required for cellular respiration; present in most organic compounds; component of water
Carbon	C	18	Forms backbone of organic molecules; can form four bonds with other atoms
Hydrogen	H	10	Present in most organic compounds; component of water
Nitrogen	N	3	Component of all proteins and nucleic acids
Calcium	Ca	1.5	Structural component of bones and teeth; important in muscle contraction, conduction of nerve impulses, and blood clotting
Phosphorus	P	1	Component of nucleic acids; structural component of bone; important in energy transfer
Potassium	K	0.4	Principal positive ion (cation) within cells; important in nerve function; affects muscle contraction
Sulfur	S	0.3	A component of most proteins
Sodium	Na	0.2	Principal positive ion in interstitial (tissue) fluid; important in fluid balance; essential for conduction of nerve impulses
Magnesium	Mg	0.1	Needed in blood and body tissues; a component of many important enzyme systems
Chlorine	Cl	0.1	Principal negative ion (anion) of interstitial fluid; important in fluid balance
Iron	Fe	Trace amount	Component of hemoglobin, myoglobin, and certain enzymes
Iodine	I	Trace amount	Component of thyroid hormones

Other elements, found in very small amounts in the body (the trace elements), include manganese (Mn), copper (Cu), zinc (Zn), cobalt (Co), fluorine (F), molybdenum (Mo), selenium (Se), and a few others.

elements, about 17 elements heavier than uranium have been made by the bombardment of elements with subatomic particles in devices known as *particle accelerators.*

About 98% of an organism's mass is composed of just six elements: oxygen, carbon, hydrogen, nitrogen, calcium, and phosphorus. Approximately 14 other elements are consistently present in living things, but in smaller quantities. Some of these, such as iodine and copper, are known as **trace elements** because they are present in such minute amounts.

Scientists have assigned each element a **chemical symbol**—usually the first letter or first and second letters of the English or Latin name of the element. For example, O is the symbol for oxygen, C for carbon, H for hydrogen, N for nitrogen, and Na for sodium (the Latin name is *natrium*). Table 2–1 lists the elements that make up a living organism and explains why each is important.

Whatever physical state matter may assume—gas, liquid, or solid—it is composed of units termed *atoms.* An **atom** is the smallest portion of an element that retains its chemical properties. The continued division of any kind of matter ultimately yields atoms; these units cannot be divided by chemical means. Most atoms are much smaller than the tiniest particle visible under a light microscope. By special scanning electron microscopy (see Chapter 4), with magnification as much as 5 million times, researchers have been able to photograph some of the larger atoms such as uranium and thorium.

ATOMIC STRUCTURE

Physicists have discovered a considerable number of subatomic particles, but for our purposes we need consider only three: protons, neutrons, and electrons. Each **proton** has one unit of a positive electrical charge; **neutrons** are uncharged particles with about the same mass as protons. Protons and neutrons make up almost all of the mass of an atom and are concentrated in the **atomic nucleus.** Each **electron** has one unit of a negative electrical charge and an extremely small mass (only about 1/1800 of the mass of a proton). The electrons, as we will see, behave as though they were spinning about in the empty space surrounding the atomic nucleus (Figure 2–2).

The Nucleus

Each kind of element has a fixed number of protons in the atomic nucleus. This number, called the **atomic number,** is written as a subscript to the left of the chemical symbol. Thus $_1H$ indicates that the hydrogen nucleus contains one proton, and $_8O$ that the oxygen nucleus contains eight protons. It

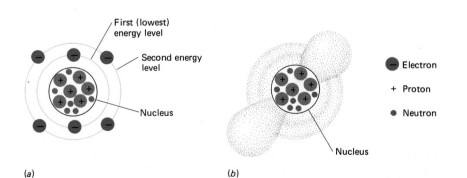

(a)

(b)

FIGURE 2–2 Two ways of representing an atom. (*a*) Bohr model of a carbon atom. Although the Bohr model is not an accurate way to depict electron configuration, it is commonly used because of its simplicity and convenience. (*b*) An electron cloud. Dots represent the probability of the electron's being in that particular location at any given moment.

FIGURE 2–3 Isotopes of carbon. Carbon-12 is the more common isotope of carbon. Its nucleus contains six protons and six neutrons, so its atomic mass is 12. Carbon-14 is a rare radioactive isotope of carbon. Because it contains eight neutrons in its nucleus rather than six, its atomic mass is 14. Both types of carbon have six protons, so both have atomic number six. Carbon-14 is often used in research to trace the fate of carbon atoms in metabolic processes.

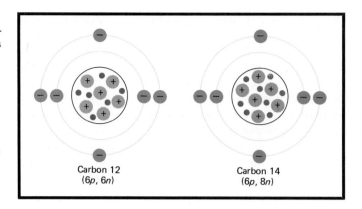

Carbon 12
(6p, 6n)

Carbon 14
(6p, 8n)

FIGURE 2–4 Anthropologists use radioisotope content to date and study fossils. The skeleton of this 11th-century inhabitant of a South African Iron Age village posed an anthropological puzzle. Physically, the man's skeleton was different from those of the other villagers, suggesting that he was not a native of the area. However, when the skeleton was analyzed for isotopes, its carbon-12–to–carbon-13 ratio was found to be similar to that of other skeletons from the same village. Since different kinds of plants incorporate different proportions of isotopes into the food produced from them, this similarity of isotope content indicates that these individuals all ate the same foods. Thus, anthropologists concluded that this man had probably spent most of his life in the village after migrating there from a distant region.

is the atomic number, the number of protons in the nucleus, that determines an atom's chemical properties. The total number of protons plus neutrons in the nucleus is the **atomic mass,** indicated by a superscript to the left of the chemical symbol. The common form of oxygen atom, with eight protons and eight neutrons in its nucleus, has an atomic number of 8 and a mass number of 16. It is indicated by the symbol $^{16}_{8}O$.

Isotopes

Atoms of the same element containing the same number of protons but different numbers of neutrons, and hence having different mass numbers, are called **isotopes.** The three isotopes of hydrogen, $^{1}_{1}H$, $^{2}_{1}H$, and $^{3}_{1}H$, contain zero, one, and two neutrons, respectively. Carbon 12 and carbon 14, two isotopes of carbon, are illustrated in Figure 2–3. Elements usually occur in nature as a mixture of isotopes.

All of the isotopes of a given element have essentially the same chemical characteristics. However, some isotopes with an excess of neutrons are unstable and tend to break down, or decay, to a more stable isotope (usually becoming a different element). Such isotopes are termed **radionuclides** (or **radioisotopes**), since they emit high-energy radiation when they decay.

Radionuclides such as ^{3}H (tritium) and ^{14}C have been extremely valuable research tools in biology and along with radionuclides of other elements are useful in medicine for both diagnosis and treatment (Figure 2–4). Despite the difference in the number of neutrons, the body treats all isotopes of a given element in a similar way. The reactions of a sugar, hormone, or drug can be followed in the body by labeling of the substance with a radionuclide such as carbon-14 or tritium. For example, the active component in marijuana (tetrahydrocannabinol) has been labeled and administered intravenously. By measurement of the amount of radioactivity in the blood and urine at successive intervals, it was determined that this compound remains in the blood and products of its metabolism remain in the urine for several weeks.

Because radiation can interfere with cell division, radioactive isotopes have been used in the treatment of cancer (a disease characterized by rapidly dividing cells). Radionuclides are also used to test thyroid gland function, to measure the rate of red blood cell production, and to study many other aspects of body function and chemistry.

Atomic Mass

The **atomic mass**[1] of an element is a number that indicates how heavy an atom of that element is compared with an atom of another element. The

---

[1]For convenience we will consider mass and weight equal, although this is not always true. Mass does not depend on the force of gravity; weight does. Thus, a person on the moon has the same mass on Earth, but because of the moon's lesser gravitational force, body weight is less there than on Earth.

mass of any single atom or molecule is exceedingly small, much too small to be conveniently expressed in terms of grams or even micrograms. Such masses are expressed in terms of the **atomic mass unit (amu),** also called the **dalton,** equal to the *approximate* mass of a proton or neutron. (The standard for comparing elements is based on assignment of an atomic mass of exactly 12 to $^{12}_{6}C$, the most common isotope of carbon.) The atomic mass for an element reflects the masses of the mixtures of isotopes that occur in nature. For example, although more than 99% of the hydrogen atoms in a naturally occurring sample have an atomic mass of 1 amu (to be precise, on the carbon = 12 scale it is 1.0000078 amu), the atomic mass of hydrogen is 1.0079 amu. This reflects the fact that a small amount of deuterium, $^{2}_{1}H$ (mass number 2), and an even smaller amount of tritium, $^{3}_{1}H$ (mass number 3), occur along with the common form of hydrogen, $^{1}_{1}H$.

Electrons and Orbitals

The space outside the atomic nucleus contains the electrons. While the mass of electrons make only a negligible contribution to the mass of an atom, electrons carry an electrical charge that has a profound impact on the chemical properties of the atom. Each electron bears a charge of -1, exactly equal but opposite to the charge on a proton. The electrons are attracted by the positive charge of the protons. Each type of atom has a characteristic number of protons and neutrons within the nucleus and a characteristic number of electrons around it, although the number and relative positions of the electrons in the atom may change during chemical reactions. In a neutral atom the number of protons in the nucleus equals the number of electrons around it. The atom as a whole has no net charge; that is, it is in a state of electrical neutrality. The positive charges of the protons equal the negative charges of the electrons.

Electrons occur in characteristic regions of space termed **orbitals.** The lowest energy orbital, the 1s orbital, is nearest the nucleus and is spherical in shape (Figure 2–5a). Other electron orbitals farther from the nucleus, the *p*, *d*, and *f* orbitals, are either spherical or dumbbell-shaped or are represented by more complex three-dimensional coordinates. Orbitals represent the places where electrons are most probably found. Electrons whirl around the nucleus, now close to it, now farther away, so that an electron cloud surrounds the nucleus. One way of illustrating an atom is to show its electron orbitals as clouds, as in Figure 2–2b. The density of the shaded areas is proportional to the probability that an electron is present there at any given moment.

Several electrons may have similar energies; these make up a **shell** and are said to be in the same **energy level.** The number of electrons in the outer energy level determines the chemical properties of atoms. The energy levels or shells of electrons in an atom can be represented by a series of concentric circles around the nucleus, as in Figure 2–2a. It is important to keep in mind, however, that electrons do *not* circle the nucleus in fixed concentric pathways. Although each orbital may contain no more than two electrons, there may be several orbitals within a given energy level or electron shell.

The way that electrons are arranged around an atom is referred to as the **electron configuration** of that atom. Electrons always fill the orbitals nearest to the nucleus before occupying those farther away. The maximum number of electrons in the innermost shell (which is a single spherical orbital) is two; the second shell has four orbitals (one spherical and three dumbbell-shaped) and thus can contain a maximum of eight electrons (Fig. 2–5). The third shell has a maximum of 18 electrons arranged in nine orbitals, and the fourth has 32 electrons in 16 orbitals. Although the third and outer shells can each contain more than eight electrons, they are most

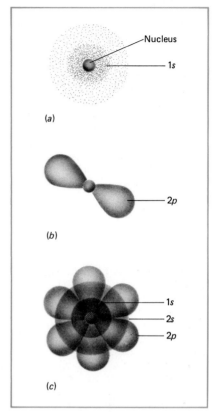

FIGURE 2–5 Representation of atomic orbitals. (*a*) The first energy level is a single spherical orbital (designated 1s) that can hold a maximum of two electrons. The electrons depicted in the diagram could be present anywhere within the dotted area. (*b*) One of the dumbbell-shaped (2p) orbitals of the second energy level. (*c*) The second energy level has four orbitals, one spherical (2s) and three dumbbell-shaped (2p). The six electrons of a carbon atom are distributed in orbitals as depicted here. Remember that the higher the energy level, the greater the average distance that the electron is located from the nucleus.

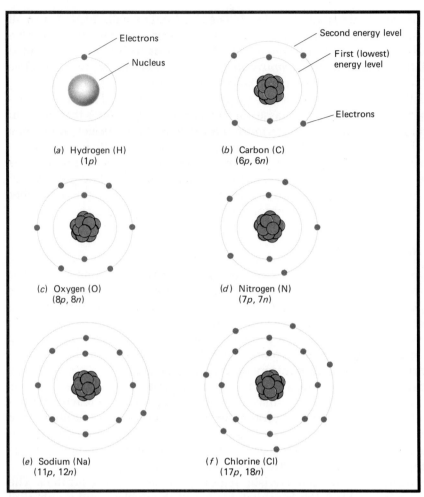

FIGURE 2–6 Bohr models of some biologically important atoms. (*a*) Hydrogen. (*b*) Carbon. (*c*) Oxygen. (*d*) Nitrogen. (*e*) Sodium. (*f*) Chlorine. Each circle represents an energy level, or electron shell. Electrons are represented by dots on the circles; *p*, proton; *n*, neutron.

stable when only eight are present. We may consider the first shell to be complete when it contains two electrons, and the other shells to be complete when they each contain eight electrons. The atomic structures of some elements important in biological systems—carbon, hydrogen, oxygen, nitrogen, sodium, and chlorine—are shown in Figure 2–6.

Keep in mind that each atom is largely empty space. The distance from an electron to the protons and neutrons in the central nucleus may be 1000 times greater than the diameter of the nucleus itself. The tendency of the negatively charged electrons to fly off in space is countered by their attraction to the atomic nucleus due to the positive charge of the protons in the nucleus.

The more distant the energy level is from the nucleus, the greater is the energy of the electrons in that level. An electron can be moved to an orbital farther from the nucleus by providing it with more energy, or an electron can give up energy and sink back to a lower energy level in an orbital nearer the central nucleus. Energy is required to move a negatively charged electron farther away from the positively charged nucleus.

When energy is added to the system, an electron can jump from one level to the next, *but it cannot stop in the space in between*. To move an electron from one level to the next the atom must absorb a discrete packet of energy known as a **quantum,** which contains just the right amount of energy needed for the transition—no more and no less. The term *quantum jump* is used in everyday language to indicate a sudden discontinuous move from one level to another.

CHEMICAL COMPOUNDS

A **chemical compound** consists of two or more different elements combined in a fixed ratio. For example, water is a chemical compound consisting of two atoms of hydrogen combined with one atom of oxygen. The properties of a chemical compound can be quite different from those of its constituent elements: At room temperature, water is usually a liquid; hydrogen and oxygen are gases.

A **chemical formula** is a shorthand method for describing the chemical composition of a compound. Chemical symbols are used to indicate the types of atoms in the molecule, and subscript numbers are used to indicate the number of each type of atom present. The chemical formula for molecular oxygen, O_2, tells us that this molecule consists of two atoms of oxygen. The chemical formula for water, H_2O, indicates that each molecule consists of two atoms of hydrogen and one atom of oxygen. (Note that when a single atom of one type is present, it is not necessary to write 1; we do *not* write H_2O_1.)

Another type of formula is the **structural formula,** which shows not only the types and numbers of atoms in a compound but also their arrangement. In each specific chemical compound the atoms are always arranged in the same way. From the chemical formula for water, H_2O, you could only guess whether the atoms were arranged H—H—O or H—O—H. The structural formula H—O—H settles the matter, indicating that the two hydrogen atoms are attached to the oxygen atom.

CHEMICAL EQUATIONS

During any moment in the life of an organism, be it a mushroom or a housefly, many complex chemical reactions are taking place. The chemical reactions that occur between atoms and compounds—for example, between methane (natural gas) and oxygen—can be described by means of chemical equations:

$$CH_4 + 2\,O_2 \longrightarrow CO_2 + 2\,H_2O + Energy$$

Methane   Oxygen   Carbon dioxide   Water

In a **chemical equation,** the **reactants** (the substances that participate in the reaction) are generally written on the left side of the equation, and the **products** (the substances formed by the reaction) are usually written on the right side. The arrow means *"yields"* and indicates the direction in which the reaction tends to proceed.

The number preceding a chemical symbol or formula indicates the number of atoms or molecules reacting. Thus, $2\,O_2$ means two molecules of oxygen, and $2\,H_2O$ means two molecules of water. The absence of a number indicates that only one atom or molecule is present.

Reactions may proceed in the reverse direction (to the left) as well as forward (to the right); at **equilibrium** the rates of the forward and reverse reactions are equal. Reversible reactions are indicated by double arrows:

$$N_2 + 3\,H_2 \rightleftharpoons 2\,NH_3$$

Nitrogen   Hydrogen   Ammonia

In this example, the arrows are drawn different lengths to indicate that when the reaction is at equilibrium there is more product than reactant.

CHEMICAL BONDS

The chemical properties of an element are determined primarily by the number and arrangement of electrons in the *outermost* energy level (electron shell). In a few elements, called the "noble gases," the outermost shell is filled. These elements are chemically inert, meaning that they do not readily combine with other elements. Two such elements are helium, with two electrons (a complete shell), and neon, with ten electrons (a complete inner shell of two and a complete second shell of eight).

The electrons in the outermost energy level of an atom are sometimes referred to as **valence electrons.** When the outer shell of an atom contains fewer than eight electrons, the atom tends to lose, gain, or share electrons to achieve an outer shell of eight (zero or two in the lightest elements).

The elements in a given compound are always present in a certain proportion by mass. This reflects the fact that atoms are combined by chemical bonds in a precise way to form the compound. A **chemical bond** is the attractive force that holds two atoms together. Each bond represents a certain amount of potential chemical energy. Bond energy is the energy necessary to break a bond. The atoms of each element form a specific number of bonds with the atoms of other elements—a number dictated by the valence electrons. The two principal types of chemical bonds are covalent bonds and ionic bonds.

Covalent Bonds

Covalent bonds involve the sharing of electrons between atoms. A compound consisting primarily of covalent bonds is called a **covalent compound.** The smallest amount of a covalent compound that retains the properties of the compound is a **molecule.**

A simple example of a covalent bond is the one joining two hydrogen atoms in a molecule of hydrogen gas, H_2 (Figure 2–7). Each atom of hydrogen has one electron, but two electrons are required to complete the first energy level. The hydrogen atoms have equal capacities to attract electrons, so neither donates an electron to the other. Instead, the two hydrogen atoms share their single electrons so that each of the two electrons is attracted simultaneously to the two protons in the two hydrogen nuclei. The two electrons thus whirl around *both* atomic nuclei and join the two atoms together.

A simple way of representing the electrons in the outer shell of an atom is to use dots placed around the chemical symbol of the element to represent the electrons. In a water molecule two hydrogen atoms are covalently bonded to an oxygen atom:

$$H\cdot\ +\ H\cdot\ +\ \cdot\ddot{\underset{\cdot\cdot}{O}}\cdot\ \longrightarrow\ H:\ddot{\underset{\cdot\cdot}{O}}:H$$

Oxygen has six valence electrons; by sharing electrons with two hydrogen atoms, it completes its outer level of eight. Each hydrogen atom obtains a complete outer level of two. (Note that in the structural formula H—O—H, each pair of shared electrons is represented by a single line. Unshared electrons are usually omitted in a structural formula.)

The carbon atom has four electrons in its outer energy level. These four electrons are available for covalent bonding:

$$\cdot\overset{\cdot}{\underset{\cdot}{C}}\cdot$$

When one carbon and four hydrogen atoms share electrons, a molecule of methane, CH_4, is formed:

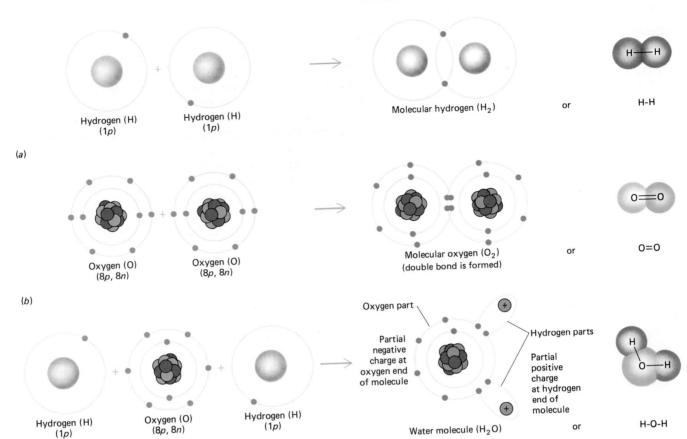

Mu

Ion

Wh
an i
lose
com
cleu
seve
com
the
life

A
are l
ions
ionic
they
elect

A
sodiu
electi
sodiu
other
stead
accep
A chle
electi
shell.
seven
charge
dium

Sodiu
(11p,

FIGURE 2–7 Formation of covalent compounds. (*a*) Two hydrogen atoms achieve stability by sharing electrons, thereby forming a molecule of hydrogen. The structural formula shown on the right is a simpler way of representing molecular hydrogen. The straight line between the hydrogen atoms represents a single covalent bond. (*b*) Two oxygen atoms share two pairs of electrons to form molecular oxygen. Note the double bond. (*c*) When two hydrogen atoms share electrons with an oxygen atom, the result is a molecule of water. Note that the electrons tend to stay closer to the nucleus of the oxygen atom than to the hydrogen nuclei. This results in a partial negative charge on the oxygen portion of the molecule and a partial positive charge at the hydrogen end. Although the water molecule as a whole is electrically neutral, it is a polar covalent compound.

Each atom shares its outer-level electrons with the other, thereby completing the first energy level of each hydrogen atom and the second energy level of the carbon atom.

The nitrogen atom has five electrons in its outer shell:

When a nitrogen atom shares electrons with three hydrogen atoms, a molecule of ammonia, NH_3, is formed:

When an electron pair is shared between two atoms, the covalent bond is termed a **single bond.** Two oxygen atoms may achieve stability by forming covalent bonds with one another. Each oxygen atom has six electrons in its outer shell. To become stable, the two atoms share two pairs of electrons, forming molecular oxygen (Figure 2–7b). When two pairs of electrons are shared in this way, the covalent bond is referred to as a **double bond.** Some atoms form **triple bonds** with one another, sharing three pairs of electrons.

FIGURE 2–11 Hydration of an ionic compound. The crystal of NaCl consists of regularly spaced ionic bonds between the Na^+ and Cl^-. When NaCl is added to water, the partial negative ends of the water molecules are attracted to the positive sodium ions and tend to pull them away from the chlorine ions. At the same time, the partial positive ends of the water molecules are attracted to the negative chloride ions, separating them from the sodium ions. When the NaCl is dissolved, each of the sodium and chlorine ions is surrounded by water molecules electrically attracted to it.

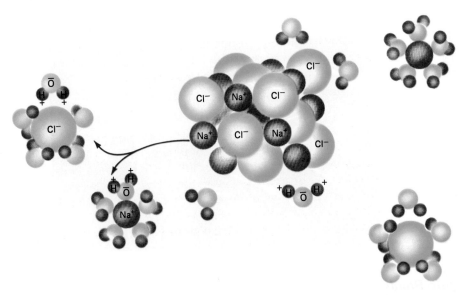

When sodium reacts with chlorine, its outermost electron is transferred completely to chlorine. The sodium ion now has 11 protons in its nucleus, 10 electrons circling the nucleus; it has a net charge of 1^+. The chlorine ion has 17 protons in its nucleus, 18 electrons circling the nucleus, and a net charge of 1^-. These ions attract each other as a result of their opposite charges. They are held together by this electrical attraction in ionic bonds to form sodium chloride,[1] common table salt.

Unlike covalently bonded atoms, compounds joined by ionic bonds, such as sodium chloride, have a tendency to **dissociate** (separate) into their individual ions when placed in water. In the solid form of an ionic compound, the constituent ions require considerable energy to be pulled apart. Water, however, is an excellent **solvent;** as a liquid it is capable of dissolving many substances. This is because of the polarity of water molecules. The localized partial positive charges (on the hydrogen atom) and partial negative charges (on the oxygen atom) on each water molecule attract the anions and cations on the surface of an ionic solid. As a result, the solid dissolves. In solution, each cation and anion of the ionic compound is surrounded by oppositely charged ends of the water molecules (Figure 2–11). This process is known as **hydration.**

$$NaCl \xrightarrow{\text{in } H_2O} Na^+ + Cl^-$$

Sodium   Sodium   Chloride
chloride   ion   ion

$$CaCl_2 \xrightarrow{\text{in } H_2O} Ca^{2+} + 2\ Cl^-$$

Calcium   Calcium   Chloride
chloride   ion   ions

$$Na_2SO_4 \xrightarrow{\text{in } H_2O} 2\ Na^+ + SO_4^{2-}$$

Sodium   Sodium   Sulfate
sulfate   ions   ion

The term *molecule* does not adequately explain the properties of ionic compounds such as NaCl. Hydrated sodium and chlorine ions do not interact with each other to the extent that "molecules" of sodium chloride can be said to exist. Likewise, if you observe a model of NaCl in its solid crystal state, you will see that each ion is actually surrounded by six ions of

---

[1] In both covalent and ionic binary compounds (*binary* denotes compounds consisting of two elements), the element having the greater attraction for the shared electrons is named second, and an -*ide* ending is added to the stem name—e.g., sodium chloride, hydrogen fluoride. The -*ide* ending is also used to indicate an anion, as in chloride (Cl^-) and hydroxide (OH^-).

opposite charge. The molecular formula NaCl indicates that sodium and chlorine ions are present in a one-to-one ratio, but in the actual crystal, no discrete molecules composed of one Na^+ ion and one Cl^- ion are present. The ions do not share electrons.

Hydrogen Bonds

Another type of bond that is extremely important in biological systems is the **hydrogen bond.** When hydrogen is combined with oxygen (or with another electronegative atom), it has a partial positive charge because its electron is positioned closer to the oxygen atom. Hydrogen bonds tend to form between an electronegative atom and a hydrogen atom that is covalently bonded to oxygen or nitrogen (Figure 2–12). The atoms involved may be in two parts of the same molecule or in two different molecules.

Hydrogen bonds are weak and are readily formed and broken. They have a specific length and orientation; this feature is very important in their role in helping determine the three-dimensional structure of large molecules such as nucleic acids and proteins. Hydrogen bonds, though relatively weak individually, occur in large numbers in the double helix of DNA and in the alpha helix of proteins (Chapter 3). The large number of bonds present compensates for the relative weakness of the individual bonds.

The water molecules in liquid water and in ice are held together in part by hydrogen bonds. The hydrogen atom of one water molecule, with its partial positive charge, is attracted to the oxygen atom of a neighboring water molecule, with its partial negative charge, forming a hydrogen bond. Each water molecule can form hydrogen bonds with a maximum of four neighboring water molecules (Figure 2–13).

Other Interactions Between Atoms

Two other types of interactions between molecules are van der Waals forces and hydrophobic interactions. These bonds, though weak, are important in maintaining the shape of many of the complex molecules in living cells. They are also important in holding together groups of nonpolar molecules, as in cell membranes.

The attractive forces between molecules, called **van der Waals forces,** occur when the molecules are very close together and are due to the interaction of their electron clouds. Van der Waals forces are weaker and less specific than the other types of interactions we have considered. They are most important when they occur in large numbers, and when the shapes of the molecules involved permit close contact between the atoms. As with hydrogen bonds, the bonding force of a single interaction is very weak. In molecules and structures that have a large number of these interactions working together, however, the binding force can be very large.

Hydrophobic (water-hating) **interactions** occur between groups of nonpolar molecules. Such groups tend to cluster together and are insoluble in water. This clustering is a result of the hydrogen bonds that hold the water molecules together, and in a sense drive the nonpolar molecules together. Hydrophobic interactions explain why oil tends to form globs when it is added to water.

MOLECULAR MASS

The molecular mass of a compound is the sum of the atomic masses of its constituent atoms; thus, the molecular mass of water, H_2O, is $(2 \times 1$ amu$) + (16$ amu$)$, or 18 atomic mass units. (Owing to the presence of isotopes, atomic mass units are not whole numbers. However, for our pur-

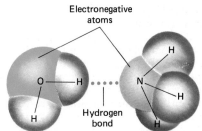

FIGURE 2–12 A hydrogen bond. The nitrogen atom of an ammonia molecule (NH_3) is joined to a hydrogen atom of a water molecule (H_2O) by a hydrogen bond. In a hydrogen bond, a hydrogen atom connected to an electronegative atom by a polar covalent bond is shared with another electronegative atom by a weak electrical attraction.

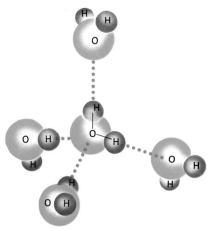

FIGURE 2–13 Hydrogen bonding of water molecules. Each water molecule tends to form hydrogen bonds with four neighboring water molecules. The hydrogen bonds are indicated by dotted lines. The covalent bonds between the hydrogen and oxygen atoms are represented by solid lines.

poses each atomic mass value has been rounded off to a whole number.) The molecular mass of the simple sugar glucose, $C_6H_{12}O_6$, which is a key compound in cellular metabolism, is (6 × 12 amu) + (12 × 1 amu) + (6 × 16 amu), or 180 atomic mass units.

The amount of a compound whose mass in grams is equivalent to its molecular mass is termed one **mole**. Thus one mole of glucose has a mass of 180 grams. A one-molar solution, represented by 1 *M*, contains one mole of the substance (e.g., 180 grams of glucose) in one liter of solution. Chemical compounds react with each other in quantitatively precise ways. For example, when glucose is burned in a fire or metabolized in a cell, one mole of glucose reacts with six moles of oxygen to form six moles of carbon dioxide (CO_2) and six moles of water.

$$C_6H_{12}O_6 + 6\ O_2 \longrightarrow 6\ CO_2 + 6\ H_2O + Energy$$

Glucose   Oxygen   Carbon   Water
dioxide

The mole is a very useful unit because we cannot do experiments with individual atoms or molecules. The very large number of units in a mole, 6.02×10^{23}, is known as *Avogadro's number*, named for the Italian physicist Amadeo Avogadro, who first calculated it. Biologists and biochemists usually deal with smaller amounts of chemical compounds—millimoles (mmoles, 1 thousandth of a mole) or micromoles (μmoles, 1 millionth of a mole).

OXIDATION-REDUCTION

Rusting—the combination of iron with oxygen—is a familiar example of oxidation and reduction.

$$4\ Fe + 3\ O_2 \longrightarrow 2\ Fe_2O_3$$

Oxidation is defined as a chemical process in which an atom, ion, or molecule loses electrons. In rusting, iron is changed from its metallic state to its iron(III) (Fe^{3+}) state; it is being oxidized.

$$4\ Fe \longrightarrow 4\ Fe^{3+} + 12\ e^-$$

The e^- is a symbol for an electron. Oxygen is simultaneously changed from its molecular state to its charged state:

$$3\ O_2 + 12\ e^- \longrightarrow 6\ O^{2-}$$

When oxygen accepts the electrons removed from the iron, it is being reduced. **Reduction** is a chemical process in which an atom, ion, or molecule gains electrons. Oxidation and reduction reactions occur simultaneously because one substance must accept the electrons that are removed from the other. Oxidation-reduction reactions are sometimes referred to as **redox reactions.**

Electrons are not easily removed from covalent compounds unless an entire atom is removed. In living cells, oxidation almost always involves the removal of a hydrogen atom from a compound; reduction often involves the addition of hydrogen (see Chapter 6).

WATER AND ITS PROPERTIES

A large part of the mass of most organisms is simply water. In human tissues the percentage of water ranges from 20% in bones to 85% in brain cells. The water content is greater in embryonic and young cells and decreases as aging occurs. About 70% of our total body weight is water; as

FIGURE 2–14 Planet Earth is sometimes referred to as the *water planet* because most of its surface is covered with water. Here, Earth is seen from Apollo II, about 98,000 nautical miles away.

FIGURE 2–15 Water has a very high surface tension because of the strength of all of its hydrogen bonds. A water strider, though more dense than water, can walk on the surface of a pond. Fine hairs at the end of its legs spread its weight over a large area, allowing its body to be supported by the surface tension of the water.

much as 95% of a jellyfish or certain plants is water. Water is not only the major component of organisms but also one of the principal environmental factors affecting them. Many organisms live within the sea or in freshwater rivers, lakes, and puddles. The physical and chemical properties of water have permitted living things to appear, to survive, and to evolve on this planet (Figure 2–14).

Water dissolves many different kinds and great quantities of compounds. Because of its solvent properties and the tendency of the atoms in certain compounds to form ions when in solution, water plays an important role in facilitating chemical reactions. Water itself is a reactant or product in many chemical reactions that occur in living tissue. Water is also the source, through plant metabolism, of the oxygen in the air we breathe, and its hydrogen atoms are incorporated into the many organic compounds in the bodies of living things. Water is also an important lubricant. It is present in body fluids wherever one organ rubs against another and in joints where one bone moves on another.

Cohesive and Adhesive Forces

Water exhibits both cohesive and adhesive forces. Water molecules have a very strong tendency to stick to each other; that is they are **cohesive.** This is due to the hydrogen bonds among the molecules. Water molecules also stick to many other kinds of substances (i.e., those substances that have charged groups of atoms or molecules on their surfaces). These **adhesive** forces explain how water makes things wet. Water has a high degree of **surface tension** because of the cohesiveness of its molecules; its molecules have a much greater attraction for each other than for molecules in the air. Thus, water molecules at the surface crowd together, producing a strong layer as they are pulled downward by the attraction of other water molecules beneath them (Figure 2–15).

Adhesive and cohesive forces account for the tendency, termed **capillary action,** of water to rise in thin tubes (Figure 2–16). Water also moves through the microscopic spaces between soil particles to the roots of plants by capillary action.

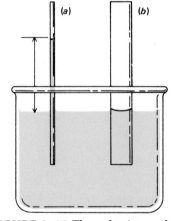

FIGURE 2–16 The cohesive and adhesive forces of water account for capillary action. (*a*) In the smaller tube, adhesive forces attract water molecules to charged groups on the surfaces of the tube. Other water molecules inside the tube are then "pulled along" by cohesive forces, which are actually caused by hydrogen bonds between the water molecules. (*b*) In the large-diameter tube, a smaller percentage of the water molecules line the glass. Because of this, the adhesive forces are not strong enough to overcome the cohesive forces of the water beneath the surface level of the container, and water in the tube rises only slightly.

Temperature Stabilization

Water has a high **specific heat;** that is, the amount of energy required to raise the temperature of water by one degree Celsius is quite large. The high specific heat of water results from the hydrogen bonding of its molecules. Raising the temperature of a substance involves adding heat energy to make its molecules move faster—to increase the kinetic energy of the molecules. Some of the hydrogen bonds holding the water molecules together must first be broken to permit the molecules to move more freely. Much of the energy added to the system is used up in breaking the hydrogen bonds, and only a portion of the heat energy is available to speed the movement of the water molecules (increase the temperature of the water). When liquid water changes to ice, a great deal of heat is liberated into the environment.

Because so much heat loss or heat input is required to lower or raise the temperature of water, the oceans and other large bodies of water have relatively constant temperatures. Thus, many organisms living in the oceans are provided with a relatively constant environmental temperature. The high water content of plants and animals living on land helps them maintain a relatively constant internal temperature. The rates of chemical reactions are greatly affected by temperature, generally doubling for each 10°C increase in temperature. The reactions of biological importance can take place only within a relatively narrow temperature range, and water helps minimize temperature fluctuations.

Because its molecules are held together by hydrogen bonds, water has a high **heat of vaporization.** More than 500 calories are required to change a gram of liquid water into a gram of water vapor. A **calorie** is a unit of heat energy (defined as 4.184 joules) that equals the amount of heat required to raise the temperature of one gram of water one degree Celsius. Since water absorbs heat as it changes from a liquid to a gas, the human body can dissipate excess heat by the evaporation of sweat, and a leaf can keep cool in the bright sunlight by evaporating water from its surface. Water's ability to conduct heat rapidly makes possible the even distribution of heat throughout the body. The properties of water are crucial in stabilizing temperatures on Earth. The quantity of water on the Earth's surface is enormous; this large mass resists both the warming effect of heat and the cooling effect of low temperatures.

Density of Water

Hydrogen bonds contribute another important property of water. Whereas most substances become more dense as the temperature decreases, water is most dense at 4°C and then begins to expand again (becoming less dense) as the temperature decreases further. Liquid water expands as it freezes because the hydrogen bonds joining the water molecules in the crystalline lattice keep the molecules far enough apart to give ice a density about 10% less than the density of water. As a result, frozen water, ice, floats upon the denser cold water (Figure 2–17). When ice has been heated enough to increase its temperature above 0°C, the hydrogen bonds between the water molecules are broken and the water molecules are free to slip closer together. The density of water is greatest at 4°C, above which water begins to expand again as the speed of its molecules increases.

This unusual property of water has been a most important factor in enabling life to appear, survive, and evolve on the Earth. If ice had a greater density than that of water, it would sink, and eventually all ponds, lakes, and even oceans would freeze solid from the bottom to the surface, making life impossible. When a body of deep water cools, it becomes covered with floating ice. The ice insulates the liquid water below it, prevent-

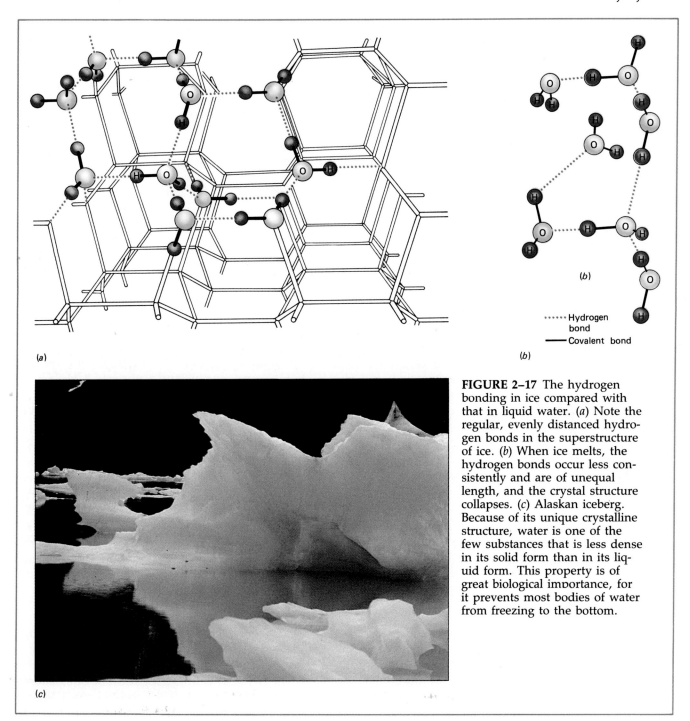

(a)

(b)

(b)

········· Hydrogen bond
————— Covalent bond

FIGURE 2–17 The hydrogen bonding in ice compared with that in liquid water. (*a*) Note the regular, evenly distanced hydrogen bonds in the superstructure of ice. (*b*) When ice melts, the hydrogen bonds occur less consistently and are of unequal length, and the crystal structure collapses. (*c*) Alaskan iceberg. Because of its unique crystalline structure, water is one of the few substances that is less dense in its solid form than in its liquid form. This property is of great biological importance, for it prevents most bodies of water from freezing to the bottom.

(c)

ing the water from freezing and permitting a variety of animals and plants to survive below the icy surface.

Ionization in Water

A further characteristic of water molecules is their slight tendency to **ionize**—that is, to dissociate into hydrogen ions (H^+) and hydroxide ions (OH^-). In pure water a very small number of water molecules form ions in this way. The tendency of water to dissociate is balanced by the tendency of hydrogen ions and hydroxide ions to reunite to form water:

$$HOH \rightleftharpoons H^+ + OH^-$$

Since water splits into one hydrogen ion and one hydroxide ion, the concentrations of hydrogen and hydroxide ions in pure water are exactly equal. Such a solution is said to be **neutral,** neither acidic nor basic (alkaline). The slight tendency of water molecules to form ions results in a concentration of hydrogen ions and of hydroxide ions of 0.0000001 (10^{-7}) moles per liter.

ACIDS AND BASES

An **acid** is a substance that dissociates in solution to yield hydrogen ions (H^+)[1] and an anion.

$$\text{Acid} \longrightarrow H^+ + \text{Anion}$$

An acid is a proton *donor.* (Remember that a hydrogen ion, or H^+, is nothing more than a proton.) A **base** is defined as a proton *acceptor.* Most bases are substances that dissociate to yield a hydroxide ion (OH^-) and a cation when dissolved in water. The part of an acid remaining after the dissociation of the H^+ is termed the **conjugate base.** The addition of a proton to a base yields its **conjugate acid.** Acids turn blue litmus paper red and have a sour taste. Hydrochloric acid (HCl) and sulfuric acid (H_2SO_4) are inorganic acids (relatively small compounds that generally do not contain carbon). Lactic acid ($CH_3CHOHCOOH$) from sour milk and acetic acid (CH_3COOH) from vinegar are two common organic acids (more complex compounds containing carbon atoms). Bases turn red litmus paper blue and feel slippery to the touch. Sodium hydroxide (NaOH) and ammonium hydroxide (NH_4OH) are inorganic bases. In later chapters we will encounter a number of organic bases such as the purine and pyrimidine bases that are components of nucleic acids.

Acids and bases dissociate when dissolved in water, releasing H^+ ions and OH^- ions, respectively. When the concentration of hydrogen ions in a solution is greater than 0.0000001 M (10^{-7} M), the solution is acidic. When the concentration of hydrogen ions is less than 10^{-7} M (and the concentration of hydroxide ions is greater than 10^{-7} M), the solution is basic, or alkaline.

pH

Since the concentration of hydrogen ions in biological fluids is usually low, it is much more convenient to express the degree of acidity or alkalinity in terms of **pH,** defined as the logarithm of the reciprocal of the hydrogen ion concentration, log [1/(H^+)]. The pH scale is thus a logarithmic one, extending from 0, the pH of a 1 M acid such as HCl, to 14, the pH of a 1 M base such as NaOH (Figure 2–18). The hydrogen ion concentration of pure water is 10^{-7} M. The logarithm of $1/10^{-7}$ is 7.0; hence, the pH of water is 7.0 (Table 2–2). At pH 7.0 the concentrations of H^+ ions and OH^- ions are exactly equal, 10^{-7} M.

---

[1]The H^+ immediately combines with water, forming a hydronium ion (H_3O^+). However, by convention H^+, rather than the more accurate H_3O^+, is used.

Table 2–2 THE RELATION OF pH TO HYDROGEN ION CONCENTRATION				
Substance	*[H+]*	*1/[H+]*	*log 1/[H+]*	*pH*
Pure water, neutral solution	0.0000001, 10^{-7}	10^7	7	7
Gastric juice	0.01, 10^{-2}	10^2	2	2
Household ammonia	0.0000000001, 10^{-11}	10^{11}	11	11

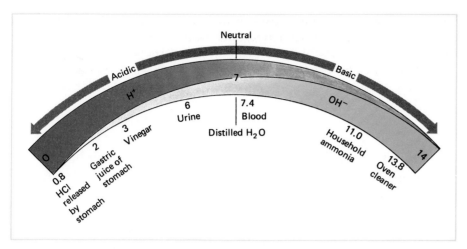

FIGURE 2–18 The pH scale. A solution with a pH of 7 is neutral because the concentrations of H^+ and OH^- are equal. The lower the pH below 7, the more H^+ ions are present, and the more acidic the solution is. As the pH increases above 7, the concentration of H^+ ions decreases and the concentration of OH^- increases, making the solution more alkaline (basic).

Since the scale is logarithmic, a solution with a pH of 6 has a hydrogen ion concentration 10 times greater than a solution with a pH of 7. A pH of 5 represents another tenfold increase in the concentration of hydrogen ions, so a solution with a pH of 5 is 10×10 or 100 times more acidic than a solution with a pH of 7. Solutions with a pH of less than 7 are acidic and contain more H^+ ions than OH^- ions. Solutions with a pH greater than 7 are alkaline, or basic, and contain more OH^- ions than H^+ ions. The contents of most animal and plant cells are neither strongly acidic nor alkaline but are an essentially neutral mixture of acidic and basic substances. Any considerable change in the pH of the cell is incompatible with life (Figure 2–19).

Buffers

Many homeostatic mechanisms operate to maintain appropriate pH values. For example, the pH of human blood is about 7.4 and must be maintained within very narrow limits. Should the blood become too acidic, coma and death may result; excessive alkalinity can result in overexcitability of the nervous system and convulsions.

A **buffer** is a substance or combination of substances that resists changes in pH when an acid or base is added. The buffer accepts or donates hydro-

FIGURE 2–19 The effects of acid rain. Sulfur oxides, emitted from fossil fuel plants and industry, and nitrogen oxides, mainly from automobile exhaust, are converted in the moist atmosphere into acids of, respectively, sulfur and nitrogen, such as sulfurous and nitrous acid. These acids are dispersed over wide areas by airflow patterns in the atmosphere. Whereas the pH of unpolluted rain averages 5.6, in some parts of the United States and Canada the pH of rain has been measured at 4.2 and even lower. Most fish species die at a pH of 4.5 to 5.0. Acid rain also affects vegetation. The roots of this spruce tree have withered and died. Even before that, the rest of the plant had for some time been suffering from nutrient deficiency and reduced efficiency of photosynthesis.

FIGURE 2–20 Buffering is used clinically as a remedy for excess stomach acid. The bubbles are CO_2 from the reaction between an acid (citric acid) and the bicarbonate ion (HCO_3) from sodium bicarbonate.

gen ions. A buffer consists of a weak acid and its conjugate base, or a weak base and its conjugate acid. One of the most common buffering systems, and one that is important in human blood, is carbonic acid and the bicarbonate ion. Bicarbonate ions are formed in the body as follows:

$$CO_2 + H_2O \rightleftharpoons H_2CO_3 \rightleftharpoons H^+ + HCO_3^-$$

Carbon dioxide   Water   Carbonic acid   Bicarbonate ion

As indicated by the arrows, the reactions are reversible.

When excess hydrogen ions are present in blood or other body fluids, bicarbonate ions combine with them to form carbonic acid, a weak acid.

$$H^+ + HCO_3^- \rightleftharpoons H_2CO_3$$

Carbonic acid

The carbonic acid is unstable and quickly breaks down into carbon dioxide and water.

Buffers also maintain a relatively constant pH when hydroxide ions are added. A buffer may release hydrogen ions, which combine with the hydroxide ions to form water.

$$OH^- + H_2CO_3 \longrightarrow HCO_3^- + H_2O$$

SALTS

When an acid and a base are mixed together, the H^+ of the acid unites with the OH^- of the base to form a molecule of water. The remainder of the acid (anion) combines with the remainder of the base (cation) to form a salt. Hydrochloric acid reacts with sodium hydroxide to form water and sodium chloride:

$$HCl + NaOH \longrightarrow H_2O + NaCl$$

A **salt** may be defined as a compound in which the hydrogen atom of an acid is replaced by some other cation. A salt contains a cation other than H^+ and an anion other than OH^-. Sodium chloride, NaCl, is a compound in which the hydrogen ion of HCl has been replaced by the cation Na^+.

When a salt, an acid, or a base is dissolved in water, its dissociated charged particles can conduct an electrical current; these substances are called **electrolytes.** Sugars, alcohols, and many other substances do not form ions when dissolved in water; they do not conduct an electrical current and are termed **nonelectrolytes.**

Cells and extracellular fluids (such as blood) of plants and animals contain a variety of dissolved salts, which include many important mineral ions. Such ions are essential for fluid balance, acid-base balance, and, in animals, nerve and muscle function, blood clotting, bone formation, and many other aspects of body function. Sodium, potassium, calcium, and magnesium are the chief cations present, and chloride, bicarbonate (HCO_3^-), phosphate (PO_4^{3-}), and sulfate (SO_4^{2-}) are important anions (Table 2–3).

The body fluids of terrestrial animals differ considerably from sea water in their total salt content. However, they resemble sea water in the kinds of salts present and in their relative abundance. The total concentration of salts in the body fluids of most invertebrate marine animals is equivalent to that in sea water, about 3.4%. Vertebrates, whether terrestrial, freshwater, or marine, have less than 1% salt in their body fluids.

Most biologists believe that life originated in the sea. The cells of those early organisms became adapted to function optimally in the presence of this pattern of salts. As larger animals evolved and developed body fluids, the pattern of salts was retained, even when some of their descendants

Table 2–3 SOME BIOLOGICALLY IMPORTANT IONS		
Name	*Formula*	*Charge*
Sodium	Na^+	1+
Potassium	K^+	1+
Hydrogen	H^+	1+
Magnesium	Mg^{2+}	2+
Calcium	Ca^{2+}	2+
Iron	Fe^{2+} or Fe^{3+}	2+ [iron(II)] or 3+ [iron(III)]
Ammonium	NH_4^+	1+
Chloride	Cl^-	1–
Iodide	I^-	1–
Carbonate	CO_3^{2-}	2–
Bicarbonate	HCO_3^-	1–
Phosphate	PO_4^{3-}	3–
Acetate	CH_3COO^-	1–
Sulfate	SO_4^{2-}	2–
Hydroxide	OH^-	1–
Nitrate	NO_3^-	1–
Nitrite	NO_2^-	1–

migrated into fresh water or onto land. Some animals have evolved kidneys and other organs, such as salt glands, that selectively retain or secrete certain ions, thus resulting in body fluids with somewhat different relative concentrations of salts. The concentration of each ion is determined by the relative rates of its uptake and excretion by the organism.

Although the concentration of salts in cells and body fluids of plants and animals is small, this amount is of great importance for normal cell function. The concentrations of the respective cations and anions are kept remarkably constant under normal conditions. Any marked change results in impaired cellular functions and ultimately in death.

■ SUMMARY

I. The chemical composition and metabolic processes of all living things are very similar; the physical and chemical principles that govern nonliving things also govern living systems.

II. An element is a substance that cannot be decomposed into simpler substances by chemical reactions.
   A. The matter of the universe is composed of 92 elements, ranging from hydrogen, the lightest, to uranium, the heaviest.
   B. Six elements—carbon, hydrogen, oxygen, nitrogen, phosphorus, and calcium—make up about 98% of an organism's content by weight.

III. Atoms are composed of a nucleus containing protons and neutrons and a cloud of electrons around the nucleus in characteristic energy levels and orbitals.
   A. Atoms of the same element that contain the same number of protons but different numbers of neutrons, and therefore have different mass numbers, are called *isotopes*.
   B. In a neutral atom, the number of protons equals the number of electrons, so the atom has no net electrical charge.

IV. Atoms are joined by chemical bonds to form larger, more complex structures called *compounds*.
   A. Covalent bonds are strong, stable bonds formed when atoms share electrons, forming molecules.

   1. Covalent bonds are nonpolar if the electrons are shared equally between the two atoms.
   2. Covalent bonds are polar if one atom has a greater affinity for electrons than the other.
   B. An ionic bond is formed when one atom donates electrons to another. An ionic compound is made up of positively charged ions (cations) and negatively charged ions (anions).
   C. Hydrogen bonds are relatively weak bonds formed when a hydrogen atom in one molecule is attracted to a highly electronegative element such as oxygen or nitrogen in another molecule or in another part of the same molecule.

V. The molecular mass of a compound is the sum of the atomic masses of its constituent atoms.

VI. Oxidation is a chemical process in which a substance loses electrons; reduction is a chemical process in which a substance gains electrons.

VII. Water accounts for a large part of the mass of most organisms.
   A. Water molecules are cohesive owing to the hydrogen bonding between the molecules; they also adhere to many kinds of substances.
   B. Water has a high degree of surface tension because of the cohesiveness of its molecules.
   C. Water has a high specific heat, which helps organ-

isms maintain a relatively constant internal temperature; this property also helps keep the oceans and other large bodies of water at a constant temperature.

D. Other important properties of water include its high heat of vaporization, its unusual density (ice is less dense than liquid water), its slight tendency to form ions, and its ability to dissolve many different kinds of compounds.

VIII. An acid is a substance that dissociates in solution to yield hydrogen ions and an anion. Acids are proton donors. Bases are proton acceptors. A base generally dissociates in solution to yield hydroxide ions.

A. The pH scale extends from 0 to 14, with 7 indicating neutrality. As the pH decreases below 7, the solution is more acidic. As a solution becomes more basic (alkaline), its pH increases from 7 toward 14.

B. An acid and its conjugate base or a base and its conjugate acid can act as a buffer to resist changes in the pH of a solution when acids or bases are added.

IX. A salt is a compound in which the hydrogen atom of an acid is replaced by some other cation. Salts provide the many mineral ions that are essential for fluid balance, nerve and muscle function, and many other body functions.

POST-TEST

1. The six elements that make up some 98% of the mass of most organisms are _____, _____, _____, _____, _____, and _____.

2. The chemical symbol for carbon is _____; for hydrogen, _____; and for oxygen, _____.

3. Elements such as cobalt, present in minute amounts in living things, are referred to as _____.

4. The three major types of subatomic particles are _____, _____, and _____.

5. Particles with a negative electric charge and an extremely small mass are termed _____.

6. The number of protons in the nucleus, called the _____ _____, is written as a subscript to the left of the chemical symbol.

7. The sum of the protons and the neutrons in the nucleus of the atom, termed the _____ _____, is indicated by a superscript to the left of the chemical symbol.

8. Atoms of the same element containing the same number of protons but different numbers of neutrons are termed _____.

9. Electrons move about the central nucleus of the atom in characteristic regions termed _____.

10. Each orbital may contain at most _____ electrons.

11. The tendency of the negatively charged electrons to fly off into space is countered by their attraction to the atomic nucleus due to the _____ charge of the protons in the nucleus.

12. The atoms of a few elements, such as _____ and _____, have a complete outermost shell of electrons; these are called _____.

13. The attraction holding two atoms together is called a _____ _____.

14. Electrically charged atoms are called _____.

15. Positively charged atoms are termed _____, and negatively charged atoms are termed _____.

Match the terms in Column A with their definitions in Column B.

Column A
16. Covalent bond
17. Hydrogen bond
18. Ionic bond
19. Molecular mass
20. Molecule
21. Products
22. Reactants
23. Valence electrons

Column B
a. Electrons in the outer orbit that determine how many electrons an atom can donate, receive, or share
b. The combination of two or more atoms joined by covalent chemical bonds
c. Substances participating in a reaction
d. Substances produced in a chemical reaction
e. Transfer of an electron from an electron donor to an acceptor and the binding together of two particles of opposite charge
f. Atoms joined by the sharing of electrons between them
g. Weak bond that holds water molecules together
h. Sum of the atomic masses of the constituent atoms

24. Water tends to rise in very-fine-bore tubes, a phenomenon termed _____ _____.

25. The amount of energy required to change one gram of liquid water to one gram of water vapor is termed the _____ _____ of water.

26. The logarithm of the reciprocal of the hydrogen ion concentration is termed _____.

27. An acid is a proton _____; a base is a proton _____.

28. A solution of weak acid and its conjugate base is called a _____.

■ REVIEW QUESTIONS

1. Distinguish between:
   a. an atom and an element
   b. a molecule and a compound
   c. an atom and an ion
2. How do isotopes of the same element differ? What is a radioisotope?
3. Contrast electrons with protons and with neutrons.
4. Compare ionic and covalent bonds, and give specific examples of each.
5. Write a chemical equation depicting the hydration of:
   a. sodium chloride
   b. calcium chloride
6. What properties of water make it an essential component of living matter?
7. How would a solution with a pH of 5 differ from one with a pH of 9? from one with a pH of 7?
8. Why are buffers important in living organisms? Give a specific example of how a buffer system works.
9. Differentiate clearly among acids, bases, and salts. What are the functions of salts in living organisms?
10. Why must oxidation and reduction occur simultaneously?
11. Describe a reversible reaction that is at equilibrium.
12. What are valence electrons? What is their significance?
13. What are hydrogen bonds? What is their significance?
14. How is each of the following determined?
    a. atomic number
    b. molecular mass

■ RECOMMENDED READINGS

Alberts, B., Bray, D., Lewis, J., Raff, M., Roberts, K., and Watson, J.D.: *Molecular Biology of the Cell.* New York, Garland Publishing Company, 1983. A complete, well-written, and easy-to-read reference text.

Baker, J.W., and Allen, G.E.: *Matter, Energy and Life: An Introduction to Chemical Concepts,* 4th ed. Reading, MA, Addison-Wesley, 1981. A presentation of the principles of thermodynamics and their application to studies of living systems.

Baum, S.J., and Scaife, C.W.: *Chemistry: A Life Science Approach,* 3rd ed. New York, Macmillan, 1987. A chemistry text emphasizing subjects of special interest to students of biology.

Bettelheim, F.A., and March, J.: *Introduction to General, Organic and Biochemistry,* 2nd ed. Philadelphia, Saunders College Publishing, 1988. A very readable reference text for those who would like to know more about the chemistry basic to life.

Darnell, J., Lodish, H., and Baltimore, D.: *Molecular Cell Biology.* New York, Scientific American Books, 1986. An up-to-date presentation of the molecular aspects of cell biology.

Frieden, E.: The chemical elements of life. *Scientific American,* January 1972, pp. 52–64. An introduction to biologically important elements with emphasis on the trace elements.

Peterson, I.: A material loss. *Science News,* September 7, 1985. A discussion of the effects of acid rain on materials.

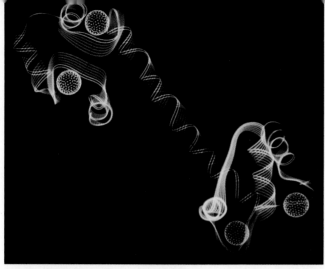

Computer representation of the protein calmodulin

3

The Chemistry of Life: Organic Compounds

■ OUTLINE

I. The versatile carbon atom
II. Isomers
III. Functional groups
IV. Polymers
V. Carbohydrates
   A. Monosaccharides
   B. Disaccharides
   C. Polysaccharides
   D. Modified and complex carbohydrates
VI. Lipids
   A. Neutral fats
   B. Phospholipids
   C. Carotenoids
   D. Steroids
VII. Proteins
   A. Amino acids
   B. Polypeptide chains are built from amino acids
   C. Protein structure—levels of organization
      1. Primary structure
      2. Secondary structure
      3. Tertiary structure
      4. Quaternary structure
   D. Protein structure determines function
VIII. Nucleic acids
   A. The nucleotide subunits of nucleic acids
   B. Other important nucleotides

■ LEARNING OBJECTIVES

After you have read this chapter you should be able to:

1. Describe the properties of carbon that make it the central component of organic compounds.
2. Distinguish among three principal types of isomers.
3. Describe the major functional groups present in organic compounds.
4. Compare the major groups of organic compounds—carbohydrates, fats, proteins, and nucleic acids—with respect to their chemical composition and function.
5. Distinguish among monosaccharides, disaccharides, and polysaccharides, and discuss the monosaccharides and polysaccharides of major importance in living things.
6. Distinguish among neutral fats, phospholipids, and steroids, and describe the composition characteristics and biological functions of each group.
7. Describe the functions and chemical structure of proteins.
8. Outline the levels of organization of protein molecules.
9. Describe the chemical structure of nucleotides and nucleic acids, and discuss the importance of these compounds in living organisms.

Most of the chemical compounds present in living organisms contain skeletons of covalently bonded carbon. These molecules are known as **organic compounds,** because at one time they were thought to be produced only by living things. Inorganic compounds are relatively small, simple compounds such as water, simple acids, bases, and salts. A few very simple compounds, including carbon dioxide and compounds containing carbonate (CO_3^{2-}), are classified as inorganic compounds even though they contain carbon.

Organic compounds are the main structural components of cells and tissues. They are the participants in and regulators of thousands of metabolic reactions, and provide energy for life processes. In this chapter, we focus on some of the major groups of organic compounds that are important in living organisms, including carbohydrates, lipids, proteins, and nucleic acids (DNA and RNA).

(a)

(b)

FIGURE 3–1 Carbon is the basis of organic compounds, of which all living things are made. (*a*) Elemental forms of carbon. An artificial diamond, a pure form of carbon, is seen at the bottom. Graphite, another form of carbon, is seen at the top. Graphite is a component of pencil lead. (*b*) Carbon accounts for more than half the dry weight of an organism. As with all living organisms, the chemistry of this porcelain crab (and that of the sea anemone on which it rests) is organized around the carbon atom.

THE VERSATILE CARBON ATOM

The chemistry of living things is organized around the element carbon (Figure 3–1). Perhaps because it can form a greater variety of molecules than any other element, carbon has emerged as the central component of organic compounds. Carbon atoms form the backbone, or principal axis, of a vast number of compounds.

Carbon's unusual properties permit formation of the large, complex molecules essential to life. A carbon atom has a total of six electrons, with two in its first energy level and four in its second energy level. With four electrons in its outer energy level, a carbon atom can form four covalent bonds with other atoms, including other carbon atoms. Carbon atoms can form very stable single covalent bonds with one another. Long chains of carbon atoms can form in this way:

—C—C—C—C—C—

Two carbon atoms can share two electron pairs with each other, forming double bonds (—C=C—). In some compounds, triple carbon-to-carbon bonds (—C≡C—) are formed. Carbon chains can be unbranched or branched. Carbon atoms can also be joined into rings (Figure 3–2). In some compounds, rings and chains are joined.

The carbon atom can form bonds with a greater number of different elements than any other type of atom. Hydrogen, oxygen, and nitrogen are atoms frequently bonded to carbon. Organic compounds consisting of only carbon and hydrogen are hydrocarbons. Although these compounds are not common in living organisms, fossil fuels are hydrocarbons; these fuels formed from organic compounds originating in organisms that lived and died millions of years ago.

The shape of a molecule is important in determining its biological properties and function. Carbon-containing molecules have a three-dimensional structure due to the tetrahedral nature of their bond angles. When a carbon atom forms four covalent single bonds with other atoms, the electron orbitals in its outer energy level become elongated and project from the carbon atom toward the corners of a tetrahedron (Figure 3–3). In this case the angle between any two of the bonds is about 109.5 degrees. This bond angle is similar in diverse organic compounds.

Generally, there is freedom of rotation around each carbon-to-carbon single bond. This property permits organic molecules to be flexible and to assume a variety of shapes, depending on the extent to which each single bond is rotated. Double and triple bonds do not permit rotation, so regions of a molecule with such bonds tend not to be flexible.

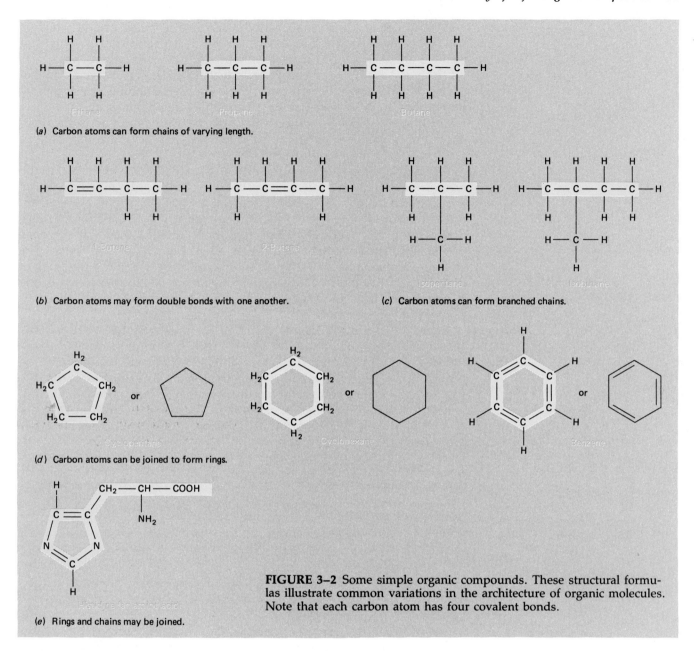

(a) Carbon atoms can form chains of varying length.

(b) Carbon atoms may form double bonds with one another.

(c) Carbon atoms can form branched chains.

(d) Carbon atoms can be joined to form rings.

(e) Rings and chains may be joined.

FIGURE 3–2 Some simple organic compounds. These structural formulas illustrate common variations in the architecture of organic molecules. Note that each carbon atom has four covalent bonds.

ISOMERS

Compounds that have the same molecular formula but different structures and thus different properties are called **isomers**. Isomers do not have identical physical or chemical properties and may have different common

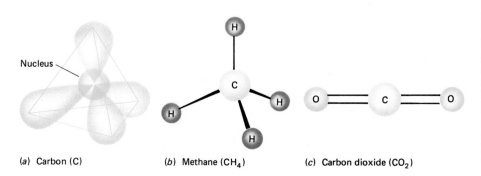

(a) Carbon (C) (b) Methane (CH_4) (c) Carbon dioxide (CO_2)

FIGURE 3–3 Carbon bonds. A carbon atom can form four covalent bonds. (a) The bonds of a carbon atom point to the four corners of a tetrahedron. This arrangement maximizes the distance between the atoms bonded to the carbon atom. (b) Methane consists of a single carbon atom bonded to four hydrogen atoms. The hydrogens are bonded symmetrically around the carbon at the points of a tetrahedron. (c) In carbon dioxide, each oxygen atom is connected to the carbon atom by a double bond. The bonds are parallel, and the molecule assumes a linear configuration.

names. Cells can distinguish between the two isomers, one of which is usually biologically active whereas the other is usually not. Three types of isomers are structural isomers, geometric isomers, and enantiomers.

Structural isomers are compounds that differ in the covalent arrangements of their atoms. For example, there are two structural isomers of the four-carbon hydrocarbon butane, one with a straight chain and the other with a branched chain (Figure 3–4a). The larger the compound is, the more structural isomers are possible. There are only 2 structural isomers of butane, but there may be up to 366,319 isomers of $C_{20}H_{42}$.

Geometric isomers are compounds that are identical with regard to the arrangement of their covalent bonds but differ in the order in which groups are arranged in space. Geometric isomers, also called *cis-trans* isomers, are present in some compounds with carbon-to-carbon double bonds. Because double bonds are not flexible like single bonds, atoms joined to the carbons of a double bond cannot rotate freely about the axis of the bonds. The *cis-trans* isomers may be drawn as shown in Figure 3–4b. The designation *cis* indicates that the two larger components are on the same side of the double bond. If they are on opposite sides of the double bond, the compound is a *trans* isomer.

Enantiomers are molecules that are mirror images of one another. Recall that the four groups bonded to a single carbon atom are arranged at the vertices of a tetrahedron. If the four bonded groups are all different, the central carbon is described as asymmetric. Figure 3–4c illustrates that the four groups can be arranged about the asymmetric carbon in two different ways that are mirror images of each other. The two molecules are enantiomers if they cannot be superimposed on one another no matter how they are rotated in space.

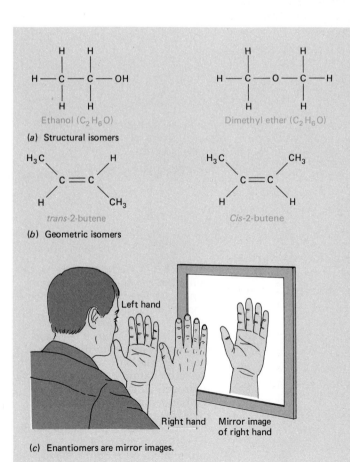

Ethanol (C_2H_6O)   Dimethyl ether (C_2H_6O)

(a) Structural isomers

trans-2-butene   *Cis*-2-butene

(b) Geometric isomers

Left hand   Right hand   Mirror image of right hand

(c) Enantiomers are mirror images.

FIGURE 3–4 Isomers. Different compounds can have the same molecular formula, but with the atoms arranged differently. (a) Structural isomers differ in the covalent arrangement of their atoms. (b) Geometric, or cis-trans, isomers have identical covalent bonds but differ in the order in which groups of atoms are arranged in space. (c) Enantiomers are molecules that are mirror images of one another.

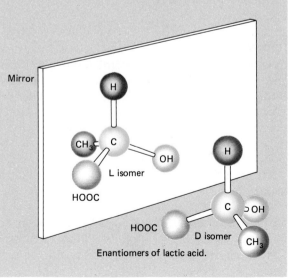

Mirror

L isomer

D isomer

Enantiomers of lactic acid.

Enantiomers are currently designated D or L based on the absolute configuration of the groups bonded to the tetrahedral carbon atom. The three-carbon compound glyceraldehyde is used as the standard for the description of all enantiomers. The D isomer of any compound is the one that has the last asymmetric carbon in the same orientation as D-glyceraldehyde. Compounds related to L-glyceraldehyde are denoted L-isomers.

When chemists synthesize organic compounds in their laboratories, a mixture that contains equal amounts of D and L-isomers is produced. In cells, only one of the two enantiomers of a compound is produced. For example, most sugars important for cells are D-sugars. Although enantiomers have similar chemical properties and identical physical properties (except for the direction in which they rotate plane-polarized light), cells can distinguish between the two isomers, and only one form is biologically active.

FUNCTIONAL GROUPS

The hydrocarbon backbone of an organic compound does not interact readily with other compounds. However, one or more of the hydrogen atoms bonded to the carbon skeleton of a hydrocarbon can be replaced by other groups of atoms. These groups of atoms, referred to as **functional groups,** readily form associations such as ionic and hydrogen bonds with other molecules. In this way functional groups help determine the types of chemical reactions in which the compound participates.

Each class of organic compounds is characterized by the presence of one or more specific functional groups. For example, as illustrated in Table 3–1, alcohols contain functional groups known as hydroxyl groups. Note that the symbol R is used to represent the remainder of the molecule of which the functional group is a part.

An important property of the functional groups found in biological molecules is their solubility in water. Positively and negatively charged functional groups are water-soluble because they associate strongly with the polar water molecule.

Bonds between carbon and hydrogen are nonpolar, so a functional group containing only a carbon-hydrogen bond such as a methyl group ($-CH_3$) is also nonpolar. Oxygen-hydrogen and nitrogen-hydrogen bonds are polar; they have a partial positive electrical charge at the hydrogen end of the bond and a partial negative electrical charge at the oxygen or nitrogen end. Thus, hydroxyl and amino groups are polar. Double bonds formed between carbon and oxygen ($C=O$) are also polar; there is a partial positive charge at the carbon end and a partial negative charge at the oxygen end. Consequently, carboxyl and aldehyde groups are polar. Functional groups that are polar interact with charged ions or with other polar groups. Compounds containing polar functional groups tend to dissolve in water because the polar groups attract water molecules.

Most compounds present in cells contain two or more different functional groups. For example, every amino acid (amino acids are molecular subunits of proteins) contains at least two functional groups—an amino group and a carboxyl group. The chemical properties of these functional groups determine the general properties of amino acids. However, many amino acids contain additional functional groups that determine the specific properties of each type of amino acid. When we know what kinds of functional groups are present in an organic compound, we can predict its chemical behavior.

Table 3–1 **SOME BIOLOGICALLY IMPORTANT FUNCTIONAL GROUPS**

Functional Group	Structural Formula	Class of Compounds Characterized By Group	Example	Description
Hydroxyl	R—OH	Alcohols	Ethanol (the alcohol contained in beverage)	Polar because electronegative oxygen attracts covalent electrons
Amino	R—NH$_2$	Amines	Amino acid	Ionic; amino group acts as base
Carboxyl	R—C(=O)—OH	Carboxylic acids (organic acids)	Amino acid	Ionic; the H can dissociate as an H$^+$ ion
Ester	R—C(=O)—O—R	Esters	Methyl acetate	Related to carboxyl group, but has hydrocarbon group in place of the OH hydrogen; polar
Carbonyl	R—C(=O)—H	Aldehydes	Formaldehyde	Carbonyl carbon bonded to at least one H atom; polar
Carbonyl	R—C(=O)—R	Ketones	Acetone	Carbonyl group bonded to two other carbons; polar
Methyl	R—CH$_3$	Component of many organic compounds	Methanol (wood alcohol)	Nonpolar
Phosphate	R—O—P(=O)(OH)—OH	Organic phosphates	Phosphate ester (as found in ATP)	Dissociated form of phosphoric acid; the phosphate ion is covalently bonded by one of its oxygen atoms to one of the carbons; ionic
Sulfhydryl	R—SH	Thiols	Cysteine	Help stabilize internal structure of proteins

POLYMERS

Many biologically important molecules such as proteins and nucleic acids are very large, consisting of thousands of atoms. Such giant molecules are known as **macromolecules,** or **polymers** (Figure 3–5). Cells produce polymers by linking together small organic compounds called **monomers.** Just as all the words in this book have been written by arrangement of the 26 letters of the alphabet in various combinations, monomers can be grouped together to form an almost infinite variety of larger molecules. The thousands of different complex organic compounds present in living things are constructed from about 40 small, simple monomers. These small compounds are linked together to form long chains of similar subunits. For example, the 20 common types of amino acid monomers can be linked together end to end in countless ways to form the polymers we know as proteins.

The synthetic process by which monomers are covalently linked is called **condensation.** Because the *equivalent* of a molecule of water is removed during the reactions that combine monomers, the term dehydration synthesis is sometimes used to describe the process. However, in biological systems, synthesis of a polymer is not simply the reverse of breakdown (which involves adding water). Synthetic processes require energy and are regulated by different enzymes (proteins that regulate chemical reactions).

Each organism is unique due to differences in monomer sequence within its DNA, the polymer that constitutes the genes. Cells and tissues within the same organism are also different due to variations in their component polymers. Muscle tissue is different from brain tissue in large part because of differences in the types and sequences of amino acids in their proteins. Ultimately this protein structure is dictated by the sequence of monomers within the DNA of the organism.

Polymers can be degraded to their component monomers by **hydrolysis** (which means "to break with water"). Bonds between monomers are broken by the addition of water. A hydrogen from the water molecule attaches to one monomer and the hydroxyl from the water attaches to the adjacent monomer. Specific examples of dehydration and hydrolysis reactions will be presented as we discuss the groups of organic compounds in more detail. The principal groups of biologically important organic compounds are summarized in Table 3–2.

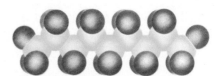

FIGURE 3–5 Monomers may be linked together to form polymers. A portion of a polyethylene polymer is shown here. The two-carbon compound ethylene (C_2H_4) is the monomer linked together to form this polymer.

CARBOHYDRATES

Sugars, starches, and celluloses are typical **carbohydrates. Sugars** and **starches** serve as fuels for cells; **celluloses** are structural components of plants. Carbohydrates contain carbon, hydrogen, and oxygen atoms in a ratio of approximately one carbon to two hydrogens to one oxygen ($(CH_2O)_n$). The term *carbohydrate*, meaning "hydrate (water) of carbon," stems from the 2:1 ratio of hydrogen to oxygen, the same ratio found in water (H_2O). Carbohydrates contain one sugar (monosaccharides), two sugar units (disaccharides), or many sugar units (polysaccharides).

Monosaccharides

Monosaccharides are simple sugars that typically contain from three to seven carbon atoms. The simplest carbohydrates are the two three-carbon sugars (trioses), glyceraldehyde and dihydroxyacetone (Figure 3–6, page 60). Ribose and deoxyribose are common pentoses, sugars that contain five carbons; they are components of nucleic acids (DNA, RNA, and related compounds). Glucose, fructose, galactose, and other sugars that consist of six carbons are called **hexoses.**

Glucose ($C_6H_{12}O_6$), the most abundant monosaccharide, is extremely important in life processes. During photosynthesis, algae and plants pro-

Table 3–2 **SOME OF THE GROUPS OF BIOLOGICALLY IMPORTANT ORGANIC COMPOUNDS**

Class of Compound	Component Elements	Description	How to Recognize	Principal Function in Living Systems
Carbohydrates	C, H, O	Contain approximately 1 C:2 H:1 O (but make allowance for loss of oxygen and hydrogen when sugar units are linked)	Count the carbons, hydrogens, and oxygens.	Cellular fuel; energy storage; structural component of plant cell walls; component of other compounds such as nucleic acids and glycoproteins
		1. Monosaccharides (simple sugars)—mainly five-carbon (pentose) molecules like ribose or six-carbon (hexose) molecules such as glucose and fructose	Look for the ring shapes: hexose or pentose	Cellular fuel; components of other compounds
		2. Disaccharides—two sugar units linked by a glycosidic bond, e.g., maltose, sucrose	Count sugar units.	Components of other compounds
		3. Polysaccharides—many sugar units linked by glycosidic bonds, e.g., glycogen, cellulose	Count sugar units.	Energy storage; structural components of plant cell walls
Lipids	C, H, O	Contain less oxygen relative to carbon and hydrogen than do carbohydrates.		Energy storage; cellular fuel, structural components of cells; thermal insulation
		1. Neutral fats. Combination of glycerol with one to three fatty acids. Monacylglycerol contains one fatty acid; diacylglycerol contains two fatty acids; triacylglycerol contains three fatty acids. If fatty acids contain double carbon-to-carbon linkages (C=C), they are unsaturated; otherwise they are saturated.	Look for glycerol at one end of molecule	Cellular fuel; energy storage.

(continued)

duce glucose from carbon dioxide and water using sunlight as an energy source. Then, during cellular respiration cells break the bonds of the glucose molecule, releasing the stored energy so that it can be used for cellular work. Glucose is also used as a component in the synthesis of other types of compounds such as amino acids and fatty acids. So central is glucose in metabolism that its concentration is carefully kept at a homeostatic (relatively constant) level in the blood of humans and other complex animals.

In the glucose molecule a hydroxyl group is bonded to each carbon except one; that carbon is double-bonded to an oxygen atom, forming a carbonyl group. In glucose the carbonyl group is at the end of the chain, so glucose is an aldehyde; if the carbonyl group is at any other position, the monosaccharide is a ketone. (By convention, the carbon skeleton of a sugar is numbered beginning with the carbon at or nearest to the carbonyl end of the open chain).

Glucose and fructose have identical molecular formulas, but their atoms are arranged differently. In fructose, a ketone, the double-bonded oxygen is linked to a carbon within the chain rather than to a terminal carbon as in glucose (which is an aldehyde). Recall that such compounds are known as *structural isomers*. Because of differences in the arrangement of their atoms, the two sugars have different chemical properties. One difference is that fructose tastes sweeter than glucose.

Table 3–2 **SOME OF THE GROUPS OF BIOLOGICALLY IMPORTANT ORGANIC COMPOUNDS**

Class of Compound	Component Elements	Description	How to Recognize	Principal Function in Living Systems
		2. Phospholipids. Composed of glycerol attached to one or two fatty acids and to an organic base containing phosphorus.	Look for glycerol and side chain containing phosphorus and nitrogen.	Components of cell membranes
		3. Steroids. Complex molecules containing carbon atoms arranged in four interlocking rings (three rings contain six carbon atoms each and the fourth ring contains five)	Look for four interlocking rings:	Some are hormones; others include cholesterol, bile salts, vitamin D.
		4. Carotenoids. Red and yellow pigments; consist of isoprene units	Look for isoprene units.	Retinal (important in photoreception) and vitamin A are formed from carotenoids.
Proteins	C, H, O, N, usually S	One or more polypeptides (chains of amino acids) coiled or folded in characteristic shapes	Look for amino acid units joined by C—N bonds.	Serve as enzymes; structural components; muscle proteins; hemoglobin
Nucleic acids	C, H, O, N, P	Backbone composed of alternating pentose and phosphate groups, from which nitrogenous bases project. DNA contains the sugar deoxyribose and the bases guanine, cytosine, adenine, and thymine. RNA contains the sugar ribose, and the bases guanine, cytosine, adenine, and uracil. Each molecular subunit, called a *nucleotide*, consists of a pentose, a phosphate, and a nitrogenous base.	Look for a pentose-phosphate backbone. DNA forms a double helix.	Storage, transmission, and expression of genetic information

Glucose and galactose differ from one another in another way. Both are hexoses and both are aldehydes. However, they differ in the arrangement of their atoms around carbon atom 4. They are mirror images.

The "stick" formulas in Figure 3–6 give a clear but somewhat unrealistic picture of the structures of some common monosaccharides. As has been discussed, molecules are not the simple two-dimensional structures depicted on a printed page. In fact, the properties of each compound depend in part on its three-dimensional structure, and three-dimensional formulas are helpful in understanding the relationship between molecular structure and biological function. Molecules of glucose and other monosaccharides in solution are not extended straight carbon chains as shown in Figure 3–7, but rather boat-shaped or chair-shaped rings formed when a covalent bond connects carbon 1 to the oxygen attached to carbon 5 or carbon 4. Glucose in solution typically exists as a ring of five carbons and one oxygen. When glucose forms a ring, two isomeric forms are possible, differing only in the orientation of an —OH group. When the hydroxyl group attached to carbon 1 is below the plane of the ring, the glucose is designated *α-glucose*; when this hydroxyl group is above the plane of the ring, the compound is designated *β-glucose*.

Disaccharides

A **disaccharide** (two sugars) consists of two monosaccharides covalently bonded to one another. The two monosaccharide units are joined by a

FIGURE 3–6 Structural formulas of some important monosaccharides (simple sugars). The monosaccharides are represented here as straight chains, called *stick formulas*. Although it is convenient to show monosaccharides in this form, they are more accurately depicted as ring structures (see Figure 3–7). Note that glucose, fructose, and galactose are structural isomers—they have the same chemical formula, $C_6H_{12}O_6$, but their atoms are arranged differently.

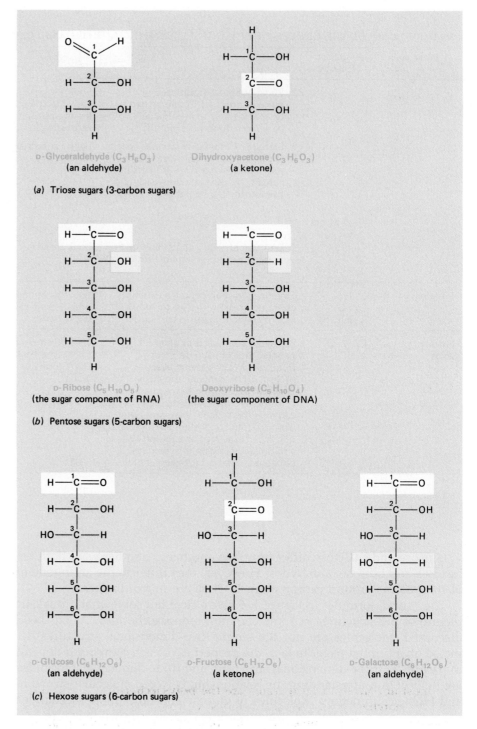

D-Glyceraldehyde ($C_3H_6O_3$)
(an aldehyde)

Dihydroxyacetone ($C_3H_6O_3$)
(a ketone)

(a) Triose sugars (3-carbon sugars)

D-Ribose ($C_5H_{10}O_5$)
(the sugar component of RNA)

Deoxyribose ($C_5H_{10}O_4$)
(the sugar component of DNA)

(b) Pentose sugars (5-carbon sugars)

D-Glucose ($C_6H_{12}O_6$)
(an aldehyde)

D-Fructose ($C_6H_{12}O_6$)
(a ketone)

D-Galactose ($C_6H_{12}O_6$)
(an aldehyde)

(c) Hexose sugars (6-carbon sugars)

glycosidic linkage, which generally forms between carbon 1 of one molecule and carbon 4 of the other molecule. The disaccharide maltose (malt sugar) consists of two covalently linked glucose units. Sucrose, the sugar we use to sweeten our foods, consists of a glucose unit combined with a fructose unit. Lactose (the sugar present in milk) is composed of one molecule of glucose and one of galactose.

A disaccharide can be hydrolyzed, that is, split by the addition of water into two monosaccharide units. During digestion maltose is hydrolyzed to form two molecules of glucose

Maltose + Water $\longrightarrow$ Glucose + Glucose

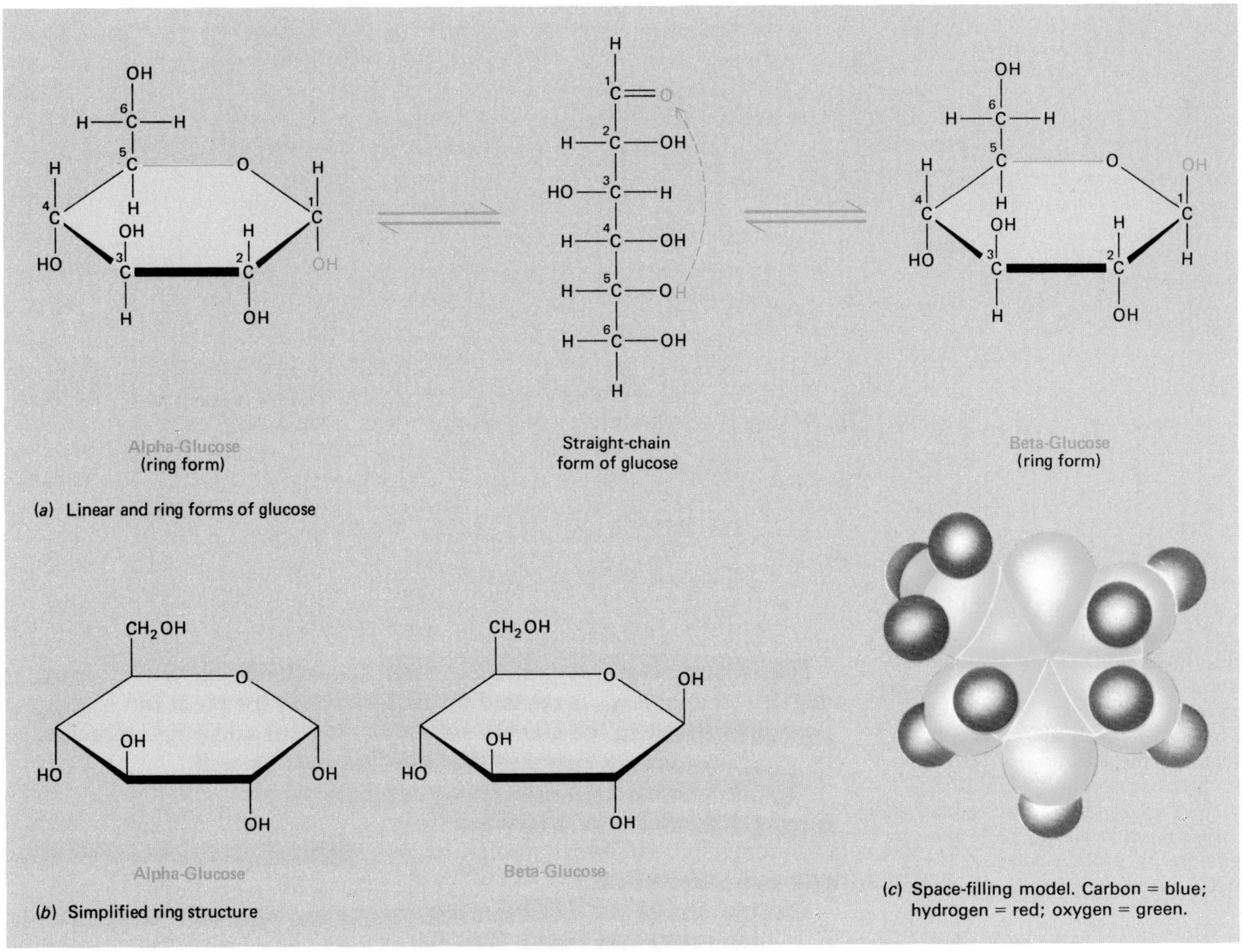

(a) Linear and ring forms of glucose

Alpha-Glucose
(ring form)

Straight-chain
form of glucose

Beta-Glucose
(ring form)

(b) Simplified ring structure

Alpha-Glucose

Beta-Glucose

(c) Space-filling model. Carbon = blue; hydrogen = red; oxygen = green.

FIGURE 3–7 Ring forms of glucose. (a) When the straight-chain form of glucose (center) dissolves in water, the molecule bends so that the —OH group on carbon 5 comes close to the =O on carbon 1. The hydrogen moves from one oxygen to the other. This permits carbon 1 to bond with the oxygen on carbon 5, producing a ring structure. Two isomeric forms are possible that differ in the orientation of the —OH group. In α-glucose, the —OH of carbon 1 is below the ring; in β-glucose, above the ring. The thick, tapered bonds in the lower portion of each ring indicate that the molecule is a three-dimensional structure. The thickest bonds represent the part of the molecule that would project out of the page toward you. (b) Simplified drawing of the ring structure of glucose. A carbon atom is assumed by convention to be present at each angle in the ring unless another atom is shown. Most hydrogen atoms have been omitted. (c) A space-filling model of a glucose molecule.

Similarly, sucrose is hydrolyzed to form glucose and fructose

$$Sucrose + Water \longrightarrow Glucose + Fructose$$

Structural formulas for the compounds in these reactions are shown in Figure 3–8.

Polysaccharides

The most abundant carbohydrates are the **polysaccharides,** a group that includes starches, glycogen, and celluloses. A polysaccharide is a macromolecule consisting of repeating units of simple sugars, usually glucose. While the precise number of sugar units present varies, typically thousands of units are present in a single molecule. The polysaccharide may be a single long chain or a branched chain. Because they are composed of different stereoisomers of glucose or because the glucose units are arranged differently, polysaccharides have very different properties.

Starch, the typical storage form of carbohydrate in plants, is a polymer consisting of glucose subunits. The monomers are joined by 1—4 linkages (Figure 3–9). Starch occurs in two forms, amylose and amylopectin. Amylose, the simpler form, is unbranched. Amylopectin, the more common form, usually consists of about 1000 units in a branched chain. Branching takes place at about every 20 to 25 units and involves a 1—6 glycosidic linkage.

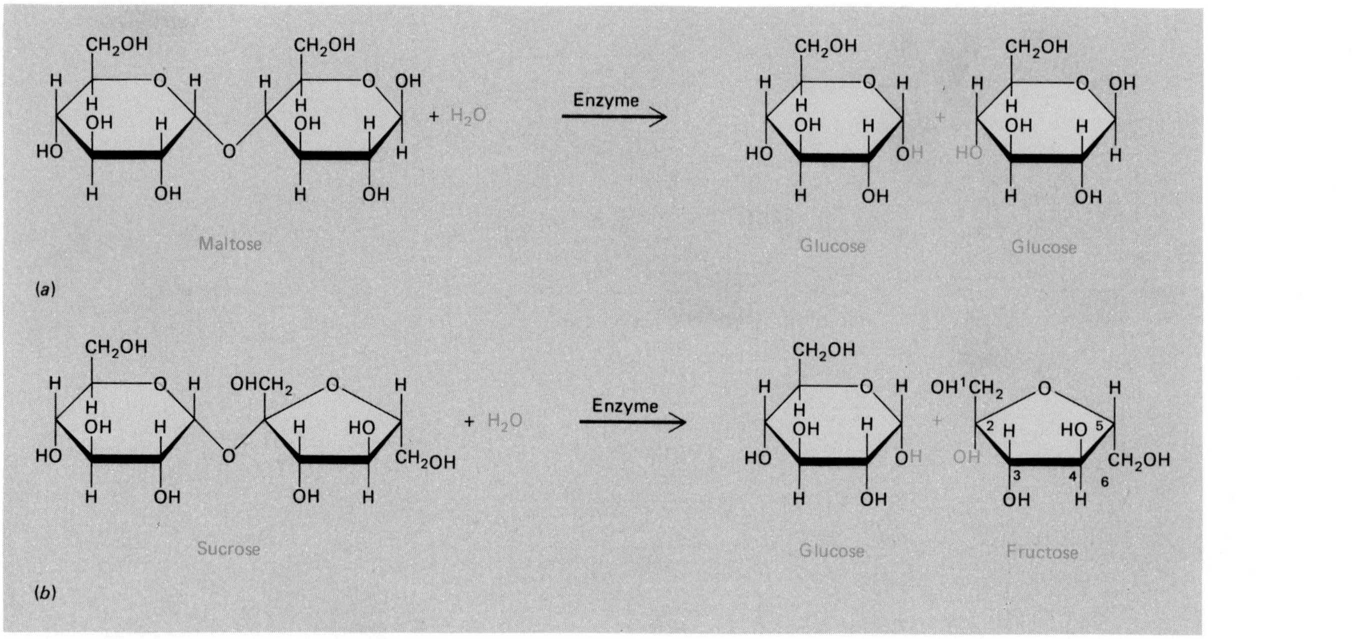

FIGURE 3–8 A disaccharide can be cleaved to yield two monosaccharide units. (*a*) Maltose may be broken down (as it is during digestion) to form two molecules of glucose. This is a hydrolysis reaction that requires the addition of water. (*b*) Sucrose can be hydrolyzed to yield a molecule of glucose and a molecule of fructose. Note that an enzyme, a protein catalyst, is needed to promote these reactions.

Plants store starch as granules within specialized organelles known as *plastids*. When energy is needed for cellular work, the plant can hydrolyze the starch, releasing the glucose subunits. Humans and other animals that eat plant foods have enzymes that can hydrolyze starch.

Glycogen (sometimes referred to as *animal starch*) is the form in which glucose is stored in animal tissues. This polysaccharide is highly branched and more water-soluble than plant starch. Glycogen is stored mainly in the liver and muscle cells.

Glucose cannot be stored as such; its small, uncharged, readily soluble molecules would leak out of the cells. The larger, less soluble starch and glycogen molecules do not readily pass through the cell membrane. Thus, instead of storing simple sugars, cells store the more complex polysaccharides such as glycogen, which can be readily hydrolyzed into simple sugars.

Carbohydrates are the most abundant group of organic compounds on Earth, and the structural carbohydrate cellulose is the most abundant carbohydrate, accounting for 50% or more of all the carbon in plants (Figure 3–10). Wood is about half cellulose, and cotton is at least 90% cellulose. Plant cells are surrounded by a strong supporting cell wall consisting mainly of cellulose. Cellulose is an insoluble polysaccharide composed of many glucose molecules joined together. The bonds joining these sugar units are different from those in starch. In starch the subunits are α-glucose and the glycosidic bonds are alpha 1—4 linkages; in cellulose β-glucose is the monomer and the linkages are beta 1—4 bonds. These bonds are not split by the enzyme that cleaves the bonds in starch. Humans do not have enzymes that can digest cellulose and so are not able to use cellulose as a nutrient. However, as will be discussed in Chapter 40, cellulose is an important component of dietary fiber and helps keep the digestive tract functioning properly.

Modified and Complex Carbohydrates

Many derivatives of monosaccharides are important biological compounds. The amino sugars glucosamine and galactosamine are compounds in which a hydroxyl group (—OH) is replaced by an amino group (—NH₂). Galactosamine is present in cartilage. Glucosamine is the molecular unit

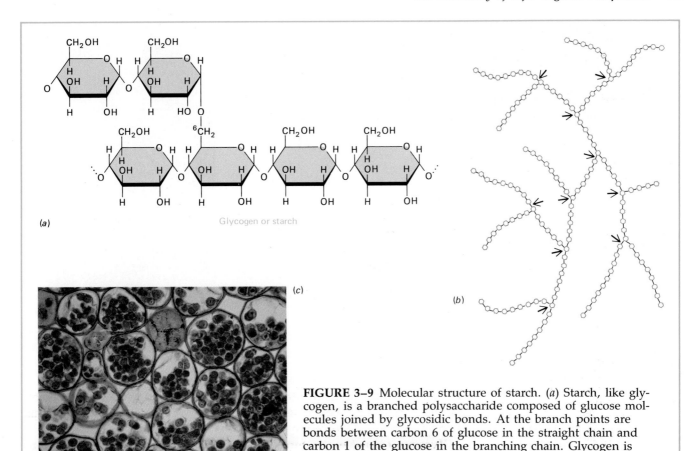

(a)

Glycogen or starch

(b)

(c)

FIGURE 3–9 Molecular structure of starch. (*a*) Starch, like glycogen, is a branched polysaccharide composed of glucose molecules joined by glycosidic bonds. At the branch points are bonds between carbon 6 of glucose in the straight chain and carbon 1 of the glucose in the branching chain. Glycogen is more highly branched than starch. (*b*) Diagram representing starch. The arrows represent the branch points. (*c*) Starch (stained purple) stored in specialized organelles called *amyloplasts*, in cells of a buttercup root (approximately ×100).

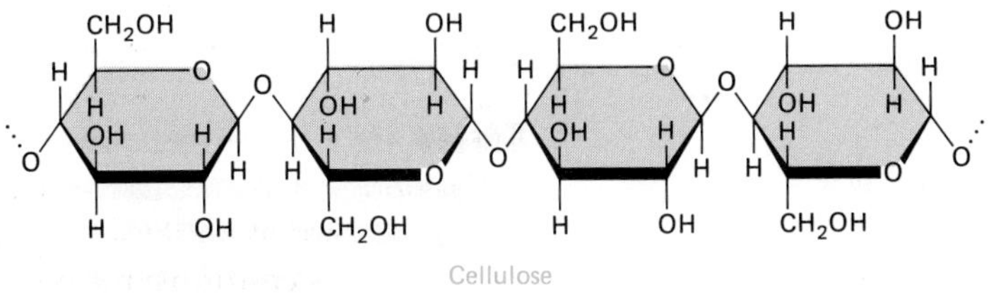

Cellulose

(a)

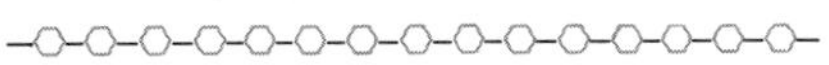

(b)

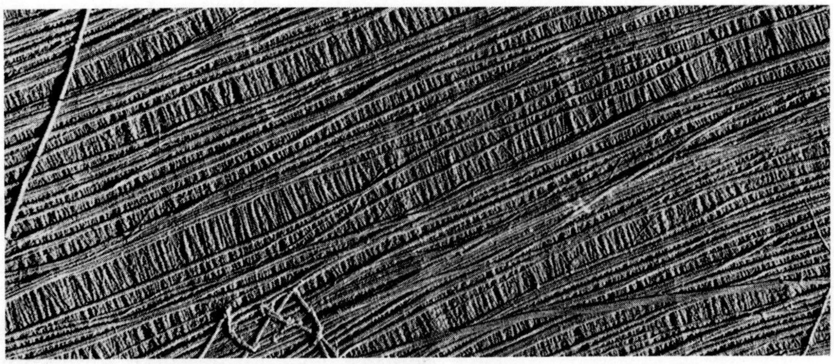

(c)

FIGURE 3–10 The structure of cellulose. (*a*) The cellulose molecule is an unbranched polysaccharide composed of approximately 10,000 glucose units joined by glycosidic bonds. (*b*) A more diagrammatic representation of cellulose structure. Each hexagon represents a glucose molecule bonded by a glycosidic bond to the adjacent glucose molecule. (*c*) An electron micrograph of cellulose fibers from the cell wall of a marine alga (approximately ×24,000).

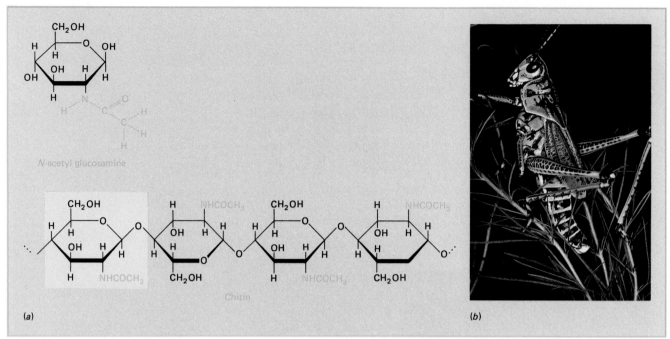

CH₂OH

N-acetyl glucosamine

CH₂OH   H   NHCOCH₃   CH₂OH   H   NHCOCH₃

Chitin

(a)

(b)

FIGURE 3–11 Chitin is a polysaccharide that rivals cellulose as the most common organic compound in the ecosphere. (*a*) The amino sugar *N*-acetyl glucosamine (NAG) is the monomer found in chitin. The polymer chitin consists of NAG subunits joined by glycosidic bonds. (*b*) Chitin is an important component of the armor-like exoskeleton (outer covering) of arthropods such as this grasshopper.

present in **chitin** the main component of the external skeletons of insects, crayfish, and other arthropods (Figure 3–11). A tough modified polysaccharide, chitin is also found in the cell walls of fungi.

Carbohydrates may also be combined with proteins to form **glycoproteins,** compounds present on the outer surface of animal cells. Most proteins secreted by cells are glycoproteins. Carbohydrates can combine with lipids to form **glycolipids,** compounds present on the surface of animal cells that are important in interactions among cells.

LIPIDS

Lipids are a heterogeneous group of compounds that have a greasy or oily consistency and are relatively insoluble in water. Like carbohydrates, lipids are composed of carbon, hydrogen, and oxygen atoms, but they have relatively less oxygen in proportion to carbon and hydrogen than do carbohydrates. Oxygen atoms are characteristic of hydrophilic (water-loving) functional groups, so lipids, with little oxygen, are much less soluble in water than most carbohydrates; in fact, lipids tend to be hydrophobic (water-hating). Among the groups of lipids especially important biologically are the neutral fats, phospholipids, steroids, carotenoids (red and yellow plant pigments), and waxes. Lipids are important biological fuels, serve as structural components of cell membranes, and some are important hormones.

Neutral Fats

The most abundant lipids in living things are the **neutral fats.** These compounds yield more than twice as much energy per gram as do carbohydrates and are an economical form for the storage of fuel reserves. Carbohydrates and proteins can be transformed by enzymes into fats and stored within the cells of adipose (fat) tissue.

A neutral fat consists of glycerol joined to one, two, or three molecules of a fatty acid. **Glycerol** is a three-carbon alcohol that contains three —OH groups (Figure 3–12). A fatty acid is a long, straight chain of carbon atoms with a carboxyl group (—COOH) at one end. About 30 different fatty acids

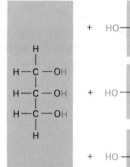

Glycerol   Fatty acid

(a)

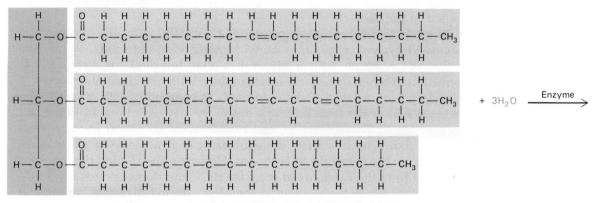

A triacylglycerol

PRODUCTS

Oleic acid

Linoleic acid

Palmitic acid

Glycerol

(b)

(c)

FIGURE 3–12 Neutral fats. (*a*) Structure of glycerol and of a fatty acid. The carboxyl (—COOH) group is present in all fatty acids. The *R* represents the remainder of the molecule, which varies with each type of fatty acid. (*b*) Hydrolysis of a triacylglycerol yields glycerol plus three fatty acids. Note that the triacylglycerol is an unsaturated fat—two of its fatty acid components contain double bonds between carbon atoms. (*c*) Honeybees on a brood comb. The comb is composed of wax secreted by special abdominal glands of the bees. It is a compound consisting of fatty acids and alcohols, and although it is classified as a lipid, it can be digested by very few animals.

are commonly found in animal lipids, and they typically have an even number of carbon atoms. For example, butyric acid, present in rancid butter, has four carbon atoms, and oleic acid, the most widely distributed fatty acid in nature, has 18 carbon atoms.

Saturated fatty acids contain the maximum possible number of hydrogen atoms, whereas **unsaturated fatty acids** contain some carbon atoms that are double-bonded with one another and are not fully saturated with hydrogens. Fatty acids with several double bonds are called *polyunsaturated fatty acids*. Fats containing unsaturated fatty acids are oils, and most of them are liquid at room temperature. Saturated fats tend to be solids at room temperature; butter and animal fat are examples. At least two fatty acids (linoleic and arachidonic) are essential nutrients, which must be included in the diet.

When a glycerol molecule combines chemically with one fatty acid, a **monoacylglycerol** (sometimes called *monoglyceride*) is formed. When two fatty acids combine with a glycerol, a **diacylglycerol** (or *diglyceride*) is formed, and when three fatty acids combine with one glycerol molecule, a **triacylglycerol** (or *triglyceride*) is formed. In combining with glycerol, the carboxyl end of the fatty acid attaches to the oxygen of one of the —OH groups, forming a covalent linkage known as an *ester bond*. In the overall reaction that produces a fat, the equivalent of a molecule of water is removed from the glycerol and fatty acid. However, the H^+ and the OH^- are removed from the reactants in separate steps and do not necessarily combine as H_2O when the reaction is complete. During digestion the neutral fats are hydrolyzed to produce fatty acids and glycerol.

Phospholipids

Phospholipids represent an important class of lipids called **amphipathic lipids,** which form cell membranes. In amphipathic molecules one end is **hydrophilic** and the other end is **hydrophobic.** A phospholipid consists of

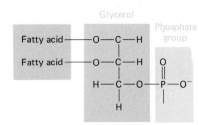

(a) Phosphatidic acid

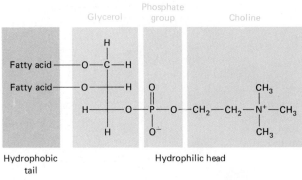

(b) Lecithin

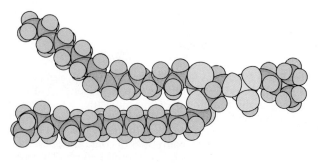

FIGURE 3–13 Phospholipids. (*a*) Many phospholipids are derivatives of phosphatidic acid, a compound consisting of glycerol chemically combined with two fatty acids and a phosphate group. (*b*) Lecithin (or phosphatidylcholine) is a phospholipid found in cell membranes. It forms when phosphatidic acid combines with the compound choline. The structural formula for lecithin is shown here along with a space-filling model.

a glycerol molecule attached to two fatty acids and to a phosphate group and linked to an organic base such as choline. Phospholipids also usually contain nitrogen in the organic base (Figure 3–13). (Note that phosphorus and nitrogen are absent in the neutral fats.)

The two ends of the phospholipid molecule differ physically as well as chemically. The fatty acid portion of the molecule is hydrophobic and not soluble in water. However, the portion composed of glycerol and the organic base is ionized and readily water-soluble. This end of the molecule is said to be *hydrophilic.* The amphipathic properties of these lipid molecules cause them to assume a certain configuration in the presence of water, with their hydrophilic water-soluble heads facing outward toward the surrounding water and their hydrophobic tails facing in the opposite direction. The cell membrane is a lipid bilayer composed of two layers of phospholipid molecules, their hydrophobic tails meeting in the middle and their hydrophilic heads oriented toward the outside of the cell membrane (Figure 3–14).

Carotenoids

The red and yellow plant pigments called **carotenoids** are classified with the lipids because they are insoluble in water and have an oily consistency. These pigments, found in the cells of all plants, play a role in photosynthesis. The carotenoid molecule consists of five carbon monomers known as *isoprene units* (Figure 3–15a).

Splitting in half a molecule of the yellow plant pigment carotene yields a molecule of vitamin A, or retinol (Figure 3–15b). Retinal, the light-sensitive chemical present in the retina of the eye, is a derivative of vitamin A. In the presence of light, retinal undergoes a chemical reaction by which light stimuli are received.

Interestingly, photoreceptors, or eyes, have evolved independently in three different lines of animals—mollusks, insects, and vertebrates. These animals have no common evolutionary ancestor equipped with eyes, yet the eyes of each of them have the same compound, retinal, involved in the process of light reception. That retinal is present in each of these types of

FIGURE 3–14 In the presence of water, lipid molecules orient themselves with their hydrophilic water-soluble heads facing outward toward the surrounding water. The hydrophobic tails face in the opposite direction. (*a*) Space-filling model of complex lipids in a bilayer. Two layers of phospholipid molecules are present with their hydrophobic tails meeting in the middle. (*b*) A lipid bilayer, such as is found in cell membranes.

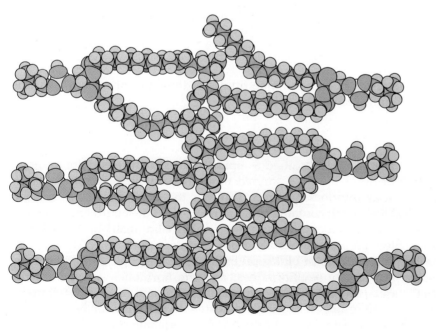

(a)

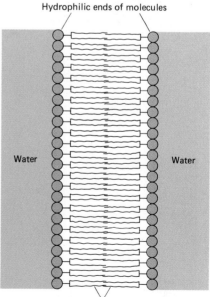

Hydrophilic ends of molecules

Water   Water

Hydrophobic ends of molecules

(b)

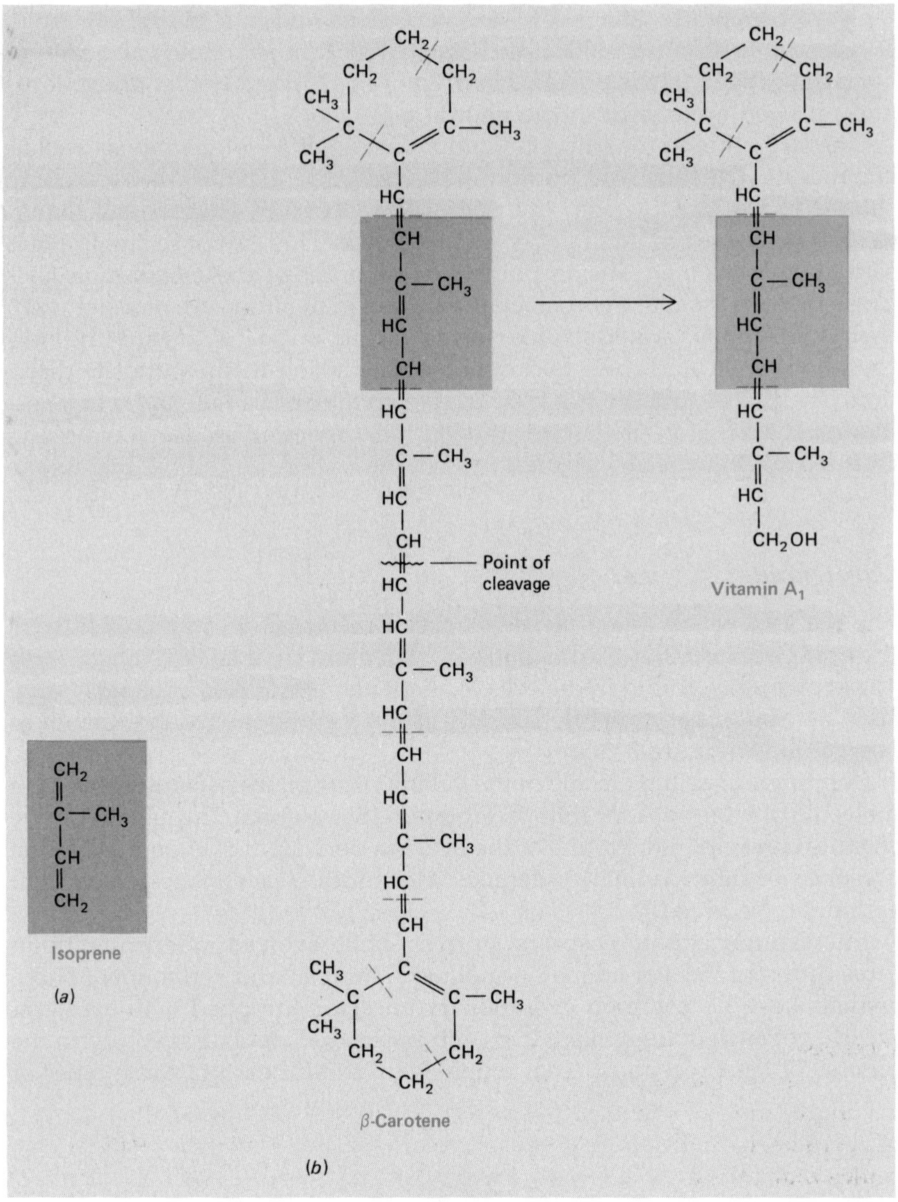

FIGURE 3–15 Carotenoids. (*a*) Isoprene is the monomer present in carotenoids. (*b*) Beta-carotene, the yellow pigment present in some plants. This carotenoid gives carrots, sweet potatoes, and other orange vegetables their color. Most animals can convert carotenoids to vitamin A. The dashed lines indicate the boundaries of the individual isoprene units within β-carotene.

eyes is the result of some unique fitness of this kind of molecule for the process of light reception.

Steroids

Although **steroids** are classified as lipids, their structure is quite different from that of other lipids. A steroid molecule contains carbon atoms arranged in four interlocking rings; three of the rings contain six carbon atoms, and the fourth contains five (Figure 3–16). The length and structure of the side chains that extend from these rings distinguish one steroid from another. Steroids are synthesized from isoprene units.

Among the steroids of biological importance are cholesterol, bile salts, the male and female sex hormones, and the hormones secreted by the adrenal cortex. Cholesterol is a structural component of animal cell membranes. Bile salts emulsify fats in the intestine so that they can be enzymatically hydrolyzed. Steroid hormones regulate certain aspects of metabolism in a variety of animals, including vertebrates, insects, and crabs.

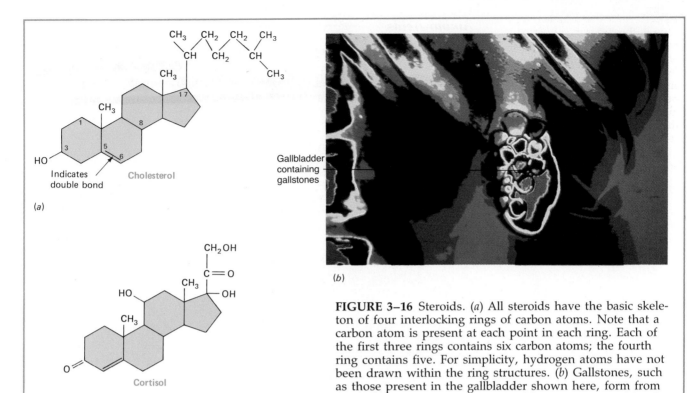

FIGURE 3–16 Steroids. (*a*) All steroids have the basic skeleton of four interlocking rings of carbon atoms. Note that a carbon atom is present at each point in each ring. Each of the first three rings contains six carbon atoms; the fourth ring contains five. For simplicity, hydrogen atoms have not been drawn within the ring structures. (*b*) Gallstones, such as those present in the gallbladder shown here, form from cholesterol. (*c*) Cortisol is a steroid hormone secreted by the adrenal glands.

PROTEINS

Proteins are of central importance in the chemistry of life. These macromolecules serve as structural components of cells and tissues, so growth and repair, as well as maintenance of the organism, depend on an adequate supply of these compounds. Some proteins serve as **enzymes,** special molecules that regulate the thousands of different chemical reactions that take place in a living system.

The protein constituents of a cell are the clue to its lifestyle. Each cell type has characteristic types, distributions, and amounts of protein that determine what the cell looks like and how it functions. A muscle cell is different from other cell types by virtue of its large content of the contractile proteins myosin and actin, which are largely responsible for its appearance as well as for its ability to contract. The protein hemoglobin, found in red blood cells, is responsible for the specialized function of oxygen transport.

Most proteins are species-specific—that is, they vary slightly in each species so that the protein complement (as determined by the instructions in the genes) is also mainly responsible for differences among species. Thus, the proteins in the cells of a dog vary somewhat from those in the cells of a fox or a coyote. The degree of difference in the proteins of two species is thought to depend on evolutionary relationships. Organisms distantly related by evolution have proteins that differ more markedly than those of closely related forms. Some proteins differ slightly even among individuals of the same species, so that each individual is biochemically unique. Only genetically identical organisms—identical twins or members of closely inbred strains of organisms—have identical proteins.

Amino Acids

A basic knowledge of protein chemistry is essential for understanding nutrition as well as other aspects of metabolism. Proteins are composed of carbon, hydrogen, oxygen, nitrogen, and usually sulfur. Atoms of these elements are arranged into molecular subunits called **amino acids.** All of

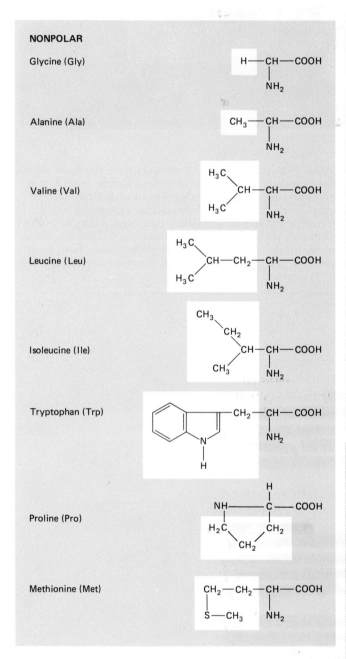

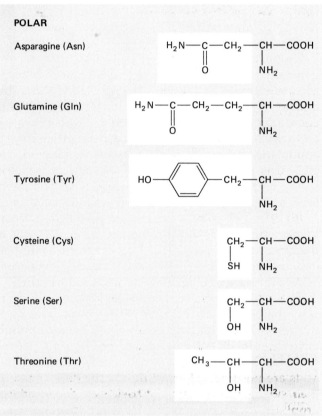

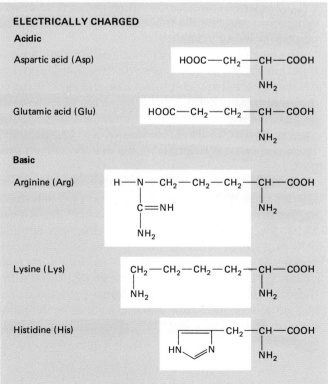

FIGURE 3–17 The amino acids commonly found in proteins. The amino acids are grouped here according to the properties of their side chains (R groups). The three-letter symbols are the conventional abbreviations for the amino acids.

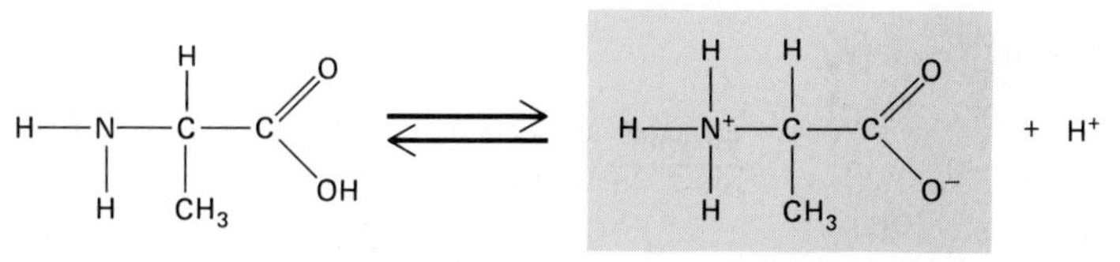

FIGURE 3–18 Ionic form of amino acids. At the pH of living cells, amino acids exist mainly as dipolar ions.

Ionized form

the twenty kinds of amino acids commonly found in proteins contain an amino group (—NH₂) and a carboxyl group (—COOH) bonded to the same carbon atom, called the **alpha carbon.** Amino acids differ in the R group or **side chains** bonded to the alpha carbon. Glycine, the simplest amino acid, has a hydrogen atom as its R group or side chain; alanine has a methyl (—CH₃) group (Figure 3–17).

Amino acids in solution at neutral pH are mainly dipolar ions. This is generally how amino acids exist at cellular pH. The amino group (—NH₂) accepts a proton and becomes —NH₃⁺, and the carboxyl group (—COOH) donates a proton and becomes dissociated —COO⁻ (Figure 3–18). As discussed earlier, a solution of an acid and its conjugate base serves as a buffer and resists changes in pH when an acid or base is added. Because of their amino and carboxyl groups, proteins in solution resist changes in acidity and alkalinity and thereby serve as important biological buffers.

The alpha carbon of an amino acid is an asymmetric carbon. Therefore, each amino acid can exist as two enantiomers, or mirror images (Figure 3–19). The two mirror images are called the L-isomer and the D-isomer. When amino acids are synthesized in the laboratory, a mixture of L and D-amino acids is produced. However, the amino acids present in living systems are almost exclusively L-isomers. One exception is a few D-amino acids present in the antibiotics produced by fungi.

The amino acids are grouped in Figure 3–17 by the properties of their side chains. Amino acids with nonpolar side chains are hydrophobic, whereas those with polar side chains are hydrophilic. Acidic amino acids have side chains that contain a carboxyl group. At cellular pH the carboxyl group is dissociated so that the R group has a negative charge. Basic amino acids are positively charged due to the dissociation of the amino group in their side chains. Acidic and basic side chains are ionic and therefore hydrophilic. In addition to the 20 common amino acids, some proteins have unusual amino acids. These rare amino acids are produced by the modification of common ones after they have become part of a protein. For example, lysine and proline may be converted to hydroxylysine and hydroxy-

FIGURE 3–19 Enantiomers of amino acids.

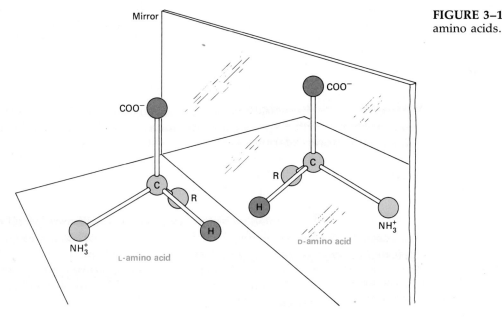

FIGURE 3–20 *a*. Formation of a dipeptide.

Glycine + Alanine → Glycylalanine (a dipeptide) + H_2O

R group, Carboxyl group, Amino group, R group, Peptide bond

(a)

Glycylalanine + Cysteine → Glycylalanylcysteine (a tripeptide) + H_2O

(b)

FIGURE 3–20 Formation of polypeptide chains. (*a*) Formation of a dipeptide. Two amino acids combine chemically to form a dipeptide. Water is produced as a byproduct during this reaction. (*b*) A third amino acid is added to the dipeptide to form a chain of three amino acids (a tripeptide, or small polypeptide). The bond between two amino acids is a peptide bond. Additional amino acids can be added to form long polypeptide chains.

proline after they have been incorporated into collagen. These amino acids can form cross links between the peptide chains that make up collagen. Such cross links are responsible for the firmness and great strength of the collagen molecule, which is a major component of cartilage, bone, and other connective tissues.

With some exceptions, plants can synthesize all of their needed amino acids from simpler substances. The cells of humans and animals generally can manufacture some, but not all, of the various kinds of biologically significant amino acids if the proper raw materials are available. Those that animals cannot synthesize but must obtain in the diet are known as **essential amino acids.** Animals differ in their biosynthetic capacities; what is an essential amino acid for one species may not be for another.

Polypeptide Chains Are Built from Amino Acids

Amino acids combine chemically with one another by bonding the carboxyl carbon of one molecule to the amino nitrogen of another (Figure 3–20). The covalent bond linking two amino acids together is called a **peptide bond.** When two amino acids combine, a **dipeptide** is formed; a longer chain of amino acids is a **polypeptide.** The complex process by which polypeptides are synthesized is discussed in Chapter 13.

A polypeptide may contain hundreds of amino acids joined in a specific linear order. A protein consists of one or more polypeptide chains. An almost infinite variety of protein molecules is possible. It should be clear that the various proteins differ from one another with respect to the number, types, and arrangement of amino acids they contain. The 20 types of amino acids found in biological proteins may be thought of as letters of a protein alphabet, each protein being a word made up of amino acid letters.

Protein Structure—Levels of Organization

The polypeptide chains making up a protein are twisted or folded to form a macromolecule with a specific **conformation,** or three-dimensional shape. This conformation determines the function of the protein. For example, the unique shape of an enzyme permits it to "recognize" and act on its substrate, the substance the enzyme regulates. The shape of a protein hormone enables it to combine with a receptor on its target cell (the cell the hormone is designed to act upon).

Proteins can be classified as fibrous or globular. In **fibrous proteins** the polypeptide chains are arranged in long sheets; in **globular proteins** the polypeptide chains are tightly folded into a compact spherical shape. Most enzymes are globular proteins. Several different levels of organization can be distinguished in the protein molecule—primary, secondary, tertiary, and quaternary (Figure 3–21).

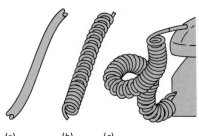

(a)   *(b)*   *(c)*

FIGURE 3–21 Protein structure. The telephone cord provides a familiar example for demonstrating (*a*) primary, (*b*) secondary, and (*c*) tertiary structure.

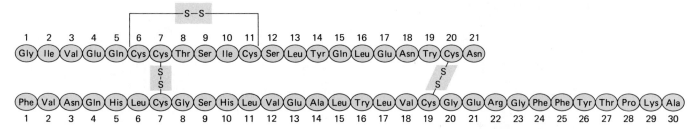

FIGURE 3–22 The primary structure of the two polypeptide chains that make up the protein insulin. The primary structure is the linear sequence of amino acids. Each oval in the diagram represents an amino acid. The letters inside the ovals are symbols for the names of the amino acids. Insulin is a very small protein.

Primary Structure

The sequence of amino acids in a polypeptide chain constitutes its **primary structure.** This sequence, as will be discussed in Chapter 13, is specified by the instructions in a gene. Using analytical methods developed in the early 1950s, investigators can determine the exact sequence of amino acids in a protein molecule. Insulin, a hormone secreted by the pancreas and used in the treatment of diabetes, was the first protein for which the exact sequence of amino acids in the polypeptide chains was ascertained. Insulin consists of 51 amino acid units in two linked chains (Figure 3–22).

Secondary Structure

The **secondary structure** of protein molecules involves the coiling of the peptide chain into a helix or some other regular conformation. The regularity is due to interactions between the atoms of the uniform backbone of the polypeptide chain. Functional groups do not play a role in forming the bonds that establish the secondary structure. Peptide chains ordinarily do not lie out flat or coil randomly, but rather undergo coiling to yield a specific three-dimensional structure. A common secondary structure in protein molecules is known as the **alpha helix.** This involves the formation of spiral coils of the polypeptide chain (Figure 3–23a). The alpha helix is a very uniform geometric structure with 3.6 amino acids occupying each turn of the helix. The helical structure is determined and maintained by the formation of hydrogen bonds between amino acids in successive turns of the spiral coil. In the alpha-helical structure the hydrogen bonding occurs between atoms within the same polypeptide chain.

The alpha helix is the basic structural unit of fibrous proteins such as wool, hair, skin, and nails. The fiber is elastic because the hydrogen bonds can be reformed. This is why human hairs can be stretched to some extent and will then snap back to their original length.

A second type of secondary structure is the **beta pleated sheet** (Figure 3–23b). Here the hydrogen bonding takes place between polypeptide chains. Each zigzag chain is fully extended, and the hydrogen bonding between them results in a sheetlike structure. Pleated sheets can also form between different regions of the same polypeptide chain (see Figure 3–26). This structure is flexible rather than elastic. Fibroin, the protein of silk, is characterized by a beta-pleated-sheet structure, and the core of many globular proteins consists of beta sheets.

Tertiary Structure

The **tertiary structure** of a protein molecule is the overall shape assumed by each polypeptide chain (Figure 3–24). This three-dimensional structure is determined by four main factors that involve interactions between R groups.

1. Hydrogen bonds between R groups of amino acid subunits in adjacent loops of the same polypeptide chain.

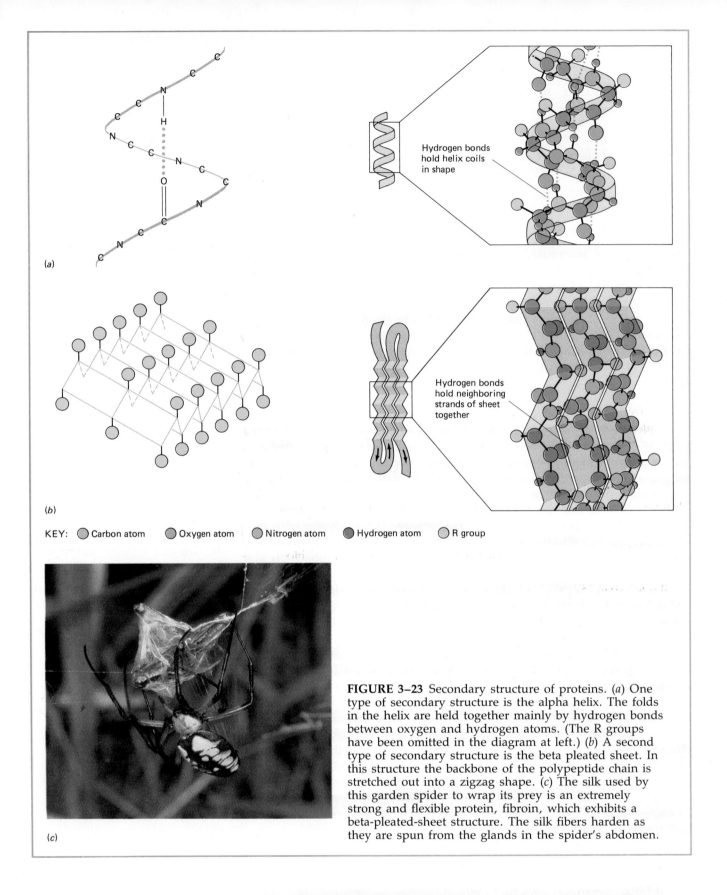

KEY: ◯ Carbon atom ◯ Oxygen atom ◯ Nitrogen atom ● Hydrogen atom ◯ R group

FIGURE 3–23 Secondary structure of proteins. (*a*) One type of secondary structure is the alpha helix. The folds in the helix are held together mainly by hydrogen bonds between oxygen and hydrogen atoms. (The R groups have been omitted in the diagram at left.) (*b*) A second type of secondary structure is the beta pleated sheet. In this structure the backbone of the polypeptide chain is stretched out into a zigzag shape. (*c*) The silk used by this garden spider to wrap its prey is an extremely strong and flexible protein, fibroin, which exhibits a beta-pleated-sheet structure. The silk fibers harden as they are spun from the glands in the spider's abdomen.

2. Ionic attraction between R groups with positive charges and those with negative charges.
3. Hydrophobic interactions resulting from the tendency of nonpolar R groups to associate in the interior of the globular structure away from the surrounding water.

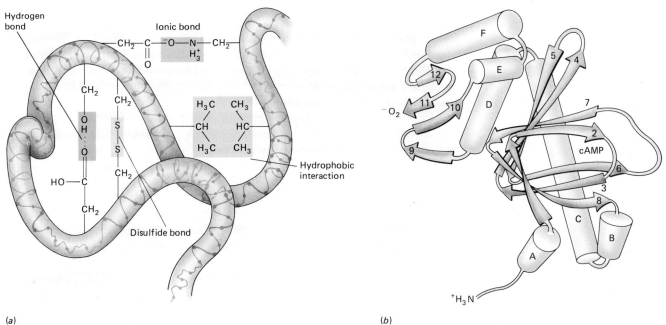

FIGURE 3–24 Tertiary structure of a protein. (*a*) The tertiary structure results from the coiling and folding of the alpha helix (or other secondary structure) into an overall globular or other shape. Hydrogen bonds, bonds between sulfur atoms, ionic attractions between R groups, and hydrophobic interactions are among the forces that hold the parts of the molecule in the designated shape. (*b*) Schematic drawing of the tertiary structure of a polypeptide that has both alpha-helical and beta-sheet secondary structure. The polypeptide is a subunit of a DNA-binding protein (CAP) from the bacterium *Escherichia coli*. The regions of the polypeptide that are in alpha-helical conformation are represented as blue tubes lettered A through F. Regions in beta conformation are represented as gray arrows numbered 1 to 12. Green lines represent connecting regions.

4. Covalent bonds known as disulfide bonds (—S—S—) link the sulfur atoms of two cysteine subunits. Disulfide bonds may link two parts of the same polypeptide chain or join two different chains.

Quaternary Structure

Proteins composed of two or more polypeptide chains have a **quaternary structure,** the arrangement assumed by the polypeptide chains, each with its own primary, secondary, and tertiary structures, to form the biologically active protein molecule. Hemoglobin, the protein in red blood cells that is responsible for oxygen transport, is an example of a globular protein with quaternary structure (Figure 3–25). Hemoglobin consists of 574 amino acids arranged in four polypeptide chains—two identical alpha and two identical beta chains. Its chemical formula is $C_{3032}H_{4816}O_{872}N_{780}S_8Fe_4$.

FIGURE 3–25 Proteins that consist of more than one polypeptide subunit assume a final quaternary shape. Hemoglobin, a globule-shaped protein containing four polypeptide subunits, is illustrated here. Its quaternary structure consists of the final shape in which the subunits combine. In hemoglobin each polypeptide encloses an iron-containing structure (shown as green discs).

Protein Structure Determines Function

The structure of a protein determines its biological activity. A single protein may have varying structure and more than one function. Many proteins are modular, consisting of two or more globular sections, called **domains,** connected by less compact regions of the polypeptide chain. Each domain may have a different function.

When a cell synthesizes a protein, the polypeptide chains spontaneously assume their three-dimensional shape. Conformation is determined by the primary structure of the polypeptide. Predicting the structure of a protein from its primary sequence of amino acids, however, is quite difficult due to the many possible combinations of folding patterns. Computer programs are being developed to predict the secondary structure of a protein from its amino acid sequence (Figure 3–26).

The biological activity of a protein can be disrupted by changes in the amino acid sequence or in the conformation of a protein. When a mutation (a chemical change in a gene) occurs, resulting in a change in the amino acid sequence of hemoglobin, the disorder known as *sickle cell anemia* may occur. The hemoglobin molecules in a person with sickle cell anemia have the amino acid valine instead of glutamic acid at position 6, that is, the sixth amino acid from the terminal end in the beta chain. The substitution of valine with an uncharged side chain for glutamate with a charged side chain makes the hemoglobin less soluble and more likely to form crystal-like structures that change the shape of the red blood cell.

Changes in the three-dimensional structure of a protein also disrupt its biological activity. When a protein is heated or treated with any of a number of chemicals, its tertiary structure becomes disordered and the coiled peptide chains unfold to give a more random conformation. This unfolding is accompanied by a loss of the biological activity of the protein—for example, its ability to act as an enzyme. Such change in shape and loss of biolog-

FIGURE 3–26 Computer-generated plot of the secondary structure of a protein predicted from its amino acid sequence. Predicting the structure of proteins from their amino acid sequence is complex, and the field is still in its infancy. This protein is 328 amino acids long. The blue line represents the amino acid chain. Sine waves represent the alpha-helical regions. Sawtooth regions represent beta-sheet structures. Strongly hydrophilic amino acids are plotted in red. Hydrophobic amino acids are plotted in yellow.

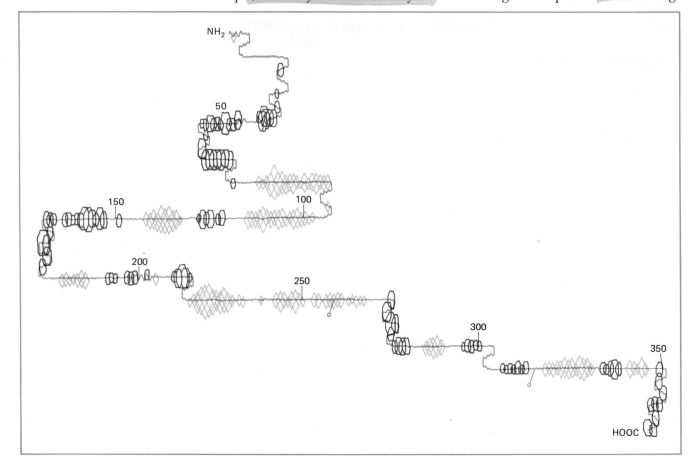

ical activity is termed **denaturation** of the protein. Denaturation generally cannot be reversed. However, under certain conditions, some proteins that have been denatured return to their original shape and biological activity when normal environmental conditions are restored.

NUCLEIC ACIDS

There are two classes of nucleic acids found in cells: **ribonucleic acids (RNA)** and **deoxyribonucleic acids (DNA).** Both function in the transmission of hereditary information and in the determination of what proteins a cell manufactures. DNA comprises the genes themselves, the hereditary material of the cell, and contains instructions for making all the proteins needed by the organism (Figure 3–27). Three types of RNA—messenger RNA, transfer RNA, and ribosomal RNA—function in the process of protein synthesis.

Like proteins, nucleic acids are large, complex molecules. They were first isolated by Miescher in 1870 from the nuclei of pus cells and their name stems from the fact that they are acidic and were first identified in nuclei.

The Nucleotide Subunits of Nucleic Acids

Nucleic acids are polymers of **nucleotides,** molecular units that consist of (1) a five-carbon sugar, either ribose or deoxyribose, (2) a phosphate group, and (3) a nitrogenous base, which may be either a double-ringed purine or a single-ringed pyrimidine (Figure 3–28). DNA contains the purines adenine (A) and guanine (G) and the pyrimidines cytosine (C) and thymine (T)

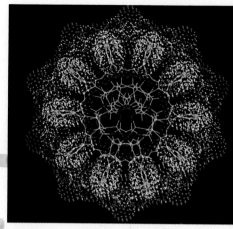

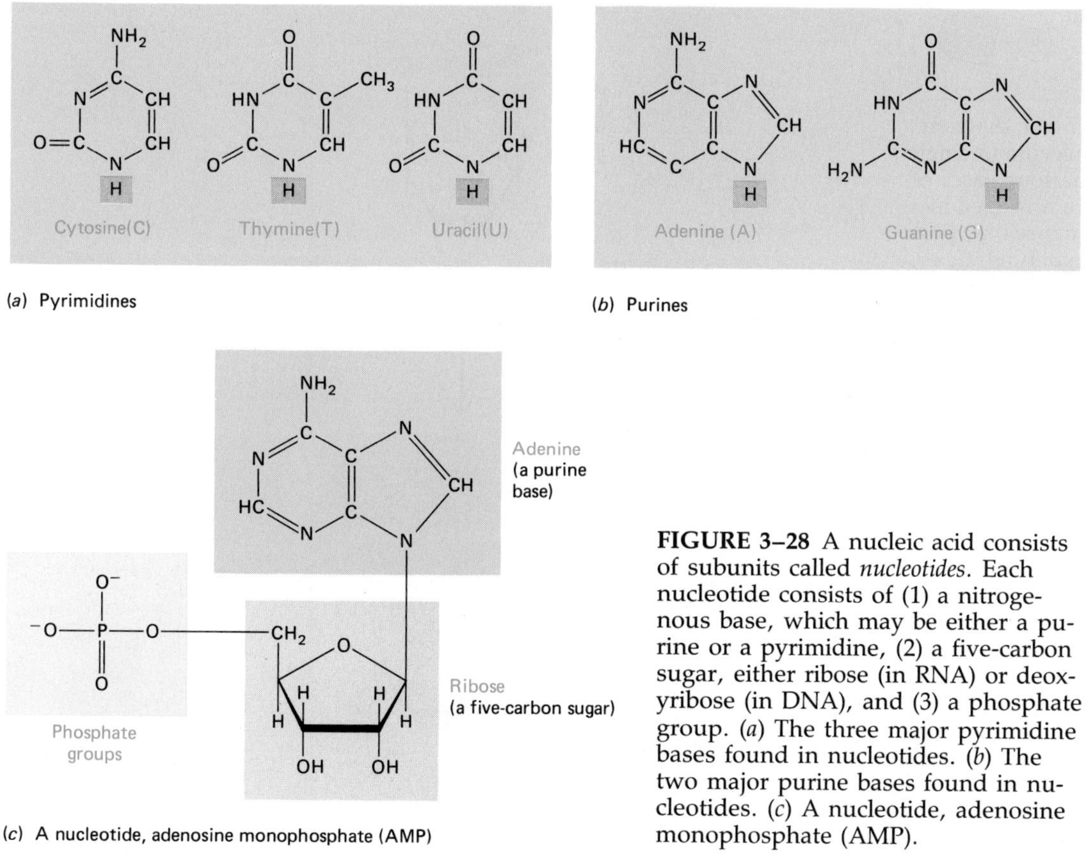

FIGURE 3–27 This unusual photograph is a computer-generated view of the DNA molecule. It simulates a view down along the axis of the molecule. The bonds appear white, and the atoms as hollow spheres of various colors; red = oxygen; blue = nitrogen; green = carbon; yellow = phosphorus; white = hydrogen.

(a) Pyrimidines

Cytosine(C)   Thymine(T)   Uracil(U)

(b) Purines

Adenine (A)   Guanine (G)

Adenine (a purine base)

Ribose (a five-carbon sugar)

Phosphate groups

(c) A nucleotide, adenosine monophosphate (AMP)

FIGURE 3–28 A nucleic acid consists of subunits called *nucleotides*. Each nucleotide consists of (1) a nitrogenous base, which may be either a purine or a pyrimidine, (2) a five-carbon sugar, either ribose (in RNA) or deoxyribose (in DNA), and (3) a phosphate group. (*a*) The three major pyrimidine bases found in nucleotides. (*b*) The two major purine bases found in nucleotides. (*c*) A nucleotide, adenosine monophosphate (AMP).

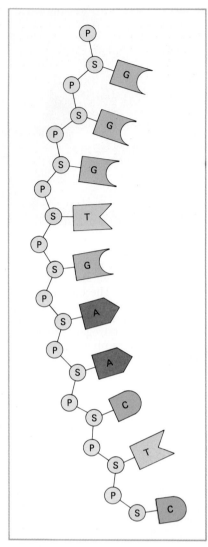

FIGURE 3–29 Schematic diagram of a nucleic acid molecule (a single strand of DNA). The four bases of each nucleic acid are arranged in various specific sequences. *P*, phosphate; *S*, sugar; *G*, guanine; *C*, cytosine; *A*, adenine; *T*, thymine.

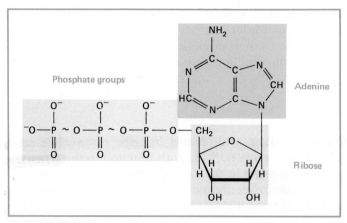

FIGURE 3–30 The structure of ATP, a nucleotide that has energy-rich bonds joining the two terminal phosphate groups to the nucleotide.

together with the sugar deoxyribose and phosphate. RNA contains the purines adenine and guanine and the pyrimidines cytosine and uracil (U), together with the sugar ribose and phosphate.

The molecules of nucleic acids are made of linear chains of nucleotides, each attached to the next by covalent bonds between the sugar molecule of one and the phosphate group of the next (Figure 3–29). These are phosphodiester bonds. As we will see in our discussion of the genetic code (Chapter 13), the specific information of the nucleic acid is coded in the

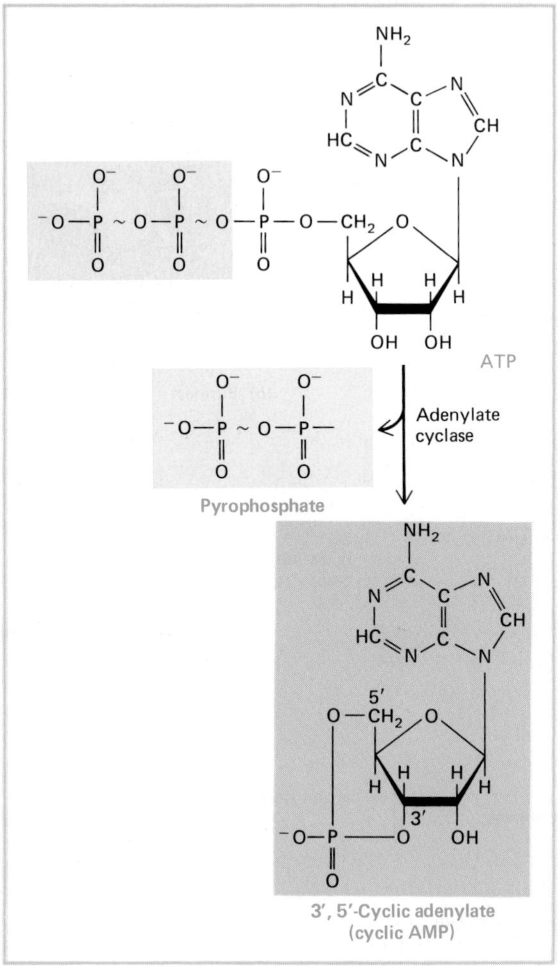

FIGURE 3–31 Formation of cyclic AMP from ATP.

ATP

Adenylate cyclase

Pyrophosphate

3′, 5′-Cyclic adenylate (cyclic AMP)

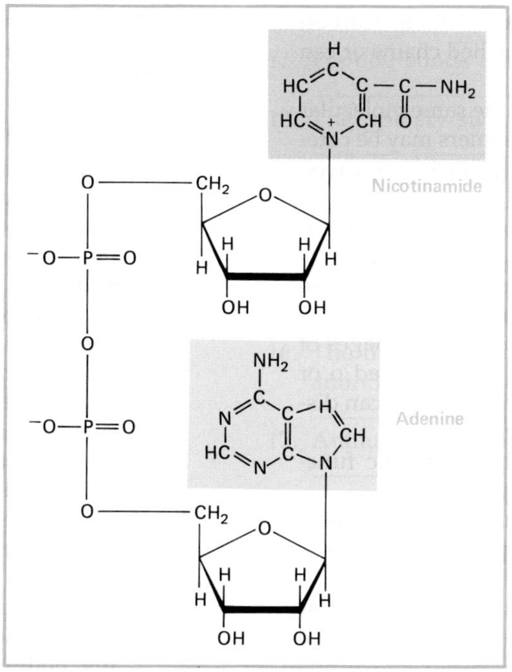

FIGURE 3–32 Structure of NAD$^+$, an important hydrogen and electron acceptor and donor. The nicotinamide portion of the molecule accepts hydrogen and is reduced in the process.

unique sequence of the four kinds of nucleotides present in the chain. The removal of the phosphate group from a nucleotide yields a compound, termed a *nucleoside*, composed of the base and sugar. DNA is composed of two nucleotide chains entwined around each other in a double helix.

Other Important Nucleotides

Besides their importance as subunits of nucleic acids, nucleotides serve other vital functions in living cells. **Adenosine triphosphate (ATP),** composed of adenine, ribose, and three phosphates (Figure 3–30), is of major importance as the energy currency of all cells. The two terminal phosphate groups are joined to the nucleotide by special "energy-rich" bonds, indicated by the ~P symbol. These bonds are called *energy-rich* because much free energy is released when they are hydrolyzed. The biologically useful energy of these bonds can be transferred to other molecules. Most of the chemical energy of the cell is stored in the energy-rich phosphate bonds of ATP, ready to be released when the phosphate group is transferred to another molecule.

A nucleotide may be converted by enzymes called **cyclases** to a cyclic form. ATP, for example, is converted to cyclic adenosine monophosphate (cyclic AMP) by the enzyme adenylate cyclase (Figure 3–31). Cyclic nucleotides play an important role in mediating the effects of hormones and in regulating various aspects of cellular function.

Cells contain several dinucleotides, which are of great importance in metabolic processes. For example, as will be discussed in Chapter 7, **nicotinamide adenine dinucleotide (NAD$^+$)** is very important as a primary electron and hydrogen acceptor and donor in biological oxidations and reductions within cells (Figure 3–32).

■ SUMMARY

I. The major groups of organic compounds are carbohydrates, lipids, proteins, and nucleic acids.

II. Chains of carbon atoms form the backbone of a large variety of organic compounds essential to life.

A. The carbon atom can form stable single covalent bonds with four other atoms, or it can form double or triple bonds with other atoms.

B. Carbon bonds with a greater number of different

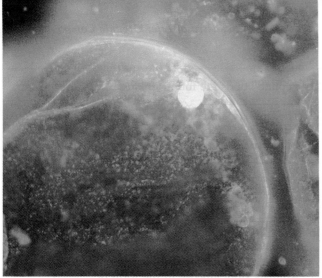
Fluorescence micrograph of tomato cell

4

Cellular Organization

■ **LEARNING OBJECTIVES**

After you have read this chapter you should be able to:

1. Discuss why the cell is considered the basic unit of life, and discuss the cell theory.
2. Explain why resolving power is an important feature of a good microscope, and discuss the differences between light and electron microscopes.
3. Discuss the general characteristics of prokaryotic and eukaryotic cells.
4. Have a good idea of size relationships among different cells and cell structures.
5. Discuss why the relationship between surface area and the volume of a cell is important in determining cell size limits.
6. Describe the structure and function of principal organelles in plant and animal cells. Be able to locate and label them on a diagram or photomicrograph.
7. Discuss the structure of the nucleus and its function in eukaryotic cells.
8. Distinguish between smooth and rough endoplasmic reticulum, and discuss the relationship between the endoplasmic reticulum and other internal membranes in the cell.
9. Explain how proteins synthesized on the endoplasmic reticulum can be processed, modified, and sorted by the Golgi complex. Discuss the different pathways a protein can take through the Golgi complex.
10. Describe the functions of lysosomes, and explain what happens when they leak.
11. Distinguish between the functions of chloroplasts and mitochondria, and explain why both organelles synthesize ATP.
12. Describe the structures of the major types of filaments that make up the cytoskeleton. Explain the importance of the cytoskeleton to the cell.
13. Discuss the structure of cilia and flagella, and explain how they function in locomotion of the cell.

All living organisms are made up of cells. Although some living systems consist of only one cell and others of several billion, even the most complex organism begins life as a single cell, the fertilized egg. In most multicellular organisms, including humans, a single cell divides to form two cells, and each new cell divides again and again, eventually forming the complex tissues, organs, and systems of the developed organism. Like the bricks of a building, cells are the building blocks of the organism.

The cell is the smallest unit of living material capable of carrying on all the activities necessary for life. It has all of the physical and chemical components needed for its own maintenance, growth, and division. When provided with essential nutrients and an appropriate environment, some cells can be kept alive and growing in laboratory glassware for many years. No cell part is capable of surviving by itself outside the cell.

THE CELL THEORY

The idea that cells are fundamental units of life is a part of the **cell theory.** Two German scientists, the botanist Matthias Schleiden in 1838 and the zoologist Theodor Schwann in 1839, were the first to point out that plants and animals are composed of groups of cells and that the cell is the basic unit of living organisms.

The cell theory was extended in 1855 by Rudolph Virchow, who stated that new cells are formed only by the division of previously existing cells. In other words, cells do not arise by spontaneous generation from nonliving matter (an idea that was rooted in the writings of Aristotle and that had persisted over many centuries). About 1880, another famous biologist, August Weismann, pointed out an important corollary to Virchow's statement, that all the cells living today can trace their ancestry back to ancient times.

In its present form, then, the cell theory includes two ideas:

1. All living things are composed of cells and cell products.
2. New cells are formed only by the division of preexisting cells.

Evidence that cells are descended from ancient cells can be seen in the fundamental similarities in the complex protein molecules found in all cells. A particular class of proteins called cytochromes is found in organisms ranging from bacteria to plants and animals. Not only are the cytochromes of all organisms similar in structure, but they also perform virtually identical functions in cells from widely different species. The fact that all cells have similar molecules of such complexity strongly suggests that modern cells are evolved from a small group of ancestral cells.

HOW CELLS ARE STUDIED

One of the most important tools for the study of cell structure is the microscope. In fact, cells were first described by Robert Hooke in 1665 when he examined a piece of cork using a microscope he had made himself. Hooke did not actually see cells in the cork; rather, he saw the walls of dead cork cells (Figure 4–1). Not until much later was it realized that the interior of the cell was an important part of the structure.

Refined versions of the *light microscope* (Figure 4–2a), which use visible light for illumination, along with the development of certain organic chemicals that specifically stain different structures in the cell, enabled biologists to discover by the early 20th century that cells contain a number of different internal structures called **organelles** (literally, "little organs"). We now know that each type of organelle in a cell performs specific functions required for the cell's existence. The development of biological stains was essential for those discoveries, since the interior of most cells is transparent in the light microscope. Most of the methods used to prepare and stain cells for observation, however, also killed the cells in the process. More

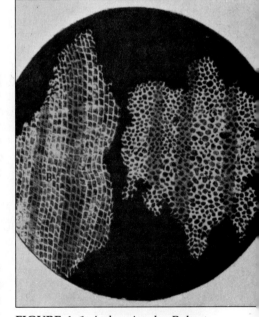

FIGURE 4–1 A drawing by Robert Hooke of the microscopic structure of a thin slice of cork. Hooke was the first to describe cells, basing his observations on the cell walls of these dead cork cells. Hooke used the term *cell* because the tissue reminded him of the small rooms that monks lived in during that period. (From the book *Micrographica*, published in 1665, in which Hooke described many of the objects that he had viewed using the compound microscope he had constructed.)

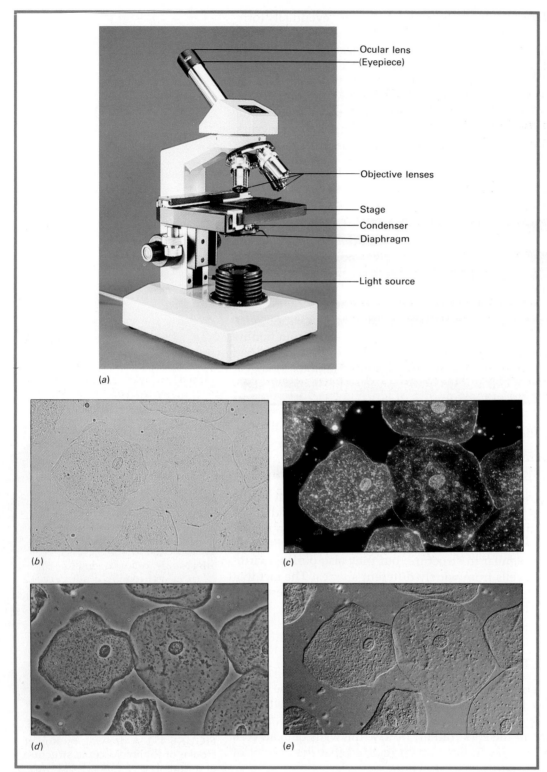

Ocular lens
(Eyepiece)

Objective lenses

Stage

Condenser

Diaphragm

Light source

(a)

(b)

(c)

(d)

(e)

FIGURE 4–2 (*a*) A student light microscope. (*b*) through (*e*) Epithelial cells using (*b*) bright field (transmitted light), (*c*) dark field, (*d*) phase contrast, and (*e*) Nomarski differential interference microscopy. The phase contrast and differential interference microscopes enhance detail by increasing the differences in optical density in different regions of the cells.

recently, sophisticated types of light microscopes have been developed that use interfering waves of light to enhance the internal structures of cells. With *phase contrast* and *Nomarski differential interference* microscopes, it is now possible to observe some of the internal structures of unstained living cells (Figure 4–2*b* and *c*). One of the most striking things that can be observed with these microscopes is that living cells contain numerous internal structures that are constantly moving and changing in shape and location.

Cells and their components are so small that ordinary light microscopes can distinguish only the gross details of many cell parts. In most cases all that can be seen clearly is the outline of a structure and its ability to be stained by certain dyes and not by others. Not until the development of the **electron microscope (EM),** which came into wide use in the 1950s, were researchers able to study the fine details, or **ultrastructure,** of cells.

Two features of a microscope determine how clearly you can view a small object. The **magnification** of the instrument is the ratio of the size of the image seen with the microscope to the actual size of the object. The best light microscopes usually magnify an object no more than 1000 times, whereas the electron microscope can magnify it 250,000 times or more. The other, even more important, feature of a microscope is its **resolving power.** Resolving power is the ability to see fine detail and is defined as the minimum distance between two points at which they can both be distinguished separately rather than being seen as a single blurred point. Resolving power depends on the quality of the lenses and the *wavelength* of the illuminating light; the shorter the wavelength, the greater the resolution. The visible light used by light microscopes has wavelengths ranging from 400 to 700 nanometers (nm); this limits the resolution (resolving power) of the light microscope to details no smaller than the diameter of a small bacterial cell.

Whereas the best light microscopes have about 500 times more resolving power than the human eye, the electron microscope has a resolving power 10,000 times greater than the eye (Figures 4–3, 4–4). This is because electrons have very short wavelengths, on the order of about 0.1 to 0.2 nm. Although this implies that the limit of resolution in the electron microscope comes close to that of the diameter of a water molecule, such resolution is difficult to achieve with biological material. It can be approached, however, when isolated molecules such as proteins or DNA are examined.

The image formed by the electron microscope cannot be seen directly. The electron beam itself consists of the charged electrons, which are focused by electromagnets in much the same way that images are focused by glass lenses in a light microscope (see Figure 4–3). For **transmission electron microscopy (TEM)** one prepares an extraordinarily *thin section* of the specimen by embedding the cells or tissue in plastic, cutting the plastic

FIGURE 4–3 Comparison of a light microscope (*left*) with transmission (*center*) and scanning electron microscopes (*right*). All three microscopes are focused by similar principles. Light rays or an electron beam focused by the condenser lens onto the specimen is magnified by the projector lens in the TEM or the eyepiece in the light microscope. The TEM image is focused onto a fluorescent screen, and the SEM image is viewed on a type of "television" screen. Lenses in the electron microscopes are actually magnets that bend the beam of electrons. The TEM diagram is inverted in order to simplify comparison with the light microscope.

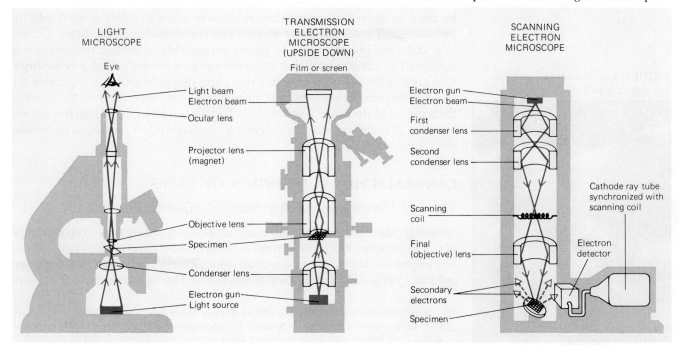

LIGHT MICROSCOPE

Eye

Light beam
Electron beam

Ocular lens

Projector lens (magnet)

Objective lens

Specimen

Condenser lens

Electron gun
Light source

TRANSMISSION ELECTRON MICROSCOPE (UPSIDE DOWN)

Film or screen

SCANNING ELECTRON MICROSCOPE

Electron gun
Electron beam

First condenser lens

Second condenser lens

Scanning coil

Final (objective) lens

Secondary electrons

Specimen

Cathode ray tube synchronized with scanning coil

Electron detector

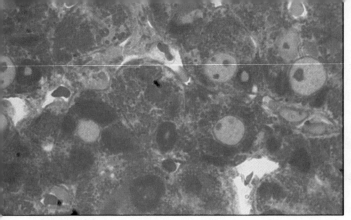

(a)

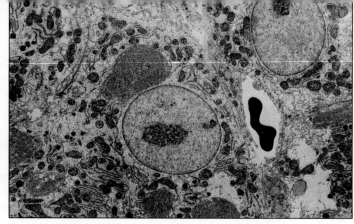

(b)

FIGURE 4–4 Comparison of a photograph taken with a modern light microscope and one taken with an electron microscope. (*a*) Rat liver cells magnified a total of 1800 times, as seen through a Nomarski interference microscope. (The magnification is greater than the theoretical limits of about 1000 times because the image has been photographed and then enlarged. Photographic enlargements do not increase the resolution of the image, so the additional magnification is sometimes referred to as *empty magnification*.) (*b*) The same cells, at about the same magnification, seen through an electron microscope. The clearer detail is a result of the greater resolving power of the electron microscope.

FIGURE 4–5 Scanning electron micrograph of budding yeast cells.

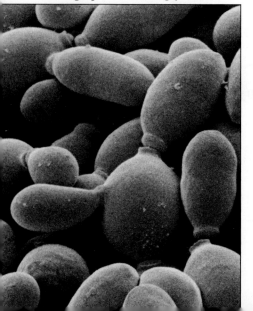

with a glass or diamond knife, and then placing the preparation on a small metal grid. The electron beam passes through the specimen and then falls on a photographic plate or a fluorescent screen. When you look at electron microscope photographs in this chapter (and elsewhere), keep in mind that they represent only a thin cross section of a cell. In order to reconstruct how something inside the cell looks in three dimensions, it is necessary to study many consecutive cross-sectional views (called serial sections) through the object.

In another type of electron microscope, the **scanning electron microscope (SEM),** the electron beam does not pass through the specimen. Instead, the specimen is coated with a thin gold film. When the electron beam strikes various points on the surface of the specimen, secondary electrons are emitted whose intensity varies with the contour of the surface. The recorded emission patterns of the secondary electrons give a three-dimensional picture of the surface of the specimen (Figure 4–5). This special kind of micrograph provides information about the shape and external features of the specimen that cannot be obtained with the transmission electron microscope.

The electron microscope is a powerful tool for studying cell structure, its primary benefit lying in its ability to provide views of the different parts of cells under different conditions. To determine the *function* of cell parts, it was necessary to develop other approaches. Researchers had to be able to purify different parts of cells so that physical and chemical methods could be used to determine what they do. There are a number of methods for purifying cell organelles that involve **cell fractionation** procedures. Generally, cells are broken apart as gently as possible and then the mixture is subjected to centrifugal force by spinning in a device called a **centrifuge.** The greater the number of rpm's (revolutions per minute), the greater the force. This permits various cell components to be separated on the basis of their different densities (Figure 4–6). Today, cell biologists often use a combination of experimental approaches to understand the function of cellular structures.

GENERAL CHARACTERISTICS OF CELLS

The Cell Membrane and Membrane Compartments

Although the microscope demonstrated striking differences from one type of cell to another, even the earliest electron microscope studies pointed out one important feature common to all cells: All cells, from bacteria to human cells, are enclosed by a **surface membrane.**[1] The surface membrane, com-

---

[1]The term *membrane* is widely used in biology to refer to any structure that is like a thin sheet. However, the cellular membranes discussed in this chapter have a unique structure consisting of a lipid bilayer and other molecular components (see Chapters 3 and 5).

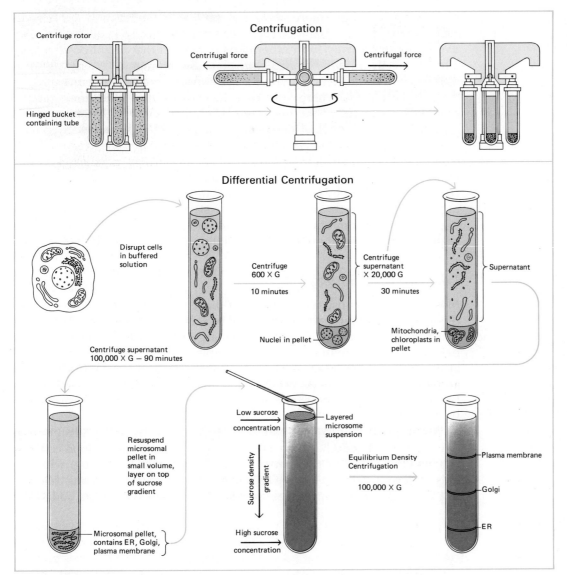

FIGURE 4–6 Cell fractionation. Cell membranes and organelles are usually separated by centrifuges, machines that can spin test tubes around. Spinning the tubes exerts a centrifugal force on its contents, which sediments particles suspended in solution (such as membranes and organelles from disrupted cells), forming a pellet at the bottom of the tube. Different cell parts have different densities, which allows us to separate them into cell fractions by centrifuging the suspension at increasing speeds **(differential centrifugation).** Membranes and the organelles from the resuspended pellets can then be further purified by **equilibrium centrifugation,** which involves layering that solution on top of a **density gradient,** usually made up of sucrose. When the density gradient is centrifuged, organelles and membranes migrate and form bands in the region of the gradient equal to their density. The purified cell fractions are then collected by puncturing the tube bottom and collecting samples of the solution.

monly known as the **plasma membrane,** has unique characteristics and plays a role in many essential functions discussed later in this and the following chapter. One essential feature of the plasma membrane is that it is a highly selective barrier. This makes the interior of the cell an enclosed compartment with a different chemical composition from the outside of the cell. Furthermore, the interior of the cell may contain other membrane-bounded compartments that form a variety of organelles.

There are a number of advantages to dividing the cell into compartments. When molecules that participate in a particular chemical reaction are concentrated in only a small part of the total cell volume, the reactants can "find each other" more easily and the rate of the reaction can be radically increased. Membrane-bounded compartments also keep certain reactive compounds away from other parts of the cell that might be adversely affected by them. Membranes also allow the storage of energy when a difference in the concentration of a substance is created on either side of the membrane; that energy can then be converted to other forms as the molecules move across the membrane from the side of high concentration to the side of low concentration. This process of **energy transduction** (discussed in Chapters 7 and 8) is the basic mechanism cells use to capture and convert energy to sustain life on Earth. Membranes in cells also serve as important work surfaces. For example, a number of chemical reactions in cells are carried out by enzymes that are bound to membranes. By organizing the enzymes that carry out successive steps of a series of reactions close together on a membrane surface, certain biological molecules that the cell requires can be made much more rapidly.

Why Are Cells So Small?

Most cells are microscopic in size, but those sizes vary over a wide range (Figure 4–7). Certain bacterial cells can be seen by a good light microscope, and some specialized animal cells are large enough to be seen with the naked eye. The human egg cell, for example, is about the size of the period at the end of this sentence. The largest cells are birds' eggs, but they are atypical in size because almost the entire mass of the egg is food reserves in the form of yolk, which is not part of the functioning structure of the cell itself.

The size and shape of cells are related to the functions the cells perform (Figure 4–8). Some cells, such as the amoeba and the white blood cell, can change their shape as they move about. Sperm cells have long, whiplike tails for locomotion, and nerve cells possess long, thin extensions that permit them to transmit messages over great distances within the body. The

FIGURE 4–7 Relative sizes of some well-known cells and their organelles on a logarithmic scale. Cells fall within broad size ranges. Prokaryotic cells vary in size from 1 to 10 μm long; eukaryotic cells (cells of plants and animals) generally fall within the range of 10 to 100 μm long. The nuclei of animal and plant cells range from about 3 to 10 μm in diameter. Mitochondria are about the size of bacteria, whereas chloroplasts are usually larger, about 5 μm long. (For an explanation of metric measurements, see Appendix C.)

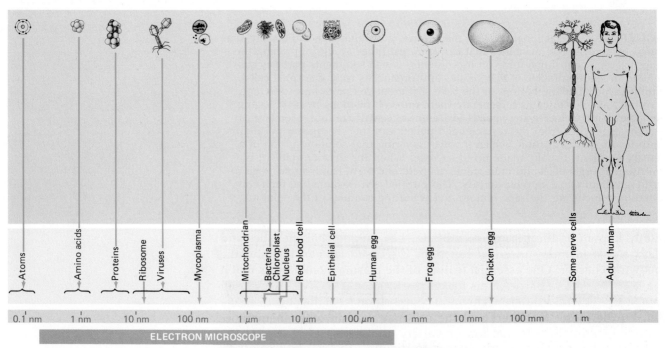

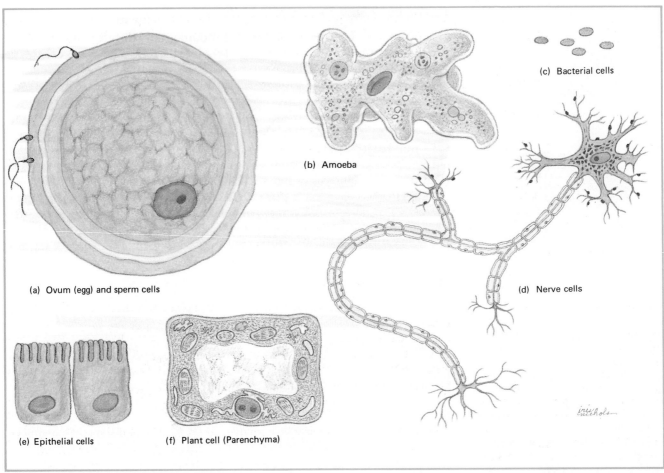

(a) Ovum (egg) and sperm cells

(b) Amoeba

(c) Bacterial cells

(d) Nerve cells

(e) Epithelial cells

(f) Plant cell (Parenchyma)

extensions on some nerve cells in the human body may be as long as 1 m. Other cells, such as epithelial cells, may be almost rectangular in shape and are stacked much like building blocks to form sheetlike structures.

Why are most cells so small? If you consider what a cell must do to grow and survive, it may be easier to understand the reasons for its small size. A cell must take in food and other materials through its plasma membrane. Once inside, these substances must move to the correct locations in the cell, where they are converted into other forms. Once the correct molecules are completed, they must again be transported to the proper sites within the cell for use. In addition, waste byproducts from various metabolic reactions must be transported out of the cell before they accumulate to toxic concentrations. In multicellular organisms, the cell must also export materials that will be used by other cells. Because cells are small, the distances molecules have to travel within them are relatively short, which speeds up many cellular functions. In addition, since essential molecules and waste products must all pass through the plasma membrane, the more surface area the cell has, the faster a given quantity of molecules can pass through it. This means that a critical factor in determining cell size is the ratio of its surface area to its volume. If you think of a cell shaped like a cube, you will see that as the length of one side is increased, the increase in the *surface area* of the cube will be proportional to the *square* of the side, but the increase in *volume* will be proportional to the *cube* of that number (Figure 4–9). The fact that the volume increases more rapidly than its surface area as the cell becomes larger places an upper limit on the size of cells. Above that size, the number of molecules required by the cell could not be transported into the cell fast enough to sustain its needs.

FIGURE 4–8 The size and shape of cells are related to the cells' functions. (*a*) An ovum (egg cell) and sperm cells. Ova are among the largest cells; sperm cells are comparatively tiny. Note the long tail (flagellum) used by the sperm cell in locomotion. By whipping its flagellum, the sperm can move toward the egg. (*b*) The amoeba changes its shape as it moves from place to place. (*c*) Bacterial cells are small, which enables them to grow and divide rapidly. (*d*) Nerve cells are specialized to transmit messages from one part of the body to another. (*e*) Epithelial cells join to form tissues that cover body surfaces and line body cavities. (*f*) The bulk of the organs of most young plants consist of parenchymal cells.

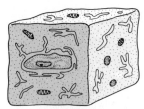

One 2-cm cube

(a)

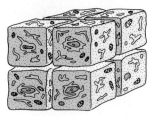

Eight 1-cm cubes

(b)

FIGURE 4–9 Eight small cells have a much greater surface area (cell membrane) in relation to their total volume than does one large cell. This concept is easier to grasp if you imagine that each of these cells is a potato. The amount of mashed potatoes you could prepare from eight small potatoes would be the same as from one large one, but which would you rather peel?

Prokaryotic and Eukaryotic Cells

Organisms can be grouped into two fundamentally different types according to the structure and complexity of their cells. **Eukaryotes** are organisms whose cells contain a membrane-bounded structure called the *nucleus*, which serves to localize the hereditary material, DNA. The cells of **prokaryotes** (meaning "before the nucleus") lack a nucleus and are generally much smaller than eukaryotic cells. Prokaryotes are single-celled organisms that belong to the kingdom Monera, which includes the bacteria and the cyanobacteria (see Chapter 24). The DNA in prokaryotic cells is usually confined to one or more nuclear regions, sometimes called **nucleoids** (Figure 4–10). Nucleoids are not enclosed by a separate membrane.

Prokaryotic cells have a **plasma membrane**, which confines the contents of the cell to an internal compartment, but they do not have distinct internal membrane systems in the form of organelles. In some prokaryotic cells the plasma membrane may be folded inward to form a complex of internal

Table 4–1 EUKARYOTIC CELL STRUCTURES AND THEIR FUNCTIONS		
Structure	**Description**	**Function**
The Cell Nucleus		
Nucleus	Large structure surrounded by double membrane; contains nucleolus and chromosomes	Control center of cell
Nucleolus	Granular body within nucleus; consists of RNA and protein	Site of ribosomal RNA synthesis; ribosome, subunit assembly
Chromosomes	Composed of a complex of DNA and protein known as chromatin; visible as rodlike structures when the cell divides	Contain genes (units of hereditary information that govern structure and activity of cell)
The Membrane System of the Cell		
Cell membrane (plasma membrane)	Membrane boundary of living cell	Contains cytoplasm; regulates movement of materials in and out of cell; helps maintain cell shape; communicates with other cells
Endoplasmic reticulum (ER)	Network of internal membranes extending through cytoplasm	Synthetic site of membrane lipids and many membrane proteins; origin of intracellular transport vesicles carrying proteins to be secreted
Smooth	Lacks ribosomes on outer surface	Lipid biosynthesis; drug detoxification
Rough	Ribosomes stud outer surface	Manufacture of many proteins destined for secretion or for incorporation into membranes
Ribosomes	Granules composed of RNA and protein; some attached to ER, some free in cytoplasm	Synthesize polypeptides
Golgi complex	Stacks of flattened membrane sacs	Modifies proteins, packages secreted proteins, sorts other proteins to vacuoles and other organelles
Lysosomes	Membranous sacs (in animals)	Contain enzymes to break down ingested materials, secretions, wastes
Vacuoles	Membranous sacs (mostly in plants, fungi, algae)	Transport and store ingested materials, wastes, water

membranes along which the energy-transforming reactions of the cell are thought to take place. Some prokaryotic cells may also have **cell walls** or **outer membranes,** which are structures that enclose the entire cell, including the plasma membrane.

The term *eukaryote* means "true nucleus" and indicates that eukaryotic cells have their DNA contained in a distinct nucleus surrounded by a nuclear membrane. They also have many types of membrane-bounded organelles that partition the cytoplasm of the cell into additional compartments. Table 4–1 summarizes the types of organelles typically found in eukaryotic cells. Some organelles may be found only in specific cells. For example, chloroplasts, structures that trap sunlight for energy conversion, are found only in cells that carry on photosynthesis. The many specialized organelles of eukaryotic cells allow them to overcome some of the problems associated with large size, so they can be considerably larger than prokaryotic cells.

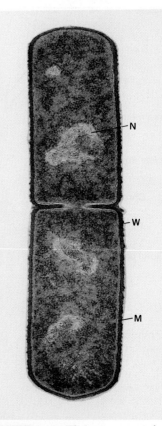

FIGURE 4–10 The structure of a prokaryotic cell that is about to complete cell division. An electron micrograph of the bacterium, *Bacillus subtilus*. This cell has a prominent cell wall (*W*) surrounding the plasma membrane (*M*). The nucleiod regions (*N*) are clearly visible.

Table 4–1 (Continued)

Structure	*Description*	*Function*
Microbodies (e.g., peroxisomes)	Membranous sacs containing a variety of enzymes	Sites of many diverse metabolic reactions
Energy-Transducing Organelles		
Mitochondria	Sacs consisting of two membranes; inner membrane is folded to form cristae	Site of most reactions of cellular respiration; transformation of energy originating from glucose or lipids into ATP
Plastids	System of three membranes; chloroplasts contain chlorophyll in internal thylakoid membranes	Chlorophyll captures light energy; ATP and other energy-rich compounds are formed and then used to convert CO_2 to glucose
The Cytoskeleton		
Microtubules	Hollow tubes made of subunits of tubulin protein	Provide structural support; have role in cell and organelle movement and cell division; components of cilia, flagella, centrioles
Microfilaments	Solid, rodlike structures consisting of actin protein	Provide structural support; play role in cell and organelle movement and cell division
Centrioles	Pair of hollow cylinders located near center of cell; each centriole consists of nine microtubule triplets (9 × 3 structure)	Mitotic spindle forms between centrioles during animal cell division; may anchor and organize microtubule formation in animal cells; absent in higher plants
Cilia	Relatively short, projections extending from surface of cell covered by plasma membrane; made of two central and nine peripheral microtubules (9 + 2 structure)	Movement of some single = celled organisms; used to move materials on surface of some tissues
Flagella	Long, projections made of two central and nine peripheral microtubules (9 + 2 structure); extend from surface of cell; are covered by plasma membrane	Cellular locomotion by sperm cells and some single-celled organisms

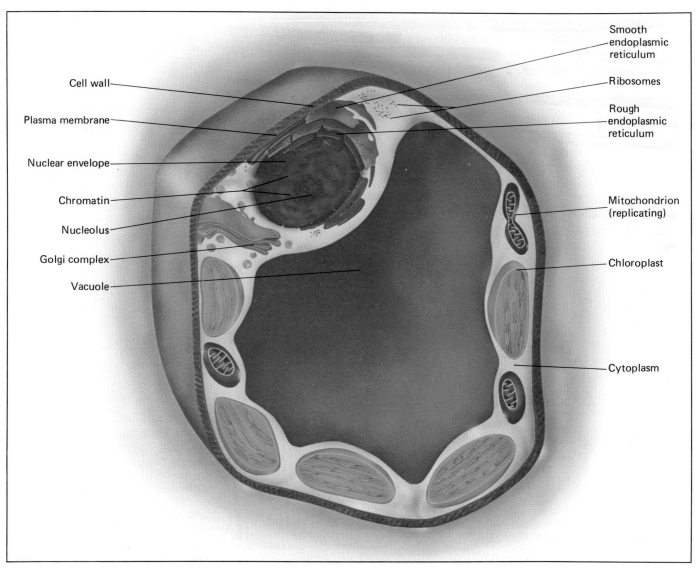

Cell wall

Plasma membrane

Nuclear envelope

Chromatin

Nucleolus

Golgi complex

Vacuole

Smooth endoplasmic reticulum

Ribosomes

Rough endoplasmic reticulum

Mitochondrion (replicating)

Chloroplast

Cytoplasm

FIGURE 4–11 Diagram of a generalized plant cell. Some plant cells will not have all the organelles shown in this diagram. Cells from photosynthetic tissues contain chloroplasts, for example, whereas root cells do not. Chloroplasts, a cell wall, and prominent vacuoles are characteristic of plant cells. Many of the other components are also found in animal cells.

Plant and Animal Cells

Although plants and animals are both eukaryotes, plant cells differ from animal cells in several ways (Figures 4–11 through 4–14):

1. Although all cells are limited by plasma membranes, plant cells are also surrounded by stiff walls containing cellulose, which prevent them from changing their shape or position.
2. Plant cells contain **plastids,** membrane-bounded structures that produce and store food material. The most familiar and abundant of these are chloroplasts.
3. Most plant cells have one large or several small conspicuous compartments called *vacuoles*, used for transporting and storing nutrients, water, and waste products.
4. Certain organelles, such as centrioles and lysosomes (discussed below), are absent from the cells of complex plants.

INSIDE THE EUKARYOTIC CELL

Early biologists believed that the cell consisted of a homogeneous jelly, which they called *protoplasm*. With the electron microscope and other modern research tools, perception of the world within the cell has been greatly expanded. We now know that the cell is highly organized and amazingly

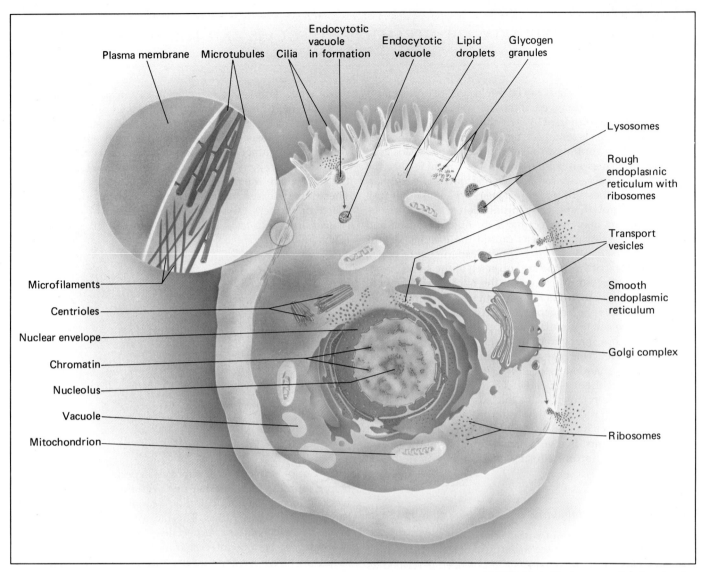

Plasma membrane   Microtubules   Cilia   Endocytotic vacuole in formation   Endocytotic vacuole   Lipid droplets   Glycogen granules

Lysosomes

Rough endoplasmic reticulum with ribosomes

Transport vesicles

Smooth endoplasmic reticulum

Golgi complex

Ribosomes

Microfilaments

Centrioles

Nuclear envelope

Chromatin

Nucleolus

Vacuole

Mitochondrion

FIGURE 4–12 A generalized animal cell. Depending on the cell type, certain organelles may be more or less prominent features.

complex (Figures 4–13 and 4–14). It has its own control center, internal transportation system, power plants, factories for making needed materials, packaging plants, and even a "self-destruct" system. Today the word *protoplasm*, if used at all, is used in a very general way. Specifically, the portion of the protoplasm outside the nucleus is called the **cytoplasm,** and the corresponding material within the nucleus is termed the **nucleoplasm.** The various organelles are suspended within the fluid component of the cytoplasm and nucleoplasm. Each of the membrane-bounded organelles forms one or more separate compartments within the cytoplasm. Together those compartments may occupy as much as half the volume of the cytoplasm in a typical animal cell.

The Cell Nucleus

The most prominent organelle within the cell is usually the **nucleus.** In most cases the nucleus is spherical or oval in shape and averages 5 μm in diameter. Due to its size and the fact that it often occupies a relatively fixed position near the center of the cell, some early investigators guessed long before experimental evidence was available that the nucleus served as the control center of the cell (see Focus on *Acetabularia*). Most cells have one nucleus, although there are a few exceptions (to be discussed in later chapters).

(Text continued on p. 96)

Focus on ACETABULARIA: THE MERMAID'S WINEGLASS AND THE SECRET OF LIFE

The role of the nucleus can be inferred from experiments in which it is removed from the cell and the consequences are examined. When the nucleus of a single-celled amoeba is removed surgically with a microneedle, the amoeba continues to live and move, but it does not grow and dies after a few days. We conclude that the nucleus is necessary for the metabolic processes (primarily the synthesis of nucleic acids and proteins) that provide for growth and cell reproduction.

But, you may object, what if the operation itself and not the loss of the nucleus caused the ensuing death? This can be decided by a controlled experiment in which two groups of amoebae are subjected to the same operative trauma, but in one group the nucleus is removed, while in the other it is not. We can insert a microneedle into some of the amoebae and push the needle around inside the cell to simulate the operation of removing the nucleus, but then withdraw the needle, leaving the nucleus inside. Amoebae treated with such a sham operation recover and subsequently grow and divide, but the amoebae without nuclei die. This permits the inference that it is the removal of the nucleus and not simply the operation that causes the death of the amoebae.

More details of nuclear function have been supplied by experiments involving *Acetabularia*, one of the largest known kinds of unicellular algae.

Introducing Acetabularia

In the imaginations of romantically inclined biologists, the little seaweed *Acetabularia* resembles a mermaid's wineglass. Less imaginatively, it has been described as looking like a little green toadstool measuring, at most, 5 to 7 cm in length (see photograph). Although it is a typical alga of tropical seas, it also occurs in some subtropical waters that are both shallow and somewhat rocky.

In the 19th century biologists discovered that this insignificant underwater plant consists of a single giant cell. Small for a seaweed, *Acetabularia* is gigantic for a cell. It consists of (1) a rootlike **holdfast,** (2) a long cylin-

drical **stalk,** and (3) at sexual maturity, a cuplike **cap.** The nucleus is found in the holdfast, about as far away from the cap as it can be. In due course, the nucleus divides by **meiosis** (a type of cell division in which the chromosome number is reduced to half the normal number), and its progeny of pronuclei swim up the stalk into the cap, where they become the pronuclei of sex cells. These are released upon maturity to swim away in search of partners. Although there are several species of *Acetabularia*, with caps of different shapes, all species function similarly.

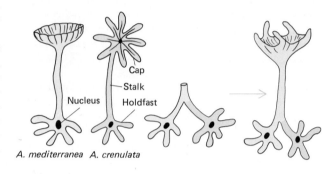

A. mediterranea   A. crenulata

Hämmerling's and Brachet's Experiments

If the cap of *Acetabularia* is removed experimentally just before reproduction, another one will grow after a few weeks. Such behavior, common among lower organisms, is called **regeneration.** This fact attracted the attention of investigators, especially Hämmerling and Brachet, who became interested in the relationship that might exist between the nucleus and the physical characteristics of the plant. Because of its great size, *Acetabularia* could be subjected to surgery impossible with smaller cells. These investigators and their colleagues performed a brilliant series of experiments that in many ways laid the foundation for much of our modern knowledge of the nucleus. In most of these experiments they employed two species of *Acetabularia*, *A. mediterranea*, which has a smooth cap, and *A. crenulata*, with a cap broken up into a series of finger-like projections.

The kind of cap that is regenerated depends on the species of *Acetabularia* used in the experiment. As you might expect, *A. crenulata* regenerates a "cren" cap, and *A. mediterranea* regenerates a "med" cap. But it is possi-

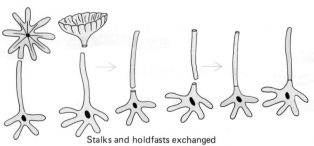

Stalks and holdfasts exchanged

ble to graft together two capless algae of different species. Through this union, they regenerate a common cap that has characteristics intermediate between those of the two species involved. Thus, it is clear that something about the lower part of the cell controls cap shape.

Stalk Exchange

It is possible to attach a section of *Acetabularia* to a holdfast that is not its own by telescoping the cell walls of the two into one another. In this way the stalks and holdfasts of different species may be intermixed.

First, we take *A. mediterranea* and *A. crenulata* and remove their caps. Then, we sever the stalks from the holdfasts. Finally, we exchange the parts.

What happens? Not, perhaps, what you would expect! The caps that regenerate are characteristic *not* of the species donating the holdfasts but of that donating the stalks!

However, if the caps are removed once again, this time the caps that regenerate are characteristic of the species that donated the holdfasts. This will continue to be the case no matter how many more times the regenerated caps are removed.

From all this we may deduce that the ultimate control of the cell is vested in the holdfast, because from now on, no matter how often the caps of these grafted plants are removed, they are always regenerated according to the species of the holdfast. However, there is a time lag before the holdfast gains the upper hand. The simplest explanation for this delay is that the holdfast produces some cytoplasmic temporary messenger substance whereby it exerts its control, and that initially the grafted stems still contain enough of that substance from their former holdfasts to regenerate a cap of the former shape. But this still leaves us with the question of what it is about the holdfast that accounts for its dictatorship. An obvious suspect is the nucleus.

Nuclear Exchange

If the nucleus is removed and the cap cut off, a new cap regenerates. *Acetabularia*, however, is usually able to regenerate only once without a nucleus. If the nucleus of an alien species is now inserted, and the cap is cut off once again, a new cap will be regenerated that is characteristic of the species of the nucleus! If more than one kind of nucleus is inserted, the regenerated cap will be intermediate in shape between those of the species that donated the nuclei.

There is only one reasonable explanation for these observations: The control of the cell exerted by the holdfast is attributable to the nucleus that is located there. This information helped provide a starting point for research on the role of the nucleic acids in the control of all cells and all cellular life.

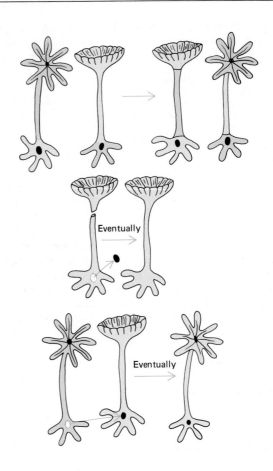

Eventually

Eventually

The Control of the Cell

Following is what is known of nuclear function based on findings in the experiments with *Acetabularia*:

1. Ultimate control of the cell is exercised by the nucleus. In the end, the form of the cap is determined by the kind of nucleus present in the holdfast. In the long run, the only thing that can successfully compete for control with a nucleus is another nucleus.
2. Some control of the form of the cap is exercised by the nonnuclear parts of the cell, presumably the cytoplasm or something in it. But since that substance can exercise control for just one regeneration, it must be limited in quantity and unable to reproduce itself without the nucleus. Perhaps it is perishable as well.
3. The source of the messenger substance must be the nucleus.

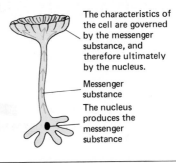

The characteristics of the cell are governed by the messenger substance, and therefore ultimately by the nucleus.

Messenger substance

The nucleus produces the messenger substance

FIGURE 4–13 The structure of an animal cell. (*a*) Electron micrograph of a human pancreas cell, whose specialized function is to secrete large amounts of a protein. Most of the structures of a typical animal cell are present. However, like most animal cells, it has certain features associated with its specialized functions. Most of the membranes in the cell are rough endoplasmic reticulum, since that is the site where the secreted protein is synthesized. The large, circular dark bodies are zymogen granules containing inactive enzymes. When released from the cell they catalyze chemical reactions such as the breaking down of peptide bonds of ingested proteins in the intestine. (*b*) A drawing based on the electron micrograph, illustrating the structures shown in this cross section. Another cell type may have many of the same membrane fractions, but other organelles may occupy most of the cytoplasm. Heart muscle cells, for example, are packed with mitochondria and would have very little endoplasmic reticulum.

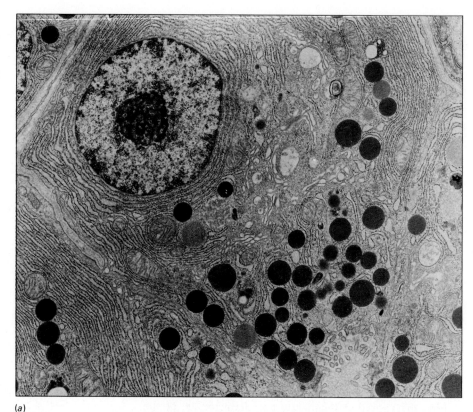

(*a*)

Chromatin Nucleolus Nucleus Smooth endoplasmic reticulum

Plasma membrane

Golgi complex

Desmosome

Ribosomes

Zymogen granules

Rough endoplasmic reticulum

Mitochondria

(*b*)

Nuclear Envelope

The nuclear envelope consists of two membranes that separate the nuclear contents from the surrounding cytoplasm (Figure 4–15). The two membranes of the envelope are fused at intervals, forming **nuclear pores,** such

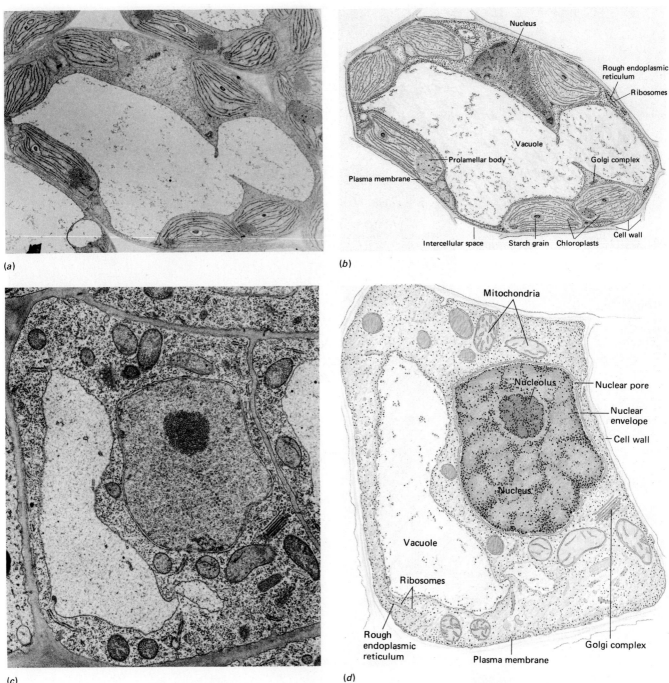

(a)

(b)

Nucleus

Rough endoplasmic reticulum

Ribosomes

Vacuole

Golgi complex

Prolamellar body

Plasma membrane

Intercellular space   Starch grain   Chloroplasts   Cell wall

(c)

(d)

Mitochondria

Nucleolus

Nuclear pore

Nuclear envelope

Cell wall

Nucleus

Vacuole

Ribosomes

Rough endoplasmic reticulum

Plasma membrane

Golgi complex

FIGURE 4–14 The structure of a plant cell. (*a*) and (*b*) Electron micrograph and drawing illustrating a cell from the leaf of a young bean plant, *Phaseolus vulgaris*, magnified approximately ×7000. The vacuole dominates most of the cross section of the cell. Prolamellar bodies are membranous regions typically seen in developing chloroplasts. (*c*) and (*d*) A root cell from *Arabidopsis thailiana*. In the root cell there are no chloroplasts, and more mitochondria are evident in the cross section.

that the inside of the nucleus communicates with the cytoplasm of the cell. Nuclear pores appear to allow the passage of materials to the cytoplasm from the interior of the nucleus and vice versa, but the process is highly selective, permitting only specific molecules to pass through these openings. Attached to the inside of the nuclear membrane is a layer of specific proteins that apparently serve as a skeletal framework for the nucleus and may play an important role in the breakdown and reassembly of the nuclear membrane during cell division.

Chromatin and Chromosomes

Almost all the DNA in a cell is located in the interior of the nucleus. The DNA molecules make up the **genes,** which contain the chemically coded instructions for producing virtually all the proteins needed by the cell. The

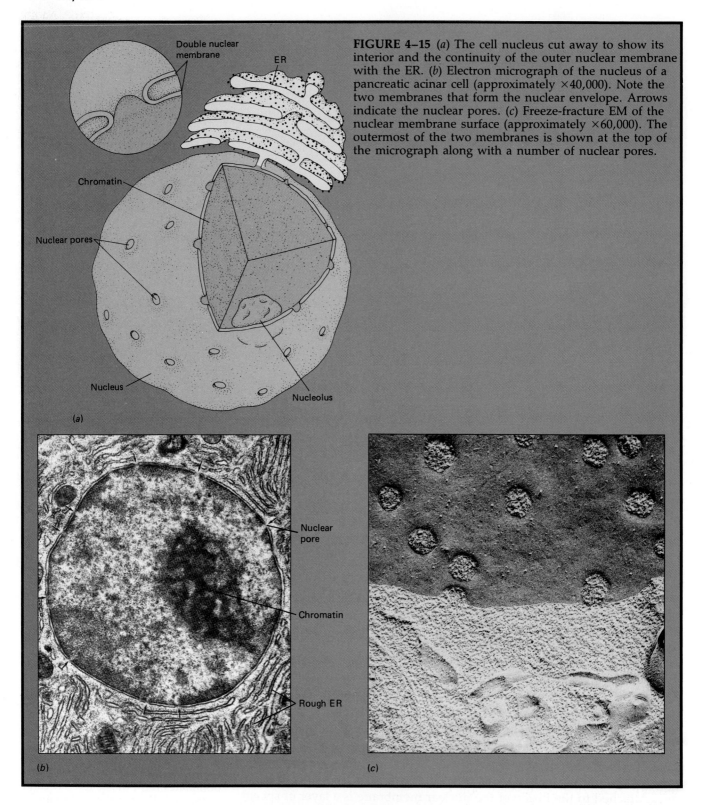

FIGURE 4–15 (*a*) The cell nucleus cut away to show its interior and the continuity of the outer nuclear membrane with the ER. (*b*) Electron micrograph of the nucleus of a pancreatic acinar cell (approximately ×40,000). Note the two membranes that form the nuclear envelope. Arrows indicate the nuclear pores. (*c*) Freeze-fracture EM of the nuclear membrane surface (approximately ×60,000). The outermost of the two membranes is shown at the top of the micrograph along with a number of nuclear pores.

nucleus controls protein synthesis (which takes place in the cytoplasm) by sending **messenger** RNA molecules, which are copies of the parts of genes that code for proteins (Chapter 13), through the nuclear membrane to small beadlike structures in the cytoplasm called *ribosomes* (see below). Once the messenger RNA is attached to the ribosomes, its information can be interpreted and protein can be synthesized.

In cells that are not in the process of dividing, the DNA is complexed with proteins in what appears to be an irregular network of granules and strands called **chromatin.** Although chromatin appears disorganized, it is not. Because DNA molecules are extremely long and thin, they are packed inside a cell in the same way that several miles of thread would be packed inside a golf ball. The chromatin is organized by arrangement into structures called **chromosomes.** As a cell divides, the chromosomes must be duplicated within the nucleus, and the two copies must be separated in such a way that no portion of either is lost or ends up in the wrong place. As the cell prepares to divide, the DNA and proteins that form each chromosome become even more tightly coiled than usual, so that the chromosomes get shorter and thicker and ultimately become visible in the microscope (Figure 4–16).

Nucleolus

In many cells the most visible structure within the nucleus is the **nucleolus** (plural, *nucleoli*), which usually stains differently from the surrounding chromatin. The nucleolus, a compact body that is *not* membrane-bounded, is the assembly site of specialized structures called *ribosomes* (discussed below).

The Internal Membrane System

Endoplasmic Reticulum and Ribosomes

If you examine the electron micrograph in Figure 4–13, one of the most prominent features is a maze of parallel internal membranes that encircle the nucleus and extend into many regions of the cytoplasm of the cell. This complex of membranes is the **endoplasmic reticulum (ER),** which can form a significant part of the total volume of the cytoplasm in certain types of cells. A higher magnification micrograph of one type of ER is shown in Figure 4–17. Remember that the electron micrograph represents only a thin cross section of the cell, so there is a tendency to interpret the photographs as depicting a series of tubes. In fact, these membranes usually consist of a series of sheets that are tightly packed and folded flat (Figure 4–17a) forming compartments within the cytoplasm. The cavity formed by the membrane sheets is called a **cisterna** (plural, *cisternae;* a Latin word meaning reservoir). In most cells the cisternae of the ER appear to be interconnected. It may be that the ER is continuous with the outer membrane of the cell nucleus (see Figure 4–15) so that the compartment formed between the two nuclear membranes is connected to the cisternae. The membranes of other organelles are not connected to the ER and appear to form distinct and separate compartments within the cytoplasm.

The ER membranes and their cisternae contain a large variety of enzymes that catalyze many different types of chemical reactions. In some cases the membranes serve as a framework for systems of enzymes on which some of the biochemical reactions take place. Other ER enzymes are located within the cisternae. The two surfaces of the membrane contain different sets of enzymes and represent regions of the cell with different synthetic capabilities, just as different regions of a factory are used to make different parts of a particular product.

Notice that in both Figures 4–13 and 4–17, one membrane face (the cytoplasmic side) is studded with dark particles, the **ribosomes,** whereas the other membrane face (the cisternal side) appears to be bare. Ribosomes are the site of protein synthesis in the cell, and many of the ribosomes found in the cell at any one time can be found bound to the ER surface. Ribosomes occur in all kinds of cells, from bacteria to complex plant and animal cells. A ribosome consists of two subunits, each composed of RNA and protein, that are combined to form the active protein-synthesizing unit (see Chapter 13). Not all proteins are synthesized on the surface of the ER

FIGURE 4–16 Scanning electron micrograph of a chromosome from a hamster cell. Just before division, the loose threads of DNA that make up the chromosomes assemble into the knotted coils you see here.

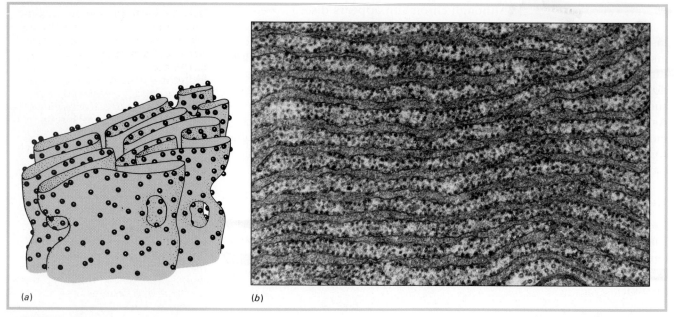

(a)   (b)

FIGURE 4–17 Rough endoplasmic reticulum (ER). (*a*) Diagram of rough ER. (*b*) Electron micrograph (magnification approximately ×70,000) of the rough ER from a secretory cell of the sea anemone *Metridium*. This form of ER consists of parallel arrays of broad, flat sacs. The outer surface (cytosolic side) of the ER membrane is studded with ribosomes; the surface on the inner compartment (called the *cisterna*) is smooth.

membranes; in fact, some are synthesized on ribosomes that are found free within the cytoplasm.

The ER plays a central role in the synthesis and assembly of proteins of ER and other membranes. Many proteins that are exported from the cell (such as digestive enzymes) or those that are destined for other organelles are formed on ribosomes that are attached to the ER membrane. The proteins are then transferred to other membranes by small **transport vesicles,**[1] which bud off the ER membrane and then insert into the target membrane. The membranes that communicate with the ER in this manner are sometimes collectively referred to as the **endomembrane system** or the **internal membrane system** (Figure 4–18). These include the nuclear and plasma membranes as well as the Golgi complex and lysosome membranes (discussed below).

There are two types of ER. **Rough ER** has ribosomes attached to it and consequently appears rough in electron micrographs. **Smooth ER** does not have ribosomes bound to it, so its outer membrane surfaces have a smooth appearance. The smooth ER is the primary site of phospholipid, sterol, and fatty acid metabolism. Smooth ER also serves an important function by localizing detoxifying enzymes that break down chemicals such as carcinogens (cancer-causing molecules) and convert them to water-soluble products that can be excreted from the body. Certain types of cells, such as liver cells, which synthesize and process much of the cholesterol and other bodily lipids and serve as the major detoxification site of the body, contain extensive amounts of smooth ER. In other cells of the body, the smooth ER may be a minor membrane component.

Signal Sequences and Membrane-Bound Ribosomes

Although ribosomes can be found both free in the cytoplasm and bound to the ER membrane, there are no apparent differences between the particles found in these two locations. In fact, it is the protein synthesized on the ribosome that appears to determine whether the ribosome is to be free or membrane-bound. Proteins that are secreted, for example, contain a se-

---

[1]Vesicles are small membrane-bounded sacs.

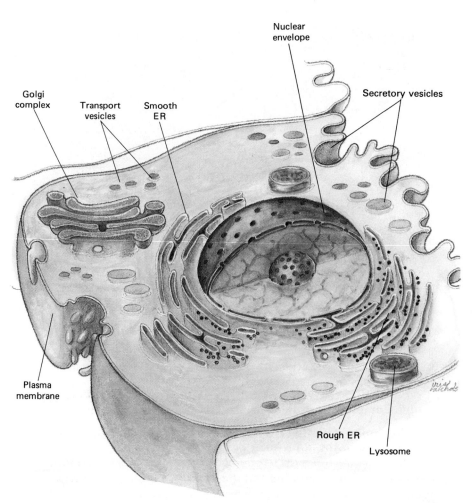

Golgi complex

Transport vesicles

Smooth ER

Nuclear envelope

Secretory vesicles

Plasma membrane

Rough ER

Lysosome

FIGURE 4–18 The endomembrane system consists of functionally different membranes that communicate with each other. Some membranes are physically connected, and others communicate through transport vesicles that bud from one membrane and fuse with another membrane in the system. Many of the membrane components originate in the ER and then progress to the cell surface or to other organelles by way of the Golgi complex. A molecule in the lumen of the ER might move via vesicles through several other compartments in the system and then pass through the plasma membrane to the outside by way of a secretory vesicle. The internal compartments of endomembranes can thus be considered counterparts to the exterior of the cell.

quence of about 20 to 24 amino acids, called the **signal sequence,** that is usually on the first part of the polypeptide made by the ribosome. If a protein lacks a signal sequence, it will be synthesized entirely in the cytoplasm on free ribosomes. A messenger RNA that codes for a secretory protein initially binds to a free ribosome in the cytoplasm. During the time that the signal sequence is being synthesized, the ribosome remains in the cytosol, but as the signal sequence emerges from the ribosome, it is recognized by a complex of molecules, termed the **signal recognition particle,** which binds to the signal sequence and directs the ribosome to a **docking protein** on the ER membrane. These events result in the binding of the ribosome to the ER (Figure 4–19). As the polypeptide chain continues to be synthesized, it is pushed through the ER membrane into the cisterna of the ER, where it will be packaged into transport vesicles for modification and secretion. In some cases the signal sequence is removed from the protein as it passes through the ER membrane; in other instances it may contribute to the structure and function of the protein and may remain intact.

Secretory proteins are not the only class of proteins that contain signal sequences and are synthesized on ER-bound ribosomes. Some proteins that are an integral part of the plasma membrane or are found in other parts of the endomembrane system may also have signal sequences. There also appear to be **stop transfer signals,** which cause certain proteins to be only partially inserted through the ER membrane; these proteins are then "stuck" in the membrane, with domains in both the ER cisterna and the cytoplasm (see Chapter 5).

FIGURE 4–19 Signal sequences. Many proteins contain targeting signals, which direct them to their correct cellular location. For example, a signal sequence on the end of a partially synthesized protein destined to be sorted and modified by the ER and Golgi complex (such as secreted proteins) is recognized by a signal recognition particle (SRP). The SRP directs the ribosome to bind to a docking protein on the ER surface, triggering the passage of the completed part of the protein across the membrane to the ER lumen. The remainder of the protein crosses the membrane as it is synthesized by the ribosome. In many cases the signal sequence is removed by a signal peptidase enzyme located on the inside of the ER membrane.

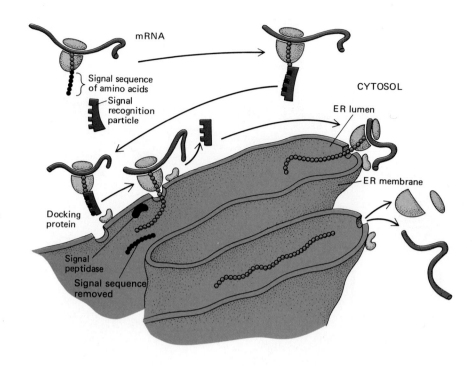

The Golgi Complex: A Protein-Packaging and -Processing Plant

The **Golgi complex** was first described in 1898 by the Italian microscopist Camillo Golgi, who found a way to specifically stain that organelle. In many cells the Golgi complex consists of stacks of platelike membranes, and may be distended in certain regions to form **vesicles,** or **sacs,** which are filled with cellular products (Figure 4–20). In a cross section view like Figure 4–20*e*, many of the ends of the sheetlike layers of Golgi membranes are distended. The arrangement of the membranes in that figure is characteristic of well-developed Golgi complexes in many types of cells. In some animal cells the Golgi complex is often located at one side of the nucleus; in other animal and plant cells there are many Golgi bodies usually consisting of separate stacks of membranes dispersed throughout the cell.

The Golgi complex functions principally as a protein-processing, -sorting, and -modifying apparatus. Most proteins that are secreted from the cell, are a part of the plasma membrane, or are routed to other organelles of the endomembrane system pass through the Golgi complex. After those proteins are synthesized on ribosomes attached to the rough ER, they are transported to the Golgi complex in small vesicles formed from the ER membrane. The vesicles fuse with the membranes of the complex that are closest to the nucleus (Figure 4–20*a* and *b*; see also Chapter 5) and the proteins then pass through the separate layers of the organelle (probably by way of membrane transport vesicles). While moving through the Golgi apparatus, the proteins are modified in different ways, resulting in the formation of complex biological molecules (Figure 4–20*c*). Often, carbohydrates are added to the protein in a stepwise process, which results in the formation of a **glycoprotein.** Some of those sugars are actually added to the protein in the rough ER and are further modified in the Golgi complex. Glycoproteins are proteins with complex branched-chain polysaccharides attached to a number of different amino acids. In other parts of the Golgi complex, molecules such as fatty acids may be covalently attached to the proteins. Each type of protein is modified in a different way, depending on complex signals that are part of the amino acid sequence of the polypeptide chain. In some cases the carbohydrates and other molecules that are added

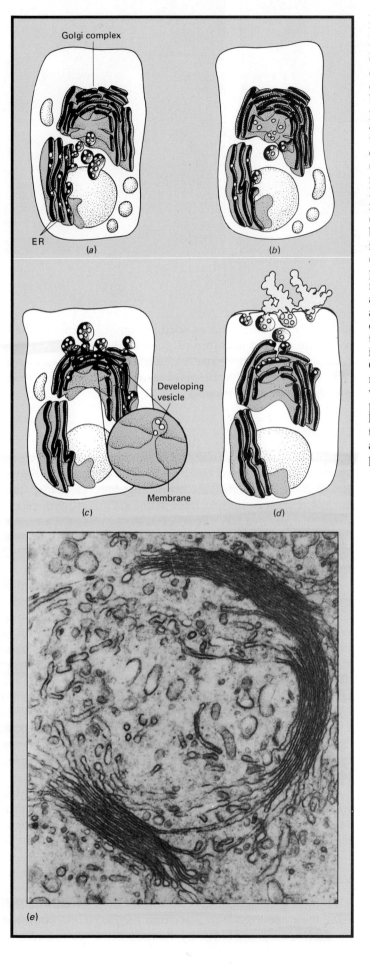

FIGURE 4–20 The Golgi complex. Diagrams (*a*) through (*d*) show the passage of proteins through the Golgi complex during the secretory cycle of a mucus-secreting goblet cell that lines the intestine. By labeling newly synthesized mucus proteins briefly with radioactive amino acids, it is possible to follow their movement in the cell at different times after their synthesis. (*a*) Immediately after synthesis, the mucus proteins are found in the ER, where they were formed on membrane-bound ribosomes. (*b*) Minutes later, some of the labeled proteins have migrated to the inner layers of the Golgi complex. (*c*) A short time later the labeled proteins can be seen at the outer face of the Golgi apparatus. Many are inside vesicles, which develop at the outer surface of the organelle. (*d*) In the final stages of secretion, labeled proteins can be seen in membrane vesicles between the Golgi complex and the plasma membrane. Some of the membrane vesicles have fused with the plasma membrane and have released their mucus contents outside the cell. (*e*) Electron micrograph of a section through the Golgi complex from the sperm cell of a ram.

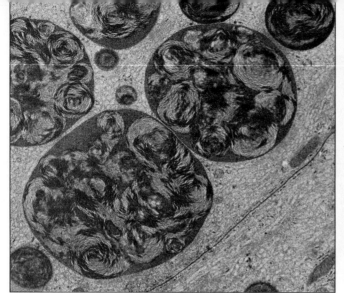

(a)

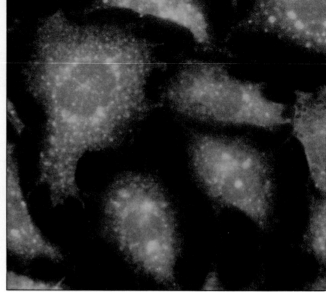

(b)

FIGURE 4–21 (*a*) Electron micrograph showing different stages of lysosome formation. Primary lysosomes bud off from the Golgi complex. After a lysosome encounters material to be digested, it is known as a *secondary lysosome*. The secondary lysosomes shown here contain various materials being digested. (*b*) Distribution of lysosomes in a cell. These cells are stained with a dye that emits a yellow-orange fluorescent light in an acid environment.

to the protein are used as sorting signals, allowing the Golgi complex to route the protein to different parts of the cell. The Golgi complex of plant cells also produces a variety of extracellular polysaccharides used as components of the cell wall. (However, cellulose is produced at the plasma membrane in most plants).

Lysosomes

Small sacs of digestive enzymes called **lysosomes** are dispersed in the cytoplasm of animal cells (Figure 4–21). The enzymes in these organelles break down complex molecules, including fats, proteins, carbohydrates, and nucleic acids, originating from both within and without the cell. About 40 different enzymes have been identified in lysosomes; most are active near the pH of 5. These enzymes originate in the Golgi complex, where they are identified and sorted to the lysosomes by unique carbohydrate signals that have been attached to the proteins.

In a cell that is short of fuel, lysosomes may break down organelles so that their components may be used as an energy source. Lysosomes are also used to degrade foreign molecules that have been ingested by cells. When a white blood cell or a scavenger cell ingests a bacterium or debris from dead cells, the foreign matter is enclosed in a vesicle composed of part of the cell membrane. One or more lysosomes then fuse with the vesicle that contains foreign matter to form a larger vesicle called a *secondary lysosome*. The powerful digestive enzymes in the lysosome come in contact with the foreign molecules and degrade them into their component parts.

When a cell dies, the lysosome membranes break down, releasing the digestive enzymes into the cytoplasm, where they break down the cell itself. This "self-destruct" system accounts for the rapid deterioration of many cells following death.

Some forms of tissue damage as well as part of the aging process may be related to "leaky" lysosomes. Rheumatoid arthritis is thought to result in part from damage done to cartilage cells in the joints by enzymes that are released from the lysosomes.

Vacuoles

Although lysosomes have been identified in almost all kinds of animal cells, their occurrence in plant and fungal cells is open to debate. Many of the functions carried out by lysosomes in animal cells are performed in

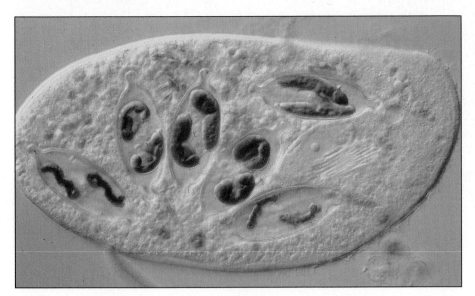

FIGURE 4–22 The protozoan *Chilodonella* (approximately ×150). Inside its body are vacuoles containing ingested diatoms (diatoms are small, photosynthetic protists). From the number of diatoms scattered about its insides, one might judge that *Chilodonella* has a rather voracious appetite.

plant and fungal cells by a large, single membrane-bounded sac referred to as the **vacuole** (see Figure 4–14). Although the terms *vacuole* and *vesicle* are sometimes used interchangeably, vacuoles are usually larger structures, sometimes produced by the merging of many vesicles.

More than half the volume of a plant cell may be occupied by a large, central vacuole containing stored food, salts, pigments, and wastes. Plants lack systems for disposing of metabolic waste products that are toxic to the cells; such waste products often aggregate and form small crystals inside the vacuole making the vacuole look almost "empty" in the EM. The vacuole may also serve as a storage compartment for inorganic compounds in plant cells and for storage molecules such as proteins in seeds. Compounds that are noxious to predators may also be stored in some plant vacuoles as a means of defense.

Vacuoles can have numerous other functions and are actually present in many types of animal cells, and most commonly in single-celled protists. Most protozoa have food or digestion vacuoles that contain food undergoing digestion (Figure 4–22), and many have contractile vacuoles, which remove excess water from the cell.

Microbodies

Microbodies are membrane-bounded organelles containing a variety of enzymes that promote an assortment of metabolic reactions. During some of those reactions, such as in the breakdown of fats, hydrogen peroxide (H_2O_2, a substance toxic to the cell) is produced. **Peroxisomes,** the type of microbody (Figure 4–23) in which these reactions occur, contain enzymes that split hydrogen peroxide, rendering it harmless. Peroxisomes in liver and kidney cells may be important in detoxifying certain compounds such as ethanol (the alcohol in alcoholic beverages).

Plant cells contain two main types of microbodies. A type of peroxisome that is found in leaves plays a part in photosynthesis (see Chapter 8). Another type of microbody, the **glyoxysome,** contains enzymes used to convert stored fats in plant seeds to sugars. The sugars are used by the young plant as an energy source and as a component needed to synthesize other compounds. Animal cells lack glyoxysomes and cannot convert fats into sugars.

FIGURE 4–23 Two peroxisomes in a leaf cell of tobacco (approximately ×58,000). Three mitochondria and portions of two chloroplasts are seen adjacent to the peroxisomes.

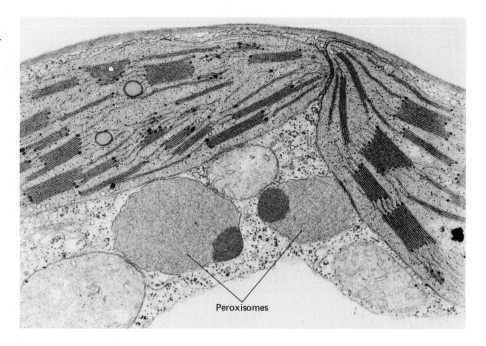

Peroxisomes

FIGURE 4–23 Two peroxisomes in a leaf cell of tobacco (approximately ×58,000). Three mitochondria and portions of two chloroplasts are seen adjacent to the peroxisomes.

Energy-Transducing Organelles

Mitochondria

Eukaryotic cells contain complex organelles called **mitochondria** (singular, *mitochondrion*). These organelles are the site of most of the chemical reactions that convert the chemical energy present in certain foods to another form of chemical energy, ATP. (You will recall from Chapter 3 that the chemical energy of ATP can be used to drive a variety of chemical reactions in the cell. The conversion of food energy to ATP, termed cellular **respiration,** is discussed in Chapter 7). Mitochondria are most numerous in cells that are very active. More than 1000 have been counted in a single liver cell, but the number varies among cell types. Mitochondria vary in size, ranging from 2 to 8 μm in length, and they are capable of changing size and shape rapidly. Mitochondria usually give rise to other mitochondria by growth and division.

Each mitochondrion is bounded by a double membrane, which creates two different compartments within the organelle (Figure 4–24; see also Chapter 7 for more detailed descriptions of structure). The **intermembrane space** is the compartment formed between the outer and inner membranes; the **matrix** is the compartment enclosed by the inner membrane. The outer membrane of the mitochondrion is smooth and somewhat like a sieve in that it allows many small molecules to pass through it. By contrast, the inner membrane is a "tight" membrane, which strictly regulates the types of molecules that can move across it; this membrane is folded repeatedly into projections, called **cristae,** which serve to increase its surface area. The matrix compartment contains enzymes that are used to break down food molecules and release their energy. The inner membrane contains a complex series of enzymes and other proteins that are involved in transforming the energy in food molecules into a different form of chemical energy stored in ATP.

Plastids

Structures known as **plastids** produce and store food materials in algae and plant cells. **Chloroplasts** are the most common type of plastid. These organelles contain the green pigments **chlorophyll a and b,** which trap light

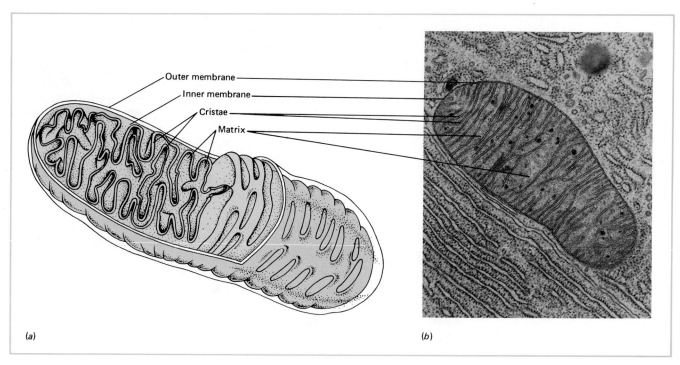

(a)

- Outer membrane
- Inner membrane
- Cristae
- Matrix

(b)

energy for photosynthesis (see Chapter 8). Chloroplasts also contain a variety of yellow and orange pigments known as **carotenoids** (see Chapter 3). A unicellular alga may have only a single large chloroplast, whereas a plant leaf cell may have as many as 20 to 100.

Chloroplasts are complex organelles, typically disc-shaped structures bounded by an inner and an outer membrane (Figure 4–25; see also Chap-

FIGURE 4–24 The mitochondrion. (*a*) Diagram of a mitochondrion cut open to show the cristae. (*b*) Electron micrograph of a typical mitochondrion from the pancreas of a bat, showing the cristae and matrix (approximately ×80,000). Note the extensive rough endoplasmic reticulum at the lower left of the micrograph and some lysosomes at the upper right.

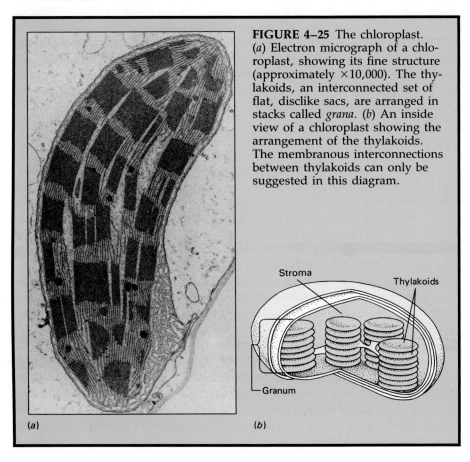

(a)

(b)

- Stroma
- Thylakoids
- Granum

FIGURE 4–25 The chloroplast. (*a*) Electron micrograph of a chloroplast, showing its fine structure (approximately ×10,000). The thylakoids, an interconnected set of flat, disclike sacs, are arranged in stacks called *grana*. (*b*) An inside view of a chloroplast showing the arrangement of the thylakoids. The membranous interconnections between thylakoids can only be suggested in this diagram.

ter 8 for more detailed descriptions of structure). The space enclosed by the inner membrane, called the **stroma,** contains enzymes responsible for producing glucose from carbon dioxide and water using energy trapped from sunlight. The inner chloroplast membrane also encloses a third system of membranes, consisting of an interconnected set of flat, disclike sacs called **thylakoids.** The thylakoid membranes form a third compartment within the chloroplasts called the **thylakoid space.** These chlorophyll-rich membranes are similar to the inner membrane of the mitochondria in that they are involved in the formation of ATP. Energy trapped from sunlight by the chlorophyll molecules is used to excite electrons which will be used to form molecules of ATP and other energy-rich compounds. The energy contained in those molecules is then used to form glucose from carbon dioxide and water.

Chloroplasts are only one of several types of organelles that can develop from precursor organelles called **proplastids.** Proplastids are found in unspecialized plant cells, particularly in growing, undeveloped tissues. Depending on the special functions that these cells will eventually have, the proplastids can mature into chloroplasts (stimulated by exposure to light); **chromoplasts,** which contain pigments that give fruits and flowers their characteristic colors; or **leucoplasts,** which are not pigmented and are primarily found in roots and tubers, where they are used to store starch.

Origin of Mitochondria and Chloroplasts

Although almost all of the DNA in eukaryotic cells resides in the nucleus, both mitochondria and chloroplasts have DNA molecules in their inner compartments. These DNA molecules specify a small number of the proteins found in these organelles. The majority of the mitochondrial and chloroplast proteins, however, are made on free ribosomes and then transported to their appropriate locations within the organelle. The existence of a separate set of DNA molecules in mitochondria and chloroplasts, along with other characteristics which are prokaryote-like, have suggested to some biologists that these organelles may have actually evolved from prokaryotic organisms which originally lived inside larger cells and gradually adapted so that they were no longer autonomous organisms. This idea has become a major part of one theory concerning how eukaryotic organisms came into existence (Chapter 20).

The Cytoskeleton

If you look closely at cells from different animal tissues, you will find striking and characteristic differences in cell shape. If you watch these cells while they are growing in the laboratory, it will also be apparent that the cells can change shape and in many cases can move about. The shapes of these cells and their ability to move are determined in large part by a complex network of protein filaments within the cell called the **cytoskeleton.** The term is somewhat misleading because it implies a static structure, whereas the cytoskeleton as a whole is highly dynamic and constantly changing. The protein filaments that make up the cytoskeletal framework were originally classified on the basis of their relative sizes. The two major types of filaments that make up the cytoskeleton in all eukaryotic cells are **microfilaments** (also known as **actin filaments**), which are 7 nm in diameter, and **microtubules,** which are 25 nm in diameter. Both microfilaments and microtubules are fibers formed from beadlike *globular* protein subunits, which can be rapidly assembled and disassembled in the cell. Although both types of filaments are major components in the cytoskeleton of the cell, they also play a role in forming other structures involved in cellular movement and organization.

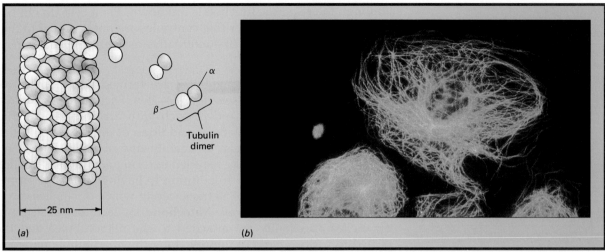

FIGURE 4–26 Microtubules.
(*a*) Structure and assembly of microtubules. Microtubules are constructed by adding dimers of alpha- and beta-tubulin to an end of the hollow cylinder. Each turn of the spiral takes 13 dimers. Disassembly occurs by removal of subunits from the ends of the filaments. (*b*) Fluorescent micrograph of a kangaroo rat epithelial cell showing the extensive distribution of microtubules. This cell was stained with fluorescent antibodies, which bind to the tubulin, permitting the microtubules to be viewed.

In many animal cells there is also a third class of filaments, **intermediate filaments,** which have a diameter of 8 to 10 nm, intermediate to that of the other two. Intermediate filaments are made from *fibrous* protein subunits and are more stable than microtubules and microfilaments. One type of intermediate filament is made of keratin, a protein similar in primary structure to the keratin of hair and fingernails.

Microtubules

Microtubules are hollow, rod-shaped structures (Figure 4–26). In addition to playing a role in the formation of the cytoskeletal structure, they are involved in the movement of chromosomes during cell division and are the major structural components of cilia and flagella, special structures used in locomotion. In order for microtubules to act as a structural framework or participate in cell movement, they must be anchored to other parts of the cell. In nondividing cells the microtubules appear to extend from a region called the **cell center** or **microtubule-organizing center.** In the cell center of almost all animal cells are two structures arranged at right angles to each other called **centrioles** (Figure 4–27). These structures are made of nine sets of three microtubules arranged to form a hollow rod. The centrioles replicate during cell division and appear to play a role in microtubule assembly,

FIGURE 4–27 Centrioles. (*a*) Electron micrograph of a pair of centrioles from monkey endothelial cells (approximately ×73,000). (*b*) A line drawing of the centrioles. Note that one centriole has been cut longitudinally and one transversely.

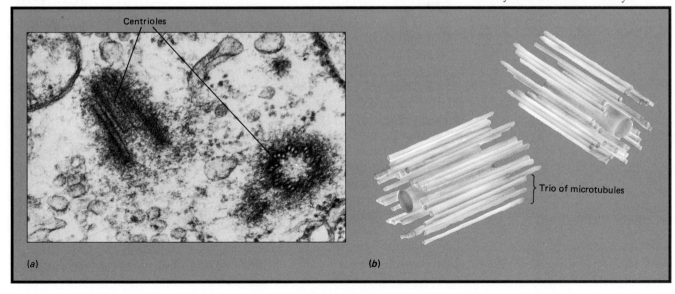

although their specific function is unknown. Plant cells do not have centrioles, which suggests either that centrioles are not essential to the microtubule assembly process or that alternative assembly mechanisms are possible.

MICROTUBULE ASSEMBLY. Microtubules are formed from dimers[1] of protein subunits called **tubulins** (see Figure 4–26*b*). Each dimer is made up of two very similar subunits, α and β, each with a molecular weight of about 55,000. Microtubules grow by the addition of dimers preferentially to one end of the tubules and can be readily disassembled by the removal of subunits, which can then be recycled to form microtubules in other parts of the cell (see Figure 4–26). In addition to having structural properties, microtubules appear to serve as tracks along which organelles can be moved to different parts of the cell. Mitochondria, secretory vesicles, and other organelles apparently are attached to the microtubules and then transported to various parts of the cell along the microtubule network by ATP-requiring proteins, which act as "motors" for the movement. One such motor protein, named *kinesin*, has been isolated and can be shown to direct the movement of isolated organelles along purified microtubules.

The ability of microtubules to be assembled and disassembled rapidly can be seen during cell division, when much of the cytoskeletal apparatus in cells appears to break down and many of the tubulin subunits are reassembled into a structure called the **spindle,** which serves as a framewok for the separation of daughter chromosomes into the progeny cells.

CILIA AND FLAGELLA. Microtubules are used in cell movements. Many cells have movable whiplike structures projecting from their surface that exhibit a beating motion. If a cell has one, or only a few, of these appendages and they are relatively long in proportion to the size of the cell, they are called **flagella** (singular, *flagellum*). If the cell has many short appendages, they are called **cilia** (singular, *cilium*). Both cilia and flagella are used by cells to move through a watery environment or to move liquids and particles across the surface of the cell. These structures are commonly found on one-celled and small multicellular organisms. In animals flagella serve as the tails of sperm cells and cilia commonly occur on the surface of cells that line internal ducts of the body (e.g., respiratory passageways).

Each cilium or flagellum consists of a slender, cylindrical stalk covered by an extension of the plasma membrane. The core of the stalk contains a group of microtubules arranged so that there are nine pairs of tubules around the circumference and two microtubules in the center (Figure 4–28). This **9 + 2 arrangement** is characteristic of all eukaryotic cilia and flagella. The microtubules move by sliding in pairs past each other. The sliding force is generated by *dynein* proteins, which are attached to the microtubules like small arms. These proteins use the energy stored in ATP in such a way that the arms on one pair of tubules are able to "walk" along the adjacent pair of tubules, causing the entire structure to bend back and forth. Thus, the microtubules on one side of a cilium or a flagellum extend farther toward the tip than those on the other side, resulting in a beating motion (Figure 4–28*d*).

At the base of each cilium and flagellum is a **basal body,** which has a "9 × 3" structure similar to that of a centriole (see Figure 4–27). The basal body appears to be the organizing structure for the cilium or flagellum when it first begins to form. However, experiments have shown that as growth proceeds the tubulin subunits are added to the tips of the microtubules rather than the base of the structure (Figure 4–29).

---

[1]A dimer is a structure formed by the combination of two monomers (similar, simpler units).

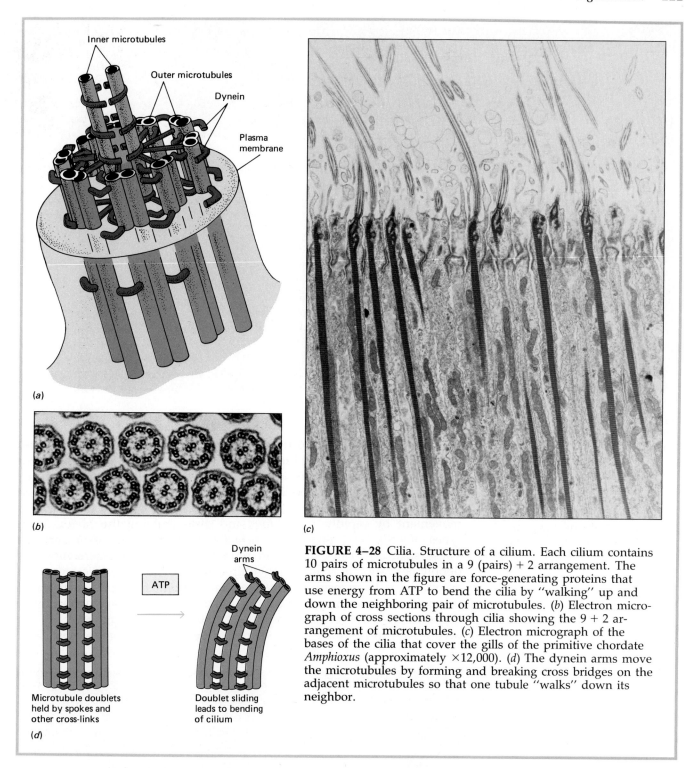

Inner microtubules

Outer microtubules

Dynein

Plasma membrane

(a)

(b)

(c)

ATP

Dynein arms

Microtubule doublets held by spokes and other cross-links

Doublet sliding leads to bending of cilium

(d)

FIGURE 4–28 Cilia. Structure of a cilium. Each cilium contains 10 pairs of microtubules in a 9 (pairs) + 2 arrangement. The arms shown in the figure are force-generating proteins that use energy from ATP to bend the cilia by "walking" up and down the neighboring pair of microtubules. (b) Electron micrograph of cross sections through cilia showing the 9 + 2 arrangement of microtubules. (c) Electron micrograph of the bases of the cilia that cover the gills of the primitive chordate *Amphioxus* (approximately ×12,000). (d) The dynein arms move the microtubules by forming and breaking cross bridges on the adjacent microtubules so that one tubule "walks" down its neighbor.

Microfilaments

Microfilaments are solid fibers composed of the protein *actin* and actin-associated proteins (Figure 4–30). In muscle cells actin is associated with another protein, myosin, to form fibers that generate the forces involved in muscle contraction (Chapter 39). Actin microfilaments perform two different types of functions in nonmuscle cells. When actin is associated with myosin, it can form contractile structures that are involved in various cell movements. Actin can also be cross-linked with other proteins to form

(a)

(b)

(c)

(d)

FIGURE 4–29 How cilia grow.
(a) A carpet of cilia lining the trachea of a rat. (b) One of the millions of cells bearing those cilia, at an early stage of their life. The new cilia project like spines from the cell surface. In (c) they are much longer, and in (d) they form a pattern like a crown or flower on the top of the cell. In the center of the radiating cilia are a group of finger-like microvilli (see Chapter 5).

bundles of fibers that provide mechanical support for various cell structures. *Stress fibers* are actin bundles that lie close to the plasma membrane of fibroblast (connective tissue) cells and appear to provide the stress or tension that causes these cells to assume a flattened shape. Actin fibers themselves cannot contract, but they appear to be capable of generating movement by rapidly assembling and disassembling the fibers. Many types of cells have finger-like *microvilli* projecting from their surfaces. These structures can extend and retract due to the polymerization and depolymerization of actin fibers within the microvilli. Actin microfilaments associated with myosin are involved in transient functions such as cell division in animals, in which contraction of a ring of actin complexed with myosin causes the constriction of the cell to form two daughter cells. This occurs after microtubules act to separate duplicated chromosomes.

FIGURE 4–30 Actin filaments. An electron micrograph of isolated actin filaments, spread on a plastic film.

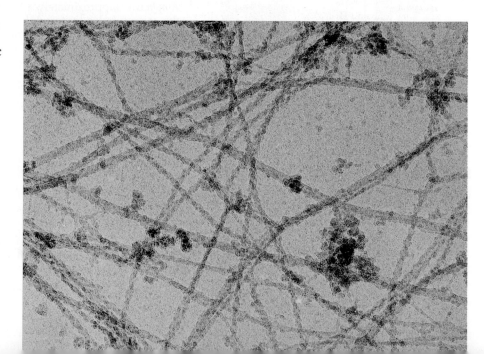

Intermediate Filaments

Intermediate filaments are very stable, tough fibers made of polypeptides that can range widely in size among different cell types and different species. These fibers are believed to help strengthen the cytoskeleton and are abundant in parts of a cell that may be subject to mechanical stress. The assembly of these filaments is probably irreversible; unpolymerized subunits are not abundant in cells. Cells may be able to regulate the length of intermediate filaments, however, by use of enzymes that break down their polypeptides into smaller fragments. It is not clear whether they are involved in cellular functions aside from their structural role.

Microtrabecular Lattice

One of the difficulties in determining the ultrastructural details of the cytoskeleton with conventional electron microscopy is that the sections of the cell are too thin to distinguish connections that may be present between the different fibrous elements. One approach to this problem is the use of a high-voltage electron microscope, which allows the viewing of thicker sections of cells without a significant loss of resolution. When cells are examined under certain conditions with that instrument, the primary filaments of the cytoskeleton appear to be connected by a network of fine protein fibers called the **microtrabecular lattice** (Figure 4–31). Other approaches, however, have failed to reveal such an extensive network through the cytoplasm. It is not clear whether this interconnecting network (which might link elements of the cytoskeleton together and provide a structural framework for the cytoplasm itself) actually exists in living cells or whether it is produced by the association of soluble proteins in the cytoplasm during the sample preparation process.

The Extracellular Matrix

Although the contents of the cell are effectively contained by the plasma membrane, most cells are also surrounded by some type of secreted coating that extends beyond the cell surface. Plant cells are surrounded by thick **cell walls** that contain multiple layers of the polysaccharide *cellulose*.

FIGURE 4–31 Elements of the cytoskeleton. (*a*) The cytoskeleton consists of networks of several types of fibers including microtubules, microfilaments (shown as stress fibers), intermediate filaments (not depicted), and perhaps microtrabecular strands. The cytoskeleton forms the shape of the cell, anchors organelles, and sometimes rapidly changes shape during cellular locomotion. (*b*) Micrograph showing a fibroblast cell. Microfilaments are stained in blue and red, microtubules in green.

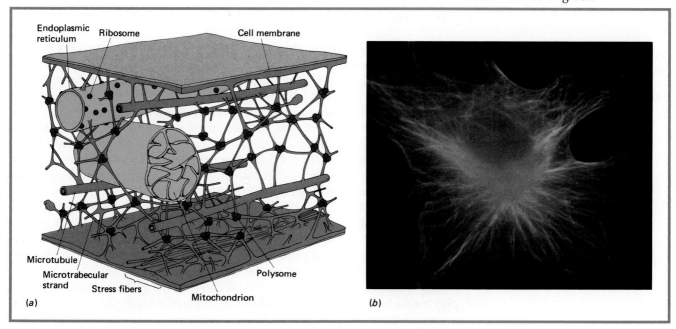

Endoplasmic reticulum Ribosome Cell membrane

Microtubule
Microtrabecular strand
Stress fibers
Polysome
Mitochondrion

(a)   (b)

FIGURE 4–32 Plant cell walls consist of multiple layers of cellulose fibers secreted from the cell. Between two adjacent plant cells is a region containing gluelike polysaccharides, called *pectins,* which cement the cells together. Growing plant cells first secrete a thin primary wall that is flexible and can stretch as the cell grows. The thicker layers of the secondary wall are secreted after the cell stops growing.

(a)

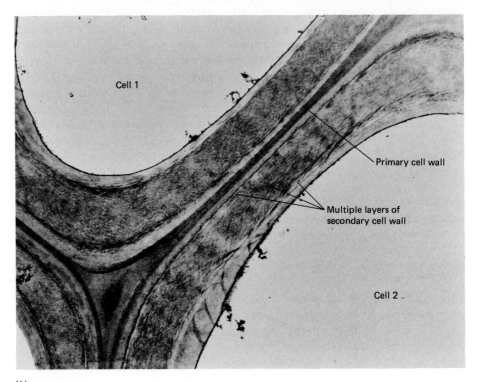

(b)

These molecules are formed into bundles of fibers that make up the bulk of the cell wall (Figure 4–32). Other polysaccharides are used in the cell wall to form cross links between the cellulose fibers. Each layer of cellulose fibers in plant cell walls runs in a different direction from the preceding layer, giving the structure great mechanical strength. Growing plant cells secrete a thin *primary cell wall,* which can stretch and expand as the cell increases its size. After the cell stops growing, either new wall material is secreted that thickens and solidifies the primary wall, or multiple layers of a *secondary cell wall* with a different composition are formed between the primary wall and the plasma membrane.

Animal cells do not have rigid cell walls, although many secrete proteins and polysaccharides that are bound to their outer surfaces and fill spaces between cells in tissues. The **glycocalyx** is a coat formed by polysaccharide side chains of lipids and proteins that are a part of the plasma membrane. Many of these molecules contain negatively charged regions, which give a negative charge to the surface of most cells. In many cases these coatings play a role in cellular contact and recognition, in addition to increasing the mechanical strength of multicellular tissues.

■ SUMMARY

I. The cell is considered the basic unit of life because it is the smallest self-sufficient unit of living material.

II. Modern cell theory states that organisms are composed of cells and products of cells. All cells arise by division of preexisting cells.

III. Biologists have learned much about cellular structure by studying cells with light and electron microscopes. The electron microscope has superior resolving power, enabling investigators to see details of cell structures not observable with conventional microscopes. Information about the function of cellular structures requires the use of cell fractionation and biochemical methods in addition to microscopic observations.

IV. All cells are surrounded by a plasma membrane that forms a cytoplasmic compartment, which contains the contents of the cell.

  A. Membrane-bounded compartments allow cells to conduct specialized activities within small areas of the cytoplasm, are used to concentrate molecules, serve as a system of energy storage, and are used to organize metabolic reactions within the cell.

  B. Cells are small so that the ratio of surface area to cell volume is favorable for rapid diffusion of molecules into the cell.

  C. Prokaryotic cells are bounded by a plasma membrane but lack a nucleus and have little or no internal membrane organization.

  D. Eukaryotic cells have a nucleus and a cytoplasm, which is organized into membrane-bounded compartments called *organelles.* Plant cells differ from animal cells in that they possess rigid cell walls, plastids, and large vacuoles; complex plant cells lack centrioles.

V. The organelles of eukaryotic cells assume many diverse functions.

  A. The nucleus, the control center of the cell, contains genetic information in the form of genes on the chromosomes.

    1. The nucleus is bounded by a double-membrane system with pores that communicate with the cytoplasm.

    2. Genetic information in the nucleus is carried by the DNA, which is complexed with protein to form a material known as *chromatin.* Chromatin complexes are organized into chromosomes, which become visible when the cell divides.

    3. The nucleolus is a region in the nucleus that is the site of ribosome RNA synthesis and ribosome assembly.

  B. The endoplasmic reticulum (ER) is a series of folded internal membranes that has many functions.

    1. Rough ER is studded along its outer walls with ribosomes, which manufacture proteins.

    2. Smooth ER is the site of lipid biosynthesis and detoxifying enzymes.

    3. Proteins synthesized on rough ER can be transferred to other membranes or secreted from the cells by transport vesicles, which are formed by membrane-budding and are then targeted to different cellular membrane locations.

  C. The Golgi complex is a series of flattened membrane sacs that process, sort, and modify proteins synthesized on the ER. It adds carbohydrates and lipids to proteins and can route proteins (by way of transport vesicles) to the plasma membrane, lysosomes, and possibly other membrane systems.

  D. Lysosomes function in intracellular digestion; they contain degradative enzymes that break down substances internalized by cells, as well as worn-out cell structures.

  E. Microbodies are membrane-bounded sacs that can contain enzymes with diverse functions. Peroxisomes are microbodies that break down hydrogen peroxide.

  F. Mitochondria are double-membrane systems in which the inner membrane is folded to form cristae.

    1. The matrix of the mitochondrion (space inside the inner membrane) is the site where energy-rich molecules derived from glucose or fatty acids are broken down, releasing chemical energy.

    2. Proteins in the inner mitochondrial membrane are involved in transforming energy released by the breakdown of glucose or fatty acids into chemical energy stored in ATP.

  G. Cells of algae and plants contain plastids; chloroplasts are triple-membrane systems containing internal thylakoid membranes, which are organized as stacks of flat, disclike structures.

    1. Thylakoid membranes contain chlorophyll, which traps energy in sunlight and plays a role in its transfer to chemical energy in the form of ATP.

    2. The stroma is the space bounded by the inner membrane. The stroma is the site of carbohydrate synthesis from carbon dioxide and water using energy from the ATP synthesized in the thylakoids.

  H. The cytoskeleton is an internal framework in animal cells made of at least three types of fiber. Much of the cytoskeleton can be rapidly disassembled and reassembled in a different form, altering the shape of the cell.

    1. Microtubules are hollow cylinders formed from subunits of the protein tubulin.

    2. Microfilaments, filaments with a smaller diameter than microtubules, are formed from subunits of the protein actin.

    3. Intermediate filaments are formed from several different types of protein.

    4. Microfilaments and microtubules can be rapidly assembled and disassembled; intermediate filaments are stable structures.

  I. Cilia and flagella are structures that project from the cell surface and are used for cell movement. They are formed from microtubules (9 + 2 structure) and covered by the plasma membrane.

  J. Microtubules are used to form centrioles and basal bodies (9 × 3 structure), which appear to be organizing centers for microtubule formation in animal cells.

  K. Actin microfilaments can generate movement by rapid polymerization and depolymerization. Actin microfilaments associated with other proteins such as myosin can slide past one another, generating force and movement.

L. Plant cells secrete cellulose and other polysaccharides to form rigid cell walls.

M. Some animal cells are covered by a glycocalyx, a coating formed from carbohydrate regions of glycoproteins and glycolipids on the surface of the cell.

■ POST-TEST

1. The ability of a microscope to reveal fine detail is known as _____ _____.

2. Proteins that are to be secreted from the cell are synthesized on _____ bound to the _____ _____.

3. The hereditary material _____ is found in the _____ of prokaryotic cells. In dividing eukaryotic cells, it is complexed with proteins in tightly coiled structures called _____. In nondividing cells, it is in the form of loose strands called _____.

4. Many proteins destined for the plasma membrane are synthesized on the _____ _____ _____. They are then transferred by _____ _____ to the _____ _____, where they are modified by the addition of _____ and _____.

5. Powerful hydrolytic enzymes contained in the _____ are released when the cell dies and digests the cellular remains.

6. Membrane-bounded organelles that break down H_2O_2 are termed _____.

7. The shelflike folds of the inner mitochondrial membrane called _____ are the site of _____ synthesis.

8. The cytoplasmic organelles involved in the synthesis and storage of carbohydrates termed _____ contain _____ different membranes.

9. Chlorophyll, which is located in the _____ membranes of chloroplasts, is used to trap energy from _____ for use in _____ synthesis.

10. The cylindrical, hollow cytoplasmic filaments, _____, play a role in controlling the shape and movement of cells.

11. The flexible framework in the cytoplasm of the cell called the *cytoskeleton* is composed of _____, _____, and _____ filaments.

12. _____ and _____ are movable, whiplike structures projecting from the cell surface. These are used to move the cell through surrounding liquid or to move liquid across the surface of the cell. The core of each is composed of 11 groups of _____ arranged with _____ in the center and _____ pairs around the circumference.

13. In addition to having a cell membrane, plant cells are surrounded by a _____ _____, formed primarily from fibers of the polysaccharide _____, which is secreted from the cell.

14. _____ are the membrane compartments in plant cells that are used for the storage of water and waste products. They may also have functions similar to _____ of animal cells.

15. Ribosome subunits are assembled in the _____ region of the _____ in eukaryotic cells.

16. The Golgi complex modifies proteins by adding complex carbohydrates to certain amino acids in the polypeptide chains to form _____.

17. Label the following diagrams of plant and animal cells (see Figures 4–11 and 4–12 for the correct labels.)

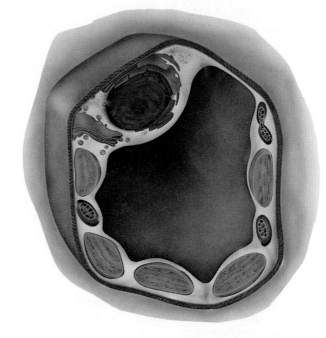

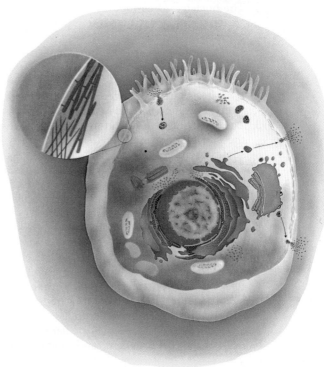

Match the subcellular organelles in Column A with their functions in Column B.

Column A
18. glyoxysomes
19. centrioles
20. chromosomes
21. cilia
22. flagella
23. Golgi complex
24. lysosomes
25. microfilaments
26. microtubules
27. mitochondria
28. nucleolus
29. nucleus
30. plastids
31. ribosomes
32. rough endoplasmic reticulum
33. smooth endoplasmic reticulum
34. vacuoles

Column B
a. Organelles that contain enzymes to convert stored fats into sugars
b. Site of lipid biosynthesis
c. Internal membranes, site of protein synthesis for plasma membrane and secreted proteins
d. RNA and protein particles involved in protein synthesis
e. Packages and processes secretory products of the cell
f. Packets of hydrolytic enzymes
g. Move materials along the cell surface
h. Long projections from the cell surface that move the cell along

i. Contains DNA and chromosomes
j. Ribosome assembly site
k. Contain the genes
l. Membrane compartment that regulates water and waste
m. Site of most reactions of cellular respiration
n. Thin fibers that provide structural support inside the cell
o. Component of cilia, flagella, centrioles
p. Pairs of cylindrical structures containing microtubules in a 9 × 3 arrangement
q. Membranous structures containing pigments

■ REVIEW QUESTIONS

1. Trace the development of the cell theory. Why is this theory important to an understanding of how living things work?
2. What are the main differences between plant and animal cells? between prokaryotic and eukaryotic cells?
3. Draw diagrams of a prokaryotic cell, a plant cell, and an animal cell. Label the organelles.
4. Sketch the membranes of chloroplasts and mitochondria. Label the membranes and their compartments. Describe the activities that take place on the different membranes and in the different compartments of these organelles.
5. What are the functions of each of the following?
   a. ribosomes
   b. endoplasmic reticulum
   c. Golgi complex
   d. lysosomes
6. Trace the path of a protein from its site of synthesis to its final destination for the following:
   a. a secretory protein
   b. a protein found inside the lysosome
   c. a protein associated with the plasma membrane
7. Describe the differences between microfilaments and microtubules. Compare their structure and the different roles they play in cell structure and function.
8. Why are lysosomes sometimes referred to as the *self-destruct system* of the cell?
9. Describe plant cell walls. How are they formed?

■ RECOMMENDED READINGS:

Alberts, B., Bray, D., Lewis, M., Raff, M., Roberts, K., and J.D. Watson. *Molecular Biology of the Cell.* Garland, 1983, Chapters, 4, 7, 8, 10, 12 and 19. A comprehensive text on cell biology, well written and well illustrated.

Avers, C.J. *Molecular Cell Biology,* Benjamin Cummings, Menlo Park, CA. 1986. An excellent presentation of the details of cell structure and function.

Darnell, J., Lodish, H., and D. Baltimore. *Molecular Cell Bi-* *ology.* Scientific American Books. 1986. Chapters 1, 5, 6, 14, 18, 19, and 20. A more recent and equally comprehensive text to Alberts et al.

de Duve, C. *A Guided Tour of the Living Cell.* Scientific American Library 1984. An engrossing, beautifully illustrated tour of the cell in the form of journeys through different membrane and organelle systems.

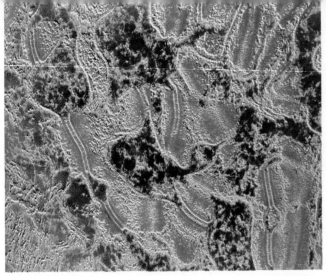

False-color TEM of desmosomes in human skin tissue

5

Biological Membranes

■ **LEARNING OBJECTIVES**

After you have read this chapter you should be able to:

1. Discuss the importance of the cell membrane to the cell, describing its various functions.
2. Describe the currently accepted model for the structure of the cell membrane.
3. Discuss the properties of the lipid bilayer and explain how they are reflected in the fluid mosaic model for membrane structure.
4. Discuss how membrane proteins associate with the lipid bilayer; discuss the different functions that membrane proteins assume.
5. Contrast the physical processes of diffusion and osmosis with the physiological process by which materials are transported across cell membranes.
6. Solve simple problems involving osmosis; for example, predict whether cells will swell or shrink under various osmotic conditions.
7. Summarize the currently accepted hypotheses concerning how small hydrophilic molecules can move across membranes.
8. Compare the different processes of facilitated diffusion and active transport; discuss the ways in which energy is supplied to active transport systems.
9. Compare endocytotic and exocytotic transport mechanisms.
10. Describe the structures and compare the functions of desmosomes, tight junctions, gap junctions, and plasmodesmata.

It is difficult to discuss the structure and function of cells and cell organelles without considering the role of membranes. In order to carry out the many chemical reactions necessary to sustain life, the cell must maintain an appropriate internal environment. This is possible because all cells are physically separated from the outside world by a limiting **plasma membrane.** The extensive internal membranes of eukaryotic cells create additional compartments containing unique environments where highly specific functions necessary for the maintenance and survival of the cell are carried out. Without membranes it would have been impossible for life on Earth to have evolved to its present complexity.

Cell membranes are not inanimate walls; they are complex and dynamic structures made from molecules that have unusual properties, permitting selective interactions between internal membrane systems within the cell and between the cell and the environment. Among the many functions of cell membranes are the regulation of movement of materials in and out of the cell, the transmission of signals and information between the environment and the interior of the cell, and the ability to act as an energy transfer and storage system.

In order to understand how membranes do these things, we need to examine what we know about the structure and composition of membranes in general. This chapter examines how materials ranging from simple to complex molecules are able to move across membranes. It also considers specialized features that lead to complex interactions between membranes of different cells. Although most of our discussion centers on the structure and functions of plasma membranes, many of the concepts discussed are applicable to internal membrane systems as well.

THE STRUCTURE OF BIOLOGICAL MEMBRANES

When you examine an electron micrograph and compare the sizes of different structures in the cell, one of the most striking things is how exceedingly thin membranes seem to be (Figure 5–1). Cell membranes are no more than 10 nm thick.

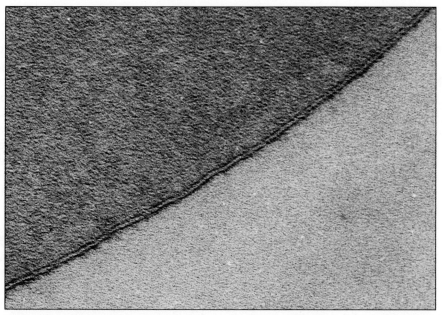

(a)

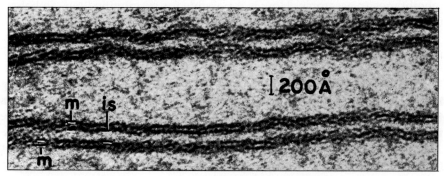

(b)

FIGURE 5–1 (a) The plasma membrane is the interface between the cell and its environment. (b) Electron micrograph (approximately ×240,000) of four regions of a plasma membrane. The black line is a size marker indicating 200 Ångstrom units or 20 nm. The dark lines represent the hydrophilic heads of the lipids, while the light zone represents the hydrophobic tails. m, membrane; is, intracellular space.

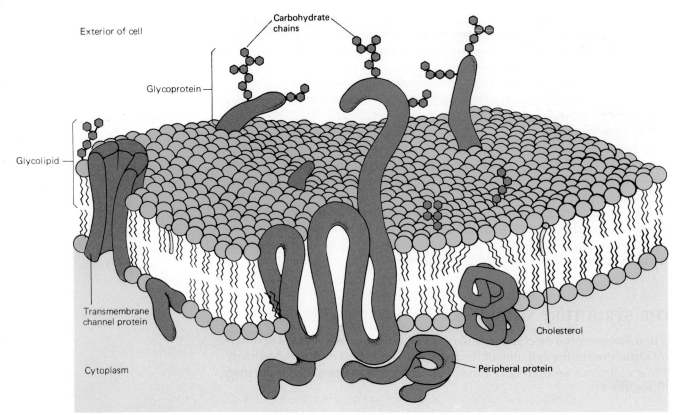

FIGURE 5–2 The fluid mosaic model of membrane structure. A representation of the plasma membrane from a eukaryotic cell, illustrating the various structures of some of the plasma membrane proteins found in human cells.

Long before the development of the electron microscope, it was known that membranes are composed of both lipids and proteins. Work by researchers in the 1920s and 1930s had provided clues that the core of the cell membrane is composed of lipids, mostly phospholipids. Furthermore, by examining the membrane of the red blood cell (which has only a plasma membrane) and comparing the surface area of the membrane with the total number of lipid molecules per cell, investigators were able to conclude that the membrane is composed of phospholipids no more than 2 molecules thick! Since many proteins have a diameter greater than 10 nm, a major problem in understanding the basic structure of membranes was to determine how the molecules that make up cell membranes can be arranged to fit in such a small space. In 1972, S. J. Singer and G. L. Nicolson proposed a model of membrane structure that represents a synthesis of the known properties of biological membranes. According to this fluid mosaic model, which has gained wide acceptance, the membrane consists of a fluid bilayer of lipid molecules (a double layer of lipid) in which the proteins are embedded (Figure 5–2).

We now know that membrane lipids have unusual properties that allow them to form bilayered structures and that these structures allow biological membranes to assemble. How is it possible for membrane lipids to behave in this way?

The Lipid Bilayer

The physical characteristics of phospholipid molecules, especially the ways in which these molecules associate with water, are what allow them to form bilayers. Recall from Chapter 3 that phospholipids are made of two fatty acid chains linked to two of the three carbons of a glycerol molecule. The two fatty acid chains on the molecule are nonpolar hydrophobic (water-hating) molecules. The third carbon of the glycerol is linked to a polar, hydrophilic (water-loving) organic molecule, which usually contains a nitrogen atom or a carbohydrate group of some type. Molecules of this type, which have distinct hydrophobic and hydrophilic regions, are called **amphipathic molecules.** All lipids that make up the core of biological membranes have amphipathic characteristics.

Since molecules such as phospholipids have one end that associates freely with water and an opposite end that does not, the most favorable conformation for them to assume when they are dispersed in water is a bilayer (Figure 5–3). The bilayer structure allows the hydrophilic headgroups of the phospholipids to associate freely with the aqueous medium and the hydrophobic fatty acid chains to be in the interior of the structure away from the water molecules.

Not all lipids are capable of forming bilayers. Triacylglycerols, for example, are so predominantly hydrophobic that they form oil droplets within the cell. Steroids, such as cholesterol, are so insoluble in water that they tend to form crystalline-like structures when they are put into water by themselves. The important features of bilayer-forming lipids are the following:

1. They have two distinct regions, one hydrophobic and one hydrophilic (together, this makes them strongly amphipathic).
2. Their shape allows them to associate with water most favorably as a bilayer structure.

An important property of phospholipid bilayers is that under appropriate conditions they behave like *liquid crystals* (Figure 5–4). Bilayers have crystal-like properties in that the lipid molecules are arranged in an orderly array with headgroups on the outside and fatty acid chains on the inside; they have liquid-like properties in that, despite the orderly arrangement of their molecules, their hydrocarbon chains are in constant motion. Thus, a molecule can move rapidly from one point to another on the same side of the bilayer. Such movement gives the bilayer the property of a *two-dimensional fluid.* Under normal conditions this means that a single phospholipid molecule can travel across the surface of a eukaryotic cell in seconds.

The fluid-like qualities of lipid bilayers also allow molecules embedded in them to move along the plane of the membrane (as long as they are not anchored in some way). This was elegantly demonstrated by David Frye and Michael Ediden with experiments in which they followed the movement of membrane proteins on the surface of two cells that had been joined together (Figure 5–5). When the plasma membranes of a mouse cell and a human cell are fused, within minutes membrane proteins from each cell migrate and become randomly distributed over the single continuous plasma membrane that surrounds the joined cells.

Biological membranes are normally fluid because the fatty acid chains in a lipid bilayer are in constant motion. Remember from Chapters 2 and 3 that molecules are free to rotate around single covalent bonds. Since most of the bonds in hydrocarbon chains are single bonds, the chains themselves can undergo very rapid, twisting motions (these motions are temperature dependent). Most biological membranes are in the liquid crystalline state; however, at low temperatures, van der Waals interactions (Chapter 3) between hydrocarbon chains lined up close to each other convert phospholipid bilayers to a solid gel state. You may be familiar with a

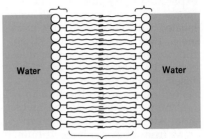

Hydrophilic head group region

Water   Water

Hydrophobic fatty acid region

FIGURE 5–3 A phospholipid bilayer. Phospholipids form bilayers in water so that the hydrophobic fatty acid chains are not exposed to the water. The headgroups of the phospholipids on each surface of the bilayer are hydrophilic and are in contact with the aqueous medium.

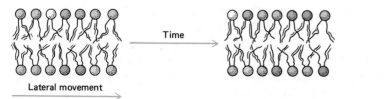

Lateral movement

Time

"Flip-flop" not allowed

FIGURE 5–4 Fluid properties of a phospholipid bilayer. Most phospholipid bilayers in cells are in a liquid-crystalline state. This means that the hydrocarbon chains of the phospholipid molecules are in constant motion, allowing each molecule to move laterally on the same side of the bilayer.

FIGURE 5–5 In an elegant series of experiments, membrane proteins of mouse cells and human cells were labeled with fluorescent dye markers in two different colors. When the plasma membranes of a mouse cell and a human cell were then fused, mouse proteins were observed migrating to the human side and human proteins to the mouse side. After a short time the proteins from both mouse and human were randomly distributed on the cell surface. This demonstration was convincing evidence that proteins in membranes are not part of a static structure like bricks in a wall, but instead are highly mobile entities in a two-dimensional fluid.

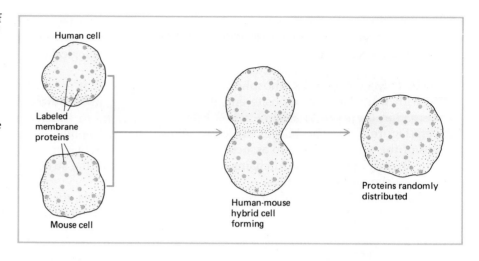

similar situation with cooking fats. Some fats are solid at room temperature, whereas others are liquid. One of the major differences between these two types of fat is the number of double bonds in their fatty acids. Kinks in the double bonds of the hydrocarbon chains of cooking oil prevent the hydrocarbon chains from coming close enough together to form van der Waals contacts, thus effectively lowering the temperature at which the oil (or the lipids of the membrane) will crystallize.

Many organisms regulate the unsaturated fatty acid composition of their membrane lipids in response to changes in temperature in order to provide an optimal fluid state for their membranes. Some molecules, such as the steroid cholesterol, fit between phospholipid molecules and act as "fluidity buffers." At low temperatures cholesterol molecules prevent hydrocarbon chains from becoming close enough to form van der Waals contacts, which would promote crystallization. At high temperatures cholesterol molecules appear to restrict the excessive motion of fatty acid chains, which might result in the membrane's becoming unstable or permeable to certain molecules. Plasma membranes of eukaryotic cells tend to have very high levels of sterols, approaching 1 sterol molecule per phospholipid molecule.

Lipid bilayers, particularly those in the liquid crystalline state, also have other important biological properties. Bilayers by themselves tend to resist forming free ends; as a result, they tend to be self-sealing and under most conditions spontaneously round up to form closed vesicles. Fluid bilayers also are flexible, allowing cell membranes to constantly change shape without breaking. Finally, under appropriate conditions lipid bilayers have the ability to fuse with other bilayers. **Membrane fusion** is an important cellular phenomenon (Figure 5–6). When a vesicle fuses with another membrane, both membrane bilayers and their compartments become continuous with each other. This allows materials to be transferred from one compartment to another or to move from a secretory vesicle to the outside of a cell by a process known as *exocytosis*. In a similar but reverse process, *endocytosis*, large molecules are brought into the cell from the outside by forming vesicles from a section of membrane. Both endocytosis and exocytosis are discussed later in this chapter.

Membrane Proteins

Until recently, it was not clear how proteins were associated with membranes. Early investigators found it difficult to accept the idea that proteins could associate with any part of membranes other than the outer surface. This led to difficulties in explaining why membranes are so thin as well as to the idea that membrane proteins must be very uniform and must as-

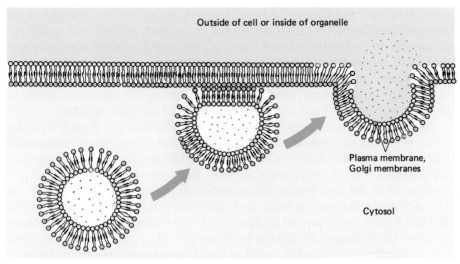

(a) Exocytosis, endomembrane transport

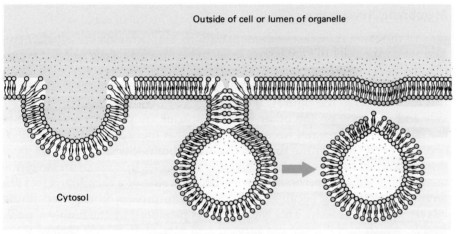

(b) Endocytosis, budding

FIGURE 5–6 Membrane fusion in endocytosis and exocytosis. Both processes involve the fusion of lipid bilayers. Exocytosis and movement of molecules between organelles involve the contact and fusion of a vesicle with a membrane, releasing the vesicle's contents into another compartment or to the outside of the cell. Endocytosis and the formation of intracellular transport vesicles originate from an invagination or a "bud" from the membrane, followed by the fusion of two regions of the membrane that come in contact with each other. Notice that in exocytosis the two cytoplasmic sides of the membrane make contact with each other to initiate membrane fusion, while in endocytosis the two noncytoplasmic layers make the first contacts. This means that endocytosis and exocytosis are not the same processes in reverse but, rather, two different types of cellular events.

sume conformations that allow them to lie like thin sheets on the membrane surface. Several pieces of evidence argued against these models. One was that membranes purified by cell fractionation contain many different proteins, which vary widely in composition and structure. Another came from physical chemical studies of membrane proteins showing that many membrane proteins are **globular;** that is, their dimensions are too large to be associated just with the membrane surface. Finally, evidence emerged that some proteins are associated with membranes in such a way that one region (or domain) of the protein is on one side of the membrane while another part of the protein is on the opposite side. Thus, the most reasonable model of membrane structure is one of a *mosaic* of proteins, many of which are mobile and extend into or completely through the lipid bilayer.

We now know that there are two major types of membrane proteins: integral membrane proteins and peripheral proteins. **Integral membrane proteins** have regions that are inserted into the hydrophobic regions of the lipid bilayer. Some of these proteins pass all the way through the membrane so that large parts of them are on either side of the membrane; these particular integral proteins are also called *transmembrane proteins.* Other integral proteins are located in large part on one side of the membrane (in the cytoplasm or protruding from the cell surface), with only a small region extending into the bilayer, where it serves as an anchor. Finally, some integral membrane proteins are almost completely buried within the hydrophobic region of the bilayer, having polypeptide chains that pass back and forth across the bilayer as many as 12 times.

Peripheral proteins, the other type of membrane protein, can be easily removed from the membrane without disrupting the structure of the bilayer. They usually bind to exposed regions of integral proteins.

Integral membrane proteins are able to insert into the lipid bilayer because the parts of them that are within the membrane have hydrophobic surfaces, compatible with the interior of the bilayer. If a membrane protein has a region with a hydrophilic surface, that region is usually found protruding from the membrane, in contact with the aqueous medium. (*Note:* The difference between soluble and membrane-bound proteins is not that one has hydrophobic amino acids and the other does not; rather, in soluble proteins the hydrophobic amino acids are buried in the interior of the protein molecule away from the water, whereas in integral membrane proteins the hydrophobic stretches of amino acids are on the surface, in contact with the fatty acid chains of the bilayer.)

Membrane Asymmetry

One of the most remarkable demonstrations that proteins are actually embedded in the lipid bilayer comes from freeze-fracture electron microscopy (Figure 5–7), which enables investigators to literally see the membrane from "inside out." When the two sides of a membrane are compared by this method (as in the figure), large numbers of particles are found on one side and very few on the other. These particles are proteins embedded in the bilayer. This does not necessarily mean that there are more proteins on one side of the membrane than on the other, but rather that most are more firmly attached to a given side. Thus, the molecules that make up biological membranes are *asymmetrically distributed*. Each side of a membrane has different characteristics because each membrane protein is oriented in the bilayer in only one way. This asymmetry is produced by the highly specific way in which membranes are formed and moved from one region of the cell to another.

FIGURE 5–7 (*a*) In the freeze-fracture method, the path of membrane cleavage is along the hydrophobic interior of the lipid bilayer, resulting in two complementary fracture faces: (1) an outwardly directed inner half-membrane presenting the P-face, from which project the majority of the membrane poteins and (2) a relatively smooth, inwardly directed outer half-membrane presenting the E-face, which shows occasional protein particles. In good fractures particles are visible on both of the inside faces of the fractured membrane, as shown in the figure. These particles are integral membrane proteins inserted in the lipid bilayer. Freeze-fractured bilayers of lipids alone do not have particles on the fracture planes. (*b*) A freeze fracture made of a membrane from a cell of the eye of a monkey. Notice the greater number of proteins on the P-face of the membrane.

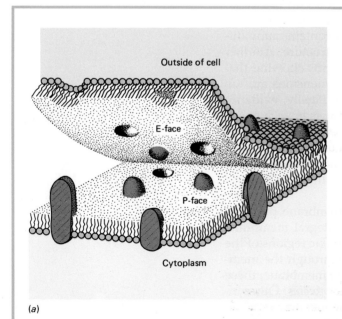

(a)

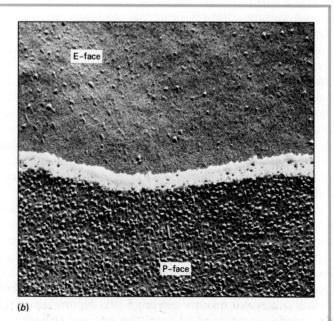

(b)

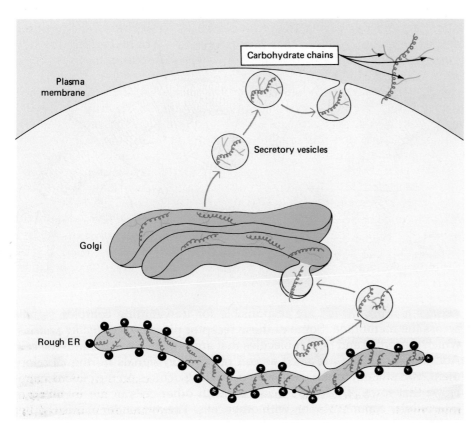

FIGURE 5–8 Transport pathway of proteins destined for the plasma membrane. Notice that the part of the protein on the surface of the cell originates in the cisternae of the ER and passes through the compartments of the Golgi complex and secretory vesicles. In a sense, the space in these compartments is equivalent to the space outside the cell. Carbohydrates are added in the ER lumen only to the parts of the proteins that pass through the lumen of the Golgi complex. Those glycosylated regions of the proteins will be located on the cell surface.

As an example, look again at Figure 5–2 and notice how the protein molecules are oriented in the plasma membrane. Carbohydrates are attached to the parts of the proteins that are exposed on the surface of the cell, but none are attached to the parts of the proteins that are exposed to the cytoplasm. This asymmetric distribution of carbohydrates is due to the way in which the glycoproteins are inserted into the membranes when they are first synthesized.

As you will recall from Chapter 4, plasma membrane proteins are formed by ribosomes on the rough endoplasmic reticulum (ER) and are inserted through the ER membrane as they are synthesized. Carbohydrates are added to the proteins in the lumen of the ER. If you follow the budding and membrane fusion that are part of the transport process (Figure 5–8), you can see that the part of the protein that protrudes into the inner compartment (cisterna) of the ER will also be exposed to the inner compartment of the Golgi complex, where the enzymes that modify carbohydrates on the protein are located. That region of the protein will remain in the inner compartment as it buds from the Golgi complex and is packaged into a secretory vesicle. When the secretory vesicle fuses with the plasma membrane, the carbohydrate-containing portion of the protein that had been oriented to the inside of the vesicle will become that part of the membrane protein that is exposed on the cell surface.

Functions of Membrane Proteins

Why should a membrane such as the plasma membrane illustrated in Figure 5–2 require so many different proteins? This diversity of proteins in a membrane is a reflection of the number of functions that take place in the membrane. Generally, proteins in the plasma membrane fall into several broad functional groups (Figure 5–9). A number of them are involved in the *membrane transport* of small molecules. Others are membrane-bound

FIGURE 5–9 Some types of intrinsic membrane proteins and their functions. (*a*) Cell adhesion proteins firmly attach membranes of adjacent cells and may serve as anchoring points for networks of internal cables. (*b*) Protein channels between two cells that allow communication of small molecules between neighboring cells. (*c*) Transport proteins allow the selective passage of essential molecules into the cell either passively by diffusion or actively, through energy dependent processes (*d*) Signal receptor/transducer proteins. Proteins which bind external signal molecules and transfer a message to the interior of the cell. (*e*) Attachment sites of soluble enzymes and cytoskeleton. Some intrinsic membrane proteins have multiple functions such as transport of specific molecules and serving as attachment sites for cytoskeleton elements and soluble enzymes. (*f*) ATP-driven pumps actively transport ions from one compartment to another as an energy storage mechanism. (*g*) Enzymes. Intrinsic membrane proteins may be enzymes with active sites located on either side or in the interior of the membrane.

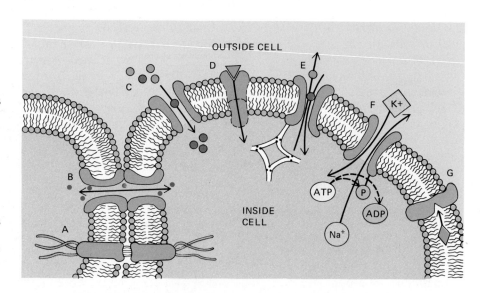

receptor proteins, which are responsible for transmitting complex signals across the membrane. Some of these receptor proteins are *binding proteins*, which recognize and bind molecules that are to be transported into the cell. Additional proteins are used as cell recognition signals during development, and still others are part of specialized structures on the plasma membrane that form physical associations with other cells or are involved in intercellular communication with other cells. The remainder of this chapter examines this functional diversity.

HOW MATERIALS PASS THROUGH MEMBRANES

Whether a membrane will permit the molecules of any given substance to pass through it depends on the structure of the membrane and the size and charge of the molecules. A membrane is said to be *permeable* to a given substance if it will permit that substance to pass through and *impermeable* if it will not permit the substance to pass through. A selectively permeable membrane allows some but not other substances to pass through it. Permeability is primarily a property of the membrane. All of the biological membranes surrounding cells, nuclei, vacuoles, mitochondria, chloroplasts, and the other subcellular organelles are selectively permeable. Water molecules, for example, can rapidly cross a fluid lipid bilayer by passing through gaps that occur as a fatty acid chain momentarily moves out of the way. Other small molecules, such as glucose and charged ions, do not pass freely through the bilayer either because of their size or because they are repulsed by a layer of electrical charges on the surface of the membrane (Table 5–1).

Table 5–1 PERMEABILITY OF THE LIPID BILAYER TO DIFFERENT SUBSTANCES		
Type of Molecule	*Example*	*Permeability*
Hydrophobic	N_2, O_2, hydrocarbons	Freely permeable
Small polar	H_2O, CO_2, glycerol, urea	Freely permeable
Large polar	Glucose, other uncharged monosaccharides, disaccharides	Not permeable
Ions/charged molecules	Amino acids, H^+, HCO_3^-, Na^+, K^+, Ca^+, Cl^-, Mg^+	Not permeable

In response to varying environmental conditions or cellular needs, the cell membrane may be a barrier to a particular substance at one time and actively promote its passage at another. By regulating chemical traffic in this way, a cell can exert some control over its own internal ionic and molecular composition, which can be very different from that on the outside. In the nonliving world, materials move passively by physical processes such as diffusion; in living organisms, materials can also be moved actively by physiological processes, including active transport, exocytosis, and endocytosis (see Table 5–2). Such active physiological processes (discussed later in this chapter) require the expenditure of energy by the cell.

Diffusion

Some substances pass into or out of cells and move about within cells by simple diffusion, a physical process based on random motion. At temperatures above absolute zero (0° Kelvin or −273°C—the point where all motion stops), all atoms and molecules possess kinetic energy, or energy of motion. The three states of matter—solid, liquid, and gas—differ with respect to the freedom of movement of their constituent molecules. The molecules of a solid are closely packed, and the forces of attraction between them allow them to vibrate but not to move around. In a liquid the molecules are farther apart; the intermolecular forces are weaker, and the molecules move about with considerable freedom. In a gas the molecules are so far apart that intermolecular forces are negligible; molecular movement is restricted only by the walls of the container that encloses the gas. This means that atoms and molecules in liquids and gases are able to move apart from each other in a kind of "random walk." This random motion is responsible for **diffusion,** the *net* movement of particles (atoms, ions, molecules, etc.) from a region of high concentration to one of lower concentration such that eventually the particles are evenly distributed (Figure 5–10). We can therefore say that diffusion involves a net movement of particles down a concentration gradient (a change in the concentration of a substance from one point to another). This does not mean that particles are prohibited from moving in the "wrong" direction (against the gradient). However, since there are initially more particles in the region of high concentration, it logically follows that more particles will move randomly from that region into the low-concentration region than from the region of low concentration into the region of high concentration. The rate of diffusion is a function of the size and shape of the molecules, their electrical charges, and the temperature. As the temperature rises, the molecules move faster and the rate of diffusion increases.

Examining a drop of water under the microscope does not reveal the motion of the water molecules; they are much too small to be seen. However, when a drop of India ink, which contains fine particles of carbon, is added to the water, the motion of the carbon particles becomes visible

FIGURE 5–10 The process of diffusion. When a small lump of sugar is dropped into a beaker of water, its molecules dissolve, as shown in (a). The sugar molecules begin to diffuse throughout the water in the container, as seen in (b) and (c). The arrows in (a) indicate net movement of sugar molecules; individual molecules move randomly in all directions. Eventually, diffusion results in an even distribution of sugar molecules throughout the water in the beaker, as shown in (d).

(a)

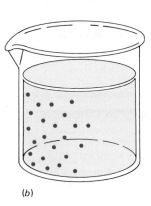

(b)

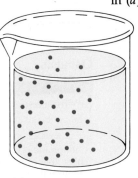

(c)

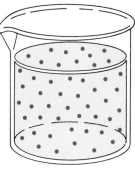

(d)

under the microscope. Each carbon particle is constantly being bumped by water molecules. The recoil from these bumps moves the carbon particle in an aimless, zigzag path. This motion of small particles—termed **Brownian movement,** after Robert Brown, an English botanist who first observed it when he looked through the microscope at pollen grains in a drop of water—provides a model of how diffusing molecules move.

As diffusion occurs, each individual molecule moves in a straight line until it bumps into something—another molecule or the side of the container. Then it rebounds and moves in another direction. An individual molecule may move as fast as several hundred meters per second, but each molecule can go only a fraction of a nanometer before it bumps into another molecule and rebounds. Thus, the progress of any given molecule in a straight line is quite slow. Molecules continue to move even when they have become uniformly distributed throughout a given space. However, as fast as some molecules move in one direction, others move in the opposite direction, so the system as a whole is at equilibrium (i.e., there is no net change). The molecules of any number of different substances will diffuse independently of each other within the same solution; ultimately all will become uniformly distributed. More commonly in biological systems, an equilibrium condition is never attained. For example, carbon dioxide is formed within the cell when sugars and other molecules are metabolized during the process of cellular respiration. Carbon dioxide readily diffuses across the plasma membrane but then is rapidly removed by the bloodstream. This limits the opportunity for the molecules to reenter the cell, so a sharp concentration gradient of carbon dioxide molecules always exists across the membrane.

Although diffusion occurs rapidly over microdistances, it takes a long time for a molecule to travel distances measured in centimeters. This overall slow rate of diffusion has important biological implications, since it limits the number of molecules that can reach a cell or an organism by simple diffusion.

Dialysis

The diffusion of a **solute** (a dissolved substance) through a selectively permeable membrane is termed **dialysis.** To demonstrate dialysis, one can fill a cellophane bag[1] with a sugar solution and immerse it in a beaker of pure water (Figure 5–11). If the cellophane membrane is permeable to sugar as well as to water, the sugar molecules will pass through it, and the concentration of sugar molecules in the water on the two sides of the membrane will eventually become equal. Subsequently, solute molecules (as well as water molecules) will continue to pass through the membrane, but there will be no net change in concentration, for the rate of movement will be equal in the two directions. Kidney dialysis is a practical application of this process; waste products, which diffuse readily across the artificial membrane used in the dialysis apparatus, can be removed from the bloodstream, but blood cells, blood proteins, and other large molecules are unable to diffuse and thus are retained.

Osmosis

Osmosis is a special kind of diffusion that involves the movement of *solvent* (in this case, water) molecules through a selectively permeable membrane.

---

[1]Cellophane is often used as an "artificial membrane." It is composed of polysaccharide molecules and can be formed into a thin sheet that will allow the passage of water molecules through it. Such membranes can be constructed with varying permeability to different solutes.

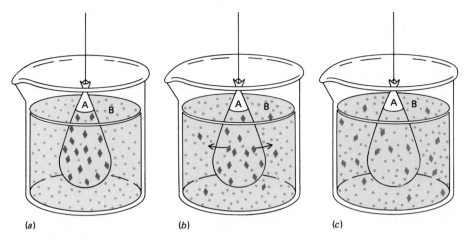

(a)   (b)   (c)

FIGURE 5–11 Dialysis. (*a*) A cellophane bag filled with a sugar solution is immersed into a beaker of water. The cellophane acts as a selectively permeable membrane, permitting passage of the sugar and water molecules, but preventing passage of larger molecules. (*b*) The arrows indicate the net movement of sugar molecules through the membrane into the water of the beaker. (*c*) Eventually the sugar becomes distributed equally between the two compartments. Although sugar molecules continue to diffuse back and forth, the net movement is zero. The same is true for the water molecules. Blue dots represent water molecules; red dots represent sugar molecules.

The water molecules pass freely in both directions, but, as in all types of diffusion, net movement is from the region where the molecules are more concentrated to the region where they are less concentrated. Most solute molecules (see Table 5–1) cannot diffuse freely through the selectively permeable cell membrane.

The principles involved in osmosis can be illustrated by use of an apparatus called a U-tube (Figure 5–12). The U-tube is divided into two sections by a selectively permeable membrane that allows solvent (water) molecules to pass freely but excludes solute molecules (sugar, salt, etc.). A water-and-solute solution is placed on one side, and pure water is placed on the other. The side containing the solute dissolved in the water has a lower concentration of water than the pure-water side because the solute molecules "dilute out" the water molecules. Therefore, there is a net movement of water molecules from the pure-water side (with a high concentration of water) to the water-and-solute side (with a lower concentration of water), as a result of which the fluid level drops on the pure-water side and rises on the water-and-solute side. Because the solute molecules do not move across the membrane, there will still be a difference in water concentration between the two sides. Net movement of water will continue, and the fluid level will continue to rise on the side containing the solute. Under weightless conditions this process could go on indefinitely, but on Earth the weight of the rising column of fluid will eventually exert enough pressure to stop further changes in fluid levels, although water molecules will continue to pass through the selectively permeable membrane in both directions.

We define the **osmotic pressure** of a solution as the tendency of water to move into that solution by osmosis. In our U-tube example, we could measure the osmotic pressure by inserting a piston on the water-and-solute side of the tube and measuring how much pressure needs to be exerted by the piston to prevent the rise of fluid on that side of the tube. A solution with a high solute concentration will have a low water concentration and a high osmotic pressure (Table 5–2); conversely, a solution with a low solute concentration will have a high concentration of water and a low osmotic pressure.

Isotonic, Hypertonic, and Hypotonic Solutions

Often, we would like to compare the relative osmotic pressures of two solutions. Dissolved in the fluid compartment of every living cell are salts, sugars, and other substances that give that fluid a certain osmotic pressure. When a cell is placed in a fluid with exactly the same osmotic pressure, there is no net movement of water molecules either into or out of the cell;

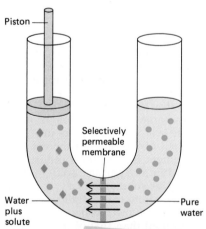

Piston

Selectively permeable membrane

Water plus solute

Pure water

FIGURE 5–12 Demonstration of osmosis. The U-tube contains pure water on the right and water (blue) plus a solute (red) on the left, separated by a selectively permeable membrane that allows water molecules, but not solute molecules, to diffuse across the membrane. The arrows indicate the direction of *net* movement of water molecules. The fluid level will rise on the left and fall on the right. The force that must be exerted by the piston in order to prevent the rise in fluid level is equal to the osmotic pressure of the solution.

Table 5–2 OSMOTIC TERMINOLOGY			
Solute Concentration in Solution A	*Solute Concentration in Solution B*	*Tonicity*	*Direction of Net Movement of Water*
Greater	Less	A hypertonic to B B hypotonic to A	B to A
Less	Greater	B hypertonic to A A hypotonic to B	A to B
Equal	Equal	Isotonic	No net movement

the cell neither swells nor shrinks. Such a fluid is said to be **isotonic** (i.e., of equal osmotic pressure) to the fluid within the cell. Normally, our blood plasma (the fluid component of blood) and all of our body fluids are isotonic to our cells; they contain a concentration of water equal to that in the cells. A solution of 0.9% sodium chloride (sometimes called *physiologic saline*) is isotonic to the cells of humans and other mammals. Human red blood cells placed in 0.9% sodium chloride will neither shrink nor swell (Figure 5–13).

If the surrounding fluid has a concentration of dissolved substances greater than the concentration within the cell, it has a higher osmotic pressure than the cell and is said to be **hypertonic** to the cell; a cell in a hypertonic solution loses water and shrinks. Human red blood cells placed in a solution of 1.3% sodium chloride will shrink (see Figure 5–13). If a cell that has a cell wall is placed in a hypertonic medium, it loses water to its surroundings, and its contents shrink away from the wall; this process is called **plasmolysis.** Plasmolysis occurs in plants when large amounts of salts or fertilizers are contained in the soil or water around them.

If the surrounding fluid contains a lower concentration of dissolved materials than the cell, it has a lower osmotic pressure and is said to be **hypotonic** to the cell; water then enters the cell and causes it to swell. Red blood cells placed in a solution of 0.6% sodium chloride will take up water, swell, and burst.

FIGURE 5–13 Osmosis and the living cell (*a*). A cell is placed in an isotonic solution. Because the concentration of solutes (and thus of water molecules) is the same in the solution as in the cell, water can pass in and out of the cell, but the net movement is zero. (*b*) A cell is placed in a hypertonic solution. This solution has a greater solute concentration (and thus a lower water concentration) than that of the cell. This results in a net movement of water out the cell (*arrows*), and the cell becomes dehydrated, shrinks, and may die. (*c*) A cell is placed in a hypotonic solution. The solution has a lower solute (and thus a greater water) concentration than that in the cell. The cell contents therefore have higher osmotic pressure than the solution. There is a net movement of water molecules into the cell (*arrows*), causing the cell to swell. The cell may even burst.

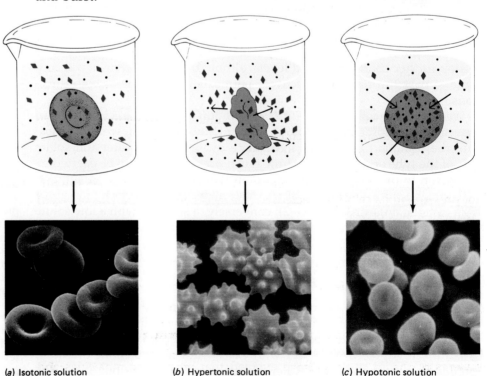

(a) Isotonic solution   (b) Hypertonic solution   (c) Hypotonic solution

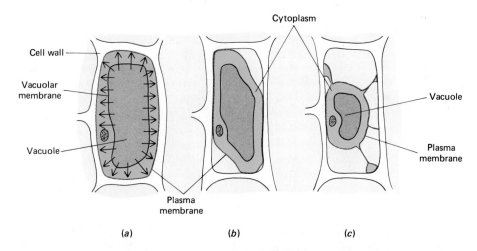

Cell wall

Vacuolar membrane

Vacuole

Cytoplasm

Plasma membrane

Vacuole

Plasma membrane

(a)   (b)   (c)

FIGURE 5–14 (*a*)Turgor pressure in a plant cell. In hypotonic surroundings, the contents of the cell fill the space within the wall. The arrows indicate the turgor pressure of the cell. (*b*), (*c*) If the cell is placed in a hypertonic medium, it loses water and its contents shrink. The cell is said to be plasmolyzed. (*d*) Section through a dividing *Escherichia coli*, a bacterium that has plasmolyzed due to a hypertonic environment.

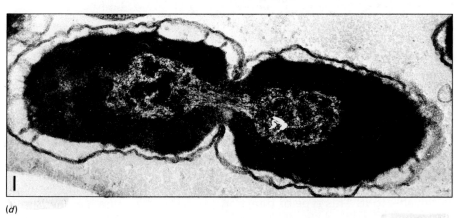

(*d*)

Turgor Pressure

The rigid cell walls of plant cells, algae, bacteria, and fungi enable these cells to withstand, without bursting, an external medium that is very dilute, containing only a very low concentration of solutes. Because of the substances dissolved in the cells' cytoplasm, the cells are hypertonic to the outside medium (the outside medium is hypotonic to the cell's cytoplasm). Water moves into the cells by osmosis, filling their central vacuoles and distending the cells. The cells swell, building up a pressure, termed **turgor pressure,** against the rigid cellulose cell walls (Figure 5–14). The cell walls can be stretched only very slightly, and a steady state is reached when the resistance to stretching of the cell walls prevents any further increase in cell size and there is no net movement of water molecules into the cells (although, of course, molecules continue to move back and forth across the cell membrane). Turgor pressure is an important factor in providing support for the body of nonwoody plants. Thus, a flower wilts when the turgor pressure in its cells has decreased (the cells have become plasmolyzed) owing to a lack of water.

Carrier-Mediated Transport of Small Molecules

The cell membrane is relatively impermeable to most of the larger polar molecules (see Table 5–1). This is advantageous because most of the compounds that are metabolized are polar, and this impermeability of the cell membrane prevents their loss by diffusion. To transport polar nutrients such as glucose and amino acids through the lipid bilayer into the cell, systems of carrier proteins have evolved that bind these molecules and

FIGURE 5–15 A model for the facilitated diffusion of glucose. The transport protein is capable of binding glucose on the outside of the cell and then changing its shape so that a channel is opened to the inside of the cell. When the glucose molecule is released on the inside, the protein reverts to its original conformation and is ready to bind the next molecule.

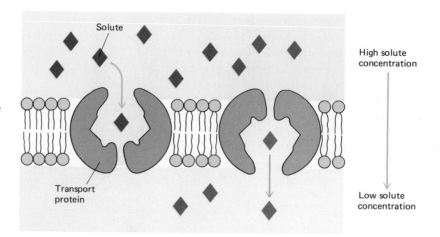

transfer them across the membrane. The transfer of solutes across the cell membrane by a carrier system is termed **carrier-mediated transport.** The energy required for mediated transport can come from two sources, facilitated diffusion and active transport.

Facilitated Diffusion

In the simpler case, the cell can use the energy stored in the concentration gradient of a substance that is normally in higher concentration outside the plasma membrane than inside the cell. Under these circumstances, the substance will diffuse into the cell as long as the cell membrane is made permeable to it. This type of transport mechanism is known as **facilitated diffusion.**

Facilitated diffusion mechanisms rely on the existence of *carrier proteins,* which combine temporarily with the solute molecule and accelerate its movement through the membrane. The carrier protein is not changed by this action; after it transports a solute molecule, it is free to bind with another solute molecule (Figure 5–15). An important example of a facilitated diffusion carrier is the glucose transporter in red blood cells. Glucose transporters are glycoproteins that account for about 2% of the total membrane protein. These cells keep the internal concentration of glucose low by immediately adding a phosphate group to entering glucose molecules, converting them to highly charged glucose phosphates that cannot pass back through the membrane.

The mechanism by which glucose transport occurs is not entirely clear. It appears that the carrier protein does not create a "hole" in the membrane for glucose to pass through; if that were the case, other related molecules or molecules smaller than glucose could also pass through the pore. It is more likely that glucose binds specifically to a region of the protein that is exposed to the outside of the cell, and that this binding changes the shape of the protein such that a channel is opened in the protein (or between several subunits of the same polypeptide chain) that allows the glucose molecule to be released on the inside of the cell. According to this model, when the glucose is released on the inside, the protein reverts to its original structure and is available to bind the next glucose molecule on the outside of the cell.

Carrier-Mediated Active Transport

Although adequate amounts of some molecules can be transported across the cell membrane by diffusion, other types of molecules are required by the cell in concentrations that are higher than the concentrations outside the cell. These molecules are moved across the cell membrane by **active transport** mechanisms. Since active transport requires that molecules be "pumped" *against* a concentration gradient (i.e., from a region of low concentration to a region of high concentration), an energy source is required. Active transport systems therefore usually involve either a source of metabolic energy generated by the cell in the form of adenosine triphosphate (ATP) or another type of stored energy derived from the hydrolysis of ATP.

One of the most striking examples of an active transport mechanism is the **sodium-potassium pump,** which is found in all animal cells (Figure 5–16). The pump is a specific protein in the plasma membrane that uses energy in the form of ATP to exchange sodium ions on the inside of the cell for potassium ions on the outside of the cell. This results in the creation of an imbalance of sodium and potassium ions on opposite sides of the membrane, so that under normal conditions the concentration of potassium ions is about 10 times greater inside the cell than outside while the concentration of sodium ions is about 10 to 15 times greater outside the cell than inside. Cells are able to use these large concentration gradients to generate an electrical potential (separation of electrical charges) across the membrane, which is the basis of the electrical impulses used in the transmission of nerve impulses (Chapter 45). These concentration gradients also store energy, which can be used to drive other active transport systems. So important is the electrochemical gradient produced by these pumps that some cells (e.g., nerve cells) expend 70% of their total energy metabolism just to drive this one transport system.

The use of electrochemical potentials for such purposes is not confined to the plasma membrane of animal cells. Plant and fungal cells also use ATP-driven plasma membrane pumps to pump protons from the cytoplasm of their cells to the outside. Removal of positively charged protons from the cytoplasm of these organisms results in a large difference in the concentration of protons, such that the outside of the cells is positively charged and the inside of the plasma membrane is negatively charged. As we will see in the following chapters, these ATP-driven proton pumps, used in the reverse direction to produce ATP in bacteria, mitochondria, and chloroplasts, are the major energy transduction system in all cells ranging from bacteria to cells in complex plants and animals.

According to the currently accepted hypothesis, sodium-potassium pumps (as well as all other ATP-driven pumps) are transmembrane proteins, which extend entirely through the membrane. By undergoing a series of conformational changes, the pumps are able to exchange sodium for potassium across the cell membrane. Unlike facilitated diffusion, at least one of the conformational changes in the pump cycle requires energy, which is released from ATP. The energy appears to be transferred to the pump proteins from ATP by the covalent bonding of one of the ATP phosphate groups to the protein, followed by removal of the phosphate later in the pump cycle.

Cotransport Systems

The electrochemical gradient generated by the sodium-potassium pump also provides sufficient energy to power the active transport of a number of other essential molecules. In these reactions the sodium-potassium concentration gradient **cotransports** the required molecule along with a sodium or a potassium ion. By linking the transport of a required molecule *against* its concentration gradient with the transport of sodium or potassium *down* its concentration gradient, the energy from ATP is indirectly used to drive the active transport of a required substance.

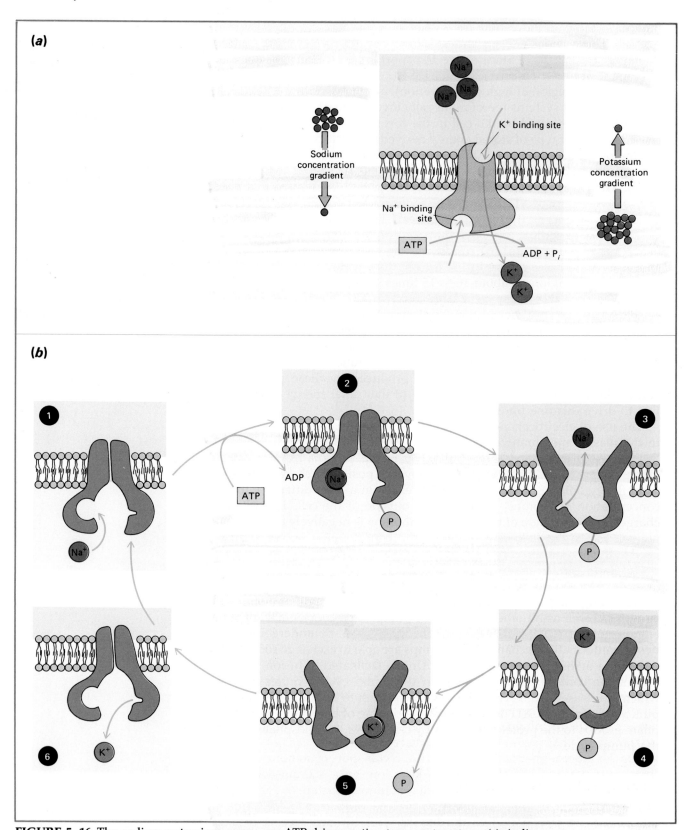

FIGURE 5–16 The sodium-potassium pump, an ATP-driven active transport system. (*a*) A diagram of the sodium-potassium ATPase in the plasma membrane. Each molecule of ATP used results in the movement of three sodium ions pumped out of the cell and two potassium ions pumped into the cell. (*b*) A model of how the sodium-potassium pump works. The binding of sodium (*1*) and the binding of a phosphate group from ATP to the protein (*2*) causes it to change its shape. This results in the (*3*) the transfer of sodium to the outside of the cell and (*4*) the formation of a potassium binding site accessible from the outside. The binding of potassium to the protein causes the phosphate to be released (*5*), resulting in the return of the protein to its original shape and releasing the potassium ion on the inside of the cell (*6*).

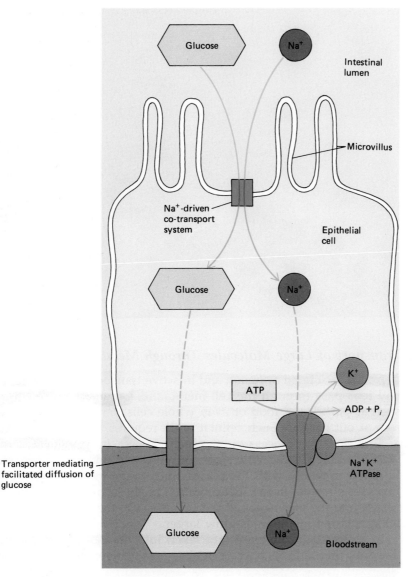

FIGURE 5–17 Coupled glucose transport systems in intestinal epithelia allow glucose to be transported through the cell from the intestine to the bloodstream. Glucose is actively transported into the cell by a sodium driven co-transport system located only on the part of the plasma membrane in contact with the intestinal lumen. The sodium gradient across the cell membrane is driven by a sodium-potassium ATPase which pumps sodium from the cytoplasm into the bloodstream, keeping the intracellular concentration low. Active transport of glucose into the cell keeps its intracellular concentration high, so that it is transported into the bloodstream by a different, facilitated diffusion transporter which is located only on the regions of the plasma membrane in contact with the bloodstream.

Integrated Multiple Transport Systems

In some cells, more than one system may work to transport a given substance. For example, the transport of glucose from the intestine to the bloodstream occurs through a thin sheet of epithelial cells that line the intestine (Figure 5–17) and have highly specialized regions, or domains, on their plasma membranes. The surface that is exposed to the intestine has many **microvilli,** finger-like protrusions that effectively increase the surface area of the membrane available for absorption. The glucose transporter protein on that region of the cell surface is part of an active transport system for glucose that is "driven" by the cotransport of sodium. The sodium concentration inside the cell is kept low by a sodium-potassium pump on the opposite surface of the cell that pumps sodium out of the cell and into the bloodstream. Because of its high concentration inside the cell, glucose can be transported to the bloodstream by facilitated diffusion.

Understanding the mechanisms behind the placement of two different transport proteins in two separate regions of the same plasma membrane is representative of the goals of cell biologists today. What are the signals that target each protein to its appropriate region on the plasma membrane? If the cell did not have a specific mechanism for handling this problem, proteins might be inserted randomly on both sides of the cell, leading to no net transport of glucose.

FIGURE 5–18 Exocytosis. A high-magnification electron micrograph of the upper surface of a secreting cell (approximately ×125,000). Secretion granules can be seen in the cytoplasm approaching the cell membrane. The filaments projecting diffusely from the cell surface are of unknown significance but may be proteins.

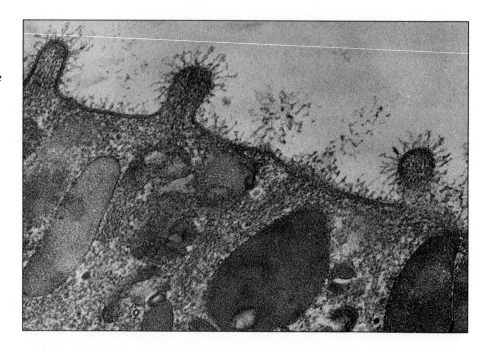

Transport of Large Molecules Through Membranes

In free or facilitated diffusion and in active transport, individual molecules and ions pass through the cell membrane. Larger quantities of material, such as particles of food or even whole cells, must sometimes be moved into or out of cells. Such cellular work requires that cells expend energy and involves the process of membrane fusion. In **exocytosis,** a cell ejects waste products or specific secretion products such as hormones by the fusion of a vesicle with the plasma membrane of the cell (Figure 5–18). Exocytosis results in the incorporation of the membrane of the secretory vesicle into the plasma membrane. This is also the primary mechanism by which plasma membranes grow larger.

In **endocytosis,** materials are taken into the cell. Several types of endocytotic mechanisms operate in biological systems. In **phagocytosis** (literally, "cell eating"), the cell ingests large solid particles such as bacteria or food (Figure 5–19). Phagocytosis is a mechanism used by protozoa and by several classes of white blood cells to ingest particles, some of which are as large as an entire bacterium. During ingestion, folds of the cell membrane enclose the particle, which has bound to the surface of the cell, and form a vacuole around it. When the membrane has encircled the particle, it fuses at the point of contact, leaving the vacuole floating freely in the cytoplasm. The vacuole then fuses with lysosomes, where the ingested material is degraded.

In the form of endocytosis known as **pinocytosis** ("cell drinking"), the cell takes in dissolved materials. Tiny droplets of fluid are trapped by folds in the plasma membrane (Figure 5–20), which pinch off into the cytoplasm as tiny vesicles. The liquid contents of these vesicles are then slowly transferred into the cytoplasm; the vesicles may themselves become progressively smaller, to the point that they appear to vanish.

In a third type of endocytosis, called **receptor-mediated endocytosis,** specific proteins or particles combine with *receptor proteins* embedded in the plasma membrane of the cell. The receptor-bound molecules then migrate into *coated pits,* which are regions on the cytoplasmic surface of the membrane coated with whisker-like structures. These coated pits form *coated vesicles* (Figure 5–21) by endocytosis. The coating on the vesicles consists of proteins, which momentarily form a basket-like structure around them. Seconds after the vesicles are released into the cytoplasm, however, the

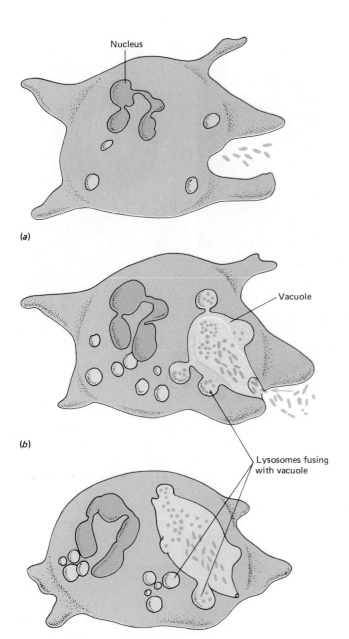

(a)

(b)

(c)

Nucleus

Vacuole

Lysosomes fusing
with vacuole

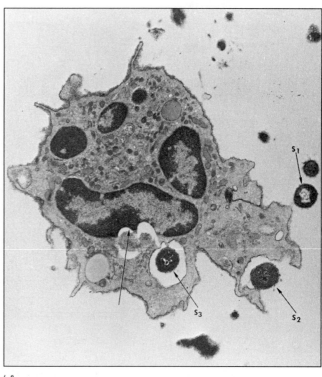

(d)

FIGURE 5–19 Phagocytosis. (*a*) In phagocytosis the cell ingests large solid particles, such as bacteria. Folds of the cell membrane surround the particle to be ingested, forming a small vacuole around it. (*b*) This vacuole then pinches off inside the cell. (*c*) Lysosomes may fuse with the vacuole and pour their potent digestive enzymes onto the digested material. (*d*) A white blood cell in the presence of *S. pyogenes* (approximately ×23,000). One bacterium (S_1) is free, one bacterium (S_2) is being phagocytized, and a third (S_3) has been phagocytized and is seen within a vacuole (phagosome). Note that near the vacuole (*see arrow*) the white blood cell's own nucleus has been partly digested.

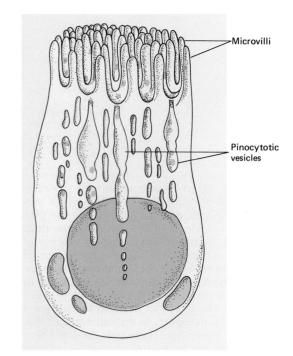

Microvilli

Pinocytotic
vesicles

FIGURE 5–20 In pinocytosis tiny particles of fluid are trapped by folds of the cell membrane, which then pinch off into the cytoplasm as little vesicles of fluid. The content of these vesicles is then slowly transferred to the cytoplasm across their membrane linings.

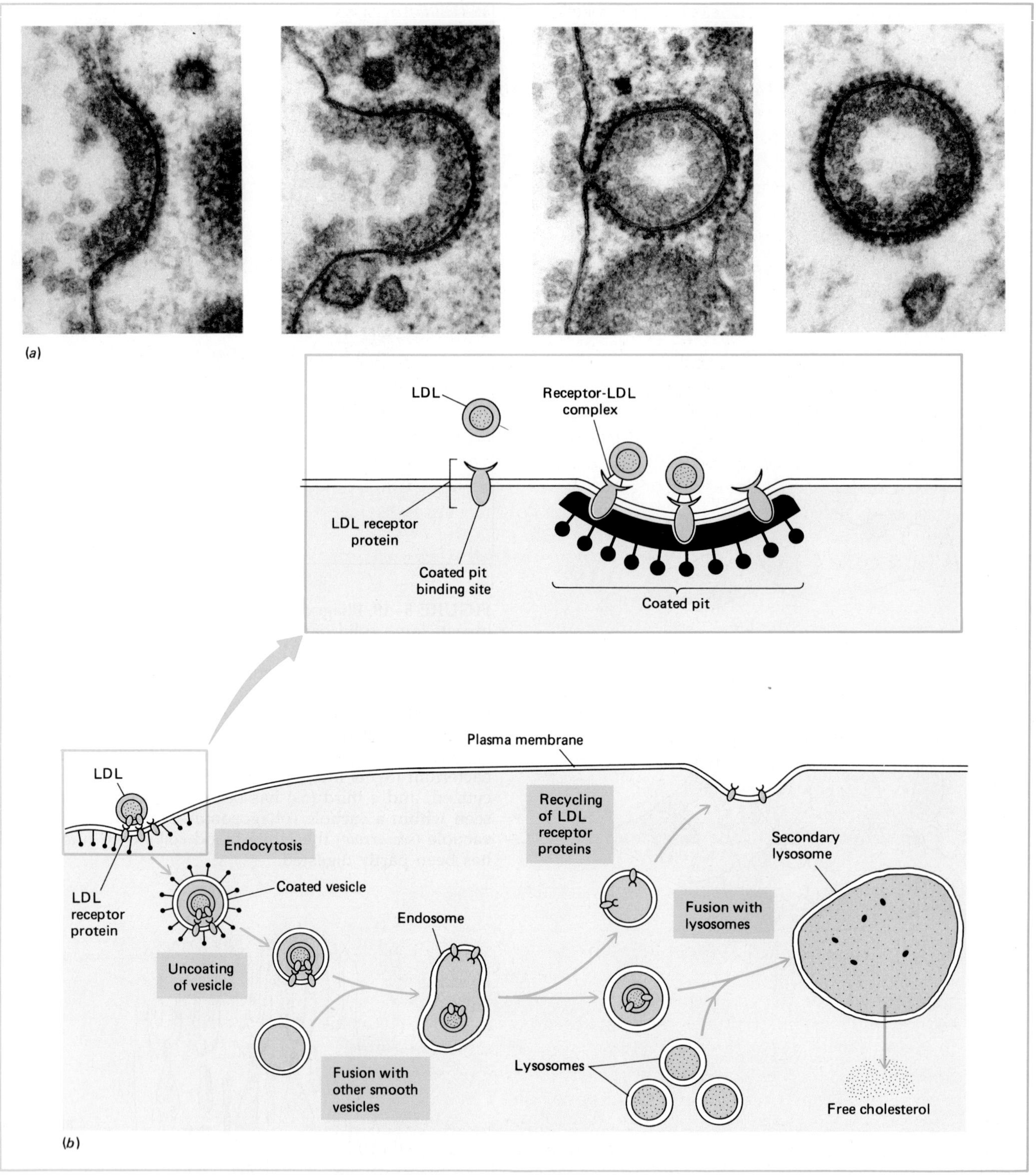

(a)

(b)

FIGURE 5–21 (a) The formation of a coated vesicle from a coated pit. (b) Low-density lipoprotein (LDL) particles, which transport cholesterol in the bloodstream, attach to specific receptor proteins on the plasma membrane. The receptor-LDL complexes move along the surface of the fluid membrane and cluster in coated pit regions of the membrane surface. Endocytosis of the coated pit results in the formation of a coated vesicle in the cytoplasm. Seconds later the coat is removed and the vesicles fuse with their counterparts to form large smooth vesicles called endosomes. The receptors and the LDL particles dissociate in the endosomes and move to different regions of the vesicles. New vesicles form from the endosomes. Those containing the receptors move to the surface and fuse with the plasma membrane, recycling the receptors to the cell surface. Vesicles containing the LDL particles fuse with lysosomes. Hydrolytic enzymes then release the cholesterol from the particles for use by the cell.

coating dissociates from them, leaving the vesicles free in the cytoplasm. The vesicles then fuse with other similar vesicles to form *endosomes,* larger vesicles in which the materials being transported are free inside and no longer attached to the membrane receptors. Endosomes themselves form two kinds of vesicles: one kind contains the receptors and can be returned to the plasma membrane; the other, which contains the ingested particles, fuses with lysosomes and is then processed by the cell.

Receptor-mediated endocytosis is an important process by which cholesterol in the bloodstream is taken up by animal cells. Much of the receptor-mediated endocytosis pathway was detailed through studies of the low-density lipoprotein (LDL) receptor by investigators Michael Brown and Joseph Goldstein, who in 1986 were awarded the Nobel prize for their pioneering work.

The recycling of the LDL receptor to the plasma membrane through vesicles illustrates a problem common to all cells that employ endocytotic and exocytotic mechanisms. In cells that are constantly involved in secretion, an equivalent amount of membrane must be returned to the interior of the cell for each vesicle that fuses with the plasma membrane; if it is not, the cell surface will keep expanding, even though the growth of the cell itself may be arrested. A similar situation exists for cells that use endocytosis. Macrophages, for example, ingest the equivalent of their entire surface membrane in about 30 minutes, requiring that an equivalent amount of recycling must occur for the cells to maintain their surface area and volume.

INTERCELLULAR CONTACTS

In multicellular organisms, cells that are in close contact with each other may develop specialized intercellular junctions on their plasma membranes. These structures allow neighboring cells to form strong connections with each other or to establish rapid communications between adjacent cells. In animals there are three common types of intercellular contacts: desmosomes, tight junctions, and gap junctions.

Desmosomes

Adjacent epithelial cells, such as those found in the upper layer of the skin, are so tightly bound to each other that strong mechanical forces are required to separate them. The structures that hold them together are button-like plaques, called **desmosomes,** that are present on the two adjacent cell surfaces (Figure 5–22). Each desmosome consists of regions of dense material associated with the inside of each cell membrane and separated by an intercellular space about 24 nm wide. The two cells are held together by protein filaments that cross the intercellular space between the desmosomes. Desmosomes are anchored on the insides of the cells to intermediate filaments, apparently coupling the intermediate filament networks of adjacent cells so that mechanical stresses are distributed throughout the tissue. The function of the desmosomes appears to be purely mechanical; they hold cells together at one point like a rivet or a spot weld. As a result, cells can form strong sheets, but substances can still pass freely through the spaces between the cell membranes.

Tight Junctions

Tight junctions are literally areas of tight connections between the membranes of adjacent cells. These connections are so tight that the spaces

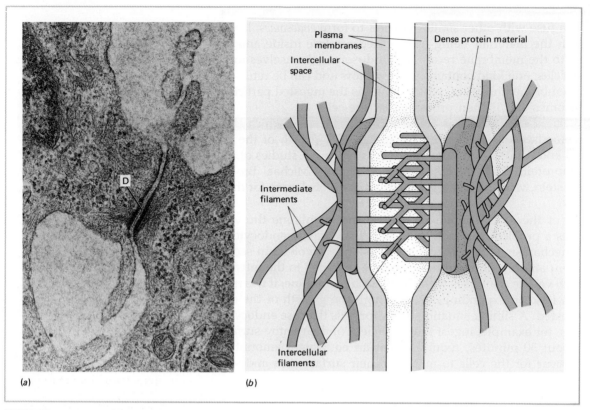

Plasma membranes

Dense protein material

Intercellular space

Intermediate filaments

Intercellular filaments

(a)

(b)

FIGURE 5–22 (*a*) Electron micrograph showing a desmosome (D) between two cells of the ovarian epithelium of a rabbit (approximately ×70,000). (*b*) Schematic drawing of the desmosome. The structure consists of discs of dense protein material on either side of the membrane, to which intermediate filaments are attached. Filaments also extend between the two regions of the cell.

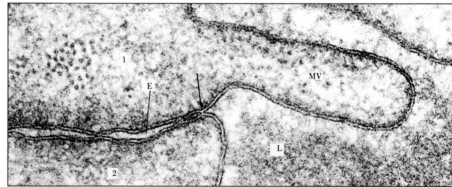

(a)

FIGURE 5–23 Tight junctions between adjacent cells. The tight junctions occur at the points of contact between two cells and would extend completely around the cells in a three-dimensional view. (*a*) Thin-section electron micrograph showing points of fusion between the plasma membranes between two cells (marked 1 and 2) lining the intestine. One junction is marked by the arrow. MV = microvillus, L = lumen (interior) of the intestine, E = extracellular space between the two cells. (*b*) Freeze-fracture electron micrograph through a region of tight junctions (TJ). The cylinders are parts of microvilli (MV) seen from inside the cell. R and F refer to the regions where the membranes contact to form the tight junctions.

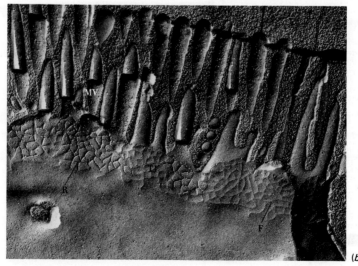

(b)

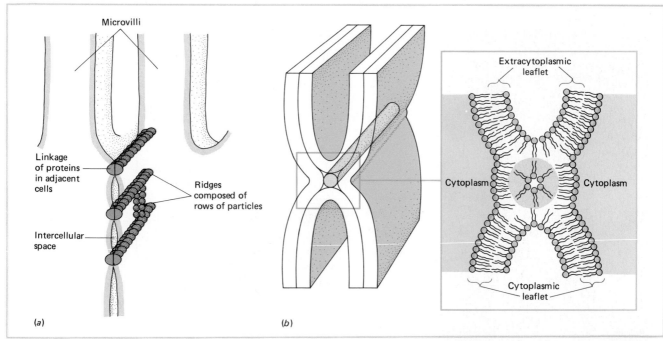

FIGURE 5–24 Two models of how tight junctions might be formed. (*a*) Junctions formed by linkage of rows of proteins between the two cells. (*b*) An alternative model in which the two outside layers of the plasma membrane bilayers are fused together to form the junction.

around the cells have completely disappeared; substances are thus prevented from passing through the layer of cells. Electron micrographs of tight junctions show that in the region of the junction the membranes of the two cells are in actual contact with each other (Figure 5–23). Several lines of evidence suggest that the membranes in the region of the tight junction may be actually *fused* together, forming a single membrane structure that completely encircles each cell in that region (Figure 5–24*b*). A layer of such cells is in effect embedded in one essentially continuous cell membrane with no gaps or intercellular spaces. Cells connected by tight junctions are used to seal off body cavities. For example, tight junctions between cells lining the intestine prevent substances from the intestine from entering the body or the bloodstream by passing around the cells. The sheet of cells thus acts as a selective barrier; food substances required by the body must be transported to the bloodstream through the cytoplasm of intestinal cells, and toxins and unwanted materials are prevented from entering the bloodstream.

Gap Junctions

A third type of intercellular connection in animal cells, the **gap junction,** is like the desmosome in that it bridges the space between cells, but the space it bridges is somewhat narrower (Figure 5–25). Gap junctions also differ in that they not only connect the membranes but also act as pores connecting the cytoplasm of adjacent cells. Gap junctions consist of hexagonal arrays of proteins forming clusters of pores, which are about 1 to 2 nm in diameter. Small molecules (e.g., ions) and small biological molecules (e.g., derivatives of ATP) are capable of passing through the pores, but larger molecules are excluded. When appropriate marker substances are injected into one of a group of cells connected by gap junctions, the marker passes rapidly into the adjacent cells but does not enter the space between the cells. Gap junctions provide for rapid chemical and electrical communications between cells. Cells in the pancreas are linked together by gap junctions in such a way that if one of a group of cells is stimulated to secrete

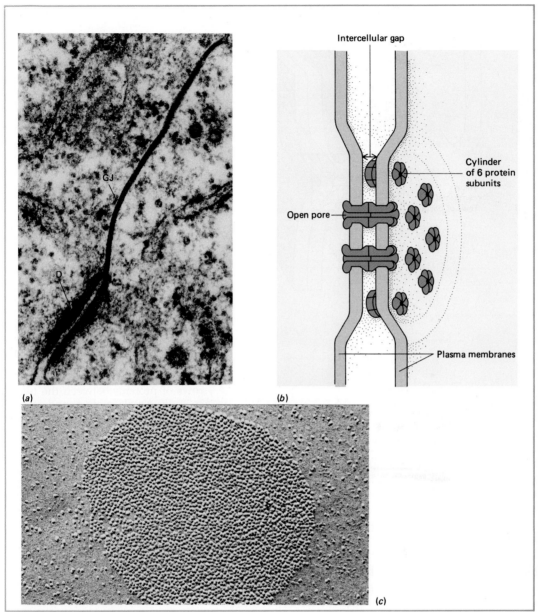

(a)

(b)

Intercellular gap

Cylinder of 6 protein subunits

Open pore

Plasma membranes

(c)

FIGURE 5–25 (a) A gap junction (*GJ*) and a desmosome (*D*) between ovarian cells of a rabbit (approximately ×180,000). Note the difference in the spaces between the two cell membranes in each structure. (b) Model of a gap junction based on electron-microscopic and x-ray diffraction data. The two membranes contain cylinders composed of six protein subunits arranged to form a pore. Two cylinders from opposite membranes are joined to form a pore which is about 1.5–2.0 nm in diameter connecting the cytoplasmic compartments of the two cells. (c) Freeze-fracture replica of the P-face (P) gap junction between two ovarian cells of a mouse, showing the numerous protein particles present (approximately ×70,000).

insulin, the signal is passed through the junctions to the other cells in the cluster, ensuring a coordinated response to the initial signal. Heart muscle cells are also linked by gap junctions in order to provide electrical coupling that synchronizes the contraction of cells.

Plasmodesmata

Plant cells are isolated from one another by rigid, thick cell walls. These walls contain channels 20 to 40 nm in diameter, called **plasmodesmata** (singular, *plasmodesma*), which are used to connect the cytoplasm of neighboring cells (Figure 5–26). The surface membranes of neighboring cells are continuous with each other through the plasmodesmata, allowing water and small molecules to pass through the openings from cell to cell. In most plasmodesmata there is a cylindrical membranous structure, called the *desmotubule*, which runs through the opening and connects the ER of the two adjacent cells.

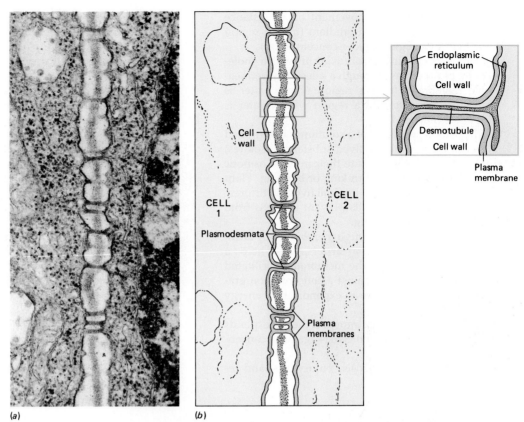

(a) (b)

FIGURE 5–26 Plasmodesmata in plant cell walls. Channels through plant cell walls form cytoplasmic connections, which allow water and small molecules to move between adjacent cells. The channels are lined with the fused plasma membranes of the two adjacent cells. In the center of most plasmodesmata are desmotubules, cylinders formed from endoplasmic reticulum membranes from both cells.

■ SUMMARY

I. Cell membranes are complex structures that (1) physically separate the interior of the cell from the outside world and (2) create compartments inside the cells of complex organisms that allow them to perform complex functions.
   A. Membranes have many different structural and functional roles.
      1. They regulate the passage of materials.
      2. They receive information that permits the cell to sense changes in its environment and respond to them.
      3. They contain specialized structures that allow specific contacts and communications with other cells.
   B. According to the fluid mosaic model of membrane structure, membranes consist of a fluid lipid bilayer in which a variety of proteins are embedded.
      1. The lipid bilayer is arranged so that the hydrophilic headgroups of the phospholipids are at the two surfaces of the structure and their hydrophobic fatty acid chains are in the middle.
      2. In almost all biological membranes the lipids of

the bilayer are in a fluid or liquid-crystalline state, which allows the molecules to move rapidly in the plane of the structure.
      3. Integral membrane proteins are embedded in the bilayer in such a way that their hydrophilic surfaces are exposed to the outside aqueous environment and their hydrophobic surfaces are in contact with the hydrophobic interior of the bilayer.
      4. Peripheral membrane proteins are associated with the surface of the bilayer and are easily removed without disrupting the structure of the membrane.
      5. Membrane proteins and lipids are asymmetrically positioned in the bilayer so that one side of the membrane has a different composition and structure from the other.

II. Biological membranes are selectively permeable; that is, they allow the passage of some substances but not others.
   A. Some ions and molecules pass through the membranes by simple diffusion.

1. Diffusion is the net movement of a substance down its concentration gradient (from a region of high to a region of low concentration).
2. Osmosis is a kind of diffusion in which molecules of water pass through a selectively permeable membrane from a region where water is more concentrated to a region where water is less concentrated.
3. The osmotic pressure of a solution is determined by the amount of dissolved substances (solutes) in solution. Cells regulate their internal osmotic pressure to prevent shrinking or bursting. Plant cells can be under high internal hydrostatic pressure, since the cell walls prevent the cells from expanding and bursting.

B. Some substances pass through membranes by facilitated diffusion, in which a carrier protein helps a molecule move through the membrane. Facilitated diffusion cannot work against a concentration gradient and does not require an expenditure of energy by the cell.

C. In carrier-mediated active transport the cell expends energy to move ions or molecules against a concentration gradient.

D. In endocytosis (phagocytosis, pinocytosis, and receptor-mediated endocytosis) materials such as food may be moved into the cell; a portion of the cell membrane envelops the material, enclosing it in a vacuole, or small vesicle, which is then released inside the cell.

E. In exocytosis, the cell ejects waste products or secretes substances such as mucus.

III. Plasma membranes of animal and plant cells contain specialized structures that allow them to have contact and communications with adjacent cells.

A. Desmosomes, tight junctions, and gap junctions are specialized structures associated with the plasma membrane of animal cells.
1. Desmosomes weld cells together to form strong tissues.
2. Tight junctions seal membranes of adjacent cells together to prevent substances from passing around cells.
3. Gap junctions are protein pores in membranes that allow communication between the cytoplasms of adjacent cells.

B. Plasmodesmata are openings in plant cell walls that allow the plasma membranes and the cytoplasm of adjacent cells to be continuous so that water and small molecules can pass through.

■ POST-TEST

1. The core of biological membranes is made of a bilayer of _____.

2. The fatty acid interior of the lipid bilayer is _____, whereas the surface of the bilayer is _____.

3. Bilayers are formed from _____ lipids— that is, lipids that have prominent _____ and _____ regions on the molecule.

4. _____ membrane proteins have domains with _____ surfaces that are associated with the interior of the bilayer. _____ membrane proteins are associated with the bilayer through interactions with other membrane components on the surface of the membrane.

5. Net movement of molecules from a region of high concentration to a region of low concentration is called _____.

6. A solution that has an equivalent concentration of solutes and hence the same osmotic pressure as the fluid inside the cell is said to be _____ to the cell's contents.

7. When a plant cell is placed in solution A, the contents inside the cell membrane can be seen to shrink away from the cell wall. Solution A is _____ to the interior of the plant cell.

8. When red blood cells are placed in solution B, the cells are seen to expand, and many of them burst. Solution B is _____ to the interior of the red blood cells.

9. When glucose enters cells by moving down its concentration gradient with the help of a carrier protein and does not require an additional source of energy, it does so by _____ _____.

10. When the glucose concentration outside the cell is lower than its concentration inside the cell, it must be moved into the cell by _____ _____.

11. The sodium-potassium pump is an _____ _____ system that pumps _____ from the inside of the cell to the outside and pumps _____ from the outside of the cell to the inside.

12. Active transport systems that derive their energy from transport of a second molecule down its concentration gradient are called _____ systems.

13. Large molecules and particles are transported into cells by an endocytotic mechanism called _____.

14. Molecules that bind to specific molecules on the cell membrane and then move into the interior of the cell through small vesicles do so by the process of _____ _____ _____.

15. _____ are specialized regions of the plasma membrane in animal cells that "spot weld" adjacent cells together.

16. _____ _____ contain specialized protein pores in the plasma membranes of animal cells that allow small molecules to pass from the cytoplasm of one cell to the cytoplasm of its neighbor.

17. _____ _____ are specialized regions of plasma membranes in animal cells that prevent substances from passing around the outsides of cells in a tissue.

18. _____ provide for water and small molecules to be transferred to adjacent plant cells.

■ REVIEW QUESTIONS

1. Make a diagram of the cell membrane showing the lipid bilayer, the various types of membrane proteins, and the carbohydrates on the outer surface of the membrane.
2. Illustrate how a transmembrane protein would be positioned in a lipid bilayer. Show which parts of the protein are hydrophilic and which parts are hydrophobic.
3. Describe the mechanisms and pathway used by cells to place carbohydrates on plasma membrane proteins. Explain why this results in the carbohydrate groups being on only one side of the lipid bilayer.
4. Discuss how molecules move as a consequence of diffusion. In what ways is the phenomenon of diffusion important to living things?
5. Differentiate clearly between osmosis and diffusion.
6. List and discuss the various functions of the plasma membrane. Discuss the nature of the proteins that carry out those functions and explain how their properties make them especially adapted for their functions.
7. Compare the transportation of compounds across the cell membrane by simple diffusion and the two major kinds of carrier-mediated transport. What additional factors and requirements are involved in the mediated transport systems?
8. Draw a diagram illustrating how membrane lipid bilayers fuse during the processes of exocytosis and endocytosis.
9. Compare the processes of phagocytosis and pinocytosis.
10. Compare the structures and functions of desmosomes and tight junctions. How are they similar? How do they differ?
11. Compare the structures and functions of gap junctions and plasmodesmata. How are they similar? How do they differ?

■ RECOMMENDED READINGS

Alberts, B., Bray, D., Lewis, J., Raff, M., Roberts, K., and Watson, J. D.: *Molecular Biology of the Cell*. New York, Garland Publishing, 1983. Chapter 6, A discussion of the structure and functions of the plasma membrane.

Avers, C. J.: *Molecular Cell Biology*. Menlo Park, CA, Benjamin Cummings, 1986. An excellent presentation of cell membrane structure and function.

Darnell, J., Lodish, H., and Baltimore, D.: *Molecular Cell Biology*. New York, Scientific American Books, 1986, Chapters 14 and 15. Comprehensive treatment of the structure and functions of the plasma membrane and transport functions.

de Duve, C.: *A Guided Tour of the Living Cell*. Scientific American Library, 1984. An illustrated tour of the cell in the form of journeys through different membrane and organelle systems.

PART 2

ENERGY IN LIVING SYSTEMS

Activities of all living cells require energy. ■ Producers capture energy from sunlight and incorporate it into their organic compounds. ■ Some of that energy is transferred to the consumers and decomposers that feed on the producers. ■ Amazingly complex sequences of reactions have evolved to capture and process the energy necessary to sustain life. ■

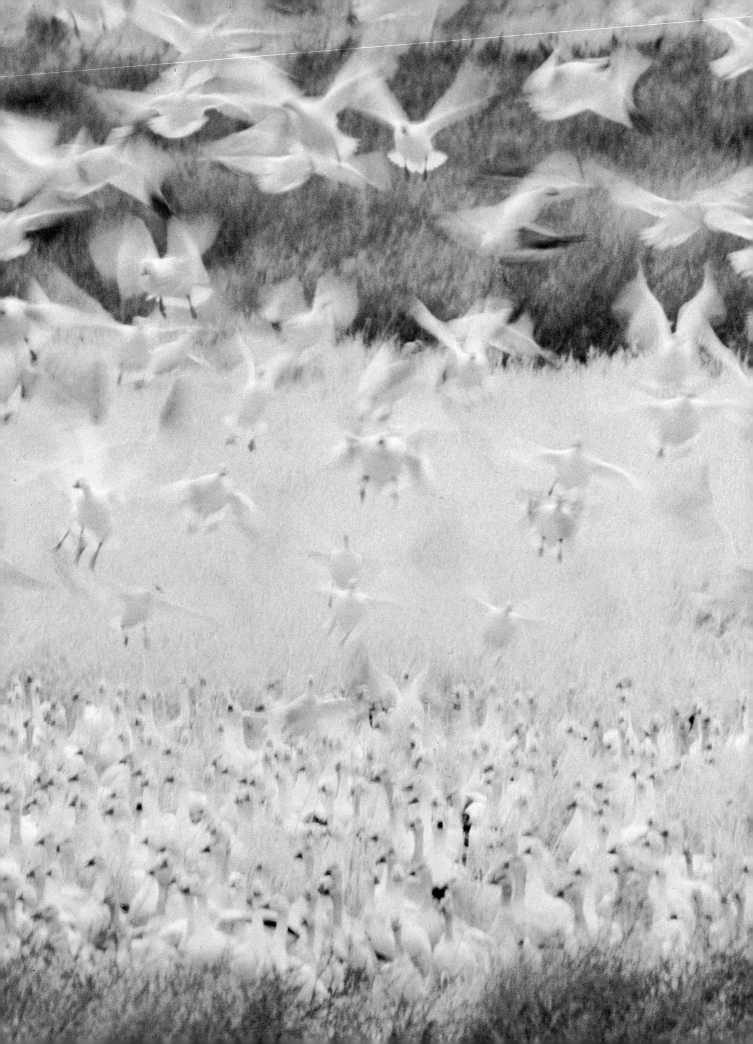

White-footed mouse (*Peromyscus leucopus*)

6

The Energy of Life

■ LEARNING OBJECTIVES

After you have read this chapter you should be able to:
1. Define the term *energy* and contrast potential and kinetic energy.
2. State the first and second laws of thermodynamics and discuss their applications to living organisms and to the ecosphere.
3. Describe the energy dynamics of a reaction that is in equilibrium.
4. Distinguish between endergonic and exergonic reactions and explain how they may be coupled so that the second law of thermodynamics is not violated.
5. Describe the chemical structure of ATP and its role in cellular metabolism.
6. Explain the function of enzymes and describe how they work.
7. Describe factors such as pH and temperature that influence enzymatic activity.
8. Compare the actions and effects of the various types of enzyme inhibitors.

Every activity of a living cell or organism requires energy. The myriad chemical reactions of cells that enable them to carry on their activities—to grow, move, maintain and repair themselves, reproduce, and respond to stimuli—together make up an organism's metabolism (Figure 6–1). We can define *metabolism* as all the chemical and energy transformations that occur within the living organism.

Because an organism has no way of creating new energy or of recycling the energy it has used, life depends on a continuous input of energy. This means there is a one-way flow of energy through any individual organism and through the ecosphere. Producers trap light energy from the sun during photosynthesis and incorporate some of that energy in the chemical

FIGURE 6–1 This bobcat is expending energy in an effort to capture the snowshoe hare. If caught and eaten, the hare will provide nutrients containing energy for future activity. For its part, the hare is expending a great deal of energy in its effort to escape becoming an energy source for the bobcat.

bonds of carbohydrates and other organic compounds. Then, a portion of that chemical energy can be transferred to the consumers, which eat the producers, and to the decomposers, which feed on them all sooner or later.

An organism must be considered an open system with respect to energy because of the one-way energy flow through it. Energy is captured, temporarily stored, and then used to perform biological work. During these processes it is converted to heat and dispersed into the environment. Because this energy cannot be reused, organisms must continually obtain fresh supplies of energy. But how do they do so? This chapter focuses on some of the principles of energy capture, storage, transfer, and use; the following two chapters explore some of the main metabolic pathways used by cells in their continuous quest for energy.

WHAT IS ENERGY?

Energy may be simply defined as the capacity to do work; more comprehensively, it can be defined as the ability to produce a change in the state or motion of matter. We can define **matter** as anything that has mass and takes up space. Here we are concerned with the ability of living systems to do biological work. It is taking enough energy to light a 75-watt bulb just to keep your brain in operation as you read these words. And at this very moment you are expending considerable amounts of energy to maintain your breathing, concentrate urine in your kidneys, digest food, circulate your blood, and maintain countless other metabolic activities. These are all forms of biological work.

Energy can exist in several different forms. These include heat, as well as electrical, mechanical, chemical, sound, and radiant energy (the energy of electromagnetic waves, such as radio waves, visible light, x rays, and gamma rays).

Potential and Kinetic Energy

Energy can be described as potential or kinetic. **Potential energy** is stored energy; it has the capacity to do work owing to its position or state. In contrast, **kinetic energy** is the energy of motion. A boulder at the top of a hill has potential energy because of its position. As the boulder rolls down the hill, the potential energy is converted to kinetic energy. It would require the input of energy to push the boulder back up the hill and restore the potential energy of its position at the top. Another example of the conversion of potential energy to kinetic energy is the release of a drawn bow (Figure 6–2). The tension in the bow and string represents stored energy; when the string is released, this potential energy is released, so that the motion of the bow propels the arrow. It would require the input of additional energy to draw the bow once again and restore the potential energy. Most of the actions of organisms involve a complex series of en-

POTENTIAL	KINETIC

FIGURE 6–2 Potential energy is stored energy. Kinetic energy is the energy of motion. Shooting a bow and arrow illustrates some of the energy transformations that take place in living systems generally. The potential chemical energy released by cellular respiration and stored temporarily in the substance adenosine triphosphate (ATP) is converted to mechanical energy, which draws the bow, and to waste heat (not shown). The energy, once again potential, that is stored in the drawn bow is released as the arrow speeds toward its target. In all of these transformations, energy is neither created nor destroyed, and in each the remaining total useful energy is less than in any of the steps that preceded it.

ergy transformations. For example, to prepare for a running event, athletes eat foods that build up their reserves of glycogen. During the event, the athletes' bodies continuously convert that energy stored in the glycogen into the kinetic energy used to run the race.

Measuring Energy

To study energy transformations, scientists must be able to measure energy. How is this done? Heat is a convenient form in which energy can be measured because all other forms of energy can be converted into heat. In fact the study of energy and its transformations has been named **thermodynamics**—that is, heat dynamics.

Although several units may be used in measuring energy, the most widely used in biological systems is the **kilocalorie (kcal).** A kilocalorie is the amount of heat required to raise the temperature of 1 kg of water from 14.5 to 15.5°C. Nutritionists use the kilocalorie in measuring the potential energy of foods and usually refer to it as a *Calorie* (with a capital C).

THE LAWS OF THERMODYNAMICS

All of the activities of our universe—from the life and death of cells to the life and death of stars—are governed by two laws of energy, the laws of thermodynamics.

The First Law of Thermodynamics

According to the **first law of thermodynamics,** known also as the *law of conservation of energy,* the energy of the universe is constant. Similarly, the total energy of any **system—** that is, of any object and its surroundings—remains constant. The term *surroundings* refers to the rest of the universe. As an object undergoes a change, it may absorb energy from its surroundings or deliver energy into its surroundings. The difference in the energy content of the object in its initial and final states must be equalled by a corresponding change in the energy content of the surroundings.

The first law of thermodynamics holds that during ordinary chemical or physical processes, energy can be transferred and changed in form but can be neither created nor destroyed. The universe is a closed system when it comes to energy. As far as we know, the energy present when it formed

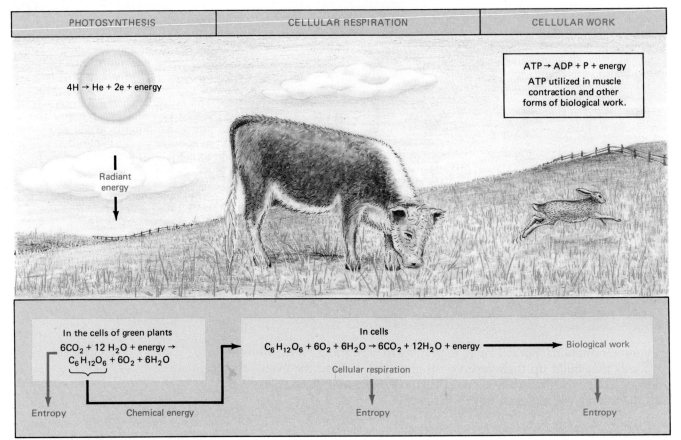

| PHOTOSYNTHESIS | CELLULAR RESPIRATION | CELLULAR WORK |

$4H \rightarrow He + 2e + energy$

$ATP \rightarrow ADP + P + energy$

ATP utilized in muscle contraction and other forms of biological work.

Radiant energy

In the cells of green plants
$6CO_2 + 12\,H_2O + energy \rightarrow$
$C_6H_{12}O_6 + 6O_2 + 6H_2O$

Entropy   Chemical energy

In cells
$C_6H_{12}O_6 + 6O_2 + 6H_2O \rightarrow 6CO_2 + 12H_2O + energy \longrightarrow$ Biological work

Cellular respiration

Entropy

Entropy

FIGURE 6–3 Three major types of processes that transform energy in the world of life are photosynthesis, cellular respiration, and biological work. (The chemical reactions shown are simplified. See Chapters 7 and 8 for more detail.)

some 20 billion years ago is all the universe can ever have; it cannot be added to or subtracted from.

Although an organism can neither make nor destroy energy, it can capture some from its environment and use it for its own needs. Organisms can also transform energy from one form to another. During photosynthesis, plant cells transform light energy to electrical energy and then to chemical energy stored in chemical bonds. Some of that chemical energy may later be transformed by some animal that eats the plant to the mechanical energy of muscle contraction or some other needed form (Figure 6–3). As these many transformations take place, some of the energy is converted to heat energy and dissipated into the environment. Although this energy can never again be used by the organism, it is not really "lost"; it still exists in the surrounding physical environment.

The Second Law of Thermodynamics

Order is an extremely unlikely state. If you were to drop a crystal vase, the fragments of glass would not spontaneously jump back into place to reconstruct the vase; in fact, it would be very difficult for you to gather all the small shards of glass and fit them together to mend the vase. Out of the multitude of possible ways in which the pieces could be assembled, only one would represent the highly ordered form that was the vase.

According to the **second law of thermodynamics,** the entropy of the universe is continuously increasing. **Entropy** may be defined as a randomized, disordered state of energy that is unavailable to do work (Figure 6–4). Entropy is a measure of randomness or disorder. The second law holds that physical and chemical processes proceed in such a way that the entropy of the system increases. In almost all energy transformations, there is

FIGURE 6–4 Entropy. (*a*) As particles leave a crystal to go into a solution, they become more disordered. The entropy of this system increases during the process. (*b*) Imagine that you shake a beaker containing two colors of marbles. A disordered arrangement (left) is much more probable than an ordered arrangement (right) in which all marbles of the same color remain together.

a loss of some energy in the form of heat to the surroundings (Figure 6–5). The energy thus lost is no longer available to do work.

Heat is actually the energy of the random motion of molecules and is the most disorganized form of energy. Heat can be made to do work only when a temperature difference causes it to flow from a warmer region to a cooler region. Temperature is the same throughout a living cell, so heat cannot be used to do biological work.

Although the total amount of energy in the universe remains constant, the energy available to do work is decreasing with time. This is because useful forms of energy are continuously degraded to heat, which is the least useful form of energy. Because entropy in the universe is continuously increasing, eventually, some billions of years in the future, all energy will be random and uniform in distribution. With only this useless form of energy, no work will be possible; the universe will have run down.

It is important to understand that the second law of thermodynamics is consistent with the first law. The total amount of energy in the universe is not decreasing with time, but the energy available to do work is being degraded to random molecular motion.

Because of the second law of thermodynamics, no process requiring energy is ever 100% efficient. Cellular energy utilization is about 55% efficient; the other 45% of the energy is lost as heat. Such biological processes are quite efficient compared with most machines made by human beings; for example, a gasoline engine is only about 17% efficient.

Because living organisms are highly organized, they are very unstable. In fact, life is a constant struggle against the second law of thermodynamics. The survival of individual organisms, as well as of ecosystems, depends on continuous energy input. Thus, producers must carry on photosynthesis, and consumers and decomposers must eat.

Energy lost
↓
Energy gained
by surroundings
Net energy change = 0

FIGURE 6–5 As a bird flies, some of its energy is lost as heat. That energy is gained by the surroundings, so the net energy change is zero.

Free Energy

The force that drives all processes in living and nonliving systems is the tendency of systems to reach the condition of maximal entropy. Energy in the form of heat is either given up or absorbed by the object to allow the system to reach the state of maximum entropy. The total heat content, or **enthalpy, H,** of a system is its total potential energy. In a chemical reaction, the enthalpy of reactants or products equals their total bond energies. When bonds are formed or broken, enthalpy is absorbed or released. Entropy and enthalpy are related by a third dimension of energy, termed **free energy.** We can think of free energy as that component of the total energy of a system that is available to do work under conditions of constant temperature and pressure. It is thus the aspect of thermodynamics of greatest interest in biology.

Entropy, represented by the letter *S*, and free energy, represented by the letter *G*, are related inversely; as entropy increases, the amount of free energy decreases. The two are related by the following equation:

$$\Delta G = \Delta H - T\Delta S$$

in which *H* is the enthalpy of the system and *T* is the absolute temperature in degrees kelvin. The symbol **Δ** means "change in."

All physical and chemical processes proceed with a decline in free energy until they reach an equilibrium at which the free energy of the system is at a minimum and the entropy is at a maximum. Free energy is useful energy in biological systems; entropy is a state of degraded, useless energy.

CHEMICAL REACTIONS AND ENERGY

A chemical reaction is a change involving the molecular structure of one or more substances; matter is changed from one substance with characteristic properties to another with new properties. During the reaction, energy is released or absorbed. For example, hydrochloric acid (HCl) reacts with the base sodium hydroxide (NaOH) to yield water (H_2O) and the salt sodium chloride (NaCl). In the process, energy is released as heat:

$$HCl + NaOH \longrightarrow H_2O + NaCl + energy\ (heat)$$

The chemical properties of HCl and NaOH are very different from those of H_2O and NaCl. Note that the number of atoms of a given element in the products is equal to the number of atoms of that element in the reactants. Atoms are neither destroyed nor created in a chemical reaction but simply change partners, as in a complex atomic square dance. This is an expression of the first law of thermodynamics.

Exothermic and Endothermic Reactions

Nearly every physical or chemical event in both living and nonliving systems is accompanied by the transfer of heat to the surroundings or by the absorption of heat from the surroundings. A process in which heat is transferred to the surroundings is **exothermic** (Figure 6–6). A process in which heat is absorbed from the surroundings is **endothermic.**

Many of the machines used in industry are heat engines that release energy. A familiar example of a heat engine is an engine driven by steam produced by the burning of coal to heat water in a boiler. In living systems heat is not a useful way of transferring or storing energy. Under conditions of constant pressure, heat can do work only when it can flow from a region of higher temperature to a region of lower temperature. Living organisms are basically isothermal (equal-temperature) systems. There is no significant temperature gradient—that is, difference in temperature—among the various parts of a cell or the various cells in a tissue. Cells cannot act as heat engines because they have no means of permitting heat to flow from a warmer to a cooler object. Furthermore, temperatures above 50°C would denature and inactivate enzymes and other cellular components.

Cells use the energy stored in the chemical bonds of complex organic molecules. However, this energy must be properly channeled if it is to perform work. Random energy release or the sudden release of large amounts of heat would be more likely to produce metabolic chaos than the order essential to living things. Endothermic and exothermic reactions in

(a)

(b)

FIGURE 6–6 An exothermic reaction. (*a*) The element bromine is in the beaker, and the element aluminum is on the watch glass. (*b*) When the aluminum is added to the bromine, the reaction is so vigorous that the aluminum melts and glows white hot.

living systems are carefully regulated and involve the transfer of very small amounts of energy.

Chemical Reactions Are Reversible

Free energy transformations that release a large amount of heat are inefficient because heat cannot be stored in living cells. Therefore, when heat is released, it must be radiated to the surrounding environment; this depletes the energy supply of the organism. Generally, when free energy transformations take place in living systems, only a small amount of free energy is released as heat at one time.

In most biochemical reactions there is little free energy difference between reactants and products. As a result, as long as external energy inputs continue to be available, most of the reactions that occur within living cells—including the important reactions of metabolism—are theoretically reversible. In fact, reversibility is characteristic of many biochemical reactions. Reversibility of biochemical reactions allows cells to control their release of free energy in accordance with their needs, and it permits many of their large biological molecules to be rebuilt or otherwise recycled for continued use in metabolic processes.

Reversibility is indicated by a double arrow: $\rightleftharpoons$. Whether a reaction will occur and whether it will proceed from right to left or left to right depends on factors such as the energy relations of the several chemicals involved, their relative concentrations, and their solubility.

Equilibrium

Suppose that over a ten-year period the population of a city remains the same. Some new folks have moved into town, but others have moved out or perhaps died; thus, the net change in the population is zero. We might say that the population in this city is in the state of *dynamic equilibrium*. In an **equilibrium** the rate of change in one direction is equal to the rate of change in the opposite direction.

Consider a reaction in terms of the numbers of each type of molecule involved. At the beginning of a reaction, only the reactant molecules may be present. These molecules move about and collide with one another with sufficient energy to react. As more and more product molecules are produced, fewer and fewer reactant molecules are left. As the product molecules increase in number, they collide more frequently, and some have sufficient energy to initiate the reverse reaction. The reaction thus proceeds in both directions simultaneously and eventually reaches an equilibrium in which the rate of the reverse reaction is about the same as the rate of the forward reaction.

For each reaction, the thermodynamic **equilibrium constant (K)** expresses the chemical equilibrium reached by the system. This is a fixed ratio of products and reactants. For the reaction

$$A + B \longrightarrow C + D,$$

$$K = \frac{[C] \times [D]}{[A] \times [B]}$$

The brackets around the letters mean "the concentration of." The equilibrium constant is different for every reaction; it is determined by the tendency of the reaction components to reach maximum entropy, or minimum free energy, for the system. A reaction with a very small K (say, 10^{-7}) hardly proceeds at all. A reaction with a very large K (say, 10^7) goes almost to completion.

When a reaction is at equilibrium, the free energy difference between the products and reactants of a chemical reaction is zero. Any change, such as a change in temperature or pressure, that affects the reacting system may cause the equilibrium to shift. The reaction may then proceed in a specific direction until once again the free energy difference is zero and a new equilibrium has been established.

When there is little free energy difference between the two sides of a chemical equation, the direction of a reaction is governed mainly by the concentrations of the reactants and products. The reaction tends to proceed in the direction that will minimize the difference in concentration between the substances on the two sides of the equation. If one of them—say, the product—is continuously removed as it is formed, the reaction will indeed proceed to completion; that is, all the reactant molecules are used up. (See Focus on Law of Mass Action.)

Spontaneous Reactions

A reaction that produces products that contain less free energy than did the original reactants tends to proceed spontaneously. A **spontaneous reaction** can occur without the addition of outside energy. Spontaneous reactions are not necessarily instantaneous; in fact, they may occur over a long period of time.

Focus On THE LAW OF MASS ACTION

The **law of mass action** states that, when all other conditions are constant, the rate of the reaction is proportional to the concentrations of the reactants. Consider a bottle of club soda. It is essentially a solution of carbon dioxide dissolved in water under high pressure (for which you can easily pay more than a dollar). Yet that bottle of club soda is not really just dissolved carbon dioxide; it is a solution of carbonic acid. To make it, carbon dioxide had to react with water in this fashion:

$$H_2O + CO_2 \longrightarrow H_2CO_3$$

As everyone knows, if the cap is left off, the opposite occurs:

$$H_2CO_3 \longrightarrow H_2O + CO_2$$

Bubbles of gas fizz to the surface; if the bottle is left open long enough, eventually no CO_2 is left. This is clearly an example of a reversible reaction. Notice that its direction depends on the pressure of the carbon dioxide gas. However, this is just another way of saying that its direction depends on the concentration of carbon dioxide gas.

When the club soda is manufactured, it is treated with high-pressure carbon dioxide. This causes large amounts of carbon dioxide to dissolve in water, producing a high concentration of the dissolved gas. The only way this concentration can be reduced is for it to react with the water to become carbonic acid. Yet this in turn produces increasing concentrations of carbonic acid, which increases the likelihood that some of those car-

bonic acid molecules will break down. Ultimately an equilibrium is reached, whereby just as many carbonic acid molecules break down as are forming. At that point the solution will contain varying concentrations of all three molecular species: carbon dioxide, water, and carbonic acid. So it will continue as long as the bottle remains unopened.

As soon as the cap is removed, carbon dioxide rushes out, and further amounts are lost by diffusion into the air, which has a very low carbon dioxide content. This reduces the concentration of carbon dioxide in the solution, rendering it less probable that molecules of carbon dioxide will collide with water molecules to make carbonic acid. In fact, the opposite occurs: The decrease in carbon dioxide means that carbonic acid already present will be converted to carbon dioxide and water, with the carbon dioxide steadily leaving the system. When equilibrium is reached, the concentrations of both carbon dioxide and carbonic acid will be considerably less than they were in the unopened bottle.

What this example illustrates is that when there is little free energy difference on the two sides of a chemical equation, the direction of its reaction will be governed mostly by the *concentrations* attained by products and reactants, with the reaction tending to proceed so as to minimize the difference in concentration between the chemical species on the two sides of the equation. If one of them is continually removed, for instance, by gaseous diffusion, evaporation, or precipitation, then even despite minimal energy differences, the reaction will nevertheless proceed to completion.

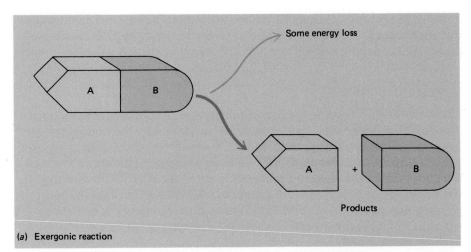

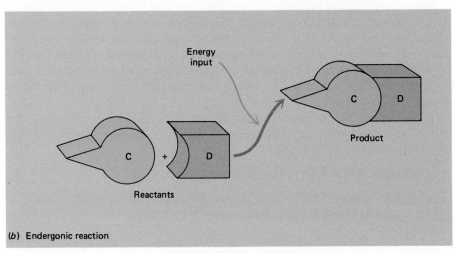

FIGURE 6–7 Exergonic and endergonic reactions. (*a*) In an exergonic reaction there is a net loss in energy. The products have less energy than was present in the reactants. (*b*) In endergonic reactions there is a net gain in energy. An endergonic reaction occurs only when energy lost from some other system is fed into the reaction. The product has more energy than was present in the reactants.

Spontaneous reactions, which release free energy and can therefore perform work, are also referred to as **exergonic reactions** (Figure 6–7). Because energy is released, the products contain less energy than the reactants. Exergonic reactions have a high equilibrium constant, K, and a negative free energy change, ∆G.

Reactions that are not spontaneous require an input of free energy and are said to be **endergonic.** In an endergonic reaction free energy is absorbed from its surroundings; as a result, the products contain more energy than the reactants. Endergonic reactions have a very low equilibrium constant and a positive free energy change (Figure 6–8).

Most exothermic (heat-releasing) reactions are also exergonic, but if the disorder of the reacting system has increased, then more free energy is released than would be indicated by the amount of heat released. This

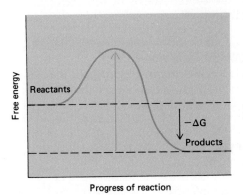

(a) Exergonic reaction

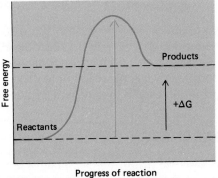

(b) Endergonic reaction

FIGURE 6–8 Energy changes in exergonic and endergonic reactions. (*a*) In exergonic reactions free energy is released, and the product has less energy than the reactants. (*b*) In endergonic reactions there is a net input of free energy, so the products contain more energy than the reactants. Note that even the exergonic reaction requires some input of energy to get started. This initial investment energy is termed *activation energy.*

increase in entropy can sometimes be sufficient to make even an endothermic (heat-absorbing) reaction or process occur spontaneously. Such is the case with the melting of ice and the dissolving of certain solids in liquids.

Coupled Reactions

Many metabolic reactions in a living organism—protein synthesis, for example—are endergonic. These reactions are driven by the energy released from exergonic reactions. Indeed, in the living cell endergonic and exergonic reactions are **coupled;** the thermodynamically favorable exergonic reaction provides the energy required to drive the thermodynamically unfavorable endergonic reaction. In such coupled systems the endergonic reaction can proceed only if the decline in the free energy of the exergonic reaction to which it is coupled is larger than the gain in free energy of the endergonic reaction.

How does a living organism, from the time it is "born" until the time it dies, employ outside energy inputs to compensate for its continuous loss of free energy? Two factors make these energy inputs available: First, organisms are part of a large universe with a vast reserve of free energy. Second, they have within themselves special structures, enzymes, and genetic information needed to direct their negatively entropic life processes.

To see how the chemical machinery of the cell is able to supply the energy to direct an endergonic reaction, consider the free energy change, ΔG, in the following reaction:

(1) $A \longrightarrow B + C \qquad \Delta G = +5$ kcal/mole

Since ΔG is positive, free energy is absorbed rather than released and the reaction is not spontaneous. By way of contrast, consider the following reaction:

(2) $C \longrightarrow D \qquad \Delta G = -8$ kcal/mole

This reaction can proceed spontaneously because it loses free energy and therefore has a *negative* ΔG.

Note that, whereas the reaction $A \rightarrow B + C$ has a positive free energy change (+5 kcal/mole), the reaction $C \rightarrow D$ has a larger negative free energy change (−8 kcal/mole). Because the free energies of reactions are additive, cells can use exergonic reactions to drive endergonic reactions. When reactions (1) and (2) occur together, they form a system that has an overall negative free energy change (−3 kcal/mole); that is, energy is released. This released energy can be used by cells for cellular work.

To sum it up:

(1) $A \longrightarrow B + C \qquad \Delta G = +5$ kcal/mole

(2) $\underline{C \longrightarrow D \qquad\quad \Delta G = -8 \text{ kcal/mole}}$

$\quad\ A \longrightarrow B + D \qquad \Delta G = -3$ kcal/mole

If D is the desired product, then this is a thermodynamically feasible way to produce it. The second reaction pulls the first one along.

Generally, for each endergonic reaction occurring in a living cell, there is a coupled exergonic reaction to drive it. Often, the exergonic chemical reaction involves the breakdown of adenosine triphosphate (ATP).

THE ENERGY CURRENCY OF THE CELL—ATP

Staying alive requires a continuous expenditure of energy. Cells must grow, maintain, and repair themselves. This requires the synthesis of proteins and other needed compounds as well as the manufacture of cellular organelles and the production of new cells. In order to maintain homeostatic conditions, cells must continuously transport ions and other substances across cell membranes. Cells also carry on mechanical work, such as contraction or the beating of cilia or flagella. What is the immediate source of energy for cellular work?

In all living cells energy is temporarily packaged within a remarkable chemical compound called **adenosine triphosphate (ATP)**, which stores large amounts of energy for very short periods of time. We may think of ATP as the energy currency of the cell. When you work you earn money, so you might say that your energy is symbolically stored in the money you earn; in the same way, the energy of the cell is stored in ATP. When you earn extra money, you might deposit some in the bank; similarly, the cell deposits energy as lipid in fat cells or as glycogen in liver and muscle. Moreover, just as you dare not make less money than you spend, so too the cell must avoid energy bankruptcy, which would mean its death. Finally, just as you (alas) do not keep what you make very long, so too the cell is forever spending its ATP.

ATP is a nucleoside triphosphate consisting of three main parts (Figure 6–9): (1) a nitrogen-containing base, adenine, which also occurs in DNA; (2) ribose, a five-carbon sugar; and (3) three inorganic phosphate groups identifiable as phosphorus atoms surrounded by oxygen atoms. Inorganic phosphate is usually designated P_i. Notice that the phosphate groups are attached to the end of the molecule in a series, rather like three passenger cars behind a locomotive. The couplings—that is, the chemical bonds attaching to the last two phosphates—also resemble those of a train in that they are readily attached and detached.

The bonds linking the phosphate groups of ATP are unstable; they can be broken by hydrolysis. All three phosphate groups are negatively charged and tend to repel one another. As a result, the phosphate bonds are relatively weak and easily broken. When ATP is hydrolyzed, the energy released is used to form new bonds that are stronger than the phosphate bonds. Energy is released during the reaction due to the chemical change to a more stable condition.

When the third phosphate is removed, the remaining molecule is **adenosine diphosphate (ADP)**. This is an exergonic reaction releasing 7.3 kcal of energy per mole of ATP that is hydrolyzed:

$$ATP + H_2O \longrightarrow ADP + P \qquad \Delta G = -7.3 \text{ kcal/mole}$$

When the two terminal phosphate groups (called a *pyrophosphate group*)

FIGURE 6–9 ATP, the energy currency of all living things. (*a*) ATP is composed of adenine (shown in blue type), ribose (red), and three phosphate groups (purple). The bonds linking the two terminal phosphate groups of ATP (shown in red) are unstable and easily hydrolyzed. They are energy-rich, with a large potential for transferring the phosphate group together with some of the energy of the bond. (*b*) A computer-generated model of ATP. The red balls are oxygen atoms; the blue, nitrogen; the green, carbon; the yellow, phosphate; and the white, hydrogen. Note the hydrogen atom attached to the last oxygen in the triphosphate group. At different pH values, this and other oxygen atoms might be bonded with hydrogen or be present in ionized form.

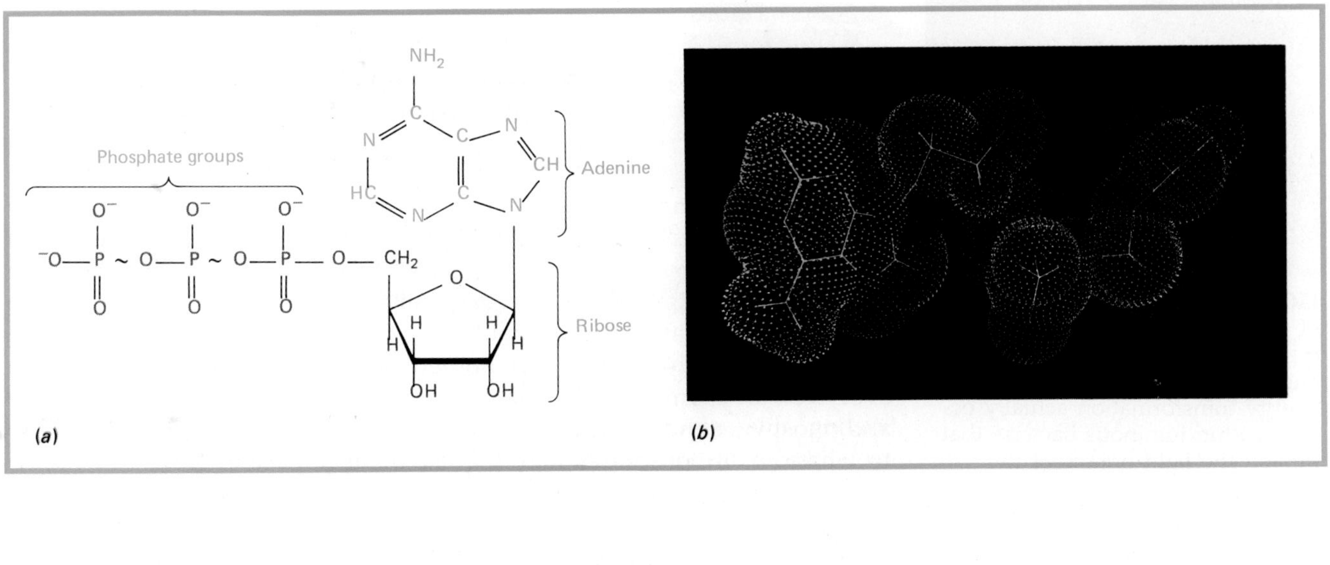

(a)

(b)

FIGURE 6–10 ATP is an important link between endergonic and exergonic reactions in living cells.

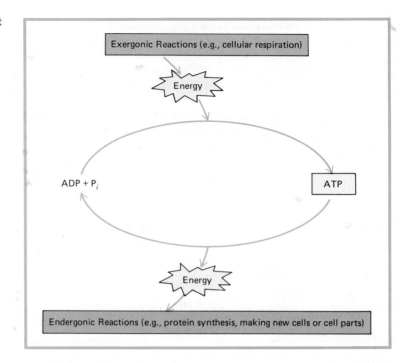

Exergonic Reactions (e.g., cellular respiration)

Energy

ADP + P$_i$

ATP

Energy

Endergonic Reactions (e.g., protein synthesis, making new cells or cell parts)

FIGURE 6–11 The chemical energy of ATP is converted to light energy in the light organs of this deep-sea angler fish, *Caulophryne jordani*. The energy transformation actually occurs within luminous bacteria that inhabit the light organs of the fish.

are removed, the molecule that remains is adenosine monophosphate (AMP).

The energy from the hydrolysis of ATP is coupled to endergonic processes within the cell with the help of specific enzymes. This process involves the transfer of a phosphate group from ATP to some other compound. The addition of a phosphate group to a molecule is referred to as **phosphorylation.** Energy-rich bonds may be designated by wavy lines (see Figure 6–9).

When a phosphate is attached to AMP, it becomes ADP, and when a phosphate is added to ADP, ATP is produced. These reactions are readily reversible.

$$AMP + P_i + energy \longrightarrow ADP$$
$$ADP + P_i + energy \longrightarrow ATP$$

As the equations indicate, energy is required to add a phosphate to either the AMP or the ADP molecule. Conversely, since energy can neither be created nor destroyed, the energy is released or transferred to another molecule when the phosphate is detached. Thus, ATP is an important link between exergonic (energy-releasing) and endergonic (energy-requiring) reactions (Figure 6–10).

ATP is formed from ADP and inorganic phosphate when nutrients are oxidized or when the radiant energy of sunlight is trapped in photosynthesis. Energy released from exergonic reactions is packaged in ATP molecules for use in endergonic reactions. This energy may be used to produce fats or glycogen, molecules stockpiled for long-term energy storage.

The cell contains a pool of ADP, ATP, and phosphate in a state of equilibrium. Large quantities of ATP cannot be stockpiled in the cell; in fact, studies suggest that a bacterial cell has no more than a 1-second supply of ATP. Thus, ATP molecules are used almost as quickly as they are produced. A human at rest uses about 45 kg of ATP each day, but the amount present in the body at any given moment is less than 1 gm. Every second in every cell an estimated 10 million molecules of ATP are made from ADP and phosphate and an equal number are hydrolyzed, yielding their energy to whatever life processes may require them (Figure 6–11).

ENZYMES: CHEMICAL REGULATORS

The principles of thermodynamics help us predict whether a reaction can occur but tell us nothing about the speed of the reaction. A glucose solution will keep indefinitely in a bottle if it is kept free of bacteria and molds. It must be subjected to high temperature, strong acids, or bases before it will break down. Living cells need glucose products and cannot wait for centuries for glucose to break down, nor can they use extreme conditions to cleave glucose molecules. In cells chemical reactions are brought about by **catalysts,** substances that affect the speed of a chemical reaction without being consumed by the reaction. Biological catalysts belong to a class of proteins called **enzymes.**

Cells require a slow, steady release of energy, and they must be able to regulate that release to meet metabolic energy requirements. Accordingly, fuel molecules are slowly oxidized and energy is extracted in small amounts during cellular respiration, which includes sequences of 30 or more reactions. In fact, most cellular metabolism proceeds by a series of steps, so that a molecule may go through as many as 20 or 30 chemical transformations before it reaches some final state. Even then, the seemingly completed molecule may be preempted by yet another chemical pathway so as to be totally transformed or consumed in the course of energy production. The changing needs of the cell require a system of flexible chemical control. The key elements of this control system are the remarkable enzymes.

Enzymes and Activation Energy

Like all catalysts, enzymes affect the rate of a reaction by lowering the energy needed to activate the reaction. Even a strongly exergonic reaction, which releases more than enough energy as it proceeds, is prevented from beginning by an energy barrier. This is because, for new chemical bonds to form, existing ones must first be broken. The energy required to overcome this barrier and start the reaction going is called **activation energy.** An enzyme greatly reduces the activation energy necessary to initiate a chemical reaction (Figure 6–12).

In a population of molecules of any kind, some molecules have a relatively high energy content, others have a lower energy content, and the energy content of the entire population of molecules conforms to a bell-shaped curve of normal distribution. Only molecules with a relatively high energy content are likely to react to form the product. To make the reaction proceed more quickly, we must lower the activation barrier. An enzyme lowers the activation energy of the reaction and allows a larger fraction of the population of molecules to react at any one time. The enzyme does this by forming an unstable intermediate complex with the **substrate,** the sub-

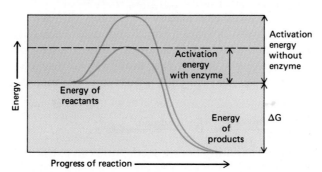

FIGURE 6–12 A catalyst such as an enzyme speeds up a reaction by lowering its activation energy. A catalyzed reaction (blue curve) proceeds more quickly than an uncatalyzed reaction (red curve) because it has a lower barrier of activation energy to overcome.

FIGURE 6–13 A bombardier beetle uses the catalyzed decomposition of hydrogen peroxide as a defense mechanism. The oxygen gas formed in the decomposition forces out water and other chemicals with explosive force. Since the reaction is very exothermic, the water comes out as steam.

stance on which it operates. This complex decomposes, forming the product and freeing the catalyst to react with a second molecule of reactant.

An enzyme can only promote a chemical reaction that could be made to proceed without it. There is nothing in the action of a catalyst of any kind that could change the operation of the second law of thermodynamics, so enzymes do not influence the direction of a chemical reaction or the final concentrations of the molecules involved. They simply speed up reactions.

Naming Enzymes

Enzymes are usually named by the addition of the suffix *-ase* to the name of the substance acted upon. For example, sucrose is split by the enzyme **sucrase** to give glucose and fructose. There are group names for enzymes that catalyze similar reactions. Lipases cleave triacylglycerols, proteinases cleave the peptide bonds in proteins, and dehydrogenases transfer hydrogens from one compound to another.

Enzymes Are Very Efficient Catalysts

The catalytic ability of some enzymes is truly phenomenal. For example, one molecule of the iron-containing enzyme **catalase** brings about the decomposition of 5 million molecules of hydrogen peroxide (H_2O_2) per minute at 0°C. Hydrogen peroxide is a poisonous substance produced as a byproduct in a number of enzyme reactions. Catalase protects cells by destroying peroxide.

Hydrogen peroxide can be split by iron atoms alone, but only at a very slow rate. It would take 300 years for an iron atom to split the same number of molecules of H_2O_2 that a molecule of catalase (containing one iron atom) splits in 1 second!

Enzymes Are Specific

Most enzymes are highly **specific,** catalyzing only a few closely related chemical reactions or, in many cases, only one particular reaction. For example, the enzyme urease, which decomposes urea to ammonia and carbon dioxide, attacks no other substrate. The enzyme sucrase splits only sucrose; it does not act on maltose or lactose. Peroxidase decomposes several different peroxides, including hydrogen peroxide.

A few enzymes are specific only in that they require that the substrate have a certain kind of chemical bond. For example, the lipase secreted by the pancreas will split the ester bonds connecting the glycerol and fatty acids of a wide variety of fats.

Enzymatic Teamwork

Enzymes usually work in teams, with the product of one enzyme-controlled reaction serving as the substrate for the next. We can picture the inside of a cell as a factory with many different assembly lines (and disassembly lines) operating simultaneously. Each of the assembly lines is composed of a number of enzymes, each of which carries out one step on the molecule, such as changing molecule A into molecule B, and then passes it along to the next enzyme, which converts molecule B into molecule C, and so on.

$$A \xrightarrow{\text{enzyme 1}} B \xrightarrow{\text{enzyme 2}} C$$

From germinating barley seeds, one can extract two enzymes that will convert starch to glucose. The first, amylase, hydrolyzes starch to maltose; the second, maltase, splits maltose to glucose. Eleven different enzymes, working consecutively, are required to convert glucose to lactic acid. The

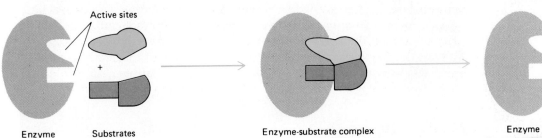

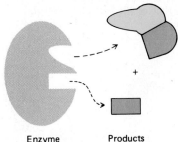

Enzyme   Substrates

Enzyme-substrate complex

Enzyme   Products

FIGURE 6–14 Lock-and-key mechanism of enzyme action. The substrate fits the active sites of the enzyme molecule much as a key fits a lock. However, in this model the lock acts on the key rather than the other way around. When the products separate from the enzyme, the enzyme is free to catalyze the production of additional products; the enzyme is not permanently changed by the reaction. In recent years the lock-and-key concept of enzyme action has undergone some modification, as shown in Figure 6–15.

same series of 11 enzymes is found in human cells, in green leaves, and in bacteria.

How Enzymes Work

Enzymes form temporary chemical compounds with their substrates. These complexes then break up, releasing the product and regenerating the original enzyme molecule for reuse.

Enzyme + substrate 1 + substrate 2 $\longrightarrow$ enzyme-substrate complex

Enzyme-substrate complex $\longrightarrow$ enzyme + product(s)

The enzyme itself is not permanently altered or consumed by the reaction.

Why does the enzyme-substrate complex break up into chemical products different from those that participated in its formation? As shown in Figure 6–14, each enzyme has one or more regions, called **active sites**, which in the case of a few enzymes have been shown to be actual indentations in the enzyme molecule. These active sites are located close to the enzyme's surface, and during the course of a reaction, substrate molecules occupying these sites are temporarily brought close together and react with one another. It is thought that when the enzyme and substrate bind together, the shape of the enzyme molecule changes slightly. This produces strain in critical bonds in the substrate molecules, causing these bonds to break. The new chemical compound thus formed has little affinity for the enzyme and moves away from it. An enzyme can be thought of as a molecular lock into which only specifically shaped molecular keys—the substrates—can fit.

Unlike a lock and key, however, the shape of the enzyme does not seem to be exactly complementary to that of the substrates. According to the *induced-fit model* of enzyme action, when the substrate combines with the enzyme, it may induce a change in the shape of the enzyme molecule. This is possible because the active sites of an enzyme are not rigid. The change in shape results in an optimum fit for the substrate-enzyme interaction and can put strain on the substrate. This stress may help bonds break, thus promoting the reaction (Figures 6–15 and 6–16).

FIGURE 6–15 Comparison of models of enzyme action. (a) The lock-and-key model. (b) The induced-fit model. Chemical reactions are favored when substrate molecules get close enough to one another to react, when they are presented to each other in the right orientation, and when their existing chemical bonds are strained. Enzymes often facilitate all three of these processes. The substrate's bonds are strained apparently because most active sites on the enzyme are a bit bigger than the substrate molecules. Accordingly, when a fit is forced on the active site, it exerts a kind of pull on the substrate, helping to pull it apart. To be sure, the fit of the enzyme and substrate must not be too poor, or they will have no affinity for one another.

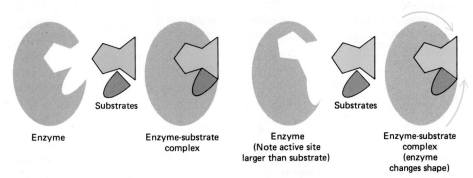

Enzyme

Enzyme-substrate complex

Substrates

Enzyme (Note active site larger than substrate)

Enzyme-substrate complex (enzyme changes shape)

Substrates

(a) Lock-and-key model

(b) Induced fit model

FIGURE 6–16 Model of the induced fit of an enzyme with its substrate.

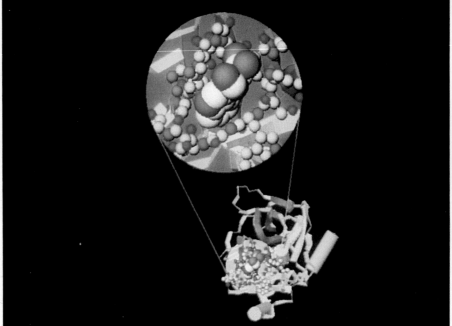

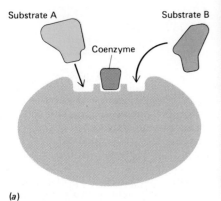

(a)

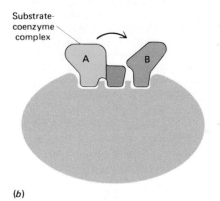

(b)

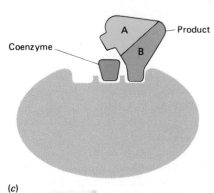

(c)

FIGURE 6–17 Coenzyme action. Some enzymes are not able to attach directly to the substrates the chemical reactions of which they catalyze. Such an enzyme employs accessory coenzymes to serve as adaptors, aiding the attachment of one or more substrates to the enzyme's active sites. First one substrate combines with the coenzyme to form a coenzyme-substrate complex. Then the coenzyme-substrate complex combines with the second substrate, forming a complex that yields the products and releases the coenzyme. Some familiar vitamins serve as coenzymes for vital enzymes of cellular metabolism.

Cofactors

Some enzymes—for example, pepsin, secreted by the stomach—consist only of protein. Other enzymes have two components, a protein referred to as the **apoenzyme** and an additional chemical component called a **cofactor.** The cofactor of some enzymes is a metal ion. In fact, most of the trace elements—elements like iron, copper, zinc, and manganese, which are required in very small amounts—function as cofactors.

An organic, nonpolypeptide compound that serves as a cofactor is called a **coenzyme.** Most vitamins are coenzymes or serve as raw materials from which coenzymes are synthesized. Neither the apoenzyme nor the cofactor alone has catalytic activity; only when the two are combined does the enzyme function (Figure 6–17).

Regulating Enzymatic Action

Enzymes regulate the chemistry of the cell, but what controls the enzymes? One mechanism of enzyme control depends simply on the amount of enzyme produced. The synthesis of each type of enzyme is directed by a specific gene. The gene, in turn, may be switched on by a signal from a hormone or by some other type of cellular product. When the gene is switched on, the enzyme is synthesized. The amount of enzyme present then influences the rate of the reaction. Up to a maximum value, the rate of an enzyme-dependent reaction increases as the concentration of the enzyme increases (Figure 6–18).

If the pH and temperature are kept constant and if an excess of substrate is present, the rate of an enzymatic reaction is directly proportional to the concentration of enzyme present (Figure 6–18b). If the enzyme concentration, pH, and temperature are kept constant, the initial rate of an enzymatic reaction is proportional to the concentration of substrate present, up to a limiting value (Figure 6–18c).

The product of one enzymatic reaction may control the activity of another enzyme, especially in a complex sequence of enzymatic reactions. For example, in the following system,

$$A \xrightarrow{\text{enzyme 1}} B \xrightarrow{\text{enzyme 2}} C \xrightarrow{\text{enzyme 3}} D \xrightarrow{\text{enzyme 4}} E$$

each step is catalyzed by a different enzyme. The final product, E, may inhibit the activity of Enzyme 1. When the concentration of E is low, the

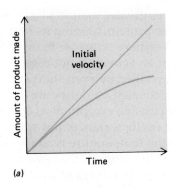
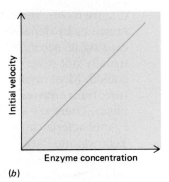
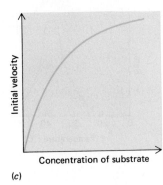

(a)

(b)

(c)

sequence of reactions proceeds rapidly. However, as the concentration of E increases, it serves as a signal for Enzyme 1 to slow down and eventually to stop functioning. Inhibition of Enzyme 1 stops this entire sequence of reactions. This type of enzyme regulation, in which the formation of a product inhibits an earlier reaction in the sequence, is called **feedback control.**

Another important method of enzymatic control depends on the activation of enzyme molecules that are present in an inactive form in the cytoplasm. In their inactive form the active sites of the enzyme are inappropriately shaped, so that the substrates do not fit. Among the factors that influence the shape (conformation) of the enzyme are acidity, alkalinity, and the concentration of certain salts.

Some enzymes, known as **allosteric enzymes,** possess a receptor site, called an **allosteric site,** on some region of the enzyme molecule other than the active site. (The word *allosteric* means "another space.") Most allosteric enzymes contain more than one polypeptide chain, each with its own active site. The allosteric site is often located where the polypeptides are joined.

If a substrate binds covalently to a site other than the active site, it may stimulate enzyme action or may inhibit it. The substance that binds to the allosteric enzyme is referred to as a **regulator,** or *modulator*.

The enzyme protein kinase is an allosteric enzyme with only one polypeptide chain; both the active site and the allosteric sites are located on this single chain. Another protein molecule serves as a regulator that binds reversibly to the allosteric site. This regulator is an inhibitor that inactivates the enzyme. In the body, protein kinase is in this inactive form most of the time (Figure 6–19). When protein kinase activity is needed, the compound cyclic AMP contacts the enzyme-regulator complex and removes the regulator, activating the protein kinase.

Enzyme + regulator ⟶ enzyme-regulator complex

Enzyme-regulator complex + cAMP ⟶ active enzyme + regulator-cAMP

Other Factors Affecting Enzyme Activity

Enzymes generally work best under certain narrowly defined conditions referred to as *optima*. These include appropriate temperature, pH, and salt concentration. Any departure from optimal conditions adversely affects enzyme activity.

Temperature

Enzymes are inactivated by high temperatures. Enzymatic reactions occur slowly or not at all at low temperatures, but the catalytic activity reappears when the temperature is raised to normal. The rates of most enzyme-controlled reactions increase with increasing temperature, within limits

FIGURE 6–18 The rate of an enzyme reaction is influenced by several factors. (*a*) Reaction rate as a function of time. As the substrate is consumed and the reaction approaches equilibrium, the reaction rate drops to zero. (*b*) Reaction rate as a function of the amount of enzyme added. Enough substrate and cofactors are added so that these do not limit the rate of the reaction. (*c*) Reaction rate as a function of the amount of substrate. Enough enzyme and cofactors are added so that these do not limit the rate of the reaction.

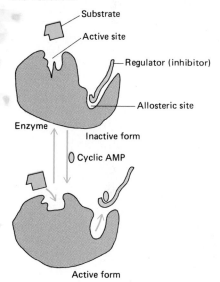
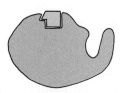

FIGURE 6–19 The enzyme protein kinase is inhibited by a regulator protein that binds reversibly to the allosteric site. When the enzyme is in this inactive form, the shape of the active site is modified so that substrate cannot combine with it. Cyclic AMP removes the regulator activating the enzyme, allowing the substrate to combine with the active site.

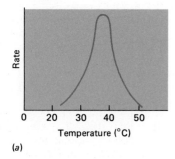

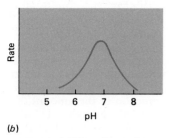

FIGURE 6–20 The effect of temperature (*a*) and pH (*b*) on the rate of enzyme-catalyzed reactions. Substrate and enzyme concentrations are constant.

(Figure 6–20). Temperatures greater than 50 to 60°C rapidly inactivate most enzymes by denaturing the protein, altering its molecular conformation by causing its secondary and tertiary structure to unwind. This inactivation is usually not reversible; that is, activity is not regained when the enzyme is cooled. Most organisms are killed by even short exposure to high temperature; their enzymes are inactivated, and they are unable to continue metabolism. There are a few remarkable exceptions to this rule: Certain species of cyanobacteria can survive in the waters of hot springs, such as the ones in Yellowstone Park, where the temperature is almost 100°C; these organisms are responsible for the brilliant colors in the terraces of the hot springs. Still other cyanobacteria live at temperatures much above that of boiling water in undersea hot springs, where the extreme pressure keeps water in its liquid form rather than as steam.

pH

The activity of an enzyme is markedly changed by any alteration in the acidity or alkalinity of the reaction medium. Full enzyme activity requires a specific number of positive and negative charges on the enzyme. Changes in pH add or remove hydrogen ions from the protein, thereby changing the number of positive and negative charges on the protein molecule and affecting its activity. Pepsin, a protein-digesting enzyme secreted by cells lining the stomach, is remarkable in that it will work only in a very acid medium—optimally at pH 2. In contrast, the pH optimum of trypsin, the protein-splitting enzyme secreted by the pancreas, is 8.5, on the alkaline side of neutrality. Many of the enzymes that operate within cells become inactive when the medium is made very acid or very alkaline. Strong acids or bases irreversibly inactivate enzymes by permanently changing their molecular conformation. Most enzymes are active only over a narrow range of pH.

Enzyme Inhibition

Most enzymes can be **inhibited** (so that their activity is decreased) or even destroyed by certain chemical agents. Enzyme inhibition may be reversible or irreversible. **Reversible** inhibitors can be competitive or noncompetitive. In **competitive** inhibition, the inhibitor competes with the normal substrate for binding to the active site of the enzyme (Figure 6–21). A competitive inhibitor usually is structurally similar to the normal substrate and so fits into the active site and combines with the enzyme. However, it is not similar enough to substitute fully for the normal substrate in the chemical reaction, and the enzyme cannot attack it to form reaction products. A competitive inhibitor occupies the active sites only temporarily and does not permanently damage the enzyme. In fact, competitive inhibition can be reversed by an increase in the substrate concentration.

In **noncompetitive** inhibition, the inhibitor binds with the enzyme at a site other than the active site. Such an inhibitor renders the enzyme inactive by altering its shape. Many important noncompetitive inhibitors are metabolic substances that regulate enzyme activity by combining reversibly with the enzyme.

Irreversible inhibitors combine with a functional group of an enzyme and permanently inactivate or destroy the enzyme. Most poisons are irreversible inhibitors. Nerve gases, for example, poison the enzyme acetylcholinesterase, which is important to the function of nerves and muscles. Cytochrome oxidase, one of the enzymes of the electron transport system (part of cellular respiration), is especially sensitive to cyanide. Death results from cyanide poisoning because cytochrome oxidase is irreversibly inhibited and can no longer transfer electrons from substrate to oxygen.

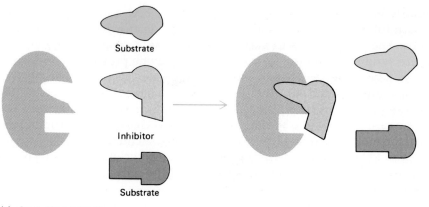

Substrate

Inhibitor

Substrate

(a) Competitive inhibition

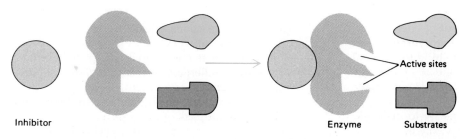

Active sites

Inhibitor

Enzyme

Substrates

Active sites not suitable for reception of substrates

(b) Noncompetitive inhibition

FIGURE 6–21 Competitive and noncompetitive inhibition. (*a*) In competitive inhibition, the inhibitor competes with the normal substrate for the active site of the enzyme. A competitive inhibitor occupies the active site only temporarily. (*b*) In noncompetitive inhibition, the inhibitor binds with the enzyme at a site other than the active site, altering the shape of the enzyme and thereby inactivating it. Noncompetitive inhibition may be reversible. Allosteric action, used by cells to control enzyme action, is a somewhat similar process (see Figure 6–19).

A number of insecticides and drugs are irreversible enzyme inhibitors. Penicillin is a good example of such a drug. This antibiotic and its chemical relatives inhibit a bacterial enzyme, transpeptidase, which is responsible for establishing some of the chemical linkages in the material of which the bacterial cell wall is composed. Unable to produce new cell walls, susceptible bacteria are prevented from multiplying effectively, as shown in Figure 6–22. Since human body cells do not possess cell walls and do not employ this enzyme, penicillin is harmless to humans, except for the occasional allergic patient.

Enzymes themselves can act as poisons if they get into the wrong compartment of the body. As little as 1 mg of crystalline trypsin injected intra-

FIGURE 6–22 Penicillin is an irreversible enzyme inhibitor. (*a*) Normal bacteria. Insert shows the new cell wall laid down between daughter cells of a dividing bacterium. (*b*) Penicillin has damaged these bacterial cell walls. The insets are magnified approximately ×54,000.

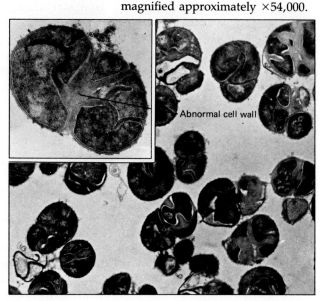

Daughter cells

New cell wall

Abnormal cell wall

(a)

(b)

venously will kill a rat. Several types of snake, bee, and scorpion venom are harmful because they contain enzymes that destroy blood cells or other tissues. The proteolytic enzymes of the pancreas, trypsin and chymotrypsin, are synthesized in the form of inactive enzyme **precursors,** molecules that are somewhat larger than the active enzyme. These are packaged in granules and secreted into the duct of the pancreas, thus protecting the pancreas itself from being digested by the enzymes it synthesizes. The enzymes are made active by other enzymes that cleave off a portion of the precursor molecule to yield the active enzyme. Acute pancreatitis, a serious, even lethal disease, occurs when the proteolytic enzymes become active while still within the pancreas and digest the cells of the pancreas and its blood vessels.

■ SUMMARY

I. Life depends on a continuous input of energy. Producers capture energy during photosynthesis and incorporate some of that energy in the chemical bonds of organic compounds. Some of this energy can then be transferred to consumers and decomposers.

II. *Energy* may be defined as the capacity to do work.
   A. Potential energy is stored energy; kinetic energy is energy of motion.
   B. A common unit used to measure energy is the kilocalorie.

III. The first law of thermodynamics states that energy can be neither created nor destroyed but can be transferred and changed in form. The second law of thermodynamics states that disorder in the universe is continuously increasing.
   A. The first law explains why organisms cannot produce energy but must borrow it continuously from somewhere else.
   B. The second law explains why no process requiring energy is ever 100% efficient; in every energy transaction, some energy is dissipated as heat. The term *entropy* refers to the energy no longer available to do work.

IV. In a chemical reaction, energy may be released or absorbed in the form of heat. An exothermic reaction releases heat. An endothermic reaction absorbs heat.

V. In an equilibrium, the rate of change in one direction is exactly the same as the rate of change in the opposite direction; the free energy difference between the reactants and products is zero.

VI. Spontaneous reactions release free energy and can therefore perform work.
   A. Reactions that release free energy are exergonic reactions; endergonic reactions require a net input of free energy.
   B. In the living cell, endergonic and exergonic reactions are coupled.

VII. ATP is the energy currency of the cell; energy is temporarily stored within its chemical bonds.
   A. ATP is formed by the phosphorylation of ADP, a process that requires a great deal of energy.
   B. ATP is a link between exergonic and endergonic reactions.

VIII. An enzyme is an organic catalyst; it greatly increases the speed of a chemical reaction without being consumed itself.
   A. An enzyme lowers the activation energy necessary to get a reaction going.
   B. Enzymes bring substrates into close contact so that they can more easily react with one another.
   C. Some enzymes consist of an apoenzyme and a cofactor. An organic cofactor is called a *coenzyme.*
   D. A cell can regulate enzymatic activity by controlling the amount of enzyme produced and by regulating conditions that influence the shape of the enzyme.
   E. Enzymes work best at specific temperatures and pH values.
   F. Most enzymes can be inhibited by certain chemical substances. Reversible inhibition may be competitive or noncompetitive.

■ POST-TEST

1. The ability to produce a change in the state or motion of matter is known as _____.
2. The energy of a particle in motion is termed _____ _____.
3. _____ is the branch of physics that deals with energy and its transformations.
4. ''Energy may be changed from one form to another but is neither created nor destroyed'' is a statement of the _____ _____ _____ _____.
5. A process in which heat is delivered to the surroundings is a(n) _____ _____.
6. A process in which heat is absorbed from the surroundings is a(n) _____ _____.
7. In thermodynamics, the term _____ is applied to a disordered state of the system.
8. ''Physical and chemical processes proceed in such a way that the entropy of the system becomes maximal'' is a statement of the _____ _____ _____ _____.

9. The _____ energy, ΔG, of a system is that part of the total energy of the system that is available to do work under conditions of constant _____ and _____.

10. A reaction can occur spontaneously only if ΔG is _____.

11. A reaction that releases energy to the system is _____.

12. To drive a reaction that requires an input of energy, some reaction that yields energy must be _____ to it.

13. ATP consists of _____, _____, and three inorganic phosphate groups.

14. The energy that is required to initiate a reaction is called _____ _____.

15. A substance that affects the rate of a chemical reaction without being consumed by the reaction is a _____.

16. _____ are biological catalysts produced by cells.

17. Enzymes and their substrates combine temporarily to form a(n) _____ _____.

18. The small portion of an enzyme molecule that combines with the substrate is the _____.

19. A(n) _____ inhibitor alters the shape of the enzyme, rendering it inactive.

■ REVIEW QUESTIONS

1. Trace the flow of energy from producer to consumer.
2. Contrast potential and kinetic energy, and give examples of each.
3. Contrast exergonic and endergonic reactions.
4. Life is sometimes described as a constant struggle against the second law of thermodynamics. Explain why this is true. How do organisms succeed in this struggle?
5. Why are coupled reactions biologically important?
6. Explain how energy is obtained from ATP. Discuss why ATP is an important link between exergonic and endergonic reactions.
7. What is activation energy? What is the relationship of a catalyst to activation energy?
8. Give the function of each of the following:
   a. active site of an enzyme
   b. coenzyme
9. Describe three factors that influence enzymatic activity.
10. Contrast competitive and noncompetitive inhibition.

■ RECOMMENDED READINGS

Alberts, B. et al.: *Molecular Biology of the Cell.* New York, Garland Publishing, Inc., 1983. A detailed, well-written account of the energy conversions that take place within chloroplasts and mitochondria.

Atkins, P.W.: *The Second Law.* San Francisco, W.H. Freeman & Co. 1984. A basic, understandable introduction to thermodynamics with an extensive section denoted to its biological implications.

Becker, W.M.: *Energy and the Living Cell: An Introduction to Bioenergetics.* New York, Harper & Row, 1977. A paperback with discussions of thermodynamics and the production and utilization of energy by living cells.

Cloud, P.: The biosphere, *Scientific American*, September 1983, pp. 176–189. A fascinating discussion of the relationship between microbial, animal, and plant life on Earth and the physical environment, as well as the energy transfer pathways that knit it all together.

Lehninger, A.L.: *Bioenergetics: The Molecular Basis of Biological Energy Transformations*, 2nd ed. Menlo Park, CA, Benjamin Cummings, 1971. A classic presentation of energy transformations in living systems; an unusually clear presentation of a rather difficult subject.

Lehninger, A.L.: *Principles of Biochemistry.* New York, Worth Publishers, 1982. An excellent text of general biochemistry with a detailed and very clear account of bioenergetics. A good source book.

Racker, E.: *A New Look at Mechanisms in Bioenergetics.* New York, Academic Press, 1976. One of the major contributors to the field presents his views of the mechanisms involved in bioenergetic processes.

Stryer, L.: *Biochemistry*, 3rd ed. San Francisco, W.H. Freeman, 1988. A well-illustrated and very readable text of general biochemistry with excellent sections on cellular energetics.

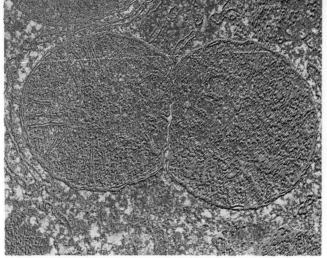
Dividing mitochondrion from rat liver cell

7

Energy-Releasing Pathways and Biosynthesis

■ OUTLINE

I. Aerobic and anaerobic catabolism
II. Oxidation-reduction reactions in metabolism
   A. Hydrogen and electron transfer
   B. Cellular respiration is a redox process
III. An overview of the reactions of cellular respiration
IV. Glycolysis
V. Anaerobic pathways
VI. Formation of acetyl coenzyme A
VII. The citric acid cycle
VIII. The electron transport system and chemiosmotic phosphorylation
   A. Electron transport
   B. The chemiosmotic model and oxidative phosphorylation
IX. Regulation of cellular respiration
X. Energy yield from glucose
XI. Catabolism of other nutrients
   A. Oxidation of amino acids
   B. Oxidation of fatty acids
XII. Biosynthetic processes
Focus on shuttles across the mitochondrial membrane

■ LEARNING OBJECTIVES

After you have read this chapter you should be able to:

1. Contrast aerobic and anaerobic pathways used by cells to extract free energy from nutrients.
2. Write a general equation illustrating hydrogen and electron transfer from a substrate to a hydrogen acceptor such as NAD^+.
3. Write a summary reaction for cellular respiration, giving the origin and fate of each substance involved.
4. List and give a brief overview of the four phases of cellular respiration, and indicate where the reactions of each phase take place in the cell.
5. Summarize the events of glycolysis, giving the key organic compounds formed and the number of carbon atoms in each; indicate the number of ATP molecules used and produced and the transactions in which hydrogen loss occurs.
6. Compare aerobic respiration with anaerobic pathways in terms of ATP formation, final hydrogen acceptor, and end products; compare alcoholic and lactate fermentation.
7. Describe the conversion of pyruvate to acetyl CoA.
8. Summarize the events of the citric acid cycle, beginning with the conversion of pyruvate to acetyl CoA; indicate the fate of carbon-oxygen segments and of hydrogens removed from the fuel molecule.
9. Summarize the operation of the electron transport system.
10. Describe chemiosmotic phosphorylation and explain how a gradient of protons is established across the inner mitochondrial membrane and the process by which the proton gradient drives ATP synthesis.
11. Indicate how the products of protein and fat metabolism feed into the same metabolic pathways that oxidize glucose.
12. Discuss the five basic principles of biosynthesis.

Every living cell must extract free energy from the organic food molecules it captures from the environment. These nutrients are broken down and some of their chemical energy is transferred to adenosine triphosphate (ATP) for later use in cellular work (Figure 7–1). The process of

FIGURE 7–1 Energy is transferred from prey to predator. Asiatic lion, *Panthera leo persica,* Gir Forest, India.

splitting larger molecules into smaller ones is an aspect of metabolism known as **catabolism,** and the individual reactions involved are called **catabolic reactions.** Cells use three different catabolic pathways to extract free energy from nutrients: aerobic cellular respiration, anaerobic cellular respiration, and fermentation.

AEROBIC AND ANAEROBIC CATABOLISM

Which catabolic pathway a cell uses to break down nutrients depends on the type of environment it inhabits. Cells that live in environments where oxygen is plentiful use the very efficient aerobic pathway, which requires molecular oxygen. Cells that inhabit the soil or polluted waters where oxygen is in short supply are adapted to use less efficient anaerobic pathways, which do not require oxygen.

Cellular respiration is the stepwise enzymatic process by which cells extract energy from glucose, fatty acids, and other organic compounds. Cellular respiration is generally aerobic; that is, it requires oxygen. During complete **aerobic respiration,** nutrients are catabolized to carbon dioxide and water. One of the most common pathways of aerobic cellular respiration involves the breakdown of the common nutrient glucose. The overall reaction for the aerobic catabolism of glucose can be summarized as follows:

$$C_6H_{12}O_6 + 6\ O_2 + 6\ H_2O \longrightarrow 6\ CO_2 + 12\ H_2O + \text{energy}$$

Cells cannot metabolize glucose to carbon dioxide and water in a single reaction. No enzyme can catalyze the direct attack by oxygen molecules on glucose. Instead, the complete catabolism of glucose occurs in a long sequence of reactions, to be examined more closely later in this chapter.

Some types of bacteria engage solely in **anaerobic respiration,** a clear advantage in soil or stagnant ponds where oxygen is in short supply. In anaerobic respiration an inorganic substance such as nitrate or sulfate is used in place of oxygen. The end products in anaerobic respiration are inorganic substances.

Certain bacteria that are adapted to anaerobic conditions use a third pathway, referred to as **fermentation.** The end products of fermentation are organic compounds. Under conditions of insufficient oxygen, human muscle cells can temporarily use a type of fermentation.

Organisms, including most plants and animals, that can survive only in an environment that provides oxygen rely on aerobic respiration for energy; these organisms are **strict aerobes.** Anaerobic bacteria, which do not require oxygen, are referred to as **anaerobes.** Some strict anaerobes are actually poisoned by oxygen. Most versatile are yeasts and certain bacteria that carry on aerobic respiration when oxygen is available but can shift to anaerobic respiration or fermentation when oxygen is in short supply; these organisms are known as **facultative anaerobes.**

OXIDATION-REDUCTION REACTIONS IN METABOLISM

Energy processing by cells involves the transfer of energy through the flow of electrons and protons. Generally, there is a sequence of oxidation-reduction reactions that take place as hydrogen or its electrons are transferred from one compound to another. Rusting—the combination of iron with oxygen—is a familiar example of oxidation and reduction:

$$4 \, Fe + 3 \, O_2 \longrightarrow 2 \, Fe_2O_3$$

Oxidation is a chemical process in which a substance loses electrons. In rusting, iron changes from its metallic state to its iron(III) (Fe^{3+}) state. We say it is being oxidized:

$$4 \, Fe \longrightarrow 4 \, Fe^{3+} + 12 \, e^-$$

The e^- means "electron." At the same time, oxygen is changing from its molecular state to its charged state:

$$3 \, O_2 + 12 \, e^- \longrightarrow 6 \, O^{2-}$$

Oxygen accepts the electrons removed from the iron and is then said to be *reduced*. **Reduction** is a chemical process in which a substance gains electrons. Electrons released during an oxidation reaction cannot exist in the free state in living cells. That is why every oxidation must be accompanied by a reduction in which the electrons are accepted by another atom or molecule. The oxidized molecule gives up energy, and the reduced molecule receives energy. Oxidation-reduction reactions, referred to as **redox reactions,** are an essential part of cellular respiration, photosynthesis, and other aspects of metabolism.

Hydrogen and Electron Transfer

Electrons are not easy to remove from covalent compounds unless an entire atom is removed. In living cells oxidation almost always involves the removal of a hydrogen atom from a compound. Reduction often involves a gain in hydrogen atoms.

When hydrogen atoms are removed from an organic compound, they take with them some of the energy that had been stored in their chemical bonds. The hydrogen, along with its energy, is transferred to a hydrogen acceptor molecule, which is generally a coenzyme. One of the most frequently encountered hydrogen acceptor coenzymes is **nicotinamide adenine dinucleotide,** more conveniently referred to as **NAD.** This coenzyme can temporarily package large amounts of free energy. Here is a generalized equation showing the transfer of hydrogen from a compound we will call X to NAD:

$$XH_2 + NAD^+ \longrightarrow \underset{\text{Oxidized}}{X} + \underset{\text{Reduced}}{NAD-H} + H^+$$

Note that the NAD is reduced when it combines with hydrogen. Before it is reduced, NAD^+ is an ion with a net charge of $+1$. When H is added, the charge is neutralized. The reduced form of the compound, **NADH,** is electrically neutral (Figure 7–2). Some of the energy stored in the bonds holding the hydrogens to molecule X has been transferred by this reaction to

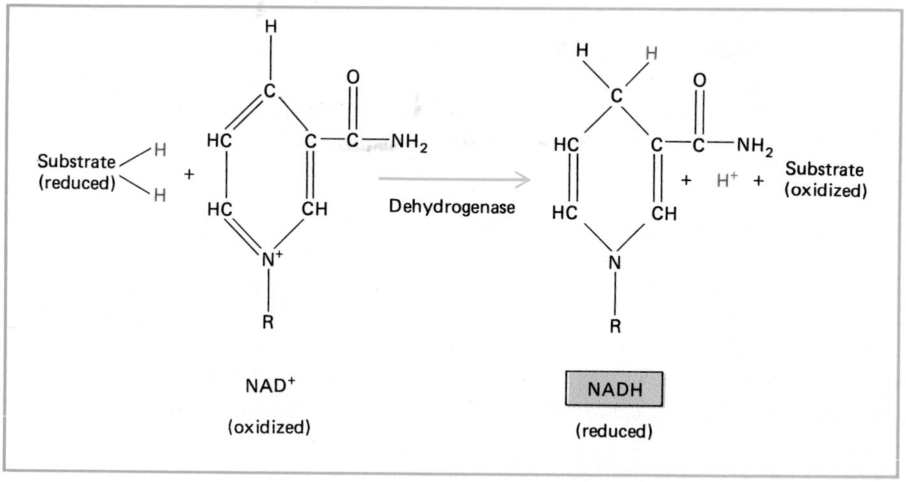

FIGURE 7–2 Reduction of NAD. The full structure of NAD is illustrated in Figure 3–32. Here, only the nicotinamide portion of NAD is shown; *R* is used to indicate the remainder of the molecule. When the substrate is oxidized, NAD is reduced. A dehydrogenase enzyme transfers a hydrogen proton and two electrons (as a hydride ion) to NAD. The NAD functions as a coenzyme. The reduced form of the coenzyme, NADH, can provide high-energy electrons for the process of ATP synthesis.

the NADH. This energy can now be used for some metabolic process, or it can be transferred through a complex series of reactions to ATP. As will be discussed, the transfer of energy in both photosynthesis and cellular respiration involves sequences of redox reactions. The acceptor compounds that make up these sequences keep energy flowing.

Other important hydrogen or electron acceptor compounds are **flavin adenine dinucleotide (FAD)** and the **cytochromes**. FAD is a nucleotide that accepts hydrogens and their electrons; the cytochromes are proteins that contain iron. The iron component accepts electrons from hydrogen and then transfers them to some other compound. Like NAD and FAD cytochromes are important in cellular respiration. Such electron acceptors provide a mechanism by which the cell can slowly and efficiently capture energy from fuel molecules (Figure 7–3).

Cellular Respiration Is a Redox Process

Cellular respiration is a redox process in which hydrogen is transferred from glucose to oxygen. Glucose is oxidized, and oxygen is reduced.

FIGURE 7–3 The total energy released by a falling object is the same whether the object is released all at once or in a series of steps. Similarly, in cellular metabolism the energy of an electron liberated from a foodstuff is the same, regardless of whether it is released all at once or gradually as it passes to successive electron acceptors. Electron acceptors function in both photosynthesis and cellular respiration. As part of a complicated scheme involving the diffusion of protons across membranes, these acceptors permit the *controlled* extraction of some of the energy generated by these processes.

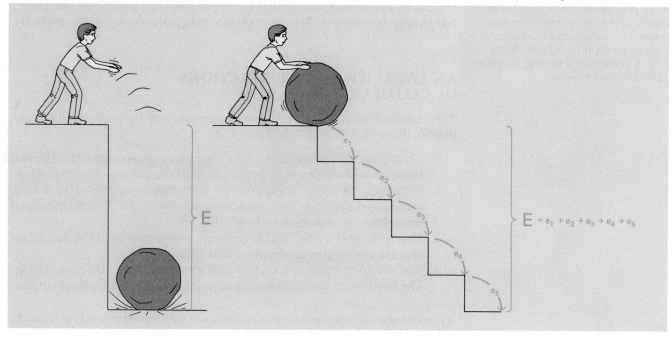

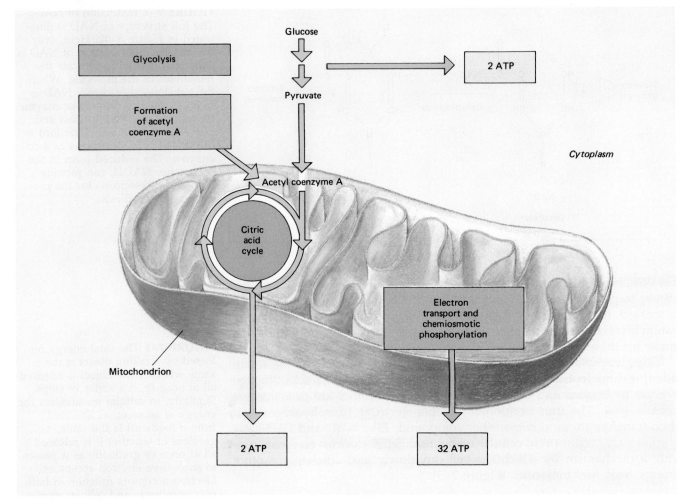

FIGURE 7–4 Four main phases in cellular respiration are (1) glycolysis, (2) the formation of acetyl coenzyme A from pyruvate, (3) the citric acid cycle, and (4) the electron transport system and chemiosmotic phosphorylation. Glycolysis occurs in the cytoplasm. Pyruvate enters a mitochondrion, where cellular respiration continues. Most ATP is synthesized during chemiosmotic phosphorylation.

$$\overbrace{C_6H_{12}O_6 + 6\ O_2 + 6\ H_2O \longrightarrow 6\ CO_2 + 12\ H_2O}^{\text{Oxidation}} + \text{energy}$$

During this process, the potential energy of electrons is reduced and chemical energy is released. This energy can then be used for ATP synthesis.

AN OVERVIEW OF THE REACTIONS OF CELLULAR RESPIRATION

The chemical reactions of cellular respiration can be grouped into four phases (Figures 7–4 and 7–5; Table 7–1):

1. Glycolysis, the conversion of the six-carbon glucose molecule to two three-carbon molecules of pyruvate[1] with the formation of two ATPs.
2. Formation of acetyl coenzyme A. Pyruvate is degraded to a two-carbon fuel molecule and combines with coenzyme A, forming acetyl coenzyme A; carbon dioxide is released.
3. The citric acid cycle, which converts acetyl coenzyme A to carbon dioxide and removes electrons and hydrogens
4. The electron transport system and chemiosmotic phosphorylation. The hydrogens and electrons removed from the fuel molecule during

---

[1]Pyruvate and many other compounds in glycolysis and the citric acid cycle exist as ions at the pH found in the cell. They sometimes associate with H^+ to form acids; for example, pyruvate forms pyruvic acid. In some textbooks these compounds are presented in the acid form.

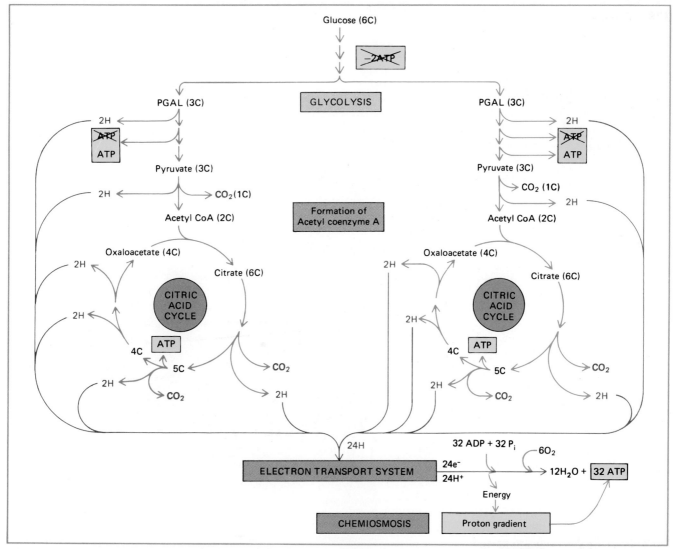

FIGURE 7–5 An overview of the reactions of cellular respiration. During glycolysis, each six-carbon glucose molecule is broken down into two molecules of a compound called *pyruvate*. The two-carbon compound acetyl coenzyme A, formed from pyruvate and coenzyme A, enters a mitochondrion. During the citric acid cycle, the two-carbon fragment from the fuel molecule is com-pletely degraded, and carbon-oxygen segments removed from the fuel molecule are released as carbon dioxide. Hydrogens removed from the fuel molecule are transferred to NAD and then pass through the electron acceptors of the electron transport system. The chemical energy released is used to synthesize ATP by chemiosmosis.

Table 7–1 SUMMARY OF CELLULAR RESPIRATION

Phase	Summary	Needed materials	End products
Glycolysis (takes place in cytoplasm)	Series of about ten reactions during which glucose is degraded to pyruvate; net profit of 2 ATPs; hydrogens released; can proceed anaerobically	Glucose, 2 ATPs, ADP, P_i	Pyruvate, ATP, hydrogen
Formation of acetyl CoA (takes place in mitochondria)	Pyruvate is degraded and combined with coenzyme A to form acetyl CoA; CO_2 is released	Pyruvate, coenzyme, A, NAD	Acetyl CoA, NADH, CO_2
Citric acid cycle (takes place in mitochondria)	Series of reactions in which fuel molecule (part of acetyl CoA) is degraded to hydrogen and carbon dioxide; aerobic	Acetyl CoA, H_2O hydrogen acceptors (e.g., NAD), ADP, P_i	CO_2, NADH, $FADH_2$, CoA, ATP
Electron transport and chemiosmosis (take place in mitochondria)	Chain of several electron transport molecules; hydrogens (or their electrons) are passed along chain; energy released is used to generate proton gradient across inner mitochondrial membrane; as protons move through ATP synthetase in membrane, ATP is synthesized; for each pair of hydrogens that enters chain, maximum of 3 ATPs can be synthesized; aerobic	Hydrogen, ADP, P_i, oxygen	ATP, water

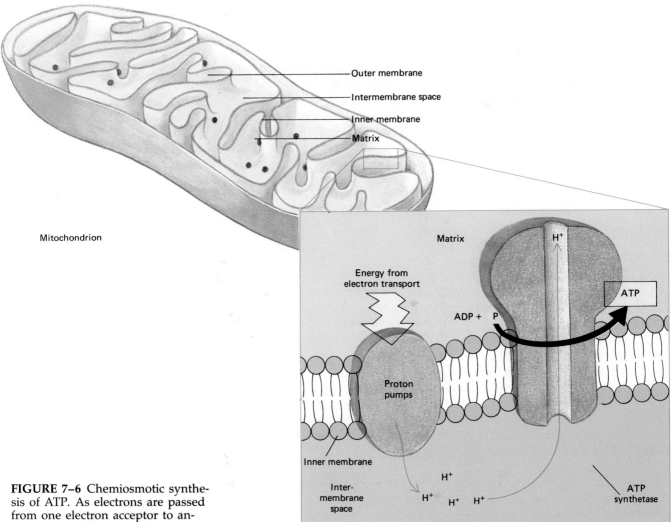

Mitochondrion

FIGURE 7–6 Chemiosmotic synthesis of ATP. As electrons are passed from one electron acceptor to another, hydrogen ions are pumped across the inner mitochondrial membrane from the matrix to the intermembrane space. The protons can flow back into the matrix only through special channels within the enzyme ATP synthetase. The energy released as the protons move down the energy gradient is used to synthesize ATP.

the preceding phases are transferred along a chain of electron acceptor compounds. Electron transport and chemiosmosis occur within the mitochondria. As the electrons are passed from one electron acceptor to another, hydrogen ions (protons) are pumped across the inner mitochondrial membrane from the matrix of the mitochondrion to the intermembrane space (Figure 7–6). The difference in the concentration of protons between the matrix and the intermembrane space is referred to as a *proton gradient*; this gradient represents potential energy. The protons can flow back into the matrix only through special channels in the inner membrane. These channels occur within the enzyme ATP synthetase. When the protons move through these channels, they move down an energy gradient, and the energy released is used to drive the synthesis of ATP. Thus, this proton gradient across the inner mitochondrial membrane couples phosphorylation (of ADP to form ATP) with oxidation.

Most reactions involved in cellular respiration can be considered one of three types: First, **dehydrogenations** are reactions in which two hydrogens (actually, two electrons plus two protons) are removed from the substrate and transferred to a coenzyme such as NAD^+ or FAD, which acts as a primary acceptor. Second, **decarboxylations** are reactions in which a carboxyl group (—COOH) is removed as a molecule of CO_2. The carbon dioxide we exhale each day is derived from decarboxylations. Third, **"make-ready" reactions** are ones in which molecules undergo rearrangements so they can subsequently undergo further dehydrogenations or decarboxylations. As we trace the reactions of cellular respiration, we will encounter many examples of these three basic types.

In Chapter 6 we considered the important basic concept of the coupling of an exergonic (energy-yielding) reaction with an endergonic (energy-requiring) reaction. Recall that, in living cells, exergonic reactions usually involve the breakdown of ATP. The free energy liberated in the hydrolysis of the high-energy phosphate bonds of ATP is used to drive the reactions that require an input of free energy. ATP is formed from ADP and P_i when fuel molecules are oxidized or when light energy is trapped in photosynthesis. The basic mechanism of energy exchange in living systems is the cyclic conversion of ATP to ADP and back.

GLYCOLYSIS

Glycolysis (literally, "splitting sugar") is the sequence of reactions that convert a glucose molecule (a six-carbon compound) to two molecules of pyruvate (a three-carbon compound) with the production of ATP. Each reaction is regulated by a specific enzyme, and there is a net gain of two ATP molecules. The reactions of glycolysis take place in the cytoplasm. Such necessary ingredients as ADP, NAD, and phosphates float freely in the cytoplasm and are used as needed. Glycolysis does not require oxygen and can proceed under aerobic or anaerobic conditions.

The reactions of glycolysis are shown in Figure 7–7. In a highly simplified model, glycolysis can be divided into two major phases. The first phase corresponds to the first four steps illustrated in Figure 7–7. During this phase, phosphate is added to the glucose molecule; the addition of phosphate to a molecule is termed **phosphorylation.** The glucose molecule is split, forming two molecules of the three-carbon compound glyceraldehyde-3-phosphate (PGAL). This transformation requires energy, so the cell must invest two molecules of ATP to initiate the oxidation of glucose.

We may summarize this portion of glycolysis as follows:

Glucose + 2 ATP $\longrightarrow\!\!\longrightarrow\!\!\longrightarrow$ 2 PGAL + 2 ADP + 2 P_i
six-carbon   three-carbon
compound   compound

Multiple arrows are used to indicate that the equation summarizes a sequence of several reactions.

In the second phase of glycolysis (Steps 5 through 9 in Figure 7–7), PGAL is oxidized with the removal of two hydrogen atoms, and certain other atoms are rearranged so that each molecule of PGAL is transformed into a molecule of pyruvate. During these reactions, enough chemical energy is released from the sugar molecule to produce four ATP molecules.

2 PGAL + 4 ADP + 4 P_i $\longrightarrow\!\!\longrightarrow\!\!\longrightarrow\!\!\longrightarrow$ 2 pyruvate + 4 H + 4 ATP
three-carbon   three-carbon
compound   compound

Note that in the first phase of glycolysis two molecules of ATP are consumed, but in the second phase four molecules of ATP are produced. Thus, glycolysis yields a net energy profit of two ATP molecules.

The two hydrogen atoms removed from each PGAL immediately combine with the hydrogen carrier molecule, NAD:

NAD^+ + 2 H $\longrightarrow$ NAD—H + H^+
oxidized NAD   reduced NAD

The fate of these hydrogen atoms will be discussed in conjunction with the electron transport system and anaerobic pathways.

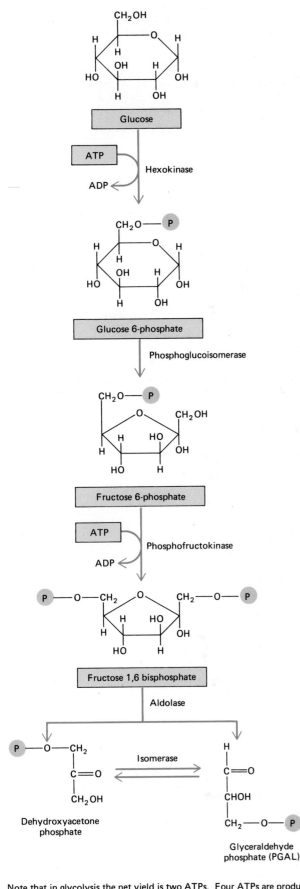

1 Glycolysis begins with a make-ready reaction in which glucose is activated. Glucose receives a phosphate group from an ATP molecule. The ATP serves as a source of both phosphate and the energy needed to attach the phosphate to the glucose molecule. (Once the ATP is spent, it becomes ADP and joins the ADP pool of the cell until turned into ATP again.) The reaction is catalyzed by the enzyme hexokinase. The phosphorylated glucose is known as glucose-6-phosphate. (Note the phosphate attached to its carbon atom 6.) Phosphorylation of the glucose makes it more chemically reactive. The electrical charge on the phosphate group also prevents the glucose from leaving the cell because the cell membrane tends to be impermeable to ions.

2 Glucose 6-phosphate undergoes rearrangement of its hydrogen and oxygen atoms, another make-ready reaction, catalyzed by an isomerase. In this reaction glucose 6-phosphate is converted to its isomer fructose-6-phosphate.

3 Next, another ATP donates a phosphate to the molecule, forming fructose-1,6-bisphosphate. So far, two ATP molecules have been invested in the process without any being produced. Phosphate groups are now bound at carbons 1 and 6 by ester bonds, and the molecule is ready to be split.

4 Fructose-1,6-bisphosphate is then split by the enzyme aldolase into two 3-carbon sugars, glyceraldehyde-3-phosphate (PGAL) and dihydroxyacetone phosphate. Dihydroxyacetone phosphate can be enzymatically converted by isomerase to its isomer glyceraldehyde-3-phosphate for further metabolism in glycolysis. Although the isomerase interconverts these sugars, only the PGAL continues in the glycolytic sequence. This shifts the equilibrium in the direction of PGAL.

Note that in glycolysis the net yield is two ATPs. Four ATPs are produced and two ATPs are consumed in the process.

FIGURE 7–7 Glycolysis. Each reaction of glycolysis is catalyzed by a specific enzyme. Note that there is a net yield of two ATP molecules.

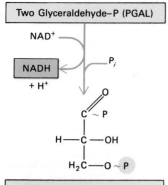

Two Glyceraldehyde–P (PGAL)

NAD⁺

NADH
+ H⁺

P$_i$

$$C \overset{O}{\sim} P$$
$$H—C—OH$$
$$H_2C—O \sim P$$

Two Bisphosphoglycerate

Phosphoglycerokinase

2 ADP

2 ATP

$$C \overset{O}{\diagdown} O^-$$
$$HC—OH$$
$$H_2C—O \sim P$$

Two 3-Phosphoglycerate

Phosphoglyceromutase

$$C \overset{O}{\diagdown} O$$
$$HC—O \sim P$$
$$H_2C—OH$$

Two 2-Phosphoglycerate

Enolase

H$_2$O

$$C \overset{O}{\diagdown} O^-$$
$$CO \sim P$$
$$CH_2$$

Two phosphoenolpyruvate

Pyruvate kinase

2 ADP

2 ATP

$$C \overset{O}{\diagdown} O^-$$
$$C = O$$
$$CH_3$$

Two pyruvate

⑤ Glyceraldehyde-3-phosphate reacts with an SH (sulfhydryl) group in the enzyme, glyceraldehyde-3-phosphate dehydrogenase, forming an H—C—OH group that can undergo dehydrogenation with NAD⁺ as hydrogen acceptor. The product of the reaction is phosphoglycerate, still bound to the enzyme. This very exergonic reaction reaction is coupled to a reaction which produces an energy-rich phsophate bond at carbon-1. In this reaction phosphoglycerate reacts with inorganic phosphate present in the cytoplasm to yield 1,3-diphosphoglycerate which is released by the enzyme. Thus, in this oxidation-reduction reaction, PGAL has been converted to a diphosphoglycerate, and the newly added phosphate group is attached with an energy-rich bond.

⑥ The energy-rich phosphate reacts with ADP to form ATP. This transfer of energy from a compound with an energy-rich phosphate is referred to as a substrate-level phosphorylation.

⑦ The 3-phosphoglycerate is rearranged to 2-phosphoglycerate by the enzymatic shift of the position of the phosphate group. This is a make-ready reaction.

⑧ Next, in an unusual reaction an energy-rich phosphate is generated by the removal of water, a dehydration reaction, rather than by the removal of hydrogen atoms (dehydrogenation). The product is phosphoenol pyruvate, or PEP. The energy-rich phosphate bond in PEP is the second such bond generated at the substrate level in the metabolism of glucose to pyruvate.

⑨ Each of the two PEP molecules can transfer its phosphate group to ADP to yield ATP and pyruvate.

FIGURE 7–8 Comparison of anaerobic with aerobic pathways. Some bacteria carry on anaerobic respiration. When oxygen is not available, yeast cells carry on alcoholic fermentation. Some bacteria carry on lactic acid (lactate) fermentation. During strenuous exercise, when there is insufficient oxygen, muscle cells shift to lactate fermentation.

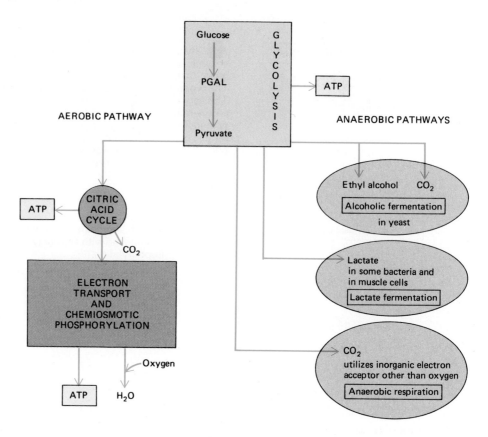

FIGURE 7–9 Yeast cells carry on alcohol fermentation.

ANAEROBIC PATHWAYS

As stated, cells use three pathways to extract free energy from nutrients: aerobic respiration, anaerobic respiration, and fermentation (Figure 7–8). In all three processes, glucose or other nutrients are oxidized, and their high-energy electrons are transferred to NAD$^+$, which becomes reduced to NADH. What happens to these electrons is different in each of the three pathways.

In aerobic respiration, the electrons removed from nutrient molecules during glycolysis and the citric acid cycle are transferred through a sequence of electron acceptors, the electron transport chain. The final electron acceptor is molecular oxygen. During anaerobic respiration, oxygen is not used as an electron acceptor. Instead, an inorganic compound, such as nitrate (NO_3) or sulfate (SO_4), serves as the final acceptor of electrons. Both anaerobic respiration and fermentation depend on the reactions of glycolysis. Recall that the net profit of two ATPs produced during glycolysis does not require the presence of oxygen.

In fermentation the final acceptor of electrons from NADH is an organic compound (rather than an inorganic one, as in anaerobic respiration). Two common types of fermentation are alcohol fermentation and lactic acid fermentation.

Yeast cells carry on **alcohol fermentation** (Figure 7–9). When deprived of oxygen, yeast cells split carbon dioxide off from pyruvate, forming a compound called *acetaldehyde* (Figure 7–10). Hydrogen from NADH + H$^+$ is transferred to acetaldehyde, producing **ethyl alcohol.** Such anaerobic reactions are the basis for the production of beer, wine, and other alcoholic beverages. Yeast cells are also used in the baking industry to produce the carbon dioxide that causes dough to rise.

Certain fungi and bacteria carry on **lactic acid fermentation.** In this alternative pathway, hydrogens removed from the fuel molecule during glycol-

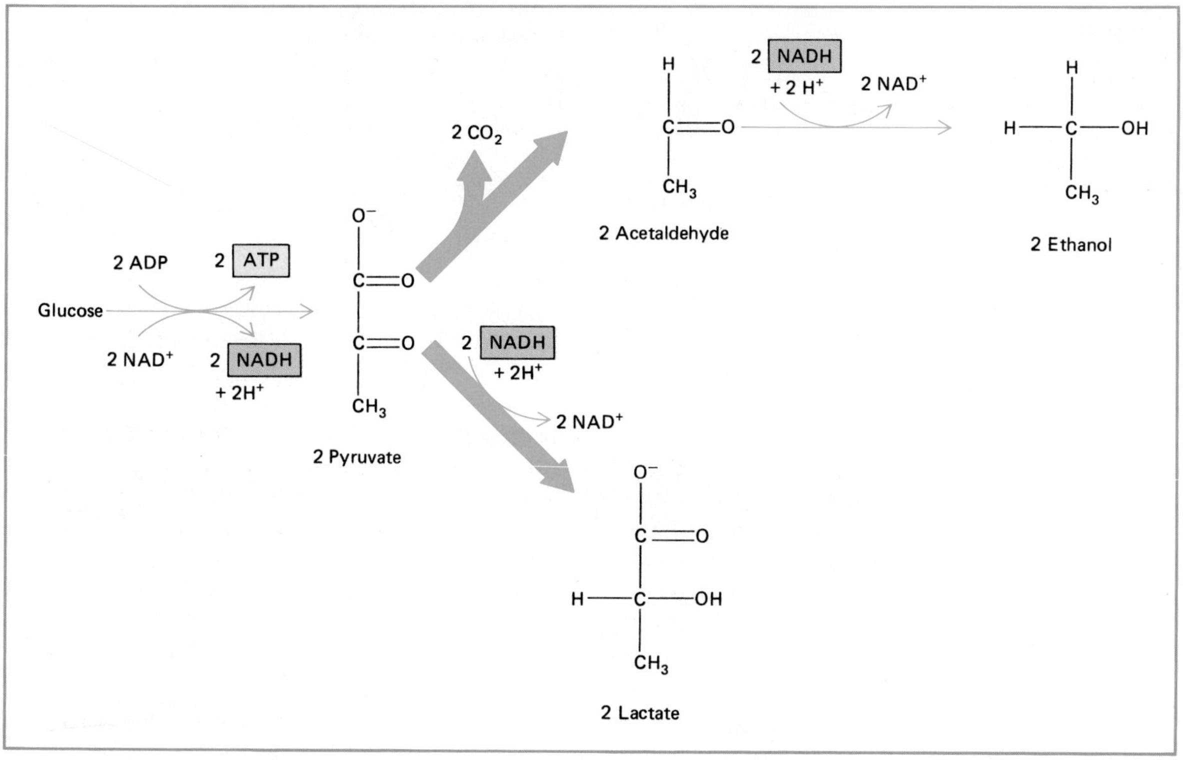

FIGURE 7–10 Fermentation. In fermentation, NADH transfers hydrogen to pyruvate, the end product of glycolysis. Thus, pyruvate serves as an electron acceptor. In alcohol fermentation, carbon dioxide is split off, and the two-carbon alcohol ethanol is the end product. In lactic acid fermentation, the final product is the three-carbon compound lactic acid. In fermentation there is a net gain of only two ATPs.

ysis are transferred to the pyruvate molecule. When hydrogen atoms are added to pyruvate, **lactate,** the ionic form of lactic acid, is formed.

Pyruvate + 2 H $\longrightarrow$ lactate

Lactate is produced when bacteria sour milk or ferment cabbage to form sauerkraut. It is also produced during muscle activity in the muscle cells of humans and other complex animals.

During strenuous physical activity, such as running, the amount of oxygen delivered to muscle cells may be insufficient to keep pace with the rapid rate of fuel oxidation. Not all of the hydrogen atoms accepted by NAD can be processed in the usual manner through the electron transport chain because there is a shortage of oxygen. In this situation muscle cells can shift from aerobic respiration to lactic acid fermentation. The hydrogens are transferred to pyruvate to form lactate. As lactate accumulates in muscle cells, it contributes to **muscle fatigue.**

Lactate acidifies the blood, which stimulates respiration, indirectly resulting in rapid breathing. The increased oxygen intake is necessary so that a portion of the lactate can be oxidized to reconvert the rest to glucose. By running faster than the circulatory system can supply oxygen to the muscles, the athlete incurs an oxygen debt that must be repaid when the exertion is over.

Anaerobic metabolism is very inefficient because the fuel is only partially oxidized. Alcohol, the end product of fermentation by yeast cells, can be burned and can even be used as automobile fuel. Obviously, it contains a great deal of energy that the yeast cells are unable to extract using anaerobic methods. Lactate, a three-carbon compound, contains even more energy than the two-carbon alcohol. In contrast, during aerobic respiration all available energy is removed because the fuel molecules are completely oxidized. A net profit of only two ATPs can be produced anaerobically from one molecule of glucose, compared with up to 38 ATPs when oxygen is

Table 7–2 THE ENERGY YIELD FROM THE COMPLETE OXIDATION OF GLUCOSE

(1) Net ATP profit from glycolysis		2 ATP*(Substrate level)
Also from glycolysis:	2 NADH₂ ⟶	4–6 ATP
(2) 2 pyruvate to 2 acetyl CoA	2 NADH₂ ⟶	6 ATP
(3) 2 acetyl CoA through citric acid cycle		2 ATP (substrate level)
	6 NADH₂ ⟶	18 ATP
	2 FADH₂ ⟶	4 ATP
(4) Total ATP profit		36–38 ATP

*These are the only 2 ATPs that can be generated anaerobically; production of all other ATPs depends on the presence of oxygen.

available (see Table 7–2). The two ATPs produced during glycolysis represent only about 5% of the total energy in a molecule of glucose. About 55% of the energy in a glucose molecule can be captured in aerobic pathways; the rest is lost as heat. This compares favorably with the finest machines we can make. In humans and other homeothermic animals (animals able to maintain a constant body temperature despite fluctuations in the temperature of their surroundings), some of the heat produced during respiration and other metabolic activities is used to maintain a constant body temperature.

The inefficiency of anaerobic metabolism necessitates a large supply of fuel. By rapidly degrading many fuel molecules, a cell can compensate somewhat for the small amount of energy that can be gained from each. To perform the same amount of work, an anaerobic cell must consume up to 20 times as much glucose or other carbohydrate as a cell metabolizing aerobically. Skeletal muscle cells, which often metabolize anaerobically for short periods, store large quantities of glucose in the form of glycogen.

FORMATION OF ACETYL COENZYME A

Pyruvate, the end product of glycolysis, contains most of the energy present in the original glucose molecule. When oxygen is not available, the cell uses a fermentation pathway. Pyruvate accepts electrons from NADH and is converted to lactic acid or ethyl alcohol. These reactions take place in the cytoplasm.

When oxygen is present, pyruvate molecules can be completely degraded during the reactions of the citric acid cycle. Pyruvate molecules move into the mitochondria, where all subsequent reactions of cellular respiration take place.

Once a pyruvate molecule enters the mitochondrion, it is converted to a compound known as **acetyl coenzyme A, (acetyl CoA),** a compound that can enter the citric acid cycle. In this complex reaction, pyruvate undergoes oxidative decarboxylation. First, a carboxyl group is removed as carbon dioxide, which diffuses out of the cell (Figure 7–11). Then the two-carbon fragment remaining is oxidized, and the hydrogens removed are accepted by NAD^+. Finally, the oxidized fragment, an acetyl group, is attached to coenzyme A. This coenzyme contains a sulfur atom, which attaches to the acetyl group by an unstable bond. The reaction is catalyzed by a multienzyme complex that contains several copies of each of three different enzymes. The overall reaction for the formation of coenzyme A is the following:

$$2 \text{ pyruvate} + 2 \text{ NAD}^+ + 2 \text{ CoA} \longrightarrow$$
$$2 \text{ acetyl CoA} + 2 \text{ NADH} + 2 \text{ H} + 2 \text{ CO}_2$$

Coenzyme A is manufactured in the cell from one of the B vitamins, panto-

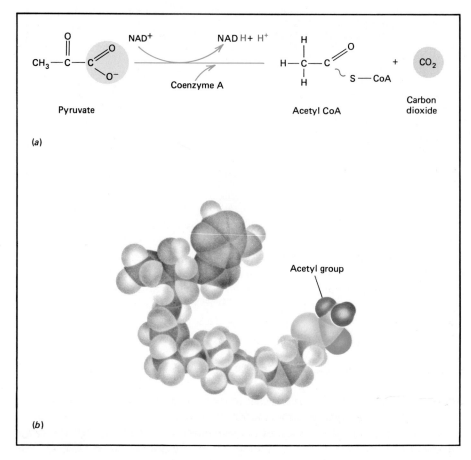

(a)

(b)

FIGURE 7–11 Formation of acetyl CoA. (*a*) Pyruvate, the end product of glycolysis, enters a mitochondrion and undergoes oxidative decarboxylation. This reaction is catalyzed by the pyruvate dehydrogenase complex. First, the carboxyl group is split off as carbon dioxide. Then the remaining two-carbon fragment is oxidized and its hydrogen transferred to NAD^+. (The hydrogens are present on an intermediate phase of the two-carbon group.) Finally, the oxidized two-carbon group, an acetyl group, is attached to coenzyme A. Coenzyme A has a sulfur atom that attaches to the acetyl group by a very unstable bond, shown as a wavy line. (*b*) A space-filling model of acetyl CoA. The acetyl group is just a small part of the coenzyme.

thenic acid. Fats and amino acids can also be degraded to acetyl CoA and can enter the respiratory pathway at this point.

Note that the original glucose molecule has now been oxidized to two acetyl groups and two CO_2 molecules. The hydrogens removed have reduced NAD^+ to NADH. Four NADH molecules have been formed, two during glycolysis and two during the oxidation of pyruvate.

THE CITRIC ACID CYCLE

The **citric acid cycle** is also known as the *tricarboxylic acid (TCA) cycle* or **Krebs cycle,** after Sir Hans Krebs, who worked it out in the 1930s. This cycle is the final common pathway in the oxidation of pyruvate, fatty acids, and the carbon chains of amino acids. The cycle takes place in the mitochondria and consists of eight steps, illustrated and described in Figures 7–12 and 7–13. Each reaction is catalyzed by a specific enzyme.

The first reaction of the cycle occurs when acetyl CoA transfers its two-carbon acetyl group to the four-carbon compound **oxaloacetate,** forming **citrate,** a six-carbon compound:

Oxaloacetate + acetyl CoA $\longrightarrow$ citrate
four-carbon   two-carbon   six-carbon
compound   compound   compound

The citrate then goes through a series of chemical transformations, losing first one, then another carboxyl group as CO_2. Most of the energy made available by the oxidative steps of the cycle is transferred to the energy-rich electrons in NADH. For each acetyl group that enters the citric acid cycle,

FIGURE 7–12 An overview of the citric acid cycle. A two-carbon acetyl group combines with the four-carbon compound oxaloacetate to form the six-carbon compound citrate. During the course of the cycle, citric acid (and its derivatives) undergoes two decarboxylations, four dehydrogenations, and several make-ready reactions. One ATP is generated at the substrate level. The four-carbon oxaloacetate is regenerated, and the cycle begins anew. Note that the CO_2 produced accounts for the two carbons that entered the cycle as part of acetyl CoA. With two turns of the cycle, six carbons—representing the six carbons from the original glucose—are split off.

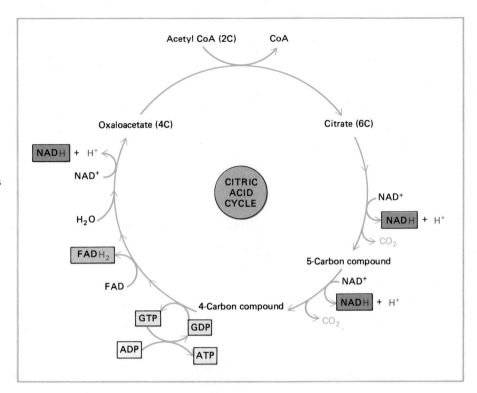

three molecules of NAD^+ are reduced to NADH. In Step six, electrons are transferred to the electron acceptor FAD rather than to NAD.

In the course of the cycle, two molecules of CO_2 and eight hydrogen atoms (eight protons and eight electrons) are removed. The CO_2 produced accounts for the two carbon atoms that entered the citric acid cycle. You may wonder why more hydrogen is generated by these reactions than entered the cycle with the fuel molecule. These hydrogens come from water molecules that are added during the reactions of the cycle. Extracting hydrogen from water requires energy, which is provided by the disruption of the chemical bonds of the fuel molecule; some of this energy is in effect stored in the hydrogen thus generated.

Because two pyruvate molecules are produced from each glucose molecule, the cycle must turn twice to process each glucose. At the end of a complete cycle, a four-carbon oxaloacetate is all that is left, and the cycle is ready for another turn. By this time the original pyruvate has lost all of its three carbons, or at least the equivalent, and may be regarded as having been completely consumed. Only one molecule of ATP is produced directly by a substrate-level phosphorylation with each turn of the cycle; how the rest of the ATP is produced is discussed below.

THE ELECTRON TRANSPORT SYSTEM AND CHEMIOSMOTIC PHOSPHORYLATION

Let us consider the fate of all the hydrogens removed from the fuel molecule during glycolysis, acetyl CoA formation, and the reactions of the citric acid cycle. The hydrogens are first transferred to primary hydrogen acceptors—the pyridine nucleotide NAD^+ or the flavin nucleotide FAD. What becomes of these hydrogens?

Electron Transport

The electron transport system is a chain of electron acceptors embedded in the inner membrane of the mitochondrion. Hydrogens are passed from

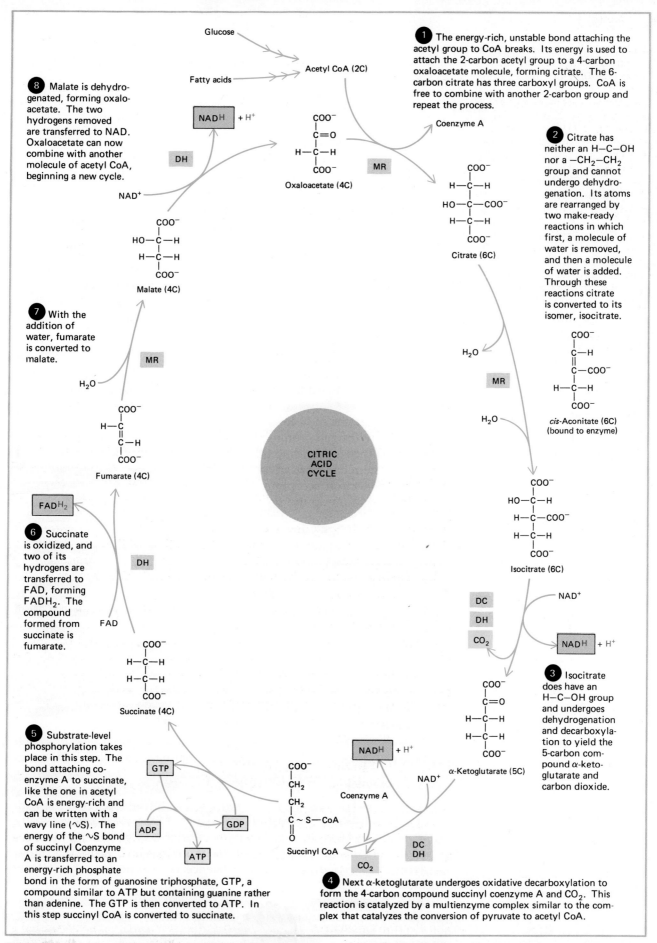

FIGURE 7–13 The citric acid cycle. During this series of reactions, acetyl CoA, produced from glucose and other organic compounds, is metabolized to yield carbon dioxide and hydrogen. The hydrogen is immediately combined with NAD or FAD and fed into the electron transport system. With each turn of the cycle, two carbons and eight hydrogens (= eight protons and eight electrons) are split off from the fuel molecule. The CO_2 produced accounts for the two carbons that entered the cycle as part of acetyl CoA.

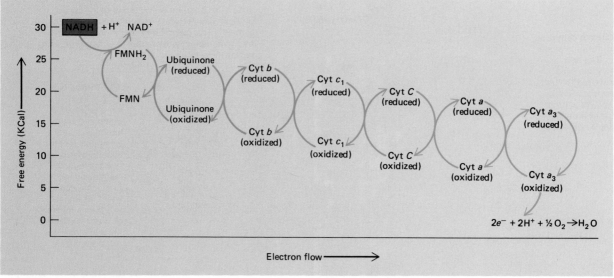

FIGURE 7–14 The electron transport chain. Hydrogens or their electrons are transferred from one electron acceptor molecule to another. The acceptor molecules are alternately oxidized and reduced as they transfer and accept hydrogen, the final electron (and hydrogen) acceptor being molecular oxygen. Water is produced as a product of these reactions. The energy released as electrons move to lower energy levels is used to transport protons across the inner mitochondrial membrane to the intermembranous space.

NADH to flavin mononucleotide (FMN), the first acceptor in the chain (Figure 7–14). As hydrogens are transferred from one to another of the electron acceptor molecules, the hydrogen protons become separated from their electrons. Hydrogens, or their electrons, pass down the "electron cascade" in a series of redox reactions.

When the hydrogen protons (H^+) separate from their electrons, they are released into the surrounding medium. The electrons entering the electron transport system have a relatively high energy content. As they pass along the chain of electron acceptors, they lose much of their energy, some of which is used to pump the protons across the inner membrane. This sets up an electrochemical gradient across the inner mitochondrial membrane that provides the energy for ATP synthesis (Figure 7–15).

The electron acceptors in the chain include FMN, ubiquinone (Q), and a group of closely related proteins called **cytochromes.** Cytochromes are characterized by a heme group, a group with four rings surrounding a central atom of iron. It is the iron that combines with the electrons from the hydrogen atoms. Cytochrome molecules accept only the electron from the hydrogen, not the entire atom. Each of the several types of cytochrome holds electrons at slightly different energy levels. Electrons are passed along from one cytochrome to the next in the chain, losing energy as they go. Finally, the last cytochrome in the chain, cytochrome a_3, passes the two electrons on to molecular oxygen. Simultaneously, the electrons reunite with protons, and the chemical union of the hydrogen and oxygen produces water.

Oxygen is the final hydrogen acceptor in the electron transport system, which explains why we require oxygen. What happens when cells are deprived of oxygen? When no oxygen is available to accept the hydrogen, the last cytochrome in the chain is stuck with it. When that occurs, each acceptor molecule in the chain may remain stuck with electrons, and the entire system may be blocked all the way back to NAD. As a result, no further ATPs can be produced by way of the electron transport system. Most cells of complex organisms cannot live long without oxygen because the amount of energy they can produce in its absence is insufficient to sustain life processes.

Lack of oxygen is not the only factor that may interfere with the electron transport system. Some poisons, including cyanide, inhibit the normal activity of the cytochrome system. Cyanide binds tightly to cytochrome a_3 so that it cannot transport electrons on to oxygen. This blocks the further passage of electrons through the chain, halting ATP production.

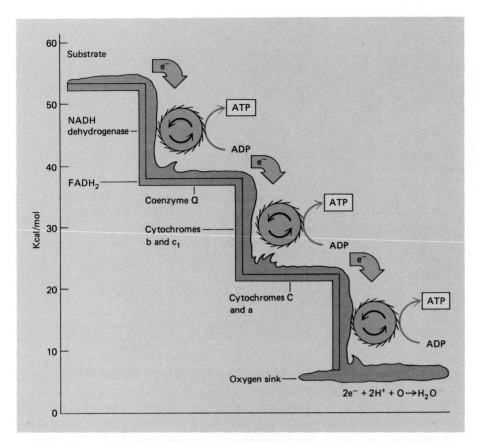

FIGURE 7–15 Electron transport may be compared to a stream of water (electrons) with three waterfalls. The flow of water (electrons) drives water wheels (the proton gradient). There are three sites in the electron transport system where energy-rich phosphate is produced. The electrons end up in the pond at the bottom of the waterfalls (the electron sink of oxygen), where they unite with protons and oxygen to yield H_2O. Electron transfer from NADH to oxygen is very exergonic, releasing about 53 kcal/mole. If released all at once, most of this energy would be lost as heat. Instead, the energy is released slowly in a series of steps, as shown here, and used to transport protons across the inner mitochondrial membrane. The membrane potential established across the membrane is the source of energy needed to synthesize ATP. For each pair of hydrogens that enters this pathway, a maximum of three ATP molecules is produced.

The Chemiosmotic Model and Oxidative Phosphorylation

As electrons are transferred along the acceptors in the electron transport chain, they are passed down an energy hill. Sufficient energy is released at three points to transport protons across the inner mitochondrial membrane and ultimately to synthesize ATP. The passage of each pair of electrons from NADH to oxygen yields three ATPs.

The flow of electrons is tightly coupled to the phosphorylation of ADP and will not occur unless phosphorylation can proceed also. This, in a sense, prevents waste, for electrons will not flow unless energy-rich phosphate can be produced. If electron flow were uncoupled from phosphorylation, there would be no production of ATP, and the energy of the electrons would be wasted as heat. Because the phosphorylation of ADP to form ATP is coupled with the oxidation of electron transport components, this entire process is referred to as **oxidative phosphorylation** (see Figure 7–15).

It had long been known that oxidative phosphorylation occurs in the mitochondria, and many experiments had shown that the transfer of electrons from NADH to oxygen results in the production of three ATP molecules; however, for a long time, just how these ATPs were synthesized remained a mystery. Then, in 1961, Peter Mitchell proposed the **chemiosmotic model,** for which he was awarded the Nobel Prize in 1978. Mitchell proposed that electron transport and ATP synthesis are coupled by a proton gradient across the mitochondrial membrane. According to this model, the stepwise transfer of electrons from NADH or $FADH_2$ through the electron carriers to oxygen results in the pumping of protons across the inner mitochondrial membrane into the space between the inner and outer mitochondrial membranes (Figure 7–16).

Protons are pumped out of the matrix by three electron transfer complexes, each associated with particular steps in the electron transport sys-

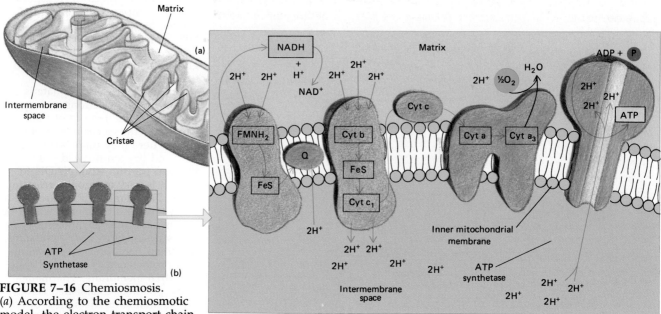

FIGURE 7–16 Chemiosmosis. (*a*) According to the chemiosmotic model, the electron transport chain in the inner mitochondrial membrane is a proton pump. (*b*) The electron acceptors in the membrane are located in three main complexes. FMN, which oxidizes NADH, is located in the first complex, the NADH dehydrogenase complex. The cytochrome b-C_1 complex consists of two cytochromes and some additional electron acceptors. The cytochrome oxidase complex includes cytochromes a and a_3. Ubiquinone (Q) and cytochrome c are mobile carriers that transfer electrons between the complexes. At three sites in the chain, the energy released during electron transport is used to transport protons (H^+) from the mitochondrial matrix to the intermembranous space, where a high concentration of protons accumulates. The protons are prevented from diffusing back into the matrix through the inner membrane everywhere except through special channels in ATP-synthetase in the membrane. The flow of the protons through the ATPase generates ATP at the expense of the free energy released as the protons pass from a region of high concentration (outside) to a region of lower concentration (inside).

tem. This process generates a membrane potential across the inner mitochondrial membrane, the medium in the intermembrane space being positively charged. The difference in the concentration of protons between the matrix and the intermembrane space represents potential energy. This potential energy results in part from a difference in pH and in part from a difference in electrical charge between the two sides of the membrane.

The inner mitochondrial membrane is impermeable to the passage of protons, which can flow back to the matrix of the mitochondrion only through special channels in the membrane. These channels occur in the enzyme **ATP synthetase**, also known as the F_0-F_1 *complex*. The ATP synthetase forms complexes called *respiratory assemblies* that project from the inner surface of the membrane and are visible by electron microscopy (Figure 7–17). The proton channels are composed of protein chains that span the membrane.

As the protons move down the energy gradient, the energy released is used by this ATP synthetase to produce ATP. The ATP synthetase acts as a

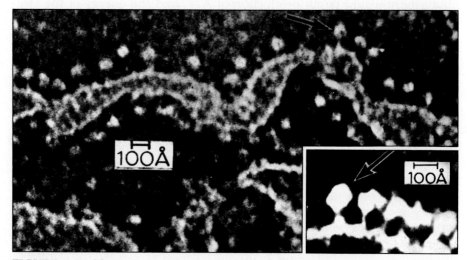

FIGURE 7–17 Electron micrograph of the inner mitochondrial membrane. The lollipop-like structures attached to the surface of the membrane are respiratory assemblies composed of complexes of the enzyme ATP synthetase.

turbine, converting one form of energy into another. The proton gradient across the inner mitochondrial membrane couples phosphorylation with oxidation.

In the process of oxidative phosphorylation, the energy of the electron transfer potential of NADH (i.e., its ability to transfer electrons to other compounds) is converted into the energy of the phosphate transfer potential of ATP (ability of ATP to transfer phosphate groups to other compounds). The electron transfer potential of a compound is reflected in its redox potential, which is expressed in volts. The redox potential of hydrogen gas, H_2, is defined as zero volts.

A compound with a negative redox potential has a lesser electron affinity and therefore more readily loses electrons than does H_2. A compound with a positive redox potential has a greater electron affinity and more readily accepts electrons than does H_2. A strong reducing agent such as NADH has a negative redox potential; a strong oxidizing agent such as O_2 has a positive redox potential. The driving force of oxidative phosphorylation is the electron transfer potential of NADH.

REGULATION OF CELLULAR RESPIRATION

Cellular respiration requires a steady input of nutrient fuel molecules and oxygen. Under normal conditions these materials are adequately provided and do not affect the rate of respiration. Instead, the rate of cellular respiration is regulated by the amount of ADP and phosphate available. In a resting muscle cell, for example, ATP synthesis continues until all the ADP has been converted to ATP. Then, when there are no more acceptors of phosphate, phosphorylation must stop. Since electron flow is tightly cou-

---

Focus on SHUTTLES ACROSS THE MITOCHONDRIAL MEMBRANE

The inner mitochondrial membrane is not permeable to NADH, which is a large molecule. The NADH produced in the cytoplasm from the dehydrogenations in glycolysis cannot diffuse into the mitochondria to transfer their electrons to the electron transport system. Unlike ATP and ADP, NADH does not have a carrier protein to transport it across the membrane. Instead, several systems have evolved to transfer the *electrons* of NADH (though not the NADH molecules themselves) into the mitochondria.

In liver, kidney, and heart cells, a special shuttle system known as the **malate-aspartate shuttle** transfers the electrons from NADH through the inner mitochondrial membrane. Once inside the matrix, malate passes the electrons to a different NAD^+, one already in the matrix. These electrons can then be passed along the electron transport system in the inner membrane, and three molecules of ATP can be produced.

In skeletal muscle, brain, and some other types of cells, another type of shuttle, the **glycerol phosphate shuttle,** operates. This shuttle requires more energy than the malate-aspartate shuttle. As a result, the electrons are at a lower energy level when they enter the electron transport chain. They are accepted by coenzyme Q, rather than by NAD^+, and so generate only two ATP molecules per pair of electrons. This is why the number of ATPs produced by the complete oxidation of a mole of glucose in skeletal muscle cells is 36 rather than 38. The NADH and $FADH_2$ produced in the oxidation of fatty acids are formed within the mitochondria and can pass electrons directly to the electron transport system.

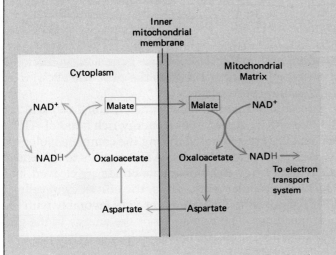

pled to phosphorylation, the flow of electrons also stops. When an energy-requiring process like muscle contraction occurs, ATP is split to yield ADP and inorganic phosphate plus energy. The ADP formed can then serve as an acceptor of phosphate and energy to become ATP once again. Oxidative phosphorylation continues until all the ADP has again been converted to ATP. Because phosphorylation is tightly coupled to electron flow, the cell possesses a system of control that can regulate the rate of ATP production and adjust it to the momentary rate of energy utilization.

Although the reactions of glycolysis take place in the cytoplasm, those of the citric acid cycle and the electron transport system occur within the mitochondria. The outer membrane of the mitochondrion engages in active transport and is thought to regulate the intake of materials. Most respiratory enzymes are associated with the inner mitochondrial membrane. It is thought that these enzymes must be exactly located with respect to one another so that hydrogens and electrons may be passed from one to another in the multitudinous and somewhat bewildering metamorphoses of cellular respiration. No assembly line yet built can match the mitochondrial disassembly line for speed and efficiency. It is a remarkable mechanism, involving thousands of elaborate transformations every second and pouring forth streams of ATP, water, and carbon dioxide as it labors along with thousands like it in every one of the billions of cells of the body.

ENERGY YIELD FROM GLUCOSE

Let us now review where biologically useful energy is released and calculate the total energy yield from the complete oxidation of glucose. Table 7–2 summarizes the arithmetic involved. In glycolysis—(1) in the table—glucose is activated by the addition of two ATP molecules and converted ultimately to 2 pyruvate + 2 NADH + 2 H$^+$ + 4 ATP, yielding a direct net profit of two ATPs. The two pyruvates are metabolized, at (2), to 2 CO_2 + 2 NADH + 2H$^+$ + 2 acetyl CoA. In the citric acid cycle, (3), the two acetyl CoA molecules are metabolized to 4 CO_2 + 6 NADH + 6 H$^+$ + 2 FADH$_2$ + 2 ATP; these are the two ATP molecules formed at the substrate level from succinyl coenzyme A.

Since the oxidation of NADH + H$^+$ in the electron transport system yields 3 ATPs per molecule, the 10 NADH molecules can yield up to 30 ATPs. In skeletal muscle and some other types of cells, the 2 NADH molecules from glycolysis yield only 2 ATPs each (see Focus on Shuttles Across the Mitochondrial Membrane), so in those cells, the total number of ATPs from NADH is only 28. The oxidation of the reduced flavin, FADH$_2$, yields two ATPs per molecule, so the two FADH$_2$ molecules yield four ATPs. Summing these at (4) in Table 7–2, we see that the complete aerobic metabolism of 1 mole (180 g) of glucose yields a maximum of 36 to 38 ATPs. Note that all but two ATP molecules were generated by reactions taking place in the mitochondria, and all but two of those generated in the mitochondria (the two formed at the substrate level in the citric acid cycle) were produced by the electron transport system.

When a mole of glucose is burned in a calorimeter, some 686 kcal are released as heat. The free energy stored in one energy-rich bond of ATP is about 7.3 kcal. When 36 ATPs are generated during the complete biological oxidation of glucose, the total energy stored in ATP amounts to 7.3 kcal × 36, or about 263 kcal. When energy-rich phosphate bonds are cleaved, this energy becomes available for cellular work. Thus, the efficiency of cellular respiration is 263/686, or about 38%. This compares very favorably with the efficiency of the finest machines. The remainder of the energy in the glucose is released as heat, used by some animals to help maintain body temperature.

CATABOLISM OF OTHER NUTRIENTS

Many organisms depend on nutrients other than glucose (or in addition to glucose) as a source of energy. Human beings and many other animals usually obtain more of their energy by oxidizing fatty acids than by oxidizing glucose. Amino acids are also used as fuel molecules. Such nutrients can be transformed into one of the metabolic intermediates that can be fed into glycolysis or the citric acid cycle (Figure 7–18).

Oxidation of Amino Acids

Amino acids are metabolized by reactions in which the amino group is first removed, a process called **deamination.** The amino group is converted to urea and excreted, but the carbon chain is metabolized and eventually enters the citric acid cycle. The sequence of reactions varies somewhat with different amino acids, but for each, a series of reactions modifies the carbon skeleton to produce a substance that is part of the citric acid cycle. Alanine, for example, undergoes deamination to become pyruvate, glutamate is converted to α-ketoglutarate, and aspartate yields oxaloacetate. Other amino acids may require several enzyme-catalyzed reactions in addition to deamination to yield a substance that is a member of the citric acid cycle. Ultimately, the carbon chains of all the amino acids are metabolized in this way.

Oxidation of Fatty Acids

Each gram of triacylglycerol contains more than twice as many kilocalories as a gram of glucose or amino acids. Fats are rich in calories because they

FIGURE 7–18 Catabolism of carbohydrates, proteins, and fats. Molecular subunits of these compounds enter glycolysis or the citric acid cycle at various points. Note that through biosynthetic pathways many of these small metabolites can be used to synthesize macromolecules. (See reverse arrows.) This diagram is greatly simplified and illustrates only a few of the principal pathways.

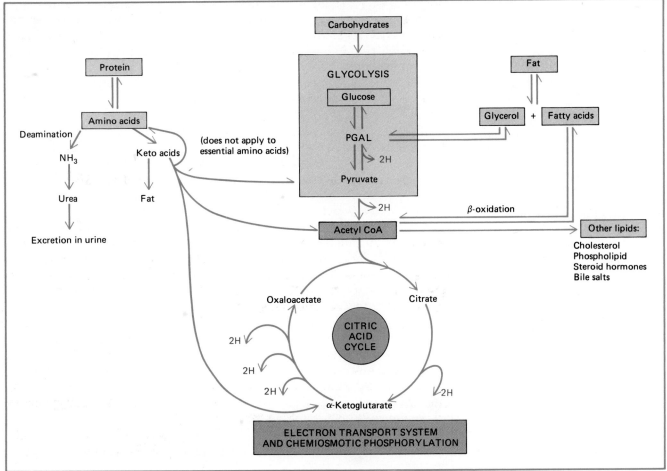

have a lot of hydrogen atoms. (Recall that electrons of hydrogen in organic compounds have a great deal of potential energy.) When completely metabolized in cellular respiration, a molecule of a six-carbon fatty acid can generate up to 44 ATPs (compared with 36 for glucose, which also has six carbons). Both the glycerol and fatty acid components of a neutral fat can be used as fuel. Phosphate is added to glycerol, converting it ultimately to PGAL or another compound that undergoes glycolysis.

Fatty acids are oxidized and split enzymatically into two-carbon compounds (acetyl groups) bound to coenzyme A; that is, they are converted to acetyl CoA. This process, which occurs in the matrix of the mitochondria, is known as *β***-oxidation** (beta oxidation). The acetyl CoA molecules enter the citric acid cycle and are converted to CO_2 with the release of hydrogen.

The long chain of a fatty acid is split into two-carbon acetyl CoA units by a cyclic sequence of reactions that include dehydrogenations and make-ready reactions but not decarboxylations. The overall process is termed *β-oxidation* because the oxygen atom is added to the *β*-carbon, the second carbon from the carboxyl group. For example, palmitate, a fatty acid with a chain of 16 carbon atoms, is first activated by enzyme-catalyzed reactions with ATP and coenzyme A to form palmityl CoA (Figure 7–19). This undergoes a dehydrogenation between the second and third carbons of the chain. The group undergoing dehydrogenation is a —CH_2—CH_2— group; the primary acceptor of the electrons and protons liberated is a flavin. The product, a long-chain molecule with a double bond between carbons 2 and 3, undergoes a make-ready reaction, the addition of a molecule of water across the double bond. The resulting molecule has an H—C—OH group at the *β*-carbon that can undergo dehydrogenation with NAD^+ as acceptor.

The product of the dehydrogenation, a long-chain fatty acid with a C=O group at the *β*-carbon (carbon 3) and with coenzyme A still attached to its carboxyl group, reacts with a second molecule of coenzyme A. The

FIGURE 7–19 Chemical reactions in the *β*-oxidation of fats. The 16-carbon compound palmitate is degraded in this process to a 14-carbon fatty acetyl CoA. This product is activated and ready to undergo FAD-linked dehydrogenation. The succeeding reactions yield an activated 12-carbon fatty acetyl CoA ready to undergo a third cycle of reactions. This process continues until the original palmitate has been cleaved to yield eight two-carbon acetyl CoA molecules.

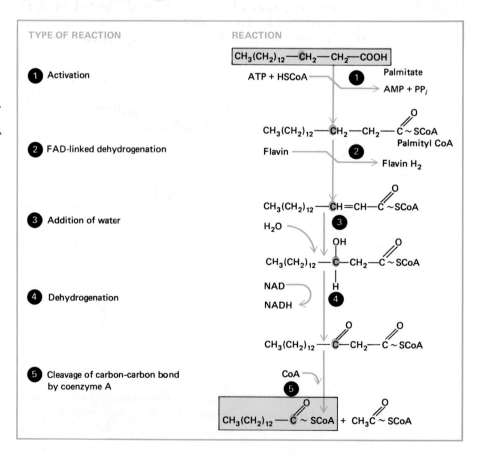

coenzyme A attacks at the C=O group and attaches to it, cleaving off carbons 1 and 2 with the original coenzyme A group attached as acetyl CoA and leaving a carbon chain that is two carbons shorter. This product has coenzyme A attached to its carboxyl group and is ready, without any further activation, to be dehydrogenated by the enzyme that uses flavin as the primary acceptor.

Four enzymes working in succession catalyze these four reactions and yield an acetyl CoA molecule plus the remaining carbon chain with a coenzyme A attached. Seven such series of dehydrogenations and make-ready reactions will split the 16-carbon chain of palmitate into eight two-carbon fragments, each with a coenzyme A group attached—eight acetyl CoA molecules. These molecules can be metabolized in the citric acid cycle.

Each cycle of β-oxidation—dehydrogenation, hydration, dehydrogenation—produces one NADH molecule and one $FADH_2$ molecule. When their electrons are passed through the electron transport system, the NADH produces three and the $FADH_2$ produces two ATPs, for a total of five ATPs. The seven cycles of β-oxidation that split palmitate to eight acetyl CoA molecules thus yield $7 \times 5 = 35$ ATPs. Each of the acetyl CoA groups yields 12 ATPs as it is metabolized in the citric acid cycle: $8 \times 12 = 96$ ATPs. Thus, the yield of ATPs, from the complete oxidation of one palmitate to CO_2 and H_2O, is $96 + 35 = 131$ ATPs. This is why fat is such a superlative energy storage molecule (and why it takes so long to lose fat by dieting!).

BIOSYNTHETIC PROCESSES

Our discussion thus far has focused on processes that break down molecules of foodstuffs and conserve their energy in the biologically useful form of energy-rich P. Cells possess a remarkable array of enzymes that catalyze a variety of biosynthetic processes, using the energy of ATP and, as raw materials, some of the five-, four-, three-, two-, and one-carbon compounds that are intermediates in the metabolism of glucose, fatty acids, amino acids, and other compounds. This **anabolic** (synthetic) aspect of metabolism is quite complex, but several basic principles can be listed:

1. In general, each cell synthesizes its own proteins, nucleic acids, lipids, polysaccharides, and other complex molecules and does not receive them preformed from other cells. Muscle glycogen, for example, is synthesized within the muscle cell and is not derived from liver glycogen.
2. Each step in the biosynthetic process is catalyzed by a separate enzyme.
3. Although certain steps in a biosynthetic sequence may proceed without the use of energy-rich phosphate, the overall synthesis of these complex molecules requires chemical energy at various points along the way.
4. The synthetic processes use as raw materials relatively few substances, among which are acetyl CoA, glycine, succinyl CoA, ribose, pyruvate, and glycerol.
5. In general, these synthetic processes are not simply the reverse of the processes by which the molecule is degraded but include one or more separate steps that differ from any step in the degradative process. These steps are controlled by different enzymes, which permits separate control mechanisms to govern the synthesis and degradation of complex molecules.
6. The biosynthetic process includes not only the formation of the macromolecular components from simple precursors but also their assembly into the several kinds of membranes that constitute the outer

FIGURE 7–20 In an adult organism the rates of synthesis and degradation are essentially equal, whereas in a growing organism the rate of synthesis must be faster than the rate of catabolism. Even in an adult organism there is a continuous turnover of molecules.

boundary of the cell and the intracellular organelles. Each cell's constituent molecules are in a dynamic state—that is, they are constantly being degraded and synthesized.

Thus, even a cell that is not growing or increasing in mass uses a considerable portion of its total energy for the chemical work of biosynthesis (Figure 7–20). A cell that is growing rapidly must allocate a correspondingly larger fraction of its total energy output to biosynthetic processes, especially the biosynthesis of protein. A rapidly growing bacterial cell may use as much as 90% of its total energy for the synthesis of proteins.

Many of the steps in biosynthetic processes involve the formation of peptide bonds, glycosidic bonds, and ester bonds, but these bonds are not formed by reactions in which water is removed. The biosynthesis of sucrose in the cane sugar plant, for example, which involves the formation of a glycosidic bond, does *not* proceed by way of the following reaction:

$$\text{Glucose} + \text{fructose} \rightleftharpoons \text{sucrose} + H_2O$$

If all reactants were present in a concentration of 1 mole/liter, it would require some 5.5 cal/mole for this reaction to go to the right. However, the concentration of glucose and fructose in the plant cell is probably less than 0.01 mole/liter, whereas the concentration of water is very high, about 55 moles/liter. Thus, the equilibrium point of the reaction under these conditions would be very far to the left.

Instead, one or more of the reactants is activated by a reaction with ATP. The terminal phosphate is enzymatically transferred to glucose with the conservation of some of the energy of the terminal phosphate bond. The glucose phosphate, with a higher energy content than free glucose, can react with fructose by way of another enzyme-catalyzed reaction to yield sucrose and inorganic phosphate.

(1) $2 \text{ glucose} + 2 \text{ ATP} \longrightarrow 2 \text{ glucose-6-P} + 2 \text{ ADP}$

(2) $\text{glucose-6-P} \rightleftharpoons \text{fructose-6-P}$

$\underline{\text{glucose-6-P} + \text{fructose-6-P} \longrightarrow \text{sucrose} + 2 \text{ P}_i}$

Sum: $2\text{ATP} + 2 \text{ glucose} \longrightarrow \text{sucrose} + 2\text{ADP} + 2 \text{ P}_i$

This reaction proceeds to the right because there is a net decrease in free energy. The 14 kcal of the 2 energy-rich P bonds are used to supply the 5.5 kcal needed to assemble the glucose and fructose into sucrose; the overall decrease in free energy is 9.5 kcal/mole. Since water is not a product of this reaction, the high concentration of water in the cell does not inhibit it.

■ SUMMARY

I. Cells use three different catabolic pathways to extract free energy from nutrients: aerobic cellular respiration, anaerobic cellular respiration, and fermentation.

II. During cellular respiration, a fuel molecule, such as glucose, is oxidized, forming carbon dioxide and water with the release of energy.

III. Cellular respiration is a redox process in which hydrogen is transferred from glucose to oxygen.

IV. The chemical reactions of cellular respiration can be grouped into four phases: glycolysis, formation of acetyl CoA, the citric acid cycle, and the electron transport system and chemiosmotic phosphorylation.

V. During glycolysis a molecule of glucose is degraded, forming two molecules of pyruvate.

A. Two ATP molecules are gained during glycolysis.

B. Four hydrogen atoms are removed from the fuel molecule.

VI. Fermentation is an anaerobic process in which the final acceptor of electrons from NADH is an organic compound. There is a net gain of only 2 ATPs per glucose molecule, compared with about 38 ATPs per glucose molecule in aerobic respiration.

A. Yeast cells carry on alcohol fermentation, in which ethyl alcohol is the final product.

B. Certain fungi and bacteria carry on lactic acid fermentation, in which hydrogen atoms are added to pyruvate, forming lactate.

VII. In aerobic cellular respiration, pyruvate molecules are combined with coenzyme A, producing acetyl CoA. Carbon dioxide is released during this process.

VIII. The four carbons remaining from the original glucose molecule are present in the two acetyl CoA molecules. Acetyl CoA enters the citric acid cycle by combining with a four-carbon compound, oxaloacetate, to form citrate, a six-carbon compound.

    A. With two turns of the citric acid cycle, the two acetyl CoAs are completely degraded.

    B. Carbon dioxide is released and hydrogens are transferred to NAD or FAD. Only one ATP is produced directly at the substrate level with each turn of the cycle.

IX. Hydrogen atoms (or their electrons) removed from fuel molecules are transferred from one electron acceptor to another down the chain of acceptor molecules that make up the electron transport system. The final acceptor in the chain is molecular oxygen, which combines with the hydrogen to form water.

    A. According to the chemiosmotic theory, energy liberated in the electron chain is used to establish a proton gradient across the inner mitochondrial membrane.

    B. The flow of protons back through the membrane (by way of the enzyme ATP synthetase) releases energy, which is used to synthesize ATP.

X. In the electron transport system, phosphorylation is tightly coupled to electron flow. The rate of ATP synthesis depends on the availability of ADP and phosphate.

XI. Organic nutrients other than glucose can be converted into appropriate compounds and fed into the glycolytic or citric acid pathway.

    A. Amino acids can be deaminated and the carbon skeleton converted to a metabolic intermediate such as pyruvate.

    B. Both the glycerol and fatty acid components of fats can be oxidized as fuel. Fatty acids are converted to acetyl coenzyme A molecules by the process of β-oxidation.

XII. The cells of living things exist in a dynamic state and are continuously building up and breaking down the many different cell constituents.

    A. In general, each cell synthesizes its own complex macromolecules, and each step in the process is catalyzed by a separate enzyme.

    B. These biosynthetic reactions are strongly endergonic and require ATP to drive them.

SUMMARY REACTIONS FOR CELLULAR RESPIRATION

Summary reaction for the complete oxidation of glucose:

$$C_6H_{12}O_6 + 6\ O_2 + 6\ H_2O \longrightarrow 6\ CO_2 + 12\ H_2O + energy$$

Summary reaction for glycolysis:

$$Glucose + 2\ ADP + 2\ P_i + 2\ NAD^+ \longrightarrow 2\ pyruvate + 2\ ATP + 2\ NADH + 2\ H^+ + 2\ H_2O$$

Summary reaction for the conversion of pyruvate to acetyl CoA:

$$2\ Pyruvate + 2\ coenzyme\ A + 2\ NAD^+ \longrightarrow 2\ acetyl\ CoA + 2\ CO_2 + 2\ NADH + 2\ H^+$$

Summary reaction for the citric acid cycle:

$$2\ acetyl\ CoA + 6\ NAD^+ + 2\ FAD + 2\ GDP + 2\ P_i + 2\ H_2O \longrightarrow 4\ CO_2 + 6\ NADH + 6\ H^+ + 2\ FADH_2 + 2\ GTP + 2\ coenzyme\ A$$

Summary reaction for the processing of hydrogens (electrons and protons) in the electron transport system:

$$NADH + H^+ + 3\ ADP + 3\ P_i + \tfrac{1}{2}\ O_2 \longrightarrow NAD^+ + 3\ ATP + H_2O$$

■ POST-TEST

1. The process of splitting larger molecules into smaller ones is an aspect of metabolism called _____ _____.

2. Yeasts and bacteria that can shift to anaerobic respiration or fermentation when oxygen is in short supply are called facultative _____.

3. A chemical process during which a substance gains electrons is called _____.

4. The pathway through which glucose is converted to pyruvate is referred to as _____.

5. The reactions of glycolysis take place within the _____.

6. During strenuous muscle activity, the pyruvate in muscle cells may accept hydrogen to become _____.

7. When deprived of oxygen, yeast cells transfer hydrogens to acetaldehyde; the product is _____.

8. A net profit of only _____ ATPs can be produced anaerobically from one molecule of glucose, compared with _____ produced in aerobic respiration.

9. The anaerobic process by which alcohol or lactate is produced as a product of glycolysis is referred to as _____.

10. Anaerobic metabolism is inefficient because the fuel molecule is only partially _____.

11. Pyruvate is decarboxylated and used to make _____.

12. During the citric acid cycle, the fuel molecule is completely oxidized and the products are _____ _____ and _____.

13. Acetyl CoA reacts with oxaloacetate to form _____.

14. Dehydrogenase enzymes remove hydrogens from fuel

molecules and transfer them to primary acceptors such as _____.

15. The carbon dioxide we exhale is derived from _____ reactions.

16. When protons move down an energy gradient in cellular respiration, energy is released and used to synthesize _____.

17. The flow of electrons in the electron transport chain is tightly coupled to _____.

18. The synthetic aspect of metabolism is referred to as _____.

■ REVIEW QUESTIONS

1. What is the specific role of oxygen in the cell? What happens when cells are deprived of oxygen?
2. How does the chemiosmotic model relate to cellular respiration? How does a proton gradient contribute to ATP synthesis?
3. Mitochondria are often referred to as the *power plants* of the cell. Justify. Be specific.
4. Why are ATPs consumed during the first steps of glycolysis?
5. What are the products of glycolysis?
6. In what form is the fuel molecule when it feeds into the citric acid cycle? What are the products of the citric acid cycle?
7. Trace the fate of hydrogens removed from the fuel molecule during glycolysis when oxygen is present.
8. Trace the fate of hydrogens removed from the fuel molecule when the amount of oxygen available in muscle cells is insufficient to support aerobic respiration.

9. Draw a mitochondrion and indicate the locations of the following:
   a. enzymes of the citric acid cycle
   b. enzymes of the electron transport system
   c. the site of the proton gradient that drives ATP production
10. Calculate how much energy is made available to the cell by the operation of glycolysis, the citric acid cycle, and the electron transport system; estimate the efficiency of each phase and of the overall process of cellular respiration.
11. Why is it advantageous that synthetic reactions are generally not the reverse of reactions in which molecules are catabolized?
12. Explain the roles of the following in cellular respiration:
    a. NAD^+
    b. cytochromes
    c. oxidative phosphorylation

■ RECOMMENDED READINGS

Alberts, Bruce, et al.: *Molecular Biology of the Cell*. New York, Garland Publishing, 1983. An excellent treatment of energy conversion in cells.

Dickerson, Richard E. Cytochrome c and the evolution of energy metabolism. *Scientific American*, March 1980, pp. 136–153.

Hinckle, P.C., and McCarty, R.E.: How cells make ATP. *Scientific American*, March 1978, pp. 104–123. An interesting presentation of the chemiosmotic theory and how it may explain both photosynthesis and oxidative phosphorylation.

Holtzman, E., and Novikoff, A.B.: *Cells and Organelles*, 3rd ed. Philadelphia, Saunders College Publishing, 1984. An integrated approach to the structural, biochemical, and physiological aspects of the cell.

Lehninger, A.L.: *Principles of Biochemistry*. New York, Worth Publishing, 1982. A standard biochemistry text with a detailed but clear account of cellular respiration. A fine source book for students.

Shulman, R.G.: NMR spectroscopy of living cells. *Scientific American*, January 1983, pp. 86–93. Spectroscopy makes it possible to study metabolic processes in intact living cells.

Stryer, L.: *Biochemistry*, 3rd ed. San Francisco, W.H. Freeman, 1988. A well-illustrated, readable text that covers the molecules of life and the concepts of cellular energetics from the ground up.

Elodea actively carrying on photosynthesis

8

Photosynthesis: Capturing Energy

■ LEARNING OBJECTIVES

After you have read this chapter you should be able to:

1. Distinguish between autotrophs and heterotrophs and between chemosynthetic and photosynthetic autotrophs.
2. Draw a diagram of the internal structure of a chloroplast, and explain how this structure facilitates the process of photosynthesis.
3. Write a summary reaction for photosynthesis, explaining the origin and fate of each substance involved.
4. Distinguish between the light-dependent and light-independent reactions of photosynthesis, and summarize the events that occur in each phase.
5. Describe the physical properties of light, and describe how the absorption of photons can activate a pigment such as chlorophyll.
6. Describe the nature of a photosystem, including the functions of antenna pigments and the reaction center.
7. Contrast cyclic and noncyclic photophosphorylation, describing each.
8. Describe how a proton gradient is established across the thylakoid membrane and how this gradient functions in ATP synthesis.
9. Describe the chemical reactions involved in the conversion of CO_2 to glucose in photosynthesis, and indicate how many molecules of ATP and NADPH are required for the process.
10. Discuss the advantages of the C_4 pathway and how this increases the effectiveness of the C_3 pathway in certain types of plants.
11. Outline the reactions involved in photorespiration.

The vast chemical industry of the biosphere has for millennia run on solar energy. Plants, blue-green algae, and certain bacteria are producers that are uniquely capable of converting solar energy into stored chemical energy by the process of **photosynthesis** (Figure 8–1). Each year these remarkable organisms produce more than 200 billion tons of food. The chemical energy stored in this food fuels the metabolic reactions that sustain life.

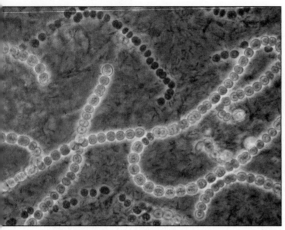

(a)

(b)

FIGURE 8–1 Photosynthetic autotrophs. These diverse producers use light energy to power the synthesis of organic compounds from carbon dioxide and water. (*a*) Filaments of *Nostoc*, a nitrogen-fixing cyanobacterium useful for enriching the soil in rice paddies. (*b*) Quaking aspen trees.

Producers are **autotrophs** ("self-nourishing"; from the Greek *auto*, "self," and *trophos*, "nourishing"), organisms that can make their food from inorganic raw materials and so are not dependent on other organisms for nourishment. Certain bacteria are **chemosynthetic autotrophs,** producers that make their organic compounds by oxidizing simple inorganic substances such as sulfur or ammonia. Chemosynthetic autotrophs do not require light as an energy source for these reactions. Most producers are **photosynthetic autotrophs,** organisms that use light as their energy source for manufacturing organic compounds from carbon dioxide and water.

Consumers and decomposers are **heterotrophs,** organisms that cannot make their own food and so must depend on other organisms for their nourishment (*heter*, "other"; *troph*, "nourishing"). These organisms must eat producers or organisms that have eaten producers. Almost all life is ultimately dependent on photosynthesis, the process in which biochemistry and ecology meet.

CHLOROPLASTS: WHERE PHOTOSYNTHESIS HAPPENS

When a bit of leaf is examined under the microscope, we can see that the green pigment, **chlorophyll,** is not uniformly distributed in the cell but is confined to small organelles, known as **chloroplasts** (Figure 8–2). In plants chloroplasts are located mainly in the cells of the **mesophyll,** a tissue inside the leaf. Each mesophyll cell has 20 to 100 chloroplasts, which can grow and divide, forming daughter chloroplasts.

By electron microscopy we can see that the chloroplast, like the mitochondrion, is bounded by an outer and an inner membrane (Figure 8–3). The inner membrane encloses a fluid-filled region called the **stroma,** which contains most of the enzymes required for the reactions of photosynthesis that do not directly require light (the reactions that convert carbon dioxide to glucose).

The inner chloroplast membrane also encloses a third system of membranes, which form an interconnected set of flat, disclike sacs called **thylakoids.** These thylakoid membranes also form a third compartment within the chloroplasts called the **thylakoid space.** In some regions, thylakoid sacs are arranged in stacks referred to as **grana.** Each granum looks something like a stack of coins, with each "coin" being a thylakoid (Figure 8–4). Some thylakoid membranes extend from one granum to another. Photosynthetic prokaryotes have no chloroplasts, but thylakoids often occur as extensions of the cell membrane and may be arranged around the periphery of the cell.

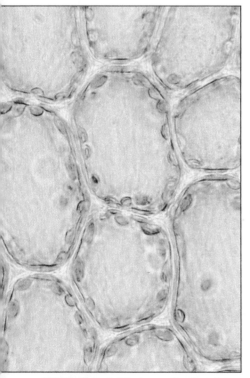

FIGURE 8–2 Photosynthesis takes place mainly within the leaves. Chloroplasts, which are present in the mesophyll cells, are the site of photosynthesis.

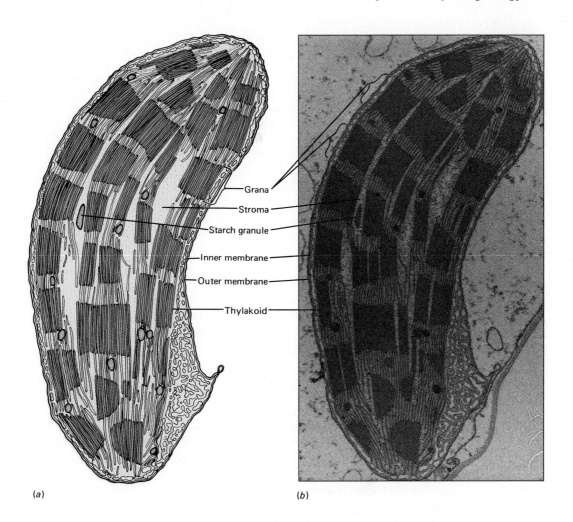

Grana

Stroma

Starch granule

Inner membrane

Outer membrane

Thylakoid

(a)   (b)

FIGURE 8–3 Structure of the chloroplast. (*a*) Diagram. (*b*) Electron micrograph of a chloroplast from a leaf cell of a timothy plant, *Phleum pratense*. Note the grana, which are stacks of thylakoids. The pigments and enzymes necessary for the light-dependent reactions of photosynthesis are associated with the thylakoid membranes.

Chlorophyll and other photosynthetic pigments and the enzymes required for the reactions of photosynthesis that require light are associated with the thylakoid membranes. These membranes, like the inner mitochondrial membrane, are involved in ATP synthesis.

AN OVERVIEW OF PHOTOSYNTHESIS

During photosynthesis, chlorophyll traps energy from sunlight and uses it to make ATP. The energy contained in the ATP is then donated to the reactions that form energy-rich carbohydrates. The principal raw materials for photosynthesis are water and carbon dioxide. Using the energy that chlorophyll molecules trap from sunlight, water is split, its oxygen liberated, and its hydrogen combined with carbon dioxide to produce carbohydrate molecules. The reactions of photosynthesis may be summarized as follows:

$$6\ CO_2 + 12\ H_2O \xrightarrow[\text{Chlorophyll}]{\text{Light energy}} C_6H_{12}O_6 + 6\ O_2 + 6\ H_2O$$

Carbon dioxide   Water   Glucose   Oxygen   Water

Photosynthesis, like cellular respiration, is a reduction-oxidation (redox) process. During respiration, organic compounds are oxidized. Hydrogens, and their electrons, split off from the fuel molecule are transferred to a

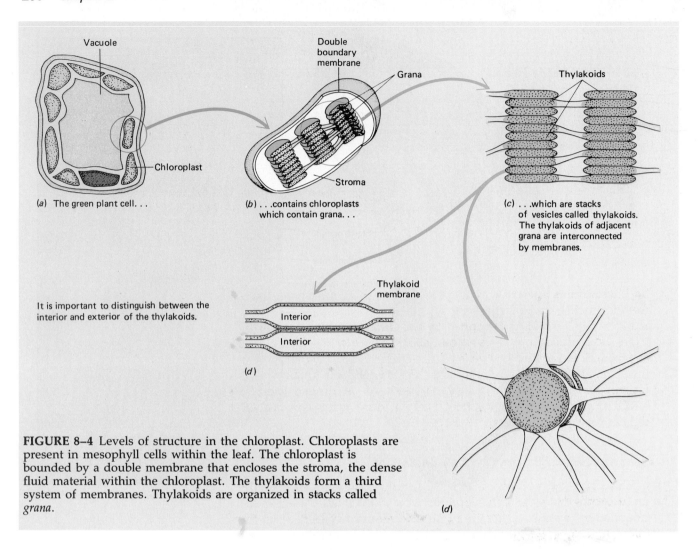

Vacuole

Chloroplast

(a) The green plant cell. . .

It is important to distinguish between the interior and exterior of the thylakoids.

Double boundary membrane

Grana

Stroma

(b) . . .contains chloroplasts which contain grana. . .

Thylakoids

(c) . . .which are stacks of vesicles called thylakoids. The thylakoids of adjacent grana are interconnected by membranes.

Thylakoid membrane

Interior

Interior

(d)

(d)

FIGURE 8–4 Levels of structure in the chloroplast. Chloroplasts are present in mesophyll cells within the leaf. The chloroplast is bounded by a double membrane that encloses the stroma, the dense fluid material within the chloroplast. The thylakoids form a third system of membranes. Thylakoids are organized in stacks called *grana*.

series of acceptor molecules and finally to molecular oxygen, forming water. Electrons lose energy as they are passed along through the chain of acceptor molecules, and that energy is used by the mitochondrion to make ATP. In photosynthesis, the direction of electron flow is reversed. Water is split, and its hydrogens and electrons are transferred through a series of electron acceptors. During this transfer, light energy raises the energy level of the electrons. This is an endergonic process in which some of the energy is used to reduce carbon dioxide, forming glucose.

The summary equation for photosynthesis describes what happens but not how it happens. The "how" is much more complex and involves many chemical reactions. For convenience, we can divide the reactions of photosynthesis into the light-dependent and the light-independent reactions.

The Light-Dependent Reactions: A Summary

The light-dependent reactions can take place only in the presence of light. During this phase of photosynthesis several important events occur:

1. Chlorophyll absorbs light energy, which is immediately converted to electrical energy. The electrical energy is represented in the flow of electrons from the chlorophyll molecule. We say that the chlorophyll becomes temporarily energized.
2. Some of the energy of the energized chlorophyll is used to make ATP

FIGURE 8–5 On sunny days the oxygen released by aquatic plants may sometimes be visible as bubbles in the water. This plant *Elodea* is actively carrying on photosynthesis, as evidenced by the oxygen bubbles.

by chemiosmosis. During this process, electrical energy is transformed to chemical energy.

3. Some of the light energy trapped by the chlorophyll is used to split water, a process known as **photolysis.** Oxygen from the water is released. Some of this oxygen is used by the plant for cellular respiration, but most of it is released into the atmosphere (Figure 8–5).

4. Hydrogen from the water combines with the hydrogen acceptor NADP, forming NADPH (reduced NADP). Here again, electrical energy is converted to chemical energy.

The Light-Independent Reactions: A Summary

Although the light-independent reactions do not *require* light, they do depend on the products of the light-dependent reactions (Figure 8–6). In the light-independent reactions, the energy of the reduced NADPH and the ATP produced during the light-dependent phase of photosynthesis is used

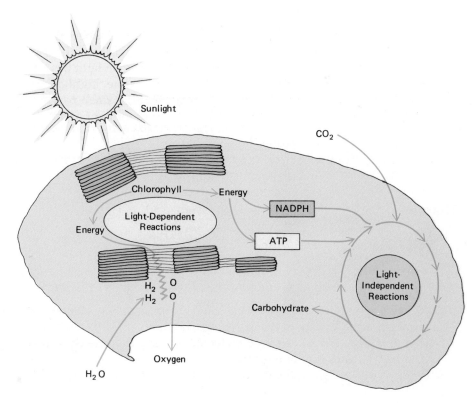

FIGURE 8–6 Summary of the light-dependent and light independent reactions of photosynthesis.

to manufacture carbohydrate; the raw materials are carbon dioxide from the air and hydrogen split off from water.

During the light-independent phase, chemical energy from the ATP and NADPH produced during the light-dependent phase is transferred to the chemical bonds of carbohydrate molecules. This form of energy packaging is most suitable for long-term storage. Some of the carbohydrate molecules produced during the light-independent phase are used as fuel molecules. Others are used to manufacture various types of organic compounds needed by the plant cells; for example, with the addition of such minerals as nitrates and sulfur from the soil, plant cells can produce proteins.

HOW THE LIGHT-DEPENDENT REACTIONS CAPTURE ENERGY

In the light-dependent reactions, the energy from sunlight is used to make ATP and to reduce the electron acceptor molecule NADP. Some of the captured energy is temporarily stored within these two compounds.

We may summarize the light-dependent reactions as follows:

$$12\ H_2O + 12\ NADP + 18\ ADP + 18\ P_i \xrightarrow[\text{Chlorophyll}]{\text{Light energy}}$$
$$6\ O_2 + 12\ NADPH + 12\ H^+ + 18\ ATP$$

Light and Atomic Excitation

Since life on our planet depends on light, it seems appropriate to discuss the nature of light and how it permits photosynthesis to occur. Light is a very small portion of a vast, continuous spectrum of radiation, the electromagnetic spectrum (Figure 8–7). All radiations in this spectrum travel in waves. At one end of the spectrum are gamma rays, which have very short wavelengths (measured in nanometers). A **wavelength** is the distance from one wave peak to the next. At the other end of the spectrum are low-frequency radiowaves, which have wavelengths so long that they are measured in kilometers.

The different colors of light (different regions of the light spectrum) are identified by their wavelengths. Within the spectrum of visible light, violet has the shortest wavelength and red the longest. Ultraviolet light has a still shorter range of wavelengths, and infrared a still longer one. Light behaves not only like a wave but also like a particle. It is made of small packets of energy called **photons,** the energy of which is different for light of different

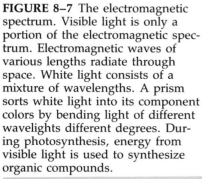

FIGURE 8–7 The electromagnetic spectrum. Visible light is only a portion of the electromagnetic spectrum. Electromagnetic waves of various lengths radiate through space. White light consists of a mixture of wavelengths. A prism sorts white light into its component colors by bending light of different wavelights different degrees. During photosynthesis, energy from visible light is used to synthesize organic compounds.

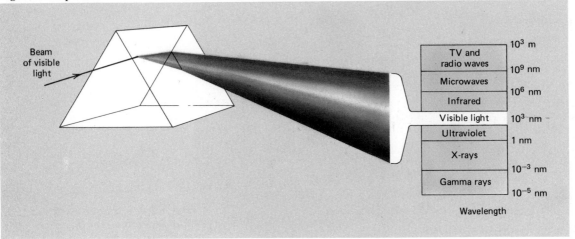

Beam of visible light

TV and radio waves	10^3 m
Microwaves	10^9 nm
Infrared	10^6 nm
Visible light	10^3 nm
Ultraviolet	1 nm
X-rays	10^{-3} nm
Gamma rays	10^{-5} nm

Wavelength

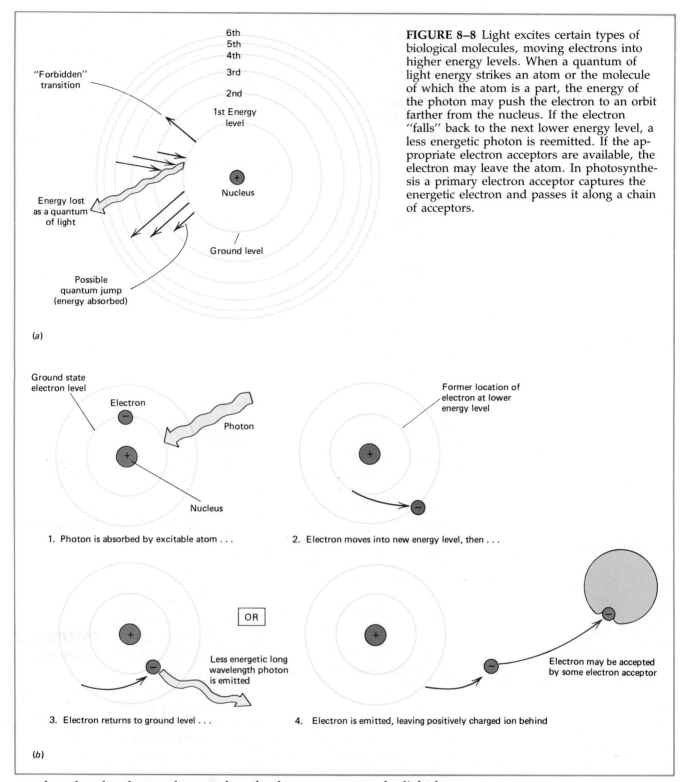

FIGURE 8–8 Light excites certain types of biological molecules, moving electrons into higher energy levels. When a quantum of light energy strikes an atom or the molecule of which the atom is a part, the energy of the photon may push the electron to an orbit farther from the nucleus. If the electron "falls" back to the next lower energy level, a less energetic photon is reemitted. If the appropriate electron acceptors are available, the electron may leave the atom. In photosynthesis a primary electron acceptor captures the energetic electron and passes it along a chain of acceptors.

wavelengths; the shorter the wavelength, the more energy the light has, and the longer the wavelength, the less energy the light has. In other words, the energy of the photon is inversely proportional to the wavelength.

Why does photosynthesis depend on visible light rather than on some other wavelength of radiation? We can only speculate. One reason may be that most of the radiation reaching our planet from the sun is within this portion of the electromagnetic spectrum. Another may be that only radiation within the visible light portion of the spectrum excites certain types of biological molecules, moving electrons into higher energy levels.

Photons interact with atoms in a variety of ways, all of which depend on the electron structure of the atom. As you should recall, an atom consists of an atomic nucleus surrounded by one or more energy levels containing electrons. In the hydrogen atom there is only a single electron, which occupies the first energy level. The lowest energy state an atom possesses is called the **ground state,** but energy can be added to an electron so that it will attain a higher energy level. When an electron is raised to a higher energy level than its ground level, the atom is said to be *excited*, or energized.

When a molecule absorbs a photon, one of its electrons is raised to a higher energy state. One of two things may then happen, depending on the atom and its surroundings:

1. The electron may soon return to its ground level (Figure 8–8; see page 204) and energy is dissipated as heat or as light of a longer wavelength than the wavelength of the absorbed light. This emission of light is **fluorescence.**
2. The excited electron may be lost, leaving the atom with a net positive charge. In this instance the electron may be accepted by a reducing agent; this is what occurs in photosynthesis.

Absorbing Light: Chlorophyll

Pigments are substances that absorb visible light. **Chlorophyll,** the main pigment used in photosynthesis, absorbs light primarily in the blue, violet, and red regions of the spectrum. Green light is not absorbed or used; it is reflected. Most plants are green because their leaves reflect most of the green light that strikes them. Different types of pigments absorb light of different wavelengths. Since the light that is absorbed is not reflected, we do not see that color.

The chlorophyll molecule is made of atoms of carbon and nitrogen joined in a complex porphyrin ring (Figure 8–9). It is strikingly similar to the heme portion of the red pigment hemoglobin in red blood cells and to cytochromes present in all cells. However, chlorophyll contains an atom of magnesium instead of an atom of iron in the center of the ring bound to two of the four nitrogen atoms. The chlorophyll molecule has a long hydrophobic tail, which holds it in the thylakoid membrane. This tail is composed of phytol, a long-chain alcohol containing a chain of 20 carbon atoms. Because of their shape, many chlorophyll molecules can be fitted together like a stack of saucers. Each thylakoid is filled with a myriad of precisely oriented chlorophyll molecules, something like the plates in a storage battery. This arrangement probably permits the generation and utilization of the minute electrical currents that power photosynthesis.

There are several kinds of chlorophyll. The most important is **chlorophyll a,** the pigment that initiates light-dependent reactions. **Chlorophyll b,** an accessory pigment, differs from chlorophyll a only in a functional group on the porphyrin; the methyl group in chlorophyll a is replaced in chlorophyll b by an aldehyde group. This difference shifts the wavelength of light absorbed by chlorophyll b, as a result of which chlorophyll b is yellow-green, whereas chlorophyll a is bright green.

Plant cells also have other accessory photosynthetic pigments, such as **carotenoids,** that are yellow and orange. Carotenoids absorb different wavelengths of light from chlorophyll and so broaden the spectrum of light that can provide energy for photosynthesis. Chlorophyll may be excited either by light or by energy passed to it from other substances that have become excited by light. Thus, when a carotenoid is excited, its energy can be transferred to chlorophyll a.

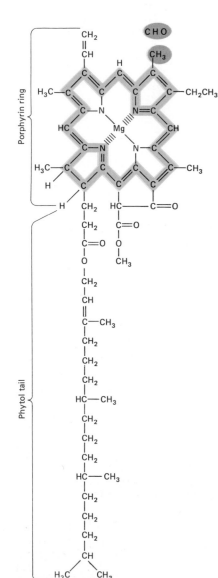

FIGURE 8–9 The structure of chlorophyll. Chlorophyll consists of a porphyrin ring and a phytol tail. The magnesium atom in the center of the ring is the part of the molecule that is excited by light. The hydrophobic phytol tail is embedded in the thylakoid membrane. The CH₃ group distinguishes chlorophyll a, which initiates light-dependent reactions, from chlorophyll b, which has an aldehyde group (CHO) in this position.

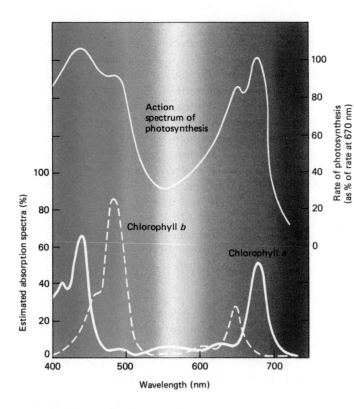

FIGURE 8–10 Absorption spectrum and action spectrum for chlorophyll. The solid line illustrates the absorption spectrum for chlorophyll a. Chlorophyll absorbs light mainly in the blue, violet, and red regions of the spectrum. The dashed line illustrates the absorption spectrum of chlorophyll b. The solid line at the top (action spectrum of photosynthesis) illustrates the effectiveness of various wavelengths of light in powering photosynthesis. Note that the action spectrum is broader due to the presence of accessory pigments, which can use additional wavelengths of light.

The Action Spectrum of Photosynthesis

An instrument known as a *spectrophotometer* is used to measure the relative abilities of different pigments to absorb different wavelengths of light. However, an **absorption spectrum** does not tell us what wavelengths are most effective in photosynthesis. The relative effectiveness of different wavelengths of light in photosynthesis is given by the **action spectrum** of photosynthesis (Figure 8–10), determined in one of the classic experiments in biology. In 1883 the German biologist T.W. Engelmann took advantage of the shape of the chloroplast in *Spirogyra*, a green alga that occurs as slimy strings in freshwater habitats, especially slow-moving or still waters (Figure 8–11*a*). *Spirogyra* wins no beauty prize in bulk, but the individual cells are exquisitely beautiful, each containing a long, spiral, emerald-green chloroplast embedded in cytoplasm. Engelmann reasoned that if these chloroplasts were exposed to a spectrum produced by a prism, photosynthesis would take place most rapidly at the points at which the chloroplast

FIGURE 8–11 Engelmann's experiment to demonstrate the wavelengths of light most effective for photosynthesis. (*a*) Filaments of *Spirogyra*, the green alga that Engelmann used in his experiment. (*b*) Engelmann illuminated a filament of *Spirogyra* with light that had been passed through a prism, producing a spectrum. In this way, different algal cells were exposed to different wavelengths of light. He used aerobic bacteria that moved toward the algal cells emitting the most oxygen. Watching through a microscope, Engelmann observed that the bacteria aggregated most densely along the cells in the blue and red portions of the spectrum. This indicated that blue and red light works most effectively for photosynthesis.

(*a*)

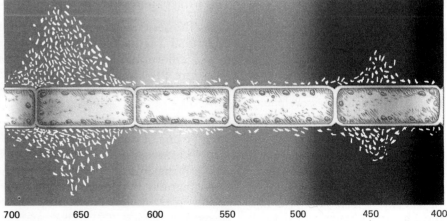

(*b*)

(a)

(b)

FIGURE 8–12 Photosynthetic autotrophs have various accessory pigments that absorb light energy and transfer it to chlorophyll a. (a) The chloroplasts of red algae and cyanobacteria contain accessory photosynthetic pigments known as *phycobilins*. (b) Carotenoids, present in most plants, are masked by chlorophyll. In the fall, when chlorophyll is metabolized, the yellow or orange colors of the carotenoids in the leaves of trees become visible.

was illuminated by the colors most readily absorbed by chlorophyll—if chlorophyll were indeed responsible for photosynthesis.

Yet how could photosynthesis be measured in those technologically unsophisticated days? Engelmann knew that photosynthesis produced oxygen, and that certain motile bacteria are attracted to areas of high oxygen concentration. He determined the action spectrum by observing which strands of *Spirogyra* bacteria swam toward: the strands located in the red and blue regions of the spectrum. The fact that the bacteria did not move toward such areas when *Spirogyra* was not present showed that bacteria are not merely attracted to any region where red or blue light is being absorbed. Because the action spectrum of photosynthesis as observed by Engelmann closely matched the absorption spectrum of chlorophyll, Engelmann was also able to conclude that it is indeed the chlorophyll in the chloroplasts (and not another compound in another organelle) that is responsible for photosynthesis (Figure 8–11). Numerous more sophisticated (but no more ingenious) studies have since amply confirmed Engelmann's conclusions.

However, the action spectrum of photosynthesis can be somewhat different from the action spectrum of pure chlorophyll, particularly in such strongly colored vegetation as red marine algae. The explanation is twofold: First, the chloroplasts contain accessory photosynthetic pigments such as carotenoids and phycobilins (accessory pigments found in red algae and cyanobacteria) in such large amounts that they mask the color of the chlorophyll and therefore absorb the green light that chlorophyll itself would reflect. Second, and more to the point, the accessory pigments transfer the energy of excitation that this green light would produce to chlorophyll molecules. The presence of such accessory pigments lets the algae use light in the green area of the spectrum more efficiently than could, for example, a chrysanthemum plant. This is an important adaptation, for it permits algae to live in deep aquatic habitats, where the red light most effective in photosynthesis has been filtered out by passage through the water.

Terrestrial plants also contain some accessory photosynthetic pigments, obvious in something like a Japanese maple or a copper beech, but visibly present in most trees only when their leaves change color in the fall. Toward the end of the growing season, the chlorophyll is metabolized and its magnesium stored in the permanent tissues of the tree, leaving only the accessory pigments in the leaves. The presence of accessory photosynthetic pigments can be demonstrated by chemical analysis of almost any leaf. It is good for us that leafy green vegetables contain carotenes; β carotene can be split in the middle to give vitamin A.

The Photosystems

The light-dependent reactions of photosynthesis begin when chlorophyll and other pigments absorb light. According to the currently accepted model, chlorophyll molecules, accessory pigments, and associated electron acceptors are organized into units called **photosystems** (Figure 8–13). There are two photosystems, each containing 200 to 300 pigment molecules.

Photosystem I contains a special form of chlorophyll a, known as **P700** because one of the peaks of its light absorption spectrum is at 700 nanometers. Photosystem II utilizes another form of chlorophyll a, **P680,** the absorption maximum of which is at a wavelength of 680 nanometers.

All pigment molecules of a photosystem apparently serve as antennae to gather solar energy. When they absorb light energy, it is passed from one pigment molecule to another until it reaches the P700 or P680 pigment molecule. These chlorophyll a molecules are located in the **reaction center.**

Only P700 or P680 is able to give up its energized electron to a primary electron acceptor. Chlorophyll is then photooxidized by the absorption of light energy, and the electron acceptor is reduced.

Up to a limiting value, the efficiency of photosynthesis increases as the intensity of light increases (Figure 8–14). At that saturation intensity, photosynthesis is occurring at a maximum rate. If light intensity is increased beyond this point, the rate of photosynthesis cannot be increased.

Cyclic Photophosphorylation

During light-dependent reactions, there are two pathways for electron flow: cyclic photophosphorylation and noncyclic photophosphorylation. The term **photophosphorylation** is used because electrons are energized by photons and then contribute their energy to the phosphorylation of ADP, thereby synthesizing ATP.

Cyclic photophosphorylation is the simplest route. Only Photosystem I is involved, and the pathway is cyclic because excited electrons that originate from P700 at the reaction center eventually return. Energized electrons are transferred along an electron transport chain within the thylakoid membrane. A series of redox reactions takes place, as in the mitochondrial electron transport system. As they are passed from one acceptor to another, the electrons lose energy, some of which is used to pump hydrogen ions across the thylakoid membrane. An ATP synthetase enzyme in the thylakoid membrane uses the energy of the proton gradient to manufacture ATP. For every two electrons that enter this pathway, one ATP molecule is synthesized by this chemiosmotic process. NADPH is not produced and oxygen is not generated. By itself, cyclic photophosphorylation could not serve as the basis of photosynthesis because, as we will see shortly, NADPH is necessary for CO_2 to be reduced to carbohydrate.

When Photosystem I absorbs two photons of light, P700 donates two electrons to an electron transport chain (Figure 8–15). Electrons are first transferred to a primary acceptor, the identity of which is not yet known. This primary acceptor transfers the electrons to a compound known as *ferredoxin.* Then electrons are passed on to a mobile electron acceptor called *plastoquinone,* which is similar to ubiquinone in the mitochondrial electron transport chain. From plastoquinone, the electrons are transferred to a cytochrome complex consisting of two cytochromes. Next in the sequence is a copper-containing compound, plastocyanin. The cycle is completed when plastocyanin transfers the now-spent electrons back to P700. The electrons in the P700 are now back in the ground state.

Cyclic photophosphorylation occurs in plant cells when there is too little NADP to accept electrons from ferredoxin. Biologists think that this process was used by ancient bacteria to produce ATP from light energy. A reaction pathway analogous to cyclic photophosphorylation in plants is present in modern photosynthetic bacteria.

Noncyclic Photophosphorylation

In **noncyclic photophosphorylation,** the more common pathway, both photosystems are used, and there is a one-way flow of electrons from water to $NADP^+$. For every two electrons that enter this pathway, there is an energy yield of two ATP molecules and one NADPH molecule.

This pathway also begins when P700 is energized and two electrons are transferred to a primary acceptor and then to ferredoxin. Then, ferredoxin transfers the electrons to the electron acceptor $NADP^+$ (Figure 8–16). When $NADP^+$ is in an oxidized acceptance condition, it is positively charged. When it accepts electrons, the electrons unite with protons to form hydrogen, so the reduced form of $NADP^+$ is NADPH. The reduction

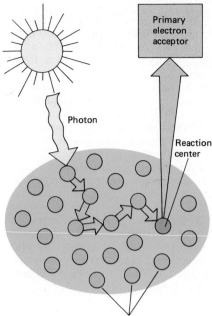

FIGURE 8–13 How a photosystem traps light energy. The many chlorophyll molecules in the unit are excited by photons and transfer their excitation energy to the specially positioned chlorophyll molecule shown in color at the reaction center.

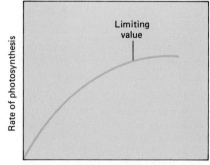

FIGURE 8–14 The rate of photosynthesis depends up to a certain level on light intensity. Only a small portion of the chlorophyll molecules in a thylakoid are excited at any one time; beyond saturation intensity, further increases in light intensity do not increase the rate of photosynthesis.

FIGURE 8–15 Cyclic photophosphorylation. When pigment molecules in Photosystem I absorb light, energy is transferred to P700, a special form of chlorophyll a. P700 gives up its energized electrons to a primary electron acceptor. Electrons are transferred in sequence to ferredoxin, plastoquionine, a cytochrome complex, plastocyanin, and finally back to chlorophyll a in the reaction center. As they are passed along this electron transport chain, electrons lose potential energy. The energy released is used to pump H^+ across the thylakoid membrane. In this way a proton gradient is established that provides energy for ATP synthesis. Note that only Photosystem I is involved. Photolysis does not occur, so neither oxygen nor NADPH is produced.

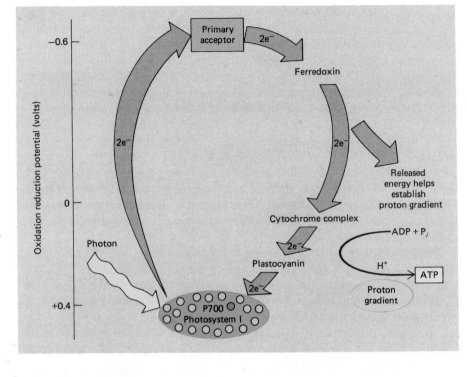

FIGURE 8–16 Noncyclic photophosphorylation. When Photosystem II is activated by absorbing photons, electrons are passed along an electron acceptor chain and are eventually donated to Photosystem I and finally to $NADP^+$. Photosystem II is responsible for the photolytic dissociation of water and the production of atmospheric oxygen. This pathway is sometimes referred to as the *Z scheme* because of its zigzag route.

of $NADP^+$ requires two electrons. Thus, two photons of light must be absorbed by Photosystem I to form one NADPH. (Ferredoxin, however, can only transfer one electron at a time.) When an electron is transferred to ferredoxin, Photosystem I becomes positively charged. It could never emit another electron, regardless of whether it became excited, unless one somehow were first restored to it; that needed electron is donated by Photosystem II.

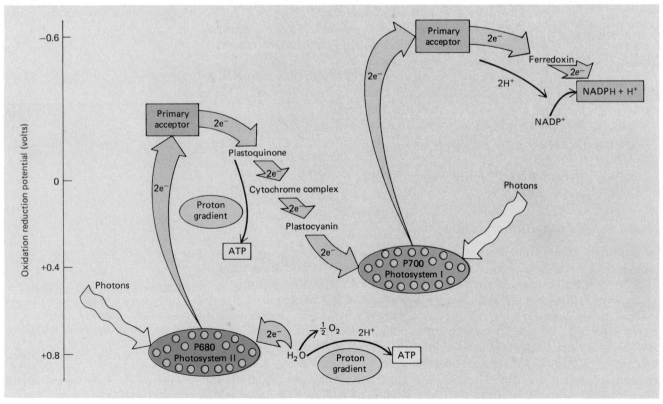

Like Photosystem I, Photosystem II is activated by a photon and gives up two electrons to a primary electron acceptor (see Figure 8–16). The electrons pass from one acceptor to another through a chain of easily oxidized and reduced acceptor molecules. As electrons are transferred along this chain of electron acceptors, they become less and less energized. Some of the energy released is used to establish a proton gradient, which leads to the synthesis of ATP. Electrons emitted from Photosystem II are eventually donated to Photosystem I.

At this point, ATP has been synthesized and electrons have been transferred to NADPH and returned to P700. However, P680, the chlorophyll a in the reaction center of Photosystem II, is now short of electrons. The missing electrons are replaced by electrons from water. When P680 absorbs light energy, it becomes positively charged and exerts a strong pull on the electrons in water molecules. Water is split by the process of photolysis into its components: electrons, protons (H^+), and oxygen. The electrons are then donated to P680; the protons are transferred to NADP, converting it to NADPH + H^+; and the oxygen is released into the atmosphere.

Chemiosmosis in Chloroplasts

Like mitochondria, chloroplasts synthesize ATP by chemiosmosis. The photosystems and electron acceptors are embedded in the thylakoid membrane. Energy released from electrons traveling through the chain of acceptors is used to pump protons from the stroma across the thylakoid membrane into the thylakoid space. These protons accumulate within the thylakoids. Because protons and hydrogen ions (H^+) are the same, this accumulation of protons causes the pH of the thylakoid interior to fall. In fact, the pH approaches 4 in bright light. This produces a pH difference of about 3.5 units across the thylakoid membrane—more than a 1000-fold difference in hydrogen ion concentration.

In accordance with the general principles of diffusion, the highly concentrated hydrogen ions inside the thylakoid tend to diffuse out. However, they are prevented from doing so because the thylakoid membrane is impermeable to them except at certain points bridged by an ATP synthetase enzyme, also known as the **CF_0-CF_1 complex**. This complex extends across the thylakoid membrane, projecting from the membrane surface both inside and outside and forming channels through which protons can leak out of the thylakoid (Figure 8–17).

The concentrated proton solution inside the thylakoid represents a low-entropy state. If the protons could move out of the thylakoid into the surrounding space, the resulting random and more or less even distribution of protons within the chloroplast would be a high-entropy state. The second law of thermodynamics identifies this situation as one in which there is a potential for useful work resulting from the change in entropy. What happens is that, as the protons pass through the CF_0-CF_1 complex, energy is released, and an enzyme catalyzes ATP synthesis. Just how this is done is the subject of active investigation.

HOW THE LIGHT-INDEPENDENT REACTIONS FIX CARBON

The light-independent reactions use the ATP and NADPH manufactured by the light-dependent reactions to reduce carbon dioxide to carbohydrate (Table 8–1). The light-independent reactions may be summarized as follows:

$$12 \text{ NADPH} + 12 \text{ H}^+ + 18 \text{ ATP} + 6 \text{ CO}_2 \longrightarrow$$
$$\text{C}_6\text{H}_{12}\text{O}_6 + 12 \text{ NADP}^+ + 18 \text{ ADP} + 18 \text{ P}_i + 6 \text{ H}_2\text{O}$$

Table 8–1 SUMMARY OF PRINCIPAL REACTIONS OF PHOTOSYNTHESIS

Reaction Series	Process	What Is Needed	End Products
Light-dependent reactions (take place in thylakoid membranes)	Energy from sunlight used to split water, manufacture ATP, and reduce NADP		
(1) Photochemical reactions	Chlorophyll energized; chlorophyll a in reaction center gives up energized electron to electron acceptor	Light energy; pigments, such as chlorophyll	Electrons
(2) Electron transport	Electrons are transported along chain of electron acceptors in thylakoid membranes; electrons reduce NADP; splitting of water provides some of H^+ that accumulates inside thylakoids	Electrons, NADP, H_2O, electron acceptors	$NADPH + H^+$, O_2, H^+
(3) Chemiosmosis	H^+ molecules are permitted to move across thylakoid membrane down proton gradient; they cross membrane through special channels in ATP synthetase (CF_0-CF_1); energy released is used to produce ATP	Proton gradient, $ADP + P_i$	ATP
Light-independent reactions (take place in stroma)	Carbon fixation; carbon dioxide is combined with organic compound	Ribulose bisphosphate, CO_2, ATP, $NADPH + H^+$	Carbohydrates, $ADP + P_i$, $NADP^+$

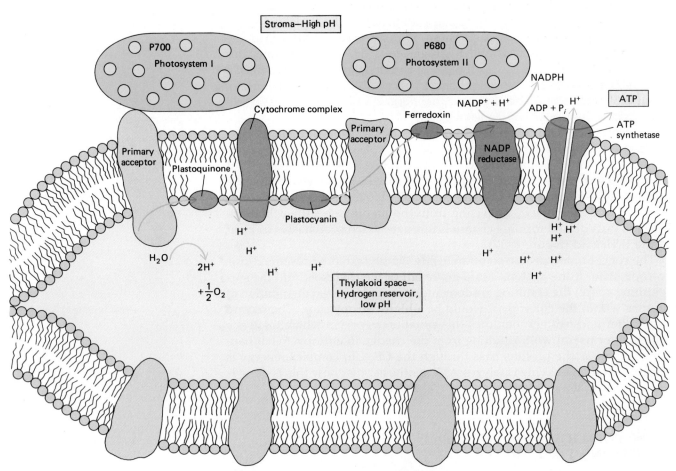

FIGURE 8–17 Electron acceptors are embedded in the thylakoid membrane, and chemiosmosis takes place across the membrane. Electron acceptors are thought to be organized in the membrane so that electrons are passed back and forth from one side of the membrane to the other. As electrons are passed along, hydrogen ions from the stroma are pumped across the thylakoid membrane into the thylakoid space. The H^+ diffuses back to the stroma along its concentration gradient through a channel within the ATP synthetase enzyme. This provides energy for ATP synthesis.

The Calvin Cycle

The light-independent reactions form a cycle known as the **Calvin cycle.** With each complete Calvin cycle, a portion of a carbohydrate molecule is produced. Two turns of the cycle are required to make just one glucose molecule.

As you read the following description of the Calvin cycle, follow the reactions illustrated in Figure 8–18. We will consider that the cycle begins with a pentose sugar phosphate called **ribulose phosphate.** In the first chemical transformation, ATP from the light-dependent reactions is used to add a second phosphate to the five-carbon skeletons of three ribulose phosphates. This chemical reaction converts them to **ribulose bisphosphate (RuBP).** A key enzyme then combines three molecules of carbon dioxide with three molecules of RuBP. This process is called **CO₂ fixation.**

Instantly, each molecule splits into two three-carbon molecules called *phosphoglycerate (PGA).* There are now six three-carbon molecules. With the energy from more ATP from the light-dependent phase, the PGA mole-

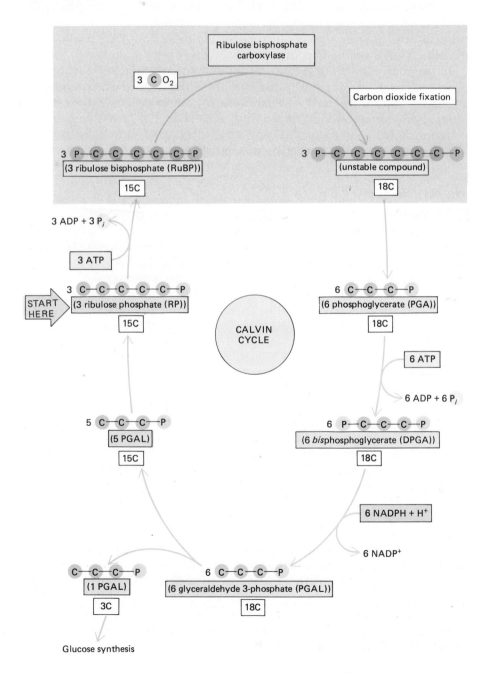

FIGURE 8–18 The Calvin cycle, or light-independent reactions. Carbon dioxide is fixed—that is, brought into chemical combination—by these reactions. For every three molecules of carbon dioxide that enter the cycle, one molecule of the three-carbon sugar PGAL is produced. To fix three molecules of carbon dioxide, nine molecules of ATP and six molecules of NADPH are required. Only the principal reactions are shown here.

cules are converted to molecules of diphosphoglycerate (DPGA). DPGA is then converted to **glyceraldehyde-3-phosphate,** known simply as **PGAL.** In this conversion hydrogen is donated by reduced NADP formed during the light-dependent reactions.

Now the plant cell is in a position to harvest the three CO_2 molecules it has put into the process, not in their original form but as carbohydrate. One of the six PGAL molecules leaves the system, to be used in carbohydrate synthesis and perhaps ultimately in the synthesis of other organic molecules, such as proteins or fats. Each of these three-carbon molecules of PGAL is essentially half a hexose (six-carbon sugar) molecule. They are joined in pairs to produce, usually, glucose or fructose. In some plants glucose and fructose are joined to produce sucrose, table sugar. This we harvest from sugar cane, sugar beets, or maple sap. The plant cell might use glucose to produce cellulose or package it as starch.

Notice that, although one of the six PGAL molecules was removed from the cycle, five of them remained. This represents 15 carbon atoms in all. Through an ingenious series of reactions, these 15 carbons and their associated atoms are rearranged into three molecules of ribulose phosphate, the very five-carbon compound with which we started. This same ribulose phosphate is now in a position to begin the process of CO_2 fixation and eventual PGAL production once again.

In summary, the inputs required for the light-independent reactions are three molecules of CO_2, hydrogen from the photolysis of water, and ATP. In the end, the three carbons from the CO_2 can be accounted for by the harvest of a three-carbon (half-hexose) molecule. The remaining organic compounds are used to synthesize the pre-existing substances with which yet another three CO_2 molecules may combine once again.

The C_4 Pathway

Most plants are referred to as *C_3 plants* because the first product of CO_2 fixation is a three-carbon compound, PGA. Some plants, referred to as *C_4 plants,* are able to fix carbon dioxide in the four-carbon compound **oxaloacetate** (Figure 8–19). These plants eventually use the Calvin cycle to produce carbohydrate, but first they incorporate CO_2 into oxaloacetate.

In addition to having mesophyll cells, C_4 plant cells have photosynthetic **bundle sheath cells** (Figure 8–19). These cells are tightly packed and form sheaths around the veins of the leaf. The mesophyll cells are located between the bundle sheath cells and the surface of the leaf. The Calvin cycle takes place within the bundle sheath cells, whereas reactions of the C_4 pathway occur in the mesophyll cells.

The C_4 pathway is present in addition to the Calvin cycle in a number of orders of plants and apparently has evolved independently several times. Plants with the C_4 pathway evidently appeared first in geographical areas with high temperatures, high light intensities, and limited amounts of water. These plants have a higher temperature optimum, have a higher light optimum, lose less water by transpiration (evaporation), and have higher rates of photosynthesis and growth than plants that use only the Calvin cycle. In fact, a major difference between C_3 and C_4 plants is that C_4 plants are not saturated with light even at the highest light intensities encountered in nature.

The essential feature of the C_4 pathway is that it concentrates carbon dioxide in the cell. Carbon dioxide enters the leaf through tiny pores, called **stomata,** that open and close in response to such factors as water content and light intensity. When the stomata are closed, the supply of carbon dioxide is greatly diminished, and in C_3 plants photosynthesis is significantly slowed. The C_4 reactions enhance carbon fixation so that photosynthesis, and ultimately growth, continue efficiently even under certain ad-

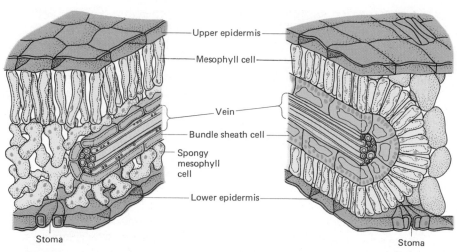

(a) Arrangement of cells in a C_3 leaf

(b) Arrangement of cells in a C_4 leaf

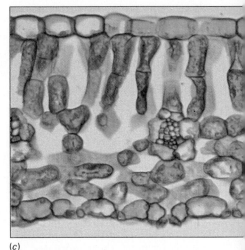

(c)

(d)

FIGURE 8–19 (a) and (b) Comparison of the leaf structures of a C_3 and a C_4 plant. In C_3 plants the Calvin cycle takes place within the mesophyll cells of the leaf. In C_4 plants, reactions that fix CO_2 into four-carbon compounds take place in the mesophyll cells. Then the Calvin cycle takes place in the bundle sheath cells, which surround the veins of the leaf. (c) Cross section of a leaf from a C_3 plant. (d) Cross section of a leaf from a C_4 plant.

verse conditions. Among the many quick-growing and aggressive plants that use the C_4 pathway are sugar cane, corn, and crabgrass. The yields of C_4 crop plants are two to three times greater than those of C_3 plants. If this pathway could be incorporated into more of our crop plants by genetic manipulation, we might well be able to greatly increase food production in some parts of the world.

In the C_4 pathway, CO_2 is added to the three-carbon compound phosphoenol pyruvate (PEP), forming oxaloacetate (which has four carbons). The reaction is catalyzed by PEP carboxylase, an enzyme that has a high affinity for carbon dioxide and binds it effectively even when carbon dioxide is at low concentration.

Oxaloacetate may be converted to other four-carbon compounds, usually malate or aspartate. The malate then passes to the bundle sheath cells, where a different enzyme catalyzes the decarboxylation of malate to yield pyruvate (which has three carbons).

$$\text{Malate} + NADP^+ \rightarrow \text{pyruvate} + CO_2 + NADPH + H^+$$

The CO_2 released condenses with rubulose 1,5-bisphosphate and enters the Calvin cycle in the usual manner (Figure 8–20). The pyruvate formed in the decarboxylation reaction returns to the mesophyll cell, where phosphoenol pyruvate is regenerated by the reaction of pyruvate with ATP. The phosphoenol pyruvate has a much higher energy of hydrolysis, (-14.8 kcal/mole) than the P of ATP (-7.3 kcal/mole). Thus, to drive the phosphorylation of pyruvate by ATP requires the energy of two ~ P, with the release of AMP. The overall reaction of the C_4 pathway is the following:

$$\begin{array}{cccc} CO_2 & + ATP \longrightarrow & AMP & + 2 P_i ++ CO_2 \\ \text{Within} & & \text{Within bundle} & \\ \text{mesophyll cells} & & \text{sheath cells} & \end{array}$$

This equation shows that two energy-rich phosphates are used up in transporting carbon dioxide to the chloroplasts of the bundle sheath cells. The role of the C_4 cycle is to increase the concentration of carbon dioxide within the bundle sheath cells so as to drive the C_3 cycle there. The operation of the C_4 cycle serves to increase the concentration of CO_2 within the bundle sheath cells some 10- to 60-fold over that in the cells of plants having only the C_3 pathway. The net reaction for the combination of the C_3 and C_4 pathways is the following:

$$6 CO_2 + 30 ATP + 12 NADPH + 12 H^+ \longrightarrow$$

$$C_6H_{12}O_6 + 30 ADP + 30 P_i + 12 NADP^+ + 6 H_2O$$

FIGURE 8–20 The C₄ pathway. (*a*) Overview of the C₄ series of reactions in photosynthesis. If, as usual, CO₂ is present in low concentration, the C₄ system readily absorbs it and in effect concentrates it for use by the C₃ system to which it is pumped. Since the C₄ system consumes some energy, ultimately made available only by photosynthesis, this system is important to the plant only at high light intensities, when the stomata can be kept open. Thus, to be effective, the C₄ system requires an abundance of both water and light. Under these conditions, it can fix more carbon than the C₃ system can fix by itself. (*b*) A more detailed diagram of the C₄ pathway showing how some of the reactions in the C₄ cycle add carbon dioxide to phosphoenol pyruvate to form oxaloacetate.

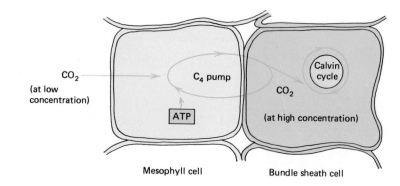

(*a*) Overview

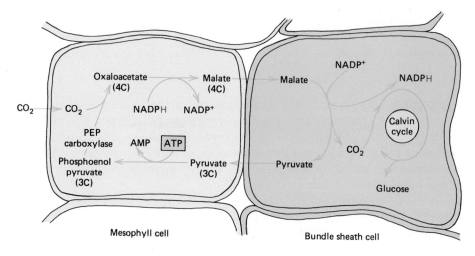

(*b*) Detailed view

The combined pathway involves the expenditure of 30 ATPs per hexose, rather than the 18 ATPs used in the absence of the C₄ pathway. The expenditure of the extra ATPs ensures a high concentration of CO₂ in the bundle sheath cells and permits them to carry on photosynthesis at a rapid rate. When light is abundant, the rate of photosynthesis is limited by the concentration of carbon dioxide; this is never very high.

At lower light intensities and temperatures, C₃ plants have the advantage. For example, winter rye, a C₃ plant, grows lavishly when crabgrass cannot. The reason is that plants with C₄ metabolism require five ATPs to fix one mole of CO₂, whereas plants with the C₃ cycle need only three ATPs.

Photorespiration

Like animal cells, plant cells carry on cellular respiration in their mitochondria, using substrates such as glucose and producing carbon dioxide. The ATP produced is used to drive the metabolic processes of the plant cells. In addition, under certain environmental conditions, many plants use oxygen and produce carbon dioxide by another process, **photorespiration.**

On bright days when the weather is hot and dry, plant cells close their stomata (the tiny pores on the leaf surface) to prevent the loss of water through the leaf. This response prevents dehydration, but it also prevents carbon dioxide from entering the leaf. As photosynthesis proceeds under these conditions, CO₂ is depleted and oxygen concentration increases. When the O₂ concentration is higher than the CO₂ concentration within the leaf, O₂ rather than CO₂ combines with the active site on ribulose bisphosphate carboxylase.

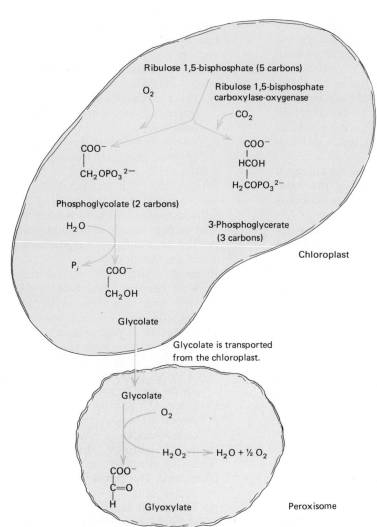

FIGURE 8–21 Formation of glyoxylate in photorespiration. Glycolate is produced within the chloroplast; oxygen is required for this process. Glycolate is then transported to a peroxisome, where it is converted to glyoxylate.

Ribulose bisphosphate carboxylase is an oxygenase as well as a carboxylase, and oxygen and carbon dioxide compete for binding to the same active site. Recall that when it functions as a carboxylase, this enzyme catalyzes the combination of CO_2 with ribulose 1,5-bisphosphate. The oxygenase function of this enzyme promotes the reaction of ribulose 1,5-bisphosphate with molecular oxygen, forming the three-carbon 3-phosphoglycerate and the two-carbon 2-phosphoglycolate (Figure 8–21). The phosphoglycolate is hydrolyzed to glycolate and its inorganic phosphate, P_i. Then the glycolate leaves the chloroplast and is metabolized further in a glyoxysome (see Chapter 4). Glycolate is converted to glyoxylate with the production of hydrogen peroxide, H_2O_2. The glyoxylate is metabolized further to carbon dioxide in the mitochondria.

Photorespiration consumes up to 50% of the carbon fixed in the Calvin cycle and so appears to be a wasteful process, with no known benefit to the plant. If photorespiration could be reduced, food supplies could be increased. Note that C_4 plants fix CO_2 under conditions that result in photorespiration in C_3 plants.

■ SUMMARY

I. Most producers are photosynthetic autotrophs, which use light as an energy source for manufacturing organic compounds from carbon dioxide and water.

II. Chloroplasts are organelles bounded by a double membrane; the inner membrane encloses the stroma.

A. In plants, chloroplasts are located mainly within mesophyll cells inside the leaf.

B. Chlorophyll and other photosynthetic pigments are found within the thylakoid membranes of chloroplasts.

C. The thylakoids are arranged in stacks called *grana*.

III. During photosynthesis, light energy is captured by chlorophyll and used to chemically combine the hydrogen from water with carbon dioxide to produce carbohydrates; oxygen is released as a byproduct.

   A. During the light-dependent reactions of photosynthesis, chlorophyll absorbs light and becomes energized. Some of the energy of the energized chlorophyll is used to make ATP; some is used to split water. Hydrogen from the water is transferred to NADP.

   B. The light-independent reactions use the energy of ATP and NADPH produced during the light-dependent reactions to manufacture carbohydrate.

IV. Light behaves as both a wave and a particle. Its particles of energy, called *photons*, can excite pigment molecules, such as chlorophyll. The resulting high-energy electrons are accepted by electron acceptor compounds.

V. Chlorophyll is a pigment that captures light energy. Its absorption spectrum is very similar to the action spectrum of photosynthesis.

VI. Chlorophyll molecules and associated electron acceptors are organized into photosystems. Only a special chlorophyll a in the reaction center actually gives up its energized electron to an electron acceptor.

VII. In cyclic photophosphorylation, electrons from Photosystem I are eventually returned to Photosystem I.

   A. A series of redox reactions takes place as energized electrons are passed along a chain of electron acceptors. Some of the energy lost is used to pump hydrogen ions across the thylakoid membrane.

   B. For every two electrons that enter the pathway, one ATP is produced. No NADPH is produced, and no oxygen is generated.

VIII. In noncyclic photophosphorylation, the electrons emitted by Photosystem II are passed through a chain of electron acceptors to $NADP^+$.

   A. In this process electrons from P680 in Photosystem II are donated to P700 in Photosystem I.

   B. Water is split and its electrons are donated to P680. Oxygen is released in the process.

IX. As electrons pass through the chain of electron acceptors in photophosphorylation, protons follow them and are pumped from the stroma into the thylakoid space. A proton gradient is set up that provides the energy for ATP synthesis.

   A. The protons can leak out of the thylakoids only through channels in the ATP synthetase (CF_0-CF_1 complex).

   B. As the protons leak through a channel within the enzyme, energy is released and used to synthesize ATP.

X. During the light-independent reactions, energy stored within ATP and reduced NADP during the light-dependent reactions is used to chemically combine carbon dioxide with hydrogen.

   A. The light-independent reactions proceed by way of the Calvin cycle.

   B. In the Calvin cycle, carbon dioxide is combined with ribulose phosphate, a five-carbon sugar. With each turn of the cycle, three carbon atoms enter the cycle. At each turn of the cycle, ribulose phosphate is regenerated.

   C. One turn of the cycle results in the synthesis of a three-carbon compound, PGAL. Two molecules of PGAL (produced with two turns of the cycle) can be used to produce one molecule of glucose.

   D. Three ATP and two NADPH molecules are consumed in the conversion of one CO_2 molecule into carbohydrate.

XI. In the C_4 pathway, the enzyme PEP carboxylase binds CO_2 effectively, even when CO_2 is at a low concentration. The initial reactions take place within mesophyll cells.

   A. The carbon dioxide is fixed in oxaloacetate, which is then converted to malate. The malate moves into a bundle sheath cell and is oxidized; the CO_2 released in the process enters the Calvin cycle.

   B. ATP is needed to phosphorylate pyruvate.

XII. In photorespiration, plant cells consume oxygen and generate carbon dioxide. This process, which decreases photosynthesis, occurs on bright, hot, dry days when plant cells close their stomata; this prevents the passage of CO_2 into the leaf.

SUMMARY EQUATIONS FOR PHOTOSYNTHESIS

The light-dependent reactions:

$$12 \text{ H}_2\text{O} + 12 \text{ NADP}^+ + 18 \text{ ADP} + 18 \text{ P}_i \longrightarrow 6 \text{ O}_2 + 12 \text{ NADPH} + 12 \text{ H}^+ + 18 \text{ ATP}$$

The light-independent reactions:

$$12 \text{ NADPH} + 12 \text{ H}^+ + 18 \text{ ATP} + 6 \text{ CO}_2 \longrightarrow \text{C}_6\text{H}_{12}\text{O}_6 + 12 \text{ NADP}^+ + 18 \text{ ADP} + 18 \text{ P}_i + 6 \text{ H}_2\text{O}$$

By canceling out the common items on opposite sides of the arrows in these two coupled equations, we obtain the simplified overall equation for photosynthesis:

$$6 \text{ CO}_2 + 12 \text{ H}_2\text{O} \longrightarrow \text{C}_6\text{H}_{12}\text{O}_6 + 6 \text{ O}_2 + 6 \text{ H}_2\text{O}$$

■ **POST-TEST**

1. Most producers are photosynthetic _____.

2. Chlorophyll is associated with the _____ membranes within organelles known as _____.

3. In photosynthesis _____ _____ and _____ are the inorganic raw materials used to produce _____; _____ is released during the process.

4. In photolysis some of the energy captured by chlorophyll is used to split _____.

5. Light is composed of particles of energy called _____.

6. The relative effectiveness of different wavelengths of light in photosynthesis is given by the _____.

7. When an electron absorbs a light quantum and moves to another orbital, the atom is said to be _____.

8. In complex plants the electron donor in the light-dependent reactions is _____.

9. In addition to having chlorophyll, most plants contain accessory photosynthetic pigments such as _____.

10. Only the special chlorophyll a in the _____ center of a photosystem actually gives up its _____ to an _____ acceptor.

11. In photophosphorylation electrons that have been energized by _____ contribute their energy to add phosphate to _____, producing _____.

12. In cyclic photophosphorylation the electrons that originate from Photosystem I are eventually transferred to _____.

13. In _____ photophosphorylation there is a one-way flow of electrons to NADP.

14. The oxygen released in photosynthesis is derived from _____.

15. The transfer of electrons through a sequence of electron acceptors provides energy to pump _____ across the thylakoid membrane.

16. As protons pass through the channels in ATP synthetase (CF_0-CF_1 complex), _____ is released and used to synthesize _____.

17. The inputs for the light-independent reactions are _____, _____, and _____

18. The process of _____ _____ involves the chemical combination of carbon dioxide with a compound such as ribulose bisphosphate.

19. _____ turns of the Calvin cycle are required to produce one glucose molecule.

20. When carbon dioxide is a limiting factor, plants that have the _____ pathway have an advantage.

21. In C_4 plants the Calvin cycle takes place in _____ cells; in these cells malate is decarboxylated, yielding _____ for photosynthesis.

22. In photorespiration ribulose 1,5-bisphosphate reacts with _____ instead of CO_2.

■ REVIEW QUESTIONS

1. What is the role of light in photosynthesis? Explain.
2. What properties of chlorophyll are significant in photosynthesis?
3. Photosynthetic prokaryotes do not have chloroplasts. How do they manage?
4. List the principal light-dependent reactions of photosynthesis.
5. What is meant by the *action spectrum* of photosynthesis?
6. What is the function of antenna chlorophylls? What is the reaction center of a photosystem?
7. How is oxygen produced during photosynthesis?
8. What features distinguish cyclic from noncyclic photophosphorylation?
9. What features of chloroplast structure are especially important in photosynthesis?
10. How is a proton gradient established across the thylakoid membrane? How does this result in ATP synthesis?
11. How are ATP and NADPH produced and used in the process of photosynthesis?
12. Summarize the events of the light-independent reactions.
13. What advantage do C_4 plants have over C_3 plants?
14. What strategies might be employed in the future to increase world food supply? Base your answer on your knowledge of photosynthesis and related processes.

■ RECOMMENDED READINGS

Barber, J. (ed.): *Topics in Photosynthesis*, (Vols. I–X). New York, Elsevier, 1976–1986. Carefully summarized research papers reporting the latest developments in our understanding of the photosynthetic process.

Clayton, R.K., and Sistron, W.: *The Photosynthetic Bacteria*. New York, Plenum Press, 1978. An interesting presentation of the evolution of the light reactions of plants and algae.

Danks, S.M., Evans, E.H., and Whittacker, P.A.: *Photosynthetic Systems: Structure, Function and Assembly*. New York, John Wiley & Sons, 1983. A well-written comparison of the mechanisms of photosynthesis in bacteria, algae, and plants.

Foyer, C.H.: *Photosynthesis*. New York, John Wiley & Sons, 1984. A comprehensive and advanced treatment of both light and dark reactions of photosynthesis, especially those occurring in higher plants.

Hatch, M.D., and Boardman, N.K.: *Photosynthesis*. New York, Academic Press, 1981. A detailed description of the chemistry and physics underlying the photosynthetic process.

Lehninger, A.L.: *Principles of Biochemistry*. New York, Worth Publishing, 1982. A standard biochemistry text with a detailed but very clear account of bioenergetics, photosynthesis, and cellular respiration.

Miller, K.R.: The photosynthetic membrane. *Scientific American*, October 1979, pp. 102–113. A description of the structure of the thylakoid membrane and how it converts light energy to chemical energy.

Stoeckenius, W.: The purple membrane of salt-loving bacteria. *Scientific American*, June 1976, pp. 38–50. A description of a curious photosynthetic mechanism that uses a pigment, rather like the visual purple of animals, instead of chlorophyll to capture the energy of light.

Stryer, L.: *Biochemistry*, 3rd ed. San Francisco, W.H. Freeman, 1988. A well-illustrated, readable text that covers the molecules of life and the concepts of cellular energetics from the ground up.

PART III

GENETICS

Every organism, even the simplest, has massive amounts of information stored in its genes. ■ Genes, which we now know to be DNA molecules, ultimately control all aspects of the life of the organism, including its metabolism, form, development, and reproduction. ■ Genetic information is responsible for the continuity of life, because genes form the essential link between generations. ■

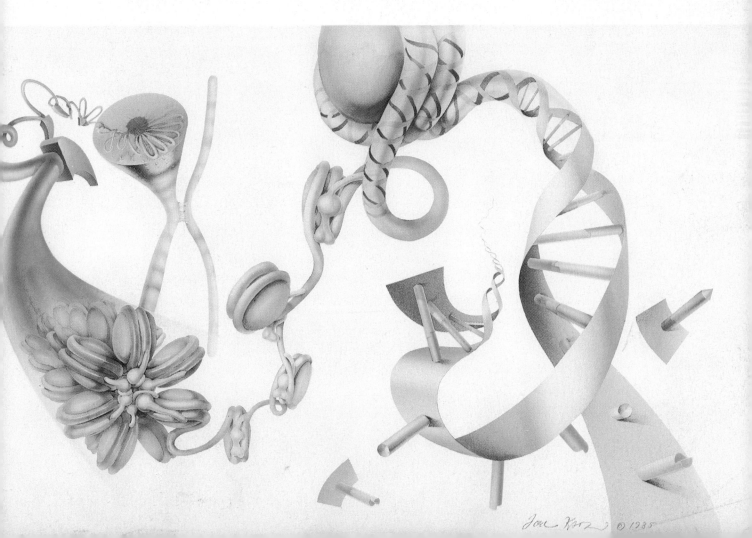

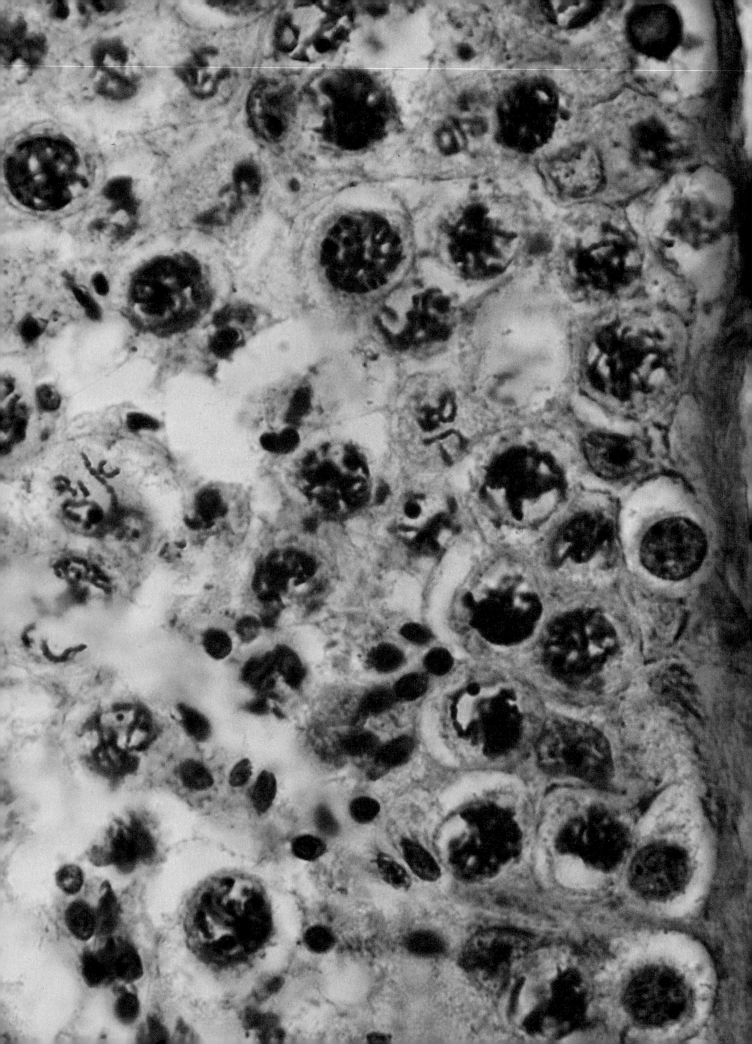

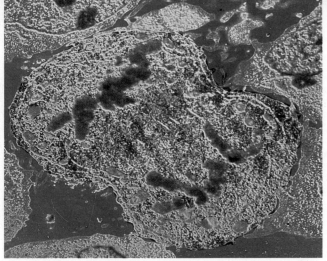

False-color electron micrograph of human lymphocyte undergoing mitotic division

9

Chromosomes, Mitosis, and Meiosis

LEARNING OBJECTIVES

After you have read this chapter you should be able to:

1. Describe the structure of a chromosome and the function of a gene.
2. Identify the stages in the cell cycle, and describe the principal events characteristic of each.
3. Explain the significance of mitosis, and describe the process.
4. Distinguish between asexual and sexual reproduction.
5. Distinguish between haploid and diploid cells, and define homologous chromosomes.
6. Contrast the events of mitosis and meiosis.
7. Compare the roles of mitosis and meiosis and of haploidy and diploidy in various generalized life cycles.

When organisms reproduce, one of the most striking things is the fact that the new generation resembles its parents. This occurs because biological information is transferred from the parents to each new individual. Human beings have been aware for centuries that "like begets like" and that one of the prime characteristics of living things is their ability to reproduce their kind. This transmission of information from parent to offspring is called **heredity**. The units of heredity are called **genes,** and the branch of biology concerned with the structure, transmission, and expression of hereditary information is called **genetics.** Since its inception at the beginning of this century, the science of genetics has advanced with great speed and currently is developing at an even more accelerated pace, largely as a result of the science of molecular genetics, established in the 1950s.

EUKARYOTIC CHROMOSOMES

The carriers of genetic information in eukaryotes are structures called **chromosomes** (Figure 9–1) contained within the cell nucleus. Although the term *chromosome* means "colored body," chromosomes are virtually color-

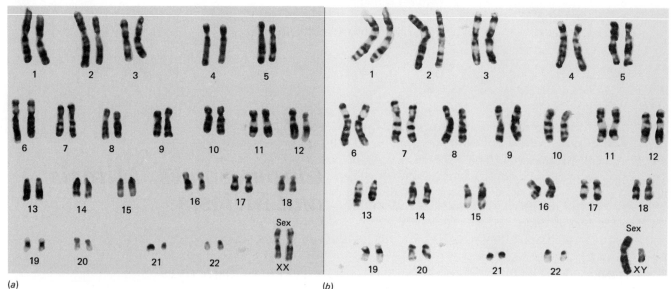

(a)   *(b)*

FIGURE 9–1 Chromosome constitutions, known as *karyotypes*, of a normal human female (*a*) and male (*b*). No significance should be attached to the apparent size differences between the female and male chromosomes; these reflect merely different degrees of chromosomal contraction in the particular cells photographed.

less; the name is a reference to their ability to be darkly stained by certain dyes. Chromosomes are made up of a complex material called **chromatin,** which consists of fibers containing about 60% protein, 35% deoxyribonucleic acid (DNA), and 5% ribonucleic acid (RNA). When the cell is not dividing, the chromatin appears as long, thin, dark-staining threads. At the time of cell division, the chromatin fibers become more and more condensed, such that the chromosomes become visible as distinct structures. The structure of chromatin is described in more detail in Chapter 12.

Genes

Each chromosome may contain hundreds or even thousands of genes. Our concept of the gene has changed considerably since the beginnings of the science of genetics, but our definitions have always centered around the gene as an informational unit involved in carrying out a function that will ultimately affect some characteristic of the organism. For example, we speak of genes controlling eye color in humans, wing length in flies, seed color in peas, and so on. Today a *gene* can be roughly defined as a part of a DNA molecule that can be copied in the form of an RNA molecule. Different kinds of RNA molecules have different specific functions, but many RNA molecules are responsible for carrying the code that specifies a particular sequence of amino acids in a polypeptide chain. This means that genes control the structure of all of the proteins of the organism, including the enzymes that catalyze every specific chemical reaction. (We discuss some of the limitations of this definition in Chapter 13.) A typical mammalian species may have about 50,000 genes, but the actual number is not known.

How Many Chromosomes?

Every individual of a given species contains a characteristic number of chromosomes in most nuclei of the body. Most cells in the body of a normal human being have exactly 46 chromosomes (see Figure 9–1). Many other species of animals and plants also have 46. It is not the *number* of chromosomes that differentiates the various species, but rather the *information* specified by the genes. A certain species of roundworm has only 2 chromosomes in each cell; some crabs have as many as 200 per cell, and some ferns have over 1200. Most species of animals and plants have chromosome numbers between 10 and 50. Numbers above and below this are not common.

THE CELL CYCLE

Once cells reach a certain size, they must stop growing or divide. Some cells, such as nerve, skeletal muscle, and red blood cells, do not normally divide once they are mature. The activities of cells that are actively growing and dividing can be described in terms of the life cycle of the cell, or the **cell cycle.** In cells that are capable of dividing, the cell cycle is the period from the beginning of one division to the beginning of the next and is customarily represented in diagrams as a circle (Figure 9–2). The length of time represented by a complete revolution of the circle is termed the **generation time, T.** The generation time can vary widely but is usually about 8 to 20 hours long in actively growing plant and animal cells.

Cell multiplication involves two main processes, mitosis and cytokinesis. **Mitosis,** a complex process involving the nucleus, ensures that each new cell contains the same number and types of chromosomes as were present in the original cell. **Cytokinesis,** which generally begins before mitosis has been completed, is the division of the cytoplasm of the cell so that two cells are formed.

Interphase

Most of the life of the cell is spent in **interphase,** the stage that occurs between successive cell divisions. This is a very active time for the cell, when it is synthesizing needed materials and growing. Most proteins and other materials are synthesized throughout interphase. The major exceptions are those molecules that must be synthesized for the cell to divide; the synthesis of these particular molecules occurs at relatively restricted times and provides us with a way of subdividing interphase. In about 1950, it was recognized that chromosomes undergo duplication during interphase and later separate and are distributed to the daughter nuclei during

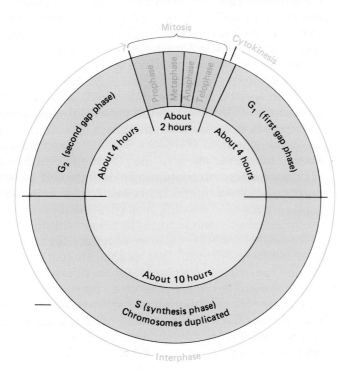

FIGURE 9–2 The cell cycle. Time intervals are relative; actual time varies with the cell type and species.

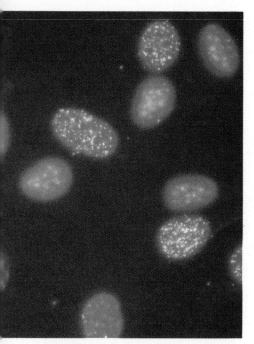

FIGURE 9–3 Photomicrograph of interphase mouse cells that have incorporated radioactive thymidine (a DNA precursor) into their DNA during the S phase. Nuclei containing newly synthesized DNA are at G_2 and can be recognized by the orange grains. Nuclei without grains are in G_1 of the cell cycle.

mitosis. The period of DNA replication during interphase serves as a major landmark, termed the *synthesis phase*, or **S phase** (Figure 9–3). Other chromosomal components, such as the chromosomal proteins, are also synthesized at this time. The duplication of the chromosomes is a complex process, to be discussed in Chapter 12.

The time between mitosis and the beginning of the S phase is termed the **G_1 phase,** or first gap phase. Nondividing cells usually remain in G_1, carrying out various metabolic activities. The G_1 phase of a cell that is actively cycling is involved with growth and, toward the end of G_1, increased activity of enzymes required for DNA synthesis. These activities make it possible for the cell to enter the S phase.

After completion of the S phase, the cell enters a second gap phase, the **G_2 phase.** At this time there is an increase in protein synthesis as the final steps in the cell's preparation for division take place. The completion of the G_2 phase is marked by the beginning of mitosis.

Mitosis

Each mitotic division is a continuous process, with each stage merging imperceptibly into the next. However, for descriptive purposes, mitosis, or the *M phase*, has been divided into four stages: prophase, metaphase, anaphase, and telophase (Figures 9–4 and 9–5). Refer to these figures as you read the description of each phase of mitosis.

Prophase

The first stage of mitosis, **prophase,** begins when the long chromatin threads begin to condense and appear as mitotic chromosomes. This condensation is accomplished mainly by a coiling process that causes the chromosomes to simultaneously become shorter and thicker. Thus, the chromosomal material is packaged such that it can ultimately be distributed to the daughter cells without tangling.

Stained with certain dyes and viewed through the light microscope, the chromosomes become visible as dark, rod-shaped bodies as prophase proceeds. At this point it becomes clear that each chromosome has been duplicated as a result of the events that occurred during the preceding S phase. Each chromosome consists of a pair of identical units, termed **sister chromatids.** Each chromatid contains a nonstaining, constricted region called the **centromere.** Sister chromatids are tightly associated in the vicinity of their centromeres (Figures 9–6 and 9–7). The chemical basis for this close association is not completely understood, although there is evidence that special kinds of DNA and special proteins that bind to DNA are involved.

If the cell is an animal cell, each of the two centrioles (see Chapter 4) will have duplicated in the previous S phase. Microtubules, composed of the protein tubulin, radiate from them, and the centriole pairs then separate and migrate toward the opposite poles of the cell. In both plant and animal cells fibers then form and begin to organize into the **mitotic spindle,** a complex structure consisting mainly of microtubules. At one time it was thought that the centrioles might be responsible for organizing the spindle; now centrioles are believed to be necessary for the formation of the basal bodies of cilia and flagella. The mechanism just described apparently provides for the orderly distribution of these organelles, thus ensuring that each of the daughter cells will receive a pair of centrioles. Centrioles are not found in the cells of higher plants, which lack flagellated sperm and do not have any other flagellated or ciliated cells. In animal cells, additional microtubules form clusters extending outward in all directions from the centrioles; these structures are called **asters** (Figure 9–8b).

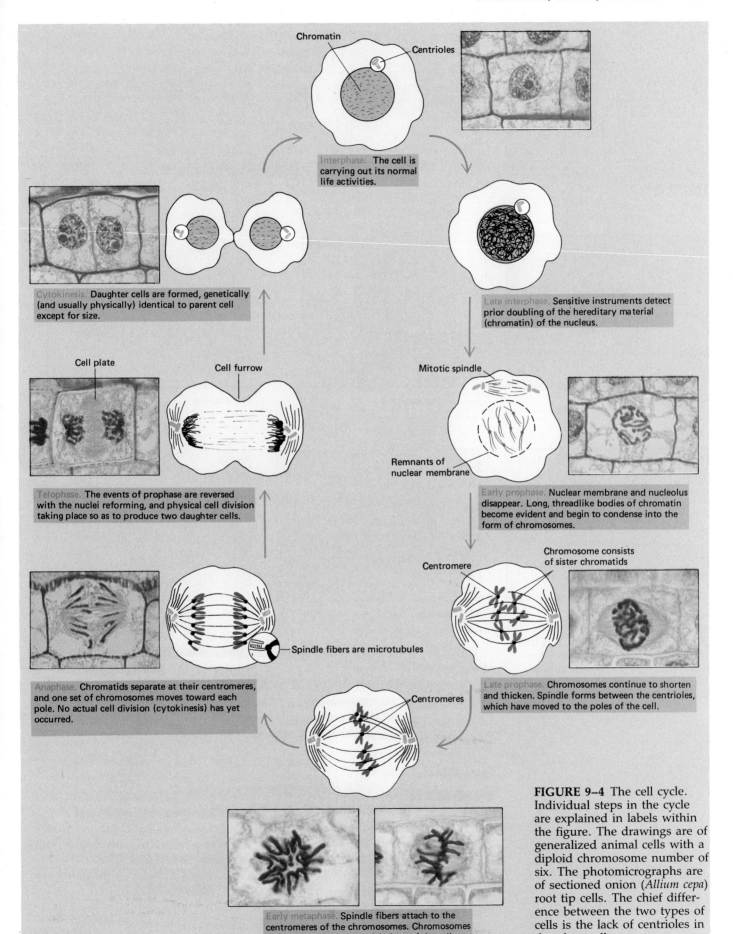

Chromatin

Centrioles

Interphase. The cell is carrying out its normal life activities.

Cytokinesis. Daughter cells are formed, genetically (and usually physically) identical to parent cell except for size.

Late interphase. Sensitive instruments detect prior doubling of the hereditary material (chromatin) of the nucleus.

Cell plate

Cell furrow

Mitotic spindle

Remnants of nuclear membrane

Telophase. The events of prophase are reversed with the nuclei reforming, and physical cell division taking place so as to produce two daughter cells.

Early prophase. Nuclear membrane and nucleolus disappear. Long, threadlike bodies of chromatin become evident and begin to condense into the form of chromosomes.

Centromere

Chromosome consists of sister chromatids

Spindle fibers are microtubules

Anaphase. Chromatids separate at their centromeres, and one set of chromosomes moves toward each pole. No actual cell division (cytokinesis) has yet occurred.

Centromeres

Late prophase. Chromosomes continue to shorten and thicken. Spindle forms between the centrioles, which have moved to the poles of the cell.

Early metaphase. Spindle fibers attach to the centromeres of the chromosomes. Chromosomes line up along the equatorial plane of the cell.

FIGURE 9–4 The cell cycle. Individual steps in the cycle are explained in labels within the figure. The drawings are of generalized animal cells with a diploid chromosome number of six. The photomicrographs are of sectioned onion (*Allium cepa*) root tip cells. The chief difference between the two types of cells is the lack of centrioles in the plant cells.

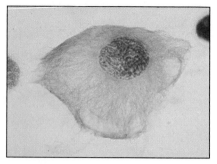

(a) Interphase

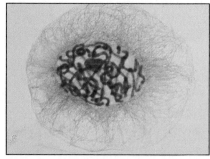

(b) Early prophase

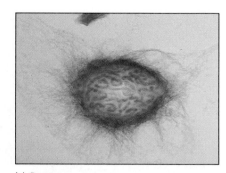

(c) Prophase

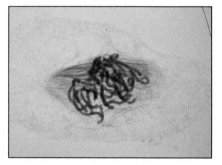

(d) Late prophase

(e) Metaphase

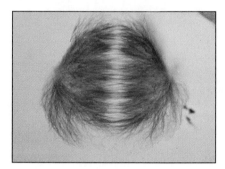

(f) Anaphase

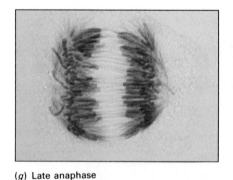

(g) Late anaphase

(h) Telophase

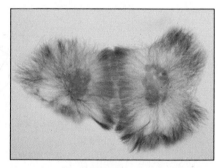

(i) Late telophase

FIGURE 9–5 Interphase and the stages of mitosis in plant cells (the blood lily, *Haemanthus*). The chromosomes have been stained and the cells flattened on microscope slides.

During prophase, the nucleolus diminishes in size and usually eventually disappears. Toward the end of prophase, the nuclear membrane breaks down, and each chromatid becomes attached to some of the spindle microtubules at its centromere (see Figure 9–7). The duplicated chromosomes then move back and forth and finally become aligned along the equator of the cell, midway between the two poles.

Metaphase

The period during which the chromosomes are lined up along the equatorial plane of the cell (termed the *metaphase plate*) constitutes **metaphase.** The mitotic spindle is complete; it is composed of numerous microtubules that extend from each pole to the equatorial region and from the centromeres to the poles (Figure 9–9). At mitotic metaphase the individual sister centromeres are attached by spindle microtubules to *opposite* poles of the cell. In animal cells the spindle microtubules end near the centrioles, in a region occupied by the **pericentriolar material** (electron-dense matter around the centrioles), but do not actually touch the centrioles themselves. In the cells of higher plants, they end near an area termed the **microtubule-organizing**

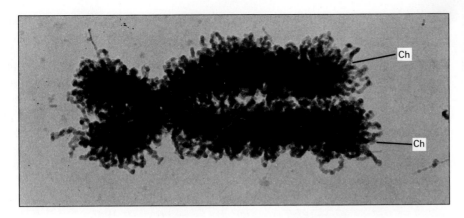

FIGURE 9–6 A human chromosome isolated and photographed at the metaphase stage of mitosis (approximately ×50,000). The chromatin fibers of the sister chromatids (*Ch*) have become slightly uncoiled.

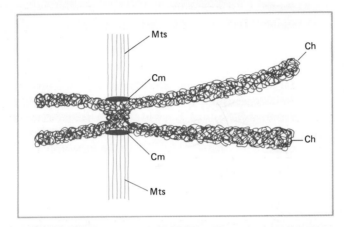

FIGURE 9–7 Structure of a mitotic metaphase chromosome. The paired structures indicated at *Cm* are the sister centromeres, which serve as microtubule (*mts*) attachment sites. Note that the sister chromatids (*Ch*) are attached in the vicinity of their centromeres.

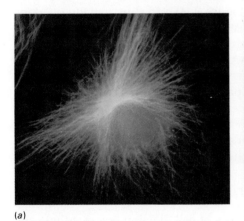 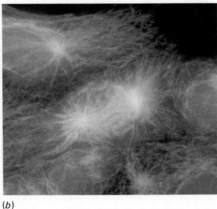

(a)   (b)

FIGURE 9–8 The transition from interphase to prophase in human cells grown in culture. The photographs in this figure and in Figure 9–9 are of cells stained with fluorescent dyes. Chromosomes are stained orange, and the microtubules are stained yellow-green. (*a*) Interphase. The microtubules are not yet organized into a mitotic spindle. (*b*) Prophase. Asters are moving toward opposite poles of the cell.

center (MTOC). The spindle is gel-like in consistency and more viscous than the surrounding cytoplasm.

During metaphase each chromatid is completely condensed and appears quite thick and discrete. Because chromosomes can be seen more clearly at metaphase than at any other stage, this is the stage at which they are usually photographed and studied for the detection of certain chromosome abnormalities (see Figure 9–1).

Anaphase

Anaphase begins as the forces that have been holding the sister chromatids together are released. *Each chromatid is now referred to as an independent chromosome.* The separated chromosomes slowly move to opposite poles. The centromeres of the chromosomes are still attached to spindle microtubules

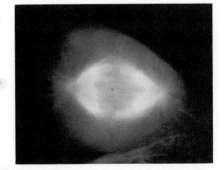

FIGURE 9–9 Metaphase. The mitotic spindle is well defined; chromosomes are lined up along the equatorial plate.

and these lead the way, with the chromosome arms trailing behind. Anaphase ends when all of the chromosomes have reached the poles.

The overall mechanism of the anaphase movement of chromosomes is still poorly understood. It has long been known that microtubules lack elastic or contractile properties. So how do the chromosomes move apart? Are they pushed or pulled, or are there other forces operating? Careful analyses of electron micrographs have yielded two main findings: First, as anaphase continues, the microtubules connecting the centromeres to the poles shorten and thus may serve to pull the chromosomes apart, perhaps through the removal of tubulin subunits from the ends of the microtubules (either in the vicinity of the MTOC or pericentriolar material or in the vicinity of the centromeres). Second, during anaphase the spindle as a whole elongates and thus may push the chromosomes apart. Lengthening of the spindle may be due to the microtubules that originate at opposite poles sliding past one another in the region of overlap at the equator.

Telophase

The final stage of mitosis, **telophase,** is characterized by a return to interphase-like conditions. The chromosomes decondense by uncoiling. A new nuclear membrane forms around each set of chromosomes, made up at least in part from recycled lipids and other components of the old nuclear membrane. The spindle microtubules disappear, and the nucleoli reappear.

Cytokinesis

Cytokinesis, the division of the cytoplasm to yield two daughter cells, generally begins during telophase (see Figure 9–4) and thus usually overlaps mitosis. The division of an animal cell is accomplished by a furrow that encircles the surface of the cell in the plane of the equator. The furrow gradually deepens and separates the cytoplasm into two daughter cells, each with a complete nucleus (Figure 9–10). In plant cells, division occurs by the formation of a **cell plate,** a partition that forms in the equatorial region of the spindle and grows laterally to the cell wall. The cell plate forms from vesicles that originate in the Golgi apparatus. Each daughter cell forms a cell membrane on its side of the cell plate, and cellulose cell walls are formed on either side of the cell plate.

The Significance of Mitosis

The remarkable regularity of the process of cell division ensures that each of the daughter nuclei will receive exactly the same number and kind of chromosomes that the parent cell had. Thus, each cell of a multicellular organism has exactly the same number and kinds of chromosomes as every other cell. If a cell receives more or fewer than the proper number of chromosomes through some malfunction of the cell division process, the resulting cell may show marked abnormalities and be unable to survive. The fact that a cell contains the genetic information needed for every characteristic of the organism explains why a single cell taken from a fully differentiated adult plant can have the potential, under suitable conditions in cell culture, to develop into an entire new plant. Similarly, at least some nuclei from differentiated frog cells can support normal development if injected into egg cells from which the nuclei have been removed (see Chapter 16).

Mitosis provides for the orderly distribution of chromosomes (and of centrioles, if present), but what about the various cytoplasmic organelles? For example, cells could not survive without mitochondria, and plant cells could not carry out photosynthesis without chloroplasts. These organelles

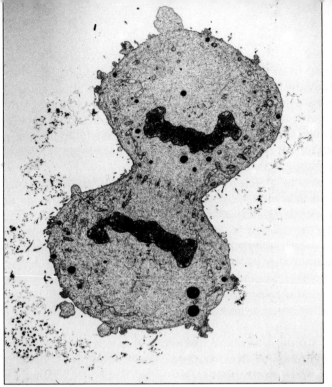

(a)

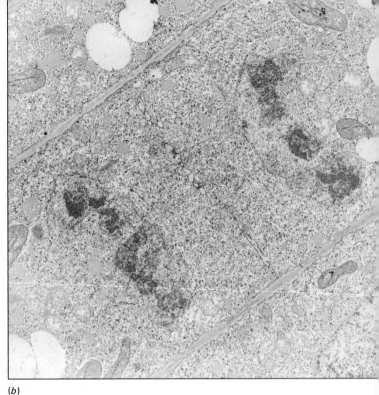

(b)

FIGURE 9–10 (*a*) Animal cell (HeLa) at midcleavage. (*b*) Telophase and cytokinesis in a root tip cell of soybean (*Glycine max*). ×19,000.

are formed by the division of previously existing mitochondria or chloroplasts. Since they are present in multiple copies in each cell, they become partitioned more or less equally between the daughter cells at cytokinesis.

Factors That Affect the Cell Cycle

The frequency of cell division varies widely in different tissues and in different species. Under optimal conditions of nutrition, temperature, and pH, the length of the cell cycle—the generation time—for any given kind of cell is constant. Under less favorable conditions, the cell cycle may be slowed; that is, the generation time is greater. However, it has not been possible experimentally to speed up the cell cycle and make cells grow faster. It appears that the length of the cell cycle for each kind of cell is the time required for the cell to carry out some precise program that has been built into it. The program has two parts, one having to do with replication of the genetic material (the chromosomes) and the other involving growth of the cell by the synthesis of various constituents.

When conditions are optimal, a prokaryotic cell can divide every 20 minutes; the generation times of eukaryotic cells are generally much longer. Cells that divide to give rise to red blood cells, cells that line the digestive tract, and cells in the reproductive layer of the skin divide frequently throughout life. In contrast, division of some of the cells in the central nervous system usually ceases in the first few months of life.

In nearly all animal cells, the production of substances that control the entrance of the cell into the S phase or the M phase depends on stimulation by growth-promoting substances present in blood. These growth factors are small proteins that appear to act specifically on some kinds of cells but not others. For example, a nerve growth factor is essential for the mitosis of certain nerve cells.

Substances that inhibit mitosis, called **chalones,** counter the action of the growth factors. The various chalones are also very specific, affecting only the type of tissue in which they are produced. For example, the chalone produced by skin cells inhibits mitosis by neighboring skin cells. Damaged skin cells are thought to synthesize less chalone, so that cells in the vicinity of a wound are released from this inhibition. They begin dividing, producing new tissue to provide for the healing of the wound. When enough healthy cells have been produced, they synthesize enough chalone to inhibit further mitotic divisions, thus turning off the wound-healing process.

Certain plant hormones are known to stimulate mitosis. Chief among these are the **cytokinins,** which act as promoters of mitosis in both normal growth and the healing of wounds.

The cell cycle can also be affected by certain drugs. **Colchicine,** a drug used to block cell division in eukaryotic cells, binds with tubulin, the major microtubule protein, and interferes with the normal function of the mitotic spindle. The spindle breaks down, preventing the chromosomes from moving to the opposite poles of the cell. As a result, the cell may end up with an extra set of chromosomes. In general, plants are relatively tolerant of extra chromosome sets; in fact, plants consisting of cells with extra sets of chromosomes tend to be larger and more vigorous than normal plants. Animals with extra chromosome sets are known, but they are relatively rare.

Antibiotics such as streptomycin and the tetracyclines prevent cell division indirectly by inhibiting protein synthesis in prokaryotic cells and thereby prolonging the G_1 phase of the cell cycle. Some drugs used in cancer therapy block one or more of the enzymes involved in DNA synthesis and cell division. Because cancer cells divide much more rapidly than most normal body cells, they are most affected by these drugs.

THE ROLE OF CELL DIVISION IN REPRODUCTION

The details of the reproductive process vary greatly among different kinds of organisms, but we can distinguish two basic types of reproduction, asexual and sexual. In **asexual** reproduction a *single parent* usually splits, buds, or fragments to give rise to two or more individuals (Figure 9–11). Because all of the cells involved are produced by mitosis, they have identical genes and therefore inherited traits that are very similar to those of the parent. Such a group of genetically identical organisms is termed a **clone.** Asexual reproduction is usually a rapid process; it permits organisms that may be well adapted to their environment to produce new generations of similarly adapted organisms.

In contrast, **sexual** reproduction generally involves *two parents*, each of which usually contributes a specialized sex cell, or **gamete.** In the case of animals and plants, the gametes are the egg cell and the sperm cell (or, in the case of higher plants, the sperm nucleus). The gametes fuse to form the fertilized egg, a single cell called a **zygote.** The egg is typically large and nonmotile, with a store of nutrients to support the growth of the embryo, which develops after the egg is fertilized. The sperm is usually small and motile and in many organisms swims actively to the egg by beating its long, whiplike tail, or **flagellum.**

Sexual reproduction has the biological advantage of making possible the recombination of the inherited traits of the two parents, so that at least a fraction of the offspring may be able to survive environmental changes or other stresses better than either parent. Of course, another fraction of the offspring may be less able to survive. Because you are now aware of the roles of chromosomes in inheritance, you have probably recognized a problem in eukaryotic sexual reproduction: How do organisms avoid producing zygotes with ever-increasing chromosome numbers? In order for us to answer this question, we need more information about the types of chromosomes found in cells.

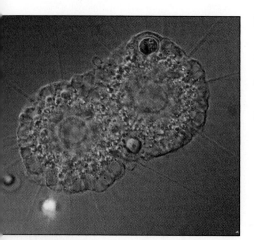

FIGURE 9–11 Amoeboid protozoan (*Actinoparys sol*) in binary fission, a form of asexual reproduction.

Chromosomes normally exist in pairs; there are typically two of each kind in the somatic (body) cells of higher plants and animals. Thus, the 46 chromosomes in human cells constitute 23 different pairs. The members of a pair, called **homologous chromosomes,** are similar in size, in shape, and in the position of their centromeres. When stained by special techniques, chromosomes often show banding patterns that are characteristic for a particular pair. In most species, chromosomes vary enough in their morphological features so that cytologists can distinguish the different homologous pairs (see Figure 9–1). The most important feature of homologous chromosomes is that they carry information for the same genetic characteristics, although not necessarily identical information. For example, members of a pair of homologous chromosomes might carry genes that specify hemoglobin structure; however, one member might have the information for normal hemoglobin while the other might specify the abnormal form of hemoglobin that is associated with sickle cell anemia (see Chapter 11).

A set containing two of each kind of chromosome is said to have the **diploid** number, or **2n** number. Gametes have only one homologue of each homologous pair; thus, they have the **haploid,** or **n,** number. In humans the diploid number is 46 and the haploid number is 23. When the sperm and egg fuse at fertilization, each gamete contributes its haploid set of chromosomes; the diploid number is thereby restored in the fertilized egg (zygote). When the zygote divides by mitosis to form the first two cells of the embryo, each cell receives a diploid set of chromosomes. In subsequent cell divisions each body cell receives a diploid set of chromosomes; thus, most body cells are diploid. However, there are exceptions. If a cell or an individual has more than the diploid chromosome number, we say that it is **polyploid.** Polyploidy is relatively rare among animals, but it is quite common among plants. In fact, about one third of all flowering plants and almost three fourths of all grasses are polyploid. As mentioned, such plants are often larger and more hardy than diploid members of the same group and may be important commercially (Figure 9–12). Modern bread wheat *(Triticum aestivum)* is a hexaploid (6n = 42) developed from three different diploid (2n = 14) species.

In order for haploid cells to be formed from diploid cells, a special kind of cell division that reduces the chromosome number is required; this process is known as **meiosis.**

FIGURE 9–12 Modern bread wheat is a hexaploid plant.

MEIOSIS

We have examined the process of mitosis, which ensures that each daughter cell receives exactly the same number and kind of chromosomes that the parent cell had. The constancy of the chromosome number in successive generations of sexually reproducing organisms is ensured by the process of **meiosis,** a special type of division that occurs during the formation of eggs and sperm in animals and of spores in plants. The term *meiosis* means "to make smaller"; the process involves two successive cell divisions during which the chromosome number is reduced to one half of its original number. Meiosis does not always immediately precede gamete formation, but it must occur at some time in the life cycle if gametes are to have haploid chromosome numbers. When two haploid gametes unite in fertilization, the fusion of their nuclei reconstitutes the normal diploid chromosome number.

How Does Meiosis Differ from Mitosis?

The events of meiosis are similar to the events of mitosis, but there are several important differences: (1) In meiosis, there are two successive nuclear and cell divisions, with the potential to yield a total of four cells. (2) Although the meiotic process consists of two cell divisions, the DNA and other chromosomal components are duplicated only once, during the interphase preceding the first meiotic division. (3) Each of the four cells produced in meiosis contains the haploid number of chromosomes—that is, only one representative of each homologous pair. (4) During meiosis, the homologous chromosomes containing genetic information from both parents are thoroughly shuffled, and one of each pair is randomly distributed to each new cell; the resulting gametes possess many different combinations of chromosomes, some of which may not have occurred in any prior generation.

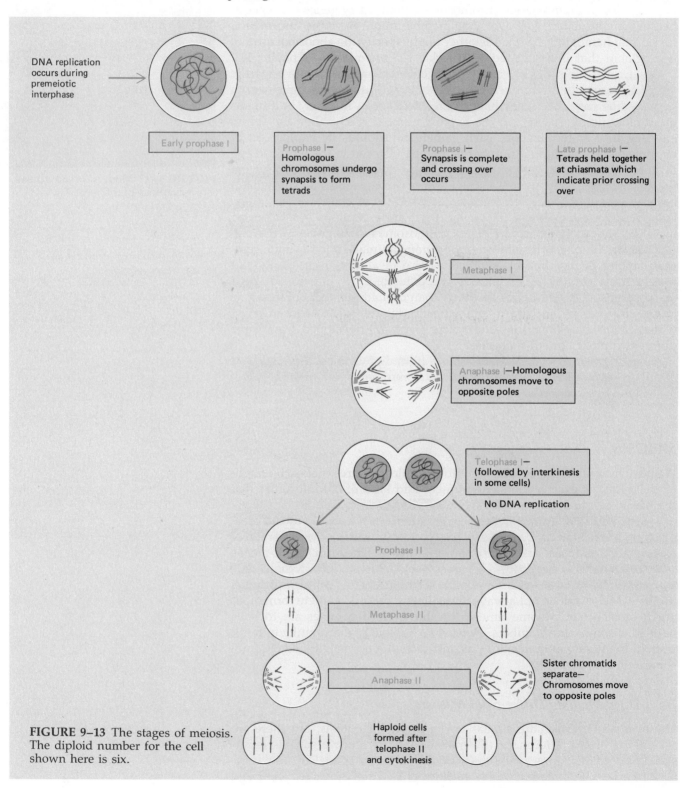

DNA replication occurs during premeiotic interphase

Early prophase I

Prophase I— Homologous chromosomes undergo synapsis to form tetrads

Prophase I— Synapsis is complete and crossing over occurs

Late prophase I— Tetrads held together at chiasmata which indicate prior crossing over

Metaphase I

Anaphase I—Homologous chromosomes move to opposite poles

Telophase I— (followed by interkinesis in some cells)

No DNA replication

Prophase II

Metaphase II

Anaphase II

Sister chromatids separate— Chromosomes move to opposite poles

Haploid cells formed after telophase II and cytokinesis

FIGURE 9–13 The stages of meiosis. The diploid number for the cell shown here is six.

The Process of Meiosis

The process of meiosis consists of two nuclear and cell divisions, designated the *first* and *second meiotic divisions*, or simply **meiosis I** and **meiosis II.** Each of these includes prophase, metaphase, anaphase, and telophase. During the first meiotic division, the members of each homologous pair of chromosomes separate and are distributed into separate cells. In the second meiotic division the chromatids that make up each chromosome separate and are distributed to the daughter cells. The following discussion describes meiosis in an animal with a diploid chromosome number of six. Refer to Figures 9–13 and 9–14 as you read.

As in mitosis, the chromosomes are duplicated during the S phase of interphase, before meiosis actually begins, and the duplicated chromosomes consist of two chromatids joined in the vicinity of their centromeres. During prophase of the first meiotic division, while the chromatids are still elongated and thin, the homologous chromosomes come to lie lengthwise side by side. This process is called **synapsis.** In our example, since the diploid number is six, there are three homologous pairs. It is customary when discussing higher organisms to refer to one member of each homologous pair as a **maternal chromosome,** because it was originally inherited from the organism's mother, and to the other as a **paternal chromosome,** because it was contributed by the father. Since each chromosome was duplicated before this time and now consists of two chromatids, synapsis results in the coming together of *four* chromatids, forming a complex known as a **tetrad.** The number of tetrads equals the haploid number of chromosomes. In our example there are 3 tetrads; in human cells there are 23 tetrads (and a total of 92 chromatids) at this stage.

During synapsis homologous chromosomes become closely associated. Electron microscopic observations reveal that a characteristic structure, known as the **synaptonemal complex,** forms between the synapsed homo-

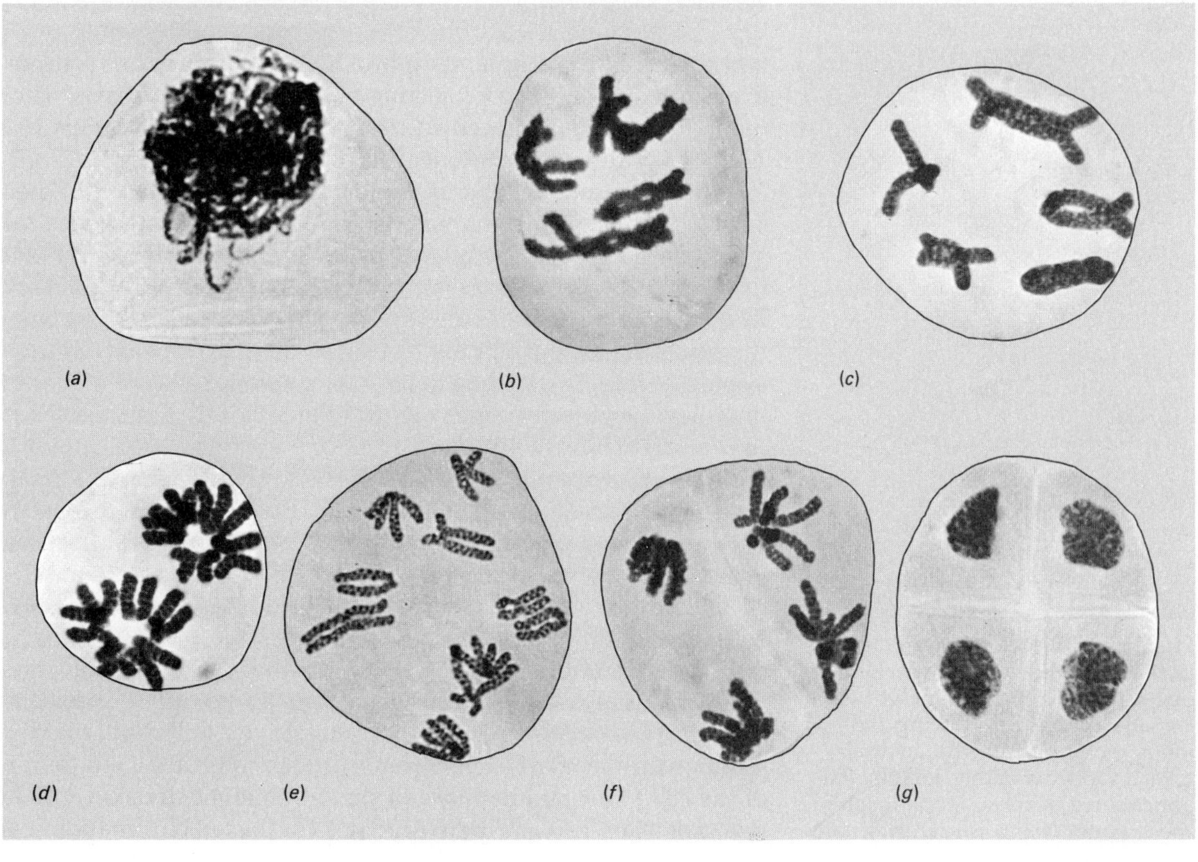

FIGURE 9–14 Photomicrographs of meiosis in the plant *Trillium erectum* (×2000). (*a*) Early prophase of the first meiotic division. (*b*) Later prophase of the first meiotic division. (*c*) Metaphase I. (*d*) Anaphase I. (*e*) Metaphase II. (*f*) Anaphase II. (*g*) Four daughter cells.

(a)   (b)   (c)

(d)   (e)   (f)   (g)

FIGURE 9–15 Electron micrograph
of a synaptonemal complex formed
between synapsing homologous
chromosomes in meiotic
Prophase I.

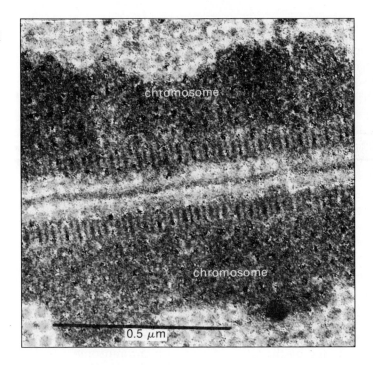

FIGURE 9–16 Photomicrograph of
part of a tetrad from a female meiotic cell (oocyte) of the newt
Triturus viridescens showing the
loops radiating from the central
thread (approximately ×1100). The
appearance of these loops inspired
their name, *lampbrush chromosomes.*

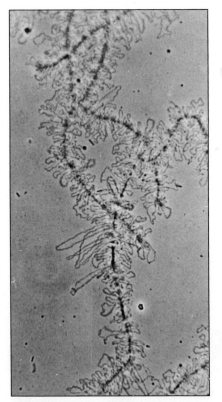

logues (Figure 9–15). Genetic material may be exchanged between homologous (nonsister) chromatids by **crossing over,** a process of breakage and
rejoining to produce new combinations of genes. The resulting **genetic
recombination** greatly enhances the amount of genetic variation among
the offspring of sexual partners.

In many species, the prophase of the first meiotic division is an extremely extended phase during which the cell grows and synthesizes nutrients. This is especially true during the formation of egg cells, since materials need to be laid down for the benefit of the future embryo. In many
types of gametes, the chromosomes assume unusual configurations during
this phase. For example, lampbrush chromosomes are composed of hundreds of pairs of loops projecting from the chromatid axis. They owe their
name to the brushes used to clean old-fashioned oil lamps (Figure 9–16).
The loops are sites of intense RNA synthesis.

While the unusual events characteristic of prophase I of meiosis are occurring, other events characteristic of mitotic prophase also take place. A
spindle of microtubules and other components forms, and in animal cells
the duplicated centrioles move to opposite poles and astral rays are
formed. The nuclear envelope disappears in late prophase, by which time
the structure of the tetrads can often be clearly seen (Figure 9–17). The
sister chromatids continue to be closely aligned along their lengths, but the
homologous chromosomes are no longer closely associated and their cen-
tromeric regions are separated from one another. At this point the homologous chromosomes are held together only at specialized regions of contact,
termed **chiasmata** (singular *chiasma*). Each chiasma represents a site at
which homologous chromatids have previously broken and rejoined
(crossing over), resulting in a cross-shaped configuration.

After the events of meiotic prophase, the tetrads are aligned on the
equator and the cell is said to be at *metaphase I.* The sister centromeres of
one chromosome are attached by spindle fibers to only one of the two
poles, and the sister centromeres of the homologous chromosome are attached to the opposite pole. (By contrast, you will recall that in mitosis the
sister centromeres of each chromosome become attached to opposite poles
of the cell.) During anaphase of the first meiotic division, the homologous
chromosomes of each pair (but not the sister chromatids) separate and

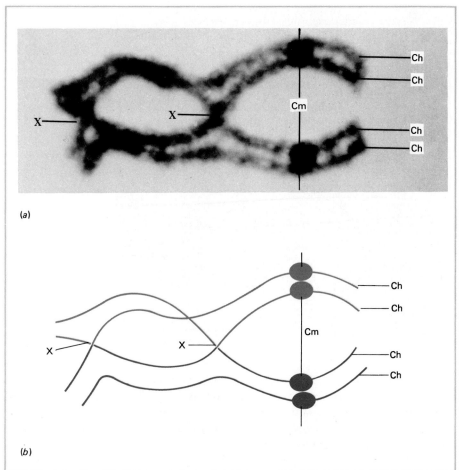

(a)

(b)

FIGURE 9–17 (*a*) Photomicrograph of a tetrad during late prophase of the first meiotic division of a salamander male meiotic cell (spermatocyte). Note the four chromatids that make up the tetrad (at Ch). The centromeres are visible at Cm and the chiasmata (resulting from crossing over) are indicated at each X. (*b*) Interpretive drawing indicating the structure of the tetrad.

move toward the poles. Each pole receives a random mixture of maternal and paternal chromosomes, but only one member of each pair is present at each pole. The sister chromatids are still united at their centromere regions. Again, this differs from mitotic anaphase, in which the sister centromeres separate and the sister chromatids pass to opposite poles. In the telophase I in our example, there are 3 duplicated chromosomes at each pole, for a total of 6 chromatids; in humans there are 23 duplicated chromosomes (46 chromatids) at each pole. During telophase, the chromatids decondense, the nuclear membrane may reorganize, and cytokinesis generally takes place.

During the interphase that follows, called **interkinesis,** there is no S phase, for no further chromosome replication takes place. In most organisms interkinesis is very brief; in some it is absent. Since the chromosomes usually remain partially condensed between divisions, the prophase of the second meiotic division is brief. Prophase II is similar to the mitotic prophase in many respects; there is no pairing of homologous chromosomes (indeed, only one homologue of each pair is present in the cell), and there is no crossing over and genetic recombination.

During metaphase II, the chromosomes line up on the equators of their cells. The first and second metaphases can be distinguished in diagrams; in the first the chromatids are arranged in bundles of four (tetrads), and in the second they are in groups of two (as in mitotic metaphase). This is not always so obvious in natural chromosomes. During anaphase II the sister chromatids, attached to spindle fibers at their centromeres, separate and move to opposite poles, just as they would at mitotic anaphase. Thus, in the telophase of the second meiotic division there is one of each kind of chromosome, the haploid number, at each pole. Nuclear membranes then

FIGURE 9–18 A simplified view of meiosis compared with mitosis. The diploid number of the cell shown here is four. (*a*) Mitosis. Note that each daughter cell has an identical set of four chromosomes (two pairs), which is the diploid number. (*b*) Meiosis. Two divisions take place, giving rise to four daughter cells. Each daughter cell has only two chromosomes, one of each pair. The chromosomes shown in blue originally came from one parent; those shown in red came from the other parent. Note that in the prophase of the first meiotic division, homologous chromosomes come together, forming tetrads.

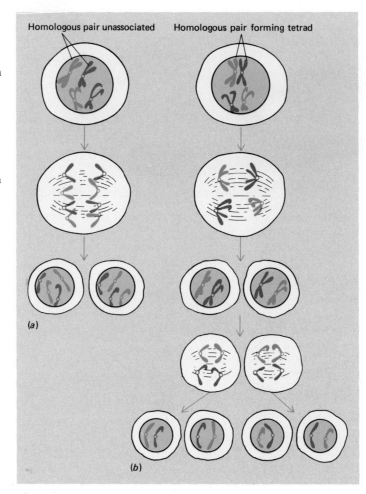

Homologous pair unassociated   Homologous pair forming tetrad

(a)

(b)

form, the chromosomes gradually elongate to form chromatin threads, and cytokinesis occurs.

Thus, the two successive divisions have the potential to yield four haploid nuclei, each containing one and only one of each kind of chromosome. Each of these cells has a different combination of genes. Figure 9–18 is a simplified comparison of mitosis and meiosis.

THE ROLE OF MITOSIS AND MEIOSIS IN SOME GENERALIZED LIFE CYCLES

Various groups of sexually reproducing eukaryotes differ with regard to the occurrence of haploidy and diploidy and the roles of meiosis and mitosis in their life cycles.

Many simple eukaryotes (including some fungi and algae) remain haploid (dividing mitotically) throughout most of their life cycles, with individuals being unicellular or multicellular. Haploid gametes (produced by mitosis) fuse to produce a diploid zygote, which undergoes meiosis immediately to restore the haploid state (Figure 9–19*a*)

Most of us are familiar with the life cycle of humans and other animals (Figure 9–19*b*). The body (somatic) cells of an individual organism multiply by mitosis and are diploid; the only haploid cells produced are the gametes. These are formed when certain **germ line** cells undergo meiosis. The formation of gametes (**gametogenesis**) in the male, termed **spermatogenesis,** results in the formation of four haploid sperm cells for each cell that enters meiosis. In contrast, female gametogenesis (**oogenesis**) results

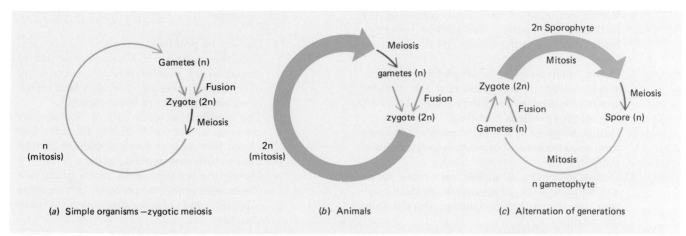

(a) Simple organisms —zygotic meiosis   (b) Animals   (c) Alternation of generations

FIGURE 9–19 Representative life cycles. (*a*) Simple eukaryote with zygotic meiosis (some fungi and algae). (*b*) Animal. (*c*) Alternation of haploid gametophyte and diploid sporophyte generations (higher plants and some algae).

in the formation of a single egg cell **(ovum)** for every cell that enters meiosis. This is accomplished by a process that apportions virtually all of the cytoplasm to only one of the two nuclei at each of the meiotic divisions. At the end of the first meiotic division one nucleus is retained; the other, called the first **polar body,** is excluded from the cell and ultimately degenerates. Similarly, at the end of the second division, one nucleus becomes the second polar body and the other nucleus survives. In this way, one haploid nucleus becomes the recipient of all of the accumulated cytoplasm and nutrients that were present in the original meiotic cell. (see Chapter 50 for a more detailed description.)

The most complex life cycles are displayed by plants and some algae (Figure 9–19*c*). These life cycles, characterized by an **alternation of generations,** consist of a multicellular diploid stage, termed the **sporophyte generation,** and a multicellular haploid stage, termed the **gametophyte generation.** Diploid sporophyte cells undergo meiosis to form haploid spores, each of which then divides mitotically to produce a multicellular haploid gametophyte. Gametophytes produce gametes by mitosis. The female and male gametes (eggs and sperm) then fuse to form a diploid zygote, which divides mitotically to produce a multicellular diploid sporophyte. In higher plants, including flowering plants, the diploid sporophyte—which includes the roots, stem, and leaves of the plant body—is the dominant form. The gametophytes are small and inconspicuous. For example, a pollen grain is actually a haploid male gametophyte in which haploid sperm nuclei are produced by mitosis.

■ SUMMARY

I. In the production of a new generation, genetic information is transferred from parent to offspring; this process is termed *heredity*. The science of genetics is the study of the structure, transmission, and expression of genes.

II. Genes are made up of DNA. DNA is complexed with protein and RNA to form the chromatin fibers that make up chromosomes.

   A. A gene is that portion of chromosomal DNA that codes for a specific RNA molecule, which may in turn code for a specific polypeptide.

   B. Each somatic cell of every diploid (2n) organism of a given species has a characteristic number of pairs of chromosomes. The two members of each pair, called *homologous chromosomes*, are similar in length, shape, and other structural features, and carry genes affecting the same characteristics of the organism.

III. The cell cycle is the period from the beginning of one division to the beginning of the next division.

   A. Interphase can be divided into the first gap phase (G_1), the chromosomal synthesis phase (S), and the second gap phase (G_2).

      1. During the G_1 phase, the cell grows and prepares for the S phase.

      2. DNA and the chromosomal proteins are synthesized during the S phase.

      3. During the G_2 phase, there is an increase in protein synthesis in preparation for cell division.

   B. During mitosis, a complete set of chromosomes is distributed to each pole of the cell, and a nuclear membrane is formed around each set.

1. During prophase, chromosomes condense, the nucleolus disappears, the nuclear envelope breaks down, and the mitotic spindle begins to form.
2. During metaphase, the duplicated chromosomes, which are composed of a pair of sister chromatids, line up along the equator of the cell; the mitotic spindle is complete.
3. During anaphase, the sister chromatids separate from one another and move to opposite poles of the cell.
4. During telophase, a nuclear membrane forms around each set of chromosomes, nucleoli reappear, the chromosomes elongate, and the spindle disappears.
   C. During cytokinesis, which generally begins in telophase and therefore overlaps mitosis, the cytoplasm divides, forming two individual cells.
IV. There are two major forms of reproduction: asexual and sexual.
   A. Offspring produced by asexual reproduction have hereditary traits that are identical to those of the single parent. These offspring constitute a clone. All of the cells involved are produced by a mitotic process.
   B. Sexual reproduction makes possible the incorporation of traits from two parents. Haploid (n) sex cells, or gametes, fuse to form a single diploid (2n) cell, the zygote.
      1. In the initial formation of haploid cells from diploid cells, a special kind of cell division termed *meiosis* is required. Although meiosis does not always immediately precede gamete formation, as a result of meiosis (at some time in the life cycle), each gamete contains one of each kind of chromosome (the haploid chromosome number).
         a. During the prophase of the first meiotic division, the members of a homologous pair of chromosomes undergo synapsis and crossing over. During crossing over, genetic material is exchanged between homologues.
         b. The members of each pair of homologous chromosomes separate during the anaphase of the first meiotic division and are distributed to different daughter cells.
         c. During the second meiotic division, the two chromatids of each homologous chromosome separate, and one is distributed to each daughter cell.
      2. When a zygote is formed, one member of each homologous pair is contributed by one parent and the other member by the other parent. Therefore, each parent contributes half of the chromosomes of the zygote.
V. Various groups of organisms differ with respect to the roles mitosis and meiosis play in their life cycles.
   A. Simple eukaryotes may be regularly haploid; the only diploid stage is the zygote, which undergoes meiosis to restore the haploid state.
   B. The somatic cells of animals are diploid; the only haploid cells are the gametes (produced by meiosis).
   C. Plants (and some algae) show alternation of generations. A diploid sporophyte forms spores by meiosis. These develop into haploid gametophytes, which produce gametes mitotically. Two haploid gametes then fuse to form a diploid zygote, which develops into a new diploid sporophyte.

■ POST-TEST

1. The tendency of individuals to resemble their parents is termed _____.
2. Chromosomes are composed of chromatin fibers, which are made up of _____, _____ and _____.
3. The period from the beginning of one cell division to the beginning of the next is termed the _____ _____.
4. DNA for the new sets of chromosomes is synthesized during the _____.
5. To facilitate description of the process, mitosis has been divided into four stages: _____, _____, _____, and _____.
6. A duplicated chromosome consists of a pair of _____ _____.
7. Each chromatid contains a constricted region, the _____, to which some of the spindle fibers attach.
8. The period during which the chromosomes are lined up on the equator of the cell constitutes _____.
9. The division of the cytoplasm to yield two daughter cells is called _____.
10. The drug _____ binds to tubulin, the major microtubule protein, thereby blocking cell division.
11. The splitting, budding, or fragmenting of a single parent to give rise to two or more offspring with hereditary traits identical to those of the parent is termed _____ reproduction.
12. A group of genetically identical individuals is called a _____.
13. The members of a pair of chromosomes are referred to as _____ chromosomes.
14. Cells that contain two complete sets of chromosomes are _____, whereas those that contain more than two complete sets are _____.
15. Gametes contain the _____ chromosome number.
16. The pairing of homologous chromosomes during prophase I is known as _____. Such an association of homologous chromosomes is termed a(n) _____.
17. The exchange of segments of homologous chromatids during meiotic prophase I is known as _____.
18. In a human with the diploid chromosome number of 46, each sperm or egg will have _____ chromosomes.
19. Alternation of haploid and diploid generations is characteristic of _____ and some algae.

■ REVIEW QUESTIONS

1. What is the chemical makeup of chromatin?
2. What is the relationship between genes and chromosomes? What are the functions of genes?
3. Two species may have the same chromosome number and yet be very different. Explain.
4. Describe the structure of a duplicated chromosome, paying special attention to the sister chromatids and the centromeres.
5. What are the stages of the cell cycle and mitosis?
6. Define the following terms:
   a. diploid
   b. haploid
   c. homologous chromosomes
7. How does meiosis differ from mitosis? Are there any points of similarity between mitosis and meiosis?
8. Compare the roles of mitosis and meiosis in organisms with different kinds of life cycles.
9. Assume that an animal has a diploid chromosome number of ten.
   a. How many chromosomes would it have in a typical body cell, such as a skin cell?
   b. How many chromosomes would be present in a cell at mitotic prophase? How many chromatids?
   c. How many tetrads would form in the prophase of the first meiotic division?
   d. How many chromosomes would be present in each gamete?

■ RECOMMENDED READINGS

Alberts, B., Bray, D., Lewis, J., Raff, M., Roberts, K., and Watson, J.D. *Molecular Biology of the Cell,* Chapters 11 and 14. New York, Garland, 1983. An extensive, detailed, and well-written discussion of cell growth and division, covering the control of cell division, the cell cycle, and the events of mitosis and meiosis.

Avers, C.J. *Genetics.* 2nd ed., Boston, Trindle, Weber, and Schmidt, 1984. A genetics text with a well-written, accurate description of mitosis and meiosis.

Pickett-Heaps, J., Tippet, D., and Porter, K. *Rethinking Mitosis,* Cell 29(1982):729–744. An advanced article about the evolution of mitosis and meiosis.

Whitehouse, H.L. *Towards an Understanding of the Mechanism of Heredity.* 3rd ed. London, St. Martin's Press, 1973. A classic text with one of the best descriptions of mitosis and meiosis.

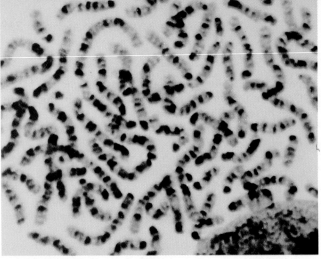

Mammal chromosomes

10

The Basic Principles of Heredity

■ LEARNING OBJECTIVES

After you have read this chapter you should be able to:

1. Define and use correctly the terms *gene, allele, locus, genotype, phenotype, dominant, recessive, homozygous, heterozygous,* and *test cross.*
2. Apply Mendel's laws to solve problems in genetics involving monohybrid and dihybrid crosses.
3. Explain how the method of progeny testing is used by commercial breeders in establishing a genetic strain that will breed true for a given trait.
4. Summarize and apply the concept of probability using the product and sum laws.
5. Solve problems in genetics involving incomplete dominance, epistasis, polygenes, multiple alleles, and X-linked traits.
6. Discuss the ways in which genes may interact to affect the appearance of a single trait; discuss how a single gene can affect many traits simultaneously.
7. Discuss the phenomena of linkage and crossing over, and solve problems involving linked genes.
8. Discuss the genetic determination of sex and the role of the Y chromosome in determining maleness in humans; compare sex determination in humans and other mammals with that in various other animals and some plants; discuss dosage compensation of X-linked genes.
9. Compare inbreeding and outbreeding, and discuss the genetic basis of hybrid vigor.

GREGOR MENDEL AND THE BEGINNINGS OF GENETICS

The basic laws of inheritance in eukaryotes were discovered by Gregor Mendel (1822–1884), an Austrian abbot who bred pea plants in his monastery garden at Brünn, in what is now Czechoslovakia (Figure 10–1). Mendel was not the first plant breeder; **hybrid** plants and animals (offspring of two genetically dissimilar parents) had been known for a long time. When Mendel began his breeding experiments, two main facts had been recognized: One was that all hybrid plants of the same kind look alike. The second was that hybrids themselves do not breed true; their offspring show a mixture of traits, some looking like their parents and some looking like their grandparents. Mendel's genius lay in his recognition of a pattern in the way the parental characteristics reappear in the offspring of hybrids. No one before had carefully categorized and counted the offspring and analyzed these regular patterns over several generations.

Just as geneticists do today, Mendel chose the organism for his experiments very carefully. The garden pea (*Pisum sativum*) had a number of advantages. Pea plants are easy to grow, and even in 1857, when Mendel began his experiments, many varieties of peas were available through commercial sources. It is impossible to study inheritance without such genetic **variation.** (If every person in the population had blue eyes, it would be impossible to study the inheritance of eye color.) Another advantage of pea plants is that it is relatively easy to conduct **controlled pollinations** with them. Pea flowers have both male and female parts. The **anthers** (pollen-producing parts of the flower) can be carefully removed to prevent self-fertilization. Pollen from a different source can then be applied to the **stigma** (receptive surface of the female parts). It is relatively easy to protect the flower from other sources of pollen because the reproductive structures are enclosed by the petals of the flower (Figure 10–2). Covering the pollinated flowers with small bags affords additional protection from the attention of pollinating insects.

Although the pea seeds were obtained from commercial sources, it was necessary for Mendel to do some important preliminary work. He spent several years developing genetically pure, or **true breeding,** lines for a number of inherited traits. During this time he apparently chose those characteristics of his pea strains that were most worthy of study, discarding or ignoring a number of others. During this time he must also have been making the initial observations that would form the basis of his theories. He eventually chose seven clearly contrasting pairs of traits: yellow versus green seeds, round versus wrinkled seeds, green versus yellow pods, tall versus short plants, inflated versus constricted pods, white seed coats versus gray seed coats, and flowers borne on the ends of the stems versus flowers appearing all along the stems.

When Mendel crossed plants from true breeding lines with two contrasting traits, such as tall plants and short plants, the plants in the first generation of offspring all looked alike and resembled one of the two parents (Figure 10–3). This first generation of offspring is termed the first **filial** (Latin for "sons and daughters") **generation,** or **F_1 generation.** The second filial generation, or **F_2 generation,** is produced by a cross of two F_1 offspring. Mendel found that the two types of individuals were present in the F_2 generation in a ratio of approximately 3:1. For example, when he crossed tall plants with short plants (these constitute the **parental,** or **P,** generation), all the members of the succeeding F_1 generation were tall. When two of these F_1-generation tall plants were either crossed or allowed to self-pollinate, the F_2 generation included 787 tall plants and 277 short plants. Clearly, in the F_1 generation the hereditary factor for shortness had been hidden by the hereditary factor for tallness.

From such results Mendel formulated his **principle of dominance,** which essentially states that in an F_1 hybrid the hereditary factor that was present in one of the parents masks expression of the hereditary factor that was present in the other parent. The hereditary factor that is expressed in the F_1 (tallness in our example) is said to be **dominant;** the trait that is hidden (shortness) is said to be **recessive.** This finding was at odds with the prevailing idea of **blending inheritance,** which implied that a hybrid should be intermediate between the two parents. Today we know that the principle of dominance does not always apply (exceptions are considered later in this chapter); nevertheless, the recognition that one hereditary factor can mask the expression of another was an important intellectual leap on Mendel's part.

FIGURE 10–1 Gregor Mendel (1822–1884).

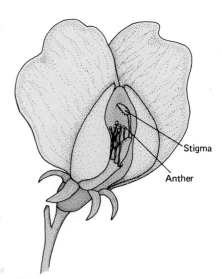

FIGURE 10–2 A cut-away drawing of a pea flower illustrating the pollen-producing anthers and the stigma, that portion of the female part of the flower that receives the pollen. Notice how the reproductive structures are enclosed by the petals.

Stigma

Anther

FIGURE 10–3 A diagram illustrating one of the crosses carried out by Gregor Mendel. Crossing a tall pea plant with a short pea plant yielded only tall offspring in the F_1 generation. However, when these offspring were self-pollinated or when two individuals were crossed, the next generation included tall and short plants in a ratio of about 3:1.

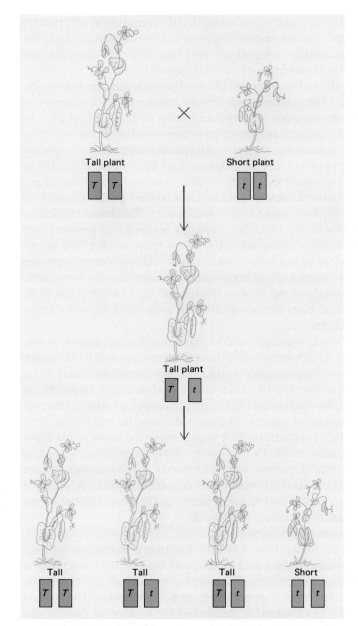

On the basis of his results, Mendel postulated that each kind of hereditary characteristic is governed by two "factors" in every individual. (Mendel's factors are equivalent to what we call **genes,** so we will use that term in our discussion.) Mendel further proposed that when gametes are formed, the genes behave like particles, becoming separated so that each sex cell (egg or sperm) contains only one member of each pair. The two genes emerge intact from this process (one does not "contaminate" the other), thus allowing for the reappearance of recessive traits in the F_2 generation. This idea, the **principle of segregation,** also ran counter to the notion of blending inheritance, according to which hereditary determinants were like fluids and became inseparably mixed once they were combined in a hybrid.

In our example the F_1-generation tall plants had two genetic factors, one for tallness (which we can designate as T) and one for shortness (which we can designate as t), but because the tall gene was dominant, these plants were tall. However, when these F_1 plants formed gametes, the gene for tallness separated (segregated) from the gene for shortness, so that half of the gametes contained a gene for tallness and the other half contained a gene for shortness. Random fertilization led to three possible combinations of factors in the offspring: one quarter with two tallness genes (TT), one quarter with two shortness genes (tt), and one half with one gene for tallness and one gene for shortness (Tt). Because both TT and Tt plants are tall, on average one could expect three quarters of the offspring to be tall (to express the dominant gene) and one quarter to be short (to express the recessive gene). Today we know that this is exactly what occurs in meiosis and the separation of homologous chromosomes and in fertilization, but it is remarkable that Mendel was able to formulate this idea at a time when mitosis and meiosis were yet to be discovered.

Mendel reported these and other findings (discussed later in this chapter) at a meeting of the Brünn Society for the Study of Natural Science and published his results in the transactions of that society in 1866. At that time biology was largely a descriptive science, and there was little interest in the application of quantitative and experimental methods such as Mendel had used. The importance of his results and his interpretations of these results was not appreciated by other biologists of the time, and his findings were neglected for nearly 35 years.

In 1900 Hugo DeVries in Holland, Karl Correns in Germany, and Erich von Tschermak in Austria independently rediscovered the laws of inheritance that had been described by Mendel. The literature search that has long been a part of scientific research led them to Mendel's paper in which these laws had been clearly stated 34 years before. They gave credit to Mendel by naming the basic laws of inheritance after him. By this time, there was much greater appreciation of the value of experimental methods in biology. The details of mitosis, meiosis, and fertilization were now known, and in 1903 W.S. Sutton pointed out the connection between Mendel's separation of genes and the separation of homologous chromosomes during meiosis. The time was right for widespread (although not universal) acceptance and extension of these ideas and their implications.

GENES AND ALLELES

Within each chromosome are many genes, each generally different from the others and each controlling the inheritance of one or more characteristics. The members of a homologous pair of chromosomes have similar genes arranged in similar order. The regularity of the mitotic process ensures that each diploid daughter cell will have two of each kind of chromosome and therefore two of each kind of gene. As the chromosomes separate during meiosis and become associated with new partners at fertilization, so, of course, must the paired genes separate and become associated with new partners. All of the phenomena of mendelian genetics depend on these simple facts. Each chromosome behaves genetically as though it were composed of a string of genes arranged in linear order. The gene controlling each specific trait occurs at a particular point in each chromosome, called a **locus** (plural, *loci*).

As stated, the inheritance of any characteristic can be studied only when there are two contrasting conditions, such as yellow versus green in Mendel's peas, normal pigmentation versus absence of pigmentation (**albinism**) in humans or other mammals (Figure 10–4), and brown versus black coat color in guinea pigs. In simple cases an individual expresses one or the other, but not both, of such contrasting conditions. Genes that govern variations of the same characteristic and that occupy corresponding loci on homologous chromosomes are termed **alleles** (Figure 10–5).

Geneticists use the term *allele* to emphasize that there are two or more alternative forms or states of the gene that can occur at corresponding specific loci on homologous chromosomes. The possible variants at each

FIGURE 10–4 Albino koala.

FIGURE 10–5 Homologous chromosomes and alleles.

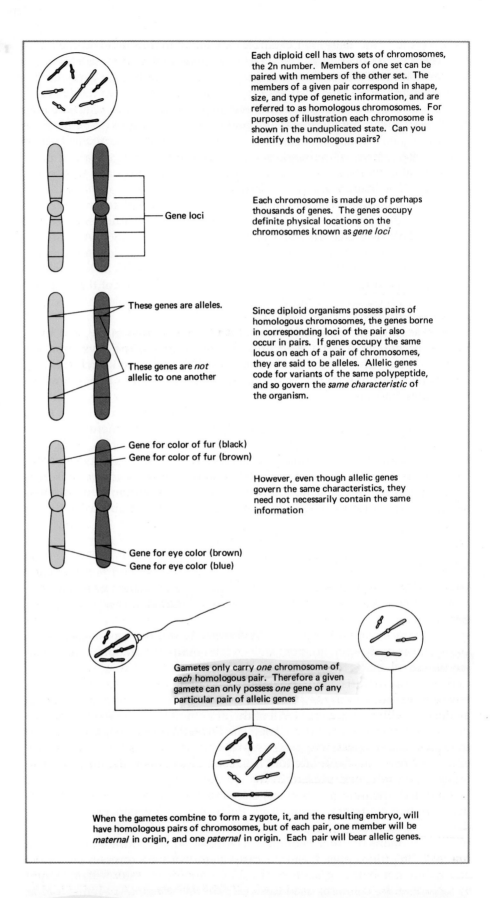

Each diploid cell has two sets of chromosomes, the 2n number. Members of one set can be paired with members of the other set. The members of a given pair correspond in shape, size, and type of genetic information, and are referred to as homologous chromosomes. For purposes of illustration each chromosome is shown in the unduplicated state. Can you identify the homologous pairs?

Gene loci

Each chromosome is made up of perhaps thousands of genes. The genes occupy definite physical locations on the chromosomes known as *gene loci*

These genes are alleles.

These genes are *not* allelic to one another

Since diploid organisms possess pairs of homologous chromosomes, the genes borne in corresponding loci of the pair also occur in pairs. If genes occupy the same locus on each of a pair of chromosomes, they are said to be alleles. Allelic genes code for variants of the same polypeptide, and so govern the *same characteristic* of the organism.

Gene for color of fur (black)
Gene for color of fur (brown)

However, even though allelic genes govern the same characteristics, they need not necessarily contain the same information

Gene for eye color (brown)
Gene for eye color (blue)

Gametes only carry *one* chromosome of *each* homologous pair. Therefore a given gamete can only possess *one* gene of any particular pair of allelic genes

When the gametes combine to form a zygote, it, and the resulting embryo, will have homologous pairs of chromosomes, but of each pair, one member will be *maternal* in origin, and one *paternal* in origin. Each pair will bear allelic genes.

locus are the alleles, each of which is assigned a single letter (or group of letters) as its symbol. It is customary to indicate a dominant allele with a capital letter and a recessive allele with a lowercase letter. The choice of the letter or letters themselves is generally determined by the first allelic variant found for that locus. For example, the allele that Mendel studied that governs the yellow color of the seed is designated Y, and the allele that is responsible for the green color is designated y. Because discovery of the yellow allele made identification of this locus possible, we commonly refer to the locus as the *yellow* locus, although pea seeds are most commonly green. The term *locus* is used to designate not only a location on a chromosome, but also a kind of "generic" gene controlling a particular *kind* of characteristic; thus, Y (yellow) and y (green) represent a specific pair of alleles of a locus involved in determining seed color in peas. To prevent confusion, the dominant allele is always designated first and the recessive allele second (Yy, never yY). Geneticists use the term *gene* sometimes to specify a locus and sometimes to specify one of the allelic variants at that locus. Usually the meaning is clear from the context.

A MONOHYBRID CROSS

The usage of genetic terms and some of the basic principles of genetics can be illustrated by examples of a simple **monohybrid cross**—that is, a cross between two individuals that differ with respect to the alleles they carry for a single locus. Our first example deals with the expected ratios in the F_2 generation, as did our previous example of Mendel's work on tall and short pea plants.

Homozygous and Heterozygous Organisms

Figure 10–6 depicts a monohybrid cross featuring a locus that governs coat color in guinea pigs. The male comes from a true breeding line of brown guinea pigs. We say that he is **homozygous** for brown because the two alleles he carries for this locus are *identical*. The black female is also from a true breeding line and is homozygous for black. What color would you expect the F_1 offspring to be? Dark brown? Spotted? It is impossible to make such a prediction without more information. In this particular case the F_1 offspring are black, but they are **heterozygous,** meaning that they carry two *different* alleles for this locus. The allele for brown coat color can be expressed only in a homozygous brown individual; it is referred to as a *recessive allele*. The allele for black coat color can be expressed in both homozygous black and heterozygous individuals; it is said to be a *dominant allele*. On the basis of this information, we can use standard notation to designate the dominant black allele as *B* and the recessive brown allele as *b*.

During meiosis in the male (bb), the two b alleles separate according to Mendel's principle of segregation so that each sperm has only one b allele. In the formation of eggs in the female (BB), the two B alleles separate so that each egg has only one B allele. The fertilization of each B egg by a b sperm results in a heterozygous animal with the alleles Bb—that is, one allele for brown coat and one for black coat. Since this is the only possible combination of alleles present in the eggs and sperm, all of the offspring will be of the same type with respect to alleles B and b.

Phenotype and Genotype

The fact that some alleles may be dominant and others recessive means that we cannot always determine the alleles carried by an organism simply by looking at it. The term used to specify the *appearance* of an individual in a given environment with respect to a certain inherited trait is known as its **phenotype.** The *genetic constitution* of that organism, most often expressed in symbols, is its **genotype.** In the cross we have been considering, the genotype of the female parent is homozygous dominant, BB, and her phe-

FIGURE 10–6 A monohybrid cross. The cross is between a homozygous brown guinea pig and a homozygous black guinea pig. The F₁ generation includes only black individuals. However, the mating of two of these offspring yields F₂ generation offspring in the ideal black:brown ratio of 3:1, indicating that the F₁ individuals are heterozygous.

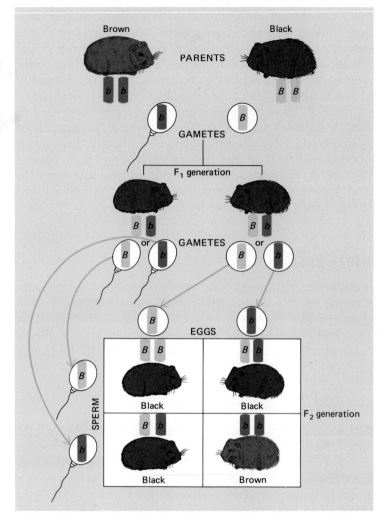

notype is black. The genotype of the male parent is homozygous recessive, bb, and his phenotype is brown. The genotype of the F₁ offspring is heterozygous, Bb, and their phenotype is black.

The phenomenon of dominance explains, in part, why an individual may resemble one parent more than the other, even if the two parents make equal contributions to their offspring's genetic constitution. In one species of animal, black coat may be dominant to brown; in another species, brown may be dominant to black. Dominance is not completely predictable and can be determined only by experiment.

During meiosis in the heterozygous (Bb) black guinea pigs, the chromosome containing the B allele becomes separated from its homologue, the chromosome containing the b allele, so that each sperm or egg contains B or b but never both. It follows that gametes containing B alleles and those containing b alleles are formed in equal numbers by heterozygous Bb individuals. There is no special attraction or repulsion between an egg and sperm containing the same kind of gene, so fertilization is a random process.

The possible combinations of eggs and sperm at fertilization may be represented in the form of a "checkerboard" devised by an early geneticist, Sir Reginald Punnett, and known as a **Punnett square** (see Figure 10–6). The types of gametes from one parent are represented across the top, and the types of gametes from the other parent are indicated along the left side; the squares are then filled in with the resulting F₂ zygote combinations. Three fourths of all F₂ offspring will be genotypically BB or Bb and phenotypically black; one fourth will be genotypically bb and phenotypically brown. The genetic mechanism responsible for the approximately 3:1 F_2 ratios (called *monohybrid F_2 phenotypic ratios*) obtained by Mendel in his pea-breeding experiments is again evident. The corresponding genotypic ratio is 1 BB:2 Bb:1 bb.

Solving Genetics Problems

In working genetics problems, it is wise to use the following procedure to avoid errors:

1. Always use standard designations for the generations. The generation with which a particular genetic experiment is begun is called the *P*, or *parental, generation*. Offspring of this generation (the "children") are called the *F_1*, or *first filial, generation*. The offspring resulting when two F_1 individuals are bred constitute the F_2, or second filial, generation (the "grandchildren"). Those resulting from the mating of F_2 individuals make up the *F_3 generation*, and so on.
2. Write down the symbols you are using for the allelic variants of each locus. Use uppercase to designate a dominant allele and lowercase to designate a recessive allele. In general, use the same letter of the alphabet to designate all of the alleles of a particular locus.
3. Determine the genotypes of the parents of each cross by making use of the following types of clues:
   a. Are they from true breeding lines? If so, they should be homozygous.
   b. Can their genotypes be reliably deduced from their phenotypes?
   c. Do the phenotypes of their offspring provide any information? See *"Focus on Deducing Genotypes"* for an example of how these determinations can be made.
4. Indicate the possible kinds of gametes formed by each of the parents.
5. Set up a Punnett square, putting the possible types of gametes from one parent down the left side and the possible types from the other parent across the top.
6. Fill in the Punnett square and read off (and sum up) the genotypic and phenotypic ratios of the offspring. Avoid confusion by consistently placing the dominant allele first and the recessive allele second in heterozygotes (Bb, never bB).

Monohybrid Test Crosses

One third of the black guinea pigs in the F_2 generation derived from the mating of homozygous black × brown parents (P generation) are themselves homozygous, BB; the other two thirds are heterozygous, Bb. Guinea pigs with the genotypes BB and Bb are alike phenotypically; they both have black coats. Since animals cannot, in general, self-fertilize, how do you think geneticists can distinguish the homozygous (BB) and heterozygous (Bb) black-coated guinea pigs? They do this by a **test cross,** in which each black guinea pig is mated with a homozygous brown (bb) guinea pig (Figure 10–7). In a test cross the two types of gametes produced by the heterozygous parent are not "hidden" in the offspring by dominant alleles coming from the other parent. Therefore, through a test cross one can deduce the genotypes of all of the classes of offspring directly from their phenotypes. If all of the offspring were black, what inference would you make about the genotype of the black parent? If any of the offspring were brown, what conclusion would you draw regarding the genotype of the black parent? Would you be more certain about one of these two inferences than the other?

FIGURE 10–7 Test crosses to determine the genotype of a black guinea pig. (*a*) If a black guinea pig is mated with a brown guinea pig and all of the offspring are black, the black parent probably has a homozygous genotype. (*b*) However, if any of the offspring are brown, the black guinea pig must be heterozygous for color.

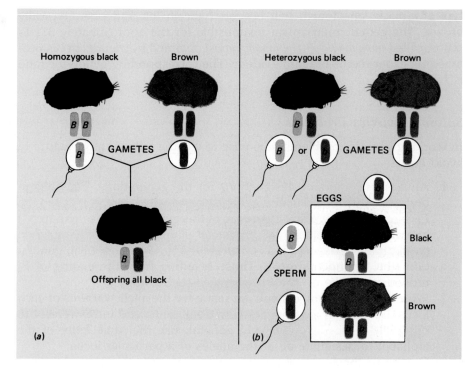

Mendel did just these sorts of experiments, breeding heterozygous tall (Tt) pea plants with homozygous recessive (tt) short ones. He predicted that the heterozygous parent would produce equal numbers of T and t gametes, whereas the homozygous short parent would produce only t gametes and that this should lead to equal numbers of tall (Tt) and short (tt) individuals among the progeny. Another value of a test cross is that it allows us to test rather directly the hypothesis that there will be 1:1 segregation of alleles in the heterozygous parent. Thus, Mendel's hypothesis not only explained the known facts, such as the monohybrid F$_2$ 3:1 phenotypic ratio, it also enabled him to predict the results of other experiments, in this case the 1:1 **monohybrid test cross** phenotypic ratio.

This sort of testing is of great importance to the commercial breeding of animals or plants when the breeder is trying to establish a strain that will breed true for a certain trait. Two bulls, for example, may look equally healthy and vigorous, yet one bull will have daughters with qualities of milk production that are distinctly superior to those of daughters of the other bull. A breeder tests the genotypes of the breeding stock by making test matings and observing the offspring; if the offspring are superior with respect to the desired trait, the parents are thereafter used regularly for breeding.

CALCULATING THE PROBABILITY OF GENETIC EVENTS

All genetic ratios are properly expressed in terms of probabilities. In the examples just discussed, we saw that in the offspring of two individuals heterozygous for the same gene pair, the ratio of the phenotype of the dominant allele to the phenotype of the recessive allele would be 3:1. A better way to express our expectations is that there are 3 chances in 4 (¾) that any particular individual offspring of two heterozygous individuals will express the phenotype of the dominant allele and 1 chance in 4 (¼) that it will express the phenotype of the recessive allele. Probabilities are always expressed as fractions (or proportions)—that is, numbers between 0 and 1. If an event is certain to occur, its probability is 1; if it is certain not to occur, its probability is 0.

Focus on DEDUCING GENOTYPES

The science of genetics resembles mathematics in that it consists of a few basic principles, which, once grasped, enable the student to solve a wide variety of problems. Very often the genotypes of the parents can be deduced from the phenotypes of their offspring. In chickens, for example, the allele for rose comb (R) is dominant to the allele for single comb (r). Suppose that a cock is mated to three different hens, as shown in the figure. The cock and hens A and C have rose combs; hen B has a single comb. Breeding the cock with hen A produces a rose-combed chick, with hen B a single-combed chick, and with hen C a single-combed chick. What type of offspring can be expected from further matings of the cock with these hens?

Since the allele for single comb, r, is recessive, all of the hens and chicks that are phenotypically single-combed must be rr. We can deduce that hen B and the offspring of hens B and C are genotypically rr. All of those individuals that are phenotypically rose-combed must have at least one R allele. The fact that the offspring of the cock and hen B was single-combed proves that the cock is heterozygous Rr, because although the single-combed chick received one r allele from its mother, it must have received the second one from its father. The fact that the offspring of the cock and hen C had a single comb proves that hen C is heterozygous, Rr. It is impossible to decide from the data given whether hen A is homozygous RR or heterozygous Rr; further breeding would be necessary to determine this. Additional matings of the cock with hen B should result in one-half rose-combed and one-half single-combed individuals; additional matings of the cock with hen C should produce three-fourths rose-combed and one-fourth single-combed chicks.

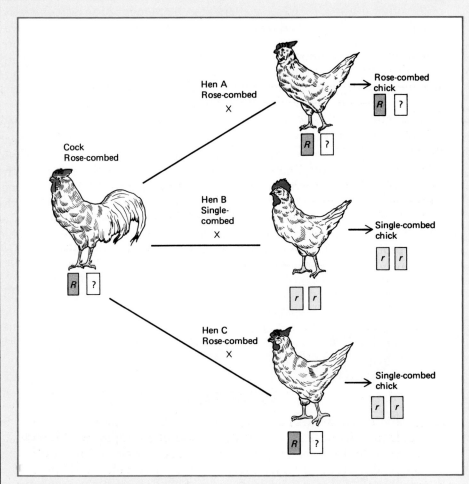

■ Deducing the parental genotypes from the phenotypes of the offspring. In chickens, the allele for rose comb (R) is dominant to the allele for single comb (r). Determine the unknown parental genotypes.

Often we wish to *combine* two or more probabilities. In order to do so, we must keep certain rules in mind; these are known as the **product law** and the **sum law.**

The Product Law

If two or more events are *independent* of each other, the probability of their both occurring is the *product* of their individual probabilities. If this seems strange to you, keep in mind that when we multiply two numbers less than 1, the product is a smaller number. Events are independent if the occurrence of one does not affect the probability of the other. For example, the probability of obtaining heads on the first toss of a coin is ½; the probability of obtaining heads on the second toss (an independent event) is also ½. The probability of obtaining heads first and second on successive tosses of the coin is the product of their individual probabilities, ½ × ½ = ¼ (or 1 chance in 4) (Figure 10–8). This product law of probability also holds for three or even more independent events. For example, the probability of choosing at random a person who is male, has blood type A, and was born in June is 0.5 × 0.4 × 0.084 = 0.0168. (The chance of being male is ½, or 0.5; of having blood type A, about ⅖ or 0.4; and of being born in June, ¹⁄₁₂, or 0.084). You may want to experiment with similar applications of the product law.

The Sum Law

Events are mutually exclusive if the occurrence of one *precludes* occurrence of the other. Mutually exclusive events can be thought of as different ways of obtaining some specified result. Naturally, if there is more than one way to obtain a result, the chances of its being obtained are improved; we therefore combine the probabilities of mutually exclusive events by summing their individual probabilities. For example, the probability that the roll of a die will produce *either* a two or a five (mutually exclusive events) is ⅙ + ⅙ = ⅓.

If we flip a coin twice, what is the probability that it will come up heads one time and tails the other time if we do not specify the order in which these events are to occur? There are two mutually exclusive ways to obtain this outcome. We could get heads the first time and tails the second; the probability of this occurring is ¼ (calculated by the product law). Alternatively, we could also get tails the first time and heads the second; the probability of this occurring is also ¼. We combine the probabilities of these mutually exclusive outcomes using the sum law: ¼ + ¼ = ½. That is, the probability of getting heads once (and only once) and tails once (and only once) on two successive tosses of the coin is ½.

FIGURE 10–8 Application of the laws of probability. When one tosses a coin twice, the probability of getting heads on the first toss is 1/2, and the probability of getting tails on the first toss is also 1/2. Identical predictions apply to the second toss. The outcome of the first toss does not affect the outcome of the second toss, so these events are independent, and we combine them by multiplying the individual probabilities (according to the product law). As shown in the Punnett square, there are four different classes of combined outcomes of two successive tosses. These are mutually exclusive, and we therefore combine them by adding them (according to the sum law). For example, heads/heads and tails/tails are mutually exclusive outcomes. If we wish to calculate the probability that we will get *either* heads/heads *or* tails/tails, we add: 1/4 + 1/4 = 1/2. These same laws of probability are used to predict genetic events.

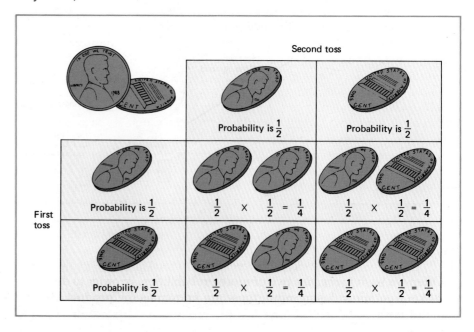

Applying the Laws of Probability

We can use the laws of probability to make a variety of calculations. For example, what are the probabilities that a family with two (and only two) children will have two girls, two boys, or one girl and one boy? The probability of having a girl first is ½, and the probability of having a girl second is also ½. These are independent events, so we combine their probabilities by multiplying: ½ × ½ = ¼. In the same way, the probability of having two boys is also ¼. In families that have both a girl and a boy, the girl can be born first or the boy can be born first. The probability that a girl will be born first is ½, and the probability that a boy will be born second is also ½. We use the product law to combine the probabilities of these two independent events: ½ × ½ = ¼. Similarly, the probability that a boy will be born first and a girl second is also ¼. These two kinds of families represent mutually exclusive outcomes, so we use the sum law to combine them: ¼ + ¼ = ½. Notice that the probabilities of the three types of families (all mutually exclusive outcomes) add up to 1.

In working with probabilities, it is important to keep in mind a point that many gamblers forget. We can say that "a tossed coin has no memory." This means that if events are truly random, then past events have no influence on the probability of the occurrence of independent future events. The color of the iris of the human eye is controlled by alleles at several loci, but alleles at one locus are primarily responsible. The allele for brown eye color, B, is usually dominant to the allele for blue, b. If two heterozygous brown-eyed people marry, what is the probability that they will have a blue-eyed child? Clearly, there is 1 chance in 4 that any child of theirs will have blue eyes. Each mating is a separate, independent event; its result is not affected by the results of any previous matings. If these two brown-eyed parents have had three brown-eyed children and are expecting their fourth child, what is the probability that the child will have blue eyes? The unwary might guess that this one *must* have blue eyes, but in fact there is still only 1 chance in 4 that the child will have blue eyes and 3 chances in 4 that the child will have brown eyes.

DIHYBRID CROSSES

Our simple monohybrid crosses each involved a pair of alleles representing a single locus. Mendel also analyzed crosses involving alleles representing two or more loci. A mating that involves individuals differing in their alleles at two loci is called a **dihybrid cross.** The principles involved and the procedure for solving problems are exactly the same for monohybrid and for dihybrid (or *polyhybrid*) crosses. In the latter, of course, the number of different types of gametes is greater, and the number of different types of possible zygotes is correspondingly larger.

When two pairs of alleles are located in nonhomologous chromosomes, each pair is inherited independently of the other; that is, each pair separates during meiosis independently of the other. An example of a dihybrid F_2 is illustrated in Figure 10–9. When a homozygous black, short-haired guinea pig (BBSS, since short hair is dominant to long hair and, as we saw earlier, black is dominant to brown) and a homozygous brown, long-haired guinea pig (bbss) are mated, the BBSS animal produces gametes that are all BS and the bbss individual produces gametes that are all bs. Each gamete contains one and only one allele for each of the two loci. The union of the BS and bs gametes yields only individuals with the genotype BbSs. All of the F_1 offspring are heterozygous for hair color and for hair length, and all are phenotypically black and short-haired.

When two of the F_1 individuals are mated, each produces four kinds of gametes with equal probability: BS, Bs, bS, and bs. Hence, the Punnett

FIGURE 10–9 A dihybrid cross. When a black, short-haired guinea pig is crossed with a brown, long-haired one, all of the offspring are black and have short hair. However, when two members of the F₁ generation are crossed, the ratio of phenotypes is 9:3:3:1. Note that the two pairs of alleles considered here segregate independently.

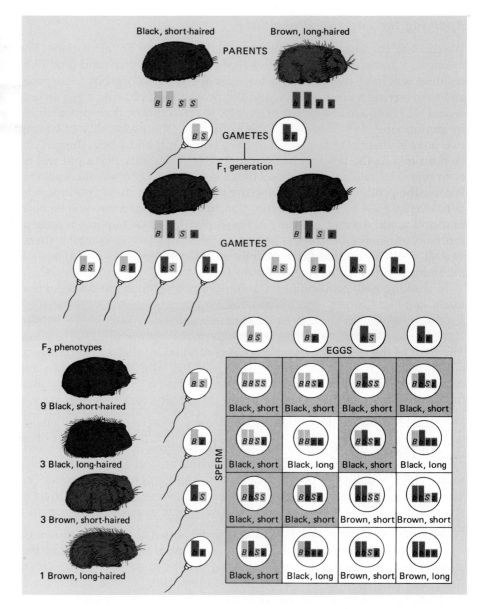

square will have 16 squares representing the zygotes, some of which will be genotypically or phenotypically alike. There are 9 chances in 16 of obtaining a black, short-haired individual; 3 chances in 16 of obtaining a black, long-haired individual; 3 chances in 16 of obtaining a brown, short-haired individual; and 1 chance in 16 of obtaining a brown, long-haired individual. This 9:3:3:1 phenotypic ratio is expected in a dihybrid F₂ if the loci are on nonhomologous chromosomes. It was on the basis of similar results that Mendel formulated his third principle of inheritance, the **principle of independent assortment** of alleles, which states that alleles of two or more different loci are distributed randomly with respect to one another during meiosis. As we shall see, independent assortment is actually a special case and does not always apply.

SUMMARY OF MENDEL'S LAWS

Now that we have learned about monohybrid and dihybrid crosses, we can state more formally how Mendel's laws apply to them. Mendel's *principle of dominance* applies to those cases where one allele masks another in a heter-

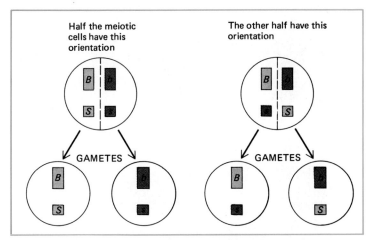

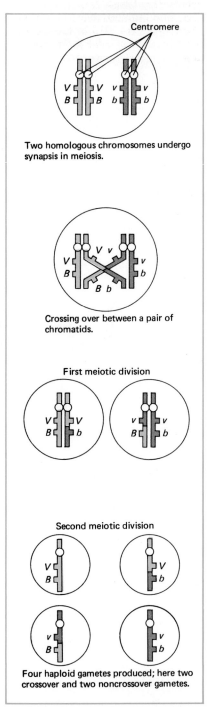

FIGURE 10–10 The meiotic basis of independent assortment. Notice that there are two independent ways that alleles of two unlinked loci can line up at meiotic metaphase I and subsequently disjoin at anaphase I. The first orientation produces half BS gametes and half bs gametes. The second orientation produces half Bs gametes and half bS gametes. Approximately half of the meiotic cells will have the first orientation and the other half will have the second, resulting in an overall 1:1:1:1 ratio for the four possible types of gametes. The sister chromatids and the second meiotic division are not shown.

ozygous individual. Mendel's *principle of segregation* may be stated as follows: Genes exist in individuals in allelic pairs, and in meiosis the alleles separate, or segregate. When the gametes are produced, each gamete has one and only one allele for each locus. This law can be best illustrated by a test cross mating between a heterozygous black guinea pig and a homozygous recessive brown guinea pig. Among the offspring there will be approximately ½ black and ½ brown guinea pigs because of the segregation of the alleles into separate gametes during meiosis.

Mendel's *principle of independent assortment* states that the members of one gene pair separate from each other in meiosis independently of the members of the other gene pairs. Each pair therefore comes to be assorted at random in the gametes. (As you can imagine, this law does not apply if the two gene pairs are located on the same pair of homologous chromosomes.) This law is illustrated by the dihybrid cross discussed previously in which the segregation of the Bb alleles is independent of the segregation of the Ss alleles. This is due to the fact that there are two different ways for nonhomologous chromosomes to be distributed during meiosis. These occur randomly, with about half of the meiotic cells having one kind of orientation and half having the opposite orientation (Figure 10–10).

LINKAGE, CROSSING OVER, AND CHROMOSOME MAPPING

Chromosomes are inherited as units, and they pair and separate during meiosis as units; thus, all of the alleles at different loci on a given chromosome tend to be inherited together. If the chromosomal units never changed, the genes on any one chromosome would always be inherited together. However, during meiosis, when the chromosomes pair and undergo synapsis (see Chapter 9), crossing over and genetic recombination occur. During **crossing over**, homologous (nonsister) chromatids may exchange entire segments of chromosomal material by a process of breakage and rejoining (Figure 10–11). A chromatid that has undergone crossing over has a new combination of genes; the term **genetic recombination** is

FIGURE 10–11 Crossing over and genetic recombination. Crossing over, the exchange of segments between chromatids of homologous chromosomes, permits the recombination of genes (for example, vB and Vb). The farther apart genes are located on a chromosome, the greater is the probability that they will be separated by an exchange of segments.

FIGURE 10–12 A cross involving linkage and crossing over. In fruitflies, the genes for vestigial versus normal wings and for black versus gray body are linked; they are located on the same chromosome. (Fruitflies are unusual in that crossing over occurs only in females and not in males. It is far more common for crossing over to occur in both sexes.)

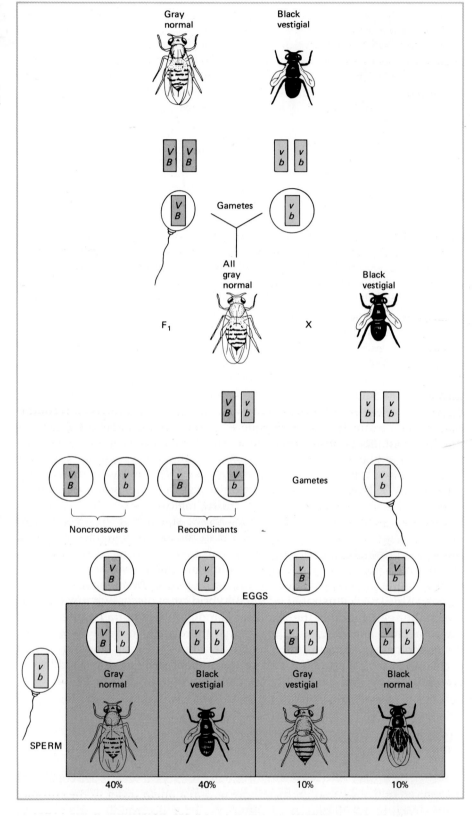

applied to the results. This exchange of segments occurs at random along the length of the paired homologous chromosomes. Several exchanges may occur at different points during a single meiotic division. In general, the greater the physical distance between any two genes in the chromosome, the greater is the likelihood that the genes will be separated by crossing over.

In fruit flies a locus controlling wing shape (the dominant allele V for normal wings and the recessive allele v for vestigial wings) and a locus controlling body color (the dominant allele B for gray and the recessive allele b for black) are located in the same pair of homologous chromosomes (Figure 10–12). They therefore tend to be inherited together and are said to be **linked.** If a homozygous BBVV fly is crossed with a homozygous bbvv fly, the F_1 flies will all have gray bodies and normal wings, and their genotype will be BbVv. However, if these F_1 flies are crossed with homozygous bbvv flies (a dihybrid test cross), the offspring will appear in a ratio that differs from the ordinary dihybrid test cross ratio.

If the loci governing these characteristics were *not* linked, their alleles would undergo *independent assortment* during meiosis. Thus, ¼ of the offspring would appear gray-bodied and normal-winged (BbVv); ¼ would be black-bodied and normal-winged (bbVv); ¼ would be gray-bodied and vestigial-winged (Bbvv); and ¼ would be black-bodied and vestigial-winged (bbvv). (Notice that the dihybrid test cross allows us to determine the genotypes of the offspring directly from their phenotypes.) If the loci *were* linked and no exchange of chromosomal segments occurred, only the **parental types**—flies with gray bodies and normal wings and flies with black bodies and vestigial wings—would appear among the offspring, and these would be present in approximately equal numbers.

However, in our example, there is an exchange between these two loci in some of the meiotic cells of the heterozygous female flies (see Figure 10–11). Because of this crossing over, some gray-bodied, vestigial-winged flies and some black-bodied, normal-winged flies will be seen among the offspring. These **recombinant types** are the flies that received a **recombinant** gamete from the heterozygous F_1 parent. A recombinant gamete is one that contains a *combination* of genes that was not present in the parental (P) generation. If the loci are linked, a recombinant gamete must carry a chromatid that has undergone crossing over so that it now contains a *new combination* of alleles for these two loci. In this instance about 20% of the gametes are recombinant gametes. Of the recombinant types, approximately 10% are gray flies with vestigial wings and 10% are black flies with normal wings. In such crosses, about 40% of the offspring are gray flies with normal wings, and another 40% are black flies with vestigial wings. These two make up the **parental,** or nonrecombinant, class of offspring.

The genetic distance between two loci in a chromosome is measured in **map units,** or recombination units, which serve as a measure of the percentage of crossing over between them. There is a rough correlation between this genetic distance and the actual physical distance along the chromosome. One can calculate the percentage of recombination by dividing the number of offspring receiving either of the two possible kinds of recombinant gametes (10 + 10) by the total number of offspring (40 + 40 + 10 + 10) and multiplying by 100. Thus, the V locus and the B locus can be said to have 20% recombination between them, or they can be said to be 20 map units apart. A 1% recombination between two loci equals a distance of 1 map unit.

The frequencies of recombination between specific loci have been measured in a number of species. All of the experimental results are consistent with the hypothesis that genes are present in a linear order in the chromosomes. Figure 10–13 illustrates the method for determining the order of genes in a chromosome.

Crossing over occurs at random, and more than one crossover between two loci in a single tetrad can occur in a given cell undergoing meiosis. We can observe only the frequency of offspring receiving recombinant gametes from the heterozygous parent, not the actual number of crossovers. In fact, the actual frequency of crossing over will be slightly more than the observed frequency of recombinant gametes. This is because the simultane-

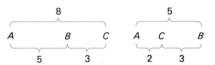

FIGURE 10–13 Genetic mapping. Gene order (i.e., which gene lies between the other two) is determined by the percentage of recombination between each of the possible pairs. In this hypothetical example, the percentage of recombination between A and B is 5% (corresponding to 5 map units) and between B and C is 3% (3 map units). If the recombination between A and C is 8% (8 map units), B must be in the middle. However, if the recombination between A and C is 2%, then C must be in the middle.

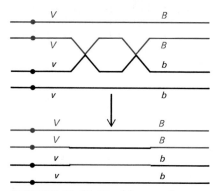

FIGURE 10–14 Double crossing over, which does not result in the formation of a recombinant gamete.

ous occurrence of two crossovers involving the same two homologous chromatids will reconstitute the original combination of genes (Figure 10–14). When two loci are relatively close together, this effect is minimized.

All of the genes in a particular chromosome tend to be inherited together and therefore are said to constitute a **linkage group.** The number of linkage groups determined by genetic tests is equal to the number of pairs of chromosomes. Through the putting together of the results of many crosses, detailed linkage maps have been developed for a number of eukaryotes, including the fruitfly (which has four pairs of chromosomes), the mouse, yeast, and *Neurospora* (a mold). (See "Focus on Genetic Mapping.") In addition, special genetic methods have made possible the development of a detailed map for *Escherichia coli,* a bacterium that has a single circular chromosome, and a number of other prokaryotes and viruses.

THE GENETIC DETERMINATION OF SEX

An exception to the general rule that the two members of a homologous pair of chromosomes are alike in size and shape is the sex chromosomes. The cells of the females of most species of animals contain two identical sex chromosomes, called **X chromosomes.** In contrast, in males the set of two sex chromosomes is composed of a single X chromosome and a smaller **Y chromosome** with which the X chromosome undergoes partial synapsis during meiosis. Human males have 22 pairs of **autosomes,** which are chromosomes other than the sex chromosomes, plus one X chromosome and one Y chromosome; females have 22 pairs of autosomes plus two X chromosomes.

The Y Chromosome

In humans and other mammals, maleness is determined in large part by the presence of the Y chromosome. Much of the evidence for this comes from studies on persons with abnormal sex chromosome constitutions (see Chapter 11). A person with an XXY constitution is a nearly normal male in external appearance but has underdeveloped gonads (Klinefelter syndrome). A person with one X but no Y chromosome has the appearance of an immature female (Turner syndrome).

In humans and other species in which the normal male has one X and one Y chromosome, half of the sperm contain an X chromosome and half contain a Y chromosome. All eggs contain one X chromosome (Figure 10–15). Fertilization of an X-bearing egg by an X-bearing sperm results in an XX female zygote; fertilization by a Y-bearing sperm results in an XY male zygote. We would expect to have equal numbers of X- and Y-bearing sperm and a 1:1 ratio of females to males. In fact, however, more males are conceived than females and more males die before birth. Even at birth the ratio is not 1:1; some 106 boys are born for every 100 girls. It is not known why this occurs, but it is assumed that the Y-bearing sperm have some competitive advantage.

An XY mechanism of sex determination is believed to operate in most species of animals, although it is not universal and many of the details may vary. The fruitfly, *Drosophila,* has XX females and XY males, but the Y is not male-determining; a fruitfly with an X chromosome and no Y chromosome has a male phenotype. In birds and butterflies the mechanism is reversed, with males being the equivalent of XX and females the equivalent of XY. In animals that are **hermaphroditic,** organs of both sexes are found in the same individual. These animals do not have sex chromosomes. Most flowering plants are hermaphrodites. When the sexes are in separate flowers but on the same plants, the plants are said to be **monoecious;** corn, wal-

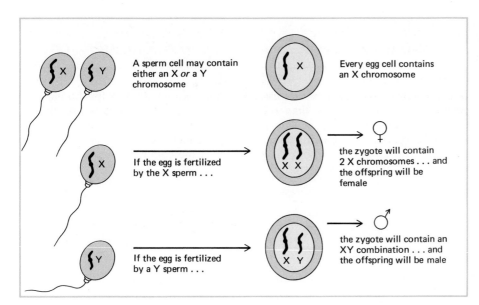

nuts, and oaks are examples of such plants. Far fewer flowering plants are **dioecious,** having male and female floral organs on separate plants. Dioecious plants with recognized sex chromosomes include the wild (not cultivated) strawberry.

X-Linked Traits

The human X chromosome contains many loci that are required in both sexes, whereas the Y chromosome contains only a few genes, principally the genes for maleness. Traits controlled by genes located in the X chromosome, such as color blindness and hemophilia, are sometimes called **sex-linked** traits. It is more appropriate, however, to refer to these as **X-linked** traits, since they follow the pattern of transmission of the X chromosome and strictly speaking are not linked to the sex of the organism per se.

A male receives from his father his Y chromosome (making him male) and from his mother a single X chromosome and therefore all of his genes for X-linked traits. A female receives one X from her mother and one X from her father. In the male, every X chromosome allele present is expressed, whether that allele was dominant or recessive in the female parent. A male is always **hemizygous** at every X-linked locus. The term *hemi* means "half"; a hemizygous male is neither homozygous nor heterozygous.

For most X-linked loci, the abnormal or uncommon allele is recessive in the female and the normal or most common allele is dominant. In a female, two recessive X-linked alleles must be present for the abnormal phenotype to be expressed, whereas in the hemizygous male a single abnormal allele will be expressed. One practical consequence of this is that although these abnormal alleles may be carried by a female, they are usually expressed only in their male offspring.

In order to be expressed in a female, a recessive X-linked allele must be present on both chromosomes; that is, the alleles must be inherited from both parents. A color-blind female, for example, must have a color-blind father and a mother who is at least heterozygous for color blindness (Figure 10–16). Such a combination is unusual. In contrast, a color-blind male need only have a mother who is heterozygous for color blindness; his father can be normal. Hence, X-linked recessive traits are generally much more common in males than in females, a fact that may explain why human male embryos are more likely to die.

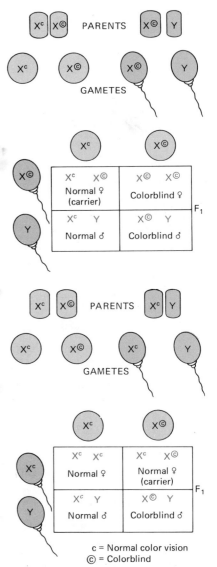

FIGURE 10–16 X-linkage. Two crosses involving color blindness, an X-linked recessive trait. Note that the Y chromosome does not carry a gene for color vision.

Focus on GENETIC MAPPING

A genetic map provides a summary of the genetic information about a species. The map positions are determined by measurement of the frequency of recombination between the genes. Genetic maps have been at least partially worked out for many organisms, including some bacteria, some fungi, corn, (*Zea mays*), the fruitfly (*Drosophila melanogaster*), the mouse, and the human. More than 1000 gene loci have been identified on the map of *Drosophila's* four chromosomes. A partial map showing some of the gene loci is illustrated here. Note that both the normal and the common mutant phenotypes are indicated.

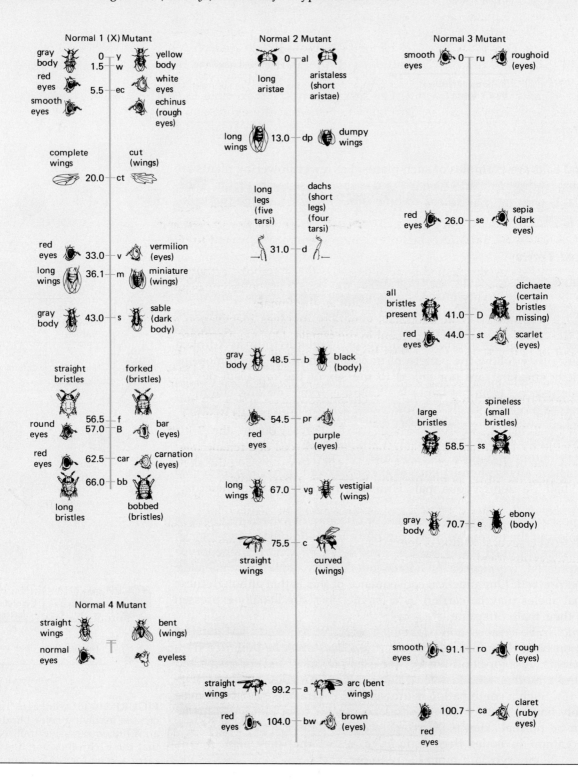

Sex-Influenced Traits

Not all of the characteristics that differ in the two sexes are X-linked. Certain **sex-influenced** traits are inherited through autosomal genes, but the *expression* of alleles at these loci can be altered or influenced by the sex of the animal. Therefore, males and females with the same genotype with respect to these loci may have different phenotypes. Pattern baldness in humans, characterized by premature loss of hair on the front and top of the head but not on the sides, is far more common among males than among females. It has been proposed that a single pair of alleles is involved. The allele B_1, which is responsible for pattern baldness, is dominant in males and recessive in females; the allele for normal hair growth can be designated B_2. Individuals with the genotype B_1B_1 show pattern baldness, regardless of sex. Persons with a B_1B_2 genotype are bald if they are male but not bald if they are female. Individuals with the genotype B_2B_2 are not bald, regardless of sex. There is evidence that the expression of most sex-influenced traits is strongly modified by sex hormones. For example, male hormones (see Chapter 49) are strongly implicated in the expression of pattern baldness.

Dosage Compensation

The X chromosome contains numerous genes that are required by both sexes, yet a normal female has two copies ("doses") for each locus, whereas a normal male has only one. Generally, a mechanism of **dosage compensation** is required to make one dose in the male equivalent to two doses in the female.

Male fruitflies accomplish this by increasing the metabolic activity of their single X chromosome. In most tissues the male X chromosome is just as active as the two X chromosomes present in the female.

Dosage compensation in mammals generally involves inactivation of one of the two X chromosomes present in the female. During interphase a dark spot of chromatin, called a **Barr body**, is visible at the edge of the nucleus of female mammalian cells (Figure 10–17). The Barr body has been found to represent one of the two X chromosomes, which has become dense and darkly staining. The other X chromosome resembles the autosomes in that during interphase it is a greatly extended thread that is not evident by light microscopy. From this and other evidence, the British geneticist Mary Lyon has suggested that in any one cell of a female mammal, only one of the two X chromosomes is active in expressing the alleles that it carries; the other is inactive and is seen as a Barr body.

Because only one X chromosome is active in any one cell, female mammals are generally **mosaic** for the expression of alleles at X-linked loci. This mosaicism is sometimes (but not always) evident in the phenotype. Mice and cats have several X-linked genes for certain coat colors. Females that are heterozygous for such genes may show patches of one coat color in the

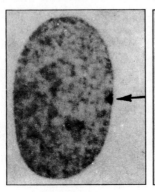 

FIGURE 10–17 A Barr body in human fibroblasts cultured from the skin of a female (approximately ×2200).

FIGURE 10–18 A calico cat. This cat has X-linked alleles for both black and yellow (or orange) pigmentation of the fur, but because of random X chromosome inactivation, black is expressed in some clones of cells and yellow (or orange) is expressed in others. Because other genes affecting fur color are also present, white patches are usually evident as well.

midst of areas of the other coat color. This phenomenon, termed **variegation,** is evident in calico and tortoise-shell cats (Figure 10–18). Early in development, when relatively few cells are present, X chromosome inactivation occurs randomly in each cell. When any one of these cells divides by mitosis, the cells of the resulting clone will all have the same active X chromosome, and therefore a patch of cells that all express the same color will develop.

THE RELATIONSHIP BETWEEN GENOTYPE AND PHENOTYPE

The relationship between a given locus and the characteristic it controls may be simple: A single pair of alleles at a locus may regulate the appearance of a single characteristic of the organism (e.g., tall versus short). Alternatively, the relationship may be more complex: A pair of alleles at a locus may participate in the control of several characteristics, or alleles at many loci may cooperate to regulate the appearance of a single characteristic. Not surprisingly, these more complex relationships are quite common. As you will learn in Chapters 13 and 14, each locus is a portion of DNA in which biological information is stored as a triplet (three-base) code in the sequence of nucleotides that compose the double helix of the DNA molecule. The information is "read out," and in a great many cases a specific protein is ultimately synthesized. The presence of a specific protein, such as an enzyme, usually provides the chemical basis for the genetic trait. Because most biologically important molecules are synthesized by complex metabolic pathways involving a number of enzymes, it is not difficult to appreciate why relationships between genes and the characteristics of the organism are complex.

We may assess the phenotype on one or many levels. It may be a morphological characteristic such as shape, size, or color. It may be a physiological characteristic or even a biochemical trait, such as the presence or absence of a specific enzyme required for the metabolism of a specific substrate. The phenotypic expression of genes may be altered by changes in the environmental conditions under which the organism develops.

Incomplete Dominance and Codominance

From studies of the inheritance of many traits in a wide variety of organisms, it is clear that one member of a pair of alleles may not be completely dominant to the other. Indeed, it is improper to use the terms *dominant* and *recessive* in such instances. For example, red and white are common flower colors in Japanese four o'clocks. Each color breeds true when these plants are self-pollinated. What flower color might we expect in the offspring of a cross between a red-flowering plant and a white-flowering one? Without knowing which is dominant, we might predict that all would have red flowers or all would have white flowers. This cross was first made by the German botanist Karl Correns (one of the rediscoverers of Mendel's work), who found that all F_1 offspring have pink flowers! Does this result in any way prove that Mendel's assumptions about inheritance are wrong? Quite the contrary, for when two of these pink-flowered plants were crossed, offspring appeared red-flowered, pink-flowered, and white-flowered in a ratio of 1:2:1 (Figure 10–19). In this instance, as in all other aspects of science, results that differ from those predicted simply prompt scientists to reexamine and modify their assumptions to account for these exceptional results. The pink-flowered plants are clearly the heterozygous individuals, and neither the red allele nor the white allele is completely dominant.

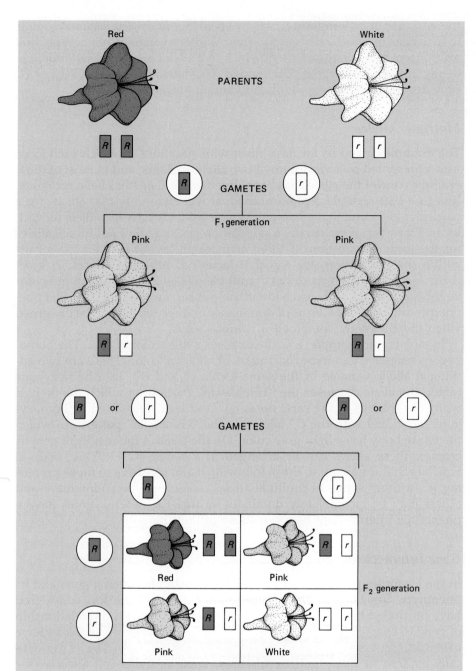

FIGURE 10–19 Incomplete dominance in Japanese four o'clocks. Red is incompletely dominant to white in some types of flowers. A plant with the genotype Rr has pink flowers.

When the heterozygote has a phenotype that is intermediate between those of its two parents, the genes are said to show **incomplete dominance.** In these crosses the genotypic and phenotypic ratios are identical.

Incomplete dominance is not unique to Japanese four o'clocks. Red- and white-flowered sweet pea plants also produce pink-flowered plants when crossed, and numerous additional examples are known in both plants and animals.

In both cattle and horses, reddish coat color is not completely dominant to white coat color. Heterozygous individuals have coats that are roan-colored—that is, reddish—but with spots of white hairs. If you saw a white mare nursing a roan-colored colt, what would you guess was the coat color of the colt's father? Because the reddish and white colors are expressed independently in the roan heterozygote, we sometimes refer to this as a case of **codominance.** Strictly speaking, *incomplete dominance* refers to instances in which the heterozygote is intermediate in phenotype, and *codominance* refers to instances in which two alleles are expressed independently in the heterozygote. The human ABO blood group (see Chapter 11) provides a classical example of codominant alleles.

Multiple Alleles

The examples given so far have dealt with situations in which each locus was represented by a maximum of two allelic variants, and in most of these examples one of the alleles has been dominant and one has been recessive. It is true that a single diploid individual will have only two alleles for a particular locus and that a haploid gamete will only have one allele for each locus. However, if we survey a population, we may find additional alleles for the locus in question. If three or more alleles for a given locus exist within the population, we say that locus has **multiple alleles.** A great many loci can be shown to have multiple alleles if the population is surveyed carefully. Usually, each identifiable allele can produce a distinct phenotype, and certain patterns of dominance and recessivity can be discerned when the alleles are combined in various ways.

In rabbits, for example, a C allele causes a fully colored coat. The homozygous recessive genotype, cc, causes albino coat color. There are two additional allelic variants of the same locus, c^h and c^{ch}. In a homozygous rabbit, the allele c^h causes the "Himalayan" pattern, in which the body is white but the tips of the ears, nose, tail, and legs are colored. A homozygous individual with the c^{ch} allele has the "chinchilla" pattern, in which the entire body has a light gray color. On the basis of the results of genetic crosses, these alleles can be arranged in a series—$C > c^{ch} > c^h > c$—in which each is dominant to those following it and recessive to those preceding it. In other series of multiple alleles, some may be codominant and others may be incompletely dominant so that the heterozygotes have a phenotype intermediate between those of their parents.

Gene Interactions

In the examples presented so far, the relationship between a gene and its phenotype has been direct, precise, and exact, and the loci considered have controlled the appearance of single traits. However, the relationship of gene to characteristic may be quite complex. Most genes probably have many different effects, a quality referred to as **pleiotropy.** This is dramatically evident in many genetic diseases, such as sickle cell anemia (see Chapter 11), in which multiple symptoms can be traced to a single pair of alleles. Albino individuals have a lack of pigment in the skin, hair, and eyes, demonstrating that a single locus can simultaneously affect a number of characteristics. In addition, virtually every characteristic of the organism is controlled by a large number of loci. We are not always aware of this because not all of these loci have been identified by the finding of allelic variants. **Epistasis** is a common type of gene interaction in which the presence of a particular allele of one gene pair determines whether certain alleles of another gene pair will be expressed. By such mechanisms several pairs of alleles may interact to affect a single trait, or one pair may inhibit or reverse the effect of another pair. More than 12 pairs of alleles interact in various ways to produce coat color in rabbits, and more than 100 pairs are concerned with eye color and shape in fruitflies.

One of the simplest types of gene interaction is illustrated by the inheritance of combs in poultry (Figure 10–20). The allele for a rose comb, R, is dominant to that for a single comb, r. Another gene pair governs the inheritance of a pea comb, P, versus a single comb, p. A single-combed fowl

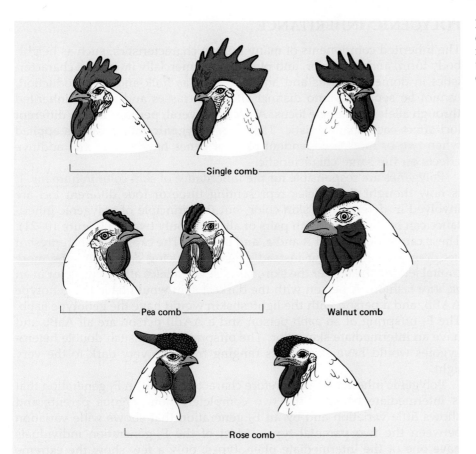

FIGURE 10–20 The different types of genetically determined combs in roosters. Two gene pairs govern the inheritance of these types of combs.

Single comb

Pea comb

Walnut comb

Rose comb

must therefore have the genotype pprr; a pea-combed fowl is either PPrr or Pprr; and a rose-combed fowl is either ppRR or ppRr. When a homozygous pea-combed fowl is mated to a homozygous rose-combed one, the offspring have neither a pea nor a rose comb, but a completely different type, called a *walnut comb*. The walnut comb phenotype is produced whenever a fowl has one or two R alleles plus one or two P alleles. What would you predict about the types of combs among the offspring of two heterozygous walnut-combed fowl, PpRr? How does this form of epistasis affect the phenotypic ratio in the F_2 that would be produced? Is it the typical mendelian 9:3:3:1 ratio?

We have already seen that coat color in guinea pigs is determined by the B and b allelic pair, with the B allele for black coat dominant to the b allele for brown coat. The expression of either phenotype, however, depends on the presence of a dominant allele at yet another locus. This allele, C, codes for the enzyme tyrosinase, which converts a colorless precursor into the pigment melanin and hence is required for the production of any kind of pigment. Thus, an animal that is homozygous recessive for this allele lacks the enzyme and produces no melanin, and is therefore a white-coated, pink-eyed albino, no matter what combination of B and b alleles may be present. When an albino guinea pig with the genotype ccBB is mated to a brown guinea pig with the genotype CCbb, the F_1 generation will be black coated, CcBb. When two such animals are mated, their offspring will appear black-coated, brown-coated, and albino in a ratio of 9:3:4. (Make a Punnett square to verify this.) In this example of epistasis, pairs of genes representing two entirely different loci interact in such a way that one dominant (C) will produce its effect regardless of whether the other is present, but the second (B) will produce its effect only in the presence of the first.

POLYGENIC INHERITANCE

The inherited components of many human characteristics, such as height, body form, and skin color, and of many commercially important characteristics in domestic plants and animals, such as milk and egg production, cannot be separated into distinct alternate classes and are not inherited through alleles at a single locus. Alleles at several, perhaps many, different loci affect each characteristic. The term **polygenic inheritance** is applied when two or more independent pairs of genes have similar and additive effects on the same characteristic.

Polygenes are responsible for the inheritance of skin color in humans. It is now thought that alleles representing three or four different loci are involved in determining skin color, but the principle of polygenic inheritance can be illustrated with pairs of alleles at only two loci (Figure 10–21). These can be designated A and a, and B and b. The capital letters represent *not* dominant alleles, but instead alleles producing dark skin—the more capital letters, the darker the skin, because the alleles affect skin color in an *additive* fashion. A person with the darkest skin would have the genotype AABB, and a person with the lightest skin would have the genotype aabb. The F_1 offspring of an aabb person and a AABB person are all AaBb and have an intermediate skin color. The offspring of two such double heterozygotes would have skin colors ranging from the very dark to the very light.

Polygenic inheritance is therefore characterized by an F_1 generation that is intermediate between the two completely homozygous parents and shows little variation and by an F_2 generation that shows wide variation between the two parental types. Most of the F_2-generation individuals have one of the intermediate phenotypes; only a few show the extreme phenotypes of the grandparents (P generation). On average, only 1 of 16 will be as dark as the very dark grandparent, and only 1 of 16 will be as

FIGURE 10–21 A simplified model to illustrate polygenic inheritance of skin color in humans. Here it is assumed that skin color is governed by alleles at two loci. The alleles producing dark skin are shown as capitals, but they are not dominant. Instead they act in an additive fashion.

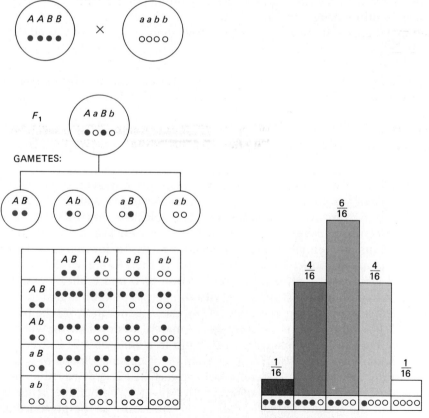

(a)

(b) Phenotypic ratios

light as the very light grandparent. The genes A and B produce about the same amount of darkening of the skin; hence, the genotypes AaBb, AAbb, and aaBB all produce similar intermediate phenotypes.

The model used here for the inheritance of skin color in humans is a rather simple example of polygenic inheritance because only two major allelic pairs were used. The inheritance of height in humans involves alleles representing ten or more loci. Because many allelic pairs are involved, and because height is modified by a variety of environmental conditions, the height of adults ranges from perhaps 125 to 215 cm. If we were to measure the height of 1000 adult American men selected at random, we would find that only a few are as tall as 215 cm or as short as 125 cm. The height of most would cluster around the mean, about 170 cm. When the number of people at each height is plotted against height in centimeters and the points are connected, the result is a bell-shaped curve, called a **normal distribution curve** (Figure 10–22). If you were to measure the height of 1000 men and women combined, what sort of curve would you expect the data to generate?

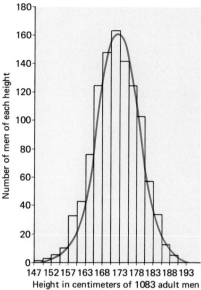

FIGURE 10–22 A normal distribution curve showing the distribution of height in 1083 adult white males. The blocks indicate the actual number of men whose heights were within the unit range. For example, there were 163 men whose heights were between 170 and 173 cm. The smooth curve is a normal curve based on the mean and standard deviation of the data. Note the bell shape of the curve.

INBREEDING, OUTBREEDING, AND HYBRID VIGOR

How do geneticists go about establishing a breed of cow that will give more milk, a strain of hens that will lay bigger eggs, or a variety of corn with more kernels per ear? By selection of the organisms that manifest the desired phenotype and use of these organisms in further matings, a true breeding strain with the commercially advantageous trait is gradually developed. Such a strain should be homozygous for all of the polygenes involved, whether they be dominant, recessive, or additive in their effects. It is clear that there is a limit to the effectiveness of breeding by selection. When a strain becomes homozygous for all of the polygenes involved, further selective breeding cannot increase the desired quality. Moreover, because of **inbreeding**—the mating of two closely related individuals—the strain may become homozygous for multiple deleterious traits as well (Figure 10–23). Certain dog breeds, for instance, are known for their susceptibility to congenital dislocation of the femur. There is evidence that human inbreeding increases the frequency of birth defects; this is why marriages of close relatives (first cousins or closer) are forbidden by law in many states.

The mating of individuals of totally unrelated strains, termed **outbreeding,** frequently leads to offspring that are much better adapted for survival

FIGURE 10–23 Inbreeding and outbreeding. Corn plants from left to right represent successive generations of inbreeding (self-pollination). There is an obvious reduction in vigor for several generations, but eventually plants become homozygous at most genetic loci, and new generations are more uniform.

than either parent. Such improvement reflects a phenomenon called **hybrid vigor.** Mongrel dogs are often hardier than highly inbred purebreds. The mule, a hybrid resulting from the mating of a horse and a donkey, is a strong, sturdy beast, better suited for many tasks than either parent. However, like many animals that are the product of two different species, the mule cannot reproduce. (The slightly different horse and donkey chromosomes cannot undergo proper synapsis during meiosis.) A large proportion of the corn, wheat, and other crops grown in the United States consists of hybrid strains. Each year the seed to grow these crops must be obtained by mating of the original strains. The hybrids are heterozygous at a great many loci and give rise, even when self-fertilized, to a wide variety of forms, none of which is as good as the original hybrid. (The seeds produced by F_1 hybrid corn plants are not normally planted, but eaten instead!)

One explanation of hybrid vigor is the following: Each of the parental strains is homozygous for certain undesirable recessive genes, but any two strains are homozygous for different undesirable genes. Each strain contains dominant genes to make up for the recessive undesirable genes of the other strain. One strain then might have the genotype AAbbCCdd, and another strain the genotype aaBBccDD. (The capital letters represent dominant genes for desirable traits, and the lowercase letters represent recessive genes for undesirable traits.) The hybrid offspring, with the genotype AaBbCcDd, would express all of the desirable and none of the undesirable traits of the two parental strains.

Sometimes a heterozygous individual will express a more extreme phenotype than either of the two kinds of homozygotes. Geneticists call this phenomenon **overdominance.** If the phenotype of the heterozygote is more desirable, we say that there is a **heterozygote advantage.** The use of this term implies that at least sometimes there is an advantage to heterozygosity for the sake of heterozygosity. In humans, individuals who are heterozygous for the recessive allele responsible for sickle cell anemia (s) and the normal dominant allele (S) appear to have increased resistance to the parasite that causes malaria, a significant advantage in areas of the world where malaria is still uncontrolled. Homozygous normal individuals (SS) appear to be less resistant to malaria. Homozygous sickle cell individuals (ss) are at a distinct disadvantage due to severe anemia and other serious effects of the sickle cell allele.

■ SUMMARY

I. Mendel's inferences from his experiments with the breeding of garden peas have been tested repeatedly in all kinds of diploid organisms and found to be true. These principles have been extended and now can be stated in a more modern form.

   A. Today we know that the genes are in chromosomes; we use the term *locus* to refer to the site a gene occupies in the chromosome.

   B. Different states of a particular gene are called *alleles*; they occupy corresponding loci on homologous chromosomes. Genes therefore exist in pairs (alleles) in diploid individuals.

   C. An individual that carries two identical alleles for a given locus is said to be *homozygous* for that locus. If the two alleles are different, that individual is said to be *heterozygous* for that locus.

   D. According to Mendel's principle of dominance, one allele (the dominant allele) may mask the expression of the other allele (the recessive allele) in a

heterozygous individual. For this reason an individual's appearance (phenotype) may be different from his or her genetic constitution (genotype). Dominance does not always apply, and alleles may be incompletely dominant or codominant.

   E. According to Mendel's principle of segregation, during meiosis, the alleles for each locus separate, or segregate, from each other as the homologous chromosomes separate. When haploid gametes are formed, each will contain one and only one allele for each locus.

   F. According to Mendel's principle of independent assortment, during meiosis, each pair of alleles will separate independently of the pairs of alleles that are located in other pairs of homologous chromosomes. Alleles of different loci therefore assort randomly into the gametes.

   G. Each chromosome behaves genetically as if it were composed of genes arranged in a linear order.

Genes that are in the same chromosome are said to be *linked* and do not undergo independent assortment. Linked genes can be recombined by the process of crossing over (breaking and rejoining of homologous chromatids), which occurs in meiotic prophase. By measuring the frequency of recombination (resulting from crossing over) between various genes, it is possible to construct a genetic map of a chromosome.

H. A cross between homozygous parents (P generation) that differ from each other with respect to their alleles at one locus is called a *monohybrid cross;* if they differ at two loci, it is a *dihybrid cross.* The first generation of offspring is heterozygous and is called the *first filial,* or F_1, *generation;* the generation produced by a cross of two F_1 individuals is the *second filial,* or F_2, *generation.* If an F_1 individual is crossed with a homozygous recessive individual, the cross is called a *test cross.*

II. Genetic ratios can be expressed in terms of probabilities.
   A. Any probability is expressed as a fraction—that is, the number of favorable events divided by the total number of events. This can range from 0 (an impossible event) to 1 (a certain event).
   B. Probabilities can be multiplied and added like any other fractions.
   C. The probability of two independent events occurring together is the product of the probabilities of each occurring separately.
   D. The probability that one or the other of two mutually exclusive events will occur is the sum of their separate probabilities.

III. The sex of humans and many other animals is determined by the X and Y sex chromosomes or their equivalents. Chromosomes that are not sex chromosomes are called *autosomes.*
   A. Normal female mammals have two X chromosomes; normal males have one X and one Y.
   B. The fertilization of an X-bearing egg by an X-bearing sperm results in a female (XX) zygote. The fertilization of an X-bearing egg by a Y-bearing sperm results in a male (XY) zygote.
   C. The Y chromosome in mammals appears to be responsible for determining male sex.

D. The X chromosome contains many important genes that are unrelated to sex determination and are required by both males and females. A male receives all of his X-linked genes from his mother. A female receives X-linked genes from both parents.
   E. In the female mammal there is dosage compensation of X-linked genes so that two copies in the female are equivalent to one copy in the male. One of the two X chromosomes is inactive in each cell and is seen as a darkly-staining Barr body at the edge of the interphase nucleus.

IV. The term *multiple alleles* is applied to three or more alleles that can occupy a single locus on the chromosome; each allele produces a specific phenotype. A *diploid* individual has any two of the alleles; a *haploid* individual or *gamete* has only one.

V. The relationship between a gene and its phenotype may be quite complex.
   A. Most genes have many different effects; this is known as *pleiotropy.*
   B. The presence of a particular allele of one gene pair may determine whether alleles of another gene pair will be expressed; this is called *epistasis.*

VI. In polygenic inheritance, two or more independent pairs of genes may have similar and additive effects on the phenotype.
   A. Many human characteristics, such as height and skin color, as well as many characteristics in other animals and plants, are inherited through polygenes.
   B. In polygenic inheritance, the F_1 generation is intermediate between the two parental types and shows little variation. The F_2 generation shows wide variation between the two parental types.

VII. Inbreeding, the mating of two closely related individuals, greatly increases the probability that an individual offspring will be homozygous for recessive genes. Outbreeding, the mating of totally unrelated individuals, increases the probability that the offspring will be heterozygous at many loci. These heterozygous individuals may be stronger and better able to survive than either parent, a phenomenon known as *hybrid vigor.*

■ POST-TEST

1. The specific site in the chromosome occupied by a given gene is termed its _____.
2. Genes governing different states of the same characteristic of the organism and occupying corresponding loci in homologous chromosomes are termed _____.
3. A cross between two organisms differing with respect to alleles of a single locus is a _____ cross.
4. An organism's genetic constitution, expressed in symbols, is called its _____.
5. The appearance of an individual with respect to a given inherited characteristic is known as its _____.
6. An allele that is expressed in the phenotype of a heterozygous individual is a _____ allele.

7. A _____ allele can only be expressed in the phenotype of a homozygous individual.
8. An organism with two identical alleles for a particular locus is said to be _____ for that locus; an organism with two different alleles for a particular locus is said to be _____ for that locus.
9. The offspring of the parental (P) generation are called the _____ _____ generation. This is abbreviated as the _____ generation.
10. The probability that two independent events will coincide is the _____ of their individual probabilities.
11. The probability that one or the other of two mutually exclusive events will occur is the _____ of their individual probabilities.

12. A probability of _____ expresses a certainty; a probability of _____ expresses an impossibility.

13. In crosses involving a pair of alleles that show incomplete dominance, the genotypic and phenotypic ratios are _____.

14. A mating of individuals that have different alleles at two loci is called a _____ cross.

15. If a particular characteristic of the organism is governed by two or more pairs of genes that have similar and additive effects, we say that characteristic is under _____ control.

16. If three or more alleles for a given locus are present in the population, we say that locus has _____ alleles.

17. The genes in a given chromosome tend to be inherited together and are said to be _____.

18. The mating of two closely related individuals, such as first cousins, is termed _____.

19. The offspring of totally unrelated parents may be better adapted for survival than either parent. This phenomenon is called _____ _____.

■ REVIEW QUESTIONS

1. Show by diagrams how the alleles in gene pairs located in different pairs of chromosomes assort independently in meiosis.

2. In peas, yellow seed color is dominant to green. State the colors of the offspring of the following crosses:
   a. homozygous yellow × green
   b. heterozygous yellow × green
   c. heterozygous yellow × homozygous yellow
   d. heterozygous yellow × heterozygous yellow

3. If two animals heterozygous for a single pair of alleles are mated and have 200 offspring, about how many would be expected to have the phenotype of the dominant allele (i.e., to look like the parents)?

4. When two long-winged flies were mated, the offspring included 77 with long wings and 24 with short wings. Is the short-winged condition dominant or recessive? What are the genotypes of the parents?

5. A blue-eyed man, both of whose parents were brown-eyed, married a brown-eyed woman whose father was blue-eyed and whose mother was brown-eyed. If eye color is inherited simply, what are the genotypes of the individuals involved?

6. Outline a breeding procedure whereby a true-breeding strain of red cattle could be established from a roan bull and a white cow.

7. What is the probability of rolling a seven with a pair of dice?

8. In rabbits, spotted coat (S) is dominant to solid color (s) and black (B) is dominant to brown (b). A brown spotted rabbit is mated to a solid black one, and all the offspring are black spotted. What are the genotypes of the parents? What would be the appearance of the F_2 generation if two of these F_1 black spotted rabbits were mated?

9. The long hair of Persian cats is recessive to the short hair of Siamese cats, but the black coat color of Persians is dominant to the brown-and-tan coat color of Siamese. If a pure black, long-haired Persian is mated to a pure brown-and-tan, short-haired Siamese, what will be the appearance of the F_1 offspring? If two of these F_1 cats are mated, what is the chance that a long-haired, brown-and-tan cat will be produced in the F_2 generation?

10. The expression of an allele called *frizzle* in fowl causes abnormalities of the feathers. As a consequence, the animal's body temperature is lowered, adversely affecting the function of many internal organs. When one gene affects many characteristics of the organism in this way, we say that gene is _____.

11. A walnut-combed rooster is mated to three hens. Hen A, which is walnut-combed, has offspring in the ratio of 3 walnut:1 rose. Hen B, which is pea-combed, has offspring in the ratio of 3 walnut:3 pea:1 rose:1 single. Hen C, which is walnut-combed, has only walnut-combed offspring. What are the genotypes of the rooster and three hens?

12. What kinds of matings result in the following phenotypic ratios?
   a. 3:1
   b. 1:1
   c. 9:3:3:1
   d. 1:1:1:1

13. The weight of the fruit in a certain variety of squash is determined by three pairs of genes: AABBCC produces fruits weighing 6 lb each, and aabbcc produces fruits weighing 3 lb each. Each gene represented by a capital letter adds $\frac{1}{2}$ lb to the weight. When a plant that produces 6-lb fruits is crossed with a plant that produces 3-lb fruits, all of the offspring produce fruits that weigh 4.5 lb each. What would be the weights of the fruits produced by the F_2 plants, if two of these F_1 plants were crossed?

14. The X-linked *barred* locus in chickens controls the pattern of the feathers, with the alleles B for barred pattern and b for no bars. If a barred male is mated to a nonbarred female, what will be the appearance of the male and female progeny? What commercial usefulness does this have?

15. A locus for coat color in cats is X-linked and has two alleles: B, which produces yellow coat, and b, which produces black coat. Heterozygous females, Bb, have a tortoise-shell pattern. What kind of offspring (and in what proportions) would be expected to result from the mating of a black male and a tortoise-shell female? Would you ever expect to see a male tortoise-shell cat?

16. Individuals of genotype AaBb were mated to individuals of genotype aabb. One thousand offspring were counted:

Aabb	474
aaBb	480
AaBb	20
aabb	26

This type of cross is known as a _____
_____. Are these loci linked? What are the
two parental classes and the two recombinant classes of
offspring? What is the percentage of recombination between these two loci? How many map units apart are
they?

17. Genes A and B are 6 map units apart and A and C are 4
map units apart. Which gene is in the middle if B and C
are 10 map units apart? Which is in the middle if B and
C are 2 map units apart?

■ RECOMMENDED READINGS

There are a number of well-written college level genetics
texts that cover the principles of transmission genetics in
eukaryotes. The following are representative examples:

Avers, C.J. *Genetics.* 2nd ed. Boston, Trindle, Weber, and
Schmidt, 1984.

Klug, W.S. and M.R. Cummings, *Concepts of Genetics*, Columbus, Ohio, Charles E. Merrill, 1983.

Snyder, L., Freifelder, D. and Hartl, D. *General Genetics,*
Boston, Jones and Bartlett, 1985.

Strickberger, A.H. *Genetics.* 3rd ed. New York, Macmillan,
1985.

Suzuki, D.T., Griffiths, A.J.F., Miller, J., and Lewonton,
R.C. *An Introduction to Genetic Analysis.* 3rd ed. New
York, W.H. Freeman, 1986.

Whitehouse, H.L., *Towards an Understanding of the Mechanism of Heredity.* 3rd ed. London, St. Martin's Press,
1973.

Family portrait

11

Human Genetics

■ LEARNING OBJECTIVES

After you have read this chapter you should be able to:

1. Discuss the various reasons that human beings are not very favorable subjects for the study of inheritance.
2. Distinguish between environmentally induced and inherited abnormalities, and between chromosome abnormalities and gene defects.
3. Describe the phenomenon of nondisjunction, and discuss its role in the production of Down syndrome, Klinefelter syndrome, and Turner syndrome.
4. Describe how amniocentesis is used in the prenatal diagnosis of human genetic abnormalities; state the relative advantages and disadvantages of amniocentesis and chorionic villus sampling.
5. Discuss the scope and implications of genetic counseling.
6. Discuss the inheritance of the ABO and Rh system human blood types.
7. Discuss the role of polygenes in the inheritance of quantitative traits in humans.
8. Discuss the impact of genetics on human society.

Quite naturally, geneticists have great interest in the study of human genetics. The trouble is that humans do not serve well as the subjects of most types of genetic research. For the study of the mode of inheritance in any species, geneticists should ideally (1) have standard stocks of genetically identical individuals—that is, **isogenic strains**—that are homozygous at virtually all of their loci, (2) conduct **controlled matings** between members of different isogenic strains, and (3) raise the offspring under carefully controlled conditions. The organisms favored in genetic studies include bacteria, fungi, fruitflies, mice, and corn. These organisms produce many offspring for each pair of parents, and the time between successive generations is short.

Of course, it would be virtually impossible, as well as unethical, to conduct such studies on humans. Members of the human species are very diverse; that is, they are heterozygous for many genes. Although some phenotypes may be a factor in mate selection, genotype is not, and in any case few persons would be willing to select a mate according to the needs of genetic researchers. Furthermore, human families are small and more than 20 to 30 years elapse between generations. Finally, few people would be willing to release their children to be raised in a controlled laboratory environment.

The study of human genetics has nevertheless been quite productive, due to the use of alternative methodologies. Early studies of human heredity usually dealt with readily identified pairs of contrasting traits and their distribution among members of a family, as illustrated by the **pedigree** shown in Figure 11–1. This method is still useful, but because human families tend to be small and information on certain family members may be lacking, it has serious limitations. Therefore, human geneticists also use methods that allow them to make inferences about the mode of inheritance of a trait based on studies of its distribution in an entire population. By applying the laws of probability to data obtained from a relatively large sample of individuals that are representative of the population, it is often possible to determine if the mode of inheritance is simple or complex, if more than one locus is involved, and so on. The methods used by population geneticists are introduced in Chapter 17.

Despite the inherent difficulties, a great deal has been learned about human inheritance, and the field is progressing rapidly. This results in part from the medical attention given to genetic diseases in humans. The extensive medical records of disease serve as a very useful data pool on which hypotheses may be based and against which they may be tested. Furthermore, some of the phenomena in human inheritance that were initially quite puzzling have been clarified by the solutions of analogous problems in the inheritance of bacteria, fungi, flies, and mice. In the field of chemical genetics, many important discoveries made originally with human material have been confirmed by experiments using bacteria or fungi. For example, the earliest evidence that genes determine the sequence of amino acids in proteins came from a comparison of normal hemoglobin molecules (the oxygen-carrying protein in the blood) with the abnormal hemoglobin found in persons suffering from sickle cell anemia, a genetic disease discussed later in this chapter.

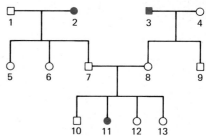

FIGURE 11–1 A pedigree for albinism (lack of pigment in the hair, skin, and eyes), which is inherited as an autosomal recessive trait. Males are indicated by squares and females by circles. Individuals showing the trait under study are indicated by color symbols; those not showing the trait are indicated by white symbols. Relationships are indicated by connecting lines; all members of the same generation are placed on the same row. Thus, 11 is an albino girl whose paternal grandmother, 2, and maternal grandfather, 3, are also albinos. All of her other relatives shown are phenotypically normal. Notice that the inheritance pattern could not be autosomal dominant or X-linked dominant because neither of 11's parents is albino. If the trait were X-linked recessive, her mother would have to be a heterozygous carrier and her father would have to be an albino (which he is not).

INHERITED HUMAN TRAITS

The development of each organ of the body is regulated by a large number of genes that interact in complex ways. The mechanisms of inheritance of many physical traits and hundreds of specific enzymes are now known. In fact, the loci of many genes have been identified, and chromosome maps, although incomplete, have been worked out for each human chromosome.

The age at which a particular gene expresses itself phenotypically may vary widely. Most characteristics develop long before birth, but some, such as hair and eye color, are not fully expressed until shortly after birth. Others, such as muscular dystrophy, become evident in early childhood. Still others, such as glaucoma and Huntington's chorea, develop only in adulthood.

BIRTH DEFECTS

A birth defect, or **congenital defect,** is simply one that is present at birth; it may or may not be inherited. Some congenital abnormalities are inherited; others are produced by environmental factors that affect the developmental process. For example, if a woman contracts the viral disease rubella (commonly known as *German measles*) during the first 3 months of pregnancy, there is a substantial risk that her offspring will show congenital malformations. Environmental factors that have been linked with birth defects are discussed in Chapter 51.

Certain abnormalities are the result of mutations involving a single locus; sickle cell anemia and albinism are examples of this type of defect. Other abnormalities, such as Down syndrome, result from an abnormal number of chromosomes.

HUMAN CYTOGENETICS

Cytogenetics is the branch of biology that uses the methods of cytology to study genetics. Many of the basic principles of genetics were discovered by experiments with simpler organisms in which it was possible to relate genetic data with cytological events. These experiments included the microscopic examination of chromosomes to determine their number and structure. Some of the organisms used in genetics, such as the fruitfly *Drosophila*, have very few chromosomes (only four pairs). In *Drosophila* salivary glands and certain other tissues, the chromosomes are large enough that their structural details are readily evident. Therefore, this organism has provided unique opportunities for correlating certain kinds of genetic changes with certain kinds of alterations in chromosome structure. Although the science of human cytogenetics is not nearly as refined, many useful determinations are still possible.

Karyotypes

The normal human karyotypes for males and females are shown in Figure 9–1. The term **karyotype** refers both to the chromosome composition of an individual and to the photomicrograph showing that composition. In karyotyping, cells from the bone marrow, blood, or skin are incubated with chemicals that stimulate mitosis. These chemicals are derived from certain plants and are called **lectins.** The cells are then treated with the drug colchicine, which arrests them at mitotic metaphase, and placed in a hypotonic solution, which causes them to swell, enabling the chromosomes to spread out so they can be readily visualized. The cells are then fixed and spread on slides and the chromosomes stained to reveal the patterns of bands, which are unique for each homologous pair. After the chromosomes have been photographed, each chromosome is cut out of the photographic print and the homologous pairs are identified and placed together. Chromosomes are identified by length, by the position of the centromere, by banding patterns, and by other morphological features such as knobs. The largest chromosome is about five times as long as the smallest one, but there are only slight size differences among some of the intermediate-sized ones (see Figure 9–1).

Chromosome Abnormalities

Polyploidy, the presence of multiples of complete chromosome sets, is common in plants but rare in animals. In fact, when it occurs in all the cells of the body, polyploidy is lethal in humans and many other animals. Triploidy (3n) is sometimes found in embryos that have been spontaneously aborted in early pregnancy. The few triploid or tetraploid (4n) individuals that have been born alive and lived for a few days have been found to be mosaic, containing a mixture of diploid and triploid cells.

Abnormalities involving the presence of an extra chromosome or the absence of a chromosome are much more common in humans; these conditions are called **aneuploidies.** An individual with an extra chromosome—that is, with three of one kind—is said to be **trisomic.** An individual lacking one member of a pair of chromosomes is said to be **monosomic.** These aneuploidies generally arise as a result of an abnormal meiotic (or mitotic) division in which chromosomes fail to separate at anaphase; this phenomenon is called **nondisjunction.** In meiosis chromosomal nondisjunction may occur during the first or second meiotic division (or both). For example, two X chromosomes that fail to separate at either the first or the second meiotic division might enter the egg nucleus, leaving the polar body with no X chromosome. Alternatively, the two joined X chromosomes might go into the polar body, leaving the egg with no X chromosome. Nondisjunction of the XY pair in the male might lead to the formation of sperm that have both an X and a Y chromosome or of sperm with neither an X nor a Y chromosome. Similarly, nondisjunction at the second meiotic division can produce sperm with two Xs or two Ys. Some of these examples of meiotic nondisjunction are illustrated in Figure 11–2. When an abnormal gamete unites with a normal one, the resulting zygote will have a chromosome abnormality that will be seen in every cell of the body. Nondisjunction during a mitotic division will lead to the establishment of a clone of abnormal cells in an otherwise normal individual; such an individual will therefore be a mosaic of normal and abnormal cells.

In some aneuploidies a part of a chromosome may break off and attach to another chromosome. Such a **translocation** may result in an abnormally long chromosome, in which there is **duplication** of some of the genes, or a short chromosome, in which there is **deletion** of some genes. Table 11–1 summarizes some disorders that are produced by aneuploidies (see also Figures 11–4 and 11–7).

FIGURE 11–2 Examples of meiotic nondisjunction of the sex chromosomes in the human male. Only the X (red) and Y (blue) chromosomes are shown. (*a*) Nondisjunction in the first meiotic division results in two XY sperm and two sperm with neither an X nor a Y. (*b*) Second-division nondisjunction of the X chromosome results in one sperm with two X chromosomes, two with one Y each, and one with no sex chromosomes. Similarly, nondisjunction of the Y results in one sperm with two Y chromosomes, two with one X each, and one with no sex chromosome. By contrast, nondisjunction in the female (not shown) results in the formation of eggs with two X chromosomes or no sex chromosomes, regardless of whether it occurs in the first or second meiotic division.

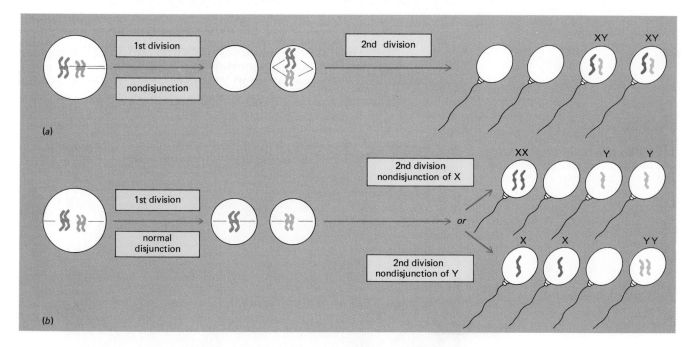

Table 11–1 SOME CHROMOSOME ABNORMALITIES

Karyotype	Common Name	Clinical Description
Trisomy 13		Multiple defects, with death by age 1 to 3 months
Trisomy 15		Multiple defects, with death by age 1 to 3 months
Trisomy 18		Ear deformities, heart defects, spasticity, and other damage; death by age 1 year
Trisomy 21	Down syndrome	Overall frequency is about 1 in 700 live births. True trisomy is usually found among children of older (age 40+) mothers, but translocation resulting in the equivalent of trisomy may occur in children of younger women. A similar, though less marked, influence is exerted by the age of the father. Trisomy 21 is characterized by an epicanthic skin fold (i.e., a fold of skin above the eye) which, although not the same as that in the Mongolian race, produces an Oriental appearance—hence the former name *mongolism* for this syndrome; varying degrees of mental retardation (usually an IQ of 70 or below), although more intelligent exceptions are known; and short stature, protruding furrowed tongue, transverse palmar crease, and cardiac deformities, all of which are common. Patients usually die by age 30 to 35 years. 50% die by age 3 or 4 years. Affected persons are unusually susceptible to respiratory infections, leukemia, and Alzheimer's disease. Females are fertile, if they live to sexual maturity; if able to reproduce, they produce Down syndrome in 50% of their offspring.
Trisomy 22		Similar to Down syndrome, but with more skeletal deformities
XO	Turner syndrome (gonadal dysgenesis)	Short stature, webbed neck, sometimes slight mental retardation; ovaries degenerate in late embryonic life, leading to rudimentary sexual characteristics; gender is female
XXY	Klinefelter syndrome	Male with slowly degenerating testes, enlarged breasts
XYY		Unusually tall male with heavy acne; some tendency to mild mental retardation
XXX		Despite triploid X chromosomes, usually fertile, fairly normal females
Short 5 (deletion of short arm of chromosome 5)	Cri-du-chat syndrome	Microcephaly, severe mental retardation; in infancy, cry resembles that of a cat; defective chromosome is heterozygous
Deletion of one arm of chromosome 21	Philadelphia chromosome	Chronic granulocytic leukemia

Down Syndrome

Cytogenetic studies have clarified the origin of **Down syndrome,** one of the most common chromosomal abnormalities in humans. Persons suffering from this syndrome have abnormalities of the face, eyelids, tongue, hands, and other parts of the body and are mentally and physically retarded (Figure 11–3). They are also unusually susceptible to certain diseases, such as leukemia and Alzheimer's disease. The term **mongolism,** which is no longer used today because of its racist connotations, was originally applied to this condition because of a characteristic fold of the eyelid in affected

FIGURE 11–3 Children participating in a music class. The child in the center has Down syndrome.

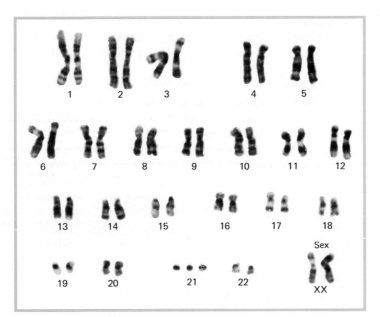

FIGURE 11–4 Karyotype of an individual with the free trisomy 21 form of Down syndrome. Note the presence of an extra chromosome 21.

persons that is superficially similar to the eyelid fold typically found in Mongolian peoples.

Cytogenetic studies have revealed that most persons with Down syndrome have 47 chromosomes (Figure 11–4). One of the smaller chromosomes, chromosome 21, is present in triplicate; affected persons are thus said to be **trisomic** for chromosome 21. Recall that ordinarily there are two of each kind of chromosome; this is the normal **disomic** condition. The presence of this extra chromosome is believed to arise by nondisjunction during meiosis.

Down syndrome occurs in only about 0.15% of all births, but it shows a marked increase in incidence with increasing maternal age (Figure 11–5); it

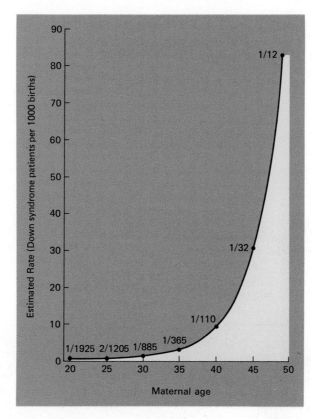

FIGURE 11–5 Relationship between maternal age and the incidence of Down syndrome.

is 100 times more likely in the offspring of mothers who are 45 years of age or older than it is in the offspring of mothers who are under 19 years of age. The occurrence of Down syndrome is affected much less by the age of the father.

Because of this striking correlation of the incidence of Down syndrome with increased maternal age, it is thought that in most (but certainly not all) cases the condition is due to nondisjunction in the mother. The reason for this is not fully understood. One possible explanation relates to differences in meiosis in human males and females. In a human female all of the cells that will ever enter meiosis do so before she is even born. They become arrested in meiotic prophase and remain in that state until she reaches puberty. After that time, one cell per month resumes meiosis. Therefore, when a woman produces an egg to be fertilized, that egg is essentially as old as she is. In contrast, in human males new cells are continually entering meiosis, and the entire process of sperm production takes only about 50 days.

In about 4% of patients with Down syndrome, only 46 chromosomes are present. However, one is abnormal: Extra genetic material from chromosome 21 has been translocated onto one of the larger chromosomes, such as chromosome 14. We will refer to the abnormal translocation chromosome as the *14/21 chromosome*. Affected persons have one chromosome 14, one 14/21 chromosome, and two copies of chromosome 21. The genetic material from chromosome 21 is thus present in triplicate. When the karyotypes of such an individual's parents are studied, either the mother or the father is usually found to have only 45 chromosomes, although he or she is generally phenotypically normal. Such a person has one chromosome 14, one 14/21 chromosome, and one chromosome 21. Although the karyotype is abnormal, there is no extra genetic material. In contrast to true, or free, **trisomy 21**, this translocation form of Down syndome can run in families (Figure 11–6), and its incidence does not increase with maternal age.

FIGURE 11–6 The translocation form of Down syndrome. In a carrier individual (either the mother or the father) most of the genetic material from a chromosome 21 has become fused to a chromosome 14, creating a 14/21 translocation chromosome. In this example the father is the carrier; he has 45 chromosomes, with one 14, one 21, and one 14/21. At anaphase I of meiosis, the chromosome 14 and the 14/21 translocation chromosome usually disjoin from one another. The chromosome 21 can go either to the pole with the chromosome 14 or to the pole with the 14/21 translocation chromosome. On average, such a carrier will produce four kinds of sperm; when they fertilize normal eggs, these sperm will produce four kinds of zygotes. One quarter of the zygotes will have only one chromosome 21 (monosomy 21), a lethal condition; one quarter will have the translocation form of Down syndrome; one quarter will be genotypically and phenotypically normal; and one quarter will be translocation carriers like the father.

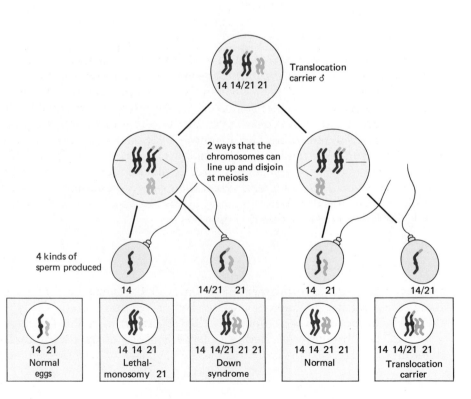

The presence of this extra chromosomal material leads to the complex physical and mental abnormalities that characterize Down syndrome. Paradoxically, although no genetic information is missing in these individuals, the extra "doses" of chromosome 21 genes bring about some type of genetic imbalance that is responsible for abnormal physical and mental development. Down syndrome is quite variable in its expression, with some individuals being far more severely affected than others. Genetic imbalances created by the addition or deletion of all or part of a chromosome typically result in multiple defects. When a disease causes multiple symptoms, we refer to it as a **syndrome;** virtually all chromosomal abnormalities fall into this category. Because the nervous system is so complicated in its development, it appears to be quite sensitive to altered gene dosages, and some form of mental retardation commonly accompanies most chromosome abnormalities.

In general, chromosome abnormalities involving the autosomes are far more devastating in their consequences than are abnormalities involving the sex chromosomes. In addition to Down syndrome, very few autosomal trisomies are known. The condition known as autosomal **monosomy,** in which only one member of a pair is present, is apparently incompatible with life, since it is not seen in live births.

By contrast, there appears to be a relative tolerance for sex chromosome abnormalities, apparently due at least in part to the phenomenon of dosage compensation discussed in Chapter 10. Recall that, according to the *single active X hypothesis,* mammals compensate for extra X chromosomal material by rendering all but one X chromosome inactive. The inactive X is seen as a Barr body, a region of darkly staining, condensed chromatin next to the membrane of the interphase nuclei.

The presence of the Barr body in the cells of normal females (but not of normal males) makes possible **nuclear sexing** to determine whether an individual is genetically female or male. As we shall see in our discussion of abnormal sex chromosome constitutions, nuclear sexing by the Barr body test has serious limitations and can therefore serve only as an initial screen. Any unusual findings should be followed up by examination of the chromosomes themselves (karyotype analysis).

Klinefelter Syndrome (XXY)

Persons with Klinefelter syndrome have 47 chromosomes, including two X chromosomes and one Y chromosome (Figure 11–7). These persons are nearly normal males, but they have small testes and produce few or no sperm. Evidence that the Y chromosome is the major determinant of the male phenotype has been substantiated by the recent discovery of a gene, called the **testis-determining factor,** on the Y chromosome. Persons with Klinefelter syndrome tend to be unusually tall and to have female-like breasts, and about half show some degree of mental retardation; most live relatively normal lives, however. When their cells are examined, they are found to have one Barr body per cell; on the basis of such a test, they would be erroneously classified as females.

Turner Syndrome (XO)

Persons with Turner syndrome have only 45 chromosomes; they are monosomic for the X chromosome and lack a Y chromosome. We write their sex chromosome constitution as *XO*, the *O* referring to the absence of a second sex chromosome. Because of the absence of the strong male-determining effect of the Y chromosome, these persons develop essentially

FIGURE 11–7 Karyotype of a person with Klinefelter syndrome. Note the presence of two X chromosomes, plus a Y chromosome. The autosomes are normal.

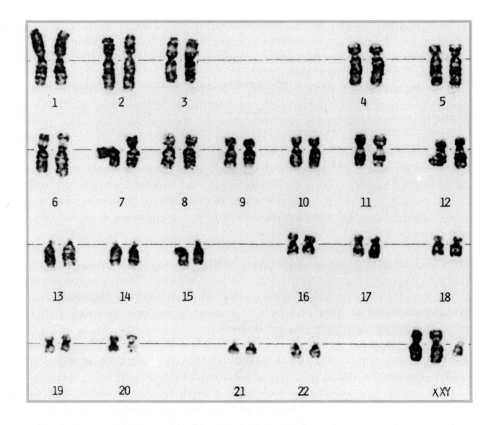

as females. However, both their internal and external genital structures are immature, and they are sterile. Examination of their cells reveals no Barr bodies, since there is no "extra" X chromosome to be inactivated.

XYY Karyotype

Persons with an X chromosome plus two Y chromosomes are phenotypically fertile males. The other characteristics of these persons (they are unusually tall, with severe acne) hardly merit the term *syndrome*; hence the designation *XYY karyotype*. Some years ago there was a widely publicized suggestion that persons with this condition are more likely than usual to display criminal tendencies and be incarcerated in penal institutions, but further studies have failed to substantiate this.

The Significance of Chromosome Abnormalities

Recognizable chromosome abnormalities are seen in less than 1% of all live births, but there is substantial evidence that the rate at conception is much higher. About 17% to 20% of all recognized pregnancies end in a spontaneous abortion ("miscarriage"). Approximately half of these spontaneously aborted embryos have major chromosome abnormalities, including autosomal trisomies (including trisomy 21), triploidy and tetraploidy, and Turner syndrome (XO), the last of which is the most common. Autosomal monosomies are exceedingly rare. It is *unlikely* that they never occur; it is far more probable that autosomal monosomy is so incompatible with life that a spontaneous abortion occurs very early, before the woman is even aware that she is pregnant. Some investigators place surprisingly high estimates (50% or more) on the rate of loss of very early embryos. It is widely assumed that chromosome abnormalities are responsible for a substantial fraction of these.

GENES AND DISEASE

Hundreds of human disorders involving enzyme defects have been found to be due to genetic mutations (Table 11–2). These disorders, sometimes referred to as **inborn errors of metabolism,** include phenylketonuria (PKU). Two other such genetic disorders associated with single gene defects (although not necessarily in genes coding for enzymes) are cystic fibrosis and sickle cell anemia. Not all human genetic diseases have a simple inheritance pattern. However, most of those that do are transmitted as autosomal recessive traits and so are expressed only in the homozygous state.

Autosomal Recessive Disorders

Phenylketonuria (PKU)

We are unable to cure any genetic disease today; the best we can hope for is successful treatment of the symptoms. Perhaps the most dramatic success to date has been in the treatment of **phenylketonuria (PKU).** Homozygous recessive individuals lack an enzyme that normally converts the amino acid phenylalanine to another amino acid, tyrosine. These persons instead convert phenylalanine into toxic products, which accumulate and damage the central nervous system. The ultimate result is severe mental retardation. A homozygous PKU infant is usually healthy at birth because its mother, who is heterozygous, produces enough enzyme to prevent phenylalanine accumulation before birth. However, during infancy and early childhood, the toxic products eventually cause irreversible damage to the central nervous system.

In the early 1950s it was found that if PKU infants are identified and placed on a special low-phenylalanine diet early enough, the symptoms can be dramatically alleviated. Biochemical tests for PKU have been developed, and screening of newborns is widespread in the United States, with more than 90% of all infants being tested. Because of these screening programs and the availability of effective treatment, more than 1000 children have been saved from severe mental retardation. By the age of about 6 years, such children are able to discontinue the diet. Although they still accumulate phenylalanine, the sensitive period is past.

It is ironic that the success of treatment of PKU in childhood presents a new problem today. If a homozygous female who was saved from mental retardation becomes pregnant, the high phenylalanine levels in her blood can damage the brain of the fetus she is carrying, even though that fetus is only heterozygous. Therefore, she must resume the diet during pregnancy, a procedure that is usually (though not always) successful in preventing the effects of **maternal PKU.** It is thus important that all PKU females be carefully monitored, not only during infancy and childhood, but throughout reproductive life as well. In the United States the PKU allele is most common among persons whose ancestors came from certain northern European countries.

Sickle Cell Anemia

Sickle cell anemia is inherited as an autosomal recessive trait. The disease is most common in persons of African descent, and about 1 in 12 black Americans is heterozygous for it. The blood cells of a person with sickle cell anemia are shaped like a sickle, or half-moon, whereas normal red blood cells are biconcave discs. The sickle cell contains hemoglobin molecules with a slightly different structure from that of hemoglobin molecules in a normal red blood cell. Sickle cell hemoglobin molecules have the amino

(Text continues on page 282)

Table 11–2 SOME IMPORTANT GENETIC DISORDERS

Name of Disorder	Mode of Inheritance	Clinical Description	Treatment, If Any	Comments
Alkaptonuria	Autosomal recessive	Pigmentation of cartilage and fibrous tissue, with eventual development of arthritis; presence of homogentisic acid causes urine to darken when it stands	Arthritis may be treated	Deficiency of enzyme homogentisic acid dehydrogenase
Childhood pseudohypertrophic muscular dystrophy	X-linked recessive	Begins in the first 3 years of life. Muscles swell, then undergo fatty degeneration. Progressive muscular deterioration leads to confinement, then to death in the early 20s	Symptomatic	Also known as Duchenne-type muscular dystrophy; extremely rare in females, but heterozygotes sometimes exhibit minor muscle function defects
Cystic fibrosis	Autosomal recessive	High level of sweat electrolytes, pulmonary disease, cirrhosis of the liver, pancreatic malfunction, and, especially, nonsecretion of digestive enzymes. No spermatogenesis in males, but females sometimes reproduce. Life expectancy is 12–16 years, with some affected persons living into their 30s or 40s. Commonest in persons of Northern European extraction	Symptomatic, with emphasis on digestive enzyme replacement and control of respiratory infections	Kills more children than diabetes, rheumatic fever, and poliomyelitis combined; exists in different degrees of severity; thick mucus interferes with lung clearance
Gangliosidosis (e.g., Tay-Sachs disease)	Autosomal recessive	Several types exist. One, Tay-Sachs disease, results from the deficiency of hexosaminidase A. All variants involve the abnormal accumulation of sphingolipids, ordinarily released from nerve and other cells by the action of whatever enzyme is deficient. Evidently several different enzymes are required. In most cases blindness, paralysis, and death occur in first few years of life.		Tay-Sachs disease is especially prevalent among Jews of Eastern European ancestry
Hemoglobinopathic disease (e.g., sickle cell anemia)	Group of autosomal recessive or incompletely dominant traits	Abnormalities of red blood cells caused by the presence of certain inappropriate amino acids at crucial locations in the hemoglobin molecule. In sickle cell anemia, for instance, one of the beta-chain amino acids is the "wrong" one, resulting in decreased solubility of hemoglobin molecules in low-oxygen environments such as tissue capillaries. This causes extreme shape distortions such as sickling, which in turn leads to premature destruction of the cell.	Varies with type of disease. Some (e.g., hereditary methemoglobinemia) may require no treatment; some cannot be treated at all. Sickle cell anemia can be treated to some degree.	These traits are similar and related but not allelic. Microcytic anemia is commonest in Mediterranean populations, sickle cell anemia in some black populations. In heterozygotes, sickle cell anemia offers some protection against malaria.

Table 11–2 (CONTINUED)

Name of Disorder	Mode of Inheritance	Clinical Description	Treatment, If Any	Comments
Hemophilia	X-linked recessive	Chronic bleeding, including bleeding into joints, with resultant arthritis; more than one variety of hemophilia exists	Treated with clotting factors	Even heterozygotes have some clotting factor deficiencies.
Lesch-Nyhan syndrome	X-linked recessive	Slowly developing paralysis accompanied by mental deficiency and self-mutilation, with patients persistently biting themselves. Gout usually develops because of deficiency of the enzyme involved in purine metabolism. The heterozygous condition is detectable (see comments).	Gout may be treated. Neurological symptoms are not treatable, and early death is inevitable.	Deficiency of a specific enzyme is to blame. Half the cells of the female carrier are enzyme-deficient. If her hair follicles are biopsied and studied, they will be found to be enzyme-negative, enzyme-positive, or mixed, but not all of them will contain the enzyme, which they normally would.
Phenylketonuria (PKU)	Autosomal recessive	Deficiency of liver phenylalanine hydroxylase leads to a chain of events beginning with excessive phenylalanine in the blood. This causes a depression in the levels of other amino acids, leading in turn to excessively light coloration and mental deficiency.	A low-phenylalanine diet minimizes symptoms. Most states have extensive PKU screening programs in which newborns are tested for excessive blood phenylalanine, or for presence of metabolic products in the urine.	Since melanin is synthesized from tyrosine, tyrosine deficiency caused by phenylalanine hydroxylase deficiency results in light coloring of skin and hair.
Red-green color-blindness				
Deutan variety	X-linked recessive	Patient can distinguish only 5 to 25 hues, as against the normal ability to see 150 + . Though visual acuity is normal, the "green" cone pigment is deficient. Subjectively, all colors are perceived as hues of blue and yellow.		Actually a series of alleles of differing degrees of severity, with the more normal dominant over the more deficient varieties. In both protan and deutan forms, heterozygous females show some color vision defects.
Protan variety	X-linked recessive, not allelic to deutan	Similar to deutan variety, but here the "red" cone pigment is missing. NOTE: Other defects of color vision associated with cone deficiencies also exist. Most of these are X-linked.		
Tyrosinase-negative oculocutaneous albinism (T − albinism)	Autosomal recessive	Absence of pigmentation due to functional absence of tyrosinase; visual acuity 20/200 or less; marked susceptibility to skin cancer	Avoidance of sunlight	Somewhat more common among blacks than whites
Tyrosinase-positive oculocutaneous albinism (T + albinism)	Autosomal recessive	Reduction of pigmentation due to malabsorption of tyrosine by body cells. If heavy pigmentation is genetically specified, some pigmentation will survive, though in some cases phenotype is virtually identical with T − . Pigmentation and visual acuity improve with age.		Highest incidence in American Indians, less in blacks, least in whites. Hybrid T + /T − persons appear normal.

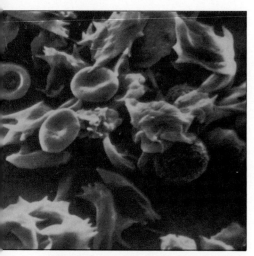

FIGURE 11–8 Red blood cells from a patient with sickle cell anemia (approximately ×4000). Note the abnormal shape of some of the cells.

acid valine instead of glutamic acid at position 6 (the sixth amino acid from the amino terminal end) in the beta chain. The substitution of an amino acid with a hydrophobic, uncharged side chain (valine) for one with a hydrophilic, charged side chain (glutamate) makes the hemoglobin less soluble, so that it tends to form crystal-like structures that change the shape of the red blood cell (Figure 11–8). This occurs in the veins after the oxygen has been released from the hemoglobin. The blood cells' abnormal sickle shape slows blood flow and blocks small blood vessels, with resulting tissue damage and painful episodes; these cells also have short life spans, leading to anemia in affected persons. Available treatments include measures to relieve pain, transfusions, and some forms of drug therapy, but these are of limited effectiveness, and children with sickle cell anemia generally lead short, painful lives.

Cystic Fibrosis

The most common autosomal recessive disorder in white children is **cystic fibrosis** (Figure 11–9). About 1 of every 20 persons in the United States is a heterozygous carrier of the gene for cystic fibrosis, which is characterized by abnormal secretions in the body. The most severe effect of this disorder is on the respiratory system, which produces abnormally viscous mucus that cannot be easily removed by the cilia that line the bronchi and thus serves as a culture medium for dangerous bacteria. These bacteria or their toxins attack the surrounding tissues, leading to the development of recurring pneumonia and other complications. The heavy mucus also occurs elsewhere in the body (e.g., in the ducts of the pancreas and liver and in the intestines), causing digestive difficulties and other effects.

Although the disease process of cystic fibrosis basically is known, how the abnormal gene produces these effects or how to cure the disease is not known. Treatment of symptoms includes administration of antibiotics to control bacterial infections and physical therapy to clear mucus from the respiratory system. Without such treatment, death would occur in infancy. With treatment, about 50% of affected persons live into their 20s; about 4 of those years are spent in a hospital. Persons with cystic fibrosis die in what for almost everyone else is the prime of life.

Tay-Sachs Disease

Tay-Sachs disease is an autosomal recessive disease of the central nervous system that results in blindness and severe mental retardation. The symptoms begin within the first year of life and result in death before the age of 5 years. Due to the absence of an enzyme, a normal membrane lipid in the brain fails to break down properly and accumulates in the cells, causing damage for which no treatment is available. The abnormal allele is especially common in the United States among Jews whose ancestors came from Eastern Europe (Ashkenazi Jews). By contrast, Jews whose ancestors came from the Mediterranean region (Sephardic Jews) have a very low frequency of the allele.

Huntington's Disease—An Autosomal Dominant Disorder

Huntington's disease, also known as *Huntington's chorea,* is due to a rare autosomal dominant allele that causes severe mental and physical deterioration, uncontrollable muscle spasms, personality changes, and ultimately insanity. There is no effective treatment. Every child of an affected individual has a 50% chance of also being affected (and of course passing the abnormal allele to half of his or her offspring). Ordinarily we would expect a dominant allele with such devastating effects to occur only as a new

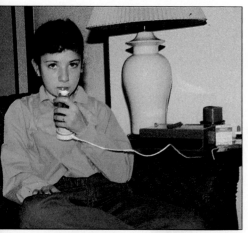

FIGURE 11–9 A child with cystic fibrosis using a nebulizer, which disperses medications into a fine mist that can then be inhaled.

mutation and not to oe transmitted to future generations. However, this disease is characterized by onset relatively late in life (usually between the ages of 35 and 50), so an individual may reproduce before knowing with certainty whether the allele is present.

The nature of the biochemical defect causing Huntington's disease is unknown. Nevertheless, new tests utilizing recombinant DNA technology (see Chapter 15) allow some persons at risk to determine whether they possess the allele. The decision to be tested is understandably a highly personal one. Certainly, the information can be very useful for those who must make decisions such as to whether to have children. However, someone who tests positive must then live with the certainty of eventually developing this devastating and incurable disease. It is to be hoped that identification of affected persons before the onset of symptoms may ultimately contribute to the development of effective treatments. If persons who have the Huntington's allele choose not to reproduce, the frequency of this allele in the population will decrease.

Hemophilia A—An X-Linked Recessive Disorder

Hemophilia A is sometimes referred to as a *disease of royalty* due to its incidence in some male descendants of Queen Victoria, but it is also found in many nonroyal pedigrees. Characterized by the lack of a blood-clotting factor, Factor VIII, it causes severe bleeding from even a slight wound. Because the mode of inheritance is X-linked recessive, affected persons are almost exclusively male, having inherited the abnormal allele on the X chromosome from their heterozyous carrier mother. Today, treatments consist of blood transfusions and administration of Factor VIII by injection. Unfortunately, both of these treatments are costly and have been associated with infection with HIV-1, the virus that causes AIDS; therefore, appropriate recombinant DNA technology (see Chapter 15) has been developed to provide a safer (and eventually less expensive) source of the clotting factor.

DIAGNOSIS OF GENETIC ABNORMALITIES

Genetic abnormalities may become manifest during early intrauterine life or not until late in adult life. Given that early detection increases the possibilities for prevention or alleviation of the effects of genetic abnormalities, efforts have been made over the years to detect such abnormalities at birth. In the past 20 years physicians have become increasingly bold about approaching the fetus while it is still in the uterus, and intrauterine diagnosis of a number of genetic abnormalities has thus become possible.

In one diagnostic technique, known as **amniocentesis,** a sample of the fluid surrounding the fetus (the amniotic fluid) is obtained (Figure 11–10). In this technique a needle is inserted through the lower abdomen of the pregnant woman and through the wall of the uterus into the uterine cavity, and the fluid is then drawn into a syringe. Although this procedure is not without risk, it is relatively safe, the positions of the fetus and the needle being determined through **ultrasound imaging** (Figure 11–11). The amniotic fluid contains living cells that have been sloughed off from the body of the fetus and hence are genetically identical to the cells of the fetus. These amniotic fluid cells can then be grown in culture in the laboratory. After 2 to 3 weeks, dividing cells from the culture can be studied for evidence of chromosomal abnormalities.

Amniocentesis is mostly performed on pregnant women over 35 years of age, whose offspring have a higher than normal incidence of Down syndrome. Many other tests have been developed to detect a number of sim-

FIGURE 11–10 The process of prenatal diagnosis of genetic disease by amniocentesis.

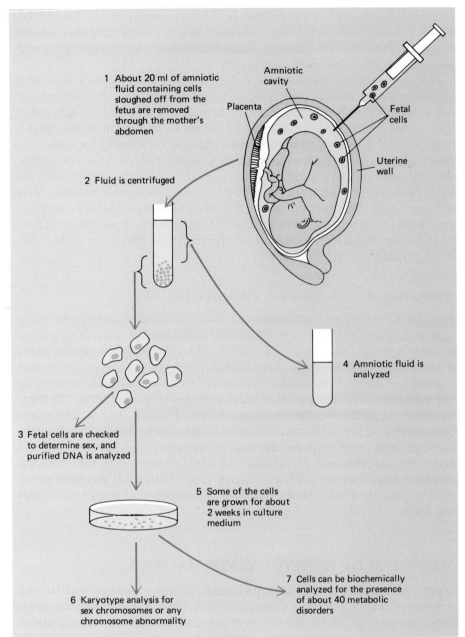

1 About 20 ml of amniotic fluid containing cells sloughed off from the fetus are removed through the mother's abdomen

Amniotic cavity

Placenta

Fetal cells

Uterine wall

2 Fluid is centrifuged

4 Amniotic fluid is analyzed

3 Fetal cells are checked to determine sex, and purified DNA is analyzed

5 Some of the cells are grown for about 2 weeks in culture medium

7 Cells can be biochemically analyzed for the presence of about 40 metabolic disorders

6 Karyotype analysis for sex chromosomes or any chromosome abnormality

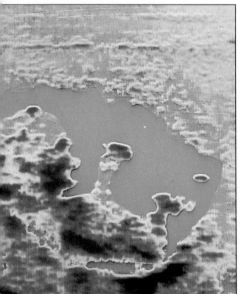

FIGURE 11–11 False-color ultrasound image of a human fetus (outlined for clarity) after 12 weeks in the womb. Ultrasound imaging uses sound waves rather than x rays to provide an image.

ply inherited genetic disorders, but these disorders are rare enough that the tests are usually only done if there is reason to suspect a problem. Enzyme deficiencies can often be detected through incubation of cells recovered from amniotic fluid with the appropriate substrate and measurement of the product; this technique has been useful in the prenatal diagnosis of disorders such as Tay-Sachs disease. The tests for a number of diseases, including sickle cell anemia and Huntington's disease, are less direct, requiring the use of recombinant DNA technology and other techniques of molecular biology (see Chapter 15). Methods for detecting many more genetic diseases are now being actively sought by researchers.

In addition, amniocentesis is useful in detecting defects in the development of the neural tube, which are relatively common (about 1 in 300 births) nongenetic congenital malformations. Such defects are associated with abnormally high levels of a normally occurring protein, alpha-fetoprotein, in the amniotic fluid.

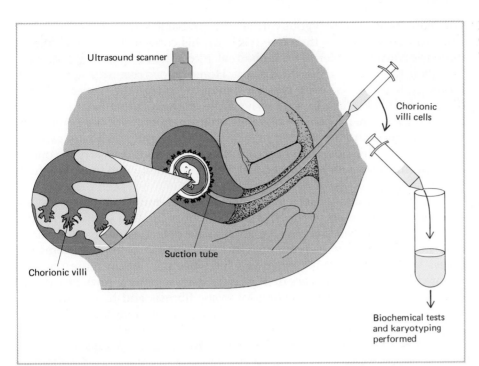

One problem with amniocentesis is that most of the conditions it detects are incurable and the results are generally not obtained until well into the second trimester, when abortion is particularly difficult both psychologically and medically. Therefore, efforts have been made to develop tests from which results can be obtained earlier in the pregnancy. One such test, **chorionic villus sampling (CVS)** (Figure 11–12), involves removing and studying cells that will form the fetal contribution to the placenta (and hence should be genetically identical to the fetal cells). CVS may be associated with a slightly higher risk of infection or miscarriage than is amniocentesis, but it has an advantage in that results usually can be obtained within the first trimester.

Although both amniocentesis and CVS can diagnose certain genetic disorders with a high degree of accuracy, there are many disorders that they cannot diagnose, and they are not foolproof; therefore, the lack of an abnormal finding on these tests is no guarantee of a normal pregnancy.

GENETIC COUNSELING

Couples who have had one abnormal child or who have a relative affected with a hereditary disease and hence are concerned about the risk of abnormality in one of their children may seek genetic counseling. Genetics clinics are available in most metropolitan centers.

Advice, of course, can be given only in terms of the *probability* that any given offspring will have a particular condition. The geneticist needs a carefully taken family history of both the man and the woman and may use tests for the detection of heterozygous carriers of certain conditions. When a disease involves only a single gene locus, probabilities can usually be easily calculated. For example, if one prospective parent is affected with a trait that is inherited as an autosomal dominant disorder, such as Huntington's disease, the probability that any given child will have the disease is 0.5.

The birth of one child who is affected with an autosomal recessive trait, such as albinism or PKU, to phenotypically normal parents establishes that both parents are heterozygous carriers; the probability that any subsequent child will be affected is therefore 0.25. (In this context, the term *carrier* is used specifically to refer to an individual who is heterozygous for a recessive allele that causes a genetic disease. Homozygous recessive individuals are not called *carriers*, even though they also "carry" the disease.) For a disease that is inherited by a recessive allele on the X chromosome, such as hemophilia A, the probability depends on the genotypes of the parents. A normal woman and an affected man will have daughters who are carriers and sons who are normal. The probability that a son of a carrier woman and a normal man will be affected is 0.5; the probability that their daughter will be a carrier is also 0.5.

It is now possible to detect the carriers of several genetic diseases, so that counseling can be provided when both husband and wife are heterozygous. For diseases that involve an enzyme defect, carriers often show only half the level of enzyme activity characteristic of normal homozygotes. Voluntary screening programs have been set up in more than 50 major cities to detect carriers of Tay-Sachs disease among Jews of Eastern European descent. Carrier testing for cystic fibrosis and hemophilia A is much more complicated and is therefore done only if there is reason, based on family history, to suspect that a person may be a carrier.

Persons heterozygous for sickle cell anemia can be readily identified with a simple blood test. They have a mixture of normal and abnormal hemoglobins in their red blood cells, with about 45% of their total hemoglobin being abnormal. Such persons, said to have **sickle cell trait,** are not ill, and their blood cells do not usually undergo sickling, although they can be made to do so when the amount of oxygen is reduced.

In conditions in which the pattern of inheritance is unknown or doubtful, an estimate of the probability of appearance of a given trait can be obtained from a table of empirical risk.[1] It is difficult to give a precise probability, because the trait may be inherited in different ways in different pedigrees. For example, retinitis pigmentosa, a disease that causes blindness due to deposition of pigment on the retina, appears to be inherited as an X-linked trait in some pedigrees and as an autosomal trait in others.

Mental deficiency, epilepsy, deafness, congenital heart disease, anencephaly, harelip, spina bifida, and hydrocephalus are conditions about which inquiries are commonly made. With such conditions the possibility must be considered that some environmental factor played a role in the appearance of the abnormality in the previous child. Did the mother have an infectious disease during pregnancy (e.g., rubella)? Was she receiving some kind of drug therapy, or was she subjected to any radiation? By dissecting the environmental contributions, the geneticist can make a better estimate of the probability of recurrence of the trait in subsequent offspring.

VARIATION IN THE HUMAN POPULATION

One does not need to be a geneticist to recognize that human beings are very diverse, and it is widely recognized that much human variation has a genetic basis. However, it is difficult to determine the genetic contribution to characteristics that are hard to assess, such as intelligence and behavior; other types of variation are more easily studied.

---

[1] *Empirical risk* simply means observed risk. By recording the instances of appearance of a trait in the offspring of affected parents, geneticists can estimate the likelihood that children of specified crosses will have the trait. Naturally it is best to *calculate* the risk if possible, but this is practical only in cases of known simple inheritance of particular diseases.

Human Blood Types

Studies on human blood have contributed much to our understanding of human variation. Blood samples are relatively easy to obtain, and blood is a complex tissue consisting of a number of cell types and extracellular molecules that can be studied.

The ABO Blood Group

The human blood types O, A, B, and AB are inherited through multiple alleles representing a single locus. Allele I^A provides the code for the synthesis of a specific glycoprotein, antigen A, which is expressed on the surface of the red blood cells. (Immunity is discussed in Chapter 43; for now we will define antigens simply as compounds capable of stimulating an immune response.) Allele I^B leads to the production of a different (but related) glycoprotein, antigen B. The allele i^O does not code for an antigen, although it represents the same locus as I^A and I^B. Allele i^O is recessive to the other two. Neither allele I^A nor allele I^B is dominant to the other; they are both expressed phenotypically and are therefore **codominant.** Persons with the genotype I^AI^A or I^Ai^O have **blood type A** (Table 11–3); those with genotype I^BI^B or I^Bi^O have **blood type B;** and those with genotype i^Oi^O have **blood type O.** When both the I^A and I^B alleles are present, both antigen A and antigen B are produced in the red blood cells; persons with this I^AI^B genotype have **blood type AB.**

The antibodies anti-A and anti-B are proteins that appear in the plasma of persons lacking the corresponding antigens on their red blood cells. (Antibodies are proteins produced by the immune system that combine with specific antigens; hence, anti-A combines with antigen A.) Because of their specificity for the corresponding antigens, these antibodies are used in standard tests to determine blood type.

Determining the blood types of the persons involved may be helpful in settling cases of disputed parentage. However, blood tests can never prove that a certain man *is* the father of a certain child; they can determine only whether he *could* be the father. Could a man with blood type AB be the father of a child with blood type O? Could a man with blood type O be the father of a child with blood type AB? Could a type-B child with a type-A mother have a type-A father or a type-O father?

More than a dozen other sets of blood types, including the Rh (discussed in the next section) group, are inherited through other loci, independently of the ABO blood types. Determining some of these types in a given person may be useful in establishing relationships that could not be made certain by ABO blood typing alone. When the answer is critical,

Table 11–3 ABO BLOOD TYPES*				Frequency in U.S. Population (%)	
Phenotype (Blood Type)	Genotypes	Antigen on RBC	Antibodies in Plasma	Western European Descent	African Descent
A	I^AI^A, I^Ai^O	A	Anti-B	45	29
B	I^BI^B, I^Bi^O	B	Anti-A	8	17
AB	I^AI^B	A, B	None	4	4
O	i^Oi^O	None	Anti-A, anti-B	43	50

*This table and the discussion of the ABO system have been simplified somewhat. Actually, some type-A persons have two type-A antigens and are designated type A_1, and some with only one antigen are termed type A_2. *RBC*, red blood cells.

tissue typing can be used, a more sophisticated type of genetic test that can come very close to proving parentage (it has greater than 99% certainty). In addition, new tests involving recombinant DNA technology (see Chapter 14) will eventually be very useful not only in resolving cases of disputed parentage, but also in identifying suspects in criminal cases.

The Rh System

Named for the rhesus monkeys in whose blood it was first found, the Rh system consists of at least eight different kinds of Rh antigens, each referred to as an **Rh factor.** By far the most important of these factors is **antigen D.** About 85% of persons living in the United States who are of Western European descent are Rh-positive, which means that they have antigen D on the surface of their red blood cells. (This is in addition to the antigens of the ABO system and other blood groups.) The 15% or so of this population who are Rh-negative have no antigen D and will produce antibodies against that antigen when they are exposed to Rh-positive blood. The allele for antigen D is dominant to the allele for the absence of antigen D. Hence, Rh-negative persons are homozygous recessive, and Rh-positive persons are heterozygous or homozygous dominant.

Although several kinds of maternal-fetal blood type incompatibilities are known, **Rh incompatibility** is probably the most important (Figure 11–13). If a woman is Rh-negative and the father of the fetus she is carrying is Rh-positive, the fetus may also be Rh-positive, having inherited the D allele from the father. Ordinarily there is no mixing of maternal and fetal blood; molecules are exchanged between these two circulatory systems by the placenta. However, late in pregnancy or during the birth process, a small quantity of blood from the fetus may pass through some defect in the placenta. The fetus's red blood cells, which bear antigen D, sensitize the mother's white blood cells, inducing them to form antibodies to antigen D. When the woman becomes pregnant again, her sensitized white blood cells produce antibodies that can cross an intact placenta and enter the fetal

FIGURE 11–13 Rh incompatibility can cause serious problems when an Rh-negative woman and an Rh-positive man produce Rh-positive offspring. (*a*) Some antigen D–bearing red blood cells leak across the placenta from the fetus into the mother's blood. (*b*) The mother produces D antibodies in response to the D antigens on the fetal red blood cells. (*c*) In her next pregnancy, some of the mother's D antibodies cross the placenta and enter the blood of her fetus, causing red blood cells to rupture and release hemoglobin into the circulation. As a result the fetus may develop erythroblastosis fetalis.

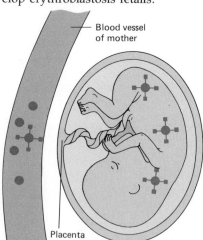

Blood vessel of mother

Placenta

A few Rh+ RBCs leak across the placenta from the fetus into the mother's blood

(a)

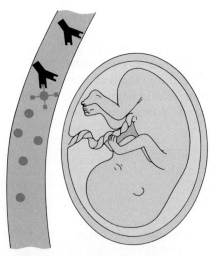

The mother produces anti-Rh antibodies in response to Rh antigen on Rh+ RBCs

(b)

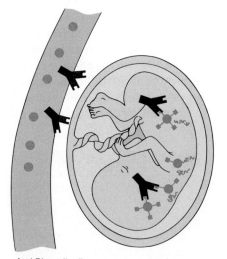

Anti-Rh antibodies cross the placenta and enter the blood of the fetus. Hemolysis of Rh+ blood occurs. The fetus may develop erythroblastosis fetalis.

(c)

● Rh– RBC of mother

Rh+ RBC of fetus with Rh antigen on surface

Anti-Rh antibody made against Rh+ RBC

Hemolysis of Rh+ RBC

blood. There they combine with the antigen D molecules on the surface of the fetal red blood cells, causing the cells to clump together. Breakdown products of the hemoglobin released into the circulation damage many organs, including the brain. In extreme cases of this disease, known as **erythroblastosis fetalis,** so many fetal red blood cells are destroyed that the fetus dies before birth.

When Rh-incompatibility problems are suspected, fetal blood can be exchanged by transfusion before birth, but this is a risky procedure. Rh-negative women are now treated just after childbirth (or at termination of pregnancy by miscarriage or abortion) with a preparation of anti-Rh antibodies known as Rho-Gam. These antibodies apparently clear the Rh-positive fetal red blood cells from the mother's blood very quickly, thus minimizing the chance that her own white blood cells will be sensitized. The antibodies that have been introduced are also soon eliminated from her body. As a result, when she becomes pregnant again her blood will not contain the anti-D that could harm her next baby.

Polygenic Inheritance in Humans

Many human characteristics are **quantitative traits**—that is, they represent some measurable quantity such as height. Such characteristics show continuous variation in the population because of the number of loci involved, which can range from a few to a great many (polygenic inheritance), and because of environmental factors, the role of which is both significant and difficult to quantify. In addition to height and skin color (see Chapter 10), the components of human intelligence that are inherited are apparently under polygenic control. It is impossible to specify what these are, given our current inability even to define intelligence in a way that will be universally accepted. However, many researchers believe that polygenic factors determine the upper limit of mental ability in persons within the normal population; how close each individual comes to that limit would then depend on a variety of environmental factors, including nutrition and experience.

Inheritance of Some Other Common Physical Characteristics

We are often curious about the inheritance of certain physical characteristics. Some of these have relatively simple modes of inheritance. For example, dark hair color is due to heavy deposits of the pigment melanin in the hair shaft. There are probably multiple alleles of the locus governing the deposition of melanin, the "darker" alleles being somewhat, but incompletely, dominant to the "lighter" alleles. Red hair color is governed by another unlinked locus with at least two incompletely dominant alleles, one coding for reddish pigment, the other for lack of such pigment. The appearance of red hair is determined by the presence of one or two "red" alleles and by partial or complete masking of the expression of these alleles by heavy melanin deposits.

Eye color is determined by the pattern of melanin distribution in the iris. There is no blue pigment, only brown or yellowish melanin. If the melanin is deposited in such a way that much of the light is reflected back from the eye, the iris will appear blue. Although inheritance of eye color is not simple, the inheritance patterns seen in most families can be explained if we assume that one locus is involved and that the alleles that govern the darker colors are dominant to the alleles that govern the lighter colors (blue, gray, hazel, etc.). It is not uncommon for two dark-eyed parents to have a light-eyed child; it is relatively rare for two light-eyed parents to have a dark-eyed child, but this can occur.

A number of other human characteristics show relatively simple inheritance patterns. For example, the allele for dimples is generally dominant to the allele for no dimples, the allele for freckling is usually dominant to that for no freckles, and curly hair is usually dominant to straight hair.

GENETICS AND SOCIETY

As we have seen, abnormal alleles that can lead to certain genetic diseases may be especially common in a particular race or ethnic group. However, this does not mean that that group is especially susceptible to genetic diseases in general; abnormal alleles are found in all populations. It is not always easy to explain why a certain allele may be present at a particularly high frequency in a certain group.

"Heterozygote advantage" in the form of resistance to malaria is the widely accepted explanation for the high frequency of the sickle cell allele among persons of African descent (see Chapter 10). Various theories have been advanced to explain the high frequencies of other genetic diseases in certain populations or ethnic groups. One hypothesis is that heterozygous carriers of a particular abnormal allele may have some kind of heterozygote advantage that causes them to have more children than noncarriers, but definitive evidence on this point is lacking for cystic fibrosis, Tay-Sachs disease, and other disorders. Sometimes history provides possible explanations. For example, in the Middle Ages Jews in Eastern Europe experienced a contraction in the size of their population due to widespread persecution (a phenomenon known as a *population bottleneck*). A coincidental high frequency of the Tay-Sachs allele in the small surviving population is one possible explanation of the high frequency (1/28) of this allele in the Ashkenazi population today.

It is often argued that medical treatment of persons affected with genetic diseases, especially those who are able to reproduce, greatly increases the frequency of abnormal alleles in the population. However, most genetic diseases that are simply inherited show an autosomal recessive inheritance pattern. Only homozygous persons will manifest these diseases phenotypically; heterozygous carriers, who are far greater in number, will be phenotypically normal. For example, if 1 in 20 persons in the United States is heterozygous for cystic fibrosis, the chance that a husband and wife will both be heterozygous is $(1/20) \times (1/20) = 1/400$. On average, 1/4 of the children of such a couple would have cystic fibrosis, so the frequency of affected individuals in the population would be about $(1/400) \times (1/4) = 1/1600$.

All human beings are carriers of something; each of us is probably heterozygous for several (3–15) very harmful alleles, any of which could cause debilitating illness or death in the homozygous state. So why aren't genetic diseases more common? Each of us has many thousands of essential genes, any of which can be mutated. It is very unlikely that the abnormal alleles that are carried by one person will also be carried by that person's mate. Of course, this possibility is more likely for a common harmful allele, such as that responsible for cystic fibrosis.

Relatives are more likely than nonrelatives to carry the same harmful alleles, having inherited them from a common ancestor. In fact, a greater than normal frequency of a genetic disease among offspring of **consanguineous matings** (matings of close relatives) is an important clue that the mode of inheritance is autosomal recessive. There is a significant risk of genetic disease among the offspring of consanguineous matings, and the overall cost of such matings to society in terms of increased frequency of affected individuals in the population can be quite high. First-cousin marriages are therefore prohibited in many states.

■ SUMMARY

I. Geneticists investigating human inheritance cannot make specific crosses of pure genetic strains; instead, they must rely on studies of populations and analysis of family pedigrees.

II. A great variety of physical and biochemical traits are inherited.

III. Studies of the karyotype (the number and kinds of chromosomes present in the nucleus) permit detection of individuals with various chromosome abnormalities.

A. Such studies can detect trisomy, in which one has an extra chromosome, and monosomy, in which one lacks one member of a pair of chromosomes.

B. The most common form of Down syndrome (trisomy 21), Klinefelter syndrome (XXY), and Turner syndrome (XO) are examples of trisomy or monosomy.

IV. Abnormal alleles of a number of loci are responsible for many inherited diseases, such as PKU, sickle cell anemia, and cystic fibrosis. Most human genetic diseases that show a simple inheritance pattern are transmitted as autosomal recessive traits.

V. Some genetic diseases can be diagnosed long before birth by amniocentesis or chorionic villus biopsy. Both chromosome abnormalities and some gene defects can be diagnosed through study of the cells obtained.

VI. Genetic counselors can advise prospective parents who have a history of genetic disease regarding the probabilities of their giving birth to affected offspring. It is possible to detect the presence of a number of harmful alleles.

VII. Variation in the human population is exemplified by human blood types, among which are the ABO blood group and the Rh system. The Rh-positive offspring of an Rh-negative mother may develop a very serious disease known as *erythroblastosis fetalis*.

VIII. The effect of genetics on society is complex.

A. The fact that a particular abnormal allele is especially common in a certain racial or ethnic group does not mean that that group has a higher frequency of abnormal alleles in general.

B. Some alleles that cause a genetic disease when they are homozygous may be advantageous in heterozygous individuals, at least in a particular environment. For example, the allele responsible for sickle cell anemia in homozygous individuals appears to confer resistance to malaria on heterozygotes.

C. Since most abnormal alleles are recessive, they are manifested phenotypically only in homozygotes, who constitute a tiny fraction of the carrier population. It has been estimated that virtually every individual in the population is heterozygous for several abnormal alleles.

■ POST-TEST

1. Standard stocks of genetically identical individuals are called _____ strains.

2. The array of chromosomes present in a given cell is called the _____.

3. An abnormality that is present and evident at birth is called a birth defect, or _____ defect.

4. An abnormality in which there is one more or one fewer than the normal number of chromosomes is called a(n) _____.

5. A person who has an extra chromosome (three of one kind) is said to be _____.

6. A person who is missing a chromosome, having only one member of a pair, is termed _____.

7. The failure of chromosomes to separate normally during cell division is called _____.

8. The transfer of a part of one chromosome to a nonhomologous chromosome is termed _____.

9. Individuals with trisomy 21, or _____ syndrome, are mentally and physically retarded and have abnormalities of the face, tongue, and eyelids.

10. An XXY individual has the disorder known as _____ syndrome; the XO individual has _____ syndrome.

11. An inherited disorder that is due to a defective or absent enzyme is called an inborn error of _____.

12. The sickle cell allele produces an altered _____ molecule, which is less soluble than usual and is more likely than normal to crystallize and deform the shape of the red blood cell.

13. In a person with _____ _____, the mucus is abnormally viscous and tends to plug the ducts of the pancreas and liver and to accumulate as pools in the lungs.

14. In the process of _____, a sample of amniotic fluid is obtained by insertion of a needle through the walls of the abdomen and uterus and into the uterine cavity.

15. _____ _____ sampling involves removal and study of cells that will make up the fetal contribution to the placenta.

16. Couples who have some reason to be concerned about the possibility of genetic abnormalities in their offspring may seek _____ _____.

■ REVIEW QUESTIONS

1. What means have been devised for overcoming the difficulties of studying human inheritance?

2. What is meant by *nondisjunction?* What human abnormalities appear to be the result of nondisjunction?

3. Are all congenital traits hereditary?
4. What are the relative advantages and disadvantages of amniocentesis and chorionic villus sampling?
5. How can carriers of certain genetic diseases be identified?
6. What is meant by *inborn errors of metabolism?* Give an example.
7. Mrs. Doe and Mrs. Roe had babies at the same hospital and at the same time. Mrs. Doe took home a girl and named her Nancy. Mrs. Roe took home a boy and named him Richard. However, she was sure that she had given birth to a girl and brought suit against the hospital. Blood tests showed that Mr. Roe was blood type O, Mrs. Roe was type AB, and Mr. and Mrs. Doe were both type B. Nancy was type A, and Richard was type O. Had an exchange occurred?
8. Imagine that you are a genetic counselor. What advice or suggestions might you give in the following situations?
   a. A couple has come for advice because the woman had a sister who died of Tay-Sachs disease.
   b. A pregnant woman has learned that as a newborn she suffered from erythroblastosis fetalis; she is concerned that she might have a similarly affected child.
   c. A young man and woman who are not related are engaged to be married. However, they have learned that the man's parents are first cousins. They are worried that they might have an increased risk of genetic defects in their own children.
   d. A young woman's paternal uncle (her father's brother) has hemophilia A. Her father is free of the disease and there has never been a case of hemophilia A in her mother's family. Should she be concerned about the possibility of hemophilia A in her own children?
   e. A 20-year-old man is seeking counseling because his father has just been diagnosed as having Huntington's disease.

■ RECOMMENDED READINGS

Some representative texts on human genetics are the following:

Hartl, D.L. *Our Uncertain Heritage: Genetics and Human Diversity.* Philadelphia, J.B. Lippincott, 1977.

Mange, A.P., and Mange, E.J. *Genetics: Human Aspects.* Philadelphia, Saunders College Publishing, 1980.

Sutton, E.P. *Human Genetics*, 4th ed. San Diego, Harcourt, Brace, Jovanovich, 1988.

Article:

Patterson, D. The Causes of Down Syndrome. *Scientific American*, 257: 52–61, 1987.

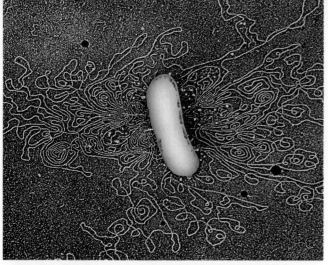
A single DNA molecule spills out in loops from a broken *E. coli* bacterium

12

DNA: The Secret of Life

■ LEARNING OBJECTIVES

After you have read this chapter you should be able to:

1. Discuss the evidence that DNA is the fundamental genetic material.
2. Discuss the evidence that led to the determination of the structure of DNA.
3. Describe the structure of a chain of DNA; compare the structures of the nucleotide subunits, and explain how they are linked together to form a DNA chain.
4. Describe the basic features of the B form of DNA; explain how the two strands of DNA are oriented with respect to each other.
5. Explain the base pairing rules for DNA, and describe how complementary bases are able to bind to each other.
6. Discuss the process of semiconservative replication.
7. Discuss the basic features of the DNA replication process.
8. Explain why DNA replication is discontinuous and bidirectional.
9. Compare the organization of DNA in chromosomes of prokaryotic cells with that in chromosomes of eukaryotic cells.
10. Discuss why the size of DNA requires that it must be highly organized in both prokaryotic and eukaryotic cells.

Why do you look so much like your parents? Because you have inherited copies of your parents' genes. But what is it about genes that causes you and your parents to be so similar? In the first half of this century, following the rediscovery of Mendel's laws, geneticists used elegant crossing experiments to learn how genes are arranged on chromosomes and how they are transmitted from generation to generation. However, two very basic questions remained unanswered: What are genes made of? How do genes work? Although the study of inheritance patterns did little to answer these questions, it contributed to an emerging set of predictions about the characteristics of the molecules that make up genes and about what genes do. Among the properties attributed to genes was the ability to store information in a stable form that can be accurately copied from generation to generation. The molecular basis of genetic changes called **muta-**

tions, which are responsible for the rare appearance of new genetic variants, also required explanation. A mutation is any change in a gene; usually a mutation will be recognized if it produces an individual with a new genetic trait. Once a mutation occurs, the new genetic characteristic is stably transmitted to succeeding generations in the same manner as the original. Thus, mutations were particularly puzzling, since they were obviously caused by sudden, but genetically stable, changes in the properties of genes.

WHAT DO GENES DO?

Through the first half of the 20th century, considerable evidence accumulated that genes had something to do with the making of proteins. The first evidence that genes and proteins are somehow related was published in 1908 by an English physician, Archibald Garrod, who theorized that certain inherited diseases in humans are caused by a block in a sequence of chemical reactions that take place in the body.

In his book, *Inborn Errors of Metabolism,* Garrod gave as an example a genetic disease called **alkaptonuria,** which is inherited as a simple mendelian trait. The condition involves a block in the metabolic reactions that break down the amino acids phenylalanine and tyrosine; the urine of affected persons turns black when exposed to air (Figure 12–1). Homogentisic acid, the substance that turns black and an intermediate in this breakdown pathway, normally is oxidized and then further converted to carbon dioxide and water. Garrod theorized that persons with alkaptonuria lack the oxidation enzyme, causing homogentisic acid to accumulate in their tissues and blood and to be excreted in their urine. Before the second edition of his book had been published in 1923, it was determined that the blood of affected persons did indeed lack the enzyme that oxidizes homogentisic acid. Garrod was right: A specific gene could be associated with the absence or presence of a specific enzyme. Despite the implications of this finding, however, little work was done in this area, primarily because errors in metabolism appeared to occur only in patients with rare diseases, which made genetic experiments impossible.

A major advance in understanding the relationship between genes and enzymes came in the early 1940s, when George Beadle and Edward Tatum developed a new approach to the problem. Most efforts until that time had centered on studying known genes and attempting to determine what biochemical reactions they affected. Experimenters did this by examining traits that were visible and already known, such as the eye-color genes of *Drosophila* and the genes that control pigments in plants. Although these studies showed that such traits were controlled by a series of biosynthetic reactions, it was not clear whether the genes acted as enzymes or whether they determined the specificities of enzymes in more complex ways.

Beadle and Tatum decided to take the opposite approach. Rather than try to identify which enzymes were affected by single genes, they decided to look for *mutations* in genes that interfere with known metabolic reactions that produce essential chemicals such as amino acids and vitamins. Beadle and Tatum used the bread mold *Neurospora* for their studies for several reasons. First, wild type[1] *Neurospora* can make all of its essential biological chemicals when it is grown on a simple minimal medium containing sugar, salts, inorganic nitrogen, and the vitamin biotin. Even a mutant that cannot make a substance such as an amino acid can still be grown if that substance is simply added to the minimal medium. Second, *Neurospora* grows primarily as a haploid organism. Thus, a mutation introduced in a

---

[1]Wild type is a term commonly applied to nonmutant strains or individuals.

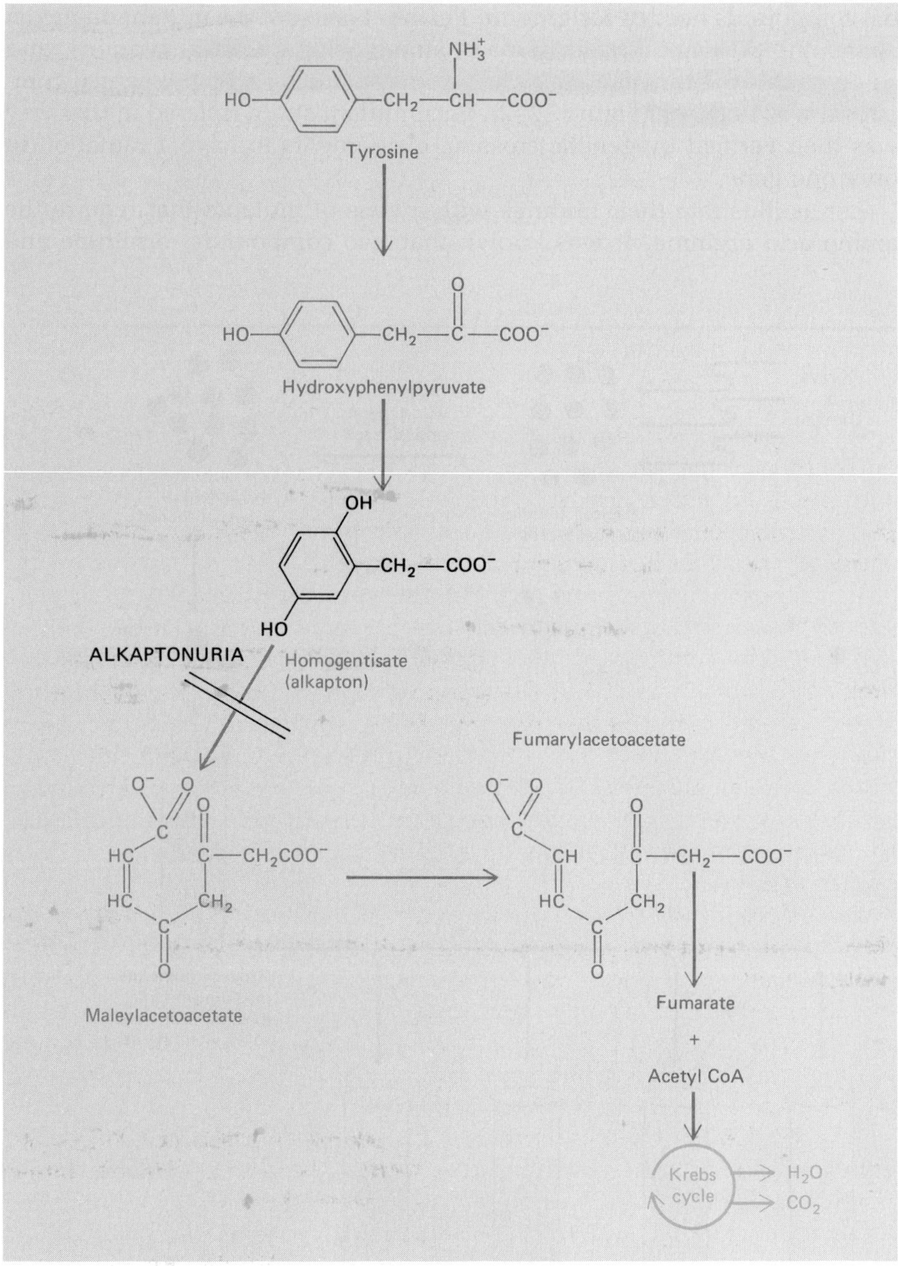

FIGURE 12–1 Pathway by which the amino acid tyrosine is catabolized. The mutation that causes alkaptonuria (black urine) produces a defect in the enzyme that normally converts homogentisic acid to maleylacetoacetate. Homogentisic acid thus accumulates in the blood and is excreted through the urine. When the homogentisic acid in the urine comes in contact with air, it oxidizes and turns black. Garrod proposed that the alkaptonuria gene causes the absence of a specific enzyme, or an "inborn error of metabolism."

gene will not be masked by an opposite, normal allele and can therefore be immediately identified. Third, *Neurospora* produces haploid spores, known as **conidia,** each of which can fuse with another haploid cell to undergo a brief sexual phase of the life cycle (see the generalized life cycles of simple organisms in Figure 9–19). The zygote that is produced undergoes meiosis to form haploid sexual spores; thus, researchers can use sexual crosses to perform genetic analyses of isolated mutants.

Beadle and Tatum searched for large numbers of mutants that were unable to synthesize biologically important molecules (e.g., amino acids, purines, pyrimidines, vitamins). They did this by exposing the haploid conidia to x rays or ultraviolet radiation, both of which were known to produce mutations. Following exposure, the conidia were grown on a **complete medium** containing all essential molecules so that mutants produced by the irradiation could survive and reproduce. If an isolated mutant were found to grow on the complete medium but not on the minimal medium, Beadle and Tatum reasoned that it was unable to produce one of the essential compounds needed for growth. Further testing of the mutant on media containing different combinations of amino acids, purines, vitamins, and so on enabled the investigators to determine exactly which essential compound was required (Figure 12–2). Each mutant strain isolated in that way was then verified by genetic crossing experiments to have a mutation in only one gene.

Let us illustrate their findings with a class of mutants that require the amino acid arginine. It was known that two compounds, ornithine and

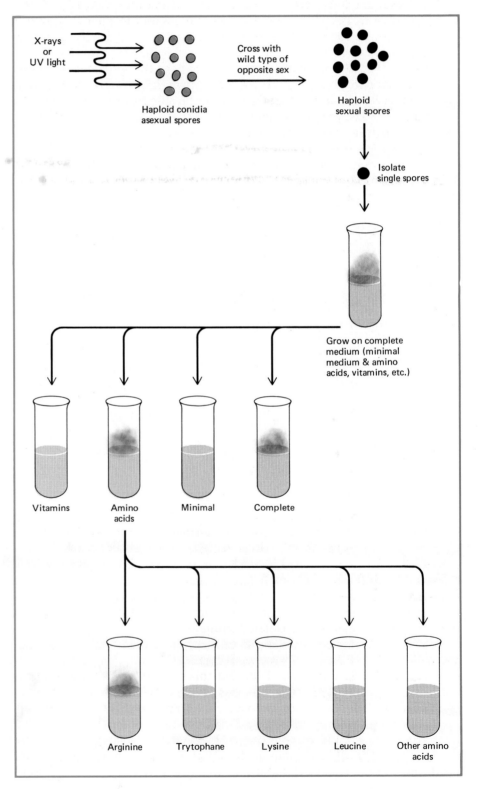

FIGURE 12–2 Isolation and identification of mutant strains of *Neurospora*. Beadle and Tatum determined the basic relationship between genes and enzymes by isolating and characterizing large numbers of mutants defective in a number of well-known metabolic pathways. Haploid asexual spores of *Neurospora* were irradiated to produce random mutations and then mated with another strain to produce haploid sexual spores. Isolated spores were then allowed to grow on a complete medium containing all the amino acids and vitamins that are normally made by *Neurospora*. Each strain was also tested on a minimal medium to determine whether a mutation was present. No growth on minimal medium indicated the presence of a biochemical mutation. Mutant strains were then tested for growth on minimal medium containing individual vitamins or amino acids. In this case the medium containing the amino acid arginine is the one that supports growth, indicating that the mutation affects some part of the arginine biosynthetic pathway.

citrulline, are precursors to arginine in its biosynthetic pathway. The order of these intermediates in the pathway is shown below. Beadle and Tatum found that whereas some of the arginine-requiring mutants could grow on ornithine, citrulline, or arginine, others could grow on only citrulline or arginine, and still others could grow only on arginine.

$$X \longrightarrow \text{ornithine} \longrightarrow \text{citrulline} \longrightarrow \text{arginine}$$

Enzyme A   Enzyme B   Enzyme C

If you examine the pathway, you can see that a mutation that inactivated the enzyme catalyzing reaction A would correspond to the first group of mutants, since after the block enzymes B and C would still be able to convert ornithine or citrulline to arginine. A mutation that inactivated the enzyme that catalyzed reaction B would produce the second class of mutants, so that citrulline could be converted to arginine but ornithine could not. Neither ornithine nor citrulline would be able to support growth in a mutant that lacked enzyme C, since both of these precursors are produced before the blocked step (Figure 12–3).

Using this approach, Beadle and Tatum were able to analyze large numbers of mutants affecting a number of metabolic pathways. They invariably found that *for each individual gene identified, only one enzyme was affected*. This

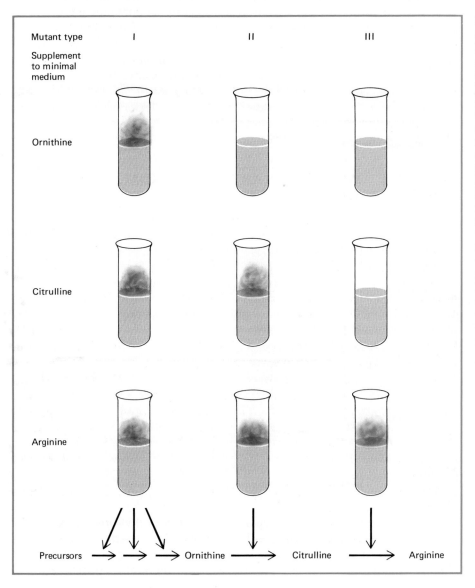

FIGURE 12–3 Associating *Neurospora* mutants with steps in a metabolic pathway. By analyzing different mutant strains that require the same amino acid or vitamin, it was possible to verify that each mutant gene affected only a single enzymatic step in the pathway. In this example, a number of different mutant strains that require the amino acid arginine were tested to see which step in the pathway was blocked. Arginine requiring strains I, II, and III were all found by genetic crossing methods to have mutations in different genes. Mutant type I grows on ornithine, citrulline, and arginine and must therefore be blocked prior to the formation of all three compounds. Mutant type II is unable to grow on ornithine but allows the conversion of citrulline to arginine. Mutant type III can grow only on arginine and must therefore be blocked at a point after the synthesis of both ornithine and citrulline.

Mutant type   I   II   III

Supplement to minimal medium

Ornithine

Citrulline

Arginine

Precursors → → → Ornithine ——→ Citrulline ——→ Arginine

one-to-one correspondence between genes and enzymes was succinctly stated as the **one gene, one enzyme hypothesis.**

The idea that a gene might encode the information for a single enzyme held for almost a decade, until it was found that many genes encode proteins that are not enzymes. In addition, some studies showed that many proteins may be constructed from two or more polypeptide chains (e.g., the alpha and beta subunits of hemoglobin), each of which may be encoded by a different gene. As a consequence of these developments, the definition was modified to state that one gene is responsible for one *polypeptide chain*. Even this definition has proved to be only partially correct, as we shall see in Chapter 13.

WHAT ARE GENES MADE OF?

We now know that genes are made of deoxyribonucleic acid, or DNA; however, when the molecular nature of the gene was first studied, most scientists were convinced that the genetic material had to be protein. Proteins are made up of over 20 different kinds of amino acids in many different combinations, which allows each type of molecule to have unique properties; in contrast, nucleic acids are made of only four nucleotides arranged in what appears to be a regular and uninteresting fashion. Beadle and Tatum's experiments had shown that genes control the production of proteins. Furthermore, protein molecules acting as enzymes were known to catalyze virtually all cellular metabolic reactions, from which all other biological molecules and cellular structures are made. Given their obvious complexity and diversity compared with other molecules, it seemed that proteins must be the stuff of which genes are made. In fact, however, genes were found to be made of DNA, despite the seemingly simple composition of DNA molecules; the amount and complexity of information in a molecule is determined not by the number of its subunits but rather by the structure of the molecule itself.

EVIDENCE THAT GENES ARE MADE FROM NUCLEIC ACIDS

Because many researchers believed that DNA could not be the hereditary material, several early clues to its role went largely unrecognized. In 1928 Frederick Griffith made a curious observation concerning two strains of *Pneumococcus* bacteria. It was already known that the smooth (S) strain, named for its formation of smooth colonies on solid growth medium, is **virulent,** or lethal; when that strain is injected into mice, the animals contract pneumonia and die. It was also known that the related rough (R) strain, which forms colonies with a rough surface, is **avirulent,** or nonlethal. Griffith found that when *heat-killed* S-strain cells and live R-strain cells were both injected into mice, the mice frequently died. Griffith was then able to isolate *living* S-strain cells from the dead mice. Because neither the heat-killed S strain nor the living R strain could be converted to the living virulent form when injected by itself, it appeared that something in the heat-killed cells was capable of converting the avirulent cells to the lethal form. This ability of some chemical substance in the dead bacteria to convert a related strain to a genetically stable, new form was termed **transformation** (Figure 12–4). Later, Avery, MacLeod, and McCarty of the Rockefeller Institute determined that the "transforming principle" was in fact DNA.

In the space of the next few years, new evidence demonstrated that the haploid nuclei of pollen grains and gametes such as sperm contained only half the amount of DNA found in somatic diploid cells of the same species. This is exactly what would be predicted from mendelian principles!

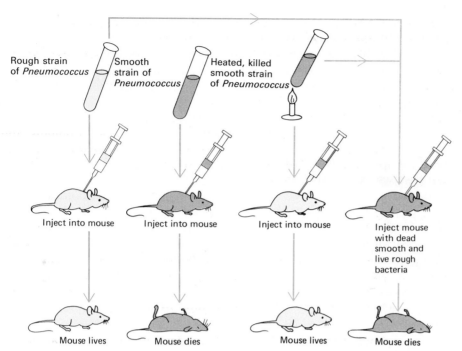

FIGURE 12–4 Frederick Griffith demonstrated the transfer of genetic information from dead, heat-killed bacteria to living bacteria of a different strain. Although neither the rough strain of *Pneumococcus* nor the heat-killed smooth strain could kill a mouse, a combination of the two did. Autopsy of the dead mouse showed the presence of living, smooth strain pneumococci. These results indicated that some substance in the heat-killed smooth strain was responsible for the transformation to virulence. Later, Avery and coworkers demonstrated that purified DNA isolated from the smooth strain restores virulence to the rough strain bacteria, establishing that the DNA carried the genetic information necessary for the bacterial transformation.

By 1952, Alfred Hershey and Martha Chase were able to show in a series of elegant experiments that viruses that infect and multiply in bacteria do so by injecting their DNA into the bacterial cells and leaving most of their protein on the outside (Figure 12–5). This result emphasized the importance of DNA in the reproduction of the virus and was seen by many to be an important indication that DNA was the hereditary material.

THE STRUCTURE OF DNA

DNA was not widely accepted as the genetic material, however, until James Watson and Francis Crick demonstrated that its structure could carry information and that it could serve as its own template for replication. The story of how its structure came to be determined is one of the most remark-

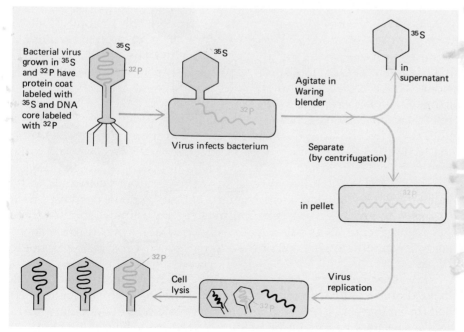

FIGURE 12–5 The Hershey-Chase experiment demonstrated that only the DNA of a bacterial virus is necessary for the reproduction of new viruses. The virus consists of a DNA core, which contains phosphorus but no sulfur atoms, surrounded by a protein coat, which contains sulfur but no phosphorus atoms. By growing the virus on medium containing the radioactive isotopes ^{32}P and ^{35}S, Hershey and Chase were able to label specifically the protein and DNA with different isotopes. They found that only the labeled DNA entered the bacterium and that the virus protein could be separated from the cells after infection without interfering with the replication of the virus. All the genetic information needed for the synthesis of new protein coats and new viral DNA was provided by the parental viral DNA.

able chapters in the history of modern biology. In order to discuss how it came about, though, we need to review what was known about the chemical structure of DNA when Watson and Crick became interested in the problem.

Nucleotides

As discussed in Chapter 3, each DNA building block is a nucleotide consisting of a pentose sugar (deoxyribose), a phosphate, and a nitrogenous base. The bases include the **purines,** adenine (A) and guanine (G), and the **pyrimidines,** thymine (T) and cytosine (C). As shown in Figure 12–6, the nucleotides are linked together by covalent **phosphodiester bonds** to form a sugar-phosphate backbone. The structure of the nucleotide subunits in DNA is identical to that of AMP (see Chapter 3), with the exception that the sugar does not contain a hydroxyl group on the 2′ carbon atom. The nitrogenous base is attached to the 1′ carbon of the sugar, and the phosphate is attached to the 5′ carbon. (The "prime" is used to distinguish individual carbon atoms in the sugar from those in the base.) In DNA, the

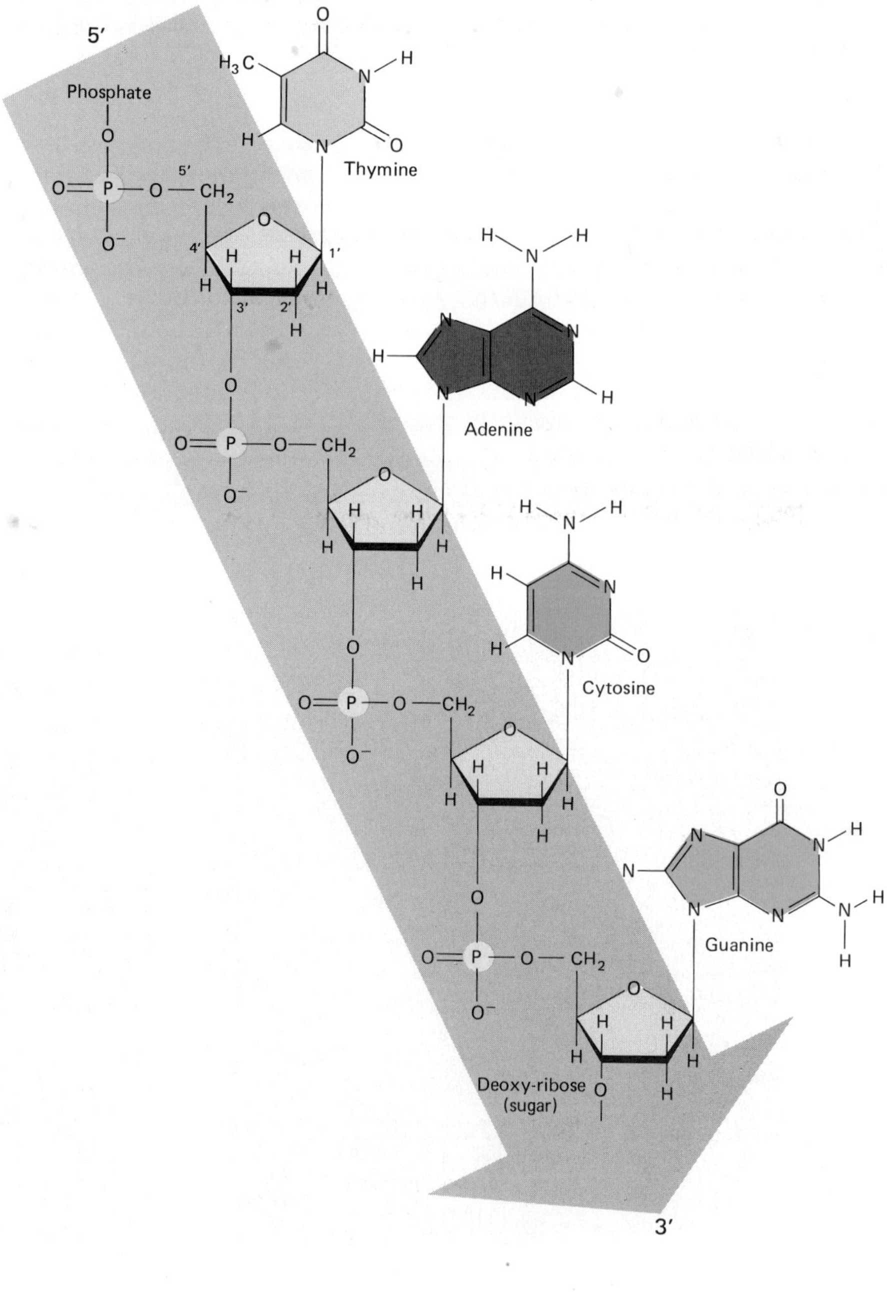

FIGURE 12–6 Chemical structure of DNA. The DNA molecule is made of deoxyribonucleotide monophosphate subunits. Each subunit is composed of a phosphate group linked to the sugar, deoxyribose at its 5′ carbon atom. Linked to the 1′ carbon of the sugar is one of four nitrogenous bases. The purine bases, adenine and guanine, have two ring structures; the pyrimidine bases, thymine and cytosine have a one ring structure. Phosphodiester bonds link the 5′ and 3′ carbon atoms of adjacent deoxyribose sugars. The schematic drawing illustrates the direction of the polynucleotide chain with the 5′ end at the top of the figure and the 3′ end at the bottom.

5′ phosphate of one nucleotide sugar is linked to the 3′ carbon of the adjacent nucleotide sugar. Linking the sugars of the nucleotides together with 5′, 3′ phosphodiester bonds, it is possible to form a polymer of indefinite length. If you look at Figure 12–6 closely, you will see that the polynucleotide chain has a direction to it. No matter how long the molecule is, one end will have a sugar with a 5′ carbon atom and the other end will have a sugar with a free 3′ carbon atom.

Chargaff's Rules

By 1950, the nucleotide composition of DNA from a number of organisms and tissues had been determined by Erwin Chargaff and his coworkers at Columbia University. Chargaff and his group developed procedures for chemically degrading the DNA to release the purine and pyrimidine bases. Researchers then separated and measured the bases to determine their relative proportions. Analysis of the DNA from different organisms yielded a simple relationship that turned out to be an important clue to its structure. No matter what organism or tissue was used, analysis of its DNA showed, in Chargaff's words, that the "ratios of purines to pyrimidines and also of adenine to thymine and of guanine to cytosine were not far from 1." In other words, in DNA molecules,

A = T and G = C

The Double Helix

Other important information about the structure of DNA came from x-ray diffraction studies on crystals of purified DNA carried out by Rosalind Franklin in the laboratory of M.H.F. Wilkens. X-ray diffraction is a powerful method for determining distances between atoms of molecules that are arranged in a regular, repeating crystalline structure (Figure 12–7). X rays have such extremely short wavelengths that they can be scattered by the electrons surrounding the atoms in a molecule. Atoms that have a dense electron cloud around them (e.g., phosphorus, oxygen) have a higher probability of deflecting electrons than do atoms with lower atomic numbers. When a crystal is exposed to an intense beam of x rays, the regular

FIGURE 12–7 X-ray diffraction photographs of suitably hydrated fibers of DNA. (*a*) Pattern obtained using the sodium salt of DNA. (*b*) Pattern obtained using the lithium salt of DNA. This pattern permits a most thorough analysis of DNA. The diagonal pattern of spots (reflections) stretching from 11 o'clock to 5 o'clock and from 1 o'clock to 7 o'clock provides evidence for the helical structure of DNA. The elongated horizontal reflections at the top and bottom of the photographs provide evidence that the purine and pyrimidine bases are stacked 0.34 nm apart and are perpendicular to the axis of the DNA molecule.

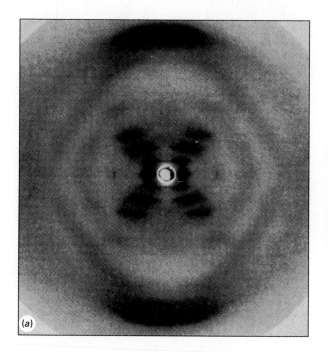
(a)

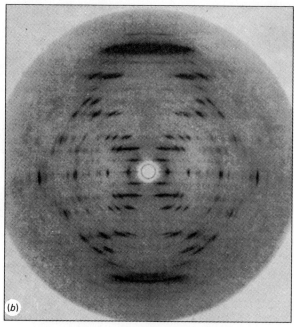
(b)

arrangement of the atoms in the crystal causes the x rays to be diffracted, or bent, in specific ways. The pattern of diffracted x rays is seen on film as dark spots. Mathematical analysis of the arrangement and distances between the spots can then be used to determine precise distances between atoms and their orientation within the molecules.

Franklin had already produced clear x-ray crystallographic films of DNA patterns when Watson and Crick began to pursue the problem of the structure of DNA. The pictures clearly showed that DNA had a type of helical structure and that there were three major types of regular, repeating patterns in the molecule with the dimensions 0.34 nm, 2.0 nm, and 3.4 nm. Franklin and Wilkins had inferred from these patterns that the nucleotide bases (which are flat molecules) are stacked one on top of the other like a group of saucers. Using this information, Watson and Crick began to build scale models of the DNA components and then fit them together to agree with the experimental data.

After a number of trials, the two worked out a model that fit the existing diffraction and chemical data (Figure 12–8). The nucleotide chains could be arranged to conform to the dimensions of the x-ray data only if each DNA molecule were to consist of *two* polynucleotide chains arranged in a coiled **double helix.** In their model, the sugar-phosphate backbone of each chain formed the outside of the helix, with the bases stacked like rungs of a ladder in the interior. To fit the data, each chain had to run in the direction

FIGURE 12–8 Molecular models of DNA. On the left is a space-filling model of the DNA double helix. On the right is a diagrammatic model of the DNA double helix with certain of its dimensions shown in nanometers (nm). The ribbons represent the sugar-phosphate backbone of each strand; the arrows indicate that the two strands extend in opposite directions.

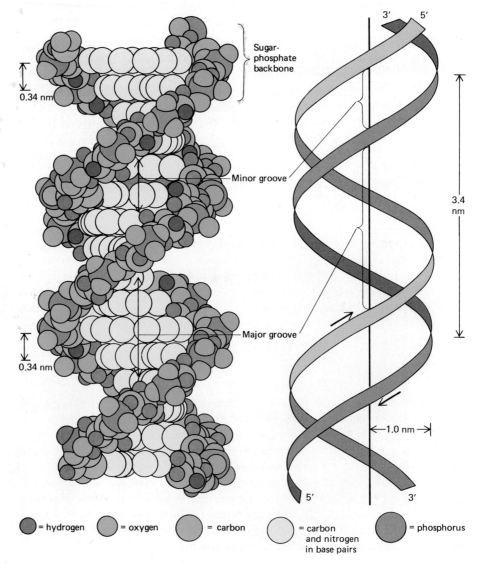

opposite that of its partner; therefore, each end of the molecule had an exposed 5′ phosphate on one strand and an exposed 3′ hydroxyl group on the other. Since the two strands ran in opposite directions, they were said to be **antiparallel** to each other.

The reasons for the 0.34-nm and 3.4-nm periodicities were readily apparent from the model: Each layer of bases was exactly 0.34 nm from the next layers above and below. Since exactly ten bases were present in each full turn of the helix, each turn was 3.4 nm high. The 2.0 nm reflections were thought to be due to the precise and constant width of the helix, which also helped explain Chargaff's rules.

Early in their attempts to determine the structure, Crick had learned that adenine and thymine could pair in solution through the formation of hydrogen bonds and that guanine and cytosine paired in a similar manner. Notice in Figure 12–6 that the pyrimidines, cytosine and thymine, contain only one ring of atoms and are smaller than the purines, guanine and adenine, which contain an additional five-membered ring. Study of the models made it clear to Watson and Crick that if each cross rung of the ladder contained one purine and one pyrimidine, the width of the helix at that point would be exactly 2.0 nm; the combination of two purines (each of which is 1.2 nm wide) would be wider and that of two pyrimidines would be narrower. Further examination of the model showed that adenine paired with thymine and guanine paired with cytosine could be arranged in such a way that the correct hydrogen bonds would form between them; the opposite combination, cytosine with adenine and guanine with thymine, would *not* lead to favorable hydrogen bonding between the two bases.

The nature of the hydrogen bonding between adenine and thymine and between guanine and cytosine is shown in Figure 12–9. Two hydrogen bonds can form between adenine and thymine; three can form between guanine and cytosine. This concept of **specific base pairing** was the explanation for Chargaff's rules. The amount of cytosine had to equal the amount of guanine, since for every cytosine in one chain there must be a paired guanine in the other chain. Similarly, for every adenine in the first chain there must be a thymine in the second chain. Thus, the sequences of bases in the two chains are **complementary,** but not identical, to each other; in other words, *the sequence of nucleotides in one chain dictates the complementary sequence of nucleotides in the other.*

The double-helix model made clear how DNA can provide genetic information. Opposite strands are complementary, so the entire molecule satisfies the criteria of Chargaff's rules. Although there are restrictions on how the bases pair with each other, the number of sequences of bases in either strand is virtually unlimited, with different sequences coding for different information. Since DNA molecules in a cell can be millions of nucleotides long, strings of base sequences can store enormous amounts of information.

DNA REPLICATION

Two immediately apparent and distinctive features of the Watson-Crick model made it seem more likely that DNA was the genetic material. One was the idea that DNA could carry coded information in its sequence of bases. The other was the mechanism by which information in DNA could be precisely copied—a process known as **DNA replication.** The importance of the replication mechanism of the genetic material was known to Watson and Crick, who noted in a classic and now famous piece of understatement at the end of their first brief paper, "It has not escaped our notice that the specific pairing we have postulated immediately suggests a possible copying mechanism for the genetic material."

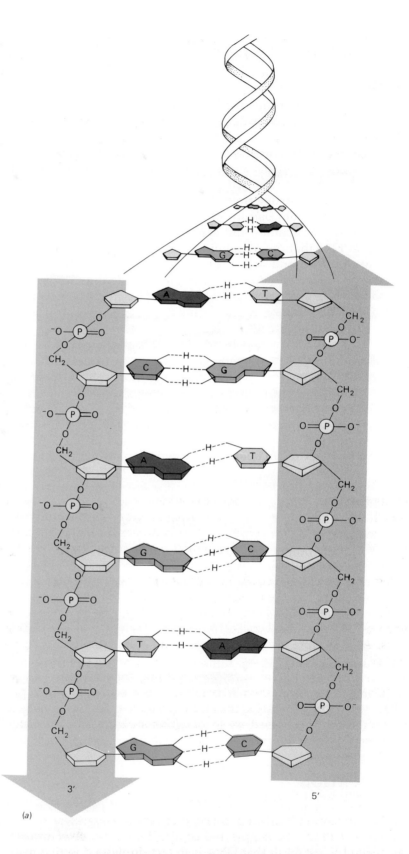

FIGURE 12–9 (*a*) Physical structure of DNA. Notice that the two sugar-phosphate chains extend in opposite directions. This orientation permits the complementary bases to pair.

(*a*)

What the model suggested was that, since the nucleotides could pair with each other in a complementary fashion, each strand of the DNA molecule could serve as a template, or pattern, for the synthesis of the opposite strand (see Figure 12–9). All that would be necessary was for the hydrogen bonds between the two strands to be broken, allowing the two chains to separate. Each half-helix could then pair with complementary nucleotides

Adenine   Thymine

Deoxyribose   Deoxyribose

Guanine   Cytosine

Deoxyribose   Deoxyribose

(b)

FIGURE 12–9 (continued) (b) Diagram of the hydrogen bonding between base pairs thymine (T) and adenine (A) (top) and guanine (G) and cytosine (C) (bottom). The AT pair has two hydrogen bonds; the GC pair has three.

to replace its missing partner. The result would be two DNA chains, each identical to the original one and each consisting of one original strand from the parent molecule and one newly synthesized complementary strand. This type of information copying is known as a **semiconservative** mechanism.

The recognition that DNA could be copied in this way revealed in turn how DNA could provide a third essential characteristic of the genetic material, the ability to mutate. It was known that mutations, or genetic changes, can arise in genes and can then be transmitted faithfully to succeeding generations. According to the double-helix model, mutations could represent a *change in the sequence of bases in the DNA*. If DNA is copied by a mechanism involving complementary base pairing, any change in the sequence of bases on one strand would result in a new sequence of complementary bases being paired during the next replication cycle. The new base sequence would then be passed on to daughter molecules by the same mechanism used to copy the original genetic material, as if no change had occurred (Figure 12–10).

Evidence for Semiconservative Replication

Although the semiconservative replication mechanism suggested by Watson and Crick was (and is) a simple and compelling model, experimental proof was needed to establish that DNA is in fact duplicated in that manner. First it was necessary to rule out several other possibilities. For example, with a *conservative* replication mechanism, both parent (or old) strands would remain together, and the two newly synthesized strands would form a double helix. With a *dispersive* mechanism, the newly synthesized DNA would contain random regions of both parental and newly synthesized strands. In order to distinguish between the semiconservative repli-

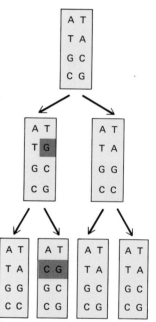

FIGURE 12–10 How a mutation could be stabilized by DNA replication. If genes are information coded in the sequence of bases on a DNA molecule, then a change (mutation) made in even one of the bases will be copied accurately by complementary base pairing. In the example, an adenine base in one of the DNA strands has been changed to guanine (this could occur by a rare error in DNA replication or by one of several other known mechanisms). When the DNA molecule is replicated again, one of the strands gives rise to a molecule that is exactly like the parent strand; the other, mutated, strand gives rise to a molecule with a new combination of bases that will be perpetuated generation after generation.

FIGURE 12–11 Meselson and Stahl's demonstration that DNA is replicated by a semiconservative mechanism. In this experiment, the bacterium *E. coli* is grown in heavy (¹⁵N) nitrogen growth medium for many generations and then transferred to light (¹⁴N) nitrogen medium. The top of the figure illustrates the predicted labeling pattern by semiconservative replication for cells grown in the heavy medium and for the first and second generations following transfer to the light medium. By isolating the DNA from cells grown at these different times and determining the density of the molecules in each generation, Meselson and Stahl were able to confirm that the DNA is replicated in a semiconservative manner.

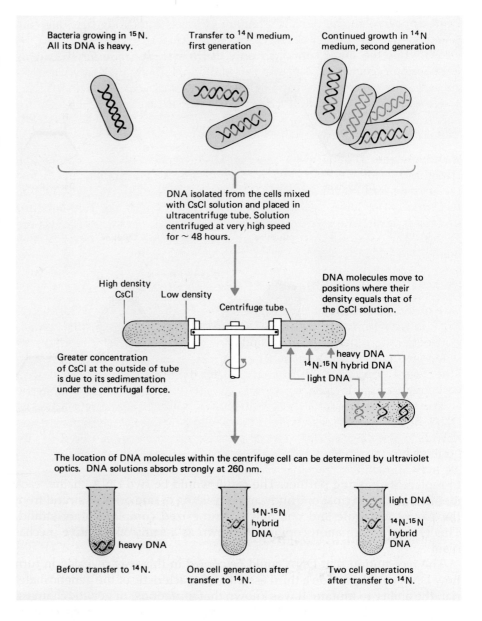

cation mechanism and other possibilities, it was necessary to distinguish between old and newly synthesized strands of DNA.

In 1957, Matthew Meselson and Franklin Stahl employed the heavy nitrogen isotope ¹⁵N (ordinary nitrogen is ¹⁴N) to label parent strands of DNA molecules in the bacterium *Escherichia coli* (Figure 12–11). Incorporation of the heavy N atoms increased the density of the DNA molecules, as demonstrated by **density gradient centrifugation.** When the labeled DNA isolated from some of the bacterial cells was centrifuged in a tube containing a density gradient formed by a heavy salt solution, the molecules became distributed in the tube according to their densities. As expected, the DNA molecules that contained heavy nitrogen accumulated in the centrifuge tube at the high-density region of the gradient. The remaining bacteria containing the ¹⁵N-labeled DNA were then transferred to medium containing the naturally abundant, lighter ¹⁴N isotope and allowed to undergo several more cell divisions. The newly synthesized DNA strands became less dense because they incorporated bases containing the ¹⁴N isotope. Molecules of DNA from cells isolated after one generation had a density indicating that they contained half as many ¹⁵N atoms as the DNA of the parental generation. After another cycle of cell division, two species of

DNA appeared in the density gradient at levels indicating that one consisted of "hybrid" DNA helices (labeled with equal amounts of ^{15}N and ^{14}N DNA) while the other contained only DNA with the naturally occurring light isotope. Each strand of the parental double-helix molecule was thus conserved in a *different* daughter molecule, exactly as predicted by the semiconservative replication model.

The Replication Process

Although the general principles of DNA replication are simple and straightforward predictions from the Watson-Crick model, the actual mechanism and machinery of DNA synthesis are quite complex. Due to the structure of DNA and the need for highly accurate replication of the original strands, the process requires a large number of proteins and enzymes. The essential features of the replication process are described below.

DNA Strands Must be Unwound During Replication

One of the properties of the double helix originally noted by Watson and Crick is that the two strands of the molecule are physically intertwined. Consequently, when the two strands are separated for replication to proceed, the ends of the helix must either rotate or twist into tighter coils. Strand separation, by itself, is a complex process. Since DNA molecules are very long, replication can work properly only if the strain on the molecule is relieved as the two strands are unwound. Unwinding is accomplished by **DNA helicase enzymes,** which travel along the helix, unwinding the strands as they move. Once the strands are separated, **helix-destabilizing proteins** bind to single-stranded DNA, preventing re-formation of the double helix until the strands are copied.

DNA Synthesis Always Proceeds in a 5'→3' Direction

The enzymes that catalyze the linking together of the nucleotide subunits are called **DNA polymerases.** These enzymes add nucleotides to the 3' end of a polynucleotide strand that is *paired* to the strand that is being copied (Figure 12–12). The DNA polymerases use **deoxyribonucleotide triphos-**

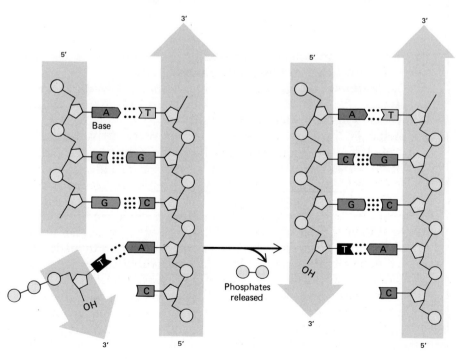

FIGURE 12–12 DNA synthesis. The building blocks for the DNA molecule are nucleoside triphosphates, which lose two of their phosphates when they are linked to the 3' carbon of the sugar at the end of the growing chain. The specificity of the polymerase enzymes that catalyze the polymerization reactions requires that the growing chain always elongates in the 5' → 3' direction.

phates as the subunits for the polymerization reactions. The subunits are similar to ATP in that they contain three phosphate groups linked to the 5'-carbon of the sugar group, plus a base. As the nucleotides are linked together, two of the phosphates are removed, releasing energy, which is used to drive the synthetic reaction. Since elongation of the new polynucleotide chain is accomplished by linkage of the 5'-phosphate group of the next nucleotide subunit to the 3'-hydroxyl sugar at the end of the growing strand, the new strand of DNA grows from its 5' end in the direction of its 3' end.

DNA Synthesis Requires a Primer

The fact that DNA polymerases can add nucleotides only to the end of an existing DNA strand raises the question of how DNA synthesis can be initiated once the two strands are separated (Figure 12–13). The answer is that a short piece (usually about five nucleotides) of an **RNA primer** is first synthesized by an aggregate of proteins called a **primosome**. This RNA primer pairs with the single-stranded DNA template at the point of initiation of replication. DNA polymerase can then initiate synthesis by adding

FIGURE 12–13 (*a*) DNA synthesis begins at a specific base sequence, termed the *origin of replication*. (*b*) Strands are separated at the origin and unwound by DNA helicase, which "walks" along the DNA molecule preceding the DNA-synthesizing enzymes. Single-stranded regions are prevented from re-forming into double strands by helix-destabilizing proteins, which bind to single-stranded DNA. The region of active DNA synthesis is associated with the "replication fork," formed at the junction of the single strands, and the double-stranded region. DNA synthesis proceeds on both single strands of the fork in a 5' → 3' direction. (*c*) As the new strands continue to grow in the first direction, unwinding and replication initiates on the other side of the origin, so that replication proceeds in both directions. (*d*) Completion of replication results in the formation of two daughter molecules, each containing one newly synthesized strand.

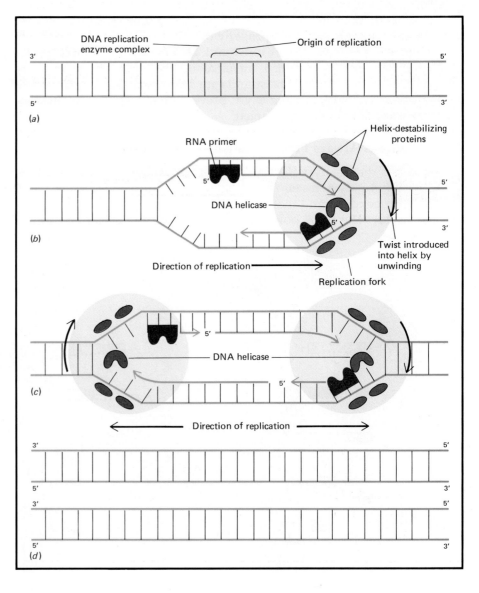

subunits to the 3' end of the RNA primer. The primer is later degraded by specific enzymes and the space filled in with DNA.

DNA Replication is Discontinuous

One of the major obstacles to understanding how DNA replication works was the fact that the complementary DNA strands run in opposite directions. Since DNA synthesis proceeds only in the direction of 5' → 3' (which means that the strand being copied is being read in a 3' → 5' direction), it would appear that only one strand could be copied at a time. We know, however, that this is not the case. DNA replication is initiated at specific sites on the DNA molecule, termed **origins of replication,** and both strands are replicated at the same time. One strand, the **leading strand,** is formed continuously following initiation of synthesis, whereas its partner strand, the **lagging strand,** is formed in short pieces, which are later linked together to make a completed molecule.

DNA is synthesized from prokaryotic and eukaryotic chromosomes at a Y-shaped structure in the molecule referred to as the **replication fork** (Figure 12–14). The lagging strand is synthesized in short (100- to 1000-nucleotide) pieces, called **Okazaki fragments** after their discoverer, Reijii Okazaki. Each Okazaki fragment is initiated by a separate primer and is then extended toward the 5' end of the previously synthesized fragment by DNA polymerase. The DNA polymerase protein in *E. coli* is a complex enzyme with several functions. As the growing fragment approaches the one synthesized previously, one part of DNA polymerase degrades the previous RNA primer, allowing other polymerases to fill in the gap between the two fragments. These are then linked together by **DNA ligase,**

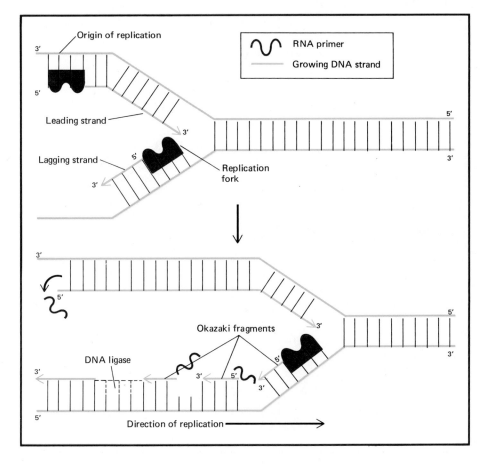

FIGURE 12–14 Discontinuous replication. Since elongation can only proceed in a 5' → 3' direction, the two strands of the replication fork are copied in different ways. The *leading strand* is synthesized continuously in a direction toward the replication fork, the *lagging strand* is synthesized in short pieces called *Okazaki fragments,* in a direction away from the replication fork. Initiation of synthesis for both strands requires an *RNA primer* since DNA can be elongated only by addition to the 3' end of an existing strand. After elongation has begun, the RNA primer is degraded and the adjoining fragments are linked together by DNA ligase.

an enzyme that uses a phosphodiester bond to link the 3′ end of one DNA fragment to the 5′ end of another.

Most DNA Synthesis is Bidirectional

When double-stranded DNA is separated, two forklike structures are created (Figure 12–15) so that the molecule is replicated in both directions from the origin of replication. In prokaryotic cells there is usually only one origin of replication on each circular chromosome, so that the two replication forks proceed around the circle, eventually meeting at the other side and completing the formation of two new DNA molecules. In eukaryotic chromosomes, each of which is composed of one linear DNA molecule, there are usually multiple origins of replication. In that case each replication fork proceeds until it meets one coming from the opposite direction, resulting in the formation of a chromosome containing two DNA double helices (Figure 12–14*b*).

FIGURE 12–15 Bidirectional replication of DNA in bacterial and eukaryotic chromosomes. (*a*) In *E. coli* the circular chromosome has only one origin of replication. DNA synthesis proceeds from that point in both directions around the circular chromosome until the two replication forks meet. (*b*) Eukaryotic chromosomal DNA contains multiple origins of replication. DNA synthesis also proceeds in both directions from those origins until the replication fork from one "replication bubble" meets the replication fork from an adjacent bubble. The photograph shows a segment of a eukaryotic chromosome that has been partially replicated. The conditions used to produce this picture also resulted in the separation of some of the newly formed DNA. *1* and *2* are the daughter double helices; *N* is a portion of the double helix not yet replicated. The double helix is separated in the regions marked *S*, illustrating its double structure.

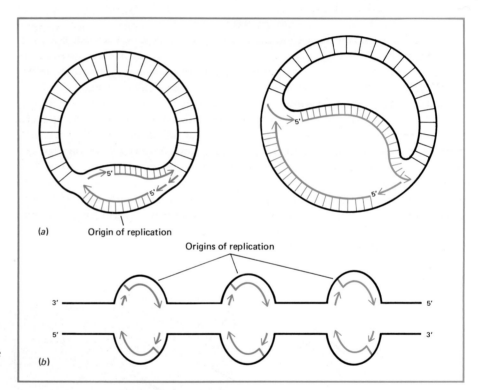

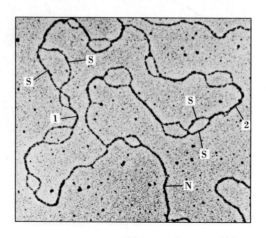

THE ORGANIZATION OF DNA IN CHROMOSOMES

Prokaryotic and eukaryotic cells differ markedly with regard to their DNA content and the manner in which the DNA molecules are organized. In bacterial cells such as *E. coli*, most, if not all, of the DNA is in the form of a single *circular*, double-stranded molecule. In contrast, almost all of the DNA in eukaryotic cells is organized in the nucleus as multiple chromosomes, which can vary markedly in size and number among different species. Each eukaryotic chromosome appears to contain a single *linear* double-stranded molecule prior to DNA replication. There are also quite large differences in DNA content between prokaryotes and eukaryotes. An *E. coli* cell normally contains about 4×10^6 base pairs of DNA in its single chromosome which, if stretched out, would be almost 1.35 mm long. This in itself represents a significant organizational problem, since the total length of the DNA in the cell is about 1000 times greater that the length of the cell itself. Consequently, the DNA molecule must be twisted and folded compactly to fit inside the bacterial cell.

A typical eukaryotic cell contains much more DNA than a bacterium. Although the nucleus of a human cell is about the size of a large bacterial cell, it contains more than 1000 times the amount of DNA found in *E. coli*. The haploid DNA content of a human cell is about 3×10^9 base pairs which, if stretched end to end, would be almost 1 m long. To fit in such a small space, the DNA is wound around proteins called **histones** to form structures called **nucleosomes.** The basic unit of the complex consists of about 140 base pairs of DNA wrapped around a disc-shaped core of eight histone protein molecules (Figure 12–16) and additional histone proteins that are bound to the section of DNA that links two neighboring nucleosomes. The nucleosomes are units of higher levels of organization that form **chromatin,** the nucleoprotein complex of which chromosomes are made. Chromatin fibers are composed of nucleosomes organized into large coiled loops held together with a set of nonhistone **scaffolding proteins** (Figure 12–17). Figure 12–18 illustrates the dense packing of DNA fibers on a histone-depleted mouse chromosome.

FIGURE 12–16 The units of histone surrounded by DNA in the chromosomes are called nucleosomes. (*a*) A model for the structure of a nucleosome. Each nucleosome bead contains a set of eight histone molecules, which form a protein core around which the double-stranded DNA is wound. The DNA wound around the histone consists of 146 nucleotide pairs; another segment of DNA about 60 nucleotide pairs long links nucleosome beads. A linker DNA segment plus one nucleosome bead together constitute a nucleosome. One type of histone (H1) covers the linker DNA between adjacent nucleosomes. This histone appears to be responsible for packing nucleosomes and may help link them to one another. (*b*) Nucleosomes from the nucleus of a chicken red blood cell. Each spherical structure and its adjacent linker is a nucleosome. Normally nucleosomes are packed more closely together, but the preparation procedure has spread them apart, revealing the DNA linkers.

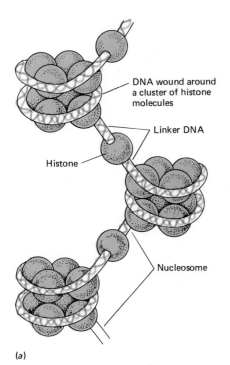

DNA wound around a cluster of histone molecules
Linker DNA
Histone
Nucleosome
(a)

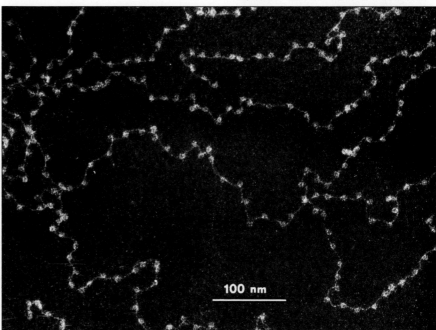

100 nm
(b)

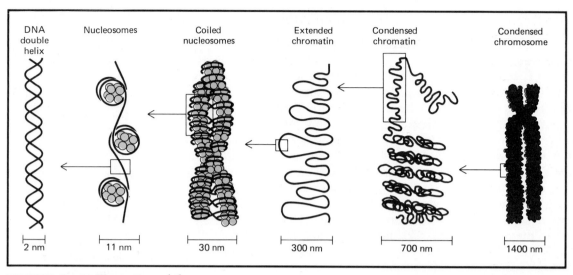

DNA double helix	Nucleosomes	Coiled nucleosomes	Extended chromatin	Condensed chromatin	Condensed chromosome
2 nm	11 nm	30 nm	300 nm	700 nm	1400 nm

FIGURE 12–17 Illustration of the levels of organization within the eukaryotic chromosome.

FIGURE 12–18 Electron micrographs of a mouse chromosome depleted of histones. (*a*) Note how densely packed the DNA fibrils are, even though they have been released from the proteins which organize them into tightly coiled structures. The dark structure running from left to right across the photograph is the scaffolding proteins. (*b*) Higher magnification of the area outlined in (*a*). The DNA is still organized in the form of loops which protrude from the scaffolding.

The size of DNA relative to the cells in both prokaryotes and eukaryotes creates additional complications for the DNA replication process. Not only is there strain on the molecules as strands are unfolded, there is also the problem of separating the daughter molecules without tangling and breaking them. Replication of circular molecules in bacteria can result in the formation of *supercoils*, which may knot and tangle the newly formed DNA molecules together. The separation of duplicated strands of eukaryotic DNA, which is in a highly condensed state in the nucleus, poses even greater problems. Suppose you wanted to duplicate a length of thread wound around a spool by making a side-by-side copy of every point along the thread. How could you separate the two strands of thread so that neither would be broken, in a space not much larger than the spool itself? Clearly there is a solution to this problem, since replicating cells do it with precision generation after generation. The DNA in both prokaryotes and eukaryotes is associated with special enzymes, called **topoisomerases,** that create breaks in the DNA molecules and then rejoin the strands, effectively untying knots that develop during replication. In eukaryotes, the untangling mechanism must also be intimately associated with the organization of the eukaryotic chromosome. The exact role that organization plays, however, remains to be discovered.

(*a*)

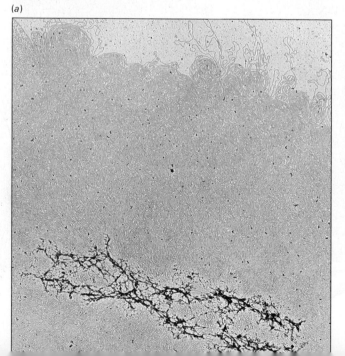

(*b*)

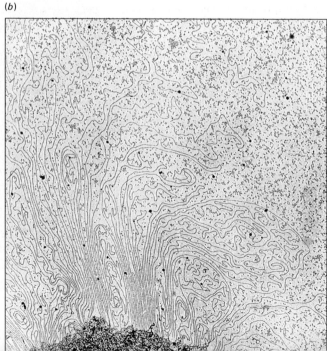

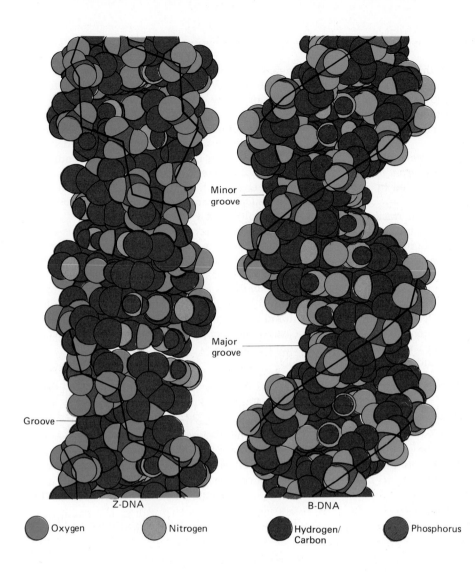

Minor
groove

Major
groove

Groove

Z-DNA

B-DNA

○ Oxygen ○ Nitrogen ● Hydrogen/ ● Phosphorus
Carbon

FIGURE 12–19 Z DNA compared to B DNA. Z-DNA forms a left-handed helix, and its major groove zigzags. Because of the zigzag and because it is so different from all previously known forms of DNA, Z DNA was named from the other end of the alphabet! Z DNA is genetically inactive, and is most likely to form only when certain sequences of bases are present in the DNA strand. It may function as part of a genetic control mechanism.

FORMS OF DNA

Several alternative forms of DNA have been discovered since Watson and Crick proposed their model of what is now known as **B DNA.** Most of the alternative forms involve relatively minor variations in the pitch and tilt of the helix, but recently a radically different form, called **Z DNA,** has come to light (Figure 12–19). This form of DNA forms a left-handed helix rather than the "standard" right-handed Watson-Crick variety. Moreover, its groove zigzags, which is one of the reasons it is called *Z DNA*. The significance of this form is beginning to be appreciated. Z DNA is quite common in certain regions of chromosomes and is thought to be involved in regulating some eukaryotic genes by blocking their activity. Evidence also suggests that potential Z DNA–forming sequences are associated with "recombination hotspots," which are regions of DNA in chromosomes that recombine at high frequency by breaking and rejoining.

DNA probably assumes other conformations in the cells. For example, there is evidence that the base sequences in some regions are arranged so that the two strands of the helix can separate, allowing each strand to fold and pair with itself, forming hairpin-like structures. Structures of this type may in fact be regions that can be recognized by special DNA-binding proteins involved in controlling the activity of genes, or they may serve as signals specifying recombination sites (Figure 12–20).

FIGURE 12–20 Computer simulation of a piece of synthetic DNA used in biophysical studies on the different structural forms of DNA.

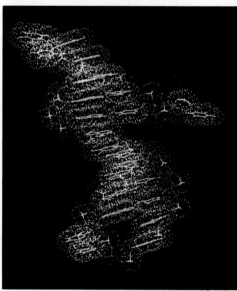

■ SUMMARY

I. Many geneticists in the first half of this century thought that genetic information was probably in the form of proteins. Proteins were known to be complex molecules, whereas nucleic acids were thought to be rather simple molecules with a limited ability to store information.

 A. The work of Garrod on inborn errors of metabolism and that of Beadle and Tatum with *Neurospora* mutants suggested that each protein is specified by a single gene.

 B. Early evidence that genes are made from nucleic acids came from "transformation" experiments in which the DNA of one strain of bacterium was found to be able to change a related bacterium to a strain with new genetic characteristics.

 C. Further evidence that nucleic acids are the essential components of genetic information came from studies of bacterial viruses that have DNA as their genetic material. These showed that when a bacterial cell becomes infected with a virus, only the DNA from the virus enters the cell and that this DNA is sufficient for the virus to reproduce and form new virus particles.

 D. The convincing evidence that DNA is the genetic material came from Watson and Crick's studies on the structure of DNA, which showed how information can be stored in the structure of the molecule and how the DNA molecules can serve as templates for their own duplication.

II. The structure of DNA

 A. DNA is formed as a polymer of nucleotides.

 1. Each nucleotide subunit contains a nitrogenous base consisting of one of the purines (adenine or guanine) or one of the pyrimidines (thymine or cytosine). The bases are covalently linked to the five-carbon sugar deoxyribose.

 2. The polymeric chain is joined together by phosphate groups, which link the 5′ carbon of the deoxyribose of one nucleotide to the 3′ deoxyribose carbon of its neighbor.

 B. Each DNA molecule is composed of two polynucleotide chains, which form a double-helix structure. The two chains extend in opposite directions, so that at each end of the DNA molecule one chain has an exposed 5′ deoxyribose carbon and the other has an exposed 3′ deoxyribose. Thus, the chains are antiparallel to each other.

 C. The deoxyribose-phosphate backbone of each chain is on the outside of the helix; the purine and pyrimidine bases are on the inside.

 D. The two chains of the helix are held together by hydrogen bonding between specific base pairs. Adenine (A) pairs with thymine (T), and guanine (G) pairs with cytosine (C).

 1. The complementary base pairing between A and T and between G and C are the basis of Chargaff's rules, which state that A = T and G = C.

 2. Since the two strands of DNA are held together by complementary base pairing, it is possible to predict the base sequence of one strand if one knows the base sequence of the other strand.

III. During DNA replication, the two strands of the double helix unwind. Each strand serves as a template for the formation of a new complementary strand.

 A. DNA replication is semiconservative—that is, each daughter double helix contains one strand from the parent molecule and one newly synthesized strand.

 B. DNA replication is a complex process requiring a number of different enzymes.

 1. The enzyme that catalyzes the formation of the new DNA strand from deoxyribonucleoside triphosphates is a DNA polymerase.

 2. Additional enzymes and other proteins are required to unwind and stabilize the separated DNA helix, to form primers, and to link together fragments of newly synthesized DNA.

 C. DNA synthesis always proceeds in a 5′ → 3′ direction. This requires that the synthesis of one DNA strand be discontinuous, in short fragments, called *Okazaki fragments*. The opposite strand is copied continuously.

 D. DNA replication is bidirectional, starting at one point, termed the *origin of replication*, and proceeding in both directions from that point. A eukaryotic chromosome may have multiple origins of replication and at any one time may be replicating at many points along its length.

IV. DNA is organized into chromosomes.

 A. Chromosomes of prokaryotic cells are usually in the form of circular DNA molecules.

 B. Eukaryotic chromosomes have several levels of organization.

 1. The DNA is associated with proteins called *histones*. The organizational unit, a beadlike structure called a *nucleosome*, consists of a coiled portion of DNA associated with histones plus an adjacent piece of linker DNA with a histone attached.

 2. The nucleosomes themselves are organized into large coiled loops held together by nonhistone scaffolding proteins.

 3. DNA molecules are much longer than the nuclei or the cells that contain them. The organization of DNA into chromosomes allows the DNA to be accurately replicated and segregated into daughter cells without tangling.

V. Several new forms of DNA have been discovered. Most of the DNA in chromosomes is in a form similar to the B form of DNA that was described by Watson and Crick. Z DNA is a "left-handed" form of DNA that may be involved in controlling the activity of certain genes and may also play some role in certain types of genetic recombination involving breaking and rejoining.

■ POST-TEST

1. Early evidence that DNA is the genetic material came from _____ experiments that showed that purified _____ is capable of changing a bacterial strain to a genetically stable new form.
2. DNA molecules are polymers composed of _____ monophosphates linked together by _____ groups, which join the _____ carbon of one nucleotide sugar with the _____ carbon of the adjacent nucleotide sugar.
3. Nucleotides found in DNA are composed of the five-carbon sugar _____. Attached to its 5' carbon is a _____ group. Attached to its 1' carbon is one of four _____ bases. The bases adenine and guanine are called _____; the bases thymine and cytosine are referred to as _____.
4. Chargaff's rules were formulated by analysis of the _____ composition of DNA from different species. These findings stated that the number of _____ bases equalled the number of _____ bases and that the number of _____ bases equalled the number of _____ bases.
5. The basic information needed by Watson and Crick to construct their model of DNA came from knowledge of the chemical structure of DNA and the _____ studies of Franklin and Wilkins.
6. Each DNA molecule consists of two anti-_____ chains arranged in a double _____.
7. The two chains of DNA are held together by base pairing between the complementary bases _____ and _____ and the complementary bases _____ and _____.
8. The process of copying a DNA molecule is termed DNA _____.
9. DNA is replicated by a _____ mechanism; the newly synthesized molecule contains one _____ strand and one _____ strand.
10. In order for DNA replication to start, the two strands must be separated at a point in the molecule known as the _____ _____ _____.
11. The newly synthesized strand of DNA is always synthesized starting at its _____ end and is elongated toward its _____ end.
12. The DNA molecules are formed from nucleotide precursors, which are _____ _____.
13. DNA replication is referred to as a *discontinuous process*, which means that one of the strands must be synthesized in short pieces, called _____ _____. The other strand of DNA is synthesized as a _____ molecule.
14. DNA is synthesized by an enzyme called _____ _____. The pieces of the lagging (discontinuous) strand are joined together by an enzyme called _____ _____.
15. The DNA in prokaryotic cells is usually in the form of a _____ molecule; the DNA in a eukaryotic chromosome is a single, _____ molecule.
16. Prokaryotic chromosomes generally have _____ origin(s) of replication, while eukaryotic chromosomes have _____ origin(s) of replication.
17. Most DNA synthesis in both prokaryotic and eukaryotic cells is _____, which means that it proceeds in both directions from the origin of replication.
18. The *E. coli* bacterial cell contains about _____ the amount of DNA that is found in a human cell. The length of DNA in an *E. coli* cell is about _____, whereas the length of DNA in a single human cell is about _____.
19. DNA in eukaryotic cells is organized into subunits called _____, which are composed of about 140 base pairs of DNA wrapped around a core of histone proteins and an adjacent region of linker DNA with an additional histone.
20. _____ fibers, which are composed of nucleosomes, are organized into large coiled _____, which are held together by nonhistone _____ _____.

■ REVIEW QUESTIONS

1. Discuss the evidence that DNA is the chemical substance of the gene. What properties of the DNA molecule allow it to serve as the genetic material? What characteristics must any molecule have if it is to serve as genetic material?
2. How did the experiments of Griffith and later Avery and coworkers point to DNA as the essential genetic material? How did the Hershey-Chase experiment establish that DNA was the genetic material? Did either of these experiments demonstrate how DNA could function as the chemical basis of genes?
3. Describe the structure of a chain of DNA. What types of subunits make up the chain? How are they linked together?
4. Explain Chargaff's rules. Does a single strand of DNA obey the rules? How do Chargaff's rules relate to the structure of DNA?
5. Describe the structure of DNA as determined by Watson and Crick. What important features of its structure show how it could serve as the genetic material?
6. Describe the mechanism by which DNA is replicated. Why is DNA replication discontinuous?
7. Compare the structure of a bacterial chromosome with that of a eukaryotic chromosome.
8. Compare the amount of DNA in a bacterial cell with that in the nucleus of a eukaryotic cell. How do the dimensions of the DNA molecules compare with the dimensions of the cell in each case?

■ RECOMMENDED READINGS

Felsenfeld, G. DNA. *Scientific American*, October 1985, pp. 58–67. An excellent, well-illustrated article on the structure and organization of DNA.

Judson, H.F. *The Eighth Day of Creation: Makers of the Revolution in Biology*. New York, Simon & Schuster, 1979. A beautifully written and fascinating account of the history of molecular biology.

Watson, J.D. *The Double Helix*. New York, Atheneum, 1968. Watson's view of the discovery of the structure of DNA. Somewhat controversial, but entertaining and insightful reading.

Watson, J.D., and Crick, F.H.C. Molecular structure of nucleic acids: A structure for deoxyribose nucleic acid. *Nature* 171: 737–738, 1953. Watson and Crick's original report—a simple, clearly written two-page paper that shook the scientific world.

Watson, J.D., Hopkins, N.H., Roberts, J.W., Steitz, J.A., and Weiner, A.M. *Molecular Biology of the Gene*, 4th ed. Menlo Park, CA, Benjamin Cummings, 1987. The most recent edition of Watson's classic text.

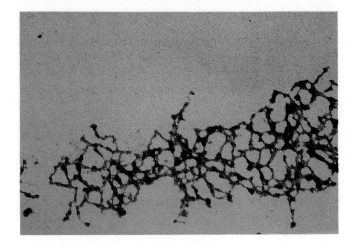

13

RNA and Protein Synthesis

■ OUTLINE

I. An overview of gene expression
II. Transcription: The synthesis of RNA
   A. The transcription process
   B. The structure of messenger RNA
III. Translation: Decoding the message
   A. The activation of amino acids
   B. The structure of transfer RNA
   C. The ribosomes
   D. Protein synthesis
   E. Polyribosomes
   F. A comparison of prokaryotic and eukaryotic protein synthesis
      1. Introns and exons
      2. Exon shuffling
IV. The genetic code
   A. Codon–amino acid specificity
   B. The second genetic code
V. What is a gene?
VI. Changes in genes: Mutations
Focus on reverse transcription, jumping genes, and pseudogenes

■ LEARNING OBJECTIVES

After you have read this chapter you should be able to:

1. Outline the mechanism by which a gene is transcribed to yield mRNA and the mechanisms by which mRNA is translated to yield a specific protein product.
2. Discuss the genetic code and its characteristics; explain why the code is said to be redundant and universal.
3. Compare the structure of RNA and DNA, and describe how transcription of a gene is initiated.
4. Discuss the role of tRNA molecules in the translation of genetic information into protein; explain what features of the structure of tRNA are important in decoding genetic information.
5. Discuss the structure of the ribosome, and explain the roles different parts of its structure play in protein synthesis.
6. Discuss the essential features of initiation, elongation, and termination in the process of protein synthesis.
7. Compare the structures of prokaryotic and eukaryotic mRNA molecules, and explain how they differ.
8. Compare the process of translation in prokaryotic and eukaryotic cells; explain how differences in mRNA structure affect the way in which mRNAs are translated.
9. Discuss the evolutionary implications for a universal genetic code.
10. Discuss the different types of mutations that affect the base sequence of DNA. Explain the different effects that can be produced by each type of mutation.

In the preceding chapter we saw how DNA molecules can be used to store information. We also saw how that information can be replicated by a cell so that it can be passed accurately to its descendants. Those basic features of DNA originally described by Watson and Crick are now known to be the same in virtually all cells, from bacteria to humans.

In the mid-1950s it was evident that the genetic information in DNA contained the code for making all of the proteins needed by the cell. However, it was impossible on the basis of the structure of DNA alone to deduce how cells could translate that information into proteins. It would be another 12 years after Watson and Crick's famous paper before we would understand how cells could accomplish such a feat. Much of that understanding came from the study of gene functions in bacteria. After the discovery of the structure of DNA, prokaryotic cells quickly became the organisms of choice for the study of those problems; they could be grown quickly and easily and seemed to contain only the minimal amount of DNA necessary for growth and reproduction.

In this chapter we will see how genetic information is decoded and used to make proteins. In the next chapter we will see how the entire process is controlled. We will first focus our attention on gene expression in prokaryotic cells, since these cells are best understood. We will then compare that process with what is known about eukaryotic cells; knowledge about these cells is increasing at a breathtaking pace due to the groundwork laid by the study of the simpler bacterial systems.

AN OVERVIEW OF GENE EXPRESSION

Two basic steps are involved in the making of proteins from genetic information in DNA. In the first step, a copy is made of the information in the DNA that is needed to make a certain protein; this copy, another form of nucleic acid, is called **messenger RNA (mRNA)** (Figure 13–1). This process is similar to DNA replication in that the mRNA strand is formed by complementary base pairing with one strand of the DNA. Since the synthesis of mRNA involves production of a copy of nucleic acid information (DNA) in the form of another nucleic acid (RNA), the process is called **transcription.**

In the second stage of the process, the transcribed information in the mRNA is converted into the amino acid sequence of a protein. Since that process involves transformation of the "nucleic acid language" in the mRNA molecule into the "amino acid language" of the protein, it is called the **translation** of the genetic code. The information contained within the mRNA is in the form of a three-base **genetic code.** A group of three bases in the mRNA that specifies an amino acid is termed a **codon.** Each codon in the mRNA specifies one of the amino acids in the protein encoded by the gene; for example, the codon that specifies the amino acid threonine is 5'-ACG-3' (Figure 13–2). Translation involves the recognition and decoding of codons in the mRNA. This is accomplished by "adapter" molecules, called **transfer RNAs (tRNAs).** Each tRNA molecule has a sequence of three bases, called the **anticodon,** which can recognize a codon on the mRNA by complementary base pairing. The amino acid specified by the anticodon is attached to another part of the tRNA molecule.

The conversion of genetic information into protein requires the linking together of amino acids in the correct order by peptide bonds. This is accomplished by **ribosomes** (see Chapter 4), complex structures composed of two different subunits each of which contains a number of proteins and

FIGURE 13–1 Transcription. Information in the DNA that codes for a protein is transcribed in the form of a molecule of messenger RNA (mRNA). The mRNA molecule is formed by complementary base pairing with the information-containing strand of DNA.

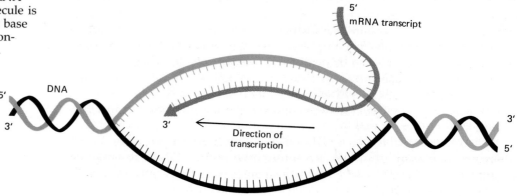

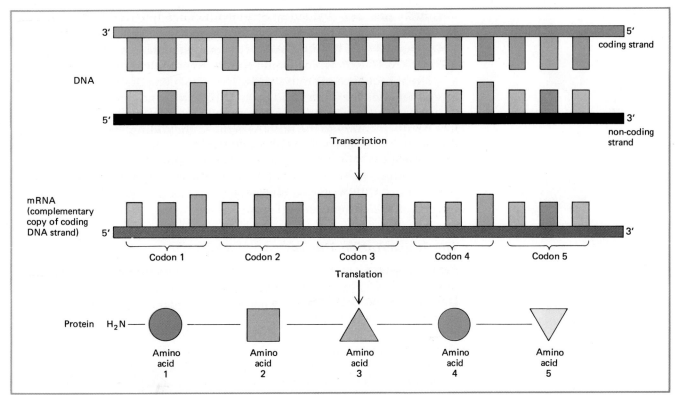

FIGURE 13–2 Genetic information in mRNA is composed of sets of three bases called codons, each of which specifies one amino acid. Messenger RNA codons are translated consecutively, specifying the sequence of amino acids in the polypeptide chain.

ribosomal RNA (rRNA). Ribosomes attach to the 5' end of the mRNA and travel along it, allowing the correct tRNAs to decode the message into a sequence of amino acids and linking them together to form a polypeptide. Polypeptides are thus formed from the information originally encoded in the DNA, by the linking together of amino acids in a linear sequence.

None of this could occur without the genetic code that directs it. The puzzle to scientists before (and after) the revelations of Watson and Crick was how the four bases in DNA could be used to govern the assembly of 20 amino acids into a vast variety of cellular proteins. When the new model of DNA forced them to consider this problem, however, it became apparent that the DNA bases could serve as a four-letter alphabet. Using three-letter combinations of the four bases (4^3), it is possible to form a total of 64 "words," more than sufficient to specify all of the naturally occurring amino acids. A decade later, the "cracking" of the genetic code was completed, verifying the existence of the three-base **triplet** code that is common to all organisms.

TRANSCRIPTION: THE SYNTHESIS OF RNA

RNA differs from DNA in that it contains the sugar ribose instead of deoxyribose and the base uracil instead of thymine (Figure 13–3). The molecules of RNA involved in protein synthesis are single-stranded, although certain regions of them may have complementary sequences that allow them to fold back to form paired regions within the single molecule. (Some forms of RNA in certain viruses do form two-stranded helices, in which case they serve the same function as DNA. The poliomyelitis virus, for example, enters the cell as a single-stranded molecule but replicates itself as two complementary strands of RNA.)

Most RNA is synthesized by **DNA-dependent RNA polymerases,** which are enzymes present in all cells. These enzymes require DNA as a template or pattern and, like DNA polymerase, use nucleoside triphosphates as substrates (Figure 13–4).

Usually only one of the strands in a protein-coding region of the DNA is transcribed (Figure 13–5a). Consider a length of DNA that contains the DNA bases 5'–ATTGCCAGA-3'. Its complementary strand would read 3'-TAACGGTCT-5', which would specify an entirely different protein with a different sequence of amino acids. Thus, only one of the DNA strands in a gene contains the amino acid coding sequence, and that is the one that is transcribed into mRNA. The DNA strand that is transcribed (and is therefore complementary to the mRNA) is called the **coding,** or **sense,** strand of the DNA. The opposite (complementary) DNA strand is the **noncoding,** or **antisense,** strand of the DNA. On a chromosome a DNA strand may serve as the coding strand at one point and as the noncoding strand at another point. Thus, mRNA may be transcribed in either direction, but on different regions of the chromosome (Figure 13–5b).

The Transcription Process

RNA polymerase starts transcription by recognizing a specific **"promoter"** base sequence at the beginning of a gene. Like DNA, an mRNA chain is always synthesized by the addition of nucleotides, one at a time, to the 3'

FIGURE 13–3 An RNA molecule. RNA is made of ribonucleotide monophosphate subunits linked together by 5'-3' bonds like DNA. Three of the nitrogenous bases— adenine, guanine, and cytosine— are the same as those found in DNA. However, instead of having thymine, RNA has the base uracil, which pairs with adenine. All four nucleotides contain the five-carbon sugar ribose, which has a hydroxyl group on the 2' carbon atom.

FIGURE 13–4 A diagram of the transcription process. The DNA helix is unwound. The exposed DNA strand on the right is being copied. Incoming ribonucleotide triphosphates pair with complementary bases on the DNA template strand. RNA polymerase cleaves two phosphates from the nucleotide and covalently links the remaining phosphate to the 3' end of the growing RNA chain. Thus, RNA is also synthesized in a 5' → 3' direction.

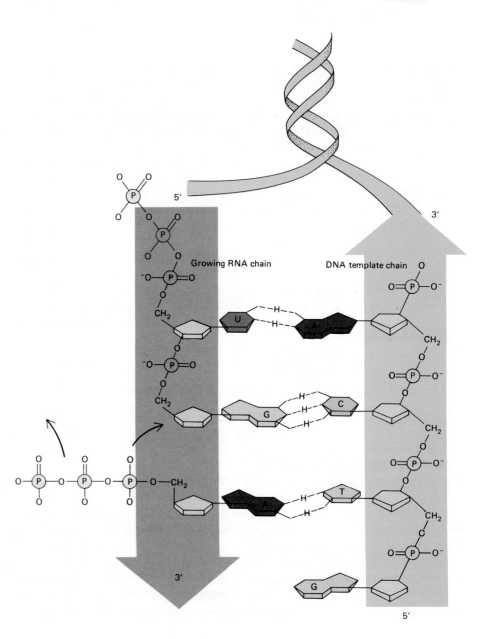

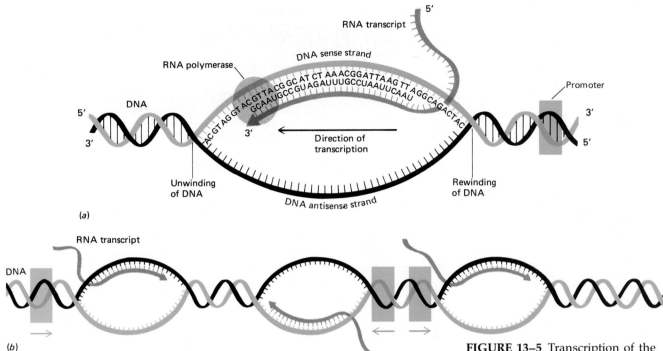

(a)

(b)

FIGURE 13–5 Transcription of the bacterial mRNA. *(a)* The RNA is synthesized in a $5' \rightarrow 3'$ direction from the sense strand (or coding strand) of the DNA molecule. Transcription starts downstream from DNA promoter sequences, which serve as the RNA polymerase recognition site. The promoter sequences are not transcribed. Transcription stops at termination sequences downstream from the coding sequences in the gene. Termination sequences signal the RNA polymerase to stop transcription and be released from the DNA. *(b)* Only one of the two strands is transcribed for a given gene, but the opposite strand may be transcribed for a neighboring gene. Each transcript is started at its own promoter.

end of the growing molecule. Unlike DNA synthesis, however, the initiation of RNA synthesis does not require a primer; instead, the 5' end of a new mRNA chain always starts with a nucleoside triphosphate (a nucleoside that has three phosphates linked to the 5'-carbon of its ribose sugar). Therefore, the mRNA has a triphosphate group at its 5' end and a free-OH group at its 3' end (Figure 13–6). In *Escherichia coli*, RNA polymerase recognizes a specific promoter sequence of bases in the DNA with the help of a polypeptide that alters the shape of the polymerase so that it can bind to those bases. Different genes may have slightly different promoter sequences, so that the cell can direct which genes are transcribed at any one time. Bacterial promoters are usually about 40 bases long and are positioned in the DNA five to eight bases before the point at which the RNA will start to be transcribed. (Generally, when we refer to a sequence of bases in a gene or the mRNA molecule copied from it, we say it is *upstream* or *downstream* of some reference point. **Upstream** means toward the 5' end of the mRNA sequence; **downstream** means toward the 3' end of the message. Therefore, the promoter sequences are upstream of the coding sequence in DNA.) Once the polymerase has recognized the correct promoter, it unwinds the helix and proceeds to transcribe only the coding strand of the DNA molecule.

The termination of transcription, like its initiation, is controlled by a set of specific base sequences at the end of the gene. These sequences act as "stop" signals for the RNA polymerase.

The Structure of Messenger RNA

The completed bacterial RNA contains more than the sequence of nucleotides that codes for the protein. RNA polymerase starts transcription of the gene well upstream of the coding sequences. As a result, mRNAs have a **5' leader sequence,** which contains recognition signals for ribosome binding so that the ribosomes can be properly positioned to translate the message. The leader sequence is followed by the **coding sequences,** which contain the actual message for the protein. At the end of the coding sequences are special termination signals that specify the end of the protein. These are followed by noncoding 3' trailing sequences, which can vary in length. In

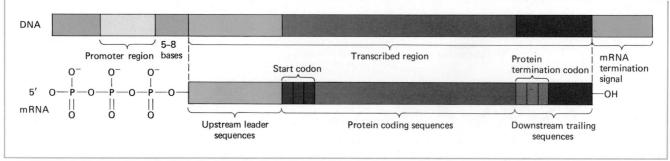

FIGURE 13–6 The structure of a bacterial mRNA molecule and the region of DNA from which it is transcribed. RNA polymerase recognizes promoter sequences on the DNA located five to eight bases upstream from the base where RNA synthesis is initiated. RNA synthesis stops when the polymerase encounters termination signals downstream from the protein-coding sequences. The nucleotide at the 5' end of the mRNA molecule has three phosphate groups attached to the 5' carbon of its ribose. The nucleotide at the 3' end of the molecule has an exposed hydroxyl group. Ribosome recognition sites are located in 5' leader sequences, which are upstream from the protein-coding sequences. Protein-coding sequences begin at a start codon following the leader sequences and end at termination codons near the 3' end of the molecule. Noncoding trailing sequences, which can vary in length, follow the protein-coding sequences.

bacterial cells one or more proteins may be encoded by a single mRNA molecule (see Chapter 14). In those organisms the lifetime of mRNA is usually measured in minutes. It is not uncommon under some growth conditions for the 5' end of the mRNA to be translated and then degraded before the 3' end of the molecule has been transcribed. Figure 13–7 illustrates how transcription and translation are coupled in *E. coli*.

TRANSLATION: DECODING THE MESSAGE

As already stated, the sequence of amino acids in a polypeptide chain is determined by the sequence of codons in mRNA in a process termed **translation.** This process adds another level of complexity to the process of information transfer, since it involves the conversion of information from a code using the four-base alphabet of nucleic acids to the 20–amino acid alphabet of proteins. Translation involves the coordinated functioning of over 100 kinds of macromolecules, including components of the ribosomes, mRNA, tRNAs, and various other factors.

The Activation of Amino Acids

Recall from Chapter 3 that amino acids are joined together by peptide bonds to form proteins. This involves linking of the amino group of one amino acid to the carboxyl group of an adjacent one. The reaction is endergonic and thus does not proceed spontaneously; the amino acids must be activated by an enzyme-mediated reaction with ATP to provide energy for peptide bond formation. Amino acid activation, a two-step process, is catalyzed by enzymes that are specific for each amino acid. The first step involves the reaction of the amino acid with ATP to form an amino acid—AMP molecule. The enzyme then catalyzes the transfer of that amino acid from the AMP to a specific tRNA to form an **aminoacyl tRNA** (Figure 13–8). The aminoacyl tRNA then serves as an "adapter," which has (1) an anticodon, which can recognize the complementary three-base codon on the mRNA that is the code for that particular amino acid and (2) a region to which the amino acid is covalently bonded. When this covalent bond is broken, sufficient energy becomes available to form the peptide bond that will link this amino acid to a growing polypeptide chain.

The fact that tRNA serves as an adapter in protein synthesis had been predicted by Francis Crick. He argued that the differences in structure between a polynucleotide chain and a polypeptide chain are so great that there is no simple way for amino acids to bind directly to a nucleotide sequence to make a protein. By attaching amino acids to small RNA adapter molecules, however, it is possible for the adapter RNA to bind to the mRNA coding sequence in a way that aligns the amino acids in the correct sequence to form the peptide chain. Notice that it is necessary for there to be not only a specific tRNA for each amino acid but also a specific enzyme that will link that amino acid to the correct tRNA.

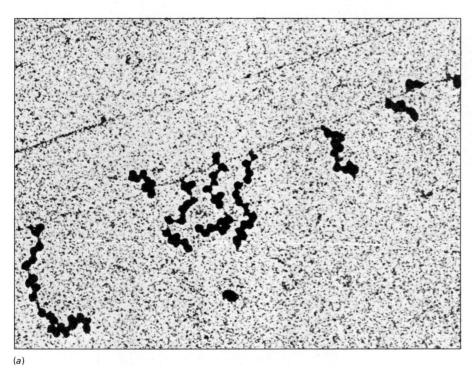

(a)

FIGURE 13–7 Coupled transcription and translation in *E. coli*. (a) Electron micrograph of two strands of DNA, one inactive and the other actively producing mRNA. The multiple ribosomes attached to the longer mRNA molecules make a polyribosome, so that protein synthesis occurs while the mRNA is being completed. (b) Diagrammatic representation of the coupled transcription and translation processes.

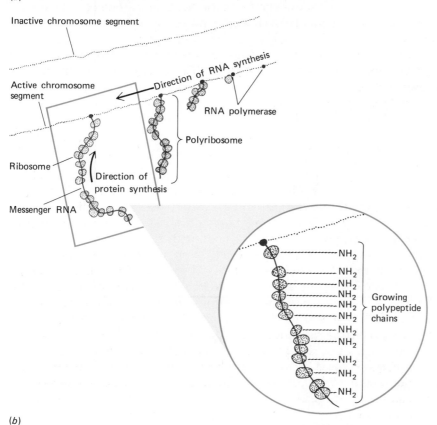

(b)

The Structure of Transfer RNA

Although tRNA molecules are considerably smaller than mRNA or rRNA molecules, they have a complex structure, with several specialized regions with specific functions. Think about the properties a tRNA molecule must have:

1. It must be recognized by a specific enzyme that adds the correct activated amino acid.

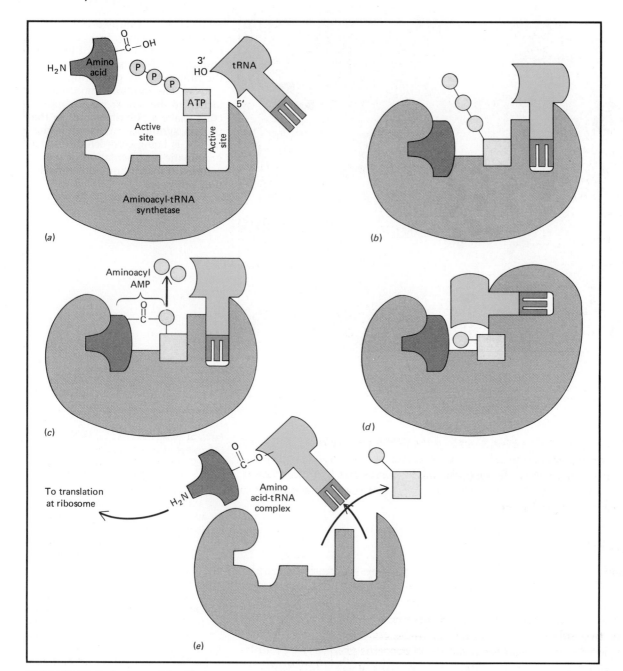

FIGURE 13–8 Amino acid activation and formation of the amino acid–tRNA complex. Aminoacyl-tRNA synthetase catalyzes two separate enzymatic reactions. The first step involves activation of the amino acid, with ATP used to form aminoacyl AMP. In the second step the activated amino acid is coupled to the 3' end of the tRNA through its carboxyl group, releasing the AMP molecule.

2. It must be recognized by ribosomes.
3. It must have a specific complementary binding sequence for the correct mRNA codon.
4. It must have a region that serves as the attachment site for the charged amino acid.

The tRNAs are polynucleotide chains about 70 nucleotides long (Figure 13–9). Each has a number of unique base sequences as well as some that are common to all. All have a basic three-dimensional structure that resembles an L-shaped molecule (Figure 13–9a). The anticodon site, which recognizes the three-base code on the mRNA, is at one end of the "L," and the amino acid–binding site is at the other end of the "L." One of the major determinants of tRNA structure is complementary base pairing within the molecule. This causes the structure to be doubled back and folded to form three or more loops of unpaired nucleotides, which are stabilized by pairing of complementary bases in the stem parts of the molecule (Figure 13–9b). The

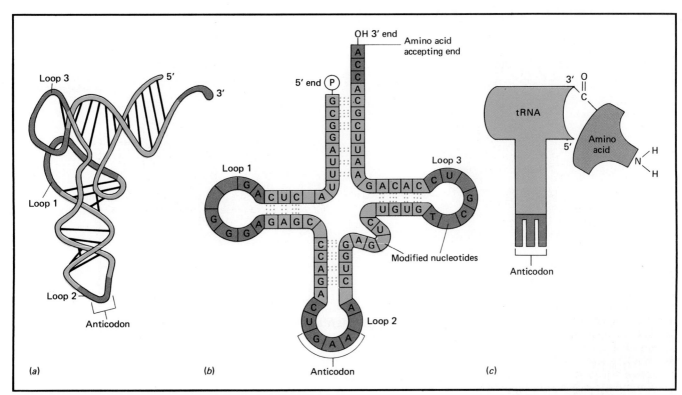

(a)   (b)   (c)

FIGURE 13–9 (a) and (b) Two representations of the structure of a typical tRNA molecule; tRNA molecules are the compounds that "read" the genetic code. A diagram of the actual shape of a tRNA molecule is shown in (a). Its three-dimensional shape is determined by intramolecular hydrogen bonds between base-paired regions, which are most clearly observed in the two-dimensional cloverleaf form depicted in (b). One loop contains the triplet anticodon that forms specific base pairs with the mRNA codon. The amino acid is attached to the terminal ribose at the 3' OH end, which has the nucleotide sequence CCA. Each tRNA has guanylic acid, G, at its 5' end and also contains several modified nucleotides. The pattern of folding results in a constant distance between anticodon and amino acid in all tRNAs examined. (c) Schematic figure illustrating how the amino acid is attached to its tRNA by its carboxyl group, leaving its amino group exposed for peptide bond formation.

amino acid–binding site is at the 3' end of the tRNA molecule. The *carboxyl* group of the amino acid is bound to the exposed 3' hydroxyl group of the terminal adenine nucleotide, leaving the *amino group* free to participate in peptide bond formation. The anticodon site is in a three-base sequence in the middle of the second loop.

The Ribosomes

The importance of ribosomes and the role of protein synthesis in cellular metabolism are exemplified by a rapidly growing *E. coli* cell, which contains some 15,000 ribosomes. These ribosomes are responsible for nearly one third of the total mass of the cell. Ribosomes from all organisms are composed of two subunits; the smaller of these contains 21 proteins and one RNA molecule, and the larger contains 35 proteins and two RNA molecules. Each subunit can be isolated in the laboratory in an intact form and then further separated into each of its RNA and protein constituents. Under certain conditions it is then possible to reassemble each subunit in a functional form by adding each component in its correct order. With this approach, and sophisticated electron microscope methods, it has been possible to determine the three-dimensional structure of the ribosome (Figure 13–10a) as well as how it is assembled in the living cell. The large subunit contains a depression on one surface into which the small subunit fits. The mRNA fits in a groove formed between the contact surfaces of the two subunits.

Within each ribosome are two pockets, the **A** and **P** binding sites for tRNA molecules (Figure 13–10b). The ribosome accepts the charged aminoacyl tRNA at the A site. Following peptide bond formation between the amino acid in the A site and the end of the growing peptide chain, the tRNA (now with the entire peptide chain attached) is moved to the P site of the ribosome, leaving the now vacant A site available for the next aminoacyl-tRNA molecule. One of the roles of the ribosome is to hold the mRNA template, the aminoacyl-tRNA precursor, and the growing peptide chain

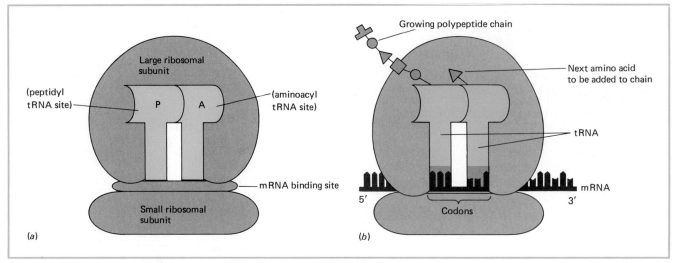

FIGURE 13–10 (*a*) A schematic model of a ribosome. (*b*) Ribosome-binding sites. The mRNA passes through a space in the ribosome formed between the two subunits. Located within the ribosome are two binding sites for tRNAs that recognize adjacent codons. The A site (aminoacyl-tRNA site) binds an aminoacyl-tRNA that will be used to add an amino acid to the growing chain. The P site (peptidyl-tRNA site) binds the tRNA that is linked to the growing polypeptide chain.

all in the correct orientation so that the genetic code can be read and the peptide bond formed.

Protein Synthesis

The process of protein synthesis is generally considered to occur in three distinct stages: **initiation, elongation,** and **termination** (Figures 13–11 and 13–12). The initiation process consists of a number of steps and requires a number of proteins, called **initiation factors.** Initiation begins with the loading of a special **initiation tRNA** onto the small ribosomal subunit. In all organisms the genetic code for the initiation of protein synthesis is AUG, which codes for the amino acid methionine (see Table 13–1). In *E. coli* there are two types of methionine tRNA that recognize the codon AUG. After methionine has been attached to the initiator tRNA, it is modified by the addition of formic acid to its amino group. Every protein in *E. coli* is synthesized with the modified amino acid, **N-formyl methionine** (Figure 13–11), at its amino terminal end. Formylated methionine is used only for the first amino acid in a polypeptide chain; if an AUG codon appears in the middle of a protein-coding sequence, a regular methionine-tRNA will be used.

Once the initiator tRNA is loaded on the small subunit, the complex binds to the special **ribosome recognition sequences** near the 5' end of the mRNA; these are upstream of the coding sequences. Binding results in alignment of the anticodon of the initiator tRNA with the AUG initiation codon of the mRNA. The large ribosomal subunit then binds to the complex, forming the completed ribosome.

The addition of other amino acids to the forming polypeptide is the process of **elongation.** The initiator tRNA is bound to the P site of the ribosome, leaving the A site unoccupied so that it can be filled by the aminoacyl tRNA specified by the next codon. Figure 13–12 outlines the events involved in elongation. The appropriate aminoacyl tRNA binds to the A site by specific base pairing of its anticodon with the complementary mRNA codon. This binding step requires energy, which in this case comes from guanosine triphosphate (GTP), an energy donor similar to ATP. The amino group of the amino acid at the A site is now aligned with the carboxyl group of the preceding amino acid at the P site. Peptide bond formation then takes place between the amino group of the new amino acid and the carboxyl group of the preceding amino acid. The reaction is catalyzed by **peptidyl transferase,** an enzyme that is tightly bound to the ribosome.

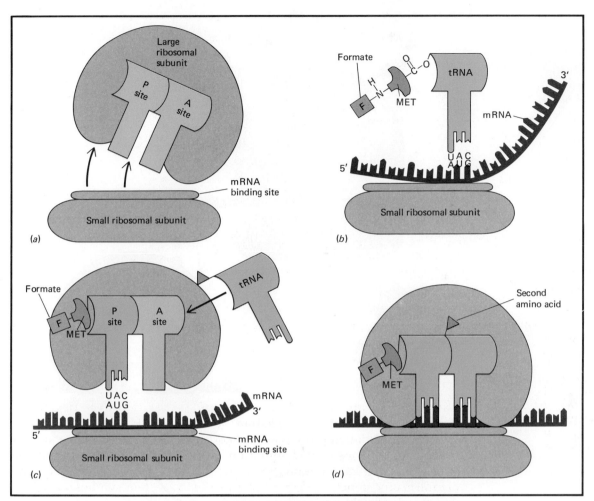

FIGURE 13–11 Initiation of protein synthesis. *(a)* Ribosomal subunits are dissociated following translation of a message. *(b)* Formation of an initiation complex. This complex consists of protein initiation factors, formylated methionine
tRNA, and the small ribosomal subunit, which bind to recognition sequences
near the 5' end of the mRNA. *(c)* The large ribosomal subunit binds to the initiation complex. Formylated methionine tRNA is bound to the P site of the completed ribosome, its anticodon paired with the initiation codon AUG. *(d)* The
second aminoacyl-tRNA recognizes the adjacent codon and binds to the A site.

In the process, the amino acid attached at the P site is released from its
tRNA and becomes attached to the aminoacyl tRNA at the A site. Remember from Chapter 3 that polypeptide chains have direction, or polarity. The
amino acid on one end has a free amino group (the amino terminal end),
and the amino acid at the other end has a free carboxyl group (the carboxyl
terminal end). If you follow the sequence through, you will see that amino
acids are always added to the carboxyl terminal end. *Protein synthesis always
proceeds from the amino terminal to the carboxyl terminal end of the growing
peptide chain.*

After the peptide bond is formed, the tRNA molecule is removed from
the P site and released into the cytoplasm, where a new amino acid can be
added to it. The growing peptide chain, *which is now attached to the tRNA in
the A site,* is then translocated to the P site, leaving the A site open for the
next tRNA–amino acid complex. This translocation process requires energy, which is again supplied in the form of GTP. In the process of translocation, the ribosome and the message move in relation to each other so that
the codon specifying the next amino acid in the polypeptide chain is posi-

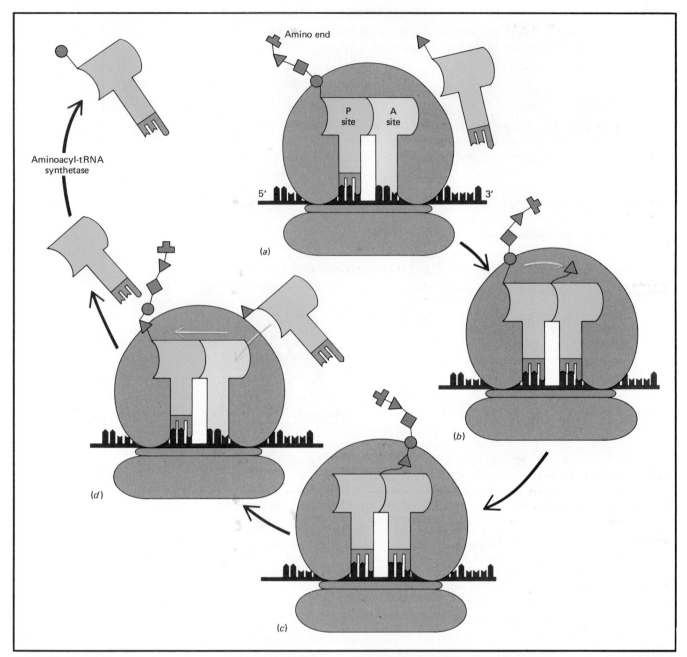

FIGURE 13–12 Peptide elongation mechanism. (*a*) The peptidyl tRNA occupies the P site. A tRNA binds to the A site by base pairing with the complementary codon at that position. (*b*) A peptide bond is formed between the amino group of the amino acid from the tRNA at the A site and the carboxyl group of the amino acid that was attached to the tRNA at the P site. (*c*) The polypeptide chain is now attached at its carboxyl end to the tRNA at the A site. (*d*) The tRNA is ejected from the P site, after which the peptidyl tRNA and its complementary mRNA codon are translocated to the P site. The A site is now ready to accept the next aminoacyl-tRNA designated by the mRNA codon occupying the A site.

tioned in the unoccupied A site. This process involves movement of the ribosome in the 3′ direction along the mRNA molecule; thus, *translation of the mRNA always proceeds in a 5′ to 3′ direction.*

Formation of each peptide bond requires only about $\frac{1}{20}$ of a second, so that an average-sized protein of about 360 amino acids is completed in about 18 seconds.

The synthesis of the peptide chain is terminated by "release factors" that recognize **termination,** or **stop, codons** at the end of the coding sequence. The three codons UAA, UGA, and UAG are special stop signals that do not code for any amino acid. Recognition of a termination codon by the release factors causes the ribosome to dissociate into its two subunits, which can then be used to form a new initiation complex with another mRNA molecule.

Polyribosomes

In *E. coli*, transcription and translation are *coupled* (see Figure 13–7). This means that ribosomes can bind to the 5' end of the growing message and initiate translation long before the message is completed. As many as 15 ribosomes may be bound to a single mRNA molecule, and these are spaced as close together as 80 nucleotides. Messenger RNA molecules that contain clusters of ribosomes are referred to as **polyribosomes** or sometimes **polysomes.**

Although a number of polypeptide chains can be actively synthesized on a single messenger at any one time, the half-life (the time it takes for half of the molecules to be degraded) of mRNA molecules in bacterial cells is only about 2 minutes. Usually, degradation of the 5' end of the message begins even before synthesis of the message is complete, and once the ribosome recognition sequences are degraded, no more ribosomes can attach to the message and initiate protein synthesis.

A Comparison of Prokaryotic and Eukaryotic Protein Synthesis

Although the basic mechanisms of transcription and translation are quite similar, there are some fundamental differences between eukaryotes and prokaryotes, particularly with regard to the characteristics of the mRNAs. Whereas bacterial mRNAs are used immediately after transcription without further processing, eukaryotic mRNA molecules undergo specific **post-transcriptional modification and processing.**

As we have seen, bacterial mRNA is translated as it is synthesized from the DNA chromosome. In contrast, eukaryotic chromosomes are confined to the nucleus of the cell, whereas protein synthesis takes place in the cytoplasm. This means that once the message is transcribed in the nucleus, the completed message must be transported through the nuclear membrane and into the cytoplasm before it can be translated. A number of modifications are made on the original transcript while it is in the nucleus before it is competent for transport and translation (Figure 13–13). Not all eukaryotic messages are modified to the same extent by this process, but the essential pathway for messenger modification is the same in all eukaryotic nuclei.

Modification of the eukaryotic message begins when the growing RNA transcript is about 20 to 30 nucleotides long. At that point enzymes add a **cap** to the 5' end of the mRNA chain. The cap is in the form of an unusual nucleotide, 7-methyl guanylate, which is guanosine monophosphate that has had a methyl group added to one of the nitrogens in the base. The capping nucleotide is also unusual in that *it is attached to the 5' end of the message backwards!* Instead of the usual 5'-3' phosphate linkage, the 5' phosphate of the cap is linked to the 5' phosphate at the end of the molecule, forming a 5'-5' linkage. A possible reason the 5' end of the message is capped in this way is to protect the message from the types of degradation that occur with unprotected bacterial messages. Eukaryotic mRNAs are in fact much more stable than prokaryotic mRNAs, with half-lives ranging from 30 minutes to as long as 24 hours. The average half-life of an mRNA molecule in a mammalian cell is about 10 hours.

A second modification of eukaryotic mRNAs occurs at the 3' end of the molecules. Near the 3' end of each completed message is a sequence of bases that serves as a **polyadenylation signal.** Within about 1 minute of

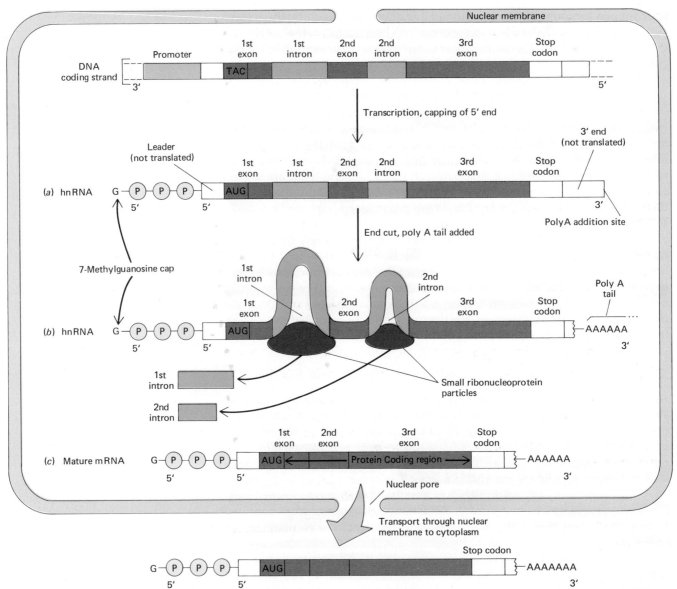

FIGURE 13–13 Transcription and processing pathway of RNA in the eukaryotic cell nucleus. A typical eukaryotic gene may have multiple exons (coding sequences) and introns (noncoding sequences). (*a*) A DNA sequence containing both exons and introns is transcribed by RNA polymerase to make the primary transcript (hnRNA). While the hnRNA is transcribed, the molecule is "capped" by the addition of a modified base to its 5' end. (*b*) The 3' end of the completed hnRNA molecule is cleaved at a sequence that designates the poly-A addition site. A poly-A tail (50–200 nucleotides long) is then added to the exposed 3' OH group. (*c*) Introns are removed from the molecule and the exons spliced together. (*d*) The mature mRNA is transported through the nuclear membrane into the cytoplasm to be used for protein synthesis.

completion of the transcript, enzymes in the nucleus recognize the polyadenylation signal and cut the mRNA molecule. This is followed by the addition to the 3' end of a **polyadenylated (or poly-A) tail,** which consists of a string of 100 to 250 adenine nucleotides. The function of this modification is not clear; perhaps it also helps stabilize the mRNA against degradation.

Introns and Exons

The final step in mRNA modification is one of the most surprising recent findings in molecular biology. Many eukaryotic genes have **interrupted**

coding sequences, caused by long sequences of bases within the protein-coding sequences of the gene that do not code for amino acids in the final protein product! The noncoding regions within the gene are called **introns** (*in*tervening sequences), as opposed to **exons** (*ex*pressed sequences), which are parts of the protein-coding sequence. The number of introns found in genes is quite variable. For example, the β-globin gene, which produces one of the components of hemoglobin, contains 2 introns; the ovalbumin gene of egg white contains 7; and the gene specifying another egg-white protein, conalbumin, contains 16. In many cases the combined lengths of the introns are much longer than the combined protein-coding exon sequences. For instance, the ovalbumin gene (an egg-white protein in chickens) contains about 7700 base pairs, whereas the exon (protein-coding) sequences together are only 1859 bases long.

When a gene containing introns is transcribed, the entire gene is copied as a large RNA transcript referred to as **heterogeneous nuclear RNA (hnRNA).** This molecule contains both exon and intron sequences. In order for the hnRNA to be made into a functional message, the introns must be cut from the molecule and the exons must then be spliced together to form a continuous protein-coding message. The splicing reactions are mediated by special base sequences within and to either side of the introns. Splicing itself can occur by several different mechanisms. In many instances the splicing involves the association of small **nuclear ribonucleo-protein complexes (snRNPs),** which bind to the introns and catalyze the excision and splicing reactions. Surprisingly, in some cases the RNA within the intron has the ability to splice itself without the use of proteins. This means that a nucleic acid is capable of acting as an enzyme! The discovery of this property has led to the recognition of a new class of biological catalysts, termed **ribozymes,** which are formed from RNA molecules rather than proteins.

Although most eukaryotic mRNAs require capping, tailing, and splicing reactions, not all mRNA molecules are modified by all three mechanisms. Some messages, for example, do not contain introns and so do not require splicing. Other eukaryotic mRNAs are capped but do not contain introns or poly-A tails.

Exon Shuffling

Why are there introns in eukaryotic genes? It seems incredible that as much as 75% of the original transcript of a gene might be useless information that has to be removed to make a working message. One theory, proposed by Walter Gilbert of Harvard University, is that introns separate coding sequences that correspond to different *structural and functional domains of proteins.* Gilbert proposed that new proteins with new functions can rapidly emerge when a new combination of exons is produced by genetic recombination between intron regions of genes that code for different proteins. This evolutionary advantage, along with the increased stability of eukaryotic mRNA molecules (which allows cells to expend less energy making mRNAs) may permit higher organisms to maintain a large load of noncoding DNA within their genes.

Analysis of the DNA and amino acid sequences from a number of eukaryotic genes has provided evidence for this theory of **exon shuffling.** The low-density lipoprotein (LDL) receptor protein (a protein found on the surface of human cells that binds to cholesterol transport molecules), for example, has a number of domains that are related to parts of several other proteins with totally different functions.

THE GENETIC CODE

We have seen that in prokaryotes there is a direct correspondence in the sequence of subunits from DNA to mRNA to protein. Thus, the three are **colinear.** Early in 1961 Crick concluded from a mathematical analysis of the coding problem that three consecutive nucleotides in a strand of mRNA must provide the code for a single amino acid in a polypeptide chain. Experimental evidence for the existence of a three-base code was obtained from the laboratory of M. Nirenberg and H. Matthaei. By constructing artificial mRNA molecules with known base sequences, Nirenberg and Matthaei were able to study which amino acids would be incorporated into protein in purified protein synthetic systems. When the synthetic mRNA polyuridylic acid (UUUUUU . . .) was added to a mixture of purified ribosomes, aminoacyl tRNAs, and essential cofactors needed to synthesize protein, only phenylalanine was incorporated into the resulting polypeptide chain. The inference that UUU is the code for phenylalanine was inescapable. Similar experiments showed that polyadenylic acid (AAAAAAAA . . .) codes for lysine and that polycytidylic acid (CCCCCCCCC . . .) codes for proline. Through the use of mixed nucleotide polymers (such as a random polymer of A and C) as artificial messengers, it became possible to assign the other nucleotide triplets to specific amino acids. However, three of the codons, UAA, UGA, and UAG, were not found to specify the synthesis of any polypeptide chain. These codons, the **stop,** or **termination, codons** mentioned earlier, are now known to be the signals that specify the end of the coding sequence for a polypeptide chain.

Taken together, these experiments led to the coding assignments of all 64 possible codons, listed in Table 13–1. Investigators were also able to demonstrate conclusively that the code is a triplet code and that it is a nonoverlapping code. In other words, codons do not overlap each other; they are read in sequence on the mRNA. Furthermore, as the mRNA code was examined in many other organisms, it became apparent that the genetic code is **universal;** the mRNA code is the same in the bacterium *E. coli* as in plants, armadillos, and humans, which suggests that the genetic code was established very early in the evolutionary history of life on Earth. As a consequence, bacterial cells can translate a human genetic message and make the same protein that would be made in a human cell. Similarly, human cells can correctly translate bacterial mRNAs to make bacterial proteins.

Recently, some minor exceptions to the universality of the genetic code have been discovered. In several single-celled protozoans, two of the stop codons, UAA and UAG, have been found to code for a single amino acid. The other exceptions are found in mitochondria, which contain their own DNA and protein-synthesis machinery for a small number of genes. These codes vary, depending on the organism; however, almost all of the coding assignments in each case are identical to the normal genetic code.

Remember that *the genetic code we define and use is an mRNA code.* The tRNA anticodon sequences as well as the DNA sequence from which the message is transcribed are complementary to the sequences shown in Table 13–1. For example, the mRNA codon for the amino acid methionine is 5′-AUG-3′. It is transcribed from the DNA base sequence 3′-TAC-5′, and the tRNA anticodon is 3′-UAC-5′.

Codon–Amino Acid Specificity

If you carefully examine the codon assignments in Table 13–1, you will see that some of the amino acids are specified by more than one codon. This **redundancy** in the code has certain characteristic patterns. Notice that the codons CCU, CCC, CCA, and CCG all code for the amino acid proline. The

Table 13–1 THE GENETIC CODE: THE SEQUENCE OF NUCLEOTIDES IN THE TRIPLET CODONS OF ᴍRNA THAT SPECIFY A GIVEN AMINO ACID

First Position (5' end)	Second Position	Third Position (3' end)			
		U	C	A	G
U	U	UUU UUC Phenylalanine		UUA UUG Leucine	
	C	UCU UCC UCA UCG Serine			
	A	UAU UAC Tyrosine		UAA UAG Stop	
	G	UGU UGC Cysteine		UGA Stop	UGG Tryptophane
C	U	CUU CUC CUA CUG Leucine			
	C	CCU CCC CCA CCG Proline			
	A	CAU CAC Histidine		CAA CAG Glutamine	
	G	CGU CGC CGA CGG Arginine			
A	U	AUU AUC AUA Isoleucine			AUG Methionine
	C	ACU ACC ACA ACG Threonine			
	A	AAU AAC AAA AAG Asparagine			
	G	AGU AGC Serine		AGA AGG Arginine	
G	U	GUU GUC GUA GUG Valine			
	C	GCU GCC GCA GCG Alanine			
	A	GAU GAC Asparagine		GAA GAG Glutamine	
	G	GGU GGC GGA GGG Glycine			

only difference among the four codons involves the nucleotide at the 3' end of the triplet. Although the code may be read three nucleotides at a time, only the first two nucleotides appear to contain specific information for proline. A similar pattern can be seen for many other amino acids. Only methionine and tryptophan have single triplet codes. All other amino acids are specified by two to six nucleotide triplets. There are 61 codons that specify amino acids; although most cells contain only about 40 different tRNA molecules, some of these tRNAs can pair with more than one codon, so all of the codons can still be used. This apparent breach of the base-pairing rules was first proposed by Francis Crick as the **wobble hypothesis.** Crick reasoned that the third nucleotide of the tRNA anticodon (which is the 5' base of that sequence) may be capable of forming hydrogen bonds with more than one base. Investigators later established this experimentally by determining the anticodon sequences of tRNA molecules and testing their specificities in artificial systems. In some cases, "wobble" is facilitated by the presence of an unusual base, **inosine,** at the third position of the tRNA anticodon. Inosine can form hydrogen bonds with A, C, or U in the third position. As a consequence, some tRNA molecules can functionally recognize as many as three separate codons specifying the same amino acid.

The Second Genetic Code

The determination of the genetic code was a remarkable feat, but it solved only a part of the translation problem. Keep in mind that the code in Table 13–1 allows one nucleic acid triplet (the codon) to recognize a complementary nucleic acid triplet (the anticodon). In order for the translation process to work there must also be a way for enzymes to accurately join each amino acid to its correct tRNA. This requires what some call a **second genetic code,** which allows the conversion of nucleic acid information to amino acid information. This bilingual code involves the amino-acyl tRNA synthetase enzymes.

Usually there is only one amino-acyl tRNA synthetase enzyme for each amino acid, but in many cases one enzyme must recognize several different tRNAs with different anticodons. For methionine and glutamine, which have one and two tRNAs, respectively, the synthetase apparently does recognize the anticodon, but for other amino acids with multiple tRNAs the anticodon is clearly not the recognition site.

Recently, researchers have found in the stem of alanine tRNA close to the amino acid binding site a single pair of bases that appear to serve as the critical recognition code for the synthetase enzyme. When the two bases were placed in the same positions in either a cysteine or a phenylalanine tRNA, the alanine tRNA synthetase linked alanine to those tRNA molecules. This result indicates that these two bases serve as the code recognized by this particular synthetase. Working out the rest of the second genetic code is an area of ongoing research.

WHAT IS A GENE?

In Chapter 12 we traced the development of ideas regarding the nature of the gene. For a time it was useful to define a gene as a sequence of nucleotides that codes for one polypeptide chain. As we have continued to find out more about how genes work, we have revised our definition of the gene. We now know that there are genes whose function is to produce RNA molecules such as rRNA and tRNA. Others make snRNAs used to modify complex mRNA molecules. Recent studies have also shown that in eukaryotic cells a single gene may be capable of producing more than one polypeptide chain by modifications in the way the mRNA is processed. We might define the gene as a **unit of transcription;** however, given the complex roles genes seem to play, it might be more accurate to define the gene in terms of its product, as a *transcribed nucleotide sequence that yields a product with a specific cellular function.*

CHANGES IN GENES: MUTATIONS

One of the first major discoveries about genes was that they can undergo changes, called **mutations.** We now know that mutations are caused by changes in the nucleotide sequence of the DNA. As we saw in Chapter 12, once the DNA sequence is changed, DNA replication copies the altered sequence just as it did the normal one, making it stable over an indefinite number of generations. The mutated gene has no greater tendency than the original gene to mutate again. Mutations provide the diversity of genetic material that makes it possible to study inheritance and the molecular nature of genes. As we shall see in later chapters, mutations also provide the variation necessary for evolution to occur within a given species.

Genes can be altered by mutation in a number of ways (Figure 13–14). The simplest type of mutation, called a **point mutation,** or **base-substitution mutation,** involves a change in only one pair of nucleotides. It is now possible to determine where a specific point mutation occurs in a gene by using recombinant DNA methods to isolate the gene and determine its sequence of bases (see Chapter 15). Often these mutations result from errors in base pairing that occurred during the replication process, such as replacement of an AT base pair with a GC, CG, or TA pair. Such a mutation may cause the altered DNA to be transcribed into an altered mRNA, which may be translated into a peptide chain with only one amino acid different from the normal sequence.

Mutations that result in the substitution of one amino acid for another are sometimes referred to as **missense mutations.** Substitution of a different amino acid into a protein in this way can have a wide range of effects. If the amino acid substitution occurs at or near the active site of the enzyme, the activity of the altered protein may be decreased or even destroyed. If a missense mutation involves a change in an amino acid that is not part of the active site or involves the substitution of a closely related amino acid (one with very similar chemical characteristics), the mutation may be *silent* and undetectable, at least if one simply examines the effects it has on the whole organism. Silent mutations occur relatively frequently; thus, the true number of mutations that occur in an organism or a species is much greater than what is actually observed.

Nonsense mutations are a kind of point mutation that can change an amino acid–specifying codon into a termination codon. A nonsense mutation in a gene usually destroys the function of the gene product; in the case of a gene specifying a protein, the part of the polypeptide chain that occurs after the termination codon will be missing.

In **frameshift mutations,** nucleotide pairs are *inserted into* or *deleted from* the molecule. Insertion or deletion of a base in a DNA sequence causes bases downstream of that site to shift their positions in the codons they would normally occupy. As a result of this shift, codons downstream of the insertion site will specify an *entirely new sequence of amino acids.* Depending on where the insertion or deletion occurs in the gene, a number of different effects can be produced. In addition to producing an entirely new polypeptide sequence immediately after the change, frameshift mutations usually result in the appearance of a stop or termination codon within a short distance of the mutation, resulting in the termination of the already altered polypeptide chain. As a consequence, most frameshift mutations in a gene specifying an enzyme result in the loss of that enzyme activity. If the enzyme is an essential one, the effect on the organism can be disastrous.

Other types of mutations may occur at the level of the chromosome due to a change in chromosome structure (see Chapter 11). These types of changes usually have a wide range of effects because they involve large numbers of genes.

One type of mutation whose mechanism of action has only recently been understood is caused by DNA sequences that "jump" into the middle of a gene. These movable sequences of DNA, called **transposons,** not only can disrupt the functions of some genes but under some conditions can also activate inactive genes (See Focus on Reverse Transcription, Jumping Genes, and Pseudogenes).

Focus on REVERSE TRANSCRIPTION, JUMPING GENES, AND PSEUDOGENES

For several decades, a central dogma of molecular biology was that the flow of genetic information always proceeds from DNA to RNA to protein. An important exception to this rule was discovered by Howard Temin in 1964. Temin found that infection with certain cancer-causing RNA tumor viruses, such as the Rous sarcoma virus, is blocked by inhibitors of DNA synthesis and by inhibitors of DNA transcription. This suggested that DNA synthesis and transcription are required for the multiplication of RNA tumor viruses and that information flows in the reverse direction in these organisms— that is, from RNA to DNA. Temin proposed that a **DNA provirus** is formed as an intermediate in the replication of RNA tumor viruses. This hypothesis required a new kind of enzyme—one that would synthesize DNA using RNA as a template. In 1970, Temin and David Baltimore discovered just such an enzyme, and in 1975 they shared the Nobel Prize for their discovery. This RNA-directed DNA polymerase, also known as **reverse transcriptase,** was found to be present in all RNA tumor viruses. (Some non-tumor-forming RNA viruses, however, replicate themselves directly without employing a DNA intermediate.)

After an infecting RNA tumor virus enters the host cell, the viral reverse transcriptase forms a DNA strand that is complementary to the viral RNA. Subsequently, a second DNA strand is synthesized. The double-stranded DNA provirus is then integrated into the host cell's DNA. The DNA of the provirus is transcribed, and viral proteins are formed in the cytoplasm as the viral mRNA is translated. Other viral RNA molecules are formed and incorporated into mature virus particles enclosed by a protein coat. Because of their reversal of the usual direction of informational flow, such viruses have become known as **retroviruses.** The AIDS virus (HIV-1) is one of the most widely known retroviruses.

Until recently, reverse transcription was thought to be associated only with retroviruses. Evidence now suggests that reverse transcription may be more common, which may explain the curious phenomena of "jumping" genes and pseudogenes.

"Jumping genes," or **mobile genetic elements,** were discovered in maize (corn) by Barbara McClintock in the 1940s. McClintock observed that certain genes appeared to be "turned off" and "turned on" spontaneously. She deduced that the mechanism involved a gene that moved from one region of a chromosome to another, where it would either activate or inactivate genes in that vicinity. It was not until the development of recombinant DNA methods (see Chapter 15) and the discovery

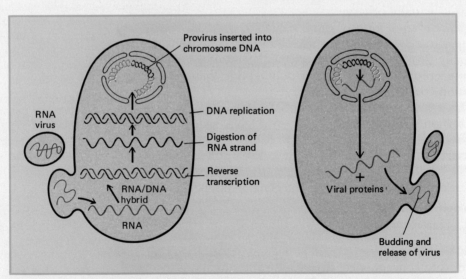

■ Events that occur during the life cycle of an RNA tumor virus (retrovirus). The RNA-containing virus particle attaches to the cell membrane, releasing the RNA into the cell. The reverse transcriptase enzyme makes a DNA copy of the RNA, after which the DNA strand is replicated to form a double-stranded molecule. The double-stranded DNA molecule is then inserted into the DNA of one of the cell's chromosomes. The viral DNA uses the cell's transcription apparatus to make RNA copies, which are released into the cytoplasm and translated to make essential proteins for the virus particles. The RNA and proteins then bud from the cell, forming new virus particles that can infect other cells.

All of the mutations discussed so far, called **spontaneous mutations,** occur infrequently as a consequence either of mistakes in DNA replication or of defects in the mitotic or meiotic segregation of chromosomes. How-

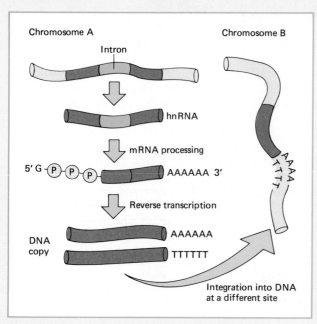

Some jumping genes in eukaryotic cells appear to move through an RNA intermediate. An intron-containing gene is transcribed and processed to form a mature mRNA with a poly-A tail. Reverse transcriptase makes a DNA copy of the intron-free mRNA, including the poly-A tail, and the DNA molecule is integrated into another region of the cell's DNA, most likely on a different chromosome. Since the promoter is not part of the mRNA transcript, the transposed gene is usually nonfunctional and becomes a silent "pseudogene," which continues to accumulate mutations. Eventually the protein-coding sequence contains numerous nonsense codons.

of jumping genes in a wide variety of organisms that this phenomenon began to be understood. In recognition of her insightful findings, Barbara McClintock was awarded the Nobel Prize in 1983.

Jumping genes, also called **transposable elements** or **transposons,** are segments of DNA that range from a length of a few hundred to several thousand bases. The elements themselves seem to require a special **transposase** enzyme in order to be incorporated into a new location within the chromosome. The longer elements may contain other genes that go along "for the ride."

Many of the transposable elements have been found to have similarities to retroviruses. Both tend to have unusual ends to their DNA sequences, and the genes themselves are remarkably similar, especially those that code for the proteins that are required for reverse transcription and integration into the chromosome.

Recently, evidence has emerged that reverse transcriptase is involved in the mechanism by which some transposons move. Experiments in Gerald Fink's laboratory at MIT have provided evidence that one type of transposition is not a result of the DNA sequence itself jumping from one location to another; instead, the *information moves through an RNA intermediate.* Fink and his colleagues used recombinant DNA methods to insert an intron into the DNA sequence of a yeast transposon as a way of identifying it. They then set up conditions that allowed them to recover and analyze the transposed sequence once it had "jumped." When the transposed

sequence appeared at a new location, the intron had been removed, just as introns are removed during the processing of normal mRNA molecules. Since the enzymes are known only to splice RNA, it appears that the transposed DNA sequence had been produced from a processed RNA copy of the original DNA, rather than from the DNA itself. This would require the RNA sequence to be converted back to DNA by reverse transcriptase activity within the yeast cells.

Other evidence of nonviral reverse transcriptase activities in cells comes from analyses of **pseudogenes,** which closely resemble certain types of normal genes in mammalian cells. Pseudogenes are DNA sequences that are almost identical to those of normal genes, except that they are riddled with mutations that prevent them from functioning in normal protein synthesis. Many pseudogenes resemble DNA copies of mRNA, but where a normal gene would contain one or more introns, the pseudogenes do not. Many pseudogene DNA sequences also end with long poly-A tails. One theory concerning the origin of pseudogenes is that they are derived from the processed mRNAs of normal genes, which were retrotranscribed into DNA by reverse transcriptase. These DNA copies were then inserted back into the chromosome, but since they lack promoter sequences they are not expressed and simply act as excess baggage, silently accumulating mutations. It is estimated that there may be hundreds or thousands of such sequences in normal human DNA.

ever, some regions of DNA are much more prone to undergo mutation than others. Such **hot spots** are often single nucleotides or short stretches of repeated nucleotides. They may consist of **unusual bases** that spontane-

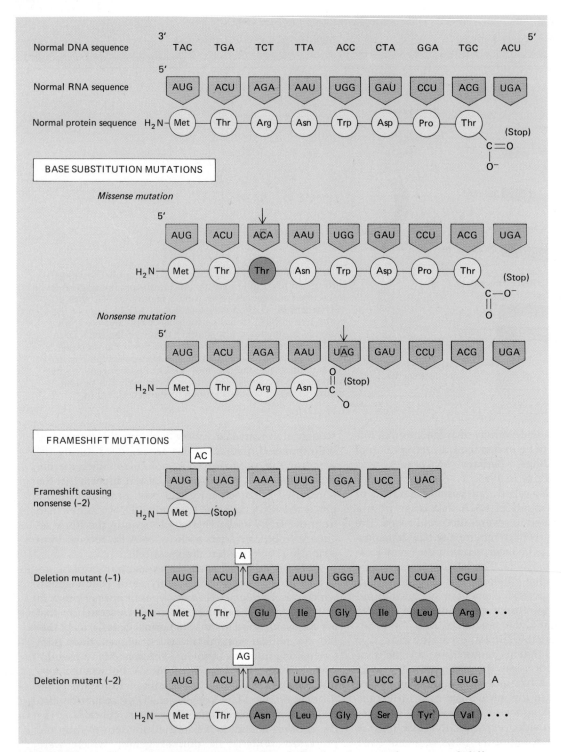

FIGURE 13–14 Types of mutations. **Base pair substitutions** can produce several different types of mutations. *Missense mutations* change a single amino acid thus producing a polypeptide that is the same length as the original. Depending on the nature of the new amino acid and its position in the polypeptide chain these may produce proteins that range from being completely functional (a *silent mutation*) to those having no activity. *Nonsense mutations*, caused by the conversion of an amino acid specifying codon to a termination codon produces a shorter, usually non-functional protein. **Frameshift mutations** result from the insertion or deletion of one or two bases and usually have more drastic effects. They cause the base sequence following the mutation to shift to a new reading frame, completely altering the structure and function of the rest of the polypeptide. Frameshifts usually create a termination codon a short distance downstream from the original mutation thus having the same effect as a nonsense mutation created by a base substitution.

ously change their structure or of short stretches of repeated nucleotides, which can cause DNA polymerase to "slip." Certain mutations can increase the mutation rate, probably by making DNA replication less precise. The rates of spontaneous mutations of different human genes range from 5×10^{-4} to 5×10^{-5} mutations per gene per generation. Since humans have a total of some 50,000 to 100,000 genes, the total mutation rate is in the order of one mutation per person per generation; in other words, each of us has some mutant gene that was not present in either of our parents. Most of these mutations are recessive or insignificant and are not noticeably expressed, usually because we are diploid and have a "normal" gene on the other chromosome.

Not all mutations occur spontaneously; many are caused by agents known as **mutagens,** including x rays, gamma rays, cosmic rays, ultraviolet rays, and other types of ionizing radiation. Some chemical mutagens react with and modify specific nucleotide bases in the DNA, leading to mistakes in complementary base pairing when the DNA molecule is replicated. Other mutagens become inserted in the DNA molecule and change the normal reading frame during replication. There is a close relationship between mutations and cancer. Many mutagens are also **carcinogens,** which produce cancer in higher organisms.

■ SUMMARY

I. The process by which information coded by genes is used to specify the sequence of amino acids in proteins involves two steps: transcription and translation.

   A. Transcription involves production of a copy of the information-containing DNA strand in the form of an RNA molecule. Messenger RNA (mRNA) molecules contain information that specifies the amino acid sequences of polypeptide chains.

   B. Translation involves conversion of the information encoded by the mRNA into a sequence of amino acids to make a protein.

      1. The information in the mRNA is in the form of a three-base genetic code.

         a. Each three-base sequence in the mRNA represents a codon, which specifies one amino acid in the polypeptide chain.

         b. Codons are arranged in a linear sequence on the mRNA molecule and are translated sequentially to give the same sequence of amino acids in the polypeptide chain.

      2. Translation of the genetic code requires adapter molecules in the form of tRNAs and complex machinery, including ribosomes, which contain rRNA molecules and many different proteins.

II. Messenger RNA is synthesized by DNA-dependent RNA polymerase enzymes.

   A. RNA is formed from ribonucleotide triphosphate precursors, which contain the sugar ribose and the bases uracil, adenine, guanine, and cytosine.

   B. RNA synthesis is initiated at sites on the DNA called *promoter regions*. RNA polymerase binds to the promoter and proceeds to unwind the DNA at that point, copying one strand of the DNA molecule.

   C. Like DNA, RNA subunits are linked together with a 5'-3' phosphate linkage. The same base-pairing rules are followed as in DNA replication, except that uracil is substituted for thymine.

   D. RNA synthesis proceeds in a 5'→3' direction, which means that the complementary DNA strand is copied in a 3'→5' direction.

III. Transfer RNAs are the "decoding" molecules in the translation process.

   A. Each tRNA molecule is specific for only one amino acid. One part of the molecule contains a three-base anticodon, which is complementary to a codon on the mRNA. Attached to one end of the tRNA molecule is the amino acid that is specified by that mRNA codon.

   B. The binding of an amino acid to its tRNA is an energy-requiring process that is catalyzed by one of a group of aminoacyl synthetase enzymes. Each enzyme is specific for a particular combination of amino acid and tRNA. The process is driven by ATP, which activates the amino acid so that it will have sufficient energy to form a peptide bond during protein synthesis.

IV. Ribosomes provide the mechanical machinery necessary to properly couple the tRNAs to their codons on the mRNA, to catalyze the formation of peptide bonds between amino acids, and to move the mRNA so that the next codon can be read.

   A. Each ribosome is made of a large and a small subunit; each subunit contains rRNA and a large number of proteins.

   B. The first stage of protein synthesis is a two-step process called *initiation*.

      1. The first step in initiation involves the binding of the small ribosomal subunit, protein initiation factors, and the initiation tRNA to the 5' region of the mRNA.

      2. The second step involves the binding of the large subunit of the ribosome to complete formation of the protein-synthesizing complex.

   C. The second stage of protein synthesis consists of a cycle of *elongation*, in which amino acids are added one by one to the growing polypeptide chain as the ribosome proceeds along the mRNA. This process involves binding of the correct tRNA to a second site on the ribosome, formation of a peptide bond between the amino group of that amino acid and the carboxyl group of the last amino acid on the growing polypeptide chain, and movement of the ribosome to the next codon on the mRNA.

   1. Elongation proceeds in a 5′→3′ direction along the mRNA.

   2. The polypeptide chain grows from its amino terminal end to its carboxy terminal end.

D. The final stage of protein synthesis, *termination*, occurs when the ribosome reaches one of three special termination, or stop, codons, which do not specify any amino acid. This triggers release of the completed polypeptide chain from its linkage to the last tRNA and causes the ribosomal subunits to fall apart so they can be recycled with a new mRNA.

E. In bacterial cells, transcription and translation are coupled. Translation of the mRNA molecule usually begins before the 3′ end of the transcript is completed.

F. A single mRNA molecule can be translated by several ribosomes; these groups of ribosomes on an mRNA are called *polyribosomes*.

V. The basic features of transcription and translation are the same in prokaryotic and eukaryotic cells. However, eukaryotic genes and their mRNA molecules are more complex than those of bacteria.

A. Eukaryotic mRNA molecules are capped at the 5′ end with a modified guanosine triphosphate. Many also have a tail of poly-A nucleotides at the 3′ end. These modifications appear to protect the molecules from degradation, giving them long lifetimes compared with those of bacterial mRNA.

B. In many eukaryotic genes the coding regions, called *exons*, are interrupted by noncoding regions, called *introns*.

C. Transcription of eukaryotic genes produces a heterogeneous nuclear RNA (hnRNA) molecule that contains both introns and exons. Following transcription, capping, and tailing, the introns are removed and the exons are spliced together to produce a continuous protein-coding sequence.

D. Introns are thought to allow recombination to occur between exons of different genes, facilitating the rapid development of genes that code for new types of proteins. This is thought to be an evolutionary advantage in complex organisms.

VI. The genetic code was determined through the construction of artificial mRNA molecules with known nucleotide sequences. Translation of the artificial messages in the test tube allowed investigators to assign 61 codons to amino acids and to classify other codons as punctuation (stop or start signals).

A. Certain codons serve as start and stop signals. The start signal for all proteins is the code AUG, which specifies the amino acid methionine. Three codons serve as stop signals for the end of a protein. Those three do not specify an amino acid.

B. The genetic code is nearly universal, which suggests that it was established early in the evolutionary history of life on Earth. The only exceptions to the code are minor variations found in mitochondria, which contain their own DNA, and in some ciliated organisms.

C. Some amino acids are specified by more than one codon; thus, the code is said to be *redundant*. In many cases one tRNA molecule recognizes more than one codon for the same amino acid. This phenomenon, known as *wobble*, is caused by the ability of the third nucleotide in some anticodons to violate base-pairing rules and pair with more than one base in the third position of a codon.

D. The genetic code is read from mRNA in only one reading frame; that is, it is read as a series of nonoverlapping triplets that specify a single sequence of amino acids.

VII. Since genes are now known to code not only for proteins but also for specific RNA molecules, a gene is now defined as a sequence of nucleotides that can be transcribed to yield a product with a specific cellular function.

VIII. Mutations can produce many effects. Types of mutations range from disruption of the structure of a chromosome to a change in only a single pair of nucleotides.

A. A point mutation can destroy the function of a protein if it alters a codon so that it specifies a different amino acid (missense mutation) or so that it becomes a termination codon (a nonsense mutation). A point mutation has minimal effects if the amino acid is not altered or the codon is changed to specify a similar amino acid in the protein.

B. Insertion or deletion of one or two base pairs in a gene invariably destroys the function of that protein, since it results in a frameshift mutation that changes the codon sequences downstream from the mutation.

C. Mutations can be produced by errors in DNA replication, by physical agents such as x rays or ultraviolet rays, or by chemical mutagens. Mutations can also occur through transposable genetic elements, or "jumping genes," which move from one part of a chromosome to another, disrupting the function of a part of the DNA.

■ POST-TEST

1. The process by which information is copied from DNA to mRNA is called _____ .

2. The process by which genetic information in mRNA is decoded to specify the amino acid sequence of a protein is called _____ .

3. An amino acid is specified in genetic code as a sequence of _____ called a(n)_____ .

4. The type of RNA molecule that "decodes" the information in a codon into an amino acid is _____ .

5. The machinery that is used to form the peptide bonds

between amino acids during translation is a(n)_____ .

6. Messenger RNA is synthesized by DNA-dependent _____ _____ from _____ _____ subunits.

7. Ribonucleotides differ from the deoxyribonucleotide subunits found in DNA in that the sugar is _____ and the base _____ is substituted for _____ .

8. The "start" signals for transcription on DNA are _____ regions, which are just before the point at which the _____ end of the mRNA will be made.

9. The part of the tRNA molecule that is complementary to the appropriate codon on the mRNA is called the _____ site.

10. An amino acid is attached by its _____ group to the _____ end of its tRNA molecule.

11. The energy for forming a peptide bond is derived from the process of _____ _____ _____ , an ATP-requiring process that results in the attachment of the amino acid to its _____ .

12. Each ribosome is made of two subunits, each of which contains _____ and _____ molecules.

13. The first stage of protein synthesis is a two-step process called _____ . The first step involves the binding of the _____ ribosomal subunit, protein _____ factors, and _____ tRNA to the _____ end of the mRNA.

14. The second stage of protein synthesis is the _____ cycle, which involves the sequential binding of the _____ specified by each codon, _____ _____ formation, and translocation of the ribosome to the next _____ on the mRNA.

15. The final stage of protein synthesis, called _____ , occurs when the ribosome reaches a(n)_____ codon. This causes the completed _____ chain to be released from the last _____ and the _____ to dissociate from the mRNA.

16. A complex consisting of a group of ribosomes bound to a single mRNA molecule is a(n) _____ .

17. Many eukaryotic genes contain noncoding sequences, called _____ , which interrupt protein-coding _____ sequences.

18. The 5′ end of a eukaryotic mRNA is _____ after it is transcribed by a modified _____ triphosphate molecule.

19. Many eukaryotic mRNAs have a modified 3′ end consisting of a(n) _____ tail.

20. The precursor to a eukaryotic mRNA molecule is referred to as a(n) _____ molecule. In order for it to become a functional mRNA, it must be _____ and _____ . Its _____ are then removed and its _____ are spliced together to make a continuous protein-coding region.

21. Mutations that change a single base pair in a gene thereby converting an amino acid–specifying codon to a termination codon, are called _____ mutations. Mutations caused by the insertion or deletion of one or two bases in a gene are called _____ mutations.

■ REVIEW QUESTIONS

1. Discuss how genetic information is transferred from the gene to the amino acid sequence of a protein.
2. Discuss the genetic code. What type of translation control signals are present within the code? Why is it a three-base code? Why is it redundant?
3. Describe tRNA molecules, and explain why they serve as "adapters," decoding the genetic code into an amino acid sequence.
4. Outline the steps involved in the translation of a protein. Describe the steps of initiation, elongation, and termination in the synthesis of a protein.
5. Discuss the phenomenon of wobble. Why is the redundancy in the genetic code important to the idea of wobble?
6. Explain how the genetic code was deciphered. What experimental procedures needed to be developed before this could be accomplished?
7. Discuss the different types of mutations and their effects.
8. Compare and contrast the formation of RNA in prokaryotic and eukaryotic cells. How do the differences affect the way in which each type of mRNA is translated?

■ RECOMMENDED READINGS

Darnell, J.E., Jr. RNA. *Scientific American*, October 1985, pp. 68–88. An excellent discussion of the role of RNA in the translation of DNA into protein and interesting speculations on how RNA itself may have been the first genetic material.

Darnell, J., Lodish, H., and Baltimore, D. *Molecular Cell Biology*. New York, Scientific American Books, 1986. A comprehensive discussion of transcription and translation in prokaryotic and eukaryotic cells.

Watson, J.D., Hopkins, N.H., Roberts, J.W., Steitz, J.A., and Weiner, A.M. *Molecular Biology of the Gene*, 4th ed. Menlo Park, CA, Benjamin Cummings, 1987. A classic text.

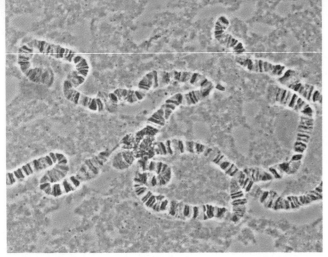

Puffed polytene chromosomes from the salivary gland of a *Drosophila* larva

14

Gene Regulation

With only a few exceptions, the cells in the human body have identical sets of genes and genetic information. After all, the ancestry of each cell can be traced back to one original fertilized egg cell. We shall see in Chapter 16 that usually none of the genes derived from that egg cell are lost during the many cell divisions that occur during the development of an adult. Yet muscle cells are strikingly different from fat cells, kidney cells, or liver cells, each of which have distinct shapes and functions and make different sets of proteins. If these cells contain the same genetic information, why are they not identical in structure and molecular composition? The reason is that only a part of that information is expressed (or active) in any given cell. Some genes are totally inactive in some cells and active in others. Other genes may be active in all cells, but their protein products may be abundant in one type of cell, yet present in only small amounts in another. This is because most genes are under some type of regulatory control, which is essential for the survival of organisms as complex as humans and as simple as bacteria.

What are the mechanisms that regulate gene activity? Some answers are emerging for eukaryotic cells. However, our clearest ideas of how gene activity is controlled come from bacterial cells. In this chapter we first examine how the expression of genes is controlled in bacterial cells. We then survey some recent advances in research on gene regulation in eukaryotes. In Chapter 16 we examine what is known about the regulation of genes in the early development of complex animals and plants. This is a type of gene regulation that allows for the development of complex organisms with many types of cells, all of which are descended from a single cell.

MECHANISMS OF GENE REGULATION IN BACTERIA

An *Escherichia coli* cell has on the order of 2000 to 4000 genes. Some of these encode proteins that are needed at all times in the life of the cell (e.g., enzymes involved in glycolysis). Other gene products are needed only when the bacterium is placed under special growth conditions. For instance, the bacteria living in the colon of an adult cow are not normally exposed to the milk sugar lactose. However, if those cells were to end up on the foot of a fly that then landed in a bucket of milk, they would have lactose available as a source of energy. This, in a way, poses a dilemma. Should a bacterial cell invest energy and materials to continuously produce lactose-metabolizing enzymes on the off-chance that it might fall into a bucket of milk? Given that the average lifetime of an actively growing *E. coli* cell is about 30 minutes, such a strategy would appear to waste energy and materials. Yet if bacterial cells did not produce those enzymes, they might one day starve in the midst of an abundant potential food resource. *E. coli* solves this problem by regulating the production of many of its metabolic enzymes in order to efficiently use available sources of carbon for energy and the synthesis of essential molecules.

Cells have two basic ways of controlling their metabolic activity: they can regulate the *activity* of certain enzymes (how effectively an enzyme molecule works), and they can control the *number* of enzyme molecules present in each cell. Some enzymes may be regulated in both ways in the same cell. An *E. coli* cell growing on glucose is estimated to need about 800 different enzymes. Some of these must be present in large amounts; others are required in only small quantities. In order for the cell as a whole to function properly, these enzymes must be under some type of efficient control.

The Operon Concept: Transcriptional Control of Protein Synthesis

One of the most intensively studied enzyme systems is that required for the metabolism of the sugar lactose. With that series of enzymes the French researchers François Jacob and Jacques Monod first demonstrated in 1961 how gene regulation could work. The sugar lactose is a disaccharide consisting of one molecule each of glucose and galactose (see Chapter 3). In order for *E. coli* to use lactose as a carbon source, the sugar must be cleaved into glucose and galactose by the enzyme **β-galactosidase**. Galactose is

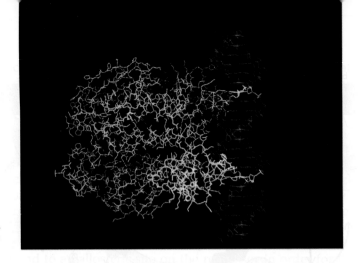

FIGURE 14–4 A computer-generated picture depicting the three-dimensional structure of active CAP binding to DNA. CAP is a dimer consisting of two identical polypeptide chains, each of which binds one molecule of cAMP. The blue lines represent the amino acids in the proteins; the green lines are those amino acids known to make contact with the DNA in the CAP-binding region. The two cAMP molecules bound to the protein are shown in red. (courtesy of Dr. Richard Ebright, Rutgers University)

The lactose operon actually has a very inefficient promoter sequence; that is, it has a low affinity for RNA polymerase even when the repressor protein is inactivated. However, a DNA sequence adjacent to the promoter site is a binding site for another protein, called the **catabolite activator protein (CAP)** (Figure 14–4). CAP, an allosteric molecule, increases the affinity of the region for RNA polymerase, allowing the enzyme to bind tightly to the DNA. In its active form CAP has **cyclic AMP (cAMP),** an altered form of adenosine monophosphate, bound to its allosteric site. As glucose is depleted from the bacterial cells, cAMP levels increase. The cAMP molecules bind to the CAP, and the resulting complex then binds to the CAP-binding site near the lactose operon promoter, thus stimulating transcription of the operon. Thus, the operon is fully active only if lactose is present and intracellular glucose levels are low.

Regulons

CAP differs from the lactose and tryptophan repressors in that it can control transcription of a number of operons involved in the metabolism of various carbon sources, such as the sugars galactose, arabinose, maltose, and lactose. A group of operons controlled by one regulator of this type is generally referred to as a **regulon** (Figure 14–5). Gene systems involved in carbon utilization are regulated in this way. A number of other multigene systems in bacteria are also controlled in this manner. For example, genes involved in nitrogen and phosphate metabolism each consist of multiple sets of operons controlled by one or more combinations of regulatory genes. Other complex multigene systems respond to changes in environmental conditions, such as rapid shifts in temperature, exposure to radiation, changes in osmotic pressure, and changes in oxygen levels. Clues to the existence of a regulon system usually come from the isolation of a mutant bacterial strain that has an altered regulatory gene. A single mutation that destroys the activity of the CAP, for example, not only prevents the cell from using lactose as a carbon source, but also prevents it from using other sugars whose metabolism is controlled by genes in the regulon.

Constitutive Genes

Many of the gene products encoded by the *E. coli* chromosome are needed only under certain environmental or nutritional conditions. As we have seen, these are the genes that are regulated at the level of transcription. They can be turned on and off as metabolic and environmental conditions change. Other gene products, such as those involved in the manufacture of ATP, are required at all times in the life of the bacterial cell. These genes, which are constantly transcribed, are termed **constitutive genes.** Although

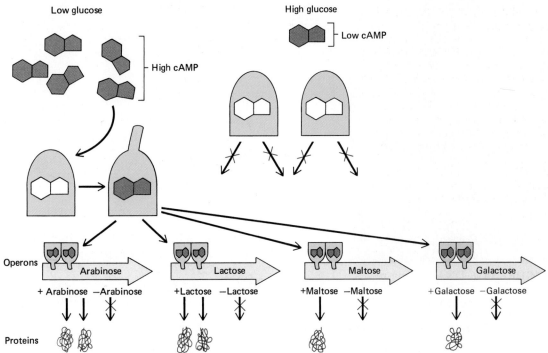

FIGURE 14–5 The carbon utilization regulon in *E. coli*. A group of operons that convert a number of different sugars to glucose are under positive control by the CAP. When glucose levels are high, cAMP levels are low, rendering CAP inactive. Under these conditions, none of the operons are active, even if the appropriate carbon source is available. When glucose levels are low, cAMP levels increase, activating the CAPs, which bind to their recognition sites in the promoter regions of all operons. If the inducer for that operon is available, its repressor will be inactivated, and transcription of the message will take place at a rapid rate.

all of the genes that encode constitutive enzymes like those of the glycolysis pathway are required to make their products continually, they are not necessarily transcribed (or translated) at the same rate. Some enzymes work more effectively or are more stable than others and consequently need to be made in fewer numbers in order to participate in the metabolic pathway. Those constitutive genes that encode proteins required in large numbers are generally transcribed more rapidly than genes for proteins required at lower levels. The rate of transcription of these genes is controlled by their promoter sequences. Genes that have efficient ("strong") promoters bind RNA polymerase more frequently and consequently transcribe more mRNA molecules than those that have inefficient ("weak") promoters.

Genes that code for repressor or activator proteins that regulate metabolic enzymes are usually constitutive and produce their protein products constantly. Since each cell usually needs relatively few molecules of any specific repressor or activator protein, the promoters for those genes tend to be relatively inactive.

How Do Regulatory Proteins Work?

Although a number of DNA-binding regulatory proteins have been identified, the question remains as to how these proteins work. How do they bind to the DNA? What base sequences do they recognize? Do they unwind the helix and bind to bases, similar to complementary base pairing, or do they recognize other parts of the DNA molecule? How can a protein bound to a DNA sequence cause another protein, such as RNA polymerase, to bind and transcribe a gene more effectively?

Many DNA-binding regulatory proteins appear to be modular proteins; that is, they have a DNA-binding domain plus other domains that can activate RNA polymerase or bind effector molecules such as allolactose or tryptophan. Both the lactose repressor and CAP have a DNA-binding region that contains two alpha-helical segments that touch the DNA (Figure 14–6). Representatives of another major class of regulatory proteins have multiple "zinc fingers," which are loops of amino acids held together by

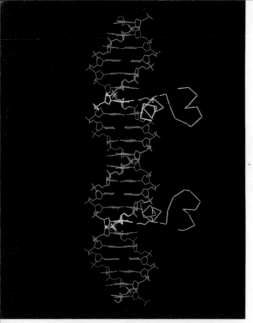

FIGURE 14–6 DNA-binding regions of regulatory proteins. (*a*) A computer model showing the two regions of CAP (see Figure 14–4) involved in binding to the CAP site on the lactose operon. Each polypeptide subunit has two alpha helical segments that contact the DNA and recognize the specific base sequences in the CAP site. The polypeptide backbones of the two DNA binding regions of the protein are shown in blue. The bases that are recognized by CAP are exactly one turn of the helix apart (courtesy Dr. Richard Ebright).

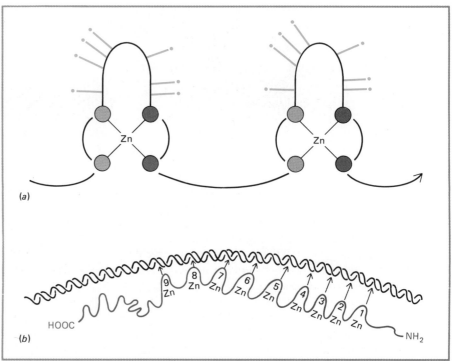

FIGURE 14–7 "Zinc-finger" DNA-binding proteins. (*a*) A diagram showing how zinc atoms cause regions of the polypeptides to form finger-like loops, which can insert into the grooves of the DNA and bind to specific base sequences. The colored circles represent amino acids that bind to the zinc atoms and form the loop. Each loop consists of about 13 amino acids. In this figure the lines projecting from the loop represent amino acids that are thought to recognize specific base sequences in the DNA. (*b*) Zinc-finger proteins that are known have from two to nine fingers. Each finger is thought to fit into a separate groove in the DNA of a control region for a specific gene.

metal ions (Figure 14–7). Researchers have been able to deduce how those proteins recognize specific DNA sequences (Figure 14–8). Both types of proteins seem to bind to the DNA by inserting either a helix or "fingers" into the grooves of the DNA molecule without unwinding the strands of

FIGURE 14–8 Computer generated models of how (*a*) CAP and (*b*) lactose repressor protein recognize their specific base sequences on the DNA. Only the amino acids from one of the DNA-binding regions of each protein are shown. The amino acids shown in blue contribute only to the structure of the binding region; their functional groups have been removed to simplify the model. The amino acids and bases that hydrogen-bond together are shown in yellow and green. Despite the similarity in their DNA binding regions, these two proteins recognize markedly different DNA base sequences. (courtesy Dr. Richard Ebright)

(*a*)

(*b*)

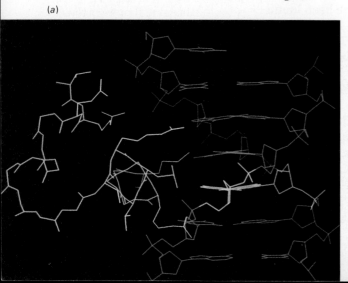

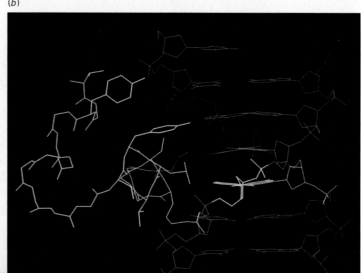

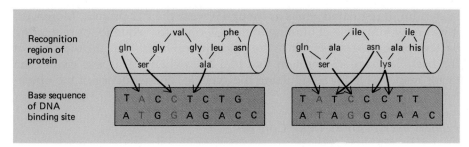

Recognition region of protein

Base sequence of DNA binding site

FIGURE 14–9 A comparison of the amino acid–base pair interactions between two related DNA-binding proteins from a bacteriophage that infects *E. coli.* Only certain amino acids in the "recognition helix" regions are used to bond with the bases. Two of the amino acids are the same in the two proteins and recognize the same bases; other amino acids differ, making the two proteins recognize specific combinations of bases.

DNA. Recall from Chapter 3 that an alpha helix of a protein is arranged such that the peptide bonds are in the interior of the helix and the functional groups of the amino acids are on the surface of the helix. This arrangement allows certain functional groups of amino acids in that part of the protein to form hydrogen bonds with specific base pairs in the DNA. The hydrogen bonds that form between the protein and the DNA differ from those used for complementary base pairing (Figure 14–9).

Mark Ptashne and his colleagues at Harvard University determined that the part of the repressor protein that fits into the groove of the DNA determines its binding specificity. They reasoned that, if an alpha helix is the part of the protein that recognizes a specific DNA sequence, changing the helix to another form would cause it to recognize a new DNA sequence, thus giving the protein the ability to regulate a different gene. Using genetic engineering methods (see Chapter 15), these researchers performed a "helix swap" (Figure 14–10) between a DNA-binding regulatory protein of bacteriophage 434 (a virus isolated from *E. coli*) and a similar protein from bacteriophage P22 (isolated from a strain of *Salmonella* bacteria). Neither of these proteins can recognize the operator sequence controlled by the other protein. The scientists modified the "recognition helix" part of the 434 protein so that it contained the amino acids normally found in the recognition helix of P22 protein. When the engineered protein was produced in a

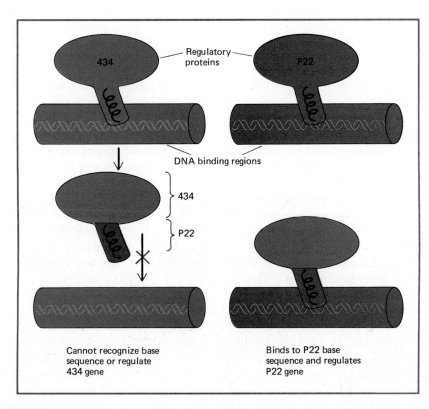

FIGURE 14–10 The "helix swap" experiment. The bacteriophage 434 protein was modified so that its DNA-binding "recognition helix" contained the amino acids found in a phage P22 regulatory protein. Even though most of the engineered molecule was identical to the 434 protein, it could no longer bind to the 434 gene. Instead, the hybrid protein bound to the P22 base sequence, and regulated that gene, establishing that the helix region is responsible for the specificity of the protein.

bacterium, it regulated genes as though it were the P22 protein. Moreover, it was now incapable of binding to the 434 operator sequence or regulating bacteriophage 434 genes. Since the "recognition helix" was the only part of the bacteriophage 434 protein that had been changed, it was clearly the factor responsible for recognizing the specific base sequence of the operator region.

Posttranscriptional Gene Control

Control of Translation

Although much of the variability in protein levels in *E. coli* is determined by regulating transcription, there are other regulatory mechanisms that operate at other levels of gene expression. Some of these systems involve the control of the *translation of mRNA.*

Since the lifetime of an mRNA molecule in a bacterial cell is very short, a molecule that is translated rapidly can produce more proteins than one that is translated slowly. Some mRNA species in *E. coli* are translated as much as 1000 times faster than others. Most of the differences appear to be due to the speed at which ribosomes can attach to the mRNA and begin translation. The structure of the 5′ end of the mRNA in the ribosome-binding region appears to be important in this type of regulation. There is a sequence of bases, usually found a short distance upstream of the initiation codon, that binds to a group of complementary bases on the 3′ end of a specific ribosomal RNA molecule. The base sequence on the mRNA apparently serves as a signal to the ribosome to bind at that point and initiate transcription. The strength of the ribosome binding (and presumably the speed at which translation is initiated) is determined by the number and position of these ribosome-binding bases. Messenger RNA molecules that have a long stretch of bases that are complementary to the rRNA are translated rapidly, whereas those with only a few complementary bases are translated more slowly.

Translation can also be controlled by proteins that bind to a sequence of bases close to the initiation codon. Ribosomal proteins, which are some of the most abundant proteins in the cell, are regulated this way (Figure 14–11). Even ribosomal proteins must be made on ribosomes; this is not a problem, since each cell is formed from a preexisting cell that contains a supply of ribosomes. However, ribosomes and their proteins are expensive to produce in terms of the total chemical resources of the cell.

Remember that the *E. coli* ribosome consists of 3 molecules of RNA and about 50 separate proteins. A completed ribosome requires a single molecule of every protein, except for one, which is present as four copies. In order to produce ribosomes efficiently, the cell has to synthesize all of these proteins and the rRNA molecules in their proper ratios. Ribosomal RNA production appears to be controlled by the rate of protein synthesis in the cell. When protein synthesis declines, a certain number of free ribosomes will not be actively synthesizing protein. This triggers a regulatory system that slows the rate of rRNA synthesis. Ribosomal proteins are regulated, in turn, by the amount of free rRNA that is available for ribosome construction. In this system a few key ribosomal proteins are the regulatory signals that repress their own synthesis and also the synthesis of several other ribosomal proteins.

Most ribosomal protein genes are located in one of several operons, each consisting of one regulatory ribosomal protein gene and several other ribosomal protein genes. Normally as the proteins are synthesized they bind to the rRNA molecules and are immediately assembled into ribosomes. When synthesis is unbalanced and some ribosomal proteins are produced in great excess, their operons continue to synthesize mRNA at normal rates, but the extra regulatory proteins inhibit the *translation* of their own messages.

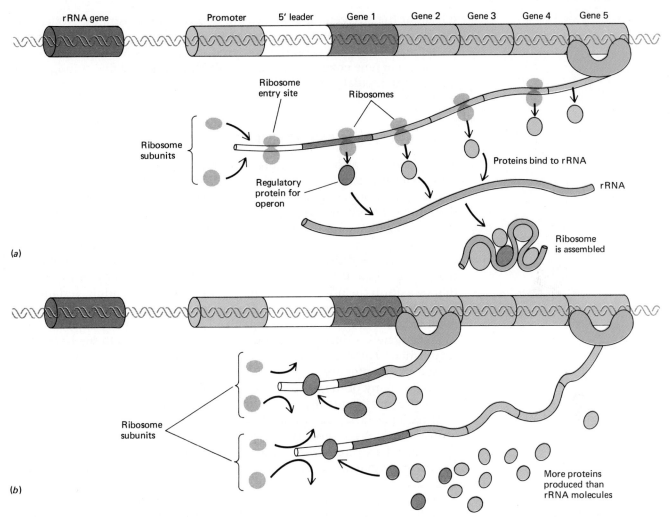

The regulatory protein blocks translation of the entire mRNA by binding to a sequence at the 5' end of the molecule called the **ribosome entry site.** That site is the only place on the mRNA where the ribosome can initiate translation. Thus, all of the ribosomal proteins encoded on that message are prevented from being synthesized until the levels of the regulatory protein drop in the cytoplasm.

The production of ribosomal proteins thus depends ultimately on the availability of rRNA. If unused rRNA is present, the regulatory proteins preferentially bind to it and participate in the assembly of new ribosomes. Under these conditions none of the ribosomal control proteins are available to block translation of their mRNAs. When the cell runs out of rRNA, the control proteins have nothing to bind to except their own mRNA, effectively blocking further translation.

Posttranslational Control

Some enzyme control mechanisms do not involve regulation of gene expression. Instead they act as switches causing one or more existing enzymes to become active or inactive. This provides a rapid way of controlling enzyme activities in response to changes in the intracellular concentrations of essential molecules such as amino acids. A common method for adjusting the rate of synthesis of a product of a metabolic pathway is through **feedback inhibition** (see Chapter 6). The end product binds to the first enzyme in the pathway at an allosteric site, temporarily inactivating the enzyme. When the first enzyme in the pathway is turned

FIGURE 14–11 The control of ribosomal protein manufacture. Ribosomal protein genes are organized in operons. One of the genes encodes a ribosomal protein that can also bind to a base sequence on the ribosome entry site at the 5' end of the mRNA. (*a*) When ribosomal proteins are produced in the same ratios as rRNA, all of the proteins bind to rRNA molecules and participate in ribosome assembly. (*b*) When ribosomal proteins from the operon are synthesized in excess of those needed to bind to rRNA, the extra regulatory proteins bind to the mRNA at the 5' ribosome entry site, preventing further translation of the messages.

off, all of the succeeding enzymes are deprived of substrates. Notice that this differs from the end-product repression of the tryptophan operon discussed earlier. In that case, the end product of the pathway prevented the formation of new enzymes. Feedback inhibition acts as a fine-tuning mechanism that regulates the activity of the existing proteins in a metabolic pathway.

EUKARYOTIC GENE CONTROL

Eukaryotic cells also must regulate the expression of their genes. Because these cells are complex, especially in multicellular organisms, their regulation is also more complex. Eukaryotic cells must respond to changes in their environment by turning on and off certain appropriate sets of genes. The organization of cells into tissues that have different functions may also require that a gene respond in one way to a given signal in one tissue and another way in a different tissue. The complex process of development also requires additional modes of regulation. During the many cell divisions that transform the original fertilized egg into the mature adult, certain cells may become committed to a particular form and function. For development to proceed in an orderly, organized way, many different genes have to be activated or inactivated, and specific regulatory patterns have to be established. The molecular and genetic basis of development will be discussed in Chapter 16.

Unlike many of the prokaryotic genes discussed so far, most eukaryotic genes are not found in operon-like clusters. However, each eukaryotic gene has specific regulatory sequences, which are essential in the control of transcription. As in prokaryotes, a number of different regulatory systems have been discovered. Many of the "housekeeping" enzymes (those essential enzymes that are needed by all cells) appear to be encoded by constitutive genes, which are expressed in all cells at all times. Some inducible genes have also been found; these respond to environmental threats or stimuli such as heavy metal ingestion, virus infection, and heat shock. Some genes appear to be inducible only at certain periods during the life of the organism; they are thought to be controlled by **temporal regulation** mechanisms. Finally, a number of genes are under the control of **tissue-specific regulation.** For example, a gene involved in the production of a particular enzyme may be regulated by one stimulus (e.g., a hormone) in muscle tissue, by an entirely different stimulus in pancreatic cells, and by a third stimulus in liver cells.

Control of Transcription

In eukaryotic as in prokaryotic cells, the transcription of both constitutive and inducible genes requires a promoter. In multicellular eukaryotes the promoter consists of a sequence of bases to which RNA polymerase binds and which is about 30 base pairs upstream from the transcription initiation site. The promoter region also contains one or more sequences of 8 to 12 bases known as **upstream promoter elements (UPE)** (Figure 14–12) within a short distance of the RNA polymerase–binding site. The strength of the promoter seems to depend on the number and type of UPEs; thus, a constitutive gene containing only one element would be weakly expressed, whereas a constitutive gene containing five or six elements would be actively transcribed.

Enhancers

Regulated eukaryotic genes require not only the promoter elements but also DNA sequences called **enhancers.** Whereas the promoter elements are required for accurate and efficient initiation of mRNA synthesis, enhancers

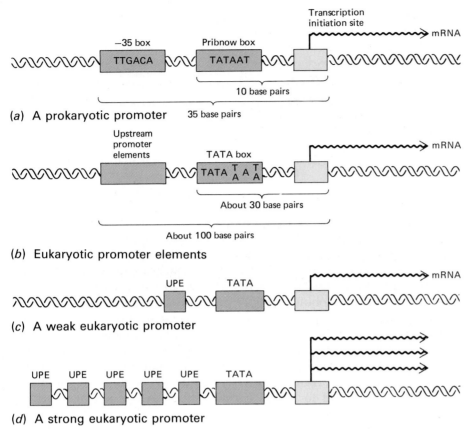

(a) A prokaryotic promoter

(b) Eukaryotic promoter elements

(c) A weak eukaryotic promoter

(d) A strong eukaryotic promoter

FIGURE 14–12 A comparison of prokaryotic and eukaryotic promoter elements. (a) Prokaryotic promoters consist of two short DNA sequences, called the Pribnow box and the −35 box, which are usually centered, respectively, 10 and 35 bases upstream from the transcription initiation site (the base in the DNA at which mRNA synthesis begins). The base sequence shown in the boxes are those most commonly found on the coding strand of DNA in those regions. The two "boxes" are RNA polymerase binding sites on the DNA. (b) Promoters in eukaryotes usually consist of a "TATA box" located 30 base pairs upstream from the transcription initiation site. The most commonly found sequence of bases on the coding strand of the TATA box is shown in the diagram (both T and A are found frequently at the positions where they are shown together). Eukaryotic promoters must also have one or more upstream promoter elements (UPEs). The strength of the promoter depends on the number of UPEs ahead of the gene. (c) A weakly expressed constitutive gene will contain only one UPE. (d) In contrast, a strongly expressed gene is likely to contain several UPEs. UPEs are thought to be binding sites for proteins that increase the rate of transcription by increasing the affinity of RNA polymerase for the promoter region.

increase the *rate* of RNA synthesis that has been initiated at a promoter. Remarkably, enhancer sequences can regulate a gene on the same DNA molecule from very long distances (up to thousands of bases away from the promoter), and they can do so from positions that are either upstream or downstream of the promoters they control (Figure 14–13). Enhancers cannot regulate a gene on a different chromosome, which indicates that the enhancer DNA sequences do not encode a regulatory protein. Furthermore, they can work in either orientation; that is, an enhancer sequence cut out of the DNA and turned to point in the opposite direction will still regulate the gene it normally controls.

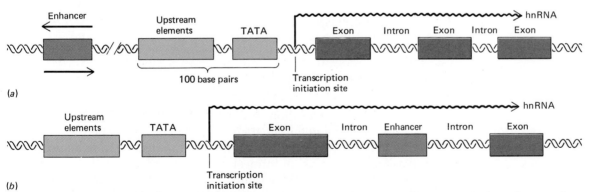

FIGURE 14–13 Enhancers are eukaryotic DNA sequences that can stimulate transcription of a gene by several orders of magnitude, at distances thousands of bases from the promoter. An enhancer can also work in either direction, upstream or downstream, from the promoter. (a) A typical eukaryotic gene controlled by an upstream enhancer. The promoter, consisting of the TATA box and UPEs, is within about 100 bases of the transcription initiation site; the enhancer sequence is several thousand bases from the promoter. (b) The internal enhancer of the immunoglobulin-G gene is positioned in an intron inside the gene. The enhancer itself is tissue-specific and activates the gene only in a very small set of cells involved in the production of antibody molecules.

FIGURE 14–14 How an enhancer might act at a large distance from the promoter. The enhancer binds transcription-activating proteins, which can contact the promoter region when the intervening DNA forms a loop.

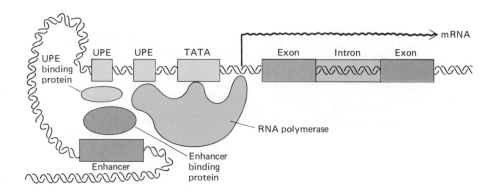

The first enhancer elements were discovered in the virus SV-40, which infects mammalian cells. Since then, enhancer elements have been found to act as regulators of a number of eukaryotic cellular genes. The first of the cellular enhancers was discovered simultaneously in the laboratories of W. Schaffner and the 1987 Nobel prize winner, M. Tonegawa, as a component of the immunoglobulin-G (IgG) heavy-chain gene (see Chapter 43). Surprisingly, this enhancer sequence was located within an intron in the middle of the gene. The discovery of this enhancer sequence was considered a breakthrough in our understanding of gene regulation, not only because enhancers were now known to be important regulatory elements in eukaryotes (and not confined to viruses) but also because the actions of enhancers were found to be *cell-specific*. Although the IgG heavy-chain gene and its enhancer are found in all cells, the enhancer regulates the production of that protein only in a small, highly specific group of cells that make only one type of antibody molecule.

Both enhancers and UPEs apparently become active when specific regulatory proteins are bound to them. For example, some zinc-finger regulatory proteins in eukaryotes bind to steroid hormones that enter the cell. When the steroid hormone is bound to the protein, the shape of the protein is modified, enabling it to bind the enhancer sequence tightly and activate transcription of the gene it controls.

One particularly puzzling property of enhancers is the fact that they work over such long distances. It is thought that the DNA between the enhancer and the promoter sequences forms a loop that allows the protein bound to the enhancer to come in contact with proteins associated with the promoter DNA sequences (Figure 14–14). When this occurs, transcription is stimulated. Alternatively, some enhancers may prevent regions of the DNA from being organized into nucleosomes (see Chapter 12), allowing those regions to be actively transcribed.

Overlapping Transcriptional Units

Certain types of proteins of eukaryotes can be found in more than one form and in more than one location in the cell. In some cases these forms are encoded by the same gene but their mRNAs are transcribed in different ways. The enzyme invertase, which cleaves the disaccharide sucrose, exists in two forms in yeast: an intracellular form and a form that is secreted from the cell into the growth medium. Both forms are encoded by the same gene (Figure 14–15); however, their mRNAs are transcribed from two different transcription "start" sites. The longer mRNA encodes the extracellular form and contains an additional "start" codon in the 5' part of the messenger RNA. When a ribosome translates the longer message, the protein is synthesized beginning with the first AUG codon, making a larger polypeptide with additional amino acids at its amino-terminal end. Those amino acids serve as a "signal sequence" designating that the protein is to

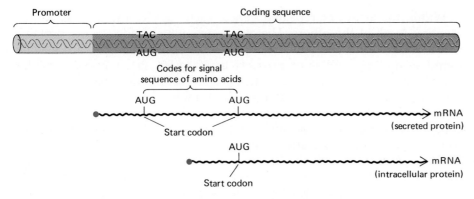

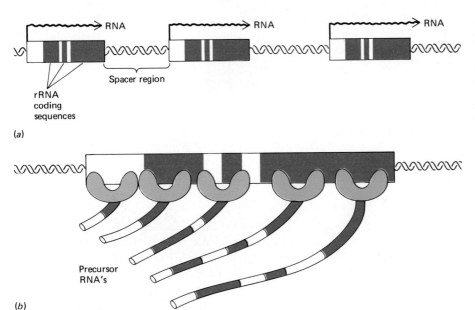

FIGURE 14–15 Overlapping transcription units. Two forms of the same protein can be made from a single gene by the initiation of transcription at two different sites. The yeast invertase gene encodes an mRNA with two start codons (AUG) in its coding region. If transcription is initiated between the two regions, a short mRNA is made that will yield, when translated, an intracellular form of the enzyme. If transcription starts at a point upstream from both AUG codons, a longer message will be made. Translation will begin at the first start codon, producing an enzyme with a "signal sequence" at the N-terminal end of the polypeptide chain that targets the protein to the endoplasmic reticulum for secretion from the cell.

be processed through the Golgi apparatus and then secreted from the cell (see Chapter 4). The smaller mRNA has only one start codon and does not contain the signal sequence when it is translated. This enzyme remains in the cytoplasm of the cell.

Other genes have tissue-specific overlapping transcriptional units. The primary hnRNA transcript of the gene for amylase (an enzyme that breaks down starch) in the mouse salivary gland is several thousand bases longer than its counterpart in the liver. This is because in the salivary gland transcription starts at a site on the DNA that is farther upstream. Although the coding portions of both mRNAs are identical after splicing and processing in both cell types, transcription in the salivary gland occurs about 100 times more frequently than transcription in the liver, resulting in the production of higher levels of the salivary amylase enzyme.

Chromosome Structure and Transcriptional Control

Repeated Gene Sequences

For certain gene products that are required in large amounts, a single gene cannot provide enough copies of its mRNA to meet the cell's needs. The requirement for high levels of those products may be met if multiple copies of genes are present in the chromosome. Genes of this type, whose products are essential for all cells, may occur as tandemly repeated gene sequences (Figure 14–16) in all cells. Other genes, which may be required by only a small group of cells, may be selectively replicated in those cells in a process called **gene amplification** (see Chapter 16).

FIGURE 14–16 Repeated gene sequences. Multiple copies of genes are required when large amounts of their products are needed by the cell. (*a*) Human ribosomal RNA genes are arranged as 200 to 300 tandemly repeated copies. Each transcription unit encodes a single copy of three of the ribosomal RNAs. (*b*) The requirement for rRNA is so great in actively growing cells that each of the units must be maximally loaded with RNA polymerases.

Within the arrays of repeated genes, each copy of the gene is almost identical to all of the others. Histone genes, which code for the proteins that associate with DNA to form nucleosomes (Chapter 12), are usually found as multiple copies of as many as 50 to 500 genes in multicellular organisms. Genes for rRNA and tRNAs also occur in multiple copies in all cells. In order to ensure that the rRNA molecules are made in equal amounts, the RNA genes are arranged as multiple transcription units, each containing one copy of the three rRNA genes. Most eukaryotic species contain around 150 to 450 such transcription units per cell. The demand for rRNA is so great in actively growing mammalian cells that, even though hundreds of copies of the genes are present, each gene must be copied simultaneously by many RNA polymerases. A single rRNA gene in these cells is usually copied simultaneously by about 100 RNA polymerase enzymes, which must be spaced no more than 80 bases apart on the DNA.

Gene Inactivation

In the higher eukaryotic organisms, only a subset of the genes present in a cell are active at any one time. The genes that are inactivated differ from cell type to cell type and in many cases seem to be in an irreversibly quiescent state. A number, but not all, of the inactive genes appear to be associated with highly compacted chromatin which can be seen as densely staining regions of chromosomes during cell division. These regions of chromatin remain tightly coiled throughout the cell cycle and even during interphase are visible as darkly staining fibers, called **heterochromatin.** Evidence suggests that the DNA of heterochromatin is not transcribed. When one of the two X chromosomes is inactivated in female mammals, most of the inactive X chromosome becomes heterochromatic and is seen as the Barr body (see Chapter 10). Active genes are associated with a more loosely packed chromatin structure, called **euchromatin** (Figure 14–17).

Posttranscriptional Control of Gene Expression

Prokaryotic mRNA has a half-life that is usually measured in minutes; eukaryotic mRNA, even when it is turned over rapidly, is far more stable. Prokaryotic mRNA is transcribed in a form that can immediately be trans-

FIGURE 14–17 The structure of the chromosome affects transcription. (*a*) An inactive region of DNA such as heterochromatin is organized into nucleosomes, which are highly condensed as tight coils. (*b*) Active genes are associated with decondensed chromatin, sometimes in response to specific inducing signals. The loosely packed chromatin increases the accessibility of RNA polymerases required for translation of the region.

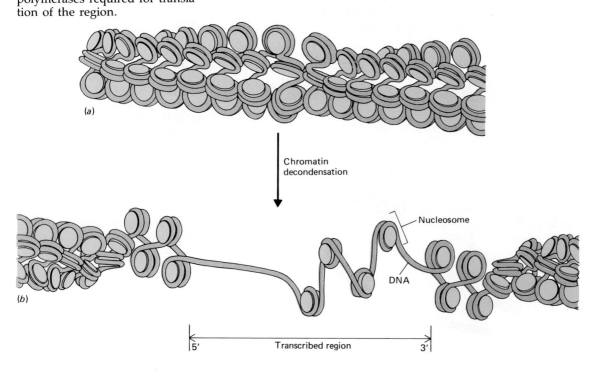

(a)

Chromatin decondensation

Nucleosome

DNA

(b)

5'   Transcribed region   3'

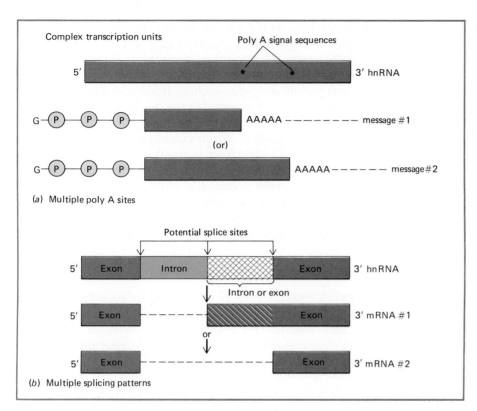

FIGURE 14–18 Differential mRNA processing. Complex transcription units can be processed in several ways to yield two or more mRNAs each of which encodes a different protein. (*a*) Multiple polyadenylation signals can be present in the hnRNA transcript; mRNAs encoding different proteins are produced, depending on which site is cleaved and polyadenylated. (*b*) Multiple splicing patterns result in the formation of different mRNAs, depending on which splicing pattern is used.

lated. In contrast, eukaryotic mRNA molecules require further modification and processing before they can be used in protein synthesis (see Chapter 13). The message is capped, spliced, and polyadenylated and then has to be transported from the nucleus to the cytoplasm to initiate translation. These events represent potential control points at which translation of the message and production of its encoded protein can be regulated.

Differential mRNA Processing

Because of the complexity of eukaryotic genes and the processing of their messages, it is possible for cells to use the same gene to produce more than one type of protein. Several forms of regulation involving mRNA processing have been discovered. In some instances, the same gene can be used to produce one type of protein in one tissue and a different type of protein in another tissue (Figure 14–18). In the thyroid gland, the calcitonin gene produces a polypeptide hormone used to retain calcium in the body; in the brain and nervous tissue, the same gene produces an entirely different polypeptide that is used as a neurotransmitter (a chemical that transmits signals between nerve cells). The two forms of the mRNA are controlled by **differential nuclear RNA processing,** involving two possible polyadenylation sites within the gene. In the thyroid gland, polyadenylation of the transcribed message at the first site results in a short transcript, which is then spliced to form a message for calcitonin. Polyadenylation at a site further downstream in nervous tissue results in the formation of a longer transcript, which is spliced differently and is then translated to produce the neurotransmitter. Other types of differential mRNA processing employ multiple splicing patterns, which generate different mRNAs depending on the location of the exon splicing sites.

Control of mRNA Stability

Another important mechanism for controlling the level of a specific protein in eukaryotes involves the stability of mature mRNA molecules in the cytoplasm of the cell. By controlling the lifetime of a particular kind of mRNA molecule, it is possible to control the number of proteins that can be translated from it. In some cases the control of messenger stability is under hormonal control. This is true for vitellogenin, for example, a protein made in the livers of certain female animals, such as frogs and chickens. After vitellogenin is synthesized, it is transported to the oviduct, where it is used in the formation of yolk proteins in the egg. The synthesis of vitellogenin is regulated by the hormone estradiol. When estradiol levels are high, the half-life of vitellogenin mRNA in the liver is about 500 hours. When cells are deprived of estradiol, the half-life of the mRNA drops rapidly to less than 3 hours. This leads to a rapid decrease in cellular vitellogenin mRNA levels and decreased synthesis of the vitellogenin protein. In addition to affecting the stability of the mRNA, the hormone seems to control the rate at which the messenger is synthesized.

Posttranslational Control of Gene Expression

Regulation of eukaryotic enzyme activity can also occur after the protein is synthesized. As in bacteria, many metabolic pathways in eukaryotes contain allosteric enzymes that are regulated through feedback inhibition. Many eukaryotic proteins are also extensively modified after they are synthesized. In proteolytic processing the proteins are synthesized as inactive precursors, which are converted to an active form by removal of a portion of the polypeptide chain. Other proteins may be regulated in part by selective degradation, which keeps their numbers constant within the cell. Enzyme activity can also be reversibly modified by the addition or removal of chemical functional groups. These types of modifications allow the cell to respond rapidly to fast-changing environmental or nutritional conditions.

Strategies of Prokaryotic and Eukaryotic Gene Regulation

Prokaryotic and eukaryotic cells have distinctly different strategies for regulating the activity of their genes. In large part this reflects the way in which these organisms make their living. Prokaryotic cells grow rapidly and are highly adapted for exploiting whatever food sources they encounter during their brief (20 to 30-minute) lifetimes. Consequently, they carry little excess baggage in the way of genes they would never use. By controlling the transcription of related genes in operons that can be rapidly turned on and off as needed, they also synthesize only those gene products that they need at any particular time. This type of regulation also requires the rapid turnover of mRNA molecules. Messages do not accumulate and continue to be translated when they are not needed.

Bacteria rarely regulate enzyme levels by degrading proteins. Once the synthesis of a protein is stopped, the rapid growth of the bacterial cell dilutes out the remaining protein molecules so rapidly that it is usually not necessary to expend the energy to break them down. Only when cells are starved or deprived of essential amino acids are proteases actively used to break down proteins no longer needed for survival in order to recycle their amino acids.

Bacterial cells exist independently, so each cell must be able to perform all of its essential functions. Since these cells grow rapidly and have relatively short lifetimes, the emphasis of bacterial gene regulation is *economy*, with genes being induced or repressed only as they are needed.

Eukaryotic cells have different regulatory requirements. In multicellular organisms, groups of cells cooperate with each other in a division of labor. Because a single gene may need to be regulated in different ways in different types of cells, eukaryotic gene regulation is relatively complex, occurring not only at the level of transcription, but also at other levels of gene expression. Eukaryotic cells also usually have long lifetimes, during which they may need to respond to many different stimuli numerous times. Rather than synthesize new enzymes each time they respond to a stimulus, these cells use preformed enzymes and proteins that can be switched rapidly from an inactive to an active state.

Much of the emphasis of gene regulation in multicellular organisms is on *specificity* in the form and function of the cells in each tissue. Each type of cell has a certain set of genes that are active and other sets of genes that may never be used once the specialized cells in that tissue are formed. Apparently the adaptive advantages of cellular cooperation in eukaryotes far outweigh the detrimental effects of carrying a load of inactive genes for generation after generation of somatic (body) cells.

■ SUMMARY

I. Most regulated genes in bacteria are organized into units, called *operons*, which may encode several proteins.
  A. Each operon is controlled by a single promoter region upstream from the protein-coding regions.
  B. A sequence of bases called the *operator* overlaps the promoter and serves as the regulatory switch controlling the operon.
    1. A repressor protein binds specifically to the operator sequence, preventing RNA polymerase from binding to the promoter and thus blocking transcription of the operon.
    2. When the repressor is not bound to the operator, RNA polymerase can bind to the promoter and transcription can proceed.
  C. An inducible operon such as the lactose operon is normally turned off. The repressor protein is synthesized in an active state and binds to the operator. If cells are exposed to lactose, a metabolite of that sugar binds to an allosteric site on the repressor protein, causing it to change its shape. The altered repressor cannot bind to the operator, and the gene is turned on.
  D. A repressible operon such as the tryptophan operon is normally turned on. The repressor protein is synthesized in an inactive state and cannot bind to the operator. A metabolite that is usually the end product of a metabolic pathway acts as a corepressor. When intracellular corepressor levels are high, one of the molecules binds to an allosteric site on the repressor, changing its shape so that it can bind to the operator and thereby turn transcription of the operon off.
  E. Repressible and inducible operons are under negative control. When the repressor protein binds to the operator, transcription of the operon is turned off.
  F. Some inducible operons are also under positive control. A separate DNA-binding protein can bind to the DNA and activate transcription of the gene.
    1. The lactose operon is activated by the CAP (catabolite activator protein), which binds to the promoter region, stimulating transcription by binding RNA polymerase tightly.
    2. CAP binding requires the presence of cAMP, levels of which increase in the absence of glucose.
  G. Groups of operons can be organized into multigene systems, known as regulons, which are controlled by a single regulatory protein. CAP activates a number of operons associated with the metabolism of carbohydrates.

II. Constitutive genes are genes that are not inducible or repressible. These genes are active at all times. Regulatory proteins such as CAP and the repressor proteins are produced constitutively. The activity of each of these genes is controlled by the strength of its promoter regions.

III. DNA-binding regulatory proteins work by recognizing and binding to specific sequences of bases in the DNA. Regions of the proteins fit into the grooves of the DNA molecules and hydrogen-bond to specific functional groups of the base pairs.

IV. Some genes are controlled after the mRNA is translated.
  A. Ribosomal protein genes are controlled at the level of translation. If more ribosomal proteins are made than there is available rRNA, one of the proteins coded for by a particular operon can bind to its own message, thus blocking translation of all the proteins coded for by that operon.
  B. The activity of key enzymes in metabolic pathways can also be controlled by feedback inhibition.

V. Eukaryotic genes usually encode only one protein. Regulation of eukaryotic genes can occur at the level of transcription, mRNA processing, translation, and the protein product.
  A. The promoter of a regulated eukaryotic gene consists of an RNA polymerase–binding site and short DNA sequences known as *upstream promoter elements* (UPEs). The strength of the promoter is determined by the number and types of UPEs within the promoter region.
  B. Inducible eukaryotic genes are controlled by enhancer elements, which can operate thousands of bases away from the promoter. They can be positioned on either side of the gene and in some cases are internal to the gene. Proteins that bind to enhancers appear to increase RNA polymerase binding to the promoter. This may require that the DNA

between the enhancer and the promoter form a loop so that the enhancer and the promoter can come into contact with each other.

C. The activity of eukaryotic genes is affected by chromosome structure.
1. Some genes whose products are required in large amounts exist as multiple copies in the chromosome. Other genes required in large amounts in only some cells may be selectively amplified by DNA replication.
2. Genes can be inactivated by changes in the chromosome structure. Densely packed regions of chromosomes called *heterochromatin* contain genes that are in an inactive state. Active genes are associated with a loosely packed chromatin structure called *euchromatin*.

D. Many eukaryotic genes are regulated after the RNA transcript is made.
1. Gene regulation can occur as a consequence of mRNA processing. In some cases a single gene can produce different forms of a protein, depending on how the hnRNA is polyadenylated or spliced.
2. Certain regulatory mechanisms increase the stability of mRNA, allowing more proteins to be formed per mRNA molecule prior to degradation.
3. Posttranslational control of eukaryotic genes can occur by feedback inhibition or by modification of the structure of the protein.

■ POST-TEST

1. Regulated prokaryotic genes are organized into clusters called _____.
2. A regulatory region that overlaps the promoter sequences and is associated with inducible and repressible operons is called the _____.
3. Repressor proteins that bind to the operator regions of operons block transcription of the genes by preventing _____ _____ from binding to the promoter. A DNA binding protein that blocks transcription is an example of _____ gene control.
4. Inducible operons are controlled by regulatory proteins that are normally in a(n) _____ state. These proteins are inactivated when a metabolite binds to a(n) _____ site, changing the _____ of the protein chain.
5. Repressible operons are controlled by regulatory proteins that are normally _____. When the _____ binds to the allosteric site of the repressor protein, it is converted to a(n) _____ repressor.
6. Positive gene control involves proteins that bind to the DNA and _____ transcription.
7. CAP is a positive controlling element that requires _____ _____ in order to bind to the CAP site of an operon.
8. When glucose is unavailable as a carbon source in bacteria, cAMP levels are _____.
9. CAP is a controlling element of a system of multiple operons called a(n) _____, which is involved in controlling carbohydrate metabolism.
10. Genes that are not under regulatory control are termed _____. These genes encode enzymes needed by cells at all times and are transcribed _____.
11. An example of a constitutive gene is the lactose _____ gene.
12. DNA-binding regulatory proteins work by inserting part of the polypeptide into one or more _____ in the DNA molecule so that amino acids can _____ bond with specific bases in the regulatory sequence.
13. Genes such as bacterial ribosomal protein genes are regulated at the level of _____.
14. Some metabolic pathways in bacterial cells are regulated by _____ _____, in which the end product blocks the activity of the first enzyme in the pathway.
15. In eukaryotic cells, the promoter region of genes consists of an RNA polymerase–binding site and adjacent _____ _____ elements. The strength of the promoter is determined by the _____ of these elements in front of the gene.
16. Inducible eukaryotic genes may be controlled by a(n) _____, which can be positioned thousands of bases away from the promoter.
17. Some single eukaryotic genes can make different forms of the protein they encode by differential _____ processing.
18. Eukaryotic genes whose products are required in large numbers may be present in _____ copies on the eukaryotic chromosome. Other genes that are required in large numbers in some cells but not in others may be _____ by selective replication of a small part of the chromosome.
19. Genes that are present in tightly coiled regions of chromosomes called _____ are _____.
20. Active genes are found in loosely packed chromatin called _____.

■ REVIEW QUESTIONS

1. Describe the structural elements of the DNA that make up the lactose operon. Explain the functions of the
   a. promoter
   b. operator
   c. CAP-binding regions
2. Discuss how the lactose operon is regulated in response

to the sugar lactose. What is required before transcription of the gene can be initiated when lactose is added to the growth medium of a bacterial culture?

3. Compare the regulation of a repressible operon, such as the tryptophan operon, to the inducible lactose operon. In what ways do the operons differ? In what ways are the structures and functions of the two operons similar?

4. Discuss the role of glucose in the positive control of the lactose operon. How is the CAP similar to the lactose repressor protein? How is it different?

5. Compare the types of bacterial genes that are associated with inducible operons, those that are associated with repressible operons, and those that are constitutive.

6. Compare the structure of a prokaryotic promoter region with known eukaryotic promoter regions. How does the regulation of inducible eukaryotic genes differ from the regulation of inducible prokaryotic genes?

7. Explain why it is necessary for certain genes in eukaryotic cells to be present in multiple copies.

8. Discuss how the structure of eukaryotic chromosomes can affect the activity of genes.

9. Discuss how a single eukaryotic gene can produce different forms of its encoded protein by differential mRNA processing.

■ RECOMMENDED READINGS

Darnell, J., Lodish, H., and Baltimore, D. *Molecular Cell Biology*. New York, Scientific American Books, 1986. A comprehensive discussion of transcription and translation in prokaryotic and eukaryotic cells.

Lewin, B. *Genes,* 3rd ed. New York, John Wiley, 1987. An excellent review of current knowledge in molecular biology and gene regulation.

Watson, J.D., Hopkins, N.H., Roberts, J.W., Steitz, J.A., and Weiner, A.M. *Molecular Biology of the Gene,* 4th ed., Vol. I. Menlo Park, CA, Benjamin Cummings, 1987. A revised edition of a classic text.

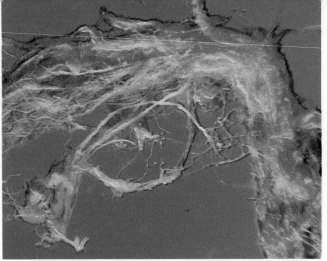

Synthesized strands of animal DNA

15

Recombinant DNA

■ LEARNING OBJECTIVES

After you have read this chapter you should be able to:

1. Explain how restriction enzymes cut DNA molecules, and discuss the role they play in recombinant DNA technology.
2. Describe a plasmid vector and explain how it can be used to amplify fragments of DNA from another organism.
3. Discuss how gene libraries are constructed and explain the difference between a genomic library and a cDNA library.
4. Explain why it is important to clone eukaryotic genes from both a genomic library and a cDNA library.
5. Explain how a gene can be cloned using DNA hybridization probes.
6. Discuss how a gene is restriction mapped.
7. Explain how the DNA sequence of a gene is determined.
8. Discuss how DNA probes are made and how they are used to diagnose genetic disorders and to identify the genetic makeup of individuals in a population.
9. Explain how genetic engineering can be used to produce important proteins and other products.
10. Discuss some of the problems encountered in using *E. coli* to produce proteins from eukaryotic cells, and explain how the use of transgenic plants and animals may solve some of those problems.

In the mid-1970s a revolution occurred in the field of biology. The development of **recombinant DNA technology** led to radical new approaches to biological research in areas ranging from molecular genetics to evolution. At the same time it led to unprecedented advances in fields such as pharmaceutics, medicine, and agriculture. One of the rapidly advancing areas of study today is **genetic engineering**—the modification of the DNA of an organism to produce new genes with new characteristics.

Recombinant DNA techniques permit the formation of new combinations of genes by allowing us to isolate a DNA fragment containing a gene from one organism and then introduce it into another. These methods allow us to place foreign DNA into cells ranging from simple bacteria to complex plants and animals. Under the right conditions the introduced

(a)

(b)

FIGURE 15–1 Recombinant DNA. (a) A scientist extracting purified recombinant DNA plasmids stained with a fluorescent compound from a centrifuge tube, with the aid of UV light. (b) An automated DNA sequencer.

DNA can be replicated (just as the DNA of the host cell is replicated) and passed on to the daughter cells when the cell divides. In this way a particular DNA sequence can be amplified or **cloned** to provide millions of identical copies, which can then be subjected to biochemical analysis.

The cloned DNA sequence can be modified in various ways in order to gain insights into how a gene or its product works. It can also be combined with parts of other genes in order to alter the ways in which it is expressed. For example, an isolated gene whose protein product is normally made in small amounts can be combined with regulatory elements from another gene; these may then cause its protein to be produced by cells in large amounts for study or for other applications.

In a way, genetic engineering at the molecular level is not new. Organisms have always had mechanisms for exchanging and modifying genes, and "experiments" in recombinant DNA and gene transfer have occurred throughout evolution as a consequence of natural interactions between different organisms. Those processes, of course, have gone on randomly, without any particular human goal. In order for genetic engineering to become effective and beneficial to humans, genes had to be modified in a controlled fashion. From primitive times humans have, in fact, practiced an extremely useful but not very efficient form of genetic engineering. By choosing plants or animals with desirable characteristics for selective breeding, it was possible to select (or engineer) organisms with new combinations of genes, using the natural reproductive processes. The development of recombinant DNA technology and molecular genetic engineering methods simply allowed a more direct approach to the creation of new combinations of DNA sequences.

RECOMBINANT DNA METHODS

Recombinant DNA technology was not developed in a short period of time. It actually began with the first studies of the genetics of bacteria and their viruses. It was only after decades of basic research and the accumulation of extensive knowledge that the current technology became feasible. Among other things, bacteria have provided us with special enzymes, known as restriction enzymes, which cut DNA molecules only in specific places. In addition, recombinant DNA molecules are most often introduced into bacterial cells so that they can be amplified or cloned, and certain aspects of the genetic systems of bacteria facilitate this process.

Restriction Enzymes: Cleaving DNA

A major breakthrough in the development of recombinant DNA technology was the discovery of bacterial enzymes called **restriction enzymes,** which are able to cut DNA molecules only at specific base sequences. One restriction enzyme may recognize and cut the base sequence 5'-GAATTC-

FIGURE 15–2 Restriction enzymes. Many restriction enzymes cut DNA at sequences of bases that are palindromic (each strand has the same base sequence, but in the opposite direction). Cutting the sequence leaves complementary "sticky" ends.

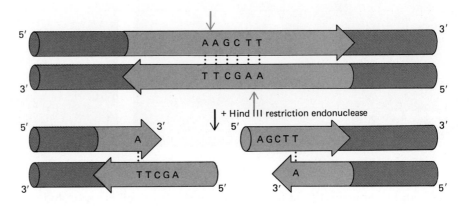

3' in a DNA molecule, while another will cut only the sequence 5'-GATC-3'. Bacteria normally use these enzymes as a defense mechanism, to attack the DNA of a virus that enters the cell. The bacteria protect their own DNA from attack by altering it in some way after it is synthesized. Purification of these enzymes enabled scientists to cut DNA from chromosomes into shorter fragments in a controlled way (Figure 15–2).

Many of the restriction enzymes that are used for recombinant DNA studies cut sequences that are **palindromic,** which means that the base sequence of one strand reads the same as its complement, but in the opposite direction. (Thus, the complement of our example, 5'-GAATTC-3', reads 3'-CTTAAG-5'.) By cutting both strands of the DNA in an asymmetric manner, these enzymes leave fragments with complementary, single-stranded ends; these are called **sticky ends** because they can pair (by hydrogen bonding) with the complementary single-stranded ends of other DNA molecules that have been cut with the same enzyme. Once two molecules have been joined together in this way, they can then be treated with **DNA ligase,** an enzyme that covalently links the two fragments to form a stable recombinant DNA molecule.

Restriction enzymes vary widely in the number of bases in the DNA sequences that they recognize, ranging from as few as four bases to as many as 23 bases. Based on probability alone, we expect the restriction sequence of a "four base cutter" to occur in a DNA molecule once on the average of every 4^4 or 256 bases, while one that recognizes six bases would cut fragments that average 4^6 or 4096 bases in size. Restriction enzymes that recognize sequences with large numbers of bases are particularly suited for studying very large DNA molecules such as those that make up entire chromosomes.

Recombinant DNA Vectors

Most recombinant DNA molecules are isolated and amplified by introducing them into cells of the bacterium *Escherichia coli*. In order to isolate a specific piece of DNA (after it has been cut by a restriction enzyme), it is first necessary to incorporate that fragment into a suitable carrier, or **vector molecule** (Figure 15–3). Viruses that infect bacteria (called **bacteriophages,** or literally "bacteria eaters") or special DNA molecules called **plasmids** are commonly used as vectors. A plasmid is a small circular DNA molecule that has its own origin of replication. These plasmids can be isolated from bacterial cells in pure form and then introduced into other cells by a method called **transformation** (Chapter 12), which involves altering the bacterial cell wall in order to make it permeable to the DNA molecules. Once a plasmid enters a cell, it is replicated and will be distributed to the daughter cells during cell division. Plasmids do not carry genes that are

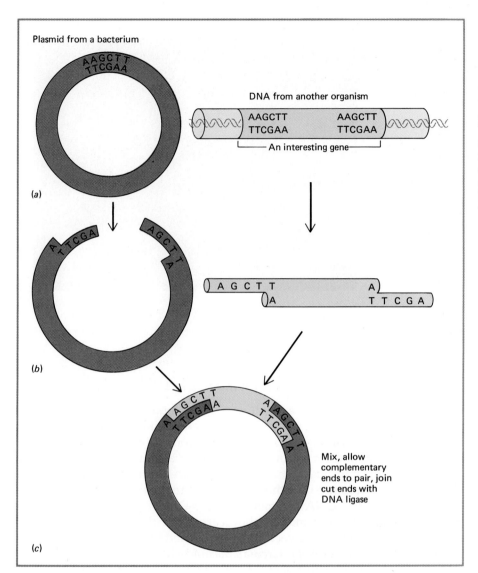

FIGURE 15–3 Creating a recombinant DNA molecule. DNA molecules from two different organisms (*a*) are cut with the same restriction enzyme, creating DNA molecules with complementary ends (*b*). In this example one molecule is a circular plasmid from a bacterium. The recombinant DNA (*c*) is constructed by mixing the two types of molecules so that their cohesive ends pair. DNA ligase then forms covalent 5'–3' phosphodiester bonds between the junctions of the two molecules.

essential to the *E. coli* cells, but they often carry genes that are useful under some environmental conditions, such as those that confer resistance to particular antibiotics.

The plasmids now used in recombinant DNA work have been extensively "engineered" to include a number of features helpful in the isolation and analysis of cloned DNA (Figure 15–4). A limiting property of any plasmid, however, is the size of the DNA fragment that it can effectively carry. DNA fragments of less than 10 kilobases (kb) can usually be inserted into plasmids for use in *E. coli*. However, many eukaryotic genes span DNA fragments that are much larger than 10 kilobases, and these require the use of bacteriophage vectors.

Recombinant DNA can also be introduced into cells of higher organisms. For example, engineered viruses are used as vectors in mammalian cells. These viruses have been disabled so that they do not kill the cells they infect; instead their DNA, and any foreign DNA they may carry, becomes incorporated into the chromosomes of the cell following infection. As we shall discuss later, other methods have been developed that do not require a biological vector. These involve techniques such as injecting the DNA directly into the cell nucleus or allowing cells to incorporate DNA that has been absorbed onto calcium phosphate crystals.

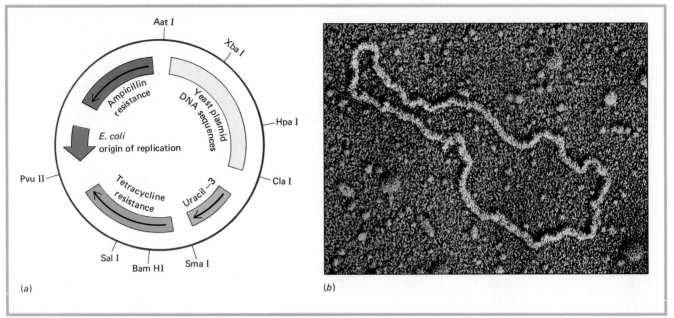

FIGURE 15-4 (a) A map of a genetically engineered plasmid vector used for both *E. coli* and the yeast *Saccharomyces cerevisiae*. This vector has been constructed from DNA fragments isolated from plasmids, *E. coli* genes and yeast genes in order to have useful features. Letters on the outer circle designate sites for those restriction enzymes that will cut the plasmid only at that one position. The plasmid has two origins of replication, for *E. coli* and yeast, allowing it to replicate independently in either type of cell. Resistance genes for the antibiotics ampicillin and tetracycline and the yeast URA-3 gene (for an enzyme involved in uracil biosynthesis) are also shown. The URA-3 gene is used to transform yeast cells lacking that particular enzyme; cells that take up the plasmid will grow on a uracil-deficient medium. (b) Electron micrograph of plasmids from *E. coli*.

Isolation of Genes Using Recombinant DNA Molecules

Since a single gene is only a small part of the total DNA in an organism, isolating the piece of DNA containing that gene is like finding a needle in a haystack. In order to do so you will need a powerful detector.

Isolating a gene from an organism such as a mouse first requires the construction of a **library,** or gene bank, from the mouse DNA (Figure 15–5). The first step is to cut the DNA with a restriction enzyme, generating a population of DNA fragments. These DNA fragments vary in size and in the genetic information they carry, but they all have identical "sticky ends." The plasmid vector DNA is treated with the same restriction enzyme, which converts the circular plasmids into linear molecules with "sticky ends" complementary to the "sticky ends" of the mouse DNA fragments. The two kinds of DNA (mouse and plasmid) are mixed together under conditions that promote hydrogen bonding of complementary bases, and the paired ends of the plasmid and mouse DNA are then joined by DNA ligase.

We now have a mixture of recombinant plasmids, with each one containing a different mouse DNA sequence. In order to find the one we are interested in, each will have to be amplified or cloned, so that there will be millions of copies to work with. This process occurs inside the cells of *E. coli*. First the recombinant plasmids are inserted into antibiotic-sensitive *E. coli* cells by transformation. This is done in such a way that there is a low ratio of plasmids to cells, so that it should be rare for a cell to receive more than one plasmid molecule, and not all of the cells will receive a plasmid. The cells are incubated on a nutrient medium that also contains antibiotics, so only those cells that have incorporated a plasmid (which contains a gene for antibiotic resistance) will grow.

The researcher may then spread a sample of the bacterial culture on solid growth medium in petri plates. If the suspension of cells is dilute enough, the cells will be widely separated. When each cell reproduces it gives rise to a **colony,** which is a clone of genetically identical cells. All of the cells of a particular colony will contain the same recombinant plasmid, so during this process a specific sequence of mouse DNA has also been cloned. The major task now is to determine which colony (out of thousands) contains the cloned fragment of interest. There are a number of ways in which this can be done.

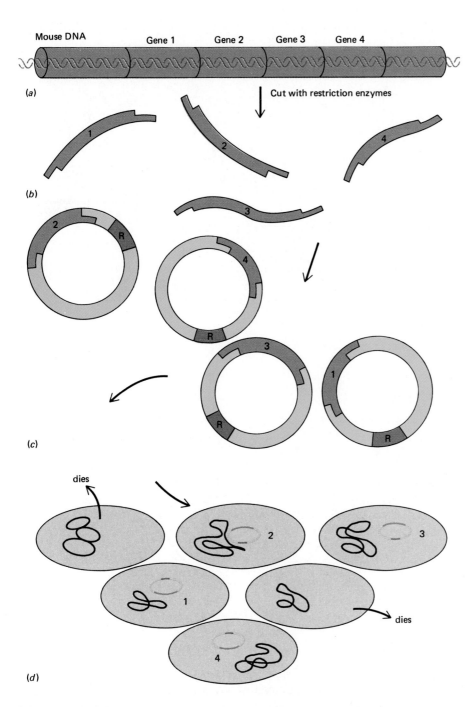

FIGURE 15–5 Construction of a genomic library from mouse DNA. DNA from mouse cells (*a*) is cut with restriction enzymes into fragments (*b*) that are ligated to a complementary restriction site in a vector molecule, which also contains an antibiotic resistance gene (*c*). The recombinant plasmids are used to transform an antibiotic-sensitive *E. coli* strain under conditions that assure each bacterial cell receives only one plasmid molecule (*d*). The cells are then grown on antibiotic-containing medium, so that only those that receive the plasmid will survive.

Genetic Probes

A common approach to the problem involves the use of a **genetic probe,** which is a radioactively labeled segment of RNA or single-stranded DNA that is complementary to the target gene. Suppose we wish to identify the gene that codes for insulin. Because we know the amino acid sequence of insulin, we could synthesize a radioactive single-stranded DNA molecule that is complementary to the DNA sequence that codes for insulin. (This is not as simple as it may sound. Remember that the genetic code is redundant, so even if you know the amino acid sequence of all or part of a protein, those amino acids could be coded for by a number of different base sequences.) The single-stranded DNA probe will **hybridize** (become attached by complementary base pairing) with the DNA sequence that codes for insulin (Figure 15–6). If the DNA from a small group of cells from a particular colony binds to the probe (Figure 15–7), then that DNA will become radioactive and can be detected by X-ray film. The rest of the cells in the colony can then be grown in quantity for further testing.

FIGURE 15–6 Identifying complementary DNA sequences by hybridization with radioactive DNA probes. Total cellular DNA contains a large number of different genes. A single gene (or a sequence of bases) in that DNA can be detected by using a cloned copy of the gene on a vector (*a*) to make complementary radioactive probe molecules (*b*). DNA isolated from the cell is denatured to produce single strands of DNA (*c*), which are then bound to the surface of a nitrocellulose filter (*d*). The filter is incubated with the radioactive probe DNA, which specifically hybridizes (pairs with) complementary regions of the bound DNA.

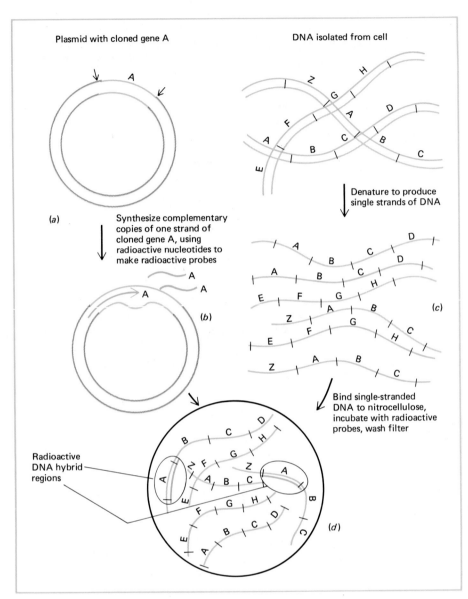

FIGURE 15–7 Isolating a gene by using a genetic probe. (*a*) *E. coli* cells containing a DNA library are spread on solid nutrient medium in dishes so that only one cell will be found in each location. Each cell gives rise to genetically identical descendants to form a colony on the medium. Each colony contains plasmids with only a single DNA fragment from the library. (*b*) To identify which colonies contain the required gene, a few cells from each colony are transferred to nitrocellulose filters which bind the DNA from the cells. (*c*) The filter is incubated with a radioactive probe DNA that is complementary to the desired gene. (*d*) DNA from cells that contain the complementary sequence to the probe will become radioactive, and can be detected by X-ray film. By examining the pattern of spots on the film, one can then return to the original culture plates and remove those colonies containing the correct plasmid.

Genomic Libraries

Populations of recombinant DNA molecules can be made in several different ways. The total DNA per cell of an organism is referred to as that organism's **genome.** If the DNA is extracted from mouse cells, as in our

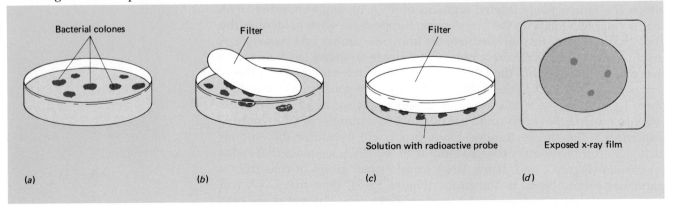

example, we would refer to it as mouse genomic DNA. A very large population of recombinant plasmids, each containing a fragment of the genome, is referred to as a **genomic library.** The library (population) should contain a complete set of mouse DNA sequences, although each individual recombinant plasmid (which is analogous to a "book" in the library) will contain only a single fragment of the total mouse genome.

cDNA Libraries

Remember, however, that many eukaryotic genes contain introns and that bacterial cells cannot remove introns from RNA. To avoid cloning those parts of a gene that do not code for proteins, libraries can also be constructed from DNA copies of the eukaryotic messenger RNA. Those copies are made by isolating messenger RNA and making DNA copies of the message using **reverse transcriptase** (Chapter 12). The complementary DNA (cDNA) copies of the message can then be inserted into the DNA of a plasmid or virus vector to form a **cDNA library** (Figure 15–8).

There are several advantages to cloning a gene from both a cDNA library and a genomic library. Analysis of the genomic DNA clone gives useful information about the structure of the gene on the chromosome and the structure of the primary hnRNA transcript. Analysis of the cDNA clone allows investigators to determine the characteristics of the protein encoded by the gene. This includes its exact amino acid sequence and the structure of the processed messenger RNA. Furthermore, since the cDNA copy of the messenger RNA does not contain intron sequences, comparison of the

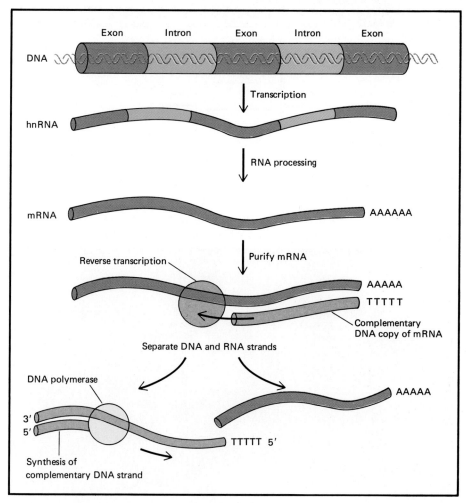

FIGURE 15–8 Constructing complementary DNA (cDNA) from a eukaryotic messenger RNA. The cDNA for constructing libraries is made by isolating messenger RNA from cells. Reverse transcriptase is used to make a complementary copy of the RNA. Synthesis of the second DNA strand is usually done by using DNA polymerase to make a complementary copy of the cDNA strand.

cDNA and genomic base sequences will reveal the locations of intron and exon coding sequences on the chromosome.

Cloned cDNA sequences are also useful in cases where it is desirable to produce a eukaryotic protein in *E. coli*. Remember that bacteria lack the enzymes to remove introns from eukaryotic mRNA transcripts. If an intron-containing human gene such as human growth hormone were to be introduced into *E. coli*, the bacterium would not be able to remove the introns from the transcribed RNA in order to make a functional mRNA for the production of its protein product. If a cDNA clone of the gene were inserted into the bacterium, however, its transcript would contain an uninterrupted coding region that could be translated into a functional protein.

Analysis of a Cloned Gene Sequence

Once a plasmid containing a desired DNA fragment has been isolated, the cloned piece of DNA can be used as a research tool for a wide variety of applications. Even if the purpose of cloning the gene is to obtain the encoded protein for some industrial or pharmaceutical process, a great deal of information must be obtained about the gene and how it functions before it can be "engineered" for a particular application.

FIGURE 15-9 Gel electrophoresis of DNA fragments. (*a*) The size of DNA fragments can be determined by electrophoresis through agarose or polyacrylamide gels. The gel material is poured as a thin slab on a glass or Plexiglas holder, and samples containing DNA fragments of different sizes are loaded in wells formed at one end of the gel. DNA molecules are negatively charged, so the molecules migrate through the gel toward the positive pole of an electrical field. The rate at which the molecules travel through the gel is inversely proportional to their molecular weight. Therefore, the smallest DNA fragments will travel the longest distance. By including DNA fragments of a known size in some of the wells of the gel, accurate molecular weights of the unknown fragments can be obtained. (*b*) A gel containing separated DNA fragments. The gel is stained with the dye, ethidium bromide, which binds to DNA and is fluorescent under UV light.

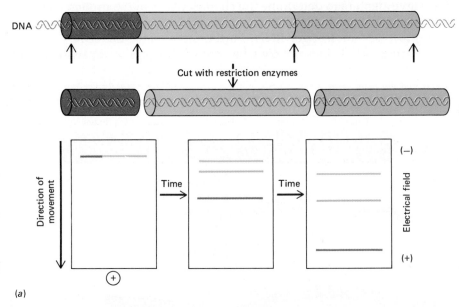

(a)

(b)

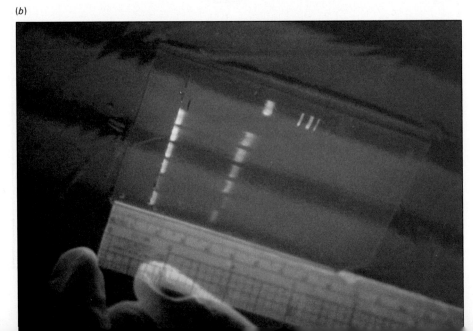

One of the first things that is done with a gene after it has been cloned is to construct a **restriction map** of the DNA fragment. Restriction mapping DNA involves identifying sites that are attacked by specific restriction enzymes, which will serve as landmarks for further studies. That information is then used to isolate (subclone) smaller DNA fragments for a variety of purposes. The mapping procedure involves cutting the DNA fragment with various combinations of restriction enzymes, and then separating the DNA fragments by **gel electrophoresis** (Figure 15–9) in order to determine the molecular weight of each cut fragment. After the size of each fragment is known, the positions of the restriction sites and the distances between them can be determined (Figure 15–10). Once the restriction map has been established, subcloned regions of the DNA fragment can be sequenced or used as DNA probes for analytical purposes.

DNA sequencing (Figure 15–11) is usually done by copying a strand of the cloned DNA in four different reaction mixtures, using a modified form of DNA polymerase. The copies are made in such a way that the newly synthesized DNA chain is broken randomly at positions corresponding to one of the four bases in each reaction mixture. The lengths of the fragments from each reaction are then determined by gel electrophoresis, allowing us to read off the sequence of bases on the cloned DNA fragment from one end to the other.

Knowing the DNA sequence in the cloned gene allows investigators to identify which parts of the DNA molecule contain the actual protein coding sequences, as well as which parts may be regulatory regions involved in gene expression. The amino acid sequence of the protein can then be read

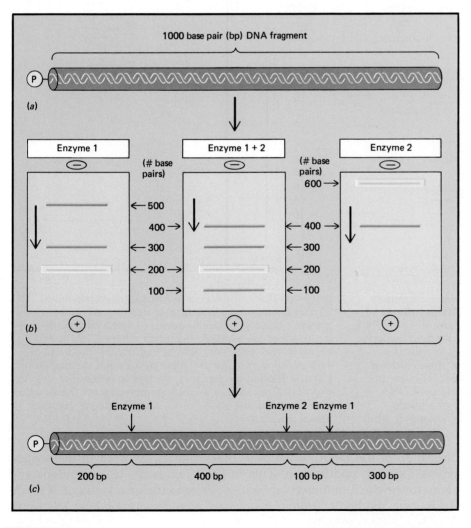

FIGURE 15–10 Restriction mapping a cloned DNA fragment. Restriction maps are constructed by determining the sizes of fragments produced when purified DNA is digested by various restriction enzymes. (*a*) A 1000 base pair DNA fragment is labeled on one end with radioactive phosphorus to provide a reference point. Samples of the DNA are cut by either or both of two different restriction enzymes. The sizes of DNA fragments and the location of the radioactive fragment are determined by gel electrophoresis (*b*). The positions of the restriction sites in the original fragment with respect to the radioactive end of the molecule are then deduced (*c*).

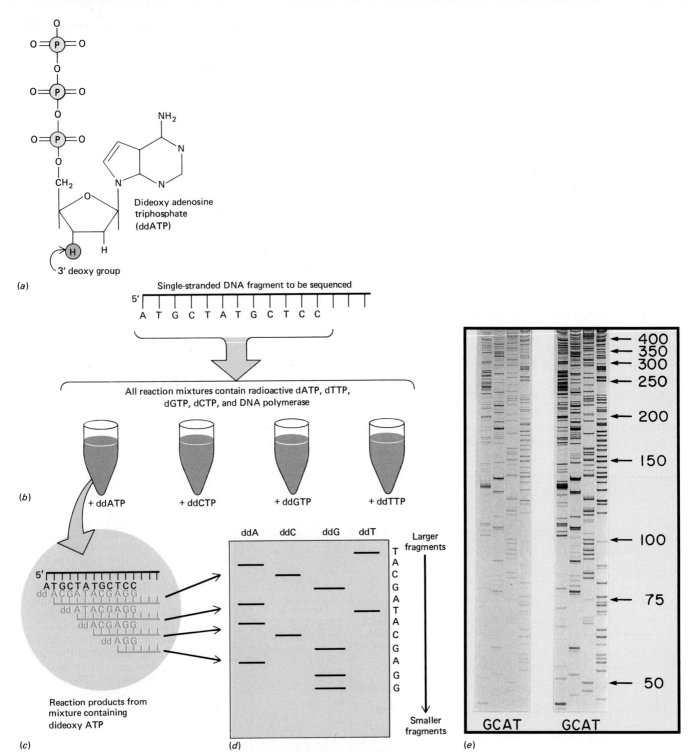

FIGURE 15–11 Sequencing DNA. The most commonly used method for sequencing DNA involves synthesis of complementary copies of a single DNA strand with DNA polymerase. (*a*) The procedure uses the random incorporation of *dideoxy*nucleotide triphosphates, modified nucleotides that lack a 3' hydroxl group, thus blocking further elongation of the new DNA chain. (*b*) Four different reaction mixtures are used to sequence a DNA fragment; each contains a small amount of a single dideoxynucleotide, such as dideoxy ATP (adenine), and larger amounts of the four normal deoxynucleotides. (*c*) The random incorporation of dideoxy ATP into the growing chain generates a series of DNA fragments ending in all of the possible positions where adenine is found in the fragment. (*d*) The radioactive products of each reaction mixture are separated by gel electrophoresis, and located by exposing the gel to X-ray film. (*e*) An exposed X-ray film of a DNA sequencing gel. Each set of four lanes represents A, C, G, and T dideoxy reaction mixes, respectively.

directly from the DNA, along with other signals involved in messenger processing and modification. This in itself represents a tremendous advance for research in molecular biology. Prior to the development of DNA sequencing methods, protein sequences were determined by laborious methods from highly purified protein samples. Although protein microsequencing technology has also advanced rapidly, in most cases cloning and sequencing the gene is easier than purifying and sequencing a particular protein. DNA sequence information is now kept in large computer databases that are available to investigators for comparing newly discovered protein sequences with those already known. By searching for DNA (and amino acid) sequences in the database, it is possible to gain a great deal of information about the function and structure of the gene product as well as the evolutionary relationships of a newly cloned gene.

Radioactive probes made from restriction fragments of cloned genes can also be used to detect related sequences in DNA or RNA from other cells. Since the probe DNA will bind only to complementary DNA sequences, it is not necessary to purify those fragments from the rest of the cellular DNA or RNA. The DNA to be probed is simply cut with restriction enzymes, and the entire collection of fragments is separated by electrophoresis. The separated fragments are then bound to nitrocellulose filters, which pick up the DNA much as a blotter picks up ink. When the DNA on the filter is incubated with the radioactive probe, any complementary fragments from the cellular DNA can be located. This type of blot hybridization is used to diagnose certain types of genetic disorders, since the radioactive probes can sometimes be used to identify alleles that are associated with certain genetic defects (see Focus on Probing for Genetic Disease).

The restriction method can also be used to examine the variability of genes within a population of individuals. Random mutations and recombination in the DNA of a population of organisms normally give rise to a number of **restriction polymorphisms** (differences in restriction enzyme sites due to changes in the base sequences in the DNA), which can be used to determine how closely related are different members of the population to each other. This is done by cutting the DNA from a number of individuals, using different restriction enzymes, and probing with a DNA sequence that naturally occurs a number of times within the organism's genome.

This "DNA fingerprinting" method is now being used by criminal investigators as a way of conclusively identifying individuals from small amounts of blood, semen, or other DNA-containing tissue left at the scene of a crime. The current methods are capable of identifying a person with a greater accuracy than traditional fingerprinting. Restriction polymorphism analysis has been found to be an especially powerful tool in the fields of population and evolutionary biology as well as diagnostic and forensic medicine, since it can be used to determine the degree of genetic relatedness between individuals of the same species.

GENETIC ENGINEERING

Recombinant DNA technology has not only provided a new and unique set of tools for examining fundamental questions about how living cells work; it has also provided new approaches to problems of applied technology in many other fields. In some cases the production of genetically engineered proteins and genetically engineered organisms has begun to have a considerable impact on our lives. The most striking of these to date have been in the fields of pharmaceutics and medicine.

Human insulin produced by *E. coli* was one of the first genetically engineered proteins to be commercially produced (Figure 15–12). Prior to the development of the altered bacterium to produce the human hormone, insulin was derived exclusively from other animals. Many diabetic persons become allergic to the insulin from those sources, since the amino acid sequences in those proteins differ slightly from those of the human hormone. The ability to produce the human protein by recombinant DNA

Focus on PROBING FOR GENETIC DISEASE

Genetic engineering techniques can be used in the characterization and detection of genetic disease. The earliest studies of this sort were carried out on sickle cell anemia, since that disease produces a thoroughly characterized protein abnormality. As we discussed in Chapter 11, the sickle cell hemoglobin differs from the normal protein by only a single amino acid at position 6 in one of the two kinds of polypeptide chains. The base sequence of the normal hemoglobin allele that codes for that amino acid and its two neighbors contains a recognition site for the restriction enzyme MstII, which reads:

—CCTNAGG—

(The letter N in the sequence means that the restriction enzyme will recognize any base in that position.)

The sickle cell hemoglobin allele differs by only one base, causing the sequence to read:

—CCTGTGG—

This change is sufficient to prevent the restriction enzyme from cutting the DNA at that point. This single base change therefore results in a restriction fragment length polymorphism, which can be detected by blot hybridization methods. By synthesizing a radioactive probe complementary to the DNA sequence on one side of the restriction site, it is possible to differentiate that specific fragment of DNA from all of the other DNA fragments in the genome. When normal human DNA is cut with MstII, a single radioactive band can be detected

on the blotting filter, and that band corresponds to a DNA molecule that is 1.15 kilobases long. Since the restriction site is abolished in individuals who have the sickle cell gene, a longer DNA fragment is detected; it is 1.35 kilobases in length. Heterozygous individuals have one copy of the normal allele and one copy of the sickle cell allele. When their DNA is analyzed, two bands can be detected by the probe, corresponding to the 1.15 and 1.35 kb fragments. Thus it is possible to distinguish individuals who are homozygous normal, homozygous recessive, and heterozygous carriers of the trait.

Actually, the DNA probe method of detection is not used to detect sickle cell anemia in adults, because the abnormal hemoglobin can be more simply detected by directly analyzing the hemoglobin proteins from the blood of individuals. If we wish to determine whether a fetus will suffer from the disease, however, it is necessary to examine DNA from cells from the amniotic fluid rather than cells from the fetal bloodstream (Chapter 11).

Few genetic diseases are as well understood as sickle cell anemia. As more abnormal genes are identified and sequenced, however, an increasing number of abnormal alleles will be detected in healthy carriers and in fetuses. Recently the genes for cystic fibrosis and Duchenne's muscular dystrophy have been cloned. Unlike the case of sickle cell anemia, no protein has yet been identified that can be associated with these diseases; consequently, use of the cloned gene sequence will be an important method for detecting parents who are heterozygous.

methods has resulted in enormous medical benefits to those individuals. Similar benefits may be gained from genetically engineered human growth hormone (Chapter 49), which is required by some children to overcome growth deficiencies. Human growth hormone could previously be obtained only from cadavers. Only small amounts were available, and there was evidence that some of the preparations were contaminated with viruses. A third promising medical application is the development of genetically engineered human blood clotting factors, which will help to alleviate a particularly tragic situation in the treatment of hemophilia (Chapter 11). During their lives, hemophiliacs normally require many treatments with purified clotting factor obtained from donated blood to stop uncontrolled bleeding. More than half of the hemophiliacs in the United States today have antibodies against viruses that cause diseases such as hepatitis and AIDS, which were present in contaminated clotting factor extracted from blood serum.

FIGURE 15–12 Scanning electron micrograph of *E. coli* cells "bulging" with human insulin.

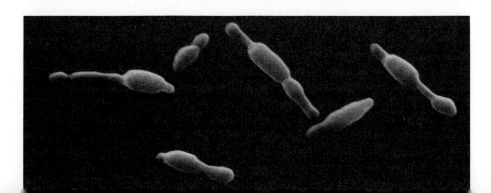

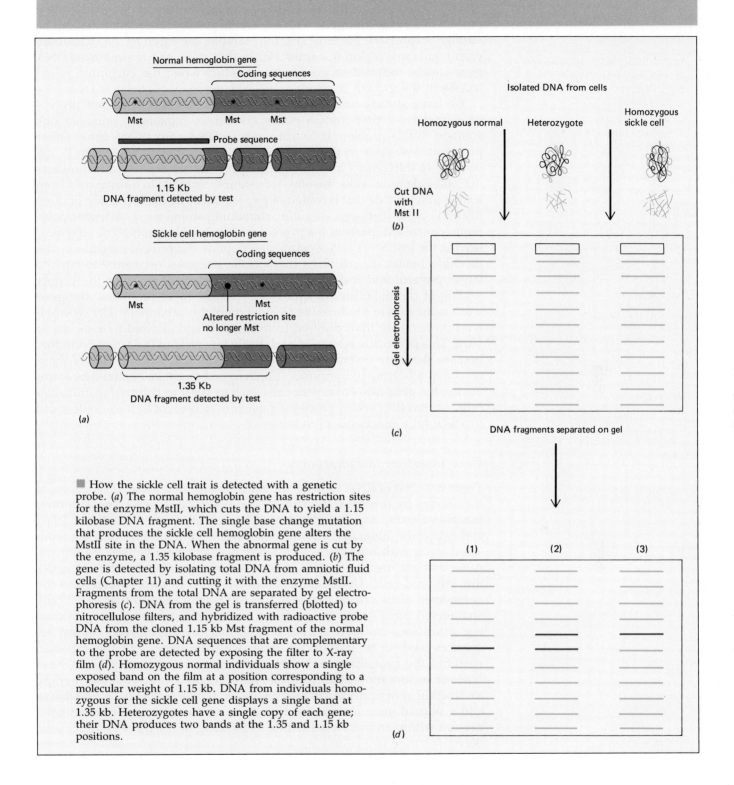

How the sickle cell trait is detected with a genetic probe. (*a*) The normal hemoglobin gene has restriction sites for the enzyme MstII, which cuts the DNA to yield a 1.15 kilobase DNA fragment. The single base change mutation that produces the sickle cell hemoglobin gene alters the MstII site in the DNA. When the abnormal gene is cut by the enzyme, a 1.35 kilobase fragment is produced. (*b*) The gene is detected by isolating total DNA from amniotic fluid cells (Chapter 11) and cutting it with the enzyme MstII. Fragments from the total DNA are separated by gel electrophoresis (*c*). DNA from the gel is transferred (blotted) to nitrocellulose filters, and hybridized with radioactive probe DNA from the cloned 1.15 kb Mst fragment of the normal hemoglobin gene. DNA sequences that are complementary to the probe are detected by exposing the filter to X-ray film (*d*). Homozygous normal individuals show a single exposed band on the film at a position corresponding to a molecular weight of 1.15 kb. DNA from individuals homozygous for the sickle cell gene displays a single band at 1.35 kb. Heterozygotes have a single copy of each gene; their DNA produces two bands at the 1.35 and 1.15 kb positions.

Expression of Recombinant DNA in Bacteria

Even if a gene has been isolated and successfully introduced into *E. coli*, the bacterium does not necessarily make its protein in large quantities. Several obstacles stand in the way of producing gene products of higher organisms in bacteria. One factor is that the gene has to be correctly associated with an appropriate set of regulatory and promoter sequences that the bacterial RNA polymerase can recognize. Recall from Chapter 13 that the regulatory regions of prokaryotic and eukaryotic genes are quite different. A usual approach to this problem is to combine the amino acid coding portion of a eukaryotic gene with a bacterial promoter sequence that can be strongly expressed. Some genes, for example, are fused to the lactose operon regulatory region (Chapter 14). In that state, the recombinant DNA gene can be induced to produce its product when the bacterium is fed lactose in the growth medium.

We have already discussed the fact that bacterial cells cannot process RNA molecules that contain eukaryotic intron sequences, and that one solution to this problem is to introduce a cDNA copy of the gene. Other problems may arise in the expression of a recombinant protein in *E. coli* because of differences in the ways the proteins are expressed in prokaryotic and eukaryotic cells. Insulin, for example, is made in human cells from a large polypeptide that is folded in a specific way by the formation of three disulfide bonds between six sulfur-containing amino acids. After the polypeptide is folded, parts of the protein are removed by proteolytic enzymes, leaving the insulin as two separate polypeptide chains held together by the disulfide bonds. *E. coli* lacks the specific enzymes necessary to cut the larger protein, and it does not have the mechanism to cause the proper folding of the molecule. In order to overcome these problems, the gene was engineered to produce the two polypeptides separately. The recombinant proteins are then purified from the cells and allowed to associate in vitro. This procedure results in a relatively low yield of the active hormone, because there are several ways that insulin can fold but only one of these results in a functional hormone. A better yield might be obtained by introducing the gene into eukaryotic cells such as yeast or cultured mammalian cells, so that the protein processing machinery present in those cells could produce fully functional proteins.

Gene Insertion in Eukaryotes

There is a class of RNA viruses, the **retroviruses,** that make DNA copies of themselves by reverse transcription (see Focus on Reverse Transcription, Jumping Genes, and Pseudogenes, Chapter 13). Sometimes those DNA copies become integrated into the host chromosomes, where they are replicated along with host DNA. Altered mouse leukemia viruses are retroviruses that are used as vectors in order to incorporate recombinant genes into cultured cells. Under certain conditions genes incorporated into the engineered virus can be expressed in the animal cells so that genetically engineered proteins can be produced. A major disadvantage of introducing genes into cultured animal cells is that the appropriate control sequences have to be introduced along with the gene, and the expected yields of the proteins encoded by the viruses are generally low. These types of vectors show some promise, however, as a means of **gene therapy** for treating genetic disorders in individuals. It may someday be possible to add a normal gene incorporated in the virus to correct a defective gene present within an individual with a debilitating or life-threatening genetic defect.

Transgenic Animals

One new approach to genetic engineering of animal proteins is to use live animals to produce the recombinant protein. A system that shows considerable promise for both research and commercial applications is the production of **transgenic animals.** This usually involves microinjecting the DNA of a particular gene into the nucleus of a recipient fertilized egg cell (Figure 15–13). The microinjected eggs are then implanted into the uterus of a female and allowed to develop. In one study of this type, the gene for growth hormone was isolated from a library of DNA from the rat and

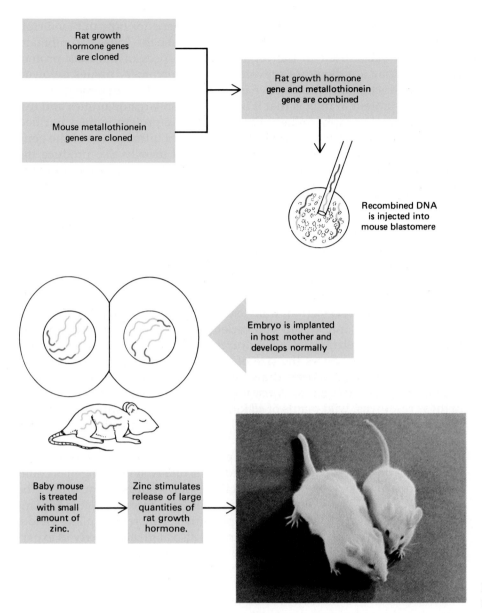

FIGURE 15–13 How to make a giant mouse.

combined with the promoter region of a mouse gene that normally produces metallothionein, a protein whose synthesis is stimulated by the presence of heavy metals such as zinc. The metallothionein regulatory sequences were used as a switch to turn the production of rat growth hormone on and off at will. After the engineered gene was injected into fertilized mouse egg cells, the eggs were implanted into the uterus of a mouse and allowed to develop into embryos. In the embryos in which the gene transplant had been successful, exposure to small amounts of zinc as they developed increased growth. One mouse, which developed from an egg that had received two copies of the growth hormone gene, grew to more than double the normal size. As might be expected, such mice are also able to transmit their increased growth capability to their offspring. These transgenic offspring have already been shown to have valuable research applications in studies involving regulation of gene expression (Chapter 16), the function of the immune system, genetic diseases, viral diseases, and the genes responsible for the development of cancer.

Transgenic animals have recently been used to develop strains that secrete important proteins in milk. The gene for tissue plasminogen activator (TPA), a protein that dissolves blood clots that cause heart attacks, has been introduced into transgenic mice, and the gene for human blood clotting factor has been similarly introduced into sheep. These recombinant genes have been fused to the regulatory sequences of the milk protein genes and they are therefore activated only in mammary tissues involved in milk production and are inactive in other tissues. The advantage of producing the protein in milk is that it is produced in large quantities and can be harvested by simply milking the animal. The protein is then purified from the milk. The animals are not harmed by the introduction of the gene and since the progeny of the transgenic animal usually also produce the recombinant protein, transgenic strains can be established simply by breeding the animals.

Engineering Plants

Plants have been selectively bred for thousands of years. The success of such efforts depends upon the presence of desirable traits in the variety of plant being selected or in closely related wild or domesticated plants whose traits can be transferred by crossbreeding. Even primitive varieties of cultivated plants may have certain traits, such as disease resistance, that could be advantageously introduced into varieties more suited to modern needs. However, many relatives of agricultural plants are rapidly becoming extinct. This greatly reduces the size of the potential gene pool from which agricultural researchers may draw. Indeed, just when their genetic resources are most needed, and just when our technologies promise access to these resources, wild plants of *all* kinds are threatened with extinction as the last available agricultural land is brought into cultivation to feed the exploding human population.

If genes could be introduced into plants from strains or species with which they do not ordinarily interbreed, or if totally synthetic genes could be introduced into them, the possibilities for improvement would be greatly increased. Much research funding has been made available to plant geneticists because of the economic potential of increased plant yields. Geneticists working with plants are perhaps also at greater liberty to experiment with new techniques than those working with animals, for manipulation of plant genes does not usually demand the same type of ethical considerations.

Unfortunately, a suitable vector for the introduction of recombinant genes into many types of plant cells has proved very difficult to find. The most widely used vector system employs the crown gall bacterium, *Agrobacterium tumefaciens*, which normally produces plant tumors (Figure 15–14). It does this by introducing a special plasmid, called the Ti (for *tumor inducing*) plasmid, into the cells of its host. The plasmid induces abnormal growth by forcing the plant cells to produce elevated levels of a plant growth hormone, cytokinin (Chapter 37). The plasmid also diverts the metabolism of the host cells to produce substances known as opines, which are simple derivatives of amino and keto acids and are the preferred food sources for the bacterium.

The Ti plasmid has been used as a vector to insert genes into plant cells. It is possible to "disarm" the plasmid so that it does not induce tumor formation. The cells into which the altered virus is introduced are essentially normal except for the genes that the experimenter has inserted. It has been shown that genes placed in the plant genome in this fashion are transmitted sexually via seeds to the next generation, but they can also be propagated asexually if desired.

A major problem with the Ti vector is that only dicotyledonous plants (Chapter 28) can be transformed by the plasmid. Unfortunately, the grain plants that are the main food source for humans belong to the group of monocotyledonous plants, and these are unaffected by Ti. Intensive re-

FIGURE 15–14 A crown gall tumor growing on a tree. The growth of this tumor is induced by a plasmid carried by *Agrobacterium tumefaciens*.

search is therefore under way to develop vector systems that will work in monocotyledonous plants. One approach has been the development of a genetic "shotgun." Microscopic metal fragments are coated with DNA, and these are then shot into plant cells in order to penetrate the cell walls. Some of the cells retain the DNA and are transformed by it. Those cells can then be cultured and used to regenerate an entire plant (Chapter 16).

An additional complication of plant genetic engineering is that a number of important plant protein genes are located on the DNA of the chloroplasts (Chapter 4). Chloroplasts are essential in photosynthesis, and photosynthesis is the basis for plant productivity. Obviously, it would be useful to develop methods for changing the portion of the chloroplast information that resides within the organelle itself. Methods of chloroplast engineering are currently the focus of intense research interest.

Recombinant DNA Regulations

People who have experienced the direct applications of recombinant DNA technology today would undoubtedly agree that those developments have been important and beneficial. In the 1970s when the new technology was introduced, however, many scientists regarded the potential misuses as being at least equally significant. There was the possibility that an organism with undesirable ecological or other effects might be produced, not by design, but by accident. Totally new strains of bacteria or other organisms, with which the world of life has no previous experience, might be difficult to control. This was recognized immediately by those who developed the recombinant DNA methods and led them to insist on stringent guidelines for making the new technology safe.

Recent history has failed to bear out these genetic worries. Experiments over the past 15 years in thousands of university and industrial laboratories have demonstrated that, at least until now, recombinant DNA manipulations can be carried out safely. One of the main concerns, the accidental release of laboratory bacterial strains containing dangerous genes into the environment, has turned out to be groundless. Laboratory strains of *E. coli* are poor competition for the wild strains in the outside world and quickly perish. In those respects, the fears of accidentally cloning or releasing a dangerous gene into the environment from the laboratory seem to be laid to rest. This does not mean, however, that *intentional* manipulations of dangerous genes are not a possibility.

Most scientists today, however, recognize the importance of recombinant DNA technology and agree that the perceived threat to mankind and the environment was overestimated. Many of the restrictive guidelines for using recombinant DNA have been removed as the safety of many of the experiments has been established. Stringent restrictions still exist, however, in certain areas of recombinant DNA research where dangers are known to exist and where there are still questions about the effects that might be produced on the environment. These restrictions are most evident in research that proposes to introduce recombinant organisms into the wild, such as agricultural strains of plants whose seeds or pollen might be spread in an uncontrolled manner. A great deal of research activity is now concentrated on determining the effects of introducing recombinant organisms in the wild; within the next few years we should be able to determine more clearly whether such dangers exist.

■ SUMMARY

I. Recombinant DNA technology is concerned with isolating and amplifying specific sequences of DNA from an organism. This is done by incorporating them into re- combinant vector DNA molecules, which can then be propagated and amplified in organisms such as *E. coli*.
A. Restriction enzymes are used to cut DNA isolated

from an organism into specific fragments in order that they may be incorporated into the vector molecules.

1. Each type of restriction enzyme recognizes and cuts DNA at a highly specific base sequence. Those sequences are usually 4 to 23 bases long, depending on the enzyme.
2. Many restriction enzymes cleave DNA sequences so that single-stranded cut ends are produced and these are complementary to each other ("sticky ends").

B. The most common recombinant DNA vectors are constructed from naturally occurring circular DNA molecules called plasmids or from bacterial viruses called bacteriophages; both of these are usually found in *E. coli.*

C. Recombinant DNA molecules are often constructed by allowing the complementary ends of a DNA fragment and a plasmid (which have both been cut with the same restriction enzyme) to associate by complementary base pairing. The DNA strands are then covalently linked by DNA ligase, thus forming the recombinant molecule.

D. Single genes are isolated from recombinant DNA libraries, which are constructed from fragments of DNA inserted into appropriate vector molecules.

1. Genomic libraries are formed from the total DNA from an organism. Genes that are present in recombinant DNA genomic libraries from eukaryotes will contain introns. Those genes can be amplified in *E. coli* along with the molecule that contains them, but the protein coding regions will not be expressed in *E. coli.*
2. cDNA libraries are formed by first making DNA copies of mRNA isolated from eukaryotic cells; these are then incorporated into recombinant DNA vectors. Since the introns have been removed from the mRNA molecules, eukaryotic genes in cDNA libraries can sometimes be expressed in *E. coli* to make a protein product.

E. Analysis of a cloned sequence can yield useful information about the gene and its protein, and can enable investigators to identify and subclone DNA fragments that can be used as molecular probes for many purposes.

1. The first step in analyzing a cloned DNA sequence is to construct a restriction map, thereby identifying sites that are cut by specific restriction enzymes. These can be used for subcloning the DNA and as landmarks for determining the orientation and location of the internal gene sequences.
2. Determining the nucleotide sequence of a cloned

DNA fragment gives information about the structure of the gene it contains and about the probable amino acid sequence of the encoded proteins.

3. Subcloned restriction fragments of a cloned gene can be used as radioactive DNA probes to identify related complementary DNA and RNA sequences from other organisms. The DNA or RNA to be identified is separated from other nucleic acids by gel electrophoresis and then blotted onto special paper. The radioactive DNA probe is then hybridized by complementary base pairing to the bound DNA, and the radioactive band or bands of DNA can be identified.

   a. DNA blotting methods are used in diagnosing genetic diseases, to determine whether persons are heterozygous for diseases such as cystic fibrosis and potentially for many other genetic disorders.
   b. DNA blotting is used to identify restriction fragment length polymorphisms, which can be used to determine the relatedness of individuals in a population.

II. Genetic engineering is a technology that uses genetic and recombinant DNA methods to devise new combinations of genes in order to produce improved pharmaceutical and agricultural products.

A. Genes isolated from one organism can be modified and expressed in other organisms ranging from *E. coli* to transgenic plants and animals. The yield of a functional protein that is coded for by an introduced gene usually depends on whether the host organism can properly process the genetic information, and on the characteristics of the particular protein product derived from the gene.

1. Expression of eukaryotic proteins in bacteria, such as *E. coli*, requires that the gene be linked to regulatory elements that the bacterium can recognize. In addition, bacterial cells do not contain many of the enzymes needed for the posttranslational processing of eukaryotic proteins, such as hormones, which often need to be modified after the protein is synthesized.
2. Expression of eukaryotic genes in eukaryotic organisms shows great promise, since the processing and modification machinery for eukaryotic proteins is already present in plant and animal cells.

   a. Production of important pharmaceutical products can be engineered in transgenic animals so that the products are secreted in milk.
   b. Genetic engineering of plants and domestic animals holds the promise of increasing the availability of food.

■ **POST-TEST**

1. Recombinant DNA methods involve the isolation of specific DNA sequences by combining a DNA fragment from one organism with other DNA molecules so that the _____ molecule can be amplified in a organism such as *E. coli.*

2. _____ enzymes are proteins that cut DNA at highly specific base sequences.

3. Many restriction enzymes that cut DNA do so at base sequences that are _____, yielding DNA fragments that have single-stranded _____ or "sticky" ends.

4. Recombinant DNA vectors used in *E. coli* are usually in the form of small, circular DNA molecules called _____ or bacterial viruses called _____.

5. A recombinant DNA plasmid is usually constructed by allowing the complementary ends of a DNA fragment and a plasmid that have been cut with the _____ restriction enzyme associate by hydrogen bonding. The ends of the molecules are then covalently linked together by _____.

6. Gene _____ are composed of plasmids or bacteriophages that contain a collection of DNA fragments that collectively represent all of the genes from an organism.

7. A _____ library is composed of the total DNA of an organism. If the DNA is from a eukaryotic cell many of the genes will contain _____ sequences.

8. A _____ library is composed of DNA fragments made from mRNA molecules that have been copied by _____ _____.

9. A first step in analyzing a DNA sequence once it has been cloned is to construct a _____ _____ of the DNA in order to locate landmark restriction sites, which can be used for further studies.

10. DNA sequencing today usually involves making copies of a cloned DNA sequence under conditions that _____ the DNA chain at points where specific bases are located.

11. Short fragments of a cloned DNA sequence can be used as radioactive DNA _____, which can be used to locate complementary nucleic fragments from other organisms by blot hybridization methods.

12. A problem with producing human proteins such as insulin in bacteria such as *E. coli* is that the bacterial cells do not contain the necessary machinery for _____ processing of the proteins.

13. Plants and animals that have been modified by the introduction of recombinant DNA are referred to as _____ strains.

■ REVIEW QUESTIONS

1. Explain how recombinant DNA molecules are constructed.

2. Explain how gene libraries are constructed and discuss the differences between a genomic library and a cDNA library.

3. Discuss how a gene is restriction mapped and how the sequence of a DNA fragment would be determined.

4. Explain how a fragment of a cloned DNA sequence can be used as a DNA probe for genetic diseases.

5. Discuss how a eukaryotic gene can be expressed in a bacterium such as *E. coli*. Explain what problems might have to be overcome in order for the protein to be produced in the bacterium. How might these problems be solved by using transgenic plants or animals?

■ RECOMMENDED READINGS

Darnell, J., Lodish, H., and Baltimore, D. *Molecular Cell Biology.* Scientific American Books, W.H. Freeman, New York, 1986. Chapters 6 and 7. Well-written accounts of the tools of molecular biology and recombinant DNA, starting with the cells and organisms used in the technology and proceeding to its methods and applications.

Drlica, K. *Understanding DNA and Gene Cloning: A Guide for the Curious.* John Wiley and Sons, New York, 1984. A coverage of the methods of recombinant DNA for those with little experience in chemistry.

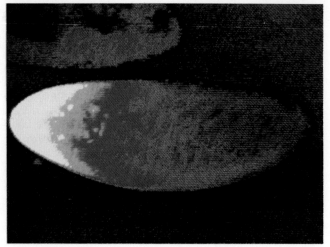

Computer-enhanced photograph of a *Drosophila* egg

16

The Molecular Genetics of Development

■ LEARNING OBJECTIVES

After you have read this chapter you should be able to:

1. Distinguish between cellular determination and cellular differentiation.
2. Discuss the meaning of morphogenesis.
3. Describe the kinds of experiments that indicate that at least some differentiated plant cells and animal nuclei are totipotent. Discuss how these findings support the idea of nuclear equivalence.
4. Discuss why the choice of a suitable organism is important in the study of the genetic control of development.
5. Indicate the features of development and genetics of *Drosophila*, *Caenorhabditis*, and the mouse that have made these organisms so valuable to researchers.
6. Distinguish between maternal effect genes, zygotic genes, and homeotic genes in *Drosophila*.
7. Define the phenomena of induction and programmed cell death.
8. Discuss how certain genes may function as genetic switches in development and provide some examples of such genes.
9. Describe how transgenic organisms are shedding light on developmental processes today.
10. Describe some examples of homeotic-like transformations in plants.
11. Discuss some of the known exceptions to the general phenomenon of nuclear equivalence.

The human body contains over 200 recognizably different types of cells (Figure 16–1). Combinations of those cells are organized into remarkably diverse and complex structures such as the eye, the hand, and the brain, each one capable of carrying out many sophisticated activities. Most remarkable of all, however, is the fact that all of the structures of the body and the different cells within them are descended from a single fertilized egg.

The process by which the progeny of an egg cell specialize and organize into a complex organism is called **development,** the study of which includes some of the most interesting and difficult problems in biology today.

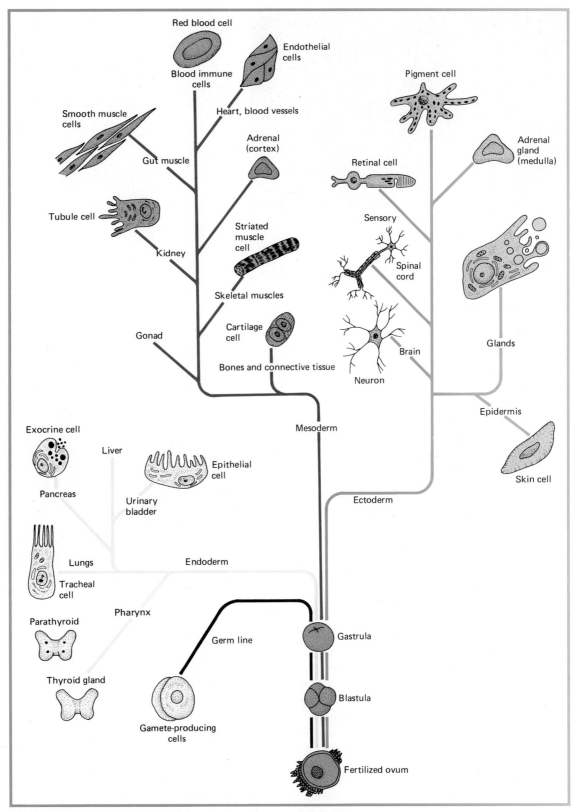

FIGURE 16–1 Lineages of differentiated cells in animals. Repeated divisions of the fertilized egg result in the formation of tissues (see Chapter 51 for a discussion of how these are formed) from which groups of specialized cells will be produced. Germ line cells (cells that produce the gametes) are set aside early in the development process. Somatic cells progress along the developmental pathways by making a series of developmental "decisions" that progressively determine the fates of different lines of cells.

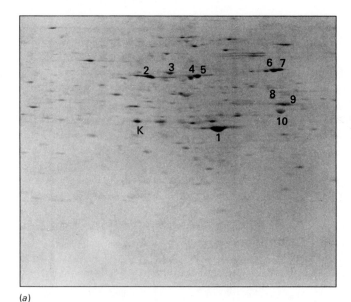

(a)

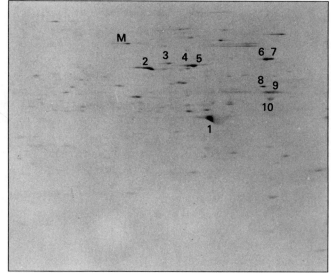

(b)

FIGURE 16–2 Proteins from different tissues of the mouse. The spots in the photographs are proteins from (*a*) kidney, (*b*) muscle, and (*c*) liver cells. The proteins were separated by two-dimensional gel electrophoresis, a method that separates the proteins in the horizontal direction by their electric charge followed by a second separation in the vertical direction by molecular weight. Several hundred proteins can be distinguished in each panel. The spots that are labeled with numbers are present in all tissues, but notice that many of them are present in different amounts from one tissue to another. The proteins that are labeled with letters are only found in that specific tissue.

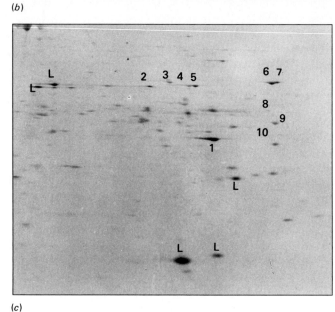

(c)

All multicellular plants and animals undergo complex patterns of development. The root cells of plants, for example, have structures and functions that are very different from those of the various types of cells located in plant leaves (Figure 4–14). Remarkable diversity can also be found at the molecular level; most strikingly, each type of plant or animal cell makes a highly specific set of proteins (Figure 16–2). In some cases, such as the protein hemoglobin in red blood cells, that one cell-specific protein may make up more than 90% of the total mass of protein in the cell. Other cells may have a complement of cell-specific proteins that occur in small amounts but still play an essential role. However, since certain proteins are required in every type of cell (all cells, for example, require certain enzymes for glycolysis), cell-specific proteins usually make up only a fraction of the total number of different kinds of proteins.

The final step leading to cell specialization during development is called **differentiation.** Differentiation, however, is not an instantaneous process. During the many cell divisions required for the fertilized egg to develop into the adult, groups of cells will become gradually committed to particular patterns of gene activity. This process of gradual commitment is called **determination.** For example, at a certain stage of development, a group of cells will be determined to become a part of the arm. Later, some of the progeny of those cells will differentiate to become skeletal muscle cells.

The details of the genetic mechanisms that control differentiation are not the only parts of the developmental puzzle. Still other complex questions emerge such as how muscle cells form a specific arm muscle, such as the biceps. This development of form is called **morphogenesis** and is also under genetic control. Like cellular differentiation, morphogenesis occurs over a series of stages, referred to as **pattern formation.** During development, cells in specific locations within the embryo become progressively organized into structures that are the beginnings of the adult body plan. Regulatory genes (Chapter 14) also play an important role in the process of pattern formation. These genes are activated by signals from other genes in a complex network of interactions necessary to specify the developmental fate of cells at different locations in the embryo. Not all of the controlling signals lie within the borders of each individual cell. Some genes (and their products) involved in cellular differentiation and pattern formation also require signals from the environment, which progressively tell the cell "where it is" and "what it should become," a process that requires that controlling signals be emitted from neighboring cells.

Until recently, little was known about how genes act in controlling development. Gene interactions in development are too complex to unravel using only traditional genetic methods. Today the tools of recombinant DNA methodology and the genetic analysis of mutant organisms with altered developmental patterns are being combined with more traditional descriptive and experimental methods to derive fresh insights into the role of gene regulation in one of the most fascinating and complex areas of biology.

THE PROCESS OF DIFFERENTIATION

One explanation for the observation that each type of differentiated cell makes a unique set of proteins might be that during development each group of cells loses the genes it does not need and retains only those genes that are required. With just a few exceptions, however, this does not seem to be true. According to the concept of **nuclear equivalence,** the nuclei of essentially all differentiated adult cells of an individual are genetically (but not necessarily metabolically) identical to each other and to the nucleus of the fertilized egg cell from which they were descended. This means that virtually all of the **somatic**[1] cells in an adult have the same genes; they are simply expressed in different tissues in different ways.

The evidence for the idea of nuclear equivalence comes from cases in which differentiated cells or their nuclei have been found to be capable of supporting normal development. Such cells or nuclei are said to be **totipotent.**

Totipotency in Plants and Animals

In plants it is possible to demonstrate that at least some differentiated cells can be induced to become the equivalent of embryonic cells (Figure 16–3). **Tissue culture** techniques are used to isolate cells from certain plants and allow them to grow in a nutrient medium. In some of the first experiments, single root cells of the carrot were induced to divide in a liquid nutrient

---

[1]Somatic cells are cells of the body and are distinguished from **germ line** cells, those cells that will ultimately give rise to a new generation. The distinction between somatic cells and germ line cells is not clear-cut in plants. In animals, however, germ line cells, whose descendants will ultimately undergo meiosis and differentiate as gametes, are generally set aside early in development.

FIGURE 16–3 Development of a carrot plant from differentiated somatic cells. Discs of phloem cells, which are specialized for nutrient transport, were isolated from carrot root tissues. When the cells were cultured in a liquid nutrient medium, clumps of dedifferentiated cells developed from individual phloem cells. These clumps (embryoids) closely resembled plant embryos in their early stages of development and then progressed to form embryonic shoots and roots. Transferring the embryonic tissue to a solid nutrient medium stimulated the tissues to form plantlets, which could then be grown into mature plants.

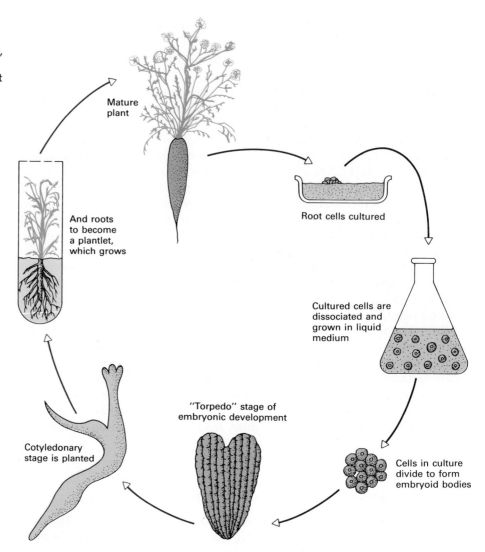

Mature plant

Root cells cultured

Cultured cells are dissociated and grown in liquid medium

And roots to become a plantlet, which grows

"Torpedo" stage of embryonic development

Cotyledonary stage is planted

Cells in culture divide to form embryoid bodies

medium and to form groups of cells called "embryoid" (embryo-like) bodies. These clumps of dividing cells could then be transferred to an agar medium, which provides nutrients plus a solid supporting structure for the developing plant cells. Upon transfer to the agar, some of the embryoid cells gave rise to roots, stems, and leaves. The resulting "plantlets" could then be transplanted to soil where they would ultimately develop into adult plants capable of producing flowers and viable seeds. The methods of plant tissue culture are now extensively used to produce genetically engineered plants, for they allow the regeneration of whole plants from cells that have incorporated recombinant DNA molecules (see Chapter 32).

Similar experiments have been attempted with animal cells, but so far it has not been possible to induce a fully differentiated somatic cell to behave like a zygote. Instead, it has been possible to test whether steps in the process of determination are reversible by transplanting the *nucleus* of a cell in a relatively late stage of development into an egg cell whose own nucleus has been destroyed (Figure 16–4). In those experiments, nuclei from amphibian cells at different stages of development were transplanted into egg cells. Some of the transplants proceeded normally through a number of developmental stages, and a few even developed into normal tadpoles. As a rule, the nuclei transplanted from cells at earlier stages were most likely to support development to the tadpole stage. As the fate of the cells became more and more determined, the probability that a transplanted nucleus could control normal development diminished rapidly. In a few

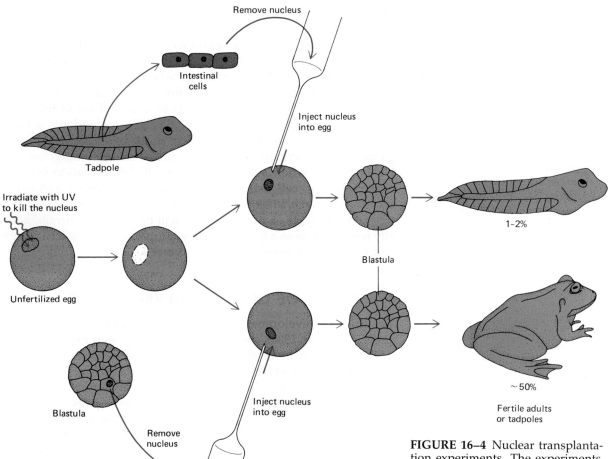

FIGURE 16–4 Nuclear transplantation experiments. The experiments of Briggs and King and later J. Gurdon showed that a nucleus from a differentiated amphibian cell could program development. This was done by injecting it into an egg whose own nucleus had been destroyed by ultraviolet radiation. The probability of success of the procedure was dependent on the developmental stage of the transplanted nucleus. If a nucleus is taken from a cell at the blastula stage of development (when cell division has produced about 1000 cells formed in the shape of a ball), there is a high probability that it will program normal development, resulting in a fertile adult. Most trials using nuclei from tadpole intestinal cells (a much later developmental stage) resulted in no growth, probably as a result of damage to the egg or the nucleus by the procedure. In a small number of trials, however, normal development would proceed until the tadpole stage, indicating that the necessary genes to program development to that point were still present.

cases, nuclei isolated from the specialized intestinal cells of a tadpole were able to direct development to another tadpole stage. This occurred infrequently, but in such experiments, success counts more than failure, and one can safely conclude that at least some nuclei of differentiated cells are in fact totipotent and have not lost any genetic material.

There are several important interpretations of these results. In the cases where specialized plant cells and animal nuclei are totipotent, it is clear that genes have not been lost as a consequence of development. That is, genes that were apparently inactive were, in fact, capable of being reactivated when the cells or nuclei were placed in a new environment. Second, even if no genetic material is lost during development, the nuclei of cells do undergo metabolic changes that make it progressively more difficult to remain in a totipotent state. This is especially true of animal nuclei, although various kinds of animals differ considerably in this regard.

Differential Gene Activity

Since genes do not appear to be regularly lost during development, it follows that the differences in the molecular composition of cells must occur by *regulating the activities of different genes*. This process of developmental gene regulation is often referred to as **differential gene activity**. During development, certain sets of genes in the cells are inactivated, and others are activated in order to produce the unique sets of proteins found in each tissue. Even expression of genes that are active in all cells can be regulated during development so that the *quantity* of each product varies from one tissue type to another. (The contractile protein actin, for example, is found

FIGURE 16–6 Developmental stages of *Drosophila* from the egg to the adult fly.

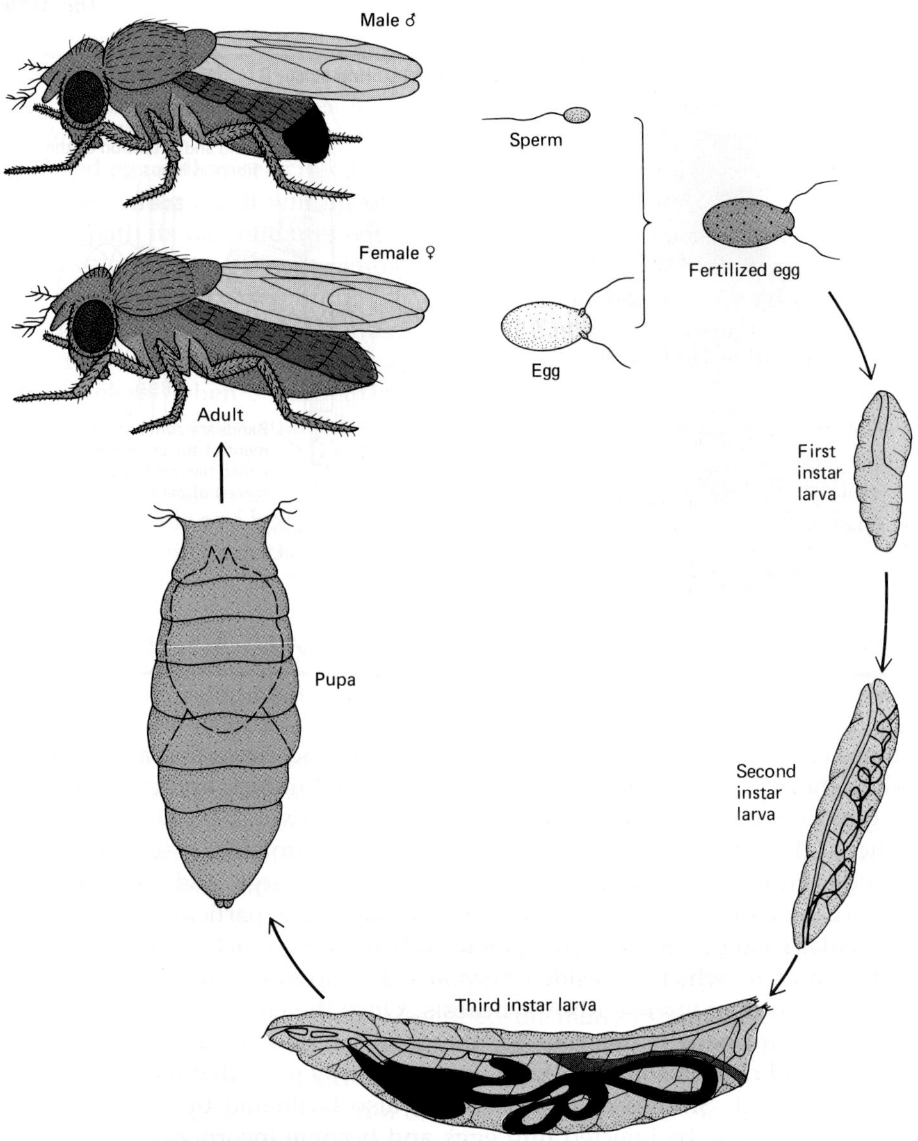

Male ♂

Sperm

Egg

Fertilized egg

First instar larva

Second instar larva

Third instar larva

Pupa

Adult

Female ♀

FIGURE 16–7 Diagram of the imaginal discs in the *Drosophila* larva illustrates the structures they give rise to in the adult fly.

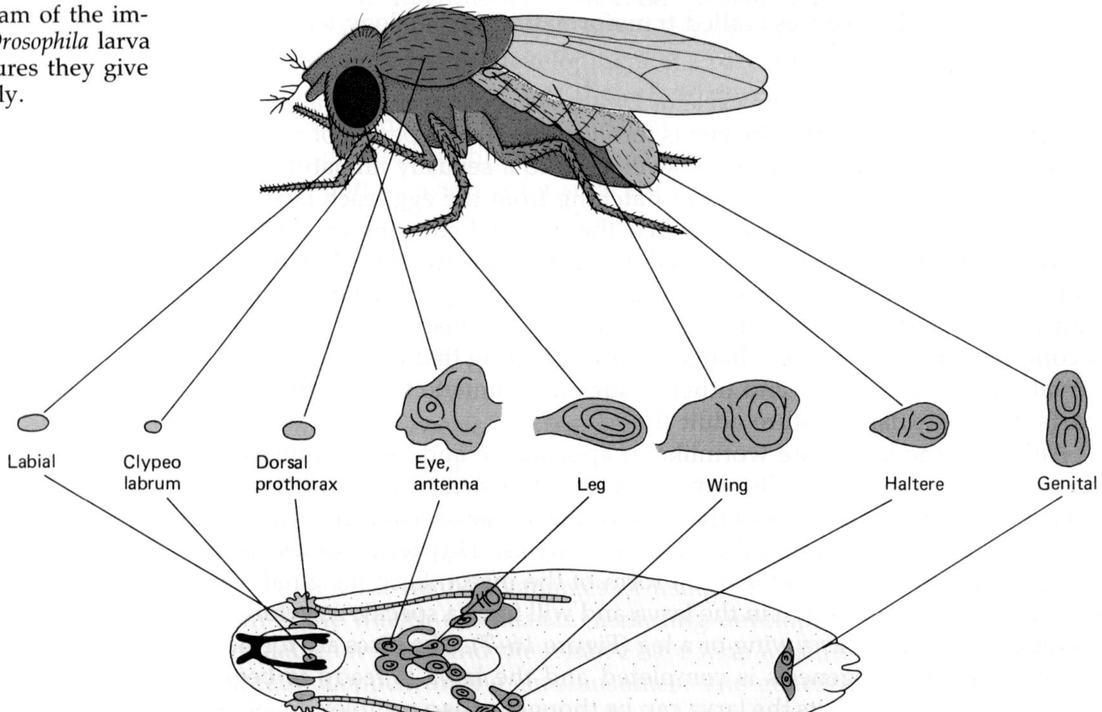

Labial

Clypeo labrum

Dorsal prothorax

Eye, antenna

Leg

Wing

Haltere

Genital

for the adult fly. The organization of the precursors of the adult structures, like the imaginal discs, is under complex genetic control. Over 50 different genes have been identified to date that specify the formation of the discs, their positions within the larva, and their ultimate functions within the adult fly. Those genes have been identified through mutations that either prevent certain discs from forming or alter their structure or ultimate fate.

GENETIC CONTROL OF DROSOPHILA DEVELOPMENT.   Many types of developmental mutants have been identified in *Drosophila*. In our discussion we pay particular attention to those that affect the segmented body plan of the organism.

Maternal Effect Genes.   The earliest stages of *Drosophila* development are controlled by maternal genes that act to organize the structure of the egg cell. As the egg develops in the ovary of the female, stores of messenger RNA (mRNA) along with yolk proteins and other cytoplasmic molecules are passed into it from the surrounding maternal cells. Therefore, all of these mRNA molecules are transcribed exclusively from genes found in the mother. The genes represented by those mRNA molecules are referred to as **maternal effect** genes. By analyzing mutants that are defective in those genes, it has been found that many of them are involved in establishing the polarity of the embryo, by designating which parts of the egg are dorsal or ventral and which are anterior or posterior (see Focus on Body Plan and Symmetry, Chapter 29). For example, maternal effect mutations are known that result in an embryo that has two heads or two posterior ends, due to the absence of specific signals in the egg. The mRNA transcripts from some of the maternal effect genes can be identified by their ability to hybridize with radioactive DNA probes from the cloned genes. Alternatively, their protein products can be identified by antibodies that specifically bind to them. In some cases, the mRNA or its protein product can be seen to form a concentration gradient in the embryo (Figure 16–8). These gradients may provide positional information that specifies the location of each nucleus or cell within the embryo. That information may then be interpreted by a cell as signals specifying the developmental path it should follow.

In many cases, the effects of maternal effect mutations can be reversed by injecting normal maternal messenger RNA into the mutant embryo. When this is done, the fly will develop normally, indicating that the gene product is needed for a short time only at the earliest stages of development.

Zygotic Genes.   Immediately after fertilization, the zygote nucleus in the *Drosophila* egg divides, thus beginning a remarkable series of 13 mitotic divisions (Figure 16–8b). Each of these divisions takes only 5 or 10 minutes, which means that the DNA in the nuclei is replicated constantly at a very rapid rate. During that time, the nuclei do not transcribe RNA. Cytokinesis does not take place, and the several thousand nuclei that are produced by those divisions remain at the center of the egg until division number eight. At that time, most of the nuclei start to migrate to the periphery of the egg. Membranes begin to form around the nuclei that are now in the periphery. Embryonic mRNA production begins, and some of the **zygotic genes** begin to be expressed. (It is customary to refer to the genes of the embryo itself as zygotic genes, even though the embryo is no longer a zygote). At that time, certain zygotic genes begin to extend the developmental program beyond the pattern established by the maternal genome. So far at least 24 zygotic **segmentation genes** have been identified that are responsible for generating a repeating pattern of segments within the embryo (Figures 16–8, 16–9, and 16–10). The segmentation genes appear to fall into three classes representing a rough hierarchy of gene action (Table 16–1). The **"gap genes"** are apparently the first sets of zygotic genes to act. These genes seem to inter-

Morphology   *Gene activity*

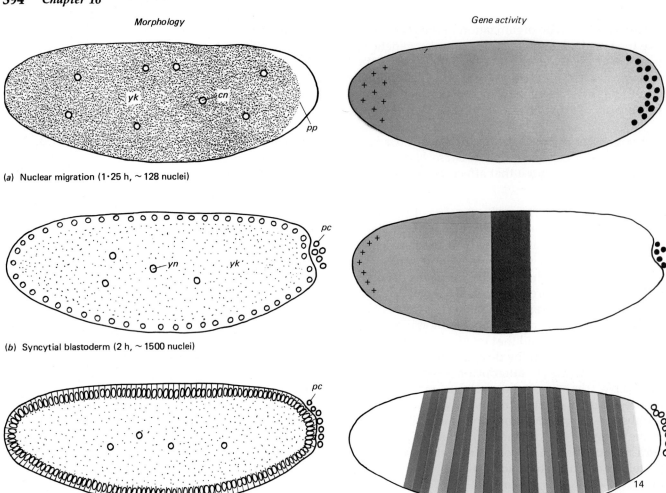

(a) Nuclear migration (1·25 h, ~ 128 nuclei)

(b) Syncytial blastoderm (2 h, ~ 1500 nuclei)

(c) Cellular blastoderm (2·5 h, ~5000 cells)

FIGURE 16–8 The early development of the *Drosophila* egg. The diagrams on the left show the structure of the embryo at different times after fertilization. The panels on the right show the patterns of activity of particular genes at each of those stages. (a) 1.25 hours (about 128 nuclei). Between the seventh and eighth nuclear divisions, the nuclei start to migrate to the periphery of the egg. The products of several maternal genes can be located in different regions of the egg. The crosses mark the location of maternal mRNA transcribed from a gene which defines the anterior (head) end of the egg. The dots represent the location of mRNA transcribed from a gene which specifies how cells located in the posterior of the embryo will develop. The pink region represents a concentration gradient of a maternal mRNA extending from the anterior to the posterior end. The protein product of the mRNA molecules appears to be part of a system of determinants that organize the early pattern of development in the embryo. (b) The pattern at 2 hours (about 1500 nuclei). Most of the nuclei have reached the perimeter of the egg and have started to make their own messenger RNA. The maternal mRNA shown in pink in the previous panel is now being transcribed from the corresponding zygotic gene by the nuclei in the anterior part of the embryo. The mRNA from a zygotic gap gene is transcribed from cells in only one segment in the middle of the embryo. (c) The pattern at 2.5 hours (about 5000 cells). Membranes start to form around nuclei located at the perimeter of the egg. Messenger RNAs from two pair rule genes can be detected as a series of stripes around the embryo. These stripes mark the prepattern used to form the segments that are found in the mature larva. The boundaries of each stripe in the prepattern are defined by two or more different genes. The genes that form the prepattern affect another set of genes that defines the actual pattern of segments in the embryo.

pret the maternal anterior–posterior information in the egg and begin the organization of the segments. A mutation in one of the gap genes usually results in an embryo having one or more missing segments. The other two classes of segmentation genes do not act on small groups of segments but rather act on all of the segments. For example, mutations in the **"pair rule"** genes delete every other segment, while mutations in the **"segment polarity"** class of genes produce segments in which one part is missing and the remaining part is duplicated as a mirror image.

(a)

FIGURE 16–9 The location of the mRNA molecules transcribed from the engrailed gene, a gene that specifies the boundaries of each segment. The dots in the panels are silver grains produced on x-ray film by radioactive DNA from the cloned engrailed gene sequence that hybridized to its complementary mRNA in the embryo. The autoradiographs are photographed through a microscope in two ways: (a) through normal optics, which cause the silver grains to appear as black dots, or (b) through dark field optics, which reverse the image.

Each gene can be shown to have distinctive times and places in the embryo where it is active. The observed pattern of expression of the maternal and zygotic genes that control segmentation indicates that cells destined to form adult structures are determined by a progressive series of developmental decisions, according to the following model. First, the head-to-tail axis and then the dorsal and ventral regions of the embryo are determined by maternal segmentation genes that are thought to form gradients of **morphogens** in the egg. (A morphogen is a hypothetical chemical agent that affects the differentiation of cells and development of form.) Zygotic segmentation genes then respond to the amounts of various morphogens at each location to control the production of a series of segments from the head to the posterior region. Then, within each segment, other genes are activated that "read" the position of the segment and "interpret"

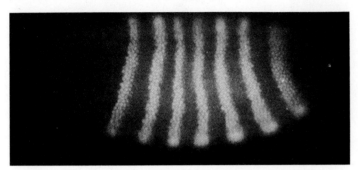

FIGURE 16–10 Locating the pattern of expression of the *fushi tarazu* gene. The activity of a gene that regulates development can also be detected by locating its protein product using fluorescent antibody molecules. The fluorescent probe molecules detect proteins only in the nucleus of each cell.

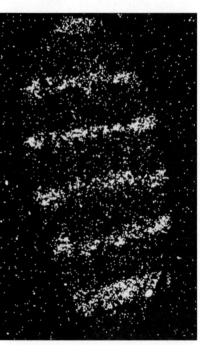

(b)

Table 16–1 **CLASSES OF GENES INVOLVED IN PATTERN FORMATION OF EMBRYONIC SEGMENTS IN *DROSOPHILA***

Type of Gene	Site of Gene Activity	Proposed Function(s) of Genes
Maternal effect genes	Maternal tissues	Initiate pattern formation by activating regulatory genes in nuclei in certain locations of embryo
Gap genes	Embryo	Cause alternate segments to be missing; some may influence activity of pair rule genes, segment polarity genes, and homeotic genes
Pair rule genes	Embryo	Cause parts of segments to be missing; some may influence activity of segment polarity genes and homeotic genes
Segment polarity genes	Embryo	Delete part of every segment; replace with mirror image of remaining structure; may influence activity of homeotic genes
Homeotic genes	Embryo	Control the identity of the segments; homeotic mutations cause parts of fly to form structures normally formed in other segments

that information in order to specify which body part that segment should become. Within each compartment, the position of each cell is further specified so that each cell now has a specific "address," which is designated by combinations of the activities of the regulatory genes.

One model for the interaction of the zygotic segmentation genes is that they act in sequence, with the gap genes acting first, then the pair rule genes, and finally the segment polarity genes. Each time a new group of genes acts, cells of a particular group become more finely restricted in the way that they will develop. As the embryo develops, it is progressively subdivided into smaller specified regions.

Homeotic Genes. The genes that actually designate the final adult structure formed by each of the imaginal discs mainly belong to a separate class of genes called **homeotic** genes. Because of their involvement in segment identity, mutations in these genes cause one body part to be substituted for

FIGURE 16–11 A homeotic mutant of *Drosophila*. (*a*) Head of normal fly and a fly with the Antennapedia mutation. (*b*) Head of fly with one type of mutation in the Antennapedia gene. This particular mutant is one of the more extreme forms of the Antennapedia gene. Most of the alleles produce only incomplete legs in place of the antennal structures.

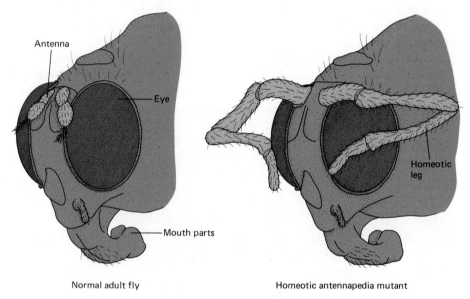

Normal adult fly   Homeotic antennapedia mutant

(*a*)

(*b*)

	1									10										20
Fly { Antp	Arg	Lys	Arg	Gly	Arg	Gln	Thr	Tyr	Thr	Arg	Tyr	Gln	Thr	Leu	Glu	Leu	Glu	Lys	Glu	Phe
ftz	Ser	Lys	Arg	Thr	Arg	Gln	Thr	Tyr	Thr	Arg	Tyr	Gln	Thr	Leu	Glu	Leu	Glu	Lys	Glu	Phe
Ubx	Arg	Arg	Arg	Gly	Arg	Gln	Thr	Tyr	Thr	Arg	Tyr	Gln	Thr	Leu	Glu	Leu	Glu	Lys	Glu	Phe
Frog	Arg	Arg	Arg	Gly	Arg	Gln	Ile	Tyr	Ser	Arg	Tyr	Gln	Thr	Leu	Glu	Leu	Glu	Lys	Glu	Phe

	21									30										40
Fly { Antp	His	Phe	Asn	Arg	Tyr	Leu	Thr	Arg	Arg	Arg	Arg	Ile	Glu	Ile	Ala	His	Ala	Leu	Cys	Leu
ftz	His	Phe	Asn	Arg	Tyr	Ile	Thr	Arg	Arg	Arg	Arg	Ile	Asp	Ile	Ala	Asn	Ala	Leu	Ser	Leu
Ubx	His	Thr	Asn	His	Tyr	Leu	Thr	Arg	Arg	Arg	Arg	Ile	Glu	Met	Ala	Tyr	Ala	Leu	Cys	Leu
Frog	Arg	Phe	Asn	Arg	Tyr	Leu	Thr	Arg	Arg	Arg	Arg	Ile	Glu	Ile	Ala	Asn	Ala	Leu	Cys	Leu

	41									50										60
Fly { Antp	Thr	Glu	Arg	Gln	Ile	Lys	Ile	Trp	Phe	Gln	Asn	Arg	Arg	Met	Lys	Trp	Lys	Lys	Glu	Asn
ftz	Ser	Glu	Arg	Gln	Ile	Lys	Ile	Trp	Phe	Gln	Asn	Arg	Arg	Met	Lys	Ser	Lys	Lys	Asp	Arg
Ubx	Thr	Glu	Arg	Gln	Ile	Lys	Ile	Trp	Phe	Gln	Asn	Arg	Arg	Met	Lys	Leu	Lys	Lys	Glu	Ile
Frog	Thr	Glu	Arg	Gln	Ile	Lys	Ile	Trp	Phe	Gln	Asn	Arg	Arg	Met	Lys	Trp	Lys	Lys	Glu	Arg

FIGURE 16–12 A comparison of the homeobox sequences of three *Drosophila* genes with the homologous sequence from an amphibian gene. About 60 amino acids are encoded by the 180 nucleotide homeobox sequence. The amino acids in the *Drosophila* sequences that are identical are shaded. The frog gene was identified and cloned by its ability to hybridize with the DNA from the *Drosophila* homeobox region. The amino acids of the frog DNA coding sequence that are identical to those of the *Drosophila* sequence are shown in color.

another and therefore produce some very peculiar changes in the adult. Among the most striking of these are the **Antennapedia** mutants, which have legs that grow from the head at a position where the antennae would normally be found (Figure 16–11). The original homeotic genes in *Drosophila* were first identified by the altered phenotypes that were produced by mutant alleles. When the DNA sequences of a number of homeotic genes were analyzed, it was discovered that there is a short DNA sequence of approximately 180 base pairs that is characteristic of many homeotic genes. This sequence has been termed the **homeobox** (Figure 16–12). Using the homeobox sequence of bases as a molecular probe made it possible to clone new homeotic genes in *Drosophila* that had not been previously identified. Surprisingly, the homeobox probe also detected homologous DNA sequences in a wide range of other organisms, including humans. This finding has generated considerable excitement because it suggests that the homeobox may be one of the first important clues for identifying and cloning genes that control development in higher organisms.

The homeobox sequences of a large number of genes have been determined. Comparisons of those DNA sequences have shown that the sequence itself is highly conserved during evolution and shows remarkable similarities between organisms as diverse as sea urchins, yeast, and humans. The first clues to the function of the proteins that are coded for by genes containing homeoboxes came from computer-generated searches. Those comparisons showed that the amino acid sequences coded for by the homeoboxes (known as homeodomains) are homologous to amino acid sequences in certain DNA-binding proteins. Most, if not all, of these "homeotic" proteins contain a helix-turn-helix sequence in the homeodomain that corresponds to a similar domain found in certain regulatory proteins in prokaryotic cells and their viruses (Chapter 14). Other evidence that the homeotic genes code for DNA-binding regulatory proteins comes from the finding that all of the *Drosophila* proteins that contain homeodomains accumulate in the nucleus. Some of the segmentation genes that act earlier in development code for a "zinc-finger" type of DNA binding regulatory protein (Chapter 14), while other segmentation gene proteins have other types of amino acid sequences that are common to several different proteins. The functions of the latter sequences are unknown, although the proteins that contain them may have closely related functions.

The evidence that many genes involved in the control of development encode proteins that bind to specific DNA sequences indicates that those proteins may act as genetic "switches" regulating the expression of other genes. Once proteins of this type have been identified, it is then possible to use the purified proteins to identify the DNA "target" sequences to which they bind. By using this approach, it should be possible to identify large parts of the regulatory pathway involved in different stages of development.

Caenorhabditis elegans

One of the simplest systems for the study of development genes is the nematode worm *Caenorhabditis elegans.* The study of this animal was begun in the 1960s by Sydney Brenner, a molecular biologist. Today it is becoming an important tool for answering basic questions about the development of individual cells within a multicellular organism. *Caenorhabditis* is a small worm that even as an adult is only 1.5 mm long and contains only about 1000 somatic cells (the exact number depends on the sex) and about 2000 germ cells. It can exist as either a **hermaphrodite** (an organism with both sexes in the same individual) or as a male. The hermaphroditic individuals are self-fertilizing, which makes it easy to obtain offspring that are homozygous for newly induced recessive mutations. The availability of males that can mate with the hermaphrodites makes it possible to do genetic crosses as well. Since its body is transparent, it is possible to follow the development of literally every somatic cell in the worm using a Nomarski differential interference microscope (Chapter 4) (Figure 16–13). As a consequence of herculean efforts by several laboratories, the lineage of each somatic cell in the adult has now been determined. The results of those studies have shown that the nematode has a very rigid developmental pattern. After fertilization, the egg undergoes repeated divisions in order to produce about 550 cells organized to form a small, sexually immature larva inside the egg case. The larva then hatches from the egg, and further cell divisions give rise to the adult worm.

The lineage of each somatic cell in the adult can be traced to a single cell in a small group of **stem** cells or **founder cells** that are formed early in development (Figures 16–14 and 16–15). If a particular founder cell is destroyed or removed, the structures that cell would normally give rise to will

FIGURE 16–13 Development of *Caenorhabditis elegans.* (*a*) A Nomarski interference micrograph of a newly hatched larva nematode. (*b*) Diagram illustrating structures in the adult hermaphrodite structure.

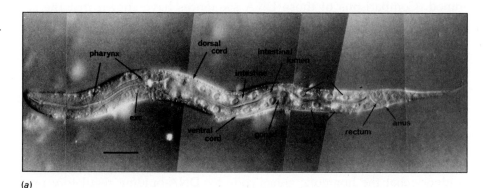

(a)

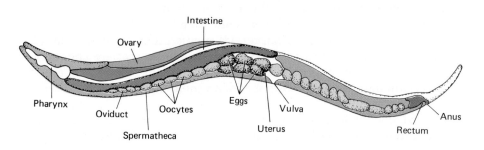

(b)

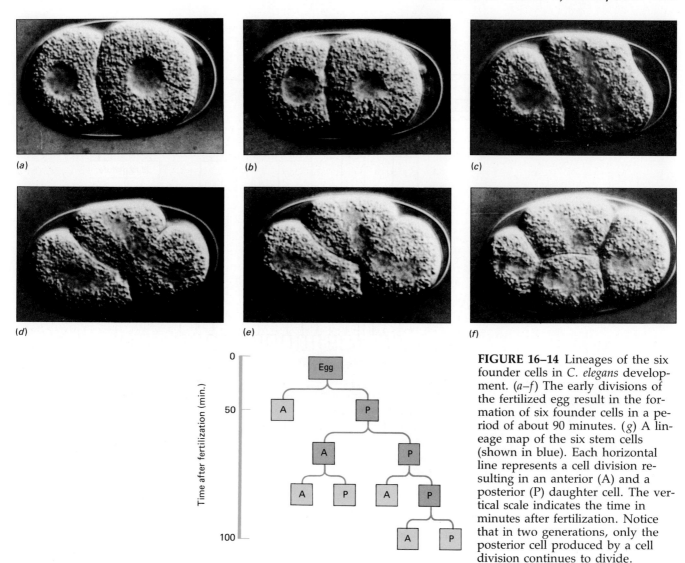

(a)

(b)

(c)

(d)

(e)

(f)

(g)

FIGURE 16–14 Lineages of the six founder cells in *C. elegans* development. (*a–f*) The early divisions of the fertilized egg result in the formation of six founder cells in a period of about 90 minutes. (*g*) A lineage map of the six stem cells (shown in blue). Each horizontal line represents a cell division resulting in an anterior (A) and a posterior (P) daughter cell. The vertical scale indicates the time in minutes after fertilization. Notice that in two generations, only the posterior cell produced by a cell division continues to divide.

be missing. An embryo with such an invariant developmental pattern is said to be highly **mosaic,** meaning that the fates of cells are largely predetermined. It was originally thought that each founder cell would give rise to only one organ. The detailed analysis of cell lineages, however, tells us that many of the structures found in the adult, such as the nervous system or the musculature, are in fact derived from more than one founder cell. Conversely, a few lineages have been identified where a nerve cell and a muscle cell are formed in the final cell division leading to the differentiation of the adult structure.

A number of mutations affecting cell lineages have been isolated, and many of these appear to have properties that would be expected of genes that are involved in control of developmental decisions.

By using microscopic laser beams that are small enough to destroy individual cells, it is possible to ask what influence one cell may have on the development of a neighbor. Consistent with the rigid pattern of cell lineages, in most cases destruction of an individual cell in *Caenorhabditis* results in the absence of all of the structures derived from that cell and the normal differentiation of all of the neighboring somatic cells. This suggests that development in each cell is regulated through its own internal program. However, there are also a few examples of **induction,** that is, cases where differentiation of a cell can be influenced by interactions with particular

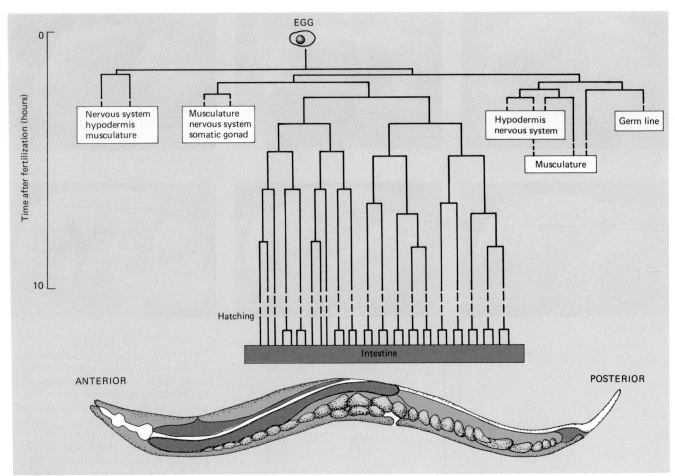

Time after fertilization (hours)

EGG

Nervous system
hypodermis
musculature

Musculature
nervous system
somatic gonad

Hypodermis
nervous system

Germ line

Musculature

Hatching

Intestine

ANTERIOR

POSTERIOR

FIGURE 16–15 A lineage map of the cells in *C. elegans* that form the intestine.

neighboring cells. One example can be found in the formation of the vulva (pl. vulvae), which is the structure through which the eggs are laid. There is a single nondividing cell, called the anchor cell, that is a part of the gonad (the structure in which the germ-line cells undergo meiosis to produce the eggs). This cell attaches to the ovary and to a point on the outer surface of the animal triggering the formation of a passage through which the eggs pass to the outside. When the anchor cell is present, cells on the surface organize to form the vulva and its opening. If the anchor cell is destroyed by a laser beam, however, the vulva does not form and the cells that would normally form the vulva remain as surface cells (Figure 16–16).

Analysis of certain cell lineage mutations has been useful in understanding such inductive interactions. For example, several mutations are known that cause more than one vulva to form. In such mutant animals, multiple vulvae will form even if the anchor cell is destroyed. Thus, the mutant cells do not require an inductive signal from an anchor cell in order to form a vulva. Evidently the gene or genes responsible for vulva formation are constitutively expressed in these mutants. Conversely, mutants lacking a vulva are also known. In some of these, the cells that would normally form the vulva appear to be unable to respond to the inducing signal from the anchor cell.

During development in *Caenorhabditis* there are a number of instances where cells are produced and die shortly thereafter. Such phenomena have been observed in other organisms as well. For example, the human hand is formed as a webbed structure, but the fingers become individualized when the cells between them die. In *Caenorhabditis*, these **programmed cell deaths** are under genetic control and a number of mutants have been isolated that alter the pattern of these deaths. Molecular analysis of the loci

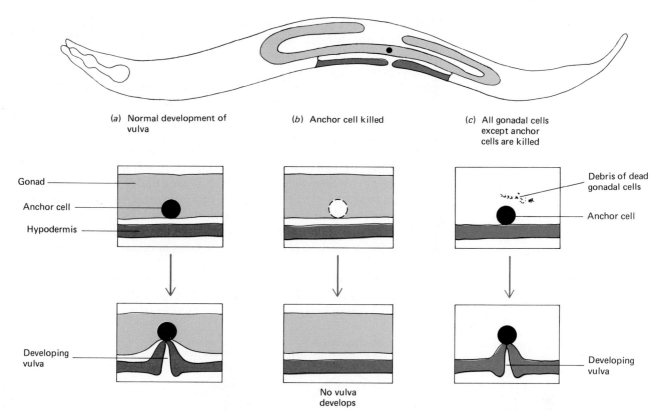

(a) Normal development of vulva

(b) Anchor cell killed

(c) All gonadal cells except anchor cells are killed

Gonad

Anchor cell

Hypodermis

Debris of dead gonadal cells

Anchor cell

Developing vulva

No vulva develops

Developing vulva

FIGURE 16–16 A single "anchor cell" induces neighboring cells to form the vulva in *C. elegans*. A schematic diagram showing how laser destruction of single cells or a group of cells can be used to demonstrate the influence of a cell on its neighbors.

that are identified by these mutations should shed considerable light on the phenomenon of programmed cell death in general.

Mutations are also known that appear to identify so-called **chronogenes,** that is, genes that are involved in developmental timing. One such locus has recessive alleles that cause certain cells to adopt fates that would ordinarily be seen later in development. Dominant alleles of the same locus cause certain cells to adopt fates that would usually be expressed earlier. Such genes appear to be good candidates for "switches" that control developmental timing. Many of these genes are now being cloned, and important information about these genes and their products should soon be available.

Developmental Genetics of the Mouse

Mammalian embryos develop in markedly different ways than the embryos of *Drosophila* and *Caenorhabditis*. The laboratory mouse is the best-studied example of early mammalian development.

EARLY MOUSE DEVELOPMENT. The early development of the mouse and other mammals is similar in many ways to human development, which is described in detail in Chapter 51. During the early developmental period, the embryo lives free in the reproductive tract of the female. It then implants in the wall of the uterus, and from that time on, its needs are provided by the mother. Consequently, mammalian eggs are very small and contain little in the way of food reserves. Almost all of the research on mouse development has concentrated on the stages leading up to implantation because up until that time the embryo is free-living and can be experimentally manipulated. During that period, a number of critical developmental decisions take place that will have a significant effect on the future organization of the embryo.

Following fertilization, a series of cell divisions gives rise to a loosely packed group of cells. It has been possible to show that all of the cells in the very early mouse embryo are equivalent. For example, at the two-cell stage of mouse embryogenesis, one of the two cells can be destroyed by pricking it with a fine needle. Implanting the remaining cell into the uterus of a foster mother in most cases leads to the development of a normal mouse. Conversely, two embryos at the eight-cell stage of development can be fused together and implanted into a foster mother, resulting in the development of a normal-sized mouse (Figure 16–17). By using two embryos with different genetic markers (such as coat color), it can be demonstrated that the resulting mouse indeed has four parents. These mice will have fur that consists of patches of different colors derived from clusters of genetically different cells. Animals that are formed in this way are called **chimeras.** (The term *chimera* is derived from the name of a mythical beast that had the head of a lion, the body of a goat, and the tail of a snake, and is used today to refer to any organism that contains two or more kinds of genetically dissimilar cells.) Chimeras have been useful in allowing the use of genetically marked cells to trace the fates of certain cells during development.

The responses of mouse embryos to these kinds of manipulations are in marked contrast to the mosaic or predetermined nature of early *Caenorhabditis* development, where the destruction of one of the founder cells

FIGURE 16–17 The procedures for the creation of chimeric mice. Embryos from two different strains of mice are removed from females. (1) The cells from the two embryos are combined in vitro. (2) The aggregated embryos continue to develop. (3) The embryo is implanted in the uterus of a foster mother. Each of the offspring has four genetically distinct parents. The foster mother, however, is not genetically related.

8-cell-stage mouse embryo whose parents are white mice

8-cell-stage mouse embryo whose parents are black mice

Zona pellucida of each egg is removed by treatment with protease

Embryos are pushed together and fuse when incubated at 37°C

Development of fused embryos continues *in vitro* to blastocyst stage

Blastocyst is transferred to pseudopregnant mouse, which acts as a foster mother

The baby mouse has four parents, but the foster mother is not one of them

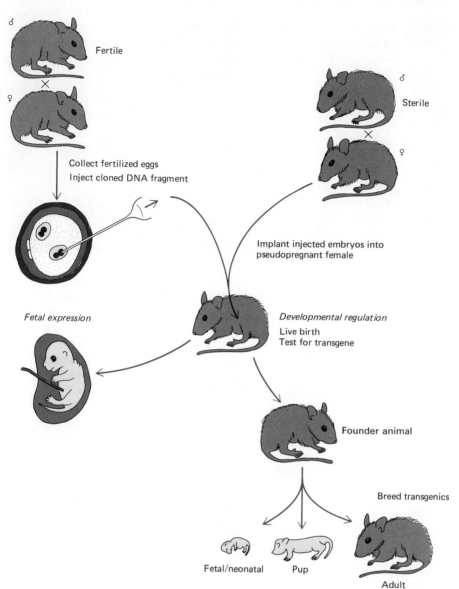

FIGURE 16–18 The production of transgenic mice. Cloned DNA fragments are injected into the pronucleus of a fertilized egg. The eggs are then surgically transferred to a foster mother. The presence of the foreign gene can be examined in the transgenic animal or the animal can be bred to establish a transgenic line of mice.

results in loss of a significant portion of the embryo. For this reason, we say that early development of the mouse (and presumably of other mammals) is highly **regulative**. This means that the early embryo acts as a self-regulating whole, which can accommodate missing or extra parts. On the other hand, it has not been possible so far to demonstrate totipotency of either cells or nuclei from later stages of mouse development.

THE USE OF TRANSGENES IN THE STUDY OF DEVELOPMENTAL REGULATION. In experiments similar to those with *Drosophila*, foreign DNA can be injected into fertilized mouse eggs and can be incorporated into the chromosomes and expressed (Figures 16–18 and 16–19). The resulting **transgenic** (Chapter 15) mice have given researchers some insights into how genes are activated during development. Scientists can identify a transgene that has been introduced into a mouse and determine whether it is active by marking the gene in several ways. Sometimes a similar gene from a different species is used, and its protein can be distinguished from the mouse protein by specific antibodies. It is also possible to construct "hybrid genes" that contain the regulatory elements of a mouse gene of interest together with part of another gene that codes for a "reporter" protein, such as an enzyme, not normally found in the mouse.

A number of developmentally controlled genes have been introduced into mice and have yielded important information about gene regulation.

FIGURE 16–19 Technique for microinjection of fertilized mouse eggs. The egg is held by suction on a holding pipet on the right. The DNA is injected into the nucleus by the glass needle (on the left) which is about 1 μm in diameter at the tip. Successful injection into the nucleus causes it to swell (lower panel).

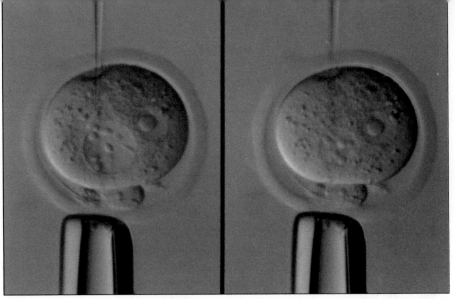

FIGURE 16–19 Technique for microinjection of fertilized mouse eggs. The egg is held by suction on a holding pipet on the right. The DNA is injected into the nucleus by the glass needle (on the left) which is about 1 μm in diameter at the tip. Successful injection into the nucleus causes it to swell (lower panel).

(a)

(b)

FIGURE 16–20 Homeotic mutations in corn. (a) Anther ear. (b) Tassel seed.

Most importantly, when developmentally controlled genes from other species such as humans or rats have been introduced into mice, they have been shown to be regulated in the same way as they normally are in the donor animal. For example, when introduced into the mouse, human genes encoding insulin, globin, and crystallin, which are normally expressed in cells of the pancreas, blood, and eye lens, respectively, are expressed only in those same tissues in the mouse. The fact that these genes are correctly expressed in their appropriate tissues indicates that the signals for tissue-specific gene expression are highly conserved through evolution. This is an exciting finding because it means that information on the regulation of genes controlling development in one organism can have valuable applications to other organisms such as humans.

Plants

Certain well-characterized plants are also being used in the study of the genetic control of development. Many of these are economically important crop plants, such as the corn plant, *Zea mays*. A number of genes with developmental effects are known in corn, including some that can be thought of as analogous to the homeotic genes of *Drosophila*. In corn the female and male flowers are borne on separate structures, with the ear carrying the female parts and the tassel carrying the male parts. When an ear is first formed, both female and male flower parts are present, but the male parts are repressed and the female parts continue to develop (Figure 16–20). Conversely, when a tassel is formed, only the male parts develop, although both male and female parts are present initially. Some alleles of the *tassel seed* locus cause the development of female flower parts on the tassels. Conversely, *anther ear* mutants have male flowers produced on the ear. Like mutations of the *Drosophila* homeotic genes, these mutations may very well identify developmental switch genes that specify alternative fates, in this case male or female structures.

Another plant that is being used increasingly to study genetics and development in plants is a member of the mustard family, *Arabidopsis* (Figure 16–21). Although *Arabidopsis* itself is of no economic importance, it has a number of advantages for research. One of these is the fact that the plant is quite small, so thousands of individuals can be grown in limited space. Chemical mutagens can be used to produce mutant strains, and a number of developmental mutants, including some that have "homeotic-like" characteristics, have been isolated. One such mutant transforms the flower petals of the plant into stamens, reproductive structures that contain functional pollen grains. The plant has a very small and simple genome, a fact

(a)

(b)

FIGURE 16–21 *Arabidopsis.* (a) *A. thaliana*, a normal flower with 4 petals, 4 sepals (hidden by the petals), six stamens, and a central ovary. (b) A homeotic mutant with only sepals and petals.

that greatly facilitates cloning of genes. In addition, cloned foreign genes can be inserted into *Arabidopsis* cells, and these can be integrated into the chromosomes and expressed. These transformed cells can be induced to differentiate into transgenic plants.

Other Molecular Mechanisms in Development

Although the concept of nuclear equivalence appears to apply to most cells in higher organisms, certain types of developmental regulation can also involve physical changes in the genome. Such changes in the structure of the genome are not common occurrences.

Genomic Rearrangements

The activity of some genes may be modified during development by different types of **genomic rearrangements** that lead to actual physical changes in the structure of the gene. In some cases, parts of genes are rearranged by recombination to make new coding sequences. This is an important mechanism for the development of the immune system (see Chapter 43).

Another type of rearrangement involves the replacement of an active gene with a copy of a "silent" gene located on a different part of the same chromosome. The baker's yeast *Saccharomyces cerevisiae* is a simple eukaryote that has two sexes or mating types called **a** and **α**. The mating type of a cell is determined by an active gene located at a position close to the middle of one of the yeast chromosomes called the mating type locus. At some distance on either side of the active gene are two silent genes called MAT a and MAT α. If a copy of the MAT a gene occupies the mating type locus, the mating type is a; if a copy of the MAT α gene occupies that site, the mating type of the cell is α. These yeast strains can switch their mating type from one form to the other as frequently as every generation (Figure 16–22) by a process that removes the gene located at the mating type locus and replaces it with a sequence of DNA copied from the silent gene that corresponds to the opposite mating type.

A somewhat similar system of gene replacement also takes place in the unicellular parasite *Trypanosoma brucei*, which causes sleeping sickness in humans and related diseases in other animals (Figure 16–23). When the parasite infects humans, it is able to defeat the immune system by constantly changing the glycoprotein molecules that are exposed on the surface of its cell. Unlike yeast, which has only two basic copies of the mating

PART IV

EVOLUTION

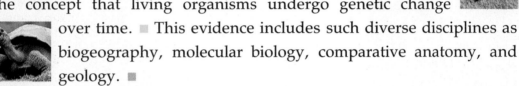

One of the basic principles of modern biology is evolution. ■ A vast assemblage of scientific evidence supports the concept that living organisms undergo genetic change over time. ■ This evidence includes such diverse disciplines as biogeography, molecular biology, comparative anatomy, and geology. ■

Darwin Lake on Isabella Island in the Galapagos

17

Darwin and Natural Selection

■ LEARNING OBJECTIVES

After you have read this chapter you should be able to:

1. Discuss the historical foundations of evolution.
2. Describe natural selection as outlined by Darwin.
3. Summarize the evidence for evolution from the fossil record.
4. Distinguish between compressions, impressions, molds, casts, and petrifactions.
5. Summarize some of the evidence supporting evolution that is obtained from the following fields: comparative anatomy, embryology, biogeography, biochemistry, and molecular biology.
6. Define and give examples of homologous and analogous organs.
7. Discuss how scientists make inferences about evolutionary relationships from the sequence of amino acids in specific proteins or the sequence of nucleotides in particular genes in organisms.

THE CONCEPT OF EVOLUTION

All of the vast diversity of life forms present on our planet evolved from one or a few simple kinds of organisms. This includes both plants and animals, microscopic bacteria and giant blue whales, living and extinct species. All of the species that exist today arose from earlier species by a process Darwin originally described as "descent with modification," or **evolution.** Evolution is a genetic change in a population of organisms. It does not refer to the changes that occur to an individual organism within its lifetime. Individuals do not evolve, but populations can.

Because evolution involves populations rather than individuals, one must consider the **gene pool,** which is all the genes present in a given species population. Another way to think of evolution, using the concept of the gene pool, is that evolution involves changes in the gene frequencies within the gene pool.

As an example, let us consider the evolution of bacterial resistance to antibiotics (Figure 17–1). The use of antibiotics like penicillin has increased dramatically since the 1940s. Once antibiotics began to be used for human and animal infections, it was believed that they would eliminate bacterial diseases. This has not occurred. Each time penicillin is used, most, but not all, of the bacteria present are killed. The survivors have a genetic resist-

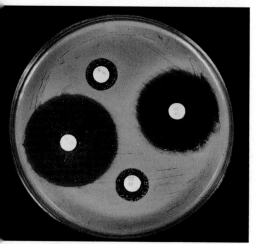

FIGURE 17–1 Antibiotic resistance in bacteria. An even coating of bacteria covers the surface of this culture dish except where an antibiotic to which the bacteria are sensitive prevents this growth. Varieties of bacteria that have developed resistance to certain antibiotics are common today, especially in hospital environments.

FIGURE 17–2 How did the giraffe get its long neck? Lamarck hypothesized that giraffes acquired longer necks by continually stretching into the trees to eat leaves unavailable to other large herbivores, and that they passed this characteristic on to their offspring. Although Lamarck's mechanism of evolution was incorrect, he was the first scientist to propose that organisms undergo evolution by natural means.

ance to penicillin, and so this trait is passed on to their progeny. This results in a larger percentage of penicillin-resistant bacteria than was present in the original population. On the practical side, physicians today recognize that antibiotics cannot be used indiscriminately, or we will be unable to control certain types of diseases because of increase in the genetic frequency of antibiotic resistance.

Since the frequency of appearance of certain genes has changed in the bacterial population, evolution has occurred. If enough of these types of changes occur over time, a new species might arise, but it is important to recognize that evolution may or may not give rise to a new species. The concept of evolution is the cornerstone of biology, for it enables us to make sense of the tremendous variety in the living world. Biologists do not question the occurrence of evolution, but the actual mechanisms that control evolution are under close study and active debate.

EARLY IDEAS ABOUT EVOLUTION

Although Charles Darwin is universally associated with evolution, ideas of evolution antedate Darwin by centuries. Aristotle (384–322 BC) saw much evidence of design and purpose in nature and arranged all of the organisms that were known to him in one "Scale of Nature" that extended from the very simple to the most complex. He visualized living organisms as being imperfect but moving toward a more perfect state. This has been interpreted by some as the germ of an idea of evolution, but Aristotle is very vague on the nature of this "movement toward the more perfect state" and certainly did not propose any notion of the origin of species.

Long before Darwin, odd fragments resembling bones, teeth, and shells (fossils) had been discovered embedded in rocks. Some of these corresponded to parts of familiar living animals, but others were strangely unlike any known form. Fossils of marine invertebrates were sometimes found in sedimentary rocks high on mountains! Leonardo da Vinci correctly interpreted these finds in the 15th century as the remains of animals that had existed in previous ages but had become extinct.

During the Renaissance, there was an increased interest in the study of nature and a movement away from simple reliance on the interpretations of early authorities. Modern scientific thought, based on observations, experiments, and vigorous inductive and deductive logic, emerged in the 17th century with the work of Francis Bacon, William Harvey, Isaac Newton, and René Descartes. Only in the 18th century did this new science begin to have much effect on interpretations of the biological world. As new continents were explored, the discovery of new species and more fossils led many to believe that the world of life as well as the physical world must be guided by natural laws.

The most thoroughly considered view of evolution before Darwin was expressed by Jean Baptiste de Lamarck in his *Philosophie Zoologique* (1809). Like most biologists of his time, Lamarck believed that all living things are endowed with a vital force that drives them to evolve toward greater complexity. He also believed that organisms could pass on traits acquired during their lifetimes to their offspring. As an example of this line of reasoning, Lamarck suggested that the long neck of the giraffe evolved when a short-necked ancestor took to browsing on the leaves of trees instead of on grass (Figure 17–2). Lamarck theorized that the ancestral giraffe, in reaching up, stretched and elongated its neck. Its offspring, inheriting the longer neck, stretched still further. As the process was repeated over many generations, the present long neck was achieved.

The mechanism for Lamarckian evolution was an "inner drive" for self-improvement, which was discredited when the mechanisms of heredity were discovered. Lamarck's contribution to science is important, however, because he was the first to propose that organisms undergo change over time as a result of some natural phenomenon rather than divine intervention. It remained for Charles Darwin (Figure 17–3) to discover the actual mechanism of evolution, by natural selection.

CHARLES DARWIN AND HIS CONTEMPORARIES

Charles Darwin (1809–1882) was sent at the age of 15 to study medicine at the University of Edinburgh. Finding the lectures intolerably dull, he transferred after 2 years to Christ College, Cambridge University, to study theology. At Cambridge he joined a circle of friends interested in natural history and through them became acquainted with Professor John Henslow, a naturalist. Shortly after leaving Cambridge, and upon Henslow's recommendation, Darwin was appointed "gentleman naturalist" on the ship *Beagle,* which was to make a 5-year cruise around the world to prepare navigation charts for the British navy (Figure 17–4). The *Beagle* left Plymouth, England, in 1831 and cruised slowly down the east coast and up the west coast of South America (Figure 17–5). While other members of the company mapped the coasts and harbors, Darwin had an opportunity to study the animals, plants, fossils, and geological formations of both coastal and inland regions, areas that had not been extensively explored. He collected and catalogued thousands of specimens of plants and animals and kept copious notes of his observations. He experienced first-hand the tremendous richness of the flora and fauna of these regions.

The *Beagle* spent some time at the Galapagos Islands, 600 miles west of Ecuador, where Darwin continued his observations and collections of the flora and fauna. He compared the animals and plants of the Galapagos with those of the South American mainland (Figure 17–6). He was particularly impressed by their similarities and wondered why the creatures of the Galapagos should resemble those from South America more than those from Africa, for example. Moreover, although there were similarities between the two groups, there were distinct differences. There were even differences in the birds (Figure 17–7) and reptiles from one island to the next! He pondered these observations and tried to develop an adequate explanation for their distribution.

The general notion in the mid-1800s was that creatures did not change significantly over time, that they looked the same as the day they were created. True, there were some troubling exceptions to this idea. For one thing, breeders could induce a great deal of variation in domesticated plants and animals in just a few generations (Figures 17–8 and 17–9). This was accomplished by selecting which traits were desirable and breeding only those individuals that possessed the desired traits, a procedure known as **artificial selection.**

Evidence found in rocks also was beginning to contradict the accepted views. A number of fossils were discovered that did not have living counterparts. Then too, geological evidence suggested that the Earth was far older than previously had been suspected. During the early 19th century, Charles Lyell developed the geological theory of **uniformitarianism.** He proposed that mountains, valleys, and other physical features of the Earth's surface were not created in their present form. Instead, they were formed over long periods of time by the slow geological processes of vulcanism, uplift, erosion, and glaciation, which still occur today. The slow pace of these geological processes indicated that the Earth was very, very old.

The ideas of Thomas Malthus were another important influence on Darwin. Malthus was a clergyman and economist who noted that populations increase in size geometrically ($2 \rightarrow 4 \rightarrow 8 \rightarrow 16 \rightarrow 32$ and so on) until checked by factors in the environment. In the case of humans, Malthus

FIGURE 17–3 Charles Darwin as a young man. This portrait was made shortly after Darwin returned to England from his voyage around the world on the *H.M.S. Beagle.* Observations made during this voyage helped him formulate the concept of evolution by natural selection.

FIGURE 17–4 A replica of the *H.M.S. Beagle,* which was made for a television presentation about Charles Darwin.

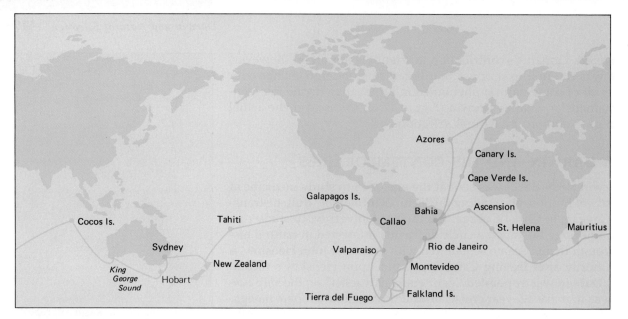

FIGURE 17–5 Voyage of the *H.M.S. Beagle.*

FIGURE 17–6 The animals and plants of the Galapagos Islands. (*a*) A marine iguana on a rocky shore. (*b*) The webbed feet of the red-footed booby (*Sula sula*) can grasp tree branches. (*c*) The blue-footed booby is a separate species (*S. nebouxii*) that is distinct from the red-footed booby. (*d*) A tree cactus (*Opuntia echios.* Other *Opuntia* in the Galapagos are not tree forms. (*e*) The Sally lightfoot crab is a subspecies that is endemic to the Galapagos. (*f*) Swallow-tailed gulls, *Creagrus furcatus,* are the only nocturnal gulls in the world. (*g*) A Hood Island saddleback tortoise (*Geochelone elephantopus hoodensis*). (*h*) This cactus, *Brachycereus nesioticus,* grows on recent lava flows. The entire genus is endemic to the Galapagos.

(a)

(b)

(c)

FIGURE 17–7 Three species of Darwin's famous Galapagos Island finches. As you can see, these are drab, unremarkable-appearing birds, which are evidently all derived from a common ancestry. Despite this common origin, the 14 known species are variously specialized for a variety of lifestyles, or ecological niches, that are elsewhere filled by birds of different species, which never had the opportunity to colonize the Galapagos Islands. The likely derivation of such different birds from a common ancestry suggested to Darwin that species originated by natural selection. (a) Cactus finch, *Geospiza scandens.* (b) A large ground finch, *Geospiza magnirostra.* This bird has an extremely heavy nutcracker-type bill adapted for eating heavy-walled seeds. (c) Woodpecker finch, *Camarhyncus pallidus.* This remarkable bird has insectivorous habits similar to those of woodpeckers but lacks the complex beak and tongue adaptations that permit woodpeckers to reach their prey. The adaptations of the woodpecker finch to this lifestyle are almost entirely behavioral. In one of the few known instances of animal tool use, this bird digs insects out of bark and crevices using cactus spines, twigs, or even dead leaves.

FIGURE 17–8 Some of the numerous dog varieties that have been produced by artificial selection. (a) West Highland. (b) Boxer. (c) Bulldog. (d) Standard Poodle. (e) Pomeranian. (f) Beagle.

(a)

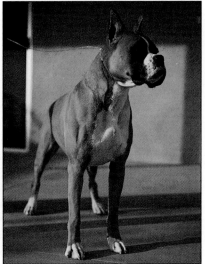

(b)

(c)

(d)

(e)

(f)

419

(a)

(b)

(c)

(d)

(e)

(f)

FIGURE 17–9 A number of common vegetables are members of the same species, *Brassica oleracea*, which includes (*a*) broccoli, (*b*) cauliflower, (*c*) kale, (*d*) brussels sprouts, (*e*) kohlrabi, and (*f*) cabbage. Selection is responsible for the tremendous variation shown within this species.

suggested that wars, famine, and pestilence served as the inevitable and necessary brakes on the growth of populations.

Darwin's years of observing the habits of animals and plants had introduced him to the struggle for existence described by Malthus. It occurred to Darwin that in this struggle favorable variations would tend to be preserved and unfavorable ones destroyed. The result of this would be adaptation of the population to the environment and, eventually, the origin of new species. Time was all that was required in order for new species to originate, and the geologists of the era, including Darwin's friend, Lyell, had supplied evidence that the Earth was indeed old enough to provide an adequate amount of time.

Darwin had at last obtained a working theory, that of "survival of the fittest." He spent the next 20 years accumulating a tremendous body of facts demonstrating that evolution had occurred and formulating his arguments for natural selection.

As Darwin was pondering his ideas, Alfred Russel Wallace, who was studying the flora and fauna of Malaysia and Indonesia, was similarly struck by the diversity of living things and the peculiarities of their distribution. Wallace arrived at a similar conclusion that evolution occurred by natural selection. In 1858, he sent a brief essay to Darwin, by then a world-renowned biologist, asking his opinion.

Darwin's friends persuaded him to present Wallace's paper along with an abstract of his own views, which he had prepared and circulated to a few friends several years earlier. Both papers were presented in July 1858 in London at a meeting of the Linnaean Society. Darwin's monumental book on the *Origin of Species by Means of Natural Selection* was published in November 1859.

NATURAL SELECTION

Darwin's mechanism of natural selection consists of four observations about the natural world. (1) Overproduction: Each species produces more offspring than will survive to maturity. (2) Variation: There is variation among the offspring. It is important to remember that the variation necessary for evolution by natural selection is genetic and can be passed on to progeny. (3) Competition: The species compete with one another for the limited resources available to them, i.e., "struggle for existence." (4) Survival to reproduce: Those offspring that possess the most favorable combination of characteristics will be most likely to survive and reproduce, i.e., "survival of the fittest." Natural selection results in the increase of "favorable" genes and the decrease of "unfavorable" genes within a population. Over time these changes may be significant enough to cause a new species to arise.

SOME EVIDENCE FOR EVOLUTION

Evolution is now supported by an enormous body of observations and experiments. In this text we can report only a small fraction of this wealth of evidence. Although biologists still do not agree completely on some aspects of the mechanism by which evolutionary changes occur, the concept that evolution has taken place is now well documented. It is consistent with all the information that has been brought to bear upon it.

Any scientific theory should lead to observations or testable predictions that, if not true, would require the theory to be modified or rejected. The concept of evolution is testable. For example, we would predict that mammalian fossils would not be found in rocks of the same age as those containing the fossils of ancestral fish but would be found in rocks laid down subsequently on top of the rock layers (strata) containing the ancestral fish. We would predict that the sequence of amino acids in human hemoglobin would be very similar to that of the chimpanzee and would show greater differences from the sequence in the hemoglobin of a horse or a whale. If such predictions prove to be untrue, the theory of evolution would be falsified and would need modification or replacement. However, all findings to date do conform with those predicted by the theory.

Paleontology

Perhaps the most direct evidence for evolution comes from the sciences of geology and paleontology. Geology deals with studies of the earth and its history. Paleontology is the science of discovery, identification, and interpretation of **fossils.**

The term *fossil* (Latin: *fossilis,* something dug up) refers not only to the parts of an animal's or plant's body that may survive, but also to any impression or trace left by previous organisms. If the body part has been trapped in sediments without being completely decomposed, the fossil is known as a **compression.** Some organic material still remains in compressions. If the pressure and heat is great during the formation of rock in which the organism is embedded, all of the organic material may be "vaporized." In this case, all that is left is an **impression** (Figure 17–10) of the original plant or animal.

(a)

(b)

(c)

FIGURE 17–10 Several types of fossils. (*a*) Impression fossil of a fern leaf. (*b*) Petrifaction of wood from the Petrified Forest National Park in Arizona. The cellular details are preserved in the fossil. (*c*) Cast fossil of a trilobite. The sediments of the sea bottom in which this arthropod lived have metamorphosed to form hard shale.

The commonest vertebrate fossils are skeletal parts. From the shapes of the bones and the positions of the bone scars that indicate points of muscle attachment, paleontologists can infer an animal's posture and style of walking, the position and size of its muscles, and the contours of its body. By a careful study of the fossil remains, paleontologists can reconstruct what an animal probably looked like in life.

In some fossils, the original hard parts, or even the soft tissues of the body, may be replaced by minerals. Iron pyrites, silica, and calcium carbonate are some of the common minerals that infiltrate buried tissues. These are known as **petrifactions** (Figure 17–10). The famous petrified forest of Arizona consists of trees that were buried and infiltrated with minerals. The muscles of a shark more than 350 million years old were so well preserved by this process that not only the individual muscle fibers, but even their cross-striations, could be observed in thin sections under the microscope.

Molds and **casts** (Figure 17–10) are produced in a different fashion. Molds are formed by the hardening of the material surrounding the buried organism, followed by the decay and removal of the tissues. The mold may subsequently be filled by minerals that harden to form casts that are replicas of the original structures.

Footprints or trails made in soft mud that subsequently hardened are a common type of fossil (Figure 17–11). From such remains, the paleontologist can infer something of the structure and locomotion of the animal that made them. Thus, fossils provide a record of animals and plants that lived earlier, some understanding of where and when they lived, and an idea of

FIGURE 17–11 Dinosaur footprints occurring in sedimentary rocks in Texas. Human feet are included for size comparison.

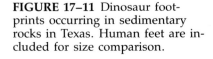

the kind of environment in which they lived. When enough fossils of organisms of different geological ages have been found, we can trace the lines of evolution that gave rise to those organisms.

Some more recent animal remains have been exceptionally well preserved by being embedded in bogs, tar, amber (Figure 17–12), or ice. The remains of some wooly mammoths deep-frozen in Siberian ice for more than 25,000 years were so well preserved that the meat was edible. Some of its DNA was preserved and could be analyzed.

The formation and preservation of a fossil require that an organism be buried under conditions that will slow the process of decay. This is most likely to occur if an organism's remains are covered quickly by fine particles of soil suspended in water. The soil particles are deposited as sediment around the animal or plant and cover it. Remains of aquatic organisms may be trapped in bogs, mud flats, sand bars, or deltas. Remains of terrestrial organisms that lived on a flood plain may also be covered by water-borne sediments or, if the organism lived in an arid region, by wind-blown sand. Animals may be trapped in a tar pit as in La Brea in Los Angeles (Figure 17–13) or be covered by volcanic ash as in Pompeii following the eruption of Mount Vesuvius.

Because of the conditions required for preservation, the fossil record is not a random sample of past life. There is bias in the record toward aquatic organisms and those living in the few terrestrial habitats conducive to fossil formation. For example, relatively few fossils of forest animals have been found. This is because plant and animal remains on the forest floor decay very rapidly, before fossilization can occur. Another reason for bias in the fossil record is that those organisms with hard body parts like bones and shells are more likely to form fossils.

To be interpreted, the sedimentary layers containing fossils must be arranged in chronological order. The layers of sedimentary rock, if they have not been disturbed, occur in the sequence of their deposition, with the more recent strata on top of the older, earlier ones. However, geological events that occur after the rocks were formed initially may change the relationship of some of the layers.

Geologists can identify specific sedimentary layers of rock by features such as their mineral content, their position in the layers, and by certain key invertebrate fossils, known as **index fossils,** that characterize a specific layer over large geographical areas. With this information, geologists can arrange strata and the fossils they contain in chronological order and identify comparable layers in widely separated localities.

FIGURE 17–12 A spider embedded in amber. It has been preserved almost perfectly for millions of years.

FIGURE 17–13 Recreation of a scene at the Rancho La Brea tar pits (now a part of Los Angeles, California) in the Pleistocene epoch. Fossilized skeletal remains of all these animals were actually found together in this locality, allowing us to form some picture of the paleo-ecology of the area. In the left foreground is one saber-toothed tiger. The giant vultures, now extinct, had a wingspan of 3 meters. In the background are mastodons.

Comparative Anatomy

Study of the details of the structure of any particular organ system in the diverse members of a given phylum reveals a basic similarity of form that is varied to some extent from one class to another. For example, a bird's wing, a dolphin's front flipper, a bat's wing, and a human arm and hand, although superficially dissimilar, are composed of a very similar arrangement of bones, muscles, and nerves (Figure 17–14). Each has a single bone, the **humerus,** in the proximal part of the limb, followed by a **radius** and **ulna,** the two bones of the forearm, then a group of **carpels** in the wrist,

FIGURE 17–14 Homology. The bird wing, bat wing, dolphin flipper, and human arm are homologous because they have a basic, underlying similarity.

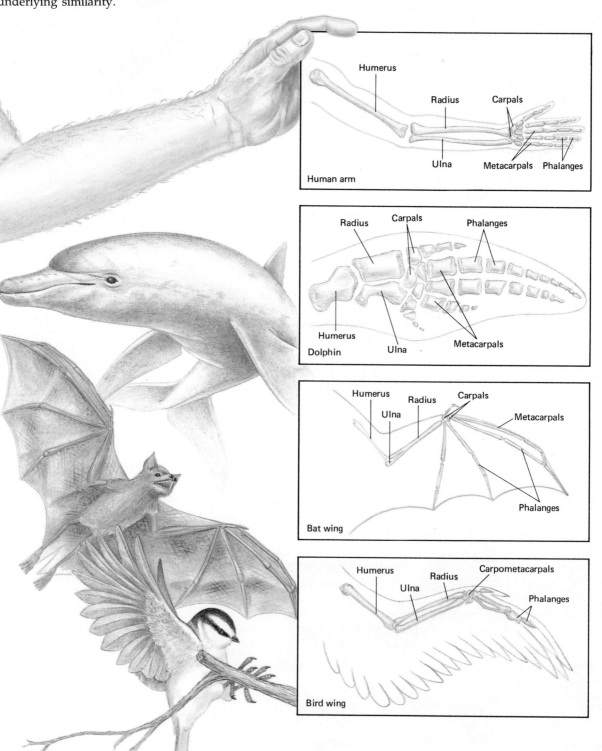

and a variable number of **digits.** This is particularly striking because wings, flippers, and the human arm are used in different ways for different functions, and there is no mechanical need for them to be so similar. Similar arrangements of parts of the forelimb are evident in the ancestral reptiles and amphibians and even in the first fishes to come out of water onto land. Darwin pointed out that such basic structural similarities in organs used in different ways are precisely the expected outcome if evolution has taken place. Organs of different organisms that have a similarity of form due to a common evolutionary origin are termed **homologous.**

With the acceptance of the theory of evolution, biologists came to realize that the homology of organs is due to their common evolutionary origin. Both bird and bat wings evolved from the forelimb of a common vertebrate ancestor. However, the flying surfaces of their wings are quite different. Feathers grow out from the posterior margin of the wings on the bird, while the flight surface of the bat's wing is essentially a webbed hand. Flight evolved independently in the two groups. Therefore, although the forelimbs are utilized as wings in both birds and bats, they are modified in quite different ways.

(a)

Organs that are not homologous but simply have similar functions in different organisms are termed **analogous** organs. For example, the lungs of mammals and the trachea (air tubes) of insects are analogous organs that have evolved to meet, in quite different ways, the common problem of obtaining oxygen. The wings of various unrelated flying animals, such as insects and vertebrates, resemble one another superficially (Figure 17–15). In more fundamental aspects, however, the wings are quite different. Vertebrate wings are modified forelimbs supported by bones, while insect wings are outgrowths of the upper wall of the thorax and are supported by chitinous veins.

The existence of homologous organs is important evidence of evolution. They are used to determine the interrelationships of living organisms. The existence of analogy is also crucial proof of evolution and adaptation. Comparisons of organisms with analogous organs indicate they have separate ancestries. Analogous organs are of evolutionary interest because they show how unrelated groups may adapt to common problems as their evolution leads to their structural and functional convergence in similar habitats.

(b)

FIGURE 17–15 Analogous structures. The wings of birds and insects, although used for similar functions, have no underlying structural similarity.

Comparative anatomy also demonstrates the existence of **vestigial organs.** Many organisms contain such organs or parts of organs that are seemingly nonfunctional and degenerate, often undersized or lacking some essential part. In the human body there are more than 100 such structures that have been viewed as vestigial, including the appendix, the coccyx (fused tail vertebrae), the wisdom teeth, and muscles that move the ears. Whales (Figure 17–16) and pythons have vestigial hind leg bones; wingless birds have vestigial wing bones; many blind, burrowing, or cave-dwelling animals may have vestigial eyes, and so on.

The occasional presence of a vestigial organ is to be expected as an ancestral species evolves and adapts to different modes of life. Some organs become much less important for survival and may end up as vestiges. When an organ loses all or much of its function, there is no longer any selective advantage in possessing it. In addition, selective pressure for getting rid of the vestigial organ altogether is weak, and so the organ tends to remain.

Embryology

The resemblance between embryos of different animals is closer than the resemblance between their adults (Figure 17–17). In fact, it is difficult to

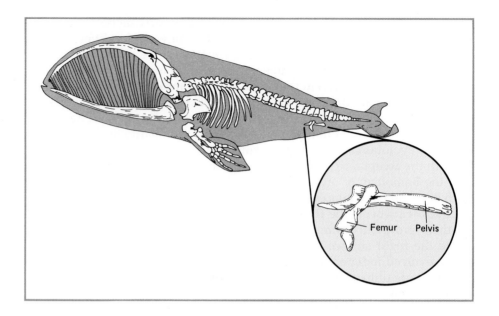

distinguish the early embryos of a fish, frog, chick, pig, or human. Segmented muscles, gill pouches, a tubular heart undivided into left and right sides, a system of aortic arches in the gill region, and many other features are found in the embryos of all vertebrates. However, none of these features persists in the adults of reptiles, birds, or mammals.

Why are these fishlike features present in the embryos of reptiles, birds, and mammals? All of these structures are necessary and functional in the developing fish. The small segmented premuscular blocks of the embryo

FIGURE 17–17 The early stages of embryonic development in several vertebrates. There are numerous structural similarities shared by the early stages, including the presence of a tail and gill pouches. (*a*) Turtle. (*b*) Mouse. (*c*) Human. (*d*) Pig. (*e*) Chick.

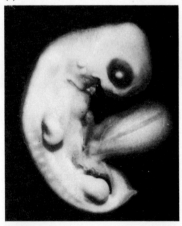

(a) Turtle

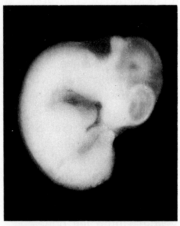

(b) Mouse

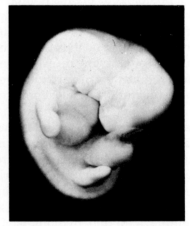

(c) Human

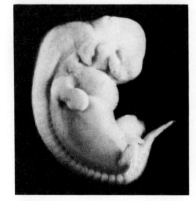

(d) Pig

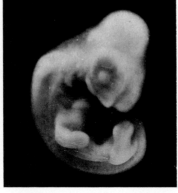

(e) Chick

give rise to the segmented muscles that are used by the adult in swimming. The gill pouches break through to the surface as gill slits. The heart remains undivided for it pumps venous blood forward to the gills that develop in association with the aortic arches. Since the higher vertebrates evolved from fish, they share the fish's basic pattern of development. The accumulation of genetic changes since the fish diverged in evolution from the higher vertebrates modifies the pattern of development of the higher vertebrate embryos.

Biogeography

The majority of plant and animal species have characteristic geographic distributions. Frequently, species are not even found everywhere that they *could* survive, as we would expect if climate and topography were the only factors determining their distribution. Central Africa, for example, has elephants, gorillas, chimpanzees, lions, and antelopes, while Brazil, with a similar climate and environmental conditions, has none of these. South America does have prehensile-tailed monkeys, sloths, and tapirs, however. The present distribution of organisms seems understandable only on the basis of evolution.

The **range** of a given species—that is, the portion of the Earth over which it is found—may be only a few square miles or, as with humans, almost the entire world. In general, closely related species do not have identical ranges, nor are their ranges far apart. They are usually adjacent, but separated by a barrier of some sort, such as a mountain or desert.

As we might expect, then, regions such as Australia and New Zealand, which have been separated from the rest of the world for a long time (Chapter 20), have a flora and fauna specific to these areas. Australia has populations of egg-laying mammals (monotremes) and pouched mammals (marsupials) found nowhere else. During the Mesozoic era (Chapter 20), Australia was isolated from the rest of the world. Its primitive mammals, therefore, never had any competition from the later-evolving placental mammals, which are thought to have competitively eliminated the monotremes and most of the marsupials everywhere else they may have existed. The original Australian mammals gave rise to a variety of forms that were able to take advantage of the different habitats available (Figure 31–34).

The kinds of animals and plants found on oceanic islands in general resemble those of the nearest mainland; yet they include some species found nowhere else. Darwin studied the flora and fauna of the Cape Verde Islands, some 400 miles west of Dakar, Africa, and of the Galapagos Islands, a comparable distance west of Ecuador. On each archipelago, the plants and terrestrial animals were indigenous, but those of the Cape Verdes resembled African species and those of the Galapagos resembled South American ones.

Darwin concluded that organisms from the neighboring continent migrated, or were carried, to the island and subsequently evolved into new species. The animals and plants found on oceanic islands are only those that could survive the trip there. There are no frogs or toads on the Galapagos, even though there are woodland spots ideally suited for such creatures, because neither the animals nor their eggs can survive exposure to sea water. There are no terrestrial mammals either, although there are many bats, as well as land and sea birds. The occurrence of these particular forms—closely related to, yet not identical with, those of the Ecuador coast—suggests strongly that evolution has modified the descendants of the first animals and plants to reach the islands.

Sometimes the range of closely related species can seem puzzling. Alligators are found only in the rivers of the southeastern United States and in the Yangtze River of China. Similarly, sassafras, tulip trees, and magnolias grow only in the eastern United States, Japan, and eastern China. Geologists infer from the various fossil layers that early in the Cenozoic era (Chapter 20), the northern hemisphere was low-lying and much flatter than it is now, and the North American continent was connected with eastern Asia by a land bridge at the Bering Strait. The climate of this region was much warmer than at present, as judged by the fossil evidence, which shows that alligators, magnolia trees, and sassafras were distributed over the entire region. Later in the Cenozoic, as the Rockies increased in height, the western part of North America became colder and drier, causing the organisms adapted to a warm, humid climate to become extinct.

Then, with the Pleistocene glaciations, the ice sheets moving from the North met the desert and mountain regions in western North America, eliminating any surviving temperate-zone plants. In the southeastern United States and eastern China, there were regions untouched by the glaciation in which the magnolia trees and alligators survived. Because the alligators and magnolia trees of the two regions have been separated for several million years, they have followed separate evolutionary pathways and are slightly different, although they are still closely related species in the same genera.

The study of the distribution of plants and animals is the science of **biogeography.** One of its basic tenets is that each species of animal and plant originated only once. The particular place where this occurred is known as the species' **center of origin.** The center of origin is not a single point but the range of the population when the new species was formed. From its center of origin, each species spreads out until halted by a barrier of some kind—physical, such as an ocean or mountain; environmental, such as an unfavorable climate; or ecological, such as the presence of organisms that compete with it for food or shelter (Figures 17–18 and 17–19).

FIGURE 17–18 Animals and plants are distributed around the world in a distinctive pattern, which reveals the existence of six major biogeographic realms. Each of these is characterized by the presence of certain unique species. These biogeographic realms are the direct outcome of the centers of origin of certain species, of their past migrations, and of the barriers they encountered.

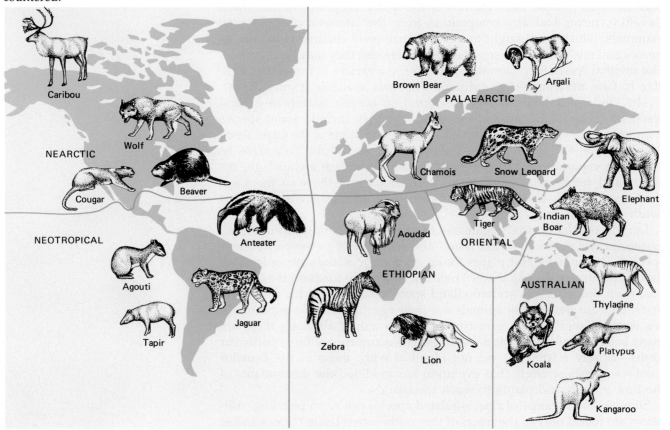

(a)

(b)

FIGURE 17–19 Structural similarities in two unrelated plant families. These plants have evolved in similar desert environments in different parts of the world. (a) A member of the spurge family in the Serengeti of Africa. (b) A member of the cactus family in Baja, Mexico.

FIGURE 17–20 Arranged in the order of their blood protein similarity to human beings, the chimpanzee (a) is most similar and the lemur (e), least. Pygmy chimpanzee is shown in (a). (b) Baboons are Old World (Eastern Hemisphere) monkeys. (c) Spider monkey, a New World monkey with a strong prehensile tail, used in swinging from tree to tree. (d) The tailless potto, an Old World primate. The large, forward-directed eyes are adapted for binocular night vision. (e) The ring-tailed lemur, one of the most primitive of living primates.

Biochemistry and Molecular Biology

It has become clear that similarities and differences in the biochemistry and molecular biology of organisms closely parallel morphological ones. Indeed, if we established evolutionary relationships based solely on biochemical and molecular characters instead of the usual morphological ones, the end result would be a very similar phylogenetic tree.

The blood serum of each species of vertebrate contains specific proteins, coded for by specific genes, whose degree of similarity can be determined by antigen-antibody reactions. When serum proteins are compared by this method, our closest "blood relations" are the great apes and then, in order, the Old World monkeys, the New World monkeys, and, finally, the tarsioids and lemurs (Figure 17–20). The biochemical relationships of a variety of organisms tested in this way correlate with and complement the relationships determined by other means. Cats, dogs, and bears have closely similar plasma proteins. Cows, sheep, goats, and deer constitute another

(a)

(b)

(d)

(c)

(e)

429

FIGURE 17–21 A diagram illustrating the differences in amino acid sequences in cytochrome *c* obtained from different species of animals and fungi. The numbers indicate how many amino acids in the cytochrome *c* of a given species differ from those of various other species. For example, the horse cytochrome *c* has 12 amino acids in it that are different from those in human cytochrome *c* but has only 3 amino acids that are different from those in pig cytochrome *c*.

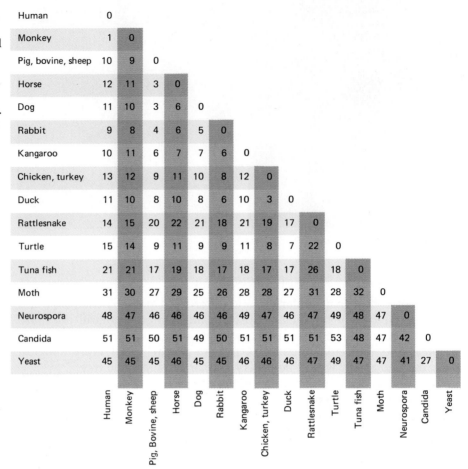

group with closely related serum proteins. Similar tests of the sera of crustaceans, insects, and mollusks have shown that the forms regarded as being closely related from anatomical or paleontological evidence have comparably similar serum proteins.

Investigations of the sequence of amino acids in proteins, such as hemoglobin or cytochrome (Figure 17–21), obtained from different species, have revealed great similarities and certain specific differences. The pattern of these differences demonstrates the nature and number of underlying mutations that must have occurred in evolution. Mutations occur, on average, at a certain rate within a given taxonomic group. From the number of alterations in the nucleotide sequence of one organism compared with another, we can estimate the age of a species or higher taxonomic group. The evolutionary relationships of organisms inferred from such biochemical studies parallel the relations inferred earlier on the basis of structural similarities.

Recent advances in molecular biology, such as the development of methods to determine the sequence of the nucleotide base pairs in DNA, have provided another means of demonstrating evolutionary relationships (Table 17–1). We would expect that species that are believed to be closely related on the basis of other evidence would also have a greater proportion of their DNA nucleotides in common than distantly related species. This has been shown in a number of investigations. The more closely species are believed to be related on the basis of other evidence, the greater is the percentage of DNA sequences that they have in common.

Darwin speculated that all forms of life are related through descent with modification from earliest organisms. This speculation has been verified as we have learned more about molecular biology. Even organisms that are

Table 17–1 DIFFERENCES IN NUCLEOTIDE SEQUENCES IN DNA AS EVIDENCE OF PHYLOGENETIC RELATIONS

Species Pairs	Percentage Differences in Nucleotide Sequences Between Pairs of Species
Human–chimpanzee	2.5
Human–gibbon	5.1
Human–Old World monkey	9.0
Human–New World monkey	15.8
Human–lemur	42.0

From Stebbins, G.L., *Darwin to DNA, Molecules to Humanity*, San Francisco, W.H. Freeman, 1982.

very remotely related, such as *Homo sapiens* and *Escherichia coli*, have some proteins such as cytochrome *c* in common. In the course of the long independent evolution of the two organisms, mutations have resulted in substitution of amino acids at various locations in the protein, but the cytochrome *c* molecules of all species are clearly similar in structure and function. Further evidence that all life is related comes from the universality of the genetic code. The genetic code has been passed along essentially unchanged through all the branches of the evolutionary tree since its origin in an extremely early form of life.

■ SUMMARY

I. Evolution is the unifying concept of biology. It enables us to make sense of the tremendous variety of life that exists in the world.

II. Charles Darwin and Alfred Wallace independently proposed essentially identical theories of evolution by natural selection.

   A. Overproduction: Each species produces more offspring than will survive to maturity.

   B. Variation: Genetic variation exists among these offspring.

   C. Competition: There is competition among these offspring for the resources needed for life, i.e., food, space, habitat.

   D. Survival to reproduce: The offspring with the most favorable combinations of genetic characteristics are most likely to survive and reproduce, passing those characters on to the next generation.

III. The concept that evolution has taken place is now well documented.

   A. Perhaps the most direct evidence for evolution comes from paleontology.

      1. Fossils are remains or traces of ancient animals and plants.

      2. There are several types of fossils. These include compressions, impressions, molds, casts, and petrifactions.

   B. Evidence supporting evolution is derived from comparative anatomy.

      1. Homologous organs have basic structural similarities, even though the organs may be used in different ways. Homologous organs indicate evolutionary ties between the organisms possessing them.

      2. Analogous organs have similar functions but are not homologous and do not indicate close evolutionary ties.

      3. The occasional presence of a vestigial organ is to be expected as an ancestral species evolves and adapts to different modes of life.

   C. Embryology provides evidence for evolution.

      1. The resemblance between embryos of different animals is closer than the resemblance between their adults.

      2. The accumulation of genetic changes since organisms diverged in evolution modifies the pattern of development of the higher vertebrate embryos.

   D. The distribution of plants and animals (biogeography) supports evolution.

      1. Areas that have been separated from the rest of the world for a long time have a flora and fauna specific to those areas.

      2. Each species originated only once (at its center of origin).

      3. From its center of origin, each species spreads out until halted by a barrier of some kind.

   E. Biochemistry and molecular biology provide compelling evidence for evolution.

      1. Blood sera of closely related vertebrates are more similar than sera of distantly related vertebrates.

      2. The sequence of amino acids in common proteins such as cytochrome or hemoglobin reveals greater similarities in closely related species.

      3. A greater proportion of the sequence of nucleotides in DNA is identical in closely related organisms.

      4. The universality of the genetic code is further evidence that all life is related.

■ POST-TEST

1. Thomas Malthus believed that populations increase in numbers: (a) arithmetically; (b) geometrically.
2. Inherent in Darwin's theory of evolution is the concept that organisms: (a) have the potential to produce more offspring than can survive; (b) can produce only limited numbers of offspring that will survive; (c) can only replace individuals that are killed off by disease, war, or famine.
3. During his cruise on the *Beagle*, Darwin started thinking about the origin of species when confronted with: (a) observation of populations of animals and plants that vary from one part of the world to the next; (b) the presence of many kinds of life on relatively young volcanic islands, such as the Galapagos, that resemble animals and plants on the coasts of adjacent continents; (c) the various kinds of fossils that he found in South America; (d) all of these.
4. The genetic constitution of an entire population of a given organism is termed its: (a) genotype; (b) gene pool; (c) karyotype.
5. An organ that appears to have little or no function, and is smaller than a similar, fully functional equivalent in the organism's ancestor or relatives, is known as a (an): (a) homologous organ; (b) analogous organ; (c) vestigial organ.
6. The blood sera of humans matches which group most closely? (a) New World monkeys; (b) Old World monkeys; (c) lemurs; (d) great apes; (e) tarsioids.
7. The wings of butterflies and bats have similar functions but are quite different. This is an example of: (a) homologous organs; (b) analogous organs; (c) vestigial organs.
8. A fossil in which the body part has been trapped in sediments without being completely decomposed is known as a: (a) compression; (b) impression; (c) petrifaction; (d) mold; (e) cast.
9. Geologists can identify specific sedimentary layers of rock by certain key invertebrate fossils, known as: (a) ancient fossils; (b) range fossils; (c) index fossils.
10. The portion of the earth over which a given species is found is its: (a) center of origin; (b) range; (c) barrier.

■ REVIEW QUESTIONS

1. How can you account for the fact that both Darwin and Wallace independently and almost simultaneously proposed essentially identical theories of evolution by natural selection?
2. Explain briefly the concept of biological evolution.
3. In what ways does Lamarck's theory of adaptation not agree with present evidence?
4. Consider the giraffe's long neck. Explain how this came about using Lamarck's theory of evolution. Then explain the giraffe using Darwin's mechanism of natural selection.
5. Why are only inherited variations important in the evolutionary process?
6. Distinguish between the different kinds of paleontological evidence used to support evolution.
7. Discuss the factors that might interfere with our obtaining a complete and unbiased picture of life in the past from a study of the fossil record.
8. Explain why marsupials are widespread in Australia and almost nonexistent elsewhere.
9. List as many vestigial structures in the human body as you can.

■ RECOMMENDED READINGS

Colbert, E.H. *Evolution of the Vertebrates.* 3rd ed. New York, John Wiley and Sons, 1980. A textbook recommended for its presentation of information regarding vertebrate fossils.

Greene, J.C. *Science, Ideology, and World View.* Berkeley, University of California Press, 1982. An interesting exposition of how evolution influences the way we view the universe and our place in it.

Darwin, C.R. *On the Origin of Species by Means of Natural Selection or the Preservation of the Favored Races in the Struggle for Life.* New York, Cambridge University Press, 1975. A readily obtainable reprint of one of the most important books of all time. Darwin's long essay is still of great significance to modern readers.

Simpson, G.G. *Splendid Isolation: The Curious Story of South American Mammals.* New Haven, Yale University Press, 1980. A charmingly written account of the evolutionary history of the mammals of South America.

Stanley, S.M. *Earth and Life Through Time.* San Francisco, W.H. Freeman, 1985. A presentation of evolution for the general public.

Stebbins, G.L. *Darwin to DNA, Molecules to Humanity.* San Francisco, W.H. Freeman, 1982. A very broad, interesting discussion of evolutionary theory and evidence.

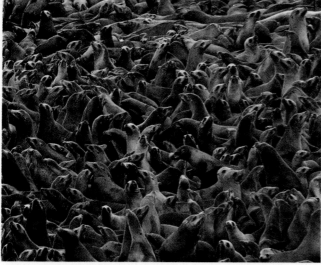

Colony of California sea lions, Año Nuevo Island

18

Population Genetics

■ LEARNING OBJECTIVES

After you have read this chapter you should be able to:

1. Distinguish between the gene pool of a population and the genotype of an individual.
2. Describe the Hardy-Weinberg law and its role in population genetics.
3. Discuss the factors that can alter the gene frequencies in populations: genetic drift, gene flow (migration), mutation, and natural selection.
4. Explain the biological cost of genetic load in heterozygote advantage.
5. Distinguish between stabilizing selection, directional selection, and disruptive selection. Describe how each plays a role in evolution.

THE GENE POOL

Evolution is the change in a population of organisms over a period of time. This change is inherited from one generation to the next. Although Darwin recognized that evolution occurred in populations, he did not understand how genetics worked in this regard. One of the most significant advances in biology since Darwin's time has been the elucidation of the genetic basis of evolution.

As mentioned briefly in the last chapter, each species possesses an isolated pool of genes ("gene pool"). This pool includes all possible alleles at each locus of each chromosome present in the individuals that make up the population. Because most species are diploid, each individual member of a population contains only two of each locus and may be either homozygous or heterozygous at each locus. Therefore, an individual within a population has only some of all the genes found in its gene pool (Figure 18–1). Moreover, the variation present in a given population indicates that each individual has a different portion of the genes that exist in the gene pool.

FIGURE 18–1 A gene pool. This drawing shows only one genetic locus (A), with three different alleles possible at that locus (A_1, A_2, A_3). Because each individual (represented by the small circles) is diploid, it will possess only two alleles for each genetic locus. The gene frequencies represented in this drawing are 0.40 for A_1, 0.35 for A_2, and 0.25 for A_3.

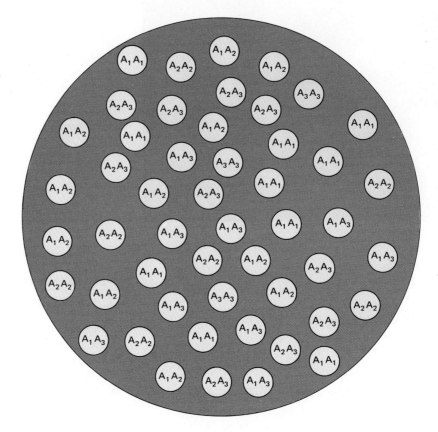

If a population is not evolving, frequencies of each allele remain constant from generation to generation. Changes in allele frequencies over successive generations indicate that evolution has occurred. This type of evolution is sometimes referred to as **microevolution,** because it involves changes taking place *within* a population. In this chapter, we examine the factors responsible for microevolution after first considering the genetics of a population that is not evolving.

THE HARDY-WEINBERG LAW

If we set a trap over a bunch of ripe bananas, we would probably catch a population of fruit flies, *Drosophila*. After we gently anesthetize and count them, we might find that we have a thousand fruit flies, 910 with gray bodies and 90 with black bodies. After releasing the fruit flies, they would mate. If we trapped and counted the next generation of fruit flies, we would find a population that is essentially the same as the previous one, with roughly 9 gray flies to every black fly. If we did this for a succession of generations, we would always get the same result.

The explanation for this stability of populations in successive generations was provided in 1908 by G.H. Hardy, an English mathematician, and W. Weinberg, a German physician, who independently pointed out that the frequencies of the members of a pair of alleles in a population are described by the expansion of the binomial equation.

Mendel's laws, as we have seen, describe the frequency of genotypes among offspring of a single mating pair. In contrast, the Hardy-Weinberg law describes the frequencies of alleles in the genotypes of an entire breeding population. The Hardy-Weinberg law shows that in large populations, sexual reproduction alone will not cause changes in the gene frequencies. Thus, knowledge of the Hardy-Weinberg law is essential for an understanding of the mechanisms of evolutionary change in sexually reproducing populations.

We now expand the fruit fly example to explain the Hardy-Weinberg law. A few simple crosses of black and gray fruit flies would reveal that the allele for gray body, *B*, is dominant over the allele for black body, *b*. Gray-bodied flies include some that are homozygous, *BB*, and some that are heterozygous, *Bb*. Obviously, all the black flies will be homozygous, *bb*. The frequency of either allele, *B* or *b*, is described by a number from zero to one. If an allele is totally absent from the population, its frequency is zero. If in the population all of the alleles at that locus are the same, then the frequency of that allele is one.

Since the gene must be either *B* or *b*, the sum of their frequencies must equal one. If we let *p* represent the frequency of the *B* allele, and *q* the frequency of the *b* allele in the population, then $p + q = 1$. When we know the value of either *p* or *q*, we can calculate the value of the other: $p = 1 - q$ and $q = 1 - p$.

The binomial equation describing the relationship of these allele frequencies in the population is $(p + q)^2$. When this is multiplied, we obtain the frequency of the offspring genotypes:

$$\underset{\text{Frequency of } BB}{p^2} + \underset{\text{Frequency of } Bb}{2pq} + \underset{\text{Frequency of } bb}{q^2}$$

From the fact that we had 90 black flies in our population of 1000, the frequency of the *bb* genotype, q^2, is 90/1000 = 0.09. Since q^2 equals 0.09, *q* is equal to the square root of 0.09, or 0.3. From our previous discussion, we know that $p = 1 - q$ or $1 - 0.3 = 0.7$

Based on this information, we can calculate the frequency of homozygous gray flies, *BB*: $p^2 = 0.7 \times 0.7 = 0.49$. The frequency of heterozygous gray flies, *Bb*, would be: $2pq = 2 \times 0.7 \times 0.3 = 0.42$. Thus, approximately 490 of the gray flies will be homozygous and 420 will be heterozygous. Note that the sum of homozygous and heterozygous gray flies equals 910, the number with which we began. Any population in which the distribution of alleles *B* and *b* conforms to the relation $p^2 + 2pq + q^2$, whatever the absolute values for *p* and *q* may be, is in genetic equilibrium (Figure 18–2). No evolution is taking place within the population.

The proportion of alleles in successive generations in a situation of genetic equilibrium will be the same, provided the following:

1. **No mutations.** In this instance, there must be no mutations of *B* or *b*.
2. **Random mating.** The three types of genotypic individuals (*BB, Bb, bb*) must not select their mates on the basis of genotype. There must be equal probabilities of mating between genotypes.
3. **Large size.** The population of individuals must be large enough so that the laws of probability function.
4. **Isolation.** There can be no exchange of genes with other populations that might have different gene frequencies.
5. **No natural selection.** If natural selection is occurring, certain alleles will be favored over others, and the gene frequencies will change.

FACTORS THAT CHANGE GENE FREQUENCIES

In studying populations in nature or in the laboratory, the Hardy-Weinberg law is used to test whether mating is random and whether evolutionary changes are taking place. If the members of a population are mating randomly, and if no other factors are affecting allele frequencies, then the frequencies of the various genotypes should be very close to those calculated with the Hardy-Weinberg formula.

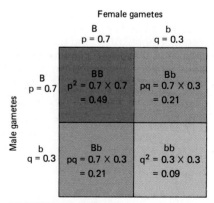

FIGURE 18–2 The random union of eggs and sperm containing B or b alleles. The frequency of the appearance of each of the possible genotypes (*BB, Bb, bb*) in the offspring is calculated by multiplying the frequencies of the alleles *B* and *b* in eggs and sperm.

However, the frequencies of the genotypes are often significantly different from those expected on the basis of the Hardy-Weinberg law. Evolution, stated in its simplest terms, represents a departure from the Hardy-Weinberg law of genetic stability. These changes in the gene pool of a population result from such phenomena as mutations, genetic drift, natural selection, and differential migration of organisms possessing particular genotypes. Without these, genetic frequencies in a large, freely interbreeding population will not change from generation to generation.

Genetic Drift

The size of a population has important effects on allele frequencies because the probability of a departure from the initial frequency by the chance mating of individuals is inversely related to population size. Let us assume that a gene pool has the alleles *A* and *a*, each with a frequency of 0.5. This can be simulated by using a bowl of 1000 beads, half of them red and half of them blue. In taking a random sample of ten beads from the bowl, it is unlikely that you will get exactly five reds and five blues. But if the sample is as large as 100, it is much more likely that you will approach 50:50.

The smaller the sample, then, the greater the probability that there will be a significant departure from the true value. If the sample size is as small as 2, the probability that both will be red is 0.5×0.5, or 0.25. The probability that both will be blue is $0.5 \times 0.5 = 0.25$. The probability that one will be red and one will be blue is only 0.5.

The effect is similar when a breeding population is very small. The probability of two individuals with the same traits mating is increased, and there is a reasonable chance that the variability of the descendant generation will deviate from that of the parent population (Figure 18–3).

If the population remains small for many generations, sampling errors accumulate, and the population's variability drifts in a random way. One allele may be eliminated by chance from the population, even if it determines a trait that is of adaptive value. Also, in such small populations, there is a strong tendency for all members of the population to become homozygous for one of its alleles. When this occurs, the allele is said to be "fixed" in the gene pool; that is, its frequency becomes 1.0.

The production of random evolutionary changes in small breeding populations is termed **genetic drift.** Genetic drift results in changes in the gene pool of a population. However, genetic drift affects gene frequencies randomly. Its direction may or may not be the same as that of other factors that change gene frequencies, such as migration, mutation, and natural selection. As a result, genetic drift may reinforce or oppose those forces.

FIGURE 18–3 The smaller the breeding population, the more likely it is that allele frequencies will change. This phenomenon is known as genetic drift.

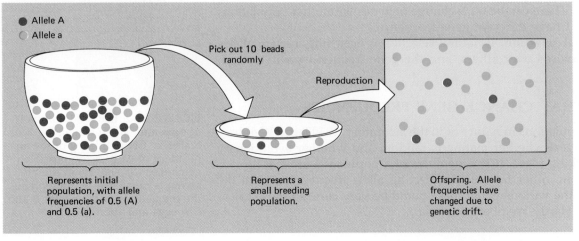

Allele A
Allele a

Pick out 10 beads randomly

Reproduction

Represents initial population, with allele frequencies of 0.5 (A) and 0.5 (a).

Represents a small breeding population.

Offspring. Allele frequencies have changed due to genetic drift.

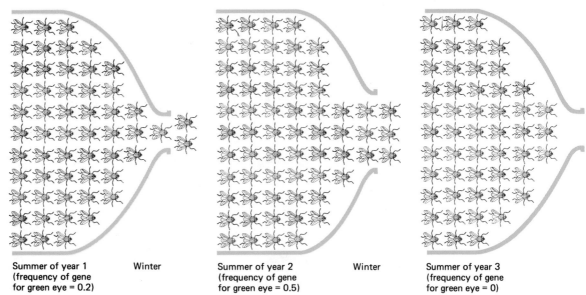

Summer of year 1
(frequency of gene
for green eye = 0.2)

Winter

Summer of year 2
(frequency of gene
for green eye = 0.5)

Winter

Summer of year 3
(frequency of gene
for green eye = 0)

FIGURE 18–4 Bottleneck effect. Since only a small population of flies survives the winter, its genotypes, not necessarily resulting from natural selection, determine the genetic frequencies of the entire succeeding summer population.

Because of fluctuations in the environment, for example, depletion of food supply or outbreak of disease, a population may periodically experience a rapid and marked decrease in the number of individuals. The population goes through a **bottleneck,** and genetic drift can affect the variability of the few survivors. As the population again increases in size, the frequencies of many alleles may be quite different from those in the population preceding the decline (Figure 18–4).

The unusual distribution of ABO blood group allele frequencies in human populations may have resulted partly from population bottlenecks during the migration of small bands of early humans. The I^B allele responsible for the B antigen has frequencies as high as 0.30 in parts of Asia and India but 0.10 or less in Europe and much of Africa (Figure 18–5). This allele is absent in Native Americans, although their ancestors migrated here from Asia.

FIGURE 18–5 The frequency of the I^B allele in different geographical locations.

Percentage
frequencies

25–30

20–25

15–20

10–15

5–10

0–5

FIGURE 18–6 Founder effect. In this example, the genetic frequencies of a population have been determined by the genotypes that happened to be possessed by its founders but were not characteristic of frequencies in the population as a whole.

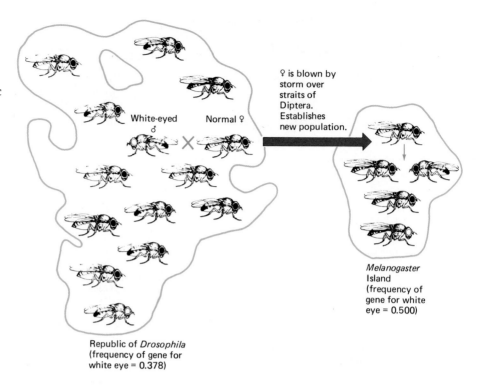

Republic of *Drosophila*
(frequency of gene for
white eye = 0.378)

White-eyed ♂ Normal ♀

♀ is blown by storm over straits of Diptera. Establishes new population.

Melanogaster Island (frequency of gene for white eye = 0.500)

Genetic drift is also important when one or a few individuals extend beyond the normal range for that species and establish a colony, as in the colonization of oceanic islands. The colonizers carry with them a small and random sample of the alleles of the gene pool from which they came, and their number is small enough initially for genetic drift to operate. The phenomenon has been termed the **founder effect** (Figure 18–6). As a result, isolated populations may have very different gene frequencies from those characteristic of the species elsewhere, and these differences may very well be random rather than adaptive.

Gene Flow

FIGURE 18–7 A member of a deme of bullfrogs. The frogs in one pond tend to mate with other frogs in the same pond.

Members of a species are not distributed uniformly throughout their range but occur in clusters that are spatially separated to some extent from other clusters. For example, the bullfrogs of one pond form a cluster separated from those in an adjacent pond (Figure 18–7). Some exchanges occur by migration between ponds, but the frogs in one pond are much more likely to mate with those in the same pond. Members of a species tend to be distributed in such local populations, termed **demes** (Greek *demos*, people).

The migration of individuals between demes causes a corresponding movement of alleles, or **gene flow,** that can have significant evolutionary consequences. As alleles "flow" from one population to another, they increase the amount of variability within the population receiving them. If the gene flow between two demes is great enough, these populations will become more similar genetically. Because gene flow has a tendency to reduce the amount of variation between two populations, it tends to counteract the effects of natural selection and genetic drift.

The amount of migration is dependent on patterns of breeding and dispersal in a species. Although the migration of certain animals like birds is apparent, many migrations are less obvious. For example, plant pollen may be carried long distances by wind or animals (Chapter 28). Seeds and fruits, which are formed after sexual reproduction has taken place, are often modified for dispersal, in some cases over long distances (Chapter

36). For example, coconut fruits may be transported hundreds, or even thousands, of miles by ocean currents.

If the amount of migration by members of a population is large, and if populations differ in their allelic frequencies, then significant changes can result. Humans have had a long period of relative isolation of population groups until recently (the past 300 years). The increase in gene flow has been significant in altering gene frequencies within various populations. For example, the United States has been a "melting pot" for racial groups, each of which originally had many attributes of a deme. American blacks have a frequency of the Rh^0 allele (one of the alleles of the Rh blood group) of 0.45. In contrast, the frequency of this allele in African blacks is 0.63. The reduced frequency in American blacks has been attributed to an influx of alleles from the white population in which the frequency of Rh^0 is very low, 0.03. Gene frequencies have changed in the white population, too, for gene flow has occurred in both directions. The corresponding change in the Rh^0 allele in white populations has not been as dramatic because the white population in the United States is larger than the black population.

Mutation

Variation is introduced into a gene pool through **mutation,** which is the source of all new alleles. Mutations result from a change in the nucleotide base pairs of the gene, from a rearrangement of genes within chromosomes so that their interactions produce different effects, or from a change in the chromosomes.

Mutations occur randomly and spontaneously. The rates of mutation are relatively stable for a particular locus, but vary by several orders of magnitude between loci within a single species, and between different species. In higher animals, a gene mutation rate of 3×10^{-5} (one gamete in 300,000) is representative. This may seem small, but the effect is significant when multiplied by the total number of loci. If we make the rather conservative estimate that humans have two copies of each of 100,000 genes (2×10^5) and an average mutation rate of 3×10^{-5}, then each of us on average carries 6 new mutant genes: $2 \times 10^5 \times 3 \times 10^{-5} = 6$. When this is multiplied in turn by the population size, it is evident that the mutation process indeed maintains a large supply of variability.

Since most mutations occur in somatic cells, they are not heritable. When the organism with the somatic mutation dies, the mutation dies with it. Some mutations, however, alter the DNA in the reproductive cells. These mutations may or may not affect the offspring, since most of the DNA in a cell is "silent" and does not code for specific polypeptides or proteins. If a mutation occurs in the DNA that codes for a polypeptide, it may still have little effect in altering the structure or function of that polypeptide. However, when the polypeptide is altered enough to change how it functions, the mutation is usually harmful.

Mutations produce random changes with respect to the direction of evolution. In a population adapting to a dry environment, mutations that are appropriate for adaptations to dry conditions are no more likely to occur than ones adapting to wet conditions or that have no relationship to the changing environment. The effect of a random change on the members of a population well adapted to its current environment is more likely to be harmful than beneficial.

Most mutations produce small changes in the phenotype that often are detectable only by sophisticated biochemical techniques. By acting against seriously abnormal phenotypes, natural selection eliminates or reduces to low frequencies the major deleterious mutations. Small mutations, even ones with slightly harmful phenotypic effects, have a better chance of being incorporated into the gene pool where at some later time they may produce traits that are helpful or adaptive for the population.

Natural Selection: Differential Reproduction

Genetic drift is random change. Gene flow and mutation may occur in a given direction, but the direction is unrelated to the nature of the environment. Only natural selection is adaptive and brings the variability in gene pools into harmony with the environment. It checks the disorganizing effects of the other forces and leads to adaptation.

Natural selection may operate at any number of different times in the life cycle of an organism. There may be nonrandom mating (which Darwin called "sexual selection"), nonrandom fecundity (that is, differences in the number of offspring produced), or nonrandom survival to reproductive age. The last is particularly common and frequently involves subtle interactions between organisms and the environment in which they live.

To study selection, biologists have developed measurements of the reproductive failure or success of one phenotype relative to other ones. The **selection coefficient,** S, is a measure on a scale of 0 to 1 of the elimination of one phenotype relative to the more successful one (0 = minimal elimination, 1 = maximal elimination). The **adaptive value,** or **fitness,** W, measures the relative success of the phenotype and is the complement $(1 - S)$ of the selection coefficient. Each phenotype is characterized by both a selection coefficient and an adaptive value, and their sum must equal 1. If 40% of one phenotype is eliminated relative to the alternative phenotype, then its selection coefficient $S = 0.4$ and its adaptive value $W = 1 - S = 0.6$.

Natural selection has two facets, for it both eliminates unfit individuals (negative selection) and favors the fit ones (positive selection). Selection is the only force known that can bring genetic variation into harmony with the environment and lead to adaptation. By eliminating the alleles with less favorable traits, selection changes the composition of the gene pool in a favorable direction and increases the probability that the favorable alleles responsible for an adaptation will come together in the same individuals.

GENETIC VARIATION

Changes in the types and frequencies of genes in gene pools, whether by natural selection or other means, are possible only if there is a source of inherited variation. It is clear that genetic variation is the raw material for evolutionary change. Without genetic variation there can be no differences in ability to reproduce and, therefore, no natural selection.

The gene pools of populations contain large reservoirs of genetic variation that have been introduced by mutation. Beneficial mutations are incorporated into the gene pool. Very harmful ones are eliminated, and neutral and mildly harmful ones may be carried in a population.

The reservoir of variability is large enough for populations to respond to selection pressures in many directions. This has been demonstrated in experiments with the fruit fly, *Drosophila.* A population is established from offspring of a single female fly fertilized in the wild. Her progeny have a certain variability and, by appropriate selection procedures, it is possible either to increase or decrease a trait such as the mean number of bristles in a certain part of the thorax and to extend the range of variability beyond that in the initial population (Figure 18–8). Since the changes occur in a few generations and can go in either direction, stored genetic variability rather than new mutations must be the source of the variation.

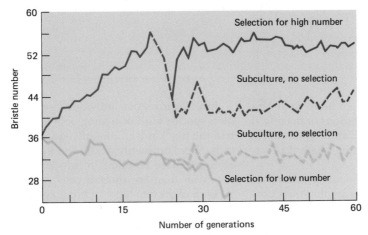

FIGURE 18–8 Genetic variation in *Drosophila melanogaster.* The original female fruit fly had 36 bristles on a part of her thorax. Investigators selected for flies with both a low number of bristles *(bottom solid green line)* and a high number of bristles *(top solid red line).* The dashed lines represent subcultures when selection was discontinued.

During sexual reproduction, recombination of alleles already present in a population further increases the population's variability, for it produces new combinations of alleles that result in new genotypes and new phenotypes. The effect of recombination can be surprisingly great. Nine different genotypes are generated in a dihybrid cross (*AaBb* × *AaBb*) involving only two genes on different loci, each with only two alleles. If we were dealing with five different genes at different loci, each with six alleles, the number of different genotypes possible would be 4,084,101! Some of the combinations that are generated may be adaptively superior, and natural selection could favor them.

Heterozygosity

Quite often the number of individuals heterozygous for a specific gene (*Aa*) are more common in the population than would be predicted by the Hardy-Weinberg law. It thus appears that the heterozygous condition, Aa, has a higher degree of fitness than either homozygote, AA or aa. This phenomenon, **heterozygote advantage,** is demonstrated in humans by the selective advantage bestowed on heterozygous carriers of the sickle cell allele (see Focus on Pleiotropy, Heterozygote Advantage, and the Sickle Cell Allele).

Individuals that are heterozygous for a number of different genes often demonstrate **hybrid vigor.** Hybrid corn, which is produced by crossing two parental strains that have a high degree of homozygosity, is valued for its consistently high productivity (Figure 18–9). The parental strains, which are much smaller and less productive than their hybrid offspring, are formed by **inbreeding,** or mating of genetically similar individuals. Homozygosity, or fixation of alleles, increases with each successive generation of inbreeding.

The reasons for heterozygote superiority are not always clear. By definition, a heterozygous individual has both dominant and recessive alleles. You may recall that most genes affect more than one trait and may be beneficial for some and deleterious for others. It may be that in heterozygous individuals the beneficial aspects of each allele are expressed in the phenotype. Or the unfavorable effects of each allele, which are evident in the homozygous condition, are not expressed in the presence of the other allele in the heterozygote.

Also, an individual with both alleles may produce two slightly different proteins that complement one another. Different proteins can be detected by techniques like gel electrophoresis (Figure 18–10). For example, if the two alleles code for slightly different versions of the same enzyme, called

FIGURE 18–9 Hybrid vigor. Hybrid corn *(center)* has many desirable characteristics not present in its two homozygous parental strains.

Focus on PLEIOTROPY, HETEROZYGOTE ADVANTAGE, AND THE SICKLE CELL ALLELE

Many genes are pleiotropic and affect more than one trait. Whether a mutant allele is deleterious or beneficial must be evaluated in the context of all its phenotypic effects in the particular environment in which it is acting.

The mutant allele for sickle cell anemia produces an altered hemoglobin, HbS, that is less soluble than normal hemoglobin, HbA, especially at low oxygen tensions. It tends to precipitate as long crystals within the red cell. This deforms or sickles the cell so that it is more likely to be destroyed in the spleen. Individuals who are homozygous for the sickle cell allele usually die at an early age.

Heterozygous individuals carry alleles for both normal and sickle cell hemoglobin. This heterozygous condition also causes the individual to be more resistant to a particularly virulent type of malaria caused by the protozoon, *Plasmodium falciparum*, than are individuals who are homozygous for the allele for normal hemoglobin. In certain parts of Africa, India, and Southern Asia where malaria is prevalent, heterozygous individuals survive in greater numbers than either homozygote (see map).

In a heterozygous individual, each allele produces its own specific kind of hemoglobin and the red cells contain the two kinds in roughly equivalent amounts. Such cells do not ordinarily sickle, and the red cells containing HbS are more resistant to infection from the malarial organism than are the red cells containing only HbA.

Each of the two types of homozygous individuals is at a disadvantage. Those homozygous for the sickling allele are likely to die of anemia, while those homozygous for the normal allele may die of malaria. The frequency of the sickle cell allele reaches 40% in the gene pools of black populations in parts of Africa.

Resistance to falciparum malaria is of no advantage in North America, and the frequency of the sickle cell allele in gene pools of American blacks has been reduced to 4 or 5%. This is due to dilution of the pool by gene flow from other populations and by selection against it.

Pleiotropic genes and those whose effects vary with different environments are very common. This suggests that a mutant allele that is harmful in a well-adapted organism living in a stable environment may become beneficial if the environment changes.

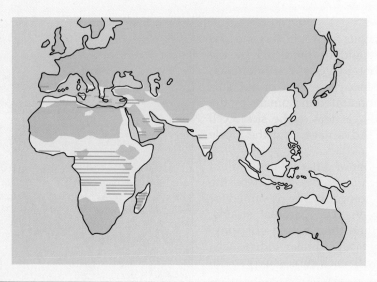

■ Map showing the distribution of sickle cell anemia (bars) compared with the distribution of falciparum malaria *(yellow region)*. The correlation strongly suggests that the resistance of heterozygous individuals to malaria has served to balance the deleterious effects of sickle cell anemia.

allozymes, this may confer a selective advantage for the individual that possesses both (Figure 18–11). Allozymes may function optimally under slightly different conditions. Presence of both allozymes allows the reaction catalyzed by the allozymes to occur over a wider range of conditions.

Heterozygosity maintains genetic variation. When heterozygous individuals are superior in fitness to either homozygote, it is unlikely that either the dominant or recessive allele will be eliminated. Sexual reproduction of the favored heterozygotes produces homozygotes of each type as well as more heterozygotes. This is the biological cost for heterozygosity. The organism must maintain a tremendous variability in its DNA, including alleles that are advantageous in the heterozygous condition but lethal

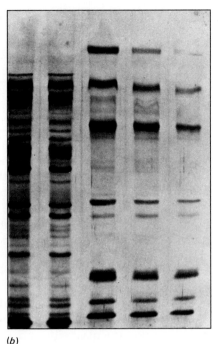

(a)

(b)

FIGURE 18–10 Comparison of proteins by gel electrophoresis. (*a*) In this apparatus, a strong electric potential is applied to a tube of gel containing a mixture of proteins. The proteins migrate in the gel in response to the electric potential, but their rate of migration depends upon their chemical structure. (*b*) When exposed to an appropriate reagent, the proteins can be observed and their patterns compared.

or deleterious when homozygous. This **genetic load** is the price a species pays for the advantage conferred by the heterozygous condition.

Polymorphism

Polymorphism is the presence of more than one allele for a given locus. A gene pool contains a tremendous reservoir of variability, much of it present at low frequency and much of it hidden. Until recently, biologists could not estimate the total amount of polymorphism in populations because they could recognize only those loci that have mutations conspicuous enough to have effects in breeding experiments. Biologists can now take a random sample of proteins from an organism and, by biochemical techniques such as electrophoresis, estimate how many exist in two or more forms as determined by different amino acid sequences. Each variety of a particular protein is coded by a different allele. Using this type of data, it is estimated that 25% of the loci in vertebrate populations are polymorphic.

Although any diploid individual may have only two alleles at any locus, the gene pool of a population may contain more than two alleles at a specific genetic locus. As an example of polymorphism in humans, the human

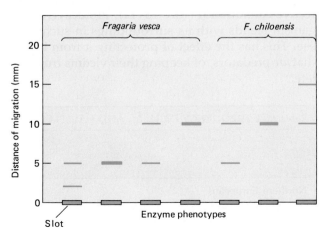

Slot

Enzyme phenotypes

FIGURE 18–11 Genetic variation in two species of wild strawberry (*Fragaria*) in California. Tissue extracts containing the enzyme peroxidase from separate individuals were placed in slots in a slab of gel (*bottom of drawing*). An electric current was applied to the gel, with the positive side at the top of the slab and the negative side at the bottom. Because peroxidase has a net negative charge, it migrated toward the positive side. Slight variations in amino acid sequence in the peroxidase molecules caused them to have slightly different negative charges and, therefore, migrate at different rates. Four different allozymes for peroxidase were found in the individuals studied. Of course, each individual can only possess a maximum of two different allozymes because strawberries are diploid. Individuals with two different allozymes for peroxidase are heterozygous, while those with only one allozyme are homozygous.

Table 18–1 FREQUENCY OF ABO BLOOD GROUPS IN HUMAN POPULATIONS				
	Frequency of Groups			
	O	*A*	*B*	*AB*
Northern Europeans	.40	.45	.10	.05
African blacks	.42	.24	.28	.06
American blacks	.67	.29	.03	.01

blood groups (A, B, AB, and O) are determined by three alleles, I^A, I^B, and i. All three alleles are present in the gene pool, but any given individual has no more than two of them.

The frequencies of the A, B, AB, and O blood types have been measured in several different human populations (Table 18.1). From these it is possible to calculate the gene frequencies for I^A, I^B, and i (Table 18.2). The frequency of the I^B allele varies most strikingly in different populations, with a 20-fold difference between the frequency of the I^B allele in African blacks and American Indians.

Of what possible selective advantage is a high degree of polymorphism? Populations with the genetic variability inherent in polymorphism maintain evolutionary plasticity and can respond to changing environmental conditions better. Polymorphism provides the diversity upon which natural selection can act.

Neutral Variation

Some of the genetic variation observed in a population may confer no selective advantage to the individuals possessing it. That is, it may not be adaptive. This **neutral variation** is very difficult to determine. It is relatively easy to prove that an allele is beneficial or deleterious, provided its effect is observable. But the variation in alleles, which is apparent by protein electrophoresis, often involving very slight differences, may or may not be neutral. These alleles may be influencing the organism in ways that are difficult to measure or assess. Also, an allele that is neutral in one environment may be beneficial or deleterious in another.

ADAPTIVE EVOLUTION

As a result of natural selection, organisms possess adaptations that increase their fitness for the particular environment in which they live. There are many examples of adaptive evolution. One is the development of **protective coloration** (Figure 18–12). Protective coloration exists when an organism blends with its surroundings in such a way as to make it hard to see. This has the effect of protecting it from its predators or, in organisms that *are* predators, of keeping their victims from noticing them until it is too late.

Table 18–2 FREQUENCY OF I^A, I^B, AND i ALLELES IN HUMAN POPULATIONS		
	Frequency of Alleles	
	I^A	I^B	i
Northern Europeans	.29	.08	.63
African blacks	.13	.23	.64
American Indians	.17	.01	.82

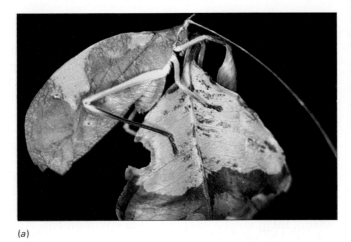

(a)

(b)

FIGURE 18–12 Camouflage conferred by protective coloration. (*a*) A leaf-mimicking katydid from Peru. (*b*) As with many ground-nesting birds, in their natural surroundings the chicks of nighthawks are almost invisible both to us and to predators.

Examples of protective form and coloration abound in nature. Some katydids resemble leaves so closely that you would never guess that they are animals—until they start to walk. The chicks of ground-nesting birds usually have feathers that blend in with the surrounding weeds and earth so that they simply cannot be discerned from a distance. Such protective coloration has been preserved and accentuated by means of natural selection.

An organism's close imitation of the appearance of an organism to which it is unrelated is called **mimicry.** One example is **Batesian mimicry,** or the resemblance of a harmless or palatable species to one that is dangerous, obnoxious, or poisonous (Figure 18–13). There are many examples. For instance, a harmless moth may resemble a bee so closely that even a biologist would hesitate to pick it up. The well-known monarch butterfly is poisonous to birds and mammals because of the toxic substances it absorbs while feeding on poisonous milkweed plants as a caterpillar. The monarch has many imitators among other butterflies, which closely resemble it in color but are entirely edible by birds. Natural selection has maintained a resemblance that gives its possessor almost as much protection as the "model," because as soon as predators learn to associate the distinctive markings of the model with its undesirable characteristics, they tend to avoid all similarly marked animals.

Protective coloration and mimicry are two examples of adaptive evolution brought about by natural selection. An organism that is well adapted to its environment has an increased chance it will survive and reproduce, passing some of its genes on to the next generation. We now examine some of the phenomena responsible for adaptive evolution of populations.

FIGURE 18–13 Batesian mimicry. (*a*) At the right is Jordan's salamander, *Plethodon jordanii*, a distasteful species. To its left is *Desmognathus imitator*, a palatable species. (*b*) (*top*) Few would want to get close enough to this insect to discover that it is actually a moth. (*bottom*) A genuinely noxious insect, the golden paper wasp.

(a)

(b)

445

Coadapted Gene Complexes

Natural selection does not operate directly on an organism's genotype. Rather, it operates on the phenotype, which is an expression of the genotype. The phenotype represents an interaction of all the alleles in the organism's genetic makeup. As we have seen, it is rare that a single gene pair has complete control over a single phenotypic trait, such as Mendel originally observed. Much more common is the interaction of several genes at different loci for the expression of a single trait. Many plant and animal characteristics are under this type of control.

Frequently, there is a collection of genes that affect the same trait or function, a **coadapted gene complex.** Due to natural selection, these genes interact compatibly. When they are present together, the organism possessing them is well adapted to its environment. Any alleles that interact unfavorably with the others have been eliminated, or their effects have been modified. The genes of a coadapted gene complex may occur in close proximity on the same chromosome. This increases the likelihood that they will be inherited together.

Selection

Natural selection changes gene frequencies in populations. It is important to recognize that the forces of natural selection do not cause the development of the "perfect" organism. Natural selection does not develop new phenotypes; rather, it "weeds out" those phenotypes that are less adapted to environmental challenges so that those that are better adapted survive and pass their genes on to their progeny.

We have learned that most traits are controlled by several different genes at different loci, i.e., polygenes. When traits are under polygenic control, such as human height, a range of phenotypes occurs with most of the population located in the median range and fewer at either extreme. This normal distribution forms a standard bell curve (Figure 18–14a).

There are three main processes of natural selection: stabilizing, directional, and disruptive selection (Figure 18–14b). These processes cause changes in the normal distribution of phenotypes in a population. Although we consider each process separately, their influences generally overlap in nature.

Stabilizing Selection

The process of natural selection that is associated with a population that is well adapted to its environment is known as **stabilizing selection.** Most

FIGURE 18–14 Different types of natural selection. (*a*) A trait, such as height, that is under polygenic control exhibits a normal distribution of phenotypes. (*b*) As a result of stabilizing selection, the curve is narrower and higher. Directional selection moves the curve in one direction. Disruptive selection results in two or more peaks.

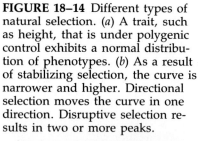
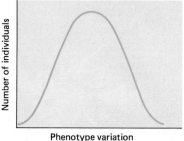

(*a*)

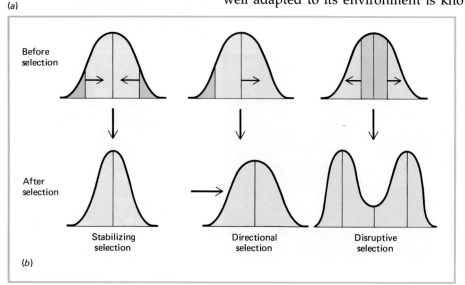

(*b*)

Focus on KIN SELECTION

Altruistic behavior, in which an individual appears to behave in such a way as to benefit others rather than itself, can be observed in the more complex social groups of animals. A particularly clear case of altruistic behavior has been observed by biologists Watts and Stokes in the mating of wild turkeys. Several differing groups of males, each with a dominance hierarchy, gather in a special mating territory and go through their displays of tail spreading, wing dragging, and gobbling in front of females who come to the area to copulate. One group attains dominance over other groups as a result of cooperation among the males within the group. The dominant male of the dominant group then copulates frequently with the females. The males who helped establish the dominant group, but have low status within it, appear to gain nothing. Close analysis, however, has shown that members of a group are brothers from the same brood. Since they share many genes with the successful male, they are indirectly perpetuating many of their genes. Altruism is closely related to kin selection, a type of natural selection where the behavior of one individual increases the likelihood of survival or reproduction of one or more genetically related individuals.

Kin selection may account for the evolution of the complex societies of social insects in which some individuals are specialized for reproduction while other close relatives do the chores of the colony. In the bee society, the workers are sterile females, and the queen functions vicariously as their reproductive organ. If the queen successfully produces offspring, a large portion of the genes shared by the queen and workers will have been passed on to the next generation, even though the workers themselves have not reproduced.

■ Prairie dog. The rodents live in large colonies in which a few act as sentries. Though the sentry places its life in grave danger when it exposes itself outside its burrow, it acts to protect its siblings and by so doing helps ensure that the genes they all have in common will be perpetuated in the population.

Another example of kin selection may be found among Florida jays. Here, nonreproducing individuals aid in the rearing of their siblings' young. Nests tended by these additional helpers as well as parents produce more young than nests with the same number of eggs overseen only by parents. By helping to care for their siblings' children, therefore, these individuals have a better chance of ensuring that at least some of their genes (the genes shared with their siblings) will be maintained in future populations.

populations are probably under the influence of stabilizing selection most of the time. In stabilizing selection phenotype extremes are selected against. In other words, those individuals with a phenotype near the mean are favored.

One of the most widely studied cases of stabilizing selection involves human birth weight, which is controlled primarily by a set of polygenes and by environmental factors. Based on extensive data from hospitals, it has been determined that infants born with intermediate weights are more likely to survive (Figure 18–15). Infants at either extreme, i.e., too small or too large, have higher rates of mortality. Stabilizing selection operates to reduce the population's variability so it is close to the weight with the minimum mortality rate.

Because stabilizing selection tends to decrease variation by favoring those individuals near the mean of the normal distribution at the expense of those at either extreme, the bell curve narrows. Actually, this decrease in variation represented by a narrower bell curve usually doesn't occur in nature because other forces act against it. Phenomena like mutation and recombination are continually adding to the variability of a population.

Stabilizing selection is particularly common in an environment that has been stable for an extended period of time. There are numerous examples

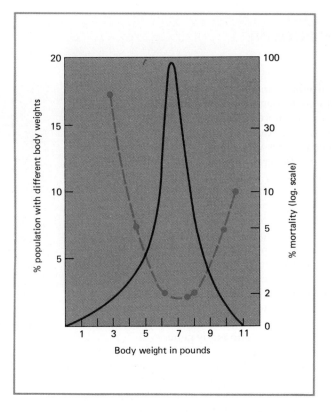

FIGURE 18–15 Human birth weight is an example of stabilizing selection. Infants with very low or very high birth weights have a higher mortality rate. Solid red line indicates number of infants at each birth weight; broken blue line indicates mortality rate at each birth weight.

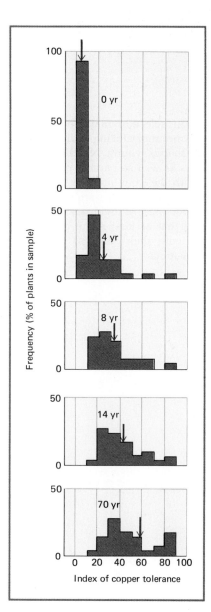

of organisms that have remained much the same for the past several million years. Based on fossil evidence, it is known that the ginkgo (Chapter 28) has not changed appreciably for approximately 200 million years. The lungfish and horseshoe crab also have not undergone evolution in millions of years. They are well adapted to the environments in which they live.

Directional Selection

If an environment changes over time, **directional selection** may favor those phenotypes at one of the extremes of the normal distribution (Figure 18–16). One phenotype may gradually replace another. Directional selection can only occur, however, if the appropriate allele (the one that is favored under the new circumstances) is already present in the population.

One classic example of directional selection is the peppered moth population studied in England (Figure 18–17). Most of the "peppered" moths *(Biston betularia)* in rural England have a black and white peppered wing color. Only a few are melanic, or all black. In industrial regions, the situa-

FIGURE 18–16 Directional selection of genes for copper tolerance in populations of *Agrostis stolonifera*, a grass. Individuals were taken from grass populations growing in areas that had been contaminated by copper (10–20 parts per million) from copper smelting areas in Lancashire, England. Year 0 represents grass growing in uncontaminated soil, while the other graphs represent grass growing in soil known to be contaminated by copper for the specified time (4 to 70 years). All plants were grown in a nutrient solution with 0.5 parts per million of copper. Root growth was measured to give an index of copper tolerance: 0 = no growth (complete inhibition) and 100 = maximum growth (no inhibition). Plant roots taken from uncontaminated soils barely grew in the experiment, while plant roots from copper-contaminated soil grew progressively better at increasing ages. The arrow indicates the mean; note how the arrow progresses in one direction with an increase in time. Copper tolerance is under polygenic control.

(a)

(b)

FIGURE 18–17 Studies of peppered moth populations in England indicate directional selection. There are two forms of the peppered moth, the black and white "peppered" form and the dark melanic form. (*a*) Protective coloration hides the peppered form on lichen-covered trees, while the dark form is obvious. Dark forms are less common because birds eat more of them. Much of rural England has lichen-covered trees that favor the peppered form. (*b*) In industrial areas, air pollution has killed the lichens that cover the tree bark in rural areas. The melanic form of peppered moth is favored here because it is less obvious than the peppered form. It is interesting that lichens are reappearing on the trees in industrial areas where the air quality has been improved. The peppered form is making a comeback (due to directional selection) in those areas.

tion is reversed and most of the moths are black, while only a few are "peppered." Moths rest during the day on tree trunks, where some are eaten by birds. Professor H.B.D. Kettlewell of Oxford University postulated in the 1950s that the peppered pattern is less conspicuous on the lichen-covered trees of rural areas, while the melanic form is less conspicuous on trees in industrial areas. (Lichens are quite susceptible to air pollution and, therefore, would not be found on tree trunks in industrial areas.) To test this, hundreds of male moths of each type were raised, marked with a spot of paint under their wings, and released in both rural and industrial areas. The survivors were recaptured after a period of time by attracting them with light or females. Significantly more melanic forms survived in the industrial areas, and more peppered forms survived in the rural areas.

Natural selection was and is shaping the gene pool of each deme of moths to local conditions. In some localities, directional selection operated toward the peppered phenotype. In other localities, selection occurred in the opposite direction toward the melanic form. Seldom, however, does a population become entirely one type. Male peppered moths fly considerable distances, so gene flow between populations helps to maintain some polymorphism.

Disruptive Selection

Sometimes extreme changes in the environment may favor two or more variant phenotypes at the expense of the mean. That is, more than one phenotype may be favored in the new environment, while the average phenotype originally present is selected against. **Disruptive selection** is a special type of directional selection in which there is a trend in several directions rather than one.

A clear example of disruptive selection involves Batesian mimicry. In some localities in Africa there are three different distasteful species of butterfly. Different females of the edible swallow tail butterfly, *Papilio dardanus*, mimic each of the distasteful models (Figure 18–18). Disruptive selection has favored varieties of the swallow tail that resemble any of the model species. The initial single population has been disrupted into three different populations that differ in their color pattern as each has mimicked a different distasteful model.

We have seen that evolution is a change in gene frequencies in the gene pool of a species population. Natural selection, originally proposed by Charles Darwin, is the most significant factor in changing gene frequencies in populations, whether it involves preserving the status quo or favoring trends in one or more directions.

(a)

(b)

(c)

species gene pool because gene flow between them is prevented. To block a chance occurrence of individuals from two different species overcoming one isolating mechanism, most species have two or more isolating mechanisms.

Isolating mechanisms that work to restrict the gene flow between species also may be found *within* a species. Each species is composed of local populations, or races, that are separated geographically and/or ecologically. This results in limited genetic exchange between populations. Sometimes local populations, in adapting to local conditions, diverge to the point where they become reproductively isolated from the rest of the species. This may lead to the formation of a new species.

Prezygotic Isolating Mechanisms

There are two groups of isolating mechanisms, prezygotic and postzygotic. **Prezygotic isolating mechanisms** prevent fertilization from ever taking place. Since male and female gametes never come into contact, an interspecific zygote never forms.

Temporal Isolation

Sometimes genetic exchange between two groups is prevented because they reproduce at different times of the day, season, or year. There are many examples of **temporal isolation.** The fruit flies, *Drosophila pseudoobscura* and *Drosophila persimilis*, have ranges that overlap to a great extent, but they do not interbreed. *D. pseudoobscura* is sexually active in the afternoon and *D. persimilis* in the morning.

Similarly, there are two species of sage, *Salvia*, with overlapping ranges in southern California. *Salvia mellifera* (black sage) flowers in early spring, while *S. apiana* (white sage) blooms in late spring and early summer.

Ecological Isolation

Although two closely related species may be found in the same geographical area, they usually live and breed in different habitats in that area. This can cause reproductive isolation between the two groups (Figure 19–2). For example, wood frogs breed in temporary woodland ponds, while bullfrogs breed in larger, permanent bodies of water. They are separated by **ecological isolation.**

Behavioral Isolation

Many animal species have distinctive courtship behaviors, so mating between species is prevented by **behavioral isolation.** Courtship is an exchange of signals between a male and a female. A male approaches a female and gives a sign or pattern of signals that may be visual, auditory, or chemical. If the female belongs to the same species, she recognizes the signals and returns her own distinctive signals. Further correct exchanges of signals eventually result in mating. If members of two different species

FIGURE 19–2 Reproductive isolating mechanisms in closely related flycatcher species in North America. Although the flycatchers are nearly identical in appearance and have overlapping ranges, they remain as distinct, reproductively isolated species. They are isolated ecologically because each species is found in a particular habitat within its range during mating. Also, they are isolated behaviorally because each species has its own characteristic song, which serves to identify it to other flycatchers of the same species. (a) Least flycatcher, *Empidonax minimus.* (b) Acadian flycatcher, *E. virescens.* (c) Traill's flycatcher, *E. trailii.*

begin courtship, one partner may not recognize one of the signals and may fail to respond. The courtship behavior will stop at that point.

Fruit flies, for instance, exhibit a definite, species-specific courting behavior (Figure 19–3). Part of the behavior is a "love song," a series of buzzes of just the right pitch and rhythm performed by the male. Differences in "love songs" keep some species of *Drosophila* apart.

Mechanical Isolation

Morphological or anatomical differences which inhibit mating between species are known as **mechanical isolation.** Sometimes members of different species will court and attempt copulation, but the structure of their genital organs is incompatible, so successful mating is prevented. The interbreeding of certain insect species is thwarted in this way.

Many flowering plants have physical differences in their flower parts that help them maintain their reproductive isolation from one another. The sage plants that were used earlier as an example of temporal isolation also have mechanical isolation. *Salvia mellifera,* which is pollinated by small bees, has a different floral structure than *S. apiana,* which is pollinated by large carpenter bees. The differences in floral structures prevent the insects from cross-pollinating the two species.

Gametic Isolation

If mating has occurred between two species, the union of gametes may still not occur. Molecular and chemical differences between species cause **gametic isolation.** The egg and sperm may simply be incompatible. In aquatic animals that release their eggs and sperm into the surrounding water simultaneously, interspecific fertilization is extremely rare. There is evidence that the egg surfaces contain specific proteins that will bind only to complimentary molecules on sperm cells of the same species.

Sometimes plant pollen doesn't germinate on the stigma of a separate species. Alternatively, the pollen may germinate and grow a pollen tube, but fertilization still doesn't occur. Different species, and even different races in the same species, have different style lengths and pollen tube lengths. That is, the pollen tube of a particular species is coordinated to grow the length of the style that is characteristic for that species. If two races or species have different flower sizes and, therefore, different style lengths, they may be incompatible. Pollen that is genetically programmed to grow a short pollen tube cannot grow the entire length of a long style. Thus, fertilization does not occur.

Postzygotic Isolating Mechanisms

When prezygotic isolating mechanisms fail, as they occasionally do, **postzygotic isolating mechanisms** may come into play. This ensures reproductive failure even though fertilization took place. In Chapter 18 we found that the offspring formed from two different populations within a species, known as hybrids, often exhibit advantages over either parent. By contrast, hybrids formed from the union of two separate species usually have numerous problems and are at a severe disadvantage.

Hybrid Inviability

Generally, the embryonic development of an interspecific (between-species) hybrid is aborted. Development is a complex process requiring the precise interaction and coordination of many genes. Apparently, the genes from parents belonging to different species do not interact properly in reg-

(a)

(b)

(c)

FIGURE 19–3 Courtship and mating in the fruit fly, *Drosophila.* (a) The male follows the female and vibrates his right wing, forming a buzzing noise. Each species of fruit fly produces its own characteristic sound and rhythm. (b) In the second stage, the male licks the genitilia of the female while continuing to vibrate. (c) Mating follows.

ulating the mechanisms for normal embryonic development. Isolation is achieved by **hybrid inviability.** For example, nearly all of the hybrids die in the embryonic stage when the eggs of a bullfrog are fertilized artificially with sperm from a leopard frog.

Plants also exhibit hybrid inviability. For example, in crosses between different species of *Iris,* the hybrid embryo develops but dies before reaching maturity as a result of breakdown of the endosperm in the seed.

Hybrid Sterility

If an interspecific hybrid develops successfully, reproduction of the hybrid may still not occur. There are several reasons why this is so. Hybrid animals may exhibit courtship behaviors incompatible with those of either parental species. As a result, they will not mate.

More often, the gametes of an interspecific hybrid are abnormal due to problems during meiosis. This is particularly true if the two species have different chromosome numbers; synapsis, pairing of homologous chromosomes, cannot occur properly. For example, a mule is the offspring of a female horse ($2n = 64$) and a male donkey ($2n = 62$). This type of union almost always results in sterile offspring ($2n = 63$) (Figure 19–4).

Many examples of **hybrid sterility** in plants have been documented. Sometimes the interaction of genes from two species causes a hybrid's anthers to develop improperly. Such male sterility has been found in hybrids between different tobacco (*Nicotiana*) species.

Hybrid Breakdown

Occasionally, an interspecific hybrid offspring develops that is fertile and produces a second, F_2, generation. However, the second generation, which develops from a cross between two hybrids or between a hybrid and one of the parent strains, has defects that prevent it from successfully reproducing. For example, **hybrid breakdown** in the F_2 generation of a cross between two sunflower species in the *Layia* genus was 80%. In other words, 80% of the F_2 were defective. Hybrid breakdown can also occur in the F_3 and later generations.

SPECIATION

There are two distinct kinds of evolution. Phyletic evolution, or **anagenesis,** refers to the conversion of an entire population over time to a form so different from the original species that it's considered a new species. That is, a sequence of species occurs over time, without an increase in

FIGURE 19–4 Hybrid sterility. Mules are interspecific hybrids formed by mating a female horse with a male donkey. Although the mule exhibits valuable characteristics of each of its parents, it is sterile.

FIGURE 19–5 Adaptive radiation. All the various mammals shown are deduced, on the basis of comparative anatomy and, to a somewhat lesser extent, the fossil record, to have evolved from the common shrewlike ancestor depicted in the center. Each of the organisms is specifically adapted to a different ecological niche.

the number of species. Diversifying speciation is known as **cladogenesis,** in which one or more new species are derived from the parent species, which continues to exist. This gives rise to two or more species from a single ancestral one. Sometimes several to many species are formed from a single ancestral species by repeated cladogenesis. This is known as **adaptive radiation** (Figure 19–5). The process of cladogenesis is not only more common than anagenesis, but it is the only process that increases biological diversity (increases the total number of species).

We are now ready to consider how entirely new species may arise from previously existing ones. A required step in the evolution of a new species is the reproductive isolation of a population from the rest of the species. When the population is sufficiently different from its ancestral species so that no genetic exchange can occur between them, even if the two populations meet, we say that speciation has occurred. There are two main types of speciation, allopatric and sympatric.

Allopatric Speciation

Speciation that occurs when one population becomes geographically separated from the rest of the species and subsequently evolves is known as geographical speciation, or **allopatric speciation.** Allopatric speciation is believed to be the most common method of speciation and has been the most important in the evolution of new species of animals.

Mechanisms of Allopatric Speciation

There are several ways this geographic isolation might occur. The Earth's surface is in a constant state of change. Rivers change their courses. Glaciers migrate. Mountain ranges form. Land bridges develop, separating previously united aquatic populations. Large lakes diminish into several smaller, geographically separated pools. It is important to recognize that what might be an imposing geographical barrier to one species may be of no consequence to another. Each species has its own methods of dispersal. For example, as a lake subsides into smaller pools, fish are usually unable to cross the land barriers between the pools and so become isolated. Birds, on the other hand, can easily fly from one pool to another. Likewise, hydrophytic plants, such as cattails, that disperse their fruits by air currents, would not be isolated by this barrier.

Alternatively, a small population may migrate and colonize a new area, away from the original species range. This colony would be geographically isolated from its parent species. The Galapagos Islands and the Hawaiian Islands represent examples of geographical areas that were colonized by individuals of a few species. From these original colonizers, the distinctive groups of unique species characteristic of each island arose (Figure 19–6).

No species is genetically uniform throughout its range. There are slight differences within a species due to adaptations to local conditions, with the populations at the periphery of the species range frequently exhibiting the most distinct differences. These differences are often enough to subdivide the species into varieties, or races. When geographical isolation occurs, it usually separates a small population at the periphery of the species range from the rest of the species. This population was already genetically different from the species members in the middle of its range. This genetic divergence is amplified by selection and random changes such as genetic drift.

Because the population is geographically isolated, there is no interbreeding with the rest of the species and, therefore, no gene flow between it and the parent species. Moreover, the isolated habitat may be different in several ways from the parent species habitat. Climate and soil factors will be distinct, and there will be a different set of biological organisms with which the isolated population must interact. As a result of these habitat differences, the isolated population will face different selective pressures. Most small populations that are faced with these new selective pressures will not speciate but will become extinct. However, some populations will survive. Over time, the isolated population that survives will adapt to the new habitat, and its gene pool will diverge from the gene pool of the original

(a)

FIGURE 19–6 (*a*) The nene (pronounced "nay-nay"), *Branta sandvicensis,* is a goose found in the Hawaiian islands. It is believed to have evolved from a small population of geese that originated in North America. (*b*) The Canadian goose, *Branta canadensis,* is believed to be a close relative of the Hawaiian goose. Although the nene is endangered, strict conservation measures have brought it back from the brink of extinction.

(b)

species. Eventually, the differences between the two will be so great that they will be unable to interbreed even if their range becomes continuous again.

Speciation is more likely to occur if the original isolated population is small. You will recall that the founder effect and genetic drift are more influential in small populations (Chapter 18). These factors tend to cause rapid genetic changes in the isolated population. The genetic divergence caused by the founder effect and genetic drift are further accentuated by the different set of selective pressures to which the population is exposed.

Occasionally, the geographical barrier may disappear after a period of isolation. A river might change its course, or a glacier might retreat. Alternatively, members of the isolated population may migrate back to the habitat of the original species. There are three possibilities when an isolated population is reunited with the parent species. If the population has diverged enough that it is unable to mate with the parent species, speciation has occurred and the divergent population is recognized as a separate and distinct species. A second possibility is that speciation has not occurred. In this case, the population will be able to mate successfully with the parent species, and gene flow will be restored.

There is a third possibility when geographical barriers are removed between an isolated population and its parent species. It may be that speciation has not quite occurred, but the isolated population is very different from the original species. This results in very limited gene flow between the two groups. Usually, natural selection causes the two groups to diverge even more in this type of situation, in a process known as **character displacement.** This prevents the two groups from directly competing, since the differences between them place them in different niches in the same environment.

There are well-documented examples of character displacement between two closely related species, as well (Figure 19–7). For example, the flowers of two species of *Solanum* found in Mexico are very similar in areas

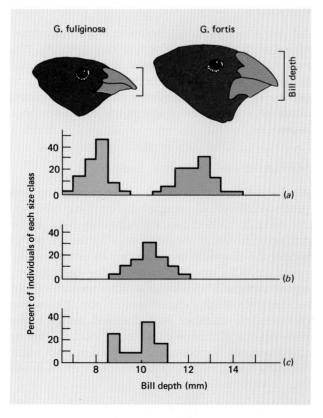

FIGURE 19–7 Character displacement in bill depth in two species of finches from the Galapagos, *Geospiza fuliginosa* and *G. fortis.* Birds with larger bill depths can crack larger seeds. (*a*) When the two species are found in the same location, *G. fuliginosa* (blue) has a smaller average bill depth than *G. fortis* (green). (*b*), (*c*) When they occur on separate islands, the average bill depths of each are similar. (*b*) is *G. fortis* and (*c*) is *G. fuliginosa.*

where the two species do not overlap. In areas where their ranges overlap, there is a noticeable difference in flower size. Because of this difference, the flowers are pollinated by different bees. In other words, character displacement reduces interspecific competition, in this case for the same kinds of bees.

Examples of Allopatric Speciation

Many examples of allopatric speciation can be traced to the barriers formed by the glaciations of the Pleistocene epoch. A western population of the Pleistocene European bear, *Ursus arctos,* was separated from the rest of the species and evolved into the cave bear, *Ursus spelaeus.* The eastern population remained as *Ursus arctos.* This reconstruction has been supported by fossil evidence.

Given the requirements for speciation we have considered, allopatric speciation may be quite rapid. Early in the 15th century, a small population of rabbits was released on Porto Santo, a small island off the coast of Portugal. There were no rabbits or other competitors and no carnivorous enemies on the island, and the rabbits thrived. By the 19th century, they were markedly different from the ancestral European stock. They were only half as large and had a different color pattern. Moreover, their lifestyle was different, as they were more nocturnal. Most significant, they could not produce offspring when bred with members of the ancestral European species. Within 400 years, a new species of rabbit had developed.

Lakes and pools of water provide the isolation for allopatric speciation of aquatic organisms that islands provide for terrestrial plants and animals. Large lakes formed by glacial melt at the end of the Pleistocene in what is now Nevada were populated by one or several species of pupfish. With the gradual demise of the large glacial lakes as a result of glacial retreat and a dryer climate, isolated pools were left. Today, there are numerous species of pupfish, but each is restricted to a single water hole.

Sympatric Speciation

Although geographical isolation is an important factor in many cases of evolution, it is not an absolute requirement. When a population forms a new species within the same geographical region as its parent species, **sympatric speciation** has occurred. The divergence of two gene pools that occurs in the same geographical range is especially common in plants, although some cases of sympatric speciation in animals have been documented.

Mechanisms of Sympatric Speciation

We have seen that hybrids formed from the union of two species rarely produce robust offspring and that these offspring are usually sterile, like the mule. Sterility is generally the case because the two parent species have different chromosome numbers. During gametogenesis, meiosis occurs to reduce the chromosome number. In order for the chromosomes to be parcelled correctly into the gametes, the homologous chromosomes pair during metaphase I. This cannot occur properly in the hybrid offspring of two species because the chromosomes aren't homologous. However, *if* the 2n chromosome number was doubled before meiosis, then the pairing of homologous chromosomes could occur. This spontaneous doubling of chromosomes has been documented in both plants and animals. It is not a common occurrence, but neither is it rare. It produces nuclei with multiple sets of chromosomes.

FIGURE 19–8 Polyploidy in day lilies *(Hemerocallis fulva)*. The tetraploid (4n) on the right is larger and more robust than the diploid (2n) on the left, from which it is derived.

Polyploidy is the possession of more than two sets of chromosomes (Figure 19–8). When it occurs in conjunction with the joining of chromosomes from two different species, it is known as **allopolyploidy,** and it can cause the hybrid to be fertile. This is because the polyploid condition provides the homologous chromosomes that can pair during meiosis. As a result, the gametes may be viable (Figure 19–9).

Allopolyploids can mate with themselves (self-fertilization) or with similar individuals. However, they are reproductively isolated from both parents because the gametes of the allopolyploid have a different number of chromosomes than those of either parent.

If a population of allopolyploids (i.e., a new species) becomes established, selective pressures will cause one of three outcomes. It is very pos-

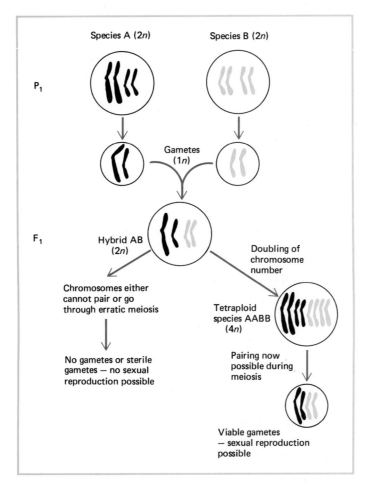

FIGURE 19–9 How a fertile allopolyploid is formed. Interspecific hybridization occurs between two species, yielding a hybrid F₁ generation. If doubling of the chromosomes does not occur, they will be unable to undergo normal meiosis, and the hybrid will be sterile *(left)*. If the chromosomes double, the hybrid will be able to undergo meiosis and will be fertile *(right)*.

FIGURE 19–10 An allopolyploid primrose, *Primula kewensis,* arose during the 20th century. The F$_1$ hybrid of *P. floribunda* (2n = 18) and *P. verticillata* (2n = 18) was a diploid perennial (2n = 18) and was sterile. Three different times it spontaneously formed a fertile branch, which was a polyploid (2n = 36) and produced seeds.

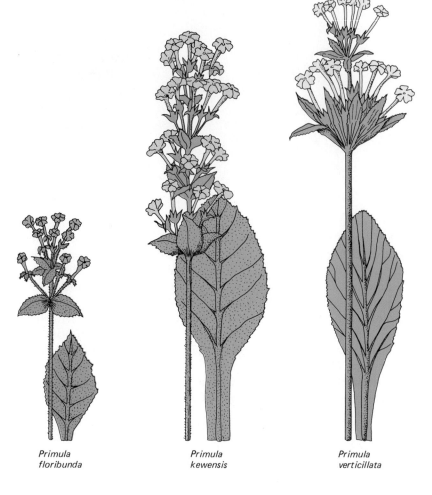

Primula floribunda

Primula kewensis

Primula verticillata

sible that the population will be unable to compete and it will become extinct. A second possibility is that the allopolyploid individuals will fill a new niche in the environment and so coexist with both parent species. A third possibility is that the new hybrid species may compete for the niche occupied by either of its parent species. If it has a combination of characters that make it more fit than the parent species for all or part of the original range of the parent, the hybrid species will replace the parent.

Although hybridization/polyploidy is extremely rare in animals, it has been a significant factor in the evolution of the flowering plants. Slightly less than one half of all flowering plants are believed to be polyploid. Most of these are allopolyploids (Figure 19–10). Moreover, hybridization/polyploidy provides a mechanism for extremely rapid speciation. A single generation is all that is needed to form a new, reproductively isolated species. Hybridization/polyploidy is believed to explain the rapid appearance of flowering plants in the fossil record and the incredible diversity in flowering plants today (more than 250,000 species).

Examples of Sympatric Speciation

Several species of hemp nettle occur in temperate parts of Europe and Asia. One of these, *Galeopsis tetrahit* (2n = 32), is a naturally occurring allopolyploid, which was formed by the hybridization of two species, *G. pubescens* (2n = 16) and *G. speciosa* (2n = 16). This speciation, which occurred in nature, was experimentally reproduced in the laboratory. *G. pubescens* and *G. speciosa* were crossed to produce F$_1$ hybrids that were mostly sterile. Nevertheless, both F$_2$ and F$_3$ generations were formed. In

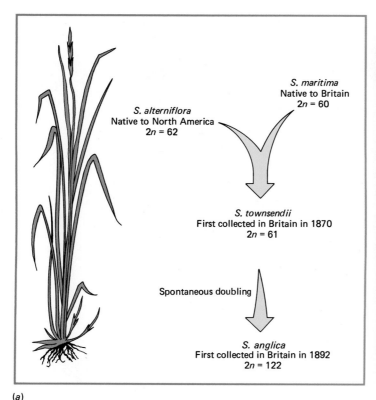

(a)

(b)

FIGURE 19–11 How the allopolyploid *Spartina anglica* was formed. (a) This plant formed in nature by the interspecific hybridization of two species of cordgrass and the subsequent doubling of the hybrid's chromosomes. (b) Close-up photo of *S. alterniflora* growing in a salt marsh.

the F_3 there was a plant with 2n = 32, which yielded fertile F_4 offspring. These artificial, allopolyploid plants had the same morphology and chromosome number as the naturally occurring *G. tetrahit*. When the experimentally produced plants were crossed with the naturally occurring *G. tetrahit*, a fertile F_1 generation was formed.

A new species of cordgrass was formed in nature by hybridization/polyploidy in the recent past (Figure 19–11). The parent species were *Spartina maritima*, which is native to Europe and has a diploid chromosome number of 60, and *Spartina alterniflora*, which is native to North America and has a diploid chromosome number of 62. *S. alterniflora* was accidentally introduced to Europe around 1860. In approximately 10 years, a new species of cordgrass appeared along the English coast. *Spartina townsendii* was formed by hybridization and had a diploid chromosome of 61. It was sterile but reproduced by vegetative propagation. Shortly after, a vigorous species of *Spartina* arose by chromosome doubling in *S. townsendii*. The new species, *S. anglica*, was a polyploid with a chromosome number of 122. It spread rapidly and soon was found on both sides of the English channel.

PACE OF EVOLUTION

In studying the fossil record, it becomes apparent that the pace of evolution varies from one group of organisms to another. In some instances it is fast, and in others it is slow. It appears that certain major groups have evolved a great deal in a relatively short amount of time, for example, mammals and flowering plants. Other organisms appear relatively unchanged, sometimes for millions of years. Mosses do not appear to have evolved much in the past 4 million years, and some organisms like the lungfish appear much the same as in 150-million-year-old fossils.

Although the pace of evolution varies, there is no question that it accelerates when there are strong selective forces. A changing, challenging environment causes rapid evolutionary change in populations.

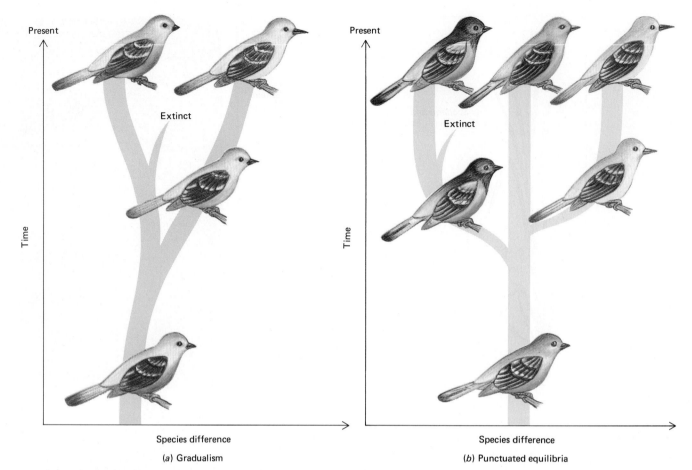

Present

Time

Species difference

(a) Gradualism

Extinct

Present

Time

Species difference

(b) Punctuated equilibria

Extinct

FIGURE 19–12 There are two theories about the pace of evolution. (*a*) In gradualism there is a slow, steady change in species over time. (*b*) In punctuated equilibria there are long periods of little evolutionary change (stasis) followed by short periods of rapid speciation.

While evolutionary biologists generally agree that natural selection is the main mechanism responsible for speciation, there is currently much debate on the timing of evolutionary change during a species' existence. Biologists fall into two groups in this debate, and each has compelling evidence. Sometimes the two groups use the same evidence, interpreting it differently to support opposing ideas. One group supports **gradualism,** a slow and steady accumulation of changes, while the other group proposes **punctuated equilibria,** in which evolution proceeds with periods of inactivity followed by very active phases (Figure 19–12). It is also possible that both may be correct: Evolutionary pace may be quite erratic, fast at times, slow at others.

Gradualism

Gradualism represents the traditional approach to evolution. It propounds that populations slowly diverge from one another by the accumulation of adaptive characteristics within a population. These adaptive characteristics accumulate as a result of different selective pressures brought on by the populations living in different environments.

If a species evolves by gradualism, there will be a number of intermediate steps, or "missing links." Proponents point to the fossil record of the evolution of the horse, with its gradual increase in size, as an example of gradualism (Figure 19–13). They further argue that there are few transitional forms in the fossil record because the fossil record is incomplete. A strong case for gradualism was recently presented in trilobite evolution (see Focus on Trilobites—Evidence of Gradualism).

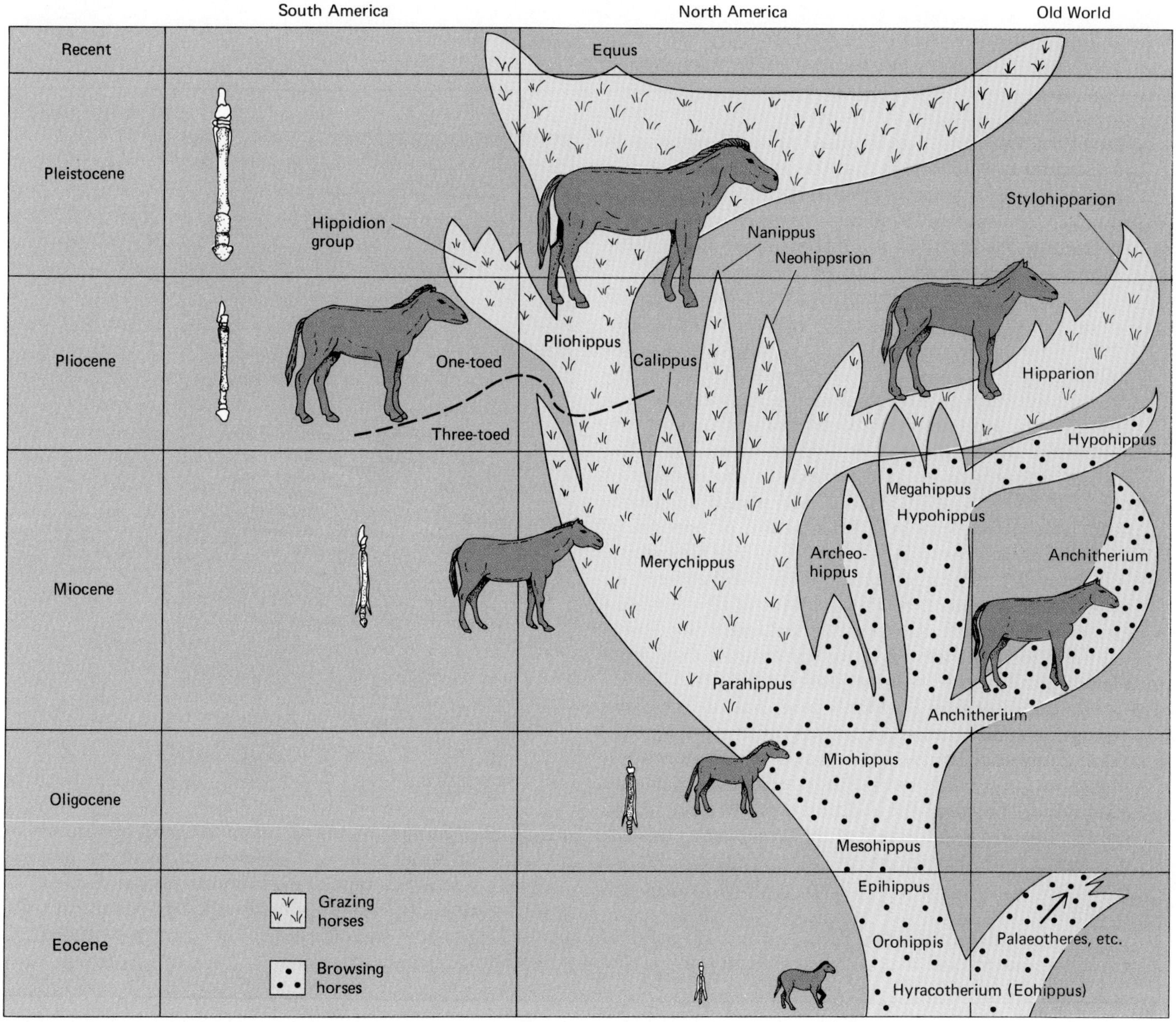

	South America	North America	Old World
Recent		Equus	
Pleistocene	Hippidion group		Stylohipparion
Pliocene	One-toed / Three-toed	Pliohippus, Calippus, Nanippus, Neohippsrion	Hipparion, Hypohippus
Miocene		Merychippus, Parahippus, Archeohippus, Megahippus, Hypohippus	Anchitherium
Oligocene		Miohippus, Mesohippus	Anchitherium
Eocene		Epihippus, Orohippis, Hyracotherium (Eohippus)	Palaeotheres, etc.

Grazing horses

Browsing horses

FIGURE 19–13 The evolution of the horse is often cited as evidence for gradualism. Note the slow increase in size over time.

The fact that there is abundant evidence in the fossil record of long periods of no change in a species (**stasis**) seems to argue against gradualism. However, proponents believe that stasis in fossils is deceptive because fossils don't show all aspects of evolutionary change. Fossils can show changes in external anatomy and skeletal structure, but such characteristics as internal anatomy, molecular changes and behavioral changes, which also represent evolution, will not be revealed by fossils.

Punctuated Equilibria

Many evolutionary biologists support punctuated equilibria, whereby evolution of new species normally proceeds in "spurts." These relatively short periods of active evolution produce new species and are followed by long periods of little or no evolutionary change (stasis). Then evolution resumes, new species form, and many old ones are out-competed and become extinct. According to punctuated equilibria, most of a species' existence is spent in stasis, and a very small percentage is spent in active evolutionary change. It is important to realize that a "short" amount of time for speciation may mean thousands of years. Such a period of time is

Focus on TRILOBITES—EVIDENCE OF GRADUALISM

One of the criticisms of gradualism made by proponents of punctuated equilibria is that the fossil record shows scant evidence of a gradual transition during the evolution from one species to another. In other words, there are few intermediate forms, which would indicate slow, progressive change.

In 1987, a paper was published in *Nature* about an exhaustive study of approximately 15,000 fossil trilobites from a 3-million-year period. Peter Sheldon studied eight lineages of the small, invertebrate marine organisms, concentrating on the number of ribs in the exoskeleton of each. He found that each lineage showed gradual change during the 3 million years, with each of the lineages showing a gradual increase in the number of ribs. There was no evidence of a long period of equilibrium (stasis) followed by a brief period of speciation.

The significance of ribs in trilobites, which are extinct, is unknown. One suggestion is that each rib covered an appendage. Another is that the addition of ribs provided extra strength. It is also possible that extra ribs had a neutral effect, but were selected for because they were pleiotropically linked to other beneficial traits.

The publication of Sheldon's work has reignited the discussion about gradualism versus punctuated equilibrium. Some supporters of punctuated equilibrium interpret Sheldon's work quite differently. They say that such minor change as the addition of a few more ribs to the trilobite exoskeleton in 3 million years is equivalent to stasis. And they point to two recent studies on bryozoa and clam fossils that clearly support punctuated equilibria.

Some scientists are of the opinion that both types of evolution may occur, that the pace of evolution may be

Area of new ribs

■ Fossil trilobites showing the area of new ribs.

steady and gradual in certain instances and abrupt in others. Even if that is the case, scientists are still going to want to know which type of evolution is most important and why some organisms appear to undergo change in a gradual manner while others have no change followed by sudden, dramatic change.

short when compared to the period of time a species exists. Evolutionary biologists that support punctuated equilibria mention that sympatric speciation and even allopatric speciation can occur in a relatively short period of time.

When rapid evolutionary change occurs, it may be due to several factors. It is possible that major changes in the environment could trigger speciation by mechanisms already discussed. Another possibility is that a major genetic change, such as a significant reorganization of chromosomes, could account for sudden speciation. This genetic change would have to alter the organism's development so that the offspring would be quite different from the parents. At this point natural selection would become involved.

Punctuated equilibria accounts for the abrupt appearance of a new species in the fossil record, with little or no record of intermediate forms. That is, proponents believe there are few transitional forms in the fossil record because there were no transitional forms during speciation. Even the fossil evidence for the evolution of the horse can be interpreted to support punctuated equilibria. A close examination of the fossil record of horse evolution shows that at each "stage" there were both larger and smaller forms. Proponents argue that the natural selection of the larger forms only gives the appearance of a gradual change in size. Finally, the fossil record shows long periods of little change. This stasis is supported by punctuated equilibria.

EXTINCTION

Extinction, the end of a lineage, occurs when the last individual of a species dies. It is a permanent loss, for once a species is extinct it can never reappear. Extinctions have occurred continually since the origin of life on Earth. By one estimate, there is only 1 species living today for every 2000 that have become extinct. Extinction is the eventual fate of all species, in the same way that death is the eventual fate of all living things.

While extinction does have a negative impact on biological diversity, it has one positive evolutionary aspect. When species become extinct, the ecological niches they occupied become vacant. As a result, those organisms still living evolve and radiate out to fill the unoccupied niches. In other words, the extinct species are replaced by new species.

During the course of life on Earth, there appear to have been two types of extinction. The continuous, low-level extinction of species, sometimes called **background extinction,** is one. The second type has occurred five or six times during Earth's history. At these times, **mass extinctions** of numerous species and higher taxa have taken place. The time periods when mass extinctions occurred may have been for millions of years, but that is a relatively short period compared with the history of life on Earth. Each period of mass extinction, which appears to have been indiscriminate in its choice of which species survived and which became extinct, was followed by a period of "mass speciation" (Figure 19–14).

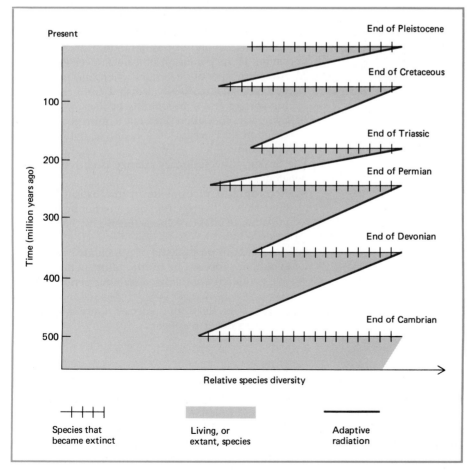

FIGURE 19–14 Mass extinctions have taken place a number of times in the Earth's history. The period following each mass extinction had a high amount of speciation as organisms evolved to fill the ecological niches left by the extinct forms.

The causes of extinction, particularly mass extinction, are not well understood. Both environmental and biological factors seem to be involved. Major changes in the climate could adversely affect plants and animals that are unable to adapt to them. Marine organisms, in particular, are adapted to a very steady, unchanging climate. If the Earth's temperature were to decrease overall by just a few degrees, many marine species would probably die. Some paleobiologists believe that climatic changes could be responsible for mass extinctions in the past.

It is also possible that mass extinctions were due to changes in the environment triggered by catastrophes. If the Earth was bombarded by a large meteorite, for example, the dust going into the atmosphere upon impact could have blocked much of the sunlight. In addition to killing many plants, this would have lowered the Earth's temperature, leading to the death of many marine organisms.

Biological factors can also trigger extinction. When a new species forms, it may be able to out-compete an older species, leading to its demise. Humans have had a profound impact on extinction. The tremendous increase in human population has caused us to spread into areas of the Earth that were previously not part of our range. The habitats of many animal and plant species are destroyed whenever humans invade an area. This can result in their extinction. Indeed, some biologists believe the Earth has entered the largest period of mass extinction in its entire history, and that this has been triggered by human activity.

■ SUMMARY

I. A species is a group of more or less distinct organisms that has the potential to interbreed with one another but not with members of different species.
II. Biological isolating mechanisms restrict the gene flow between species and sometimes between different populations within a species.
   A. Prezygotic isolating mechanisms prevent fertilization from taking place.
      1. Temporal isolation is due to the two groups reproducing at different times of the day, season, or year.
      2. Ecological isolation is caused by habitat differences between two closely related species living in the same geographical area.
      3. Distinctive courtship behaviors prevents mating between species (behavioral isolation).
      4. Mechanical isolation is due to morphological or anatomical differences in the reproductive structures of plants and animals.
      5. Molecular and chemical differences may cause gamete incompatibility between species (gametic isolation).
   B. Postzygotic isolating mechanisms assure reproductive failure even though fertilization has taken place.

1. Hybrid inviability is abortion of the hybrid embryo.
2. Hybrid sterility prevents hybrids from reproducing.
3. Hybrid breakdown prevents hybrids from reproducing beyond one generation.
III. Speciation is the evolution of a new species.
   A. Allopatric speciation occurs when one population becomes geographically isolated from the rest of the species and subsequently evolves.
   B. Sympatric speciation does not require geographical isolation and occurs as a result of hybridization/polyploidy.
IV. The timing of evolutionary change is currently being debated.
   A. According to proponents of gradualism, populations slowly diverge from one another by the accumulation of adaptive characteristics within a population.
   B. According to proponents of punctuated equilibria, evolution proceeds in spurts. Short periods of active evolution are followed by long periods of stasis.
V. Extinction is the death of a species. Once a species is extinct it can never reappear.

■ POST-TEST

1. A species is a group of organisms with a common: (a) coadapted gene complex; (b) gene pool; (c) postzygotic isolating mechanism; (d) appearance.
2. When two closely related species that are found in the same geographical range reproduce at different times

of the year, this is known as: (a) temporal isolation; (b) ecological isolation; (c) hybrid inviability; (d) behavioral isolation; (e) hybrid breakdown.
3. If two different species have reproductive structures that prevent mating, they fail to reproduce due to:

(a) gametic isolation; (b) hybrid inviability; (c) mechanical isolation; (d) ecological isolation; (e) hybrid sterility.

4. Which of the following is *not* an example of a prezygotic isolating mechanism? (a) ecological isolation; (b) temporal isolation; (c) behavioral isolation; (d) hybrid sterility; (e) gametic isolation.

5. The most important method of speciation in animal evolution is: (a) allopatric speciation; (b) sympatric speciation.

6. Several to many species formed from a single ancestral species is known as: (a) reproductive isolation; (b) subspecies; (c) hybrid breakdown; (d) anagenesis; (e) adaptive radiation.

7. The divergence of two closely related groups in the same geographical area so that their differences place them in different niches in the same environment is known as: (a) anagenesis; (b) hybrid sterility; (c) character displacement; (d) hybridization; (e) polyploidy.

8. An individual that possesses multiple sets of chromosomes, in which one or more of those sets came from a different species is known as a(an): (a) allopolyploid; (b) character displacement; (c) cladogenesis; (d) hybridization; (e) anagenesis.

9. The fact that the fossil record shows few transitional forms during speciation is used to support: (a) gradualism; (b) punctuated equilibria.

10. This type of extinction is believed to have occurred during five or six periods of the Earth's history: (a) background extinction; (b) mass extinction.

■ REVIEW QUESTIONS

1. Give an example of each of the following:
   a. temporal isolation
   b. ecological isolation
   c. behavioral isolation
   d. mechanical isolation
   e. gametic isolation
2. When prezygotic isolating mechanisms fail, postzygotic isolating mechanisms may come into play. Describe the three types of postzygotic isolating mechanisms and give an example of each.
3. Give at least five geographical barriers that might lead to allopatric speciation.
4. Why is speciation more likely to occur if the original isolated population is small?
5. When does character displacement occur? Why does it occur?

6. Explain how hybridization and polyploidy can cause a new species to form in as little time as one generation.
7. If you were in a debate and had to support gradualism, what would you say? What would you say if you were supporting punctuated equilibrium?
8. What role does extinction play in evolution?
9. Is (a) anagenesis or cladogenesis? What about (b)? What about (c)?

(● = species; e = extinction; dots in a vertical line represent the same species.)

■ RECOMMENDED READINGS

Ambrose, E.J. *The Nature and Origin of the Biological World.* New York, John Wiley, 1982. A textbook on evolution that discusses the major controversial topics in the field.

Grant, V. *Plant Speciation.* 2nd ed. New York, Columbia University Press, 1981. An authoritative and comprehensive review of plant speciation.

Mayr, E. *Animal Species and Evolution.* Cambridge, Harvard University Press, 1963. A classic on animal evolution.

Stanley, S. *Macroevolution: Pattern and Process.* San Francisco, W.H. Freeman, 1979. An exposition of punctuated equilibrium and the question of whether it is important in speciation and evolution.

Primordial rock formations on Año Nuevo Island, California

20

The Evolutionary History of Life

■ LEARNING OBJECTIVES

After you have read this chapter you should be able to:

1. Describe the conditions on early Earth.
2. Outline the major steps that are believed to have occurred in the origin of life.
3. Briefly describe the geological features and distinguishing plant and animal life for the Precambrian, Paleozoic, Mesozoic, and Cenozoic eras.
4. Define macroevolution and discuss it in the context of unusual features, evolutionary trends, adaptive radiation, and extinction.
5. Describe the modern synthesis of evolution.

ORIGIN OF LIFE

The last three chapters have been concerned with how life evolved, but we have not dealt with a fundamental question involving biological evolution: How did life begin? The hypothesis generally accepted by scientists is that life developed from nonliving matter. This process, called **chemical evolution,** involved several stages. First, small organic molecules were synthesized. Over time they accumulated. Large macromolecules like proteins and nucleic acids were assembled from smaller molecules. The macromolecules acted on one another, collecting into more complicated assemblages that could eventually metabolize and replicate. These assemblages developed into cell-like structures which, ultimately, became the first true cells.

After the first cells originated, they evolved over several billion years into the rich biological diversity on our planet today. It is believed that life originated on Earth only once and that this occurred under environmental circumstances quite different from those on Earth today. And so, to understand the origin of life, we must consider the conditions of early Earth. Although we will never be certain of the exact conditions on Earth when life arose, scientific evidence from a number of sources can provide us with valuable clues.

Conditions on Early Earth

The formation of the Earth and the rest of our solar system is tied to the formation of the universe. It is believed that the universe was not always spread out the way it is today. Between 10 and 20 billion years ago, the universe was a dense compaction that exploded (the Big Bang), hurling dust, debris, and gases into space. This material has been hurling outward ever since, so that the universe is continually expanding. As the materials cooled, atoms of different elements formed, particularly hydrogen and helium. The cooling and compression of this matter ultimately formed the stars and planets.

Our sun is a second- or third-generation star that formed 5 or 10 billion years ago. As the solar matter compressed by gravitational forces, it ignited, producing a tremendous amount of heat. This heat triggered the formation of other elements from hydrogen and helium. Some of this matter was ejected from the sun and coalesced with debris, dust, and gases encircling the sun. Thus, the planets formed.

The Earth is approximately 4.6 billion years old. The matter making up early Earth compacted as a result of gravitational forces, with the heaviest elements, nickel and iron, forming the center core, the medium-weight elements forming the mantle, and the lighter elements remaining near the surface. The first atmosphere, composed largely of the lightest elements, hydrogen and helium, was lost from the Earth because the Earth's weak gravitational forces couldn't hold it.

The Earth is believed to have been cold originally. As gravitational compaction continued, heat built up. This was increased by energy from radioactive decay. This heat occasionally escaped in hot springs and volcanoes, which also produced gases. These gases formed the second atmosphere of early Earth. It was a strongly reducing atmosphere with little or no free oxygen present. The gases included carbon dioxide (CO_2), water vapor (H_2O), carbon monoxide (CO), hydrogen (H_2), and nitrogen (N_2). It is also possible that the early atmosphere contained some ammonia (NH_3), hydrogen sulfide (H_2S), and methane (CH_4), although these reduced molecules may have been rapidly broken down by ultraviolet radiation from the sun. As the temperature of the Earth slowly cooled, water vapor condensed and torrential rains fell, forming the oceans. The falling rain eroded the earth's surface, adding minerals to the oceans, making them "salty."

There are four requirements for chemical evolution. First, life could only have evolved in the absence of free oxygen. Oxygen is very reactive and would have broken down the organic molecules that are a necessary step in the origin of life. The Earth's atmosphere was strongly reducing, however, so any free oxygen would have formed oxides with other elements. A second requirement for the origin of life would be energy. Early Earth was a place of high energy, with violent thunderstorms, volcanoes, and intense radiation, including ultraviolet radiation from the sun (Figure 20–1). More ultraviolet radiation was probably produced by the "young" sun than is produced today, and the Earth had no protective ozone layer to block much of this radiation. Third, the chemicals that would be the building blocks for chemical evolution must be present. These included water, dissolved inorganic minerals (present as ions), and the gases present in the early atmosphere. A final requirement would have been time: time for molecules to accumulate and react. The age of the Earth provides adequate time for chemical evolution. The Earth is approximately 4.6 billion years old, and there is geological evidence of the appearance of simple life forms 3.5 billion years ago.

FIGURE 20–1 Conditions on early Earth would have been inhospitable for most of today's life forms. The strongly reducing atmosphere lacked oxygen. Volcanoes erupted, spewing gases that contributed to the atmosphere. Violent thunderstorms produced torrential rainfall that eroded the land.

Origin of Organic Molecules

Since organic molecules are the building materials for living organisms, it is reasonable to consider how they might have originated. The concept that simple organic molecules like sugars, nucleotides, and amino acids could form spontaneously from nonliving raw materials was first hypothesized in the 1920s by two scientists working independently, Oparin, a Russian biochemist, and Haldane, a Scottish physiologist and geneticist. Their hypothesis was tested in the 1950s by Urey and Miller, who designed an apparatus that simulated conditions then believed to be prevalent on early Earth (Figure 20–2). The atmosphere they started with was rich in H_2, CH_4, H_2O, and NH_3. They exposed this atmosphere to an electric discharge, which simulated lightning. Their analysis of the chemicals produced in a week revealed that amino acids and other building blocks had been synthe-

FIGURE 20–2 Stanley Miller and Harold Urey used an apparatus similar to this to replicate what they believed were the conditions of early Earth. An electric spark was produced in the upper right flask to simulate lightning. The gases present in the flask reacted together, forming a number of basic organic compounds.

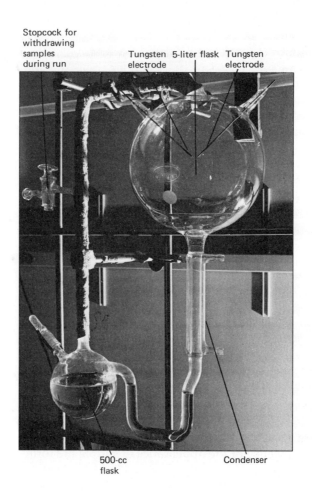

Stopcock for withdrawing samples during run

Tungsten electrode 5-liter flask Tungsten electrode

500-cc flask

Condenser

sized. We now believe that the Earth's early atmosphere was not rich in methane or ammonia, but similar experiments using different combinations of gases have produced a wide variety of organic molecules, including nucleotide bases of RNA and DNA.

Oparin envisioned that the organic molecules would, over vast spans of time, accumulate in the shallow seas, as a "sea of organic soup." Under such conditions, he believed that larger organic molecules (polymers) would form by the union of smaller ones (monomers). Based on scientific evidence accumulated since Oparin's time, most scientists believe that polymerization to form proteins, nucleic acids, and other large organic molecules would not have occurred under such conditions. For one thing, many polymerization reactions involve **dehydration synthesis,** in which two molecules are joined by the removal of water. It is unlikely that a water-producing reaction would occur in the water without enzymes. Also, it is doubtful that the concentration of organic monomers would have reached high enough levels in the oceans to stimulate their polymerization.

It is more likely that organic polymers were synthesized and accumulated on rock or clay surfaces. Clay is particularly intriguing as a site for polymerization because it contains zinc and iron ions which might possibly assist as catalysts. Also, clay binds the exact forms of sugars and amino acids that are found in living organisms. Other amino acids and sugars may be produced but do not bind to clay. To test whether polymers could form under these conditions, Fox heated a mixture of dry amino acids and obtained polypeptides. He called the product of this spontaneous polymerization a **proteinoid.**

After polymers are produced, could they assemble into more complex structures? Scientists have worked with several different **protobionts,** spontaneous assemblages of organic polymers. They have been able to make protobionts that resemble simple life forms in several ways, helping us to envision how complex nonliving molecules took that giant leap and became living cells. Protobionts often divide in half after they have "grown." Their internal environment is chemically different from the external environment. And some of them show the rudiments of metabolism. They are amazingly organized, considering their relatively simple composition.

One type of protobiont, the **microsphere,** was formed by adding water to proteinoids (Figure 20–3). Microspheres are spherical and have osmotic properties. Some of them produce an electric potential across their surface, reminiscent of membrane potentials in cells. Microspheres can also absorb materials from their surroundings and respond to changes in osmotic concentration as though they were surrounded by membranes, even though they contain no lipid. **Liposomes** are protobionts made from lipids. In water they form a spherical structure surrounded by a lipid bilayer similar

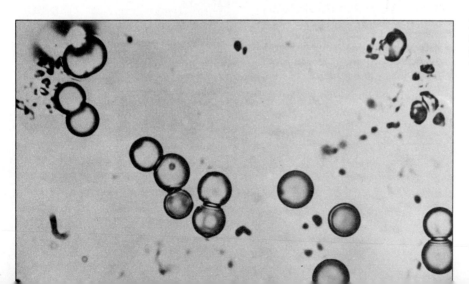

FIGURE 20–3 Proteinoid microspheres are tiny spheres (1–2 μm in diameter) that exhibit some of the properties of life.

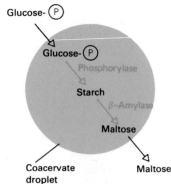

FIGURE 20–4 Coacervates are capable of very simple metabolic pathways. A coacervate containing phosphorylase and amylase was able to absorb glucose 1-phosphate from the medium and convert it to maltose, which was detected in the medium.

FIGURE 20–5 Microfossils of early cells. (*a*) This tubular, filamentous microfossil is from sediments in western Australia and is 3.5 billion years old. (*b*) These prokaryotes are from the Gunflint Iron Formation in Ontario and existed 2 billion years ago.

(*a*)

(*b*)

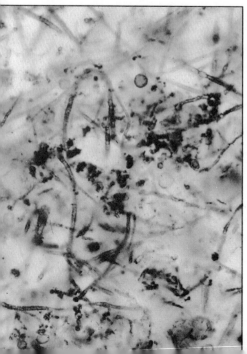

in structure to cell membranes. A final example of a protobiont is the **coacervate.** Oparin formed coacervates from relatively complex mixtures of polypeptides, nucleic acids, and polysaccharides. Coacervates are capable of very simple metabolism (Figure 20–4). When he made a coacervate out of short-chain RNAs and the enzyme responsible for replicating nucleic acids, and placed it in a medium that contained nucleotide triphosphates, the coacervates "grew," replicated, and divided.

The First Cells

Studying protobionts can help us appreciate that relatively simple "precells" can exhibit some of the properties of life. However, it is a major step from molecular aggregates such as protobionts to living cells. Yet fossil evidence indicates that prokaryotic cells were thriving 3.5 billion years ago.

Unquestionably, the first cells to evolve were prokaryotic (Figure 20–5). Australian and South African rocks have yielded microscopic fossils of prokaryotic cells 3.4 to 3.5 billion years old. **Stromatolites** are another type of fossil evidence of the Earth's earliest cells. These column-like rocks are composed of many minute layers of prokaryotic cells, usually cyanobacteria. Living stromatolites are still found in hot springs and in shallow pools of fresh and salt water. Over time, sediment collects around the cells and gradually becomes mineralized. Meanwhile, a new layer of living cells grows over the older, dead cells. Stromatolites are found in a number of places in the world, including the Canadian Great Slave Lake and the Gunflint Iron Formations along Lake Superior in the United States. Some of them are extremely ancient. One group in Western Australia, for example, is several billion years old. There are still living colonies that form stromatolites in Yellowstone National Park and in Shark Bay, Australia (Figure 20–6).

We have said that the origin of cells from macromolecular assemblages was a major step in the origin of life. Actually, it was probably a series of small steps. Two crucial parts of that process would have been the origin of molecular reproduction and the origin of metabolism.

Molecular Reproduction

Polynucleotides (RNA and DNA) can form spontaneously on clay in much the same way as polypeptides. It is generally believed that RNA was the first "information" molecule to evolve in the progression toward the first cell. Proteins and DNA came later. One of the surprising properties of RNA is that it often has catalytic properties (Figure 20–7). Catalytic RNAs,

FIGURE 20–6 Stromatolites at Shark Bay in Western Australia that are approximately 2000 years old. These formations are composed of mats of cyanobacteria and minerals like calcium carbonate. Some fossil stromatolites are 3.5 billion years old.

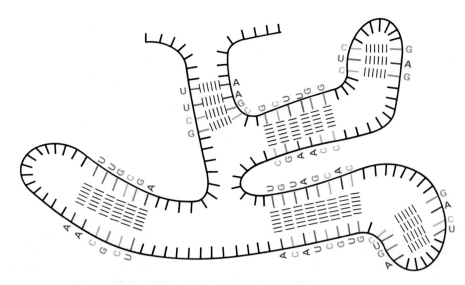

FIGURE 20–7 Single-stranded RNA can form base pairs with itself, creating a precise conformation that may have catalytic properties. The order of the nucleotides determines the ultimate shape of the molecule.

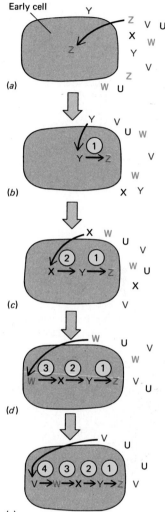

(a)

(b)

(c)

(d)

(e)

or **ribozymes,** function like enzymes in this regard. They are used in present-day cells to help process RNA into final products: rRNA, tRNA, and mRNA. Before the evolution of true cells, it is possible that this RNA catalyzed the formation of more RNA in the clays or shallow rock pools. If one adds RNA strands to a test tube containing RNA nucleotides, replication occurs without enzymes. This reaction is increased if zinc is added as a catalyst. You will recall that zinc is bound to clay.

RNA can also direct protein synthesis. Some of the single-stranded RNA molecules fold back on themselves due to the interaction of the nucleotides composing the strand. Sometimes the conformation of the folded molecule is such that it weakly binds to an amino acid. If amino acids are held together closely by RNA molecules, they may bond together, forming a polypeptide.

In living cells, information is transferred from DNA to RNA to proteins. We have considered how RNA and proteins might have evolved. The final step in the evolution of informational molecules would have been to incorporate DNA into the information transfer system. Because DNA is a double helix, it is more stable and less reactive than RNA. There would still be a need for RNA, however, because DNA is not catalytic.

There are several more steps involved before a true, living cell could be formed from macromolecular aggregations. At the present time, we have very little knowledge about how these might have occurred. For example, how did the genetic code originate? This step must have occurred very early in the origin of life because virtually all living organisms possess the same code. Also, how did a membrane made of lipid and protein envelope the macromolecular assemblage, permitting the accumulation of some molecules and the exclusion of others?

Metabolism

Metabolism, all the biochemical reactions performed by a living organism, involves major sequences of reactions which occur in a step-by-step fashion. It is considered likely that metabolism also arose step by step. Horowitz postulated in 1945 that an organism would acquire, by successive gene mutations, the enzymes needed for metabolic pathways. However, these enzymes would be formed in the *reverse* order of the sequence in which they are ultimately used for normal metabolism.

For example, let us suppose that our first primitive organism required an organic compound, Z, for its growth (Figure 20–8). This substance, Z, and a vast variety of other organic compounds, Y, X, W, V, U, and so forth,

FIGURE 20–8 Evolution of metabolic pathways probably proceeded backwards. (*a*) Compound Z, required by the primitive cell, was obtained from the environment as long as it was in abundance. (*b*) A mutation produced an enzyme 1 that enabled the cell to use compound Y to make compound Z. Compound Y was obtained from the environment as long as it was available. (*c*) A new mutation produced an enzyme 2 that could convert X to Y. The cell continued to make enzyme 1 that converted Y to Z. (*d*) A new enzyme was added to the pathway, converting compound W to X. (*e*) The final metabolic pathway, starting with compound V and ending with the desired end product, Z, involved the production of 4 enzymes.

were present in the environment. They had been spontaneously synthesized previously. The organism would be able to survive as long as the supply of compound Z lasted. If a mutation occurred for a new enzyme enabling the organism to synthesize Z from compound Y, the organism with this mutation would be able to survive when the supply of compound Z was exhausted. A second mutation that established an enzyme for catalyzing a reaction in which substance Y could be made from substance X would again have survival value when the supply of Y was exhausted. Similar mutations would have set up enzymes enabling the organism to use successively simpler substances, W, V, U, and so on.

Heterotrophy and Autotrophy

The first cells were almost certainly anaerobic prokaryotes. Some of the earliest cells may have been **heterotrophic,** obtaining the organic molecules they needed for energy from the environment as opposed to synthesizing them. They probably consumed many types of organic molecules that had spontaneously formed: sugars, nucleotides, and amino acids, to name a few. They obtained the energy needed to support life by fermenting these organic compounds. Fermentation is, of course, an anaerobic process.

Molecular biology is playing an increasingly important role in elucidating characteristics of the first cells. In 1988, Lake made a careful comparison of ribosomal RNA from numerous organisms. His analysis showed that all living organisms have a common ancestor that was most likely a prokaryote that metabolized sulfur and lived in hot springs (Figure 20–9). This contrasts sharply with the previous view that the last common ancestor was a heterotroph and shows the power of molecular techniques in evolutionary biology. When interpreted with care, molecular evidence can aid in answering difficult evolutionary questions.

Before the supply of spontaneously generated organic molecules was exhausted, mutations may have occurred that gave organisms possessing them a distinct selective advantage. These cells could obtain energy from a new source, sunlight. They were able to store the radiant energy in the form of a chemical, adenosine triphosphate. Probably later, they were able to expand this process further, storing radiant energy as chemical energy in organic molecules such as sugars. These photosynthetic organisms did not require the energy-rich organic compounds which were of limited availability from the environment.

FIGURE 20–9 Aerial view of a hot sulfur spring. The last common ancestor of all living organisms was probably a prokaryote that lived in hot springs and metabolized sulfur. Such environments today are populated by numerous bacteria and cyanobacteria.

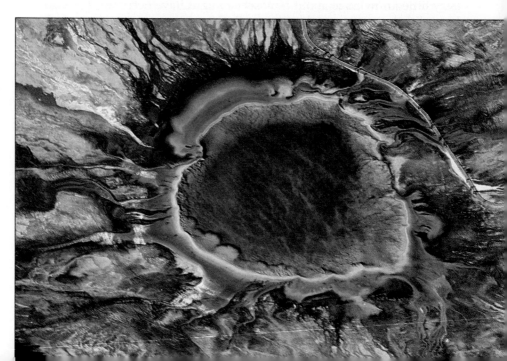

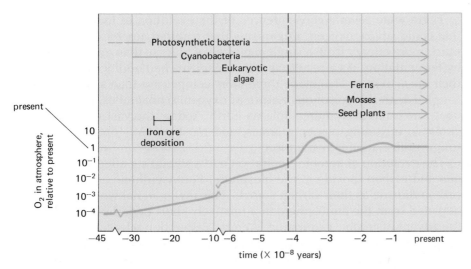

FIGURE 20–10 The increase in oxygen over time is shown, along with the organisms that were responsible for its evolution. Oxygen concentration is given as a fraction of the present concentration and is plotted on a log scale. The vertical dashed line represents the early Silurian period when it is believed that the level of oxygen in the atmosphere reached 10% of its present level, a point at which the ozone layer started to form.

Photosynthesis requires not only light energy, but also a source of hydrogen, which is used to reduce carbon dioxide when organic molecules are synthesized (Chapter 8). Most likely the first photosynthetic **autotrophs** used the energy of sunlight to split hydrogen-rich molecules like hydrogen sulfide, H_2S, releasing elemental sulfur in the process. Indeed, the green sulfur bacteria and the purple sulfur bacteria still use H_2S. A third group of bacteria, the purple nonsulfur bacteria, use other organic molecules or hydrogen gas as a hydrogen source.

The first photosynthetic autotrophs to split water in order to obtain hydrogen were the cyanobacteria. Water is quite abundant on earth, and the selective advantage that splitting water bestowed on them caused the cyanobacteria to thrive. In the process of splitting water, oxygen was released as a gas, O_2. Initially, the oxygen released from photosynthesis oxidized minerals in the ocean and the Earth's crust. Over time, more oxygen was released than could be utilized by these **sinks,** and oxygen began to accumulate in the oceans and atmosphere (Figure 20–10).

The timing of the events just described has been estimated based on geological and fossil evidence. The first autotrophs probably evolved about 3.4 billion years ago. Rocks from that period contain traces of chlorophyll. Cyanobacteria appeared between 2.5 and 2.7 billion years ago. Evidence such as the stromatolites discussed previously is used to date their appearance. By 2 billion years ago, the cyanobacteria had produced enough oxygen to begin to change the atmosphere in a significant manner.

Evolution of Aerobes

The increase in atmospheric oxygen had a profound effect on the Earth and on life. First, oxygen in the upper atmosphere reacted to form **ozone,** O_3 (Figure 20–11). Ozone blanketed the Earth, preventing much of the sun's ultraviolet radiation from penetrating to the Earth's surface. It enabled living organisms to live closer to the surface in aquatic environments and even on land! Because the energy in ultraviolet radiation had been used to form spontaneously generated organic molecules, their synthesis decreased. Obligate anaerobes were poisoned by the oxygen, and many species undoubtedly perished. Some anaerobes, however, evolved ways to neutralize the oxygen so it could not harm them. Some organisms even evolved ways to *use* the oxygen becoming so prevalent in their environment. As a result, some organisms evolved the capacity to use oxygen to extract energy from food. Aerobic respiration was tacked onto the existing process of glycolysis. Like other types of metabolism, it probably evolved in a step-by-step fashion.

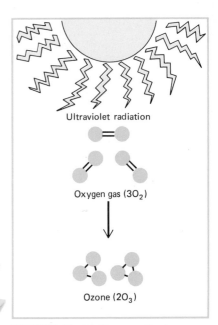

FIGURE 20–11 Ozone, O_3, is formed in the upper atmosphere when ultraviolet radiation from the sun breaks the double bonds of oxygen molecules.

There were several consequences of the evolution of living organisms that could utilize oxygen. Organisms that respired aerobically could gain much more energy from a single molecule of glucose than anaerobes could by fermentation. As a result, aerobic organisms had additional energy for their life activities. This made them more competitive than anaerobes and, coupled with the poisonous nature of oxygen to anaerobes, forced anaerobes into a relatively minor role on Earth. Today, the vast majority of organisms, including plants, animals, and most protists, prokaryotes, and fungi, utilize aerobic respiration.

The evolution of aerobic respiration had a stabilizing effect on both oxygen and carbon dioxide in the biosphere. Photosynthetic organisms used carbon dioxide as their carbon source. This raw material would have been depleted from the atmosphere in a relatively short period of time without the advent of aerobic respiration. Aerobic respiration released carbon dioxide as a waste product from the complete breakdown of organic molecules. Carbon thus started cycling in the biosphere, moving from the abiotic environment to photosynthetic organisms to heterotrophs that ate the plants. Carbon was then released back into the abiotic environment as carbon dioxide by respiration, and the cycle continued. In like manner, oxygen was produced in photosynthesis and utilized in aerobic respiration.

Origin of Eukaryotic Cells

It is logical to think of the ancestors of modern organisms as being very simple. Among modern organisms, the very simplest forms of cellular life are prokaryotes. That is one reason biologists think the earliest cells were prokaryotic. You will recall that prokaryotic cells lack nuclear membranes as well as other membranous organelles such as mitochondria, endoplasmic reticulum, chloroplasts, and the Golgi complex (Chapter 4).

Eukaryotes appeared in the fossil record 1.5 billion years ago. How did eukaryotic cells arise from prokaryotes? The **endosymbiont theory** suggests that mitochondria, chloroplasts, and perhaps even centrioles and flagella may have originated from symbiotic relationships between two prokaryotic organisms (Figure 20–12). Thus, chloroplasts are viewed as former photosynthetic bacteria (but generally not cyanobacteria) and mitochondria as former bacteria (or photosynthetic bacteria that lost the ability to photosynthesize). These endosymbionts were originally ingested by the host cell but not digested. They survived and reproduced along with the host cell so that future generations of the host also contained endosymbionts. The two organisms developed a mutualistic relationship, and eventually the endosymbiont lost the ability to exist outside its host.

This theory stipulates that each of these partners brought to the relationship something the other lacked. For example, mitochondria provided the ability to employ oxidative metabolism, which was lacking in the original host cell; chloroplasts provided the ability to use a simple carbon source (carbon dioxide); and spiral bacteria provided the ability to move by eventually becoming flagella. The host cell provided a safe habitat and raw materials or nutrients.

The principal evidence in favor of the endosymbiont theory is that mitochondria and chloroplasts possess *some*, but not *all*, of their own genetic apparatus. They have their own DNA (as a circular chromosome much like prokaryotes) and their own ribosomes (which resemble prokaryotic ribosomes rather than eukaryotic ribosomes). They have some of the machinery for protein synthesis, including tRNA molecules, and are able to conduct protein synthesis on a limited scale. Mitochondria and chloroplasts are both self-replicating, which means they divide independently of the cell in which they reside. Further, it is possible to poison them with an antibiotic that affects bacteria but not eukaryotic cells. Mitochondria and

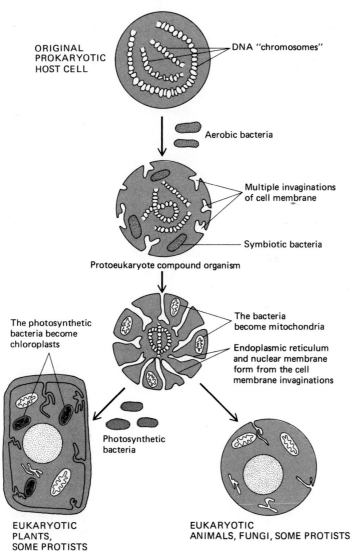

ORIGINAL
PROKARYOTIC
HOST CELL

DNA "chromosomes"

Aerobic bacteria

Multiple invaginations
of cell membrane

Symbiotic bacteria

Protoeukaryote compound organism

The bacteria
become mitochondria

The photosynthetic
bacteria become
chloroplasts

Endoplasmic reticulum
and nuclear membrane
form from the cell
membrane invaginations

Photosynthetic
bacteria

EUKARYOTIC
PLANTS,
SOME PROTISTS

EUKARYOTIC
ANIMALS, FUNGI, SOME PROTISTS

FIGURE 20–12 The endosymbiotic theory of the origin of the eukaryotes.

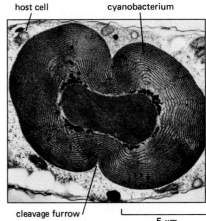

host cell   cyanobacterium

cleavage furrow   5 μm

FIGURE 20–13 The flagellate, *Cyanophora paradoxa*, contains a cyanobacterial endosymbiont.

chloroplasts are enveloped by a double membrane. The outer membrane is envisioned as having developed from the invagination of the host cell's plasma membrane, while the inner membrane developed from the endosymbiont's plasma membrane.

There are a number of endosymbiotic relationships today (Figure 20–13). Many corals have algae living within their cells. This is one of the reasons that coral reefs are so productive. In the gut of the termite lives a protozoon, *Myxotricha paradoxa*, with several different endosymbionts, including spirochete bacteria that are attached to the protozoon and function as flagella. Also, the colonial tunicate, *Diplosoma virens*, has photosynthetic prokaryotes living within its cells. This relationship is particularly intriguing because the prokaryote is a chloroxybacterium rather than a cyanobacterium. Chloroxybacteria were recently discovered and have the same pigment system—chlorophylls *a* and *b* and carotenoids—as most chloroplasts (Figure 20–14).

The endosymbiont theory is not the final answer to how eukaryotic cells evolved from prokaryotes. It does not explain how the genetic material in the nucleus came to be surrounded by a membrane, for example. And the evidence supporting the evolution of motile structures like flagella and cilia from prokaryotes is weak. There are no traces of genetic material in flagella. Further, the 9 + 2 arrangement of the microtubules in flagella has not been found in any prokaryote to date.

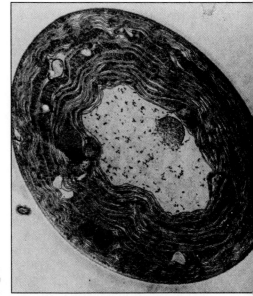

FIGURE 20–14 Ultrastructure of *Prochloron*, a chloroxybacterium that possesses the pigments found in higher plants.

HISTORY OF LIFE

The sediments of the Earth's crust consist of five major rock strata, each subdivided into minor strata, lying one on top of the other. These sheets of rock were formed by the accumulation of mud and sand at the bottom of oceans, seas, and lakes. Each contains certain characteristic fossils that serve to identify deposits made at approximately the same time in different parts of the world. Geological time has been divided into **eras,** which are subdivided into **periods,** which in turn are composed of **epochs** (Table 20–1 and Figure 20–15). Between the major eras, and serving to distinguish them, there were widespread geological disturbances, which raised or lowered vast regions of the Earth's surface and created or eliminated shallow inland seas. These disturbances altered the distribution of sea and land organisms and may have triggered the mass extinction of many life forms. The raising and lowering of portions of the Earth's crust result from the slow movements of the enormous plates that compose the crust (see Focus on Continental Drift).

Precambrian Life

The richest deposits of fossils date from the beginning of the "explosion of life" that occurred during the Cambrian period, some 570 million years ago. However, there is evidence that life existed long before the Cambrian period. Signs of **Precambrian** life date from the **Archean era,** which began about 3.5 billion years ago.

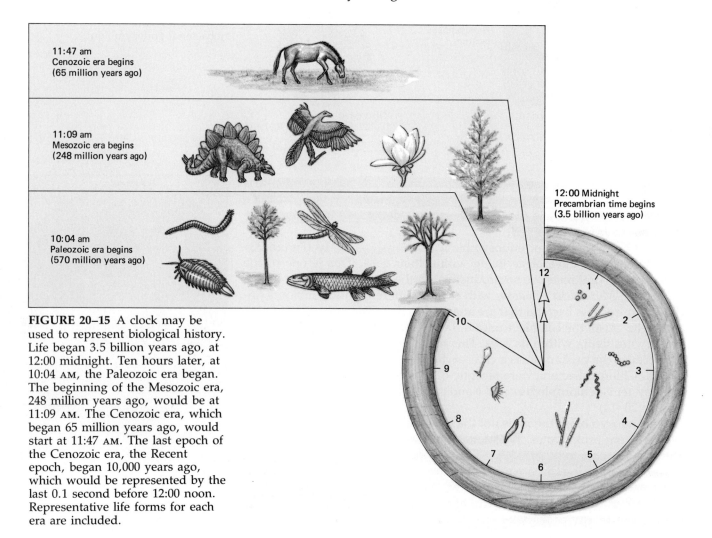

FIGURE 20–15 A clock may be used to represent biological history. Life began 3.5 billion years ago, at 12:00 midnight. Ten hours later, at 10:04 AM, the Paleozoic era began. The beginning of the Mesozoic era, 248 million years ago, would be at 11:09 AM. The Cenozoic era, which began 65 million years ago, would start at 11:47 AM. The last epoch of the Cenozoic era, the Recent epoch, began 10,000 years ago, which would be represented by the last 0.1 second before 12:00 noon. Representative life forms for each era are included.

11:47 am
Cenozoic era begins
(65 million years ago)

11:09 am
Mesozoic era begins
(248 million years ago)

10:04 am
Paleozoic era begins
(570 million years ago)

12:00 Midnight
Precambrian time begins
(3.5 billion years ago)

Table 20–1 SOME IMPORTANT BIOLOGICAL EVENTS IN GEOLOGICAL TIME*

Era	Period	Epoch	Time from Beginning of Period to Present (millions of years)**	Geological Conditions	Plants and Microorganisms	Animals
Cenozoic (Age of Mammals)	Quaternary	Recent	0.01	End of last Ice Age; warmer climate	Decline of woody plants; rise of herbaceous plants	Age of *Homo sapiens*
		Pleistocene	2.0	Four Ice Ages; glaciers in Northern Hemisphere; uplift of Sierras	Extinction of many species	Extinction of many large mammals
	Tertiary	Pliocene	5	Uplift and mountain-building; volcanoes; climate much cooler	Development of grasslands; decline of forests; flowering plants	Large carnivores; many grazing mammals; first known human-like primates
		Miocene	25	Climate drier, cooler; mountain formation	Flowering plants continue to diversify	Many forms of mammals evolve
		Oligocene	38	Rise of Alps and Himalayas; most land low; volcanic activity in Rockies	Spread of forests; flowering plants, rise of monocotyledons	Apes evolve; all present mammal families are represented
		Eocene	55	Climate warmer	Gymnosperms and flowering plants dominant	Beginning of Age of Mammals; modern birds
		Paleocene	65	Climate mild to cool; continental seas disappear		Evolution of primate mammals
Mesozoic (Age of Reptiles)	Cretaceous		144	Continents separated; formation of Rockies; other continents low; large inland seas and swamps	Rise of flowering plants; gymnosperms decline	Dinosaurs reach peak, then become extinct; toothed birds become extinct; first modern birds; primitive mammals
	Jurassic		213	Climate mild; continents low; inland seas; formation of mountains; continental drift continues	Gymnosperms common	Large, specialized dinosaurs; first toothed birds; insectivorous marsupials
	Triassic		248	Many mountains form; widespread deserts; continental drift begins	Gymnosperms dominate	First dinosaurs; egg-laying mammals
Paleozoic (Age of Ancient Life)	Permian		286	Continents merge as Pangaea, glaciers; formation of Appalachians; continents rise	Conifers diversify; cycads evolve	Modern insects appear; mammal-like reptiles; extinction of many Paleozoic invertebrates
	Carboniferous		360	Lands low; great coal swamps; climate warm and humid; later cooler	Forests of ferns, club mosses, horsetails, and gymnosperms	First reptiles; spread of ancient amphibians; many insect forms; ancient sharks abundant
	Devonian		408	Glaciers; inland seas	Terrestrial plants well established; first forests; gymnosperms appear; bryophytes appear	Age of Fishes; amphibians appear; wingless insects appear; many trilobites
	Silurian		438	Continents mainly flat; flooding	Vascular plants appear; algae dominant in aquatic environment	Fish evolve; terrestrial arthropods
	Ordovician		505	Sea covers continents; climate warm	Marine algae dominant; terrestrial plants first appear	Invertebrates dominant; first fish appear
	Cambrian		570	Climate mild; lands low; oldest rocks with abundant fossils	Algae dominant in aquatic environment	Age of marine invertebrates; most modern phyla represented
(Precambrian) Proterozoic			1500	Planet cooled; glaciers; formation of Earth's crust; mountains form	Primitive algae and fungi, marine protozoans	Toward end, marine invertebrates
Archean			3.5 billion years ago		Evidence of first prokaryotic cells	
Origin of the earth			4.6 billion years ago			
Origin of the universe			15–20 billion years ago			

*You may want to study this table starting from the bottom and working your way up through time.
**Based on Harland et al., *A Geologic Time Scale*, Cambridge, Cambridge University Press, 1982.

(a)

(b)

FIGURE 20–16 These Precambrian fossils were found in the Ediacaran Hills of South Australia. (*a*) *Spriggina,* a segmented worm, was approximately 4 cm long and is visible on the right hand side of the ruler. (*b*) This unidentified fossil organism lived in shallow marine waters.

The Archean era began with the formation of the Earth's crust, when rocks and mountains were already in existence and the processes of erosion and sedimentation had begun. Because the rocks of the Archean era are very deeply buried in most parts of the world, they are considered to be the most ancient. However, Archean rocks are exposed at the bottom of the Grand Canyon and along the shores of Lake Superior.

The Archean era lasted 2 billion years and was characterized by widespread volcanic activity and giant upheavals that raised mountains. The heat, pressure, and churning associated with these movements probably destroyed most of whatever fossils may have been formed, but some evidence of life still remains. This evidence consists of traces of graphite or pure carbon, which may be the transformed remains of primitive life. These remains are especially abundant in what were the oceans and seas of that era. Fossils of what appear to be cyanobacteria have been recovered from several Archean formations.

The second era, the **Proterozoic era,** which began approximately 1.5 billion years ago, was thought to be almost a billion years in length. It was characterized by the deposition of large quantities of sediment, reflecting massive erosion and perhaps glaciation. The fossils found in the later Proterozoic rocks show clear-cut examples of some major groups of fungi, protists, plants, and animals. One source of rich deposits of Precambrian fossils has been South Australia. The forms of life found there include jellyfish, corals, segmented worms, and two animals with no resemblance to any other known fossil or living form (Figure 20–16). These fossils are from very late in Precambrian time. Except for an arbitrary geological boundary, they might well be considered early Cambrian.

The Paleozoic Era

The oldest subdivision of the **Paleozoic era,** the **Cambrian period,** is represented by rocks rich in fossils. All the present-day animal phyla, except the chordates, are present, at least in marine sediments. There were arachnidlike forms, some of whose descendants (such as the horseshoe crab) exist almost unchanged today. The sea floor was covered with simple sponges, corals, crinoid echinoderms growing on stalks, snails, bivalves, primitive cephalopods, brachiopods, and trilobites.

Except for the chordates, the major types of body plans were established so early in the history of the eukaryotes that very little further change of a basic nature is seen. This probably indicates that, by the early Cambrian period, animal forms had reached a degree of adaptation that allowed them to exploit the Earth and adapt to changes in the environment with only limited modifications in their body plans.

Anglaspis

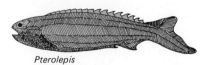

Pterolepis

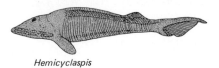

Hemicyclaspis

FIGURE 20–17 Three fossil ostracoderms, which were primitive jawless fishes.

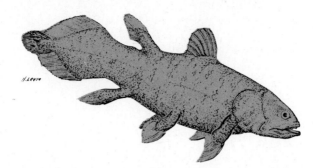

FIGURE 20–18 *Latimeria,* a modern-day coelacanth. A photograph of a coelacanth is shown in Figure 31–21.

According to geologists, the continents were gradually flooded during the Cambrian period. In the **Ordovician period,** this submergence reached its maximum, so that much of what is now land was covered by shallow seas. Inhabiting the seas were giant cephalopods, squid or nautilus-like animals with straight shells 5 to 7 meters long and 30 cm in diameter. The first traces of the early vertebrates, the jawless, bony-armored fish called **ostracoderms,** are also found in Ordovician rocks (Figure 20–17).

Two life forms of great biological significance appeared in the **Silurian period,** the land plants and the air-breathing animals. The first known land plants resembled ferns rather than mosses. These primitive vascular plants reproduced by spores. The evolution of terrestrial plants allowed terrestrial animals to evolve. Plants provided food and shelter for the first land animals. The only air-breathing land animals that have been discovered in Silurian rocks were arachnids that resembled scorpions.

A great variety of fishes appeared in the **Devonian period.** In fact, the Devonian is frequently called the "Age of Fishes." Unlike the ostracoderms, the Devonian fishes typically had jaws, an adaptation that enables a vertebrate to chew and bite. Appearing in Devonian deposits are sharks and the three main types of bony fish: lungfishes, lobe-finned fishes, and the ray-finned fishes. A few lungfishes have survived to the present. The ray-finned fishes later gave rise to the major modern orders of fishes. The lobe-finned fishes, some of which are considered ancestral to the land vertebrates, were thought to have become extinct by the end of the Mesozoic era. However, in 1939 the first living **coelacanth** was discovered off the coast of Madagascar (Figure 20–18).

Upper Devonian sediments contain fossil remains of salamander-like ancient amphibians that were often very large, with short necks and heavy, muscular tails (Figure 20–19). These creatures, whose skulls were encased in bony armor, were quite similar in many respects to the lobe-finned fishes.

The early vascular plants diversified during the Devonian period. Ferns, club mosses, horsetails, and seed ferns all flourished. The Devonian was the first period characterized by forests (Figure 20–20). Wingless insects and millipedes are believed to have originated in the late Devonian.

The **Carboniferous period** is named for the great swamp forests whose remains persist today as major coal deposits. The land during this time was covered with low swamps filled with horsetails, club mosses, ferns, seed ferns, and gymnosperms. The first reptiles, the **cotylosaurs,** appeared in the Carboniferous period, flourished in the Permian, and became extinct

FIGURE 20–19 *Eryops* was a primitive amphibian.

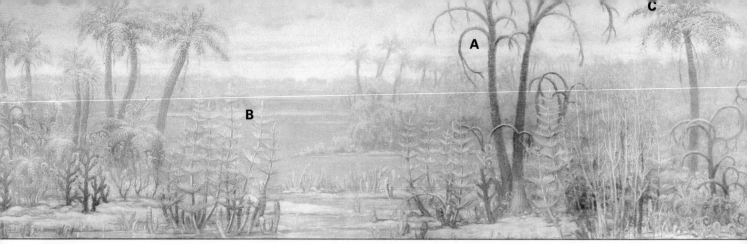

FIGURE 20–20 A restoration of a Middle Devonian forest in the eastern United States. *(A)* An early club moss. *(B)* An early horsetail. *(C)* An early tree fern.

early in the Mesozoic era. *Seymouria*, a cotylosaur, was typical (Figure 20–21). Two important groups of winged insects occurred for the first time in the Carboniferous: cockroaches and dragonflies (Figure 20–22). The dragonflies ranged in size from smaller than today's dragonflies to some with a wingspan of 75 cm.

The final period of the Paleozoic, the **Permian period,** was characterized by great changes in climate and topography. At the end of the Permian, a general folding of the Earth's crust raised a great mountain chain from Nova Scotia to Alabama. These mountains were originally higher than the present Rockies. Other mountain ranges formed in Europe at this time. A glaciation, spreading from the Antarctic, covered most of the Southern Hemisphere, extending almost to the equator in Brazil and Africa.

Many Paleozoic forms of life may have been unable to adapt to the climatic and geological changes and became extinct. Even many marine

FIGURE 20–21 *Seymouria* was one of the first reptiles. Its heavier skeleton represents a transitional form between the amphibians and reptiles.

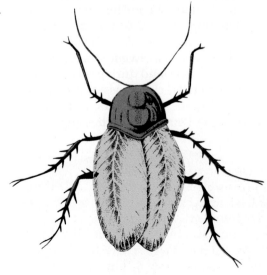

FIGURE 20–22 A primitive cockroach from the Carboniferous period.

FIGURE 20–23 *Cynognathus*, a Permian reptile that had many mammalian characteristics.

forms became extinct, perhaps owing to cooler water temperatures. Most of the plants that were dominant in the Carboniferous period became extinct during the Permian. The seed plants became dominant, with the diversification of conifers and the appearance of cycads.

During the late Carboniferous and early Permian, a group of reptiles appeared that are believed to be the ancestors of mammals. These were the carnivorous reptiles that were more slender and lizard-like than the cotylosaurs. In the latter part of the Permian, mammalian-like reptiles called **therapsids** appeared. One of these, *Cynognathus*, the "dog-jawed" reptile, was a slender, light-built animal with a skull intermediate between that of a reptile and a mammal (Figure 20–23). Its teeth, instead of being conical and all alike, as reptilian teeth are, were differentiated into incisors, canines, and molars. In the absence of information about the animal's soft parts, whether it had scales or hair, whether or not it was homeothermic (warmblooded), and whether it suckled its young, it is called a reptile.

The Mesozoic Era

The **Mesozoic era** is dated as beginning about 248 million years ago and lasting some 183 million years. It is divided into the **Triassic, Jurassic,** and **Cretaceous periods.** The outstanding feature of the Mesozoic era was the origin, differentiation, and final extinction of a large variety of reptiles. For this reason, the Mesozoic is commonly called the "Age of Reptiles." From a botanical viewpoint, the Mesozoic was dominated by gymnosperms until the mid-Cretaceous, when the flowering plants replaced them.

The most primitive reptilian line present in the Mesozoic includes the ancient cotylosaurs and the turtles. Turtles are first seen in Permian strata. They have the most complicated body armor of any land animal. With this protection, both marine and land forms have survived with few structural changes since before the time of the dinosaurs. Most of the snakes and lizards found in Mesozoic formations are also similar to their present-day descendants. The marine lizards of the Cretaceous period are a notable exception to this (Figure 20–24). They attained a length of 13 meters, with a long tail useful in swimming.

Of all the reptilian branches, the **dinosaurs** are the most famous. There were two main groups of dinosaurs. The **saurischians** were fast, twolegged forms ranging from the size of a dog to the ultimate representative of this group, the gigantic carnivore of the Cretaceous, *Tyrannosaurus* (Figure 20–25). Other saurischians had a plant diet and a four-legged gait. Some of these were among the largest animals that ever lived: *Brontosaurus*, with a length of 21 meters; *Diplodocus*, with a length of 29 meters; and *Brachiosaurus*, with an estimated weight in excess of 50 tons.

The other group of dinosaurs, the **ornithischians,** were entirely herbivorous. Although some of them walked upright, the majority had a fourlegged gait. Some had no front teeth and may have possessed a stout, horny, birdlike beak. In some forms this was broad and ducklike (hence the name "duck-billed dinosaurs"). Webbed feet were characteristic of this type. Other species had great armor plates, possibly as protection against

FIGURE 20–24 *Tylosaurus*, a large marine reptile, belongs to a group that was ancestral to modern lizards.

FIGURE 20–25 *Tyrannosaurus,* the largest of the flesh-eating dinosaurs, reached a length of 15 meters and a height of 6 meters. Its head was as much as 2 meters long and was equipped with many sharp teeth, whose edges were serrated like the blades of steak knives. The long tail was probably used as a counterweight to the immense head.

the carnivorous saurischians. *Ankylosaurus,* dubbed the "reptilian tank," had a broad, flat body covered with armor plates and large, laterally projecting spines (Figure 20–26).

Two other groups of Mesozoic reptiles were the marine **plesiosaurs** and **ichthyosaurs** (Figure 20–27). The extremely long neck of the plesiosaurs took up over half of their total length of 15 meters. The ichthyosaurs had a body form superficially like that of a fish or porpoise, with a short neck, large dorsal fin, and shark-type tail.

FIGURE 20–26 *Ankylosaurus,* a heavily armored ornithischian. One would not guess from the external appearance of this creature that the ornithischians were so named for their birdlike pelvis.

FIGURE 20–27 Two types of marine reptiles. (*a*) *Plesiosaurus.* (*b*) *Ichthyosaurus.* Note the similarity of *Plesiosaurus* to animals such as modern seals and the similarity of *Ichthyosaurus* to modern porpoises. During their long reign as the dominant animals on Earth, the reptiles radiated into almost every conceivable environment.

Although the reptiles were the dominant animals of the Mesozoic, many other important organisms occur in the same formations. Most of the modern orders of insects appeared during that era. Snails and bivalves increased in number and diversity. Sea urchins reached their peak. Mammals first appeared in the Triassic, and birds first appear in Jurassic formations. During the early Triassic, the most abundant plants were gymnosperms (Figure 20–28). By the end of the Cretaceous, many flowering plants resembling present-day species had appeared and were the dominant vegetation.

Excellent bird fossils, some even showing the outlines of feathers, have been preserved from the Jurassic. *Archaeopteryx* is the classic example (Figure 20–29). This animal was about the size of a crow, had rather feeble wings, jawbones armed with teeth, and a long reptilian tail covered with feathers. Increasingly, *Archaeopteryx* is interpreted as a representative of a rather rare group of reptiles, one branch of which gave rise to the birds, but not as a bird itself. True birds do occur in the Cretaceous rocks, some of them apparently even older than *Archaeopteryx*.

At the end of the Cretaceous period, a great many animals abruptly became extinct. Most gymnosperms, with the exception of conifers, also perished. Changes in climate may have been a factor in their demise. While

FIGURE 20–28 Reconstruction of plants from the Mesozoic. Smaller ferns are dwarfed by the gymnosperm trees: conifers, cycads (resembling palms), and ginkgoes *(upper left)*.

Focus on CONTINENTAL DRIFT

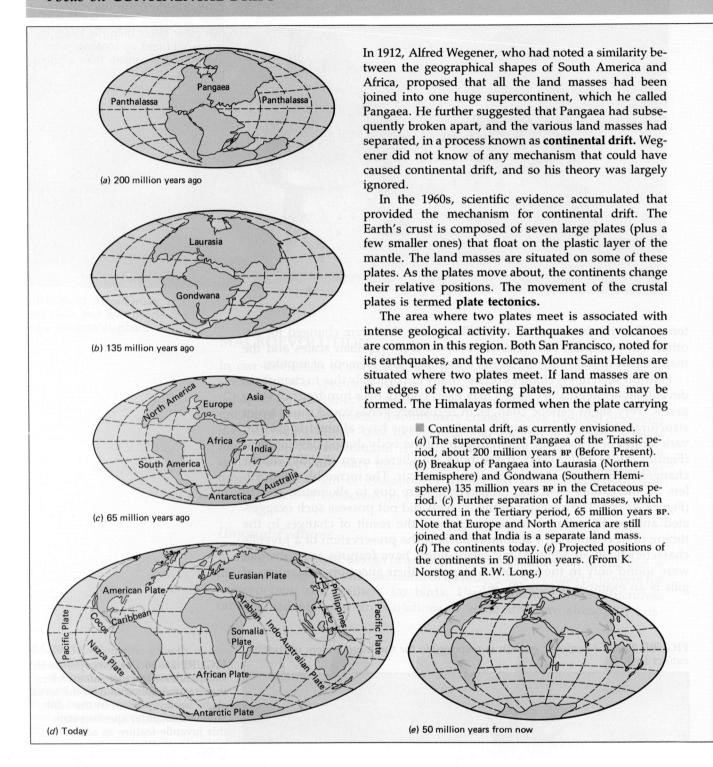

(a) 200 million years ago

(b) 135 million years ago

(c) 65 million years ago

(d) Today

(e) 50 million years from now

In 1912, Alfred Wegener, who had noted a similarity between the geographical shapes of South America and Africa, proposed that all the land masses had been joined into one huge supercontinent, which he called Pangaea. He further suggested that Pangaea had subsequently broken apart, and the various land masses had separated, in a process known as **continental drift.** Wegener did not know of any mechanism that could have caused continental drift, and so his theory was largely ignored.

In the 1960s, scientific evidence accumulated that provided the mechanism for continental drift. The Earth's crust is composed of seven large plates (plus a few smaller ones) that float on the plastic layer of the mantle. The land masses are situated on some of these plates. As the plates move about, the continents change their relative positions. The movement of the crustal plates is termed **plate tectonics.**

The area where two plates meet is associated with intense geological activity. Earthquakes and volcanoes are common in this region. Both San Francisco, noted for its earthquakes, and the volcano Mount Saint Helens are situated where two plates meet. If land masses are on the edges of two meeting plates, mountains may be formed. The Himalayas formed when the plate carrying

▪ Continental drift, as currently envisioned. (a) The supercontinent Pangaea of the Triassic period, about 200 million years BP (Before Present). (b) Breakup of Pangaea into Laurasia (Northern Hemisphere) and Gondwana (Southern Hemisphere) 135 million years BP in the Cretaceous period. (c) Further separation of land masses, which occurred in the Tertiary period, 65 million years BP. Note that Europe and North America are still joined and that India is a separate land mass. (d) The continents today. (e) Projected positions of the continents in 50 million years. (From K. Norstog and R.W. Long.)

Evolutionary Trends

The fossil record contains a number of examples of changes in a lineage that appear to exhibit an overall trend (Figure 20–37). For example, in the evolution of flowers, the number of floral parts like stamens and carpels tends to become reduced. In the mammal lineage, there is an evolutionary trend toward an increase in brain size relative to body size, as well as an overall trend toward an increase in body size.

India rammed into the plate carrying Asia. When two plates grind together, one of them is sometimes buried under the other, in a process known as subduction. When two plates move apart, a ridge of lava forms between them that continually expands as the plates move farther apart. The Atlantic Ocean is getting larger because of the buildup of lava along the mid-Atlantic ridge, where two plates are separating.

Knowledge that the continents were at one time connected and have since drifted apart is useful in explaining the geographical distribution of plants and animals, or biogeography (Chapter 17). Likewise, continental drift has played a major role in the evolution of different life forms. When Pangaea originally formed, about 250 million years ago, it brought together plants and animals that had evolved separately from one another, leading to competition and possible extinctions. Marine life was adversely affected, largely because, with the continents joined as one large mass, there would have been less coastline. Coastal areas are shallower and, therefore, have high concentrations of marine organisms.

Pangaea separated into several land masses approximately 180 million years ago. As the continents began to drift apart, populations became geographically isolated in different environmental conditions, the ideal setting for evolution.

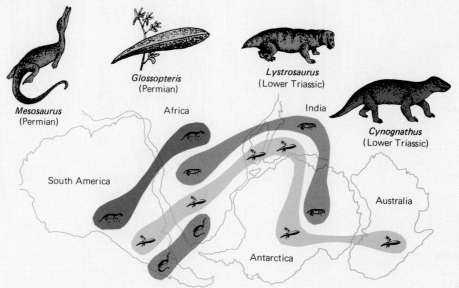

The distribution of the same fossils of different animal and plant species on four continents suggests that the continents were once joined.

Evolutionary trends do not appear to indicate gradualism (Chapter 19). The fossil record usually indicates periods of change which reflect the evolutionary trend, followed by periods of stasis. At any rate, trends can only apply in a general sense because there are many individual exceptions to

FIGURE 20–37 The titanotheres, an extinct group of mammals, exhibited an evolutionary trend toward increased size. The increase in horn size is an example of allometry.

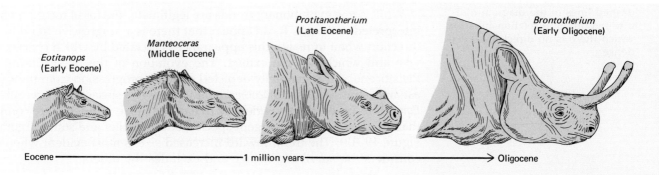

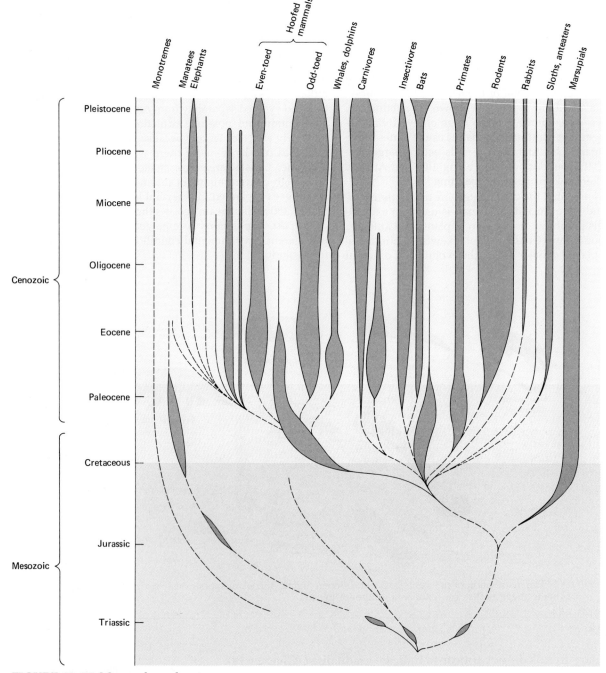

FIGURE 20–38 Mammals underwent adaptive radiation at the end of the Cretaceous. The decline of the reptiles at this time is believed to have allowed the mammals to radiate into many groups in a relatively short period of time. The dashed lines in the diagram indicate hypothetical relationships for which there is no direct fossil evidence.

the overall trend. While elephant evolution has revealed a trend in the direction of increased size, a number of extinct elephants were smaller than their ancestors.

While some evolutionary trends are legitimate, the fossil record must be interpreted with care. It can appear that there is a progressive trend in one direction when in reality this appearance is dictated by which species survive and which become extinct. The evolution of the horse during the Pleistocene as it is generally depicted shows a progression from the smallest, ancestral species, *Hyracotherium,* to the largest species, the present-day *Equus.* However, the evolutionary tree of the horse is much more complex than that, with many divergent species, some smaller and some larger (see Figure 19–13). The trend toward increased size is most evident when one considers the surviving lineage and excludes the others.

One can think of **species selection** in much the same way that natural selection applies to individuals in a population. In species selection, the species that survive the longest in evolutionary history, and produce the largest number of new species, have the greatest effect on an evolutionary trend. But the *direction* of the trend is a consequence of environmental pressure. If environmental conditions change, the trend may reverse direction or simply stop.

Adaptive Radiation

Once a novel feature evolves that represents an evolutionary advancement, adaptive radiation may occur, producing many species with that advancement. However, care must be taken in interpreting a cause-and-effect relationship between an advancement and adaptive radiation. It is tempting to take the simplistic approach and state, for example, that the evolution of the flower triggered adaptive radiation of thousands of species of flowering plants. Flowering plants exhibited rapid adaptive radiation, perhaps due to the evolution of a more competitive method of sexual reproduction (the flower). However, adaptive radiation in the angiosperms may be a consequence of other advancements that the flowering plants possessed instead of, or in addition to, flowers.

Adaptive radiation appears to be more common during periods of major environmental change, but it is difficult to determine if these changes trigger adaptive radiation. It is possible that major environmental change has an indirect effect on adaptive radiation by increasing the rate of extinction (Chapter 19). Extinction creates empty niches, which are available for adaptive radiation. Mammals had evolved millions of years before they underwent adaptive radiation. The radiation of mammals is thought to be a result of the extinction of the dinosaurs (Figure 20–38).

DOES MICROEVOLUTION LEAD TO MACROEVOLUTION?

We have seen that such mechanisms as differential reproduction through natural selection, mutation, genetic drift, and migration explain the evolutionary changes within a population, i.e., microevolution (Chapter 18). But can these mechanisms be used to explain the evolution of genera, orders, and higher taxa? Does natural selection account for macroevolution? Can population genetics help explain major evolutionary events?

The concepts of evolution presented in Chapters 17 to 19 represent the **modern synthesis** of evolution. Most biologists believe that the modern synthesis is adequate to explain macroevolution. That is, given enough time, the same processes that lead to speciation will produce new genera, new families, new orders, new classes, and new phyla. However, some biologists believe that adaptation may not be as significant in the evolution of higher taxa as it is in microevolution.

It is always possible that other mechanisms will be discovered to have an important role in macroevolutionary events. If this occurs, the mechanisms involved in evolution will have to be reevaluated in light of the new evidence. This is what makes the study of evolution such an intellectually stimulating pursuit.

■ SUMMARY

I. Life began from nonliving matter by chemical evolution.
   A. The four requirements for chemical evolution are: (1) absence of oxygen; (2) energy; (3) chemical building blocks; (4) sufficient time.
   B. The sequence of events in chemical evolution were: (1) the origin and accumulation of small organic molecules; (2) the assembly of macromolecules; (3) the formation of macromolecular assemblages; (4) the formation of cells.

II. The first cells were anaerobic and prokaryotic.
   A. The oldest cells in the fossil record are 3.4 to 3.5 billion years old.
   B. The last common ancestor of all living organisms on Earth was probably a sulfur-metabolizing prokaryote that inhabited hot springs.
   C. The evolution of photosynthesis ultimately changed early life because it generated oxygen.
   D. Aerobic organisms evolved the ability to use oxygen in respiration.
   E. Mitochondria, chloroplasts, and possibly flagella evolved from prokaryote endosymbionts.

III. The Earth's history is divided into eras, periods, and epochs.
   A. Life began and evolved into different groups of animals, plants, protists, and fungi during the Precambrian.
   B. During the Paleozoic era, land plants evolved to the gymnosperms, and fish and amphibians flourished.
   C. The Mesozoic era was characterized by the evolution of flowering plants and reptiles. Insects flourished. Birds and early mammals appeared.
   D. In the Cenozoic era, which includes the present time, there was a diversification of flowering plants and mammals, including humans and their ancestors.

IV. Macroevolution is the evolution of taxa above the species level.
   A. It includes the origin of unusual features, evolutionary trends, adaptive radiation, and extinction.
   B. The modern synthesis of evolution is probably adequate to explain macroevolution.

■ POST-TEST

1. Which of the following gases was not part of the Earth's atmosphere before living organisms evolved? (a) carbon dioxide; (b) oxygen; (c) water vapor; (d) hydrogen; (e) nitrogen.

2. Which of the following was not a requirement for chemical evolution? (a) energy; (b) time; (c) absence of oxygen; (d) chemical building blocks; (e) ammonia and methane gases.

3. Although Oparin envisioned life as originating in a "sea of organic soup," it is more likely that it evolved here: (a) clay surfaces; (b) freshwater lakes; (c) glacier surfaces.

4. Protobionts formed from assemblages of polypeptides, nucleic acids, and polysaccharides: (a) microspheres; (b) coacervates; (c) proteinoids; (d) stromatolites; (e) liposomes.

5. What kind of evidence did Lake use to determine the last common ancestor of life? (a) fossils; (b) geological evidence; (c) radiocarbon dating; (d) molecular evidence; (e) comparative morphology.

6. The first autotrophs probably used sunlight to split which hydrogen-rich molecules? (a) water; (b) ammonia; (c) carbon dioxide; (d) hydrogen sulfide; (e) ozone.

7. Evidence in support of the endosymbiont theory is weakest for which organelle? (a) chloroplasts; (b) flagella; (c) mitochondria; (d) none of these.

8. Which of the following is in the correct chronological order, starting with the oldest?
   (a) Precambrian, Mesozoic, Cenozoic, Paleozoic
   (b) Cenozoic, Mesozoic, Paleozoic, Precambrian
   (c) Precambrian, Cenozoic, Mesozoic, Paleozoic
   (d) Mesozoic, Precambrian, Cenozoic, Paleozoic
   (e) Precambrian, Paleozoic, Mesozoic, Cenozoic.

9. Which of the following was not important in the Cenozoic era? (a) mammals; (b) birds; (c) insects; (d) flowering plants; (e) dinosaurs.

10. A "new" structure that is a variation of some structure already in existence is called a(an): (a) allometric growth; (b) autotroph; (c) protobiont; (d) preadaptation; (e) paedomorphosis.

11. The incredibly large antlers of the extinct Irish elk were due to: (a) paedomorphosis; (b) allometric growth; (c) species selection; (d) preadaptations; (e) adaptive radiation.

12. Which of the following may occur where two tectonic plates meet? (a) earthquakes; (b) subduction; (c) volcanism; (d) mountain formation; (e) all of these.

■ REVIEW QUESTIONS

1. If chemical evolution occurred once, why can't it occur again?

2. What are the four requirements for chemical evolution, and why is each essential?

3. Briefly describe one of the contributions of each of these scientists: a. Oparin; b. Urey and Miller; c. Fox; d. Lake.

4. Which informational macromolecule probably evolved first and why?

5. Give some of the evidence used to support the endosymbiont theory.

6. In macroevolution, how are novel changes in structure related to development?

7. Give an example of each of the following: a. preadaptation; b. allometric growth; c. paedomorphosis.

8. Does microevolution lead to macroevolution? Why or why not?

■ RECOMMENDED READINGS

Foster, A., and Gifford, E. *Comparative Morphology of Vascular Plants.* 2nd ed. San Francisco, W.H. Freeman, 1974. A classic textbook that gives a detailed account of ancient plant structure.

Futuyma, D. *Evolutionary Biology.* 2nd ed. Sunderland, Massachusetts, Sinauer Associates, Inc., 1986. Probably the best in-depth treatment of evolution available.

Lewin, R. *Thread of Life.* Washington, D.C., Smithsonian Books, 1982. A visually striking discussion of evolution from the origin of life to the present.

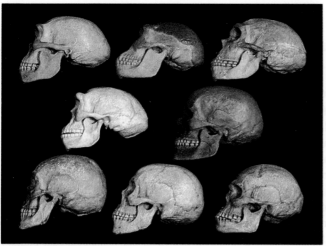

Evolution of human skulls

The Evolution of Primates

■ LEARNING OBJECTIVES

After you have read this chapter you should be able to:

1. Explain why primates have adaptations for an arboreal existence, even though many primates live on the ground, and describe the morphological adaptations primates possess for an arboreal existence.
2. List the two suborders in the order Primates, give several distinguishing features of each, and describe representative examples of each.
3. Describe skeletal and skull differences between apes and hominids.
4. Give a brief description of each of these hominids: australopithecines, *Homo habilis*, *Homo erectus*, *Homo sapiens*.
5. Explain what cultural evolution is and how it has had an impact on the human experience.

PRIMATE EVOLUTION

Most people have an interest in their roots. To many of us, this means trying to discover the immediate ancestors of our great grandparents. In this chapter we examine what we might call our "deep roots," as we trace human ancestry back some 65 million years to the earliest primates. For nearly a century after Darwin's *Origin of Species*, fossil evidence on human ancestry was rather sparse and unsatisfactory. However, research over the last four or five decades, especially in East Africa, has provided us with some reasonable answers to the question, "Where did we come from?"

Humans and other primates are mammals, members of the class Mammalia. As you know, mammals are **homeothermal** (warm-blooded) animals that produce body hair and feed their young milk from mammary glands. Most mammals are **viviparous** and bear their young alive as opposed to laying eggs. Although mammals evolved from reptiles over 210 million years ago, they were of secondary importance at that time. It was the "Age of Reptiles," and reptiles were the dominant animals in almost every habitat on Earth.

There were three main lines of mammals during the Mesozoic era: (1) the multituberculates, which may have given rise to monotremes like the duck-billed platypus; (2) the marsupials, which were the ancestors of modern-day kangaroos and opossums; and (3) small shrewlike placental mammals that ate insects and lived a nocturnal existence in the trees.

advanced anthropoids in a number of ways. Their snouts are shortened and the eyes point forward completely. When the tarsier sits or climbs upright, its head is positioned on the vertebral column at an angle, enabling it to face forward instead of directly upward.

Anthropoids

The anthropoids evolved from a group of prosimians during the Oligocene, approximately 36 million years ago (Figure 21–2). This took place in Africa or, possibly, Asia. From there the anthropoids quickly spread throughout Europe, Asia, and Africa. They branched into two main groups, the New World monkeys and the Old World monkeys. It is not clear how the New World monkeys got to South America, as Africa and South America had already split by continental drift (Chapter 20). Originally, it was believed that the New World monkeys had evolved from a separate prosimian line. Various types of comparative biochemistry, including amino acid sequencing of proteins, have indicated that the New World and Old World monkeys share a common ancestor. At any rate, the New World and Old World monkeys have been separated for millions of years and have evolved along different paths. The Old World monkeys were ancestral to the apes and **hominids,** a group composed of humans and their ancestors. One of the earliest apelike anthropoids was discovered in Egypt and named *Aegyptopithecus.* It was a small arboreal ape that existed approximately 35 million years ago.

During the Miocene epoch, the apes and Old World monkeys diversified. Fossils of an early forest ape, *Dryopithecus,* are of special interest because it may have given rise to modern apes as well as the human line. Although the forest ape was arboreal, it apparently spent part of the time on the ground. It lacked the long forearms characteristic of the apes today and had a sloping cranium with bony ridges above the eyes. The dryopithecines were distributed widely across Europe, Africa, and Asia. As the climate gradually cooled and became drier, their distribution became more limited.

By the beginning of the Pliocene epoch, approximately 5 million years ago, the apes that developed from the dryopithecines were restricted primarily to tropical rain forests. Unfortunately, the moist conditions of the tropics preclude the formation of many fossils, so our knowledge of ape evolution is sketchy. However, sometime between 5 and 7 million years ago, a new form of primate arose, the hominid.

Monkeys

Monkeys and apes are generally larger than prosimians. Most are active during the day, or **diurnal,** as compared with the nocturnal prosimians. Like the prosimians, they are generally tree dwellers, although their diet is more varied than that of the prosimians. Different groups eat leaves, fruits, buds, insects, and even small vertebrates. Probably the most significant difference between the prosimians and the apes and monkeys is the size of the brain. The cerebral cortex, in particular, is more developed in monkeys and apes.

The New World monkeys are arboreal and possess long, slender limbs for easy movement in the trees (Figure 21–4). Many have a **prehensile** tail that is capable of wrapping around branches. Some of them have a smaller thumb, and in certain cases the thumb is totally absent. Their facial anatomy is different from that of the Old World monkeys, as they have flattened noses with the nostrils opening to the side. They live in groups and exhibit social behavior. New World monkeys are restricted to Central and

(a)

(b)

FIGURE 21–4 New World and Old World monkeys. (*a*) Most New World monkeys have prehensile tails that can function almost as effectively as another limb. Shown here is a red howler monkey in Peru. (*b*) Baboons are ground-dwelling Old World monkeys. These yellow baboons are feeding in the Amboseli National Park in Kenya.

South America. Examples include howler monkeys, squirrel monkeys, and spider monkeys.

Many Old World monkeys are arboreal, although some, like the baboons and macaques, are ground dwellers (Figure 21–4). The ground dwellers, which are **quadrupedal** and walk on all fours, evolved from arboreal monkeys, however. None of them has a prehensile tail, and some even lack tails. They have a thumb that is fully opposable. Unlike the New World monkeys, their nostrils are directed downward and are closer together. Old World monkeys are larger than New World monkeys. They are social animals and are distributed in tropical parts of Africa and Asia.

Hominoids

Apes and humans share a number of features and are collectively called **hominoids.** There are four genera of apes classified into two families (Figure 21–5). The gibbons (*Hylobates*) are sometimes known as lesser apes and are placed in a separate family, Hylobatidae. The family Pongidae includes the orangutans (*Pongo*), gorillas (*Gorilla*), and chimpanzees (*Pan*).

Gibbons are well adapted for an arboreal existence. They are natural acrobats, and can **brachiate,** or swing, with their weight supported by one arm at a time. Orangutans are also tree dwellers, but both gorillas and chimpanzees have adapted to life on the ground. They have retained elongated forearms typical of tree-dwelling primates but use these to assist in quadrupedal walking, sometimes known as **knuckle walking** because of the way they fold their digits when moving. Apes, like humans, lack tails. They are larger than monkeys, with the exception of the gibbon. Their social organization, particularly in the gorillas and chimpanzees, is more complex. It is believed this is due, in part, to their larger brains.

Antigen-antibody tests of similarities in serum proteins show that, of all the primates, gorillas and chimpanzees have serum proteins most nearly like those of the human. The amino acid sequence in the chimpanzee's hemoglobin is identical with that of the human; that of the gorilla and rhesus monkey differ from the human's in 2 and 15 amino acids, respectively. Molecular studies of DNA sequences indicate the chimpanzee is our nearest living relative among the apes (Figure 21–6).

(a)

(b)

(c)

(d)

FIGURE 21–5 The apes. (*a*) Gibbons are extremely acrobatic and often move through the trees by brachiation. (*b*) Orangutans are solitary apes that seldom leave the protection of the trees. (*c*) A gorilla knuckle-walking. (*d*) Chimpanzees live in groups and have a complex social behavior.

HOMINID EVOLUTION

The **hominid** line separated from the ape line approximately 3.5 to 4 million years ago. General trends in human evolution are evident from the fossil record, but we do not have enough evidence to make specific conclusions. There are simply too few early hominid fossils, and the ones we have are represented by only a few bones. Moreover, it is impossible to determine many aspects about early hominid biology or appearance or behavior from fossilized bones. Nevertheless, it is evident that early hominids evolved a **bipedal** (two-footed) posture before their brains enlarged. This is an example of **mosaic evolution,** two traits evolving independently of one another and at different rates. In order to understand the evolutionary progression from the earliest hominids to modern humans, we must examine some of the characteristics of the skeleton and skull.

As compared to the ape skeleton, the human skeleton possesses distinct differences that reflect our ability to stand erect and walk on two feet (Figure 21–7). These differences also reflect the change in habitat for early hominids, from an arboreal existence in the forest to spending at least part of the time on the ground. The skeletal differences include a greater curvature of the spine to allow for better balance and weight distribution. The

502

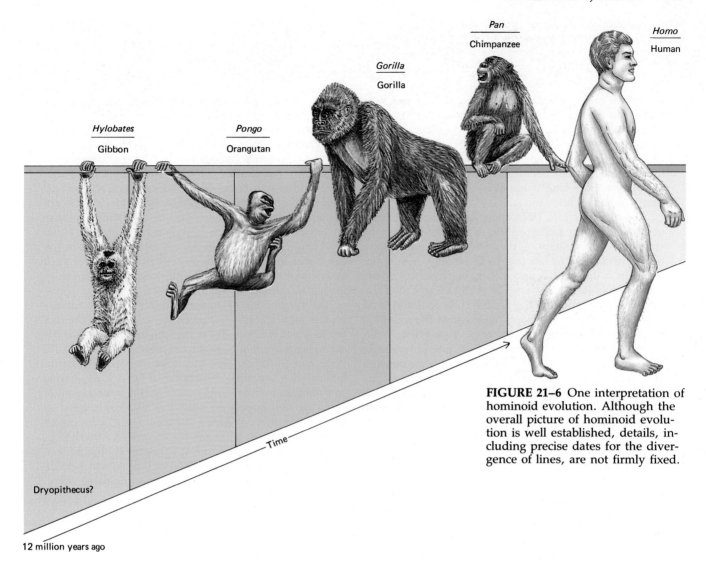

Hylobates
Gibbon

Pongo
Orangutan

Gorilla
Gorilla

Pan
Chimpanzee

Homo
Human

Time

Dryopithecus?

12 million years ago

FIGURE 21–6 One interpretation of hominoid evolution. Although the overall picture of hominoid evolution is well established, details, including precise dates for the divergence of lines, are not firmly fixed.

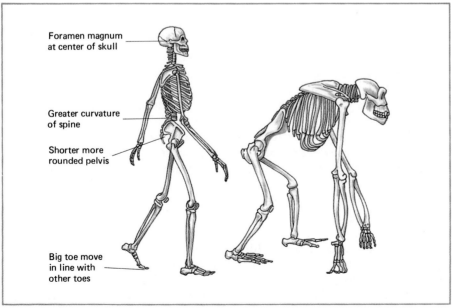

Foramen magnum at center of skull

Greater curvature of spine

Shorter more rounded pelvis

Big toe move in line with other toes

FIGURE 21–7 Comparison of gorilla and human skeletons. Note the skeletal adaptations for bipedalism in humans.

Focus on MOLECULAR CLOCKS

One of the most powerful new tools in determining how long it has been since two organisms had a last common ancestor is molecular biology. Each gene coding for a specific protein has a specific rate of change, or mutation, over time. By comparing the nucleotide sequences in the same gene for two different species, it is possible to estimate how long it has been since they diverged from a common ancestor.

As you may recall, DNA codes for the specific order of amino acids in a polypeptide, which determines the shape, or conformation, of the molecule. A substitution of one nucleotide for another in DNA may or may not result in a different amino acid being inserted in the polypeptide. Recall that a triplet of DNA codes for a single amino acid and that the genetic code common to all organisms specifies that some amino acids are coded for by several possible triplet sequences. Also, even if a different amino acid is substituted in the polypeptide chain, the conformation of the molecule may or may not change. That is, a mutation in DNA does not necessarily cause changes in the polypeptide it codes for. Such changes in DNA are neutral mutations because they do not affect the organism's phenotype. Neutral mutations are important in molecular clocks because natural selection has no effect on them.

When molecular evidence is used to calculate a last common ancestor, care must be taken. The rate of mutation is generally the same for different species, but there are exceptions. Primates, in particular, have been shown to have a lower mutation rate than other organisms, although the reasons for this are not clear.

When molecular biologists first started using molecular clocks, paleoanthropologists were skeptical. However, evidence from molecular clocks has correlated well with the dating of fossil evidence. Molecular clocks cannot be calculated for organisms that are extinct, because we have no DNA from such organisms to analyze. However, molecular clocks have been used to determine the last common ancestor when no fossil evidence for such an ancestor was known. This information provides clues for locating fossils of such ancestors.

The sequences of mitochondrial DNA in various human populations were recently compared by the molecular clock method. These data suggest all humans living today had a last common ancestor that lived 200,000 years ago, in Africa. This ancestor was popularly identified as "Eve" because mitochondrial DNA is a maternal contribution to the next generation. (Eggs contain mitochondria, but sperm do not.) Eve was determined to have originated in Africa because there were much greater differences in the mDNA of various African populations than in other groups. Presumably, the African populations had a longer period of time in which to diverge from one another because Eve originated there.

pelvis is shorter and more rounded, providing a better attachment of muscles used for upright walking. The hole in the base of the skull for the spinal cord, called the **foramen magnum,** is located at the back of the skull in apes. In contrast, the foramen magnum in humans is located in the middle of the bottom of the skull to allow the head to be positioned for erect walking. An increase in the length of the legs relative to the arms, and movement of the big toe so it is in line with the rest of the toes, further adapted the early hominids for bipedalism.

Another major trend in human evolution is an increase in the size of the brain relative to the size of the body (Figure 21–8). In addition, the ape skull possesses prominent bony ridges above the eye sockets, while these **supraorbital ridges** are lacking in human skulls. Human faces are flatter than those of apes, and the jaws are different. The arrangement of teeth in the ape jaw is somewhat rectangular as compared to a rounded or U-shape in humans. Apes have larger teeth than humans, and their canines are especially large.

Early Hominids

Human evolution occurred in Africa. The earliest hominids belong to the genus *Australopithecus*, or "Southern ape," and appeared approximately 3.8 million years ago. The actual number of species assigned to this genus is under debate. It is very difficult to decide whether differences in the relatively few skeletal fragments that have been discovered indicate individual variation within a species or separate species. Most biologists recognize between two and four species of australopithecines.

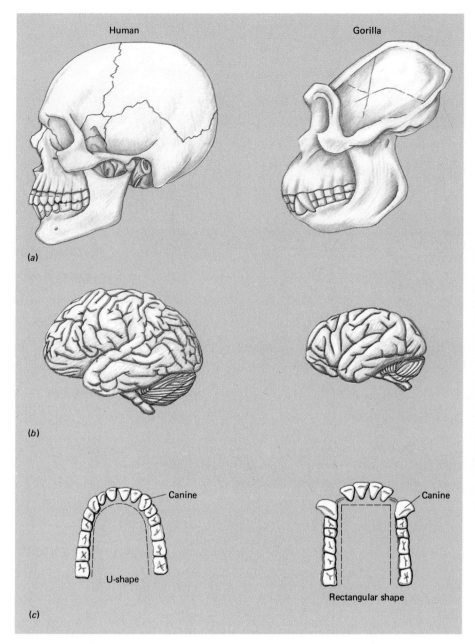

Human

Gorilla

(a)

(b)

Canine

Canine

U-shape

Rectangular shape

(c)

FIGURE 21–8 Comparison of features of the ape and human head. (*a*) The ape skull has pronounced supraorbital ridges. Note how the human skull is flatter in the front and has a more pronounced chin. (*b*) The human brain, particularly the cerebrum, is larger than that of an ape. (*c*) The human jaw is structured so that the teeth are arranged in a U-shape. Human canines are reduced in size compared to ape canines.

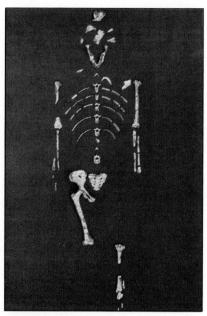

FIGURE 21–9 The skeletal remains of Lucy, a hominid approximately 3.5 million years old.

FIGURE 21–10 Three hominids (*Australopithecus afarensis*) walked across ash scattered by a volcanic eruption over 3.6 million years ago in Africa. Their footprints were compacted by a rain shower shortly thereafter. The footprints, which were discovered in 1976 by Mary Leakey and her associates, indicate *A. afarensis* had a bipedal gait.

The most ancient hominids are assigned to the species *A. afarensis*. Several fossils of skeletal remains of *A. afarensis* have been discovered, including a remarkably complete skeleton named Lucy (Figure 21–9). In addition, in 1976 fossil footprints were discovered of three individuals who walked over 3.6 million years ago (Figure 21–10). The footprints plus pelvis, leg, and foot bones indicate that the development of an upright posture and bipedalism occurred early in human evolution. *A. afarensis* was a small hominid, approximately 3 feet tall. Its face projected forward, and its ape-like skull covered a small brain. The cranial capacity was 450–500 cubic centimeters compared to a modern human cranial capacity of 1400 cc (Figure 21–11). Even when differences in body size are taken into account, *A. afarensis* still possessed a small brain. Its **dentition,** the number and arrangement of teeth, was primitive and included long canines.

Many scientists think *A. afarensis* evolved into the more advanced australopithecine, *A. africanus*, which appeared approximately 3 million years ago. The first *A. africanus* fossil was discovered in South Africa in 1924, and

Hominid	Date of appearance (millions of years)	Mean cranial capacity (cc)	Skull
Australopithecus afarensis	3.8	450–500	
Australopithecus africanus	3.0	494	
Australopithecus robustus	2.0	500	
Homo habilis	1.8	650cc	
Homo erectus	1.5	950	
Homo sapiens	0.2	1400	

since then a number of others have been found. This rather small hominid walked erect and possessed hands and teeth that were distinctly human-like. Based on characteristics of the teeth, it is believed that *A. africanus* ate both plants and animals. Like *A. afarensis*, its brain was small, approximately 500 cc. Two larger forms of *Australopithecus* have been identified, *A. robustus* and *A. boisei*, but it is generally agreed that neither is in the direct line to humans.

Focus on TOOL-MAKING: A HUMAN ACTIVITY?

One of the established notions in paleoanthropology is that the ability to make tools is a strictly human characteristic. Indeed, many authorities have considered tool-making an indication that an early hominid belonged in the *Homo* genus. While it is recognized that chimpanzees and other primates can occasionally use items from their environment as tools, the conscious effort to make tools has become ingrained as defining the beginning of human culture. We have felt that humans can make tools and nonhumans cannot for two reasons. First, our superior brains give us the analytical ability to design and fashion tools. Second, certain characteristics of the human hand enable us to hold objects with the precision required in tool-making.

It is possible that we may have to rethink our views about what, if anything, constitutes a specific human activity. In 1988, several fossil hand bones that had been recovered from a cave in South Africa were examined by Randall Susman. A number of fossil skull fragments had already been studied from the cave and identified as belonging to *Australopithecus robustus*, an early hominid from an extinct line that did not give rise to humans. Although *A. robustus* was decidedly apelike in overall appearance and brain size, the hand bones are distinctly modern. They indicate that *A. robustus* would have had the manual dexterity required to fashion stone tools. Additional evidence on the cave floor includes some bone and stone artifacts that could be interpreted as crude tools used by *A. robustus*, a vegetarian, to dig and chop plants. Susman concluded that *A. robustus* may have used tools and that the reason for its extinction was certainly not its inability to use tools. Susman further concluded that the early success of the *Homo* line could not be attributed to its ability to make and use tools.

Not all scholars of human evolution are happy with Susman's conclusions. Some point out that the cave also had a few skeletal fragments of *Homo habilis* or *H. erectus* and that these bones and the scattered tools could belong to that early human.

The final resolution of whether or not *A. robustus* had a human-like hand and used tools may not occur until additional fossil evidence is unearthed. Regardless of the outcome, Susman has succeeded in cautioning paleoanthropologists to be very careful in interpreting evidence about early humans. It will no longer be safe to assume that all stone tools and artifacts are evidence of early human activity.

Homo habilis

The first hominid to have enough human features to be placed in the same genus as modern humans is *Homo habilis. H. habilis* had a larger cranial capacity, an average of 650 cc, than the australopithecines (Figure 21–11). This early human appeared approximately 1.8 million years ago and persisted for over half a million years. Fossils of *H. habilis* have been found in numerous areas in Africa. These sites contain the first primitive tools, stones that had been chipped to make sharp edges for cutting or scraping. Although other primates occasionally use tools, *H. habilis* represents the first to consciously design them (see Focus on Tool-making for a decidedly different view).

The relationship between the australopithecines and *H. habilis* is not clear. Using physical characteristics as evidence, some biologists believe that the australopithecines were ancestors of *H. habilis*. Others think that *H. habilis* and *A. africanus* were contemporaries for much of their existence and that *H. habilis* was in a direct line to humans, but *A. africanus* was not (Figure 21–12). Hopefully, discoveries of additional fossils will help clarify their relationship.

Homo erectus

There is more agreement on interpretation of the fossils classified as *Homo erectus*, as numerous fossils have been found. *H. erectus* evolved in Africa as did the other hominids, but migrated into Europe and Asia. For this reason, the oldest fossils of *H. erectus*, 1.5 million years old, are found in Africa, and the later ones are more widely distributed in the Old World.

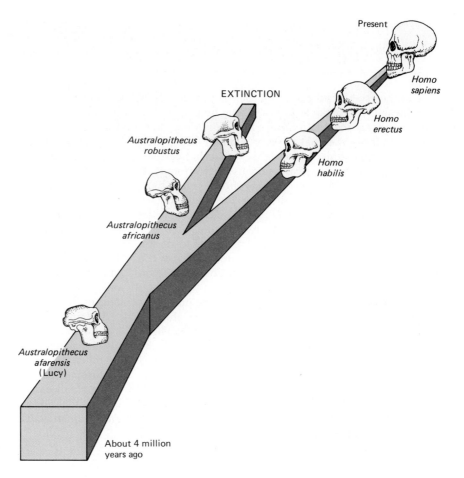

FIGURE 21–12 One possible representation of the human lineage. Paleoanthropologists are not in complete agreement about the details of our lineage, but many feel the current evidence supports the evolution of *H. habilis* from *A. afarensis*. In this representation *A. africanus* and *A. robustus* are not in a direct line to *H. sapiens*.

The Peking man and Java man discovered in Asia were later examples of *H. erectus*, which existed until approximately 200,000 years ago.

H. erectus was taller than *H. habilis*, bipedal, and fully erect. During the course of its existence, its brain got progressively larger, evolving from a cranial capacity of 850 cc to between 1000 and 1200 cc (Figure 21–11). Their skulls, although larger, did not possess totally modern features. They retained the heavy supraorbital ridges and projecting faces that are more characteristic of the apes (Figure 21–13).

The increase in mental faculties associated with an increase in brain size enabled these early humans to make more advanced stone tools, including hand axes and other tools that have been interpreted as choppers, borers, and scrapers. Their intelligence enabled them to survive in areas that were cold. *H. erectus* wore clothing, built fires, and lived in caves or shelters. It is not known for sure whether they were hunters or scavengers. To date, no weapons have been unearthed at their sites.

Homo sapiens

Humans having features modern enough to classify them within our species appeared approximately 200,000 years ago. Their brains continued to enlarge, developing from 850 cc in earliest individuals to the current cranial capacity of 1400 cc (Figure 21–11).

One of the earliest groups of *H. sapiens* was the Neanderthals. They were first discovered in the Neander Valley in Germany but had widespread distribution. These early humans had a short, sturdy build. Their brains were slightly larger than that of modern *H. sapiens*, but their faces still projected slightly, with less pronounced chins, and they still had heavy brow ridges (Figure 21–14).

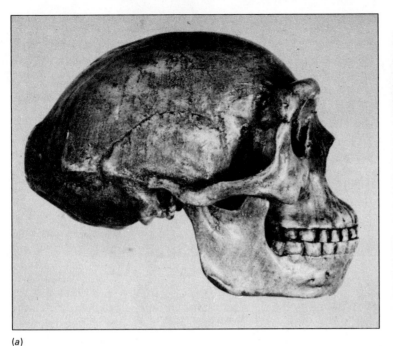

(a)

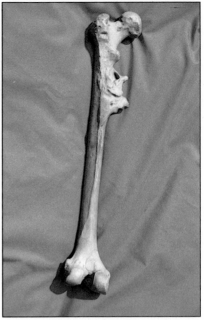

(b)

FIGURE 21–13 *Homo erectus.* (a) A replica of the skull. Note the massive bony ridges over the eyes and the receding forehead and protruding jaws. (b) A view of the femur of *Homo erectus*, discovered by Eugene Dubois. The well-developed linea aspera (a long ridge that serves as a point of attachment for many hip muscles) indicates an erect posture for this hominid, as is reflected in its scientific name. The advanced bony tumor on the femur is possible evidence that sick *Homo erectus* hominids were cared for by well ones.

Their tools, including spear points, were more sophisticated than those of *H. erectus*. Studies of Neanderthal sites indicate their culture included hunting for large animals. The existence of skeletons that were old or had healed fractures demonstrates they cared for the elderly and sick, which is an advanced example of social cooperation. They apparently had rituals, possibly of religious significance, and buried their dead. The presence of food, weapons, and flowers in the graves indicates they had the abstract concept of an afterlife.

The disappearance of the Neanderthals is a mystery. Other groups of *H. sapiens* with more modern features coexisted with the Neanderthals. It is possible that the Neanderthals interbred with these humans, diluting their features beyond recognition, or perhaps the other humans "outcompeted" them. It is also possible that the Neanderthals could not adapt to the climate changes of the Pleistocene and that their disappearance is unrelated to the presence of other humans.

H. sapiens with thoroughly modern features existed from 40,000 years ago and possibly earlier. It is not known whether these humans are derived from an isolated population of Neanderthals. Their skulls lacked heavy brow ridges and possessed a distinct chin. The Cro-Magnon culture in France and Spain exemplifies these humans. Their weapons and tools were complex and often made of materials other than stone, including bone, ivory, and wood. They made stone blades that were very sharp.

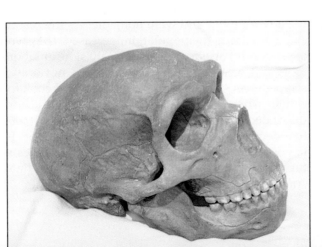

FIGURE 21–14 Neanderthal skull. Note the very heavy supraorbital ridge and the protruding face. The Neanderthal brain size was greater than that of modern humans.

FIGURE 21–15 The Cro-Magnon people painted animals on cave walls in Europe. These are some of the earliest representations of human art and have been interpreted as having a religious significance, possibly for guaranteeing a successful hunt.

They developed art, possibly for ritualistic purposes, including cave paintings, engraving, and sculpture (Figure 21–15). The existence of a variety of complex tools and art is an indication that they may have possessed language abilities, used to transmit their culture to younger generations.

Studies of mitochondrial DNA of different geographical populations of humans today indicate that the first modern *H. sapiens* developed in Africa during the early part of the late Pleistocene. Similar studies of nuclear DNA (in the beta-globulin cluster) and Y chromosome DNA also support the African origin. More fossil evidence will be needed to determine the exact time and place for the last common ancestor of humans. Once evolved, these modern humans migrated extensively over the Earth. Some crossed the Bering land bridge into North America. Others traveled across ocean waters to reach Australia. If modern *H. sapiens* originated in Africa, then it would appear that the Neanderthal *H. sapiens* was not in a direct line to modern *H. sapiens*.

CULTURAL EVOLUTION

Genetically speaking, humans are not very different from other primates. Most of our genes are shared with gorillas and chimpanzees. However, humans do possess a greater intelligence and have been able to maximize this intelligence through **cultural evolution.** Cultural evolution is the progressive addition of knowledge to the human experience. Human culture is dynamic; it is modified as we obtain new knowledge (Figure 21–16). Cultural evolution is generally divided into three stages: (1) the development of hunter/gatherer societies; (2) the development of agriculture; and (3) the Industrial Revolution.

Early humans that were hunter/gatherers relied on what was available in the environment and were nomadic. As the resources in a given area were exhausted or as the population increased, they would migrate to a different area. These societies required a division of labor and the ability to make tools and weapons, which are needed not only to kill game, but also to scrape hides, dig up roots and tubers, and cook food. Although we are not certain when hunting was incorporated into human society, we do know that it declined in importance approximately 15,000 years ago, possibly due to a decrease in large animals, triggered in part by a change in climate. A few isolated groups of hunter/gatherer societies survived to the 20th century, including the Mountain Lapps of Scandinavia and the Bushmen of Australia.

FIGURE 21-16 The progressive improvement of stone tools is evidence of cultural evolution. (*a*) Hand axes appeared approximately 1.5 million years ago. This primitive hand ax was found at the Olduvai Gorge in Africa. (*b*) This particular tool, an Acheulean hand ax from Europe, is more advanced than the earliest ones discovered in Africa. (*c*) Several Neanderthal Mousterian tools. These tools represent more advanced examples of stone tools. Each is specialized for a particular task. (*d*) Stone blades were fashioned by Cro-Magnon humans. Note that the length of the blade is longer than its width. These are examples of the most advanced tools made from stone.

(*a*)

(*b*)

(*c*)

(*d*)

Development of Agriculture

Evidence that humans had begun to cultivate crops approximately 10,000 years ago includes the presence of agricultural tools and plant material at archaeological sites. Agriculture, keeping animals as well as cultivating plants, resulted in a more dependable food supply. It appears from recent archaeological evidence that agriculture arose in several steps. Although there is a lot of variation from one site to another, plant cultivation usually occurred first in combination with hunting. Animal domestication followed at a later period. Agriculture, in turn, often led to more permanent dwellings since considerable time was invested in growing crops in one area. Often, villages and cities grew up around the farmlands, but relating the advent of agriculture to the establishment of villages and towns is complicated by recent discoveries. For example, Abu Hureyra in Syria was a village founded *before* agriculture arose. The people subsisted on the rich plant life of the area and migrating herds of gazelle. Once people turned to agriculture, however, they seldom went back to hunting/gathering to provide food.

Archaeological evidence indicates that agriculture developed independently in several different regions. There were three main centers of agriculture and several minor ones. Each of the main centers was associated with cultivation of a cereal crop, although other foods were grown as well. Cereals are grasses, which are members of the monocot group of flowering plants (Chapter 28). The cereals associated with the three main centers of agriculture are wheat, corn, and rice.

Wheat was cultivated in the semiarid regions along the eastern edge of the Mediterranean. Other crops that originated in this area include peas, lentils, grapes, and olives. Central and South America were the sites of the maize, or corn, culture. Squash, chili peppers, beans, and potatoes were also cultivated there. In the Far East, in Southern China, there is evidence for the early cultivation of rice and other crops like soybeans. The actual date for the domestication of rice is unknown because rice is cultivated in a wetter environment, conditions that prevent preservation of archaeological evidence.

Corn, wheat, and rice are all propagated by seed, which requires fairly sophisticated agricultural practices. It has been suggested that cultivation of plants that could be propagated vegetatively occurred earlier. Plants that are cultivated in this manner, such as bananas, yams, potatoes, and manioc, do not preserve as well as grains because of their high water content. For that reason, we may have no evidence of their cultivation prior to the cereal crops.

Other advances in agriculture include the domestication of animals, which were kept to supply food, milk, and hides. In the Old World they were also used to prepare fields for planting. Another major advance in agriculture was the use of irrigation, which dates to 7000 years ago in the Near East.

Producing food agriculturally was more time-consuming than obtaining food by hunting and gathering. However, it was also more productive. In hunter/gatherer societies, everyone shares the responsibility of obtaining

FIGURE 21–17 The human population is increasing geometrically, as Malthus predicted (see Chapter 17).

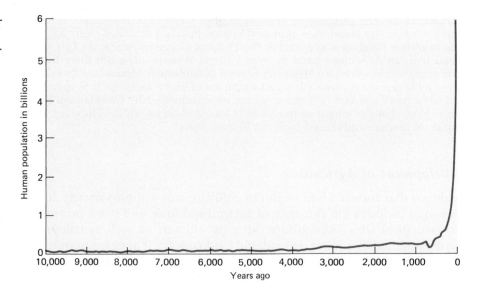

food. In agricultural societies fewer people are needed to provide food for everyone. This freed some people to pursue other endeavors, including religion, art, and various crafts.

The Impact of Cultural Evolution Today

Cultural evolution has had a profound effect on human society and on other life forms. The Industrial Revolution, which began in the 18th century, resulted in the concentration of people in urban areas where centers of manufacturing are located. Advances in agriculture encouraged this, as fewer and fewer people were needed to provide food for everyone. The spread of industrialization has increased the demand for natural resources to supply the raw materials for industry. The human population has increased so dramatically that some biologists fear the Earth cannot support our numbers (Figure 21–17). As it is, millions of people are malnourished or undernourished. Almost all the arable land on Earth is under cultivation.

Cultural evolution has resulted in large-scale disruption and degradation of the environment. Tropical rain forests and other natural environments are rapidly being eliminated. Soil, water, and air pollution occurs in many places. Desertification is increasing as plant cover is removed from marginal, arid lands so they can be cultivated. Many plant and animal species cannot adapt to the rapid changes humans are causing in the environment and are perishing. The decrease in biological diversity due to extinction is alarming biologists and others.

On a positive note, we are aware of the negative effects we have caused on the environment. And we have the intelligence to modify our behavior to improve these conditions. By educating younger generations, we can help them develop environmental sensitivity. Then, cultural evolution may be our salvation instead of our destruction.

■ SUMMARY

I. Primates evolved from small, arboreal, shrewlike mammals.
   A. Primates are adapted for an arboreal existence by the following: the presence of five digits, including an opposable thumb; long, slender limbs that move freely at the hips and shoulders; and eyes located in front of the head.
   B. Primates are divided into two suborders, the prosimians and the anthropoids.
      1. Prosimians include lemurs, tarsiers, and lorises.
      2. Anthropoids include monkeys, apes, and humans.
II. Anthropoids evolved from prosimian ancestors during the Oligocene.

A. The early anthropoids branched into two groups, the New World monkeys and the Old World monkeys.
B. Apes evolved from the Old World monkey lineage.
C. There are four genera of apes: gibbons; orangutans; gorillas; and chimpanzees.

III. The hominid line separated from the ape line approximately 3.5 to 4 million years ago.
   A. The earliest hominids belong to the genus *Australopithecus*. The australopithecines walked on two feet, a human feature.
   B. *Homo habilis* was an early hominid that had some human features the australopithecines lacked, including a slightly larger brain. *H. habilis* fashioned tools from stone.
   C. *Homo erectus* had a larger brain than *H. habilis,* made

more sophisticated tools, and discovered how to use fire.
   D. *Homo sapiens* appeared approximately 200,000 years ago.
      1. The brain continued to enlarge during its evolution.
      2. It is likely that modern *Homo sapiens* evolved from a common African ancestor.

IV. Cultural evolution is the progressive addition of knowledge to the human experience.
   A. It is made possible by an evolutionary increase in brain size in humans.
   B. Two of the most significant advances in cultural evolution were the development of agriculture and the Industrial Revolution.

■ POST-TEST

1. Which of the following is an adaptation for dwelling in trees? (a) bipedalism; (b) opposable thumbs; (c) large brains; (d) short, thickened limbs with limited mobility; (e) eyes on the sides of the head.
2. Tarsiers and lemurs are examples of: (a) hominids; (b) anthropoids; (c) New World monkeys; (d) prosimians; (e) hominoids.
3. The anthropoids evolved from a group of: (a) prosimians; (b) hominids; (c) shrewlike mammals; (d) hominoids; (e) Old World monkeys.
4. One of the earliest apelike anthropoids that evolved 35 million years ago was: (a) *Australopithecus*; (b) *Pan*; (c) *Homo*; (d) *Aegyptopithecus*; (e) *Pongo*.
5. Which of the following features is more characteristic of humans than of apes? (a) large supraorbital ridges; (b) brachiation; (c) foramen magnum located toward

back of skull; (d) opposable big toe; (e) small canines.
6. Which of the following includes both the apes and the humans but not the monkeys? (a) hominids; (b) hominoids; (c) anthropoids; (d) prosimians.
7. The first hominid to walk erect on two feet was: (a) *Homo sapiens*; (b) *Homo erectus*; (c) *Australopithecus*; (d) *Homo habilis*.
8. The earliest hominid to be placed in the genus *Homo*: (a) *H. erectus*; (b) *H. sapiens*; (c) *H. habilis*.
9. Both molecular and fossil evidence have recently been used to identify the origin of modern humans as: (a) Europe; (b) Asia; (c) Africa; (d) Australia; (e) North America.
10. Which of the following centers of agriculture was the origin of corn, chili peppers, beans, and potatoes? (a) Near East; (b) Far East; (c) New World.

■ REVIEW QUESTIONS

1. Distinguish between each of the following:
   a. mammals and primates
   b. prosimians and anthropoids
   c. New World monkeys and Old World monkeys
   d. hominoids and hominids
   e. hominids and humans
2. Tell three different adaptations primates have for an arboreal existence.
3. Give at least three ways an ape skull differs from a human skull.
4. Give at least three ways an ape skeleton differs from a human skeleton.
5. Cite one anatomical feature and one behavioral feature that distinguishes each of the following from its immediate ancestor:

   a. *Australopithecus afarensis*
   b. *Homo habilis*
   c. *Homo erectus*
   d. *Homo sapiens*—Neanderthal
   e. *Homo sapiens*—modern
6. Draw two possible family trees of the hominids listed in question 5.
7. How did the origin of agriculture impact human development?
8. How is cultural evolution related to biological evolution? (Hint: The evolution of what biological characteristic contributed to cultural evolution?)
9. How has cultural evolution helped humans?
10. How has cultural evolution affected the rest of the Earth besides humans?

■ RECOMMENDED READINGS

Heiser, C. *Seed to Civilization: The Story of Food.* 2nd ed. San Francisco, W.H. Freeman Company, 1981. A fascinating account of the origin of agriculture and important food crops.

Klein, R. *The Green World: An Introduction to Plants and People.* 2nd ed. New York, Harper & Row, 1987. Includes a detailed description of the origin of agriculture, along

with some of the environmental problems that are the result of cultural evolution.

Lewin, R. *Thread of Life.* Washington, D.C., Smithsonian Books, 1982. Includes a presentation of human evolution.

Weaver, K. The search for our ancestors. *National Geographic*, November 1985, pp. 560–623. The story of human evolution written in spell-binding fashion.

PART V

THE DIVERSITY OF LIFE

Millions of different organisms inhabit our planet. ■ To make some order of this diversity, biologists have developed a system of classification in which each organism is assigned to one of five kingdoms. ■

Anemone nematocysts

22

The Classification of Living Things

None of the millions of organisms that inhabit our planet comes with a label giving its name and evolutionary or ecological relationship with other organisms. This task has been left to human invention. In order to distinguish among various organisms and to communicate with one another about them, biologists have classified and named them. The science of classifying and naming organisms is known as **taxonomy,** and the biologists who specialize in classification are taxonomists.

Although designing a classification system must come in part from human imagination, we will see that taxonomists need not function arbitrarily. Numerous similarities and differences in appearance, lifestyle, and origin can be discerned among life forms. Assigning relative weight to these characteristics gives challenge and controversy to the work of the taxonomist.

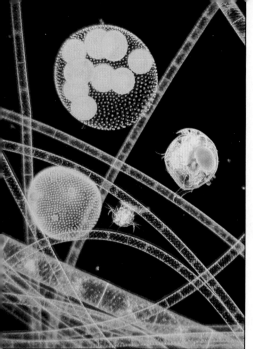

FIGURE 22–1 Photomicrograph of a concentrated sample of marine microorganisms (plankton) from the Rhode Island Sound (×225). Such variety may be bewildering even to the modern graduate student; to Linnaeus at a time when the microscope was just being developed, the array of microorganisms being viewed for the first time seemed beyond hope of systematic classification. However, a modern planktonologist could identify and completely classify each one of the microorganisms (mostly diatoms) shown here.

THE DEVELOPMENT OF TAXONOMY

How would you use what you already know about living things if you wanted to assign them to categories? You have the advantage of already knowing the common names of many organisms. Would you place insects, bats, and birds in one category because they all have wings and fly? And would you, perhaps, place squid, whales, fish, penguins, and Olympic backstroke champions in another category because they all swim? Or would you classify organisms according to a culinary scheme, placing lobsters and tuna in the same part of the menu, perhaps, as "seafood"?

All of these schemes might be valid, depending on the purpose you might have in attempting to classify life. Similar methods have been used throughout history. Animals, for example, were classified by St. Augustine in the fourth century as useful, harmful, or superfluous—to human beings. Anthropologists have discovered that some cultures still employ a similar system of classification. Among organisms they find useful, the Australian aborigines make sophisticated distinctions. Often these are much the same distinctions that biologists would make. However, among organisms they consider useless, aborigines draw only the most obvious distinctions.

In Renaissance times scholars began to abandon narrowly utilitarian schemes of classification and sought categories that might be inherent in the organisms themselves. These were originally thought to reflect natural groupings arrayed around an ideal type, an archetype in the mind of a creator, and were arranged roughly in an order that proceeded from the simple to the complex. Out of the many classification schemes that were proposed, the system designed by Carolus Linnaeus in the mid-18th century (described briefly in Chapter 1) has survived with some modification to the present day. Linnaeus probably intended to design a static system of classification, for he had no theory of evolution in mind when he set it up. Neither did he or his colleagues have any conception of the vast number of living and extinct organisms that would later be discovered (Figure 22–1). Yet it is remarkable how flexible and adaptable to the influx of new biological knowledge and theory his system has proved to be. There are very few 18th-century inventions that survive today in a form that would still be recognizable by their originators.

THE BINOMIAL SYSTEM OF CLASSIFICATION

Linnaeus developed a **binomial system** of nomenclature, a system based on a unique two-part name for each organism. The first part of the name designates the genus and the second part, the specific epithet. This is a descriptive word expressing some quality of the organism. The specific epithet is always used together with the full or abbreviated generic name preceding it. In each scientific name, the genus is given first and is capitalized, whereas the specific epithet is given second and is not capitalized.

The **species** is the basic unit of classification. Recall that a species is a group of organisms with structural, functional, and developmental similarities that breed with one another to produce fertile offspring, and do not interbreed with members of other species under natural conditions. Members of a species share a common evolutionary ancestry. Closely related species are grouped together in the next higher unit of classification, the **genus** (plural, *genera*).

In accordance with the binomial system, the scientific name of the domestic dog, *Canis familiaris*, applies to all varieties of tame dogs—collies, German Shepherds, cocker spaniels, Chihuahuas, and so on. All of these belong to the same species and all are able to interbreed. Related species of the same genus are *Canis lupus,* the wolf; *Canis latrans,* the coyote; and

Canis aureus, the golden jackal. The cat, which belongs to a different genus, is named *Felis catus.*

The use of Latin rather than a modern language in naming organisms is a carryover from the days when Latin was the international language of scholars. Why do we continue to use Latin, rather than common, names for plants and animals? Why call a sugar maple *Acer* (maple) *saccharum* (sugar)? The main reason is to be accurate and avoid confusion, for in some parts of America this same tree is called either hard maple or rock maple. The tree generally called white pine is *Pinus strobus,* but some biologists refer to *Pinus flexilis* and *Pinus glabra* as white pines; still other biologists call *Pinus strobus* northern pine, soft pine, or Weymouth pine. There are many instances of confusing common names, but these examples should make it clear that exact scientific names are necessary for unambiguous communication.

TAXONOMIC CLASSIFICATION IS HIERARCHICAL

Just as species may be grouped together to form a genus, a number of related genera constitutes a **family,** and families may be grouped into **orders,** orders into **classes,** and classes into **divisions** for plants or fungi, or into **phyla** for animals or protists. For example, the family Felidae includes all catlike animals—genus *Felis,* the house cat; genus *Panthera,* the leopard genus; and three or four other genera (Figure 22–2). Family Felidae, along with family Ursidae (bears) and several other families of animals that eat mainly meat, is placed in order Carnivora (Table 22–1). Order Carnivora, order Primates (the order to which humans belong), order Rodentia (rodents), and several other orders belong to Class Mammalia (mammals), a class of animals with hair, mammary glands that produce milk for the young, and differentiation of teeth into several types. Class Mammalia, class Aves (birds), class Reptilia (reptiles), and several other classes are grouped in phylum Chordata, which in turn belongs to kingdom Animalia. For a description of human classification, see Focus on Why You Are *Homo sapiens.*

A **taxon** (plural, *taxa*) is a particular grouping, for example a particular species, genus, or phylum, defined by the classification system. For example, the phylum Chordata is a taxon that contains several classes, including Mammalia and Amphibia. Mammalia is a taxon that includes many different orders.

Subspecies

The species is the fundamental unit of classification, but not the smallest in use. Geographically distinct populations within a species often display certain consistent characteristics that serve to distinguish them from other populations of the same species. If they interbreed, however, they are not truly separate species but are termed instead **subspecies,** or **varieties.** For small organisms such as bacteria, the term **strain** is used.

Although members of different subspecies do not ordinarily differ very much from one another, these differences may be sufficient to affect their behavior, biochemistry, or other characteristics important in biological research. This can cause problems for scientists attempting to duplicate or extend one another's research findings when they use the same species, but different subspecies, as their experimental organisms. Deer mice, many species of oaks, and numerous other kinds of organisms occur in a number of varieties.

Although subspecies are usually distinguishable from one another by experts, they may grade imperceptibly into one another at the borders of their geographical ranges where there is opportunity to interbreed freely.

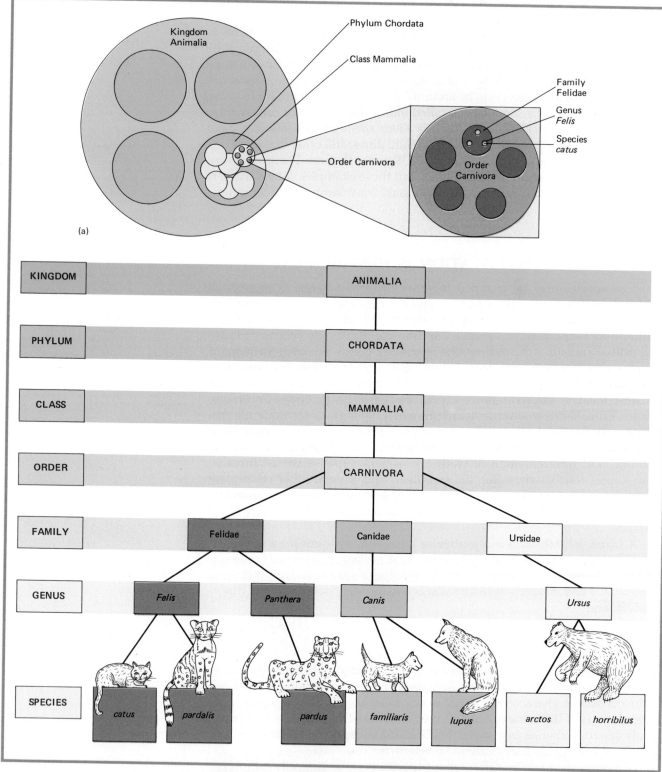

FIGURE 22–2 The principal categories used in classifying an organism. (*a*) The domestic cat is used to illustrate the hierarchical nature of our taxonomic system. (*b*) Three of the several families of order Carnivora are illustrated here. Family Felidae, the cats, is made up of 4 or 5 genera depending on the taxonomic scheme followed. Genus Felis includes *Felis catus,* the domestic cat, and several other species such as *Felis pardalis,* the ocelot. Family Canidae—wolves, foxes, jackals, and dogs—includes 12 genera. Genus Canis includes *Canis familiaris,* the domestic dog, and *Canis lupus,* the timber wolf. Family Ursidae, the bears, is made up of 6 genera.

Table 22–1 CLASSIFICATION OF THE DOMESTIC CAT, THE HUMAN BEING, AND CORN

Category	Classification of Cat	Classification of Human Being	Classification of Corn
Kingdom	Animalia	Animalia	Plantae
Phylum (Division)	Chordata	Chordata	Tracheophyta
Subphylum (Subdivision)	Vertebrata	Vertebrata	Spermatophytina
Class	Mammalia	Mammalia	Angiospermae
Order	Carnivora	Primates	Commelinales
Family	Felidae	Hominidae	Poaceae
Genus	*Felis*	*Homo*	*Zea*
Species	*catus*	*sapiens*	*mays*

Some of these subspecies may be in the process of becoming reproductively isolated and will, in the course of time, become clear-cut species (Figure 22–3). Thus they provide an opportunity for field studies of gene pools and of the speciation process.

Splitting and Lumping

Many organisms fall into easily recognizable, apparently natural groups, and their classification presents no obvious difficulty. Others, though, appear to lie on the borderline between two groups, having some characteristics in common with each. These organisms are difficult to assign to one group or the other.

The number and inclusiveness of the principal groups vary according to the basis used for classification and the judgment of the taxonomist making the decisions. Some taxonomists like to group organisms into already existing units; they are referred to as **"lumpers."** Others prefer to establish separate categories for forms that do not fall naturally into one of the existing classifications; they are called **splitters."** "Lumpers" consider that there are 10 animal phyla and four plant divisions, whereas "splitters" recognize up to 33 animal phyla and up to 12 plant divisions.

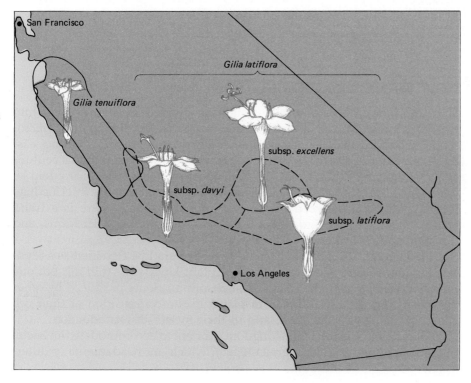

FIGURE 22–3 The ranges and distinguishing features of the flowers of the subspecies of the California wildflower, *Gilia latiflora*, and of *G. tenuiflora*, a closely related subspecies. From their similarities, it seems probable that *G. tenuiflora* was originally a subspecies of *G. latiflora*. Since it now overlaps geographically with *G. latiflora* without interbreeding, it is now a distinct, separate species. For simplicity, three other subspecies of *G. latiflora* with narrower distributions have been omitted. (The full name of a subspecies includes genus, species, and subspecies designations, e.g., *Gilia latiflora excellens*.)

Focus on WHY YOU ARE *Homo sapiens*

1. At present there are five generally recognized kingdoms of organisms and two domains. Since human cells have discrete nuclei surrounded by nuclear membranes, you belong to the domain Eukaryota. Your cells lack chloroplasts and cell walls, and you are a multicellular heterotroph, with highly differentiated tissues and organ systems. That makes you a member of the kingdom Animalia.

2. What kind of an animal are you? You possess a spinal column composed of bony vertebrae that has largely replaced a cartilaginous rod you had as an embryo, the notochord. At that time you also had structures that had you been a fish would have developed into gill slits. You have a dorsal nerve cord and brain, both of which still retain remnants of their embryonic cavities. These traits mark you as a chordate and a vertebrate—that is, you belong to the phylum Chordata (because you either have or have had a notochord), and to the subphylum Vertebrata (because you have vertebrae that replaced the notochord).

3. Among the vertebrates there are several classes: cartilaginous fish, bony fish, jawless fish, amphibia, reptiles, mammals, and birds. You are homeothermic (warmblooded) and so must be either a bird or a mammal. Lacking feathers and having teeth and (if you are female) the potential for nursing your young, you are a mammal. If you are male, do not be concerned; even if you cannot nurse, having hair is enough.

4. Within the mammals there are three subclasses: the Prototheria, the Metatheria, and the Eutheria. The Prototheria are confined either to zoos or to the Australian continent and its environs. In the present day they include the duck-billed platypus and the spiny anteater, both of which, in addition to other unusual traits, lay eggs. The Metatheria, most of which are also from Australia, usually carry their still-embryonic young around in a pouch and totally lack a placenta (an organ of exchange between mother and developing embryo). If you did not hatch from an egg or spend your infancy in a pouch, you can be confident of your status as a eutherian.

5. A number of orders exist within the Eutheria. The insectivores, for instance, include the moles and shrews, the Chiroptera are the bats, and the Carnivora include the dogs, cats, and ferrets, among others. Your opposable thumbs, frontally directed eyes, flat fingernails, and several other characteristics identify you as a primate, along with monkeys, apes, and tarsiers.

6. Primates include a number of families. You and the New World monkeys are obviously very different—they have prehensile tails, for instance, which you and all Old World monkeys and apes lack; indeed, you and the apes lack tails altogether. Your posture is upright, you have long legs and short arms, and not much body hair. You are blessed with your very own family that has no other modern occupants: the Hominidae.

7. Within the Hominidae anthropologists distinguish several species, all but one of which are known only as fossils. *Australopithecus* is one of these. If you are alive, you do not belong to any of those extinct genera, but to the genus *Homo*.

8. Again, the genus *Homo* has only one living species—*sapiens*. Since many taxonomists insist that the species name always includes the genus name, please think of yourself as *Homo sapiens*.

HOW MANY KINGDOMS?

Since the time of Aristotle, biologists have divided the living world into two kingdoms, **Plantae** and **Animalia.** After the development of microscopes it became increasingly obvious that many organisms could not easily be assigned to either the plant or animal kingdom. More than a century ago a German biologist, Ernst Haeckel, suggested that a third kingdom be established to include all the single-celled organisms that are intermediate in many respects between plants and animals. Today, there is a trend among biologists to include all of the algae, even multicellular forms, and the slime molds in kingdom **Protista.**

In 1969, R. H. Whitaker suggested that the **Fungi** be classified as a separate kingdom rather than as part of the plant kingdom. After all, not one fungus is photosynthetic. Fungi must absorb nutrients produced by other organisms. Fungi also differ from plants in the composition of their cell walls, in their body structure, and in their modes of reproduction.

Biologists have established kingdom **Monera** to accommodate the bacteria and cyanobacteria (blue-green algae), which are fundamentally differ-

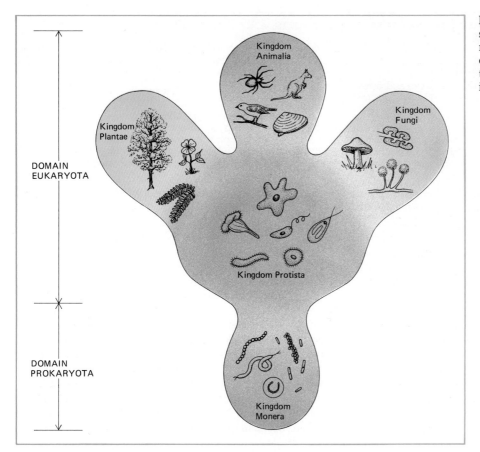

FIGURE 22–4 The five-kingdom system of classification. The monerans—bacteria and cyanobacteria—constitute domain Prokaryota. Protists, plants, fungi, and animals are included in the domain Eukaryota.

ent from all other organisms in that they lack distinct nuclei and other membranous organelles. Most biologists now recognize the five kingdoms illustrated in Figure 22–4 and described in Table 22–2.

THE DOMAINS

Taxonomic categories larger than kingdoms have been proposed by some biologists. The monerans—the bacteria and cyanobacteria—differ profoundly from all other organisms: They lack a nucleus and they have no mitochondria, chloroplasts, lysosomes, or other membranous organelles characteristic of nonmonerans. To emphasize this gulf between monerans and all other living organisms, taxonomists have suggested that monerans be classified in a domain named **Prokaryota,** based on the Greek words meaning *before* and *nucleus.* All other organisms would be assigned to domain **Eukaryota,** meaning *true nucleus.* The names of these domains reflect the concept that the prokaryotes evolved first and eventually gave rise to the eukaryotes.

SYSTEMATICS: RECONSTRUCTING PHYLOGENY

Modern classification is based on evolutionary relationships. The study of evolutionary relationships among organisms is called **systematics.** A systematist seeks to reconstruct the evolutionary relationships, or **phylogeny,** of organisms. Once these relationships are defined, the classification of organisms can be based on common ancestry.

Table 22–2 **THE FIVE KINGDOMS: MONERA, PROTISTA, FUNGI, PLANTS, AND ANIMALS**

Kingdom	Characteristics	Ecological Role
Monera	Prokaryotes (lack distinct nuclei and other membranous organelles); single-celled; microscopic	
Bacteria	Cells walls composed of peptidoglycan (a substance derived from amino acids and sugars); many secrete a capsule made of a polysaccharide material. In pathogenic bacteria, the capsule may protect against defenses of host. Cells may be spherical (cocci), rod-shaped (bacilli), or coiled (spirilla).	Decomposers; some chemosynthetic autotrophs; important in recyling nitrogen and other elements. A few are photosynthetic, usually employing hydrogen sulfide as hydrogen source. Some pathogenic (cause disease); some used in industrial processes.
Cyanobacteria	Specifically adapted for photosynthesis; use water as a hydrogen source. Chlorophyll and associated enzymes organized along layers of membranes in cytoplasm. Some can fix nitrogen.	Producers; blooms (population explosion) associated with water pollution
Protista	Eukaryotes; mainly unicellular or colonial	
Protozoa	Microscopic; unicellular; depend upon diffusion to support many of their metabolic activities	Important part of zooplankton; near base of many food chains. Some are pathogenic (e.g., malaria is caused by a protozoan).
Eukaryotic algae	Some difficult to differentiate from the protozoa. Some have brown pigment in addition to chlorophyll.	Very important producers, especially in marine and fresh-water ecosystems
Slime molds	Protozoan characteristics during part of life cycle; fungal traits during remainder	
Fungi	Eukaryotes; plantlike but lack chlorophyll and cannot carry on photosynthesis	Decomposers, probably to an even greater extent than bacteria. Some are pathogenic (e.g., athlete's foot is caused by a fungus).
Molds, yeasts, mildew, mushrooms, rust	Body composed of threadlike hyphae; rarely, discrete cells. Hyphae may form tangled masses called mycelia, which infiltrate whatever the fungus is eating or inhabiting. Mycelium is often invisible, as in mushrooms.	Some used as food (yeast used in making bread and alcoholic beverages); some used to make industrial chemicals or antibiotics; responsible for much spoilage and crop loss
Plantae	Multicellular eukaryotes; adapted for photosynthesis; photosynthetic cells have chloroplasts. All plants have reproductive tissues or organs and pass through distinct developmental stages and alternations of generations. Cell walls of cellulose; cells often have large central vacuole. Indeterminate growth; often no fixed body size or exact shape.	Other organisms depend upon plants to produce foodstuffs and molecular oxygen.
Animalia	Multicellular eukaryotic heterotrophs, many of which exhibit advanced tissue differentiation and complex organ systems. Lack cell walls. Able as a rule to move about by muscle contraction; extremely and quickly responsive to stimuli, with specialized nervous tissue to coordinate responses; determinate growth.	Almost the sole consuming organisms in the biosphere, some being specialized herbivores, carnivores, and detrivores (eating dead organisms or organic material such as dead leaves).

Monophyletic Versus Polyphyletic

A population of organisms has not only a dimension in space—its range—but also a dimension in time. The population extends backward in time, merging with other species populations much like branches of a tree (Figure 22–5). Species have various degrees of evolutionary relationship with one another, depending on the length of time that has elapsed since their populations diverged—the point at which they have common ancestry. If all of the subgroups within any taxonomic group have a common ancestry, the grouping is referred to as **monophyletic.** A taxon containing a common ancestor and all the species descended from it is called a **clade.**

Many taxa in current use are **polyphyletic,** consisting of several evolutionary lines and not including a common ancestor. For example, mammals are thought to have evolved from at least three different groups of Triassic reptiles.

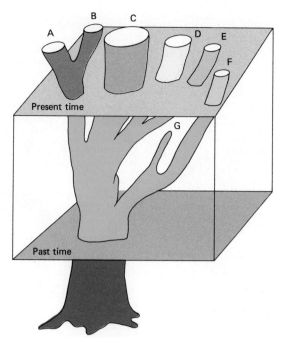

FIGURE 22–5 The evolutionary relationships of six hypothetical monophyletic species. Circular cross sections of the branches represent the species at the present time. Junctions of the branches represent points of common ancestry. Species G is extinct. Which groupings of these species might be considered a genus? Modern taxonomists, working solely with living forms, compare them on the basis of their modern similarities and differences. Paleontologists investigating the fossil record may confirm the findings of the taxonomists if all the organisms classified together taxonomically appear to share their similarities as far back in the fossil record as they can be traced. Many of the groups classified together today on the basis of their structural similarities are thought to share a common, though remote, origin and therefore also share genetic similarities.

Homologous Structures

There is general, although not complete, agreement as to what constitutes a species, but the grouping of species into higher taxonomic groups is more subjective and difficult to define. For example, in Figure 22–5, should species A and B be placed within a single genus or do they represent two distinct genera? If species C and E are distinct genera, should D be part of either of these genera? Similar difficulties exist in determining the assignments to families, orders, classes, and phyla. Most biologists base their judgments about the degree of relationship of organisms on the extent of similarity between living species, and, when available, on the fossil record.

In evaluating similarities, biologists look primarily for *homologous structures* in different organisms. The presence of homologous structures in different organisms implies that *divergent evolution* has occurred from a common ancestor (Figure 22–6). In contrast, similarities among *analogous structures* result not from shared ancestry but from *convergent evolution*. This sometimes occurs when unrelated or distantly related organisms adapt to similar environmental conditions.

Primitive and Derived Characters

Closely similar organisms are thought to be closely related, and less similar organisms are viewed as being more remotely related. However, the choice of the similarities used is extremely important. How does the systematist interpret the significance of these similarities? In making decisions about taxonomic relations, the biologist first examines the characteristics that are common to the largest group of organisms and interprets them as indicating the most remote common ancestry. These **primitive characters** have remained essentially unchanged. **Derived characters** are characteristics that have evolved more recently. Such characters suggest a more recent common ancestor. A feature viewed as a derived character in a large group may be seen as a primitive character in a smaller taxon. More recent common ancestry is indicated by classification in smaller and smaller taxonomic groups.

For example, the three small bones in the middle ear are useful in identifying a branching point between mammals and reptiles. The evolution of this character was a unique event in all mammals, and only mammals have

FIGURE 22–6 Convergent and divergent evolution. In divergent evolution, an ancestral group, for example primitive mammals, branches and gives rise to two or more lines of evolution which may lead far from the initial ancestral design. The presence of homologous structures suggests divergent evolution. In convergent evolution, distantly related groups such as birds and bats may come to resemble one another in certain respects as they evolve to fit similar modes of life. For example, the wings of the bird and the bat are analogous structures that indicate similar adaptations but do not imply common ancestry.

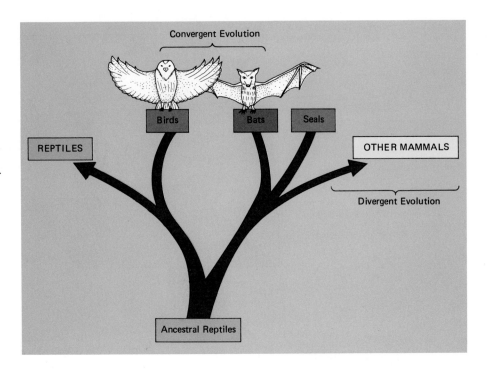

this derived character. However, when comparing one mammal with another, the three ear bones are a primitive character because all mammals have inherited them. They have no value in distinguishing mammalian groups. A search must be made for other derived characters that can be used in establishing branching points among the different groups of mammals.

Choosing Taxonomic Criteria

Both fish and porpoises have streamlined body forms, but this is viewed as an analogous adaptation and is less important than the derived characteristics they share with other organisms. For this reason they are not classified together. The porpoise shares important derived characteristics with mammals such as humans—the ability to breathe air, nurse young, maintain a constant body temperature, and grow hair. Thus, the porpoise is classified as a mammal and is viewed as descending from a mammalian ancestor.

Although the porpoise has more in common with us than it does with a fish, there are some characteristics that all three kinds of animals share. Among these are a notochord (skeletal rod) and rudimentary gill slits in the embryo stage, and a dorsal tubular nerve cord. These shared primitive characteristics serve as a basis for classification and indicate a common ancestry. This ancestry is more remote between the porpoise and the fish than between the porpoise and human beings. Therefore, fish, humans, and porpoises are grouped together in the large category called phylum Chordata, and humans and porpoises are also classified together in Class Mammalia, a smaller category indicating their closer relationship (Figure 22–7).

Deciding the appropriate weights for various traits in determining taxonomic categories is not always simple, even in apparently straightforward instances. What, for example, are the most important invariable characteristics of a bird? We might list feathers, beak, wings, absence of teeth, the egg-laying trait, and warmbloodedness. Yet some mammals (the monotremes) also have beaks, lack teeth, and lay eggs (Figure 22–8). Should we classify them as birds?

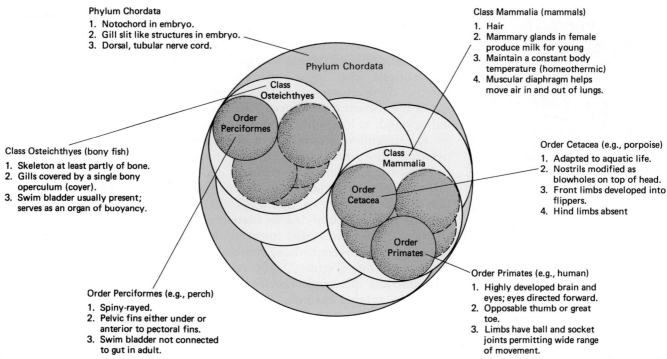

Phylum Chordata
1. Notochord in embryo.
2. Gill slit like structures in embryo.
3. Dorsal, tubular nerve cord.

Class Mammalia (mammals)
1. Hair
2. Mammary glands in female produce milk for young
3. Maintain a constant body temperature (homeothermic)
4. Muscular diaphragm helps move air in and out of lungs.

Class Osteichthyes (bony fish)
1. Skeleton at least partly of bone.
2. Gills covered by a single bony operculum (cover).
3. Swim bladder usually present; serves as an organ of buoyancy.

Order Cetacea (e.g., porpoise)
1. Adapted to aquatic life.
2. Nostrils modified as blowholes on top of head.
3. Front limbs developed into flippers.
4. Hind limbs absent

Order Perciformes (e.g., perch)
1. Spiny-rayed.
2. Pelvic fins either under or anterior to pectoral fins.
3. Swim bladder not connected to gut in adult.

Order Primates (e.g., human)
1. Highly developed brain and eyes; eyes directed forward.
2. Opposable thumb or great toe.
3. Limbs have ball and socket joints permitting wide range of movement.

FIGURE 22–7 Shared derived characteristics. Members of class Osteichthyes (bony fish) and class Mammalia (the mammals) share many more characteristics with one another and with the members of the other classes of phylum Chordata than they do with members of any other class of any other phylum. For example, a perch has more in common with a monkey than with a sea star or clam. Members of various orders of the same class share more characteristics than members of orders that belong to different classes. Thus, a porpoise has more shared derived characters in common with a human being than with a perch. This indicates a more recent common ancestry for the porpoise and the human.

No mammal, however, has feathers. Is this trait absolutely diagnostic of birds? According to the conventional taxonomic wisdom, the presence of feathers could be used to decide what is and is not a bird. This applies only to modern birds, however. Some extinct reptiles may have been covered with feathers, while not being birds in any meaningful sense. Moreover, as Figure 22–9 shows, there are even some modern exceptions!

Usually, organisms are classified on the basis of a combination of traits rather than on one perhaps superficial trait such as the ability to live in

FIGURE 22–8 A few mammals share important characteristics with birds. (*a*) The duck-billed platypus, a monotreme, lays eggs, has a beak and lacks teeth. Should we classify it as a bird? (*b*) Australian pelicans (*Pelecanus conspicillatus*).

(a)

(b)

FIGURE 22–9 The possession of feathers is one of the key characteristics of birds, but as demonstrated by this naked chicken, it is possible for a bird to lack feathers entirely. Naked chickens arose as a mutation in the 1950s. Investigators considered breeding naked hens in order to save the feed that would be used metabolically to produce feathers. Unfortunately, the uninsulated hens shivered so much that they used up more energy in muscular contraction than they would have used to produce feathers.

water. The significance of these combinations is determined inductively, that is, by an integration and interpretation of data. Such induction is a necessary first step in all science. The taxonomist proposes, for example, that birds should all have beaks, feathers, no teeth, and so on; this is really a hypothesis. Then he or she reexamines the living world and observes whether there are organisms that might reasonably be called birds that do not fit the current definition of "birdness." If not, the definition is permitted to stand, at least until someone discovers too many exceptions to it. If so, the definition is modified or abandoned. Sometimes, the taxonomist persuades the world that the apparent exception—the bat, for instance—resembles a bird only superficially, and should not be considered one.

Taxonomy is a dynamic science that proceeds by the constant reevaluation of data, hypotheses, and theoretical constructs. As new data are discovered and old subjected to reinterpretation, the ideas of taxonomists change. During the 1980s, for example, a group of organisms, the Lorcifera, were discovered whose combination of traits did not fit those of any existing phylum (Figure 22–10). A new phylum was established just to accommodate this single species.

To take another example, as evolutionary conceptions about the origin of cellular life have changed, biologists have considered creating a new, sixth kingdom for the Archaebacteria, a subgroup of bacteria with unusual habitats. Although these bacteria do not visibly differ very much from others, they branched off very early (archae is from the Greek for "ancient") as one of three cell lineages (archaebacteria, eubacteria, and eukaryotes) that evolved from a universal ancestor. Each of the three cell lineages has a distinct ribosomal RNA structure.

Molecular Biology: New Taxonomic Tools

Evolution of new species is not always signaled by obvious structural changes. For example, two distinct species of fruitflies may appear indistinguishable. Some of their macromolecules, however, are different. Variations in the structure of specific macromolecules among species, just like differences in anatomic structure, result from mutation. Macromolecules that are functionally similar in two different types of organisms are considered homologous if their subunit structure is similar. Recently developed methods that enable biologists to compare the nucleotide sequences of various nucleic acids and the amino acid sequences of proteins have become extremely important taxonomic tools. Comparison of molecular structure is

FIGURE 22–10 Only one new animal phylum—that for a group of microscopic marine animals—has been added to the animal kingdom since 1900. The recent discovery of the marine animal Lorcifera has resulted in the proposal that a second new phylum be added. Lorcifera larvae propel themselves with a pair of appendages attached to the body by a ball-and-socket joint. The tiny adults, about 230 μm long, lack appendages for swimming. Both larvae and adults have head spines and a flexible, retractable tubelike mouth. These animals live between grains of shell gravel in the ocean bottom. Are there many other unique organisms yet to be discovered and classified?

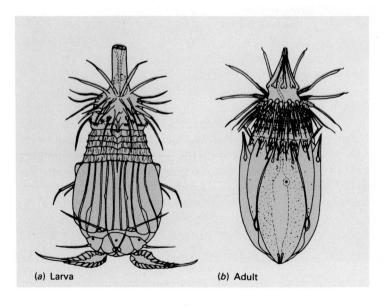

(a) Larva   (b) Adult

an objective measure of evolutionary relationships, a measure that can be quantified.

Molecular Clocks

The number of differences in nucleotide sequence of DNA in two groups reflects the time since the groups branched off from a common ancestor. Many mutations are thought to be neutral; they do not significantly affect the function of the protein for which they code. Thus, they do not affect the organism's phenotype. Such mutations are not affected by natural selection, and tend to accumulate randomly.

Because the basic function of most genes does not generally change during evolution, we can expect that neutral mutations will occur at a constant rate. In addition, DNA evolves at a similar rate in different species. Whatever changes in the rate of evolution do occur will average out over the span of evolutionary time. This constancy in DNA evolution enables biologists to use specific genes as **molecular clocks**. Such clocks can be used to date the divergence of two groups from a common ancestor.

Protein evolution also provides molecular clocks. Although various proteins evolve at different rates, a given protein apparently evolves at a constant rate. The degree of difference in amino acid sequence of their proteins reflects the time that has passed since the groups diverged.

Protein Similarities

By comparing amino acid sequences of proteins we can gain some idea of degree of relatedness of two organisms; the greater the correspondence in their amino acid sequences, the more closely they are thought to be related. In Chapter 7 we discussed the respiratory protein cytochrome c present in all aerobic organisms. Human and chimpanzee cytochrome c molecules have identical amino acid sequences. In another primate, the rhesus monkey, one of the 104 amino acids in the sequence is different. However, cytochrome c of the dog, a nonprimate, differs from human cytochrome c by 13 amino acids.

Protein similarity can be ascertained by serological techniques that involve the immunological comparison of proteins. Much of the original development of this technique was done shortly after the turn of the century by George Nuttall, who injected rabbits with the blood serum of other organisms. The rabbits developed antibodies to proteins in the alien serum that acted as antigens. (An antigen is a macromolecule recognized by the body as foreign; antibodies are proteins that have the capacity to act against specific antigens. These concepts are discussed in greater detail in Chapter 43.) When another specimen of the experimental serum was mixed with the blood serum of the rabbits, the antigen was bound to the antibody and a cloudy precipitate formed (Figure 22–11).

Antisera such as those from the immunized rabbit may cross-react with antigens that are similar but not identical to the antigens used in eliciting the antibodies. If a rabbit has been immunized to guinea pig serum, for example, some (but less) precipitation would occur if its antiserum were then mixed with the serum of a chinchilla. This result indicates a similarity of the blood proteins, reflecting that both the guinea pig and chinchilla are rodents belonging to the same family. For a rodent whose exact taxonomic position was controversial, its degree of relationship to the guinea pig could be estimated by assessing the degree to which its serum cross-reacted with the anti-guinea pig serum. This can be done by using instruments that measure the density of the antibody-antigen precipitate or by using an agar-well technique shown in Figure 22–12. By similarly comparing the unknown animal immunologically with other rodents such as rats,

FIGURE 22–11 When blood serum from a guinea pig is injected into a rabbit, the rabbit makes antibodies (special proteins) to the foreign proteins (antigens) in the guinea pig blood. Later, rabbit blood serum containing these antibodies is mixed with blood serum from other test animals. The more similar the test animal's proteins are to the guinea pig proteins, the greater will be the reaction. Thus, the density of the antibody-antigen precipitate can be used to estimate how closely related the animals are.

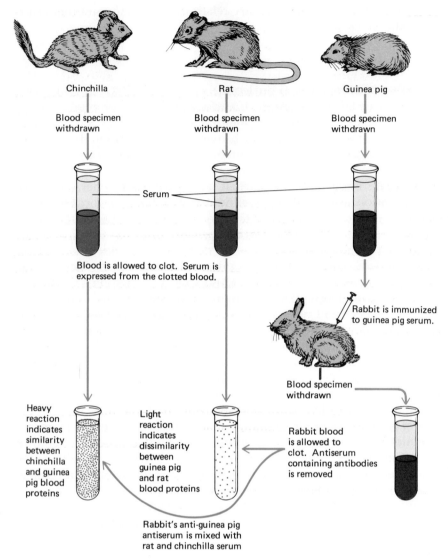

Chinchilla

Rat

Guinea pig

Blood specimen withdrawn

Blood specimen withdrawn

Blood specimen withdrawn

Serum

Blood is allowed to clot. Serum is expressed from the clotted blood.

Rabbit is immunized to guinea pig serum.

Blood specimen withdrawn

Heavy reaction indicates similarity between chinchilla and guinea pig blood proteins

Light reaction indicates dissimilarity between guinea pig and rat blood proteins

Rabbit blood is allowed to clot. Antiserum containing antibodies is removed

Rabbit's anti-guinea pig antiserum is mixed with rat and chinchilla serum

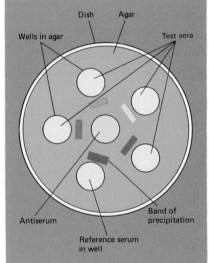

Dish   Agar

Wells in agar

Test sera

Antiserum

Band of precipitation

Reference serum in well

FIGURE 22–12 The agar-well technique for testing the degree of relationship of various organisms. Blood serum from a rabbit is placed in the center well of an agar dish. Blood serum from a variety of test animals is placed in surrounding wells. The extent of precipitation reflects how similar the proteins in the animal's blood is to that of the reference animal (guinea pig) and is thus an indication of the degree of relatedness of the animals.

hamsters, and squirrels, its place within the order Rodentia could be determined.

Such techniques, however, cannot be used as ultimate arbiters of taxonomy. Serological, and to a lesser extent, electrophoretic techniques are often handicapped because they are applied to mixtures of unknown or poorly characterized proteins, so that which protein is cross-reacting may not be clear. Only the tedious and expensive amino acid sequencing techniques are really reliable as the basis for exact taxonomic determination, but even they do not necessarily settle all controversy.

Nucleic Acid Similarities

Among related species the DNA sequences for the same structural genes are very similar. Detailed restriction maps within large homologous regions of chromosomes of related organisms are also very similar. For example, the beta-like globin region of several primates has been mapped. Even though the gorilla diverged from humans more than 6 million years ago, 65 out of the 70 restriction sites are identical.

Recently, Morris Goodman of Wayne State University School of Medicine and his coworkers determined the nucleotide sequence of a portion of

DNA from each of three species of primates. From their analysis of the 7000-nucleotide sequence, the investigators inferred a common ancestral gene. The simplest branching pattern that would account for the results suggests that the gorilla first split off from the common ancestor of the chimpanzee and human. Later the chimpanzee and human lines diverged.

The genomes of mammals contain thousands of copies of a segment known as alu-DNA. The repetitive sequences of this DNA account for up to 10% of the human genome. Biologists speculate that alu-DNA plays a role in the initiation of DNA synthesis. Even between closely related species there are differences in alu-DNA, and these differences are thought to reflect evolutionary changes.

DNA hybridization techniques (Chapter 15) are used to compare DNA from different organisms. Short segments of DNA can be isolated from two different organisms. This DNA is denatured so that it separates into complementary strands. Then, the DNA fragments from the two species are permitted to recombine to form hybrid DNA. The extent of hybridization depends on the number of base pairs that correspond in the DNA of the two species. To determine the extent of pairing, the hybrid DNA is heated until the two strands separate. The higher the temperature necessary for separation of the strands, the more bases were paired, and thus, the more similar is the DNA of the two species.

Biologists can also infer how closely two species are related by comparing the variations among ribosomal nucleotide sequences. Using a statistical model for comparing such sequences, James A. Lake has recently proposed a major change in the evolutionary tree. Lake suggests that the ancestral cell resembled sulfur-metabolizing bacteria that inhabit hot springs. On the basis of his computer analyses, he claims that a branch of prokaryotes known as eocytes are more closely related to eukaryotes than to other prokaryotes. Eocytes are sulfur-metabolizing, thermophilic bacteria. He proposes calling eocytes and eukaryotes the *karyotes* and the rest of the bacteria and cyanobacteria, *prokaryotes*.

APPROACHES TO TAXONOMY

In constructing a phylogenetic tree, taxonomists consider branch points that indicate the time at which a particular group of organisms evolved. They also consider the extent of divergence between branches, or how different two groups have become since they originated from a common ancestor and embarked upon different evolutionary pathways. Which of these bits of evolutionary data is utilized more in classifying a group of organisms depends upon one's approach to taxonomy. Three major approaches are phenetics, cladistics, and classical evolutionary taxonomy.

Phenetics

Pheneticists argue that we cannot be sure that our view of phylogeny is correct and therefore should not base classification on phylogeny. These taxonomists do not try to reconstruct evolutionary history. The **phenetic** (phenotypic) system is a numerical taxonomy based on phenotypic similarities. In this system organisms are grouped according to the number of characteristics they share, without trying to determine whether their similarities are homologous. Pheneticists argue that it is not important to try to sort homologous and analogous characteristics, because overall the number of similarities that two organisms have in common will reflect the degree of homology.

A taxonomist who follows the phenetic system would explain that porpoises are classified along with humans as mammals rather than fish be-

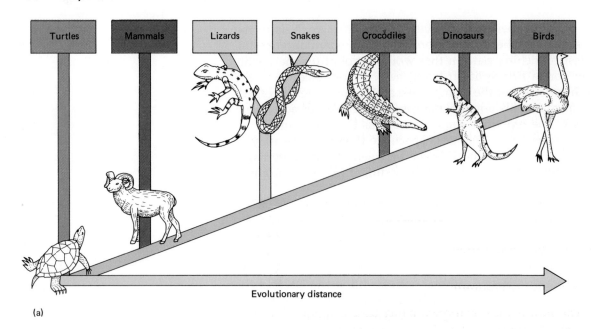

Evolutionary distance

(a)

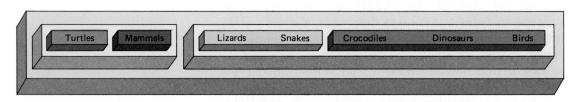

(b)

FIGURE 22–13 According to the cladistic approach, birds and reptiles are classified together. (*a*) A simple cladogram. Unlike a traditional phylogenetic tree, the cladogram is not concerned with evolutionary time and instead focuses on relative degree of relationship. (*b*) Nested sets of taxonomic boxes can be constructed from the cladogram. Each box contains a taxon. Each taxon must contain a common ancestor so the taxa are nested.

cause they share more characteristics with mammals. Pheneticists assign numbers to many arbitrarily chosen traits (more than 100) that are given equal weight. They designate these traits as present (+) or absent (−) in the organisms of a particular category. This information is fed into a computer, which indicates which groups have the most traits in common. A pure phenetics approach is not very popular among taxonomists because it does not work very well. However, phenetics has made an important contribution; taxonomists have found the phenetic emphasis on quantitative comparisons useful.

Cladistics

The **cladistic** approach to taxonomy emphasizes phylogeny, focusing upon how long ago one group branched off from another. Cladists insist that taxa be monophyletic; each taxon should contain a common ancestor and its descendants. Currently recognized polyphyletic taxa should be divided. Cladists do not concern themselves with divergence. The significance or magnitude of specific adaptational differences between the descendants of a common ancestor are not important. A cladist would say that porpoises cannot be classified with fish because they evolved much later in time than fish. In fact, the whales appeared on earth millions of years after the fish.

Cladists develop branching taxonomic patterns called cladograms. A cladogram consists of a series of branches. Each branch represents the splitting of two new groups from a common ancestor. Cladists use carefully defined objective criteria. However, some taxonomists criticize cladists because they ignore later evolutionary changes that take place in groups that have split.

Consider the evolutionary grouping of turtles, lizards, snakes, dinosaurs, crocodiles, and birds. Birds, along with dinosaurs, are thought to be descendants of reptiles that shared a common ancestor with the modern crocodiles and alligators. Crocodiles and birds, then, might be considered sister groups, which some cladists might place in the same class. This decision might be made without considering that crocodiles appear to have much more in common adaptationally and ecologically with lizards, snakes, and turtles (all of which descended from a different reptilian ancestor) than they do with birds.

Cladists would classify birds and reptiles in a nested series of monophyletic taxa (Figure 22–13). Birds and reptiles would be assigned to a single taxon because they share a common ancestor. Within that taxon, perhaps called a class, birds and crocodilians might be grouped together in one order because they have a common ancestor. Lizards and snakes would be assigned to a different order because they share a different common ancestor. Turtles would be placed in a third order. In this way cladists develop phylogenetic systems of classification.

Classical Evolutionary Taxonomy

More traditional **classical evolutionary taxonomy** uses a system of phylogenetic classification, and presents evolutionary relationships in a phylogenetic tree. Classical taxonomists consider both evolutionary branching and the extent of divergence that has occurred in a lineage since it branched from a stem group (Figure 22–14). Classical evolutionary taxonomy is the most widely accepted approach and is the one used in this book.

A taxonomist using this approach to classification might explain that porpoises are mammals rather than fish because they share many characteristics with other mammals, and because these characteristics can be traced to a common ancestor. Organisms are classified in the same cate-

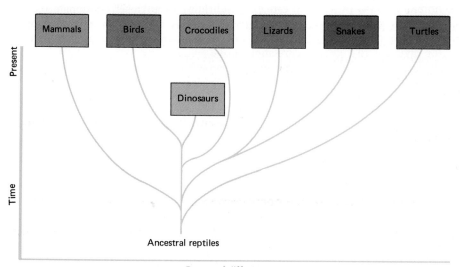

(a)

FIGURE 22–14 Classical evolutionary taxonomists consider both branching and extent of divergence that has occurred since branching occurred. (a) The branching points (numbered) and degrees of difference in the evolution of the major groups of reptiles. Turtles, snakes, lizards, and crocodiles are most similar, but birds, dinosaurs, and crocodiles are most closely related because they branched most recently from a common ancestor. (b) Hierarchical classification of these groups by the classical method.

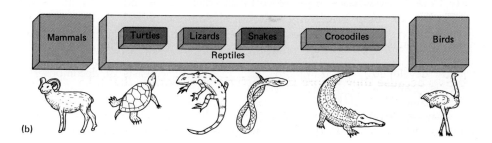

(b)

gory according to their shared characteristics only if those traits are derived from a demonstrable common ancestor. The significance of the adaptations possessed by related organisms is also considered. If, for example, egg-laying mammals could be shown to have a very different ancestry from the placental mammals, the classical taxonomist might erect a separate class to accommodate them. On the other hand, common ancestry, though necessary for inclusion in the same category, would not by itself be sufficient grounds for inclusion.

A classical taxonomist would almost surely classify birds and crocodiles separately, for example, even though they share a common ancestor. The birds would be placed in class Aves on the basis of their characteristic feathers, warmbloodedness, and other features that indicate extensive divergence since branching from the early reptilian stock. In short, the classical taxonomist would be a splitter more than a lumper. The shared characters of turtles, lizards, snakes, crocodiles, and dinosaurs would be emphasized and these animals would be assigned to class Reptilia. All of these animals have horny scales and are coldblooded.

■ SUMMARY

I. The science of classifying and naming organisms is taxonomy.

II. The modern system of scientific taxonomy is based on the binomial system first used consistently by Linnaeus.
   A. In this system the basic unit of classification is the species.
   B. Each organism is given a two-part name consisting of the genus name and the species name. For example, the scientific name for the human is *Homo sapiens* and that for the domestic cat is *Felis catus.*

III. The hierarchical system of classification currently used includes domain, kingdom, phylum (or division, in plants and protists), class, order, family, genus, and species. Subspecies may be listed after species.

IV. The five-kingdom classification in current use recognizes the kingdoms Monera, Protista, Fungi, Plantae, and Animalia.

V. The Monera are assigned to domain Prokaryota because they lack distinct nuclei and other membrane-bounded organelles. The Protista, Fungi, Plantae, and Animalia, which all have discrete nuclei and membrane-bounded organelles, are assigned to domain Eukaryota.

VI. Modern classification is based on evolutionary relationships, or phylogeny.
   A. All of the organisms in a monophyletic taxon have a common ancestor; the organisms in a polyphyletic taxon may have evolved from different ancestors.
   B. Homologous structures imply divergent evolution from a common ancestor.
   C. Shared primitive characters suggest common ancestry; shared derived characters indicate more recent common ancestry.
   D. The human, porpoise, and fish are all classified in the same phylum, the chordates, on the basis of shared primitive characters; the human and porpoise are also classified in the same class, a smaller category, indicating their closer relationship.
   E. Comparison of DNA and protein structure provides a powerful tool for confirming evolutionary relationships.

VII. Three main approaches to taxonomy are phenetics, cladistics, and classical evolutionary taxonomy.
   A. The phenetic system is a numerical taxonomy based on phenotypic similarities; organisms are grouped according to the number of characteristics they share without trying to determine whether their similarities are homologous.
   B. The cladistic approach emphasizes phylogeny, focusing on how long ago one group branched off from another; cladists insist that taxa be monophyletic.
   C. Classical evolutionary taxonomy considers both evolutionary branching and the extent of divergence; this is the most widely accepted approach.

■ POST-TEST

1. The science of classifying and naming organisms is _____ .

2. In the binomial system of nomenclature developed by _____ each organism is given a _____ and a _____ name.

3. In the hierarchy of taxonomic classification a number of related genera constitute a _____ .

4. The kingdom that includes the algae is _____ .

5. Kingdom _____ consists of decomposers such as yeasts.

6. The members of a monophyletic group have a common _____ .

7. The presence of _____ structures in differ-

ent organisms suggests that divergent evolution has occurred.

8. The porpoise and the human both have the ability to nurse their young, whereas the less closely related fish does not. The ability to nurse their young is a shared _____ character.

9. The constancy in DNA and _____ evolution permits biologists to use these macromolecules as molecular _____.

10. The _____ system is a numerical taxonomy based on phenotypic similarities.

11. Taxonomists who follow the _____ school of taxonomy might classify crocodiles and birds in the same group.

12. A system of phylogenetic classification is used by _____ _____ taxonomists.

A complete classification of the human would be as follows:
13. Domain _____
14. Kingdom _____
15. Phylum _____
16. Subphylum _____
17. Class _____
18. Subclass _____
19. Order _____
20. Family _____
21. Genus _____
22. Species _____

■ REVIEW QUESTIONS

1. How would you define a species? a class? a phylum? a division?
2. Distinguish between "natural" and "artificial" systems of classifying organisms.
3. What are the advantages of a "five-kingdom" system over a "two-kingdom" one? List some organisms that are easily assigned to a kingdom. What organisms are especially difficult to assign a place in the taxonomic hierarchy?
4. What taxonomic problems was the five-kingdom scheme intended to solve? Has it created new problems?
5. Outline the complete taxonomic system devised by Linnaeus.
6. Why are there some difficulties in attempting to use the concept of a species?

7. How can shared derived characteristics be used to determine relationships among organisms?
8. In which kingdom would you clasify each of the following?
   a. an oak tree
   b. an amoeba
   c. *Escherichia coli*
   d. a tapeworm
9. Compare the phenetic, cladistic and phylogenetic approaches to taxonomy.
10. Of what use to a taxonomist would be knowledge of the amino acid sequences of the proteins of various organisms?

■ RECOMMENDED READINGS

Corliss, J.O. Consequences of creating new kingdoms of organisms. *Bioscience*, Vol. 33, May 1983. The objections of a holdout against the five-kingdom scheme of taxonomy.

Dobzhansky, T., et al. *Evolution*. San Francisco, W. H. Freeman, 1977. An introduction to evolution that includes chapters on taxonomy.

Eldredge, N., and J. Cracraft. *Phylogenetic Patterns and the Evolutionary Process*. New York, Columbia University Press, 1980. A discussion of the interaction between taxonomy and evolutionary theory.

Gardner, E. J. *History of Biology*. Minneapolis, Burgess Publishing Company, 1965. Chapter 8 summarizes the work of Linnaeus and that of his lesser known contemporaries and predecessors in the establishment of our system of classification.

Gould, S. J. *The Panda's Thumb: More Reflections in Natural History*. New York, Norton, 1980.

Krogmann, D. W. Cyanobacteria (blue-green algae)—their evolution and relation to other photosynthetic organisms, *Bioscience*, Vol. 31, No. 2, February 1981. A good example of the application of modern taxonomic techniques.

Leedale, G. F. How many are the kingdoms of organisms?

Taxon, Vol. 23, 1974. Discusses problems with assigning organisms to discrete kingdoms.

Lewin, R. Molecular clocks scrutinized. *Science*, May 3, 1985. A summary of what is known about molecular clocks.

Margulis, L., and K. V. Schwartz. *Five Kingdoms. An Illustrated Guide to the Phyla of Life on Earth*. San Francisco, W.H. Freeman, 1982. The tremendous diversity of living things, beautifully illustrated.

Mayr, E. Biological classification: toward a synthesis of opposing methodologies. *Science*, Vol. 214, October 1981. A discussion of phenetics, cladistics, and evolutionary classification suggesting that all three methods be utilized in taxonomy.

Moore, R. T. Proposal for the recognition of super ranks. *Taxon*, 23:650–652, 1974. A suggestion for using the term dominium to comprise the (1) viruses, (2) prokaryota, and (3) eukaryota.

Palleroni, N. J. The taxonomy of bacteria, *Bioscience*, Vol. 33, No. 6, June 1983. Modern approaches to microbial taxonomy are discussed following a discussion of the history of bacterial classification systems.

Sibley, C. G., and J. F. Ahlquist. Reconstructing bird phylogeny by comparing DNAs. *Scientific American*, February, 1986. An interesting account of a modern taxonomic method.

Whittaker, R. H. New concepts of kingdoms of organisms. *Science,* Vol. 163, 1969. A proposal for classifying living things according to a five-kingdom system.

Whittaker, R. H. On the broad classification of organisms. *Quarterly Review of Biology,* 34:210, 1959. The earliest rumble of what was to become the five-kingdom revolution.

Whittaker, R. H. and L. Margulis. Protist classification and the kingdoms of organisms. *Biosystems,* Vol. 10, 1978.

Woese, C. R., and G. E. Fox. Phylogenetic structure of the Prokaryote domain: the primary kingdoms. *Proceedings of the National Academy of Sciences,* 74 (11):5088–5090, 1977. Significant for its proposal of both the domain concept and the suggestion that the domain Prokaryota should contain two kingdoms, the eubacteria and the methanogens.

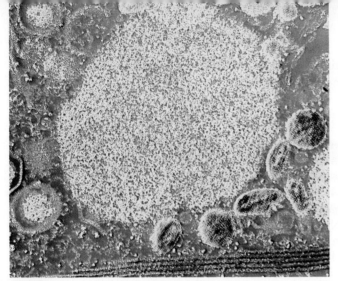
False-color TEM of cell affected with myxomatosis virus

23

Viruses

■ LEARNING OBJECTIVES

After you have read this chapter you should be able to:

1. Compare a virus with a free-living cell.
2. Describe the structure of a virus, or draw and label a virus.
3. Describe bacteriophages, and explain why they have been significant in the development of knowledge about viruses.
4. Trace the steps that take place in the process of viral infection.
5. Contrast a lytic infection with a lysogenic infection.
6. Give the basis of species and tissue specificity of viral infection.
7. Identify two viral infections of plants, and describe viroids.
8. Describe the infection of an animal cell by a virus and compare acute, chronic, latent, and slow virus infections.
9. Summarize what is currently known of the relationship between viruses and cancer.
10. Speculate upon the evolutionary origin of viruses.

Viruses lie on the threshold between life and nonlife. They are not cellular, cannot move about on their own, and cannot carry on metabolic activities independently. All cellular forms of life contain both DNA and RNA, but a virus contains *either* DNA *or* RNA, not both. Most viruses lack ribosomes and enzymes necessary for protein synthesis. They can reproduce, but only within the complex environment of the living cells that they infect. In a sense, viruses come alive only when they infect a cell. They are superbly adapted for their parasitic mode of life.

Because they are not cellular and cannot carry on metabolic activities on their own, viruses are not classified in any of the five kingdoms of living things. Furthermore, no system of virus classification has yet been agreed upon because so little is known about their evolutionary relationships. Although a system has been proposed for dividing them into families and genera, it is not universally accepted. At present viruses are usually grouped on the basis of four main criteria: (1) size; (2) shape; (3) presence or absence of an outer envelope; and (4) the type of nucleic acid—DNA or RNA—they contain, and whether it is single-stranded or double-stranded.

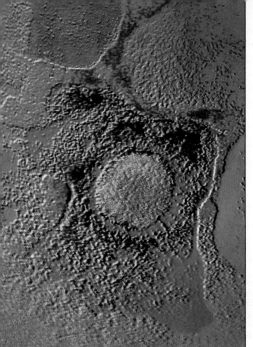

FIGURE 23–1 False-color transmission electron micrograph (TEM) of a single virion (virus particle) of cytomegalovirus, a member of the herpesvirus group. This virus causes a condition among newborns called cytomegalic disease, characterized by jaundice and an enlarged liver and spleen. It is transmitted to the infant by the mother either while in the uterus or during birth. The virus was named cytomegalo, meaning large cell, because it causes each cell it infects to have a swollen appearance. The central protein capsid is shown in pale green. The viral DNA is contained within the capsid. Surrounding the capsid is a large lipoprotein envelope shown in orange (×73,500 at 35mm size).

They are also sometimes classified according to the types of diseases they cause or their mode of transmission.

STRUCTURE OF A VIRUS

A **virus** is a tiny particle consisting of a nucleic acid core surrounded by a protein coat called a **capsid.** The term **virion** refers to a single, infective virus particle. Some viruses are surrounded by an outer membranous **envelope** containing proteins, lipids, carbohydrates, and traces of metals. There are DNA viruses (Figure 23–1) and RNA viruses. Whether the virus contains DNA or RNA, that type of nucleic acid serves as its genetic material, or genome. The viral genome may consist of fewer than five genes or as many as several hundred. However, viruses never have tens of thousands of genes like the cells of more complex organisms.

Only the largest virus, the smallpox virus, can be seen with the light microscope. Most viruses are much smaller than bacteria, and indeed some are scarcely larger than a large molecule of protein. Individual particles of all but the smallpox virus are less than 0.25 micrometers in diameter and can be photographed only with an electron microscope. However, *accumulations* of viruses growing in the cytoplasm of an infected cell are visible with a light microscope.

The shape of a virus is determined by the organization of protein subunits, called **capsomeres,** that make up the capsid. Viruses are generally either helical or polyhedral in shape, or a complex combination of both (Figure 23–2). Helical viruses, such as the tobacco mosaic virus, appear as long rods or threads; their capsid is a hollow cylinder with helical structure. Polyhedral viruses are somewhat spherical in shape. The plant virus known as bushy stunt virus is a polyhedral virus that lacks an outer envelope. Another polyhedral virus, the influenza virus, is surrounded by an outer envelope with glycoprotein spikes that aid in adhering to the host cell. The poliovirus is a polyhedral virus with 20 triangular faces and 12 corners.

Unlike cells, viruses can be crystallized. Later, if the inert crystals are put back into the appropriate host cells, they can multiply and produce the symptoms of disease.

BACTERIOPHAGES

Among the most complex viruses are those that infect bacteria. These viruses are known as **bacteriophages** (bacteria eaters), or simply **phages** (Figure 23–2d). The most common bacteriophage structure consists of a long nucleic acid molecule coiled within a polyhedral head. Most, but not all, utilize DNA as their genetic material. Many phages have a tail attached to the head. Fibers extending from the tail may be used to attach to a bacterium.

There are many varieties of phages, and they are usually species-specific (or strain-specific), meaning that one type of phage generally attacks only one species (or strain) of bacteria. Because phages can be easily cultured within living bacteria in the laboratory, most of our knowledge of viruses has come from studying these bacterial viruses. It has now been shown that phages can serve as excellent models for the viruses that infect animal cells (see Focus on Culturing Viruses).

Virulent, or **lytic,** bacteriophages cause lytic infections. After these viruses multiply, they **lyse** (destroy) the host cell. **Temperate,** or **lysogenic,** viruses do not kill their host cells during the lysogenic cycle. However, these viruses may revert to a lytic cycle and then destroy their host. Some temperate viruses integrate their nucleic acid into the DNA of the host cell and multiply whenever the host cell DNA replicates.

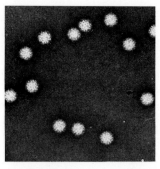

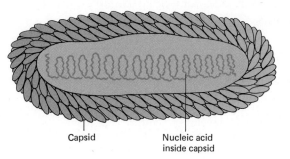

Capsid   Nucleic acid inside capsid

(a) Tobacco mosaic virus, a helical virus (not enveloped)

FIGURE 23–2 Viral structure. A virus consists of a nucleic acid core surrounded by a capsid (protein coat). Some viruses are surrounded by an outer membranous envelope. Viruses are generally either helical or polyhedral in shape, or a complex combination of both. (a) Tobacco mosaic virus has a helical capsid and appears rod-shaped (approximately ×140,000). (b) Bushy stunt virus has a polyhedral capsid. (approximately ×260,000). (c) Influenza virus is a polyhedral virus surrounded by a phospholipid envelope (×200,000). (d) This bacteriophage, known as T₄, is a complex combination of helical and polyhedral shapes (approximately ×275,000).

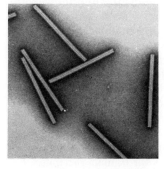

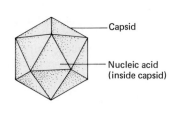

Capsid

Nucleic acid (inside capsid)

(b) Bushy stunt virus, a polyhedral virus (not enveloped)

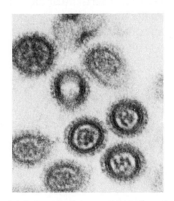

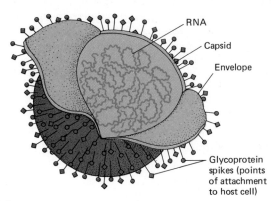

RNA

Capsid

Envelope

Glycoprotein spikes (points of attachment to host cell)

(c) Influenza virus, a polyhedral enveloped virus

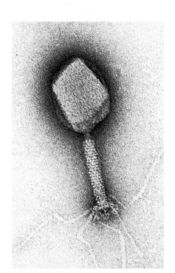

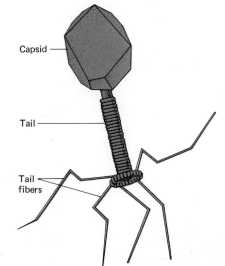

Capsid

Tail

Tail fibers

(d) T₄ bacteriophage, a polyhedral and helical virus (not enveloped)

.0001 μm

FIGURE 23–3 An osmotically shocked bacteriophage (a virus that attacks bacteria). Its single molecule of DNA has been released from the phage coat by breaking the coat protein.

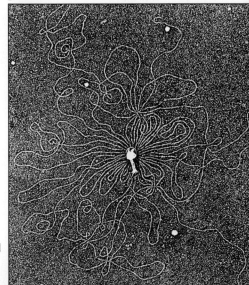

Focus on CULTURING VIRUSES

Since viruses can multiply only when they have infected living cells, they cannot be cultured on a nonliving medium. One of the first, and still very useful, methods for culturing animal viruses was culturing them in developing chick embryos. A small piece of egg shell is removed about a week or two after fertilization of the egg, and the material containing the virus is injected through the opening. The virus can be injected into the embryo itself, or the virus can be injected onto one of the membranes surrounding the embryo—usually the yolk sac or chorioallantoic membrane (see figure). The opening in the shell can then be sealed with paraffin wax, and the egg incubated at 36°C. The virus multiplies within the living cells and can later be separated from the host cells by centrifugation. Cultivation of viruses in developing chick embryos has been used in production of virus for various vaccines including smallpox, influenza, and yellow fever vaccines. This method is also used for immunological and other research studies.

Currently the most widely used method for culturing viruses is tissue culture. Almost any type of animal cell can now be cultured in an appropriate culture medium in glass or plastic dishes. Appropriate viruses for the particular tissue obligingly infect these cells. Some vaccines are now prepared from viruses grown in tissue culture. This is advantageous to persons who may be aller-

gic to eggs, and therefore to the vaccines prepared from viruses cultured in chick eggs. Viruses grown in tissue culture may induce characteristic changes in the tissue culture cells just as they do within the body, and so provide an important model for studying viral infection and its effects on cells.

What types of cells are used for tissue culture? Cells can be obtained from normal animal tissues (mouse, hamster, chicken, monkey, or human). Unfortunately, such cells can only be subcultured a few times before they die. Cell cultures obtained from embryonic tissue are capable of a greater number of cell divisions in culture than cells derived from adult tissue. Cells obtained from malignant tissue appear to be capable of an unlimited number of cell divisions. Cancer cells taken from a woman, Henrietta Lacks, who died of cancer in 1952 have continued to divide in Petri dishes throughout the world. These cells, referred to as HeLa cells, have been widely used in research. Malignant cells are not used, however, for producing viruses for human vaccines.

Some viruses cannot be cultured in developing chick embryos or in laboratory glassware. These must be propagated in living animals, usually mice, guinea pigs, rabbits, or primates. An advantage of animal inoculation is that researchers can study typical symptoms of the infection as they develop.

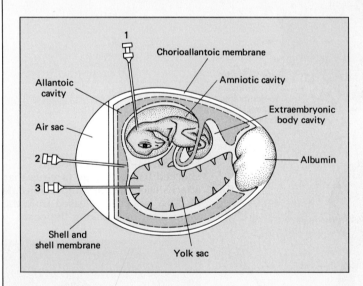

■ Diagrammatic section through a developing chick embryo from 10 to 12 days old, indicating how viruses can be injected into (1) the head of the embryo, (2) the alloantoic cavity, and (3) the yolk sac.

VIRAL REPLICATION IN LYTIC INFECTIONS

Outside a living cell, a virus has no metabolic activity and cannot reproduce itself. When a virus infects a susceptible host cell, it uses the host cell's metabolic machinery to replicate its nucleic acid and produce its specific proteins.

Several steps in the process of viral infection are common to almost all bacteriophages (Figure 23–4):

1. **Attachment** to the surface of the host cell. The virus attaches to spe-

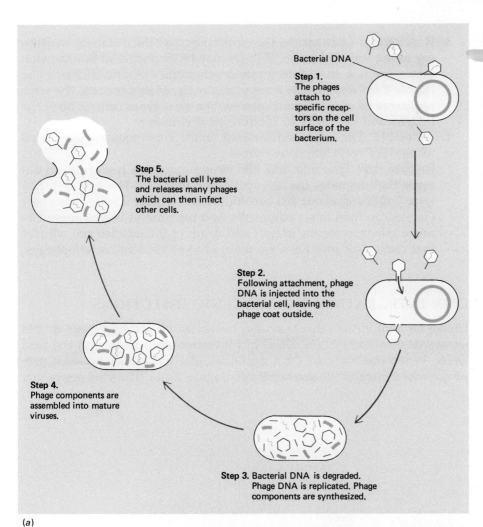

Bacterial DNA

Step 1.
The phages attach to specific receptors on the cell surface of the bacterium.

Step 5.
The bacterial cell lyses and releases many phages which can then infect other cells.

Step 2.
Following attachment, phage DNA is injected into the bacterial cell, leaving the phage coat outside.

Step 4.
Phage components are assembled into mature viruses.

Step 3. Bacterial DNA is degraded. Phage DNA is replicated. Phage components are synthesized.

(a)

FIGURE 23–4 The sequence of events in a lytic infection. (*a*) The phage attaches to the surface of the host cell. It then injects its nucleic acid into the cell. The bacterial DNA is degraded, and the phage DNA is replicated. Phage components are synthesized and assembled, producing new phages. The bacterial cell bursts open, releasing phages, which can then infect other cells. (*b*) Phage infecting *Escherichia coli*, a bacterium. Many phages are attached to the cell wall. The head, tail, and base plate of most virus particles are clearly visible. The tail core extending from the base plate to the cell wall is hollow and acts like a hypodermic needle in injecting viral DNA into the cell. The break in the bacterial cell wall is an artifact produced during preparation for viewing under the electron microscope.

(b)

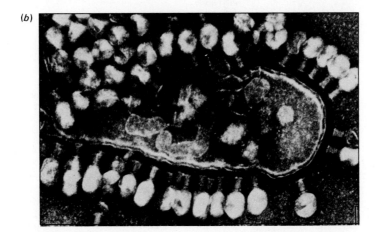

cific receptor sites on the host cell wall. Since each bacterial species (and sometimes each strain within a species) has different receptor sites, a virus will attach to only a specific species (or strain). A laboratory technique called bacteriophage typing makes use of this discriminatory ability of viruses to distinguish among various strains of bacteria.

2. **Penetration.** After the virus has attached to the cell surface, its nucleic acid is injected through the cell membrane and into the cytoplasm of the host cell. The capsid of a phage remains on the outside. Most viruses that infect animal cells, in contrast, enter the host cell intact by phagocytosis.

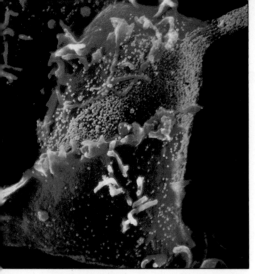

FIGURE 23–5 An infected human cell has burst open and is releasing new viruses (blue).

3. **Replication.** Once inside, the virus takes over the metabolic machinery of the cell. The bacterial DNA may be degraded so that the viral genes are free to dictate future biochemical operations. Using the host cell's ribosomes, its energy, and many of its enzymes, the virus replicates its own macromolecules. The viral genes contain all of the information necessary to produce new viruses.

4. **Assembly.** The newly synthesized viral components are assembled to produce complete new viruses.

5. **Release.** In a lytic infection, the virus produces a **lysozyme,** an enzyme that degrades the cell wall of the host cell. The host cell then lyses, releasing about 100 bacteriophages (Figure 23–5). These new viruses can then infect other cells, and the process begins anew. Because infection results in lysis and death of the infected cell, viruses that cause lytic infections are referred to as virulent bacteriophages.

DNA INTEGRATION IN LYSOGENIC INFECTIONS

Unlike virulent viruses that lyse their host cells, temperate viruses do not always destroy their hosts. They can integrate their DNA into the host DNA. When the bacterial DNA replicates, the viral DNA (called a **prophage** when integrated) also replicates (Figure 23–6). The viral genes that

FIGURE 23–6 The sequence of events in a lysogenic infection. A temperate virus can integrate its DNA into the host DNA. The integrated DNA, called a prophage, replicates when the bacterial DNA replicates. Lysogenic conversion may occur in which the bacterial cells exhibit new properties. Under certain conditions, the prophage may become lytic and begin a lytic cycle.

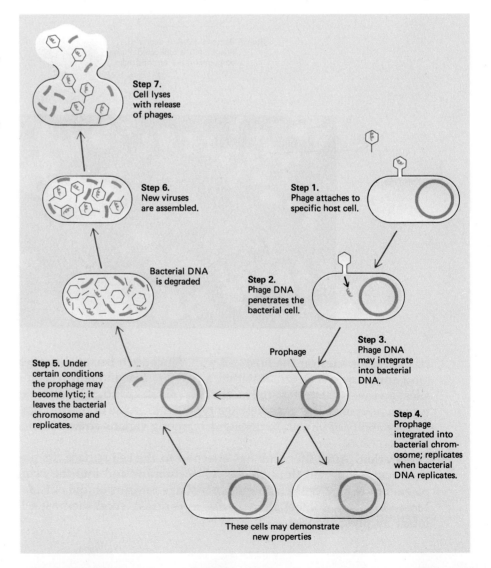

Step 7. Cell lyses with release of phages.

Step 6. New viruses are assembled.

Step 1. Phage attaches to specific host cell.

Bacterial DNA is degraded

Step 2. Phage DNA penetrates the bacterial cell.

Prophage

Step 3. Phage DNA may integrate into bacterial DNA.

Step 5. Under certain conditions the prophage may become lytic; it leaves the bacterial chromosome and replicates.

Step 4. Prophage integrated into bacterial chromosome; replicates when bacterial DNA replicates.

These cells may demonstrate new properties

code for viral structural proteins may be repressed indefinitely. The bacterial (host) cell, on the other hand, may behave almost normally. Host cells carrying prophages are referred to as lysogenic.

In some cases bacterial cells containing temperate viruses may exhibit new properties. This is called **lysogenic conversion.** For example, the bacterium that causes diphtheria produces the toxin responsible for the disease symptoms only when inhabited by a specific phage. In fact, the toxin is actually encoded by the phage. In the same way, a phage is responsible for producing the toxin associated with scarlet fever; only when the streptococci are lysogenic can they cause scarlet fever. **Clostridium botulinum** bacteria synthesize the toxin that causes botulism only when they are lysogenic for certain phages.

Certain external conditions can cause the phage nucleic acid to enter a lytic phase, releasing new phages. When a lysogenic cell does lyse, the phages released may contain some bacterial DNA in place of their own genetic material. When it infects a new bacterium, such a phage can introduce this DNA into the genome of the host. Known as **transduction,** this process permits genetic recombination in the new host cell (Figure 23–7).

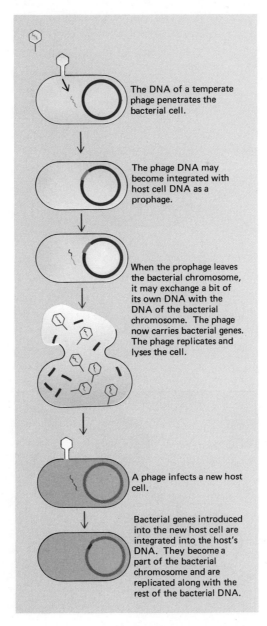

FIGURE 23–7 Transduction. A phage can transfer bacterial DNA from one bacterium to another.

The DNA of a temperate phage penetrates the bacterial cell.

The phage DNA may become integrated with host cell DNA as a prophage.

When the prophage leaves the bacterial chromosome, it may exchange a bit of its own DNA with the DNA of the bacterial chromosome. The phage now carries bacterial genes. The phage replicates and lyses the cell.

A phage infects a new host cell.

Bacterial genes introduced into the new host cell are integrated into the host's DNA. They become a part of the bacterial chromosome and are replicated along with the rest of the bacterial DNA.

This ability of some viruses to transfer DNA from one cell to another is taken advantage of in recombinant DNA studies (Chapter 14).

ANOTHER FORM OF COEXISTENCE

A few bacterial viruses (as well as some animal viruses) release new viruses slowly without destroying the host cell (Figure 23–8). In such cases the host cell carries on its own metabolic activities, although some of its energy is used to produce new viruses. As they are assembled, mature viruses migrate to the cell membrane. They appear to exit from the host cell in a process that may be the reverse of penetration.

FIGURE 23–8 Coexistence with viruses. (*a*) In some viral infections, assembly of new viruses and their release may occur slowly without destroying the host cell. A few types of phages coexist with their host cells in this way. (*b*) Virus particles budding from the surface of a human cell. Each new virus has incorporated some of the cell membrane of the host cell; however, the host is able to repair its membrane after the budding is complete.

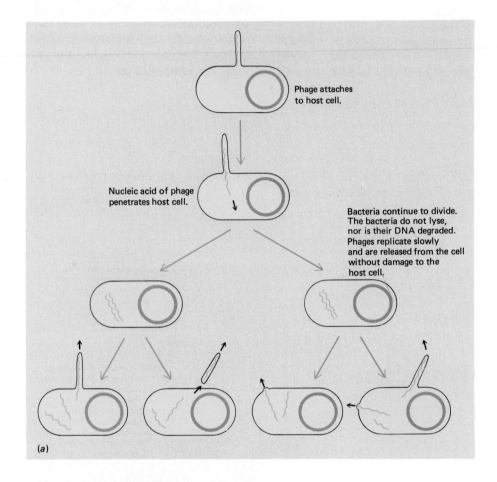

Phage attaches to host cell.

Nucleic acid of phage penetrates host cell.

Bacteria continue to divide. The bacteria do not lyse, nor is their DNA degraded. Phages replicate slowly and are released from the cell without damage to the host cell.

(a)

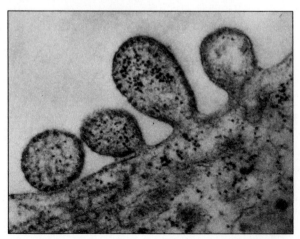

(b)

PLANT VIRUSES AND VIROIDS

In 1892, the Russian botanist Iwanowski found that tobacco mosaic disease (Figure 23–9)—so called because the infected tobacco leaves have a spotted appearance—could be transmitted to healthy plants by daubing their leaves with the sap of diseased plants. The sap was infective even after it had been passed through filters fine enough to remove all bacteria.

Tobacco mosaic virus consists only of an RNA core surrounded by a protein capsid; it is elongate and lacks an outer envelope. After infecting a host cell, this virus attaches to the host's ribosomes and is translated as though it were mRNA. For this reason tobacco mosaic virus is referred to as a plus-strand RNA virus. Many types of plant viruses belong to this group, including those that cause mosaic diseases such as alfalfa mosaic disease and those that cause stunt diseases such as tomato bushy stunt.

Plant viruses cause serious agricultural losses. Since cures are not known for most viral diseases of plants, agricultural scientists have focused their efforts on prevention. Virus-resistant strains of important crop plants are being developed. A common practice is to burn plants that have been infected.

Viral diseases are spread among plants by insects. They can also be inherited by way of infected seeds or by asexual propagation. Once a plant is infected, the virus can spread through the plant body by passing through the cytoplasmic connections (plasmodesmata) that penetrate the cell walls between adjacent cells.

Plants may also be infected by **viroids,** infective agents that are even smaller and simpler than viruses. Each viroid consists of a very short strand of RNA (only 250 to 400 nucleotides). The amount of RNA present may be sufficient to code for a single medium-sized protein. No proteins are associated with them and they have no protective coat (Figure 23–10). Evidence suggests that the viroid does not code for any proteins. Host enzymes are used to replicate the viroid's RNA. Viroids are generally found within the cell nucleus, and they cause disease either by interfering with intron splicing or by interfering with the regulation of the host's genes.

All viroids that have been identified infect plants. Viroids have been linked to several diseases of complex plants, including potato spindle-tuber disease and a disease that causes stunting of chrysanthemums.

VIRUSES THAT INFECT ANIMALS

Hundreds of different viruses infect humans and other animals. Viruses have been identified that can infect almost every type of cell. Most viruses cannot survive very long outside a living host cell, and so their survival depends on their being transmitted from one cell to another or from animal to animal.

Infection By Animal Viruses

The type of receptor molecules on the surface of a virus determines what type of cell the virus can infect. While most viruses lack specialized structures for attachment to a host cell, their surfaces do have receptor molecules that aid in chemical attachment. Some viruses, such as the adenoviruses, have fibers that project from the capsid and are thought to help the virus adhere to receptor sites on the host cell. Other viruses, such as those that cause herpes, influenza, and rabies, are surrounded by a lipoprotein envelope with projecting glycoprotein spikes that serve as receptors.

FIGURE 23–9 Tobacco plant infected with tobacco mosaic virus. The virus produces a yellow and green mottling, or mosaic pattern. Young leaves are especially vulnerable; however, the disease tends to reduce crop yields rather than killing the plants outright.

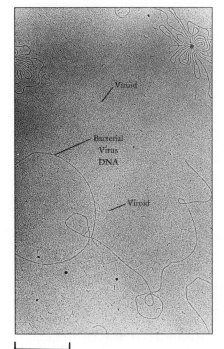

0.25 μm

FIGURE 23–10 A viroid is a rod-like structure consisting of a single-stranded circular molecule of RNA. This electron micrograph compares the size of a viroid with molecules of bacteriophage DNA.

545

Receptor sites vary with each species and sometimes with each type of tissue. Thus, some viruses can infect only humans, because their receptors can combine only with receptor sites found on human cell surfaces. The measles virus and poxviruses can infect many types of tissues because their receptors can combine with receptor sites on a variety of cells. However, poliovirus receptors can attach only to certain tissue cells—spinal cord, throat, and intestinal cells.

When an unenveloped animal virus binds to a receptor site on the surface of a host cell, the cell engulfs it and transports it into the cytoplasm in a process resembling phagocytosis. An enveloped virus fuses with the host cell membrane and passes into the cytoplasm, where the envelope and capsid are removed.

Like other viruses, those that infect animal cells must replicate and produce new virus particles. Viral nucleic acid is replicated and viral proteins are synthesized that inhibit the host DNA, RNA, and protein synthesis. Viral DNA and protein synthesis is similar to the process by which the host cell would normally carry out its own DNA and protein synthesis. In most RNA viruses, transcription takes place with the help of an RNA polymerase. However, the virus (HIV) that causes AIDS and some tumor-causing viruses use a DNA polymerase called **reverse transcriptase** (see Chapter 13, Focus on Reverse Transcription, Jumping Genes, and Pseudogenes). Reverse transcriptase catalyzes the synthesis of a complementary DNA strand, using the viral RNA as a template. The DNA then acts as the template for production of its complement so that a double-stranded viral DNA molecule is produced. This DNA is then used to synthesize copies of the viral RNA.

Structural proteins needed by the virus are synthesized and the capsid is produced. After new virus particles are assembled, they exit the host cell. Viruses that do not have an outer envelope exit by cell lysis; the plasma membrane ruptures, releasing the viral particles. Enveloped viruses receive their lipoprotein envelopes as they pass through the plasma membrane (or, in some types, the nuclear membrane). They are released slowly by a process called budding, and these viruses do not destroy the host cell when they exit.

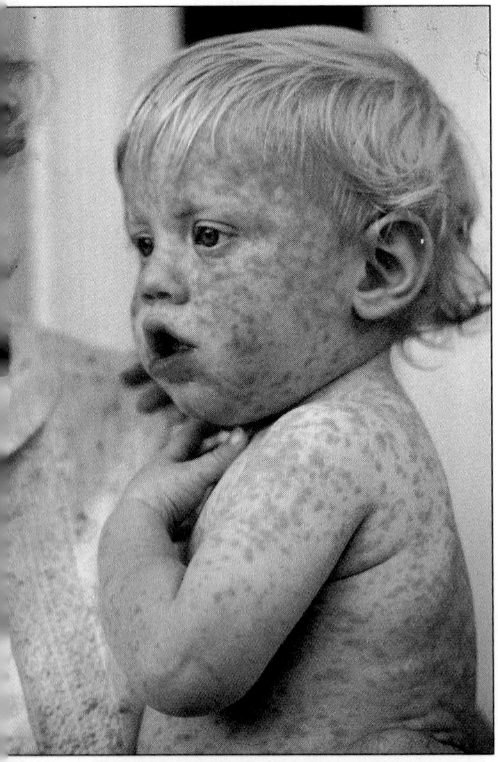

Viral proteins synthesized within the host cell may damage the cell in a variety of ways. Such proteins may alter the permeability of the cell membrane, or may inhibit synthesis of host nucleic acids or proteins. Viruses sometimes damage or kill their host cells by their sheer numbers. A poliovirus may produce 100,000 new viruses within a single host cell!

One way that cells react to viral infection is by the production of **interferons,** proteins that interfere with viral replication. Interferon is released from infected cells and helps to protect uninfected cells in the area. Interferon and other responses to virus infections are discussed in Chapter 43.

Viral Diseases in Animals

Improvements in nutrition and hygiene and the development of effective vaccines have virtually eradicated some viral diseases such as smallpox and poliomyelitis in developed countries. However, other viral diseases such as hepatitis B and AIDS remain as serious causes of disease and death.

Animal diseases caused by viruses include hog cholera, foot-and-mouth disease, canine distemper, swine influenza, and Rous sarcoma in fowl. Humans are prone to a variety of viral diseases, including chickenpox, herpes simplex (one type of which is genital herpes), herpes zoster, mumps, rubella (German measles), rubeola (measles), rabies, warts, influenza, and AIDS (Figure 23–11; Table 23–1). Indeed, it has been estimated that each of us suffers from two to six viral infections each year. Fortunately, most of these are relatively benign forms such as the common cold.

FIGURE 23–11 Rubella (German measles) is caused by an RNA virus spread by close contact. When contracted during pregnancy, it can cause birth defects. Immunity appears to be lifelong following infection. Vaccination has greatly decreased the incidence of this disease.

Table 23–1 **ANIMAL VIRUSES**

Group	Diseases Caused	Characteristics
DNA Viruses		
Poxviruses	Smallpox, cowpox, and economically important diseases of domestic fowl	Large complex, oval-shaped viruses that replicate in the cytoplasm of the host cell
Herpesviruses	Herpes simplex Type 1 (cold sores); herpes simplex Type 2 (genital herpes, a sexually transmitted disease); herpes zoster (chickenpox and shingles). The Epstein-Barr virus has been linked with infectious mononucleosis and Burkitt's lymphoma.	Medium to large, enveloped viruses; frequently cause latent infections; some cause tumors
Adenoviruses	About 40 types known to infect human respiratory and intestinal tracts; common cause of sore throat, tonsillitis, and conjunctivitis; other varieties infect other animals	Medium-sized viruses
Papovaviruses	Human warts and some degenerative brain diseases; cancer in animals other than humans	Small viruses
Parvoviruses	Infections in dogs, swine, arthropods, rodents	Very small viruses; some contain single-stranded DNA; some require a helper virus in order to multiply
RNA Viruses		
Picornaviruses	About 70 types infect humans including polioviruses; enteroviruses infect intestine; rhinoviruses infect respiratory tract and are main cause of human colds; coxsackievirus and echovirus cause aseptic meningitis.	Diverse group of small viruses
Togaviruses	Rubella, yellow fever, equine encephalitis	Large, diverse group of medium-sized, enveloped viruses; many transmitted by arthropods
Myxoviruses	Influenza in humans and other animals	Medium-sized viruses that often exhibit projecting spikes
Paramyxoviruses	Rubeola, mumps, distemper in dogs	Resemble myxoviruses but somewhat larger
Reoviruses	Vomiting and diarrhea in children	Contain double-stranded RNA

We are most familiar with **acute** viral infections, in which the disease is short-lived. However, some viruses cause **latent** infections, in which the viruses remain quietly in the body for years before becoming active. Sometimes, symptoms reappear periodically. After the initial infection, the herpes virus that causes cold sores can infect certain ganglia (groups of nerve cell bodies), where it may remain for years without causing symptoms. However, during times of stress or ill health, the virus may be activated and cause cold sores once again. Environmental factors such as exposure to the sun may also activate the virus. During the latent period the virus usually cannot be detected. It is during the active phase that the herpes virus can be transmitted through physical contact with the active sore.

In **chronic** viral infections, the virus can be shown to be present even though the carrier may not exhibit symptoms. Infected individuals can also transmit the disease to others even when no symptoms are present. **Slow** virus infections generally cause slow, progressive degeneration of the tissues involved. Such infections often lead to death. Multiple sclerosis is suspected of being a disease of this type.

Viruses and Cancer

Both RNA and DNA viruses are known to cause cancer in animals. For example, the Rous sarcoma virus is known to cause cancer in domestic fowl. The nucleic acid of these cancer-causing viruses becomes integrated into the DNA of the host cells, which are transformed into cancer cells. This viral nucleic acid probably codes for enzymes that change important proteins in the host cell.

Some viruses that cause cancer have one or a few genes, called **oncogenes,** that are responsible for their ability to transform host cells into cancer cells. Oncogenes have also been found in normal cells of most species, and studies indicate that activation of these cellular oncogenes can transform normal cells to cancer cells. Oncogenes appear to code for many different kinds of proteins, including cellular growth factors, membrane receptors, and protein kinases. Changes in the cell that maintain these genes in an active state of transcription appear to promote cancer. Such changes apparently result from infection by a retrovirus, from mutation, and from currently unknown factors. Some viruses that lack oncogenes may work by activating existing cellular oncogenes.

RNA viruses that have been associated with cancer are retroviruses that contain reverse transcriptase. A class of human retroviruses (the HTLV viruses) has been linked to certain leukemias. This HTLV virus enters a T lymphocyte (a type of white blood cell) and triggers a chain of events that leads to leukemia. A related virus (now called HIV) has been identified as the causative agent of AIDS.

No tumor-causing DNA virus has been isolated from human tumors, but evidence links several DNA viruses with human cancers. One type of herpesvirus, the **Epstein-Barr virus,** is thought to cause **Burkitt's lymphoma,** a cancer of the lymphatic system. The Epstein-Barr virus apparently infects almost all human beings. In some it causes infectious mononucleosis (popularly referred to as mono). In Central Africa this virus has been linked to Burkitt's lymphoma. The reason for this dramatic difference in virulence is not known. However, since Burkitt's lymphoma is endemic in areas where malaria is prevalent, the possibility exists that there is an interaction between the agents that cause the two illnesses. The same virus has been implicated in the development of nasopharyngeal carcinoma among persons of Chinese ancestry.

Herpes simplex virus type 2 has been linked with cervical cancer, and hepatitis B virus has been associated with liver cancer. Viruses may also play a role in Hodgkin's disease, breast cancer, and Kaposi's sarcoma, a disease common among AIDS patients.

Prions

Viroids had been suspected of causing certain animal diseases including scrapie, a progressive neurological disease of sheep. Recently, however, a subviral pathogen called a **prion** has been implicated. The prion (a coined name for a proteinlike infectious particle)—even smaller than the viroid—appears to consist only of a glycoprotein. The glycoprotein contains at least one polypeptide about 250 amino acids long, and no nucleic acid component is present. Prions polymerize in infected tissue, forming rods. Prions are thought to be associated with two rare central nervous system diseases in humans—Creutzfeldt-Jakob disease and kuru.

Treating Viral Infections

As will be discussed in Chapter 43, the body has powerful weapons—immune mechanisms—with which it defends itself against viral invasion. Vaccines against viruses have been developed that take advantage of the body's immune mechanisms. Such vaccines help prevent viral infection. However, modern medicine does not yet have much to offer one who has been infected by a virus. Effective antiviral drugs have proved difficult to develop.

Antibiotics kill bacteria, but not viruses. Antibiotics inhibit enzymes involved in the synthesis of needed molecules in bacteria. Viruses lack their own distinct metabolic machinery and so provide far less of a target for drugs to attack. For this reason antiviral therapy is still in its infancy.

Among the antiviral drugs that have been developed thus far, acyclovir is considered the most successful. Acyclovir is a nucleoside analogue that mimics the structure of the nucleoside components of DNA or RNA. This drug inhibits replication by certain herpes viruses, so it is used in the treatment of genital herpes. Although it does not cure the disease, it speeds healing when an attack occurs. A few other antiviral drugs have been developed and others are being tested, including a few that show promise for the treatment of AIDS.

THE ORIGIN OF VIRUSES

What is the evolutionary origin of the viruses? One hypothesis is that the ancestors of modern viruses were primitive, free-living heterotrophs that evolved in the primordial sea. These early "organisms" fed upon the organic nutrients that surrounded them. By the time these nutrients were depleted, some organisms had become autotrophs, while others had evolved the enzyme systems necessary to derive energy by feeding on the autotrophs. The early viruses developed neither of these life styles; instead, they adapted to a parasitic mode of life.

Another hypothesis was that viruses evolved from cellular ancestors, becoming highly specialized as parasites. During their evolution, they "lost" their cellular components—all but the nucleus.

The hypothesis currently thought most likely is that viruses are bits of nucleic acid that "escaped" from cellular organisms. According to this view some viruses may trace their origin to animal cells, others to plant cells, and still others to bacterial cells. Their multiple origin might explain the specificity with which they infect different types of of organisms. Perhaps viruses infect only those species closely related to the organisms from which they originated. This hypothesis is supported by the genetic similarity between virus and host cell—a closer similarity than exists among various types of viruses.

■ SUMMARY

I. A virus is a tiny particle or virion consisting of a core of DNA or RNA surrounded by a capsid (protein coat).
   A. Viruses are much smaller than bacteria.
   B. Viruses are helical, polyhedral, or a combination of both shapes.

II. Bacteriophages are viruses that infect bacteria.
   A. Some phages are virulent, or lytic.
   B. Other phages are temperate, or lysogenic.

III. Viral infection includes the following processes: attachment to the host cell, penetration, replication, assembly, and release.
   A. In a lytic infection, the virus produces a lysozyme, which causes the host cell to lyse, releasing the new viruses.
   B. In lysogenic infections, temperate viruses integrate their DNA into the host DNA.
      1. Such nucleic acid integration may confer new properties on the host cell.
      2. Phages released from lysogenic cells may contain a portion of bacterial DNA, which can lead to transduction in a new host cell; this process has important applications in recombinant DNA experiments.

   3. In some viral infections, the host cell is permitted to continue its metabolic activities, and viruses are slowly assembled and released with minimal damage to the host cell.

IV. Plant viruses cause serious agricultural losses. Plants may also be infected by viroids, which are smaller and simpler than viruses.

V. Unenveloped animal viruses enter the host cell by a process similar to phagocytosis; enveloped viruses fuse to the host cell membrane and then pass into the cell.
   A. Capsids are removed within the host cell, and the virions replicate and produce new virions.
   B. Animal cells produce interferons in response to viral infection.
   C. Human viral infections may be acute, chronic, latent, or slow.
   D. Viruses cause cancers in many types of animals, and there is evidence that they cause certain human cancers. For example, the Epstein-Barr virus is thought to cause Burkitt's lymphoma.

VI. Viruses are thought to have had a multiple evolutionary origin. According to this view they are bits of nucleic acid that escaped from cellular organisms.

■ POST-TEST

1. The core of a virus consists of _____ or _____, but never both.
2. Bacteriophages are viruses that infect _____.
3. The five main steps in bacteriophage infection are _____, _____, _____, _____, and _____.
4. A lysozyme is an enzyme produced by the _____, which degrades the _____ of the host.
5. The part of a bacteriophage that actually enters the host cell is its _____ _____.
6. In _____ _____, bacterial cells containing temperate viruses exhibit new properties.
7. Virulent viruses cause _____ infections.
8. Lysogenic viruses are also known as _____ viruses.
9. Lysogenic phages can transfer nucleic acid from one virus to another, resulting in genetic recombination; this process is known as _____.
10. _____ are proteins produced by host cells that interfere with viral replication.
11. In a _____ infection, viruses may rest quietly in the body for years before becoming active.
12. Oncogenes are responsible for the ability of some viruses to transform cells into _____ _____.
13. _____ are even smaller and simpler than viruses; each consists of a very short strand of RNA without any sort of protective _____.
14. The portion of viral DNA integrated into the host DNA is referred to as a _____.
15. The type of infection in which the host cell bursts open and releases many viruses is a _____ infection.

■ REVIEW QUESTIONS

1. Why are viruses often looked upon as being on the threshold between nonlife and life? What characteristics does a virus share with a living cell? What characteristics of life are lacking in a virus?
2. Draw a diagram of a virus, and label its parts.
3. What is a bacteriophage? Why have phages been important in the development of knowledge about viruses?
4. List the steps in the process of viral infection, and briefly describe each step.
5. How do lysogenic infections differ from lytic infections?
6. What limits a virus in the number of species (or tissue types) that it can infect?
7. What is a latent viral infection? a slow infection?
8. What is the possible relationship between the Epstein-Barr virus and human cancer? Explain.
9. Define the following terms:
   a. capsid
   b. virion
   c. temperate virus
   d. lysozyme
   e. transduction

■ RECOMMENDED READINGS

Butler, P. J. G., and A. Klug. The assembly of a virus. *Scientific American*, November 1978. An account of how tobacco mosaic virions assemble themselves.

Diener, T. O. The viroid—A subviral pathogen. *American Scientist*, September–October 1983. A discussion of the origin, structure, and pathogenesis of viroids with speculation about prions.

Gallo, R. C. The first human retrovirus. *Scientific American*, December 1986, pp. 88–98. The discovery of the first human retrovirus laid the groundwork for identifying the related virus that causes AIDS.

Hirsch, M. S., and J. C. Kaplan. Antiviral therapy. *Scientific American*, April 1987, pp. 76–85. A discussion of the problems encountered in developing antiviral drugs and a description of some new antiviral drugs.

Hogle, J. M., M. Chow, and D. J. Filman. The structure of poliovirus. *Scientific American*, March 1987, pp. 42–49. The poliovirus has become a model for investigating the molecular links between viral form and function.

Johnson, H. T., and C. McArthur. Myelopathies and retroviral infections. *Annals of Neurology*, Vol. 21, No. 2, February 1987. A discussion of pathologic changes in the spinal cord resulting from infection with various retroviruses, including HIV (the virus that causes AIDS).

Prusiner, S. B. Prions. *Scientific American*, October 1984. An examination of the tiny infective agents composed only of glycoprotein.

Simon, K., H. Garoff, and A. Helenius. How an animal virus gets into and out of its host cell. *Scientific American*, February 1982, pp. 58–70. An excellent article detailing the process of viral infection in animal cells.

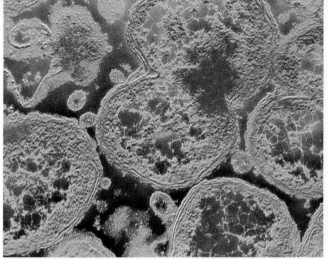

False-color TEM of *Neisseria meningitis*, the bacterium that causes cerebrospinal meningitis

24

Kingdom Monera

■ LEARNING OBJECTIVES

After you have read this chapter you should be able to:

1. Describe the distinguishing characteristics of members of kingdom Monera.
2. Compare the cyanobacteria with other bacteria.
3. Summarize the ecological importance of the cyanobacteria.
4. Describe the structure of a bacterial cell, emphasizing the cell wall and DNA.
5. Characterize bacteria as heterotrophs or autotrophs, and compare bacterial photosynthesis with plant photosynthesis.
6. Distinguish between facultative anaerobes and obligate anaerobes.
7. Summarize the three mechanisms of genetic recombination (transformation, conjugation, and transduction) that take place in bacteria.
8. Describe the three main shapes of the eubacteria, and give examples of gram-positive and gram-negative eubacteria.
9. Characterize each of the following groups of bacteria: myxobacteria, spirochetes, actinomycetes, mycoplasmas, rickettsias, and chlamydias.
10. Give two important ecological roles of bacteria.

It is sobering to reflect that for most of human history bacteria were completely unknown. Neither Aristotle nor King Solomon had ever heard of them. Yet today we realize that life on Earth would be impossible without them, and that they have been the cause of most human disease and many deaths. Careful study of bacteria has led to much of our knowledge of the chemical basis of heredity, and a very large part of our general knowledge of cell biology. Genetically engineered bacteria are now producing materials such as human insulin in commercial quantities that would be impractical or impossible to manufacture in any other way, and there is promise of much more to come (Figure 24–1).

SCIENTISTS AND BACTERIA

It is likely that the first person to see bacteria was Antony van Leeuwenhoek (1632–1723), who was a draper in Delft, Holland. With hand lenses he ground himself, he examined almost everything at hand—pond water,

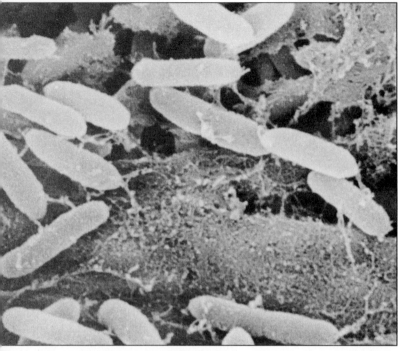

(a)

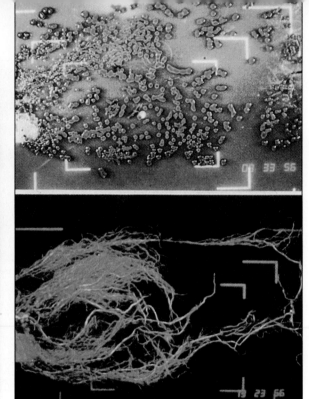

(b)

FIGURE 24–1 (*a*) Genetically engineered bacteria have great potential in agriculture, industry, and medicine. *Pseudomonas fluorescens*, the rod-shaped bacterium shown here on corn roots, can be given a toxin-producing gene from the unrelated bacterium *Bacillus thuringiensis*, which might enable it to attack insect pests feeding on the corn roots. (*B. thuringiensis* is an important insect pathogen in nature which is widely used as a biological insecticide.) (*b*) It is important, however, to take precautions against possible ecological damage caused by genetically engineered bacteria. For this reason we must be able to track the occurrence of these altered organisms in the environment. Genetically altered *P. fluorescens* bacteria were given two genes taken originally from *E. coli* intestinal bacteria. One of these permits *P. fluorescens* to consume the sugar lactose and the other enables it to produce a bright blue pigment. If *P. fluorescens* bacteria grow on a medium containing lactose and produces the blue pigment (top), it must be a genetically altered organism. If not, it is the wild-type strain (bottom).

sea water, vinegar, pepper solutions (he wanted to find out what made peppercorns hot), feces, saliva, semen, and many other things. He described the objects he saw in letters to the Royal Society of London. In one letter written in 1683, his description of the size, shape, and characteristic motion of certain organisms he had observed leaves no doubt that they were bacteria. But it was left to others to discover the significance of bacteria.

The extensive research of Louis Pasteur in the 1870s and 1880s revealed the importance of bacteria as agents of disease and decay. This stimulated work by Robert Koch, Joseph Lister, and others, and the science of bacteriology blossomed rapidly in the latter part of the 19th century. Pasteur's studies of the "diseases" of souring wine and beer showed that they were caused by microorganisms that entered the wine or beer from the air and brought about undesirable fermentations, yielding products other than alcohol. By gently heating (a process now known as **pasteurization**) the grape juice or beer mash to kill the undesirable organisms and only afterward seeding the cooled juice with yeast, these "diseases" could be prevented.

Another of Pasteur's contributions to bacteriology was his unequivocal demonstration that bacteria cannot arise by spontaneous generation (see Chapter 1). After his study of wine, Pasteur was asked by the French government to investigate a disease of silkworms. When Pasteur found that this, too, was caused by microorganisms, he reasoned that many animal and plant diseases might be caused by the invasion of "germs." During his investigation of chicken cholera and of anthrax (a disease of sheep and cattle), he came upon a method of treatment, that of immunization, which greatly reduced the death rate from these diseases.

Lord Lister, an English surgeon, was one of the first to understand the significance of Pasteur's discoveries and to apply the germ theory to surgical procedures. He initiated antiseptic techniques by dipping all his operating instruments into carbolic acid and by spraying the scene of the operation with that germicide. The result was a marked decline in the number of deaths following surgery.

Table 24–1 SOME BACTERIA THAT INFECT HUMANS

Bacterium	Characteristics	Importance
Eubacteria		
Staphylococcus aureus	Cocci that often form clusters; gram-positive	Can live harmoniously as part of normal microbial community. Opportunistic; can cause boils. Also toxin is a major cause of food poisoning.
Streptococcus pyogenes	Cocci that form pairs and chains; gram-positive	Causes "strep throat," ear infections, scarlet fever. Induces rheumatic fever
Streptococcus pneumoniae	Cocci that form pairs or chains; gram-positive	Causes pneumonococcal pneumonia and meningitis
Clostridium tetani	Slender, gram-positive bacilli; strictly anaerobic; form spores	Causes tetanus (lockjaw); potent toxin affects nervous system
Clostridium botulinum	Large gram-positive bacilli; anaerobic; form spores	A soil organism that causes botulism; potent toxin affects nervous system
Neisseria gonorrhoeae	Gram-negative cocci that form pairs (diplococci); adhere to cells via pili	Causes gonorrhea
Escherichia coli	Gram-negative bacilli; facultative anaerobes	Lives as part of normal intestinal microbial community; opportunistic strains among them can cause diarrhea, urinary tract infections, and meningitis
Salmonella	Gram-negative bacilli	One species causes food poisoning (diarrhea, vomiting, fever); another species can cause typhoid fever; a third species causes infections of the blood
Hemophilus influenzae	Gram-negative small rods	Causes infections of upper respiratory tract and ear; can cause meningitis
Rickettsias		
Rickettsia rickettsii	Short rod-shaped; obligate intracellular parasite	Can cause Rocky Mountain spotted fever; transmitted by tick from dog or rodent
Spirochetes		
Treponema pallidum	Very slender, tightly coiled spirals; move via axial filament	Causes syphilis
Actinomycetes		
Mycobacterium tuberculosis	Slender, irregular rods	Causes tuberculosis of lungs and other tissues
Mycobacterium leprae	Slender, irregular rods	Causes Hansen's disease (leprosy)
Chlamydias		
Chlamydia trachomatis	Gram-negative cocci; obligate parasites	Causes trachoma (the leading cause of blindness); causes a sexually transmitted disease (lymphogranuloma venereum)

THE ORIGINAL EARTHLINGS?

The Earth's oldest fossils, bacteria resembling contemporary photosynthetic cyanobacteria, were found in rocks dated by geologists at 3.5 billion years old. In contrast, the oldest eukaryotic fossils are dated at about 800 million years old (Figure 24–2). It is frequently suggested that cyanobacteria were responsible for the original development of the Earth's atmosphere that made extensive eukaryotic life possible. According to the endosymbiotic theory (Chapter 25), eukaryotes originated from a union of bacteria-like ancestors.

All modern prokaryotes are assigned to their own kingdom, the Monera. On the whole the Monera are distinguished by their negative traits and their simplicity. Their cell volume is only about one-thousandth that of

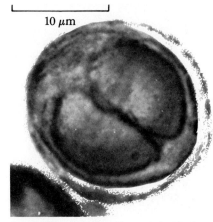

10 µm

FIGURE 24–2 Fossil cyanobacteria, dated at approximately 850 million years ago. This four-celled colony of cyanobacteria surrounded by a sheath or capsule was discovered in a thin sheath of ancient rock in Australia.

small eukaryotic cells, and their length only about one-tenth. Most are unicellular organisms, but some form colonies or filaments containing specialized cells and so are believed by some to be multicellular organisms. The cytoplasm contains ribosomes but usually lacks membrane-bound organelles typical of eukaryotic cells. Thus there are no mitochondria, endoplasmic reticulum, Golgi complex, or lysosomes (Figure 24–3). The genetic material of a prokaryote is contained in a single circular DNA molecule that lies in the cytoplasm, not surrounded by a nuclear membrane. The DNA duplicates before the cell divides asexually by a simple splitting that avoids the complexity of mitosis. The simplicity of their reproduction may be the reason that bacteria can multiply with amazing speed.

Most prokaryotic cells have a cell wall surrounding the cell membrane, but its structure and composition differ from those of eukaryotic cell walls. Some prokaryotes have flagella, but unlike the flagella of eukaryotes (which are composed of two central hollow microtubules surrounded by nine pairs of similar microtubules often connected to other internal tubular structures), these are solid structures made of long filaments of protein. None of the prokaryotes have cilia. However, some (the spirochetes) have batteries of internal filaments that permit the cell to move about and swim. Some species of spirochetes even live attached to other cells, functioning like flagella and permitting the host cell to travel through such places as termites' guts.

Despite the relative simplicity of prokaryotes, the diversity of the microbial world is astonishing. The development of the electron microscope revealed that the blue-green algae, formerly considered "simple plants," were in essence bacteria. As a result, they are now called **cyanobacteria.** Even more fundamental, perhaps, was the discovery that certain odd microbes, now termed the **Archaebacteria,** have molecular traits that seem to set them apart from all other cellular life, including the remainder of the bacteria. Reflecting this new perception, it has been proposed that the bacterial kingdom Monera should be split into two kingdoms or subkingdoms, with the Archaebacteria in one and the remaining prokaryotes in the other.

THE ARCHAEBACTERIA

Under a microscope all bacteria appear fundamentally similar. Biochemically, however, the Archaebacteria are very different from other bacteria. One of their most striking distinguishing features is the absence of pepti-

FIGURE 24–3 Much simpler in their structure than eukaryotes, prokaryotes are mainly distinguished by the things that they lack. The much-folded bacterial chromosome, for instance, floats free in the cytoplasm without a surrounding membrane, so that it must be called a **nucleoid** rather than a nucleus. There are no mitochondria or chloroplasts. Most of the functions performed by elaborate systems of internal membranes in eukaryotes are carried out by the plasma membrane of bacteria. This may be surrounded, however, by several layers of material lacking in eukaryotes. The cell wall maintains turgidity and cellular shape, additional membranes may help to confine protons used in chemiosmosis, and a capsule serves for defense.

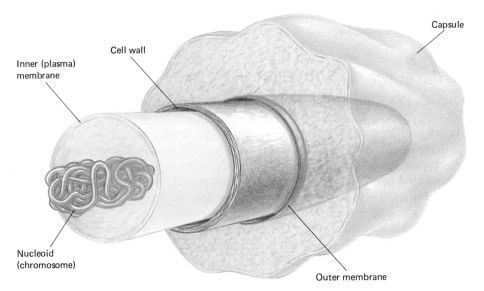

Capsule

Cell wall

Inner (plasma) membrane

Nucleoid (chromosome)

Outer membrane

Focus on CULTURING BACTERIA IN THE LABORATORY

One of the most significant advances in the heroic age of bacteriology was the invention of methods to grow bacteria in pure culture. In nature, many species of bacteria are intermingled, forming mixed communities that are difficult or impossible to study scientifically. In order to perform meaningful experiments, it is necessary to produce a pure culture consisting of a single species. In the laboratory, bacteria can be cultured on a medium of agar (a solidifying agent extracted from seaweed) that is enriched with appropriate nutrients. The medium can be poured into a test tube or into a Petri dish, a flat glass or plastic dish with a cover, as shown in the figure. The agar medium is liquid when hot, but it gels to form a solid as it cools. Bacteria cannot move very much on a solid medium.

To obtain a pure culture of bacteria, a drop of saliva, earth, or some other likely material can be spread or streaked out on the surface of the agar medium with a wire loop or bent glass rod. This thins out the bacteria so that they are separated from one another. The medium containing the bacteria is then incubated at an appropriate temperature for several hours or days. Isolated bacteria multiply, each giving rise to a colony of daughter cells. When about 10 million to 100 million cells are present, the colony is visible to the naked eye as a small, raised, often circular area on the agar medium. The color, shape, and consistency of the colony vary with the species.

Some of the members of a colony can be carefully streaked onto a fresh dish of agar with a sterile wire loop. After incubation, this Petri dish should contain only colonies formed by the desired species. In this way, any bacteria contaminating the original colony can be avoided. The culture is now pure; it contains only a single kind of bacteria.

■ The preparation of bacterial cultures in the laboratory must be performed with exacting but not difficult technique. Using a sterilized wire loop, the technician transfers a drop of broth containing bacteria to the surface of solid medium contained in a Petri dish, holding the cover in such a way as to prevent contamination from the atmosphere. The bacteria are carefully streaked over the surface of the culture medium.

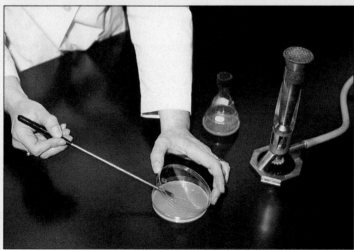

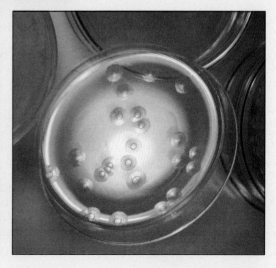

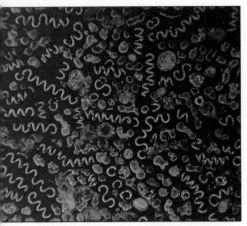

(a)

(b)

FIGURE 24–4 Some living cyano-bacteria, magnified about 60 times. (*a*) *Anabaena spiroides* is the spiral-shaped organism and *Microcystis aeruginosa,* the irregularly shaped one. Both of these species are often toxic. (*b*) Aerial photograph of a hot spring containing halobacteria capable of withstanding conditions hostile to all other known life.

doglycan (page 561) in the cell wall. There are other important (though quite technical) differences in proteins and cell chemistry that set the Archaebacteria apart from other bacteria. Some of these biochemical differences may account for the fact that many Archaebacteria inhabit extreme environments, such as deep-sea hot springs whose temperature may exceed 100°C.

The Archaebacteria include three groups:

1. **Extreme halophiles.** The halobacteria can live only in extremely salty environments such as salt ponds. They are sometimes found in brines used to cure fish, making their presence known by forming red patches. Despite the high salinity of their surroundings, the osmotic pressure of the cytoplasm of halophiles usually lies within the range that is normal for most bacteria. Some of the halophiles are capable of photosynthesis in which the energy of sunlight is captured by the purple pigment **bacteriorhodopsin.**

Bacteriorhodopsin is the sole known kind of protein found in the purple patches of the plasma membrane that lend their color to these bacteria. It is basically a protein consisting of seven subunits that spans the membrane in such a way that a channel is formed in the membrane. Bacteriorhodopsin chemically resembles the rhodopsin found in animal eyes, and like rhodopsin, is excited by light. The energy from a single photon of light suffices to pump two protons from the interior of the cell to the exterior. This pumping is done by the bacteriorhodopsin itself. An electrical and chemical gradient is thus established across the plasma membrane, which is utilized by the cell to generate ATP (Chapter 8).

2. **Methanogens.** These anaerobes produce methane from carbon dioxide and hydrogen. They inhabit sewage and swamps and are common in the digestive tracts of humans and other animals. In such habitats, organic material decomposes under extremely anaerobic conditions. The methanogens are probably the commonest of the archaebacteria.
3. **Thermoacidophiles.** These bacteria normally grow in hot, acidic environments. Some are found in hot sulfur springs.

The biochemical and metabolic differences between the Archaebacteria and other bacteria suggest that these groups may have diverged from each other long ago—relatively early in the history of life. It is argued that many of the extreme conditions to which the modern Archaebacteria are adapted resemble conditions thought to have existed on the primitive Earth—conditions that have persisted in such places as hot springs ever since.

TYPICAL BACTERIA

The bacteria comprise an extremely wide variety of organisms. Here are not only some microorganisms capable of photosynthesis, but some that photosynthesize using hydrogen sulfide rather than water as a hydrogen source. Others keep all life going by bringing the nitrogen of the atmosphere into chemical combination. This diversity seems to demand recognition in the way we discuss bacteria and perhaps in the way we classify them. Our systems of classification may not be "natural," however, in that they probably do not accurately reflect the evolutionary relationships among many of the bacteria.

The Cyanobacteria

The cyanobacteria (formerly known as the blue-green algae) are found in ponds, lakes, swimming pools, and moist soil, as well as on dead logs and the bark of trees. Some also occur in the oceans, and a few species inhabit hot springs. A few types are unicellular, and all are microscopic, but most occur as large globular colonies or as long filaments united by extracellular materials. Some species show a division of labor among members of a colony. In certain species some cells are specialized for fixing nitrogen (i.e., incorporating atmospheric nitrogen into chemical compounds); other cells are specialized to reproduce; and still others may attach the colony to the substrate.

Most cyanobacteria are **photosynthetic autotrophs.** They contain chlorophyll *a*, which is also found in photosynthetic eukaryotes. They have several varieties of accessory pigments, including carotenoids, which are present in some eukaryotes and some bacteria. **Phycocyanin,** a blue pigment, is found only in cyanobacteria and red algae. Some have a red pigment, **phycoerythrin,** that is also present in red algae (this has been suggested as evidence for an evolutionary relationship between the two groups). Chlorophyll and the accessory pigments, when they occur, are not enclosed in plastids as they are in algae and in plant cells, but are dispersed along membranes in the periphery of the cell or stacked in the cytoplasm.

As far as is known, cyanobacteria reproduce only asexually, usually by fission. Colonies may fragment, and then cells of each of the separated parts reproduce to form new colonies. In some species cells are capable of turning into thick-walled **endospores** that are highly resistant to adverse conditions. They may remain dormant for many years until conditions are favorable for them to germinate and give rise to new colonies of cells.

Structure of Cyanobacteria

Like other monerans, cyanobacteria lack a nuclear membrane and other membranous organelles such as mitochondria and chloroplasts. However, unlike other bacteria, cyanobacteria have internal membranes called **photosynthetic lamellae,** which contain chlorophyll and enzymes needed for photosynthesis (Figure 24–5).

The tough cell wall does not contain cellulose, but consists of other polysaccharides linked with polypeptides. Many of the cyanobacteria secrete a sticky gelatinous substance, which may form a sheath around the cell wall. The gelatinous material often contains pigments and may also contain toxins that prevent fish and other organisms from feeding on them.

Despite their name, only about half of the "blue-green algae" are actually blue-green. The coloration is modified by the photosynthetic pigments within the cell, producing brown, black, purple, yellow, blue, green, or even red individuals. The Red Sea acquired its name from red cyanobacte-

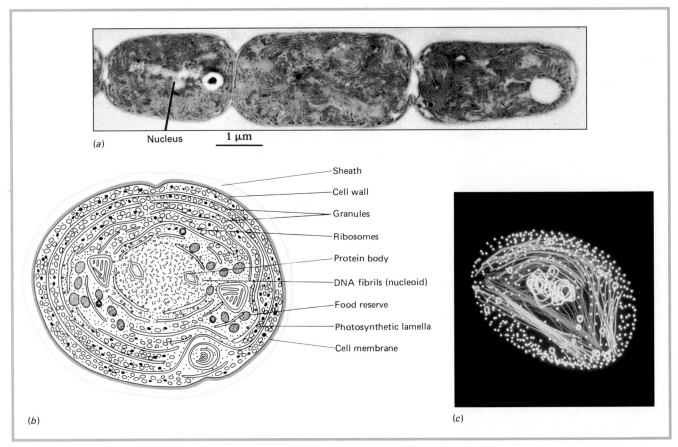

(a) Nucleus   1 μm

Sheath
Cell wall
Granules
Ribosomes
Protein body
DNA fibrils (nucleoid)
Food reserve
Photosynthetic lamella
Cell membrane

(b)   (c)

FIGURE 24–5 Structure of cyanobacteria. (*a*) Electron micrograph of *Anabaena*. The pale area is nucleic acid, but no nuclear membrane is present. There are membranous lamellae in the cytoplasm, however, which function in photosynthesis somewhat as do the thylakoid membranes in plant chloroplasts. Gas vacuoles help to keep the organisms afloat in the sunlit layers of water. (*b*) Diagram of the structure of a cyanobacterium. (*c*) Computer-generated reconstruction of a unicellular cyanobacterium.

ria, which sometimes occur there in such great numbers that they color the water red.

Cyanobacteria do not have flagella, but some of the filamentous species are capable of a curious back-and-forth oscillating movement. Others have a slow, gliding motion.

Significance of Cyanobacteria

As producers, cyanobacteria provide oxygen and organic material for other organisms. Many species can **fix nitrogen**—that is, incorporate atmospheric nitrogen into inorganic compounds that can be used by plants. This process enriches soil and sea bed. Owing to the presence of cyanobacteria in the rice paddies of Southeast Asia, rice can be grown on the same land for many years without adding nitrogen fertilizer. In some regions cyanobacteria are now added to the soil to increase crop yield. Some (*Spirulina*) have been used directly as human food. Since these organisms require nothing but sunlight, alkaline water, nitrogen, and carbon dioxide from the atmosphere for their growth, they might serve as an extremely economical and surprisingly rich source of dietary protein.

Cyanobacteria form symbiotic relationships with many organisms including protists, fungi, and some plants. Together with fungi they form some kinds of lichens (discussed in Chapter 26). Those that form cooperative relationships within the cells of other organisms, known as **endosymbiosis,** usually lack a cell wall and function quite like chloroplasts within the host cell, producing food for their symbiotic partner. One such recently discovered endosymbiote, *Prochloron* (Figure 24–6), lives in the cloacal cavity of a species of tunicate (a type of marine invertebrate, Chapter 31). *Prochloron* is more than a mere curiosity, for it possesses both chlorophyll *a* and chlorophyll *b*, much like the chloroplasts of plants and protists. It has

FIGURE 24–6 *Prochloron*, a recently discovered endosymbiotic cyanobacterium, lives in the cells of the cloacas of these Australian tunicates, producing their green color.

been suggested that *Prochloron* is a surviving member of the now otherwise extinct group of photosynthetic prokaryotes that gave rise to plant chloroplasts.

Some cyanobacteria can tolerate extremes of salinity, temperature, and pH that would kill true algae and many other organisms. In fact, cyanobacteria not only are able to thrive in polluted lakes and ponds but often become the dominant species under such conditions. Fish find them largely inedible, and cyanobacteria may reproduce so extensively that they form "blooms" (population explosions) in the water. The bloom may cause the water to become extremely turbid, limiting the penetration of sunlight. Many of the cyanobacteria die as a result of the crowding and shading. Their decomposition by heterotrophic bacteria consumes large amounts of oxygen from the water and may result in fish kills. Some cyanobacteria also produce toxic metabolic products that kill fish or any animal that takes in the water (Figure 24–7).

The Remaining Bacteria

The eubacteria are the classic bacteria studied by the pioneer microbiologists. Most bacteria belong to this group, and when bacteria are casually mentioned, eubacteria are usually what is meant. Early investigators

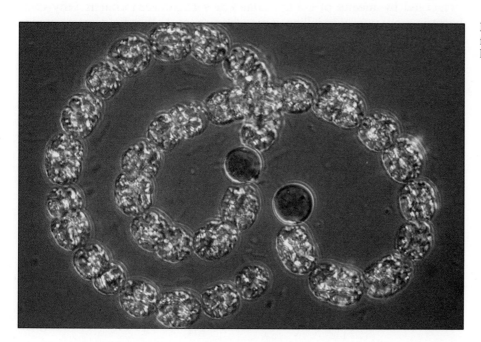

FIGURE 24–7 *Anabaena circinalis*, a filamentous cyanobacterium with heterocysts.

quickly realized that bacteria are present almost universally, being abundant in air, in liquids such as milk, and in and on the bodies of plants and animals, both living and dead. In fact, relatively few places in the world are devoid of bacteria, for they can be found in fresh and salt water, as far down as several meters deep in the soil and in deep gravel aquifers, in the ice of glaciers, and even in oil deposits far underground.

When conditions are especially adverse, bacteria of many species can form naturally dehydrated endospores and remain in a state of suspended animation—sometimes for years—until environmental conditions are favorable. Bits of rock and ice drilled from depths of 430 m in Antarctica contained bacteria that had lain dormant at very low temperatures for at least 10,000 years! When these bacteria were brought to more favorable temperatures, they became activated and resumed normal metabolic activities.

Structure of Typical Eubacteria

Most bacterial species exist as single-celled forms, but some are found as colonies or as filaments of joined cells. The cell membrane, the active barrier between the cell and the external environment, governs the passage of molecules moving into and out of the cell. Enzymes needed for the operation of the electron transport system (which in eukaryotic cells are found in mitochondria) are attached to the cell membrane.

The cell wall surrounding the cell membrane provides a strong, rigid framework that supports the cell, maintains its shape, and, probably most important, keeps it from bursting because of osmotic pressure. (Most bacteria seem to be adapted to hypotonic surroundings.) The great strength of the bacterial cell wall may be attributed to the properties of **peptidoglycan,** a macromolecule found only in prokaryotes. Peptidoglycan consists of two unusual types of sugar linked with short peptides. The peptides contain two amino acids found only in bacterial cell walls. The sugars and peptides are linked to form a single macromolecule that surrounds the entire cell membrane.

Normally bacteria cannot survive without their cell walls. When wall-less forms are created experimentally, they must be maintained in isotonic solutions to keep them from bursting. Cell walls are of little help when the bacterium is confronted with a hypertonic environment, as found in food preserved by means of a high sugar or salt content. That is why most bacteria grow poorly in jellies, jams, salt fish, and other foods preserved in this way.

Almost 100 years ago the Danish physician Christian Gram developed a procedure that is still used to divide bacteria into two groups based on their staining properties. Bacteria that absorb and retain crystal violet stain during laboratory staining procedures are referred to as **gram-positive,** whereas those that do not retain the stain are **gram-negative.** The cell walls of gram-positive bacteria are very thick and consist primarily of peptidoglycan. Examples of gram-positive bacteria are *Streptococcus* and *Staphylococcus.* The envelope of a gram-negative bacterial cell consists of three layers: an inner cell membrane, a thin peptidoglycan layer, and a thick outer layer of lipoprotein and lipopolysaccharide, a lipid-polysaccharide complex (Figure 24–8). *Escherichia coli* and *Salmonella* are among the gram-negative bacteria.

The differences in composition of the cell wall of gram-positive and gram-negative bacteria are of great practical importance in our dealings with them. Each type of cell wall confers certain advantages on the bacteria that possess it, and such traits are not necessarily advantageous to us! The thick peptidoglycan layer of the gram-positive bacterium makes its cell stronger and less likely to break as a result of physical stress. However, the

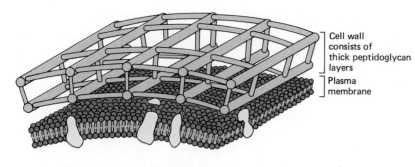

Cell wall
consists of
thick peptidoglycan
layers

Plasma
membrane

Gram positive wall

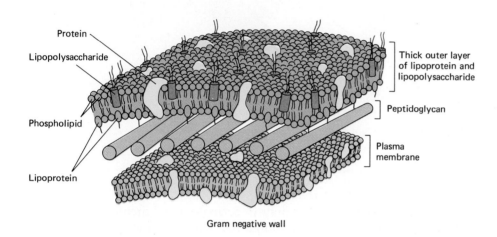

Protein

Lipopolysaccharide

Thick outer layer
of lipoprotein and
lipopolysaccharide

Peptidoglycan

Phospholipid

Plasma
membrane

Lipoprotein

Gram negative wall

FIGURE 24–8 A simplified, schematic representation of the peptidoglycan layer in gram-positive and gram-negative bacterial cell walls. The gram-negative bacterium has a second lipid bilayer outside the cell wall which resembles the plasma membrane.

gram-positive cell wall is more susceptible to attack by lysozyme, a naturally occurring hydrolytic enzyme (present, for example, in human tears). The peptidoglycan layer of gram-negative bacteria is protected by the semipermeable outer membrane, which lysozyme does not penetrate.

The antibiotic penicillin interferes with peptidoglycan synthesis, ultimately resulting in a fragile cell wall that cannot effectively protect the cell (Figure 24–9). Penicillin works most effectively against gram-positive bacteria. Nongrowing cells and cells that produce the enzyme penicillinase (also known as beta-lactamase) are not affected by penicillin.

A few species of bacteria produce a capsule or slime layer that surrounds the cell wall. Such a capsule may provide added protection for the cell against phagocytosis both by other microorganisms (in free-living species) and by their host's white blood cells (in the case of pathogenic bacteria). Some kinds of bacteria that normally have capsules also exist in unencapsulated strains maintained in laboratory culture. Normally virulent *Streptococcus pneumoniae* bacteria, for instance, are harmless if they have no capsules. As you may recall, this fact is of considerable significance in the history of molecular genetics (Chapter 12).

The dense cytoplasm of the bacterial cell contains ribosomes and storage granules that hold glycogen, lipid, or phosphate compounds. Although the membranous organelles of eukaryotic cells are absent, in some bacterial cells the cell membrane is elaborately folded inwardly. Such complex extensions of the cell membrane are called **mesosomes**. Mesosomes may function in a variety of metabolic processes and are definitely used in cell division. Respiratory enzymes *may* sometimes be associated with these membranes, and in some bacteria cellular respiration may take place there. During cell division the mesosome appears to be involved in the formation of a septum (wall) between the two new daughter cells. Other internal membranes include those mentioned in connection with photosynthesis in some bacteria, and some occurring in nitrogen-fixing bacteria.

FIGURE 24–9 Bacterium treated with a beta-lactamase–inhibiting antibiotic such as penicillin. The antibiotic interferes with the production of new cell walls. As a result, when the bacterium next divides, the new cells are not able to separate or to function normally. Arrows point to abnormal or abortive walls between daughter cells.

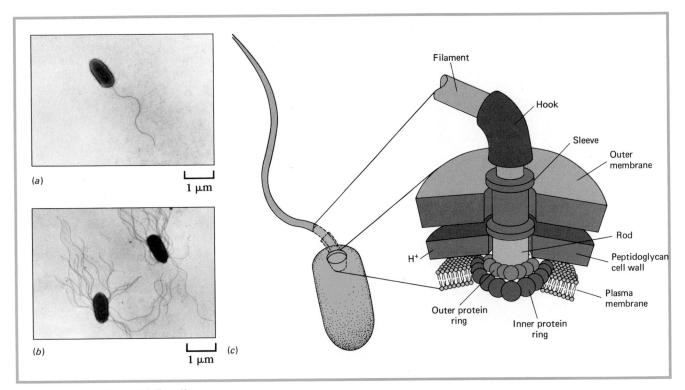

(a)

1 μm

(b)

1 μm

(c)

Filament

Hook

Sleeve

Outer membrane

Rod

Peptidoglycan cell wall

Plasma membrane

H⁺

Outer protein ring

Inner protein ring

FIGURE 24–10 Bacterial flagella. (a) A single flagellum at the end of the bacterium *Pseudomonas aeruginosa* (approximately × 5000). (b) In the bacterium *Proteus mirabilis,* flagella project from many surfaces of the cell (approximately × 3500). (c) Structure of bacterial flagellum. Diffusion of protons powers protein "motor" that spins the flagellum like a propeller.

Bacterial DNA is found mainly in a single long, circular molecule referred to as a chromosome (although histones and other proteins are not associated with it as they are in the eukaryotic chromosome). When stretched out to its full length, the bacterial chromosome is about 1000 times longer than the cell itself. A small amount of genetic information may be present as smaller DNA molecules, called **plasmids,** which replicate independently of the chromosome. Bacterial plasmids often bear genes involved in resistance to antibiotics, and are often considered to be intracellular genetic parasites or mutualists comparable to viruses.

Many types of bacteria propel themselves by means of whiplike flagella (Figure 24–10). Some species have a single flagellum or a bundle of them at one end of the cell, while others have flagella distributed over the entire surface. Their arrangement is characteristic of the species and therefore useful in identification.

As we have seen, bacterial flagella are distinctive in that they consist of a single fibril, whereas the flagella of eukaryotes are composed of 11 microtubules. At the base of a bacterial flagellum is a complex structure that produces a rotary motion, pushing the cell much as a ship is pushed along by its propeller. In this way some bacteria can travel as much as 2000 times their own length in an hour. (Imagine a man 2 m tall swimming 4 km per hour in a viscous syrup simply by twirling a long whip!)

Many gram-negative bacteria have hundreds of hairlike appendages known as **pili** (Figure 24–11). These structures are organelles of attachment that help the bacteria adhere to certain surfaces, such as the cells they will infect. Pili (so-called F-pili) are also sometimes involved in the transmission of plasmids (Chapter 15) between bacteria.

Bacterial Metabolism

A bacterial cell contains about 5000 different chemical compounds. What each of these does, how they interact, and how the bacterium synthesizes them from the nutrients it takes in are complex biochemical problems that have absorbed researchers for years. Much of the knowledge that has been

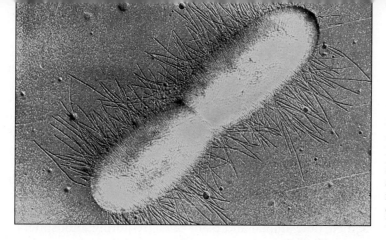

FIGURE 24–11 False color transmission electron micrograph of the bacterium *Escherichia coli*. The fine hairlike structures are probably sex pili, although other kinds of pili are also used by harmful bacteria to attach to host tissues as a preliminary to infection. This bacterium is dividing into two daughter cells ($\times$ 14,700).

gained from studying these mechanisms in bacterial cells has been successfully applied to cells of humans and other organisms, for there is surprising uniformity in basic biochemical processes.

Most bacteria are **heterotrophs,** obtaining preformed organic compounds from other organisms. And the majority of heterotrophic bacteria are free-living **saprobes,** organisms that get their nourishment from dead organic matter. Other species of heterotrophic bacteria live in symbiosis with other organisms. These symbionts may be **commensals,** which neither help nor harm their hosts (Figure 24–12). A few are parasites, which live at the expense of their host and cause diseases in plants and animals. Such disease-causing organisms are known as **pathogens.** Many symbiotic bacteria inhabit the human skin, digestive tract, and other areas of the body. They are part of the normal community of microorganisms that colonize our bodies. Most are commensals that normally do us no harm.

More than one group of photosynthetic bacteria exists. Altogether, in fact, there are four: the green sulfur bacteria, the purple sulfur bacteria, the green nonsulfur bacteria, and the purple nonsulfur bacteria, these last belonging to the archaebacteria. Some bacterial photosynthesis differs in two important ways from photosynthesis carried on by algae, plants, or cyanobacteria. First, the chlorophyll present in these bacteria absorbs light most strongly in the near-infrared portion of the light spectrum rather than in the visible light range. This enables them to carry on photosynthesis in red light that would appear very dim to human eyes.

Second, noncyanobacterial photosynthesis by bacteria does not produce oxygen because water is not used as a hydrogen donor. Instead, the sulfur bacteria use sulfur compounds such as hydrogen sulfide (H_2S) as hydrogen donors. (Remember, we are not now discussing the photosynthesis of cyanobacteria, which resembles that of plants.) Photosynthetic sulfur bacteria produce free sulfur as a waste product, somewhat as cyanobacteria and plants produce oxygen. Accumulation of such sulfur has produced important deposits that have been commercially mined.

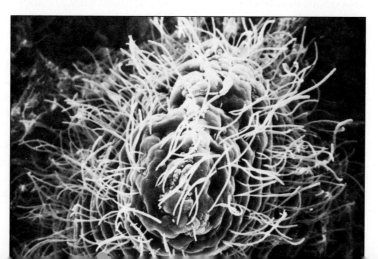

FIGURE 24–12 Numerous bacterial colonies attached to the intestinal lining of a dog (approximately $\times$ 550). Although by no means all such attachments result in infection, attachment is frequently the first step in host invasion.

Some bacteria are **chemosynthetic autotrophs,** or **chemoautotrophs.** Rather than carrying on photosynthesis, they produce their own food from simple inorganic ingredients with energy obtained from oxidizing inorganic compounds. Chemosynthetic bacteria absorb carbon dioxide, water, and simple nitrogen compounds from their surroundings. From these, they manufacture complex organic substances with energy they obtain from oxidation of ammonia to nitrites and nitrates, from the oxidation of sulfur or iron compounds, or from oxidation of gaseous hydrogen. Some of these bacteria play an important role in the nitrogen cycle (Chapter 56).

Whether they are heterotrophs or autotrophs, most bacterial cells are aerobic (like animal and plant cells), requiring atmospheric oxygen for cellular respiration. Some bacteria are **facultative anaerobes,** meaning they can use oxygen for cellular respiration if it is available, but can carry on metabolism anaerobically when necessary. Other bacteria are **obligate (strict) anaerobes,** which carry on energy-yielding metabolism only anaerobically. Some of these bacteria grow more slowly in the presence of oxygen; a few are actually killed by even low concentrations of oxygen gas.

In anaerobic metabolism, bacteria obtain their energy by the anaerobic degradation of carbohydrates or amino acids. In the process they accumulate a variety of partially oxidized intermediates such as ethanol, glycerol, and lactic acid. The anaerobic metabolism of carbohydrates is termed **fermentation,** and the anaerobic metabolism of proteins and amino acids is called **putrefaction.** The foul smells associated with the decay of food, wastes, or corpses are due to nitrogen- and sulfur-containing compounds formed during putrefaction.

Bacterial Reproduction

Bacteria generally reproduce asexually by **transverse binary fission,** in which the cell develops a transverse cell wall and then divides into two daughter cells. After the circular bacterial "chromosome" has been replicated, the transverse wall is formed by an ingrowth of both the cell membrane and the cell wall. If the newly formed cell wall does not separate completely into two walls, a chain of bacteria may be formed. The distribution of the two copies of the single chromosome is facilitated by the mesosome, which is really an elaborate connection between each chromosome and the cell membrane. The duplication of the chromosome and the division of the cell often get out of phase, so that a bacterial cell may have from one to four or even more identical chromosomes.

Bacterial cell division can occur with remarkable speed, and some species grown in an appropriately fortified and aerated culture medium can divide every 20 minutes. At this rate, if nothing interfered, one bacterium would give rise to some 250,000 bacteria within 6 hours. This explains why the entrance of only a few pathogenic bacteria into a human being can result so quickly in the symptoms of disease. Fortunately, bacteria cannot reproduce at this rate for very long, because they are soon checked by lack of food or by the accumulation of waste products. Since pathogenic bacteria have their nutritional and other needs met by the host, some other defense mechanisms must deal with the invaders (Chapter 43). Upon infection, however, reptiles, birds, and mammals are able to reduce their plasma content of iron, an essential bacterial nutrient. These vertebrates also increase their body temperature (running a fever) when infected. The increased temperature seems to interfere with the ability of bacteria to absorb any remaining iron from the body fluids.

Although complex sexual reproduction involving fusion of gametes does not occur in monerans, genetic material is sometimes exchanged between individuals. Such genetic recombination can take place by three different mechanisms: transformation, conjugation, and transduction. In

transformation, fragments of DNA released by a broken cell are taken in by another bacterial cell. This mechanism has been used experimentally to show that genes can be transferred from one bacterium to another and that DNA is the chemical basis of heredity.

In **conjugation,** two cells of different mating types (the equivalent of sexes?) come together and genetic material is transferred from one to another (Figure 24–13). Conjugation has been most extensively studied in the bacterium *Escherichia coli,* of which there are F^+ strains and F^- strains. F^+ individuals contain a plasmid known as the F factor, which is capable of organizing special hollow pili, the **sex pili,** or better, the **F pili.** These pili serve as conjugation bridges that pass from the F^+ to the F^- cell. The F pili are long and narrow, and have an axial hole through which fragments of DNA may pass from one bacterium to the other. Many biologists view plasmids as genetic parasites or commensals comparable to viruses, rather than as "accessory bacterial chromosomes." By this view, the plasmid induces the pilus as a means of infecting new host cells, and the transfer of other genes may incidentally occur. Most strains even of *E. coli* never develop sex pili, however, and the phenomenon is probably best viewed as a curiosity of great research value and of use in genetic engineering, but with little significance in nature.

In the third process of gene transfer, **transduction,** bacterial genes are carried from one bacterial cell into another within a bacteriophage, as we saw in Chapter 14.

Spore Formation

When the environment of a bacterium becomes extremely unfavorable, such as when it becomes very dry, many species become dormant. In such cases the cell loses water, shrinks slightly, and remains quiescent until water is again available. Other species form endospores—dormant, resting cells—to survive in extremely dry, hot, or frozen environments, or when food is scarce. Endospores are not exactly comparable to the spores of fungi and plants, and endospore formation is not really a kind of reproduction in bacteria, since only one endospore is formed per cell. The total number of individuals does not increase as a result of endospore production. The outer coat of the endospore develops inside the cell (Figure 24–14), surrounding the DNA and a small amount of cytoplasm. Some endospores are so strong that they can survive an hour or more of boiling, or centuries of freezing. When the environmental conditions are again suitable for growth, the endospore can absorb water, break out of its inner wall, and become an active, growing bacterial cell again.

Some Important Groups of Bacteria

Bacteria are difficult to classify because their evolutionary relationships to one another are not known, although newer biochemical criteria are being increasingly applied to this problem. Taxonomic ranks higher than the genus have limited significance. According to the current standard work on bacterial classification (*Bergey's Manual of Systematic Bacteriology,* 1984), bacteria are classified in 33 groups called **sections.** They are grouped on the basis of common shape and structure, biochemical properties, genetic characteristics, nutritional needs, habitat, and sensitivities to specific drugs.

In the following discussion, the several groups of bacteria are each given phylum rank. Many bacteriologists would rather call them classes or some other name. However they are ranked, though, there is general agreement on the *names* of the groups, which are made up of member species that do have enough in common to be placed together. Groups that have already been discussed (such as the cyanobacteria) are omitted.

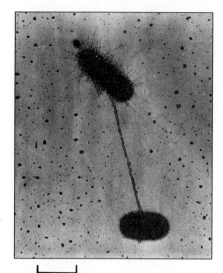

$\vdash\!\!\!-\!\!\!\dashv$
2μm

FIGURE 24–13 F-pilus connecting *Escherichia coli* bacteria. Plasmid DNA is transferred during this process. Bacterial viruses have attached to the pilus and are visible as tiny bumps.

FIGURE 24–14 Endospore within a cell of *Clostridium*. Each cell contains only one endospore, a resistant, dehydrated remnant of cytoplasm plus the nucleoid.

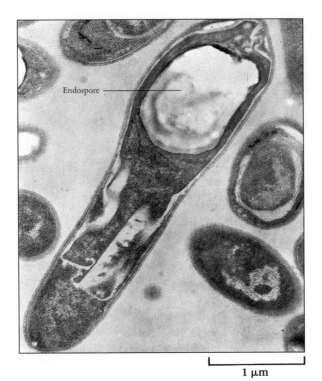

Endospore

1 μm

FIGURE 24–15 Three characteristic bacterial shapes: (*a*) Cocci. (*b*) Bacilli. (*c*) Spirilla (*Spiroplasma*).

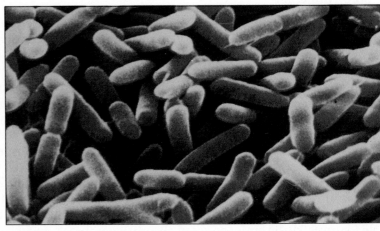

(*a*)   (*b*)

(*c*)

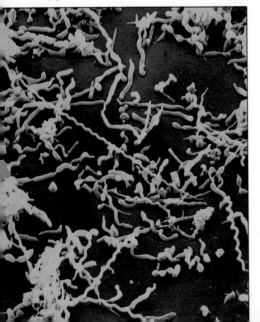

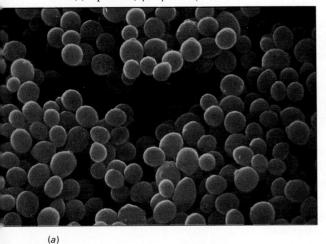

EUBACTERIA.   The **eubacteria,** or true bacteria, are a diverse group that nevertheless conform to certain general characteristics. They are often classified on the basis of their staining properties, type of metabolism, and shape. The three main shapes—as shown in Figure 24–15—are spherical, rod-shaped, and spiral. Spherical bacteria, known as **cocci** (singular **coccus**), occur singly in some species, in groups of two in others, in long chains (e.g., streptococci), or in irregular clumps that look like bunches of grapes (e.g., staphylococci). Rod-shaped bacteria, called **bacilli** (singular, **bacillus**), may occur as single rods or as long chains of rods. Diphtheria, typhoid fever, and tuberculosis are all caused by bacilli. Spiral bacteria are known as **spirilla** (singular, **spirillum**). Short, incomplete spiral-shaped bacteria are known as **comma bacteria** or **vibrios.** There is no reason to think these represent "natural" groups in the evolutionary sense, but the scheme is nevertheless useful and practical.

Most eubacteria are not pathogenic. Rather, they are harmless saprobes that are ecologically important as decomposers. The lactic acid bacteria are gram-positive bacteria that produce lactic acid as the main end-product of their fermentation of sugars. Among these are the rod-shaped **lactobacilli,** which may be found in decomposing plant material, milk, yogurt, and other dairy products. They are commonly present in animals and are among the normal inhabitants of the human mouth and vagina.

Streptococci, also gram-positive bacteria, are found in the mouth as well as in the digestive tract. Among the harmful species of streptococci are those that cause "strep throat," scarlet fever, wound infections, and skin, ear, and other infections. **Staphylococci** are gram-positive bacteria that normally live in the nose and on the skin. **Opportunistic** species such as these can cause disease when the immunity of the host is lowered. *Staphylococcus aureus* may cause boils and skin infections or may infect wounds. Certain strains (varieties) of *Staphylococcus aureus* cause food poisoning, and some are thought to cause toxic shock syndrome. The **clostridia** are a notorious group of anaerobic gram-positive eubacteria. One species causes tetanus; another causes gas gangrene; and *Clostridium botulinum,* an inhabitant of the soil, can cause botulism, an often fatal type of food poisoning.

The gram-negative bacteria exhibit differences in shape, structure, and metabolic processes. Among them are the **azotobacteria,** which have the ability to fix nitrogen. Others are chemoautotrophs, which, as described earlier, obtain energy from oxidizing inorganic compounds. Many important pathogens are gram-negative eubacteria. The gram-negative coccus **Neisseria gonorrhoeae,** for instance, causes gonorrhea. *Hemophilus influenzae* is a gram-negative bacillus that can cause infections of the respiratory tract and ear, as well as meningitis. The **enterobacteria** are a group of gram-negative rods that include free-living saprobes, plant pathogens, and a variety of symbionts that inhabit humans. *Escherichia coli,* a member of this group, inhabits the intestines of humans and other animals as part of the normal microbe population. Some strains of the enterobacterium *Salmonella* cause food poisoning.

MYXOBACTERIA.   The **myxobacteria** are unicellular short rods resembling the bacilli, but they lack a rigid cell wall. They excrete slime; when they are cultured in a Petri dish, their growth is marked by a spreading layer of slime. These bacteria glide or creep along.

Most myxobacteria are saprobes that break down organic matter in the soil, manure, or rotting wood that is home to them. Some species can break down complex substrates such as cellulose and bacterial cell walls. A few of these prey on other bacteria.

In some species, reproduction is more complex than in other bacteria. Cells swarm together to form masses, which may develop into reproductive structures called **fruiting bodies** (Figure 24–16). During this process, many bacterial cells are changed into **resting cells** (equivalent to spores)

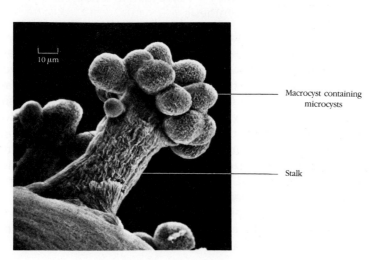

10 μm

Macrocyst containing microcysts

Stalk

FIGURE 24–16 Myxobacteria aggregate at a certain time in their life cycle and form a fruiting body. Protective resting cells, called microcysts, are formed that are very resistant to heat and drying. The fruiting body shown here was formed by *Chondromyces crocatus.*

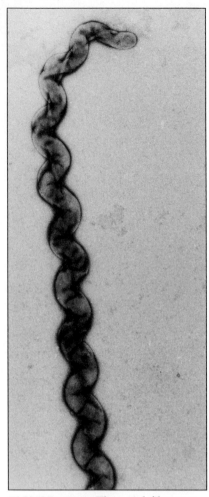

FIGURE 24–17 The axial filament is clearly visible in this photomicrograph of *Leptospira*.

FIGURE 24–18 Scanning electron micrograph of *Mycoplasma pneumoniae*, a pathogenic microorganism. Notice irregular, almost filamentous shapes of these prokaryotes (× 10,000).

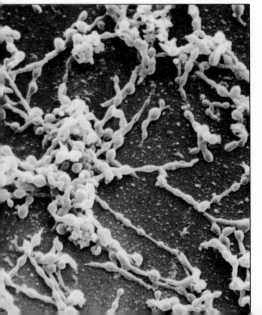

and are packaged within a protective wall, forming a cyst. When conditions are favorable, the cyst breaks open, and the resting cells become active. Although eukaryotic slime molds (Chapter 27) cannot be closely related to myxobacteria, their life cycles are somewhat similar to that of myxobacteria.

SPIROCHETES. **Spirochetes** are slender, spiral-shaped bacteria with flexible cell walls. They move by means of a unique structure called an **axial filament** or **periplasmic flagellum**. Some species inhabit freshwater and marine habitats, others form commensal associations, and a few are parasitic. The spirochete of greatest medical importance is *Treponema pallidum*, the pathogen that causes syphilis (discussed in Chapter 50).

ACTINOMYCETES. **Actinomycetes** resemble molds (which are fungi) in that their cells remain together to form branching filaments. Many produce moldlike spores, called **conidia**. However, they produce cell wall peptidoglycans, lack nuclear membranes, and have other prokaryote characteristics.

The actinomycetes perform much of the decomposition of organic materials in soil. Most members of this group are saprobes, and some are anaerobic. Several species of the group known as **Streptomyces** produce antibiotics; streptomycin, erythromycin, chloramphenicol, and the tetracyclines are among the antibiotic drugs derived from these bacteria. In fact, most known antibiotics, both medically usable and not, are of actinomycete origin.

Some actinomycetes do cause disease. *Mycobacterium tuberculosis* is the bacillus that causes human tuberculosis. *Mycobacterium leprae* causes Hansen's disease (leprosy). Other actinomycetes can cause serious lung disease or generalized infections in humans and animals.

MYCOPLASMAS. The **mycoplasmas** are tiny bacteria bounded by a pliable cell membrane, but lacking a typical bacterial cell wall. Some are so small that, like viruses, they pass through bacteriological filters. In fact, mycoplasmas are smaller than some viruses. Mycoplasmas may be the simplest form of life capable of independent growth and metabolism.

Mycoplasmas may be aerobic or anaerobic, depending on the species. Some live in soil, others in sewage. Still others are parasitic on plants or animals. Some species of *Mycoplasma* inhabit human mucous membranes but do not generally cause disease. One species causes a mild type of bacterial pneumonia in humans that until recently was thought to be caused by a virus (Figure 24–18). This disease accounts for approximately 20% of all cases of pneumonia in some areas. Since the mycoplasmas lack a cell wall, they are resistant to penicillin and other antibiotics that act on the cell walls. However, they are sensitive to tetracycline and other antibiotics that inhibit protein synthesis.

RICKETTSIAS. **Rickettsias** are small bacteria that cannot carry on metabolism independently and so are obligate intracellular parasites (meaning that they must live as parasites in order to survive). It has proved almost impossible to culture rickettsias on nonliving media. Most parasitize certain arthropods such as fleas, lice, ticks, and mites without causing specific disease in them. Diseases caused by the few species known to be pathogenic to humans (and other animals) are transmitted by arthropod **vectors** (carriers), through bites or contact with their excretions. Among these are typhus, Rocky Mountain spotted fever, and Q fever. In fact, Howard Ricketts, who discovered the rickettsia, died in Mexico in 1910 of typhus fever while studying the organisms that cause it.

Rickettsias are typically short gram-negative rods with rigid cell walls. They are not motile and do not form spores. Reproduction takes place by binary fission.

CHLAMYDIAS. **Chlamydias** differ from rickettsias in that they are spherical rather than rod-shaped. In addition, they do not depend upon arthropod vectors for transmission. Because they are obligate intracellular parasites, chlamydias were considered for many years to be large viruses. However, they are now classified as bacteria because they contain both DNA and RNA, possess ribosomes, synthesize their own proteins and nucleic acids, and are sensitive to a wide range of antibiotics. Although they do contain many enzymes and can carry on some metabolic processes, chlamydias are completely dependent on the host cell for ATP. In other words, they are energy parasites.

Studies indicate that chlamydias infect almost every species of bird and mammal. Perhaps 10% to 20% of the human population of the world is infected. Interestingly, though, individuals may be infected for many years without apparent harm. Sometimes chlamydias do cause acute infectious diseases. **Trachoma,** the leading cause of blindness in the world, is caused by a strain of *Chlamydia.* These bacteria also cause a contagious venereal disease (lymphogranuloma venereum) and psittacosis, a disease transmitted to humans by birds; and they may be responsible for many cases of pelvic inflammatory disease, infections of the urethra, and other urogenitally associated infections.

Ecological Importance of Bacteria

Bacteria play an essential role as decomposers in our biosphere. Without bacteria (and the fungi), all available carbon, nitrogen, phosphorus, and sulfur would eventually be tied up in the wastes and dead bodies of plants and animals. Life would soon cease because of the lack of raw materials for the synthesis of new cellular components.

Some bacteria can fix nitrogen from the air. They bring nitrogen gas into chemical combination by reducing it to ammonia. Other bacteria convert ammonia to nitrites and nitrates (Figure 24–20). Some nitrogen-fixing bacteria are free-living. Others form symbiotic relationships with certain plants. Some grow as nodules on the roots of beans, peas, clover, and other leguminous plants; these bacteria are so efficient at their task that

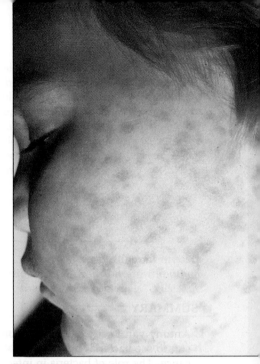

FIGURE 24–19 Patient suffering with Rocky Mountain spotted fever.

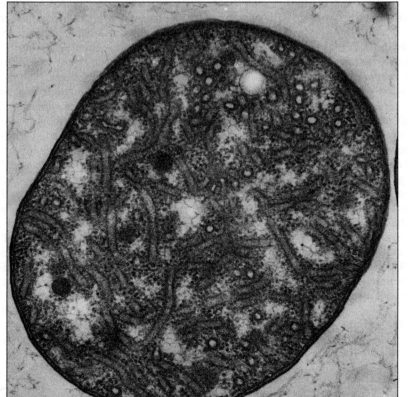

FIGURE 24–20 Some prokaryotes *do* have internal membranes! An electron micrograph of a thin section of a marine nitrifying bacterium, *Nitrococcus mobilis,* that oxidizes nitrites to nitrates. The micrograph shows both longitudinal and cross sections of the curious tubular membranes present in these cells. These membranes contain the enzymes of the electron transport system. Ribosomes show as small, dense dots. Most prokaryote internal membranes surrounding cavities are actually more or less elaborate inpocketings of the plasma membrane.

569

Light micrograph of diatoms

25

The Protist Kingdom

■ LEARNING OBJECTIVES

After you have read this chapter you should be able to:

1. Characterize the common features of the Kingdom Protista.
2. Discuss in general terms the diversity inherent in this kingdom, including modes of nutrition, morphologies, and methods of reproduction.
3. Describe some of the evolutionary relationships among the various protists.
4. Summarize current theories on the origin of eukaryotic cells, and multicellularity within the protists.
5. Discuss the evolutionary relationships of certain protists with the other kingdoms.
6. Briefly describe representative protozoa phyla.
7. Briefly characterize representative groups of algae.
8. Briefly discuss representative fungal-like protists.

FIGURE 25–1 Various protists. The protists are an extremely diverse group of organisms. Some are photosynthetic, while others are heterotrophic.

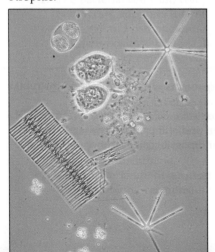

CHARACTERISTICS OF THE KINGDOM PROTISTA

The Kingdom Protista consists of a vast assemblage of eukaryotic organisms whose very diversity makes them difficult to characterize (Figure 25–1). Protistologists estimate that there are as many as 200,000 extant and extinct species of **protists.** The major feature they possess, eukaryotic cellular structure, is shared with animals, plants, and fungi. Eukaryotic cells have true nuclei and other membrane-bounded organelles such as mitochondria and plastids. Their nuclei divide by meiosis and mitosis, although there are variations in the exact process. This characteristic makes the separation between the protists and the Kingdom Monera quite distinct, however.

Size varies considerably within this kingdom, from single-celled protozoa to kelps, giant brown algae that can reach 60 meters in length. Most of the protists are microscopic, single-celled organisms. However, some have

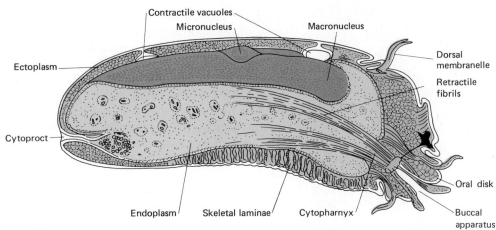

Contractile vacuoles — Micronucleus — Macronucleus

Ectoplasm —

Dorsal membranelle

Retractile fibrils

Cytoproct —

Endoplasm — Skeletal laminae — Cytopharynx

Oral disk

Buccal apparatus

FIGURE 25–2 Even though composed of a single cell, *Epidinium ecaudatum*, a ciliated protozoon, possesses a strikingly complex internal organization. *Epidinium* is a commensal in the rumen (stomach) of cows.

a colonial organization, some are **coenocytic** (multinucleate but not multicellular), and some are multicellular. Most multicellular protists have relatively simple body forms, without specialized tissues.

The word "protist" comes from Greek meaning "the very first." Protists are considered to be simple eukaryotic organisms. However, the cellular organization of the single-celled protists is more complex than that of individual plant, animal, or fungal cells (Figure 25–2). The multicellular eukaryotes have different cells, tissues, organs, and organ systems to perform the various functions of a living organism. The single-celled protists accomplish the same functions within one cell. For example, water regulation in these organisms is often controlled by special organelles, **contractile vacuoles** (Figure 25–3). Because freshwater protists continually take in water by osmosis, the contractile vacuole is needed to remove excess water. Other protists solve the water "problem" with rigid cell walls that limit absorption of water by osmosis.

Methods of obtaining nutrients are quite variable. The autotrophic protists have chlorophyll and photosynthesize like plants. Some of the heterotrophic protists obtain their food by absorption, like the fungi, while others resemble animals and ingest their food. Some of them can switch their mode of nutrition, being autotrophic at certain times and heterotrophic at others. Most protists have aerobic respiration, utilizing mitochondria to metabolize their food.

FIGURE 25–3 A micrograph of a stained *Paramecium caudatum*. Note the two large contractile vacuoles.

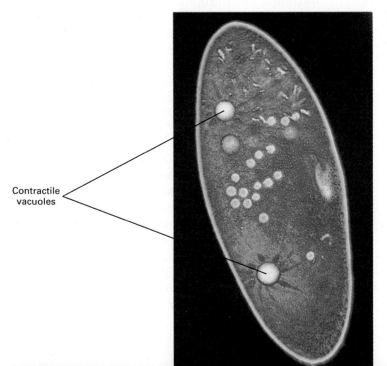

Contractile vacuoles

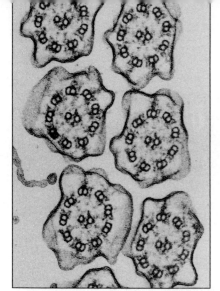

FIGURE 25–4 Electron micrograph of cross-section through several cilia showing the 9 + 2 arrangement of the microtubules which is characteristic of eukaryotes.

Many protists are free-living, while some form symbiotic associations with other organisms. These associations range from mutualism, where both partners benefit, to parasitism, with some protists being important pathogens of plants or animals.

Most of the protists are aquatic, living in oceans, freshwater ponds, lakes, and streams. They make up the **plankton,** the floating microscopic organisms that are the base of the food chain in aquatic ecosystems. Other aquatic protists attach to rocks and other surfaces in the water. The terrestrial protists are restricted to damp places like soil and leaf litter. Even the parasitic protists live in the wet environments of plant and animal body fluids.

Reproduction is also quite varied. All protists can reproduce asexually. Many also reproduce sexually with meiosis and **syngamy,** the union of gametes. However, most protists do not develop multicellular sex organs, nor do they form embryos like many higher organisms.

The protists have various means of locomotion and most are motile at some point in their life cycle, although some are nonmotile. Movement may be accomplished by amoeboid motion, by flexing individual cells, or by waving cilia or flagella. Many protists use a combination of two or more means of locomotion, e.g., flagellar and amoeboid. Their cilia and flagella are quite distinct from those of prokaryotes because protists, like all eukaryotes, possess a 9 + 2 arrangement of microtubules [i.e., nine outer doublet microtubules encircling two single microtubules (Figure 25–4).

The taxonomy of the Kingdom Protista is currently under study, with ultrastructure, biochemistry, and molecular biology adding critical information about the various groups of protists. To recognize natural relationships within the protists, some protistologists believe as many as fifty phyla are needed. Consideration of all protist groups is beyond the scope of this text, but we will discuss representative groups (Table 25–1).

A SURVEY OF REPRESENTATIVE GROUPS

When the five-kingdom system of classification was proposed by Robert Whittaker in 1969, only unicellular organisms were placed in the Protist Kingdom. The boundaries of this kingdom have been expanding since that time, although there is no universal acceptance among biologists about what constitutes a "protist." We have interpreted the Protist Kingdom broadly, including heterotrophic protists (the protozoa), autotrophic protists (the algae), and fungal-like protists (the slime molds and water molds). Although these groups may superficially resemble animals, plants, or fungi in certain respects, they should not be considered as simpler animals, plants, or fungi. At any rate, the Protist Kingdom is not a natural assemblage of organisms. Organisms are placed within it for convenience. If natural, phylogenetic relationships were the sole means of classifying organisms into kingdoms, there would be many more than five kingdoms.

Protozoa

The name **protozoa** originally was given to animal-like organisms that were not multicellular. Unicellularity does not imply simplicity, however, and many of the protozoa are structurally complex. Protozoa are animal-like in that most of them ingest their food.

The protozoa do not represent a natural grouping. Traditionally, they have been divided into four groups—flagellates, ciliates, amoebas, and spore formers. However, the flagellates and amoebas are more closely allied to each other than to other groups, and many of the spore formers are distinctly unrelated to each other. In 1980 the Society of Protozoologists

Table 25–1 **A COMPARISON OF REPRESENTATIVE PHYLA IN THE PROTIST KINGDOM**

Common Name	Phylum	Morphology	Locomotion	Photosynthetic Pigments	Special Features
Flagellates	Sarcomastigophora	Single cell, some colonial	One to many flagella; some amoeboid	—	Symbiotic forms often highly specialized
Amoeboid protozoa	Sarcomastigophora	Single cell, no definite shape	Pseudopods	—	Some with elaborate shells
Ciliates	Ciliophora	Single cell	Cilia	—	Macro/micro-nuclei
Sporozoa	Apicomplexa	Single cell	None	—	All parasitic; develop resistant spores
Dinoflagellates	Dinoflagellata	Single cell, some colonial	Two flagella	Chlorophylls *a* and *c* Carotenoids, including fucoxanthin	Many covered with cellulose plates
Diatoms	Bacillariophyta	Single cell, some colonial	Most nonmotile; some move by gliding over secreted slime	Chlorophylls *a* and *c* Carotenoids, including fucoxanthin	Silica in cell wall
Euglenoids	Euglenophyta	Single cell	Two flagella (one of them very short)	Chlorophylls *a* and *b* Carotenoids	Flexible pellicle
Green algae	Chlorophyta	Single cell, colonial, siphonous, multicellular	Most flagellated at some stage in life; some nonmotile	Chlorophylls *a* and *b* Carotenoids	Reproduction highly variable
Red algae	Rhodophyta	Most multicellular, some single cell	None	Chlorophyll *a* Carotenoids Phycocyanin Phycoerythrin	Some reef builders
Brown algae	Phaeophyta	Multicellular	Two flagella on reproductive cells	Chlorophylls *a* and *c* Carotenoids, including fucoxanthin	Differentiation of body into blade, stipe, and holdfast
Plasmodial slime molds	Myxomycota	Multinucleate plasmodium	Streaming cytoplasm, flagellated or amoeboid reproductive cells	—	Reproduce by spores formed in sporangia
Cellular slime molds	Acrasiomycota	Vegetative—single cell Reproductive—multicellular (slug)	Pseudopods (for single cells) Cytoplasmic streaming (for multicellular)	—	Aggregation of cells signaled by cyclic AMP
Water molds	Oomycota	Coenocytic mycelium	Biflagellate zoospores	—	Cellulose and/or chitin in cell walls

proposed organizing the protozoa into seven phyla, reflecting natural relationships. It must be stressed that their relationships are continually being evaluated as additional evidence becomes available.

Phylum Sarcomastigophora

Members of this phylum move by means of flagella or pseudopodia or both. Some of the flagellates form pseudopods and some of the amoebae have flagellated stages. These protozoa have predominantly one type of nucleus and do not form spores. Reproduction is primarily asexual, but gametes are sometimes formed.

FLAGELLATES.   Flagellates have spherical or elongate bodies, a single central nucleus, and one to many whiplike **flagella** that enable them to move.

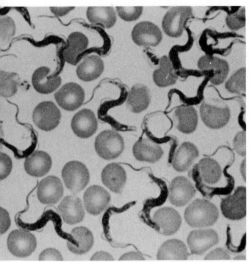

FIGURE 25–5 Rat blood infected with a parasitic flagellate, *Trypanosoma brucei,* visible as dark, wavy bodies among the red blood cells. Similar trypanosomes infect the central nervous system of human beings, causing sleeping sickness.

FIGURE 25–6 A stalked colony of *Codosiga botrytis,* a typical choanoflagellate.

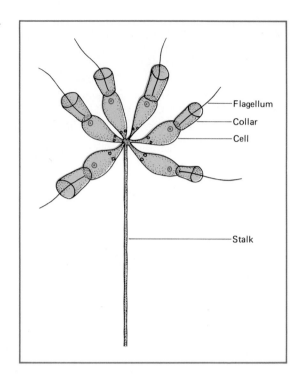

Flagellum
Collar
Cell
Stalk

Flagellates move rapidly, pulling themselves forward by lashing one or more flagella that are usually located at the anterior end. Some flagellates are amoeboid and engulf food by forming pseudopods. Others have a definite "mouth" or **oral groove,** a "gullet" or **cytopharnyx,** and specialized organelles for processing food.

The zooflagellates are heterotropic and obtain their food by ingesting living or dead organisms, or absorbing nutrients from dead or decomposing organic matter. They may be free-living or symbionts. Those with the largest number of flagella and the most specialized bodies live in the intestines of termites. They possess the enzymes to digest wood, and both the termite and the flagellates obtain their nutrients from this source. Some of the parasitic flagellates cause disease (Figure 25–5). One of these is *Trypanosoma cruzi,* the cause of African sleeping sickness (see Focus on African Sleeping Sickness).

The choanoflagellates (Figure 25–6) are of special interest because their resemblance to cells in the sponges is striking. Most biologists believe the choanoflagellates are related to the sponges, and probably to all animals. These sedentary flagellates are attached to a substrate by a stalk, and their single flagellum is surrounded by a delicate collar of cytoplasm.

AMOEBOID PROTOZOA. Many members of this group have no definite body shape. Their single cells change form as they move. These organisms reproduce asexually by cell division. Sexual reproduction has not been reported. A typical example is the amoeba, which moves by pushing out temporary cytoplasmic projections called **pseudopods** from the surface of the body (Figure 25–7). More cytoplasm flows into the pseudopods, enlarging them until all the cytoplasm has entered and the organism as a whole has moved.

Pseudopods are also used to capture food, two or more of them moving out to surround and engulf a bit of debris or another microorganism (Figure 25–8). The food that has been engulfed is surrounded by a food vacuole and digested by enzymes added by lysosomes. The digested materials are absorbed from the food vacuole, which gradually shrinks as it becomes empty. Any indigestible remnants are expelled from the body and left behind as the amoeba moves along.

FIGURE 25–7 *Chaos carolinense,* a giant amoeba ingesting a colonial green alga. Note the pseudopods extending to surround the prey.

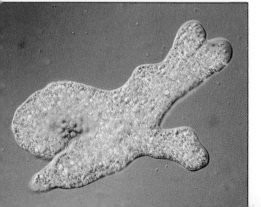

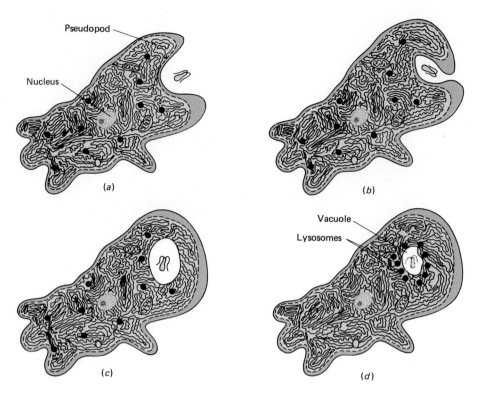

Pseudopod

Nucleus

(a)

(b)

Vacuole

Lysosomes

(c)

(d)

FIGURE 25–8 Amoeboid movement in feeding. The lysosomes release enzymes that digest the prey.

The parasitic members of amoeboid protozoa include the species that causes serious amebic dysentery in humans. Some amoebas, like *Acanthamoeba,* are usually free living but can produce opportunistic infections in the eyes of contact lens users.

Some members of this group produce calcified shells, or **tests.** For example, the oceans contain untold trillions of foraminiferans, which secrete chalky, many-chambered shells (tests) with pores through which pseudopods can be extended (Figure 25–9). The pseudopods form a sticky, interconnected net that entangles its prey (Figure 25–10). Dead foraminiferans sink to the bottom of the ocean, where their shells form a grey mud that is gradually transformed into chalk. With geological uplifting these chalk formations can become part of the land, like the white cliffs of Dover. Foraminifera are used as indicators of geophysical changes in the environment. The direction of coiling in their tests varies with environmental conditions.

Other ameboid protozoa, the actinopods, secrete elaborate and beautiful skeletons made of silica (Figure 25–11). These skeletons become mud on

FIGURE 25–9 The test of a foraminiferan, *Poneroplia perfusus.*

FIGURE 25–10 Foraminiferans secrete a shell, or test. Cytoplasm is extruded through the pores, forming a layer outside.

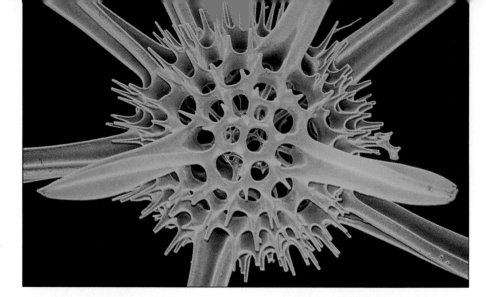

FIGURE 25–11 Marine actinopods, also called radiolarians, secrete elaborate siliceous tests which are symmetrical.

the ocean floor and eventually are compressed into siliceous, sedimentary rock. Some actinopods have long, filamentous **filopodia,** sometimes provided with rodlike skeletal elements that protrude through pores in their skeletons. Diatoms and other prey become entangled in these filopodia and are digested outside the main body of the actinopod. Cytoplasmic streaming carries the products of digestion back within the shell. Many actinopods have symbiotic algae that provide their host with the products of photosynthesis.

Phylum Ciliophora

The ciliates have a definite but somewhat flexible shape due to the presence of a flexible outer **pellicle.** In *Paramecium* the surface of the cell is covered with several thousand fine cilia that extend through pores in the pellicle and permit movement (Figure 25–12). The cilia beat with an oblique stroke so that the animal revolves as it swims. The coordination of the ciliary beating is so precise that the organism not only can go forward but can back up and turn around. Near their surface, many ciliates possess numerous small **trichocysts,** organelles that can discharge filaments be-

Focus on AFRICAN SLEEPING SICKNESS

African sleeping sickness is a serious disease caused by several species of the flagellated protozoa *Trypanosoma*. These organisms are transmitted from one mammalian host to another by the bite of the bloodsucking tsetse fly. Large areas of Africa have been effectively closed to human settlement due to the disease that the trypanosomes induce in cattle and humans. Hoofed game animals serve as epidemiological reservoirs of these microorganisms, which produce few symptoms of illness in their wild hosts. However, domestic cattle waste away and humans suffer from a fatal encephalitis when infected.

The parasites possess sophisticated countermeasures against the immune systems of their hosts. Even when the primary immune response (Chapter 43) destroys

most of the parasites, a few resistant individuals persist whose antigens differ from those of the original population. This forces the host to mount a fresh primary immune response. Meanwhile, the parasites are multiplying. Repeated cycles of reinfection, accompanied by eventual invasion of the central nervous system, in time destroy a nonresistant host.

Trypanosomes accomplish this immunological feat by means of a library of some two hundred transposable genes. Occasionally, a member of this genetic archive is copied onto a section of a chromosome equipped with a promoter that causes a new gene to be expressed instead of the old one that it replaced. The result is a new coat of antigens with which the host's immune system has had no previous experience.

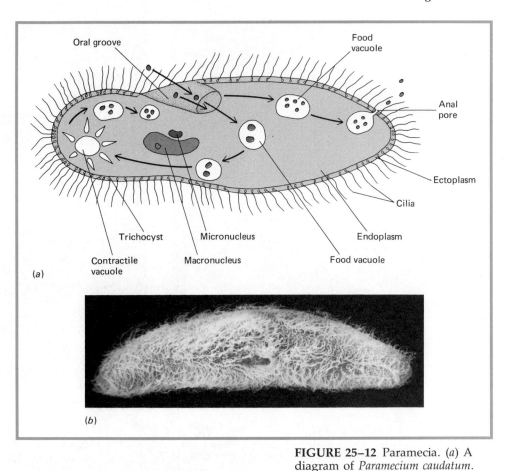

(a)

(b)

FIGURE 25–12 Paramecia. (*a*) A diagram of *Paramecium caudatum*. (*b*) A scanning electron micrograph of *Paramecium multimicronucleatum*. Note the oral groove.

lieved to aid in trapping and holding prey (Figure 25–13). Most ciliates ingest their food (Figure 25–14). Although none are photosynthetic, some have symbiotic algae living within their cells.

Ciliates differ from other protozoa in having at least two nuclei per cell, often one or more **micronuclei** that function in reproduction, and a larger **macronucleus** that controls cell metabolism and growth.

Most ciliates are capable of a sexual phenomenon called **conjugation** (Figure 25–15). *Paramecium* and other ciliates may have two, and as many as eight, mating types. During conjugation in *Paramecium* two individuals of different mating types press their oral surfaces together. Within each individual the macronucleus disintegrates and the micronucleus under-

FIGURE 25–13 *Paramecium* discharging its trichocysts, specialized organelles which produce a substance that hardens into entangling threads when the organism is disturbed.

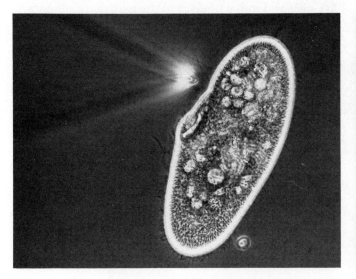

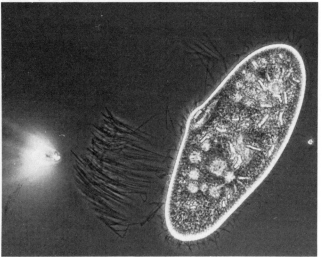

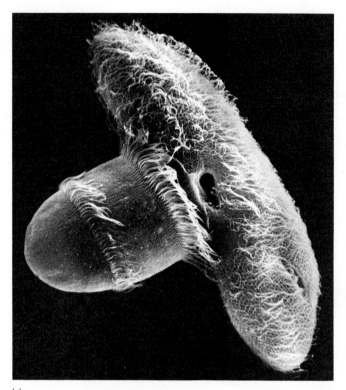

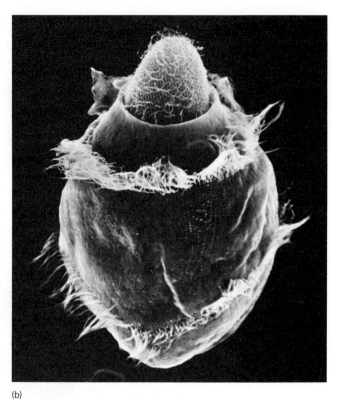

(a)

(b)

FIGURE 25–14 A heterotrophic protist devouring another protist. (*a*) *Didinium,* a ciliate, about to devour a much larger *Paramecium,* also a ciliate. (*b*) *Didinium* has almost completely engulfed *Paramecium.* Only the tip of *Paramecium* is protruding from *Didinium.*

goes meiosis, forming four haploid nuclei. Three of these degenerate, leaving one. This nucleus then divides mitotically, and one of the two identical haploid nuclei remains within the cell. The other nucleus crosses through the oral region into the other organism and fuses with the haploid nucleus already there. Thus each conjugation yields two cross "fertilizations." This leads to two new cells, both of which are genetically identical but different from the "parents" (or, more properly, the preconjugant cells). Actual cell division need not follow immediately. Cytokinesis in ciliates is a complex process involving more than simply splitting in half because of the presence of complex organelles that must be reorganized and replicated.

Not all ciliates are motile. Some forms are stalked and others, such as *Stentor,* while capable of some swimming, are more likely to remain attached to the substrate at one spot (Figure 25–16). Strong cilia set up currents in the surrounding water to bring food to them.

Suctoria are among the most aberrant of the ciliates (Figure 25–17). Young individuals have cilia and move freely, but the adults are sedentary,

FIGURE 25–15 Apparent recognition behavior of two ciliates of the species *Euplotes crassus* preparatory to conjugation. The organisms appear to embrace one another with fused bundles of cilia known as cirri.

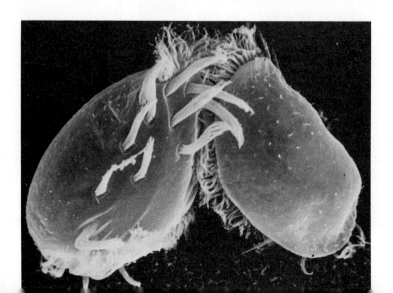

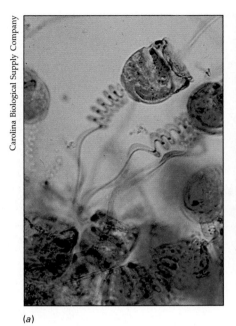

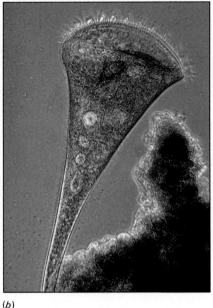

Carolina Biological Supply Company

(a)   (b)

FIGURE 25–16 Ciliates. (*a*) A group of *Vorticella*, a ciliate with a contractile stalk. (*b*) *Stentor*. Note the numerous cilia that direct food particles into its "mouth," or "gullet."

lack cilia, and have stalks by which they are attached to the substrate. Each bears a group of delicate cytoplasmic tentacles, some of which are pointed to pierce and absorb prey. Other tentacles are tipped with rounded adhesive knobs to catch and hold prey. The tentacles also secrete a toxic material that may paralyze the prey.

Phylum Apicomplexa

The sporozoa in the Phylum Apicomplexa are a large group of parasitic protozoa, some causing serious diseases such as malaria in humans. Sporozoa have neither organelles for locomotion nor contractile vacuoles. They do move, however, by flexing. At some stage in their life many develop a resistant **spore,** which is the infective agent for the next host. They often spend part of their life in one host and part in a different species.

The sporozoon causing malaria, *Plasmodium*, enters the human bloodstream when a parasitized mosquito bites a human (Figure 25–18). The parasites first enter liver cells and then red blood cells, multiplying there safe from the host's immune system. When the infected cell bursts, many new parasites are released. The released parasites infect new red blood cells and the process is repeated. The simultaneous bursting of millions of red cells causes a typical malarial chill followed by fever, as toxic substances are released and penetrate other organs of the body.

If a second, nonparasitized mosquito bites the infected human, it will suck up some parasites along with blood. A complicated process of sexual reproduction then occurs within the mosquito's stomach, and new parasites develop, some of which migrate into the mosquito's salivary glands to infect the next person bitten. This sexual reproduction does not occur within humans. For this reason elimination of the mosquito hosts, though very difficult, would eradicate the disease.

Malarial parasites elude the immune system of the host not only by hiding in its cells, but also by excreting a coating of protein equipped with decoy antigens that is readily attacked by host antibodies. From time to time the parasites cast off this coat, attached antibodies and all, and produce a new one. Hosts never do develop antibodies to the vital plasmodial antigens on the cell membrane beneath the protein coat.

FIGURE 25–17 A suctorian, *Paracineta*, attached to a filamentous alga, *Spongomorpha*. Note the absence of cilia and the presence of tentacles on the suctorian. *Paracineta* lives in a lorica, a surrounding case that is separate from and encloses the cell.

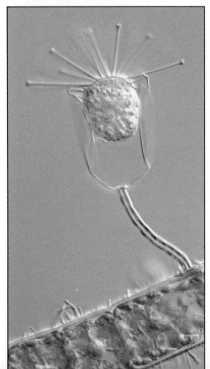

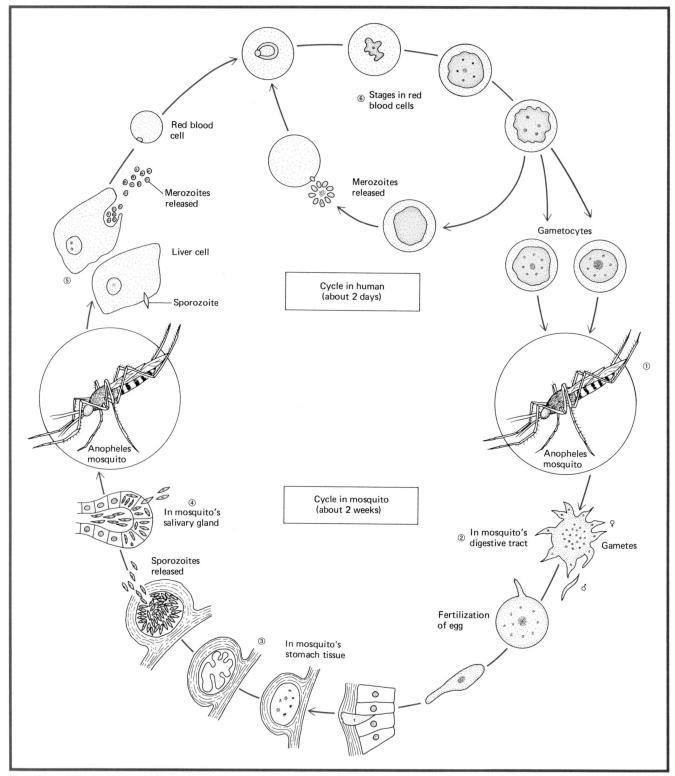

FIGURE 25–18 *Plasmodium* causes malaria in humans and other mammals. (*1*) A female *Anopheles* mosquito bites an infected person, obtaining gametocytes. (*2*) In the mosquito's digestive canal the gametocytes develop into gametes and fertilization occurs. (*3*) The zygote becomes embedded in the mosquito's stomach lining and produces sporozoites, which are released and migrate to the salivary gland. (*4*) The mosquito bites an uninfected human and transmits sporozoites to the human's blood stream. (*5*) The sporozoites enter the liver cells and divide to produce merozoites that infect red blood cells. (*6*) In the blood cells merozoites divide to form more merozoites, which reinfect more red blood cells. Or the merozoites form gametocytes. The gametocytes can be transmitted to the next mosquito that bites that human, and the process repeats itself.

Algae

The algae represent a diverse group of organisms that are mostly photosynthetic. They range in size from single-celled, microscopic forms to large, multicellular seaweeds. Unlike plants, the algae lack a cuticle (a waxy covering) and are, therefore, restricted to damp or wet environments when actively growing. Most algae do not have multicellular **gametangia** in which gametes are produced; their gametangia are formed from single cells. In addition to chlorophyll *a* and yellow and orange **carotenoids,** they possess a variety of other pigments. Classification into phyla, or divisions, is largely by pigment composition and energy storage products. Other characteristics used to classify algae include cell wall composition, number and placement of flagella, and chloroplast morphology.

Phylum Dinoflagellata

One of the most unusual groups of protists are the dinoflagellates. Most of them are unicellular, although a few colonial forms exist. Their cells are often covered with a shell of interlocking cellulose plates, with some having silicates impregnated on them (Figure 25–19). Each has two flagella. One of the flagella is wrapped around a transverse groove in the center of the cell like a belt; the other is located in a longitudinal groove perpendicular to the first and projects beyond the cell. The undulation of these flagella propels the dinoflagellate through the water like a spinning top. Indeed, their name is derived from the Greek "dinos" meaning whirling. Some dinoflagellates are bioluminescent. In tropical oceans, ships or large fish moving through the water at night cause them to light up the water with an eerie glow.

Most of the dinoflagellates are photosynthetic, possessing chlorophylls *a* and *c* in addition to carotenoids (Figure 25–20). Their energy storage products are usually oils or polysaccharides. However, a number of them are colorless, some ingesting other microorganisms for food.

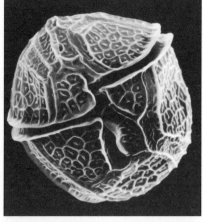

(a)

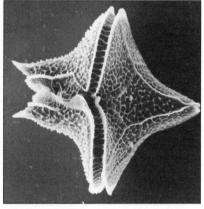

(b)

FIGURE 25–19 Scanning electron micrographs of some dinoflagellates. Note the plates that encase the single-celled body. The two flagella are located in grooves. (a) *Gonyaulax.* (b) *Protoperidinium.* (c) *Ceratium.*

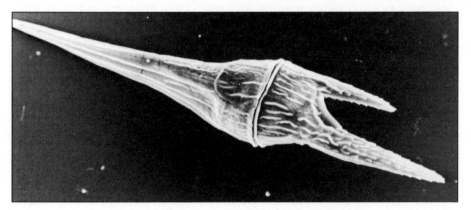

(c)

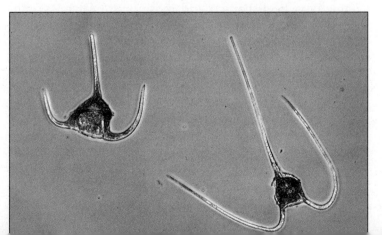

FIGURE 25–20 Two freshwater dinoflagellates of the genus *Ceratium.* These dinoflagellates are photosynthetic and possess chlorophylls a and c and carotenoids.

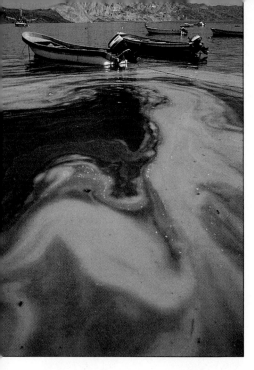

FIGURE 25–21 Red tide in Tampa Bay. The oddly patterned cloudiness in the water is produced by countless billions of dinoflagellates.

Many of the dinoflagellates are endosymbionts. The photosynthetic endosymbionts often reside in marine invertebrates such as jellyfish, corals, and mollusks. These autotrophic dinoflagellates lack cellulose plates and flagella; in this form they are called **zooxanthellae.** They photosynthesize and provide food for their mutualistic partner. Their contribution to the productivity of coral reefs alone is substantial. Some dinoflagellates that reside in other organisms lack pigmentation and do not photosynthesize. These heterotrophs are parasitic on their hosts.

The variety of life styles in the dinoflagellates demonstrates the difficulties in neatly categorizing many protists. Are the dinoflagellates algae or are they protozoa? Or should they be separated into different groups based on such features as pigmentation (or lack of it) and method of obtaining food?

Reproduction of the dinoflagellates is primarily asexual, by longitudinal cell division. A few genera have been reported to reproduce sexually. The nucleus of dinoflagellates is unusual because the chromosomes are permanently condensed and always evident. The nuclear membrane remains intact during mitosis and meiosis. Also, most dinoflagellates have spindles located *outside* the nucleus. The chromosomes do not make direct contact with the spindle microtubules. Based on the uniqueness of their chromosome morphology and mitosis, the dinoflagellates are believed to have no close extant relatives.

Ecologically, the dinoflagellates are one of the most important groups of producers in aquatic ecosystems. Most species are marine. A few of the dinoflagellates are known to have occasional population explosions, or blooms. These blooms frequently color the water orange, red, or brown and are known as **red tides** (Figure 25–21). Some of the species that form red tides produce a toxin that affects the nervous system of fish, leading to massive fish kills. Other species that produce a toxin are eaten by mollusks, which apparently are not harmed. However, the toxin accumulates within their tissue, making them poisonous to humans. It is not known what environmental conditions initiate blooms, but they are more common in the warm waters of late summer.

Phylum Bacillariophyta

Most members of the Phylum Bacillariophyta **(diatoms)** are unicellular, although there are a few colonial forms. Their cell walls are composed of two halves that overlap where they fit together, much like a Petri dish (Figure 25–22). Silica is impregnated in the cell wall, and this glasslike material forms striking, intricate patterns. Indeed, the patterns of ridges, lines, and pores etched into the cell walls are used to classify the diatoms (Figure 25–23). There are two basic groups of diatoms, those with radial symmetry (wheel-shaped) and those with bilateral symmetry. Although most diatoms are part of the floating plankton, those that grow on rocks and other surfaces move by gliding. This movement is facilitated by the secretion of a material from a small groove along the shell.

Most diatoms are photosynthetic and contain chlorophylls *a* and *c* in addition to carotenoids. Their pigment composition gives them a yellow or brown color. Food reserves are stored as oils or carbohydrates.

Diatoms primarily reproduce asexually by cell division. When a cell divides, the two halves of the cell wall separate and each becomes the larger half for a new cell. Therefore, some of the cells get progressively smaller with each succeeding generation. When the diatoms are a fraction of their

FIGURE 25–22 Scanning electron micrograph of several diatoms. The frustule (shell) is built like a pill-box, with the lid overlapping the base.

original size, sexual reproduction is triggered, with the production of gametes that shed their cell walls. Sexual reproduction restores the diatom to its original size, because the resulting zygote grows substantially before producing a new shell, and the process starts over again.

Diatoms are common in both fresh and ocean waters, but are especially abundant in cooler marine waters. They are major producers in aquatic ecosystems because of their unbelievable numbers. When diatoms die, their cell walls trickle down and accumulate in what becomes sedimentary rock. After millions of years, some of these deposits have been exposed on land by geological upheaval. This diatomaceous earth is mined and has commercial value as a filtering, insulating, and soundproofing material.

FIGURE 25–23 The structure of diatoms viewed through different methods of microscopy. (*a*) and (*b*) Scanning electron micrographs of a diatom. Note the raphe (groove) and elaborate series of wall perforations. (*c*) and (*d*) Diatoms viewed by a technique called interference optics microscopy.

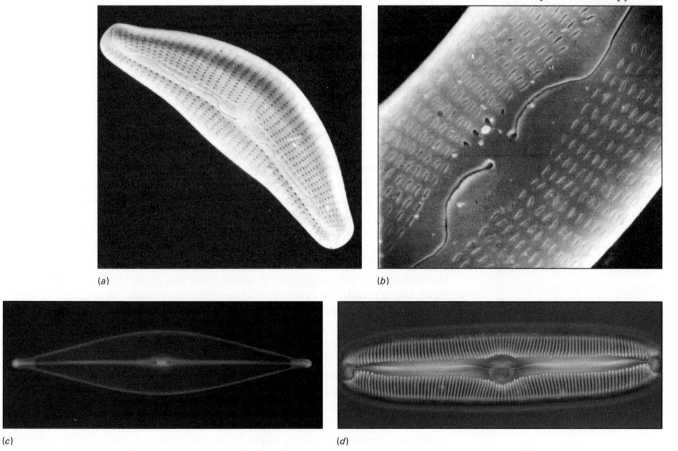

(*a*)

(*b*)

(*c*)

(*d*)

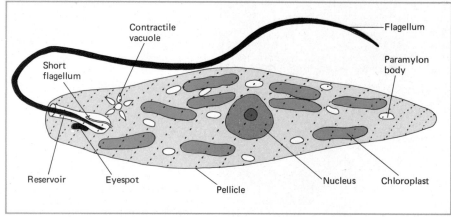

(a)

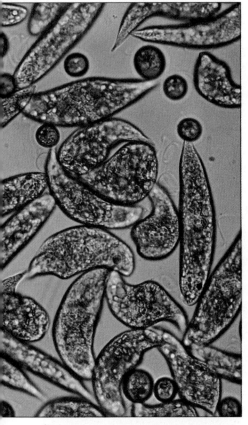

(b)

FIGURE 25–24 *Euglena.* (a) This protist has both plant-like and animal-like traits. It has at various times been classified in the plant kingdom (with the algae) and in the animal kingdom (when protozoa were considered to be animals). (b) Living euglenoids. Note the red eyespots.

Phylum Euglenophyta

All euglenoids are unicellular flagellates (Figure 25–24). They generally possess two flagella, one that is long and whiplike and one that is so short it doesn't protrude outside the cell. Their shape continually changes because their outer covering is a flexible, proteinaceous pellicle rather than a rigid cell wall. The euglenoids reproduce asexually by longitudinal cell division. None of them has ever been observed to reproduce sexually.

The euglenoids are included in our discussion of plantlike protists because many of them contain chloroplasts and photosynthesize. They have chlorophylls *a* and *b* and carotenoids, identical to those of green algae and plants. Their food is stored as paramylon, a polysaccharide. Some of the photosynthetic forms lose their chlorophyll when grown in the dark and obtain their nutrients heterotrophically by ingesting organic matter. Other species of euglenoids are always colorless and heterotrophic.

Because of their similarities to other flagellates, many biologists put euglenoids in the Phylum Sarcomastigophora. Once again, the distinction between the animal-like and plantlike protists is difficult to discern. Organisms like *Euglena* make the necessity for a protist kingdom obvious. When all living organisms were placed in either the plant or the animal kingdom, *Euglena* and other euglenoids didn't fit into either. Although the euglenoids have the same pigmentation as the green algae and plants, they are not believed to be related to either of them.

Euglenoids inhabit freshwater ponds and puddles, particularly those with large amounts of organic material. For that reason they are used as indicator species of pollution. If a body of water has large numbers of euglenoids, it is probably polluted. Marine waters and mud flats are also inhabited by some euglenoids.

Phylum Chlorophyta

If one had to pick a single word to describe the green algae (Phylum Chlorophyta), it would be "variety." These protists exhibit an amazing number of morphologies and methods of reproduction. Their body forms range from single cells to colonial forms to coenocytic, **siphonous** algae to multicellular filaments and sheets (Figure 25–25). The multicellular forms do not have tissue differentiation, however. Most are flagellated during at least part of their life history, although there are some that are totally nonmotile.

Although the green algae are structurally very diverse, they are biochemically very uniform. They are photosynthetic, with chlorophylls *a* and *b* and carotenoids present in chloroplasts of a wide variety of shapes. Starch is the main food reserve. Most possess cell walls with cellulose,

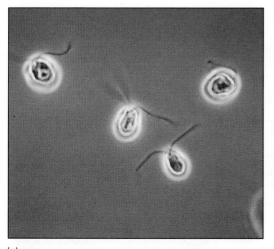

(a)

(b)

(c)

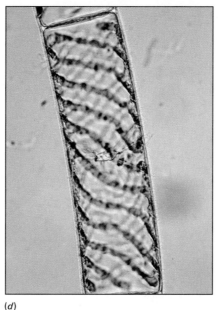

(d)

(e)

FIGURE 25–25 Diversity in the green algae. (*a*) *Chlamydomonas* is a biflagellate unicellular organism. (*b*) An example of a colonial green alga is *Volvox*. A daughter colony can be observed leaving the mother colony through a rent in its walls. (*c*) Siphonous green algae, like *Codium*, are coenocytic. (*d*) *Spirogyra* is a multicellular green alga with a filamentous body form. Note the spiral-shaped chloroplasts. (*e*) Some multicellular green algae are sheet-like. The thin, leaf-like form has given *Ulva* its common name of "sea lettuce."

although some lack walls and some are covered with scales. Many of the green algae are symbionts with other organisms. Some live in body cells of invertebrates. Others have evolved with certain fungi, growing together as a dual organism, the lichen (Chapter 26).

Green algae share a number of characteristics in common with plants. Biochemically, their pigmentation, storage products, and cell walls are identical. Because of these and other similarities, it is generally accepted that plants evolved from green, algal-like ancestors. Taxonomy of the green algae is currently under study. Although many of the simpler forms share much in common with other protistan flagellates, the relationships between different lines of green algae within the group are not clear. Research advances in ultrastructure and biochemistry are providing insight into this very diverse group.

Reproduction is as varied as morphology in the green algae. Both sexual and asexual reproduction occur in the group. Asexual reproduction may be by cell division for unicells or by fragmentation for multicellular forms. Many green algae produce spores asexually by mitosis. If these spores are flagellated and motile, they are called **zoospores.** The sexual cycles of green algae are often quite complex. Sexuality involves formation of gametes in single-celled gametangia. If the two flagellated gametes that fuse are iden-

FIGURE 25–26 Life cycle of *Chlamydomonas*. *Chlamydomonas* is a haploid green alga that has two strains, + and −, which are visually indistinguishable. Both strains reproduce asexually by cell division. At times, the + and − strains behave as gametes and fuse, forming a diploid zygote that develops into a zygospore. Meiosis occurs within the zygospore and four haploid cells emerge, two + and two −. This is an example of isogamous sexual reproduction.

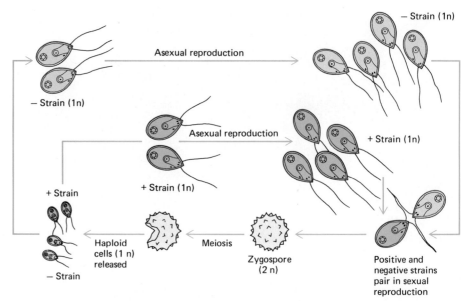

FIGURE 25–27 *Spirogyra,* a haploid organism, has a sexual phenomenon known as conjugation. (*a*) Filaments of two different strains line up, and conjugation tubes form between cells of the two filaments. (*b*), (*c*) The contents of one cell pass into the other cell through the conjugation tube. (*d*) Each zygote develops into a zygospore, which is thick-walled and dormant for a period of time. Meiosis occurs when the zygospore germinates, restoring the haploid condition.

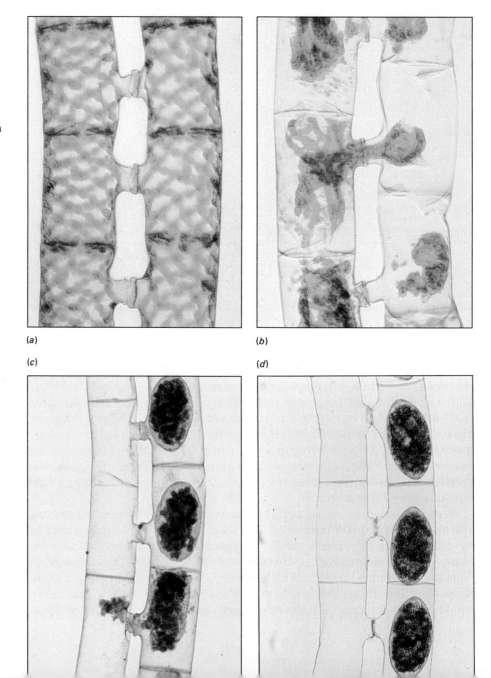

(*a*)

(*b*)

(*c*)

(*d*)

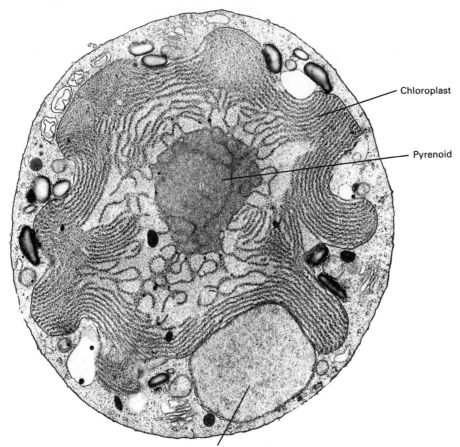

Chloroplast

Pyrenoid

Nucleus

FIGURE 25–28 Electron micrograph of a section through a one-celled red alga, *Porphyridium cruentum* (×14,850). A single stellate chloroplast with a central pyrenoid (associated with starch deposition) occupies much of the cell volume.

(a)

(b)

FIGURE 25–29 Most red algae are multicellular, many with complex filamentous bodies (a) *Odonthalia.* (b) *Plumaria.*

tical, reproduction is said to be **isogamous** (Figure 25–26). **Anisogamous** reproduction involves the fusion of two flagellated gametes of different sizes. Some green algae are **oogamous** and produce a nonmotile egg and a flagellated male gamete. Finally, some green algae exchange genetic information by conjugation (Figure 25–27).

There are both aquatic and terrestrial forms of green algae. The aquatic green algae primarily inhabit fresh water, but there are a number of marine species. Green algae that inhabit the land are restricted to damp soil, tree bark, and other moist places. They are important ecologically as the base of the food chain, particularly in freshwater habitats.

Phylum Rhodophyta

The vast majority of red algae in the Phylum Rhodophyta are multicellular, although a few unicellular species occur (Figure 25–28). The multicellular body form is commonly composed of complex, interwoven filaments that are delicate and feathery, although a few are flattened sheets of cells (Figure 25–29). Most of them attach to rocks or other substrates by a root-like **holdfast.** The chloroplasts of red algae contain phycoerythrin, a red pigment, and phycocyanin, a blue pigment, in addition to chlorophyll *a* and carotenoids. Their storage product is floridean starch, a carbohydrate similar to glycogen. The cyanobacteria have the same pigment composition as the red algae, supporting the hypothesis that cyanobacteria endosymbionts evolved into chloroplasts in the red algae.

The cell walls of red algae often contain mucilaginous polysaccharides that have commercial value. For example, agar is extracted from certain red algae and used in baking and as a culture medium for microorganisms. A second polysaccharide extracted from red algae is carrageenan, which is

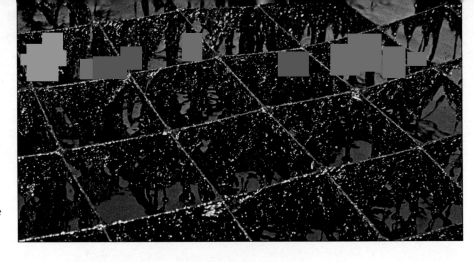

FIGURE 25–30 *Porphyra* (nori), a red alga used as food, is grown in seaweed beds such as this one in Japan. The nets are arranged so the algae are exposed during low tide and submerged during high tide.

FIGURE 25–31 Coralline algae are ecologically important in the formation of reefs. Their bodies are impregnated with calcium carbonate.

FIGURE 25–32 *Laminaria*, a typical brown alga. Note the blade, stipe, and holdfast.

used to stabilize emulsions in puddings, laxatives, and toothpastes. Red algae are important human food, particularly in oriental countries (Figure 25–30).

Reproduction in the red algae has been studied in detail for only a few species, but it is amazingly complex, with an alternation of sexual and asexual generations. Sexual reproduction is common, but at no stage in the life history of red algae are there flagellated cells. Besides lacking flagella, the red algae lack centrioles. These characteristics and several others may link the red algae to the fungi, but additional studies are needed.

The red algae are primarily found in warm tropical oceans, although a few freshwater and soil species occur. Some of the red algae incorporate calcium carbonate into their cell walls from the ocean waters. These coralline red algae are very important in building "coral" reefs, possibly more important than coral animals (Figure 25–31).

Phylum Phaeophyta

The Phylum Phaeophyta, or brown algae, are the giants of the Protist Kingdom. All are multicellular, ranging in size from several centimeters to approximately 60 meters in length. Their body forms may be tufts, "ropes," or thick, flattened, branching forms. The largest brown algae, called kelps, are tough and leathery in appearance and have considerable differentiation into leaf-like **blades,** stem-like **stipes,** and anchoring holdfasts (Figure 25–32). They often have gas-filled floats to increase buoyancy. It is important to remember that the blades, stipes, and holdfasts of brown algae are not homologous to the leaves, stems, and roots of plants. Brown algae and green plants arose from different unicellular ancestors. These structures are the result of convergent evolution.

Brown algae are photosynthetic and possess chlorophylls *a* and *c* and carotenoids in their chloroplasts. A special yellow-brown carotenoid, fucoxanthin, is found only in brown algae, dinoflagellates, and diatoms. The main food storage reserve in brown algae is a carbohydrate called laminarin.

Brown algae are commercially important for several reasons. They have a polysaccharide, algin, in their cell walls, possibly to help cement the cell walls together. It is used as a thickening agent in ice cream, marshmallows, and cosmetics. Brown algae are an important source of food for humans, especially in oriental countries, and they are rich sources of minerals, especially iodine.

Reproduction is varied and complicated in the brown algae. They reproduce sexually, and most have a well-defined alternation of generations, spending a portion of their lives as haploid organisms and a portion as diploid organisms. Their reproductive cells, zoospores and gametes, are flagellated.

Brown algae are common in cooler marine waters, especially along rocky coastlines. There they can be found mainly in the intertidal zone or relatively shallow waters. The kelps forms extensive underwater "forests" and are essential in that ecosystem as the primary food producer (Figure 25–33). These beds also provide habitats for many marine invertebrates, fish, and mammals. The diversity of life in these underwater forests is astounding, and all of it is supported by the brown algae.

There is an extensive colony of floating brown algae in an area of the central Atlantic Ocean called the Sargasso Sea. The brown alga, *Sargassum*, supports a diverse community here, providing food and shelter.

Fungal-like Protists

Some of the fungal-like protists superficially resemble the fungi because they are nonphotosynthetic and the body form is often threadlike hyphae. However, they are not fungi for several reasons. Many of these protists produce flagellated cells, which the fungi lack. They also have centrioles and produce cellulose as a major component of many of their cell walls, unlike the fungi.

Phylum Myxomycota

The vegetative (feeding) stage of plasmodial slime molds (Phylum Myxomycota) is most unusual. It is a naked mass of protoplasm that is often brightly colored (Figure 25–34). This **plasmodium** contains many nuclei that are usually diploid, but it is not divided into separate cells. The cytoplasm streams over damp, decaying logs and leaf litter, often forming a network of channels to cover a larger surface area. As it creeps along, it ingests bacteria, yeasts, spores, and decaying organic matter much as an amoeba does.

When the food supply dwindles or there is insufficient moisture, the plasmodium crawls to an exposed surface and reproduction is initiated. Usually stalked structures of intricate complexity and beauty form from the drying plasmodium (Figure 25–35). These structures, called **sporangia,** form cell walls (of cellulose and/or chitin) around each nucleus. Within the sporangium, haploid spores are formed by meiosis. These spores are extremely resistant to adverse environmental conditions. When conditions are favorable, they crack open, and a haploid reproductive cell emerges from each. It may be a one-celled biflagellate or an amoeboid cell, depend-

FIGURE 25–33 A kelp bed off the coast of California. These underwater forests are ecologically important, supporting large numbers of aquatic organisms.

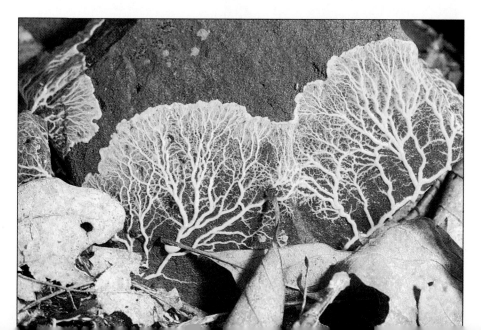

FIGURE 25–34 The plasmodium of *Physarum* is colored bright yellow. This naked mass of protoplasm is multinucleate and feeds on bacteria and other microorganisms.

(a)

(b)

FIGURE 25–35 The reproductive structures of plasmodial slime molds are often stalked sporangia. (a) *Stemonitis.* (b) *Physaram.*

ing on how wet it is. These two forms, the flagellated swarm cell and the myxamoeba, act as gametes. Eventually two of them fuse, and the diploid zygote divides by mitosis without cytoplasmic division to form a multinucleate plasmodium.

Phylum Acrasiomycota

Although organisms in the Phylum Acrasiomycota are called cellular slime molds, their resemblance to the plasmodial slime molds is superficial. Indeed, they have much closer affinities with the amoebae. During their

FIGURE 25–36 The life cycle of the cellular slime mold, *Dictyostelium discoideum.* (1) Mature fruiting body releasing spores. (2) Each spore opens to liberate an amoeba-like one-celled organism that eats, grows, and reproduces by cell division. (3) After their food supply is depleted, the cells stream together. (4) An aggregation of cells. (5) The aggregation organizes into a slug-shaped, multicellular organism. (6) The "slug" migrates for a period of time before forming a stalked fruiting body (7, 8, and 9). Cells making up the anterior third of the slug differentiate into stalk cells, while those in the posterior two thirds form the spores.

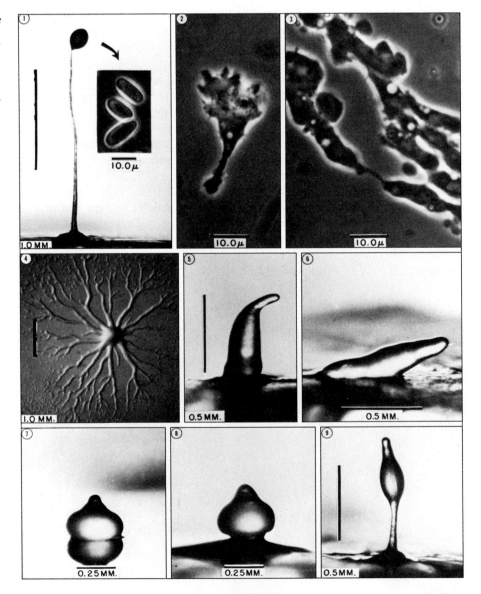

vegetative (feeding) stage, the individual amoeboid cells behave as separate, solitary organisms; they creep over rotting logs and soil or swim in fresh water, ingesting bacteria and other particles of food. Each cell has a haploid nucleus.

When moisture or food becomes inadequate, the cells send out a chemical signal, cyclic AMP (cyclic 3′,5′-adenosine monophosphate), which causes them to aggregate by the hundreds or thousands for reproduction. During this stage the cells creep about as one unit, called a **pseudoplasmodium** or "slug." Each cell of the slug retains its plasma membrane and individual identity. Eventually, the slug settles and constructs a stalked structure. The stalk forms from the cells in the anterior third of the slug. The posterior portion of the slug forms a rounded structure at the tip, within which spores differentiate. Each spore may grow into an individual amoeboid cell, and the cycle repeats itself (Figure 25–36). This reproductive cycle is asexual, although sexual reproduction has been observed occasionally. There are no flagellated stages for most of the cellular slime molds.

Phylum Oomycota

The water molds, Phylum Oomycota, were once classified as true fungi because of a superficial resemblance in morphology. Both have a body, termed a **mycelium,** that grows over a substrate, digesting it and then absorbing the predigested nutrients. The thread-like **hyphae** that make up the vegetative mycelium are coenocytic. Because there are no cross walls, the body is like one giant multinucleate cell. The cell wall may be composed of cellulose (like plants) or chitin (like fungi) or both.

When food is plentiful and environmental conditions are good, water molds reproduce asexually (Figure 25–37). A hyphal tip swells and a cross wall is formed, separating it from the mycelium. Within this structure, tiny biflagellate zoospores are formed. Each zoospore is capable of developing into a new mycelium. When environmental conditions worsen, sexual reproduction is initiated. After fusion of male and female nuclei, an **oospore** develops. Water molds often overwinter in this stage (Figure 25–38).

Some of the water molds have played infamous roles in human history. For example, the Irish potato famine of the nineteenth century was caused by a water mold, *Phytophthora*, that causes late blight of potatoes. Due to

FIGURE 25–37 When *Saprolegnia*, a common water mold, reproduces asexually, it forms sporangia. Within the magnified sporangium shown here, biflagellated zoospores have formed.

FIGURE 25–38 The life cycle of a typical water mold. Oomycetes reproduce both asexually and sexually.

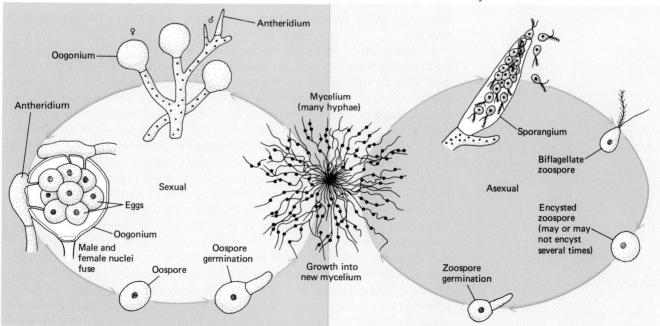

several rainy, cool summers in the 1840's, the fungus multiplied unchecked. Potato tubers rotted in the fields. Since potatoes were the staple of the Irish peasant's diet, many people starved. Estimates of the number of deaths resulting from this disease range from a quarter million to more than one million. A mass migration out of Ireland to such places as the United States also ensued.

EVOLUTION AND THE PROTISTS

Most of the protists have not left extensive fossil records because their bodies are too soft to leave permanent traces. A few with hardened cell walls or shells have abundant fossils—for example, the diatoms and foraminiferans. Therefore, evolutionary theories involving the protists are based primarily on comparisons of living organisms. Some of the most useful data for evolutionary interpretations are ultrastructure studies of cell organelles by electron microscopy. Cell biochemistry also provides important information.

Origin of Advanced Features

The protists are considered to be the first eukaryotic cells. The may have originated as early as 1.5 to 2 billion years ago. There is compelling evidence that several cell organelles, such as mitochondria and chloroplasts, arose from endosymbiotic relationships between various prokaryotic organisms. For example, chloroplasts are believed to have evolved from a symbiotic relationship between a photosynthetic prokaryote and another, larger single-celled organism, while aerobic bacteria may have been the origin of mitochondria. In each case the relationship between the endosymbiont and its host was mutually beneficial (Chapter 20).

The presence in chloroplasts of a double membrane, chloroplast DNA, and chloroplast ribosomes supports this view. Mitochondria also possess these features. Both chloroplasts and mitochondria are self-replicating, which means that they divide independently of the nucleus. If chloroplasts and mitochondria evolved in this manner, they have developed a complex relationship with the rest of the cell during their long association. For example, genes involved in photosynthesis are located in both the chloroplast genome and the nucleus. Chloroplasts, mitochondria, and other cellular organelles function interdependently; none can exist as separate entities.

Multicellularity also arose in the Protist Kingdom. The green algae, red algae, and brown algae all have considerable numbers of multicellular species. Because of profound differences among the three groups, however, it is believed that multicellularity arose independently several times. That is, these groups of algae are believed to have different ancestors. Most biologists are of the opinion that the green, red, and brown algae are the outcomes of three independent endosymbiont evolutionary lineages. For example, the green algae (and euglenoids and plants) possess chlorophylls *a* and *b*. Do any living prokaryotes possess the same pigment composition? In 1975 a bacterium, *Prochloron*, that possesses chlorophylls *a* and *b* was discovered living as an endosymbiont in marine animals. It is possible that an ancient organism like *Prochloron* gave rise to chloroplasts in green algae and/or euglenoids. Likewise, the cyanobacteria have been implicated as the ancestors of the red algae because of similarities in pigment composition. Cyanobacteria and red algae are the only organisms to contain chlorophyll *a*, phycoerythrin, and phycocyanin. Only very recently, in 1983, was a living prokaryote discovered with pigments similar to those of the brown algae and diatoms. Organisms similar to *Heliobacterium* may have been the ancestors of these groups.

Evolutionary Relationships

Protistologists use their research data to help determine evolutionary relationships among the various groups of protists. For example, the flagellates are widely regarded as the most primitive of the protozoa. It is also generally accepted that the amoeboid protozoa evolved from the flagellates. The presence of intermediate stages between the two groups (flagellates that produce amoeboid forms and amoeboid protozoa that have flagellated stages) adds support to this view. The sporozoa, although nonmotile, may have also originated from the flagellates. The origin of the ciliates is less certain, although the resemblance between cilia and flagella is obvious.

How are protists related to the other eukaryotic kingdoms? Plants, animals, and fungi are believed to have their ancestry in the Protist Kingdom. The green algae are regarded as the ancestors of plants, in part because of identical pigments and storage products. The choanoflagellates bear a striking resemblance to the sponges, and some biologists have suggested that they may be the ancestors of all animals. Some other flagellate is probably the ancestor of animals if the choanoflagellates are not. The fungi are believed to have protistan ancestors, but their lineage is less certain. The fungi and red algae both lack motile cells and share similarities in aspects of their sexual reproduction. The resemblance is strongest between the red algae and the higher fungi, opening the question about the evolutionary lineage of the lower fungi.

■ SUMMARY

I. The Kingdom Protista is composed of "simple" eukaryotic organisms.
  A. Although most protists are unicellular, their cell structure is more complex than those of animal or plant cells.
  B. Most protists are aquatic.
II. There is incredible diversity in the Protist Kingdom.
  A. Protists range in size from microscopic, single cells to multicellular organisms 60 meters in length.
  B. Different protists obtain their nutrients autotrophically or heterotrophically.
  C. Protists may be free-living or endosymbiotic, with relationships ranging from mutualism to parasitism.
  D. Many protists can reproduce both sexually and asexually. Others can reproduce only asexually.
  E. Protists have various means of locomotion, including flagella, pseudopods, and cilia. Some are nonmotile.
III. Various phyla within the Protist Kingdom show the diversity in this group of organisms.
  A. Protozoa are the heterotrophic protists.
    1. The sarcomastigophorans include the zooflagellates and the amoeboid protozoa, which move by pseudopods.
    2. The ciliates move by cilia, have a micronucleus and macronucleus, and undergo complex cell division.
    3. The sporozoa are parasitic protozoa that produce spores and are nonmotile. A sporozoon causes malaria.
  B. Algae are autotrophic protists.
    1. Dinoflagellates are mostly unicellular, biflagellate, photosynthetic organisms of tremendous ecological importance.

2. Diatoms are major producers in aquatic ecosystems. These are mostly single-celled, with silica impregnated in their cell walls.
    3. Euglenoids are single-celled, flagellated protists with pigmentation like green algae and the higher plants. They are not believed to be close relatives of either.
    4. Green algae exhibit a wide diversity in size, complexity, and reproduction. They are believed to be the ancestors of plants.
    5. Red algae are mostly multicellular, and lack motile cells. They have features which suggest to some that they may be the ancestors of the fungi. Other biologists believe that certain fungi may be the ancestors of the red algae.
    6. All brown algae are multicellular and produce flagellated cells during reproduction.
  C. Fungal-like protists were originally classified with the fungi, but have features that are clearly protistan.
    1. The vegetative body of the plasmodial slime molds is a multinucleate plasmodium. Reproduction is by spores.
    2. The cellular slime molds live vegetatively as individual, single-celled organisms. They aggregate for reproduction.
    3. The water molds have a coenocytic mycelium, and reproduce asexually by biflagellate zoospores and sexually by oospores.
IV. The probable consensus of evolutionary theorists places the flagellates at the base of the protist family tree. The flagellates are also thought to have given rise to animals, and green algae to plants. The origin of the fungi is unclear.

■ POST-TEST

1. In freshwater protozoa the _____ _____ pumps excess water out of the cell.
2. _____ is the union of gametes in sexual reproduction.
3. Cilia and flagella of protists have a 9 + 2 arrangement of _____.
4. Amoeboid sarcomastigophorans move by _____.
5. The _____ are possibly the ancestors of sponges.
6. Some actinopods have long _____ that protrude through pores in their skeletons.
7. Ciliates have a micronucleus that functions in _____ and a macronucleus that functions in _____.
8. The ciliates often display a sexual phenomenon called _____.
9. The _____ are a group of parasitic protozoa that form spores at some stage in their life.
10. Dinoflagellates are photosynthetic, biflagellate, and often covered by _____ plates.
11. A dinoflagellate bloom is known as a _____ _____.
12. The _____ are photosynthetic protists with cell walls composed of two halves that fit together like a Petri dish.
13. Chlorophylls *a* and *b* and carotenoids are found in green algae, _____, and plants.
14. _____ sexual reproduction is the fusion of two flagellated gametes of different sizes.
15. The multicellular bodies of _____ algae are differentiated into blades, stipes, holdfasts, and gas-filled floats.
16. The vegetative stage of the myxomycetes is a multinucleate _____.
17. The _____ slime molds behave as single-celled organisms until reproduction, when they aggregate.
18. The _____ _____ reproduce asexually by biflagellate zoospores and sexually by oospores.

■ REVIEW QUESTIONS

1. What are the characteristics of a typical protist? Why are protists so difficult to characterize?
2. Explain why the taxonomic position of *Euglena* is ambiguous.
3. Why are the protists so important to humans? How do they affect our lives? How are they important ecologically?
4. Some biologists still classify the algae as plants. Why could algae be considered plants? Why do most biologists classify them as protists rather than plants?
5. Plasmodial slime molds reproduce sexually, while cellular slime molds do not. Are there advantages for each organism in having the type of reproduction it does? Why do you think the cellular slime molds do not reproduce sexually?
6. Some biologists argue that multicellularity arose only once, while others feel it arose several times. Cite several reasons why each group holds its particular view.
7. Why aren't protozoa considered to be animals in this text?

■ RECOMMENDED READINGS

Donelson, J.E., and M.J. Turner. How the trypanosome changes its coat, *Scientific American*, February 1985. An explanation of how the flagellate that causes African sleeping sickness eludes the immune system in humans.

Godson, G.N. Molecular approaches to malaria vaccines, *Scientific American*, May 1985. A general explanation of the life cycle of the malarial sporozoon.

Hickman, C.P., L.S. Roberts, and F.M. Hickman. *Integrated Principles of Zoology*, 8th Edition, St. Louis, Times Mirror Mosby, 1988. A good review of the protozoa, with more emphasis on their biology than on the characteristics of each group.

Lee, J.J., S.H. Hutner, and E.C. Bovee. editors. *Illustrated Guide to the Protozoa*, Society of Protozoologists, P.O. Box 368, Lawrence, Kansas 66044, 1985. Some beautiful line drawings and electron micrographs of most of the protozoa.

Margulis, L. *Symbiosis in Cell Evolution*. San Francisco, W.H. Freeman and Company, 1981. The endosymbiont theory of the origin of eukaryotic cells is presented in depth.

Scagel, R.F., R.J. Bandoni, J.R. Maze, G.E. Rouse, W.B. Schofield, and J.R. Stein. *Nonvascular Plants: An Evolutionary Survey*. Belmont, California, Wadsworth Publishing Company, 1982. A comprehensive review of plant-like and fungal-like protists, including detailed discussions of their possible evolutionary development.

Scarlet waxy-cap mushrooms

26

Kingdom Fungi

■ **LEARNING OBJECTIVES**

After you have read this chapter you should be able to:

1. Describe the distinguishing characteristics of the Kingdom Fungi.
2. Contrast the structure of a yeast with that of a mold, and describe the body plan of a mold.
3. Trace the fate of a fungal spore that lands on an appropriate substrate such as an overripe peach, and describe conditions that permit fungal growth.
4. Give distinguishing characteristics for each of the three divisions of fungi and of the Deuteromycota, and give examples of each group.
5. Trace the black bread mold *Rhizopus nigricans* (a zygomycete) through the stages of its life cycle.
6. Trace a member of Ascomycetes (such as *Neurospora crassa*) through the stages of its life cycle.
7. Trace a member of Basidiomycetes (such as a mushroom) through the stages of its life cycle.
8. Describe the role of fungi as decomposers, and discuss the special ecological roles of lichens and mycorrhizae.
9. Summarize the economic significance of the fungi.
10. Identify several fungal diseases of plants and three human fungal diseases.

The tasty mushroom—delight of the gourmet—has much in common with the black mold that forms on stale bread and the mildew that collects on damp shower curtains. All of these life forms belong to the Kingdom Fungi, a diverse group of more than 100,000 known species. Although they vary strikingly in size and shape, all of the fungi are eukaryotes. Their cells are encased in cell walls during at least some stage in the life cycle, a characteristic that accounted in part for their original classification in the plant kingdom. However, fungi lack chlorophyll, a basic characteristic of plants. Fungi are heterotrophs that *absorb* their food through the cell wall and cell membrane. Fungi are nonmotile. They reproduce by means of spores, which may be produced sexually or asexually.

THEIR PLACE IN THE ECOSPHERE

Fungi make an important contribution to the ecological balance of our world. Like bacteria, most fungi are decomposers. These fungi are **saprobes** that absorb nutrients from wastes and dead organisms. Instead of

FIGURE 26–1 A parasitic fungus. After the ant ingests a spore of the fungus, the fungus begins to grow inside its body, absorbing nutrients from it. Eventually the fungus affects the nervous system of the ant, causing it to climb to a leaf high in the tree. There the ant dies, and the fungus develops a specialized reproductive structure known as a sporocarp. Altering the behavior of the host assures that the spores of the fungus are spread over a wide area. The effect of parasites on the behavior of host organisms is a rather new and interesting subject of biological research.

taking food inside its body and then digesting it as an animal would, a fungus digests food outside its body by secreting strong hydrolytic enzymes onto the food. In this way complex organic compounds are broken down into simpler compounds that the fungus can absorb. When fungi degrade wastes and corpses in this way, carbon, nitrogen, and mineral components of organic compounds are released and these elements can be recycled. For example, as decomposition occurs, carbon dioxide is released into the atmosphere and minerals are returned to the soil. Without this continuous decomposition, essential nutrients would soon become locked up in huge mounds of dead animals, feces, branches, logs, and leaves. These nutrients would be unavailable for use by new generations of organisms.

Fungi that are saprobes are also important in industrial fermentations. Brewing beer and making wine depend on the action of saprobic fungi.

While some fungi are saprobes, others form symbiotic relationships with other organisms. A symbiotic relationship is an intimate relationship between two or more organisms of different species. Some fungi are parasites, organisms that live in or on another organism and that are harmful to their host. Parasitic fungi absorb food from the living bodies of their hosts (Figure 26–1). Such fungi cause disease in humans and other animals, and are the most important disease-causing organisms of plants. Their activities cause billions of dollars in agricultural damage yearly.

Some types of fungi form mutualistic symbiotic relationships with other organisms. In mutualism both symbiotic partners benefit from the relationship. At the same time a mutualistic fungus absorbs nutrients from its host, it makes some contribution to its host's well-being. **Mycorrhizae** are mutualistic symbiotic relationships between fungi and the roots of complex plants. As these fungi break down organic compounds in the soil, mineral components are released and become available to the plant. A **lichen** consists of a fungus in intimate association with an alga or cyanobacterium. These compound organisms are important members of new ecologic communities and are important in soil formation (Figure 26–2). These mutualistic relationships will be discussed further in a later section.

BODY PLAN OF A FUNGUS: YEASTS AND MOLDS

The body structures of fungi vary in complexity, ranging from the single-celled yeasts to the multicellular molds, a term used loosely to include the mildews, rusts and smuts, mushrooms, and puffballs. In most fungi the rigid cell wall encasing each cell is composed in part of chitin, which is also a component of the external skeletons of insects and other arthropods.

FIGURE 26–2 The lichen *Cladonia leporina* has a mossy appearance. Because they are able to inhabit bare rock, lichens are often the pioneer organisms in a new community.

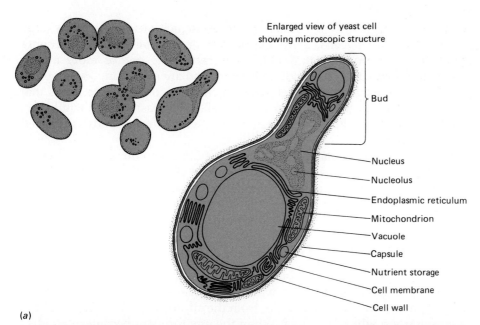

Enlarged view of yeast cell
showing microscopic structure

Bud

Nucleus

Nucleolus

Endoplasmic reticulum

Mitochondrion

Vacuole

Capsule

Nutrient storage

Cell membrane

Cell wall

(a)

FIGURE 26–3 Yeasts are unicellular fungi that reproduce asexually mainly by budding. They may also reproduce sexually. (*a*) Budding cells of the common bread yeast. (*b*) Photomicrograph of yeast cells, *Saccharomyces cerevisiae*, commonly known as baker's yeast, growing on potato dextrose agar.

(b)

Recall that chitin consists of subunits of a nitrogen-containing sugar, gluco-samine. Chitin is far more resistant to breakdown by microbes than is the cellulose of which plant cell walls are composed.

Yeasts are unicellular fungi that reproduce asexually mainly by budding (Figure 26–3) but also by fission, and sexually through spore formation. Each bud that separates from the mother yeast cell can grow into a new yeast. In general, yeast cells are larger than most bacteria. Some group together to form colonies. The yeasts are not classified as a single taxo-nomic group because many different fungi can be induced to form a yeast stage. Most yeasts, however, are found among the Ascomycota.

Most fungi are filamentous molds. A mold consists of long, branched, threadlike strings (or filaments) of cells called **hyphae** (singular, hypha). Hyphae form a tangled mass or tissue-like aggregation known as a **myce-lium** (Figure 26–4). The cobweb-like mold sometimes seen on bread con-sists of the mycelia of mold colonies. What are not seen are the extensive mycelia that grow down into the substance of the bread. The color of the

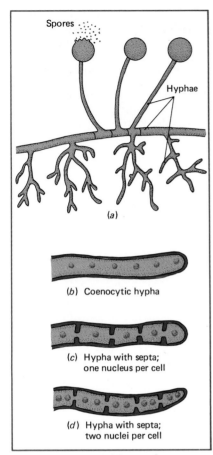

FIGURE 26–4 Molds. (*a*) Structure of a mold. (*b*) Coenocytic hypha. (*c*) Hypha divided into cells by septa; each cell has one nucleus. In some classes the septa are perforated, permitting cytoplasm to stream from one cell to another. (*d*) Septate hypha in which each cell has two nuclei.

FIGURE 26–5 Fungi can thrive in a wide range of environmental conditions; even refrigerated foods, such as these oranges, are not immune to fungal invasion.

mold comes from the reproductive spores, which are produced in large numbers on the mycelia. Some hyphae are divided by walls, called **septa** (singular, septum), into individual cells containing one or more nuclei; others are **coenocytic,** undivided by septa, and are something like an elongated, multinucleated giant cell. Septa generally contain large pores that permit organelles to flow from cell to cell. Cytoplasm flows within the hypha, providing a kind of circulation. The whole fungus body is referred to as a **thallus.**

Many fungi, particularly those that cause disease in humans, are **dimorphic**—that is, they have two forms: They can change from the yeast form to the mold form in response to changes in temperature, nutrients, or other environmental factors.

METABOLISM AND GROWTH

Fungi grow best in dark, moist habitats, but they are found universally wherever organic material is available. Moisture is necessary for their growth, and they can obtain water from the atmosphere as well as from the medium upon which they live. When the environment becomes very dry, fungi survive by going into a resting stage or by producing spores that are resistant to desiccation. Although the optimum pH for most species is about 5.6, some fungi can tolerate and grow in pH ranging from 2 to 9. Many fungi are less sensitive to high osmotic pressures than bacteria, and can grow in concentrated salt solutions or sugar solutions such as jelly that discourage or prevent bacterial growth (Figure 26–5). Fungi may also thrive over a wide temperature range. Even refrigerated food is not immune to fungal invasion.

When a fungal spore comes into contact with an appropriate substrate, perhaps an overripe peach that has fallen to the ground, it germinates and begins to grow (Figure 26–6). A threadlike hypha emerges from the tiny spore. Soon a tangled mat of hyphae infiltrates the peach, while other hyphae extend upward into the air. Cells of the hyphae secrete digestive enzymes into the peach, degrading its organic compounds to small molecules that the fungus can absorb. Fungi are very efficient at converting nutrients into new cell material. If adequate amounts of nutrients are available, fungi are able to store them in the mycelium.

REPRODUCTION

Fungal reproduction occurs in a variety of ways: asexually by fission, by budding, or, most commonly, by spore formation; or sexually by means that are characteristic for each group. Two types of reproductive structures are found in fungi. Sporangia produce spores, whereas gametangia are associated with the production of gametes. These reproductive structures are separated from the hyphae by complete septa.

Spores are usually produced on hyphae that project up into the air (aerial hyphae) above the food source. This arrangement permits the spores to be blown by the wind and distributed to new areas. Fungal spores are nonmotile cells dispersed by wind or by animals. In some fungi such as mushrooms, the aerial hyphae form large complex reproductive structures in which spores are produced. These structures are called **sporocarps** (or fruiting bodies). The familiar part of a mushroom or toadstool is a large sporocarp. We do not normally see the bulk of the organism, a nearly invisible network of hyphae buried out of sight in the rotting material it invades.

Spores may be produced either sexually or asexually. Unlike animal and vascular plant cells, fungal cells contain haploid nuclei; only the zygote is

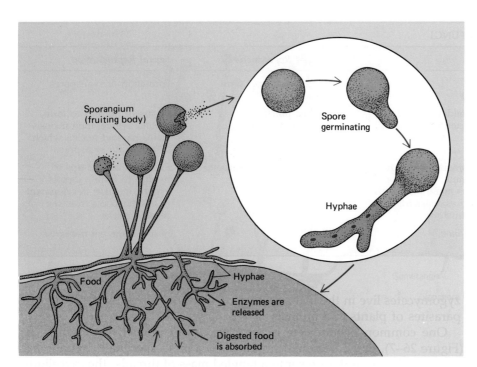

FIGURE 26–6 Germination and growth of a typical mold.

Sporangium (fruiting body)

Spore germinating

Hyphae

Food

Hyphae

Enzymes are released

Digested food is absorbed

diploid. In sexual reproduction, fungi often carry out some type of conjugation. Hyphae of two genetically different mating types come together and fuse, forming a diploid zygote. In two divisions of fungi, the genetically different nuclei do not fuse immediately, but remain separate within the fungal cytoplasm for most of the fungus's life. A hypha that has two genetically distinct nuclei within it is referred to as a **heterokaryotic** hypha. If the nuclei within a hypha are genetically similar, the hypha is **homokaryotic.** Hyphae that contain two genetically distinct nuclei within each cell are **dikaryotic.** Hyphae that contain only one nucleus per cell are **monokaryotic.**

CLASSIFYING FUNGI

The classification of fungi is based mainly on the characteristics of the sexual spores and fruiting bodies. Authorities do not agree on how to classify these diverse organisms, but the current trend is to assign them to three divisions (equivalent to phyla in animal taxonomy): division Zygomycota, division Ascomycota, and division Basidiomycota. The slime molds and the water molds (Oomycota), groups that have traditionally been classified as fungi, are now considered to be protists. Table 26–1 summarizes the classification of the fungi.

The fungi were once thought to have evolved from the algae, but many biologists now think they can be traced back to unicellular eukaryotes that are now extinct. Fungi are thought to be among the oldest eukaryotes. The three living divisions were represented in the Carboniferous period, 300 million years ago.

DIVISION ZYGOMYCOTA

The members of division **Zygomycota** are referred to as zygomycetes. They produce sexual spores, called **zygospores,** that remain dormant for a time. Their hyphae are coenocytic, that is, they lack septa. However, septa do form to separate the hyphae from sporangia or gametangia. Many

(a)

(b)

(c)

(d)

FIGURE 26–12 Some members of Basidiomycetes. (*a*) *Gyrodon meruloides*. (*b*) *Tremella mesenterica*, also called witch's butter. (*c*) Coral fungus, *Tremellodendron* sp. (*d*) Brown bracket fungi showing pattern.

basidiospores develop on the *outside* of the basidium, whereas ascospores develop *within* the ascus. The basidiospores are released, and when they come in contact with the proper environment, they develop into new mycelia. The hyphae are divided into compartments by septa. As in ascomycetes, the septa are perforated and allow cytoplasmic streaming.

The vegetative (feeding) body of the cultivated mushroom, *Agaricus campestris*, consists of a mass of white, branching, threadlike hyphae that occurs mostly below ground. Compact masses of hyphae, called buttons, develop along the mycelium. The button grows into the structure we ordinarily call a mushroom, which consists of a stalk and cap. More formally, the mushroom is referred to as a sporocarp, or **basidiocarp.** The lower surface of the cap consists of many thin perpendicular plates called **gills,** extending radially from the stalk to the edge of the cap. The basidia develop on the surfaces of these gills (Figure 26–13).

A typical basidiomycete life cycle includes three phases (Figure 26–14). Each individual fungus produces millions of basidiospores, and each basidiospore has the potential, should it happen upon an appropriate environment, to give rise to a new **primary mycelium.** Hyphae of this mycelium consist of monokaryotic (having a single nucleus) cells. When in the course of its growth such a hypha encounters another hypha of a different

FIGURE 26–13 A view of a basidiomycete, *Lepista nuda*, showing gills and basidium.

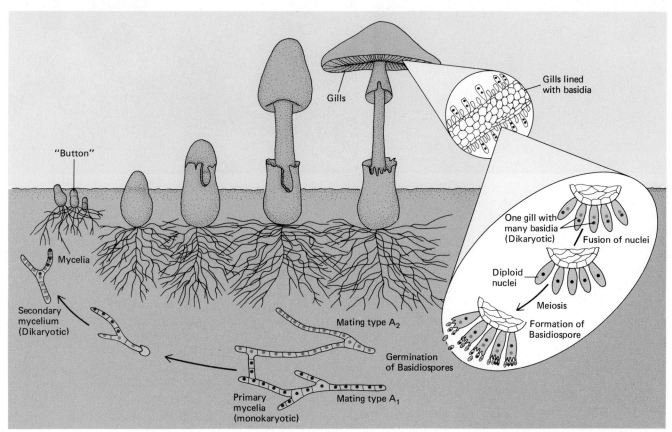

FIGURE 26–14 Life cycle of a basidiomycete. A mushroom develops from the mycelium, a mass of white, branching threads found underground. A compact button appears and grows into a fruiting body, or mushroom. On the undersurface of the fruiting body are gills, thin perpendicular plates extending radially from the stem. Basidia develop on the surface of these gills and produce basidiospores, which are shed. If these spores reach a suitable environment, they give rise to new mycelia.

mating type, the two hyphae fuse. However, the two haploid nuclei remain separate. In this way a secondary mycelium with dikaryotic hyphae is produced, in which each cell contains two haploid nuclei. Because each nucleus is of a different mating type, the hyphae are heterokaryotic.

The heterokaryotic hyphae of the mycelium grow extensively and eventually form compact masses, which are the mushrooms or basidiocarps. Each basidiocarp actually consists of intertwined hyphae matted together. On the gills of the mushroom the nuclei of the cells at the tips of the hyphae undergo nuclear fusion, forming diploid zygotes. These are the only diploid cells that form in the life history of a basidiomycete. Meiosis then occurs, forming four haploid nuclei. These nuclei move to the outer edge of the basidium. The cell wall then forms finger-like extensions into which the nuclei and some cytoplasm move. Basidiospores develop containing these nuclei. Behind the basidiospores the finger-like extensions come together, forming a cross wall and separating the basidiospores from the rest of the fungus by a delicate stalk. When the stalk breaks, the basidiospores are released.

DIVISION DEUTEROMYCOTA (IMPERFECT FUNGI)

About 25,000 species of fungi have been assigned to a group referred to as the **deuteromycetes** (Figure 26–15). They are also known as **imperfect fungi** because a sexual stage characteristic of the other fungi has not been observed during their life cycle. (Should further study reveal a sexual stage, these species will be reassigned to a different class.) Most deuteromycetes reproduce only by means of conidia and so are closely related to the ascomycetes; a few appear to be related to the basidiomycetes.

The unique flavor of cheeses such as Roquefort and Camembert is produced by the action of members of the genus *Penicillium*. The mold *P.*

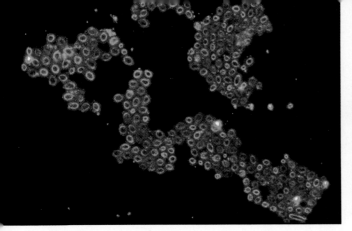

(a)

(b)

FIGURE 26–15 Deuteromycetes.
(a) *Aspergillus* from moldy bread.
(b) *Candida albicans*, a fungus that causes thrush.

roquefortii is found in caves near the French village of Roquefort; only cheeses produced in this area can be called Roquefort cheese. Another *Penicillium* species produces the antibiotic penicillin. *Aspergillus tamarii* and other imperfect fungi species are used in the Orient to produce soy sauce by fermenting soybeans.

Certain deuteromycetes cause ringworm and athlete's foot. Another member of this group, *Candida albicans*, causes thrush and other infections of mucous membranes, and still others cause systemic fungus infections.

LICHENS

Although a **lichen** looks like an individual plant, it is actually a symbiotic combination of a **phototroph** (an organism that carries on photosynthesis) and a fungus. The phototrophic component is usually either a green alga or a cyanobacterium, and the fungus is most often an ascomycete (Figure 26–16a). In some lichens from tropical regions, the fungus partner is a basidiomycete. The phototrophs found in lichens are also found as free-living species in nature, but the fungal component of the lichen is generally found only as part of the lichen. In the laboratory the fungal and algal components can be separated and grown separately in appropriate culture media. The alga grows more rapidly when separated, but the fungus grows slowly and requires many complex carbohydrates. Generally, the fungus does not produce fruiting bodies when separated in this way. The phototroph and fungus can be reassembled as a lichen, but only if they are placed in a culture medium under conditions incapable of supporting either of them independently.

What is the nature of this partnership? In the past the lichen has been considered a definitive example of mutualism, a symbiotic relationship that is beneficial to both species. The phototroph carries on photosynthesis, producing food for both members of the lichen. However, it is unclear how the phototroph benefits from the relationship. It has been suggested that the phototroph obtains water and minerals from the fungus as well as protection, mainly against desiccation. Some investigators have suggested that the lichen partnership is not really a case of mutualism but one of controlled parasitism of the phototroph by the fungus.

There are some 20,000 species of lichens (Figure 26–16b–d). Resistant to extremes of temperature and moisture, lichens grow everywhere that life can be supported at all except in polluted, industrial cities. They exist farther north than any plants of the Arctic region and are equally at home in the steaming equatorial jungle. They grow on tree trunks, mountain peaks, and bare rock. In fact, they are often the first organisms to inhabit bare rocky areas and play an important role in the formation of soil. Lichens gradually etch the rocks to which they cling, facilitating disintegration of rocks by wind and rain.

The reindeer mosses of Arctic regions are lichens that serve as the main source of food for the reindeer and caribou of the region. Some lichens produce colored pigments. One of them, orchil, is used to dye woolens, and another, litmus, is widely used in chemistry laboratories as an acid-base indicator.

Lichens vary greatly in size. Some are almost invisible; others, like the reindeer mosses, may cover miles of land with a growth that is ankle deep. Growth proceeds slowly; the radius of a lichen may increase by less than a millimeter each year. Some mature lichens are thought to be thousands of years old.

Lichens absorb minerals mainly from the air and from rainwater, although some may be absorbed directly from their substrate. They have no means of excreting the elements they absorb, and perhaps for this reason they are very sensitive to toxic compounds. Lichen growth has been used as an indicator of air pollution, especially sulfur dioxide. Absorption of such toxic compounds results in damage to the chlorophyll. Return of lichens to an area indicates a reduction in air pollution.

When a lichen dries out, photosynthesis stops, and the lichen enters a state of suspended animation in which it can tolerate severely adverse conditions such as great extremes of temperature. Lichens reproduce mainly by asexual means, usually by fragmentation. Generally, bits of the thallus break off and land on a suitable substrate, where they establish themselves as new lichens. Special dispersal units containing cells of both partners are released by some lichens. In others, the alga reproduces asexually by mitosis, while the fungus produces ascospores. The spores may be carried off by the wind and find an appropriate algal partner only by chance.

FIGURE 26–16 Lichens. (*a*) A fungal hypha encircling a single algal cell. (*b*) Lichen ascocarps. (*c*) A lichen from northern Michigan. (*d*) A multicolored lichen from Bylot Island, in northeastern Canada.

(a)

(b)

(c)

(d)

(a)

(b)

FIGURE 26–17 Mycorrhizae on the roots of (*a*) *Clonothus*. (*b*) Soybeans respond to mycorrhizae: left, a control plant; the other two plants have two types of mycorrhizae.

MYCORRHIZAE: FUNGUS-ROOTS

Mycorrhizae (fungus-roots) are symbiotic relationships between fungi and the roots of higher plants (Figure 26–17). Such relationships occur in more than 90% of all families of higher plants. The fungus benefits the plant by decomposing organic material in the soil, thus making the minerals available to the plant. The roots supply sugars, amino acids, and some other organic substances that may be used by the fungus. Basidiomycetes are the fungal partners in mycorrhizae of trees and other woody plants. Zygomycetes are the fungal partners of nonwoody plants.

The importance of mycorrhizae first became evident when horticulturalists observed that orchids do not grow unless they are colonized by an appropriate fungus. Similarly, it has been shown that many forest trees die from malnutrition when transplanted to nutrient-rich grassland soils that lack appropriate fungi. When forest soil that contains the appropriate fungi or their spores is added to the soil around these trees, they quickly assume a normal growth pattern.

ECONOMIC IMPORTANCE OF FUNGI

The vital ecological role of fungi as decomposers has already been discussed. It is well to remember that without these organisms, life on earth eventually would become impossible. The same powerful digestive enzymes that enable fungi to decompose wastes and dead organisms also permit them to reduce wood, fiber, and food to their components with great efficiency. Various molds produce incalculable damage to stored goods and building materials each year (Figure 26–18). Bracket fungi cause enormous losses by bringing about the decay of wood, both in living trees and in stored lumber. The timber destroyed each year by these basidiomycetes approaches in value that destroyed by forest fires.

Fungi for Food

Among the basidiomycetes, there are some 200 kinds of edible mushrooms and about 70 species of poisonous ones, sometimes called toadstools. Edible mushrooms can be cultivated commercially: more than 60 thousand metric tons are produced each year in the United States alone. The morels,

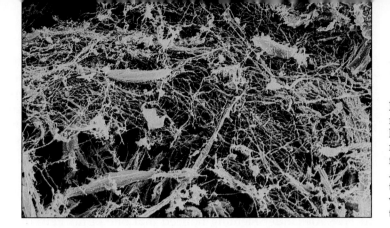

FIGURE 26–18 False color scanning electron micrograph of mycelium of dry rot in a piece of domestic plywood. Dry rot is a fungus that infects damp timber. As the mycelium grows, it destroys the cellulose in the wood and weakens its structure. Magnification × 49 at 35 mm size.

which are gathered and eaten like mushrooms, and truffles, which produce underground ascocarps, are ascomycetes. These delights of the gourmet are now being cultivated as mycorrhizae on the roots of tree seedlings.

Edible and poisonous mushrooms can look very much alike and may even belong to the same genus. There is no simple way to distinguish edible from poisonous mushrooms (Figure 26–19); they must be identified by an expert. For humans, the most toxic substances in mushrooms are certain cyclopeptides (amatoxins and phallotoxins). One of these cyclopeptides strongly inhibits messenger RNA synthesis in animal cells. Some of the most poisonous mushrooms belong to the genus *Amanita*. Toxic species of this genus have been appropriately called such names as "destroying angel" *(Amanita virosa)* and "death angel" *(Amanita phalloides)*; ingestion of a single cap can kill a healthy adult human.

Ingestion of certain species of mushrooms causes intoxication and hallucinations. The sacred mushrooms of the Aztecs, *Conocybe* and *Psilocybe*, are still used in religious ceremonies by Central American Indians and others for their hallucinogenic properties. The chemical ingredient psilocybin, chemically related to lysergic acid diethylamide (LSD), is responsible for the trancelike state and colorful visions experienced by those who eat these mushrooms.

The ability of yeasts to produce ethyl alcohol and carbon dioxide from glucose in the absence of oxygen is of great economic importance. The yeasts used in brewing beer and in baking are cultivated strains carefully kept to prevent contamination. Beer is made by fermenting grain, usually barley, flavored with hops (the dried conelike fruits of the female *Humulus lupulus* plant, a member of the mulberry family). During germination, the grain plant embryo degrades its starchy food supply to simple sugars, which are then fermented by the yeast.

During the process of making bread, carbon dioxide produced by the yeast becomes trapped in the dough as bubbles, which cause the dough to rise and give leavened bread its light texture. Both the carbon dioxide and the alcohol produced by the yeast are driven off during baking.

FIGURE 26–19 Edible and poisonous mushrooms look very much alike. (*a*) An edible mushroom, *Cantharellus cibarius*. (*b*) *Amanita muscaria*, or fly amanita, is probably the best known of the poisonous mushrooms. Its colorful yellow cap is covered by white scales. This species is not as toxic as others in the same genus and has been used in some cultures as a hallucinogen. (*c*) *Amanita phalloides*, the death angel. About 2 ounces (50 g) of this mushroom could kill a 68-kg man.

(a)

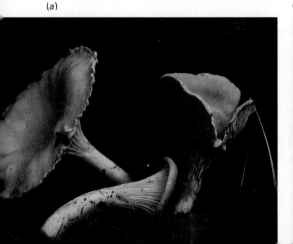

(b)

(c)

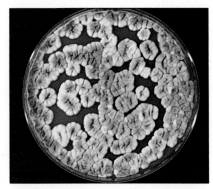

FIGURE 26–20 Colonies of *Penicillium chrysogenum* growing in a culture medium. Since Fleming's time, *P. chrysogenum* has become the fungus species most widely used to produce the antibiotic penicillin.

As noted earlier, many cheeses owe their unique flavors to specific fungi used in their production, and authentic Chinese soy sauce is produced only with the help of a deuteromycete that slowly ferments boiled soybeans. Soy sauce provides other foods with more than its special flavor; it also adds vital amino acids from both the soybeans and the fungi themselves to the low-protein rice diet. Fungi have been used in many cultures to improve the nutrient quality of the diet.

Fungi for Drugs and Useful Chemicals

In 1928 Alexander Fleming noticed that one of his Petri dishes containing staphylococci bacteria was contaminated by mold. The bacteria were not growing in the vicinity of the mold, leading Fleming to the conclusion that the mold was releasing some substance harmful to them. Within a decade of Fleming's discovery, penicillin produced by the deuteromycete *Penicillium notatum* was purified and used in treating bacterial infections (Figure 26–20). Penicillin is still the most widely used and most effective antibiotic. Another fungus, *Penicillium griseofulvicum*, produces the antibiotic griseofulvin, which is used clinically to inhibit the growth of fungi. Cyclosporine, the drug used to suppress immune responses in patients receiving organ transplants, is derived from two strains of deuteromycetes.

The ascomycete *Claviceps purpurea* infects the flowers of rye plants and other cereals. It produces a structure called an ergot where a seed would normally form in the grain head (Figure 26–21). When livestock eat this grain or when humans eat bread made from the infected rye, they may be poisoned by the very toxic substances in the ergot. These substances often cause nervous spasms, convulsions, psychotic delusions, and even gangrene. This condition, called ergotism, was known as St. Anthony's fire during the Middle Ages, when it occurred often. In 994 A.D. an epidemic of St. Anthony's fire caused more than 40,000 deaths. In 1722, the cavalry of Czar Peter the Great was felled by ergotism on the eve of the battle for the conquest of Turkey. This was one of several recorded times that a fungus changed the course of history. Lysergic acid, one of the constituents of ergot, is an intermediate in the synthesis of lysergic acid diethylamide (LSD). Some of the compounds produced by ergot are now used clinically in small quantities as drugs to induce labor, to stop uterine bleeding, to treat high blood pressure, and to relieve one type of migraine headache.

Fungi can be used as biological control agents to prevent damage by many insect pests. They can replace some very toxic biocides that are environmentally damaging. Other fungi are used commercially to produce citric acid and other chemicals of high quality.

Fungus Diseases of Plants

Fungi are responsible for many serious plant diseases, including epidemic diseases that spread rapidly and often result in complete crop failure, causing great economic loss and human suffering in some cases. All plants are apparently susceptible to some fungal infection. Damage may be localized in certain tissues or structures of the plant, or the disease may be systemic, affecting the entire plant. Fungus infections may cause stunting of plant structures or of the entire plant; they may cause growths like warts; or they may kill the plant.

Generally, a plant becomes infected after germ tubes of hyphae enter through pores (stomata) in the leaf or stem or through wounds in the plant body. As the fungal mycelia grow, they may remain mainly between the plant cells or may penetrate the cells. Parasitic fungi often produce special hyphal branches called **haustoria,** which penetrate the host cells and ob-

Carolina Biological Supply Company

FIGURE 26–21 The ascomycete *Claviceps purpurea* infects the flowers of cereals. It produces a structure called an ergot where a seed would normally form in the grain head.

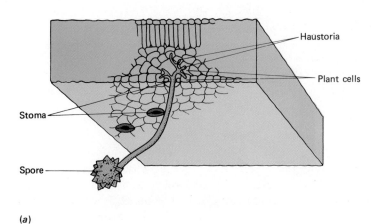

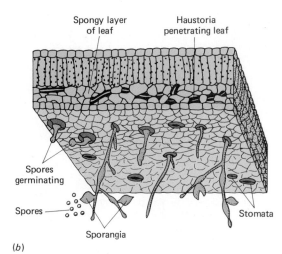

(a)

(b)

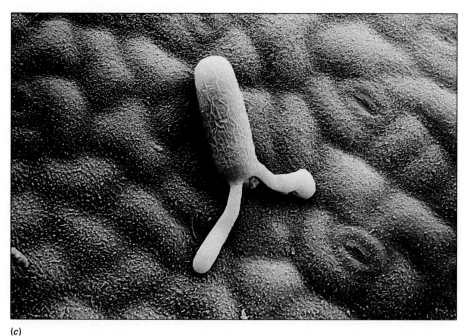

(c)

FIGURE 26–22 How plants become infected with fungi. (*a*) Germ tubes of hyphae enter through stomata or through wounds. Haustoria penetrate host cells and obtain nourishment from the cytoplasm. (*b*) The fungus may develop sporangia beneath the outer epidermis of the leaf. (*c*) False color scanning electron micrograph of a pathogenic fungus *Erisyphe pisi*, also known as *E. polygoni*, cause of powdery mildew on grasses such as cereals, clover, and peas. The infection begins with a conidium (spore) germinating a short hypha, which attaches to the leaf surface. The germinating conidium shown here is attaching to the leaf surface of a pea plant. If the plant is susceptible the fungus produces haustoria that penetrate the host cells and obtain nourishment from the cytoplasm. Magnification × 200 at 35 mm size.

tain nourishment from the cytoplasm (Figure 26–22). When the fungus is ready to reproduce, it may develop sporangia beneath the outer epidermis of the plant. Eventually, the epidermis ruptures, releasing the fungal spores into the air.

Some important plant diseases caused by ascomycetes are chestnut blight, Dutch elm disease, apple scab, and brown rot, which attacks cherries, peaches, plums, and apricots (Figure 26–23). Basidiomycetes include some 700 species of smuts and 6000 species of rusts that attack the various cereals—corn, wheat, oats, and other grains. In general, each species of smut is restricted to a single host species. Some of these parasites, such as the stem rust of wheat and the white pine blister rust, have complex life cycles that involve two or more different plants, during which several kinds of spores are produced. The white pine blister rust must infect a gooseberry or a red currant plant before it can infect another pine. The wheat rust must infect an American barberry plant[1] at one stage in its life

---

[1]The Japanese barberry plant commonly used today in hedges or as a decorative shrub, however, is resistant to infection by rust; only the American variety figures in the wheat rust life cycle.

(a)

(b)

FIGURE 26–23 Fungi are responsible for many serious plant diseases. (*a*) Apple-cedar rust on apple leaves. The mycelia cause cluster cups (aecia) on the undersides of the leaves. (*b*) Black knot of plum.

cycle. Since this has been known, the eradication of American barberry plants in wheat-growing regions has effectively reduced infection with wheat rust. However, the eradication must be complete, for a single barberry bush can support enough wheat rust organisms to infect hundreds of acres of wheat.

Fungus Diseases of Animals

Although the skin and mucous membranes of healthy animals present effective barriers to fungal penetration, some fungi cause disease in humans and other animals (Figure 26–24). Some of these cause superficial infections in which the fungi infect only the skin, hair, or nails. Others cause systemic infections, in which fungi infect deep tissues and internal organs and may spread through many regions of the body.

Ringworm and athlete's foot are examples of superficial fungus infections. Candidiasis is an infection of mucous membranes of the mouth or vagina and is among the most common fungal infections. Histoplasmosis is a serious human systemic fungus infection caused by a fungus that sporulates abundantly in soil containing bird droppings; a person who inhales the spores may then develop the infection. Most pathogenic fungi are opportunists that cause infections only when the body's immunity is lowered.

FIGURE 26–24 Some fungi infect humans and other animals. (*a*) A fungus *Microsporum canis* carried by pets that can cause fungus infections in humans, as shown in (*b*).

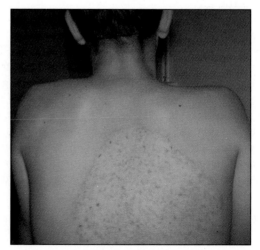

(a)

(b)

■ SUMMARY

I. Fungi are eukaryotes with cell walls.
   A. Fungi lack chlorophyll and so are heterotrophic, absorbing their food through the cell wall and cell membrane.
   B. They reproduce by means of spores, which may be produced sexually or asexually.
II. Fungi function ecologically as decomposers that produce carbon dioxide and break down organic compounds.
III. A fungus may be unicellular, the yeast form, or multicellular, the mold form.
   A. The thallus of a mold consists of long, branched hyphae, which form a mycelium.
   B. In the zygomycetes, the hyphae are coenocytic (undivided by septa).
   C. In other fungi, perforated septa are present.
IV. When a fungal spore comes into contact with an appropriate substrate, it germinates and begins to grow.
   A. Some hyphae infiltrate the substrate, digesting its organic compounds with enzymes.
   B. Spores are produced on aerial hyphae.
V. Members of division Zygomycota produce sexual resting spores called zygospores. They have coenocytic hyphae. The black bread mold *Rhizopus nigricans* is a common member of this group.
VI. Members of division Ascomycota produce asexual spores called conidia at the tips of conidiophores. Sexual (haploid) spores called ascospores are produced in asci. This group includes yeasts, morels, truffles, red and brown molds, and many others.
VII. Members of division Basidiomycota produce basidiospores on the outside of a basidium. Basidia develop on the surface of gills in mushrooms. This group includes mushrooms, toadstools, rusts, and smuts.
VIII. The deuteromycetes are the imperfect fungi, species for which a sexual stage has not been observed. Most reproduce by conidia. Members of this group include *Aspergillus tamarii*, used to produce soy sauce, and *Candida albicans*, which can cause human fungal infections.
IX. A lichen is a symbiotic combination of a fungus and a phototroph in which the fungus benefits from the photosynthetic activity of the phototroph. Lichens play an important role in soil formation.
X. Mycorrhizae are symbiotic relationships between fungi and the roots of higher plants. The fungus decomposes organic material, making minerals available to the plant. The plant may secrete organic compounds needed by the fungus.
XI. Fungi are of both positive and negative economic importance.
   A. Mushrooms, morels, and truffles are used as food; yeasts produce ethyl alcohol and so are vital in production of wines and beer, and are also used to make bread; certain fungi are used to produce cheeses and soy sauce.
   B. Fungi are used to make penicillin and other antibiotics; ergot is used to produce certain drugs; fungi can be used as biological control agents; they can be used to make citric acid and many other industrial chemicals.
XII. Fungi cause many plant diseases including potato blight, wheat rust, Dutch elm disease, and chestnut blight; they cause human diseases such as ringworm, athlete's foot, histoplasmosis, and candidiasis.

■ POST-TEST

1. Ecologically, fungi serve as _____.
2. Yeasts reproduce asexually mainly by _____, but also by _____.
3. A mold consists of threadlike strings of cells called _____, which form a tangled mass called a _____.
4. The familiar portion of a mushroom is actually a large _____.
5. *Rhizopus* and other members of the zygomycetes form sexual resting spores called _____.
6. The term **heterothallic** means that there are two _____.
7. In ascomycetes (sac fungi), asexual reproduction involves formation of spores called _____, which are pinched off at the tips of _____.
8. Sexual reproduction in ascomycetes involves production of spores known as _____ within structures called _____.
9. The type of sexual spore produced by a mushroom is a _____.
10. Basidia develop on the surface of perpendicular plates called _____.
11. The deuteromycetes are known as imperfect fungi because their _____ stage has not been observed.
12. A _____ consists of a phototroph and a fungus that are intimately related.
13. Mycorrhizae are fungi that form symbiotic relationships with the _____ of complex _____.
14. When the ascomycete *Claviceps purpurea* infects the flowers of cereals, it produces a structure called an _____ that is toxic.
15. Haustoria are hyphae produced by parasitic fungi that can _____.
16. Wheat rust must infect an American _____ at one stage in its life cycle.

■ REVIEW QUESTIONS

1. How does a fungus differ from an alga? What are the distinguishing features of a fungus?
2. How does the body plan of a yeast differ from that of a mold? Describe the body structure of a mold.

3. What is the difference between a hypha and a mycelium? between an ascus and a basidium?
4. Describe the life cycle of a typical mushroom.
5. What is the ecological importance of fungi? of lichens? of mycorrhizae?
6. What measures can you suggest to prevent bread from becoming moldy?
7. Draw diagrams to illustrate the life cycle of the black bread mold *Rhizopus nigricans*.

8. Briefly describe three important fungal diseases of plants, and three fungal diseases of humans.
9. What strategies do you think could be used to prevent fungal disasters such as the Irish potato famine?
10. What conditions might permit opportunistic fungi to cause disease?

■ RECOMMENDED READINGS

Ahmadjian, V. The nature of lichens. *Natural History*, March 1982. A beautifully illustrated summary of the algal-fungal partnership.

Anagnostakis, S. Biological control of chestnut blight. *Science*, Vol. 215, January 1982. An interesting account of the history of chestnut blight fungus in the United States and of virus-like agents that may be useful in controlling it.

Ross, I.K. *Biology of the Fungi: Their Development, Regulation, and Associations*. New York, McGraw-Hill Book Co., 1979. A good reference text on the fungi.

Strobel, G.A., and G.N. Lanier. Dutch elm disease. *Scientific American*, August 1981. An account of Dutch elm disease, a deadly fungal infection of elm trees, and of new biological techniques used to attack the fungus and the beetles that spread it.

Wernick, R. From ewe's milk and a bit of mold: a fromage fit for a Charlemagne. *Smithsonian*, Vol. 13, No. 11, February 1983. The story of the production of Roquefort cheese.

Sporangia on silver fern (*Cyanthea dealbata*)

27

The Plant Kingdom: Seedless Plants

■ **LEARNING OBJECTIVES**

After you have read this chapter you should be able to:

1. Discuss the environmental challenges faced by land plants and relate adaptations which land plants evolved to meet these challenges.
2. Name the group of organisms from which the land plants evolved and give several pieces of evidence used to support this evolutionary theory.
3. Summarize the features possessed by the bryophytes that distinguish them from green algae.
4. Discuss the advancements the ferns and fern allies have over the mosses and liverworts.
5. Diagram a generalized plant life cycle, clearly showing alternation of generations.
6. Compare the generalized plant life cycle for homosporous plants with one for heterosporous plants.

THE PLANT KINGDOM

The Plant Kingdom comprises thousands of different species that live in every conceivable habitat, from the frozen Arctic tundra to lush tropical rain forests. Plants range in size from minute, almost microscopic duckweeds to massive giant sequoias. Although plants exhibit an incredible diversity in size, habit, and form, they are believed to have evolved from common ancestors.

There are four major groups of plants living today (Table 27–1). The mosses and other bryophytes lack a vascular, or conducting, system and are therefore restricted in size. The other three groups of plants possess vascular tissues, **xylem** for water and mineral conduction and **phloem** for food conduction. Ferns and their allies are vascular plants that reproduce by spores. The gymnosperms and angiosperms are both vascular plant groups whose primary means of reproduction is by seeds. Gymnosperms are naked seed plants, whose seeds are often produced in a cone, while angiosperms produce seeds enclosed within a fruit.

Origin of Land Plants

Land plants evolved from ancient green algae (Figures 27–1 and 27–2). For several reasons, plant biologists believe that the green algae gave rise to the land plants. The green algae share a number of biochemical and metabolic traits with plants. Both contain the same photosynthetic pigments, chloro-

617

Table 27–1 **THE PLANT KINGDOM**

I. Nonvascular plants with a dominant gametophyte generation
        Division Bryophyta (bryophytes)
           Class Bryopsida (mosses)
           Class Hepatopsida (liverworts)
           Class Anthoceropsida (hornworts)
II. Vascular plants with a dominant sporophyte generation
  A. Seedless plants
        Division Pterophyta (ferns)
        Division Psilophyta (whisk ferns)
        Division Sphenophyta (horsetails)
        Division Lycophyta (club mosses)
  B. Seed plants
    1. Naked seed plants
        Division Coniferophyta (conifers)
        Division Cycadophyta (cycads)
        Division Ginkgophyta (ginkgo)
        Division Gnetophyta (gnetophytes)
    2. Seeds enclosed within a fruit
        Division Magnoliophyta (angiosperms)
           Class Magnoliopsida (dicots)
           Class Liliopsida (monocots)

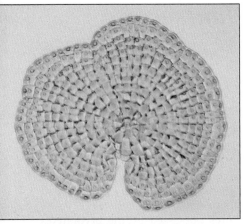

FIGURE 27–1 The green alga *Coleochaete* resembles a group of green algae which may have been the ancestors of the land plants. It is considered to be quite primitive because it shares a number of features with its extinct ancestors.

phylls *a* and *b*, carotenes, and xanthophylls. Also, both store their excess carbohydrates as starch. Cellulose is a major component of the cell walls of both. Finally, certain details of cell division, including the formation of a cell plate, are shared by plants and many green algae.

Characteristics of Plants

Plants are multicellular organisms that photosynthesize to obtain their energy. Plants use the green pigment chlorophyll to absorb radiant energy, which is then converted to the chemical energy found in carbohydrates. In addition to chlorophylls *a* and *b*, all plants have **xanthophylls** (yellow pigments) and **carotenes** (orange pigments).

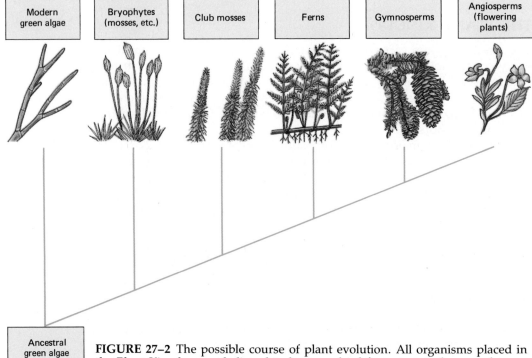

FIGURE 27–2 The possible course of plant evolution. All organisms placed in the Plant Kingdom are believed to have evolved from ancestral green algae.

One of the most important adaptations plants must have in order to survive on land is a waxy covering, the **cuticle,** over their aerial parts. The cuticle is essential for existence on land because it helps prevent desiccation, or drying out, of plant tissues by evaporation. Because plants are rooted in the ground, they cannot move to wetter areas during dry spells. Therefore, a cuticle is critical. Algae do not possess a cuticle, but land plants do.

Algae and photosynthetic plants that live in aquatic environments are bathed by water containing dissolved materials, including dissolved carbon dioxide, or carbonate. Carbonate moves by diffusion into algal cells and is used as the raw material for photosynthesis. Land plants obtain their carbon from the atmosphere as carbon dioxide. This gas must be accessible to the chloroplasts inside green plant cells. Because the external surfaces of plant stems and leaves are covered by the waxy cuticle, however, gas exchange through the cuticle between the atmosphere and the inside of cells is negligible. To overcome this problem, land plants have tiny openings, or **stomata,** in their surface tissues, which permit the gas exchange essential for photosynthesis.

Plants have multicellular sex organs, or **gametangia,** each of which possesses a sterile layer of cells surrounding the gametes. Each female organ, the **archegonium,** produces a single egg. Sperm are produced in the male sex organ, the **antheridium.** Presumably, the gametes produced within these organs are protected from desiccation by the outer jacket of sterile cells. Algae lack such organs.

There is a final difference between the land plants and the algae. In plants the fertilized egg develops into a multicellular **embryo** *within* the female gametangium. Thus, during its development the embryo is protected. In algae, development of the fertilized egg occurs independent of the gametangium. In some algae the gametes are released before fertilization, while in others the fertilized egg is released.

The Life Cycle of Plants

Plants have a clearly defined **alternation of generations;** that is, they spend part of their lives in the haploid stage and part in the diploid stage (Figure 27–3). The haploid portion of the life cycle is called the **gametophyte generation** because it gives rise to haploid gametes by mitosis. The diploid portion of the life cycle is the **sporophyte generation,** which gives rise to **spores** immediately following meiosis.

The haploid gametophytic plant produces male sex organs, or antheridia, in which sperm form. Female gametangia, or archegonia, are also formed by the gametophytic plant. Each archegonium bears a single egg. The sperm get to the archegonia by a variety of ways, and swim down the neck of the archegonium. One sperm will fuse with the egg. This process, known as **fertilization,** results in a fertilized egg, or **zygote.**

Because the zygote is diploid, it is the first stage in the sporophyte generation. The zygote divides by mitosis and develops into a multicellular embryo, which is supported and protected by the gametophytic plant. Eventually, the embryo matures into the sporophyte plant. The sporophyte plant has special cells that are capable of dividing by meiosis. These cells, called **spore mother cells,** undergo meiotic division and form haploid spores. All spores produced by plants are the result of meiosis. This is in contrast to algae and fungi, which may produce spores by both meiosis and mitosis.

Because the spores are haploid, they represent the first stage in the gametophyte generation. Each spore is capable of growing by mitosis into a multicellular gametophytic plant, and the cycle continues as previously

FIGURE 27–3 The basic plant life cycle. All land plants have modifications of this cycle. Note that plants alternate generations, spending part of their life in the haploid (gametophyte) stage and part in the diploid (sporophyte) stage.

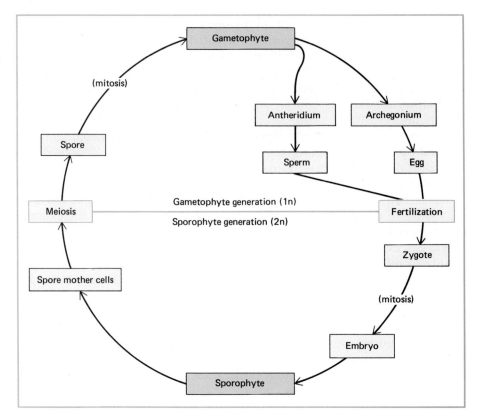

discussed. Thus, plants have an alternation of generations, alternating between a haploid (gametophyte) stage and a diploid (sporophyte) stage.

DIVISION BRYOPHYTA: MOSSES AND OTHER BRYOPHYTES

Mosses and other bryophytes are the only land plants that lack vascular tissues. Because they have no means of transporting water, food, and essential minerals for extensive distances, they are restricted in size. Also, they require a moist environment for active growth and reproduction, although some bryophytes are tolerant of dry areas.

The bryophytes are sometimes divided into three classes: the mosses, the liverworts, and the hornworts. These three groups of plants may or may not be closely related. Their life cycles are similar, however.

Class Bryopsida: The Mosses

Moss plants usually live in dense colonies or beds. Each individual plant has tiny rootlike structures, or **rhizoids,** that anchor the plant to the soil. Each plant also has an upright "stem," which bears "leaves." Because mosses lack specialized vascular tissues, they cannot be said to possess true roots, stems, or leaves. However, some moss species do have water-conducting cells and food-conducting cells, although they are not as specialized as in the vascular plants.

The leafy green plant (Figure 27–4) is the gametophyte generation of mosses. It bears its gametangia at the top of the plant. Many moss plants have separate sexes; these mosses have male plants that bear antheridia, and female plants that bear archegonia. Other mosses produce antheridia and archegonia on the same plant.

FIGURE 27–4 The leafy green gametophyte plants of mosses grow in dense clusters. These are male moss plants with multiple antheridia located at the top of each plant.

In order for fertilization to occur, the sperm produced in the antheridia must fertilize the egg held within the archegonium. Sperm transport requires some imagination to envision, particularly if archegonia and antheridia are located on separate plants at some distance from one another. How do the sperm get to the archegonia?

As you may recall, the first land animals to evolve from fish were the amphibians. Although they adapted to life on land, they still depend on water to accomplish fertilization and must return to water to reproduce. The mosses resemble the amphibians in this respect: Although they have adapted to life on land, they need water to accomplish fertilization. This requirement of water for fertilization is considered to be a primitive characteristic retained from their algal ancestors.

The sperm may be transported from antheridia to archegonia by splashing rain droplets. A rain drop lands on the top of a male gametophytic plant and sperm are released into it from the antheridia. When another rain drop lands on the male plant, it may splash the sperm-laden droplet into the air and onto the top of a nearby female plant. Also, insects may touch the sperm-laden fluid and inadvertently carry it for considerable distances. Once the sperm land on the female plant, one will swim down the neck of the archegonium and fuse with the egg.

The zygote formed as a result of fertilization grows into a multicellular embryo by mitosis and matures into the moss sporophyte. This sporophyte plant grows out of the top of the female gametophyte, and it is attached to and nutritionally dependent upon it (Figure 27–5). Although it is initially green in color and photosynthetic, it becomes a golden brown at maturity and is composed of three main parts: a **foot,** which anchors the sporophyte to the gametophyte; a **seta,** or stalk; and a **capsule,** which contains spore mother cells.

These spore mother cells undergo meiosis to form haploid spores. When the spores are mature, the capsule opens by various mechanisms to release the spores. These microscopic cells are carried by wind or rain to other places. If a spore lands in a suitable spot, it germinates and grows into a filamentous thread of green cells called a **protonema.** The protonema, which looks like a filamentous green alga, forms buds. Each bud grows into a leafy green gametophyte plant, and the life cycle continues as already described (Figure 27–6).

The gametophyte generation is considered to be the dominant generation in mosses because it is capable of living independently of the sporophyte. By contrast, the sporophyte generation in mosses is at all times attached to and dependent on the gametophyte plant. A dominant game-

FIGURE 27–5 Haircap moss. The moss sporophyte generation grows out of the top of the gametophyte plant. Each sporophyte is attached to and dependent on the gametophyte plant for nourishment. Spores are produced by meiosis within the capsule at the tip of each sporophyte.

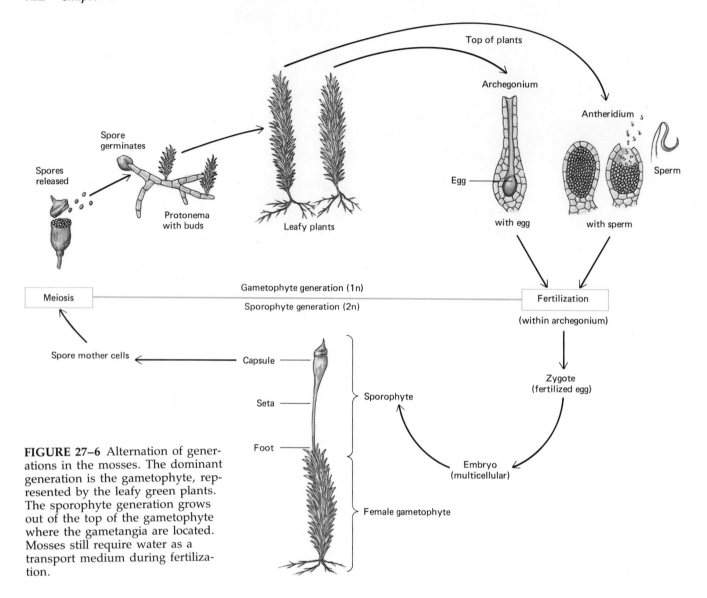

Top of plants

Archegonium

Antheridium

Spore germinates

Spores released

Protonema with buds

Leafy plants

Egg

Sperm

with egg

with sperm

Gametophyte generation (1n)

Meiosis

Sporophyte generation (2n)

Fertilization

(within archegonium)

Spore mother cells

Capsule

Seta

Sporophyte

Zygote (fertilized egg)

Foot

Embryo (multicellular)

Female gametophyte

FIGURE 27–6 Alternation of generations in the mosses. The dominant generation is the gametophyte, represented by the leafy green plants. The sporophyte generation grows out of the top of the gametophyte where the gametangia are located. Mosses still require water as a transport medium during fertilization.

tophyte generation is considered to be a primitive characteristic. Many algae also have alternation of generations with a dominant gametophyte.

Although all land plants evolved from green algal ancestors, the mosses are not in a direct path to the higher plants. They represent an evolutionary sideline that developed from the algae or possibly from vascular plants.

FIGURE 27–7 Peat mosses grow in boggy areas. As older plants die, new ones grow over them. The acidic nature of the peat bog prevents rapid decay and the dead organic material accumulates. Peat bogs can be many meters thick.

FIGURE 27–8 Many liverworts have a thallus body form characterized by flattened, ribbon-like lobes. The thallus is the gametophyte plant. Note the moon-shaped gemmae cups on the thallus. These gemmae cups contain gemmae, asexual structures which can grow into new thalli.

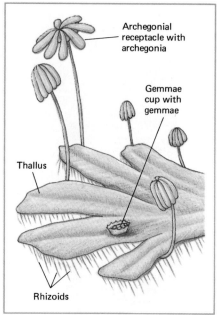

FIGURE 27–9 Some liverworts have male and female structures on separate plants. This thallus has stalked structures which terminate in archegonial receptacles with archegonia.

Mosses make up an inconspicuous but significant part of their environment. They play an important role in forming soil, especially in primary succession (see Chapter 56). Because they grow tightly packed together in dense colonies, they hold the soil in place and help to prevent soil erosion. They provide food for animals, especially birds and mammals.

Commercially, the most important mosses are the peat mosses in the genus *Sphagnum.* One of the distinctive features of *Sphagnum* is the presence of large ''empty'' cells in the ''leaves,'' which apparently function to hold water. This feature makes peat mosses particularly beneficial as a soil conditioner. For example, when added to sandy soils, they help to hold and retain moisture in the soil. In other countries peat moss is often collected, dried, and burned for fuel (Figure 27–7).

The name ''moss'' is often commonly used for plants that have no affinities to the mosses. For example, reindeer moss is a lichen that is a dominant form of vegetation in the Arctic tundra. Spanish moss is a flowering plant, and club moss is a relative of ferns.

Class Hepatopsida: The Liverworts

The morphology of some liverworts is quite different from that of mosses. Their body form is often a flattened, leaf-like **thallus,** which is lobed (Figure 27–8). They were named liverworts because the lobes of their thalli superficially resemble the lobes of the human liver. In medieval times many people believed in the Doctrine of Signatures, that a ''signature'' on each plant gave a clue to its use. Because of their resemblance to the human liver, liverworts were considered to have medicinal value in treating liver ailments, a belief that has not been supported by modern medical research.

On the underside of the liverwort thallus are rootlike rhizoids, which anchor the plant to the soil. As with other bryophytes, the liverworts lack vascular tissue. They are small and generally inconspicuous plants that are restricted largely to damp environments.

Many liverworts have a leafy appearance rather than a lobed thallus. Some of these leafy liverworts are superficially very similar to mosses, with ''leaves,'' ''stems,'' and rhizoids.

Liverworts reproduce both sexually and asexually. Their sexual reproduction involves the production of archegonia (Figure 27–9) and antheridia (Figure 27–10) on the haploid thallus. Their life cycle is basically the same

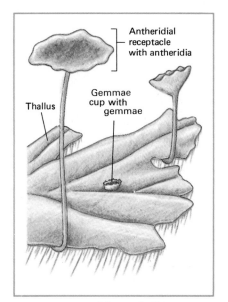

FIGURE 27–10 The male liverwort thallus produces antheridial receptacles on stalks. Numerous antheridia are embedded in the antheridial receptacle.

623

FIGURE 27–11 Aspects of the liverwort life cycle. (*a*) After fertilization of the egg within an archegonium on the archegonial receptacle, the diploid sporophyte generation develops. As in the mosses, the sporophyte generation is always attached to and dependent on the gametophyte plant. (*b*) The liverwort sporophyte has the same basic structure as the moss sporophyte, with its foot, stalk, and capsule. Meiosis occurs in the capsule, producing haploid spores. Note the elastic elaters that help to scatter the spores when the capsule cracks open. Elaters are hydroscopic, absorbing moisture from the atmosphere. Changes in humidity cause them to twist or coil, scattering spores in the process.

as that of the mosses, although some of the structures look quite different. The sporophyte generation, which is usually somewhat spherical, is attached to the gametophyte plant (Figure 27–11).

One of the ways that liverworts reproduce asexually is by forming tiny balls of tissue called **gemmae**. These gemmae are borne in a saucer-shaped structure, the **gemmae cup,** directly on the liverwort thallus. Splashing raindrops and small animals aid in the dispersal of gemmae. When gemmae land in a suitable place, each can grow into a liverwort thallus.

A second way that liverworts reproduce asexually is by thallus branching and growth. As the thallus continues to grow, the individual lobes elongate and form longer extensions of the thallus. When the older part of the thallus that originally attached the individual lobes dies, each extended lobe becomes a separate plant. Both of these mechanisms of reproduction, gemmae and thallus branching, are asexual because they do not involve fusion of gametes or meiosis.

Class Anthoceropsida: The Hornworts

The hornworts are a small group of plants whose gametophytes superficially resemble the thalloid liverworts. Hornworts may or may not be closely related to other bryophytes. For example, their cell structure, particularly the presence of a single large chloroplast in each cell, is reminiscent of certain algae. Mosses and liverworts, on the other hand, are like all other plants because they have many disk-shaped chloroplasts per cell.

Archegonia and antheridia are embedded in the gametophyte thallus. After fertilization, the sporophyte generation projects out of the gametophyte tissue, forming a spike or "horn" (Figure 27–12). Meiosis occurs within the sporangium, and spores are formed. A single gametophytic plant often produces a number of sporophytes.

FIGURE 27–12 *Anthoceros*, a typical hornwort. The leafy green thallus, which is 1–2 cm in diameter, is the gametophyte generation. After fertilization the sporophyte projects out of the thallus.

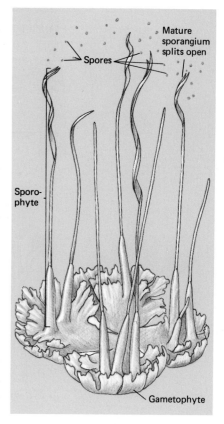

SEEDLESS VASCULAR PLANTS: FERNS AND THEIR ALLIES

The ferns represent an ancient group of plants that is still successful today. They are especially common in temperate woodlands and tropical rain forests, where they are found in the greatest variety. The three groups of plants that are considered to be allies of the ferns because of similarities in life cycles are the whisk ferns, club mosses (lycopods), and horsetails. According to the fossil record, the lycopods are the most ancient group of seedless vascular plants. The fern allies were of considerable importance in past ages. Fossil evidence indicates that these plants were often immense trees. Most are extinct today (Figure 27–13) except for a few smaller representatives of the ancient groups.

The major advancement of the ferns and their allies over mosses and other bryophytes is the presence of specialized vascular tissues. This system of conduction enables vascular plants to achieve larger sizes than mosses because water, dissolved minerals, and food can be transported to all parts of the plant. Although the ferns in temperate areas are relatively small plants, there are tree ferns in the tropics that grow to heights of 60 feet. The ferns and fern allies all have stems with vascular tissues. Most have roots and leaves as well.

The evolution of the leaf as the main organ of photosynthesis has been studied extensively. There are two basic types of leaves (Figure 27–14). The **microphyll** presumably evolved from extensions of stem tissue. Microphylls are usually small and possess one vascular strand.

FIGURE 27–13 The earliest vascular plants to colonize the land evolved approximately 400 million years ago. (*a*) *Rhynia* is a leafless plant which probably lived in marshes. (*b*) *Psilophyton* evolved somewhat later (approximately 375 million years ago). Plants like *Psilophyton* were probably the ancestors of the ferns. (*c*) *Asteroxylon* was an early club moss. All three of these plants are extinct.

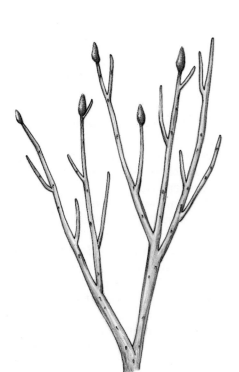

(a)

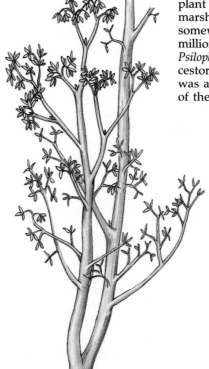

(b)

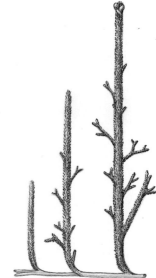

(c)

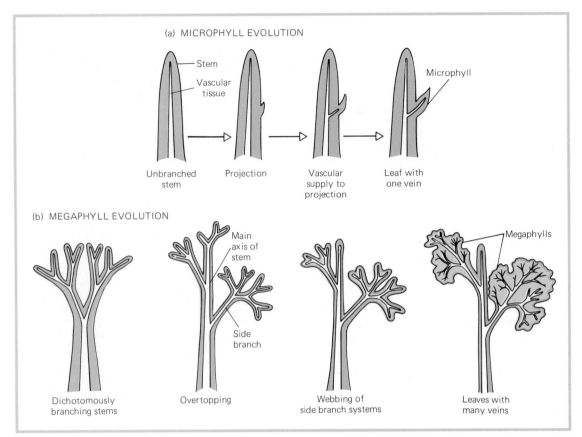

FIGURE 27–14 There are two kinds of leaves. (*a*) Microphylls are generally believed to have originated as outgrowths of stem tissue that developed a single vascular strand later. Both club mosses and horsetails have microphylls. (*b*) Megaphylls, which are more complex and have multiple veins, probably evolved from a fusion of side branches. Ferns, gymnosperms, and angiosperms have megaphylls.

Megaphylls are leaves that apparently evolved from stem branches, which gradually filled in with additional tissue, forming most leaves as we know them today. Megaphylls possess more than one vascular strand, as would be expected if they evolved from several branches.

Division Pterophyta: The Ferns

The ferns represent one of the oldest groups of vascular plants. Fossil ferns have been studied that are 360 million years old. Most of the ferns are terrestrial, although a few have returned to aquatic habitats. Although they range from the tropics to the Arctic Circle, most species are found in the moist tropics (Figure 27–15). In temperate areas, ferns commonly inhabit moist woodlands and stream banks.

The life cycle of ferns involves a clearly defined alternation of generations (Figure 27–16). The fern plant that is grown as a house plant represents the diploid, or sporophyte, generation. Its body is composed of a horizontal underground stem, called a **rhizome,** which bears roots and leaves, or **fronds.** As each young frond first emerges from the ground, it is tightly coiled and resembles the top of a violin. For this reason, it is called a **fiddlehead.** As the fiddleheads grow and expand, they unroll to form the fronds. Fern fronds are often compound and dissected, forming beautifully complex leaves. The fronds, roots, and rhizome are considered to be true plant organs because of the presence of vascular tissue in each.

Since the conspicuous plant body of the fern is the sporophyte generation, it forms spores. Spore production usually occurs on the fronds. Certain areas on the fronds develop **sporangia,** or spore cases, in which spore mother cells are formed. The sporangia are frequently borne in clusters, called **sori,** on the fronds. Within the sporangia, spore mother cells undergo meiosis to form haploid spores. When these spores are disseminated

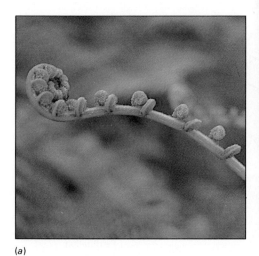
(a)

(b)

(c)

(d)

FIGURE 27–15 Representative ferns. (a) Branch of tree fern showing characteristic "fiddlehead" growth pattern of new leaves. (b) Cinnamon fern, showing brown reproductive spikes and green sterile leaves (the spikes bear the sporangia). (c) Staghorn fern, *Platycerium bifurcatum,* native to Australian rain forests, although widely cultivated elsewhere. In nature the staghorn fern is an epiphyte and grows attached to tree trunks. (d) Tree fern, *Cyathea.* These plants, which are native to tropical rain forests, are found in New Zealand, South Africa, and South America.

and land in a suitable place, they may germinate and grow by mitosis into a mature gametophyte plant.

The gametophyte plant of ferns bears no resemblance to the sporophyte generation, yet it is a fern. It is a tiny, green, often heart-shaped structure that grows flat against the ground (Figure 27–17). Called a **prothallus,** it lacks vascular tissue and has tiny rootlike rhizoids that anchor it to the ground. The prothallus produces both archegonia and antheridia on its underside. The archegonia are located near the notch of the prothallus, and each contains a single egg. Numerous sperm are produced in the antheridia, which are found scattered among the rhizoids.

Although ferns are considered to be advanced over the mosses because of the presence of vascular tissue, they have still retained the primitive requirement of water for fertilization. If a thin film of water is on the ground underneath the prothallus, it provides the transport medium in which the flagellated sperm swim to the neck of the archegonia. After one of the sperm unites with the egg in an archegonium, a diploid zygote grows by mitosis into a multicellular embryo. At this stage in its life, the sporophyte embryo is attached to and dependent upon the gametophyte. As the embryo matures into the sporophyte fern plant, the prothallus withers and dies.

The fern life cycle has a clearly defined alternation of generations between the diploid plant with its rhizome, roots, and fronds, and the haploid prothallus. The sporophyte generation is considered to be dominant not only because it is larger in size than the gametophyte, but also because it persists for an extended period of time, whereas the gametophyte dies soon after reproducing.

A trend observed in more advanced land plants is that the sporophyte generation becomes increasingly dominant and less dependent on water for fertilization. Most algae have a dominant gametophyte generation, with its corresponding production of flagellated gametes that can swim in the water. Production of nonmotile spores becomes more advantageous than flagellated gametes for land plants.

Division Psilophyta: The Whisk Ferns

The few extant species of whisk ferns are found mainly in the tropics. Although whisk ferns do not closely resemble the ferns in appearance, they are considered to be fern allies because of similarities in their life cycles. Their morphology has been under close scrutiny in recent years,

627

FIGURE 27–16 The fern life cycle. Note the clearly defined alternation of generations between the haploid and diploid stages.

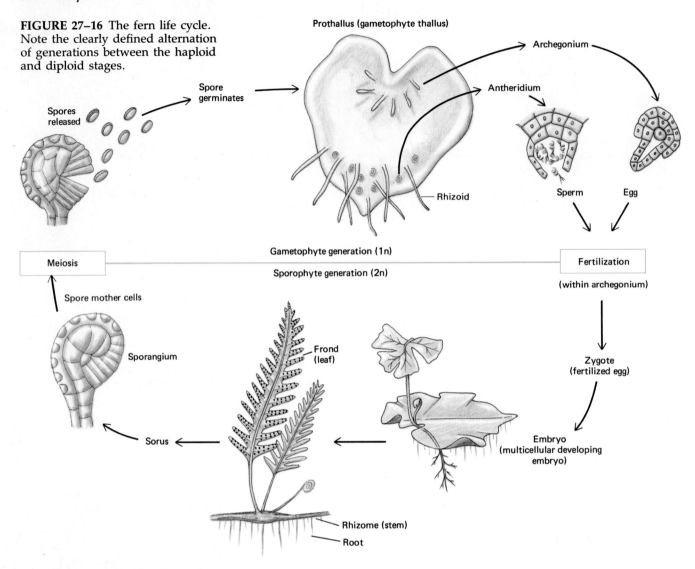

(a)

FIGURE 27–17 The prothallus is the gametophyte generation of ferns. (a) The cells of the prothallus contain chloroplasts while the rhizoids underneath are nonphotosynthetic. Archegonia and antheridia have not yet developed. (b) Close-up of a mature prothallus showing antheridia. Dark cells within each antheridium will mature into sperm.

(b)

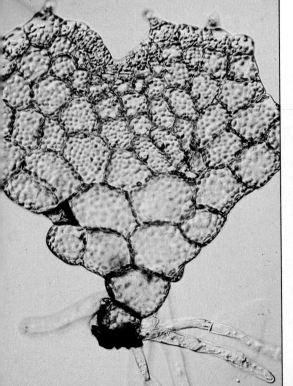

and botanists are divided over interpretation of their structures. One group considers them to be very primitive vascular plants, while another considers them to be advanced, highly specialized ferns.

Psilotum nudum (Figure 27–18) is a representative whisk fern. It lacks true roots or leaves, but does have a vascular stem. *Psilotum* has both a horizontal, underground rhizome and vertical, aboveground stems. Whenever the stem branches, it always branches in halves. This **dichotomous** branching is considered to be a primitive characteristic. The stem is green and the main organ of photosynthesis. The sporangia are borne naked on the stems, and they produce spores that can germinate to form haploid prothalli. The prothalli of whisk ferns were originally difficult to locate because they grow underground. They are nonphotosynthetic due to their subterranean location and apparently have a mutualistic relationship with a fungus, which aids in their nourishment.

Division Sphenophyta: The Horsetails

These plants were more important millions of years ago, when they were dominant land plants. The ancient horsetails are still significant to us today because they contributed largely to the vast coal deposits that we are currently using (see Focus on Ancient Plants and Coal Formation). The few extant plants are small, but very distinctive (Figure 27–19). Horsetails often grow in wet, marshy habitats. In pioneering days they were referred to as scouring rushes and were used to scrub out pots and pans along the stream banks. The hollow, jointed stems are impregnated with silica, which gives them a gritty feeling.

Horsetails have true roots, stems, and leaves. The leaves are fused in whorls at each **node,** and are greatly reduced in size. The green stem is the main organ of photosynthesis. The plants were named horsetails because certain vegetative (nonreproductive) stems have whorls of branches giving the appearance of a bushy horse tail.

When certain branches of horsetails become reproductive, each bears a terminal conelike **strobilus.** The strobilus is composed of small, reduced branches that bear sporangia. The horsetail life cycle is similar to the fern life cycle.

FIGURE 27–18 The growth habit of *Psilotum nudum.* The stem is the main organ of photosynthesis since leaves are absent. This ancient-looking vascular plant bears sporangia directly on the stems.

FIGURE 27–19 *Equisetum* species with different growth habits. (*a*) *Equisetum telematia,* a horsetail, with a wide distribution in Eurasia, Africa, and North America. It has unbranched, nongreen, nonphotosynthetic "fertile" shoots bearing cone-like strobili and separate, highly branched, green, photosynthetic "sterile" shoots. Both types of shoots arise from an underground rhizome. (*b*) A dense growth of *Equisetum* on a stream bank.

(a)

(b)

Focus on ANCIENT PLANTS AND COAL FORMATION

The industrial society in which we live depends on energy from fossil fuels. One of the most important fuels is coal, which is used to provide electricity to some homes. Coal is also used to help produce machines and other items made of steel and iron. Although coal is mined from the earth as a mineral, it is not a mineral like gold or aluminum. Coal is organic, formed from ancient plant material.

Much of the coal we use today was formed from the prehistoric remains of primitive land plants, particularly those of the Carboniferous period, approximately 300 million years ago. Five main groups of plants contributed to coal formation. Three of them were seedless vascular plants—the club mosses, horsetails, and ferns. The other two important groups of coal formers were seed plants, the seed ferns (now extinct) and primitive gymnosperms.

It is hard to imagine that the small, relatively inconspicuous club mosses, ferns, and horsetails of today could have been so significant in forming the vast beds of coal in the Earth. However, the extinct members of these groups that existed during the Carboniferous period were giants by comparison, and they formed vast forests of trees.

The climate during the Carboniferous period was warm and mild. Plants could grow year round because of the favorable weather conditions. The forests of these plants occurred in low-lying areas that were periodically flooded when the sea level rose. When the sea level receded, these plants would become established again.

When these large plants died or were blown over during storms, they were incompletely decomposed because they were covered by the swampy waters. The anaerobic conditions of the water prevented wood-rotting fungi from decomposing the plants, and anaerobic bacteria do not decompose wood rapidly. Over time, the partially decomposed plant material accumulated and consolidated.

Layers of sediment formed over the plant material when the sea level rose and flooded the low-lying swamps. With time, heat and pressure built up in these accumulated layers and converted the plant material to coal and the sediment layers to sedimentary rock.

Much later, geologic upheavals raised the layers of coal and sedimentary rock. For example, coal is found in seams (layers) in the Appalachian Mountains. The various grades of coal (lignite, bituminous, and anthracite) formed as a result of different temperatures and pressures to which they were exposed.

■ The plants of the Carboniferous Period included giant ferns, horsetails, and club mosses. The early gymnosperms were also present. Recent work suggests that Carboniferous swamps were more open than depicted here.

Division Lycophyta: The Club Mosses

One of the problems with using common names in biology is vividly portrayed by this group of plants. The most common names for the Division Lycophyta are "club mosses" and "ground pines," yet these plants are not closely related to either mosses or pines. Their affinities lie with the ferns.

(a)

(b)

FIGURE 27–20 Two species of club mosses. Although the club mosses may superficially resemble mosses, they are fern allies. The sporophyte plant has reduced, scale-like leaves which are evergreen. Spores are produced in sporangia clustered in a cone-like strobilus.

Like horsetails, club mosses were dominant plants millions of years ago. Species that are now extinct often attained great size. These large trees were major contributors to the coal deposits on our Earth (see Focus on Ancient Plants and Coal Formation).

The club mosses today are small, attractive plants most common in woodlands (Figure 27–20). They possess true roots, stems (both rhizomes and erect, aboveground stems), and small, scalelike leaves (microphylls). Sporangia are borne in conelike strobili at the tips of stems. The plants are evergreen and are often fashioned as Christmas wreaths and other holiday decorations. In some areas they are endangered by overuse.

In the alternation of generations that we have considered thus far, plants produced one type of spore as a result of meiosis. This condition, known as **homospory,** is found in the bryophytes, horsetails, whisk ferns, and most ferns and club mosses (Figure 27–21). However, certain ferns

FIGURE 27–21 A *Lycopodium* strobilus. This club moss is homosporous and produces one type of spore. The bryophytes, horsetails, whisk ferns, and most ferns are homosporous.

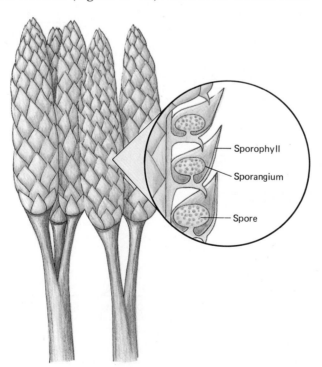

Sporophyll

Sporangium

Spore

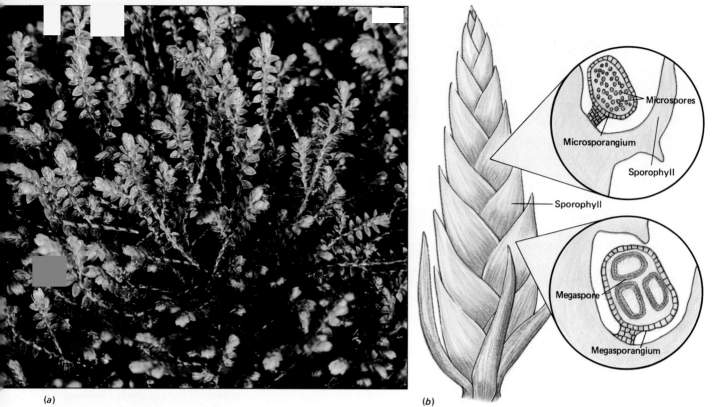

(a)

(b)

FIGURE 27–22 *Selaginella*. (*a*) The growth habit of this lycopod. (*b*) A *Selaginella* strobilus. This plant is heterosporous and produces both microspores and megaspores. Microspores develop into male gametophytes that produce sperm. Megaspores develop into female gametophytes that produce eggs. Heterospory is considered an advanced condition. It is found in certain ferns and club mosses, and in all the gymnosperms and angiosperms.

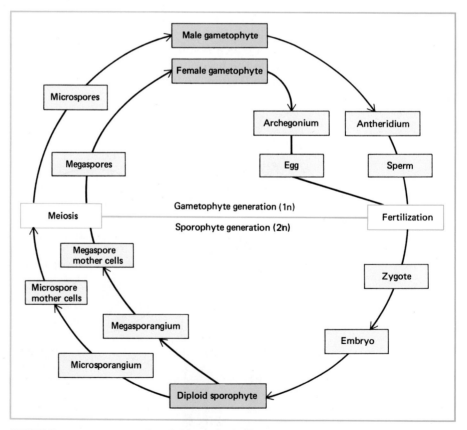

FIGURE 27–23 A generalized life cycle for heterosporous plants. These plants produce two types of spores, microspores and megaspores.

and club mosses are **heterosporous;** they produce two different types of spores as a result of meiosis. Heterospory is a significant development in plant evolution because it is found in the two most successful groups of plants on the Earth today, the gymnosperms and the angiosperms.

Selaginella is a lycopod that is heterosporous (Figure 27–22). Its strobilus bears two kinds of sporangia. One sporangium, the **microsporangium,** bears **microspore mother cells** that undergo meiosis to form tiny, haploid **microspores.** Each microspore develops into a male gametophyte, which produces sperm.

The other type of sporangium in the *Selaginella* strobilus is a **megasporangium,** within which are found **megaspore mother cells.** When megaspore mother cells undergo meiosis, they form haploid **megaspores,** each of which can develop into a female gametophyte that produces eggs. Refer to Figure 27–23 to help visualize this type of life cycle. The development of male and female gametophytes from microspores and megaspores, respectively, occurs within the sporangium. The male and female gametophytes are not truly free-living.

■ SUMMARY

I. Terrestrial plants evolved from green algal ancestors.
  A. Both have similar biochemical characteristics: the same pigments, cell wall components, and carbohydrate storage material.
  B. Both have similarities in fundamental processes like cell division.
II. The migration of plants from aquatic ecosystems to the land involved a number of anatomical, physiological, and reproductive adaptations.
  A. Land plants developed a waxy cuticle to protect against water loss.
  B. Land plants developed stomata for gas exchange necessary for photosynthesis.
  C. A trend in land plants is toward a larger, more dominant sporophyte generation. The gametophyte generation becomes less dominant in more advanced plants.
  D. Plants produce multicellular gametangia with a sterile jacket of cells surrounding the gametes. The male antheridia produce sperm and the female archegonia produce eggs.

III. Plants have alternation of generations, spending part of their life cycle in the haploid (gametophyte) stage and part in the diploid (sporophyte) stage.
  A. The gametophyte generation produces haploid gametes.
  B. These gametes fuse in a process known as fertilization, which requires water as a transport medium. The first stage in the sporophyte generation, the zygote, is the result of fertilization.
  C. The zygote develops into an embryo, which is protected and nourished by the gametophyte plant.
  D. The sporophyte plant has spore mother cells that can undergo meiosis, producing haploid spores. These spores represent the first stage in the gametophyte generation.
IV. Mosses and other bryophytes have several advancements over the green algae, including possession of a cuticle, stomata, and multicellular gametangia.
V. Ferns and fern allies have several advancements over the bryophytes, including the possession of vascular tissue and a dominant sporophyte generation.

■ POST-TEST

1. The bryophytes lack a _____ system and are therefore restricted in size.
2. Mosses and ferns reproduce by _____.
3. The waxy layer that covers aerial parts of land plants is the _____.
4. The land plants evolved from the _____ _____.
5. Land plants store their carbohydrate reserves as _____.
6. The openings in plants that allow gas exchange for photosynthesis are called _____.
7. The female gametangium, or _____, produces an egg.
8. The male gametangium, or _____, produces sperm.
9. The fusion of gametes is called _____ and results in a diploid fertilized egg, or _____.

10. Meiosis of spore mother cells results in the formation of _____.
11. Plants have _____ of _____ in which they spend part of their life in the haploid stage and part in the diploid stage.
12. The leafy green moss plant is the _____ generation.
13. The flattened leaf-like body form of many liverworts is called a _____.
14. Clusters of sporangia, termed _____, are often found on fern fronds.
15. The club mosses are most closely allied to the _____.
16. _____ have hollow, jointed stems that are impregnated with silica.
17. Certain lower vascular plants are _____ and produce two kinds of spores.

FIGURE 28–1 Dawn redwood with autumn coloration. This plant is deciduous, unlike most gymnosperms. The dawn redwood was originally identified from fossils and thought to be extinct. It was found alive in China in this century.

FIGURE 28–2 The majestic sequoias are an example of the division Coniferophyta, gymnosperms that produce their seeds in cones. Some of the redwoods are the world's tallest plants.

times of plenty and saved for times of need. Few other human foods could be stored as conveniently or for as long.

The two groups of seed plants are the **gymnosperms** and the **angiosperms.** The word *gymnosperm* is adapted from a Greek word meaning "naked seed." These plants produce seeds that are totally exposed or borne on the scales of cones. Pine, spruce, fir, and other conifers are examples of gymnosperms. The Greek from which the term *angiosperm* is derived translates as "seed enclosed in a vessel or case." Angiosperms, or flowering plants, produce their seeds within a fruit. Angiosperms include such diverse plants as corn, oaks, water lilies, cacti, and buttercups.

Both gymnosperms and flowering plants possess vascular tissue, xylem for conduction of water and dissolved minerals, and phloem for conduction of food. Both have alternation of generations, spending a portion of their lives in the diploid (sporophyte) stage and a portion in the haploid (gametophyte) stage. The gametophyte generation, however, is significantly reduced. They both are heterosporous, producing two types of spores, **microspores** and **megaspores.** Refer to Chapter 27 for a review of these advances.

GYMNOSPERMS: NAKED SEED PLANTS

The gymnosperms include some of the most interesting plants in the plant kingdom (Figure 28–1). A number of record holders are in this group. For example, the world's largest organism (in terms of sheer bulk) is the General Sherman, a giant sequoia. A redwood is the world's tallest tree, measuring almost 380 feet in height (Figure 28–2). The oldest living organism (i.e., with the longest life span) is the bristlecone pine. One living bristlecone pine in Nevada has been dated by tree ring analysis as 4900 years old.

The gymnosperms are usually classified into four divisions. The largest group is the division Coniferophyta, the **conifers.** Pine and other conifers are woody plants that bear their seeds in cones. Two divisions of gymnosperms represent evolutionary remnants of groups that were more significant in the past, the Ginkgophyta and the Cycadophyta. A final group of gymnosperms, the Gnetophyta, is a collection of some very unusual plants that share certain advancements not found in the other gymnosperms.

The fossil record indicates that seed plants apparently evolved independently several times. Seeds appeared during the Devonian period over 360 million years ago. The exact pathway of gymnosperm evolution is unclear. It is not even certain that all gymnosperms have an immediate common ancestor. The first appearance of conifers in the fossil record is in the Upper Carboniferous period, approximately 300 million years ago.

Division Coniferophyta: The Conifers

The conifers include pines, spruces, hemlocks, and firs. They are woody trees or shrubs; there are no herbaceous conifers. They are attractive plants even during winter because most are evergreen (Figure 28–3). Only a few, such as the larch and the bald cypress, are **deciduous** and shed their leaves, called **needles,** at the end of the growing season (Figure 28–4). Most conifers are **monoecious,** having separate male and female reproductive parts on the same plant. These reproductive parts are generally borne in **cones.**

Conifers occupy vast areas of the Earth today. They range from the Arctic to the tropics and are the dominant vegetation in the taiga, the vast forested regions of Canada, Northern Europe, and Siberia. In addition, they are important in the southern hemisphere, particularly in areas of South America, Australia, and Malaysia. Ecologically, they contribute food and shelter to animals. Their roots hold the soil in place and help prevent

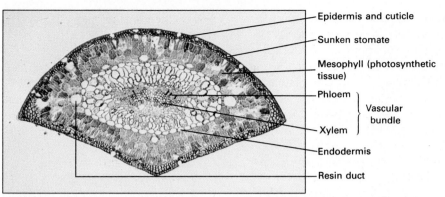

FIGURE 28–4 Cross section of a pine needle. The thick, waxy cuticle and sunken stomata (for gas exchange) are two anatomical adaptations that enable the pine tree to retain its needles throughout the winter.

Labels (clockwise from top):
Epidermis and cuticle
Sunken stomate
Mesophyll (photosynthetic tissue)
Phloem
Vascular bundle
Xylem
Endodermis
Resin duct

FIGURE 28–3 Coniferous gymnosperms are dominant plants in northern latitudes. Their evergreen needles have special adaptations for surviving in cold, i.e., physiologically arid, environments.

FIGURE 28–5 The life cycle of pine. One major evolutionary advancement of gymnosperms over the lower vascular plants is their wind-borne pollen. Pines and other gymnosperms are not dependent on water as a transport medium for sperm.

erosion. Humans use conifers for lumber (for building materials as well as paper products) and various substances like turpentine and resins. Because of their attractive appearance, there is a large business in growing conifers for landscape design and for Christmas trees.

The pine tree represents the typical conifer life cycle (Figure 28–5). The tree is the sporophyte generation and therefore produces spores. Pine is heterosporous and produces microspores and megaspores in separate

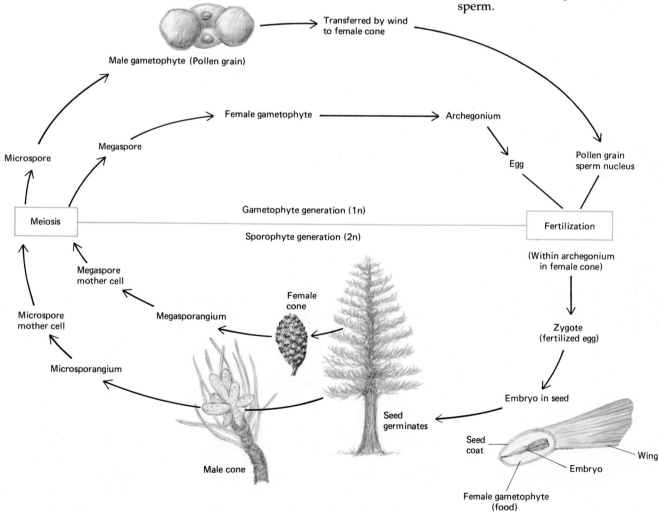

637

FIGURE 28–6 The male cones of pine produce copious amounts of pollen in the spring. These cones are about twice their natural size.

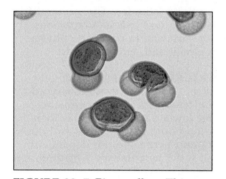

FIGURE 28–7 Pine pollen. This micrograph shows the air bladders that give pollen its buoyancy. Pine pollen is carried by wind to the female cones.

cones. The familiar woody pine cones are the female cones. They are usually located on the upper branches of the tree and bear seeds after fertilization has occurred.

The male cones are smaller than female cones and are generally produced on the lower branches in the spring (Figure 28–6). Each cone is composed of overlapping leaflike structures called **sporophylls.** At the base of each sporophyll is a sac, or microsporangium, which contains numerous microspore mother cells. These cells undergo meiosis to form haploid microspores. Each microspore then develops into the male gametophyte generation, which is extremely reduced in gymnosperms. The immature male gametophyte is also called a **pollen grain** (Figure 28–7). The pollen grains are shed from the male cones in great numbers, and some are carried by wind currents to the immature female cones.

The woody bracts of the female cones (Figure 28–8) have megasporangia at their bases. Within each megasporangium, meiosis of a megaspore mother cell produces four haploid megaspores. One of these develops into the female gametophyte, which produces an egg within each of several archegonia. The pollen grain grows a tube which digests its way through the female gametophyte tissue to the egg within the archegonium. Then, a cell within the pollen grain divides to form two nonflagellated sperm (also called sperm nuclei). One of these fuses with the egg to form the zygote which grows into the young pine embryo, forming part of the seed (Figure 28–9).

The haploid female gametophyte tissue surrounding the developing embryo becomes the nutritive tissue in the mature seed. The food and embryo are surrounded by a tough protective seed coat. The mature seed

FIGURE 28–8 Female Scotch pine cones. These mature cones have opened to shed their seeds.

has a papery wing that enables it be be carried by wind currents. When the female cone opens, the seeds are dispersed.

There is a large time lapse between the appearance of cones on a tree and the maturation of pine seeds. When **pollination,** the transfer of pollen to the female cones, occurs in the spring, the female is immature. Meiosis of the megaspore mother cells has not occurred at this time. During the following year, the female tissue matures and eggs are formed within archegonia. The pollen slowly grows a tube through the female tissues to the archegonia. **Fertilization,** the union of the egg and sperm nucleus, occurs during the spring of the following year. Seed maturation takes several months more, although some seeds remain within the female cones for a number of years before being shed.

There are several key points to remember about the pine life cycle. The sporophyte generation is dominant, and the gametophyte has decreased in size to microscopic structures in the cones. Although the female gametophyte produces archegonia, the male gametophyte is so reduced (initially, it is a single cell) that it doesn't produce antheridia. Also, the gametophyte is dependent on the sporophyte for nourishment.

A major advancement in the pine life cycle is elimination of water as a transport medium for the sperm. The pollen is carried to the female cones by air currents. Nonflagellated sperm accomplish fertilization by moving through a pollen tube to the egg. Therefore, the gymnosperms are the first land plants whose reproduction is totally adapted for life on land.

Division Cycadophyta: The Cycads

The **cycads** were a very important plant group in the prehistoric past. The Triassic period, which occurred approximately 248 million years ago, is sometimes referred to as the "Age of Cycads." The few remaining extant cycads are tropical plants with a palmlike appearance (Figure 28–10).

Cycad reproduction is similar to pine reproduction except the cycads are **dioecious** and therefore have male and female reproductive structures on separate plants. Their ovule and seed structure is most like the earliest seeds. They also have retained the primitive feature of motile sperm. These flagellated sperm are a vestige, however, since the cycads produce pollen that is carried by air or, possibly, insects. They do not need water to accomplish fertilization.

Division Ginkgophyta: Ginkgo

There is only one living representative of this plant division, the ginkgo, or maidenhair, tree (Figure 28–11). It is native to China, where it has been under cultivation for centuries. It has not been found in the wild, and it is likely that it would have become extinct had it not been cultivated in Chinese monasteries. The ginkgo represents the oldest genus of living trees. Fossil ginkgoes 200 million years old have been discovered that are nearly identical to the modern-day ginkgo.

The ginkgo is common in North America today, particularly in cities, because it is somewhat resistant to air pollution. The leaves are deciduous. The plant is dioecious, with separate male and female trees. Like the cycads, the ginkgo has flagellated sperm, a vestige that is not required, since the ginkgo produces air-borne pollen. The seeds are naked and not found within cones. Only the female trees produce seeds, a fact you should remember if you ever wish to plant ginkgoes. As the seeds mature, they produce a disgusting odor, which makes the female trees undesirable. Some cities and towns have passed ordinances making it unlawful to plant female ginkgoes.

FIGURE 28–9 Bisected gymnosperm seed. Seeds contain an embryo (in middle of seed) and food for the developing plant (surrounding the embryo). The nutritive tissue in gymnosperm seeds is haploid female gametophyte tissue.

FIGURE 28–10 A cycad growing in South Africa. Cycads are tropical gymnosperms with a palmlike appearance. Note the immense seed cones on this plant.

(a)

(b)

FIGURE 28–11 *Ginkgo biloba,* the ginkgo or maidenhair tree. *(a)* The ginkgo has separate male and female trees. Only the females bear seeds. *(b)* The unusual leaves of the ginkgo resemble the maidenhair fern, hence its common name. Note the naked, fleshy seeds.

Division Gnetophyta: The Gnetophytes

This amazingly diverse group of gymnosperms is composed of three genera that share a number of features that make them clearly more advanced than the rest of the gymnosperms. The **gnetophytes** have more efficient water-conducting cells called vessels in their xylem (see Chapter 32). Angiosperms have vessels in their xylem, but gymnosperms, with the exception of the gnetophytes, do not. Also, the cone clusters produced by some of the gnetophytes resemble flower clusters.

The genus *Gnetum* contains tropical vines and trees with leaves that resemble the angiosperms (Figure 28–12). *Ephedra* species are shrubs found in deserts and other dry regions (Figure 28–13). They resemble the horsetails in appearance and are commonly called joint firs. An Asiatic *Ephedra* is the source of the antihistamine ephedrine. The final gnetophyte genus, *Welwitschia,* contains a single species found in African deserts (Figure 28–14). The majority of its body grows underground. The above-ground stem forms a shallow disk, up to three feet in diameter, from which two ribbon-like leaves are produced. These leaves continue to grow throughout the plant's life, but the ends of the leaves are usually broken and torn by the wind. When *Welwitschia* reproduces, it forms cones around the edge of the disklike stem.

FIGURE 28–12 The leaves of *Gnetum* are similar to certain flowering plants. The gnetophytes have a number of advanced features which other gymnosperms lack.

FIGURE 28–13 *Ephedra.* Species native to desert areas in the Southwestern United States were used by pioneers to make a beverage. A common name for this plant is Mormon tea.

640

FIGURE 28–14 The most bizarre gymnosperm in the world is *Welwitschia*, native to an African desert. Although the plant only produces two leaves, the winds tear them so that it appears to have many.

DIVISION MAGNOLIOPHYTA: ANGIOSPERMS

The **angiosperms,** or flowering plants, are the most successful plants on Earth today, surpassing even the gymnosperms in importance. They have adapted to almost every habitat, except Antarctica, and, with over 250,000 species, are the dominant plants on Earth (Figure 28–15). Flowering plants reproduce sexually by forming flowers, fruits, and seeds. They possess vessels in xylem tissue (Chapter 32), and their fertilization process is unique.

Flowering plants are extremely important to humans. Our survival as a species literally depends upon the flowering plants. All of our major food crops are angiosperms, including rice, wheat, and corn. We use angiosperms to provide us with fibers like cotton and medicines like digitalis. Woody flowering plants, such as oak, cherry, and walnut, provide us with valuable lumber. Plant products as diverse as rubber, tobacco, coffee, and aromatic oils for perfumes come from flowering plants. Economic botany is the subdiscipline that deals with plants of economic importance.

Monocots and Dicots

The division Magnoliophyta is divided into two classes, Liliopsida **(monocots)** and Magnoliopsida **(dicots).** Monocots include palms, grasses, orchids, and lilies. Examples of the dicots are oaks, roses, cacti, and sunflowers. Table 28–1 provides a comparison of some of the features of the two groups.

The monocots are herbaceous plants with leaves that may be narrow and have parallel venation. The flower parts occur in multiples of three. Monocot seeds have a single **cotyledon** (embryonic seed leaf), and **endosperm** (nutritive tissue) is usually present in the mature seed.

Dicots may be herbaceous (for example, the tomato) or woody (for example, the hickory). Their leaves are variable but often broader than monocot leaves, with netted venation. Flower parts occur in multiples of four or five. Two cotyledons are present in seeds of dicots, and endosperm is usually absent in the mature seed.

Sexual Reproduction in Angiosperms

The organ of sexual reproduction in the angiosperms is the flower (Figure 28–16). Flowers have four main parts (sepals, petals, stamens, and carpels) that are arranged in whorls. A flower that has all four parts is said to be **complete,** whereas an **incomplete** flower lacks one or more of these. Although all four parts are important in the reproductive process, only the stamens (male part) and carpels (female part) participate directly in repro-

FIGURE 28–15 Flowering plant diversity. (a) Maple flowers are wind-pollinated. (b) Cactus plants have a number of adaptations, such as fleshy stems for water storage that enables them to survive the harsh desert environment. (c) Hibiscus flowers have an unusual arrangement of male and female parts. It has been suggested that hibiscus coevolved to facilitate bird pollination. (d) The Indian pipe does not contain chlorophyll or photosynthesize and it lives underground except for its flowers. (e) Angiosperms like this water lilly are adapted to wet environments. (f) This willow catkin is actually a cluster of wind-pollinated flowers. (g) It is obvious how the leopard lily got its common name. (h) All four floral parts (sepals, petals, stamens, carpels) can be seen in this trillium.

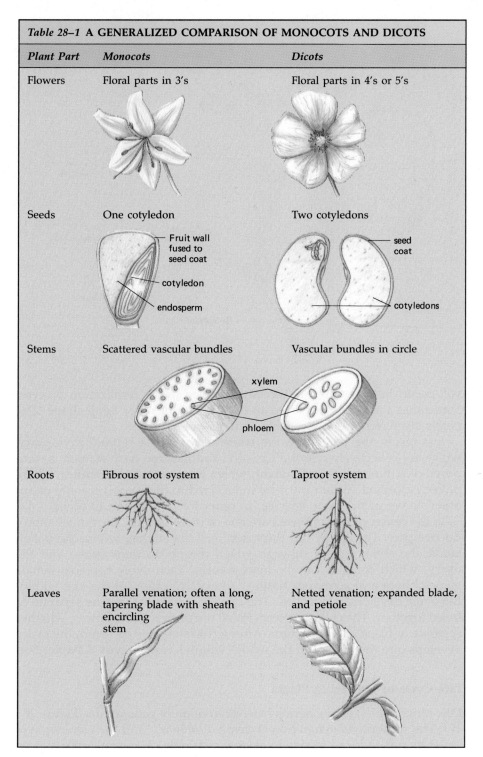

Table 28–1 **A GENERALIZED COMPARISON OF MONOCOTS AND DICOTS**

Plant Part	Monocots	Dicots
Flowers	Floral parts in 3's	Floral parts in 4's or 5's
Seeds	One cotyledon	Two cotyledons
Stems	Scattered vascular bundles	Vascular bundles in circle
Roots	Fibrous root system	Taproot system
Leaves	Parallel venation; often a long, tapering blade with sheath encircling stem	Netted venation; expanded blade, and petiole

duction. A flower that has both stamens and carpels is said to be **perfect,** whereas an **imperfect** flower has stamens or carpels, but not both.

The **sepals** make up the outermost whorl. They are leaflike in appearance and often green. The sepals cover and protect the flower parts when the flower is a bud. As the blossom opens from the bud, the sepals fold back to reveal the more conspicuous petals. The collective term for all the sepals of a flower is the **calyx.**

The **petals** are also leaflike in appearance, although they are frequently brightly colored. They play an important role in ensuring that pollination

FIGURE 28–16 Diagram of a "typical" flower. This cutaway view shows the details of basic floral structure.

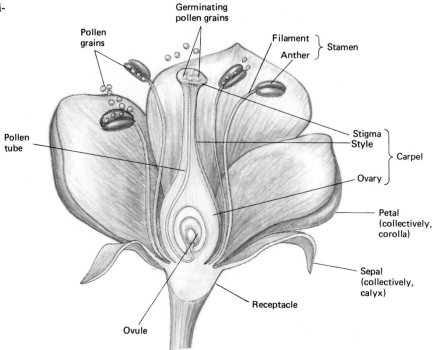

will occur, but they are not directly involved in the fertilization process. Sometimes the petals are fused to form a tube or other floral shape. The petals are referred to collectively as the **corolla.**

Just inside the petals are the **stamens,** the male reproductive parts, which are collectively referred to as the **androecium.** Each stamen is composed of a thin stalk, the **filament,** which terminates in the **anther,** where pollen is formed. Pollen must be transferred to the carpel, usually of another flower of the same species, in order for fertilization to occur.

In the center of most flowers are one or more **carpels,** the female reproductive part. The carpel has three sections: the **stigma,** where the pollen lands; the **style,** a neck through which the pollen must grow; and the **ovary,** which contains one or more **ovules.** Each ovule contains female gametophyte and accessory tissues. After fertilization of the egg within, the ovule develops into a seed. The carpels of a flower may be separate or fused together. The collective term for all the carpels of the flower, whether separate or fused, is **gynoecium.** After fertilization has occurred, the ovary develops into the **fruit,** and the ovules within become **seeds** (Chapter 36).

Life Cycle of Flowering Plants

Like other plants, angiosperms have alternation of generations (Figure 28–17). The sporophyte generation is clearly dominant, and the gametophyte generation is reduced in size to several cells only. The gametophyte is so reduced that there are no archegonia or antheridia. Like the gymnosperms and certain other vascular plants, flowering plants are heterosporous and produce two kinds of spores.

Each ovule within the ovary contains a megaspore mother cell which undergoes meiosis, producing four haploid megaspores. Three of these disintegrate and one develops into the female gametophyte generation, which is also called the **embryo sac.** The embryo sac contains eight nuclei, including one egg and two **polar nuclei.** The egg and the polar nuclei are involved in fertilization.

The anther contains microspore mother cells, which undergo meiosis to form haploid microspores. These microspores each develop into the male gametophyte generation, a pollen grain. Pollen is transferred to the stigma

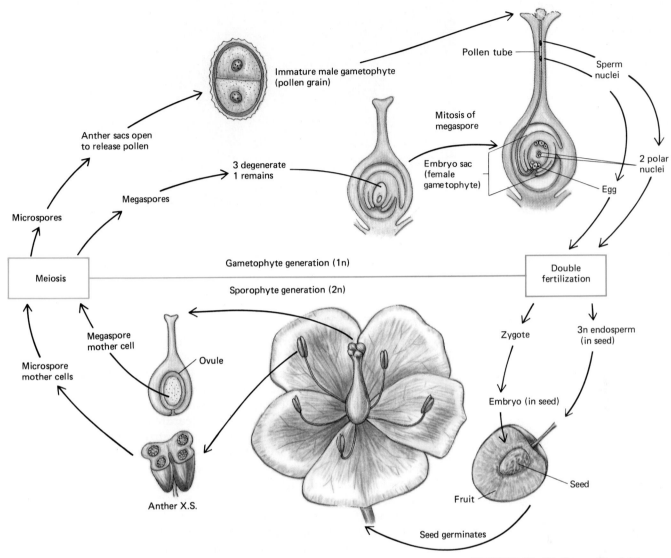

Immature male gametophyte (pollen grain)

Pollen tube

Sperm nuclei

Anther sacs open to release pollen

Mitosis of megaspore

3 degenerate 1 remains

Embryo sac (female gametophyte)

2 polar nuclei

Egg

Megaspores

Microspores

Meiosis

Gametophyte generation (1n)

Double fertilization

Sporophyte generation (2n)

Megaspore mother cell

Ovule

Zygote

3n endosperm (in seed)

Microspore mother cells

Embryo (in seed)

Anther X.S.

Seed

Fruit

Seed germinates

FIGURE 28–17 Generalized life cycle of a typical flowering plant. The most significant feature of the angiosperm life cycle is double fertilization, which is found nowhere else in the living world.

of the carpel and, if compatible, grows a thin tube down the style and into the ovary. A cell within the pollen grain divides to form two sperm nuclei (nonflagellated male gametes are little more than nuclei with genetic material). Both sperm nuclei are involved in fertilization.

Something happens in angiosperm sexual reproduction that does not occur anywhere else in the living world. When the sperm nuclei enter the embryo sac, *both* of them participate in fertilization. One sperm nucleus fuses with the egg, forming the zygote that develops into a plant embryo in the seed. The second sperm nucleus fuses with the two haploid polar nuclei, forming a **triploid** (3n) cell that develops into endosperm in the seed. This process, involving two separate cell fusions, is called **double fertilization** and is unique to the angiosperms.

As a result of double fertilization, each seed contains (1) a young plant, (2) food or nutritive tissue (the endosperm), and (3) a seed coat. In monocots the endosperm persists and is the main source of food in the mature seed. In most dicots the endosperm is used by the developing embryo, which subsequently stores food in its cotyledons.

As the seed develops, the ovary wall surrounding it enlarges and develops into the fruit. In some instances, other tissues associated with the ovary also enlarge to form the fruit. Fruits serve to protect the developing seeds from desiccation during maturation. Also, fruits often aid in the dispersal of seeds (Chapter 36). For example, dandelion fruits have feathery

FIGURE 28–18 Bees and other insects are common pollinators of flowering plants.

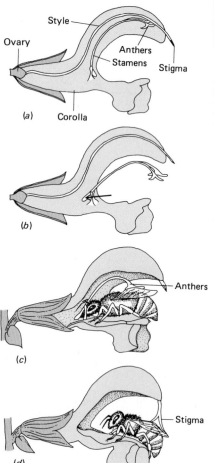

FIGURE 28–19 Flowers of sage *(Salvia)* showing how pollen is transferred to pollinators (bees). Other sage species have modifications for bird pollination. *(a)* Longitudinal section of a flower shows stamens attached to the tubular corolla by a hinge bearing a basal projection. *(b)* When the basal projection is pushed *(arrow)*, the stamens tip downward. *(c)* A bee entering the flower to reach the nectar at its base pushes against the projection, causing stamens to swing down and deposit pollen on the bee's abdomen. *(d)* When a bee alights on a flower whose stigma is exposed and receptive to pollen, the pollen-dusted part of the bee contacts the stigma and pollinates it.

plumes that enable the entire fruit to be carried by air currents. Once it lands in a suitable place, it germinates and develops into a mature sporophyte plant, and the life cycle continues as described.

Pollination Mechanisms

Before fertilization can occur, the pollen must be transferred from the anther to the stigma. Flowering plants have evolved a number of mechanisms to assure this transfer. Some of these involve animals, including insects, birds, and bats. Biologists who study pollination have developed the concept of pollination syndromes, that unrelated species of plants that have the same agent of pollination share similar features. Flowers that are pollinated by animals have various methods to attract animals, including showy petals, nectar (a sugary solution used as an attractant for pollinators), pollen (a protein-rich food for pollinators), and scent. As the animal moves from flower to flower, it inadvertently carries pollen.

Insects are often involved in pollen transfer (Figures 28–18 and 28–19). Plants that are pollinated by insects often have blue or yellow petals. The insect eye does not perceive color in the same manner as the human eye. Insects see very well in the blue and yellow range of visible light but do not see red well. Consequently, flowers that are pollinated by insects are not usually red. Insects can also see in the ultraviolet range of the electromagnetic spectrum, an area that is invisible to the human eye (Figure 28–20). Many flowers have dramatic ultraviolet markings that are invisible to us

FIGURE 28–20 Many insect-pollinated flowers have ultraviolet markings that are invisible to humans but very conspicuous to insects. *(a)* A flower as seen by the human eye. *(b)* The same flower viewed with a filter that transmits ultraviolet radiation indicates how the insect eye perceives it. The ultraviolet markings draw attention to the center of the flower where the pollen and nectar are located.

but direct the insect to the center of the flower where the pollen or nectar is located.

Insects have a well-developed sense of smell. Many insect-pollinated flowers have a strong scent which is often pleasant, but not always. For example, the carrion plant, which is pollinated by flies, has petals that are dappled with a reddish-brown color (like dried blood) and smells like rotting flesh. Flies move from one flower to another looking for a place to deposit their eggs and accomplish pollination at the same time (Figure 28–21).

Birds such as hummingbirds are effective and important pollinators (Figure 28–22). Flowers pollinated by birds are usually red to yellow. Birds see well in this region of visible light. In Europe there are no natural bird pollinators and there are also no naturally occurring plants with red flowers. Birds do not have a strong sense of smell; consequently, bird-pollinated flowers usually lack any scent.

Bats, which feed at night and do not see very well, are frequent pollinators in the tropics (Figure 28–23). Bat-pollinated flowers have dusky, dull-colored petals. These plants do produce a strong scent, usually of fermented fruit. Bats are attracted to the flowers by their scent and lap up the nectar. As they move from flower to flower, pollen is transferred. Other animals, including snails and small rodents, sometimes pollinate plants.

Some plants have evolved a pollination mechanism that relies on wind rather than animals (Figure 28–24). Flowering plants that are wind-pollinated produce many, often inconspicuous, flowers. They do not use their energy to produce large, colorful petals or scent or nectar. Wind pollination is a "hit or miss" affair, and the likelihood of pollen landing on the stigma of the same species of flower is slim. Wind-pollinated plants therefore produce copious amounts of pollen. Wind-pollinated plants include grasses, ragweed, and maples.

There are a number of examples of obligate relationships occurring between an animal pollinator and the plant it pollinates (Figure 28–25). For example, a yucca found in the Southwest can only be pollinated by one species of moth. The female moth lays her eggs in the flower ovary. Both the yucca and the moth would become extinct if something should happen to the other. Neither would be able to reproduce successfully.

Evolutionary Advancements of the Flowering Plants

Why are the angiosperms so successful today? Certainly, seed production as the primary means of reproduction and dispersal is significant. However, the angiosperms have a number of other advanced features besides highly successful reproduction. Their vascular tissues are very efficient at

FIGURE 28–21 *Stapelia variegata.* This desert angiosperm is sometimes called the carrion plant due to its coloration and bad scent. It is pollinated by flies.

FIGURE 28–22 A ruby-throated hummingbird pollinating a trumpet vine flower.

FIGURE 28–23 A greater short-nosed fruit bat pollinating a banana plant. The pollen grains on the bat's fur will be carried to the next plant, where cross pollination will occur.

(a)

(b)

FIGURE 28–24 Wind is an agent of pollination for a number of flowering plants. *(a)* Many angiosperm trees are wind-pollinated. Box elder flowers (in clusters) lack petals. *(b)* Grass flowers have pendulous stamens. Pollen produced in these stamens gets windborne easily.

conduction. Flowering plants have vessels in their xylem and sieve tube elements in their phloem (see Chapter 32). Angiosperm leaves, with their broad, expanded blades, are structured for maximum efficiency in photosynthesis (see Chapter 33). Abscission of these leaves during cold or dry spells is also an advantage that has enabled some angiosperms to expand into habitats that would otherwise be too harsh for survival. Angiosperm roots are often modified for food or water storage (see Chapter 35).

Probably most important, however, is the overall adaptability of the sporophyte generation. This adaptability is evident in the diversity of the angiosperms. The cactus is marvelously adapted for desert environments. The water lily is equally well adapted for wet environments. The flowering plants are successful because they readily adapt to new habitats and changing environments.

FIGURE 28–25 A number of orchids have petals that resemble insects. This orchid deludes male wasps, who believe it is a female wasp. As the males fly from flower to flower, attempting to copulate, pollination occurs.

■ SUMMARY

I. Seeds represent an evolutionary advancement over spores.
  A. Each seed contains a well-developed plant embryo and a food supply.
  B. The gymnosperms and the flowering plants reproduce by seeds.
II. The gymnosperms are the naked seed plants.
  A. They have several advances over the ferns, including production of wind-borne pollen.

B. There are four divisions of gymnosperms.
  1. The conifers are the largest group of gymnosperms. They are woody plants that bear needle leaves and produce their seeds in cones.
  2. The cycads are palmlike in appearance but reproduce in a manner similar to pines. There are relatively few extant members of this once large division.
  3. The ginkgo is the only living species in its divi-

Focus on POLLEN AND HAY FEVER

If you suffer from hay fever, you are not alone. Millions of people endure the sneezing and itchy, watery eyes associated with this condition. Everyone knows that one of the causes of hay fever is pollen, but many blame any plant in bloom when they are suffering. For this reason, roses and goldenrod are often unjustly accused.

Hay fever is caused by certain wind-pollinated plants. Plants that produce large, colorful petals are pollinated by animals and do not cause hay fever because their pollen does not get into the air in appreciable quantities. Wind-pollinated plants, on the other hand, must produce copious amounts of pollen to ensure that at least some of it lands on the stigmas for successful reproduction. Not all wind-pollinated plants cause an allergic reaction. For example, the conifers are all wind-pollinated, yet allergies to conifers are rare.

People with allergies can suffer at different times during the growing season, depending on which plants are pollinating and whether they are sensitized to those plants. In early spring, the trees pollinate before their leaves are fully developed. (Can you think of why pollination at this time would be advantageous for the tree?) Trees that cause allergic reactions in humans include oaks, ashes, walnuts, maples, and elms. If you suffer in late spring and early summer, you are probably allergic to the grasses, for example, bluegrass, timothy, and redtop. It is interesting that most of our major grass crops (e.g., corn, rice, and wheat) do not cause allergies in

■ Scanning electron micrograph of ragweed pollen, which is the most common pollen allergen.

humans. In late summer and early fall, people are allergic to different plants, depending on their geographical location. Ragweed is the culprit in the East, while saltbush and Russian thistle are problems in the West.

Most people are born with some resistance to pollen allergies, but many become sensitized by repeated contact. For that reason, a move to a different geographic location often temporarily halts the suffering. The biology of the allergic reaction is explained in Chapter 43.

■ Wind-pollinated plants produce copious amounts of pollen.

sion. It is a deciduous, dioecious tree. Female ginkgoes produce fleshy seeds directly on branches.
  4. The gnetophytes share a number of advances over the rest of the gymnosperms, including vessels in their xylem.

III. The flowering plants produce seeds enclosed within a fruit.
  A. The flower is their organ of sexual reproduction.
  B. There are two classes of angiosperms.
    1. The monocots have floral parts in multiples of

three, and their seeds contain one cotyledon. The nutritive tissue in their mature seeds is endosperm.

2. The dicots have floral parts in multiples of four or five, and their seeds contain two cotyledons. The nutritive tissue in their mature seeds is usually in the cotyledons.

C. The angiosperms have several advanced features.
1. Double fertilization, which results in the formation of a zygote and endosperm tissue, is unique to the angiosperms.
2. They possess vessels in the xylem.
3. Angiosperms may be pollinated by wind or animals.

■ POST-TEST

1. _____ are better than spores for reproduction because they have an embryo and food tissue.
2. Plants that shed their leaves at the end of the growing season are _____ .
3. Most conifers are _____ and have separate male and female reproductive parts on the same plant.
4. Although conifers bear their seeds in cones, they are considered the naked seed plants because their seeds are not enclosed in a _____ .
5. The male gametophyte generation of pine is called _____ .
6. The nutritive tissue in the pine seed is _____ _____ .
7. _____ is the transfer of pollen from the male to the female reproductive structure.
8. Flagellated sperm are found as vestiges in two gymnosperm groups, the _____ and the _____ .

9. This class of angiosperms, the _____ , includes the palms, grasses, and orchids.
10. The _____ is a nutritive tissue formed as a result of double fertilization.
11. The _____ is composed of the stigma, style, and ovary.
12. A flower that lacks stamens is said to be both _____ and _____ .
13. After fertilization, the _____ develops into the fruit and the _____ develops into the seed.
14. The female gametophyte generation in angiosperms is also called the _____ _____ .
15. Plants with blue petals, nectar, and a strong scent are most likely pollinated by _____ .
16. Plants with reduced or absent petals, no nectar, no scent, and copious amounts of pollen are pollinated by _____ .

■ REVIEW QUESTIONS

1. Why are seeds such a significant evolutionary development?
2. List several ways the conifers are advanced over the ferns.
3. What features do the cycads, ginkgo, and gnetophytes share with the conifers?
4. How are the angiosperms different from the gymnosperms?

5. Diagram a flower and label the following parts: calyx, corolla, stamen, carpel, sepal, petal, anther, filament, stigma, style, ovary, ovule.
6. What are the two classes of angiosperms, and how can one distinguish between them?
7. How does pollination occur in the gymnosperms? In the angiosperms?

■ RECOMMENDED READINGS

Raven, P. H., Evert, R. F., and Eichhorn, S. E. *Biology of Plants*. 4th ed. New York, Worth Publishers, Inc., 1986. An excellent general botany textbook, especially useful for current interpretations of plant structures.

Ray, P. M., Steeves, T. A., and Fultz, S. A. *Botany*. Philadelphia, Saunders College Publishing, 1983. One of the most comprehensive general botany textbooks available, particularly good in plant evolution.

K. R. Stern. *Introductory Plant Biology*. 4th ed. Dubuque, Iowa, Wm. C. Brown Publishers, 1988. This reference is written in an enjoyable style. Readers learn botany almost effortlessly.

Ballerina sea anemone

29

The Animal Kingdom: Animals Without a Coelom

■ LEARNING OBJECTIVES

After you have read this chapter you should be able to:

1. List the characteristics common to most animals; using these characteristics, develop a brief definition of an animal.
2. Identify an ecological role of animals and discuss their distribution; compare the advantages and disadvantages of life in the sea and of life in fresh water and on land.
3. Relate the animal phyla on the basis of symmetry, type of body cavity, and pattern of embryonic development, e.g., protostome versus deuterostome.
4. Identify the distinguishing characteristics of phyla Porifera, Cnidaria, Ctenophora, Platyhelminthes, Nemertinea, Nematoda, and Rotifera; compare the level of organization of each of these phyla.
5. Classify a given animal in the appropriate phylum (from among those listed in Objective 4), and at the option of your instructor, identify the class to which it belongs.
6. Trace the life cycle of the following parasites: *Ascaris*, tapeworm, hookworm, and trichina worm. Identify several adaptations that these animals possess for their parasitic life style.
7. Explain the adaptive advantages of each of the following characteristics: bilateral symmetry, cephalization, a motile larva, digestive cavity with two openings, hermaphroditism.

More than a million species of animals have been described, and perhaps several million more remain to be identified. Most members of the kingdom Animalia are classified in about 35 different phyla. The animals most familiar to us—dogs, birds, fish, frogs, snakes—are **vertebrates,** animals with backbones. However, vertebrates account for only about 5% of the species of the animal kingdom. The majority of animals are the less familiar **invertebrates,** animals without backbones. The invertebrates include such diverse forms as sponges, jellyfish, worms, and insects.

WHAT IS AN ANIMAL?

We have no difficulty identifying a horse as an animal and an oak tree as a plant, but many marine animals that live attached to rocks or docks are often mistaken for plants. Early naturalists thought that sponges were plants because they did not move from place to place. There are so many diverse animal forms that exceptions can be found to almost any definition of an animal. Still, there are some characteristics that describe at least most animals (Figure 29–1):

1. All animals are multicellular.
2. Animals are eukaryotes.
3. The cells of an animal exhibit a division of labor. In all but the simplest animals, cells are organized to form tissues, and tissues are organized to form organs. In most animal phyla, specialized body systems carry on specific functions.
4. Animals are exclusively heterotrophic; they ingest their food first and then digest it inside the body, usually within a digestive system.
5. Most animals are capable of locomotion at some time during their life cycle. However, there are some animals—for example, the sponges—that are **sessile** (firmly attached to substrate) as adults.

FIGURE 29–1 Despite their diversity, most members of the animal kingdom share several distinct traits. (*a*) Some marine animals, like this feather star from the South China Sea, do not move about from place to place and are sometimes mistaken for plants. (*b*) The nearly transparent body of this freshwater crustacean, *Simocephalus vetulus*, or water flea, shows a complex of organ systems. (*c*) As heterotrophs, all animals must feed either on producers or on other animals that eat producers. Those that feed on other animals are often highly motile and complex in their behavior. Shown here is a fishing spider feeding on a small fish. The numerous black dots on the dorsal (back) surface of the spider are actually eyes. Though these eyes do not allow the spider to see especially sharp images, they do serve to make the animal an extremely efficient hunter.

(a)

(b)

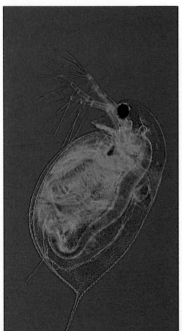

(c)

6. Most animals have well-developed sensory and nervous systems and respond to external stimuli with adaptive behavior.
7. Most animals reproduce sexually, with large, nonmotile eggs and small flagellated sperm. Sperm and egg unite to form a fertilized egg, or **zygote,** which goes through a series of embryonic stages before developing into a larva or immature form.

THEIR PLACE IN THE ENVIRONMENT

As consumers, all animals are ultimately dependent upon producers for their raw materials, energy, and oxygen. Animals are also dependent upon decomposers for recycling nutrients.

Animals are distributed in virtually every environment of the earth. Animals probably evolved in the Precambrian seas and most animal phyla still inhabit the sea. Of the three environments—salt water, fresh water, and land—the sea is the most hospitable. Sea water is isotonic to the tissue fluids of most marine animals, so there is little problem in maintaining fluid and salt balance. The buoyancy of sea water supports its inhabitants, and the temperature is relatively constant, owing to the large volume of water. **Plankton** (the organisms that are suspended in the water and float with the movement of the water) consists of tiny animals, plants, and protists that provide a ready source of food.

There are certain disadvantages to life in the sea. For example, the environment is continuously in motion. Although this motion brings nutrients to animals and washes their wastes away, animals must be able to cope with the constant churning and currents that might sweep them away. Fish and marine mammals are strong enough swimmers so that they can direct their movements and location effectively. However, most invertebrates, owing to their body structure, are unable to swim strongly enough and so have other adaptations. Some are sessile, attaching to some stable structure like a rock, so that they are not wafted about with the tides and currents. Others cling to the substratum or burrow in the sand and silt that cover the sea bottom. Many invertebrates have adapted by maintaining a small body size and becoming part of the plankton. Even while they are tossed about, their food supply continues to surround them, and so they survive successfully. The brackish water of the estuarine environment, where fresh water meets the sea, presents its own special difficulties; the lower and changing salinities restrict the types of organisms that can thrive there.

Fresh water offers a much less constant environment than sea water. Oxygen content and temperature vary, and turbidity (due to sediment suspended in the water) and even water volume fluctuate. Fresh water is hypotonic to the tissue fluids of animals, so water tends to diffuse into the animal; therefore, freshwater animals must have some mechanism for pumping out excess water while retaining salts. This osmoregulation requires an expenditure of energy. Another disadvantage is that freshwater environments generally contain less food than marine environments. For these reasons, far fewer animals make their homes in fresh water than in the sea.

Terrestrial life is even more difficult. Dehydration is a serious threat, because water is constantly lost by evaporation and is often difficult to replace. The many adaptations in terrestrial animals addressing this problem are discussed in Chapters 30 and 31. Only a few animal groups, most notably representatives of the arthropods (insects, spiders, and some related forms) and the higher vertebrates, have successfully made their homes on land.

FIGURE 29–2 Types of body symmetry in animals. (*a*) In radial symmetry, multiple planes can be drawn through the central axis; each divides the organism into two mirror images. (*b*) Most animals are bilaterally symmetrical. A midsagittal cut (lengthwise vertical cut through the midline) divides the animal into right and left halves. The head end of the animal is generally its anterior end, and the opposite end is its posterior end. The back of the animal is its dorsal surface; the belly surface is ventral. The diagram also illustrates various ways in which the body can be sectioned (cut) in order to study its internal structure. Many cross sections and sagittal sections are used in illustrations throughout this book to show relationships among tissues and organs.

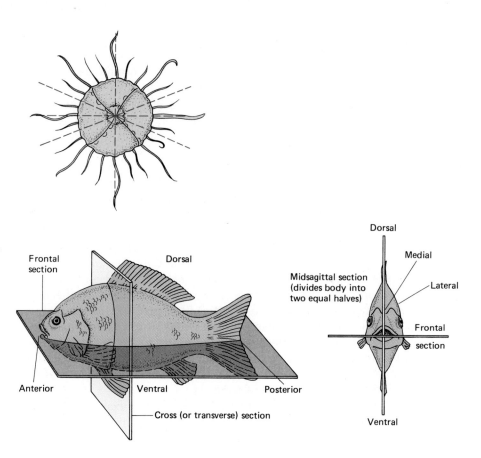

ANIMAL RELATIONSHIPS

Most biologists agree that the evolutionary origin of animals is obscure. Animals are thought to have arisen from the protists, probably from the flagellates. Although the relationships among the various phyla are a matter of conjecture, a few of the more widely held hypotheses are presented in this section.

The animal kingdom may be divided into two large groups, or subkingdoms: **Parazoa,** which consists of sponges, and **Eumetazoa,** which includes all the other animals. This distinction is made because the sponges are so different from all other animals that most biologists think that they are not directly ancestral to any other animal phylum.

Members of Eumetazoa are often further classified on the basis of body symmetry. Two phyla, the cnidarians, (jellyfish and relatives) and the ctenophores (comb jellies) are radially symmetrical and are included in the branch **Radiata.** In **radial symmetry,** similar structures are regularly arranged as spokes, or radii, from a central axis; any imaginary plane passing through the central axis divides the organism into two mirror images. All of the other animals are bilaterally symmetrical (at least in their larval stages) and belong to the branch **Bilateria.** In **bilateral symmetry,** the body is divided into roughly identical right and left halves when sliced down the middle. (See Figure 29–2 and Focus on Body Plan and Symmetry.)

Acoelomates and Coelomates

A widely held system for relating the animal phyla to one another is based upon the type of body cavity, or **coelom.** In the simplest eumetazoans (cnidarians and platyhelminths) the body is essentially a double-walled sac surrounding a digestive cavity with a single opening to the outside—the mouth. There is no body cavity, so these animals are referred to as acoe-

Focus on BODY PLAN AND SYMMETRY

Most animals exhibit **bilateral symmetry,** a type of symmetry in which the body can be divided through only one plane (which goes through the midline of the body) to produce roughly equivalent right and left halves that are mirror images. Cnidarians (jellyfish, sea anemones, and their relatives) and adult echinoderms (sea stars, sea urchins, and their relatives) exhibit radial symmetry. In them, similar body parts are arranged around a central body axis. Radial symmetry is considered an adaptation for a sessile life style, for it enables the organism to receive stimuli equally from all directions in the environment.

Bilateral symmetry is considered an adaptation to motility. The front, or **anterior,** end of the animal generally has a head where sense organs are concentrated; this end receives most environmental stimuli. The **posterior,** or rear, end of the animal may be equipped with a tail for swimming or may just follow along.

In order to locate body structures, it is helpful to define some basic terms and directions (see Figure 29–2). The back surface of an animal is its **dorsal** surface; the belly side is its **ventral** surface. (In animals that stand on two limbs, such as humans, the term **posterior** refers to the dorsal surface, and **anterior,** to the ventral surface.)

A structure is said to be **medial** if it refers to the midline of the body and **lateral** if it is toward one side of the body. For example, in a human, the ear is lateral to the nose. The terms **cephalic** and **rostral** (and **superior,** in humans) refer to the head end of the body; the term **caudal** refers to structures closer to the tail. (In human anatomy, the term **inferior** is used to refer to structures located relatively lower in the body.)

A bilaterally symmetrical organism has three axes, each at right angles to the other two: an anterior–posterior axis extending from head to tail; a dorsoventral axis extending from back to belly; and a left–right axis extending from side to side. We can distinguish three planes (flat surfaces that divide the body into specific parts): The **midsagittal** plane (or section) divides the body into equal right and left halves; this plane passes from anterior to posterior and from dorsal to ventral. Any section or plane (cut) parallel to the midsagittal plane is parasagittal (or simply **sagittal**) and divides the body into unequal right and left parts. A **frontal** (or horizontal) plane, or section, divides the body into anterior and posterior parts. A **transverse section,** or **cross section,** cuts at right angles to the body axis.

lomates (Figure 29–3). Others have a body cavity derived from the blastocoel, a cavity within the early embryo; these animals are known as **pseudocoelomates.** Animals with a true coelom are **coelomates.** In order to understand what a true coelom is, we must digress briefly into the animal's embryonic origins.

The structures of most animals develop from three embryonic tissue layers, called **germ layers.** The outer layer, called the **ectoderm,** gives rise

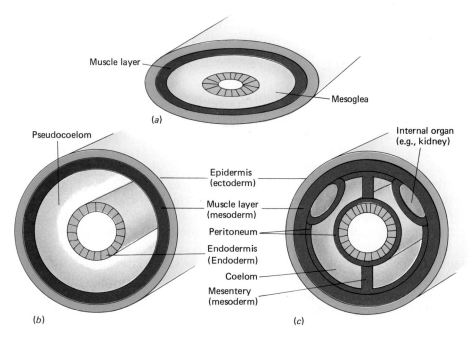

FIGURE 29–3 Three basic animal body plans are illustrated by these cross sections. (*a*) An acoelomate animal has no body cavity. (*b*) A pseudocoelomate animal has a body cavity that develops between the mesoderm and endoderm. The term *body cavity* refers to the space between the body wall and the internal organs. (*c*) In a coelomate animal, the body cavity, called a coelom, is completely lined with tissue derived from mesoderm.

FIGURE 29–4 Proposed evolutionary relationships are illustrated by this phylogenetic tree indicating acoelomate, pseudocoelomate, and coelomate phyla. The flatworms and nemerteans have a solid body and so are referred to as acoelomate. Nematodes (roundworms) and rotifers do have a body cavity, but it develops from the embryonic cavity, called a blastocoel, and is located between the mesoderm and endoderm. All other bilateral animals and the echinoderms have a true coelom, a body cavity lined by mesoderm.

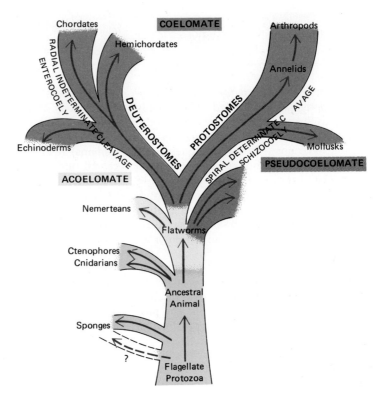

to the outer covering of the body and to the nervous system. The inner layer, or **endoderm,** lines the digestive tract. **Mesoderm,** the middle layer, extends between the ectoderm and endoderm and gives rise to most of the other body structures, including the muscles, bones, and circulatory system. The development of the germ layers is described in more detail in Chapter 51.

Complex animals usually have a tube-within-a-tube body plan: The inner tube, the digestive tract, is lined with tissue derived from endoderm and is open at each end—the mouth and the anus. The outer tube or body wall is covered with tissue derived from ectoderm. Between the two tubes is a second cavity, the body cavity. If the body cavity develops between the mesoderm and endoderm, it is called a **pseudocoelom.** If it forms within the mesoderm and is completely lined by mesoderm, the body cavity is a true **coelom.** Advantages of having a coelom are discussed in Chapter 30. The phylogenetic tree shown in Figure 29–4 indicates the relationships of the major phyla of animals based on the type of coelom they possess.

Protostomes and Deuterostomes

A different family tree is shown in Figure 29–5. In this important phylogenetic scheme, the complex animals are divided into two groups—the protostomes and deuterostomes—based on the pattern of embryonic development. These groups reflect two main lines of evolution. Early during embryonic development, a group of cells moves inward to form an opening called the **blastopore.** In most mollusks, annelids, and arthropods, this opening develops into the mouth; these animals comprise the **protostomes** (meaning "first, the mouth"). (Some taxonomists include the flatworms and pseudocoelomates as protostomes.) In echinoderms (e.g., the sea star) and chordates (the phylum that includes the vertebrates), the blastopore develops into the anus; the opening that develops into the mouth forms later in development. These animals are the **deuterostomes** ("second, the mouth").

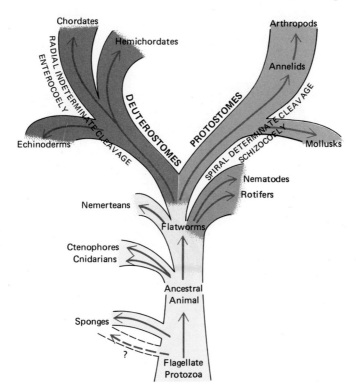

FIGURE 29–5 A phylogenetic tree based on protostome–deuterostome characteristics. In the protostomes—the mollusks, annelids, and arthropods—the blastopore develops into the mouth, cleavage is generally spiral and determinate, and the coelom develops within the mesoderm when the mesoderm splits. In the deuterostomes—the echinoderms and chordates—the blastopore develops into the anus, and the mouth develops from a second opening. Deuterostomes typically have radial, indeterminate cleavage, and the coelom develops from outpocketings of the gut.

Another difference in the development of protostomes and deuterostomes is the pattern of **cleavage,** that is, the first several cell divisions of the embryo. In protostomes, the early cell divisions are oblique to the polar axis, resulting in a spiral arrangement of cells; any one cell is located between the two cells above or below it (Figure 29–6). This pattern of division is known as **spiral cleavage.** In **radial cleavage,** characteristic of the deuterostomes, the early divisions are either parallel or at right angles to the polar axis; the cells are located directly above or below one another.

In the protostomes, the fate of each embryonic cell is fixed very early. For example, if the first four cells of an annelid embryo are separated, each cell develops into only a fixed quarter of the larva; this is referred to as **determinate cleavage.** In deuterostomes, cleavage is **indeterminate:** If the first four cells of a sea star embryo, for instance, are separated, each cell is capable of forming a complete, though small, larva. Each cell of the early embryo has the potential to develop into an entire organism.

FIGURE 29–6 Types of cleavage in embryonic development. (*a*) Spiral cleavage is characteristic of protostomes. Note the spiral arrangement of the cells. (*b*) In the radial cleavage characteristic of deuterostomes, the early divisions are either parallel or at right angles to the polar axis so that the cells are stacked in layers.

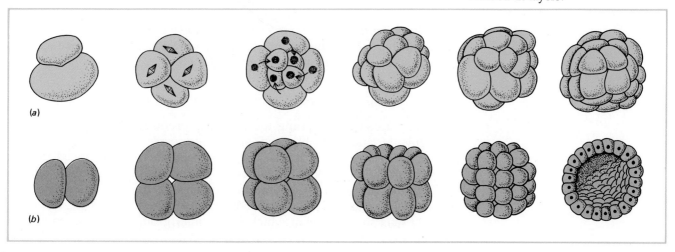

FIGURE 29–7 The coelom originates in the embryo as a block (or
blocks) of mesoderm that split off
from each side of the embryonic
gut. In schizocoely, the mesoderm
(red) splits; this split widens into
a cavity that becomes the coelom.
In enterocoely, the mesoderm outpockets from the gut, forming
pouches. The cavity within these
pouches becomes the coelom. Ectoderm is shown in blue, endoderm
in yellow.

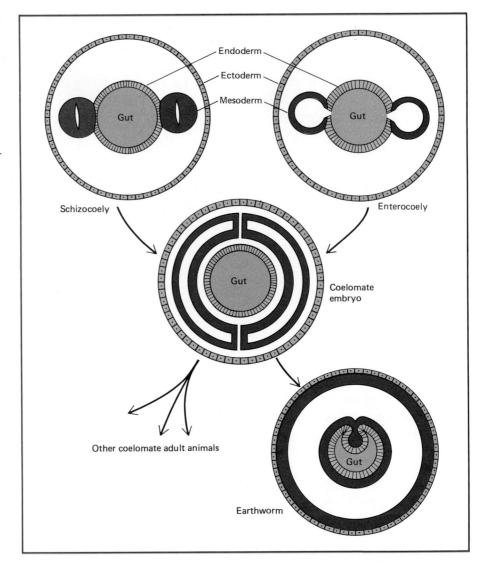

Still another difference between protostome and deuterostome development is the manner in which the coelom is formed. In protostomes, the
mesoderm splits, and the split widens into a cavity that becomes the coelom (Figure 29–7). This method of coelom formation is known as
schizocoely, and for this reason the protostomes are sometimes called
schizocoelomates. In deuterostomes, the mesoderm usually forms as "outpocketings" of the developing gut. These outpocketings eventually separate and form pouches; the cavity within these pouches becomes the coelom. This type of coelom formation is called **enterocoely,** and these animals
are sometimes referred to as **enterocoelomates.**

Simple Versus Complex

In discussing and comparing the many different groups of animals, it is
convenient to use terms like *lower* and *higher, simple* and *complex,* and *primitive* and *advanced.* Such terms as higher, complex, or advanced do not
imply that these animals are better or more nearly perfect than others.
Rather, they are used in a comparative sense to describe their hypothesized evolutionary relationships. For example, the terms *higher* and *lower*
usually refer to the level at which a particular group has diverged from a
main line of evolution. It is customary, for instance, to refer to sponges and
cnidarians as lower invertebrates because they are thought to have origi-

nated near the base of the phylogenetic tree of the animal kingdom. However, neither sponges nor cnidarians are primitive in all morphological or physiological characteristics. Each has become highly specialized to its own particular life style. Furthermore, the terms *lower* and *higher* do not necessarily imply that the higher groups have evolved directly from or through the lower groups.

PHYLUM PORIFERA: THE SPONGES

About 5000 species of **sponges** have been identified and assigned to phylum **Porifera.** These simple, multicellular animals occupy aquatic, mainly marine, habitats. Living sponges may be drab or bright green, orange, red, or purple (Figure 29–8). They are usually slimy to the touch and may have an unpleasant odor. Sponges range in size from 1 to 200 cm in height and vary in shape from flat, encrusting growths to balls, cups, fans, or vases.

Sponges are usually thought to have evolved from certain flagellate protozoans, perhaps from a hollow, free-swimming colonial flagellate. Sponge larvae resemble such flagellate colonies. In the sense that they apparently did not give rise to any other animal group, sponges seem to represent a dead end in evolution. Of course, sponges themselves continue to change as they are subjected to continual selective pressures from the environment.

Sponges are divided into three main classes on the basis of the type of skeleton they secrete: Members of the class **Calcispongiae** secrete a skeleton composed of small calcium carbonate spikes, or **spicules.** Members of the class **Hexactinellida,** the glass sponges, have a skeleton of six-rayed siliceous spicules. Sponges that belong to the class **Demospongia** have a skeleton of **spongin** (a protein material) fibers or of siliceous spicules that do not have six rays. Bath sponges consist of the dried spongin skeleton of demospongians.

Body Plan of a Sponge

Porifera, meaning "to bear pores," aptly describes the sponge body, which resembles a sac perforated with tiny holes. In a simple sponge, water enters through these pores, passes into the central cavity, or **spongocoel,** and finally flows out through the sponge's open end, the **osculum.** Water is kept moving by the action of flagellated cells that line the spongocoel. Each of these **collar cells,** or **choanocytes,** is equipped with a tiny collar that surrounds the base of the flagellum. The collar is actually an extension of the cell membrane and consists of microvilli. The choanocytes of some complex sponges can pump a volume of water equal to the volume of the sponge each minute! In some types of sponges, the body wall is extensively folded, and there are complicated systems of canals (Figure 29–9). Sponges are the only animals that have choanocytes. These cells are strikingly similar to the choanoflagellates, an order of flagellate protozoans. Many biologists think that the choanoflagellates gave rise to the sponges and also to the Eumetazoa.

Most sponges are asymmetrical, but some exhibit radial symmetry. Although a sponge is multicellular, its cells are loosely associated and do not form definite tissues. There is a division of labor, with certain cells specialized to perform particular functions such as nutrition, support, or reproduction. The epidermal cells that make up the outer layer of the sponge are capable of contraction (Figure 29–10). The choanocytes, which make up the inner layer, create the water current that brings food and oxygen to the cells and carries away carbon dioxide and other wastes; they also trap and phagocytize food particles. Between the outer and inner cellular layers of

(a)

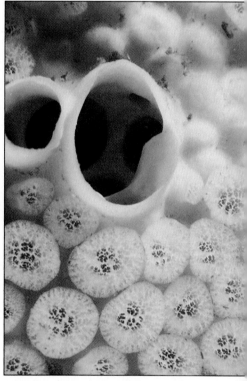

(b)

FIGURE 29–8 Sponges vary widely in size, shape, and color. (*a*) Tube sponge, a member of class Demospongiae. (*b*) View of the osculum of a sulphur sponge, *Cliona celata*.

659

FIGURE 29–9 Three types of body plans in the sponge. (*a*) Simple unfolded or asconoid type of sponge. This type of sponge cannot grow very large because its surface area is limited and the rate of water flow is too slow. (*b*) In the syconoid type of sponge, the body wall is folded, forming finger-like processes. This folding increases the surface area so that there are more choanocytes to circulate water. There is also a decrease in the size of the spongocoel so that there is less water that must be moved out through the osculum. (*c*) The most complex sponge body plan, the leuconoid type, is also the most efficient for circulating water, and most sponges have this intricately folded structure. Numerous channels replace the single, large spongocoel characteristic of the simpler types.

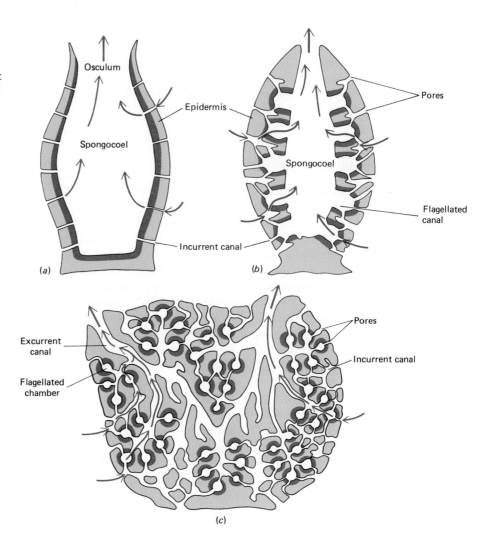

the sponge body is a gelatin-like layer supported by skeletal spicules. Amoeba-like cells, aptly called **amoebocytes,** wander about in this layer, and some of these cells secrete the spicules.

Life Style of a Sponge

Although larval sponges are flagellated and able to swim about, the adult sponge remains attached to some solid object on the sea bottom and is incapable of locomotion. Since sponges cannot swim in search of food, they are adapted for trapping and eating whatever food the sea water brings to them. Sponges are filter feeders. As water circulates through the body, food is trapped along the sticky collars of the choanocytes. Food particles pass down to the base of the collar, where they are taken by phagocytosis and either digested within the choanocyte or transferred to an amoebocyte for digestion. The amoebocytes help distribute food to other cells of the sponge. Direct absorption of nutrients from one cell to another may also help in distributing materials. Undigested food is simply eliminated into the water.

Oxygen from the water diffuses throughout the sponge. Respiration and excretion are carried on by each individual cell. Each cell of the sponge body is irritable and can react to stimuli. However, there are no sensory cells or nerve cells that would enable the animal to react as a whole. Behavior appears limited to the basic metabolic necessities such as procuring food and regulating the flow of water through the body. Pores and the osculum may be closed by the contraction of surrounding cells.

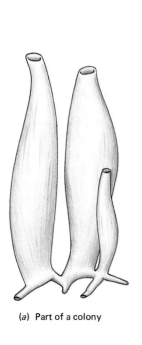

(a) Part of a colony

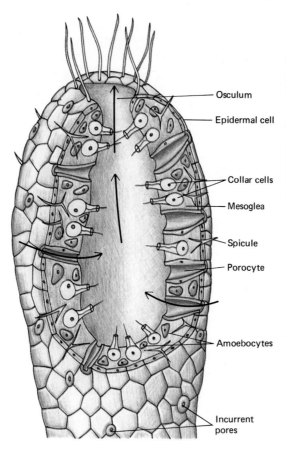

Osculum

Epidermal cell

Collar cells

Mesoglea

Spicule

Porocyte

Amoebocytes

Incurrent pores

(b)

(c)

FIGURE 29–10 Sponge structure. (a) Part of a sponge colony. (b) A simple sponge cut open to expose its cellular organization. (c) A photomicrograph of the spicules of a sponge (*Grantia*) (approximately ×100).

Sponges can reproduce asexually. A small fragment or bud may break free from the parent sponge and give rise to a new sponge or may remain to form a colony with the parent sponge. Sponges also reproduce sexually. Most sponges are **hermaphroditic,** meaning that the same individual can produce both egg and sperm. Some of the amoebocytes develop into sperm cells, others into egg cells. Hermaphroditic sponges can cross-fertilize other sponges, however. Fertilization and early development take place within the jelly-like layer. The flagellated larva moves out into the spongocoel and leaves the parent along with the stream of excurrent water. After swimming about for a day or two as part of the plankton, the larva finds a solid object, attaches to it, and settles down to a sessile life.

Sponges possess a remarkable ability to repair themselves when injured and to regenerate lost parts. When the cells of a sponge are separated from one another, they reaggregate to form a complete sponge again. When clusters of cells are isolated from one another in separate containers of seawater, each cluster will reorganize and regenerate to form a new sponge. If the disaggregated cells of two different sponge species are mixed together, the cells will sort themselves out and reorganize separate sponges of the original species.

PHYLUM CNIDARIA

Most of the 10,000 or so species of the phylum **Cnidaria**[1] are marine. They are grouped in three classes (Table 29–1): **Hydrozoa,** which includes the hydras, the hydroids such as *Obelia,* and the Portuguese man-of-war; **Scy-**

---

[1]This phylum was formerly known as Coelenterata, a name derived from the fact that the body cavity serves as the digestive cavity: *coel* = hollow; *enteron* = gut.

Table 29–1 CLASSES OF PHYLUM CNIDARIA

Class and Representative Animals	Characteristics
Hydrozoa *Hydra* *Obelia* Portuguese man-of-war	Mainly marine, but some freshwater species; both polyp and medusa stage in many species (polyp form only in *Hydra*); formation of colonies by polyps in some cases
Scyphozoa Jellyfish (e.g., *Cyanea*)	Marine; inhabit mainly coastal waters; free-swimming jellyfish most prominent forms; polyp stage restricted to small larval stage
Anthozoa Sea anemones Corals Sea fans	Marine; solitary or colonial polyps; no medusa stage; gastrovascular cavity divided by partitions into chambers increasing area for digestion; sessile

FIGURE 29–11 Some representatives of phylum Cnidaria. Note the two basic types of body form. Photographs of representative cnidarians. (*a*) A member of class Hydrozoa, *Obelia* forms a colony of polyps. (*b*) *Cyanea capillata*, a lion's mane jellyfish, a member of class Scyphozoa. The bell of this jellyfish can measure 2 meters; its tentacles trail over 10 meters. (*c*) A Portuguese man-of-war (*Physalia*) with a fish it has captured. (*d*) Polyps form the coral *Montastrea cavernosa* extended for feeding. (*e*) A sea anemone (*Urticina*) eating an orange sun star.

phozoa, which includes the jellyfish; and **Anthozoa,** which includes the sea anemones, true corals, and alcyonarians (sea fans, sea whips, and precious corals) (Figure 29–11).

All of the cnidarians have stinging cells, called **cnidocytes,** from which they get their name (Cnidaria is from a Greek word meaning "sea nettles"). The cnidarian body is radially symmetrical and is organized as a hollow sac with the mouth and surrounding tentacles located at one end. The mouth leads into the digestive cavity, called the **gastrovascular cavity.** The mouth

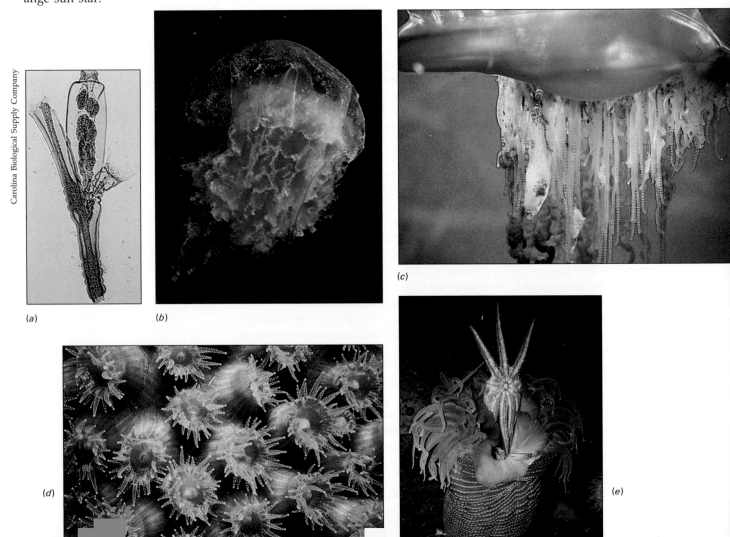

Carolina Biological Supply Company

(*a*)   (*b*)   (*c*)   (*d*)   (*e*)

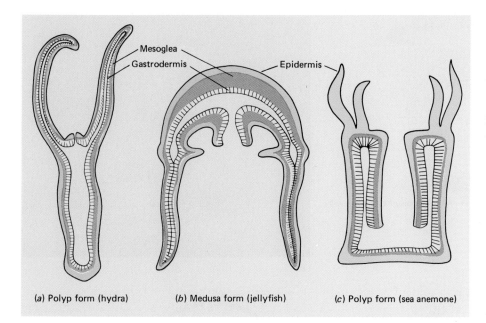

(*a*) Polyp form (hydra)   (*b*) Medusa form (jellyfish)   (*c*) Polyp form (sea anemone)

FIGURE 29–12 The polyp and medusa body forms characteristic of phylum Cnidaria are structurally similar. (*a*) The polyp form as seen in *Hydra*. (*b*) The medusa form is basically an upside-down polyp. (*c*) In the anthozoan polyp, the gastrovascular cavity is characteristically divided into chambers by vertical partitions.

is the only opening into the gastrovascular cavity and so must serve for both the ingestion of food and egestion of wastes.

Much more highly organized than the sponge, a cnidarian has two definite tissue layers. The outer **epidermis** and the inner **gastrodermis** are composed of several types of epidermal cells. These layers are separated by a gelatin-like **mesoglea.**

Cnidarians have two body shapes, the polyp and the medusa (Figure 29–12). The **polyp** form, represented by *Hydra*, resembles an upside-down, slightly elongated jellyfish. Some cnidarians have the polyp shape during their larval stage and later develop into the **medusa** (jellyfish) form. Though many cnidarians live a solitary existence, others group into colonies. Some colonies—for example, the Portuguese man-of-war—consist of both polyp and medusa forms.

Class Hydrozoa

As the most primitive class of the cnidarians, Hydrozoa are thought by some evolutionists to have given rise to both of the other classes. The cnidarian body plan is typified by a tiny animal, *Hydra*, found in freshwater ponds and appearing to the naked eye like a bit of frayed string (Figure 29–13). This animal is named after the multiheaded monster of Greek mythology with the remarkable ability to grow two new heads for each head cut off. The cnidarian hydra also has an impressive ability to regenerate: When it is cut into several pieces, each piece may grow all the missing parts and become a whole animal.

The hydra's body, seldom more than 1 cm long, consists of two layers of cells enclosing a central gastrovascular cavity. The outer **epidermis** serves as a protective layer; the inner **gastrodermis** is primarily a digestive epithelium. The bases of cells in both layers are elongated into contractile muscle fibers; those of the epidermis run lengthwise, and those in the gastrodermis run circularly. By the contraction of one or the other, the hydra can shorten, lengthen, and bend its body. Throughout its life, the animal lives attached to a rock, twig, or leaf by a pedal disk of cells at its base. At the outer end is the mouth, connecting the gastrovascular cavity with the outside and surrounded by a circlet of tentacles. Each tentacle may be as much as one and a half times as long as the body itself. The tentacles, composed of an outer epidermis and an inner gastrodermis, may be hollow or solid.

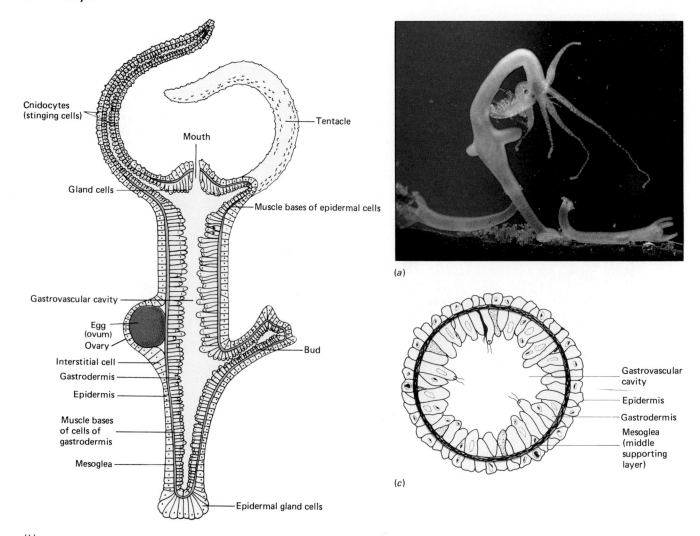

(a)

(b)

(c)

FIGURE 29–13 *Hydra* body structure. (*a*) A brown hydra, *Hydra oligactis,* capturing a small crustacean. Note the buds present on the hydra's body. One bud has already detached as a separate animal. (*b*) This *Hydra* is cut longitudinally to show its internal structure. Asexual reproduction by budding is represented on the right; sexual reproduction is represented by the ovary on the left. Male hydras develop testes that produce sperm. (*c*) Cross section through the body of a *Hydra*.

The cnidarians are unique in producing "thread capsules," or **nematocysts,** (Figure 29–14) within cnidocytes (stinging cells) in the epidermis. The nematocysts, when appropriately stimulated, can release a coiled, hollow thread. Some types of nematocyst threads are sticky; others are long and coil around the prey; a third type is tipped with a barb or spine and can inject a protein toxin that paralyzes the prey.

Each cnidocyte has a small projecting trigger (cnidocil) on its outer surface that responds to touch and to chemicals dissolved in the water ("taste") and causes the nematocyst to fire its thread. A nematocyst can be used only once; when it has been discharged, it is released from the cnidocyte and replaced by a new one, produced by a new cnidocyte. The tentacles encircle the prey and stuff it through the mouth into the gastrovascular cavity, where digestion begins. The partially digested fragments are taken up by pseudopods of the gastrodermis cells, and digestion is completed within food vacuoles in those cells.

Respiration and excretion occur by diffusion, for the body of a hydra is small enough that no cell is far from the surface. The motion of the body as it stretches and shortens circulates the contents of the gastrovascular cavity, and some of the gastrodermis cells have flagella whose beating aids in circulation. The hydra has no true circulatory organs.

The first true nerve cells in the animal kingdom are found in the cnidarians. These animals have many nerve cells that form irregular **nerve nets** connecting the sensory cells in the body wall with muscle and gland cells. The coordination achieved thereby is of the simplest sort; there is no aggre-

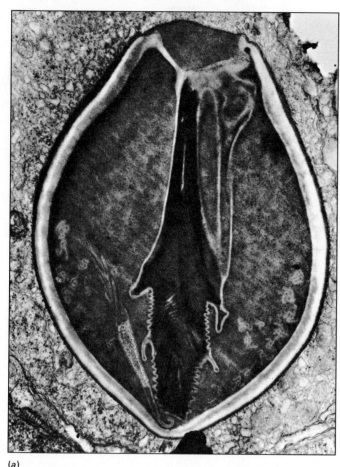

(a)

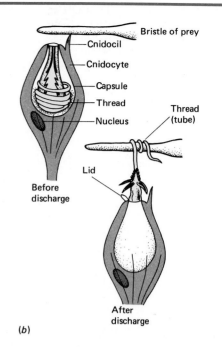

Bristle of prey
Cnidocil
Cnidocyte
Capsule
Thread
Nucleus

Before discharge

Thread (tube)
Lid

After discharge

(b)

FIGURE 29–14 Nematocysts, the thread capsules within cnidarian cnidocytes. (*a*) Electron micrograph of an undischarged nematocyst of *Hydra* (sagittal section). (*b*) Discharge of a nematocyst. When an object comes in contact with the cnidocil, the nematocyst discharges, ejecting a thread that may entangle the prey or secrete a toxic substance immobilizing the prey.

gation of nerve cells to form a brain or nerve cord, and an impulse set up in one part of the body passes in all directions more or less equally.

Hydras reproduce asexually by budding during periods when environmental conditions are optimal but are stimulated to form sexual forms, males and females, in the fall, or when the pond water becomes stagnant. Females develop an ovary, which produces a single egg; males form a testis that produces sperm. After fertilization the egg becomes covered with a shell, leaves the parent, and remains within the protective shell throughout the winter.

Many of the marine cnidarians form colonies consisting of hundreds or thousands of individuals. A colony begins with a single individual that reproduces by budding. However, instead of separating from the parent, the bud remains attached and continues to form additional buds. Several types of individuals may arise in the same colony, some specialized for feeding, some for reproduction, and others for defense. The existence of two or more different kinds of individuals within the same species is known as **polymorphism.**

The Portuguese man-of-war, *Physalia*, superficially resembles a jellyfish but is actually a colony of polyps and medusas. The individuals display remarkable polymorphism. A modified medusa acts as a float for the colony in the form of a gas-filled sac colored a vivid iridescent purple. The long tentacles of this animal, which may hang down several meters below the float, are equipped with cnidocytes. These are capable of paralyzing a large fish and can wound a human swimmer severely.

FIGURE 29–15 Life cycle of *Obelia*, a colonial marine hydrozoan. Note the specialization of individual members of the polyp colony.

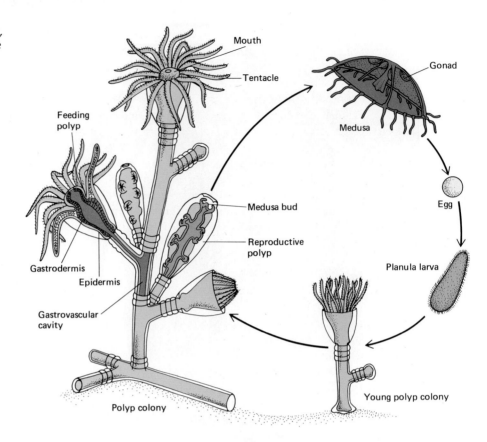

Some of the marine cnidarians are remarkable for an alternation of sexual and asexual generations analogous to that in plants. This alternation of generations differs from that of plants in that both sexual and asexual forms are diploid. Only sperm and egg are haploid. The cnidarian life cycle is illustrated by the colonial marine hydrozoan *Obelia* (Figure 29–15). In this polyp colony, the asexual generation consists of two types of polyps: those specialized for feeding and those for reproduction. Free-swimming male and female medusae bud off from the reproductive polyps. These medusae eventually produce sperm and eggs, and fertilization takes place. The zygote develops into a ciliated swimming larva called a **planula.** The larva attaches to some solid object and begins to form a new generation of polyps by asexual reproduction.

Class Scyphozoa (Jellyfish)

Among the jellyfish the medusa is the more prominent body form. It is like an upside-down hydra with a thick viscous mesoglea that gives firmness to the body. In scyphozoans, the polyp stage is restricted to a small larval stage. The largest jellyfish, *Cyanea*, may be more than 2 meters in diameter and have tentacles 30 meters long. These orange and blue monsters, among the largest of the invertebrate animals, are a real danger to swimmers in the North Atlantic Ocean.

Class Anthozoa (Corals)

The sea anemones and corals have no free-swimming medusa stage, and the polyps may be either individual or colonial forms. These animals have a small ciliated larva, which may swim to a new location before attaching to develop into a polyp.

Anthozoans differ from hydrozoans in that the gastrovascular cavity is divided by a series of vertical partitions into a number of chambers, and the surface epidermis is turned in at the mouth to line the pharynx (Figure 29–16). The partitions in the gastrovascular cavity increase the surface area for digestion, so that an anemone can digest an animal as large as a crab or fish. Although corals can capture prey, many tropical species depend for nutrition mainly on photosynthetic dinoflagellates that live within their cells.

In warm shallow seas, almost every square meter of the bottom is covered with coral or anemones, most of them brightly colored. The extravagant reefs and atolls of the South Pacific are the remains of billions of microscopic, cup-shaped calcareous skeletons, secreted during past ages by coral colonies and by coralline plants. Living colonies occur only in the uppermost regions of such reefs, adding their own calcareous skeletons to the mass.

PHYLUM CTENOPHORA (COMB JELLIES)

The **ctenophores,** or comb jellies, are a phylum of about 50 marine species. They are fragile, luminescent animals that may be as small as a pea or larger than a tomato. They are biradially symmetrical, and their body plan is somewhat similar to that of a medusa. The body consists of two cell layers separated by a thick jelly-like mesoglea.

The outer surface of a ctenophore is covered with eight rows of cilia, resembling combs (Figure 29–17). The coordinated beating of the cilia in these combs moves the animal through the water. At the upper pole of the body is a sense organ containing a mass of limestone particles balanced on four tufts of cilia connected to sense cells. When the body turns, these particles bear more heavily on the lower cilia, stimulating the sense cells. This causes the cilia in certain of the combs to beat faster and bring the body back to its normal position. Nerve fibers extending from the sense organ to the cilia control the beating. If these fibers are cut, the beating of the cilia below the incision is disorganized.

Ctenophores have only two tentacles, and most species lack the stinging cells characteristic of the cnidarians. However, their tentacles are equipped with adhesive glue cells (called colloblasts), which trap their prey.

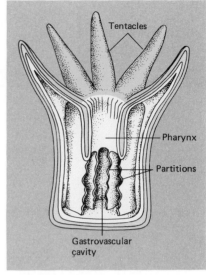

FIGURE 29–16 Structure of an anthozoan polyp. Longitudinal section.

FIGURE 29–17 Ctenophore structure. (*a*) The ctenophore *Pleurobrachia* is shown in side view with the body cut open to expose internal structures. (*b*) Top view of the ctenophore shown in (*a*). (*c*) The ctenophore *Boreo*.

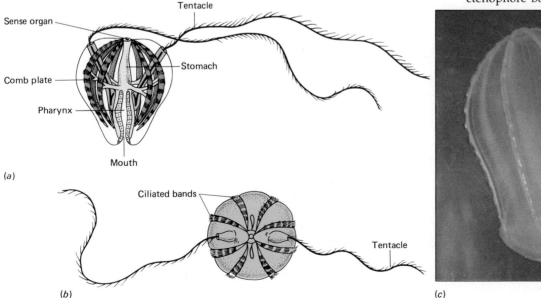

Table 29–2 **CLASSES OF PHYLUM PLATYHELMINTHES**	
Class and Representative Animals	*Characteristics*
Turbellaria Planarians (e.g., *Dugesia*)	Free-living flatworms; mainly marine, some terrestrial forms living in mud; body covered by ciliated epidermis; usually carnivorous forms that prey on tiny invertebrates or on dead organisms
Trematoda Flukes (e.g., schistosomes)	All parasites with a wide range of vertebrate and invertebrate hosts; may require intermediate hosts; suckers for attachment to host
Cestoda Tapeworms	Parasites of vertebrates; complex life cycle with one or two intermediate hosts; larval host may be invertebrate; tapeworms have suckers and sometimes hooks on scolex for attachment to host; eggs produced within proglottids, which are shed; no digestive system

PHYLUM PLATYHELMINTHES (FLATWORMS)

Members of phylum **Platyhelminthes,** the **flatworms,** are flat, elongated, legless animals. They exhibit bilateral symmetry and are the simplest members of the Bilateria. Some zoologists classify them as acoelomate protostomes, while others just refer to them as acoelomates and reserve the term *protostomes* for mollusks, annelids, and arthropods (animals with a complete digestive tract). The three classes of **Platyhelminthes** are **Turbellaria,** the free-living flatworms, including *Planaria* and its relatives; **Trematoda,** the flukes, which are either internal or external parasites; and **Cestoda,** the tapeworms, the adults of which are intestinal parasites of vertebrates (Table 29–2).

Some important characteristics of this phylum follow:

1. **Bilateral symmetry and cephalization.** Along with their bilateral symmetry, flatworms have a definite anterior end and a posterior end. This is a great advantage because an animal with a front end generally moves in a forward direction. With a concentration of sense organs in the part of the body that first meets the environment, an animal is able to detect an enemy quickly enough to escape; it is also more likely to see or smell prey quickly enough to capture it. A rudimentary head, the beginnings of **cephalization,** is evident in flatworms.

2. **Three definite tissue layers.** In addition to an outer epidermis, derived from ectoderm, and an inner endodermis, derived from endoderm, the flatworm has a middle tissue layer, derived from mesoderm, which comprises most of the body.

3. **Well-developed organs.** The flatworms are the simplest animals that have well-developed organs, functional structures made of two or more kinds of tissue. Among their organs is a muscular pharynx for taking in food, eyespots and other sensory organs in the head, a simple brain, and complex reproductive organs.

4. **A simple nervous system.** The simple brain consists of two masses of nervous tissue, called **ganglia,** in the head region. The ganglia are connected to two nerve cords that extend the length of the body. A series of nerves connects the cords like the rungs of a ladder; this type of system is sometimes called a ladder-type nervous system.

5. **Excretory structures** called **protonephridia** ending in specialized collecting cells, called flame cells.

6. A **gastrovascular cavity** in most species. It is often extensively branched and has only one opening, the mouth, usually located on the middle of the ventral surface.

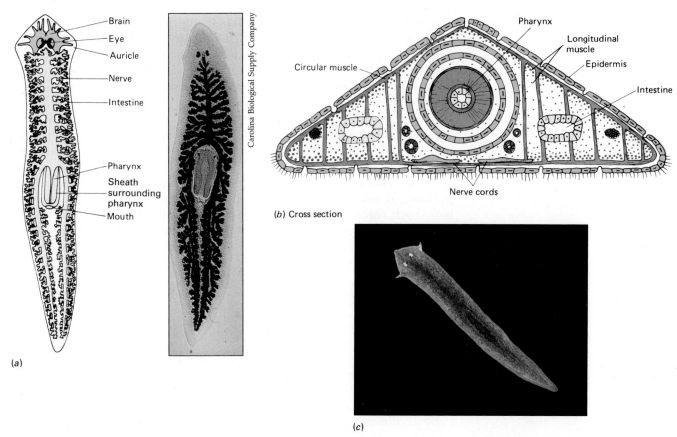

(a)

(b) Cross section

(c)

FIGURE 29–18 The common planarian, *Dugesia*. (*a*) A stained specimen compared with a line drawing. (*b*) Cross section through the body of a planarian. (*c*) A living *Dugesia*.

Both groups of parasitic flatworms—the flukes and tapeworms—are highly adapted to their parasitic life style. The epithelial layers of these animals are resistant to the digestive enzymes secreted by their hosts. They have suckers or hooks for holding on to the host. Many have complicated life cycles that enable them to change hosts (since an individual host eventually dies). To further ensure survival of the species, these worms produce large numbers of eggs. Other adaptations include the loss of unneeded structures such as sense organs or a digestive system.

Class Turbellaria

Members of the class Turbellaria are free-living, mainly marine, flatworms. **Planarians** are turbellarian flatworms found in ponds and quiet streams all over the world. The common American planarian *Dugesia* is about 15 mm long, with what appear to be crossed eyes and flapping ears called **auricles** (Figure 29–18). The auricles actually serve as organs of smell.

Planarians are carnivorous, trapping small animals in a mucous secretion. The digestive system consists of a single opening (the mouth), a pharynx, and a branched intestine. A planarian can project its **pharynx** (the first portion of the digestive tube) outward through its mouth, to suck up small pieces of the prey. Extracellular digestion takes place in the intestine by enzymes secreted by gland cells. Digestion is completed after the nutrients have been absorbed into individual cells. Undigested food is eliminated through the mouth. The lengthy intestine (actually a highly branched gastrovascular cavity) helps to distribute food to all parts of the body, so that each cell is within range of diffusion. Flatworms can survive without food for months, gradually digesting their own tissues and growing smaller as time passes. Some flatworms confiscate intact nematocysts from the hydras they eat; they incorporate them into their own epidermis and use them for defense.

A planarian's flattened body ensures that gases can reach all the cells by diffusion. There are no specialized respiratory or circulatory structures. Although some excretion takes place by diffusion, an excretory system is present. It consists of two excretory tubes that extend the length of the body and give off branches called **protonephridia** throughout their length. Each of these tubules ends in a **flame cell,** a collecting cell equipped with cilia. The beating of the cilia channels water containing wastes into the system of tubules. Planarians are capable of learning; memory is not localized within the brain but appears to be retained throughout the nervous system.

Planarians can reproduce either asexually or sexually. In asexual reproduction, an individual constricts in the middle and divides into two planarians. Each regenerates its missing parts. Sexually, these animals are hermaphroditic. During the warm months of the year, each is equipped with a complete set of male and female organs. Two planarians come together in copulation and exchange sperm cells so that their eggs are cross-fertilized.

Class Trematoda (Flukes)

Although they are parasites, the **flukes,** members of class **Trematoda,** are structurally like the free-living flatworms. They differ in having one or more suckers with which to cling to the host and in their lack of a ciliated epidermis (Figure 29–19). The organs of digestion, excretion, and coordination are like those of the other flatworms, but the mouth is anterior rather than ventral. The reproductive organs are extremely complex.

Some flukes that are parasitic in human beings are the blood flukes, widespread in China, Japan, and Egypt, and the liver flukes, common in

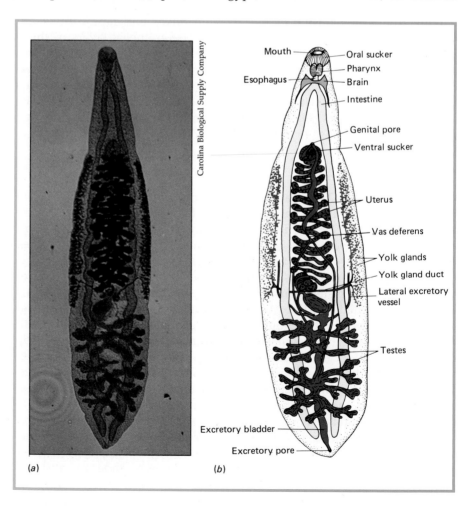

FIGURE 29–19 Flukes are adapted for a parasitic mode of life. (*a*) The human liver fluke, *Clonorchis sinensis.* (*b*) Internal structure of a fluke.

Carolina Biological Supply Company

Mouth
Esophagus
Oral sucker
Pharynx
Brain
Intestine
Genital pore
Ventral sucker
Uterus
Vas deferens
Yolk glands
Yolk gland duct
Lateral excretory vessel
Testes
Excretory bladder
Excretory pore

(*a*)　　　(*b*)

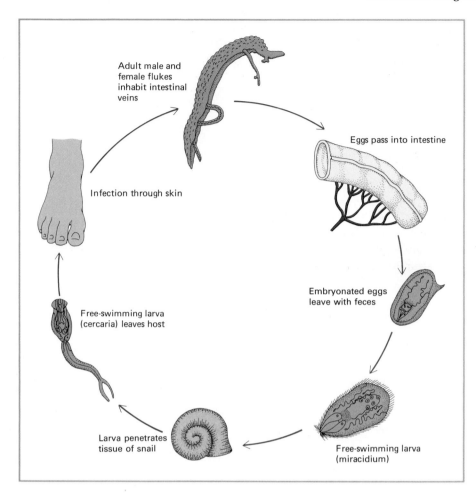

Adult male and
female flukes
inhabit intestinal
veins

Infection through skin

Eggs pass into intestine

Free-swimming larva
(cercaria) leaves host

Embryonated eggs
leave with feces

Larva penetrates
tissue of snail

Free-swimming larva
(miracidium)

FIGURE 29–20 Life cycle of a blood fluke, a schistosome. The adult male has a long canal that holds the female during fertilization.

FIGURE 29–21 False-color scanning electron micrograph of the head (scolex) of the small tapeworm *Acanthrocirrus retrisrostris*, taken in its larval, encysted form from the body of its barnacle host (*Balanus balanoides*). The tapeworm reaches maturity in the intestines of wading birds that eat barnacles. The photograph shows the piston-like rostellum, which can be withdrawn into the head or thrust out and buried in the host's tissue. Beneath the rostellum, two of the four powerful suckers are visible (×80 at 35-mm size).

China, Japan, and Korea. Blood flukes of the genus *Schistosoma* infect about 200 million people who live in tropical and temperate areas. Both blood flukes and liver flukes go through complicated life cycles, involving a number of different forms, alternation of sexual and asexual generations, and parasitism on one or more intermediate hosts, such as snails and fishes (Figure 29–20). When dams are built, marshy areas are created, which provide habitats for the aquatic snails that serve as intermediate hosts in the fluke life cycle.

Class Cestoda (Tapeworms)

Adult members of the more than 1000 different species of the class Cestoda live as parasites in the intestines of probably every kind of vertebrate, including humans. Tapeworms are long, flat, ribbon-like animals strikingly specialized for their parasitic mode of life. Among their many adaptations are suckers and sometimes hooks on the head, or **scolex,** which enable the parasite to maintain its attachment to the host's intestine (Figure 29–21). Their reproductive adaptations and abilities are extraordinary. The body of the tapeworm consists of a long chain of segments called **proglottids.** Each segment is an entire reproductive machine equipped with both male and female organs and containing as many as 100,000 eggs. Since an adult tapeworm may possess as many as 2000 segments, its reproductive potential is staggering. A single tapeworm may produce as many as 600 million eggs in a year. Proglottids farthest from the tapeworm's head contain the ripest eggs; these segments are shed daily, leaving the host's body with the feces.

Tapeworms lack certain organs. They absorb their food directly through

FIGURE 29–22 Life cycle of the beef tapeworm, a parasitic flatworm.

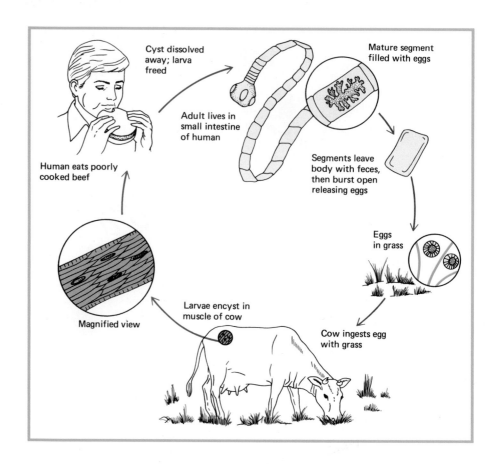

Cyst dissolved away; larva freed

Mature segment filled with eggs

Adult lives in small intestine of human

Segments leave body with feces, then burst open releasing eggs

Human eats poorly cooked beef

Eggs in grass

Magnified view

Larvae encyst in muscle of cow

Cow ingests egg with grass

their body wall from the host's intestine and have no mouth and no digestive system of their own. Some tapeworms have rather complex life cycles, spending their larval stage within the body of an intermediate host and their adult life within the body of a different, final host. As an example, let us consider the life cycle of the beef tapeworm, so named because human beings become infected when they eat poorly cooked beef containing the larvae (Figure 29–22).

The microscopic tapeworm larva spends part of its life cycle encysted within the muscle tissue of beef. When a human being ingests infected meat, the digestive juices break down the cyst, releasing the larva. The larva attaches itself to the intestinal lining and within a few weeks matures into an adult tapeworm, which may grow to a length of 50 feet. The parasite reproduces sexually within the human intestine and sheds proglottids filled with ripe eggs. Once established within a human host, the tapeworm makes itself very much at home and may remain there for the remainder of its life, as long as 10 years. A person infected with a tapeworm may suffer pain or discomfort, increased appetite, weight loss, and other symptoms or may be totally unaware of its presence.

In order for the life cycle of the tapeworm to continue, its eggs must be ingested by an **intermediate host,** in this case, a cow. (This requisite explains why we are not completely overrun by tapeworms, and why the tapeworm must produce millions of eggs to ensure that at least a few will survive.) When a cow eats grass or other foods contaminated with human feces, eggs may be ingested. The eggs hatch in the cow's intestine, and the larvae make their way into muscle. There they encyst and remain until released by a **final host,** perhaps a human eating rare steak.

Two other tapeworms that infect humans are the pork tapeworm and the fish tapeworm. The pork tapeworm infects persons who eat poorly cooked, infected pork, and the fish tapeworm is contracted by ingesting raw, or poorly cooked, infected fish. Like most parasites, tapeworms tend to be species-specific; that is, each can infect only certain specific species. For example, the beef tapeworm can spend its adult life in no other host than a human being.

PHYLUM NEMERTINEA (PROBOSCIS WORMS)

The phylum **Nemertinea** is a relatively small group of animals (about 550 species) that is considered an evolutionary landmark because its members are the simplest animals to possess definite organ systems (Figure 29–23). None of them is parasitic, and none is of economic importance. Almost all of them are marine, although a few inhabit fresh water or damp soil. They have long narrow bodies, either cylindrical or flattened, varying in length from 5 cm to more than 20 m. Some of them are a vivid orange, red, or green, with black or colored stripes.

Their most remarkable organ—the **proboscis,** from which they get their common name, proboscis worms—is a long, hollow, muscular tube, which can be everted from the anterior end of the body for use in seizing food or in defense. The proboscis secretes mucus, which is helpful in catching and retaining prey. In certain species, the proboscis is equipped with a hard point at its tip and poison-secreting glands at the base of this point. The outward movement of the proboscis is accomplished by the pressure of the surrounding muscular walls on the contained fluid; a separate muscle inside the proboscis retracts it.

An important advance displayed by the nemerteans is its **tube-within-a-tube-body plan.** The digestive tract is a complete tube, with a mouth at one end for taking in food and an anus at the other for eliminating undigested food. This is in contrast to the cnidarians and planarians, whose food enters and wastes leave by the same opening.

A second advance exhibited by the nemerteans is the separation of digestive and circulatory functions. These animals are the most primitive organisms to have a separate circulatory system. It is a rudimentary system consisting simply of muscular tubes—the blood vessels—extending the length of the body and connected by transverse vessels. Some of these primitive forms have red blood cells filled with hemoglobin, the same red pigment that transports oxygen in human blood. Nemerteans have no

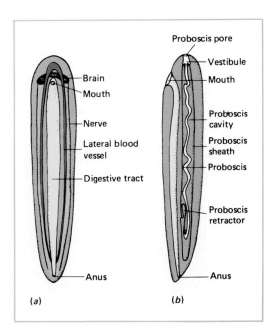

FIGURE 29–23 Structure of a typical proboscis worm or nemertean. (*a*) Dorsal view of the digestive, circulatory, and nervous systems. (*b*) Lateral view of the digestive tract and proboscis. Note the complete digestive tract that extends from mouth to anus, giving this animal a tube-within-a-tube body plan.

heart to pump the blood; the blood is circulated through the vessels by movements of the body and contractions of the muscular blood vessels.

The nervous system is more highly developed than in the flatworm. A "brain" at the anterior end of the body consists of two groups of nerve cells (ganglia) connected by a ring of nerves extending around the sheath of the proboscis. Two nerve cords extend posteriorly from the brain.

PHYLUM NEMATODA (ROUNDWORMS)

Phylum **Nematoda,** the **roundworms,** are of great ecological importance because of their role as consumers of organic matter. They promote nutrient recycling by enhancing bacterial and fungal activity in the soil. By feeding on bacteria and fungi, they eliminate excess individuals and help maintain healthy bacterial and fungal populations. Nematodes are also enormously important parasites. Their parasitic members inhabit almost every species of plant and animal. Among the 30 or so human parasites belonging to phylum Nematoda are the hookworms, the intestinal roundworm *Ascaris,* pinworms, trichina worms, and filaria worms. More than 12,000 species have been named, and perhaps hundreds of thousands of additional species remain to be identified. Nematodes are widely distributed in the soil, the sea, and fresh water. A spadeful of soil may contain more than a million of these tiny white worms, which thrash around, coiling and uncoiling.

The elongate, cylindrical, threadlike nematode body is pointed at both ends and covered with a tough **cuticle,** which is molted as they grow (Figure 29–24). Secreted by the underlying epidermis, the cuticle enables nematodes to resist desiccation, permitting them to inhabit dry soils and even deserts. Beneath the epidermis is a layer of longitudinal muscles. No circular muscles are present in the body wall. Nematodes are the most primitive animals to have a body cavity, but it is not a true coelom. Because the body cavity is not completely lined with mesoderm, it is referred to as a **pseudocoelom** (Figure 29–25).

The fluid-filled pseudocoelom is a hydrostatic skeleton that transmits the force of muscle contraction to the enclosed fluid. Unlike many other animals, nematodes lack circular muscles to antagonize the action of their longitudinal muscles. Instead, the cuticle serves that function. The nematode moves by contracting its longitudinal muscles on one side, thus deforming the cuticle on that side. When the muscles relax, the cuticle snaps back into place. Characteristic nematode movement is by a thrashing motion.

Like the proboscis worms, the nematodes exhibit bilateral symmetry, a complete digestive tract, three distinct tissue layers, and definite organ systems. However, they lack circulatory structures. The sexes are usually separate, and the male is smaller than the female. The characteristics of

FIGURE 29–24 A free-living nematode among the cyanobacteria *Oscillatoria,* which it eats.

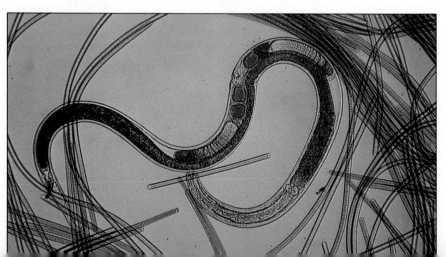

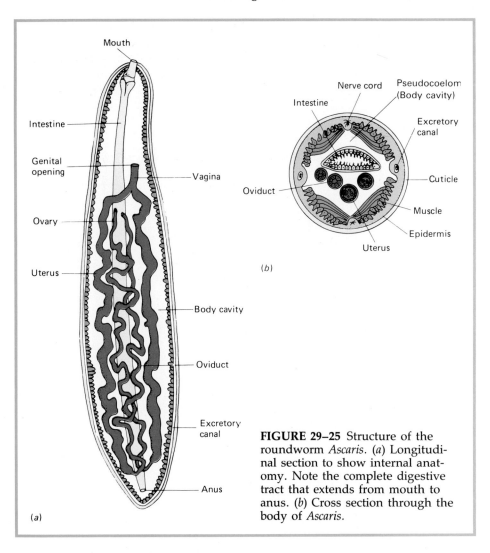

FIGURE 29–25 Structure of the roundworm *Ascaris*. (*a*) Longitudinal section to show internal anatomy. Note the complete digestive tract that extends from mouth to anus. (*b*) Cross section through the body of *Ascaris*.

nematodes and other lower invertebrate phyla are summarized in Table 29–3.

Ascaris: **A Parasitic Roundworm**

A common intestinal parasite of human beings, **Ascaris** is a pinkish worm about 25 cm long. *Ascaris* spends its adult life in the human intestine, where it makes its living by sucking in partly digested food. Like the tapeworm, it must devote a great deal of effort to reproduction in order to ensure survival of its species. The sexes are separate, and copulation takes place within the host. A mature female may lay as many as 200,000 eggs a day.

Ascaris eggs leave the human body with the feces and, where sanitation is poor (that is, in most of the world), find their way onto the soil. In many parts of the world, human wastes are utilized as fertilizer—a practice that encourages the survival of *Ascaris* and many other human parasites. People are infected when they ingest *Ascaris* eggs. The eggs hatch in the intestine, and the larvae then take an incredible journey through the body before settling in the small intestine. The larvae burrow through the intestinal wall into blood vessels or lymph vessels. Then they are carried through the heart to the lungs, where they break through into the air sacs and move up the air passageways to the throat. Finally, the larvae pass through the stomach and into the intestine, where they settle and feed on partly di-

Table 29–3 COMPARISON OF SOME LOWER INVERTEBRATE PHYLA

Phylum and Representative Animals	Level of Organization	Symmetry	Digestion	Circulation	Gas Exchange	Waste Disposal	Nervous System	Reproduction	Other Characteristics
Porifera (pore bearers) Sponges	Multicellular but tissues loosely arranged	Radial or none	Intracellular	Diffusion	Diffusion	Diffusion	Irritability of cytoplasm	Asexual, by budding; sexual, most are hermaphroditic; larvae swim by cilia; adults incapable of locomotion	Filter feeders; skeleton of chalk, glass, or spongin (a protein material)
Cnidaria Hydra Jellyfish Coral	Tissues	Radial	Gastrovascular cavity with only one opening; intra- and extracellular digestion	Diffusion	Diffusion	Diffusion	Nerve net; no centralization of nerve tissue	Asexual by budding; sexual, sexes separate	Have cnidocytes (stinging cells) along their tentacles
Platyhelminthes (flatworms) Planarians Flukes Tapeworms	Organs	Bilateral; rudimentary head	Digestive tract with only one opening	Diffusion	Diffusion	Protonephridia; flame cells and ducts	Simple brain; two nerve cords; ladder type system; simple sense organs	Asexual, by fission; sexual, hermaphroditic, but cross-fertilization in some species	Three definite tissue layers; no body cavity; many parasitic
Nemertinea Proboscis worms	Organ systems	Bilateral	Complete digestive tract with mouth and anus	At least two pulsating longitudinal blood vessels; no heart; blood cells with hemoglobin	Diffusion	Two lateral excretory canals with flame cells	Simple brain; two nerve cords; cross nerves; simple sense organs	Asexual, by fragmentation; sexual, sexes separate	No body cavity; proboscis for defense and capturing prey
Nematoda (roundworms) Ascarids Hookworms Nematodes	Organ systems	Bilateral	Complete digestive tract with mouth and anus	Diffusion	Diffusion	Excretory canals	Simple brain; dorsal and ventral nerve cords; simple sense organs	Sexual, sexes separate	Have pseudocoelom (space between internal organs and body wall); many parasitic

gested food. During their migration, the larvae can damage the lungs and other tissues. Sometimes the worms perforate the intestine and cause serious infection (peritonitis).

Other Parasitic Roundworms

The life cycle of a human **hookworm** (Figure 29–26) is somewhat similar to that of *Ascaris*. Only one host is required. Adult worms, which are less than 1.5 cm (0.5 inch) long, live in the human intestine and lay eggs, which pass out of the body with the feces. Larvae hatch and feed upon bacteria in the soil. After a period of maturation, they become infective. When a potential host walks barefoot on soil containing the microscopic larvae, or otherwise comes in contact with it, the larvae bore through the skin and enter the blood. They migrate through the body before finding their way to the intestine where they mature.

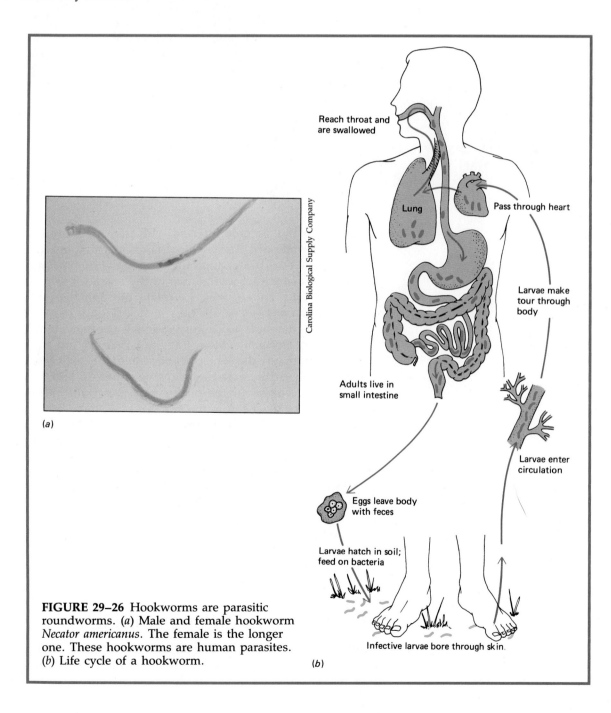

Carolina Biological Supply Company

(a)

Reach throat and are swallowed

Lung

Pass through heart

Larvae make tour through body

Adults live in small intestine

Larvae enter circulation

Eggs leave body with feces

Larvae hatch in soil; feed on bacteria

Infective larvae bore through skin

FIGURE 29–26 Hookworms are parasitic roundworms. (*a*) Male and female hookworm *Necator americanus*. The female is the longer one. These hookworms are human parasites. (*b*) Life cycle of a hookworm.

(b)

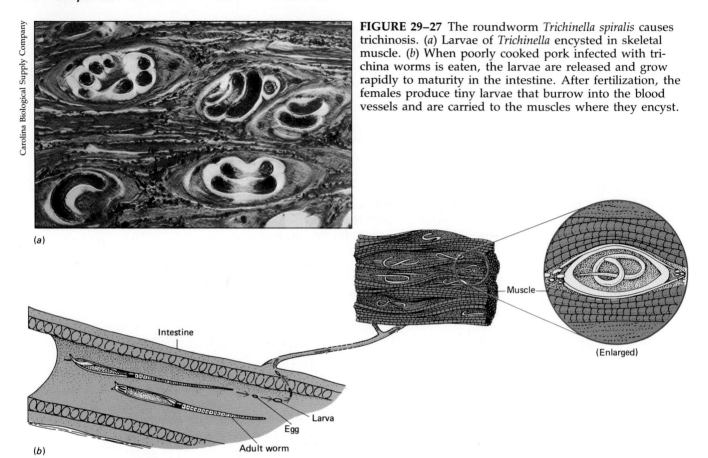

Carolina Biological Supply Company

(a)

(b)

FIGURE 29–27 The roundworm *Trichinella spiralis* causes trichinosis. (*a*) Larvae of *Trichinella* encysted in skeletal muscle. (*b*) When poorly cooked pork infected with trichina worms is eaten, the larvae are released and grow rapidly to maturity in the intestine. After fertilization, the females produce tiny larvae that burrow into the blood vessels and are carried to the muscles where they encyst.

Muscle

(Enlarged)

Intestine

Larva

Egg

Adult worm

Human beings become infected with **trichina worms** by eating poorly cooked, infected pork. The trichina parasite is adapted to live inside many animals (pigs, rats, bears, and others), and the human being is an accidental host. Adult trichina worms live in the small intestine of the host. The females produce larvae, which migrate through the body, making their way to skeletal muscle, where they encyst (Figure 29–27). Continuance of the life cycle depends upon ingestion by another animal. Since human beings are not normally eaten, trichina larvae are not liberated from their cysts and eventually die. The cysts, however, become calcified and permanently remain in the muscles, causing stiffness and discomfort. Other symptoms are caused by the presence of the adults and by the migrating larvae. No cure has been found for this infection.

Pinworms are the most common worms found in children. Adult worms, less than 1.3 cm (0.5 inch) long, live in the large intestine. Female pinworms often migrate to the anal region at night to deposit their eggs. Irritation and itching caused by this practice induce scratching, which serves to spread the tiny eggs. Eggs may be further distributed in the air and are in this way scattered throughout the house. The original host or other members of the household may be infected by ingesting the eggs. Eating with dirty hands facilitates the process. Mild infections may go unnoticed, but those more serious may result in injury to the intestinal wall, discomfort, and irritation.

PHYLUM ROTIFERA (WHEEL ANIMALS)

Among the more obscure invertebrates are the "wheel animals" of the phylum **Rotifera.** These aquatic, microscopic worms, although no larger than many protozoans, are multicellular. Rotifers have a characteristic

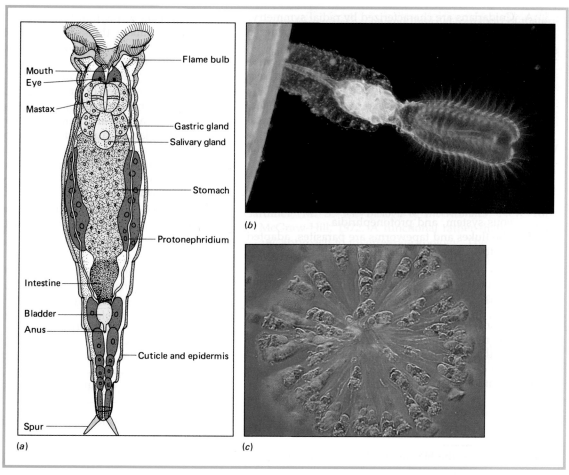

Mouth
Eye
Mastax
Flame bulb
Gastric gland
Salivary gland
Stomach
Protonephridium
Intestine
Bladder
Anus
Cuticle and epidermis
Spur

(a)

(b)

(c)

FIGURE 29–28 Wheel animals. (*a*) Structure of a rotifer. (*b*) A solitary rotifer, *Stephanboceros*, with cilia extended for feeding. (*c*) A colonial rotifer with an internal skeleton composed of silica. The lightweight skeleton helps keep the colony afloat as it travels with other plankton.

crown of cilia on the anterior end, which gives the appearance of a spinning wheel (Figure 29–28). They have a complete digestive tract, including a **mastax,** a muscular organ for grinding food; a pseudocoelom; an excretory system made up of flame cells and a bladder; and a nervous system with a "brain" and sense organs.

Rotifers are "cell constant" animals: Each member of a given species is composed of exactly the same number of cells; indeed, each part of the body is made of a precisely fixed number of cells arranged in a characteristic pattern. Cell division ceases with embryonic development, and mitosis cannot subsequently be induced; growth and repair are impossible. One of the challenging problems of biological research is the nature of the difference between such nondividing cells and the dividing cells of other animals. Do rotifers never develop cancer?

■ SUMMARY

I. Animals are eukaryotic, multicellular, heterotrophic organisms whose cells exhibit a division of labor. They generally are capable of locomotion at some time during their life cycle, can reproduce sexually, and can respond adaptively to external stimuli.

II. Animals are consumers that inhabit the sea, fresh water, and the land.

III. Animals may be classified as Parazoa or Eumetazoa; as Radiata or Bilateria; as acoelomates, pseudocoelomates, or coelomates; or as protostomes or deuterostomes.

IV. Phylum Porifera consists of the sponges.
   A. Sponges are divided into three main classes on the basis of the type of skeleton they secrete.
   B. The sponge body is a sac with tiny holes through which water enters, a central spongocoel, and an osculum.
   C. Metabolic processes in sponges depend upon diffusion of materials between the cells and the watery habitat that bathes them.

V. Phylum Cnidaria includes the hydras, jellyfish, and corals.

fening, enabling the body to withstand the pull of gravity; it provides protection against desiccation; it serves as a coat of armor to protect the animal against predators; and it serves as a point of attachment for muscles.

Reproduction on land poses still another problem. Aquatic forms generally shed their gametes in the water, and fertilization occurs there. The surrounding water serves as an effective shock absorber, protecting the delicate embryos as they develop. Some land animals, including most amphibians, return to the water for reproduction, and the larval forms—tadpoles—develop in the water. Earthworms, snails, insects, reptiles, birds, and mammals engage in internal fertilization. They transfer sperm from the body of the male directly into the body of the female by copulation. The sperm are surrounded by a watery medium or semen. The fertilized egg either is covered by some sort of tough, protective shell secreted around it by the female, or it develops within the body of the mother.

PHYLUM MOLLUSCA

Mollusks are among the best known of the invertebrates—almost everyone has walked along the shore collecting their shells. With more than 50,000 living species and 35,000 fossil species, Phylum **Mollusca** is second only to the arthropods in number of species. This phylum includes clams, oysters, octopuses, snails, slugs, and the largest of all the invertebrates, the giant squid, which may achieve a weight of several tons. Although most mollusks are marine, there are snails and clams that live in fresh water and many species of snails and slugs that inhabit the land. Representative mollusks are illustrated in Figure 30–2, and the major classes are listed in Table 30–1.

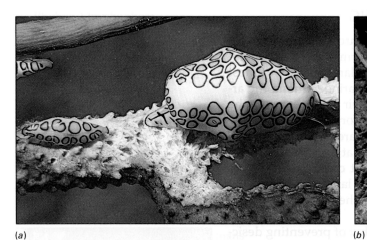

(a)

(b)

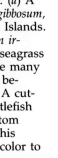

(c)

FIGURE 30–2 There are many beautiful forms of mollusks. (*a*) A flamingo tongue, *Cyphoma gibbosum*, photographed in the Virgin Islands. (*b*) A bay scallop, *Argopecten irradians*, photographed in a seagrass bed in Tampa Bay. Note the many blue eyes looking out from between the hinged shell. (*c*) A cuttlefish (*Sepia officinalis*). Cuttlefish may remain on the sea bottom waiting for passing prey. This cephalopod can change its color to simulate its background.

Table 30–1 **MAJOR CLASSES OF PHYLUM MOLLUSCA**	
Class and Representative Animals	*Characteristics*
Polyplacophora Chitons	Primitive marine animals with segmented shells; shell consists of 8 separate transverse plates; head reduced; broad foot used for locomotion
Gastropoda Snails Slugs Nudibranchs	Marine, freshwater, or terrestrial; body and shell is coiled; well-developed head with tentacles and eyes
Bivalvia Clams Oysters Mussels	Marine and freshwater; body laterally compressed; two shells hinged dorsally; hatchet-shaped foot; filter feeders
Cephalopoda Squids Octopods	Marine; fast-swimming, predatory; foot divided into tentacles, usually bearing suckers; well-developed eyes

Although mollusks vary widely in outward appearance, most share certain basic characteristics (Figure 30–3):

1. a soft body, usually covered by a dorsal shell
2. a broad, flat muscular **foot,** located ventrally, which can be used for locomotion
3. a **visceral mass,** located above the foot, that contains most of the organs
4. a **mantle,** a heavy fold of tissue covering the visceral mass and usually containing glands that secrete the shell. The mantle generally overhangs the visceral mass, forming a mantle cavity, which often contains gills.
5. a rasplike structure called the **radula,** which is a belt of teeth within the digestive system. (The radula is not present in clams and other bivalves.)

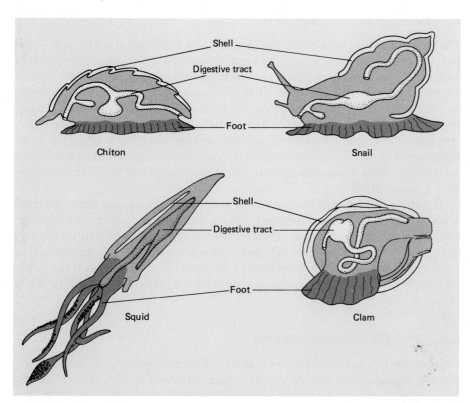

FIGURE 30–3 Variations in the basic molluskan body plan in chitons, snails, clams, and squids. Note how the foot, shell, and digestive tract have changed their positions in the evolution of the several classes.

All of the organ systems typical of complex animals are present in the mollusks. The digestive system is a tube, sometimes coiled, consisting of a mouth, buccal cavity, esophagus, stomach, intestine, and anus. The radula, located within the buccal cavity, can be projected out of the mouth and used to scrape particles of food from the surface of rocks or the ocean floor. Sometimes the radula is used to drill a hole in another animal's shell or to break off pieces of a plant.

The open circulatory system characteristic of most mollusks is well developed. The heart pumps blood into a single blood vessel, the **aorta,** which may branch into other vessels. Eventually, blood flows into a network of large sinuses, where the tissues are bathed directly; this network makes up the **hemocoel** (blood cavity). This system is referred to as an **open circulatory system,** in which the blood does not remain within a circuit of blood vessels but bathes the tissues directly. From the sinuses, blood drains into vessels that conduct it to the gills where oxygenation takes place. From the gills, the blood returns to the heart. Thus, blood flow in a mollusk follows the pattern

Heart → aorta → smaller blood vessels → sinuses → gills → heart

Open circulatory systems are not very efficient. Blood pressure tends to be low, and tissues are not very efficiently oxygenated. However, because most mollusks are sluggish animals with low metabolic rates, this type of circulatory system is adequate. In the active cephalopods (the class that includes the squids and octopods), the circulatory system is closed; the blood remains within a circuit of blood vessels.

The sexes are usually separate, with fertilization taking place in the surrounding water. Most marine mollusks pass through one or more larval stage. The first larval stage is typically a **trochophore larva,** a free-swimming, ciliated, top-shaped larva characteristic of mollusks and annelids (Figure 30–4). In most of the mollusk classes, the trochophore larva develops into a **veliger larva,** which has a shell and foot. The veliger larva is unique to the phylum Mollusca.

Relationship to the Annelids

The striking similarities in the development of mollusks and annelids—the process of spiral cleavage and the appearance of a trochophore larva—had suggested that these two phyla were related in evolutionary origin and had a common coelomate ancestor. This view was supported by the discovery in 1952 of specimens of a primitive mollusk, *Neopilina,* in material dredged from a deep trench in the Pacific Ocean off Costa Rica. Since this discovery, specimens belonging to several related species have been collected from deep water in many parts of the world. *Neopilina* and its relatives have been assigned to class **Monoplacophora** (meaning one plate) because they possess a single shell.

The most remarkable feature of the monoplacophorans is the segmental serial repetition of certain internal organs, a condition known as **metamerism.** They have five pairs of retractor muscles, six pairs of nephridia, and five pairs of gills. This fact has been interpreted by some zoologists as evidence of the segmental character of their ancestors. Moreover, they may be closely related to the annelids, which have a basically metameric body plan.

Class Polyplacophora: Chitons

Chitons, which are members of the class **Polyplacophora** (meaning many plates), are sluggish marine animals with flattened bodies (Figure 30–5). Their most distinctive feature is a shell composed of eight separate, but

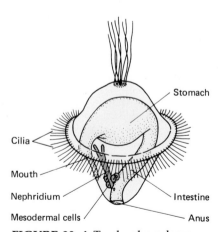

Cilia
Mouth
Nephridium
Mesodermal cells
Stomach
Intestine
Anus

FIGURE 30–4 Trochophore larva, the first larval stage of a marine mollusk. This type of larva is also characteristic of annelids. Note the characteristic ring of ciliated cells just above the mouth.

FIGURE 30–5 Chitons are sluggish marine animals with shells composed of eight overlapping plates. These lined chitons, *Tonicella lineatus*, are from coastal waters off the Pacific Northwest.

overlapping, transverse plates. The head is reduced in this class, and there are no eyes or tentacles. Chitons inhabit rocky intertidal zones, feeding on algae and other small organisms, which they scrape off rocks and shells with the radula. The broad, flat foot not only functions in locomotion but also helps the animal adhere firmly to rocks. The mantle can also be pressed firmly against the substratum, and the chiton can lift the inner edge of the mantle to create a partial vacuum. The suction developed enables the animal to adhere powerfully to its perch.

Class Gastropoda: Snails and Their Relatives

The class **Gastropoda** is the largest and numerically the most successful group of mollusks (Figure 30–6). In fact, this class, which includes the

(a)

(c)

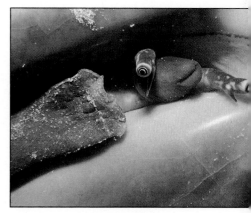

(b)

FIGURE 30–6 Representative gastropods. (a) The tulip snail *Fasciolaria tulipa*). (b) A conch, *Strombus gigas*, showing foot, mouth, and eyes. This mollusk grazes on algae. Its large, mobile proboscis sweeps across the bottom of the sea like a vacuum cleaner. (c) A thick-horned aeolid nudibranch, *Hermissenda crassicornis*.

687

FIGURE 30–7 Embryonic torsion in the gastropod *Acmaea* (a limpet). As the bilateral larva develops, the visceral mass twists 180 degrees relative to the head.

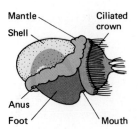
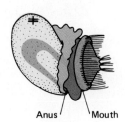
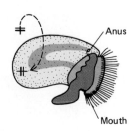

snails and their relatives, is the second largest class in the animal kingdom—second only to the insects. Gastropods inhabit a wide variety of habitats, including the seas, brackish water, fresh water, and many terrestrial areas, but they are most numerous and diverse in marine waters. Most land snails do not have gills; instead, the mantle is highly vascularized and functions as a lung. These snails are described as **pulmonate.**

We usually think of snails as having a single, spirally coiled shell, and many do. Yet many other gastropods, such as limpets and abalones, have shells like flattened dunce caps; others, such as garden slugs and the marine snails known as **nudibranchs,** have no shell at all (Figure 30–6c).

Gastropods characteristically have a well-developed head with tentacles. Two simple eyes are usually located on stalks that extend from the head. The broad flat foot is used for creeping. The most unique feature of this group is **torsion,** a twisting of the visceral mass. As the bilateral larva develops, the body twists permanently 180 degrees relative to the head. As a result, the digestive tract becomes somewhat U-shaped, and the anus comes to lie above the head (Figure 30–7). Subsequent growth is dorsal and usually in a spiral coil. The twist limits space in the body, and typically the gill, kidney, and gonad are absent on one side. The **viscera** (internal organs) of the slugs that lack shells also undergo torsion during development.

Class Bivalvia: Clams, Oysters, and Their Relatives

The soft body of members of the class **Bivalvia** is laterally compressed and completely enclosed by two shells hinged dorsally and opening ventrally (Figure 30–8). This arrangement allows the hatchet-shaped foot to protrude ventrally for locomotion. Apertures are also present for flow of water into and out of the mantle cavity. Extensions of the mantle, called siphons, permit bivalves to obtain water relatively free of sediment. There are both an **incurrent siphon** for water intake and an **excurrent siphon** for water

FIGURE 30–8 Internal anatomy of a clam.

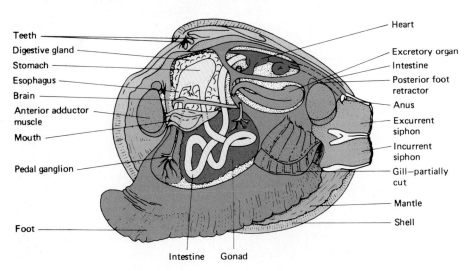
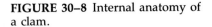

output. Large, strong muscles attached to the shell enable the animal to close its shell.

The inner pearly layer of the bivalve shell is secreted in thin sheets by the epithelial cells of the mantle. Composed of calcium carbonate, it is known as mother-of-pearl and is valued for making jewelry and buttons. When a bit of foreign matter lodges between the shell and the epithelium, the epithelial cells are stimulated to secrete concentric layers of calcium carbonate around the intruding particle; this is how a pearl is formed.

Some bivalves, such as oysters, attach permanently to the substratum. Others, like clams, burrow slowly through rock or wood, seeking protected dwellings. (The shipworm, *Teredo,* which damages dock pilings and other marine installations, is just looking for a home.) Finally, some bivalves, such as scallops, swim with amazing speed by clapping their two shells together by the contraction of a large **adductor muscle** (the part of the scallop that is eaten by humans).

Clams and oysters are filter feeders. They obtain food by straining the sea water brought in over the gills by the siphon. The water is kept in motion by the beating of cilia on the surface of the gills. This stream of water carries food particles trapped in the mucus that is secreted by the gills to the mouth. An average oyster filters about 3 liters of sea water per hour. Because bivalves are filter feeders, they have no need for a radula, and indeed they are the only group of mollusks that lack this structure.

Most bivalves have two distinct sexes. Gametes are usually discharged into the water, where fertilization takes place. In some marine and nearly all freshwater bivalves, sperm are shed into the water and fertilize the eggs within the mantle cavity of the female. In these species, the female also broods her young within the mantle cavity. Development takes place among the gill filaments. In marine bivalves, a trochophore larva typically develops, which then develops further into a veliger larva with shell and foot. Larvae of some freshwater species spend several weeks as parasites on the gills of fishes.

Class Cephalopoda: Squids, Octopods, and Their Relatives

In contrast to most other mollusks, members of the class **Cephalopoda** (meaning head–feet) are active, predatory animals. They are fast-swimming organisms, adapted for an entirely different life style than their filter-feeding relatives. Some biologists consider cephalopods the most advanced of the invertebrates.

The octopus has no shell, and the shell of the squid is reduced to a small "pen" in the mantle. *Nautilus* has a flat, coiled shell consisting of many chambers built up year by year; each year the animal lives in the newest and largest chamber of the series. By secreting a gas resembling air into the other chambers, the *Nautilus* is able to float.

The cephalopod foot is divided into tentacles—ten in squids, eight in octopods. The tentacles, or arms, surround the central mouth of the large head (Figure 30–9). Cephalopods have large, well-developed eyes that form images. Although they develop differently, the eyes are structurally much like vertebrate eyes and function in much the same way.

The tentacles of squids and octopods are covered with suckers for seizing and holding prey. In addition to a radula, the mouth is equipped with two strong, horny beaks used to kill prey and tear it to bits. The mantle is thick and muscular and fitted with a funnel. By filling the mantle cavity with water and ejecting it through the funnel, the animal can attain rapid jet propulsion in the opposite direction.

Besides its speed, the cephalopod has developed two other important mechanisms that enable it to escape from its predators, which include the

FIGURE 30–9 *Octopus macropus,* a cephalopod. The octopod lives in a den among the rocks; it may wait near the entrance of its den to seize a passing crustacean, fish, or snail.

whales and moray eels. One is its ability to confuse the enemy by rapidly changing colors. By expanding and contracting pigment cells—**chromatophores**—in its skin, the cephalopod can display an impressive variety of mottled colors. Another defense mechanism is the **ink sac,** which produces a thick black liquid. This liquid is released in a dark cloud when the animal is alarmed; while its enemy pauses, temporarily blinded and confused, the cephalopod easily escapes. The ink has been shown to paralyze the chemical receptors of some predators.

The octopus feeds on crabs and other arthropods, catching and killing them with a poisonous secretion of its salivary glands. During the day, the octopus usually hides among the rocks; in the evening, it emerges to hunt for food. Its motion is incredibly fluid, giving little hint of the considerable strength in its eight arms.

Small octopods survive well in aquaria and have been studied extensively. They have a relatively high degree of intelligence and can make associations among stimuli. Their very adaptable behavior resembles more closely that of the vertebrates than the more stereotyped patterns of behavior seen in other invertebrates.

PHYLUM ANNELIDA

Phylum **Annelida,** the segmented worms, includes earthworms, leeches, and many marine and freshwater worms. The 10,000 or so species are divided into three main classes (Table 30–2 and Figure 30–10). The term *Annelida* means ringed and refers to the series of rings, or segments, that make up the annelid body. Both the body wall and the internal organs are segmented. The segments are separated from one another by transverse partitions, the **septa.** The bilaterally symmetrical, tubular body may consist of about 100 segments. Some structures, such as the digestive tract and

Table 30–2 CLASSES OF PHYLUM ANNELIDA

Class and Representative Animals	Characteristics
Polychaeta Sandworms Tubeworms	Mainly marine; each segment bears a pair of parapodia with many setae; well-developed head; separate sexes; trochophore larva
Oligochaeta Earthworm	Terrestrial and freshwater worms; few setae per segment; lack well-developed head; hermaphroditic
Hirudinea Leeches	Most are blood-sucking parasites that inhabit fresh water; lack appendages and setae; prominent muscular suckers

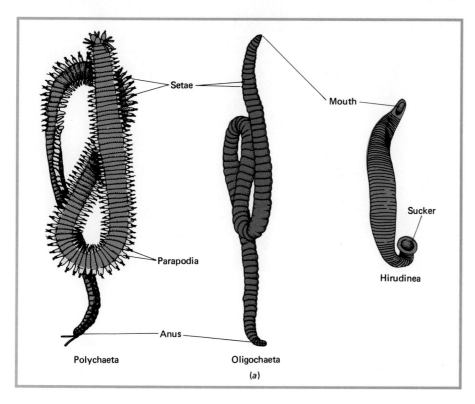

certain nerves, extend the length of the body, passing through successive segments. Other structures are repeated in each segment (Figure 30–11). Segmentation is an advantage because not only is the coelom divided into segments, but each segment has its own muscles, enabling the animal to elongate one part of its body while shortening another part. The annelid's hydrostatic skeleton is discussed in Chapter 39. In the annelid, the individual segments are almost all alike, but in many segmented animals—the arthropods and chordates—different segments and groups of segments are specialized to perform certain functions. In some groups the specialization may be so pronounced that the basic segmentation of the body plan may be obscured.

Bristle-like structures called **setae** aid in locomotion. Annelids have a well-developed coelom, a closed circulatory system, and a complete digestive tract extending from mouth to anus. Respiration takes place through the skin or by gills. Typically, a pair of excretory structures called **metanephridia** is found in each segment. The nervous system generally consists of a simple brain composed of a pair of ganglia and a double ventral nerve cord. A pair of ganglia and lateral nerves are repeated in each segment.

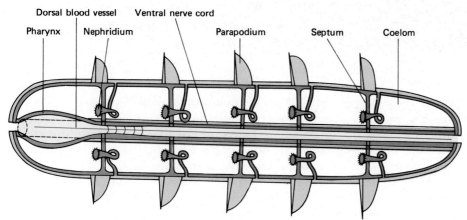

FIGURE 30–11 Metamerism in a generalized annelid. The body is segmented, and there is serial repetition of body parts (metamerism).

(a)

(b)

FIGURE 30–12 Polychaete annelids. (*a*) The large West Indian fireworm, *Hermodice carunculata,* feeds on corals and sea anemones. (*b*) The Christmas tree worm *(Spiro branchus giganteus)* photographed in a Florida coral reef. This polychaete bores a protective retreat within living or dead coral or shells. The prostomium has developed to form a crown consisting of several processes, called radioles, that close together when the worm withdraws into its tube.

Class Polychaeta

The class **Polychaeta** includes marine worms, which swim freely in the sea, burrow in the mud near the shore, or live in tubes formed by cementing bits of shell and sand together with mucus and other secretions from the body wall. Each body segment has a pair of paddle-shaped appendages called **parapodia** (singular, parapodium) that extend laterally and function in locomotion (Figure 30–12). These fleshy structures bear many stiff setae (the name *Polychaeta* means many bristles). Most polychaetes have a well-developed head or **prostomium** bearing eyes and antennae. The prostomium may also be equipped with tentacles, bristles, and **palps** (feelers). Polychaetes develop from free-swimming trochophore larvae similar to those of mollusks.

Many polychaetes have evolved behavioral patterns that ensure fertilization. By responding to certain rhythmic variations, or cycles, in the environment, nearly all of the females and males of a given species release their gametes into the water at the same time. More than 90% of reef-dwelling *Palolo* worms of the South Pacific shed their eggs and sperm within a single 2-hour period on one night of the year. In this animal the seasonal rhythm limits the reproductive period to November; the lunar rhythm, to a day during the last quarter of the moon when the tide is unusually low; and the diurnal rhythm, to a few hours just after complete darkness. The posterior half of the *Palolo* worm, loaded with gametes, actually breaks off from the rest, swims backward to the surface, and eventually bursts, releasing the eggs or sperm so that fertilization may occur. Local islanders eagerly await this annual event when they can gather up great numbers of the swarming polychaetes and broil them for dinner.

Class Oligochaeta: The Earthworms

The 3000 or so species of the class **Oligochaeta** are found almost exclusively in fresh water and in moist terrestrial habitats. These worms lack parapodia, have few bristles per segment (the name *Oligochaeta* means few bristles), and lack a well-developed head. All oligochaetes are **hermaphroditic,** meaning that male and female reproductive systems are present in the same individual. Since the earthworm is among the most familiar of all invertebrates, let us examine this animal in more detail.

Lumbricus terrestris, the common earthworm, is about 20 cm long. Its body is divided into more than 100 segments separated externally by grooves and internally by septa. The mouth is located in the first segment, the anus in the last. The earthworm's body is protected from desiccation by a thin, transparent cuticle, secreted by the cells of the epidermis (Figure 30–13). Mucus secreted by the glandular cells of the epidermis forms an additional protective layer over the body surface.

The body wall contains an outer layer of circular muscles and an inner layer of longitudinal muscles. The earthworm moves forward by contracting its circular muscles to elongate the body, grasping the ground or walls of the burrow with its setae, and then contracting its longitudinal muscles to draw the posterior end forward. Locomotion proceeds in waves. The muscles work against the hydrostatic skeleton provided by the coelomic fluid within the coelom of each segment. (For a more detailed discussion, see Chapter 39.) Each segment except the first bears four pairs of setae supplied with tiny muscles that can move each seta in and out and change its angle. Thus in locomotion, the earthworm's body is extended, anchored by the setae, and then contracted.

An earthworm literally eats its way through the soil, ingesting its own weight in soil and decaying vegetation every 24 hours. During this process the soil is turned and aerated, and nitrogenous wastes from the earthworm enrich it. This is why earthworms are vital to the formation and maintenance of fertile soil. The earthworm's soil meal, containing nutritious decaying vegetation, is processed in the complex digestive system. Food is swallowed through the muscular **pharynx,** and passes through the **esophagus** to the **stomach.** The stomach consists of two parts: a thin-walled **crop** where food is stored and a thick-walled muscular **gizzard** where food is ground to bits. The rest of the digestive system is a long, straight **intestine,** where food is digested and absorbed. The surface area of the intestine is increased by a dorsal, longitudinal fold called the **typhlosole.** Wastes pass out of the intestine to the exterior through the **anus.**

The efficient, closed circulatory system consists of two main blood vessels that extend longitudinally. The dorsal blood vessel, just above the digestive tract, collects blood from numerous segmental vessels. It contracts, pumping blood anteriorly. In the region of the esophagus, five pairs of blood vessels propel blood from the dorsal to the ventral blood vessel. Located just below the digestive tract, the ventral blood vessel conveys blood both posteriorly and anteriorly. Small blood vessels branch from it and deliver blood to the various structures in each segment as well as to the body wall. Within these structures blood flows through very tiny blood vessels (capillaries) before returning to the dorsal blood vessel.

Gas exchange takes place through the moist skin, and oxygen is usually transported by the respiratory pigment hemoglobin present in the blood plasma. The excretory system consists of paired organs, the metanephridia, repeated in almost every segment of the body. Each metanephridium consists of a ciliated funnel (nephrostome) opening into the next anterior coelomic cavity and connected by a tube to the outside of

FIGURE 30–13 Internal structure of an earthworm (an oligochaete). (*a*) Diagrammatic longitudinal section of the anterior portion. (*b*) Cross section.

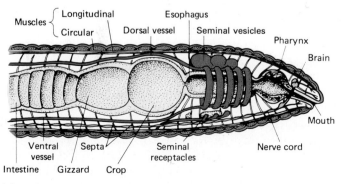

(*a*)

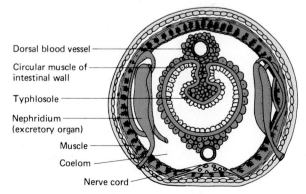

(*b*)

the body (Figure 30–13). Wastes are removed from the coelomic cavity partly by the beating of the cilia and partly by currents set up by the contraction of muscles in the body wall. The tube of the metanephridium is surrounded by a capillary network which reabsorbs usable materials from the coelomic fluid in the tube.

The metanephridia, open at both ends, are quite different from the protonephridia of the flatworms, which are blind tubules opening only to the exterior. In higher invertebrates, the adults typically have metanephridia, but larval forms usually have protonephridia. These protonephridia generally have a single long flagellum rather than a tuft of cilia. This developmental pattern of excretory structures is often used to support the concept that complex invertebrates (as well as vertebrates) evolved from forms similar to the lower invertebrates.

The nervous system consists of a pair of **cerebral ganglia** that serve as a brain, just above the pharynx in the third segment, and a **subpharyngeal ganglion,** just below the pharynx in the fourth segment. A ring of nerve fibers connects the ganglia. From the lower ganglion a double ventral nerve cord extends beneath the digestive tract to the posterior end of the body. In each segment along the nerve cord, there is a pair of fused **segmental ganglia,** from which nerves extend laterally to the muscles and other structures of that segment. The segmental ganglia coordinate the contraction of the muscles of the body wall, so that the worm can creep along. The nerve cord contains a few giant axons that transmit nerve impulses more rapidly than ordinary fibers. When danger threatens, these stimulate the muscles to contract and draw the worm quickly back into its burrow.

The subpharyngeal ganglion is the main center controlling movement and vital reflexes. It exerts control over the other ganglia in the chain. When the subpharyngeal ganglion is destroyed, all movement stops. When the brain is removed, the subpharyngeal ganglion is able to continue to control movement, but the worm is no longer able to adjust its actions to conditions in the environment. Earthworm responses are limited to reflex (preprogrammed, stereotyped) actions. However, in the laboratory earthworms can be conditioned to perform simple acts such as contracting when exposed to bright light or vibration. Living a subterranean life, the earthworm has no need for well-developed sense organs.

Like other oligochaetes, earthworms are hermaphroditic. During copulation, two worms, heading in opposite directions, press their ventral surfaces together (Figure 30–14). These surfaces become glued together by thick mucous secretions of the **clitellum,** a thickened ring of epidermis in segments 32 to 37. Sperm from each worm pass posteriorly to its clitellum and are stored in the **seminal receptacles** of the other worm. The worms then separate. A few days later the clitellum secretes a membranous **cocoon** containing an albuminous fluid. As the cocoon is slipped over the worm's head, eggs are laid into it from the female pores, and sperm are added as the cocoon passes over the seminal receptacles. When the cocoon is free, its openings constrict so that a spindle-shaped capsule is formed,

FIGURE 30–14 Two earthworms, genus *Lumbricus,* copulating. These animals are hermaphroditic, but cross-fertilize one another.

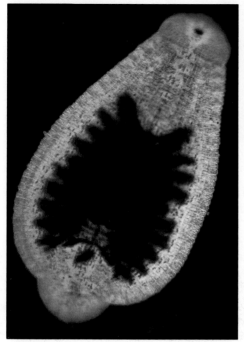

(a)

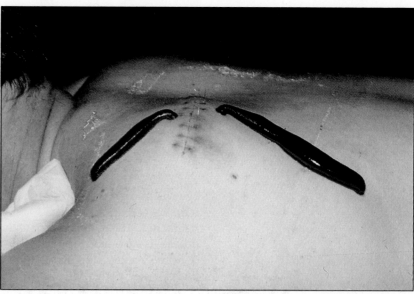

(b)

FIGURE 30–15 (a) Class Hirudinea is represented by this leech, *Helobdella stagnalis*. The dark area in its swollen body is recently ingested blood. (b) The medicinal leech, *Hirudo medicinalis*, is used to treat hematoma, an accumulation of blood within body tissues that results from injury or disease. The leech attaches its sucker near the site of injury, makes an incision, and deposits an anticoagulant called hirudin. Hirudin prevents the blood from clotting and dissolves already-existing clots.

and the eggs develop into tiny worms within the cocoon. This complex reproductive pattern is an adaptation to terrestrial life.

Class Hirudinea: The Leeches

Most **leeches,** which are members of the class **Hirudinea,** are blood-sucking parasites that inhabit fresh water. They differ from other annelids in having neither setae nor appendages. Prominent muscular suckers are present at both anterior and posterior ends for clinging to their prey. Most leeches attach themselves to a vertebrate host, bite through the skin of the host, and suck out a quantity of blood, which is stored in pouches in the digestive tract. An anticoagulant (hirudin), secreted by glands in its crop, ensures the leech a full meal of blood. Their meals may be infrequent, but they can store enough food from one meal to last a long time. The so-called medicinal leech *(Hirudo medicinalis),* a freshwater worm about 10 cm long, was used by physicians for bloodletting in the 17th and 18th centuries (Figure 30–15). This was based on the misconception that fevers and disease were caused by an excess of blood. Interestingly, leeches have found a place in modern medicine. They are used to reduce congestion from bleeding that results from damaged blood vessels (Figure 30–15).

PHYLUM ONYCHOPHORA

Only about 70 living species of **Onychophorans** are known, but this group of animals is considered important as a possible link between the annelids and the arthropods. In fact, some zoologists classify onychophorans as annelids, and others classify them with the arthropods. These wormlike animals inhabit humid tropical areas such as rain forests.

Peripatus, a caterpillarlike creature, is the best known of the onychophorans (Figure 30–16). About 5 to 8 cm long, *Peripatus* has a thin, soft cuticle covering its elongated muscular body. Its many pairs of unjointed legs bear claws. Like annelids, it is internally segmented, and many of its organs are duplicated serially. However, its jaws are derived from append-

FIGURE 30–16 The velvet worm, *Macroperipatus tarquatus,* an onychophoran from moist forested regions of Trinidad. This animal has features of both arthropods and annelids. Note the soft, sluglike body and the presence of a series of jointed legs.

ages as in arthropods, and like arthropods, it has an open circulatory system. Its coelom is reduced, for much of the body cavity is occupied by a hemocoel. Its respiratory system, which consists of air tubes (tracheal tubes), is also arthropod-like.

Some zoologists believe that terrestrial onychophorans gave rise to the insects, millipedes, and centipedes; others think that this group branched off the annelid–arthropod trunk after the annelids but before the arthropods. In either case, the onycophorans are interesting creatures with a curious mixture of annelid and arthropod characteristics.

PHYLUM ARTHROPODA

The animals that make up phylum **Arthropoda** are, without doubt, the most biologically successful of all animals. There are more of them (about 800,000 described species); they live in a greater variety of habitats; and they can eat a greater variety of foods than the members of any other phylum (Figure 30–17). Among their most important characteristics are the following:

1. *Paired, jointed appendages,* from which they get their name (arthropod means jointed foot). These appendages function as swimming paddles, walking legs, mouth parts, or accessory reproductive organs for transferring sperm.
2. A hard, armorlike *exoskeleton,* composed of chitin, that covers the entire body and appendages. The exoskeleton provides protection against excessive loss of moisture as well as predators and gives support to the underlying soft tissues. Distinct muscle bundles attach to the inner surface of the exoskeleton. These act upon a system of levers that permit the extension and flexion of parts at the joints. The exoskeleton has certain disadvantages, however. Body movement is somewhat restricted, and in order to grow, the arthropod must shed this outer shell periodically and grow another larger one, a process that leaves it temporarily vulnerable to predators.
3. A *segmented body,* like that of the annelid. In some arthropod classes, however, segments become fused together or lost during development. Segments may become fused into groups, known as **tagmata** (singular, tagma), that perform specific functions.
4. An *open circulatory system* with a dorsal heart. A hemocoel (blood cavity) occupies most of the body cavity, and the coelom is small and is filled chiefly by the organs of the reproductive system.

The Arthropod Body Plan

The bodies of most arthropods are divided into three regions: the **head,** composed of four to six segments; the **thorax;** and the **abdomen,** both of which consist of a variable number of segments. In contrast to most anne-

(a)

(b)

(c)

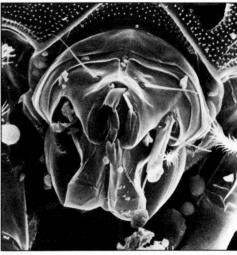

(d)

FIGURE 30–17 The arthropods are considered the most highly successful animals. (*a*) Starry-eyed hermit crab photographed in the Virgin Islands. (*b*) Florida dragonfly. (*c*) A pair of horseshoe crabs, *Limulus polyphemus*, mating. (*d*) Hay mite.

lids, each arthropod has a fixed number of segments, which remains the same throughout life. The incredible range of variations in body plan and in the shape of the jointed appendages in the numerous species almost defies description.

The nervous system of the more primitive arthropods, like that of the annelids, consists of a ventral nerve cord connecting segmental ganglia. In the more complex arthropods, the successive ganglia usually fuse together. Arthropods have a variety of well-developed sense organs: complicated eyes, such as the compound eyes of insects; organs of hearing; **antennae** sensitive to touch and chemicals; and cells sensitive to touch on the surface of the body.

The open circulatory system includes a dorsal, tubular heart that pumps blood into a dorsal artery, and sometimes several other arteries. From the arteries blood flows into large sinuses, which collectively make up the hemocoel. Blood in the hemocoel bathes the tissues directly. No capillaries or veins are present. Eventually blood finds its way back into the heart through openings, referred to as **ostia,** in its walls.

Most of the aquatic arthropods have a system of gills for gas exchange. The land forms, in contrast, typically have a system of fine, branching air tubes called **tracheae** that conduct air to the internal organs. The digestive system typically is a simple tube similar to that of the earthworm. Both ends are lined with a waxy cuticle similar to the outer layer of the exoskeleton. Excretory structures vary somewhat from class to class.

Major Groups of Arthropods

There is much disagreement concerning arthropod classification. Historically, living arthropods have been classified in two groups. Those without antennae and with clawlike chelicerae have been assigned to subphylum

FIGURE 30–18 The trilobites are extinct marine arthropods that are considered the most primitive members of the phylum. (*a*) Dorsal view of a trilobite from the Ordovician period. (*b*) Ventral view of the same trilobite.

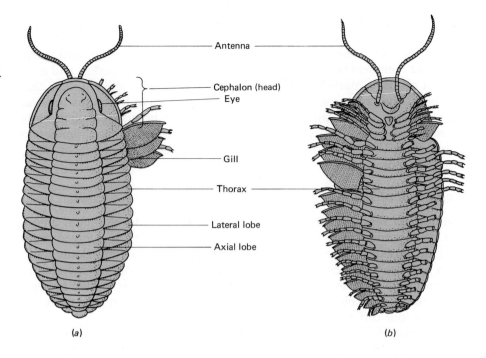

Antenna
Cephalon (head)
Eye
Gill
Thorax
Lateral lobe
Axial lobe

(a)   (b)

Covered by a hard, segmented shell, the trilobite body was a flattened oval divided into three parts: an anterior head of four fused segments bearing a pair of antennae and a pair of compound eyes; a thorax consisting of a variable number of segments; and a posterior abdomen composed of several fused segments (Figure 30–18). At right angles to these divisions, two dorsal grooves extended the length of the animal, dividing the body into a median lobe and two lateral lobes. (The name trilobite derives from this division of the body into three longitudinal parts.) Each segment of the body had a pair of segmented **biramous** (two-branched) **appendages;** each appendage consisted of an inner walking leg and an outer branch bearing gills.

It is remarkable that fossil evidence has yielded information not only about the structure of the adult but also about the developmental stages of the trilobites. These animals went through three larval periods; during each one, the larvae underwent several molts. As molts occurred, additional segments were added to the body, and the body structure became more complex.

Subphylum Chelicerata

In members of subphylum **Chelicerata,** the body consists of a **cephalothorax** and an abdomen. Chelicerates have no antennae and no chewing mandibles (see Table 30–3). Instead, the first pair of appendages, which are located immediately anterior to the mouth, are the chelicerae, used to manipulate food and pass it to the mouth. The second pair of appendages, called **pedipalps,** are modified to perform different functions in various groups. Posterior to the pedipalps there are usually four pairs of legs.

Class Merostomata: Horseshoe Crabs

Almost all members of the class **Merostomata** are extinct. The only living merostomes, the **horseshoe crabs** in the subclass Xiphosura, have survived essentially unchanged for 350 million years or more. *Limulus polyphemus* is the species common in North America. As its common name describes, this animal is horseshoe-shaped. The long spikelike tail that extends posteriorly is used in locomotion, not for defense or offense. Horseshoe crabs feed on mollusks, worms, and other invertebrates that they find on the ocean floor.

Class Arachnida: Spiders, Scorpions, Ticks, and Mites

The 60,000 or so species of class **Arachnida** include the spiders, scorpions, mites, ticks, and harvestmen or daddy long-legs (Figure 30–19). The arachnid body consists of a cephalothorax (composed of fused head and thorax) and abdomen. Arachnids have six pairs of jointed appendages. In spiders, the first pair, the chelicerae, are fanglike structures used to penetrate prey and suck out its body fluids; in some species, the chelicerae are used to inject poison into the prey. The second pair of appendages, the pedipalps, are used by spiders to hold and chew food and in some species are modified as sense organs for tasting the food. The other four pairs of appendages are used for walking. Most arachnids are carnivorous and prey upon insects and other small arthropods.

Gas exchange in arachnids takes place either by tracheal tubes or by book lungs, or by both. Each **book lung** consists of 15 to 20 plates, like pages of a book, that contain tiny blood vessels. Air enters the body through abdominal slits and circulates between the plates. As air passes over the blood vessels, oxygen and carbon dioxide are exchanged. An arachnid may have as many as four pairs of book lungs.

Spiders have **silk glands** in their abdomen; these glands secrete an elastic protein that is liquid as it emerges from their spinnerets. It hardens as it is drawn out and is used to construct webs for building nests, encasing

FIGURE 30–19 Class Arachnida includes the spiders, scorpions, mites, ticks, and harvestmen. (*a*) A marbled scorpion, *Lychas marmoreus*. (*b*) The red-spotted crab spider, *Misumena vatia*, is a voracious predator. (*c*) The red water mite is an active swimmer. The larval stages of water mites may be parasitic on aquatic insects or on the gills. (*d*) A tick, *Dermacentor andersoni* (approximately ×40).

(a)

(b)

(c)

(d)

FIGURE 30–20 A banded argiope spider wrapping a grasshopper in its web.

eggs in a cocoon, and in some species, for trapping prey (Figure 30–20). Many spiders lay down a silken dragline as they venture forth. This serves as a safety line and is also a means of communication between members of a species. From a dragline a spider can determine the sex and maturity level of the spinner.

Although spiders do have poison glands for capturing prey, only a few have poison that is toxic to humans. The most widely distributed poisonous spider in the United States is the black widow. Its poison is a neurotoxin that interferes with transmission of messages from nerves to muscles. Although painful, spider bites cause fewer than five fatalities per year in the United States; these usually occur in children.

Mites and ticks are among the greatest arthropod nuisances. They eat our crops, infect our livestock and pets, and inhabit our own bodies. Many live unnoticed, owing to their small size, but others cause disease. Certain mites cause mange in dogs and other domestic animals. Chiggers (red bugs), the larval form of red mites, attach themselves to the skin and secrete an irritating digestive fluid that may cause itchy red welts. Larger than mites, ticks are ectoparasites on dogs and other domestic animals. They can transmit diseases such as Rocky Mountain spotted fever, Texas cattle fever, relapsing fever, and Lyme disease.

Subphylum Crustacea: Lobsters, Crabs, Shrimp, and Their Relatives

Crustaceans—members of the subphylum **Crustacea**—are vital members of marine food chains, serving as primary consumers of algae and other producers, and serving as food for the many carnivores that inhabit the oceans. Countless billions of microscopic crustaceans swarm in the ocean and form the food (krill) of many fish and other marine forms such as whales.

As discussed earlier, crustaceans are characterized by having mandibles, biramous appendages, and two pairs of antennae (Figure 30–21). Their antennae serve as sensory organs for touch and taste. The mandibles, which are the third pair of appendages, are located on each side of the ventral mouth and are used for biting and grinding food. Posterior to the mandibles are two pairs of appendages, the first and second **maxillae,** used for manipulating and holding food. Several other pairs of appendages are present. Usually five pairs are modified for walking. Others may be specialized for swimming, sperm transmission, carrying eggs and young, or sensation.

As the only class of arthropods that are primarily aquatic, crustaceans generally have gills for gas exchange. Two large **antennal glands** (also referred to as green glands) located in the head remove metabolic wastes from the blood and body fluids and excrete them through ducts opening at the base of each antenna. The nervous system is somewhat similar to the annelid nervous system but is proportionately larger, and ganglia are fused and large. Most adult members of the class have compound eyes (discussed in Chapter 48). Among the other sense organs present are **statocysts** for detecting the pull of gravity.

Crustaceans characteristically have separate sexes. During copulation, the male uses specialized appendages to transfer sperm into the female. The fertilized eggs are usually brooded. The newly hatched animals pass by successive molts through a series of larval stages and finally reach the body form characteristic of the adult. The lobster, for example, molts seven times during the first summer; at each molt it gets larger and resembles the adult more. After it becomes a small adult, additional molts provide for growth. The process of molting is discussed in Chapter 39.

(a)

(b)

(c)

(d)

FIGURE 30–21 Crustaceans. (a) Broken-back shrimp from Monterey Bay. (b) Gooseneck barnacles, *Pollicipes polymerus*, are stalked barnacles that occur in large numbers on intertidal rocks along the west coast of the United States. (c) A single barnacle, *Balanus nubilis*, filtering water for food. (d) Crab.

The barnacles are the only sessile crustaceans. They differ markedly in their external anatomy from other members of the class, and it was only in 1830, when the larval stages were investigated, that the relationship between the barnacles and other crustaceans was recognized. The barnacles are exclusively marine and secrete complex calcareous cups within which the animal lives. The larvae of barnacles are free-swimming forms that go through several molts and eventually become sessile and develop into the adult form. Barnacles were described by the 19th century naturalist Louis Agassiz as "nothing more than a little shrimplike animal standing on its head in a limestone house and kicking food into its mouth."

The largest order of crustaceans, **Decapoda,** contains some 8500 species of lobsters, crayfish, crabs, and shrimp. Most decapods are marine, but a few, such as the crayfish, certain shrimp, and a few crabs, live in fresh water. The crustaceans in general and the decapods in particular show striking specialization and differentiation of parts in the various regions of the animal. The segments of trilobites and perhaps of the earliest crustaceans bore appendages that were very similar. In the lobster, no 2 of the 19 pairs of appendages are identical, and the appendages in the different parts of the body differ markedly in form and function (Figure 30–22).

The six segments of the lobster's head and the eight segments of the thorax are fused into a cephalothorax, covered on its top and sides by a shield, the **carapace.** The carapace is composed of chitin impregnated with calcium salts. The two pairs of antennae are the sites of chemoreceptors and tactile sense organs; the second pair of antennae are especially long.

703

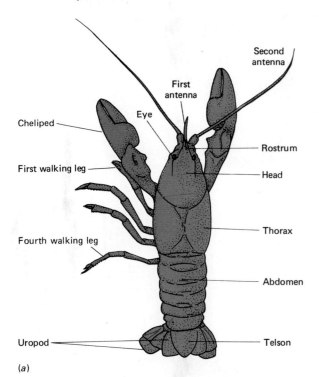

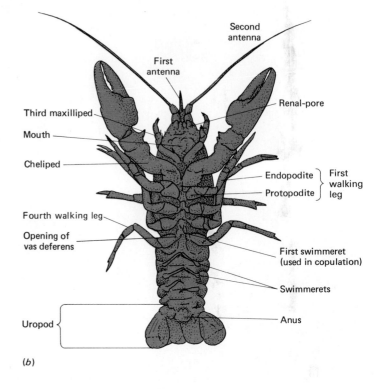

FIGURE 30–22 Anatomy of a lobster. (a) Dorsal view; (b) ventral view.

The mandibles are short and heavy, with opposing surfaces used in grinding and biting food. Behind the mandibles are two pairs of accessory feeding appendages, the first and second maxillae. The appendages of the first three segments of the thorax are the **maxillipeds,** which aid in chopping up food and passing it to the mouth. The fourth segment of the thorax has a pair of large **chelipeds,** or pinching claws. The last four thoracic segments have pairs of **walking legs.** The appendages of the first abdominal segment are part of the reproductive system and function in the male as sperm-transferring structures. On the following four abdominal segments are paired **swimmerets,** small paddle-like structures used by some decapods for swimming and by the females of all species for holding eggs. Each branch of the sixth abdominal appendages, which are called **uropods,** consists of a large flattened structure. Together with the flattened **telson,** the posterior end of the abdomen, they form a fan-shaped tail fin used for swimming backwards.

FIGURE 30–23 Some insect adaptations include (a) Stick insect. (b) Thirteen-year *Cicada,* molting. This insect requires 13 years to mature. The nymphs live in the soil feeding on roots. (c) A maybug, *Melolontha melolontha,* preparing for a landing.

(a)

(b)

(c)

Subphylum Uniramia

The insects, centipedes, and millipedes are grouped together in subphylum **Uniramia** because they all possess uniramous (unbranched) appendages. They also bear only a single pair of antennae rather than two pairs as in crustaceans.

Class Insecta

With more than 750,000 described species, the class **Insecta** is the most successful group of animals on our planet in terms of diversity and number of species, as well as number of individuals (see Figure 30–23 and 30–24). Insects are primarily terrestrial animals, but some species live in fresh

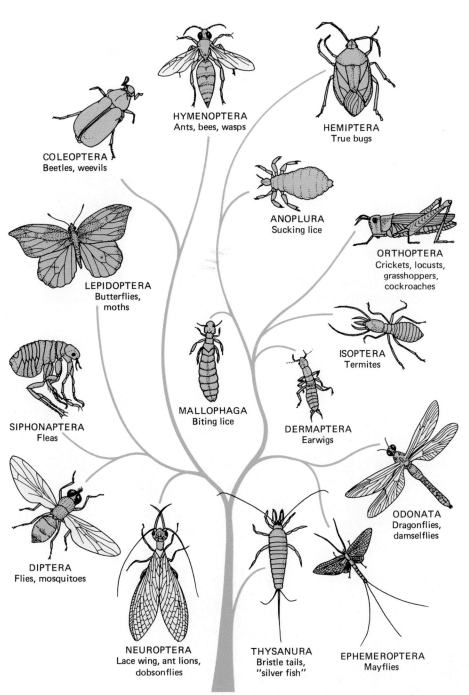

FIGURE 30–24 Representatives of some of the important orders of the class Insecta.

COLEOPTERA
Beetles, weevils

HYMENOPTERA
Ants, bees, wasps

HEMIPTERA
True bugs

ANOPLURA
Sucking lice

ORTHOPTERA
Crickets, locusts,
grasshoppers,
cockroaches

LEPIDOPTERA
Butterflies,
moths

ISOPTERA
Termites

SIPHONAPTERA
Fleas

MALLOPHAGA
Biting lice

DERMAPTERA
Earwigs

ODONATA
Dragonflies,
damselflies

DIPTERA
Flies, mosquitoes

NEUROPTERA
Lace wing, ant lions,
dobsonflies

THYSANURA
Bristle tails,
"silver fish"

EPHEMEROPTERA
Mayflies

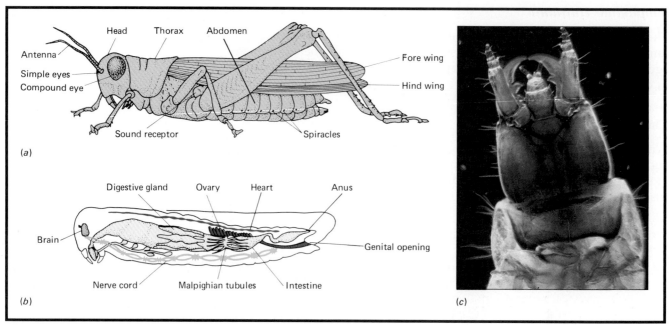

FIGURE 30–25 Insect body structure. (a) External anatomy of the grasshopper. Note the three pairs of segmented legs. (b) Internal anatomy of the grasshopper. (c) Head of the larva of a water scavenger beetle, an aquatic beetle. Can you identify the mandibles and antennae?

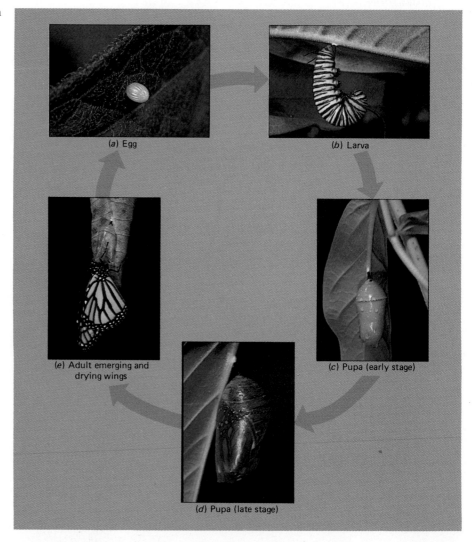

FIGURE 30–26 The life cycle of a monarch butterfly.

water, a few are truly marine, and others inhabit the shore between the tides.

An insect may be described as an **articulated** (jointed), **tracheated** (having tracheal tubes for gas exchange) **hexapod** (having six feet). The insect body consists of three distinct parts—head, thorax, and abdomen (Figure 30–25). Three pairs of legs emerge from the adult thorax, and usually two pairs of wings. One pair of antennae protrudes from the head, and the sense organs include both simple and compound eyes. A complex set of mouth parts is present; these may be adapted for piercing, chewing, sucking, or lapping. Excretion is accomplished by two to many slender **Malpighian tubules,** which receive metabolic wastes from the blood and, after concentrating them, discharge them into the intestine.

The sexes are separate, and fertilization takes place internally. During development there are several molts. In some orders there are several developmental stages called nymphal stages and gradual metamorphosis (change in body form) to the adult form (see Focus on The Principal Orders of Insects). In others there is a complete metamorphosis with four distinct stages in the life cycle: egg, larva, pupa, and adult (Figure 30–26).

Certain species of bees, ants, and termites exist as colonies or societies made up of several different types of individuals, each adapted for some particular function (Figure 30–27). The members of some insect societies communicate with each other by "dances" and by chemicals called pheromones. Social insects and their communication are discussed in Chapter 52.

SECRETS OF INSECT SUCCESS. There are more species of insects than of all other classes of animals combined. What they lack in size, insects make up in sheer numbers. It has been calculated that if all the insects in the world could be weighed, they would weigh more than all of the remaining terrestrial animals. Because they have an extraordinary ability to adapt to changes in the environment, it has been predicted that these curious creatures may eventually inherit the earth.

What are the secrets of insect success? One important factor is their body plan, which can be modified and specialized in so many ways that insects have been able to adapt to an incredible number of life styles. They have filled almost every variety of ecological niche. One of the main secrets of their success is the ability to fly. Unlike other invertebrates, which creep slowly along (or under) the ground, the insects fly rapidly through the air. Their wings and small size facilitate their wide distribution.

The insect body is well protected by the tough exoskeleton, which also helps to prevent water loss by evaporation. Other protective mechanisms include mimicry, protective coloration, and aggressive behavior. Metamor-

FIGURE 30–27 A royal cell of the termite *Nasutitermes* sp., from Peru. The queen, with an enlarged abdomen, occupies the center of the chamber. Most of the individuals are workers. A few soldiers with "squirt gun" heads and reduced mandibles can be seen.

Focus on THE PRINCIPAL ORDERS OF INSECTS

Order and Examples	Mouth Parts	Wings	Other Characteristics
*Ametabolous Insects**			
Thysanura Silverfish Bristletails	Chewing	None	Long antennae; 3 "tails" extend from posterior tip of abdomen; run fast; live in dead leaves and wood, or in houses where they eat the starch in books and clothing
Collembola Springtails	Chewing	None	Abdominal structure for jumping; live in soil, dead leaves, rotting wood
Hemimetabolous Insects			
Odonata Dragonflies Damselflies	Chewing	2 pairs; long, narrow, membranous	Predators; large compound eyes; aquatic nymph
Ephemeroptera Mayflies	Chewing	2 pairs; membranous; forewings larger than hindwings	Small antennae; vestigial mouth parts in adult; 2 or 3 "tails" extending from tip of abdomen; nymph aquatic
Orthoptera Grasshoppers Crickets Roaches	Chewing	2 pairs or none; leathery forewings, membranous hindwings	Most herbivorous, some cause crop damage; praying mantis eats other insects.
Isoptera Termites	Chewing	2 pairs or none; wings shed by sexual forms after mating	Social insects, form large colonies; main diet wood; can be very destructive
Dermaptera Earwigs	Chewing	2 pairs; forewings very short; hindwings large, membranous	Forceps-like appendage on tip of abdomen; nocturnal
Anoplura Sucking lice	Piercing and sucking	None	Ectoparasites of birds and mammals; head louse and crab louse are human parasites; vectors of typhus fever

*Insects may be divided into 3 groups based on their pattern of development. Ametabolous insects do not undergo metamorphosis (egg→immature form→adult). Hemimetabolous insects exhibit incomplete metamorphosis (egg→nymph [resembles adult in many ways but lacks functional wings and reproductive structures] →adult). Holometabolous insects undergo complete metamorphosis (egg→larva→pupa→adult). The wormlike larva, which is very different from the adult, hatches from the egg (Fig. 30–26). Typically, an insect spends most of its life as a larva. Eventually, the larva stops feeding, molts, and enters a pupal stage usually within a protective cocoon or underground burrow. The pupa does not feed and cannot defend itself. Its energy is spent remodeling its body form, so that when it emerges it is equipped with functional wings and reproductive organs.

■ Silverfish, *Lepisma saccharina*, adult and young. Silverfish are primitive ametabolous insects.

phosis divides the insect life cycle into different stages, a strategy that has the advantage of placing larval forms into their own niches so that they do not have to compete with adults for food or habitats.

IMPACT OF INSECTS ON HUMANS.   Not all insects compete with us for food or merely cause us to scratch, swell up, or recoil from their presence. Bees, wasps, mosquitos, and many other insects pollinate flowers of many crops

Order and Examples	Mouth Parts	Wings	Other Characteristics
Hemiptera (true bugs) Chinch bugs Bedbugs	Piercing and sucking	2 pairs; hindwings membranous	Mouth parts form beak; only order of insects properly called bugs
Homoptera Aphids Leaf hoppers Cicadas Scale insects	Piercing and sucking	Usually 2 pairs; membranous	Mouth parts form sucking beak; base of beak near thorax; some very destructive to plants
Holometabolous Insects			
Neuroptera Ant lions Dobson flies	Chewing	2 pairs; membranous	Larvae are predators.
Lepidoptera Moths Butterflies	Sucking	2 pairs, covered with overlapping scales	Larvae, called caterpillars, have chewing mouthparts; eat plants; adults that feed suck flower nectar.
Diptera (true flies) Houseflies Mosquitos Gnats Fruitflies	Usually piercing and sucking	Forewings functional; hindwings small, knoblike halteres	Larvae are maggots or wigglers; may be damaging to domestic animals and food; adults often transmit disease.
Siphonaptera Fleas	Piercing and sucking	None	Legs adapted for jumping; lack compound eyes; ectoparasites on birds and mammals; vectors of bubonic plague and typhus
Coleoptera Beetles Weevils	Chewing	Usually 2 pairs; forewings modified as heavy, protective coverings	Largest order of insects (more than 300,000 species); majority herbivorous; some aquatic
Hymenoptera Ants Bees Wasps	Chewing but modified for lapping or sucking in some forms	2 pairs or none; transparent when present	Some are social insects; some sting.

■ An aphid, *Aphis nerii*, giving birth to a live young. Aphids are hemimetabolous insects.

■ An io caterpillar. This larva will undergo complete metamorphosis before becoming an adult (see Figure 31–26).

and fruit trees. Some destroy harmful insects. For example, dragonflies eat mosquitos; some organic farmers even purchase lady beetles, so adept are they at ridding plants of aphids and other insect pests. Insects are important members of many food chains. Many birds, mammals, amphibians, reptiles, and even some fish depend upon insects for dinner. Many beetles and the maggots of flies are detritus feeders that break down dead plants and animals and their wastes, permitting nutrients to be recycled.

Many insect products are useful to us. Bees produce several thousand tons of honey used commercially each year, as well as a large amount of beeswax used in making candles, lubricants, chewing gum, and other products. Shellac is made from lac, a substance given off by certain scale insects that feed on the sap of trees. And the labor of silkworms provides us with millions of pounds of silk annually.

On the negative side, billions of dollars worth of crops are destroyed each year by insect pests. Whole buildings may be destroyed by termites, and clothing damaged by moths. Fire ants not only inflict painful stings but cause farmers serious economic loss because of their large mounds, which damage mowers and other farm equipment. Such mounds also reduce grazing land because livestock quickly learn to avoid them.

Blood-sucking flies, screwworms, lice, fleas, and other insects annoy and cause disease in both humans and domestic animals. Mosquitos are vectors of malaria, yellow fever, and filariasis. Body lice may carry the typhus rickettsia, and houseflies sometimes transmit typhoid fever and dysentery. Tsetse flies transmit African sleeping sickness, and fleas may be vectors of bubonic plague.

Classes Chilopoda and Diplopoda: Centipedes and Millipedes

Members of class **Chilopoda** are called **centipedes** (meaning hundred-legged), and members of class **Diplopoda** are known as **millipedes** (meaning thousand-legged). These animals are all terrestrial and are typically found beneath stones or wood in the soil in both temperate and tropical regions.

Centipedes and millipedes are similar in having a head and an elongated trunk with many segments, each bearing legs (Figure 30–28). The centipedes have one pair of legs on each segment behind the head. Most centipedes do not have enough legs to merit their name—the most common number being 30 or so—although in a few species, the number of legs is 100 or more. The legs of centipedes are long, enabling them to run rapidly. Centipedes are carnivorous and feed upon other animals, mostly insects,

FIGURE 30–28 Chilopods and diplopods have uniramous appendages. (*a*) Centipede, a member of the class Chilopoda. Centipedes have one pair of appendages per segment. (*b*) Millipede, a member of the class Diplopoda. Millipedes have two pairs of appendages per segment.

Carolina Biological Supply Company

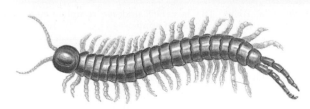

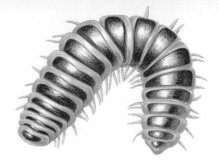

(*a*)

(*b*)

but the larger centipedes have been known to eat snakes, mice, and frogs. The prey is captured and killed with poison claws located just behind the head on the first trunk segment. A pair of poison glands at the base of the claws empty into ducts that open at the tip of the pointed, fanglike claw.

The distinguishing feature of the millipedes is the presence of **diplo-segments**—doubled trunk segments—resulting from the fusion of two original segments. Each double segment has two pairs of legs and contains two pairs of ganglia. The most anterior three or four segments have only a single pair of legs. The body of the millipede is cylindrical, whereas the body of the centipede tends to be flattened. Diplopods are not as agile as chilopods, and most species can crawl only slowly over the ground, though they can powerfully force their way through earth and rotting wood. Millipedes are generally herbivorous and feed on both living and decomposing vegetation. In both chilopods and diplopods, eyes may be completely lacking, or the animal may have simple eyes called **ocelli.** A few species of centipedes have eyes that are similar to the compound eyes of insects.

■ SUMMARY

I. The coelomate protostomes include the mollusks, annelids, and arthropods, as well as some minor phyla.

II. The coelom provides space for many organs to develop and function and permits the digestive tract to move independently of body movements; coelomic fluid helps transport materials and bathes cells that line the coelom.

III. In terrestrial animals, the body covering must prevent fluid loss, some sort of skeleton must be present to withstand the pull of gravity, and reproductive adaptations must be made, such as internal fertilization, development within the mother's body, or shells that prevent desiccation of the developing embryo.

IV. Mollusks are soft-bodied animals usually covered by a shell; they possess a ventral foot for locomotion and a mantle that covers the visceral mass.

A. Class Polyplacophora includes the sluggish marine chitons, which have segmented shells.

B. Class Gastropoda, the largest and most successful group of mollusks, includes the snails, slugs, and whelks. In gastropods, the body is twisted, and the shell (when present) is coiled.

C. Class Bivalvia includes the clams and oysters, animals enclosed by two shells, hinged dorsally.

D. Class Cephalopoda includes the squids and octopods, which are active predatory animals; the foot is divided into tentacles that surround the mouth located in the large head.

V. Phylum Annelida, the segmented worms, includes many aquatic worms, earthworms, and leeches.

A. Annelids have conspicuously long bodies that are segmented both internally and externally; their large compartmentalized coelom serves as a hydrostatic skeleton.

B. Class Polychaeta consists of marine worms characterized by bristled parapodia, used for locomotion.

C. Class Oligochaeta, which includes the earthworms, contains segmented worms characterized

by a few setae per segment. The body is divided into more than 100 segments separated internally by septa.

D. Class Hirudinea, which includes the leeches, is composed of animals that lack setae and appendages; they are equipped with suckers for sucking blood.

VI. Phylum Onychophora includes animals with both annelid and arthropod characteristics.

VII. Phylum Arthropoda is composed of animals with jointed appendages and an armor-like exoskeleton.

A. The trilobites are extinct marine arthropods that were covered by a hard, segmented shell.

B. Subphylum Chelicerata includes class Merostomata (the horseshoe crabs) and class Arachnida (spiders, mites, and their relatives).

1. In the chelicerates the first pair of appendages are chelicerae, used to manipulate food. Chelicerates have no antennae and no mandibles.

2. The arachnid body consists of a cephalothorax and abdomen; there are six pairs of jointed appendages, of which four pairs serve as legs.

C. Subphylum Crustacea includes the lobsters, crabs, and barnacles. The body consists of cephalothorax and abdomen; often, five pairs of walking legs are present. Crustaceans have two pairs of antennae and mandibles for chewing.

D. Subphylum Uniramia includes class Insecta, class Chilopoda, and class Diplopoda; members of this subphylum have unbranched appendages and a single pair of antennae.

1. An insect is an articulated, tracheated hexapod; its body consists of head, thorax, and abdomen. Insects are the most ecologically successful group of animals.

2. The centipedes have one pair of legs per body segment, whereas the millipedes have two pairs of legs per body segment.

■ POST-TEST

1. The _____ of arthropods and mollusks provides protection and serves as a point of attachment for muscles.
2. The molluskan _____ is a belt of teeth within the digestive system.
3. A hemocoel is characteristic of animals with an _____ circulatory system.
4. The first larval stage of a mollusk is a free-swimming _____ larva.
5. Segmental, serial repetition of structures within an animal is known as _____.
6. In pulmonate snails, the highly vascularized mantle functions as a _____.
7. Nudibranchs are mollusks that lack a _____.
8. Setae are bristle-like structures that function in _____.
9. In hermaphroditic animals, male and female reproductive systems are _____.
10. The earthworm typhlosole functions to _____.
11. The earthworm brain consists of a pair of _____.
12. Onychophorans are considered a possible link between the annelids and the _____.
13. Animals with an exoskeleton and paired, jointed appendages are _____.
14. The mandibles of a crustacean are used for _____.
15. Antennal (green) glands in the crustacean are _____ organs.

16. The only sessile crustaceans are the _____.
17. Chelipeds are large _____.
18. Malpighian tubules are _____ organs.

Select the appropriate animal in Column B for the description given in Column A:

Column A	Column B
19. An articulated, tracheated hexapod	a. Trilobite
20. An animal that uses book lungs	b. Insect
21. An animal with two pairs of legs per segment	c. Spider
22. A primitive arthropod with a hard, segmented shell	d. Millipede
	e. None of the above

Select the appropriate animal subphylum or class in Column B for the description given in Column A.

Column A	Column B
23. Largest and most successful group of crustaceans	a. Polychaetes
24. An arthropod subphylum whose members have unbranched appendages	b. Chelicerates
	c. Gastropods
25. Arthropods with no antennae or mandibles	d. Uniramia
	e. None of the above

■ REVIEW QUESTIONS

1. In what ways are mollusks and annelids alike? In what ways are they different?
2. Give two distinguishing characteristics for each:
   a. mollusks
   b. annelids
   c. arthropods
3. What are the advantages of each of the following?
   a. presence of a coelom
   b. the arthropod exoskeleton
   c. segmentation (metamerism)
4. Contrast the life styles of a gastropod and a cephalopod. Identify adaptations possessed by each for its particular life style.

5. What is a trochophore larva? a veliger larva?
6. Describe some of the adaptations that have contributed to insect success.
7. Distinguish between insects and spiders.
8. What are the distinguishing features of each of the arthropod subphyla?
9. What are the distinguishing features of the principal arthropod classes?
10. Identify animals that belong to each class of arthropods.

■ RECOMMENDED READINGS

Brownell, P.H. Prey detection by the sand scorpion. *Scientific American*, Vol. 251, No. 6, December 1984. The sand scorpion does not see or hear the insects it feeds on. Instead it uses receptors on its legs that are exquisitely sensitive to disturbances of the sand.

Burgess, J.W. Social spiders, *Scientific American*, March 1976. A few species of spiders interact socially and build large communal webs.

Camhi, J.M. The escape system of the cockroach. *Scientific American*, December 1980. A study of the mechanisms by which a roach rapidly escapes from predators.

Foot, J. Squid swarm. *Natural History*, Vol. 91, No. 4, April 1982. An account of the schools of squid that gather off the California coast to spawn.

Gosline, J.M., and Demont, M.E. Jet-propelled swimming in squids. *Scientific American*, Vol. 252, No. 1, January 1985. As the squid swims it takes up and expels water by contracting radial and circular muscles in its mantle wall.

Heinrich, B. The regulation of temperature in the honeybee swarm. *Scientific American*, June 1981. A discussion of thermoregulation in a swarm of bees.

Jackson, R.R. A web-building jumping spider. *Scientific*

American, Vol. 253, No. 3, September 1985. Unlike most other jumping spiders, the Australian species *Portia fimbriata* builds webs. This predatory spider hunts other spiders.

Miller, J.A. A brain for all seasons. *Science News,* Vol. 123, No. 17, April 23, 1983. A study of brain development during metamorphosis of a moth.

Miller, J.A. A skunk of a beetle. *Science News,* Vol. 115, May 19, 1979. An account of the defensive spray of a bombardier beetle.

Merritt, R.W., and Wallace, J.B. Filter-feeding insects. *Scientific American,* April 1981. An account of insects that hatch underwater and gather food with nets and brushes, and a discussion of their ecological importance.

Moore, J. Parasites that change the behavior of their host. *Scientific American,* Vol. 250, No. 5, May 1984. Certain parasites, such as thorny-headed worms which infect pill bugs, make the host more vulnerable to predation by their next host.

Nijhout H.F. The color patterns of butterflies and moths. *Scientific American,* November 1981. A study of the development of the more than 100,000 different wing patterns of butterflies and moths.

Reid, R.G.B., and Bernard, F.R. Gutless bivalves. *Science,* Vol. 208, May 1980. A description of a new species of bivalve that lacks internal digestive organs.

Richardson, J.R. Brachiopods. *Scientific American,* Vol. 255, No. 3, September 1986. One class of these clamlike animals survives by searching out environments suited to an unchanging form; the other class survives by adapting its form or behavior to the local environment.

Roper, C.F.E., and Boss, K.J. The giant squid. *Scientific American,* April 1982. A discussion of the anatomy and ecology of the giant squid.

Tangley, L. Tracing the roots of a gypsy. *Science News,* Vol. 122, No. 17, October 23, 1982. In a search for biological control mechanisms, a team of scientists traveled to China, where the gypsy moth, a major threat to forests and ornamental trees, may have originated.

Ward, P., Greenwald, L., and Greenwald, O.E. The buoyancy of the chambered nautilus. *Scientific American,* October 1980. An account of the mechanisms by which the chambered nautilus divides its shell into compartments and removes water, thereby gaining mobility.

West, S. Moon history in a seashell. *Science News,* Vol. 114, No. 25, December 16, 1978. A discussion of a theory linking the development of the chambered nautilus with lunar months.

Wicksten, M.K. Decorator crabs. *Scientific American,* February 1980. A discussion of species of spider crabs that camouflage themselves with materials that they attach to their exoskeletons.

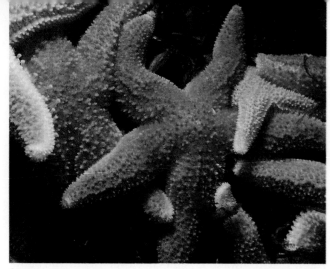

Sea star (*Asterias vulgaris*)

31

The Animal Kingdom: The Deuterostomes

■ LEARNING OBJECTIVES

After you have read this chapter you should be able to:

1. Discuss the relationship of the echinoderms and chordates, giving specific reasons for grouping them together.
2. Describe the distinguishing characteristics of the echinoderms.
3. Describe and give examples of each of the five main classes of echinoderms.
4. List the subphyla of phylum Chordata and describe the characteristics that they have in common; describe the characteristics of subphylum Vertebrata.
5. Distinguish among the classes of vertebrates and assign a given vertebrate to the correct class.
6. Trace the evolution of vertebrates according to current theory.
7. Identify adaptations that reptiles and other terrestrial vertebrates have made to life on land.
8. Contrast monotremes, marsupials, and placental mammals and give examples of members that belong to each group.
9. Identify the major orders of placental mammals and give an example of an animal that belongs to each.

It may seem strange to group the echinoderms—the sea stars and sand dollars—with the chordates, the phylum to which we belong. However, echinoderms and chordates are thought to have evolved from a common ancestor. Both groups are deuterostomes, so they share similarities in their patterns of development. As discussed in Chapter 29, deuterostomes are characterized by radial and indeterminate cleavage, origin of the mesoderm as enterocoelous pouches, formation of the coelom from cavities within the mesodermal outpocketings, and formation of the mouth from a second opening that develops in the embryo. The characteristics of some of the higher animal phyla are summarized in Table 31–1.

Table 31–1 COMPARISON OF CHARACTERISTICS OF SOME HIGHER ANIMAL PHYLA*

Phylum	Body Symmetry	Gas Exchange	Waste Disposal	Nervous System	Circulation	Reproduction	Other Characteristics
Mollusca Clams Snails Squids	Bilateral	Gills and mantle	Metanephridia	Three pairs of ganglia; simple sense organs	Open system	Sexual; sexes separate; fertilization in water	Soft-bodied; usually have shell and ventral foot for locomotion
Annelida (segmented worms) Earthworms Leeches Marine worms	Bilateral	Diffusion through moist skin; oxygen circulated by blood	Pair of metanephridia in each segment	Simple brain; ventral nerve cord; simple sense organs	Closed system	Sexual; hermaphroditic but cross-fertilize	Earthworms till soil
Arthropoda (joint-footed animals) Crustaceans Insects Spiders	Bilateral	Tracheae in insects; gills in crustaceans; book lungs or tracheae in spider group	Malpighian tubules in insects; antennal (green) glands in crustaceans	Simple brain; ventral nerve cord; well-developed sense organs	Open system	Sexual; sexes separate	Hard exoskeleton; most diverse and numerous group of animals
Echinodermata (spiny-skinned animals) Sea stars Sea urchins Sand dollars	Embryo: bilateral; adult: modified radial	Skin gills	Diffusion	Nerve rings; no brain	Open system; reduced	Sexual; sexes almost always separate	Water vascular system; tube feet
Chordata Tunicates Lancelets Vertebrates	Bilateral	Gills or lungs	Kidneys and other organs	Dorsal nerve cord with brain at anterior end	Closed system; ventral heart	Sexual; sexes separate	(1) Notochord; (2) dorsal, tubular nerve cord; (3) pharyngeal gill slits

*Members of these phyla are at the organ system level of organization and have a complete digestive tract.

(a)

(b)

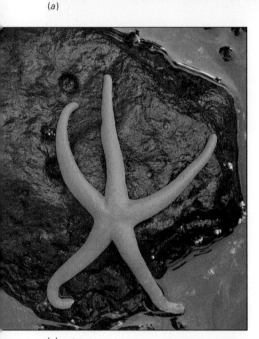

(c)

FIGURE 31–1 Some representative echinoderms. (*a*) Northern basket star, *Gorgonocephalus arcticus,* on a sponge. (*b*) Blue spotted sea urchin, *Astropyga radiata.* (*c*) Pacific Henricia sea star, *Henricia laeviuscula.*

PHYLUM ECHINODERMATA

All of the members of phylum **Echinodermata** inhabit the sea. About 6000 living and 20,000 extinct species have been identified. The living species are divided into five principal classes (Figure 31–1): Class **Crinoidea** includes the sea lilies and feather stars; class **Asteroidea,** the sea stars; class **Ophiuroidea,** the brittle stars; class **Echinoidea,** the sea urchins and sand dollars; and class **Holothuroidea,** the sea cucumbers.

The echinoderms are in many ways unique in the animal kingdom. According to some biologists, their bilaterally symmetrical, ciliated, free-swimming larvae suggest that at one time the adults were also bilaterally symmetrical and free-swimming. Then, during their evolutionary history, they became more sedentary. As they evolved, they developed striking adaptations such as the water vascular system and radial symmetry. Many echinoderms are pentaradial, meaning that the body is arranged in five parts around a central disk where the mouth is located. Other biologists think that the earliest echinoderms were sessile and later gave rise to the free-moving groups.

One of the unique characteristics of the Echinoderms is the endoskeleton, which consists of small calcareous plates (composed of $CaCO_3$), typically bearing spines that project outward; the name Echinodermata, meaning spiny-skinned, reflects this trait. The endoskeleton is covered by a thin, ciliated epidermis.

Also unique in echinoderms is the **water vascular system,** a network of canals through which sea water circulates. Branches of this system lead to numerous tiny **tube feet,** which extend when filled with fluid. The tube feet serve in locomotion and obtaining food, and in some forms, in gas exchange. The water vascular system serves as a hydrostatic skeleton for the tube feet. To extend a foot, a rounded muscular sac, or **ampulla,** at the upper end of the foot contracts, forcing water through a valve into the tube of the foot. At the bottom of the foot is a suction structure that adheres to the substratum. The foot can be withdrawn by contraction of muscles in its walls, which forces water back into the ampulla (Figure 31–2).

Echinoderms have a well-developed coelom, in which the various internal organs are located. The complete digestive system is the most prominent body system. There are a variety of respiratory structures in the various classes, including dermal gills in the sea stars and respiratory trees in sea cucumbers. Only a rudimentary circulatory system is present, and there are no specialized excretory structures. The nervous system is simple, usually consisting of **nerve rings** about the mouth with radiating nerves. There is no brain. The sexes are usually separate. Eggs and sperm are released into the water, where fertilization takes place externally.

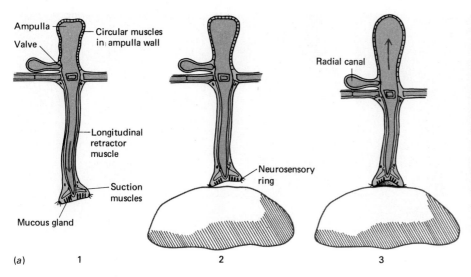

(a) 1 2 3

(b)

FIGURE 31–2 Tube feet in echinoderms. (*a*) Longitudinal section through the tube foot of a sea urchin. Sequence shows how the foot works to anchor the animal to a substrate. (*1*) When the valve to the radial canal is closed, there is a fixed volume of water in the tube foot. When the muscles in the ampulla contract, water is forced into the lower part of the foot, which elongates. (*2*) When the foot comes in contact with the substrate, the center of the sucker withdraws, creating a near-vacuum, or suction on the substrate. A secretion from the mucous glands aids in adhesion. (*3*) After adhesion of the sucker, the longitudinal muscles of the foot contract, shortening the foot and forcing fluid back into the ampulla. The valve to the radial canal remains closed as long as the hydrostatic pressure in the tube foot remains greater than the pressure within the canal. (*b*) Tube feet of a sea star.

Class Crinoidea: Feather Stars and Sea Lilies

Class Crinoidea, the most primitive class of living echinoderms, includes the feather stars and the sea lilies. The feather stars are free-swimming crinoids, though they often remain in the same location for long periods of time. Sea lilies are sessile and remain attached to the ocean floor by a stalk. Although there are relatively few living species, a great many extinct crinoids are known.

Crinoids differ from other echinoderms in that the oral (mouth) surface is turned upward, and a number of branched, feathery arms extend upward (Figure 31–3). In all other classes, the mouth is located on the ventral surface. The crinoids are filter feeders. Their tube feet, abundant on the feathery arms, are coated with mucus that traps microscopic organisms.

Class Asteroidea: Sea Stars

Sea stars, or starfish, are members of the class Asteroidea. The body of a sea star consists of a central disk from which radiate 5 to 20 or more **arms,** or **rays** (Figure 31–4). In the center of the underside of the disk is the mouth. The endoskeleton consists of a series of calcareous plates that permit some movement in the arms. Around the base of the delicate skin gills used in gas exchange are tiny pincer-like spines called **pedicellariae;** operated by muscles, these keep the surface of the animal free of debris (Figure 31–5)

FIGURE 31–3 A feather star from the Caribbean Sea.

(a)

(b)

FIGURE 31–4 Sea stars. (*a*) A sunflower star (*Pycnopodia helianthoides*) releasing eggs. (*b*) A painted sea star (*Orthasterias koehleri*) attacking a clam.

FIGURE 31–5 (*a*) The sea star *Asterias* viewed from above with the arms in various stages of dissection. (*1*) Upper surface with a magnified detail showing the features of the surface. The end is turned up to show the tube feet on the lower surface. (*2*) Arm shown in cross section. (*3*) Upper body wall of arm removed. (*4*) The upper body wall and digestive glands have been removed, and the ampullae and ambulacral plates are shown in magnified view. (*5*) All of the internal organs, except for the retractor muscles, have been removed to show the inner surface of the lower body wall. (*b*) Photomicrograph showing spines and pedicellaria on the surface of an echinoderm.

The undersurface of each arm is equipped with hundreds of pairs of tube feet. The cavities of the tube feet are all connected by radial canals in the arms; these in turn are connected by a circular canal in the central disk. The circular canal is connected by an axial stone canal to a button-shaped plate, called the **madreporite** on the upper (aboral) surface of the central disk. As many as 250 tiny pores in the madreporite permit sea water to enter the water vascular system.

Most sea stars are carnivorous and feed upon crustaceans, mollusks, annelids, and even other echinoderms. Occasionally they catch and eat a small fish. To attack a clam or other shell fish, the sea star mounts it, assuming a humped position as it straddles the edge opposite the hinge. Then, with its tube feet attached to the two shells, it begins to pull. The sea star uses many of its tube feet at a time but can change and use new groups as active tube feet get tired. By applying a steady pull on both shells over a long period of time, the sea star succeeds in tiring the powerful muscles of the clam so that they are forced to relax, opening the shell.

To begin its meal, the sea star projects its stomach out through its mouth and into the soft body of its prey. Digestive enzymes are secreted into the clam so that it is partly digested while still in its own shell. The soft parts of the clam are digested to the consistency of a thick soup and pass into the sea star body for further digestion by enzymes secreted from glands located in each arm. The sea star's water vascular system does not enable it

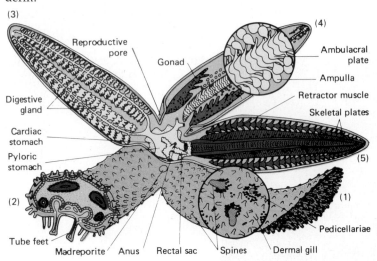

(a)

(b)

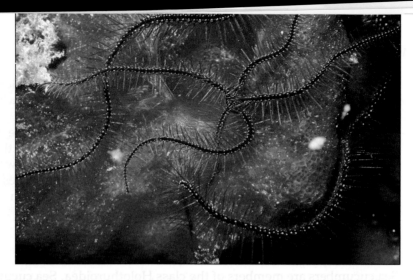

FIGURE 31–6 Brittle stars living on the surface of a sponge.

to move rapidly, but since it usually preys upon slow-moving or stationary clams and oysters, the speed of attack is not as critical as for most other predators.

The blood circulatory system in sea stars is poorly developed and probably of little help in circulating materials. Instead, this function is assumed by the coelomic fluid, which fills the large coelom and bathes the internal tissues. Metabolic wastes pass to the outside by diffusion. The nervous system consists of a ring of nervous tissue encircling the mouth and a nerve cord extending from this into each arm; there is no aggregation of nerve cells that could be called a brain.

Class Ophiuroidea: Basket Stars and Brittle Stars

Basket stars and brittle stars (serpent stars) are members of the class Ophiuroidea. They resemble asteroids in that their bodies also consist of a central disk with arms, but the arms are long and slender and more sharply set off from the central disk (Figure 31–6). Ophiuroids can move more rapidly than asteroids, using their arms to perform rowing or even swimming movements. The tube feet are not used in locomotion and are thought to serve a sensory function, perhaps that of smell or taste. Tube feet are also used to collect and handle food.

Class Echinoidea: Sea Urchins and Sand Dollars

The class Echinoidea includes the sea urchins and the sand dollars. Echinoids lack arms, and their skeletal plates are flattened and fused, forming a solid shell called a **test**. The sea urchin body is covered with spines (Figure 31–7), which in some species can penetrate flesh and are difficult to

FIGURE 31–7 (*a*) A sea urchin scavenging for food in the Red Sea. (*b*) Close-up of a poison sac on a sea urchin's spine.

(b)

(a)

Focus on SOME ORDERS OF LIVING PLACENTAL MAMMALS

Order Insectivora: moles, hedgehogs, and shrews. These are nocturnal insect-eating animals, considered to be the most primitive placental mammals and the ones closest to the ancestors of all the placentals. The shrew is the smallest living mammal; some weigh less then 5 g.

Order Chiroptera: bats. These mammals are adapted for flying; a fold of skin extends from the elongated fingers to the body and legs, forming a wing. Bats are guided in flight by a sort of biologic sonar: They emit high-frequency squeaks and are guided by the echoes from obstructions. These animals eat insects and fruit or suck the blood of other animals. Blood-feeding bats may transmit diseases such as yellow fever and paralytic rabies.

Order Carnivora: cats, dogs, wolves, foxes, bears, otters, mink, weasels, skunks, seals, walruses, and sea lions. Carnivores are flesh-eaters, with sharp, pointed canine teeth and shearing molars. In many species the canines are used to kill the prey. Carnivores have a keen sense of smell and exhibit complex social interactions. Its members are among the fastest, strongest, and smartest of animals. Limbs of the seals, walruses, and sea lions are modified as flippers for swimming. However, these animals are not completely adapted to life in the water; they come ashore to mate and bear their offspring.

Order Edentata: sloths, anteaters, and armadillos. In these animals, the teeth are reduced to molars without enamel in the front part of the jaws, or no teeth are present. Sloths are sluggish animals that hang upside down from branches. They are often protectively colored by green algae that grow on their skin. Armadillos are protected by bony plates; they eat insects and small invertebrates.

Order Rodentia: Squirrels, beavers, rats, mice, hamsters, porcupines, and guinea pigs. These are gnawing mammals with chisel-like incisors that grow continually. As they gnaw, the teeth are worn down. The rodents are one of the most successful orders of mammals; about 3000 species have been described.

Order Lagomorpha: rabbits, hares, and pikas. Like the rodents, the lagomorphs have chisel-like incisors with enamel. Their long hind legs are adapted for jumping, and many have long ears.

Order Primates: lemurs, monkeys, apes, and humans. These mammals have highly developed brains and eyes, nails instead of claws, opposable great toes or thumbs, and eyes directed forward. Most species of primates are arboreal (tree-dwelling) and are thought to have evolved from the tree-dwelling insectivores. Primates may be divided into the prosimians, which include the lemurs, lorises, and tarsiers, and the anthropoids, which include monkeys, apes, and humans. (Primate evolution is discussed in Chapter 21.)

Order Perissodactyla: horses, zebras, tapirs, and rhinoceroses. These are herbivorous hoofed mammals with an odd number of digits per foot, usually one or three toes. (Hoofed mammals are often referred to as ungulates.) The teeth are adapted for chewing. These are usually large animals with long legs.

Order Artiodactyla: cattle, sheep, pigs, deer, and giraffes. These herbivorous hoofed mammals have an even number of digits per foot. Most have two toes, but some have four. Many have antlers or horns on the head. Most are ruminants that chew a cud and have a series of stomachs in which bacteria that digest cellulose are incubated; this contributes greatly to their success as herbivores.

Order Proboscidea: elephants. These animals have a long, muscular trunk (proboscis) that is very flexible. Thick, loose skin is characteristic. The two upper incisors are elongated as tusks. Most are enormous with large heads and broad ears; the legs are like pillars. These are the largest land animals, weighing as much as 7 tons. This order includes the extinct mastodons and wooly mammoths.

Order Sirenia: sea cows and manatees. These are herbivorous aquatic mammals with finlike forelimbs and no hind limbs. They are probably the basis for most tales about mermaids.

Order Cetacea: whales, dolphins, and porpoises. These mammals have become well adapted for their aquatic life style. They have fish-shaped bodies with broad, paddle-like forelimbs (flippers). Posterior limbs are absent. Many have a thick layer of fat called blubber covering the body. These very intelligent animals mate and bear their young in the water, and the young are suckled like those of other mammals. The blue whale is the largest living animal and probably the largest animal that has ever existed.

■ Representative members of several orders of placental mammals. (a) *Trachops cirrhosus*, a frog-eating bat from Panama (family Chiroptera). After the rodents, bats are the largest order of mammals. (b) *Ursus maritimus*, a polar bear, photographed in the Kane Basin in the Arctic (family Carnivora). (c) *Erethizon dorsatum*, a porcupine (family Rodentia). Porcupine females bear only one offspring in a season. For mammals, young porcupines are unusually able to care for themselves; at the age of 2 days they are able to climb trees and find food. (d) *Leontopatheus rosalia*, the golden lion tamarin monkey. Only 150 of these primates survive in their native coastal rainforest in Brazil, which has been reduced to 2% of its original acreage. Thus, even though these animals have been able to breed well in captivity, their future as a species remains uncertain. (e) Two giraffes, *Giraffa camelopardalis* (family Artiodactyla), taking a cooling drink at an oasis in South Africa. Special vascular adaptations prevent the dangerous rise in cerebral blood pressure that would otherwise develop when the giraffe lowers its neck to drink. (f) A common dolphin, *Delphinus delphis*, photographed in the Sea of Cortez, Mexico.

(a)

(b)

(c)

(d)

(e)

(f)

FIGURE 31–35 A vertebrate family tree.

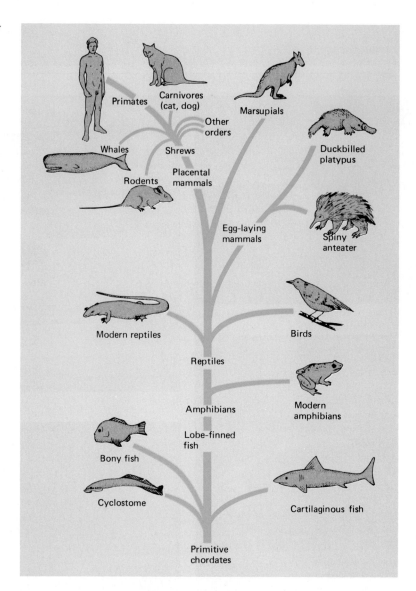

interact with other members of the group within a few minutes of birth. There are about 17 living orders of placental mammals. A brief summary of some of these is given in Focus on Some Orders of Living Placental Mammals. The probable family tree of the vertebrates is illustrated in Figure 31–35.

■ SUMMARY

I. The echinoderms and chordates are thought to be related because they are both deuterostomes and therefore share many developmental characteristics.

II. Phylum Echinodermata includes marine animals with spiny skins, a water vascular system, and tube feet; the larvae have bilateral body symmetry; most of the adults exhibit pentaradial symmetry.

   A. Class Crinoidea includes the sea lilies and feather stars; in these animals, the oral surface is turned upward; some are sessile.

   B. Class Asteroidea is made up of the sea stars, animals with a central disk from which radiate five or more arms.

   C. Class Ophiuroidea includes the brittle stars, which resemble asteroids but have longer, more slender arms, set off more sharply from the central disk.

   D. Class Echinoidea includes the sea urchins and sand dollars, animals that lack arms; they have a solid shell, and their body is covered with spines.

   E. Class Holothuroidea consists of sea cucumbers, animals with elongated flexible bodies; the mouth is surrounded by a circle of modified tube feet that serve as tentacles.

III. Phylum Chordata consists of three subphyla: Urochordata, Cephalochordata, and Vertebrata. At some time in its life cycle, a chordate has a notochord, a dorsal tubular nerve cord, and pharyngeal gill slits.

A. Subphylum Urochordata comprises the tunicates, which are sessile, filter-feeding marine animals that have tunics made of cellulose.

B. Subphylum Cephalochordata consists of the lancelets, small, segmented fishlike animals that exhibit all three chordate characteristics.

C. Subphylum Vertebrata includes animals with a vertebral column that forms the chief skeletal axis of the body. Vertebrates also have a cranium that is part of the endoskeleton, pronounced cephalization, differentiated brain, muscles attached to the endoskeleton for movement, and two pairs of appendages.

1. Class Agnatha, the jawless fish, includes the lamprey eels and hagfishes.

2. Descendants of the ostracoderms (agnathans that are the earliest known fossil chordates) are thought to have evolved jaws and paired appendages and to have given rise to the modern jawed fishes.

3. Class Chondrichthyes, the cartilaginous fish, consists of the sharks, rays, and skates.

4. Class Osteichthyes, the body fish, includes about 20,000 species of freshwater and saltwater fishes. The osteichthyes and chondrichthyes are thought to have evolved from placoderm ancestors at about the same time. Most modern bony fish are ray-finned fishes with swim bladders.

5. Modern amphibians include the salamanders, frogs and toads, and wormlike caecilians.

   a. Most amphibians return to the water to reproduce; frog embryos develop into tadpoles, which undergo metamorphosis to become adults.

   b. Amphibians use their moist skin as well as lungs for gas exchange; they have a three-chambered heart with systemic and pulmonary circulations; and they have mucous glands in the skin.

6. Class Reptilia includes turtles, lizards, snakes, and alligators.

   a. Reptiles are true terrestrial animals.

   b. Fertilization is internal; most reptiles secrete a leathery protective shell around the egg; the embryo develops an amnion and other extraembryonic membranes, which protect it and keep it moist.

   c. A reptile has a dry skin with horny scales, lungs with many chambers, and a three-chambered heart (with some division of oxygen-rich and oxygen-poor blood) and excretes uric acid.

   d. Reptiles dominated the earth during the Mesozoic era; then, during the Cretaceous period, most of them, including all of the dinosaurs, became extinct.

7. Birds (class Aves) have many adaptations for flight, including feathers, wings, and light hollow bones containing air spaces; birds have a four-chambered heart, very efficient lungs, a high metabolic rate, and a constant body temperature; they excrete solid wastes (uric acid).

   a. Birds have a well-developed nervous system and excellent vision and hearing.

   b. Birds have developed the voice and communicate with simple calls and complex songs.

8. Mammals have hair, mammary glands, differentiated teeth, and maintain a constant body temperature. They have a highly developed nervous system and a muscular diaphragm.

   a. Monotremes, mammals that lay eggs, include the duck-billed platypus and the spiny anteater.

   b. Marsupials are pouched mammals such as kangaroos and opossums. The young are born in an immature stage and complete their development in the marsupium, where they are nourished with milk from the mammary glands.

   c. Placental mammals are characterized by an organ of exchange, the placenta, that develops between the embryo and the mother; this organ supplies oxygen and nutrients to the fetus and enables it to complete development within the uterus. There are about 17 living orders of placental mammals.

■ POST-TEST

1. Adult _____ have pentaradial symmetry.

2. The _____ _____ system and _____ feet are unique to echinoderms.

3. Echinoids (e.g., sea urchins) lack _____, and their skeletal plates form a solid _____ .

4. The three distinguishing characteristics of a chordate are a _____, a dorsal _____ _____, and pharyngeal _____ _____ .

5. _____ are sessile, marine chordates often mistaken for sponges.

6. Vertebrates are distinguished from all other animals in having a _____ _____; anterior

to this structure a _____ encloses and protects the brain.

7. Fish belong to superclass _____; amphibians, reptiles, birds, and mammals belong to superclass _____ .

8. _____ scales are characteristic of sharks.

9. In sharks, the _____ receives digestive wastes, urine, and gametes.

10. The shark's spiral valve slows the passage of _____, permitting more time for _____ .

11. Modern fishes are thought to have descended from the _____ fishes; the lobe-finned fishes are

credited with being the ancestors of the _____.
12. The operculum covers the _____.
13. The labyrinthodonts are thought to have been the first successful _____.
14. The amnion is an adaptation to _____ life; it secretes a fluid that _____.
15. The only homeothermic animals are the _____ and the _____.
16. Monotremes are mammals that _____.

Match the answer in Column B with the description in Column A; there may be more than one answer for each question.

Column A
17. Have amnion
18. Have hair
19. Have four-chambered heart (two atria and two ventricles)
20. Have tube feet
21. Body covered with hard, dry, horny scales
22. Bones contain air spaces; no teeth
23. Have pharyngeal gill slits at some time in life cycle

Column B
a. Bony fish
b. Amphibians
c. Reptiles
d. Birds
e. Mammals
f. None of the above
g. All of the above (a–e)

Column A
24. Agnathan with circular sucking disk
25. Earliest known species of bird
26. A cephalochordate
27. A lobe-finned fish
28. Stem reptiles

Column B
a. Cotylosaurs
b. Lamprey
c. Archaeopteryx
d. Coelacanth
e. Amphioxus

■ REVIEW QUESTIONS

1. Why are echinoderms thought to be more closely related to chordates than to other phyla?
2. What are the three principal distinguishing characteristics of a chordate? How are these evident in a tunicate larva? in an adult tunicate? in a lancelet? in a human?
3. What characteristics distinguish the vertebrates from the rest of the chordates?
4. How do lampreys and hagfishes differ from other fishes? Of what economic importance are agnathans?
5. What is the function of gills? In general terms, how do they work? Why do you suppose aquatic mammals do not possess them?
6. Compare the skins of sharks, frogs, snakes, and mammals.
7. Give the location and function of each of the following:
   a. swim bladder
   b. placenta
   c. operculum
   d. amnion
   e. marsupium
8. Give the phylum, subphylum, class, (and order if you can) for each of the following animals:
   a. human being
   b. turtle
   c. lamprey eel
   d. *Branchiostoma* (amphioxus)
   e. dogfish shark
   f. whale
   g. frog
   h. pelican
   i. bat
9. Why are monotremes considered to be more primitive than other mammals? Some paleontologists consider them to be therapsid reptiles rather than mammals. Give arguments for and against this position.
10. Which vertebrate groups maintain a constant body temperature? How do they accomplish this? Why is this advantageous?
11. Which are more specialized animals, birds or mammals? Explain your answer.
12. According to current evolutionary theory, give the significance of each of the following:
   a. coelacanths
   b. placoderms
   c. labyrinthodonts
   d. *Seymouria*
   e. therapsids
   f. *Archaeopteryx*

■ RECOMMENDED READINGS

Adding flesh to bare therapsid bones. *Science News*, Vol. 119, No. 25, June 1981. A brief account of the therapsid reptiles thought to be our ancestors.

Alldredge, A. Appendicularians. *Scientific American*, July 1976. A description of a group of tunicates that build a house of mucus and filter food particles out of seawater.

Alldredge, A. L., and Madin, L. P. Pelagic tunicates: unique herbivores in the marine plankton. *Bioscience*, Vol. 32, No. 8, September 1982. An account of the unique adaptations of pelagic tunicates.

Austad, S. N. The adaptable opossum. *Scientific American*, Vol. 258, No. 2, February 1988, pp. 98–104. The Virginia opossum can adjust the sex ratios of its progeny, an efficient reproductive strategy that helps it adapt quickly to environmental changes.

Bramble, D. M., and Carrier, D. R. Running and breathing in mammals. *Science*, Vol. 219, No. 4582, January 1983. A physiological study of the synchronization of locomotion and breathing in running mammals.

Crews, D., and Garstka, W. R. The ecological physiology of a garter snake. *Scientific American*, Vol. 247, No. 5, November 1982. An account of the physiological and behavioral adaptations of the red-sided garter snake to its harsh environment.

Degabriele, R. The physiology of the koala. *Scientific American*, July 1980. A study of the survival adaptations of the koala, a marsupial that eats eucalyptus leaves, seldom drinks, and uses no shelter.

Eastman, J.T., and DeVries, A.L. Antarctic fishes. *Scientific American*, Vol. 255, No. 5, November 1986. Most species of fish died out when the Antarctic Ocean became icy cold, but fishes in the suborder Notothenioidei survive by making biological antifreezes and conserving energy.

Griffiths, M. The platypus. *Scientific American*, Vol. 258, No. 5, May 1988, pp. 84–91. Everything you might want to know about this interesting monotreme; the platypus has mechanoreptors and electroreceptors on its beak for detecting prey.

Gwinner, E. Internal rhythms in bird migration. *Scientific American*, Vol. 254, No. 4, April 1986. Migratory birds have a biological clock that tells them when to begin and end their flight. This clock also helps them find their destinations.

Miller, J. A. Strike. *Science News*, Vol. 120, August 29, 1981. An account of rattlesnakes and vipers striking rodent prey and an exploration of the evolutionary implications of this behavior.

Mansour, T. E. Chemotherapy of parsitic worms: new biochemical strategies. *Science*, Vol. 205, August 1979. A review of the mechanisms by which various drugs act on parasitic worm infestations.

Mossman, D. J., and Sarjeant, W. A. S. The footprints of extinct animals. *Scientific American*, Vol. 248, No. 1, January 1983. An account of vertebrate evolution with emphasis on information gained from animal tracks.

Murray, J. D. How the leopard gets its spots. *Scientific American*, Vol. 258, No. 3, March 1988. A single pattern-formation mechanism may underlie the wide variety of animal coat markings found in nature.

Myers, C. W., and Daly, J. W. Dart-poison frogs. *Scientific American*, Vol. 248, No. 2, February 1983. An account of poisonous frogs and the toxins they release.

Nelson, C.H., and Johnson, K.R. Whales and walruses as tillers of the sea floor. *Scientific American*, February 1987. As gray whales and Pacific walruses gather food from the bottom of the seas, they produce pits and furrows to an extent that rivals the disturbances caused by geologic processes.

Newman, E. A., and Hartline, P. H. The infrared "vision" of snakes. *Scientific American*, March 1982. Snakes of two families can detect and localize sources of infrared radiation.

Orr, R. T. *Vertebrate Biology*. 5th ed. Philadelphia, Saunders College Publishing, 1982. This introduction to vertebrate biology focuses on biological processes.

Romer, A. S., and Parsons, T. S. *The Vertebrate Body*. 6th ed. Philadelphia, Saunders College Publishing, 1986. A well-respected, classic textbook that takes a comparative approach to life processes in vertebrates.

Vaughan, T. A. *Mammalogy*. 3rd ed. Philadelphia, Saunders College Publishing, 1986. An introduction to the mammals; a systematic approach.

Welty, J. C. *The Life of Birds*. 4th ed. Philadelphia, Saunders College Publishing, 1988. An introduction to the biology of birds.

Wursig, B. Dolphins. *Scientific American*, March 1979. An interesting description of dolphin behavior and learning ability.

Zapol, W. M. Diving adaptations in the Weddell seal. *Scientific American*, June 1987. The Weddell seal can swim deeper and hold its breath longer than most other mammals. This adaptation is now thought to be at least in part the result of collapsible lungs and a spleen that functions as a scuba tank.

PART VI

STRUCTURES AND LIFE PROCESSES IN PLANTS

Plants are eukaryotic organisms that obtain chemical energy by photosynthesis, converting the energy of sunlight into the chemical energy of organic compounds. ▪ As a result of photosynthesis, plants serve as the base of the food chain for other living organisms. ▪ Plus, atmospheric oxygen is continually replenished by photosynthesis. ▪

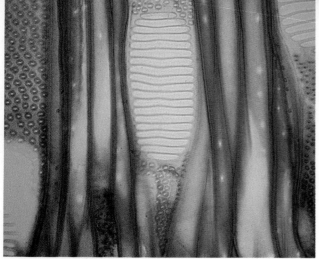

Sclariform perforation plates of a red alder vessel element

32

Growth and Differentiation

■ **LEARNING OBJECTIVES**

After you have read this chapter you should be able to:

1. Discuss what is involved in "growth" of plants.
2. Trace the stages in embryo development in angiosperms.
3. Distinguish between primary and secondary growth in plants.
4. Distinguish between apical meristems and lateral meristems.
5. Distinguish between determinate and indeterminate growth and give two examples of each.
6. Describe the following simple tissues: parenchyma, collenchyma, and sclerenchyma.
7. Characterize the vascular tissue system of plants, the xylem and phloem.
8. Summarize the dermal tissue system of plants, the epidermis and periderm.
9. Discuss the plant body, including the basic features of leaves, stems, and roots.
10. Relate how plant development is different from development in animals.

Plants, like animals, are complex multicellular organisms. When they reproduce sexually, a fertilized egg, or zygote, results from the union of haploid male and female reproductive cells. This single-celled zygote grows into a multicellular embryo. The mature plant that develops from the embryo is composed of millions of cells that are specialized into tissues and organs.

EMBRYONIC DEVELOPMENT

Seed plants produce a young plant embryo complete with nutrients in a compact package, the seed (Chapter 28). The seed develops after fertilization, the union of the egg and the sperm nucleus that forms a zygote, or fertilized egg. Mitotic divisions of the zygote to form a multicellular embryo progress in an orderly, predictable fashion that is essentially the same for both dicots and monocots (Figure 32–1). The following description is for dicot embryonic development.

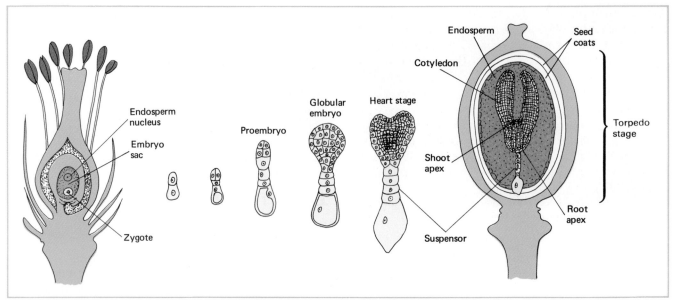

FIGURE 32–1 Embryonic development in dicots. With the first division of the zygote, polarity is established: The bottom cell develops into the suspensor, and the top cell develops into the plant embryo. In the final drawing, the embryo is still immature. In most dicots the endosperm is gone in the mature seed, its reserves having been used for further growth and development of the embryo.

The two cells formed as a result of the first division of the zygote establish polarity in the embryo. The bottom cell develops into the **suspensor,** a multicellular structure that anchors the embryo and aids in nutrient uptake from the endosperm. The top cell grows into the embryo proper. Initially, the top cell divides to form a chain of cells, the **proembryo.** As mitosis continues, a multicellular sphere of cells develops, the **globular embryo.** Tissue differentiation begins during this stage. When the two cotyledons begin to form, the embryo resembles a heart: this is called the **heart stage.** As the embryo elongates, the **torpedo stage** develops, which continues to grow into the mature embryo.

Like all other aspects of plant growth and development, the embryonic stages (proembryo, globular embryo, heart stage, and torpedo stage) are under genetic control (Chapter 16). It is possible to culture entire, multicellular plants from single cells using plant tissue culture methods. Under certain conditions, the genes that control the development of the embryo are expressed in plant tissue culture, and all the stages can be observed in their normal progression. Haploid embryos grown from pollen cells in culture may also go through the embryonic stages, demonstrating that the diploid condition is not a requirement for embryo development.

SEED GERMINATION

When the seed is mature, it often will not **germinate** immediately. A number of factors influence whether a seed will resume growth. Many of these are environmental factors, including water, oxygen, temperature, and sometimes light requirements. No seed will germinate unless it has imbibed, or absorbed, water. A watery medium in cells is necessary for active metabolism. When a seed germinates, its metabolic machinery is turned on, with numerous materials being synthesized and degraded. Therefore, water is an absolute requirement for germination. Also, there is a very high energy requirement associated with seed germination and growth. Because plants have the same aerobic respiratory pathway as animals, oxygen is needed for plant development during germination.

Another environmental factor that affects germination is temperature. Each plant species has an optimum temperature for seed germination, although germination will occur over a range of temperatures. For most plants, the optimum germination temperature is 25–30°C. Some plant seeds, such as apples, require exposure to prolonged periods of cold before their seeds will germinate. Finally, certain plants also have a light requirement for germination, especially those with tiny seeds (Chapter 36).

Some of the environmental factors that affect germination ensure the survival of the plant. If plant seeds germinated at extremely low temperatures, the young plants would not likely survive. Further, the requirement of a prolonged cold period ensures the seed will germinate in the spring rather than the winter. The light requirement ensures that a tiny seed will germinate only if it is close to the surface of the soil. If such a seed germinated several inches below the soil surface, it would not have enough food reserves to grow to the surface.

Even when external factors are optimal, certain seeds will not germinate because of internal factors. Many plant seeds are **dormant** either because they are immature and the embryo must develop further or because inhibitors are present. These inhibitors, such as abscisic acid, may be leached out of the seed by rain (Chapter 37). Once again, this ensures the survival of the plant. For example, desert annuals often have high levels of abscisic acid in their seeds which is leached out only when rainfall is sufficient to support the plant's growth after germination.

If the proper combination of external and internal factors is not present, the seed will not germinate. How long can a seed remain dormant? There have been stories of seeds germinating after thousands of years when archaeologists excavated the tombs of the Pharaohs. These accounts have not been verified. However, a scientific experiment was conducted at Michigan State University starting in 1879, when a variety of seeds were enclosed in jars and buried. Periodically, some of the jars were removed, and an attempt was made to germinate the seeds. As late as 1980, 101 years later, some of the seeds still germinated.

The first part of the plant to emerge when the seed germinates is the **radicle,** or embryonic root. As the root grows, it forces its way through the soil, encountering considerable friction. The delicate cells at the tip of the root are protected by a layer of cells known as the **root cap.** Plant stem tips are not covered by a cap of cells, but they have different ways to protect the delicate tip as it grows through the soil to the surface. The stem of the bean seedling is curved over, forming a hook, so the tip is actually pulled up through the soil (Figure 32–2). Corn and other monocots have a special sheath of cells, the **coleoptile,** surrounding the shoot (Figure 32–3). The coleoptile pushes up through the soil, and the more delicate shoot then grows up through the coleoptile sheath.

PLANT GROWTH AFTER GERMINATION

The seedling that emerges from the seed continues to grow into an adult plant. Certain parts of the plant grow throughout the life of the plant. This **indeterminate growth** is characteristic of stems and roots. Theoretically, these parts of the plant could continue to elongate forever. Other parts of the plant have **determinate growth** and discontinue growth after reaching a certain size. The size of these structures varies from species to species depending on the plant's genetic programming and environmental conditions which may limit the growth of the plant. Leaves and flowers are examples of structures that exhibit determinate growth.

Although "growth" is a term that everyone knows, it is a complex phenomenon involving three different processes. First, an increase in the number of cells, i.e., cell division, is essential to growth. However, an increase in cell number without a corresponding increase in cell size would contribute little to overall growth of the plant. Cell elongation is the second

FIGURE 32–2 Seed germination and growth of the young bean plant. Note the hook in the stem of the young seedling, which protects the delicate stem tip as it moves up through the soil. Once the shoot has emerged from the soil, the hook straightens. The stem and roots elongate by growth at their tips. As the stored food in the cotyledons is used by the developing plant, the cotyledons shrivel and fall off the stem. At this point the young plant meets all of its nutritional needs by photosynthesis.

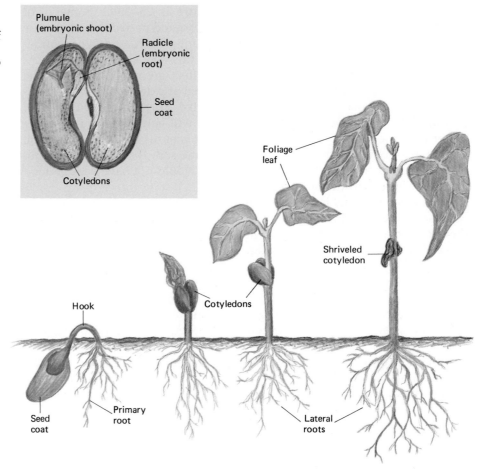

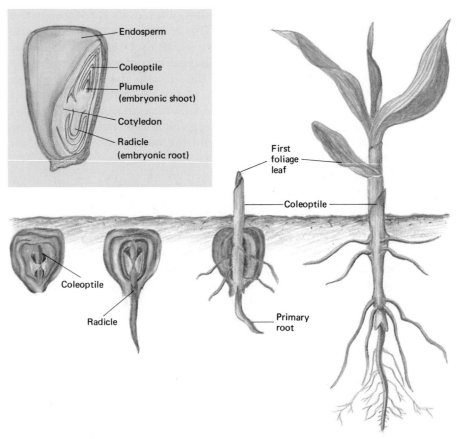

FIGURE 32–3 Seed germination and growth of the young corn plant. Note the coleoptile, a sheath of cells which emerges first from the soil. The delicate shoot tip grows up through the middle of the coleoptile.

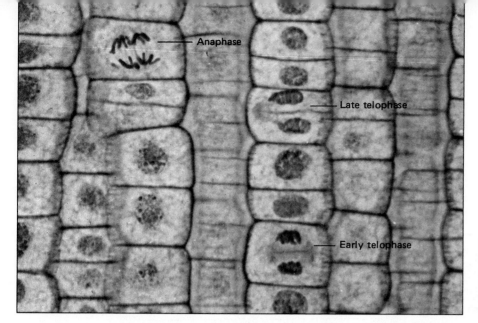

FIGURE 32–4 Meristematic cells retain the ability to divide, giving rise to all other plant tissues. This tissue from the onion (*Allium cepa*) root tip shows cells in several stages of mitosis.

process associated with growth. Finally, cells must **differentiate,** or specialize, to perform the various functions required in a complex, multicellular organism. In differentiation, cells that are genetically identical develop differences in structure and physiology, enabling them to perform different activities.

One difference between plants and animals is the location of growth. When a young animal is growing, all parts of its body grow, although not at the same rate. However, plant growth is localized into areas, called **meristems,** which are composed of cells that remain unspecialized and retain the ability to divide by mitosis (Figure 32–4).

Primary Growth and Apical Meristems

There are two kinds of growth in plants. One is **primary growth,** which is an increase in the length of the plant. The other is **secondary growth,** an increase in the girth, or width, of the plant. All plants have primary growth, but only woody plants have secondary growth. A plant with primary growth only is said to be **herbaceous.** Primary growth occurs because of the activity of **apical meristems,** meristematic areas at the tips of both stems and roots (Figure 32–5).

As discussed previously, the very end of the root is a protective layer of cells called the root cap (Figure 32–6). However, directly behind the root cap is the **area of cell division.** A microscopic examination of the area of cell division reveals the meristematic cells (Figure 32–4). These cells are very small and "boxy" in shape. They remain small because they are continually dividing.

Further back from the tip of the root, just behind the area of cell division, the cells are no longer dividing, but they are enlarging. This is the **area of cell elongation.** Here some differentiation also occurs, and immature tissues become evident. The immature tissues continue to develop and differentiate into the mature tissues of the adult plant.

Further back from the root tip, behind the area of cell elongation, the cells have completely differentiated and are fully mature. A number of specialized tissues within the root are evident in the **area of cell maturation.** For example, root hairs, extensions of epidermal cells, may be observed here.

The apical meristem of stems is essentially the same as the root apical meristem, although its appearance is quite different (Figures 32–7 and 32–8). The stem apical meristem has **leaf primordia** (embryonic leaves) and

FIGURE 32–5 Elongation in plants occurs near the tips. (*a*) A recently germinated pea (*Pisum sativum*) root has ink marks placed 1 mm apart. (*b,c*) Several hours later, the marks closest to the tip of the root have grown apart. The marks closest to the seed have remained approximately 1 mm apart, indicating little elongation further from the tip.

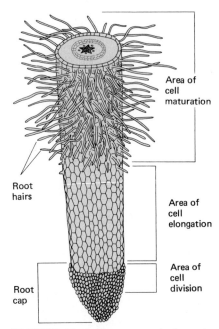

FIGURE 32–6 The root apical meristem. Just behind the root cap is the area of cell division, where mitosis occurs. Further from the tip is the area of cell elongation, where cells enlarge and begin to differentiate. The area of cell maturation has fully mature, differentiated cells. Note the root hairs in this area.

Area of cell maturation

Root hairs

Area of cell elongation

Area of cell division

Root cap

bud primordia (embryonic buds) emerging from it. A dome of tiny meristematic cells is located in the center at the very tip of the stem. Further from the tip of the stem, the immature cells elongate and start to differentiate. The immature tissues continue to develop into mature tissues that are located further back from the stem tip. The three areas (cell division, elongation, and maturation) are present in stem tips, although they are not as obvious as in the root.

Secondary Growth and Lateral Meristems

Plants with secondary growth have stems and roots that increase in girth. Woody trees and shrubs all have secondary growth in addition to primary growth. That is, these plants increase in length by primary growth and in girth by secondary growth. This increase in width is due to the activity of **lateral meristems,** which are located on the sides of the stem and root. Actually, two lateral meristems are responsible for secondary growth, the **vascular cambium** and the **cork cambium** (Figure 32–9).

The vascular cambium is a layer of meristematic cells that forms a ring, or cylinder, around the stem and root trunk. It is located exactly between the wood and bark of the plant, and its cells divide to form more wood and more bark (the inner bark, to be more precise). The cork cambium is composed of patches of meristematic cells located in the outer bark region. Cells of the cork cambium divide to form the tissues of the outer bark. A more comprehensive discussion of secondary growth is given in Chapters 34 and 35.

Plants with secondary growth are **perennial,** living year after year. Plants that are herbaceous and have only primary growth have no persistent above-ground parts. Many of them, e.g., corn and rice, are **annuals** and grow, reproduce, and die in one season. Some herbaceous plants, e.g., carrots and beets, are **biennials,** and take two years to complete growth and reproduction before dying. Herbaceous plants that are perennials, e.g., rhubarb and asparagus, live year after year, but die back each

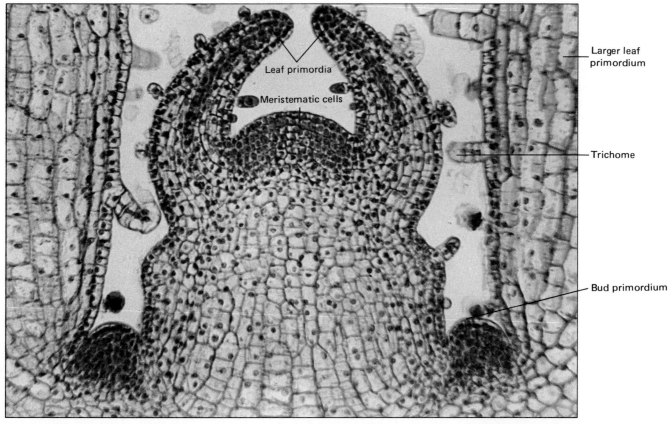

Larger leaf primordium

Trichome

Bud primordium

Leaf primordia

Meristematic cells

FIGURE 32–7 The dicot stem apical meristem. Note the leaf primordia and bud primordia.

winter. Their body parts in the soil do not die but remain dormant during the winter and send out new growth each spring.

Why do certain plants have secondary growth and make wood and bark? Secondary growth in plants confers the advantage of a longer life span than most plants without secondary growth have. Individual cells do not live forever. Typical plant cells live approximately three years, although there is a lot of individual variation. Plants with primary growth only do not have a way to replace older tissues in the stem and root. Although their tips are continually producing new cells, the older parts eventually die. Plants that have secondary growth produce new stem and root tissues to replace the older parts throughout the length of the plant, not just at the tips. Therefore, plants with secondary growth have an extended life span, sometimes for thousands of years!

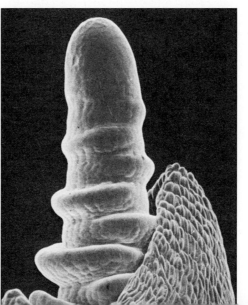

FIGURE 32–8 Scanning electron micrograph of the stem apical meristem of Arawa wheat (×200). The apex consists of cells that divide to form the shoot. The leaves are initiated just down from the apex. The leaves first appear as a ridge, but with successive cell divisions and expansions, the size of the ridge increases, and the ridge takes on the more familiar shape of a leaf.

FIGURE 32–9 In secondary growth, plants increase in girth as a result of the activity of two lateral meristems. The vascular cambium produces secondary vascular tissues, the wood and inner bark. The cork cambium produces the outer bark tissues that replace the epidermis in the secondary plant body.

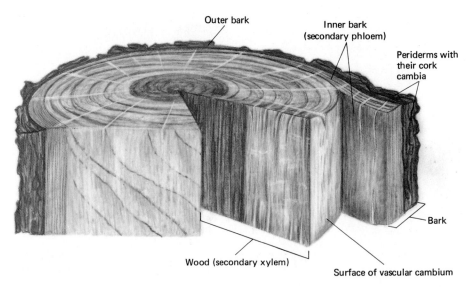

Outer bark

Inner bark (secondary phloem)

Periderms with their cork cambia

Bark

Wood (secondary xylem)

Surface of vascular cambium

CELLS AND TISSUES

As growth occurs, some cells become specialized and develop into **tissues** that make up the plant body. A tissue is a group of cells that is a structural and functional unit. The tissues of different plant organs are interconnected throughout the plant. All parts of the plant have three tissue systems. The dermal tissue system provides a covering for the plant body. The vascular tissue system is responsible for conduction of various substances in the plant, including water, dissolved minerals, and food. Finally, the ground tissue system makes up the rest of the plant body. It is composed of various cell types with a variety of functions (Table 32–1). Some plant tissues are composed of only one cell type (simple tissues), while other plant tissues have two or more cell types (complex tissues).

Parenchyma

If one were to think of a "typical" plant cell, it would be **parenchyma** (Figures 32–10, 32–11, and 32–12). Parenchyma, a simple plant tissue, is found throughout the plant. Its cells are relatively unspecialized, especially when compared with some of the tissues that will be considered in this chapter. These living cells perform a number of important metabolic func-

FIGURE 32–10 Parenchyma cells from the stamen hairs of the spiderwort (*Tradescantia virginiana*). These represent living, relatively unspecialized cells. The large vacuole contains pigmented material and occupies most of the cell. Note the nucleus and cytoplasmic strands.

Table 32–1 A SUMMARY OF PLANT CELL TYPES

Cell Type		Function	Location
Parenchyma		Secretion Storage Photosynthesis	Throughout the plant body
Collenchyma		Support	Just under stem epidermis Along leaf veins
Sclerenchyma		Support	Throughout the plant body Common in stems and certain leaves
Tracheids		Conduction of water and minerals Also provide support	Xylem
Vessel elements		Conduction of water and minerals Also provide support	Xylem
Sieve tube members		Conduction of food	Phloem
Companion cells		Aids sieve tube members in food conduction	Phloem

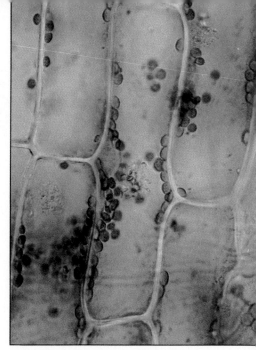

FIGURE 32–11 Some parenchyma cells contain chloroplasts. The primary function of these cells is photosynthesis.

tions for plants. Parenchyma cells often contain chloroplasts and are responsible for photosynthesis in the plant. In addition, parenchyma cells often store various materials, including food (visible as starch grains or oil droplets) and salts (visible as crystals). Secretions also may be produced by parenchyma cells. The various functions of parenchyma cells require living protoplasm.

Like all plant cells, each parenchyma cell is enclosed by a cell wall (see Chapter 4). This wall often contains layers and provides structural support for the plant. All plant cells have a **primary cell wall.** Many plant cells, as they mature, deposit additional cell wall material *inside* the primary wall, i.e., between the primary wall and the plasma membrane. This **secondary cell wall** serves to reinforce the primary wall. Parenchyma cells typically have primary walls only.

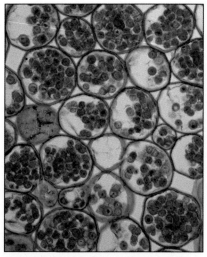

FIGURE 32–12 Parenchyma cells often function in storage. These parenchyma cells are from the cortex of a *Ranunculus* root. Note the starch grains filling the cells.

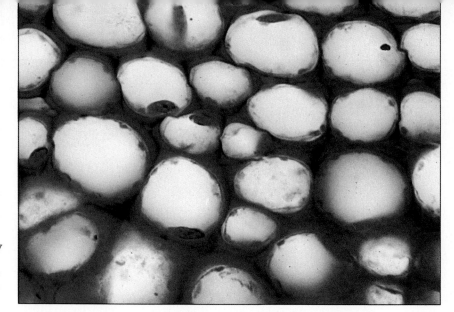

FIGURE 32–13 Collenchyma cells provide support. Note the unevenly thickened cell walls that are especially thick in the corners. Note the nuclei evident in several cells, indicating they are living at maturity.

Collenchyma

One of the simple plant tissues specialized for structural support in plants is **collenchyma** (Figure 32–13). Support is a crucial function in plants. Because plants lack a skeletal system, support of their body parts is provided by their individual cells, especially strengthening tissues such as collenchyma.

Collenchyma cells are living at maturity. Their primary walls are unevenly thickened, being especially thick in the corners. Collenchyma is not located throughout the plant. It is found in long strands, often just under the epidermis in stems and along leaf veins. In addition to providing support, collenchyma is an extremely flexible tissue.

Sclerenchyma

A second simple plant tissue specialized for structural support is **sclerenchyma.** Sclerenchyma cells have both primary and secondary cell walls. (The word *sclerenchyma* is derived from a Greek term meaning "hard.") Their walls are not only strong, or hard, but they become extremely thick, so thick that the cell's living material dies (Figure 32–14). Therefore, at

FIGURE 32–14 Sclerenchyma cells produce both primary and secondary walls. The cells walls are extremely thick and hard, providing structural support. These sclerenchyma cells are from a cherry pit.

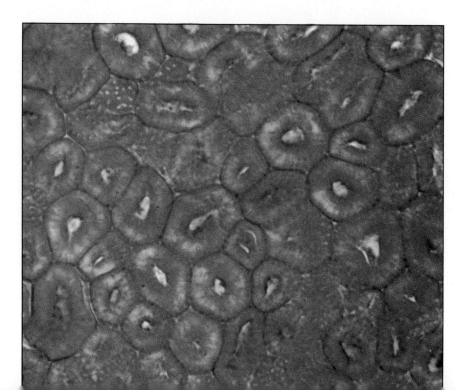

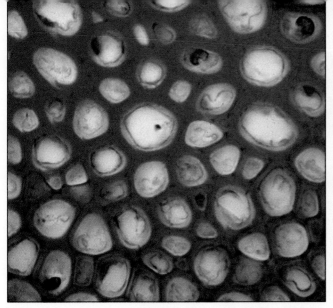

(a)

(b)

FIGURE 32–15 Fibers are a type of sclerenchyma, (a) in cross section and (b) in longitudinal section. These long, tapering cells have extremely thick cell walls and are dead at functional maturity. The fibers in (a) still have some cell contents visible.

functional maturity, when the sclerenchyma is providing support for the plant body, its cells are dead. Sclerenchyma may be located in several areas of the plant.

One type of sclerenchyma is the **fiber** (Figure 32–15). Fibers are long, tapered cells that are often located in patches or clumps. In cross section, one can appreciate the thickness of their walls and understand why the cells die.

Xylem

Xylem is a complex plant tissue that is composed of four different cell types in flowering plants. Its primary function is to conduct water and dissolved minerals from the roots to the stems and leaves. A secondary function of xylem is to provide structural support. Xylem is located throughout the plant. The xylem of the root is continuous with the stem xylem, and stem xylem is continuous with leaf xylem. Two of the four cell types found in xylem actually conduct; these are the **tracheids** and **vessel elements** (Figures 32–16 and 32–17). In addition to these cells, xylem also contains parenchyma and fiber cells.

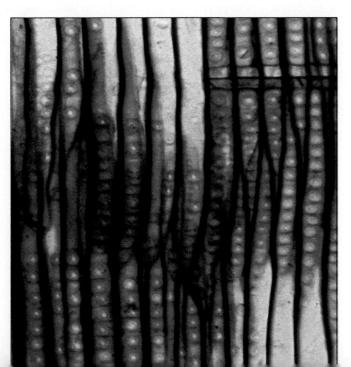

FIGURE 32–16 Tracheids in longitudinal section. These cells, which occur in clumps, transport water and dissolved minerals. Water passes readily from tracheid to tracheid through the pits, thin places in the wall.

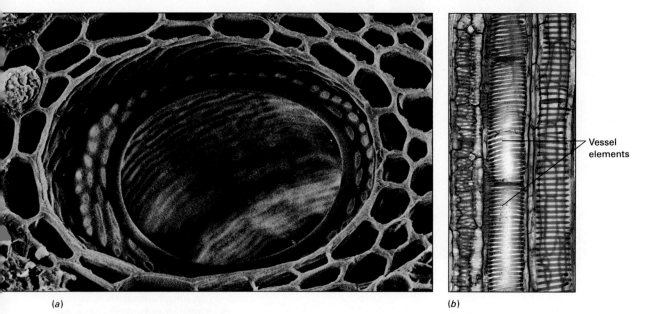

(a)

(b)

Vessel
elements

FIGURE 32–17 Vessel elements (*a*) in cross section, and (*b*) in longitudinal section. These cells are more efficient than tracheids in conducting water. Note how they are stacked end-on-end.

Tracheids and vessel elements are marvelously specialized for conduction. Both are dead at maturity. Tracheids are long, tapering cells that are hollow. Only their cell walls remain. Tracheids are located in patches, or clusters. Water is conducted up through tracheids, passing from one tracheid into another through thin places in the walls, called **pits.** Under the microscope, pits often appear to be holes in the wall, but in this case appearances are misleading.

Vessel elements are considered to be more advanced than tracheids. The cell diameter of vessel elements is wider than tracheids, so they are more efficient at water conduction. These cells are hollow, and their end walls have holes, or perforations. Vessel elements are stacked end-on-end. Water is conducted readily from one vessel element into the next. A stack of vessel elements is called a **vessel.** Like tracheids, vessel elements also have pits.

Phloem

Conduction of food throughout the plant is accomplished by the **phloem.** Like xylem, phloem is a complex tissue. In flowering plants, it is composed of four cell types, **sieve tube members, companion cells,** fibers, and parenchyma (Figure 32–18). The fibers are frequently quite extensive in phloem, providing additional support for the plant body.

FIGURE 32–18 Phloem tissue in cross section. Note the sieve plate, the end wall of the sieve tube member. The smaller cells are companion cells.

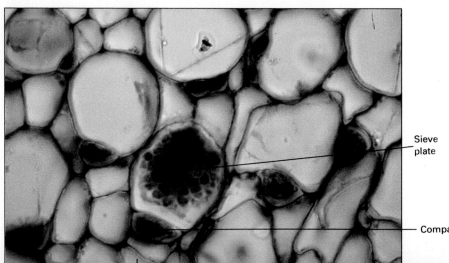

Sieve
plate

Companion cel

Food is conducted in solution through the sieve tube members, which are some of the most highly specialized cells in the living world. Sieve tube members are stacked end-on-end to form sieve tubes. Their end walls have a series of holes, called **sieve plates.** Cytoplasmic connections run from one sieve tube member into the next. They are living at maturity, but during maturation many cell organelles disintegrate, including the nucleus, vacuole, and ribosomes.

Few eukaryotic cells can function without nuclei. One example of such cells in mammals is the red blood cell. However, red blood cells can only function for a very limited period of time (approximately 120 days in humans), presumably because they lack nuclear control. Sieve tube members typically live for less than a year, although there are notable exceptions. Certain palms have sieve tube members that have remained alive approximately 100 years! It is not clear how these cells can function as long as they do.

Adjacent to each sieve tube member is a companion cell. The companion cell is a living cell, complete with nucleus and other organelles. There are numerous cytoplasmic connections between companion cells and sieve tube members. Although the companion cell does not conduct food, it plays an essential role in phloem transport (Chapter 34).

Epidermis

The dermal tissue system provides a protective covering over plant parts. In plants with primary growth, the dermal covering is a single layer of cells, the **epidermis** (Figure 32–19). The epidermis is a complex tissue that is composed of several types of cells. Most of the cells in the epidermis are parenchyma. Their cell walls are thicker to the outside of the plant for protection. In addition, epidermal parenchyma do not contain chloroplasts. Their transparent nature allows light energy to penetrate into interior tissues of the stem and leaf where photosynthesis does occur.

One of the most important requirements of the above-ground parts of the plant (i.e., stems and leaves) is protection against desiccation. The epidermal cells secrete a waxy layer, the **cuticle,** over their outer walls. This wax greatly restricts the loss of water from plant surfaces. The root epidermis does not produce a cuticle, since roots must be permeable to water in order to absorb it from the soil.

The cuticle is very efficient at preventing water loss through epidermal cells. It also provides a barrier against gases. Since photosynthetic tissues are inside the epidermis in both stems and leaves, there must be some mechanism for the gas exchange that is necessary for photosynthesis. This is accomplished by **stomates,** tiny pores formed in the epidermis by two rounded cells, called **guard cells** (Figure 32–19). A number of gases pass through the stomates by diffusion, including carbon dioxide, oxygen, and

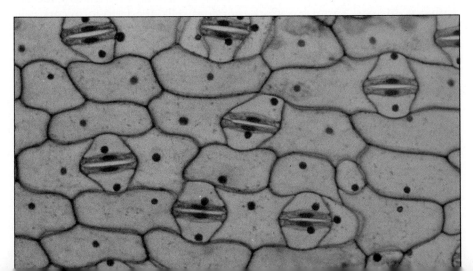

FIGURE 32–19 *Tradescantia* leaf epidermis (×350). Note the pink-colored guard cells that form openings for gas exchange.

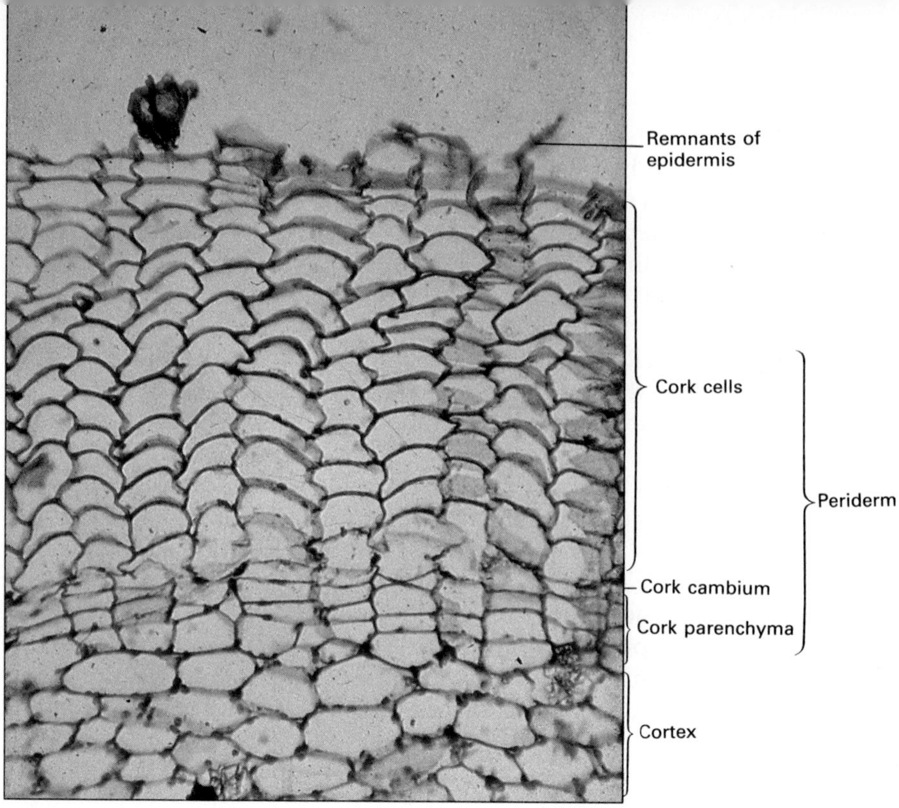

Remnants of epidermis

Cork cells

Periderm

Cork cambium

Cork parenchyma

Cortex

FIGURE 32–20 Periderm is the secondary plant body replacement for epidermis. Formed by the cork cambium, it makes up the outer bark of woody stems and roots. The cells of periderm are always arranged in vertical stacks.

water vapor. Stomates are generally open during the day when photosynthesis is occurring. They close during the night to conserve water in the plant (Chapter 33).

The epidermis also may contain special outgrowths, or hairs, termed **trichomes** (Figure 32–7). Trichomes have a number of different functions. Roots hairs are epidermal cell extensions that increase the surface area of the root that comes into contact with the soil for more effective water absorption (Chapter 35). Plants that can tolerate salty environments may have trichomes specialized for salt removal. Research indicates that the presence of trichomes on the aerial portions of desert plants may increase reflection of light off the plants, thereby cooling the internal tissues and decreasing water loss.

Periderm

Plants that increase in width by secondary growth must produce a dermal covering to replace the epidermis, which gets split apart as the plant expands. The **periderm,** a complex tissue, is the functional replacement of the epidermis (Chapter 34). It is several to many cells thick and forms the outer bark of the stem and root (Figure 32–20).

Periderm is continually being formed by a lateral meristem located within it, the cork cambium. Cork cambium cells divide to form cork cells to the outside and cork parenchyma to the inside. Cork cells are dead at maturity. Their walls are heavily coated with a waterproof substance to help reduce water loss. Cork parenchyma cells function primarily for storage.

THE PLANT BODY

All the cell and tissue types just discussed are organized into the plant body. Plants have a root system and a shoot system (Figure 32–21). The root system is generally the below-ground portion. The above-ground portion, the shoot system, is made up of a stem, which bears leaves, flowers,

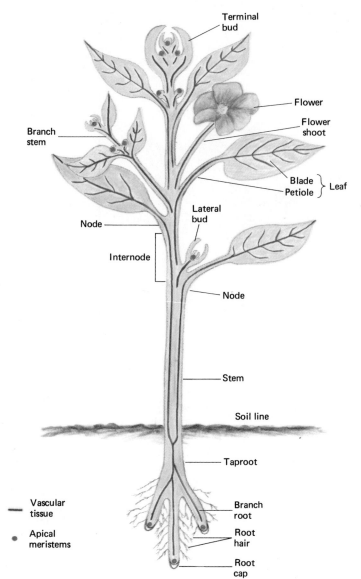

Terminal bud

Branch stem

Flower

Flower shoot

Blade
Petiole } Leaf

Node

Lateral bud

Internode

Node

Stem

Soil line

Taproot

Vascular tissue

Apical meristems

Branch root

Root hair

Root cap

FIGURE 32–21 The primary plant body of a typical herbaceous dicot plant. Note the continuity of the vascular tissues from one plant organ into the next. Primary growth occurs in meristems at the tips of plants. There is also meristematic tissue in the bud primordia. When a bud grows into a branch, this area becomes the apical meristem for the developing shoot.

and fruits. Roots, stems, leaves, flowers, and fruits are considered to be organs because they are composed of several different tissues. Some plant tissues are continuous throughout the length of the plant (e.g., vascular tissue), while others may be localized in certain organs (e.g., fibers).

There are several types of root systems (Figure 32–22). Plants with a **tap root** system have one primary root with smaller roots branching off it. The dandelion root is a good example of a tap root system. The **fibrous root** system has several main roots developing from the end of the stem. Smaller roots branch off these roots. Crabgrass and other grasses have fibrous root systems. Some plants have their roots modified for storage. These **storage roots** may be modified taproots (e.g., carrot) or fibrous roots (e.g., sweet potato).

One way that most stems can be distinguished from roots is that stems bear leaves. The area on the stem where leaves attach is the **node,** while the region of the stem between two successive nodes is the **internode.** Stems have **buds,** the **terminal bud** being the embryonic shoot at the tip of the stem. When the apical meristem is not actively growing, it is covered and protected by **bud scales,** which are modified leaves. Plants also have **lateral buds** located in the **axils** of leaves. The axil is the area on the stem directly above where the leaf attaches to the stem. When buds grow, they form stems that bear leaves or flowers.

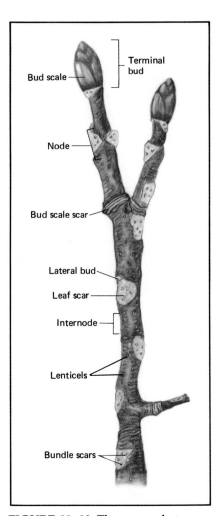

(a)

(b)

FIGURE 32–22 Root systems in plants. (*a*) The fibrous root system is characteristic of monocots. (*b*) The taproot system is common in many dicots. Both fibrous roots and taproots may be modified for food storage.

FIGURE 32–23 The external structure of a woody twig of horse chestnut, *Aesculus hippocastanum.* One can determine the age of a woody twig by counting the number of bud scale scars (don't count side branches). How old is this twig?

FIGURE 32–24 A simple leaf. Most leaves have a broad, expanded blade and a petiole.

A woody twig that has shed its leaves can be used to demonstrate stem structures (Figure 32–23). The terminal bud is covered by bud scales that protect the embryonic tissues during dormancy. When the plant resumes growth, the bud scales fall off, leaving a **bud scale scar.** Since plants form terminal buds once a year (at the end of the growing season), counting the number of bud scale scars indicates the age of the twig. The **leaf scar** shows where the leaf was attached to the plant. The vascular tissue that runs from the stem out into the leaf forms **bundle scars** within the leaf scar. Directly above the leaf scar is where the lateral bud may be found. Finally, the bark of a woody twig has tiny marks on it. These marks are the **lenticels,** sites of gas exchange in the woody stem.

Most leaves are composed of two parts. The broad, expanded portion is the **blade,** and the stalk that attaches the blade to the stem is the **petiole** (Figure 32–24). Leaves may be **simple** or **compound** (Figure 32–25). Sometimes a beginning student has difficulty telling whether a leaf is compound or really a small stem bearing several simple leaves. One easy way to tell if a plant has compound leaves is to look for lateral buds. The lateral buds form at the base of the leaf, whether it is simple or compound, never at the

Leaflet

Lateral
bud

(a)

(b)

FIGURE 32–25 Compound leaves.
(a) The leaf of Virginia creeper (*Parthenocissus quinquefolia*) is palmately compound. (b) The ash (*Fraxinus americana*) has pinnately compound leaves.

base of **leaflets.** Also, compound leaves lie in a single plane, whereas simple leaf arrangement on a stem is never in one plane.

Leaves can be arranged on a stem (Figure 32–26) several ways. Plants with **alternate** leaf arrangement have one leaf at each node. Two leaves at each node is **opposite** leaf arrangement. And three or more leaves per node is **whorled.**

Leaves are varied in their vein patterns in the blade (Figure 32–27). Monocots have **parallel** venation, while dicots have **netted** venation. Netted venation can be **palmate,** where several major veins radiate out from one point, or **pinnate,** where the major veins branch off along the entire length of the main vein.

DIFFERENTIATION IN PLANTS

Plants are extremely complex organisms composed of millions of cells organized into tissues, organs, and systems. Like animals, plants develop from a single cell, the zygote. However, differentiation of plant cells and tissues, which occurs during the entire life of the plant, is under different controls than animal development.

Of course, the ultimate control of plant differentiation is genetic. If the genes required for development of a particular characteristic are not present, that characteristic cannot develop. Location of the cell during early development also has a profound effect on what that cell will ultimately become. All this is true for animals as well, but in plants other nongenetic factors can have a tremendous influence on gene expression.

One nongenetic factor affecting plant growth and development, including differentiation, is the influence of other plant tissues and organs. Much of this control is mediated by **hormones,** substances produced in one part of the plant and transported to another, where they elicit some type of

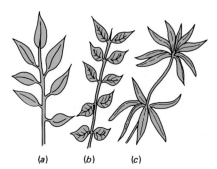

(a) (b) (c)

FIGURE 32–26 Leaf arrangement on the stem may be (a) alternate, (b) opposite, or (c) whorled, depending on the number of leaves at each node.

(a)   (b)   (c)

FIGURE 32–27 Venation patterns in leaves. (*a*) Parallel venation is characteristic of monocot leaves. (*b,c*) Netted venation is characteristic of dicot leaves. Diagram (*b*) is pinnately netted and (*c*) is palmately netted.

response (Chapter 37). Unlike animal hormones, each plant hormone affects a wide variety of growth responses in the plant throughout its lifetime. These hormones interact with one another in both stimulatory and antagonistic ways.

The external environment is a very important nongenetic factor affecting all aspects of plant growth and development. It plays a profound role for plants, but is much less important for animals. This should not be surprising when one considers that plants, being sessile, cannot respond to their environment by departing, as animals can. All aspects of plant growth are intimately connected with environmental cues. For example, in many plants, the initiation of flowering is controlled by differences in day length and darkness that occur with the changing seasons. The environment modifies and, in some cases, controls plant growth and gene expression.

Genetic control of both plant and animal development has already been considered in Chapter 16. Much of our current understanding of plant cell, tissue, and organ differentiation has come from experimental studies involving cell and tissue cultures. Plant biologists attempted unsuccessfully to grow isolated plant cells in culture beginning in the early 1900s. Initially, plant cells could be kept alive in a chemically defined, sterile medium, but they wouldn't divide. It was discovered that addition of certain natural materials like coconut milk induced the cells to divide in culture. Of course, coconut milk has a complex composition, so the division-inducing substance was not chemically identified for some time (Chapter 37). By the late 1950s, plant cells from a variety of sources could be cultured successfully, dividing to produce a mass of undifferentiated cells, or **callus.**

In 1958, F.C. Steward, a plant physiologist at Cornell University, succeeded in generating an entire carrot plant from a single cell in the carrot root. This demonstrated conclusively that each cell has the genetic blueprint for the entire organism. His work also showed that it is possible to grow an entire plant from a single cell, providing one can induce expression of the proper genes at the proper times (Chapter 16).

Since Steward's pioneering work, many plants have been successfully cultured using a variety of cell sources (Figure 32–28). Plants have been

(a)

(b)

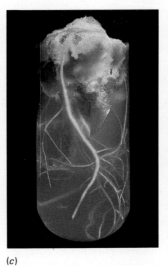
(c)

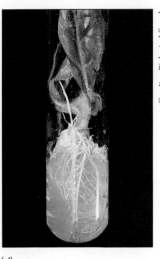
(d)

Carolina Biological Supply

regenerated from different tissues (e.g., pith), organ explants (e.g., root apical meristems, young embryos), and cells. The generation of plants from haploid pollen cells has proved to be particularly valuable because every gene in a haploid plant is expressed.

It is possible to remove the cell walls from individual plant cells. Such **protoplasts** can then be induced to fuse with other protoplasts (Figure 32–29). After fusion, a new plant can sometimes be regenerated from the protoplast. For example, tomato and potato protoplasts have been experimentally induced to fuse, and the hybrid plant (a topato?) subsequently developed. One disadvantage of protoplast fusion is that the results of the fusion cannot be predicted. In the case of the tomato–potato fusion, the regenerated plants had stems like potatoes (no fruits) and below-ground parts like tomatoes (no tubers).

Cell and tissue culture techniques can obviously be used to help elucidate many fundamental questions involving growth and development in plants. The practical potential of these techniques should also be obvious. Using tissue culture, it is possible to regenerate large numbers of genetically identical plants from cells of a single, genetically superior plant. It is possible to alter genetic composition while in cell culture and then have these changes expressed during regeneration. Research involving cell and tissue cultures is one of the exciting and fruitful areas of biological research today.

FIGURE 32–28 Propagating tobacco by tissue culture. (*a*) A fragment of undifferentiated tissue from the center of a tobacco stem is placed in a culture medium. A complete plant can be formed from the tissue fragment because each cell of the fragment contains all the genetic information for the entire organism. Different kinds of hormones in culture media will produce different growth responses. (*b*) Placed on a callus initiation medium, cells begin to proliferate, and undifferentiated tissue (callus) forms. (*c*) The callus produces roots on a root initiation medium. (*d*) By changing the relative levels of several plant hormones, shoots can be initiated. Plants grown by tissue culture techniques can be transferred to soil and grown normally.

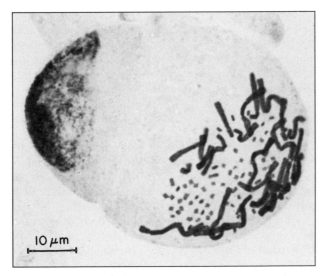
10 μm

FIGURE 32–29 The hybrid cell undergoing mitosis was derived by protoplast fusion between soybean (*Glycine max*) and vetch (*Vicia hajastana*). The large chromosomes are derived from the vetch and the small ones from the soybean.

Focus on SOME EXPERIMENTAL METHODS IN EMBRYOGENESIS

During the past 50 years, plant biologists have developed increasingly sophisticated methods to study embryo development in plants. Initially, biologists concentrated on descriptive aspects of embryogenesis, involving light microscopy studies of different stages in the developmental process. Even today, important contributions are being made in descriptive embryology.

Autoradiography is one technique that is used to help determine where specific chemical events are occurring in the embryo during its development. For example, to monitor RNA in developing cells, one can incubate them with radioactively labelled precursors of RNA. After incubation, the cells are stained, sectioned, and placed on slides. The slides are covered with a photographic substance and left in the dark. During this time, the radiation emitted from the radioactively labelled RNA exposes the photographic substance. After development of the slide, the exact location of radioactivity in the cell can be determined.

Analysis of the proteins produced during different stages of embryo development is done in several ways. Polyacrylamide gel electrophoresis is an effective way to separate individual proteins from one another. In electrophoresis, the proteins are placed into slots cut into a slab of gel. Electricity is applied across the gel for a period of time. Different proteins have different charges, determined by their amino acid composition. As a result of these charge differences, as well as differences in molecular weight, the proteins will migrate at different rates of speed across the gel. The proteins may be visualized by using a chemical that changes color in the presence of protein or by autoradiography (if the proteins have first been labelled with radioactive amino acids).

Tissue culture techniques have had a great impact on experimental approaches to embryology. It is much easier to study certain aspects of embryo development in a tissue culture system than in an intact ovule. For example, the development of haploid embryos from pollen cells provides data that intact diploid embryos within the seed never could. Manipulation of levels of plant hormones in tissue culture has shed some light on their roles in embryology, but many questions remain. For example, it is unclear how the same plant hormones can have different effects on the embryo at different stages.

Gene expression during development is of great interest to plant biologists. But how does one go about studying which genes affect different stages in embryo development? The corn, or maize, plant is probably the most useful organism to use for several reasons. First, because maize is one of our most important crop plants, it has been studied a great deal. Much is already known about different aspects of its growth and development, as well as its genetics and physiology. Also, there are a number of maize mutants that have abnormalities in their embryonic development. Because most mutants have abnormalities in their endosperm development as well as their embryos, it is relatively easy to identify them by looking for unusual characteristics in the endosperm. It is useful to determine how normal development is interrupted in mutants. For example, it is now known that different genes direct at least some of the development of root and shoot meristems in embryos. This was determined by studying an embryo that has abnormal development in the shoot meristem but normal root meristem development.

Embryo development in plants will provide scientists with a fruitful area of study in the future. With the development of sophisticated molecular techniques, plant biologists have the tools to determine more information than was even dreamed possible a short time ago.

■ SUMMARY

I. Plant embryos develop in the seed in an orderly, predictable fashion.

II. Seed germination is affected by a number of factors.
   A. External environmental factors that may affect seed germination include requirements for oxygen, water, temperature, and light.
   B. Internal factors affecting seed germination include immature embryos and the presence of inhibitors.

III. Plant growth is localized in regions, or meristems, and involves cell division, cell elongation, and differentiation.

IV. Plants have two kinds of growth: primary growth, an increase in length; and secondary growth, an increase in width of the plant.

V. Plant cells are organized into tissues.
   A. Parenchyma tissue is composed of relatively unspecialized, living cells.

B. Collenchyma tissue is composed of living cells that help support the plant.

C. Sclerenchyma tissue is composed of dead cells that help support the plant.

D. Xylem is a complex tissue that functions to conduct water and dissolved minerals.

E. Phloem is a complex tissue that functions to conduct food.

F. The epidermis covers the plant body and functions primarily for protection.

G. The periderm covers the plant body in plants with secondary growth. Its primary function is for protection.

VI. The plant body is organized into a root system and a shoot system.
   A. The shoot system is composed of the stem and leaves.

B. Although separate organs (roots, stems, leaves, flowers, and fruits) exist in the plant, many tissues are integrated throughout the plant body, providing continuity from organ to organ.

VII. Plant development is controlled not only by genetic factors but also by external factors.

A. Other plant tissues and organs exert a profound influence on plant development.

B. Many environmental factors determine gene expression and affect plant development.

■ POST-TEST

1. Stems and roots have _____ growth because they grow throughout the life of the plant.
2. Primary growth, the increase in the length of the plant, is found in localized areas of the plant, the _____ meristems.
3. The _____ _____ is a protective covering over the root tip.
4. Stem apical meristems differ from root apical meristems in bearing embryonic structures, _____ _____ and _____ _____.
5. The two lateral meristems responsible for secondary growth are the _____ _____ and the cork cambium.
6. Plants that complete their life cycle in 1 year are called _____, those that complete it in 2 years are _____, and those that live year after year are _____.
7. Storage, secretion, and photosynthesis are the functions of _____.
8. The two tissues specialized for support are _____ and sclerenchyma.
9. Conduction of water and minerals in the xylem occurs in _____ and vessel elements.
10. Conduction of food in the sieve tube members of the phloem is aided by _____ cells.
11. The outer covering of plants with primary growth is _____, while plants with secondary growth are covered by the _____.
12. Plants like grasses have a _____ root system.
13. Dormant terminal buds are covered by _____ _____.
14. Herbaceous stems have _____ for gas exchange, while woody stems have _____.
15. A single leaf composed of several leaflets is said to be _____.
16. Leaf arrangement on the stem may be alternate, _____, or whorled.
17. Monocots have _____ leaf venation and dicots have _____ leaf venation.

■ REVIEW QUESTIONS

1. Put the following stages of embryonic development in order and briefly describe each: torpedo stage, globular stage, proembryo, heart stage.
2. How is growth in plants different from growth in animals?
3. What factors influence the germination of seeds? Are these factors advantageous for plants?
4. What benefit does secondary growth confer on plants (as compared to plants with primary growth only)?
5. How is plant development different from animal development?
6. Why do you think plant development is so sensitive to nongenetic factors such as the presence of other cells and tissues and environmental influences?
7. How does tissue culture genetics compare with the usual approach to plant breeding? What are the advantages and disadvantages of each?

■ RECOMMENDED READINGS

Raghaven, V., *Embryogenesis in Angiosperms.* Cambridge, Cambridge University Press, 1986. An excellent review of all aspects of embryogenesis in angiosperms, especially good in reviewing experimental methods employed.

Raven, P.H., Evert, R.F., and Eichhorn, S.E. *Biology of Plants.* 4th ed. New York, Worth Publishers, Inc., 1986. An excellent general botany textbook, especially useful for current interpretations of plant structures.

Ray, P.M., Steeves, T.A., and Fultz, S.A. *Botany.* Philadel-phia, Saunders College Publishing, 1983. One of the most comprehensive general botany textbooks available, particularly good in experimental botany.

Sheridan, W.F., and Clark, J.K. Maize embryogeny: A promising experimental system. *Trends in Genetics.* Vol. 3, 1987, pp. 3–6. A paper extolling the virtues of corn for embryology studies.

Weier, T.E., Stocking, C.R., Barbour, M.G., and Rost, T.L. *Botany.* New York, John Wiley, 1982. This book is particularly strong in coverage of plant structures.

Lupine in the Sierra Nevada

33

Leaves and Photosynthesis

■ LEARNING OBJECTIVES

After you have read this chapter you should be able to:

1. Describe the major tissues of the leaf.
2. Relate leaf structure to its function of photosynthesis.
3. Discuss transpiration and its effects on the plant.
4. Relate the physiological changes that accompany stomatal opening and closing.
5. Compare leaf anatomy in dicot and monocot leaves.
6. Discuss leaf abscission: why it occurs and the physiological and anatomical changes that precede it.
7. List several modified leaves and give the function of each.

PHOTOSYNTHESIS AND LEAF ANATOMY

The primary function of leaves is to collect radiant energy and convert it to a form that can be used by the plant. This process, photosynthesis, has been examined in detail in Chapter 8. In photosynthesis, plants are able to take relatively simple molecules, carbon dioxide and water, and convert them into sugar. Oxygen is given off as a waste product. During this process, radiant energy is converted to chemical energy, the energy bonding the sugar molecules together. There are two metabolic uses for the sugar formed during photosynthesis. It may be broken down by respiration, releasing the chemical energy stored in its bonds for other cellular purposes. Also, the sugar molecules provide the cell with basic building materials. The cell modifies the sugar molecules, converting them into a number of other important compounds.

Major Tissues of the Leaf

In Chapter 32, the basic morphology of leaves was considered. Most leaves have a broad, flattened blade that is very efficient in collecting radiant energy. The leaf is a plant organ because it is composed of several different tissues. These tissues are organized in a way that optimizes their main function, photosynthesis (Figures 33–1 and 33–2).

Because the blade has top and bottom sides, the leaf has two epidermal layers, the **upper epidermis** and the **lower epidermis.** The cells making up this outer covering of the leaf are living parenchyma cells (Chapter 32). Lacking chloroplasts, they are relatively transparent. One interesting feature of leaf epidermal cells is that the cell wall on the outside of the leaf is

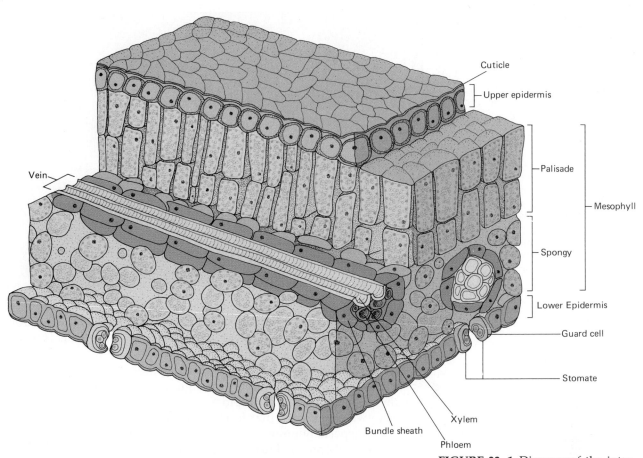

FIGURE 33–1 Diagram of the internal arrangement of tissues in a typical leaf blade. The photosynthetic tissue, the mesophyll, is often arranged into palisades and spongy layers. Veins branch throughout the mesophyll. The blade is covered by an upper and lower epidermis.

thicker than that on the inside. This may afford the plant additional protection from injury or water loss.

Because leaves have such a large surface area exposed to the atmosphere, water loss by evaporation from the surface is unavoidable. The very feature that makes leaves so efficient in collecting the sun's rays is its undoing in water relations. However, the epidermal cells secrete a noncellular waxy layer, the **cuticle,** which serves to reduce water loss. The cuticle varies in thickness in different plants. Generally, the upper epidermis has a thicker cuticle than the lower epidermis.

The epidermis of most leaves is covered with various trichomes (Figure 33–3). Indeed, some leaves with large numbers of trichomes feel quite

FIGURE 33–2 Scanning electron micrograph of a leaf cross section.

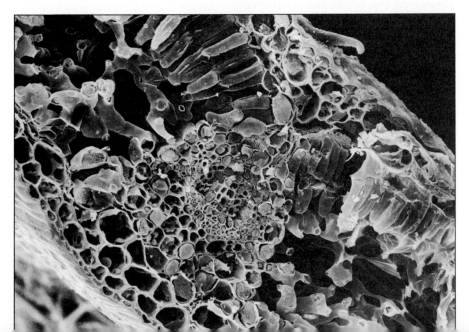

FIGURE 33–3 Scanning electron micrograph of a nettle leaf. The leaf epidermis is often covered with trichomes that may limit the transpiration of water, discourage herbivores, sting, or perform other functions.

fuzzy. Leaf trichomes frequently aid in the reduction of water loss from the leaf surface by maintaining a layer of moist air next to the leaf. This reduces evaporation from the leaf's surface.

The leaf epidermis is covered with tiny pores, or **stomates,** flanked by specialized cells in the epidermis, the **guard cells** (Figure 33–4). They are usually the only cells in the epidermis that have chloroplasts. Each pore and the two guard cells that form it are called a **stomatal apparatus.** Stomates are especially numerous on the lower epidermis and in many cases are located *only* on the lower surface. This adaptation reduces water loss because the stomates are shielded from direct sunlight.

The photosynthetic tissue of the leaf, the **mesophyll,** is sandwiched between the upper and lower epidermis. The word *mesophyll* comes from Greek, meaning "the middle of the leaf." Mesophyll cells are parenchyma cells with chloroplasts. They are very loosely arranged with lots of air spaces between them. Quite often, the mesophyll is divided into two specific areas. Toward the upper epidermis the cells are stacked into a **palisades** layer, while in the lower portion the cells are more loosely and more irregularly arranged in the **spongy** layer.

The veins of a leaf run through the mesophyll. Branching is extensive, so that no mesophyll cell is very far from a vein (Figure 33–5). Each vein contains two types of vascular tissue (Chapter 32). The **xylem** is located on the upper half of the vein, toward the upper epidermis. The **phloem** is always on the lower side of the vein.

FIGURE 33–4 The epidermis of a lily leaf. Note the puzzle-shaped epidermal cells, which are relatively transparent. Each stomatal pore is flanked by two bean-shaped guard cells.

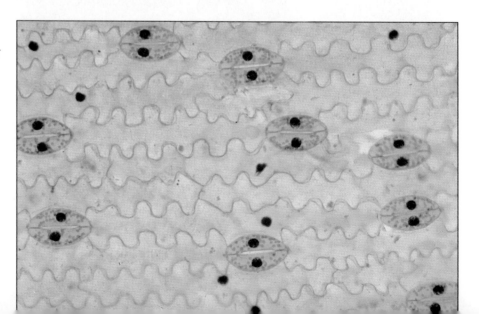

Veins are usually surrounded by one or more layers of nonvascular cells, the **bundle sheath.** Bundle sheaths are composed of parenchyma or sclerenchyma cells. Frequently the bundle sheath has columns, or **extensions,** that extend through the mesophyll to the upper and lower epidermis. Bundle sheath extensions may be composed of parenchyma, collenchyma, or sclerenchyma.

Structure–Function Relationship in Leaves

The transparent epidermis allows the light to penetrate to the center of the leaf where the photosynthetic tissue, the mesophyll, is located. Carbon dioxide diffuses into the interior of the leaf from the atmosphere through the stomates. Water is obtained from the soil and transported to the leaf in the xylem. The loose arrangement of the mesophyll tissue, with its moist cell surfaces and air spaces between cells, allows for rapid diffusion of both carbon dioxide and water into the chloroplasts. Also, the oxygen produced in photosynthesis diffuses rapidly out of the mesophyll cells and passes into the atmosphere through the stomates. Both carbon dioxide and oxygen are dissolved in the water film at the cell surface before going in or out. The veins supply water to the photosynthetic tissue (in the xylem) and carry the sugar produced in photosynthesis to other parts of the plant (in the phloem). The bundle sheath and bundle sheath extensions provide additional support to prevent the leaf, which is structurally weak owing to the large amount of air space in the mesophyll, from collapsing under its own mass.

Plants with C_4 photosynthesis have a characteristic arrangement of photosynthetic bundle sheath cells and mesophyll cells around each vein. These plants have enzymes for the C_4 pathway located in the mesophyll cells and enzymes for the C_3 pathway of photosynthesis in the bundle sheath cells (Chapter 8).

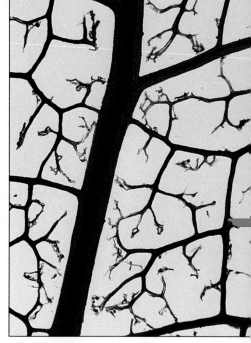

FIGURE 33–5 Portion of a dicot leaf with the mesophyll removed so the veins become more prominent. Note the extensive branching of the veins.

Transpiration

Despite leaf adaptations, such as the cuticle to control water loss, approximately 99% of the water a plant absorbs from the soil is lost by evaporation from the leaves and stem. The loss of water vapor from land plants is called **transpiration.**

The cuticle is effective in reducing water loss. It is estimated that only 1 to 3% of water lost from a plant passes directly through the cuticle. Most of the water loss occurs through the stomates. The numerous pores which are so effective in gas exchange for photosynthesis are also openings through which water can escape. Also, the loose arrangement of the mesophyll cells provides a large amount of internal air within the leaf in which water can evaporate.

A number of environmental factors influence the transpiration rate. For example, with higher temperatures more water is lost from plant surfaces. Wind also increases the transpiration rate. A high relative humidity decreases transpiration because the air is already saturated, or nearly so, with water vapor. Light increases the transpiration rate, in part because it causes the stomates to open.

Transpiration may seem like a wasteful process. However, there may be some benefits to the large amount of water plants lose by transpiration. First, transpiration, like sweating in animals, has a cooling effect on the plant. When water passes from a liquid state to a vapor, it absorbs a great deal of heat. As the water molecules leave the plant, this heat is carried with them. It is possible that the cooling effect of transpiration prevents overheating of the plant, particularly in direct sunlight. On a hot summer

FIGURE 33–6 Temporary wilting in pumpkin (*Cucurbita pepo*) leaves (*a*) in the late afternoon of a hot day and (*b*) the following morning. During the night, the plants recovered by absorbing water from the soil while transpiration was negligible.

(*a*)

(*b*)

day, the internal temperature of leaves is lower than that of the surrounding air.

A second possible benefit of transpiration is that it provides the plant with essential minerals. Remember that the water a plant transpires is absorbed from the soil. It is not pure water, but rather a very dilute solution of dissolved mineral salts. Many of these minerals are required for the plant's growth. It has been suggested that transpiration enables a plant to take in enough essential minerals and that plants could not satisfy their mineral requirement if they did not transpire.

There is no doubt, however, that under certain circumstances transpiration can be harmful to a plant. On hot summer days, plants frequently lose more water by transpiration than is replaced from the soil. Their cells experience a loss of **turgor,** and the plant wilts (Figure 33–6). If a plant is able to recover overnight, as a result of negligible transpiration (closed stomates) while water is still being absorbed from the soil, the plant is said to have experienced temporary wilting. Most plants recover from temporary wilting with no ill effects. In cases of prolonged drought, the soil may not contain sufficient moisture to permit recovery from wilting. A plant that cannot recover overnight is said to be permanently wilted and will die unless water is supplied immediately.

It appears that transpiration is a "mixed blessing" for plants. It has possible benefits and potential hazards. At any rate, transpiration is unavoidable in plants because they have stomates.

Guttation

There are times when some plants excrete water as a liquid (Figure 33–7). This process, known as **guttation,** occurs when transpiration is negligible and available soil moisture is high. Guttation frequently occurs at night

FIGURE 33–7 Guttation in wild strawberry (*Fragaria* sp.) plants.

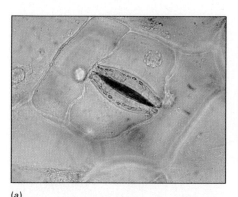

(a)

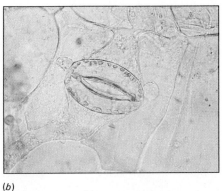

(b)

FIGURE 33–8 Stomatal opening in *Zebrina*. (*a*) Closed. (*b*) Open.

because the stomates are closed, but water continues to move into the roots by osmosis (Chapter 35). Many leaves have special openings through which the water is literally forced out. People sometimes wrongly attribute the early morning droplets of water produced on leaves by guttation to dew, which is water condensation from the air.

Stomatal Opening and Closing

The stomates are open during the day when photosynthesis is occurring and closed at night when photosynthesis is shut down (see Focus on Photosynthesis in Desert Plants for an interesting exception). This opening and closing of the pores is caused by changes in the shape of the guard cells. When water moves into guard cells from surrounding cells, they become turgid (swell) and bend outward at the center, creating the pore. When water leaves the guard cells, they become flaccid and collapse against one another, closing the pore (Figure 33–8).

The opening and closing of guard cells is triggered by an environmental factor, daylight or darkness. Other factors are also involved, including carbon dioxide concentration. A low concentration of CO_2 induces the stomates to open even in the dark. Another factor that affects stomatal opening and closing is severe water stress. During prolonged drought, plant stomates will remain closed even during the day. This mechanism is under hormonal control (Chapter 37).

The opening and closing of stomates also appears to be under the control of an internal clock that approximates the 24-hour cycle. Plants placed in darkness continue to open and close their stomates at more or less the same times even in the absence of environmental cues like light and darkness. This internal biological clock is known as a **circadian rhythm.** Other examples of circadian rhythms are given in Chapter 37.

The data from numerous experiments and observations suggest that the stomates of plants open and close by the **potassium ion (K$^+$) mechanism** (Figure 33–9). The appearance of light triggers an influx of potassium ions into the guard cells from surrounding cells of the epidermis. This movement of potassium, which occurs through ion channels in the membrane by active transport and requires ATP, has been experimentally measured and verified (Chapter 5). The increase of K$^+$ ions in the guard cells lowers the relative concentration of water in those cells. Therefore, water passes into the guard cells from surrounding epidermal cells by osmosis. This in turn changes the shape of the guard cells, and the pore opens.

The guard cells remain electrically neutral during the influx of positively charged K$^+$ ions because of the concurrent movement of other ions. H$^+$ ions produced in the guard cells from ionized organic acids are pumped into the surrounding epidermal cells. Also, Cl$^-$ ions are brought into the guard cells with the K$^+$ ions.

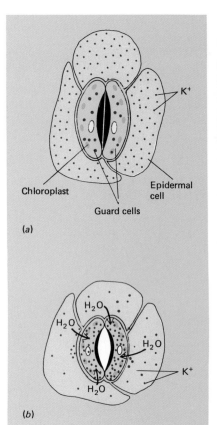

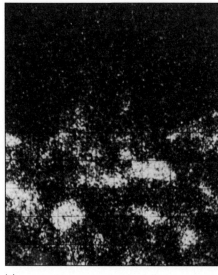

(c)

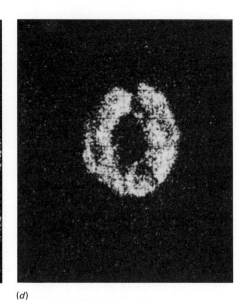

(d)

FIGURE 33–9 Movement of K$^+$ ions into and out of the guard cells affects stomatal opening and closing. (a) When the stomate is closed, K$^+$ ions are randomly distributed throughout the epidermis. (b) With the influx of K$^+$ ions into the guard cells from surrounding epidermal cells, water moves into the guard cells by osmosis. This changes the shape of the guard cells, and the pore appears. (c) Radiograph of a strip of epidermis with a closed stomate. The white spots indicate K$^+$ ions. Note how they are more or less randomly distributed. (d) Radiograph of the same strip of epidermis with the stomate open. Note how the K$^+$ ions are concentrated within the guard cells.

Focus on PHOTOSYNTHESIS IN DESERT PLANTS

Plants living in dry, or xeric, conditions have a number of special anatomical and physiological adaptations that enable them to survive. Many of them have succulent tissues for water storage. That is, they possess large amounts of parenchyma cells with large cell vacuoles to hold the water. They also have stomates which open at night and close during the day. This reduces water loss from transpiration by a considerable amount. But closed stomates during the day mean that gas exchange for photosynthesis cannot occur. C$_3$ plants must fix carbon during the day, using sunlight to produce ATP and NADPH + H$^+$, which then power the dark reactions to make sugar.

Many succulents have evolved a special photosynthetic pathway called Crassulacean Acid Metabolism, or CAM. The name comes from the stonecrop plant family, the Crassulaceae, which possess this pathway. Other unrelated plants, including the pineapple and most cacti, have it as well. CAM plants fix CO$_2$ during the night (when stomates are open) into malic acid, which is stored in the vacuole. During the day (when stomates are closed and gas exchange cannot occur between the plant and the atmosphere), the malic acid is decarboxylated to yield CO$_2$ again. Now the CO$_2$ is available to be fixed into sugar by combining with ribulose bisphosphate, by the usual C$_3$ photosynthetic pathway.

The pathway of carbon in CAM plants may sound familiar to you. It is very similar to the C$_4$ pathway discussed in Chapter 8. There are important differences, however. You will recall that C$_4$ plants initially fix carbon dioxide into four carbon organic acids in leaf mesophyll cells. The acids are later decarboxylated to produce CO$_2$, which is fixed by the C$_3$ pathway in the bundle sheath cells. In other words, the C$_4$ and C$_3$ pathways occur in different cells within the leaf of a C$_4$ plant.

In CAM plants, the initial fixation of carbon dioxide occurs at night. Decarboxylation of the 4-carbon malic acid and subsequent production of sugar from CO$_2$ by the C$_3$ pathway occur during the day. That is, the C$_4$ and C$_3$ pathways occur at different times within the same cell of a CAM plant.

The CAM pathway is found in over 25 different plant families. It is most common in angiosperms, especially dicots, but occurs in several monocots, *Welwitschia* (a gymnosperm), and some ferns. The CAM pathway is the result of convergent evolution because it apparently evolved independently in different groups of unrelated plants. It is a very successful adaptation to xeric conditions. Plants with CAM photosynthesis can survive in deserts where C$_3$ and C$_4$ plants cannot.

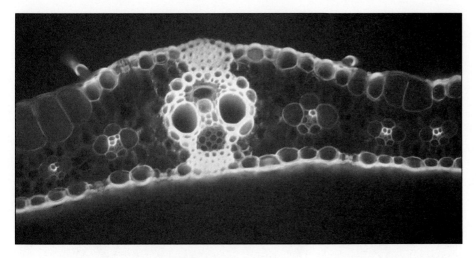

FIGURE 33–10 Photomicrograph of a cross section of a corn leaf stained by a special technique that, among other things, causes phloem to fluoresce pink (approximately ×60). Corn is a monocot. Note how parallel venation causes the veins to appear evenly spaced. Also, except for the midvein, the veins are more or less the same size.

Like other plant responses to light, the stomates have varying sensitivities to light of different colors (i.e., wavelengths). The action spectrum for stomatal opening shows greatest activity in the blue and, to a lesser extent, in the red regions. Also, dim blue light induces stomatal opening while dim, red light does not. Any physiological response to light must involve a **photoreceptor,** a pigment that absorbs the light prior to induction of the biological response. These and other data suggest that the photoreceptor for stomatal opening and closing is a yellow flavoprotein located on the tonoplast (vacuolar membrane) in the guard cells.

Light also affects stomatal opening indirectly by inducing photosynthesis. This reduces the internal concentration of CO_2 in the leaf, which triggers opening of the stomates.

In the late afternoon or early evening, the stomates close by a reversal of the process. The K^+ ions are pumped out of the guard cells into the surrounding epidermal cells. Water leaves the guard cells by osmosis, the cells collapse, and the pore closes.

Comparison of Dicot and Monocot Leaves

Plants as diverse as beans and maples are dicots. Their leaves are usually composed of a broad, flattened blade and a petiole. Venation is netted. Monocots such as lilies and corn often have long, narrow leaves that wrap around the stem in a sheath, rather than a petiole. Parallel venation is characteristic of monocot leaves (Figure 33–10).

Internal anatomy of dicots and monocots is different as well. Dicot leaves typically have two distinct regions in the mesophyll, the palisades and spongy layers. Mesophyll in many monocot leaves is not differentiated into palisades and spongy tissue. Because dicots have netted venation, a cross section of a dicot blade often shows veins in cross section as well as lengthwise views. In cross section, the parallel venation pattern of monocot leaves produces evenly spaced veins, all of uniform size except the midvein.

The upper epidermis of some monocot leaves has large, thin-walled cells, **bulliform cells,** located on either side of the midvein (Figure 33–11). These cells may be involved in the folding inward of the leaf during drought. When water is plentiful, the bulliform cells are turgid and the leaf is open. When the bulliform cells lose water, the leaf folds inward, reducing transpiration.

There are also differences between the guard cells in monocot and dicot leaves (Figure 33–12). Dicots have guard cells that are shaped like beans.

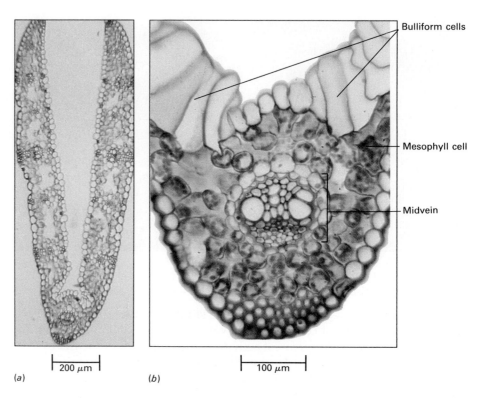

FIGURE 33–11 Cross section of bluegrass, a monocot. (*a*) The overall view of the leaf. Note the absence of distinct regions of palisades and spongy mesophyll. Also evident is the evenly spaced parallel venation characteristic of monocots. (*b*) Higher magnification of the midvein region showing the bulliform cells. These bulliform cells are partially expanded. When the bulliform cells are fully turgid, the leaf blade is expanded rather than folded up.

(*a*) 200 μm (*b*) 100 μm

Bulliform cells

Mesophyll cell

Midvein

The epidermal cells surrounding them are not noticeably different from other epidermal cells. Monocot leaves, on the other hand, have guard cells shaped like dumbbells. Each guard cell is associated with a special epidermal cell called a **subsidiary cell.**

FIGURE 33–12 Variation in guard cells. (*a*) Monocot guard cells are narrow in the center and thicker at each end. Each monocot guard cell is associated with a special cell in the epidermis, a subsidiary cell. (*b*) Dicot guard cells are bean-shaped.

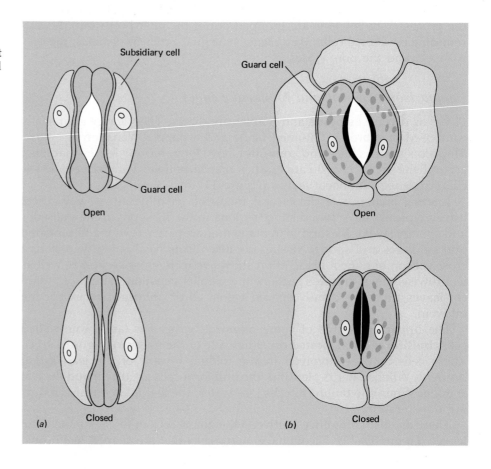

Subsidiary cell

Guard cell

Guard cell

Open

Open

Closed

Closed

(*a*)

(*b*)

LEAF ABSCISSION

In temperate climates, the leaves turn color and **abscise,** or fall off, when winter approaches. Most woody plants with broad leaves must shed their leaves in order to survive the low temperatures of winter. You have seen that leaves lose a tremendous amount of water by transpiration. During the winter, water relations become critical for plants. As the ground chills, absorption of water by the roots is inhibited. If the ground freezes, there can be no absorption. If a plant were to maintain its broad leaves during the winter, it would continue to lose water by transpiration but would be unable to replace water by absorption from the soil.

In the lower temperatures of winter, the plant's metabolism, including its photosynthetic machinery, slows down a great deal. Therefore, plants have little need of leaves in winter. Abscission is a complex process that involves many physiological changes in the plant. All of it is orchestrated by changing levels of plant hormones (Chapter 37).

As autumn approaches, the plant reabsorbs many of the essential minerals located in the leaves. Nitrogen, phosphorus, and possibly potassium move into the woody tissues from the leaves. The level of sugar in the leaves rises. Chlorophyll is broken down, and some of the accessory pigments in the chloroplast, carotenes and xanthophylls, become evident. These pigments were always present in the leaf but were masked by the chlorophyll. In addition, other water-soluble pigments may be synthesized in the vacuole. The various combinations of these pigments are responsible for the brilliant colors found in autumn landscapes in temperate climates.

The area where the leaf petiole detaches from the stem is structurally different from surrounding tissues (Figure 33–13). This difference may be seen in early summer in some plants. This area, called the **abscission zone,** is composed primarily of thin-walled parenchyma cells and is anatomically weak. There are few fibers. On the stem side of the abscission zone, a protective layer of cork cells develops. These cells have a waxy material called **suberin,** which is impermeable to water, impregnated in their walls.

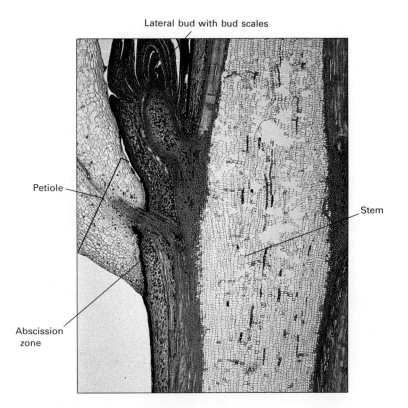

Lateral bud with bud scales

Petiole

Stem

Abscission zone

FIGURE 33–13 A longitudinal section through a maple branch, showing the base of the petiole. Note the abscission zone where the leaf will abscise from the stem. A lateral bud with its protective bud scales is evident above the petiole.

FIGURE 33–14 The stem of a cactus functions both for photosynthesis and water storage. The leaves of cacti are modified into spines for protection.

As fall approaches, enzymes dissolve the middle lamella, which is the "cement" holding the cells together, in the abscission zone. By this time, there is nothing holding the leaf to the stem but a few xylem cells. A sudden breeze is enough to make the final break, and the leaf detaches. The protective layer remains, sealing off the area and forming the leaf scar (Chapter 32).

LEAF MODIFICATIONS

Although photosynthesis is the main function of leaves, there are a number of special modifications that certain leaves have for functions other than photosynthesis.

Some plants have leaves specialized for protection. **Spines,** which are hard and pointed, may be found on plants like cacti (Figure 33–14). (In the cactus, the main organ of photosynthesis is the stem rather than the leaf.) While leaves may be modified as spines, stems are sometimes modified as **thorns,** which serve the same function.

Many vines have **tendrils** (Figure 33–15). Tendrils, which are usually specialized leaves, are for grasping and holding onto other structures. Because vines are climbing stems that do not support their own weight, tendrils are needed to keep the vine attached to the structure on which it's growing.

The winter buds of a dormant woody plant are covered by protective **bud scales** (Chapter 32). Bud scales are modified leaves. They protect the delicate meristematic tissue of the shoot from injury and desiccation.

Leaves may be modified for storage of water or food. For example, the **bulb** is a short stem to which large, fleshy leaves are attached. Onions and tulips form bulbs, which grow underground. Many desert plants have fleshy, succulent leaves for water storage. These leaves are usually green and function for photosynthesis as well.

Some of the most bizarre examples of modified leaves are those of insectivorous plants. These plants have their leaves modified to trap animal prey, usually insects. Most insectivorous plants grow in poor soil that is deficient in certain essential minerals. These plants obtain some nutrients by digesting insects and other small animals.

Some insectivorous plants have passive traps. For example, the leaves of the pitcher plant are shaped so that rainwater collects within, forming a reservoir (Figure 33–16). An insect that is attracted to the pitcher by its scent or nectar may lean over the edge and fall in. It is prevented from crawling out by a row of stiff spines that points downward around the lip of the pitcher. The insect eventually drowns. Enzymes produced by the plant digest part of the insect's body. This material is absorbed by the

FIGURE 33–15 Tendrils may be modified leaves or stems. These pea tendrils are modified leaves. Tendrils grasp onto objects and aid the plant in climbing.

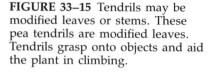

(a)

(b)

Carolina Biological Supply Company

FIGURE 33–16 The pitcher plant has leaves modified to form a pitcher that collects water, drowning its prey. (a) Growth habit of a common pitcher plant, *Sarracenia purpurea*. (b) A cut-away view of a pitcher reveals accumulated insect bodies and debris.

plant. In the tropics, pitcher plants may be large enough to hold 1 liter or more of water.

The Venus flytrap is an example of an insectivorous plant with active traps (Figure 33–17). The leaves resemble tiny bear traps. Each side of the leaf has three small hairs located on it. If an insect alights and brushes against two of the hairs, the trap springs shut with amazing rapidity (Chapter 37). After the insect has died and been digested, the trap reopens, and the indigestible remains fall off.

Plants living in unusual environments often have evolved specialized adaptations for that environment. The flower pot plant is a small tropical plant which is an **epiphyte.** It lives high in the forest canopy attached to a tree, but it does not parasitize the tree. This plant has leaves specialized to form a "flower pot" (Figure 33–18). Rainwater and various minerals that are leached out of the leaves above collect in the pot. A root runs into the flower pot leaves and absorbs water and dissolved minerals high above the ground floor.

FIGURE 33–17 The modified leaves of a Venus flytrap snap shut on its prey. Venus flytraps are found in bogs in North and South Carolina. (a) Growth habit of Venus flytrap, *Dionaea muscipula*, with opened traps. (b) A closed trap and insect prey. After enzymes produced by the plant digest as much of the insect as is digestible, the trap will reopen.

(a)

(b)

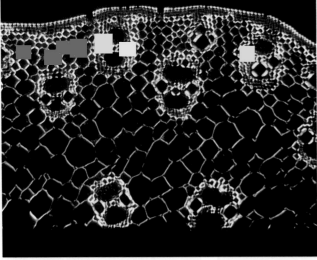

A corn stem viewed under polarized light

34

Stems and Plant Transport

OUTLINE

 I. Functions of stems
 II. Structure of stems with primary growth
    A. Dicot stems
    B. Monocot stems
    C. Structure–function relationships in primary stems
 III. Structure of stems with secondary growth
    A. Development of secondary growth
       1. Vascular cambium
       2. Cork cambium
    B. Trees and terminology
 IV. Transport in plants
    A. Movement of water and minerals in xylem
       1. Water potential
       2. Root pressure
       3. Tension–cohesion
    B. Movement of sugars in phloem
       1. Pathway of movement
       2. Mechanism of movement—pressure flow
Focus on tree ring analysis

LEARNING OBJECTIVES

After you have read this chapter you should be able to:

1. Compare the structure of primary dicot stems and monocot stems.
2. Outline how a stem develops from primary growth to secondary growth.
3. Discuss the structure of woody stems and relate the common terms (like wood and bark) that are associated with woody plants to their anatomy.
4. List several functions of stems and discuss how the structure of stems relates to their function.
5. Describe the pathway of water movement in plants.
6. Discuss root pressure and tension–cohesion as mechanisms to explain the rise of water in xylem.
7. Describe the pathway of food transport in plants.
8. Discuss the pressure flow theory of sugar transport in phloem.

FUNCTIONS OF STEMS

The basic plant body has three parts. Roots serve to anchor the plant and aid in absorption of materials from the soil. Leaves are for photosynthesis, converting radiant energy into the chemical energy of sugar. Stems are usually above-ground structures that link the roots to the leaves.

Stems have three main functions. They support the leaves and reproductive structures. Because leaves must absorb sunlight for photosynthesis, most stems grow upright with the leaves arranged on the stem for maximum light absorption of each leaf. A second function of stems is conduction. Stems conduct water and dissolved minerals that were absorbed from the soil by the roots to the leaves and other plant structures. Stems also conduct the food produced in the leaves by photosynthesis to the roots and other parts of the plant. It should be emphasized, however, that stems are not the only plant organ to conduct materials. The vascular system is continuous throughout all parts of the plant, and conduction occurs in roots, stems, leaves, and reproductive structures. The third main function of many stems is to produce new living tissue. Stems continue to grow throughout the life of the plant, by both primary and secondary growth

FIGURE 34–1 Two ancient baobab trees from Madagascar. The baobab stores large volumes of water and starch in its stem tissues.

(Chapter 32).In addition to the main functions of support, conduction, and production of new stem tissue, there are a number of stems that are modified for asexual reproduction (Chapter 36). Finally, some stems are specialized for photosynthesis and storage (Figure 34–1).

STRUCTURE OF STEMS WITH PRIMARY GROWTH

In Chapter 32 you learned that plants may have two different types of growth. Primary growth is the increase in the length of the plant and is localized at the tips of plants, the apical meristems. Secondary growth is the increase in the girth of the plant and is due to the activity of lateral meristems located on the sides of the plant. All plants have primary growth, while only some plants have secondary growth.

Plant stems with only primary growth are herbaceous. Although primary stems all have the same basic tissues, the arrangement of tissues in the stem varies considerably. We consider primary stem structure in the two groups of angiosperms, the dicots and the monocots.

Dicot Stems

The outer covering of primary stems is the **epidermis**. Inside the epidermis is a layer that is several cells thick, the **cortex**. The cortex is a complex tissue that may contain parenchyma, collenchyma, and sclerenchyma (Chapter 32).

The vascular tissue of dicots is located in patches that are arranged in a circle (Figure 34–2). Each patch, or vascular bundle, contains both **xylem** and **phloem**. The xylem is usually located on the inner side of the vascular bundle, while the phloem is usually located to the outside. Sandwiched between the xylem and the phloem is a single layer of cells, the **vascular cambium**. Frequently, there is a patch of fibers directly outside the phloem, the **phloem fiber cap**. Although the vascular bundles are arranged in a circle in cross section, it is important to remember that they run as long strands throughout the length of the stem, continuous with the vascular tissues of the root and leaves.

The inside of the dicot stem is **pith**, a tissue composed of large, thin-walled parenchyma cells. Due to the arrangement of the vascular tissues in bundles, there is no distinct separation of cortex and pith between the bundles. These areas of the stem are usually referred to as **pith rays**.

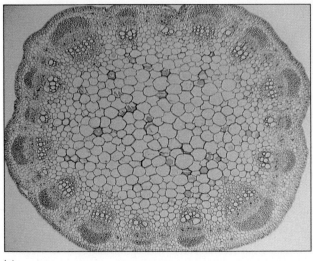

(a)

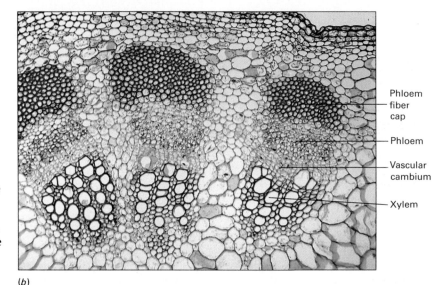

Phloem
fiber
cap

Phloem

Vascular
cambium

Xylem

FIGURE 34-2 Primary growth in a dicot stem. (*a*) Cross section of a *Helianthus* (sunflower) stem showing the arrangement of tissues. The vascular bundles are arranged in a circle. (*b*) Close-up of the vascular bundles in *Helianthus*. The xylem is to the inside and the phloem to the outside of the bundle. Each vascular bundle is "capped" by a batch of fibers for additional support.

(b)

Monocot Stems

Monocot stems are also covered with an epidermis. As in dicot stems, the vascular tissue runs in strands throughout the length of the stem. However, these vascular bundles are not arranged in a circle as in dicots. Rather, they are scattered throughout the stem (Figure 34–3). Therefore, the monocot stem does not have distinct areas of cortex and pith. Instead, the parenchyma tissue in which the vascular tissues are embedded is called **ground tissue** or, sometimes, **ground parenchyma**.

Each vascular bundle in monocot stems contains xylem to the inside and phloem to the outside. There is no vascular cambium in monocot stems, however. Monocots have primary tissues only and do not produce wood and bark. Although some monocots attain considerable size, e.g. palm trees, most never produce secondary tissues (Figure 34–4).

Structure–Function Relationships in Primary Stems

The epidermis in primary stems serves for protection. It is covered by the cuticle, a waxy layer that also covers the leaf epidermis (Chapter 33). The stem cuticle serves to reduce water loss.

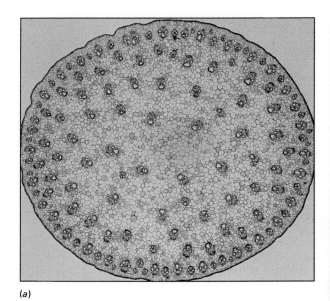

(a)

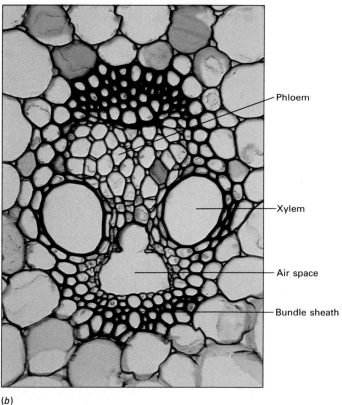

Phloem

Xylem

Air space

Bundle sheath

(b)

FIGURE 34–3 Arrangement of stem tissues in *Zea mays* (corn), a monocot. (*a*) Cross section of stem showing the scattered vascular bundles. (*b*) Close-up of one of the bundles. The air space is where the first xylem elements were formed. The entire bundle is enclosed in a sheath of sclerenchyma for additional support.

As might be expected from the various types of cells it contains, the cortex in dicot stems has several functions. If the stem is green, photosynthesis occurs in the cortex in parenchyma cells. Parenchyma also serves as storage tissue. Starch grains and crystals are frequently evident in cortex parenchyma. The collenchyma and sclerenchyma in the cortex provide strength and support for the stem.

The vascular tissues function not only for conduction but also for support. The xylem transports water and dissolved minerals, while the phloem transports food. Fibers may be found in both xylem and phloem, although they are usually more extensive in the phloem. These fibers add considerable strength to the stem body. The vascular cambium is responsible for secondary growth, which is considered later.

Finally, the parenchyma cells that make up the pith in the center of dicot stems function primarily for storage. Although monocots lack a distinct cortex and pith, their ground tissue performs the same functions.

FIGURE 34–4 Palms are monocots which attain considerable size but do not have secondary growth. Their increase in girth is due to the division of primary cells rather than division of a vascular cambium.

STRUCTURE OF STEMS WITH SECONDARY GROWTH

Secondary growth occurs in plants as a result of the activity of two lateral meristems (Chapter 32). Cells in the **vascular cambium** divide and produce secondary xylem and secondary phloem, which become the functional replacements of primary xylem and phloem. The second lateral meristem, the **cork cambium**, divides to produce cork cells and cork parenchyma. The cork cambium and the tissues it produces are collectively referred to as **periderm**, which functions as a replacement for the epidermis.

Development of Secondary Growth

In angiosperms, only certain dicots have secondary growth. Table 34–1 summarizes the anatomical relationships of various tissues in a dicot stem.

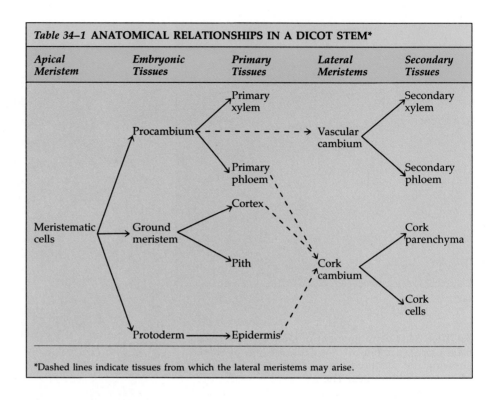

Table 34-1 **ANATOMICAL RELATIONSHIPS IN A DICOT STEM***

*Dashed lines indicate tissues from which the lateral meristems may arise.

Woody plants such as apple, hickory, and maple are examples of dicots with secondary growth. The gymnosperms also have secondary growth. Examples of gymnosperms are pine, juniper, and spruce.

Plants with secondary growth develop from plants that have primary growth. That is, a woody plant continues to increase in length at the tips of the branches and roots. However, the older parts of the plant further back from the tips develop secondary tissues.

Vascular Cambium

In the primary dicot stem, we have seen that the vascular cambium is a thin layer of cells sandwiched between the xylem and phloem in the vascular bundles (Figure 34–5). It does not initially form a solid ring of cells because the vascular bundles are separated by pith rays. When production of secondary tissues is induced in the primary stem, the vascular cambium forms an uninterrupted ring of cells. It is able to do so because certain cells in each pith ray dedifferentiate and become meristematic again. These cells connect to the vascular cambium cells in each vascular bundle, forming a ring of vascular cambium.

When a cell in the vascular cambium divides, one of the daughter cells remains meristematic, i.e., part of the vascular cambium. The other cell may divide again several times, but it eventually develops into mature secondary tissue. Cells in the vascular cambium divide tangentially to form tissues in two directions. That is, the cells formed from the dividing vascular cambium are sometimes located *inside* the ring of vascular cambium and sometimes *outside* it (Figure 34–6).

The tissue formed to the inside of the vascular cambium is secondary xylem (Figure 34–7). It is also known as wood. The tissue formed to the outside of the vascular cambium is secondary phloem. Secondary phloem makes up the inner bark of a woody plant. The vascular cambium is sandwiched in between the wood and the inner bark. Besides dividing tangentially to produce secondary xylem and phloem, the vascular cambium divides radially to form more vascular cambium as the plant increases in girth.

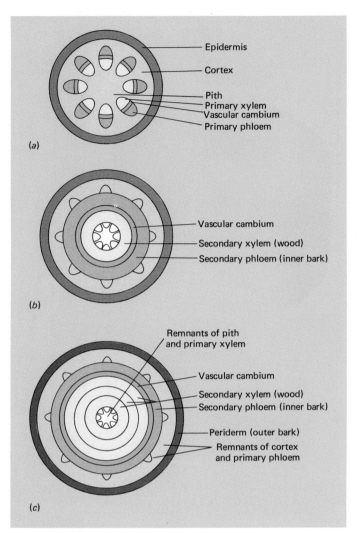

FIGURE 34–5 Development of secondary growth in the dicot stem is shown in cross section. (*a*) A primary dicot stem. The vascular cambium is sandwiched between the primary xylem and primary phloem in each vascular bundle. (*b*) The vascular cambium forms a solid ring of cells and begins to divide, forming secondary xylem on the inside and secondary phloem on the outside. Note that the primary xylem and primary phloem are split apart in this process. (*c*) A 3-year-old woody stem. The vascular cambium produces more secondary xylem than phloem. The original primary tissues are no longer functional. The epidermis has been replaced by tissues formed by a second lateral meristem, the cork cambium.

Plants with secondary growth can attain tremendous girth. While the secondary xylem and phloem accomplish vertical transport throughout the plant, there must also be a way to accomplish *horizontal* transport. Lateral movement of materials is accomplished by **rays**, which are composed of chains of parenchyma cells. The rays are continuous from the xylem to the phloem. Various materials, including water, minerals, and food, are transported laterally in the rays.

What happens to the original primary tissues of the stem once secondary growth commences? As the stem begins to increase in thickness, owing to the activity of the vascular cambium, the orientation of the original primary tissues to each other changes. The secondary tissues, secondary xylem and phloem, are laid down between the primary xylem and primary phloem within each vascular bundle. Therefore, as secondary growth continues, the primary xylem and phloem in each vascular bundle get split apart (Figure 34–8).

The primary tissues remaining inside the ring of secondary tissue, i.e., the pith and primary xylem, are under pressure from the changes in growth and soon get crushed beyond recognition. The primary phloem, cortex, and epidermis are also subjected to the pressures produced by secondary growth and get split apart and sloughed off.

Secondary tissues replace the primary tissues in function. The secondary xylem conducts water and dissolved minerals in the woody plant. It contains the same types of cells found in the primary xylem (Chapter 32). The arrangement of the different cell types in secondary xylem creates

FIGURE 34–6 Radial section view of a dividing vascular cambium cell. Note that the vascular cambium divides in two directions, forming secondary xylem to the inside and secondary phloem to the outside. These cells differentiate to form the mature cell types associated with xylem and phloem. As secondary xylem accumulates, the vascular cambium moves outward, and the woody stem increases in width.

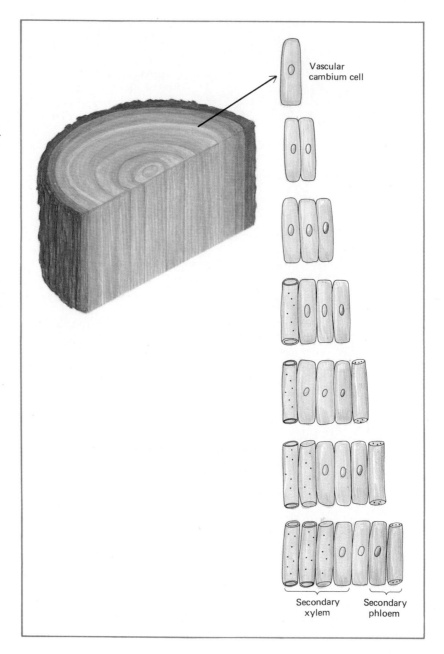

Vascular cambium cell

Secondary xylem

Secondary phloem

distinctive wood characteristics for each species. Secondary phloem conducts food. The same types of cells as in primary phloem are also found in secondary phloem, although there are usually more fibers in secondary phloem (Chapter 32).

Cork Cambium

The cork cambium arises from parenchyma cells in the cortex, epidermis, or phloem. These cells dedifferentiate and become meristematic. Unlike the vascular cambium, the cork cambium does *not* form a continuous ring of dividing cells. Instead, it forms patches of meristematic cells that cut into successively deeper layers of tissue.

Like the vascular cambium, the cork cambium divides to form new tissue both to its inside and outside (Figure 32–20). It forms cork cells to the outside (Chapter 32). Because these cells are dead at maturity and have heavily suberized walls, any primary tissues located external to the cork

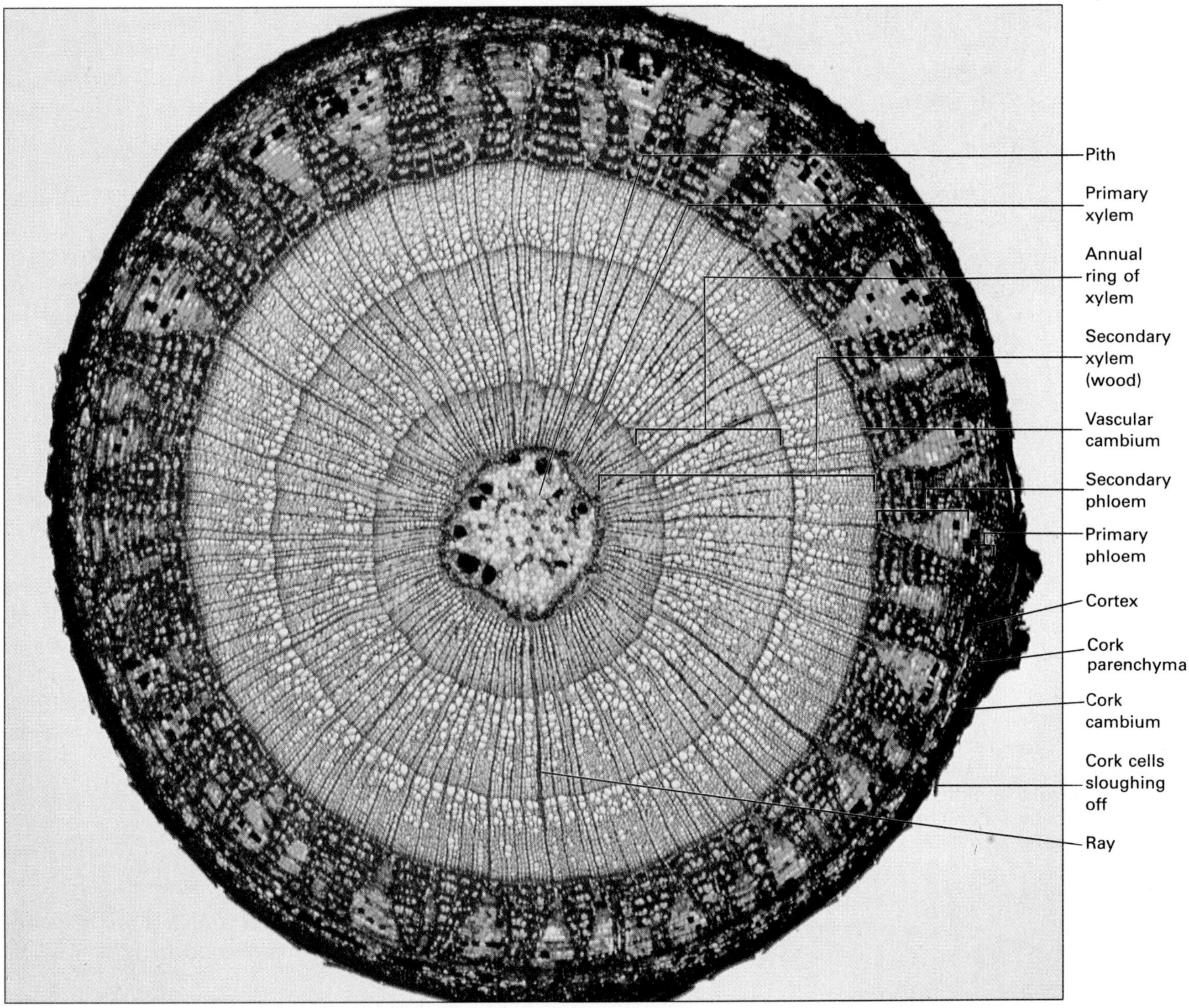
Carolina Biological Supply Company
Pith
Primary xylem
Annual ring of xylem
Secondary xylem (wood)
Vascular cambium
Secondary phloem
Primary phloem
Cortex
Cork parenchyma
Cork cambium
Cork cells sloughing off
Ray

FIGURE 34–7 Cross section of a 3-year-old *Tilia* (basswood) stem.

die. The cork cambium forms cork parenchyma to the inside. The cork parenchyma is only one to several cells thick, which is much less than the thickness of the cork cell layer.

The periderm and any tissues external to it form the outer portion of the bark (Figure 32–9). The thickness, patterns, and texture of bark vary considerably from species to species. These differences are due to varying activity of the cork cambium, which is under genetic control. The cork cells produced by the cork cambium protect the plant against mechanical injury, fire, temperature extremes, and water loss.

Trees and Terminology

If you've ever examined different types of lumber, you may have noticed that some trees have wood with two different colors (Figure 34–9). The older wood in the center of the tree is **heartwood**. It has a brownish-red color, in contrast to the newer wood closer to the bark, the **sapwood**. A microscopic examination of heartwood reveals the difference. The vessels and tracheids of heartwood are plugged up with various materials. Therefore, heartwood cannot function in conduction. The functional xylem is the sapwood. Heartwood is denser than sapwood and does provide mechani-

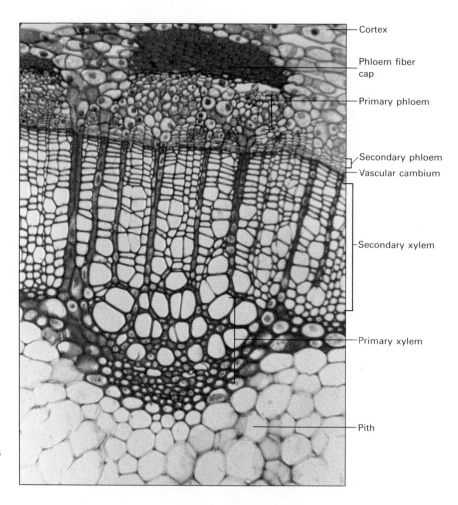

FIGURE 34–8 Cross section through part of a magnolia stem showing a vascular bundle that has been split apart by secondary growth.

cal support. There is some evidence that heartwood is also more resistant to decay. The materials plugging the cells of heartwood include various pigments, tannins, gums, and resins.

Almost everyone has heard of **hardwood** and **softwood.** Botanically speaking, hardwood is the wood of angiosperms, while softwood is the wood of gymnosperms. Pine and other gymnosperms typically have wood that lacks fibers and vessel elements. The conducting cell in gymnosperm

FIGURE 34–9 Cross section through a tree trunk revealing the darker heartwood in the center of the tree and the lighter sapwood. The sapwood is the functioning xylem, conducting water and dissolved minerals.

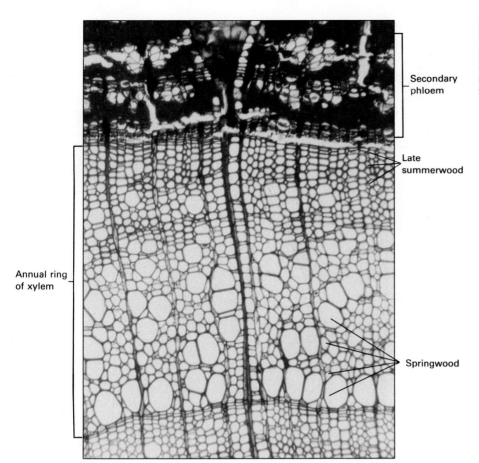

FIGURE 34–10 Portion of *Tilia* (basswood) stem, in cross section, showing one complete annual ring. Note the differences between the springwood and the later summerwood.

wood is the tracheid. These anatomical differences generally make gymnosperm wood softer than angiosperm wood, although there is a lot of variation from species to species.

It is possible to determine the age of a woody stem by counting the layers of wood, or **annual rings**. Examination of these annual rings with a magnifying lens would reveal that there is no ring, or line, separating one year's growth from the next. The appearance of a ring is due to differences in size and wall thickness between secondary xylem formed in the preceding year and that formed in the following year.

In the spring, when water is plentiful, the wood formed by the vascular cambium has thin-walled, large-diameter vessels and tracheids. It is appropriately called **springwood**. As summer progresses and water becomes less plentiful, the wood formed has thicker walls and smaller-diameter vessels and tracheids, i.e., **late summerwood**. It is the difference between the late summerwood of the preceding year and the springwood of the following year that gives the appearance of rings (Figure 34–10).

Plants that grow in temperate climates where there is a growing period and a period of dormancy (i.e., winter) exhibit annual rings. In the tropics, environmental conditions determine the presence or absence of rings, but they aren't reliable in determining the ages of trees. A lot of information about climate in past times can be gleaned from study of ancient tree rings (see Focus on Tree Ring Analysis).

The wood of a tree has quite different appearances depending on the angle at which it is cut (Figure 34–11). These differences are apparent both macroscopically and microscopically. In cross section, the annual rings appear as concentric rings, while the rays appear as straight lines coming out from the center of the wood. In a tangential section, which is a longitudinal section cut perpendicular to the radius, the annual rings appear as

FIGURE 34–11 A block of wood showing cross, radial, and tangential sections. The appearance of the rays and annual rings is distinctive for each section.

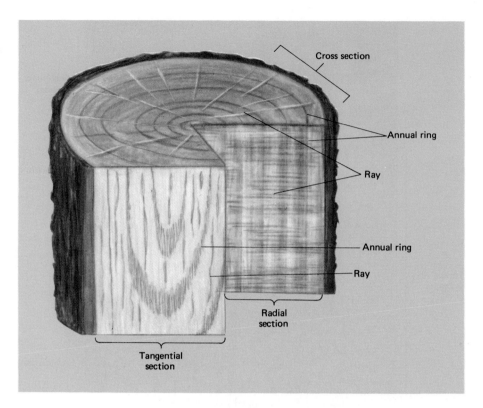

Focus on **TREE RING ANALYSIS**

In temperate areas, the age of a tree can be determined by counting the number of tree rings. Other useful information can be obtained by analyzing tree rings as well. For example, the size of each ring varies depending on environmental conditions, including precipitation and temperature. Sometimes the variation in tree rings can be attributed to one factor, and similar patterns appear in the rings of many tree species over a large geographical area. For example, trees in the Southwest of the United States have similar ring patterns due to variation in the amount of annual precipitation. Years with an adequate amount of precipitation produce larger layers of growth, while years of drought produce much smaller layers.

It is possible to study the sequence of rings back in time several thousand years. First a master chronology, a complete sample of rings dating back as far as possible, is developed. One starts with an old tree that is currently

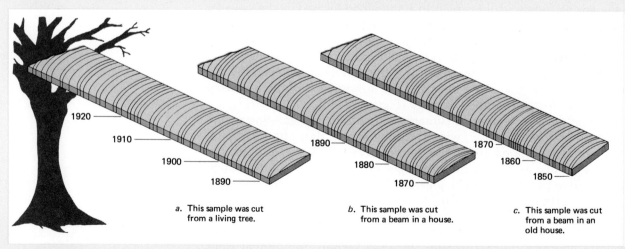

a. This sample was cut from a living tree.

b. This sample was cut from a beam in a house.

c. This sample was cut from a beam in an old house.

■ Tree ring dating. A master chronology is developed using progressively older pieces of wood from the same geographical area. By matching the rings of a wood sample of un-known age to the master chronology, the age of the sample can be accurately determined.

vertical lines that often come together in a V-shape and the rays appear as specks or very short lines. A radial section, which is a longitudinal section cut through the center (along the radius), has annual rings that look like lines running the length of the wood and rays that look like horizontal strips.

As the woody stem increases in width over the years, the branches that it bears grow along with it as long as they're alive. When a branch dies, it no longer continues to grow with the stem. In time the stem grows out and surrounds it. The basal portion of an embedded branch is called a **knot**. It is possible for the knot to contain bark as well as wood. The presence of knots in wood weakens its commercial value, except for ornamental purposes. Some plants (knotty pine) are valued for their high production of knots.

TRANSPORT IN PLANTS

The movement of food, water, and minerals within a multicellular plant is called **translocation**. Water and minerals are transported in xylem, while dissolved food is transported in phloem. Translocation in plants does not resemble the movement of materials in animals because nothing *circulates*. The materials being translocated in the xylem travel in one direction only. While movement in different phloem cells can be in several directions, it cannot be said to circulate.

living. The older rings toward the center of the living tree can be matched with the youngest rings toward the outside of a dead tree or even a piece of wood from a house. By using older and older sections of wood, even those found in prehistoric dwellings, and overlapping their matching ring sequences, one obtains a master chronology of the area.

Tree ring analysis, or dendrochronology, has been used extensively in several unrelated fields. Astronomers have correlated growth patterns over the years with cycles of sunspots. This work was pioneered by the American astronomer, Andrew Douglass, during the early years of the 20th century. Tree ring analysis has been extremely useful in dating prehistoric sites of native Americans in the Southwest. For example, the Cliff Palace in the Mesa Verde National Park dates back to 1073 AD. Tree ring analysis indicates that an extended drought forced the original inhabitants to abandon their home. Climatologists use tree ring data to study climate patterns in the past. Tree rings are also being analyzed for other disciplines, including ecology (to study succession) and environmental science (to study the effects of pollution).

■ The Cliff Palace in the Mesa Verde National Park in Colorado has been dated using tree ring analysis.

Movement of Water and Minerals in Xylem

The water and dissolved minerals, which form approximately a 0.1% solution, move within the tracheids and vessel elements, which you will recall are hollow, dead cells (Chapter 32). The movement of water in the xylem is the most rapid of any movement in plants. On a hot summer day, water has been measured moving upward in the xylem at 2 feet per minute.

Water initially moves horizontally into the roots from the soil, passing through several tissues until it reaches the xylem (Chapter 35). Once the water is in the tracheids and vessel elements, it travels upward through a continuous network of xylem from root to stem to leaf. The dissolved minerals are carried along passively in the water. The plant does not expend energy to transport water, which moves as a result of natural physical processes. How does water move to the tops of plants? Clearly, it is either pushed up from the bottom of the plant, or it is pulled up from the top of the plant. Both mechanisms exist in plants.

Water Potential

In order to understand the movement of water, it will be helpful to consider **water potential**, which is the free energy of water. The water potential of pure water is set at 0 bars by convention because it cannot be measured directly (the bar is a metric unit of pressure). It is possible to measure the differences in energy of water molecules in different situations, however. When substances are dissolved in water, the free energy of water decreases. This means that solutes that are dissolved in water will lower the water potential to a negative number. Water potential is a measure of the tendency of water to diffuse or evaporate. Water moves from an area of higher water potential to an area of lower, or more negative, water potential.

Root Pressure

Water moves into the roots of plants by osmosis (Chapter 5). The direction of water flow in osmosis can be explained by using the concept of water potential. In osmosis, water always moves from an area of higher to an area of lower water potential. The water potential for the soil varies, depending on how much water is in the soil. When a soil is very dry, its water potential is very low (i.e., very negative). When a soil has an average amount of water, its water potential is higher, although it is still a negative number because dissolved minerals are present in very dilute concentrations.

The water potential in root cells is also negative due to the presence of dissolved minerals, sugars, and other osmotically active substances. Roots contain more dissolved materials than the soil water, unless the soil is very dry. This means that under normal conditions the water potential of the root is more negative than the water potential of the soil, and water moves into the root.

In the root pressure mechanism, water moves into the roots by osmosis because of the difference in water potential between the soil and the roots. This accumulation of water creates a pressure in the root that forces the water up the xylem.

Root pressure is a real phenomenon in plants. Guttation is an example of this process (Chapter 33). However, plant physiologists have measured root pressure and found that it is not strong enough to explain the rise of water to the tops of the tallest trees. Root pressure exerts an influence in smaller plants, but it clearly does not cause water to rise hundreds of feet in the tallest plants.

Tension–Cohesion

A second possible explanation for the rise of water in plants is that a tension is created at the top of the plant to pull the water up. This process works much like a person sucking a liquid up through a straw. The tension created at the top of the plant is the evaporation-pull of transpiration (Chapter 33). Once again, this can be explained in terms of water potential. The atmosphere has an extremely negative water potential. There is a gradient in water potentials from the least negative (in the soil) up through the plant to the most negative (the atmosphere). This literally pulls the water up through the plant (Figure 34–12).

This pulling of water is only possible as long as there is a solid, unbroken column of water in the xylem. Water has a tendency to form an unbroken column because of the cohesiveness of water molecules. Water molecules are strongly attracted to one another due to hydrogen bonding (Chapter 2). Also, the adhesion of water to the walls of the xylem cells is an important factor in maintaining an unbroken column of water.

To summarize, in the tension–cohesion mechanism, a tension is created at the top of the plant by transpiration. This tension pulls the water up to the top of the plant. The cohesive and adhesive properties of water enable it to form a solid column, which can be pulled.

Is tension–cohesion powerful enough to explain the rise of water in the tallest plants? Plant physiologists have calculated that the tension created by transpiration is strong enough to pull water up 500 feet. Since the tallest trees on earth are approximately 350 feet high, tension–cohesion easily accounts for their water transport.

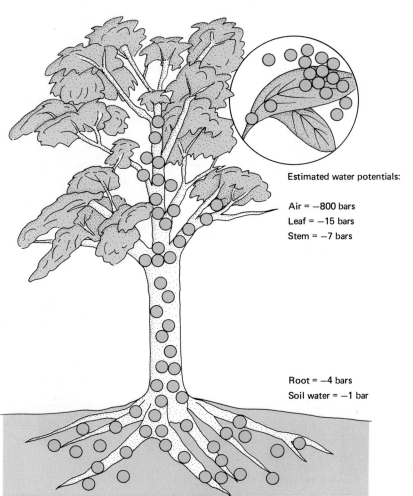

Estimated water potentials:

Air = −800 bars
Leaf = −15 bars
Stem = −7 bars

Root = −4 bars
Soil water = −1 bar

FIGURE 34–12 The evaporation-pull of transpiration is driven by the gradient in water potentials from the soil through the plant to the air. Water moves from higher to lower (more negative) water potentials.

Movement of Sugars in Phloem

The sugar produced by photosynthesis is often stored temporarily as starch in the chloroplasts. It is converted into the disaccharide, sucrose, before being loaded into the phloem for transport to the rest of the plant. Sucrose, or common table sugar, is the predominant form of food carried in the phloem. Movement of materials in the phloem is rapid. Although not as rapid as xylem transport, phloem transport has been measured at approximately 1 inch per minute.

Pathway of Movement

Movement within the phloem tissue may be up or down in the plant. Sucrose may be transported from its place of manufacture (the leaf) to a place of storage (the root, fruit, or seed). It may also be transported from the leaf or root to actively growing regions like the root or shoot apical meristems, where it would be quickly utilized. Phloem transport is due to a gradient that is established between the *source*, where the sugar is loaded into the phloem, and the *sink*, where the sugar is removed from the phloem.

Mechanism of Movement—Pressure Flow

At the source—the leaf—the dissolved sucrose moves from the mesophyll cells where it was manufactured and is actively loaded into the companion cells. The active loading requires ATP energy, which probably works through a proton pump, forming a gradient of hydrogen ions. This mechanism is supported by the changes in pH that have been observed in sugar loading. Presumably the sugar accompanies the flow of H^+ ions back across the membrane. Once the sugar is in the companion cell, it readily moves into the sieve tube elements through the many cytoplasmic connections between the two cells (Figure 34–13). The increase in dissolved sugars in the sieve tube element decreases the water potential of that cell. As a result, water moves by osmosis into the sieve tubes, creating a pressure. This pressure pushes the sugar solution through the phloem much as water is forced through a hose.

At its destination, the sink, sugar is actively unloaded from the sieve tube elements, with ATP being required. With a loss in sugar, the water potential in the sieve tube elements at the sink increases. Therefore, water moves out of the sieve tubes by osmosis.

It is the sugar gradient, i.e., the difference in sugar concentrations between the source and the sink, that causes transport in the phloem. The

FIGURE 34–13 Proposed model of sugar loading into phloem in the leaf. The energy of ATP is used to pump H^+ ions out of the companion cell. The return of H^+ ions through a protein channel in the companion cell plasma membrane is accompanied by sucrose. Once the sucrose is loaded into the companion cell, it moves through cytoplasmic connections into the sieve tube element.

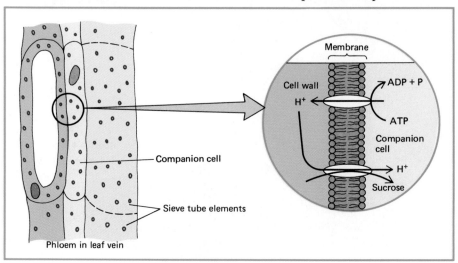

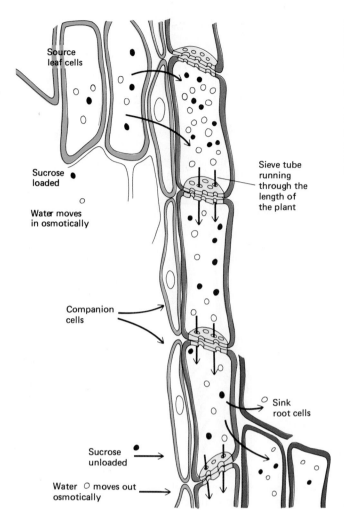

Source leaf cells

Sucrose loaded

Water moves in osmotically

Sieve tube running through the length of the plant

Companion cells

Sink root cells

Sucrose unloaded

Water moves out osmotically

FIGURE 34–14 Pressure flow mechanism for phloem transport. Sugar is actively loaded into the sieve tube element at the source. As a result, water moves osmotically into the sieve tube element. At the sink the sugar is actively unloaded and water leaves the sieve tube element by osmosis. The gradient of sugar from source to sink causes pressure flow through the sieve tube toward the sink.

actual flow of sugar solution in the phloem does not require energy. However, both loading and unloading sugar at the source and sink require energy derived from ATP (Figure 34–14).

Study of phloem transport in plants is difficult. The cells are under hydrostatic pressure, so cutting into the phloem to observe it causes the contents of the sieve tube elements to be sucked against one end wall. Much useful information about phloem transport has been obtained using radioactive tracers (Figure 34–15). Aphids, small wingless insects that insert their mouthparts into phloem sieve tubes for feeding, have also been a useful tool in phloem research (Figure 34–16). The pressure flow mechanism adequately explains current data on phloem transport. However, there is still a lot to be learned about this complex process.

■ SUMMARY

I. The primary functions of stems are to support, conduct, and produce new stem tissue.

II. Stems with primary growth have an epidermis, vascular tissue, and cortex and pith, or ground tissue.

   A. Dicot stems have the vascular bundles arranged in a circle and have a distinct cortex and pith.

   B. Monocot stems have scattered vascular bundles and ground tissue instead of a distinct cortex and pith.

III. Secondary growth occurs in some dicots and all gymnosperms.

   A. The vascular cambium produces secondary xylem (wood) to the inside and secondary phloem (inner bark) to the outside.

   B. The cork cambium produces cork parenchyma to the inside and cork cells to the outside.

IV. Water and dissolved minerals move upward in the xylem from the root to the stem to the leaves.

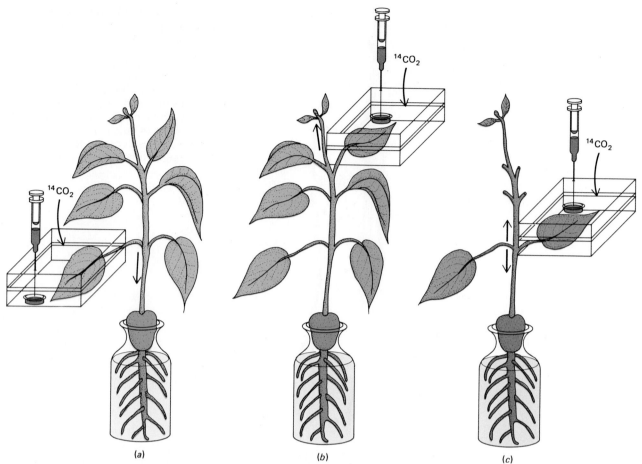

(a)   (b)   (c)

FIGURE 34–15 Source-to-sink translocation of radioactively labeled sugars in the phloem. When a leaf is fed $^{14}CO_2$, radioactive carbon is fixed into the sugar produced by photosynthesis. The movement of the sugar can be followed using autoradiographic techniques. After exposure to the labeled CO_2, the stem tissue is freeze-dried, sliced into thin sections, and placed on photographic film. The part of the film in contact with radioactive substances is exposed. After development, the exact location of the radioactive sugars can be determined. (a) Lower leaf given $^{14}CO_2$, translocation downward. (b) Upper leaf given $^{14}CO_2$, translocation upward. (c) Upper leaves removed and $^{14}CO_2$ fed to a lower leaf, with some translocation in both directions.

FIGURE 34–16 Mature aphid feeding on a lower side of a branch of basswood *(Tilia americana)*. The aphid is about 6 mm long. (a) The aphid's feeding apparatus (stylet) is inserted into the phloem. The pressure in the punctured phloem drives the sugar solution through the stylet and into the aphid's digestive system. (b) A microscopic view of *Tilia* phloem, showing the aphid stylet has penetrated a sieve tube element.

(a)

(b)

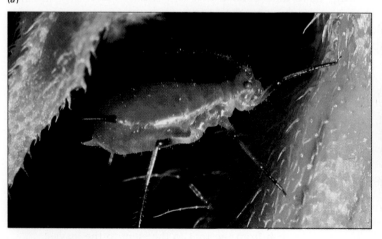

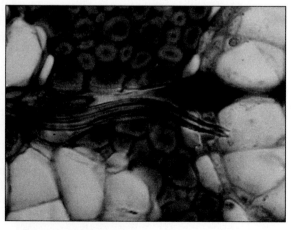

A. Root pressure, caused by the movement of water into the root from the soil due to differences in water potential, can explain the rise of water in small plants. Guttation is a consequence of root pressure.

B. Tension–cohesion causes the rise of water in even the largest plants.
  1. The evaporation-pull of transpiration causes a tension at the top of the plant. This is due to a gradient in water potentials from the soil up through the plant to the atmosphere.
  2. A solid, unbroken column of water is pulled up through the plant as a result of the cohesive and adhesive nature of water.

V. Dissolved food is transported up or down in the phloem.

A. Sucrose is the predominant form of food transported in the phloem.

B. Movement of materials in the phloem is caused by pressure flow.
  1. Sugar is actively loaded into the sieve tubes at the source. This requires ATP. As a result, water moves into the sieve tubes by osmosis.
  2. Sugar is actively unloaded from the sieve tubes at the sink. This requires ATP. As a result, water leaves the sieve tubes by osmosis.
  3. Transport of materials between the source and sink is driven by mass flow, or the pressure created by the additional water entering the phloem at the source.

■ POST-TEST

1. Vascular tissue arranged in a circle is characteristic of the primary stems of _____.
2. Primary _____ stems lack a distinct pith and cortex.
3. The two lateral meristems responsible for secondary growth are the _____ _____ and the _____ _____.
4. The cork cambium and the tissues it produces are collectively called the _____ and make up the outer region of the bark.
5. Horizontal transport of materials is accomplished by _____ in plants with secondary growth.
6. In plants with secondary growth, the _____ serves as a functional replacement for the epidermis.
7. Botanically speaking, the wood of a tree is _____.
8. The _____ of a tree is the inner layers of wood that are nonfunctional and usually pigmented.
9. An annual ring of wood is formed by the difference between the _____ of the preceding year and the _____ of the following year.
10. The basal portion of an embedded branch is called a _____.
11. _____ is the science of tree ring analysis.
12. The movement of food, water, and minerals within a plant is called _____.
13. The free energy of water in a particular situation is referred to as its _____ _____.
14. The water potential of pure water is set by convention at _____ bars.
15. The presence of dissolved solutes in water _____ the water potential.
16. This mechanism of water movement, _____ _____, is not strong enough to explain the rise of water to the tops of the tallest trees.
17. In the tension–cohesion mechanism, the tension is created at the top of the plant by the evaporation-pull of _____.
18. The area of the plant where sugar is loaded into the phloem is known as the _____.

■ REVIEW QUESTIONS

1. List several functions of stems and describe the tissue(s) responsible for each function.
2. When secondary growth commences, it is said that certain cells dedifferentiate and become meristematic. Could a tracheid ever do this? Why or why not?
3. What happens to the primary tissues of the stem when secondary growth occurs?
4. What is water potential? How can it be used to explain the movement of water in osmosis? in transpiration?
5. Explain the tension–cohesion mechanism of water transport. Make sure you consider both the "tension" and "cohesion" aspects of the mechanism.
6. Describe the pressure flow mechanism of sugar movement in the phloem, including the activities at the source and sink.

■ RECOMMENDED READINGS

Galston, A.W., Davies, P.J., and Satter, R.L., *The Life of the Green Plant*. 3rd ed. Englewood Cliffs, New Jersey, Prentice-Hall, Inc., 1980. Contains an excellent explanation of conduction in both xylem and phloem.

Raven, P.H., Evert, R.F., and Eichhorn, S.E., *Biology of Plants*. 4th ed. New York, Worth Publishers, Inc., 1986. An excellent general botany textbook, especially useful for current interpretations of plant structures.

Tippo, O. and Stern, W.L., *Humanistic Botany*. New York, W.W. Norton and Company, 1977. This book is an enjoyable presentation of the human relevance of plants. The history and utilization of dendrochronology is thoroughly covered.

Radish seedlings

35

Roots and Mineral Nutrition

▪ LEARNING OBJECTIVES

After you have read this chapter you should be able to:

1. Compare the structures of primary dicot and monocot roots.
2. Outline how a root develops from primary growth to secondary growth.
3. List several functions of roots and discuss how their structure relates to their functions.
4. Describe several variations in roots that perform unusual functions.
5. Discuss the pathway of water movement in roots.
6. List the five components of soil and give the ecological significance of each.
7. Describe the factors involved in soil formation.
8. Outline the criteria an element must satisfy in order to be considered essential for plant growth.
9. List the 16 elements that are essential for plant growth.
10. Give a physiological role of each essential element in plants.

FIGURE 35–1 Prop roots in corn (*Zea mays*). These adventitious roots provide additional support.

FUNCTIONS OF ROOTS

Roots are essential plant organs. They serve to anchor the plant, and they absorb water and dissolved minerals from the soil. These materials are then transported throughout the plant. Many roots also serve as storage organs. Excess sugars produced in the leaves by photosynthesis are transported to the roots for storage until they are needed. The roots use some of this sugar for their own respiratory needs. Most of the sugar that is stored in the root is transported via the phloem to other parts of the plant when needed. Some plants, like beets and sweet potatoes, have roots that are enormously swollen with food storage tissues.

Other roots are modified for additional functions besides anchorage, absorption, conduction, and storage. Some roots are produced at unusual places on the plants. These **adventitious roots** are frequently aerial. **Prop roots,** which are more common in monocots than dicots, provide additional support for the plant (Figures 35–1 and 35–2). They frequently arise from the main stem of the plant. Corn is an example of a plant that produces prop roots.

Other aerial roots have additional functions. For example, plants adapted to wet environments where the soil is flooded often have roots modified for aeration. Although roots live in the soil, they require oxygen for respiration. A flooded soil is depleted of oxygen, so aerial roots assist in getting oxygen to the below-ground roots. Black mangrove and bald cypress are examples of plants with aerial roots for aeration (Figure 35–3).

Epiphytes are plants that grow attached to other plants. Many epiphytes have unusual root modifications. For example, certain epiphytic orchids have photosynthetic roots. Epiphytic roots may absorb moisture as well.

Plants that produce bulbs often have **contractile roots,** which contract and pull the bulb and stem deeper into the ground (Figure 35–4). Contractile roots are common in monocots, but certain dicots and ferns possess them also.

STRUCTURE OF ROOTS WITH PRIMARY GROWTH

All primary roots have certain tissues also found in stems, such as epidermis, cortex, xylem, and phloem (Chapter 34). Roots also have several tissues and structures not found in stems, including a root cap and root hairs. Each root tip is covered by a root cap, a protective layer many cells thick covering the delicate root apical meristem. Root hairs are extensions of epidermal cells located in the area of maturation near the root tip (Chapter 32). Although stems and leaves may have various types of hairs, they are distinct from root hairs in structure and function.

FIGURE 35–2 The banyan tree has prop roots that develop from branches.

FIGURE 35–3 Many plants living in swampy areas have special roots for aeration. Cypress "knees" may provide oxygen for roots that are buried in anaerobic mud. Cypress trees growing in normal soil do not produce knees.

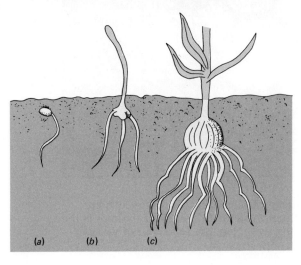

FIGURE 35–4 Plants that produce bulbs often have contractile roots. (a) Germination of the seed. (b) During the first growing season, contractile roots do not pull the bulb (c) appreciably deeper in the soil. During succeeding seasons, contractile roots pull the bulb deeper and deeper until it reaches a depth of temperature stability.

805

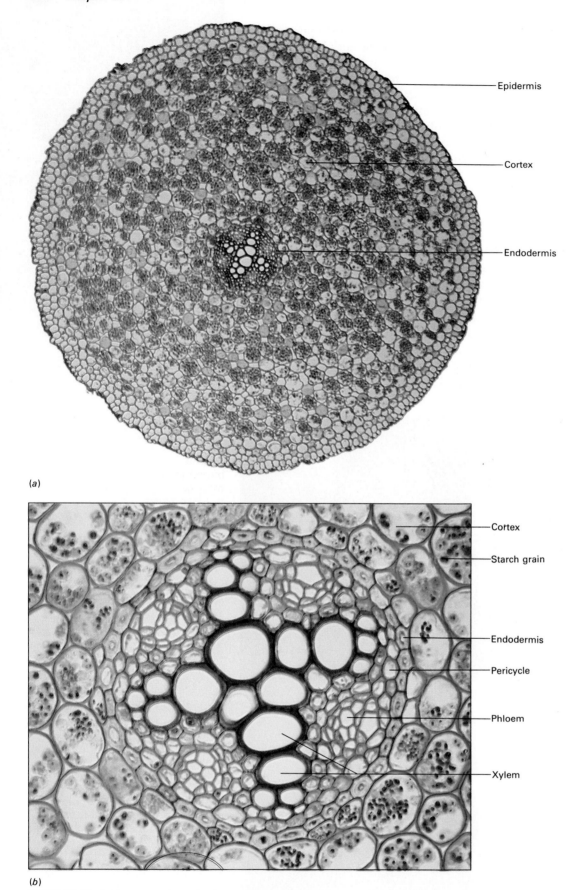

Epidermis

Cortex

Endodermis

(a)

Cortex

Starch grain

Endodermis

Pericycle

Phloem

Xylem

(b)

FIGURE 35–5 Cross section of a buttercup (*Ranunculus*) root. Buttercups are dicots with primary growth. (*a*) Entire root. Note that the bulk of the root is the cortex. (*b*) A close-up of the center of the root. Note the solid core of vascular tissues.

There is considerable variation in the internal arrangement of tissues in both dicot and monocot roots. We will examine a representative root for each.

Dicot Roots

Like other parts of the plant, dicot roots are covered by a single layer of protective tissue, the **epidermis** (Figure 35–5). Unlike the epidermis of aerial parts of the plant, however, the root epidermis does not secrete a waxy cuticle, which would impede the absorption of water from the soil. The root hairs are another modification enabling the root to absorb more water from the soil, as they greatly increase the surface area of the root in contact with the moist soil (Figures 35–6 and 35–7). Root hairs are short-lived and never develop into multicellular root branches. Branches of roots develop in a different manner.

The root **cortex** is not composed of the variety of cell types found in stem cortex (Chapter 34). Rather, parenchyma with lots of intercellular spaces makes up the bulk of the cortex. Usually there is no collenchyma in roots, although some sclerenchyma develops as the root ages. The inner layer of the cortex, the **endodermis,** is different from the rest of the cortex. Endodermal cells fit snugly against each other. Each has a special bandlike region on its radial and transverse walls, the **Casparian strip** (Figure 35–8).

FIGURE 35–6 Root hairs on a radish seedling. Each delicate hair is a single cell extension of the root epidermis. Root hairs increase the surface area of the root in contact with the soil.

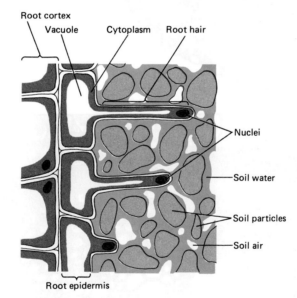

Root cortex
Vacuole Cytoplasm Root hair
Nuclei
Soil water
Soil particles
Soil air
Root epidermis

FIGURE 35–7 Stages in root hair development. The nucleus migrates into the root hair.

FIGURE 35–8 A few cells of the endodermis. Note the Casparian strip around the radial and transverse walls. The endodermis controls water uptake by the root.

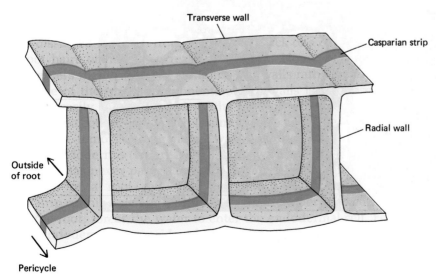

Transverse wall
Casparian strip
Radial wall
Outside of root
Pericycle

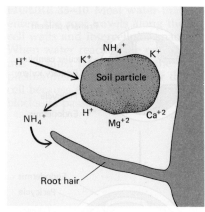

FIGURE 35–13 Cation exchange in the soil. The negatively charged soil particles bind various positively charged ions, or cations. It is possible for hydrogen ions to replace or be exchanged for these cations, freeing them for absorption by the root. The hydrogen ions for this exchange come from the soil and from the root.

A final factor involved in soil formation is the **topography,** or surface features of a region. For example, soil formation depends on whether the soil is located on a mountain top where significant erosion might occur, or in a valley where sedimentation might increase the rate of soil formation.

Soil Composition

Soil is a complex substance composed of five materials: inorganic minerals, organic matter, soil organisms, soil atmosphere, and soil water. The inorganic minerals that come from weathered rock form the basic soil material. The texture of a soil is determined by the amounts of inorganic soil particles of different sizes. The large particles are **sand** (0.02 to 2 mm diameter), the medium are **silt** (0.002 to 0.02 mm diameter), and the small are **clay** (less than 0.002 mm diameter). The clay component is very important in determining many characteristics of the soil, in part because each clay particle has negative charges on its outer surface. Clay tends to attract and bind positively charged mineral ions, preventing them from being leached out of the soil (Figure 35–13).

Different soils have different combinations of soil particle sizes. A loamy soil, which is good for agriculture, has approximately 40% each of sand and silt and about 20% of clay. Soils with larger proportions of sand are not as desirable for plant growth because they don't hold water and mineral ions well. Soils with a larger proportion of clay tend to get compacted, robbing the soil of spaces that can be filled by water and air.

The organic matter of a soil is the remains of dead plants, animals, and microorganisms. Organic matter is decomposed by the microorganisms

FIGURE 35–14 Essential elements like nitrogen, potassium, and calcium cycle from the soil into living organisms and back into the soil.

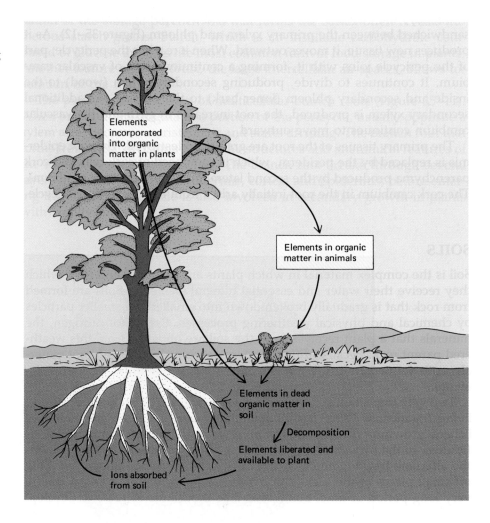

Elements incorporated into organic matter in plants

Elements in organic matter in animals

Elements in dead organic matter in soil

Decomposition

Elements liberated and available to plant

Ions absorbed from soil

that inhabit the soil. In the process, essential mineral elements are released into the soil and may be reabsorbed by plants (Figure 35–14). Organic matter is also important because it alters certain soil characteristics. It increases the water holding capacity of the soil, acting much like a sponge. For this reason it is an important additive, especially for sandy soils. The partly decayed organic portion of the soil is referred to as **humus.**

The organisms living in the soil form a complex community in numbers too vast to comprehend. A single teaspoon of good agricultural soil may contain millions of living organisms. Soil organisms include bacteria, fungi, algae, protozoa, worms, insects, and larger plants and animals (Figure 35–15). The bacteria and fungi are particularly essential in decompos-

FIGURE 35–15 There is an incredible diversity of organisms living in the soil. In addition to plant roots, the soil contains bacteria, algae, protozoa, fungi, worms, insects, and a variety of other organisms.

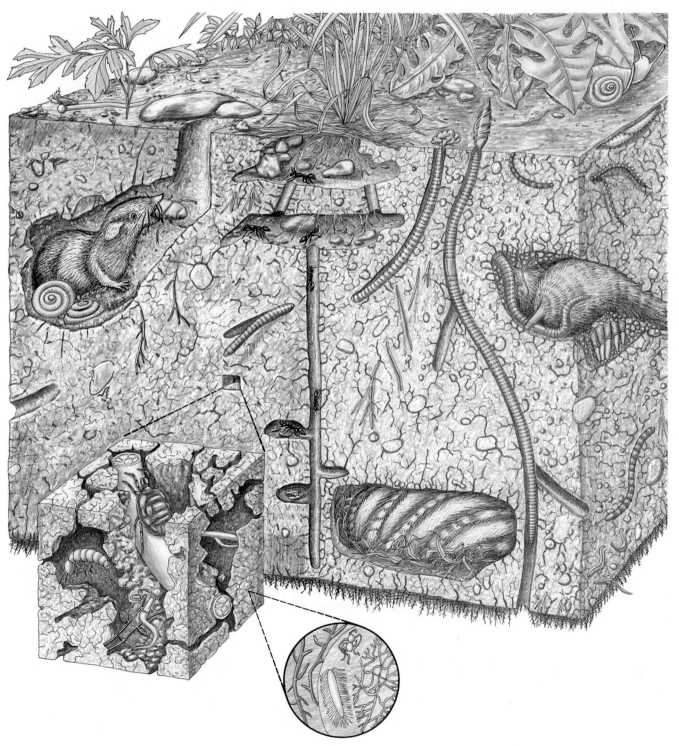

FIGURE 35–16 Cultivated citrus species show a marked dependency on a mycorrhizal association for adequate growth. Shown here are orange seedlings after six months, with mycorrhizae (left) and without mycorrhizae (right). Applications of calcium nitrate and ammonium nitrate were given to the two sets of plants, but only those with mycorrhizae absorbed these nitrates effectively, aiding in the plants' nutrition and creating a marked difference in growth response.

ing dead organic material. They are also important in nutrient cycles. For example, most steps in the nitrogen cycle involve microorganisms. Mycorrhizae, symbiotic associations between fungi and the roots of most plants, aid in uptake of mineral ions. The hyphae come into contact with a large amount of soil, and absorb minerals and transfer them to the plant. Food produced by photosynthesis is transferred to the fungus. Plant growth is often enhanced by mycorrhizal associations (Figure 35–16).

Soil is not a solid. Approximately 30 to 60% of the volume of soil is occupied by space between soil particles. These spaces are filled with varying proportions of soil air and soil water. Soil air has a slightly different composition of gases than atmospheric air. Generally, the level of oxygen is lower and the level of carbon dioxide is higher in soil air. This is due to the respiratory activities of the living organisms in the soil. Both air and water are necessary in soil for good plant growth. Plant roots need oxygen for respiration. Plants living in swamps and marshes have special adaptations enabling them to survive in anaerobic soil, including large air spaces within the plant tissues and special roots for aeration.

Soil Conservation

The formation of soil from rock on land is a gradual process. Similarly, soil is gradually worn away, or eroded, from land surfaces. This is a natural process. Two environmental factors that promote erosion are water and wind. Rainfall loosens soil particles, which can then be carried away by running water. Wind is particularly effective in removing soil from bare, dry land. The Dust Bowl in the 1930s was a vivid example of the effect of wind on erosion. Soil from Colorado was blown as far as several hundred miles off the Atlantic coast. Although erosion is a natural process, its effects are much reduced when there is sufficient plant cover. Plant roots are very effective at holding the soil in place.

Another environmental factor that affects soil fertility is fire. Fire can have an adverse effect on the soil for several reasons. First, it removes plant cover from the soil, making it more susceptible to erosion. If a fire is sufficiently hot, it can burn off the nitrogen and carbon from the soil, thereby reducing its fertility. It should be mentioned, however, that fire is an essential component of certain ecosystems, such as chaparral, many grasslands, and several types of pine forest.

MINERAL NUTRITION IN PLANTS

There are more than 90 naturally occurring elements in the Earth. Over 60 of these have been found in plant tissues, including elements as common as carbon and as rare as gold. Not all of these elements are considered essential for plant growth, however.

How do biologists determine whether an element is essential? One of the most useful methods is **hydroponics,** growing plants in aerated water with dissolved mineral salts. It is impossible to conduct mineral nutrition experiments by growing plants in soil, which is too complex and contains too many elements. However, one can grow plants in a solution of water and all known required mineral salts. If it is suspected that a particular element is essential for plant growth, plants are grown in a nutrient solution that contains all known essential elements except the one in question. If plants grown in the absence of that element are unable to develop normally or complete their life cycle, the element may be essential. Additional criteria are used to determine whether an element is essential. The element must be shown to have a direct effect on the metabolism of the plant. Also, the element must be demonstrated to be essential for a wide variety of plant species.

Plant physiologists sometimes measure a plant's uptake using radioactive isotopes. Once an element has been demonstrated by hydroponics to be essential, the use of autoradiography can help locate the element within plant tissues (Chapter 32).

Focus on COMMERCIAL HYDROPONICS

Hydroponics, the practice of growing plants in an aerated solution of chemically defined mineral salts, has been used by scientists to determine which elements are essential. Initially, entrepreneurs hailed hydroponics as the scientific way to grow plants in places where soil was poor or unavailable. However, the expenses involved in commercially growing produce for human consumption prevented hydroponics from becoming more than a curiosity. Recent technical improvements have revived the interest in commercial hydroponics.

There are several places where hydroponics has great potential. It is being tried experimentally in desert countries in the Middle East, where the soil is too arid to support cultivation and water is unavailable for irrigation. When plants are grown hydroponically in greenhouses, little water is used compared with traditional agriculture. Hydroponics is also being tried in termperate latitudes, particularly for winter crops.

There are several advantages to hydroponics. First, it is possible to grow these crops in areas where pathogens and pests are completely absent. This means that the crops aren't exposed to pesticides. Also, hydroponics can be used to grow crops near their area of use, saving on transportation costs.

The main disadvantage of hydroponics is the expense. The plants must be supplied with nutrient solution, which must be continually monitored and adjusted. Heating and lighting costs are high. Aeration of the roots was a major expense, although recent developments like the nutrient film technique have cut costs considerably. In the nutrient film technique, the plants are grown in plastic trenches through which a film of nutrient solution is run. In this way, the roots get adequate aeration. The nutrient solution is saved and reused on the plants, cutting down on water and mineral costs.

Although we will probably never replace traditional agriculture with hydroponics, it has been shown to be a viable alternative in certain situations. As new techniques are developed, hydroponics may become even more common.

■ Hydroponically grown lettuce. There is no soil in the plastic container, which serves only as a float to keep this and other lettuce plants (shown in rows in the background) from sinking into the nutrient solution (kept covered with white plastic film except for holes in the pots). Although soil is not used, these plants yield excellent salad greens.

FIGURE 35–17 Tobacco plants illustrating the effects of deficiencies of specific elements. The plant in the center (Ck.) received all the essential elements. The others were supplied with all essential elements except the one indicated on the label. All plants are the same age and variety. Some of them exhibit chlorosis (breakdown of chlorophyll) and necrosis (death of tissue).

Essential Elements

Sixteen elements have been demonstrated to be essential for plant growth (Figure 35–17). Nine of these are required in fairly large quantities (greater than 0.05% dry weight) and are therefore known as **macronutrients:** carbon, hydrogen, oxygen, nitrogen, phosphorus, potassium, sulfur, calcium, and magnesium. The remaining seven **micronutrients** are needed in trace amounts for normal plant growth and development: iron, boron, manganese, copper, molybdenum, chlorine, and zinc.

Four of the sixteen elements are obtained from water or gases in the atmosphere. Carbon is taken from carbon dioxide by photosynthesis. Oxygen is obtained from oxygen gas and water. Water also supplies hydrogen to the plant. Plants get their nitrogen from the soil as ions of nitrogen salts, nitrate (NO_3^-) and ammonium (NH_4^+), but it is "fixed" into that form from nitrogen gas (N_2) by various microorganisms in the soil. The remaining twelve essential elements are obtained from the soil as dissolved mineral ions. Their ultimate source is the parent rock from which the soil was formed.

Some of the roles of the essential elements are summarized in Table 35–2. Carbon, hydrogen, and oxygen are found in all biologically important molecules, including lipids, carbohydrates, nucleic acids, and proteins. Nitrogen is part of proteins, nucleic acids, and chlorophyll. Phosphorus is critical for plants because it is found in nucleic acids, phospholipids (an essential part of membranes), and energy transfer molecules like ATP. The middle lamella, the cementing layer of the plant cell wall, contains calcium. Calcium has also been implicated in a number of physiological roles in plants, including membrane permeability. Magnesium is part of the chlorophyll molecule. Sulfur is essential because it is found in certain amino acids and vitamins.

Potassium, which is required in fairly substantial amounts by plants, is not found in a specific compound or group of compounds. Rather, it remains as free K^+ ions in the plant cells. It has a very important role in

Table 35–2 **FUNCTIONS OF ESSENTIAL ELEMENTS**

Element	Major Functions
Carbon	Structural—in carbohydrates, lipids, proteins, and nucleic acids
Hydrogen	Structural—in carbohydrates, lipids, proteins, and nucleic acids
Oxygen	Structural—in carbohydrates, lipids, proteins, and nucleic acids
Nitrogen	Structural—in proteins, nucleic acids, chlorophyll, certain coenzymes
Phosphorus	Structural—in nucleic acids, phospholipids, ATP (energy transfer compound)
Calcium	Structural—in middle lamella of cell walls Physiological—role in membrane permeability
Magnesium	Structural—in chlorophyll Physiological—enzyme activator in carbohydrate metabolism
Sulfur	Structural—in certain amino acids and vitamins
Potassium	Physiological—osmosis and ionic balance, e.g., opening and closing of stomates; enzyme activator
Chlorine	Physiological—ionic balance; involved in light reactions of photosynthesis
Iron	Physiological—part of enzymes involved in photosynthesis and respiration
Manganese	Physiological—part of enzymes involved in respiration and nitrogen metabolism
Copper	Physiological—part of enzymes involved in photosynthesis
Zinc	Physiological—part of enzymes involved in respiration and nitrogen metabolism
Molybdenum	Physiological—part of enzymes involved in nitrogen metabolism
Boron	Physiological—exact role unclear; involved in membrane transport and calcium utilization

maintaining the turgidity of cells because it is osmotically active. Its role in the opening and closing of stomates through its effect on osmosis in the guard cells has already been discussed (Chapter 33). Another element that has a role in turgor balance of cells is chlorine. The Cl^- ion is present in very minute amounts in plants, but research indicates that it is essential for photosynthesis in addition to its osmotic role.

Five of the micronutrients (iron, manganese, copper, zinc, and molybdenum) are involved in various enzymes, often as enzyme activators. Potassium is also involved in certain enzymatic reactions. The role of boron in plants is unclear. Recent experiments have suggested that boron is involved in membrane transport. It also appears to affect calcium utilization.

Besides the sixteen essential elements, several additional elements have been shown to be essential for specific plants. Nickel is involved in enzymatic reactions in legumes such as peas and beans. Sodium is probably essential for saline plants, sugar beets, and bluegrass. Silicon enhances the growth of various grasses. After further evaluation, one or more of these may be added to the list of essential elements.

Fertilizers

In a balanced ecosystem the minerals removed from the soil by plants are returned when the plants or the animals that eat them die and decompose. However, the agricultural practices of humans prevent this cycle from occurring. The removal of crops from the land gradually depletes the soil of certain key elements. Likewise, homeowners mow their laws and remove the clippings, preventing decomposition and cycling of minerals that were in the grass blades. Plant growth is limited by the essential material (water, sunlight, or some essential element) that is in shortest supply. This is sometimes called the **concept of limiting factors.** In order to sustain productivity of agricultural soils, fertilizers are periodically added to replace those minerals that are limiting factors.

There are two main types of fertilizer, organic and inorganic. Organic fertilizers come from natural sources such as cow manure and ground corncobs. Green manure, an organic fertilizer, is a crop that is planted in the soil and deliberately plowed under to decompose rather than being harvested. Frequently, this crop is a plant that has nitrogen fixation in its roots, thereby increasing the amount of nitrogen in the soil. Organic fertilizers have several advantages over inorganic fertilizers. First, they increase the amount of organic material in the soil, which improves the water-holding capacity of the soil. Organic fertilizers also release the minerals they contain gradually, as decomposers break down the organic material. The addition of organic material to the soil changes the biota in the soil. In ways that are not clear to scientists, this sometimes suppresses microorganisms that could cause plant disease.

Inorganic fertilizer is manufactured. Its exact chemical composition is known. Most inorganic fertilizers contain three main elements (nitrogen, phosphorus, and potassium) that are usually the limiting factors in plant growth. The numbers on fertilizer bags (e.g., 10,20,20) tell the relative concentrations of each of the three elements (N,P,K). An advantage of inorganic fertilizers over organic is that one knows precisely what is being applied to the soil. By varying the relative concentrations of nitrogen, phosphorus, and potassium, different growth responses can be induced in plants. For example, if one were growing a lettuce crop, it would be best to use a fertilizer with a high nitrogen content, because that stimulates vigorous vegetative growth rather than reproduction. However, if one were growing tomatoes and applied a fertilizer with a high nitrogen content, the production of tomatoes would be quite low. Although the plants would grow vigorously, they would form few flowers and, therefore, few tomato fruits.

Obviously, there are advantages for each type of fertilizer. The chemical elements supplied by each are identical, however. Nitrogen from commercial, inorganic fertilizer is the same as nitrogen from organic fertilizer.

■ SUMMARY

I. Anchorage, absorption, conduction, and storage are the functions of roots.
II. Roots may be modified for several functions, including support, aeration, and photosynthesis.
III. Primary roots have an epidermis, cortex, endodermis, pericycle, xylem, and phloem.
  A. The epidermis protects the root.
  B. The cortex contains storage tissue.
  C. The endodermis controls water uptake by the root.
  D. The pericycle is the origin of branch roots.
  E. The xylem conducts water and dissolved minerals.
  F. The phloem conducts food.
IV. There are some differences between monocot and dicot roots.
  A. Monocot roots often have a pith.
  B. Dicot roots with secondary growth have a vascular cambium.
V. Secondary roots have wood and bark just like stems with secondary growth.
VI. Soil is the complex material in which plants root.

A. Factors influencing soil formation include parent rock, climate, living organisms, time, and topography.
B. Soil is composed of inorganic minerals, organic material, living organisms, soil air, and soil water.
C. Erosion is a natural process that may be accelerated under certain circumstances.
VII. Plants require certain elements for normal growth.
  A. Nine elements are macronutrients: carbon, oxygen, hydrogen, nitrogen, potassium, phosphorus, sulfur, magnesium, and calcium.
  B. Seven elements are micronutrients: iron, boron, manganese, copper, zinc, molybdenum, and chlorine.
  C. These elements are part of the structure of biological molecules, are important in the ionic balance of cells, and are involved in enzyme reactions.
  D. Some of the essential elements may be added to the soil as organic or inorganic fertilizer.

■ POST-TEST

1. _____ roots are produced at unusual places on the plant.
2. Plants with bulbs often have _____ roots that pull the bulb deeper into the ground.
3. The waterproof region around the radial and transverse walls of endodermis cells is the _____ _____.
4. The _____ is the origin of branch roots.
5. The center of the dicot root is _____.
6. The center of a monocot root is _____.
7. Minerals may pass through a membrane against the concentration gradient by _____ _____.
8. The largest inorganic soil particles are _____, the medium-sized particles are _____, and the smallest particles are _____.
9. Growing plants in aerated water with dissolved mineral salts is known as _____.
10. _____ are essential elements required in fairly large quantities.
11. Although more than 60 elements have been found in plant tissues, only _____ of them are essential for plant growth.
12. _____ is an essential element found in phospholipids, nucleic acids, and energy transfer molecules like ATP.
13. _____ and chlorine are osmotically active and have a role in maintaining the turgidity of cells.
14. Green manure is an example of a/an _____ fertilizer.
15. The three elements that most often limit plant growth are potassium, phosphorus, and _____.

■ REVIEW QUESTIONS

1. Trace the pathway of water into the root from the soil.
2. Are minerals absorbed in the same manner as water? If not, how are they absorbed by the root?
3. How does a root with primary growth develop secondary tissues?
4. List the five components of soil and tell why each is important to plants.
5. Explain how chemical and physical weathering processes convert rock into soil particles.
6. How does fire affect soil fertility?
7. What criteria have biologists used to determine which elements are essential for plant growth?
8. Give the advantages of both organic and inorganic fertilizers.

■ RECOMMENDED READINGS

Galston, A.W., P.J. Davies, and R.L. Satter. *The Life of the Green Plant*, Third Edition. Prentice-Hall, Inc., Englewood Cliffs, New Jersey, 1980. Contains an excellent explanation of plant mineral nutrition and hydroponics.

Raven, P.H., R.F. Evert, and S.E. Eichhorn. *Biology of Plants*, Fourth Edition. Worth Publishers, Inc., New York, 1986. An excellent general botany textbook, especially useful for current interpretations of plant structures.

Lupine pods explode, dispersing seeds

36

Reproduction in Flowering Plants

SEXUAL VERSUS ASEXUAL REPRODUCTION

Many angiosperms are able to reproduce both sexually and asexually. Sexual reproduction involves the formation of flowers and, after fertilization, fruits and seeds (Chapter 28). More specifically, however, sexual reproduction is the fusion of haploid gametes, the egg and the sperm nucleus. The union of these cells is called fertilization and occurs within the ovary of the flower. The offspring of sexual reproduction exhibit a great deal of individual variation. This is due in part to the recombination of chromosomes that occurs during meiosis and to the union of dissimilar gametes, often from two different parents. Sexual reproduction has several advantages for the plant. It makes it possible for new combinations of genes to occur that might make the plant better suited to its habitat. Also, the fruits and seeds of many plants have various mechanisms for dispersal, making it possible for the plant to extend its range.

Asexual reproduction in angiosperms does not usually involve the formation of flowers, fruits, and seeds. Instead, vegetative structures (roots, stems, and leaves) form offspring. In asexual reproduction there is one parent rather than two, and the offspring are formed by mitosis. This means the offspring are genetically identical to the parent and to each other. Asexual reproduction is ideal for producing large numbers of identical offspring from a desirable parent. Humans take advantage of asexual propagation techniques to produce large numbers of plants from a single plant that has a desirable combination of genetic characteristics (Figure 36–1).

ASEXUAL REPRODUCTION IN FLOWERING PLANTS

Angiosperms have evolved many methods of asexual reproduction. Most of these involve modified vegetative parts. In particular, there are a number of asexual structures that are modified stems.

Stems

The **rhizome** is a horizontal, underground stem that may or may not be fleshy for storing food. Although rhizomes may resemble roots, they are really stems, as shown by the presence of scale-like leaves, buds, nodes, and internodes. Rhizomes frequently branch in different directions (Figures 36–2 and 36–3). Over time, the old portion of the rhizome dies, eventually separating the two branches into distinct plants. The iris and many grasses are examples of plants that have rhizomes. Humans propagate plants with rhizomes by dividing or cutting the rhizome into smaller pieces, each with a bud. Each piece is capable of growing into an entire plant.

Another underground stem is the **tuber,** which is greatly enlarged for food storage. White potatoes and *Caladium* are examples of plants that produce tubers (Figure 36–4). The "eyes" of the white potato are actually lateral buds, evidence that the tuber is an underground stem rather than a

FIGURE 36–1 Hybrid orchid. To retain its desirable combination of horticultural characteristics, this plant must be propagated asexually.

FIGURE 36–2 Rhizomes growing out from the base of a plant. New shoots arise from buds that develop along the rhizome.

(a)

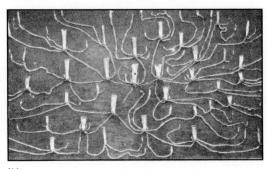

(b)

FIGURE 36–3 Cattails (*Typha latifolia*) (*a*) The number of plants that can arise by rhizomes from one plant in a single growing season can be quite large. (*b*) The leaves and roots have been cut back, showing the rhizome branching and growth. The plant in the center (with the star) was the original shoot.

root. Humans propagate tubers by cutting them into pieces, each with a lateral bud. When planted, each will grow into a plant.

The **bulb** is a shortened underground stem to which fleshy storage leaves are attached (Figure 36–5). Bulbs are globose, or round, and are covered by paper-like bulb scales. They frequently form small daughter bulbs that are initially attached to the mother bulb. Humans separate these daughter bulbs to increase the number of plants, but this process also occurs in nature. The contractile roots (Chapter 35) of some daughter bulbs contract and eventually pull the daughter bulb away from its parent. Lilies, tulips, onions, and daffodils form bulbs.

An underground stem that superficially resembles the bulb is the **corm** (Figure 36–5). The storage organ in the corm is the much thickened stem, rather than leaves as in the bulb. The entire corm is stem tissue that is covered with papery scales. These scales are modified leaves that are attached to the corm at nodes. Lateral buds frequently arise on the corm. Plants that produce corms include the crocus, gladiolus, and cyclamen.

Stolons, or runners, are horizontal stems that run above ground (Figure 36–6). They are characterized by having long internodes. Adventitious buds develop along the stolon, and each bud gives rise to a new plant. The strawberry is an example of a plant that produces stolons.

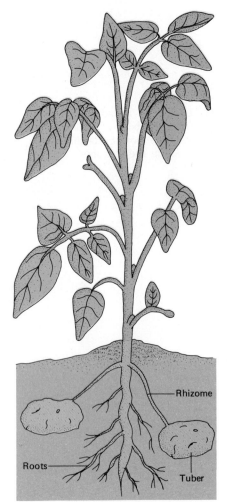

FIGURE 36–4 Potatoes are seldom grown from seed. Instead, a tuber is cut into pieces, each with an "eye." The plant that grows forms rhizomes, which enlarge at the ends into tubers.

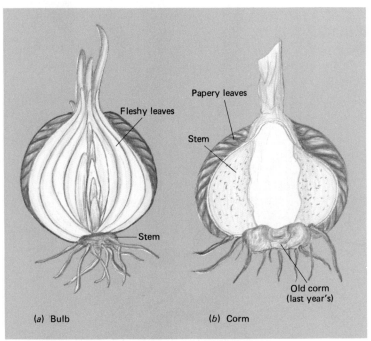

FIGURE 36–5 Bulbs and corms. (*a*) The bulb is an underground stem to which are attached overlapping, fleshy leaves. (*b*) The entire corm is stem tissue, in contrast to the bulb.

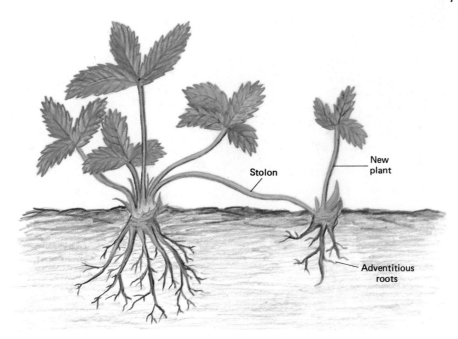

Stolon

New plant

Adventitious roots

FIGURE 36–6 The wild strawberry (*Fragaria virginiana*) reproduces asexually by forming stolons, or runners.

Leaves

Some plants are capable of forming plantlets along their leaf margins. *Kalanchoe*, commonly called "mother of thousands," has meristematic tissue in the leaf that gives rise to an individual plant at each notch in the leaf (Figure 36–7). When these plants attain a certain size, they drop to the ground and grow.

Roots

Generally, buds originate on stems rather than roots. Some roots produce **suckers,** which are aboveground stems that develop from adventitious buds on the roots. Each sucker grows roots at the base of the stem (Figure 36–8). Examples of plants that form root suckers include the black locust, pear, apple, cherry, red raspberry, and blackberry. It is possible to separate the suckers from the parent plant. Some weeds are able to produce consid-

FIGURE 36–7 The "mother of thousands" (*Kalanchoe*) produces young plants along the margins of the leaves. When the young plants attain a certain size, they drop off and root in the ground.

FIGURE 36–8 A grove of aspen trees in Utah. Quite often the entire grove is descended from a single plant that reproduced asexually by forming adventitious buds on the roots. These buds developed into suckers, each of which became a separate tree. Because the entire grove is genetically identical, their responses to the environment are uniform. In spring they break dormancy simultaneously, and in the fall their leaves turn color at the same time.

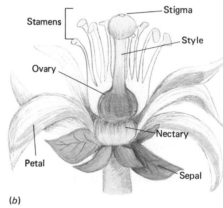

FIGURE 36–9 The head of a dandelion. Each individual plumed structure is a fruit that contains a seed. The seeds were produced asexually, by apomixis. The fruits are dispersed by wind currents.

(a)

Stamens

Stigma

Style

Ovary

Nectary

Petal

Sepal

(b)

FIGURE 36–10 Development of orange fruits. (*a*) Orange flowers. Note that several of the flowers have already lost their petals and stamens, revealing the stigma, style, and ovary of the female structure. (*b*) Diagram of an orange flower. Note the nectary in the diagram and in photo (*a*). (*c*) Maturing ovaries. The stigma and style have dropped off several ovaries already. (*d*) Longitudinal section through a maturing ovary. Note the abscission zone at the base of the style, where it will separate from the ovary. (*e*) Immature fruits. (*f*) Mature fruits. The orange is a modified berry. (*g*) Cross section through an orange. Each section of an orange is a single carpel. The ovary that develops into the fruit is composed of several fused carpels.

(c)

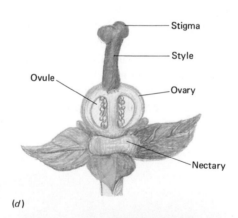

Stigma

Style

Ovule

Ovary

Nectary

(d)

(e)

(f)

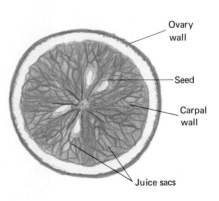

Ovary wall

Seed

Carpal wall

Juice sacs

(g)

erable numbers of plants by this method. These plants are difficult to control, because pulling the plant out of the soil seldom removes all the roots. In response to wounding, the roots produce more and more adventitious buds, which can be a considerable nuisance.

Apomixis

Sometimes plants produce seeds and fruits without meiosis, fusion of gametes, and the other aspects of sexual reproduction. When this occurs, it is known as **apomixis.** For example, the embryo may develop from a diploid cell in the ovule rather than from the diploid zygote that forms from the union of two haploid gametes. The seeds produced by apomixis are a form of asexual reproduction because the embryo is genetically identical to the original parent. Example of plants that reproduce by apomixis include the dandelion, citrus plants, blackberry, and certain grasses (Figure 36–9).

SEXUAL REPRODUCTION IN FLOWERING PLANTS

The life cycle of angiosperms, including details about the flower, were considered in Chapter 28. After fertilization has occurred within the ovule in the ovary, the ovule develops into a seed and the ovary surrounding it develops into a **fruit** (Figure 36–10). Therefore, a fruit can be defined as a mature, ripened ovary. There are several types of fruits, which vary in structure due to variations in the flowers from which they were formed. We will consider a few representative types.

Types of Fruits

There are four basic types of fruits (Table 36–1). A **simple fruit** develops from a single ovary of a single flower. Most fruits are simple fruits. At maturity simple fruits may be fleshy or dry. Two examples of fleshy fruits are the berry and the drupe (Figure 36–11). The **berry** is a fleshy fruit that has soft tissues throughout. Using this definition, a tomato is a berry, as are grapes and bananas. A **drupe** is a simple, fleshy fruit that has a hard, stony pit surrounding the seed. Examples of drupes include peaches, plums, and avocados.

Many simple fruits are dry at maturity. These fruits fall into two main categories. Some are **dehiscent,** and split open at maturity (Figure 36–12). The milkweed **follicle** is an example of a simple, dry, dehiscent fruit that

(a)

(b)

FIGURE 36–11 Simple fruits that are fleshy at maturity. (*a*) The tomato is an example of a berry. Note that the fruit is soft throughout. (*b*) The peach is a drupe. The hard, stony pit is part of the ovary. The single seed is inside the pit.

Table 36–1 SOME TYPES OF FRUITS*

I. Simple fruit
  A. Fleshy
    1. Berry
    2. Drupe
  B. Dry
    1. Dehiscent
      a. Follicle
      b. Legume
      c. Capsule
    2. Indehiscent
      a. Grain
      b. Achene
II. Aggregate fruit
III. Multiple fruit
IV. Accessory fruit

*The number of specific fruit types is great. This table includes only those types that are discussed in the text.

(a)

(b)

(c)

FIGURE 36–12 Simple, dry, dehiscent fruits. (*a*) The milkweed follicle splits open along one seam. (*b*) The bean fruit is a legume, which splits open along two seams at maturity. (*c*) The capsule splits open along multiple seams or pores at maturity. These iris fruits dehisce along three seams.

FIGURE 36–13 Simple, dry, indehiscent fruits. (*a*) The corn fruit is a grain. In grains the fruit wall is fused to the seed coat. (*b*) The sunflower fruit is an achene. The seed coat is not fused to the fruit wall. Therefore, it is possible to peel off the fruit wall, separating it from the seed.

splits open along one seam or suture to release the seeds. The **legume** is a fruit that splits open along two seams or sutures. Pea pods are examples of legumes. So are green beans, although they are harvested before the fruit has dried out and split open. A **capsule** splits open along multiple seams or pores. Poppy and cotton fruits are capsules.

Other simple, dry fruits do not split open at maturity. These fruits are **indehiscent** (Figure 36–13). The **grain** is an example of a simple, dry, indehiscent fruit. Each grain contains one seed; the seed coat is fused to the fruit wall, so it appears that the grain is a seed rather than a fruit. Kernels of corn and wheat are actually fruits of this type. The **achene** is a similar fruit in that it is simple, dry, indehiscent, and contains a single seed. However, the seed coat is not fused to the fruit wall in achenes. Rather, the single seed is attached to the fruit wall at one point only. Therefore, one can separate the achene from its seed. The sunflower fruit is an example of an achene. One can peel off the fruit wall to reveal the seed within.

Aggregate fruits are a second main type of fruit. An aggregate fruit is formed from a single flower that contains many separate carpels. After fertilization each ovary from each individual carpel enlarges. As they enlarge, they fuse to form a single fruit. The raspberry and blackberry are examples of aggregate fruits (Figure 36–14).

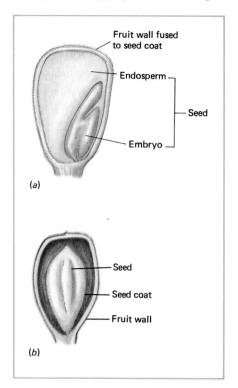

Fruit wall fused to seed coat

Endosperm

Seed

Embryo

(a)

Seed

Seed coat

Fruit wall

(b)

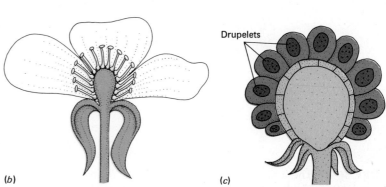

(b)

(c)

Drupelets

FIGURE 36–14 Raspberries and blackberries are examples of aggregate fruits. (*a*) Developing fruits. (*b*) Cutaway view of a raspberry flower, showing the many separate carpels in the center of the flower. (*c*) Longitudinal section through a raspberry fruit, which is an aggregate of tiny drupes.

(a)

FIGURE 36–15 The pineapple is a multiple fruit, formed from the ovaries of many separate flowers.

A third type is the **multiple fruit,** which is formed from the ovaries of many flowers. Because these flowers grow in close proximity, the ovary from each fuses with nearby ovaries as it enlarges and develops after fertilization. Pineapples and osage oranges are multiple fruits (Figure 36–15).

Accessory fruits are the fourth type. They are different from the other types in that other plant tissues, in addition to ovary tissue, make up the fruit. For example, the major edible portion of the strawberry is the red, fleshy **receptacle,** which is the terminal part of the flower stalk (Figure 36–16). Apples and pears are also accessory fruits. The outer part of each fruit is the enlarged **floral tube** that surrounds the ovary (Figure 36–17).

FIGURE 36–16 The strawberry is three fruits in one! It is an accessory fruit because the major part of it is tissue other than ovary tissue. It is an aggregate fruit because it develops from a single flower that has many separate carpels. Finally, each speck on the strawberry is a tiny achene that develops from one of the separate carpels.

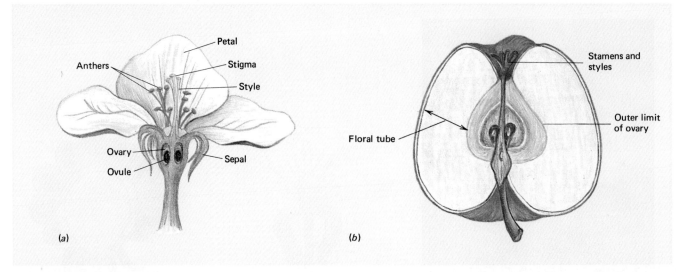

FIGURE 36–17 The apple is an accessory fruit. (*a*) Note the floral tube surrounding the ovary in the apple flower. This tube becomes the major edible portion of the apple. (*b*) Longitudinal section through an apple, showing the floral tube and ovary tissue.

Fruit and Seed Dispersal

The angiosperms have evolved a number of ways to disperse their seeds. This has given some of them the opportunity to expand their range. If the seed is carried to an environment that is suitable for growth, the seed will germinate and the plant will become established in that habitat. In some cases the seed is the actual agent of dispersal. In others it is the fruit. The tumbleweed is an example of the entire plant being the agent of dispersal, as it detaches and blows across the ground. As it bumps along, seeds fall out (Figure 36–18).

The wind is responsible for seed and fruit dispersal in many plants (Figure 36–19). Plants that have winged fruits, such as the maple, are adapted for wind dispersal. Light, feathery plumes on the fruit or seed allow it to be transported, often considerable distances. The dandelion fruit and milkweed seed have this type of adaptation.

Some plants have evolved special structures that aid in animal dispersal of their seeds and fruits (Figure 36–19). The spines and barbs of the cocklebur and similar fruits catch in the fur of animals. Fleshy, edible fruits are also adapted for animal dispersal. As they are eaten, the seeds are often swallowed. Because of their thick seed coats, they are not digested. Rather, they pass through the animal's digestive tract and are deposited in the animal's feces some distance away from the plant.

FIGURE 36–18 Tumbleweeds (*Salsola kali*) blown against a fence. As they blow across the land, their seeds fall out and are thus dispersed over great distances.

(a)

(b)

(c)

(d)

(e)

The coconut is a good example of a fruit that is adapted for dispersal by water (Figure 36–19). It has air spaces and corky floats that make it buoyant. It is capable of being carried by ocean currents for thousands of miles. When it washes ashore, it germinates and grows into a coconut palm tree.

Some fruits accomplish dispersal without relying on wind, animals, or water. These fruits are often explosive, forcibly discharging their seeds. Pressures due to differences in turgor or to drying out cause them to burst open suddenly, scattering the seeds for considerable distances (Figure 36–20).

ENVIRONMENTAL CUES THAT INDUCE SEXUAL REPRODUCTION

The initiation of sexual reproduction is often under environmental control, particularly in temperate latitudes. This is important for the plant's survival, because the timing of sexual reproduction is critical to reproductive success. Plants must be able to flower and form fruits and seeds before dormancy is induced by the onset of winter. A number of plants can detect changes in the relative amounts of daylight and darkness that accompany the changing seasons, and flower in response to these changes. These plants vary in their response to the duration and timing of light and dark, but the overall mechanism of detection is the same. Other plants have temperature requirements that induce sexual reproduction.

FIGURE 36–19 Methods of seed and fruit dispersal. (*a*) The feathery plumes of a milkweed seed make it buoyant for dispersal by wind. (*b*) The fruits of sugar maple have wings for wind dispersal. (*c*) Spiny barbs of the common sandspur get caught in animal fur. When it falls off, it is usually far from its original location. (*d*) Fleshy fruits are eaten by animals such as this meadow vole. The seeds are frequently swallowed whole and pass unharmed through the animal's digestive tract. (*e*) Coconuts are adapted for water dispersal. When it washes ashore, the coconut germinates, often thousands of miles from its original home.

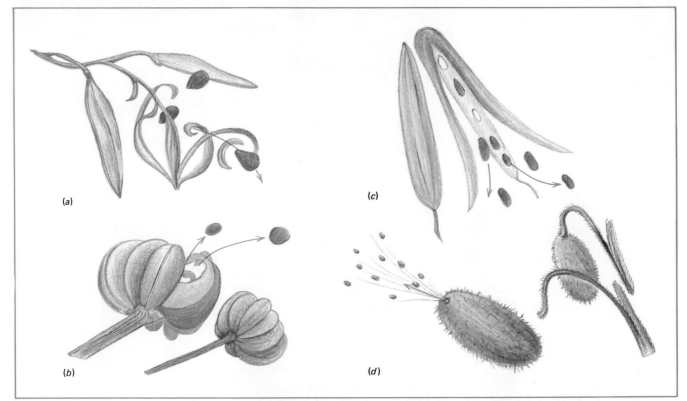

FIGURE 36–20 Forcibly discharged seeds. (*a*) Touch-me-not (*Impatiens*). Differences in turgor pressure between inner and outer walls of the fruit cause it to split apart and curl up, ejecting the seeds in the process. (*b*) Sandbox tree (*Hura crepitans*). As the fruit dries, the segments split apart and sling the seeds out. (*c*) Bitter cress (*Cardamine*). As the fruit dries, stresses are created that cause it to snap apart explosively. (*d*) Squirting cucumber (*Ecballium*). When the fruit separates from the stalk, the seeds are squirted from a basal opening.

FIGURE 36–21 The chrysanthemum is a short-day plant. The plant on the left received 8 hours of daylight and 16 hours of darkness. The plant on the right received 16 hours of daylight and 8 hours of darkness.

Light and Photoperiodism

Photoperiodism is the response of a plant to the relative lengths of daylight and darkness. Flowering is one of several physiological activities that are photoperiodic in some plants. For example, if one were to plant biloxi soybeans at two-week intervals from early May to August, they would all flower at the same time in September, regardless of size or age.

Plants can be placed into three main groups on the basis of how photoperiodism affects their flowering. **Short-day plants** were initially defined as plants that flower when exposed to some critical day length or less. However, the important factor in initiation of flowering in short-day plants is the long, uninterrupted period of darkness rather than the short period of daylight. In other words, short-day plants flower when the night length is equal to or greater than some critical length. Common short-day plants are the chrysanthemum and the poinsettia (Figure 36–21). These plants typically flower in late summer or fall.

Long-day plants were initially defined as being able to flower when the day length is equal to or greater than some critical amount. However, a more accurate definition would be that long-day plants require some critical period of darkness or less. Plants that flower in late spring or summer, such as clover, black-eyed Susan, and lettuce, are long-day plants. The critical day length (or night length) varies from species to species. Two different plants could have the same critical day length, but one could be a short-day plant and the other could be a long-day plant.

Some plants do not initiate flowering in response to changing amounts of daylight and darkness. These **day-neutral plants** have some other type of stimulus, either external or internal, that causes them to flower. The tomato, dandelion, string bean, and pansy are day-neutral plants.

In order for plants, or any living organism, to have a biological response to light, there must be something in that organism that perceives the light. There are often different **photoreceptors** for different physiological responses. What is the photoreceptor for photoperiodism? That is, what determines whether it is daylight or darkness in plants prior to the initiation of flowering?

Phytochrome

The photoreceptor involved in photoperiodism and a number of other light-initiated physiological responses of plants is a blue-green, proteinaceous pigment called **phytochrome.** Phytochrome, which is universal in vascular plants, has two forms. It can readily convert from one form to the other upon absorption of light of specific wavelengths. One form, designated P_R (for red-absorbing phytochrome), absorbs red light (660 nm) strongly. In the process the conformation of the molecule changes to the second form of phytochrome, P_{FR}. This form of phytochrome is so designated because it absorbs red light of longer wavelengths than P_R, described as far-red light (730 nm). When P_{FR} absorbs far-red light, it reverts back to the original form, P_R. The P_{FR} form of phytochrome is less stable than the P_R form and reverts spontaneously, albeit slowly, to P_R in the dark. The physiologically active form of phytochrome is P_{FR}.

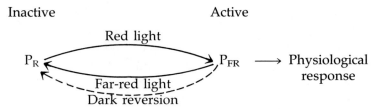

But what does a pigment that absorbs red light and far-red light have to do with daylight? The sun's light is composed of the entire spectrum of visible light in addition to ultraviolet and infrared. However, sunlight has more red light than far-red light. Therefore, the phytochrome in a plant exposed to the sunlight will be a mixture of both P_R and P_{FR}, with P_{FR} predominating. During the night the P_{FR} will revert back to P_R.

Phytochrome and Photoperiodism

In short-day plants the active form of phytochrome, P_{FR}, *inhibits* flowering. In order to flower, these plants need long nights. The long period of darkness allows the P_{FR} to revert back to P_R so the plant has some minimum time during the 24-hour period with *no* P_{FR} present. This initiates flowering.

Biologists have experimented with short-day plants by growing them under a short-day/long-night regime, but interrupting the night with a short burst of red light. Exposure to red light for as brief a period as 10 minutes in the middle of the night will prevent flowering in short-day plants (Figure 36–22). This effect occurs because the brief exposure to red light converts some of the phytochrome to the P_{FR} form. Therefore, the plant does not have a sufficient period of time at night without any P_{FR}.

The effect that a short period of red light in the middle of the night has on short-day plants is reversible. That is, if a short-day plant is grown

FIGURE 36–22 Photoperiodic response of a short-day plant (top row) and a long-day plant (bottom row) to different periods of light and dark. Note that the short-day plant does not flower when exposed to 8 hours of daylight and 16 hours of darkness interrupted with a brief flash of light. This same treatment induces the long-day plant to flower.

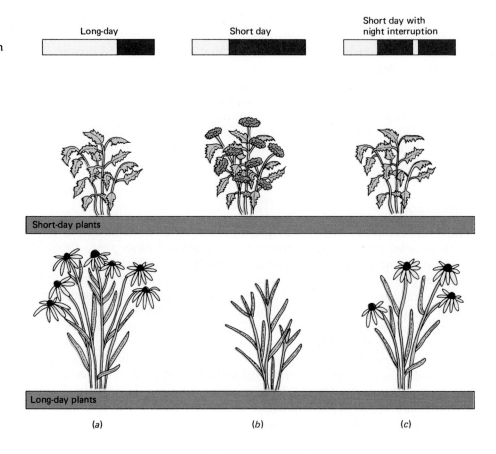

under conditions of short days and long nights, with a brief flash of red light followed by a brief flash of far-red light, that plant will flower. Based on our understanding of the photoreversible nature of phytochrome, this observation is easy to explain. Short-day plants need long nights to allow for dark reversion of P_{FR} to P_R in order to induce flowering. A brief flash of red light in the middle of the night converts P_R into P_{FR}. However, if this is followed by a period of far-red light, the P_{FR} that was formed is converted back into P_R. Therefore, flowering occurs.

In long-day plants the active form of phytochrome, P_{FR}, *induces* flowering. Long-day plants that are exposed to a long-day/short-night regime flower. The long days cause these plants to produce predominantly P_{FR}. During the short nights P_{FR} is changed to P_R, but because the night is short the plant has very little time with no P_{FR} present during a 24-hour period. Hence, it flowers.

Plant biologists are puzzled by the observation that the active form inhibits flowering in short-day plants and induces flowering in long-day plants. Why different plants respond so differently to P_{FR} is not known at this time. Biologists are also seeking the exact mechanism of phytochrome action. That is, once it has absorbed light and changed into another form, what happens next? Does this somehow trigger the production of a hormone, which then triggers flowering? Or does phytochrome work by affecting membranes or enzymes or gene expression? Does it work by some combination of these methods? Regardless of its mode of action, the universal presence of phytochrome in vascular plants attests to its importance.

Other Roles for Phytochrome

Phytochrome has been implicated in a number of physiological responses besides flowering. For example, it is involved in the light requirement that some seeds have for germination (Chapter 32). Proof that phytochrome is

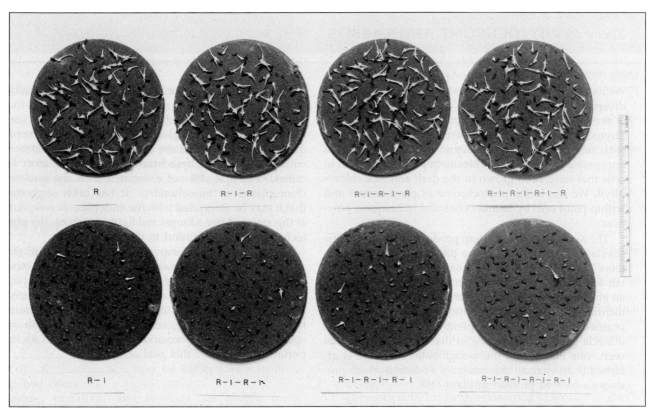

FIGURE 36–23 The control of lettuce seed germination by red (R) and far-red (I) light. Seeds are moistened and then exposed to red light (for 1 minute each exposure) and far-red light (for 4 minutes each exposure) in the sequences indicated. If the last exposure is red light, most of the seeds germinate. If the last exposure is far-red light, they remain dormant.

the photoreceptor for this response is evident in its photoreversibility when exposed to red or far-red light. Seeds with a light requirement must be exposed to red light. Exposure to red light converts P_R to P_{FR}, and germination occurs. However, if the seeds are exposed to a brief period of red light followed by a brief period of far-red, they will not germinate because P_{FR} is converted back to P_R, the inactive form. Experiments on the photoreversible nature of phytochrome have been conducted in which seeds are exposed to alternating forms of light many, many times. Regardless of how many light treatments one gives the seeds, they always respond to the *last* treatment. If the last treatment is red light, the seeds will germinate. If the last treatment is far-red light, the seeds remain dormant (Figure 36–23).

Other physiological functions under the influence of phytochrome include sleep movements in leaves (Chapter 37), shoot dormancy, leaf abscission, and pigment formation in flowers, fruits, and leaves. If phytochrome has been implicated in such diverse physiological responses, then light is required to initiate these responses. The importance of light in various plant functions besides photosynthesis cannot be overemphasized. Timing of daylight and darkness is a key way for plants, as well as animals, to measure the change in time from one season to the next. This measurement is crucial for survival, particularly in environments where the climate fluctuates.

Temperature and Vernalization

In certain plants the temperature has an effect on flowering. The promotion of flowering by treatment with cold is known as **vernalization.** The part of the plant that must be exposed to cold varies. For some plants the moist seeds must be exposed to a period of several weeks of cold. For other plants the young, recently germinated seedlings have the cold requirement. Some plants have an absolute requirement for the cold period. That

Schefflera grown in a testtube

37

Plant Hormones and Responses

■ LEARNING OBJECTIVES

After you have read this chapter you should be able to:

1. Distinguish between a tropism and a turgor movement.
2. Distinguish between phototropism, geotropism, and thigmotropism.
3. List several different ways each of the following hormones affect plant growth and development: auxin, gibberellins, cytokinins, ethylene, and abscisic acid.
4. Relate the steps involved in the acid growth hypothesis regarding how auxin induces cell elongation.
5. Explain the effect of gibberellin on gene activity in germinating barley seeds.
6. Give an example of a physiological response in plants that may be due to varying ratios of several hormones rather than one specific hormone.

FIGURE 37–1 Plant growth in the direction of light, phototropism, demonstrates that plants respond to their environment.

PLANT MOVEMENTS AND GROWTH RESPONSES

Because most plants are firmly rooted in the ground, it is generally assumed that they are incapable of self-directed movements. However, there are a variety of growth movements and responses found in plants. Some of these are very gradual and others are quite rapid and spectacular, as when the Venus flytrap snaps its leaf shut in less than 0.1 second after being stimulated (Chapter 33).

Many aspects of plant growth and development use environmental cues to determine when, whether, and to what extent they will happen. We have already examined several of these. The germination of seeds is influenced by water, oxygen, temperature, and light (Chapters 32 and 36). Light affects such diverse physiological responses as photosynthesis (Chapter 8), stomatal opening and closing (Chapter 33), and flowering (Chapter 36). And flowering is promoted in many plants by periods of low temperature (Chapter 36).

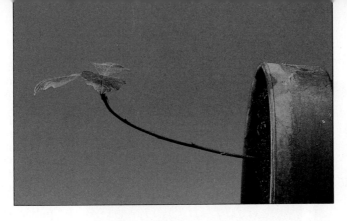

FIGURE 37–2 Stems exhibit negative gravitropism. When placed on its side, this white oak stem responded by growing upward, against the direction of gravity. The response is evident in 15 hours or less.

Tropisms

Plant growth in response to an external stimulus is known as a **tropism.** Tropisms may be positive or negative depending on whether the plant grows toward or away from the stimulus. **Phototropism** is the growth of a plant due to the direction of light (Figure 37–1). Most stems exhibit positive phototropism and bend toward light. A growth in response to the direction of gravity is **gravitropism** (Figure 37–2). Stems generally exhibit negative gravitropism, while roots exhibit positive gravitropism. **Thigmotropism** is growth in response to a mechanical stimulus, such as contact with a solid object. The twining or curling growth of tendrils is an example of thigmotropism (Figure 37–3). There are also tropisms in plants due to other stimuli in the environment such as water, temperature, chemicals, and oxygen.

Turgor Movements

Mimosa pudica, the sensitive plant, dramatically folds its leaves in response to an external stimulus (Figure 37–4). The stimulus may be mechanical, electrical, chemical, or thermal. It is possible that this unusual behavior protects the plant from predators.

When a *Mimosa* leaf is stimulated by touching, an electrical signal moves down the leaf to special cells in an organ at the base of the petiole, the **pulvinus.** While it is recognized that plants such as *Mimosa* can use electrical signals for intercellular communication, the actual mechanism of the transmission is imperfectly understood at this time. When the electrical signal reaches the pulvinus cells, it triggers a loss of turgor in those cells as water, potassium ions, and tannins leave. The sudden change in turgor is responsible for the leaf movement. The tannins, which are normally stored in the vacuole, impart a bad taste to the tissue, and some researchers have suggested this as a further mechanism of *Mimosa* to avoid predation.

The closure of the Venus flytrap leaf (Chapter 33) is similar in its mechanism to that of *Mimosa* (Figure 33–17). An electrical signal, which moves much more rapidly than in *Mimosa*, induces a movement of water out of certain motor cells. A movement of ions is also associated with the water movement, but it is not known which ions are involved.

Circadian Rhythms

Plants, animals, and microorganisms appear to have an internal timer, or biological clock, that approximates a 24-hour cycle. These internal cycles are known as **circadian rhythms** (obtained from Latin words meaning "around" and "day"). Circadian rhythms usually are between 20 and 30 hours. In nature the rising and setting of the sun resets the clock each day. Phytochrome (Chapter 36) has been implicated as the photoreceptor involved in resetting the biological clock for many plants.

FIGURE 37–3 The twining motion of a tendril is an example of thigmotropism. Here an adult *Heliconius cydno* made the mistake of roosting on a live *Passiflora* tendril overnight.

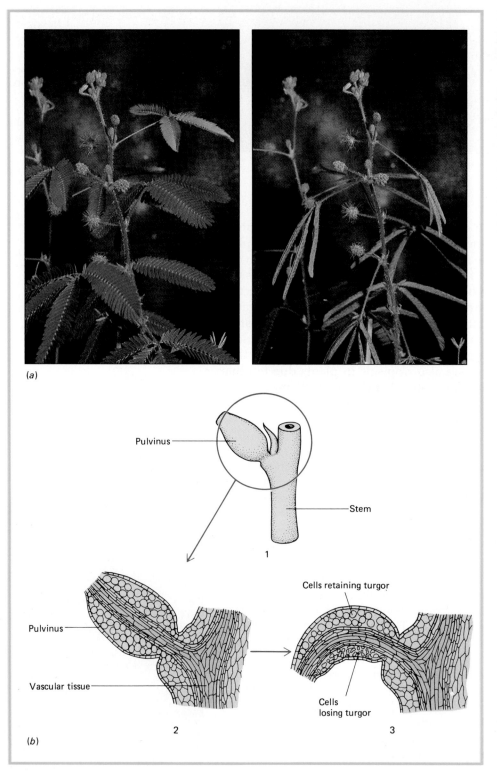

(a)

Pulvinus

Stem

1

Pulvinus

Vascular tissue

Cells retaining turgor

Cells
losing turgor

2   3

(b)

FIGURE 37–4 Turgor movements in the "sensitive plant," *Mimosa pudica*. (a) Left, *Mimosa pudica* before being disturbed. Right, the plant several seconds after being touched. Note how the leaves have folded and drooped. (b) (1) The base of the petiole, showing the pulvinus. (2) Section through the pulvinus, showing cells when leaf is undisturbed. (3) Section through the pulvinus, showing loss of turgor that produces the folding of the leaves.

One example of circadian rhythms in plants is the opening and closing of stomates that occurs independently of light and darkness (Chapter 33). Plants placed in continual darkness for extended periods continue to open and close their stomates. Another example of circadian rhythms in plants is the sleep movements observed in the common bean and other plants (Figure 37–5). During the day the leaves are horizontal for optimum light absorption. At night the leaves fold down or up, perpendicular to their daytime orientation. By connecting the leaf to a pen on a rotating disk, the

(a)

(b)

FIGURE 37–5 Sleep movements in *Oxalis*. (a) Leaf position during the day. (b) Leaf position at night.

movement can be measured and timed (Figure 37–6). Results from studies such as this indicate that the plant actually "anticipates" sunrise and sunset, as the movements begin *before* the sun rises or sets. These sleep movements occur independently of the 24-hour cycle in nature. If bean plants are placed in continual darkness or continual light, the movements continue, although on an approximately 23-hour cycle.

HORMONAL REGULATION OF PLANT GROWTH AND DEVELOPMENT

Plants, like animals, use **hormones** to regulate their development and growth. Plant hormones are organic compounds produced in one part of the plant and transported to another part, where they elicit a physiological response. Hormones are effective in extremely small amounts. For that reason their study is very challenging. In plants the study of hormones is even more difficult because each plant hormone elicits many different responses. Also, the effects of different hormones overlap, so that it is difficult to determine which hormone, if any, is the primary cause of the response. Finally, plant hormones may be stimulatory or inhibitory depending on their concentrations.

There are five groups of plant hormones: auxins, gibberellins, cytokinins, ethylene, and abscisic acid. Together they control the growth of the plant at various stages of development.

FIGURE 37–6 A record of sleep movements can be made by attaching a leaf to a rotating drum. The plant is maintained in continuous dim light but will continue to exhibit sleep movements.

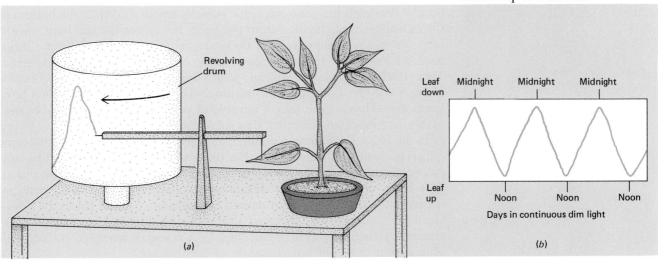
(a)

(b)

FIGURE 37–7 Darwin's experiment with coleoptiles of canary grass seedlings. (Upper row) Some plants were uncovered, some were covered only at the tip, some had the tip removed, and some were covered everywhere but at the tip. (Lower row) After exposure to light coming from one direction, the uncovered plants and the plants with uncovered tips (right) grew toward the light. The plants with covered tips (left center) or tips removed (right center) did not bend toward light. Darwin and his son concluded that the tip is sensitive to light and produces some "influence" that moves down the plant and causes the bending.

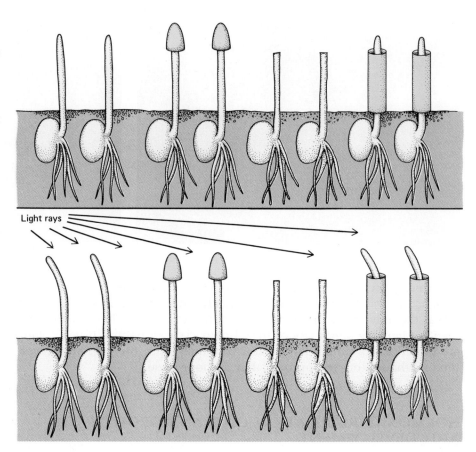

Light rays

Auxins

Although he is known mostly for originating the concept of natural selection to explain evolution, Charles Darwin was a gifted naturalist who experimented on a wide number of plants and animals. Darwin and his son, Francis, were interested in phototropism, the growth of plants toward light. In the 1880s they experimented with newly germinated canary grass seedlings (Figure 37–7). The first part of the seedling to emerge from the soil is the coleoptile (Chapter 32). When they exposed coleoptiles to unidirectional light, the coleoptiles bent toward the light. The bending occurred close to, but not at the very tip of, the coleoptile. They tried to influence this bending in several ways. For example, they covered the tip of the coleoptile as soon as it emerged from the soil. The plants treated in this manner did not bend! Likewise, bending did not occur when the coleoptile tip was removed (i.e., the coleoptile was decapitated). When the bottom of the coleoptile was shielded from the light, the coleoptile still bent toward light. From these experiments, the Darwins concluded that "some influence is transmitted from the upper to the lower part, causing it to bend." This conclusion fits the definition of a hormone exactly. Thus, Charles Darwin was the first person to produce data suggesting that plants have hormones. However, it took a number of years before the techniques necessary to extract and identify this substance were available.

In the 1920s Frits Went, a young Dutch scientist, isolated the phototropic hormone from oat coleoptiles. He removed the oat coleoptile tips and placed them on tiny blocks of agar for a period of time. When he put an agar block on the side of a decapitated coleoptile in the dark, bending occurred (Figure 37–8). Went named this substance **auxin** (from the Greek word for "enlarge" or "increase"). The purification and elucidation of its chemical structure was accomplished by a research team led by Kenneth Thimann at the California Institute of Technology.

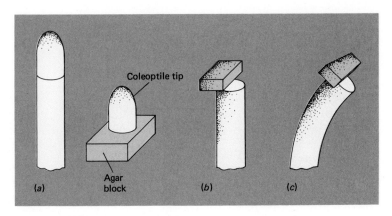

FIGURE 37–8 Frits Went's experiment. (*a*) Coleoptile tips were placed on agar blocks for a period of time. (*b*) The agar block was transferred to a decapitated coleoptile. It was placed off-center, and the coleoptile was left in continuous darkness. (*c*) The coleoptile bent, indicating that a chemical had been transferred from the original coleoptile tip to the agar block to the decapitated coleoptile.

Chemistry of Auxins

Auxins are any compounds that stimulate phototropic curvature in oat coleoptiles. The main auxin found in plants is indoleacetic acid, or IAA (Table 37–1). Its structure is similar to that of the amino acid, tryptophan, from which it is synthesized. A number of synthetic auxins have been

Table 37–1 THE FIVE PLANT HORMONES

Hormone	Chemical Structure	Site of Production	Method of Translocation
Auxin		Shoot apical meristem, young leaves, seeds	Polar transport in parenchyma cells
Gibberellin		Young leaves, root and shoot apical meristems, embryo in seed	Unknown
Cytokinin		Roots	Xylem
Ethylene		Stem nodes, ripening fruit, senescing tissue	Unknown (diffusion?)
Abscisic acid		Older leaves, root cap, stem	Vascular tissue

FIGURE 37–9 The acid-growth hypothesis. Auxin activates a proton pump in the plasma membrane. This pumps hydrogen ions out of the cell to the cell wall, changing its pH. The lowered pH of the cell wall activates enzymes that break the cross links holding the cellulose microfibrils together. The pressure created by increasing turgor then allows the wall to expand.

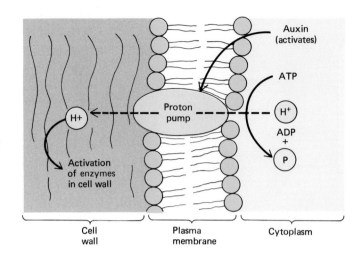

made with similar structures. IAA is synthesized in the shoot apical meristem, young leaves, and seeds. It is not translocated in either the xylem or the phloem. It moves through the plant within the parenchyma cells, at a rate that is too fast to be accounted for by diffusion. The movement of auxin is called **polar transport** because it is always unidirectional, or polar, from the top of the shoot toward the roots. Polar transport requires energy and is not due to the influence of gravity. If a section of stem is inverted, the auxin still moves toward the root end of the plant.

Functions of Auxins

Auxin causes cell elongation in plants. Recall that cell elongation occurs in apical meristems just behind the area of cell division (Chapter 32). Auxin apparently exerts this effect by changing the cell walls so they can expand. According to the **acid-growth hypothesis,** auxin triggers a proton pump in the plasma membrane (Figure 37–9). This causes a flow of H^+ ions through to the cell wall, acidifying it and activating certain enzymes that break

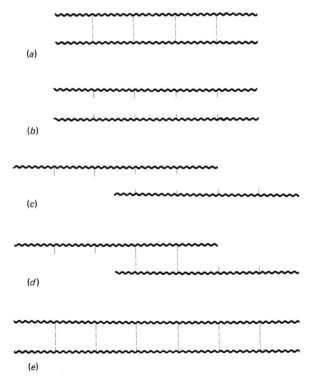

FIGURE 37–10 How wall expansion associated with cell elongation may occur. (*a*) The cell wall is composed of cellulose microfibrils held in place by polysaccharide cross links. (*b*) Enzymes activated by a lowered pH break the cross links. (*c*) The pressure created by increasing turgor causes the wall to stretch. (*d*) New polysaccharide cross links are formed, holding the wall in its new position. (*e*) New cell wall materials are synthesized, completing the expanded wall.

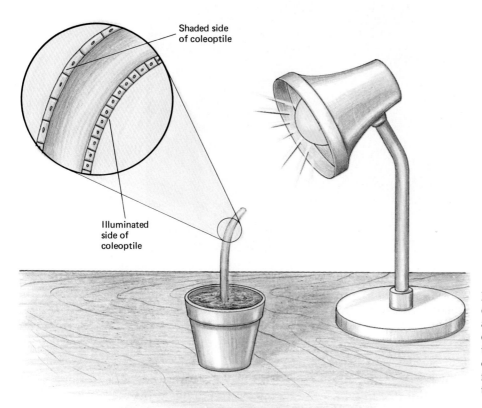

Shaded side
of coleoptile

Illuminated
side of
coleoptile

FIGURE 37–11 Phototropism is due to the unequal distribution of auxin. Auxin travels down the side of the stem or coleoptile *away* from the light, causing the cells on the darkened side to elongate. Therefore, the stem or coleoptile bends toward light.

bonds between cell wall molecules (Figure 37–10). As a result, the wall becomes flexible and can stretch as water accumulates in the vacuole.

Phototropism can be explained by auxin's effect on cell elongation. When a plant is exposed to a unidirectional source of light, the auxin migrates to the dark side of the stem before being transported down the stem. As a result, the cells on the dark side of the stem elongate more than the cells exposed to light, and the stem bends (Figure 37–11).

Auxin also influences gravitropism, although the mechanism is incompletely understood at this time. To complicate matters, other hormones have been implicated in gravitropism in addition to IAA. Auxin is believed to be the hormone involved in thigmotropism and the other tropisms as well.

Certain plants tend to branch out very little. Growth in these plants occurs from the apical meristem, rather than from the lateral meristems. Such plants are said to exhibit **apical dominance.** In plants with strong apical dominance, it appears that auxin inhibits the development of lateral buds (Figure 37–12). When the apical meristem is pinched off, the auxin source is removed and lateral buds develop into branches.

IAA produced by seeds stimulates the development of the fruit. When auxin is applied to flowers in which fertilization has not been allowed to occur, the ovary enlarges and develops into a seedless fruit. Seedless tomatoes have been developed in this manner. Auxin is not the only hormone involved in fruit development, however.

Synthetic auxins have several other commercial applications. NAA (naphthalenacetic acid) is used to stimulate root development on stem cuttings for asexual propagation, particularly of woody plants (Figure 37–13). A synthetic auxin, 2,4-D (2,4-dichlorophenoxyacetic acid), is used as a selective herbicide. It is applied at high concentrations and causes exaggerated growth in some plant parts and growth inhibition in others. For reasons that are not understood at the present time, monocots are less sensitive to the concentration of 2,4-D that is applied. Therefore, an appli-

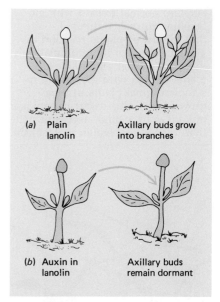

(a) Plain
lanolin

Axillary buds grow
into branches

(b) Auxin in
lanolin

Axillary buds
remain dormant

FIGURE 37–12 Auxin inhibits the development of lateral buds. Both (*a*) and (*b*) have had the tip (source of auxin) removed. (*a*) The control has some lanolin applied to the cut tip. Because there is no auxin to diffuse down the stem, the lateral buds develop. (*b*) When auxin is applied in lanolin, the lateral buds do not develop.

FIGURE 37–13 African violet leaf cuttings. (Left) Cuttings treated with NAA, a synthetic auxin, formed many adventitious roots. (Right) Cuttings placed in water (control) did not form roots in the same time period.

cation of 2,4-D to a lawn or a field of corn will kill the broad-leaved weeds (dicots) and not harm the grass or corn (monocots).

Gibberellins

In the 1920s a Japanese plant scientist, E. Kurosawa, was working on a disease of rice in which the young rice seedlings grow extremely tall and spindly, fall over, and die. The cause of the disease was discovered to be a fungus in the genus *Gibberella* that produces a substance, named **gibberellin,** causing the symptoms. It wasn't until after World War II that scientists in Europe and North America learned of the exciting work done by the Japanese. The first gibberellin discovered in plants was isolated from bean seeds in 1960. Gibberellins are involved in many normal functions of plants. In the case of the "foolish seedling disease of rice," the symptoms were caused by an abnormally high gibberellin concentration in the plant tissue.

Chemistry of Gibberellins

Gibberellins have a complex chemical structure composed of five rings (Table 37–1). More than 70 naturally occurring gibberellins have been discovered; they have the same basic structure but differ in number of double bonds and location of chemical groups. These structural differences are important, however. Some gibberellins have pronounced effects on plant growth, while others are inactive. Gibberellins are produced in the root and stem apical meristems, the young leaves, and the embryo in the seed. It is not known how translocation of gibberellins occurs.

Functions of Gibberellins

As in the foolish seedling disease of rice, gibberellins promote stem elongation in many plants. When a gibberellin is applied to a plant this elongation may be spectacular, particularly in plants that normally have very short stems. Single gene dwarf mutants of corn and peas will grow to a normal height when treated with gibberellins (Figure 37–14). Gibberellins are also involved in the rapid stem elongation that occurs when many plants initiate flowering. This phenomenon is known as **bolting** (Figure 37–15). Gibberellins cause stem elongation by inducing both cell division and cell elon-

FIGURE 37–14 Effect of gibberellin on normal and dwarf corn plants. From left to right: dwarf, untreated; dwarf treated with gibberellin; normal treated with gibberellin; normal, untreated. Note that the dwarf plants respond to gibberellin much more dramatically than the normal plants. Dwarf plants treated with gibberellin resemble normal plants in their growth rate. This dwarf variety is a mutant with a single recessive gene that impairs gibberellin metabolism.

FIGURE 37–15 Bolting in cyclamen; this condition was caused by treatment with gibberellin. (Left) Untreated control. (Right) Treated with gibberellin.

gation. The mechanism of cell elongation is different from that caused by auxin, however.

Gibberellins are involved in several reproductive processes in plants. They stimulate flowering, particularly in long-day plants. In addition, they can substitute for the cold requirement that biennials have before the initiation of flowering (Chapter 36). If gibberellins are applied to biennials during their first year of growth, flowering occurs without the cold period. Gibberellins, like auxins, affect the development of fruits. Commercially, gibberellins are applied to several varieties of grapes to produce larger berries (Figure 37–16).

Gibberellins are involved in the germination process in many plants. The embryo in the seed produces gibberellins that trigger other physiological responses involved in germination. In plants with light or cold requirements for seed germination, gibberellins can be substituted for the specific environmental requirement.

Gibberellins have an important role in the production of enzymes in germinating cereal seeds. The mechanism of action has been studied in detail in germinating barley seeds. The young plant embryo produces gibberellins, which stimulate the seed to synthesize digestive enzymes. These enzymes digest the stored starch in the endosperm, making it available to the young plant as sugar (Figure 37–17).

FIGURE 37–16 Effect of gibberellin on the growth of grapes. (a) Untreated grapes (control). (b) Grapes were treated with gibberellin, producing larger berries.

(a)

(b)

FIGURE 37–17 Mobilization of insoluble storage reserves during germination of a grass seed such as barley. Diagrams *a* to *e* depict conditions within the seed at the times that appear directly above on the graph. (*a*) Seed seen in longitudinal section, showing release of gibberellin (GA) from embryo into endosperm after the seed is wetted. (*b*) Aleurone cells respond to GA by producing digestive enzymes (shown by black dots), excreting these into the starchy endosperm. (*c*) Enzymes break down starch and other molecules in endosperm, releasing soluble nutrients (color dots), which the embryo's cotyledon absorbs (*d*) and delivers to the shoot and root for growth. As coleoptile appears above ground its tip opens, and the enclosed first foliage leaf emerges (*d*). (*e*) By the time storage reserves are depleted, the seedling's first foliage leaf has expanded and begun photosynthesis. Although the formation of enzymes to mobilize insoluble storage reserves occurs in seeds generally, control of enzyme formation by GA seems to be found largely or exclusively in seeds of grasses.

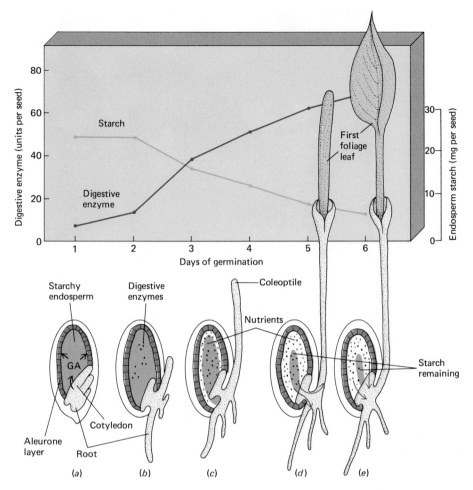

Cytokinins

In the 1940s and 1950s a number of researchers were trying to determine substances that might induce plant cells to divide in tissue culture (Chapter 32). Folke Skoog and others at the University of Wisconsin discovered that cells would not divide without some substance that was transported in the vascular tissue of plants. This active substance was also found in coconut milk and autoclaved herring sperm DNA. Finally, in 1956 the active substance was isolated from herring sperm and called a **cytokinin** because it induces cell division, or cytokinesis. In 1963 the first naturally occurring cytokinin was identified from corn and named zeatin. Since that time several similar molecules have been extracted from other plants.

Chemistry of Cytokinins

Cytokinins have an intriguing structure. They are similar to a purine, adenine (Table 37–1). Cytokinins are found as part of certain transfer-RNA molecules not only in plants, but also in animals and microorganisms. In plants they are produced in the roots and transported in the xylem to all parts of the plant.

Functions of Cytokinins

Cytokinins promote cell division and differentiation in intact plants. They are a required ingredient of plant tissue culture media (Chapter 32) and must be present in order to induce mitosis. In tissue culture, cytokinins interact with auxin during **organogenesis,** the formation of organs. For

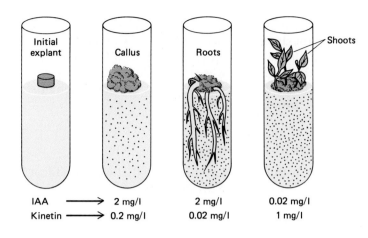

FIGURE 37–18 Growth responses of tobacco tissue culture to auxin and cytokinin. The initial explant is a small piece of sterile tissue from the pith of a tobacco stem, which is placed on a nutrient agar medium as shown at left. After several weeks, the kinds of growth illustrated occur on media supplemented with the indicated levels of auxin and cytokinin.

example, in tobacco tissue culture a high ratio of cytokinins to auxin induces shoot formation, while a low ratio of cytokinins to auxin induces root formation (Figure 37–18).

Cytokinins and auxin also interact in the control of apical dominance. Here their relationship is antagonistic, as auxin inhibits the growth of lateral buds while cytokinin promotes their growth (Figure 37–19). The situation is reversed in roots, with auxin promoting the growth of branch roots and cytokinins inhibiting it.

One very interesting effect of cytokinins on plant cells is to delay their **senescence,** or aging. Plant cells, like all living cells, go through a natural aging process. This process is accelerated in plant parts that are cut, such as cut flowers. Cytokinins somehow promote cells to maintain their normal levels of protein and nucleic acids, thus delaying the rapid aging associated with cut plant parts. It is believed that plants must have a continual supply of cytokinins from the roots. Cut flowers, of course, lose their source of cytokinins. Commercially, cytokinins are sprayed on cut flowers to prevent their rapid senescence.

Ethylene

The effects of **ethylene** on plants had been noted in the 1800s, long before it was recognized as a natural plant hormone. Before electricity was invented, a mixture of various gases called coal gas was used to illuminate homes and street lights. It was noted that plants growing near street lights

FIGURE 37–19 Apical dominance and the effect of auxin and cytokinin. (*a*) When the apical meristem is intact, auxin produced there inhibits the growth of the lateral buds. (*b*) When the apical meristem is removed, the lateral buds grow into branches. (*c*) If a paste of lanolin and auxin is applied to the cut stump immediately after the apical meristem is removed, the lateral buds are inhibited. (*d*) When cytokinins are added to lateral buds of a plant treated as in (*c*), the lateral buds grow. Cytokinin can overcome the apical dominance exerted by auxin.

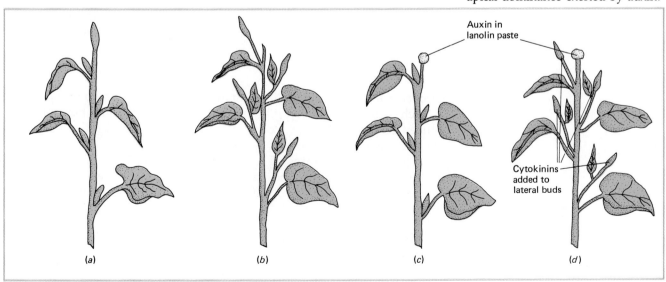

were altered in several ways. The trees shed their leaves early, flowers faded quickly and their petals fell off, and newly sprouted seedlings grew horizontally rather than erect. In 1901 a Russian plant physiologist determined that ethylene was the ingredient in coal gas that caused these effects, but it wasn't until 1934 that scientists demonstrated that ethylene was also produced by plants.

Chemistry of Ethylene

Ethylene is the only plant hormone that is a gas (Table 37–1). It is colorless and smells like ether. Ethylene is produced in several places in plants. It is produced in the nodes of stems, in ripening fruits, and in senescing tissues such as leaves.

Functions of Ethylene

Ethylene has a major role in many aspects of senescence, including the ripening process in fruits. A number of physiological changes occur during fruit ripening. Fruits often change color, as chlorophyll is degraded and other pigments are synthesized. Starch and acids stored in the fruit are converted to sugars, giving the fruit a sweet taste. The fruit cell walls are partly broken down, making the fruit tissue softer. And flavors characteristic of the particular fruit are synthesized. Ethylene triggers these physiological changes. Further, ethylene has a "domino effect." As a fruit ripens, it produces ethylene, which triggers an acceleration of the ripening process. This induces the fruit to produce more ethylene, which further accelerates ripening. The expression "one rotten apple spoils the lot" is true. A rotten apple is one that is overripe. This apple produces large quantities of ethylene, which diffuse and then trigger the ripening process in nearby apples. Ethylene is used commercially to promote the uniform ripening of bananas. Bananas are picked while green and shipped to their destination. There they are exposed to ethylene before delivery to stores. Thus, they ripen uniformly in the stores.

Another effect of ethylene is also related to senescence. Ethylene has been implicated as the hormone that induces leaf abscission (Chapter 33). However, abscission is actually under the control of two plant hormones that are antagonistic toward one another, ethylene and auxin. As the leaf ages and autumn approaches, the level of auxin in the leaf decreases. This initiates several changes in the abscission zone. Concurrently, cells in the abscission zone begin producing ethylene, which triggers other actions. To further complicate the process, it is possible that cytokinins may be involved. Cytokinins, like auxin, decrease in amount as leaf tissue ages.

Abscisic acid

Abscisic acid was discovered simultaneously in 1963 by two independent research teams. P.F. Wareing in England was working on a hormone that induced bud dormancy in woody plants, and F.T. Addicott in California was working on a hormone that promoted the abscission of cotton fruits. Later, when the structures of both hormones were found to be identical, the hormone was given one name, abscisic acid. This was unfortunate because abscisic acid is involved in dormancy, but not in abscission.

Chemistry of Abscisic Acid

Abscisic acid is a single compound rather than a family of compounds. It is a six-carbon ring with a number of substitutions (Table 37–1). It is produced in the leaf, root cap, and stem and is transported in the vascular tissue. Abscisic acid, or ABA, is also high in seeds and fruits, but it is not clear whether it is synthesized or transported there.

Functions of Abscisic Acid

Abscisic acid is sometimes referred to as the "stress hormone." It promotes changes in plant tissues that are exposed to unfavorable conditions, or stressed. The effect of ABA on plants suffering from water stress is best understood. Abscisic acid increases dramatically in the leaves of plants that are exposed to severe drought conditions. The high level of ABA in the leaves triggers the outflow of potassium ions from the guard cells. This induces water to leave the guard cells by osmosis, and the guard cells collapse (Chapter 33). The closing of stomates in water-stressed plants saves a large amount of water that is normally transpired through the stomates, increasing the plant's likelihood of survival. When water is restored to the plant, the stomates do not open immediately. The level of ABA in the leaf cells must decrease before that can occur.

The onset of winter could also be considered a type of stress on the plant. Woody plants cease growth and prepare for winter by forming protective coverings of bud scales over their terminal buds (Chapter 32). These adaptations are promoted by abscisic acid.

Another winter adaptation that involves abscisic acid is dormancy in seeds. If seeds germinated in the autumn, the delicate seedlings would be killed by the first frost. Many seeds have high levels of ABA in their tissues and are, therefore, unable to germinate (Figure 37–20).

The evidence that abscisic acid is the only hormone involved in both plant and seed dormancy is not conclusive. The addition of gibberellin reverses the effects of dormancy. In seeds the level of ABA decreases during the winter, while the level of gibberellin increases. Cytokinins have also been implicated. Once again we see that a single physiological activity in plants may be controlled by the interaction of several hormones (Table 37–2). The actual response may be due to changing ratios of hormones instead of the effect of a single hormone.

FIGURE 37–20 Blooming desert annuals. Sufficient rain has recently fallen to leach abscisic acid from their seeds, so germination occurred. The moisture now available is also adequate to sustain growth throughout their life cycle.

Table 37–2 **SOME OF THE INTERACTIONS BETWEEN PLANT HORMONES DURING VARIOUS ASPECTS OF PLANT GROWTH**

Physiological Activity	*Auxin*	*Gibberellin*	*Cytokinin*	*Ethylene*	*Abscisic Acid*	*Other Factors for Some Plants*
Seed germination		Promotes	?		Inhibits	Cold requirement, light requirement
Growth of seedling into mature plant	Cell elongation, organogenesis[1]	Cell division and elongation	Cell division and differentiation, organogenesis[1]			
Apical dominance	Inhibits lateral bud development		Promotes lateral bud development	?		
Initiation of reproduction (flowering)		Stimulates flowering in some plants[2]	?			Cold requirement, photoperiod requirement
Fruit development and ripening	Development	Development		Promotes ripening		Light requirement (for pigment formation)
Leaf abscission	Inhibits		Inhibits	Promotes		Light requirement
Winter dormancy of plant		Breaks	?		Promotes	Light requirement
Seed dormancy		Breaks	?		Promotes	

[1]In plant tissue culture.
[2]Gibberellin cannot be considered as *the* flowering hormone. There is evidence for a flowering hormone that has not yet been isolated and characterized.

■ SUMMARY

I. Plants can grow in response to external stimuli.
   A. Phototropism is the growth of a plant due to the direction of light.
   B. Gravitropism is the growth of a plant due to the influence of gravity.
   C. Thigmotropism is the growth of a plant in response to contact with a solid object.
II. Some plants respond to external stimuli by changes in turgor in special cells, i.e., turgor movements.
   A. The sensitive plant, *Mimosa pudica,* dramatically folds its leaves in response to various stimuli.
   B. The closure of Venus flytrap leaves is an example of turgor movements.
III. Circadian rhythms, which are regular rhythms in growth or activities of the plant, approximate the 24 hour day and are reset by the biological clock.
IV. Hormones regulate plant growth and development and are effective in small amounts.
   A. The functions of hormones overlap.

B. Many effects of hormones may be due to the ratios of concentrations of several hormones rather than the effect of a single hormone.
V. There are five classes of plant hormones.
   A. Auxins are involved in cell elongation, phototropism, gravitropism, apical dominance, and fruit development.
   B. Gibberellins are involved in stem elongation, flowering, and the germination of seeds.
   C. Cytokinins promote cell division and differentiation, delay senescence, and interact with auxins in apical dominance.
   D. Ethylene has a role in the ripening of fruits and leaf abscission. Ethylene is involved in many aspects of plant senescence.
   E. Abscisic acid is the stress hormone. It is involved in stomatal closure due to water stress, and bud and seed dormancy.

■ POST-TEST

1. _____ is the growth of a plant due to the direction of light.
2. Plant roots generally exhibit _____ gravitropism.
3. The twining of tendrils is an example of _____ .
4. The _____ is an organ at the base of the petiole in *Mimosa* that can undergo rapid changes in turgor, causing dramatic movements of the leaves.
5. Sleep movements observed in beans are an example of _____ _____ .
6. _____ are organic compounds produced in one part of the plant and transported to another, where they cause a positive or negative effect.

7. The movement of auxin, called _____ _____, is unidirectional from the top of the shoot to the roots.

8. According to the _____-_____ hypothesis of cell elongation in plants, auxin triggers a proton pump in the plasma membrane that acidifies the wall, activating enzymes that break bonds between cell wall molecules. Therefore, the wall can be stretched.

9. A synthetic _____, 2,4-D, is used as a selective herbicide.

10. Research on a disease of rice provided the first clues about _____.

11. Gibberellins have an important role in the production of _____ that digest starch in germinating seeds.

12. _____ interact with auxin during organogenesis in tissue culture.

13. The relationship between cytokinins and auxins in apical dominance is _____.

14. _____ delay senescence, while _____ promotes it.

15. The only plant hormone that is a gas is _____.

16. The stress hormone is _____ _____.

17. Abscisic acid promotes the _____ of woody twigs.

■ REVIEW QUESTIONS

1. Why might some plants have sleep movements?
2. Of what value is dormancy in seeds?
3. How is auxin involved in phototropism?
4. If sections of coleoptiles are placed in an acidic solution, they elongate as if auxin were present. Explain why this elongation occurs.
5. Could there be any significance to the large number of gibberellins found in plants?
6. Discuss the hormones that are involved in each of the following: germination of seeds; growth and development of the plant; ripening of fruits; abscission of leaves; and dormancy of seeds.

■ RECOMMENDED READINGS

Galston, A.W., P.J. Davies, and R.L. Satter. *The Life of the Green Plant*, Third Edition. Prentice-Hall, Inc., Englewood Cliffs, New Jersey, 1980. Contains a detailed account of plant hormones.

Raven, P.H., R.F. Evert, and S.E. Eichhorn. *Biology of Plants*, Fourth Edition. Worth Publishers, Inc., New York, 1986. An excellent general botany textbook, with good explanations of plant movements and growth responses.

Weier, T.E., C.R. Stocking, M.G. Barbour, and T.L. Rost. *Botany*. John Wiley, New York, 1982. Another textbook with excellent coverage of plant hormones.

PART VII

STRUCTURES AND LIFE PROCESSES IN ANIMALS

The body of a complex animal is a remarkable mechanism composed of cells, tissues, and organ systems. ■ Ten principal organ systems perform together to maintain life and function. ■ Each system is responsible for a specific group of activities. ■

Bengal tiger (*Panthera tigris*)

38

The Animal Body: Tissues, Organs, and Organ Systems

■ LEARNING OBJECTIVES

After you have read this chapter you should be able to:

1. Discuss the advantages and disadvantages of the multicellular condition.
2. Define tissue, organ, and organ system.
3. Compare the four principal kinds of animal tissue—epithelial, connective, muscular, and nervous tissues—and give their respective functions.
4. Compare the main types of epithelial tissue and locate each in the body.
5. Compare the main types of connective tissue and give their functions.
6. Compare the three types of muscle tissues and their functions.
7. Cite the functions of nervous tissue and distinguish between neurons and glial cells.
8. List the organ systems characteristic of complex animals, describe their functions, and discuss how each system is involved in maintaining the constancy of the internal environment.

Animals, like plants and most fungi, are **multicellular,** that is, they are composed of many cells (Figure 38–1). These organisms are not merely colonies of similar cells but are composed of a number of different types of cells, each with a characteristic size, shape, structure, and function. In most animals, cells are organized into tissues, tissues into organs, and organs into organ systems.

WHY ARE SO MANY LIVING THINGS MULTICELLULAR?

Animals can be large because they are multicellular. We do not see amoebas as large as whales slithering around because it is inefficient for a cell to become very large (Chapter 4). Recall that when its size approaches the limits of efficiency, a cell divides to form two cells. In unicellular organisms, cell division results in the production of two new individuals, whereas in multicellular organisms, the two new cells may remain associated to form part of the whole organism.

(a)

(b)

FIGURE 38–1 All animals are multicellular; a large animal such as the bull elephant from western Kenya (*a*) is composed of more cells than the much smaller anole lizard.

Human beings, elephants, and other complex animals are composed of millions of cells, rather than one giant cell. The number of cells, not their individual size, is responsible for the different sizes of various organisms. The cells of an earthworm, a human being, or an elephant are about the same size; the elephant is larger because its genes are programmed to provide for a larger number of cells.

With multicellularity also comes the specialization of cells. In a unicellular organism, such as a bacterium or a flagellate, the single cell must carry on all the activities necessary for the life of the organism. An organism composed of many cells can assign specific tasks to different cells. For example, some groups of cells are specialized to transport materials, whereas others contract to enable the organism to move. How do these cells associate and perform such specialized functions? To answer this we examine the tissues, organs, and organ systems of complex animals.

ANIMAL TISSUES

A **tissue** consists of a few types of closely associated cells that are adapted to carry out specific functions. Animal tissues may be classified as epithelial, connective, muscular, or nervous. Each kind of tissue is composed of cells with a characteristic size, shape, and arrangement.

Epithelial Tissues

Epithelial tissue (also called **epithelium**) consists of cells fitted tightly together, forming a continuous layer or sheet of cells covering a body surface or lining a cavity within the body. One surface of the sheet is attached to the underlying tissue by a noncellular **basement membrane** composed of tiny fibers and of nonliving polysaccharide material produced by the epithelial cells. In addition to the outer layer of the skin, the linings of the digestive and respiratory tracts and the lining of the kidney tubules are examples of epithelial tissues.

Epithelial tissues function in protection, absorption, secretion, or sensation. The epithelial layer of the skin covers the entire body and protects it from a variety of deleterious effects of the environment, including mechanical injury, harmful chemicals, bacteria, and fluid loss. The epithelial tissue lining the digestive tract absorbs nutrients and water into the body. Other epithelial cells are organized into glands, adapted for the secretion of cell products.

Everything that enters the body or leaves it must cross one or more layers of epithelium. Food that is taken into the mouth and swallowed is not really "inside" the body: This occurs only when the substance is absorbed through the epithelium of the gut and enters the blood. The permeability of the various epithelia regulates to a large extent the exchange of substances between the different parts of the body and between the organism and the external environment.

Many epithelial membranes are subjected to continuous wear and tear. As outer cells are sloughed off, they must be replaced by new ones from below. Such epithelial tissues generally have a rapid rate of cell division so that new cells are continuously produced to take the place of those lost.

Three types of epithelial cells can be distinguished on the basis of their shape (see Table 38–1): **Squamous** epithelial cells are thin, flattened cells shaped like pancakes or flagstones. **Cuboidal** epithelial cells are short cylinders that in side view are cube-shaped, resembling dice. Actually, each cell has a complex shape, usually that of an eight-sided polyhedron. **Columnar** epithelial cells look like tiny columns or cylinders when viewed from the side. The nucleus is usually located near the base of the cell. Viewed from above, or in cross section, these cells appear hexagonal in shape. Columnar epithelial cells may have cilia on their free surface that beat in a coordinated way, moving materials in one direction. Most of the respiratory tract is lined with ciliated epithelium; the ciliary beating helps to move particles of dust and other foreign material away from the lungs.

Epithelial tissue may be **simple**—that is, composed of one layer of cells—or **stratified**—composed of two or more layers (see Table 38–1). Simple epithelium is usually located in areas where materials must diffuse through the tissue or where substances are secreted, excreted, or absorbed. Stratified epithelial tissue is located in regions where protection is the main function. Stratified squamous epithelium is found in the skin and lining the mouth and esophagus of humans and other vertebrates. A third arrangement of epithelial cells is **pseudostratified** epithelium, so named because its cells falsely appear to be layered. All of the cells really do rest on a basement membrane, but not every cell is tall enough to reach the free surface of the tissue. This may give the false impression that there are two or more cell layers. Some of the respiratory passageways are lined with pseudostratified epithelium equipped with cilia.

Table 38–1 illustrates the main types of epithelial tissue and indicates where they are located in the body, as well as describing their functions.

The linings of the body cavities and the linings of the blood and lymph vessels are derivatives of **mesenchyme,** a generalized embryonic tissue that gives rise to connective tissues rather than epithelial germ layers. Structurally, however, they are in all respects typical epithelial cells. To distinguish them from the true epithelia, the linings of blood and lymph vessels are termed **endothelium.**

A layer of cells specialized to receive stimuli is called **sensory epithelium.** The olfactory epithelium in the lining of the nose, for example, contains neurons that respond to the presence of certain chemicals in inhaled air. These cells are responsible for the sense of smell.

Table 38–1 EPITHELIAL TISSUES

Type of Tissue	Main Locations	Functions	Description and Comments
Simple squamous epithelium Nuclei	Air sacs of lungs, lining of blood vessels	Passage of materials where little or no protection is needed and where diffusion is major form of transport	Cells are flat and arranged as single layer
Simple cuboidal epithelium Nuclei of cuboidal epithelial cells Lumen of tubule 	Lining of kidney tubules, gland ducts	Secretion and absorption	Single layer of cells; from the side each cell looks like short cylinder; sometimes have microvilli for absorption
Simple columnar epithelium Goblet cell Nuclei of columnar cells 	Lining of much of digestive tract, upper part of respiratory tract	Secretion, especially mucus; absorption, protection, movement of mucous layer	Single layer of columnar cells, often with nuclei located in base of each cell almost in row; sometimes with enclosed secretory vesicles (goblet cells), highly developed Golgi complex, and cilia

Table 38–1 (CONTINUED)

Type of Tissue	Main Locations	Functions	Description and Comments
Stratified squamous epithelium 	Skin, mouth lining, vaginal lining	Protection only; little or no absorption or transit of materials; outer layer continuously sloughed off and replaced from below.	Several layers of cells, with only the lower ones columnar and metabolically active. Division of lower cells causes older ones to be pushed upward toward surface.
Pseudostratified epithelium 	Some respiratory passages, ducts of many glands, sometimes ciliated	Secretion, protection, movement of mucus	Comparable in many ways to columnar epithelium, except that not all cells are the same height. Thus, though all cells contact the same basement membrane, the tissue appears stratified. Nuclei not in line. Ciliated, mucus-secreting, or with microvilli cells contact the same basement membrane, the tissue appears stratified. Nuclei not in line. Ciliated, mucus-secreting, or with microvilli

A **gland** consists of one or more epithelial cells specialized to produce and secrete a product such as sweat, milk, mucus, wax, saliva, hormones, or enzymes (Figure 38–2). The epithelial tissue lining the cavities and passageways of the body typically contain specialized mucus-secreting cells called **goblet-cells.** The mucus lubricates these surfaces and facilitates the movement of materials.

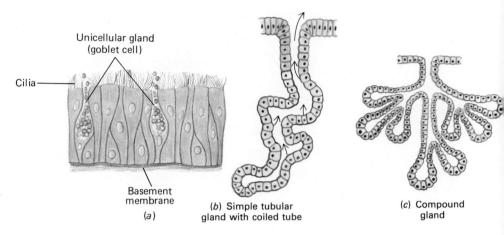

Unicellular gland
(goblet cell)

Cilia

Basement
membrane

(a)

(b) Simple tubular
gland with coiled tube

(c) Compound
gland

FIGURE 38–2 A gland consists of one or more epithelial cells. (a) Goblet cells are unicellular glands that secrete mucus. (b) Sweat glands are simple tubular glands with coiled tubes similar to the one shown here. The walls of the gland are constructed of simple cuboidal epithelium. (c) The parotid salivary glands are compound glands like the one shown here.

Connective Tissues

The main function of **connective tissues** is to join together the other tissues of the body. Connective tissues also support the body and its structures and protect underlying organs. In addition, almost every organ in the body has a supporting framework of connective tissue, called **stroma.** The epithelial components of the organ are supported and cushioned by the stroma.

There are many kinds of connective tissues and many systems for classifying them. Some of the main types of connective tissue are (1) loose and dense connective tissues, (2) elastic connective tissue, (3) reticular connective tissue, (4) adipose tissue, (5) cartilage, (6) bone, and (7) blood, lymph, and tissues that produce blood cells. These tissues vary widely in the details of their structure and in the specific functions they perform (Table 38–2; page 862).

Connective tissues contain relatively few cells embedded in an extensive **intercellular substance** consisting of threadlike microscopic **fibers** scattered throughout a matrix secreted by the cells. The **matrix** is a thin gel composed of polysaccharides. The cells of different kinds of connective tissues differ in their shape and structure and in the kind of matrix they secrete. The nature and function of each kind of connective tissue are determined in part by the structure and properties of the intercellular substance. Thus, to some extent, connective tissue cells perform their respective functions indirectly by secreting the matrix, which does the actual connecting and supporting.

Fibers and Cells of Connective Tissue

Connective tissue contains three types of fibers: collagen, elastic, and reticular. **Collagen fibers,** the most numerous type, extend in all directions and are composed of bundles of smaller parallel **fibrils** (Figure 38–3). They are flexible but resist tension, stretching only very slightly in response to a pull. With a great-enough force they will break. **Elastic fibers** branch and fuse to form networks; they can be stretched by a force and then will return to their original size and shape when the force is removed. They are not composed of bundles of fibrils but do contain microfibrils evident by electron microscopy. **Reticular fibers** are very small, branched fibers that form delicate networks not visible in ordinary stained slides. The reticular fibers become apparent when a tissue is stained with silver.

Collagen and reticular fibers contain the protein **collagen** (rich in the amino acids glycine, proline, and hydroxyproline). Collagen is a very tough material, and these fibers impart great strength to structures in which they occur. (Meat is tough because of its collagen content.) The

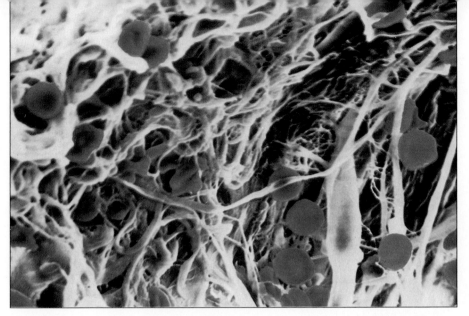

FIGURE 38–3 False-color scanning electron micrograph of human connective tissue (approximately ×440), showing collagen fibers which appear as an irregular mass of yellow strands. Red blood cells (erythrocytes) are interspersed between the fibers. Connective tissue provides structural and metabolic support for other tissues.

tensile strength of collagen fibers has been compared to that of steel. When treated with hot water, collagen is converted into the soluble protein gelatin. Because there is so much connective tissue in the body, about one third of all mammalian protein is collagen.

Connective tissue also contains several types of cells. **Fibroblasts** are connective tissue cells that produce the protein and carbohydrate complexes of the matrix as well as the fibers. The fibroblasts release specific protein components that arrange themselves to form the characteristic fibers. Fibroblasts are especially active in developing tissue and in healing wounds. As tissues mature, the numbers of fibroblasts decrease, and they become less active. Mature, inactive fibroblasts are referred to as **fibrocytes.**

Pericytes are undifferentiated (unspecialized) cells located along the outer walls of small blood vessels (capillaries) that run through connective tissues. It is thought that pericytes give rise to other types of cells when necessary. For example, when an injury occurs, pericytes are thought to multiply and give rise to fibroblasts that can produce the components needed to heal the wound.

Macrophages, the scavenger cells of the body, are also common in connective tissues. They wander through the tissues, cleaning up cellular debris and phagocytizing foreign matter including bacteria (Figure 38–4). Among the other types of cells seen in connective tissues are mast cells, which release histamine during allergic reactions; adipose (fat) cells; and plasma cells, which produce antibodies.

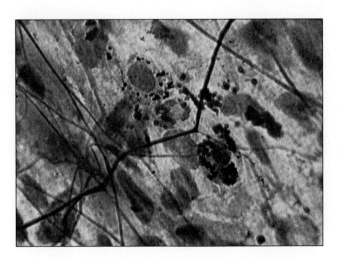

FIGURE 38–4 Loose ordinary connective tissue in the mesentery of a rabbit which had been injected with india ink (approximately ×1000). The macrophages have ingested the ink particles. Fibroblasts (the elongated cells) are also present. Note that the collagen is stained pink and the elastic fibers black.

Table 38–2 **CONNECTIVE TISSUES**

Type of Tissue	Main Locations	Functions	Description and Comments
Loose (areolar) connective tissue Collagen fibers / Nuclei of fibroblasts	Everywhere support must be combined with elasticity, e.g., subcutaneous layer	Support; reservoir for fluid and salts	Fibers produced by fibroblast cells embedded in semifluid matrix and mixed with miscellaneous group of other cells
Dense connective tissue	Tendons, strong attachments between organs; dermis of skin	Support; transmission of mechanical forces	Bundles of interwoven collagen fibers interdigitated with rows of fibroblast cells
Elastic connective tissue	Structures that must both expand and return to their original size, such as lung tissue and large arteries; ligaments	Confers elasticity	Branching elastic fibers interspersed with fibroblasts
Reticular connective tissue	Framework of liver, lymph nodes, spleen	Support	Consists of interlacing reticular fibers

Table 38–2 (CONTINUED)

Type of Tissue		Main Locations	Functions	Description and Comments
Adipose tissue		Subcutaneous layer; pads around certain internal organs	Food storage; insulation; support of such organs as mammary glands, kidneys	Fat cells are star-shaped at first; fat droplets accumulate until typical ring-shaped cells are produced.
Cartilage	Chondrocytes — Lacuna — Intercellular substance	Supporting skeleton in sharks, rays, and some other vertebrates; in other vertebrates, forms ends of bones; supporting rings in walls of some respiratory tubes; tip of nose; external ear	Flexible support and reduction of friction in bearing surfaces	Cells (chondrocytes) separated from one another by the gristly intercellular substance; occupy little spaces in it
Bone	Lacunae — Haversian canal — Matrix	Forms skeletal structure in most vertebrates	Support, protection of internal organs; calcium reservoir; skeletal muscles attach to bones	Osteocytes located in lacunae; in compact bone, lacunae arranged in concentric circles about haversian canals
Blood		Within heart and blood vessels of circulatory system	Transports oxygen, nutrients, wastes, and other materials	Consists of cells dispersed in a fluid intercellular substance

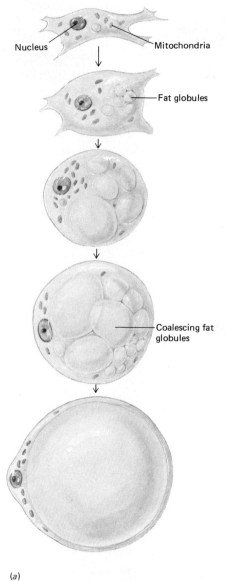

Nucleus

Mitochondria

Fat globules

Coalescing fat globules

(a)

(b)

FIGURE 38–5 Storage of fat in a fat cell. (*a*) As more and more fat droplets accumulate in the cytoplasm, they coalesce to form a very large globule of fat. Such a fat globule may occupy most of the cell, pushing the cytoplasm and the organelles to the periphery. (*b*) Photomicrograph of adipose tissue (approximately ×100). The fat droplets were dissolved by chemicals used to prepare the tissue, leaving large spaces. Because of these spaces, the cells tend to collapse and no longer appear round.

Loose and Dense Connective Tissues

Loose connective tissue (also called **areolar** tissue) is the most widely distributed connective tissue in the body. It is found as a thin filling between body parts and serves as a reservoir for fluid and salts. Nerves, blood vessels, and muscles are wrapped in this tissue. Together with adipose tissue, loose connective tissue forms the subcutaneous (that is, below the skin) layer, the layer that attaches skin to the muscles and other structures beneath. Loose connective tissue consists of fibers strewn in all directions through a semifluid matrix. Its flexibility permits the parts it connects to move.

Dense connective tissue is very strong, though somewhat less flexible than loose connective tissue. Collagen fibers predominate. In **irregular** dense connective tissue, the collagen fibers are arranged in bundles distributed in all directions through the tissue. This type of tissue is found in the lower layer (dermis) of the skin. In **regular** dense connective tissue, the collagen bundles are arranged in a definite pattern, making the tissue greatly resistant to stress. Tendons, the cable-like cords that connect muscles to bones, consist of this tissue.

Elastic and Reticular Connective Tissues

Elastic connective tissue consists mainly of bundles of parallel elastic fibers. It is found in ligaments, the bands of tissue that connect bones to one another. Structures that must expand and then return to their original size, like the walls of the large arteries and lung tissue, contain elastic connective tissue. **Reticular connective tissue** is composed mainly of interlacing reticular fibers. It forms a supporting stroma in many organs, including the liver, spleen, and lymph nodes.

Adipose Tissue

Adipose tissue is rich in fat cells, which store fat and release it when fuel is needed for cellular respiration. It is found in the subcutaneous layer and in tissue that cushions internal organs. An immature fat cell is somewhat star-shaped. As fat droplets accumulate within the cytoplasm, the cell assumes a more rounded appearance (Figure 38–5). Fat droplets eventually merge with one another until finally a single large drop of fat is present. This large drop occupies most of the volume of the mature fat-storing cell. The cytoplasm and its organelles are pushed to the cell edges, where a bulge is typically created by the nucleus. A cross section of such a fat cell looks like a ring with a single stone. (Cytoplasm forms the ring, and the nucleus, the stone.)

When you study a section of adipose tissue through a microscope, it may remind you of chicken wire (Figure 38–5*b*). The "wire" is represented by the rings of cytoplasm, and the large spaces indicate where fat drops existed before they were dissolved by chemicals used to prepare the tissue. The empty spaces may cause the cells to collapse, resulting in a wrinkled appearance.

Cartilage and Bone

The supporting skeleton of vertebrates is composed of cartilage or bone. **Cartilage** is the supporting skeleton in the embryonic stages in all vertebrates, but it is largely replaced in the adult by bone in all but the sharks and rays. The supporting structure of the external ear, the supporting rings in the walls of the respiratory passageways, and the tip of the nose in humans are examples of structures composed of cartilage. Cartilage is firm,

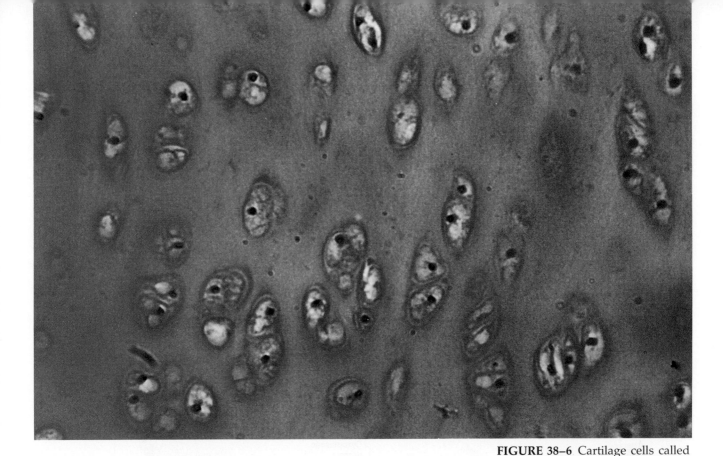

yet elastic. Cartilage cells called **chondrocytes** secrete this hard, rubbery matrix around themselves and also secrete collagen fibers, which become embedded in the matrix and strengthen it. Chondrocytes eventually come to lie singly or in groups of two or four in small cavities called **lacunae** in the matrix (Figure 38–6). Cartilage cells in the matrix remain alive. Cartilage tissue lacks nerves, lymph vessels, and blood vessels. Chondrocytes are nourished by diffusion of nutrients and oxygen through the matrix.

Bone is the major vertebrate skeletal tissue. It is similar to cartilage in that it consists mostly of matrix material containing lacunae and inhabited by the cells that secrete and maintain the matrix (Figure 38–7). Unlike cartilage, however, bone is a highly vascular tissue with a substantial blood supply. Diffusion alone would never suffice for the nourishment of the bone cells, called **osteocytes,** because the matrix consists not only of collagen, mucopolysaccharides, and other organic materials but also of the mineral apatite, a complex calcium phosphate. Diffusion through such a substance would be impracticably slow. Thus, the osteocytes of bone communicate with one another and with capillaries by tiny channels, **canaliculi,** which contain fine extensions of the cells themselves. Because it is important that no bone cell be located very far from the nearest blood vessel, the osteocytes are arranged around central capillaries in concentric layers called **lamellae,** which form spindle-shaped units known as **osteons.** The capillaries, as well as nerves, run through central microscopic channels known as **haversian canals.**

Bone also contains large multinucleated cells called **osteoclasts,** which can dissolve and remove the bony substance, as can the osteocytes themselves. The shape and internal architecture of the bone can gradually change in response to normal growth processes and to physical stress. The calcium salts of bone render the matrix very hard, and the collagen prevents the bony matrix from being overly brittle. Bones are amazingly light and strong. Most bones have a large **marrow cavity** in the center; this may contain yellow marrow, which is mostly fat, or red marrow, the connective tissue in which red and some white blood cells are made.

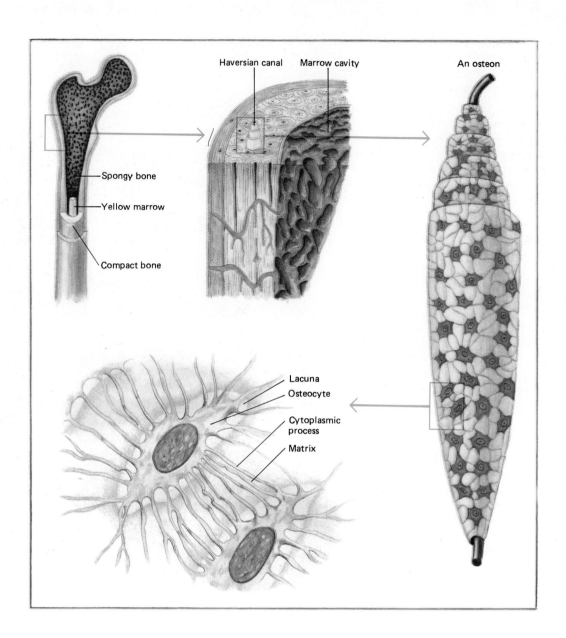

Haversian canal   Marrow cavity

An osteon

Spongy bone

Yellow marrow

Compact bone

Lacuna
Osteocyte
Cytoplasmic process
Matrix

FIGURE 38–7 Compact bone is made up of units called osteons. Blood vessels and nerves run through the haversian canal within each osteon. In bone the matrix is rigid and hard. Bone cells become trapped within lacunae but communicate with one another by way of cytoplasmic processes that extend through tiny canals.

Blood

In mammals, **blood** is composed of red and white cells and platelets, suspended within **plasma,** the liquid, noncellular part of the blood. Plasma transports many kinds of substances from one part of the body to another. Some of these substances are simply dissolved in the plasma, whereas others are bound to proteins such as albumins. Most biologists classify blood with connective tissues.

The **red blood cells** (erythrocytes) of humans and other vertebrates contain the red respiratory pigment **hemoglobin,** which can combine easily and reversibly with oxygen. Oxygen, combined as oxyhemoglobin, is transported to the cells of the body by the red blood cells. The red cells of most mammals are flattened biconcave disks that lack a nucleus (Table 38–2); those of other vertebrates are oval and have a nucleus. In many

Focus on NEOPLASMS—UNWELCOME TISSUES

A **neoplasm** (new growth) is an abnormal mass of cells. Neoplasms, or **tumors,** can develop in many species of animals and plants. A benign ("kind") tumor tends to grow slowly, and its cells stay together. Because benign tumors form discrete masses, often surrounded by connective tissue capsules, they can usually be removed surgically. Unless a benign neoplasm develops in a place where it interferes with the function of a vital organ, it is not lethal.

A malignant ("wicked") neoplasm, or **cancer,** usually grows much more rapidly than a benign tumor. Neoplasms that develop from connective tissues or muscle are referred to as sarcomas, and those that originate in epithelial tissue are called carcinomas. Unlike the cells of benign tumors, cancer cells do not retain the typical structural features of the cells from which they originate.

How normal cells may be transformed to cancer cells is discussed in Focus on Oncogenes and Cancer (Chapter 16). When a transformed cell multiplies, all the cells derived from it are also abnormal. Two basic defects in behavior that characterize most cancer cells are rapid multiplication and abnormal relations with neighboring cells. While normal cells respect one another's boundaries and form tissues in an orderly, organized manner, cancer cells grow helter-skelter upon one another and infiltrate normal tissues (see figure). Apparently they are no longer able to receive or respond appropriately to signals from surrounding cells.

Studies indicate that many neoplasms grow to only a few millimeters in diameter and then enter a dormant stage, which may last for months or even years. At some point, cells of the neoplasm release a chemical substance that stimulates nearby blood vessels to develop new capillaries that grow out toward the neoplasm and infiltrate it. Once a blood supply is ensured, the neoplasm grows rapidly and can become life threatening.

Death from cancer is almost always caused by **metastasis,** which is a migration of cancer cells through blood or lymph channels to distant parts of the body. Once there, they multiply, forming new malignant neoplasms, which may interfere with normal function in the tissues being invaded. Cancer often spreads so rapidly and extensively that surgeons are unable to locate all the malignant masses.

Why some persons are more susceptible to cancer than others remains a mystery. Some researchers think that cancer cells form daily in everyone but that in most persons the immune system (the system that provides protection from disease organisms and other foreign invaders) is capable of destroying them. According to this theory, cancer is a failure of the immune system. Another suggestion is that different persons have different levels of tolerance to environmental irritants. As many as 90% of cancer cases are thought to be triggered by environmental factors. Certain personality factors and styles of coping with stressful life events have been hypothetically linked with higher risk for cancer. For

(a)

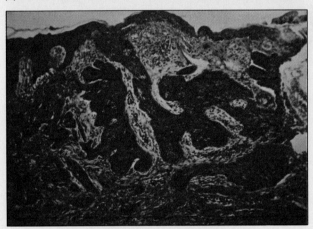

(b)

■ Normal skin (*a*) compared with cancerous tissue (*b*). Note the disruption of the normal tissue structure by the invasion of the neoplasm.

example, studies suggest that individuals who deny and repress anger, and who respond to negative life events with chronic hopelessness and depression are at higher risk for developing cancer.

Cancer is the second most common cause of death in the United States and, despite advances in its treatment, there is no miracle cure on the horizon. Fewer than 50% of cancer patients survive five years from the time cancer is first diagnosed. Currently, the key to survival is early diagnosis and treatment with a combination of surgery, radiation therapy, and drugs that suppress mitosis (chemotherapy). Since cancer is actually an entire family of closely related diseases (there are more than 100 distinct varieties), it may be that there is no single cure. Most investigators agree, however, that a greater understanding of basic control mechanisms and communication systems of cells is necessary before effective cures can be developed.

invertebrate animals, the oxygen-carrying pigments are not located within a cell but are dissolved in the plasma, coloring it red or blue.

Human blood contains five different kinds of **white blood cells,** each with distinct size, shape, structure, and functions. None of the white blood cells contain hemoglobin, but some can move around by amoeboid motion and slip through the walls of the blood vessels, entering the tissues of the body to engulf bacteria and other foreign particles. The white cells constitute an important line of defense against disease bacteria.

Platelets are not whole cells but are small fragments broken off from large cells located in the bone marrow. In complex vertebrates they play a key role in the clotting of blood (Chapter 42).

Muscle Tissue

The movements of most animals result from the contraction of the elongated, cylindrical, or spindle-shaped cells of **muscle tissue.** Each muscle cell, usually referred to as a **fiber** because of its length, contains many small, longitudinal, parallel contractile fibers called **myofibrils.** The proteins **myosin** and **actin** are the chief components of myofibrils. Muscle cells perform only mechanical work by contracting, getting shorter or thicker; they cannot exert a push.

Three types of muscle tissue are found in vertebrates (Figure 38–8). **Cardiac muscle** is present in the walls of the heart. **Smooth muscle** occurs in

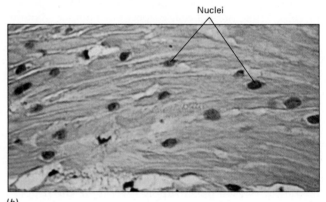

(b)

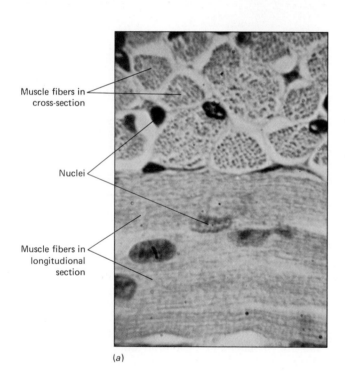

Muscle fibers in cross-section

Nuclei

Muscle fibers in longitudional section

(a)

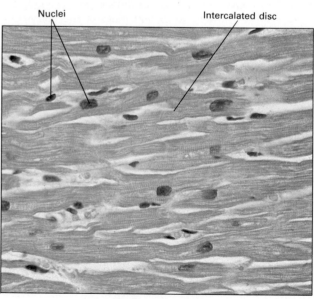

Nuclei

Intercalated disc

(c)

FIGURE 38–8 Muscle tissue. (*a*) Skeletal muscle is striated, voluntary muscle (magnification approximately ×1000). (*b*) Smooth muscle tissue lacks striations and is involuntary (magnification approximately ×450). (*c*) Cardiac muscle tissue is striated, has branched fibers, and is involuntary. The special junctions between cardiac muscle cells are called intercalated disks.

Table 38–3 **THE TYPES OF MUSCLE TISSUES**

	Skeletal	*Smooth*	*Cardiac*
Location	Attached to skeleton	Walls of stomach, intestines, etc.	Walls of heart
Type of control	Voluntary	Involuntary	Involuntary
Shape of fibers	Elongated, cylindrical, blunt ends	Elongated, spindle-shaped, pointed ends	Elongated, cylindrical fibers that branch and fuse
Striations	Present	Absent	Present
Number of nuclei per fiber	Many	One	One or two
Position of nuclei	Peripheral	Central	Central
Speed of contraction	Most rapid	Slowest	Intermediate
Ability to remain contracted	Least	Greatest	Intermediate

(a) Skeletal muscle fibers

(b) Smooth muscle fibers

(c) Cardiac muscle fibers

the walls of the digestive tract, uterus, blood vessels, and certain other internal organs. **Skeletal muscle** makes up the large muscle masses attached to the bones of the body. Skeletal muscle fibers are among the exceptions to the rule that cells have only one nucleus; each skeletal muscle fiber has many nuclei. The nuclei of skeletal muscle fibers are also unusual in their position: They lie peripherally, just under the cell membrane. This is thought to be an adaptation to increase the efficiency of contraction. The entire central part of the skeletal muscle fiber is occupied by the contractile units, the myofibrils. Skeletal muscle cells may be as long as 2 or 3 cm.

By light microscopy, both skeletal and cardiac fibers are seen to have alternate light and dark transverse stripes, or **striations.** These microscopic stripes are involved in the contraction process, for they change their relative sizes during contraction (see Figure 39–15): The dark stripes remain essentially constant, but the light stripes decrease in width. Striated muscle fibers can contract rapidly but cannot remain contracted. A striated muscle fiber must relax and rest momentarily before it can contract again. Skeletal muscle fibers are generally under voluntary control, whereas cardiac and smooth muscle fibers are not usually regulated at will. Table 38–3 summarizes the distinguishing features of the three kinds of muscle. In the bodies of invertebrates, the distribution of muscle types may be quite different from that in vertebrates, and one or more types may be missing completely.

Nervous Tissue

Nervous tissue is composed of **neurons,** cells specialized for conducting electrochemical nerve impulses, and **glial cells,** cells that support and nourish the neurons (Figure 38–9). Certain neurons receive signals from

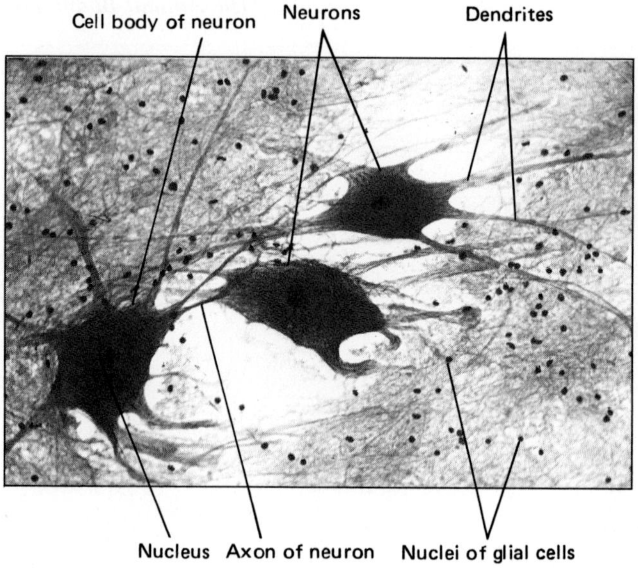

FIGURE 38–9 Nervous tissue consists of neurons and glial cells (approximately ×50).

the external or internal environment and transmit them to the spinal cord and brain; other nerve cells process and store the information. This is the cellular basis for the complex functions of consciousness, memory, thought, and directed movement.

Neurons come in many shapes and sizes, but typically each has an enlarged **cell body,** which contains the nucleus, and from which two kinds of thin hairlike extensions project. **Dendrites** are fibers specialized for receiving impulses either from environmental stimuli or from another cell. The single **axon** is specialized to conduct impulses away from the cell body. Axons usually are long and smooth but may give off an occasional branch; they typically end in a group of fine branches. Axons range in length from a millimeter or two to more than a meter. Those extending from the spinal cord down the arm or leg in the human may be a meter or more in length. Neurons communicate at junctions called **synapses;** thus, they are functionally connected and can pass impulses for long distances through the body. A **nerve** consists of a great many fibers bound together by connective tissue.

ANIMAL ORGANS AND ORGAN SYSTEMS

Complex animals have a great variety of organs. Although an animal organ may be composed mainly of one type of tissue, other types are needed to provide support, protection, and a blood supply and to allow transmission of nerve impulses. For example, the heart consists mainly of cardiac muscle tissue, but it is lined and covered by epithelium, contains blood vessels composed of connective tissue, and is regulated by nervous tissue.

The organ systems of complex animals include the integumentary, skeletal, muscle, nervous, circulatory, digestive, respiratory, urinary (excretory), endocrine, and reproductive systems (Figure 38–10 on pages 872 and 873). See Table 38–4 for a summary of their principal organs and functions. In the digestive system, for example, organs include the mouth, esophagus, stomach, small and large intestines, liver, pancreas, and salivary glands. This system functions to process food, reducing it to its simple components. The digestive system transfers the products of digestion into the blood for transport to all of the body's cells.

Table 38–4 **THE ORGAN SYSTEMS OF A MAMMAL AND THEIR FUNCTIONS**

System	Components	Functions	Homeostatic Ability
Integumentary	Skin, hair, nails, sweat glands	Covers and protects body	Sweat glands help control body temperature; as barrier, the skin helps maintain steady state
Skeletal	Bones, cartilage, ligaments	Supports body, protects, provides for movement and locomotion, calcium depot	Helps maintain constant calcium level in blood
Muscular	Organs mainly of skeletal muscle; cardiac muscle; smooth muscle	Moves parts of skeleton, locomotion; movement of internal materials	Ensures such vital functions as nutrition through body movements; smooth muscle maintains blood pressure; cardiac muscle circulates the blood
Digestive	Mouth, esophagus, stomach, intestines, liver, pancreas	Ingests and digests foods, absorbs them into blood	Maintains adequate supplies of fuel molecules and building materials
Circulatory	Heart, blood vessels, blood; lymph and lymph structures	Transports materials from one part of body to another; defends body against disease	Transports oxygen, nutrients, hormones; removes wastes; maintains water and ionic balance of tissues
Respiratory	Lungs, trachea, and other air passageways	Exchange of gases between blood and external environment	Maintains adequate blood oxygen content and helps regulate blood pH; eliminates carbon dioxide
Urinary	Kidney, bladder, and associated ducts	Eliminates metabolic wastes; removes substances present in excess from blood	Regulates blood chemistry in conjunction with endocrine system
Nervous	Nerves and sense organs, brain and spinal cord	Receives stimuli from external and internal environment, conducts impulses, integrates activities of other systems	Principal regulatory system
Endocrine	Pituitary, adrenal, thyroid, and other ductless glands	Regulates body chemistry and many body functions	In conjunction with nervous system, regulates metabolic activities and blood levels of various substances
Reproductive	Testes, ovaries, and associated structures	Provides for continuation of species	Passes on genetic endowment of individual; maintains secondary sexual characteristics

■ SUMMARY

I. Multicellular organisms are composed of a number of cell types, each specialized and adapted to carry out specific functions.
   A. A tissue is an aggregation of similarly specialized cells, and their intercellular substance, that associate to perform a specific function or group of functions.
   B. Several types of tissue may be united to form an organ; several organs may be organized as an organ system.
II. Multicellular organisms are able to achieve a much larger size than is possible in single-celled ones. In a multicellular organism, cells can specialize to perform specific functions.
III. Animal tissues are classified as epithelial, connective, muscular, or nervous.
   A. Epithelial tissue may form a continuous layer or sheet of cells covering a body surface or lining a body cavity; some epithelial tissue is specialized to form glands.
      1. Epithelial tissue functions in protection, absorption, secretion, or sensation.
      2. Epithelial cells may be squamous, cuboidal, or columnar in shape.
      3. Epithelial tissue may be simple, stratified, or pseudostratified. (Features of each are summarized in Table 38–1.)
   B. Connective tissue joins together other tissues of the body, supports the body and its organs, and protects underlying organs.
      1. Connective tissue consists of cells, such as fibroblasts and macrophages, and the intercellular substance secreted by the cells.
      2. Some types of connective tissue are loose and dense connective tissue, elastic connective tissue, reticular connective tissue, adipose tissue,

FIGURE 38–10 The principal organ systems of the human body.

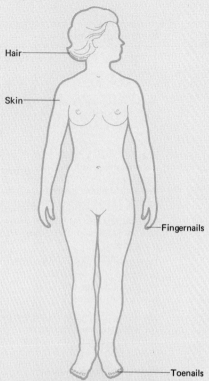

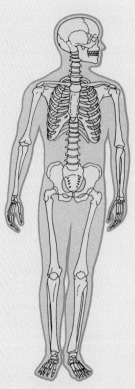

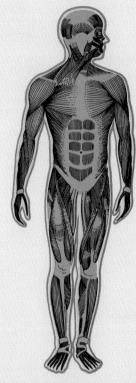

(1) The integumentary system consists of the skin and the structures such as nails and hair that are derived from it. This system protects the body, helps to regulate body temperature, and receives stimuli such as pressure, pain, and temperature.

(2) The skeletal system consists of bones and cartilage. This system helps to support and protect the body.

(3) The muscular system consists of the large skeletal muscles that enable us to move, as well as the cardiac muscle of the heart and the smooth muscle of the internal organs.

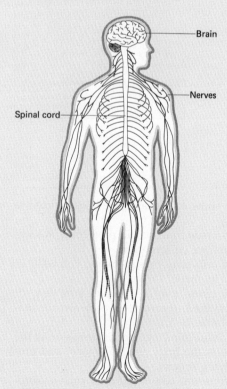

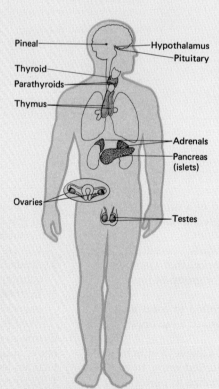

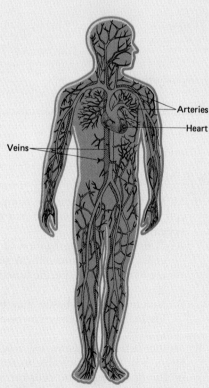

(4) The nervous system consists of the brain, spinal cord, sense organs, and nerves. This is the principal regulatory system.

(5) The endocrine system consists of the ductless glands that release hormones. It works with the nervous system in regulating metabolic activities.

(6a) The circulatory system includes the heart and blood vessels. This system serves as the transportation system of the body.

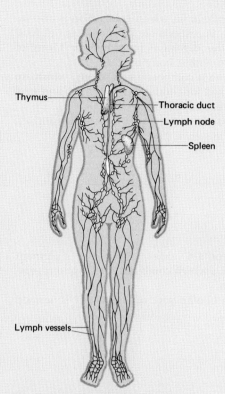

Thymus

Thoracic duct

Lymph node

Spleen

Lymph vessels

(6b) The lymphatic system is a subsystem of the circulatory system; it returns excess tissue fluid to the blood and defends the body against disease.

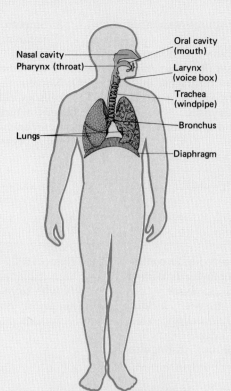

Nasal cavity

Pharynx (throat)

Oral cavity (mouth)

Larynx (voice box)

Trachea (windpipe)

Bronchus

Lungs

Diaphragm

(7) The respiratory system. Consisting of the lungs and air passageways, this system supplies oxygen to the blood and excretes carbon dioxide.

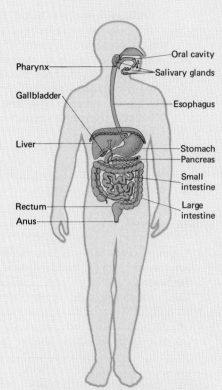

Pharynx

Oral cavity

Salivary glands

Gallbladder

Esophagus

Liver

Stomach

Pancreas

Small intestine

Rectum

Large intestine

Anus

(8) The digestive system consists of the digestive tract and glands that secrete digestive juices into the digestive tract. This system mechanically and enzymatically breaks down food and eliminates wastes.

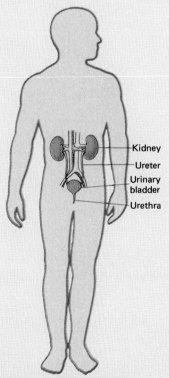

Kidney

Ureter

Urinary bladder

Urethra

(9) The urinary system is the main excretory system of the body, and helps to regulate blood chemistry. The kidneys remove wastes and excess materials from the blood and produce urine.

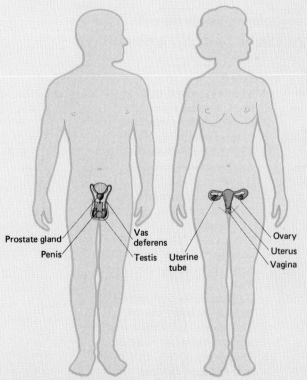

Prostate gland

Vas deferens

Penis

Testis

Uterine tube

Ovary

Uterus

Vagina

(10) Male and female reproductive systems. Each reproductive system consists of gonads and associated structures. The reproductive system maintains the sexual characteristics, and perpetuates the species.

cartilage, bone, and blood. (Features of each are summarized in Table 38–2.)

C. Muscle tissue is composed of cells specialized to contract. Each cell is an elongated fiber containing many small longitudinal, parallel contractile units called myofibrils. The chief components of myofibrils are the proteins actin and myosin.

1. Skeletal muscle is striated and is under voluntary control.
2. Cardiac muscle is striated; its contraction is involuntary.
3. Smooth muscle, in which contraction is also involuntary, is responsible for movement of food through the digestive tract and for forms of movement within body organs.

D. Nervous tissue is composed of neurons, which are cells specialized for conducting impulses, and glial cells, which are supporting cells.

IV. Complex animals have a great variety of organs and ten principal organ systems. (See Table 38–4.)

■ POST-TEST

1. A group of cells fitting tightly together to form a continuous sheet covering a body surface or lining a cavity of the body is termed an _____ tissue.
2. The functions of epithelial tissues include _____, _____, _____, and _____.
3. On the basis of their shape, we can distinguish _____, _____, and _____ epithelia.
4. The mammalian respiratory tract is lined with _____.
5. The outer layer of the skin is composed of _____.
6. Epithelial cells specialized to produce and secrete a product are called _____.
7. The supporting connective tissue framework of an organ is called _____.
8. _____ _____ are small branched fibers forming delicate networks in tissues.

19. Erythrocytes contain the respiratory pigment _____, which combines readily and reversibly with oxygen.
20. The liquid, noncellular part of the blood is termed _____.

Match the terms in Column A with their definitions in Column B.

Column A	Column B
9. Canaliculi	a. Undifferentiated cells on outer walls of capillaries
10. Chondrocytes	b. Scavenger cells that clean up cellular debris
11. Collagen	c. The ground substance of connective tissue; intercellular substance
12. Fibroblasts	d. A unique protein present in connective tissues, secreted by fibroblasts
13. Glial cells	e. Connective tissue cells that produce and secrete the proteins and other components of the matrix
14. Macrophages	f. Cartilage cells that secrete a flexible, rubbery matrix
15. Matrix	g. Tiny channels in bone containing extensions of bone cells
16. Myofibrils	h. Fragments of cells that play a role in blood clotting
17. Pericytes	i. Small, longitudinal, parallel contractile fibers in muscle cells
18. Platelets	j. Supporting cells present in nervous tissue

■ REVIEW QUESTIONS

1. What sort of arguments can you muster for the position that multicellular organisms have an advantage over unicellular organisms?
2. What are the functions of epithelial tissues? How are the cells adapted to carry out these functions?
3. What is the structure of bone? of adipose tissue? of loose connective tissue? How are each of these adapted to carry out its special functions?
4. Compare the properties of the three types of muscle.
5. Discuss the structure of a nerve cell and how this adapts it for its function.
6. Name the kinds of tissues you would expect to find in the following organs: the lung, the heart, the intestines, and the salivary glands.
7. List the principal organ systems found in a complex animal and give the functions of each.

■ RECOMMENDED READINGS

Caplan, A.I. Cartilage. *Scientific American*, October 1984, pp. 84–94. Cartilage's basic properties of strength and resilience are explained in terms of the tissue's molecular structure.

Kessel, R.G., and Kardon, R.H. *Tissues and Organs: A Text-Atlas of Scanning Electron Microscopy*. San Francisco, W.H. Freeman Co. Publisher, 1979. A striking collection of scanning electron micrographs.

National Geographic Society Book Service. *The Incredible Machine*. Washington, D.C., National Geographic, 1986. A beautiful and informative introduction to the human body. Features the incredible art of renowned photographer Lennart Nilsson.

Solomon, E.P., and Davis, P.W. *Human Anatomy and Physiology*. Philadelphia, Saunders College Publishing, 1983. A very readable presentation of human anatomy and physiology.

Hunting cheetah

39

Protection, Support, and Movement: Skin, Skeleton, and Muscle

■ **OUTLINE**

■ **LEARNING OBJECTIVES**

After you have studied this chapter you should be able to:

1. Describe the external epithelium of invertebrates and summarize its functions.
2. Compare vertebrate skin with the external epithelium of invertebrates and identify the principal derivatives of vertebrate skin.
3. Compare different types of skeletal systems, including the hydrostatic skeleton, exoskeleton, and endoskeleton; give advantages of the vertebrate endoskeleton as compared to the arthropod exoskeleton.
4. Identify the main divisions of the human skeleton and the bones that make up each division.
5. Describe the structure of a typical long bone.
6. Summarize bone development, differentiating between endochondral and membranous bone development.
7. Describe the gross and microscopic structure of skeletal muscle.
8. List in sequence the events that take place in muscle contraction.
9. Compare the roles of glycogen, creatine phosphate, and ATP in providing energy for muscle contraction.
10. Describe the antagonistic action of muscles.
11. Describe the functional relationship between skeletal and muscle tissues.

Some animals run, some jump, some fly. Others remain rooted to one spot, sweeping their surroundings with tentacles. Many contain internal circulating fluids, pumped by hearts and contained by hollow vessels that maintain their pressure by gentle squeezing. Digestive systems push food along with peristaltic writhings. And in all these cases, each action is powered by muscle, a specialized tissue that, however varied its effects, has but one action: It can contract.

In many animals, the muscles responsible for locomotion are anchored to the skeleton, which then serves to transmit forces. The skeleton also functions to support the body and to protect the delicate organs within. Epithelial coverings of invertebrates and skin of vertebrates protect the tissue beneath and may also be specialized to perform a variety of other functions such as secretion or respiration. In this chapter we discuss skin, skeleton, and muscle—systems that are closely interrelated in function and significance.

OUTER COVERINGS: PROTECTION

Epithelium is the tissue that covers all external and internal surfaces of the animal body. The outer epithelium forms a protective shield around the body.

External Epithelium of Invertebrates

In invertebrates epithelial tissue is simple, rather than stratified, and the external epithelium is generally cuboidal or a low columnar type. The external epithelium has a protective function and may also be specialized for secretion or gas exchange. Epithelial cells may be modified as sensory cells that are selectively sensitive to light, chemical stimuli, or mechanical stimuli such as contact or pressure.

In many species, the epithelium contains secretory cells that produce a protective cuticle or secrete lubricants or adhesives. In some species, these cells release odorous secretions used for communication among members of the species or for marking trails. Others produce poisonous secretions used for offense or defense. In earthworms, a lubricating, mucous secretion serves as a moist slime for diffusion of gases across the body wall. This lubricating secretion also reduces friction as the earthworm pushes its way through the soil.

In some species, an epithelial secretion may be limited to a particular region of the body surface. In the gastropod mollusk, for example, a mucous secretion is released from the foot to produce a slime track through which the snail glides (Figure 39–1). Epithelial cells in the basal region of *Hydra* allows this animal to temporarily cling to some object. Insects are able to walk upside down across a ceiling owing to the gummy substance secreted by gland cells in the epithelium of the terminal segments of their legs.

In some insects, for example weaver ants, epithelial secretions are released as fine threads of considerable strength and are used to construct nests. The spinning glands of spiders develop as invaginations of epidermal cells. In Lepidopteran insects, silk is synthesized from amino acids in silk-forming glands.

The Vertebrate Skin

In many fish, in the African ant-eating pangolin (a mammal), and in some reptiles, the skin is developed into a set of scales formidable enough to be considered armor. Even in human beings, the skin has considerable strength. Human skin (Figure 39–2) includes a variety of structures, including fingernails and toenails, hair of various types, sweat glands, oil (sebaceous) glands, and several types of sensory receptors responsible for our ability to feel pressure, temperature, and pain.

Human skin and the skin of other mammals contains mammary glands specialized in females for secretion of milk. Oil glands present in human skin empty via short ducts into hair follicles (see Figure 39–3). They secrete a substance called **sebum,** a complex mixture of fats, waxes, and hydrocarbons. In humans, these glands are especially numerous on the face and scalp. The oil secreted keeps the hair moist and pliable and prevents the skin from drying and cracking. (It is excessive sebum produced at puberty

FIGURE 39–1 The foot of the gastropod mollusk like this striped snail (*Helicella candicans*) releases a mucous secretion producing a slime track through which the animal glides.

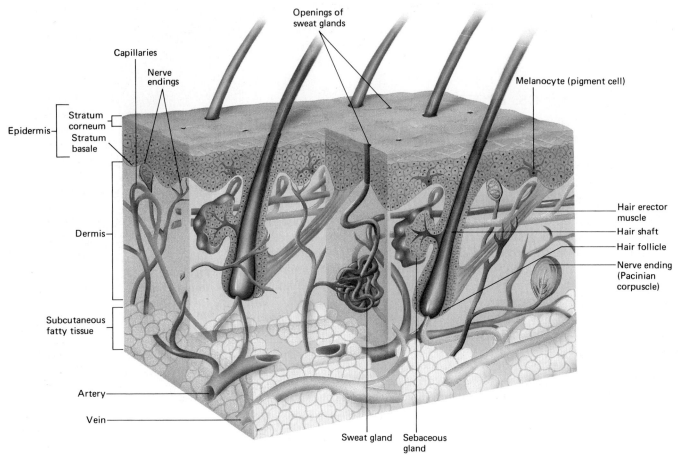

Openings of sweat glands

Capillaries

Nerve endings

Melanocyte (pigment cell)

Epidermis

Stratum corneum
Stratum basale

Hair erector muscle

Hair shaft

Hair follicle

Nerve ending (Pacinian corpuscle)

Dermis

Subcutaneous fatty tissue

Artery

Vein

Sweat gland   Sebaceous gland

FIGURE 39–2 The structure of mammalian skin.

in response to increased levels of sex hormones that fills the glands and follicles, producing the too familiar inflammation called **acne.**)

In a human being, the skin functions as a thermostatically controlled radiator, regulating the elimination of heat from the body (Chapter 1). About 2.5 million sweat glands secrete sweat, and its evaporation from the surface of the skin lowers the body temperature.

The skin in some other vertebrates varies considerably from ours (Figure 39–4). Instead of hairs, birds have feathers, which nevertheless form in a manner comparable to hairs and provide even more effective insulation than fur. Among the poikilothermous vertebrates, one finds epidermal scales (as in reptiles), naked skin covered with mucus (as in many amphibians and fish), and skin with bony or toothlike scales. Some skin, such as that of certain tropical frogs, is even provided with poison glands. Skin and its derivatives are often brilliantly colored in connection with courtship rituals, territorial displays, and various kinds of communication. The human blush pales alongside the spectacular displays of such animals as peacocks.

The Epidermis

The outer layer of skin, the **epidermis,** is the interface between the delicate tissues within and the hostile universe. The epidermis consists of several strata, the lowest of which is the **stratum basale,** and the outermost, the **stratum corneum** (see Figure 39–2). In the stratum basale, cells continuously divide, and the new cells are pushed upward by yet other cells being produced below them. As the epidermal cells move upward in the skin, they mature. In almost all vertebrates, there are no capillaries in the epider-

Sebaceous glands

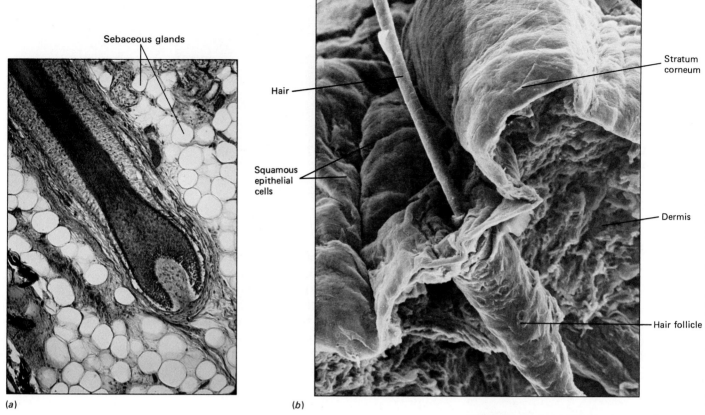

Hair

Squamous
epithelial
cells

Stratum
corneum

Dermis

Hair follicle

(a)

(b)

FIGURE 39–3 (*a*) Photomicrograph of a hair follicle. (*b*) Scanning electron micrograph (approximately ×250) of human skin with hair follicle.

mis, and so the maturing cells are progressively deprived of more and more nourishment and become ever less active metabolically.

As they move upward, epidermal cells manufacture the distinctive skin protein, **keratin,** an elaborately coiled protein that confers on the skin considerable mechanical strength combined with flexibility, for the coils are capable of stretching much like springs. Keratin is quite insoluble and serves as an excellent body surface sealant. When fully mature, epidermal cells are also dead—as dead and as waterproof as shingles. Like shingles, the cells of stratum corneum, the outermost layer, continuously wear off and must be continuously replaced.

FIGURE 39–4 The feathers of birds and scales of reptiles are modifications of the skin. (*a*) Mallard duck. (*b*) Common iguana.

(a)

(b)

FIGURE 39–5 Arctic mammals, such as this harbor seal, have thick layers of subcutaneous fat that serve as insulation.

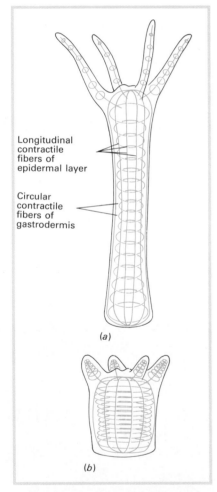

Longitudinal contractile fibers of epidermal layer

Circular contractile fibers of gastrodermis

(a)

(b)

FIGURE 39–6 Movement in *Hydra.* The longitudinally arranged cells are antagonistic to the cells arranged around the body axis. (*a*) Contraction of the circular muscles elongates the body. (*b*) Contraction of the longitudinal muscles shortens the body.

The Dermis

Underneath the epidermis lies the **dermis** (see Figure 39–2), which consists of a dense, fibrous connective tissue composed principally of the protein collagen. The major part of each sweat gland is embedded in the dermis, and the hair follicles reach down into it. The dermis also contains blood vessels, which nourish the skin, and sense organs concerned with touch. Mammalian skin rests on a layer of jelly-like subcutaneous tissue, composed mainly of fat that insulates us from unfavorable outside temperature extremes (Figure 39–5).

SKELETONS: LOCOMOTION AND SUPPORT

A muscle must have something to act upon, something by means of which its contractions can be transmitted to leg, body, or wing. In some of the simplest animals, this is no more than the glutinous, jelly-like substance of the body itself or perhaps a fluid-filled body cavity. More complex animals, however, require a true skeleton to receive, transmit, and transform the simple movement of their muscular tissues.

In a few instances this skeleton is internal—plates or shafts of calcium-impregnated tissue. But in most cases the skeleton is not a living tissue at all but a lifeless deposit atop the epidermis—a shell, or exoskeleton. In addition to its function in locomotion, the skeleton serves to support the body and to protect the internal organs.

Hydrostatic Skeletons

Imagine an elongated balloon full of water. If one were to pull on it, it would lengthen, but it would also lengthen if it were squeezed. Conversely, it would shorten if the ends were pushed. In *Hydra* and other cnidarians, cells of the two body layers are capable of contraction.

The contractile cells in the outer epidermal layer are arranged longitudinally, whereas the contractile cells of the inner layer (the gastrodermis) are arranged circularly around the central body axis (Figure 39–6). These two groups of cells work in **antagonistic** fashion. What one can do, the other can undo. When the epidermal, longitudinal layer contracts, the hydra shortens, and because of the fluid present in its gastrovascular cavity, force is transmitted so that it thickens as well. On the other hand, when the inner, circular layer contracts, the hydra thins, but its fluid contents force it also to lengthen.

The hydra is mechanically little more than a simple bag of fluid. Its fluid interior acts as a hydrostatic skeleton, since it transmits force when the contractile cells contract against it. Hydrostatic skeletons permit only crude mass movements of the body or its appendages. Delicacy is difficult because, in a fluid, force tends to be transmitted equally in all directions and hence throughout the entire fluid-filled body of the animal. It is not easy for the hydra to thicken one part of its body, for example, while thinning another.

The annelid worms have a more sophisticated hydrostatic skeleton that permits more versatile body movements than those of *Hydra*. The body of an earthworm, to take a familiar example, may protrude from its burrow on damp evenings to feed on bits of decayed vegetation on the surface. Its posterior end may then protrude in order to defecate the familiar worm castings. This practice is not without its hazards, for if the hungry worm waits there too long, an equally hungry bird is likely to find it. If the bird is quick enough, the story ends right there. If not, the giant nerve axons of the worm's ventral, solid nerve cord swiftly transmit impulses that stimulate the longitudinal muscles and inhibit the circular muscles. Abruptly the longitudinal muscles contract, pulling the body of the worm toward the safety of its burrow. If the worst occurs and the bird obtains a firm hold, the worm holds on too, with its swollen, contracted anterior now fitting the burrow like a cork in a bottleneck. If the bird releases its hold, the worm will rapidly crawl down the burrow to safety. But how?

If the worm is progressing anterior end first, it must protrude a thinned portion of its body into the burrow ahead. Then, while anchored posteriorly by a thickened portion of itself, the worm must cause its anterior end to swell. Having thus gripped the burrow ahead, the worm releases its posterior grip, and by longitudinal muscle contraction, drags the whole body toward the anchored anterior end. It repeats this process again and again (Figure 39–7).

All this is made possible by the transverse partitions, or **septa,** that subdivide the body cavity of the worm. These septa isolate portions of the body cavity and its contained fluid, permitting the hydrostatic skeletons of each segment to be largely independent of one another. Thus, the contraction of the circular muscle in the elongating anterior end need not interfere

FIGURE 39–7 Annelid locomotion. Can you infer the segments in which longitudinal or circular muscle is active in each stage? The worm is aided in anchoring itself by bristle-like setae.

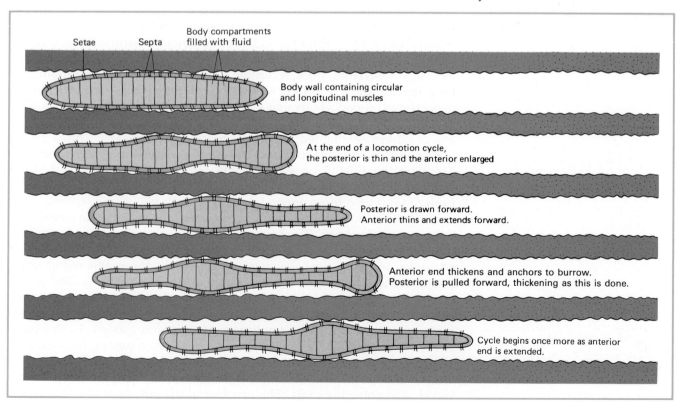

Setae Septa Body compartments filled with fluid

Body wall containing circular and longitudinal muscles

At the end of a locomotion cycle, the posterior is thin and the anterior enlarged

Posterior is drawn forward. Anterior thins and extends forward.

Anterior end thickens and anchors to burrow. Posterior is pulled forward, thickening as this is done.

Cycle begins once more as anterior end is extended.

with the action of the longitudinal muscle in the segments of the still-anchored posterior.

For animals that do more than drag themselves along on their bellies, the hydrostatic skeleton is insufficient. Yet some examples of it occur to a small extent in higher invertebrates and even in vertebrates. Among mollusks, for example, the feet of bivalves are extended and anchored by a hydrostatic blood pressure mechanism not too different from that used by the earthworm. The multitudinous tube feet of echinoderms, such as the sea star and sea urchin, are moved by an ingenious version of the hydrostatic skeleton, and even in man, the penis becomes erect and stiff because of the turgidity of pressurized blood in its cavernous spaces.

External Skeletons

Although there are others, the two major groups of animals with external skeletons are the mollusks and the arthropods. In both mollusks and arthropods, the shell is a nonliving product of the cells of the epidermis, but it differs substantially in function between the two phyla.

The Mollusk Exoskeleton

In mollusks, the exoskeleton basically provides protection, with its major muscle attachments serving the skeleton rather than the skeleton serving the muscles. Thus, the common clam has a pair of muscles whose major function is to hold the two valves of the shell tightly shut against the onslaughts of seastar and chowder maker.

The Arthropod Exoskeleton

In arthropods, however, the skeleton serves not only to protect but also to transmit forces in ways fully comparable to those found in the skeletons of vertebrates. Whereas in mollusks the shell is primarily an emergency retreat, with the bulk of the body nakedly and succulently exposed at other times, in arthropods the exoskeleton covers every bit of the body. It even extends inward as far as the stomach on one end and for a considerable distance inward past the anus on the other. Though the arthropod exoskeleton is a continuous one-piece sheath, it varies greatly in thickness and flexibility, with large, thick, inflexible plates separated from one another by thin, flexible joints arranged segmentally. Enough joints are provided to make the arthropod's body just as flexible as that of many vertebrates. This exoskeleton is also extensively modified to form specialized tools or weapons or otherwise adapted to a vast variety of life styles.

The chief disadvantage of the arthropod exoskeleton is also profound: The rigid exoskeleton prevents growth. To overcome this disadvantage, arthropods must **molt**—that is, cast off their old integument (Figure 39–9) from time to time to accommodate new growth.

To understand molting, we must examine the tissue structure of the arthropod exoskeleton. The living **epidermis** of an arthropod consists of a thin single layer of more or less cuboidal cells interspersed with a variety of glandular and sometimes sensory cells. Directly above this living layer, and in contact with it, is the nonliving **cuticle** (Figure 39–8). The cuticle consists of a mixture of protein and chitin. **Chitin** is a kind of polysaccharide composed of linked, chemically modified glucose units called glucosamine, which, as the name implies, contain amino groups. In the outer portion of the cuticle, cross linkages are established among the chains of chitin by the action of tanning agents secreted by epidermal glands.

The outer portion of the cuticle is the **epicuticle**, composed of proteins, waxes, and sometimes oils. The function of the epicuticle is to retard the

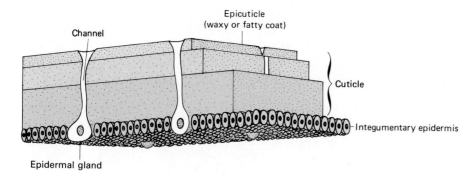

FIGURE 39–8 Structure of the arthropod exoskeleton.

evaporation of water, so it is particularly likely to be present and well developed in the terrestrial arthropods, such as insects. If the greasy epicuticle is experimentally removed from a cockroach—little more than thorough wiping is required—the insect dies from dehydration in a few hours. In most aquatic and a few terrestrial arthropods, the cuticle is impregnated with calcium salts, producing an exoskeleton almost as hard as a mollusk shell. In the joint membranes, these hard layers are reduced or absent for the sake of flexibility.

The developmental stages between moltings are known as **instars.** Somewhere near the end of an instar, an endocrine gland produces a hormone, **ecdysone,** which initiates molting, or **ecdysis.** In response to ecdysone, the epidermis secretes both a new cuticle and enzymes that attack the old cuticle at its base. It is not clear why the new cuticle is not digested also, but when the old one has been sufficiently loosened and dissolved, it splits open along predetermined seams, and the animal slowly and painstakingly wriggles and pulls every appendage and every other detail of its complicated anatomy out of the old shell. Even the internal linings of mouth and anus must be detached. At last the exhausted animal lies still and recuperates. Eventually it drinks or swallows air to stretch its new suit of hardening chitin in preparation for future tissue growth.

Ecdysis obviously produces a temporarily very weakened animal, unable to defend itself and almost unable to move, with a soft integument open to attack by any predator (Figure 39–9). Accordingly, this process is usually carried out in some very sheltered location. During the ensuing hours or days, epidermal glands secrete the tanning agents that harden the cuticle, and in some arthropods channels in the cuticle carry other glandular secretions to the surface, which becomes the new epicuticle.

Internal Skeletons

Endoskeletons, or internal skeletons, are extensively developed only in echinoderms and chordates. Echinoderms have spicules and plates of calcium salts embedded in the tissues of the body wall. These serve mainly for support and protection, in some animals (for example, sea urchins) forming what amounts to an internal shell (Figure 39–10). It is the vertebrates that employ the internal skeleton for its full range of potential—for support and for protection, but primarily for the transmission of forces.

Composed of living tissue, the endoskeleton grows in pace with the growth of the animal as a whole, eliminating the need for molting. It also permits the animal to grow, potentially, to great size. Compare the largest land vertebrates—elephants and dinosaurs—to the largest land arthropods—beetles a few inches long. If beetles grew to the size of horses, their external armor would weigh so much that it would probably collapse or at least prevent the unfortunate animal from moving. So much for the giant insects of the horror movies!

FIGURE 39–9 A cicada molting. This insect requires 13 years to mature.

(a)

(b)

FIGURE 39–10 The echinoderm endoskeleton serves mainly for support and protection. (*a*) The endoskeleton of this sea urchin is composed of spicules and plates of nonliving calcium salts embedded in tissues of the body wall. (*b*) A living slate pencil sea urchin (*Heterocentrotus mammillaths*).

The endoskeleton probably also permits a greater variety of possible motions than does an exoskeleton. In humans, complex motions are produced by an equally complex interaction of many muscles. But there simply is not room for a great many muscles inside the armor of an arthropod limb. Indeed, in some arthropods, especially the spiders, the hydrostatic action of the body fluid is just as important as the intrinsic musculature in producing limb movement.

Many vertebrates possess bones that humans lack, such as the skeleton of the gill arches of fish. Careful studies of the embryos of humans and other mammals have shown, however, that a number of elements of the skull originate embryonically in the same way as do the gill arches of fishes. The tiny middle ear bones—malleus, incus, and stapes—are examples of such elements.

The Human Skeleton

The human skeleton has two main divisions. The **axial skeleton,** located along the central axis of the body, consists of the skull, vertebral column, ribs, and sternum (breastbone). The **appendicular skeleton** consists of the bones of the appendages, upper and lower extremities (arms and legs), plus the bones making up the **girdles** that connect the appendages to the axial skeleton—the shoulder girdle and most of the pelvic girdle (Figure 39–11).

The skull, the bony framework of the head, consists of the cranial and facial bones. The 8 cranial bones enclose the brain, and 14 bones make up the facial portion of the skull. Several cranial bones that are single in the adult human result from the fusion of two or more bones that were originally separate in the embryo or even in the newborn (Figure 32–9).

The human spine, or vertebral column, supports the body and bears its weight. It consists of 24 **vertebrae** and two fused bones, the sacrum and coccyx. The regions of the vertebral column are the **cervical** (neck), composed of 7 vertebrae; the **thoracic** (chest), which consists of 12 vertebrae; the **lumbar** (back), composed of 5 vertebrae; the **sacral** (pelvic), consisting of 5 fused vertebrae; and the **coccygeal,** also consisting of fused vertebrae.

Although they differ in size and shape in different regions of the vertebral column, a typical vertebra consists of a bony central portion that bears most of the body weight, the **centrum,** and a dorsal ring of bone, the neural arch, which surrounds and protects the delicate spinal cord. Vertebrae may also have various projections for the attachment of ribs and muscles and for articulating (joining) with neighboring vertebrae. The first vertebra, the **atlas** (named for the mythical Greek who held the world on his shoulders), has rounded depressions on its upper surface into which fit two projections from the base of the skull. Since there are no ribs in the human neck,

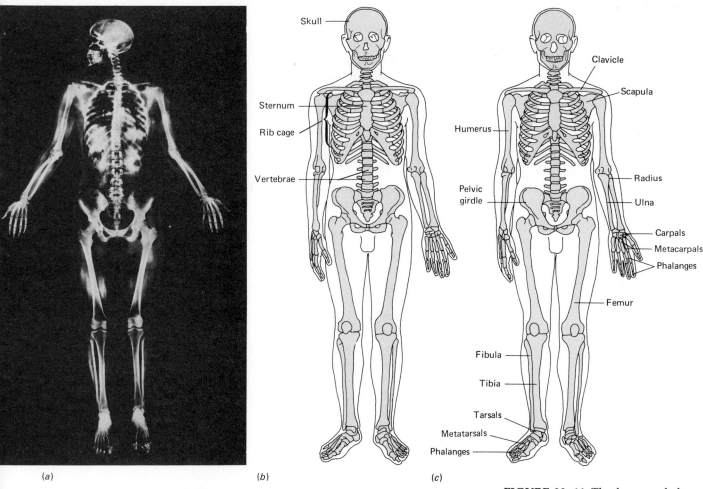

(a)   (b)   (c)

FIGURE 39–11 The human skeleton. (*a*) Compounded series of radiographs of the whole body of a young adult female. (*b*) Bones of the axial skeleton, anterior view. (*c*) Bones of the appendicular skeleton, anterior view.

the cervical vertebrae lack the little facets whereby ribs are attached to the vertebrae of the thorax.

The rib cage is a bony basket formed by the sternum (breastbone), thoracic vertebrae, and 12 pairs of ribs. It protects the internal organs of the chest, including the heart and lungs, and supports the chest wall, preventing it from collapsing as the diaphragm contracts with each breath. Each pair of ribs is attached posteriorly to a separate vertebra. Of the 12 pairs of ribs in the human, the first 7 are attached ventrally to the sternum (breastbone), the next 3 are attached indirectly by cartilages, and the last 2, called "floating ribs," have no attachments to the sternum.

The **pectoral girdle** (shoulder girdle) consists of the two collarbones, or **clavicles,** and the two shoulderblades, or **scapulas.** The **pelvic girdle** consists of a pair of large bones, each composed of three fused hipbones. Whereas the pelvic girdle is securely fused to the vertebral column, the pectoral girdle is loosely and flexibly attached to it by muscles.

The human extremities are comparatively generalized, each terminating in five **digits**—the fingers and toes, whereas the more specialized appendages of other animals may be characterized by four digits (as in the pig), three (as in the rhinoceros), two (as in the camel), or one (as in the horse).

Great apes and humans do have a highly specialized feature—the opposable thumb. (In addition, great apes have an opposable big toe; and in humans, though it is not opposable, the big toe is similar enough in structure to the thumb that it can be used as a surgical substitute.) The opposable thumb can readily be wrapped around objects such as a tree limb in climbing, but it is especially useful in grasping and manipulating objects. It can be opposed to each finger singly or to all of them collectively. The muscles that move the thumb are almost as powerful as those of all the other fingers put together.

In humans, the upper appendages (which contain our thumbs) are not used for locomotion as in other mammals, including the great apes. Our hands have been emancipated. The combination of opposable thumbs and upright posture enables us to use our hands to shape and build, to effect changes in our environment to a greater extent than any other organism on Earth.

Structure of a Typical Long Bone

The radius, one of the two bones of the forearm, is a typical long bone (Figure 39–12). It has numerous muscle attachments, arranged so that the bone operates as a lever that amplifies the motion they generate. Muscles cannot shorten enough, by themselves, to produce large movements of the body parts to which they are attached.

Like other bones, the radius is covered by a connective tissue membrane, the **periosteum,** capable of laying down fresh layers of bone and thus increasing the diameter of the bone. The main shaft of a long bone is known as its **diaphysis.** The expanded ends of the bone are called **epiphyses.** In children, a disk of cartilage, the **metaphysis,** is found between the epiphyses and the diaphysis. The metaphyses are growth centers that disappear at maturity, becoming vague **epiphyseal lines.** Within the long bone, there is a central **marrow cavity** filled with a fatty connective tissue known as yellow bone marrow. The marrow cavity is lined with a thin membrane, the endosteum.

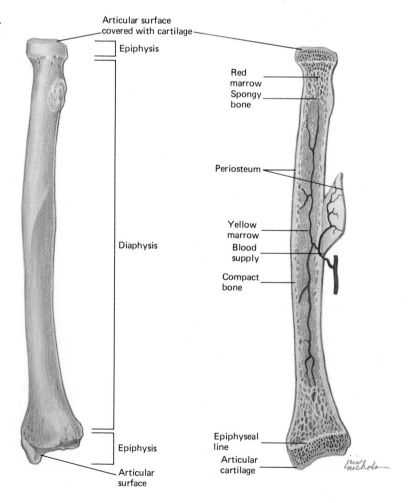

FIGURE 39–12 Anatomy of a bone. (*a*) Structure of a typical long bone. (*b*) Internal structure of a long bone.

At its joint surfaces, the outer layer of a bone consists of articular cartilages, which serve as low-friction bearings for the joints (Figure 38–9). The joints are enclosed in joint capsules full of a lubricant, the **synovial fluid.**

The radius has a thin, outer shell of **compact bone,** which is very dense and hard and is found primarily near the surfaces of a bone, where great strength is needed. Recall from Chapter 38 that compact bone consists of interlocking, spindle-shaped units called **osteons,** or **haversian systems** (Figure 38–7). Within an osteon, osteocytes, the mature bone cells, are found in small cavities called **lacunae.** The lacunae are arranged in concentric circles around central **haversian canals.** Blood vessels that nourish the bone tissue pass through the haversian canals. Threadlike extensions of the cytoplasm of the osteocytes extend through narrow channels called **canaliculi.** These cellular extensions connect the osteocytes.

Interior to the thin shell of compact bone is a somewhat spongy filling of **spongy bone,** also called **cancellous bone** which despite its loose structure provides most of the mechanical strength of the bone. Spongy bone consists of a meshwork of thin strands of bone. The spaces within the spongy bone are filled with bone marrow.

Bone Development

During fetal development, bones form in two ways. Long bones, such as the radius, develop from cartilage replicas, in a process called **endochondral bone** development. The flat bones of the skull, for example the frontal bone, the vertebrae, and some other bones, develop from a noncartilage connective tissue scaffold (Figure 39–13). This is known as **intramembranous bone** development.

Osteoblasts are bone-building cells. They secrete the protein collagen, which forms the strong, elastic fibers of bone. A complex calcium phosphate called **apatite** is present in the tissue fluid. This compound automatically crystallizes around the collagen fibers, forming the hard matrix of bone. As the matrix forms around the osteoblasts, they become isolated within the small spaces called lacunae. The trapped osteoblasts are referred to as osteocytes.

FIGURE 39–13 Mature (*a*) and fetal (*b*) skulls. Compare these skulls, noting the differences in proportions between the two. In the fetal specimen the frontal bones have not yet fused to produce the single adult frontal bone. The same is true of the lower jaw. The skull of an infant also has a **fontanel** (*arrow*) or gap, between the skull bones at the superior surface of the skull; the fontanel closes after birth.

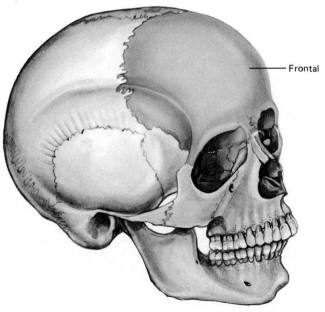

Frontal

(a)

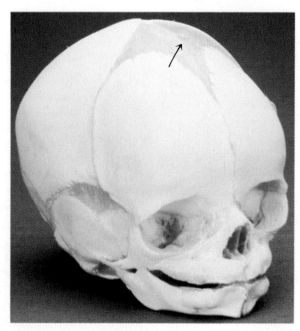

(b)

Bones are modeled during growth and remodeled continuously throughout life in response to physical stresses and other changing demands. As muscles develop in response to physical activity, the bones to which they are attached thicken and become stronger. As bones grow, bone tissue must be removed from the interior, especially from the walls of the marrow cavity. This process keeps bones from getting too heavy. **Osteoclasts** are the cells that break down bone in a process referred to as bone resorption. These bone-breaker cells are very large cells that move about, secreting enzymes that digest bone. Osteoclasts and osteoblasts work side by side to shape bones and to form the precise grain needed in the finished bone.

As initially laid down, the collagen fibers crisscross in various directions. But to achieve full mechanical strength, they must attain a definite grain. This requires that the fine structure of bone be rebuilt even if the bone as a whole is not to be remodeled to any great extent.

The mechanical stresses to which a bone is habitually subjected determine the direction in which its grain should run. These forces cause the existing apatite to generate electrical signals somewhat as a crystal in a phonograph arm generates electrical signals in response to the vibrations of a needle in the record grooves. In accordance with the electrical fields thus set up, osteoclasts digest the collagen fibers. Crystallized apatite apparently cannot exist in the body apart from collagen, and so it dissolves in the surrounding fluids.

As fast as the former bone is removed, fresh osteoblasts busily deposit new bone in its place. This new bone, however, has a definite grain and is organized into spindle-shaped osteons. It is believed that most bone is completely made over as many as ten times during the course of an average lifetime.

MUSCLES: MOVEMENT

Locomotion, manipulation, circulation of blood, and the propulsion of food through the digestive tract all require some way of generating mechanical forces and motion. The muscles serve as motors of the body, making possible all these actions and much more. The three types of muscle—skeletal, smooth, and cardiac—each specialized for its particular task, were described in Chapter 38.

A Muscular Zoo

The humblest of animals has the ability to move. The simple cnidarian *Hydra* has only two layers of cells, both consisting of epithelial tissue. Yet these cells are muscular as well as epithelial; their elongated bases contain contractile strands. In other cnidarians (some anthozoans and some scyphozoans), distinct muscle cells specialized for contraction are present, and in some instances these are grouped together in conspicuous bands of muscle. In flatworms, muscle occurs as a specialized tissue, organized into definite layers. Arthropods have muscles that are typically cross-striated like vertebrate skeletal muscle. See Focus on Flight Muscles of Insects.

Muscle Structure

What we commonly think of as a muscle—the biceps in your arms, for example—consists of thousands of individual cells, each wrapped in connective tissue. Because muscle cells are elongated in shape, they are often referred to as **fibers.** Muscle fibers are arranged in bundles known as **fascicles,** which are wrapped by connective tissue.

Focus on FLIGHT MUSCLES OF INSECTS

The only invertebrate phylum that possesses striated muscle to any great extent is the arthropods. In some cases, arthropod striated muscle may be more highly differentiated even than that of vertebrates. This is seen best in the **flight muscles** of insects where many muscle contractions result from a single nervous input.

In most flying insects, the flight muscles are attached not directly to the wings but to the flexible portions of the exoskeleton that articulate with the wings. Each contraction of the muscles produces a dimpling of the exoskeleton in association with a downstroke and, depending on the exact arrangement of the muscles, sometimes on the upstroke as well. When the dimple springs back into its resting position, the muscles attached to it are stretched. The stretching initiates another contraction immediately, and the cycle is repeated. The deformation of the cuticle is transmitted as a force to the wings and they beat—so fast that we may perceive the sound as a musical note. In the common blowfly, for instance, the wings may beat at a frequency of 120 cycles *per second.* Yet in that same blowfly, the neurons that innervate those furiously contracting flight muscles are delivering impulses to them at the astonishingly low frequency of three per second. It seems very likely that the mechanical properties of the musculoskeletal arrangement are what provide the stimuli for contraction by stretching the muscle fibers at the resonant frequency of the system. But the nerve impulses are needed to maintain it.

Insect flight muscle in action has a very high metabolic rate, perhaps the highest of any tissue anywhere. Accordingly, it contains more mitochondria than any known variety of muscle, and it is elaborately infiltrated with tiny air-filled tracheae that carry oxygen directly to each cell (see figure). Many insects have special adaptations to rid the body of the excess heat produced by the flight muscles. The rapidly flying sphinx moth, for example, has what amounts to a radiator in its abdomen, a great blood vessel that carries heat from the thorax, where it is generated, and emits it into the cool of the night.

Flight muscles must be kept at operating temperature if they are to function. You have probably noticed the constant twitching of the wings of such insects as wasps even when they are crawling instead of flying. Probably this behavior is necessary to keep the temperature of the flight muscles high for instant combat readiness. You may also have noticed that the bodies of many moths are quite furry. The fur of these insects (more properly called **pilus**) serves the same function as fur in a mammal—to conserve body heat. When the moth awakens and prepares for flight, it shivers its flight muscles at a low frequency to warm them up, constricting its abdominal blood vessel to keep the heat in its thorax. Gradually the frequency of the shivering increases until, at a critical moment, the moth spreads its wings and hums off into the darkness.

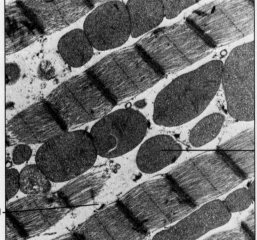

Insect flight muscle, such as that of the bumblebee shown here, may be the most powerful muscle found in any organism. Oxygen is brought directly to the muscle by the tracheal tubes, which convey air into the muscle cell itself. Note the prominent striations and the tremendously convoluted internal membranes of the many mitochondria.

Each muscle fiber is a spindle-shaped cell with many nuclei (Figure 39–14). The cell membrane, known in a muscle cell as the **sarcolemma,** has multiple inward extensions that form a set of **T-tubules** (transverse tubules). The cytoplasm of a muscle fiber is referred to as **sarcoplasm,** and the endoplasmic reticulum as **sarcoplasmic reticulum.**

Threadlike structures called **myofibrils** run lengthwise through the muscle fiber. The myofibrils are composed of two types of even tinier struc-

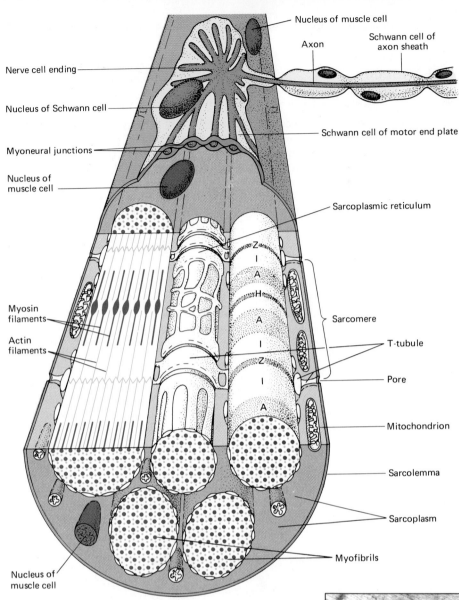

Nucleus of muscle cell

Axon

Schwann cell of axon sheath

Nerve cell ending

Nucleus of Schwann cell

Schwann cell of motor end plate

Myoneural junctions

Nucleus of muscle cell

Sarcoplasmic reticulum

Z
I
A
H
A
I
Z
I
A

Sarcomere

Myosin filaments

Actin filaments

T-tubule

Pore

Mitochondrion

Sarcolemma

Sarcoplasm

Nucleus of muscle cell

Myofibrils

(a)

FIGURE 39–14 Skeletal muscle structure. (a) The structure of a skeletal muscle cell. Notice the eccentric placement of the nuclei, just under the sarcolemma. (b) Light photomicrograph showing striations (approximately ×200). (c) Electron micrograph (approximately ×30,000). Note that striations persist at this much higher magnification. Black line indicates 1 micrometer. GLY, glycogen; MY, myosin filaments; ACT, actin filaments; M, mitochondrion; TS, transverse tubule or T-system; A, H, I, and Z, the zones and bands in the muscle tissue.

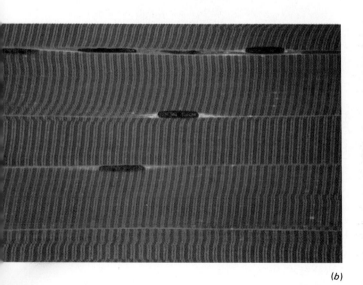

(b)

(c)

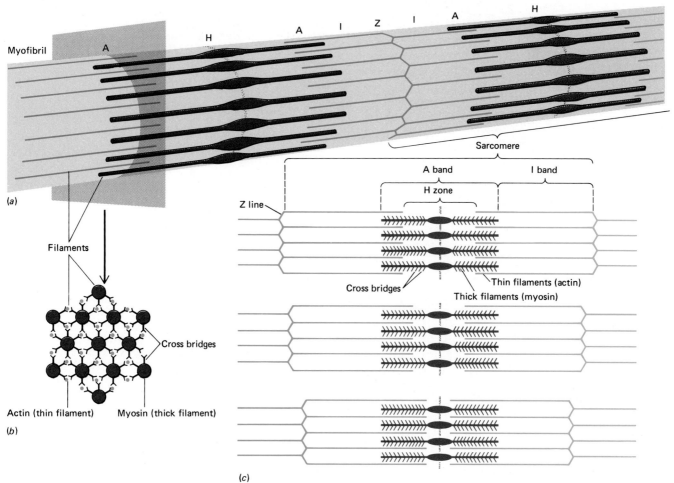

FIGURE 39–15 A myofibril stripped of the accompanying membranes. (*a*) The Z lines mark the ends of the sarcomeres. (*b*) Cross section of myofibril shown in (*a*). (*c*) Filaments slide past each other during contraction. Notice the way the filaments overlap. It is the regular pattern of overlapping filaments that gives rise to the striated appearance of skeletal and cardiac muscle. In the top drawing of *c* the myofibril is relaxed. In the middle drawing, the filaments have slid toward each other, increasing the amount of overlap and shortening the muscle cell by shortening its sarcomeres. At bottom, maximum contraction has occurred; the sarcomere has shortened considerably. Letters represent zones along the myofibril.

tures, the **myofilaments.** The thick myofilaments, called **myosin filaments,** consist mainly of the protein myosin, while the thin **actin filaments** consist mostly of the protein actin. Myosin and actin filaments are arranged lengthwise in the muscle fibers so that they overlap. Their overlapping produces a pattern of bands, or **striations,** characteristic of striated muscle. These are designated by specific letters, as indicated in Figure 39–15. A **sarcomere** is a unit of thick and thin filaments. Sarcomeres are joined at their ends by an interweaving of filaments called the Z line.

Muscle Contraction

The typical pull of a muscle results from the shortening of its cells, which in turn results from the actin and myosin filaments actively pulling themselves past and between one another (see Figure 39–15). Each myosin filament consists of about 200 molecules of the protein myosin in a parallel arrangement. A rounded head extends from each rod-shaped myosin molecule. The head of the myosin molecule bears a binding site that is complementary to binding sites on the actin filament. Each actin filament contains 300 to 400 rounded actin molecules arranged in two chains plus other regulatory proteins. Each sarcomere is capable of independent contraction. When many sarcomeres contract together, they produce the contraction of the muscle as a whole.

During muscle contraction, the actin filaments are pulled inward between the myosin filaments. As this occurs, the muscle shortens. We can

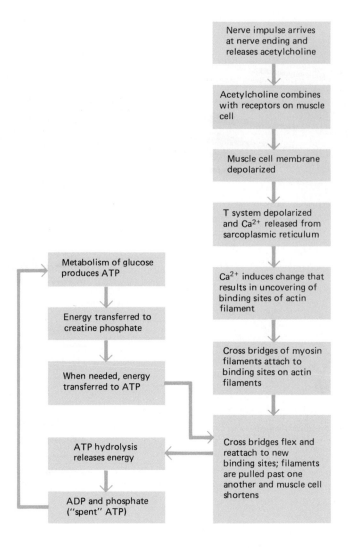

FIGURE 39–16 Summary of the events of muscular contraction.

summarize the process of muscle contraction as follows (Figure 39–16):

1. When a nerve impulse (a neural message) passes down a motor neuron (a nerve cell that stimulates a muscle) and arrives at the junction between the neuron and muscle the neuron releases a compound known as acetylcholine.
2. The acetylcholine diffuses across the junction (myoneural cleft) between the neuron and the muscle fiber and combines with receptors on the surface of the muscle fiber (Figure 39–17).
3. In response, the sarcolemma undergoes an electrical change called **depolarization,** which we study in more detail as it occurs in nerve cells. (Depolarization is unique in muscle cells because it is not confined to the surface membrane of the cell but actually travels *into* the cell along the T tubules, which are inward extensions of the cell membrane.) Depolarization may initiate an impulse (an electric current) that spreads over the sarcolemma. The electric current generated is known as an **action potential.** Excess acetylcholine is broken down by the enzyme **cholinesterase.**
4. The impulse spreads through the T tubules and stimulates protein channels in the sarcoplasmic reticulum to open, allowing calcium ions to move out of storage and flow into the sarcoplasm.
5. The calcium induces a process that uncovers binding sites on the actin filaments.
6. The ends of the myosin molecules that make up the thick filaments are called **cross bridges** because they bridge the gaps between thick and thin filaments. These cross bridges attach to the binding sites on

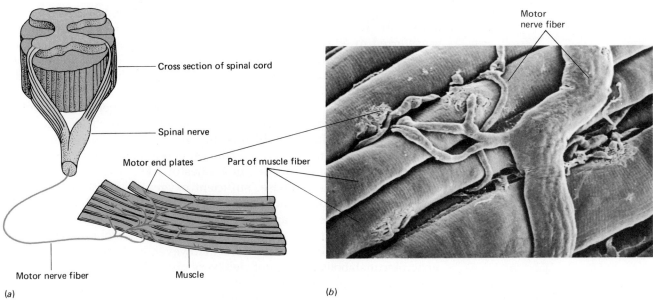

FIGURE 39–17 Motor units. (*a*) A motor unit typically includes many more muscle fibers than appear here, averaging about 150 muscle fibers each, but some units have less than a dozen fibers, while others have several hundred. (*b*) Scanning electron micrograph of some of the cells in a motor unit (×900). Note how the large neuron branches send subdivisions to each cell in the motor unit.

the actin filaments. This process is powered by energy from ATP molecules.

7. The cross bridges now release their hold on the first set of binding sites (also using ATP) and reach for the second. The process is repeated with the third, and so on (Figure 39–15). This series of stepping motions actively pulls the myosin and actin filaments past one another with an apparent tendency to combine along as great a length as possible. As this process continues, the muscle shortens.

8. Relaxation of the muscle occurs when the calcium is pumped back into the sarcoplasmic reticulum.

Even when we are not moving, our muscles are in a state of partial contraction known as **muscle tone.** Stimulated by messages from nerve cells, some muscle fibers are contracted at any given moment. Muscle tone is an unconscious process that helps keep muscles prepared for action. When the motor nerve to a muscle is cut, the muscle becomes limp (completely relaxed), or flaccid.

Energy For Muscle Contraction

Muscle cells are often called upon to perform strenuously, and must be provided with large amounts of energy. The immediate source of energy necessary for muscle contraction is ATP. ATP is necessary both for the pull exerted by the cross bridges and for their release from each active site, as they engage, in hand-over-hand fashion, in their tug of war on the thin filaments. Rigor mortis, the temporary but very marked muscular rigidity appearing after death, results from ATP depletion following the cessation of cellular respiration that occurs at death.[1]

Sufficient energy can be stored in the energy-rich bonds of ATP molecules for only the first few seconds of strenuous activity. The main phosphorylated compound in vertebrate muscle cells is not ATP—but an en-

---

[1]Rigor mortis does not persist indefinitely, however, for the entire contractile apparatus of the muscles degenerates eventually, restoring pliability. The phenomenon is temperature-dependent, so given the prevailing temperature, a medical examiner can estimate the time of death of a cadaver from its degree of rigor mortis. Perhaps it should be said that rigor mortis is not by itself muscular contraction; it only tends to freeze the corpse in its position at the time of death. Thus, tales of corpses sitting, pointing to their murderers, and otherwise carrying on posthumously may be entertaining but have no factual basis.

ergy storage compound known as **creatine phosphate,** which can be stockpiled. As it is needed, the energy stored in creatine phosphate is transferred to ATP. However, the supply of creatine phosphate does not last very long either during vigorous exercise. As ATP and creatine phosphate stores are depleted, muscle cells must replenish their supply of these energy-rich compounds.

Fuel is stored in muscle fibers in the form of glycogen, a large polysaccharide formed from hundreds of glucose units. Stored glycogen is degraded, yielding glucose, which is then broken down in cellular respiration. When sufficient oxygen is available, enough energy is captured from the glucose to produce needed quantities of ATP.

During strenuous exercise, sufficient oxygen may not be available to meet the needs of the rapidly metabolizing muscle cells. Under these conditions, muscle cells are capable of breaking down fuel molecules anaerobically (without oxygen) for short periods of time. Anaerobic metabolism is a very rapid method of generating ATP. However, as discussed in Chapter 7, anaerobic metabolism does not yield very much ATP. The depletion of ATP results in weaker contractions and muscle fatigue. The waste product lactic acid is produced during anaerobic breakdown of glucose. Lactic acid buildup contributes to muscle fatigue. During muscle exertion, an oxygen debt develops, which is paid back during the period of rapid breathing that typically follows strenuous exercise.

The energy conversion of muscular contraction is not very efficient. Only about 30% of the chemical energy of the glucose fuel is actually converted to mechanical work. The remaining energy is accounted for as heat, produced mainly by frictional forces within the muscle cell. This is why we get hot when we work hard physically and also why we shiver when we are cold: The muscle contraction involved in shivering is one way of producing heat to warm the body.

Muscle Action

Skeletal muscles produce movements by pulling on tendons, tough cords of connective tissue that anchor muscles to bone. Tendons, in turn, pull on bones. Most muscles pass across a joint and are attached to the bones that form the joint. When the muscle contracts, it draws one bone toward or away from the bone with which it articulates.

Muscles can only pull; they cannot push. Muscles act **antagonistically** to one another; the movement produced by one can be reversed by another. The biceps muscle, for example, permits you to flex your arm, whereas the triceps muscle allows you to extend it once again (Figure 39–18). Thus, the biceps and triceps can work antagonistically to one another.

The muscle that contracts to produce a particular action is known as the **agonist.** The muscle that produces the opposite movement is referred to as the **antagonist.** When the agonist is contracting, the antagonist is relaxed. Generally, movements are accomplished by groups of muscles working together so that there may be several agonists and several antagonists in any action. Note that muscles that are agonists in one movement may be antagonists in another. The superficial muscles of the human body are shown in Figures 39–19 and 39–20.

Variations in Muscle Response

The three types of muscle differ in the ways they respond. Smooth muscle (Figure 38–8) often contracts in response to simple stretching, and its contraction tends to be lengthy and sustained. It is well adapted to performing such tasks as the regulation of blood pressure by sustained contraction of

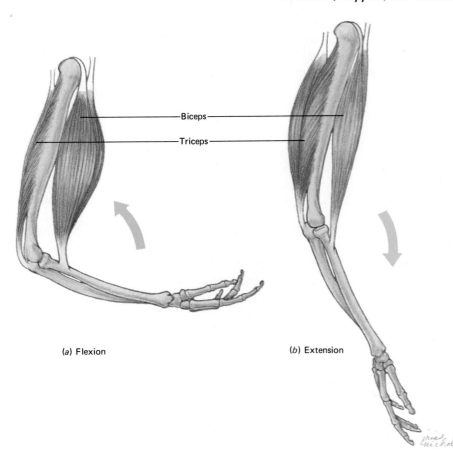

(a) Flexion

(b) Extension

FIGURE 39–18 The antagonistic arrangement of the biceps and triceps muscles.

the walls of the arterioles. Although smooth muscle contracts slowly, it shortens much more than striated muscle does. Though not well suited for running or flying, smooth muscle squeezes superlatively.

Cardiac muscle contracts abruptly and rhythmically, propelling blood with each contraction. Sustained contraction of cardiac muscle would be disastrous! Skeletal muscle, when stimulated by a single brief stimulus, contracts with a quick, single contraction called a **simple twitch.** Ordinarily, simple twitches do not occur except in laboratory experiments. In the normal animal, skeletal muscle receives a series of separate stimuli very close together. These produce not a series of simple twitches, however, but a single, smooth, sustained contraction called **tetanus.** Depending upon the identity and number of muscle cells tetanically contracting, we thread a needle or haul a rope.

Not all muscular activities are the same, however. Dancing or, even more so, typing requires quick response rather than the long, sustained effort that might be appropriate in hauling a rope. In many animals, entire muscles are specialized for quick or slow responses. In chickens, for instance, the white breast muscles are efficient for quick responses, since flight is an escape mechanism for chickens. On the other hand, chickens walk about on the ground all day, so the dark leg and thigh meat is composed of muscle specialized for more sustained activity.

There is no human equivalent of light and dark meat. However, we do possess individual muscle cells that are specialized for either fast or slow response, which can be distinguished microscopically with the appropriate staining (Figure 39–21). The proportions of **slow-twitch** and **fast-twitch** fibers vary from muscle to muscle in the same person and also differ among persons. It has long been believed that the relative proportions of

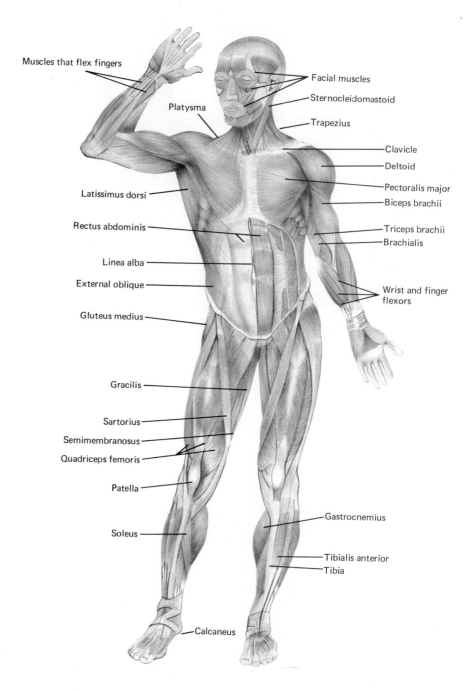

Muscles that flex fingers

Facial muscles

Sternocleidomastoid

Platysma

Trapezius

Clavicle

Deltoid

Pectoralis major

Latissimus dorsi

Biceps brachii

Rectus abdominis

Triceps brachii

Brachialis

Linea alba

Wrist and finger flexors

External oblique

Gluteus medius

Gracilis

Sartorius

Semimembranosus

Quadriceps femoris

Patella

Gastrocnemius

Soleus

Tibialis anterior

Tibia

Calcaneus

FIGURE 39–19 Superficial muscles of the human body, anterior view.

the two determined the kind of athletic activity at which one might have the greatest potential proficiency and also that this proportion was genetically determined. Recent evidence, however, indicates that the proportions of the two kinds of fibers in human muscle can be changed by appropriate training, at least to some degree.

■ SUMMARY

I. In invertebrates, epithelial tissue may contain secretory cells that produce a protective cuticle; secrete lubricants or adhesives; produce odorous or poisonous secretions; or produce threads for nests or webs. Invertebrate epithelium may be specialized for sensory or respiratory functions.

II. Human skin includes nails, hair, sweat glands, oil glands, and sensory receptors.
   A. Cells in the stratum basale of the epidermis continuously divide; as they are pushed upward toward the skin surface, these cells mature, produce keratin, and eventually die.

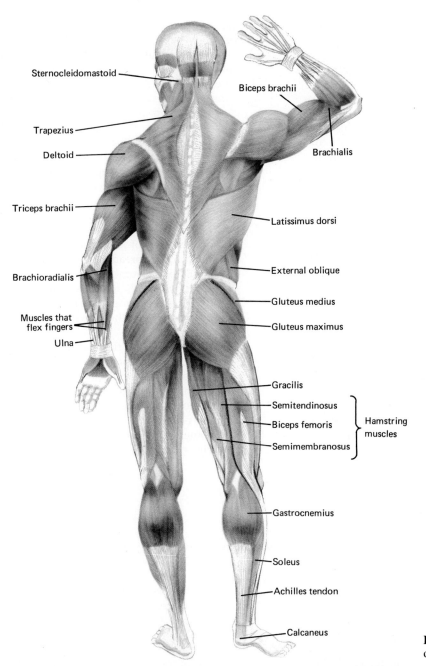

Sternocleidomastoid

Trapezius

Deltoid

Triceps brachii

Brachioradialis

Muscles that
flex fingers

Ulna

Biceps brachii

Brachialis

Latissimus dorsi

External oblique

Gluteus medius

Gluteus maximus

Gracilis

Semitendinosus

Biceps femoris

Semimembranosus

} Hamstring
muscles

Gastrocnemius

Soleus

Achilles tendon

Calcaneus

FIGURE 39–20 Superficial muscles of the human body, posterior view.

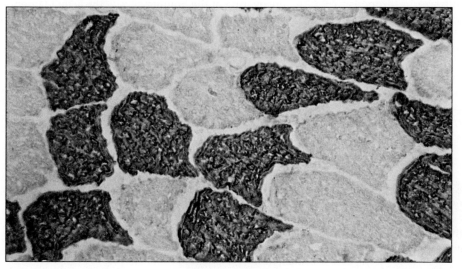

FIGURE 39–21 Slow (light-colored) and fast (dark) striated muscle fibers in cross section (approximately ×640). The fast fibers have been stained by a technique that identifies a particular kind of quick-acting tropomyosin, a protein that is probably responsible for the characteristic differences in the two kinds of muscle. The white meat of chickens and turkeys is composed of fast-twitch fibers, and the dark, of slow-twitch fibers, but in human beings both kinds of fibers are found in all muscles.

B. The dermis, which consists of dense, fibrous connective tissue, rests on a layer of subcutaneous tissue composed largely of fat.

III. The skeleton transmits mechanical forces generated by muscle and also supports and protects the body.

A. *Hydra* and many other invertebrates have a hydrostatic skeleton in which fluid is used to transmit forces generated by contractile cells or muscle. In *Hydra* the circular and longitudinal layers of contractile cells form an antagonistic relationship.

B. Exoskeletons are characteristic of mollusks and arthropods. The arthropod skeleton, composed mainly of chitin, is jointed for flexibility. This nonliving skeleton prevents growth, making it necessary for arthropods to molt periodically.

IV. Endoskeletons, found in echinoderms and chordates, are composed of living tissue and therefore are capable of growth.

A. The human skeleton consists of an axial portion and an appendicular portion.

B. The radius, a typical long bone, consists of a thin outer shell of compact bone surrounding the inner cancellous bone. Within the long bone is a central marrow cavity.

C. Long bones such as the radius develop from cartilage replicas; this is endochrondral bone formation. Other bones, such as the flat bones of the skull, develop from a noncartilage connective tissue replica; this is membranous bone development.

V. All animals have the ability to move. Specialized muscle tissue is found in most invertebrate phyla and in all

of the vertebrates. As muscle contracts (shortens), it moves body parts by pulling on them.

A. A muscle such as the biceps consists of hundreds of muscle fibers.

B. The striations of skeletal muscle fibers reflect the interdigitations of their actin and myosin filaments. A unit of actin and myosin filaments makes up a sarcomere.

C. During muscle contraction, the actin filaments are pulled inward between the myosin filaments.

1. Muscle contraction begins when a motor neuron releases acetylcholine into the myoneural cleft.

2. The acetylcholine combines with receptors on the surface of the muscle fiber.

3. This results in depolarization of the sarcolemma and initiation of an action potential.

4. The impulse spreads through the T tubules and stimulates calcium release.

5. Calcium initiates a process that uncovers the binding sites of the actin filaments.

6. Cross bridges of the myosin filaments attach to the binding sites.

7. The cross bridges flex and reattach to new binding sites so that the filaments are pulled past one another and the muscle shortens.

D. ATP is the immediate source of energy for muscle contraction, but muscle tissue has another energy storage compound, creatine phosphate. Glycogen is the fuel stored in muscle fibers.

E. Muscles act antagonistically to one another.

■ POST-TEST

1. The vertebrate skin consists of two main layers, the outer _____ and the inner _____.

2. The cells of the stratum _____ of the epidermis are dead and almost waterproof.

3. The protein _____ confers mechanical strength, flexibility, and waterproofing on the skin.

4. _____ skeletons have the principal or even sole function of transmitting muscular force.

5. Since an exoskeleton tends to limit size, arthropods must _____ from time to time in order to grow.

6. The internal skeletons of echinoderms and chordates are known as _____.

7. The radius has a thin outer shell of _____ bone and a spongy filling of _____ bone.

8. Synovial fluid serves as a _____ in _____.

9. The two types of myofilaments in muscle tissue are _____ filaments and _____ filaments.

10. Unscramble this list of the events of muscle contraction into the correct sequence:
a. calcium release
b. T-system depolarization
c. acetylcholine release
d. nerve impulse
e. uncovering of the binding sites of the actin filaments
f. cross bridges flex
g. cross bridges release binding sites

11. Creatine phosphate's function is _____ in the muscle cell.

12. Fuel is stored in muscle cells in the form of the polysaccharide _____; the immediate source of energy for muscle contraction is _____.

■ REVIEW QUESTIONS

1. Compare vertebrate skin with the external epithelium of invertebrates.

2. What properties does keratin confer on human skin?

3. What is a hydrostatic skeleton? Which functions does it perform?

4. How do the septa in the annelid worm contribute to the flexibility of its hydroskeleton?

5. What are the disadvantages of an exoskeleton? Compare the arthropod exoskeleton to the vertebrate endoskeleton.

6. Describe the process of arthropod molting.
7. Describe the divisions of the human skeleton.
8. Draw a typical long bone such as the radius and label its parts.
9. Contrast the functions of osteoblasts and osteoclasts. Why is it important that bones be continuously remodeled?
10. Compare the two types of myofilaments in muscle tissue. What is a sarcomere?
11. Outline the sequence of events that causes a muscle cell to contract, beginning with the stimulation of its nerve and including cross-bridge action.
12. What is the role of ATP in muscle contraction? What is the function of creatine phosphate? of glycogen?
13. What is the role of agonists? of antagonists? Why is it important that a muscle can switch roles?

■ RECOMMENDED READINGS

Austin, P.R., et al. Chitin: new facets of research. *Science* Vol. 212, pp. 749–753, May 15, 1981. The practical uses of chitin, the most widely distributed animal skeleton carbohydrate, may eventually rival those of cellulose, the major plant skeletal carbohydrate.

Buller, A.J., and Buller, N.P. *The Contractile Behavior of Skeletal Muscle* (booklet). Burlington, N.C., Carolina Biological Supply Company, 1978. Emphasis is on classical physiology of contraction and stimuli.

Clark, R.B. *Dynamics in Metazoan Evolution: The Origin of the Coelom and Segments.* Oxford, Clarendon Press, 1964. Despite the title, this is basically a discussion of locomotory adaptations in lower animals with emphasis on the hydrostatic skeleton.

Cole, R.P. Myoglobin function in exercising skeletal muscle. *Science* Vol. 216, pp. 523–525, April 30, 1982. Its function long a mystery, muscle hemoglobin at last yields up some of its secrets. Though the details remain unknown, the substance is shown to be necessary for normal muscular oxygen consumption.

Gray, J. *Animal Locomotion.* London, Weidenfeld and Nicholson, 1968. The hydrostatic skeleton and the role of the musculoskeletal system in higher animal locomotion.

Hadley, N.F. The arthropod cuticle. *Scientific American*, Vol. 255, No. 1, pp. 104–112, July 1986. The arthropod cuticle is largely responsible for the adaptive success of the arthropods. The author discusses the properties that permit the cuticle to provide protection and support.

Harrington, W.F. *Muscle Contraction* (booklet). Burlington, N.C., Carolina Biological Supply Company, 1981. Contains hard-to-find and valuable criticisms of the cross bridge theory of muscle contraction.

Heinrich, B., and Bartholomew, G.A. Temperature control in flying moths. *Scientific American*, pp. 87–95, June 1972. A description of the mechanisms of temperature regulation in moths and of the relationship between temperature and flight.

Huxley, A. *Reflections on Muscle.* Princeton, Princeton University Press, 1980. The originator of the sliding filament theory of muscular contraction discusses the history and prospects of the scientific understanding of contraction.

Lazarides, E., and Revel, J.P. The molecular basis of cell movements. *Scientific American*, pp. 100–112, May 1978. The role of microfilaments in cell movement.

Lowenstam, H.A. Minerals formed by organisms. *Science* Vol. 211, pp. 1126–1130, March 13, 1981. A fine comparative study of skeletal systems.

Luttgens, K., and Wells, K.F. *Kinesiology: Scientific Basis of Human Motion.* 7th ed. Philadelphia, Saunders College Publishing, 1982. How skeleton, muscles and nervous system interact to permit and produce the multitude of motions of which the human body is capable.

Montagna, W., and Parakkel, P.F. *The Structure and Function of the Skin.* 3rd ed. New York, Academic Press, 1974.

Morey, E.R. Spaceflight and bone turnover. *Bioscience*, pp. 168–172, 1984. Spaceflight may become practical only when the demineralization of bone that it produces is stopped. This will require an extension of our fundamental knowledge of bone mineral turnover mechanisms and their control.

Neville, C. *The Biology of the Arthropod Cuticle* (booklet). Burlington, N.C., Carolina Biological Supply Company. A beautiful, short summary.

A red slug (*Arion rufus*) lubricates its food with a stream of yellow mucus

40

Processing Food

■ LEARNING OBJECTIVES

After you have read this chapter you should be able to:

1. Describe, in general terms, the following steps in processing food: ingestion, digestion, absorption, and elimination.
2. Compare the food habits of herbivores, carnivores, and omnivores and compare adaptations that they possess for their particular mode of nutrition.
3. Compare the nutritional life styles of parasites, commensals, and mutualistic partners.
4. Compare the types of digestive systems found in cnidarians and flatworms with the digestive system of an earthworm.
5. Trace a bite of food through each structure of the human digestive tract and describe the changes that take place en route.
6. Describe the four layers of the wall of the vertebrate digestive tract.
7. Describe the types of teeth found in mammals and give their functions.
8. Draw and label a tooth and give the function of each structure.
9. Describe the accessory digestive glands of terrestrial vertebrates, and how each promote digestion.
10. Describe the anatomic features of the small intestine that increase its surface area, and discuss their advantages.
11. Trace the step-by-step digestion of a carbohydrate, a lipid, and a protein.
12. Draw a diagram of an intestinal villus and label its parts.
13. Describe the absorption of glucose, amino acids, and fat.

Nutrients are the substances present in food that are needed by organisms as building blocks for growth and repair of the body, as ingredients for making compounds needed for metabolic processes, and as an energy source for running the machinery of the body. Obtaining nutrients is so important that many organisms have undergone adaptations in response to their nutritional needs and the means by which they obtain and process food. An organism's body plan and its life style are adapted to its particular mode of obtaining food.

All animals are **heterotrophs,** organisms that must obtain their energy and nourishment from the organic molecules manufactured by other organisms. Because heterotrophs eat the macromolecules made by other organisms, they must break down these molecules and refashion them for their own needs. For example, humans cannot incorporate the proteins in steak directly into their own muscles. The body must first break down the steak proteins into their component amino acids and deliver these to the muscle cells. Then the cells must arrange these components into human muscle proteins.

After foods are selected and obtained, they are **ingested,** that is, taken into the digestive tract. In a human, ingestion includes taking the food into the mouth and swallowing it. Once the food is inside the organism it must be **digested** or broken down before its nutrients can be utilized by the animal. Typically, large pieces of food are first mechanically broken down into smaller ones. Then, the macromolecules in the food are enzymatically degraded into small molecules that can be absorbed and utilized.

In sponges and other animals with intracellular digestion, nutrients are absorbed from the phagosome (the vesicle enclosing the food) directly into the cytoplasm. In animals that are equipped with digestive tracts, **absorption** involves the passage of nutrients through the cells lining the digestive tube and into the blood or other body fluids. Nutrients are then distributed to all parts of the organism and utilized for metabolic activities within each cell. Undigested and unabsorbed food is ejected from the body, a process referred to as **egestion** in simpler forms and as **elimination** in more complex animals.

MODES OF NUTRITION

Animals are consumers, ultimately dependent upon plants for their food, energy, and oxygen. Some animals are **herbivores,** or **primary consumers,** which eat exclusively, or mainly, plant materials. Herbivores may be consumed by flesh-eating **carnivores,** which also may consume one another. **Omnivores** eat both plants and animals. Some animals are **symbionts,** organisms that form intimate nutritional relationships with members of other species.

Adaptations of Herbivores

Herbivores eat only algae or plants. They may restrict their diet to a particular plant part—the leaves, for example, or perhaps the roots, seeds, or nectar. A herbivore may eat only one or a few species of plants. Many aquatic herbivores are filter feeders. As large volumes of water flow by, their filtering system traps tiny plants or protists in a mucous secretion. Cilia then sweep the mucus containing the captured food to the mouth.

Terrestrial plants contain a great deal of supporting material, including cell walls made of cellulose. Animals cannot digest cellulose, so it is somewhat of a problem for them to obtain nutrients from the plant material they eat. The adaptations utilized to exploit plant food sources are strikingly varied (Figure 40–1). Some herbivorous insects have piercing and sucking mouthparts so that they can pierce through the tough cell walls and suck the sap or nectar within the plant cells. Other herbivores simply eat great quantities of food. Grasshoppers, locusts, elephants, and cows, for example, all spend a major part of their lives eating. Most of what they eat is not efficiently digested and moves out of the body as waste, almost unchanged. However, by eating large enough quantities, sufficient material is digested and absorbed to provide the nourishment necessary to sustain their life processes.

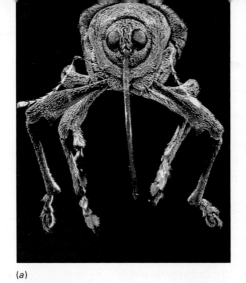

(a)

(b)

(c)

(d)

FIGURE 40–1 Adaptations of herbivores. (*a*) An acorn weevil. The impressively long "snout" of this little beetle is used both for feeding and to make a hole in the acorn through which an egg is deposited. When it has hatched, the larva feeds on the contents of the acorn seed. (*b*) The parrot has a powerful beak for cracking nuts and seeds. Unlike most birds, the parrot can use one foot to manipulate its food and feed itself. (*c*) The rhinoceros can use its horn to uproot and overturn small trees and bushes; it then eats the leaves. Members of some species use their lips to break off grass. (*d*) Camels live on seeds, dried leaves, and whatever desert plants they can find. This animal can eat sharp cactus thorns without injury because the lining of its mouth is very tough.

Many herbivores are equipped with jaws and teeth or toothlike structures for ingesting food. The teeth of herbivorous mammals include wide molars for grinding plant food, which often have enamel specially adapted to heavy wear. Herbivores also have longer and more elaborate digestive tracts than carnivores, so that food can be retained for digestion for a long time.

A common and very interesting adaptation of herbivores is a symbiotic relationship with microorganisms that inhabit their digestive tracts. These microorganisms break down cellulose cell walls to allow the digestive enzymes access to the nutrients within the cell. A small amount of the cellulose may be completely hydrolyzed to glucose subunits. Termites, cows, and horses are among the herbivores that enjoy such symbiotic relationships. Many vertebrate herbivores have a specialized section of the digestive tract called the **cecum** which houses bacteria capable of digesting cellulose.

In ruminant mammals (including sheep, cattle, and deer), the stomach is divided into fermentation chambers (Figure 40–2). When a cow eats grass, very little chewing occurs before swallowing. The food passes through the esophagus into the **rumen** and **reticulum**, where it is mixed and churned. Microorganisms that inhabit these chambers secrete enzymes that break down cellulose cell walls. Some of the cellulose is degraded into sugars, which are then used by the bacteria themselves. The bacteria produce fatty acids during their metabolism, some of which are absorbed by the animal's rumen, and are in fact its main energy source.

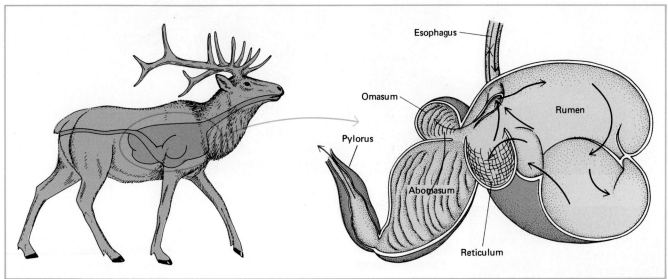

FIGURE 40–2 The ruminant stomach is divided into chambers. Food passes from the esophagus to the rumen and reticulum, where microorganisms digest cellulose. The partly digested food is pushed back into the animal's mouth where it is rechewed. After reswallowing, the partly digested food moves into the omasum and finally into the abomasum, or true stomach.

Food that is not sufficiently chewed, called the "cud," is regurgitated back up into the cow's mouth, where it is chewed again. When it is re-swallowed, it enters the **omasum,** where it is mechanically churned. Partly digested food and the microorganisms mixed with it then enter the **abomasum,** or true stomach. Like the stomach of other vertebrates, the abomasum produces digestive enzymes. Both the food and the symbiotic bacteria are digested in the abomasum so that many of the nutrients absorbed by the bacteria are recovered by the cow. In addition to digesting cellulose, these symbiotic bacteria produce vitamins and amino acids that can be used by the host.

Adaptations of Carnivores

Carnivores eat meat, and some eat large amounts of bones as well. As predators their first problem is to find and capture their prey. Even simple invertebrate carnivores have adaptations for this purpose. For example, *Hydra* and its relatives have long tentacles equipped with stinging cells called cnidocytes. Should an organism brush by a cnidocyte, a long thread-like lasso springs forward, entangling and perhaps paralyzing the prey. The tentacles then deliver the meal into the waiting mouth. Dinner is captured without the hydra's ever having to move from its perch.

Vertebrate carnivores have a variety of interesting adaptations for catching prey (Figure 40–3). The fast-moving tongue of the frog captures many a fly, while the long, quick legs, sharp teeth, and claws of the lion enable it to catch and kill gazelles. Carnivorous mammals have well-developed canine teeth for stabbing during combat; their molars are modified for shredding meat into small chunks that can be swallowed easily. The digestive juices of the stomach break down proteins, and because meat is more easily digested than plant food, carnivores' digestive tracts are shorter.

Omnivores

Omnivores, such as bears and humans, include both plant material and meat in their diet. They obtain food by a wide variety of mechanisms. Most aquatic filter-feeding organisms ingest both tiny plants and animals. Earthworms take in large amounts of soil containing both animal and plant material; they utilize the organic material for food and egest the rest. Omnivores are generally equipped to distinguish among a wide range of smells and tastes, which enable them to select various foods.

(a)

(b)

(c)

FIGURE 40–3 Adaptations of carnivores. (*a*) The long-nose butterfly fish (*Forcipiger longirostris*) has a mouth adapted for picking small worms and crustaceans from tight spots in coral reefs. (*b*) With lightning speed the Burmese python strikes at its prey, then suffocates it before consuming it whole. (*c*) With its wide field of vision and fast reflexes, the California mantis is a very able carnivore.

Symbionts

A symbiont is an organism that lives in intimate association with a member of another species. One or both of the organisms usually derive nutritional benefit from the association. Three types of symbionts are parasites, commensals, and mutualistic partners.

A **parasite** lives on or in the body of another living organism, the **host** species, from which it obtains its nourishment (Figure 40–4). **Ectoparasites,** such as fleas and ticks, live on the host's body; **endoparasites,** such as tapeworms and hookworms, live inside the host. Whether the parasite nourishes itself from food ingested by its host or by sucking the host's blood, it is strictly a freeloader. An effective parasite does not kill its host before it provides itself with passage to its next host. For example, a tapeworm might live within the intestine of its host for many years.

A **commensal** is an organism that derives benefit from its host without either doing harm to or benefiting the host. Commensalism is especially common in the ocean. Practically every worm burrow and hermit crab shell contain some uninvited guests that take advantage of the shelter and abundant food supplied by the host.

Mutualistic partners are two species of organisms that live together for their mutual benefit. They may be unable to survive separately. A classic example of mutualism is the flagellate protozoan that lives in the intestine of the termite. The termite eats wood but cannot produce the enzymes necessary to digest it. Its mutualistic partner, the flagellate, cannot chew wood and cannot survive outside the termite's gut, but it does produce the enzymes necessary to digest cellulose. Termites cannot survive without these intestinal inhabitants. Newly hatched termites lick the anus of another termite to obtain a supply of flagellates, a practice which requires these insects to live together in social groups.

INVERTEBRATE DIGESTIVE SYSTEMS

Sponges, the simplest animals, are filter feeders. They obtain food by filtering microscopic plants and animals out of sea water. Individual cells (both choanocytes and amoebocytes) phagocytize the food particles. Digestion is intracellular within a food vacuole (phagosome), and nutrients can be transferred to other cells. Wastes are egested into the water that continuously circulates through the sponge body.

Cnidarians, such as *Hydra* and jellyfish, capture small aquatic animals with the help of their cnidocytes and tentacles. The mouth opens into a large gastrovascular cavity lined by cells that secrete enzymes that hydrolyze proteins to peptides (Figure 40–5). Flagella on the gastrodermal cells mix the food. Digestion continues intracellularly within food vacuoles, and the products diffuse to the epidermal cells. Undigested food is ejected through the mouth by contraction of the body.

Free-living flatworms generally feed on small invertebrates such as crustaceans, rotifers, and annelid worms. Planarians begin to digest their prey even before ingesting it, by extending the pharynx through the mouth and secreting digestive enzymes onto the prey. Once ingested, the food is pumped into the branched intestine by waves of muscular contraction (peristalsis). Extracellular digestion proceeds as intestinal cells secrete digestive enzymes. Partly digested food fragments are then phagocytized by cells of the intestinal lining, and digestion is completed intracellularly within food vacuoles. The highly branched intestine facilitates the distribution of digested food, but there is no separate circulatory system. Like cnidarians, flatworms have no anal opening, so undigested wastes are eliminated through the mouth.

In most other invertebrates, and in all vertebrates, the digestive tract is a complete tube with an opening at each end. Food enters through the mouth, and undigested food is eliminated through the anus. Peristaltic movements push the food in one direction, so that more food can be taken in while previously eaten food is being digested and absorbed farther down the tract. Various parts of the tube are specialized to perform specific functions. For example, the digestive tract of the earthworm includes a mouth, a muscular pharynx that secretes a mucous material to lubricate food particles, an esophagus, a thin-walled crop where food is stored, a thick muscular gizzard where food is ground against small stones, and a long, straight intestine in which extracellular digestion occurs. The intestine terminates in an anus through which food wastes are eliminated. Some invertebrates—certain worms, insects, mollusks, crustaceans, and sea urchins—have hard, toothed mouthparts that can tear off and chew bits of food.

VERTEBRATE DIGESTIVE SYSTEMS

The vertebrate digestive tract is a complete tube extending from mouth to anus (Figure 40–6). In simple vertebrates, the digestive tract, or **gut**, may be a rather simple tube. In more complex vertebrates, however, various regions of the gut have specialized structures and functions. Food succes-

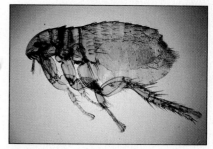

(a)

(b)

FIGURE 40–4 Parasitic symbionts. (*a*) The flea is a well-known ectoparasite. Its body shape is adapted for slipping through fur, and its long hind legs are built for jumping. Note the claws, which help the flea hold on to hairs on the host's body. (*b*) A most interesting form of predation is practiced by the brachonid wasp. The adult wasp lays her eggs under the skin of an insect like this saddleback caterpillar of the flannel moth. When the eggs hatch, the larvae feed on the body fluids of the caterpillar until ready to spin cocoons. This caterpillar is almost covered by cocoons and has little chance of developing into an adult moth. The wasps are beginning to emerge from their cocoons.

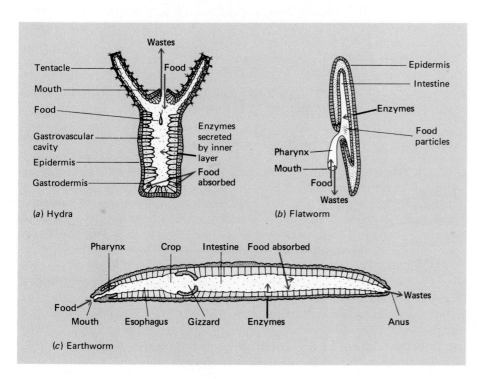

(a) Hydra

(b) Flatworm

(c) Earthworm

FIGURE 40–5 Food processing in several types of invertebrates. The *hydra* (*a*) and the flatworm (*b*) each have a digestive tract with a single opening that serves as both mouth and anus. (*c*) The earthworm, like most complex animals, has a complete digestive tract extending from mouth at one end of the body to anus at the other end.

FIGURE 40–6 Like all vertebrates, the salamander has a complete digestive tract. The liver and pancreas are digestive glands that secrete digestive juices into the digestive tract.

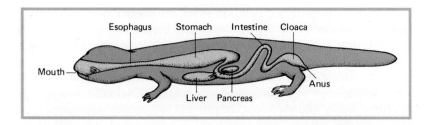

FIGURE 40–6 Like all vertebrates, the salamander has a complete digestive tract. The liver and pancreas are digestive glands that secrete digestive juices into the digestive tract.

sively passes through these parts of the digestive tract: mouth, pharynx (throat), esophagus, stomach, small intestine, large intestine, and anus. All vertebrates have accessory glands that secrete digestive juices into the digestive tract. These include the liver and the pancreas and, in terrestrial vertebrates, the salivary glands (Figure 40–7).

Wall of the Digestive Tract

The wall of the mammalian digestive tract, from the esophagus to the rectum, has a similar structure and is composed of the same four layers. From the **lumen** (inner space) outward, they are the mucosa, submucosa, muscularis, and adventitia (Figure 40–8). The **mucosa** lines the digestive tract.

FIGURE 40–7 The human digestive system. Note the complete digestive tract, a long, coiled tube extending from mouth to anus. Locate the three types of accessory glands.

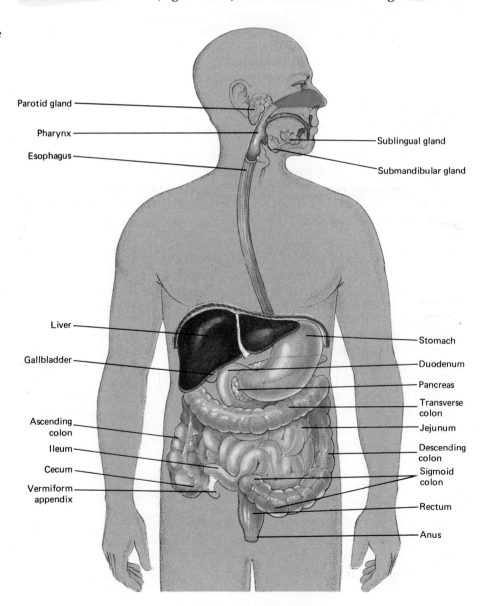

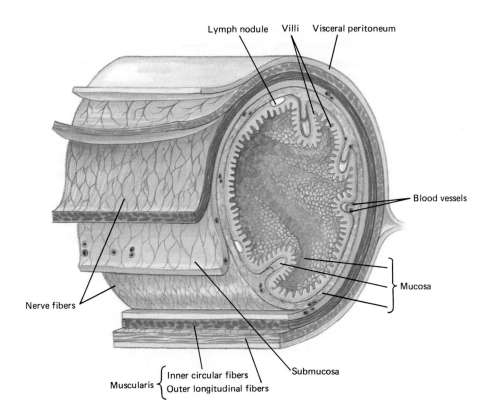

Lymph nodule Villi Visceral peritoneum

Blood vessels

Mucosa

Submucosa

Nerve fibers

Muscularis { Inner circular fibers
Outer longitudinal fibers

FIGURE 40–8 Section through the wall of the human small intestine illustrating the mucosa, submucosa, muscularis, and visceral peritoneum.

It consists of epithelial tissue resting upon a layer of connective tissue. **Goblet cells** in the epithelial tissue secrete mucus, which protects and lubricates the inner surface of the digestive tract. The multicellular glands of the digestive tract are formed as inpocketings of the mucosa. In the stomach and intestine, the mucosa is greatly folded to increase the secreting and absorbing surface of the digestive tube.

The **submucosa,** made up of connective tissue, binds the mucosa to the muscle layer beneath. The submucosa is rich in blood vessels, lymph vessels, and nerves. Along most of the digestive tract, the **muscularis** consists of two layers of smooth muscle, an inner one, which has muscle fibers arranged circularly, and an outer layer with fibers arranged longitudinally. Localized contractions of these muscles help to mechanically break down food and to mix it with digestive juices. Rhythmic waves of contraction of these muscles push food along through the digestive tract in the process of **peristalsis** (see Figure 40–13).

The **adventitia** is the outer connective tissue coat of the digestive tract. Below the level of the diaphragm it is covered by a layer of squamous epithelium and is called the **visceral peritoneum.** By various folds it is connected to the **parietal peritoneum,** a sheet of connective tissues that lines the walls of the abdominal and pelvic cavities. Between the visceral and parietal peritoneums, there is a potential space, the **peritoneal cavity.** Inflammation of the peritoneum, called **peritonitis,** can be very serious because infection can spread along the peritoneum to most of the abdominal organs.

Inside the Mouth

Food is ingested into the mouth, and both mechanical and chemical digestion begin there. In humans, the mouth, and especially the tongue and teeth, also aid in speech. The fleshy, sensitive lips that surround the opening of the mouth help guide food into the mouth. The mouth cavity is supported by jaws and is bounded on the sides by teeth, gums, and

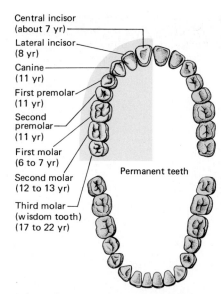

Central incisor
(about 7 yr)

Lateral incisor
(8 yr)

Canine
(11 yr)

First premolar
(11 yr)

Second
premolar
(11 yr)

First molar
(6 to 7 yr)

Second molar
(12 to 13 yr)

Third molar
(wisdom tooth)
(17 to 22 yr)

Permanent teeth

FIGURE 40–9 Human permanent teeth. Approximate time of eruption is shown in parentheses. One quadrant of the jaw has been shaded.

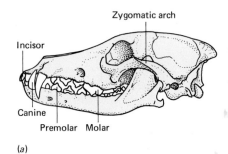

Zygomatic arch

Incisor

Canine

Premolar Molar

(a)

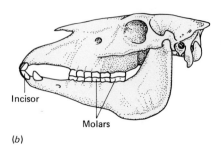

Incisor

Molars

(b)

FIGURE 40–10 Comparison of the teeth of carnivore and herbivore. (a) Skull of a coyote. (b) Skull of a domestic horse.

cheeks, and on the bottom by the tongue. Its roof, the **palate,** is a shelf that separates the mouth cavity from the nasal cavity. The anterior bony portion is the **hard palate,** and the posterior fleshy part, the **soft palate.**

A muscular, mobile **tongue** is characteristic of most terrestrial vertebrates. In many amphibians and reptiles, and in some birds, the tongue is used to help capture food. You may have seen the long tongue of a frog dart outward with lightning speed to catch an insect on its sticky tip. Such catapulting tongues are attached anteriorly and are free in the back. The mammalian tongue manipulates food, pushing it between the teeth to be chewed and then shaping it into a mass, called a **bolus,** which is then swallowed.

In mammals, the **taste buds** are concentrated in the **papillae,** tiny projections on the tongue's surface. Sensory cells in the taste buds respond to different chemicals, enabling humans to distinguish four primary tastes: sweet, sour, salty, and bitter.

The Teeth

The **teeth** are used to bite, tear, crush, and grind food. Unlike the simple, pointed, and conical teeth of fish, amphibians, and reptiles, the teeth of mammals vary in size and shape and are specialized to perform specific functions (Figure 40–9). The chisel-shaped **incisors** are used for biting and are especially large in gnawing animals such as mice, rats, squirrels, and beavers. In herbivores, the lower incisors are well developed for cutting off grass. Upper incisors do not develop in ruminants. These animals crop grass by pulling it with their tongue and upper lip across the cutting edge of the lower incisors.

The four long, pointed **canines** (one in each quadrant of the jaw just lateral to the incisors) are used for stabbing and tearing food. Carnivores such as wolves, dogs, and lions have prominent canine teeth, sometimes referred to as fangs (Figure 40–10). (In fact, these teeth are called canines because they are so large in dogs.) Male baboons and some other animals use canines in dominance displays and in defense.

Humans have eight **premolars** and 12 **molars,** arranged with two premolars and three molars on each side of the upper and lower jaw (see Figure 40–9). These teeth have flattened surfaces for crushing and grinding food. Herbivores have large, flat molars for grinding plant material. In humans, the third molars, called **wisdom teeth,** frequently fail to erupt or, if they do emerge from the gum, are often crooked and useless. This is usually interpreted as reflecting a trend in evolution of modern humans toward a shortening of the jaws, with a resulting crowding of the teeth, which leaves inadequate space for the last molar.

Although shaped somewhat differently among species, all mammalian teeth have the same basic structure. The part of the tooth projecting above the gum is the **crown** (Figure 40–11). One or more **roots** are hidden beneath the gumline, and the somewhat constricted junction between the crown and root is the **neck.** Each root is embedded in a socket of the jawbone.

In the crown and upper neck region, a hard, outer **enamel** covers the tooth. Enamel is the hardest substance in the body. Beneath the enamel is a thick layer of **dentin,** which makes up most of the tooth and resembles bone in composition and hardness. In the neck and root region, a calcified connective tissue called **cementum** covers the dentin and attaches it, via connective tissue, to its socket.

Beneath the dentin is a **pulp cavity** filled with **pulp,** a soft connective tissue containing blood vessels, lymph vessels, and nerves. Narrow extensions of the pulp cavity, called **root canals,** pass through the roots of the

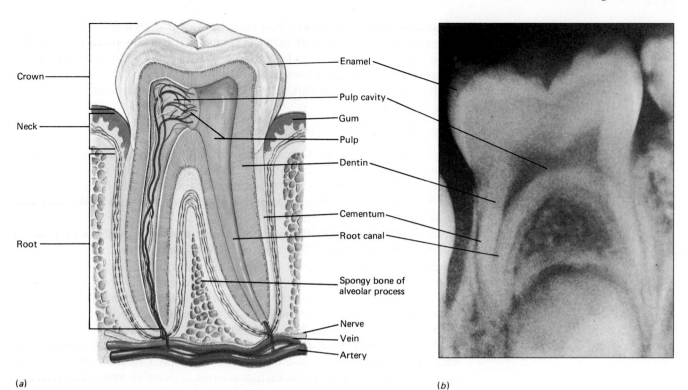

Crown

Neck

Root

Enamel

Pulp cavity

Gum

Pulp

Dentin

Cementum

Root canal

Spongy bone of
alveolar process

Nerve

Vein

Artery

(a)

(b)

FIGURE 40–11 Structure of a
tooth. (*a*) Sagittal section through a
lower human molar. (*b*) X-ray of a
healthy tooth.

tooth. Blood vessels, lymph vessels, and nerves reach the pulp cavity by
way of the root canals.

When allowed to accumulate, bacteria that inhabit the mouth form dental plaque on the surfaces of the teeth, particularly near the gum line.
Carbohydrates (especially sucrose) deposited on the teeth are fermented
by the bacteria of the plaque. Organic acids produced during fermentation demineralize the outer layers of the teeth and cause **dental caries,** or
cavities.

The Salivary Glands

Aquatic animals have plenty of water available to moisten food and aid in
swallowing. Terrestrial vertebrates possess **salivary glands,** which produce
saliva to perform these functions. In mammals, there are three main pairs
of salivary glands: the parotid, submandibular, and sublingual glands. The
parotid glands, the largest salivary glands, are located in the tissue below
and in front of the ears (Figure 40–7). These glands often become infected
and swell when a person has the mumps. The **submandibular glands** lie
below the jaw, and the **sublingual glands** lie under the tongue. Secretions
from these glands are delivered into the mouth cavity through tiny ducts.

Saliva consists of a thin, watery component containing the digestive
enzyme **salivary amylase** and a mucous component that lubricates the passage of the bolus during swallowing. Saliva also contains salts and substances that kill bacteria. Salivary amylase begins the digestion of carbohydrates by hydrolyzing starch to maltose. Saliva is normally slightly acidic,
with a pH of about 6.7, and amylase works best at this pH. After it reaches
the stomach, the bolus is penetrated by the very acidic gastric juice, and
the amylase is then inactivated.

Humans secrete about one liter of saliva daily. Secretion is regulated by
control centers in the brain that send messages to the glands by way of
nerves. Feeling food in the mouth or tasting it stimulates these control
centers. Even smelling, seeing, or thinking about food may stimulate an

increase in saliva secretion. Food such as sour pickles and lemons are the strongest stimuli. Very little saliva is secreted during sleep, and should the body become dehydrated, salivation slows or stops altogether. The dry feeling in the mouth that results is one of the stimuli that indicates thirst, motivating us to ingest fluids and thus restore homeostasis.

Through the Pharynx and Esophagus

During swallowing, food passes from the mouth cavity into the pharynx (throat region), the muscular cavity where the digestive and respiratory systems cross. The **esophagus** is a muscular tube extending from the pharynx to the stomach. It passes between the lungs behind the heart and penetrates the diaphragm.

The movement of food from the mouth to the stomach is aided by a series of reflex actions. The first part of swallowing is under voluntary control. The tongue is raised against the roof of the mouth, and the bolus of food between the tongue and palate is pushed into the pharynx by a wavelike movement of the tongue (Figure 40–12). When swallowing begins, breathing is stopped momentarily by a reflex mechanism that prevents food from passing into the respiratory passageways.

Several openings in the pharynx close by reflex action before food reaches the pharynx. This ensures that the food will pass only into the esophagus. The hard bump in the ventral midline of the neck, perhaps known to you as the Adam's apple, is the **larynx.** Contraction of muscles raises the larynx so that its opening (the glottis) is sealed off against a flap of tissue called the **epiglottis.** This action prevents food from entering the respiratory passageway. You can observe the raising of the larynx by watching someone swallow: It bobs upward with each swallow.

Reflex movements propel the bolus through the pharynx and into the esophagus. As the bolus enters the esophagus, a peristaltic wave pushes it downward toward the stomach (Figure 40–13). This journey takes only about 10 seconds. Gravity is not necessary to pull food through the esophagus. Astronauts at zero gravity are able to swallow, and even if you are standing on your head, food will reach the stomach!

The opening from the esophagus to the stomach is controlled by a portion of circular muscle that acts as a sphincter. When a peristaltic wave reaches the lower portion of the esophagus, the ring of muscle relaxes, permitting the bolus to enter the stomach. When the sphincter fails to close during digestion, the highly acidic gastric juice may splash up into the esophagus and cause a burning sensation. This sensation is known as heartburn because it is experienced in the region over the heart.

FIGURE 40–12 Position of the tongue and epiglottis during (a) breathing and (b), (c) swallowing. In (b), note how a bolus is pushed from the mouth into the pharynx by the tongue to initiate swallowing.

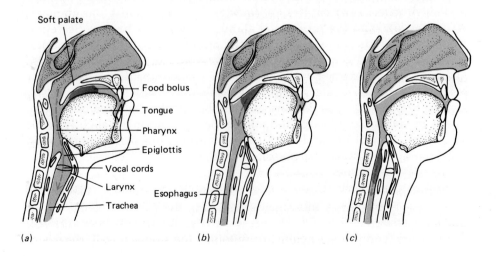

Soft palate

Food bolus
Tongue
Pharynx
Epiglottis
Vocal cords
Larynx
Trachea
Esophagus

(a)   (b)   (c)

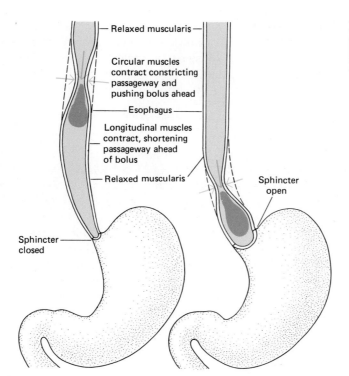

- Relaxed muscularis
- Circular muscles contract constricting passageway and pushing bolus ahead
- Esophagus
- Longitudinal muscles contract, shortening passageway ahead of bolus
- Relaxed muscularis
- Sphincter closed
- Sphincter open

FIGURE 40–13 Peristalsis. Food is moved through the digestive tract by waves of muscular contraction known as peristalsis.

In the Stomach

The human **stomach** is a thick-walled, muscular sac on the left side of the body just beneath the lower ribs. The muscular layers of the stomach wall are very thick and include a diagonal layer of fibers in addition to the circular and longitudinal fibers found elsewhere in the wall of the digestive tract (Figure 40–14). When empty, the stomach is collapsed and shaped somewhat like a hotdog. As a meal is eaten, the stomach stretches, assuming the shape of a football; its capacity is about one liter. The lining of the empty stomach appears wrinkled because of prominent folds of the mucosa called **rugae.** As the stomach fills with food, the rugae gradually flatten out.

The stomach is lined with simple columnar epithelium that secretes large amounts of mucus. Millions of microscopic **gastric glands** extend deep down into the mucosa. They secrete the **gastric juice. Parietal cells** of the gastric glands produce hydrochloric acid, and **chief cells** produce a protein, **pepsinogen,** the inactive precursor of the enzyme pepsin. The gastric juice is highly acidic, with a pH of about 0.8. However, the gastric juice mixes with mucus and food so that the final pH of the gastric contents is about 2. This is sufficiently acidic to kill most of the bacteria that enter the stomach with the food (see Focus on Peptic Ulcers).

Pepsinogen is converted to active **pepsin** by the removal of a portion of the molecule. This reaction is catalyzed by hydrochloric acid (HCl) and by pepsin itself. The HCl also provides an optimum pH for pepsin action. Pepsin, the principal enzyme of the gastric juice, initiates the digestion of proteins. One of its most important actions is to digest the protein collagen found in the connective tissue of meat. As collagen is broken down by pepsin, the protein within the muscle cells becomes more accessible to the digestive juices and can be digested.

The activities of the stomach are regulated by both the nervous and endocrine systems. When food is contemplated, smelled, viewed, or tasted, the brain sends messages stimulating the gastric glands. By the time food arrives in the stomach, gastric juices have already been released. Then, when food presses against receptors in the stomach wall, the gastric

FIGURE 40–14 From the esophagus, food enters the stomach, where it is mechanically and enzymatically digested. (*a*) Structure of the stomach. (*b*) The stomach lining and gastric glands.

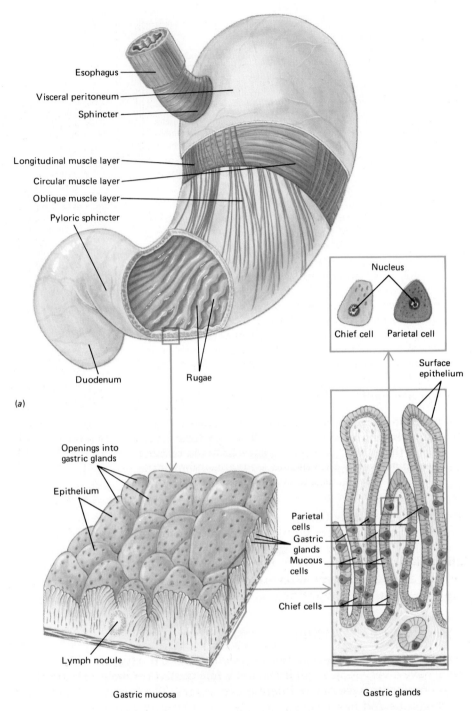

glands are further stimulated. Stretching the stomach with food also stimulates the stomach mucosa to release a hormone called **gastrin.** This hormone is absorbed into the blood and transported to the gastric glands, where it stimulates release of gastric juice. The presence of partly digested proteins, caffeine, or moderate amounts of alcohol in the stomach also stimulates secretion of gastrin.

After a meal, food may remain in the stomach for more than 4 hours. As it is churned, mashed, and digested by gastric juice, the food is converted into a soupy mixture called **chyme.** Peristaltic waves slowly push the chyme along toward the exit of the stomach. Only water, salts, and lipid-soluble substances such as alcohol are absorbed from the stomach. The exit of the stomach is generally kept closed by contraction of a ring of muscle,

Focus on PEPTIC ULCERS

One of the wonders of physiology is that gastric juice does not normally digest the stomach wall itself. Several protective mechanisms prevent this from happening. Cells of the gastric mucosa secrete an alkaline mucus that coats the stomach wall and also neutralizes the acidity of the gastric juice along the lining. In addition, the epithelial cells of the lining fit tightly together, preventing gastric juice from leaking between them and onto the tissue beneath. Should some of the epithelial cells be damaged, they are quickly replaced. In fact, the lifespan of an epithelial cell in the gastric mucosa is only about 3 days. About a half million of these cells are shed and replaced every minute.

Still, these mechanisms sometimes malfunction or prove inadequate, and a small bit of the stomach lining is digested, leaving an open sore or **peptic ulcer.** Substances such as alcohol and aspirin reduce the resistance of the stomach mucosa to digestion by gastric juice. Peptic ulcers occur more often in the duodenum than in the stomach. They also sometimes occur in the lower part of the esophagus.

Peptic ulcers may bleed, leading to anemia. If the ulcer extends into the muscularis, large blood vessels may also be damaged, resulting in hemorrhage. A **perforated ulcer** is one that extends all the way through the wall of the stomach or other affected organ. The opening created may allow bacteria and food to pass through to the peritoneum, leading to peritonitis and shock. Perforation is the main cause of death from ulcers.

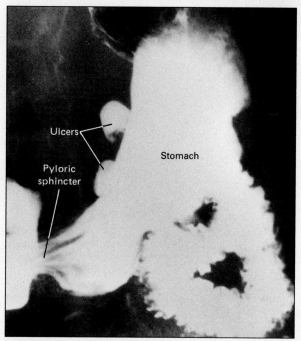

■ An x-ray of an ulcer in the wall of the stomach. The stomach and intestine have been filled with a contrast medium, making them appear white. This fluid also fills the cavities of the ulcers.

the **pyloric sphincter.** When digestion in the stomach is completed, the pyloric sphincter relaxes so that chyme can be pushed, a few milliliters at a time, into the small intestine.

Inside the Small Intestine

The length of the small intestine is correlated with the type of diet. Herbivores have a very long small intestine, while carnivores have a shorter one, and omnivores have one of intermediate length. The frog larva (tadpole) is herbivorous and has a long small intestine, but the carnivorous adult frog has a much shorter one, relative to its body size.

The human small intestine is a coiled tube about 2.6 m long by 4 cm in diameter. Curved like the letter c, the first 21 cm or so of the small intestine is the **duodenum** (see Figure 40–17). As the small intestine turns downward, it is called the **jejunum.** The jejunum extends for about 0.9 m before becoming the **ileum.** The duodenum is held in place by connective tissue ligaments that attach to the liver, stomach, and dorsal body wall. The rest of the small intestine (and most of the large intestine) is loosely anchored to the dorsal body wall by a thin transparent membrane called the **mesentery.**

The inner lining of the small intestine is not smooth like the inside of a water pipe. Instead, the inner surface is intricately folded in three ways: First, the mucous membrane is thrown into visible **circular folds.** Then, the mucous membrane is pushed up into millions of microscopic fingerlike projections called **villi** (Figures 40–15 and 40–16). Finally, the intestinal

FIGURE 40–15 The surface of the small intestine is studded with villi and tiny openings into the intestinal glands. Here, some of the villi have been opened to show the blood and lymph vessels within.

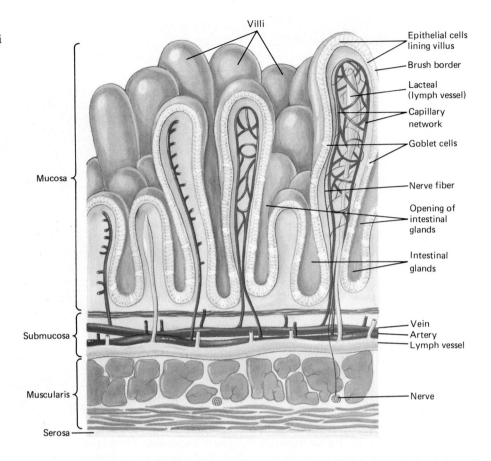

surface is further increased by thousands of **microvilli,** which are folds in the cell membrane of the exposed borders of the epithelial cells (Figure 40–17). When viewed with an electron microscope, the microvilli give the epithelial lining a fuzzy appearance, termed a **brush border.** Together, the circular folds, villi, and microvilli increase the surface area of the small intestine so extensively that if the lining could be completely unfolded and spread out, its surface would approximate the size of a tennis court!

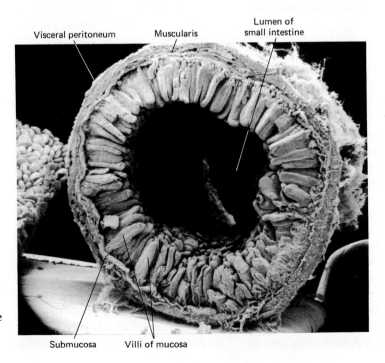

FIGURE 40–16 Scanning electron micrograph of a cross section of the small intestine (approximately ×30).

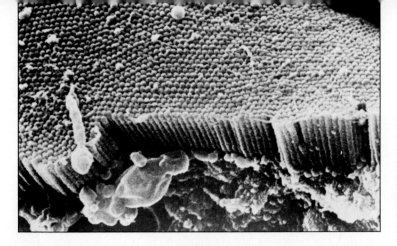

FIGURE 40–17 Scanning electron micrograph (approximately ×14,000) of the surface of an epithelial cell from the lining of the small intestine showing microvilli. The epithelium has been cut vertically to allow the microvilli to be viewed from the side as well as from above.

Most of the enzymatic digestion of food takes place in the duodenum. Bile from the liver and pancreatic juice from the pancreas are secreted into the duodenum. As we will see, these secretions are important in digestion. In addition, millions of tiny **intestinal glands** in the intestinal mucosa secrete intestinal juice, which serves as a medium for digestion and absorption of nutrients. The intestinal epithelial cells produce several enzymes that catalyze the final steps in digestion.

As elsewhere in the digestive tract, contractions in the small intestine produce both mixing movements and peristaltic waves. Several hours are required for chyme to be propelled through the length of the small intestine and into the large intestine. Movement and digestion in the small intestine are regulated by neural messages and by hormones. When acidic chyme from the stomach comes in contact with the mucosa of the duodenum, a hormone called **secretin** is released from the duodenal mucosa. Secretin stimulates both the pancreas and the liver to release some of their secretions. The presence of fatty acids or partly digested proteins in the duodenum also stimulates the duodenal mucosa to release a hormone known as **cholecystokinin,** or **CCK.** This hormone stimulates the pancreas and gallbladder (see Table 40–5) and is thought to affect the appetite control centers in the brain.

The Pancreas

The pancreas and liver are large accessory digestive glands that develop in the embryo as outgrowths of the digestive tract. The **pancreas** is an elongated gland that lies in the abdominal cavity between the stomach and duodenum (Figure 40–18). The cells that secrete the pancreatic enzymes are arranged in units called **acini,** which look like clusters of grapes. The ducts leading from the acini not only conduct the enzymes but also secrete a sodium bicarbonate solution, which makes the pancreatic juice somewhat alkaline. The pancreatic enzymes include (1) the proteolytic enzymes **trypsin, chymotrypsin,** and **carboxypeptidase;** (2) **pancreatic lipase,** which hydrolyzes neutral fats; (3) **pancreatic amylase,** which degrades almost all carbohydrates except cellulose to disaccharides; (4) an esterase that splits cholesterol esters; and (5) ribonuclease and deoxyribonuclease, which split RNA and DNA into free nucleotides (Table 40–1).

All of the proteolytic enzymes are secreted as inactive precursors. Trypsin is activated in the duodenum when it comes in contact with the enzyme **enterokinase,** which is secreted by the intestinal mucosa. Enterokinase splits off a portion of the precursor molecule trypsinogen to yield the active enzyme trypsin and an inactive molecular fragment. The activated trypsin then activates the other proteases. To further protect itself from digestion by the proteases it secretes, the pancreas produces an internal trypsin inhibitor, which inactivates any trypsin that might become activated in the pancreas. If the pancreas is damaged (as in alcoholism) or if the duct is

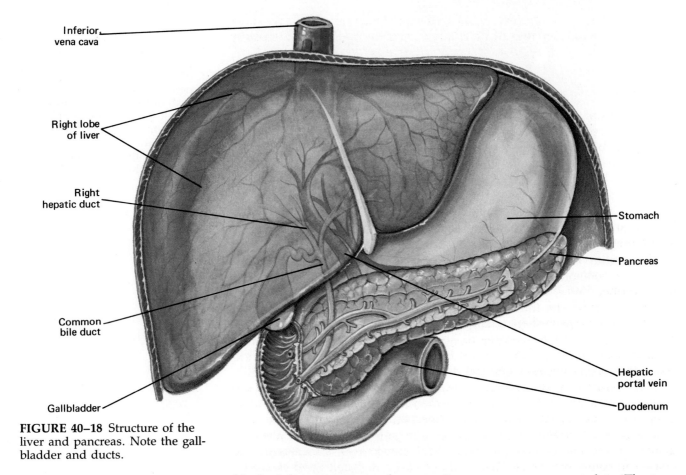

Inferior
vena cava

Right lobe
of liver

Right
hepatic duct

Common
bile duct

Gallbladder

Stomach

Pancreas

Hepatic
portal vein

Duodenum

FIGURE 40–18 Structure of the
liver and pancreas. Note the gallbladder and ducts.

blocked, large amounts of pancreatic enzymes may accumulate. The trypsin inhibitor system then may be overwhelmed and the proteases may
digest the tissues of the pancreas. This can result in **acute pancreatitis,**
which is often fatal.

Table 40–1 ENZYMES IMPORTANT IN DIGESTION				
Enzyme	*Source*	*Optimum pH*	*Substrate*	*Product*
Salivary amylase	Saliva	Neutral	a-Glycosidic bonds of starch and glycogen	Maltose
Pepsin	Stomach	Acid	Peptide bonds within chain and adjacent to tyrosine or phenylalanine	Peptides
Rennin	Stomach	Acid	Peptide bonds in casein	Coagulated casein
Trypsin	Pancreas	Alkaline	Peptide bonds within chain adjacent to lysine or arginine	Peptides
Chymotrypsin	Pancreas	Alkaline	Peptide bonds within chain adjacent to tyrosine, phenylalanine, or trytophan	Peptides
Lipase	Pancreas	Alkaline	Ester bonds of fats	Glycerol, fatty acids, mono- and diacylglycerols
Amylase	Pancreas	Alkaline	a-Glycosidic bonds of starch and glycogen	Maltose
Ribonuclease	Pancreas	Alkaline	Phosphate esters of RNA	Nucleotides
Deoxyribonuclease	Pancreas	Alkaline	Phosphate esters of DNA	Nucleotides
Carboxypeptidase	Intestinal glands	Alkaline	Peptide bond adjacent to free carboxyl end	Free amino acids
Aminopeptidase	Intestinal glands	Alkaline	Peptide bond adjacent to free amino end	Free amino acids
Enterokinase	Intestinal glands	Alkaline	Trypsinogen	Trypsin
Maltase	Intestinal glands	Alkaline	Maltose	Glucose
Sucrase	Intestinal glands	Alkaline	Sucrose	Glucose and fructose
Lactase	Intestinal glands	Alkaline	Lactose	Glucose and galactose

The pancreas is an endocrine, as well as an exocrine, gland. Its endocrine component, the **islets of Langerhans,** secretes the hormones insulin and glucagon, which regulate the concentration of glucose in the blood.

The Liver

The **liver** is one of the largest and functionally most complex organs in the body. Each liver cell can carry on hundreds of metabolic activities. The liver (1) secretes **bile,** important in fat digestion; (2) removes nutrients from the blood; (3) converts glucose to glycogen for storage and glycogen to glucose as needed; (4) stores iron and certain vitamins; (5) converts amino acids to keto acids and urea; (6) manufactures many proteins found in the blood; (7) detoxifies many drugs and poisons that enter the body; (8) phagocytizes bacteria and worn-out red blood cells; and (9) performs countless functions in the metabolism of amino acids, fats, and carbohydrates.

Liver cells continuously secrete small amounts of bile, which pass through a system of ducts into the **common bile duct.** The common bile duct empties into the duodenum, but the exit from the duct is usually closed by a sphincter muscle. When the sphincter is constricted, bile is shunted into the pear-shaped **gallbladder** for storage. When fat enters the duodenum, it stimulates release of the hormone CCK from the intestinal mucosa. CCK stimulates the gallbladder to contract and relaxes the sphincter so that bile is released into the duodenum.

Bile consists of water, bile salts, bile pigments, cholesterol, salts, and lecithin (a phospholipid). **Bile salts** are made by the liver from cholesterol. They act as detergents, emulsifying (mechanically breaking down into droplets) the fats in the intestine. When large fat globules are mechanically dispersed into many small globules, their surface area is increased, and lipase can come into contact with individual fat molecules and cleave off fatty acids. (When the bile duct is obstructed and bile salts are absent from the intestine, both digestion and absorption of fats are impaired, causing much of the fat eaten to be wastefully eliminated in the feces.) Bile salts are conserved by the body. They are reabsorbed in the lower part of the intestine and transported back to the liver by the blood to be secreted once again.

Cholesterol is synthesized in the liver; its concentration in the bile reflects the amount of lipid in the diet. Cholesterol is rather insoluble in water but combines with bile salts and lecithin to form soluble molecular aggregates called **micelles.** Under certain abnormal conditions, cholesterol precipitates and produces hard little pellets called **gallstones** (Figure 40–19). Persons on high-fat diets over a period of years tend to develop gallstones more readily than do those on low-fat diets.

The color of bile results from the presence of bile pigments (green, yellow, orange, or red in different animal species). Bile pigments are formed from the heme portion of hemoglobin by enzymatic processes in the liver. In the intestine, bile pigments are metabolized further by bacterial enzymes and become a brownish color. These brown pigments are responsible for the color of feces. Sometime the excretion of bile pigment is prevented by some obstruction of the bile duct such as gallstones. The pigments then accumulate in the blood and tissues, imparting a yellowish tinge to the skin, a condition known as **jaundice.** Absence of the pigments from the intestinal contents leaves the feces pale like whitish clay.

Enzymatic Digestion

As chyme is moved through the digestive tract by peristaltic action, mixing contractions, and motion of the villi, enzymes come into contact with the nutrients and digest them.

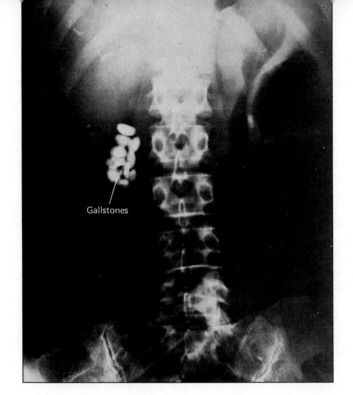

FIGURE 40–19 X-ray showing clacified gallstones in gallbladder.

Carbohydrate Digestion

Polysaccharides such as starch and glycogen are an important part of the food ingested by humans and most other animals. The glucose units of these large molecules are joined by glycosidic bonds linking carbon 4 (or 6) of one glucose molecule with carbon 1 of the adjacent glucose molecule. These bonds are hydrolyzed by **amylases,** enzymes that digest polysaccharides to the disaccharide maltose (Table 40–2). Although the amylases of

Table 40–2 **SUMMARY OF CARBOHYDRATE DIGESTION**

Location	Source of Enzyme	Digestive Process*
Mouth	Salivary glands	Polysaccharides (e.g., starch) $\xrightarrow{\text{salivary amylase}}$ Maltose + Dextrin
Stomach		Action continues until salivary amylase is inactivated by acidic pH
Small intestine Lumen	Pancreas	Undigested polysaccharides and dextrins $\xrightarrow{\text{pancreatic amylase}}$ Maltose
Brush border	Intestine	Disaccharides hydrolyzed to monosaccharides as follows:
		Maltose (malt sugar) $\xrightarrow{\text{maltase}}$ Glucose + Glucose
		Sucrose (table sugar) $\xrightarrow{\text{sucrase}}$ Glucose + Fructose
		Lactose (milk sugar) $\xrightarrow{\text{lactase}}$ Glucose + Galactose

*◯ = monosaccharide (complete structures are shown in Chapter 3).

Table 40–3 SUMMARY OF PROTEIN DIGESTION

Location	Source of Enzyme	Digestive Process*
Stomach	Stomach (gastric glands)	Protein $\xrightarrow{\text{pepsin}}$ Polypeptides
Small intestine Lumen	Pancreas	Polypeptides $\xrightarrow[\text{chymotrypsin}]{\text{trypsin,}}$ Tripeptides + Dipeptides A—A—A—A—A A—A—A A—A | A—A—A—A—A
		Dipeptides $\xrightarrow{\text{carboxypeptidase}}$ Free amino acids A—A A A A

Brush borders (and within cytoplasm of epithelial cells)	Small intestine	Tripeptides + Dipeptides $\xrightarrow{\text{peptidases}}$ Free amino acids
		A—A—A A—A A A
		                           A
		                     A A
		                       A

*A = amino acid units or, when standing alone, a free amino acid.

the digestive tract can split the α-glycosidic bonds present in starch and glycogen, they cannot split the β-glycosidic bonds present in cellulose. In most vertebrates, amylase is secreted only by the pancreas. However, in humans and certain other mammals, amylase is also secreted by the salivary glands.

Amylases cannot split the bond between the two glucose units of maltose. Enzymes produced by the cells lining the small intestine break down disaccharides such as maltose to monosaccharides. These enzymes are found in the brush border of the epithelial cells and are thought to catalyze hydrolysis while the disaccharides are being absorbed through the epithelium. **Maltase,** for example, splits maltose into two glucose molecules.

Protein Digestion

Several kinds of proteolytic enzymes are secreted into the digestive tract (Table 40–3). Each is specific for peptide bonds in a specific location in a polypeptide chain. Three main groups are exopeptidases, endopeptidases, and dipeptidases. **Exopeptidases** split the peptide bond joining the terminal amino acids to the peptide chain. For example, carboxypeptidase cleaves the peptide bond joining the amino acid with the free terminal carboxyl group to the peptide chain. Aminopeptidase splits off the amino acid with a free terminal amino group.

Endopeptidases cleave only peptide bonds *within* a peptide chain. Pepsin, trypsin, and chymotrypsin are endopeptidases. They differ in their requirements for specific amino acids adjacent to the bond to be split (Figure 40–20). These endopeptidases split peptide chains into smaller fragments, which are then cleaved further by exopeptidases. The combined action of the endopeptidases and exopeptidases results in splitting of the protein molecules to dipeptides. **Dipeptidases** in the brush borders of the duodenum then split these to free amino acids. The free amino acids, and some dipeptides and tripeptides, are absorbed through the epithelial cells lining the villi and enter the blood.

Lipid Digestion

Lipids are usually ingested as large masses of triacylglycerols (also referred to as triglycerides). They are digested largely within the duodenum by pancreatic lipase (Table 40–4). Like other proteins, lipase is water-soluble,

FIGURE 40–20 Formula of a peptide indicating point of attack of pepsin (*P*), trypsin (*T*), chymotrypsin (*C*), amino peptidase (*AP*), and carboxypeptidase (*CP*).

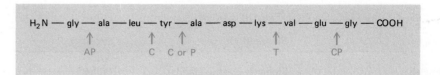

but its substrates are not. Thus, the enzyme can attack only those molecules of fat at the surface of a mass of fat. The bile salts are detergents that reduce the surface tension of fats, breaking the large masses of fat into smaller droplets. This greatly increases the surface area of fat exposed to the action of lipase and so increases the rate of lipid digestion.

Conditions in the intestine are usually not optimal for the complete hydrolysis of lipids to glycerol and fatty acids. The products of lipid digestion therefore include monoacylglycerols (monoglycerides) and diacylglycerols (diglycerides), as well as glycerol and fatty acids. Undigested triacylglycerols remain as well, and some of these are absorbed without digestion.

Control of Digestive Juice Secretion

Most digestive enzymes are produced only when food is present in the digestive tract. The quantity of each enzyme secreted reflects the amount needed for digestion of the food materials present. The salivary glands are controlled entirely by the nervous system, but secretion of other digestive juices is regulated by both nervous and endocrine mechanisms. For example, gastric juice is secreted in response both to neural messages and to the hormone gastrin. Both mechanisms are triggered by the presence of food in the stomach. Table 40–5 summarizes the actions of the principal hormones of the digestive system. Several other substances are suspected of being digestive system hormones, but just what they do and how they do it are not yet clear.

Absorption

After the digestive enzymes have split the large molecules of protein, polysaccharides, lipids, and nucleic acids into their constituent subunits, the products are absorbed through the wall of the intestine. These ingested

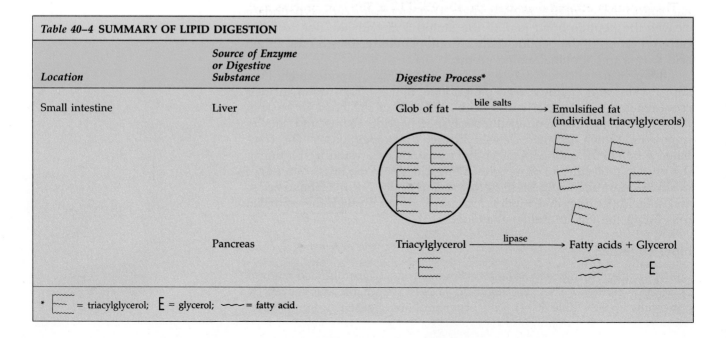

Table 40–4 SUMMARY OF LIPID DIGESTION

Location	Source of Enzyme or Digestive Substance	Digestive Process*
Small intestine	Liver	Glob of fat ——bile salts——> Emulsified fat (individual triacylglycerols)
	Pancreas	Triacylglycerol ——lipase——> Fatty acids + Glycerol

* ⌇ = triacylglycerol; E = glycerol; ⌇⌇ = fatty acid.

Table 40–5 HORMONES FOR THE DIGESTIVE TRACT				
Hormone	*Source*	*Target Tissue*	*Actions*	*Factors That Stimulate Release*
Gastrin	Stomach (mucosa)	Stomach (gastric glands)	Stimulates gastric glands to secrete pepsinogen	Distention of the stomach by food; certain substances such as partially digested proteins and caffeine
Secretin	Duodenum (mucosa)	Pancreas	Stimulates release of alkaline component of pancreatic juice	Acidic chyme acting on mucosa of duodenum
		Liver	Increases rate of bile secretion	
Cholecystokinin (CCK)	Duodenum (mucosa)	Pancreas	Stimulates release of digestive enzymes	Presence of fatty acids and partially digested proteins in duodenum
		Gallbladder	Stimulates contraction and emptying	
Gastrin inhibitory peptide	Duodenum (mucosa)	Stomach	Decreases stomach motor activity, thus slowing emptying	Presence of fat or carbohydrate in duodenum

materials, however, account for only a small part of the total amount of fluid absorbed daily (about 1.5 liters of a total of approximately 9 liters). The rest consists of mucus and digestive juices released by the digestive system itself.

Most substances are absorbed through the villi in the wall of the small intestine (Figure 40–15). Each villus consists of a single layer of epithelial cells covering a network of blood capillaries and a central lymph vessel called a **lacteal.**

Absorption occurs in part by simple diffusion, in part by facilitated diffusion, and in part by active transport. Glucose and amino acids are absorbed by active transport. Absorption of these nutrients is coupled with the active transport of sodium. Fructose is absorbed by facilitated diffusion.

After nutrients such as amino acids are transported into the epithelial cells lining the villi, they accumulate within the cells and then diffuse into the blood of the intestinal capillaries. Amino acids and glucose are transported to the liver by the **hepatic portal vein.** In the liver this vein gives rise to a vast network of sinusoids (tiny blood vessels similar to capillaries), which allow the nutrient-rich blood to course slowly through the liver tissue. This gives the liver cells opportunity to remove nutrients and certain toxic substances from the circulation.

The products of lipid digestion are absorbed by a different process and different route (Figure 40–21). Fatty acids and monoacylglycerols combine with bile salts to form soluble complexes called micelles. This greatly facilitates absorption, because the micelles transport the fatty substances to the brush borders. When the micelles come into contact with the epithelial cells of the villi, the monoacylglycerols and fatty acids (both soluble in the lipid of the cell membrane) diffuse into the cell, leaving the rest of the micelle behind to combine with new fatty acids and monoacylglycerols.

In the epithelial cells, free fatty acids and glycerol are assembled once again into triacylglycerols by the endoplasmic reticulum. These triacylglycerols then are packaged into globules with absorbed cholesterol and phospholipids and covered with a thin coat of protein. Such protein-covered fat globules are called **chylomicrons.** They pass out of the epithelial cell and into the lacteal of the villus. The chylomicrons are transported by the lymph and eventually are emptied with the lymph into the blood. About 90% of absorbed fat enters the blood circulation in this indirect way. The rest, mainly short-chain fatty acids such as those in butter, are absorbed directly into the blood. After a meal rich in fats, the great number of chylomicrons in the blood may give the plasma a turbid, milky appearance for a few hours.

FIGURE 40–21 Overview of the process of lipid absorption by an epithelial cell lining the intestine.

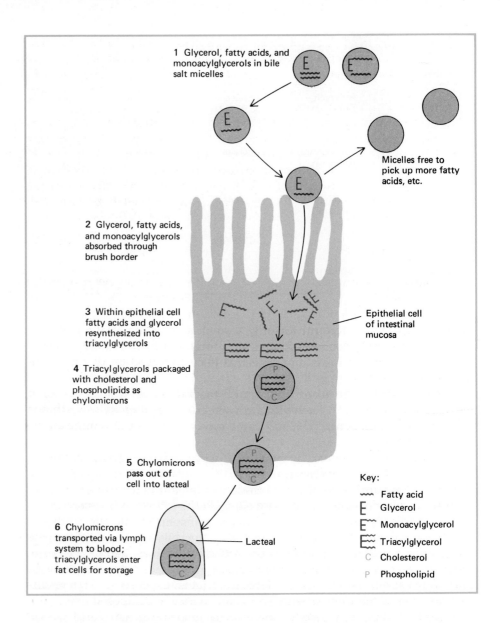

1 Glycerol, fatty acids, and monoacylglycerols in bile salt micelles

Micelles free to pick up more fatty acids, etc.

2 Glycerol, fatty acids, and monoacylglycerols absorbed through brush border

3 Within epithelial cell fatty acids and glycerol resynthesized into triacylglycerols

Epithelial cell of intestinal mucosa

4 Triacylglycerols packaged with cholesterol and phospholipids as chylomicrons

5 Chylomicrons pass out of cell into lacteal

6 Chylomicrons transported via lymph system to blood; triacylglycerols enter fat cells for storage

Lacteal

Key:

〜〜 Fatty acid
E Glycerol
E〜 Monoacylglycerol
E≡ Triacylglycerol
C Cholesterol
P Phospholipid

Most of the nutrients in the chyme are absorbed by the time the chyme reaches the end of the small intestine. What is left of the chyme (mainly waste) passes through a sphincter, the **ileocecal valve,** and into the large intestine.

Through the Large Intestine

Approximately 9 hours elapse from the time food is ingested until its remnants reach the large intestine. From 1 to 3 days or even longer may be required for the slow journey through the large intestine. The large intestine functions in the following ways:

1. It absorbs sodium and water from the chyme. Sodium is absorbed by active transport, and water follows by osmosis. The chyme is slowly solidified into the consistency of normal feces.
2. It incubates bacteria. The movements of the large intestine are sluggish, giving bacteria time to grow and reproduce there. Some kinds of intestinal bacteria are mutualistic partners with their human hosts. They produce certain vitamins (vitamin K, thiamine, riboflavin, and vitamin B_{12}) in exchange for a place to live and the remnants of their

host's last meal. The presence of harmless bacteria in the intestine inhibits the growth of pathogenic varieties. Should the normal ecology of the large intestine be disturbed, however, as sometimes happens when a person takes certain antibiotics, harmful bacteria may multiply and cause disease.

3. It eliminates wastes. Undigested and unabsorbed food, as well as cells that are sloughed off from the intestinal mucosa, are eliminated from the body by the large intestine in the form of feces. A distinction should be made between elimination and excretion. **Elimination** is the process of getting rid of digestive wastes, materials that never left the digestive tract and so never participated in metabolic activities. **Excretion** refers to the process of getting rid of metabolic wastes, and is mainly the function of the kidneys. The large intestine does, however, excrete bile pigments.

The large intestine is shorter in length but larger in diameter than the small intestine. Its regions include the cecum, ascending colon, transverse colon, descending colon, sigmoid colon, rectum, and anus, the opening for elimination of feces (see Figure 40–7). The small intestine empties into the side of the ascending colon about 7 cm from its end. This leaves a blind pouch, the **cecum,** which hangs down below the junction of small and large intestines. The **appendix,** a worm-shaped blind tube about the size of the little finger, hangs down from the end of the cecum. The presence of the cecum and appendix is something of a mystery, as they have no known function in human beings. They were probably larger in our remote ancestors and did function in them in the digestion of plant foods. Herbivores such as rabbits and guinea pigs have a large, functional cecum, containing bacteria that digest cellulose.

Both mixing and peristaltic movements occur in the large intestine, but both are ordinarily slower and more sluggish than those in the small intestine. Periodically, usually after eating, more vigorous peristaltic movements force the contents along. When a mass of fecal material reaches the weak sphincter at the entrance to the rectum, it relaxes, allowing feces to enter the rectum. Distention of the rectum stimulates nerves in its walls and brings about the impulse to **defecate,** that is, expel feces. This results in relaxation of the internal anal sphincter, which is composed of smooth (involuntary) muscle. However, the external anal sphincter, which is composed of skeletal muscle, remains contracted until voluntarily relaxed. Thus, defecation is a reflex action that can be voluntarily inhibited by keeping the external sphincter muscle contracted.

The feces of a healthy human contains about 75% water by weight. The solid portion consists of about 30% bacteria, both live and dead. The rest is cellulose and other undigested and unabsorbed remnants of food, dead cells, salt, and bile pigments.

If the lining of the colon is irritated, as in certain infections, motility of the large intestine may be increased while absorption is decreased. The intestinal contents pass rapidly through the colon, and only a small amount of water is absorbed from them. This condition, known as **diarrhea,** results in frequent defecation and watery feces. Prolonged diarrhea can result in loss of water and of needed electrolytes such as sodium and potassium. Dehydration, especially in infants, may be very serious, or even fatal. The opposite condition, **constipation,** results when the contents pass through the colon too slowly, so that too much water is removed from them. The feces then become excessively hard and dry. Constipation may be caused by a diet containing too little fiber.

Cancer of the colon is one of the most common types of cancer in the United States and many other industrialized countries (Figure 40–22). This

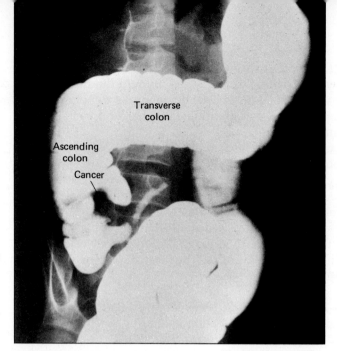

FIGURE 40–22 X-ray of the large intestine of a patient with cancer of the colon. The lumen of the large intestine has been filled with a suspension of barium sulfate, which makes irregularities in the wall visible. The cancer is evident as a mass that projects into the lumen.

type of cancer is thought to be related to diet, since the disease is more common in people whose diets are very low in fiber. It has been suggested that a low-fiber diet results in less frequent defecation, allowing prolonged contact between the mucosa of the colon and carcinogenic substances in foods.

■ SUMMARY

1. Animals are heterotrophs and consumers.
   A. Herbivores, the primary consumers, eat algae or plants.
   B. Carnivores consume meat and are usually predators.
   C. Omnivores, such as bears and humans, eat both meat and plant material.
   D. Symbionts are members of two different species that live in intimate association with one another. Three types of symbionts are parasites, commensals, and mutualistic partners.
II. Digestion of food is accomplished in a variety of ways in invertebrates.
   A. Sponges have no digestive system; digestion is carried on intracellularly.
   B. Cnidarians and flatworms have digestive systems with only one opening, which serves as both mouth and anus.
   C. In more complex invertebrates the digestive tract is a complete tube with an opening at each end.
III. The vertebrate digestive system is a complete tube extending from mouth to anus.
   A. The wall of the digestive tract from the lumen outward consists of mucosa, submucosa, muscularis, and adventitia.
   B. Ingestion takes place through the mouth, and both mechanical breakdown and chemical digestion begin there.
      1. Each tooth consists mainly of dentin covered by enamel in the crown region and by cementum in the root.
      2. Three pairs of salivary glands produce saliva in terrestrial vertebrates. Saliva moistens food, aids in swallowing, and contains salivary amylase, which initiates carbohydrate digestion.
   C. During swallowing, food passes from the mouth cavity through the pharynx and into the esophagus. Peristaltic action moves it through the esophagus and into the stomach.
   D. In the stomach, food is mechanically broken down. There, the enzyme pepsin in the gastric juice initiates protein digestion. The food is reduced to a soupy mixture called chyme.
   E. Most enzymatic digestion of food takes place in the duodenum, which receives secretions from both the liver and pancreas.
      1. The liver secretes bile, which emulsifies fat.
      2. The pancreatic juice contains proteolytic enzymes, lipase, pancreatic amylase, and other enzymes.
   F. Polysaccharides are digested to maltose by salivary and pancreatic amylases. Maltase in the brush border of the intestine splits maltose into glucose, the main product of carbohydrate digestion.
   G. Proteins are split by pepsin in the stomach and by proteolytic enzymes in the pancreatic juice. The dipeptides produced are then split by dipeptidases in the brush borders of the duodenum. Free amino acids are the end products of protein digestion.
   H. Lipids are emulsified by bile salts and then hydrolyzed by lipase in the pancreatic juice.
   I. Secretion of digestive enzymes is regulated by nerves and hormones.
   J. Most digested nutrients are absorbed through the

villi of the small intestine. Monosaccharides and amino acids enter the blood; glycerol, fatty acids, and monoacylglycerols enter the lymph.

K. The large intestine absorbs sodium and water from the intestinal contents, incubates bacteria, and eliminates wastes.

■ POST-TEST

1. Once food has been ingested, it must be _____ and then _____ through the lining of the digestive tract.

2. Vertebrate herbivores generally have a specialized section of the digestive tract called the _____ in which live _____.

3. When a ruminant animal chews its cud, it is actually _____.

4. Animals that eat mainly meat are _____; those that include both meat and plant material in their diets are _____.

5. A flea is an _____ parasite; a tapeworm is an _____ parasite.

6. The inner lining of the digestive tract, also called the _____, contains cells that secrete mucus.

7. The rhythmic waves of contraction that move food through the digestive tract are referred to as _____.

8. _____ are teeth used for biting; they are especially large in gnawing animals.

9. The parotid glands are the largest _____.

10. Food passing through the pharynx next enters the _____; food leaving the stomach next enters the _____.

11. The prominent folds in the stomach lining are called _____.

12. The gastric juice is secreted by _____ _____ in the mucosa of the _____.

13. The common bile duct conducts bile into the _____.

Select the most appropriate answer from Column B for each entry in Column A. The same answer may be used more than once, and each entry may have more than one answer.

Column A	Column B
14. Hormone that stimulates release of gastric juice	a. Pepsin
15. Initiates digestion of proteins	b. Bile
16. Splits maltose	c. Dipeptidases
17. Secreted by pancreas	d. Gastrin
18. Secreted by epithelial cells lining small intestine	e. Amylase
19. Emulsifies fats	f. Maltase
20. Chylomicrons are protein-covered globules of _____.	

■ REVIEW QUESTIONS

1. Contrast adaptations in herbivores and carnivores that help each group to be successful in its particular nutritional life style.
2. Give examples of parasites, commensals, and mutualistic partners, and describe how each relates to its host.
3. Compare food processing in a sponge, a hydra, a flatworm, and an earthworm.
4. Trace a bit of lettuce through the human digestive system, listing in sequence each structure through which it would pass.
5. Trace the ingredients of a hamburger sandwich through the human digestive system, indicating all changes that occur en route. (Hint: The hamburger contains both protein and fat; the bun contains starch, a polysaccharide.)
6. What prevents the stomach from being digested by gastric juice? What happens when these protective mechanisms fail?
7. How does the absorption of fat differ from the absorption of glucose?
8. Why is it advantageous that the inner lining of the digestive tract is not smooth like the inside of a water pipe? What structures increase its surface area?
9. Give four functions of the liver.
10. How are the movements and secretion of the digestive system regulated? Give specific examples.

■ RECOMMENDED READINGS

Davenport, H.W. Why the stomach does not digest itself. *Scientific American*, January 1972, pp. 86–93. An interesting discussion of the features of the stomach that protect it from digestion.

Kessel, R.G., and Kardon, R.H. *Tissue and Organs: A Text-Atlas of Scanning Electron Microscopy*. San Francisco, Freeman, 1979. Chapter 9, which focuses on the digestive system, includes many fascinating electron micrographs.

Moog, F. The lining of the small intestine, *Scientific American*, November 1981, pp. 154–176. The cells lining the small intestine are covered by a membrane that actively digests foods and speeds nutrients into the blood.

Solomon, E.P., and Davis, P.W. *Human Anatomy and Physiology*. Philadelphia, Saunders College Publishing, 1983. Chapter 21 presents a readable discussion of the structure and function of the digestive system.

A diet high in saturated fats and cholesterol raises the blood-cholesterol level by as much as 25%. On the other hand, ingestion of polyunsaturated fats tends to decrease the blood cholesterol level. For these reasons, many people now cook with vegetable oils rather than butter and lard, drink skim milk rather than whole milk, eat ice milk instead of ice cream, and use margarine instead of butter.

Lipid Storage

Recall from the last chapter that lipids are absorbed into the lymph circulation and are transported as tiny droplets called chylomicrons. A **chylomicron** consists of triacylglycerols, cholesterol, phospholipids, and a small amount of protein present as a covering around the lipid droplet. From the lymph circulation the chylomicrons enter the blood. As they pass through the capillaries of the adipose tissue and sinusoids of the liver, most of the chylomicrons are removed.

Cholesterol is stored primarily in the liver cells. Phospholipids are degraded to yield fatty acids, which are reassembled into triacylglycerols within the liver and fat cells. These triacylglycerols, along with those from the chylomicrons, are stored in the fat and liver cells. The fat cells of adipose tissue can store large amounts of triacylglycerols, up to 95% of their volume (see Figure 38–5). For this reason adipose tissue is often referred to as the fat depot of the body.

The cells of adipose tissue are very active metabolically. Triacylglycerols are constantly being synthesized and degraded. This occurs so rapidly that the triacylglycerol content in the fat cells is entirely changed within a three-week period. Thus, the fat molecules stored in your fat cells are not the same ones that were there last month. Sad to say, the total amount of fat is probably much the same! (See Focus on Obesity.)

Using Fat as Fuel

The main function of adipose tissue is to store triacylglycerols until they are used as fuel by the cells. Fats are the body's principal energy storage form because they are a very concentrated source of energy and because they are not water-soluble. The metabolism of a gram of fat to carbon dioxide and water yields up to 9 kcal of energy, more than twice as much as the metabolism of a gram of protein or carbohydrate. Most cells can use fatty acids as a source of energy. Between meals, in fact, most cells shift their metabolism so that fatty acids are oxidized in preference to glucose. This shift spares glucose for the brain cells, which are less able to use fatty acids as fuel.

Normally, the fasting level of free fatty acids in the blood is about 15 mg/ 100 ml. As cells remove fatty acids from the blood and the concentration decreases, fats are mobilized from the fat cells. Fat cells degrade triacylglycerol and release free fatty acids into the blood. These fatty acids combine with a certain type of albumin—a plasma protein—and are transported in this way to the liver.

Fatty acids are metabolized by **β-oxidation** (beta-oxidation; Chapter 7), which takes place mainly in the liver cells. Some of the acetyl coenzyme A formed is oxidized in the citric acid cycle, by the liver cells, but some is transported to other cells (Figure 41–3). Before these molecules can leave the liver cells, they must be converted to a form that can diffuse freely out of the liver cells and be transported in the blood. Two acetyl coenzyme A molecules are combined and the coenzyme A portions of the molecule are split off, forming one of three types of **ketone bodies** (acetoacetic acid, β-hydroxybutyric acid, and acetone). Ketone bodies are taken up from the

Focus on OBESITY

At present obesity is one of the most important nutritional problems in the United States. A person who has an excessive accumulation of body fat, and who is more than 20% overweight, is obese. About 23% of adult males, 30% of adult females, and 10% of children in the United States fit into this category. Obesity predisposes to a number of diseases, including cardiovascular disease and diabetes, and thus also decreases life expectancy.

Obesity can result from an increase in the size of fat cells (hypertrophic obesity) or from an increase in number of fat cells (hyperplastic obesity), or both. The number of fat cells in the adult body appears to be determined mainly by the amount of fat stored during infancy and childhood. Overfed infants may develop up to three times more fat cells than those fed a diet more balanced in calories. Such a person may also develop a higher setting of the brain center that controls food intake. When an overweight person goes on a diet, fat is mobilized and the fat cells decrease in size, but they do not disappear; should the person begin to overeat once again, the waiting fat cells simply fill up like tiny balloons with new fat. Persons who become overweight during middle age or old age probably do not develop new fat cells. Those they have, just become larger as they store more fat.

Humans of normal weight regulate their body weight about a specific set point that represents their characteristic weight. In obesity it is thought that the set point is elevated so that the body regulates around the abnormally high weight. When an obese person restricts calorie intake, the body is thought to lower its metabolic rate and become more efficient in digesting and absorbing food in an effort to stay at the familiar set point. This explains why it is so difficult to lose weight and to maintain the weight loss once it is achieved. Exercise has been shown to be an important component of a weight loss program because exercise raises the metabolic rate. When an obese person loses weight and maintains the weight loss for a period of time (perhaps six months to two years) the body is thought to lower its set point so that eventually maintaining the weight loss becomes less difficult.

Most overweight people overeat owing to a combination of poor eating habits and psychological factors. The only cure for obesity is to adjust food intake to meet energy needs. To lose weight, energy intake must be less than energy output. The body will then draw on its fat stores for the needed kilocalories, and body weight will decrease. This is best done by a combination of increased exercise and decreased total caloric intake.

Despite all of the claims by proponents of fad reducing diets, the most effective weight management programs are those that advocate a permanent shift to a more balanced life style. Such programs include instruction in behavior modification to help patients achieve life style changes, including exercise programs and a shift to more healthful eating patterns. Dietary treatment consists of a well-balanced diet (1000 to 1500 kilocalories for the mildly obese) that provides the bulk of the calories in the form of complex carbohydrates.

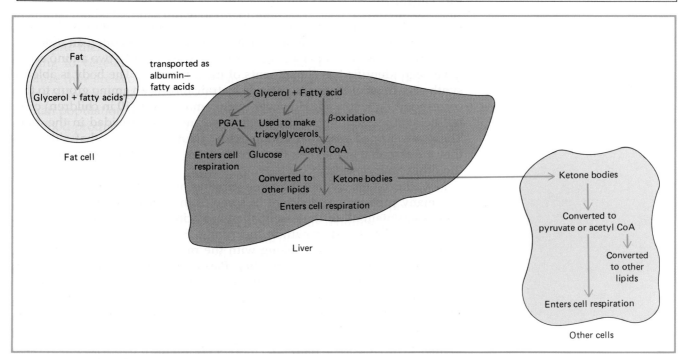

FIGURE 41–3 Overview of lipid metabolism.

Table 41–2 SOME IMPORTANT MINERALS AND THEIR FUNCTIONS

Mineral	Functions	Comments
Calcium	Component of bone and teeth; essential for normal blood clotting; needed for normal muscle and nerve function	Good sources: milk and other dairy products, green leafy vegetables. Bones serve as calcium reservoir.
Phosphorus	As calcium phosphate, an important structural component of bone; essential in energy transfer and storage (component of ATP) and in many other metabolic processes; component of DNA and RNA	Performs more functions than any other mineral; absorption impaired by excessive intake of antacids
Sulfur	As component of many proteins (e.g., insulin), essential for normal metabolic activity	Sources: high-protein foods such as meat, fish, legumes, nuts
Potassium	Principal positive ion within cells; influences muscle contraction and nerve excitability	Occurs in many foods
Sodium	Principal positive ion in interstitial fluid; important in fluid balance; essential for conduction of nerve impulses	Occurs naturally in foods; sodium chloride (table salt) added as seasoning; too much ingested in average American diet; in excessive amounts, may lead to high blood pressure
Chlorine	Principal negative ion of interstitial fluid; important in fluid balance and in acid–base balance	Occurs naturally in foods; ingested as sodium chloride
Copper	Component of enzyme needed for melanin synthesis; component of many other enzymes; essential for hemoglobin synthesis	Sources: liver, eggs, fish, whole wheat flour, beans
Iodine	Component of thyroid hormones (hormones that stimulate metabolic rate)	Sources: seafoods, iodized salt, vegetables grown in iodine-rich soils. Deficiency results in goiter (abnormal enlargement of thyroid gland).
Cobalt	As component of vitamin B_{12}, essential for red blood cell production	Best sources are meat and dairy products. Strict vegetarians may become deficient in this mineral.
Manganese	Necessary to activate arginase, an enzyme essential for urea formation; activates many other enzymes	Poorly absorbed from intestine; found in whole-grain cereals, egg yolks, green vegetables
Magnesium	Appropriate balance between magnesium and calcium ions needed for normal muscle and nerve function; component of many coenzymes	Occurs in many foods
Iron	Component of hemoglobin, myoglobin, important respiratory enzymes (cytochromes), and other enzymes essential to oxygen transport and cellular respiration	Mineral most likely to be deficient in diet. Good sources: meat (especially liver), nuts, egg yolk, legumes. Deficiency results in anemia.
Fluorine	Component of bones and teeth; makes teeth resistant to decay	In areas where it does not occur naturally, fluorine may be added to municipal water supplies (fluoridation). Excess causes tooth mottling.
Zinc	Component of at least 70 enzymes, including carbonic anhydrase; components of some peptidases, and thus important in protein digestion; may be important in wound healing	Occurs in many foods

for muscle contraction. Calcium deficiency and malabsorption, vitamin D deficiency, and decrease in estrogen are factors implicated in the development of **osteoporosis** (meaning "porous bone"). In osteoporosis, the most prevalent bone disease, the osteoblasts become less active and there is a decrease in bone mass. This makes the bones brittle. Osteoporosis is common among postmenopausal women and in the elderly. Risk for this disease can be decreased by a calcium-rich diet and by weight-bearing exercises such as walking and bicycling, which stimulate formation of new bone tissue.

VITAMINS

The discovery of vitamins and the analysis of their properties and functions in metabolism have been among the most notable achievements in science since the turn of this century. **Vitamins** are a group of unrelated

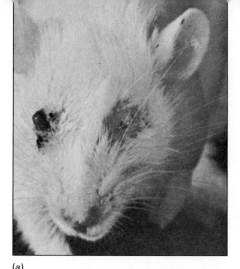 

(a) (b)

FIGURE 41-7 Effects of vitamin A deficiency. (*a*) This rat is suffering from a typical eye disorder produced by lack of vitamin A. (*b*) The eyes have been restored to normal by feeding the rat 3 IU (about 0.001 mg) of vitamin A daily.

organic compounds that are required in the diet in very small amounts, and are essential for normal metabolism and good health. Some vitamins help to regulate metabolism by serving as part of coenzymes. For example, niacin (one of the B complex vitamins) is a component of the coenzyme NAD, which is essential to cellular respiration and many other biological processes.

About 20 vitamins are important in human nutrition. They are classified into two groups: the fat-soluble vitamins and the water-soluble vitamins. The **fat-soluble vitamins,** found in association with lipids, are vitamins A, D, E, and K. The **water-soluble vitamins** include vitamin C and those belonging to the B complex.

Although most plants and animals use many of the same vitamins for their metabolic activities, vitamin requirements are not identical. Plants synthesize all of the vitamins they need. Whereas humans require vitamin C in the diet, most animals can synthesize vitamin C from glucose and so do not require a dietary source. Some animals do not require certain vitamins for their metabolic processes. For example, essential roles for vitamins D and K have not been demonstrated in invertebrates.

Vitamin deficiency results in predictable metabolic disorders and clinical symptoms (Figures 41–7, 41–8, and 41–9). For example, vitamin A defi-

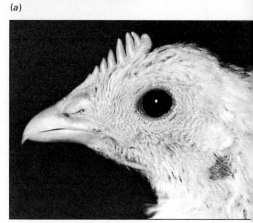

(a)

(b)

FIGURE 41-8 Effects of deficiency of pantothenic acid, a B-complex vitamin. (*a*) Chick after being fed a diet deficient in pantothenic acid. The eyelids, corners of the mouth, and adjacent skin are inflamed. The growth of feathers is retarded, and the feathers are rough. (*b*) The same chick after three weeks on a diet with pantothenic acid; the lesions are completely cured.

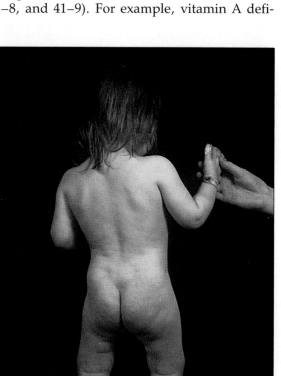

FIGURE 41-9 A child with rickets. A deficiency of vitamin D decreases the body's ability to absorb and use calcium and phosphorus and produces soft, malformed bones. These are most clearly evident in the ribs and in wrists and ankles. Note the bowed legs.

Table 41–3 **THE VITAMINS**

Vitamins and U.S. RDA*	Actions	Effect of Deficiency
Fat-soluble		
Vitamin A, Retinol 5000 IU†	Converted to retinal, a necessary component of retinal pigments, essential for normal vision; essential for normal growth and integrity of epithelial tissue; promotes normal growth of bones and teeth by regulating activity of bone cells	Failure of growth; night blindness; atrophy of epithelium; epithelium subject to infection; scaly skin
Vitamin D, Calciferol 400 IU	Promotes calcium absorption from digestive tract; essential to normal growth and maintenance of bone	Bone deformities; rickets in children; osteomalacia in adults
Vitamin E, Tocopherols 30 IU	Inhibits oxidation of unsaturated fatty acids and vitamin A that help form cell and organelle membranes; precise biochemical role not known	Increased catabolism of unsaturated fatty acids, so that not enough are available for maintenance of cell membranes and other membranous organelles; prevents normal growth
Vitamin K probably about 1 mg	Essential for blood clotting	Prolonged blood clotting time
Water-soluble		
Vitamin C (ascorbic acid) 60 mg	Needed for synthesis of collagen and other intercellular substances; formation of bone matrix and tooth dentin, intercellular cement; needed for metabolism of several amino acids; may help body withstand injury from burns and bacterial toxins	Scurvy (wounds heal very slowly and scars become weak and split open; capillaries become fragile; bone does not grow or heal properly)
B-complex Vitamins		
B_1, Thiamine 1.5 mg	Derivative acts as coenzyme in many enzyme systems; important in carbohydrate and amino acid metabolism	Beriberi (weakened heart muscle, enlarged right side of heart, nervous system and digestive tract disorders)
B_2, Riboflavin 1.7 mg	Used to make coenzymes (e.g., FAD) essential in cellular respiration	Dermatitis, inflammation and cracking at corners of mouth; mental depression
Niacin 20 mg	Component of important coenzymes (NAD and NADP) essential to cellular respiration	Pellagra (dermatitis, diarrhea, mental symptoms, muscular weakness, fatigue)
B_6, Pyridoxine 2 mg	Derivative is coenzyme in many reactions in amino acid metabolism	Dermatitis, digestive tract disturbances, convulsions
Pantothenic acid 10 mg	Constituent of coenzyme A (important in cellular metabolism)	Deficiency extremely rare
Folic acid 0.4 mg	Coenzyme needed for reactions involved in nucleic acid synthesis and for maturation of red blood cells	A type of anemia
Biotin 0.3 mg	Coenzyme needed for carbon dioxide fixation	
Vitamin B_{12} 6 mg	Coenzyme important in nucleic acid metabolism	Pernicious anemia

*RDA is the recommended dietary allowance, established by the Food and Nutrition Board of the National Research Council, to maintain good nutrition for healthy persons.
†International Unit: the amount that produces a specific biological effect and is internationally accepted as a measure of the activity of the substance.

ciency results in night blindness, vitamin D deficiency in rickets, and vitamin K deficiency in prolonged clotting time. Table 41–3 lists the vitamins, their actions, and the effects of deficiency.

Vitamin production is a multimillion-dollar industry. Yet there is disagreement on whether healthy individuals should take vitamin supplements. Some authorities think that people who eat a balanced diet do not require vitamin pills. Others argue that most of us do not eat enough fresh fruits and vegetables and should take vitamin supplements. There is also controversy regarding the effects of megadoses (massive doses) of individual vitamins such as vitamin C (often taken to prevent or minimize colds) or vitamin E (frequently taken as a protection against vascular disease and

Sources	Comments
Liver, fish-liver oils, egg; yellow and green vegetables	Can be formed from provitamin carotene (a yellow or red pigment); sometimes called anti-infection vitamin because it helps maintain epithelial membranes; excessive amounts harmful
Liver, fish-liver oils, egg yolk, fortified milk, butter, margarine	Two types: D_2, a synthetic form; D_3, formed by action of ultraviolet rays from sun upon a cholesterol compound in the skin; excessive amounts harmful
Oils made from cereals, seeds, liver, eggs, fish	
Normally supplied by intestinal bacteria; green leafy vegetables	Antibiotics may kill bacteria; then supplements needed in surgical patients
Citrus fruits, strawberries, tomatoes	Possible role in preventing common cold or in the development of acquired immunity(?); harmful in very excessive dose
Liver, yeast, cereals, meat, green leafy vegetables	Deficiency common in alcoholics
Liver, cheese, milk, eggs, green leafy vegetables	
Liver, meat, fish, cereals, legumes, whole-grain and enriched breads	
Liver, meat, cereals, legumes	
Widespread in foods	
Produced by intestinal bacteria; liver, cereals, dark green leafy vegetables	
Produced by intestinal bacteria; liver, chocolate, egg yolk	
Liver, meat, fish	Contains cobalt; intrinsic factor secreted by gastric mucosa needed for absorption

other processes associated with aging). To date, research studies have generated conflicting results.

Too much vitamin A or D can be harmful. Moderate overdoses of the water-soluble vitamins are excreted in the urine, but the fat-soluble vitamins are not easily excreted and may accumulate in the body tissues. Overdoses of vitamin A may result in skin ailments, retarded growth, enlargement of the liver and spleen, and painful swelling of the long bones. An excess of vitamin D can cause weight loss, mineral loss from the bones, and calcification of soft tissues, including the heart and blood vessels. Very high doses of vitamin D ingested by pregnant women have been linked to a form of mental retardation in the offspring.

WATER

Water, which makes up about two thirds of the human body and about 98% of a jellyfish, is an essential component of every cell. It is the medium in which the other chemicals of the body are dissolved and in which all chemical reactions occur. Water also serves as an active participant in many chemical reactions. For example, in digestion a water molecule is required for each sugar, amino acid, or fatty acid unit split from a carbohydrate, protein, or fat. Water is also used to transport materials within cells and from one place in the body to another. It is the fluid part of blood, lymph, urine, and sweat. It helps distribute and regulate body heat and, as perspiration, cools the body surface. (The properties of water are discussed in Chapter 2.)

Although the exact amount varies widely with individual activities and climate, an average of about 2.4 liters of water is lost from the body daily. This loss must be replaced promptly. Humans can live for several weeks without food, but only a few days without water. Much of the daily requirement for water can be satisfied by eating foods, because all foods contain some water. Certain fruits and vegetables contain as much as 95% water.

Water is one of the main factors that limits the distribution of animal populations. Most animals have mechanisms for controlling the water content of their bodies, but different animals have different tolerances for both water loss and water gain. Some small invertebrates can withstand long periods of dehydration by forming waterproof cysts around themselves. When water in the environment is again available, they emerge none the worse for the experience. Certain desert vertebrates live indefinitely without drinking water by obtaining it from the foods they eat and from the oxidation of their food. Problems of water balance in aquatic animals are discussed in Chapter 45.

■ SUMMARY

I. Metabolism is the sum of all chemical and energy transformations that take place in the organism.
   A. Anabolism is the synthetic phase of metabolism in which large molecules are synthesized from smaller ones.
   B. Catabolism is the phase of metabolism in which larger molecules are split into smaller ones.
II. Basal metabolic rate (BMR) reflects the amount of energy an organism must expend just to survive. Total metabolic rate is the sum of BMR and the energy needed to carry on daily activities.
III. When energy input equals energy output, body weight remains constant.
   A. When energy input exceeds output, the excess is stored in fat, and body weight increases.
   B. When energy input is less than output, the body draws on its fuel reserves (fat), and body weight decreases.
IV. Glucose, the end-product of carbohydrate digestion, is used by the cells mainly as fuel for cellular respiration.
V. The liver maintains a relatively constant concentration of glucose in the blood, with a fasting level equal to 90 mg/100 ml.
   A. When the concentration of glucose in the blood exceeds steady-state conditions, the liver removes glucose from the blood and converts it to glycogen for storage (the process of glycogenesis).
   B. As glucose is removed from the blood by the cells, its concentration falls below the steady state. Liver cells then convert glycogen back to glucose (the process of glycogenolysis) and return glucose to the blood.
   C. When glycogen stores are depleted, liver cells convert amino acids and glycerol to glucose (the process of gluconeogenesis).
VI. Fat is stored in adipose tissue. When the concentration of glucose in the blood falls below the steady state, fat is mobilized and can be used as an energy source.
   A. Fatty acids are degraded via β-oxidation to molecules of acetyl coenzyme A.
   B. Acetyl coenzyme A can be used as fuel by the liver cells or converted to ketone bodies for transport to other cells.
   C. Acetyl coenzyme A can be used to synthesize steroid hormones, cholesterol, or the fatty acids of phospholipids and triacylglycerols.
VII. Amino acids are used mainly in anabolism.
   A. Amino acids are used to make structural proteins, enzymes, functional proteins such as hemoglobin, and nucleic acids.

B. Amino acids are deaminated in the liver, and the resulting keto acids can be used as an energy source, or converted to glucose or fatty acids for storage as glycogen or fat.

VIII. Minerals required by the body include iron (a component of hemoglobin), iodine (a component of thyroid hormones), calcium and phosphorus (components of bones and teeth), and sodium and chlorine (needed for maintaining fluid balance).

IX. Many vitamins serve as part of coenzymes. Their actions, sources, and the effects of various vitamin deficiencies are listed in Table 41–3.

X. Water is a vital component of all organisms. Homeostasis depends upon fluid balance.

■ POST-TEST

1. The building or synthetic phases of metabolism are termed _____; the breaking-down aspects of metabolism are called _____.
2. The rate at which the body uses energy under resting conditions is the _____ _____ rate.
3. The total metabolic rate is the sum of the BMR and the _____.
4. When energy input equals energy output, body weight _____; when energy input exceeds energy output, body weight _____.
5. The principal sources of energy in the human diet are _____ and _____.
6. Glucose is utilized by the cells mainly as _____.
7. Glycogenesis is the process of _____.
8. During glycogenolysis, glycogen is _____.
9. Liver cells convert certain amino acids and other nutrients to glucose during the process of _____.
10. Cholesterol is stored mainly in the _____ cells; triacylglycerols are stored mainly in _____ cells.
11. Ketone bodies are produced during the metabolism of _____; they are taken up by the cells, are converted back to _____ _____, and then may be used as fuel.

12. In ketosis the pH of the blood and tissues becomes_____.
13. Essential amino acids are amino acids that must be _____.
14. Complete proteins contain _____.
15. Black beans and rice are considered a nutritious meal because together these foods contain all of the _____ _____ _____ _____.
16. During deamination of an amino acid, the _____ is split off the _____ skeleton.
17. The fat-soluble vitamins are vitamins _____, _____, _____, and _____.

Select the most appropriate substance from Column B for the description in Column A.

Column A	Column B
18. Component of thyroid hormones	a. Vitamin B$_{12}$
19. Component of hemoglobin	b. Iodine
20. Component of retinal pigments	c. Vitamin K
21. Deficiency results in pernicious anemia	d. Iron
22. Essential for blood clotting	e. Vitamin A

■ REVIEW QUESTIONS

1. What types of nutrients are used as fuel for cellular respiration? What other types of nutrients are required for a balanced diet?
2. Give two examples of catabolic reactions and two of anabolic reactions.
3. What is the difference between basal metabolic rate and total metabolic rate? What could you do to increase your own total metabolic rate?
4. What happens when energy input exceeds energy output? When energy input equals energy output?
5. How does the liver help to maintain a constant concentration of glucose in the blood?
6. When and why is gluconeogenesis important?
7. What are the products of β-oxidation? What is the fate of these products?
8. What function do ketone bodies serve? What is ketosis?
9. What is the fate of excess amino acids? Explain.
10. Why is each of the following nutreints required by the body?
    a. essential amino acids

    b. calcium
    c. iodine
    d. iron
    e. vitamin A
    f. vitamin D
11. What happens when each of the following is deficient in the diet?
    a. essential amino acids
    b. vitamin D
    c. vitamin K
    d. cholesterol
12. What types of foods contain complete proteins? What types contain unsaturated fatty acids?
13. What would be the approximate caloric content and the specific types of nutrients needed for an active young man? Describe how the diet of each of the following persons should differ from this:
    a. a 10-year-old boy
    b. a 65-year-old man
    c. a pregnant woman

■ **RECOMMENDED READINGS**

Beddington, J.R., and R.M. May. The harvesting of inter-acting species in a natural ecosystem. *Scientific American,* November 1982, pp. 62–69. Harvesting a biological resource such as krill affects the whales and other animals that normally feed upon it.

Brown, M.S., and J.L. Goldstein. How LDL receptors influence cholesterol and atherosclerosis. *Scientific American,* November 1984, pp. 58–66. Some Americans have too few LDL receptors, which normally remove particles carrying cholesterol from the circulation. Absence of these receptors puts individuals at high risk for atherosclerosis and heart attacks.

Hartbarger, J.C. and N.J. Hartbarger. *Eating for the Eighties: A Complete Guide to Vegetarian Nutrition.* Philadelphia, W.B. Saunders, 1981.

Hinman, C.W. Potential new crops. *Scientific American,* July 1986, pp. 33–37. A number of new plants show promise as sources of food; some are approaching commercial production.

Krause, M.B., and L.K. Mahan. *Food, Nutrition, and Diet Therapy,* 7th ed. Philadelphia, W.B. Saunders, 1984. A comprehensive discussion of the science of nutrition and its application to the maintenance of health.

Kretchmer, N., and W. van B. Robertson. *Human Nutrition.* San Francisco, W.H. Freeman, 1978. An interesting collection of articles from *Scientific American* covering all levels of nutrition.

Swaminathan, M.S. Rice. *Scientific American,* Vol. 250, no. 1, pp. 80–93. This member of the grass family is one of three on which the human species largely subsists.

False color scanning electron micrograph of human red blood cells.

42

Internal Transport

■ **LEARNING OBJECTIVES**

After you have read this chapter you should be able to:

1. List seven functions of a circulatory system.
2. Compare invertebrates that can function with no specialized circulatory system with those that require a circulatory system.
3. Compare open and closed circulatory systems and give examples of invertebrates that have an open system.
4. List and briefly describe the principal components of human blood, giving the function of each component.
5. Describe the life cycle of a red blood cell.
6. Describe and identify in illustrations five types of leukocytes, and give the function of each type.
7. Summarize the events involved in blood clotting.
8. Compare the structures and functions of arteries, arterioles, capillaries, sinusoids, and veins.
9. Compare the hearts of a fish, amphibian, reptile, bird, and mammal.
10. Describe the external and internal structure of the human heart.
11. Describe the heart's conduction system and cardiac muscle.
12. Briefly describe the events of the cardiac cycle, and relate normal heart sounds to the events of this cycle.
13. Define cardiac output, describe how it is regulated, and identify factors that affect it.
14. Explain the physiological basis for arterial pulse and for blood pressure.
15. Trace a drop of blood from one organ or body part to another through each part of the heart, and through the pulmonary and systemic circulations.
16. Describe atherosclerosis, giving its complications and risk factors.
17. List the functions of the lymphatic system, and describe how this system operates to maintain fluid balance.

Most animal cells require a continuous supply of nutrients and oxygen, and continuous removal of waste products. Very small organisms living in a watery environment can accomplish this by simple diffusion and do not need specialized circulatory structures. In larger organisms, diffusion cannot supply enough raw materials to all of the cells, and other mechanisms are required to transport materials to and from cells not in direct contact with the water in which the animal lives. In complex animals, whether aquatic or terrestrial, specialized structures are present to accomplish internal transport. These structures make up the **circulatory system.**

A circulatory system typically consists of the following:

1. **blood,** a fluid connective tissue consisting of cells and cell fragments dispersed in fluid
2. a pumping device, usually called a **heart**
3. a system of blood vessels or spaces through which the blood circulates

In the **closed circulatory system** present in many animals, blood passes through a continuous network of blood vessels. The arthropods and most mollusks have an **open circulatory system,** so named because the heart pumps blood into vessels that have open ends. Blood spills out of them, filling the body cavity and bathing the cells. Blood passes back into the circulatory system through openings within the heart (or through open-ended blood vessels that lead to the heart). Movement of blood through an open system is not as rapid or as efficient as through a closed system.

A circulatory system may perform the following functions:

1. Transport nutrients from the digestive system and from storage depots to all of the body cells
2. Transport oxygen from respiratory structures (skin, gills, lungs) to the cells of the body, and carbon dioxide from the cells to the respiratory structures
3. Transport wastes from each cell to the excretory organs
4. Transport hormones from endocrine glands to target tissues
5. Help maintain fluid balance
6. Defend the organism against invading microorganisms
7. Help regulate body temperature in homeothermic ("warm-blooded") animals

INVERTEBRATES WITH NO CIRCULATORY SYSTEM

No specialized circulatory structures are present in sponges, cnidarians, ctenophores, flatworms, or nematodes. In aquatic forms, the tissues are bathed in water laden with oxygen and nutrients. Wastes simply diffuse into the water and are washed away. Because the bodies of these invertebrates are only a few cells thick, diffusion is an effective mechanism for distributing materials to and from their cells.

In cnidarians, the central gastrovascular cavity serves as both a digestive organ and a circulatory organ (Figure 42–1). The animal's tentacles capture prey and stuff it through the mouth and into the cavity, where digestion occurs. The digested nutrients then pass into the cells lining the cavity, and through them to cells of the outer layer. Movement of the animal's body, as it stretches and contracts, stirs up the contents of the central cavity and aids circulation.

In the flatworm planaria, the branched intestine brings nutrients to all regions of the body. Nutrients diffuse into the tissue fluid of the mesenchyme, and then to cells of the mesenchyme and the outer layer of the body. As in cnidarians, circulation is aided by contractions of the muscles of the body wall, which agitate the fluid in the intestine and the tissue

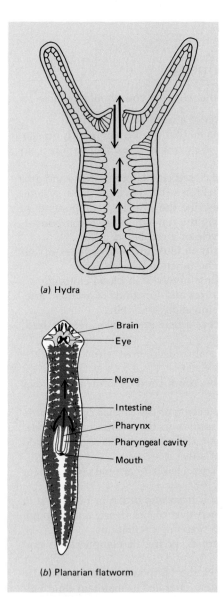

(a) Hydra

Brain
Eye

Nerve

Intestine
Pharynx
Pharyngeal cavity
Mouth

(b) Planarian flatworm

FIGURE 42–1 Some invertebrates with no circulatory system. (a) In *hydra* and other cnidarians, the gastrovascular cavity serves a circulatory function, permitting nutrients to come in contact with the body cells. (b) In planarian flatworms, the branched intestine conducts food to all of the regions of the body.

fluid. Oxygen distribution and waste removal from individual cells depend mainly upon diffusion.

Fluid in the pseudocoelom of nematodes and other pseudocoelomate animals helps to circulate materials. Nutrients, oxygen, and wastes dissolve in this fluid and diffuse through it to and from the individual cells of the body. Body movements of the animal result in movement of the fluid, facilitating distribution of these materials to all parts of the body.

INVERTEBRATES WITH OPEN CIRCULATORY SYSTEMS

In mollusks (except for the cephalopods) and in arthropods, the circulatory system is open. In many mollusks the heart is surrounded by a pericardial cavity. The heart typically consists of three chambers: two **atria,** which receive blood (often referred to as hemolymph in animals with open circulatory systems) from the gills, and a **ventricle,** which pumps this oxygen-rich blood to the tissues. Blood vessels leading from the heart open into large spaces called sinuses, enabling the blood to bathe the body cells. Such blood-filled spaces form a **hemocoel** (blood cavity). From the hemocoel, blood passes into vessels that lead to the gills; there blood is recharged with oxygen and passes into blood vessels that return it to the heart. In the clam, the ventricle pumps blood both forward and backward (Figure 42–2).

Some mollusks, as well as arthropods, have a blood pigment, **hemocyanin,** that contains copper. Hemocyanin transports oxygen and imparts a bluish color to the blood of these animals (the original bluebloods!).

In insects and other arthropods, a tubular heart pumps blood into blood vessels (arteries) that eventually deliver blood to the sinuses of the hemocoel. Blood then circulates through the hemocoel, eventually finding its way back to the pericardial cavity surrounding the heart. It enters the heart through tiny openings called **ostia,** which are equipped with valves to prevent backflow. Some insects have accessory "hearts" that help pump blood through the extremities, particularly the wings. Circulation of the blood is faster during muscular movement. Thus, when an animal is active and most in need of nutrients for fuel, its own movement ensures effective circulation. In insects, oxygen is transported directly to the cells by the tracheae of the respiratory system, rather than by the circulatory system.

INVERTEBRATES WITH CLOSED CIRCULATORY SYSTEMS

A rudimentary closed circulatory system is found in the proboscis worms (phylum Nemertinea; Figure 42–3). This system consists of a complete network of blood vessels, including at least two vessels that extend the length of the body. These communicate anteriorly and posteriorly by means of connecting blood sinuses. Branches from the main vessels extend into the tissues. No heart is present; instead, blood flow depends upon movements of the animal and upon contractions in the walls of the large blood vessels. Blood may move in either direction. Blood cells containing colored pigments are found in this group.

Earthworms and other annelids have a complex closed circulatory system. Two main blood vessels extend lengthwise in the body. The ventral vessel conducts blood posteriorly, while the dorsal vessel conducts blood anteriorly. Dorsal and ventral vessels are connected by lateral vessels in every segment. Branches of the lateral vessels deliver blood to the skin, where it is oxygenated, and to the various tissues and organs. In the anterior part of the worm are five pairs of contractile blood vessels (sometimes referred to as hearts) that connect dorsal and ventral vessels (Figure 42–3). Contractions of these paired vessels and of the dorsal vessel, as well as contraction of the muscles of the body wall, circulate the blood. Earth-

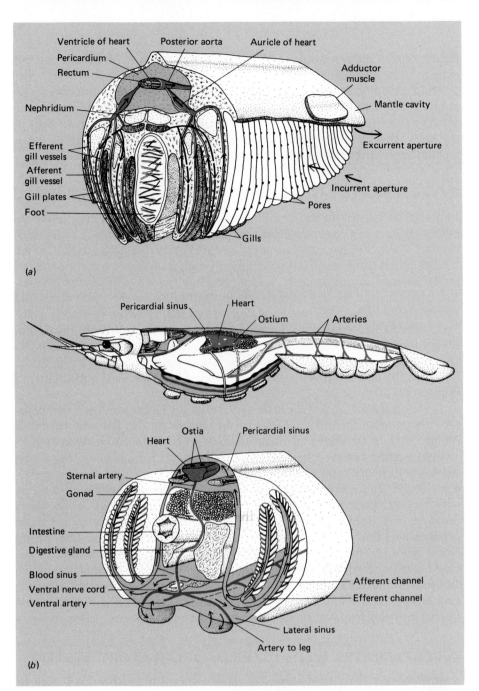

(a)

(b)

FIGURE 42–2 Mollusks and arthropods have an open circulatory system. (a) Circulatory system of the clam. (b) Circulatory system of the crayfish. Lateral view and cross section.

FIGURE 42–3 Examples of a closed circulatory system. (a) A rudimentary closed circulatory system is present in nemertines. (b) Earthworms have a complex closed circulatory system with five pairs of contractile blood vessels that deliver blood from the dorsal vessel to the ventral vessel.

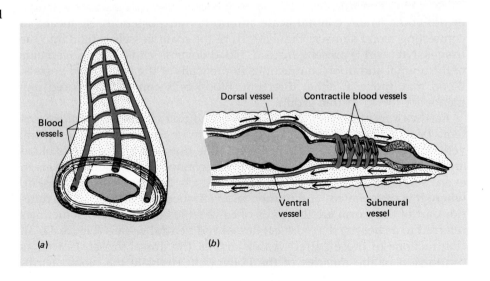

(a)

(b)

948

worms have hemoglobin, the same red pigment that transports oxygen in vertebrate blood; however, their hemoglobin is not within red blood cells but is dissolved in the blood plasma. The circulatory system of the earthworm apparently does little to remove wastes from the body cells. This function is delegated instead to the coelomic fluid, which transports waste to the nephridia.

Although other mollusks have an open circulatory system, the fast-moving cephalopods (squid, octopus) require a more efficient means of internal transport. They have a closed system made even more effective by the presence of accessory "hearts" at the base of the gills, which speed the passage of blood through the gills.

The circulatory system of the sea cucumbers (holothuroids) is the most highly developed system of any of the echinoderms. Its vessels parallel the tubes of the water vascular system, and it appears to transport both nutrients and oxygen.

Invertebrate chordates have a closed circulatory system, usually with a ventral heart. An exception is the cephalochordate amphioxus, which has no heart and must depend upon certain pulsating blood vessels to move its blood.

THE VERTEBRATE CIRCULATORY SYSTEM

The circulatory systems of all vertebrates are fundamentally similar, from fishes, frogs, and reptiles to birds and human beings. All have a muscular heart that pumps blood into a closed system of blood vessels (Figure 42–4). **Arteries** are blood vessels that branch and rebranch, carrying blood away from the heart and to the various organs of the body. The tiniest branches, **arterioles,** deliver blood to the **capillaries,** vessels with walls so thin that nutrients and oxygen can diffuse through them to the individual cells of the body. Wastes from the cells diffuse into the blood through the thin walls of the capillaries and are carried to the excretory organs. After passing through a network of capillaries, blood flows into **veins,** the vessels that conduct the blood back to the heart. Because this basic plan is similar in all vertebrates, much can be learned about the human circulatory system by dissecting an animal such as a shark or a frog.

Blood

In humans the total circulating blood volume is about 8% of the body weight—5.6 liters (6 quarts) in a 70-kg (154-lb) person. This is about the amount of oil in the crankcase of most cars! Although blood appears to be a homogeneous crimson fluid as it pours from a wound, it is composed of a pale yellowish fluid, called **plasma,** in which red blood cells, white blood cells, and blood platelets are suspended. About 55% of the blood is plasma; the remaining 45% is made up of blood cells and platelets. The loss of water in profuse sweating may reduce the plasma volume to 50% of the blood, and drinking large quantities of fluid may increase it to 60%. Because cells and platelets are heavier than plasma, the two may be separated from plasma by the process of centrifugation. Plasma does not separate from blood cells in the body because the blood is constantly mixed as it circulates in the blood vessels.

Plasma

Plasma is composed of water (about 92%), proteins (about 7%), salts, and a variety of materials being transported, such as dissolved gases, nutrients, wastes, and hormones. Plasma is in dynamic equilibrium with the intersti-

FIGURE 42–6 Principal varieties of formed elements (blood cells and platelets) in the circulating blood.

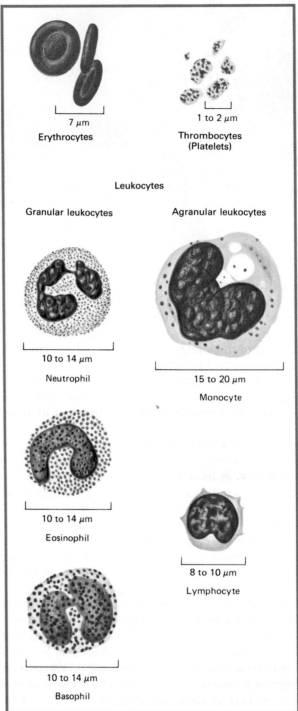

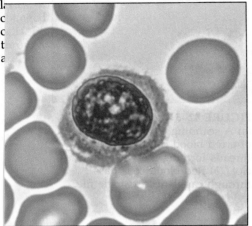

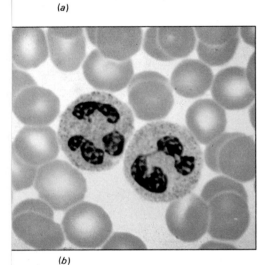

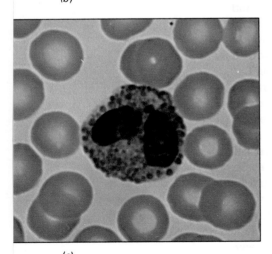

(a)

(b)

(c)

FIGURE 42–7 Photomicrographs of circulating blood cells. (*a*) A lymphocyte surrounded by red blood cells. (*b*) Two neutrophils. Note the lobed nucleus. (*c*)) An eosinophil.

seeking out and ingesting bacteria. They also phagocytize the remains of dead tissue cells, a clean-up task that must be performed after injury or infection. **Eosinophils** have large granules that stain bright red with eosin, an acidic dye. These cells increase in number during allergic reactions and during parasitic infections such as tapeworm infections. **Basophils** exhibit deep blue granules when stained with basic dyes. Like eosinophils, these cells are thought to play a role in allergic reactions. Basophils contain large amounts of the chemical histamine, which they release in injured tissues and in allergic responses. Because they contain the anticlotting chemical **heparin,** basophils may play a role in preventing blood from clotting inappropriately within the blood vessels.

Table 42–1 CELLULAR COMPONENTS OF BLOOD

	Normal Range	Function	Pathology
Red blood cells	Male: 4.2–5.4 million/μl Female: 3.6–5.0 million/μl	Oxygen transport; carbon dioxide transport	Too few: anemia Too many: polycythemia
Platelets	150,000–400,000/μl	Essential for clotting	Clotting malfunctions; bleeding; easy bruising
White blood cells (total)	5000–10,000/μl		
Neutrophils	About 60% of WBCs	Phagocytosis	Too many: may be due to bacterial infection, inflammation, leukemia (myelogenous)
Eosinophils	1–3% of WBCs	Some role in allergic response	Too many: may result from allergic reaction, parasitic infection
Basophils	1% of WBCs	May play role in prevention of clotting in body	
Lymphocytes	25–35% of WBCs	Produce antibodies; destroy foreign cells	Atypical lymphocytes present in infectious mononucleosis; too many may be due to leukemia (lymphocytic), certain viral infections
Monocytes	6% of WBCs	Differentiate to form macrophages	May increase in monocytic leukemia, tuberculosis, fungal infections

Agranular leukocytes lack large distinctive granules, and their nuclei are rounded or kidney-shaped. Two types of agranular leukocytes are lymphocytes and monocytes. Some **lymphocytes** are specialized to produce antibodies, while others attack foreign invaders such as bacteria or viruses directly. Just how they manage these feats is discussed in the next chapter.

Monocytes are the largest WBCs, reaching 20 μm in diameter. They are manufactured in the bone marrow. After spending about 24 hours in the blood, a monocyte leaves the circulation. Development is completed within the tissues, where the monocyte greatly enlarges and becomes a **macrophage,** a giant scavenger cell. All of the tissue macrophages develop in this way. Macrophages voraciously engulf bacteria, dead cells, and any debris littering the tissues.

In human blood there are normally about 7000 WBCs per cubic millimeter of blood (only one for every 700 red blood cells). During bacterial infections the number may rise sharply, so that a **white blood cell count** is a useful diagnostic tool. The proportion of each kind of WBC is determined by a **differential** WBC count. The normal distribution of leukocytes is indicated in Table 42–1.

Leukemia is a form of cancer in which any one of the kinds of white cells multiplies rapidly within the bone marrow. Many of these cells do not mature, and their large numbers crowd out developing red blood cells and platelets, leading to anemia and impaired clotting. A common cause of death from leukemia is internal hemorrhaging, especially in the brain. Another frequent cause of death is infection because, although there may be a dramatic rise in the white cell count, the cells are immature and abnormal, and unable to defend the body against disease organisms. Although no cure for leukemia has been discovered, radiation treatment and therapy with antimitotic drugs can induce partial or complete remissions lasting as long as 15 years in some patients.

Platelets and Blood Clotting

In most vertebrates other than mammals, the blood contains small, oval cells called **thrombocytes,** which have nuclei. In mammals thrombocytes are tiny, spherical, or disk-shaped bits of cytoplasm that lack a nucleus.

They are usually referred to as blood **platelets.** About 300,000 platelets per µl are present in human blood. Platelets are formed from bits of cytoplasm that are pinched off from very large cells (megakaryocytes) in the bone marrow. Thus, a platelet is not a whole cell but a fragment of cytoplasm enclosed by a membrane.

Platelets play an important role in **hemostasis** (the control of bleeding). When a blood vessel is cut, it constricts, reducing loss of blood. Platelets stick to the rough, cut edges of the vessel, physically patching the break in the wall. As platelets begin to gather they release ADP, which attracts other platelets. Within about five minutes after injury a complete platelet patch, or temporary clot, has formed.

At the same time that the temporary clot is formed, a stronger, more permanent clot begins to develop. More than 30 different chemical substances interact in this very complex process. The series of reactions that leads to clotting is triggered when one of the clotting factors in the blood is activated by contact with the injured tissue. In **hemophiliacs** (persons with "bleeder's disease") one of the clotting factors is absent as a result of an

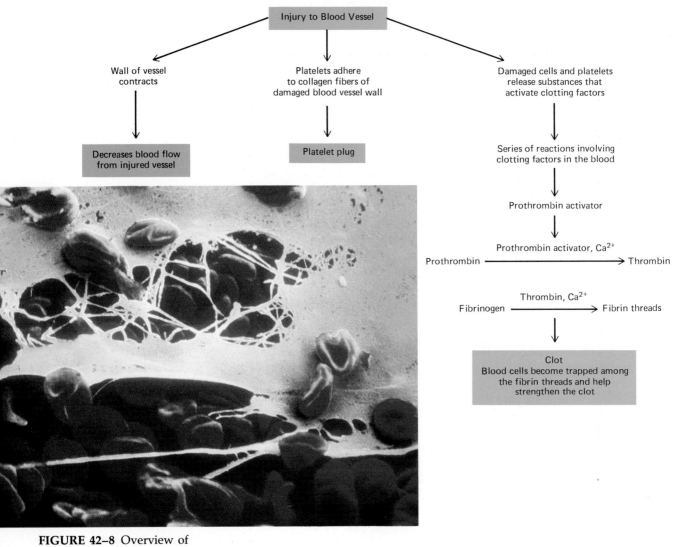

FIGURE 42–8 Overview of blood clotting. Scanning electron micrograph of part of a blood clot, showing red blood cells enmeshed in a network of fibrin.

inherited genetic mutation. In a very simplified way the clotting process can be summarized as shown in Figure 42–8.

Prothrombin, a globulin manufactured in the liver, requires vitamin K for its production. In the presence of clotting factors, calcium ions, and compounds released from platelets, prothrombin is converted to **thrombin.** Then thrombin catalyzes the conversion of the soluble plasma protein fibrinogen to an insoluble protein, **fibrin.** Once formed, fibrin polymerizes, producing long threads that stick to the damaged surface of the blood vessel and form the webbing of the clot. These threads trap blood cells and platelets, which help to strengthen the clot.

The Blood Vessels

The circulatory system of a vertebrate includes three main types of blood vessels: arteries, veins, and capillaries (Figure 42–9). A blood vessel wall, like the wall of the heart, has three layers (Figure 42–10). The innermost layer **(tunica intima),** which lines the blood vessel, consists mainly of **endothelium,** a tissue that resembles squamous epithelium. The middle layer (tunica media) consists of connective tissue and smooth muscle cells, and the outer layer (tunica adventitia) is composed of connective tissue rich in elastic and collagen fibers.

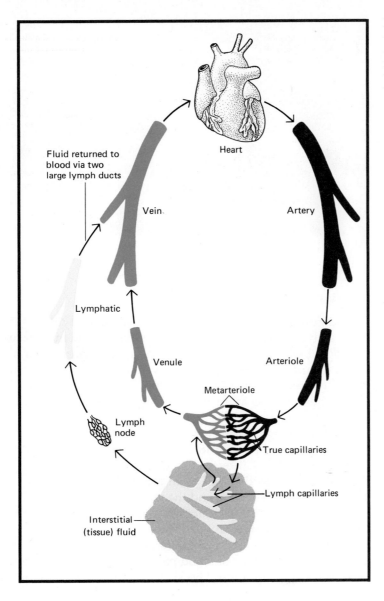

FIGURE 42–9 Types of blood vessels and their relationship to one another. Lymphatic vessels return interstitial fluid to the blood by way of ducts that lead into large veins in the shoulder region.

FIGURE 42–10 Blood vessel structure. (*a*) Comparison of the walls of an artery, vein, and capillary. (*b*) Scanning electron micrograph of a branch of the hepatic artery nourished by small capillaries. Notice the elliptical depressions in the artery wall. This preparation was made by injecting the blood vessels with a special plastic, followed by treatment with a corrosive that removed all surrounding tissues. The nuclei of the flattened cells that lined the artery in life made the oval depressions.

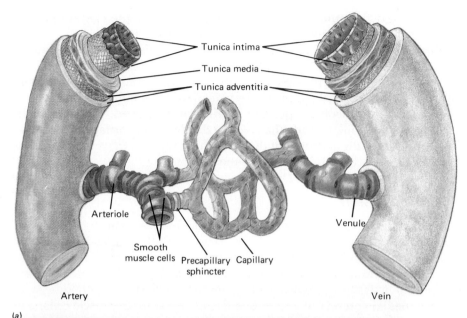

Tunica intima
Tunica media
Tunica adventitia

Arteriole

Smooth muscle cells
Precapillary sphincter
Capillary
Venule

Artery

Vein

(a)

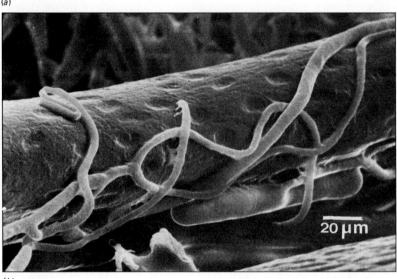

20 µm

(b)

Arteries

In addition to the endothelial lining, the inner layer of most arteries contains a layer of elastic tissue, the **internal elastic membrane,** which gives additional strength to the walls. In large arteries the middle layer is the thickest, and it contains several layers of elastic fibers that enable the arteries to stretch as they fill with blood delivered to them with each heartbeat. These elastic arteries branch into smaller distributing arteries that deliver blood to specific organs.

Within an organ or tissue, a distributing artery branches to form very small arteries called **arterioles,** which are important in determining the amount of blood distributed to a tissue and in maintaining blood pressure. Smooth muscle in the wall of an arteriole can contract or relax, changing the diameter of the vessel and the volume of blood that can pass through it. Arteriolar contraction produces **vasoconstriction;** relaxation causes **vasodilatation.** Such changes in blood flow are under control of the nervous system; these changes help to maintain appropriate blood pressure and to determine the amount of blood passing to a particular organ. For example, during resting conditions the skeletal muscles receive only about 15% of

Table 42–2 **BLOOD FLOW TO REGIONS OF THE HUMAN BODY UNDER BASAL CONDITIONS AND DURING STRENUOUS EXERCISE**

	Basal Conditions*		Exercise	
	ml/min	*% of Total*	*ml/min*	*% of Total*
Brain	700	14	750	4.2
Heart	200	4	750	4.2
Bronchi	100	2	200	1.1
Kidneys	1100	22	600	3.3
Liver	1350	27	600	3.3
Via portal vein	1050	21		
Via hepatic artery	300	6		
Skeletal muscles	750	15	12,500	70.3
Bone	250	5	250	1.4
Skin	300	6	1,900	10.7
Thyroid gland	50	1	50	0.3
Adrenal glands	25	0.5	25	0.2
Other tissue	175	3.5	175	1.0
	5000	100	17,800	100

*Data from Guyton, A. C. *Function of the Human Body,* 4th ed. Philadelphia, W. B. Saunders Co., 1974. Based on data compiled by Dr. L. A. Sapirstein.

the blood, but during exercise the muscles receive about 70% of the blood (Table 42–2).

Capillaries

Arterioles deliver blood into **capillaries,** microscopic vessels with walls just one cell thick. Only the walls of the capillaries are thin enough to permit the exchange of nutrients, gases, and wastes between blood and tissues. Capillary walls consist of a single layer of cells, the endothelium, that is continuous with the endothelial lining of the artery and vein on either side. Each capillary is only about 1 mm (0.04 inch) long; yet there are so many of these tiny vessels that almost every cell in the body is within two or three cells of a capillary—close enough for oxygen and nutrients to diffuse from the blood to every cell. The number of capillaries in the body is almost beyond calculation. In tissues with a high metabolic rate they are very close together. One investigator places the number of capillaries in muscle at about 240,000 per square centimeter.

The amount of blood that the capillaries in the body could hold is so great that you would need about 40% more blood to fill them all completely. Actually, at any moment only about 5% of your blood can be found within capillaries. Thus at any time most of these tiny vessels are not filled with blood. During periods of intense activity of a particular organ, most of its capillary networks fill with blood (Figure 42–11).

Capillaries are somewhat ''leaky'' because some of their endothelial cells have tiny pores, and because others overlap slightly. As blood passes through a capillary, some of the plasma passes through its walls and out

FIGURE 42–11 Changes in blood flow through a capillary bed as the tissue becomes active. (*a*) When the tissue is inactive, only the metarterioles are open. (*b*) When the tissue becomes active, the decreased oxygen tension in the tissue brings about a relaxation of the precapillary sphincters, and the capillaries open. This increases the blood supply and the delivery of oxygen to the active tissue.

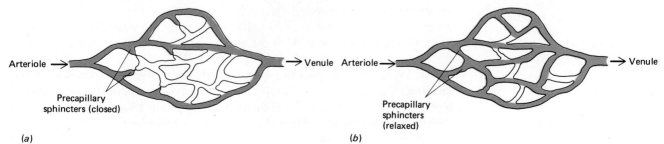

(a)   (b)

FIGURE 42–12 Nutrients, oxygen, and other materials diffuse out of the blood and through the tissue fluid that bathes the cells. Carbon dioxide and other waste products diffuse out of the cells and enter the blood through the capillary wall.

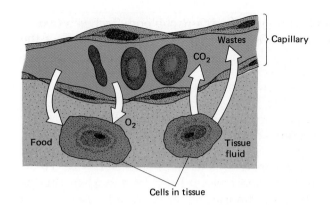

into the tissues. This fluid, which bathes the tissues, is called **tissue fluid,** or **interstitial fluid.** It may be laden with nutrients and oxygen, which pass out of the blood by diffusion (Figure 42–12).

The small vessels that directly link arterioles with **venules** (small veins) are **metarterioles.** The so-called **true capillaries** branch off from the metarterioles and then rejoin them (Figure 42–8). True capillaries also interconnect with one another. Wherever a capillary branches from a metarteriole, a smooth muscle cell called a **precapillary sphincter** is present. These sphincters can open or close to regulate passage of blood. Precapillary sphincters open and close continuously, directing blood first to one and then to another section of tissue. These sphincters also (along with the smooth muscle in the walls of arteries and arterioles) regulate the blood supply to each organ and its subdivisions. Such mechanisms ensure that blood flow meets the changing needs of the body as a whole, as well as the metabolic needs of the tissue being serviced. For example, during exercise, the increased metabolic rate of muscle cells demands a greater blood supply. Arterioles serving the muscle dilate, permitting a 10-fold increase in the amount of blood delivered to these cells.

Veins

In mammals about 50% of the blood may be found within the veins at any given moment. Vein walls are much thinner and less elastic than those of arteries. By the time blood flows through capillaries and into veins, its pressure is quite low. What, then, keeps the blood flowing through the veins? Breathing and other forms of muscular activity contribute to venous blood flow. When muscles contract, the veins within them are compressed; this helps to push the blood along (Figure 42–13). Most veins larger than 2 mm (0.08 inch) in diameter that conduct blood against the force of gravity

FIGURE 42–13 The action of skeletal muscles in moving blood through the veins. (*a*) Resting condition. (*b*) Muscles contract and bulge, compressing veins and forcing blood toward the heart. The lower valve prevents backflow. (*c*) Muscles relax, and the vein expands and fills with blood from below. The upper valve prevents backflow.

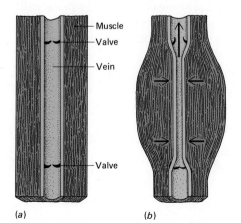

(a)   (b)   (c)

are equipped with **valves** that prevent backflow of blood. Such valves usually consist of two **cusps** formed by inward extensions of the wall of the vein.

When a person stands still for a long period of time, blood tends to accumulate in the veins of the legs. Excessive pooling of blood may stretch the veins so that the cusps of their valves no longer meet. This is especially likely to occur in persons whose occupations require them to stand for long hours each day. **Varicose veins** are likely to result from stretched veins, particularly in overweight persons or in those who have inherited weak vein walls. A varicose vein appears dilated, twisted, and elongated. **Hemorrhoids** are varicose veins in the anal region. They develop when venous pressure in that area is elevated abnormally, as in chronic constipation (due to straining) or during pregnancy, when the increased size of the uterus results in added pressure on the veins of the legs and the lower abdomen.

The Heart

In vertebrates, the heart consists of one or two chambers called **atria,** which receive blood returning from the tissues, and one or two chambers called **ventricles,** which pump blood into the arteries. Additional chambers are present in some animals.

Comparison of Vertebrate Hearts

Surveying the vertebrates, we find that the structure of the heart and circulatory system reflects evolutionary development. Fish, the earliest vertebrates, have a simple heart; amphibians, reptiles, and birds and mammals have hearts of increasingly complex structure (Figure 42–14).

FISH. Because it has only one atrium and one ventricle, the fish heart is usually described as a two-chambered heart. Actually, two accessory chambers are present. A thin-walled **sinus venosus** receives blood returning from the tissues and pumps it into the atrium. The atrium then contracts, sending blood into the ventricle. The ventricle in turn pumps the blood into an elastic **conus arteriosus,** which does not contract. These four compartments are separated by valves that prevent blood from flowing backward. From the conus, blood flows into a large artery, the ventral aorta, which branches to distribute blood to the gills. Because blood must

FIGURE 42–14 The evolution of the vertebrate heart. (*a*) In the fish heart there is one atrium and one ventricle. (*b*) The amphibian heart consists of two atria and one ventricle. (*c*) The reptilian heart has two atria and two ventricles, but the wall separating the ventricles is incomplete so that blood from the right and left chambers mix to some extent. (*d*) Birds and mammals have two atria and two ventricles, and blood rich in oxygen is kept completely separate from oxygen-poor blood.

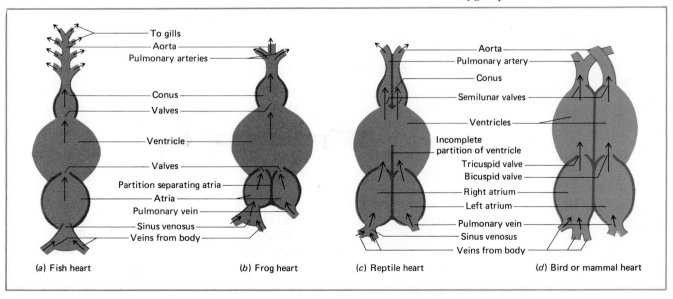

(a) Fish heart   (b) Frog heart   (c) Reptile heart   (d) Bird or mammal heart

pass through the capillaries of the gills before flowing to the other tissues of the body, blood pressure is low through most of the system. This low-pressure circulatory system permits only a low rate of metabolism in the fish.

AMPHIBIANS.   The three-chambered amphibian heart consists of two atria and a ventricle. A thin-walled sinus venosus collects blood returning from the veins and pumps it into the right atrium. Blood returning from the lungs passes directly into the left atrium. Both atria pump blood into the single ventricle. In the frog heart, oxygenated and deoxygenated blood are kept somewhat separate. Deoxygenated blood is pumped out of the ventricle first and passes into the tubular conus arteriosus, which has a spiral fold that helps to separate the blood. Much of the deoxygenated blood is directed to the lungs and skin, where it can be charged with oxygen. Oxygenated blood is delivered into arteries, which conduct it to the various tissues of the body.

REPTILES.   In reptiles the heart consists of two atria and two ventricles. In all reptiles except the crocodiles, however, the wall between the ventricles is incomplete, so that some mixing of oxygenated and deoxygenated blood does occur. Mixing is minimized by the timing of contractions of the left and right side of the heart and by pressure differences.

BIRDS AND MAMMALS.   The hearts of birds and mammals have completely separate right and left sides. The wall between the ventricles is complete, preventing the mixture of oxygenated blood in the left side with deoxygenated blood in the right side. The conus has split and become the base of the aorta and pulmonary artery. No sinus venosus is present as a separate chamber (although a vestige remains as the sinoatrial node, or pacemaker, described later in the chapter).

Complete separation of right and left hearts makes it necessary for blood to pass through the heart twice each time it makes a tour of the body. As a result, blood in the aorta of birds and mammals contains more oxygen than that in the aorta of the lower vertebrates. Hence the tissues of the body receive more oxygen, a higher metabolic rate can be maintained, and the homeothermic condition is possible. Birds and mammals can maintain a constant, high body temperature even in cold surroundings.

The pattern of blood flow in birds and mammals may be summarized as follows:

> veins → right atrium → right ventricle →
> one of the pulmonary arteries → capillaries in the lung →
> one of the pulmonary veins → left atrium → left ventricle →
> →aorta

Structure of the Human Heart

In an average lifetime of 70 years, the human heart beats about 2.5 billion times, pumping about 180 million liters of blood. Yet this remarkable organ is only about the size of a fist and weighs only about 400 g (less than a pound). The heart is a hollow, muscular organ located in the chest cavity directly under the breastbone (Figure 42–15). Enclosing it is a tough connective tissue sac, the **pericardium.** The inner surface of the pericardium and the outer surface of the heart are covered by a smooth layer of epithelium-type cells. Between these two surfaces is a small **pericardial cavity** filled with fluid, which reduces friction to a minimum as the heart beats.

The right atrium and ventricle are separated from the left atrium and ventricle by a wall, or **septum** (Figure 42–16). Between the atria the wall is known as the **interatrial septum;** between the ventricles it is the **interventricular septum.** On the interatrial septum a shallow depression, the **fossa**

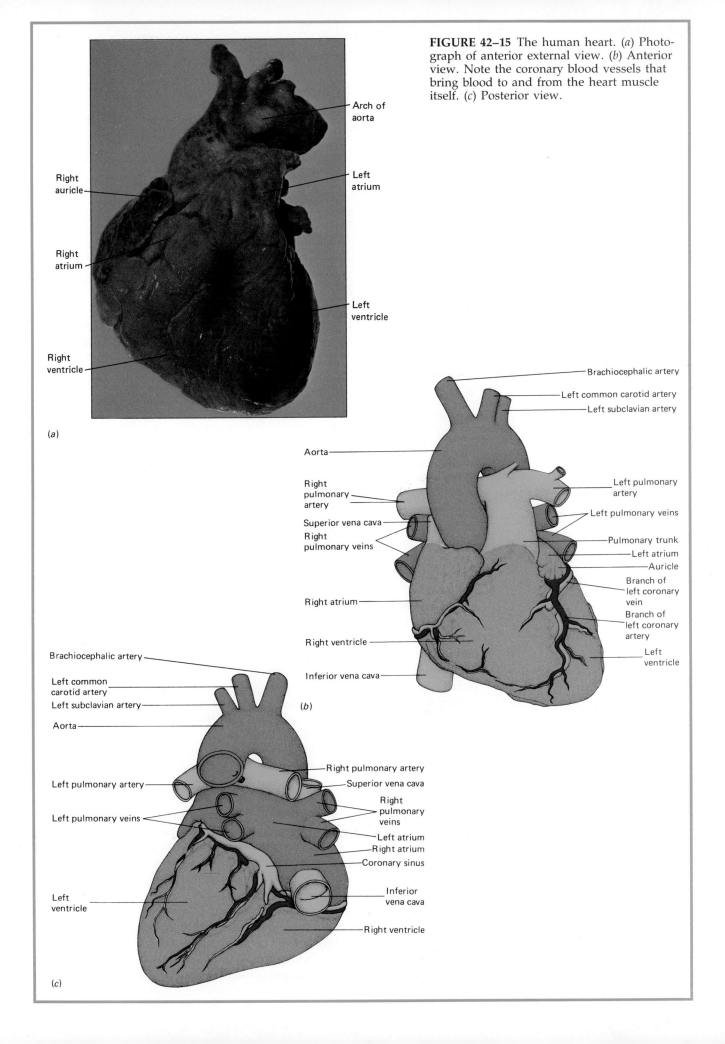

FIGURE 42–15 The human heart. (*a*) Photograph of anterior external view. (*b*) Anterior view. Note the coronary blood vessels that bring blood to and from the heart muscle itself. (*c*) Posterior view.

Arch of aorta

Right auricle

Right atrium

Right ventricle

Left atrium

Left ventricle

(*a*)

Brachiocephalic artery

Left common carotid artery

Left subclavian artery

Aorta

Right pulmonary artery

Superior vena cava

Right pulmonary veins

Right atrium

Right ventricle

Inferior vena cava

Left pulmonary artery

Left pulmonary veins

Pulmonary trunk

Left atrium

Auricle

Branch of left coronary vein

Branch of left coronary artery

Left ventricle

(*b*)

Brachiocephalic artery

Left common carotid artery

Left subclavian artery

Aorta

Left pulmonary artery

Left pulmonary veins

Left ventricle

Right pulmonary artery

Superior vena cava

Right pulmonary veins

Left atrium

Right atrium

Coronary sinus

Inferior vena cava

Right ventricle

(*c*)

FIGURE 42–16 Section through the human heart showing chambers, valves, and connecting blood vessels.

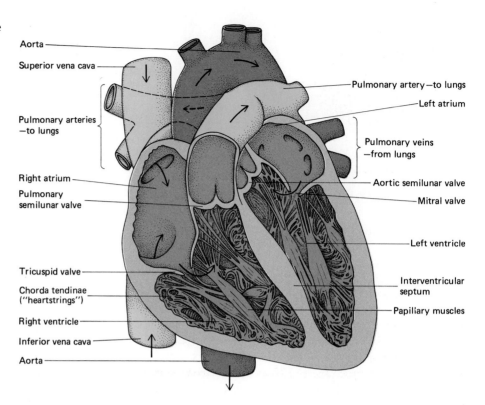

Aorta

Superior vena cava

Pulmonary arteries —to lungs

Right atrium

Pulmonary semilunar valve

Tricuspid valve

Chorda tendinae ("heartstrings")

Right ventricle

Inferior vena cava

Aorta

Pulmonary artery—to lungs

Left atrium

Pulmonary veins —from lungs

Aortic semilunar valve

Mitral valve

Left ventricle

Interventricular septum

Papillary muscles

ovalis, marks the place where an opening, the **foramen ovale,** was located in the fetal heart. In the fetus, the foramen ovale permits the blood to move directly from right to left atrium so that very little blood passes to the as yet nonfunctional lungs. At the upper surface of each atrium lies a small, muscular pouch called an **auricle.**

To prevent blood from flowing backward, the heart is equipped with valves that close automatically. The valve between the right atrium and ventricle is called the right **atrioventricular (AV) valve** (also known as the tricuspid valve). The left AV valve is referred to as the **mitral valve.** The AV valves are held in place by stout cords, or "heart-strings," the **chordae tendineae.** These cords attach the valves to the **papillary muscles** that project from the walls of the ventricles. When blood returning from the tissue fills the atria, blood pressure upon the AV valves forces them to open into the ventricles. Blood then fills the ventricles. As the ventricles contract, blood is forced back against the AV valves, pushing them closed. However, contraction of the papillary muscles and tensing of the chordae tendineae prevent them from opening backward into the atria. These valves are like swinging doors that can open in only one direction.

Semilunar valves (named for their flaps, which are shaped like half-moons) guard the exits from the heart. The semilunar valve between the left ventricle and the aorta is known as the **aortic valve,** and the one between the right ventricle and the pulmonary artery as the **pulmonary valve.** When blood passes out of the ventricle, the flaps of the semilunar valve are pushed aside and offer no resistance to blood flow. But when the ventricles are relaxing and filling with blood from the atria, the blood pressure in the arteries is higher than that in the ventricles. Blood then fills the pouches of the valves, stretching them across the artery so that blood cannot flow back into the ventricle (Figure 42–17).

Valve deformities are sometimes present at birth or may result from certain diseases such as rheumatic fever or syphilis. As a consequence of inflammation and scarring, valves may be thickened so that the passageway for blood is narrowed. Sometimes the valve tissues are eroded so that the flaps cannot close tightly, causing blood to leak backward and reducing

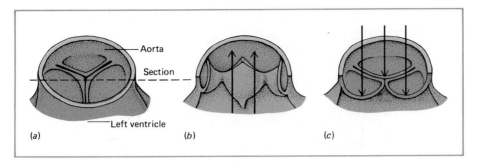

(a)   (b)   (c)

FIGURE 42–17 The operation of the semilunar valves. (a) Arrangement of the three pouches of the semilunar valves. The aorta has been cut across just above its point of attachment to the ventricle to expose the valves. (b) When the ventricle contracts, the expelled blood (*arrows*) pushes the pouches aside and passes aside into the aorta. (c) When the ventricle relaxes, blood from the aorta fills the pouches (*arrows*), causing them to extend across the cavity and prevent the leakage of blood back into the heart.

the efficiency of the heartbeat. Diseased valves can now be surgically replaced with artificial valves.

The wall of the heart is composed mainly of cardiac muscle attached to a framework of collagen fibers. At their ends cardiac muscle cells are joined by dense bands called **intercalated disks** (Figure 42–18). Each disk is a type of gap junction (see Chapter 5) in which two cells overlap slightly. This type of junction is of great physiological importance because it offers very little resistance to the passage of an action potential. Ions move easily through the gap junctions, allowing the entire atrial (or ventricular) muscle mass to contract as one giant cell. Because it acts as a single unit, cardiac muscle is sometimes referred to as a **functional syncytium.** The atrial syncytium is separated from the ventricular syncytium by fibrous connective tissue.

How the Heart Works

Horror films not infrequently feature a scene in which a heart cut out from the body of its owner continues to beat. Scriptwriters of these tales actually have some factual basis for their gruesome fantasies, for when carefully removed from the body, the heart does continue to beat for many hours if kept in a nutritive, oxygenated fluid. This is possible because the contractions of cardiac muscle begin within the muscle itself and can occur independently of any nerve supply.

FIGURE 42–18 Cardiac muscle. (a) Cardiac muscle as seen with the light microscope (approximately ×900). (b) An electron micrograph of cardiac muscle; A, a band; m, m line; Z, z line; M, mitochondrion; ID, intercalated disc.

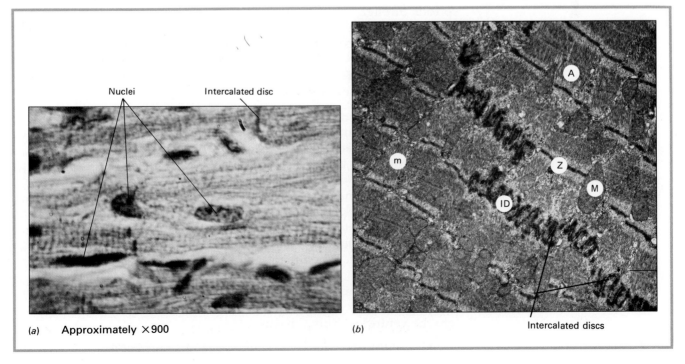

(a)   **Approximately ×900**   (b)

FIGURE 42–19 The conduction system of the heart.

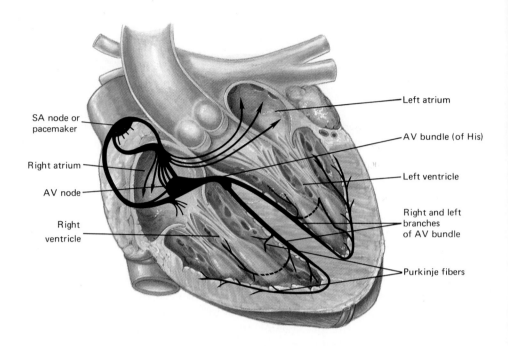

SA node or pacemaker

Right atrium

AV node

Right ventricle

Left atrium

AV bundle (of His)

Left ventricle

Right and left branches of AV bundle

Purkinje fibers

To ensure that the heart beats in a regular and effective rhythm, there is a specialized conduction system. Each beat is initiated by the **pacemaker,** called the **sinoatrial (SA) node** (Figure 42–19). This is a small mass of specialized cardiac muscle in the posterior wall of the right atrium near the opening of the superior vena cava. Ends of the SA node fibers fuse with surrounding ordinary atrial muscle fibers so that the action potential spreads through the atria, producing atrial contraction.

One group of atrial muscle fibers conducts the action potential directly to the **atrioventricular (AV) node** located in the right atrium along the lower part of the septum. Here transmission is delayed briefly, permitting the atria to complete their contraction before the ventricles begin to contract. From the AV node the action potential spreads into specialized muscle fibers called **Purkinje fibers.** These large fibers make up the **atrioventricular (AV) bundle.** The AV bundle then divides, sending branches into each ventricle. When an impulse reaches the ends of the Purkinje fibers it spreads through the ordinary cardiac muscle fibers of the ventricles.

Each minute the heart beats about 70 times. One complete heart beat takes about 0.8 second and is referred to as a **cardiac cycle.** That portion of the cycle in which contraction occurs is known as **systole;** the period of relaxation is **diastole.** Figure 42–20 shows the sequence of events that occur during one cardiac cycle.

HEART SOUNDS.   When you listen to the heartbeat with a stethoscope you can hear two main heart sounds, lub-dup, which repeat rhythmically. The first heart sound, **lub,** is low-pitched, not very loud, and of fairly long duration. It is caused mainly by the closing of the AV valves and marks the beginning of ventricular systole. The lub sound is quickly followed by the higher-pitched, louder, sharper, and shorter **dup** sound. Heard almost as a quick snap, the dup marks the closing of the semilunar valves and the beginning of ventricular diastole.

The quality of these sounds tells a discerning physician much about the state of the valves. When the semilunar valves are injured, a soft hissing noise ("lub-shhh") is heard in place of the normal sound. This is known as a **heart murmur** and may be caused by any injury that has affected the valves so that they do not close tightly, permitting blood to flow backward into the ventricles during diastole.

ELECTROCARDIOGRAMS.   As each wave of contraction spreads through the

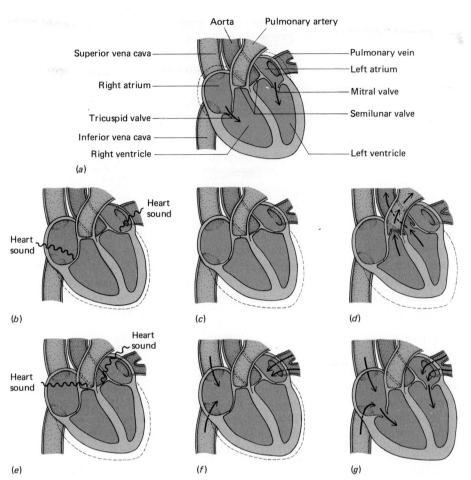

FIGURE 42–20 The cardiac cycle. Arrows indicate the direction of blood flow; dotted lines indicate the change in size as contraction occurs. (*a*) Atrial systole: Atria contract, and blood is pushed through the open tricuspid (right AV) and mitral valves into the ventricles. The semilunar valves are closed. (*b*) Beginning of ventricular systole: Ventricles begin to contract, and pressure within the ventricles increases and closes the tricuspid and mitral valves, causing the first heart sound. (*c*) Period of rising pressure. (*d*) The semilunar valves open when the pressure within the ventricle exceeds that in the arteries, and blood spurts into the aorta and pulmonary artery. (*e*) Beginning of ventricular diastole: When the pressure in the relaxing ventricles drops below that in the arteries, the semilunar valves snap shut, causing the second heart sound. (*f*) Period of falling pressure: Blood flows from the veins into the relaxed atria. (*g*) The tricuspid and mitral valves open when the pressure in the ventricles falls below that in the atria; blood then flows into the ventricles.

heart, electrical currents spread into the tissues surrounding the heart and onto the body surface. By placing electrodes on the body surface on opposite sides of the heart, the electrical activity can be amplified and recorded by either an oscilloscope or an electrocardiograph. The written record produced is called an **electrocardiogram,** or ECG (Figure 42–21). Malfunctioning of the heart causes abnormal action currents, which in turn produce an abnormal ECG.

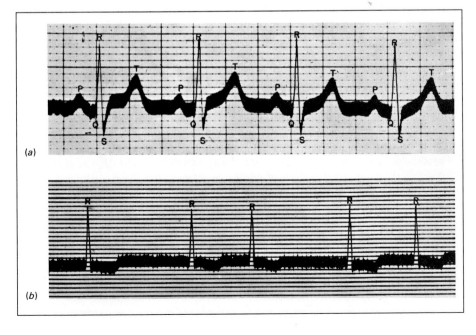

FIGURE 42–21 Electrocardiograms. (*a*) Tracing from a normal heart. The P wave corresponds to the contraction of the atria, the QRS complex to the contraction of the ventricle, and the T wave to the relaxation of the ventricle. (*b*) Tracing from a patient with atrial fibrillation. The individual muscle fibers of the atrium twitch rapidly and independently. There is no regular atrial contraction and no P wave. The ventricles beat independently and irregularly, causing the QRS wave to appear at irregular intervals.

CARDIAC OUTPUT.   The volume of blood pumped by one ventricle during one beat is called the **stroke volume.** By multiplying the stroke volume by the number of times the ventricle beats in one minute, the cardiac output can be computed. In other words, the **cardiac output** is the volume of blood pumped by one ventricle in one minute. For example, in a resting adult the heart may beat about 72 times per minute and pump about 70 ml of blood with each contraction.

Cardiac output = Stroke volume × Heart rate

(number of ventricular
contractions per minute)

$$= \frac{70 \text{ ml}}{\text{stroke}} \times \frac{72 \text{ strokes}}{\text{minutes}}$$

= 5040 ml/min (or a little more than 5 liters/min)

The cardiac output varies dramatically with the changing needs of the body. During stress or heavy exercise, the normal heart can increase its cardiac output four- to fivefold, so that up to 25 liters of blood can be pumped per minute. Cardiac output varies with changes in either stroke volume or heart rate.

Regulation of Stroke Volume.   Stroke volume depends mainly upon venous return, the amount of blood delivered to the heart by the veins. According to **Starling's law of the heart,** the greater the amount of blood delivered to the heart by the veins, the more blood the heart pumps (within physiological limits). When extra amounts of blood fill the heart chambers, the cardiac muscle fibers are stretched to a greater extent and contract with greater force, pumping a larger volume of blood into the arteries. This increase in stroke volume increases the cardiac output (Figure 42–22).

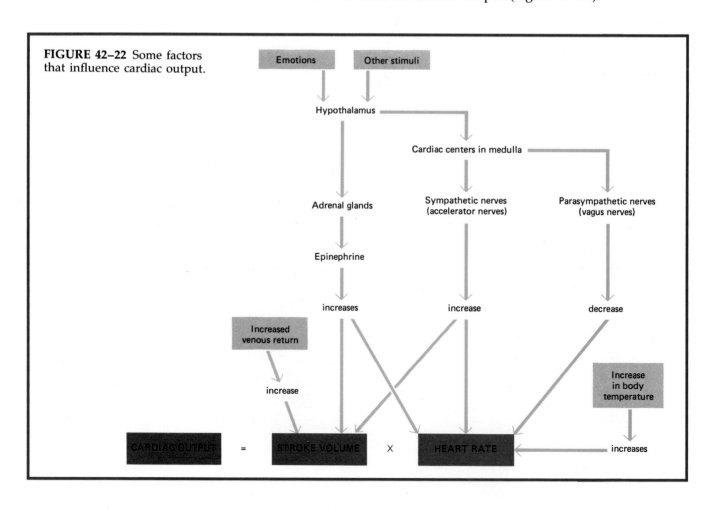

FIGURE 42–22 Some factors that influence cardiac output.

The release of **norepinephrine** by sympathetic nerves also increases the force of contraction of the cardiac muscle fibers. **Epinephrine** released by the adrenal glands during stress has a similar effect on the heart muscle. When the force of contraction increases, the stroke volume increases, and this in turn increases cardiac output.

Regulation of Heart Rate. Although the heart is capable of beating independently of its control systems, its rate is, in fact, carefully regulated by the nervous system. Sensory receptors in the walls of certain blood vessels and heart chambers are sensitive to changes in blood pressure. When stimulated they send messages to **cardiac centers** in the medulla of the brain. These cardiac centers maintain control over two sets of autonomic nerves that pass to the SA node. Sympathetic nerves release norepinephrine, which speeds the heart rate and increases the strength of contraction. Parasympathetic nerves release **acetylcholine,** which slows the heart and decreases the force of each contraction.

Hormones also influence heart rate. During stress, the adrenal glands release epinephrine and norepinephrine, which speed the heart. An elevated body temperature can greatly increase heart rate; during fever, the heart may beat more than 100 times per minute. As you might expect, heart rate decreases when body temperature is lowered. This is why a patient's temperature may be deliberately lowered during heart surgery.

Pulse and Blood Pressure

You have probably felt your pulse by placing a finger over the radial artery in your wrist or over the carotid artery in the neck. Arterial **pulse** is the alternate expansion and recoil of an artery. Each time the left ventricle pumps blood into the aorta, the elastic wall of the aorta expands to accommodate the blood. This expansion moves down the aorta and the arteries that branch from it in a wave (Figure 42–23). When the wave passes, the elastic arterial wall snaps back to its normal size. Every time the heart contracts, a pulse wave begins, so the number of pulsations you count in an artery per minute indicates the number of heartbeats.

Blood pressure is the force exerted by the blood against the inner walls of the blood vessels. It is determined by the blood flow and the resistance to that flow. **Blood flow** depends directly upon the pumping action of the heart. When cardiac output increases, blood flow increases, causing a rise

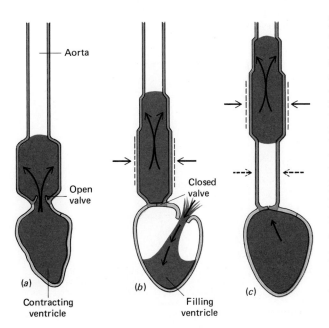

(a) Aorta — Open valve — Contracting ventricle

(b) Closed valve — Filling ventricle

(c)

FIGURE 42–23 The movement of blood from the ventricle through the elastic arteries. For simplicity, only one ventricle and artery are shown, and the amount of stretching of the arterial wall is exaggerated. (*a*) As the ventricle contracts, blood is forced through the semilunar valves, and the adjacent wall of the aorta is stretched. (*b*) As the ventricle relaxes and begins to fill for the next stroke, the semilunar valve closes and the expanded part of the aorta contracts, causing the adjacent part of the aorta to expand as it is filled with blood. (*c*) The pulse wave of expansion and contraction is transmitted to the next adjoining section of the aorta.

in blood pressure. When cardiac output decreases, blood flow decreases, causing a fall in blood pressure. The volume of blood flowing through the system also affects blood pressure. If blood volume is reduced by hemorrhage or by chronic bleeding, the blood pressure drops. On the other hand, an increase in blood volume results in an increase in blood pressure. For example, a high dietary intake of salt causes retention of water. This may result in an increase of blood volume and lead to higher blood pressure.

Blood flow is impeded by resistance; when the resistance to flow increases, blood pressure rises. **Peripheral resistance** is the resistance to blood flow caused by viscosity and by friction between the blood and the wall of the blood vessel. In the blood of a healthy person, viscosity remains fairly constant and is only a minor factor influencing changes in blood pressure. More important is the friction between the blood and the wall of the blood vessel. The length and diameter of a blood vessel determine the surface area of the vessel in contact with the blood. The length of a blood vessel does not change, but the diameter, especially of an arteriole, does. A small change in the diameter of a blood vessel causes a big change in blood pressure. The resistance is inversely proportional to the fourth power of the vessel radius. Thus if the radius is doubled, the resistance would be dramatically reduced to one sixteenth of its former value, and the flow would increase 16-fold.

Blood pressure in arteries rises during systole and falls during diastole (Figure 42–24). Normal blood pressure for a young adult is about 120/80 millimeters of mercury, abbreviated mm Hg, as measured by the sphygmomanometer (Figure 42–25). Systolic pressure is indicated by the numerator, diastolic by the denominator.

When the diastolic pressure consistently measures more than 95 mm Hg, the patient may be suffering from high blood pressure, or **hypertension.** In hypertension, there is usually increased vascular resistance, especially in the arterioles and small arteries. Workload of the heart is increased because it must pump against this increased resistance. As a result, the left ventricle increases in size and may begin to deteriorate in function. Heredity, obesity, and possibly high dietary salt intake are thought to be important factors in the development of hypertension.

As you might guess, blood pressure is greatest in the large arteries and decreases as blood flows through the smaller arteries and capillaries (Figure 42–26). By the time blood enters the veins, its pressure is very low, even approaching zero. Flow rate can be maintained in veins at low pressure because they are low-resistance vessels. Their diameter is larger than

FIGURE 42–24 The changes in blood pressure in an artery as the heart beats.

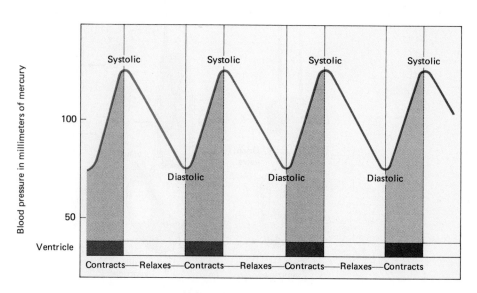

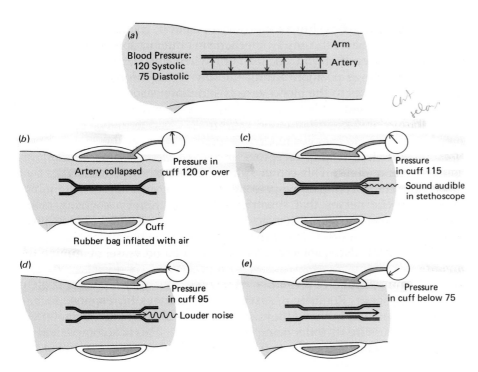

FIGURE 42–25 The measurement of blood pressure by the sphygmomanometer in a young healthy adult. (*a*) Blood pushes against the arterial walls. (*b*) When the pressure in the rubber cuff is 120 mm Hg, the artery is collapsed and blood cannot flow through. (*c*) With the pressure in the cuff just below 120 mm Hg, the artery is collapsed except for a small period during systole, when a small amount of blood squirts through, producing a sound audible in the stethoscope. (*d*) With the pressure in the cuff at 95 mm Hg, the artery is open for a longer time during systole, a greater amount of blood passes through, and a louder noise is produced. (*e*) When the pressure in the cuff drops below 75 mm Hg, the artery is open continuously, blood passes through continuously, and no noise is heard.

that of corresponding arteries, and there is little smooth muscle in their walls. As described earlier, flow of blood through veins depends upon several factors, including muscular movement, which compresses veins, and valves, which prevent backflow.

When a person stands perfectly still for a long time, as when a soldier stands at attention, blood tends to pool in the veins. This is so because when fully distended with blood, the veins can accept no more blood from the capillaries. Pressure in the capillaries increases, and large amounts of plasma are forced out of the circulation through the thin capillary walls. Within just a few minutes, as much as 20% of the blood volume can be lost from the circulation in this way—with drastic effect. Arterial blood pressure falls dramatically, so that blood flow to the brain is reduced. Sometimes the resulting lack of oxygen in the brain causes fainting, a protective response aimed at increasing blood supply to the brain. Lifting a person who has fainted to an upright position can result in circulatory shock and even death.

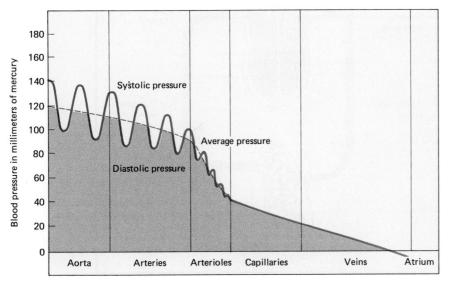

FIGURE 42–26 Blood pressure in different types of blood vessels of the body. The systolic and diastolic variations in arterial blood pressures are shown. Note that the venous pressure drops below zero (below atmospheric pressure) near the heart.

Each time you get up from a horizontal position, changes occur in your blood pressure. Several complex mechanisms interact to maintain normal blood pressure so that you do not faint when you get out of bed each morning or change position during the day. When blood pressure falls, sympathetic nerves to the blood vessels stimulate vasoconstriction so that pressure rises again.

The **baroreceptors** present in the walls of certain arteries and in the heart wall are sensitive to changes in blood pressure. When an increase in blood pressure stretches the baroreceptors, messages are sent to the **vasomotor center** in the medulla. This center stimulates parasympathetic nerves that slow the heart, lowering blood pressure. Also, the vasomotor center inhibits sympathetic nerves that constrict arterioles, thereby lowering blood pressure. These neural reflexes act continuously to maintain blood pressure.

Hormones are also involved in regulating blood pressure. The **angiotensins** are a group of hormones that act as powerful vasoconstrictors. The enzyme **renin** stimulates formation of angiotensins from a plasma protein (angiotensinogen). Renin is released by the kidneys in response to low blood pressure within the kidneys. The kidneys also act *indirectly* to maintain blood pressure by influencing blood volume. This is accomplished by hormonal regulation of the rate at which salt and water are excreted.

The Pattern of Circulation

One of the main jobs of the circulation is to bring oxygen to all of the cells of the body. In humans, as in other mammals and in birds, blood is charged with oxygen in the lungs. Then it is returned to the heart to be

FIGURE 42–27 Highly simplified diagram showing the pattern of circulation through the systemic and pulmonary circuits. Red represents oxygenated blood; blue represents deoxygenated blood.

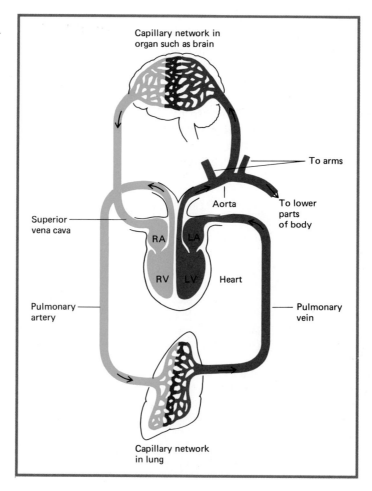

pumped out into the arteries that deliver it to the other tissues and organs of the body. There is a double circuit of blood vessels—(1) the **pulmonary circulation,** which connects the heart and lungs, and (2) the **systemic circulation,** which connects the heart with all of the tissues of the body. This general pattern of circulation may be traced in Figure 42–27.

Pulmonary Circulation

Blood from the tissues returns to the right atrium of the heart partly depleted of its oxygen supply. This oxygen-poor blood, loaded with carbon dioxide, is pumped by the right ventricle into the pulmonary circulation. As it emerges from the heart, the large **pulmonary trunk** branches to form the two **pulmonary arteries,** one going to each lung. These are the only arteries in the body that carry oxygen-poor blood. In the lungs the pulmonary arteries branch into smaller and smaller vessels, which finally give rise to extensive networks of **pulmonary capillaries** that bring blood to all of the air sacs of the lung. As blood circulates through the pulmonary capillaries, carbon dioxide diffuses out of the blood and into the air sacs. Oxygen from the air sacs diffuses into the blood so that, by the time blood enters the **pulmonary veins** leading back to the left atrium of the heart, it is charged with oxygen. Pulmonary veins are the only veins in the body that carry blood rich in oxygen.

In summary, blood flows through the pulmonary circulation in the following sequence:

right atrium → right ventricle → pulmonary artery →
pulmonary capillaries (in lung) → pulmonary vein → left atrium

Systemic Circulation

Blood entering the systemic circulation is pumped by the left ventricle into the **aorta,** the largest artery of the body. Arteries that branch off from the aorta conduct blood to all of the regions of the body. Some of the principal branches include the **coronary arteries** to the heart wall itself, the **carotid arteries** to the brain, the **subclavian arteries** to the shoulder region, the **mesenteric artery** to the intestine, the **renal arteries** to the kidneys, and the **iliac arteries** to the legs (Figure 42–28). Each of these arteries gives rise to smaller branches, which in turn give rise to smaller and smaller vessels, somewhat like branches of a tree that divide until they form tiny twigs. Eventually blood flows into the capillary network within each tissue or organ.

Blood returning from the capillary networks within the brain passes through the **jugular veins.** Blood from the shoulders and arms drains into the **subclavian veins.** These veins and others returning blood from the upper portion of the body merge to form the **superior vena cava,** a very large vein that empties blood into the right atrium. **Renal veins** from the kidneys, **iliac veins** from the lower limbs, **hepatic veins** from the liver, and other veins from the lower portion of the body return blood to the **inferior vena cava,** which delivers blood to the right atrium.

As an example of blood circulation through the systemic circuit, let us trace a drop of blood from the heart to the right leg and back to the heart:

left atrium → left ventricle → aorta → right common iliac
artery → smaller arteries in leg → capillaries in right
leg → small vein in leg → right common iliac vein →
inferior vena cava → right atrium → right ventricle
→into pulmonary circulation

CORONARY CIRCULATION. The heart muscle is not nourished by the blood within its chambers, because its walls are too thick for nutrients and oxy-

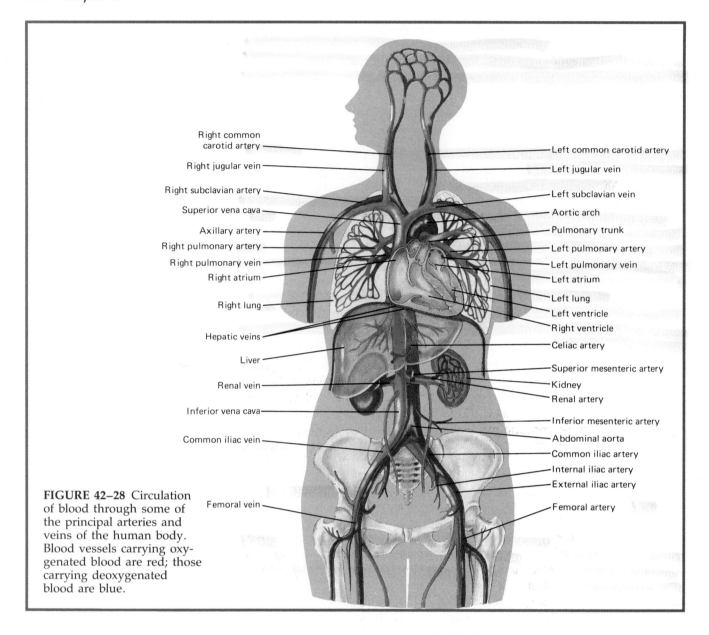

FIGURE 42–28 Circulation of blood through some of the principal arteries and veins of the human body. Blood vessels carrying oxygenated blood are red; those carrying deoxygenated blood are blue.

Right common carotid artery
Right jugular vein
Right subclavian artery
Superior vena cava
Axillary artery
Right pulmonary artery
Right pulmonary vein
Right atrium
Right lung
Hepatic veins
Liver
Renal vein
Inferior vena cava
Common iliac vein
Femoral vein

Left common carotid artery
Left jugular vein
Left subclavian vein
Aortic arch
Pulmonary trunk
Left pulmonary artery
Left pulmonary vein
Left atrium
Left lung
Left ventricle
Right ventricle
Celiac artery
Superior mesenteric artery
Kidney
Renal artery
Inferior mesenteric artery
Abdominal aorta
Common iliac artery
Internal iliac artery
External iliac artery
Femoral artery

gen to diffuse through them to all of the cells. Instead, the cardiac muscle is supplied by coronary arteries branching from the aorta at the point where that vessel leaves the heart. These arteries branch, giving rise to a network of blood vessels within the wall of the heart. Nutrients and gases are exchanged through the coronary capillaries. Blood from these capillaries flows into coronary veins, which join to form a large vein, the **coronary sinus.** The coronary sinus empties directly into the right atrium; it does not join either of the venae cava.

When one of the coronary arteries is blocked, the cells in the area of the heart muscle served by that artery are deprived of oxygen and nutrients, and die. The affected muscle stops contracting; if sufficient cardiac muscle is affected, the heart may stop beating entirely. This is a common cause of "heart attack" and is often the result of atherosclerosis (see Focus on Cardiovascular Disease).

CIRCULATION TO THE BRAIN.   Four arteries—two internal carotid arteries and two vertebral arteries (branches of the subclavian arteries)—deliver blood to the brain. At the base of the brain, branches of these arteries form an arterial circuit called the **circle of Willis.** In the event that one of the arteries serving the brain becomes blocked or injured in some way, this arterial circuit helps ensure blood delivery to the brain cells via other vessels. Blood from the brain returns to the superior vena cava by way of the internal jugular veins at either side of the neck.

HEPATIC PORTAL SYSTEM. Blood almost always travels from artery to capillary to vein. An exception to this sequence is made by the **hepatic portal system,** which delivers blood rich in nutrients to the liver. Blood is conducted to the small intestine by the superior mesenteric artery. Then, as it flows through capillaries within the wall of the intestine, blood picks up glucose, amino acids, and other nutrients. This blood passes into the mesenteric vein and then into the **hepatic portal vein.** Instead of going directly back to the heart (as most veins would), the hepatic portal vein delivers nutrients to the liver. Within the liver, this vein gives rise to an extensive network of tiny blood sinuses. As blood courses through the hepatic sinuses, liver cells remove nutrients and store them. Eventually liver sinuses merge to form hepatic veins, which deliver blood to the inferior vena cava. The hepatic portal vein contains blood that, although it is laden with food materials, has already given up some of its oxygen to the cells of the intestinal wall. Oxygenated blood is supplied to the liver by the hepatic artery.

The Lymphatic System

In addition to the blood circulatory system, vertebrates have a subsystem of the circulatory system known as the **lymphatic system** (Figure 42–29). The lymphatic system has three important functions: (1) to collect and return interstitial fluid to the blood, (2) to defend the body against disease organisms by way of immune mechanisms, and (3) to absorb lipids from the digestive tract. In this section we will focus upon the first function. Immunity is discussed in Chapter 43 and lipid absorption is discussed in Chapter 40.

Design of the Lymphatic System

The lymphatic system consists of (1) an extensive network of lymphatic vessels that conduct **lymph,** the clear, watery fluid formed from interstitial fluid, and (2) **lymph tissue,** a type of connective tissue that has large numbers of lymphocytes. Lymph tissue is organized into small masses of tissue called **lymph nodes** and **lymph nodules.** Tonsils and adenoids are examples of lymph tissue. The **thymus gland** and **spleen,** which consist mainly of lymph tissue, are also part of the lymphatic system.

Tiny "dead-end" capillaries of the lymphatic system extend into almost all of the tissues of the body (Figure 42–30). Lymph capillaries converge to form larger **lymphatics** (lymph veins). There are no lymph arteries.

Interstitial fluid enters lymph capillaries and then is conveyed into lymphatics. At certain locations the lymphatics empty into lymph nodes, where lymph is filtered. Bacteria and other harmful matter are removed from the lymph. The lymph then flows into lymphatics leaving the lymph node. Lymphatics from all over the body conduct lymph toward the shoulder region. These vessels join the circulatory system at the base of the subclavian veins by way of ducts—the **thoracic duct** on the left side, and the **right lymphatic duct** on the right.

Some nonmammalian vertebrates such as the frog have lymph "hearts," which pulsate and squeeze lymph along. However, in mammals the walls of the lymph vessels themselves pulsate, pushing lymph along. Valves within the lymph vessels prevent the lymph from flowing backward. Lymph flow is increased when anything compresses body tissue, for this has the effect of squeezing the lymph vessels. When muscles contract or when arteries pulsate, for example, pressure on the lymph vessels enhances the flow of lymph. The rate at which lymph flows is slow and

FIGURE 42–29 The lymphatic system. Note that while the lymphatic vessels extend into most tissues of the body, the lymph nodes are clustered in certain regions. The right lymphatic duct drains lymph from the upper right quadrant of the body. The thoracic duct drains lymph from other regions of the body.

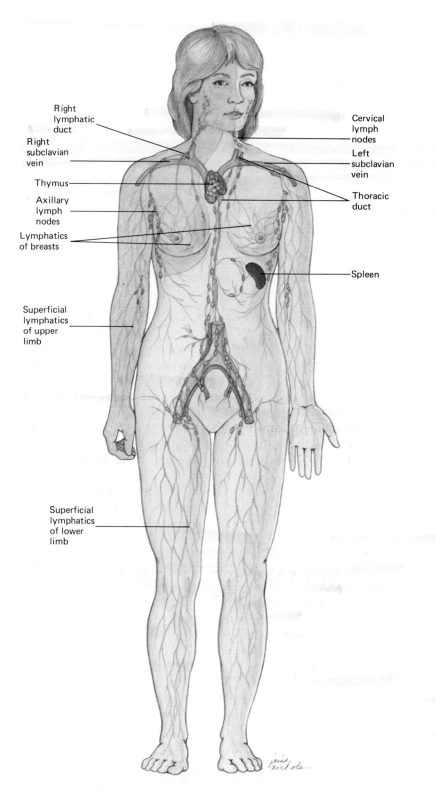

Right lymphatic duct

Right subclavian vein

Thymus

Axillary lymph nodes

Lymphatics of breasts

Superficial lymphatics of upper limb

Superficial lymphatics of lower limb

Cervical lymph nodes

Left subclavian vein

Thoracic duct

Spleen

variable, but the total lymph flow is about 100 ml per hour—very much slower than the 5 liters per minute of blood flowing in the vascular system.

Role in Fluid Homeostasis

Recall that when blood enters a capillary network it is under rather high pressure, so that some plasma is forced out of the capillaries and into the tissues. Once it leaves the blood vessels, this fluid is called interstitial fluid, or tissue fluid. It is somewhat similar to plasma but contains no red blood

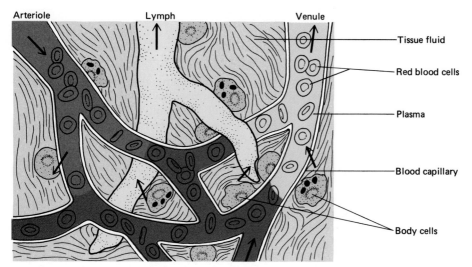

FIGURE 42–30 The relation of lymph capillaries to blood capillaries and tissue cells. Note that blood capillaries are connected to vessels at both ends, whereas lymph capillaries, outlined in color, are "dead-end streets" and contain no erythrocytes. The arrows indicate direction of flow.

cells or platelets, and only a few white blood cells. Its protein content is about one fourth of that found in plasma. This is because proteins are too large to pass easily through capillary walls. Smaller molecules dissolved in the plasma do pass out with the fluid leaving the blood vessels. Thus, interstitial fluid contains glucose, amino acids, other nutrients, and oxygen, as well as a variety of salts. This nourishing fluid bathes all the cells of the body.

The main force pushing plasma out of the blood is hydrostatic pressure, that is, the pressure exerted by the blood on the capillary wall, which is caused by the beating of the heart (Figure 42–31). At the venous ends of the capillaries the hydrostatic pressure is much lower. Here the principal force is the osmotic pressure of the blood, which acts to draw fluid back into the capillary. However, osmotic pressure is not entirely effective. Protein does not return effectively into the venous capillaries and instead

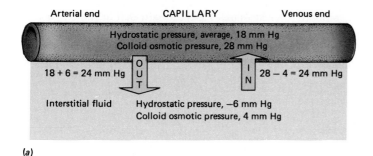

(a)

FIGURE 42–31 Materials are exchanged between blood and tissue fluid through the thin walls of the capillaries. (*a*) Hydrostatic and osmotic pressures are responsible for the exchange of materials between blood and tissue fluid. (*b*) The path by which water flows in and out of capillaries to the tissue fluid.

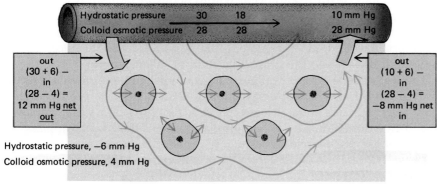

(b)

Focus on CARDIOVASCULAR DISEASE

Cardiovascular disease is the number one cause of death in the United States and in most other industrial societies. Most often death results from some complication of **atherosclerosis*** (hardening of the arteries as a result of lipid deposition). Although atherosclerosis can affect almost any artery, the disease most often develops in the aorta and in the coronary and cerebral arteries. When it occurs in the cerebral arteries it can lead to a **cerebrovascular accident (CVA),** commonly referred to as a stroke.

Although there is apparently no single cause of atherosclerosis, several major risk factors have been identified:

1. Elevated levels of cholesterol in the blood, often associated with diets rich in total calories, total fats, saturated fats, and cholesterol.
2. Hypertension. The higher the blood pressure, the greater the risk.
3. Cigarette smoking. The risk of developing atherosclerosis is two to six times greater in smokers than nonsmokers and is directly proportional to the number of cigarettes smoked daily.
4. Diabetes mellitus, an endocrine disorder in which glucose is not metabolized normally.

The risk of developing atherosclerosis also increases with age. Estrogen hormones are thought to offer some protection in women until after menopause, when the concentration of these hormones decreases. Other suggested risk factors that are currently being studied are obesity, hereditary predisposition, lack of exercise, stress and behavior patterns, and dietary factors such as excessive intake of salt or refined sugar.

In atherosclerosis, lipids are deposited in the smooth muscle cells of the arterial wall. Cells in the arterial wall

*Atherosclerosis is the most common form of arteriosclerosis, any disorder in which arteries lose their elasticity.

proliferate and the inner lining thickens. More lipid, especially cholesterol from low-density lipoproteins, accumulates in the wall. Eventually calcium is deposited there, contributing to the slow formation of hard plaque. As the plaque develops, arteries lose their ability to stretch when they fill with blood, and they become progressively occluded (blocked), as shown in the figure. As the artery narrows, less blood can pass through to reach the tissues served by that vessel and the tissue may become **ischemic** (lacking in blood). Under these conditions the tissue is deprived of an adequate oxygen supply.

When a coronary artery becomes narrowed, **ischemic heart disease** can occur. Sufficient oxygen may reach the heart tissue during normal activity, but the increased need for oxygen during exercise or emotional stress results in the pain known as **angina pectoris.** Persons with this condition often carry nitroglycerin pills with them for use during an attack. This drug dilates veins so that venous return is reduced. Cardiac output is lowered so that the heart is not working so hard and requires less oxygen. Nitroglycerin also dilates the coronary arteries slightly, allowing more blood to reach the heart muscle.

Myocardial infarction (MI) is the very serious, often fatal, form of ischemic heart disease that often results from a sudden decrease in coronary blood supply. The portion of cardiac muscle deprived of oxygen dies within a few minutes and is then referred to as an **infarct.** The term myocardial infarction is used as a synonym for heart attack. MI is the leading cause of death and disability in the United States. Just what triggers the sudden decrease in blood supply that causes MI is a matter of some debate. It is thought that in some cases an episode of ischemia triggers a fatal arrhythmia such as **ventricular fibrillation,** a condition in which the ventricles contract very rapidly without actually pumping blood. In other cases, a **thrombus** (clot) may form in a diseased

tends to accumulate in the interstitial fluid, and not as much fluid is returned to the circulation as escapes. These potential problems are so serious that fluid balance in the body would be significantly disturbed within a few hours and death would occur within about 24 hours if it were not for the lymphatic system.

The lymphatic system preserves fluid balance by collecting about 10% of the interstitial fluid and the protein that accumulates in it. Once it enters the lymph capillaries, the interstitial fluid is called lymph.

The walls of the lymph capillaries are composed of endothelial cells that overlap slightly. When interstitial fluid accumulates, it presses against these cells, pushing them inward like tiny swinging doors that can swing in only one direction. As fluid accumulates within the lymph capillary, the cell doors are pushed closed.

Obstruction of the lymphatic vessels can lead to **edema,** a swelling that results from excessive accumulation of interstitial fluid. Lymphatic vessels can be blocked as a result of injury, inflammation, surgery, or parasitic

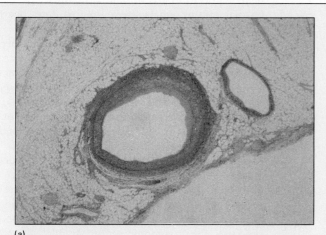

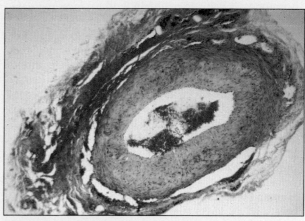

(a)

(b)

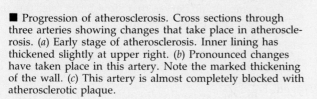

■ Progression of atherosclerosis. Cross sections through three arteries showing changes that take place in atherosclerosis. (a) Early stage of atherosclerosis. Inner lining has thickened slightly at upper right. (b) Pronounced changes have taken place in this artery. Note the marked thickening of the wall. (c) This artery is almost completely blocked with atherosclerotic plaque.

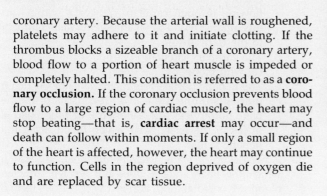

coronary artery. Because the arterial wall is roughened, platelets may adhere to it and initiate clotting. If the thrombus blocks a sizeable branch of a coronary artery, blood flow to a portion of heart muscle is impeded or completely halted. This condition is referred to as a **coronary occlusion.** If the coronary occlusion prevents blood flow to a large region of cardiac muscle, the heart may stop beating—that is, **cardiac arrest** may occur—and death can follow within moments. If only a small region of the heart is affected, however, the heart may continue to function. Cells in the region deprived of oxygen die and are replaced by scar tissue.

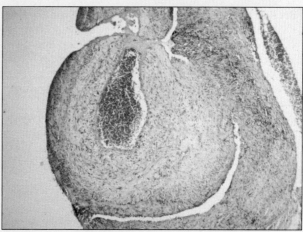

(c)

infection. For example, when a breast is removed (mastectomy) because of cancer, lymph nodes in the underarm region often are also removed in an effort to prevent the spread of cancer cells. The patient's arm may swell greatly because of the disrupted lymph circulation. Fortunately, new lymph vessels develop within a few months and the swelling slowly subsides.

Filariasis, a parasitic infection caused by a larval nematode that is transmitted to humans by mosquitoes, also disrupts lymph flow. The adult worms live in the lymph veins, blocking lymph flow. Interstitial fluid then accumulates, causing tremendous swelling. The term **elephantiasis** has been used to describe the swollen legs sometimes caused by this disease, because they resemble the huge limbs of an elephant (Figure 42–32).

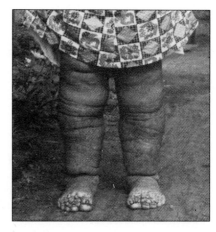

FIGURE 42–32 Lymphatic drainage is blocked in the limbs of this individual because of the parasitic infection known as filariasis. The condition characterized by such swollen limbs is elephantiasis.

■ SUMMARY

I. The circulatory system in most animals transports nutrients, oxygen and carbon dioxide, wastes, and hormones; it helps to maintain fluid balance, regulate body temperature in homeothermic animals, and defend the organism against invading microorganisms.

II. Sponges, cnidarians, ctenophores, flatworms, and nematodes have no specialized circulatory structures.

III. Arthropods and most mollusks have an open circulatory system in which blood flows into a hemocoel.

IV. Other invertebrates and all vertebrates have a closed circulatory system in which blood flows through a continuous network of blood vessels.

V. In the vertebrate circulatory system a muscular heart pumps blood into a system of arteries, capillaries, and veins.

VI. Human blood consists of red blood cells, white blood cells, and platelets suspended in plasma.
   A. Plasma proteins include fibrinogen, globulins, albumin, and lipoproteins.
   B. Red blood cells transport oxygen and aid in carbon dioxide transport.
   C. White blood cells defend the body against disease-causing organisms.
   D. Platelets function in hemostasis.

VII. Arteries carry blood away from the heart; veins return blood to the heart.
   A. Arterioles constrict and dilate to regulate blood pressure and distribution of blood to the tissues.
   B. Capillaries are the exchange vessels with thin walls through which materials pass back and forth between the blood and tissues.
   C. Veins are equipped with valves that prevent backflow of blood.

VIII. In vertebrates the heart consists of one or two atria, which receive blood, and one or two ventricles, which pump blood into the arteries.
   A. The four-chambered hearts of birds and mammals have completely separate right and left sides, which separate oxygenated blood from deoxygenated blood.
   B. The heart is enclosed by a pericardium and is equipped with valves that prevent backflow of blood.
   C. The sinoatrial node initiates each heartbeat. A specialized conduction system ensures that the heart will beat in a coordinated manner.
   D. Cardiac output equals stroke volume times heart rate.
      1. Stroke volume depends upon venous return and upon neural messages and hormones.
      2. Heart rate is regulated mainly by the nervous system but is influenced by hormones, body temperature, and certain other factors.

IX. Pulse is the alternate expansion and recoil of an artery; blood pressure is the force exerted by the blood against the inner walls of the blood vessel.
   A. Blood pressure is greatest in the arteries and decreases as blood flows through the capillaries.
   B. Baroreceptors sensitive to changes in blood pressure send messages to the vasomotor center in the medulla. When informed of an increase in blood pressure, the vasomotor center stimulates parasympathetic nerves that slow heart rate, thereby reducing blood pressure.
   C. Angiotensins raise blood pressure.

X. The pulmonary circulation connects heart and lungs; the systemic circulation connects the heart with all of the tissues.
   A. In the systemic circulation the left ventricle pumps blood into the aorta, which branches into arteries leading to all of the body organs; after flowing through capillary networks within various organs, blood flows into veins that conduct it to the right atrium.
   B. The coronary circulation supplies the heart muscle with blood.
   C. The hepatic portal system circulates blood rich in nutrients through the liver; it consists of the hepatic portal vein and a network of tiny blood sinuses in the liver, from which blood flows into hepatic veins.

XI. Atherosclerosis leads to ischemic heart disease in which the heart muscle does not receive sufficient blood. Myocardial infarction is a very serious form of ischemic heart disease.

XII. The lymphatic system collects interstitial fluid and returns it to the blood. It plays an important role in homeostasis of fluids.

■ POST-TEST

Select the most appropriate answer or answers from Column B for the description in Column A.

Column A	Column B
1. No circulatory system	a. Insect
2. Open circulatory system	b. Bird
3. Closed circulatory system	c. Flatworm
4. Heart with two atria and two ventricles	d. Earthworm
	e. Cnidarian
	f. Clam

5. Hemocyanin is a pigment that transports _____.

6. When the proteins involved in blood clotting are removed from plasma, the remaining fluid is called _____.

7. The _____ _____ fraction of plasma proteins contains many types of antibodies.

8. A deficiency in hemoglobin is referred to as _____.

Select the most appropriate term in Column B to fit the description in Column A.

Column A	Column B
9. Transport oxygen	a. Platelets
10. Principal phagocytic cells in blood	b. Red blood cells
	c. Macrophages
11. Release histamine	d. Basophils
12. Develop from monocytes	e. Neutrophils
13. Initiate clotting	

14. Prothrombin requires vitamin _____ for its production.
15. In the presence of thrombin, fibrinogen is converted to the insoluble protein _____.
16. _____ are blood vessels important in maintaining appropriate blood pressure.
17. Nutrients and other materials are exchanged through the walls of _____.
18. In birds and mammals, blood leaving the right ventricle enters one of the _____ _____.
19. The _____ valves guard the exits of the heart.
20. The portion of the cardiac cycle during which the heart contracts is referred to as _____; the relaxation phase is _____.
21. The cardiac output is the _____.
22. According to Starling's law of the heart, the greater the amount of blood delivered to the heart by the veins, _____.
23. The force exerted by the blood against the inner walls of the blood vessels is known as _____.
24. Blood pressure is determined by _____ _____ and by _____.
25. The largest artery in the human body is the _____.
26. The carotid arteries deliver blood to the _____.
27. The renal veins deliver blood from the _____; the hepatic veins deliver blood from the _____.
28. The term myocardial infarction (MI) is used as a synonym for _____ _____.
29. Baroreceptors are sensitive to changes in _____ _____.

30. The angiotensins are a group of hormones that are powerful _____.
31. Label the diagram below.

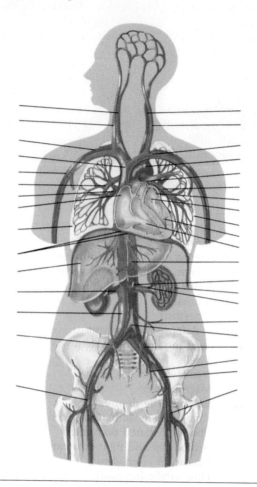

■ REVIEW QUESTIONS

1. List five functions of the circulatory system.
2. Compare the manner in which nutrients and oxygen are transported to the body cells in a hydra, a planarian, an earthworm, a clam, and a frog.
3. Contrast an open with a closed circulatory system.
4. List the functions of the main groups of plasma proteins.
5. Contrast the structure and functions of red and white blood cells.
6. What is anemia? What are the causes of this condition?
7. Summarize the process by which blood clots.
8. Compare the functions of arteries, capillaries, and veins. Why are arterioles important in maintaining homeostasis?
9. Compare the hearts of a fish, frog, reptile, and bird.
10. Define cardiac output, and describe factors which influence it.

11. What is the relationship between blood pressure and peripheral resistance? What mechanisms regulate blood pressure?
12. Trace the path of a red blood cell from the inferior vena cava to the aorta, and from the jugular vein to the kidney.
13. What is the function of the hepatic portal system? How does its sequence of blood vessels differ from that in most other circulatory routes?
14. List four risk factors associated with the development of atherosclerosis. How is atherosclerosis linked to ischemic heart disease? Describe myocardial infarction.
15. How does the lymph system help maintain fluid balance? What is the relationship between plasma, interstitial fluid, and lymph?

■ RECOMMENDED READINGS

Cantin, M., and Genest, J., The heart as an endocrine gland. *Scientific American*, February 1986. In addition to its pumping function, the heart produces a hormone that helps regulate blood pressure and volume.

Doolittle, R.F. Fibrinogen and fibrin. *Scientific American*, December 1981, pp. 126–135. The clotting process can be understood in terms of the structure of fibrinogen.

Silberner, J. Risky business. *Science News,* January 25, 1986. A discussion of gene markers for atherosclerosis.

Zucker, M.B. The functioning of blood platelets. *Scientific American*, June 1980, pp. 86–103. The interactions of substances in blood platelets and substances in blood plasma and in tissue play complex roles in health and disease.

leave the circulation and enter the tissues. As fluid leaves the circulation, it also brings with it needed oxygen and nutrients.

Although inflammation is often a local response, sometimes the entire body is involved. **Fever** is a common clinical symptom of widespread inflammatory response. A peptide called interleukin 1, released by macrophages, somehow resets the body's thermostat in the hypothalamus. Substances known as prostaglandins are also involved in this resetting process. Fever interferes with viral activity and decreases circulating levels of iron. This makes it difficult for microorganisms to obtain needed iron supplies, placing the invaders at a metabolic disadvantage.

Phagocytosis

One of the main functions of inflammation appears to be increased phagocytosis. A phagocyte ingests a bacterium or other invading organism by flowing around it like an amoeba and engulfing it. As it ingests the organism, the cell wraps it within membrane pinched off from the cell membrane. The vesicle containing the bacterium is called a **phagosome.** One or more lysosomes adhere to the phagosome membrane and fuse with it. Under a microscope the lysosomes look like granules, but when they fuse with the phagosome the granules seem to disappear. For this reason, the process is referred to as degranulation. The lysosome releases potent digestive enzymes onto the captured bacterium, and the phagosome membrane releases hydrogen peroxide onto the invader. These substances destroy the bacterium and break down its macromolecules to small, harmless compounds that can be released or even utilized by the phagocyte.

After a neutrophil phagocytizes 20 or so bacteria it becomes inactivated (perhaps by leaking lysosomal enzymes) and dies. A macrophage can phagocytize about 100 bacteria during its lifespan. Can bacteria counteract the body's attack? Certain bacteria are able to release enzymes that destroy the membranes of the lysosomes. The powerful lysosomal enzymes then spill out into the cytoplasm and may destroy the phagocyte. Other bacteria, such as those that cause tuberculosis, possess cell walls or capsules that resist the action of lysosomal enzymes.

Some macrophages wander through the tissue phagocytizing foreign matter and bacteria; when it is appropriate, they release antiviral agents (Figure 43–3). Others stay in one place and destroy bacteria that pass by. For example, air sacs in the lungs contain large numbers of tissue macrophages that destroy foreign matter entering with inhaled air.

Specific Defense Mechanisms

Nonspecific defense mechanisms destroy pathogens and prevent the spread of infection while the specific defense mechanisms are being mobilized. Several days are required to activate specific immune responses, but once in gear, these mechanisms are extremely effective. There are two main types of specific immunity: **cell-mediated immunity,** in which lymphocytes attack the invading pathogen directly, and **antibody-mediated immunity,** in which lymphocytes produce specific antibodies designed to destroy the pathogen. Several types of cells are important in specific immunity.

Cells of the Immune System

As discussed in Chapter 42, several types of mononuclear blood cells can be distinguished under the light microscope. The larger mononuclear cells (10 to 18 micrometers in diameter) containing a kidney-shaped nucleus and some light granules are monocytes. Recall that monocytes develop into

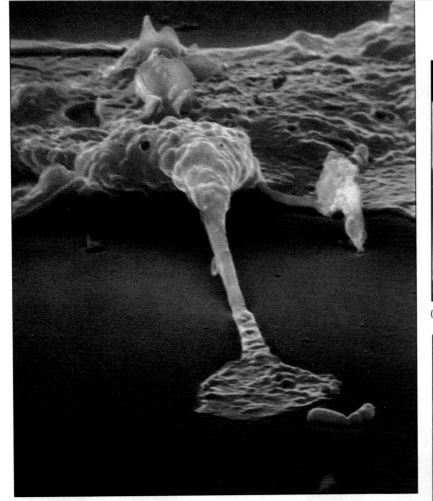

(b)

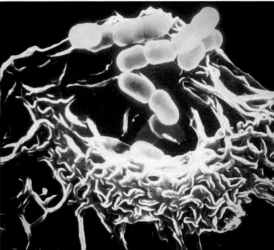

(a)

(c)

FIGURE 43–3 The macrophage is an incredibly efficient warrior. (*a*) A macrophage (grey) extends a pseudopod toward an invading *E. coli* bacterium (green) that is already multiplying. (*b*) The bacterium is trapped within the engulfing pseudopod. (*c*) The macrophage sucks the trapped bacteria in along with its own cell membrane. The macrophage's cell membrane will seal over the bacteria, and powerful lysosomal enzymes will destroy them.

macrophages, large phagocytic cells that engulf and destroy bacteria and other invaders.

The small (6 to 10 micrometers in diameter) mononuclear blood cells are lymphocytes, the main warriors in specific immune responses. The trillion or so lymphocytes are stationed strategically in the lymphatic tissue throughout the body. The agranular mononuclear lymphocytes with a high nuclear to cytoplasmic ratio are **T lymphocytes,** referred to as **T cells,** and **B lymphocytes,** or **B cells.** One population of mononuclear cells has some granules as well as a relatively low nuclear to cytoplasmic ratio. These are called large granular lymphocytes (LGL). The LGLs include the NK cell mentioned earlier, which is important for killing virally infected cells and tumor cells.

T CELLS. Like macrophages and B cells, T cells originate from stem cells in the bone marrow (Figure 43–4). On their way to the lymph tissues, the future T cells stop off in the thymus gland for processing. The T in T lymphocytes stands for thymus-derived; somehow the thymus gland influences the differentiation of lymphocytes capable of immunological response.

T lymphocytes are responsible for cellular immunity (Figure 43–5). On their cell surfaces, T cells have antigen-binding molecules referred to as **T-cell receptors.** There are millions of different T cells, each with specific antibodies. The immune system has the potential to produce millions of different antibodies, each programmed to respond to a different antigenic determinant. Two principal theories explaining this diversity have been debated. The **germ line theory** proposed that we are born with a different gene for the production of every possible antibody variable region. One criticism of this theory has been that a mammal has only about a million genes, and only a small number of these could be devoted to coding antibodies. Thus, we do not have sufficient numbers of genes to account for the millions of different antibodies.

The **somatic mutation theory** held that we inherit only a few types of genes for making antibody variable regions and that mutations occur within the DNA of the lymphocytes, producing thousands of lymphocytes with slightly different variable regions. Each somatically generated V (variable) region gene could combine with a C (constant) region gene to form the immunoglobulin molecule. Thus, each variety of lymphocyte would be programmed to produce certain specific antibodies. These mutations might occur in the thymus or bursa equivalent.

Recent research indicates that the correct explanation is probably a combination of these theories. The ability to make many different antibodies is inherited, but this diversity is probably increased by mutation as well as by recombination. However, how can we hypothesize somatic development of diversity at one end of the molecule (V region) and maintenance of constant amino acid chains at the other (C region)? In 1965 Dreyer and Bennett proposed that two genes, not one, code for a single immunoglobulin chain.

More recent recombinant DNA techniques clearly describe the genes for immunoglobulin production. Several V region genes code for any V region. These genes have spaces between them as well as additional genes for joining V regions to C regions. These spacers and joining sequences increase the chance for recombinational events during transcription. This process occurs in the production of both light and heavy chains. Then, the two chains associate to form the completed antibody.

Thus, generation of immunoglobulin diversity is caused by multiple germ line V genes, recombinational events, and to a lesser extent point somatic mutations. The result to the individual is the existence of thousands of V regions (on B cell surfaces and in serum) that are able to bind to most foreign matter.

Cell-Mediated Immunity

The T cells and monocyte/macrophage system are responsible for cell-mediated immunity. These cells are especially effective in attacking viruses, fungi, and the types of bacteria that live within host cells. How do the T cells know which cells to attack? Once a pathogen invades a body cell, the host cell's macromolecules may be altered. The immune system then regards that cell as foreign, and T cells destroy it. Cytotoxic T cells also destroy cancer cells, and, unfortunately, the cells of transplanted organs.

As in antibody-mediated immunity, activated helper T cells are necessary for a response. Recall that T cells have receptors on their surfaces capable of reacting with antigens on the surfaces of invading cells or altered host cells. As with B cells, only the variety of lymphocyte able to react to the specific antigen presented—that is, the competent lymphocyte—becomes sensitized. However, as discussed earlier, a T cell cannot recognize an antigen if it is presented alone. The antigen must be presented to the T cell as part of a complex with HLA that the T cell recognizes as "self."

Once stimulated, T cells increase in size, proliferate, and give rise to a sizable clone of cytotoxic T cells and memory cells (Figure 43–10). Cytotoxic T cells make up the cellular infantry; they leave the lymph nodes and make

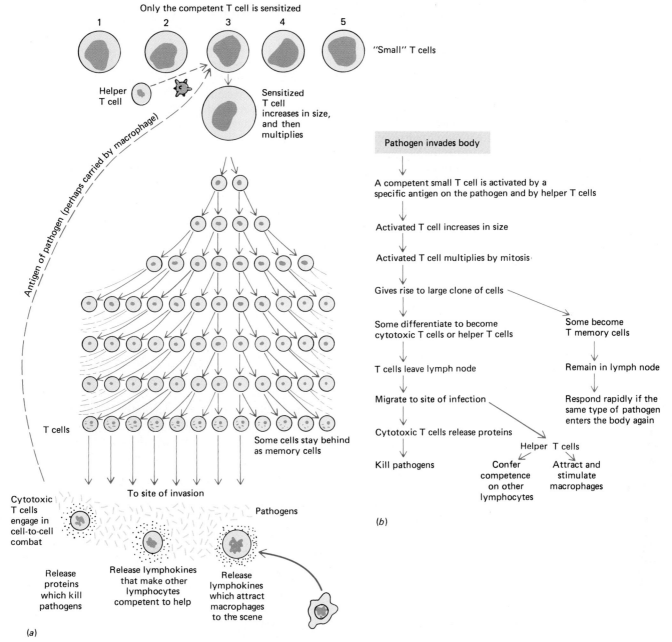

Only the competent T cell is sensitized

1 2 3 4 5 "Small" T cells

Helper T cell

Sensitized T cell increases in size, and then multiplies

Antigen of pathogen (perhaps carried by macrophage)

T cells

Some cells stay behind as memory cells

To site of invasion

Cytotoxic T cells engage in cell-to-cell combat

Pathogens

Release proteins which kill pathogens

Release lymphokines that make other lymphocytes competent to help

Release lymphokines which attract macrophages to the scene

(a)

Pathogen invades body

A competent small T cell is activated by a specific antigen on the pathogen and by helper T cells

Activated T cell increases in size

Activated T cell multiplies by mitosis

Gives rise to large clone of cells

Some differentiate to become cytotoxic T cells or helper T cells

Some become T memory cells

T cells leave lymph node

Remain in lymph node

Migrate to site of infection

Respond rapidly if the same type of pathogen enters the body again

Cytotoxic T cells release proteins

Helper T cells

Kill pathogens

Confer competence on other lymphocytes

Attract and stimulate macrophages

(b)

their way to the infected area. These killer cells can destroy a target cell within seconds after contact. Then, the T cell disengages itself from its victim cell and seeks out a new target cell.

After a cytotoxic T cell combines with antigen on the surface of the target cell, it secretes a powerful group of proteins known as perforins. The T cell releases granules containing these cytotoxic proteins at the site of cell contact, producing lesions in the target cell and rapid lysis. Under some conditions, cytotoxic T cells release soluble proteins called **lymphotoxins,** which are especially toxic for cancer cells.

T cells and macrophages at the site of infection secrete interleukins, interferons, and a variety of other substances that help regulate immune function. Some interleukins confer competence upon other lymphocytes in the area, increasing the ranks of cytotoxic T cells. Other interleukins enhance the inflammatory reaction, attracting great numbers of macrophages to the site of infection. Interleukins and gamma interferon stimulate macrophages, making them more active and effective at destroying pathogens.

FIGURE 43–10 Cell-mediated immunity. When activated by an antigen and by a helper T cell, a competent T cell gives rise to a large clone of cells. Many of these differentiate to become cytotoxic T cells, which migrate to the site of infection. There, they release proteins known as perforins and lymphotoxins that destroy invading pathogens.

First recognized in 1981, **acquired immunodeficiency syndrome (AIDS)** is a deadly disease that is spreading through the population at an alarming rate. More than 40,000 cases of AIDS have been reported in the United States, and the U.S. Public Health Service estimates that about 270,000 individuals will have had AIDS by 1991.

AIDS results from infection with a retrovirus identified as human immunodeficiency virus (HIV). The virus rapidly infects helper T cells, resulting in irreversible defects in immunity. Recall that helper T cells stimulate the proliferation of B and T cells, stimulate macrophages, and perform other functions that enhance the immune response. When the helper T cell population is depressed, the ability to resist infection is severely impaired. AIDS victims die within several months to about 5 years of rare forms of cancer, pneumonia, and other opportunistic infections that pose little threat to individuals with fully functioning immune systems.

About 90% of patients dying of AIDS develop a neuropsychological disorder known as AIDS dementia complex, which results from direct infection of the central nervous system by the retrovirus. AIDS dementia complex is characterized by progressive cognitive, motor, and behavioral dysfunction that typically ends in coma and death.

Current evidence indicates that AIDS is transmitted mainly by semen during sexual intercourse with an infected person or by direct exposure to infected blood or blood products. Those most at risk are homosexual and bisexual men (72% of cases) and intravenous drug users (20%). Individuals such as hemophiliacs who require frequent blood transfusions are also at risk, as are infants born to mothers with AIDS. Effective blood-screening procedures have been developed to safeguard blood bank supplies, so that risk of infection from blood transfusion has been greatly reduced. Use of condoms during sexual intercourse provides some protection against the virus. AIDS is not spread by casual contact. People do not contract the disease by hugging, kissing, sharing a drink, or using the same bathroom facilities. Close friends and family members who live with AIDS patients are not more likely to get the disease.

It is estimated that more than 2 million persons in the United States have antibody to HIV but have no symptoms of the disease. Apparently, not everyone exposed to HIV actually contracts the disease. AIDS may be an opportunistic infection that causes disease in individuals with inadequate immune function. Susceptibility to HIV may depend on a combination of genetic, environmental, and psychosocial factors. The latter include personality variables and coping styles that influence susceptibility to environmental stressors.

Many persons exposed to HIV develop the AIDS related complex (ARC); its symptoms include night sweats, fever, swollen lymph glands, and weight loss. Patients with ARC may eventually develop AIDS, but the extent of that risk is not yet known.

Research laboratories throughout the world are searching for drugs that will successfully combat the AIDS virus. One drug currently being tested (azidothymidine or AZT) blocks HIV replication. AZT blocks the action of reverse transcriptase, the enzyme needed by the retrovirus for incorporation into the host cell's DNA. Because the AIDS virus often infects the central nervous system, an effective drug will have to cross the blood-brain barrier. Research directed at developing a vaccine against AIDS has not yet been successful because the retrovirus mutates rapidly, giving rise to many viral strains.

While immunologists work to develop a successful vaccine, massive educational programs are being developed that aim at slowing the spread of AIDS. Spreading the word that having multiple sexual partners increases the risk for AIDS and teaching sexually active individuals the importance of "safe" sex may help to slow the epidemic. Some have suggested that public health facilities offer free condoms to those who are sexually active and free sterile hypodermic needles to those addicted to drugs. The cost of these measures would be far less than the cost of medical care for increasing numbers of AIDS patients and the toll in human suffering.

■ HIV virus particles (blue) that cause AIDS attack a helper T cell. HIV seriously impairs the immune system by rapidly destroying helper T cells.

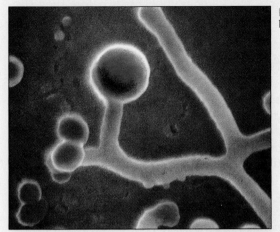

■ HIV virus particles budding from the ends of branched microvilli. Magnification about ×400,000.

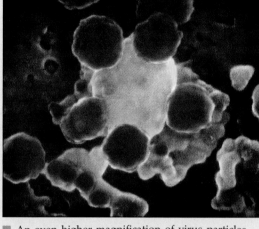

■ An even higher magnification of virus particles budding from a "bleb" (a cytoplasmic extension broader than a microvillus). Note the pentagonal symmetry often evident in other biological branching and flowering structures.

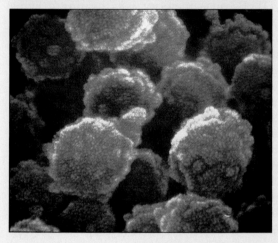

■ HIV virus particles at extremely high magnification (about ×1,200,000). The surfaces are grainy and the outlines slightly blurred because the preparation was coated with coarse-grained heavy metal salt (palladium).

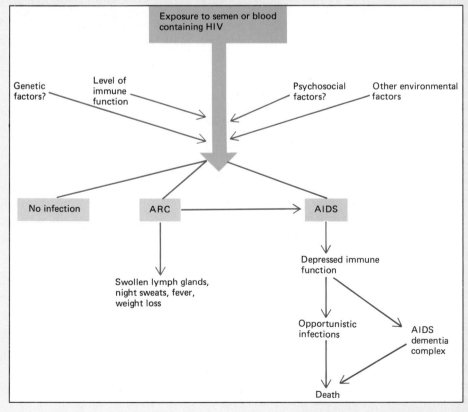

Exposure to semen or blood containing HIV

Genetic factors?

Level of immune function

Psychosocial factors?

Other environmental factors

No infection   ARC   AIDS

Swollen lymph glands, night sweats, fever, weight loss

Depressed immune function

Opportunistic infections

AIDS dementia complex

Death

■ Exposure to semen or blood that contains HIV can lead to AIDS related complex (ARC) or to AIDS. Many factors apparently determine whether a person exposed to AIDS virus will contract the disease. Some individuals do not suffer infection. However, the risk increases with multiple exposures.

Suppressor T cells are stimulated by antigen. These cells help regulate both T cells and B cells. Suppressor T cells multiply more slowly than cytotoxic T cells, so more than a week generally elapses before they suppress an immune response.

Primary and Secondary Responses

The first exposure to an antigen stimulates a **primary response.** Injection of an antigen into an immunocompetent animal causes specific antibodies to appear in the blood plasma in 3 to 14 days. After injection of the antigen there is a brief **latent period,** during which the antigen is recognized and appropriate lymphocytes begin to form clones. Then there is a **logarithmic phase,** during which the antibody concentration rises logarithmically for several days until it reaches a peak (Figure 43–11). IgM is the principal antibody synthesized. Finally, there is a **decline phase,** during which the antibody concentration decreases to a very low level.

A second injection of the same antigen, even years later, evokes a **secondary response** (Figure 43–11). Because memory cells bearing a living record of the encounter with the antigen persist throughout an individual's life, the secondary response is generally much more rapid than the primary response, with a shorter latent period. The amount of antigen necessary to evoke a secondary response is much less than that needed for a primary response. More antibodies are produced than in a primary response, and the decline phase is slower. In a secondary response the predominant antibody is IgG. The **affinity,** or strength of fit, of the antibody also increases following secondary exposure.

The body's ability to launch a rapid, effective response during a second encounter with an antigen explains why we do not usually suffer from the same disease several times. Most persons get measles or chicken pox, for example, only once. When exposed a second time, the immune system destroys the pathogens before they have time to establish themselves and cause symptoms of the disease. Booster shots of vaccine are given in order to elicit a secondary response, reinforcing the immunological memory of the disease-producing antigens.

You may wonder, then, how a person can get flu or a cold more than once. Unfortunately, there are many varieties of these diseases, each

FIGURE 43–11 Primary and secondary responses of antibody formation to successive doses of antigens. Antigen 1 was injected at day 0, and the immune response was assessed by measuring antibody levels to the antigen. At week 4, the primary response had subsided. Antigen 1 was injected again along with a new protein, antigen 2. Note that the secondary response to antigen 1 was greater and more rapid than the primary response. A primary response was made to the newly encountered antigen 2.

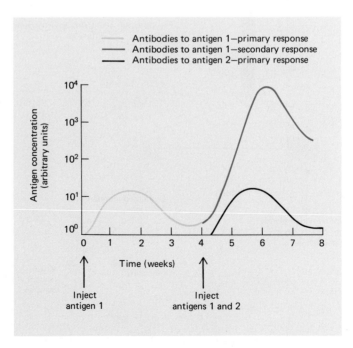

caused by a virus with slightly different antigens. For example, more than 100 different viruses cause the common cold, and new varieties of "cold" and flu virus evolve continuously by mutation (a survival mechanism for them), which may result in changes in their surface antigens. Even a slight change may prevent recognition by memory cells. Because the immune system is so specific, each different antigen is treated by the body as a new immunological challenge.

Active and Passive Immunity

We have been considering **active immunity,** immunity developed following exposure to antigens. After you have had measles as a young child, for example, you develop immunity that protects you from contracting measles again. Active immunity can be *naturally* or *artificially* induced (Table 43–1). If someone with measles sneezes near you and you contract the disease, you naturally develop active immunity. Active immunity can also be artificially induced by **immunization,** that is, by injection of a vaccine. In this case, the body launches an immune response against the antigens contained in the measles vaccine and develops memory cells, so that future encounters with the same pathogen will be dealt with swiftly.

Effective vaccines can be prepared in a number of ways. A virus may be attenuated (weakened) by successive passage through cells of nonhuman hosts. In the process, mutations occur that adapt the pathogen to the nonhuman host, so that it can no longer cause disease in humans. This is how polio vaccine, smallpox vaccine, and measles vaccine are produced. Whooping cough and typhoid fever vaccines are made from killed pathogens that still have the necessary antigens to stimulate an immune response. Tetanus and botulism vaccines are made from toxins secreted by the respective pathogens. The toxin is altered so that it can no longer destroy tissues, but its antigens are still intact. When any of these vaccines is introduced into the body, the immune system actively develops clones, produces antibodies, and develops memory cells.

In **passive immunity,** an individual is given antibodies actively produced by another organism. The serum or gamma globulin containing these antibodies can be obtained from humans or animals. Animal sera are less desirable because their nonhuman proteins can themselves act as antigens, stimulating an immune response that may result in a clinical illness termed serum sickness.

Passive immunity is borrowed immunity, and its effects are not lasting. It is used to boost the body's defense temporarily against a particular dis-

Table 43–1 ACTIVE AND PASSIVE IMMUNITY			
Type of Immunity	*When Developed*	*Development of Memory Cells*	*Duration of Immunity*
Active			
Naturally induced	Pathogens enter the body through natural encounter (e.g., person with measles sneezes on you)	Yes	Many years
Artificially induced	After immunization	Yes	Many years
Passive			
Naturally induced	After transfer of antibodies from mother to developing baby	No	Few months
Artificially induced	After injection with gamma globulin	No	Few months

ease. For example, during the Vietnam War, in areas where hepatitis was widespread, soldiers were injected with gamma globulin containing antibodies to the hepatitis pathogen. Such injections of gamma globulin offer protection for only a few months. Because the body has not actively launched an immune response, it has no memory cells and cannot produce antibodies to the pathogen. Once the injected antibodies wear out, the immunity disappears.

Pregnant women confer natural passive immunity upon their developing babies by manufacturing antibodies for them. These maternal antibodies, of the IgG class, pass through the placenta (the organ of exchange between mother and developing child) and provide the fetus and newborn infant with a defense system until its own immune system matures. Babies who are breast-fed continue to receive immunoglobulins, particularly IgA, in their milk. These immunoglobulins provide considerable immunity to the pathogens responsible for gastrointestinal infection, and perhaps to other pathogens as well.

How the Body Defends Itself Against Cancer

Some immunologists think that a few normal cells are transformed into cancer cells every day in each of us in response to viruses, hormones, radiation, or carcinogens in the environment. Because they are abnormal cells, some of their surface proteins are different from those of normal body cells. Such proteins act as antigens, stimulating an immune response. According to the **theory of immune surveillance,** the body's immune system destroys these abnormal cells whenever they arise. Only when these mechanisms fail do abnormal cells divide rapidly, resulting in cancer.

Every component of the immune system helps defend against cancer cells. Tumor cells exhibit abnormal surface antigens that induce both cellular and antibody-mediated responses. T cells produce interleukins, which attract macrophages and NK cells and activate them. The T cells also produce interferons, which have an antitumor effect. The macrophages themselves produce factors, including TNF (tumor necrosis factor), that inhibit tumor growth.

Cytotoxic T cells, macrophages, and natural killer cells attack cancer cells (Figure 43–12). Natural killer cells are capable of killing tumor cells or virally infected cells upon first exposure to the foreign antigen. Patients with advanced cancer are thought to have lower natural killer cell activity than normal persons.

What prevents killer T cells, macrophages, and natural killer cells from effectively destroying cancer cells in some persons? The immune system cells may fail to recognize the cancer cells as foreign, or they may recognize them but be unable to destroy them. Sometimes the presence of cancer cells stimulates B cells to produce IgG antibodies that combine with antigens on the surfaces of the cancer cells. These **blocking antibodies** may block the T cells so that they are unable to adhere to the surface of the cancer cells and destroy them. For some unknown reason, the blocking antibodies are not able to activate the complement system that would destroy the cancer cells. Interestingly, the presence of antibodies in this case is harmful.

An exciting approach in cancer research involves the production of **monoclonal antibodies.** In this procedure, mice are injected with antigens from human cancer cells. After the mice have produced antibodies to the cancer cells, their spleens are removed and cells containing the antibodies are extracted from this tissue. These cells are fused with cancer cells from other mice. Because of the apparently unlimited ability of cancer cells to divide, these fused hybrid cells will continue to divide indefinitely. Researchers select hybrid cells that are manufacturing the specific antibodies

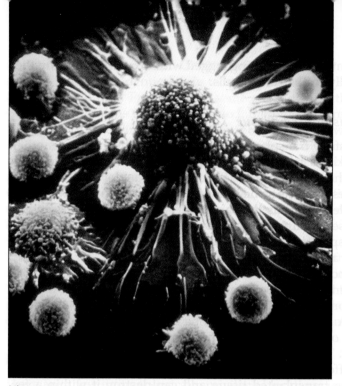

(a)

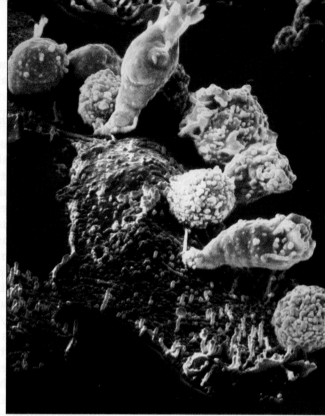

(b)

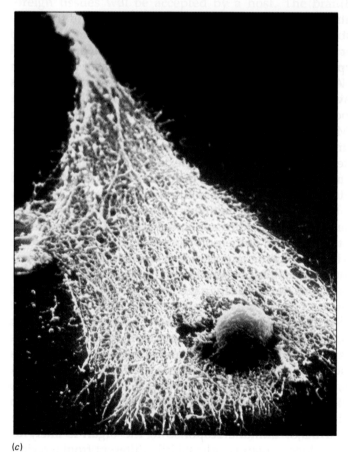

(c)

FIGURE 43–12 Cytotoxic T cells defend the body against cancer cells. (*a*) An army of cytotoxic T cells surround a large cancer cell. The T cells recognize the cancer cell as nonself because it displays altered antigens on its surface. (*b*) Some of the cytotoxic T cells elongate as they chemically attack the cancer cell, breaking down its cell membrane. (*c*) The cancer cell has been destroyed; only a collapsed fibrous cytoskeleton remains.

needed, and then clone them in a separate cell culture. Cells of this clone produce large amounts of the specific antibodies needed—hence the name monoclonal antibodies. Such antibodies can be injected into the very same cancer patients whose cancer cells were used to stimulate their production, and are highly specific for destroying the cancer cells. (Monoclonal antibodies specific for a single antigenic determinant can now be produced.) In

ability is delayed. If the antigens are present continuously, the ability to respond to them will be postponed indefinitely. However, this same animal will respond quite normally to the presence of other antigens, will form antibodies to them, and will reject grafts from other strains of mice.

Immunological tolerance can be induced in this manner only in fetal or neonatal animals. If the thymus gland of a newborn animal is removed, the animal's lymph nodes remain small and the animal is deficient in cellular immunity. An animal treated in this way will accept tissue grafts from other animals that differ from it genetically.

Autoimmune Disease

Sometimes self-tolerance appears to break down and the body reacts immunologically against its own tissues, causing an **autoimmune disease.** Some of the diseases that result from such failures in self-tolerance are rheumatoid arthritis, multiple sclerosis, Graves' disease, myasthenia gravis, systemic lupus erythematosus (SLE), insulin-dependent diabetes (Type I), and perhaps infectious mononucleosis.

Myasthenia gravis is an autoimmune disease in which function at the neuromuscular junctions is impaired. Affected persons (more than 15,000 in the United States) experience muscle weakness and are easily fatigued. Gradually, victims lose muscle control. Sometimes respiratory muscles are affected to a life-threatening extent. Most myasthenia gravis patients have a circulating antibody that combines with acetylcholine receptors in the motor end plates. This interaction blocks the receptors and can damage or even destroy them.

What causes the production of abnormal antibodies in myasthenia gravis and in other autoimmune diseases? No one really knows. Some investigators have suggested genetic predisposition, perhaps involving HLA types; others speculate that prior damage to the tissue is involved. Some studies suggest that a viral infection in the involved tissue previously stimulated the body to manufacture antibodies against the infected cells. Then, after the virus has been destroyed, the body continues to manufacture harmful antibodies capable of attacking the body cells—even though they are no longer infected. A combination of these factors may be responsible.

Allergic Reactions

The immune system normally functions to defend the body against pathogens and to preserve homeostasis, but sometimes the system malfunctions. **Allergy** is a state of altered immune response that is harmful to the body. Allergic persons have a tendency to manufacture antibodies against mild antigens, called **allergens,** that do not stimulate a response in nonallergic individuals. In many kinds of allergic reactions, distinctive IgE immunoglobulins called **reagins** are produced. About 15% of the population of the United States are plagued by an allergic disorder such as allergic asthma or hayfever. There appears to be an inherited tendency to these disorders.

Let us examine a common allergic reaction—a hayfever response to ragweed pollen (Figure 43–14). When an allergic person inhales the microscopic pollen, allergens stimulate the release of IgE from sensitized plasma cells in the nasal passages. The IgE attaches to receptors on the membranes of mast cells, large connective tissue cells filled with distinctive granules. Each mast cell has thousands of receptors to which the IgE may attach. Each IgE molecule attaches to a mast cell receptor by its C region end, leaving the V region end of the immunoglobulin free to combine with the ragweed pollen allergen.

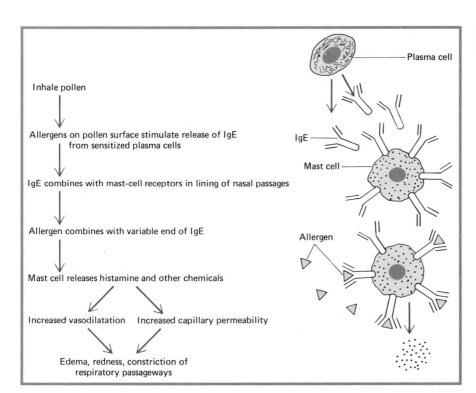

FIGURE 43–14 A common type of allergic response.

Inhale pollen

Allergens on pollen surface stimulate release of IgE from sensitized plasma cells

IgE combines with mast-cell receptors in lining of nasal passages

Allergen combines with variable end of IgE

Mast cell releases histamine and other chemicals

Increased vasodilatation   Increased capillary permeability

Edema, redness, constriction of respiratory passageways

Plasma cell

IgE

Mast cell

Allergen

When the allergen combines with IgE antibody, the mast cell rapidly releases its granules (Figure 43–15). When exposed to extracellular fluid, the granules release histamine, serotonin, and other chemicals that cause inflammation. These substances produce dilation of blood vessels and increased capillary permeability, leading to edema and redness. Such physiological responses cause the victims' nasal passages to become swollen and irritated. Their noses run, they sneeze, their eyes water, and they feel generally uncomfortable.

In **allergic asthma,** an allergen-IgE response occurs in the bronchioles of the lungs. Mast cells release SRS-A (slow-reacting substance of anaphylaxis), which causes smooth muscle to constrict. The airways in the lungs sometimes constrict for several hours, making breathing difficult.

Certain foods or drugs act as allergens in some persons, causing a reaction in the walls of the gastrointestinal tract that leads to discomfort and diarrhea. The allergen may be absorbed and cause mast cells to release granules elsewhere in the body. When the allergen-IgE reaction takes place in the skin, the histamine released by mast cells causes the swollen red welts known as **hives.**

Systemic anaphylaxis is a dangerous kind of allergic reaction that can occur when a person develops an allergy to a specific drug such as penicillin, or to compounds present in the venom injected by a stinging insect. Within minutes after the substance enters the body, a widespread allergic reaction takes place. Large amounts of histamine are released into the circulation, causing extreme vasodilation and permeability. So much plasma may be lost from the blood that circulatory shock and death can occur within a few minutes.

The symptoms of allergic reactions are often treated with **antihistamines,** drugs that block the effects of histamine. These drugs compete for the same receptor sites on cells targeted by histamine. When the antihistamine combines with the receptor, it prevents the histamine from combining and thus prevents its harmful effects. Antihistamines are useful clinically in relieving the symptoms of hives and hayfever. They are not

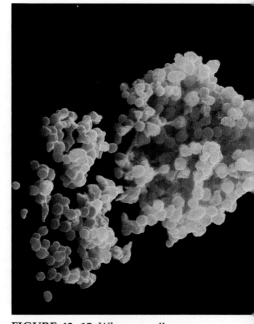

FIGURE 43–15 When an allergen combines with IgE bound to a mast cell receptor, the mast cell explodes, releasing granules filled with histamine and other chemicals that cause the symptoms of an allergic response.

completely effective, however, because mast cells release substances other than histamine that cause allergic symptoms.

In serious allergic disorders patients are sometimes given **desensitization therapy.** Very small amounts of the very antigen to which they are allergic are either injected or administered in the form of drops daily over a period of months or years. This stimulates production of IgG antibodies against the antigen. When the patient encounters the allergen, the IgG immunoglobulins combine with the allergen, blocking its receptors so that the IgE cannot combine with it. In this way a less harmful immune response is substituted for the allergic reaction. Desensitizing injections of the antigen are also thought to stimulate suppressor T cell activity.

■ SUMMARY

I. Immune responses depend upon the ability of an organism to distinguish between self and nonself.

II. Most invertebrates are capable only of nonspecific responses such as phagocytosis and the inflammatory response.

III. Vertebrates can launch both nonspecific and specific responses.

A. Nonspecific defense mechanisms that prevent entrance of pathogens include the skin, acid secretions in the stomach, and the mucous lining of the respiratory passageways.

B. Should pathogens succeed in breaking through the first line of defense, other nonspecific defense mechanisms are activated to destroy the invading pathogens.

1. When pathogens invade tissues, they trigger an inflammatory response, which brings needed phagocytic cells and antibodies to the infected area.

2. Neutrophils and macrophages phagocytize and destroy bacteria.

C. Specific immune responses include antibody-mediated immunity and cell-mediated immunity. Both T cells and B cells respond to antigens.

D. In antibody-mediated immunity, competent B cells are activated when specific antigens are presented by a macrophage and when exposed to interleukins secreted by helper T cells. B cells multiply, giving rise to clones of cells. Some differentiate to become plasma cells, which secrete specific antibodies.

E. Antibodies are highly specific proteins called immunoglobulins. They are produced in response to specific antigens. Antibodies are grouped in five classes according to their structure.

F. Antibody combines with a specific antigen to form an antigen-antibody complex, which may inactivate the pathogen, stimulate phagocytosis, or activate the complement system. The complement system increases the inflammatory response and phagocytosis; some complement proteins digest portions of the pathogen cell.

G. In cell-mediated immunity, specific T cells are activated by the presence of specific antigens and by helper T cells; these activated T cells multiply, giving rise to a clone of cells.

1. Some T cells differentiate to become cytotoxic T cells, which migrate to the site of infection and chemically destroy pathogens.

2. Some sensitized T cells remain in the lymph nodes as memory cells; others become helper T cells or supressor T cells.

H. Second exposure to an antigen evokes a secondary immune response, which is more rapid and more intense than the primary response.

I. Active immunity develops as a result of exposure to antigens; it may occur naturally or may be artificially induced by immunization. Passive immunity develops when an individual is injected with antibodies produced by another person or animal, and is temporary.

J. According to the theory of immune surveillance, the immune system destroys abnormal cells whenever they arise; diseases such as cancer develop when this immune mechanism fails to operate effectively.

K. Transplanted tissues possess protein markers known as major histocompatibility complex that stimulate graft rejection, an immune response (launched mainly by T cells) that destroys the transplant.

L. Immunological tolerance to foreign tissues can be induced experimentally under certain conditions.

M. In autoimmune diseases the body reacts immunologically against its own tissues.

N. In an allergic response, an allergen can stimulate production of IgE antibody, which combines with the receptors on mast cells; the mast cells then release histamine and other substances, causing inflammation and other symptoms of allergy.

■ POSTTEST

1. An antigen is a substance capable of stimulating a(n) _____ _____.

2. Specific proteins produced in response to specific antigens are called _____.

3. When infected by viruses, some cells respond by producing proteins called _____ .

4. The clinical characteristics of inflammation are _____ , _____ , _____ , and _____ .

5. T lymphocytes are thought to originate in the _____ _____ ; they are processed in the _____ ; and then proliferate in the _____ tissues.

6. Lymphokines are released by _____ .

7. When the body is invaded by the same pathogen a second time, the immune response can be launched more rapidly owing to the presence of _____ cells.

8. The cells that produce antibodies are _____ cells.

9. An antigenic determinant gives the antigen molecule a specific configuration that can be "recognized" by an _____ .

10. The _____ confers immunological competence upon T cells.

11. The complement system is activated when an _____ complex is formed.

12. In opsonization, complement proteins _____ .

13. Although artificially induced, immunization is a form of _____ immunity.

14. An individual injected with antibodies produced by another organism is receiving _____ immunity.

15. In humans the major histocompatibility complex is called the _____ group.

16. An autograft consists of tissue transplanted from _____ .

17. In graft rejection the host launches an effective _____ _____ against _____ tissue.

18. Cornea transplants are highly successful because the cornea is an immunologically _____ site.

19. An _____ is a mild antigen that does not stimulate a response in an individual who is not _____ .

20. In a typical allergic reaction mast cells secrete _____ and other compounds that cause _____ .

◼ REVIEW QUESTIONS

1. How does the body distinguish between self and nonself? Are invertebrates capable of making this distinction?

2. Contrast specific and nonspecific defense mechanisms. Which type confronts invading pathogens immediately? How do the two systems work together?

3. How does inflammation help to restore homeostasis?

4. Give two specific ways in which cell-mediated and antibody-mediated immune responses are similar, and three ways in which they are different.

5. Describe three ways in which antibodies work to destroy pathogens.

6. John is immunized against measles. Jack contracts measles from a playmate in nursery school before his mother gets around to having him immunized. Compare the immune responses in the two children. Five years later, John and Jack are playing together when Judy, who is coming down with measles, sneezes on both of them. Compare the immune responses in Jack and in John.

7. Why is passive immunity temporary?

8. What is immunological tolerance?

9. What is graft rejection? What is the immunological basis for it?

10. List the immunological events that take place in a common type of allergic reaction such as hayfever.

11. Explain the theory of immunosurveillance. What happens when immunosurveillance fails?

12. What is an autoimmune disease? Give two examples.

13. What public policy decisions would you recommend that might help slow the spread of AIDS?

◼ RECOMMENDED READINGS

Bolotin, Carol. Drug as hero. *Science 85,* June 1985, 68–71. The story of cyclosporin, a drug that selectively inhibits the rejection of transplanted organs.

Buisseret, P.D. Allergy. *Scientific American,* August 1982, 86–95. A discussion of the cellular and biochemical changes that occur during an allergic response.

Cohen, I.R. The self, the world and autoimmunity. *Scientific American,* April 1988, 52–60. Self-recognition is important for health as well as for certain diseases.

Darnell, J., H. Lodish, and D. Baltimore. *Molecular Cell Biology.* Scientific American Books, New York, 1986. Chapter 24 clearly presents the basics of immunology.

Edelson, R.L., and J.M. Fink. The immunologic function of skin. *Scientific American,* June 1985, 46–53. Specialized cells in the skin play interacting roles in the response to foreign invaders.

Gallo, R.C. The AIDS virus. *Scientific American,* January 1987, 46–56. An overview of the discovery of the AIDS virus and a description of the pathogen.

Jaroff, L. Stop that germ. *TIME,* May 23, 1988, 56–64. An excellent summary of immune function, with a focus on recent discoveries in immunology.

Kennedy, R.C., J.L. Melnick, and G.R. Dreesman. Antiidiotypes and immunity. *Scientific American,* July 1986, 48–52.

Laurence, J. The immune system in AIDS. *Scientific American,* December 1985. A discussion of the effects of AIDS on the immune system and of how we may be able to stop this disease.

Lerner, R.A. Synthetic vaccines. *Scientific American,* February 1983, 66–74. A report on experiments on the preparation of synthetic vaccines.

Marrack, P., and J. Kappler. The T cell and its receptor. *Scientific American,* February 1986. The surface proteins

body surface also have a relatively low metabolic rate, so that only small quantities of oxygen are needed.

How does an animal such as the earthworm exchange gases through its body surface? Movements of the worm and air currents ventilate the air so that fresh oxygen-rich air is brought in contact with the body surface. Gland cells in the epidermis secrete mucus, which keeps the body surface moist. Oxygen from tiny air pockets in the loose soil that the earthworm inhabits dissolves in the mucus and then diffuses through the body wall. The oxygen diffuses into blood circulating in a network of capillaries just beneath the outer cell layer. Oxygen is transported by the blood to all of the cells of the earthworm body, diffusing from the blood into the cells. Carbon dioxide from the body cells diffuses into the blood and is transported to the body surface, from which it diffuses out into the environment.

When it rains, water fills the air pockets in the soil. Water holds less oxygen than air, so the oxygen available may not be sufficient at such times. This is why earthworms often come to the surface of the soil after it rains—to come in contact with the air above the ground.

Tracheal Tubes

In insects and some other arthropods (e.g., chilopods, diplopods, some mites, and some spiders) the respiratory system consists of a network of **tracheal tubes** (Figure 44–3). Air enters the tracheal tubes through a series of tiny openings called **spiracles** along the body surface. The maximum number of spiracles in an insect is 20—two thoracic pairs and eight abdominal pairs—but the number and position vary in different species. In large or active insects, air moves into and out of the spiracles by movements of the body or by rhythmic movements of the tracheal tubes. For example, the grasshopper draws air into its body through the first four pairs of spiracles when the abdomen expands. Then the abdomen contracts, forcing air out through the last six pairs of spiracles.

Once inside the body, the air passes through the branching tracheal tubes, which extend to all parts of the animal. The tracheal tubes terminate in microscopic, fluid-filled tracheoles, and gases are exchanged between these tracheoles and the body cells. Although hemolymph fills the body cavity in these organisms, it does not function in the transport of gases. The tracheal system itself conducts air deep within the insect body, near

FIGURE 44–3 Tracheal tubes are characteristic of insects and some other invertebrates. (*a*) A scanning electron micrograph of a mole cricket trachea (approximately ×1300). The corrugations are a very long spiral around the tube, which may strengthen the tracheal wall somewhat as does the spring within the plastic hoses of many vacuum cleaners and hair dryers. The tracheal wall is composed of chitin. (*b*) Each tracheal tube and its branches conduct oxygen to the body cells of the insect.

(a)

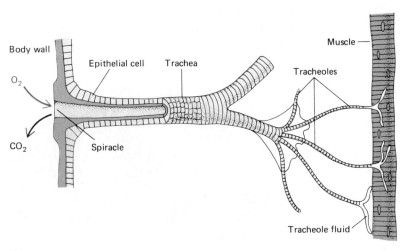

(b)

enough to each cell so that gas exchange can take place directly between the respiratory system and the body cells. In some insects the branching tracheal tubes end in elastic air sacs, which expand and contract, aiding gas exchange.

Gills

Gills are respiratory structures found mainly in aquatic animals, because gills are supported in water, but tend to collapse in air. They are moist, thin structures that extend out from the body surface (Figure 44–4). In many animals the outer surface of the gills is exposed to water, while the inner side is in close contact with networks of blood vessels.

Sea stars and sea urchins have **dermal gills** that project from the body wall. The beating of the cilia of the epidermal cells ventilates the gills by moving a stream of water over them. Gases are exchanged through the gills between the water and the coelomic fluid inside the body. Various types of gills are found in some annelids, aquatic mollusks, crustaceans, fish, and amphibians. Molluskan gills are folded, providing a large surface for respiration. In bivalve mollusks and in simple chordates, the gills are adapted for trapping and sorting food. Rhythmic beating of cilia draws water over the gill area, and food is filtered out of the water as gases are exchanged.

In chordates the gills are usually internal. A series of slits perforates the pharynx, and the gills are located along the edges of these gill slits. In bony fish, the fragile gills are protected by an external bony plate, the **operculum.** Movements of the operculum help to pump oxygenated water in through the mouth. The water flows over the gills and then exits through the gill slits.

Each gill in the bony fish consists of many **filaments,** which provide an extensive surface for gas exchange (Figure 44–5). The filaments extend out into the water, which continuously flows over them. A capillary network delivers blood to the gill filaments, facilitating diffusion of oxygen and carbon dioxide between blood and water. The very impressive efficiency of this system depends on the flow of blood in a direction opposite to the movement of the water. This arrangement, referred to as a **countercurrent exchange system,** maximizes the difference in oxygen concentration between blood and water.

If blood and water flowed in the *same* direction, the difference between the oxygen concentrations in blood (low) and water (high) would be very large initially and very small at the end. The oxygen concentration in the water would decrease as the oxygen concentration in the blood increased. When the concentrations of oxygen in the two fluids became equal, net diffusion of oxygen would stop. A great deal of oxygen would remain in the water.

In the countercurrent exchange system, however, blood low in oxygen comes in contact with water that is partly oxygen-depleted. Then, as the blood becomes more and more oxygen-rich, it comes in contact with water with a progressively higher concentration of oxygen. In this way a high rate of diffusion is maintained, ensuring that a very high percentage (more than 80%) of the available oxygen in the water diffuses into the blood.

Oxygen and carbon dioxide do not interfere with one another's diffusion, and they diffuse in opposite directions at the same time. This is because oxygen is more concentrated outside the gills than within, but carbon dioxide is more concentrated inside the gills than outside. Thus, the same countercurrent exchange mechanism that ensures efficient influx of oxygen works in reverse fashion to ensure equally efficient outgo of carbon dioxide.

FIGURE 44–4 The gills of the salmon provide an extensive surface for gas exchange.

FIGURE 44–5 Function of the fish gill. (*a*) Gills are an efficient breathing apparatus for the mature Siamese fighting fish (*Betta splenden*), while it builds an oxygen-rich bubble nest for its eggs. (*b*) The gills are located under a bony plate, the operculum, which has been removed in this side view. The gills occupy the opercular chamber, and form the lateral wall of the pharyngeal cavity. (*c*) The fish continuously pumps water through its mouth and over the gill arches. In inspiration, the fish enlarges its pharyngeal cavity, seen here from the top. This produces a negative (−) pressure in the pharyngeal cavity and water enters through the mouth. (*d*) When the fish exhales, it closes the mouth and the oral valve behind the lips, so that water is forced between the gills and out. Muscular action produces the high positive (+++) pressure in the pharyngeal cavity that is necessary to force the water between the gills. (*e*) Each gill consists of a cartilaginous gill arch to which two rows of leaflike gill filaments are attached. Blood circulates through the gill filaments as water passes among them. (*f*) Each gill filament has many even smaller extensions called secondary lamellae, which further increase the surface area. The lamellae contain capillaries that receive blood depleted of oxygen. The blood flows through the capillaries in the direction *opposite* to that of the water washing over the lamellae. As blood flows through the capillaries, it picks up oxygen from the water. In this countercurrent exchange, the blood is charged with oxygen very efficiently.

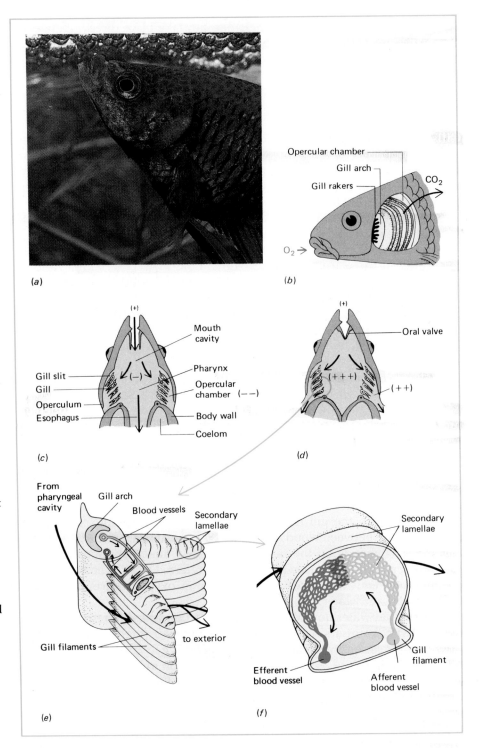

Lungs

Lungs are broadly defined as respiratory structures that develop as ingrowths of the body surface or from the wall of a body cavity such as the pharynx (Figure 44–6). Arachnids and some small mollusks (particularly terrestrial snails and slugs, but also some others) have lungs that depend almost entirely on diffusion for gas exchange. The book lungs of spiders are enclosed in an inpocketing of the abdominal wall. These lungs consist of a series of parallel, thin plates filled with blood. The plates are separated by air spaces that are connected to the outside environment through a spiracle.

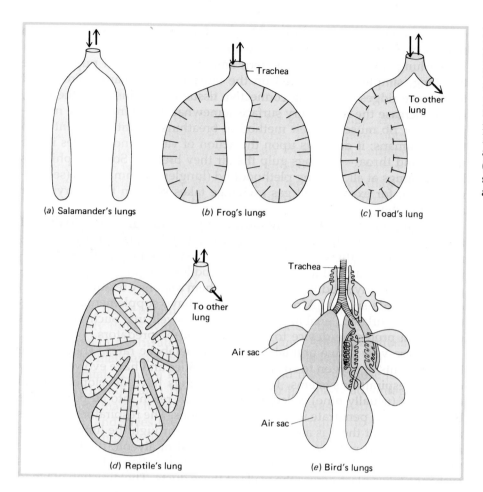

(a) Salamander's lungs

(b) Frog's lungs

(c) Toad's lung

Trachea

To other lung

(d) Reptile's lung

Trachea

Air sac

Air sac

To other lung

(e) Bird's lungs

FIGURE 44–6 The structure of the lung varies in the different vertebrate classes. Note the progressive increase in surface area for gas exchange. Salamander lungs are simple sacs, frog lungs have small ridges in the lung wall that help increase surface area, and birds have an elaborate system of lungs and air sacs. Mammalian lungs (Figure 44–7) have millions of air sacs that increase the surface available for gas exchange.

Larger mollusks and vertebrates with lungs have some means of forcefully moving air across the lung surface, that is, of ventilating the lung. This helps to ensure adequate oxygenation in active animals.

As far as can be ascertained from the evidence of fossils, crossopterygian fish, thought to be the ancestors of the amphibians, had lungs somewhat similar to those of modern lungfish. Three genera of lungfish are known today. They live in the headwaters of the Nile, in the Amazon, and in certain Australian rivers. Streams inhabited by the lungfish dry up during seasonal droughts; during these dry periods lungfish remain in the mud of the streambed, exchanging gas by means of their lungs. These fish are also equipped with gills, which they use when swimming. The African lungfish uses both its gills and lungs all year around, and cannot survive if deprived of air.

Remains of crossopterygian fish occur extensively in the fossil record. Those found in Devonian strata are believed to be similar to the ancestors of amphibians. Numerous amphibian fossils also occur in adjacent ancient strata. The geological evidence suggests that periodic droughts occurred in Devonian times. Some paleontologists hypothesize that lungs may have evolved originally as an adaptation to these conditions, and that lungs or lunglike structures may have been present in all early bony fish.

Most modern fish have no lungs, but almost all of them do possess homologous swim bladders (Chapter 31). Although it is primarily a hydrostatic organ that permits the fish to control its own density, the swim bladder may also store oxygen. Like the vertebrate lung, the swim bladder develops as an outpocketing of the pharynx. A swim bladder that retains its connection to the pharynx can serve as an *accessory respiratory organ.* When the oxygen level in the pond or lake is low, the fish may come to the

copper-containing proteins found in many species of mollusks and arthropods. These do not have a heme (porphyrin) group. When oxygen is combined with the copper, the compound appears blue. Without oxygen, it is colorless. Hemocyanins are dispersed in the blood rather than confined within cells.

THE HUMAN RESPIRATORY SYSTEM

FIGURE 44–7 The human respiratory system. The paired lungs are located in the thoracic cavity. The muscular diaphragm forms the floor of the thoracic cavity, separating it from the abdominal cavity beneath. An internal view of one lung illustrates its extensive system of air passageways. The microscopic air sacs are shown in later figures.

The respiratory system in humans and other air-breathing vertebrates includes the lungs and the system of tubes through which air reaches them (Figure 44–7). Air enters the body through the **external nares,** or nostrils, which open into the **nasal cavities.** The paired nasal cavities are separated by a wall of cartilage, the **nasal septum.** Three projections called **conchae** extend from the lateral wall of each nasal cavity, increasing the surface area over which air travels as it moves through the nasal cavities. The **olfactory epithelium,** the organ of smell, is located in the roof of the nasal cavity.

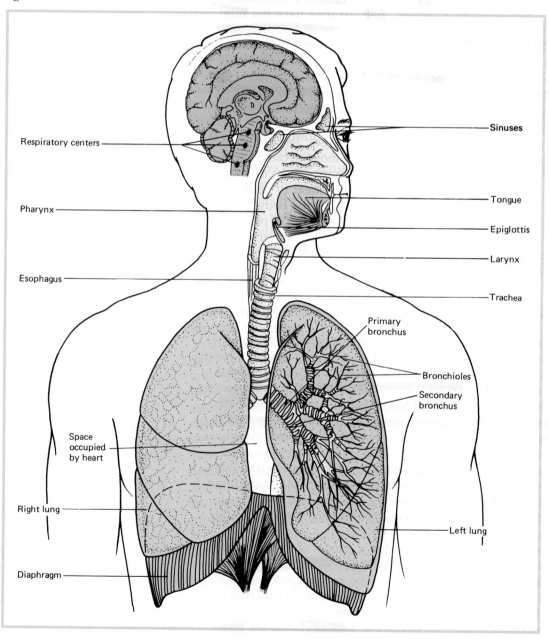

Respiratory centers

Pharynx

Esophagus

Space occupied by heart

Right lung

Diaphragm

Sinuses

Tongue

Epiglottis

Larynx

Trachea

Primary bronchus

Bronchioles

Secondary bronchus

Left lung

The nasal cavities are lined with ciliated epithelium complete with mucus cells. These cells produce more than 400 ml of mucus a day, providing a thin layer of mucus that traps dirt particles in the inhaled air. The beating of cilia moves a continuous band of mucus along toward the throat, from which it is swallowed together with saliva. In this way, dirt particles are delivered to the digestive system, which disposes of them, and the delicate lower part of the respiratory system is protected from contact with such irritating foreign matter.

The nasal cavities are connected to the **sinuses,** small cavities in the bones of the skull. Mucus produced by the epithelial lining of the sinuses drains into the nose. When the capillaries in the sinuses or nasal cavities dilate during infection or allergic response, fluid accumulates in these tissues, which then swell. This, along with secretion of extra amounts of mucus, produces the "stopped-up" feeling that accompanies colds or hayfever.

Air passes by way of the internal nares to the pharynx. Whether breathing begins through the nose or the mouth, air finds its way into the pharynx. Nose breathing is more desirable, however, because as air passes through the nose it is filtered, moistened, and brought to body temperature.

An opening in the floor of the pharynx leads into the larynx, which is evident externally as the "Adam's apple." In most mammals the larynx contains **vocal cords,** folds of epithelium that vibrate and produce sounds as air passes over them. Muscles adjust the tension of the cords to produce sounds of varying pitch.

The **epiglottis** is a flap of tissue that automatically seals off the larynx during swallowing so that neither food nor water can enter the airway. Occasionally this automatic mechanism fails, and food goes down into the larynx. When any foreign matter comes in contact with the larynx, a cough reflex is initiated. Coughing expels the foreign matter from the respiratory system. If coughing does not push out the food or material that has entered, the larynx may be blocked, which results in choking (see Focus on Choking).

From the larynx air passes into the trachea (the windpipe). Like the larynx, the trachea is kept from collapsing by rings of cartilage in its wall. These are necessary because air pressure in the trachea is less than atmospheric pressure during inspiration. At the level of the first rib the trachea branches into two cartilaginous bronchi, one extending to each lung. Inside the lung, each bronchus branches, giving rise to smaller bronchi, which, after several generations of branching, give rise ultimately to the tiny bronchioles. These branch repeatedly into smaller and smaller passageways, leading eventually into clusters of alveoli.

The very thin walls of the alveoli (only one cell thick) permit gases to diffuse easily (Figure 44–8). A thin film of detergent-like phospholipid known as **surfactant** coats the alveoli and facilitates their dilation by lowering the surface tension. Each alveolus is enveloped in a network of capillaries. Thus only two membranes separate the air in the alveolus from the blood: the epithelium of the alveolar wall and the endothelium of the capillary.

Both the trachea and bronchi are lined by ciliated epithelium, which contains mucus cells (Figure 44–9). Particles of dust and bacteria are trapped in the film of mucus covering these cells. The beating of cilia moves a stream of dirty mucus back up to the pharynx, from which it is swallowed. This mechanism keeps foreign matter out of the lungs, serving as a cilia-propelled mucus escalator.

Neither mucus nor ciliated cells are present in the smallest bronchioles or in the air sacs. Foreign particles such as those from tobacco smoke that find their way into the air sacs may remain there indefinitely, or may be

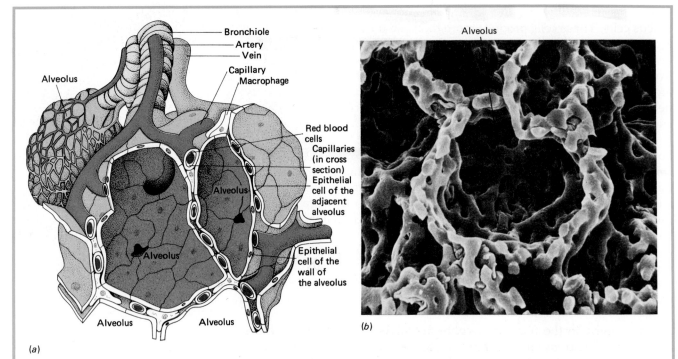

Bronchiole
Artery
Vein
Capillary
Macrophage

Alveolus

Red blood cells
Capillaries (in cross section)
Epithelial cell of the adjacent alveolus

Alveolus

Epithelial cell of the wall of the alveolus

Alveolus

Alveolus   Alveolus

(a)

(b)

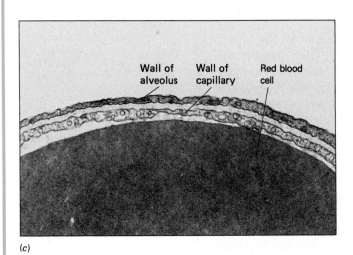

Wall of alveolus   Wall of capillary   Red blood cell

(c)

FIGURE 44–8 Structure of the alveolus. (*a*) The alveolar wall consists of extremely thin squamous epithelium. Each alveolus is enmeshed by a network of capillaries, facilitating free exchange of gases between the alveolus and the blood. (*b*) Scanning electron micrograph showing the capillary network surrounding a portion of one alveolus, seen in the center (approximately ×850). The vascular system has been injected to provide a clear, three-dimensional picture. Such casts are prepared by injecting a resin into the pulmonary artery. The resin fills all branches of the vessel; after it sets, the cellular and fibrous elements of the lung, including the blood vessels, are digested to leave only casts of the circulatory system. (*c*) Oxygen in the lung diffuses the short distance through the thin wall of the alveolus and then through the capillary wall to reach the blood. A portion of a red blood cell (large dark structure) is visible within the capillary.

phagocytized by macrophages. Such macrophages may accumulate in the lymph nodes of the lungs, permanently blackening them.

The lungs are large, paired spongy organs that occupy the thoracic cavity. Each lung, as well as the cavity in which the lung rests, is covered by a thin sheet of smooth epithelium called the **pleura.** The potential space between the pleura covering the lung and the pleura lining the chest cavity is the **pleural cavity.** A film of fluid over the pleura keeps the membrane moist and enables the lung to move in the chest cavity during breathing with little friction. Inflammation of the pleura, called **pleurisy,** results in the secretion of fluid into the pleural cavity, causing considerable pain during breathing.

The thoracic cavity is closed and has no communication with the outside atmosphere or with any other body cavity. It is bounded on the top and sides by the chest wall, which contains the ribs, and on the bottom by the strong, dome-shaped diaphragm.

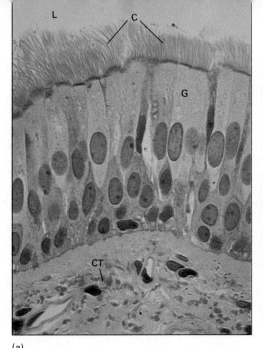

(a)

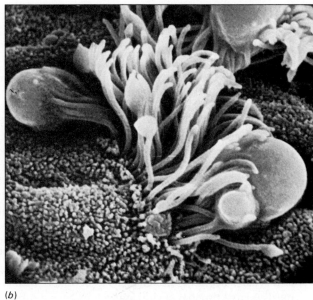

(b)

Inside, each lung consists of bronchioles, alveoli, and great networks of capillaries, all supported by connective tissue rich in elastic fibers. Lymph tissue and nerves are also present. The surface area available for gas exchange in the lung is tremendous—more than 50 times the area of the skin, or an area that approximates the size of a tennis court.

The Mechanics of Breathing

Breathing is the mechanical process of taking air into the lungs—**inspiration**—and letting it out again—**expiration.** Oxygen continuously moves from the air in the alveoli into the blood, while carbon dioxide constantly moves from the blood into the alveoli. If oxygen is to be continuously available, the air in the alveoli must be continuously replaced with fresh air. In the resting adult the normal breathing cycle of inspiration and expiration is repeated about 12 times each minute.

In humans and other mammals the ribs, chest muscles, and diaphragm are easily movable, and the volume of the chest cavity can be increased or decreased at will. During inspiration, certain chest muscles contract, drawing the front ends of the ribs upward and outward, an action made possible by the hingelike connection of the ribs with the vertebrae (Figure 44–10). The diaphragm contracts, becoming less convex in shape, which also has the effect of enlarging the chest cavity. As the chest expands, the film of fluid on the pleura pulls the membranous walls of the lungs outward,

FIGURE 44–9 The airway is lined with ciliated epithelium. (*a*) Pseudostratified columnar epithelium of a bronchus (approximately ×4000). *L*, lumen of bronchus; *C*, cilia; *G*, goblet cell; *CT*, connective tissue. (*b*) Ciliated epithelial cells from a hamster trachea, viewed from above by a scanning electron microscope that magnifies them about 7000 times. The round bodies are mucus droplets.

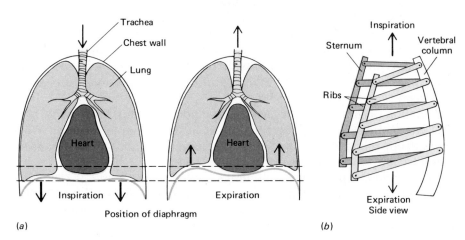

(a)

(b)

FIGURE 44–10 The mechanics of breathing. (*a*) Changes in the position of the diaphragm in expiration and inspiration result in changes in the volume of the chest cavity. (*b*) Changes in the position of the rib cage in expiration and inspiration. The elevation of the front ends of the ribs by the chest muscles causes an increase in the front-to-back dimension of the chest, and a corresponding increase in the volume of the chest cavity. When the volume of the chest cavity is increased, air moves into the lungs.

Focus on CHOKING

Choking kills an estimated 8,000 to 10,000 people per year in the United States. Many of these have long suffered from some degree of paralysis or other malfunction of the muscles involved in swallowing, often without consciously realizing it. Swallowing is a very complex process in which the mouth, pharynx, esophagus, and vocal cords must be coordinated with great precision. Functional muscular disorders of this mechanism can originate in a variety of ways—as birth defects, for example, or from brain tumors or vascular accidents involving the swallowing center of the brainstem.

Choking is more likely to occur in restaurants, where social interactions and unfamiliar surroundings are likely to distract a person's attention from swallowing and where alcohol is more likely to be taken with the meal. A large number of choking victims have a substantial blood alcohol content upon autopsy, which suggests the possibility that in them a marginally effective swallowing reflex has been further and fatally compromised by the effects of alcohol on the brain.

Anyone who begins to choke and gasp during a meal should be asked if he or she can speak. If not, as indicated by shaking the head or other gestures, the person is probably suffering a laryngeal obstruction rather than a coronary heart attack. If you are present during such an episode, be aware that you can take certain steps that can save the person's life: As a first step, deliver a strong blow to the victim's back with the open hand. If this fails, stand behind the victim, bring your arms around the person's waist, and clasp your hands just about the beltline. Your thumbs should be facing inward against the victim's body. Then squeeze abruptly and strongly in an upward direction. In most instances the residual air in the lungs will pop the obstruction out like a cork from a bottle. This is called the abdominal thrust, or **Heimlich maneuver.**

If the Heimlich maneuver must be performed with the victim lying down, place the victim face up. Kneel astride the victim's hips and, with one of your hands on top of the other, place the heel of your bottom hand on the abdomen slightly above the navel but below the rib cage. Press into the victim's abdomen with a quick upward thrust. This may be repeated if necessary. If there is no response within 15 or 20 seconds, it may be necessary to start cardiopulmonary resuscitation, or CPR. (See Focus on Cardiopulmonary Resuscitation [CPR].)

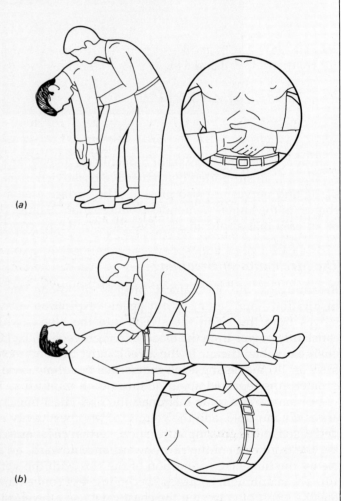

(a)

(b)

along with the chest walls. This increases the space within each lung. The air molecules in the lung now have more space in which to move about, and the pressure of the air in the lung decreases to 2 or 3 mm Hg below atmospheric pressure. Air from outside the body then rushes in through the air passageways and fills the lungs until the two pressures are equal once again.

Expiration occurs when the diaphragm and chest muscles relax. When these chest muscles relax, the ribs return to their original position. The simultaneous relaxation of the diaphragm permits the abdominal organs to push it back up to its initial convex shape. The volume of the chest cavity decreases, and the pressure in the lungs consequently increases (to 2 or 3 mm Hg above atmospheric pressure). The distended elastic air sacs return to their normal size, expelling the air that was inhaled and bringing the pressure back to atmospheric levels. Thus, in inspiration, the millions of

Table 44–2 COMPOSITION OF INHALED AIR COMPARED WITH THAT OF EXHALED AIR			
	% Oxygen (O_2)	**% Carbon Dioxide (CO_2)**	**% Nitrogen (N_2)**
Inhaled air (atmospheric air)	20.9	0.04	79
Exhaled air (alveolar air)	14.0	5.60	79
As indicated, the body uses up about one-third of the inhaled oxygen. The amount of CO_2 increases more than 100-fold because it is produced during cellular respiration.			

tiny air sacs fill with air like so many balloons, while during expiration, the air rushes out of the alveoli, partially deflating the balloons.

The Quantity of Air Respired

The amount of air moved into and out of the lungs with each normal resting breath is called the **tidal volume.** The normal tidal volume of a young adult male is about 500 ml. The **vital capacity** is the maximum amount of air a person can exhale after filling the lungs to the maximum extent.

Vital capacity is greater than tidal volume. This means that the lungs are not completely emptied of stale air and filled with fresh air with each breath. For this reason, alveolar air contains less oxygen and more carbon dioxide than does atmospheric air (Table 44–2). Expired air has had less than one fourth of its oxygen removed and can be breathed over again—a good thing for those in need of mouth-to-mouth resuscitation!

Exchange of Gases in the Lungs

After the lungs are ventilated, the oxygen in the alveoli must pass into the pulmonary capillaries, and carbon dioxide from the blood in the pulmonary capillaries must pass in the reverse direction. Since there is normally a greater concentration of oxygen in the lung alveoli than in the blood within the pulmonary capillaries, oxygen diffuses from the alveoli into the capillaries (Figure 44–11; Table 44–2). On the other hand, the concentration of carbon dioxide is greater in the pulmonary capillaries than in the alveoli, so the carbon dioxide diffuses in the opposite direction—from blood to alveolar air.

The factor that determines the direction and rate of diffusion is the pressure or tension of the particular gas. According to Dalton's law of partial pressures, in a mixture of gases the total pressure of the mixture is the sum of the pressures of the individual gases. Each gas exerts, independently of the others, the same pressure it would exert if it were present alone. The pressure of earth's atmosphere, that is, the **barometric pressure,** at sea level is able to support a column of mercury (Hg) 760 mm high. Because the atmosphere is made up of about 21% oxygen, oxygen's share of that pressure is $0.21 \times 760 = 160$ mm Hg. Thus, 160 mm Hg is the partial pressure of oxygen, abbreviated P_{O_2}.

Blood passes through the lung capillaries too rapidly to become completely equilibrated with the alveolar air. The partial pressure of oxygen in arterial blood is about 100 mm Hg. The P_{O_2} in the tissues ranges from 0 to 40 mm Hg, so that oxygen diffuses out of the capillaries and into the tissues. Not all of the oxygen leaves the blood, however. The blood passes through the tissue capillaries too rapidly for equilibrium to be reached; thus, the partial pressure of oxygen in venous blood returning to the lungs is about 40 mm Hg.

FIGURE 44–11 Gas exchange. (*a*) Exchange of gases between air sacs and capillaries in the lung. The concentration of oxygen is greater in the alveoli than in the pulmonary capillaries, so oxygen moves from the alveoli into the blood. Carbon dioxide is more concentrated in the blood than in the alveoli, so it moves out of the capillaries and into the alveoli. (*b*) Exchange of gases between the capillary and body cells. Here, oxygen is more concentrated in the blood than in the cells, so it moves out of the capillary into the cells. Carbon dioxide is more concentrated in the cells, and so it diffuses out of the cells and moves into the blood.

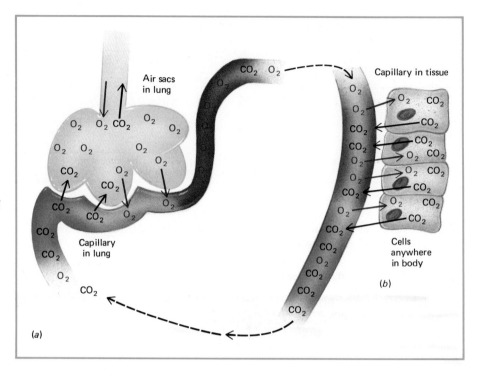

The continuous metabolism of glucose and other substrates in the cells results in the continuous production of carbon dioxide and utilization of oxygen. Consequently, the concentration of oxygen in the cells is lower than that in the capillaries entering the tissues, and the concentration of carbon dioxide is higher in the cells than in the capillaries. Thus, as blood circulates through capillaries of a tissue such as brain or muscle, oxygen moves by diffusion from the capillaries to the cells, and carbon dioxide moves from the cells into the blood.

Throughout the system, from lungs to blood to tissues, oxygen moves from a region of higher concentration to one of lower concentration: Oxygen moves from the air to the blood, and then to the tissue fluid, and is finally used in the cells. Carbon dioxide also moves from a region of greater to one of lesser concentration. It diffuses from the cells where it is produced through the tissue fluid and blood to the lungs and then out of the body.

Oxygen Transport

At rest, the cells of the human body utilize about 250 ml of oxygen per minute, or about 300 liters every 24 hours. With exercise or work this rate may increase as much as 10- or 15-fold. If oxygen were simply dissolved in plasma, blood would have to circulate through the body at a rate of 180 liters per minute to supply enough oxygen to the cells at rest. This is because, as we have seen, oxygen is not very soluble in blood plasma. Actually, the blood of a human at rest circulates at about 5 liters per minute and supplies all of the oxygen the cells need. Why are only 5 liters per minute rather than 180 liters per minute required?

The answer is hemoglobin, the respiratory pigment in red blood cells. Hemoglobin transports about 97% of the oxygen. Only about 3% is dissolved in the plasma. Plasma in equilibrium with alveolar air can take up in solution only 0.25 ml of oxygen per 100 ml, but the properties of hemoglobin enable whole blood to carry some 20 ml of oxygen per 100 ml. The protein portion of hemoglobin is composed of four peptide chains, typically two α and two β chains, to which are attached four **heme** (porphyrin) **rings.** An iron atom is bound in the center of each heme ring.

Hemoglobin has the remarkable property of forming a loose chemical union with oxygen. An oxygen molecule may attach to each of four iron atoms. In the lung (or gill), oxygen diffuses into the erythrocyte and combines with hemoglobin (Hb) to form oxyhemoglobin (HbO_2):

$$Hb + O_2 \rightleftharpoons HbO_2$$

Hemoglobin would, of course, be of little value to the body if it could only take up oxygen. It must also *release* the oxygen where needed. The reaction goes to the right in the lungs, forming oxyhemoglobin, and to the left in the tissues, releasing oxygen. Oxyhemoglobin is bright scarlet, giving arterial blood its color; reduced hemoglobin is purple, giving venous blood a darker hue.

The ability of oxygen to combine with hemoglobin and to be released from oxyhemoglobin is influenced by several factors, including the concentration of oxygen, the concentration of carbon dioxide, the pH, and the temperature. The **oxygen-hemoglobin dissociation curves** shown in Figure 44–12 illustrate the fact that as oxygen concentration increases, there is a progressive increase in the amount of hemoglobin that is combined with oxygen. This is known as the **percent saturation** of the hemoglobin. The percent saturation is highest in the pulmonary capillaries where the concentration of oxygen is greatest. In the capillaries of the tissues where there is less oxygen, the oxyhemoglobin dissociates, releasing oxygen. There, the percent saturation of hemoglobin is correspondingly less.

The extent to which oxyhemoglobin dissociates is determined mainly by oxygen concentration, but is also influenced by carbon dioxide. Carbon dioxide reacts with water in the plasma to form carbonic acid, H_2CO_3. An increase in the carbon dioxide concentration increases the acidity and lowers the pH of the blood. Oxyhemoglobin dissociates more readily in a more acidic environment. Lactate released from active muscles also lowers the pH of the blood and has a similar effect on the oxygen-hemoglobin dissociation curve. Displacement of the oxyhemoglobin dissociation curve by a change in pH is known as the **Bohr effect.**

Some carbon dioxide is transported by the hemoglobin molecule. Although it attaches to the hemoglobin molecule in a different way and at a different site than oxygen, the attachment of a carbon dioxide molecule causes the release of an oxygen molecule from the hemoglobin. Thus carbon dioxide concentration affects the oxygen-hemoglobin dissociation

FIGURE 44–12 Oxygen dissociation curves. These curves show that as oxygen concentration increases, there is a progressive increase in the amount of hemoglobin that is combined with oxygen. The curves also show how carbon dioxide affects the dissociation of oxyhemoglobin. Look at the vertical axis, labeled *percentage saturation*. If the blood contains 20% O_2 by volume, which is one fourth of the amount it could contain, it is said to be 25% saturated. The left curve shows what happens if no carbon dioxide is present. The middle curve shows the situation when the P_{CO_2} is 40, which is typical of arterial blood. The right curve indicates a P_{CO_2} of 90, which is quite unhealthy. Now find the location on the horizontal axis where the partial pressure of oxygen is 40, and follow the line up through the curves. Notice how the saturation of hemoglobin with oxygen differs among the three curves, even though the partial pressure of oxygen is the same.

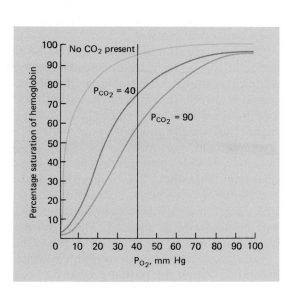

curve in two ways. This results in an extremely efficient transport system. In the capillaries of the lungs (or gills in fishes), carbon dioxide concentration is relatively low and oxygen concentration is high, so oxygen saturates a very high percentage of hemoglobin. In the capillaries of the tissues, carbon dioxide concentration is high and oxygen concentration is low, so oxygen is released from the hemoglobin.

Carbon Dioxide Transport

When carbon dioxide enters the blood, a small percentage of it becomes dissolved in the plasma. Most of it enters the red blood cells, where an enzyme called **carbonic anhydrase** catalyzes the following reaction:

$$CO_2 + H_2O \xrightarrow{\text{carbonic anhydrase}} H_2CO_3 \longrightarrow H^+ + HCO_3^-$$

This reaction takes place slowly in the plasma, but the carbonic anhydrase in the red blood cells accelerates the rate of the reaction by about 5000 times. (It also accelerates the reverse reaction in the lungs by the same factor.)

Most of the hydrogen ions released from the carbonic acid combine with hemoglobin, which is a strong buffer. Many of the bicarbonate ions diffuse into the plasma. Chloride ions diffuse into the red blood cells to replace the bicarbonate ions, a process known as the **chloride shift.**

Some of the carbon dioxide that enters the red blood cell combines with hemoglobin. The bond between the hemoglobin and carbon dioxide is very weak; the reaction, therefore, is readily reversible. By far, most—about 70%—of the carbon dioxide is transported as the bicarbonate ion. About 23% is transported in combination with hemoglobin, and about 7% is dissolved in the plasma.

Any condition (such as pneumonia) that interferes with the removal of carbon dioxide by the lungs leads to an increased concentration of carbon dioxide in the form of carbonic acid and bicarbonate ions in the blood. This condition is called **respiratory acidosis.** Although the pH of the blood is not actually acidic in this state, it is lower than normal.

Regulation of Respiration

Respiratory centers are groups of neurons and synapses that receive information relevant to respiration, evaluate it, and send messages to the respiratory muscles. Such respiratory centers have been located in the medulla and pons of the brain. They fire rhythmically so that at rest we breathe 12 to 14 times per minute.

The basic rhythm of respiration can be altered in response to changing needs of the body. Carbon dioxide concentration is the most important chemical stimulus for regulating the rate of respiration. Recall that an increase in carbon dioxide concentration results in an increase in hydrogen ions from carbonic acid. Specialized nerve endings called **chemoreceptors** within the medulla and within the walls of the aorta and carotid arteries are sensitive to changes in hydrogen ion concentration. Even a slight increase in carbon dioxide concentration (as might occur, for example, during exercise) stimulates these chemoreceptors and causes an increase in the rate and depth of respiration. As carbon dioxide is removed by the lungs, the hydrogen ion concentration in the blood and other body fluids decreases, and homeostasis is restored. Then, since the respiratory centers are no longer stimulated, the rate and depth of breathing return to normal.

Substantial decreases in oxygen concentration can also affect the rate of breathing. When the partial pressure of oxygen falls markedly, the aortic and carotid chemoreceptors are stimulated and send messages to the respiratory centers to increase the rate of respiration. It is interesting that oxygen concentration does not affect the respiratory centers directly, and that in healthy persons living at sea level, oxygen concentration generally does not play an important part in regulating respiration.

Although breathing is an involuntary process, the action of the respiratory centers can be consciously influenced for a short time by either stimulating or inhibiting them. For example, you can inhibit respiration by holding your breath. You cannot hold your breath indefinitely, though, because eventually you feel a strong impulse to breathe. Even if you were able to ignore this, you would soon pass out and breathing would resume.

Individuals who have stopped breathing because of drowning, smoke inhalation, electric shock, or cardiac arrest can sometimes be sustained by mouth-to-mouth resuscitation until their own breathing reflexes can be initiated again. Cardiopulmonary resuscitation (CPR) is a method for aiding victims who have suffered respiratory or cardiac arrest or both. For an overview of the procedure, see Focus on Cardiopulmonary Resuscitation (CPR).

Hyperventilation

Underwater swimmers and some Asiatic pearl divers voluntarily **hyperventilate** before going under water. By taking a series of deep inhalations and exhalations, you can markedly reduce the carbon dioxide content of the alveolar air and of the blood. As a result, it takes longer before the impulse to breathe becomes irresistible.

Focus on CARDIOPULMONARY RESUSCITATION (CPR)

Cardiopulmonary resuscitation, or **CPR,** is a method for aiding victims of accidents or heart attacks who have suffered cardiac arrest and respiratory arrest. It should not be used if the victim has a pulse or is able to breathe. It must be started immediately, because irreversible brain damage may occur within about 3 minutes of respiratory arrest. Here are its ABCs:

Airway Clear airway by extending victim's neck. This is sometimes sufficient to permit breathing to begin again.
Breathing Use mouth-to-mouth resuscitation.
Circulation Attempt to restore circulation by using external cardiac compression.

The procedure for CPR may be summarized as follows:

I. Establish the unresponsiveness of the victim.
II. Procedure for mouth-to-mouth resuscitation:
1. Place the victim on his or her back on a firm surface.
2. Clear the throat and mouth, and tilt the head back so that the chin points outward. Make sure that the tongue is not blocking the airway. Pull the tongue forward if necessary.
3. Pinch the nostrils shut and forcefully exhale into the victim's mouth. Be careful, especially in children, not to overinflate the lungs.
4. Remove your mouth and listen for air rushing out of the lungs.

5. Repeat about 12 times per minute. Do not interrupt for more than 5 seconds.
III. Procedure for external cardiac compression:
1. Place the heel of one hand on the lower third of the victim's breastbone. Keep your fingertips lifted off the chest. (In infants, two fingers should be used for cardiac compression; in children, use only the heel of the hand.)
2. Place the heel of the other hand at a right angle to and on top of the first hand.
3. Apply firm pressure downward so that the breastbone moves about 4 to 5 cm (1.6 to 2 in) toward the spine. Downward pressure must be about 5.4 to 9 kg (12 to 20 lb) with adults (less with children). Excessive pressure can fracture the sternum or ribs, resulting in punctured lungs or a lacerated liver. This rhythmic pressure can often keep blood moving through the heart and great vessels of the thoracic cavity in sufficient quantities to sustain life.
4. Relax your hands between compressions to allow the chest to expand.
5. Repeat at the rate of at least 60 compressions per minute. (For infants or young children, 80 to 100 compressions per minute are appropriate.) Fifteen compressions should be applied, then two breaths, in a ratio of 15:2.

When hyperventilation is continued for a long period, the person may feel dizzy and sometimes becomes unconscious. This is because a certain concentration of carbon dioxide is needed in the blood to maintain normal blood pressure. (This mechanism operates by way of the vasoconstrictor center in the brain, which maintains the muscle tone of blood vessel walls.) Furthermore, if divers hold their breath too long, the low concentration of oxygen may result in unconsciousness and drowning.

High Flying and Low Diving

The barometric pressure decreases at progressively higher altitudes. Since the concentration of oxygen in the air remains at 21%, the partial pressure of oxygen decreases along with the barometric pressure. At an altitude of 6000 m (19,500 ft), the barometric pressure is about 350 mm Hg, the partial pressure of oxygen is about 75 mm Hg, and the hemoglobin in arterial blood is about 70% saturated with oxygen. At 10,000 m, the barometric pressure is about 225 mm Hg, the partial pressure of oxygen is 50 mm Hg, and arterial oxygen saturation is only 20%. Thus, getting sufficient oxygen from the air becomes an ever-increasing problem at higher altitudes.

When a person moves to a high altitude, the body adjusts over a period of time by producing a greater number of red blood cells. In a person breathing pure oxygen at 10,000 m, the oxygen would have a partial pressure of 225 mm Hg, and the hemoglobin would be almost fully saturated with oxygen. Above 13,000 m, however, the barometric pressure is so low that even breathing pure oxygen would not permit complete oxygen saturation of arterial hemoglobin.

A person can remain conscious only until the arterial oxygen saturation falls to 40% to 50%. This level is reached at about 7000 m when the person is breathing air, or 14,500 m when pure oxygen is used. All high-flying jets have cabins that are airtight and pressurized to the equivalent of the barometric pressure at an altitude of about 2000 m.

Hypoxia, a deficiency of oxygen, results in drowsiness, mental fatigue, headache, and sometimes euphoria. The ability to think and to make judgments is impaired, and there is a loss of ability to perform tasks requiring coordination. If a jet were flying at about 11,700 m and underwent sudden decompression, the pilot would lose consciousness in about 30 seconds and become comatose in about 1 minute.

In addition to the problems of hypoxia, a rapid decrease in barometric pressure can cause **decompression sickness** (the bends). Whenever the barometric pressure drops below the total pressure of all gases dissolved in the blood and other body fluids, the dissolved gases tend to come out of solution into a gaseous state and form bubbles. A familiar example of such bubbling occurs each time you uncap a bottle of soda, thus reducing the pressure in the bottle. The carbon dioxide is released from solution and bubbles out into the air. In the body, it is nitrogen that causes the problem because it has such a low solubility in blood and tissues. Nitrogen comes out of solution and forms bubbles that may block capillaries, interfering with blood flow. Other tissues may also be damaged. The clinical effects of decompression sickness are pain, dizziness, paralysis, unconsciousness, and even death.

Decompression sickness is even more common in deep-sea diving than in high-altitude flying. As a diver descends, the surrounding pressure increases tremendously—1 atmosphere for each 10 m. To prevent the collapse of the lungs, a diver must be supplied with air under pressure, thereby exposing the lungs to very high alveolar gas pressures (Figure 44–13).

At sea level an adult human has about 1 liter of nitrogen dissolved in the body, with about half of that in the fat and a half in the body fluids. After a

FIGURE 44–13 A diver carries tanks containing compressed air in order to expose the lungs to high alveolar gas pressures. This prevents lung collapse at depths where the surrounding pressure is very high.

diver's body has been saturated with nitrogen at a depth of 100 m, the body fluids contain about 10 liters of nitrogen. To prevent this nitrogen from rapidly bubbling out of solution and causing decompression sickness, the diver must be brought to the surface gradually, with stops at certain levels on the way up. This allows the nitrogen to be expelled slowly through the lungs.

Some mammals can spend rather long periods of time in the ocean depths without coming up for air; see Focus on Adaptations of Diving Mammals.

Oxygen poisoning can also result from inhaling compressed air. One of its worst effects is convulsions, which can be lethal to divers deep beneath the surface. Just why oxygen is toxic is not understood, but several possibilities have been suggested. One hypothesis is that high concentrations of oxygen may inactivate some of the essential respiratory enzymes. There is also evidence that high concentrations of oxygen decrease blood flow through the brain by 25% to 50%. Still another suggestion is that too much oxygen in the cell may cause the production of oxidizing free radicals. (A free radical is a particle with an unpaired electron and is generally very

Focus on ADAPTATIONS OF DIVING MAMMALS

The Weddell seal can swim under the ice at a depth of 500 meters for more than an hour without coming up for air. The bottle-nosed whale can remain in the ocean depths for as long as two hours. Porpoises, whales, seals, beavers, mink, and several other air-breathing mammals have adaptations that permit them to dive for food or disappear below the water surface for several minutes to elude their enemies.

Diving mammals have about twice the volume of blood, relative to their body weight, as nondivers. Many have high concentrations of myoglobin, an oxygen-binding pigment similar to hemoglobin found in muscles. They do not, however, have larger lungs than those in nondiving mammals.

When a mammal dives to its limit, a group of physiological mechanisms known collectively as the **diving reflex** are activated. Breathing stops. Bradycardia (slowing of the heart rate) occurs. The heart rate may decrease to one tenth of the normal rate, reducing the body's consumption of oxygen and energy. Blood is redistributed, with the lion's share going to the brain and heart, the organs that can least withstand anoxia. Skin, muscles, digestive organs, and other internal organs can survive with less oxygen, and so receive less blood while an animal is submerged. Muscles shift from aerobic to anaerobic metabolism.

Diving mammals do not take in extra air before a dive. In fact, seals exhale before they dive, and the lungs of whales are compressed during diving. These adaptations are thought to reduce the chance of decompression sickness, because with less air in the lungs there is less nitrogen in the blood to dissolve during the dive.

The diving reflex is present to some extent in human beings, where it may act as a protective mechanism during birth when an infant may be deprived of oxygen for several minutes. Many cases of near-drownings have

been documented in which the victim had been submerged for several minutes (as long as 45 minutes) in very cold water before being rescued and resuscitated. In many of these survivors there was no apparent brain damage. The shock of the icy water slows the heart rate, increases blood pressure, and shunts the blood to the internal organs of the body that most need oxygen (blood flow in the arms and legs decreases). Metabolic rate decreases so that less oxygen is required.

■ Mother and baby spotted dolphins (*Tursiops truncatus*).

(a)

(b)

FIGURE 44–14 Air pollution contributes to respiratory disorders. (a) Industry spews tons of pollutants into the atmosphere. (b) Asbestos fibers are a serious health threat. This macrophage has engulfed several fibers, but will probably die from this toxic meal, leaving the asbestos fibers lodged in the lung tissue.

FIGURE 44–15 Lung tissue. (a) Normal lung tissue. (b) Lung tissue with accumulated carbon particles. Despite the body's defenses, when we inhale smoky, polluted air, especially over a long period of time, dirt particles do enter the lung tissue and remain lodged there.

reactive.) Free radicals may oxidize certain essential cell components, thereby damaging the cell's metabolic machinery.

Effects of Smoking and Air Pollution

A human being breathes about 20,000 times each day, inhaling about 35 pounds of air—six times the amount of food and drink consumed. Many of us breathe dirty urban air laden with particulates, carbon monoxide, sulfur oxides, and other harmful substances (Figure 44–14 and Chapter Opener). How do these air pollutants affect the respiratory system?

Mucus cells in the epithelial lining of the respiratory passageways react to the irritation of inhaled pollutants by secreting increased amounts of mucus. The ciliated cells, damaged by the pollutants, are unable to effectively clear the mucus and trapped particles from the airways. The body attempts to clear the airways by coughing. (Smoker's cough is a well-known example.) The coughing causes increased irritation, and the respiratory system becomes susceptible to viral, bacterial, and fungal infections. Particles of dirt also accumulate in the lungs over a period of years (Figure 44–15).

One of the body's fastest responses to inhaling dirty air is **bronchial constriction.** The bronchial tubes narrow, providing increased opportunity

(a)

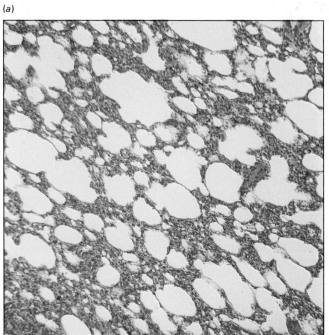

(b)

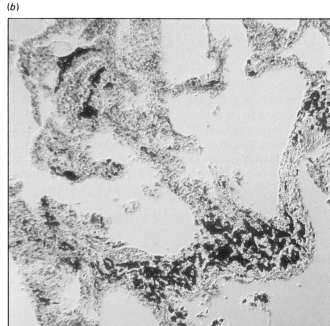

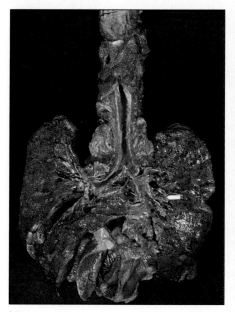
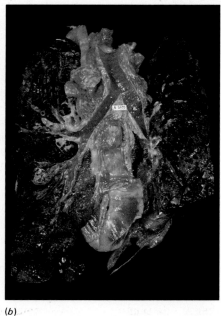

(a)

(b)

FIGURE 44–16 A diseased lung compared to a healthy lung. (*a*) Normal human lungs and major bronchi. (*b*) Human lungs and heart showing effects of cigarette smoking.

for the inhaled particles to be trapped along the sticky mucous lining of the airway. Unfortunately, the narrowed bronchial tubes allow less air to pass through to the lungs, so that oxygen available to the body cells is decreased. Chain smokers and those who live in polluted industrial areas experience almost continuous bronchial constriction.

Chronic bronchitis and emphysema are **chronic obstructive pulmonary diseases** that have been linked to smoking and air pollution. More than 75% of patients with chronic bronchitis have a history of heavy cigarette smoking. Advanced emphysema is virtually unknown in those who have never smoked regularly.

In **chronic bronchitis,** the bronchioles secrete too much mucus and are constricted and inflamed. Respiratory cilia cannot clear the passages of the mucus and particles that partly clog them. Affected persons are short of breath and often cough up mucus. They are susceptible to frequent episodes of **acute bronchitis,** a temporary inflammation usually caused by viral or bacterial infection. Persons with chronic bronchitis often develop emphysema.

In **emphysema,** obstruction of the bronchioles results in increased airway resistance and increased work in breathing, especially during expiration. The alveolar walls break down so that several air sacs become joined to form larger and less elastic alveoli (Figure 44–16). The number of capillaries in the lungs is reduced. Because the air sacs are not as elastic as they should be, air is not expelled effectively, and stale air accumulates in the air sacs. This decreases the amount of oxygen available to combine with hemoglobin in the blood. To compensate, the right ventricle of the heart pumps harder and becomes enlarged. Emphysema patients often die of heart failure.

Cigarette smoking is also the main cause of **lung cancer.** About 80% of deaths from lung cancer would not occur if people did not smoke cigarettes. (See Focus on Facts About Smoking.) In fact, cigarette smoking is the major single cause of all cancer deaths in the United States. More than 10 of the compounds in the tar of tobacco smoke have been shown to cause cancer in animals. These carcinogenic compounds irritate the cells lining the respiratory passages and alter their metabolic balance. Normal cells are transformed into abnormal cancerous ones, which multiply rapidly and invade surrounding tissue.

Focus on FACTS ABOUT SMOKING

- The life of a 30-year-old who smokes 15 cigarettes a day is shortened by an average of more than 5 years.
- If you smoke more than one pack per day, you are about 20 times more likely to develop lung cancer than a nonsmoker. According to the American Cancer Society, cigarette smoking causes more than 75% of all lung cancer deaths.
- If you smoke, you are more likely to develop atherosclerosis, and you double your chances of dying from cardiovascular disease.
- If you smoke, you are 20 times more likely to develop chronic bronchitis and emphysema than a nonsmoker.
- If you smoke, you are seven times more likely to develop peptic ulcers (especially malignant ulcers) than a nonsmoker.
- If you smoke, you have about 5% less oxygen circulating in your blood (because of carbon monoxide–hemoglobin) than a nonsmoker.
- If you smoke when you are pregnant, your baby will weigh about 6 ounces less at birth, and there is double the risk of miscarriage, stillbirth, and infant death than if you did not smoke.
- Workers who smoke one or more packs of cigarettes per day are absent from their jobs because of illness 33% more often than nonsmokers.
- Risks increase with the number of cigarettes smoked, inhaling, smoking down to a short stub, and use of nonfilter or high-tar, high-nicotine cigarettes. Cigar and pipe smokers have lower risks than cigarette smokers because they do not inhale as much. Cigarette smokers who switch to cigars and continue to inhale actually increase their risks.
- Nonsmokers confined in living rooms, offices, automobiles, or other places with smokers are adversely affected by the smoke. For example, when parents of infants smoke, the infant has double the risk of contracting pneumonia or bronchitis in its first year of life.
- When smokers quit smoking, their risk of dying from chronic pulmonary disease, cardiovascular disease, or cancer decreases. (Precise changes in risk figures depend upon the number of years the person smoked, the number of cigarettes smoked per day, the age of starting to smoke, and the number of years since quitting.)
- If everyone in the United States stopped smoking, more than 300,000 lives would be saved each year.

■ SUMMARY

I. In organisms less than 1 mm thick, gas exchange can occur effectively by diffusion, but in larger and more complex forms, specialized respiratory structures are required.
   A. In nudibranch mollusks, most annelids, small arthropods, and some vertebrates, gas exchange occurs through the entire body surface.
   B. In insects and some other arthropods, the respiratory system consists of a network of tracheal tubes.
   C. Gills are respiratory structures characteristic of aquatic animals.
      1. In echinoderms, simple dermal gills project from the body wall.
      2. In chordates, gills are usually internal, located along the edges of the gill slits.
      3. In bony fish, a countercurrent exchange system promotes diffusion of oxygen into the blood and diffusion of carbon dioxide out of the blood and into the water.
   D. Large mollusks and terrestrial vertebrates have lungs with some means of ventilating them.
   E. Most modern fish do not have lungs but possess homologous swim bladders.
II. Gas exchange with air is more efficient than gas exchange with water because air contains more oxygen than water and because oxygen diffuses more rapidly through air than through water.
III. Respiratory pigments greatly increase the capacity of blood to transport oxygen.
IV. The human respiratory system includes the lungs and a system of tubes through which air reaches them. A breath of air passes through the nose, pharynx, larynx, trachea, bronchus, bronchioles, and alveoli.
   A. During breathing, the diaphragm and rib muscles contract, expanding the chest cavity. The membranous walls of the lungs move outward along with the chest walls, decreasing the pressure within the lungs. Air from outside the body rushes in through the air passageways and fills the lungs until the pressure once more equals atmospheric pressure.
   B. Tidal volume is the amount of air moved into and out of the lungs with each normal breath. Vital capacity is the maximum volume that can be exhaled after the lungs are filled to the maximum extent.
   C. Oxygen and carbon dioxide are exchanged between alveoli and blood by diffusion.
   D. About 97% of the oxygen in the blood is transported as oxyhemoglobin.
      1. As oxygen concentration increases, there is a progressive increase in the amount of hemoglobin that combines with oxygen.
      2. Owing to the Bohr effect, oxyhemoglobin dissociates more readily as carbon dioxide concentration increases.
   E. About 70% of the carbon dioxide in the blood is transported as bicarbonate ions.
   F. Respiratory centers located in the medulla and pons control the basic rhythm of respiration.
      1. The respiratory centers are stimulated by chemoreceptors sensitive to an increase in hy-

drogen ions, which results from increased carbon dioxide concentration.

2. The respiratory system can also be stimulated by signals from chemoreceptors sensitive to very low oxygen concentration.

G. Hyperventilation reduces the concentration of carbon dioxide in the alveolar air and in the blood.

H. As altitude increases, barometric pressure decreases and less oxygen enters the blood. This situation can lead to hypoxia. A rapid decrease in barometric pressure can cause decompression sickness.

I. Inhaling polluted air results in bronchial constriction, increased mucus secretion, damaged ciliated cells, and coughing. It eventually can lead to chronic bronchitis, emphysema, or lung cancer.

▪ POST-TEST

1. Specialized respiratory structures must have _____ walls so that _____ easily occurs; they must be _____ so that gases can be dissolved; and they are typically richly supplied with _____ _____ to ensure transport of gases.

2. In insects, air enters a network of _____ tubes through openings called _____.

3. The operculum is a bony plate that protects the _____ in _____ _____.

4. Respiratory structures that develop as ingrowths of the body surface are called _____.

5. A fish can rise or sink, or maintain a particular level in the water without muscular effort, by regulating the amount of air in its _____ _____.

6. In birds, the lungs have several extensions referred to as _____ _____.

7. In the mammalian respiratory system, inhaled air passing through the larynx next enters the _____, and then passes into a _____.

8. In the mammalian respiratory system, gas exchange takes place through the thin walls of the _____.

9. In mammals, the floor of the thoracic cavity is formed by the _____.

10. The chief problem with breathing air is _____ _____.

Select the most appropriate answer in Column B for each description in Column A.

Column A
11. Seals off larynx during swallowing
12. Cavities in bones of skull
13. Initiates cough reflex
14. Covers lung
15. Coated with surfactant

Column B
a. Sinuses
b. Larynx
c. Alveoli
d. Pleura
e. Epiglottis

16. The maximum amount of air a person can exhale after filling the lungs to the maximum extent is the _____ _____.

17. The extent to which oxyhemoglobin dissociates is determined mainly by the _____ concentration.

18. An increase in carbon dioxide concentration lowers the blood pH and results in greater dissociation of _____; this is known as the _____ _____.

19. Most carbon dioxide is transported in the blood as _____.

20. Hypoxia is a deficiency in _____.

21. Bronchial constriction is one of the body's most rapid responses to _____.

22. In _____, the alveolar walls break down so that several air sacs join to form larger, less elastic alveoli.

23. The main cause of lung cancer is _____.

24. Label the following diagram. (Refer to Figure 44–7 as necessary.)

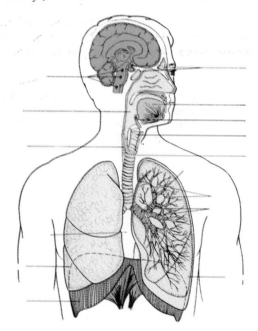

▪ REVIEW QUESTIONS

1. Why are specialized respiratory structures necessary in a tadpole but not in a flatworm?

2. Compare ventilation in a sea star with ventilation in a human.

3. Compare gas exchange in the following animals:
   a. earthworm   c. fish
   b. grasshopper   d. frog

4. Why are lungs more suited for an air-breathing verte-

brate and gills more effective in a fish? Why are lungs internal?

5. What respiration problem is solved by respiratory pigments? How?
6. Why does alveolar air differ in composition from atmospheric air?
7. What physiological mechanisms bring about an increase in rate and depth of breathing during exercise? Why is such an increase necessary?
8. What is the advantage of having lungs with millions of alveoli rather than lungs consisting of simple sacs, as in the mud puppies?
9. In what way might it be an advantage for a fish to have lungs as well as gills? What function do the "lungs" of most modern fish serve?
10. How does the countercurrent exchange system increase the efficiency of gas exchange between gills and blood?
11. What mechanisms does the human respiratory system have for getting rid of inhaled dirt? What happens when so much dirty air is inhaled that these mechanisms cannot function effectively?
12. What factors affect the dissociation of oxyhemoglobin?
13. What happens to a deep-sea diver who surfaces too quickly?

■ RECOMMENDED READINGS

Feder, M.E., and W.W. Burggren. Skin breathing in vertebrates. *Scientific American,* November 1985, 126–142. An interesting account describing how skin functions in gas exchange in some vertebrates.

Kanwisher, J.W., and S.H. Ridgway. The physiological ecology of whales and porpoises. *Scientific American,* June 1983, 111–119. A discussion of adaptations for deep diving.

Perutz, M.F. Hemoglobin structure and respiratory transport. *Scientific American,* December 1978, 92–125. Nobel prize winner Perutz describes how hemoglobin changes its shape to facilitate oxygen binding and release.

Randall, D.J., A.P. Burggren, A.P. Farrell, and M.S. Haswell. *The Evolution of Air Breathing in Vertebrates.* Cambridge University Press, New York, 1981. A discussion of structure and function of respiratory adaptations in terrestrial vertebrates.

Zapol, W.M. Diving adaptations of the Weddell seal. *Scientific American,* June 1987, 100–105. Physiological adaptations enable the seal to swim deeper and hold its breath longer than most other mammals.

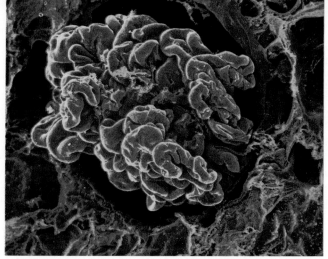

Over a million capillary tufts (glomeruli) filter blood in the kidneys

45

Fluid Balance and Disposal of Metabolic Wastes

■ LEARNING OBJECTIVES

After you have read this chapter you should be able to:

1. Identify the principal types of metabolic wastes (including three types of nitrogenous wastes), and describe how each is produced and how each is excreted.
2. Compare the excretory organs of a flatworm, a crustacean, an insect, and a vertebrate, explaining briefly how each functions.
3. Compare the adaptations of marine bony fish and cartilaginous fish that enable them to carry on osmoregulation.
4. Compare the osmoregulatory problem of a freshwater fish with that of a marine fish, and describe how the freshwater fish solves the problem.
5. Describe the adaptations that enable each of the following animals to solve their osmoregulatory problems: amphibians; marine birds and reptiles; and a dehydrated human being.
6. Label on a diagram the organs of the human urinary system, and give the functions of each.
7. Label on a diagram the principal parts of the nephron, and give the function of each structure. Include the following structures: Bowman's capsule, glomerulus, proximal convoluted tubule, loop of Henle, distal convoluted tubule, collecting duct, afferent arteriole, and efferent arteriole.
8. Summarize the process of urine formation, comparing filtration and reabsorption.
9. Trace a drop of filtrate from Bowman's capsule to its release from the body as urine.
10. Relate the countercurrent mechanism in the kidney to the process of urine concentration.
11. Describe the effects of each of the following on the volume and composition of urine: ADH; low fluid intake; alcohol intake; and eating a bag of potato chips.
12. Summarize the functions of the kidneys in maintaining homeostasis.

Water shapes the basic nature of life and its distribution on earth. Most organisms consist mainly of water, which is the medium in which metabolic reactions take place. The simplest forms of life are small

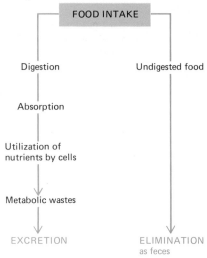

FOOD INTAKE

Digestion | Undigested food

Absorption

Utilization of nutrients by cells

Metabolic wastes

EXCRETION | ELIMINATION as feces

FIGURE 45–1 Excretion contrasted with elimination.

organisms that live in the sea. They obtain their food and oxygen directly from the sea water that surrounds them, and they release waste products into it. Larger, more complex animals and most terrestrial animals have their own internal sea—the blood and interstitial fluid—to bathe their cells and to transport and dissolve nutrients, gases, and waste products.

Terrestrial animals have a continuous stream of water flowing through them. Water is taken in with food and drink and is formed in metabolic reactions. Some of this water becomes part of the blood, which transports materials throughout the organism. Then, interstitial fluid is formed from the blood plasma to bathe all of the cells of the body. Excess water evaporates from the body surface or is excreted by specialized structures.

The water content of the body, as well as the concentration and distribution of ions, must be precisely regulated, by a process called **osmoregulation.** Freshwater protozoans have contractile vacuoles that periodically pump out the excess water that is osmotically drawn into the cell. In many organisms, the same structures that rid the body of excess water and ions often are also adapted to rid the body of metabolic wastes. These organs make up the **excretory system.**

As cells carry on metabolic activities, waste materials are generated. If permitted to accumulate, such waste products would reach toxic concentrations and threaten homeostasis. Therefore, waste must be continuously removed from the body. **Excretion is the process of removing metabolic wastes from the body.** The term *excretion* is sometimes confused with the term **elimination** (Figure 45–1). Undigested and unabsorbed food materials are **eliminated** from the body in the feces. Such substances never actually participated in the organism's chemical metabolism or entered body cells but merely passed through the digestive system.

FUNCTIONS OF THE EXCRETORY SYSTEM

Typically, an excretory system helps maintain homeostasis in three ways: (1) It excretes metabolic wastes; (2) it carries on osmoregulation (regulating both the fluid and salt content of the body); and (3) it regulates the concentrations of most of the constituents of body fluids. To carry out these functions, an excretory organ collects fluid, generally from the blood or interstitial fluid. It then adjusts the composition of this fluid by **reabsorbing** substances the body needs. Finally, the adjusted excretory product (urine, for example) must be expelled from the body.

WASTE PRODUCTS

The principal waste products generated by metabolic activities in most animals are water, carbon dioxide, and nitrogenous wastes. Carbon dioxide is mainly excreted by gills, lungs, or other respiratory surfaces. Excretory organs such as kidneys excrete most of the water and nitrogenous wastes.

Nitrogenous wastes include ammonia, uric acid, and urea. The first step in the catabolism of amino acids is **deamination,** a process in which the amino group is removed and converted to **ammonia** (Figure 45–2). Aquatic invertebrates and freshwater vertebrates excrete nitrogen mainly as ammonia. The water that surrounds these animals dilutes the ammonia and quickly washes it away. Ammonia is highly toxic and cannot be permitted to accumulate in the body or in the organism's surroundings. In many organisms it is quickly converted to some less toxic nitrogenous waste product such as uric acid or urea.

Uric acid is produced from ammonia and also when nucleotides from nucleic acids are metabolized. Uric acid is poorly soluble and precipitates from a supersaturated solution, forming crystals. Thus, uric acid can be

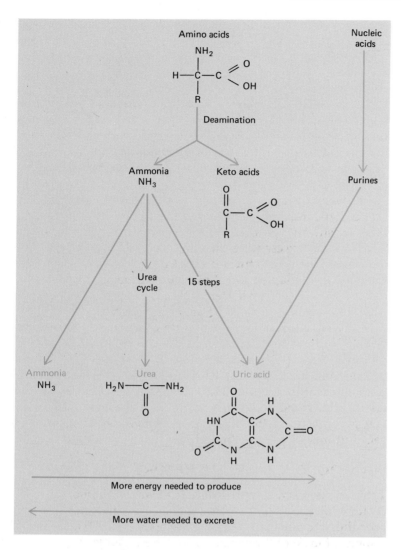

FIGURE 45–2 Nitrogenous wastes are formed by deamination of amino acids and by metabolism of nucleic acids. Ammonia is the first metabolic product of deamination. In many animals, ammonia is converted to urea via the urea cycle (see Figure 45–3). Other animals convert ammonia to uric acid. Energy is required to convert ammonia to urea and uric acid, but less water is required to excrete these wastes.

excreted as a crystalline paste with little fluid loss. Uric acid excretion is an important adaptation for conserving water in many terrestrial animals such as insects, pulmonate snails, certain reptiles, and birds. In birds, the absence of a urinary bladder and the frequent excretion of uric acid wastes as part of the feces contributes to the light body weight that facilitates flight. Because it is nontoxic, excretion of uric acid is an adaptive advantage for species whose young begin their development enclosed in eggs.

Urea is the principal nitrogenous waste product of amphibians and mammals. It is produced mainly in the liver. The sequence of reactions by which the urea molecule is synthesized from ammonia and carbon dioxide is known as the **urea cycle** (Figure 45–3). These reactions require specific enzymes and the input of energy by the cells. However, urea is far less toxic than ammonia and so can accumulate in higher concentrations without causing tissue damage. Thus, urea can be excreted in more concentrated form. Because urea is highly soluble, it is excreted dissolved in water, and more water is needed to excrete urea than to excrete uric acid.

WASTE DISPOSAL AND OSMOREGULATION IN INVERTEBRATES

Sponges and cnidarians have no specialized excretory structures. Their wastes pass by diffusion from the intracellular fluid to the external environment. For this reason, certain environments such as coral reefs are espe-

FIGURE 45-3 The urea cycle is a sequence of enzymatic reactions by which the amino groups from amino acids are converted to urea. Carbon dioxide and ammonia are combined to form carbamoyl phosphate, which then combines with ornithine to form citrulline. Aspartate (the ionic form of the amino acid aspartic acid) then reacts with citrulline to form the intermediate argininosuccinate. This intermediate compound is cleaved to yield arginine and fumarate. Thus, the amino group of aspartate is transferred to arginine. The arginine is hydrolyzed by the enzyme arginase to yield urea and ornithine, which can be used in the next cycle. The energy to drive the cycle and synthesize urea is provided by the two ATPs used in the synthesis of carbamoyl phosphate and the ATP used in the synthesis of argininosuccinate.

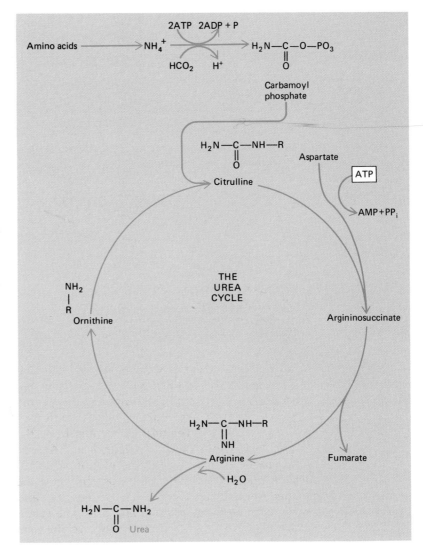

cially prone to damage when changes in water current or stagnation occur. The vast majority of sponges and cnidarians are marine organisms living in an isotonic environment. They have no special problems of excess water intake. There are a few freshwater sponges and a few freshwater cnidarians such as *Hydra* that live in a medium that is very hypotonic to their intracellular fluid. How these organisms prevent inflow of excess water or how they pump it out remains a mystery.

Nephridial Organs

Nephridial organs are a common type of excretory organ found among invertebrates. They consist of simple or branching tubes that open to the outside of the body through nephridial pores. Flatworms and nemertines are the simplest animals with specialized excretory organs. Although metabolic wastes do pass by diffusion through the body surface, these animals also have nephridial organs that consist of tubules with enlarged blind ends containing cilia. These organs are known as **protonephridia** and have **flame cells**. A branching system of excretory ducts connects the protonephridia with the outside (Figure 45–4). The flame cells lie in the fluid that bathes the body cells, and wastes diffuse into the flame cells and from there into the excretory ducts. The beating of the cilia (which when seen under the microscope suggests a flickering flame) presumably moves fluid in the ducts out toward the excretory pores.

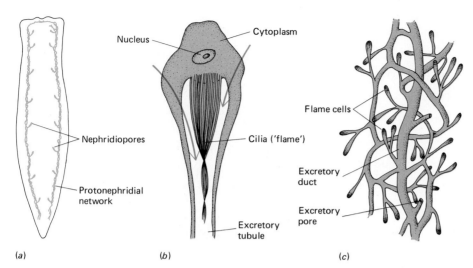

FIGURE 45–4 Nephridial organs are common among invertebrates. (*a*) The excretory organs of a typical flatworm. (*b*) A single flame cell. (*c*) Wastes collected by flame cells pass through excretory ducts to and out of the animal through nephridiopores.

The chief role of the flame cells is to regulate the water content. In fact, the number of protonephridia in a planarian is adjusted to the salinity of the environment. Planaria grown in slightly salty water develop fewer flame cells, but the number quickly increases if the concentration of salt in the environment is reduced.

Earthworms have nephridial organs called **metanephridia** in each segment of their bodies. In contrast to the flame cells of flatworms, the metanephridium is a tubule open at both ends. The inner end opens into the coelom as a ciliated funnel (Figure 45–5). Around each tubule is a network of capillaries, which permits the removal of wastes from the blood. Fluid from the coelom passes into the metanephridium. As this fluid passes through the long, looped tubule, its composition is adjusted. Water and substances such as glucose and salts are reabsorbed into the coelom and ultimately returned to the blood. Wastes are concentrated and pass out of the body as part of the very dilute, copious urine. The earthworm excretes urine at a rate of about 60% of its total body weight each day.

Molluskan nephridial organs are sometimes called "kidneys." Coelomic fluid enters the metanephridium through the ciliated nephrostome. As it passes along the metanephridium, some substances are selectively reabsorbed and others are secreted into the fluid. In aquatic forms, the resulting urine is expelled into the mantle cavity.

Antennal Glands

Antennal glands, also called **green glands,** are the principal excretory organs of crustaceans. A pair of these structures is located in the head, typically at the base of the antennae. Each gland consists of a coelomic sac, a greenish glandular chamber with folded walls, an excretory tubule, and an exit duct that is enlarged to form a bladder in some species (Figure 45–6). Fluid from the blood is filtered into the coelomic sac, and its composition is adjusted as it passes through the excretory organ. Needed materials are reabsorbed into the blood. Wastes can also be actively secreted from the blood into the filtrate within the excretory organ.

Malpighian Tubules

The excretory system of insects consists of organs called **Malpighian tubules** (Figure 45–7). There may be from two to several hundred tubules, depending upon the species. Malpighian tubules have blind ends that lie in the hemocoel, bathed in blood. Their cells transfer wastes by diffusion or

FIGURE 45–5 The excretory structures of the earthworm are a series of paired metanephridia. Each consists of a ciliated funnel opening into the coelom, a coiled tubule, and a nephridiopore opening to the outside. (*a*) Cross section. (*b*) Three-dimensional internal view, showing parts of three segments.

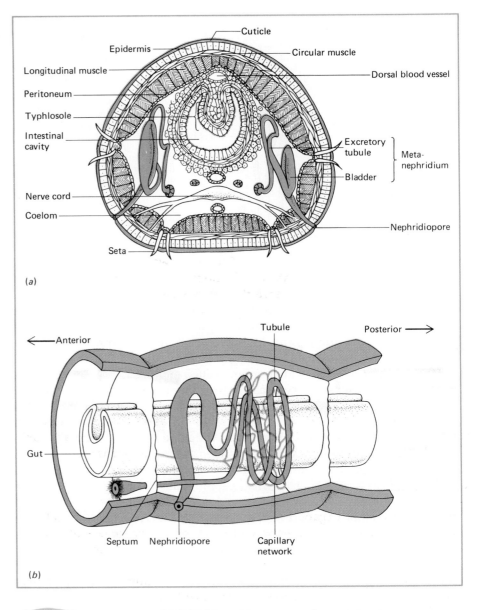

(a)

(b)

FIGURE 45–6 Antennal glands are the principal excretory structures of crustaceans. Fluid from the hemolymph is filtered into the coelomic sac, and its composition is adjusted as it passes through the excretory ducts. The exit duct may be enlarged to form a bladder.

active transport from the blood to the cavity of the tubule. Each tubule has a muscular wall, and its slow, writhing movements assist the passage of wastes down its lumen to the gut cavity. The Malpighian tubules empty into the intestine. Water is reabsorbed into the hemocoel both from the

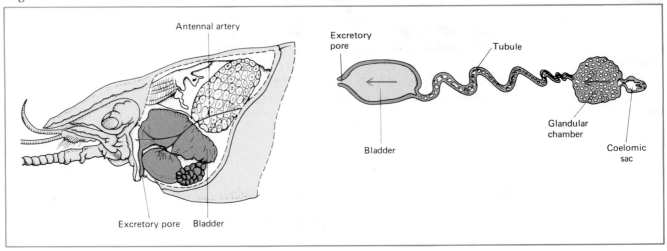

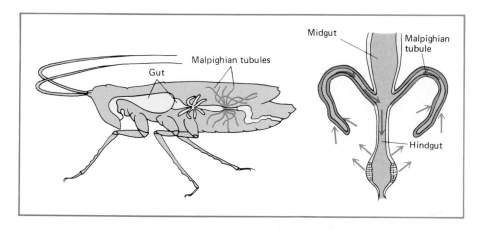

tubule and from the digestive tract. The major waste product, uric acid, precipitates as the water is reabsorbed, and is excreted as fairly dry pellets. This adaptation helps to conserve the insect's body fluids.

OSMOREGULATION AND METABOLIC WASTE DISPOSAL IN VERTEBRATES

In most vertebrates, not only the urinary system but also the skin, lungs or gills, and digestive system function to some extent in metabolic waste disposal and fluid balance. In addition, some vertebrates have special salt glands, which excrete salt. By removing water, salts, and potentially toxic waste materials, these organs help to maintain homeostasis.

The Vertebrate Kidney

The main excretory organs in vertebrates are **kidneys,** which excrete nitrogenous wastes, water, a variety of salts, and some other substances. Together with the urinary bladder and associated ducts, they form the **urinary system.**

The vertebrate kidney is composed of functional units called **nephrons,** which are described in a later section. Most vertebrate kidneys function by filtration, reabsorption, and secretion. Blood is filtered, and the initial filtrate that enters the nephron contains all of the substances present in the blood except large compounds such as proteins. (Blood cells and platelets are, of course, too large to be filtered out of the blood.) As the filtrate passes through the coiled tubules of the nephron, needed materials such as glucose, amino acids, salts, and water are selectively reabsorbed into the blood. Some substances are secreted from the blood into the filtrate. Thus, the composition of the filtrate is slowly adjusted, and the urine that is finally excreted consists of water, metabolic waste products, and materials such as salts.

Osmoregulation

Aquatic vertebrates have a continuous problem of osmoregulation. The marine bony fishes have blood and body fluids that are hypotonic to sea water. They tend to lose water osmotically and, like Coleridge's "Ancient Mariner," are in danger of becoming dehydrated even though surrounded by water. Many bony fish compensate by drinking sea water constantly. They retain the water and excrete salt by the action of specialized cells in their gills. Only a small volume of urine is produced, and the kidneys of such fish are generally very small (Figure 45–8).

FIGURE 45–8 Osmoregulation in marine fish and sharks. (*a*) Marine fish live in a hypertonic medium and so they lose water by osmosis. To compensate, a fish drinks salt water, excretes the salt, and produces very little urine. (*b*) The shark solves its osmotic problem differently. It accumulates urea in high enough concentration to become hypertonic to the surrounding medium. As a result, some water enters the body by osmosis. A large quantity of hypertonic urine is excreted.

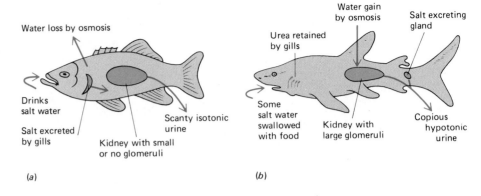

(a)   (b)

Marine chondrichthyes (sharks and rays) solve their osmotic problem in a different way (Figure 45–8). They accumulate urea and retain it in a concentration high enough to render their blood and interstitial fluid slightly hypertonic to sea water. Some water enters the body osmotically through the gills. Their tissues are adapted to function at concentrations of urea that would be toxic to most other animals. Such animals are able to excrete a hypotonic urine despite the high salinity of their environment. Excess salts are excreted both by the kidneys and by a special rectal gland in the posterior region of the intestine.

Freshwater fish are in continuous danger of becoming water-logged due to the osmotic inward flow of water. Although they are covered with a mucous secretion that retards the passage of water into the body, water does enter through the gills. The kidneys of these fish selectively reabsorb salts but not water, and they excrete a copious, dilute urine (Figure 45–9). Water entry, though, is only part of the problem of osmoregulation in freshwater fish. These animals also tend to lose salts to the surrounding fresh water. To compensate, special cells in the gills actively transport salt from the water into the body. Salts also enter the body as part of the animal's food.

Most amphibians are at least semiaquatic, and their mechanisms of osmoregulation are similar to those of freshwater fish. They produce a large amount of dilute urine. A frog can lose through its urine and skin an amount of water equivalent to one third of its body weight in a day. Loss of salt, both through the skin and in the urine, is compensated for by active transport of salt by special cells in the skin.

Certain reptiles and marine birds have salt glands in the head that can excrete the salt that enters the body with the sea water they drink (Figure 45–10). Salt glands are usually inactive and only begin to function in response to osmotic stress. Thus, when sea water or salty food is ingested, the salt glands excrete a fluid laden with sodium and chloride.

Whales, dolphins, and other marine mammals ingest sea water along with their food. Their kidneys produce a very concentrated urine, much more salty than sea water (Figure 45–10). This is an important physiological adaptation, especially for marine carnivores. The high-protein diet of these animals results in production of large amounts of urea, which must be excreted in the urine.

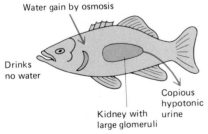

FIGURE 45–9 Freshwater fish live in a hypotonic medium, so water continuously enters the body by osmosis. These fish excrete a large quantity of dilute urine.

FLUID BALANCE AND EXCRETION IN HUMANS

In human beings, as in many other terrestrial vertebrates, the kidneys, skin, lungs, and digestive system all function to some extent in waste disposal (Figure 45–11). The lungs excrete water and carbon dioxide. The liver excretes bile salts and bile pigments (the breakdown products of hemoglobin), which pass into the intestine and then pass out of the body with the

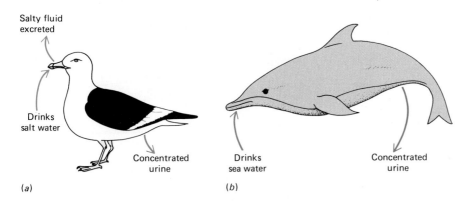

FIGURE 45–10 Osmoregulation in marine birds and mammals. (*a*) Marine birds have salt glands in the head that can excrete the excess salt that enters the body with the sea water they drink. They excrete a concentrated urine. (*b*) Marine mammals drink sea water along with their food. Their kidneys produce a very salty urine.

feces. Though primarily concerned with the regulation of body temperature, the sweat glands also excrete 5–10% of all metabolic wastes. Sweat contains the same substances (salts, urea, and water) as those in urine but is much more dilute, having only about an eighth as much solid matter. The volume of perspiration varies, ranging from about 500 ml on a cool day to as much as 2 or 3 liters on a hot one. While doing hard work at high temperatures, a human may excrete from 3 to 4 liters of sweat in an hour!

The Human Kidney and Its Ducts

The human kidneys are paired, bean-shaped structures about 10 cm long, each about the overall size of a fist. They are located just below the diaphragm in the abdominal cavity (Figure 45–12). The outer portion of the kidney is called the **renal cortex;** the inner portion is the **renal medulla.** As urine is produced, it flows into the **renal pelvis,** a funnel-shaped chamber, which is continuous with the **ureter,** the duct conveying urine from the kidney. The ureter and blood vessels connect with the kidney at its **hilus,** an indentation on its medial border.

FIGURE 45–11 In many terrestrial vertebrates, the kidneys, lungs, skin, and digestive system all participate in disposal of metabolic wastes. (*a*) The kidney conserves water by reabsorbing it. (*b*) Disposal of metabolic wastes in humans and other terrestrial mammals. To conserve water, a small amount of hypertonic urine is usually produced. Nitrogenous wastes are produced by the liver and transported to the kidneys. All cells produce carbon dioxide and some water during cellular respiration.

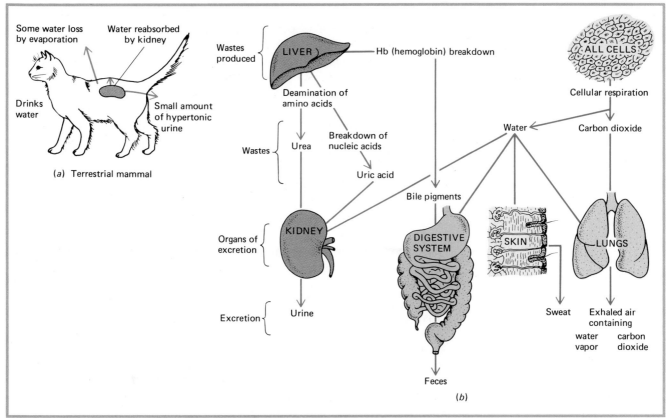

FIGURE 45–12 The human urinary system. (*a*) Urine is produced in the kidneys, then conveyed by the ureters to the urinary bladder for temporary storage. The urethra conducts urine from the bladder to the outside of the body. (*b*) Structure of the kidney. When urine is produced, it flows into the renal pelvis and leaves the kidney through the ureter. The renal artery delivers blood to the kidney; the renal vein drains blood from the kidney. (*c*) The location of the two main types of nephrons.

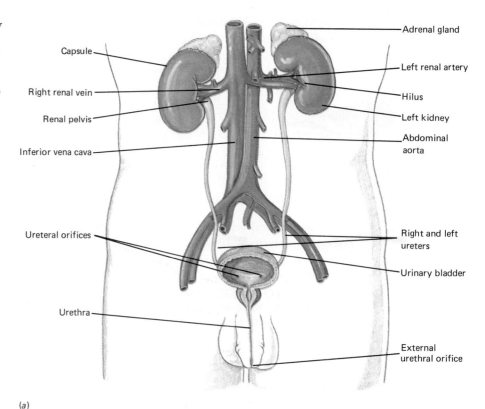

(*a*)

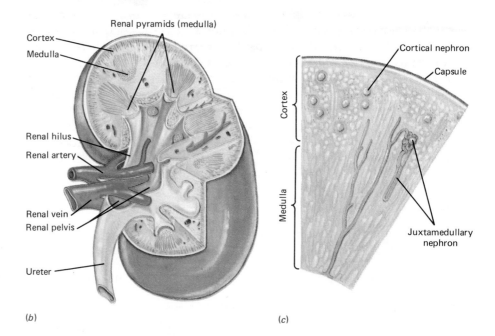

(*b*)   (*c*)

The kidneys produce urine from the water, salts, wastes, and other materials that they filter from the blood. By adjusting the amount of water, salts, and other materials excreted, the kidneys help to maintain the body's internal chemical balance. Indeed, they are indispensable homeostatic organs.

The urine, excreted by the kidney in a continuous trickle, collects in the renal pelvis. From the pelvis, urine, moved along by peristaltic contractions, passes through one of the paired **ureters.** Each ureter, about 25 cm long, connects its kidney with the **urinary bladder,** a hollow, muscular organ located in the lower, anterior portion of the pelvic cavity. The smooth muscle and special transitional epithelium of the urinary bladder enable it to stretch to accommodate up to 800 ml (more than a pint and a half) of urine. As the volume of urine in the bladder increases, the distension of its muscular walls stimulates nerve endings to send impulses to the brain, producing the sensation of fullness. Impulses may be sent back to the bladder, causing **micturition** (urination), the expulsion of urine from the bladder.

The **urethra** is a duct leading from the urinary bladder to the outside of the body. In the female, the urethra is short and transports only urine. In the male, however, the urethra is lengthy and passes through the penis. Semen as well as urine is transported through the male urethra (although at different times). The length of the male urethra discourages bacterial invasion of the bladder, so that such infections are less common in males than in females.

Bladder control depends upon the learned ability to facilitate or inhibit the reflex action that causes micturition. Humans can learn to empty the bladder voluntarily at a convenient time even before it is full. Similarly, micturition can be inhibited for some time even after the bladder is full. Such voluntary control cannot be exerted by an immature nervous system, and most babies are unable to develop complete urinary control until about 2 years of age.

The Nephron

Each kidney contains more than a million functional units called nephrons that regulate the composition of the blood and excrete wastes. A nephron consists of two main parts: a **renal corpuscle** and a **renal tubule** (Figure 45–13). The renal corpuscle is made up of a double-walled, hollow cup of cells, called **Bowman's capsule,** and a spherical tuft of capillaries, the **glomerulus,** which projects into the capsule. The renal tubule consists of three main regions: the **proximal** (near) **convoluted tubule, loop of Henle,** and the **distal** (far) **convoluted tubule.** Each distal convoluted tubule delivers its content into a **collecting duct.** Many distal convoluted tubules empty into each collecting duct.

The inner wall of Bowman's capsule consists of specialized epithelial cells called **podocytes.** The podocytes possess elongated foot processes, which cover the surfaces of most of the glomerular capillaries. Spaces between the "toes" of these foot processes are called **slit pores.**

Urine Formation

Urine is produced by a combination of three processes: filtration, reabsorption, and tubular secretion.

Filtration

Filtration occurs at the junction between the glomerular capillaries and the wall of Bowman's capsule (Figure 45–14). Blood is delivered to the kidneys by the renal arteries. Branches of the renal arteries give rise to afferent arterioles. (An afferent arteriole conducts blood into a structure; an efferent arteriole conducts blood out of a structure.) In the kidney, an **afferent arteriole** delivers blood into the capillaries of the glomerulus. The blood then drains from the glomerulus by a much narrower **efferent arteriole.** Constriction of the efferent arteriole produces high hydrostatic pressure in the glomerulus. This hydrostatic pressure forces fluid to leave the capillaries of the glomerulus. The fluid passes through the slit pores and enters the urinary pipeline. Once it enters Bowman's capsule, this fluid is referred to as the **glomerular filtrate.**

The amount of fluid that enters Bowman's capsule depends on the **effective filtration pressure,** the combination of mechanical and osmotic forces

FIGURE 45–13 Each kidney is composed of more than a million microscopic nephrons. (*a*) Diagrammatic view of the basic structure of a nephron. (*b*) Detailed view of a renal corpuscle. Note that the distal convoluted tubule is actually adjacent to the afferent arteriole. (*c*) Low-power scanning electron micrograph of a portion of the kidney cortex, showing glomeruli and much of the kidney's circulatory system. Urine forms by filtration from the blood in the glomeruli and by adjustment of the filtrate as it passes through the tubules that drain the glomeruli. (For a magnified view of a single glomerulus see chapter opening photograph.)

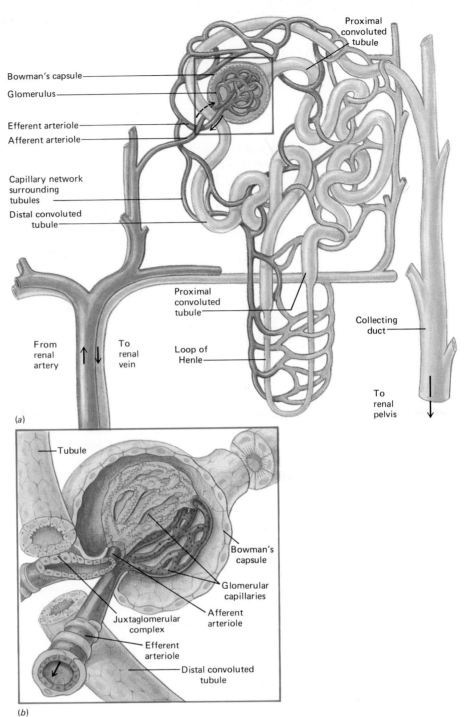

(*a*)

(*b*)

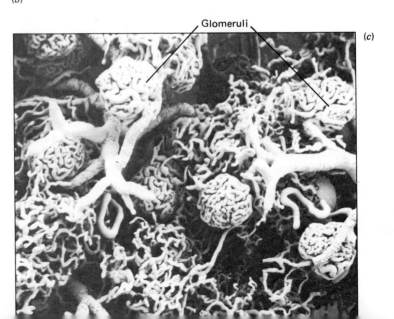

(*c*)

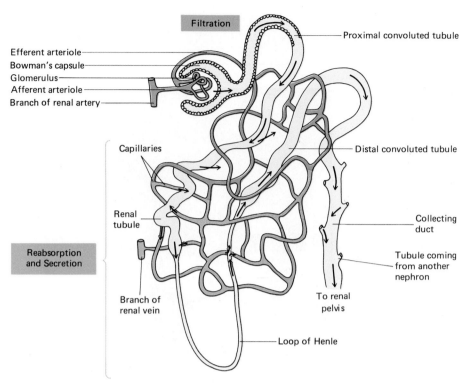

Filtration

Efferent arteriole
Bowman's capsule
Glomerulus
Afferent arteriole
Branch of renal artery

Proximal convoluted tubule

Capillaries

Distal convoluted tubule

Renal tubule

Collecting duct

Reabsorption and Secretion

Tubule coming from another nephron

Branch of renal vein

To renal pelvis

Loop of Henle

(a)

FIGURE 45–14 Nephron function. (*a*) Diagram of a nephron showing where filtration, reabsorption, and secretion take place. The dashed, colored arrows indicate reabsorption of water, glucose, salts, and other substances in the capillaries. (*b*) Diagram of a nephron showing the pressure gradients that move fluids from blood to glomerular filtrate (filtration pressure). Substances are reabsorbed, largely in the proximal convoluted tubules, by processes of active transport. (*c*) The total fluid movement in all of the nephrons.

Afferent arteriole   Efferent arteriole

95 mm Hg   18 mm Hg

70 mm Hg

Bowman's capsule

32 mm Hg Colloidal osmotic pressure

Net filtration

14 mm Hg

Proteins Water, Salts

Distal convoluted tubule

Proximal convoluted tubule

10 mm Hg

Collecting tubule

(b)

Blood flow 1200 ml/min   Blood flow 1075 ml/min

Glomerular filtrate flow 125 ml/min

Blood flow 1199 ml/min to vein

Urine flow 1 ml/min

(c)

that determines filtration. The main force *favoring* filtration is the hydrostatic pressure of the blood in the glomerulus. Most of the *resistance* to this push results from (1) the resistance of the capillary wall and the wall of Bowman's capsule to the passage of material, (2) the hydrostatic pressure of the fluid already in the lumen of Bowman's capsule, and (3) the difference in osmotic pressure between the blood and the filtrate.

Filtration determines the composition of urine only to a minor extent because, except for holding back most of the plasma proteins, filtration is not a selective process. Except for a small amount of albumin, the large plasma proteins remain in the blood with the blood cells and platelets. However, smaller materials dissolved in the plasma—such as glucose,

Focus on KIDNEY DISEASE, DIALYSIS, AND TRANSPLANT

Kidney disease ranks fourth among major diseases in the United States. Kidney function can be impaired by infections, by poisoning caused by substances such as mercury or carbon tetrachloride, by lesions, tumors, kidney stone formation, shock, or by many circulatory diseases. One of the most common kidney diseases is *glomerulonephritis,* which is actually a large number of related chronic diseases in which the glomeruli are damaged. The damage is thought to result from an autoimmune response.

In chronic kidney disease there is a progressive loss of renal function that may eventually reach the stage of **kidney failure.** In kidney failure there is a decrease in the glomerular filtration rate and the kidneys are unable to maintain homeostasis of the blood. Homeostatic balance of water, sodium, potassium, calcium, and other salts is no longer possible, and nitrogenous wastes are not excreted. Retention of water causes edema and, as the concentration of hydrogen ions increases, acidosis develops. Nitrogenous wastes accumulate in the blood and tissues, a condition referred to as **uremia.** If untreated, the acidosis and uremia can cause coma and eventually death. Chronic kidney failure can be treated by kidney dialysis or by kidney transplant.

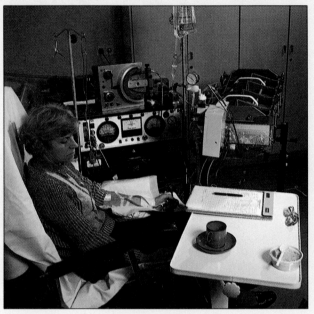

■ A patient undergoing dialysis in a hospital setting.

Kidney Dialysis Is Used to Treat Patients with Kidney Failure

Dialysis is the process of separating solutes in a solution by diffusion across a semipermeable membrane. A kidney dialysis machine can be used to restore appropriate solute balance to a patient whose kidneys are not functioning.

In extracorporeal dialysis, a plastic tube is surgically inserted into both an artery and a vein in the patient's arm or leg. These tubes can then be connected to a circuit of plastic tubing from a dialysis machine. The patient's blood flows through the tubing, which is immersed in a solution containing most of the normal blood plasma constituents in their homeostatic proportions. The walls of the plastic tubing constitute a semipermeable membrane. Since the dialysis fluid contains no wastes, nitrogenous wastes such as urea pass from the patient's blood through minute pores in the tubing and into the surrounding solution. As the blood circulates repeatedly through the tubing in the machine, dialysis continues, eventually adjusting most of the values of the patient's blood chemistry to normal ranges. Although much improved by recent engineering advances, machine dialysis is very expensive ($20,000 to $30,000

per year), clumsy, and inconvenient, and may produce serious side effects such as osteoporosis.

A different dialysis technique, continuous ambulatory peritoneal dialysis (CAPD), makes use of the fact that the peritoneum (the lining of the abdominal cavity) is a differentially permeable membrane. A plastic bag containing dialysis fluid is attached to the patient's abdominal cavity, and the fluid is allowed to run into the abdominal cavity. After about 30 minutes, the fluid is withdrawn into the bag and discarded. This process is repeated about three times each day. This type of dialysis is much more convenient but poses the threat of peritonitis, should bacteria enter the body cavity with the dialysis fluid.

Kidney Transplant Is More Effective than Dialysis

Long-term use of dialysis is not as desirable for the patient as would be a functioning kidney. With a successful kidney transplant, a patient can live a more normal life with far less long-term expense. At present more than two thirds of kidney transplants are successful for several years, although physicians must routinely treat the problems of graft rejection (discussed in Chapter 43). There are several recipients of kidney transplants who have survived for more than 20 years.

amino acids, sodium, potassium, chloride, bicarbonate, other salts, and urea—are filtered out of the blood and become part of the filtrate.

The total volume of blood passing through the kidneys is about 1200 ml per minute or about one fourth of the entire cardiac output. The plasma passing through the glomerulus loses about 20% of its volume to the glomerular filtrate; the rest leaves the glomerulus through the efferent arteriole. The normal glomerular filtration rate amounts to about 180 liters (about 45 gallons) each 24 hours. This is four and a half times the amount of fluid in the entire body! Common sense tells us that urine could not be excreted at that rate. Within a few moments, dehydration would become a life-threatening problem.

Reabsorption

The threat to homeostasis posed by the vast amounts of fluid filtered by the kidneys is avoided by **reabsorption.** About 99% of the filtrate is reabsorbed into the blood through the renal tubules, leaving only about 1.5 liters to be excreted as urine. Reabsorption permits precise regulation of blood chemistry by the kidneys. Needed substances such as glucose and amino acids are returned to the blood, while wastes and excess salts and other materials remain in the filtrate and are excreted in the urine. Each day the tubules reabsorb more than 178 liters of water, 1200 g (2.5 lb) of salt, and about 250 g (0.5 lb) of glucose. Most of this, of course, is reabsorbed many times over.

After leaving the glomerulus, the efferent arteriole delivers blood to a second capillary network in the nephron. This capillary network surrounds the renal tubule and receives materials returned to the blood by the tubule. Blood from these capillaries then flows into small veins, which eventually merge to form the large renal vein draining each kidney.

The simple epithelial cells lining the renal tubule are well adapted for reabsorbing materials. They have abundant microvilli, which increase the surface area for reabsorption (and give the inner lining a "brush border" appearance; see Figure 45–15). These cells also contain large numbers of mitochondria needed to provide the energy for running the cellular pumps that actively transport materials.

About 65% of the filtrate is reabsorbed as it passes through the proximal convoluted tubule. Glucose, amino acids, vitamins, and other substances of nutritional value are reabsorbed there, as are many ions, including sodium, chloride, bicarbonate, and potassium. Some of these ions are actively transported; others follow by diffusion. Reabsorption continues as the filtrate passes through the loop of Henle and the distal convoluted tubule. Then the filtrate is further concentrated as it passes through the collecting duct leading to the renal pelvis.

Normally, substances that are useful to the body, such as glucose or amino acids, are reabsorbed from the renal tubules. If the concentration of a particular substance in the blood is high, however, the tubules may not be able to reabsorb all of it. The maximum rate at which a substance can be

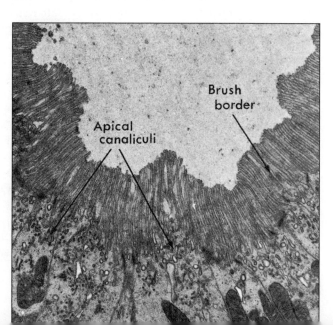

FIGURE 45–15 An electron micrograph of the cells of the proximal tubule of the kidney showing the brush order of the lumen (approximately ×6000).

Brush border

Apical canaliculi

reabsorbed is called its **tubular transport maximum,** or **Tm.** For example, the Tm for glucose averages 320 mg/min for an adult human being. Normally, the tubular load of glucose is only about 125 mg/min, so almost all of it is reabsorbed. However, if glucose is filtered in excess of the Tm, that excess will not be reabsorbed but will instead pass into the urine.

Each substance that has a Tm also has a **renal threshold** concentration in the plasma. When a substance exceeds its renal threshold, the portion not reabsorbed is excreted in the urine. In a person with diabetes mellitus, the concentration of glucose in the blood exceeds its threshold level (about 150 mg glucose per 100 ml of blood), so glucose is excreted in the urine. Its presence in the urine is evidence of this disorder.

Secretion

Some substances, especially potassium, hydrogen, and ammonium ions, are secreted from the blood into the filtrate. Some drugs, such as penicillin, are also removed from the blood by secretion. Secretion occurs mainly in the region of the distal convoluted tubule.

Secretion of hydrogen ions is an important homeostatic mechanism for regulating the pH of the blood. Secretion of hydrogen ions takes place through the formation of carbonic acid. Carbon dioxide, which diffuses from the blood into the cells of the distal tubules and collecting ducts, combines with water to form carbonic acid. This acid then dissociates hydrogen ions and bicarbonate ions. When the blood becomes too acidic, more hydrogen ions are secreted into the urine.

The secretion of potassium is also very important. When potassium concentration is too high, nerve impulses are not effectively transmitted and the strength of muscle contraction decreases. In fact, a high potassium concentration can weaken the heart and lead to death from heart failure. When the concentration of potassium ions is too high, potassium ions are secreted from the blood into the renal tubules, and the ions are excreted in the urine. Secretion results from a direct effect of the potassium ions on the tubules. A high potassium ion concentration in the blood also stimulates the adrenal cortex to increase its output of the hormone aldosterone. This hormone further stimulates secretion of potassium.

How Urine Is Concentrated

When fluid intake is high, a large volume of dilute urine is excreted. Wastes must be excreted even when fluid intake is low, but to conserve fluid, a small volume of concentrated urine is produced. The ability of the kidneys to produce either concentrated or dilute urine depends on a high salt concentration in the interstitial fluid in the medulla of the kidney. This high salt concentration is maintained by salt reabsorption from various regions of the renal tubule and a **countercurrent mechanism.**

The salts and urea in the glomerular filtrate are used to increase the osmotic concentration of the interstitial fluid. This causes water to pass from the filtrate into the interstitial fluid. From there, water is reabsorbed into the blood.

There are two types of nephrons: the cortical nephrons and the more internal juxtamedullary nephrons. The **juxtamedullary nephrons** have a very long loop of Henle that extends deep into the medulla. This loop consists of a descending loop, into which filtrate flows, and an ascending loop, through which the filtrate passes on its way to the distal convoluted tubule. The loop of Henle is specialized to produce a high concentration of sodium chloride in the medulla. A highly hypertonic interstitial fluid is set up near the bottom of the loop.

The walls of the descending portion of the loop of Henle are relatively permeable to water but relatively impermeable to salt and urea. As the

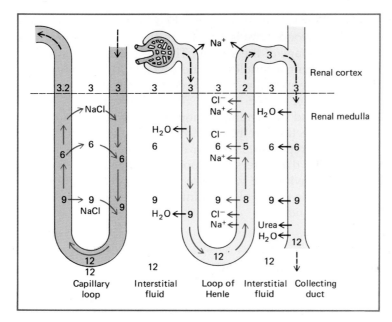

FIGURE 45–16 Countercurrent mechanism. The walls of the descending loop of Henle are relatively permeable to water but impermeable to salt and urea. Water passes out of the descending loop, leaving a more concentrated filtrate inside. At the turn of the loop, the walls of the tubule become more permeable to salt and less permeable to water. Salt moves out into the interstitial fluid. Thus, salt concentration is greater at the bottom of the loop. Higher in the ascending loop, chloride pumps transport chloride out into the interstitial fluid, and sodium follows. Urea moves out into the interstitial fluid through the collecting ducts. This countercurrent mechanism establishes a very hypertonic interstitial fluid near the renal pelvis that draws water osmotically from the filtrate in the collecting ducts. The solute concentration is given in milliosmols per liter.

filtrate passes down the loop of Henle, water passes out by osmosis, leaving a more concentrated filtrate in the loop of Henle (Figure 45–16). Water diffuses out as a result of a high concentration of sodium chloride and urea in the interstitial fluid (discussed later).

At the turn of the loop of Henle, the walls of the tubule become more permeable to salt and less permeable to water. As the concentrated filtrate moves up the ascending portion of the loop of Henle, salt moves out into the interstitial fluid. Thus, salt is more concentrated at the bottom of the loop. Higher in the ascending part of the loop of Henle, pumps are present that transport chloride out into the interstitial fluid. The negatively charged chloride ions attract positively charged sodium ions, which follow them.

The collecting ducts are permeable to urea, allowing the concentrated urea in the filtrate to diffuse out into the interstitial fluid. This results in a high urea concentration in the interstitial fluid, which is responsible for water passing out of the descending part of the loop of Henle.

Because water passes out of the descending loop, there is a high concentration of salt at the bottom of the loop. However, since salt (but not water) is removed in the ascending loop, by the time the filtrate flows through the distal convoluted tubule, it is isotonic to blood or even hypotonic to it. As the filtrate moves down the collecting duct, water continues to pass by osmosis into the interstitial fluid where it is collected by blood vessels.

Note that there is a *counterflow* of fluid through the two limbs of the loop of Henle. Filtrate passing down through the descending portion of the loop is flowing in a direction opposite to filtrate moving upward through the ascending loop. The filtrate is concentrated as it moves down the descending portion of the loop and diluted as it moves up the ascending part of the loop. This **countercurrent mechanism** helps maintain a high salt concentration in the interstitial fluid of the medulla. The hypertonic interstitial fluid draws water osmotically from the filtrate in the collecting ducts.

The collecting ducts are routed so that they pass through the zone of hypertonic interstitial fluid on their way to the renal pelvis. This loss of water from the ducts concentrates the urine to such an extent that it can be hypertonic to blood. *A hypertonic urine has a low concentration of water and so conserves water.* The urine becomes most hypertonic in thirst (in fact, thirst is a signal that the fluid content of the blood is low). In thirst, the permeability of the collecting duct walls becomes very high due to the action of the hormone ADH (discussed in a later section).

The water that diffuses from the filtrate into the interstitial fluid is removed by blood vessels known as the **vasa recta** and carried off in the venous drainage of the kidney. The vasa recta are long, looped extensions of the efferent arterioles of the juxtamedullary nephrons. They extend deep into the medulla, only to return "hairpin fashion" to the cortical venous drainage of the kidney. Blood flows in opposite directions in the ascending and descending regions of the vasa recta just as filtrate flows in opposite directions in the ascending and descending portions of the loop of Henle. This represents another countercurrent flow in which much of the salt and urea that enters the blood leaves again; the solute concentration of the blood leaving the vasa recta is only slightly higher than that of blood entering. This mechanism maintains the high solute concentration of the interstitial fluid.

COMPOSITION OF URINE

By the time the filtrate reaches the renal pelvis, its composition has been precisely adjusted. Useful materials have been returned to the blood by reabsorption. Wastes and unneeded materials that became part of the filtrate by filtration or secretion have been retained by the renal tubules. The adjusted filtrate, called **urine,** is composed of about 96% water, 2.5% nitrogenous wastes (mainly urea), 1.5% salts, and traces of other substances.

The composition of the urine yields many clues to body function or malfunction. **Urinalysis,** the physical, chemical, and microscopic examination of urine, is a very important diagnostic tool and is used to monitor many disorders such as diabetes mellitus.

FLUID HOMEOSTASIS

If salt and water intake were always the same, and if an organism were always exposed to the same conditions (such as heat and humidity) that influence water loss, the kidneys would not have to vary their output. However, both the ingestion of water and its loss vary greatly from time to time, so that in order to maintain homeostasis, the kidney output must be appropriately regulated.

Regulation of Urine Volume

The rate of reabsorption of water from the collecting ducts varies. The permeability of these ducts to water is increased by **antidiuretic hormone (ADH),** a hormone synthesized in the hypothalamus and released by the posterior lobe of the pituitary gland.

When fluid intake is low, the body begins to become dehydrated (Figure 45–17). The concentration of salts dissolved in the blood becomes greater, causing an increase in osmotic pressure of the blood. Specialized receptors (osmoreceptors) in the brain are sensitive to such change. The posterior lobe of the pituitary responds with increased ADH secretion. As a result, the walls of the collecting ducts become more permeable and more water is reabsorbed. In this way, more water is conserved for the body, blood volume increases, and conditions are restored to normal. Thus, the more ADH that is secreted, the less water is lost from the body. ADH promotes a small volume of concentrated urine.

On the other hand, intake of large amounts of fluid results in dilution of the blood and a fall in osmotic pressure. Release of ADH decreases, lessening the amount of water reabsorbed from the collecting ducts. A large amount of dilute urine is then produced.

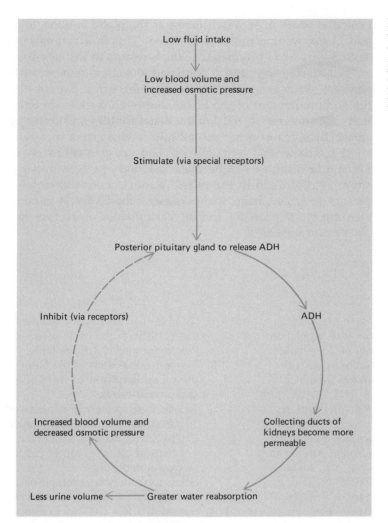

FIGURE 45–17 Regulation of urine volume. When fluid is low, blood volume decreases and osmotic pressure increases. The posterior lobe of the pituitary gland releases the hormone ADH, which stimulates increased water reabsorption.

ADH release is increased during sleep and is reduced by some diuretic agents, such as alcoholic beverages, that increase urine volume. (Caffeine is also a diuretic but works by a different mechanism.) In the disorder **diabetes insipidus,** the body cannot make sufficient ADH, or the kidneys lack receptors for ADH. Large volumes of urine are produced, and the sufferer must drink immense quantities of water to compensate for this fluid loss. Diabetes insipidus can now be treated with injections of ADH or with an ADH nasal spray. (Diabetes insipidus should not be confused with diabetes mellitus, a disorder of carbohydrate metabolism; see Chapter 49. Diabetes mellitus also increases urine flow, and with some thought you can explain why. The clue is that the increased glucose content of the urine increases its tonicity.)

Regulation of Sodium Reabsorption

Fluid balance is linked closely with sodium balance because sodium is the most abundant extracellular ion, accounting for about 90% of all positive ions outside the cells. Recall that when salt concentration increases, water is drawn to the region osmotically. An increase in dietary salt intake thus increases the osmotic pressure of the blood. Special receptors (the thirst center) in the hypothalamus sense this increase in osmotic pressure and cause the sensation of thirst and the desire to drink fluids.

When the thirst drive is activated, an animal drinks water. This water intake dilutes the sodium, restoring osmotic balance of the blood. However, the addition of water increases the blood volume, and now the normal blood volume must be restored. This is accomplished in part by a decrease in ADH production, which results in increased water excretion.

Sodium concentration is regulated to some extent by the hormone **aldosterone,** which is secreted by the adrenal cortex. Even a slight decrease in the sodium content of the blood causes an increase in aldosterone secretion. Aldosterone stimulates the distal tubules and collecting ducts to increase their reabsorption of sodium. Aldosterone secretion is also regulated by renin and angiotensin. The enzyme **renin** is secreted by the kidneys in response to a decrease in blood pressure or in sodium or potassium concentration in the blood. Renin converts angiotensinogen in the plasma to angiotensin, which causes blood vessels to constrict, thus increasing blood pressure and increasing aldosterone secretion. Sodium reabsorption increases.

■ SUMMARY

I. Three functions of the excretory system are excretion of metabolic wastes, osmoregulation, and regulation of concentrations of the constituents of body fluids.

II. Principal metabolic wastes in most animals are water, carbon dioxide, and nitrogenous wastes such as ammonia, urea, and uric acid.

III. Various mechanisms for waste disposal and osmoregulation are found in invertebrates.
   A. Nephridial organs consist of tubes that open to the outside of the body through nephridial pores. The protonephridia of flatworms and the metanephridia of earthworms are nephridial organs.
   B. Antennal glands, found in crustaceans, consist of a coelomic sac, a greenish glandular chamber, an excretory tubule, and an exit duct. Fluid filtered into the coelomic sac is adjusted as it passes through the excretory organ.
   C. The Malpighian tubules of insects have blind ends that lie in the hemocoel. Wastes are transferred from the blood to the tubule by diffusion and active transport. Water is reabsorbed into the hemocoel, and uric acid is excreted into the intestine.

IV. The vertebrate kidney functions in excretion and osmoregulation and is vital in maintaining homeostasis.

V. Aquatic vertebrates have a continuous problem of osmoregulation.
   A. Marine bony fish lose water osmotically. They compensate by drinking sea water and excreting salt through their gills; only a small volume of urine is produced.
   B. Marine cartilaginous fish retain large amounts of urea, which enables them to take in water osmotically through the gills. This water can be used to excrete a hypotonic urine.
   C. Freshwater fish take in water osmotically. They excrete a large volume of dilute urine.
   D. Other aquatic vertebrates have specific adaptations for dealing with osmoregulation. Marine birds and reptiles, for example, have salt glands that excrete excess salt.

VI. The urinary system is the principal excretory system in human beings and other vertebrates.
   A. The kidneys produce urine, which passes through the ureters to the urinary bladder for storage. During micturition the urine passes through the urethra to the outside of the body.
   B. The functional units of the kidneys are the nephrons. Each nephron consists of a renal corpuscle and a renal tubule.
      1. The renal corpuscle consists of a glomerulus surrounded by Bowman's capsule.
      2. The renal tubule consists of the proximal convoluted tubule, the loop of Henle, and the distal convoluted tubule.
   C. Urine formation is accomplished by filtration of plasma, reabsorption of needed materials, and secretion of a few substances into the renal tubule.
      1. Plasma is filtered out of the glomerular capillaries and into Bowman's capsule, but macromolecules such as protein are retained in the blood. Filtration is not a selective process. Thus, needed materials such as glucose and amino acids become part of the filtrate, as well as wastes and excess substances.
      2. About 99% of the filtrate is reabsorbed from the renal tubules into the blood. This is a very selective process that returns needed materials to the blood and adjusts the composition of the filtrate.
      3. Potassium ions, hydrogen ions, ammonia, and certain drugs are secreted from the blood into the renal tubule.
      4. Countercurrent exchange establishes a very hypertonic interstitial fluid which draws water osmotically from the filtrate in the collecting ducts. This permits concentration of urine in the collecting ducts so that urine hypertonic to the blood can be produced.
   D. The adjusted filtrate, called urine, consists of water, nitrogenous wastes, salts, and a variety of other substances.
   E. Urine volume is regulated by the hormone ADH, which is released in appropriate amounts by the posterior lobe of the pituitary. ADH increases the permeability of the collecting ducts so that more water is reabsorbed.

■ POST-TEST

1. The process of removing metabolic wastes from the body is _____.
2. The principal nitrogenous waste product of insects and birds is _____ _____.
3. The principal nitrogenous waste product of amphibians and mammals is _____.
4. Flatworms have excretory structures called _____, which are characterized by _____ cells.
5. Earthworms have _____ in each of their body segments.
6. The principal excretory organs of crustaceans are _____ _____.
7. The excretory system of insects consists of organs called _____ _____.
8. The vertebrate kidney consists of functional units called _____.

For each group, select the most appropriate answer from Column B for each description in Column A.

Column A	Column B
9. Drink water constantly; small kidneys produce small urine volume	a. Freshwater fish
10. Accumulate urea; excrete hypotonic urine; excrete sodium	b. Marine bony fish
11. Kidneys produce concentrated urine, saltier than surrounding water	c. Marine mammals
12. Excrete large volume of dilute urine; transport salt into body	d. Marine chondrichthyes (sharks)
13. Have salt glands in head that excrete excess salt	e. Marine birds

Column A	Column B
14. Outer portion of human kidney	a. Cortex
15. Delivers urine to bladder	b. Medulla
16. Funnel-shaped chamber that receives urine from collecting ducts.	c. Ureter
	d. Urethra
	e. Renal pelvis

Column A	Column B
17. Site of filtration	a. Collecting duct
18. Receives filtrate from Bowman's capsule	b. Proximal convoluted tubule
19. Site of most renal secretion	c. Distal convoluted tubule
	d. Bowman's capsule
	e. Ureter

20. The glomerulus consists of a tuft of _____ that project into _____.
21. Blood is delivered to the glomerulus by an _____ _____ and leaves the glomerulus in an _____ _____.
22. Once the fluid that has left the glomerular capillaries enters Bowman's capsule, it is known as the _____ _____.
23. The tubular transport maximum is the maximum rate at which a substance can be _____.
24. When a substance exceeds its renal threshold concentration, the portion not reabsorbed is _____.
25. A _____ urine is low in water and so conserves water.
26. Antidiuretic hormone (ADH) increases the permeability of the _____ _____ so that more water is _____ and the volume of urine is _____ (increased/decreased).
27. A person with diabetes insipidus does not produce sufficient _____ and as a result _____.

■ REVIEW QUESTIONS

1. Compare mechanisms of metabolic waste disposal in the following:
   a. earthworms   d. insects
   b. flatworms   e. vertebrates
   c. crayfish
2. Define osmoregulation. What type of osmoregulatory problem is faced by marine fish? How do they solve it?
3. How do freshwater fish solve their osmoregulatory problems? How is this accomplished by marine birds?
4. In the human urinary system, name the structure associated with each of the following:
   a. filtration   d. temporary storage of urine
   b. absorption   e. conduction of urine out of
   c. urea formation   the body
5. What substances are present in normal human urine?
6. Compare the following nitrogenous wastes:
   a. ammonia   b. uric acid   c. urea
   Explain how each is formed.
7. Contrast filtration, reabsorption, and secretion.
8. Trace a molecule of urea from its formation in the liver to its excretion in the urine.
9. Why is glucose normally not present in urine? Why is it present in the urine of an individual with diabetes mellitus?
10. How is urine volume regulated? Explain.
11. In what ways are the kidneys vital homeostatic organs?

■ RECOMMENDED READINGS

Dantzler, W.H. Renal Adaptations of Desert Vertebrates. *Bioscience*, February 1982, Vol. 32, No. 2, pp 108–112. Adaptations in renal physiology are integrated with other physiological and behavioral mechanisms for desert survival.

Heatwole, H. Adaptations of Marine Snakes, *American Scientist*, Sept-Oct 1978, pp 594–604. Among the adaptations discussed are those that permit several groups of snakes to inhabit the marine environment by maintaining fluid balance.

Pollie, R. Comprehending Kidney Disease, *Science News*, 2, October 1982, p 218. A simple theory is offered to explain a complex syndrome.

Smith, H.W. *From Fish to Philosopher*, 2nd ed. Boston, Little, Brown, & Co., 1961. A fascinating summary of the evolution of the vertebrate kidney.

Solomon, E.P., and P.W. Davis. *Human Anatomy and Physiology*, Philadelphia, Saunders College Publishing, 1983. Chapters 23 and 24 focus on the human urinary system and fluid balance.

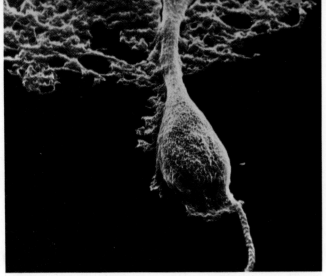

A single neuron with branching dendrites (top) and a rope-like axon

Neural Control: Neurons

■ LEARNING OBJECTIVES

After you have read this chapter you should be able to:

1. Trace the flow of information through the nervous system.
2. Describe the principal divisions of the vertebrate nervous system.
3. Compare the functions of neurons and neuroglia.
4. Draw a diagram of a neuron, giving the name and function of each structure.
5. Compare a nerve with a ganglion.
6. Explain how the resting potential of a neuron is maintained.
7. Contrast local changes in potential with an action potential.
8. Describe the effects of calcium imbalance, local anesthetics, and such agents as DDT on neuron excitability.
9. Describe synaptic transmission, and explain how it regulates the direction of neural transmission.
10. Identify the neurotransmitters described in the chapter, and give an example of where each is secreted.
11. Draw diagrams to illustrate convergence, divergence, facilitation, and reverberating circuits.
12. Define and describe the process of neural integration.
13. Draw a reflex pathway consisting of three neurons, label each structure, and indicate the direction of information flow; relate reflex action to the processes of reception, transmission, integration, and actual response.

Vibrations from approaching footsteps provoke an earthworm to retreat quickly into its burrow. When a crayfish is hungry, it seeks out and devours a wiggling worm. A child learns to look both ways before crossing a busy street, and a college student learns the principles of biology or calculus. All of these activities, and countless others, are made possible by the nervous system. In complex animals the endocrine system works with the nervous system in regulating many behaviors and physiological processes. The endocrine system generally provides a relatively slow and long-lasting regulation, whereas the nervous system permits a very rapid response.

Changes within the body or in the outside world that can be detected by an organism are termed **stimuli.** The ability of an organism to survive and to maintain homeostasis depends largely upon how effectively it can respond to stimuli in its internal or external environment.

INFORMATION FLOW THROUGH THE NERVOUS SYSTEM

The nervous system is bombarded with thousands of stimuli each day. It receives information, transmits messages, sorts and interprets incoming data, and then sends the proper messages so that the responses will be coordinated and homeostatic. Even the simplest response to a stimulus generally requires a sequence of information flow through the nervous system that includes **reception** of information, **transmission of impulses** to the brain or spinal cord, **integration,** transmission of impulses from the brain or spinal cord, and **response by an effector**—usually a muscle or gland.

Imagine that you are very hungry and an obliging friend sets a delicious steak dinner before you. You cannot lift the first forkful to your mouth until reception, transmission, integration, and response by an effector have taken place (Figure 46–1). First, you must detect the food—the stimulus. At least two types of **receptors** (your eyes and olfactory epithelium in your nose) receive the information. Second, these messages must be sent to your brain, informing you that they have received a stimulus. **Afferent (sensory) neurons** transmit this information in the form of neural impulses *from* the sense organs *to* the brain.

When a decision to eat the food has been made, **efferent (motor) neurons** transmit the message from the brain to the appropriate effector cells—in this case, certain muscle fibers in your arm and hand. The last process in this sequence is the actual contraction of the muscle fibers necessary to carry out the response. Now, finally, you spear the food with the fork and lift it into your mouth. Your ability to perform all of these actions depends on the function of the neuron.

ORGANIZATION OF THE VERTEBRATE NERVOUS SYSTEM

In vertebrates the **central nervous system (CNS)** consists of a complex brain that is continuous with a single, dorsal, tubular nerve cord (spinal cord). The central nervous system integrates all incoming information and determines appropriate responses. Within the CNS, afferent neurons **synapse** (that is, make functional connections) with **association neurons,** also called **interneurons.** Each association neuron may synapse with many other association neurons. At these synapses, incoming neural messages are sorted out and interpreted.

The **peripheral nervous system (PNS)** consists of the sensory receptors and the nerves lying outside the brain and spinal cord. Twelve pairs of cranial nerves (ten in fish and amphibians) and several pairs of spinal nerves (31 pairs in humans) link the CNS with various parts of the body. Afferent neurons in these nerves continually inform the CNS about stimuli that impinge upon the body. Efferent neurons then transmit the impulses from the CNS to appropriate effector cells—muscles and glands—which make the adjustments necessary to preserve homeostasis.

For convenience, the PNS can be subdivided into somatic and autonomic systems. The **somatic system** consists of the receptors and nerves concerned with stimuli in the outside world. The **autonomic system** consists of the receptors and nerves operating to regulate the internal environment. In the autonomic system there are two types of efferent nerves: sympathetic and parasympathetic. These divisions of the vertebrate nervous system are discussed in Chapter 47.

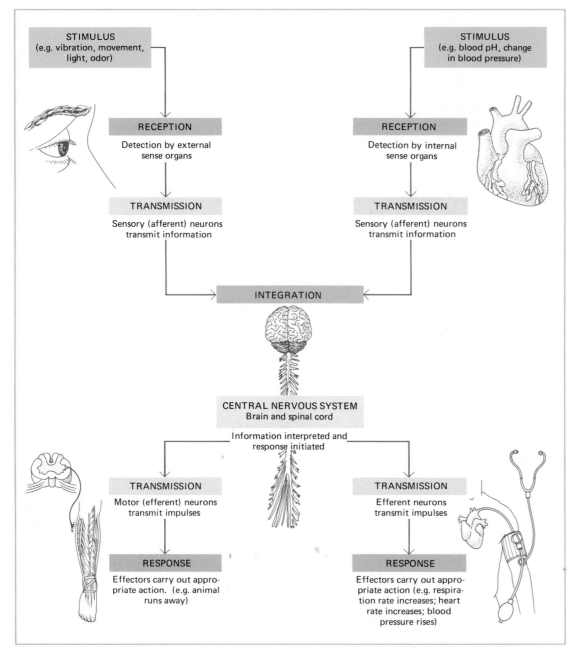

FIGURE 46–1 Flow of information through the nervous system.

STRUCTURE OF THE NEURON

The **neuron,** or nerve cell, is the structural and functional unit of the nervous system. The specialized supporting cells of nervous tissue are called **neuroglia.** A typical **multipolar neuron** consists of a **cell body**, which has bushy cytoplasmic extensions called **dendrites,** and a single, long cytoplasmic extension, the **axon** (Figure 46–2). Other types of neurons are illustrated in Figures 46–3 and 46–4.

Cell Body

The cell body contains the nucleus and many of the other organelles as well as the bulk of the cytoplasm. It is concerned with the metabolic maintenance and growth of the neuron. The cell body contains **Nissl substance,** which consists of deeply staining regions of cytoplasm rich in rough endoplasmic reticulum and free ribosomes. This is the site of protein synthesis.

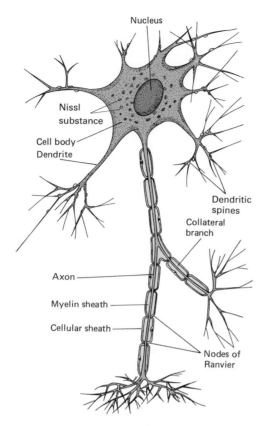

FIGURE 46–2 Structure of a multipolar neuron. The axon of this neuron is myelinated, and so the myelin sheath is shown as well as the cellular sheath.

Nucleus

Nissl substance

Cell body

Dendrite

Dendritic spines

Collateral branch

Axon

Myelin sheath

Cellular sheath

Nodes of Ranvier

Microtubules and microfilaments are distributed throughout the cytoplasm. The microtubules help to maintain the shape of the neuron, especially of the dendrites and axon, and play a role in transporting materials through the axon. The short microfilaments, composed of actin, are also thought to function in transporting materials.

Dendrites

Dendrites are highly branched extensions of the cytoplasm that project from the cell body. Dendrites and the surface of the cell body are special-

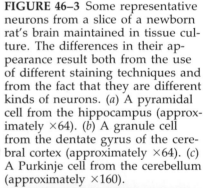

FIGURE 46–3 Some representative neurons from a slice of a newborn rat's brain maintained in tissue culture. The differences in their appearance result both from the use of different staining techniques and from the fact that they are different kinds of neurons. (*a*) A pyramidal cell from the hippocampus (approximately ×64). (*b*) A granule cell from the dentate gyrus of the cerebral cortex (approximately ×64). (*c*) A Purkinje cell from the cerebellum (approximately ×160).

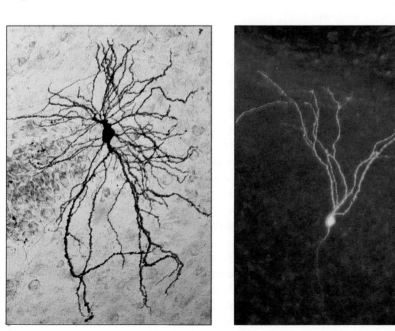

(*a*)

(*b*)

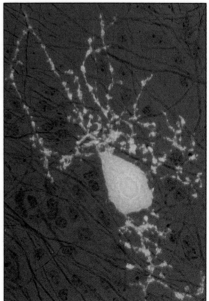

(*c*)

FIGURE 46–4 Types of neurons classified according to the number of extensions. (*a*) A multipolar neuron has many short dendrites and one long axon. Motor neurons are of this type. (*b*) A bipolar neuron has one axon and one dendrite. Bipolar neurons are found in the retina of the eye, in the olfactory nerve, and in the nerves coming from the inner ear. (*c*) A unipolar neuron has a short process extending from the cell body. This process branches into two long processes. Since it functions like a dendrite, the distal process is generally referred to as a dendrite. Sensory neurons are unipolar.

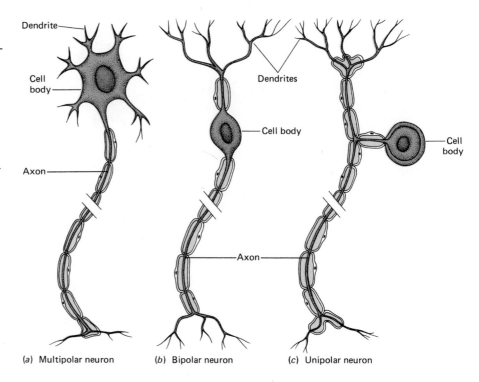

(*a*) Multipolar neuron   (*b*) Bipolar neuron   (*c*) Unipolar neuron

ized to receive stimuli. Their surfaces are dotted with thousands of tiny **dendritic spines,** which are the sites of specialized junctions with other neurons.

The Axon

The axon arises from a thickened area of the cell body, **the axon hillock.** It transmits neural impulses from the cell body to another neuron or to an effector cell. Although microscopic in diameter, the axon may be very long. A motor neuron in the leg of a giraffe may be barely 0.1 mm in diameter, but it spans a distance of several meters extending from the spinal cord to the toe. Perhaps because of its impressive length, the axon is often referred to as a **nerve fiber.**

At its distal end the axon branches extensively. At the ends of these branches are tiny enlargements called **synaptic knobs.** These knobs release a substance (a neurotransmitter) that transmits messages from one neuron to another. Branches called **collaterals** may arise along the length of the axon, permitting extensive interconnections among neurons.

Axons of some peripheral neurons are enveloped by a **cellular sheath,** or **neurilemma,** composed of supporting cells called **Schwann cells.** The cellular sheath is important in regeneration of injured nerves (see Focus on Regeneration of an Injured Neuron). In forming the cellular sheath, Schwann cells line up along the axon and wrap themselves around it. On some axons the Schwann cells produce an insulating inner sheath called the **myelin sheath.** This covering forms when the Schwann cell membrane becomes wrapped around the axon several times (Figures 46–5 and 46–6). Myelin is a white, lipid-rich substance that makes up the cell membrane of the Schwann cell. This fatty material is an excellent electrical insulator and speeds the conduction of nerve impulses (see "Saltatory Conduction," later in this chapter). Between adjacent Schwann cells there are gaps called **nodes of Ranvier.** At these points, which are from 50 to 1500 μm apart, the axon is not insulated with myelin. Almost all axons more than 2 μm in diameter are myelinated, that is, possess myelin sheaths. Those with a smaller diameter are generally unmyelinated.

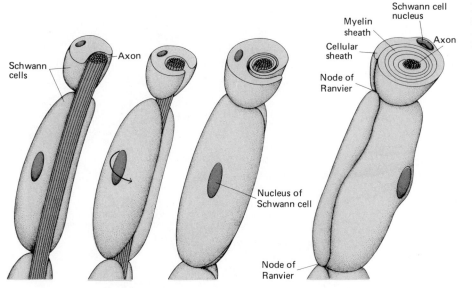

FIGURE 46–5 Formation of the myelin sheath around the axon of a peripheral neuron. A Schwann cell wraps its cell membrane around the axon many times to form the insulating myelin sheath. The rest of the Schwann cell remains outside the myelin sheath, forming the cellular sheath.

Other axons, including those within the CNS, have no neurilemma. Their myelin sheaths are formed by another type of neuroglial cells (oligodendrocytes) rather than by Schwann cells. Certain areas of the brain and spinal cord are composed principally of myelinated axons. The myelin imparts a whitish color to these areas, so that they are referred to as white matter.

When myelin is destroyed, nerve function is impaired. **Multiple sclerosis** is a neurological disease in which patches of myelin deteriorate at irregular intervals along neurons in the CNS. The myelin is replaced by a type of scar tissue, and the affected neurons are not able to conduct impulses. This leads to impaired neural function, including loss of coordination, tremor, and paralysis of affected body parts.

NERVES AND GANGLIA

The nerves observed in gross anatomical dissection consist of hundreds or even thousands of axons wrapped in connective tissue (Figure 46–7). A nerve can be compared to a telephone cable. The individual axons correspond to the wires that run through the cable, and the sheaths and connective tissue coverings correspond to the insulation. Within the CNS, bun-

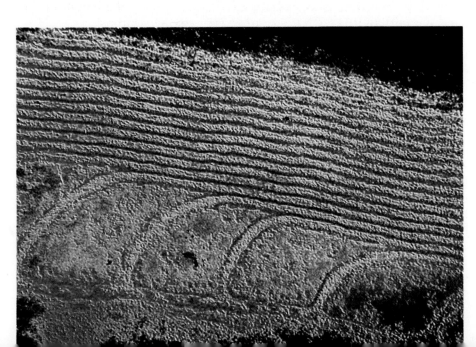

FIGURE 46–6 False-color transmission electron micrograph showing the myelin sheath surrounding the human auditory nerve. In longitudinal section, concentric layers of the myelin sheath appear as orange bands at top of image. (approximately ×88,000)

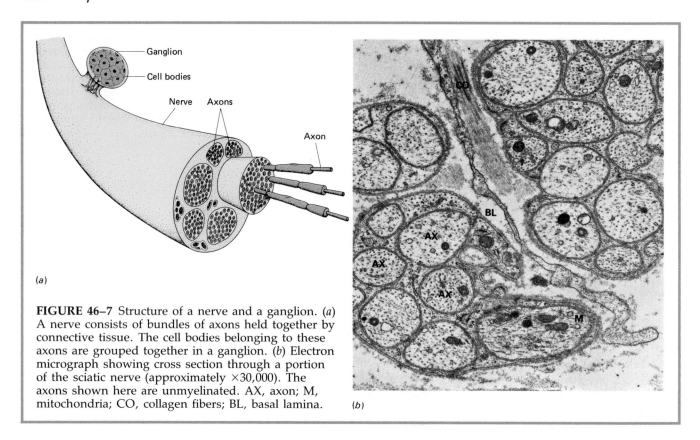

FIGURE 46–7 Structure of a nerve and a ganglion. (*a*) A nerve consists of bundles of axons held together by connective tissue. The cell bodies belonging to these axons are grouped together in a ganglion. (*b*) Electron micrograph showing cross section through a portion of the sciatic nerve (approximately ×30,000). The axons shown here are unmyelinated. AX, axon; M, mitochondria; CO, collagen fibers; BL, basal lamina.

dles of axons are referred to as **tracts** or **pathways,** rather than nerves. The cell bodies of neurons are usually grouped together in masses called **ganglia.**

TRANSMISSION OF A NEURAL IMPULSE

Once a receptor has been stimulated, the information must be conducted through a sequence of neurons. Transmission of a nerve impulse down the length of a neuron is an electrochemical process that depends upon changes in ion distribution. Transmission from one neuron to another across a synapse is generally a chemical phenomenon involving the secretion of neurotransmitter by the axon and the action of chemoreceptors in the dendrite.

The Resting Potential

In a **resting neuron**—one not transmitting an impulse—the inner surface of the plasma membrane is negatively charged compared with the interstitial fluid surrounding it (Figure 46–8). The resting neuron is said to be **electrically polarized,** that is, the inside of the membrane and the interstitial fluid outside are oppositely charged. When electric charges are separated in this way, they have the potential to do work should they be permitted to come together. The difference in electric potential between the two sides of the membrane may be expressed in millivolts (mV). (A millivolt is a thousandth of a volt and is a unit for measuring electrical potential.)

The **resting potential** of a neuron is about 70 mV. By convention this is expressed as −70 mV because the inner surface of the plasma membrane is negatively charged relative to the interstitial fluid. The resting membrane can be measured by placing one electrode, insulated except at the tip, inside the cell and a second electrode on the outside surface. The two elec-

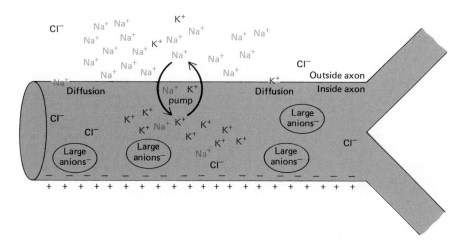

FIGURE 46–8 Segment of an axon of a resting (nonconducting) neuron. Sodium-potassium pumps in the cell membrane actively pump sodium out of the cell and pump potassium in. Sodium is unable to diffuse back to any extent, but potassium does diffuse out along its concentration gradient. Negatively charged proteins and other large anions are present in the cell. Because of the unequal distribution of ions, the inside of the axon is negatively charged compared to the tissue fluid outside.

trodes are connected with an instrument such as a galvanometer, which measures current by electromagnetic action. If both electrodes are placed on the outside surface of the neuron, no potential difference between them is registered; all points on the outside are at the same potential.

What is responsible for the resting potential? It results from the presence of a slight excess of *negative* ions *inside* the plasma membrane and a slight excess of *positive* ions *outside* the plasma membrane. This imbalance in ion distribution is brought about by several factors. The plasma membrane of the neuron has very efficient sodium-potassium pumps that actively transport sodium out of the cell and potassium into the cell. These pumps work against a concentration gradient and an electrochemical gradient. Operating these pumps requires a great deal of energy—more than one third of all of the ATP energy used by a resting cell. About three sodium ions are pumped out of the neuron for every two potassium ions that are pumped in. Thus, more positive ions are pumped out than in.

Ions also cross the neuron membrane through ion-specific channels that may be open or closed. When the channels are open, ions can move through them from an area of higher to an area of lower concentration. In the resting neuron the membrane is much more permeable to potassium than to sodium because the potassium channels are open, whereas the sodium channels are closed. As a result, sodium ions pumped out of the neuron cannot easily pass back into the cell, but potassium ions pumped into the neuron are able to diffuse out.

Potassium ions leak out through the membrane along a concentration gradient until the positive charge outside the membrane reaches a level that repels the outflow of additional positively charged potassium ions. A steady state is reached when the potassium outflow equals the inward flow of sodium ions. At this point the resting potential is about −70 mV.

Contributing to the overall ionic situation are large numbers of negatively charged proteins and organic phosphates within the neuron that are too big to diffuse out. The plasma membrane is permeable to negatively charged chloride ions, but because of the positively charged ions that accumulate outside the membrane, chloride ions are attracted to the outside and tend to accumulate there.

The resting potential is due mainly to the presence of large protein anions inside the cell and to the outward diffusion of potassium ions along their concentration gradient. The conditions for this diffusion, however, are set by the action of the sodium-potassium pumps.

Local Changes in Potential

An electrical, chemical, or mechanical stimulus may alter the resting potential by increasing the permeability of the plasma membrane to sodium, or

Focus on REGENERATION OF AN INJURED NEURON

When an axon is separated from its cell body by a cut, it soon degenerates. A hollow tube of Schwann cells remains, but myelin eventually disappears. As long as the cell body of the neuron has not been injured, however, it is capable of regenerating a new axon. Sprouting begins within a few days following cutting (see accompanying figure). The growing axon enters the old sheath tube and proceeds along it to its final destination. Axons can grow in the absence of sheaths if some conduit is provided for them. They can, for example, be made to grow within sections of blood vessels or extremely fine plastic tubes. The length of time required for regeneration depends on how far the nerve has to grow and may require as long as 2 years. When cuts occur within the spinal cord or brain, regeneration, if it occurs at all, is very feeble. It is thought that growth of new sprouts in the CNS is prevented by scar tissue formed by neuroglial cells at the site of injury.

It is remarkable that (if not blocked by scar tissue or other barrier) each regenerating axon of a cut peripheral nerve finds its way back to its former point of termination, whether this is a specific connection in the central nervous system or a specific muscle or sense organ in the periphery. If, during the early stages of development of an amphibian, an extra limb bud is transplanted next to the normally developing limb, both will grow to maturity. The extra limb then moves synchronously with the normal one. Anatomical examination reveals that the nerve that innervates the normal limb sends out branches to the extra one. Clearly, the extra limb exerts some stimulating influence on the growing nerves to produce more branches, and some directive influence as well.

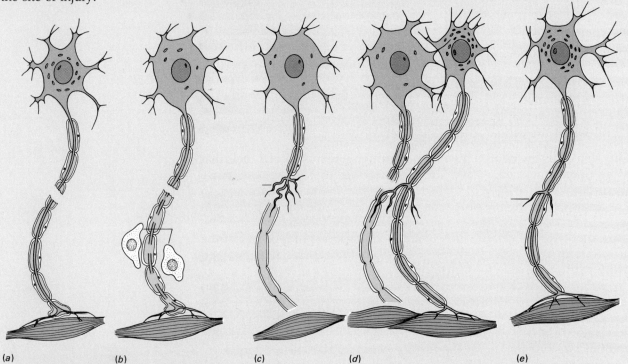

(a) (b) (c) (d) (e)

Regeneration of an injured neuron. (*a*) A neuron is severed. (*b*) The part of the axon that has been separated from its nucleus degenerates. Its myelin sheath also degenerates, and macrophages phagocytize the debris. (*c*) The tip of the severed axon begins to sprout, and one or more sprouts may find their way into the empty cellular sheath, which has remained intact. The sprout grows slowly and becomes myelinated. (*d*) Sometimes an adjacent undamaged neuron may send a collateral sprout into the cellular sheath of the damaged neuron. (*e*) Eventually the neuron may regenerate completely, so function is fully restored. Unused sprouts degenerate.

by making the neuron more negative relative to the interstitial fluid. Such local responses are called **graded potentials** because they vary in magnitude, depending on the strength of the stimulus applied.

Some neurophysiologists think that the **excitatory stimuli** open sodium gates leading into the sodium channels. This permits sodium ions to rush down the concentration gradient into the cell. The passage of positive sodium ions into the cell causes the membrane potential to become less negative for a brief moment, and so is known as **depolarization.** In contrast, **inhibitory stimuli** may hyperpolarize the membrane, that is, increase the resting potential. This occurs because of an increase in permeability of the membrane to potassium, which permits potassium ions to flow out of the neuron. With additional positively charged potassium ions outside the membrane, the neuron becomes more negative relative to the outside than when it was at rest.

Local changes in potential can cause a flow of electric current. The greater the change in potential, the greater the flow of current. Such a local current flow can function as a signal only over a very short distance, because it fades out within a few millimeters of its point of origin. As we will see, however, graded potentials can be added together, resulting in action potentials.

The Action Potential

The membrane of a neuron can depolarize by as much as 15 mV (which changes the resting potential to about −55 mV) without initiating an impulse. When the extent of depolarization reaches about −55 mV, a critical point called the **threshold level** or firing level is reached. At this point the resulting depolarization is self-propagating; that is, it spreads down the axon as a wave of depolarization without fading. The electrochemical gradient established by this spreading wave of depolarization is called a **nerve impulse** or **action potential** (Figure 46–9).

When the threshold level is reached, an almost explosive action occurs as the action potential is produced. The movement of Na^+ through the neuron membrane quickly causes the membrane potential to shoot to about +35 mV, so that there is a momentary reversal in polarity. The sharp rise and fall of the action potential is referred to as a **spike.** Figure 46–10 illustrates an action potential.

The action potential is an electric current of sufficient strength to induce collapse of the resting potential in the adjacent area of the membrane. The impulse moves along the axon at a velocity and amplitude that are constant for each type of neuron. The neuron is said to obey an **all-or-none law,** since there are no variations in intensity of the action potential: Either the neuron fires completely, or it does not fire at all.

As the wave of depolarization moves along the axon, the normal polarized state is quickly reestablished behind it. By the time the action potential moves a few millimeters along the axon, the membrane over which it has just passed begins to repolarize. The sodium gates close, so that the membrane becomes impermeable to sodium. Potassium gates in the membrane then open, permitting potassium to leave. The accumulation of potassium ions outside the membrane results in repolarization. Repolarization occurs very rapidly, but redistribution of sodium and potassium ions back to the resting condition requires more time. Resting conditions are slowly reestablished as the sodium-potassium pumps actively transport excess sodium out of the neuron.

During the millisecond or so during which it is depolarized, the axon membrane is in an **absolute refractory period,** when it cannot transmit an action potential no matter how great a stimulus is applied. Then, for two or three additional milliseconds, while the resting conditions are being reestablished, the axon is said to be in a **relative refractory period.** During this time another potential can be generated if the stimulus is stronger than the normal threshold stimulus. Even with the limits imposed by their refractory periods, neurons can transmit several hundred impulses per second!

The electrical and chemical processes involved in the transmission of a nerve impulse are similar in many ways to those involved in muscle contraction. Compared with a contracting muscle, however, a transmitting

Stimulus

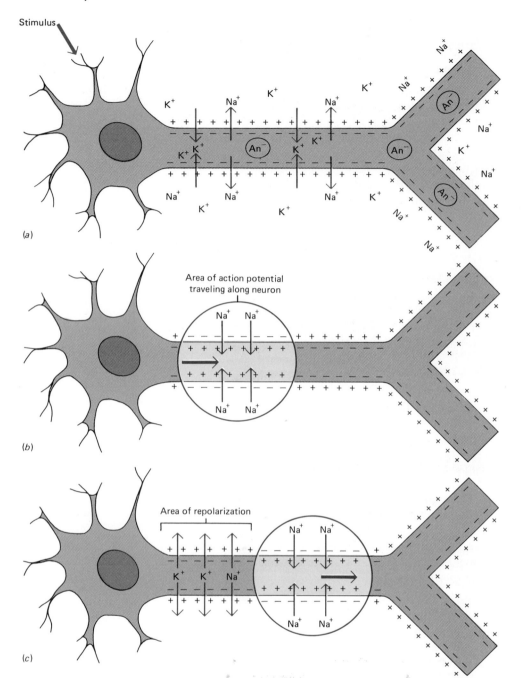

(a)

(b)

(c)

FIGURE 46–9 Transmission of an impulse along an axon. (*a*) The dendrites (or cell body) of a neuron are stimulated sufficiently to depolarize the membrane to firing level. The axon is shown still in the resting state and has a resting potential. (*b*) and (*c*) An impulse is transmitted as a wave of depolarization that travels down the axon. At the region of depolarization, sodium ions diffuse into the cell. As the impulse progresses along the axon, repolarization occurs quickly behind it.

nerve expends little energy; the heat produced by one gram of nerve stimulated for one minute is equivalent to the energy liberated by the oxidation of 10^{-6} grams of glycogen. This means that if a nerve contained only 1% glycogen to serve as fuel, it could be stimulated continuously for a week or more without exhausting the supply. Nerve fibers are practically incapable of being fatigued as long as an adequate supply of oxygen is available. Whatever "mental fatigue" may be, it is not due to the exhaustion of the energy supply of nerve fibers!

Saltatory Conduction

The smooth, progressive impulse transmission just described is characteristic of unmyelinated neurons. In myelinated neurons the myelin insulates the axon except at the nodes of Ranvier, where the membrane makes direct contact with the interstitial fluid. Depolarization skips along the axon from

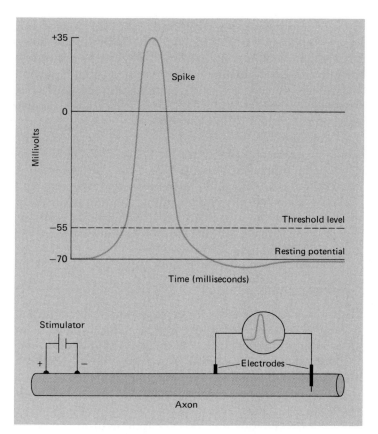

FIGURE 46–10 An action potential recorded with one electrode inside the cell and one just outside the plasma membrane. When the axon depolarizes to about −55 millivolts, an action potential is generated. (The numerical values are included as representative examples. These values vary for different nerve cells.)

one node of Ranvier to the next, jumping electrically over insulated regions (Figure 46–11). The ion activity at the node opens voltage-sensitive sodium channels, resulting in a current that passes to and depolarizes the next node along the axon. Known as **saltatory conduction,** this type of impulse transmission is more rapid than the continuous type.

Saltatory conduction requires less energy than continuous conduction. Only the nodes depolarize, so fewer sodium and potassium ions are displaced, and the cell does not have to work as hard to reestablish resting conditions each time an impulse is transmitted.

Substances That Affect Excitability

Substances that increase the permeability of the membrane to sodium make the neuron more excitable than normal. For example, calcium balance is essential to normal neural function. When insufficient concentra-

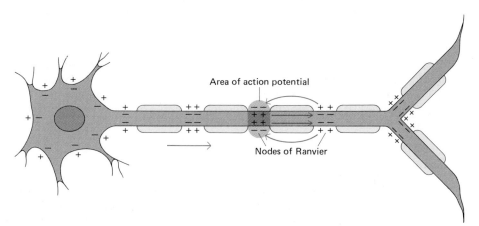

FIGURE 46–11 Saltatory conduction. In a myelinated axon the impulse leaps along from one node of Ranvier to the next.

tions of calcium ions are present, the sodium gates apparently do not close completely between action potentials. Sodium ions then leak into the cells, lowering the resting potential and bringing the neuron closer to firing. The neuron fires more easily—sometimes even spontaneously. As a result, the muscle innervated by the neuron may go into spasm, a condition known as low-calcium tetany. When calcium ions are too numerous, neurons are less excitable and more difficult to fire.

Local anesthetics such as procaine and cocaine are thought to decrease the permeability of the neuron to sodium. Excitability may be so reduced that the neuron cannot propagate an action potential through the anesthetized region. DDT and other chlorinated hydrocarbon biocides interfere with the action of the sodium pump. When nerves are poisoned by these chemicals, they are unable to transmit impulses. Fortunately for us, insects are poisoned by much lower concentrations of these biocides than those harmful to humans.

Synaptic Transmission

Recall that a **synapse** is a junction between two neurons. The neurons are generally separated by a tiny gap from 2.0 to more than 20 nm wide (less than a millionth of an inch), called the **synaptic cleft.** A neuron that ends at a specific synapse is referred to as a **presynaptic neuron,** whereas a neuron that begins at a synapse is a **postsynaptic neuron.** A neuron may be postsynaptic with respect to one synapse and presynaptic with respect to another.

Some presynaptic and postsynaptic neurons come very close together (within 2 nm of one another) and form low-resistance gap junctions. Such junctions permit an impulse to be electrically transmitted directly from one cell to another. Electrical synapses of this sort are found in cnidarian nerve nets and in parts of the nervous systems of earthworms, crayfish, and fish.

At most synapses, however, a gap of more than 20 nm separates the two plasma membranes, and the impulse is transmitted by special substances called **neurotransmitters.** When an impulse reaches the synaptic knobs at the end of a presynaptic axon, it stimulates the release of neurotransmitter into the synaptic cleft. This chemical messenger rapidly diffuses across the narrow synaptic cleft and affects the permeability of the membrane of the postsynaptic neuron.

The synaptic knobs continuously synthesize neurotransmitter and store it in little vesicles. Mitochondria in the synaptic knobs provide the ATP required for this synthesis. The enzymes needed are produced in the cell body and move down the axon to the synaptic knobs. Each time an action potential reaches a synaptic knob, calcium ions pass into the cell. This apparently induces several hundred vesicles to fuse with the membrane and to release their contents into the synaptic cleft (Figures 46–12 and 46–13). After diffusing across the synaptic cleft, the neurotransmitter combines with specific receptors on the dendrites or cell bodies of postsynaptic neurons. This opens gates in the membrane, and the resulting redistribution of ions affects the electrical potential of the membrane, either depolarizing or hyperpolarizing it.

In order for repolarization to occur quickly, any excess neurotransmitter must be removed. It is either inactivated by enzymes or reabsorbed into the synaptic vesicles.

Excitatory and Inhibitory Signals

If the effect of a neurotransmitter is to partially depolarize the membrane, the change in potential is called an **excitatory postsynaptic potential,** or **EPSP.** On the other hand, if the effect of the neurotransmitter is to hyper-

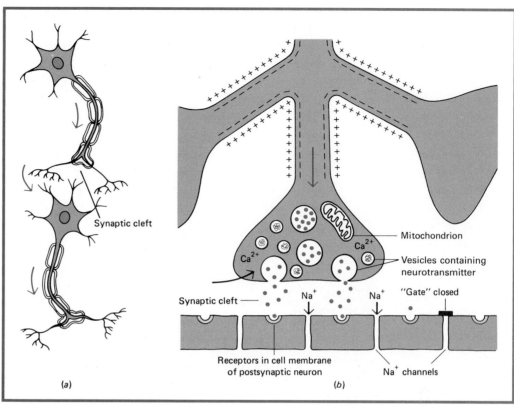

Synaptic cleft

Mitochondrion

Vesicles containing
neurotransmitter

Ca^{2+} Ca^{2+}

Synaptic cleft Na^+ Na^+ "Gate" closed

Receptors in cell membrane
of postsynaptic neuron Na^+ channels

(a)   (b)

polarize the postsynaptic membrane, the change in potential is called an **inhibitory postsynaptic potential,** or **IPSP.**

One EPSP by itself is usually too weak to trigger an action potential in the postsynaptic neuron. Its effect is subliminal, that is, below threshold level. However, EPSPs may be added together, a process known as **summation. Temporal summation** occurs when repeated stimuli cause new EPSPs to develop before previous ones have faded. By adding several EPSPs the neuron may be brought to threshold level. In **spatial summation,** several synaptic knobs release neurotransmitter at the same time, so that the postsynaptic neuron is stimulated at several points simultaneously. The added effects of this stimulation can also bring the postsynaptic neuron to the critical firing level.

Neurotransmitters

More than 60 different substances are now known (or suspected) to be neurotransmitters, and a number of neuropeptides have been identified that modulate the effect of neurotransmitters (Table 46–1). Many neurons secrete at least two neurotransmitters, which are independently regulated. Often a neurotransmitter and a neuropeptide are released simultaneously. The neuropeptide exerts a longer lasting effect. A postsynaptic neuron can have receptors for more than one type of neurotransmitter. Some of its receptors may be excitatory and some inhibitory.

The two neurotransmitters that have been investigated most extensively are acetylcholine and norepinephrine. **Acetylcholine** triggers muscle contraction. It is released not only by motor neurons that innervate skeletal muscle but also by some neurons in the autonomic system and by some neurons in the brain. Cells that release acetylcholine are referred to as **cholinergic neurons.** Acetylcholine has an excitatory effect on skeletal muscle but an inhibitory effect on cardiac muscle. (Whether a neurotransmitter excites or inhibits is at least in part a property of the postsynaptic

FIGURE 46–12 Transmission of an impulse between neurons, or from a neuron to an effector. (*a*) In most synapses the wave of depolarization is unable to jump across the synaptic cleft between the two neurons. (*b*) The problem is solved by chemical transmission across synapses. When an impulse reaches the synaptic knobs at the end of a presynaptic neuron, calcium ions enter the synaptic knob from the interstitial fluid. The calcium ions apparently cause the synaptic vesicles to fuse with the membrane and release neurotransmitter into the synaptic cleft. The neurotransmitter diffuses across the synaptic cleft and may combine with receptors in the membrane of the postsynaptic neuron. This may trigger an impulse in the postsynaptic neuron. It is thought that when neurotransmitter combines with the postsynaptic receptors, the permeability of the postsynaptic membrane to certain ion changes. This can result in either depolarization or hyperpolarization. For example, in depolarization sodium gates open, permitting sodium to rush into the axon.

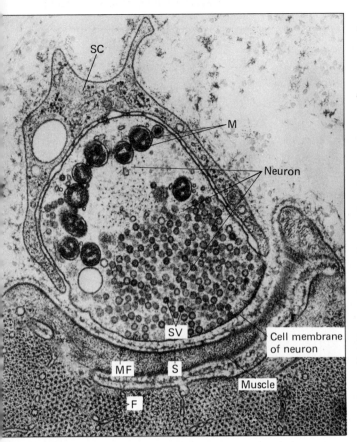

(a)

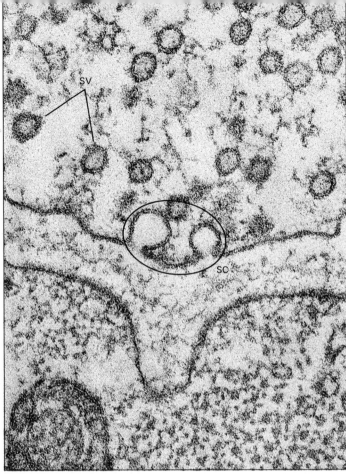

(b)

FIGURE 46–13 Transmission across a synapse. (*a*) Electron micrograph of a synaptic knob filled with synaptic vesicles (approximately ×20,000). This is a motor neuron synapsing with a muscle fiber. SC, Schwann cell; M, mitochondria; SV, synaptic vesicle; S, synaptic cleft; MF, membrane of muscle fiber; F, filaments of muscle. (*b*) Electron micrograph of a synaptic knob and cleft (approximately ×125,000). A portion of the neuron is shown above the cleft; the muscle is shown below the cleft. Two synaptic vesicles (*circled area*) are merging with the cell membrane of knob and discharging neurotransmitter into the cleft. SV, synaptic vesicles; SC, synaptic cleft.

receptors with which it combines.) After acetylcholine is released into a synaptic cleft and combines with receptors on the postsynaptic neuron, the remaining molecules must be removed to prevent repeated stimulation of the muscle. The enzyme **cholinesterase** catalyzes the breakdown of acetylcholine into choline and acetate.

Nerve gases and organophosphate biocides inactivate cholinesterase, so that the amount of acetylcholine in the synaptic cleft increases with each successive nerve impulse. This causes repetitive stimulation of the muscle fiber and may lead to life-threatening muscle spasms. Should the muscles of the larynx go into spasm, for example, a person may die of asphyxiation.

Norepinephrine is released by sympathetic neurons as well as by many neurons in the brain and spinal cord. Neurons that release norepinephrine are called **adrenergic neurons.** Norepinephrine and the neurotransmitters epinephrine and dopamine belong to a class of compounds known as **catecholamines.** After their release from synaptic knobs, catecholamines are removed mainly by re-uptake into the vesicles in the synaptic knobs. Some are inactivated by enzymes such as monoamine oxidase (MAO). Catecholamines affect mood, and many drugs that modify mood do so by altering the levels of these substances in the brain.

Direction and Speed of Conduction

In the laboratory it can be demonstrated that an impulse can move in both directions along a single axon. However, in the body an impulse generally stops when it reaches the dendrites, because there is no neurotransmitter present there to conduct it across a synapse. This limitation imposed by the location of the neurotransmitters makes neural transmission unidirectional. Neural pathways thus function as one-way streets, with the usual

Table 46–1 SOME NEUROTRANSMITTERS

Substance	Origin	Comments
Acetylcholine	Myoneural (muscle–nerve) junctions; preganglionic autonomic endings;* postganglionic parasympathetic nerve endings; parts of brain	Inactivated by cholinesterase
Norepinephrine	Postganglionic sympathetic endings; reticular activating system; areas of cerebral cortex, cerebellum, and spinal cord	Reabsorbed by vesicles in synaptic knob; inactivated by MAO (monoamine oxidase); norepinephrine level in the brain affects mood
Dopamine	Limbic system; cerebral cortex; basal ganglia; hypothalamus	Thought to affect motor function; may be involved in pathogenesis of schizophrenia;† amount reduced in Parkinson's disease
Serotonin (5-HT, 5-hydroxytryptamine)	Limbic system; hypothalamus; cerebellum; spinal cord	May play a role in sleep; LSD antagonizes serotonin; thought to be inhibitory
Epinephrine	Hypothalamus; thalamus; spinal cord	Identical to the hormone released by the adrenal glands
GABA (γ-aminobutyric acid)	Spinal cord; cerebral cortex; Purkinje cells in cerebellum	Acts as inhibitor in brain and spinal cord; may play role in pain perception
Glycine	Released by neurons mediating inhibition in spinal cord	Acts as an inhibitor
Endorphins	CNS and pituitary gland	Neuropeptides that have morphinelike properties and suppress pain; may help regulate cell growth; linked to learning and memory
Enkephalins	Brain and gastrointestinal tract	Neuropeptides thought to inhibit pain impulses by inhibiting release of substance P (see below); bind to same receptors in brain as morphine
Substance P	Brain and spinal cord, sensory nerves, intestine	Transmits pain impulses from pain receptors into CNS

*These and other structures listed in this table are discussed in Chapter 47.
†Studies suggest that the brains of schizophrenics have more dopamine receptors than those of nonschizophrenics.

direction of transmission from the axon of the presynaptic neuron across the synapse to the dendrite or cell body of the postsynaptic neuron.

Compared with the speed of an electric current or the speed of light, the rate of nerve impulse travel is rather slow. Still, some neurons can transmit impulses at a rate of more than 120 meters per second. The rate of conduction of impulses increases as the diameter of the axon increases. Such an increase in diameter decreases the internal electrical resistance along the length of the axon, permitting sodium ions to spread rapidly when they enter the axon. In some invertebrates, giant axons that can transmit impulses very rapidly are employed to conduct danger signals. In vertebrates, the presence of the myelin sheath[1] permits rapid saltatory conduction of impulses along myelinated neurons. The largest, most heavily myelinated neurons conduct impulses most rapidly.

When considering speed of conduction through a sequence of neurons, the number of chemical synapses must be taken into account. Each time an impulse is conducted from one neuron to another, there is a slight synaptic delay (about 0.5 msec). Synaptic delay is due to the time required for the release of neurotransmitter, its diffusion, its binding to postsynaptic membrane receptors, and the generation of the action potential in the postsynaptic neuron.

---

[1]By separating the nodes, where depolarization is possible, the myelin sheath forces an electric current to flow between a depolarized node and the next that has not yet been depolarized. This current immediately produces depolarization of that node. Since current flow is faster than depolarization, the fewer and farther apart the nodes are, the more quickly the nerve impulse is conducted.

INTEGRATION

Neural integration is the process of adding and subtracting incoming signals and determining an appropriate response. Each neuron synapses with hundreds of other neurons. It is the job of the dendrites and cell body of every neuron to integrate the numerous messages that they are continually receiving. Hundreds of EPSPs and IPSPs may be tabulated before an impulse is actually transmitted. When sufficient excitatory neurotransmitter predominates, the neuron is brought to the threshold level and an action potential is generated. Such an arrangement permits the neuron and the entire nervous system far greater flexibility and a wider range of response than would be possible if every EPSP generated an action potential.

Every neuron acts as an integrator, sorting through the thousands of bits of information continually bombarding it. Since more than 90% of the neurons in the body are located in the brain and spinal cord, most neural integration takes place there.

ORGANIZATION OF NEURAL CIRCUITS

Neurons are organized into specific pathways, or **circuits.** Within a neural circuit many presynaptic neurons may converge upon a single postsynaptic neuron. In **convergence,** the postsynaptic neuron is controlled by signals from two or more presynaptic neurons (Figure 46–14a). An association neuron in the spinal cord, for instance, may receive information from sensory neurons entering the cord, from neurons originating at other levels of the spinal cord, and even from neurons bringing information from the brain. Information from all of these converging neurons must be integrated before an action potential is generated in the association neuron and an appropriate motor neuron is stimulated.

In **divergence,** a single presynaptic neuron stimulates many postsynaptic neurons (Figure 46–14b). Each presynaptic neuron may synapse with 25,000 or more different postsynaptic neurons. In **facilitation,** the neuron is brought close to threshold level by EPSPs from various presynaptic neurons but is not yet at the threshold level. The neuron can be easily excited by a new EPSP. Figure 46–15 illustrates facilitation: Neither neuron A nor neuron B can by itself fire neuron 2 or 3. However, when either neuron A or B is fired, neurons 2 and 3 are facilitated. Then when the other presynaptic neuron fires, the postsynaptic neuron receives sufficient neurotransmitter to generate an action potential.

FIGURE 46–14 Organization of neural circuits. (a) Convergence of neural input. Several presynaptic neurons synapse with one postsynaptic neuron. This organization in a neural circuit permits one neuron to receive signals from many sources. (b) Divergence of neural output. A single presynaptic neuron synapses with several postsynaptic neurons. This organization allows one neuron to communicate with many others.

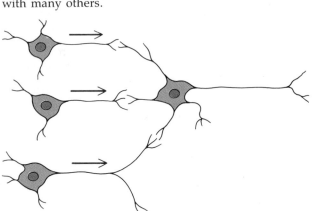

(a)

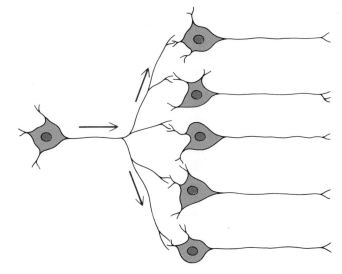

(b)

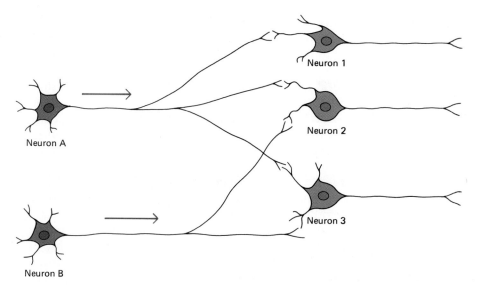

FIGURE 46–15 Facilitation. Neither neuron *A* nor neuron *B* can itself fire neuron *2* or *3*. However, stimulation by either *A* or *B* does depolarize the neuron toward the threshold level (if the stimulation is excitatory). This facilitates the post-synaptic neuron, so if another presynaptic neuron stimulates it, the threshold level may be reached and an action potential generated.

A very important type of neural circuit is the **reverberating circuit.** This is a neural pathway arranged so that a neuron collateral synapses with an association neuron (Figure 46–16). The association neuron synapses with a neuron in the sequence that can send new impulses again through the circuit. New impulses can be generated again and again until the synapses involved become fatigued (owing to depletion of neurotransmitter), or until they are stopped by some sort of inhibition. Reverberating circuits are thought to be important in rhythmic breathing, in maintaining alertness, and perhaps in short-term memory.

REFLEX ACTION

A **reflex action** is a stereotyped, automatic response to a given stimulus that depends only on the anatomic relationships of the neurons involved. Reflexes are functional units of the nervous system, and many physiological mechanisms depend upon reflex actions. A reflex typically involves part of the body rather than the whole. Constriction of the pupil in response to bright light is an example. Breathing, heart rate, salivation, and regulation of blood pressure and temperature are other examples of reflex actions. A change in body temperature, for instance, acts as a stimulus,

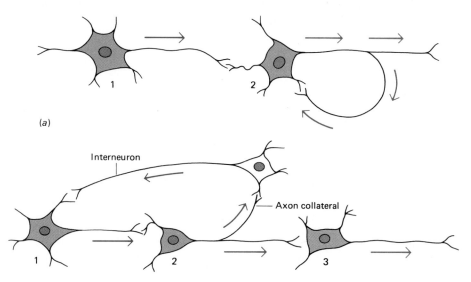

FIGURE 46–16 Reverberating circuits. (*a*) A simple reverberating circuit in which an axon collateral of the second neuron turns back upon its own dendrites, so the neuron continues to stimulate itself. (*b*) In this neural circuit an axon collateral of the second neuron synapses with an interneuron. The interneuron synapses with the first neuron in the sequence. New impulses are triggered again and again in the first neuron, causing reverberation.

causing the temperature-regulating center in the brain to mobilize homeostatic mechanisms that bring body temperature back to normal. Many responses to external stimuli, such as withdrawing from painful stimuli, are also reflex actions.

The simple knee jerk, or patellar reflex, is an example of a very simple type of reflex requiring a chain of only two sets of neurons. Because only one group of synapses is involved, this type of reflex is called a **monosynaptic reflex.** Yet even this simple reflex action requires the sequence of information flow through the nervous system discussed earlier—reception, transmission, integration, and response by an effector. In the knee jerk the receptors are muscle spindles that respond to stretch stimuli when the tendon is tapped suddenly. Afferent neurons transmit the impulses to the spinal cord, where integration takes place at the synapses between afferent and efferent neurons. An efferent neuron then transmits the impulse to the effector cells. The muscle fibers contract, resulting in a sudden straightening of the leg. The patellar reflex is important clinically because it is exaggerated when certain spinal tracts are damaged, and blocked when certain peripheral nerves or reflex centers in the spinal cord are damaged.

Withdrawal reflexes are **polysynaptic,** requiring participation of three sets of neurons (Figure 46–17). For example, an accidental burn on your finger would cause you to jerk your hand away from the painful stimulus even before you became aware of the pain. The pain receptors (dendrites of sensory neurons) in your fingers send messages through afferent neurons to the spinal cord. There each neuron synapses with an association neuron. Integration takes place and impulses are sent via appropriate efferent neurons to muscles in the arm and hand, instructing them to contract, jerking the hand away from the harmful stimulus. At the same time that the association neuron sends a message to the motor neuron, it may also dispatch a message up the neurons in the spinal cord to the brain. Very quickly you become conscious of your plight and can make the decision to hold your hand under cold water. This is not part of the reflex action, however.

That the brain is not essential to many reflex actions can be demonstrated by an experiment often performed in college physiology laboratories. The brain of a frog is destroyed, creating a "spinal" animal. Then a piece of acid-soaked paper is applied to the animal's back. No matter how many times the piece of paper is placed on the skin, one leg will invariably move upward and flick it away. This experiment also demonstrates that reflex actions are stereotyped and automatic. A frog with a brain might make the response two or three times, but eventually would try a different response—perhaps hop away.

Some reflex actions—for example, the pupil reflex—do involve the brain, but only the so-called lower parts, which are functionally similar to

FIGURE 46–17 The withdrawal reflex involves a chain of three neurons. A sensory neuron transmits the message from the receptor to the central nervous system, where it synapses with an association neuron. Then an appropriate motor neuron (shown in red) transmits an impulse to the muscles that move the hand away from the flame (the response).

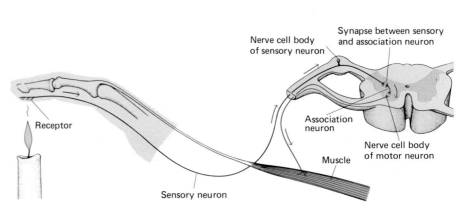

the spinal cord and have nothing to do with conscious thought. Some reflex actions can be consciously inhibited or facilitated. An example is the reflex that voids the urinary bladder when it fills with urine. In babies, urination is a reflex action; when the bladder fills with urine and is stretched to a critical point, a sphincter muscle relaxes and urine passes out of the body. In early childhood we learn to facilitate the reflex consciously, stimulating it before bladder pressure reaches a critical level. We also learn to inhibit the reflex consciously if the bladder becomes full at an inconvenient time or place.

Much of the behavior in a relatively simple animal such as a sea star can be explained in terms of reflex actions. The sea star can extend and retract its tube feet and make postural movements associated with ambulation. The extension and retraction of the tube feet are unoriented reflex responses. The apparent coordination of all the tube feet, retracting in unison when a wave washes over the animal, is simply the sum of individual responses to a common stimulus. Numerous aspects of crustacean behavior are primarily reflex responses—the withdrawal of the eye stalk, the opening and closing of the claws, and movements associated with escape, defense, feeding, and copulation.

Although we might, in theory, construct an animal capable of many responses simply out of reflexes, reflexes in complex animals account for only a small part of total behavior. Reflex actions and behavior are discussed further in Chapter 52.

■ SUMMARY

I. Information flow through the nervous system begins with reception. Information is then transmitted to the CNS via afferent neurons. Integration takes place within the CNS, and appropriate nerve impulses are then transmitted by efferent neurons to the effectors that carry out the response.

II. The vertebrate nervous system is divided into central nervous system (CNS) and peripheral nervous system (PNS).
  A. The CNS consists of brain and spinal cord.
  B. The PNS consists of sensory receptors and nerves; it may be divided into somatic and autonomic systems.

III. The neuron is the structural and functional unit of the nervous system.
  A. A typical multipolar neuron consists of a cell body from which project many branched dendrites and a single, long axon.
  B. The axon is surrounded by a neurilemma. Many axons are also enveloped in a myelin sheath.

IV. A nerve consists of hundreds of axons wrapped in connective tissue; a ganglion is a mass of cell bodies.

V. Transmission of a neural impulse is an electrochemical mechanism.
  A. A neuron that is not transmitting an impulse has a resting potential.
    1. The inner surface of the plasma membrane is negatively charged compared to the outside.
    2. Sodium-potassium pumps continuously transport sodium ions out of the neuron, and transport potassium ions in.
    3. Potassium ions are able to leak out more readily than sodium ions are able to leak in.
  B. Excitatory stimuli are thought to open sodium gates in the plasma membrane. This permits sodium to enter the cell and depolarize the membrane. Inhibitory stimuli hyperpolarize the membrane.
  C. When the extent of depolarization reaches threshold level, an action potential may be generated.
    1. The action potential is a wave of depolarization that spreads along the axon.
    2. The action potential obeys an all-or-none law.
    3. As the action potential moves down the axon, repolarization occurs very quickly behind it.
  D. Saltatory conduction takes place in myelinated neurons. In this type of transmission, depolarization skips along the axon from one node of Ranvier to the next.
  E. Excitability of a neuron can be affected by calcium concentration and by certain substances such as local anesthetics and biocides.
  F. Synaptic transmission generally depends upon release of a neurotransmitter from vesicles in the synaptic knobs of the presynaptic neuron.
    1. The neurotransmitter diffuses across the synaptic cleft and combines with specific receptors on the postsynaptic neuron.
    2. This may cause an excitatory postsynaptic potential (EPSP) or an inhibitory postsynaptic potential (IPSP).
    3. Temporal or spatial summation may bring the postsynaptic neuron to the threshold level.
  G. The largest, most heavily myelinated neurons conduct impulses most rapidly.

VI. Neural integration is the process of adding and subtracting EPSPs and IPSPs and determining an appropriate response.

VII. Complex neural pathways are possible because of such neuronal associations as convergence, divergence, and facilitation.

VIII. A simple reflex action includes reception of a stimulus, transmission of impulses to the CNS via an afferent neuron, integration within the CNS, transmission of impulses via a motor neuron to an effector, and response by the effector.

■ POST-TEST

1. Afferent neurons transmit information from the _____ _____ to the _____.

2. Functional connections between neurons are called _____.

3. Changes in the environment that can be detected by an organism are termed _____.

4. The peripheral nervous system consists of sensory structures and _____.

5. A nerve cell is properly referred to as a _____; the supporting cells of nervous tissue are called _____.

6. Dendrites are specialized to _____; the axon functions to _____ from the _____ to a _____.

7. The cellular sheath is important in _____.

8. In some peripheral neurons, Schwann cells produce both a _____ _____ and a _____ _____.

9. A ganglion consists of a mass of _____.

10. The _____ _____ of a neuron is due mainly to the outward diffusion of potassium ions along their concentration gradient.

11. _____ stimuli are thought to open sodium gates, thereby permitting sodium to rush into the cell.

12. The passage of sodium ions into the neuron _____ the cell membrane.

13. A wave of depolarization that travels down the axon is called a nerve impulse or _____.

14. Because there is no variation in the intensity of an action potential, the neuron is said to obey an _____ _____ law.

15. In saltatory conduction, depolarization skips along the axon from _____.

16. When insufficient calcium ions are present, the _____ _____ is lowered and the neuron fires _____ (*more/less*) easily.

17. When an impulse reaches the synaptic knobs, it stimulates the release of _____.

18. The neurotransmitter that stimulates muscle contraction is _____.

19. Adrenergic neurons release the neurotransmitter _____.

20. In _____, a single presynaptic neuron stimulates many postsynaptic neurons.

21. A stereotyped, automatic response to a given stimulus that depends only on the anatomic relationships of the neurons involved is called a _____ _____.

22. In a typical withdrawal reflex, pain receptors send messages through _____ neurons to the _____ _____, where _____ takes place.

23. Label the following diagram. (Refer to Figure 46–2 as necessary.)

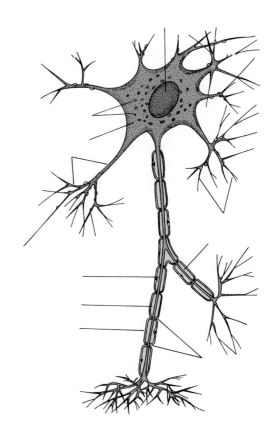

■ REVIEW QUESTIONS

1. Distinguish between a neuron and a nerve.

2. Imagine that a very unfriendly-looking monster suddenly appears before you. What processes must take place within your nervous system before you can make your escape?

3. Contrast the functions of an afferent neuron and an efferent neuron.

4. Describe the functions of the following:
   a. myelin
   b. a ganglion
   c. neuroglia
   d. dendrites
   e. axon

5. What is meant by the resting potential of a neuron?

How do sodium-potassium pumps contribute to the resting potential?

6. What is an action potential? What is responsible for it?
7. What is the all-or-none law?
8. Contrast saltatory conduction with conduction in an unmyelinated neuron.
9. How is neural function affected by the presence of too much calcium? too little calcium?
10. Describe the functions of the following substances:

a. acetylcholine
b. cholinesterase
c. norepinephrine

11. Contrast convergence and divergence.
12. What is summation? Describe facilitation.
13. Draw a diagram illustrating a withdrawal reflex, and label each neuron involved. Indicate the direction of neural transmission.

■ RECOMMENDED READINGS

Bloom, F.E. Neuropeptides. *Scientific American*, October 1981. An account of the discovery and actions of neuropeptides, which help regulate bodily activities, in some cases acting as both neurotransmitters and hormones.

Fine, A. Transplantation in the central nervous system. *Scientific American*, August 1986. Neurons transplanted from embryos can establish functional connections in the adult brain and spinal cord.

Gottlieb, D.I. Gabaergic neurons. *Scientific American*, February 1988. GABA is an inhibitory neurotransmitter in the brains of all mammals.

Keynes, R.D. Ion channels in the nerve-cell membrane. *Scientific American*, March 1979. A discussion of the generation of a nerve impulse by the flow of sodium and potassium ions through channels in the neuron membrane.

Llinas, R.R. Calcium in synaptic transmission. *Scientific American*, October 1982. The role of calcium is studied in the giant synapse of a squid.

Morell, P., and W.T. Norton. Myelin. *Scientific American*, May 1980. A description of myelin and its functions.

Patterson, P., D. Potter, and E. Furshpan. The chemical differentiation of nerve cells. *Scientific American*, July 1978. A discussion of the differentiation of adrenergic and cholinergic nerve cells.

Schwartz, J.H. The transport of substances in nerve cells. *Scientific American*, April 1980. A discussion of the movement of substances long distances between the cell body and the neuron endings.

Wurtman, R.J. Nutrients that modify brain function. *Scientific American*, April 1982. Increasing the level of neurotransmitter precursors in the blood amplifies signals from some nerve cells. They may eventually be used clinically.

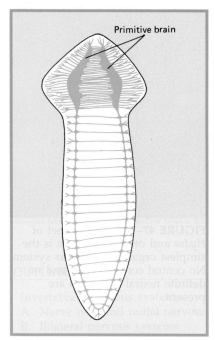

FIGURE 47–2 Planarian flatworms have a ladder-type nervous system. Cerebral ganglia in the head region serve as a simple brain and, to some extent, control the rest of the nervous system.

impulses toward a central nervous system, and **efferent** nerves, which transmit impulses away from the central nervous system and to the effector cells. Various parts of the central nervous system are usually specialized also to perform specific functions, so that distinct structural and functional regions can be identified.

4. Increased number of association neurons and complexity of synaptic contacts. This permits much greater integration of incoming messages, provides a greater range of responses, and allows far more precision in responses.

5. Cephalization, or formation of a head. A bilaterally symmetric animal generally moves in a forward direction, so concentration of sense organs at the front end of the body is important for detection of an enemy quickly enough to escape or for seeing or smelling food in time for its capture. Response can be more rapid if these sense organs are linked by short pathways to decision-making nerve cells nearby. Therefore, nerve cells are also usually concentrated in the head region, comprising a definite brain.

In planarian flatworms, the head region contains concentrations of nerve cells referred to as **cerebral ganglia** (Figure 47–2). These serve as a primitive "brain" and exert some measure of control over the rest of the nervous system. Two ventral longitudinal nerve cords extend from the ganglia to the posterior end of the body. Transverse nerves connect the brain with the eyespots and anterior end of the body. This arrangement is referred to as a **ladder-type** nervous system.

In annelids and arthropods, there is also typically a pair of ventrally located longitudinal nerve cords (Figures 47–3 and 47–4). The cell bodies of the nerve cells are massed into pairs of ganglia located in *each* body segment. Afferent and efferent neurons are located in lateral nerves that link the ganglia with muscles and other body structures.

When the earthworm brain is removed, the animal can move almost as well as before, but when it bumps into an obstacle, it persists in futile

FIGURE 47–3 The nervous system of the earthworm is typical of those found in other annelids. (*a*) Lateral view. The cell bodies of the neurons are located in ganglia found in each body segment. They are connected by the paired ventral longitudinal nerve cords. (*b*) Cross section through the body of the earthworm.

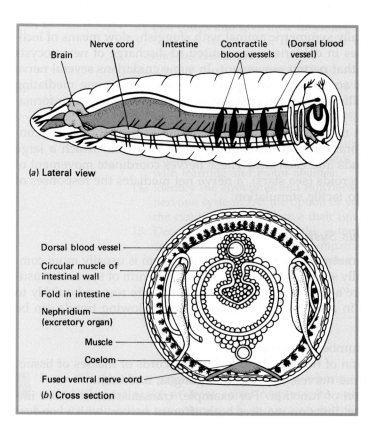

(a) Lateral view

Brain
Nerve cord
Intestine
Contractile blood vessels
(Dorsal blood vessel)

Dorsal blood vessel
Circular muscle of intestinal wall
Fold in intestine
Nephridium (excretory organ)
Muscle
Coelom
Fused ventral nerve cord

(b) Cross section

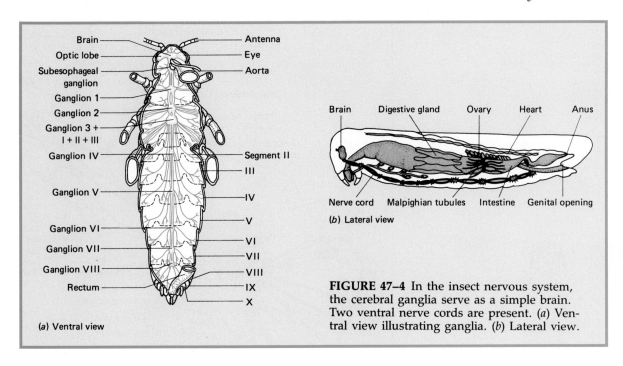

FIGURE 47–4 In the insect nervous system, the cerebral ganglia serve as a simple brain. Two ventral nerve cords are present. (*a*) Ventral view illustrating ganglia. (*b*) Lateral view.

efforts to move forward instead of turning aside. The brain is therefore necessary for adaptive movements; it enables the earthworm to respond appropriately to environmental change.

In complex arthropods, especially insects, some of the abdominal ganglia move anteriorly in the course of embryonic development and fuse with the thoracic ganglia. In some arthropods, the cerebral ganglia are somewhat brainlike in that specific functional regions have been identified in them (Figure 47–4).

In mollusks there are typically at least three pairs of ganglia: **cerebral ganglia,** found dorsal to the esophagus (which serve as a coordinating center for complex reflexes and which also have a motor function); **visceral ganglia,** located among the organs (which control shell opening and closing); and **pedal ganglia,** located in the foot (which control the movement of the foot). The visceral and pedal ganglia are connected to the cerebral ganglia by nerve cords.

In cephalopods, such as the octopus, there is a tendency toward concentration of the nerve cells in a central region (Figure 47–5). All the ganglia are massed in the **circumesophageal ring,** which contains about 168 million nerve cell bodies. With this complex brain, it is no wonder that the octopus is considered to be among the most intelligent of the invertebrates. Octopuses have considerable learning abilities and can be taught quite complex tasks.

THE VERTEBRATE BRAIN

In the early embryo, the brain and spinal cord differentiate from a single tube of tissue, the **neural tube.** Anteriorly, the tube expands and differentiates into the structures of the brain. Posteriorly, the tube develops into the spinal cord. Brain and spinal cord remain continuous and their cavities communicate. As the brain begins to differentiate, three primary swellings develop in the anterior end of the neural tube. These give rise to the **forebrain, midbrain,** and **hindbrain** (Figure 47–6). As indicated in Table 47–1, the forebrain further subdivides to form the **telencephalon** and **diencephalon.** The telencephalon gives rise to the cerebrum, and the diencephalon to

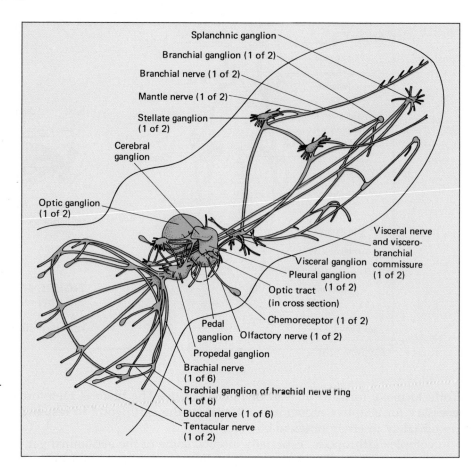

FIGURE 47–5 The cephalopod nervous system. Several ganglia, including the cerebral, optic, and pedal ganglia, make up the "brain." These structures contain millions of nerve cell bodies.

the thalamus and hypothalamus. The hindbrain subdivides to form the **metencephalon,** which gives rise to the cerebellum and pons, and the **myelencephalon,** which gives rise to the medulla. The medulla, pons, and midbrain make up the **brain stem,** the elongated portion of the brain that looks like a stalk holding up the cerebrum.

The most posterior part of the brain, the **medulla,** is continuous with the spinal cord. Its cavity, the **fourth ventricle,** communicates with the **central canal** of the spinal cord and with the cerebral aqueduct, a channel that runs through the midbrain. The cerebral aqueduct connects the fourth ventricle with the **third ventricle** (located within the diencephalon). In turn, the third ventricle is connected with the **lateral ventricles,** also called the **first** and **second ventricles,** within the cerebrum by way of a channel known as the **interventricular foramen.**

FIGURE 47–6 Early in the development of the vertebrate embryo, the anterior end of the neural tube differentiates into the forebrain, midbrain, and hindbrain. These primary divisions subdivide and then give rise to specific structures of the adult brain (see Table 47–1).

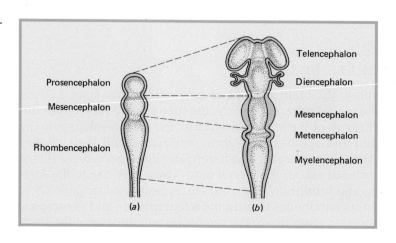

Table 47–1 **DIFFERENTIATION OF CNS STRUCTURES**

Early Embryonic Divisions	Subdivisions	Derivatives in Adult	Cavity
Brain			
Forebrain (prosencephalon)	Telencephalon	Cerebrum	Lateral ventricles (first and second ventricles)
	Diencephalon	Thalamus, hypothalamus, epiphysis (pineal body)	Third ventricle
Midbrain (mesencephalon)	Midbrain	Optic lobes in fish and amphibians; superior and inferior colliculi	Cerebral aqueduct
Hindbrain (rhombencephalon)	Metencephalon Myelencephalon	Cerebellum, pons Medulla oblongata	Fourth ventricle
Spinal cord		Spinal cord	Central canal

As illustrated in Figure 47–7, all vertebrates from fish to mammals have the same basic brain structure. Certain parts of the brain are specialized to perform specific functions, and some regions, such as the cerebellum and cerebrum, are vastly more complex in the higher vertebrates.

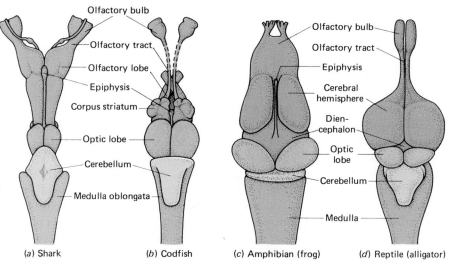

(a) Shark (b) Codfish (c) Amphibian (frog) (d) Reptile (alligator)

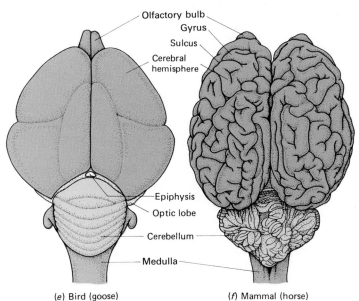

(e) Bird (goose) (f) Mammal (horse)

FIGURE 47–7 Comparison of the brains of members of six vertebrate classes indicates basic similarities and evolutionary trends. Note that different parts of the brain may be specialized in the various groups. For example, the large olfactory lobes in the shark brain (a) are essential to this predator's highly developed sense of smell. During the course of evolution, the cerebrum and cerebellum have become larger and more complex. In the mammal (f), the cerebrum is the most prominent part of the brain; the cerebral cortex, the thin outer layer of the cerebrum, is highly convoluted (folded), which greatly increases its surface area.

The Hindbrain

The walls of the medulla are thick and made up largely of nerve tracts that connect the spinal cord with various parts of the brain. In complex vertebrates, the medulla contains discrete nuclei that serve as **vital centers** regulating respiration, heart beat, and blood pressure. Other reflex centers in the medulla regulate such activities as swallowing, coughing, and vomiting.

The size and shape of the **cerebellum** vary greatly among the vertebrate classes. Development of the cerebellum in different animals is correlated roughly with the extent and complexity of muscular activity. In some fish, birds, and mammals, the cerebellum is highly developed, whereas it tends to be small in cyclostomes, amphibians, and reptiles. The cerebellum coordinates muscle activity and is responsible for muscle tone, posture, and equilibrium. Injury or removal of the cerebellum results not in paralysis but in impairment of muscle coordination. A bird without a cerebellum is unable to fly, and its wings thrash about jerkily. When the human cerebellum is injured by a blow or by disease, muscular movements are uncoordinated. Any activity requiring delicate coordination, such as threading a needle, is very difficult, if not impossible.

In mammals, a large mass of fibers known as the **pons** forms a bulge on the anterior surface of the brain stem. The pons is a bridge connecting the spinal cord and medulla with upper parts of the brain. It contains nuclei that relay impulses from the cerebrum to the cerebellum and also contains centers that help regulate respiration.

The Midbrain

In fish and amphibians, the midbrain is the most prominent part of the brain. In these animals, the midbrain is the main association area. It receives incoming sensory information, integrates it, and sends decisions to appropriate motor nerves. The dorsal portion of the midbrain is differentiated to some degree in these lower vertebrates. For example, the **optic lobes,** specialized for visual interpretations, are part of the midbrain. In reptiles, birds, and mammals, many of the functions of the optic lobes are assumed by the cerebrum. In mammals, the midbrain consists of the **superior colliculi,** which are centers for visual reflexes such as pupil constriction, and the **inferior colliculi,** which are centers for certain auditory reflexes. The mammalian midbrain also contains the **red nucleus,** a center that integrates information regarding muscle tone and posture.

The Forebrain

The forebrain consists of the diencephalon and telencephalon. The diencephalon contains the thalamus and hypothalamus (Figure 47–7). In all vertebrate classes, the **thalamus** is a relay center for motor and sensory messages. In mammals all sensory messages (except those from the olfactory receptors) are delivered to the thalamus before being relayed to the sensory areas of the cerebrum.

Below the thalamus, the **hypothalamus** forms the floor of the third ventricle. The hypothalamus is the principal integration center for the regulation of the viscera (internal organs). It provides input to centers in the medulla and spinal cord that regulate activities such as heart rate, respiration, and digestive system function. The hypothalamus also links the nervous and endocrine systems. In fact, the pituitary gland (an important endocrine gland) hangs down from the hypothalamus. **Releasing hormones** produced by the hypothalamus regulate the secretion of several hormones produced by the anterior lobe of the pituitary gland. In reptiles, birds, and mammals, body temperature is controlled by the hypothalamus. The hypothalamus also regulates appetite and water balance and is involved in emotional and sexual responses.

The telencephalon differentiates to form the cerebrum, and, in most vertebrate groups, the **olfactory bulbs.** The olfactory bulbs are concerned with the chemical sense of smell, the dominant sense in most aquatic and terrestrial vertebrates. In fact, much of brain development in vertebrates appears to be focused upon the integration of olfactory information. In fish and amphibians, the cerebrum is almost entirely devoted to the integration of such incoming sensory information.

Birds are an exception among the vertebrates in that their sense of smell is generally poorly developed. In them, however, a part of the cerebrum called the **corpus striatum** is highly developed. This structure is thought to control the innate, stereotyped, yet complex action patterns characteristic of birds. Just above the corpus striatum is a region thought to govern learning in birds.

In most vertebrates, the cerebrum is divided into right and left hemispheres. Most of the cerebrum is made of **white matter** consisting mainly of axons connecting various parts of the brain. In mammals and most reptiles, there is a layer of **gray matter** called the **cerebral cortex** that makes up the outer portion of cerebral tissue. Certain reptiles possess a different type of cortex, not found in lower vertebrates, known as the **neopallium;** it serves as an association area, a region that links sensory and motor functions and is responsible for higher functions such as learning. The neopallium is much more extensive in mammals, making up the bulk of the cerebrum,[1] which becomes the most prominent part of the brain. In mammalian embryonic development, in fact, the cerebrum expands and grows backwards, covering many of the other brain structures.

In mammals, the cerebrum is responsible for many of the functions that are performed by other parts of the brain in lower vertebrates. In particular, it has many complex association functions lacking in reptiles, amphibians, and fish. In small or simple mammals, the cerebral cortex may be smooth. However, in large complex mammals, the surface area is greatly expanded by numerous folds called **convolutions,** or **gyri.** The furrows between them are called **sulci** when shallow and **fissures** when deep.

THE HUMAN CENTRAL NERVOUS SYSTEM

As in other vertebrates, the human central nervous system (CNS) consists of the brain and spinal cord (Figure 47–8). These soft, fragile organs are carefully protected. Both are encased in bone and wrapped in three layers of connective tissue collectively termed the **meninges.** The three meningeal layers are the tough, outer **dura mater,** the middle **arachnoid,** and the thin, vascular **pia mater** that adheres closely to the tissue of the brain and spinal cord (Figure 47–9). **Meningitis** is a disease in which these coverings become infected and inflamed.

Between the arachnoid and the pia mater is the **subarachnoid space,** which contains **cerebrospinal fluid.** This shock-absorbing fluid cushions the brain and spinal cord and prevents them from bouncing against the bones of the vertebrae or skull with every movement. Cerebrospinal fluid also circulates through the ventricles of the brain. It is produced by special networks of capillaries called the **choroid plexuses** that project from the pia mater into the ventricles. After circulating through the CNS, cerebrospinal fluid is reabsorbed into the blood.

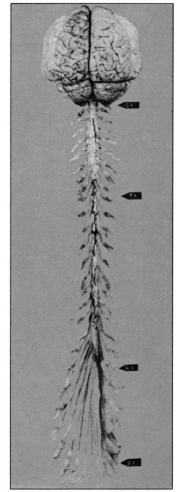

FIGURE 47–8 Photograph of human brain and spinal cord. The roots of the spinal nerves are still attached. Note the group of nerves that extend caudally from the lower region of the cord. Because they resemble a horse's tail, they are referred to as the cauda equina. These nerves have been left undisturbed on the right but have been fanned out on the left.

---

[1]In humans about 90% of the cerebral cortex is neopallium and consists of six distinct cell layers.

Table 47–2 **DIVISIONS OF THE HUMAN BRAIN**

Division	Description	Functions
Medulla	Most inferior portion of the brain stem; continuous with spinal cord. Its white matter consists of nerve tracts passing between spinal cord and various parts of the brain; its gray matter consists of nuclei. Its cavity is the fourth ventricle.	Contains vital centers (within its reticular formation) that regulate heartbeat, respiration and blood pressure; contains reflex centers that control swallowing, coughing, sneezing, and vomiting; relays messages to other parts of the brain
Pons	Consists mainly of nerve tracts passing between the medulla and other parts of the brain; forms a bulge on anterior surface of brain stem; contains a respiratory center	Serves as a link to connect and integrate various parts of the brain; helps regulate respiration
Midbrain	Just superior to the pons; contains red nucleus; cavity is the cerebral aqueduct; posteriorly consists of superior and inferior colliculi	Superior colliculi mediate visual reflexes; inferior colliculi mediate auditory reflexes. Red nucleus integrates information regarding muscle tone and posture.
Diencephalon Thalamus	Located on each side of the third ventricle; consists of two masses of gray matter partly covered by white matter; contains many important nuclei	Main relay center conducting information between spinal cord and cerebrum. Incoming messages are sorted and partially interpreted within thalamic nuclei before being relayed to appropriate centers in the cerebrum.
Hypothalamus	Forms ventral floor of third ventricle; contains many nuclei. Optic chiasms mark crossing of the optic nerves. The pituitary stalk connects pituitary gland to hypothalamus.	Contains centers for control of body temperature, appetite, and fluid balance; secretes releasing hormones that regulate pituitary gland; helps control autonomic functions; involved in some emotional and sexual responses
Cerebellum	Second largest part of the brain; consists of two lateral cerebellar hemispheres; superior to the fourth ventricle	Responsible for smooth, coordinated movement; maintains posture and muscle tone; helps maintain equilibrium
Cerebrum	Largest, most prominent part of brain. Longitudinal fissure divides cerebrum into right and left hemispheres, each containing a lateral ventricle and each divided into six lobes; frontal, parietal, occipital, temporal, limbic, and insular.	Center of intellect, memory, language, and consciousness; receives and interprets sensory information from all of the organs; controls motor functions
Cerebral cortex	Convoluted, outer layer of gray matter, functionally divided into three areas: (1) Motor areas (2) Sensory areas (3) Association areas	Control voluntary movement and certain types of involuntary movement Receive incoming sensory information from eyes, ears, touch and pressure receptors, and other sense organs. Sensory association areas interpret sensory information. Responsible for thought, learning, language, judgment, and personality; store memories; connect sensory and motor areas
White matter	Consists of association fibers that interconnect neurons within the same hemisphere; fibers that interconnect the two hemispheres (e.g., corpus callosum), and fibers that are part of ascending and descending tracts. Basal ganglia are located within the white matter.	Link various areas of the brain

sensation of light; their removal causes blindness. The centers for hearing are located in the lateral **temporal lobes** of the brain above the ear; stimulation by a blow causes a sensation of noise. Although removal of both auditory areas causes deafness, removal of one does not cause deafness in one ear but rather produces a decrease in the auditory acuity of both ears.

A fissure called the **central sulcus** crosses the top of each hemisphere from medial to lateral edge. This partially separates the **primary motor areas** in the **frontal lobes** controlling the skeletal muscles from the **parietal lobes** just behind the furrow. The parietal lobes are responsible for the sensations of heat, cold, touch, and pressure that result from stimulation of sense organs in the skin. In both motor and sensory areas, there is a further specialization along the furrow from the top of the brain to the side. Neu-

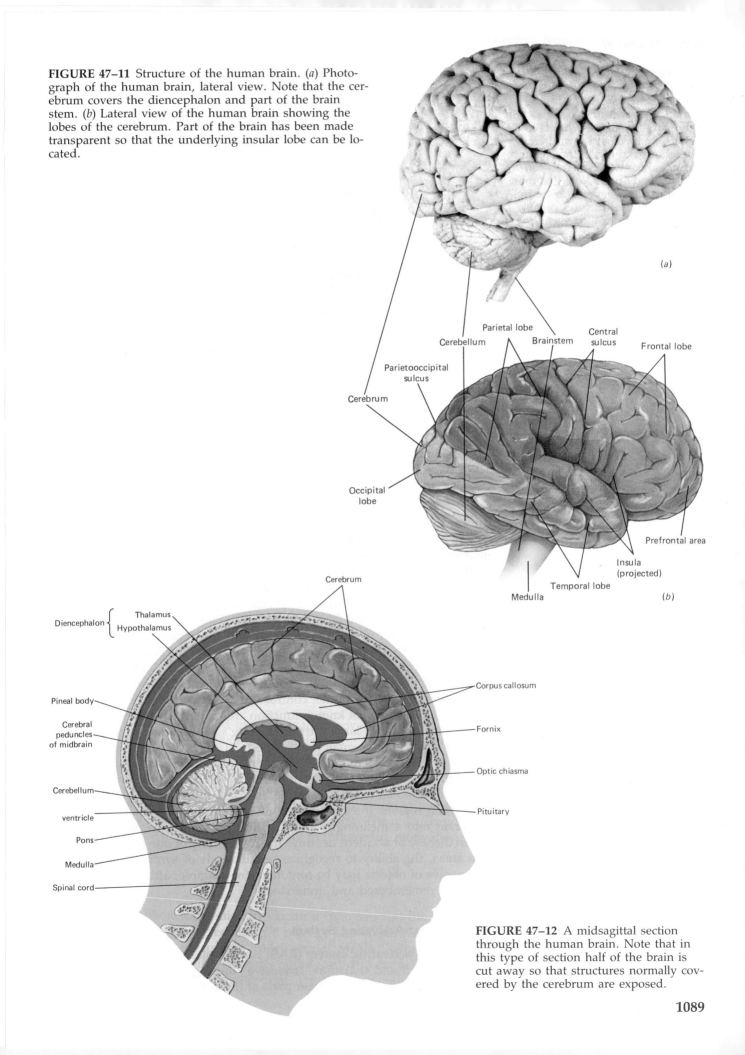

FIGURE 47–11 Structure of the human brain. (*a*) Photograph of the human brain, lateral view. Note that the cerebrum covers the diencephalon and part of the brain stem. (*b*) Lateral view of the human brain showing the lobes of the cerebrum. Part of the brain has been made transparent so that the underlying insular lobe can be located.

Parietal lobe

Central sulcus

Cerebellum

Brainstem

Frontal lobe

Parietooccipital sulcus

Cerebrum

Occipital lobe

Prefrontal area

Insula (projected)

Medulla

Temporal lobe

(*a*)

(*b*)

Cerebrum

Thalamus

Diencephalon { Hypothalamus

Corpus callosum

Pineal body

Fornix

Cerebral peduncles of midbrain

Optic chiasma

Cerebellum

Pituitary

ventricle

Pons

Medulla

Spinal cord

FIGURE 47–12 A midsagittal section through the human brain. Note that in this type of section half of the brain is cut away so that structures normally covered by the cerebrum are exposed.

1089

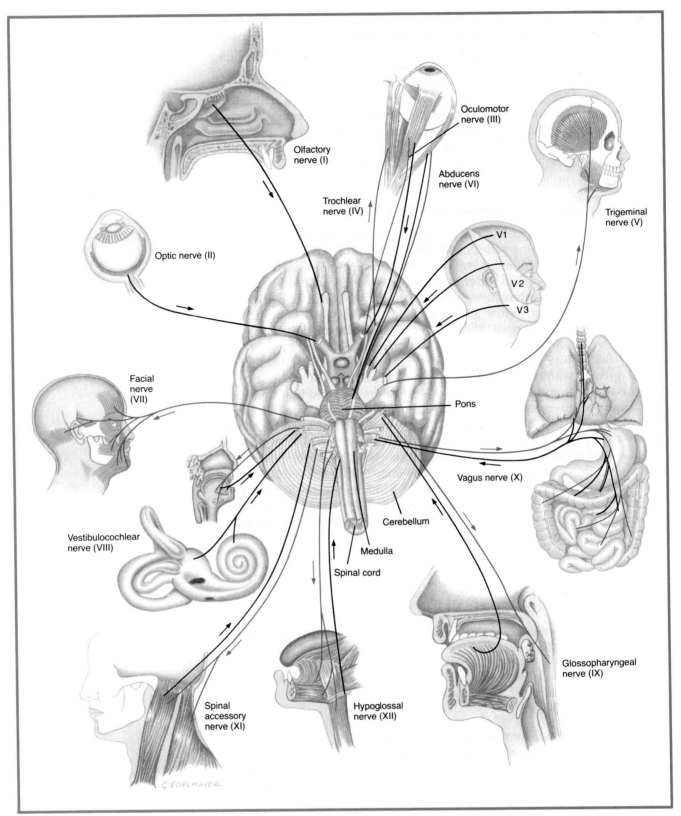

FIGURE 47–17 Ventral (basal) view of the human brain showing emergence of the cranial nerves. (Black indicates sensory fibers; color indicates motor fibers.)

The Spinal Nerves

In human beings, 31 symmetric pairs of **spinal nerves** emerge from the spinal cord (Figure 47–18). All the spinal nerves are mixed nerves, containing both motor and sensory neurons. Each nerve innervates the receptors and effectors of one region of the body. Each spinal nerve has two **roots,**

Table 47–3 **THE CRANIAL NERVES OF MAMMALS**

Number	Name	Origin of Sensory Fibers	Effector Innervated by Motor Fibers
I	Olfactory	Olfactory epithelium of nose (smell)	None
II	Optic	Retina of eye (vision)	None
III	Oculomotor	Proprioceptors* of eyeball muscles (muscle sense)	Muscles that move eyeball (with IV and VI); muscles that change shape of lens; muscles that constrict pupil
IV	Trochlear	Proprioceptors of eyeball muscles	Other muscles that move eyeball
V	Trigeminal	Teeth and skin of face	Some of muscles used in chewing
VI	Abducens	Proprioceptors of eyeball muscles	Other muscles that move eyeball
VII	Facial	Taste buds of anterior part of tongue	Muscles used for facial expression; submaxillary and sublingual salivary glands
VIII	Auditory (vestibulocochlear)	Cochlea (hearing) and semicircular canals (senses of movement, balance, and rotation)	None
IX	Glossopharyngeal	Taste buds of posterior third of tongue, lining of pharynx	Parotid salivary gland; muscles of pharynx used in swallowing
X	Vagus	Nerve endings in many of the internal organs: lungs, stomach, aorta, larynx	Parasympathetic fibers to heart, stomach, small intestine, larynx, esophagus, and other organs
XI	Spinal accessory	Muscles of shoulder	Muscles of neck, shoulder
XII	Hypoglossal	Muscles of tongue	Muscles of tongue

*Proprioceptors are receptors located in muscles, tendons, or joints that provide information about body position and movement.

points of attachment with the spinal cord. All of the sensory neurons enter the cord through the **dorsal root;** all motor fibers leave the cord through the **ventral root** (Figure 47–19). Just before the dorsal root joins the spinal cord, it is marked by a swelling called the **spinal ganglion,** or **dorsal root ganglion,** which consists of the cell bodies of the sensory neurons. Cell bodies of the motor neurons are located within the gray matter of the cord. Dorsal and ventral roots unite, forming the spinal nerve.

If the dorsal root is severed, the part of the body innervated by that nerve suffers complete loss of sensation without paralysis of muscles. If the ventral root is cut, there is complete paralysis of muscles innervated by that nerve, but the senses of touch, pressure, temperature, pain, and kinesthesis (muscle sense) are not impaired.

Peripheral to the junction of the dorsal and ventral roots, each spinal nerve divides into branches. The dorsal branch serves the skin and muscles of the back. The ventral branch serves the skin and muscles of the sides and ventral part of the body. The autonomic branch innervates the viscera. The ventral branches of several spinal nerves form tangled networks called **plexuses** (Figure 47–18). Within a plexus, the fibers of a spinal nerve may separate and then regroup with fibers that originated in other nerves. Thus, nerves emerging from a plexus consist of neurons that originated in

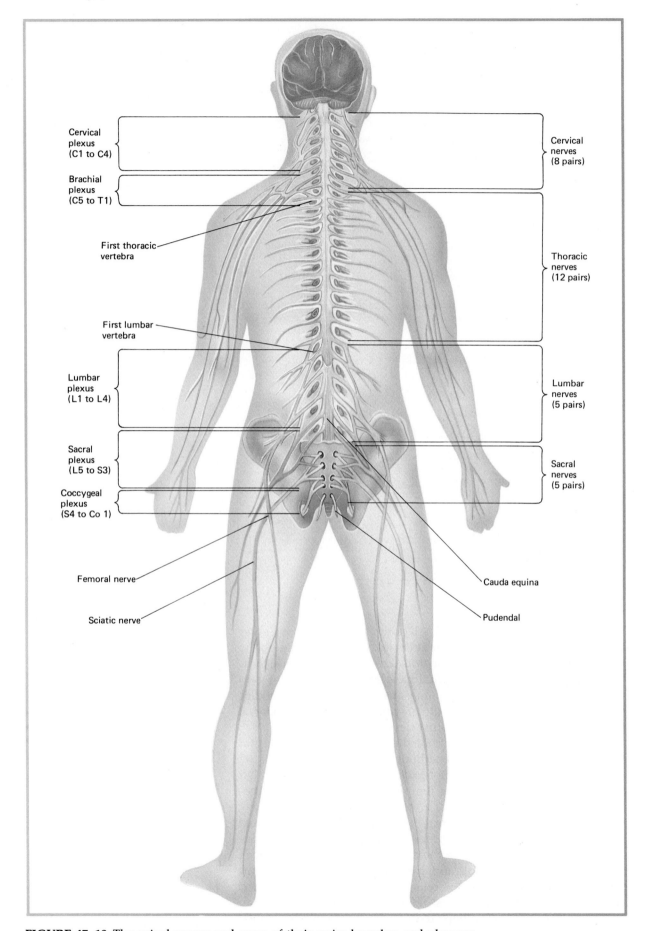

Cervical
plexus
(C1 to C4)

Brachial
plexus
(C5 to T1)

First thoracic
vertebra

First lumbar
vertebra

Lumbar
plexus
(L1 to L4)

Sacral
plexus
(L5 to S3)

Coccygeal
plexus
(S4 to Co 1)

Femoral nerve

Sciatic nerve

Cervical
nerves
(8 pairs)

Thoracic
nerves
(12 pairs)

Lumbar
nerves
(5 pairs)

Sacral
nerves
(5 pairs)

Cauda equina

Pudendal

FIGURE 47–18 The spinal nerves and some of their major branches and plexuses.

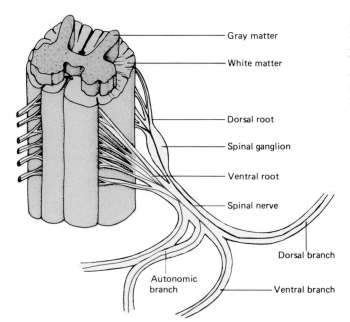

Gray matter

White matter

Dorsal root

Spinal ganglion

Ventral root

Spinal nerve

Dorsal branch

Autonomic branch

Ventral branch

FIGURE 47–19 Dorsal and ventral roots emerge from the spinal cord and join to form a spinal nerve. The spinal nerve divides into several branches. (Black indicates sensory fibers; color indicates motor fibers.)

several different spinal nerves. Among the principal plexuses are the cervical plexus, the brachial plexus, the lumbar plexus, and the sacral plexus.

The Autonomic System

The **autonomic system** helps to maintain a steady state within the internal environment of the body. For example, this system helps to maintain a constant body temperature and to regulate the rate of the heartbeat. Receptors within the viscera relay information through afferent nerves to the CNS, and the impulses are transmitted along efferent neurons to the appropriate muscles and glands.

The efferent portion of the autonomic system is subdivided into sympathetic and parasympathetic systems. Many organs are innervated by both (Figures 47–20 and 47–21). In general, the **sympathetic system** mobilizes energy and enables the body to respond to stress (Table 47–4). Its nerves speed the heart's rate and force of contraction, increase blood pressure, increase blood sugar concentration, and reroute blood circulation when required so that skeletal and cardiac muscles receive the additional amounts of blood needed to support their maximum effort.

The **parasympathetic system** is most active in ordinary, restful situations. After a stressful episode, it decreases the heart rate, decreases blood pressure, and stimulates the digestive system to process food. The parasympathetic system is dominant during relaxation or calm, quiet activities. The sympathetic and parasympathetic systems work together to orchestrate the numerous complex activities continuously taking place within the body (Table 47–5).

Instead of utilizing a single efferent neuron, as in the somatic system, the autonomic system uses a relay of two neurons between the CNS and the effector. The first neuron, called the **preganglionic neuron,** has a cell body and dendrites within the CNS. Its axon, part of a peripheral nerve, ends by synapsing with a **postganglionic neuron.** The dendrites and cell body of the postganglionic neuron are located within a ganglion outside the CNS. Its axon terminates near or on the effector. The sympathetic ganglia are paired, and there is a chain of them on each side of the spinal cord from the neck to the abdomen, the **paravertebral sympathetic ganglion chain.** Some sympathetic preganglionic neurons do not end in these ganglia but instead pass on to ganglia located in the abdomen close to the aorta

(Text continues on page 1102)

FIGURE 47–20 Dual innervation of the heart and stomach by sympathetic and parasympathetic nerves. (Sympathetic nerves are shown in color; postganglionic fibers are shown as dotted lines.)

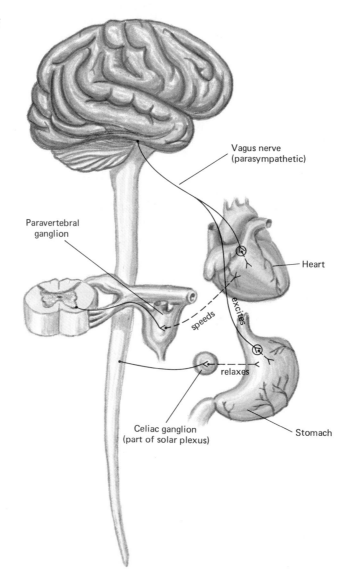

Characteristic	Sympathetic System	Parasympathetic System
General effect	Prepares body to cope with stressful situations	Restores body to resting state after stressful situation; actively maintains normal configuration of body functions
Extent of effect	Widespread throughout body	Localized
Transmitter substance released at synapse with effector	Norepinephrine (usually)	Acetylcholine
Duration of effect	Lasting	Brief
Outflow from CNS	Thoracolumbar levels of spinal cord	Craniosacral levels (from brain and spinal cord)
Location of ganglia	Chain and collateral ganglia	Terminal ganglia
Number of postganglionic fibers with which each preganglionic fiber synapses	Many	Few

Table 47–4 COMPARISON OF SYMPATHETIC WITH PARASYMPATHETIC SYSTEM

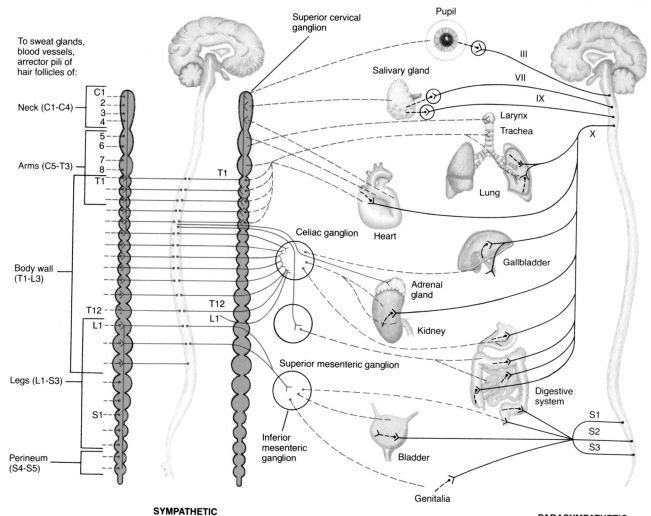

To sweat glands, blood vessels, arrector pili of hair follicles of:

Neck (C1–C4)

Arms (C5–T3)

Body wall (T1–L3)

Legs (L1–S3)

Perineum (S4–S5)

Superior cervical ganglion

Celiac ganglion

Superior mesenteric ganglion

Inferior mesenteric ganglion

Pupil

Salivary gland

Larynx
Trachea

Lung

Heart

Gallbladder

Adrenal gland

Kidney

Digestive system

Bladder

Genitalia

SYMPATHETIC

PARASYMPATHETIC

FIGURE 47–21 Sympathetic and parasympathetic nervous systems. For clarity, peripheral and visceral nerves of the sympathetic system are shown on separate sides of the cord. Complex as it appears, this diagram has been greatly simplified. (Colored lines represent sympathetic nerves, black lines represent parasympathetic nerves, and dotted lines represent postganglionic nerves.) See Table 47–5 for specific action of the nerves.

Table 47–5 **COMPARISON OF SYMPATHETIC AND PARASYMPATHETIC ACTIONS ON SELECTED EFFECTORS***

Effector	*Sympathetic Action*	*Parasympathetic Action*
Heart	Increases rate and strength of contraction	Decreases rate; no direct effect on strength of contraction
Bronchial tubes	Dilates	Constricts
Iris of eye	Dilates pupil	Constricts pupil
Sex organs	Constricts blood vessels; ejaculation	Dilates blood vessels; erection
Blood vessels	Generally constricts	No innervation for many
Sweat glands	Stimulates	No innervation
Intestine	Inhibits motility	Stimulates motility and secretion
Liver	Stimulates glycogenolysis (conversion of glycogen to glucose)	No effect
Adipose tissue	Stimulates free fatty acid release from fat cells	No effect
Adrenal medulla	Stimulates secretion of epinephrine and norepinephrine	No effect
Salivary glands	Stimulates thick, viscous secretion	Stimulates profuse, watery secretion

*Refer to Figure 47–21 as you study this table. Note that many other examples could be added to this list.

1101

Table 47–6 EFFECTS OF SOME COMMONLY USED DRUGS

Name of Drug	Effect on Mood	Actions on Body	Dangers Associated with Abuse
Barbiturates (e.g., Nembutal, Seconal)	Sedative-hypnotic*; "downers"	Inhibit impulse conduction in RAS: depress CNS, skeletal muscle, and heart; depress respiration; lower blood pressure; cause decrease in REM sleep	Tolerance, physical dependence, death from overdose, especially in combination with alcohol
Methaqualone (e.g., Quaalude, Sopor)	Hypnotic	Depresses CNS; depresses certain polysynaptic spinal reflexes	Tolerance, physical dependence, convulsions, death
Meprobamate (e.g., Equanil, Miltown; "minor tranquilizers")	Antianxiety drug**; induces calmness	Causes decreases in REM sleep; relaxes skeletal muscle; depresses CNS	Tolerance, physical dependence; coma and death from overdose
Valium, Librium ("mild tranquilizers")	Reduce anxiety	May reduce rate of impulse firing in limbic system; relax skeletal muscle	Minor EEG abnormalities with chronic use; very large doses cause physical dependence
Phenothiazines (chlorpromazine; "major tranquilizers")	Antipsychotic; highly effective in controlling symptoms of psychotic patients	Affect levels of catecholamines in brain (block dopamine receptors, inhibit uptake of norepinephrine, dopamine, and serotonin); depress neurons in RAS and basal ganglia	Prolonged intake may result in Parkinson-like symptoms
Antidepressant drugs (e.g., Elavil)	Elevate mood; relieve depression	Block uptake of norepinephrine, so more is available to stimulate nervous system	Central and peripheral neurological disturbances; incoordination; interfere with normal cardiovascular function
Alcohol	Euphoria; relaxation; release of inhibitions	Depresses CNS; impairs vision, coordination, judgment; lengthens reaction time	Physical dependence; damage to pancreas; liver cirrhosis; possible brain damage
Narcotic analgesics (e.g. morphine, heroin)	Euphoria; reduction of pain	Depress CNS; depress reflexes; constrict pupils; impair coordination; block release of substance P from pain-transmitting neurons	Tolerance; physical dependence; convulsions; death from overdose

*Sedatives reduce anxiety; hypnotics induce sleep.
**Antianxiety drugs reduce anxiety but are less likely to cause drowsiness than the more potent sedative-hypnotics.

and its major branches. These ganglia are known as **collateral ganglia.** Parasympathetic preganglionic neurons synapse with postganglionic neurons in **terminal ganglia** located near or within the walls of the organs they innervate.

The sympathetic and parasympathetic systems also differ in the neurotransmitters they release at the synapse with the effector. Both preganglionic and postganglionic parasympathetic neurons secrete acetylcholine. Sympathetic postganglionic neurons release norepinephrine (although their preganglionic neurons secrete acetylcholine). Table 47–4 compares the sympathetic and parasympathetic systems.

The autonomic system got its name from the original belief that it was independent of the CNS, that is, autonomous. Physiologists have shown that this is not so and that the hypothalamus and many other parts of the CNS help to regulate the autonomic system. Although the autonomic system usually functions automatically, its activities can be consciously influenced. **Biofeedback** provides a person with visual or auditory evidence concerning the status of an autonomic body function; for example, a tone

Name of Drug	Effect on Mood	Actions on Body	Dangers Associated with Abuse
Cocaine	Euphoria; excitation followed by depression	CNS stimulation followed by depression; autonomic stimulation; dilates pupils; local anesthesia; inhibits reuptake of norepinephrine	Mental impairment; convulsions; hallucinations; unconsciousness; death from overdose
Amphetamines (e.g., Dexedrine)	Euphoria; stimulant; hyperactivity; "uppers," "pep pills"	Stimulate release of dopamine and norepinephrine; block reuptake of norepinephrine and dopamine into neurons; inhibit monoamine oxidase (MAO); enhance flow of impulses in RAS: increase heart rate; raise blood pressure; dilate pupils	Tolerance; possible physical dependence; hallucinations; death from overdose
Caffeine	Increases mental alertness; decreases fatigue and drowsiness	Acts on cerebral cortex; relaxes smooth muscle; stimulates cardiac and skeletal muscle; increases urine volume (diuretic effect)	Very large doses stimulate centers in the medulla (may slow the heart); toxic doses may cause convulsions
Nicotine	Psychological effect of lessening tension	Stimulates sympathetic nervous system; combines with receptors in postsynaptic neurons of autonomic system; effect similar to that of acetylcholine, but large amounts result in blocking transmission; stimulates synthesis of lipid in arterial wall	Tolerance; physical dependence; stimulates development of atherosclerosis
LSD (lysergic acid diethylamide)	Overexcitation; sensory distortions; hallucinations	Alters levels of transmitters in brain (may inhibit serotonin and increase norepinephrine); potent CNS stimulator; dilates pupils sometimes unequally; increases heart rate; raises blood pressure	Irrational behavior
Marijuana	Euphoria	Impairs coordination; impairs depth perception and alters sense of timing; inflames eyes; causes peripheral vasodilation; exact mode of action unknown	In large doses, sensory distortions, hallucinations; evidence of lowered sperm counts and testosterone (male hormone) levels

may be sounded when blood pressure reaches a desirable level. Using such techniques, subjects have learned to control certain autonomic activities such as brain wave pattern, heart rate, blood pressure, and blood sugar level. Even certain abnormal heart rhythms can be consciously modified.

EFFECTS OF DRUGS ON THE NERVOUS SYSTEM

Among the most widely used drugs in the United States today are tranquilizers, sedatives, alcohol, stimulants, and antidepressants—all drugs that affect mood. Many of these drugs act by altering the levels of neurotransmitters within the brain. For example, amphetamines increase the amount of norepinephrine within the RAS, thus stimulating the CNS. Table 47–6 describes the effects of several types of commonly used and abused drugs.

Many mood drugs, including alcohol, are taken in order to induce a feeling of **euphoria**, or well-being. Habitual or prolonged use of almost any mood drug may result in **psychological dependence**, in which the user becomes emotionally dependent upon the drug. When deprived of it, the user may become irritable and feel unable to carry out normal activities.

Some drugs induce **tolerance** when they are taken continuously for several weeks. This means that increasingly large amounts are required in order to obtain the desired effect. Tolerance is thought to occur when the liver cells are stimulated to produce larger quantities of the enzymes that metabolize and inactivate the drug. Use of some drugs, such as heroin, tobacco, alcohol, or barbiturates, may also result in **physical addiction,** in which physiological changes take place in body cells, making the user dependent upon the drug. When the drug is withheld, the addict suffers physical illness and characteristic **withdrawal symptoms.**

Physical addiction can also occur because certain drugs, such as morphine, have components similar to substances that body cells normally manufacture on their own. The continued use of such a drug, followed by sudden withdrawal, causes potentially dangerous physiological effects because the body's natural production of these substances has been depressed. It may be some time before homeostasis is reestablished.

■ SUMMARY

I. Among invertebrates, nerve nets and radial nervous systems are typical of radially symmetric animals, and bilateral nervous systems are characteristic of bilaterally symmetric animals.
   A. A nerve net consists of nerve cells scattered throughout the body; no CNS is present. Response of these animals to stimuli is generally slow and imprecise.
   B. Echinoderms typically have a nerve ring and nerves that extend into various parts of the body.
   C. In a bilateral nervous system, there is a concentration of nerve cells to form nerves, nerve cords, ganglia, and (in complex forms) a brain. There is also an increase in numbers of neurons, especially of the association neurons. This permits greater precision and a wider range of responses.

II. In the vertebrate embryo, the brain and spinal cord arise from the neural tube. The anterior end of the tube differentiates into forebrain, midbrain, and hindbrain.
   A. The hindbrain subdivides into the metencephalon and myelencephalon.
      1. The myelencephalon develops into the medulla, which contains the vital centers and other reflex centers.
      2. The metencephalon gives rise to the cerebellum and pons.
         a. The cerebellum is responsible for muscle tone, posture, and equilibrium.
         b. The pons connects various parts of the brain.
   B. The midbrain is the largest part of the brain in fish and amphibians. It is their main association area, linking sensory input and motor output. In reptiles, birds, and mammals, the midbrain has a lesser function and is used as a center for certain visual and auditory reflexes. It also contains the red nucleus that integrates information about muscle tone and posture.
   C. The forebrain differentiates to form the diencephalon and telencephalon.
      1. The diencephalon develops into thalamus and hypothalamus.
         a. The thalamus is a relay center for motor and sensory information.
         b. The hypothalamus controls autonomic functions, links nervous and endocrine systems, controls temperature, appetite, and fluid balance, and is involved in some emotional and sexual responses.
      2. The telencephalon develops into the cerebrum and olfactory bulbs.
         a. In fish and amphibians, the cerebrum functions mainly to integrate incoming sensory information.
         b. In birds, the corpus striatum controls stereotypical but complex behavior patterns. Another part of the cerebrum is thought to govern learning.
         c. The simplest animals possessing a neopallium are certain reptiles. The neopallium is present in birds and mammals and accounts for about 90% of the cerebral cortex in the human brain. In mammals, the cerebrum has complex association functions.

III. The human brain and spinal cord are protected by bone and by three meninges and are cushioned by cerebrospinal fluid.
   A. The spinal cord consists of ascending tracts, which transmit information to the brain, and descending tracts, which transmit information from the brain. Its gray matter consists of many nuclei that serve as reflex centers.
   B. The human cerebral cortex consists of motor areas, which control voluntary movement; sensory areas, which receive incoming sensory information; and association areas, which link sensory and motor areas and also are responsible for learning, language, thought, and judgment.
      1. The reticular activating system is responsible for maintaining consciousness.
      2. The limbic system affects the emotional aspects of behavior, motivation, sexual behavior, autonomic responses, and biological rhythms.

3. Alpha wave patterns are characteristic of relaxed states, beta wave patterns of heightened mental activity, and delta waves of non-REM sleep.
4. Metabolic rate slows during non-REM sleep. REM sleep is characterized by dreaming.
5. Short-term memory may depend upon reverberating circuits in the brain. Mechanisms of long-term memory are not understood.
6. Environmental experience can cause physical and chemical changes in the brain.

IV. The peripheral nervous system consists of sensory receptors and nerves, including the cranial and spinal nerves and their branches.

V. The autonomic system regulates the internal activities of the body.
A. The sympathetic system enables the body to respond to stressful situations.
B. The parasympathetic system influences organs to conserve and restore energy.

VI. Many drugs alter mood by increasing or decreasing the concentrations of specific neurotransmitters within the brain.

■ POST-TEST

1. The simplest organized nervous system is the _____ _____ found in *Hydra* and other cnidarians.
2. _____ nerves transmit impulses toward a central nervous system, whereas _____ nerves transmit impulses away from the CNS.
3. The cerebral ganglia in a flatworm serve as a simple _____.
4. In addition to cerebral ganglia, mollusks typically have _____ ganglia located among the organs and _____ ganglia located in the foot.
5. In vertebrates, the embryonic neural tube expands anteriorly to form the _____ and develops posteriorly into the _____ _____.
6. The medulla, pons, and midbrain make up the _____ _____.
7. The fourth ventricle, located within the _____, communicates with the _____ _____ of the spinal cord.

For each group, select the most appropriate answer from Column B for the description given in Column A.

Column A
8. Most prominent part of amphibian brain
9. Links nervous and endocrine systems
10. Most prominent part of mammalian brain
11. Coordinates muscle activity
12. Contains vital centers

Column B
a. Cerebellum
b. Cerebrum
c. Midbrain
d. Medulla
e. Hypothalamus

Column A
13. Controls innate, complex action patterns in birds
14. Convey voluntary motor impulses from cerebrum down spinal cord
15. Shallow furrows between gyri
16. Makes up most of human cerebral cortex
17. Layers of connective tissue that protects CNS

Column B
a. Neopallium
b. Corpus striatum
c. Meninges
d. Pyramidal tracts
e. Sulci

Column A
18. Contains primary motor areas
19. Action system concerned with emotional behavior and with reward and punishment
20. Contains the visual centers
21. Maintains wakefulness

Column B
a. Limbic system
b. RAS
c. Frontal lobe
d. Temporal lobe
e. None of the above

22. As you answer these questions, your brain should be emitting _____ waves.
23. Dreaming takes place during _____ sleep.
24. Cranial nerve X, the _____ nerve, innervates the _____.
25. The sympathetic nervous system mobilizes _____ and helps the body respond to _____.
26. Sensory nerves enter the spinal cord through the _____ root.
27. Some drugs induce tolerance, which means that _____.
28. Label the following diagram. (Refer to Figure 47–12 as necessary.)

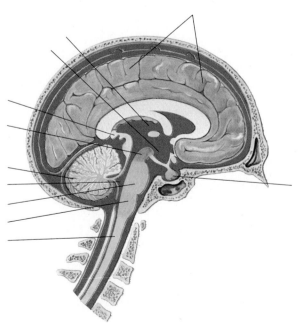

■ **REVIEW QUESTIONS**

1. Compare the nervous system of a hydra with that of a planarian flatworm.
2. Describe characteristics of the bilateral nervous system. What are some advantages of the bilateral system?
3. Compare the flatworm nervous system with that of a vertebrate. In each case, how does the nervous system impact on the animal's life style?
4. Identify the parts of the adult vertebrate brain derived from the embryonic forebrain, midbrain, and hindbrain.
5. What are the functions of each of the following structures in the human brain: medulla, midbrain, cerebellum, thalamus, hypothalamus, and cerebrum.
6. Compare the fish midbrain and cerebrum with that of the mammal.
7. Describe the protective coverings of the human CNS, and give the function of the cerebrospinal fluid.
8. Cite experimental evidence supporting the view that environmental experience can alter the brain.
9. Contrast the structure and function of the sympathetic system with that of the parasympathetic system.
10. Describe how each of the following drugs affects the CNS: a. alcohol; b. phenothiazines; c. barbiturates; d. amphetamines.

■ **RECOMMENDED READINGS**

Black, I.B., et al. Biochemistry of information storage in the nervous system. *Science*, Vol. 256, June 5, 1987, pp. 1263–1268. Use of molecular approaches has defined new mechanisms that store information in the mammalian nervous system.

Bower, B. Million-cell memories. *Science News*, Vol. 130, No. 20, November 15, 1986, pp. 313–315. Large parts of the brain may participate in a simple memory.

Constantine-Paton, M., Law, M. The development of maps and stripes in the brain. *Scientific American*, Vol. 247, No. 6, December 1982, pp. 62–70. An account of studies of the visual system and its organization.

Goldstein, G.W., Betz, A.L. The blood-brain barrier. *Scientific American*, Vol. 255, No. 3, September 1986, pp. 74–83. A discussion of how brain capillaries regulate the entrance of materials into the brain.

Kandel, E.R., Schwartz, J.H. Molecular biology of learning: modulation of transmitter release. *Science*, Vol. 218, October 1982. A review that focuses on the biochemistry of learning in a marine mollusk.

Mishkin, M., Appenzeller, T. The anatomy of memory. *Scientific American*, Vol. 256, No. 6, June 1987, pp. 80–89. Deep structures in the brain may interact with perceptual pathways in outer layers of the brain to transform sensory stimuli into memories.

Morrison, A.R. A window on the sleeping brain. *Scientific American*, Vol. 248, No. 4, April 1983, pp. 94–101. A study of REM sleep and accompanying paralysis.

Wurtman, R.J. Alzheimer's disease. *Scientific American*, Vol. 252, No. 1, pp. 62–74. Alzheimer's disease affects more than 100,000 Americans a year. Models of this progressive dementia are presented.

Wurtman, R.J. Nutrients that modify brain function. *Scientific American*, April 1982, pp. 50–59. An exciting discussion of the effect of such nutrients as choline, tryptophane, and tyrosine on brain function.

Bull frog (*Rana catesbeiana*)

48

Sense Organs

■ **LEARNING OBJECTIVES**

After you have read this chapter you should be able to:

1. Distinguish among exteroceptors, proprioceptors, and interoceptors, and explain the importance of each group.
2. Name the five types of receptors that are classified according to the types of energy to which they respond. Give examples of specific sense organs of each type.
3. Describe how a sense organ functions, including definitions of energy transduction, receptor potential, and adaptation in your answer.
4. Describe how the following mechanoreceptors work: tactile receptors, statocysts, lateral line organs, and proprioceptors.
5. Compare the function of the saccule and utricle with that of the semicircular canals in maintaining equilibrium.
6. Trace the path taken by sound waves through the structures of the ear, and explain how the organ of Corti functions as an auditory receptor.
7. Describe the receptors of taste and smell.
8. Describe the ways thermoreceptors are advantageous in various types of animals.
9. Contrast simple eyes, compound eyes, and the mammalian eye.
10. Label the structures of the mammalian eye on a diagram, and give the functions of each of the accessory structures.
11. Identify two types of photoreceptors in the human retina, and compare their functions, taking into account the role of rhodopsin.

    Sense organs link organisms with the outside world and enable them to receive information about their environment. The kinds of sense organs an animal has determine just how it perceives the world. We humans live in a world of rich colors, numerous shapes, and varied sounds. But we cannot hear the high-pitched whistles audible to dogs and cats or the ultrasonic echoes by which bats navigate (Figure 48–1). Nor do we ordinarily recognize our friends by their distinctive odors. And although vision is our dominant and most refined sense, we are blind to the ultraviolet hues that light up the world for insects.

FIGURE 48–1 Bats navigate by ultrasonic echoes inaudible to the human ear. They emit high-pitched clicking sounds and use the echos to locate objects.

WHAT IS A SENSE ORGAN?

A **sense organ** is a specialized structure consisting of one or more **receptor cells** and, sometimes, **accessory cells;** it detects changes in the environment and transmits that information into the nervous system. For example, rod and cone cells located in the retina are the receptor cells of the human eye that respond to light. Accessory structures (cornea, lens, iris, and ciliary muscles) enhance the versatility of the sense organ but in some instances may limit its performance. The lens, for example, enhances the ability of the eye to see by adjusting the focus of the light rays. On the other hand, the lens filters out ultraviolet light before it reaches the retina, so we cannot see it. The cells in the retina can respond to light of this wavelength, however.

Receptor cells may be either neuron endings or specialized cells that are in close contact with neurons. Human taste buds, for example, are modified epithelial cells connected to one or more neurons.

HOW SENSE ORGANS ARE CLASSIFIED

Sense organs can be classified according to the location of the stimuli affecting them. **Exteroceptors** receive stimuli from the outside environment so that an animal can know and explore the world, search for food, find and attract a mate, find shelter, recognize friends, detect enemies, and learn. **Proprioceptors** are sense organs within muscles, tendons, and joints that enable the animal to perceive the position of its arms, legs, head, and other body parts, along with the orientation of its body as a whole. With the help of our proprioceptors we humans can get dressed or eat in the dark.

Interoceptors are sense organs *within* body organs that detect changes in pH, osmotic pressure, body temperature, and the chemical composition of the blood. We are usually not conscious of messages sent to the central nervous system (CNS) by these receptors as they work continuously to maintain homeostasis. We become aware of their activity when they enable us to perceive such diverse internal conditions as thirst, hunger, nausea, pain, and orgasm. Interoceptors are described in conjunction with discussions of blood pressure, temperature regulation, respiration, and other specific body functions.

Another way that sense organs can be classified is according to the type of energy to which they respond (Table 48–1). **Mechanoreceptors** respond to mechanical energy—touch, pressure, gravity, stretching, or movement. **Chemoreceptors** respond to certain chemical stimuli, while **photoreceptors**

Table 48–1 CLASSIFICATION OF RECEPTORS BY STIMULI TO WHICH THEY RESPOND

Type of Receptor	Examples	Effective Stimulus
Mechanoreceptors	Tactile receptors	Touch, pressure
	Statocysts in invertebrates	Gravity
	Lateral line organs in fish	Waves, currents in water
	Proprioceptors	Movement and body position
	Muscle spindles	Detect muscle contraction
	Golgi tendon organs	Stretch of a tendon
	Joint receptors	Detect movement in ligaments
	Halteres in flies	Strains in cuticle
	Labyrinth of vertebrate ear	
	Saccule and utricle	Gravity; linear acceleration
	Semicircular canals	Angular acceleration
	Cochlea	Pressure waves (sound)
Chemoreceptors	Taste hair of fly, taste buds, olfactory epithelium	Specific chemical compounds
Thermoreceptors	Temperature receptors in blood-sucking insects and ticks; pit organs of pit vipers; nerve endings and receptors in skin and tongue of many animals	Heat
Electroreceptors	Organs in skin of some fish	Electric currents in water
Photoreceptors	Ommatidia of arthropods	Light energy
	Retina of vertebrates	Light energy

detect light energy. **Thermoreceptors** respond to heat or cold. Some fish have well-developed **electroreceptors,** which detect electrical energy.

Traditionally, humans are said to have five senses: touch, smell, taste, sight, and hearing. Balance is now also recognized as a sense, and touch is viewed as a compound sense that involves detection of pressure, pain, and temperature. In this chapter we also consider some proprioceptors that enable us to sense muscle tension and joint position.

HOW SENSE ORGANS WORK

Receptor cells absorb energy, **transduce** (convert) that energy into electrical energy, and produce a *receptor potential* (Figure 48–2). In its capacity as a detector or sensor, a receptor receives a small amount of energy from the environment. Each kind of receptor is especially sensitive to one particular form of energy. Rods and cones in the retina absorb the energy of photons. Temperature receptors respond to radiant energy transferred by radiation, conduction, or convection. Electricity is detected by the energy of electrons. Taste buds and olfactory cells detect the change in energy accompanying the binding of specific molecules to their chemical receptors.

Receptor cells are remarkably sensitive to appropriate stimuli. The rods and cones of the eye, for example, are stimulated by an extremely faint beam of light, whereas only a very strong light can stimulate the optic nerve directly. The negligible amount of vinegar that can be tasted, or the amount of vanilla that can be smelled, would have no effect if applied directly to a nerve fiber.

The various kinds of environmental energy act as triggers for the receptor cells to perform biological work. These relationships are best exemplified by a very simple sense organ, the tactile hair of an insect. This hair plus its associated cells constitutes a complete sense organ, but only the

SENSORY CODING AND SENSATION

The stimulation of any sense organ initiates what might be considered a coded message, composed of action potentials transmitted by the nerve fibers and decoded in the brain. Impulses from the sense organ may differ in any of several ways: (1) the total number of fibers transmitting, (2) the specific fibers carrying action potentials, (3) the total number of action potentials passing over a given fiber, (4) the frequency of the action potentials passing over a given fiber, or (5) the time relations between action potentials in specific fibers. These are the possibilities in the "code" sent along the nerve fiber; how the sense organ initiates different codes and how the brain analyzes and interprets them to produce various sensations are not yet understood.

All action potentials are qualitatively the same. Light of the wavelength 400 nanometers (blue), sugar molecules (sweet), and sound waves of 440 hertz (A above middle C) all cause action potentials to be sent to the brain via the appropriate nerves; these action potentials are identical. How can the organism assess its environment accurately? The qualitative differentiation of stimuli must depend either upon the sense organ itself, upon the brain, or upon both. In fact, it depends upon both. Primarily, our ability to discriminate red from green, hot from cold, or red from cold depends on particular sense organs and their individual sensitive cells being connected to specific cells in particular parts of the brain.

The frequency of the repetitive action potential codes the intensity of the stimulus. Since each receptor normally responds to only one category of stimuli (i.e., light, sound, taste, and so forth), a message arriving in the central nervous system along this nerve is interpreted as meaning that a particular stimulus occurred. Interpretation of the message and, in the case of humans, of the quality of sensation depends upon which central association neurons receive the message. Sensation, when it occurs, takes place in the brain. Rods and cones do not see; only the combination of rods, cones, and centers in the brain see. Furthermore, many sensory messages never give rise to sensations. For example, chemoreceptors in the carotid sinus and the hypothalamus sense internal changes in the body but never stir our consciousness.

Since only those nerve impulses that reach the brain can result in sensations, any blocking of the impulse along the nerve fibers by an anesthetic has the same effect as removing the original stimulus entirely. The sense organs, of course, will continue to initiate impulses that can be detected by the proper electrical apparatus, but the anesthetic prevents them from reaching their destination.

Spatial localization of stimuli impinging on the body, especially mechanical and pain stimuli, also depends upon the destination of specific nerves in the brain. The importance of the brain in localization and making sensations possible is emphasized by the phenomenon of referred pain. A well-known example of referred pain is that often experienced by persons suffering from heart pains; such persons may complain of pain in the shoulder, upper chest, or left arm. Actually, the stimuli originate in the heart, but the nerve impulses terminate in the same part of the brain as do impulses genuinely originating in the shoulder, chest, or arm.

Cross-fiber patterning, another method of coding information, is probably the one used in olfactory organs. An olfactory organ does not contain a specific receptor for each of the thousands of individual odors that can be recognized. Instead, there is evidence that a limited number of categories of receptors exists. Several receptors are thought to react to each odor. The brain probably makes a statistical analysis of the pattern of responding receptors and from this pattern infers the odor.

The **temporal pattern** of action potentials generated in a single neuron may serve as a code for different stimuli. The single taste receptors of flies, for example, generate action potentials at an even, regular frequency when the stimulus is salt but generate irregular frequencies when the stimulus is acid.

In invertebrates it is usual for the axon of a sensory neuron to extend all the way to the CNS without synapsing. In these circumstances, the message generated at the periphery arrives unaltered. However, in the compound eye of the arthropod, as in many vertebrate sense organs, many interneurons are interposed between the receptor and the CNS. The vertebrate retina and the olfactory bulb have an exceptionally complicated neural circuitry. As a consequence of all these synaptic connections, the original message is altered and may lose or gain some of its information.

MECHANORECEPTORS

Mechanoreceptors respond to touch, pressure, gravity, stretch, or movement. Some of these sense organs are concerned with enabling an organism to maintain its primary body attitude with respect to gravity (for us, head up and feet down; for a dog, dorsal side up and ventral side down; for a tree sloth, ventral side up and dorsal side down).

Mechanoreceptors are also concerned with maintaining postural relations (i.e., the position of one part of the body with respect to another). This information is essential for all forms of locomotion and for all coordinated and skilled movements, from spinning a cocoon to completing a reverse one-and-a-half dive with twist. In addition, mechanoreceptors provide information about the shape, texture, weight, and topographical relations of objects in the external environment. Finally, mechanoreceptors affect the operation of some internal organs. For example, they supply information about the presence of food in the stomach, feces in the rectum, urine in the bladder, or a fetus in the uterus.

Tactile Receptors

The simplest mechanoreceptors are free nerve endings in the skin that are directly stimulated by contact with any object on the body surface. Somewhat more complex are the tactile receptors that lie at the base of a hair or bristle (Figure 48–4). They are stimulated indirectly when the hair is bent or displaced. A receptor potential then develops, and a few action potentials may be generated. Because this type of receptor responds only when the hair is moving, it is known as a **phasic** receptor. Even though the hair may be maintained in a displaced position, the receptor is not stimulated unless there is motion. Such tactile hairs are found in many invertebrates as well as vertebrates and are involved in orientation to gravity, in postural orientation, and in the reception of vibrations in air and water, as well as in contacts with other objects.

The remarkable tactile sensitivity of human skin, especially on the fingertips and lips, is due to a large number of diverse sense organs (Figure 48–5). By making a careful point-by-point survey of a small area of skin, using a stiff bristle to test for touch, a hot or cold metal stylus to test for temperature, and a needle to test for pain, it has been found that receptors for each of these sensations are located at different spots. By comparing the distribution of the different types of sense organs and the types of sensations produced, it has been found that the free nerve endings are responsible for pain perception, that a variety of tiny sense organs (e.g., Meissner's corpuscles, Ruffini's end organs, and Merkel's disks) are re-

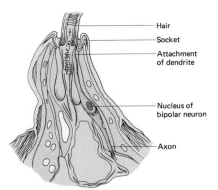

Hair
Socket
Attachment of dendrite

Nucleus of bipolar neuron

Axon

FIGURE 48–4 A tactile hair from a caterpillar, showing the attachment of the dendrite of the bipolar neuron (the mechanoreceptor) at the point where the shaft of the hair enters the socket.

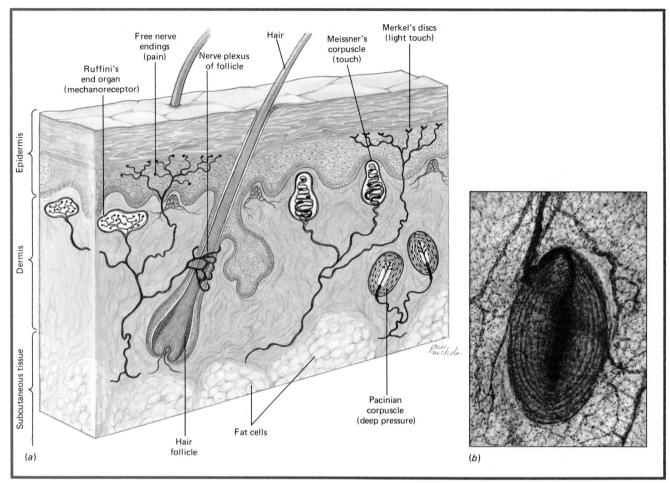

FIGURE 48–5 Sense organs within the human skin. (*a*) Diagrammatic section through the human skin showing the types of sense organs present. The free nerve endings respond to pain; tactile hairs, Merkel's disks, Ruffini's end organs, and Meissner's corpuscles respond to touch; pacinian corpuscles respond to deep pressure. (*b*) Pacinian corpuscle, a deep pressure receptor.

sponsible for touch, and that Pacinian corpuscles mediate the sensation of deep pressure. The receptors responsible for sensations of cold and warmth are not known.

The Pacinian corpuscle has been particularly well studied. The bare nerve ending is surrounded by lamellae (connective tissue layers) interspersed with fluid. Compression causes displacement of the lamellae, which provides the deformation stimulating the axon. Even though the displacement is maintained under steady compression, the receptor potential rapidly falls to zero and action potentials cease—an excellent example of adaptation. The Pacinian corpuscle is a phasic receptor responding to velocity (rapid movement of the tissue).

Gravity Receptors: Statocysts

All organisms are oriented in a characteristic way with respect to gravity. When displaced from this normal position, they quickly adjust the body to reassume it. To accomplish this, receptors must continually send information regarding the position and movements of the body to the CNS.

Many invertebrates have specialized sense organs called **statocysts** that serve as gravity receptors. A statocyst is basically an infolding of the epidermis lined with receptor cells that have hairs (Figure 48–6). The cavity contains a **statolith** (sometimes more than one), which is a tiny granule of loose sand grains or calcium carbonate. The particles are held together by an adhesive material secreted by cells of the statocyst. Normally the particles are pulled downward by gravity and stimulate the hair cells. When the position of the statolith changes, the hairs of the receptor cells are bent.

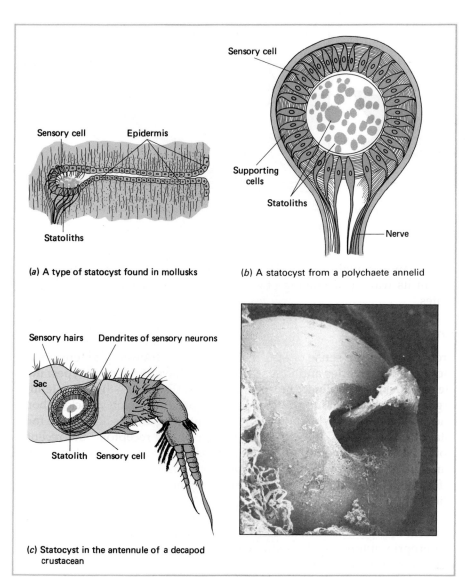

(a) A type of statocyst found in mollusks

(b) A statocyst from a polychaete annelid

(c) Statocyst in the antennule of a decapod crustacean

FIGURE 48–6 Many invertebrates have statocysts, which serve as sensors of gravitational force.

This mechanical displacement results in receptor potentials and action potentials that inform the CNS of the change in position. By "knowing" which hair cells are firing, the animal knows where down is and so can correct any abnormal orientation.

In a classic experiment on crayfish, the function of the statocyst was demonstrated by substituting iron filings for sand grains in the statocyst. The force of gravity was overcome by holding magnets above the animals. The iron filings were attracted upward toward the magnets, and the crayfish began to swim upside down in response to the new information provided by their gravity receptors.

Lateral Line Organs

Lateral line organs are found in fishes and in aquatic and larval amphibians. Typically this sense organ consists of a long canal running the length of the body and continuing into the head on both sides of the animal (Figure 48–7). The canals are lined with receptor cells that have hairs. Above the hairs and enclosing their tips is a mass of gelatinous material, called a **cupula,** secreted by the receptor cells.

The receptor cells are thought to respond to waves, currents, or disturbances in the water. The water moves the cupula and causes the hairs to bend. This results in messages being dispatched to the CNS. The lateral

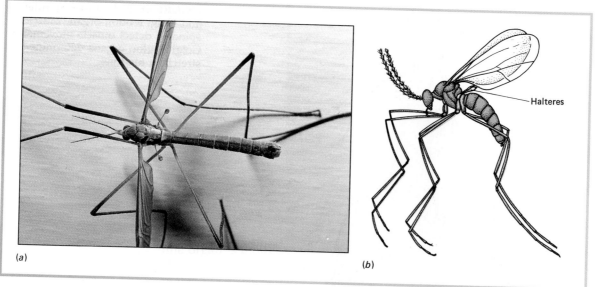

(a)

(b)

—Halteres

FIGURE 48–9 In dipterans, the hind wings are reduced to halteres, small stalks expanded at the distal end. (*a*) "Daddy long legs" (crane fly) showing halteres. (*b*) The halteres of a gnat.

a thin stalk (Figure 48–9) and resembles an Indian club. The base is folded and articulated in a complicated fashion and is equipped with about 418 mechanoreceptors. These respond to strains produced in the cuticle by gyroscopic torque produced by the beating of the halteres. These oscillating masses generate forces at the base of the stalk as the whole fly rotates. They probably do not act as stabilizing gyroscopes of the sort placed in ships to offset their movement. Their action is indirect in that their mechanoreceptors signal the CNS to make the necessary corrections in flight.

The Labyrinth of the Vertebrate Ear

When we think of the ear, we think of hearing. However, fishes do not use their ears for hearing. In fact, in all vertebrates the basic function of the ear is to help maintain equilibrium. Typically the ear also contains gravity receptors. Although many vertebrates do not have outer or middle ears, all of them have inner ears.

The inner ear consists of a complicated group of interconnected canals and sacs, often referred to as the **labyrinth.** In mammals the labyrinth consists of two saclike chambers, the **saccule** and **utricle,** and three **semicircular canals,** as well as a snail-shaped structure known as the **cochlea.**

Collectively, the saccule, utricle, and semicircular canals are referred to as the **vestibular apparatus** (Figure 48–10). Destruction of the vestibular apparatus leads to a considerable loss of the sense of equilibrium. A pigeon in which these organs have been destroyed is unable to fly but in time can relearn how to maintain equilibrium using visual stimuli. Equilibrium in the human depends not only upon stimuli from the organs in the inner ear but also upon the sense of vision, stimuli from the proprioceptors, and stimuli from cells sensitive to pressure in the soles of the feet.

The saccule and utricle house gravity detectors in the form of small calcium carbonate ear stones called **otoliths** (Figure 48–11). The sensory cells of these structures are similar to those of the lateral line organ. They consist of groups of hair cells surrounded at their tips by a gelatinous cupula. The receptor cells in the saccule and utricle lie in different planes. Normally, the pull of gravity causes the otoliths to press against particular hair cells, stimulating them to initiate impulses sent to the brain by way of sensory nerve fibers at their bases. When the head is tilted, or in **linear acceleration** (change in speed when the body is moving in a straight line), the otoliths press upon the hairs of other cells and stimulate them. This

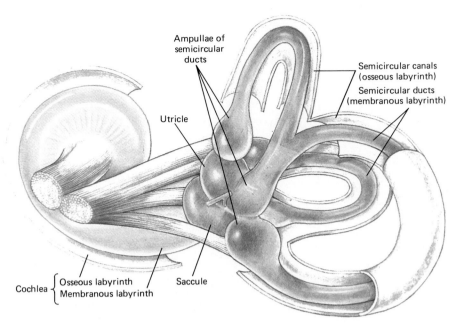

Ampullae of
semicircular
ducts

Semicircular canals
(osseous labyrinth)

Semicircular ducts
(membranous labyrinth)

Utricle

Cochlea { Osseous labyrinth
Membranous labyrinth

Saccule

FIGURE 48–10 The human inner ear with the membranous labyrinth exposed. Because this is a posterior view, the utricle and saccule can be seen. Note that the membranous labyrinth is shown in color only within the semicircular canals and is not colored in the cochlea.

enables the animal to perceive the direction of gravity and of linear acceleration or deceleration when the head is in any position.

Information about turning movements, referred to as **angular acceleration,** is furnished by the three semicircular canals. Each of these is connected with the utricle and lies in a plane at right angles to the other two like the sides of a box where they meet at a corner. Each canal is a hollow ring filled with fluid called **endolymph.** At one of the openings of each canal into the utricle is a small, bulblike enlargement, the **ampulla.** Within each ampulla is a clump of hair cells called a **crista,** similar to those in the utricle and saccule but lacking otoliths. These receptor cells are stimulated by movements of the endolymph in the canals (Figure 48–12).

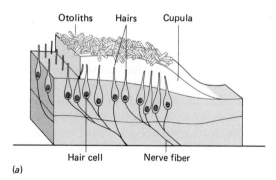

Otoliths   Hairs   Cupula

Hair cell   Nerve fiber

(a)

FIGURE 48–11 Otoliths are present in the succule and utricle. Compare the positions of the otoliths and hairs in (*a*) with those in (*b*). Changes in head position cause the force of gravity to distort the cupula, which in turn distorts the hairs of the hair cells; the hair cells respond by sending impulses down the vestibular nerve (part of the auditory nerve) to the brain.

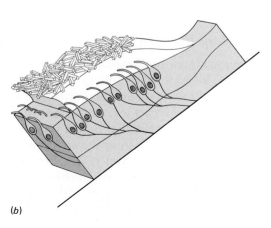

(b)

FIGURE 48–12 How movement of the endolymph within the semicircular ducts of the ampulla distorts the cupula. The hair cells of the cupula then are bent, reporting any change to the brain via the vestibular nerve.

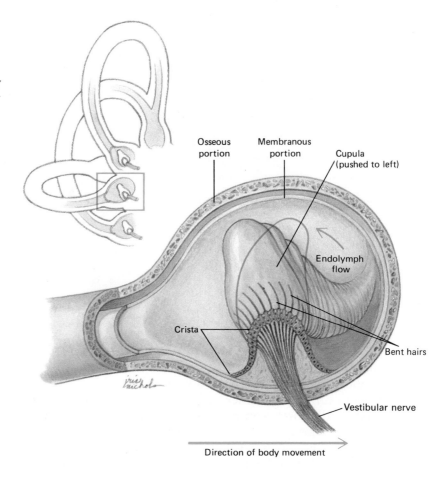

When the head is turned, there is a lag in the movement of the fluid within the canals, so that the hair cells move in relation to the fluid and are stimulated by its flow. This stimulation produces not only the consciousness of rotation but also certain reflex movements in response to it—movements of the eyes and head in a direction opposite to the original rotation. Since the three canals are located in three different planes, a movement of the head in any direction stimulates the movement of the fluid in at least one of the canals.

We humans are used to movements in the horizontal plane, which stimulate certain semicircular canals, but are unused to vertical movements parallel to the long axis of the body. Movements such as the motion of an elevator, or of a ship pitching in a rough sea, stimulate the semicircular canals in an unusual way, and may cause sea sickness or motion sickness, with their resulting nausea or vomiting. When a person so affected lies down, the movement stimulates the semicircular canals in a different way, and nausea is less likely to occur.

Auditory Receptors

Some fish hear by means of receptors in the utricle, but hearing does not seem to be important to them. Hearing is an important sense in tetrapods, however. Both birds and mammals have a highly developed sense of hearing based in the cochlea, a structure in the inner ear that contains mechanoreceptor hair cells that detect pressure waves.

Part of the labyrinth of the inner ear, the cochlea is a spiral tube coiled two and a half turns and resembling a snail's shell (Figure 48–13). If the cochlea were uncoiled, as in Figure 48–14c, it would be seen to consist of three canals separated from each other by thin membranes and coming almost to a point at the apex. Two of these canals, or ducts, the **vestibular**

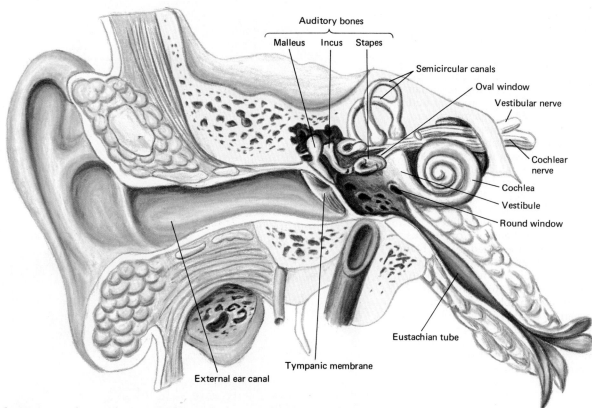

Auditory bones
Malleus Incus Stapes
Semicircular canals
Oval window
Vestibular nerve
Cochlear nerve
Cochlea
Vestibule
Round window
Eustachian tube
Tympanic membrane
External ear canal

FIGURE 48–13 The anatomy of the human ear.

canal (also known as the scala vestibuli) and the **tympanic canal** (scala tympani), are connected with one another at the apex of the cochlea and are filled with a fluid known as **perilymph.** The middle canal, the **cochlear duct** (scala media), is filled with endolymph and contains the actual auditory receptor, the **organ of Corti.**

Each organ of Corti contains about 24,000 hair cells arranged in rows extending the entire length of the coiled cochlea. Each cell is equipped with hairlike projections extending into the cochlear duct. These cells rest upon the **basilar membrane,** which separates the cochlea from the tympanic canal. Overhanging the hair cells is another membrane, the **tectorial membrane,** attached along one edge to the membrane on which the hair cells rest, with its other edge free. The hair cells initiate impulses in the fibers of the cochlear (auditory) nerve.

In terrestrial vertebrates, accessory structures change sound waves in the air to pressure waves in the cochlear fluid. In the human ear, for example, sound waves pass through the **external auditory meatus** (the canal of the outer ear) and set the **eardrum** (the membrane separating outer ear and middle ear) vibrating. These vibrations are transmitted across the middle ear by three tiny bones, the **malleus** (hammer), **incus** (anvil), and **stapes** (stirrup) (so called because of their shapes). The hammer is in contact with the eardrum, and the stirrup is in contact with the membrane at the opening of the inner ear called the **oval window.** The vibrations pass through the oval window to the fluid in the vestibular canal.

Since liquids cannot be compressed, the oval window could not cause movement of the fluid in the vestibular duct unless there were an escape valve for the pressure. This is provided by the **round window** at the end of the tympanic canal. The pressure wave presses upon the membranes separating the three ducts, is transmitted to the tympanic canal, and causes a bulging of the round window. The movements of the basilar membrane produced by these pulsations are believed to rub the hair cells of the organs of Corti against the overlying tectorial membrane, thus stimulating them and initiating nerve impulses in the dendrites of the cochlear nerve lying at the base of each hair cell.

FIGURE 48–14 The cochlea is the part of the inner ear concerned with hearing. (*a*) Cross section through the cochlea showing the organ of Corti resting on the basilar membrane and covered by the tectorial membrane. (*b*) Scanning elec-electron micrograph of guinea pig organ of Corti showing inner hair cells, *IHC*, and three rows of outer hair cells, *OHC 1* (magnification ×1,790). *N*, nucleus; *S*, stereocilia; *TM*, tectorial membrane; *SM*, scala media (cochlear duct); *ST*, scala tympani (tympanic duct); *black arrows*, basilar membrane; *HC*, *DC*, *PC*, various supporting cells. (*c*) Diagram of the cochlea uncoiled and drawn out in a straight line. (*d*) How the organ of Corti works. Vibrations transmitted by the hammer, anvil, and stirrup set the fluid in the vestibular canal in motion; these vibrations are transmitted to the basilar membrane and the organ of Corti. The hair cells of the organ of Corti are the receptor cells for hearing. The hair cells are innervated by the cochlear nerve, a branch of the auditory nerve.

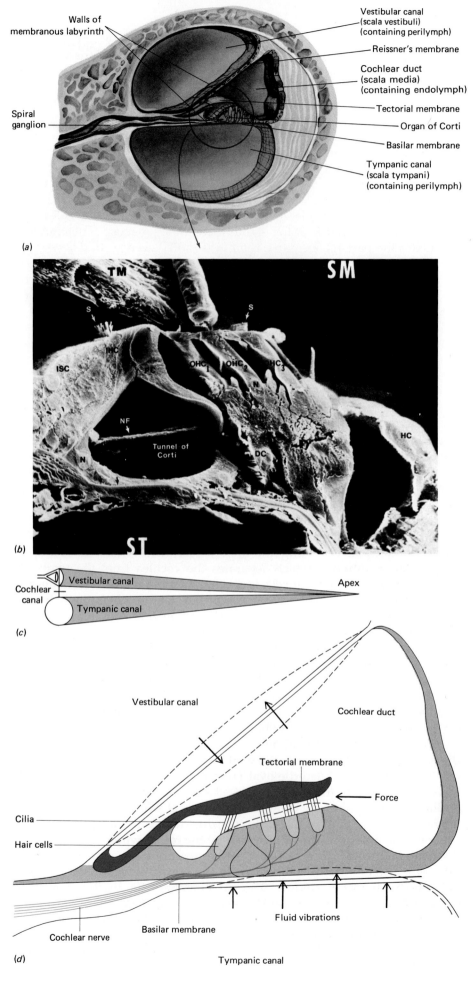

Walls of membranous labyrinth

Spiral ganglion

Vestibular canal (scala vestibuli) (containing perilymph)

Reissner's membrane

Cochlear duct (scala media) (containing endolymph)

Tectorial membrane

Organ of Corti

Basilar membrane

Tympanic canal (scala tympani) (containing perilymph)

(*a*)

TM SM S S IHC ISC PC OHC₁ OHC₂ HC₃ N NF N Tunnel of Corti DC HC ST

(*b*)

Cochlear canal Vestibular canal Tympanic canal Apex

(*c*)

Vestibular canal Cochlear duct

Tectorial membrane

Force

Cilia

Hair cells

Cochlear nerve Basilar membrane Fluid vibrations Tympanic canal

(*d*)

Since sounds differ in pitch, loudness, and tone quality, any theory of hearing must account for the ability to differentiate between these characteristics of sound. Pitch depends upon frequency. Low-frequency vibrations result in the sensation of low pitch, whereas high-frequency vibrations result in the sensation of high pitch. Up to about 4000 cycles per second, the nerve impulses produced have the same frequency as the sounds that cause them. At frequencies lower than 60 cycles per second, the entire basilar membrane vibrates. Action potentials are set up in the cochlear nerve that reflect the rhythm of the sound, and this informs the brain about the pitch. Frequencies greater than 60 cycles per second result in the basilar membrane vibrating unequally along its length. Sounds of a given frequency set up resonance waves in the fluid in the cochlea that cause a particular section of the basilar membrane to vibrate. The vibration stimulates the particular group of hair cells in that section. In this way the brain infers the pitch of a sound by taking note of the particular hair cells that are stimulated. Thus, the brain may recognize particular pitches by the frequency of the nerve impulses reaching it, as well as by the identity of the nerve fibers conducting the impulses.

Loud sounds cause resonance waves of greater amplitude and lead to a more intense stimulation of the hair cells and to the initiation of a greater number of impulses per second passing over the auditory nerve to the brain. Variations in the quality of sound, such as are evident when an oboe, a cornet, and a violin play the same note, depend upon the number and kinds of **overtones,** or **harmonics,** present. These stimulate different hair cells in addition to the main stimulation common to all three. Thus, differences in quality are recognized by the *pattern* of the hair cells stimulated. Careful histological work has shown that the nerve fibers from each particular part of the cochlea are connected to particular parts of the auditory area of the brain, so that certain brain cells are responsible for the perception of sensations of high tones and others, for low tones.

The human ear is equipped to register sound frequencies between about 20 and 20,000 cycles per second, although there are great individual differences. Some animals—dogs, for example—can hear sounds of much higher frequencies. The human ear is more sensitive to sounds between 1000 and 4000 cycles per second than to higher or lower ones. Within this range, the ear is extremely sensitive. In fact, when the energy of audible sound waves is compared with the energy of visible light waves, the ear is 10 times more sensitive than the eye.

The normal human ear is extremely efficient. Any further increase in sensitivity would probably be useless, since it would pick up the random movement of air molecules, which would result in a constant hiss or buzzing. Similarly, if the eye were more sensitive, a steady light would appear to flicker because the eye would be sensitive to the individual photons (light particles) impinging upon it.

There is little fatigue connected with hearing. Even though the ear is constantly assailed by noises, it retains its acuity and fatigue disappears after a few minutes. When one ear is stimulated for some time by a loud noise, the other ear also shows fatigue (i.e., loses acuity), indicating, not unexpectedly, that some of the fatigue is in the brain rather than in the ear itself.

Deafness may be caused by injuries or malformations of either the sound-transmitting mechanisms of the outer, middle, or inner ears, or of the sound-perceiving mechanism of the latter. The external ear may become obstructed by wax secreted by the glands in its wall; the middle-ear bones may become fused after an infection; or, more rarely, the inner ear or auditory nerve may be injured by a local inflammation or by the fever accompanying some disease.

When the ear is subjected to intense sound, the organ of Corti is injured. This was demonstrated by an experiment in which guinea pigs were exposed to continuous pure tones for a period of several weeks. When their cochleas were examined microscopically after death, it was found that the guinea pigs subjected to high-pitched tones suffered injury only in the lower part of the cochlea, while those subjected to low-pitched tones suffered injury only in the upper part of the cochlea. Members of rock bands, boilermakers, and other workers subjected to loud, high-pitched noises over a period of years frequently become deaf to high tones because the cells near the base of the organ of Corti become injured.

CHEMORECEPTORS: TASTE AND SMELL

Throughout the animal kingdom many feeding, social, sexual, and reproductive activities are initiated, regulated, or influenced in some way by specific chemical cues in the environment. Insects, for example, use many chemicals in communication, defense from predators, and recognizing specific foods. Many vertebrates employ chemical secretions to mark territory, attract their sexual partners, and defend themselves. Chemoreception is also used to help carnivores track prey and to help the intended prey elude the carnivores.

The sensitivity of chemoreceptors varies greatly. Some, such as those in the skin of a frog, may be gross and nonspecific. A frog is thus unable to differentiate between certain stimuli and will scratch its back when dilute acid or concentrated solutions of inorganic salts are applied to the skin. Free nerve endings are the chemoreceptors involved. This common chemical sense is widely distributed among aquatic animals. Among mammals it is restricted to moist areas of the body. Recall how your eye smarts and waters in the presence of ammonia fumes or a peeled onion, and how a broken blister stings if touched by a nonphysiological solution.

Two highly sensitive chemoreceptive systems are the senses of **taste** and **smell** (olfaction). These are easily distinguishable in humans and other terrestrial organisms. However, in aquatic organisms, especially members of lower phyla, it becomes increasingly difficult to decide what is taste and what is olfaction.

The Sense of Taste in Insects

One of the most thoroughly studied organs of taste is the taste hair of the fly (Figure 48–15). The terminal segments of the legs and the mouthparts of flies, moths, butterflies, and a number of other insects are equipped with very sensitive hairs. In the fly, each one of these contains four taste receptors and a tactile receptor. All are primary neurons. One taste receptor is more or less specific to sugars, one to water, and two to salts. If water is placed on one hair of a thirsty fly, action potentials generated by the water cell pass directly to the CNS and cause the fly to respond by extending its retractable proboscis and drinking. Similarly, sugar on a particular hair stimulates the sugar receptor and causes feeding. Salt causes the fly to reject the solution.

The Human Sense of Taste

The organs of taste in higher vertebrates are budlike structures known as **taste buds,** which are located in the mouth predominantly on the tongue. In humans they are found mainly in tiny elevations, or papillae, on the surface of the tongue. There is a rapid turnover of taste bud cells; every 10 to 30 hours the cells are completely replaced.

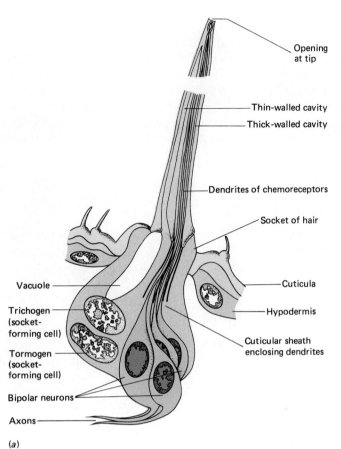

Opening at tip

Thin-walled cavity

Thick-walled cavity

Dendrites of chemoreceptors

Socket of hair

Cuticula

Vacuole

Hypodermis

Trichogen (socket-forming cell)

Cuticular sheath enclosing dendrites

Tormogen (socket-forming cell)

Bipolar neurons

Axons

(a)

FIGURE 48–15 A chemoreceptive (taste-sensing) hair of the blowfly. Four of the five neurons are shown.

A taste bud is an oval body that consists of an epithelial capsule containing several taste receptors. Each taste receptor is an epithelial cell with a border of microvilli at its free surface (Figure 48–16). A hairlike projection extends to the external surface of the taste bud through an opening called the taste pore. The connections between the taste receptors and the nerve cells that innervate them are complicated, with each taste receptor being innervated by more than one neuron. Furthermore, some neurons are connected with one taste cell, while others are connected with many. This complexity of connections renders interpretation of taste-sensory physiology difficult.

Traditionally four basic tastes are recognized: sweet, sour, salty, and bitter. Although the greatest sensitivity to each of these tastes is restricted to a given area of the human tongue (Figure 48–17), not all papillae are restricted in their sensitivity to a single category of taste. Some, indeed, are responsive specifically to salty, bitter, or sweet taste, but the majority respond to two or more categories of taste. Nor is a single taste bud restricted in its sensitivity to a single type of chemical. A single taste receptor may respond to more than one category of taste. Thus, the detection and processing of information in the taste organs of the tongue are very complex. Taste discrimination probably depends on a code that consists of cross-fiber patterning; that is, each receptor responds to more than one kind of chemical, but no two respond exactly alike, so that the total pattern of messages going to the brain is different for different solutions.

Flavor does not depend on the perception of taste alone. It is actually compounded of taste, smell, texture, and temperature. Smell affects flavor because odors pass from the mouth to the nasal chamber via the internal nares. No doubt you have observed that when you have a cold, food seems to have little "taste." Actually, the taste buds are not affected, but the blockage of nasal passages severely reduces the participation of olfactory reception in the composite sensation of flavor.

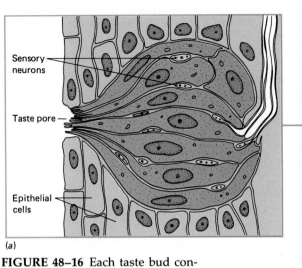

(a)

FIGURE 48–16 Each taste bud consists of an epithelial capsule containing several taste receptors. (a) Diagram of a taste bud. (b) Electron micrograph of papillae of the tongue containing taste buds. The taste buds are the oval bodies (approximately ×400).

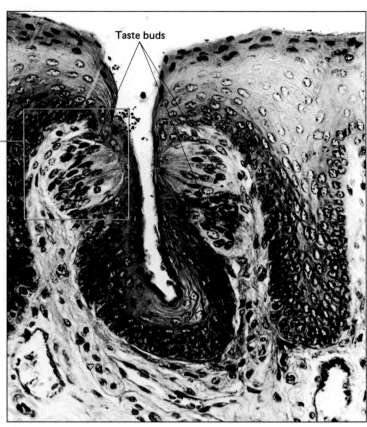

(b)

FIGURE 48–17 The surface of the tongue, showing distribution of taste buds sensitive to sweet, bitter, sour, and salt.

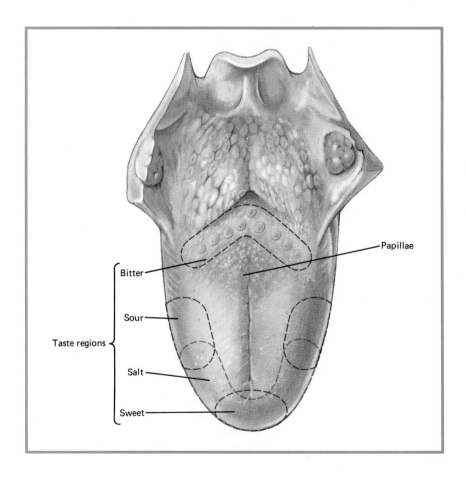

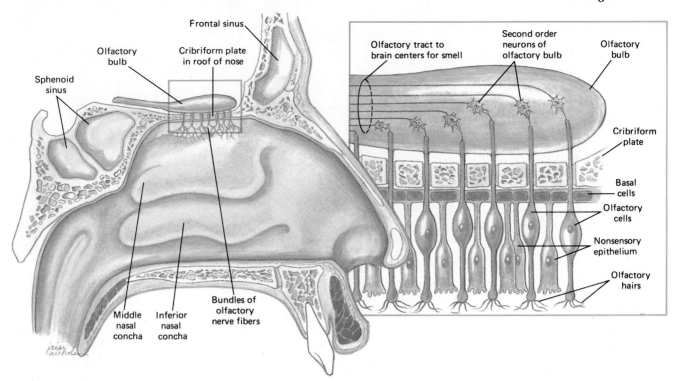

FIGURE 48–18 Location and structure of the olfactory epithelium. Note that the receptor cells are located in the epithelium itself.

The Sense of Smell

In terrestrial vertebrates, olfaction occurs in the nasal epithelium. In humans, the **olfactory epithelium** is located in the roof of the nasal cavity (Figure 48–18). This epithelium contains about 20 million specialized olfactory cells with axons that extend upward as the fibers of the olfactory nerves. These fibers penetrate the exceedingly thin cribriform plate of the ethmoid bone on the cranial floor through many sievelike pores. The end of each olfactory cell on the epithelial surface bears several olfactory hairs that are believed to react to odors (chemicals) in the air.

Unlike the taste buds, which are sensitive to only a few chemical sensations, the olfactory epithelium is thought to react to as many as 50. Mixtures of these primary smell sensations produce the broad spectrum of odors that we are capable of perceiving. The olfactory organs respond to remarkably small amounts of a substance. For example, ionone, the synthetic substitute for the odor of violets, can be detected by most people when it is present in a concentration of only one part in more than 30 billion parts of air.

Despite its sensitivity, smell is perhaps the sense that adapts most quickly. The olfactory receptors adapt about 50% in the first second or so after stimulation, so that even offensively odorous air may seem odorless after only a few minutes. Part of this adaptation is thought to take place in the CNS.

THERMORECEPTORS

Heat is another form of radiant energy to which all living things respond. Although not much is known about their specific thermoreceptors, many invertebrates are sensitive to gradations in temperature. In very small animals, the CNS itself may respond to temperature change. In others, free nerve endings in the integument may be responsible for detection of heat.

FIGURE 48–19 A yellow eye-lash viper showing a pit organ, a sensory structure located between each eye and nostril. The pit organ can detect the heat from a warm-blooded animal up to a distance of 1 to 2 meters.

Mosquitoes and other blood-sucking insects and ticks must use thermoreception in their search for a homeothermic host. Some have temperature receptors on their antennae sensitive to changes of less than 0.5°C.

Certain snakes (pit vipers and boas) use thermoreceptors to locate warm-blooded prey. Pits in the heads of pit vipers are thermoreceptors that can detect heat generated by a small animal as far as a half meter away (Figure 48–19). The snake orients its head so that the pits on both sides of its head detect the same amount of heat. This position indicates that the snake is facing its prey directly, and presumably improves the snake's "aim" when striking. Each pit organ consists of a cavity 1 to 5 mm in diameter in the head of the snake. In rattlesnakes, a very thin membrane is suspended across the cavity. About 7000 heat-sensitive axons (of the trigeminal nerve) are located along the membrane surface. When warmed, each of these axons can send action potentials to the snake's brain.

In mammals, free nerve endings and specialized receptors in the skin and tongue detect temperature changes. Thermoreceptors in the hypothalamus of the brain detect internal changes in temperature and receive and integrate information from thermoreceptors on the body surface. The hypothalamus then initiates homeostatic mechanisms that ensure a constant body temperature.

ELECTRORECEPTORS

Electric organs are found in a few species of rays and bony fishes. Some of these animals have electroreceptors on the body that are linked with the neurons supplying the lateral line organs. These receptors are used to detect electric currents in the water. In species that produce a weak current, the electric organs are used to help in orientation. This is particularly useful in murky water where visibility and olfaction is poor. The fish is informed of an object in the surrounding water when the electrical conductivity of the object distorts the flow-pattern of the electric current. Some animals with electric organs can deliver a powerful shock to stun prey or defend themselves.

PHOTORECEPTORS

Just as matter consists of atoms, light consists of units called **photons.** The energy content of one photon is defined as one **quantum.** In photoreception, quanta of light energy striking the light-sensitive cells trigger the receptor cell to transmit a nerve impulse. Light energy is absorbed by certain pigments. **Rhodopsins** are the photosensitive pigments found in the eyes of cephalopod mollusks, arthropods, and vertebrates.

Eyespots and Simple Eyes

Cells are sensitive to radiant energy, including light, but this general sensitivity (found even among protozoa) is not considered photoreception. Some protozoa have **eyespots,** which are more sensitive to light than the rest of the cell surface, but on the evolutionary scale, the first true light-sensitive organs are found in certain cnidarians and in flatworms. Their photoreceptor organs, comparable to simple eyes, are called **ocelli.** In planarian flatworms, they are bowl-shaped structures containing black pigment (Figure 48–20). At the bottom of the pigment are clusters of light-sensitive cells. The pigment shades these light-sensitive cells from all light except that coming from above and slightly to the front. This arrangement enables the planarian to detect the direction of the source of light. These photoreceptors can also distinguish light intensity. Planarians have other light-sensitive cells over the body surface and therefore can continue to react to light even after their ocelli have been destroyed. However, their responses became random and slow.

Animals with eyespots or very simple eyes can detect light, but they cannot see objects. Effective image formation, called **vision,** requires a more complex eye, usually with a lens that can focus an image on the light-sensitive cells. A necessary first step in the evolution from photoreceptor to true eye, therefore, was the development of a lens to concentrate light on a group of photoreceptors. As better lens systems evolved, the photoreceptors were able to form images, and an eye in the strict sense of the word evolved. Two fundamentally different types of complex eyes are the camera eye of some cephalopods (squids and octopuses) and vertebrates, and the compound eye of the arthropods (Figure 48–21). (The vertebrate eye and the cephalopod eyes are analogous structures; that is, they evolved independently of one another and are functionally but not structurally similar.)

Compound Eyes

Compound eyes are found in crustaceans and insects. Not only do these eyes look different from vertebrate eyes, but they also see differently (Figures 48–21 and 48–22). The surface of a compound eye appears faceted. Each **facet** is the convex cornea of one of its visual units called an **ommatidium.** The number of ommatidia varies in different species. For example, in the eye of certain crustaceans there are only 20, whereas in the eye of a dragonfly there are as many as 28,000.

Each ommatidium consists of a cornea, a lens, and a light-sensitive **rhabdome,** the central core of the ommatidium. The rhabdome is surrounded by retinular cells, which transmit the sensory stimulus (Figure 48–23). A sheath of pigmented cells envelops each ommatidium. Arthropod eyes usually adapt to different intensities of light. In nocturnal and crepuscular (dark- or dusk-loving) insects and many crustaceans, pigment is capable of migrating proximally and distally. When the pigment is in the proximal position, each ommatidium is shielded from its neighbor, and only light entering directly along its axis can stimulate the receptors. When the pigment is in the distal position, light striking at any angle may pass through several ommatidia and stimulate many retinal units. As a result, in dim light, sensitivity of the eye is increased, and in bright light, the eye is protected from excessive stimulation. Pigment migration is under neural control in insects and under hormonal control in crustaceans. In some species it follows a daily rhythm.

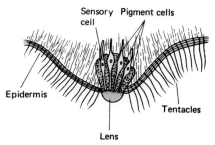

(a) Simple invertebrate eye (Ocellus from Jellyfish)

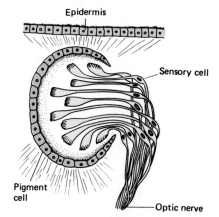

(b) Ocellus of planarian worm

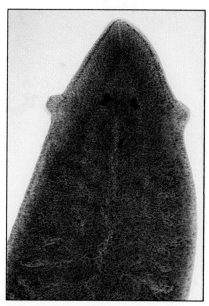

(c)

FIGURE 48–20 Simple invertebrate eyes. (a) Ocellus from bell of jellyfish. (b) Ocellus of planarian worm. (c) Planarian worm (*Planaria agilis*) showing eyes.

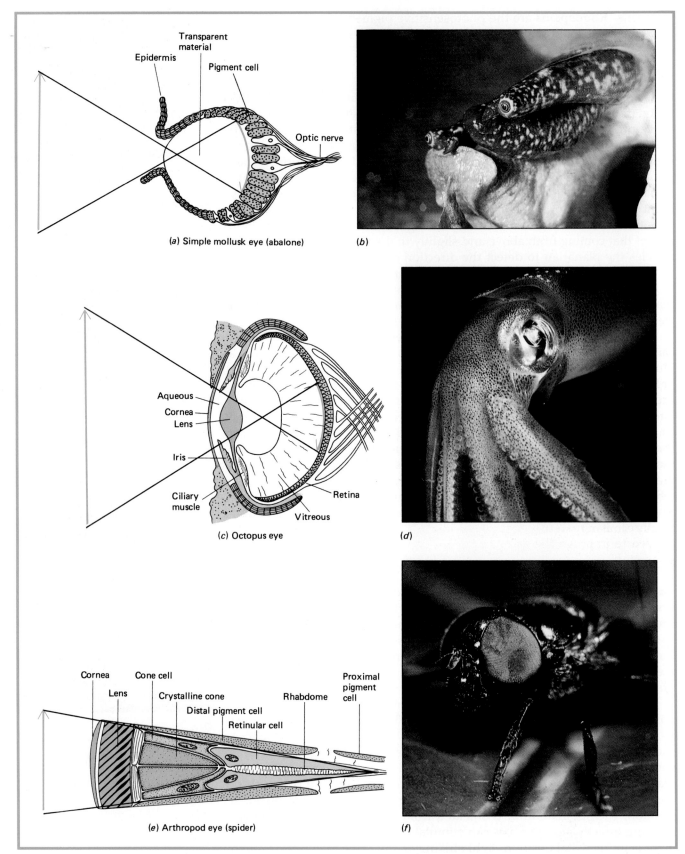

FIGURE 48–21 Comparison of different types of eyes. (a) Some mollusks have a simple eye that lacks a lens and works like a pinhole camera. (b) The eyes of a conch. (c) Some cephalopods have eyes that work like vertebrate eyes. An adjustable lens is present. (d) The eye of a common squid (*Loligo vulgaris*). (e) A unit from the compound eye of an insect or crustacean. This type of eye registers changes in light and shade so that the insect can detect movement. The compound eyes of most of these organisms do not actually form images. (f) The eyes of a fly.

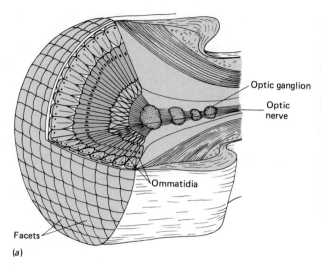

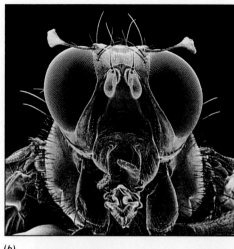

(a)

(b)

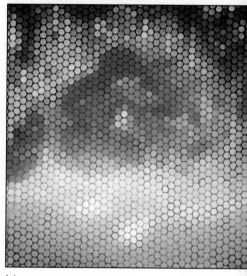

(c)

Compound eyes do not perceive form well. Although the lens system of each ommatidium is adequate to focus a small inverted image on the retinular cells, there is no evidence that such images are actually perceived as images by the organism. However, all the ommatidia together do produce a composite image. Each ommatidium, in gathering a point of light from a narrow sector of the visual field, is in fact sampling a mean intensity from that sector. All of these points of light taken together form a **mosaic** picture. To appreciate the nature of this mosaic picture, we need only look at a newspaper photograph through a magnifying glass; it is a mosaic of many dots of different intensities. The clearness and definition of the picture depends upon how many dots there are per unit area—the more dots, the better the picture. So it is with the compound eye. The image as perceived by the animal is probably much better in quality than might be suspected from the structure of the compound eye. The nervous system of an insect is apparently capable of image-processing similar to that employed to improve the quality of photographs sent to the earth by robot spacecraft.

Although the compound eye can form only coarse images, it compensates for this by being able to follow **flicker** to higher frequencies. Flies are able to detect flickers up to about 265 per second. In contrast, the human eye can detect flickers of only 45 to 53 per second; because flickering lights

FIGURE 48–22 (*a*) The compound eye. (*b*) The Mediterranean fruit fly (*Ceratitis capitata*). (*c*) A bee's eye view of poppies.

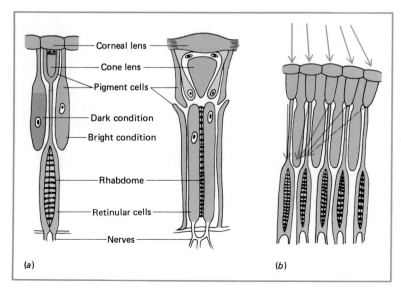

(a)

(b)

FIGURE 48–23 (*a*) Insect ommatidia, showing a diurnal type (*left*) and a nocturnal type (*right*). In the diurnal type, the pigment is shown in two positions: adapted for very dark conditions on the left side and for relatively bright conditions on the right. (*b*) Nocturnal type of eye adapted for dark conditions, showing how light can be concentrated upon one rhabdome from several lenses. If the pigment moved downward, light from peripheral lenses would be screened out.

(a)

(b)

FIGURE 48–24 Insects see ultraviolet light and their world appears differently colored from ours. (a) *Nedelia trilobata* appears to be uniformly yellow to the human eye. (b) When viewed with UV film, the flowers have darker areas that represent light-absorbing centers.

fuse above these values, we see motion pictures as smooth movement and the ordinary 60-cycle light in the room as steady. To an insect, both motion pictures and room lighting must flicker horribly. However, because the insect has such a high critical flicker fusion rate, any movement of prey or enemy is immediately detected by one of the eye units. Hence the compound eye is peculiarly well suited to the arthropod's way of life.

Compound eyes are different from our eyes in two other respects. They are sensitive to different wavelengths of light ranging from the red into the ultraviolet (UV), and they are able to analyze the **plane of polarization** of light. Accordingly, an insect can see UV light well, and its world of color is much different from ours. Since different flowers deflect UV light to different degrees, two flowers that appear identically colored to us may appear strikingly different to insects. How the world appears to an insect with UV vision can be appreciated by viewing the landscape through a television camera with a UV light-transmitting lens (Figure 48–24). A sky that appears equally blue to us in all quadrants reveals quite different patterns to an insect, because the insect's eyes can detect differences in the plane of polarization of the light in various parts of the sky. Honeybees and some other arthropods employ this ability as a navigational aid.

The Mammalian Eye

The eye of a squid or octopus is rather like a simple box camera equipped with "slow" (requiring much light for development) black-and-white film, whereas the mammalian eye is like a 35-millimeter camera loaded with extremely sensitive color film. The analogy between the mammalian eye and a camera is an especially apt one. The eye (Figure 48–25) has a **lens** that can be focused for different distances; a diaphragm, the **iris**, which regulates the size of the light opening, the **pupil;** and a light-sensitive

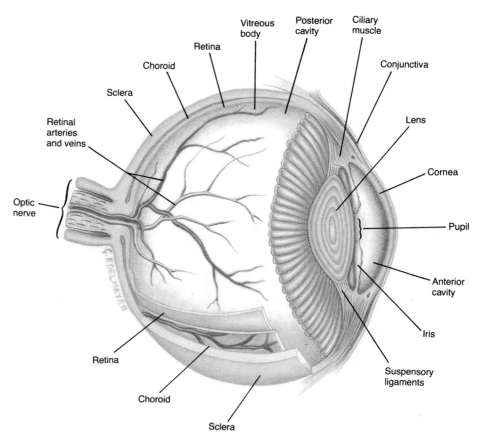

FIGURE 48–25 Structure of the human eye.

retina located at the rear of the eye, corresponding to the film of the camera. Next to the retina is a sheet of cells filled with black pigment that absorbs extra light and prevents internally reflected light from blurring the image (cameras are also painted black on the inside). This sheet, called the **choroid,** also contains the blood vessels that nourish the retina.

The outer coat of the eyeball, called the **sclera,** is a tough, opaque, curved sheet of connective tissue that protects the inner structures and helps to maintain the rigidity of the eyeball. On the front surface of the eye this sheet becomes the thinner, transparent **cornea,** through which light enters.

The lens is a transparent, elastic ball located just behind the iris. It bends the light rays coming in and brings them to a focus on the retina. The lens is aided by the curved surface of the cornea and by the refractive properties of the liquids inside the eyeball. The **anterior cavity** between the cornea and the lens is filled with a watery substance, the **aqueous fluid.** The larger **posterior cavity** between the lens and the retina is filled with a more viscous fluid, the **vitreous body.** Both fluids are important in maintaining the shape of the eyeball.

At its anterior margin, the choroid is thick and projects medially into the eyeball, forming the **ciliary body.** The ciliary body consists of ciliary processes and the ciliary muscle. The ciliary processes are glandlike folds that project toward the lens and secrete the aqueous fluid. The eye has the power of **accommodation,** meaning it can change focus for near or far vision by changing the curvature of the lens. This is made possible by the stretching and relaxing of the lens by the **ciliary muscle,** a part of the ciliary body that is attached to the lens by the suspensory ligaments. When the eye is at rest, the ciliary muscle is relaxed and the ligaments are tense. This pulls the lens into a flattened (ovoid) shape, focusing the eye for far vision. When the ciliary muscle contracts, tension on the suspensory ligaments lessens; the elastic lens then assumes a rounder shape for near vision.

As people grow older, the lens enlarges and becomes less elastic and thereby less able to accommodate for near vision. When this occurs, spectacles with one portion ground for distance vision and one portion ground for near vision (bifocals) may be worn to accomplish what the eye can no longer do.

The amount of light entering the eye is regulated by the **iris,** a ring of smooth muscle that appears as blue, green, or brown, depending on the amount and nature of pigment present. The iris is composed of two mutually antagonistic sets of muscle fibers; one that is arranged circularly and contracts to decrease the size of the pupil and one that is arranged radially and contracts to increase the size of the pupil. The response of these muscles to changes in light intensity is not instantaneous but requires from 10 to 30 seconds. Thus, when a person steps from a light to a dark area, some time is needed for the eyes to adapt to the dark, and when a person steps from a dark room to a brightly lighted area, the eyes are dazzled until the size of the pupil is decreased. The retina of the eye (soon to be discussed) is also able to adapt to changes in light intensity.

Each eye has six muscles stretching from the surface of the eyeball to various points in the bony socket. These muscles enable the eye as a whole to move and be oriented in a given direction. Cranial nerves innervate the muscles in such a way that the eyes normally move together and focus on the same area.

The light-sensitive part of the vertebrate eye is the retina, which lines the posterior two thirds of the eyeball covering the choroid. The retina contains an abundance of receptor cells called, according to their shape, **rods** and **cones.** In the human eye, there are about 125 million rods and 6.5 million cones. In addition, the retina contains many sensory and connector neurons and their axons. Curiously, the sensitive cells are at the *back* of the

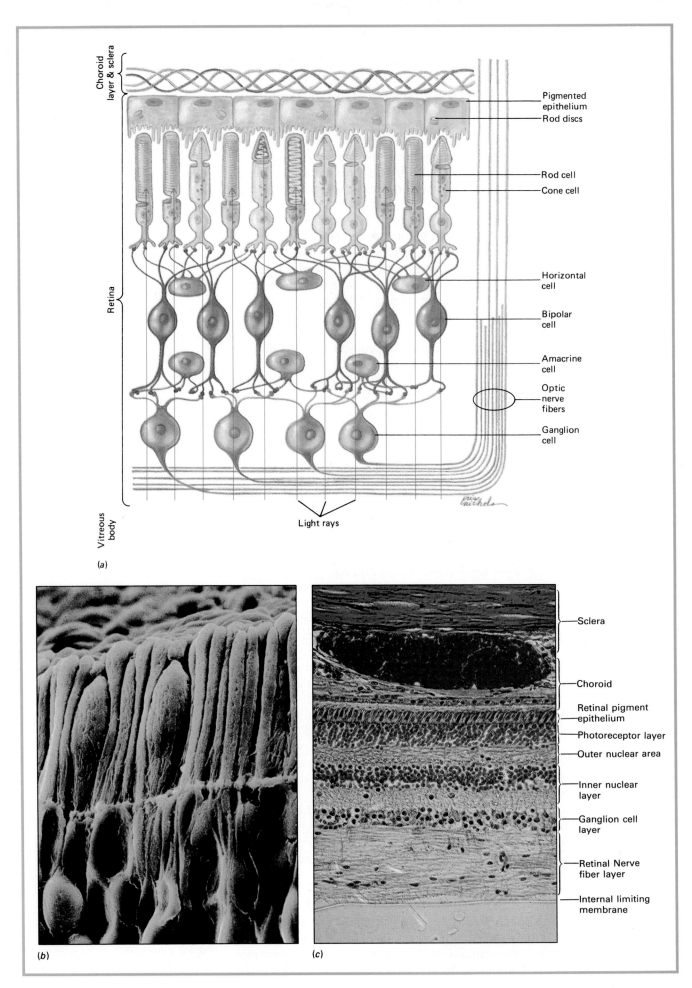

Choroid layer & sclera

Pigmented epithelium

Rod discs

Rod cell

Cone cell

Horizontal cell

Bipolar cell

Amacrine cell

Optic nerve fibers

Ganglion cell

Retina

Vitreous body

Light rays

(a)

(b)

(c)

Sclera

Choroid

Retinal pigment epithelium

Photoreceptor layer

Outer nuclear area

Inner nuclear layer

Ganglion cell layer

Retinal Nerve fiber layer

Internal limiting membrane

1134

FIGURE 48–26 (*opposite*) The retina. (*a*) Neuronal connections in the retina. The elaborate interconnections among the various layers of cells allow them to interact and to influence one another in a number of ways. (*b*) Rods (elongated structures) and two cones (shorter, thicker, structures) make up the top row in this micrograph (magnification ×10,000). The elongated rods permit us to see shape and movement, whereas the shorter cones allow us to view our world in color. (*c*) Photomicrograph of a portion of the sclera, choroid, and retina. Magnification about ×300.

retina; to reach them, light must pass through several layers of connecting neurons (Figure 48–26).

At a point in the back of the eye, the individual axons of the sensory neurons unite to form the **optic nerve,** which then passes out of the eyeball. Here there are no rods and cones. This area is called the "blind spot," since images falling on it cannot be perceived. Its existence can be demonstrated by closing the left eye and focusing the right eye on the X in Figure 48–27. Starting with the page about 13 cm from the eye, move it away until the circle disappears. At that position the image of the circle is falling on the blind spot and so is not perceived.

In the center of the retina, directly in line with the center of the cornea and lens, is the region of keenest vision, a small depressed area called the **fovea.** Here are concentrated the light-sensitive cones, responsible for bright-light vision, for the perception of fine detail, and for color vision.

The other light-sensitive cells, the rods, are more numerous in the periphery of the retina, away from the fovea. These function in twilight or dim light and are insensitive to colors. We are not ordinarily aware that only those objects more or less directly in front of our eyes can be perceived in color, but this can be demonstrated by a simple experiment. Close one eye and focus the other on some point straight ahead. As a colored object is gradually brought into view from the side, you will be aware of its presence and of its size and shape before you are aware of its color. Only when the object is brought closer to the direct line of vision, so that its image falls on a part of the retina containing cones, can its color be determined. The rods are actually more sensitive in dim light than are the cones. Since the rods are located not in the center but in the periphery of the retina, it is a curious fact that you can see an object better in dim light if you look at it not directly (when its image would fall on the cones in the center of the retina) but slightly to one side of it, so that its image falls on the rods in the periphery of the retina.

The Chemistry of Vision

The substance primarily responsible for the ability to see is **rhodopsin** (also called visual purple), found only in the rod cells, together with a series of closely related pigments found in the cone cells. Rhodopsin consists of a large protein known as an opsin, chemically joined with a carotenoid called **retinal,** which is made from vitamin A. Two isomers of retinal exist: the 11-*cis* form and the all-*trans* form.

FIGURE 48–27 Demonstration of the blind spot on the retina. See text for details.

FIGURE 48–28 The visual cycle. When light strikes rhodopsin, it breaks down, depolarizing the rod cell which contains it. This produces an impulse. See text for further explanation.

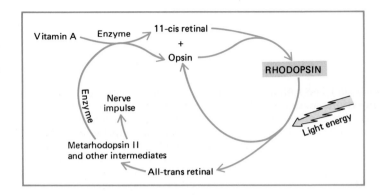

When light strikes rhodopsin, it transforms **11-*cis* retinal** to **all-*trans* retinal.** This change in shape causes the breakdown of rhodopsin into its components, opsin and retinal. During this process, rhodopsin is converted into a series of intermediate compounds. One of these intermediates is **metarhodopsin II,** which is thought to be the key compound leading to the generation of action potentials in ganglion cells, and ultimately transmission of neural impulses to the brain. The all-*trans* retinal is converted back to the 11-*cis* form by an enzyme. Then the retinal combines with opsin to produce rhodopsin once again. This sequence of reactions is known as the visual cycle (Figure 48–28).

It has been shown that a single quantum of light can be absorbed by a single molecule of rhodopsin and lead to the excitation of a single rod. When the eye is exposed to a flash of light lasting only a millionth of a second, it sees an image of light that persists for nearly a tenth of a second. This is the length of time that the retina remains stimulated following a flash. This persistence of images in the retina enables your eye to fuse the successive flickering images on a motion picture screen, so that what is actually a rapid succession of still pictures is perceived as moving persons and objects.

The ability to see an exceedingly faint light depends on the amount of rhodopsin present in the retinal rods. This in turn depends on the relative rates of synthesis and breakdown of rhodopsin. In bright light, much of the rhodopsin is broken down to free retinal and opsin. The synthesis of rhodopsin is a relatively slow process, and the concentration of rhodopsin in the retina is never very great so long as the eye is exposed to bright light. When the eye is suitably shielded from light, the rhodopsin breakdown does not occur and its concentration gradually builds up until essentially all of the opsin has been converted to rhodopsin. The sensitivity of the eye to light, a function of the amount of rhodopsin present, can increase 1 million-fold if the eye is dark-adapted for as long as an hour.

Color Vision

The chemistry of the cones and of color vision is less well understood. There are three different types of cones and three different cone pigments. One cone pigment, **iodopsin,** is composed of retinal and an opsin different from that found in the rods. The cones are considerably less sensitive to light than the rods and cannot provide vision in dim light. The prime function of the cones is to perceive colors.

The three different types of cones respond respectively to blue, green, and red light. This has been substantiated recently by the extraction of three kinds of color receptors—red, green, and blue—from human and other primate retinas. Each type can respond to light with a considerable range of wavelengths. The green cones, for example, can respond to light

Focus on DEFECTS IN VISION

The most common defects of the human eye are near-sightedness (myopia), farsightedness (hypermetropia), and astigmatism. In the normal eye, as shown in (*a*), the shape of the eyeball is such that the retina is the proper distance behind the lens for the light rays to converge in the fovea. In a **nearsighted** eye, illustrated in (*b*), the eyeball is too long and the retina is too far from the lens. The light rays converge at a point in front of the retina, and are again diverging when they reach it, resulting in a blurred image.

In a **farsighted** eye, the eyeball is too short and the retina too close to the lens (part (*c*) of the figure). Light

rays strike the retina before they have converged, again resulting in a blurred image. Concave lenses correct for the nearsighted condition by bringing the light rays to a focus at a point farther back, and convex lenses correct for the farsighted condition by causing the light rays to converge farther forward.

In **astigmatism,** the cornea is curved unequally in different planes, so that the light rays in one plane are focused at a different point from those in another plane, as shown in (*d*). To correct for astigmatism, lenses must be ground unequally to compensate for the unequal curvature of the cornea.

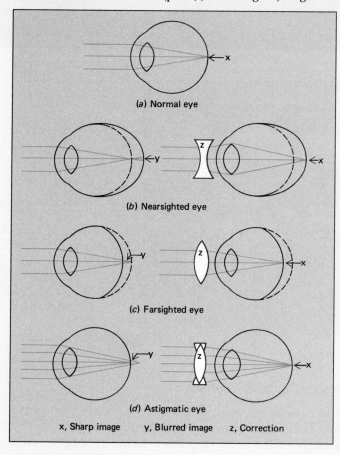

(*a*) Normal eye

(*b*) Nearsighted eye

(*c*) Farsighted eye

(*d*) Astigmatic eye

x, Sharp image   y, Blurred image   z, Correction

■ Common abnormalities of the eye that result in defects in vision. (*a*) Normal eye, in which parallel rays coming from a point in space are focused as a point on the retina. (*b*) Nearsighted eye, in which the eyeball is elongated so that parallel light rays are brought to a focus in front of the retina (on dotted line, which represents the position of the retina in a normal eye) and so form a blurred image on the retina. This is corrected by placing a concave lens in front of the eye, which diverges the light rays, making it possible for the eye to focus these rays on the retina. (*c*) Farsighted eye, in which the eyeball is shortened and light rays are focused behind the retina. A convex lens converges the light rays so that the eye focuses them onto the retina. (*d*) Astigmatic eye, in which light rays passing through one part of the eye are focused on the retina, while light rays passing through another area of the lens are not focused on the retina. This is a result of the unequal curvature of the lens or cornea. A cylindrical lens corrects this by bending light rays going through only certain parts of the eye.

---

of any wavelength from 450 to 675 nanometers (i.e., blue, green, yellow, orange, and red light), but they respond to green light more strongly than to any of the others. Intermediate colors other than blue, green, and red are perceived by the simultaneous stimulation of two or more types of cones. The "red" cones are actually more sensitive to yellow light than to red, but they respond to red before any of the others do and therefore behave as red receptors. By a comparison of the rate at which various receptors respond, the brain is able to detect light colors of intermediate frequency. Color blindness results when one or more of the three types of cones is absent. This is an inherited sex-linked condition (see Chapter 11).

(a)

(b)

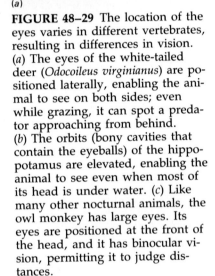

(c)

FIGURE 48–29 The location of the eyes varies in different vertebrates, resulting in differences in vision. (*a*) The eyes of the white-tailed deer (*Odocoileus virginianus*) are positioned laterally, enabling the animal to see on both sides; even while grazing, it can spot a predator approaching from behind. (*b*) The orbits (bony cavities that contain the eyeballs) of the hippopotamus are elevated, enabling the animal to see even when most of its head is under water. (*c*) Like many other nocturnal animals, the owl monkey has large eyes. Its eyes are positioned at the front of the head, and it has binocular vision, permitting it to judge distances.

Binocular Vision and Depth Perception

The position of the eyes in the head of humans and certain other higher vertebrates permits both eyes to be focused on the same object (Figure 48–29). This **binocular vision** is an important factor in judging distance and depth. To focus on a near object, the eyes must converge (turn inward so that the animal becomes slightly cross-eyed). The proprioceptors in the eye muscles causing this convergence are stimulated by this contraction to send impulses to the brain. Thus, part of judgment of distance and depth depends upon impulses originating when the sensory fibers in those muscles are stimulated. In addition, the eyes, being some distance apart (a little over 5 cm in humans), see things from slightly different angles and therefore get slightly different views of a close object. The images of a given object that the two eyes perceive differ most for a near object and least for a distant object. By comparing the differences, the brain is able to infer distances. Depth perception is also made possible by the differential size of near and far objects on the retina, by perspective, by overlap and shadow, by distance over the horizon, and by the increasing dimness of distant objects.

■ **SUMMARY**

I. A sense organ consists of one or more receptor cells and sometimes accessory cells. Receptor cells may be neuron endings or specialized cells in close contact with neurons.

II. Exteroceptors are sense organs that receive information from the outside world. Proprioceptors are sense organs within muscles, tendons, and joints, which enable the animal to perceive orientation of the body and position of its parts. Interoceptors are sense organs within body organs.

III. Sense organs also may be classified on the basis of the type of energy to which they respond. Thus, there are mechanoreceptors, chemoreceptors, photoreceptors, thermoreceptors, and electroreceptors.

IV. Receptor cells absorb energy, transduce that energy into electrical energy, and produce receptor potentials.

V. Adaptation of a receptor to a continuous stimulus results in diminished perception. For this reason, adaptation to an unpleasant odor or noise occurs after a few moments.

VI. Mechanoreceptors respond to touch, pressure, gravity, stretch, or movement.
  A. The tactile receptors in the skin are mechanoreceptors that respond to mechanical displacement of hairs or of the receptor cells themselves.
  B. Statocysts are gravity receptors found in many invertebrates.
    1. When the position of the statolith within the statocyst changes, hairs of receptor cells are bent.
    2. Messages sent to the CNS inform an animal of which hairs have been stimulated; from this the animal can determine where "down" is and can correct for any abnormal orientation.
  C. Lateral line organs supplement vision in fish and some amphibians by informing the animal of moving objects or objects in its path.
  D. Muscle spindles, Golgi tendon organs, and joint receptors are proprioceptors that respond continuously to tension and movement in the muscles and joints.
  E. The saccule and utricle of the vertebrate ear contain otoliths that change position when the head is tilted or when the body is moving forward. The hair cells stimulated by the otoliths send impulses to the brain, enabling the animal to perceive the direction of gravity.
  F. The semicircular canals of the vertebrate ear inform the brain about turning movements. Their cristae are stimulated by movements of the endolymph.
  G. The organ of Corti within the cochlea is the auditory receptor in birds and mammals.
    1. Sound waves pass through the external auditory meatus, cause the eardrum to vibrate, and are transmitted through the middle ear by the hammer, anvil, and stirrup.
    2. Vibrations pass through the oval window to fluid within the vestibular duct. Pressure waves press upon the membranes separating the three ducts of the cochlea.
    3. Movements of the basilar membrane rub the hair cells of the organ of Corti against the overlying tectorial membrane, thus stimulating them.
    4. Nerve impulses are initiated in the dendrites of the auditory neurons lying at the base of each hair cell.

VII. Chemoreceptors include receptors for taste and smell.
  A. Taste receptors are specialized epithelial cells located in taste buds.
  B. The olfactory epithelium contains specialized olfactory cells with axons that extend upward as fibers of the olfactory nerves.

VIII. Thermoreceptors are important in warm-blooded animals in providing cues about body temperature. In some invertebrates they are used to locate a warm-blooded host.

IX. Photoreceptors in very simple eyes detect light, but these eyes do not form images effectively. Effective image formation and interpretation is called vision.
  A. In the human eye, light enters through the cornea, is focused by the lens, and sensed as an image by the retina. The iris regulates the amount of light that can enter.
  B. When light strikes rhodopsin in the rod cells, a chemical change in retinal occurs that breaks down the rhodopsin, triggering depolarization of the rod cell.
  C. The rods form images in black and white, whereas the cones function in color vision.
  D. The compound eye found in insects and crustaceans consists of ommatidia, which collectively form a mosaic image.

■ **POST-TEST**

1. A sense organ consists of one or more _____ cells and, sometimes, _____ cells.
2. Exteroceptors are sense organs that _____; proprioceptors enable an animal to perceive _____, along with the _____ of the body as a whole.
3. _____ detect light energy; _____ respond to touch, gravity, or movement.
4. Receptor cells absorb _____, transduce this energy into _____ energy, and produce a _____ _____.
5. The diminishing response of a receptor to a continued, constant stimulus is called _____.
6. Statocysts serve as _____ receptors; their

action depends upon mechanical displacement of receptor cell hairs by change in position of a _____.

7. The lateral line organ of fishes is thought to supplement _____; its canals are lined with receptor cells that have _____; the receptor cells secrete a mass of gelatinous material called a _____.

8. Three main types of vertebrate proprioceptors are _____ _____, which detect muscle movement; _____ _____ organs, which determine stretch in tendons; and _____ receptors, which detect movement in ligaments.

9. Phasic receptors respond only to _____; tonic sense organs respond as long as the _____ is present.

10. _____ are balancing organs that stabilize flight in flies.

11. The basic function of the vertebrate ear is to help maintain _____.

12. The inner ear consists of interconnected canals and sacs called the _____; in jawed vertebrates this structure consists of two saclike chambers, the _____ and _____, and three _____ canals, as well as the cochlea.

13. The rocks in your head (within the saccule and utricle) called _____ are actually _____ detectors.

14. Each semicircular canal is filled with fluid called _____; at one of the openings of each canal into the utricle is a small enlargement, the _____.

15. The cochlea, located in the _____ ear, contains mechanoreceptor hair cells that detect _____ waves.

16. The actual auditory receptor is the organ of _____; it is located within the _____ canal.

17. The senses of taste and smell depend upon _____.

18. The photosensitive pigments in the eyes of vertebrates and arthropods are _____.

Select the most appropriate answer in column B for each description in Column A.

Column A
19. Light-sensitive part of human eye
20. Regulates size of pupil
21. Perceive color
22. Region of keenest vision
23. Visual unit in compound eye

Column B
a. Ommatidium
b. Cones
c. Retina
d. Iris
e. Fovea

24. Label the diagrams below. (Refer to Figures 48–13 and 48–22 as necessary.)

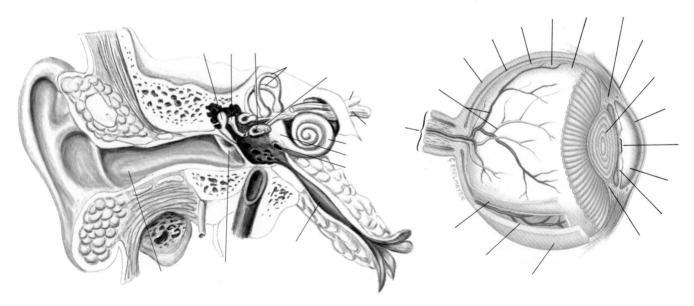

■ REVIEW QUESTIONS

1. What are mechanoreceptors? Give three examples.
2. What is a proprioceptor? What is its function in the mammalian body?
3. How do anesthetics reduce or eliminate the sensation of pain?
4. Draw a diagram of the human eye, labeling all parts. How are rods and cones distributed in the retina?
5. Discuss the mechanism by which photoreceptors are stimulated by light. What is the function of rhodopsin? How is it regenerated?
6. Describe the anatomical abnormality that produces each of the following visual defects:
   a. myopia
   b. hypermetropia
   c. astigmatism
7. How does the human eye adjust to near and far vision and to bright and dim lights?
8. Draw a diagram of the human ear, labeling all parts.
9. Discuss the mechanism by which the sensory cells of the ear are stimulated by sound waves.

10. What are otoliths and what is their role in maintaining equilibrium?

11. Contrast the function of the insect's compound eye with that of the vertebrate eye.

■ **RECOMMENDED READINGS**

Abu-Mostafa, Y.S., and Psaltis, D. Optical neural computers. *Scientific American*, Vol. 256, No. 3, March 1987, pp. 88–94. The arrangement of neurons in the brain can be used as a model for building a computer that can solve problems, such as recognizing patterns, that involve memorizing all possible solutions.

Hudspeth, A.J. The hair cells of the inner ear. *Scientific American*, Vol. 248, No. 1, January 1983, pp. 54–64. A description of the mechanism by which hair cells in the inner ear respond to convey information about acoustic tones and acceleration. Some of this author's conclusions are controversial.

Levine, J.S., and MacNichol, E.F. Jr. Color vision in fishes. *Scientific American*, February 1982, pp. 140–149. A discussion of retinal pigments in fishes and their relationship to the evolution of the eye.

Miller, J.A. Colorful views of vision. *Science News*, Vol. 120, October 3, 1981. An account of a technique for mapping components of the visual system.

Newman, E.A., and Hartline, P.H. The infrared 'vision' of snakes. *Scientific American*, March 1982, pp. 116–124. Snakes of two families can detect and localize sources of infrared radiation.

Poggio, T., and Koch, C. Synapses that compute motion. *Scientific American*, Vol. 256, No. 5, pp. 46–52. Studies of cells in the eye that interpret movement may help clarify mechanisms involved in other neural processes.

Schnapf, J.L., and Baylor, D.A. How photoreceptor cells respond to light. *Scientific American*, Vol. 256, No. 4, April 1987, pp. 40–47. How a single photoreceptor cell in the eye registers the absorption of a single photon.

Treisman, A. Features and objects in visual processing. *Scientific American*, Vol. 255, No. 5, November 1986, pp. 114–125. We perceive meaningful wholes visually by automatically extracting features from a scene and assembling them into objects.

Anemone shrimp *(Perislemmes yucatansis)*

49

Animal Hormones: Endocrine Regulation

■ OUTLINE

I. The chemical nature of hormones
II. Mechanisms of hormone action
   A. Activation of genes
   B. Action through second messengers
   C. Prostaglandins
III. Regulation of hormone secretion
IV. Invertebrate hormones
   A. Endocrine regulation of reproductive development in cephalopods
   B. Color change in crustaceans
   C. Hormonal control of insect development
V. Vertebrate hormones
   A. Endocrine disorders
   B. The hypothalamus and pituitary gland
   C. Growth and development
     1. Growth hormone
       a. Regulation of growth hormone secretion
       b. Abnormal growth
     2. Thyroid hormones
       a. Regulation of secretion
       b. Thyroid hormones and growth disorder
   D. Regulation of glucose concentration in the blood: insulin and glucagon
     1. Regulation of insulin and glucagon secretion
     2. Diabetes mellitus
     3. Hypoglycemia
   E. Response to stress
     1. Hormones of the adrenal medulla
     2. The effects of cortisol
Focus on identification of endocrine tissues and hormones

■ LEARNING OBJECTIVES

After you have read this chapter you should be able to:

1. Define the terms *hormone* and *endocrine gland* and identify sources of hormones other than endocrine glands.
2. Compare two mechanisms of hormone action: activation of genes and second messengers such as cyclic AMP.
3. Describe the regulation of hormone secretion by negative feedback mechanisms and draw diagrams illustrating the regulation of secretion of each of the hormones discussed in this chapter.
4. Cite the general sources and actions of invertebrate hormones.
5. Summarize the interaction of hormones that control development in insects.
6. Locate the principal vertebrate endocrine glands, list the hormones secreted by each, and summarize their actions.
7. Support the following concept: The hypothalamus is the link between nervous and endocrine systems. (Include a description of the mechanisms by which the hypothalamus influences the anterior and posterior lobes of the pituitary gland.)
8. Relate the actions of growth hormone and thyroid hormones to growth; explain the consequences of hyposecretion and hypersecretion of these hormones.
9. Compare the actions of insulin and glucagon in regulating the concentration of glucose in the blood; summarize the physiological problems associated with diabetes mellitus and hypoglycemia.
10. Compare the roles of the adrenal medulla and the adrenal cortex in helping the body adapt to stress.

The endocrine system works closely with the nervous system to maintain the steady state of the body. The long-term adjustments of metabolism, growth, and reproduction are generally under endocrine control. Hormones play a major role in regulating the concentrations of glucose, sodium, potassium, calcium, and water in the blood and interstitial fluid. The endocrine system is also important in the body's adaptation to stress.

An **endocrine gland** is a ductless gland that secretes one or more hormones, which are transported by the blood. Endocrine glands can be distinguished from exocrine glands, such as sweat glands or salivary glands, which release their secretions into ducts for transport to some body surface or cavity.

Traditionally, animal hormones have been defined as chemical messengers produced by endocrine glands and transported by the blood to target tissues, where they stimulate a change in some physiological activity. In recent years, **endocrinology,** the branch of biology dealing with endocrine activity, has broadened its scope to include chemical messengers not secreted by discrete endocrine glands. Some hormones, for example, are produced by neurons. (See Focus on Identification of Endocrine Tissues and Hormones.) Hormones can be classified in the following groups:

1. Hormones secreted by endocrine glands.
2. Hormones secreted by neurons (Figure 49–1). These are **neurohormones,** and the cells that secrete them are **neurosecretory cells.** Antidiuretic hormone (ADH) and oxytocin, produced by specialized neurons in the hypothalamus, are examples of vertebrate neurohormones. Many neurohormones are secreted by various invertebrates. Some endocrinologists consider neurotransmitters to be hormones because they qualify as chemical messengers.
3. Chemical messengers released by individual cells.

Almost all cells produce substances by which they communicate with other cells. Growth factors, prostaglandins, and histamine are examples. Some of the hormones in this group simply diffuse through the interstitial fluid and into their target tissues.

In order to include chemical messengers produced by neurons and other cells as hormones, we must broaden the traditional definition of a hormone. A more contemporary definition follows: A **hormone** is a chemical messenger produced by one type of cell that has a specific regulatory effect on the activity of another type of cell. As a group, hormones are very

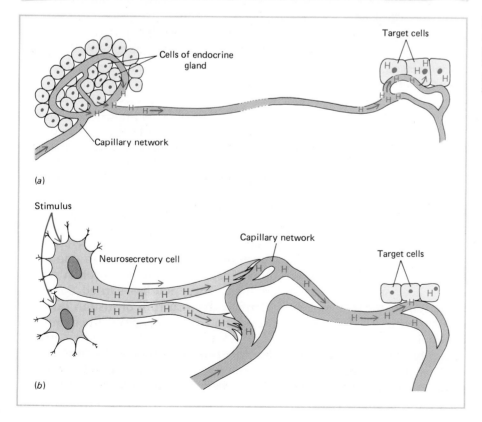

(a)

(b)

FIGURE 49–1 Hormones may be secreted by the cells of endocrine glands, as in (*a*), or by neurosecretory cells, as in (*b*). Hormones are generally transported by the blood and are taken up by target cells.

Focus on IDENTIFICATION OF ENDOCRINE TISSUES AND HORMONES

How are endocrine tissues identified? How can an investigator prove that any particular hormone exists? Traditionally, there are three main experimental procedures used to make such determinations:

1. The suspected endocrine tissue is removed. If it is an endocrine tissue, removal should produce deficiency symptoms in the experimental animals.
2. The tissue is then replaced by transplanting similar tissue from another animal, often to a different location within the body cavity. The changes induced by removing the tissue should be reversed by replacing it.
3. The suspected hormone is extracted from the endocrine tissue of one animal and injected into an experimental animal from which the suspected endocrine tissue has been removed. Deficiency symptoms should be relieved by replacing the suspected hormone.

Most of the known hormones and endocrine glands were established by these criteria. Today these techniques are supplemented by investigation with the electron microscope and by sophisticated biochemical and physiological procedures.

In 1849 the German physiologist Berthold performed the first clear-cut and successful experiments in endocrinology. He removed the testes from cockerels (young roosters) and observed that the comb became atrophic. He then transplanted a testis into some of the birds and observed that the combs then grew back to normal size. He postulated that these male sex glands secrete a substance into the bloodstream that is essential for the normal male secondary sex characteristics.

important in maintaining steady states within the body and they are effective at very low concentrations.

Pheromones are chemicals produced by an animal for communication with other animals of the same species. Since pheromones are generally produced by exocrine glands and do not regulate metabolic activities within the animal that produces them, most biologists do not classify them as hormones. Their role in regulating behavior is discussed in Chapter 52.

THE CHEMICAL NATURE OF HORMONES

Although hormones are chemically diverse, they belong to the protein family (amines, peptides, proteins) or they are lipids (derived from fatty acids; or steroids). Epinephrine and norepinephrine released by the adrenal glands (as well as by certain neurons) of vertebrates are amines derived from the amino acid tyrosine (Figure 49–2). The thyroid hormones are also derived from tyrosine. Oxytocin and antidiuretic hormone, produced by neurosecretory cells in the hypothalamus, are short peptides composed of nine amino acids. Seven of the amino acids are identical in the two hormones; yet the substances have quite different actions. One of the pigment-concentrating hormones found in crustaceans is an octapeptide.

Glucagon, secretin, adrenocorticotrophic hormone (ACTH), and calcitonin are somewhat longer peptides with about 30 amino acids in the chain. Insulin, secreted by the beta cells of the islets of Langerhans in the pancreas, consists of two peptide chains joined by disulfide bonds. The pioneering work of Fred Sanger and his colleagues at the University of Cambridge determined the sequence of the 21 amino acids in one chain and the 30 amino acids in the other. Insulin was the first protein whose amino acid sequence was determined. These studies earned Sanger a Nobel prize, for they not only clarified the structure of insulin but also established methods by which the amino acid sequence of other proteins could be determined.

Insulin is synthesized in the beta cells of the pancreas as a single peptide composed of 84 amino acids. This compound, called **proinsulin,** undergoes folding; three disulfide bonds are formed; and a connecting peptide, containing 33 amino acids, is removed from the center of the original chain

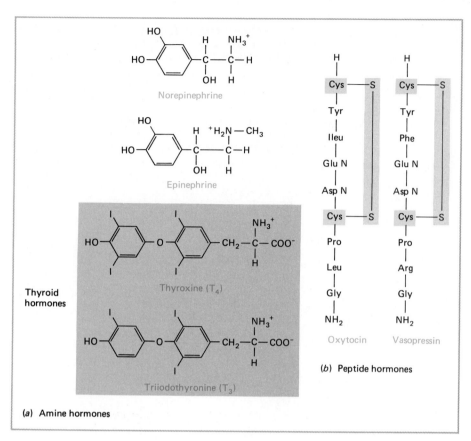

FIGURE 49–2 (*a*) Some hormones derived from amino acids. Note the presence of iodine in the thyroid hormones. (*b*) Peptide hormones produced in the hypothalamus and secreted by the posterior pituitary. Note that they differ by only two amino acids.

by enzymatic cleavage. This leaves two peptide chains joined by disulfide bridges (Figure 49–3). Proinsulin is an example of a **prohormone.** Several other peptide hormones are synthesized as prohormones and undergo partial cleavage to yield smaller peptides with full hormonal activity. Growth hormone, thyroid-stimulating hormone, and the gonadotropic hormones, all secreted by the anterior lobe of the pituitary gland, are proteins with molecular weights of 25,000 or more.

Prostaglandins and the juvenile hormones of insects are hormones derived from fatty acids (Figure 49–4). The molting hormone (also called ec-

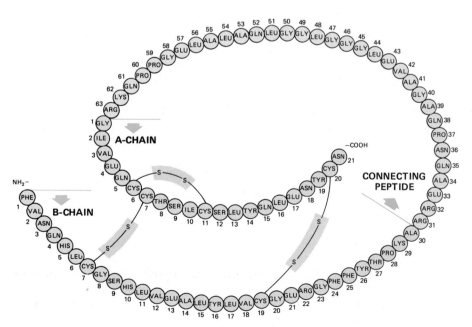

FIGURE 49–3 The molecular structure of proinsulin, a single polypeptide chain, and its conversion to insulin, composed of two peptide chains. This is achieved by the formation of three disulfide bonds and the removal of a section of the peptide chain between peptide A and peptide B.

FIGURE 49–4 Hormones derived from fatty acids and steroids. (*a*) Juvenile hormone and prostaglandins are derived from fatty acids. (*b*) Steroid hormones.

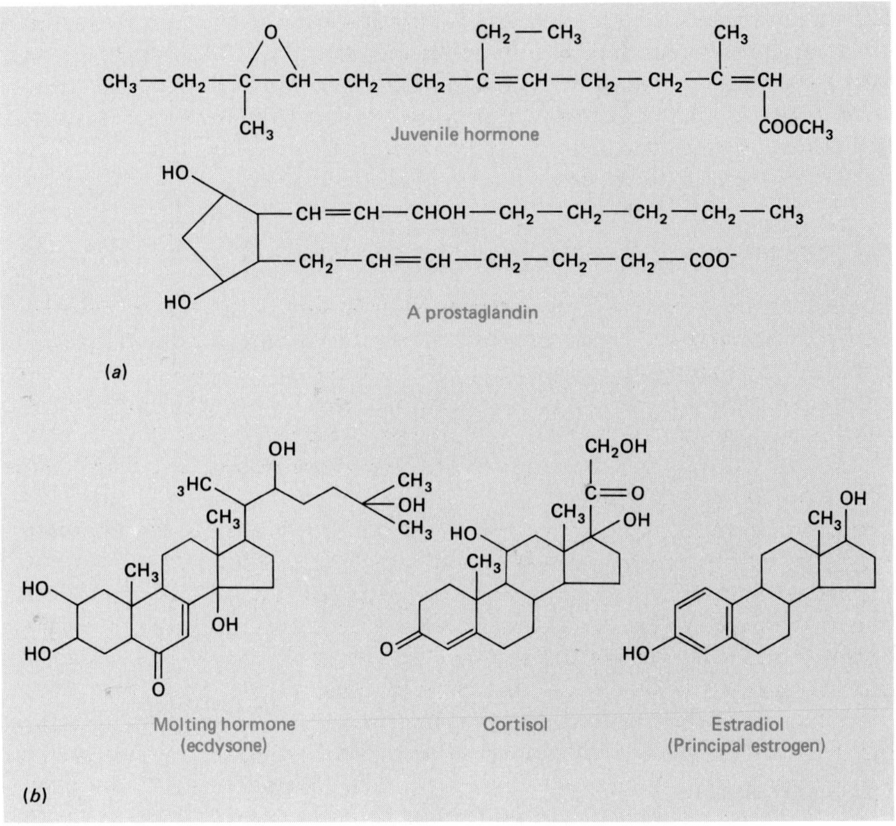

(*a*)

(*b*)

Molting hormone (ecdysone)   Cortisol   Estradiol (Principal estrogen)

FIGURE 49–5 The sequence of reactions by which testosterone (the principal male sex hormone), the female sex hormones estradiol, glucocorticoids, and mineralocorticoids are synthesized from progesterone (another female sex hormone). Progesterone is synthesized from cholesterol. Note that the methyl (CH$_3$) groups attached to each line extending upward have been omitted. Multiple arrows indicate intermediate steps that have been omitted from the diagram.

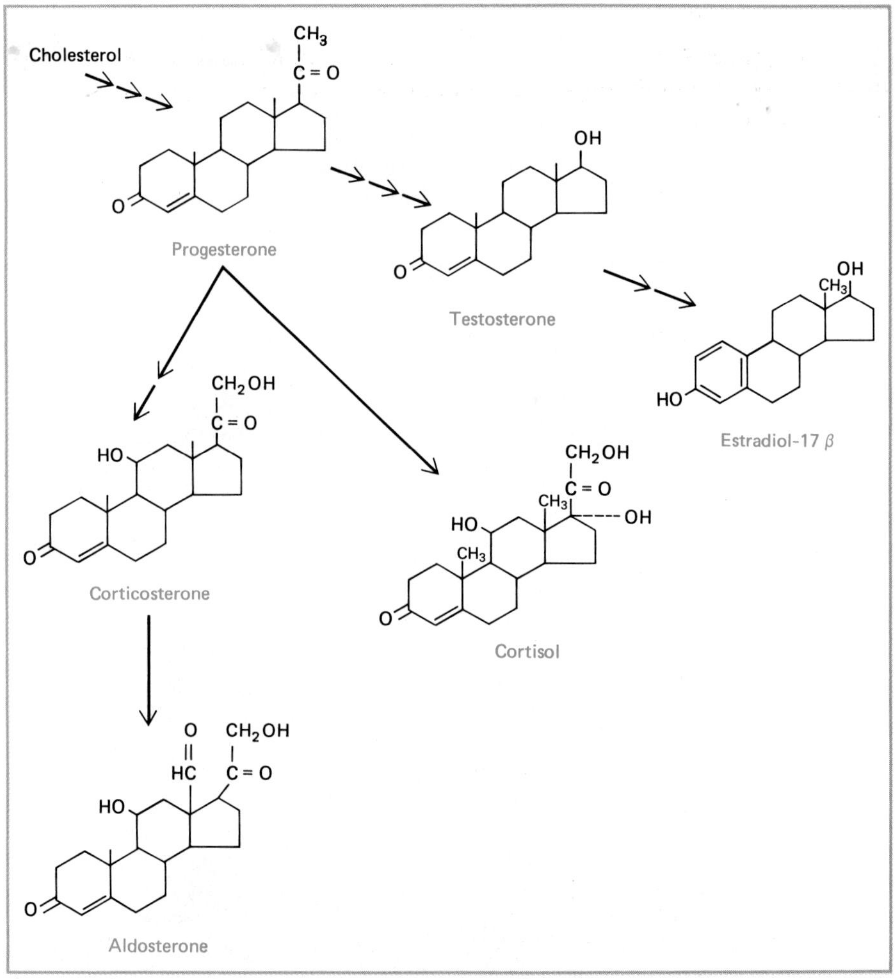

Cholesterol

Progesterone

Testosterone

Estradiol-17 β

Corticosterone

Cortisol

Aldosterone

dysone) of insects is a steroid with 27 carbons arranged in four rings and a tail as in the cholesterol molecule. In vertebrates, the adrenal cortex, testis, ovary, and placenta secrete steroids synthesized from cholesterol. Progesterone, one of the female sex hormones, is the precursor of the steroid hormones produced by the adrenal cortex, of the male hormone testosterone, and of the female hormone estradiol (Figure 49–5).

MECHANISMS OF HORMONE ACTION

In vertebrates, most endocrine glands secrete small amounts of their hormones continuously. Thus, although present in minute amounts, about 30 different hormones are circulating in the blood at all times. Some hormones are transported bound to plasma proteins. Hormone molecules are continuously removed from the circulation by target tissues. They are also removed by the liver, which inactivates some hormones, and by the kidneys, which excrete them.

Hormone molecules diffuse from the blood into the interstitial fluid and are taken up by their target tissue. How does the target tissue recognize its hormone? Specialized receptor proteins in the target tissue cells specifically bind the hormone. The receptor protein may be compared to a lock and the hormones to different keys. Only the hormone that fits the lock—the specific receptor—can influence the metabolic machinery of the cell.

Most cells have receptors for more than one type of hormone, as several different hormones may be involved in regulating their metabolic activities. Some hormones act synergistically on their target tissues; that is, the presence of one enhances the effect of another.

Activation of Genes

Steroid hormones are relatively small, lipid-soluble molecules that pass through the cell membrane of a target cell, through the cytoplasm, and into the nucleus (Figure 49–6). Specific protein receptors in the nucleus com-

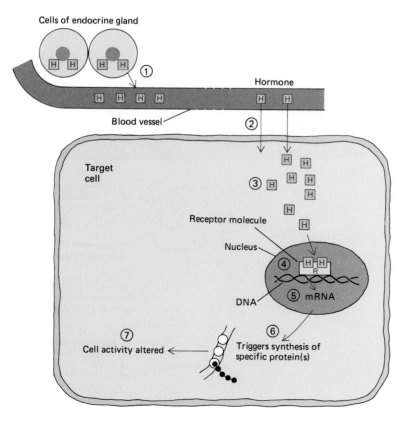

FIGURE 49–6 Activation of genes by steroid hormones. (*1*) Steroid hormone is secreted by endocrine gland and transported by the blood to a target cell. (*2*) Steroid hormones are small lipid-soluble molecules that pass freely through the cell membrane. (*3*) The hormone passes through the cytoplasm to the nucleus. (*4*) Inside the nucleus, the hormone combines with a receptor and the steroid hormone–receptor complex combines with a protein associated with the DNA. (*5*) This activates specific genes, leading to mRNA transcription and (*6*) the synthesis of specific proteins. The proteins produce the response recognized as the hormone's action (*7*).

bine with the hormone to form a hormone-receptor complex. This hormone-receptor complex then combines with a protein associated with the DNA. This combination activates certain genes and leads to synthesis of the messenger RNAs coding for specific proteins. Although they are not steroids, thyroid hormones are also small, hydrophobic molecules that pass through the plasma membrane and bind to receptors within the nucleus.

Action Through Second Messengers

Some protein hormones do not enter the target cell but instead combine with receptors in the cell membrane of the target cell. The hormonal message is then relayed to the appropriate site within the cell by a second messenger. In the 1960s, Earl Sutherland identified **cyclic AMP** (adenosine monophosphate or cAMP) as a hormone intermediary. He suggested that when certain hormones combine with a receptor, the membrane-bound enzyme adenylcyclase is activated. Adenylcyclase catalyzes the conversion of ATP to cyclic AMP (Figures 49–7 and 49–8). Cyclic AMP then acts as a second messenger and stimulates specific cyclic AMP-dependent protein kinases. These enzymes add phosphate groups to proteins and either activate or inhibit their activity. Membrane permeability may then be altered, or protein synthesis may be stimulated.

The specific action initiated by cyclic AMP depends upon the types of enzyme systems and other molecules present in the cell. This explains how the hormone can promote different responses in different kinds of cells. Among the hormones that act by increasing cyclic AMP are adrenocorticotropic hormone (ACTH), thyroid stimulating hormone (TSH), calcitonin, parathyroid hormone, epinephrine, and the gonadotropic hormones (See Table 49–1).

Some hormones (and a number of neurotransmitters) bind to specific phospholipid membrane receptors triggering specific responses but do not use cyclic AMP as the second messenger. Many of these chemical messengers increase the concentration of calcium ions in the cytoplasm. The second messenger in many cases is the compound inositol triphosphate,

FIGURE 49–7 Second-messenger mechanism of hormone action. After they are (*1*) secreted by an endocrine gland and transported by the blood to a target cell, many hormones (especially large protein hormones) (*2*) combine with receptors in the cell membrane of the target cell. (*3*) The hormone–receptor complex activates an enzyme, adenylcyclase, located on the inner surface of the membrane. The adenylcyclase catalyzes the conversion of ATP to cyclic AMP, a second messenger. (*4*) Cyclic AMP then activates protein kinases which (*5*) phosphorylate specific proteins. (*6*) The activity of the cell is altered. See Figure 49–8 for a specific example of hormone action via cyclic AMP.

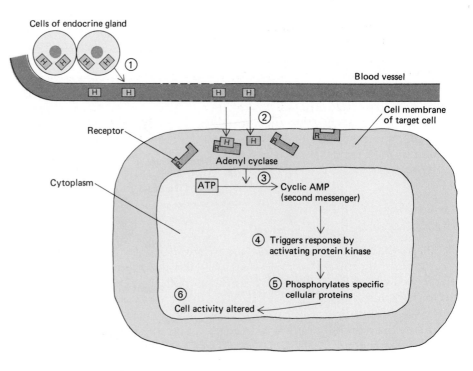

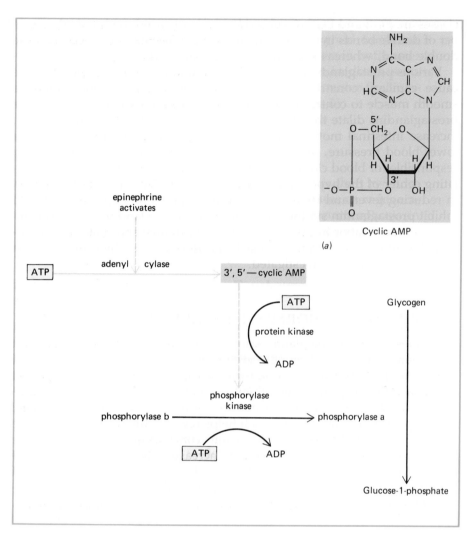

FIGURE 49–8 Hormone action via the second messenger cyclic AMP. (*a*) Structure of cyclic AMP (adenosine 3, 5-monophosphate). (*b*) The sequence of enzymatic events by which a hormone such as epinephrine or glucagon affects cell activity. Epinephrine activates adenylcyclase, bringing about the synthesis of cyclic AMP. This in turn activates a protein kinase that phosphorylates the enzyme phosphorylase kinase. Phosphorylase kinase then activates phosphorylase, bringing about the cleavage of glycogen and the secretion of glucose.

which is split off from the phospholipid. This second messenger causes the release of calcium ions from binding sites in the mitochondria and endoplasmic reticulum. This second messenger mechanism can produce various effects in the cell, including activation of protein kinases independent of cyclic AMP. Note that calcium ions can be viewed as a third messenger in this sequence of metabolic events.

Many of the actions of calcium in the cell are mediated by a calcium-binding protein known as **calmodulin.** The calmodulin-calcium complex binds to certain enzymes, activating them. Some of the cellular processes regulated by this complex include membrane phosphorylation, neurotransmitter release, and microtubule disassembly. Among the hormones that act by increasing the concentration of calcium ions in the cell are oxytocin and in cells with certain receptors, epinephrine, norepinephrine, and ADH.

Prostaglandins

Prostaglandins are derivatives of polyunsaturated fatty acids, each with a five-carbon ring in its structure. Prostaglandins have many hormone-like effects and may help to regulate the action of other hormones by stimulating or inhibiting the formation of cyclic AMP.

Prostaglandins are synthesized and released by many different tissues, including the seminal vesicles, the lungs, liver, and digestive tract. They are classified on the basis of their chemical structure. Among the major

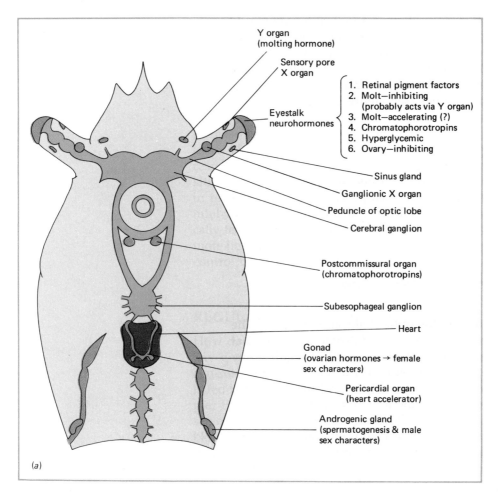

Y organ
(molting hormone)

Sensory pore
X organ

Eyestalk
neurohormones

1. Retinal pigment factors
2. Molt—inhibiting
   (probably acts via Y organ)
3. Molt—accelerating (?)
4. Chromatophorotropins
5. Hyperglycemic
6. Ovary—inhibiting

Sinus gland

Ganglionic X organ

Peduncle of optic lobe

Cerebral ganglion

Postcommissural organ
(chromatophorotropins)

Subesophageal ganglion

Heart

Gonad
(ovarian hormones → female
sex characters)

Pericardial organ
(heart accelerator)

Androgenic gland
(spermatogenesis & male
sex characters)

(a)

(b)

FIGURE 49–11 Color change in crustaceans. (*a*) Locations of the most extensively studied endocrine glands of a crustacean. (*b*) Protective coloration in the ghost crab (*Ocypode quadrata*).

many activities, including molting, migration of retinal pigment, reproduction, heart rate, and metabolism (Figure 49–11). One of the most interesting—and novel—activities regulated by hormones is color change.

Pigment cells of crustaceans are located in the integument beneath the exoskeleton. Pigment may be black, yellow, red, white, or even blue. Color changes are produced by changes in the distribution of pigment granules within the cells. When the pigment is concentrated near the center of a cell, its color is only minimally visible; but when it is dispersed throughout the cell, the color shows to advantage. These pigments provide protective coloration. By appropriate condensation or dispersal of specific types of pigments, a crustacean can approximate the color of its background. Distribution of pigment, that is, color display, is regulated by hormones produced by neurosecretory cells.

Hormonal Control of Insect Development

Like crustaceans, insects have endocrine glands as well as neurosecretory cells. The various hormones interact with one another to regulate growth and development, including molting and morphogenesis. Hormones also help regulate metabolism and reproduction.

Hormonal control of development in insects is complex and varies among the many species. Generally, some environmental factor (e.g., temperature change) affects neurosecretory cells in the brain. Once activated, these cells produce a hormone referred to as **brain hormone (BH)** (or ecdysiotropin), which is transported down axons and stored in the **corpora cardiaca** (Figure 49–12). When released from the corpora cardiaca, BH stimulates the **prothoracic glands,** endocrine glands in the prothorax, to

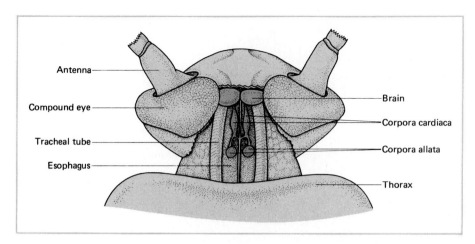

FIGURE 49–12 Dissection of an insect head shows the paired corpora allata and corpora cardiaca, as well as the brain.

produce **molting hormone,** also called **ecdysone.** Molting hormone stimulates growth and molting.

In the immature insect, endocrine glands called **corpora allata** secrete **juvenile hormone.** This hormone suppresses metamorphosis at each larval molt so that the insect increases in size but retains its immature state (Figure 49–13). After the molt, the insect is still in a larval stage. When the concentration of juvenile hormone decreases, metamorphosis occurs, and the insect is transformed into a pupa (Chapter 30). In the absence of juvenile hormone, the pupa molts and becomes an adult. The secretory activity of the corpora allata is regulated by the nervous system, and the amount of juvenile hormone decreases with successive molts.

VERTEBRATE HORMONES

In vertebrates, hormones regulate such diverse activities as growth, metabolic rate, utilization of nutrients by cells, and reproduction. They are largely responsible for regulating fluid balance and blood homeostasis, and

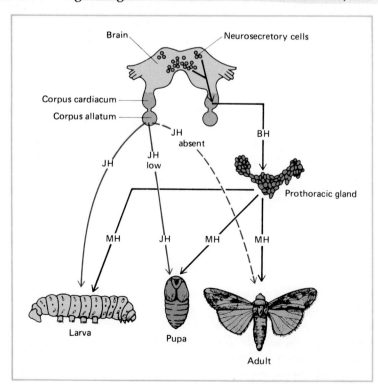

FIGURE 49–13 The neural and endocrine control of growth and molting in a moth. Neurosecretory cells in the brain secrete a hormone, BH, which stimulates the prothoracic glands to secrete molting hormone (*MH*). In the immature insect, the corpora allata secrete juvenile hormone (*JH*), which suppresses metamorphosis at each larval molt. Metamorphosis to the adult form occurs when molting hormone acts in the absence of juvenile hormone.

FIGURE 49–14 Location of the principal endocrine glands in the human male and female.

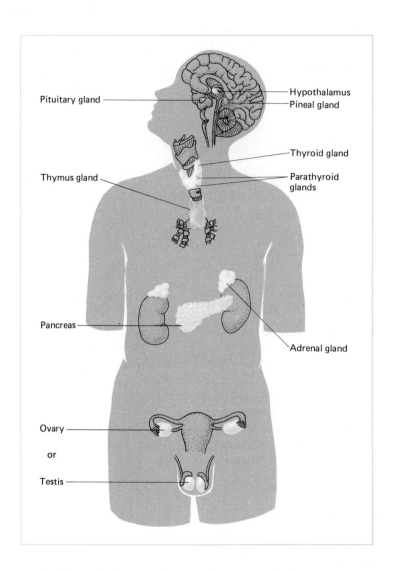

Pituitary gland

Hypothalamus

Pineal gland

Thyroid gland

Thymus gland

Parathyroid glands

Pancreas

Adrenal gland

Ovary

or

Testis

they help the body to cope with stress. The principal human endocrine glands are illustrated in Figure 49–14. Most vertebrates possess similar endocrine glands. Table 49–1 gives the physiological actions and sources of some of the major vertebrate hormones.

Endocrine Disorders

When a disorder or disease process affects an endocrine gland, the rate of secretion may become abnormal. If **hyposecretion** (reduced output) occurs, target cells are deprived of needed stimulation. If **hypersecretion** (abnormal increase in output) occurs, the target cells may be overstimulated. In some endocrine disorders, an appropriate amount of hormone is secreted, but target cells lack receptors (or the receptors may not function properly), so that the cell may not be able to take up the hormone. Any of these abnormalities leads to predictable metabolic malfunctions and clinical symptoms (see Table 49–2).

The Hypothalamus and Pituitary Gland

Hormonal activity is controlled directly or indirectly by the hypothalamus, which links the nervous and endocrine systems. In response to input from other areas of the brain and from hormones in the blood, neurons of the hypothalamus secrete hormones that regulate the release of hormones from the pituitary gland. The hypothalamus produces two hormones, **ADH** and **oxytocin,** which pass down the axons of neurons into the poste-

Table 49–1 SOME ENDOCRINE GLANDS AND THEIR HORMONES*

Endocrine Gland and Hormone	Target Tissue	Principal Actions
Hypothalamus Releasing and release-inhibiting hormones	Anterior lobe of pituitary gland	Stimulates or inhibits secretion of specific hormones
Hypothalamus (production) **Posterior lobe of pituitary (storage and release)** Oxytocin	Uterus	Stimulates contraction
	Mammary glands	Stimulates ejection of milk into ducts
Antidiuretic hormone (vasopressin)	Kidneys (collecting ducts)	Stimulates reabsorption of water; conserves water
Anterior lobe of pituitary Growth hormone	General	Stimulates growth by promoting protein synthesis
Prolactin	Mammary glands	Stimulates milk production
Thyroid-stimulating hormone	Thyroid gland	Stimulates secretion of thyroid hormones; stimulates increase in size of thyroid gland
Adrenocorticotropic hormone	Adrenal cortex	Stimulates secretion of adrenal cortical hormones
Gonadotropic hormones (follicle-stimulating hormone, FSH; luteinizing hormone, LH)	Gonads	Stimulate gonad function and growth
Thyroid gland Thyroxine (T_4) and triiodothyronine (T_3)	General	Stimulate metabolic rate; essential to normal growth and development
Calcitonin	Bone	Lowers blood calcium level by inhibiting bone breakdown by osteoclasts
Parathyroid glands Parathyroid hormone	Bone, kidneys, digestive tract	Increases blood calcium level by stimulating bone breakdown; stimulates calcium reabsorption by kidneys; activates vitamin D
Islets of Langerhans of pancreas Insulin	General	Lowers glucose concentration in the blood by facilitating glucose uptake and utilization by cells; stimulates glycogenesis; stimulates fat storage and protein synthesis
Glucagon	Liver, adipose tissue	Raises glucose concentration in the blood by stimulating glycogenolysis and gluconeogenesis; mobilizes fat
Adrenal medulla Epinephrine and norepinephrine	Muscle, cardiac muscle, blood vessels, liver, adipose tissue	Help body cope with stress; increase heart rate, blood pressure, metabolic rate; re-route blood; mobilize fat; raise blood sugar level
Adrenal cortex Mineralocorticoids (aldosterone)	Kidney tubules	Maintain sodium and phosphate balance
Glucocorticoids (cortisol)	General	Help body adapt to long-term stress; raise blood glucose level; mobilize fat
Pineal gland Melatonin	Gonads, pigment cells, other tissues(?)	Influences reproductive processes in hamsters and other animals; pigmentation in some vertebrates; may control biorhythms in some animals; may help control onset of puberty in humans
Ovary† Estrogens (Estradiol)	General; uterus	Develop and maintain sex characteristics in female; stimulate growth of uterine lining
Progesterone	Uterus; breast	Stimulates development of uterine lining
Testis‡ Testosterone	General; reproductive structures	Develops and maintains sex characteristics of males; promotes spermatogenesis; responsible for adolescent growth spurt

*The ovaries and testes and their hormones are discussed in Chapter 50. The digestive hormones are described in Chapter 40.
†For more detailed description see Table 50–2.
‡For more detailed description see Table 50–1.

Table 49–2 **CONSEQUENCES OF ENDOCRINE MALFUNCTION**

Hormone	Hyposecretion	Hypersecretion
Growth hormone	Pituitary dwarf	Gigantism if malfunction occurs in childhood; acromegaly in adult
Thyroid hormones	Cretinism (in children); myxedema, a condition of pronounced adult hypothyroidism (BMR is reduced by about 40%; patient feels tired all of the time and may be mentally slow); goiter, enlargement of the thyroid gland (see Figure)	Hyperthyroidism; increased metabolic rate, nervousness, irritability
Parathyroid hormone	Spontaneous discharge of nerves; spasms; tetany; death	Weak, brittle bones; kidney stones
Insulin	Diabetes mellitus	Hypoglycemia
Adrenocortical hormones	Addison's disease (body cannot synthesize sufficient glucose by gluconeogenesis; patient is unable to cope with stress; sodium loss in urine may lead to shock)	Cushing's disease (edema gives face a full-moon appearance; fat is deposited about trunk; blood glucose level rises; immune responses are depressed)

■ Goiter resulting from iodine deficiency.

rior lobe of the pituitary gland (Figure 49–15). These neurosecretions are stored within the tips of the axons in the posterior lobe until the neurons are stimulated. They are then released from the posterior lobe and enter the circulation.

FIGURE 49–15 The hormones secreted by the posterior lobe of the pituitary are actually manufactured in cells of the hypothalamus. The axons of these neurons extend down into the posterior lobe of the pituitary. The hormones are packaged in granules that flow through these axons and are stored in their ends. The hormone is secreted into the interstitial fluid as needed and transported by the circulatory system.

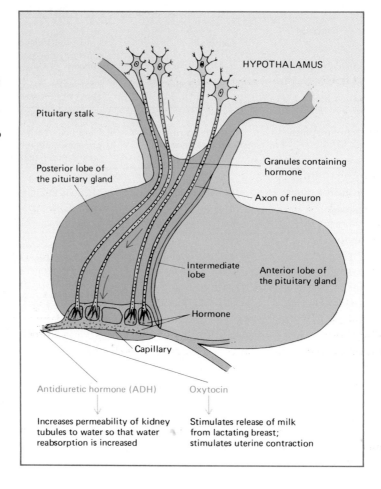

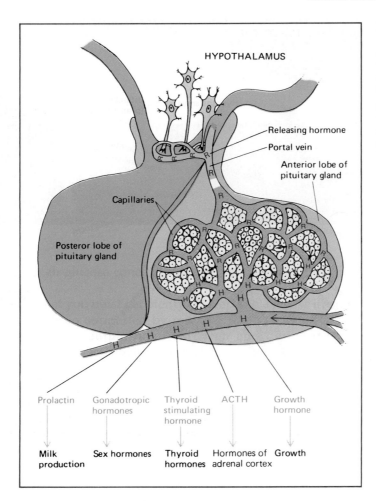

FIGURE 49–16 The hypothalamus secretes several specific releasing and release-inhibiting hormones, which reach the anterior lobe of the pituitary gland by way of portal veins. Each releasing hormone stimulates the synthesis of a particular hormone by the cells of the anterior lobe. (*R*, releasing hormone; *H*, hormone.)

The hypothalamus regulates the anterior lobe of the pituitary gland by secreting several **releasing** and **release-inhibiting hormones,** each more or less specific for one type of cell in the pituitary. These neurohormones are released by the hypothalamus, enter capillaries, and pass through special portal veins that connect the hypothalamus with the anterior lobe of the pituitary. (This portal vein, like the hepatic portal vein, does not deliver blood to a larger vein directly but connects two sets of capillaries.) Within the anterior lobe of the pituitary, the portal veins divide into a second set of capillaries. The hypothalamic releasing and release-inhibiting hormones pass through the walls of these capillaries into the tissue of the anterior lobe where they regulate production and secretion of the hormones produced by the anterior lobe of the pituitary (Figure 49–16).

Only about the size of a pea, the **pituitary gland** is a remarkable organ that secretes at least nine hormones. These hormones regulate the activities of several other endocrine glands and influence a wide range of physiological processes (Figure 49–17). The **posterior lobe** of the pituitary gland releases the hormones ADH and oxytocin produced by the hypothalamus. The **anterior lobe** produces growth hormone, prolactin, and several **tropic hormones,** hormones that regulate other endocrine glands. (The anterior lobe also produces a peptide called β-lipotropin, the precursor of endorphins and enkephalins, which are substances that function in pain perception).

The pituitary of some vertebrates contains an intermediate lobe that secretes **melanocyte stimulating hormone (MSH).** This hormone regulates skin pigmentation. In humans, the intermediate lobe is poorly developed, and the function of MSH is not known.

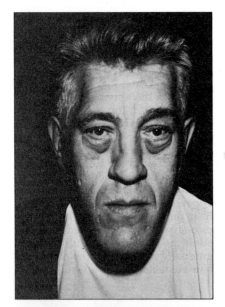

mones are also necessary for cellular differentiation. The development of tadpoles into adult frogs cannot take place without thyroxine; the hormone appears to selectively regulate the synthesis of proteins. Thyroid hormones stimulate growth by promoting protein synthesis and enhancing the effect of growth hormone.

REGULATION OF SECRETION. The secretion of thyroid hormone is regulated by a feedback system between the anterior pituitary and the thyroid (Figure 49–21). The pituitary secretes **thyroid-stimulating hormone (TSH),** which acts by way of cyclic AMP to promote synthesis and secretion of thyroid hormones and to increase the size of the thyroid gland itself. When the concentration of thyroid hormones in the blood rises above normal, the cells in the anterior pituitary are inhibited and the release of TSH is decreased.

Too much thyroid hormone in the blood may also affect the hypothalamus, inhibiting secretion of TSH-releasing hormone. However, the hypothalamus is thought to exert its regulatory effects mainly in certain stressful situations, such as exposure to extreme weather conditions.

THYROID HORMONES AND GROWTH DISORDER. Extreme hypothyroidism during infancy and childhood results in low metabolic rate and retarded mental and physical development, a condition known as **cretinism.** A cretin dwarf is very different from a pituitary dwarf. When diagnosed and treated early enough, the effects of cretinism can be prevented by clinical administration of thyroid hormone.

An adult may also suffer from hypothyroidism. When there is almost no thyroid function, the basal metabolic rate (BMR) is reduced by about 40% and the condition known as **myxedema** develops. An individual with myxedema feels tired and may be mentally slow.

Hyperthyroidism does not result in abnormal growth but does increase metabolic rate by up to 60% or more. Individuals with hyperthyroidism tend to be nervous, irritable, and emotionally unstable. Hyperthyroidism often results from Graves' disease (toxic goiter), in which the thyroid gland increases in both size and activity. Graves' disease can be treated with drugs, by surgery, or by injecting radioactive iodine which is taken up by the thyroid and destroys some of the thyroid tissue.

FIGURE 49–20 A case of acromegaly. Instead of producing growth in height, the hypersecretion of GH in the adult causes a thickening of connective tissue. Note the enlarged nose, jaw, and cheekbones.

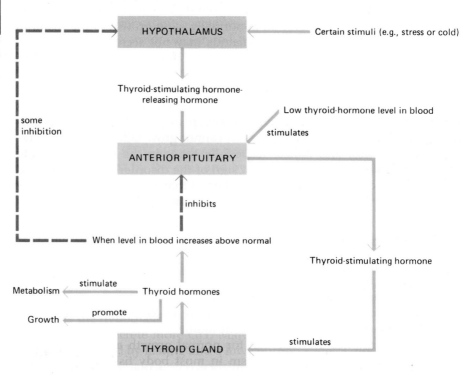

FIGURE 49–21 Regulation of thyroid hormone secretion. Green arrows indicate stimulation; red arrows indicate inhibition.

Any abnormal enlargement of the thyroid gland is termed a **goiter** and may be associated with either hypo- or hypersecretion (see illustration in Table 49–2). One cause of goiter is Graves' disease; another is **endemic goiter** caused by dietary iodine deficiency. Without iodine, the gland cannot make thyroid hormones, so their concentrations in the blood decrease. In compensation, the anterior lobe of the pituitary secretes large amounts of TSH and the thyroid gland enlarges, sometimes to gigantic proportions. However, enlargement of the gland cannot increase production of the hormones because iodine is still a missing ingredient. Thanks to iodized salt, simple goiter is no longer common in the United States. In other parts of the world, however, an estimated 200 million persons still suffer from this easily preventable disorder.

Regulation of Glucose Concentration in the Blood: Insulin and Glucagon

More than a million small clusters of cells known as the **islets of Langerhans** are scattered throughout the pancreas. About 70% of the islet cells are **beta cells,** which produce the hormone **insulin. Alpha cells** secrete the hormone **glucagon.** Both insulin and glucagon are proteins.

Insulin exerts widespread influence on metabolism, but its principal effect is to facilitate the uptake of glucose into skeletal muscle and fat cells. By stimulating cells to take up glucose from the blood, insulin lowers the concentration of glucose in the blood (Figure 49–22). Liver cells are very permeable to glucose and do not require insulin to take it up from the blood. However, insulin increases the amount of glucokinase, an enzyme that uses ATP to phosphorylate glucose so that it cannot diffuse out of the liver cells. In this way insulin acts to trap glucose inside liver cells. The hormone also stimulates glycogenesis (glycogen formation) and storage of glycogen in the liver. Insulin stimulates protein synthesis and promotes fat storage.

Glucagon acts antagonistically to insulin. Its principal action is to increase the concentration of glucose in the blood. It does this by stimulating glycogenolysis (conversion of glycogen to glucose) in the liver, and by stimulating gluconeogenesis (production of glucose from other nutrients). Glucagon mobilizes fatty acids and amino acids as well as glucose.

The insulin-glucagon system is a powerful, fast-acting mechanism for keeping the concentration of glucose in the blood within a narrow range. Brain cells are especially dependent upon a continuous supply of glucose because they are normally unable to utilize other nutrients as fuel.

Regulation of Insulin and Glucagon Secretion

Secretion of insulin and glucagon is controlled by the concentration of blood sugar. After a meal, when the blood glucose concentration rises as a result of intestinal absorption, the beta cells are stimulated to increase insulin secretion (Figure 49–22). Then, as the cells remove glucose from the blood, decreasing its concentration, insulin secretion decreases.

When a person has not eaten for several hours, the concentration of glucose in the blood begins to fall. When it falls from its normal fasting level of about 90 mg/100 ml to about 70 mg/100 ml, the alpha cells of the islets secrete glucagon. Glucose is mobilized from the liver stores, and the blood sugar concentration returns to normal.

Diabetes Mellitus

One of the most serious disorders of carbohydrate metabolism is **diabetes mellitus.** Although many of the symptoms of diabetes can be controlled,

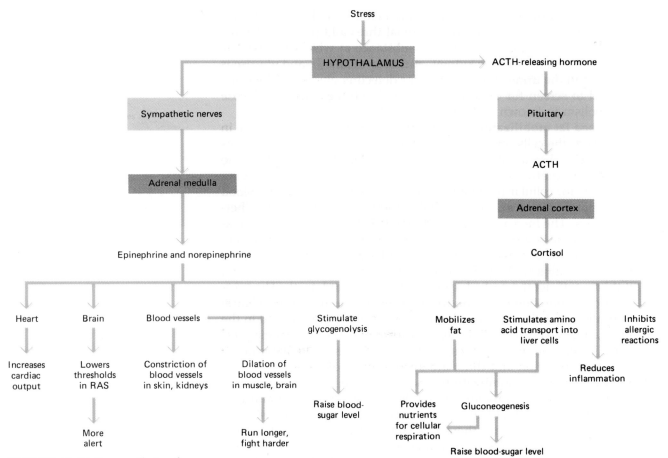

FIGURE 49–23 Some effects of stress.

epinephrine (adrenaline) and **norepinephrine** (noradrenaline). Recall that norepinephrine is also secreted as a neurotransmitter by many neurons. The hormone secreted by the adrenal medulla has much longer-lasting effects (about ten times longer) because it is removed from the blood more slowly than it is removed from synapses. Still, the nervous system and the adrenal glands reinforce one another's action during stress or excitement (such as participation in a competitive athletic event).

During sudden stress, the brain sends messages through sympathetic nerves to the adrenal medulla. The medulla initiates an **alarm reaction** which enables a person to think more quickly, fight harder, or run faster than usual. Metabolic rate increases as much as 100%. The adrenal medullary hormones stimulate the heart to beat faster, raise blood pressure, and lower thresholds in the RAS (reticular activating system) of the brain (Figure 49–23). These hormones also enlarge airways, permitting more effective breathing. In fact, epinephrine and related drugs are used clinically to relieve nasal congestion and asthma.

Hormones of the adrenal medulla cause blood to be rerouted to those organs essential for emergency action. Blood vessels to the skin, digestive organs, and kidneys are constricted, while those that deliver blood to the brain, skeletal muscles, and heart are dilated. Constriction of blood vessels in the skin has the added advantage of decreasing blood loss in case of hemorrhage. It also explains the sudden paling that comes with fear or rage. Epinephrine and norepinephrine also raise the concentration of glucose in the blood by stimulating glycogenolysis in the liver. They raise fatty acid concentrations in the blood by mobilizing fat stores from adipose tissue. These actions provide additional fuel molecules for the rapidly metabolizing muscle cells.

The Effects of Cortisol

Cortisol (also known as hydrocortisone) is one of the principal steroid hormones produced by the adrenal cortex. It belongs to a group of hormones known as **glucocorticoids.** Cortisol reinforces the actions of epinephrine and norepinephrine by making nutrients available for gluconeogenesis; it increases the transport of amino acids into liver cells and increases the liver enzymes necessary to convert amino acid skeletons to glucose. These actions contribute to the formation of large amounts of glycogen in the liver and help to ensure adequate fuel supplies for the cells, especially when the body is under stress.

Cortisol also reduces inflammation by inhibiting the synthesis of prostaglandins and decreasing the permeability of capillary walls, thereby reducing swelling. Cortisol is thought to help stabilize lysosome membranes so that their potent enzymes do not leak out and damage tissues. Because of these actions, cortisol and related glucocorticoids are used clinically to reduce inflammation in allergic reactions, arthritis, and certain types of cancer and with caution in infections.

Any type of stress stimulates the hypothalamus to secrete **corticotropin releasing factor (CRF)** (Figure 49–23). This hormone stimulates certain cells in the anterior lobe of the pituitary to secrete **adrenocorticotropic hormone (ACTH)** (also called corticotropin). In turn, this tropic hormone stimulates the adrenal cortex to grow and to increase its output of cortisol. So potent is the effect of ACTH that it can result in as much as a 20-fold increase in cortisol secretion within minutes. Cortisol has direct negative feedback effects on both the hypothalamus and the pituitary. Hyposecretion of the adrenal cortex can result in Addison's disease, a condition in which the body becomes unable to cope with stress; hypersecretion results in Cushing's disease, in which immune response is depressed (See Table 49–2).

During prolonged stress, such as might be caused by a chronic disease or an unhappy marriage, the blood levels of adrenal hormones, especially of cortisol, remain elevated. Blood pressure may remain abnormally high, and metabolism in general remains geared to help the body cope with stress. Such chronic stress is thought to be harmful because of the side effects of long-term elevated levels of adrenal hormones. Chronic high blood pressure may contribute to heart disease, and increased levels of fat in the blood may promote atherosclerosis. High levels of cortisol over a long period of time can interfere with normal immune responses, increasing susceptibility to infection. Such effects are seen when large doses of steroids are administered clinically to human patients and when experimental animals receive injections of large amounts of cortisol. Among the diseases that have been linked to excessive amounts of cortisol (or the closely related steroids used clinically) are ulcers, hypertension, atherosclerosis, and possibly diabetes mellitus.

■ SUMMARY

I. Hormones are chemical messengers that help to regulate homeostasis; many hormones are produced by endocrine glands and transported to their target tissues by the blood.

II. Hormones may be steroids, proteins, or derivatives of fatty acids or amino acids.

III. Special receptors in the cells of target tissues bind specific hormones.

    A. Steroid hormones pass through the plasma membrane of a target cell and combine with specific receptors in the nucleus. These hormone-receptor complexes are thought to activate specific genes.

    B. Many protein hormones combine with receptors in the cell membrane of the target cell and act by way of a second messenger such as cyclic AMP.

    C. Prostaglandins may help regulate hormone action by regulating cyclic AMP formation.

IV. Hormone secretion is regulated by negative feedback control mechanisms.

V. Many invertebrate hormones are secreted by neurons

rather than by endocrine glands. They help to regulate regeneration, molting, metamorphosis, reproduction, and metabolism.

A. The optic glands of mollusks are endocrine organs that stimulate gonadal development.

B. Distribution of pigment in crustaceans is regulated by hormones secreted by neurosecretory cells.

C. Hormones control development in insects.
1. When stimulated by some environmental factor, neurosecretory cells in the insect brain secrete a hormone known as BH.
2. BH is stored in the corpora cardiaca; when released, it stimulates the prothoracic glands to produce molting hormone (ecdysone), which stimulates growth and molting.
3. In the immature insect, the corpora allata secrete juvenile hormone, which suppresses metamorphosis at each larval molt.
4. The amount of juvenile hormone decreases with successive molts; when its concentration becomes low, the larva develops into a pupa; in its absence the pupa changes into an adult.

VI. In vertebrates the hypothalamus is the link between the nervous and endocrine systems.

A. The hypothalamus regulates the anterior lobe of the pituitary gland by producing several releasing and release-inhibiting hormones that regulate secretion of pituitary hormones.

B. The hypothalamus produces ADH and oxytocin, the hormones released by the posterior lobe of the pituitary.

C. Among the hormones that influence growth and development are growth hormone (GH), secreted by the anterior pituitary, and thyroid hormones, secreted by the thyroid gland.
1. Both GH and the thyroid hormones promote protein synthesis.
2. Hypersecretion of GH during childhood results in gigantism; hyposecretion may result in a pituitary dwarf. Hyposecretion of thyroid hormones during infancy and childhood may result in cretinism.

D. Glucose concentration in the blood is regulated mainly by insulin and glucagon.
1. Insulin lowers the concentration of glucose in the blood by stimulating uptake of glucose by skeletal muscle and fat cells, and by promoting glycogenesis.
2. Glucagon raises the blood sugar level by stimulating glycogenolysis and gluconeogenesis.
3. In diabetes mellitus, cells are unable to utilize glucose properly and turn to fat and protein for fuel.

E. The adrenal medulla and the adrenal cortex secrete hormones that help the body cope with stress.
1. The adrenal medulla secretes epinephrine and norepinephrine, which increase heart rate, metabolic rate, and strength of muscle contraction, and reroute the blood to the organs that require more blood in time of stress.
2. The adrenal cortex releases cortisol, which promotes gluconeogenesis in the liver, thereby raising the glucose concentration; this hormone provides fuel needed for the increased metabolic activities stimulated by the adrenal medullary hormones.

■ POST-TEST

1. The tissues influenced by a specific hormone are referred to as its _____ _____.
2. Neurosecretory cells are neurons that secrete _____.
3. A hormone may be defined as a _____.
4. Many protein hormones combine with receptors in the _____ of the target cell; the hormonal message is relayed by a second messenger such as _____ _____.
5. Hormone secretion is regulated by _____ _____ control mechanisms.
6. The insect hormone known as BH stimulates the prothoracic glands to secrete _____ hormone.
7. Releasing and release-inhibiting hormones that regulate pituitary gland activity are secreted by the _____.
8. Somatomedins mediate _____ responses.

Select the most appropriate answer in Column B for each description in Column A; the same answer may be used more than once.

Column A
9. Stimulates protein synthesis
10. A steroid hormone
11. Increases heart rate; raises blood pressure
12. Lowers blood glucose level
13. Helps body deal with long-term stress
14. Hyposecretion may cause cretinism
15. Used clinically to reduce inflammation

Column B
a. Norepinephrine
b. Growth hormone
c. Cortisol
d. Glucagon
e. None of the above

16. ACTH stimulates the adrenal _____ to increase secretion of _____.
17. Acromegaly results from _____ secretion of _____.
18. The _____ hormones stimulate the rate of metabolism.
19. In diabetes mellitus, there are a decreased utilization of _____ and an increased utilization of fat and _____.
20. When the concentration of thyroid hormones in the blood rises above normal levels, the _____ is inhibited and slows its release of _____.

■ REVIEW QUESTIONS

1. Why do some biologists advocate changing the traditional definition of a hormone? Do you think neurotransmitters should be included as hormones? How about growth factors?
2. How are most hormones transported? How do their target tissues recognize them?
3. Draw a diagram illustrating how a steroid hormone can activate a gene.
4. What is the role of cyclic AMP in hormone action?
5. Why are prostaglandins of interest to both endocrinologists and drug companies?
6. Draw a diagram illustrating how negative feedback regulates the secretion of the following hormones:
   a. parathyroid hormone
   b. cortisol
   c. thyroid hormones
7. Draw a diagram illustrating the hormonal control of development in insects.
8. Locate each of the following endocrine glands in the human body:
   a. pituitary
   b. thyroid
   c. parathyroids
   d. islets of Langerhans
   e. adrenal glands
9. List the hormones secreted by each of the following glands, and give their actions:
   a. posterior lobe of the pituitary
   b. anterior lobe of the pituitary
   c. thyroid
   d. islets of Langerhans
   e. adrenal medulla
   f. adrenal cortex
10. Explain how the hypothalamus influences endocrine activity.
11. Compare the actions of growth hormone and thyroid hormones in stimulating growth?
12. Compare the abnormalities of growth associated with hyposecretion of hypersecretion of GH and thyroid hormones.
13. Contrast the actions of insulin and glucagon in regulating the concentration of glucose in the blood.
14. What physiological disturbances are associated with diabetes mellitus?
15. Imagine that you are a full-time student and also hold a very demanding job that you do not enjoy. In addition, you are in a relationship that is no longer satisfying. How do your adrenal glands help your body cope with this chronic stress?

■ RECOMMENDED READINGS

Bloom, F. Neuropeptides. *Scientific American*, October 1981. Neuropeptides are sometimes considered hormones, sometimes neurotransmitters.

Cantin, M., and Genest, J. The heart as an endocrine gland. *Scientific American*, Vol. 255, No. 5, November 1986, pp. 76–81. The atria secrete a recently discovered hormone that helps regulate blood pressure and volume.

Carmichael, S.W., and Winkler, H. The adrenal chromaffin cell. *Scientific American*, Vol. 253, No. 2, August 1985, pp. 40–49. Chromaffin cells offer insights into the functioning of other secretory cells, especially neurons.

Crews, D. The hormonal control of behavior in a lizard. *Scientific American*, August 1979. Sexual behavior is controlled by the interaction of the brain and hormones from the gonads.

Edwards, D.D. Diabetes autoimmunity seen, stopped. *Science News*, Vol. 132, November 7, 1987. Reports of two clinical trials suggest that early use of an immune system suppressor could stop destruction of insulin-producing cells in diabetics.

Guillemin, R., and Burges, R. Hormones of the hypothalamus. *Scientific American*, November 1972. Hormones released by the hypothalamus control the pituitary.

Hadley, M.E. *Endocrinology*. New Jersey, Prentice Hall, 1984. A basic introduction to endocrinology.

Solomon, E.P., and Davis, P.W. *Human Anatomy and Physiology*. Philadelphia, Saunders College Publishing, 1983. Chapter 14 focuses on endocrine regulation in humans.

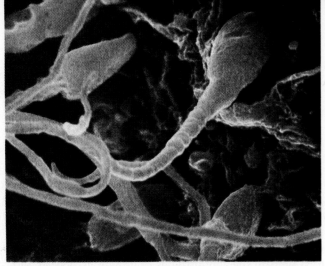

Sperm and egg

50

Reproduction

■ LEARNING OBJECTIVES

After you have read this chapter you should be able to:

1. Compare asexual and sexual reproduction, giving two examples of asexual reproduction.
2. Describe adaptive advantages of the following reproductive styles: metagenesis, parthenogenesis, and hermaphroditism.
3. Trace the passage of sperm cells through the male reproductive system from their origin in the seminiferous tubules until they leave the body in the semen.
4. Label the structures of the male reproductive system on a diagram and give the functions of each.
5. Label the structures of the female reproductive system on a diagram and give the functions of each.
6. Trace the development of an ovum and its passage through the female reproductive system until it is fertilized.
7. Describe the actions of testosterone and of the gonadotropic hormones in the male.
8. Describe the hormonal regulation of the menstrual cycle and identify the time of important events of the cycle such as ovulation and menstruation.
9. Describe the physiological changes that occur during sexual response in the male and the female.
10. Describe the process of human fertilization.
11. Compare the methods of birth control in Table 50–3 with respect to mode of action, effectiveness, advantages, and disadvantages.

If there is one feature of a living system that is unique, it is the ability to reproduce and perpetuate its species. The survival of each species requires that its members produce new individuals to replace those that die.

At the molecular level, reproduction is a function of the unique capacity of nucleic acids to replicate themselves. At the level of the whole organism, reproduction ranges from the simple fission of bacteria or protists to the incredibly complex structural, functional, and behavioral processes of reproduction in higher animals. Reproduction in numerous organisms has been discussed throughout this book. In this chapter we summarize some major features of animal reproductive processes and then focus upon human reproduction.

ASEXUAL REPRODUCTION

In **asexual reproduction** a single parent splits, buds, or fragments to give rise to two or more offspring that have hereditary traits identical with those of the parent. Sponges and cnidarians can reproduce by **budding,** in which a small part of the parent's body separates from the rest and develops into a new individual (Figure 50–1). It may split away from the parent and establish an independent existence, or it may remain attached and become a more or less independent member of the colony.

Salamanders, lizards, sea stars, and crabs can grow a new tail, leg, arm, or certain other organs if the original one is lost. In some cases, this ability to regenerate a part becomes a method of reproduction. The body of the parent may break into several pieces; each piece then regenerates the missing parts and develops into a whole animal. Such reproduction by **fragmentation** is common among flatworms. Similarly, oyster farmers learned long ago that when they tried to kill sea stars by chopping them in half and throwing the pieces back into the sea, the number of sea stars preying on the oyster bed doubled! A sea star can, in fact, regenerate an entire new individual from a single arm.

SEXUAL REPRODUCTION

Most animals reproduce by sexual reproduction, which generally involves two parents. Each contributes a specialized **gamete** (an egg or sperm) and these fuse to form the fertilized egg, or **zygote.** The egg is typically large and nonmotile, with a store of nutrients to support the development of the embryo. The sperm is usually small and motile, adapted to propel itself by beating its long, whiplike tail.

Sexual reproduction has the biological advantage of promoting genetic variety among the members of a species, because the offspring is the product of genes contributed by both parents, rather than a genetic copy of a single individual. By making possible the genetic recombination of the inherited traits of two parents, sexual reproduction gives rise to offspring that may be better able to survive than either parent. Advantageous genes can spread quickly through a population that reproduces sexually. This permits the rapid and effective spread of adaptations that enable a species to survive in an ever-changing environment.

Fertilization, the fusion of sperm and egg, may take place inside the body—**internal fertilization**—or outside—**external fertilization** (Figure 50–2). Most aquatic animals practice external fertilization. Mating partners usually release eggs and sperm into the water simultaneously. Many gametes are lost—to predators, for example—but so many are released that a sufficient number of sperm and egg cells meet and unite to perpetuate the species.

In internal fertilization, matters are left less to chance. The male generally delivers sperm cells directly into the body of the female. Her moist tissues provide the watery medium required for movement of sperm. Most terrestrial animals, as well as a few fish and some other aquatic animals, practice internal fertilization.

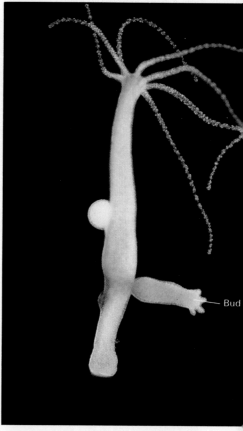

Bud

FIGURE 50–1 *Hydra* is one of many organisms that reproduces asexually by budding. A part of the body grows outward and then separates and develops into a new individual. The portion of the parent body that buds is not specialized exclusively for performing a reproductive function.

(a)

(b)

(c)

FIGURE 50–2 External fertilization and internal fertilization. (*a*) External fertilization is illustrated by these spawning frogs (*Rana temporaria*). Most amphibians must return to water for mating. The female lays a mass of eggs, while the male mounts her and simultaneously deposits his sperm in the water. (*b*) Internal fertilization is practiced by desert tortoises. (*c*) A butterfly, *Heliconius ismenius,* ovipositing (depositing eggs) on a passionflower plant. These eggs were fertilized internally.

REPRODUCTIVE SYSTEMS

Although the details of the reproductive process vary from one animal to another, some generalizations about reproductive systems may be helpful in understanding the variations (Figure 50–3). The basic components of the male reproductive system are the male gonad, or **testis,** in which sperm are produced, and the **sperm duct** for the transport of sperm to the exterior of the body. Parts of the sperm duct, or areas adjacent to it, may be modified for specific functions. A part of the duct may be specialized for sperm storage; such a structure is often called a **seminal vesicle.** (The human organ termed the seminal vesicle, however, does *not* store sperm.) There also may be glandular areas for the production of semen, which serves as a vehicle for the sperm and may also activate, nourish, and protect them. The terminal part of the sperm duct may open onto or into a copulatory organ, a **penis,** which in internal fertilization provides for the transfer of sperm to the female.

The basic parts of the female system are the female gonad, or **ovary,** and the **oviduct,** a tube for the transport of eggs. Like the sperm duct, the oviduct may be modified in any of several ways in different species. It may have a glandular portion for the secretion of an egg shell, a case, or a cocoon. A section of the oviduct, a **seminal receptacle,** may be modified for the storage of sperm following their transfer from the male. Another portion of the oviduct may be modified as a **uterus** for egg storage or for the development of the fertilized egg within the body of the female. The terminal portion of the oviduct may be adapted as a **vagina** for receiving the male copulatory organ.

The male and female systems of most animals do not possess all these modifications, and these structures cannot be arranged in any kind of progressive sequence of evolutionary changes. Some otherwise simple animals have extremely complex reproductive systems. This is especially likely to be true of parasites. The presence or absence of some specific adaptation of the reproductive system is correlated with the circumstances of reproduction—whether the animal lives in the sea, in fresh water, or on land; whether fertilization is external or internal; whether the eggs are liberated singly into the water to develop or are deposited onto the ocean floor or river bottom within envelopes; and whether or not the developing individual passes through a larval stage.

Most species of fishes utilize a simple form of reproduction in which the female sheds into the water enormous numbers of eggs, which may or may not be fertilized by sperm from some passing male. Thereafter, both male

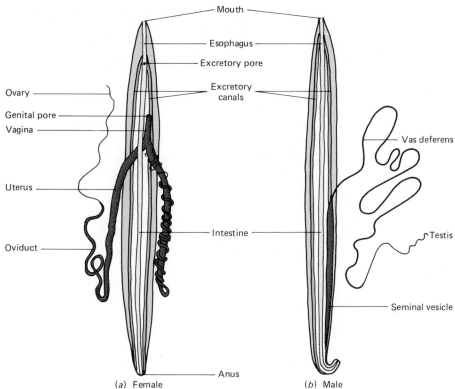

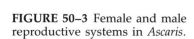

FIGURE 50–3 Female and male reproductive systems in *Ascaris.*

(a) Female   (b) Male

FIGURE 50–4 Reproductive variations. (*a*) Metagenesis in the hydrozoa *Pennaria tiarelle.* Magnified approximately 3.8 times, this closeup of two polyps shows one bearing two male medusae. Parthenogenesis takes place in (*b*) the greenfly (*Hemiptera aphidoidea*) and (*c*) this *Noteus,* a rotifer, with an egg. (*d*) Hermaphroditic copulation in earthworms.

(a)

(b)

and female fishes ignore the eggs. In such fishes, the uterus is missing and the male does not have a copulatory organ. In many species, however, the male undertakes much responsibility for the care of the eggs and young. In the sea catfish, for instance, the fertilized eggs are deposited in a clump that the male picks up and carries in his mouth during most of their incubation (presumably fasting and yielding not to temptation!).

SOME REPRODUCTIVE VARIATIONS

Animals show tremendous diversity in their methods of reproduction (Figure 50–4). Even members of the same class may differ markedly in their reproductive processes.

Metagenesis

Metagenesis is an alternation of asexual and sexual generations. For example, in the hydrozoan *Obelia,* a polyp generation gives rise asexually by budding to a generation of medusas (Figure 29–15). The motile medusas

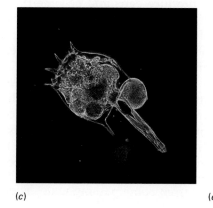

(c)

(d)

produce gametes and reproduce sexually, giving rise to a new generation of polyps. Thus, there is an alternation of generations—polyp, medusa, polyp, medusa, and so on. Both generations consist of diploid organisms.

Parthenogenesis

Parthenogenesis (virgin development) is a form of reproduction in which an unfertilized egg develops into an adult animal. Parthenogenesis is common among rotifers, some gastropod mollusks, some crustaceans, insects, especially honeybees and wasps, and some reptiles. Some species of arthropods (and even some vertebrates—lizards) consist entirely of females that reproduce in this way. More commonly, parthenogenesis occurs for several generations, after which males develop, produce sperm, and mate with the females to fertilize their eggs. In some species, parthenogenesis is advantageous in maintaining the social order; in others, it appears to be an adaptation for survival in times of stress or when there is a serious decrease in population.

A special form of parthenogenesis occurs in honeybees. The queen honeybee is inseminated by a male during the "nuptial flight." The sperm she receives are stored in a little pouch connected with her genital tract but closed off by a muscular valve. As the queen lays eggs, she can either open this valve, permitting the sperm to escape and fertilize the eggs, or keep the valve closed, so that the eggs develop without fertilization. Generally, fertilization occurs in the fall, and the fertilized eggs are quiescent during the winter. The fertilized eggs become females (queens and workers); the unfertilized eggs become males (drones). Some species of wasps alternately produce a parthenogenetic generation and a generation developing from fertilized eggs.

Hermaphroditism

Many invertebrate species are **hermaphroditic,** meaning that a single individual produces both eggs and sperm. A few, such as the parasitic tapeworms, are capable of self-fertilization. Although this form of reproduction is still classified as sexual (since both eggs and sperm are involved), it is an exception to the important generalization that sexual reproduction involves two different individuals. Hermaphroditism of this type in tapeworms is an adaptation suited to the tapeworm's mode of life, in which only a single tapeworm may infect a host but must nonetheless be able to reproduce in order for the species to survive.

Most hermaphrodites do not reproduce by self-fertilization. Rather, as in earthworms, two animals copulate, and each inseminates the other. In some hermaphroditic species, self-fertilization is prevented by the development of testes and ovaries at different times. In the clam *Mercenaria mercenaria*, 98% of the population are males when they first reach maturity; later in their lives, they produce eggs and become functional females. A few continue to produce only sperm and remain functional males. In the American oyster *Crassostrea virginica*, sperm are mainly produced when the animal first matures. The next year, the animal produces eggs, and so on, in a regular annual alternation of sexes.

Certain fish exhibit a somewhat similar reproductive pattern in which sex is related to dominance. In one species of wrasse, the dominant fish is always male, and he lords his position over a harem of females. If he is removed or dies, one of the remaining female fish reverses sex and becomes the new patriarch. In another species of wrasse, the opposite situation exists: The dominant fish is always female, and the males change sex to achieve dominance if she is removed.

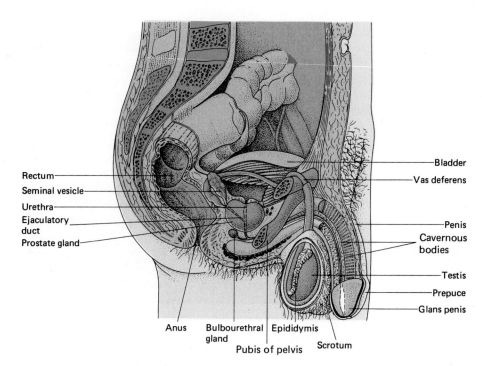

FIGURE 50–5 Anatomy of the human male reproductive system. The scrotum, penis, and pelvic region are shown in sagittal section to illustrate their internal structures.

Rectum
Seminal vesicle
Urethra
Ejaculatory duct
Prostate gland

Bladder
Vas deferens
Penis
Cavernous bodies
Testis
Prepuce
Glans penis

Anus
Bulbourethral gland
Epididymis
Pubis of pelvis
Scrotum

HUMAN REPRODUCTION: THE MALE

The reproductive responsibility of the human male, as in other mammals, is to produce sperm cells and deliver them into the female reproductive tract. When a sperm combines with an egg, it contributes half the genetic endowment of the offspring and determines its sex. The male reproductive system is illustrated in Figure 50–5. Male structures include the testes (which produce sperm and the hormone testosterone), the scrotum (which contains the testes), conducting tubes (which transport sperm from the testes to the outside of the body), accessory glands (which produce semen), and the penis (copulatory organ).

Sperm Production—The Testes

In humans and other vertebrates, **spermatogenesis,** the process of sperm cell production, occurs in the paired male gonads, or **testes.** More specifically, spermatogenesis takes place within the walls of a vast tangle of hollow tubules, the **seminiferous tubules,** within each testis (Figure 50–6). The seminiferous tubules are partially lined with primitive, undifferentiated stem cells called **spermatogonia.** These cells give rise to sperm cells (Figure 50–7).

During embryonic development and childhood, the spermatogonia divide by mitosis, producing more spermatogonia. At adolescence, some spermatogonia continue to divide by mitosis, but about half of them enlarge and become **primary spermatocytes.** These cells undergo meiosis. (You may want to review the discussion of meiosis in Chapter 9.) In many animals, gamete production occurs only in the spring or fall, but humans have no special breeding season. In the human adult male, spermatogenesis proceeds continuously and millions of sperm are produced daily.

Each primary spermatocyte undergoes a first meiotic division to produce **secondary spermatocytes** (Figure 50–8). In the second meiotic division, each secondary spermatocyte gives rise to two spermatids. Four sperma-

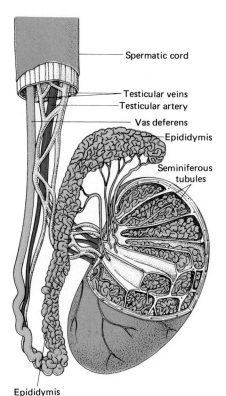

Spermatic cord

Testicular veins
Testicular artery
Vas deferens
Epididymis
Seminiferous tubules

Epididymis

FIGURE 50–6 Structure of the testis, epididymis, and spermatic cord. The testis is shown in sagittal section to illustrate the arrangement of the seminiferous tubules.

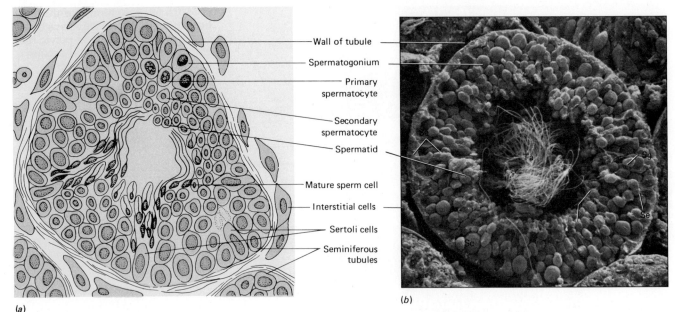

(a)

(b)

FIGURE 50–7 Structure of a semi-niferous tubule showing developing sperm cells in various stages of spermatogenesis. (*a*) Identify the sequence of sperm cell differentiation. Note the Sertoli cells and the interstitial cells. (*b*) A scanning electron micrograph of a transverse section through a seminiferous tubule (×580). *Se,* Sertoli cell; *Sc,* primary spermatocyte; *Sg,* spermatogonium.

Labels (a): Wall of tubule · Spermatogonium · Primary spermatocyte · Secondary spermatocyte · Spermatid · Mature sperm cell · Interstitial cells · Sertoli cells · Seminiferous tubules

tids are produced from the original primary spermatocyte. The spermatid is a fairly large cell with much more cytoplasm than is present in a mature sperm.

The haploid spermatid differentiates into a mature sperm by a rather complicated process. The nucleus shrinks, and part of the Golgi complex becomes the **acrosome,** which produces enzymes that help the sperm penetrate the egg (Figure 50–9). The two centrioles nestle in a small depression on the nuclear surface. One of these, the distal centriole, gives rise to the flagellum of the sperm. Although somewhat longer than most, the sperm flagellum is typical of eukaryote flagella, with the usual 9 + 2 arrangement of microtubules. Finally, most of the cytoplasm is discarded to be phagocytized by the large nutritive Sertoli cells present within the seminiferous tubules.

Sperm of different mammals vary somewhat. For example, the rat sperm has a hooklike head. Invertebrate sperm can vary radically from mammalian sperm. Crustacean sperm have no flagella at all and cannot propel themselves effectively. When they contact an egg, however, they grasp its surface by means of three pointed projections, like grappling hooks. Their middle piece is built like a spring, uncoiling to push the nucleus through the outer membrane of the egg. The intestinal parasitic nematode *Ascaris* has an amoeboid sperm (Figure 50–10).

Sperm cells cannot develop at body temperature. Although the testes develop within the abdominal cavity of the male embryo, about 2 months before birth they descend into the **scrotum,** a skin-covered sac suspended from the groin. The scrotum serves as a cooling unit, maintaining sperm at about 2°C below body temperature. In rare cases where the testes fail to descend, the seminiferous tubules eventually degenerate and the male becomes **sterile,** that is, unable to father offspring.

The scrotum is an outpocketing of the pelvic cavity and is connected to it by the **inguinal canals.** As the testes descend, they pull their blood vessels, nerves, and conducting tubes after them. These structures, encased by the cremaster muscle and by layers of connective tissue, constitute the **spermatic cord.** The inguinal region remains a weak place in the abdominal wall. Straining the abdominal muscles by lifting heavy objects may result in tearing of the inguinal tissue; a loop of intestine can then bulge into the scrotum through such a tear. This is an **inguinal hernia.**

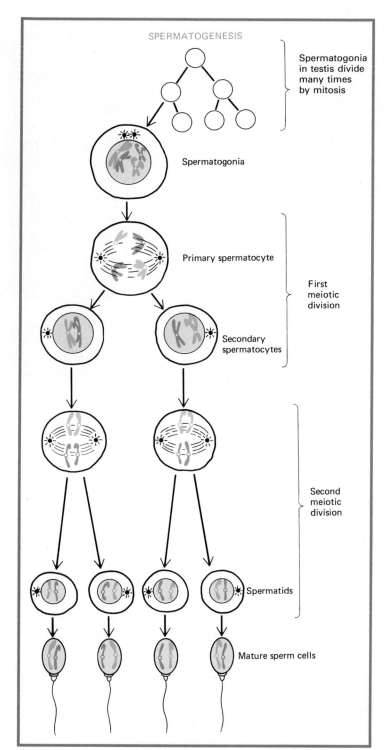

SPERMATOGENESIS

Spermatogonia in testis divide many times by mitosis

Spermatogonia

Primary spermatocyte

First meiotic division

Secondary spermatocytes

Second meiotic division

Spermatids

Mature sperm cells

FIGURE 50–8 Spermatogenesis. The primary spermatocyte undergoes meiosis, giving rise to four spermatids. The spermatids differentiate, becoming mature sperm cells.

Sperm Transport

Sperm cells leave the seminiferous tubules of the testes through a series of small tubules (vasa efferentia) that pass from the testis and empty into a larger tube, the epididymis. The **epididymis** of each testis is a complexly coiled tube in which sperm complete their maturation and are stored.

From each epididymis, sperm pass into a sperm duct, the **vas deferens,** which extends from the scrotum through the inguinal canal and into the pelvic cavity. Each vas deferens empties into a short **ejaculatory duct,**

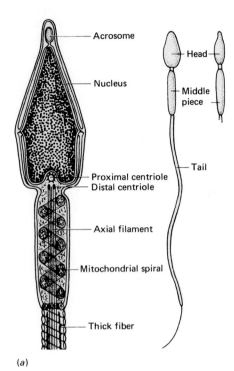

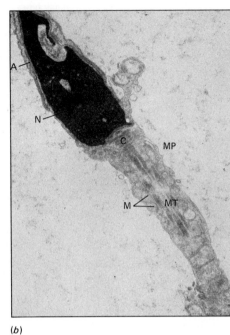

(a)

(b)

(c)

FIGURE 50–9 Sperm cell structure. (a) A mammalian sperm. *Left*, a head and middle piece. Structures shown would be visible through an electron microscope. *Middle and right diagrams*, top and side views of a sperm. (b) Electron micrograph of a human sperm cell (approximately ×37,500). *A*, acrosome; *N*, nucleus; *MP*, midpiece; *M*, mitochondria; *MT*, microtubules; *C*, centrioles. (c) Scanning electron micrograph of human sperm (approximately ×2400).

which passes through the prostate gland and then opens into the urethra. The single **urethra,** which at different times conducts either urine or semen, passes through the penis to the outside of the body.

Semen Production

As sperm are transported through the conducting tubes, they are mixed with secretions from the accessory glands. About 3.5 ml of **semen** is ejaculated during sexual climax. Semen consists of about 400 million sperm cells suspended in the secretions of the seminal vesicles, prostate gland, bulbo-urethral glands, and small glands in the walls of the ducts.

The paired **seminal vesicles** empty into the vasa deferentia, and the single **prostate gland** releases its alkaline secretion into the urethra. During

FIGURE 50–10 The intestinal parasite *Ascaris* has an unusual amoeboid sperm cell. It lacks a flagellum and moves by means of pseudopods. Visible here are the cytoskeleton, mitochondria, and dark granules of unknown function that may store food.

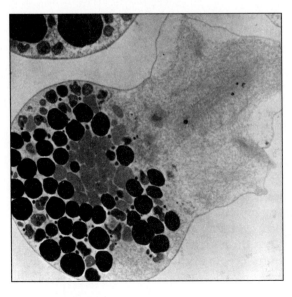

sexual arousal, the paired **bulbourethral glands** (also called Cowper's glands) release a few drops of alkaline fluid, which can neutralize the acidity of the urethra and aid in lubrication.

The Penis

The **penis** is an erectile copulatory organ adapted to deliver sperm into the female reproductive tract. It consists of a long **shaft** that enlarges to form an expanded tip, the **glans.** Part of the loose-fitting skin of the penis folds down and covers the proximal portion of the glans, forming a cuff called the **prepuce,** or **foreskin.** In the operation termed **circumcision** (commonly performed on male babies for either hygienic or religious reasons), the foreskin is removed.

Under the skin, the penis consists of three parallel columns of **erectile tissue,** sometimes called **cavernous bodies** (corpora cavernosa) (Figure 50–11). One of these columns surrounds the portion of the urethra that passes through the penis. The erectile tissue consists of large blood vessels called venous sinusoids. When the male is sexually stimulated, nerve impulses cause the arteries of the penis to dilate. Blood rushes into the vessels of the erectile tissue, causing the tissue to swell. This compresses veins that conduct blood away from the penis, slowing the outflow of blood through them. Thus, more blood enters the penis than can leave, causing the erectile tissue to become further engorged with blood. The penis thus becomes erect, that is, longer, larger in circumference, and firm. Though the human penis does not contain any bone, penis bones do occur in some other mammals, such as bats.

Reproductive Hormones in the Male

At about age 10, the hypothalamus begins to mature in its function of regulating sex hormones. It secretes a releasing hormone which stimulates

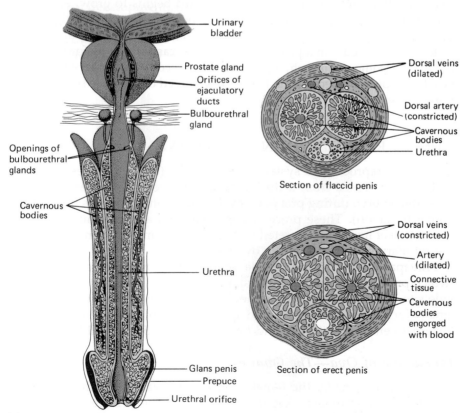

(a)

(b)

FIGURE 50–11 Internal structure of the penis. (*a*) Longitudinal section through the prostate gland and penis. (*b*) Cross section through flaccid and erect penis. Note that the erectile tissues of the corpora cavernosa (cavernous bodies) are engorged with blood in the erect penis.

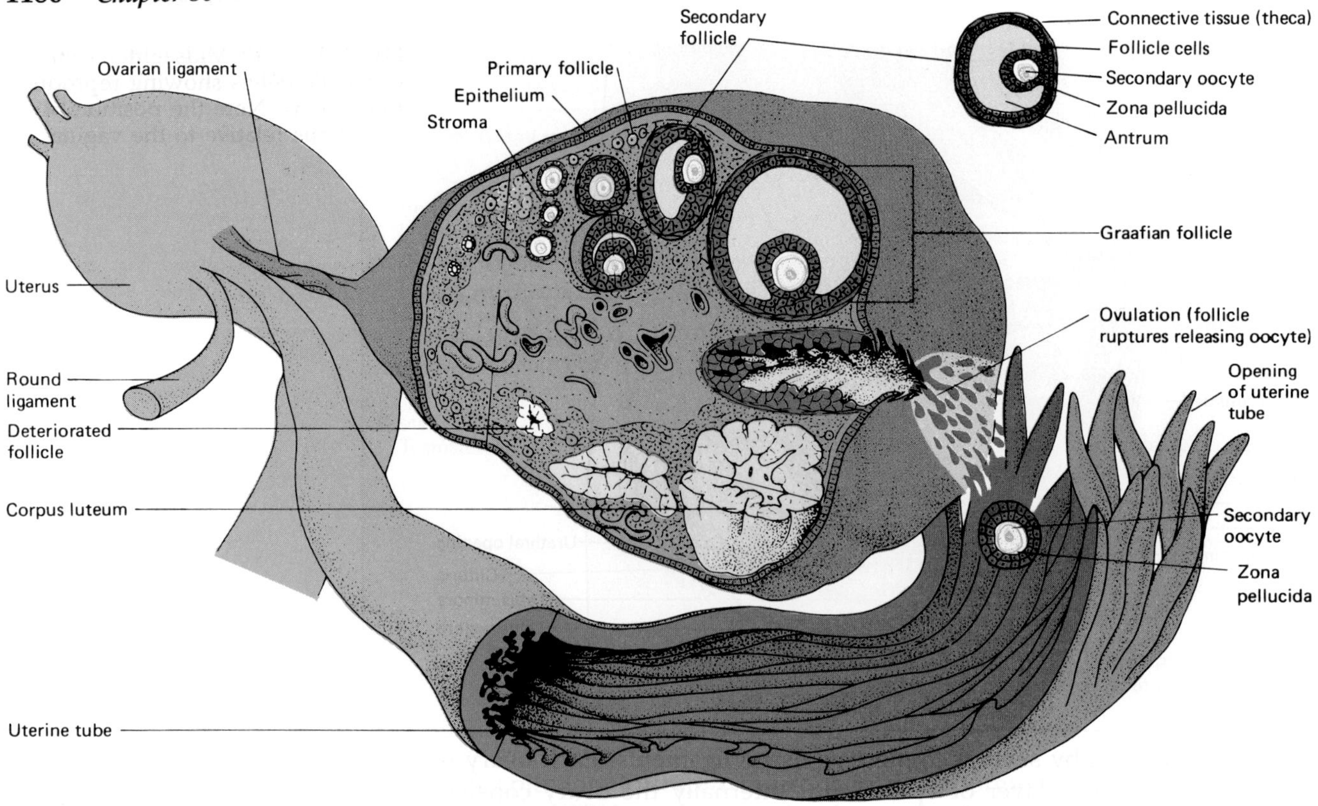

FIGURE 50–14 Microscopic structure of the ovary. Follicles in various stages of development are scattered throughout the ovary.

With the onset of puberty, a few of the follicles develop each month in response to FSH secreted by the anterior pituitary. As the follicle grows, the primary oocyte completes its first meiotic division, producing two cells that are markedly disproportionate in size (Figure 50–15). The smaller one, the **first polar body,** may later divide to form two polar bodies, but these eventually disintegrate. The larger cell, the **secondary oocyte,** proceeds to the second meiotic division but halts in metaphase until it is fertilized. When meiosis does continue, the second meiotic division gives rise to a single ovum and a second polar body. The polar bodies are small and apparently serve to dispose of unneeded chromosomes with a minimum amount of cytoplasm.

Recall that in the male large numbers of sperm are needed and in spermatogenesis each primary spermatocyte gives rise to four sperm. In contrast, only a few ova are needed during the reproductive life of a female and, accordingly, each primary oocyte generates only one ovum.

As an oocyte develops, it becomes separated from its surrounding follicle cells by a thick membrane, the **zona pellucida.** The follicle cells themselves proliferate so that the follicle grows in size. As the follicle develops, follicle cells secrete fluid, which collects in a space (antrum) created between them (Figure 50–16). Connective tissue from the stroma surrounds the follicle cells; this tissue contains cells that secrete the steroid sex hormone estradiol.

As a follicle matures, it moves closer to the surface of the ovary, eventually resembling a fluid-filled blister on the ovarian surface (Figure 50–16b). Such mature follicles are called **Graafian follicles.** Normally only one follicle fully matures each month. Several others may develop for about a week and then deteriorate. These remain in the ovary as atretic (degenerate) follicles.

At **ovulation,** the mature ovum is ejected through the wall of the ovary and into the pelvic cavity. The portion of the follicle that remains in the ovary develops into the **corpus luteum,** a temporary endocrine gland.

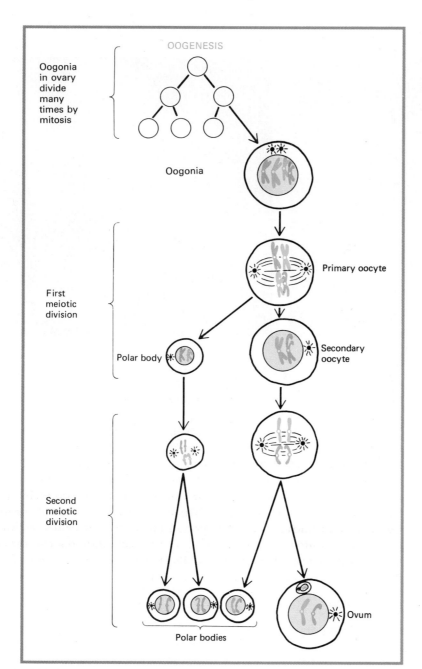

FIGURE 50–15 Oogenesis. Only one functional ovum is produced from each primary oocyte. The other cells produced are polar bodies that degenerate.

OOGENESIS

Oogonia in ovary divide many times by mitosis

Oogonia

First meiotic division

Primary oocyte

Polar body

Secondary oocyte

Second meiotic division

Polar bodies

Ovum

Ovum Transport—The Uterine Tubes

Almost immediately after ovulation, the ovum passes into the funnel-shaped opening of the **uterine tube** (also called fallopian tube or oviduct). Peristaltic contractions of the muscular wall of the uterine tube and beating of the cilia in its lining help to move the ovum along toward the uterus. Fertilization takes place within the uterine tube. If fertilization does not occur, the ovum degenerates there.

The Uterus

The uterine tubes empty into the upper corners of the pear-shaped uterus (see Figure 50–13). About the size of a fist, the **uterus** occupies a central position in the pelvic cavity. This organ has thick walls of smooth muscle and a mucous lining, the **endometrium,** which thickens each month in preparation for possible pregnancy. If an ovum is fertilized, the tiny em-

Follicle cells Ovary surface

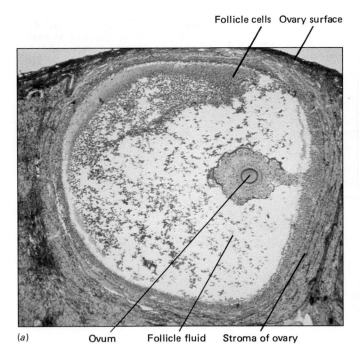

(a) Ovum Follicle fluid Stroma of ovary

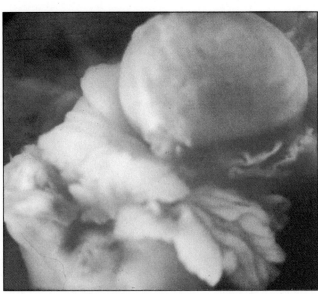

(b)

FIGURE 50–16 Follicle development. (a) A stained section through a developing follicle. The ovum is surrounded by a layer of follicle cells that will be released along with it. These follicle cells become the corona radiata, a layer that acts as a barrier to sperm cells and that may help ensure that the egg is fertilized by only one of the many sperm that approach it. (b) The mature follicle develops just under the surface of the ovary, producing a tightly stretched fluid-filled blister (shown here from the outside) that will eventually burst, releasing the ovum.

bryo finds its way into the uterus and is implanted in the endometrium. Here it grows and develops, sustained by nutrients and oxygen delivered by surrounding maternal blood vessels. If fertilization does not occur during the monthly cycle, the endometrium sloughs off and is discharged. This process is called **menstruation.**

The lower portion of the uterus, called the **cervix,** projects slightly into the vagina. The cervix is a common site of cancer in women. Detection is usually possible by the routine Papanicolaou test (Pap smear) in which a few cells are scraped from the cervix during a regular gynecological examination and studied microscopically. About 50% of cervical cancer is now detected in this way at very early stages of malignancy, when the patient can be cured.

The Vagina

The **vagina** is an elastic, muscular tube that extends from the uterus to the exterior of the body. The vagina serves as a receptacle for sperm during sexual intercourse and as part of the birth canal when development of the fetus is complete.

External Genital Structures

The external female sex organs, collectively known as the **vulva,** include liplike folds, the **labia minora,** which surround the opening to the vagina and the opening to the urethra (Figure 50–17). External to the delicate labia minora are the heavier, hairy **labia majora.** Anteriorly the labia minora merge to form the prepuce of the **clitoris,** a very small erectile structure comparable to the male glans penis. Like the penis, the clitoris contains erectile tissue that becomes engorged with blood during sexual excitement. Rich in nerve endings, the clitoris serves as a center of sexual sensation in the female.

The **mons pubis** is the mound of fatty tissue just above the clitoris at the junction of the thighs and torso. At puberty it becomes covered by coarse pubic hair. The **hymen** is a thin ring of tissue that may partially block the entrance to the vagina. It is often ruptured during a woman's first sexual intercourse. However, the hymen may be destroyed by strenuous exercise

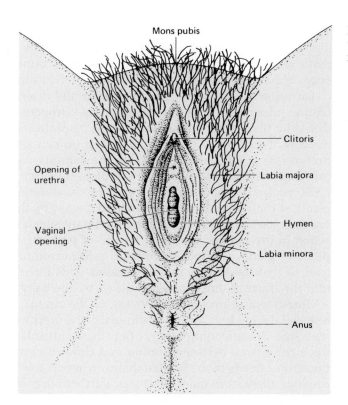

FIGURE 50–17 The vulva, the external genital structures of the female.

Mons pubis

Clitoris

Opening of urethra

Labia majora

Hymen

Vaginal opening

Labia minora

Anus

during childhood or by the use of tampons inserted into the vagina to absorb the menstrual flow.

The Breasts

The breasts, which contain the mammary glands, overlie the pectoral muscles and are attached to them by connective tissue. Fibrous bands of tissue called **ligaments of Cooper** firmly connect the breasts to the skin. Each breast is composed of 15 to 20 lobes of glandular tissue, further subdivided into lobules made of connective tissue in which gland cells are embedded. The secretory cells are arranged in little grapelike clusters called **alveoli** (Figure 50–18). Ducts from each cluster unite to form a single duct from

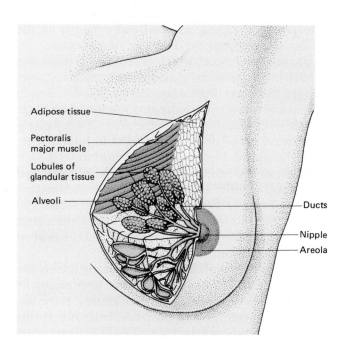

FIGURE 50–18 The mature human female breast.

Adipose tissue

Pectoralis major muscle

Lobules of glandular tissue

Alveoli

Ducts

Nipple

Areola

each lobe, so that there are 15 to 20 tiny openings on the surface of each nipple. The amount of adipose tissue around the lobe of the glandular tissue determines the size of the breasts and accounts for their soft consistency. The size of the breasts does not affect their capacity to produce milk.

Lactation is the production of milk for the nourishment of the young. During pregnancy, high concentrations of estrogens and progesterone produced by the corpus luteum and by the placenta stimulate the glands and ducts of the breast to develop, resulting in increased breast size. For the first couple of days after childbirth, the mammary glands produce a fluid called **colostrum,** which contains protein and lactose but little fat. After birth, prolactin, a hormone secreted by the anterior lobe of the pituitary gland, stimulates milk production, and usually by the third day after delivery milk itself is produced. When the infant suckles at the breast, a reflex action in the mother results in release of prolactin and oxytocin from the pituitary gland. Oxytocin stimulates cells surrounding the alveoli to contract so that the alveoli are compressed. This forces milk from the alveoli into the ducts where it can be sucked by the baby.

Breastfeeding promotes recovery of the uterus, because the oxytocin released during breastfeeding stimulates the uterus to contract to nonpregnant size. Breastfeeding a baby offers many advantages, including promoting a close bond between mother and child. Breast milk is tailored to the nutritional needs of the human infant, whereas cow's milk is more likely to produce allergies to dairy products. Furthermore, breastfed babies receive antibodies in the colostrum and mother's milk that are thought to play a protective role, resulting in a lower incidence of infantile diarrhea and even of respiratory infection during the second 6 months of life.

Hormonal Control of the Menstrual Cycle

As a female approaches puberty, the anterior pituitary secretes the gonadotropic hormones FSH and LH, which signal the ovaries to become active. Interaction of FSH and LH with estrogen and progesterone from the ovaries regulates the **menstrual cycle,** the monthly chain of events that prepares the body for possible pregnancy. The menstrual cycle runs its course every month from puberty until **menopause,** the end of a women's reproductive (though not sexually active) life.

Although there is wide variation, a typical menstrual cycle is 28 days long (Figure 50–19). The first day of the cycle is marked by menstruation, the monthly discharge through the vagina of blood and tissue from the endometrium. Ovulation occurs on about the 14th day of the cycle.

During the **menstrual phase** of the cycle, which lasts about 5 days, the pituitary gland releases FSH in response to gonadotropin releasing hormone (GnRH) from the hypothalamus. FSH stimulates a few follicles to develop in the ovary. Cells in the connective tissue surrounding the follicle cells secrete the steroid hormone β-estradiol during the **preovulatory phase** of the menstrual cycle. (Estradiol is the principal estrogen in humans; estrogens are a group of closely related 18-carbon steroid hormones.)

Estradiol stimulates growth of the endometrium, which thickens and develops new blood vessels and glands. The sharp rise in the concentration of estradiol in the blood stimulates the anterior lobe of the pituitary to secrete LH. Together LH and FSH stimulate ovulation. LH then stimulates the portion of the follicle that remains in the ovary after the ovum has been ejected to develop into a corpus luteum.

The corpus luteum produces both estradiol and **progesterone** during the **postovulatory phase.** These hormones stimulate the uterus to continue its preparation for pregnancy. Progesterone stimulates tiny glands in the endometrium to secrete a fluid rich in nutrients. If the ovum is fertilized, this nutritive fluid nourishes the early embryo when it arrives in the uterus on

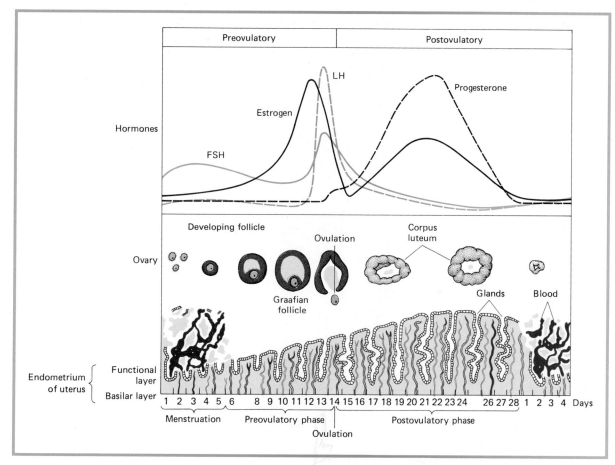

FIGURE 50–19 The menstrual cycle. The events that take place within the pituitary, ovary, and uterus are precisely synchronized. When fertilization does not occur, the cycle repeats itself about every 28 days. Compare this illustration with Figure 50–20.

about its 4th day of development (Figure 50-20). On about the 7th day after fertilization, the embryo begins to implant in the thick endometrium. Membranes that develop around the embryo secrete **human chorionic gonadotropin (HCG),** a hormone that signals the mother's corpus luteum to continue to function.

If the ovum is not fertilized, the corpus luteum begins to degenerate, and the concentrations of progesterone and estradiol in the blood fall dramatically. Small arteries in the endometrium constrict, and the tissue they supply becomes ischemic (oxygen-deprived). As cells die and damaged arteries rupture and bleed, menstruation begins once again.

Table 50–2 lists the actions of the pituitary and ovarian reproductive hormones. Note that, like testosterone in the male, estradiol is responsible for the growth of the sex organs at puberty, for body growth, and for development of secondary sexual characteristics. In the female these include breast development, broadening of the pelvis, and the characteristic development and distribution of muscle and fat responsible for the shape of the female body.

Some testosterone is produced in the female, mainly in the adrenal cortex. This hormone is largely responsible for the adolescent growth spurt and for development of pubic and axillary hair in the female (as well as in the male). In the male some estradiol is produced in the testis.

Details of the hormonal interaction that regulates the menstrual cycle are not fully known. The releasing hormone, GnRH, secreted by the hypothalamus in bursts, stimulates the secretion of FSH and LH from the anterior lobe of the pituitary (Figure 50–21). FSH stimulates the early maturation of the follicles in the ovary, and FSH and LH together stimulate the final maturation of a follicle. The rise in estradiol secreted by the developing

FIGURE 50–20 The menstrual cycle is interrupted when pregnancy occurs. The corpus luteum does not degenerate, and menstruation does not take place. Instead, the wall of the uterus remains thickened so that the embryo can develop within it.

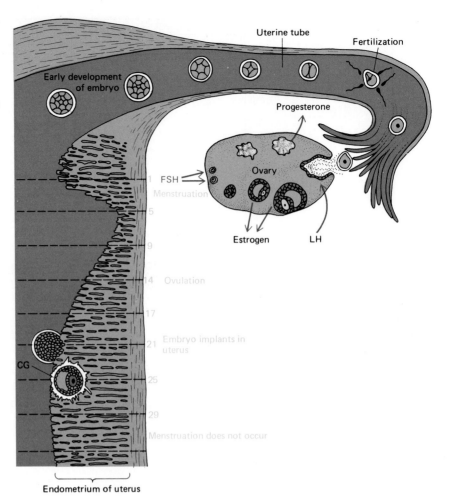

Table 50–2 PRINCIPAL REPRODUCTIVE HORMONES AND THEIR ACTIONS IN FEMALES

Endocrine Gland and Hormone	Principal Target Tissue	Principal Actions
Anterior pituitary Follicle-stimulating hormone (FSH)	Ovary	Stimulates development of follicles; with LH, stimulates secretion of estradiol and ovulation
Luteinizing hormone (LH)	Ovary	Stimulates ovulation and development of corpus luteum
Prolactin	Breast	Stimulates milk production (after breast has been prepared by estradiol and progesterone)
Ovary Estradiol	General	Growth of body and sex organs at puberty; development of secondary sex characteristics (breast development, broadening of pelvis, distribution of fat and muscle)
	Reproductive structures	Maturation; monthly preparation of the endometrium for pregnancy; makes cervical mucus thinner and more alkaline
Progesterone	Uterus	Completes preparation of and maintains endometrium for pregnancy
	Breasts	Stimulates development of mammary glands

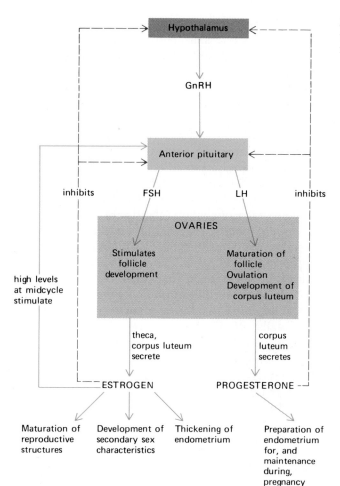

FIGURE 50–21 Hormones that regulate the menstrual cycle. Red arrows indicate inhibition.

follicles causes the burst in LH secretion that stimulates ovulation and formation of the corpus luteum. The high constant levels of estradiol and progesterone maintained by the corpus luteum have a negative feedback effect on FSH and LH secretion. The feedback effects of estradiol and progesterone are thought to act directly on the pituitary gland. Once the corpus luteum begins to degenerate, estrogen and progesterone levels fall, and the secretion of FSH and LH then increase once again.

Estradiol–progesterone imbalance has been suggested as the basis of **premenstrual syndrome (PMS),** a condition experienced by some women several hours to 10 days before menstruation and ending a few hours after onset of menstruation. Symptoms include fatigue, anxiety, depression, irritability, headache, edema, and skin eruptions.

PHYSIOLOGY OF SEXUAL RESPONSE

Human reproduction is accomplished sexually by the union of ovum and sperm. During copulation, also called **coitus** or sexual intercourse in humans, the male deposits semen into the upper end of the vagina. The complex structures of the male and female reproductive systems and the complex physiological, endocrine, and psychological phenomena associated with sexual activity are adaptations that promote the successful union of sperm and ovum and the subsequent development and nurturing of the resulting embryo.

Sexual stimulation results in two basic physiological responses: vasocongestion and increased muscle tension. **Vasocongestion** occurs as blood flow is increased to the genital structures and to certain other tissues such as the breasts and skin. During vasocongestion, erectile tissues within the penis and clitoris, as well as in other areas of the body, become engorged with blood.

Sexual response can be divided into four phases: **excitement, plateau, orgasm,** and **resolution.** Sexual excitement can result from psychological or physical stimulation, or usually both. Such stimulation, often referred to as sexual foreplay, promotes vasocongestion and increased muscle tension. Before the penis can enter the vagina and function in coitus, it must be erect; accordingly, penile erection is the first male response to sexual excitement. In the female, vaginal lubrication is the first response to effective sexual stimulation; the lubricating fluid is produced in the vaginal wall owing to vasocongestion (glands are not present in the vagina). During the excitement phase, the vagina lengthens and expands in preparation for receiving the penis; the clitoris and breasts become vasocongested, and the nipples become erect.

If erotic stimulation continues, sexual excitement heightens to the plateau phase. Vasocongestion and muscle tension increase markedly. In the female the inner two thirds of the vagina continue to expand and lengthen. The walls of the outer one third of the vagina become greatly vasocongested so that the vaginal entrance becomes somewhat constricted. In this vasoconstricted state, the outer one third of the vagina is referred to as the orgasmic platform. In the male the penis increases in circumference. In both sexes, blood pressure increases and heart rate and breathing are accelerated.

Coitus may begin during the excitement or plateau phase. During coitus the penis is moved back and forth in the vagina by movements referred to as pelvic thrusts. Physical and psychological sensations resulting from this friction (and from the entire intimate experience between the partners) lead to orgasm, the climax of sexual excitement.

Although lasting only a few seconds, orgasm is the phase of maximum sexual tension and its release. In the male, orgasm is marked by **ejaculation** of semen. Contractions of the vas deferens propel sperm into the ejaculatory ducts. The accessory glands contract, adding their secretions; then, contractions of the ejaculatory ducts, urethra, and certain muscles of the pelvic floor eject the semen from the penis. Muscular contractions continue at about 0.8-second intervals for several seconds. After the first few contractions, their intensity decreases, and they become less regular and less frequent.

In the female, stimulation of the clitoris is important in heightening the sexual excitement that leads to orgasm. Orgasm is marked by rhythmic contractions of the pelvic muscles and the orgasmic platform, starting at approximately 0.8-second intervals and recurring 5 to 12 times. After the first 3 to 6 contractions, their intensity decreases, and the time interval between them increases. No fluid ejaculation accompanies orgasm in the female. In both sexes, heart rate and respiration more than double, and blood pressure rises markedly both just before and during orgasm.

Orgasm is followed by the resolution phase, during which muscle relaxation and detumescence (the subsiding of swelling) restore the body to its normal state. In most males, there is a refractory period during which physiological response to sexual stimulation does not occur. Duration of the refractory period varies in different individuals and also in different situations. Most women are able to repeat the sexual cycle immediately, reaching orgasm again if appropriately stimulated.

Erection of the penis is necessary for effective coitus. Chronic inability to sustain an erection is termed **erectile dysfunction** (formerly referred to as impotence). This disorder is often associated with psychological issues.

FERTILIZATION

Fertilization is the fusion of sperm and ovum to produce a zygote. The process of fertilization and the subsequent establishment of pregnancy together are referred to as **conception.** When conditions in the vagina and cervix are favorable, sperm begin to arrive at the site of fertilization in the upper uterine tube within 5 minutes after ejaculation. Contractions of the uterus and uterine tubes (perhaps stimulated by prostaglandins in the semen) are thought to help transport the sperm. The sperms' own motility is probably important in approaching and fertilizing the ovum.

If only one sperm is needed to fertilize an ovum, why are millions involved in each act of coitus? For one thing, sperm movement is undirected, so that many lose their way. Others die as a result of unfavorable pH or phagocytosis by leukocytes and macrophages in the female tract. Only a few thousand succeed in traversing the correct uterine tube and reaching the vicinity of the ovum. Additionally, large numbers of sperm may be necessary to penetrate the covering of follicle cells, the **corona radiata,** that surrounds the ovum. Each sperm is thought to release small amounts of enzymes from its acrosome that help break down the cement-like substance holding the follicle cells together.

Once one sperm has entered the ovum, there is a rapid electrical change, followed by a slower chemical change in the cell membrane of the ovum that prevents the entrance of other sperm. As the fertilizing sperm enters the ovum, it usually loses its tail (Figure 50-22). Sperm entry stimulates the ovum to complete its second meiotic division. The head of the haploid sperm then swells to form the **male pronucleus** and fuses with the female pronucleus, forming the diploid nucleus of the zygote (Figure 50–23). The process of fertilization is described in more detail in Chapter 51.

After ejaculation into the female reproductive tract, sperm remain alive and retain their ability to fertilize an ovum for about 48 hours. The ovum itself remains fertile for only about 24 hours after ovulation. Thus, there are only about three days during each menstrual cycle (days 12 to 15 in a very regular 28-day cycle) when sexual intercourse is likely to result in fertilization.

In view of the many factors working against fertilization, it may seem remarkable that it ever occurs! Yet the frequency of coitus and the large number of sperm deposited at each ejaculation enable the human species not only to maintain itself but to increase its numbers at an alarming rate.

STERILITY

Men with fewer than 20 million sperm per milliliter of semen are usually considered to be sterile. When a couple's attempts to produce a child are unsuccessful, a sperm count and analysis may be performed in a clinical laboratory. Sometimes semen is found to contain large numbers of abnormal sperm or occasionally none at all. In about one fourth of the cases of mumps in adult males, the testes become inflamed; in some of these cases the spermatogonia die, resulting in permanent sterility. Low sperm counts also have been linked to cigarette smoking, and studies show that men who smoke tobacco are also more likely than nonsmokers to produce abnormal sperm. Exposure to chemicals such as DDT and PCBs may also result in low sperm count and sterility.

One cause of female sterility is scarring of the uterine tubes. Inflammation of the uterine tubes, frequently caused by gonorrhea (see Focus on Sexually Transmitted Diseases), may result in scarring, which blocks the tubes so that the ovum cannot pass to the uterus. Sometimes partial constriction of the uterine tube results in tubal pregnancy, in which the em-

FIGURE 50–22 Fertilization. (*a*) Each sperm is thought to release a small amount of enzyme that helps to disperse the follicle cells surrounding the ovum. (*b*) After a sperm cell enters the ovum, the ovum completes its second meiotic division, producing an ovum and a polar body. (*c*) Pronuclei of sperm and ovum combine, producing a zygote with the diploid number of chromosomes. (*d*) A scanning electron micrograph of a sperm cell fertilizing a hamster ovum.

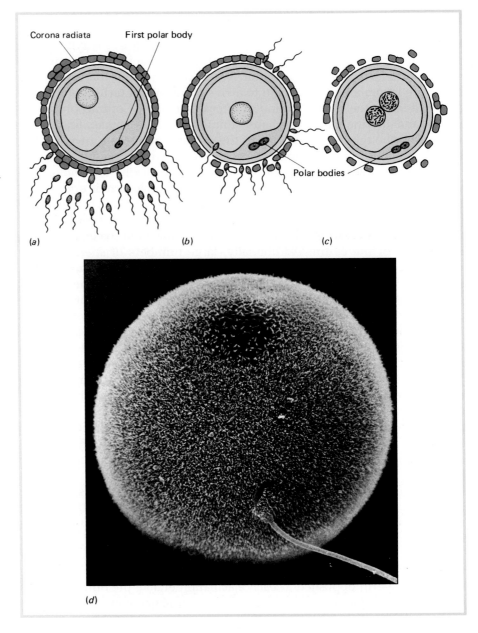

Corona radiata   First polar body

Polar bodies

(*a*)   (*b*)   (*c*)

(*d*)

bryo begins to develop in the wall of the uterine tube because it cannot progress to the uterus. Uterine tubes are not adapted to bear the burden of a developing embryo; thus, the uterine tube and the embryo it contains must be surgically removed before it ruptures and endangers the life of the mother.

FIGURE 50–23 Light micrograph of human ovum surrounded by sperm (approximately ×600).

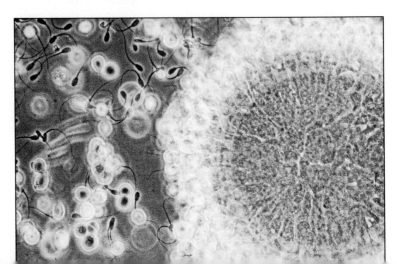

Women with blocked uterine tubes usually can produce ova and can foster embryonic development in the uterus. However, they need clinical assistance in getting the ovum from the ovary to the uterus. The ovum can be removed from the ovary, fertilized with the husband's sperm in laboratory glassware (in vitro fertilization), and then placed in the woman's uterus, where it may develop normally (see Focus on Novel Origins).

BIRTH CONTROL

Populations of various animal species are limited under natural conditions by a variety of mechanisms, even apart from such things as disease and food shortage. Under crowded conditions, parental neglect, feeding of only the strong offspring, and cannibalism are three ways in which species control their populations. In some species spontaneous abortion, genetic deterioration, or death by stress occurs.

Among human populations, some use of abortion as a means of birth control has been found among every population studied. Even infanticide has been practiced, especially in primitive societies. For centuries, human beings have searched for more desirable methods of birth control. Highly effective contraceptive (literally, "against conception") methods are now available, but all have side effects, inconveniences, or other disadvantages (see Table 50–3). The ideal contraceptive has not yet been developed.

Oral Contraceptives: The Pill

More than 80 million women worldwide use **oral contraceptives** (more than 8 million in the United States alone). The most common preparations consist of a combination of synthetic progestin and synthetic estrogen. (Natural hormones are destroyed by the liver almost immediately, but the synthetic ones resist destruction.) Starting on the 5th day of the menstrual cycle, a woman takes one pill each day for about 3 weeks. She then stops taking the pill, and menstruation begins about 3 days later. Ovulation does not occur if she resumes the pills on the 5th day of her cycle. When taken correctly, these pills are about 99.9% effective in preventing pregnancy.

Most oral contraceptives prevent pregnancy by preventing ovulation. By maintaining postovulatory levels of ovarian hormones in the blood, the pituitary gland is inhibited and does not produce the surge of FSH and LH that stimulates ovulation. Some advantages of oral contraceptives are summarized in Table 50–3.

The Intrauterine Device (IUD)

The **intrauterine device (IUD)** is a small plastic loop or coil that must be inserted into the lumen of the uterus by a medical professional. Once in place, it can be left in the uterus indefinitely or until the woman wishes to conceive. Some types of IUDs are about 99% effective. The mode of action of the IUD is not well understood, but it is thought that the IUD sets up a minor local inflammation in the uterus, attracting macrophages, which destroy the tiny embryo before it implants. IUDs that contain copper may interfere with both migration of sperm and implantation of the embryo. Because of side effects such as pelvic inflammatory disease and increased risk of infertility, IUDs are being prescribed less frequently, and are no longer manufactured in the United States.

Other Common Methods

Other common contraceptive methods are listed and compared in Table 50–3. Two highly effective methods that present physical barriers to sperm are the contraceptive diaphragm and the condom. The **contraceptive dia-**

Focus on SOME SEXUALLY TRANSMITTED DISEASES

Sexually transmitted diseases, or STD (commonly referred to as venereal diseases), are, next to the common cold, the most communicable diseases in the world. **AIDS,** described by President Reagan as "Public Enemy Number One," is discussed in Chapter 43 in the Focus on AIDS.

Gonorrhea

About 1 million patients with **gonorrhea** are treated each year in the United States, making it the most prevalent communicable disease except for the common cold. One reason for its frequency is that there are many strains of the causative organism. Although a person may develop immunity to one strain after infection, reinfection may occur following exposure to another strain.

Gonorrhea is caused by a bacterium, *Neisseria gonorrhoeae*. Infection most often results from direct sexual contact, but it may rarely be transmitted indirectly by contact with towels or toilet seats bearing the infective bacteria.

From 2 to 31 days after exposure, the bacteria produce exotoxin (antigens), which may cause redness and swelling at the site of infection. In the male the infection usually begins in the urethra. Symptoms include frequent urination, accompanied by a severe burning sensation, and a profuse discharge from the penis. Infection may spread to the prostate, seminal vesicles, and epididymides. Abscesses may develop in the epididymides, causing extensive destruction that may lead to sterility.

In about 60% of infected women no symptoms occur, so that treatment is sought only after the woman learns that a sexual partner is infected. The urethra and cervix provide favorable environments for development of the bacteria. If untreated, the infection may spread to the uterine tubes. This type of infection tends to heal, leaving dense scar tissue, which may effectively close the walls of the uterine tubes and result in sterility. In both sexes, untreated gonorrhea may spread to other parts of the body, including heart valves, meninges, and joints. Joint involvement causes a type of arthritis.

Diagnostic criteria include clinical symptoms and positive identification of *N. gonorrhoeae* bacteria in cultures of body secretions. Penicillin is the drug of choice, but tetracyclines and other antibiotics have been used effectively. Strains of gonorrhea bacteria that are resistant to penicillin have been known since about 1976, and the tetracycline spectinomycin has been used as a back-up drug since that time. Recently, however, a strain of gonorrhea resistant to both of these drugs has been reported.

Syphilis

Although not as common as gonorrhea, **syphilis** is a more serious disease, causing death in 5% to 10% of those infected. It is caused by a slender, corkscrew-shaped spirochete, *Treponema pallidum*, and is contracted almost exclusively by direct sexual contact. *T. pallidum* enters the body through a tiny break in the skin or mucous membrane; within 24 hours it invades the lymph or circulatory system and spreads throughout the body.

The course of the disease may be divided into four stages: primary, secondary, latent, and advanced. The primary stage begins about 3 weeks after infection (although the time may vary considerably) with the appearance of a **primary chancre,** a small, painless ulcer at the site of infection, usually the penis in the male or the vulva, vagina, or cervix in the female (see figure). The primary chancre is highly infectious, and sexual contact at this time will almost certainly transmit the disease to an uninfected partner. In many cases the chancre goes unnoticed and spontaneously heals within about a month.

About 2 months after initial infection the secondary stage occurs, sometimes marked by an influenza-like

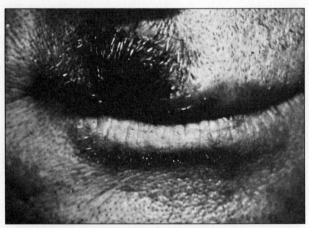

■ Clinical symptoms of syphilis. Primary syphilitic chancre. This is usually the first symptom.

phragm mechanically blocks the passage of sperm from the vagina into the cervix. Its effectiveness is increased by covering it with spermicidal jelly or cream; the diaphragm is inserted just prior to coitus (Figure 50–24). The most commonly used male contraceptive device, the **condom** holds the semen so that sperm cannot enter the female tract. The condom provides some protection against AIDS and other sexually transmitted diseases.

syndrome accompanied by a mild rash. Sometimes annular (circular) scaling lesions develop. Mucous membranes of the mouth and genital structures may exhibit lesions containing large populations of spirochetes. The secondary stage is the most highly contagious because the many lesions literally teem with spirochetes.

If untreated, the disease enters a latent (hidden) stage, which may last for 20 years or longer. Although there may be no obvious symptoms at this time, the spirochetes invade various organs of the body.

About a third of untreated patients recover spontaneously. Another third retain the disease in the latent form. The remaining third develop lesions, usually in the cardiovascular system, and less frequently in the CNS or elsewhere. Lesions known as **gummas** are characteristic of advanced syphilis; they occur mainly in skin, liver, bone, and spleen. Involvement of the cardiovascular system is the principal cause of death from syphilis.

In the early primary stage, syphilis can be diagnosed by darkfield examination, which involves identifying the spirochete (using microscopic techniques) in fluid collected from a lesion. Within 2 weeks after the appearance of the primary chancre, blood tests may be utilized for diagnoses. One of the most common tests in current use is the VDRL (Venereal Disease Research Laboratory) serological test, which is based on the presence of an antibody-like substance, **reagin,** that appears in the blood in response to the infection.

As with gonorrhea, penicillin is the drug of choice. When treated in its early stages, syphilis can be completely cured. Late syphilis is more difficult to cure, and lesions may have already caused permanent damage to various organs.

Genital Herpes

Genital herpes is an increasingly common disease caused by herpes simplex virus type 2. (This is not to be confused with the herpes simplex type 1 virus, which causes ordinary cold sores.) Symptoms usually appear within a week after infection. First infections are often very mild, however, and may even go unnoticed. When they occur, symptoms take the form of tiny, painful blisters on the genital structures. The blisters may break down into ulcers. Fever, enlarged lymph nodes, and other influenza-like symptoms may appear. The ulcers usually heal within a few days or weeks.

In some patients the disease leaves permanently; in others it recurs periodically (although usually with decreasing severity) for many years. It is thought that the virus may be harbored in the dorsal root ganglia, which receive sensory fibers from the reproductive structures.

There is no cure for genital herpes, although some drugs, such as acyclovir, lessen the duration and severity of initial infections, and oral forms of the drug show promise in preventing recurrence. Pain medications are often prescribed, and sulfa drugs are used to treat secondary bacterial infections that may develop. Because a link has been shown between genital herpes and cervical cancer, women with this disease should have annual Pap smears.

Trichomoniasis

Trichomoniasis is a common infection of the genital tract caused by the flagellate protozoan *Trichomonas vaginalis.* It can easily be transmitted by dirty toilet seats and towels as well as sexual contact. This disease affects about 20% of women during their reproductive years. Symptoms include vaginal itch and vaginal discharge in women. Males may not have symptoms.

Pelvic Inflammatory Disease

Pelvic inflammatory disease (PID) is an infection of the cervix, uterus, uterine tubes, or ovaries. It results most commonly from infection transmitted by intercourse but also from abortion procedures performed under unsanitary conditions. Patients with IUDs are especially susceptible. PID, now the most common sexually transmitted disease in the United States, is often caused by chlamydial infections. PID may now be the most common cause of sterility in women, since it leaves women sterile in more than 15% of cases.

Nongonococcal Urethritis

Nongonococcal urethritis (NGU), also known as nonspecific urethritis, is an inflammation of the urethra that is not caused by the bacterium *Neisseria gonorrhoeae.* NGU is frequently caused by *Chlamydia* bacteria and responds to treatment with tetracycline. Symptoms of this disease include swelling and narrowing of the urethra so that urination becomes difficult, burning pain accompanying urination, and sometimes a pus-containing discharge.

Sterilization

Short of abstinence, the safest method of birth control is **sterilization,** which renders an individual permanently incapable of parenting offspring. About 75% of sterilization procedures are currently performed on men.

Table 50-3 BIRTH CONTROL METHODS

Method	Failure Rate*	Mode of Action	Advantages	Disadvantages
Oral contraceptives	0.3; 5	Prevents ovulation; may also affect endometrium and cervical mucus and prevent implantation	Highly effective; sexual freedom; regular menstrual cycle	Minor discomfort in some women; possible thromboembolism; hypertension, heart disease in some users; possible increased risk of infertility; should not be used by women who smoke
Intrauterine device (IUD)	1; 5	Not known; probably stimulates inflammatory response	Provides continuous protection; highly effective	Cramps; increased menstrual flow; spontaneous expulsion; prescribed less frequently due to increased risk of pelvic inflammatory disease and infertility; no longer manufactured in United States
Spermicides (sponges, foams, jellies, creams)	3; 20	Chemically kill sperm	No side effects (?) Vaginal sponges are effective in vagina for up to 24 hours after insertion. Sponges also act as physical barriers to sperm cells.	Some evidence linking spermicides to birth defects
Contraceptive diaphragm (with jelly)**	3; 14	Diaphragm mechanically blocks entrance to cervix; jelly is spermicidal	No side effects	Must be prescribed (and fitted) by physician; must be inserted prior to coitus and left in place for several hours after intercourse
Condom	2.6; 10	Mechanical; prevents sperm from entering vagina	No side effects; some protection against STD, including AIDS	Interruption of foreplay to fit; slightly decreased sensation for male; could break
Rhythm†	13; 21	Abstinence during fertile period	No side effects (?)	Not very reliable
Douche	40	Flush semen from vagina	No side effects	Not reliable; sperm beyond reach of douche in seconds
Withdrawal (coitus interruptus)	9; 22	Male withdraws penis from vagina prior to ejaculation	No side effects	Not reliable; contrary to powerful drives present when an orgasm is approached. Sperm present in fluid secreted before ejaculation may be sufficient for conception.
Sterilization Tubal ligation	0.04	Prevents ovum from leaving uterine tube	Most reliable method	Usually not reversible
Vasectomy	0.15	Prevents sperm from leaving scrotum	Most reliable method	Usually not reversible
Chance (no contraception)	About 90			

*The lower figure is the failure rate of the method; the higher figure is the rate of method failure plus failure of the user to utilize the method correctly. Based on number of failures per 100 women who use the method per year in the United States.
**Failure rate is lower when used together with spermicidal foam.
†There are several variations of the rhythm method. For those who use the calendar method alone, the failure rate is about 35. However, by taking the body temperature daily and keeping careful records (temperature rises after ovulation), the failure rate can be reduced. Also, by keeping a daily record of the type of vaginal secretion, changes in cervical mucus can be noted and used to determine time of ovulation. This type of rhythm contraception is also slightly more effective. When women use the temperature or mucus method and have intercourse *only* more than 48 hours *after* ovulation, the failure rate can be reduced to about 7.

The most common method of male sterilization is **vasectomy** (Figure 50–25), in which each vas deferens is severed and tied off. Vasectomies can be performed using a local anesthetic in a physician's office. A small incision is made on each side of the scrotum. Each vas deferens is then separated from the other structures of the spermatic cord and is cut; its ends are tied or clipped so that they cannot grow back together.

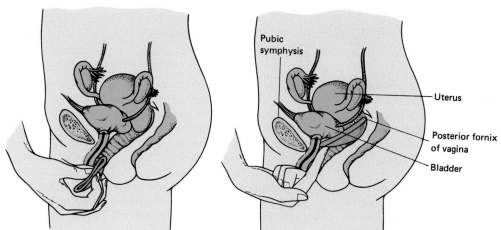

Pubic symphysis

Uterus

Posterior fornix of vagina

Bladder

FIGURE 50–24 Insertion of a contraceptive diaphragm.

Testosterone secretion is not affected by vasectomy, so this procedure in no way affects masculinity. Sperm continue to be produced, though at a much slower rate. Instead of being ejaculated, however, they eventually die and are destroyed by macrophages in the testis or epididymis. Because sperm account for very little of the semen volume, there is no noticeable change in the amount of semen ejaculated. In cases where a reversal of the procedure has been requested, surgeons have had a success rate of about 30%. The low success rate may be because some sterilized men eventually develop antibodies against their own sperm, making the sperm nonviable.

Female sterilization generally involves **tubal ligation,** cutting and tying of the uterine tubes (Figure 50–25). As in the male, hormone balance and sexual performance are not affected.

Abortion

Abortion is the termination of pregnancy, resulting in the death of the embryo or fetus. An estimated 40 million abortions are deliberately performed each year worldwide (more than a million in the United States). Three kinds of abortions may be distinguished: spontaneous abortions, therapeutic abortions, and those undertaken as a means of birth control. **Spontaneous abortions** (popularly called miscarriages) occur without intervention and often are nature's way of destroying a defective embryo. **Therapeutic abortions** are performed in order to maintain the health of the mother or when there is reason to suspect that the embryo is grossly abnormal. The third type of abortion, the type performed as a means of birth control, is the most controversial.

Most first-trimester abortions (those performed during the first 3 months of pregnancy) and some later ones are performed using a suction method. After the cervix is dilated, a suction aspirator is inserted in the uterus, and the embryo and other products of conception are evacuated. In another method, dilation and curettage ("D&C"), the endometrium is scraped out. (D&Cs may also be performed to remove polyps from the uterus and for diagnosis and treatment of a number of other uterine disorders.)

Later in pregnancy, abortions are performed using saline injections. The amniotic fluid surrounding the embryo is removed with a needle and replaced with a solution containing salt and prostaglandins. The fetus dies within 1 to 2 hours; labor begins several hours later.

When abortion is performed during the first trimester by skilled medical personnel, the mortality rate of the mother is about 1.9 per 100,000. After the first trimester, this rate rises to 12.5 per 100,000 (Table 50–4). The death

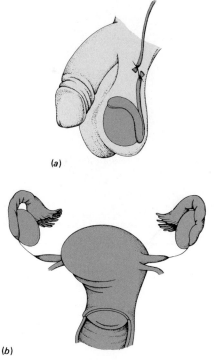

(a)

(b)

FIGURE 50–25 Sterilization. (*a*) In vasectomy, the vas deferens (sperm duct) on each side is cut and tied. (*b*) In tubal ligation, each uterine tube is clamped, or cut and tied, so that ovum and sperm can no longer meet.

Focus on NOVEL ORIGINS

About 10,000 children born each year are products of **artificial insemination.** Usually this procedure is sought when the male partner of a couple desiring a child is sterile or carries a genetic defect. Although the sperm donor remains anonymous to the couple involved, his genetic qualifications are screened by physicians.

In vitro fertilization is a technique by which an ovum is removed from a woman's ovary, fertilized in a test tube, and then reimplanted in her uterus. Such a procedure may be attempted if a woman's fallopian tubes are blocked or if they have been surgically removed. With the help of this technique a healthy baby was born in England in 1978 to a couple who had tried unsuccessfully for several years to have a child. Since that time, thousands of children have been conceived within laboratory glassware.

Another novel procedure is **host mothering.** In this procedure, a tiny embryo is removed from its natural mother and implanted into a female substitute. The foster mother can support the developing embryo either until birth or temporarily until it is implanted again into the original mother or into another host. This technique has already proved useful to animal breeders. For example, embryos from prize sheep can be temporarily implanted into rabbits for easy shipping by air, and then reimplanted into a foster mother sheep, perhaps of inferior quality. Host mothering also has the advantage of allowing an animal of superior quality to produce more offspring than would be naturally possible. In one recent series of experiments mouse embryos were frozen for up to 8 days and then successfully transplanted into host mothers. Host mothering may someday be popular with women who can produce embryos but are unable to carry them to term.

Someday society may have to deal with **cloning** (not yet a reality in the case of humans). In this process the nucleus would be removed from an ovum and replaced with the nucleus of a cell from a person who wished to produce a human copy of himself. Theoretically, any cell nucleus could be used, even a white blood cell nucleus. The fertilized ovum would then be placed into a human uterus for incubation; the resulting baby would be an identical, though younger, twin to the individual whose nucleus was used.

■ A newborn bongo and its surrogate mother, an eland. As a young embryo, the bongo was transplanted into the eland's uterus, where it implanted and developed. Bongos are a rare and elusive species inhabiting dense forests in Africa. The larger and more common elands, members of the same genus, inhabit open areas.

Table 50–4 DEATHS IN THE UNITED STATES FROM PREGNANCY AND CHILDBIRTH AND FROM VARIOUS BIRTH CONTROL METHODS	
	Death Rate per 100,000 Women
Pregnancy and childbirth	9
Oral contraceptive	4
IUD	0.5
Legal abortions	
First trimester	1.9
After first trimester	12.5
Illegal abortion performed by medically untrained individuals	about 100

rate from illegal abortions performed by medically untrained individuals is about 100 per 100,000. In contrast to these figures, the death rate from pregnancy and childbirth is about 9 per 100,000.

■ SUMMARY

I. In asexual reproduction, a single parent endows its offspring with a set of genes identical with its own. In sexual reproduction, each of two parents contributes a gamete containing half of the offspring's genetic endowment.

II. Unusual variations of reproduction include metagenesis, parthenogenesis, and hermaphroditism.

III. Human reproductive processes include gametogenesis, cyclic changes in the female body in preparation for pregnancy, sexual intercourse, fertilization, pregnancy, and lactation.

A. The male reproductive system includes the testes, which produce sperm and testosterone; a series of conducting tubes; accessory glands; and the penis.

1. The testes, housed in the scrotum, contain the seminiferous tubules, where the sperm are produced, and the interstitial cells, which secrete testosterone.

2. Sperm complete their maturation and are stored in the epididymis and may also be stored in the vas deferens.

3. During ejaculation, sperm pass from the vas deferens to the ejaculatory duct and then into the urethra, which passes through the penis.

4. Human semen contains about 400 million sperm suspended in the secretions of the seminal vesicles and prostate gland.

5. The penis consists of three columns of erectile tissue; when this tissue becomes engorged with blood, the penis becomes erect.

B. The female reproductive system includes the ovaries, which produce ova and the hormones estradiol and progesterone, the uterine tubes, the vagina, the vulva, and the breasts.

1. After ovulation, the ovum enters the uterine tube, where it may be fertilized.

2. The uterus serves as an incubator for the developing embryo.

3. The vagina is the lower part of the birth canal; it receives the penis during coitus.

4. The clitoris is the center of sexual sensation in the female.

C. The gonadotropic hormones FSH and LH stimulate sperm production and testosterone secretion. Testerone is responsible for establishing and maintaining primary and secondary sex characteristics in the male.

D. The first day of menstrual bleeding marks the first day of the menstrual cycle. Ovulation occurs at about day 14 in a typical 28-day menstrual cycle. Events of the menstrual cycle are coordinated by the gonadotropic and ovarian hormones.

1. FSH stimulates follicle development; FSH and LH together stimulate ovulation; LH promotes development of the corpus luteum.

2. The developing follicles release estradiol, which stimulates development of the endometrium and is responsible for the secondary female sex characteristics.

3. The corpus luteum secretes progesterone, which stimulates final preparation of the uterus for possible pregnancy.

E. Vasocongestion and increased muscle tension are two basic physiological responses to sexual stimulation. The cycle of sexual response includes the excitement stage, plateau stage, orgasm, and resolution.

F. Fertilization restores the diploid chromosome number, determines the sex of the offspring, and triggers development.

G. Effective methods of birth control include oral contraceptives, intrauterine devices, contraceptive diaphragms, condoms, and sterilization.

■ POST-TEST

1. The type of reproduction in which an animal divides into several pieces and then each piece develops into an entire new animal is called _____.

2. In metagenesis there is an alternation of _____.

3. Parthenogenesis is a type of reproduction in which an unfertilized egg _____.

4. An individual that can produce both eggs and sperm is described as _____.

5. A sex cell (either egg or sperm) is properly called a _____; a fertilized egg is a _____.

6. An adult who is unable to parent offspring is said to be _____.

For each group, select the most appropriate answer from Column B for each description in Column A.

Column A	Column B
7. Sperm produced here	a. Seminiferous tubules
8. Produce testosterone	b. Prostrate gland
9. Columns of erectile tissue	c. Interstitial cells
10. Secretes alkaline fluid into urethra	d. Cavernous bodies
11. Sac that holds the testes	e. None of the above

Column A

12. Produces gametes
13. Thickens each month in preparation for pregnancy
14. Lower portion of uterus
15. Fertilization takes place here

Column B

a. Uterine tube
b. Ovary
c. Cervix
d. Endometrium
e. None of the above

Column A

16. Produces FSH
17. Produces progesterone
18. Center of sexual sensation
19. Extends from uterus to exterior of body

Column B

a. Hymen
b. Corpus luteum
c. Anterior lobe of pituitary
d. Clitoris
e. None of the above

Column A

20. Responsible for secondary sexual characteristics in female
21. Responsible for secondary sexual characteristics in male
22. Produced by pituitary
23. Stimulates glands in endometrium to develop

Column B

a. Testosterone
b. Estradiol
c. LH
d. Progesterone
e. None of the above

Column A

24. Prevents ovulation
25. Prevent sperm from entering vagina
26. Procedure in which vas deferens is severed
27. Blocks passage of sperm from vagina into uterus

Column B

a. IUD
b. Contraceptive diaphragm
c. Oral contraceptive
d. Condom
e. None of the above

28. A _____ abortion is performed in order to maintain the mother's health or when the embryo is thought to be grossly abnormal.
29. Tubal ligation is a common method of _____ _____.
30. The menstrual cycle runs its course every month from puberty until _____.
31. Label the following diagrams. (Refer to Figures 50–5 and 50–13 in the text as necessary.)

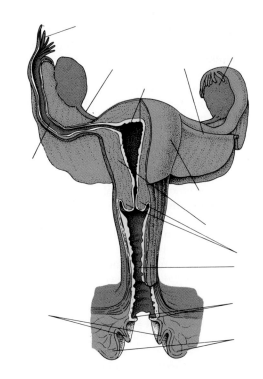

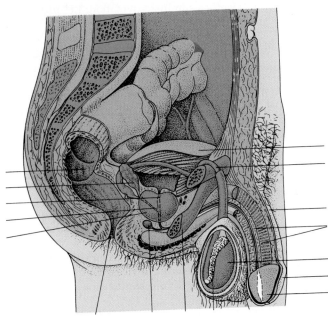

■ REVIEW QUESTIONS

1. Compare asexual with sexual reproduction, and give specific examples of asexual reproduction.
2. What are the advantages of the following adaptions: (1) metagenesis (2) parthenogenesis. Give an example of each.
3. Can you think of any biological advantages of hermaphroditism? Explain.
4. Compare the functions of the ovaries and the testes.
5. Trace the passage of sperm from a seminiferous tubule through the male reproductive system until it leaves the male body during ejaculation. Assuming that ejaculation takes place within the vagina, trace the journey of the sperm until it meets the ovum.
6. What are the actions of testosterone? of estradiol? of progesterone?

7. What is the function of the corpus luteum? Which hormone stimulates its development?
8. What are the actions of FSH and LH in the female?
9. Why are so many sperm produced in the male and so few ova produced in the female?
10. Which methods of birth control are most effective? least effective?
11. Draw a diagram of the principal events of the menstrual cycle, including ovulation and menstruation. Indicate on which days of the cycle sexual intercourse would most likely result in pregnancy.
12. Distinguish between the following terms: erectile dysfunction, sterility, and castration.
13. Give the physiological basis for penile erection.

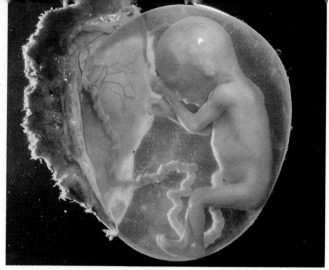

Human fetus at 16 weeks, 16 cm (6.4 inches).

51

Development

■ OUTLINE

■ LEARNING OBJECTIVES

After you have read this chapter you should be able to:

1. Relate the preformation theory and the theory of epigenesis to current concepts of development.
2. Compare the roles of cell proliferation, growth, morphogenesis, and cellular differentiation in the development of an organism.
3. Summarize the functions of fertilization, and describe the four processes involved.
4. Describe the principal characteristics of each of the early stages of development: zygote, cleavage, morula, blastula, gastrula.
5. Contrast cleavage in the sea star (or *Amphioxus*), in the amphibian, and in the bird.
6. Summarize the fate of each of the germ layers and compare gastrulation in *Amphioxus*, the amphibian, and the bird.
7. Define organogenesis and trace the early development of the nervous system.
8. Trace the development of the extraembryonic membranes and placenta, giving the functions of each membrane.
9. Summarize the general course of human development.
10. Identify the principal events of each stage of labor.
11. Contrast postnatal with prenatal life, describing several adaptations that the neonate must make in order to live independently.
12. List specific steps that a pregnant woman can take to promote the well-being of her developing child, and describe how the embryo can be affected by nutrients, drugs, cigarette smoking, pathogens, and ionizing radiation.
13. Trace the stages of the human life cycle.
14. Identify anatomical and physiological changes that occur with aging, and discuss current theories of aging.
15. Discuss genetic and nongenetic factors that interact to regulate development, relating experiments discussed in this chapter.

Development includes all of the changes that take place during the entire life of an organism from conception to death. In this chapter we will focus mainly on development of the embryo, but we also touch on

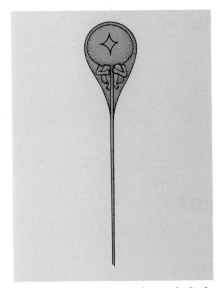

FIGURE 51–1 The preformed "little man" within the sperm as visualized by 17th-century scientists. (After Hartseeker's drawing from "Essay de Dioptrique," Paris, 1694.)

growth and maturation of the organism after birth and on the aging process.

Many scientists of the 17th century thought that the egg cell contained a completely formed, though miniature, human being. They believed that all the parts were already present, so that the embryo had only to grow in size. This concept is known as the **preformation theory.**

By the end of the 17th century two competing groups of preformationists emerged. One group, the ovists, thought that the preformed organism resided within the egg; the opposing group, the spermists, were cetain that the "little man" was housed in the sperm. Using their crude microscopes, some investigators even imagined that they could see a completely formed tiny human being within the head of the sperm (Figure 51–1).

Some scientists carried the theory to an extreme form, arguing that every woman contained within her body a miniature of every individual who would ever descend from her. Her children, grandchildren, great grandchildren, and so on were thought to be preformed, each within the reproductive cells of the other. Some investigators of that time even computed mathematically how many generations could fit, one within the other's gametes. They concluded that when all these generations had lived and died, the human species would end. Jan Swammerdam, a renowned preformationist, felt that this concept explained original sin. He wrote, "In nature there is no generation, but only propagation, the growth of parts. Thus, original sin is explained, for all men were contained in the organs of Adam and Eve. When their stock of eggs is finished the human race will cease to be." The idea of preformation was not restricted in its application to human beings alone. All plant and animal species were included. For almost 200 years this theory was seriously debated by scientists and philosophers.

An opposing view, the **theory of epigenesis,** gained experimental support as better techniques for investigation were developed. This theory held that the embryo develops from a formless zygote and that the structures of the body take shape in an orderly sequence, developing their characterisitic forms only as they emerge.

Today we know that development is largely epigenetic. No microscopic organism waits preformed in either gamete. Development proceeds from one cell to billions, from a formless mass of cells to an intricate, highly specialized and organized organism. However, a spark of truth can be found in the preformationist view. Although the "little man" itself is not to be found within the zygote, its blueprint *is* there, precisely encoded in the form of chemical specifications within the DNA of the genes.

DEVELOPMENTAL PROCESSES

How does a microscopic, unspecialized zygote give rise to the blood, bones, brain, and all the other complex structures of an organism? As we will see, development is a balanced combination of several processes: cell proliferation (mitosis and cytokinesis), growth, morphogenesis, and cellular differentiation.

The single-celled zygote undergoes mitosis and cytokinesis, forming two cells; then each of these cells divides, giving rise to four cells. The process is repeated again and again, producing the billions of cells of the adult organism. Growth occurs both by the increase in the number of cells, and by the increase of size of these cells. An orderly pattern of cellular proliferation and growth provides the cellular building blocks of the organism. But these processes alone would produce only a formless heap of cells.

Cells must arrange themselves into specific structures and appropriate body forms. The precise and complicated cellular movements that bring about the form of a multicellular organism with its intricate pattern of tissues and organs are termed **morphogenesis.**

Not only must cells be arranged into specific structures, but they must also perform varied and specialized functions. Cells must be specialized as well as organized. In order to function in different fashions, body structures must be made of different components. During early development, cells begin to change from their initial structure and function. They also become different from one another, specializing biochemically and structurally to perform specific tasks. More than 200 distinct types of cells can be found in the adult vertebrate body. The process by which cells become specialized is known as **cellular differentiation.**

As you read the following sections on development, bear in mind that cell proliferation, growth, morphogenesis, and cellular differentiation are intimately interrelated. The pattern of early development is basically similar for all animals.

FERTILIZATION

Fertilization is the union of sperm and ovum to produce a **zygote,** or fertilized egg (Figure 51–2). Fertilization serves three functions: (1) The diploid number of chromosomes is restored as the sperm contributes its haploid set of chromosomes to the haploid ovum. (2) In mammals and many other animals, sex of the offspring is determined. (3) Fertilization provides the stimulation necessary to initiate the reactions in the egg that permit development to take place.

Fertilization involves four steps. First, the sperm must contact the egg and recognition must occur. Second, the sperm enters the egg. Third, the sperm and egg nuclei fuse. Finally, the egg is activated and development begins.

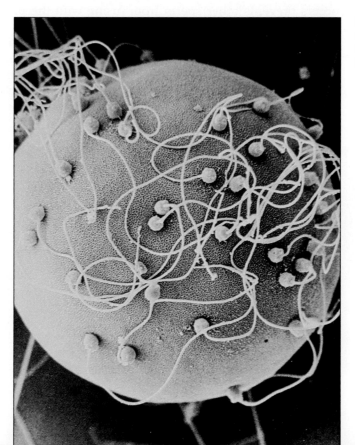

FIGURE 51–2 Sperm swarm around an egg of a surf clam in this scanning electron micrograph (approximately ×2100).

Contact and Recognition

The egg is surrounded by a very thin **vitelline membrane** and, outside of this, by a thick glycoprotein layer called the **jelly coat** (zona pellucida in mammals). In a few species (e.g., some cnidarians) the egg (or a surrounding structure) secretes a chemotactic substance that attracts sperm of the same species. However, in most species chemical attraction of sperm by the egg has not been shown, and meeting of the gametes may be largely a matter of chance.

The activation of sperm by the egg, the **acrosome reaction,** has been most studied in marine invertebrates. When the sperm contacts the jelly coat surrounding the egg, the acrosome (a structure at the head of the sperm) releases enzymes that digest a path through the jelly coat to the vitelline membrane of the egg. Actin molecules join to form actin filaments, which permit the acrosome to extend outward and contact the vitelline membrane of the egg. Recognition that the sperm and egg are of the same species occurs at this time.

If the species are the same, a species-specific protein known as **bindin,** located on the acrosome, adheres to a species-specific bindin receptor located on the vitelline membrane.

Regulation of Sperm Entry

Once acrosome-vitelline membrane fusion occurs, enzymes dissolve a bit of the vitelline membrane in the area of the sperm head. The plasma membrane of the egg is covered with microvilli. Several microvilli elongate to surround the head of the sperm, forming a **fertilization cone.** The sperm is then drawn into the egg by contraction of the fertilization cone. As this occurs, the plasma membranes of sperm and egg fuse (Figure 51–3).

As soon as one sperm enters the egg, two reactions occur that prevent additional sperm from entering. Within a few seconds a fast block to polyspermy occurs. This incomplete block involves an electrical change in the plasma membrane of the egg. Ion channels in the plasma membrane open, permitting sodium ions to pass into the cell. The depolarization that occurs prevents other sperm from fusing with the plasma membrane.

A second mechanism preventing entrance of more than one sperm, referred to as the slow block to polyspermy, is the **cortical reaction.** Depolarization of the egg plasma membrane results in calcium ion release from thousands of cortical granules present beneath the membrane. The cortical granules then release enzymes by exocytosis into the area between the plasma membrane and the vitelline membrane. Some of these enzymes

FIGURE 51–3 Fertilization. (*a*) When the first sperm makes contact with the egg's plasma membrane, microvilli elongate and surround the head of the sperm. This fertilization cone draws the sperm into the cytoplasm of the egg. (*b*) This computer reconstruction shows the fusion of sperm and egg plasma membranes that marks the instant of fertilization.

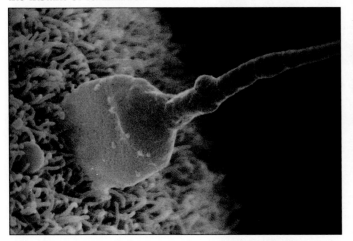

(a)

(b)

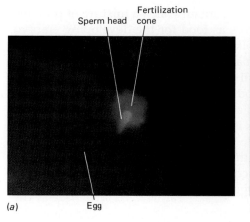

Sperm head | Fertilization cone

(a) Egg

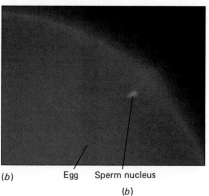

(b) Egg | Sperm nucleus

(b)

FIGURE 51–4 These light micrographs of fertilization were obtained using special fluorescent staining techniques. (*a*) The sperm nucleus (blue) is shown as it first enters the egg. It is surrounded by a crown of egg microfilaments (red). (*b*) The sperm is shown here well on its way into the egg. The microfilaments have begun to disperse.

dissolve the protein linking the vitelline membrane to the plasma membrane. Other substances released by the cortical granules result in osmotic passage of water into the space between the plasma membrane and vitelline membrane. This elevates the vitelline membrane, and it then becomes the fertilization membrane, a hardened membrane that prevents entry of sperm. This slow block requires one to several minutes, but it is a complete block. (In mammals, a fertilization membrane does not form, but the enzymes released alter the zona pellucida sperm receptors so that no additional sperm can bind to them.) In some species polyspermy does occur, but only the first sperm that enters fertilizes the egg.

Fusion of the Genetic Material

After the sperm nucleus enters the egg through the fertilization cone, it is thought to be guided toward the egg nucleus by a system of microtubules that form within the egg (Figure 51–4). The sperm nucleus swells to form the male pronucleus. The nucleus of the ovum becomes the female pronucleus, and then the haploid pronuclei fuse, forming the diploid nucleus of the zygote.

Activation of the Egg

Release of calcium ions into the egg cytoplasm is necessary for the cortical reaction, and it also triggers metabolic changes within the cell. There is an increase in oxygen use by the cell as certain compounds within the egg cell are oxidized. Within a few minutes after sperm entry there is a burst of protein synthesis.

In some species sperm penetration initiates rearrangement of the cytoplasm. For example, in the amphibian egg some of the superficial cytoplasm containing dark granules shifts, exposing the underlying lighter-colored cytoplasm. This cytoplasm, which appears gray, is referred to as the gray crescent; it marks the region where gastrulation, an early developmental process, begins in the amphibian embryo.

In some embryos, the egg can be artificially activated without sperm penetration by swabbing it with blood and pricking the cell membrane with a needle, by calcium injection, or by diverse other treatments. These eggs develop parthenogenetically (that is, without being fertilized).

Although it appears to be a relatively simple cell, the zygote has the potential to give rise to all the cell types of the complete individual (Figure 51–2). Since the sperm cell is quite tiny in comparison with the egg, the bulk of the zygote cytoplasm comes from the ovum. However, it contains the chromosomes contributed by both sperm and egg.

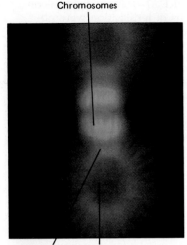

Chromosomes

Spindle fibers   Centriole

FIGURE 51–5 The first mitotic division of the new organism. The centrioles are derived from the sperm, but the microtubules (in green) of the mitotic spindle are part of the egg's dowry. The brilliantly blue-fluorescing chromosomes were provided by both.

The eggs of many kinds of animals contain yolk—a mixture of proteins, phospholipids, and fats—that serves as food for the developing embryo. The amount and distribution of yolk vary among different animal groups. Whereas bird eggs contain great quantities of yolk, those of mammals and some invertebrates have only small amounts. In eggs with large amounts of yolk, this metabolically inert food substance is concentrated at one pole of the cell; such eggs are described as **telolecithal**. The pole of the cell that contains the largest amount of yolk is known as the **vegetal** pole; the opposite, more metabolically active pole is the **animal pole**. In yolk-poor eggs the yolk is uniformly dispersed through the cytoplasm; such eggs are **isolecithal**. However, using other criteria, animal and vegetal poles can be distinguished even in yolk-poor eggs.

CLEAVAGE: FROM ONE CELL TO MANY

Shortly after fertilization the zygote undergoes a series of rapid mitoses, collectively referred to as **cleavage**. The zygote undergoes mitosis and divides to form a two-cell embryo (Figures 51–5 and 51–6). Then, each of these cells undergoes mitosis and divides, bringing the number of cells to four. Repeated divisions continue to increase the number of cells, or **blastomeres,** making up the embryo. At about the 32-cell stage the embryo is referred to as a **morula.** The cells of the morula continue to multiply, eventually forming a hollow ball of several hundred cells, the **blastula.** The cells of the blastula surround a fluid-filled cavity, the blastocoele.

What Cleavage Accomplishes

During cleavage the cells do not grow, so the mass of cells produced is not larger than the original zygote. The principal effect of early cleavage, then, is to partition the zygote into many small cells that serve as basic building units. Their small size allows the cells to move about with relative ease, arranging themselves into the patterns necessary for further development. Each cell moves by amoeboid motion, probably guided by the proteins of its own cell coat and those of other cell surfaces. Specific properties conferred upon cell membranes by these surface proteins are important in helping cells to "recognize" one another and therefore in determining which ones will adhere to form tissues.

Patterns of Cleavage

The amount of yolk in the egg affects the pattern of cleavage. In the isolecithal eggs characteristic of most invertebrates and simple chordates, there are relatively small amounts of uniformly distributed yolk. Cleavage in these eggs is described as **holoblastic** because the entire egg divides, producing cells of roughly the same size.

In radial cleavage, the first division passes through both animal and vegetal poles and splits the egg into two equal cells. The second cleavage division passes through both poles of the egg at right angles to the first, and separates the two cells into four equal cells. The third division is horizontal, at right angles to the other two, and separates the four blastomeres into eight—four above and four below the line of cleavage. This holoblastic pattern of radial cleavage is seen in the sea star and in *Amphioxus* (Figures 51–6 and 51–7).

You may recall from Chapter 29 that the radial cleavage described above is characteristic of deuterostomes. Some protostomes have a pattern of early cell division known as **spiral cleavage** (Figure 51–8). In spiral cleav-

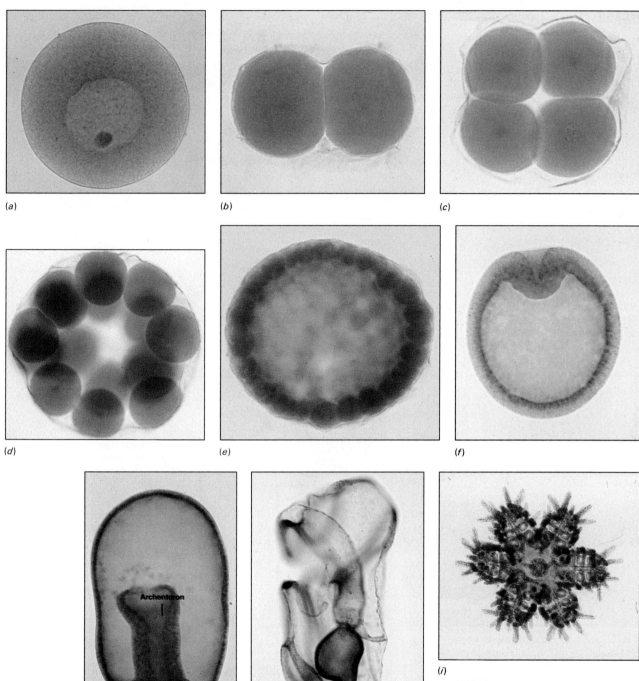

(a)

(b)

(c)

(d)

(e)

(f)

Archenteron

(g)

(h)

(i)

FIGURE 51–6 Development of a seastar. (*a*) Unfertilized seastar egg. (*b*) Two-cell stage. (*c*) Top view of four-cell stage. (*d*) Sixteen-cell stage. (*e*) Cross section through 64-cell blastula. (*f*) Section through early gastrula. (*g*) Section through middle gastrula. (*h*) Seastar larva. (*i*) Young seastar. In this type of cleavage the entire egg becomes partitioned into cells. The blastopore is the opening into the inner cavity, the archenteron. All views are side views with the animal pole at the top, except (*c*) and (*i*) which are top views.

age, the early cell divisions after the first two are oblique to the polar axis, resulting in a spiral arrangement of cells; each cell is located between the two cells below it.

The telolecithal eggs of bony fishes and amphibians contain a large amount of yolk concentrated toward the vegetal pole. The cleavage divisions in the vegetal hemisphere are slowed by the presence of the inert yolk. As a result the blastula consists of many small cells in the animal hemisphere and fewer, but larger, cells in the vegetal hemisphere (Figure 51–9). For this reason the lower wall is much thicker than the upper one and the blastocoel is displaced upward.

FIGURE 51–7 Isolecithal cleavage and gastrulation in *Amphioxus* viewed from the side. *A*, Mature egg with polar body. *B–E*, Two-, four-, eight-, and 16-cell stages. *F*, Embryo at 32-cell stage cut open to show the blastocoel. *G*, Blastula. *H*, Blastula cut open. *I*, Early gastrula showing beginning of invagination at vegetal pole (*arrow*). *J*, Late gastrula. Invagination is completed and blastopore has formed.

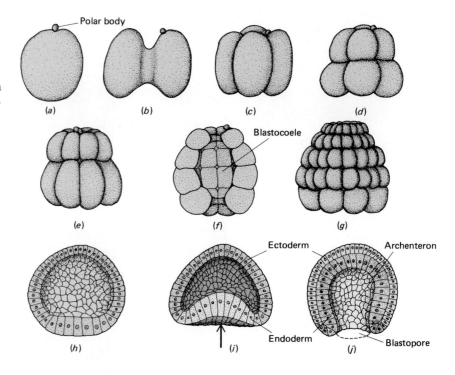

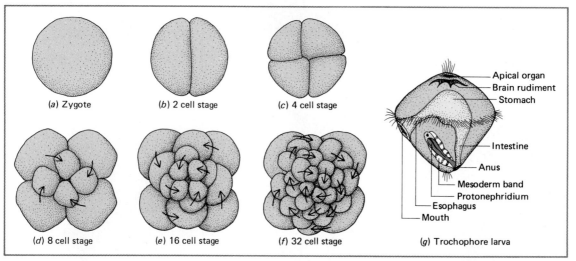

FIGURE 51–8 Development in annelids. (*a*) through (*f*) are top views of the animal pole. The successive cleavage divisions occur in a spiral pattern as indicated. (*g*) A typical trochophore larva. The upper half of the trochophore develops into the extreme anterior end of the adult worm; all the rest of the adult body develops from the lower half.

The telolecithal eggs of reptiles, birds, and some fishes have a very large amount of yolk and only a small amount of cytoplasm concentrated at the animal pole. In such eggs, cleavage is said to be **meroblastic** because cell division takes place only in the **blastodisc,** the small disc of cytoplasm at the animal pole (Figure 51–10). In birds and some reptiles, the blastomeres form two layers, an upper epiblast and below that a thin layer of flat cells,

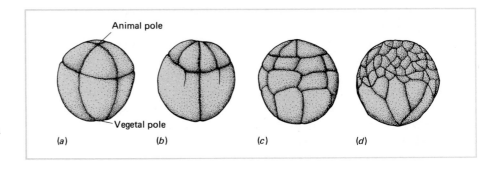

FIGURE 51–9 Successive stages in telolecithal cleavage in the frog, viewed from the side.

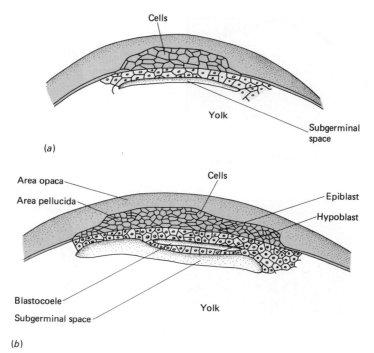

Cells

Yolk

Subgerminal space

(a)

Area opaca

Area pellucida

Cells

Epiblast

Hypoblast

Blastocoele

Subgerminal space

Yolk

(b)

FIGURE 51–10 Successive stages in the cleavage of a hen's egg. (*a*) Cleavage is restricted to a small disk of cytoplasm on the upper surface of the egg yolk called the blastodisc. A subgerminal space appears beneath the blastodisc, separating it from the unsegmented yolk. (*b*) The blastodisc cleaves into an upper epiblast and a lower hypoblast separated by the blastocoel.

the hypoblast. The epiblast and hypoblast are separated from each other by the blastocoel, and are separated from the underlying yolk by another cavity, the subgerminal space. The subgerminal space appears only under the central portion of the blastodisc. This central portion, called the area pellucida, appears somewhat transparent. The more opaque part of the blastodisc that rests directly on the yolk is the area opaca.

GASTRULATION: FORMATION OF GERM LAYERS

The process by which the blastocyst becomes a three-layered embryo, called a **gastrula,** is termed **gastrulation.** During gastrulation, the cells arrange themselves into three distinct **germ layers,** or embryonic tissue layers—the ectoderm, mesoderm, and endoderm.

Fate of the Germ Layers

The cells lining the hollow center of the gastrula make up the **endoderm,** which eventually lines the digestive tract and its outgrowths such as the liver, pancreas, and lungs. The outer wall of the gastrula consists of the germ layer known as **ectoderm.** The ectoderm eventually forms the outer layer of the skin and gives rise to the nervous system and sense organs.

A third layer of cells, the **mesoderm,** proliferates between the ectoderm and endoderm. The mesoderm gives rise to the skeletal tissues, muscle, circulatory system, excretory system, and reproductive system (Table 51–1). The complex morphogenetic movements of cells and their ability to form the germ layers depend upon instructions from the genes of the embryo.

Patterns of Gastrulation

Gastrulation in the sea star and in *Amphioxus* are illustrated in Figures 51–6 and 51–7, respectively. Gastrulation begins when the blastoderm at the vegetal pole flattens and then bends inward. The cells of a section of one wall of the blastula move inward, a process referred to as invagination. The invaginated wall eventually meets the opposite wall, obliterating the original blastocoel. In this way the embryo is converted into a double-walled

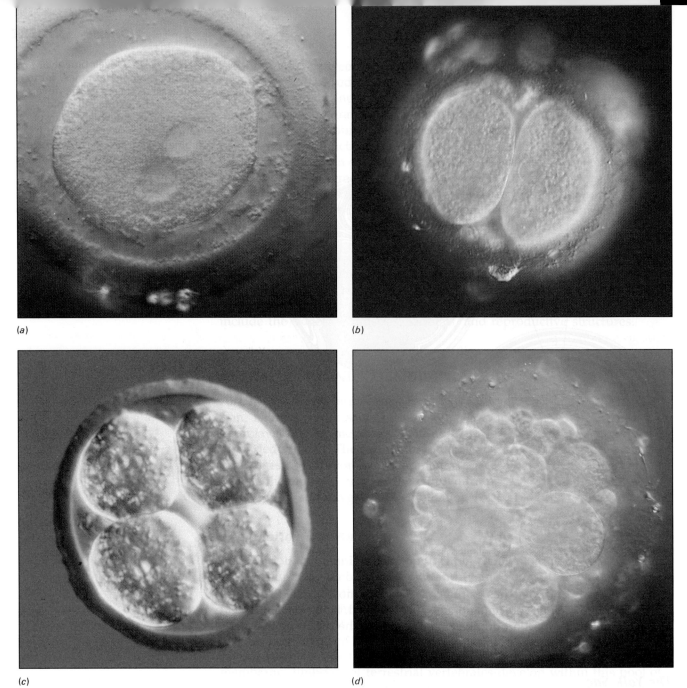

(a)

(b)

(c)

(d)

FIGURE 51–16 Early human development (approximately ×250). (a) Human zygote. This single cell contains the genetic instructions for producing a complete human being. (b) Two-cell stage. (c) Eight-cell stage. (d) Cleavage continues, giving rise to a cluster of cells called the morula.

vessels) as the embryonic circulation develops. Enzymes released by the invading embryonic cells destroy some of the endometrial tissue, including many small blood vessels. Small amounts of blood oozing from these damaged vessels form pools about the chorionic villi.

As the human embryo grows, the region on the ventral side from which the folds of the amnion, yolk sac, and allantois grew becomes relatively smaller, and the edges of the amniotic folds come together to form a tube that encloses the other membranes. This tube, the **umbilical cord,** connects the embryo with the placenta (Figures 51–15 and 51–18). In addition to the yolk sac and allantois, the umbilical cord contains the two umbilical arteries and the umbilical vein. The **umbilical arteries** connect the embryo with a vast network of capillaries developing within the villi; blood from the villi returns to the embryo through the **umbilical vein.**

The placenta actually consists of the portion of the chorion that develops villi, together with the uterine tissue underlying the villi, which contains maternal capillaries and small pools of maternal blood. The blood of the fetus in the capillaries of the chorionic villi comes in close contact with the mother's blood in the tissues between the villi. However, they are always separated by a membrane through which substances may diffuse or be actively transported. *Maternal and fetal blood do not normally mix in the placenta or any other place.*

Several hormones are produced by the placenta. From the time the embryo first begins to implant itself, its trophoblastic cells release **human chorionic gonadotropin (HCG),** which signals the corpus luteum that pregnancy has begun. In response, the corpus luteum increases in size and releases large amounts of progesterone and estrogen, which in turn stimulate continued development of endometrium and placenta. Without HCG, the corpus luteum would degenerate and the embryo would be aborted and flushed out with the menstrual flow. In such a case the woman would probably not even know that she had been pregnant. When the corpus luteum is removed before about the eleventh week of pregnancy, the embryo is spontaneously aborted. After that time, however, the placenta itself produces enough progesterone and estrogens to maintain pregnancy.

HUMAN DEVELOPMENT

In this section we will focus on human development as an example of development in placental mammals. Prenatal (before birth) human development normally requires 280 days from the time of the mother's last menstrual period to the birth of the baby, or 266 days from the time of conception (Table 51–2).

Table 51–2 **SOME IMPORTANT DEVELOPMENTAL EVENTS IN THE HUMAN EMBRYO**

Time from Fertilization	Event
24 hours	Embryo reaches two-cell stage
3 days	Morula reaches uterus
7 days	Blastocyst begins to implant
2.5 weeks	Notochord and neural plate are formed; tissue that will give rise to heart is differentiating; blood cells are forming in yolk sac and chorion.
3.5 weeks	Neural tube forming; primordial eye and ear visible; pharyngeal pouches forming; liver bud differentiating; respiratory system and thyroid gland just beginning to develop; heart tubes fuse, bend, and begin to beat; blood vessels are laid down.
4 weeks	Limb buds appear; three primary vesicles of brain formed.
2 months	Muscles differentiating; embryo capable of movement. Gonad distinguishable as testis or ovary. Bones begin to ossify. Central cortex differentiating. Principal blood vessels assume final positions.
3 months	Sex can be determined by external inspection. Notochord degenerates. Lymph glands develop.
4 months	Face begins to look human. Lobes of cerebrum differentiate. Eye, ear, and nose look more "normal."
Third trimester	Lanugo appears, then later is shed; neurons become myelinated; tremendous growth of body
266 days (from conception)	Birth

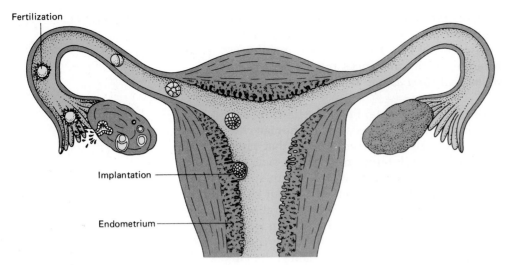

Fertilization

Implantation

Endometrium

FIGURE 51–17 Cleavage takes place as the embryo is moved along through the uterine tube to the uterus.

Early Development

By about 24 hours after fertilization, the human zygote has completed meiosis and divided to become a two-cell embryo. Each of the cells of the two-cell-stage embryo undergoes mitotic division, bringing the number of cells to four. Repeated divisions continue as the embryo is pushed along the uterine tube by ciliary action and muscular contraction. By the time the embryo reaches the uterus, on about the fifth day of development, it is in the morula stage (Figures 51–16 and 51–17).

When the embryo enters the uterus, the membrane that has surrounded it, the **zona pellucida,** is dissolved. Now the embryo is bathed in a nutritive fluid secreted by the glands of the uterus. Nourished in this manner, the embryo continues its development for two or three days, floating free in the uterine cavity. During this period its cells arrange themselves to form a blastula, which in mammals is called a **blastocyst** (Figure 51–18a). The outer layer of cells, the **trophoblast,** eventually forms the chorion and amnion that surround the embryo. A little cluster of cells, the **inner cell mass,** projects into the cavity of the blastocyst. These cells give rise to the embryo proper.

Twins

Occasionally, the cells of the two-cell embryo may separate and each cell may develop into a complete organism. Alternately, the inner cell mass may subdivide to form two independent groups of cells, each of which develops independently. Since these cells have identical sets of genes, the individuals formed are exactly alike—**identical twins.** Very rarely, the two inner cell masses are not completely separated and give rise to **conjoined** (Siamese) **twins. Fraternal twins** develop when a woman ovulates two eggs and each is fertilized by a different sperm. Each zygote has its own distinctive genetic endowment, so the individuals produced are not identical. Triplets (and other multiple births) may similarly be either identical or fraternal. In the United States, twins are born once in about 88 births, triplets once in 88^2 (or 7,744), and quadruplets once in 88^3 (or 681,472).

Implantation

On about the seventh day of development the embryo begins to **implant** in the endometrium of the uterus (Figure 51–18a). The trophoblast cells in contact with the uterine lining secrete enzymes that erode an area just large enough to accommodate the tiny embryo. Slowly the embryo works its

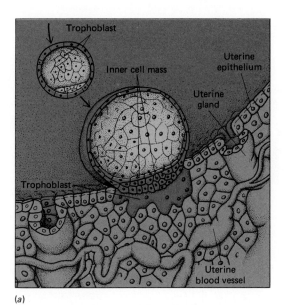

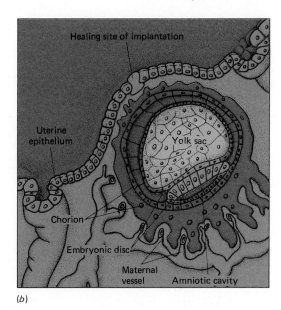

(a)

(b)

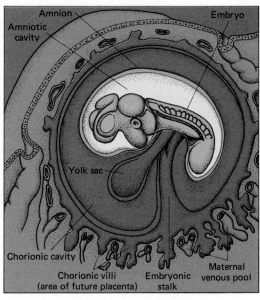

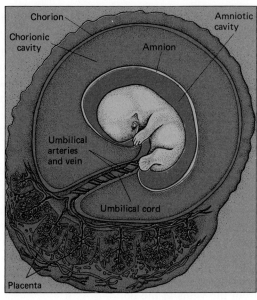

(c)

(d)

FIGURE 51–18 Implantation and development of the early human embryo. (*a*) About 7 days after fertilization the blastocyst drifts to an appropriate site along the uterine wall and begins to implant itself. The cells of the trophoblast proliferate and invade the endometrium. (*b*) About 10 days after fertilization. The chorion has formed from the trophoblast. (*c*) By 25 days intimate relationships have been established between the embryo and the maternal blood vessels. Oxygen and nutrients from the maternal blood are now satisfying the embryo's needs. Note the specialized region of the chorion that will soon become the placenta. The embryonic stalk will become part of the umbilical cord. (*d*) At about 45 days the embryo and its membranes together are about the size of a Ping-Pong ball, and the mother still may be unaware of her pregnancy. The amnion filled with amniotic fluid surrounds and cushions the embryo. The yolk sac has been incorporated into the umbilical cord. Blood circulation has been established through the umbilical cord to the placenta.

way down into the underlying connective and vascular tissues. The opening through which the blastocyst enters the lining of the uterus is closed, first by a blood clot, and eventually by the overgrowth of regenerated epithelial cells. All further development of the embryo takes place *within* the endometrium of the uterus (Figure 51–19).

FIGURE 51–19 Photograph of implanted blastocyst at 12 days after fertilization.

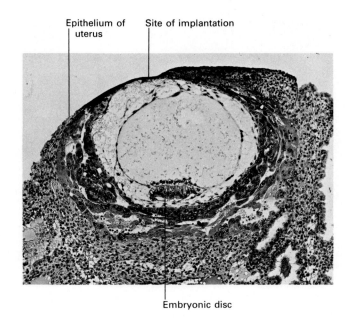

Epithelium of uterus   Site of implantation

Embryonic disc

During implantation, enzymes destroy some tiny maternal capillaries in the wall of the uterus. Blood from these capillaries comes in direct contact with the trophoblast of the embryo, temporarily providing a rich source of nutrition.

Implantation is completed by the ninth day of development. This would correspond to about the twenty-third day of a woman's menstrual cycle, so that despite all the developmental activities taking place in her uterus, a woman would probably not even know that she was pregnant at this time.

The Beginnings of Differentiation and Organogenesis

In the human embryo, the first visible sign of differentiation occurs during gastrulation, when one group of cells of the inner cell mass becomes flattened and the remaining cells become columnar. The flattened cells develop into the endoderm, whereas the columnar cells become ectoderm.

During the second and third weeks of development, gastrulation occurs in the manner described for the bird embryo. Next the notochord begins to form and then induces formation of the neural plate. The neural tube develops and the forebrain, midbrain, and hindbrain are all established by the fifth week of development. A week or so later, the forebrain begins to grow outward, forming the rudiments of the cerebral hemispheres.

During the first month of development, the heart begins to take shape and to beat—about 60 times each minute (Figure 51–20). In the pharyngeal

FIGURE 51–20 Ventral views of successive stages in the development of the heart. The heart forms from the fusion of two blood vessels. At first it is upside down, with the end that receives blood from the veins at the bottom. As it develops, the heart twists and turns in position to carry the atrium to a position above the ventricle, and the conus and sinus disappear. The chambers divide to form the four-chambered heart. The heart begins to beat spontaneously without nervous stimulation.

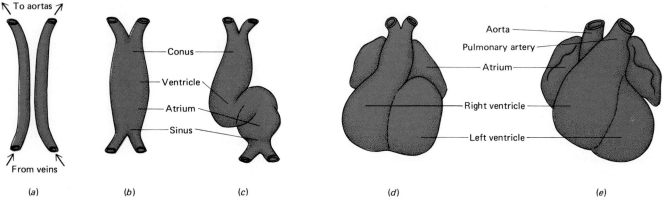

To aortas

From veins

Conus

Ventricle

Atrium

Sinus

Aorta

Pulmonary artery

Atrium

Right ventricle

Left ventricle

(a)   (b)   (c)   (d)   (e)

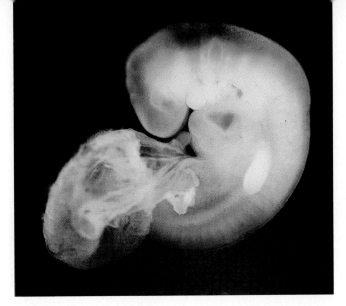

FIGURE 51–21 Human embryo at 29 days, about 7 mm (0.3 inch) long. Note the slender tail and developing limb buds. The tail will regress during later development. The heart can be seen below the head near the mouth of the embryo. Branchial arches appear as ''double chins.''

region, the pharyngeal pouches, branchial grooves, and branchial arches form. In the floor of the pharynx at the level of the fourth pharyngeal pouches a tube of cells grows downward, forming the primordial trachea, which gives rise to the lung buds. The digestive system also gives rise to outgrowths that will develop into the liver, gallbladder, and pancreas. Near the end of the first month, the limb buds begin to differentiate; these structures give rise to arms and legs (Figures 51–21 through 51–23).

All of the organs continue to develop during the second month. A thin tail becomes prominent during the fifth week (see Figure 51–21), but fails to keep pace with the rapid growth of the rest of the body and so becomes inconspicuous by the end of the second month. Muscles develop, and the embryo is capable of some movement (Table 51–2). The brain begins to send impulses to regulate the functions of some organs, and a few simple reflexes are evident. After the first two months of development the embryo is referred to as a **fetus.**

By the end of the **first trimester** (first three months of development), the fetus is recognizably human (Figures 51–24 and 51–25). The external genital structures have differentiated, indicating the sex of the fetus. Ears and eyes approach their final positions. Some of the skeleton becomes distinct, and the notochord degenerates. The fetus performs breathing movements, pumping amniotic fluid into and out of its lungs, and even carries on sucking movements. By the end of the third month, the fetus is almost 56 mm (about 2.2 in.) long and weighs about 14 g (0.5 oz).

Later Development

During the second trimester (Figure 51–25), the fetal heart (now beating about 150 times per minute) can be heard with a stethoscope. The fetus moves freely through the amniotic cavity; during the fifth month, the mother usually becomes aware of fetal movements (''quickening'').

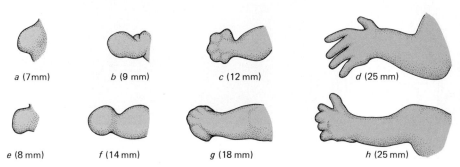

a (7mm)   *b* (9 mm)   *c* (12 mm)   *d* (25 mm)

e (8 mm)   *f* (14 mm)   *g* (18 mm)   *h* (25 mm)

FIGURE 51–22 Stages in the development of the human arm (*upper row*) and leg (*lower row*) between the fifth and eighth weeks. The number of millimeters indicated in parentheses refers to the length of the whole embryo from crown to rump at that stage of limb development.

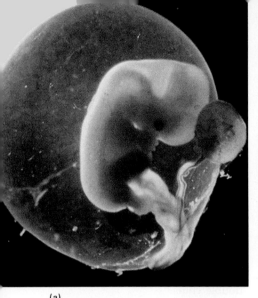

(a)

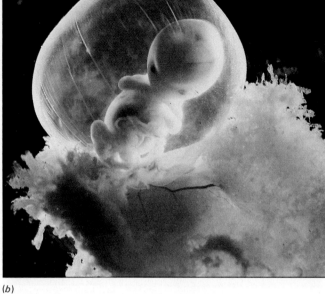

(b)

FIGURE 51–23 Photographs of developing human embryos. (a) Human embryo at 5½ weeks, 1 cm (0.4 inch) long. Limb buds have lengthened and the eyes have become prominent. (b) In its seventh week of development, the embryo is 2 cm (0.8 inch) long. The dark red object inside the embryo is the liver.

During the final trimester, the fetus grows rapidly, and final differentiation of tissues and organs occurs. At the beginning of the sixth month the skin has a wrinkled appearance, perhaps because it is growing faster than the underlying connective tissue. If born prematurely at this age the fetus attempts to breathe and is able to move and cry, but almost always dies because its brain and lungs are not sufficiently developed to sustain vital functions such as rhythmic breathing and the regulation of body temperature.

During the seventh month the cerebrum grows rapidly and develops convolutions. The grasp and sucking reflexes are evident, and the fetus may suck its thumb. Most of the body is covered by a downy hair called **lanugo,** which is usually shed before birth. Occasionally the lanugo is not shed until a few days after birth.

FIGURE 51–24 The human fetus at 10 weeks of development. (a) Note position in the uterine wall. (b) Photograph at 10 weeks.

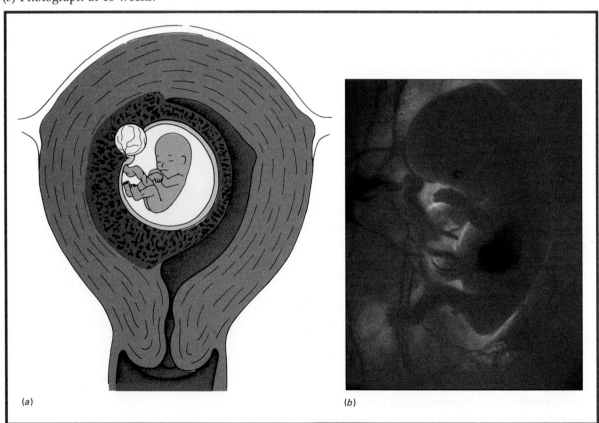

(a)

(b)

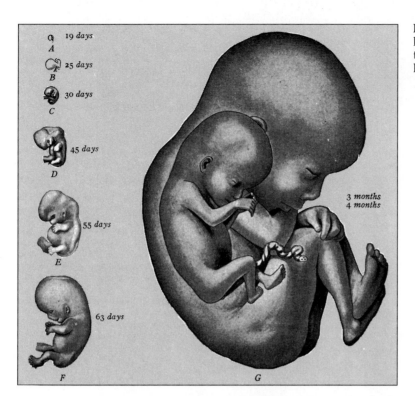

FIGURE 51–25 A graded series of human embryos. Note the characteristic position of the arms and legs in the 4-month fetus.

During the last months of prenatal life, a protective creamlike substance, the **vernix,** covers the skin, and hair begins to grow on the scalp. At birth the average full-term baby weighs about 3000 g (7 lb) and measures about 52 cm (20 in.) in total length (or about 35 cm from crown to rump).

The Birth Process

The human **gestation period,** the duration of pregnancy, is normally 280 days, from the time of the last menstrual period to the birth of the baby (or 266 days from conception). Babies born as early as 28 weeks or as late as 45 weeks after the last menstrual period may survive. The factors that actually initiate the process of birth, or **parturition,** after the period of gestation is complete are not well understood. Childbirth begins with a long series of involuntary contractions of the uterus, experienced as the contractions of **labor.**

Labor may be divided into three stages. During the first stage, which typically lasts about 12 hours, the contractions of the uterus move the fetus down toward the cervix. The cervix dilates and becomes effaced, losing its normal shape and flattening so that the fetal head can pass through. During the first stage of labor the amnion usually ruptures, releasing about a liter of amniotic fluid, which flows out through the vagina. In the second stage, which normally lasts between 20 minutes and an hour, the fetus passes through the cervix and vagina and is born, or "delivered" (Figure 51–26). With each uterine contraction the woman holds her breath and bears down so that the fetus is expelled from the uterus by the combined forces of uterine contractions plus the contractions of the muscles of the abdominal wall.

After the baby is born, the contractions of the uterus squeeze much of the fetal blood from the placenta back into the infant. After the pulsations in the umbilical cord cease, the cord is tied and cut, severing the child from the mother. The stump of the cord gradually shrivels until nothing remains but the depressed scar, the **navel.**

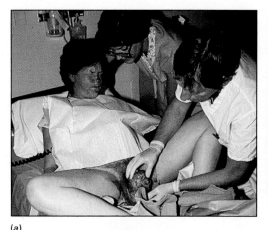

(a)

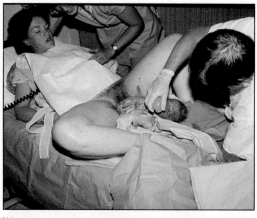

(b)

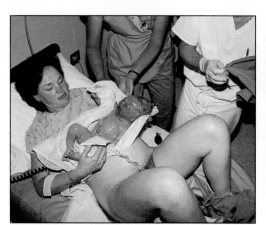

(c)

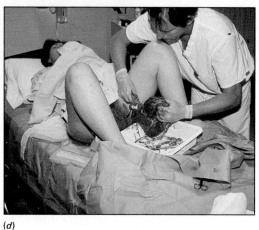

(d)

FIGURE 51–26 Birth of a baby. In about 95% of all human births the baby descends through the cervix and vagina in the head-down position. (*a*) The mother bears down hard with her abdominal muscles, helping to push the baby out. When the head fully appears, the physician or midwife can gently grasp it and guide the baby's entrance into the outside world. (*b*) Once the head has emerged, the rest of the body usually follows readily. The physician gently aspirates the mouth and pharynx to clear the upper airway of any amniotic fluid, mucus, or blood. At this time the neonate usually takes its first breath. (*c*) The baby, still attached to the placenta by its umbilical cord, is presented to its mother. (*d*) During the third stage of labor the placenta is delivered.

During the third stage of labor, which lasts 10 or 15 minutes after the birth of the child, the placenta and the fetal membranes are loosened from the lining of the uterus by another series of contractions and expelled. At this stage they are called collectively the **afterbirth.** In humans and certain other mammals in which the placenta forms a very tight connection with the uterine lining, the expulsion of the placenta is accompanied by some loss of blood. In other mammals in which the connection between the fetal membranes and uterine wall is not close, the placenta can pull away from the uterine wall without causing bleeding. Following birth, the size of the uterus decreases, and its lining is rapidly restored.

During labor an obstetrician may administer drugs such as oxytocin or prostaglandins to increase the contractions of the uterus, or may assist with special forceps or other techniques. In some women, the aperture between the pelvic bones through which the vagina passes is too small to permit the passage of the baby, so the child must be delivered by **cesarean section,** an operation in which an incision is made in the abdominal wall and uterus.

Adaptations to Birth

Great changes take place within a short time after a baby is born. During prenatal life, the fetus received both food and oxygen from the placenta. Now the newborn's own digestive and respiratory systems must function. Correlated with these changes are several major changes in the circulatory system.

Normally the **neonate** (newborn infant) begins to breathe within a few seconds of birth, and cries within a half a minute. If anesthetics have been given to the mother, however, the fetus may also have been anesthetized, and breathing and other activities may be depressed. Some infants may not begin breathing until several minutes have passed. This is one of the reasons for the current trend toward natural childbirth, in which as little medication as possible is used.

The neonate's first breath is thought to be initiated by the accumulation of carbon dioxide in the blood after the umbilical cord is cut. This stimulates the respiratory centers in the medulla. The resulting expansion of the lungs enlarges its blood vessels (which previously were partially collapsed). Blood from the right ventricle flows in increasing amounts through the pulmonary vessels instead of through the arterial duct. Before birth, the arterial duct connected the pulmonary artery and aorta allowing blood to bypass the lungs during fetal life.

ENVIRONMENTAL INFLUENCES UPON THE EMBRYO

Although it is obvious that a baby's growth and development are influenced by the food he eats, the air he breathes, the disease organisms that infect him, and the chemicals or drugs to which he is exposed, it is less obvious that *prenatal* development is also affected by these environmental influences. Life before birth is, in fact, even more sensitive to environmental changes than it is in the fully formed baby.

About 5% of all babies born alive, or 175,000 babies per year, arrive with a defect of clinical significance. Such birth defects account for about 15% of deaths among newborns. Birth defects may be caused by genetic or environmental factors or a combination of the two. Genetic factors were discussed in Part 3. In this section we will examine some environmental conditions that affect the well-being of the embryo.

Substances or conditions that intrude upon the developing embryo from the outside environment may cause significant damage at one period in development, yet appear to be harmless during a later stage. Timing is important. Each developing structure has a critical period during which it is most susceptible to unfavorable conditions. Generally this critical period occurs early in the development of the structure, when interference with cell movements or divisions may prevent formation of normal shape or size, resulting in permanent malformation. Since most structures form during the first three months of embryonic life, the embryo is most susceptible to environmental conditions during this early period. During a substantial portion of this time the mother may not even realize that she is pregnant and may therefore take no special precautions to minimize potentially dangerous influences.

Anything that circulates in the maternal blood—nutrients, drugs, or even gases—may find its way into the blood of the fetus. Some drugs, as well as other types of agents, are **teratogens;** they can interfere with normal development (Figure 51–27). Table 51–3 describes some of the environmental influences upon development.

Recent advances in medicine have enabled physicians to diagnose some defects while the embryo is in the uterus. In some cases treatment is possible before birth. **Amniocentesis,** discussed in Chapter 12, is one technique used to detect certain defects. Figure 51–28 is a **sonogram,** discussed in

(a)

(b)

FIGURE 51–27 Thalidomide administered to the marmoset (*Callithrix jacchus*) produces a pattern of developmental defects similar to those found in humans. (*a*) Control marmoset fetus obtained from an untreated mother on day 125 of gestation. (*b*) Fetus (same age as control) of marmoset treated with 25 mg/kg thalidomide from days 38 to 52 of gestation. The drug suppresses limb formation, perhaps by interfering with the function of cholinergic nerves.

Table 51–3 ENVIRONMENTAL INFLUENCES ON THE HUMAN EMBRYO		
Factor	*Example and Effect*	*Comment*
Nutrition	Severe protein malnutrition doubles the risk of birth defects. Fewer brain cells are produced, and learning ability may be permanently affected.	Growth rate is mainly determined by rate of net protein synthesis by embryo's cells.
Excessive amounts of vitamins	Vitamin D is essential, but excessive amounts may result in a form of mental retardation. Too much vitamin A and K may also be harmful.	Vitamin supplements are normally prescribed for pregnant women, but some women mistakenly reason that if one vitamin pill is beneficial, four or five might be even better.
Drugs	Many drugs affect the development of the fetus. Even aspirin has been shown to inhibit growth of human fetal cells (especially kidney cells) cultured in the laboratory. It may also inhibit prostaglandins, which are concentrated in growing tissue.	Common prescription drugs are generally taken in amounts based on the body weight of the mother, which may be hundreds or thousands of times too much for the tiny embryo.
Alcohol	When a woman drinks heavily during pregnancy, the baby may be born with fetal alcohol syndrome, that is, deformed and mentally and physically retarded. Low birth weight and structural abnormalities have been associated with as little as two drinks a day. It is thought that some cases of hyperactivity and learning disabilities may be caused by alcohol intake of the pregnant mother.	Fetal alcohol syndrome is thought to be one of the leading causes of mental retardation in the United States.
Heroin	Maternal heroin use is associated with a high mortality rate and high prematurity rate.	Infants that survive are born addicted to the drug and must be treated for weeks or months.
Methadone	Methadone use by mother results in addiction also.	
Thalidomide	Thalidomide, marketed as a mild sedative, was responsible for more than 7000 grossly deformed babies born in the late 1950s in 20 different countries. Principal defect was **phocomelia,** a condition in which babies are born with extremely short limbs, often with no fingers or toes.	Interferes with cellular metabolism; most hazardous when taken during the fourth to sixth weeks, when limbs are developing.
Cigarette smoking	Cigarette smoking reduces the amount of oxygen available to the fetus because both maternal and fetal hemoglobin are combined with carbon monoxide. May slow growth and can cause subtle forms of damage. (In extreme form carbon monoxide poisoning causes gross defects such as hydrocephaly.)	Mothers who smoke deliver babies with lower than average birth weights. They also have a higher incidence of spontaneous abortions, stillbirths, and neonatal deaths. Studies also indicate a possible link between maternal smoking and slower intellectual development in the offspring.
Pathogens	Rubella (German measles) virus crosses the placenta and infects the embryo; interferes with normal metabolism and cell movements. Causes a syndrome that involves blinding cataracts, deafness, heart malformations, and mental retardation. The risk is greatest (about 50%) when a pregnant woman contracts rubella during the first month of pregnancy; it declines with each succeeding month.	A rubella epidemic in the United States in 1963–65 caused about 20,000 fetal deaths and 30,000 infants born with gross defects.
	Syphilis is transmitted to the fetus in about 40% of infected women. Fetus may die or be born with defects and congenital syphilis.	Pregnant women are routinely tested for syphilis during prenatal examinations.
Ionizing radiation	When mother is subjected to x rays or other forms of radiation during pregnancy, infant has higher risk of birth defects and leukemia.	Radiation was one of the earliest teratogens to be recognized.

Chapter 11, a photograph taken of the embryo by using ultrasound. Such previews are helpful in diagnosing defects and also in determining the position of the fetus, and whether a multiple birth is pending.

THE HUMAN LIFE CYCLE

Development begins at conception and continues through the stages of the human life cycle until death (see Table 51–4). We have examined briefly the development of the embryo and fetus, the birth process, and the adjustments it requires of the neonate. The **neonatal period** is usually considered

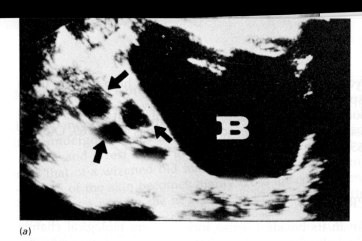

(a)

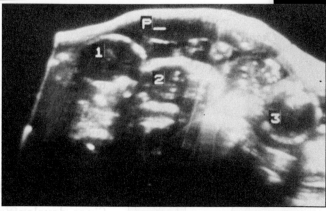

(b)

to extend from birth to the end of the first month of extrauterine life. **Infancy** follows the neonatal period and lasts until the rapidly developing infant can assume an erect posture (i.e., can walk), usually between 10 and 14 months of age. Some regard infancy as extending to the end of the first 2 years. **Childhood,** also a period of rapid growth and development, continues from infancy until adolescence.

Adolescence is the time of development between puberty and adulthood. During adolescence a young person experiences the physical and physiological changes that result in physical and reproductive maturity (Figures 51–29 and 51–30). This is also a time of profound psychological

FIGURE 51–28 Ultrasonic techniques can be used to monitor follicle maturation and ovulation, as well as to give the physician information about the fetus. (a) Sonogram taken with ultrasound techniques, showing three follicles of equal maturation in the left ovary of a human. (b) Triplets in the same patient at 16 weeks of pregnancy. P, placenta. Such previews are valuable to the physician in diagnosing defects and predicting multiple births.

Table 51–4 STAGES IN THE HUMAN LIFE CYCLE

Stage	Time Period	Characteristics
Embryo	Conception to end of eighth week of prenatal development	Development proceeds from single-celled zygote to embryo that is about 30 mm long, weighs 1 g, and has rudiments of all its organs
Fetus	Beginning of ninth week of prenatal development to birth	Period of rapid growth, morphogenesis, and cellular differentiation, changing tiny parasite to physiologically independent organism
Neonate	Birth to 4 weeks of age	Neonate must make vital physiological adjustments to independent life: it must now process its own food, excrete its wastes, obtain oxygen, and make appropriate circulatory changes
Infant	End of fourth week to 2 years of age (sometimes, ability to walk is considered end of infancy)	Rapid growth; deciduous teeth begin to erupt; nervous system develops (myelinization), making coordinated activities possible; language skills begin to develop
Child	Two years to puberty	Rapid growth; deciduous teeth erupt, are slowly shed and replaced by permanent teeth; development of muscular coordination; development of language skills and other intellectual abilities
Adolescent	Puberty (approximately ages 11–14) to adult	Growth spurt; primary and secondary sexual characteristics develop; development of motor skills; development of intellectual abilities; psychological changes as adolescent approaches adulthood
Young adult	End of adolescence (approximately age 20) to about age 40	Peak of physical development reached; individual assumes adult responsibilities that may include marriage, fulfilling reproductive potential, and establishing career; after age 30, physiological changes associated with aging begin
Middle-aged adult	Age 40 to about age 65	Physiological aging continues, leading to menopause in women and physical changes associated with aging in both sexes (e.g., graying hair, decline in athletic abilities, wrinkling skin); this is period of adjustment for many as they begin to face their own mortality
Old adult	Age 65 to death	Period of senescence (growing old); physiological aging continues; maintaining homeostasis more difficult when body is challenged by stress; death often results from failure of cardiovascular or immune system

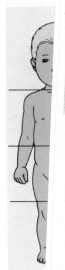

FIGUI
head,
child

FIG
grov
syst
mer

Associative Learning

In **associative learning,** an animal's behavior changes because of an association it makes between two stimuli. There are two types of associative learning: classical conditioning and operant (instrumental) conditioning.

Classical Conditioning

In **classical conditioning,** a normal response to a stimulus becomes associated with a new stimulus, which then is also capable of eliciting the response. The eminent Russian physiologist Ivan Pavlov studied this type of learning by carefully measuring saliva secretion when dogs were presented food. The food in this case is the normal or **unconditioned stimulus** (UCS) and the act of salivating is the **unconditioned response** (UCR). When a neutral stimulus such as a ringing bell is presented to the dog prior to the food, the dog makes an association between the neutral stimulus (called the **conditioned stimulus,** or CS) and the food (UCS). After the UCS and CS are paired for a sufficient number of times, the salivating response becomes a **conditioned response** (CR). The dog now salivates when the CS is presented. The CR may vary slightly from the UCR, but it is basically the same (in this case, both are salivation).

Each time the conditioned stimulus (the bell) is followed by the unconditioned stimulus (the food), the conditioned response (salivation) is *reinforced,* and for this reason the UCS is referred to as a **reinforcer.** The CS must always precede the UCS because, once a relationship between the two is established, the CS serves as a signal that the UCS will occur. The dog now salivates in response to the bell and thus has learned to give an old response to a new stimulus.

Because of their historical import, Pavlov's experiments with dogs are frequently cited as a paradigm (example or model) for classical conditioning. Another example is an aquarium fish that learns to swim to the surface for food (CR) when the glass is tapped (CS). The UCS is the odor (or sight) of the food and the UCR is the fish's response to this odor or sight (i.e., swimming to the surface). Thus tapping and odor (or sight) have been linked by classical conditioning. No doubt classical conditioning of this sort occurs commonly among vertebrates. As soon as a fishing vessel enters the harbor, gulls resting on the dock become excited and begin flying around, as they have associated the shrimp boat (CS) with the presence of offal (UCS). The UCR in this case is their normal response to the presence of food. Because the UCS and CS have been paired, their response to the presence of the fishing vessel is the CR.

In both sensitization and classical conditioning, an animal's response to a stimulus depends on another stimulus, but in classical conditioning the associated stimulus is specific: the fish swims to the surface of the aquarium in response to a stimulus specifically associated with food. In contrast, after being repeatedly fed, a sensitized fish may approach *any* stimulus, even nonfood objects that only remotely resemble food.

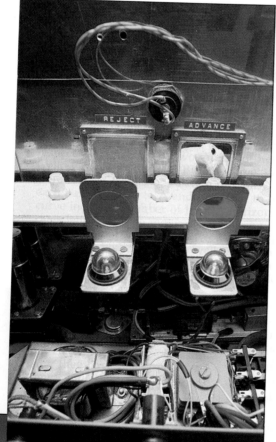

FIGURE 52–14 To demonstrate operant conditioning, an animal is placed in a chamber that contains a button which, when pressed, releases food into the chamber. The animal makes an association between pressing the button and receiving the food; thereafter, every time it presses the button, this behavior is reinforced with food.

Operant Conditioning

In **operant (instrumental) conditioning,** a type of learning extensively studied in the United States by B.F. Skinner and his colleagues, the sequence of events is contingent (i.e., dependent) upon the behavior of the animal. The animal is rewarded or punished after performing a behavior it discovers by chance. (For this reason, this type of learning has also been called "trial and error learning.") In a typical experiment, an animal moves about in a cage (Figure 52–14) and eventually presses a button, without knowing beforehand that pressing the button will release food into the

FIGURE 52–15 Operant conditioning may occur in nature when predators, such as this gray fox, discover an area where prey is plentiful. They make an association between visiting that area and capturing prey (the reward).

cage. The animal makes an association between the button and the food (the reward). Thereafter, whenever it presses the button, this behavior (pressing the button) is reinforced with the food.

In classical conditioning the reinforcer is the UCS, because this is the stimulus that actually reinforces the learned behavior (the CR). In operant conditioning, the reinforcer is the reward (e.g., the food) that follows when the animal behaves in a certain way (e.g., pressing the button).

Examples of operant conditioning in nature are not difficult to imagine. A fox randomly searches the forest for prey and chances upon two areas: one where rodents are plentiful and easily caught and another where vicious dogs always drive off intruders like foxes (Figure 52–15). Very quickly the fox learns to frequent the area with the rodents and to avoid the dogs, just as experimental animals learn to press certain buttons and not others.

There are many examples of operant conditioning in invertebrates. Larval grain beetles can learn mazes, and, rather interestingly, this learning survives through metamorphosis into the adult stage. Cockroaches can be conditioned to avoid shock, even when their heads have been removed. This form of learning (passive avoidance) can be demonstrated by delivering an electric shock to the leg when the roach lowers its leg into a saline solution. Roaches that have been shocked lower their leg into the solution fewer times than do controls.

In both classical and instrumental conditioning the learned response must be periodically reinforced or it will not persist. In fact, periodic reinforcement is more effective than continuous reinforcement in maintaining an operant response. If the response is insufficiently reinforced, it diminishes. This is known as **extinction.**

Latent Learning

In **latent learning,** sometimes called ''exploratory learning,'' an animal stores information about its environment that later can influence its behavior. Rats permitted to explore a maze without any apparent reward nonetheless store certain information about the maze (Figure 52–16). When later they are trained to run the maze for a reward (food), they learn to do this more quickly than rats not previously exposed to the maze. Obviously some type of learning occurred when the rats explored the maze. If, as has been suggested, latent learning is a variant of operant conditioning, the rat must have been rewarded in some way for learning some of the maze's features while it was exploring the maze. Perhaps the learning that accompanies exploration is reinforced by the reduced anxiety that comes with becoming familiar with one's surroundings; or maybe the mere process of observing and remembering features about one's surroundings, because of a ''natural'' curiosity, is a sufficient reinforcer.

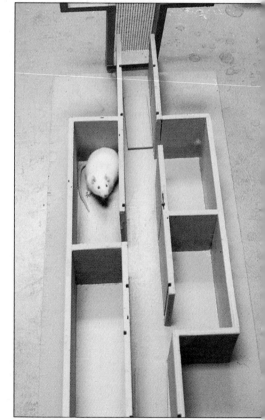

FIGURE 52–16 Rats that have previously been placed in a maze learn more quickly to escape the maze for a reward than do rats placed in the maze for the first time. Therefore, the first group must have ''learned'' the maze, even though it is not clear what the reward was. This type of learning is known as latent learning.

FIGURE 52–17 Insight learning, and in this case, simple tool use. Confronted with the problem of reaching food hanging from the ceiling, the chimpanzee stacks boxes until it can climb and reach the food. Many other examples of apparent insight are known from the behavior of these animals.

Insight Learning

Insight learning is perhaps the most difficult type of learning to demonstrate conclusively in animals, even though we take it for granted in ourselves. Insight learning is the intuitive ("insightful") solution to a problem that has not been previously attempted. One criterion for insight is what has been called the "aha experience" (one "sees the light!"). An animal may be familiar with the separate parts of a problem that are necessary for its solution; but it solves a problem through insight only when it recognizes the relationships, causes and effects of a particular contingency or occurrence. If it is capable of doing this, it is capable of solving problems it has never encountered before.

More than seventy years ago the psychologist Wolfgang Köhler offered evidence for reasoning (or insight) in chimpanzees. He observed that they obtained bananas suspended from the ceiling and out of reach by first stacking boxes to make a platform (Figure 52–17). They then climbed on top of this structure to reach the bananas.

Even this impressive performance may not demonstrate conclusively that animals reason. Critics point out that all of the manipulations observed in these chimps—stacking and climbing on boxes, striking objects with sticks, etc.—are part of the repertoire of play behaviors in chimps. They may have climbed on the stacked boxes while playing. In other words, we may not have to use insight learning to explain their success, since they may not have seen obtaining the banana as a problem to be solved at all! Although there are other observations of primate behavior that suggest insight learning in animals, the issue has not been settled.

Imprinting

Imprinting is a very specialized type of learning. A young waterfowl, such as a gosling, forms a learned attachment to its mother during the first few days of its life. This behavior, known as imprinting, is adaptive since it insures that the gosling will follow its mother at this critical time of its life (Figure 52–18). Its chances of surviving increase because of the mother's protection. Imprinting does not occur in birds that remain helpless in the nest until they mature and are ready to leave. An ability to imprint in these species would serve no useful function.

Imprinting differs from most types of learning in that it normally can occur only at a specific time during an animal's life. This **critical period** is usually shortly after hatching (about 12 hours), and it generally lasts for

FIGURE 52–18 Imprinting. When young, these geese formed an attachment to a human being that has remained firm into their maturity.

just a few days. At this time a young bird will imprint on an adult of its own species if given a choice. In the absence of an adult bird it will imprint on any moving object, including a person. It has been shown experimentally that birds imprint more firmly on nonliving structures if these structures have moving parts and sound-producing devices. Also, birds imprint more firmly on an object if they must crawl over obstacles in their path in order to follow it.

Imprinting shares some features with song learning in some birds. Many young birds learn to sing their own species' song correctly only if they hear it at a critical period during their life. Some scientists have suggested that children learn languages more easily than adults because they are in a critical period for language acquisition comparable to the critical period for song learning in birds.

BEHAVIORAL ECOLOGY

Natural selection tends to produce animals that are maximally efficient at propagating their genes, and this is accomplished by making the best choices in mate selection, territory defense, and foraging. A topic that currently interests a number of behavioral ecologists is **optimal foraging,** which is theoretically the most efficient way for an animal to obtain food.

Optimal foraging may be illustrated by the following example, which is fictitious for the sake of simplicity. Suppose there exists a monkey that eats only one species of fruit and that this fruit is contained in a husk that is either soft, leathery, or hard. It takes 1 minute to remove a soft husk, 5 minutes to remove a leathery husk, and 10 minutes to remove a hard husk. If the forest contains equal numbers of hard-husked, leathery-husked, and soft-husked fruits, if these three forms are randomly distributed, and if a monkey moves randomly and eats every fruit it encounters, then at the end of a foraging bout it should have eaten about the same number of each husk form.

But is it necessarily the best use of the monkey's time and energy to eat all three husk types? Since removing hard husks takes so long, perhaps it is more efficient to pass up these fruits and eat only the soft-husked and leathery-husked fruits. Or maybe the best procedure is to eat only the soft-husked fruits.

These three possible ways of feeding are examples of **feeding strategies.** There are two other feeding strategies: select only the hard-husked fruits, or select the hard-husked fruits and the leathery-husked fruits and pass up the soft-husked fruits. These two strategies are clearly less efficient ways to forage, since they require a greater expenditure of time and energy for the same amount of food. The term **strategy** used in this context does not imply planning as it does in human behavior, but refers to the way an animal locates, procures, and handles food.

Assuming that once husked, all three forms can be eaten in the same brief time, we may ask which of the following strategies is most adaptive, that is, which one permits the monkey to forage optimally:

Strategy I: select the soft-husked fruits whenever encountered and pass up the other two kinds
Strategy II: select the soft-husked and leathery-husked fruits whenever encountered, and pass up the hard-husked fruits
Strategy III: select all three kinds whenever encountered

Strategy I has its trade-offs. Since only every third fruit is soft-husked, the monkey must travel three times as far to get the same number of fruits

B. In conflict situations, animals may direct their attack behavior toward a substitute object (redirected behavior) or may behave in a way that appears irrelevant or inappropriate (displacement behavior).

C. Behaviors such as FAPs may acquire a communication function during evolution, a process known as ritualization, during which they become incorporated into displays.

D. The application of ethological methods to the study of human behaviors is known as human ethology.

IV. Most behaviors are regarded as having both genetic ("nature") and environmental ("nurture") components.

A. The "blueprint" of a behavior is the DNA that is responsible for the development of an organism. The ways an organism responds to external stimuli are limited by the genotype, but can be altered within these limitations by the environment (learning).

B. The genetic and environmental components of behavior can be dealt with at the population level in terms of heritability: a measure that compares the amount of phenotypic variability that is due strictly to the variation in genes with the total amount of phenotypic variability.

V. Learning, in which a behavior is altered by a previous experience, may be classified as follows.

A. In habituation, an animal's unlearned response to a constant or repeated stimulation wanes.

B. In sensitization, an animal's response to a stimulus becomes more likely as the stimulus is repeated.

C. In classical conditioning, an animal makes an association between a neutral stimulus (conditioned stimulus) and a response (conditioned response). Prior to conditioning, the response (then called an unconditioned response) was elicited only by the normal stimulus (unconditioned stimulus).

D. In operant conditioning, the sequence of events depends on the behavior of the animal, which is rewarded or punished after performing a behavior it discovers by chance.

E. Other types of learning include latent learning, in which the animal stores information about the environment that later can influence its behavior; and insight learning, the intuitive solution to a problem that has previously not been attempted.

F. In imprinting, an animal forms a learned attachment to an object (usually its mother) during a critical period of its life, normally the first few days.

VI. Behavioral ecology is concerned with topics such as optimal foraging, which is theoretically the most efficient strategy for an animal to obtain food.

VII. The study of animal migration includes (1) the stimulus to migrate and (2) the mechanisms that allow organisms to orient and migrate.

VIII. Social behavior deals with the interactions between animals.

IX. Animal communication involves the transmission of signals but does not utilize (as far as is known) symbolic language in the human sense.

A. Animals often transmit signals involuntarily as a result of their physiological state.

B. Pheromones are chemical signals that convey information between members of a species.

X. Dominance hierarchies result in the suppression of aggressive behavior.

XI. Organisms often inhabit a home range, from which they seldom or never depart. This range, or some portion of it, may be defended from members of the same (or occasionally different) species.

A. Defended areas are called *territories*, and the defensive behavior is *territoriality*.

B. Often, territorial defense is carried out by display behavior rather than actual fighting.

XII. Courtship behavior ensures that the male is a member of the same species, and it permits the female to assess the quality of the male.

A. A pair bond is a stable relationship between a male and a female that ensures cooperation behavior in mating and rearing the young.

B. Parental care increases the probability that the offspring will survive. A high investment in parenting is less advantageous to the male than to the female.

XIII. Play gives the young animal a chance to practice adult patterns of behavior.

XIV. Sociobiology focuses on the evolution of social behavior through natural selection.

■ POST-TEST

1. _____ is the study of species-typical behaviors in natural environments from the point of view of adaptation and evolution.

2. The long-term evolutionary causes of behavior are called _____ factors.

3. In the honey bee's dance, the direction of the food source, relative to the sun, is indicated by the angle of the _____ with respect to the vertical axis.

4. The stimulus that elicits a FAP is called a _____ stimulus.

5. The genotype in worker bees for both uncapping and removing dead larvae from cells is _____.

6. In the type of learning known as _____,

there is an increased probability than an animal will respond to a stimulus that has been presented to it.

7. In Pavlov's classical conditioning experiments, the bell is the _____ stimulus.

8. Optimal foraging theory takes into account variables such as traveling time, _____ time, and eating time.

9. _____ are visual signals that include movements, postures, and facial expressions.

10. The ability of an animal in unfamiliar surroundings to find its goal without relying on familiar landmarks is called _____.

11. An important difference between human and animal

communication is that animal communication is not generally _____ .

12. _____ are chemical signals that convey information between members of a species.

13. An arrangement of members of a population by status is called a _____ _____ .

14. The geographical area that members of a population seldom leave is the _____ _____ .

15. Territoriality tends to reduce _____ and control _____ growth.

16. A _____ _____ is a stable relationship between animals of the opposite sex that ensures cooperative behavior in mating and rearing the young.

17. According to sociobiology, an organism and its adaptations are ways that its genes have of _____ _____ .

■ REVIEW QUESTIONS

1. Discuss possible proximate and ultimate factors for urine marking in dogs.
2. By considering the defining characteristics of FAPs, argue for or against the position that sucking behavior in a newborn baby is a FAP.
3. In what ways do behavior patterns change through ritualization?
4. Distinguish between a sign stimulus and an innate releasing mechanism.
5. Discuss how the concept of heritability can be applied to the breeding of animals that have a specific behavioral trait.
6. Give an example of operant conditioning in a wild animal, and explain why this learned behavior would be due to operant rather than to classical conditioning.
7. Give an example of classical conditioning in invertebrates and explain why it is classified as such by comparing it to Pavlov's experiments on dogs.
8. Explain, using a graph, the relationship between traveling time, handling time, and food abundance with respect to optimal foraging.
9. Discuss circannual rhythms and their relevance to the annual stimulus for migration in birds.

■ RECOMMENDED READINGS

Bonner, J.T. *The Evolution of Culture in Animals*. Princeton, Princeton University Press, 1980. The title tells the contents but fails to convey the charm of this scholarly little book.

Borgia, G. "Sexual selection in bowerbirds," *Scientific American*, June 1986. Males compete fiercely for mating sites; the females mating choice is based on architectural adornment.

Burgess, J.W. "Social Spiders," *Scientific American*, March 1976. A few species of spiders interact socially and build large communal webs.

Camhi, J.M. "The Escape System of the Cockroach," *Scientific American*, December 1980. A study of the mechanisms by which a roach rapidly escapes from predators.

Crews, D., and W.R. Garstka. "The Ecological Physiology of a Garter Snake," *Scientific American*, November 1982. An account of the physiological and behavioral adaptations of the red-sided garter snake to its harsh environment.

Gallup, G.G., Jr. "Self-Awareness in Primates," *American Scientist* July–August 1979. Chimpanzees and the like may, in a sense, be persons.

Gould, J.L. *Ethology*. New York, W.W. Norton, 1982. The best and most up-to-date general text so far.

Gould, S.J. "The Guano Ring," *Natural History*, January 1982. An easily studied and interpreted example of territoriality.

Grier, J.W. *Biology of Animal Behavior*. St. Louis, Times Mirror/Mosby, 1984. A well-written introduction to animal behavior integrating the structure and function of behavior with ethology, comparative psychology, and neurobiology.

Heinrich, B. "The Regulation of Temperature in the Honeybee Swarm," *Scientific American*, June 1981. A discussion of thermoregulation in a swarm of bees.

Ligon, J.D., and S.H. Ligon. "The Cooperative Breeding Behavior of the Green Woodhoopoe," *Scientific American*, July 1982. Kin selection and altruism in a vertebrate society.

Matthews, G.V.T. Orientation and Position-Finding by Birds (booklet). Burlington, N.C., Carolina Biological Supply Company, 1974. Though written before the magnetic senses of animals were at all understood, this is a good summary of research on the use of visual cues by birds in navigation.

Partridge, B.L. "The Structure and Function of Fish Schools," *Scientific American*, June 1982. Schooling confuses predators and is coordinated by lateral line sensory input.

Von Frisch, K. *Animal Architecture*. New York, Harcourt Brace Jovanovitch, 1974. The construction projects of animals, from spider webs to anthills, not forgetting birds' nests.

Wicksten, M.K. "Decorator Crabs," *Scientific American*, February 1980. A discussion of a species of spider crabs that camouflage themselves by attaching materials to their exoskeletons.

Wilson, E.O. *Sociobiology, the New Synthesis*. Cambridge, Mass., N.H. Belknap Press, 1975. An important (to say the least) summary of Wilson's proposed mechanisms of the evolution of altruism and his views on kin selection.

Wursig, B. "Dolphins," *Scientific American*, March 1979. An interesting description of dolphin behavior and learning ability.

PART VIII
ECOLOGY

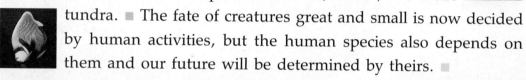There is no isolation in the biosphere. ■ The creatures of the deepest ocean abyss are linked by atmosphere and water to those of the tropical rain forest, desert, and arctic tundra. ■ The fate of creatures great and small is now decided by human activities, but the human species also depends on them and our future will be determined by theirs. ■

Sunrise on Marco Island, Florida

53

The Planetary Environment

■ **LEARNING OBJECTIVES**

After you have read this chapter you should be able to:
1. List the main determinants of the earth's climate.
2. Summarize the factors that compose the planetary heat budget.
3. Discuss the role of convection and the Coriolis effect in the production of global air and water flow patterns.
4. Summarize the hydrologic cycle.
5. Give the main causes of precipitation.
6. Summarize the processes of soil formation, giving the main varieties of soil and their primary characteristics.

THE SUBSTANCE OF ECOLOGY

Ecology is at least three things—a concept, a science, and, in our day, a political movement. The concept of ecology was first popularized in the 19th century by Ernst Haeckel, who also disseminated its name—*eco* from the Greek for "household," and *logy* from the Greek word for "study," or "body of instruction." Nature, viewed from the standpoint of ecology, is something like a great estate in which the nonliving environment and living things interact in an immense web of relationships.

Ecologists study those relationships, not only among organisms, but between single organisms and their environment. Ecology can be global and generalized, or very specific, depending on the viewpoint of the scientist. Thus an ecologist could work with species diversity in an oak forest or determine the mineral and light requirements of a single species of oak tree. Ecology is perhaps the broadest of the biological sciences, with explicit links to behavior, physiology, biochemistry, genetics, and indeed every biological discipline.

The universality of ecology brings subjects into view that are not traditionally part of biology. Geology and earth science are extremely impor-

FIGURE 53–1 The ecosphere. No significant loss or gain of matter occurs on the Earth as a whole. All materials utilized by living things must be continuously recycled within this vast, closed system.

tant, for the planetary environment must be taken into account. And since human beings are organisms, all of their activities have a bearing on ecology. Even economics and politics have profound ecological implications. We shall explore as many of these relationships as possible in the ensuing chapters, but it seems logical to begin with a discussion of the great planetary incubator upon which all living things are dependent—the Earth itself.

THE PLANETARY ENVIRONMENT

Almost completely isolated from all but sunlight, our planet has often been compared to a vast spaceship, one whose life-support system is life itself. The living things aboard this great spacecraft produce oxygen, cleanse its air, adjust its gases, transfer energy, and recycle waste products with great efficiency. Yet none of these things would be possible without the nonliving environment of the world, the plumbing of our spaceship Earth. Since the science of ecology deals with the nonliving environment as well as the living, it is fitting to call this, the greatest of all systems of life, the **ecosphere** (Figure 53–1).

The Earth is composed of the **biosphere,** the totality of its living inhabitants; the **hydrosphere,** its supply of water (both frozen and liquid); its gaseous **atmosphere;** and, bulking by far the greatest, its **lithosphere,** that is, rocks and their derivatives. The nonliving environment is influenced by the living biosphere, but all of the constituents of the biosphere came ultimately from the solid crust of the planet, the lithosphere.

The influence of the living biosphere on the nonliving environment can be illustrated by the producers, without whose photosynthesis (Figure 53–2) conditions on the Earth would resemble those that exist currently on the planet Venus. Surface temperatures would range in hundreds of degrees Celsius, there would be practically no oxygen in the atmosphere, and possibly no water. The composition of the surface of the lithosphere also would be profoundly different from what we know.

Mineral substances used by living things, such as phosphorus or iodine, obviously originated in the lithosphere, and so, less obviously, does the global supply of carbon dioxide and water. These substances are continually emitted by volcanoes. Ultimately, the oxygen that primary producers make from water is a derivative of those substances, and hence of the lithosphere. Minerals are continuously rendered soluble by living things, leached from the lithosphere by percolating water, incorporated temporarily into the bodies of organisms, and often deposited as their remains in sedimentary rocks in the ocean depths. From there they may be uplifted to form continents, or subducted into the Earth's interior. Vast deposits of phosphate rock, of calcium carbonate, and of silica represent eons of accumulated skeletons of once-living things. Spottily recycled by geological processes to the Earth's surface, these minerals may eventually reenter the biosphere.

Sunlight

Fusion power, sought as the energy basis of future civilization, is really the oldest of all energy sources, for it has powered the Earth since its creation. In the sun's core, highly compressed, heated hydrogen atoms are forced to fuse with one another, producing helium. Every second, an estimated 596 million metric tons of hydrogen become 532 metric tons of helium. The difference, some 44 million metric tons of mass, does not simply disappear but is converted into energy, released mainly as heat and light. Virtually all of this energy is emitted to outer space in the form of radiation—light,

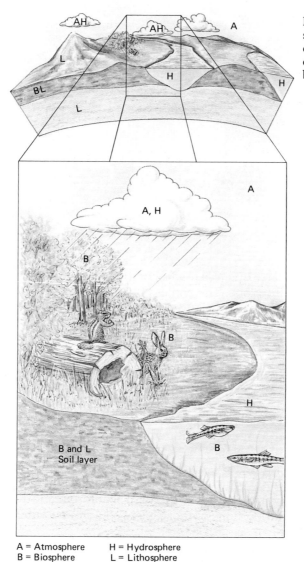

A = Atmosphere   H = Hydrosphere
B = Biosphere   L = Lithosphere

infrared radiation, and other varieties. An infinitesimal portion of this energy strikes the Earth's atmosphere, and a minute part of this tiny trickle of energy operates the biosphere. All living things ultimately operate on solar power, but solar energy also directly influences the nonliving environment, particularly the hydrosphere and atmosphere.

The "antennae" that collect solar radiation to power the life processes of the Earth's organisms are the mostly green leaves and photosynthetic tissues of plants and certain microorganisms. Only the sunlight actually reaching photosynthetic tissues can be used for photosynthesis. Thus very little of the sunlight falling on large areas of mountainous regions, extreme deserts, and similar places nearly void of plant life is transformed into biotic energy.

Even when photosynthetic organisms are present, they seldom cover an area completely. By far most of what is absorbed simply heats the rocky surface, the water, and the atmosphere. Even that which is absorbed by primary producers is eventually released as heat by one metabolic process or another (unless the organisms incorporating it are geologically buried to form coal or oil deposits). Almost all delays in the flow of heat energy are temporary. All of the absorbed heat, plus whatever is released by nuclear reactions in the depths of the Earth, is ultimately radiated away.

The slightest imbalance between absorption and output has far-reaching

FIGURE 53–3 The amount of energy per unit area that the Earth's surface receives is determined by the angle at which the sun's rays strike the Earth. This angle varies from place to place due to the spherical shape of the Earth and the inclination of its axis. In this diagram, which represents the situation in the month of June, solar radiation strikes the Earth perpendicularly in the Northern Hemisphere but very obliquely in the Southern Hemisphere. This difference produces summer conditions in the Northern Hemisphere and winter in the Southern Hemisphere, both at the same time.

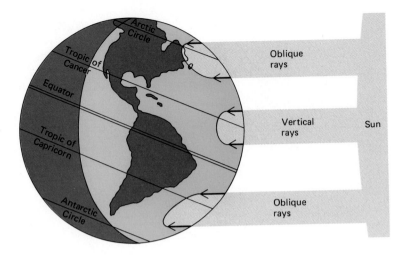

consequences to raise or lower the surface and atmospheric temperature of the globe. The endless fluctuations of what we call normal climate depend upon that balance, as do droughts, floods, and ice ages.

The Heat Budget of the Earth

In the daytime, 23% of the solar radiation that falls upon the Earth is immediately reflected by clouds, and another 3% by surfaces, especially snow, ice, and oceans. Yet another 14% is lost by diffusion and direct reradiation of light. The remaining 60% is absorbed but is eventually lost by radiation of long-wave infrared energy into space, not only during the day but at night. Notice that these figures add up to 100%. If they did not, the temperature of the Earth would fall or rise until energy output exactly equalled input once again. In fact, the atmosphere keeps the Earth's surface much warmer than it otherwise would be, and relatively minor changes in the proportions of those gases could cause substantial further warming, as we shall see in Chapter 57.

These figures are planetary averages and may vary substantially because of local conditions. For example, high clouds (even of dust or soot) increase energy reflection. Low clouds and even water vapor increase energy absorption, as does carbon dioxide. The most significant local variation, however, is produced by a combination of the Earth's roughly spherical shape and the oblique inclination of its axis of rotation. These factors produce great variation in **insolation,** the energy delivered by sunlight.

Lightfall and Insolation

The fall of light on the face of the Earth is determined by two main factors: distance from the sun and the inclination of the Earth's axis. While distance from the sun varies during the year, its effect upon climate and season is tiny compared to that produced by the inclination of the Earth's axis.

Since this inclination is always the same, 23 1/2 degrees, during half of the year the Northern Hemisphere is inclined *toward* the sun, and during the other half, *away* from the sun (Figure 53–3). Naturally, the orientation of the Southern Hemisphere is just the opposite at these times (see also Figure 53–4). The principal difference this inclination makes is seen in the angles at which the sun's rays strike the Earth at any one time.

Owing to the size of the sun's radiating surface, rays of light emitting from it are approximately parallel to one another. On the average, they fall most nearly in a vertical fashion near the Equator but more and more obliquely as one approaches the poles. This means that the energy content of the light is spread out over a larger area of surface near the poles, for there

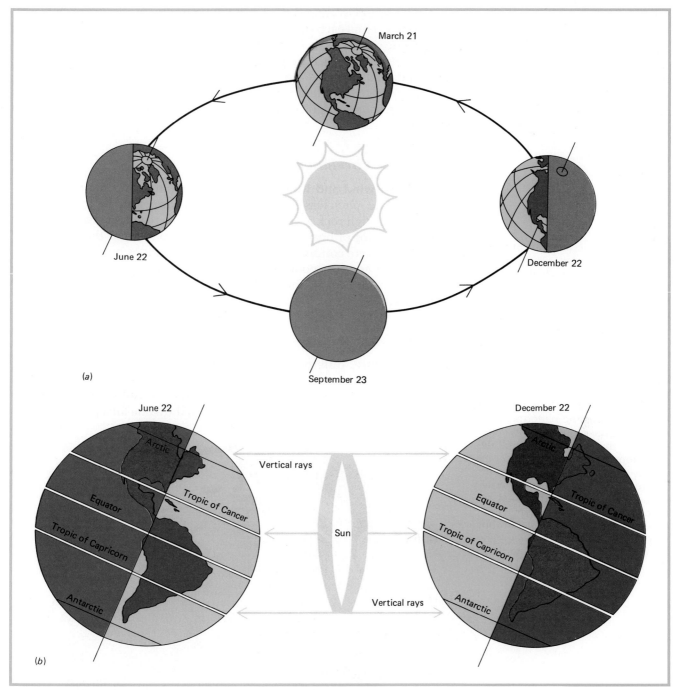

(a)

(b)

FIGURE 53–4 As the Earth travels around the sun, the inclination of its axis remains the same. As a result, the sun's rays strike the Northern and Southern Hemispheres obliquely at different times of the year. Notice that the absolute angle of the sun's rays at the equator is more nearly the same at all seasons.

the rays are almost parallel to the surface of the planet. This reduces the light's ability to heat the surface. Also, rays of light entering the atmosphere obliquely near the poles must usually pass through more air than those falling near the Equator, allowing more of their energy to be reflected out to space. Even if the axis of the Earth were not inclined at all, these factors would be sufficient to produce great climatic variations at different latitudes. To sum up, climatic differences between the various parts of the Earth are fundamentally determined by the spherical shape of the Earth, which, in turn, determines the angle with which the sunlight strikes the Earth at different latitudes.

Figure 53–4 displays the relationship of the Earth's surface to the sun's rays in the month of June. Notice that at this time they actually fall vertically on the surface of the Earth at the latitude of the Tropic of Cancer. Six months later, in December, they will fall vertically on the Tropic of Capricorn instead. Thus each of these nonequatorial latitudes experiences tropical sunshine once a year. By contrast, the seasons in equatorial regions are likely to be distinguished by precipitation, so that wet and dry, rather than warm and cold, seasons are the rule. At sea level, these areas have an endless summer. The temperate and polar zones have pronounced cool and warm seasons, opposite from one another in the Northern and Southern Hemispheres.

Atmospheric Circulation

The action of wind and water affect the atmosphere and hydrosphere by making climate much less extreme over the surface of the Earth. The transfer of heat through rock is very slow, since rock is really an insulator rather than a conductor of heat. The poles would be colder and the Equator far hotter if heat were conducted between the two regions only by rock. Both air and water are also poor conductors of heat, but since they are fluid, they can transfer heat by their motion. Such heat transfer is usually called **convection.**

Convective Heat Transfer

In large measure, differences in temperature drive the circulation of the atmosphere. The heated surface of the Equator heats the air that is in contact with it, causing this air to expand. As it rises, it cools and sinks again. Much of it recirculates almost immediately to the same areas it has left, but

FIGURE 53–5 Because the sun heats the Earth's surface unequally, convective forces create atmospheric movement. The greatest solar energy input occurs at the Equator, heating air most strongly in that area. The air rises and travels poleward, but is cooled in the process so that much of it descends again around 30 degrees latitude in both hemispheres. The rising air produces a prevailing high-pressure area in most equatorial regions. At higher latitudes the patterns of movement are more complex. For example, the small circles at 50 degrees represent the storm-generating jet streams. Heavy rains occur where air rises and cools, especially in the tropics. Where the dried air descends deserts may occur. Polar deserts have little precipitation but may be less extreme than temperate or tropical deserts due to the low rate of evaporation that prevails there.

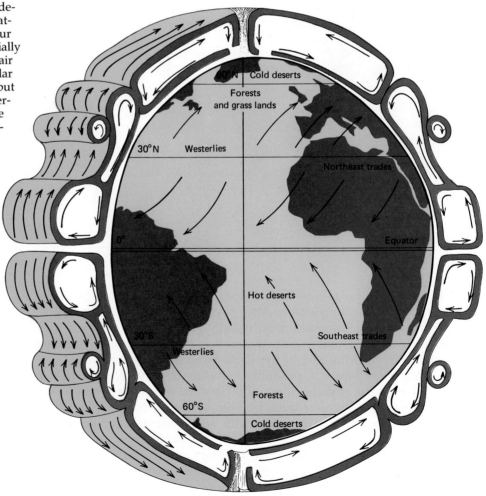

most of the remainder flows poleward, where eventually it is chilled. Similar local convection and poleward flow occur at higher latitudes as well (Figure 53–5). As air cools by contact with the polar ground and ocean, it sinks to flow toward the Equator, generally beneath the sheets of warm air flowing poleward at the same time. Also, seasonal heating of the polar regions and midlatitudes prevents the stagnation of air, so that constant fluxes of air masses do battle over the temperate zones. This produces much more variable weather than is seen at the Equator. Most important, this constant flux of air transfers heat from the Equator toward the poles and, in returning, cools the land over which it passes. This constant turnover does not equalize temperatures over the surface of the Earth but does moderate them. The atmosphere circulates upward from the Equator and for the most part returns to the Equator without substantial detour, although it is dryer.

The Coriolis Effect

The Earth's rotation also contributes indirectly to air movement (and to water movement as well). A person standing exactly at the North Pole would hardly become dizzy despite the rotation of the Earth, since he would rotate only once a day. Now imagine hiking (or sledding) a kilometer south of the pole. Should he desire to do so, even wearing snowshoes he could easily keep up with the Earth's rotation.

By similar thinking, a sedentary Soviet citizen of Leningrad is moving rotationally at a speed of 830 kilometers per hour (kmph) (Figure 53–6), while at the same time a Kampalan farmer in Africa is zipping along at 1660 kmph owing to the fact that he is sitting on the equator and therefore must traverse a much larger circle in the course of a day.

If a spacecraft were to orbit from Leningrad to Kampala at such a speed that it would take an hour to arrive, its path would simultaneously be affected by another speed, an eastward velocity of 830 kmph imparted by the earth's rotation *at the latitude of Leningrad.* As the craft traveled, the

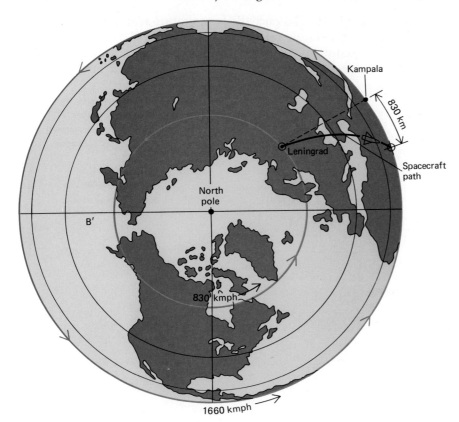

FIGURE 53–6 The Coriolis effect causes the path of a spacecraft to curve with respect to the earth's surface. The absolute speed of the Earth's rotation varies with the distance from the poles. Because the entire globe must turn around once each day, a point on the Equator must travel much further during that time than a point near one of the poles.

eastward speed of the Earth beneath it would steadily increase while the missile's speed remained the same. Since the velocity difference between Kampala and Leningrad is 830 kmph, the ship would land 830 km *West* of Kampala! Less spectacular but nevertheless significant deviations of this sort can be detected even in the case of long-range artillery fire. A missile or artillery shell travels in a straight line, but curves with respect to the surface of the Earth because the Earth is rotating below it.

This effect, called the **Coriolis effect,** is a deflection of the path of objects traveling north or south caused by the rotation of the Earth. It is most marked for massive objects like artillery shells, but it influences substances as light as air if the distances traveled are long enough and if their north-south speed is low. The poleward movement of air masses fulfills these criteria. Thus, as a consequence of the Coriolis effect, air masses curve in an eastward direction as they travel poleward, and a prevailing west wind can be observed in many areas, especially the tropics. On the other hand, the cooled polar air flows back toward the Equator in a westward direction. These flows interact with topography and with the distribution of oceans and currents to produce consistent weather trends that in turn produce the great climatic zones of the Earth.

Water

Not only does water comprise most of the contents of living things, but the properties of water determine many of the conditions on the Earth that are favorable for life. Water covers 70% of the Earth's surface, most of it residing in the oceans (Figure 53–7). The very magnitude of such vast masses of water gives this remarkable substance broad ecological significance above even this. The distinctive physical ecology of water is closely related to its remarkable heat and density relationships, its surface tension, and its ability to act as a solvent.

Water and Heat

If energy is added to a substance the temperature will rise, but temperature and energy are not simply related. An example will make this clear. Suppose a bathtub contains 80 liters of water at 60° C. Clearly there is more heat energy in the bathtub than in a cup of water (0.2 liter) at the same

FIGURE 53–7 Most of the water in the hydrosphere resides in the oceans. Most of the rest is tied up in glaciers and icecaps, with only a tiny portion remaining as liquid fresh water.

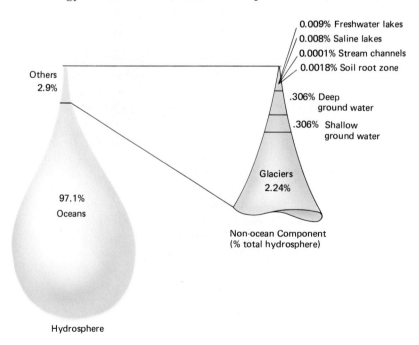

0.009% Freshwater lakes
0.008% Saline lakes
0.0001% Stream channels
0.0018% Soil root zone
.306% Deep ground water
.306% Shallow ground water
Glaciers 2.24%
Others 2.9%
97.1% Oceans
Non-ocean Component (% total hydrosphere)
Hydrosphere

temperature. In fact, there is far more heat energy in that tub at 60° C than in a cup of boiling water at 100° C. Considering a different aspect of this matter, which will cool down faster—a 1-liter balloon full of air at 60° C or a 1-liter cannonball of iron at 60° C? It obviously will take the cannonball longer, showing that the two substances differ in their heat energy content, even though their volumes are the same.

We may express all this more precisely using the concept of **specific heat**—the energy needed to raise or lower a unit of mass of a given substance by 1° C. It takes much less heat to raise the temperature of air than of iron. The surprising fact is, however, that more heat is needed to raise the temperature of water than of iron! Water has a remarkably high specific heat. As a consequence, water increases or decreases in temperature more slowly than almost any substance—rock, for instance, and especially air— even if all three are exposed to the same amount of insolation. As a result, water moderates the climate of adjacent land areas, and aquatic habitats experience less temperature variation than terrestrial ones do.

The average temperature of the Earth lies close to the mean of those governing the three physical states of water, so that common local variations can easily change it to and from a liquid, solid, or vaporized state. Unlike almost all other substances, water's greatest density is not as a solid, but as a liquid at 4° C. Below that temperature it expands and becomes *less* dense. This is why ice floats. When water freezes, it releases a great deal of heat. This does not ordinarily raise its temperature, but it does help to slow further freezing. Similarly, water absorbs a great deal of heat when it evaporates and conversely releases a great deal of heat when it condenses again.

The heat released by water upon condensation *can* raise the temperature of the air with which water vapor is usually mixed. This air, being warmed, rises even higher in the atmosphere, thus inducing turbulence and sometimes creating storms.

The Hydrologic Cycle

Since the average temperatures of the Earth's surface favor the evaporation of water, water constantly enters the atmosphere, mostly from the seas. Any cooling of the air containing water vapor may result in precipitation. This has the effect of transporting water inland, where it supports terrestrial ecosystems. Rising moist air masses carry water vapor steadily upward. As the air enters regions of decreasing pressure, it expands and therefore cools.

The amount of water vapor that air can contain at any given temperature is its **saturation value.** The lower the temperature, the lower the saturation value. Whether cooled by this or some other mechanism, if the air is sufficiently humid eventually the temperature may fall so low that the **dew point** is attained, the temperature at which water will condense from the air. Especially if accompanied by turbulence and condensation nuclei such as dust, this cooling produces **precipitation,** that is, snow, hail, or rain.[1] Most precipitation occurs over the oceans, but that which is carried onto the land sustains almost all terrestrial life. The amount of precipitation, the rate of its removal by evaporation or runoff, and its seasonal distribution profoundly influence the distribution of terrestrial life. Eventually, water returns to the sea, although it may be delayed for months in snowfields or for millenia in glaciers. The global recycling of water is called the **hydrologic cycle** (Figure 53–8).

---

[1] While not technically precipitation, dew forms from atmospheric moisture and may sometimes (in foggy coastal deserts or some mountainous areas) locally contribute more water to the community than precipitation does.

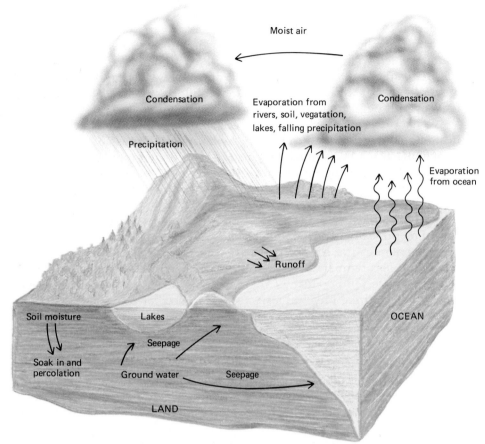

Moist air

Condensation

Evaporation from rivers, soil, vegatation, lakes, falling precipitation

Condensation

Precipitation

Evaporation from ocean

Runoff

Soil moisture

Lakes

OCEAN

Soak in and percolation

Seepage

Ground water

Seepage

LAND

FIGURE 53–8 The hydrologic cycle. Water vapor carries precipitation over land and sea, with temporary stays in freshwater areas and the bodies of living things. It all returns to the sea, to be recycled yet again.

Worldwide, about a billion tons of water evaporate and another billion are precipitated every minute. Imbalances in this equation produce such disasters as ice ages. It is sobering to realize that the causes of ice ages are not well understood and may depend upon such seemingly minor upsets in the usual order of planetary business as a few more or less sunspots than normal, a bit more or less carbon dioxide in the atmosphere, or changes in the number of soot and dust particles. It is even more sobering to realize that we have produced important changes in some of these parameters by the activity of our industrial civilization without really knowing what effect those changes may have.

Oceanic Currents

Almost everyone has seen waves raised by the action of wind on a lake or ocean. Persistent prevailing winds, given long enough to act, can produce deep mass movement of ocean water known as currents.

The oceans are not uniformly distributed over the globe (Figure 53–9). There is clearly more water in the Southern Hemisphere than in the Northern. Even more important, the potential circumpolar flow of water in the Southern Hemisphere is almost unimpeded by land masses. In the Northern Hemisphere, circumpolar movement also takes place, but in a westerly direction. In reality these movements are part of larger, circular current patterns that occur in the Pacific and Atlantic Oceans, moving clockwise in the Northern Hemisphere and counterclockwise in the Southern (Figure 53–10). The ultimate explanation for these circular current patterns (called **gyres**) is the Coriolis movement of the prevailing winds that drive them.

The great Humboldt current, for instance, drives northward along the west coast of South America. This cold current cools the South American

FIGURE 53–9 The Northern and Southern Hemispheres have greatly differing proportions of land and water, with far more water occurring in the Southern Hemisphere. Ocean currents are more free to flow longitudinally in the Southern Hemisphere.

continent almost to the Equator. Quite opposite in its action, the Gulf Stream carries warm Caribbean waters northeast, transfering heat and minerals, and greatly moderating the coastal European climate.

Climate and Precipitation

Average temperature, temperature extremes, precipitation, the seasonal distribution of precipitation, day length, and season length are biologically the most important dimensions of climate. Combinations of these climatic factors produce the biomes such as tundra, desert, rain forest, and prairie (Chapter 54). We have seen what determines day length, insolation, and, to a large degree, temperature. Differences in precipitation are more subtly determined and depend upon a combination of all other climatic factors.

The heavy-rainfall areas of the tropics result mainly from the equatorial upwelling of moisture-laden air. High water surface temperatures encourage the evaporation of vast quantities of water, and prevailing winds blow

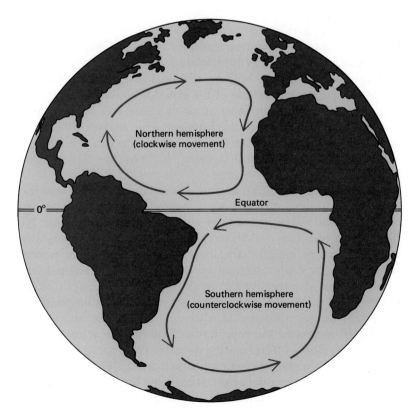

FIGURE 53–10 The basic pattern of the main ocean currents results in part from the action of winds and in part from the Coriolis effect. The main ocean current flow—clockwise in the Northern Hemisphere and counterclockwise in the Southern Hemisphere—results partly from the Coriolis effect, which also causes the difference in direction.

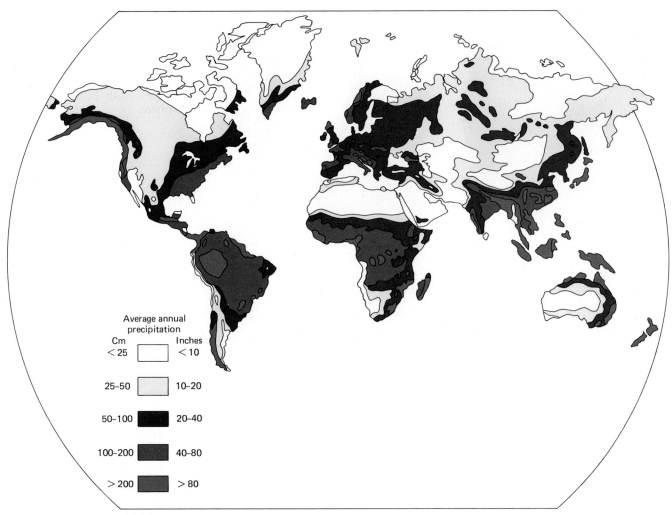

Average annual precipitation

Cm		Inches
< 25		< 10
25–50		10–20
50–100		20–40
100–200		40–80
> 200		> 80

FIGURE 53–11 Hemispheric map of average precipitation. Notice the light precipitation at Northern latitudes, in the centers of many continents, and to the leeward of mountain ranges.

great packets of the resulting moist air over land masses. Convective heating of the air by the heavily insolated ground causes heavy and frequent storms. The air, driven high by the heat of condensation, eventually returns to earth on either side of the equator between the Tropics of Cancer and Capricorn. By then most of its moisture has been squeezed from it, and the dry air returns to the Equator. This makes little biological difference over the ocean, but the lack of moisture in this returning air produces some of the great tropical deserts, such as the Sahara Desert (Figure 53–11).

Air also dries from long journeys over land masses. Near the windward coasts of continents rainfall may be heavy, but not all the precipitation runs off the ground surfaces; some of it reevaporates into the air. Nevertheless, there is a continuing deficit, so that in the temperate zones deep continental interiors are usually dry, being situated far from potential oceanic sources for the replenishment of water in the air that passes over them.

Moisture is also removed from air by topographic uplift of air masses. If, as occurs on the North American west coast, prevailing winds blow onto a mountain range, moisture is squeezed out of the air, especially on the windward slopes of the mountains. Downwind, a low-precipitation **rain shadow** develops, often creating deserts. Thus some of the regional differences in worldwide precipitation result from the drying of air as it is returned to more equatorial areas, some from long travel over continents and some from cooling produced by mountainous regions.

FIGURE 53–12 Capp's glacier heads on Mount Torbert (elevation 11,413 feet) in Alaska.

Glaciation

Glaciation (Figure 53–12), one of the more spectacular results of precipitation, has had consequences of great import for living things of the past, present, and probably future. As glaciers advance they push living things ahead of them; as they retreat, the ranges of organisms can expand. When glaciers become continental in size, water becomes sequestered in them, ocean levels fall, and coastlines expand. When the glaciers melt once more, oceans rise again, as does the land formerly depressed by their great weight. Other effects, many of them very poorly understood, produce worldwide climatic changes.

Glaciers form when snow accumulates faster than it melts. Low temperatures will not by themselves produce glaciers; high rates of precipitation are also required. Accumulated snow compacts itself into a material called firn, and firn compacts into ice (Figure 53–13). The ice, when it becomes

FIGURE 53–13 Glaciers form when snow builds up in the zone of accumulation, where it becomes firn and then ice. The ice flows slowly downhill, but melts at some point, the zone of ablation, at a rate equaling its flow. A **terminal moraine** of rock and mud forms at the terminus or snout of the glacier. (*a*) Alpine glaciers form from the accumulation of snow and firn in old watercourses. Typically several glaciers flow into one valley. (*b*) Continental ice caps are immense glaciers that bury and erode most topographical features beneath them. When these melt, the absence of their weight may cause the land beneath them to rise.

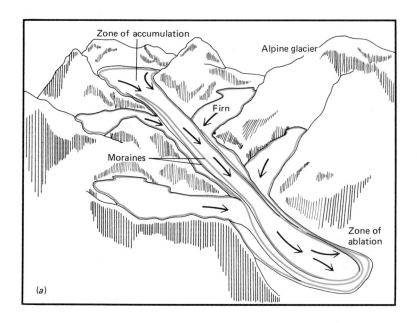

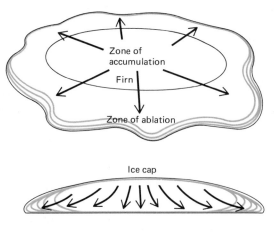

(b)

FIGURE 53–14 Maximum extent of Pleistocene glaciers of North America. All plant and animal life in the area either died or was forced to move, if it could. As the glaciers melted, survivors repopulated the formerly glaciated areas.

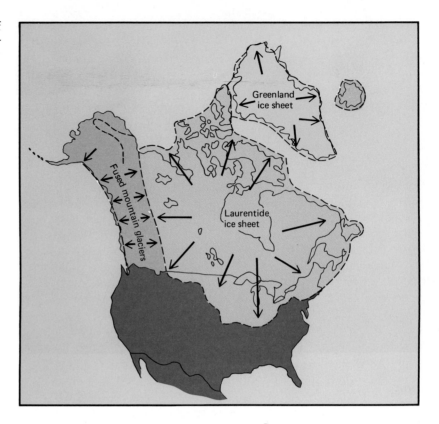

thick enough, fractures along crystal planes and moves along the fractured surfaces. En masse, it behaves almost like a viscous liquid and can flow like a very slow river or move as a gradually expanding sheet of ice.

Alpine glaciers form in mountainous regions, often in somewhat sheltered locations where snow melting is slow. As they flow down the mountainside, they tear rock and debris from the rock beneath, to either side, and behind them. The dislodged material then scrapes along the rock, further eroding it and often cutting grooves. If the snout of the glacier is melting appreciably, rocks and finely ground rock flour are deposited in the form of a terminal moraine. If the glacier retreats, its former course is covered with such debris. Most modern glacial soils are geologically young, with ages measured in thousands of years, and in many cases soluble minerals have not yet been released from them by weathering. Consequently, plant communities growing on glacial soils are often deficient in some minerals. Animal members of local food chains, including humans, may also be affected. Iodine deficiency leading to simple goiter is marked in formerly glaciated regions of the world.

In some cases glaciers run together to form icefields, and in others true ice caps develop, such as those of Greenland and Antarctica. A series of glaciations has occurred in the Pleistocene geological era (Figure 53–14), the most recent of which ended or at least paused perhaps 10,000 to 13,000 years ago. During these ice ages, vast sheets of ice covered large portions of the continents of the Northern Hemisphere, with attendant boreal forest and tundra communities (Chapter 54) pushing far into what are now temperate or even subtropical zones. Organisms, especially plants, that were trapped by the encroaching ice became extinct, while others adjusted their ranges, moving back and forth with advances and retreats of the glaciers. Some idea of the scale of these glaciations may be gained from their modern results. In North America, for example, the Great Lakes originated as accumulations of glacial meltwater, and Long Island, New York, is almost entirely a terminal moraine containing rocks that lay originally as far north as Canada.

We are unable to predict the glaciation of the future. At the present time glaciers seem to be shrinking, but it is not certain what the cause of this may be or whether this trend will continue. It is entirely possible that the ice caps of Greenland and Antarctica may melt away. It is less probable that the continental glaciers may return. Additional factors, especially human activities, are also likely to have a bearing on the fate of the glaciers (Chapter 57).

Microclimate

There are many exceptions to the overall climatic conditions in any habitat. Differences in elevation, in the steepness and directional orientation of slopes, or in exposure to prevailing winds may produce local variations in climate known as **microclimates,** and these may be quite different from their overall surroundings. It is really the microclimate that is most important to an organism, for the microclimate of its habitat is the climate it actually experiences and with which it must cope. Sometimes it is possible for animals to migrate to a local area of favorable microclimate (Figure 53–15) or even to substantially modify their own microclimate to dimensions more favorable to their existence. For instance, the temperature within a forest is usually lower and the relative humidity higher than that outside the forest, and the temperature and humidity beneath the litter of the forest floor can be expected to differ still more. By burrowing, desert-dwelling organisms survive in climatic conditions that would kill them in

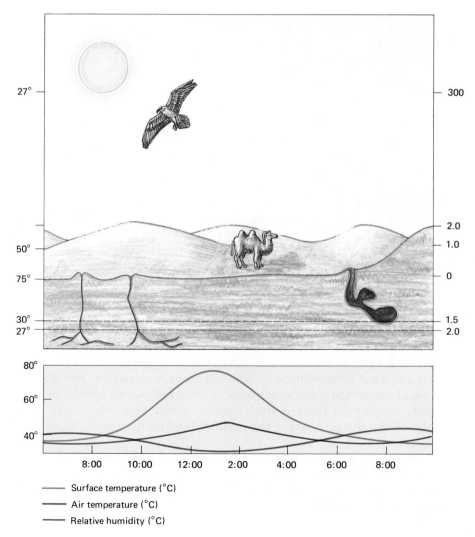

FIGURE 53–15 Substantial vertical differences in microclimate in a desert habitat. The sunlit surface is a dangerous 75°C, but no more than a meter above it the temperature is fully 25°C lower. Soaring birds live at an equable 27°C with little energy expenditure, and so do burrowing organisms with even less.

Surface temperature (°C)

Air temperature (°C)

Relative humidity (°C)

FIGURE 53–16 A cushion of moss campion. These little plants grow in tundra and alpine habitats, producing an internal habitat of still air and moderate temperatures despite the wind and cold outside.

minutes on the surface. The cooler daytime microclimate in their burrows permits them to survive until night when the macroclimate is more equable, allowing them to leave their retreats to forage or hunt.

Perhaps the most striking example of microclimate is produced by such alpine plants as moss campion (Figure 53–16). These little angiosperms grow in a low, ovoid cushion whose streamlined shape encourages the drying alpine wind to pass over them. At the same time they are able to absorb light and solar heat which is retained in the still air within. Since wind cannot blow away accumulating litter and dead plant parts within the cushion, these accumulate, allowing the retention of as much as a teaspoonful of water that the plants otherwise would not have. In the daytime, the temperature of the interior of one of these cushion plants can be several degrees warmer than that of its immediate surroundings. Small animals such as insects may shelter in these miniature warm forests, away from the dry, ultraviolet-suffused world of cold and wind outside.

Soil

Living things depend on the soil and also help to produce it, but soil originates primarily from the erosion of parent rock that underlies the soil. Once soil forms, it continues to develop and differentiate under the action of living things of all kingdoms. Minerals enter the biosphere from the soil or from water that has leached them from the soil.

As rain falls on soil, the unabsorbed **surface runoff** washes some of the particles away in the all-too-familiar process of soil loss, **erosion.** The presence of plants and their roots cushions the force of the rainfall, greatly reducing the erosion rate to a point where in many natural ecosystems it can hardly be detected.

Soil Water

Considerable water may soak into the ground, depending on the topography and the mineral content of the soil. When water finally reaches an impermeable layer of bedrock, it fills the spaces between the soil particles, forming the **zone of saturation** in which there is no air at all. The limit of the zone of saturation is the **water table,** from which water usually may be pumped. As long as there is underground water flow, the water table roughly follows the contours of the ground. It is not necessarily level (Figure 53–17).

Above the zone of saturation lies the **zone of aeration,** which, as its name implies, contains varying amounts of air. The lowest portion of the zone of aeration contains considerable **capillary water,** which is water held against the force of gravity by capillary action. Above this only **hygroscopic**

Precipitation

Organic additions and soil faunal activity

Weathered bedrock

Runoff (surface water)

Groundwater flow

Joints

Solid bedrock

Water table

Mineral particle

Air

Water

Below water table zone of saturation

Hygroscopic water

Capillary fringe water

Run-in gravitational water zone of aeration

FIGURE 53–17 The relations of soil and water. Much rainfall simply runs off the surface, but some soaks into the aerated upper layers of the soil. Some remains with the soil water as capillary fringe water or hygroscopic water—held in the first instance by physical and the second by chemical forces. Some runs down to the surface of the water table and flows more or less horizontally thereafter until it reaches an outlet such as a spring. Some can even enter bedrock and flow through crevices and joints. Below the water table the soil contains little or no air.

water is found, water that is closely held to the soil particles by physicochemical forces. It may take quite some time for water to cease percolating through the soil, however, and while it is doing so considerable water is available in all zones. All plants can take advantage of percolating water and capillary water, but most cannot absorb hygroscopic water. If a plant (usually a tree) has a long enough taproot to reach the water table, it is immune to drought while the water table lasts. Shallowly rooted plants are more drought-sensitive.

Water also can move upward in the soil as the water table rises, or it can rise by capillary action. Minerals tend to be leached from the soil by downward movement of water, but they also may be deposited (sometimes in toxic amounts) by upward movement combined with evaporation (Figure 53–18).

FIGURE 53–18 Soil development is affected by both climate and living things. The decay of dead plant litter produces a humus-rich topsoil that is redistributed downward through the activities of burrowing animals such as earthworms. Precipitation leaches minerals from the soil and also moves very fine particles such as silt. If rainfall is moderate, much of this may be redeposited in lower but nearby soil layers. During hot, dry periods capillary rise of water and evaporation may redeposit materials in upper layers.

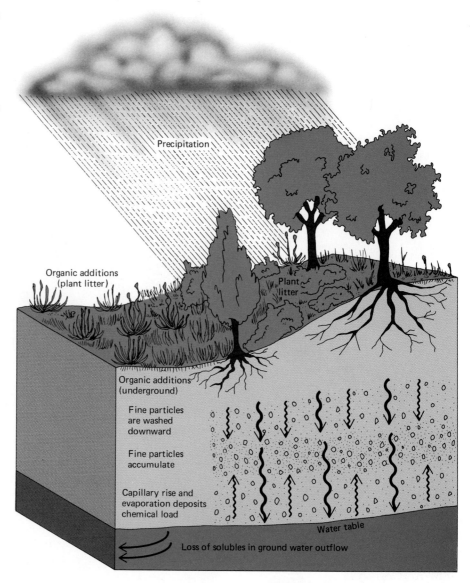

Precipitation

Organic additions (plant litter)

Plant litter

Organic additions (underground)

Fine particles are washed downward

Fine particles accumulate

Capillary rise and evaporation deposits chemical load

Water table

Loss of solubles in ground water outflow

Soil Layers

Shed leaves, dead roots and stems, and woody parts of plants accumulate in the upper portions of the soil, decaying to form humus. The organic-rich uppermost layer of many soils is called the **O-horizon** (Figure 53–19).

The formation of humus is greatly assisted by detritus-feeding burrowing organisms such as earthworms and termites. These, plus ants, continually carry soil and humus to and from the surface in the course of their activities, churning and mixing it. Although the negative economic impact of termites is obvious, their ecological importance is not always appreciated. The average colony of Formosan termites may contain as many as 7 million termites, weigh 60 pounds collectively, and consume 3 pounds of wood per day—converting it to carbon dioxide, water, and minerals far more rapidly than microorganisms alone ever could.

Although ants and termites are very important, particularly in tropical ecosystems, earthworms are probably the best known of all soil-dwelling organisms. These annelids continuously eat soil, digesting and living upon its humus. Their digestive systems are chemically specialized for the task, being provided with chitinase, and unusual cellulase enzymes that attack the largest organic components of undecomposed litter. As the worms tunnel in the soil they aerate it and deposit nitrogenous waste products in its

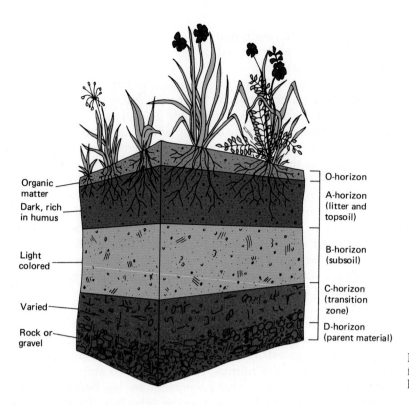

Organic matter

Dark, rich in humus

Light colored

Varied

Rock or gravel

O-horizon

A-horizon (litter and topsoil)

B-horizon (subsoil)

C-horizon (transition zone)

D-horizon (parent material)

FIGURE 53–19 A typical soil profile, showing five main layers or horizons. Not all soils have all five.

deeper layers. Periodically, the worms surface to defecate, leaving little piles of processed soil on the surface.

Some idea of the ceaseless activity of these little excavators can be gained by their burial of surface features such as rocks and even tombstones. Some species are able to grasp dead leaves and other surface litter in their mouths on nocturnal trips to the surface, dragging this food into their burrows and eating it there. This probably has an important effect on soil nitrogen distribution. Earthworms eventually die and they themselves decay. When they do, ammonia formed from their body proteins may add significantly to the fixed nitrogen supply of the deeper layers of soil and, as their corpses blend with lower layers of soil, the O-horizon is extended downwards.

In cool climates, as water percolates through the soil it tends to pick up the finest particles and deposit them in deeper layers, causing a gradual accumulation of the finest particles in the deeper layers. In soils derived from granitic rocks, these fine sediments are clay, so that clay layers tend to develop in deeper layers of well-watered soils. The coarser particles left in the upper layers may form a sandy or even gravelly soil where erosion is extreme. Minerals leached out of surface soil or decaying litter may accumulate in the clay; sand will not hold them. Since clay particles are small and closely packed, hygroscopic water tends to fill all the spaces between them so that little air is present. This, together with the relative hardness of dry clay, may effectively prevent the penetration of plant roots through part or all of such soils. In many cases an almost concrete-hard layer called **hardpan** will develop. The granular, relatively nutrient-poor layers beneath the O-horizon are called the **A-horizon;** the clay-rich layers, the **B-horizon.** Composed of relatively unaltered deep material adjacent to the bedrock and often in the saturation zone, the **C-horizon** underlies all the others.

Climate and Soil

This well-differentiated soil structure is not universal but develops mainly in well-watered temperate areas where it produces soils known as **podzols,** a word of Russian origin that means "ash-grey." There are several variants of podzol soils, each caused by differences in precipitation, temperature, and bedrock material.

In climates with somewhat less precipitation, water may evaporate or be lost by plant transpiration as fast as (or faster than) it percolates through the soil. In such cases the water table (if any) may be very deep and a layer of mineral-rich subsoil may develop where minerals are deposited by evaporating water. These minerals do not necessarily originate directly by leaching from the upper layers of soil but may be pulled up by capillary action from the deepest portions. This process, **calcification,** is common in prairie or steppe soils. Such soils are usually rich in humus, for the grasses that live there may be very productive if there is sufficient rainfall to support them. In some cases, however, very low precipitation may produce soils that are actually saline, especially in desert habitats. In such areas only wild plants or crops specially adapted to saline conditions can grow, and some such areas are totally devoid of life. Owing to their low productivity, desert soils may have little or no organic content and therefore no O-horizon.

Perhaps the most extreme effect of precipitation on the development of soil is **laterization,** a process most likely to occur in the humid tropics, especially rainforest habitats. Very rapid decay processes practically prevent the formation of humus, so there may be almost no O-horizon in the soil. Any minerals that may be present are leached out by heavy rainfall. Even silica tends to be lost, leaving little but aluminum and iron compounds. The stony, aluminum-rich laterite that is produced gives this process its name. Some laterites are bricklike and stony in texture and have actually been used as building materials in Southeast Asia and elsewhere in the tropics.

Lateritic soils are so mineral-poor that one would think them to be biologically unproductive. Lush tropical ecosystems seem to belie this, but tropical rain forests live on the soil more than in it. Most of the minerals that in other ecosystems reside in the soil are here found only in the tissues of the local vegetation. Crop plants are typically unable to recycle minerals rapidly. As a consequence, when rain forests are cut down and burned to make room for grazing or agriculture, all the nutrients they contain are immediately lost. Since the soil contains few nutrients, crops and pastures soon fail. Fortunately, there are some exceptions.

FIGURE 53–20 Traditional agriculture such as these Filipino rice terraces, often recycles materials that more "modern" systems lose and must continually replace.

The most extreme lateritic soils are believed to be geologically very old, some of them possibly dating from pre-Cambrian times. There would seem to have been ample time for mineral loss in such ancient soils. Younger, volcanic tropical soils such as those found in the Phillipines may still retain much mineral content available for crops.

Traditional methods of tropical agriculture are often designed to retain nutrients instead of allowing them to be washed away. Hillside terracing produces platforms that retain silt and nutrients that otherwise would wash down the slopes (Figure 53–20). Diked rice paddies have the same effect in low-lying areas. In fact, with nutrients recycled by fertilization with human and animal wastes, and the deliberate encouragement of nitrogen-fixing organisms, rice paddies serve almost as complete artificial ecosystems. Their internal methods of recycling minerals serve as substitutes for the far more elaborate natural mechanisms that they replace.

Local Recycling

It is clear that soil is as much a product of life as skeletons or wood. Minerals are absorbed by roots, incorporated into plant and animal bodies, released by decay, and eventually or immediately taken up again to be reincorporated into living things (Figure 53–21). Yet these processes are not perfect; there is always some leakage away from a community of living things. The nature of the leakage varies with the element, but the tendency is almost always to lose some elements to the oceans or the atmosphere. Fortunately even these are not necessarily final resting places but may serve to transport the substances back to where they can once again be incorporated into living things, as will be discussed in the next chapter.

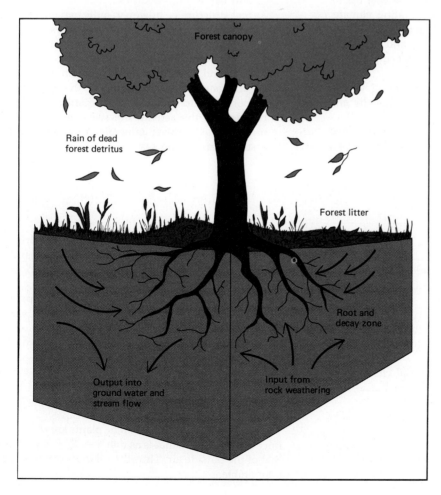

FIGURE 53–21 A forest nutrient cycle. Much of the mineral material lost in the form of shed leaves, decayed roots, and the like may be reabsorbed by plants, including the plants that lost them in the first place. Some is lost by leaching into inaccessible ground water, but some is gained by weathering from parent material.

■ SUMMARY

I. Ecology is the study of the interactions between living organisms and their environment, and among living things.

II. The unique planetary environment of the Earth makes life possible.

A. Sunlight is the major (almost the sole) source of energy available to the biosphere.

1. The heat budget of the Earth balances heat loss by radiation against absorption. Its climate depends in large measure on the set of conditions under which these are equal.

2. Local climate, however, is determined largely by the angle at which the local solar input falls to the surface. Most of the time this is greatest and least variable in equatorial regions, whose climate therefore is hotter and less variable than that of temperate and polar areas.

B. Global climate is moderated by heat transfer via air and ocean.

1. Atmospheric heat transfer by convection from Equator to poles produces both a poleward movement of warm air and a countervailing motion of cool air. These meet in the temperate zone, producing unstable weather.

2. The Coriolis effect diverts air and water currents longitudinally. The Coriolis effect results from the rotation of the Earth.

C. Water is a virtually unique substance that transfers heat efficiently, but whose high specific heat resists changes in temperature, thus moderating climate.

1. The hydrologic cycle results from the evaporation of water from both land and sea and its subsequent precipitation when the air that contains it is cooled.

2. Oceanic currents result from prevailing winds. They redistribute mineral nutrients globally and often profoundly affect local climate.

D. Precipitation is greatest where air heavily saturated with water vapor has a chance to cool to the dew point. Accordingly, humid areas tend to be located where air passes over the ocean and then is forced upward by mountainous topography or where such air is otherwise cooled. Deserts develop in the rain shadows of mountain ranges or in continental interiors.

E. Glaciation results when snowfall locally exceeds the rate of its melting. Glaciers flow slowly over rock, which they vigorously erode. Past ice ages have greatly influenced the present distribution of natural communities.

F. Microclimate is the local variation of climate that is actually experienced by living organisms. By choosing favorable microclimates, or even by creating them, organisms can often live in areas whose macroclimate is hostile to life.

G. Soil is important in determining the distribution of living organisms. Its development depends largely on local climate, especially precipitation.

1. Water occurs in the soil. The most temporary form is percolation water, but hygroscopic and capillary water are more long lasting. Plants are usually able to utilize only percolation and capillary water unless their roots reach the water table.

2. Some soils are strongly layered, especially the temperate climate soils known as chernozem. These soils typically possess a topmost layer of humus-rich material, underlain by sand or clay, and below that, bedrock.

3. Soil structure is, however, quite variable. In tropical areas, for instance, there may be essentially no humus and little else except clay. The most extreme of such soils are called laterites.

H. The mineral nutrients needed by communities of living things must be extracted from the soil. Once incorporated into the bodies of organisms, these are recycled by the activities of scavengers and decay organisms. Whatever minerals are lost find a final resting place in the sea.

■ POST-TEST

1. Ecology treats the interaction of living things with one another and with their _____ .

2. The energy that operates the world of life originates almost exclusively in _____ _____ .

3. The prevailing temperature of the Earth is determined by the heat balance of the Earth, that is, the relationship between gain from the sun and loss by _____ .

4. The warmth and constancy of equatorial climates results mostly from the direct _____ with which sunlight falls upon their surface.

5. Heat is transmitted from the Equator to the poles by _____ of atmosphere and hydrosphere.

6. The high _____ _____ of water moderates climates of areas adjacent to large bodies of water.

7. The _____ _____ , which results from an interaction between motion, inertia, and the rotation of the earth, displaces the paths of atmospheric and oceanic currents to the East and West.

8. Precipitation will probably occur when the temperature of humid atmosphere falls below its _____ .

9. Mountain ranges may produce downwind arid _____ _____ .

10. Glaciation occurs when the rate of snowfall exceeds its rate of _____ .

11. Recently glaciated soils tend to be low in readily available _____ _____ .

12. The hollow stems of certain arctic plants are able to produce a temperature that is as much as 10° C higher than that of the outside air. Thus their tissues are exposed to a favorable _____ .

13. _____ soils have a well-differentiated structure consisting of several layers known as _____.

14. Extreme and long-continued weathering processes often produce _____ tropical soils that consist mostly of clay with little or no humus or available nutrients.

■ REVIEW QUESTIONS

1. What does the word *ecology* actually mean?
2. What proportion of the sun's light that strikes the Earth's surface is employed by plants for photosynthesis?
3. What determines the temperature of the Earth? How might this balance be disturbed?
4. What basic forces determine the circulation of the Earth's atmosphere? Can you describe its general directions?
5. Give the properties of water that are of the greatest significance in its role in determining the Earth's climate.
6. Summarize the hydrologic cycle.
7. What forces produce the main oceanic currents? What is their role in determining the climate of the Earth?
8. What conditions produce precipitation? What are some of the factors that produce areas of precipitation extremes such as rain forests and deserts?
9. What factors produce glaciers? What has their role been in the production of some of the major habitats that exist today?
10. What is meant by microclimate? Give some examples of how living things seek or produce favorable microclimates in which to live.
11. How does soil form? What are the main components of soil?
12. What is meant by soil horizons? How does the horizon structure of soil vary with climate?
13. How does water accumulate in and under the ground?
14. How do local associations of living things recycle mineral nutrients? What is the importance of this?

■ RECOMMENDED READINGS

Foster, R.J. *Physical Geology*. Columbus, Ohio, Charles E. Merrill, 1983. Introduction to the geological processes that determine the basic structure of the Earth upon which organisms must live.

Gabler, R.E., et. al. *Essentials of Physical Geography*, 3d ed. Philadelphia, Saunders College Publishing, 1987. A synthesis of geology, economics, and ecology.

Lutgens, F.K., and E.J. Tarbuck. *The Atmosphere: An Introduction to Meteorology*. Englewood Cliffs, N.J., Prentice-Hall, 1986. The operation and causes of climate and weather.

Turk, J., and A. Turk. *Environmental Science*, 4th ed. Philadelphia, Saunders College Publishing, 1988. Ecology with a human emphasis. Includes an extensive discussion of the natural resources essential to civilization and the geologic processes that produce them or influence them.

tween the areas where plants and animals can live exists between the aquatic and terrestrial environments. This contrast reflects the substantial differences in the physical and chemical properties of water and air. These properties determine the basic and distinctive adaptations of the organisms that live in each habitat—sometimes in surprising ways.

Thermoregulation, to take one example, is easier for an animal living in air because air does not conduct heat away from the body as efficiently as water does. When the surroundings are hot, the evaporation of water can be used to cool a terrestrial organism. Yet temperatures are usually more extreme in air than in water.

Skeletal support affords another example. Aquatic organisms require less skeletal support than terrestrial ones because of the greater density of the fluid medium supporting the watery tissues of living things. Organisms such as jellyfish and whales could not conceivably live outside of the water. Aquatic organisms also can grow large. Consider the largest known insects—several inches long. Yet the largest known aquatic arthropods, certain Japanese crabs, span almost a dozen feet.

Of great importance to plants, water can carry nutrients directly to all parts of organisms immersed in it, so that aquatic plants do not usually have vascular tissues. But water also absorbs light heavily in comparison to air, therefore transmitting it poorly. This limits photosynthesis to the shallows. Finally, water (especially warm water) is unable to hold large amounts of oxygen in solution, so that water-breathing organisms must get along with far less of this vital element than terrestrial creatures.

TERRESTRIAL LIFE ZONES: BIOMES

The distribution of terrestrial communities is governed fundamentally by climate. Climate is shaped not only by temperature, but also by precipitation. Climate, in turn, affects soil development, another important governing factor (see Chapter 53).

Biogeographical realms are regions made up of entire continents or large parts of a continent separated by major geographical barriers and characterized by the presence of certain types of animals and plants. Within these biogeographical realms are **biomes**—large, relatively distinct, community units, which are also characterized by the kinds of plants and animals present in them. The boundaries of biomes are established by a complex interaction of climate, physical factors, and other factors produced by their living inhabitants.

Climate determines biomes more than any other factor (Figure 54–1). The more polar communities, for example, are colder and have a shorter growing season. It is this that determines the distribution of the tundra community (discussed below) and, to a lesser extent, the taiga as well. Away from the polar communities, however, precipitation becomes a very important differential factor, producing the temperate and tropical communities of desert, steppe, prairie, and deciduous woodland, in increasing order of precipitation.

Part of the effect of precipitation is direct—cacti are obviously better adapted to dryness than cattail plants! Part of it, however, is indirect—differences in soil structure and nutrient availability, for instance. Yet these, in turn, depend on precipitation. In the tropics, extremely high levels of precipitation produce the rain forest, whereas less precipitation produces the distinctive savannah and tropical desert areas.

Some of these life zones are mimicked to some extent by the altitudinal distribution of organisms in mountainous areas. The base of a tropical mountain, for instance, might be clothed in rain forest. Higher and cooler altitudes might possess deciduous trees reminiscent of those in northern deciduous woodlands. Above that, one might find evergreen forests, and

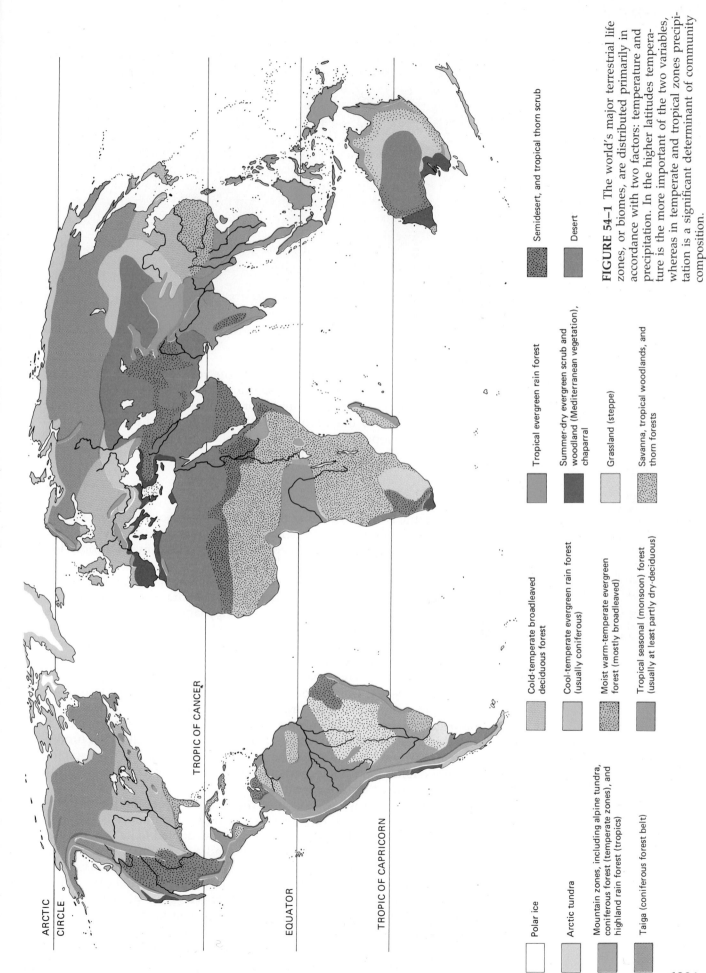

Polar ice

Arctic tundra

Mountain zones, including alpine tundra, coniferous forest (temperate zones), and highland rain forest (tropics)

Taiga (coniferous forest belt)

Cold-temperate broadleaved deciduous forest

Cool-temperate evergreen rain forest (usually coniferous)

Moist warm-temperate evergreen forest (mostly broadleaved)

Tropical seasonal (monsoon) forest (usually at least partly dry-deciduous)

Tropical evergreen rain forest

Summer-dry evergreen scrub and woodland (Mediterranean vegetation), chaparral

Grassland (steppe)

Savanna, tropical woodlands, and thorn forests

Semidesert, and tropical thorn scrub

Desert

FIGURE 54–1 The world's major terrestrial life zones, or biomes, are distributed primarily in accordance with two factors: temperature and precipitation. In the higher latitudes temperature is the more important of the two variables, whereas in temperate and tropical zones precipitation is a significant determinant of community composition.

ARCTIC CIRCLE

TROPIC OF CANCER

EQUATOR

TROPIC OF CAPRICORN

FIGURE 54–2 Altitude zonation at a northern temperate latitude. Notice the coniferous forest eventually giving way to bare rock and arctic-type ice caps. This photo was taken in Glacier Park, Montana; the valley and the jagged faces of the rock were formed by the movement of glaciers through the area.

above the evergreens, a kind of tundra. At the very top, a permanent ice or snow cap might be found, similar to the nearly lifeless polar land areas.

Communities found in areas of hilly or mountainous topography (Figure 54–2) reflect not only the latitude, but the altitude and the direction faced by the various slopes, for the angle of insolation is more direct on slopes facing toward the Equator and more oblique on those facing away.

Tundra

In the extreme North, wherever the snow melts seasonally, there exists a distinctive circumpolar **tundra** community (Figure 54–3). (The Southern Hemisphere has no equivalent because it has no land in the proper latitudes.)

For the most part, the land of the tundra has low relief and drainage is slow. That, combined with the low rates of evaporation that result from cold annual temperatures, produces a swampy landscape of broad shallow lakes, sluggish streams, and bogs. In addition, tundra has a layer (varying in depth and in thickness) of permanently frozen ground, the **permafrost,** which interferes with subterranean drainage. Frozen mammals, especially extinct mammoths, have been unearthed from the permafrost with the meat still edible—at least by dogs—after tens of thousands of years. Permafrost can be spotty in its distribution. Where the permafrost is overlain by substantial amounts of unfrozen soil, forest habitats are more likely to be present. Although there is little precipitation over much of the tundra, low temperatures, flat topography, and the permafrost layer conspire to keep the habitat wet. Some parts of the tundra, however, have so little precipitation that they are essentially arctic deserts.

The very short growing season of the tundra affects the life of everything in it. The great natural stress to which this community is subjected results in low species diversity, but the species adapted to its extreme conditions often exist in great numbers. The tundra is dominated by reindeer moss (actually a lichen), grasses, grasslike sedges, and annuals. There are no readily recognizable trees or shrubs except in very sheltered localities, although dwarf willows and other dwarfed trees do grow widely. The need to complete an entire life cycle in a span of weeks produces frantic rates of growth and development in annuals, made possible in large part by the great length of each summer day. In many places the sun does not set at all

(a)

(b)

(c)

FIGURE 54–3 (*a*) An example of tundra vegetation in Alaska. (*b*) The alpine tundra vegetation of the mountainous western United States is not the same as the northern tundra and is adapted to somewhat different conditions, although it often contains some of the same species. (*c*) Alpine tundra, Alaska. Low-growing willows, sedges, and other recumbent vegetation make up a dry tundra in a mountainous area that is surrounded at lower altitudes by boreal forest. Sea level tundra does not occur for hundreds of miles northward.

for a considerable number of days in midsummer. Many arctic flowering plants have parabolic flowers designed like solar collectors to focus the sun's heat on developing ovules, thus providing a warm haven for pollinating insects.

Most modern animal life (Figure 54–4) of the tundra is small, consisting of ratlike lemmings, weasels, arctic foxes, snowshoe hares, ptarmigan, snowy owls, hawks, and the like. The immense musk-oxen, caribou, and reindeer are modern exceptions, although wooly rhinoceros and mammoths also roamed the prehistoric tundra of Europe, Asia, and North America. There are no reptiles or amphibians. One might also expect few or no insects, but insects are able to survive over winter as eggs or pupae and occur in great numbers. Mosquitoes are particularly numerous.

Tundra soils tend to be geologically young, since most of them have only recently been glaciated. For this reason they are usually nutrient-poor, with distinctly different vegetation often growing in the vicinity of animal roosts and dens and their associated droppings. The O-horizon may be poorly developed, for although the low temperatures prevailing over most of the year inhibit decay, there is also relatively little primary production, hence little organic litter.

An incomplete equivalent of the tundra community occurs on a few mountains (Figure 54–3), but the thin dry air, the high levels of ultraviolet

(a)

(b)

(c)

(d)

FIGURE 54–4 The tundra biome contains a variety of species, although fewer than more southerly habitats. (*a*) Alaskan tundra in the fall. Like the trees of temperate deciduous forests, many of the low plants of the tundra, such as bearberry (*b*), change their foliage colors. The brilliant hues probably attract birds and mammals, which disperse the seeds by eating the fruit. Wildlife of the tundra includes (*c*) the snowy owl, which, like most owls, is a nocturnal predator, and (*d*) the arctic fox. The color of this animal varies according to the season. During the long winter its coat is almost entirely white.

light exposure, higher levels of precipitation (especially snowfall), and the usual absence of a permafrost layer make these mountain "tundras" not fully comparable to their sea-level prototype. Also, their species diversity tends to be even lower than that of the tundra proper.

One reason for this particularly low diversity is that mountaintop tundras are almost like islands, often separated from one another and the tundra proper by hundreds of miles of other types of habitat. In fact, a mountaintop tundra usually contains species of the tundra proper only if it actually *was* a part of the tundra proper during past glacial times. Many high tropical mountains in South America and Africa tend to possess an *equivalent* of the tundra community; they are populated with species that are ecologically equivalent to those of the tundra proper and sometimes even more striking in their adaptations but that do not occur on the northern tundra.

Owing to low rates of photosynthesis, all tundra regenerates very slowly after disturbance. Even casual use by hikers can be enough to destroy mountaintop tundra, and long-lasting damage (likely to persist for hundreds of years) has been done to large portions of the arctic tundra by oil exploration and military use.

Boreal Forest

Like the tundra, the northern boreal forest community is also circumpolar, dominated by coniferous evergreens (Figure 54–5) but with a soil possessing a much thicker O-horizon than the tundra. Permafrost is usually ab-

FIGURE 54–5 Taiga, or boreal forest.

FIGURE 54–6 Temperate coniferous rain forest. Note epiphytic vegetation.

sent. This community has an acid, mineral-poor soil characterized by a great depth of partly decomposed pine and spruce needles (or larch, in the harsher areas). The growing season of the boreal forest is somewhat longer than that of the tundra, but it is short compared to that of the temperate zone, where so many varieties of agricultural plants have been developed. Deciduous angiosperms such as aspen or birch may form striking assemblages in burned or logged-over areas, but, overall, conifers clearly dominate. The probable reason is that the growing season is so short that photosynthesis is needed year round. Yet water cannot be absorbed by roots from frozen soil, or even cold soil (approximately 5°C).

It can be quite difficult to draw a line between tundra and boreal forest. The southernmost tundra is interspersed with patches of evergreen trees, usually where the permafrost is deeply buried or locally absent. If the soil is waterlogged (as it often is) a peat swamp or **muskeg** habitat develops, dominated by spindly, stunted black spruce trees. These are replaced by spindly, stunted white spruce trees in dryer or higher locales.

Conifers have many drought-resistant adaptations (such as their needle-like leaves with minimal surface area for water loss) which suit them well to withstand the physiological drought of the northern winter months while permitting them to photosynthesize even then. Although most of the boreal forest is not well suited to agriculture because of its short growing season and mineral-poor soil, the boreal forest yields vast quantities of lumber and pulpwood, plus furs and other forest products.

The animal life of the boreal forest consists of some larger species such as caribou (which migrate into the area from the tundra in winter), wolves, bears, and moose. However, most of the animal life is medium-sized to small, such as rodents, rabbits, and fur-bearing predators, like lynx, sable, and fisher. Most species of birds are seasonally abundant but migrate away in the winter. Insects are abundant, but amphibians and reptiles are sparsely represented except in the southern extensions.

FIGURE 54–7 Krummholz vegetation on Pennsylvania Mountain in the Mosquito range of the Colorado Rockies.

Temperate Coniferous Forests

The boreal forest extends quite far south on the western coast of North America, reaching as far as northern California. Here, the short growing season of the subarctic has an equivalent, produced by the heavy seasonal rains characteristic of the West Coast. The extensive drought that follows the rains every year is much like the tundra's physiological drought of winter. Like the tundra, the boreal forest has a mountain equivalent often found at the middle altitudes, and some mature coniferous forests occur in the North American Southwest.

On the northwest coast of North America there exists a coniferous **temperate rain forest** (Figure 54–6), which is unlike the tropical rain forest but which contains many fascinating ecologically equivalent species whose adaptations parallel those of the tropical rain forest. Like the tropical rain forest soon to be discussed, the temperate rain forest has relatively nutrient-poor soils, although their organic content may be high. Also like the tropical rain forest, it is rich in epiphytic vegetation—plants that grow nonparasitically on other plants. These are mainly club mosses, lichens, and ferns instead of the bromeliads and orchids characteristic of the tropics. Still another parallel between the two types of rain forest is the dense pattern of growth which shades out underbrush and produces an open forest floor obstructed only by the dispersed trunks of large trees.

The upper limit of tree growth on mountains is known as the **timberline.** Usually it is located just below a tundra area. Here a stunted, miniature forest, the **krummholz** (Figure 54–7), may occur. The krummholz is usually composed of intricately entangled dwarfed conifers resembling bonsai trees (which are horticulturally produced by deliberate harsh treatment). Below the krummholz large trees may occur but are often distorted into a characteristic **flagged** shape in which the branches are mostly confined to the downwind side of the trunk. This results from the accumulation of ice on the upwind side which kills the branches.

Temperate Communities

Below the boreal forest, communities do not extend evenly around the world because of the influence of precipitation. In temperate latitudes rainfall varies greatly with longitude. In general, continental interiors tend to be dry, but this results from a variety of causes. Permanent high-pressure

FIGURE 54–8 Seasonal changes in a temperate deciduous forest. (*a*) Dense, green hardwood foliage during summer. (*b*) Color changes in foliage during fall.

(a)

(b)

areas, such as those over the Sahara Desert, may nudge moist air masses aside. Air passing over a large land mass also may dry out without having the opportunity to be recharged with fresh moisture. The climate of the North American continent, however, is dominated by rain shadows cast by mountain ranges, especially in the West.

As prevailing westerly winds push against the bases of the Sierra Nevada and Cascade Mountains, masses of moist air from the Pacific Ocean are forced upward, subjected to cooling, and precipitate much of their moisture, as discussed in the preceding chapter. Thus, the westerly slopes of the mountains are so well watered that a kind of rain forest develops. Considerable rain falls in the upper reaches of the eastern slopes also, but by the time the air has sunk back to lower altitudes, most of the available moisture has been wrung from it and its relative humidity is very low indeed.

Temperate zone communities are differentiated from one another mainly by precipitation, which produces differences in soil profile and mineral content that indirectly influence community composition. Apart from temperate rain forest, deciduous woodland is the most humid of all temperate terrestrial life zones. More highly mineralized soils are found in prairie and desert areas, which are dominated by grasses or drought-resistant vegetation.

FIGURE 54–9 A broad-leaved evergreen subtropical forest, near Miami, Florida.

Deciduous Forests

Where temperate zone precipitation ranges from about 70 to 150 cm annually, **temperate deciduous forests** tend to develop. Those of the northeastern and middle eastern United States serve as an example. These habitats are dominated by broad-leaved hardwood trees that lose their foliage annually (Figure 54–8), and broad-leaved evergreen trees make up an increasing proportion of the trees in this life zone as one proceeds southward (Figure 54–9).

Typically, the soil of a deciduous forest consists of a deep clay-rich B-horizon layer, a well-developed O-horizon, and an A-horizon that is rich in humus. As organic materials in the humus decay, mineral ions are released. If they are not immediately absorbed by the roots of the living trees, these ions are washed into the clay, where they may be retained.

Deep plowing, which mixes the mineral-bearing clay with humus, produces a fair agricultural soil, making all of it accessible to crop roots. Erosion resulting from the lack of a proper root structure in the soil can be serious, as can both the loss of minerals by run-off and the consumption of crop plants at some geographically distant point.

Deciduous forests were among the first communities to be converted to agricultural use, yet they have survived surprisingly well into the 20th century. In Europe and Asia many habitats that were originally forested have been cultivated for thousands of years by traditional methods without substantial loss in fertility. However, American farmers of the 18th and 19th centuries, perceiving land as a limitless resource, often abandoned the wise soil conservation practices of their forebears and allowed erosion and other forms of soil depletion to destroy their land.

In many areas, especially New England, the highly glaciated mountainous terrain had only a thin layer of soil, which was swiftly exhausted. Ironically, the abandonment of the land that resulted from poor agricultural productivity permitted succession to restore much of the original forest, although doubtless in stunted form. Nevertheless, there now is probably more wildlife habitat in the eastern United States today than in the continuously cultivated countryside of western Europe.

The deciduous woodland originally contained such large mammalian fauna as puma, wolves, deer, bison, bears, and other species now extinct,

FIGURE 54–10 Western grassland of the North American plains.

plus many small mammals and birds. Both reptiles and amphibians abounded, together with a denser and more varied insect life than exists today.

Temperate Grasslands

The North American midwest affords us an excellent example of a temperate grassland. Here there are few trees, except for those that grow near watercourses, but grass grows in great profusion in the thick wind- and glacier-deposited soil. Summers are hot, winters are cold, and rainfall is often uncertain.

It is difficult to say where the eastern deciduous forest originally left off and where the grassland began (Figure 54–10), because North American Indians often burned forests and grasslands by accident, for agriculture, or to clear land for traveling and hunting. Repeated burning encourages grassland and discourages the development or persistence of forests. It is likely, however, that the eastern deciduous forest originally extended through Illinois, into Missouri, and westward.

As rainfall decreased with longitude, there was a greater tendency for minerals to accumulate in a marked horizon just below the topsoil instead of washing out of the soil. The soil had considerable humus content and a well-developed O-horizon, for many prairie grasses are similar to deciduous plants in that the above-ground portions of the plants die off each winter, while the roots survive underground. Also, many prairie grasses are sod-forming grasses, meaning that their roots form a continuous underground mat like those we see in lawns surrounding residences and other buildings, only much thicker and tougher—so much so that a wooden plow has difficulty breaking it apart.

The development of the steel plow in postcolonial times, and later the development of the steam tractor, spelled the doom of the original prairie. This biome was so well suited to agriculture that little of it now persists except in such places as abandoned railroad rights-of-way. Almost nowhere can we see even an approximation of what our ancestors saw as they settled the Midwest. It is not surprising that the American Midwest,

FIGURE 54–11 A sagebrush (*Artemisia tridentata*) environment in Nevada. This type of growth is typical of cold deserts.

1298

(a)

(c)

(b)

(d)

FIGURE 54–12 The chaparral biome is surprisingly varied. A sclerophyllous evergreen vegetation develops here where hot, dry summers alternate with rainy winters. Deep-rooted species possess small, leathery leaves (sclerophylls) that lose a minimum of transpired water to the dry air. The seemingly slight difference of the direction faced by the slopes in (a) makes a substantial difference in the nature of the community that develops on them, with low oaks growing on the moister slope. (b) A scrub juniper and oak community in Arizona. (c) The yucca is a plant that grows in many environments, including chaparral and moist deserts. It spends a considerable amount of its energy budget to create the large flowering stalk. Because all vegetation here is widely spaced, the tall stalk is needed for visibility to its pollinator, the yucca moth. (d) The protective coloration of this horned lizard is an important adaptation to life in a region where ground cover is scarce.

the Ukraine, and other such areas became the breadbaskets of the world. These habitats are ideal for such grasses as the cereal grains.

Formerly, certain species of grass grew as tall as a man on horseback, and the land was covered with herds of grazing animals, mainly bison. The principal predators were wolves, although in sparser, drier areas their place was taken by coyotes. Smaller fauna included prairie dogs (actually rodents) and their predators (foxes, black-footed ferrets, and various raptorial birds), grouse, reptiles (such as snakes and lizards), and great numbers of insects.

Steppes, or shortgrass prairies, are grassland habitats with somewhat greater precipitation than the moister deserts. These have a thinner coating of grass than more humid prairies, a higher proportion of bunch grasses, and some bare soil that may occasionally show through. Native grasses are drought-resistant, but this may not be true of introduced cereals or grazing grasses. A combination of drought, injudicious cultivation, and overgrazing resulted in widespread destruction of cultivated steppes both in the United States in the 1930s (the well-known "dust bowl") and in similar areas of the Soviet Union more recently. Steppes may grade imperceptibly into moist desert habitats dominated by sagebrush (Figure 54–11) in cooler areas or cacti (or their ecological equivalents) in warmer areas.

Chaparral

Some temperate habitats have mild winters combined with very dry summers. Such **mediterranean climates,** as they are called, occur not only in the mediterranean area of the world but also in California (Figure 54–12), Australia, and elsewhere. During the rainy season the habitat may be lush and green, but grasses and annuals die or lie dormant during the hot, dry summer. Trees and shrubs often have **sclerophyllous** leaves—hard, small, leathery leaves that resist water loss. Frequent fires occur in this habitat, so

(a)

(b)

(c)

(d)

(e)

(f)

FIGURE 54–13 Inhabitants of hot deserts are often strikingly adapted to the demands of their environment. (*a*) The moister deserts of North America frequently contain large cacti such as this giant saguaro. (*b*) Flowering barrel cactus. Much of the body of this plant is devoted to water storage tissue. In all true cacti, leaves are absent or reduced to the status of spines. Photosynthesis is carried out by the stem alone, which minimizes the surface area subject to evaporation and serves to conserve water. Spines discourage herbivores that might consume the cactus both for its food value and its water content. (*c*) A gecko from the central Namib Desert (Africa). Its webbed feet act like snowshoes, enabling the lizard to run rapidly on the surface of the sand. (*d*) Desert cottontail. (*e*) Gopher snake. A number of desert snakes travel by "sidewinding" due to uncertain purchase of sand. (*f*) Joshua trees.

that many perennial and annual plants are specifically fire-adapted, often reproducing best following a fire. Similar communities occur in the various parts of the world where mediterranean-type climates occur, but in the North American Southwest this community is known as **chaparral.** Chaparral is usually dominated by shrubs, but may contain drought-resistant pine or oak trees.

Deserts

Deserts are arid areas with sparse to almost nonexistent plant life. The low water content of the desert atmosphere leads to temperature extremes of heat and cold, so that a 100°F change of temperature in a single day is sometimes possible. Deserts vary greatly, depending upon the amount of precipitation they receive. Less rainfall than occurs in the chaparral produces the desert of the Hollywood western (Figure 54–13)—in North

(a)

(b)

(c)

(d)

(e)

FIGURE 54–14 The tropical rain forest. (*a*) A broad view of a rain forest on one of the Hawaiian Islands. It is possible that some trees in this photograph are species introduced by settlers and traders who began to arrive in the mid-19th century. (*b*) and (*c*) Tropical rain forest trees typically possess elaborate systems of buttress roots that support them in the shallow, often wet soil. In very high rainfall areas, their leaves may have elongated "drip tips" which drain water off rapidly, thus discouraging the growth of tiny epiphytes, the epiphylls, which would otherwise develop on the leaves and cut off some sunlight from them. (*d*) Other larger epiphytes, such as this bromeliad, grow on trunks and branches. (*e*) Saving the investment of materials in skeletal tissues of the trunk, a strangler fig climbs up the trunk of a palm tree which may eventually be killed.

America consisting of picturesque large and small cacti, yucca plants, contorted Joshua trees, and widely scattered bunchgrass or its ecological equivalent. Still less rainfall produces a Sahara-like desert with widely scattered plant life and, in places, virtually no life at all.

Desert animals tend to be small and of cryptic habits, remaining under cover or returning to shelter periodically during the heat of the day. At night they come out to forage. In addition to specialized insects, there are many specialized desert reptiles: lizards, tortoises, and snakes, especially venomous snakes such as the American sidewinder rattlesnake and similarly adapted Old World vipers.

Mammals include such rodents as the American kangaroo rat, which does not have to drink water but can subsist solely on the water content of its food plus metabolically generated water. In American deserts there are also jack-rabbits, and in Australian deserts, ecologically equivalent kangaroos. Carnivores like the African fennec fox and some raptorial birds, especially owls, live on the rodents and rabbits. A few larger herbivores, such as antelopes, may also be found.

Desert plant adaptations are, if anything, even more striking than those of the animals. While all plants must perform photosynthesis, the stomata that make gas exchange possible are the weakest point in the plant's water conservation system. On the other hand, the evaporation of water from leaf mesophyll also cools the leaf. On the whole, perennial desert plants tend to have reduced leaves or even none, as in the new world cactuses and similar old world *Euphorbias* (see Chapter 17). Still others shed their leaves for most of the year, budding only in the brief moist season. Desert plants are noted for **allelopathy,** an adaptation in which toxic substances secreted by roots or found in shed leaves inhibit the establishment of competing plants nearby. Many desert plants are provided with defensive spines, thorns, or toxins to resist the heavy grazing pressure they experience in this food-short and water-short environment.

A few deserts are so dry that virtually no plant life occurs in them, as is the case in portions of the African Namib Desert and the Atacama-Sechura Desert of Chile and Peru. In the case of these two coastal deserts, however, heavy fog provides condensation that animals and some plants are often able to utilize. The animals usually emerge at night or early morning and lick the dew from their bodies. Condensation may also form on plant leaves, then drip off to moisten the roots. In a few cases the leaves can directly absorb water. The almost entirely heterotrophic communities of these deserts must live on detritus blown in from the outside or washed up on the beaches.

The Tropics

Tropical and subtropical habitats (Figure 54–14) are at least as varied as temperate ones, and like temperate life zones they are determined mainly by precipitation. Thus there are not only tropical forests, but also grasslands and deserts. In the tropics the seasonal distribution of rainfall is especially important. There are grasslands that would be rain forests except that almost all of their rainfall occurs during two months of the year. Lush vegetation could scarcely persist for 10 months unwatered, especially where evaporation is encouraged by high temperatures.

Tropical biomes are often located in areas with ancient, mineral-poor soil that has been extensively leached by high precipitation, producing a soil poor in both nutrient and organic content. In areas with less rainfall, the tropical equivalents of grasslands and deserts occur. The tropical rain forest is very productive despite the scarcity of mineral nutrients in its soil.

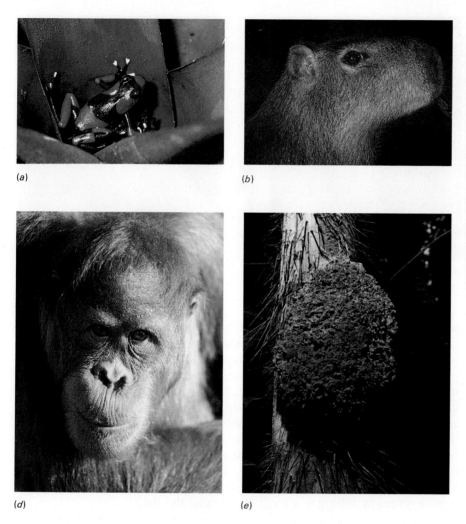

(a)

(b)

(d)

(e)

(c)

FIGURE 54–15 Some animals of the rain forest. (*a*) An arrow-poison frog from Colombia. The bright colors of this frog are an aposematic trait (Chapter 19) which warn that it is inedible. Human beings, however, utilize the frog's poison to coat the tips of the arrows they use for hunting. (*b*) A capybara, the largest rodent in South America. These animals are amphibious, spending large amounts of time in the water. (*c*) Three-toed sloth. (*d*) Orang-utan. (*e*) Tropical termite nest.

Tropical Rain Forests

Of all the life zones in the world, the coral reef and the tropical rain forest are unexcelled in species diversity and variety (Figure 54–15). No one species dominates the tropical rain forest; one could travel for a quarter mile without encountering two members of the same species of tree, especially in subclimax communities.

Despite what you may have seen in Tarzan movies, the vegetation of tropical rain forests is not dense at ground level except near stream banks and in areas that are recovering from lumbering, fire, or agricultural use. The continuous canopy of leaves overhead produces a dark habitat with an extremely humid microclimate. A large part of the precipitation received by the tropical rain forest is locally recycled water that comes from the transpiration of the forest's own trees. If trees are removed over an extensive area, the rainfall may be so reduced that reestablishment of the original forest becomes unlikely.

A fully developed rain forest has at least three distinct stories of vegetation. The topmost story consists of the crowns of occasional very tall trees. It is exposed entirely to the sunlight. The middle story forms a continuous canopy of leaves that lets in very little sunlight for the support of the sparse understory. The understory itself consists of both plants specialized for life there and the seedlings of taller trees.

All stories of vegetation support extensive epiphytic communities of smaller plants that grow in crotches, on bark, or even on the leaves of their

(b)

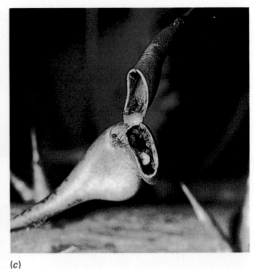
(c)

(a)

FIGURE 54–16 (*a*) Tropical (Costa Rica) acacia tree. (*b*) The thorns harbor ants that feed on extrafloral nectaries. The energy cost of the nectar is worth while to the host plant since the ants defend it from leaf-eating animals. (*c*) An ant brood living inside an acacia thorn.

hosts. The epiphytes are not parasitic in the usual sense, but their numbers can become so great as to break branches or interfere with photosynthesis. Most of the mineral nutrient turnover of the rain forest probably takes place among epiphytes and the growing leaves of the forest canopy, where most animal biomass is concentrated, rather than on the forest floor. Since little light penetrates to the understory, many of the plants living there are adapted to climb upon already established host trees rather than invest their meager photosynthetic resources in the dead cellulose tissues of their own trunks. Tropical vines as thick as a man's thigh abound. This adaptation can be seen in extreme form in several species of **strangler tree** (see Figure 54–14). The strangler fig, for instance, overgrows the trunk of the host, eventually killing it after it has served the fig.

Even rainforest epiphytes have a serious water problem. No matter how hard it may rain, the rain tends to run off the host tree branches immediately. For this reason, cacti are excellent potential rain forest epiphytes and a number of species are common there. Other epiphytes, such as the staghorn fern, possess specialized leaves or other body parts that retain pockets of spongy soil which absorb water. Still others, especially orchids, have specialized spongy tissues surrounding their roots which not only store water but serve as a home for nitrogen-fixing cyanobacteria and other symbiotic algae. Certain large epiphytes known as bromeliads store as much as a gallon of rainwater in their leaf cups and absorb this water between rains by means of tiny scales on the bases of their leaves which function like root hairs. Mosquito larvae, other insects, and even specialized species of crabs and frogs live in this odd aerial swamp.

Epiphytes also have a problem with mineral nutrition: How can they absorb anything from the soil when they are perched sometimes hundreds of feet above it? The nutrient cycles of the rain forest canopy are only now being worked out in detail, but epiphytes apparently eke out their mineral nutrition by a variety of means. Nitrogen and phosphorus may be absorbed from animal droppings; indeed the main reason for the existence of the bromeliad cups may be to attract animals whose excrement or corpses serve as sources of nitrogen.

Epiphytes of the rain forest possess means of reabsorbing minerals which are unparalleled elsewhere. They are, for example, able to remove the tiny amounts of minerals contained in rainwater and compete vigorously for even this sparse source of mineral nutrition.

Like the orchids previously described, many other epiphytes also possess symbiotic nitrogen-fixing microorganisms. Recent evidence even indicates that epiphytes may contribute to the nutrition of their host tree, for host tree roots growing directly from the branches (known as **adventitious roots**) often extensively penetrate the epiphyte community, from which they presumably absorb minerals.

The trees of the tropical rain forest are usually evergreen (though not coniferous), but there are exceptions. Their roots are often shallow, forming a mat 2 or 3 feet thick on the surface of the soil, missing hardly a particle of mineral nutrient released from leaves and litter by decay processes. Despite their poor anchorage, swollen bases or flying buttresses hold the trees upright and aid in the extensive distribution of the shallow roots.

Since the temperature is high year round, decay organisms, ably assisted by detritus-feeding ants and termites, decompose organic litter before it can become humus. At the same time, the high rate of rainfall leaches nutrients rapidly from the soil. Nutrients are extracted from decomposing material by highly developed mycorrhizae. They are then transferred to the roots of living plants before they have a chance to enter the soil. The absorptive mechanisms of the community are so efficient that the run-off water often has a lower mineral content than the rain that falls there.

Ants are especially prominent in the tropical rain forest and throughout the tropics generally. In temperate and even subtropical areas, these insects usually nest in sunny areas, because for full activity their bodies must be warmed by the sun. In the high and relatively unvarying temperatures of the rain forest, however, they need no sunlight and so nest and rove everywhere. Some, the so-called army ants, build no permanent nests but bivouac nightly throughout their nomadic existence.

Perhaps the most fascinating tropical ants are those that have achieved a symbiotic association with certain species of *Acacia* trees and shrubs. The trees bear extremely large thorns with spongy interiors where the ants nest. The trees even bear extrafloral nectaries and protein-rich bodies that provide nourishment for their ant guests. The ants repay the favor, as it were, by enthusiastically attacking any creature brash enough to attack, feed upon, or even touch their tree home. The best known species of such ants are those that nest in hollow *Acacia* thorns. Most such *Acacia* species are veld or even desert species, but the protective ant symbiosis also occurs in rain forest habitats, particularly in association with tree genera other than *Acacia*, Figure 54–16.

Rain forest animals include the most abundant and varied insect, reptile, and amphibian fauna on the face of the earth. Birds, too, are varied and often brilliantly colored. Much mammalian life is arboreal (sloths, monkeys), although some large ground-dwelling forms, including even elephants, are found there. Because rain forest soils are not conducive to fossil formation, little is known of the inhabitants of rain forests of the geologic past. Even today, much remains unknown about the wonderfully varied and intricately interrelated life of the present rain forest.

Explosive human population growth in tropical countries may spell the end of most or all rain forests by the end of the century. It is believed that many rain forest organisms will be rendered extinct in this way before they have even been scientifically described. A surprisingly large number of useful drugs, such as quinine, has been discovered among the exotic chemical compounds used by tropical plants as chemical defenses against grazers and parasites. One wonders how many other potentially useful species of plant and animal life have already vanished. The total ecological impact of rain forest destruction is unknown at present, but it is likely that the burning or decay of felled trees will contribute substantially to the world atmospheric carbon dioxide content while removing one of the largest carbon dioxide fixation areas of active photosynthesis.

FIGURE 54–17 The savanna biome of eastern Africa. These grasslands formerly supported large herds of grazing animals and their predators, which are swiftly vanishing under pressure from pastoral and agricultural land use.

The local ecological impact is likely to be immediate and substantial. Once the protective mat of roots has been removed from the soil, erosion is likely to become an even more severe problem than in temperate climates and often will be accompanied by accelerated laterization (see Chapter 53). Since the mineral content of the rain forest community resides mostly in its vegetation, rain forest soils tend to be poor. Even if left alone, areas of destruction may not succeed to mature forest again if the soil is excessively depleted. Moreover, if the area destroyed is extensive, the rather inefficiently dispersed seeds of the forest species may not be able to make their way to much of it.

FIGURE 54–18 Inhabitants of the savanna. (*a*) Cape Buffalo grazing in Uganda. (*b*) Lions killing a zebra. In the open spaces of the savannah (shown here during the dry season), hunting in pairs or groups is advantageous in tiring or trapping swift-moving prey. (*c*) What is not consumed by the hunters is left for the scavengers, which often double as predators themselves. Here hyenas compete with buzzards, which are strikingly adapted as scavengers. Their long, almost bald necks allow them to reach inside carcasses and make a last effort to pick the body clean.

(*a*)

(*b*)

(*c*)

Savannah

The **savannah** (or **veld**) life zone is a tropical grassland or very open woodland, depending on one's viewpoint (Figure 54–17). Widely dispersed trees such as *Acacia*, bristling with thorns that provide protection against herbivores, grow amid long grasses. The greatest herbivore biomass in the modern world occurs in the African savannah. Here live the great herds of antelope, giraffe, zebra, and the like. Large predators, such as lions and hyenas, kill and scavenge the herbivores. Savannah is produced naturally either by low rainfall or by sharply seasonal rainfall. In areas of seasonally varying rainfall, the herds and their predators (Figure 54–18) may migrate annually, much as caribou migrate between tundra and taiga in the north.

Tropical grasslands are being rapidly converted to range for cattle and other animals. Severe overgrazing in places has converted marginal savannah to actual desert. On the other hand, lumbering and grazing have converted much bordering rain forest to savannah. In both cases large amounts of original wildlife habitat are being subjected to probably irreversible destruction.

AQUATIC HABITATS

The world's surface is mostly water, and of the portion that is liquid, most is salt water. In water (as on land) mineral nutrients are collectively a significant limiting factor, but temperature is less important, and water itself is not an important limiting factor. In most terrestrial habitats there is no great shortage of light, the floor of the rain forest being the most important exception. But this is not true in aquatic habitats. Water greatly interferes with the penetration of light, so floating photosynthesizers must remain near the surface, and vegetation attached to the bottom can grow only in the shallowest zones.

The aquatic habitat contains three main ecological categories of organisms: free-floating plankton, strongly swimming nekton, and bottom-dwelling benthos. All categories are constrained by the small amount of light penetrating the water. This limits productivity to the euphotic zones—upper or shallow waters of aquatic life zones in which photosynthesis is practical. The benthos may be further subdivided into intertidal, subtidal, and very deep-dwelling abyssal communities. Of these only the abyssal community is without its own primary producers and subsists on the leftovers that wash or fall into it from the lighted areas of the water.

The most fundamental division in aquatic ecology is probably between fresh and salt water. The physiological adaptations required to permit osmoregulation in freshwater animals is a major strain on an organism's resources. Some animals, such as echinoderms, are never found in fresh water at all.[1]

The primary ecological subdivisions of aquatic organisms are the plankton, the nekton, and the benthos. The **plankton** (Figures 54–19 and 54–20) consist of organisms (usually small or microscopic) that are relatively feeble swimmers and thus, for the most part, are carried about at the mercy of currents and waves. They are unable to swim far horizontally, but some species are capable of large diurnal migrations in a vertical direction and occupy different strata of water at different times of the day or sometimes at different seasons. Plankton may be subdivided into two major catego-

---

[1]Such organisms, adapted to a very narrow range of salinity, are known as **stenohaline;** those that can live in a variety of salinities are **euryhaline.** Some fish and crustaceans are euryhaline as adults but stenohaline as larvae. Others, such as salmon and eels, live in fresh water during part of their life cycles and salt during another part, migrating between the two.

FIGURE 54–19 Mixed marine zooplankton, crustaceans (including copepods).

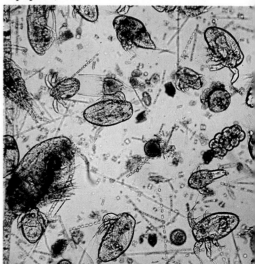

FIGURE 54–20 A baleen whale siphoning the surface of the open ocean for plankton. The bird nearby has been attracted by fish stirred up near the surface of the water by the motion of the whale.

ries: the phytoplankton and the zooplankton. The **phytoplankton** are primary producers. Some are cyanobacteria and free-floating algae of several types, diatoms usually predominating. The **zooplankton** include planktonic protists and animals, including the larval stages of many organisms that are large as adults.

The **benthos** consist of bottom-dwelling creatures, whether sessile and fixed to one spot (as barnacles are), burrowing (as in the case of many worms and echinoderms), or simply walking about on the bottom (such as lobsters).

The **nekton** consist of the larger, more strongly swimming organisms such as fish and whales, creatures likely to reside at the ends of aquatic food chains and creatures of economic and commercial importance to human beings.

Fresh Water

Freshwater habitats include streams, lakes, and ponds. Even temporary ponds have a typical assemblage of inhabitants: protists and algae, mosquito larvae, toad tadpoles, fairy shrimp, which hatch from eggs that lie dormant during dry spells in the bottom, and water strider insects (see Figure 2–15), which live atop the surface film while the water lasts (they can fly away when it dries up). The kinds of organisms found in stream communities vary greatly, mostly in accordance with the strength of the current. In fast streams the water is often cold and highly oxygenated, and the inhabitants may have adaptations such as suckers that keep them from being swept away. Large, slow-moving streams resemble lakes ecologically, and it is on lake habitats that we will focus our attention.

A typical lake shore is inhabited by emergent "air-breathing" vegetation, such as cattails and water lilies (Figure 54–21). The lake shore plus several other concentric communities of deeper-dwelling, entirely aquatic plants constitute the **littoral zone,** the shallow water area around the margins of a lake, pond, or watercourse. It is the most highly productive zone of the lake. A shallow lake may consist entirely of a littoral zone. Algae, particularly filamentous algae and diatoms, may exceed the biomass of the higher plants in the littoral zone. The littoral zone contains frogs and their tadpoles, turtles, annelid worms, crayfish and other crustaceans, insect larvae, and many nongame fish, such as perch or carp, plus some sport fish, such as bass (Figure 54–21). Here, too, at least in the quieter areas, one finds surface film dwellers, sometimes collectively called the **neuston,** such as water striders and whirligig beetles.

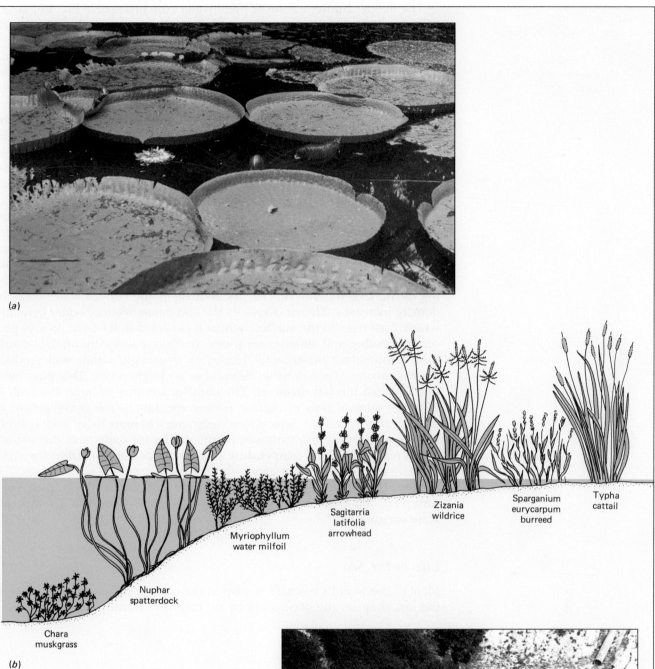

(a)

(b)

Chara
muskgrass

Nuphar
spatterdock

Myriophyllum
water milfoil

Sagitarria
latifolia
arrowhead

Zizania
wildrice

Sparganium
eurycarpum
burreed

Typha
cattail

(c)

FIGURE 54–21 (*a*) A small freshwater pond provides an example of both an ecosystem and the littoral zone. These great tropical Victoria lilies support a rich array of invertebrates, consumed by one another, by larger invertebrates, and by fishes. (*b*) Zonation of vegetation about ponds and along river banks. Note the changes in vegetation with water depth. (*c*) Bass spawning ground. Each circular crater is a bass "nest." These fish use the shallows of the littoral zone for breeding and could easily be destroyed by the destruction of this zone—for example, by lakeside "development."

The deeper **limnetic zone** of open water contains sparse life, but larger fish spend most of their time here, although they may visit the littoral zone to feed and breed. Owing to its depth, less vegetation grows here.

The deepest zone, the **profundal zone,** is out of reach of effective quantities of light for photosynthesis. Primary producers cannot prosper here because it is below their **compensation point,** the point at which the products of photosynthesis balance the rate of catabolism. Nevertheless, much food drifts into this zone from adjacent zones or falls into it from the lighted waters above. When dead plants and animals reach the profundal zone, decay bacteria liberate the minerals their bodies contain, but these cannot be effectively recycled because primary producers do not exist there. In consequence, the profundal habitat tends to be both mineral-rich and anaerobic, with few forms of higher life occupying it.

This marked layering is accentuated by the marked **thermal stratification** characteristic of large lakes, especially in temperate zones. Cool water remains at the bottom in the summer, being separated from the warmer water above by a marked and abrupt temperature transition, the **thermocline.** In temperate zone lakes, falling temperatures in the fall cause a mixing of the lake waters. This occurs because as the surface water cools its density increases. Thus it displaces the less dense warmer water beneath, which then rises to the surface, where it is cooled in its turn. As this process of cooling and sinking continues, the lake eventually reaches a uniform temperature throughout. Hereafter, even light winds will produce further mixing, since density differences no longer exist. This phenomenon is called the **fall turnover.** The sudden presence of large amounts of nutritive mineral ions in surface waters encourages the development of high algal populations, which form temporary **blooms** then, and again in the spring. The spring turnover occurs as the ice melts and the surface water reaches 4°C (its temperature of greatest density). During the summer, thermal stratification again occurs.

Since there is little seasonal temperature variation in the tropics, such turnover, and the thermocline responsible for it, is mostly a temperate zone occurrence.

Life in the Sea

Most of the world's biomass resides in the oceans; however, as we saw in the last chapter, the oceans are by no means uniformly productive.

Plankton

Marine plankton are mostly small or microscopic forms: tiny crustaceans, diatoms and other protists, jellyfish, swimming tunicates, ocean sunfish, arrow worms, and the larval forms of many bottom-dwelling organisms. One can see from this that "plankton" is not a taxonomic classification. Plankton are at the base of most marine food chains, but not all plankton eaters are small. The baleen whale, the largest organism ever known to have lived, eats plankton directly, straining it out of the water through the whalebone in its mouth.

The marine benthos include detritus feeders (such as many worms and bivalve mollusks), many crustaceans (such as crabs, lobsters, and some shrimp), sessile creatures (such as sea anemones, corals, and tunicates), and even some fish (such as flatfish, rays, and moray eels). Aquatic plants equipped with holdfasts are also part of the shallow-water benthos, but they do not extend past the euphotic zone.

Marine nekton tend to depend ultimately on plankton and sometimes on benthic organisms as well, for it is there that the food chains of most fish (including most commercially valuable fish) begin. It is obvious that the

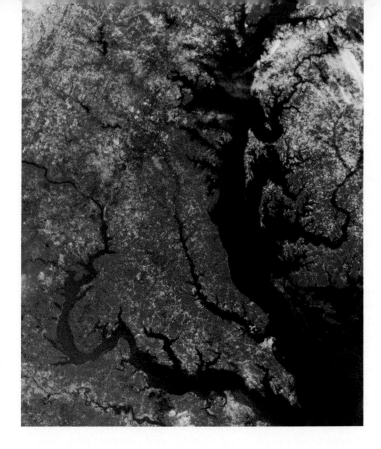

FIGURE 54–22 Chesapeake Bay is a major estuary surrounded by minor, tributary estuaries. The Potomac River enters the main estuary at the bottom of the picture. Such environments are extensively threatened by coastal development, overfishing, and many kinds of pollution.

best fishing tends to be near land, in the relatively shallow and well-fertilized continental shelf areas.

Marine Life Zones

Although marine and freshwater life zones are comparable in many ways, their shallowest and deepest areas differ substantially. The depths of even the deepest lakes do not approach those of the oceanic abysses, which are extremely deep areas that extend miles below the sunlit surface.

SALT MARSHES AND ESTUARIES. Where the sea meets the land there may be one of several kinds of ecosystems: a rocky shore, a sandy beach, an intertidal mud flat, or a tidal estuary containing salt marshes (Figure 54–22). An **estuary** is a coastal body of water, partly surrounded by land, with access to the open sea and a large supply of fresh water from rivers. It usually contains **salt marshes,** very shallow to swampy areas filled with mud, which are dominated by grasses (Figure 54–23). Its salinity fluctuates be-

FIGURE 54–23 The estuarine environment. (*a*) Brigantine salt marsh estuary in New Jersey. (*b*) The glasswort, *Salicornia virginia,* is an estuary plant that grows just above the high tide mark in marshy areas. It is adapted to salty soil conditions by its ability to accumulate comparable concentrations of salt in its own tissues.

(a)

(b)

tween that of sea water and that of fresh water. Many estuaries undergo marked variations in temperature, salinity, and other physical properties in the course of a year. To survive there, organisms must have a wide range of tolerance to these changes.

The waters of estuaries are among the most fertile in the world, often having a much greater productivity than the adjacent sea or the fresh water up the river. This high productivity is brought about by (1) the action of the tides, which promote a rapid circulation of nutrients and help remove waste products, (2) the importation of nutrients into the estuarine ecosystem from the land drained by rivers and creeks that run into the estuary, and (3) the presence of many kinds of plants, which provide an extensive photosynthetic carpet and whose roots and stems also mechanically trap much potential food material. As leaves and plants die, they decay, forming the basis of many detritus food chains. The majority of commercially important fin and shell fish spend their larval stages in estuaries among the protecting roots and tangle of decaying stems.

Estuaries and salt marshes have often appeared to uninformed people to be worthless. In consequence, they have been used as dumps for the castoffs of industrial civilization and have become severely polluted. More recently they have been "filled" with dredged bottom material to form artificial land for residential and industrial development. A large part of the total productivity of the marine environment has been lost in this way.

THE INTERTIDAL ZONE.   The sea exhibits zonation comparable to that of freshwater habitats but profoundly influenced by tides and currents to an extent

FIGURE 54–24 Zonation in the sea.

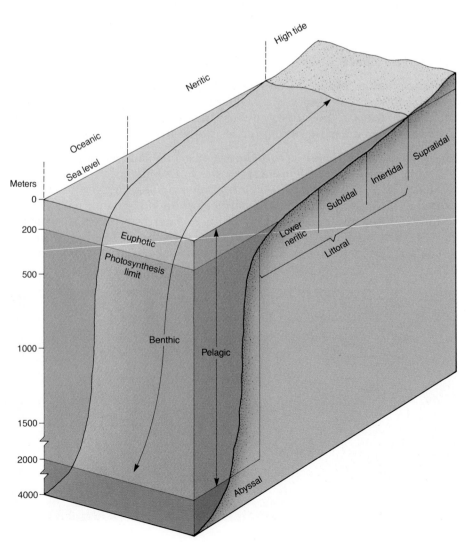

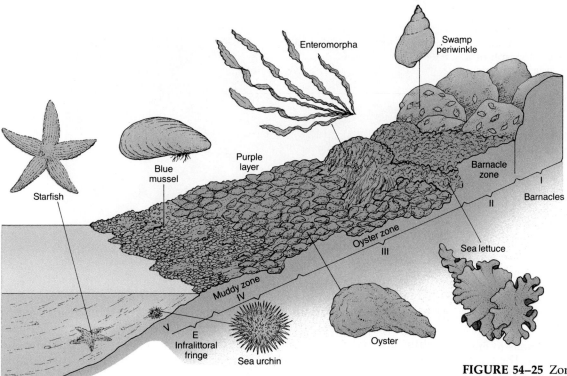

FIGURE 54–25 Zonation along a rocky shore of the mid-Atlantic North American coastline. I. Bare rock with some black (brown) algae and swamp periwinkle. II. Barnacle zone. III. Oyster zone, which also includes sea lettuce and purple (red) algae. IV. Muddy zone with mussel beds. V. Infralittoral zone with sea stars and other typical animals.

unknown in freshwater habitats (Figure 54–24). The gravitational pulls of both sun and moon produce two tides a day throughout the ocean, but the height of those tides varies seasonally and by the influence of the local topography. The area between low and high tide is the **intertidal zone** (Figures 54–25 and 54–26). The high levels of light and nutrients, together with an abundance of oxygen, make the intertidal zone a potentially excellent habitat. Yet it is a very stressful one. If an intertidal beach is sandy, the

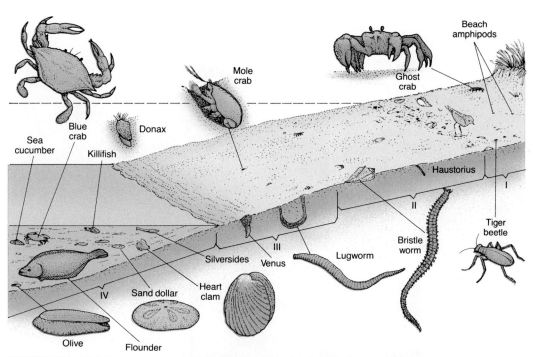

FIGURE 54–26 Life zones on a sandy beach. The gradation, less marked than that of a rocky shore, is nevertheless evident. I. Supratidal zone: ghost crabs and sand fleas. II. Flat beach zone: ghost shrimp, bristle worms, clams. III. Intertidal zone: clams, lugworms, sand ''crabs'' (which follow the retreating or advancing waters). IV. Subtidal zone.

1313

FIGURE 54–27 Intertidal life zonation. Several clearly defined areas can be seen in this photograph of a rocky beach at low tide in the vicinity of Botany Beach, British Columbia. The barnacles covering the higher rocks show up as white, and the mussels, on lower rocks, as dark blue or black. Pink tones are contributed by coralline algae, with green and brown (kelp) algae in the very lowest areas.

FIGURE 54–28 Adaptations to life in the intertidal zone. (*a*) The seaweeds clinging to these rocks have very little hard structure to their bodies, yet have strong holdfasts that enable them to remain in place against the onslaught of the tides. (*b*) Holdfast of bull kelp, *Diurvillea antarctica* (*c*) These Bahamian snails shelter from the wave action by clinging tightly to a cavity in the rocks. Their shells are so shaped as to offer minimal resistance to the rushing seawater. (*d*) Life underwater in a tidal pool: dog whelk, rock barnacle, obelia kelp, rockweed, sea anemone, and blue mussel.

inhabitants must contend with a constantly shifting environment that threatens to engulf them and gives them scant protection against wave action (Figure 54–26). Consequently, most sand-dwelling organisms are continuous and active burrowers. They are, however, able to follow the tides up and down the beach, and so do not usually have any notable adaptations to drying and exposure.

On the other hand, a rocky shore (Figure 54–25) provides a fine anchorage but is exposed to wave action when submerged and to drying (to say nothing of seasonal heating and freezing) when exposed to the air. A typical rocky shore denizen will have some way of sealing in moisture (perhaps by closing its shell, if it has one), plus powerful means of anchorage to the rocks (Figure 54–27). Mussels, for example, have horny, threadlike anchors, and barnacles have a special cement gland. Rocky shore intertidal plants usually have thick, gummy polysaccharide coats, which dry out slowly when exposed, and flexible bodies not easily broken by wave action. Some members of this community hide in burrows or crevices at

(a)

(b)

(c)

(d)

FIGURE 54–29 This offshore caye (or sand island) is surrounded by an area of shallow water. Notice that submerged vegetation grows thickly only in the shallows. That is because light penetrates water rather poorly; the deeper the water, the sparser is its bottom vegetation.

low tide, and some small semiterrestrial crustaceans run about the splash line, following it up and down the beach (Figure 54–28).

THE SUBTIDAL ZONE.   The **subtidal zone** is below the lowest tide but still shallow enough for vigorous photosynthesis (Figure 54–29). Largely protected from wave action, this area supports a varied community of echinoderms (such as sea stars), fish, burrowing worms, eelgrass, and the like. Samples of this community can be seen in **tidal pools,** which are temporary marine ponds left behind on rocky shores by the receding tide. Shore birds find the subtidal and intertidal zones rich hunting and fishing grounds. (The subtidal zone can be considered the shallowest portion of the neritic zone [below].)

THE NERITIC (CONTINENTAL SHELF) ZONE.   Nekton and larger benthic organisms are mostly confined to the shallower **neritic** waters of less than 200 feet deep because that is where their food is. Not only is there considerable vegetation on the bottom, but as you can see from Figure 56–2, this is also where the chlorophyll content of the water itself is high, although not as high as in an estuary. That means that little commercial fishing is worthwhile more than a hundred miles offshore, a fact that is not without significance in determining a nation's "territorial waters."

DEEP WATERS.   Some 88% of the ocean is more than a mile deep, far below any algal compensation depth and therefore, for the most part, an aquatic desert. Most of the life that exists under the tremendous pressures and darkness of the abyss depends upon whatever food filters into its habitat from the upper lighted regions. The principal exceptions are the deep sea oases located near abyssal hot springs (**thermal vents**), to be discussed in Chapter 56.

Animals of the abyss are strikingly adapted to darkness and scarcity of food (Figure 54–30). Many have illuminated organs, enabling them to see

FIGURE 54–30 A deep-sea fish. Most fish that live at great ocean depths have weak or vestigial eyes. Many have luminous organs enabling them to locate one another (it is thought) for mating or social display. Inside the nearly transparent body of this particular fish are the remains of small crustaceans that may have drifted down from the productive surface waters of the ocean.

Focus on Coral Reefs

The most wonderfully varied marine environment is the coral reef community. Such reefs are confined to warm waters (although coral is not), much as most rain forests occur in warm climates. Like the rain forest, they are highly productive and highly diverse communities, often occurring in nutrient-poor habitats.

Coral reefs start in shallow water, for example on the sides of a volcanic island near the surface. Wave and weather action eventually erode the island to water level, but the coral does not erode in the same way. Being alive, it is capable of compensatory growth. The result can be a circular reef of coral, an **atoll,** surrounding a lagoon of quiet water. Sometimes the bottom sinks, or the ocean level rises slowly enough for the growth of the coral organisms to keep up with it. If this process continues long enough, a living coral reef may persist, perched upon layer after layer of its smothered forebears. Sometimes coral reefs grow as barriers, comparable to offshore sand bars, as in the great barrier reef of Australia.

The upper living portions of coral reefs must grow in the euphotic zone. Many coral reefs are composed principally of red coralline algae, and even coral animals, usually anthozoan coelenterates, have hordes of intracellular symbiotic dinoflagellates (**zooxanthellae**) living in their tissues. Although species of coral without zooxanthellae do exist, only species with zooxanthellae are able to build reefs. Other reef-dwelling organisms, such as the giant clam, *Tridacna*, may also possess symbiotic algae. Not only do the coral polyps capture food with stinging tentacles, but thanks to their symbiotic partners, benefit from photosynthesis as well. A coral polyp is almost the marine equivalent of a lichen, and some biologists consider corals "honorary plants," for their tissues may contain even more zooxanthellae than animal cells. Viewed as plants, their primary productivity compares with that of rain forests and estuaries.

The shallow location also means that the reef is exposed to turbulent wave action. The disadvantage of this is that organisms are easily damaged by wave action and may be stranded occasionally at low tide. The advantage, however, is that the surf provides filter-feeders with an energy subsidy, much as is the case in the intertidal littoral zone.

The waters in which coral reefs are found are often nutrient-poor. Parts of the South Pacific Ocean, for example, tend to be unproductive for this reason. How coral reefs, productive as they are, overcome this disadvantage is not fully known. However, efficient local recycling mechanisms (similar to those of rain forests) appear to play a part, as do, it is suspected, the zooxanthellae. (The zooxanthellae themselves benefit from a much higher carbon dioxide availability in the tissues of their hosts than they would find in the warm water outside.)

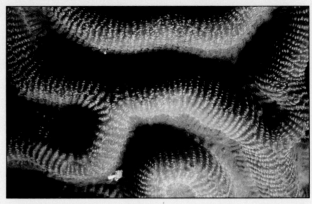
(b)

■ The complexity and productivity of the coral reef rivals or exceeds that of any terrestrial environment. In many ways the coral reef is the aquatic equivalent of the tropical rain forest. (a) Pink sea fan coral growing at right. (b) Brain coral (*Colpophyllia natans*) with symbiotic algae. (c) Giant clam. (d) Glassy sweepers. (e) Symbiosis between zoanthid anemone and sponge (red). (f) Blue seastar (*Linckia*). (g) Clownfish with sea anemone. (h) Plumeworm with coral. (i) Queen anglefish (*Holacanthus ciliaris*).

(a)

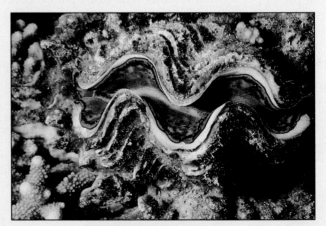
(c)

(d)

(g)

Coral reefs contain hundreds of species of fishes and invertebrates which occur nowhere else. Many are brilliantly colored or poisonous or possess other striking and bizarre adaptations. The multitude of ecological niches and relationships occurring there seems comparable only to the tropical rain forest among terrestrial ecosystems. As in the rain forest, however, competition is intense, particularly competition for light and space to grow.

For instance, corals of the genus *Podillopora* possess a striking defense mechanism that protects them from coral browsers such as the crown-of-thorns starfish. Tiny commensal crabs live among the branches of the coral. When a starfish crawls onto the coral, the crabs swarm over it, nipping off its tube feet and eventually killing it if it does not promptly retreat! This relationship is a close ecological equivalent of that between *Acacia* and ants, which was discussed earlier in this chapter.

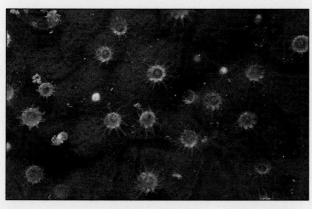

(e)

(h)

(f)

(i)

one another for mating. Some abyssal squid even emit a luminous cloud of ink which may distract predators. A great many abyssal organisms are predators (there is little other choice!) and live in dispersed populations. Some are remarkably adapted killing machines prepared to take quick and maximum advantage of any rare opportunity to feed. Their scattered population suggests that it might be hard for them to find mates, but studies to confirm this must await the development of ways to observe the ocean bottom over long periods. One species of deep-sea anglerfish hangs onto her mate once she has found him, for he permanently attaches to her body and grows into it—the perfect male career parasite!

LIFE ZONE INTERACTION

Not one of the biomes or life zones we have discussed exists in isolation. All interact. We have seen that it is profitable, for example, to plant bushes around a trout stream or pond, for not only do these bushes help to keep the water cool, but also insects living on the bushes fall into the water, where the fish devour them. To take another example, when parts of the Amazon rain forest flood annually, fish leave the stream beds and range widely over the forest floor, where they have been shown to play a role in dispersing the seeds of many species of land plants.

The final example comes from the polar zones, whose waters may be much more productive than the surrounding land. In Antarctica, for instance, there is hardly any terrestrial community of organisms, but there are many seabirds, seals, and other semiaquatic air breathers. These are supported exclusively by the ocean. Their waste products, cast-off feathers, and the like, if deposited on land, support what lichens and insects may occur there. Polar bears and seals perform similar roles on the shores of Arctic seas.

Inhabitants of various biomes may interact over wide distances, even global distances in the case of migratory birds and fish. Migratory birds commonly spend critical parts of their life cycles in entirely different countries, which can make their conservation difficult. It does little good, for instance, to protect a songbird in one country if the inhabitants of the next put it in the cooking pot as soon as it lands in their neighborhood. This concept of large-scale interaction makes ecology difficult for many people to grasp or apply.

■ SUMMARY

I. Living things and associations of living things are restricted in their occurrence to habitats to which their adaptations suit them.
II. For this reason, living things make up certain characteristic geographic assemblages. Such terrestrial life zones are known as biomes.
  A. Characterized by a permanently frozen layer of subsoil, the permafrost, the tundra is the northernmost of the terrestrial biomes. It possesses low-growing vegetation adapted to cold and a very short growing season.
  B. The boreal forest habitats lie south of the tundra and stretch across both North America and Eurasia. The boreal forest is dominated by coniferous trees.
  C. Temperate coniferous forests exist where a pronounced dry season gives the evergreen needle-

leaved trees a competitive edge. Some of these habitats do, however, have high seasonal rainfall and can even qualify as temperate rain forest.
  D. Temperate communities are extremely varied, including deciduous forest, prairie, chaparral, and desert regions. Differences in precipitation produce the differentiation of temperate communities from one another.
    1. Deciduous forests are dominated by broadleaved trees that for the most part lose all their leaves seasonally. The soil is well differentiated into several horizons. Much modern temperate zone agricultural land was originally deciduous forest.
    2. Moderate precipitation produces the temperate grasslands. Typically these possess a deep, min-

eral-rich soil and are well suited to the growing of grain crops in modern times.

 3. Chaparral is an assemblage of small-leaved shrubs and trees which develops in a few areas of modest rainfall and mild winters.

 4. Deserts are produced by low rates of precipitation and possess communities whose organisms have noteworthy water-conserving adaptations. Desert soils are often saline.

E. Tropical biomes include the tropical rain forest, the savannah, and deserts, plus intermediate communities. As is the case with temperate communities, most of the differences among tropical life zones result from variations in precipitation.

 1. Tropical rain forests are produced by very high rainfall evenly distributed throughout the year. They have high species diversity and are noted for three stories of forest foliage, many epiphytes, and often a lateritic soil.

 2. Many tropical grasslands are similar to open forests, with scattered trees interspersed with grassy areas. These are called savannah or veld.

III. Aquatic life zones differ from terrestrial largely because of the differences in physical properties of air and water. One of the most striking limiting factors in aquatic environments is the availability of light for photosynthesis. Oxygen also is in short supply. However, water's density provides support for the tissues and bodies of aquatic organisms, which reduces their need for skeletal structures.

A. Freshwater communities are differentiated on the basis of water depth. The marginal littoral zone contains emergent vegetation and heavy growth of algae. The limnetic zone is intermediate in depth, and the deepest profundal zone holds little life. Temperate zone lakes and ponds exhibit thermal stratification. The thermocline delimits the warmer water above from the deep, cold water. Annual spring and fall turnover remixes these layers.

B. The main marine habitats are similar in principle to those of fresh water but are modified by tidal action. They include salt marshes and estuaries, the intertidal zone, the subtidal and continental shelf areas, and the abyss.

 1. Aquatic life is ecologically divided into the plankton, nekton, and benthos. In large measure, the free-floating microscopic plankton support aquatic communities. Owing to the availability of mineral nutrients near continental margins, shallower waters are the most productive of plankton.

 2. Marine life zones

  a. The very productive estuary community receives a high input of nutrients from the adjacent land and serves as an important nursery area for the young stages of many aquatic organisms.

  b. Organisms of the intertidal zone possess adaptations that enable them to resist wave action. Sandy shore inhabitants usually burrow; rocky shore organisms cling to rocks. Because oxygen, light, and food are readily available, intertidal communities are abundant and productive.

  c. The subtidal and continental shelf zones are characterized by rich plankton and benthic vegetation.

  d. Coral reef–building organisms employ symbiotic photosynthetic organisms to produce food directly for the use of the coral polyps. The species diversity of coral reefs is unexcelled in aquatic habitats.

  e. The abyssal zone is without primary producers (except in the case of hot spring oasis communities), so that its animal inhabitants are exclusively predators or scavengers subsisting on input from the areas of the ocean capable of primary production.

■ POST-TEST

1. The tundra typically has a _____ rate of precipitation, a short growing season, and a permanently frozen underground layer of _____.

2. _____ _____ are the dominant vegetation of the boreal forest.

3. In the temperate zone, the deciding factor in producing forest versus grassland is usually _____.

4. In temperate deciduous woodlands, minerals leached from decomposing humus may accumulate in a layer of _____.

5. Epiphytic vegetation must be adapted to a shortage of _____ and _____ _____.

6. Species diversity is exceptionally _____ in tropical rain forest and marine _____ _____ habitats.

7. The _____ is a tropical habitat in which widely spaced trees are interspersed with grassland.

8. Compared with terrestrial habitats, aquatic environments are less variable in _____ and have less available _____ and _____.

9. Because they are shallow and abundantly provided with mineral nutrients, _____ are highly productive habitats that shelter larval stages of many marine organisms.

10. Temperate zone lakes are thermally stratified, with warm and cold layers separated by a transitional _____. This stratification is lost at the time of the _____ _____.

11. Emergent vegetation grows in the _____ zone of freshwater lakes and streams.

12. Organisms living in a sandy beach usually escape wave action by _____.

13. The lightless _____ community is almost entirely heterotrophic, living on an input of dead organisms from other marine habitats.

■ REVIEW QUESTIONS

1. What factors produce the major terrestrial biomes?
2. List the terrestrial biomes and give the location or locations of each.
3. What physical and chemical properties of water are most important in determining the adaptations of the organisms found in aquatic habitats?
4. What are the major freshwater life zones found in a large pond or lake?
5. What is a thermocline? What is its ecological significance?

6. What are plankton? What is their role in aquatic ecology?
7. How do the inhabitants of a rocky beach differ from those of a sandy beach? What characteristic adaptations tend to occur in the two habitats?
8. What is meant by "compensation point"? What is its significance to benthic communities?
9. Describe the ecological structure of a coral reef.

■ RECOMMENDED READINGS

Also consult the general ecology works listed at the ends of Chapters 53 and 55.

Coker, R.E. *Streams, Lakes and Ponds.* New York, Harper and Row, 1968 (originally published 1954). Despite its age, still accurate and easily the most readable introduction to freshwater habitats. Includes a practical approach to water pollution.

Cole, G.A. *Textbook of Limnology.* 3d ed. St. Louis, C. V. Mosby, 1983. Ecology of fresh waters.

Forsyth, A., and K. Miyata. *Tropical Nature.* New York, Charles Scribner's sons, 1984. Beautifully written introduction to tropical rain forests. At once scholarly and anecdotal.

Mitsch, W.J., and J.G. Gosselink. *Wetlands.* New York, Van Nostrand Reinhold Company, 1986. Complete discussion of every aspect of freshwater and coastal wetland ecology.

Teal, J., and M. Teal. *Life and Death of the Salt Marsh.* Boston, Little, Brown, 1969. Popular, very readable and personal summary of coastal wetland ecology with a plea for wetlands conservation.

Zwinger, A.H., and B. Willard. *Land Above the Trees: A Guide to American Alpine Tundra.* New York, Harper and Row, 1986 (originally published 1972). Unique and beautifully written guide to the tundra most accessible to Americans.

Variation within a population: noontime crowd on Wall St., New York City

55

Populations

■ **LEARNING OBJECTIVES**

After you have read this chapter you should be able to:
1. Define *population* and summarize the main properties of a population of organisms.
2. Summarize the main types of population distribution.
3. Give the main factors that produce population change.
4. Define *environmental resistance* and give its role in determining population growth and size.
5. Summarize the concept of limiting factors and give their role in environmental resistance.
6. Distinguish the effects of density-dependent limiting factors from those of density-independent limiting factors.
7. Give examples of density-dependent limiting factors.
8. Name, describe, and draw the main kinds of growth curves.
9. Define *K-strategies* and *R-strategies*, and give an example of each.

One of the most important subjects in ecology is the study of populations. For one thing, every thinking person should be concerned with the dynamics of the swiftly growing human population. Biologists have, in addition, profound concerns with the populations of other organisms, not only because of their possible extinction, but also in connection with evolution, population genetics, and normal functioning of ecosystems.

WHAT POPULATION IS

At first glance it may seem that population biology does not require separate consideration. After all, organisms are individuals and individuals make up populations; populations in turn make up communities. Yet populations have properties that individuals do not possess. Individuals do not exhibit birth rate, death rate, sex ratio, or median age. These properties have been recognized and much of the groundwork for their study has been laid by such people as insurance actuaries, but biologists apply them, sometimes in special forms, to the study of all organisms and their mutual influence upon one another's numbers.

On the other hand, populations have properties that communities also lack. The most important of these is genetics. Since ecological communities do not share a common gene pool, there is no obvious way that natural selection or evolution can produce change on the level of communities as such, although these things do occur in the populations of the species that make up communities, thus affecting communities indirectly.

But to think clearly about populations we must begin with as exact a definition of the term as possible. A **population** is a group of organisms *of the same species* living in a particular geographic area. Yet by itself, a bald population figure tells us relatively little. A thousand mice in a square mile is a very different matter from a thousand mice in an acre! Also, within a ten-square-mile area containing 10,000 mice there might be an occasional acre with few or none.

Density

Population density is the number of individuals of a species per unit of habitat area. Habitats vary in the population density of any species of organism that they can support, and this density may also vary from year to year. Miller and his colleagues counted red grouse in northwest Scotland at two locations only 2.5 km apart. At one location the population neither increased nor declined significantly, but at the other it almost doubled in 1965 and 1966 and then declined to its initial density once again. The reason was a change in the habitat. The area where the population density increased had been experimentally burned. Young plant growth produced by this was beneficial to the grouse. So population density is not an inherent property of a species; it is determined in large part by external factors.

Distribution Patterns

It is obvious that organisms do not occur haphazardly or even exist in all locations where they theoretically could live. Even within a population, distribution may be spotty (Figure 55–1). Sometimes this spottiness is due to random factors. Of the three major types of distribution, **random distribution** seems the most likely to occur but is not common or easy to observe, partly because important environmental factors affecting distribution usually do not occur at random. Flour beetle larvae in a container of flour are randomly distributed, for example, but their environment is unusually uniform.

Uniform distribution is somewhat more common. A field of soybeans or wheat, for example, exhibits very regular and even population distribution—so much so that it is obviously artificial. Uniform distribution sometimes does occur naturally, however. When competition between individuals is severe, when they wage warfare on one another by allelopathy (see Chapter 54), and in some kinds of territorial behavior, individuals share space to a roughly equal extent.

Perhaps most common, **clumped distribution** often results from the presence of family groups and pairs in animals, and from inefficient seed dispersal or asexually propagated clones in plants. An entire grove of aspen trees, for example, may originate from a single seed. Clumping may also be advantageous in itself; social animals derive much benefit from mutual aid and kin selection.

(a)

(b)

(c)

FIGURE 55–1 (*a*) Near random distribution of Mexican poppies in a field. (*b*) Uniform distribution of desert plants, probably due to their production of toxic substances that inhibit the growth of plants nearby (allelopathy). (*c*) A combination of social behavior and locally concentrated resources such as this water hole produces clumped distribution, as with these zebra.

POPULATION GROWTH

Whether a population grows, declines, or remains stable depends on the balance of two factors: the rate of arrival of organisms and their rate of departure. The arrival rate is determined by the rate at which organisms immigrate into a habitat along with the rate at which they are born (or hatched, sprouted, or otherwise produced). The rate of departure depends on the rate of emigration and of death. Although there are exceptions, especially in the case of migratory species, immigration and emigration rarely make a substantial contribution to population growth. The important factors are **natality rate,** the rate at which organisms reproduce, and **mortality rate,** the rate at which they die.

Population *growth* is a special case of population *change.* To determine the speed of this change, we must also take into account the time involved, that is, the change in time. To express these changes we will employ the Greek letter *delta,* printed as a triangle in the following equation:

$$\frac{\Delta N}{\Delta t} = b - d \tag{1}$$

where b is the natality, d the mortality, ΔN the change in population, and Δt the time interval.

A modification of this equation tells us at what rate the population is growing at a particular time, that is, its **instantaneous growth rate.** Equation (2) describes **per capita growth rate,** that is, how rapidly each individual is reproducing. (In equation 2, the expression $b - d$ is replaced by r.) Ordinarily only females are taken into account, since in the case of most animals, only females produce young (some modification is necessary in the case of many plants or asexually reproducing organisms). To determine the per capital natality rate and mortality rate, these values are divided by the size of the female segment of the population at the *beginning* of the time period under consideration. If r is the per capita growth rate, and p is the initial population size, then

$$\frac{\Delta N/p}{\Delta t} = r \qquad (2)$$

and the instantaneous growth rate[1] can be expressed as

$$\frac{dN}{dt} = rN \qquad (3)$$

Notice that, since $r = b - d$, if the organisms are born faster than they die, r is positive and population size will increase. If they die faster than they are born, r is negative and population size will decrease. If r is equal to zero, births and deaths match, and population size is stable despite continued reproduction.

r tells us the rate at which a population is actually increasing, and it can be determined by measurement. Also important, however, is the maximum rate at which the organisms *could* reproduce under ideal conditions. This theoretical maximum, the **innate capacity to increase** or **biotic potential,** depends on such factors as litter size and the minimum age at which females bear young. Sometimes it can be estimated by experiment or reliable observation.

Mathematically, the biotic potential is expressed as r_m. The actual growth rate may be equal to the biotic potential in the initial stages of population growth of a rapidly reproducing species, but as a rule population growth is constrained to a lower value than biotic potential in various ways. As we shall see, in the end population stops growing altogether no matter what the biotic potential may be.

Biotic Potential

Many factors can determine the biotic potential of a population. Asexual reproduction frequently results in extremely high biotic potentials, as can be seen in the division of bacteria and protozoa, or the reproduction of many plants by underground or underwater roots or stems (as in the case of water hyacinths). Yet high biotic potential also can occur in sexually reproducing organisms. Consider the vast numbers of eggs that can be produced by a tapeworm or laid by a single female fly.

For mammals and birds, high biotic potentials often occur in rodents and songbirds with large litter or clutch sizes. Low biotic potentials may occur in carnivores such as cheetahs or golden eagles, but there are many exceptions. Large herbivores such as horses or elephants tend to have low biotic potential, as can some exceptional organisms with peculiar adaptations, such as the 17-year cicada. Biotic potential depends not only upon the number of offspring produced at a time, but also upon how frequently they are produced and how long the reproductive life of the organism may be.

---

[1]*Instantaneous growth rate* is the rate of growth at a particular instant, rather than average growth rate during the entire period of study.

Mortality and Survivorship

Mortality is the inverse of biotic potential, so that mortality rate opposes birth rate. The population growth rate depends upon the balance of birth rate and death rate, its rate of increase or decrease depending on the number of individuals that survive.

As an example, let us consider the survivorship of human beings in the United States in 1980. As the graph in Figure 55–2 shows, there was an early hazardous time of life for newborn infants and young children, especially males, resulting principally from premature birth, congenital defects, and undeveloped immunity to infectious disease. Once this inital period was safely passed, however, few people died until well into their thirties. The mortality rate increased substantially in middle age, with a great acceleration in the fifties and thereafter.

Such a survivorship curve is typical of long-lived organisms whose young have a high likelihood of survival. In the case mentioned, the curve reflects an almost complete freedom from predation, excellent care of the young, and a combination of medical and social measures conducive to the reduction of infant mortality, although not necessarily to the reduction of mortality at later ages. Although similar survivorship curves have been found in some large and formidable animals, it is not characteristic of most species. More will be said about human population traits in Chapter 57.

Another type of survivorship is exhibited by the herring gull. These birds were studied on Kent Island, Maine, in the 1930s, with the results summarized in Figure 55–3.

Paying close attention to the scale of the horizontal axis, notice that the vast preponderance of mortality occurs almost immediately despite the protection and care given to the helpless chicks by the parent birds. Much of this reflects predation or attack and destruction by other herring gulls, but some occurs as a result of such factors as inclement weather, infectious disease, or starvation following the death of the parent.

Once the chicks become fully fledged and independent, their survivorship increases dramatically. Notice, however, that mortality occurs at about the same rate (as reflected in the slope of the graph) throughout the remaining life of the birds. This constancy probably results from essentially random events causing death year after year with little age bias. As a result, few or no herring gulls in a state of nature die from causes comparable to the degenerative diseases of "old age" that cause death in most human Americans. Partly for this reason the maximum ages observed for animals kept in protected captivity are not necessarily indicative of what happens in a natural state.

Another type of survivorship curve is well demonstrated by a tropical palm tree (Figure 55–4) occurring in the upland forests of Puerto Rico (as reported by Leigh van Valen), but similar curves can be drawn for many animals such as oysters, whose larvae are widely scattered and dependent upon chance events to permit their survival.

Most of the seeds produced by this palm tree fail to sprout, owing partly to lack of dispersal, partly to predation by seed-eating animals, and partly to attack by specialized insects, fungi, and the like. Even once germinated, seedlings and young trees have great difficulty becoming established in tropical forests owing to competition from larger trees. Only when the trees are big enough to obtain abundant sun is their survival reasonably assured. From then on their mortality is extremely low until, at about age 150, they begin to die at a great rate, presumably from causes analogous to those of human "old age."

Putting the three curves together and idealizing them a great deal, we obtain a family of curves (Figure 55–5) describing the ideal survivorships of organisms with highly effective protection of their young, intermediate protection, and little or no protection. However, it is difficult to find examples that perfectly fit each of these curves.

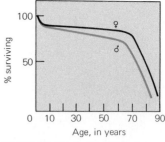

FIGURE 55–2 Survivorship curves for Americans, based on the 1980 census data.

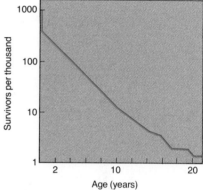

FIGURE 55–3 Survivorship curves for a herring gull population. Data collected from Kent Island, Maine, 1934–1939, as reported by Paynter. Baby gulls were banded to establish identity.

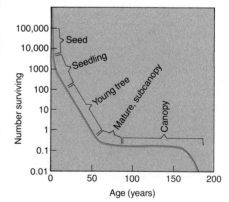

FIGURE 55–4 Survivorship of a species of palm tree. This species (*Euterpe globosa*) grows in Puerto Rican upland forests. During the period of comparative youth shown here, the mortality rate holds constant at about 10 to 15% per year, dropping at maturity. Around 150 years of age, the mortality rate greatly increases, with none surviving past their mid-170s. (From Van Valen, L. Life, death, and energy of a tree. *Biotropica* 7:263, 1975.)

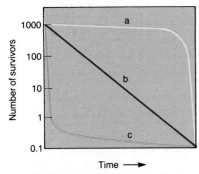

FIGURE 55-5 Three theoretical survivorship curves describing two extremes in life span strategy (*a*) and (*c*) and a third intermediate state (*b*).

It now becomes clear why human population growth has become so marked. It is not because of our high biotic potential (which is comparable to that of elephants or 17-year cicadas), but because of our astonishingly great survivorship, particularly at young ages and during our reproductive years.

GROWTH CURVES AND STEADY STATES

The rate of natural increase (*r*) would produce a steady population growth indefinitely in an environment of unlimited resources. Clearly no such environment exists, but population growth will follow the rate of natural increase for a time in the initial stages of population growth, when the resources of the environment are large in comparison with the population. One often observes this in bacterial or protist cultures or even in the case of some insects.

However, the rate of natural increase is not the *same* as the rate of population growth in such a case. The reason is partly that in rapidly multiplying organisms, parents are still reproducing even after offspring have begun to reproduce. Mostly, however, it is simply that offspring have offspring in their turn. Thus even among humans one child does not usually yield one grandchild, but two, three, four, or more. Unrestricted population growth therefore resembles compound interest in which money breeds more money. Although it cannot remain unrestricted indefinitely, in its initial stages it can be overwhelmingly fast.

Figure 55-6 shows the growth of a bacterial culture. The graph shows an initial **lag phase** of low initial growth, which steepens into a **logarithmic stage,** whose shape is approximately described by Equation 1, above. During the stage of logarithmic growth, the time it takes the population to double, known as **doubling time,** will decrease. Eventually, however, logarithmic growth levels off and the population enters a steady state from which it may then decline. Such a growth curve is called an **S-shaped curve.** The value, *K*, at which growth stops represents the collective action of all the components of environmental resistance. *If it is sustained, K* is the **carrying capacity** of the environment for this particular organism. If it is not sustained, carrying capacity has been temporarily exceeded and *K* represents only the maximum population that can be attained for the time being. For instance, fish and game management programs often artificially hatch and rear very large numbers of sport fish or game birds which are released just before their season officially begins. Often, those individuals that escape the hunters or fishers soon die and do not permanently increase the population.

The logarithmic portion of a population growth curve can be approximately described by a modified growth equation called the **logistic equation.** It describes what determines the growth rate at any one time.

$$\frac{\Delta N}{\Delta t} = rN\frac{K - N}{K} \tag{4}$$

It can easily be demonstrated that while the number of organisms (*N*) is small, the rate of growth will be high. But if *N* begins to approach *K*, the growth rate falls nearly to zero despite high population densities.

The logarithmic portion of the S-shaped growth curve is unable to persist indefinitely, but while population is still actively growing, its graph resembles the letter J. In such a J-shaped curve, environmental resistance becomes most significant toward the end of population growth. In such cases it often reflects little more than the sudden depletion of nutrients, which abruptly halts population growth and then produces a decline. However, environmental resistance can have other components, such as

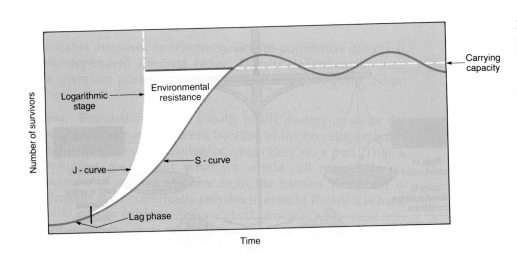

FIGURE 55–6 A typical, simple growth curve such as can be observed initially in a bacterial culture. The J-shaped curve of population growth changes to an S-shaped curve as the result of increasing (but often poorly understood) environmental limitations. When the carrying capacity of the environment is reached, population tends to fluctuate narrowly around it, usually because of both reduced natality and increased mortality.

the accumulation of toxic waste products. If environmental resistance gradually increases from an early stage of population growth, it produces a **sigmoid curve** whose *K*-value is low and whose slope is less than that of a J-shaped curve.

Both J-shaped and sigmoid curves are easier to observe in a laboratory experiment than under natural conditions. Reasons include (1) Natural populations are usually freer to disperse, (2) complex limiting factors may come into play at various times, and (3) natural populations are rarely observed in their early stages of growth. Such curves can be observed, however, among organisms newly introduced to favorable habitats without significant natural enemies and among the human population of the world at the present time.

Populations do not always stabilize somewhere around the highest point they attain, but may experience a **population crash,** an abrupt decline from high population density to very low density. Such an abrupt change, which can even result in local extinction, is observed commonly in bacterial cultures or other populations, the upper limits of whose resources are nonrenewable or inflexibly fixed. When the vital resources are all consumed, the organisms all die—to the last individual.

Sometimes, even when resources are renewable, they may take longer to regenerate than the remaining members of the population can survive. In the case of one arctic island, for example, a particularly bad winter buried all available forage under snow and ice so that a reindeer population of thousands almost completely starved, leaving just one individual alive. It is instructive that this last case was due in part to an unpredictable density-independent limiting factor—severe weather. However, it almost certainly would have left more animals alive if their population density were not so close to the carrying capacity of the habitat.

Environmental Resistance

Population growth is ultimately stopped by an increase in mortality and sometimes by a decrease in natality. Both mortality and natality have causes. These causes are specific but often very difficult to identify. The important factor, however, is their summation. Most causes appear to be environmental, which has led to the term **environmental resistance** (Figure 55–7) to denote the limitations on the ability of the environment to support populations of organisms. Such resistance can arise from resource shortage, such as the lack of some critical plant nutrients, food, or water. Some resistance comes from disease and some from predation or competition from members of the same or other species. Notice that this last may result

FIG
tion
fact
thos
Wh
envi
tion
men
pote
crea

SPACE. Competition for resources also includes competition for living space. Part of the reason for this is that living space contains other resources, especially food and concealing cover in the case of animals, or perhaps scarce water or minerals in the case of plants. It seems logically to follow from this that highly territorial animals should arrange their territories so as to partition the identifiable resources of their environment evenly among them. But a clear-cut relationship between territoriality and environmental resources is hard to demonstrate. Yet obviously there is an indirect relationship, which implies that space itself can be a limiting factor.

Space is at a premium for many benthic, sessile marine organisms. Favored marine habitats such as oceanic hot spring oases, rocky intertidal zones, or continental shelves tend to be crowded with sedentary inhabitants. This is especially true of firm, rocky substrates to which such organisms as barnacles, mussels, and anemones attach themselves permanently or temporarily. Probably for this reason, most such organisms have planktonic larvae capable of being widely dispersed to whatever suitable habitat might be available to them. So scarce is space that one finds barnacles growing on other hard-surfaced marine organisms such as mussels, crabs, whales, and even turtles (Figure 55–8). Not surprisingly, many marine organisms possess adaptations that discourage such colonization. Most echinoderms, for example, possess clawlike pedicellariae (Chapter 31) covering the exposed surface of their bodies. Often these pedicellariae are even provided with poison glands with which to dispatch interloping larvae.

Numerous studies have demonstrated the crucial importance of space in the intertidal and subtidal zones. In a typical experiment, a coastal area of the Pacific Ocean supported 15 species of animals. Some of these were predatory, including a chief predator—a kind of starfish judged to be especially effective in its gluttonous role. The starfish was particularly catholic in its taste, consuming a wide variety of prey organisms. The experimenters removed all specimens of this species of starfish from the study area. Under the greatly reduced predation, all other species were limited only by direct competition for space. Seven of the original species became locally extinct, crowded out by faster-growing colleagues, showing that the predator was not the only or even the most important potential limiting factor. In fact, from the point of view of one of the extinguished prey organisms, the presence of the starfish was a mixed advantage. When one limiting factor was removed, others became significant, *but limiting factors continued to exist.*

Although space seems a clear-cut limiting factor in sedentary organisms such as barnacles or plants, there are few experimental demonstrations of

FIGURE 55–8 Living space is scarce on the sea bottom. These barnacles will eventually be shed when the host crab molts.

its significance for mobile animals such as mammals. The reason is that it is hard to separate limitations of space from consequent limitations of the resources that the space contains. But one such experiment was performed a number of years ago by John B. Calhoun and his colleagues.

Calhoun obtained a room of modest size which he converted into something of a rat paradise, provided with abundant supplies of food, water, and nesting boxes in which young could be raised. A single pair of laboratory rats was introduced and allowed to breed, as were their descendants. The rat population initially increased at a great rate, but eventually reached an ultimate high value from which it declined, in the end to zero.

In the course of the study Calhoun observed numerous disruptions in social behavior, courtship, and other reproductive behavior which probably resulted from excessive interaction forced by the abnormally high population density. In other studies of crowded animals investigators have found evidence of adrenal steroid insufficiency, presumably also resulting from excessive interaction. For Calhoun's rats, crowding did more to reduce natality than to increase mortality. One possible reason may have been maternal stress.

It is not certain, however, that lack of space in itself produced these results. Shortages of nesting sites, inbreeding, and even undetected infectious disease could have been responsible. Also, the study was of doubtful general applicability. In nature, rats are free to disperse and are not adapted to conditions of high population density such as those under which organisms such as cockroaches and human beings seem to thrive. On the other hand, although a rat's choice of habitat is broad, it is not unlimited. Field studies in such habitats as garbage dumps should be rewarding if unpleasant for the researchers.

However, there have been some more recent interesting studies of urban squirrels yielding results reminiscent of Calhoun's. Urban squirrels frequently subsist under conditions like those imposed on the rats. They are often deliberately fed by human residents and live in isolated habitat islands—parks—in cities from which they cannot readily disperse. Investigators have observed both higher population densities and some evidence of stress-induced disease not seen in squirrels living under less cramped conditions.

PREDATION. Predation has obvious potential as a limiting factor for the population of a prey organism, but this potential is not usually realized. Apparently prey and predator organisms are mutually adapted to one another. Were it not so, one or both would swiftly become extinct.

Extinction is a real possibility in the case of a totally defenseless prey organism. There have even been historical examples, such as the Mauritian dodo bird, now a metaphor for extinction. On the other hand, a highly inefficient predator would not get enough to eat, while an excessively efficient one would destroy its prey species and, lacking alternative foods, would swiftly become extinct itself. The endangered everglades kite eats nothing but the large apple snails that live in the Florida Everglades. As the Everglades shrink or are degraded by surrounding development, there are fewer and fewer apple snails. Not amazingly, there are fewer and fewer everglades kites, too, despite the protection that is extended to these birds. In this instance the population of the predator depends on that of the prey, and not the reverse; both the kites and the snails have presumably cohabited the everglades at least since glacial times.

A more classic example is that of the snowshoe hare and Canada lynx (Figure 55–9). The Hudson Bay Company has kept continuous records of the number of hare and lynx pelts bought from trappers since colonial times. Fluctuations in the number of hare skins are consistently seen to be followed, after a brief time lag, by similar fluctuations in the number of

FIGURE 55–9 Changes in the relative abundance of the lynx and the snowshoe hare, as indicated by the number of pelts received by the Hudson Bay Company. This is the classic case (but by no means the only case) of cyclical oscillation in population density and of dependence of predator upon prey.

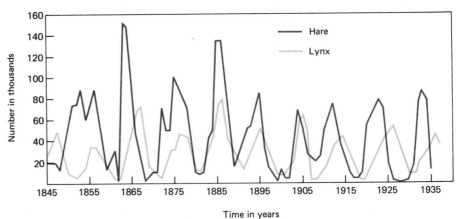

lynx skins. It is hard to view the lynx as controlling the numbers of hares in this instance; rather, the predator population seems to be controlled by that of the prey. Evidently, this is true in most instances of predator-prey interaction, although the relationship is by no means always this clear.

As you can see from the other graphs that are included, the hare-lynx pattern is not the only prey-predator relationship that has been observed. Even between hare and lynx, no cyclical relationship is evident in some environments where both exist. Moreover, between tawny owls and voles, their chief rodent prey, there is also no obvious relationship, the owl population remaining relatively constant despite vole population fluctuations. It seems clear that vole supply is not the usual limiting factor governing owl population; food isn't everything in life!

To illustrate further the variation one finds in the relation of prey and predator numbers, let us conclude the story of the Isle Royale moose, which has not continued to be one of wild population oscillations accompanied by mass starvation and death. In the winter of 1949 a small group of timber wolves wandered across the ice to the island to find themselves surrounded by one of their favorite prey organisms.

Since then,* the populations of both wolves and moose have remained nearly constant, with about two dozen wolves per 800 moose. Wolves, which are pack-hunting predators, have some difficulty killing adult healthy moose. They tend to pick out the very young, the old, and the unhealthy, killing one every three days on the average. This rate of predation approximately balances the moose rate of natural increase in this environment, resulting in not only a stable but a healthy moose population. Probably no moose ever dies of old age on Isle Royale.

Although cases of excessively efficient predators are rare, unfortunately one of the few such species is a familiar one—*Homo sapiens.* We have been able to survive the extinction of many prey species, apparently beginning in prehistoric times, because of our very large repertoire of alternative food and feeding strategies based on culture and technology unknown among other organisms. It is likely that the woolly mammoth, the giant ground sloth, and many others were destroyed by glacial-era Americans, Europeans, and perhaps Australians and Africans. Similar events continue today, probably at an even greater rate of occurrence. The far-too-numerous modern examples of extinction produced by human predation include the passenger pigeon, Steller's sea cow, and many others. In most instances, as prey species become rare, predators find it increasingly difficult to find them, and usually some prey survive in out-of-the-way places. However, people are often able to search out wild game in every available refuge and kill every last one of them.

In situations involving nonhuman predators, however, such extreme results can usually be observed only in simplified laboratory systems. For instance, one case of predator efficiency was studied under laboratory conditions by Gause (Figure 55–10). The familiar ciliate protozoan *Paramecium*

---

*Recently, the wolf population has declined drastically due to introduced disease.

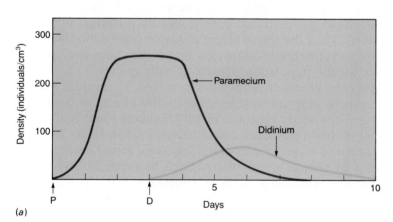

(a)

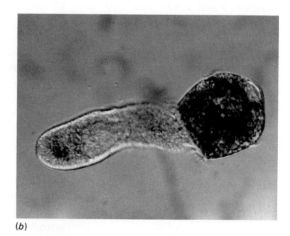

(b)

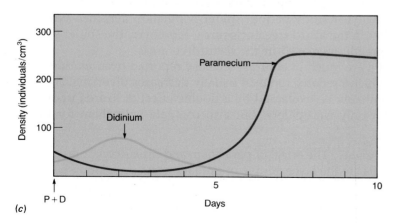

(c)

FIGURE 55–10 Predator-prey interactions between *Paramecium* and *Didinium.* (a) *Paramecium* and *Didinium* are introduced into a culture containing no refuges for *Paramecium.*
(b) *Didinium* consuming a *Paramecium.* (c) *Paramecium* and *Didinium* are introduced into a culture vessel containing detritus (sediment) to provide refuges for *Paramecium.* (Modified from original illustrations by Gause.)

(a)

(b)

FIGURE 55–11 (*a*) Prickly pear cactus, native to the New World, invaded Australia, rendering millions of acres of grazing land unusable. (*b*) The cactus was controlled successfully by the introduction of *Cactoblastis* moths and other natural enemies. Moth larvae here consume the pulp of a prickly pear pad.

is preyed upon by another smaller ciliate, *Didinium*, which penetrates the pellicle of its prey with a kind of beak and sucks out the living cytoplasm. When Gause introduced *Didinium* into a culture of *Paramecium*, the initially abundant prey species swiftly became extinct in the culture, followed shortly by its predator. Gause reasoned that if the *Paramecium* could hide from their formidable predators, some might survive. Accordingly, he provided a culture with much sediment on the bottom, among the particles of which the *Paramecium* might find refuge. This strategy was successful in protecting some members of the prey species, even though they became so hard to find that the predators starved. In the absence of *Didinium*, the remaining *Paramecium* underwent a population explosion shortly thereafter. It would seem that under simplified laboratory conditions the ecological relationship between these two species was most unstable. Perhaps in a more complex natural community containing broader choices of prey for the *Didinium* and more defense resources to be employed by *Paramecium* against predators, the populations of both would be more stable.

The most dramatic instances of predator-based population homeostasis have been observed in the case of pest organisms introduced into a new habitat without their natural predators. The resulting abnormally high population densities of such organisms (usually plants or insects) are what makes them pests. In such cases predators do appear to be density-dependent limiting factors. One reason may be that the more dense the population of the prey organism, the more easily the prey can be found.

American prickly pear cactus introduced to Australia, European "klamath weed" introduced to North America, African cottony scale insects, and others have become notable nuisances, even menaces. For instance, vast areas of Australia became infested with prickly pear, rendering them unsuitable as wildlife habitat or as rangeland. Biologists introduced the American *Cactoblastis* moth, which devastated the cactus and reduced the cactus plague to tolerable proportions. But it is important to note that the cactus did *not* become extinct in Australia. Both it and its predator[1] persist at moderate densities. Thus the effect of the predator was to moderate the population of the prey; one might almost say that the predator-prey system was homeostatic for the population density of both species.

These examples are, however, somewhat contrived. In complex natural communities it can be very difficult to separate the effects of predation from other factors. In a study reported in 1987, Schoener and Spiller noticed reports that on tropical islands where lizards exist there are few spiders. More spiders and more species of spiders were found on lizardless islands. Deciding to subject these observations to experimental study, Schoener and Spiller staked out plots of vegetation (mainly sea grape shrubs) and enclosed some of them with lizard-proof screen enclosures. All lizards were removed from some of the plots (which must have taken some doing) and web-building spiders were introduced into them all. At the end of a two-year study period, spider population densities averaged 2.5 times higher in the lizard-free enclosures. Moreover, the spider species *diversity* had greatly increased in the lizard-free areas.

These results are much in contrast to those reported from marine communities, in which predator removal usually decreases diversity. This difference can probably be explained by a double effect of lizards upon spiders. They not only eat spiders, but also compete with them for insect prey.

PARASITISM AND DISEASE. The effect of parasitism and disease on population is somewhat similar to the last example of predation. As population becomes

---

[1]Even a herbivorous organism such as *Cactoblastis* is classified ecologically as a predator if it kills its host organism.

more dense, the distance between organisms decreases, they meet one another more frequently, and the chance of their transmitting disease to one another by contact or by animal vectors becomes greater. If the disease is fatal or markedly reduces the chances of its host's reproducing, population will diminish to the point where transmission becomes unlikely. If the disease has not killed off all the host organisms, population density should be constant thereafter.

However, host organisms usually possess means to resist disease, as was discussed in Chapter 43. As levels of host immunity increase, the pool of susceptible individuals decreases. Eventually so few have the disease and so few contract it, that the disease ceases to be an important limiting factor. Moreover, the host organism may develop genetically based resistance to it by evolutionary mechanisms, so that the disease becomes a minor rather than a life-threatening one.

European hares introduced in colonial times overran parts of the Australian continent, competing seriously with domestic animals and wildlife for forage and other essentials, and doing other ecological damage. Hunting, trapping, poisoning, and huge fences failed to stop their spread. Eventually Australians deliberately imported from Latin America the virus that produces the rabbit disease myxomatosis. This disease is transmitted by parasitic insects, especially mosquitoes. The mosquitoes sought out every rabbit they could, sick or well, infecting all and sundry in the process. The rabbit population melted away initially before the onslaughts of this plague. However, a few rabbits were able to develop resistance, and also the virus seemed to become less virulent. Rabbits still persist in Australia, although now in more moderate numbers.

The results were different in the case of the American chestnut. This tree was formerly dominant over a vast area of North America and was economically as well as ecologically important, providing food for wildlife and some domestic animals and decay-resistant timber for fences, railroad ties, and other items. Today this fine tree has no economic importance whatever. A chestnut blight disease destroyed almost all American chestnut trees over a period of a few years. The exact source of the disease is unknown, but it is believed to have originated in the Orient. A very few trees persist, but do not seem to be able to reproduce successfully. Almost all of them, in fact, live in a state of perpetual ill health. Although vigorous efforts to save it are being made, it is all too likely that the American chestnut will eventually become genetically, as well as commercially, extinct.

MUTUALISM. Microorganisms and other organisms can increase population densities by means of relationships that are beneficial to both members of a partnership between organisms of different species. Such a relationship is called **mutualism.** We do not refer to obligatory mutualism such as exists between termites and their intestinal protozoa but to facultative relationships such as exist between treehoppers and ants. Although not obviously a *limiting* factor, mutualism can be density-dependent, as we shall now see.

Some years ago Catherine Bristow studied the ironweed treehoppers of an abandoned field in New Jersey. These little homopterans suck phloem sap from plants, especially shrubs and trees, providing them with abundant carbohydrate but little protein or amino acids. To obtain the amino acids they must ingest far more carbohydrate than they can use. The excess sugar-rich fluid is defecated in the form of honeydew. A number of animals will consume this sweet material, including human beings, bees, and (especially) ants.

Numerous predators such as ladybugs feast upon the leafhoppers themselves, especially the nymphs. As a result, only perhaps 5 to 10% of the eggs laid by a female leafhopper reach maturity. In view of this it is odd that leafhoppers tend to aggregate, sometimes forming insect herds that not only stay together but move together on branches and trunks of trees. It would seem that such aggregations would only attract predators to the defenseless insects and that they would be safer if widely dispersed. However, aggregations of leafhoppers also produce large quantities of honeydew, and the honeydew attracts ants.

Bristow discovered that female ironweed leafhoppers tend to deposit their eggs at such a time that when they hatch the nymphs will probably attract *Myrmica* ants. These formidable insects not only consume honeydew, but attack any predators rash enough to show an interest in the leafhoppers. As a result of the ants care, the survival rate of the leafhoppers more than doubles. Under ant "domestication" individual leafhoppers grow about 20% larger than they do without the ants, reaching maturity very quickly—in only about 80% of the time otherwise required.

BEHAVIORAL MECHANISMS.   Population densities are not always regulated by brutal malthusian mechanisms. Sometimes behavioral mechanisms act to regulate population densities short of the theoretical maximum population density, or **carrying capacity** of the environment.

Perhaps the best example of such a mechanism is the ordinary honey bee, *Apis mellifera*. Having originated in the tropics, these social insects spread northward by a combination of two behavioral adaptations: nesting in cavities such as hollow trees, and communal thermoregulation. The cavity provides shelter, especially during the winter. This by itself would not be enough, but by clustering together the bees achieve a collective temperature homeostasis comparable to that of a single mammal or bird.

Studies by Southwick indicate that to survive very cold temperatures (as low as $-25°$ C) a minimum of 2000 bees is required. The behavior is complex and involves more than simply huddling and pooling metabolic heat. By opening and closing ventilation channels into their collective mass, for example, the bees adjust heat loss so as to maintain not only a high but an even "body" temperature. Should the temperature rise excessively, the wax of the comb will melt, producing a horrid sticky disaster. However, bees are also able to thermoregulate at high temperatures by importing water into the hive, scattering droplets throughout it, and fanning air through it with their wings. The evaporation of water cools the hive much as the evaporation of water from moist body surfaces cools vertebrates. Presumably a minimum colony size is required for this also.

Maximum colony sizes are related to the size of the nesting cavity, the availability of food, and other factors. Maximum colony size is controlled pheromonally. Instead of the rude dominance mechanisms employed by many bumblebees and wasps to suppress worker reproduction, the queen honeybee secretes an exocrine **queen substance** which is licked from her body by her attendants and is quickly passed thereafter from one worker to another by regurgitation **(trophallaxis)** until it permeates the hive. The continuing presence of queen substance suppresses the ovarian function of the workers and also serves as assurance that a queen is, indeed, present. If the queen dies, the sudden absence of the queen substance is noticed by the workers and they immediately begin to rear royal larvae to replace her.

But if the population of the colony exceeds 15,000 to 20,000, this dilutes the queen substance among the many workers without eliminating it. Their response to this, under the appropriate conditions, is to begin rearing royal larvae preparatory to swarming. Once queen cells have been started, the worker bees stop feeding the queen. She gradually ceases to lay eggs and begins to wander over the brood comb, emitting occasional "piping" noises. Any mature queens still in their brood cells respond with "quacking" sounds. This indicates that the colony could survive without the old queen and serves as a signal to swarm.

(a)   (b)   (c)

FIGURE 55–12 The population and many other traits of a honeybee colony are socially controlled, largely by social hormones (pheromones). (*a*) The queen is licked by workers. The queen substance, a pheromone, spreads throughout the colony by the exchange of stomach contents. (*b*) Bees habitually feed and share food with one another. In this case the queen is being fed. (*c*) The egg about to be laid by the queen, here shown inspecting a brood cell, is one of thousands produced in her lifetime.

When the weather is favorable, foraging workers cease to look for nectar and pollen, and most activity within the hive slows or stops. Some unknown factor triggers a sudden evacuation, and half or more of the colony, accompanying the old queen, suddenly swarms out of the nest to form a cluster nearby. Scout bees from the swarm may locate another suitable cavity and thus begin a new colony. Meanwhile, back at the old colony, one of the young queens emerges from her brood cell, stings her unemerged sisters to death, and in due course flies out to mate. Then she settles down to her lifelong career as a professional egg-laying machine.

Behavioral mechanisms of population regulation among vertebrates are less clear-cut and mostly related to territoriality. Among many songbirds, for example, a male that is unable to establish a territory cannot attract a mate or establish a nest (Figure 55–12). In the case of migratory songbirds, the first males to arrive in suitable habitat establish territories immediately and defend them. Despite that, newcomers settle nearby, and under this increased pressure the original territories shrink to the accompaniment of much squabbling (Figure 55–13).

There is, however, a minimum size past which territories are not compressible. A domiciled male may lose his territory but cannot afford to let it shrink too far, probably because the resources of a very small territory would not support a family. What of the males that do not gain territories? They are still around, skulking inconspicuously about and awaiting a chance to replace any territorial male that is killed. In experiments in which such domiciled males were deliberately shot, it was observed that they were swiftly replaced from a seemingly inexhaustible supply of floating males. Thus, in some birds, territory functions to regulate population (although there is no reason to think it originated for this reason).

In many other animals also, territoriality functions to control population. However, the exact nature of the relationship between territoriality and population is related to the limiting factors of the environment, which could be food supply (as it probably is with most songbirds), or whatever shortage is most critical.

In pinnipeds (seals, sea lions, and walruses) the limiting factor is not usually food, but the fact that they cannot bear and raise young in the

FIGURE 55–13 Conflict between territorial male sharptailed grouse.

water. In the arctic, for example, they must breed either on the ice or on remote beaches not accessible to predators such as wolves and polar bears, and such sites are rare.

Among the beach-breeding species, the males strive for dominance in a multitude of bloody brawls, with only the winner able to copulate with females and he only while his strength holds out. When, after some days or weeks of dominance, he becomes exhausted, another male defeats and replaces him. The females, for their part, breed successfully only if they are able to find suitable beaches, and these are often as crowded as Fort Lauderdale during spring break. Mothers must leave their young periodically to fish and have no assurance that the infant will still be there when they return.

This example illustrates the difficulty in assessing the relationship between behavior and population control. We have already seen (Chapters 10, 19, and 20) that mating systems are of great genetic importance. Is population control the main adaptive function of the pinniped system, or does it have a more important genetic function?

INTERACTION OF LIMITING FACTORS.   As originally conceived, limiting factors were supposed to be independent in function. Thus, an excess of one element in an agricultural fertilizer could not make up for a lack of another. While there is much truth to that view, it is an oversimplification. What determines the northern limit of citrus tree culture in North America? It seems obvious that minimum annual temperature is the crucial factor. Yet a well-watered tree can withstand lower temperatures than one that is stressed by drought. At least two potential limiting factors actually interact to determine the range of citrus. Despite this, when the minimum temperature is low enough, it is quite true that no amount of water or anything else can make up for it.

GROWTH AND DISPERSAL STRATEGIES

It appears that all species possess adaptations that function to disperse their organisms away from their point of origin. The reason seems to be that such adaptations give an organism a competitive advantage. All animals and plants die sooner or later. This creates vacancies within their range, vacancies that would never be filled if other species members did not tend to disperse themselves. An organism with efficient dispersal adaptations would be likely to reach the vacant spot sooner than one with inefficient dispersal adaptations. Natural selection should favor efficient dispersion even in isolated populations that never actually expand (i.e., island populations of flightless birds). In many cases, however, dispersal mechanisms may make it possible for an organism to expand its range or to quickly colonize new and suitable habitats that become available within its range. It is common, for instance, for a grassland or forest to be destroyed by fire. This removes the dominant vegetation and makes room for the arrival of new inhabitants that are specifically adapted to pioneering.

R-Strategies

Early stages of community history may be harsh habitats, but not to organisms specifically adapted to those conditions. Such organisms live initially among abundant resources, but eventually competition from other organisms will limit or expunge their population. It is advantageous for these creatures to emphasize effective dispersal mechanisms so they may settle in a habitat whose resident community may have been recently destroyed. Once established, it is also desirable for them to multiply rapidly so as to get ahead of any competitors that may also arrive. They also need to be able to thrive under microclimatic and other conditions that prevail in a disturbed habitat. Please recall the logistic equation in which *r* stood for the rate of natural increase. Because such organisms have a high rate of natural increase, they are known as **r-strategists.** Some of the best examples are common weeds.

Dandelions don't make plans or prepare strategies, of course, but if they did they could hardly improve on the package of adaptations they possess as typical *r*-strategists. *Taraxacum officinale* was originally a European weed (or crop, depending on one's viewpoint), but this could hardly be guessed from its now nearly universal distribution. Almost everyone knows how they are propagated—by a parachute-like fruit containing little seeds that are scattered widely by any strong breeze or wind (Figure 55–14). A 5-mile-per-hour breeze is strong enough for flight. In temperate climates any suitable habitat will be colonized as soon as it becomes available.

The seeds of some varieties of dandelions are produced by an almost asexual process that is anything but typical for flowering plants. Although dandelions do produce pollen, in the most common varieties the grains do not germinate, or else their nuclei fail to fertilize the ovules. Instead, the ovules are diploid and require no fertilization. They simply grow into a seed whose genetic composition is presumably identical to that of the parent plant. Upon germination, the new plant follows this with another less bizarre form of asexual reproduction. The mature dandelion plant has a long taproot that splits longitudinally to form two, four, and so on, so that natural clones of plants spread out from the original progenitor in genetically identical clumps. Moreover, the flat, spreading habit of growth preserves the dandelion from the lawnmower, while its very rapid production of seed makes reproduction likely between mowings.

Why does the dandelion emphasize asexual reproduction? It has several advantages for an *r*-strategist. In the first place, no insect vectors or even other dandelion plants are required. Both may be in short supply in seriously disturbed habitats in the early stages of their resettlement. Second, it is fast. Third, as long as the particular kind of habitat for which they are adapted is common, it preserves without change a genotype ideally suited to that habitat. If people ever stop maintaining lawns, grassy highway medians, and pastures, however, asexual reproduction could become a liability for the dandelion.

FIGURE 55–14 Dandelions. Many varieties of this common invader of disturbed habitat reproduce asexually. Even seeds are often produced asexually, to be distributed by wind and air currents.

K-Strategies

Unlike the organisms of early successional stages, those of climax communities live under stable conditions. Their populations do not ordinarily become extinct even locally because they are the best fitted organisms for survival under the prevailing conditions; that is why their community undergoes no further change. Competition from established conspecifics or from other species is severe, opportunities for invasion of new habitats are limited, and lifespans are likely to be long. Instead of the frenetic emphasis on reproduction and competition one finds among the *r*-strategists, these organisms seem more restrained and genteel. Since their populations tend to be stable and close to what can be supported by the carrying capacity of their habitat, these organisms are called **k-strategists.**

k-Strategists can be found among animals as well as plants. Owls, for instance, are important predators in forest ecosystems and specialized species also occur in tundra, prairie, and desert areas. On the whole they are rather sedentary animals, traveling long distances only when forced to do so by food shortages.

Tawny owls were studied by H.N. Southern, who discovered that they pair-bond for life, both living and hunting on adjacent, well-defined territories. Southern found that tawny owls regulate their reproduction in accordance with the resources (especially food) present in their territories. Even in an average year, 30% of the birds do not breed. If food supplies are more limited than initially indicated, many of those that breed fail to incubate their eggs. Rarely do the owls lay the maximum number of eggs, and often they delay breeding until late in the season when the rodent populations on which they depend have become large. In a bad year, not one baby owl may survive to adulthood and even in a good year a quarter of the hatchlings die.

In one population of tawny owls there were 50 adults, with an average of eleven that died annually. With such a high death rate, the average couple could produce no more than one surviving offspring every other year if the population were to remain stable. Actually there is a slight excess and as many as seven young owls must seek their fortune elsewhere each year. Despite this, their population is regulated behaviorally on the whole; crude starvation does not usually intervene.

POPULATION CONTROL AS AN ADAPTATION

The monarch butterfly is noteworthy for its migratory behavior and its beautiful coloration. It is also notably poisonous. The larvae often eat milkweed, a plant that contains a class of naturally occurring pesticides known as **cardiac glycosides.** Monarch caterpillars belong to a small and select group of milkweed pests that are resistant to these chemicals. The glycosides do accumulate in the caterpillar's body, however, and remain in it through pupation and metamorphosis. The eclosed butterfly usually is not only poisonous but exceedingly nauseating to any predator (usually a bird) that might chance to eat it.

If the bird survives its first monarch meal, it will probably never attempt another one. Such ingestion-associated aversive learning is remarkably swift, a single trial usually sufficient to teach a lesson that lasts a lifetime. The occasional monarch that falls victim to a naive, inexperienced young bird produces immunity to predation from that particular bird for the entire population. The showy and distinctive monarch coloration has become associated in classical pavlovian fashion with the illness that the butterfly induces. Most insectivorous birds are sight-hunting predators that will henceforth avoid any butterfly that looks like a monarch—even if it is *not* a monarch!

Some (evidently!) tasty butterflies that are not monarchs resemble monarchs very closely (Figure 55–15). Sight-hunting entomologists have long known that a butterfly closely similar to the monarch exists, and more recently it has been determined that these, the viceroys, are perfectly edible (at least by birds) and are readily and repeatedly eaten by birds that have never consumed genuine monarchs. Ordinarily, however, there is little consumption of both monarchs and their mimics by birds, probably because monarchs are common enough that most potential bird predators happen across monarchs early in their careers.

The viceroys need not be poisonous to benefit from their shared warning coloration, but they had better not become too numerous. Their system of mimicry works only if a potential predator has encountered the real thing previously. Even after the monarch has reproduced, it lives for a long time (at least a long time for a butterfly—perhaps two years). This long life serves to increase the number of monarchs, even though the monarch senior citizens will produce no further young during that time. The more monarchs there are, the more likely it is that new-fledged birds will be taught their nauseating lesson! This is not true of the viceroys. If they were to become too numerous or live too long, the birds would learn the wrong

FIGURE 55–15 One can hardly tell a viceroy butterfly (upper insect) from a monarch, and neither can insectivorous birds. The poisonous, brightly colored monarch thus protects the innocuous viceroy.

lesson by being likely to initially encounter viceroys instead of monarchs. Thus the viceroys must reproduce at a low rate and die quickly thereafter.

The important point is that among these butterflies population density is the result of inherited factors. It is clearly determined by a direct genetic mechanism and by other associated traits that are part of the total adaptational package of this organism.

■ SUMMARY

I. Populations of organisms have certain properties that individual organisms lack, such as birth rate and age ranges.
   A. Population density is the number of individuals of a species per unit of habitat area.
   B. Population distribution may be random, clumped, or, rarely, uniform.
II. Population growth results from the difference between the rate of arrival and the rate of departure of organisms. Usually the most significant forms of arrival and departure are birth and death.
   A. Biotic potential is the theoretical maximum rate at which a population can grow.
   B. Mortality is the death rate of organisms in a population. Those that survive make up the actual population.
   C. Environmental limitations prevent unbridled population growth. The totality of all such limitations constitutes environmental resistance.
   D. Limiting factors are the specific inhibitors of population growth. Generally a limiting factor is an essential in potentially short supply.
      1. Density-independent limiting factors are conditions or events that depress population without tending to stabilize it.
      2. Density-dependent limiting factors are most effective at high population densities. For this reason they tend to stabilize population size or density.
         a. Resource depletion is a common limiting factor. Examples include a lack of food or nesting sites needed for continued population growth.
         b. With resource depletion, competition for whatever resources remain intensifies and becomes a limiting factor itself.
      c. Predation is sometimes a limiting factor on the population of a prey organism, but more commonly the population of a predator is governed by that of its prey.
      d. Parasitism and disease are effective density-dependent limiting factors because they are transmitted more readily among members of a dense population than among members of a dispersed one.
      e. Mutualistic, cooperative relationships tend to increase the population density of the species that participate in them.
      f. Population is regulated to some extent and in some species by behavioral mechanisms, especially territoriality. It is not suggested that population control is the reason that territoriality exists.
III. Population growth curves take a variety of forms, for example, an S-form. The S-shaped curve shows an initial lag phase of population growth, followed by a logarithmic phase, followed by a static or exhaustion phase. A J-shaped curve resembles the logarithmic portion of the S-curve and occurs in the essential absence of limiting factors.
IV. It is essential for all organisms to possess some means of population dispersal, but the particular strategy employed depends on the adaptations of the organism.
   A. An R-strategy is employed by pioneer, "weed" organisms. It emphasizes wide dispersal, large numbers of seed or offspring, and often, genetic uniformity.
   B. A K-strategy is characteristic of inhabitants of mature communities whose replacement is slow.
V. Population strategies are part of the total adaptation package of a species.

■ POST-TEST

1. Clumped distribution of population often results from an uneven distribution of _____ within a habitat.
2. A graph of population growth approaches a vertical line in the _____ stage of increase.
3. Population is potentially able to grow very rapidly because offspring are able to produce offspring of their own. This resembles _____ _____ in finance.
4. Large herbivores such as elephants tend to have a _____ biotic potential.
5. For most species, the highest mortality occurs _____ in life.
6. Rapid human population growth results not so much from great _____ _____ as from high _____.
7. A rapidly growing bacterial culture exhibits an initial _____ _____ of low population growth.
8. The classic definition of a limiting factor is that it is whatever essential needed by an organism is in the _____ supply.
9. Calhoun observed that an experimental population of rats declined, evidently due to a shortage of _____.

10. In studies of rocky shore marine communities, it is usually seen that the removal of predators _____ the diversity of the community.

11. As a rule, predator populations are controlled by the abundance of their _____, rather than the reverse.

12. On the whole, disease is a population-_____ limiting factor.

13. In a hive of honeybees, population size is controlled by a pheromone known as _____ _____ .

14. Among animals, would cockroaches or blue whales be better examples of an *r*-strategy of dispersal and population growth?

■ REVIEW QUESTIONS

1. Why is population increase from reproduction usually more significant than that resulting directly from immigration?

2. Describe an S-shaped population growth curve.

3. What is a limiting factor? Give examples. Can a surplus of one limiting factor compensate for the lack or shortage of another?

4. What is the logistic equation? What is its significance?

5. Suppose that it is proposed that the number of ring-neck pheasants available to hunters should be increased by (a) killing as many foxes (thought to prey upon the birds) as possible and (b) by feeding the birds in the winter. Would either approach be successful, in your opinion? Why, or why not?

6. Two examples of population control by behavioral mechanisms are given in this chapter: honeybees and territorial birds. Bear in mind that natural selection can act only on an individual organism and not, as far as we can tell, on a group. Viewed in this light, do you think that natural selection maintains these two mechanisms of population control, or are they incidental effects of behavior that is selected for some other reasons?

7. Some plants produce two kinds of seeds: one that usually comes to rest near the parent plant and one with a fruit that produces wide dispersal. The first type is produced by self-fertilization and the second by cross-fertilization. Why might the difference exist?

8. Similarly, many species of aphids produce vast numbers of wingless forms by parthenogenesis (a form of nearly asexual reproduction in which females produce female young without fertilization) and a few winged individuals that reproduce themselves sexually. Can you explain the significance of this?

■ RECOMMENDED READING

Begon, M., J.L. Harper and C.R. Townsend. *Ecology: Individuals, Populations and Communities.* Sunderland, Mass., Sinauer Associates, 1986. A population-oriented (but clear and understandable) introduction to general ecology.

Denizens of a forest floor community

56

Communities

■ **LEARNING OBJECTIVES**

After you have read this chapter you should be able to:

1. Define *community* and *ecosystem*, give the salient characteristics of each, and give an example of each.
2. Characterize producers, consumers, and decomposers, and give the function of each category of organism in a community.
3. Summarize the concept of limiting factors, and describe their relationship to the ecological niche.
4. Define *ecological niche* and give examples.
5. Summarize the concept of competitive exclusion in relation to niche differentiation.
6. Summarize the carbon, nitrogen, and phosphorus cycles.
7. Summarize the concept of a food chain and compare it to that of a food web.
8. Describe a typical food pyramid with emphasis on the biomass to be expected at different trophic levels.
9. Give reasons for the reduction of biomass with trophic level, and summarize at least two practical consequences of this reaction.
10. Define *gross primary productivity, net primary productivity, turnover,* and *standing crop.*
11. Summarize the main determinants of community diversity.
12. Outline the concept of ecological succession.
13. Contrast autogenic with allogenic succession.
14. Summarize the facilitation and inhibition hypotheses of succession.
15. Summarize recent changes in ecological thinking with respect to the nature of communities, the nature of succession, community complexity, and niche differentiation.

    As we have seen, the science of ecology encompasses the study of all the relationships among living things and between each of them and its nonliving environment. These relationships can be considered from two main viewpoints. First, one may view all the interrelationships of a single species of organism, viewing its ecology as a set of adaptations. This

FIGURE 56–1 Each organism is a member of a community, often specifically adapted for its life in a given community. The study of a particular organism and its relationships with the environment is called autecology. (*a*) The water requirements of the African "living stone" plant are minimal; it is specifically adapted for life in a dry habitat. (*b*) How different are the needs of water lilies! Their autecology excludes them from the desert communities inhabited by "living stones," although they might be capable of some degree of interaction in a desert oasis. (*c*) The total community of which many organisms are part is the object of the discipline of synecology. This complex coral reef community contains many sub-communities, and is itself part of the marine habitat. (*d*) Sponges are part of a sea bed community and form one of their own. Each sponge is a community in itself, often harboring a variety of commensals such as little crabs.

approach is known as **autecology.** An expression such as "the ecology of honeybees" implies an autoecological viewpoint. One may also view these relationships as a whole—as they exist within a biological community. Such an approach sees ecology as a property of a community of organisms. It is known as **synecology.** The approach of this chapter will be synecological, focusing primarily on the relationships organisms have with one another and examining the communities of organisms established by the intricate webwork of these relationships.

COMMUNITIES AND ECOSYSTEMS

The term *community* has a far broader sense in biology than in everyday speech. For the biologist, a **community** is an association of organisms of different species living together with some degree of mutual interdependence. Thus you, your dog, and the fleas on your dog are all members of the same community! One might add cockroaches, silverfish, dandelions, grass, maple trees, and much more to the list.

(*b*)

(*a*)

(*c*)

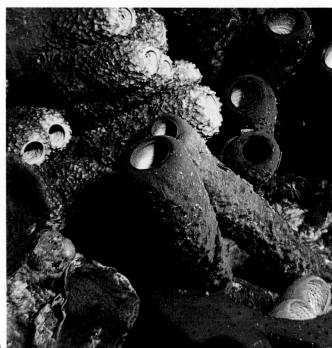

(*d*)

FIGURE 56–2 Ecosystem interaction. Rain falling on the land leaches mineral plant nutrients from the rocks and soil, while decay processes liberate these nutrients from the bodies of dead land organisms. Watercourses eventually wash them into the sea. These nutrients are also required by aquatic organisms, which grow at highest densities near the coast where the nutrient content of the sea water is highest. This photograph was prepared by using a special instrument that is sensitive to the chlorophyll content of microorganisms in the water. Note that chlorophyll levels are low in most of the open sea, which is, for the most part, little more than an aquatic desert of very low productivity.

CHLOROPHYLL <u>a</u> + PHAEOPIGMENTS <u>a</u> (MG/M³)

Organisms form integrated communities of varying sizes (Figure 56–1). Since communities are rarely completely isolated, they really do interact with one another—even when that interaction is not readily apparent. Furthermore, communities are nested within one another like Chinese boxes; there are communities within communities. A forest is a community, but so is a rotting log in that forest, containing fungus, insects, and even mice. So, too, is the association of microorganisms living within the gut of a termite in that log. On the other end of the scale, a great biological province containing a forest is also a community, as is the entire living world of which that province, in turn, forms a part.

Living things also have a nonliving environment that is essential to their lives. Food, minerals, air, water, and sunlight are just as much a part of a honeybee's ecology as the flowers which it pollinates and from which it takes nectar.

The nonliving environment plus the living communities it contains make up an **ecosystem.** Thus a lake plus its various communities of swimming, bottom dwelling, and floating inhabitants constitute an ecosystem. Although an ecosystem typically is fairly self-sufficient, like communities, ecosystems do not have precise boundaries. All ultimately exchange materials with one another, and all influence one another. Thus the fate of a far-off tropical rain forest may profoundly influence weather patterns in a temperate zone, and the pesticides deposited on a prairie may affect the invertebrates and birds of a distant river delta (Figure 56–2).

COMMUNITY ROLES

The organisms of a community fall into categories and perform the roles of **producer, consumer,** and **decomposer.** Most communities contain representatives of all three. All three interact extensively with one another and all three (Figure 56–3) are usually present in stable communities.

(a)

(b)

(c)

(d)

(e)

(f)

FIGURE 56–3 Producers, consumers, and decomposers. (a) The trees and epiphytic vegetation of the tropical rain forest account for most of the primary production of food within this ecosystem. (b) The three-toed sloth is a primary consumer; it eats mostly leaves. The low energy content of this fare may account for the sloth's proverbial sluggishness. (c) Sloths are a common prey for the Harpy eagle, a secondary consumer or predator. (d) An-other style of secondary consumption is exemplified by the king vulture, a scavenger or detritus feeder. (e) Other detritus feeders, tropical termites, often build immense nests in trees while consuming dead vegetable matter elsewhere. (f) Finally, decomposers such as this cup fungus, reduce all living things to their mineral constituents, plus carbon dioxide and water.

Focus on LIFE WITHOUT THE SUN

In the mid 1970s an oceanographic expedition studied the Galapagos Rift, a deep cleft in the ocean floor off the coast of Ecuador. The expedition revealed a series of extremely hot springs on the floor of the abyss in which seawater apparently had penetrated to be heated by the hot rocks below. During its sojourn within the earth the water had also been charged with mineral compounds, including hydrogen sulfide, H_2S. Certain bacteria, called chemosynthetic autotrophs, extract the energy they need for building carbon dioxide into carbohydrates not from light by photosynthesis but from the oxidation of inorganic compounds. Some of these bacteria can cause hydrogen sulfide to react with oxygen, producing water and sulfur or even sulfate. These reactions are exergonic and provide the energy to operate a version of the light-independent reactions of photosynthesis.

At the tremendous depths of the Galapagos Rift there is no light for photosynthesis and therefore little plankton except what rains down dead from the lighted surface layers. But the hot springs support a rich and bizarre life in great contrast to the surrounding lightless desert of the abyssal floor. Many of the species in these oases of life were new to science; some could not even be assigned with confidence, at that time, to known phyla. Some had astonishing adaptations. One species of clam, for example, is unique among invertebrates in possessing red blood cells that contain hemoglobin. The mystery is, what do these species live on?

The basis of the food chains in these aquatic oases is the chemically autotrophic bacteria we have mentioned, which are remarkable in being able to survive and multiply in water so hot (exceeding 200° C) that it would not even remain in liquid form were it not under such extreme pressure. These bacteria function as primary producers. Some of the energy that activates these living communities originated not only in the thermonuclear reactions of the sun but also in the heat released by slow radioactive decay in the depths of the earth.

The local ocean floor is dotted with the remains of extinct communities of this sort, which died out when the hot springs ceased to flow. Where new hot springs occur, however, swimming larval forms are apparently able to re-establish the communities in new locations. One would hardly think that brine, far hotter than the normal boiling point of water and charged with noxious gases and other chemicals, could be a necessity to any living thing. There is hardly a more dramatic example of a limiting factor.

(a)

(b)

■ The inhabitants of the Galapagos Rift. (a) Scanning electron micrograph (×5200) of chemically autotrophic bacteria, which are the base of the food chain in hydrothermal vent systems. (b) Chemically autotrophic bacteria living in the tissues of these beard worms extract hydrogen sulfide and carbon dioxide from water to manufacture organic compounds. Because beard worms lack digestive systems, they depend on the organic compounds provided by the endosymbiotic bacteria, along with materials filtered from the surrounding water and digested extracellularly. Also visible in the photograph are some filter-feeding mollusks.

Producers

Producers manufacture complex organic molecules from simple inorganic substances, usually using the energy of sunlight to do so. In other words, they are autotrophs. By incorporating their chemical output into their own bodies, producers make their bodies or body parts a food resource potentially available to other organisms. Plants are the most significant producers, but in aquatic environments plantlike protists, cyanobacteria, and sometimes sulfur bacteria (Chapter 24) are also important. Sunlight is the source of energy that powers almost all life processes on the face of the earth. However, deep-sea hot spring oasis communities depend upon chemoautotrophic bacteria (see accompanying box) as producers, although such assemblages are probably of minor overall ecological importance.

Consumers

Animals, animal-like protists, and a very few predatory bacteria and fungi make up the consumers. Consumers use the bodies of other organisms, including those of other consumers or even of decomposers, as a source of food energy and body building materials. Some are **primary consumers,** which usually means that they are exclusively herbivorous. Cattle and deer are examples. **Secondary consumers** include "meat eaters," the **predators** or **carnivores,** which consume other animals exclusively. Lions and tigers are excellent examples. Still others consume a variety of organisms, whether plant or animal. These are **omnivores,** examples being bears, pigs,

---

Focus on SYMBIOSIS

The most delicate and precise adaptations organisms must have are probably adaptations to one another. This is evident even in such ecological relationships as exist between predator and prey, but the kind of mutual relationship called symbiosis is probably the most exacting of them all.

Symbiosis, which means "living together," is a term that encompasses a spectrum of relationships. At the one extreme are the **parasitic** relationships, where one member of the association benefits at the expense and to the harm of the other. At the other extreme we find mutualism in which *both* parties benefit, equally as far as one can judge. Somewhere in the middle lies **commensalism** in which the benefits of the relationship are one-sided, yet there is little harm to the other party (Fig. A).

How would you describe the relationship of the pilot fish to the shark (Fig. B)? These fish cling to the body of the shark host, feeding on its scraps that it probably would not eat anyway. They might do a little harm to the shark in slowing it down, or causing it to expend

more energy in swimming, yet it is possible that they aid it in subtle ways as well, perhaps by sensing potential enemies.

Algae live inside the body cells of *Hydra viridissima*. The hydra subsists on the organic products of the alga's photosynthesis. Is it a parasite? Perhaps the nitrogenous wastes of the hydra's metabolism make an important contribution to the nutrition of the algae (Fig. C).

Cleaning symbiosis is an especially interesting case of mutual adaptation. A tiny, brilliantly colored wrasse, which could be swallowed at a gulp by the much larger fish (such as a grouper) it services, swims about them with impunity (Fig. D). The host may even open its mouth so that the little cleaner may pick decaying food from its teeth, or devour parasites deep in the pharynx. The bright color of the wrasse serves as a negative releaser (see Chapter 48) which inhibits predation from its fish hosts. Indeed so strong is the host's inhibition that the relationship serves as the basis of a kind of mimicry in which a brightly colored imitator of the wrasse swims up to the host fish. When the host presents itself for grooming, the imposter steals, perhaps, into its mouth.

■ A. This Caribbean shrimp (genus *Pereclemines*) might receive some leftover food from the host anemone, but, primarily, it is given protection from predators. The anemone probably neither benefits nor is harmed by the relationship.

■ B. A shark with several pilotfish attached.

and humans. Many animals, however, do not fit readily into one of these categories, since they are able to modify their food preferences to some degree when the need arises.

Much variation occurs in the life styles of consumers, especially such consumers as predators and parasites. These depend upon other consumers for their sustenance, so that sophisticated adaptations may be needed. The consumees may resist being consumed and possess adaptations for their self-preservation.

Parasitism is one of the most effective of such special life style adaptations. Parasitism differs from predation in that while a parasite, like a predator, attacks a living victim (called a **host**), it does not immediately (and may never) kill the host organism. It has been said that a predator lives off capital, while a parasite lives off interest. Some parasite-like predators do gradually kill the host. These are **parasitoids,** such as certain wasp larvae

Quickly grasping a mouthful of succulent flesh, the mimic darts off before the startled host can react.

Perhaps the most interesting relationships, at least from an evolutionary point of view, are the frankly parasitic ones. Often parasites lack body parts that their nonparasitic relatives do possess. An example is the wingless fly (*Braula*) shown (Fig. E) riding on the back of a bee host. What use are wings to the fly? The bee does all the flying; why carry excess baggage?

Yet such adaptations do severely limit many parasites in their choice of a host. Bird lice, for example, are best transmitted from generation to generation of their hosts by nest contact, but this tends to minimize the chance that they will be transmitted to other species of bird. Through this and many other adaptations, parasites become not only host-dependent but host-specific as well. If hosts evolve a resistance to parasites, then the parasites may well have to coevolve, thus keeping up with the new "design features" of their hosts. In fact, evolutionary taxonomists have used the more easily inter-

■ D. A cleaner fish with a grouper.

preted features of the related species of parasites to give them clues as to the possible evolution of their hosts. If two species of parasites seem closely related, the reasoning goes, so must be their hosts.

■ C. A budding *Hydra* with symbiotic algae.

■ E. A scanning electron micrograph (falsely colored to show contrasts) of a wingless fly on the back of a bee.

FIGURE 56–4 An adult fly lays her eggs on a gypsy moth larva.

that infest spiders or caterpillars for extended periods (Figure 56–4). The parasitoid adaptation is actually widespread among insects. Some parasitoids may have hyperparasitoids that live on them! It is estimated that fully 25% of all animal species are parasitoids, most of these being insects.

Finally, we come to the scavengers, also called **detritus feeders,** especially important in aquatic habitats. There, these burrow in bottom muck, consuming the organic matter that collects there. Earthworms are a terrestrial equivalent, as are termites and fly maggots. Detritus feeders work together with decomposers to destroy corpses and waste products, as discussed below.

An earthworm, for example, actually eats its way through the earth, digesting much of the organic matter the soil may contain, reducing its particle size by "chewing" it in a kind of gizzard. Earthworms even possess enzymes capable of digesting cellulose. When this partially digested soil is eventually defecated by the worm, the minerals that were originally incorporated into such organic materials as DNA have now been excreted in the form of simple inorganic compounds like phosphate and ammonia, readily available for absorption by plants. The partial destruction of cell walls and other resistant structures in the humus renders it more susceptible to further attack by other detritus feeders and decomposers. Earthworms also aerate the soil and redistribute its minerals and organic matter by their extensive tunneling.

Decomposers

The process of mineral liberation can take place even without detritus feeders, although often less rapidly. Dead wood is invaded first by sugar-metabolizing fungi that consume such readily available simple carbohydrates as glucose or maltose. Then, when these carbohydrates are exhausted, cellulose-digesting fungi complete the digestion of the wood, often aided by termites and bacteria. The fungus itself may be eaten by a variety of fungivores like millipedes, whose waste products consequently contain minerals liberated from the fungus. Thus the complex food chains within the soil constantly break down organic matter, liberating its minerals for recycling into plant and animal bodies once more.

Ecosystems contain a balanced representation of all major ecological categories of organisms—consumers, decomposers, and producers. It is not hard to appreciate the role of the producers, which provide both the food and oxygen for all life. Even the decomposers are necessary to the survival of any community. Without them, not only would corpses and waste products accumulate indefinitely, but elements such as carbon, nitrogen, and phosphorus contained in various organisms would become permanently locked up and unavailable for use by new generations of living things.

ECOLOGICAL NICHES

Every organism has an **ecological niche,** the totality of its adaptations, its use of resources, and the lifestyle to which it is fitted (Figure 56–5). Not just a habitat, an ecological niche defines an organism's way of life—its profession, as it were. Thus the ecological niche of a dog flea is not merely the dog, but the flea's special version of a parasitic way of life, its life history, and much else. Each of the factors involved is a **dimension** of the flea's ecological niche. A flea lives on the surface of a host's skin. A mite living in the dog's hair follicles therefore has a different ecological niche, even though the two parasites share the same dog. A great many factors determine an organism's ecological niche, most of them doubtless unknown and at present unknowable. A complete description of an organism's ecological niche would involve numerous dimensions.

We must try to distinguish between the role an organism *could* play in the community from the role that it actually does fulfill. For example, one might be capable of becoming a doctor or a lawyer, but few people manage to be both. A person's actual life style is chosen from among many possibilities. Similarly, the ecological niche of an organism may be potentially far broader than it is in actuality. Put differently, an organism is usually capable of utilizing much more of its environment's resources than it actually does use.

The potential ecological niche of an organism is its **fundamental niche,** but various factors such as competition with other species may exclude it from part of its fundamental niche. Thus, the life style that an organism actually pursues and the resources that it actually utilizes comprise its **realized niche.**

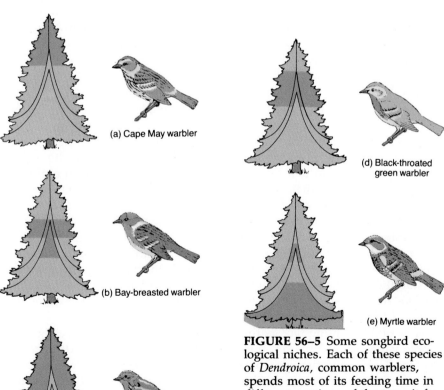

(a) Cape May warbler

(b) Bay-breasted warbler

(c) Blackburnian warbler

(d) Black-throated green warbler

(e) Myrtle warbler

FIGURE 56–5 Some songbird ecological niches. Each of these species of *Dendroica*, common warblers, spends most of its feeding time in different portions of the trees it frequents and also consumes somewhat different insect food. The colored regions indicate where each species spends at least half its foraging time. (After MacArthur.)

(a)

(b)

FIGURE 56–6 (*a*) Green Carolina anole. (*b*) Brown Cuban anole.

An example may help to make this clear. The little Carolina anole lizard is native to Florida. In former years these lizards were widespread. However, an introduced but related species of Cuban anole became common several years ago, especially in urban areas. Suddenly the Carolina anoles became rare—driven out, apparently, by the competition from the larger Cuban lizards. Careful investigation disclosed, however, that the Carolina anoles were still around but were now confined largely to the foliated crowns of trees, where they were less easily seen.

From this it became clear that the habitat portion of the Carolina anole's fundamental niche included both trunks and crowns of trees, exterior house walls, and much else. The Cuban anoles were able to drive them out from all but the tree crowns, so that their realized niche became much smaller as a result of the environmental competition. Since all natural communities consist of numerous species, many of which compete to some extent, the complex interaction among them produces the realized niche of each.

Limiting Factors

The factors that actually determine a realized niche can be extremely difficult to identify. For this reason the concept of the ecological niche is largely theoretical, although some of its dimensions often can be experimentally determined. Whatever environmental variable tends to restrict the realized niche of an organism is a **limiting factor** (also discussed from a different viewpoint in Chapter 55). But what factors do actually determine the realized niche of a creature?

An organisms's life style is basically determined by its total package of adaptations, whether structural, physiological, or behavioral. General and specific adaptations also determine the tolerance an organism has for environmental extremes. It is obvious that if any feature of its environment lies outside the bounds of its tolerance, then the organism cannot live there. No one would expect to find a cactus living in a pond, but water lilies would hardly be surprising there.

Most of the limiting factors that have proved practical to investigate are simple variables such as soil mineral content, temperature extremes, and the like. These investigations have disclosed that any factor whose supply exceeds the tolerance of an organism or which is present in quantities smaller than the minimum requirements of an organism limits the occurrence of that organism in a community. By their interaction, such factors help to define an organism's realized niche.

We have already briefly considered the concept of limiting factors in connection with population growth. Limiting factors also determine the distribution of a species and its realized niche. The concept of the limiting factor was originated in the 19th century by the pioneering agricultural chemist Justus von Liebig, who propounded what is now called **the law of the minimum.** As later amended in 1913 by V. E. Shelford, the law of the minimum holds that the growth of each organism is limited by whatever factor essential to it is in shortest supply or is present in harmful excess. This consideration applies throughout the life cycle of an organism. For instance, although adult blue crabs can live in almost fresh water, they cannot become permanently established in such areas because their larvae cannot tolerate fresh water. Similarly, the ring-necked pheasant, a popular game bird, has been widely introduced in North America but does not survive in the southern United States. The adult birds do well, but the eggs cannot develop properly in the high temperatures that prevail there.

However, we now understand that von Liebig viewed limiting factors much too narrowly. He understood, rightly, that an excess of one limiting factor cannot make up for the deficiency of another. Nevertheless, environmental stresses interact in a complex manner so that minima or near-minima can influence one another or interact. Therefore, especially when an organism is near the limit of its tolerance of some conditions, the interaction of potential limiting factors is synergistic, collectively limiting distribution more stringently than any one of them could do by itself.

Niche Width

Some organisms have very narrow ranges of tolerance to environmental factors; others can survive within much broader limits. Ecologists use the prefixes *steno-* (meaning "narrow") and *eury-* (meaning "wide") to indicate the ranges of tolerance to a given factor that an organism may have.

A eurythermic organism, for example, will tolerate wide variations in temperature. The housefly is eurythermic and can tolerate temperatures ranging from 5° to 45° C. The contrasting extreme adaptation to cold of the Antarctic fish *Trematomus bernacchi* is remarkable. It is so stenothermic that it can tolerate temperatures only between −2° C and +2° C. At 1.9° C this fish is immobile from heat prostration! Just the same, this very narrowly adapted and highly specialized fish can live where all or almost all more broadly adapted fish cannot.

As we have seen in the case of the anole lizards, competition can be one of the chief determinants of actual niche realization. The initial evidence for this came from a series of experiments by the Russian biologist A. F. Gause. In one study Gause grew two species of the ciliate protozoan *Paramecium* together, *P. aurelia* and the larger *P. caudatum* (Figure 56–7). When grown separately, each species quickly increased its population to a high level, which it maintained for some time thereafter. When grown together, however, only *P. aurelia* thrived. *P. caudatum* dwindled and eventually died out. Under different sets of culture conditions, however, *P. caudatum* may prevail. Gause interpreted this to mean that although one set of conditions favored one species, a different set favored the other. Nevertheless, because both species were similar, given time one or the other would eventually triumph at the other's complete cost. As he put it, "As a result of competition two similar species scarcely occupy similar niches, but displace each other in such a manner that each takes possession of certain peculiar kinds of food and modes of life in which it has an advantage over its competitor."

This kind of thinking eventually gave rise to the principle of **competitive exclusion,** which holds that evolutionary forces tend to drive organisms that may be closely similar ecologically into differing clusters of adaptations, so that their ecological niches become progressively differentiated.

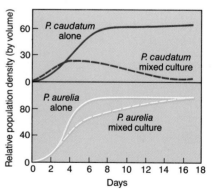

FIGURE 56–7 Competition among *Paramecium.* The solid-line curves show how the population of each species of *Paramecium* grows in a single-species environment; the dotted curves show how they grow when in competition with each other. (After G. F. Gause, *Science*, Vol 79, 1934.)

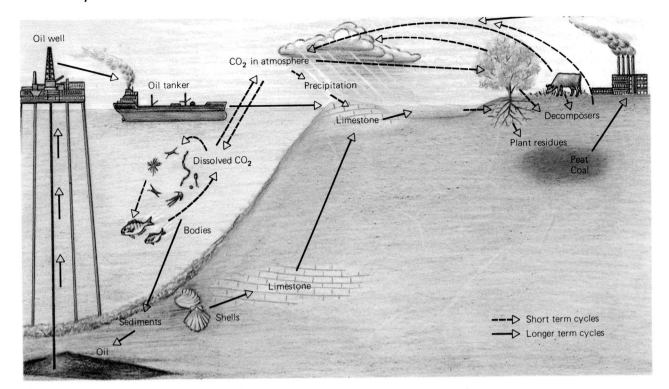

FIGURE 56–8 Simplified diagram of the carbon cycle. The dashed arrows indicate rapid processes; the solid arrows denote relatively slow ones, for example, the geological uplift that carries limestone to the surface, where its carbonate dissolves in water in the form of bicarbonate ion and is utilized by aquatic plants as a carbon source.

COMMUNITY AND ECOSYSTEM CYCLES

Since the earth and biosphere are really a closed system, materials utilized by organisms cannot be lost, although they can end up in locations that are outside the reach of organisms. Usually, however, materials are reused and often recycled both within and among communities. If a substance has a common gaseous compound or state, as in the case of nitrogen and carbon, it is recycled with relative ease. If all its important compounds are nongaseous, as in the case of phosphorus, only local recycling within the community is effective.

The Carbon Cycle

Since organic life is based upon the properties of the carbon atom, carbon must be available to living things. Most carbon enters the atmosphere in the form of carbon dioxide emitted from such sources as industrial combustion and volcanoes. The accumulation of this gas in the atmosphere would produce a disastrous rise in temperature comparable to that prevailing on the planet Venus if carbon were not also removed from the atmosphere. By photosynthesis, plants and cyanobacteria remove carbon dioxide from the air and **fix** it, that is, incorporate it into complex chemical compounds (Figure 56–9). These compounds then are usually used for fuel by the producer that first made them or by a consumer or decomposer that has the opportunity, thus liberating the carbon dioxide once again.

With this continuous input plus continuous recycling, the total living mass of the earth's organisms should increase indefinitely. This has not happened because carbon also can be carried out of the reach of living things by some of their own life activities. The flotation oils of diatoms, for instance, have apparently given rise to the underground deposits of petroleum and natural gas that have accumulated in the geological past, and vast coal beds have been formed from the bodies of ancient trees that did not decay fully before they were buried.

It is believed, however, that most of the carbon dioxide that leaves the atmosphere is incorporated permanently into the shells of marine organisms—shells that can form seabed deposits thousands of feet thick. Almost the only way such carbon can reenter the atmosphere is by subduction (see Chapter 20) and associated volcanic action.

Short-term fluctuations in atmospheric carbon dioxide content probably can be compensated for by simple chemical buffering in sea water, but over the long term, increase in carbon deposit by biological processes is far more important and permanent. A true negative feedback loop, biological carbon fixation and deposition should keep the atmospheric carbon dioxide content approximately constant indefinitely. However, biological compensation requires thousands or even millions of years to be effective. Substantial emissions of carbon dioxide by increased volcanic action or human activity could produce great climatic changes that would be permanent on a human time scale.

The Nitrogen Cycle

We and other consumers ultimately depend upon photosynthetic autotrophs for chemically combined or **fixed nitrogen,** utilized in the form of proteins and nucleic acids. Plants depend upon both chemosynthetic autotrophs and decomposers for recycling the nitrogen upon which their survival depends (Figure 56–9).

At first glance it would appear that there is no danger of nitrogen starvation. After all, the surrounding ocean of atmosphere contains about 80% nitrogen. But the element nitrogen occurs combined in stable diatomic molecules (N_2) and zealously resists any tendency to unite with any other common substance. Biologically, molecular nitrogen is a nearly inert gas. The overall reaction that breaks up diatomic nitrogen and combines the

FIGURE 56–9 The nitrogen cycle, including the effects of human activities. We have deliberately *not* simplified this diagram so as to convey some impression of the multitude of pathways whereby this vital substance travels throughout the biosphere.

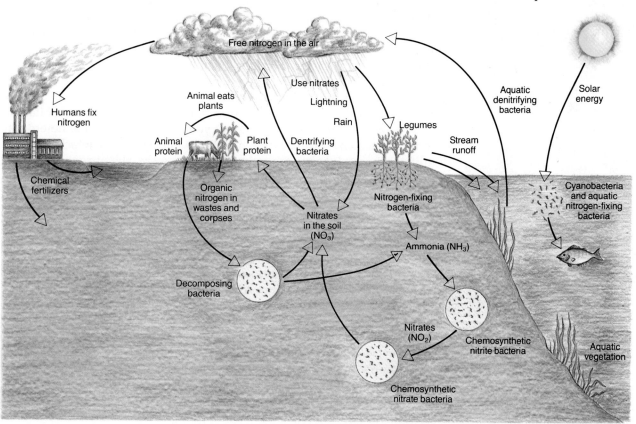

nitrogen with such elements as oxygen or hydrogen is strongly endergonic. Photosynthesis might seem an excellent source of the required energy, and so it is, but not at first hand. The oxygen liberated by photosynthesis destroys **nitrogenase,** the enzyme that welds nitrogen into chemical combination, usually as part of an ammonium or nitrate ion.

Most plants, as we have seen in Chapter 35, require nitrogen in the form of nitrate, NO_3, or ammonium, NH_3, mainly to manufacture amino and nucleic acids. Nitrate is produced by chemautotrophic bacteria known as **nitrifying** bacteria, which oxidize nitrite, NO_2, to nitrate, NO_3. The oxidation furnishes the bacteria with energy. Nitrite, in its turn, is produced from ammonia by other bacteria that oxidize ammonia as *their* source of energy. Ammonia itself is produced by still other bacteria that reduce amino acids to ammonia, water, and carbon dioxide. Amino acids are the products of decay of plant and animal proteins such as occur in corpses and wastes. Certain bacteria also convert the waste products urea or uric acid to ammonia.

On the other hand, **denitrifying bacteria** produce free nitrogen gas from fixed nitrogen as the ultimate result of their work, reversing the action of the nitrifying and nitrogen-fixing bacteria. Denitrifying bacteria are anaerobic. They live chiefly in the deepest reaches of the soil near the water table, a nearly oxygen-free environment. They employ nitrate or nitrite as electron acceptors in place of oxygen. The action of the denitrifying bacteria would produce a net deficit of chemically combined nitrogen in the biosphere were it not for the action of the nitrogen-fixing bacteria.

Although considerable nitrogen is fixed by combustion, volcanic action, and lightning discharges, most nitrogen fixation is carried out by nitrogen-fixing microorganisms, which must employ the enzyme nitrogenase to do so. Since nitrogenase can function only in the absence of oxygen, the microorganisms that employ it must insulate the enzyme from oxygen by some means. Some nitrogen-fixing microorganisms live beneath layers of oxygen-excluding slime on the roots of a number of higher plants. But the most important nitrogen-fixing microorganisms live in special swellings, or **nodules,** on the roots of leguminous plants such as beans or peas (Figure 56–10) and some other woody plants in several families (the actinorhizal plants). It is estimated that in terrestrial environments these mutualistic bacteria can fix nitrogen 100 times faster than other, less vigorous nitrogen-fixing organisms, employing the energy released from carbohydrate provided by the photosynthesis of their host plant.

A large part of the nitrogen fixation that goes on in aquatic habitats is done by cyanobacteria. In filamentous cyanobacteria occasional heterocyst cells occur, which explicitly function to fix nitrogen and which are evidently not themselves photosynthetic. (If they were, the oxygen they generated would inactivate their nitrogenase.) Some water ferns (Figure 56–11) have cavities in which such cyanobacteria live, somewhat as nitrogen-fixing bacteria live in root nodules. Cyanobacteria also fix nitrogen in symbiotic association with cycads and some other terrestrial plants, and in certain lichens.

The reduction of nitrogen gas to ammonia by nitrogenase is a remarkable accomplishment of biological industry, achieved without the tremendous heat, pressure, and energy consumption of the commercial processes developed by technology. Even so, it takes the consumption of 12 grams of glucose or the equivalent to fix a single gram of nitrogen biologically.

In summary, the four main groups of microorganisms involved in the nitrogen cycle perform, respectively, the following functions: (1) they produce ammonia from decaying proteins, urea, or uric acid; (2) they oxidize ammonia—ultimately to nitrates; (3) they defix nitrogen by liberating it from nitrate; and (4) they fix molecular nitrogen directly and turn it into ammonia. Ecologists believe that virtually all nitrogen now in the earth's atmosphere has been fixed and liberated many times by these organisms.

FIGURE 56–10 Root nodules of a pea plant. Mutualistic *Rhizobium* bacteria live in these nodules, living on energy derived from sugars provided by their legume host. The bacteria efficiently fix nitrogen, some of which is utilized by the host plant. The ultimate death and decay of both partners enriches the soil with the nitrogen they have brought into chemical combination.

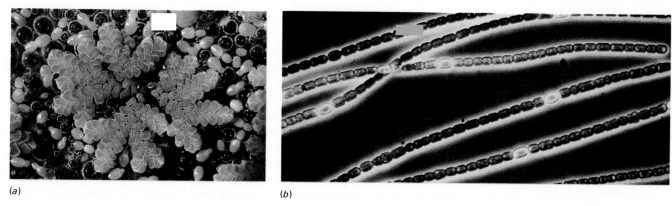

(a)

(b)

FIGURE 56–11 (*a*) Azolla, a water fern that harbors nitrogen-fixing cyanobacteria. (*b*) Strand of cyanobacteria (*Anabaena*), including distinctive heterocysts.

In light of this, it is remarkable that at any one time there are probably only a few kilograms of the enzyme nitrogenase on the entire planet. Those few kilograms have sustained the biosphere for millennia.

The Phosphorus Cycle

As water runs over rocks, it gradually wears away the surface and carries off a variety of minerals, either in solution or in suspension. Phosphate minerals are among the most important of those that are lost from the land in this way. Because the most common phosphorus minerals are quite insoluble, there is no quick way that they can be returned to the land. Instead, they tend to be deposited permanently on the sea floor (Figure 56–12), although geological processes of uplift may some day expose these

FIGURE 56–12 The phosphorus cycle. Recycling of phosphorus is rendered difficult by the fact that no biologically important compound or state of phosphorus is gaseous.

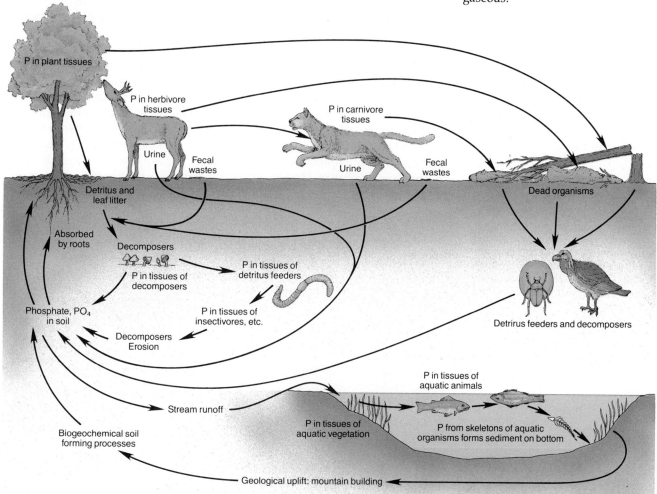

sediments on new land surfaces, from which they can once again be eroded.

Phosphorus in the soil of plant communities is taken in by plant roots in the form of inorganic phosphates. Once in the plant body, they are converted into a variety of organic phosphate intermediates in the metabolism of carbohydrates, fats, and other compounds. Some of these are incorporated into nucleic acids and related compounds. Animals obtain most of their phosphate as inorganic or organic compounds in the food they eat, although in some localities the drinking water may contain substantial amounts of inorganic phosphate.

Very little phosphorus is lost from natural communities, but few communities today can reasonably be considered to be in a "natural state." If there were some way that phosphorus could be returned to communities via the atmosphere (as with nitrogen), its loss would not be of critical importance. As it is, for practical purposes, phosphorus, once lost, is lost for good, for it ends up and remains in the sea for millenia. Moreover, when phosphorus loss is accelerated from soil by such practices as the clear-cutting of timber, or as a result of erosion from agricultural or residential land use, the excessive phosphorus can produce ecological damage to the aquatic communities through which it passes.

Phosphorus enters aquatic communities through absorption by algae and plants (Figure 56–13), which are in turn consumed by large organisms and by microorganisms and other plankton. These, then, are eaten by various fin and shell fish, which may also feed upon one another. A small portion of the fish and invertebrates are eaten by sea birds, which may defecate on their own roosting places. **Guano,** the manure of sea birds, contains large amounts of phosphorus and nitrate. The phosphate contained in guano thus finds its way back to the land, and some of it may be able to enter terrestrial food chains in this way, although the amounts involved are obviously trivial.

The bulk of phosphorus transport in the food chain is in the opposite direction. Corn grown in Iowa may be used to fatten cattle in an Illinois feedlot. Beef from these cattle may be consumed by people living far away— for instance, in New York City. Part of the phosphorus absorbed from the Iowa soil by the roots of the corn plants ends up in feedlot wastes, which probably eventually wash into the Mississippi River. More ends up flushed down toilets in places like Manhattan. To replace this steady loss, Iowa farmers add phosphate fertilizer to their fields. That fertilizer is made (more than likely) in Florida from deposits of phosphate rock that will

FIGURE 56–13 In the open sea and large lakes, organisms from the upper productive waters eventually die and sink to the bottom. Here, phosphate and other minerals may be released by scavenging food chains. Phosphate-rich water may upwell to the surface, but bones and other phosphate-rich solids remove some of the phosphate by becoming a permanent part of the bottom sediments.

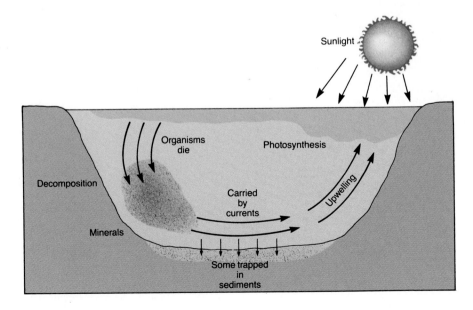

probably be exhausted somewhere around the year 2000, at present rates of use. We have thus made phosphorus into a nonrenewable resource.

FOOD WEBS, CHAINS, AND ECOLOGICAL ENERGY FLOW

Erasmus Darwin, Charles Darwin's grandfather, is said to have been fond of repeating the phrase "eat or be eaten." Whatever the social truths of the aphorism may be, to be biologically true it must be amended to "eat *and* be eaten"! Virtually everything eats plants—directly or indirectly. The herbivores or omnivores that feed upon them directly are themselves eaten by predators or fed upon by parasites. The predators and parasites, however, are as desirable for food as herbivores and may be eaten by other predators and parasites. When any of the preceding die or produce waste products, their corpses or wastes may be eaten by detritus feeders, scavengers, or decomposers, all of which can be eaten in their turn.

Food Webs

A list of the eaters and eaten describes a **food chain,** or at least it would if each of them had just one choice of food. However, there are rather few such highly restricted food chains in the world of life, since few organisms other than parasites eat just one kind of other organism. More typically, the flow of energy and materials through communities takes place in accordance with a range of choices of food on the part of each organism involved. In a community of reasonable complexity, numerous alternative pathways are possible. One could best express these as a diagram resembling a web or network.

Thus a **food net** or **food web** is a more realistic model of the flow of energy and materials through communities (Figure 56–14). Food nets represent an attempt to describe the numerous alternative food energy–transfer pathways in a typical community, most of whose members have more than one choice of food. Obviously, the greater the number of differ-

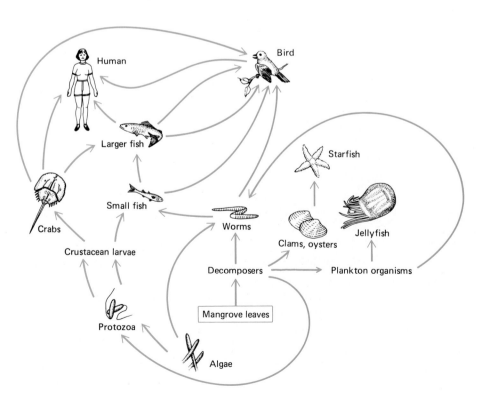

FIGURE 56–14 A greatly simplified diagram of a marine food web. Arrows point from each organism to its consumer or consumers. Inclusion of alternative food choices would greatly increase the interconnections.

ent species residing in a community, the greater the complexity of the food net and the greater the number of alternative pathways by which food energy may travel throughout the community.

Food Pyramids

Food performs two main functions in an organism's body. First, the substances it contains, after appropriate processing, can be used as building materials from which cells and tissues are constructed or with which they are repaired. Second, it can be used as a source of energy to run life processes. That energy is almost always originally solar energy that was first captured by the primary producers. In the end, all creatures must make use of the sun's energy, for according to the first law of thermodynamics, energy cannot be created; it must be reused. But according to the second law of thermodynamics, with each reuse there is a substantial dispersal of energy; it cannot be recycled. The amount of energy available to a particular group of organisms is reflected in their **biomass,** their total mass in the community.

Organisms may be classified roughly according to how closely their feeding habits and adaptations place them to the community autotrophs. Autotrophs that are the bases of food chains are also known as the **primary producers,** since they produce all the available food in the first place. The primary producers themselves constitute the first **trophic level** ("trophic" means "feeding") and usually possess the highest biomass and the greatest numbers of all trophic levels. The herbivores that feed directly upon them are known as **primary consumers.** Those eating the primary consumers are the **secondary consumers,** and so on.

Each successive level is usually much reduced in both biomass and numbers, as compared to its predecessor. A diagram that reflects these relationships graphically resembles a pyramid (Figure 56–15) and is known as a **food pyramid.**

These three categories represent the first three (and usually the only three) trophic levels, with the primary producers occupying the first trophic level, the primary consumers occupying the second, and so on. It is rare to find more than four trophic levels, at least in terrestrial communities.

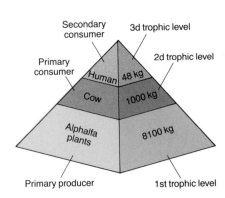

(a)

(b)

FIGURE 56–15 Trophic levels. (a) A trophic pyramid, much idealized. Numbers are illustrative only and are not intended to be exact. (b) All three trophic levels are represented in this Colorado cattle drive.

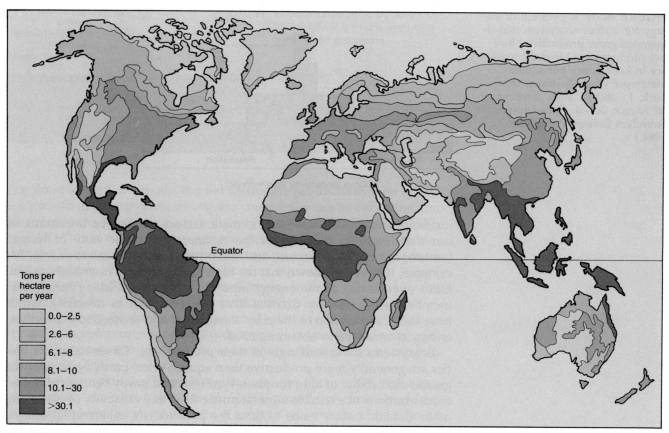

Tons per
hectare
per year

- 0.0–2.5
- 2.6–6
- 6.1–8
- 8.1–10
- 10.1–30
- >30.1

Equator

FIGURE 56–16 Worldwide terrestrial productivity. Temperature, rainfall, and insolation determine the productivity of land ecosystems. Wet tropical regions rank the highest, with deserts and arctic regions the lowest.

Productivity

It is of both theoretical and practical importance to estimate how much food is available to the trophic levels of a community. In almost all cases, this is ultimately determined by how well the plants fix carbon dioxide, for that partly determines how well they grow. However, in order to produce growth, carbon compounds must remain in the plant body, which is not always the case. Like animals, plants respire oxygen, release energy from food compounds for their own metabolic use, and release carbon dioxide back into the atmosphere. A plant usually keeps 25% of the carbon that it fixes.

The productivity of a given community, depending as it does upon that of its plants, varies greatly depending upon such factors as solar energy influx, other energy influx, the presence of climatic or other limiting factors, and human intervention.

The **gross primary productivity** of an ecosystem is the *rate* at which organic matter is produced during photosynthesis. The **net primary productivity** represents the *rate* at which this organic matter is actually incorporated into plant bodies so as to produce growth (Figure 56–17). Only the net primary productivity is available for the nutrition of heterotrophs, and of this only a portion is actually utilized by them. Both gross and net primary productivity may be expressed in terms of calories fixed photosynthetically per year, grams of carbon incorporated into tissue per year, or biomass (the crude total mass of living tissue in all organisms) increase per year. Expressed as a formula,

Net primary productivity = gross primary productivity − respiration

What determines productivity? A number of factors may interact. The influx of solar energy, which probably accounts for the high productivity of many tropical ecosystems, is exceedingly important. Availability of mineral

Focus on MICROCOSMS

A favorite illustration of ecosystems has always been a balanced aquarium, that is, an aquarium containing fish and plants, along with decomposing bacteria. If properly set up, it should be possible for the inhabitants of such an aquarium to survive indefinitely, even if it is totally sealed off from the outside world. Unfortunately, whenever this is actually tried, everything usually dies in short order. This result is of more than academic interest owing to the development of space flight. Spacecraft sent on journeys lasting years cannot be expected to carry all the food and oxygen they will need. It is obvious that they should be balanced terrariums, growing their own food, recycling their own wastes, and producing their own oxygen by photosynthesis. However, in the Soviet Bios experiments in which humans were sealed inside completely closed systems similar to projected spacecraft, the walls became covered with green slime and the humans contracted severe diarrhea, a situation not to be desired halfway to Mars.

In 1977 Dr. Joe Hanson of NASA was able to develop stable, sealed ecosystems containing shrimp and algae. Engineering Research Associates of Tucson, Arizona, was able to develop mass-produced versions of Hanson's systems that are now commercially available. By extensive experiment, they were able to develop a controlled mixture of as many as 100 species of organisms that worked well (most of these were microorganisms). For example, in addition to shrimp and algae it was also necessary to include *Nitrosomonas* bacteria to convert the toxic ammonia excreted by shrimp into nitrite. Nitrite, also toxic, is in turn converted to nitrate by *Nitrobacter*. The nitrate can be utilized as a nitrogen source by the algae. The entire system, called a **microcosm,** is supplied inside a sealed glass sphere resembling a paperweight. These decorative little ecosystems may tell us much about the management of spacecraft, including our spaceship Earth.

■ An ecosphere, an aquarium-like microcosm. (From Yensen N. P. Ecosystems in glass. *Carolina Tips*, April 1, 1988. Courtesy of Ecosphere Associates, Ltd.)

species of trees than one would expect in so large a state—only a fraction of the national average. Ecuador, at the opposite extreme, is about the size of the state of Colorado yet has more than 1300 species of birds—nearly twice as many as the United States and Canada combined.

4. Diversity bears a definite, although complex, relationship to ecological succession, as will soon be discussed.

5. Diversity is often higher at the margins of distinctive habitats than in their centers, because the margins contain all or most of the ecological niches of the habitats that they border. This is known as the **edge effect.**

6. Diversity is reduced when any one species of organism enjoys a decided position of dominance within a community so that it is able to appropriate a disproportionate share of available resources, thus crowding out many other species.

7. Diversity is greatly affected by biotic history. An area recently vacated by glaciers, for instance, will have a low diversity because few

species will as yet have had a chance to enter it and become established. A long-established stable area might have high diversity even if it is in other ways a poor habitat.

ECOLOGICAL SUCCESSION

Communities of organisms do not spring into existence full-blown but develop gradually through a series of stages until they reach a state of maturity. The process of community development is called **succession.**

Primary Succession

Primary succession begins in a habitat not substantially influenced by previous biotic communities, often on a bare rock surface such as cooled volcanic lava (Figure 56–22), rock scraped clean by glacial action, or even the wall of an abandoned building. Although the details may vary, in such a succession one might first observe a community of lichens, followed by mosses and drought-resistant ferns, followed in turn by tough grasses and herbs. These might be replaced by low shrubs (if sufficient soil has accumulated), and then by forest trees in several distinct stages. Each such succession is called a **sere,** and its succeeding communities are **seral stages.**

Secondary Succession

Secondary succession takes place in a habitat already substantially modified by a preexisting community (Figure 56–22). A moderate forest fire, the systematic logging of a forest, and an abandoned old field are common examples. In primary succession a sere resulting in a mature forest would probably go through a shrubby grassland stage similar to the old field at about the midpoint of its progress toward the climax community. In other instances, such as in the case of a forest fire, a habitat is produced that is unlike any that occurs in primary succession.

Some organisms are specifically adapted to the early stages of secondary succession. For instance, some plants such as fireweed (Figure 56–23) or

FIGURE 56–22 Primary succession. This view shows small plants growing on recently cooled volcanic lava in Hawaii. (Charles Seaborn.)

FIGURE 56–23 Secondary succession in the boreal forest. The northern forests of Canada, Siberia, and Alaska are fundamentally coniferous communities, but when they are disturbed by logging or fire, one of the first plants to grow is fireweed (*a*), followed by shrubs such as alders. The midsuccession trees (*b*) are usually hardwoods, particularly aspens, such as the brilliant yellow trees shown here. In time, conifers will displace the aspens.

(a)

(b)

certain conifers reproduce and grow especially well right after fires. Very long-lived seeds of such species as dog fennel may also persist for tens to hundreds of years in the soil, only to sprout when and if conditions favorable for their growth occur. This is sometimes seen even in urban areas when an old building or pavement is removed from the ground and an almost pure stand of the persistent species sprouts suddenly from the newly exposed soil.

Why does succession take place? In some cases, succession occurs mostly as a result of the influence of physical and chemical changes originating outside the community, such as when a pond fills with silt and eventually becomes land. This is **allogenic succession.** Less easy to understand, **autogenic succession** occurs because of the way living organisms modify the habitat and community of which they are a part by their own actions.

Allogenic Succession

A salt marsh develops initially from a tidal mudflat, rich in detritus. Just what triggers the successional process is not known with certainty, but it appears to be related to these factors:

1. Progressive shallowing due to the ongoing deposit of silt by tide and current, together with geologic forces that change the elevation of the shoreline.
2. Growth of algal mats on the regularly submerged surface of the mud which may stabilize the sediments sufficiently for the roots of vascular plants to gain a foothold.
3. Accumulation of chemically combined nitrogen and phosphorus by erosion from nearby land, by pollution, or by the activities of nitrogen-fixing cyanobacteria (Cyanobacteria are probably the most important nitrogen-fixing organisms in aquatic habitats.)

The pioneer community may be dominated by mangroves, *Spartina* grass, or a mixed group of specialized grasses. Oddly, it tends to develop all at once over a broad area. The species involved must meet a strict set of qualifications: They must be very salt-tolerant, especially in higher areas where salt tends to become most concentrated by evaporation; they must be able to grow rapidly into an unstable substratum and fix it in place; and they must be able to reproduce under these conditions.

Red mangroves, for example, propagate viviparously; seedlings sprout from germinating seeds while the fruit remains on the parent tree. Such seedlings are able to take root immediately upon arrival in a favorable habitat. Mangroves also remove much of the salt from the water their roots absorb by a sort of reverse osmosis powered by transpiration. Also, they possess salt glands whereby excess salt is eliminated. Finally, they employ prop roots to hold them in place firmly in the bottom mud. *Spartina* grass also has salt-extruding glands and, like many grasses, spreads readily by subsurface stolons to form what is almost a littoral sod.

Once a pioneer community is in place it may well be stable for decades. If succession occurs at all, it is likely to begin in the part of the marsh nearest land. The initial displacement of *Spartina* arises from several factors. For one thing, the roots of the plants trap additional silt between them, leading to further accretion of soil. Litter (that is, dead plant parts) accumulates not only from the death of *Spartina* itself but also from material brought in from outside the community by tide and wind. The litter interferes with the reproduction of *Spartina* and the accumulated silt retains rainwater so that the salinity, which inhibits competition from less stringently adapted plants, is reduced.

(a)

(b)

(c)

FIGURE 56–24 Reproduction in the red mangrove. (*a*) Red mangrove tree. (*b*) Flowers of red mangrove. (*c*) The red mangrove is viviparous. These elongated structures are actually seedlings about to be released from the fruit in which they have germinated and from which they have been nourished for several weeks.

Once invading plants begin to be established, their shade inhibits *Spartina* growth in their immediate vicinity and this results in their further spread. Once begun, the allogenic conditions favoring the replacement of *Spartina* often take place faster than successor organisms can take advantage of them.

Ultimately, coastal marshes can become converted into completely terrestrial woodland communities. In the vicinity of Biscayne Bay, Florida, for example, areas now far inland are underlain by peaty deposits containing remnants of mangrove, mahogany, and other plants known to grow in or near coastal marshes.

Autogenic Succession

There are two main explanations of autogenic succession, not mutually exclusive. According to the first of these views to have been proposed, each community prepares the way for its successor. The other and later opinion holds that successor communities more usually inhibit the species occurring in their preceding seral stages. It is also true that random events may influence succession, especially in determining just which species happen to arrive at a particular stage.

Consider a bare rock surface. Few habitats are less hospitable. Its temperature may approach 90° C in the sunlight, and, unless it is actually raining, it may be totally devoid of moisture. Whatever minerals are present in the rock are locked up in its crystalline structure, unavailable to living things. Few animals could do more than briefly rest on such a surface and in any case would find nothing there to eat.

Microorganisms are the first pioneers. In an aquatic habitat a film of bacterial slime may develop first on such a surface, but algae, fungi, or lichens are the usual terrestrial pioneers. In one study, ground rock was mixed with nutrient agar. When *Penicillium* fungus was grown on the medium, much more of the mineral content of the rock was rendered soluble than in the absence of this organism. Usually, however, fungus does not occur by itself unless some kind of organic matter is available to it, for ordinarily fungus is a decomposer—a heterotroph.

That is less true of lichens, however. These compound organisms consist of both fungi and algae and are able to live not only on the surface of many rocks, but in the case of porous ones, beneath the surface in a sheltered and somewhat moister microhabitat. Evidently the fungal component of the lichen secretes digestive juices that are able to render the mineral content of the rock soluble. Upon the death of the fungus these minerals may become available to the algae. Meanwhile the fungus parasitizes the algae, which act as the primary producers within this ultimately simple microecosystem. Both fungi and algae are very resistant to desiccation and merely cease to metabolize when water is not available. Yet they quickly resume their lives when moisture becomes available. Some species of lichen can absorb their own weight in water within moments of moistening.

As generations of lichen pass, several important cumulative changes occur. First, the biomass of the community increases, so that even more solubilized minerals are stored in the living tissue of the community. Second, fine particles of rock become detached from the surface or even within the rock itself. Third, as lichens die, their decomposing remains form humus, which mixes with the rock "flour" to form a primitive soil. Fourth, water is absorbed by the lichens whenever it is available and is retained in the tissues of the lichens and in the new, thin soil layer for longer periods of time than ever before. As all of these changes occur, an increasing diversity of tiny animals makes its home in the lichens and soil, especially desiccation-resistant forms such as tardigrades.

These changes have the important effect of mitigating the harshness of the conditions under which the community must live, making it possible for moss plants to grow there. In fact, since the moss plants are able to grow faster than lichens, they tend to replace any lichens that die. The higher productivity possible to moss leads to a greater accumulation of biomass and ultimately, of humus. This leads to further community change, probably through increased moisture retention.

Trends in Succession

Certain trends often can be made out during the course of succession. In the first place, community productivity goes up (Figure 56–25). Secondly, biomass tends to increase. (After all, how much biomass can there be in lichens growing on a rock surface?) Third, species diversity of all kinds

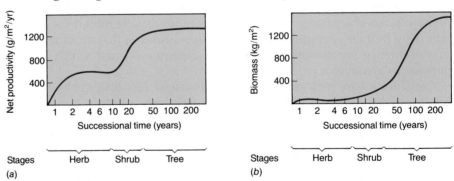

FIGURE 56–25 (*a*) Net primary productivity increased in the course of succession on an area of oak-pine forest in Brookhaven, New York. The reason for the "break" in the curve is that the community was invaded initially by herbs and later by larger woody plants. (*b*) Accompanying changes in biomass. While herbs were dominant, little increase in biomass was observed. Later, with woody plant invasion, there was considerable increase. Most of the woody tissue of such plants is, however, not living. (After Whittaker, *Communities and Ecosystems*, 2nd ed. New York, MacMillan, 1975. After B. Holt and G. M. Woodwell.)

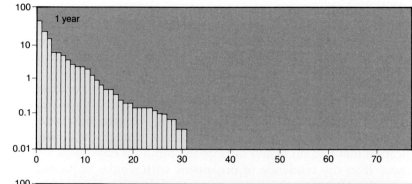

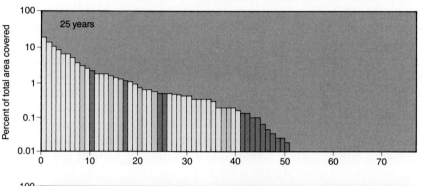

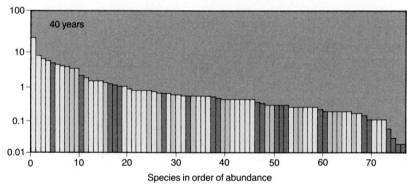

Species in order of abundance

FIGURE 56–26 Changes in species diversity during secondary succession. The gray bars represent herbs, the colored bars represent trees, and the white bars represent shrubs. Not only did the species make-up of this old field change over the years, but the variety increased. (After May, R. M. The evolution of ecological systems. *Scientific American*, September, 1978. Copyright 1978 by Scientific American, Inc. All rights reserved.)

usually increases for most of the course of the succession (Figure 56–26). (In the end, diversity may decrease as a result of the competitive dominance of a few or even one species.)

NEW IDEAS IN ECOLOGY

Great changes have occurred in the past 20 years in the understanding and even the acceptance of the basic concepts of ecology by biologists. The areas that have seen the greatest changes are (1) the nature of communities, (2) the nature of succession, (3) the role of complexity in communities, and (4) niche differentiation.

The Nature of Communities

F. E. Clements, a biologist who was professionally active during the first third of this century, was struck by the overall uniformity of large tracts of vegetation in the world, for example, boreal forest or chaparral. He also noted that even though the species composition of a community living in a particular habitat might be different from that of a community in a climatically similar habitat elsewhere in the world, overall the two were usually closely analogous. He saw communities (which he called "associations") as something like compound organisms—"superorganisms" whose member species cooperated with one another in a manner similar to the cooperation of the parts of the body of an individual organism. This body went through certain stages of development, like that of an organism, and eventually reached an adult state. The developmental process was succession, and the adult state was the climax community.

This cooperative view of the community stressed the interaction of its members. Its opponents held that these interactions were less important than climate, soil, or even chance in producing communities; indeed, the very concept of a community was questionable. It might be a category of classification that had no reality outside the minds of ecologists, reflecting little more than the tendency of organisms with similar requirements to live in similar places.

One way to distinguish between these two views is to census the species that make up communities. If most species are organized into discrete groups, then it should be possible to recognize clusters of associations by statistical analysis. Another prediction holds that the boundaries between communities should be fairly sharp, and few species should be shared by adjacent, distinguishable communities. When studies informed by these proposals were performed, the results were at best controversial. They do not seem to support the interactive concept of a community. If anything, they favor an individualistic rather than a mutualistic view of the community.

The nature of the community is of more practical importance than it might appear. If organisms are highly interdependent, then the slightest disturbance of a community will tend to be propagated throughout it by a kind of ripple effect. If individualism is the norm, then what happens to one species will often have little or no effect on others.

The Nature of Succession

What causes succession to take place? If the community is a superorganism, then succession is inevitable and should be largely unvarying in all its stages wherever it takes place within a given biogeographical province. Moreover, if the climax community is the "goal" to which succession proceeds, then it is reasonable to suppose that each community in a sere prepares the way for its successor. This view of succession is known as the **facilitation** hypothesis.

Actually, many cases of facilitation are known. For example, when Alaskan glaciers retreat, the mineral-poor clay soil that is often left behind is colonized first by plants that harbor vigorous nitrogen-fixing bacteria. The accumulation of fixed nitrogen in the soil permits invasion by other plants that require this nutrient. However, other cases are known in which organisms that arrive by chance in the new habitat secrete allelopathic substances or otherwise wage war on the established inhabitants. This results in community change by competition rather than cooperation.

In still other cases one sees no real succession at all, but rather the ongoing, cyclic replacement of one group of species in repeating fashion.

The idea of a climax community has come under particularly strong attack. Although the idea of an indefinitely stable, unchanging community seems unassailable, the truth is that we cannot be sure that any community will not change. Over geologically significant periods of time communities certainly do change; consider what a different place the state of Wisconsin was during the last glaciation. Who is to say that it will not one day become such a place again? Would any of its communities remain unchanged? According to this line of argument, it is merely a matter of some communities changing more slowly than others. The very slowest of them appear to us to be climax communities.

Community Complexity

Communities vary greatly in their complexity, partly, as we have seen, in response to stress (Figure 56–27), and seemingly in response to climate also. Yet odd facts cast some doubt on these generalizations. For instance, species diversity is often very great in certain desert habitats (e.g., the Namib desert), which seem to be as stressful an environment as the subarctic.

It has been suggested that the diversity of the tropical rain forest is due not so much to its stability or to any lack of environmental stress as to its history. It is believed that the tropical rain forest is a very ancient habitat that has undergone few climatic changes in the entire history of the earth, despite the glaciations that have afflicted the temperate and arctic zones. During this time, it is argued, many species have differentiated, have experienced few lasting barriers to their dispersal and intermingling, and have experienced few or no abrupt climatic changes that might lead to their extinction. Community complexity, in other words, results from long-term community stability.

It has classically been thought to be the other way around, that is, that community stability results from community complexity. The higher the diversity, it is argued, the less critically important any one component should be. With a multitude of possible interactions (such as food chains) within the community, it is unlikely that any single disturbance could affect enough components of the system to make a significant difference in its functioning. Thus, the almost complete loss of the very important American Chestnut tree to the chestnut blight fungus has had very little ecological impact on the complex Appalachian woodlands of which it used to be a part.

On the other hand, a single day of frost can devastate a southern lawn composed almost entirely of a cold-sensitive variety of grass. By this thinking, as one species after another is lost from a community, a critical point of simplification may ultimately be reached where it has no further resistance to stress and may utterly collapse.

The greatest objection to this view arises from mathematical models, which indicate that more complex systems are actually less resistant to stress than less complex ones, at least if the assemblage of species in them is random so that their interdependence is minimal. According to modern views of community structure, this is a reasonable assumption. However, the relationship between community stability and its complexity has not as yet been thoroughly tested experimentally.

FIGURE 56–27 The Dolly Sods area of West Virginia, a community subject to considerable natural stress. Branches are missing from one side of many coniferous trees, mostly because of harsh prevailing winds and the development of rime ice at low temperatures. Very few plant species are able to grow in this area of sparse soil and harsh climate. The stress is not entirely natural, however, for the area was denuded of much of its original topsoil by disastrous fires in past years.

Niche Differentiation

Despite the serious reservations about competitive exclusion that many biologists express, most seem to feel that there is much truth in the concept. Indeed, the ecological niches of community members *are* often mutually compressed in such a way that community members compete minimally with one another.

From Gause's and other experiments it *seems* clear that if two species of organisms have identical or nearly identical sets of requirements, one is bound, if only as a result of chance factors, to possess or to gain an advantage over the other. In time the competition will prove fatal to one. This concept of competitive exclusion is sometimes summarized as the so-called "4-C law"—Complete Competitors Cannot Coexist, as Garrett Hardin put it in 1960. But is it really true?

It is easy to discover apparent contradictions to the competitive exclusion principle. In Florida, for instance, one sees many instances of the coexistence in the wild of native fish and exotic cichlid fish that certainly seem to compete completely. Field botanists have often remarked that seemingly closely competitive species are often collected together. However, it is possible that with further careful study the realized niches of these organisms may be found to differ significantly in some way not yet understood.

In 1945 the ornithologist David Lack studied the life styles of the closely similar cormorants *Phalacrocorax carbo* and *P. aristotelis*. Both species of sea bird occur together, yet both are cliff nesters and fish eaters and so seem to be a counterexample of competitive exclusion. Lack was able to show, however, that *P. carbo* nests mainly on broad, flat ledges and feeds mostly in shallow estuaries and harbors. *P. aristotelis* nests on narrow ledges and feeds further out to sea. It has been suggested, however, that this is not a true case of niche differentiation. To some ecologists it looks very much like *P. carbo* is forcing *P. aristotelis* out of the choicest habitat and feeding areas. Perhaps *P. carbo* will one day bring *P. aristotelis* to extinction. If this happens, it will be an example of competitive exclusion conforming to Gause's views, but as yet it has not happened.

As you may recall from Chapter 1, an unfalsifiable hypothesis is one which cannot be proved false even if it is false. Competitive exclusion may be just such an unfalsifiable hypothesis, or it may turn out to be valid only in certain special cases. It is easy to claim that difficulties may disappear with further study, but this is not the same as showing the principle of competitive exclusion to be true.

Ecology as a Science

Science is more than the systematic collection of data. Science seeks to explain the significance of data, to formulate hypotheses, to test hypotheses, and to assemble everything into a total picture based on principles. The physical sciences were historically the first to be successfully approached in this way and are still producing dramatic new views of the universe. Biology as applied on the level of the organism and the cell is also far advanced in explanatory power. As yet, however, ecology seems more comparable to the "soft" sciences of sociology and psychology than to the physical sciences. This is doubtless due to the complexity and subtlety of its subject matter.

It is the great complexity of the subject matter of ecology that has thus far defeated most of our attempts to produce sweeping generalizations. We find all too often that our models and paradigms are contradicted by the evidence, but we find ourselves unable to replace them with better. The assured understanding of ecological principles that will allow us to confidently apply the science to practical affairs still seems to lie in the future.

■ **SUMMARY**

I. A biological community consists of a group of organisms that interact and live together. A reasonably complete and independent community makes up an ecosystem.

II. The major community roles are those of producer, consumer, and decomposer.
   A. Producers are the photosynthetic organisms that are the basis of most food chains.
   B. Consumers are almost exclusively animals. They feed upon other organisms.
   C. Decomposers recycle the components of corpses and wastes by feeding on them.

III. Organisms play particular parts in ecosystems. The distinctive life style and adaptational package of an organism is its ecological niche.
   A. The nature of the limiting factors for an organism is determined by that organism's ecological niche.
   B. Organisms usually are able to exploit more resources and play a broader role in the life of their community than what is actually the case. They are constrained by the competition of other organisms.

IV. All vital substances (although not energy) are continually recycled through ecosystems, becoming continually available to new generations of organisms.
   A. For example, carbon is converted to carbon dioxide by the metabolic processes of most organisms but is reincorporated into complex compounds (fixed) by photosynthesis. The rate of entry of carbon into the biosphere is roughly balanced by its loss as insoluble compounds on ocean beds.
   B. Although the atmosphere is mostly nitrogen, the high energy cost of fixing this substance in chemical combination makes it a potential limiting factor. Although nitrogen compounds are liberated from organic substances by decay, these compounds are degraded by some microorganisms. The imbalance is redressed by symbiotic and free-living nitrogen-fixing microorganisms, principally bacteria and cyanobacteria.
   C. Because it has no biologically important gaseous compounds, phosphorus is not readily recycled but tends to be deposited in sea beds, from which it is only slowly and unreliably recycled by geological processes.

V. The trophic relationships of a community may be expressed as food chains, or more realistically, as food webs, which show the multitude of alternative pathways that energy may take among the consumers and decomposers of a community.
   A. Gross primary productivity of an ecosystem is the rate at which organic matter is produced during photosynthesis. Net primary productivity expresses the rate at which some of this matter is incorporated into plant bodies. Net primary productivity is less than gross primary productivity because of the losses resulting from plant metabolism.
   B. Food pyramids express the progressive reduction in biomass and numbers of organisms found in successive trophic levels. However, when turnover is rapid, inverted pyramids may be observed.

VI. Community complexity is related to a variety of factors. Expressed in terms of diversity, it is often high when the number of potential ecological niches is great, when a community is not isolated or severely stressed, in the later stages of succession, at the edges of adjacent communities, and in communities with a long history.

VII. Communities succeed one another in orderly fashion.
   A. Primary succession occurs in a habitat that is initially devoid of life.
   B. Secondary succession may go through some of the same stages as primary but begins with a preexisting community, such as an abandoned pasture.
   C. Autogenic successions occur as a result of the actions of the organisms within the changing community. Allogenic successions result primarily from the action of outside influences.

VIII. The science of ecology is by no means in a finished or definitive state, probably because of the great complexity of its subject matter.
   A. It is increasingly believed that community composition results from an individualistic, almost random action on the part of its member species.
   B. There is less tendency today to view succession as a cooperative process of facilitation; instead it is seen as the result of competition in which members of successor communities attack and inhibit their predecessors. It is also increasingly believed that random processes are very important in succession and that it is not a goal-oriented or inevitable process. There is also a tendency to view with skepticism the concept of a totally stable climax community.
   C. Classically community stability and complexity have been thought to be directly related. Many ecologists now see them as being, if anything, inversely related.
   D. Niche differentiation is a very active area of ecological research. The 4-Cs law—Complete Competitors Cannot Coexist—is no longer taken for granted in all quarters.

■ **POST-TEST**

1. Ecologically speaking, yeast would be classified as a _____ .

2. Maggots, however, would be considered to be _____ _____ .

3. _____ usually do not kill the organism on which they feed.

4. Symbiotic organisms that are beneficial to one another are known as _____ .

5. The _____ _____ is the totality of an organism's adaptations, use of resources, and the life style to which it is fitted.

6. The _____ _____ is the ideal ecological niche that an organism could potentially occupy.

7. The interaction of limiting factors plus competition from other species helps to determine an organism's _____ _____ .

8. A species of crab that can tolerate a broad range of salinity of the water in which it lives is a stenohaline/euryhaline organism.

9. "Complete Competitors Cannot Coexist" is a statement of the principle of _____ _____ .

10. Carbon dioxide originally enters the atmosphere mainly by _____ action.

11. Nitrite is oxidized to nitrate by _____ bacteria.

12. Most nitrogen fixation is performed by microorganisms, especially mutualistic _____ and _____ .

13. Persistent pesticides and certain other substances are concentrated in the bodies of predators by a process known as _____ _____ .

14. Net primary productivity equals gross primary productivity minus _____ .

15. Standing crop refers to the current _____ _____ in a community.

16. Inverted pyramids of numbers are most likely to occur in the case of _____ and _____ organisms.

17. Isolated communities are likely to be _____ diverse than those that are adjacent to large areas of suitable habitat.

18. Community diversity tends to _____ in later stages of succession.

19. Ecological succession progresses by stages collectively known as a _____ .

20. A primary succession in which lichens prepare the habitat for succeeding communities is an example of _____ succession.

21. The preceding is also an example of succession produced by _____ .

22. F.E. Clements, an early ecologist, viewed communities as _____ .

23. It is now thought that community structure results from _____ interaction of its members.

■ REVIEW QUESTIONS

1. What is meant by autecology? Synecology?

2. Can you distinguish a community from an ecosystem? Could one speak of a deciduous forest as a community? What about a rotting log within that forest? Which would be best referred to as an ecosystem?

3. How might one distinguish between a decomposer and a detritus feeder?

4. What is a parasitoid?

5. What is an ecological niche? A fundamental niche? Why is a realized niche usually "narrower" than a fundamental niche?

6. Can you give examples of limiting factors not mentioned in the textbook?

7. Would you expect to find more stenothermic organisms in a tropical rain forest or a tundra habitat? Why?

8. Who was A.F. Gause and what important ecological concept did he originate?

9. Describe the carbon and phosphorus cycles. What are their fundamental differences?

10. Why are nitrogen-fixing microorganisms essential to the continuance of life on the Earth?

11. Why is the concept of a food web preferable to that of a food chain in most cases?

12. What is a pyramid of numbers? How might it differ from a pyramid of biomass?

13. What accounts for the existence of inverted pyramids?

14. What is net primary productivity?

15. What is meant by standing crop?

16. Why are predators especially susceptible to the biological magnification of persistent pesticides?

17. The Florida Everglades have been transected by a number of highways that are difficult for the Florida "panther" to cross. However, there is still ample habitat for the panthers that remain. Despite this, the number of panthers continues to decrease. Propose an explanation related to pyramids of biomass (of course there are other explanations).

18. Give two hypotheses purporting to explain the process of ecological succession.

19. Give at least two important ecological concepts or generalizations that have been called into question in recent years.

■ RECOMMENDED READINGS

Crawley, M.J. (Editor). *Plant Ecology.* Oxford, Blackwell Scientific Publications, 1986. Plant ecology is the basis of almost all community ecology, and here is a thoroughly modern discussion of plant ecology from an up-to-date ecological viewpoint.

Krebs, C. *Ecology, the Experimental Analysis of Distribution and Abundance,* 3d ed. New York, Harper and Row, 1985. A community-oriented approach to ecology.

Mitsch, W.J., and J.G. Gosselink. *Wetlands.* Van Nostrand Reinhold, New York, 1986. A theoretical and practical analysis of this important ecosystem type with a view to its conservation.

Traditional agriculture on an Amish farm in Pennsylvania

57

Human Ecology

■ **OUTLINE**

I. The ecology of agriculture
   A. Agricultural communities
   B. Germ plasm resources
   C. New systems of agriculture
   D. Forestry
II. Pesticides
   A. Problems with using pesticides
   B. DDT and other chlorinated hydrocarbons
   C. Reducing pesticide use
   D. Alternatives to pesticides
III. Waste disposal and pollution
   A. Water pollution
      1. Mechanisms of water pollution
      2. Control of water pollution
   B. Air pollution
      1. Ecological effects of air pollution
      2. Meteorology of air pollution
      3. Control of air pollution
      4. Global air pollution
         a. Carbon dioxide and climate change
         b. Ozone layer destruction
   C. Radioactive pollution
   D. Solid waste disposal
IV. Recycling
V. Energy options
VI. Extinction
VII. Overpopulation
   A. Human population growth
   B. When is a nation overpopulated?
VIII. The future
Focus on the last winter

■ **LEARNING OBJECTIVES**

After you have read this chapter you should be able to:

1. Review the development and impact of modern human life styles upon the ecosystems of the earth.
2. Contrast an agricultural community with a typical natural community, relating the differences to ecological instability.
3. Summarize the direct and indirect ecological impact of modern agriculture and forestry practices and outline new or proposed methods of agriculture.
4. Describe two problems associated with pesticide use and two problems specifically associated with the use of persistent pesticides (chlorinated hydrocarbons).
5. Summarize the principal alternatives to chemical biocides in agricultural pest control and be able to briefly discuss the advantages and disadvantages of each.
6. Summarize the role of germ plasm resources in the development of new varieties of crop plants.
7. Summarize the sources of water pollution and describe their ecological effects.
8. Describe methods of water pollution control.
9. Discuss the principal ecological effects and climatic implications of air pollution, together with means whereby these effects might be controlled.
10. Describe the principal methods of solid-waste disposal and discuss the relevance of recycling to waste disposal.
11. Discuss nuclear power, fusion power, and solar power as energy options.
12. Summarize the process of extinction, listing factors that contribute to the decline and extinction of endangered species and providing an example of each.
13. Relate human overpopulation to specific environmental problems and explain how it is possible for humans to temporarily expand the carrying capacity of their habitat.

Human beings have an ecology much as do other animals and plants—an ecology so unusual that it merits special study, not merely because we ourselves are human beings but because of the great effect the human species has upon the rest of the biosphere. Although one rarely speaks of a species as a geological or ecological disaster, that is the category into which *Homo sapiens* seems to fit. No organism in the history of life has had a greater impact on the planet than we have had. In a few short generations we have utterly transformed the face of the earth and greatly accelerated the extinction of organisms to a rate unparalleled since the demise of the dinosaurs.

THE ECOLOGY OF AGRICULTURE

Agriculture, the original highly distinctive innovation of human ecology (Figure 57–1), has produced much of the ecological impact of the human race on its environment. Technological advances in agriculture have accelerated this impact. Practices such as scientific crop rotation, use of chemical fertilizers and pesticides, mechanical tillage, and mechanically driven irrigation systems did not exist 200 years ago.

What has made most of them widespread and practical is the ready availability of fossil fuel energy. This energy chemically fixes nitrogen in fertilizer factories, turns the wheels of tractors, and mines water from the ground. Indeed, widespread agriculture itself is an innovation. Ancient primitive peoples did not employ it and thus apparently had little more ecological impact than any other species of large predatory animal. Today, no animal can even be compared to *Homo sapiens* in ecological importance.

Agricultural Communities

Agriculture may be viewed as the establishment of artificial, greatly simplified communities of organisms yielding a product that is directly or indirectly convertible into human biomass. Such communities are unstable, are very simple, cannot be perpetuated without human intervention, are sparse wildlife habitats, and have poorly developed means of recycling nutrients.

An agricultural community is unstable because it can be maintained only by constant human intervention and energy input. Maize (corn), for example, can reproduce only with human aid. Often, too, agricultural varieties can resist their natural enemies only with human assistance.

In a natural state, host plants (those attacked by pests) are interspersed with others that a particular pest species will not eat. Moreover, wild plants are usually genetically diverse and often possess natural pest-control adaptations, such as poisonous alkaloids. Plant breeders usually breed these traits out, making the varieties they produce much more susceptible to insect attack.

Furthermore, at any one time agriculture tends to be dominated by the latest varieties of crop plants, so that any threat to a single variety is potentially able to wipe out an entire major crop for one or more years. (This happened several years ago with a dominant variety of corn that was uniquely susceptible to a new strain of fungal disease, and is happening now to some species of coconut palms.)

An agricultural community is simpler than any natural ecosystem, which means that the plants of an agricultural community usually belong to a single species. This **monoculture,** as it is called, greatly simplifies the community, and since the plant that is being monocultured usually is consumed by human beings, the community that results is highly and desirably productive from a human viewpoint (Figure 57–2). However, that very monoculture renders the agricultural community unstable and highly susceptible to insect pests or diseases, which can spread from one host organism to another with little hindrance.

In the unnatural communities of modern monocultural agriculture, therefore, unnatural pest-control methods have become necessary. Inevita-

FIGURE 57–1 Primitive slash-and-burn agriculture.
(a) The forest trees are felled and (b) then burned, releasing mineral nutrients into the soil. (c) Crops are planted, among which is the dietary staple, tapioca root. As shown in (d) through (g), this is harvested, grated, soaked, cooked, and rolled out and dried for future consumption. (h) A clearing becomes a village, which is abandoned when the soil has become depleted, so that the rain forest regenerates.

Archaeological evidence indicates that most agriculture, including that of early Europeans, was originally of this type.

FIGURE 57–2 Monoculture. This vast field of strawberries consists not only of a single species but even of a single variety of that species of plant.

FIGURE 57–3 Mechanized versus traditional agriculture. (*a*) The extraordinary productivity of Western agriculture today is due to mechanization and chemical treatment, most of it connected with weed and pest suppression. However, modern farming methods have ecological side effects, including accelerated soil erosion and pollution of groundwater and surface runoff alike with fertilizers and pesticides. (*b*) Although less productive per acre, traditional agriculture may better preserve those acres for the use of future generations. The future of agricultural science may lie in the improvement of such methods; indeed we may yet see them used again in the developed world.

bly, these methods have substantial ecological impact, often extending far outside the boundaries of the areas where they are applied. Human management also greatly reduces competition among species. Farmers go to great lengths (Figure 57–3) to suppress undesired species (that is, weeds and much wildlife, together with invertebrate pests) that eat or compete with cultivated varieties or interfere with their harvest.

The chemical cycles of an agricultural community are often incomplete. Cultivated species usually are consumed at a spot remote from where they are grown. Their substance is ordinarily not returned to the soil but escapes from the terrestrial ecosystem through a sewage disposal plant or garbage dump. This interruption of nutrient cycles causes the soil to deteriorate unless artificially fertilized.

Obviously, an agricultural community is not and cannot be the same as the natural community it has replaced. The demands of agriculture do not permit this. Farmland is therefore not a suitable habitat for most of the wild species that formerly occupied it. This realization has led to the develop-

(*a*)

(*b*)

ment of game parks and preserves in many countries. Yet even on farm-land some compromise is possible, and with enlightened management, agricultural lands can yield a modest return of wild game or at least serve as homes for wild species. Understandably, however, many farmers feel that they cannot spare the space that otherwise could be employed more profitably, and many also view patches of natural habitat as refuges from which weed species and pests can readily spread.

In the heavily populated and economically less developed countries, the situation is far worse. Hungry peasants encroach continually on wild game preserves and expunge most surviving large wild animals from the land that is supposed to be reserved for wildlife habitat. This is sometimes abetted by money supplied by developed nations for illegal trophies or by the ready market that exists for exported meat, lumber, and other products in the developed nations. On the other hand, tourism from the developed nations often provides the major motivation for whatever wildlife preservation is practiced.

Germ Plasm Resources

The gene pool of cultivated organisms is a subset of the gene pool possessed by their wild ancestors and relatives. Crop improvement depends upon the use of genes already occurring in primitive cultivated or wild varieties of crop plants. Most crop plants originated in what are now third-world nations (Figure 57–4).

The **germ plasm** is the total genome of an organism. Since the germ plasm of a cultivated variety of plant is a subset of the genome of its wild ancestral population, plant breeders have created cultivated varieties mostly by eliminating traits advantageous to the wild forms, but undesirable to us. As mentioned earlier, these might include the production of

FIGURE 57–4 Areas where plants and animals were probably first domesticated. Some species that may have been domesticated independently in two or more areas are shown in both places.

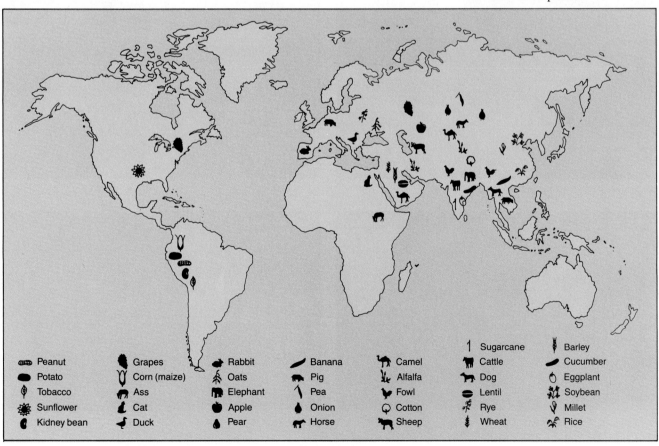

poisons, for instance, that may have functioned originally as natural pesticides. Until recently breeders were dependent on naturally occurring mutations to disable genes they considered undesirable, or to produce alleles with desirable characteristics. Although genetic engineering is capable of transplanting genes from one organism to another, it has not yet progressed to the point of designing genes to order. We must use *available* genes—recapturing, it is hoped, a larger sample of the original gene pool from which present varieties were derived. This means that the improvement of existing species of cultivated plants requires the discovery of wild relatives with potentially useful traits, or primitive cultivated varieties. To discover them we must explore the underdeveloped world, which is increasingly reluctant to let us do so.

New varieties of crop plants are profitable to those who develop them. They are also costly to produce, especially if they cannot be propagated asexually. The needed trained manpower and other resources do not usually exist in underdeveloped nations, so they must purchase expensive varieties that have been developed by others using germ plasm that was originally part of their own natural resource base. On the other hand, food production in many underdeveloped countries has greatly increased as a result of new varieties and techniques that were originated in or with substantial aid from the developed nations.

Moreover, owing to the great species diversity of tropical ecosystems, nations with warmer climates are the most likely sources of entirely new potential crop plants. These plants are likely to be lost by the destruction of tropical wild communities, especially the rain forests (as discussed in Chapter 54).

FIGURE 57–5 Destruction of tropical rainforests for subsistence logging in Ecuador (*a*) and Malaysia (*b*). Forests, such as these in Costa Rica, are also cleared for charcoal production (*c*) and to create pastureland (*d*).

(a)

(b)

(c)

(d)

Developed countries contribute to this destruction indirectly by wasteful demand for cheap timber, beef, and certain kinds of agricultural produce. To sell such products for badly needed foreign currency, underdeveloped countries often convert their rain forests and other natural habitats into tree farms, cropland, or pasture at a rate much faster than they otherwise would (Figure 57-5).

New Systems of Agriculture

Famine and malnutrition are among the gravest global problems of our time, as the news media regularly remind us. American agriculture, although by no means problem-free, often has seemed exempt from disastrous crop losses and continues to be one of the most productive agricultural systems anywhere today. Yet even in the United States, although great increases in food production have been achieved, the costs have been substantial in erosion, wildland destruction, and energy consumption. These problems are even more acute in the underdeveloped world.

Perhaps agriculture needs to be reinvented. Many "high technology" alternatives, such as hydroponic farming, have been proposed and put into limited practice, but on the whole these have proved to be even more expensive and energy-consuming than conventional agriculture.

One recent and innovative proposal is cellulose farming. Most terrestrial primary production goes into the formation of inedible cellulose. Technological advances (including biotechnology) may enable us to convert this otherwise wasted cellulose to simple sugars at low cost. These sugars then could be used as industrial feedstocks for the direct propagation of edible plant and animal tissues in factory tissue or organ culture. This also would change the season-dependence of agriculture, since trees and other perennial cellulose sources can be harvested at any time of the year.

There is no telling whether such ideas will prove practical. However, there is an approach that we can be sure is practical and energy-efficient. If the diet of developed countries could be modified to include more vegetable food, it would not only be more healthful but less ecologically wasteful, since it takes less land to raise vegetable than animal food.

A further solution is limiting population growth, which would result in fewer mouths to feed. While it is true that the improved systems of tropical agriculture known as "green revolution" techniques have greatly (if temporarily) increased the production of food in many formerly starving countries, the originators of the improved crop varieties that helped to make this possible saw their work as giving the world time to get population growth under control.

Forestry

The deforestation of all countries has proceeded with alarming speed since the Industrial Revolution and, in some localities, since ancient times. One reads, for instance, of the Biblical cedars of Lebanon, and the modern Lebanese flag even depicts one. Yet little is left of the great forests whose timber was used in the construction of the palace and temple of King Solomon. This deforestation resulted not only from simply cutting down trees, but also from heavy grazing by sheep and goats, which prevented the growth of seedlings.

More recently, firewood and wood charcoal fueled the early Industrial Revolution. To this day the greatest use of forest products worldwide is for fuel. In countries that have limited or no fossil fuel resources, this often leads to dramatic deforestation. So acute is deforestation in modern India, for example, that the major fuel used over vast areas is dried cattle dung.

FIGURE 57-6 A fir-tree seedling nursery. After the trees reach a certain height, they are transferred to another location where they are planted in rows like a crop, periodically thinned, and weed trees removed. Little wildlife inhabits such farms.

There are many other uses for timber as well. In the 19th century forests were extensively and wantonly cut to provide wood for construction timber and paper pulp, with fires often completing the destruction that logging had begun. Such practices continue today, especially in the tropics, but even in parts of the United States.

Many nations have embarked on extensive reforestation programs. However, the trees planted often are not native and do not fit in well with the local ecology. Additionally, these replanted trees are usually managed as a crop, so ecologically this **silviculture**, as it is called, resembles agriculture. Tree farms are not forest communities; usually they are single-species aggregations (Figure 57–6) with little wildlife and with no climax species. They must be managed like any other crop, which may include pesticide treatment and even the use of artificial fertilizers. Douglas fir seedlings, for instance, do not grow well in the shade of mature trees and must be planted in areas from which all mature trees have been removed. To take another example, southern longleaf pine woodlots must be burned periodically to remove the fire-sensitive seedlings of hardwood trees that otherwise would eventually replace them. After burning or clear-cutting, mineral nutrients that otherwise would be retained in the community are lost at an accelerated rate, so much so that in some studies the watercourses draining the area actually suffered from eutrophic overfertilization (see later in the chapter).

PESTICIDES

The characteristics that make pesticides economically desirable often make them ecologically dangerous. Chief among these are persistence and effectiveness against a broad spectrum of pest organisms. Pesticides often are most damaging to the predatory organisms that otherwise keep the pest in check. Yet pesticides seem necessary for modern agriculture, as we have seen.

The first-generation pesticides—inorganic chemicals such as sulfur, lead, arsenic, and mercury—were used to repel or kill pests for hundreds of years. Many of these substances are quite toxic, and they sometimes accumulated in the soil in amounts that inhibited even plant growth. Pests also became resistant to them (a case of natural selection), so that over time they lost their effectiveness. These problems led to some use of natural plant insecticides, such as nicotine (derived from tobacco leaves) and the pyrethins (derived from certain chrysanthemum-like plants).

The era of second-generation pesticides began about 1940 with the discovery of DDT. Pesticides of this type are synthetic organic compounds that can be classified into three groups: (1) DDT and related chlorinated hydrocarbons (Chlordane, Dieldrin, Mirex), (2) organophosphates (Malathion, Parathion), and (3) carbamates (Sevin, Temik). These highly effective organic pesticides have been developed and put into widespread use by everyone from farmers and health departments to suburban homeowners.

The chlorinated hydrocarbons interfere with nerve action by antagonizing the sodium pumps in nerve cell membranes. Most other organic pesticides are anticholinesterases, which render the enzyme cholinesterase incompetent (see Chapters 39 and 46). Anticholinesterases are, in general, more toxic than chlorinated hydrocarbons. Humans are less susceptible to the action of both classes of pesticides than insects, partly because of their large body size and partly because of differences in physiology. Nevertheless, serious accidental poisonings, especially with potent anticholinesterase pesticides, occur regularly.

(a)

(b)

(c)

FIGURE 57–7 Biocides, necessary to control insect pests (*a*), become concentrated as they are passed along the food chain. When predators consume poisoned insects, the concentration of pesticides increases. (*b*) Ultimate predators, such as hawks, bald eagles, peregrine falcons and pelicans that feed on contaminated birds and fish are prime victims. (*c*) Pesticides interfere with egg production and embryonic development of these predatory birds, causing thin-shelled, fragile eggs that rarely hatch in the wild, as in the case of these brown pelican eggs.

Problems with Using Pesticides

Most widely used pesticides are broad-spectrum poisons that kill nonpest species as well as pests. Another major drawback of pesticides is that many insects become resistant to them. Although initially the pesticide may cause a rapid decline in the pest population, resistant mutants gradually replace the susceptible insects. By 1980 more than 200 DDT-resistant pests were known. Often, a pesticide-resistant pest population evolves rapidly, while the natural predators of these insects do not become resistant. This occurs partly because predators reproduce more slowly than pests, partly because there are fewer of them, and partly because of biological magnification (Figure 57–7).

With their natural enemies devastated, the pest population increases in numbers and may become a greater threat than before the pesticides were used. Also, previously unobtrusive insects, whose populations were held in check by the now absent predators, may now become economically significant pests for the first time.

DDT and Other Chlorinated Hydrocarbons

The organophosphate and carbamate pesticides are **biodegradable,** which means that they decompose several weeks or months after being sprayed. DDT and other chlorinated hydrocarbons are **persistent pesticides.** They are not readily biodegradable. An additional drawback of chlorinated hydrocarbon pesticides is that they do not remain confined to the areas where they are sprayed. Food chains, air movements, and, to a lesser extent, water currents distribute persistent pesticides globally, so there is now probably no organism on land or in the sea whose tissues are completely free from them. Traces of DDT have been found even in the fat of penguins in Antarctica, thousands of miles from the nearest place where DDT was deliberately employed.

Reducing Pesticide Use

Partly because DDT does pose a serious threat to ecosystems, its use in the United States was banned in 1973. However, in the early 1980s it was still being manufactured in the United States for export to many other countries! And other countries are still manufacturing and using it. Several closely related chlorinated hydrocarbon pesticides have also been banned, but others continue to be used widely. The use of DDT is officially sanctioned by a number of international organizations because of its usefulness in checking the populations of mosquitoes that spread malaria, especially in the underdeveloped world. But even without considering some of the deeper issues involved in such use, it is clear that DDT is becoming steadily less effective as a mosquito pesticide as these insects develop increasing resistance to its action.

Although the problems connected with pesticide use are reduced somewhat when biodegradable compounds such as the carbamates or organophosphates are substituted, there is no known pesticide currently in widespread use that is free from undesirable ecological consequences. The carbamate Sevin, for instance, is almost completely harmless to mammals but is instant death for bees, and bees are vital for pollination of food and wild plants. Since all pesticides kill more than the pests against which they are directed, in the end pesticide problems can be solved only by avoiding their use. Yet this seems less and less likely to happen. In fact, we now use more than twice the amount of pesticides used in 1962, the year Rachel Carson's famous book on the subject, *Silent Spring*, was published.

Unfortunately, the very characteristics of pesticides that make them ecologically dangerous make them economically attractive to agriculturalists. A broad-spectrum pesticide also kills many varieties of *pests* with a single application, and if it is persistent it does not need to be applied as often. On the whole (and for the short term) pesticide use contributes substantially to agricultural production. There is also no reasonable doubt that public health has benefited from pesticide use, especially in underdeveloped countries.

In the long run, however, consequences of population growth made possible by these measures may outweigh the good that has been accomplished. We can avoid using pesticides, to be realistic, only if we are content to produce less food. We can produce less food only if we have fewer mouths to feed. With fewer mouths to feed, we could, perhaps, confine human settlement to the parts of the globe that are most ecologically suited to settlement—for example, by avoiding malarial swamps. Population growth and the political problems that result from it are among the most significant factors keeping us from attempting ecologically sound solutions to environmental problems.

Alternatives to Pesticides

There are some alternatives to the use of pesticides short of completely abandoning crops to the insect world, although none may permit as high a level of agricultural production as our overpopulated world demands. Some methods are effective by themselves; others help reduce the need for pesticides. Among these alternatives are the use of natural enemies, the release of sterile males, and such practices as careful crop rotation to break pest life cycles by depriving them regularly of their preferred food plant.

Third-generation pesticides may hold the greatest promise. These are synthetic versions of naturally occurring pest pheromones or hormones which, if applied carefully, can disrupt the behavior (particularly reproductive behavior) or physiology of the target organism without damaging other harmless or desirable organisms.

Biological control agents of various types can sometimes suppress the growth of pest populations, but they are most likely to be successful where the pest organism has been introduced to a habitat to which it is not native. Often, in such a case, the pest has neglected to bring along its natural enemies. *If* those enemies were among the chief limiting factors on the pest organism's multiplication in its original habitat, introducing the missing natural enemies may control the pest. Of course, the natural enemies themselves must first be investigated and shown to be harmless before we attempt to use them.

The sterile male technique has been employed against the screw-worm fly, an important pest of the cattle industry (Figure 57–8). Millions of screw-worm flies were artificially reared, then sterilized by exposure to ionizing radiation. When released in huge numbers, these sterile flies successfully competed with the relatively few fertile ones for mates.

The female screw-worm fly mates only once in her lifetime. As a result of many unproductive matings, the screw-worm fly population dramatically declined, and in some areas, especially islands, this pest has been completely eradicated. However, in places that are not geographically isolated, there is a constant reinvasion of fertile flies from adjoining areas that makes eradication impossible. Some populations of screw-worm flies have also developed a behavioral resistance to the sterile male technique. In these, the females are able to detect and discriminate against the sterile males.

FIGURE 57–8 Biological control, a superior alternative to the use of biocides against screw-worm flies, which cause tremendous damage in livestock. The male flies are sterilized and then released to mate with wild females. Since the females mate only once, they will never reproduce.

WASTE DISPOSAL AND POLLUTION

Waste may be defined as any product of our civilization that is usually discarded rather than used, or a formerly useful product no longer used for its original purpose or for any other. The disposal of wastes has become an acute problem, not only because of the quantity of such wastes, but also because of their kind. Many are not readily degraded by natural mechanisms even in small quantities. Careless waste disposal is practiced because it is advantageous to the polluter, forcing others to bear the costs, economic and otherwise.

Pollution is a reduction in the quality of the environment by the addition of materials (or heat) not normally found there, or in harmful quantities if they are normally present. Put differently, pollution exists when wastes or other substances have a significantly damaging effect upon public health, property, ecosystems, or aesthetic values.

Polluting costs us a great deal. Why, then, does it exist? Most of the reasons are economic. Here are some of them:

1. By avoiding the costs of waste disposal, the polluter compels society to bear either the costs of his pollution or the costs of correcting it.
2. By passing on the cost of clean-up to future generations (as in the case of dumping toxic chemicals) polluters avoid having to pay for it themselves.
3. By providing convenience of use through disposable packaging, manufacturers increase sales of many goods, while compelling the public to bear the costs, ecological and otherwise, of that disposal.

(What would your garbage collection bill amount to if the 60% to 80% of it that is packaging did not need to be thrown out?)

4. By deliberately designing goods to wear out or become unstylish, and by otherwise encouraging wasteful consumption, we increase the flow of goods and services through the economy.

Water Pollution

Water pollution is a first-class health menace due to its potential for transmitting infectious disease. Purification of drinking water, desirable as it is, does not address the ecological damage done by water pollution by sewage and other organic wastes, which reduces the oxygen content of natural waters and promotes cultural eutrophication (explained below).

Water pollution was one of the first varieties of pollution to attract widespread concern and efforts at correction, mainly because of its obvious role in the transmission of disease by drinking water (Figure 57–9). This has led to an effort to separate drinking water from that used for waste disposal, or failing this, the purification of drinking water. Later, the ecological damage of water pollution began to be understood, leading to the use of more effective control measures.

At present, the two major sources of water pollution are industrial waste and municipal sewage (which actually may be largely composed of industrial waste). Industry accounts for most water pollution in the United States. Usually far more concentrated than municipal sewage, industrial waste produces up to 12 times more pollution per gallon of effluent than municipal wastes do. But what they lack in concentration, municipal

FIGURE 57–9 Map of locations of cholera cases in London as recorded by Dr. John Snow in 1854. Since these cases tended to cluster around the location of a single public pump, Snow thought that they might have resulted from contamination of water drawn by this pump. This and other studies provided early impetus for the provision of reliable sources of clean water. The ecological hazards associated with water pollution began to be appreciated much later. (Redrawn after Newman and Matzke.)

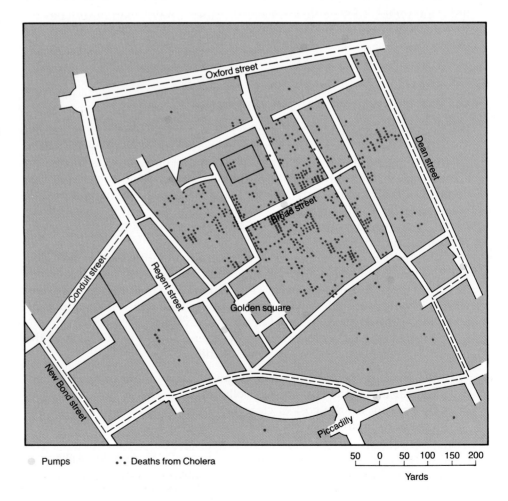

Pumps   Deaths from Cholera

50 0 50 100 150 200

Yards

wastes tend to make up in volume, particularly taking into account street drainage and surface runoff after rain and snow.

Liquid or dissolved waste is often organic, but a wide variety of chemicals (whose nature and control are beyond the scope of this book) also can be polluting. The organic component of water pollution is probably the most important, and historically was the first whose control was attempted.

Organic wastes provide a rich source of nutrients for decay bacteria and fungi. Hence feces, blood from slaughterhouses, oxygen-demanding wastes from paper mills, and peelings from vegetable-processing plants (among many other things) stimulate the growth of bacteria whose metabolism rapidly removes oxygen from the water. Industrial wastes may also contain large amounts of sediment, chemically combined nitrogen, phosphorus, carbon dioxide, methane, hydrogen sulfide, and smaller amounts of miscellaneous chemicals, heavy-metal ions, and even pesticides.

Mechanisms of Water Pollution

A polluted environment is a demanding one. Not surprisingly, few organisms can tolerate it. Yet those able to exist in it often attain astronomical numbers and very large biomass. Most aquatic organisms, including, unfortunately, many of those we hold most desirable, are sensitive to the effects of pollution and are destroyed by it.

When organic wastes are dumped into a stream, a predictable sequence of events occurs (Figure 57–10). Near the source of pollution, surprisingly, numbers of fish and other organisms may persist because the organic wastes have not yet had time to decay and the dissolved oxygen level is high. In severe instances, somewhat farther away, conditions worsen, with bacteria and fungi that degrade organic sewage using up so much oxygen that anaerobic conditions prevail. Still farther away, the polluted water has begun to purify itself: Organic materials start to disappear, oxygen diffuses into the water from the air, and, except for cultural eutrophication (discussed below), something like a normal community is established.

Small quantities of potential pollutants may enter natural ecosystems continually without producing actual pollution. If quantities of decaying organic material are small, they are merely recycled as part of the normal action of decay organisms and detritus feeders in the community. The immense quantities of wastes produced by large urban concentrations of human beings, industries, and modern agriculture are too much for the usual decay mechanisms to handle, and it is these that result in the degradation of the ecosystems that are forced to receive them.

Eutrophication is the typical successional process that occurs in geologically aging aquatic habitats. A shallow, warm lake or watercourse, whose biomass and waters are high in mineral nutrients, is said to be eutrophic. Geologically young lakes and watercourses are said to be **oligotrophic,** meaning "little food" because they usually have little plant life and therefore little of the primary production on which food chains must be based. Fish and other aquatic animals that live in them typically require low temperatures and high oxygen concentrations and usually have to obtain their food from sources outside the aquatic habitat such as insects that fall into the water. The total biomass of an oligotrophic aquatic community is usually low, but species diversity is high and many of its species (for instance trout) are highly valued.

The reason that a geologically older lake gradually becomes eutrophic is that silt and nutrients accumulate in it from nearby watersheds. The term eutrophic means "good food"; but whether the food is truly good or not, it

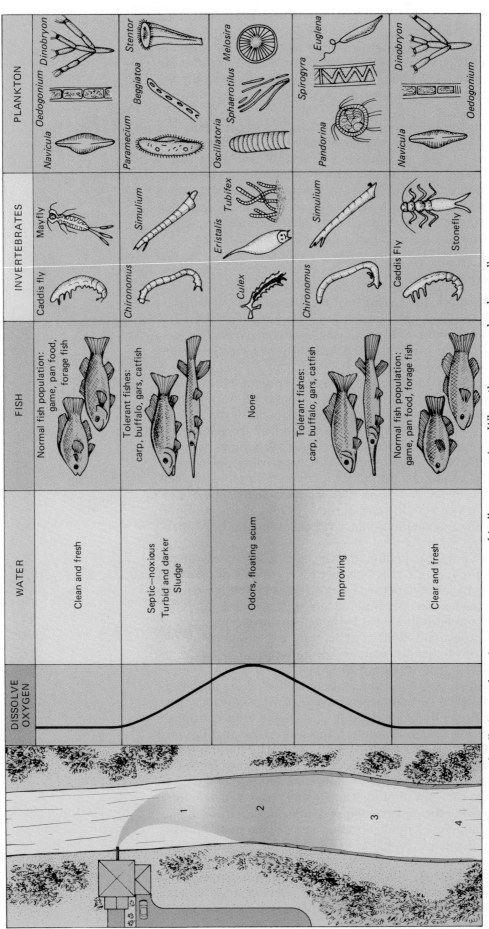

FIGURE 57–10 Zonation of pollution ecology in a stream receiving large amounts of untreated sewage. As the amount of oxygen dissolved in the water decreases, fishes disappear. Only organisms able to obtain oxygen from the surface, tolerate low oxygen tensions, or respire anaerobically can survive. When the sewage has been all decomposed by bacteria, the species of plants and animals present in the stream return approximately to normal, although eutrophic changes may persist. (After Odum.)

does become more plentiful. The water warms and the species composition changes to fish like bass or carp, but since community productivity improves, more fish are available, although of fewer kinds.

The accelerated release of minerals from pollutants produces an exaggerated form of eutrophication called **cultural eutrophication.** Cultural eutrophication can change the characteristic form of submerged and emergent vegetation undesirably, producing swampy tangles of growth (Figure 57–11) through which fish can hardly navigate. Dense blooms of cyanobacteria can occur, crowding out normal and more desirable algae and protists. At night or on cloudy days the plant life consumes more oxygen than it produces. Finally, the time inevitably comes when some event kills off a large part of the staggering plant biomass. Its decay then reduces the oxygen content of the water below that which will support most aquatic animal life, and massive fish kills result.

About half the population of the United States depends upon rivers for drinking water (the rest use groundwater). Although legislation requires industries and municipalities to treat their wastes before dumping them in streams or rivers, the purification process is never 100% effective. Thus many pollutants, including known and suspected carcinogens, are present in the drinking water of many communities. Eventually, many river pollutants (like air and land pollutants) find their way to the oceans—the ultimate sinks for most pollution. There they may continue to damage aquatic life, with only a small portion recycled biologically to the land.

FIGURE 57–11 Cultural eutrophication in an estuary. A shallow portion of Tampa Bay, seen here at low tide. Excessive nutritional enrichment of the habitat by sewage and other wastes has encouraged the excessive growth of *Ulva*, forming this continuous "lawn" of algae.

Control of Water Pollution

Natural ecosystems are sometimes capable of considerable natural pollution control. Often grossly polluted, marshlands are capable of considerable natural sewage treatment if not overloaded. As the wastes travel slowly through the sunlit shallows, a multitude of photosynthetic algae, protists, and cyanobacteria thrive upon their high mineral content, releasing oxygen that is employed by aerobic decay bacteria to swiftly decompose the organic wastes. Instead of causing aquatic eutrophication, excess mineral nutrients may be absorbed by the bodies of the grass, rushes, or other emergent vegetation of the marsh.

Most modern methods of water pollution control attempt to utilize the same processes of decomposition and decay that occur naturally, but confine them to the interior of a sewage disposal plant. Probably the method in most widespread use today, the **aerobic activated sludge process,** temporarily stores raw sewage in a holding tank. Decomposer bacteria are then added, along with much of the sludge left over from the last batch. Air is then introduced to the tank to provide oxygen for aerobic decomposition.

When the process is complete, any remaining sludge is allowed to settle and it, along with the decay bacteria it contains, is partially recycled to the next batch of sewage to be treated. Leftover sludge is digested anaerobically and what remains of it is either incinerated or, unfortunately, just dumped somewhere. This process yields an effluent from which perhaps 95% of the oxygen-demanding wastes have been removed. Unless it is treated further, however, the minerals such as phosphate that were originally present in the sewage are still in the effluent, plus whatever was released from the organic matter by decay bacteria.

The biggest difficulty with water pollution control techniques is that they are not used effectively. Overall, the nation's water quality has not improved over the past ten years; indeed, it has worsened. This results largely from population growth, which tends to counteract the effect of any control measures that might be employed. It also results from the low priority that our society seems to assign to pollution control. Clearly, much more widespread use of pollution control plants is needed, together with the development of new, inexpensive technologies for those plants to use.

Air Pollution

Air pollution is produced principally by combustion, either in automobile engines or in industrial and power generation plants. It is responsible for photochemical smog and the extremely damaging acid rain. Air pollution is locally aggravated by air stagnation and the development of inversion layers (to be discussed below), which prevent its ready dispersion. Although control technologies exist, they are, as in the case of water pollution, insufficiently employed.

Have you ever considered the plight of a fish in a polluted lake? It is exposed to a continuous dose of toxic substances every moment, every day, from which it usually has no escape. Air pollution puts us in the position of the fish, for each of us must breathe about 20,000 times every day. With each breath we inhale a wide variety of toxic gases and particles emitted by automobiles, power plants, industries, and cigarette smokers. The United States alone dumps more than 220 million tons of pollutants into the air annually.

The pollutants that account for most air pollution are carbon monoxide, sulfur oxides, nitrogen oxides, hydrocarbons, and particulates. Their relative contributions to air pollution are indicated in Figure 57–12. These pollutants interact to form secondary pollutants and smog and also produce the acids (chiefly nitric and sulfuric) that are responsible for acid precipitation. Cigarette smoke is also an important indoor air pollutant. It contains several other pollutants and represents a highly concentrated source of air contamination for all who must associate with smokers. Although the total volume of polluted air produced by smokers does not begin to compete with the amount spewed forth by industry, the health consequences are more severe, for the air polluted by the smoker is the air closest to us, producing very concentrated contamination. At least industrial pollution tends to be diluted by the time it wafts our way. Nevertheless, the effects of industrial and other gross pollution on public health and on natural communities are substantial.

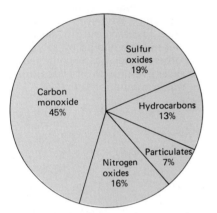

FIGURE 57–12 The average make-up of air pollution.

Ecological Effects of Air Pollution

Air pollution has many sources, from the smokestacks of power plants to the exhaust pipe of the family car. Our consumptive life style ensures a multitude of sources of such air pollution, especially in manufacturing areas and in the vicinity of cities. In Chicago air pollution at times reduces the amount of sunlight by 40%, and in New York City, by 25%. When sunlight is reduced, photosynthesis is also diminished. In addition, trees and other plants absorb great quantities of pollutants directly from the air.

Damage occurs mostly in the photosynthetic tissue (mesophyll) of the leaf (Figure 57–13). The surfaces of mesophyll cells, which are moist to facilitate gas exchange in photosynthesis, are vulnerable to attack by toxic substances in the air. Fluoride particulates, photochemical smog (generated from automobile emissions by a chemically complex atmospheric process), sulfur dioxide, and even very fine soot can kill vegetation. Some species of plants are more susceptible to certain pollutants than other species. At a concentration of less than one part per million, the common pollutant nitrogen dioxide reduces the growth of tomato plants by 30%.

In Los Angeles County and areas more than 60 miles away from it, forests are being damaged by photochemical smog. Photosynthesis is reduced by 60% when smog concentrations are 0.25 part per million. Such

(a)

(b)

(c)

FIGURE 57–13 Effects of air pollution (sulfur dioxide) on white birch leaf (a). Effects of acid rain on (b) German pine forest and (c) ornamental architecture of the Field Museum, Chicago.

decreases in photosynthesis also slow the flow of resins under the bark, rendering the trees susceptible to plant disease and insect pests.

Acid precipitation damages both terrestrial and aquatic communities in several ways. In the first place, rainwater or snow of low pH produces aquatic pollution in soft water lakes and streams, often sufficient to destroy most aquatic life. Even extremely remote areas (such as inland China) are seriously affected. Secondly, the high nitrate content of this water can contribute to cultural eutrophication. Acid precipitation directly damages plant roots and indirectly weakens vegetation by nitrate overnutrition (see Chapter 35). At the same time, other mineral nutrients such as calcium and magnesium are leached out of the soil and lost to the plants.

With the present increase in air pollution, great damage to vegetation and aquatic habitats is occurring worldwide. Air pollution respects no political boundaries. Within the United States federal legislation and control are clearly most appropriate. In the last analysis, air pollution can only effectively be regulated by diplomatic or other international measures.

Meteorology of Air Pollution

Atmospheric inversion is the cause of most air pollution episodes (Figure 57–14). In such cases weather conditions form a lid of warm air above cooler, polluted air. Although inversions do not actually increase pollution, they seal the pollutants below and prevent them from being dispersed. In Los Angeles County, where conditions favor them, inversion layers form when warm air moves in from the deserts to the east and lies over the mountains that surround Los Angeles. Beneath this warm air is a layer of cooler air that has moved in from the sea. Such inversions also form, although less regularly, in virtually every area of the country. Simple air stagnation, which can occur anywhere, is almost as bad, for then pollution is not dispersed by normal wind action.

It is that very wind action, although helpful in removing polluted air, that unfortunately results in acid rain possibly hundreds of miles from where the pollutants originated. It is common for acid rain to be caused by air pollution in other nations, and this has caused international disputes

1393

Focus on **THE LAST WINTER**

There has never been any doubt in the minds of informed people that nuclear war would be unimaginably disastrous, but for many years little attention was paid to the probable ecological consequences of nuclear exchange. Scientists assembled at a two-day conference in the fall of 1983 agreed that of all possible ecological damage, the consequences of nuclear war would be by far the most serious, exceeding immediate injuries and death and producing a horrendous climax to the environmental despoliation of our civilization.

The biological concentration of such isotopes as strontium-90 has been studied since the 1950s, and it is obvious that even a limited nuclear exchange would produce widespread environmental pollution by radioactive fallout. It has also been clear for some time that much of the protective atmospheric ozone layer would be destroyed by nuclear-generated nitrogen oxides. What is new is an appreciation of what the smoke, soot, and pulverized earth produced by nuclear explosions would do to the terrestial climate.

■ The first atomic bomb detonated, at Trinity Site on White Sands Missile Range, July 16, 1945. This began the process that led to the unprecedented power of humans to render all life on earth extinct. (U.S. Army, White Sands Missile Range.)

chickens. In turn, such animals were often slaughtered and eaten—a good example of recycling wastes.

Today's American consumer discards large quantities of paper (newspapers, paper bags, cups, plates, cartons, and other packaging materials) and substantial amounts of edible food scraps, as well as nonreturnable glass and plastic bottles, steel and aluminum cans, and wood and garden refuse, little of which is recycled (Figure 57–16). In addition to these municipal wastes, agricultural activities generate more than 1.8 billion metric tons (a metric ton is about 2205 pounds) of wastes each year, mainly manure, by no means all of it recycled. Mining and industrial wastes add to the problem.

Although some rural communities still dispose of their solid wastes in open dumps, and many coastal cities practice ocean dumping, the two principal means of solid waste disposal today are the sanitary landfill and incineration. The old-fashioned dump, with its swarming flies, roaches, and rats, is a manifest public health hazard. Dumps, if they catch fire, also can be dangerous sources of air pollution, and, once ablaze, their fires can be difficult to extinguish. Finally, the garbage they contain represents a kind of food subsidy supporting huge populations of scavengers such as

Widespread dust has entered the earth's atmosphere before. There is some evidence (and more speculation) that dust stirred up by asteroidal bombardment of the earth may have been responsible for widespread extinctions of life that are observable in the fossil record. Even within historic times large volcanic eruptions have placed enough ash in the atmosphere to produce marked, although temporary, climatic changes. The 1815 eruption of the Indonesian volcano Tambora ejected some 25 cubic miles of debris, much of which did not fall to earth immediately. In 1816 it produced a disastrous year for agriculture, the "year without a summer," in which there were three killing frosts during the New England growing season and widespread hardship throughout the Northern Hemisphere.

It would appear that the dust and soot produced by even a "moderate" nuclear exchange could be expected to almost completely obscure sunlight over the entire Northern Hemisphere and probably the Southern Hemisphere as well. Even in summer, prevailing temperatures would immediately fall below freezing, dropping as low as perhaps −15 to −20° C. The cold would persist, freezing natural bodies of water, in many cases to the bottom. Since the oceans would remain relatively warm for a considerable time, the resulting marked temperature difference between land and sea would produce storms of unparalleled violence. Darkness and cold lasting for months would cause most animals and plants to die, and probably most of them to become extinct, according to the conferees. This would be especially true of tropical life, but even temperate zone species would be decimated, particularly if the exchange occurred in summer.

Although some ecological recovery could be expected within a few years, conventional agriculture would be impossible, not only because of climatic disruption, but also because of the destruction of the industrial-based needs of agriculture such as fertilizer plants, fuel delivery, and availability of agricultural machinery and spare parts. Starving people would hunt down any surviving animals that they could but eventually would themselves starve. Moreover, radiation sickness and other disease states, brought about by widespread chemical pollution resulting from the burning of synthetic material, could be expected to weaken even the survivors. The authors of a report summarizing the conference's conclusions emphasized that survivors, at least in the Northern Hemisphere, would face extreme cold, water shortages, lack of food and fuel, heavy burdens of radiation and pollutants, disease, and psychological stress—all in twilight or in darkness. The authors also predicted that most tropical plants and animals would be rendered extinct, as would most temperate zone vertebrates.

The matter has been restudied since 1983 by several governmental and other agencies; although these studies slightly moderate the original conclusions, on the whole they stand. Even a moderate reduction in the temperature of the earth would almost certainly produce famine and suffering on an unprecedented scale, especially in the Northern Hemisphere. The recovery of industrial civilization might not be quite as unlikely as the original conferees believed, but it would still be by no means assured.

gulls. The gulls and rats associated with dumps can do direct ecological damage by preying on other plants and animals for miles around.

Ocean dumping is, in some ways, worse than that practiced on land. Many of the materials dumped out of sight of the shoreline are synthetic, and many of these, especially plastics, are not degraded rapidly or at all by decomposer organisms. By their persistence in the marine environment they damage many organisms, usually by entangling them or damaging their digestive systems.

The sanitary landfill is a slight improvement over a dump. Its advantage is gained by burying garbage daily. After waste is dumped in a sanitary landfill, it may be further compacted by bulldozers. Each day a layer of soil is pushed over the garbage to discourage flies and rats. Using the sanitary landfill method, abandoned strip mines have been filled and eventually reclaimed, and artificial mountains have been constructed for skiing in the Midwestern plains.

Not all land should be filled, however. Marshes and lakes, for example, are ecologically vital wetlands, yet they often are favorite sites for landfill. Landfills, like dumps, can pollute groundwater as contaminants leak into the ground. This is a particularly acute problem where highly toxic indus-

FIGURE 57–16 Composition of municipal trash in the United States.

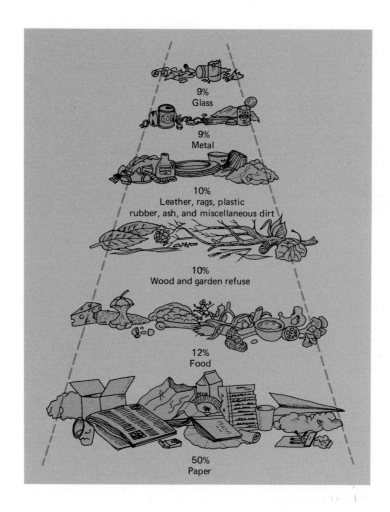

9%
Glass

9%
Metal

10%
Leather, rags, plastic
rubber, ash, and miscellaneous dirt

10%
Wood and garden refuse

12%
Food

50%
Paper

trial wastes have been buried, but even ordinary municipal waste is a significant source of groundwater pollution. Another disadvantage of sanitary landfills is the limited potential uses of filled land. If used as a building site, settling may cause walls and foundations to crack. Methane gas resulting from anaerobic decomposition in the depths of the fill may seep into buildings and constitute an explosion hazard. For these reasons filled land is best used for parks and other recreational purposes.

Many communities burn their garbage rather than dump it. In a modern incinerator the trash is burned in a carefully engineered furnace, but some air pollution still results. The heat produced by the fire may be used to boil water and generate steam that can be sold for industrial use.

RECYCLING

While technology exists for recycling most solid wastes (and this would greatly alleviate problems currently caused by their disposal), less than 10% of consumer goods are recycled. About 20% of the paper used in the United States each year is recycled. If we increased that to 50% (the amount recycled in Japan), we could save about 100 million trees per year! Enough energy would be saved to supply 750,000 homes with electricity. Why don't we practice recycling to a greater extent?

Perhaps the main reason recycling has not become more popular is that in our society recycling is more expensive for the polluter than disposal is. Manufacturers find it cheaper to consume the energy needed to produce goods from virgin materials than to hire the labor necessary to recycle.

An aluminum can is worth about one cent in the United States. For that penny, most people would rather throw the can in the trash than worry about recycling it. However, it takes about 20 times more energy to manufacture a can from the ore than from recycled scrap metal. If *all* of the costs of manufacturing and then disposing of the can were considered, the value of recycling would be more readily appreciated.

Virgin resources have been more plentiful in the United States than in many countries, and we have, in our abundance, felt free to waste them. Tax laws are probably an even greater impediment to recycling. As these laws currently stand, they encourage use of virgin materials over recycling. Legislative incentives that look to the future (before important resources are depleted) are likely to be of the most help in promoting recycling.

ENERGY OPTIONS

Human agriculture and industry are energy-dependent. Our technological civilization would be impossible without a constant large energy input. Thus far that energy has been obtained principally from burning fossil fuels. This depletes the supply of such fuels, increases carbon dioxide in the atmosphere, and may prevent fuel use for important purposes in the future. Of all substitute power generation technologies, solar power (Figure 57–17) may have the greatest potential.

Just as there could be no life on earth without the energy of the sun, there could be no modern society without the energy harnessed by human beings. One may credibly make the case that increases in energy procurement and production, from human and animal muscle power, through wind and water power, to modern electricity generation have produced most of the increase of our population. Yet as population has grown and technology has expanded, it has become increasingly apparent that we have an energy problem. Simply stated, the problem results from too many people consuming too many goods that require energy input for use or for manufacture. The effect is a rapidly advancing depletion of oil, natural gas, and even coal resources.

New technologies are being developed that will enable us to partially replace traditional energy sources with new ones, but at present we still

FIGURE 57–17 "Solar One," a solar electric plant in the Mojave Desert of California. Over 1800 heliostats (the banked mirrors surrounding the tower base) focus the sun's heat on a boiler at the top of the central tower. Thus far, solar power has proved most practical for low-power uses such as water heating and small-scale electricity generation, but in some locales it might prove usable for power production on an industrial scale.

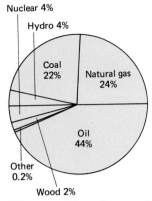

FIGURE 57-18 The mixture of energy sources currently utilized in the United States consists mostly of fossil fuels.

depend on fossil fuels (mainly petroleum) for about 90% of our energy needs (Figure 57–18), including large amounts used in food processing, shipment, and production. These fuels also produce almost 100% of our air pollution and, as we have seen, may give rise to the extremely significant atmospheric greenhouse effect. When they are gone, moreover, they will be unavailable for higher uses such as the production of commercial chemicals.

Two important energy options now being developed are nuclear power and solar power. Thirty years ago nuclear experts hailed the age of nuclear power with the claim that nuclear power would be "too cheap to meter." These words could not have been less true. With changes in the economy and massive cost overruns, nuclear power has become very expensive. Questions about environmental impact, availability of uranium ore, and safety have seriously threatened the nuclear power industry. (Only two nuclear plants planned for construction in the United States since 1978 have not been later canceled, and these two are unlikely ever to be completed.) An average of 12 mishaps per day occurs in nuclear plants, mainly due to equipment problems or failures, and the problem of safe disposal of nuclear wastes has never been solved satisfactorily.

The nuclear power plants currently in use are fission plants. A controlled chain reaction is established in some suitable material, usually a uranium isotope or plutonium. The heat that is produced boils water (or in the future, perhaps some other fluid), producing steam that drives a turbine and generator.

Fusion-type nuclear power depends on the fusion of atoms of light elements such as deuterium, tritium, and lithium to produce the heat. Nuclear fusion may prove less harmful than fission, but the technology is still in its developmental stages, and no one knows when or even if this energy option will become a reality. Meanwhile, it is probable that fission-type nuclear power plants will continue to account for a small percentage (perhaps 5%) of the world energy budget.

Many experts consider solar power a safe and viable alternative to nuclear power. The energy is free and abundant, it causes no air pollution and few environmental hazards, and the technology promises to be less expensive than alternatives. Solar energy can be used in buildings through passive designs maximizing the use of natural sunlight and through the installation of solar collectors, which trap and store heat.

Technology is available for converting sunlight into direct-current electricity; thus far it is expensive, although there has been great recent progress in this area. Photovoltaic cells, tiny cells similar to the silicon semiconductor chips used in calculators, can convert up to 20% of the sunlight striking their surface into electricity. Mass production techniques may greatly reduce the cost of photovoltaic cells in the future, making their household use affordable. Even if (as seems likely) this technology is not adequate for industrial use, it has the potential to greatly reduce the demand for fossil-fueled electric power from residences, leaving more to be used by industry and stretching fossil fuel supplies.

Other solar technologies include the following:

1. *Brine ponds*, which trap heat at high temperatures. That heat then can be converted into electricity or used directly.
2. *Reflector fields*, which are arrays of mirrors focused on an object to be heated. Such high-temperature processes as smelting can sometimes be carried out in this way, giving this method the greatest industrial potential of them all.
3. *Home solar heating*, in which fluid-filled panels or passive techniques that involve careful siting and design take advantage of seasonal changes in insolation.

FIGURE 57–19 Differences in energy consumption between the United States and Europe. At midnight, when most people are asleep, the United States is still brightly lit, as can be seen in this satellite photograph (*a*). (*b*) In contrast, most of Europe is dark, so that much less energy is wasted.

Other energy options include hydropower, wind, tide, geothermal, and ocean thermal power. With the exception of hydropower, these sources are not economically feasible at present and are not expected to contribute more than about 1% of the annual energy budget by the year 2000. Energy experts project that we will still be tightly locked into fossil fuel use for our energy needs as we enter the 21st century. In their 800-page report on the energy situation from 1985 to 2010, the Committee on Nuclear and Alternative Energy Systems concluded that the highest priority of the United States' national energy policy should be directed toward avoiding waste (Figure 57–19) and in general, reducing the growth of energy *demand*, thus minimizing both the need for new generating technologies and the pollution generated by existing ones.

EXTINCTION

By its unparalleled use of the earth's resources, the human species has caused extinction of other species in numbers comparable to the great extinctions of the geological past. Most of these have resulted directly or indirectly from habitat destruction.

Although extinctions have always occurred, human beings have raised the rate at which they occur perhaps a thousand-fold. The species most vulnerable to extinction appear to be large, predatory, and migratory, such as the polar bear. However, extinction also pursues animals with other combinations of characteristics: those that require large tracts of wilderness or solitude, live in very specialized or restricted habitats, compete with human beings in any respect, or yield economically valuable products. Extinctions also result from importing alien species against which an endangered organism has no effective defense. Even environmental pollution has been involved in some extinctions.

(a)

(b)

(c)

FIGURE 57–20 Some endangered species. (*a*) *Eschscholzia ramosa*, (*b*) Himalayan snow leopards, (*c*) golden toad of the Monteverde cloud forest preserve in Costa Rica, (*d*) giant panda.

(d)

While it is difficult to be certain, habitat destruction seems to account for most extinctions and endangerments of species. The extinction of the passenger pigeon, for example, apparently resulted from the destruction of the beech forest where it nested as much as from the market hunting that is usually blamed. Another nearly extinct species, the ivory-billed woodpecker, required large virgin tracts of cypress forest. Today little of that habitat is left, and only a recently discovered relict population of this once-common bird persists in a remote area of Cuba.

It follows that the best hope of preserving any species is preserving its habitat, especially the ecosystem of which it is a part. Preserving endangered species in zoos and similar facilities is commendable as a last resort but is unlikely to succeed for many species. In the first place, confined animals must adapt to an artificial environment, becoming, in effect, semidomesticated. It is by no means certain that they or their descendants would ever be competent under natural conditions if and when they could be reintroduced to their original habitat. Moreover, inbreeding and genetic drift among small confined populations with a limited gene pool may endanger them by genetic disease, as has already been demonstrated in some zoo animals and small wild populations of endangered species.

1404

Extinctions caused by humans (whether from hunting, habitat destruction, or other causes) result from the tendency of our species, like any other species, to multiply to the limits of environmental resistance. The resultant conversion of other life to human biomass, plus our formidable technology-aided competition with them for resources or space, has caused the destruction of many species that are not directly and obviously valuable to us. Much of the destruction has resulted more from thoughtlessness than from real need. But how can we prevent habitat destruction and the resulting extinction of most species with which we come in contact if our population continues to increase? Surely we may yet discover that species now considered "useless" have potential for domestication or may play a vital role in the ecosystem to which we ourselves belong. When that time comes, many of these species, especially tropical rain forest organisms, will almost certainly already be extinct (Figure 57–20).

OVERPOPULATION

The appropriate human population density is one that can be sustained in stable balance with the resources of our habitat. Although limiting factors can be expected to stop human population growth one day, no one can be certain at this time just what they will be. It is likely that this limitation will be accompanied by widespread and intense human misery connected with such events as starvation and warfare. Since it is certain that population growth cannot be sustained indefinitely, humane methods of limiting it are desirable.

Fifty years ago many communities dumped their raw sewage into the nearest river or bay. They could depend on natural bacterial decomposition to break it down with little negative effect upon the ecology of the waterway. As communities grew, however, more people produced more sewage than nature could handle. As water become increasingly polluted, communities had to invest in expensive sewage treatment systems. Technology, in other words, had to take over nature's functions.

More people mean more than just increased sewage, however. More people need more food, clothing, houses, schools, roads, automobiles, television sets, energy, and the material goods our society holds so dear. Each of these can be translated into increased pressure on the environment. For example, the need for more food requires the use of more pesticides and more chemical fertilizers, which results in damage to the soil and increased land and water pollution. Production of chemical fertilizers requires the use of more petroleum, contributing to energy shortages. At the same time, more people need housing, stores, schools, and roads, so more land is taken out of agricultural production, leaving farmers to try to raise more food on less land.

The one third of the world's population that lives in the developed countries consumes 85% of the earth's resources and is responsible for most of the stress placed on the environment. The other two thirds of humanity aspire to consume at our level. It has been estimated that the maximum world population that could be supported at the United States' level of affluence is less than one billion. The environmental impact of enriching even this billion to current United States norms would involve increased industrial pollution, increased erosion of agricultural land, further depletion of natural resources, and much more. Such a goal is totally unrealistic, however, because at the current growth rate world population will reach 8 billion by the year 2100! Almost certainly, most human beings living then will subsist in unprecedented conditions of poverty.

Overpopulation is thus one of the most pressing problems of our time—a problem from which we in the United States are not insulated. Aside from the moral issues involved, the National Security Council has described population increases around the world as a threat to our national security. How much greater a threat it must be to the third-world nations in which population growth is now the greatest—and to their neighbors.

Human Population Growth

It is instructive to view human population growth in the light of the principles put forth in Chapter 55. Beginning perhaps 14,000 years ago up through the Middle Ages, the human species was essentially in its lag phase of population growth. Disease and food shortages served as powerful environmental resistance. But about the time of the Industrial Revolution our population entered the logarithmic phase of its growth and it is now increasing by about 200,000 persons per day. Today, in some nations human population is doubling every 15 years, or even more rapidly.

The demographic events of the past 250 years are shown in Figure 57–21. Notice that among developed countries the birth rate has declined over the long run. The death rate has also declined, perhaps mostly due to the control of infectious disease. Since the birth rate and death rate have mostly marched in step, the population of such countries as Sweden has increased only modestly, and in some, such as West Germany, it even appears to be in a state of slight decline. The birth rate in less developed countries has also fallen dramatically, but modern medicine has brought mortality (still high by the standards of the developed nations) down even faster, so that their population has increased greatly despite the decline in the birth rate.

Current and past population growth affects age distribution in a nation's population. In the graph of the United States population shown in Figure 57–22, notice that there are very few individuals in the uppermost age ranges. Mostly because of the post–World War II baby boom, there is a bulge in the population profile at around the late teens to early thirties.

Sweden, shown in the second graph, has a stable population history. Owing to widespread use of birth control methods and ready availability of excellent medical care for the entire population, about as many Swedes are born each year as die. Thus Sweden has achieved zero population growth. However, the average Swede lives to quite an advanced age, so that the numbers of people in all age ranges are about the same up to about age 70.

Mexico, shown in the third graph, typifies heavy population growth in the Third World. In such a nation children have always been perceived as an economic asset—extra hands to work subsistence farms, and eventually, providers of care to aged parents. Until recently, many children died early from infectious disease, providing powerful economic motivation for widespread resistance to population control.

But here there is a conflict of economic interests between individuals and their society. In all three graphs we have stippled the age ranges in

FIGURE 57–21 Birth and death rates in the developed and underdeveloped world for the past 250 years.

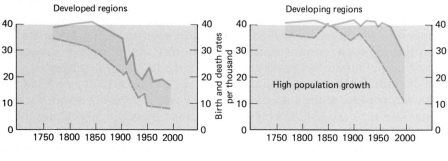

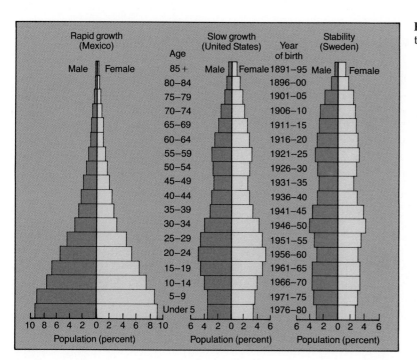

FIGURE 57–22 Population age distribution in three countries.

which people consume resources without (on the whole) making substantial economic contributions. A demand for schooling, pediatric care, nursery care, and much else is generated by the children of Mexico. Yet the resources of Mexican society are insufficient to meet this demand. Moreover, the growth rate of the Mexican population results in the doubling of the need for all of society's services every 15 years. Even an affluent country like Sweden or the United States could not readily double all its roads, schools, hospitals, sewage disposal plants, fire departments, and apartment buildings every 15 years. And if it did, think of the ecological consequences. However, the developing countries cannot even run fast enough to stay where they are, although understandably, they aspire to the same standards of affluence that the developed nations enjoy.

What has produced the demographic differences between the developed and less-developed nations? Evidently, the citizens of the developed nations ceased some time ago to regard children as an economic asset. The slowing of population growth that resulted is called the **demographic transition.** In the United States, for example, it costs in excess of $60,000 to raise a child to economic independence (not counting the cost of a college education), almost none of which is returned to the parent. Upwardly mobile people therefore tend to limit their family size. Most developed countries went through a difficult time of social transition in the 19th and early 20th centuries, when they accumulated the capital necessary to establish the industrial economies in which children are now no longer an economic asset. Not so the less-developed nations. With capital consumed as fast as it is generated by the urgent needs of their populations, they are not likely to achieve industrial economies or undergo the kind of demographic transition that has placed the developed countries in their present position.

When Is a Nation Overpopulated?

One may readily obtain agreement in the United States that other countries, especially less developed nations, are overpopulated. A glance at Figure 57–23 should help to dispel this misunderstanding. Parts of Western Europe and North America are as densely populated as any part of India or China! As suburban sprawl stretches out, merging one city into

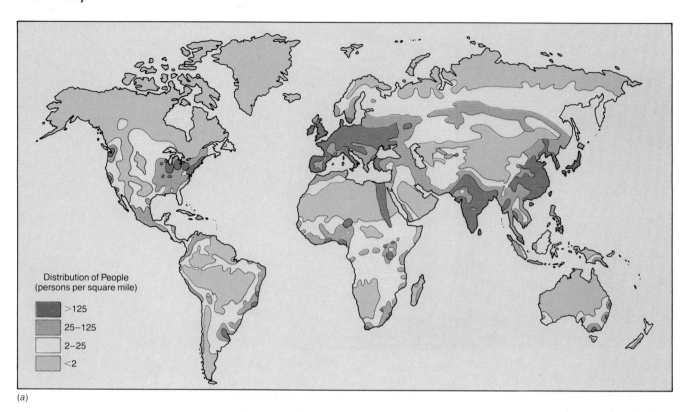

Distribution of People
(persons per square mile)

- >125
- 25–125
- 2–25
- <2

(a)

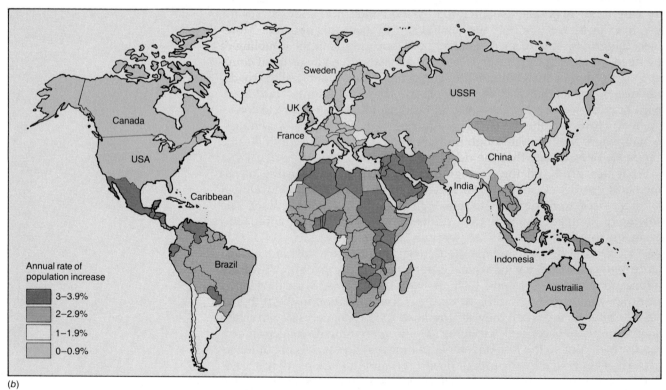

Annual rate of
population increase

- 3–3.9%
- 2–2.9%
- 1–1.9%
- 0–0.9%

(b)

FIGURE 57–23 (*a*) Although a great deal of the world seems sparsely populated, heavy population densities occur in East Asia, South Asia, Europe, and parts of North America. (*b*) Population *growth* rates, however, are low in most developed nations.

another, as farms are sold and the cost of food rises, as resources dwindle and we experience fuel and other shortages, as water pollution and air pollution produce increasing blight, many individuals no longer need to be convinced that overpopulation is a reality, a present reality, in developed nations too. Only their relatively high level of affluence and ability to import needed goods and materials maintains the high standard of living that tends to obscure their overcrowding.

In the long run the ideal population for a nation, continent, or planet is one that can be stably sustained. It is clear that our resources will not

FIGURE 57–24 Overpopulation produces pressures that extend even into remote areas. As Brazilian population expands into the Amazon wilderness, greatly aided by the trans-Amazon highway shown here, the greatest of tropical rain forest ecosystems will be lost not only to Brazil but to the rest of humanity as well.

sustain the population of even lightly peopled countries at their present population densities indefinitely. We have reached these unprecedented population densities by means of a temporary expansion of the carrying capacity of the habitat by technological means.

Thus far these expansions have depended on the consumption of nonrenewable resources and our passing some of the costs to the ecosystem. Since nonrenewable resources will dwindle and finally cease, and since the ability of the ecosystem to absorb pollution is limited, expansion cannot continue indefinitely. Populations dependent on present-day mechanical technology will eventually be drastically reduced. By that time more resources will have been consumed and the resulting population will be not only fewer in number, but materially poorer, living in a biotically impoverished world as well.

However, the human population now is growing (on the whole) without seeming resource limitation because we have discovered technological means to partially counteract infectious disease, parasitism, and predation as limiting factors for our population and have also found ways to greatly increase the resources potentially available for the production of human biotonnage. These means include, for example, high-energy-input agriculture. Until now these newly available resources have been great in comparison to our natural rate of increase. We have been less successful, however, in dealing with the waste products of our industrial civilization and the potential depletion of the energy subsidy that makes human population growth possible. So far, these potential limiting factors have not taken effect to a great extent. As we have already seen, a population that has stopped growing is not necessarily a stable one, let alone a happy one. How happy might our population be if its growth were to be curtailed by malnutrition, pollution, lack of shelter, or some other factor? Perhaps it would be wise to begin to think of humane alternatives to these before less pleasant options are forced upon us.

THE FUTURE

Ideally, population control will be attained through planning, not through violence or disease and their attendant suffering. However, although many environmental problems are rooted in human overpopulation, their severity is linked to our choice of lifestyle. Many ecologically harmful practices that are both wasteful and irrational could be altered. What seems to have been lacking is the will to do so.

As living organisms, we share much in common with the fate of other life forms on the planet. We differ from other organisms, however, in our capacity to reflect on and probe our biological identities and to reflect on the meaning of life. This talent is the key to any hope of ensuring our, and the ecosphere's, continued survival.

■ SUMMARY

I. Modern agriculture requires substantial energy and technologic input and has great environmental impact.
   A. In agriculture, people replace diverse natural communities with those consisting of one or two specially bred species.
      1. These artificial communities are unstable, simple, and usually lacking in some important ecological mineral cycles.
      2. Some wildlife can coexist with agricultural communities.
   B. Monoculture of agricultural species promotes attack by pests and disease. This is typically combatted by using pesticides.
   C. Silviculture resembles agriculture in that single-species tree farms are cultivated, rather than forest communities.
   D. Until genetic engineering techniques are more fully developed, we are in danger of a potential shortage of new genetic varieties of crop plants.
II. Most pesticides are broad-spectrum poisons that kill nonpest species as well as pests. Many pest populations become resistant to pesticides, whereas predator populations are often eradicated.
   A. Persistent pesticides are not biodegradable. They become widely distributed and then concentrated in predators by biological magnification.
   B. Alternatives to pesticide use exist and should be employed where possible.
III. Pollution exists when wastes degrade the environment's suitability for any of the organisms that would normally inhabit it, or when damage to property, aesthetic values, or public health results.
IV. Severe aquatic pollution by organic substances produces a characteristic zonation. Eventually, much of the pollution is rendered harmless by natural processes, but only after a portion of the habitat has been degraded. Most methods of organic pollution control compress natural processes of decomposition and decay into the confines of a sewage disposal plant.
V. The decomposition of sewage in natural waters can be imitated within a pollution control plant, producing a product (effluent) of low organic content, which may, however, require still further treatment to remove excess mineral nutrients.
VI. With every breath we inhale various pollutants deposited in the air and elsewhere by motor vehicles, industries, power plants, and other sources.
   A. Trees and other plants are damaged by air pollution. Widespread acid rain, caused by air pollution, menaces both terrestrial and aquatic habitats.
   B. An inversion layer may develop when a layer of warm air acts as a lid, sealing pollutants beneath so that they cannot be dispersed.
VII. Radioactive materials cannot easily be disposed of, and they persist and are passed extensively through food chains. At present most such radioactive pollution has resulted from atmospheric tests of nuclear weapons, and a bit less from nuclear power plant accidents. The ecological effects of nuclear war would be catastrophic.
VIII. Sanitary landfills and incineration are two main ways of solid waste disposal.
IX. Technology is available for recycling most solid wastes, yet less than 10% of consumer goods are recycled.
X. The nuclear power industry has been threatened by serious questions regarding its safety and environmental impact. Solar power is considered a safe alternative for the future but one whose industrial potential may be limited.
XI. Species may become endangered or extinct by a variety of processes—overexploitation, persecution, environmental pollution, or the introduction of exotic species—but habitat destruction is the most important.
XII. Increases in population translate into increased needs for food, shelter, clothing, material goods, and educational, medical, and social services. These increases result in additional environmental stress.
   A. We have temporarily extended the carrying capacity of the Earth for human beings by our technology. This expansion depends on consumption of nonrenewable resources and the passage of some of the costs to the environment in the form of pollution.
   B. In the author's view, a nation may be considered overpopulated if its total population is too great to be sustained permanently by its resources. By this criterion, virtually all areas of the modern world are overpopulated at the present time.

■ POST-TEST

1. Agricultural communities are generally simpler/more complex than natural communities.
2. Predators do not develop resistance to pesticides as quickly as pests do because they are often _____ and _____ at a slower rate.

3. Pesticides that are _____ decompose a few weeks or months after they have been applied.

4. DDT and other chlorinated hydrocarbon biocides are soluble/insoluble in water and very soluble/insoluble in body fat.

5. Owing to _____ _____, persistent pesticides, such as DDT, often tend to accumulate in the _____ levels of trophic pyramids.

6. _____ is a reduction in the quality of the environment by the addition of materials not normally found there or harmful quantities of otherwise harmless substances.

7. In the area of most recent pollution of a stream, the organic content is high/low, but the dissolved _____ level is still high.

8. In an atmospheric inversion a layer of _____ air forms a lid above one of _____ polluted air.

9. Chlorinated hydrocarbon pollution of the atmosphere is likely to cause a substantial reduction in the _____ _____ of the atmosphere that normally protects the earth's surface from excessive _____ _____.

10. In human beings, radioactive strontium becomes incorporated into _____ and _____.

11. The bulk of extinctions now occurring appear to be due to _____ _____.

12. In a rapidly expanding human population, most of the people are _____.

■ REVIEW QUESTIONS

1. What are some of the reasons that agricultural communities are unstable compared with natural communities?

2. Why are pest populations more likely to become pesticide-resistant than predator populations?

3. Give some economic and social reasons for pollution.

4. What disadvantages do you see to the use of natural enemies of pest organisms in place of pesticides?

5. What happens to the communities within aquatic ecosystems in extreme instances of pollution by organic wastes?

6. What is an inversion layer? The greenhouse effect?

7. Why are many persistent pesticides and radioactive substances subject to biological magnification? Explain this process.

8. What kinds of animals are most prone to extinction by human agency? Is there reason to be concerned about this, or is it something we can live with?

9. Which contributes more, in your opinion, to our ecological crisis: overpopulation or wasteful consumption? Show how each contributes to the major ecological problems outlined in this chapter.

10. When fossil fuels are exhausted, which means of power generation do you think will be most widely employed? What might their drawbacks be? The effect on our standard of living?

■ RECOMMENDED READINGS

Altieri, M., D. Letourneau, and J.R. Davis. Developing sustainable agroecosystems. *Bioscience*, 33(1):45–49, January, 1983. Energy-intensive agricultural ecosystems are not practical in the long run and must be replaced with alternative, indefinitely stable technologies.

Anonymous. *The International Book of the Forest*. New York, Simon and Schuster, 1981. Although tainted by the viewpoint of the commercial forest industries and to some extent an apologia for them, if its bias is kept in mind, this book is an excellent source of information not readily available elsewhere, not only about forests themselves but about their economic importance.

Bastian, R.K., and J. Benforado. Waste treatment: Doing what comes naturally. *Technology Review*, 86(2):58–66, February-March, 1983. Mother Nature's way of waste treatment is best, especially if one can make it happen in a sewage disposal facility.

Beck, M., et al. The bitter politics of acid rain. *Newsweek*, April 25, 1983, pp. 36–37. A neat, brief summation of the political and economic reasons for acid rain, and be it said, much else besides.

Botkin, D.B., and J.A. Wiens. Mono Lake: Solving an environmental dilemma. *The World & I*, May, 1988. The conflict between the need for water generated by the growth of a California metropolis and the need to preserve a unique ecosystem.

Eckholm, E. Human wants and misused lands. *Natural History* 91(6):33–48, June, 1982. How does one reconcile human needs for land with the land's needs?

Ehrlich, P. and A. Ehrlich. *Extinction*. New York, Random House, 1981. A powerful and compellingly reasoned compilation of arguments for the preservation of species. These authors have also written many other excellent titles in the environmental sciences, such as *The End of Affluence* and *The Population Bomb*.

Forman, R.T.T., and M. Godron. *Landscape Ecology*. New York, John Wiley and Sons, 1986. The integration of communities of organisms with human land use, with an extensive discussion of responsible ecological management.

Graham, F. Jr. *Since Silent Spring*. Boston, Houghton Mifflin, 1970. Rather than refer you to the original book, *Silent Spring*, we thought it better to list a somewhat updated version.

Gwatkin, D.R., and S.K. Brandel. Life expectancy and population growth in the third world. *Scientific American* 246(5):57–65, May, 1982. A decrease in death rate is the basic motor of population growth in underdeveloped countries.

Harwell, P., *Nuclear Winter: The Human and Environmental Consequences of Nuclear War*. New York, Springer Verlag, 1988. An authoritative analysis of the consequences of nuclear war for humans and the environment.

Heinrich, B. *Bumblebee Economics*. Cambridge, Mass., Harvard University Press, 1979. The life of bumblebee society considered from the standpoint of energy economy. Human societies also have energy-dependent economies.

Kinkead, E. Tennessee small fry. *The New Yorker*, January 8, 1979, pp. 52–55. Portrait of an endangered species—the snail darter fish.

Lockeretz, W. The lessons of the dust bowl. *American Scientist*, 66:560–569, September-October, 1978. The principal lesson in this paper appears to be that we have not learned our lesson about "the most severe man-made environmental problem the United States has ever seen."

Moran, J.M., M.D. Morgan, and J.H. Wiersma. *Introduction to Environmental Science*. San Francisco, Freeman, 1980. A fine introductory textbook written as if the reader had some intelligence.

Radcliffe, B., and L.P. Gerlach. The ecology movement after ten years. *Natural History*, January, 1981, pp. 12–18. Among educated, ecologically aware persons, environmental issues are still important, as to a lesser extent they are among the general population.

Revelle, R. Carbon dioxide and world climate. *Scientific American*, 247(2):35–43, August, 1982. As a result of fossil fuel use and the clearing and burning of forests, the global atmospheric content of carbon dioxide is increasing. What will this do to us?

Stommel, H., and E. Stommel. The year without a summer. *Scientific American*, June, 1979, pp. 176–186. A natural environmental mishap, produced by the explosion of the volcano Krakatoa, illustrates the fragility of the heat balance of the earth's atmosphere.

Turk, J. *Introduction to Environmental Studies*. Philadelphia, W.B. Saunders, 1980. A fine, truly college-level general textbook, notable for presenting more than one side to most questions.

Ward, G.M., T.M. Sutherland, and J.M. Sutherland. Animals as an energy source in third world agriculture. *Science*, 208:570–575, 9 May, 1980. A model of clear thinking and careful discussion whose implications reach far beyond the issues immediately addressed by the authors. Appropriate technology is often traditional technology. But even traditional technology could be improved.

Westoff, C.F. Marriage and fertility in the developed countries. *Scientific American*, 239(6):51–55, December, 1978. What are the consequences of population growth changes in the developed countries, which may actually begin to decline before the year 2015?

UNDERSTANDING BIOLOGICAL TERMS

Your task of mastering new terms will be greatly simplified if you learn to dissect each new word. Many terms can be divided into a prefix, the part of the word that precedes the main root, the word root itself, and often a suffix, a word ending that may add to or modify the meaning of the root. As you progress in your study of biology, you will learn to recognize the more common prefixes, word roots, and suffixes. Such recognition will help you analyze new terms so that you can more readily determine their meaning, and will also help you remember them.

Prefixes

a-, ab- from, away, apart (abduct, lead away, move away from the midline of the body)

a-, an- un-, -less, lack, not (asymmetrical, not symmetrical)

ad- (also **af-, ag-, an-, ap-**) to, toward (adduct, move toward the midline of the body)

allo- different (allometric growth, different rates of growth for different parts of the body during development)

ambi- both sides (ambidextrous, able to use either hand)

andro- male (androecium, the male portion of a flower)

aniso- unequal (anisogamy, sexual reproduction in which the gametes are different in size)

ante- forward, before (anteflexion, bending forward)

anti- against (antibodies, proteins that have the capacity to react against foreign substances in the body)

auto- self (autotrophic, organisms that manufacture their own food)

bi- two (biennial, a plant that takes two years to complete its life cycle)

bio- life (biology, the study of life)

circum-, circ- around (circumcision, a cutting around)

co-, con- with, together (congenital, existing with or before birth)

contra- against (contraception, against conception)

cyt- cell (cytology, the study of cells)

di- two (disaccharide, a compound made of two sugar molecules chemically combined)

dis- apart (dissect, cut apart)

ecto- outside (ectoplasm, outer layer of cytoplasm)

end-, endo- within, inner (endoplasmic reticulum, a network of membranes found within the cytoplasm)

epi- on, upon (epidermis, upon the dermis)

ex-, e-, ef- out from, out of (extension, a straightening out)

extra- outside, beyond (extraembryonic membrane, a membrane, such as the amnion, that protects the embryo)

gravi- heavy (gravitropism, growth of a plant in response to gravity)

hemi- half (cerebral hemisphere, lateral half of the cerebrum)

hetero- other, different (heterozygous, unlike members of a gene pair)

homo-, hom- same (homologous, corresponding in structure; homozygous, identical members of a gene pair)

hyper- excessive, above normal (hypersecretion, excessive secretion)

hypo-, under, below, deficient (hypotonic, a solution whose osmotic pressure is less that that of an isotonic solution)

in-, im- not (incomplete flower, a flower that does not have one or more of the four main parts)

inter- between, among (interstitial, situated between parts)

intra- within (intracellular, within the cell)

iso- equal, like (isotonic, equal strength)

macro- large (macronucleus, a large, polyploid nucleus found in ciliates)

mal- bad, abnormal (malnutrition, poor nutrition)

mega- large, great (megakaryocyte, giant cell of bone marrow)

meso- middle (mesoderm, middle tissue layer of the animal embryo)

meta- after, beyond (metaphase, the stage of mitosis after prophase)

micro- small (microscope, instrument for viewing small objects)

mono- one (monocot, a group of flowering plants with one cotyledon, or seed leaf, in the seed)

oligo- small, few, scant (oligotrophic lake, a lake deficient in nutrients and organisms)

oo- egg (oocyte, developing egg cell)

paedo- child (paedomorphosis, the preservation of a juvenile characteristic in an adult)

para- near, beside, beyond (paracentral, near the center)

peri- around (pericardial membrane, membrane that surrounds the heart)

photo- light (phototropism, growth of a plant in response to the direction of light)

poly- many, much, multiple, complex (polysaccharide, a carbohydrate composed of many simple sugars)

post- after, behind (postnatal, after birth)

pre- before (prenatal, before birth)

pseudo- false (pseudopod, a temporary protrusion of a cell, i.e., "false foot")

retro- backward (retroperitoneal, located behind the peritoneum)

semi- half (semilunar, half-moon)

sub- under (subcutaneous tissue, tissue immediately under the skin)

super-, supra- above (suprarenal, above the kidney)

sym- with, together (sympatric speciation, evolution of a new species within the same geographical region as the parent species)

syn- with, together (syndrome, a group of symptoms which occur together and characterize a disease)

trans- across, beyond (transport, carry across)

Suffixes

-able, -ible able (viable, able to live)

-ad used in anatomy to form adverbs of direction (cephalad, toward the head)

-asis, -asia, -esis condition or state of (hemostasis, stopping of bleeding)

-cide kill, destroy (biocide, substance that kills living things)

-emia condition of blood (anemia, without enough blood)

-gen something produced or generated, or something that produces or generates (pathogen, something that can cause disease)

-gram record, write (electrocardiogram, a record of the electrical activity of the heart)

-graph record, write (electrocardiograph, an instrument for recording the electrical activity of the heart)

-ic of or pertaining to (ophthalmic, of or pertaining to the eye)

-itis inflammation of (appendicitis, inflammation of the appendix)

-logy study or science of (cytology, study of cells)

-oid like, in the form of (thyroid, in the form of a shield)

-oma tumor (carcinoma, a malignant tumor)

-osis indicates disease (psychosis, a mental disease)

-pathy disease (dermopathy, disease of the skin)

-phyll leaf (mesophyll, the middle tissue of the leaf)

-scope instrument for viewing or observing (microscope, instrument for viewing small objects)

Some Common Word Roots

abscis cut off (abscission, the falling off of leaves or other plant parts)

angi, angio vessel (angiosperm, plants that produce seeds enclosed within a fruit or "vessel")

apic tip, apex (apical meristem, area of cell division located at the tips of plant stems and roots)

arthr joint (arthropods, invertebrate animals with jointed legs and segmented bodies)

aux grow, enlarge (auxin, a plant hormone involved in growth and development)

blast a formative cell, germ layer (osteoblast, cell that gives rise to bone cells)

brachi arm (brachial artery, blood vessel that supplies the arm)

bry grow, swell (embryo, an organism in the early stages of development)

cardi heart (cardiac, pertaining to the heart)

carot carrot (carotene, a yellow, orange, or red pigment in plants)

cephal head (cephalad, toward the head)

cerebr brain (cerebral, pertaining to the brain)

cervic, cervix neck (cervical, pertaining to the neck)

chlor green (chlorophyll, a green pigment found in plants)

chondr cartilage (chondrocyte, a cartilage cell)

chrom color (chromosome, deeply staining body in nucleus)

cili small hair (cilium, a short, fine cytoplasmic hair projecting from the surface of a cell)

coleo a sheath (coleoptile, a protective sheath that encircles the stem in grass seeds)

conjug joined together (conjugation, a sexual phenomenon in certain protists)

cran skull (cranial, pertaining to the skull)

decid falling off (deciduous, a plant that sheds its leaves at the end of the growing season)

dehis split (dehiscent fruit, a fruit that splits open at maturity)

derm skin (dermatology, study of the skin)

ecol dwelling, house (ecology, the study of organisms in relation to their environment, i.e., "their house")

enter intestine (enterobacteria, bacteria that inhabit humans, particularly in the digestive tract)

evol to unroll (evolution, descent of complex organisms from simpler ancestors)

fil a thread (filament, the thin stalk of the stamen in flowers)

gamet a wife or husband (gametangium, the part of a plant or protist that produces reproductive cells)

gastr stomach (gastrointestinal tract, the digestive tract)

gen generate, produce (gene, a hereditary factor)

glyc, glyco sweet, sugar (glycogen, storage form of glucose)

gon seed (gonad, an organ that produces gametes)

gutt a drop (guttation, loss of water as liquid drops from plants)

gymn naked (gymnosperm, a plant that produces seeds not enclosed within a fruit, i.e., "naked")

hem blood (hemoglobin, the pigment of red blood cells)

hepat liver (hepatic, of or pertaining to the liver)

hist tissue (histology, study of tissues)

hom, homo same, unchanging, steady (homeostasis, reaching a steady state)

hydr water (hydrolysis, a breakdown reaction involving water)

leuk white (leukocyte, white blood cell)

menin membrane (meninges, the three membranes that envelop the brain and spinal cord)

morph form (morphogenesis, development of body form)

my, myo muscle (myocardium, muscle layer of the heart)

myc a fungus (mycelium, the vegetative body of fungi)

nephr kidney (nephron, microscopic unit of the kidney)

neur, nerv nerve (neuromuscular, involving both the nerves and muscles)

occiput back part of the head (occipital, back region of the head)

ost bone (osteology, study of bones)

path disease (pathologist, one who studies disease processes)

ped, pod foot (biped, organism with two feet)

pell skin (pellicle, a flexible covering over the body of certain protists)

phag eat (phagocytosis, process by which certain cells ingest particles and foreign matter)

phil love (hydrophilic, a substance that attracts water)

phloe bark of a tree (phloem, food-conducting tissue in plants which corresponds to bark in woody plants)

phyt plant (xerophyte, a plant adapted to xeric, or dry, conditions)

plankt wandering (plankton, microscopic aquatic protists that float or drift passively)

rhiz root (rhizome, a horizontal, underground stem that superficially resembles a root)

scler hard (sclerenchyma, cells that provide strength and support in the plant body)

sipho a tube (siphonous, a tubular type of body form found in certain algae)

som body (chromosome, deeply staining body in the nucleus)

sor heap (sorus, a cluster or "heap" of sporangia in ferns)

spor seed (spore, a reproductive cell that gives rise to individual offspring in plants and protists)

stom mouth, opening (stomate, a small pore in the epidermis of plants)

thigm touch (thigmotropism, plant growth in response to touch)

thromb clot (thrombus, a clot within the body)

tropi turn (thigmotropism, growth of a plant in response to contact with a solid object, as when a tendril "turns" or wraps around a wire fence)

visc pertaining to an internal organ or body cavity (viscera, internal organs)

xanth yellow (xanthophyll, a yellowish pigment found in plants)

xyl wood (xylem, water-conducting tissue in plants, the "wood" of woody plants)

zoo animal (zoology, the science of animals)

Appendix B:

ABBREVIATIONS

The biological sciences employ a great many abbreviations, and with good reason. Many technical terms in biology and biological chemistry are both long and difficult to pronounce. Yet it can be difficult for beginners, when confronted with something like NADPH or COPD, to understand the reference. Here are some of the common abbreviations employed in biology for your ready reference.

ACTH The pituitary hormone, AdrenoCorticoTropic Hormone, that governs the secretion of adrenal cortical hormones by the adrenal cortex.

ADH The pituitary hormone, Anti-Diuretic Hormone, also called vasopressin, that governs the reabsorption of water by the kidney, and thus the amount of urine that is produced. ADH is anti-diuretic: The more ADH present, the less urine produced.

ADP Adenosine Di Phosphate. A form of the almost universal energy transfer substance of the cell, ATP, in its "discharged" state. Even ADP, however, contains some transferable energy which is employed in certain biochemical reactions.

AIDS Acquired Immune Deficiency Syndrome. Caused by a virus, AIDS is an "acquired" condition, distinguishable from hereditary defects of the immune system with which the sufferer may have been born.

AMP Adenosine Mono Phosphate. The basic "stem" of the energy transfer substance ATP; AMP contains no readily transferable energy.

ATP Adenosine Tri Phosphate. The almost universal energy transfer substance of the cell. Typically, ATP donates its terminal phosphate group to some substance, along with much of the energy formerly contained in the bond that held it to the remainder of the ATP molecule. ADP is left over, and in due course, is recycled.

ATPase An enzyme or protein that breaks down ATP, usually producing ADP. Most ATPases perform some function such as active transport or contraction which utilizes the energy that is discharged in the breakdown of ATP. One important class of ATPase, the F_o-F_1 and CF_o-CF_1 complexes of the mitochondria and chloroplasts, operate in reverse. These actually *produce* ATP, using the energy of a proton gradient to add the terminal phosphate to the ATP molecule.

AV node Atrio-Ventricular node. A mass of tissue located near the junction of the heart atria and ventricles that is responsible for the coordination of the contraction of the ventricles with that of the atria.

B lymphocyte The lymphocyte responsible for the production of antibody-mediated immunity. Originally named for the Bursa of Fabricius, a structure in which these cells mature in birds.

BMR Basal Metabolic Rate. The rate of energy production, indirectly measured by the rate of consumption. Called "basal" because it is the basic metabolic cost of living apart from any special exertion.

cAMP Cyclic Adenosine MonoPhosphate. Most commonly, A "second messenger" substance produced within the cell in response to a hormone. It is made from ATP.

CNS Central Nervous System.

COPD Chronic Obstructive Pulmonary Disease.

CP Creatine Phosphate. An important energy storage substance in vertebrate muscle.

DNA DeoxyriboNucleic Acid. The fundamental substance in which hereditary information is stored.

EM Electron Microscope.

ER Endoplasmic Reticulum.

FAD (also FADH, etc.) Flavine Adenine Dinucleotide. A hydrogen acceptor and donor involved in cellular metabolism and in photosynthesis.

FSH Follicle Stimulating Hormone. A hormone secreted by the anterior pituitary that was originally discovered in female mammals, though it also occurs in males. Stimulates the growth of the ovarian follicles and the semeniferous tubules.

GABA Gamma Amino Butyric Acid. A neurotransmitter that acts as an inhibitor in the brain and spinal cord.

GH Growth Hormone, otherwise known as somatotropin.

GTP Guanosine TriPhosphate. An energy-rich compound sometimes used by cells in place of ATP.

Hb Hemoglobin.

HCG Human Chorionic Gonadotropin. A peptide hormone produced by the placenta that helps to maintain pregnancy. Similar hormones occur in other mammals, but they are known simply as CG.

HDL High-Density Lipoprotein. Blood particle of relatively high density that transports fats.

HLA Human Leukocyte Antigen. A major complex of antigens found on human body cells that governs graft compatibility, which is governed by specific genes, the HLA genes.

Ig Immunoglobin, as in IgA, IgG, IgM, etc.

LDH Lactic DeHydrogenase enzyme.

LH Luteinizing Hormone. Although this anterior pituitary hormone also occurs and functions in males, it was discovered in females and named for its action in them, namely, to convert an ovarian follicle into a progesterone-secreting corpus Luteum. In males it promotes growth and function of the cells that secrete testosterone.

MAO Monoamine Oxidase. An enzyme that destroys epinephrine-like neurotransmitter substances.

MHC Major Histocompatibility Complex. Group of genes governing the tissue antigens, which in turn determine graft compatibility in humans.

mRNA Messenger RiboNucleic Acid. A form of RNA (see below) that carries the genetic message from DNA to the cytoplasm, where it is translated into peptides or proteins.

MSH Melanocyte Stimulating Hormone. An anterior pituitary hormone that darkens the skin by stepping up the activity of melanocytes.

NAD$^+$ (also NADH + H$^+$) Nicotine Adenine Dinucleotide. A coenzyme widely used by cells as a hydrogen acceptor or donor in a variety of metabolic reactions. NAD$^+$ is generally involved in catabolic reactions, including respiration.

NADP$^+$ (also **NADPH + H$^+$**) Nicotine Adenine Dinucleotide Phosphate. A coenzyme widely used by cells as a hydrogen acceptor or donor in a variety of metabolic reactions. NADP$^+$ is generally involved in anabolic reactions, including photosynthesis.

PNS Peripheral Nervous System.

RAS Reticular Activating System. The system of the brain important in maintaining a conscious state.

RBC Red Blood Cell, erythrocyte.

RNA RiboNucleic Acid. Nucleic acid responsible for the expression of genetic information. There are three known varieties: mRNA, tRNA, and rRNA.

rRNA Ribosomal RiboNucleic Acid. RNA that forms part of the structure of the ribosome.

SA node Sino-Atrial Node of the heart; responsible for the initiation and timing of the heartbeat.

SEM Scanning Electron Microscope.

STD Sexually Transmitted Disease.

T lymphocyte Lymphocyte with a wide variety of functions, primarily concerned with cell-mediated immunity. Named "T" for the fact that these cells mature in the thymus gland.

TEM Transmission Electron Microscope

tRNA Transfer RiboNucleic Acid. A variety of RNA responsible for carrying (transferring) amino acids to the ribosomal protein construction site.

WBC White Blood Cell, leukocyte.

Post-test Answers

CHAPTER 1

1. metabolism   2. homeostasis   3. stimulus   4. locomotion   5. asexual reproduction   6. respond
7. c   8. f   9. b   10. a   11. g   12. h   13. d
14. e   15. plants, algae (some bacteria); animals; fungi, bacteria   16. prokaryotes   17. cyanobacteria, bacteria
18. Protista   19. hypothesis   20. theory

CHAPTER 2

1. carbon, oxygen, hydrogen, nitrogen, phosphorus, calcium   2. C; H; O   3. trace elements   4. electrons, protons, neutrons   5. electrons   6. atomic number
7. mass number   8. isotopes   9. orbitals   10. two
11. positive   12. helium, neon; noble gases
13. chemical bond   14. ions   15. cations; anions
16. f   17. g   18. e   19. h   20. b   21. d
22. c   23. a   24. capillary action   25. specific heat
26. pH   27. donor; acceptor   28. buffer

CHAPTER 3

1. c   2. d   3. b   4. e   5. e   6. d   7. a
8. f   9. b   10. c   11. amino acids   12. amino acid sequence in its polypeptide chains   13. glucose, fructose   14. Cellulose   15. hydrophobic   16. carboxyl, amino   17. glycogen

CHAPTER 4

1. resolving power   2. ribosomes, rough endoplasmic reticulum   3. DNA, nucleoid chromosomes, chromatin
4. rough endoplasmic reticulum, transport vesicles, Golgi complex, carbohydrates, lipids   5. lysosomes
6. peroxisomes   7. cristae, ATP   8. chloroplasts, three   9. thylakoid, sunlight, carbohydrate (or glucose)
10. microtubules   11. microtubules, microfilaments, intermediate filaments   12. Cilia, flagella, microtubules, two, nine   13. cell wall, cellulose   14. Vacuoles, lysosomes   15. nucleolus, nucleus   16. glycoproteins   17. see figure in text   18. a   19. p
20. k   21. g   22. h   23. e   24. f   25. n
26. o   27. m   28. j   29. i   30. q   31. d
32. c   33. b   34. 1

CHAPTER 5

1. phospholipids (or lipids)   2. hydrophobic, hydrophilic   3. amphipathic, hydrophobic, hydrophilic
4. Integral, hydrophobic, peripheral   5. diffusion

6. isotonic   7. hypertonic   8. hypotonic   9. facilitated diffusion   10. active transport (carrier-mediated)
11. active transport, sodium, potassium   12. cotransport   13. phagocytosis   14. receptor mediated endocytosis   15. Desmosomes   16. Gap junctions
17. Tight junctions   18. Plasmodesmata

CHAPTER 6

1. energy   2. kinetic energy   3. thermodynamics
4. first law of thermodynamics   5. exothermic reaction
6. endergonic reaction   7. entropy   8. second law of thermodynamics   9. free energy; temperature, pressure   10. negative   11. exergonic   12. coupled
13. adenine, ribose   14. activation energy   15. catalyst   16. enzymes   17. enzyme-substrate complex
18. active site   19. noncompetitive

CHAPTER 7

1. catabolism   2. anaerobes   3. reduction   4. glycolysis   5. cytoplasm   6. lactic acid   7. ethyl alcohol   8. 2 ATPs; 36-38   9. fermentation   10. oxidized   11. carbon dioxide   12. carbon dioxide, water
13. citrate (citric acid)   14. NAD   15. decarboxylation
16. ATP   17. ATP synthesis (phosphorylation of ADP)
18. anabolism.

CHAPTER 8

1. autotrophs   2. thylakoid; chloroplasts   3. carbon dioxide, water; glucose (carbohydrate); oxygen
4. water   5. photons   6. action spectrum   7. energized   8. water   9. carotenoids   10. reaction; electron; electron   11. light (photons); ADP; ATP
12. photosystem I   13. noncyclic   14. water
15. hydrogens   16. energy; ATP   17. ATP, NADPH, carbon dioxide   18. carbon fixation   19. Two
20. C_4 21. bundle sheath cells; carbon dioxide
22. oxygen

CHAPTER 9

1. heredity   2. DNA, protein, RNA   3. cell cycle
4. S phase   5. prophase, metaphase, anaphase, telophase   6. sister chromatids   7. centromere
8. metaphase   9. cytokinesis   10. colchicine
11. asexual   12. clone   13. homologous   14. diploid, polyploid   15. haploid   16. synapsis, tetrad
17. crossing over   18. 23   19. plants

CHAPTER 10

1. locus   2. alleles   3. monohybrid   4. genotype
5. phenotype   6. dominant   7. recessive   8. homozygous   9. first filial, F_1 10. product   11. sum
12. 1, 0   13. identical   14. dihybrid   15. polygenic
16. multiple   17. linked   18. consanguineous
19. hybrid vigor

CHAPTER 11

1. isogenic   2. karyotype   3. congenital   4. aneuploidy   5. trisomic   6. monosomic   7. nondisjunction   8. translocation   9. Down   10. Klinefelter, Turner   11. metabolism   12. hemoglobin   13. cystic fibrosis   14. amniocentesis   15. Chorionic villus
16. genetic counseling

CHAPTER 12

1. transformation, DNA   2. deoxynucleotide, phosphate, 3', 5',   3. deoxyribose, phosphate, nitrogenous, purines, pyrimidines   4. base, adenine, thymine, guanine, cytosine   5. X-ray diffraction   6. parallel, helix
7. adenine, thymine, guanine, cytosine   8. replication
9. semiconservative, parent, newly synthesized (or daughter)   10. origin of replication   11. 5', 3'
12. deoxyribonucleotide triphosphates   13. Okazaki fragments, continuous   14. DNA polymerase, DNA ligase   15. circular, linear   16. one, multiple
17. bidirectional   18. 1/1000, 1.3 mm, 1 meter
19. nucleosomes   20. Chromatin, loops, scaffolding proteins.

CHAPTER 12

Challenge Question: Since DNA synthesis proceeds in a $5' \rightarrow 3'$ direction, if DNA replication were not discontinuous, only one strand of the DNA molecule (that part of the DNA which could be read in a $3' \rightarrow 5'$ direction) could be replicated while the DNA strands were being separated, the opposite strand would have to be completely unwound before replication could begin. This would result in the formation of one double stranded helix and a single stranded molecule which would have to be copied after replication of the first was complete.

CHAPTER 13

1. Transcription   2. Translation   3. 3 bases, codon
4. tRNA   5. ribosome   6. RNA polymerase, ribonucleotide triphosphate   7. ribose, uracil, thymine
8. promoter, 5'   9. anticodon   10. carboxyl, 3'
11. amino acid activation, tRNA   12. RNA, protein
13. initiation., small, initiation, initiator (or formyl methionine), 5'.   14. elongation, tRNA, peptide bond, codon   15. termination, termination or stop, polypeptide, tRNA, ribosomal subunits.   16. polyribosome
17. introns, exon   18. capped, guanosine   19. polyadenylated   20. hnRNA. capped, tailed. introns, exons.
21. nonsense, frameshift.

CHAPTER 14

1. operons.,   2. operator,   3. RNA polymerase, negative   4. active. allosteric, conformation,   5. inactive, co-repressor, active   6. activate,   7. cyclic AMP
8. high,   9. regulon,   10. constitutive, constantly
11. repressor   12. grooves, hydrogen   13. translation,   14. feedback inhibition,   15. upstream promoter, number   16. enhancer,   17. mRNA,
18. multiple, amplified   19. heterochromatin, inactive
20. euchromatin

CHAPTER 15

1. recombinant   2. Restriction,   3. palindromic, complementary,   4. plasmids, bacteriophages.   5. same, DNA ligase   6. libraries   7. genomic, intron
8. cDNA, reverse transcriptase.   9. restriction map
10. break   11. probes   12. post-translational
13. transgenic

CHAPTER 16

1. differentiated, determination   2. morphogenesis, pattern formation   3. nuclear equivalence
4. totipotent   5. differential gene activity   6. polytene   7. maternal effect   8. zygotic   9. morphogens   10. homeotic   11. DNA   12. mosaic
13. induction   14. regulative   15. transgenic
16. homeotic   17. genomic rearrangements
18. Gene amplification

CHAPTER 17

1. b   2. a   3. d   4. b   5. c   6. d   7. b
8. a   9. c   10. b

CHAPTER 18

1. c   2. d   3. c   4. a   5. b   6. b   7. a
8. a   9. a   10. d

CHAPTER 19

1. b   2. a   3. c   4. d   5. a   6. e   7. c
8. a   9. b   10. b

CHAPTER 20

1. b   2. e   3. a   4. b   5. d   6. d   7. b
8. e   9. e   10. d   11. b   12. e

CHAPTER 21

1. b   2. d   3. a   4. d   5. e   6. b   7. c
8. c   9. c   10. c

CHAPTER 22

1. taxonomy   2. Linnaeus; genus, species   3. family
4. Protista   5. Fungi   6. ancestor   7. homologous

8. derived   9. protein; clocks   10. phenetic
11. cladistic   12. classical evolutionary
13. Eukaryota   14. Animalia   15. Chordata
16. Vertebrata   17. Mammalia   18. Eutheria
19. Primates   20. Hominidae   21. Homo
22. Homo sapiens

CHAPTER 23

1. DNA, RNA   2. bacteria   3. attachment; penetration; replication; assembly; release   4. virus; cell wall
5. nucleic acid   6. lysogenic conversion   7. lytic
8. temperate   9. transduction   10. Interferons
11. latent   12. host (or normal); cancer cells
13. Viroids; coat   14. prophage   15. lytic

CHAPTER 24

1. Monera   2. nuclear membrane, organelles   3. producers   4. cell wall   5. gram-positive   6. mesosome   7. Plasmids   8. pili   9. saprobes   10. Facultative anaerobes   11. conjugation   12. spherical, rod-shaped, spiral   13. cocci, bacilli, spirilla
14. clostridia   15. spirochete   16. chlamydias
17. cell wall

CHAPTER 25

1. contractile vacuole   2. syngamy   3. microtubules
4. pseudopods   5. choanoflagellates   6. filopodia
7. reproduction; metabolism   8. conjugation   9. sporozoa   10. cellulose   11. red tide   12. diatoms
13. euglenoids   14. anisogamous   15. brown
16. plasmodium   17. cellular   18. water molds

CHAPTER 26

1. decomposers   2. budding; fission   3. hyphae; mycelium   4. fruiting body   5. zygospores
6. mating types   7. conidia; conidiophores   8. ascospores; asci   9. basidiospore   10. gills   11. sexual
12. lichen   13. roots; plants   14. ergot   15. penetrate plant cells to obtain nourishment   16. barberry plant

CHAPTER 27

1. vascular   2. spores   3. cuticle   4. green algae
5. starch   6. stomata   7. archegonium   8. antheridium   9. fertilization; zygote   10. spores   11. alternation of generations   12. gametophyte   13. thallus   14. sori   15. ferns   16. horsetails
17. heterosporous

CHAPTER 28

1. seeds   2. deciduous   3. monoecious   4. fruit
5. pollen   6. female gametophyte   7. pollination
8. cycads; ginkgoes (any order)   9. monocots
10. endosperm   11. carpel   12. incomplete; imperfect (any order)   13. ovary; ovule   14. embryo sac
15. insects   16. wind

CHAPTER 29

1. invertebrates   2. sessile   3. body cavity   4. ectoderm   5. mesoderm   6. mouth   7. cnidarians (also ctenophores and adult echinoderms)   8. one quarter   9. sponges   10. spicules   11. spongocoel; osculum   12. polyp, medusa   13. Cnidaria; Anthozoa   14. flatworms   15. smell   16. eating poorly cooked pork that contains the larvae   17. a complete digestive tract (tube-within-a-tube body plan), a separate circulatory system   18. b   19. c   20. d   21. e
22. a   23. c   24. b   25. e   26. a   27. d

CHAPTER 30

1. exoskeleton   2. radula   3. open   4. trochophore
5. metamerism   6. lung   7. shell   8. locomotion
9. present in the same animal   10. increase the surface area of the intestine   11. cerebral ganglia   12. arthropods   13. arthropods   14. biting and grinding food   15. excretory   16. barnacles   17. pinching claws   18. excretory   19. b   20. c   21. d   22. a
23. e   24. d   25. b

CHAPTER 31

1. echinoderms   2. water vascular; tube   3. arms (rays); test (shell)   4. notochord; tubular nerve cord; gill slits   5. Tunicates   6. vertebral column; cranium
7. Pisces, Tetrapoda   8. Placoid   9. cloaca
10. food; digestion   11. ray-finned; tetrapods (land vertebrates)   12. gills   13. tetrapods (land vertebrates)
14. terrestrial; keeps the embryo moist and acts as a shock absorber   15. birds, mammals   16. lay eggs
17. c, d, e   18. e   19. d, e   20. f   21. c   22. d
23. a–e   24. b   25. c   26. e   27. d   28. a

CHAPTER 32

1. indeterminate   2. apical   3. root cap   4. leaf primordia; bud primordia (any order)   5. vascular cambium   6. annuals; biennials; perennials   7. parenchyma   8. collenchyma   9. tracheids   10. companion   11. epidermis; periderm   12. fibrous   13. bud scales   14. stomates; lenticels   15. compound
16. opposite   17. parallel; netted

CHAPTER 33

1. mesophyll   2. guard cells   3. xylem   4. transpiration   5. circadian rhythms   6. potassium
7. photoreceptor   8. dicots   9. bulliform   10. abscission   11. suberin   12. tendrils   13. storage

CHAPTER 34

1. dicots   2. monocot   3. cork cambium; vascular cambium (any order)   4. periderm   5. rays
6. periderm   7. secondary xylem   8. heartwood
9. summerwood; springwood   10. knot   11. dendrochronology   12. translocation   13. water potential
14. zero   15. decreases or lowers   16. root pressure
17. transpiration   18. source

CHAPTER 35

1. adventitious   2. contractile   3. casparian strip
4. pericycle   5. xylem   6. pith   7. active transport
8. sand; silt; clay   9. hydroponics   10. macronutrients   11. sixteen   12. phosphorus   13. potassium
14. inorganic   15. nitrogen

CHAPTER 36

1. rhizome   2. tuber   3. stolon   4. asexual
5. fruit   6. berries   7. peach (or cherry or apricot or plum)   8. dehiscent   9. two   10. aggregate; multiple   11. accessory   12. wind   13. animals
14. photoperiodism   15. darkness   16. red
17. vernalization

CHAPTER 37

1. phototropism   2. positive   3. thigmotropism
4. pulvinus   5. circadian rhythms   6. hormones
7. polar transport   8. acid-growth   9. auxin
10. gibberellins   11. enzymes   12. cytokinins
13. antagonistic   14. cytokinins; ethylene   15. ethylene   16. abscisic acid   17. dormancy

CHAPTER 38

1. epithelial tissue   2. protection, absorption, secretion, sensation   3. squamous, cuboidal, columnar
4. pseudostratified ciliated epithelium   5. stratified squamous epithelium   6. glands   7. stroma
8. Reticular fibers   9. g   10. f   11. d   12. e
13. j   14. b   15. c   16. i   17. a   18. h
19. hemoglobin   20. plasma

CHAPTER 39

1. epidermis; dermis   2. corneum   3. keratin
4. Hydrostatic   5. molt   6. endoskeletons   7. compact; cancellous   8. lubricant; joints   9. actin (thin), myosin (thick)   10. d; c; b; a; e; f; g.   11. storing energy   12. glycogen; ATP.

CHAPTER 40

1. digested; absorbed   2. cecum; bacteria that digest cellulose   3. rechewing partially digested food
4. carnivores; omnivores   5. ecto; endo   6. mucosa
7. peristalsis   8. Incisors   9. salivary glands
10. esophagus; duodenum (small intestine)   11. rugae
12. gastric glands; stomach   13. duodenum   14. d
15. a   16. f   17. e   18. c,f   19. b   20. fat.

CHAPTER 41

1. anabolism; catabolism   2. basal metabolic   3. energy used to carry on daily activities   4. remains the same; increases   5. sugars, starches   6. fuel for cellular respiration   7. glycogen synthesis (from glucose)
8. reconverted to glucose   9. gluconeogenesis
10. liver; fat   11. fats; acetyl coenzyme A   12. acidic

13. ingested in the diet   14. all of the essential amino acids in sufficient quantity to support growth   15. essential amino acids   16. amino group; carbon
17. A,D,E, K   18. b   19. d   20. e   21. a
22. c.

CHAPTER 42

1. c, e   2. a, f   3. b, d   4. b   5. oxygen
6. serum   7. gamma globulin   8. anemia   9. b
10. e   11. d   12. c   13. a   14. K   15. fibrin
16. Arterioles   17. capillaries   18. pulmonary arteries
19. semilunar   20. systole; diastole   21. volume of blood pumped by one ventricle in one minute   22. the more blood the heart pumps   23. blood pressure
24. blood flow; resistance to blood flow   25. aorta
26. brain   27. kidneys; liver   28. heart attack
29. blood pressure   30. vasoconstrictors

CHAPTER 43

1. immune response   2. antibodies (or immunoglobulins)   3. interferons   4. redness, heat, swelling, pain
5. bone marrow; thymus; lymph   6. lymphocytes
7. memory   8. plasma cells (differentiated B cells)
9. antibody   10. thymus   11. antigen-antibody
12. coat pathogens   13. active   14. passive
15. HLA.   16. one location to another in the same organism   17. immune response; transplanted
18. privileged   19. allergen; allergic   20. histamine; inflammation.

CHAPTER 44

1. thin; diffusion; moist; blood vessels   2. tracheal; spiracles   3. gills; bony fish   4. lungs   5. swim bladder   6. air sacs   7. trachea; bronchus   8. alveoli   9. diaphragm   10. water loss   11. e   12. a
13. b   14. d   15. c   16. vital capacity   17. oxygen   18. oxyhemoglobin; Bohr effect   19. bicarbonate ions   20. oxygen   21. inhaling dirty air   22. emphysema   23. cigarette smoking

CHAPTER 45

1. excretion   2. uric acid   3. urea   4. protonephridia; flame   5. metanephridia   6. antennal (green) glands   7. Malpighian tubules   8. nephrons
9. b   10. d   11. c   12. a   13. e   14. a   15. c
16. e   17. d   18. b   19. c   20. capillaries; Bowman's capsule   21. afferent arteriole; efferent arteriole
22. glomerular filtrate   23. reabsorbed   24. excreted in the urine   25. hypertonic   26. collecting ducts; reabsorbed; decreased   27. ADH; suffers tremendous fluid loss in the urine

CHAPTER 46

1. sense organs; central nervous system (CNS)   2. synapses   3. stimuli   4. nerves   5. neuron; neuroglia
6. receive stimuli; transmit neural impulses; cell body; synapse   7. regeneration of injured axons   8. cellu-

lar sheath, myelin sheath   **9.** cell bodies   **10.** resting potential   **11.** Excitatory   **12.** depolarizes **13.** action potential   **14.** all-or-none   **15.** one node of Ranvier to the next   **16.** resting potential; more **17.** neurotransmitter   **18.** acetylcholine   **19.** norepinephrine   **20.** divergence   **21.** reflex action   **22.** afferent; spinal cord; integration

CHAPTER 47

1. nerve net   **2.** Afferent; efferent   **3.** brain   **4.** visceral; pedal   **5.** brain; spinal cord   **6.** brain stem **7.** medulla; spinal canal   **8.** c   **9.** e   **10.** b   **11.** a **12.** d   **13.** b   **14.** d   **15.** e   **16.** a   **17.** c **18.** c   **19.** a   **20.** e   **21.** b   **22.** beta   **23.** REM **24.** vagus; internal organs of chest and upper abdomen **25.** energy; stress   **26.** dorsal   **27.** increasingly larger amounts are needed to obtain the desired effect.

CHAPTER 48

1. receptor; accessory   **2.** receive stimuli from the outside world; position of body parts; orientation   **3.** Photoreceptors; mechanoreceptors   **4.** energy; electrical; receptor potential   **5.** adaptation   **6.** gravity; statolith **7.** vision; hairs; cupula   **8.** muscle spindles; Golgi tendon; joint   **9.** motion; stimulus   **10.** Halteres **11.** equilibrium   **12.** labyrinth; saccule, utricle; semicircular   **13.** otoliths; gravity   **14.** endolymph; ampulla **15.** inner; sound   **16.** Corti; cochlear   **17.** chemoreceptors   **18.** rhodopsins   **19.** c   **20.** d   **21.** b **22.** e   **23.** a

CHAPTER 49

1. target tissues   **2.** hormones   **3.** chemical messenger produced by one type of cell that has a specific regulatory effect on the activity of another type of cell **4.** cell membrane; cyclic AMP   **5.** negative feedback **6.** molting   **7.** hypothalamus   **8.** growth   **9.** b **10.** c   **11.** a   **12.** e   **13.** c   **14.** e   **15.** c **16.** cortex; cortisol   **17.** hyper; GH   **18.** thyroid **19.** glucose; protein   **20.** anterior lobe of the pituitary; TSH

CHAPTER 50

1. fragmentation   **2.** sexual and asexual generations **3.** develops into a new individual   **4.** hermaphroditic **5.** gamete; zygote   **6.** sterile   **7.** a   **8.** c   **9.** d **10.** b   **11.** e   **12.** b   **13.** d   **14.** c   **15.** a **16.** c   **17.** b   **18.** d   **19.** e   **20.** b   **21.** a **22.** c   **23.** d   **24.** c   **25.** d   **26.** e   **27.** b **28.** therapeutic   **29.** female sterilization   **30.** menopause

CHAPTER 51

1. morphogenesis; cellular differentiation   **2.** acrosome; vitelline   **3.** fertilization cone   **4.** electrical; cortical **5.** cleavage   **6.** holoblastic   **7.** meroblastic; blastodisc **8.** gastrulation   **9.** ectoderm; endoderm   **10.** primi-

tive streak   **11.** neural plate   **12.** brain, spinal cord **13.** amnion (amniotic fluid)   **14.** placenta   **15.** inner cell mass; embryo   **16.** implant; uterus (endometrium) **17.** fetus   **18.** gestation   **19.** newborn infant **20.** birth   **21.** determines the future differentiation of nearby cell groups   **22.** frog   **23.** development **24.** aging

CHAPTER 52

1. Ethology   **2.** ultimate   **3.** straight run   **4.** sign **5.** *uurr*   **6.** sensitization   **7.** conditioned   **8.** handling   **9.** displays   **10.** navigation   **11.** symbolic **12.** Pheromones   **13.** dominance hierarchy **14.** home range   **15.** conflict; population   **16.** pair bond   **17.** making more genes

CHAPTER 53

1. environment   **2.** the sun   **3.** radiation   **4.** angle **5.** convection   **6.** specific heat   **7.** Coriolis effect **8.** dew point   **9.** rain shadows   **10.** melting **11.** mineral nutrients   **12.** microclimate   **13.** Chernozem; horizons   **14.** laterite

CHAPTER 54

1. low; permafrost   **2.** Coniferous evergreens **3.** precipitation   **4.** clay   **5.** water; mineral nutrients **6.** high; coral reef   **7.** savannah (veld)   **8.** temperature; oxygen; light   **9.** estuaries   **10.** thermocline; fall overturn   **11.** littoral   **12.** burrowing   **13.** abyssal

CHAPTER 55

1. resources   **2.** logarithmic   **3.** compound interest **4.** low   **5.** early   **6.** biotic potential; survivorship **7.** lag phase   **8.** shortest   **9.** space   **10.** increases **11.** prey   **12.** dependent   **13.** queen substance **14.** cockroaches

CHAPTER 56

1. decomposer   **2.** detritus feeders   **3.** Parasites **4.** mutualists   **5.** ecological niche   **6.** fundamental niche   **7.** realized niche   **8.** euryhaline   **9.** competitive exclusion   **10.** volcanic   **11.** nitrifying **12.** bacteria; cyanobacteria   **13.** biological magnification **14.** respiration   **15.** plant biomass   **16.** decomposer; parasitic   **17.** less   **18.** increase   **19.** sere **20.** autogenic   **21.** facilitation   **22.** superorganisms **23.** individualistic

CHAPTER 57

1. simpler   **2.** larger; reproduce   **3.** biodegradable **4.** insoluble; soluble   **5.** biological magnification; upper **6.** Pollution   **7.** high; oxygen   **8.** warm; cooler **9.** ozone layer; ultraviolet light   **10.** teeth; bones **11.** habitat loss   **12.** children

Glossary

abdomen (ab'doh-men) (1) In mammals the region of the body between the diaphragm and the rim of the pelvis; (2) in arthropods, the posterior-most major division of the body.

abiogenesis (ab"ee-oh-jen'e-sis) The spontaneous generation of life; the origin of living things from inanimate objects.

abortion (uh-bor'shun) Expulsion of an embryo or fetus before it is capable of surviving.

abscisic acid (ab-sis'ik) A plant hormone involved in responses to stress and in dormancy.

abscission (ab-sizh'en) The normal (usually seasonal) falling off of leaves or other plant parts, such as fruits or flowers.

abscission layer A special layer of thin-walled cells, loosely joined together, extending across the base of the petiole, thus weakening the base of the leaf and finally permitting the leaf to fall.

abscission zone The area at the base of the petiole where the leaf will break away from the stem.

absorption (ab-sorp'shun) The taking up of a substance, as by the skin, mucous surfaces, or lining of the digestive tract.

absorption spectrum A measure of the amount of energy at specific wavelengths that has been absorbed as light passes through a substance. Each type of molecule has a characteristic absorption spectrum.

accessory fruit A fruit composed primarily of tissue other than ovary tissue. Apples and pears are accessory fruits.

acclimatization (a-kly'muh-tih-zay'shun) Gradual physiological changes in an organism in response to slow, relatively long-lasting changes in the environment.

acetylcholine (ah"see-til-koh'leen) A compound of choline and acetic acid; the neurotransmitter employed by cholinergic nerves.

acetyl Co A (ah-see'til) A key intermediate compound in metabolism; consists of an acetyl group covalently bonded to coenzyme A.

achene (a-keen') A simple, dry, indehiscent fruit with one seed. The fruit wall is separate from the seed coat. Sunflower fruits are achenes.

acid A substance that releases hydrogen ions (protons) in water. Acids have a sour taste, turn blue litmus paper red, and unite with bases to form salts.

acid-growth hypothesis The proposed mechanism by which auxin induces cell elongation.

acoelomate organisms (a-seel'oh-mate) Organisms that lack a body cavity (coelom).

acromegaly (ak"roh-meg'ah-lee) A condition characterized by overgrowth of the extremities of the skeleton, nose, jaws, fingers, and toes. It may be produced by excessive secretion of growth hormone from the pituitary.

acrosome (ak'roh-sohm) A cap-like structure covering the head of the spermatozoan (sperm cell).

actin (ak'tin) A protein component of muscle fiber that, together with the protein myosin, is responsible for the ability of muscles to contract.

action potential The electrical activity developed in a muscle or nerve cell during activity; a neural impulse.

activation energy The energy required to initiate a chemical reaction.

active site Area of an enzyme that accepts a substrate and catalyzes its reaction with another.

active transport Energy-requiring transport of a molecule across a membrane from a region of low concentration to a region of high concentration.

adaptation (1) The ability of an organism to adjust to its environment; (2) decline in the response of a receptor subjected to repeated or prolonged stimulation.

adaptive radiation The evolution from an unspecialized ancestor of several to many species.

adaptive value Also known as fitness, a measure of the relative success of the phenotype; adaptive value (w) is the complement of the selection coefficient (i.e., $1 - S$).

adenine (ad'eh-neen) A purine (nitrogenous base) that is a component of nucleic acids and of nucleotides important in energy transfer—adenosine triphosphate (ATP), adenosine diphosphate (ADP), and adenylic acid (AMP).

adenosine triphosphate (a-den'oh-seen) An organic compound containing adenine, ribose, and three phosphate groups; of prime importance for energy transfers in biological systems.

adipose (ad'i-pohs) Pertaining to fat; tissue in which fat is stored, or the fat itself.

adrenal glands (ah-dree'nul) Paired endocrine glands, one located just superior to each kidney.

adrenaline (uh-dren'ah-lin) See epinephrine.

adrenergic neuron (ad-ren-er'jik) A neuron that releases norepinephrine or epinephrine as a neurotransmitter.

adventitious root (ad"ven-tish'us) A root that arises in an unusual position on the plant.

aerobic (air-oh'bik) Growing or metabolizing only in the presence of molecular oxygen.

aerobic metabolism Metabolism requiring oxygen.

afferent (af'fur-ent) Structure which conducts fluid or impulses toward an organ or structure, e.g. afferent neurons conduct impulses to the central nervous system.

agglutination (ah-gloo"tih-nay'shun) The collection into clumps of cells or particles distributed in a fluid.

aggregate fruit A fruit that develops from a single flower with many separate carpels, such as a raspberry.

agnathans (ag-na'thanz) Jawless fishes; class of vertebrates, including lampreys, hagfishes, and many extinct forms.

albinism (al'bih-niz-em) A hereditary inability to form pigment resulting in abnormally light coloration.

albumin (al-bew'min) A class of protein found in animal tissues; a fraction of plasma proteins.

aldosterone (al-dos'tur-ohn) Steroid hormone produced by the vertebrate adrenal cortex which governs the excretion or retention of sodium and potassium ions.

algae (al'gee) Single-celled or simple multicellular photosynthetic organisms; important producers.

alkali (al'kuh-lie) A substance which, when dissolved in water, produces a pH greater than 7; also called a base.

allantois (a-lan'toe-iss) One of the extraembryonic membranes of reptiles, birds, and mammals; a pouch growing out of the posterior part of the digestive system.

alleles (al-leels') Genes governing variations of the same characteristic that occupy corresponding positions on homologous chromosomes; alternative forms of a single gene.

allergy A hypersensitivity to some substance in the environment, manifested as hay fever, skin rash, asthma, food allergies, etc.

allometric growth Varied rates of growth for different parts of the body during development.

allopatric speciation (al-oh-pa'trik) Speciation that occurs when one population becomes geographically separated from the rest of the species and subsequently evolves.

allopolyploid (al"oh-pol'ee-ploid) A polyploid formed by joining one or more sets of chromosomes from each of two different species.

allosteric site (al-oh-steer'ik) A site located on an enzyme molecule which enables a substance, other than the normal substrate, to bind to the molecule, and to change the shape of the molecule and the activity of the enzyme.

allozyme (al'loh-zime) One of two or more slightly different versions of the same enzyme. The presence of allozymes may confer a selective advantage on the organism that possesses them.

alternation of generations A type of life cycle characteristic of plants. Plants spend part of their life in the haploid (gametophyte) stage and part in the diploid (sporophyte) stage.

alveolus (al-vee'o-lus) (1) An air sac of the lung through which gas exchange with the blood takes place; (2) a saclike unit of some glands; (3) a tooth socket.

ameboid motion (uh-mee'boid) The movement of a cell by means of the slow oozing of its cellular contents.

amensalism (ay-men'sal-izm) A relationship between two species whereby the first species is adversely affected by the second, but the second species is unaffected by the presence of the first.

amino acid (uh-mee'no) An organic compound containing an amino group (—NH_2) and a carboxyl group (—COOH). Amino acids may be linked together to form the peptide chains of protein molecules.

amino acid activation The linking of an amino acid to AMP, the first part of the process by which an amino acid is incorporated into a protein.

aminoacyl tRNA (uh-mee"no-as'seel) Molecule consisting of an amino acid covalently linked to a tRNA.

amniocentesis (am"nee-oh-sen-tee'sis) Sampling of the amniotic fluid surrounding a fetus in order to obtain information about its development and genetic makeup.

amnion (am'nee-on) An extraembryonic membrane that forms a fluid-filled sac for the protection of the developing embryo.

amphipathic molecule (am"fih-pa'thik) Molecule that contains both hydrophobic and hydrophilic regions.

amylase (am'ih-lase) Starch-digesting enzyme of animals, e.g., human salivary amylase or pancreatic amylase.

anabolism (an-ab'oh-lizm) Chemical reaction in which simpler substances are combined to form more complex substances, resulting in the storage of energy, the production of new cellular materials, and growth.

anaerobic (an"air-oh'bik) Growing or metabolizing only in the absence of molecular oxygen.

anagenesis (an"ah-jen'eh-sis) The conversion of an entire population over time to a new species; also known as phyletic evolution.

analogous Similar in function or appearance but not in origin or development.

anaphase (an'uh-faze) The stage of mitosis between metaphase and telophase in which the chromatids separate and move toward opposite ends of the cell.

anaphylaxis (an"uh-fih-lak'sis) An acute allergic reaction following sensitization to a foreign protein or other substance.

androecium (an-dree'see-um) The male portion of a flower.

androgen (an'dro-jen) Any substance that possesses masculinizing properties, such as sex hormones.

anemia (uh-nee'mee-uh) A deficiency of hemoglobin or red blood cells.

aneuploidy (an'-you-ploy-dee) Any chromosomal aberration in which there are either extra or missing copies of certain chromosomes.

angiosperm (an'jee-oh-sperm") The traditional name for plants having flowers, fruits, and seeds.

angiotensin (an-jee-o-ten'sin) A compound formed by the action of renin and found in blood; stimulates aldosterone secretion by the adrenal cortex.

anhydrase (an-high'drase) An enzyme that chemically removes water from a substance, e.g., carbonic anhydrase, which catalyzes the formation of carbon dioxide from carbonic acid.

anion (an'eye-on) A negatively charged ion such as Cl^-.

anisogamy (an"eye-sog'uh-mee)   Reproductive process involving motile gametes of similar form but dissimilar size as in certain algae.

annelids (an'eh-lids)   Segmented worms with true coeloms, such as earthworms.

annual   A plant that completes its entire life cycle in one growing season.

annual ring   A layer of wood (secondary xylem) formed in woody plants growing in temperate areas. One layer is usually formed per year.

anther (an'thur)   The part of the stamen in flowers that produces microspores and, ultimately, pollen.

antheridium (an"thur-id'ee-im)   The male gametangium in certain land plants. Antheridia produce sperm.

antherozoid (an'ther-oh-zoid")   The motile fertilizing cell of fungi.

anthocyanins (an"thi-sigh'ah-ninz)   A class of pigments of blue, red, and violet flowers.

anthropoid (an'thro-poid)   A suborder of primates that includes monkeys, apes, and humans.

antibiotic (an"ty-by-ot'ik)   Substance produced by microorganisms that has the capacity, in dilute solutions, to inhibit the growth of or destroy bacteria and other microorganisms; used largely in the treatment of infectious diseases in humans, animals, and plants.

antibodies (an-tih-bod'ees)   Protein compounds (immunoglobulins) produced by plasma cells in response to specific antigens and having the capacity to react against the antigens.

anticodon (an'ty-koh"don)   A sequence of three nucleotides in transfer RNA that is complementary to, and combines with, the three nucleotide codon on messenger RNA, thus helping to specify the addition of a particular amino acid to the end of a growing peptide.

antidiuretic hormone (an"ty-dy-uh-ret'ik)   A hormone secreted by the posterior lobe of the pituitary which controls the rate of water reabsorption by the kidney.

antigen (an'tih-jen)   Any substance capable of stimulating an immune response; usually a protein or large carbohydrate that is foreign to the body.

antitoxin (an"ty-tok'sin)   An antibody produced in response to the presence of a toxin (usually protein) released by a bacterium.

anus (ay'nus)   The distal end and outlet of the digestive tract.

aorta (ay-or'tah)   The largest and main systemic artery of the body; arises from the left ventricle and branches to distribute blood to all parts of the body; main artery leaving the heart in vertebrates.

apical dominance (ape'ih-kl)   The inhibition of lateral buds by the apical meristem.

apical meristem (mehr'ih-stem)   An area of cell division located at the tips of plant stems and roots. Apical meristems produce primary tissues.

apoenzyme (ap"oh-en'zime)   Protein portion of an enzyme; requires the presence of a specific coenzyme to become a complete functional enzyme.

apomixis (ap"uh-mix'us)   A type of reproduction in which fruits and seeds are formed asexually, such as dandelions.

arachnids (ah-rack'nids)   Eight-legged arthropods such as scorpions and spiders.

archaebacteria (ar"kuh-bak-teer'ee-uh)   Biochemically unique bacteria thought to be similar to the ancient ancestors of all bacteria.

archegonium (ar"ke-go'nee-um)   The female organ of certain land plants (such as mosses) in which an egg is produced.

archenteron (ar-ken'ter-on)   The central cavity of the gastrula stage of embryonic development, which is lined with endoderm; primitive digestive system.

arteries   Thick-walled blood vessels that carry blood away from the heart and toward the body organs.

arteriole (ar-teer'ee-ole)   A very small artery.

arthritis (ar-thry'tis)   Inflammation of a joint.

arthropod (ar'throh-pod)   An invertebrate, such as an insect or crustacean, that has jointed legs.

artificial selection   Selection by humans of traits that are desirable in other plants or animals, and breeding only those individuals that possess the desired traits.

ascospores (ass'koh-sporz)   Set of spores, usually eight, contained in a special spore case.

asexual reproduction   Reproduction in which only one parent participates, and in which the genetic makeup of parent and offspring is the same. (See also Reproduction.)

asthma (az'muh)   A disease characterized by airway constriction and often leading to difficulty in breathing.

astigmatism (ah-stig'muh-tizm)   A defect of vision resulting from irregularity in the curvature of the cornea or lens.

atherosclerosis (ath"ur-oh-skle-row'sis)   A progressive disease in which muscle cells and lipid deposits accumulate in the inner lining of arteries, leading eventually to impaired circulation and heart disease.

atom   The smallest quantity of an element that can retain the chemical properties of that element; composed of an atomic nucleus containing protons and neutrons, together with electrons that circle the nucleus in specific orbitals.

atomic orbital   The space surrounding the atomic nucleus in which an electron will usually be found.

atrioventricular valve (ay"tree-oh-ven-trik'you-lur)   A valve between each atrium and its ventricle that prevents backflow of blood.

atrium (of the heart) (ay'tree-um)   The chamber on each side of the heart that receives blood from the veins.

auricle (of the heart) (or'ih-kl)   Small accessory pouch attached to the atrium of the mammalian heart.

autoimmune disease (aw"toh-ih-mune')   A disease in which the body produces antibodies against its own cells or tissues.

autonomic nervous system (aw-tuh-nom'ik)   The portion of the peripheral nervous system that controls the visceral functions of the body, e.g., regulates smooth muscle, cardiac muscle, and glands, thereby helping to maintain homeostasis.

autosome (aw'toh-sohm)   A chromosome other than the sex (X and Y) chromosomes.

autotrophy (aw'-toh-troh"fee) Obtaining organic molecules by synthesizing them from inorganic material.

auxin (awk'sin) A plant hormone involved in various aspects of plant growth and development.

avicularia (ah-vik"you-lair'ee-uh) Members of a colony of bryozoa specialized for defense that resemble the head of a bird.

axon (ax'on) The long, tubular extension of the neuron that transmits nerve impulses away from the cell body.

bacillus (ba-sil'us) A rod-shaped bacterium.

background extinction The continuous, low-level extinction of species that has been evident throughout much of the history of life. Compare with mass extinction.

bacteria (bak-teer'ee-uh) Unicellular, prokaryotic microorganisms belonging to Kingdom Monera. Most are decomposers, but some are parasites or autotrophs.

bacteriophage (bac-teer'-ee-oh-fayj) Virus that can infect a bacterium (literally, ''bacteria eater'').

balanced polymorphism (pol"ee-mor'fizm) An equilibrium mixture of organisms homozygous and heterozygous for a given gene that is maintained by opposing forces of natural selection.

baroreceptors (bare'oh-ree-sep"torz) Receptors within certain blood vessels that are stimulated by pressure changes.

Barr body A condensed and inactivated X-chromosome appearing as a distinctive dense spot in the nucleus of certain cells of female mammals.

basal body (bay'sl) Structure that is similar to a centriole in its arrangement of microtubules and other components, and which is involved in the organization and anchorage of a cilium or flagellum.

basal metabolic rate The amount of energy expended by the body just to keep alive, when no food is being digested and no voluntary muscular work is being done.

base A compound that releases hydroxyl ions (OH—) when dissolved in water; turns red litmus paper blue.

basement membrane A connective tissue sheet that underlies vertebrate epithelium.

basidium (ba-sid'ee-um) The clublike spore-producing organ of certain fungi.

Batesian mimicry (bate'see-un mim'ih-kree) The resemblance of a harmless or palatable species to one that is dangerous, unpalatable, or poisonous.

behavioral isolation A prezygotic isolating mechanism in which gamete exchange between two groups is prevented because each group possesses its own characteristic courtship behavior.

benthos (ben'thos) The flora and fauna of the bottom of oceans or lakes.

berry A simple, fleshy fruit in which the fruit wall is soft throughout. Tomatoes, bananas, and grapes are berries.

bicuspid (by-kus'pid) Having two points or cusps (as bicuspid teeth), or two flaps (as the mitral valve of the heart).

biennial (by-en'ee-ul) A plant that takes two years (i.e., two growing seasons) to complete its life cycle.

bile Digestive juice produced by the liver.

binary fission (by'nare-ee fish'un) Equal division of a cell or organism into two; usually a variety of asexual reproduction.

binomial nomenclature (by-nome'ee-ul) System of naming organisms by the combination of the names of genus and species.

biodegradable (by"oh-de-gray'dih-bl) Substance that can be eliminated from the environment by the action of decay organisms or detritus feeders; for example, paper.

biogenesis (by-oh-jen'eh-sis) The generalization that all living things come only from preexisting living things.

biogeography The study of the distribution of organisms.

biological clocks Means by which activities of plants or animals are adapted to regularly recurring changes.

biological oxidation Process in which electrons removed from an atom or molecule are transferred through the electron transport system of the mitochondrion.

biomass The total weight of all the organisms in a particular habitat.

biome (by'ohm) A large, easily differentiated community unit arising as a result of complex interactions of climate, other physical factors, and biotic factors.

bipedal Walking on two feet.

biosphere The entire zone of air, land, and water at the surface of the earth that is occupied by living things.

biotic potential Inherent power of a population to increase in numbers when the age ratio is stable and all environmental conditions are optimal.

birefringence (by"re-frin'jens) Property of a substance in solution to refract light differently in different planes.

bivalent (by-vale'ent) Having a chemical valence of two.

blastocoele (blas'toh-seel) The fluid-filled cavity of the blastula.

blastomere (blas'toh-meer) One cell of a blastula; a cell formed during early cleavage of a vertebrate embryo.

blastopore (blas'toh-pore) Primitive opening into the body cavity of an early embryo which may become the mouth or anus of the adult organism.

blastula (blas'tew-lah) Usually a spherical structure produced by cleavage of a fertilized ovum; consists of a single layer of cells surrounding a fluid-filled cavity.

blastocyst (blas'toh-sist) The blastula stage in the development of the mammalian embryo; a spherical mass consisting of a single layer of cells, the trophoblast, from which a small cluster, the inner cell mass, projects into a central cavity.

B lymphocyte (lim'foh-site) A type of white blood cell responsible for antibody-mediated immunity. When stimulated, B lymphocytes differentiate to become plasma cells that produce antibodies; also called B cells.

bottleneck Genetic drift that may result from the sudden decrease in a population due to environmental factors.

Bowman's capsule Double-walled, hollow sac of cells

that surrounds the glomerulus at the end of each kidney tubule.

brachiate To swing, arm to arm, from one branch to another.

brachiopods (bray'kee-oh-pods) Marine organisms possessing a pair of shells and, internally, a pair of coiled arms with ciliated tentacles.

branchial (brang'kee-ul) Pertaining to gills or the gill region.

bronchiole (bronk'ee-ole) Tiny air duct of the lung that branches from a bronchus; divides to form air sacs (alveoli).

bronchus (bronk'us) pl. **bronchi** (bronk'eye) One of the branches of the trachea and its immediate branches within the lung.

Brownian movement Motion of small particles in solution or suspension resulting from their being bumped by water molecules.

brush border The many fine hairlike processes (microvilli) extending from the free surface of certain epithelial cells, such as the cells of the proximal convoluted tubules of the mammalian kidney.

bryophytes (bry'oh-fites) Members of the plant kingdom comprising mosses, liverworts, and hornworts.

bryozoans (bry"uh-zoh'uns) Minute water animals that usually form fixed, branching, mosslike colonies; sometimes mistaken for seaweed; other species form colonies that appear as thin, lacy encrustations on rocks.

bud An undeveloped shoot that can develop into flowers, stems, or leaves. Buds may be terminal (at the tip of the stem) or lateral (on the side of the stem).

budding Asexual reproduction in which a small part of the parent's body separates from the rest and develops into a new individual, eventually either taking up an independent existence or becoming a more or less independent member of the colony or cell.

bud primordium (pry-mor'dee-um) An embryonic bud, evident in the stem apical meristem.

bud scale A modified leaf that covers and protects winter buds.

buffers Substances in a solution that tend to lessen the change in hydrogen ion concentration (pH) that otherwise would be produced by adding acids or bases.

bulb A globose, fleshy, underground bud. A bulb is a short stem with fleshy leaves, e.g., onion.

bulbourethral glands (bul-boh-yoo-ree'thrul) In the male, paired glands that secrete fluid into the urethra during sexual excitement; also called Cowper's glands.

bulliform cell (bool-ee'form) A large, thin-walled cell found in the epidermis of some monocots; aids in the folding and unfolding of leaves associated with periods of drought.

bundle sheath A ring of cells surrounding the vascular bundle in dicot and monocot leaves.

calcitonin (kal-sih-toh'nin) A hormone secreted by the thyroid gland that rapidly lowers the calcium content in the blood.

callus (kal'us) Undifferentiated tissue in plant tissue culture.

calorie (kal'oh-ree) A unit of heat. The calorie used in the study of metabolism is the large calorie or kilocalorie and is defined as the amount of heat required to raise the temperature of a kilogram of water one degree Celsius.

Calvin cycle Cyclic series of reactions occurring in the light-independent phase of photosynthesis that fixes carbon dioxide and produces glucose.

calyx (kay'liks) The collective term for the sepals of a flower.

cambium (kam'bee-um) A layer of meristematic tissue that produces lateral (secondary) growth in plants.

cancellous bone (kan'se-lus) Spongy bone; has a lattice-like structure.

capillaries (kap'i-lare-eez) Microscopic blood vessels occurring in the tissues which permit exchange of materials between tissues and blood. Lymph capillaries also drain away excess tissue fluid.

capsule (1) The portion of the moss sporophyte that contains spores; (2) a simple, dry, dehiscent fruit that opens along many seams or pores to release seeds, such as cotton fruits.

carbohydrate Compound containing carbon, hydrogen, and oxygen, in the ratio of $1C:2H:1O$, such as sugars, starches, and cellulose.

carboxyl (kar-bok'sil) A monovalent radical characteristic of organic acids; group of atoms in a molecule arranged as —COOH.

carcinogen (kar'sin-oh-jen, kar-sin'oh-gen) A substance that causes cancer or accelerates its development.

cardiac (kar'dee-ak) Pertaining to the heart.

cardiac muscle Distinctive involuntary but striated type of muscle occurring only in the vertebrate heart.

carnivore (kar'ni-vor) An animal that primarily eats flesh.

carotene (kare'oh-teen) Yellow to orange-red pigments found in carrots, sweet potatoes, leafy vegetables, etc.; can be converted in the animal body to vitamin A.

carotenoid A carotene or xanthophyll.

carpel (kar'pul) The female structure in flowers that bears ovules.

carrier-mediated transport Any form of transport across a membrane that utilizes a transport protein with a binding site for a specific substance.

carrying capacity The ability of the habitat to support a population of organisms.

cartilage (kar'til-ij) Flexible skeletal tissue of vertebrates.

Casparian strip (kas-pare'ee-un) A band of waterproof material around the walls of the cells of the root endodermis.

cast Fossil formed by a mold being filled in by minerals that harden.

catalyst (kat'ah-list) A substance that regulates the speed at which a chemical reaction occurs without affecting the end point of the reaction and without being used up as a result of the reaction.

cation (kat'eye-un) An ion bearing a positive charge.

cecum (see'kum) The first part of the large intestine; a blind sac opening into the ileum, the colon, and the vermiform appendix.

cell The basic structural and functional unit of the

body, consisting of a complex of organelles bounded by a unit membrane, and usually microscopic in size.

cell cycle Cyclic series of events in the life of a dividing eukaryotic cell consisting of M (mitosis), cytokinesis, and the stages of interphase, which are the G_1 (first gap), S (DNA synthesis), and G_2 (second gap) phases.

cell plate Forming cell wall that separates the two daughter cells produced by plant mitosis.

cellulose (sel'yoo-lohs) A complex polysaccharide that comprises the cell walls of plants.

center of origin The particular place where a species originated or evolved.

centriole (sen'tree-ohl) One of a pair of small dark-staining organelles lying near the nucleus in the cytoplasm of animal cells.

centromere (sen'tro-meer) Specialized constricted region of a chromatid that serves as the site of spindle fiber attachment during cell division; sister chromatids are joined in the vicinity of their centromeres.

cercaria (sur-kar'ee-ah) The final free-swimming larval stage of a trematode (fluke) parasite.

cercopithecoid (sur"ko-pith'e-koid) An Old World monkey; has a tail but does not use it as a limb.

cerebellum (ser-eh-bel'um) The deeply convoluted subdivision of the brain lying beneath the cerebrum which is concerned with the coordination of muscular movements; second largest part of the brain.

cerebral cortex (ser-ee'brul kor'tex) The outer layer of the cerebrum composed of grey matter and consisting mainly of nerve cell bodies.

cerebrum (ser-ee'brum) Largest subdivision of the brain; functions as the center for learning, voluntary movement, and interpretation of sensation.

chaparral (shap"uh-ral') Distinctive vegetation type characteristic of certain areas with Mediterranean climates; dominated by drought-resistant shrubs and small trees with small leaves.

chelicera (kee-lis'er-ah) Pincers; the first pair of appendages in arthropods, located immediately anterior to the mouth, and used to manipulate food into the mouth.

chemical evolution The origin of life from nonliving matter.

chemoautotrophs (kee"moh-aw'toh-trofes) Autotrophic organisms that depend on the energy of pre-existing chemical compounds in their environment rather than the energy of sunlight.

chemoreceptor (kee"moh-ree-sep'tor) A sense organ or sensory cell that responds to chemical stimuli.

chemotropism (kee"moh-tro'pizm) A growth response to a chemical stimulus.

chiasma (ky-az'muh) pl. **chiasmata** A site in a tetrad where homologous (non-sister) chromatids have undergone exchange by breakage and rejoining.

chimera (ky-mee'rah) An individual organism whose body consists of a mixture of cells derived from different zygotes.

chitin (ky'tin) An insoluble, horny protein-polysaccharide that forms the exoskeleton of arthropods and the cell walls of many fungi.

chlorenchyma (klor"en-ky'mah) Closely packed photosynthetic plant cells, usually found in leaves.

chlorophyll (klor'oh-fil) One of a group of light-trapping green pigments found in most photosynthetic organisms.

chloroplast (klor'oh-plast) A chlorophyll-bearing intracellular organelle of plant cells; site of photosynthesis.

choanocyte (koh-an'oh-sight) A unique cell having a flagellum surrounded by a thin cytoplasmic collar; characteristic of sponges and one group of protists.

cholinergic neuron (koh"in-air'jik) A nerve cell that produces acetylcholine as a transmitter substance.

chordates (kor'dates) Phylum of animals which possess, at some time in their lives, a cartilagenous dorsal skeletal structure called a notochord.

chorion (kor'ee-on) An extraembryonic membrane in reptiles, birds, and mammals that forms an outer cover around the embryo, and in mammals contributes to the formation of the placenta.

chorionic villus sampling (CVS) (kor"ee-on'ik) Study of extra-embryonic cells that are genetically identical to the cells of the embryo, making it possible to determine the genetic makeup of the embryo.

chromatid (kroh'mah-tid) One of the two halves of a replicated chromosome.

chromatin (kroh'mah-tin) The complex of DNA protein and RNA which makes up eukaryotic chromosomes.

chromomere (kro'moh-meer) One of a linear series of beadlike structures composing a chromosome.

chromosomes (kro'moh-soms) Filamentous or rod-shaped bodies in the cell nucleus that contain the hereditary units, the genes.

chrysalis (krih'suh-lis) Pupa of a butterfly or moth; immobile stage of life cycle between larval (caterpillar) form and adult winged insect.

cilium (sil'ee-um) pl. **cilia** One of many short, hairlike structures that project from the surface of some cells and are used for locomotion or movement of materials across the cell surface; structurally like flagella, including a cylinder of 9 doublet microtubules and 2 central single microtubules, all covered by a plasma membrane.

circadian rhythm (sir-kay'dee-un) An internal rhythm that approximates the 24-hour day. Circadian rhythms are found in plants, animals, and other organisms.

cisterna (sis-tur'nah) A hollow or cavity; especially the space within the endoplasmic reticulum.

citric acid cycle Aerobic series of chemical reactions in which food molecules are completely degraded to carbon dioxide and water with the release of metabolic energy; also known as the Krebs cycle.

cladogenesis (klad"oh-jen'eh-sis) Diversifying speciation in which one or more new species is derived from the parent species, which continues to exist.

class A taxonomic grouping of related, similar orders of living organisms.

classical conditioning A type of learning in which a normal response to a stimulus becomes associated with a new stimulus, after which the new stimulus elicits the response.

clay The smallest inorganic particles of soil; compare with salt and sand.

cleavage First of several cell divisions of an embryo which forms a multicellular blastula.

climax community The final, stable, and mature community in a successional series.

cline Continuous series of differences in structure or function exhibited by the members of a species along a line extending from one part of their range to another.

clitoris (klit-ó-ris) A small, erectile body at the anterior part of the vulva, which is homologous to the male penis.

cloaca (klow-a′ka) An exit chamber in lower vertebrates which receives digestive wastes and urine; may also serve as exit for reproductive system.

clone A population of cells descended by mitotic division from a single ancestral cell, or a population of genetically identical organisms asexually propagated from a single ancestor.

cnidarians (nye-dare′ee-uns) Phylum of animals possessing stinging cells called cnidoblasts; characterized by a single body opening, two tissue layers, and radial symmetry, such as the *Hydra*.

coacervate (koh-as′sir-vate) A protobiont formed from relatively complex mixtures of polypeptides, nucleic acids, and polysaccharides.

coadapted gene complex A collection of genes that affect the same trait or function. Due to natural selection, these genes interact compatibly.

cocci (kok′eye) Spherical bacterial cells, usually less than 1 micrometer in diameter.

cochlea (kok′lee-ah) The structure of the inner ear of mammals that contains the auditory receptors.

codominance (koh″dom′in-ints) Condition in which both alleles of a locus are expressed in a heterozygote.

codon (koh′don) A trio of mRNA bases that specifies an amino acid or a signal to terminate the polypeptide.

coelom (see′lum) Body cavity which forms within the mesoderm and is lined by mesoderm.

coenocyte (see′no-site) Type of fungal hypha consisting of a giant elongated multinucleated cell.

coenocytic (see″no-sit′ik) Mass of protoplasm formed by divisions of the nucleus but not the cytoplasm.

coenzyme (koh-en′zime) A nonprotein substance that is required for a particular enzymatic reaction to occur; participates in the reaction by donating or accepting some reactant; loosely bound to enzyme. Most of the vitamins function as coenzymes.

cofactor A nonprotein substance needed by an enzyme for normal action; cofactors are metal ions; others are coenzymes.

coleoptile (kol-ee-op′tile) A protective sheath that encloses the stem in grass seeds.

colinearity (koh-lin-ee-air′ih-tee) The one-to-one serial correspondence between the linear sequence of the nucleotide codons, the RNA, and the linear sequence of amino acids in the polypeptide coded for by that sequence.

collagen (kol′ah-jen) Protein in connective tissue fibers that is converted to gelatin by boiling.

collenchyma (kol-en′kih-mah) Living cells with moderately but unevenly thickened walls. Collenchyma cells help support the primary plant body.

colloid (kol′oid) A substance composed of very small, insoluble and nondiffusable particles larger than molecules but small enough to remain suspended in fluid without settling to the bottom.

commensal (kum-men′sul) An organism that lives in intimate association with another species without harming or benefiting its host.

community An assemblage of populations that live in a defined area or habitat, which can either be very large or quite small. The organisms constituting the community interact in various ways with one another.

companion cell A cell in the phloem of plants that is responsible for loading and unloading sugar into the sieve tube member for conduction.

competitive inhibition Interference with enzyme action by an abnormal substrate that is not permanently bound to the active site and that competes with the normal substrate for the site.

complete flower A flower that possesses all four main parts: sepals, petals, stamens, and carpels.

compound eye An eye, such as that of an insect, composed of many smaller, light-sensitive units called ommatidia.

compression A fossil in which the organism was trapped in sediments without being completely decomposed.

cone (1) In botany, a reproductive structure in many gymnosperms that produces either microspores or megaspores. (2) In zoology, the conical photoreceptive cell of the retina that is particularly sensitive to bright light, and, by distinguishing light of various wave lengths, mediates color vision.

conifers (kon′ih-furs) Gymnosperms bearing needlelike leaves that are well adapted to withstand heat and cold.

conjugation A sexual phenomenon in certain protists that involves exchange or fusion of a cell with another cell. The term is also is used sometimes for DNA exchange in bacteria.

connective tissue Vertebrate tissue consisting mostly of a matrix composed of cell products in which the cells are embedded, e.g., bone.

contractile root A root that contracts and pulls the plant deeper into the soil.

contractile vacuole (vak′yoo-ohl) A vacuole that expands, filling with water, and periodically contracts, ejecting the water from the cell.

convection The transfer of heat within liquids and gases that results from currents arising as a result of heat-induced local differences in density.

convergent evolution The independent evolution of similar structures that carry on similar functions, in two or more organisms of widely different, unrelated ancestry.

copulation Sexual union; act of physical joining of two animals during which sperm cells are transferred from one to the other.

cork Cells produced by the cork cambium. Cork is dead at maturity and functions for protection.

cork cambium (kam'bee-um) Lateral meristem in plants that produces cork cells and cork parenchyma. Cork cambium and the tissues it produces make up the outer bark on a woody plant.

cork parenchyma (par-en'kih-mah) One or more layers of parenchyma cells produced by the cork cambium.

corm A short, thickened underground stem specialized for food storage and asexual reproduction, as exists in the gladiolus.

cornea (kor'nee-ah) Transparent anterior covering of the eye.

corolla (kor-ohl'ah) A collective term for the petals of a flower.

corpus allatum (kor'pus al-lah'tum) An endocrine gland located in the head of insects just behind the brain. It secretes juvenile hormone.

corpus callosum (kah-loh'sum) A large bundle of fibers interconnecting the two cerebral hemispheres.

corpus luteum (loo-tee'um) Pocket of endocrine tissue in the ovary that is derived from the follicle cells; secretes the hormone progesterone.

cortex (kor'tex) The outer layer of an organ; in plants, the tissue beneath the epidermis in many nonwoody plants.

co-transport The active transport of a substance from a region of low to high concentration by coupling its transport with the transport of a substance down its concentration gradient.

cotyledon (kot"i-lee'dun) The seed leaf of the embryo of a plant, which may contain stored food for germination.

covalent bond Chemical bond involving one or more shared pairs of electrons.

cretinism (kree'tin-izm) A chronic condition due to congenital lack of thyroid secretion; results in retarded physical and mental development.

cristae (kris'tee) Shelf-like or finger-like wrinkles in the inner membrane of a mitochondrion.

crossing over The breaking and rejoining of homologous (non-sister) chromatids during early meiotic prophase I, resulting in an exchange of genetic material.

ctenophores (teen'oh-forz) Marine animals ("comb jellies") whose bodies consist of two layers of cells enclosing a gelatinous mass. The outer surface is covered with eight rows of cilia resembling combs, by which the animal moves through the water.

cultural evolution The progressive addition of knowledge to the human experience.

cuticle (kew'tih-kl) A waxy covering over the epidermis of the above-ground portion of land plants that reduces water loss from plant surfaces.

cutin (kew'tin) A waxy, waterproof material that prevents the loss of water from a leaf surface.

cyanobacteria (sy-an'oh-bak-teer'ee-uh) Prokaryotic photosynthetic microorganisms that possess chlorophyll and produce oxygen by the photolysis of water. Formerly known as blue-green algae.

cycads (sih'kads) Members of a class of woody seed plants (the Cycadopsida) which live mainly in tropical and semitropical regions and have either short, tuber-ous, underground stems, or erect, cylindrical above-ground stems.

cyclic AMP (cAMP) A form of adenosine monophosphate in which the phosphate is part of a ring-shaped structure; acts as a regulatory molecule and second messenger in organisms ranging from bacteria to humans.

cyclosis (sy-kloh'sis) The streaming motion of the living material in cells.

cytochromes (sy'toh-kromz) The iron-containing heme proteins of the electron transport system that are alternately oxidized and reduced in biological oxidation.

cytokinesis (sy"toh-kih-nee'sis) Stage of cell division in which the cytoplasm is divided to form two daughter cells.

cytokinin (sy"toh-kih'nin) A plant hormone involved in various aspects of plant growth and development.

cytoplasm (sy'toh-plazm) General cellular contents exclusive of the nucleus.

cytoskeleton Internal structure of microfilaments, intermediate filaments, and microtubules that gives shape and mechanical strength to animal cells.

cytotoxic T cells T lymphocytes that destroy cancer cells and other pathogenic cells on contact. Also known as killer T cells.

dalton A unit of molecular weight; the weight of one hydrogen atom. Named for English chemist and physicist John Dalton.

day-neutral plant A plant that does not flower in response to variations in day length which occur with changing seasons.

deamination (dee-am-ih-nay'shun) Removal of an amino group (—NH$_2$) from an amino acid or other organic compound.

deciduous (de-sid'yoo-us) Falling off at a certain time or stage of growth, such as some leaves, antlers, and the wings of some insects.

decomposers Microorganisms of decay.

dehiscent fruit (dih-his'sent) A simple, dry fruit that splits open along one or more seams at maturity; compare with indehiscent fruit.

dehydrogenation (dee-hy"dro-jen-ay'shun) A form of oxidation in which hydrogen atoms are removed from a molecule.

deme An interbreeding population within a species.

denature (dee-nay'ture) To alter the physical properties and three-dimensional structure of a protein, nucleic acid, or other macromolecule by mild treatment that does not break the primary structure, but which does destroy its activity.

dendrite (den'drite) Nerve fiber, typically branched, that conducts a nerve impulse toward the cell body.

denitrifying bacteria (dee-ny'tri-fy-ing) Bacteria that convert ammonia to nitrogen gas.

dentition The number and arrangement of teeth in the mouth.

deoxyribose Pentose lacking an OH group on carbon-2.

depolarization (dee-pol"ar-ih-zay'shun) Change in electric charge of a plasma membrane that produces the action potential.

dermis (dur'mis) The layer of dense connective tissue beneath the epidermis in the skin of vertebrates.

desmosomes (dez'moh-somz) Buttonlike plaques, present on the two opposing cell surfaces and separated by the intercellular space, that serve to hold the cells together.

determination The progressive limitation of a cell line's potential fate during development.

detritus feeders (deh-try'tus) Organisms, other than microorganisms, that feed on dead, decaying organisms or their fragments.

deuterostome (doo'ter-oh-stome) A division of coelomate animals that includes the echinoderms and chordates; characterized by radial cleavage, entercolelous formation of the coelom, and development of the anus from the blastopore.

diapause (dy'uh-paws) Inactive state of an insect during the pupal stage.

diastole (dy-as'toh-lee) Relaxation of the heart muscle, especially that of the ventricle, during which the lumen becomes filled with blood.

dichotomous A type of branching in which the branches or veins always branch exactly in half.

dicotyledon (dy-kot-ih-lee'dun) A flowering plant with embryos having two seed leaves, or cotyledons; also known as a dicot.

differentiation Development toward a more mature state; a process changing a relatively unspecialized cell to a more specialized cell.

diffusion The net movement of molecules from a region of high concentration to one of lower concentration, brought about by their kinetic energy.

dihybrid cross (dy-hy'brid) A genetic cross which takes into account the behavior of two distinct pairs of genes.

dikaryotic cells (dy-kare-ee-ot'ik) Cells having two nuclei.

dimorphic (dy-mor'fik) Having two forms, as in the alternation of generations of some plants.

dinoflagellates (dy"noh-flaj'eh-lates) Single-celled algae, surrounded by a shell of thick interlocking cellulose plates.

dioecious (dy-ee'shus) Having male and female reproductice structures on separate plants; compare with monoecious.

diploid (dip'loid) A chromosome number twice that found in gametes; containing two sets of chromosomes.

directional selection The gradual replacement of one phenotype with another due to environmental change which favors those phenotypes at one of the extremes of the normal distribution.

disaccharide (dy-sak'ah-ride) A two-unit sugar, such as sucrose, which consists of a glucose and a fructose subunit.

displacement activity An innate, stereotyped response that appears to be inappropriate or irrelevant to the situation.

disruptive selection A special type of directional selection in which changes in the environment favor two or more variant phenotypes at the expense of the mean.

distal Remote; farther from the point of reference.

diurnal (dy-ur'nl) Active during the daytime.

DNA Deoxyribonucleic acid; present in chromosomes; contains genetic information coded in specific sequences of its constituent nucleotides.

DNA replication The synthesis of DNA by using a complementary strand of DNA as a template.

dominant allele (al-leel') Gene allele which is always expressed when it is present, regardless of whether it is homozygous or heterozygous.

dorsal (dor'sl) Pertaining to the back of an animal.

double fertilization A process in the angiosperm reproductive cycle where there are two fertilizations: one results in the formation of a young plant; the second results in the formation of the endosperm.

Down syndrome An inherited defect in which individuals have abnormalities of the face, eyelids, tongue, and other parts of the body, and are retarded in both their physical and mental development; results from a trisomy of chromosome 21.

drupe (droop) A simple, fleshy fruit in which the inner wall of the fruit is hard and stony. Peaches and cherries are drupes.

duodenum (doo"uh-dee'num) Portion of the small intestine directly adjacent to the stomach.

ecdysone (ek'dih-sone) The hormone that induces molting (ecdysis) in arthropods.

echinoderms (eh-kine'oh-derms) Spiny-skinned marine animals such as starfish, sea urchins, and sea cucumbers.

ecological isolation A prezygotic isolating mechanism in which gamete exchange is prevented between two groups that are located in the same geographic area because they live and reproduce in two different ecological habitats.

ecological niche The status of an organism within a community or ecosystem; depends on the organism's structural adaptations, physiological responses, and behavior.

ecology (ee-kol'uh-jee) The study of the interrelations between living things and their environment, both physical and biotic.

ecosystem (ee'koh-sis-tem) A natural unit of living and nonliving parts that interact to produce a stable system in which the exchange of materials between living and nonliving is recycled.

ecotone (ee'koh-tone) A fairly broad transition region between adjacent biomes; contains some organisms from each of the two biomes plus some that are characteristic of, and perhaps restricted to, the ecotone.

ectoderm (ek'toh-derm) The outer of the three embryonic germ layers of the gastrula; gives rise to the skin and nervous system.

edaphic factors (ee-daf'ik) Factors in the soil that influence the distribution and numbers of plants and animals.

effector (1) A muscle or gland that contracts or secretes in direct response to nerve impulses; (2) activator of an allosteric enzyme.

efferent (ef'fur-ent) Pertaining to a structure that leads away from another structure or organ, such as the efferent arteriole of the kidney nephron.

ejaculation (ee-jak'yoo-lay-shun) A sudden expulsion, as in the ejection of semen.

electrocardiogram (eh-lek-troh-kar'dee-oh-gram) **(ECG, EKG)** A graphic record made by an electrocardiograph machine of the electrical activity of the heart.

electrochemical potential The potential energy possessed by a system in which there is a difference in electrical charge, as well as a concentration gradient of ions across a membrane.

electroencephalogram (eh-lek-troh-en-sef'uh-loh-gram) **(EEG)** A graphic record of changes in electrical potential associated with activity of the cerebral cortex.

electrolyte (ee-lek'troh-lite) An element or compound that, when melted or dissolved in solvent (usually water), dissociates into ions and is able to conduct an electric current.

electron A negatively charged subatomic particle located at some distance from the atomic nucleus.

electron transport system A series of about 6 chemical reactions during which hydrogens or their electrons are passed along from one acceptor molecule to another, with the release of energy.

element Chemically, one of the 100 or so types of matter, natural or man-made, composed of atoms, all of which have the same number of protons in the atomic nucleus and the same number of electrons circling in orbits.

elimination The ejection of waste products, especially undigested food remnants, from the digestive tract (not to be confused with excretion).

embryo (em'bree-oh) A young organism before it emerges from the egg, seed, or body of its mother; the developing human organism until the end of the second month, after which it is referred to as a fetus.

embryo sac The female gametophyte generation in angiosperms.

endergonic reactions (end"er-gon'ik) Nonspontaneous reactions requiring a net input of free energy.

endocrine glands (en'doh-crin) Glands that secrete products into the blood or tissue fluid instead of into ducts.

endocytosis (en"doh-sy-toh'sis) The active transport of substances into a cell by the formation of invaginated regions of the plasma membrane which pinch off and become cytoplasmic vesicles.

endoderm (en'doh-derm) The inner germ layer of the gastrula lining the archenteron; becomes the digestive tract and its outgrowths—the liver, lungs, and pancreas.

endodermis (en"doh-der'mis) The innermost layer of the cortex in the plant root. Endodermis cells have a Casparian strip running around radial and transverse walls.

endogenous (en-doj'eh-nus) Produced within the body, or due to internal causes.

endolymph (en'doh-limf) The fluid of the membranous labyrinth of the ear.

endomembrane system Membranes within a eukaryotic cell which are either in physical contact with each other or which exchange components through the budding and fusion of vesicles.

endometrium (en"doh-mee'tree-um) Uterine lining.

endoplasmic reticulum (en"doh-plaz'mik reh-tik'yoo-lum) **(ER)** Structure composed of numerous internal membranes within eukaryotic cells.

endorphins (en-dor'finz) Polypeptide transmitter substance of certain brain and visceral neurons whose action is mimicked by opiate alkaloids.

endoskeleton (en"doh-skel'eh-ton) Bony and cartilaginous supporting structures within the body that provide support from within.

endosperm (en'doh-sperm) The nutritive tissue that is found at some point in all angiosperm seeds.

endosymbiont hypothesis (en"doh-sim'bee-ont) Idea that certain organelles such as mitochondria and chloroplasts originated as symbiotic prokaryotes that lived inside larger cells.

endothelium (en-doh-theel'ee-um) The simple epithelial tissue that lines the cavities of the heart, blood and lymph vessels, and the eye.

endothermic (en"doh-ther'mik) The ability of animals to use metabolic energy to maintain a constant body temperature despite variations in environmental temperature.

enhancers Regulatory elements that can be located large distances away from the actual coding regions of a gene.

enkephalins (en-kef'ah-linz) Polypeptide transmitter substances employed by certain brain neurons that seem to function in pain perception and whose action is, like endorphins, mimicked by opiate alkaloids.

enterocoely (en'tur-oh-seel'ee) A type of coelom formation in which the mesoderm forms outpockets that separate to form pouches which, in turn, become the coelom.

entropy (en'trop-ee) Disorderliness; a quantitative measure of randomness or disorder, symbolized by S.

environmental resistance The sum of the physical and biological factors that prevent a species from reproducing at its maximum rate.

enzyme (en'zime) A protein catalyst produced within a living organism that accelerates specific chemical reactions.

eosinophil (ee-oh-sin'oh-fil) A type of white blood cell whose cytoplasmic granules accept acidic stains.

epicotyl (ep'ih-kot"il) The part of the axis of a plant embryo or seedling above the point of attachment of the cotyledons.

epidermis (ep-ih-dur'mis) An outer layer of cells covering the body of plants and animals; functions primarily for protection.

epididymis (ep-ih-did'ih-mis) pl. **epididymides** A coiled tube that receives sperm from the testes and conveys it to the vas deferens.

epigenesis (ep-ih-jen'eh-sis) The theory that development proceeds from a structureless cell by the successive formation and addition of new parts that do not preexist in the fertilized egg.

epinephrine (ep-ih-nef'rin) The chief hormone of the

adrenal medulla; stimulates the sympathetic nervous system.

epiphyte (ep'ih-fite) A plant that is self-nourishing but that grows upon another plant, using it for position and support.

epistasis (ep"ih-sta'sis) Condition in which certain alleles at one locus can alter the expression of alleles at a different locus.

epithelial tissue (ep-ih-theel'ee-al) The type of tissue that covers body surfaces, lines body cavities, and forms glands; also called epithelium.

epoch Geological time that is a subdivision of a period.

era In geology, one of five main divisions of geological time. Eras are divided into periods.

erythrocyte (er-eeth'roh-site) Vertebrate red blood cell.

esophagus (ee-sof'ah-gus) The muscular tube extending from the pharynx to the stomach.

estradiol (es"-trah-di'-ol) The most potent naturally occurring estrogen in humans; principle female sex hormone.

estrogens (es'troh-jens) Female sex hormones produced by the ovarian follicle; promote the development and maintenance of female reproductive structures and of secondary sex characteristics.

estrus (es'trus) The recurrent period of heat or sexual receptivity occurring around ovulation in female mammals having estrous cycles.

ethology (ee-thol'oh-jee) The study of the whole range of animal behavior under natural conditions.

ethylene (eth'ih-leen) A plant hormone involved in various aspects of plant senescence.

eubacteria (yoo'bak-teer"ee-ah) Bacteria other than the archaebacteria.

euchromatin (yoo-croh'-mah-tin) Loosely coiled chromatin which is generally capable of transcription.

Euglena (yoo-glee'nah) An alga-like flagellate protist with chloroplasts.

euglenoids (yoo-glee'noids) A group of protists that resemble *Euglena*.

eukaryote (yoo"kare'ee-ote) Organism whose cells possess nuclei surrounded by membranes.

eustachian tube (yoo-stay'shee-un) The auditory tube passing between the middle ear cavity and the pharynx in vertebrates; permits the equalization of pressure on the tympanic membrane.

eutherian (yoo-theer'ee-an) Pertaining to placental mammals in which a well-formed placenta is present and the young are born at a relatively advanced stage of development; includes all living mammals except monotremes and marsupials.

eutrophication (yoo"troh-fih-kay'shun) Process of progressive nutritional enrichment of bodies of water by natural processes or pollution.

evolution Genetic change in a population of organisms.

excretion (ek-skree'shun) The discharge from the body of a waste product of metabolism (not to be confused with the elimination of undigested food materials).

exergonic (ex-er-gon'ik) A reaction characterized by the release of energy.

exocrine glands (ex'oh-crin) Glands that excrete their products through ducts in the epithelium onto the surface of the skin, such as sweat glands and sebaceous glands.

exocytosis (ex"oh-sy-toh'sis) Export of materials from the cell by fusion of cytoplasmic vesicles with the plasma membrane.

exogenous (ek-sodj'eh-nus) Due to, or produced by, an external cause; not arising within the body.

exon (1) A protein-coding region of a eukaryotic gene; (2) the RNA transcribed from such a region (see intron).

exoskeleton (ex"oh-skel'eh-ton) An external skeleton, such as the shell of mollusks or outer covering of arthropods; provides protection and sites of attachment for muscles.

exotoxin (ex"oh-tok'sin) An extremely potent poison secreted by a bacterial cell to the outside environment.

extension A straightening out; especially used of muscular movement by which a flexed part is made straight.

extensor A muscle that serves to extend or straighten a limb.

exteroceptor (ex'tur-oh-sep"tor) One of the sense organs that receives sensory stimuli from the outside world, such as the eyes or touch receptors.

extinction The death of a species. Extinction occurs when the last individual of a species dies.

F₁ generation (first filial) The first generation of hybrid offspring resulting from a genetic cross.

F₂ generation (second filial) The offspring of the F₁ generation.

facilitated diffusion The passage of ions and molecules, bound to specific carrier proteins, across a cell membrane down their concentration gradients. Facilitated diffusion does not require a special source of energy.

facilitation (1) The promotion or hastening of any natural process; the reverse of inhibition; (2) the effect of a nerve impulse on a postsynaptic neuron which brings the postsynaptic neuron closer to firing.

family In taxonomy, the taxonomic grouping of related, similar genera.

fatty acid One of many organic acids largely composed of long chains of carbon and hydrogen atoms that occur as components of fat.

feces (fee'seez) Solidified waste products eliminated by the digestive tract of an organism.

feedback control System in which the accumulation of the product of a reaction leads to a decrease in its rate of production, or a deficiency of the product leads to an increase in its rate of production.

fermentation Anaerobic respiration that utilizes organic compounds both as electron donors and acceptors.

fertilization Union of male and female gametes.

fetus The unborn offspring after it has largely completed its embryonic development; from the third month of pregnancy to birth in humans.

fiber In plants, a type of sclerenchyma. Fibers are long, tapered cells with thick walls.

fibrin Blood protein that forms clots.

fibroblasts (fy'broh-blasts) Connective tissue cells that secrete intercellular substance.

fibrous root system   A root system in plants that has several main roots without a dominant root.

fiddlehead   A young fern frond; fiddleheads are tightly coiled and resemble the top of a violin.

filament   In botany, the thin stalk of the stamen.

fission (fish′un)   Process of asexual reproduction in which an organism divides into two approximately equal parts.

fitness   See adaptive value.

fixed action patterns (FAP)   Innate, stereotyped behaviors that are characteristic of the species.

flagellum (flah-jel′um)   Long, whiplike movable structure of cells that is used in locomotion or in moving fluid past the tissue of which the cell is a part. Eukaryote flagellae are composed of two single microtubules surrounded by 9 double microtubules, but prokaryote flagellae are filaments rotated by special structures located in the plasma membrane and cell wall.

flexor (flek′sor)   A muscle that serves to bend a limb.

floral tube   A cup formed by the fused basal parts of stamens, petals, and sepals, and which surrounds the ovary in certain flowers.

fluid mosaic model   The modern picture of the plasma membrane (and other cell membranes) in which protein molecules float in a phospholipid bilayer.

follicle (fol′i-kl)   (1) A simple, dry, dehiscent fruit that splits open along one seam to liberate the seeds; (2) a small sac of cells in the mammalian ovary that contains a maturing egg.

food chain   A sequence of organisms through which energy is transferred from its ultimate source in a plant; each organism eats the preceding and is eaten by the following member of the sequence.

food web   Complex feeding relationships in a community of organisms.

foot   In botany, the basal portion of the moss sporophyte that serves to anchor it to the gametophyte.

foramen magnum (for-ay′men)   The opening in the vertebrate skull through which the spinal cord passes.

foramen ovale (for-ay′men oh-vah′lay)   The oval window between the right and left atria, present in the fetus, by means of which blood entering the right atrium may enter the aorta without passing through the lung.

fossil   Parts of an ancient organism, or traces left by previous life.

founder effect   Genetic drift that results from a small population colonizing a new area.

frameshift mutation   Mutation that results when nucleotides are inserted into or deleted from the DNA in numbers that are not multiples of three, thereby altering the reading frame so that all of the codons downstream from the mutation will be changed.

frond   A leaf of a fern.

fruit   In angiosperms, a mature, ripened ovary. Fruits contain seeds and usually serve for seed protection and dispersal.

fucoxanthin (few″koh-zan′thin)   The brown pigment found in diatoms, brown algae, and dinoflagellates.

fungus pl. **fungi**   A plantlike eukaryote incapable of photosynthesis. Most fungi are decomposers; a few are parasitic.

G₁ phase   Gap phase in interphase of the cell cycle before DNA synthesis begins.

gametangium (gam″uh-tan′gee-um)   Special multicellular or unicellular structure of plants, protists, and fungi in which gametes are formed. In fungi, these form at the tips of hyphae of opposite mating types.

gamete (gam′eet)   A cell that functions in sexual reproduction; an egg or sperm whose union, in sexual reproduction, initiates the development of a new individual.

gametic isolation (gam-ee′tik)   A prezygotic isolating mechanism in which gamete exchange between two groups cannot occur because of chemical differences between the gametes of two different species.

gametophyte (gam-ee′toh-fite)   The haploid or gamete-producing stage in the life cycle of a plant.

ganglion (gang′glee-on)   A knotlike mass of the cell bodies of neurons located outside the central nervous system.

gap junction   Structure consisting of specialized regions of the plasma membranes of two adjacent cells containing numerous pores that allow passage of certain molecules and ions between them.

gastrodermis (gas″troh-der′mis)   The tissue lining the gut cavity that is responsible for digestion and absorption in certain phyla such as cnidaria.

gastrula (gas′troo-lah)   Early stage of embryonic development during which the three germ layers form.

gemma (jem′mah)   Vegetative bud in bryophytes that develops asexually into a new plant.

gemmae cup   Saucer-shaped structure found on certain bryophytes; it contains gemmae.

gene   A discrete unit of hereditary information that usually specifies a protein. It consists of DNA and is located on the chromosomes.

gene amplification   Process by which multiple copies of a gene are produced by selective replication, thus allowing for increased synthesis of the gene product.

gene flow   The movement of alleles between local populations, or demes, due to the migration of individuals. Gene flow can have significant evolutionary consequences.

gene pool   All the genes present in a species population.

generation time   The average length of the cell cycle in a group of cells.

genetic code   Code consisting of groups of 3 bases in mRNA which specifies individual amino acids or translation start and stop signals.

genetic drift   A random change in gene frequency in a small breeding population.

genetic load   Those alleles that are advantageous in the heterozygous condition but lethal or deleterious in the homozygous condition.

genome (jee′nome)   A complete set of hereditary factors contained in one haploid assortment of chromosomes.

genotype (jeen′oh-type, jen′oh-type)   The complete genetic makeup of an organism.

genus (jee′nus) A rank in taxonomic classification above the species.

geotropism (jee″oh-troh′pizm) A growth response toward or away from the earth; the influence of gravity on growth.

germ cells Cells within the body that give rise to gametes.

germ layer Primitive embryonic tissue layer; endoderm, mesoderm, or ectoderm.

germ line In animals, the line of cells that will ultimately undergo meiosis to form gametes.

germination The resumption of growth in seeds or spores.

gibberellin (jib″ur-el′lin) A plant hormone involved in various aspects of plant growth and development.

gill The respiratory organ of aquatic animals, usually a thin-walled projection from the body surface or from some part of the digestive tract.

gizzard A portion of the digestive tract specialized for mechanical digestion in some animals.

gland Body structure specialized for secretion.

glial cell (glee′ul) Supporting cell of central nervous tissue, comparable to a connective tissue cell elsewhere in the body, but originating from ectoderm and incapable of producing an extracellular matrix.

globulin (glob′yoo-lin) One of a class of proteins in blood plasma, some of which (gamma-globulins) function as antibodies.

glomerulus (glom-air′yoo-lus) A tuft of minute blood vessels; specifically, the knot of capillaries at the proximal end of a kidney tubule.

glucagon (gloo′kah-gahn) A pancreatic hormone that stimulates glycogenolysis, increasing the concentration of glucose in the blood.

glycocalyx (gly″koh-kay′lix) A coating or layer of oligosaccharides on the outside of an animal cell.

glycogen (gly′koh-jen) A polysaccharide formed from glucose and stored primarily in liver and (to a lesser extent) muscle tissue; the major carbohydrate stored in animal cells.

glycolysis (gly-kol′ih-sis) The metabolic conversion of glucose into pyruvate or lactate with the production of ATP. Glycolysis is a metabolic pathway present in all living cells.

glycoprotein (gly″koh-proh′teen) A protein with covalently attached carbohydrates.

glyoxysome (gly-ox′ih-sohm) Microbody that contains a large array of enzymes; prominent in the fatty endosperm tissue of germinating plant seeds.

goblet cell Mucus-secreting cell common in the intestinal lining.

Golgi body (goal′jee) Also **Golgi complex** or **Golgi apparatus** One of numerous cell organelles found in the cytoplasm of eukaryotic cells. Its membranes modify and sort products of the endoplasmic reticulum.

gonad (goh′nad) A gamete-producing gland; an ovary or testis.

gonadotropins (go-nad″oh-troh′pins) Hormones produced by the anterior pituitary gland and the embryo that stimulate the function of the testes and ovaries.

gradualism The idea that evolutionary change of a species is due to a slow, steady accumulation of changes over time.

grain A simple, dry, indehiscent fruit in which the fruit wall is fused to the seed coat, making it impossible to separate the fruit from the seed. Corn kernels are grains.

grana (gran′ah) Small bodies within chloroplasts that contain alternate layers of chlorophyll, protein, and lipid, and are the functional units of photosynthesis.

gravitropism (grav″ih-troh′pizm) Growth of a plant in response to gravity.

greenhouse effect The warming of the earth resulting from the retention of atmospheric heat caused by the action of certain gases, especially carbon dioxide.

ground meristem A primary meristem (embryonic tissue) that gives rise to cortex and pith in plants.

ground parenchyma Ground tissue; one of three fundamental tissues (the others being epidermis and vascular tissue) found in plants.

guanine (gwan′een) See nucleic acid.

guard cell A cell in the epidermis of plant stems and leaves. Two guard cells form a pore for gas exchange, collectively called a stomate.

guttation (gut-tay′shun) The appearance of water droplets on leaves, forced out through leaf pores by root pressure.

gymnosperms (jim′noh-sperms) Seed plants in which the seeds are not enclosed in an ovary; gymnosperms frequently bear their seeds in cones.

gynoecium (ji-nee′see-um) The female portion (all the carpels) of a flower.

habituation (hab-it″yoo-ay′shun) The process by which organisms become accustomed to a stimulus and cease to respond to it.

hair follicle (fol′i-kul) An epithelial ingrowth of the epidermis into the dermis that surrounds a hair.

haploid (hap′loyd) The chromosome number characteristic of gametes or spores; half the diploid number. In plants, the chromosome number of body cells of the gametophyte generation.

Hardy-Weinberg law The principle that, regardless of dominance or recessiveness, the relative frequencies of allelic genes do not change from generation to generation.

Haversian canals (ha-vur′zee-un) Channels extending through the matrix of bone and containing blood vessels and nerves.

heartwood The inner wood of many woody plants that contains various pigments, tannins, gums, and resins.

helper T cells T lymphocytes that facilitate the ability of B lymphocytes to form an antibody-producing clone in response to an antigen.

hemizygous genes (hem″ih-zy′gus) Genes carried on a heterokaryotic sex chromosome.

hemoglobin (hee′-moh-gloh″bin) The red, iron-containing protein pigment of erythrocytes that transports oxygen and carbon dioxide, and aids in regulation of pH.

hemolysis (hee-mol′ih-sis) The destruction of red blood cells and the resultant escape of hemoglobin.

hemophilia (hee″moh-feel′-ee-ah) ''Bleeder's disease'';

hereditary disease in which blood does not clot properly; caused by a globulin deficiency affecting the production of thromboplastin.

hepatic (heh-pat'ik) Pertaining to the liver.

herbaceous (er-bay'shus) A plant with soft tissues only.

herbivore (erb'i-vore) Animals which only consume plants or algae for their nutritional requirements.

heritability A measure that compares the amount of phenotypic variability that is due strictly to the variation in genes, with the total amount of phenotypic variability in the population.

hermaphrodite (her-maf'roh-dite) An organism which can produce both male and female gametes.

heterochromatin (het″ur-oh-kroh'mah-tin) Highly coiled and compacted chromatin in an inactive state.

heterocysts (het'ur-oh-sists″) Nitrogenase-containing cells of certain cyanobacteria.

heterogamy (het″ur-og'ah-mee) Reproduction involving the union of two gametes that differ in size and structure, such as the egg and sperm.

heterogeneous nuclear RNA (het″ur-oh-jeen'us) (RNA) RNA precursor to mRNA in eukarotes; contains both exons and introns; also called pre-mRNA.

heterosis See hybrid vigor.

heterospory (het″ur-os'pur-ee) Production of two types of spores in plants, microspores, and megaspores.

heterothallic (het-ur-oh-thal'ik) Pertaining to an organism having two mating types; only by combining a plus strain and a minus strain can sexual reproduction occur.

heterotrophs (het'ur-oh-trofes) Organisms that cannot synthesize their own food from inorganic materials and therefore must live either at the expense of autotrophs or upon decaying matter.

heterozygote advantage A phenomenon in which the heterozygous condition confers some special advantage on an individual that either homozygous condition does not, i.e., Aa has a higher degree of fitness than AA or aa.

heteroxygous alleles (het-ur-oh-zye'gus) Genetically mixed. Usually this term refers to a single pair of allelic genes that are unlike each other.

hibernation The dormant state of decreased metabolism in which certain animals pass the winter.

histamine (his'tah-meen) Substance released from mast cells that is involved in allergic and inflammatory reactions.

histones (his'tones) Small, positively charged (basic) proteins in the nucleus which bind to the negatively charged DNA to form the nucleosomes.

holdfast The structure for attachment to solid surfaces found in multicellular algae.

homeobox (home'ee-oh box) Specific DNA sequence found in many of the animal genes that are involved in controlling the development of the body plan.

homeostasis (home″ee-oh-stay'sis) The balanced internal environment of the body; the automatic tendency of an organism to maintain such a steady state.

homeothermic (home″ee-oh-thur'mic) Using heat generated by metabolic activities to maintain a constant body temperature. "Warm blooded." Animals with constant body temperature; examples are birds and mammals. Also called endothermic.

homeotic gene (home″ee-ot'ik) A gene that controls the formation of specific structures during development. Such genes were originally identified through mutants in which one body part was substituted for another.

hominid (hah'min-id) Any of a group of both extinct and living humans.

hominoids The apes and hominids.

homologous chromosomes (hom-ol'ah-gus) Chromosomes that are similar in morphology and genetic constitution. In humans there are 23 pairs of homologous chromosomes, each containing one member from the mother and one member from the father.

homology (hom-ol'oh-jee) Similarity in basic structural plan and development, which is assumed to reflect a common genetic ancestry.

homospory (hoh-mos'pur-ee) Production of one type of spore in plants. The spore gives rise to a bisexual gametophyte.

homozygous (hoh″moh-zy'gus) Possessing a pair of identical alleles.

hormone A chemical messenger produced by an endocrine gland or by certain cells. In animals, hormones are usually transported in the blood and regulate some aspect of metabolism.

humoral immunity (hew'mor-ul) Antibody-mediated immunity residing in the body fluids that is based on immunoglobulin proteins produced by B-lymphocytes.

humus (hew'mus) Organic matter in various stages of decomposition in the soil; a dark mold of decayed vegetable tissue that gives soil a brown or black color.

hybrid breakdown A postzygotic isolating mechanism in which, although the interspecific hybrid is fertile and produces a second (F_2) generation, the F_2 has defects that prevent it from successfully reproducing.

hybrid inviability A postzygotic isolating mechanism in which the embryonic development of an interspecific hybrid is aborted.

hybrid sterility A postzygotic isolating mechanism in which the hybrid cannot reproduce successfully.

hybrid vigor Genetic superiority due to the presence of heterozygosity in a number of different genes.

hybridization Interbreeding between different species.

hydrocarbons Organic compounds composed solely of hydrogen and carbon.

hydrogen bond A weak bond between two atoms formed when a hydrogen atom is shared between two atoms, one of which is usually oxygen; of primary importance in the structure of nucleic acids and proteins.

hydrolysis (hy-drol'ih-sis) The splitting of a compound into parts by the addition of water between certain of its bonds, the hydroxyl group being incorporated into one fragment, and the hydrogen atom into the other.

hydrophilic Attracted to water.

hydrophobic In biology, the property of repelling water molecules.

hydroponics Growing plants in an aerated solution of dissolved inorganic minerals.

hydroxyl (hy-drok'sil) An ion or chemical group with the formula —OH.

hypertonic Having a greater concentration of solute molecules and a lower concentration of solvent (water) molecules, and hence an osmotic pressure greater than that of the solution with which it is compared.

hypha (hy'fah) One of the filaments composing the mycelium of a fungus.

hypocotyl (hy'poh-kah"tl) The part of the axis of a plant embryo or seedling below the point of attachment of the cotyledons.

hypothalamus (hy-poh-thal'uh-mus) Part of the brain that functions in regulating the pituitary gland, the autonomic system, emotional responses, body temperature, water balance, and appetite; located below the thalamus.

hypotonic Having an osmotic pressure or solute content less than that of some standard of comparison.

ileum (il'ee-um) The last portion of the small intestine, extending from the jejunum to the cecum.

immune reaction The production of antibodies or T cells in response to antigens.

immunoglobulins (im-yoon"oh-glob'yoo-lins) Antibodies; highly specific proteins manufactured in response to specific antigens. See also antibodies.

immunological tolerance (im-yoon"uh-loj'ih-kl) The ability of an organism to accept cells transplanted from a genetically distinct organism.

imperfect flower A flower that lacks either stamens or carpels.

implantation (im"plan-tay'shun) (1) The insertion of a part or tissue in a new site in the body; (2) the attachment of the developing embryo to the epithelial lining (endometrium) of the uterus.

impression Fossil in which pressure and heat have destroyed all traces of organic material.

imprinting A form of rapid learning by which a young bird or mammal forms a strong social attachment to an individual (usually a parent) or object within a few hours after hatching or birth.

inbreeding Mating of genetically similar individuals. Homozygosity increases with each successive generation of inbreeding.

incomplete dominance Condition in which neither member of a pair of contrasting alleles is completely expressed when the other is present.

incomplete flower A flower lacking one or more of the four main parts: sepals, petals, stamens, and/or carpel.

indehiscent fruit (in"dih-his'ent) A simple, dry fruit that does not split open at maturity; compare with dehiscent fruit.

independent assortment The mutually independent inheritance of genes that are located on separate chromosome pairs.

indeterminate cleavage Early cell division in deuterostomes that produces blastomeres, any of which is potentially capable of developing into any body part.

index fossil Certain key invertebrate fossils that are found in the same sedimentary layers in different geographical areas. Index fossils help geologists identify comparable layers in widely separate localities.

induced fit Hypothesis that the active site of certain enzymes becomes conformed to the shape of the substrate molecule.

induction Condition in which development of one group of cells is influenced by another group of cells.

inflammation The response of body tissues to injury or infection, characterized clinically by heat, swelling, redness, and pain, and physiologically by increased vasodilatation and capillary permeability.

innate behaviors Behaviors that are inherited and typical of the species.

innate releasing mechanism The hypothetical neural circuits that enable an animal to perceive sign stimuli and respond to them with appropriate muscular contractions.

insertion In biology, the more moveable point of attachment of a skeletal muscle to a bone.

insight learning The intuitive solution to a problem that has not previously been attempted.

instinct A genetically determined pattern of behavior or responses that is not based on the individual's previous experience.

insulin (in'suh-lin) A small protein hormone secreted by the pancreas that lowers blood glucose content.

integral proteins Proteins that span or penetrate the lipid bilayer of cellular membranes.

integration In biology, the process of sorting and interpreting neural impulses leading to an appropriate response.

integumentary system (in-teg"yoo-men'tur-ee) The body's covering, including the skin and its nails, glands, hair, and other associated structures.

intercalated discs (in-tur'kah-lay-ted) Dense bands (representing specialized junctions) between cardiac muscle cells.

interferon (in"tur-feer'on) A protein produced by animal cells when challenged by a virus. Important in immune responses, it prevents viral reproduction and enables cells of the same species to resist a variety of viruses.

intermediate filaments Cytoplasmic fibers that are part of the cytoskeletal network and intermediate in size between microtubules and microfilaments.

interneuron (in"tur-noor'on) A nerve cell that carries impulses to other nerve cells, and is not directly associated with either an effector or a sense receptor.

internode The section of a stem between two nodes.

interoceptor (in"tur-oh-sep'tor) A sense organ within the body organs that transmits information regarding chemical composition, pH, osmotic pressure, or temperature.

interphase The period in the life cycle of a cell in which there is no visible mitotic division; period between mitotic divisions.

interstitial cells (in"tur-stih'shul) Cells located in groups between the seminiferous tubules in the testes; produce the male hormone testosterone.

interstitial fluid The fluid that occupies the spaces between the cells of a tissue.

intertidal zone The zone of a shoreline between the high tide mark and the low tide mark.

intron A non-protein-coding region of a eukaryotic gene and also of the RNA transcribed from such a region. Introns do not appear in mature mRNA (see exon).

invagination (in-vaj"ih-nay'shun) The infolding of one part within another, specifically a process of gastrulation in which one region folds in to form a double-layered cup.

inversion, chromosomal Turning a segment of a chromosome end for end and attaching it to the same chromosome.

ion An atom or a group of atoms bearing an electric charge, either positive (cation) or negative (anion).

iris Circular, muscular structure that separates the chambers of the vertebrate eye and serves to limit the entrance of light.

ischemia (is-kee'mee-ah) Deficiency of blood to a body part due to functional constriction or actual obstruction of a blood vessel.

islets of Langerhans (eye'lets of lahng'er-hanz) The endocrine portion of the pancreas. Alpha cells of the islets release glucagon; beta cells secrete insulin. These hormones regulate blood-sugar level.

isogamy (eye-sog'ah-mee) Reproduction resulting from the union of two gametes that are similar in size and structure.

isomer (eye'soh-mur) One of two or more chemical compounds having the same chemical formula but a different structural formula, such as glucose and fructose.

isotonic (eye"soh-ton'ik) Having identical concentrations of solute and solvent molecules, and hence the same osmotic pressure as the solution with which it is compared.

isotope (eye'suh-tope) An alternate form of an element with a different number of neutrons but the same number of protons and electrons.

isozymes (eye'soh-zimes) Different molecular forms of proteins with almost the same enzymatic activity.

jejunum (jeh-joo'num) The middle portion of the small intestine.

juvenile hormone An arthropod hormone that preserves juvenile morphology during a molt. Without it, metamorphosis toward the adult form takes place.

karyokinesis (kare"ee-oh-kin-ee'sis) The four-stage division of the nucleus in which nuclear material is distributed equally during mitosis and meiosis.

karyotype (kare'ee-oh-type) The chromosomal constitution of an individual. Representations of the karyotype are generally prepared by photographing the chromosomes and arranging the homologous pairs according to size and centromere position.

keratin (kare'ah-tin) A horny, water-insoluble protein found in the epidermis of vertebrates and in nails, feathers, hair, and horns.

ketone bodies (kee'tone) Substances such as acetone produced during fat metabolism; incompletely oxidized fatty acids which are toxic in high concentrations.

ketosis (kee-toh'sis) Abnormal condition marked by an elevated concentration of ketone bodies in the body

fluids and tissues; a complication of diabetes mellitus and starvation.

kilocalorie See calorie.

kinesis (kih-nee'sis) The activity of an organism in response to a stimulus; the direction of the response is not controlled by the direction of the stimulus (in contrast to a taxis).

kinetochore (kin-eh'toh-kore) Portion of the chromosome centromere to which mitotic spindle fibers attach.

kinesthesis (kin"es-thee'sis) Sense that gives us our awareness of the position and movement of various parts of the body.

kingdoms The broadest category of classification commonly used; (however, kingdoms are sometimes grouped in the domains Prokaryota and Eukaryota).

kinins (kin'ins) (1) Group of polypeptides produced in blood and tissues that act on blood vessels, smooth muscles, and certain nerve endings; e.g., bradykinin or kallidin; (2) one of a group of compounds containing adenine that stimulates cell division and growth of plant cells in tissue culture.

Krebs cycle See citric acid cycle.

labia (lay'bee-uh) The liplike borders of the vulva.

labyrinthodont (lab"ih-rin'thoh-dant) A member of a subclass of extinct amphibians in which the enamel of the teeth is folded in a characteristic, complex fashion. Labyrinthodonts are considered ancestral to terrestrial vertebrates.

lactation (lak-tay'shun) (1) The production or release of milk from the breast; (2) the period during which the child is nourished from the breast.

lacteal (lak'tee-al) One of the many lymphatic vessels in the intestinal villi that absorb fat.

lagging strand Strand of DNA that is synthesized in short pieces. Since its net direction of synthesis at the replication fork is 3' to 5', these short pieces are then covalently joined by DNA ligase. See also leading strand.

lamella (lah-mel'ah) A thin leaf or plate, as of bone.

larva An immature free-living form in the life history of some animals in which it may be unlike the parent.

larynx (lare'inks) The organ at the upper end of the trachea that contains the vocal cords.

latent learning Learning in which an animal stores (without apparent reward) information about its environment that can later influence its behavior.

latent period An interval, lasting about 0.01 second, between the application of a stimulus and the beginning of the visible shortening of a muscle.

lateral meristem An area of cell division located on the side of the plant. There are two lateral meristems, the vascular cambium and the cork cambium.

leader sequence Non-coding sequence of nucleotides in mRNA that is transcribed from the region that precedes (is upstream to) the coding region.

leading strand Strand of DNA that is synthesized continuously. See also lagging strand.

leaf primordium (pry-mor'dee-um) An embryonic leaf, evident in the stem apical meristem.

learning A change in the behavior of an animal that results from experiences during its lifetime.

legume (leg′yoom) A simple, dry, dehiscent fruit that splits open along two seams to release its seeds.

lenticels (len′tih-sels) Masses of cells that rupture the epidermis and form porous swellings in stems, facilitating the exchange of gases.

leukemia (loo-kee′mee-uh) A progressive, malignant disease of the blood-forming tissues, characterized by uncontrolled multiplication of, and abnormal development of, white blood cells and their precursors.

leukocytes (loo′koh-sites) White blood cells; colorless cells that defend the body against disease-causing organisms; exhibit ameboid movement.

leukoplasts (loo′koh-plasts) Colorless plastids that act as centers for the storage of materials in the cytoplasm of certain kinds of plant cells.

lichens (ly′kenz) Plant-like compound organisms composed of symbiotic algae and fungi.

ligament (lig′uh-ment) A connective tissue cable or strap that connects bones to each other or holds other organs in place.

lignin (lig′nin) The substance responsible for the hard, woody nature of plant stems and roots.

limbic system An action system of the brain that plays a role in emotional responses, motivation, autonomic function, and sexual response.

linkage The tendency for a group of genes located on the same chromosome to be inherited together in successive generations.

lipase (lip′ase) Fat-digesting enzyme.

lipid Any of a group of organic compounds which are insoluble in water but soluble in fat solvents; lipids serve as a storage form of fuel and an important component of cell membranes.

liposome (lip′uh-sohm) A protobiont made from lipids.

lithosphere (lith′oh-sfeer) Portion of the planet that is composed of rock.

littoral (lit′or-ul) The region of shallow water near the shore between the high and low tide marks.

locus The particular point on the chromosome at which the gene for a given trait occurs.

long-day plant A plant that flowers in response to long days (and short nights); compare with short-day plant.

loop of Henle (hen′lee) The U-shaped loop of a mammalian kidney tubule which extends down into the medulla.

lower epidermis The outermost layer covering the bottom of leaf blades.

lumen (loo′men) The cavity or channel within a tube or tubular organ, such as a blood vessel or the digestive tract.

luteinizing hormone (loo′tin-iz-ing) **(LH)** Anterior pituitary lobe hormone that stimulates ovulation and development of the corpus luteum; stimulates the production of progesterone; in males, stimulates testosterone secretion.

lymph (limf) The colorless fluid that is derived from blood plasma and resembles it closely in composition; contains white cells.

lymph nodes A mass of lymph tissue surrounded by a connective tissue capsule; manufactures lymphocytes and filters lymph.

lymphocyte (limf′oh-site) White blood cell with nongranular cytoplasm that is responsible for immune responses.

lysis (ly′sis) The process of disintegration of a cell or some other structure.

lysosome (ly′soh-some) Intracellular organelles present in many animal cells; contain a variety of hydrolytic enzymes which act when the lysosome ruptures or fuses with another vesicle. Lysosomes function in development and in phagocytosis.

macroevolution (mak″roh-eh-voh-loo′shun) Large-scale evolutionary change; evolutionary change involving higher taxa, such as genera and orders, i.e., above the level of species.

macromolecule Very large molecules such as a protein or nucleic acid.

macronucleus A large polyploid nucleus found, along with one or several micronuclei, in ciliates. The macronucleus regulates in cell metabolism and growth.

macronutrient An essential element that is required in fairly large amounts for normal plant growth.

macrophage (mak′roh-faje) A large phagocytic cell capable of ingesting and digesting bacteria and cellular debris.

malpighian tubule (mal-pig′ee-an) The excretory organ of many arthropods.

mandible (man′dih-bl) The lower jaw of vertebrates or the main external mouthparts of certain arthropods such as insects.

mantle Thin outside layer of mollusk body that is usually responsible for the production of the shell.

marsupials (mar-soo′pee-uls) A subclass of mammals, the Metatheria, characterized by the possession of an abdominal pouch in which the young are carried for some time after being born in a very undeveloped condition.

mass extinction The extinction of numerous species and higher taxa during a relatively short period of geological time. Compare with background extinction.

mast cell A type of cell found in connective tissue; contains histamine and is important in allergic reactions.

matrix (may′triks or mat′riks) (1) Nonliving material secreted by and surrounding connective tissue cells; frequently contains a thick, interlacing matted network of microscopic fibers; (2) the interior of the compartment formed by the inner membrane of mitochondria.

mechanical isolation A prezygotic isolating mechanism in which gamete exchange between two groups is prevented by morphological or anatomical differences between them.

mechanoreceptor (meh-kan′-oh-ree-sep″tor) A sense cell or sense organ that perceives mechanical stimuli such as those of touch, pressure, hearing, and balance.

medulla (meh-dul′uh) (1) The inner part of an organ, such as the medulla of the kidney; (2) the most posterior part of the brain, lying next to the spinal cord.

medulla oblongata (ob-long-gah′tah) Lowest portion of

the brain; contains vital centers that control heartbeat, respiration, and blood pressure.

medusa A jellyfish; a free-swimming, umbrella-shaped stage in the life cycle of certain cnidarians.

megaphyll (meg′uh-fil) A large leaf that contains multiple vascular strands. Megaphylls are found in ferns, gymnosperms, and angiosperms. (Compare with microphyll.)

megasporangium (meg″ah-spor-an′jee-um) A spore sac containing megaspores.

megaspore (meg′ah-spor) The haploid spore in heterosporous plants that gives rise to a female gametophyte.

megaspore mother cell A diploid cell in the megasporangium that undergoes meiosis to form megaspores.

meiosis (my-oh′sis) Division of the cell nucleus that produces haploid cells; produces gametes in animals and spores in plants.

melanin (mel′ah-nin) A dark brown to black pigment common in the integument of many animals and sometimes found in other organs; usually occurs within special cells called melanocytes.

meninges (men-in′jeez) sing. **meninx** The three membranes that envelop the brain and spinal cord: the dura mater, arachnoid, and pia mater.

menopause The period (usually from 45 to 55 years of age) in women when the recurring menstrual cycle ceases.

menstruation (men-stroo-ay′shun) The monthly discharge of blood and degenerated uterine lining in the human female; marks the beginning of each menstrual cycle.

meristem (mer′ih-stem) A localized area of mitosis and growth in the plant body.

mesenchyme (mes′en-kime) A loose, often jelly-like connective tissue containing undifferentiated cells, found in the embryos of vertebrates and the adults of some invertebrates.

mesentery (mes′en-tare″ee) Internal membrane of vertebrates and echinoderms which holds the viscera in place.

mesoderm (mez′oh-derm) The middle layer of the three basic tissue layers that develop in the early embryo; gives rise to connective tissue, muscle, bone, blood vessels, kidneys, and many other structures; lies between the ectoderm and the endoderm.

mesoglea (mes″oh-glee′ah) A gelatinous matrix located between the ectoderm and endoderm of cnidarians.

mesophyll (mez′oh-fil) Photosynthetic cells in the interior of a leaf.

mesophytes (mes′oh-fites) Common land plants that live in a climate with an average amount of moisture.

mesosome (mes′oh-some) Invaginated complex of membranes in some bacteria, thought to function in cell respiration.

mesothelium (mez-oh-theel′ee-um) The simple squamous cell layer of epithelium (derived from mesoderm) that covers the surface of the serous membranes (peritoneum, pericardium, pleura).

messenger RNA (mRNA) RNA that has been transcribed from DNA that specifies the amino acid sequence of a protein in eukaryotes and prokaryotes.

metabolism The sum of all the physical and chemical processes by which living organized substance is produced and maintained; the transformations by which energy and matter are made available for use by the organism.

metamorphosis (met″ah-mor′fuh-sis) An abrupt transition from one developmental stage to another, such as from a larva to an adult.

metaphase (met′ah-faze) The middle stage of mitosis and meiosis during which the chromosomes line up in the equatorial plate.

metastasis (me-tas′tuh-sis) The transfer of a disease such as cancer from one organ or part of the body to another not directly connected to it.

metazoan (met″ah-zoh′an) Multicellular animals; all animals except protozoans.

microbodies Membrane-bounded eukaryote cellular structures containing enzymes.

microevolution Changes in gene frequencies that occur within a population over successive generations.

microfilaments Tiny rodlike structures with contractile properties that make up part of the internal skeletal framework of the cell. Microfilaments are composed of actin.

micronucleus One or more smaller nuclei found, along with the macronucleus, in ciliates. The micronucleus is involved in reproduction.

micronutrient An essential element that is required in trace amounts for normal plant growth.

microphyll A small leaf that contains one vascular strand; microphylls are found in horsetails and club mosses. (Compare with megaphyll.)

microsphere A protobiont formed by adding water to proteinoids.

microsporangia (my″kroh-spor-an′jee-ah) Small "pollen" sacs that contain microspore mother cells which divide by meiosis to form microspores.

microspore The haploid spore in heterosporous plants that gives rise to a male gametophyte.

microspore mother cell A diploid cell in the microsporangium that undergoes meiosis to form microspores.

microtubules (my-kroh-too′bewls) Hollow, cytoplasmic cylinders, composed mainly of tubulin protein, that comprise such organelles as flagella and centrioles, and serve as a skeletal component of the cell.

microvilli (my-kroh-vil′ee) Minute projections of the cell membrane which increase the surface area of the cell; found mainly in cells concerned with absorption or secretion, such as those lining the intestine or kidney tubules.

mimicry (mim′ik-ree) An adaptation for survival in which an organism resembles some other living or nonliving object.

mineralocorticoids (min″ur-al-oh-kor′tih-koidz) Hormones produced by the adrenal cortex that regulate mineral metabolism and, indirectly, fluid balance. The principal mineralocorticoid is aldosterone.

miracidium (mir″ah-sid′ee-um) The first larval stage of parasitic flukes.

mitochondria (my″toh-kon′dree-ah) Spherical or elongate intracellular organelles that contain the electron transport system and certain other enzymes; site of oxidative phosphorylation. Sometimes referred to as the powerhouses of the cell.

mitosis (my-toh′sis) Division of the cell nucleus resulting in the distribution of a complete set of chromosomes to each daughter cell; cytokinesis (actual division of the cell itself) usually occurs during the telophase stage of mitosis. Mitosis consists of four phases: prophase, metaphase, anaphase, and telophase.

mitotic spindle (my-tot′ik) Structure consisting mainly of microtubules that provides the framework for chromosome movement during cell division.

mitral valve (my′trul) The bicuspid heart valve located between the left atrium and left ventricle.

mold Fossil formed by the hardening of material surrounding a buried organism and the following decay and removal of the tissues.

mole The amount of a chemical compound whose mass in grams is equivalent to its molecular weight, the sum of the atomic weights of its constituent atoms.

molecule The smallest particle of a covalently bonded element or compound that has the composition and properties of a larger part of the substance.

molting The shedding and replacement of an outer covering such as hair, feathers, or exoskeleton.

monerans (moh-nair′unz) The simplest prokaryotic microorganisms, the bacteria and blue-green algae; forms lacking true nuclei or plastids, and in which sexual reproduction is rare or absent.

monocot (mon′oh-kot) One of the two classes of angiosperms, or flowering plants. Monocots get their name (mono = one) from having one cotyledon in the seed.

monocyte (mon′oh-site) A white blood cell; a large phagocytic, nongranular leukocyte that enters the tissues and differentiates into a macrophage.

monoecious (mon-ee′shus) Having separate male and female reproductive parts on the same plant; compare with dioecious.

monomer (mon′oh-mer) A simple molecule of a compound of relatively low molecular weight which can be linked with others to form a polymer.

monosaccharide (mon-oh-sak′ah-ride) A simple hexose sugar; one that cannot be degraded by hydrolysis to a simpler sugar.

monosomy (mon′uh-soh″mee) Abnormal condition in which one member of a specific chromosome pair is absent.

monotremes (mon′oh-treems) Egg-laying mammals such as the duck-billed platypus of Australia.

morphogenesis (mor-foh-jen′eh-sis) The development of the form and structures of the body and its parts by precise movements of its cells.

morula (mor′yoo-lah) An early embryo consisting of a solid ball of large cells.

mosaic evolution Two traits within a species that evolved independently of one another and at different rates.

motor unit All the skeletal muscle fibers that are stimulated by a single motor neuron.

mucosa (mew-koh′suh) A mucous membrane, especially in the lining of the digestive and respiratory tracts.

mucus (mew′cus) A viscous secretion that serves to lubricate body parts and to trap particles of dirt or other contaminants. (The adjectival form is spelled mucous.)

Müllerian mimicry Mimicry of one species by another, where both species are dangerous to some predator.

multiple alleles (al′leels) Three or more alternative allelic genes which govern the same trait, as in the ABO series of blood types.

multiple fruit A fruit that develops from many ovaries of many separate flowers to form one fruit. Pineapples are multiple fruits.

muscle An organ which by contraction produces movement.

mutagen (mew′tah-jen) Any agent that is capable of producing mutations.

mutation A change in the nucleotide base pairs of a gene, or a rearrangement of genes within chromosomes so that their interactions produce different effects; a change in the chromosomes themselves.

mutualism An association whereby two organisms of different species each gain from being together and are often unable to survive separately.

mycelium (my-seel′ee-um) The vegetative body of fungi and certain protists (water molds); consists of a branched network of hyphae.

mycorrhizae (my″kor-rye′zee) Mutualistic associations of fungi that aid and plant roots that aid in the absorption of materials.

myelin (my′eh-lin) The fatty material that forms a sheath around the axons of nerve cells in the central nervous system and in certain peripheral nerves.

myelin sheath The fatty insulating covering around axons of certain neurons.

myocardium (my-oh-kar′dee-um) The middle and thickest layer of the heart wall, composed of cardiac muscle.

myofibrils (my-oh-fy′bril) Tiny threadlike organelles found in the cytoplasm of striated and cardiac muscle that are responsible for contractions of the cell.

myofilament (my-oh-fil′uh-munt) One of the filaments making up the myofibril; the structural unit of muscle proteins in a muscle cell.

myoglobin (my′oh-gloh″bin) A hemoglobin-like oxygen transferring protein found in muscle.

myoneural cleft (my-oh-new′rul) The space, corresponding to a synaptic cleft, between a motor neuron ending and a muscle cell.

myosin (my′oh-sin) A protein which, together with actin, is responsible for muscle contraction.

nares (nare′eez) The openings of the nasal cavities. External nares open to the body surface; internal nares, to the pharynx.

navigation   The process in which an animal finds a goal without relying on landmarks with which it is familiar.

needle   The leaf of a conifer such as pine.

nekton (nek'ton)   Collective term for free-swimming aquatic animals that are essentially independent of water movements.

nematocyst (nem-at'oh-sist)   A minute stinging structure found within cnidocytes (stinging cells) in cnidarians; used for anchorage, defense, and capturing prey.

neoteny (nee-ot'eh-nee)   Sexual maturity in a larval or otherwise immature stage.

nephridium (neh-frid'ee-um)   The excretory organ of the earthworm and other annelids which consists of a ciliated funnel, opening into the next anterior coelomic cavity and connected by a tube to the outside of the body.

nephron (nef'ron)   The functional, microscopic unit of the vertebrate kidney.

neritic zone (ner-ih'tik)   Zone of shallow water adjacent to the seacoast.

nerve   A large bundle of axons (or dendrites) wrapped in connective tissue that conveys impulses between the central nervous system and some other part of the body.

net primary productivity   Total carbon fixation by a plant community, minus the carbon lost by respiration.

neural crest (noor'ul)   Group of cells along the neural tube that migrate and form structures of the peripheral nervous system and certain other structures.

neuroglia (noor-og'lee-ah)   Connecting and supporting cells in the central nervous system.

neurohumor (noor"oh-hew'mor)   A substance secreted by the tip of a neuron that is able to activate a neighboring neuron or muscle.

neuron (noor'on)   A nerve cell; a conducting cell of the nervous system which typically consists of a cell body, dendrites, and an axon.

neurosecretion   The production of hormones by nerve cells.

neurotransmitter   Substance used by neurons to transmit impulses across a synapse. Also called transmitter substance.

neutral variation   Genetic variation in a population that may confer no selective advantage to the organisms possessing it.

neutrons (noo'tronz)   Electrically uncharged particles of matter existing along with protons in the atomic nucleus of all elements except the mass 1 isotope of hydrogen.

neutrophil (new'truh-fil)   A type of granular leukocyte.

nitrogen fixation   The ability of certain microorganisms to bring atmospheric nitrogen into chemical combination.

node   The area on a plant stem where the leaves attach.

noncompetitive inhibitor   A substance that permanently destroys the ability of an enzyme to function.

nondisjunction   Abnormal separation of homologous chromosomes or sister chromatids caused by their failure to disjoin (move apart) properly during cell division.

nonpolar covalent bond   Chemical bond in which electrons are shared equally among the participating atoms and which does not produce any electrical charge within the molecule.

nonsense mutation   Mutation that results in an amino acid specifying codon being changed to a termination codon. When the abnormal mRNA is translated, the resulting protein usually is truncated and non-functional.

norepinephrine (nor-ep-ih-nef'rin)   A neurotransmitter substance that is also a hormone secreted by the adrenal medulla.

notochord (no'toe-kord)   The flexible, longitudinal rod in the anteroposterior axis that serves as an internal skeleton in the embryos of all chordates and in the adults of some.

nuclear envelope   The double membrane system that surrounds the cell nucleus of eukaryotes.

nuclear pores   Structures in the nuclear envelope that appear to allow passage of certain molecules between the cytoplasm and the nucleus.

nucleic acid (noo-klay'ik)   DNA or RNA. A polymer composed of nucleotides that contain the purine bases adenine and guanine and/or the pyrimidine bases cytosine, thymine, and uracil.

nucleolus (new-klee'-oh-lus)   Specialized structure in the nucleus formed from regions of several chromosomes; site of ribosome synthesis.

nucleoside (new'klee-oh-side)   Molecule consisting of a nitrogenous base and a pentose sugar.

nucleosome (new'klee-oh-sohm)   Repeating unit of chromatin structure consisting of a length of DNA wound around a complex of eight histone molecules (two of each of four different types) plus a DNA linker region associated with a fifth histone protein.

nucleotide (noo'klee-oh-tide)   A molecule composed of a phosphate group, a 5-carbon sugar (ribose or deoxyribose) and a nitrogenous base (purine or pyrmidine); one of the subunits into which nucleic acids are split by the action of nucleases.

nucleus (new'klee-us)   (1) That portion of an atom that contains the protons and neutrons; the core; (2) a cellular organelle containing DNA and serving as the control center of the cell; (3) a mass of nerve cell bodies in the central nervous system.

nutrients   The chemical substances in food that are used by the body as components for synthesizing needed materials and for fuel.

nymph   A juvenile insect that often resembles the adult stage and that will become an adult without an intervening pupal stage.

obligate anaerobe (ob'lih-gate an'air-obe)   An anaerobic organism that is killed by oxygen.

ocellus (oh-sell'us)   A simple light receptor found in many different types of invertebrate animals.

olfaction (ol-fak'shun)   The act of smelling.

ommatidium (om"ah-tid'ee-um)   One of the light-detecting units of a compound eye, consisting of a cornea, lens, and rhabdome.

omnivores (om'nih-vore)   An animal which consumes both plant and animal material.

oncogene (on′koh-jeen) Any of a number of genes that usually play an essential role in cell growth or division, and that cause the formation of a cancer cell when mutated; also known as cellular oncogenes.

oncotic pressure (on-kot′ik) The osmotic pressure exerted by colloids in a solution; in microcirculatory fluid dynamics it operates to counterbalance capillary blood pressure.

ontogeny (on-toj′uh-nee) The complete developmental history of the individual organism.

Onychophora (on-ih-kof′or-ah) Rare, tropical, caterpillarlike animals, structurally intermediate between annelids and arthropods, possessing an annelidlike excretory system, an insectlike respiratory system, and claw-tipped short legs.

oocytes (oh′uh-sites) Cells that give rise to egg cells (ova) by meiosis.

oogamy (oh-og′ah-me) The fertilization of a large nonmotile female gamete by a small, motile male gamete.

oogenesis (oh″oh-jen′eh-sis) Production of female gametes (eggs).

oogonium (oh″oh-goh′nee-um) The primordial cell from which the ovarian egg arises; undergoes growth to become a primary oocyte.

operant (instrumental) conditioning A type of learning in which an animal is rewarded or punished for performing a behavior it discovers by chance.

operator site The site on DNA to which repressor molecules are bound, thereby inhibiting the synthesis of mRNA by the genes in the adjacent operon; adjacent to the structural genes in the operon.

operon (op′er-on) In prokaryotes, a group of structural genes that are transcribed as a single message plus their associated regulatory elements. An operon is controlled by a single repressor.

optimal foraging The theory that animals feed in a manner that maximizes benefits and/or minimizes costs.

orbital Any one of the permissible patterns of motion of an electron in an atom or molecule.

order In taxonomic classification, a group of related, similar families.

organ A differentiated part of the body made up of tissues and adapted to perform a specific function or group of functions, such as the heart or liver.

organelle One of the specialized structures within the cell, such as the mitochondria, Golgi complex, ribosomes, or contractile vacuole.

organizer A part of an embryo that influences some other part and directs its histological and morphological differentiation.

organogenesis The development of organs.

osmoregulation (oz″moh-reg-yoo-lay′shun) The active regulation of the osmotic pressure of body fluids so that they do not become excessively dilute or excessively concentrated.

osmosis (oz-moh-sis) Diffusion of water (the principal solvent in biological systems) through a selectively permeable membrane from a region of higher concentration of water to a region of lower concentration of water.

osmotic pressure A measure of the solute concentration of a solution.

osteoblast (os′tee-oh-blast) Primitive bone cells that give rise to mature bone.

osteocyte (os′tee-oh-site) A mature bone cell; an osteoblast that has become embedded within the bone matrix and occupies a lacuna.

osteon (os′tee-on) Spindle-shaped unit of bone composed of concentric layers of osteocytes; Haversian system of bone.

outbreeding The mating of individuals of unrelated strains.

ovary (oh′var-ee) (1) In animals, one of the paired female gonads; responsible for producing eggs and sex hormones; (2) in flowering plants, the base of the carpel that contains ovules. Ovaries develop into fruits after fertilization.

oviduct (oh′vih-dukt) Tube that carries ova from the ovary to the uterus, cloaca, or body exterior.

oviparous (oh-vip′ur-us) Bearing young in the egg-stage of development; egg-laying.

ovoviviparous (oh″voh-vih-vip′ur-us) A type of development in which the young hatch from eggs laid outside the mother's body.

ovulation (ov-yoo-lay′shun) The release of a mature egg from the ovary.

ovule (ov′yool) The part (i.e., megasporangium) which develops into the seed after fertilization.

ovum The female gamete, or egg. See also meiosis.

oxidation The loss of electrons, or in organic chemistry, the loss of hydrogen from a compound.

oxidative phosphorylation (fos″for-ih-lay′shun) The production of ATP using energy derived from the transfer of electrons in the electron transport system of the mitochondria.

oxygen debt The accumulation of lactic acid in muscles during strenuous exercise, followed by rapid breathing that provides the extra oxygen necessary to metabolize the lactic acid.

oxytocin (ok-see-toh′sin) A hormone produced by the hypothalamus and released by the posterior lobe of the pituitary; causes the uterus to contract and stimulates the release of milk from the lactating breast.

pacemaker, of the heart A specialized node of tissue in the wall of the right atrium in which the impulses regulating the heartbeat originate.

paedomorphosis (pee″doh-mor-foh′sis) Evolutionary retention of characteristics in the adult juvenile.

palate (pal′ut) Horizontal partition separating the nasal and oral cavities; the roof of the mouth.

palisade mesophyll The vertically stacked photosynthetic cells near the upper epidermis in dicot leaves.

pancreas (pan′kree-us) Large digestive gland located in the vertebrate abdominal cavity. The pancreas produces a digestive secretion containing lipase, peptidase, amylase, and nuclease enzymes; also serves as an endocrine gland, secreting the hormones insulin and glucagon into the bloodstream.

papilla (pa-pil′ah) pl. **papillae** A small nipplelike projection or elevation, such as the papilla at the base of each hair follicle.

paramylum (par"uh-my'lum) Carbohydrate storage compound present in euglenoids; chemically distinct from both starch and glycogen.

parapodia (par"uh-poh'dee-ah) Paired, thickly bristled paddles extending laterally from each segment of polychaete worms.

parasitism An intimate living relationship between organisms of two different species in which one benefits and the other is harmed.

parasympathetic A division of the autonomic nervous system concerned primarily with the control of the internal organs; functions to conserve or restore energy.

parathyroids Small, pea-sized glands closely adjacent to the thyroid gland; their secretion regulates calcium and phosphate metabolism.

parenchyma (par-en'kih-mah) Plant cells that are relatively unspecialized, are thin-walled, may contain chlorophyll, and are typically rather loosely packed; they function in photosynthesis and in the storage of nutrients.

parotid glands (pah-rot'id) The largest of the three main pairs of salivary glands.

parthenogenesis (par"theh-noh-jen'eh-sis) The development of an unfertilized egg into an adult organism; common among honey bees, wasps, and certain other arthropods.

passive immunity Temporary immunity derived from the immunoglobulins of another organism.

pathogen (path'oh-jen) An organism capable of producing disease.

pelagic (pel-aj'ik) An organism that inhabits open water, as in mid-ocean.

pellicle (pel'ih-kl) A flexible, proteinaceous covering over the body of certain protists.

penis The male sexual organ of copulation in reptiles and mammals.

pepsin (pep'sin) The chief enzyme of gastric juice; hydrolyzes proteins, particularly collagen.

peptide (pep'tide) A compound consisting of a chain of amino acid groups. A dipeptide consists of two amino acids, an octapeptide of eight, a polypeptide of many.

peptide bond A distinctive covalent carbon-to-nitrogen bond that links amino acids in peptides and proteins.

peptidoglycan (pep"tid-oh-gly'kan) A modified protein or peptide possessing an attached carbohydrate.

perennial (pur-en'ee-ul) A plant that grows year after year. Perennials may be wooly or herbaceous.

perfect flower A flower that has both stamens and carpels.

pericardium (pare-ih-kar'dee-um) The fibrous sac that surrounds the heart and roots of the great blood vessels; also forms the outer layer of the heart wall.

pericycle (pehr'eh-sy"kl) A layer of meristemmatic cells in roots that gives rise to branch roots.

periderm (pehr'ih-durm) Layers of cells covering the surface of woody stems and roots (i.e., the outer bark). Anatomically, the periderm is composed of cork cells, cork cambium, and cork parenchyma, along with traces of primary tissues.

period In geology, geological time that is a subdivision of an era. Each period is divided into epochs.

peripheral nervous system The nerves and ganglia that lie outside the central nervous system.

peripheral proteins Proteins associated with the surface of biological membranes.

peripheral resistance The hindrance to blood flow caused by friction between blood and the wall of the blood vessel; plays an important role in determining blood pressure.

peristalsis (pehr"ih-stal'sis) Powerful, rhythmic waves of muscular contraction and relaxation in the walls of hollow tubular organs, such as the ureter or parts of the digestive tract, that serve to move the contents through the tube.

peritoneum (pehr-ih-tuh-nee'um) The membrane lining the abdominal and pelvic cavities (the parietal peritoneum) and the membrane forming the outer layer of the stomach and intestine (the visceral peritoneum).

peroxisomes (pehr-ox'ih-somz) Lysosome-like vesicles containing enzymes that produce or degrade hydrogen peroxide.

petals The colored cluster of modified leaves that constitute the next-to-outermost portion of a flower.

petiole (pet'ee-ohl) The stalk by which a leaf is attached to a stem.

petrifaction A fossil in which the soft tissues of the organism are replaced by minerals such as iron pyrites, silica, and calcium.

pH The negative logarithm of the hydrogen ion concentration by which the degree of acidity or alkalinity of a fluid may be expressed.

phagocyte (fag'oh-site) A cell such as a macrophage or neutrophil that ingests microorganisms and foreign particles.

phagocytosis (fag"oh-sy-toh'sis) Literally, "cell eating"; a type of endocytosis by which certain cells engulf food particles, microorganisms, foreign matter, or other cells.

pharynx (far'inks) That part of the digestive tract from which the gill pouches or slits develop; in higher vertebrates it is bounded anteriorly by the mouth and nasal cavities, and posteriorly by the esophagus and larynx; the throat region in humans.

phenotype (fee'noh-type) The physical or chemical expression of an organism's genes (see also genotype).

phenylketonuria (fee"nl-kee"toh-noor'ee-ah) An inherited disease in which there is a deficiency of the enzyme that normally converts phenylalanine to tyrosine; if untreated, results in mental retardation.

pheromone (feer'oh-mone) A substance secreted by one organism to the external environment that influences the development or behavior of other members of the same species.

phloem (floh'em) Vascular tissue that conducts food in plants.

phloem fiber cap A cap of fibers formed just outside the phloem in vascular bundles of many dicot stems.

phospholipids (fos"foh-lip'idz) Fatlike substances in which a nitrogen- and phosphorus-containing group occurs in place of one of the fatty acids. Phospholipids comprise most of the plasma and internal membranes of cells.

phosphorylation (fos"for-ih-lay'shun) The introduction of a phosphate group into an organic molecule.

photic zone (foh'tik) Zone of aquatic habitats lying near enough the surface so that sufficient light is present for photosynthesis to take place.

photolysis (foh-tol'uh-sis) The photochemical dissolution of water in the light-dependent reactions of photosynthesis.

photon (foh'ton) A particle of electro-magnetic radiation; one quantum of radiant energy.

photoperiodism (foh"toh-peer'ee-od-izm) The physiological response of animals and plants to variations of light and darkness.

photophosphorylation (foh"toh-fos-for-ih-lay'shun) The production of ATP in photosynthesis.

photopic vision (foh-top'ik) Vision in daylight.

photoreceptor (foh'toh-ree-sep"tor) A sense organ specialized to detect light; a pigment that absorbs light before triggering a physiological response.

photorespiration (foh"toh-res-pur-ay'shun) The production of carbon dioxide and consumption of oxygen during photosynthesis at high light intensities by C-3 plants.

photosynthesis (foh"toh-sin'thuh-sis) The production of organic materials, especially glucose, from carbon dioxide and water by the energy of light. Photosynthesis is practiced by plants, some protists, and several kinds of bacteria.

photosystem A group of chlorophyll and other molecules located in the thylakoid membrane (in photoautotrophic eukaryotes) which emits electrons in response to light.

phototropism (foh"toh-troh'pizm) The growth response of an organism to the direction of light.

phycocyanin (fy"koh-sy'ah-nin) A blue chromoprotein found in cyanobacteria.

phylogeny (fy-loj'en-ee) The complete evolutionary history of a group of organisms.

phylum (fy'lum) A taxonomic grouping of related, similar classes; a category beneath the kingdom and above the class. Phyla are used in classifying animals; divisions are used for plants.

phytochrome (fy'toh-krome) A blue-green, proteinaceous pigment that is the photoreceptor for a wide variety of physiological responses, including initiation of flowering in certain plants.

phytoplankton (fy"toh-plank'tun) Microscopic floating algae and plants which are distributed throughout oceans or lakes; autotrophic plankton.

pia mater (pee'a may'ter) The inner membrane covering the brain and spinal cord; the innermost of the meninges.

pili (pil'ee) sing. **pilus** Hair or hairlike structures; especially, the external hairlike filaments of bacteria.

pinna (pin'ah) The prominent external flap of the mammalian ear.

pinocytosis (pin"oh-sy-toh'sis) Cell drinking; the engulfing and absorption of droplets of liquids by cells.

pith Large, thin-walled parenchyma cells found as the innermost tissue in many plants.

pith ray Parenchyma cells that extend out from the pith to the cortex between the vascular bundles in dicot stems.

pituitary (pit-oo'ih-tehr"ee) Small gland located at the base of the hypothalamus; secretes a variety of hormones influencing a wide range of physiological processes; also called the hypophysis.

placenta (plah-sen'tah) The partly fetal and partly maternal organ whereby materials are exchanged between fetus and mother in the uterus of eutherian mammals.

placoderms (plak'oh-durms) Earliest of the jawed fishes, known only from fossils; believed to be ancestral to both bony and cartilaginous fishes.

plankton Free-floating, mainly microscopic aquatic organisms found in the upper layers of the water; includes phytoplankton, which are photosynthetic organisms, and zooplankton, which are heterotrophic organisms.

planula (plan'yoo-lah) A ciliated cnidarian larval form.

plasma cells Cells that secrete antibodies; differentiated B-lymphocytes.

plasma, of blood The fluid portion of blood consisting of a pale yellowish fluid containing proteins, salts, and other substances, and in which the blood cells and platelets are suspended.

plasma membrane A living, functional part of the cell through which all nutrients entering the cell and all waste products or secretions leaving it must pass; the surface membrane of the cell which acts as a selective barrier to passage of molecules and ions into the cell.

plasmids (plaz'midz) Small circular DNA molecules that carry genes separate from the main bacterial chromosome.

plasmodium (plaz-moh'dee-um) (1) Multinucleate, ameboid mass of living matter that comprises the vegetative phase of the life cycle of some slime molds; (2) a single-celled organism that reproduces by spore formation and causes malaria.

plasmolysis (plaz-mol'ih-sis) The shrinkage of cytoplasm and the pulling away of the plasma membrane from the cell wall when a plant cell (or other walledcell) loses water, usually after being placed in a hypertonic environment.

plastids (plas'tidz) A family of membrane-bounded organelles occurring in photosynthetic eukaryote cells; examples are chloroplasts and amyloplasts.

platelets (playt'lets) Cell fragments in the blood that function in clotting; also called thrombocytes.

pleiotropic gene (ply"oh-troh'pik) A gene that affects a number of different characteristics in a given individual.

pleura (ploor'uh) The membrane that lines the thoracic cavity and envelops each lung.

plexus (plek'sus) A network of intersecting or interconnected nerves, blood vessels, or lymphatics.

ploidy (ploy'dee) Relating to the number of sets of chromosomes in a cell.

plumule (ploom'yool) The embryonic shoot of a seed plant.

poikilothermic (poi"kil-oh-thur'mik) Having a body temperature that fluctuates with that of the environment; "cold-blooded"; also called ectothermic.

point mutation Mutation involving a single nucleotide change in DNA.

polar body Small cell that consists almost entirely of a nucleus that is formed during oogenesis.

polar covalent bond A chemical bond established by electron sharing which produces some difference in the charge of the ends of the molecule.

polar nucleus One of two haploid cells that fuse with a sperm nucleus during double fertilization in angiosperms.

polar transport The unidirectional movement of the plant hormone, auxin, from the stem tip to the roots.

pollen The mass of microspores of seed plants which produce haploid nuclei capable of fertilization.

pollination In seed plants, the transfer of pollen from the male to the female part of the plant.

polyadenylation (pol″ee-ah-den-uh-lay′shun) Part of eukaryotic mRNA processing in which multiple adenine bases (a poly A tail) are added to the 3′ end of the molecule.

polygenes (pol′ee-jeens″) Two or more pairs of genes that affect the same trait in an additive fashion.

polygenic inheritance (pol″ee-jen′ik) Inheritance in which several independently assorted or loosely linked non-allelic genes modify the intensity of a trait, or contribute to the phenotype in additive fashion.

polymer (pol′ih-mer) A molecule built up from repeating units of the same general type, such as a protein, nucleic acid, or polysaccharide.

polymorphism (pol″ee-mor′fizm) (1) The existence of two or more phenotypically different individuals within the same species; (2) the presence of more than one allele for a given locus.

polypeptide A chain of many amino acids linked by peptide bonds.

polyploidy (pol′ee-ploy″dee) Possession of more than two sets of chromosomes per nucleus.

polyps (pol′ips) (1) Hydra-like animals; the sessile stage of the life cycle of certain cnidarians.

polyribosomes A complex consisting of a number of ribosomes attached to an mRNA molecule during translation; also known as polysomes.

polysaccharide (pol-ee-sak′ah-ride) A carbohydrate consisting of many monosaccharide units; examples are starch, glycogen, and cellulose.

polyunsaturated fat (pol″ee-un-sat′yur-ay-ted) A fat containing fatty acids that have double bonds and are not fully saturated with hydrogen.

pons (ponz) The convex white bulge that is the part of the brainstem between the medulla and midbrain; connects various parts of the brain.

portal system A circulatory pathway in which blood flows from a vein draining one region to a second capillary bed in another organ, rather than directly to the heart; examples are the renal and hepatic portal systems.

postganglionic neuron (post″gang-glee-on′ik) A neuron located distal to a ganglion.

postzygotic isolating mechanism (post″zy-got′ik) A mechanism that restricts gene flow between species

and ensures reproductive failure even though fertilization took place.

potassium (K⁺) ion mechanism Mechanism by which plants open and close their stomates. The influx of potassium ions into the guard cells causes water to move in by osmosis, changing the shape of the guard cells and opening the pore.

preadaptation An evolutionary change in an existing biological structure that enables it to have a different function.

predation Relationship in which a species kills and devours other animals.

prehensile (pree″hen′sil) Adapted for grasping by wrapping around an object, as in a prehensile tail.

prezygotic isolating mechanism (pree″zy-got′ik) A mechanism that restricts gene flow between species by preventing mating from taking place.

primary growth An increase in the length of a plant. This growth occurs at the tips of the stems and roots due to the activity of apical meristems.

primary succession Ecological succession taking place in an environment which has not previously supported a community.

primitive streak A longitudinal thickened, cellular region that develops on the embryonic disc of the eggs of reptiles, birds, and mammals; it is comparable to the lips of the blastopore.

producers Organisms, such as plants, that produce food materials from simple inorganic substances.

progesterone (pro-jes′ter-ohn) The hormone produced by the corpus luteum of the ovary and by the placenta; acts with estradiol to regulate estrous and menstrual cycles, and to maintain pregnancy.

proglottid (pro-glah′tid) One of the body segments of a tapeworm.

prokaryote (pro-kar′ee-ote) Organisms that lack membrane-bounded nuclei, and other membrane-bounded organelles; the bacteria and blue-green algae.

promoter A recognition signal encoded in DNA that functions to initiate transcription.

promoter site Sites along DNA to which RNA polymerase attaches to begin transcription.

prophage (pro′faj) A latent stage of a bacteriophage in which the viral genome is inserted into the host chromosome.

prophase The first stage in mitosis, during which the chromatin threads condense, distinct chromosomes become evident, and a spindle forms.

prop root An adventitious root that arises from the stem and provides additional support for plants.

prosimian (pro″sim′ee-un) A suborder of primates that includes the lemurs, lorises, and tarsiers.

prostaglandins (pros″tah-glan′dinz) Hormone-like unsaturated fatty acids that produce a wide variety of local effects; synthesized by most cells of the body.

prostate (pros′tate) The largest accessory sex gland of male mammals; it surrounds the urethra at the point where the vasa deferentia join it, and it secretes a large portion of the seminal fluid.

protective coloration The coloring of an organism so

that it blends into its surroundings in such a way that it is difficult to see.

protein A large, complex organic compound composed of chemically linked amino acid subunits; contains carbon, hydrogen, oxygen, nitrogen, and sulfur; proteins are the principal structural constituents of cells.

proteinoid Polypeptides obtained by heating dry amino acids; proteinoids may have formed during chemical evolution.

prothallus (pro-thal'us) The heart-shaped, haploid gametophyte plant found in ferns, whisk ferns, club mosses, and horsetails.

protist (proh'tist) One of a vast assemblage of eukaryotic organisms, primarily single-celled or simple multicellular, mostly aquatic.

protobiont (proh"toh-by'ont) Assemblages of organic polymers that spontaneously form during certain conditions. Protobionts may have been involved in chemical evolution.

protoctists (proh-tok'tists) Members of a proposed taxonomic category that includes the protists and the plantlike algae.

protocooperation (proh"toh-co-op'ur-ay-shun) Relationship in which each of two populations benefits by the presence of the other, but can survive in its absence.

proton A basic physical particle present in the nuclei of all atoms that has a positive electric charge and a mass of 1 (similar to that of a neutron); a hydrogen ion consists of a single proton.

protonema (proh"toh-nee'mah) A filament of haploid cells that grows from a moss spore; each protonema develops buds that grow into haploid moss plants.

protonephridium (proh"toh-nef-rid'ee-um) The flame-cell excretory organs of lower invertebrates and of some larval higher animals.

protoplasm Obsolete term for cellular contents.

protoplast Plant, fungal, or bacterial cell without its cell wall. Protoplasts are produced by enzymatically digesting the cell wall.

protostome (proh'toh-stome) Major division of the animal kingdom in which the blastopore develops into the mouth, and the anus forms secondarily; includes the annelids, arthropods, and mollusks.

protozoa (proh"toh-zoh'ə) Single-celled, animal-like protists, including amoebas, ciliates, flagellates, and sporozoans.

proximal Relatively near the body center.

proximate factors The immediate conditions or mechanisms that cause particular behaviors.

provirus (pro"vy'rus) A part of a virus, consisting of nucleic acids only, which has been inserted into a host genome.

pseudocoelom (soo"doh-see'lom) A body cavity between the mesoderm and endoderm; derived from the blastocoele.

pseudoplasmodium The aggregation of cells for reproduction in cellular slime molds.

pseudopod A temporary extension of an ameboid cell, which the cell uses for feeding and locomotion.

pseudostratified columnar epithelium An epithelial tissue that appears layered, but all of whose cells are attached to the same basement membrane.

pulse The rhythmic expansion of an artery that may be felt with the finger. It is due to blood ejected with each cardiac contraction, and thus is felt in time with the heartbeat.

pulvinus (pul-vy'nus) A special structure at the base of the petiole that functions in leaf movement by changes in turgor.

punctuated equilibrium The concept that evolution proceeds with periods of inactivity (i.e., periods of little or no change within a species) followed by very active phases, so that major adaptations or clusters of adaptations appear suddenly in the fossil record.

pupa (pew'pah) A stage in the development of an insect, between the larva and the imago (adult); a form that neither moves nor feeds, and may be in a cocoon.

purines (pure'eenz) Nitrogenous bases with carbon and nitrogen atoms in two interlocking rings; components of nucleic acids ATP, NAD, and other biologically active substances. Examples are adenine and guanine.

pyrenoid (py'reh-noid) Starch-containing granular bodies seen in the chromatophores of certain protozoa.

pyrimidines (pyr-im'ih-deenz) Nitrogenous bases composed of a single ring of carbon and nitrogen atoms; components of nucleic acids. Examples are thymine, cytosine, uracil.

quadrupedal Walking on all fours.

quantum A unit of radiant energy; the amount of energy emitted or absorbed by atoms or molecules.

radicle (rad'ih-kl) The embryonic root of a seed plant.

radula (rad'yoo-lah) A rasplike structure in the alimentary tract of chitons, snails, squids, and certain other mollusks.

range The portion of the earth in which a particular species occurs.

ray A chain of parenchyma cells that functions for lateral transport of food, water, and minerals in woody plants.

reabsorption The selective removal of certain substances from the glomerular filtrate by the cells of the tubules of the kidney and their return into the blood.

recapitulation Any tendency for embryos in the course of development to repeat the sequence of stages in the embryonic development of their evolutionary ancestors.

receptacle In botany, the end of a flower stalk where the floral parts are attached.

receptor (1) A specialized neural structure that is excited by a specific type of stimulus; (2) a site on the cell surface specialized to combine with a specific substance such as a hormone or transmitter substance.

recessive genes Genes not expressed in the heterozygous state.

recombinant DNA Any DNA molecule made by combining genes from different organisms.

redirected behaviors Innate, stereotyped behaviors that are directed toward a substitute object.

redox reactions (ree'dox) Chemical reactions in which

one substance is oxidized and another reduced; involves the transfer of one or more electrons from one reactant to another.

red tide A population explosion, or bloom, of dinoflagellates.

reduction In chemistry, the gain of electrons by a substance, or the chemical addition of hydrogen; the opposite of oxidation.

reflex An inborn, automatic, involuntary response to a given stimulus which is determined by the anatomic relations of the involved neurons; generally functions to restore homeostasis.

refractory period The brief period of time that must elapse after the response of a neuron or muscle fiber, during which it cannot respond to another stimulus.

regulator genes Special genes that provide codes for the synthesis of repressor or activator proteins.

releaser A stimulus that triggers an unlearned behavior; a communication signal between members of a species.

renal (ree'nl) Pertaining to the kidney.

renal corpuscle The complex formed by a glomerulus and the surrounding Bowman's capsule; filtration, the first step in urine formation, occurs here.

renin (reh'nin) A proteolytic enzyme released by the kidney in response to ischemia or lowered pulse pressure, which changes angiotensinogen into angiotensin, leading to an increase in blood pressure.

repressor The protein substance produced by a regulator gene that represses protein synthesis in a specific gene.

reproductive isolation The reproductive barriers that prevent a species from interbreeding with another species. As a result, each species' gene pool is isolated from other species.

respiration 1. Cellular respiration is the process by which cells utilize oxygen, produce carbon dioxide, and conserve the energy of food molecules in biologically useful forms, such as ATP. 2. Organismic respiration is the act or function of gas exchange.

resting potential The membrane potential (difference in electric charge) of an inactive neuron (about 70 millivolts).

reticular activating system (reh-tik'yoo-lur) **(RAS)** A diffuse network of neurons in the brainstem responsible for maintaining consciousness.

reticulum (reh-tik'you-lum) A network of fibrils or filaments, either within a cell or in the intercellular matrix.

retina The innermost of the three layers of the eyeball, which is continuous with the optic nerve and contains the light-sensitive rod and cone cells.

retrovirus (ret'roh-vy"rus) An RNA virus that produces a DNA intermediate in its host cell.

reverse transcriptase Enzyme produced by retroviruses to enable the transcription of DNA from the viral RNA in the host cell.

rhizoids (ry'zoids) Colorless, hairlike absorptive filaments analogous to roots that extend from the base of the stem of mosses, liverworts, and fern prothallia.

rhizome (ry'zome) A horizontal underground stem that gives rise to above-ground leaves.

rhodophytes (roh'doh-fites) The division of red algae, found almost entirely in the oceans.

rhodopsin (roh-dop'sin) Visual purple; a light-sensitive pigment found in the rod cells of the vertebrate eye and also employed for photosynthesis by certain bacteria.

ribonucleic acid (RNA) A family of single-stranded nucleic acids that function mainly in protein synthesis.

ribosomes (ry'boh-sohms) Organelles that are part of the protein synthesis machinery; consist of a larger and a smaller subunit each composed of ribosomal RNA (rRNA) and ribosomal proteins.

RNA polymerase Family of enzymes that catalyze the synthesis of RNA molecules from DNA templates.

ribozyme (ry'boh-zime) A molecule of RNA that has catalytic ability.

rickettsia (rih-ket'see-uh) A type of disease organism intermediate in size and complexity between a virus and a bacterium; parasitic within cells of insects and ticks; transmitted to humans by the bite of an infected insect or tick.

ritualization The modification of a behavior pattern, through evolution, to serve a communicative function.

rod The rod-shaped light-sensitive cells of the retina, which are particularly sensitive to dim light and mediate black and white vision.

root cap A covering of cells over the root tip that protects the delicate meristematic tissue directly behind it.

root hair An extension of an epidermal cell in roots. Root hairs increase the absorptive capacity of roots.

root pressure The positive pressure of the sap in the roots of plants, generated by the hypertonicity of the sap with respect to the water in the surrounding wall.

rough ER Major division of the endoplasmic reticulum that contains ribosomes and functions in protein synthesis.

rugae (roo'jee) Folds, such as those in the lining of the stomach.

SA node The sinoatrial node of the heart in which the impulse triggering the heartbeat originates; the pacemaker of the heart.

sand Inorganic soil particles that are larger than clay or silt.

saprobic nutrition (sap-roh'bik) A type of heterotrophic nutrition in which organisms absorb their required nutrients from nonliving organic material.

sapwood The outermost, youngest wood in woody plants. Sapwood conducts water and dissolved minerals; compare with heartwood.

sarcolemma (sar"koh-lem'mah) The muscle cell plasma membrane.

sarcomere (sar'koh-meer) A segment of a striated muscle cell located between adjacent Z-lines that serves as a unit of contraction.

sarcoplasm (sar'koh-plazm) The cytoplasm of a muscle cell.

sarcoplasmic reticulum System of vesicles in a skeletal or cardiac muscle cell which surrounds the myofibrils and releases calcium in muscle contraction; a modified endoplasmic reticulum.

saturated fatty acid A fatty acid with no double bonds between adjacent carbon atoms. It is completely saturated with hydrogen.

savannah A tropical or subtropical grassland containing scattered trees.

schizocoel (skiz'oh-seel) A body cavity formed by the splitting of embryonic mesoderm into two layers.

sclera, of the eye (skler'ah) The firm, fibrous outer coat of the eyeball.

sclerenchyma (skler-en'kim-uh) Cells that provide strength and support in the plant body. Sclerenchyma cells are dead at maturity and have extremely thick walls.

scrotum (skroh'tum) The external sac of skin found in most male mammals that contains the testes and their accessory organs.

sebaceous glands (seb-ay'shus) Glands in the skin that secrete sebum, an oily material that lubricates the skin surface.

secondary growth An increase in the width of a plant due to the activity of lateral meristems, vascular cambium, and cork cambium.

secondary response A rapid production of antibodies induced by a second injection of antigen several days, weeks, or even months after the primary injection.

seed A plant reproductive body that is composed of a young, multicellular plant and nutritive tissue (food).

selection coefficient A measure (s) on a scale of 0 to 1 of the elimination of one phenotype relative to the more successful one. (0 = minimal elimination, 1 = maximal elimination)

selectively permeable membrane A membrane that allows some substances to cross it more easily than others. Biological membranes are generally permeable to water but restrict the passage of many solutes.

semen Fluid composed of sperm suspended in various glandular secretions that is ejaculated from the penis during orgasm.

semicircular canals The passages in the vertebrate inner ear which contain structures that control the sense of dynamic equilibrium.

semilunar valve (sem-eye-loo'nar) A valve with flaps shaped like half-moons such as the aortic valve and the pulmonary artery valve.

seminal vesicles 1. In mammals, glandular sacs which secrete a component of the seminal fluid. 2. In some invertebrates, structures that store sperm.

seminiferous tubules (sem"in-if'ur-us) Tiny coiled ducts within the testis in which spermatogenesis occurs.

senescence The aging process.

sensitization An increased response by an animal to a stimulus that has been presented before.

sepals (see'puls) The outermost parts of a flower, usually leaflike in appearance, that protect the flower as a bud.

sere (seer) A sequence of communities that replace one another in succession in a given area; the transitory communities are called seral stages. The series ends with a climax community typical of the climate in that part of the world.

serotonin (seer"oh-tone'in) An amino acid-like neurotransmitter employed by certain neurons.

Sertoli cells (sur-tole'ee) Supporting cells of the tubules of the vertebrate testis.

serum Light yellow liquid left after clotting of blood has occurred.

sessile (ses'sile) Permanently attached to one location. Coral animals, for example, are sessile.

seta (seet'ah) A thin stalk; for example, the stalk of the moss sporophyte.

sex-linked characteristics Genetic traits carried on a sex chromosome, such as hemophilia in humans.

sex-linked genes Genes borne on a sex chromosome. In mammals almost all sex-linked genes are borne on the X-chromosome.

sexual dimorphism (dy-mor'fizm) Difference in body proportions, coloring, or other characteristics in the two sexes of a species.

short-day plant A plant that flowers in response to short days and long nights; compare with long-day plants.

sieve tube member The cell that conducts food in the phloem of plants.

signal recognition particle (SRP) A complex macromolecule in the cytoplasm of eukaryotic cells that binds to ribosomes engaged in translating a protein with a signal sequence, thus allowing the ribosome to bind to the ER membrane.

signal sequence A sequence of amino acids in a protein that indicates that it is to be translocated across the endoplasmic reticulum.

sign stimulus The critical stimulus properties that stimulate an innate response in an animal.

silt Medium-sized inorganic soil particles; compare with clay and sand.

simple fruit A fruit that develops from a single ovary of a single flower.

sinoatrial node See SA node.

sinus A cavity or channel such as the air cavities in the cranial bones or the dilated channels for venous blood in the cranium.

siphonous (sy'fun-us) A type of body form that is tubular and coenocytic; found in certain algae.

skeletal muscle Voluntary or striated muscle of vertebrates, so-called because it usually is directly or indirectly attached to some part of the skeleton.

smooth ER Portion of the endoplasmic reticulum that has no ribosomes and thus appears smooth; produces steroids and other lipids.

sodium-potassium pump Cellular active transport mechanism that transports sodium out of, and potassium into, cells.

solute (sol'yoot) The dissolved substance in a solution.

solvent A liquid substance, such as water, in which other materials may be dissolved.

somatic cell A cell of the body not involved in sexual reproduction.

somatic nervous system That part of the nervous system that keeps the body in adjustment with the external environment; includes the sensory receptors on the body surface and within the muscles, and the nerves that link them with the central nervous system.

somites (soh'mites) Paired, blocklike masses of mesoderm, arranged in a longitudinal series alongside the neural tube of the embryo, forming the vertebral column and dorsal muscles.

sorus (soh'rus) A cluster of sporangia (in the ferns).

speciation Evolution of a new species.

species A group of organisms with similar structural and functional characteristics that in nature breed only with each other and have a close common ancestry; a group of organisms with a common gene pool.

specific heat The amount of heat required to raise 1 gram of a substance 1°C.

sperm The motile, haploid male reproductive cell of animals and some plants and protists; spermatozoon.

spermatid (spur'mah-tid) An immature sperm cell.

spermatocyte (spur-mat'oh-site) A cell formed during spermatogenesis which gives rise to spermatids, and ultimately to mature sperm cells.

spermatogenesis (spur"mah-toh-jen'eh-sis) The production of sperm by meiosis.

spermatozoa (spur-mah-toh-zoh'uh) Mature sperm cells.

sperm nucleus A nonflagellated male reproductive cell. Sperm nuclei are produced by angiosperms and gymnosperms.

S-phase Phase in interphase of cell cycle during which DNA and other chromosomal components are synthesized.

sphincter (sfink'tur) A group of circularly arranged muscle fibers, the contractions of which close an opening, such as the pyloric sphincter at the end of the stomach.

spindle The intracellular apparatus, composed of microtubules, that separates chromosomes in cell division of eukaryotes.

spine A leaf that is modified for protection, such as a cactus spine; compare with thorn.

spiracle (speer'ih-kl) An opening for gas exchange, such as the opening on the body surface of a trachea in insects.

spiral cleavage Distinctive spiral pattern of blastomere production in an early protostome embryo.

spleen Abdominal organ located just below the diaphragm that removes worn-out blood cells and bacteria from the blood and plays a role in immunity.

spongy mesophyll (mes'oh-phil) The irregularly arranged, photosynthetic tissue closest to the lower epidermis in leaves.

sporangium (spor-ran'jee-um) A spore case, found in plants and certain protists and fungi.

spore A reproductive cell that gives rise to individual offspring in plants, algae, fungi, and certain protozoa.

spore mother cell A diploid cell that undergoes meiosis to form haploid spores in plants.

sporophyll (spor'oh-fil) A leaf-like structure that bears spores.

sporophyte (spor'oh-fite) The diploid portion of a plant life cycle which produces spores by meiosis.

springwood The large, thin-walled conducting cells produced in the xylem of woody plants in the spring when water is plentiful; compare with summerwood.

squamous (skway'mus) Flat and scalelike, as in squamous epithelium.

stabilizing selection Natural selection that acts against extreme phenotypes and favors intermediate variants; associated with a population well-adapted to its environment.

stamen (stay'men) The male part of flowers that produces pollen.

statocyst (stat'oh-sist) A cellular cyst containing one or more granules; used in a variety of animals to sense gravity and motion.

stele (steel) Vascular cylinder of stem and root; term for the pericycle and the tissues within it: xylem, phloem, and parenchyma.

steroids (steer'oids) Complex molecules containing carbon atoms arranged in four interlocking rings, three of which contain six carbon atoms each and the fourth of which contains five; the male and female sex hormones and the adrenal cortical hormones of vertebrates are examples, as is the insect hormone ecdysone. Steroids are chemical derivatives of cholesterol.

stigma (1) That portion of the carpel where the pollen lands prior to fertilization; (2) sometimes used to mean an eyespot.

stimulus A physical or chemical change in the internal or external environment of an organism potentially capable of provoking a response.

stipe A short stalk or stemlike structure that is a part of the body of certain brown algae.

stolon An above-ground, horizontal stem with long internodes. Stolons often form buds that develop into separate plants.

stomatal apparatus A collective term for the stomate (pore) and the cells associated with it.

stomate Small pore flanked by specialized cells (i.e., guard cells) which are located in the epidermis of land plants; stomata allow for gas exchange necessary for photosynthesis.

stoneworts Multicellular green algae found in freshwater ponds; resemble miniature trees, with structures that superficially look like, and serve the functions of, roots, stems, leaves, and seeds, though they are not anatomically like their counterparts in higher plants.

stop codon See termination codon.

stratum basale (strat'um bah-say'lee) The deepest layer of the epidermis, consisting of cells that continuously divide.

striated muscle See skeletal muscle.

strobilus (stroh'bil-us) A conelike structure that bears sporangia.

stroke volume The volume of blood pumped by one ventricle during one contraction.

stroma The fluid region of the chloroplast which surrounds the grana and thylakoids.

stromatolite (stroh-mat'oh-lites) A column-like rock that is composed of many minute layers of prokaryotic cells, usually cyanobacteria. Some stromatolites are over 3 billion years old.

style The neck connecting the stigma to the ovary of a carpel.

suberin (soo'ber-in) A waterproof material found in plants which occurs in the covering of leaf scars, in cork cells, and in the Casparian strip of endodermal cells.

submucosa (sub-myoo-koh'suh) The layer of connective tissue that attaches the mucous membrane to the tissue below.

subsidiary cell A cell associated with the guard cell in monocot leaves.

substrate A substance on which an enzyme acts; a reactant in an enzymatically catalyzed reaction.

sucker A shoot formed from an adventitious bud that develops on certain roots.

sulcus (sul'kus) pl. **sulci** A groove, trench, or depression, especially one occurring on the surface of the brain separating the gyri.

summerwood Smaller, thick-walled conducting cells formed in the secondary xylem of woody plants during the summer when water is not as plentiful as it was in spring; compare with springwood.

suppressor T cells T lymphocytes that suppress the immune reaction.

supraorbital ridges (soop"rah-or'bit-ul) Prominent bony ridges above the eye sockets. Ape skulls have supraorbital ridges; human skulls do not.

suspensor In plant embryo development, a multicellular structure that anchors the embryo and aids in nutrient absorption from the endosperm.

symbiosis (sim-bee-oh'sis) An intimate relationship between two or more organisms of different species (see parasitism, commensal, endosymbiont hypothesis, mutualism).

sympathetic nervous system A subdivision of the autonomic nervous system; its general effect is to mobilize energy, especially during stress situations; prepares the body for fight-or-flight response.

sympatric speciation (sim-pat'rik) The evolution of a new species within the same geographical region as the parent species.

symplast (sim'plast) In plants, the pathway traversed by water through the cells of the plant root rather than between them.

synapse (sin'aps) The junction between two neurons or between a neuron and an effector.

synapsis (sin-ap'sis) The pairing and union side-by-side of homologous chromosomes during Prophase I of meiosis.

syncytium (sin-sit'ee-um) A multinucleate mass of cytoplasm produced by the merging of cells.

syngamy (sin'gah-mee) Sexual reproduction; the union of the gametes in fertilization.

systole (sis'tuh-lee) The contraction phase of the cardiac cycle.

taiga (tie'gah) Northern coniferous forest biome found primarily in Canada, northern Europe, and Siberia.

tap root A root system in plants that has one main root with smaller roots branching off it.

taxis (tak'sis) An orientation movement of a motile organism response to a stimulus.

taxonomy (tax-on'ah-mee) The science of naming, describing, and classifying organisms.

T cell Lymphocyte that is processed in the thymus. T cells have a wide variety of immune functions but are primarily responsible for cell-mediated immunity.

tectorial membrane (tek-tor'ee-ul) The roof membrane of the organ of Corti in the cochlea of the ear.

telophase (teel'oh-faze or tel'oh-faze) The last stage of mitosis and meiosis when, having reached the poles, the chromosomes become decondensed and a nuclear envelope forms around each group.

temperate forest Distinctive forest community of the temperate zone, usually dominated by deciduous angiosperm trees.

temporal isolation A prezygotic isolating mechanism in which genetic exchange is prevented between two groups because they reproduce at different times of the day, season, or year.

tendon A connective tissue structure that joins a muscle to another muscle, or a muscle to a bone. Tendons transmit the force generated by a muscle.

tendril A leaf or stem that is modified for holding or attaching onto objects.

termination codon Any codon in mRNA that does not code for an amino acid (UAA, UAG and UGA). This stops the translation of a peptide at that point. Also called stop codon.

territoriality Behavior pattern in which one organism (usually a male) delineates a territory of his own and defends it against intrusion by other members of the same species and sex.

tertiary structure (tur'she-air"ee) The three-dimensional shape of a protein that forms spontaneously as a result of interactions of side chains.

testis (tes'tis) The male gonad that produces spermatozoa; in humans and certain other mammals the testes are situated in the scrotal sac.

testosterone (tes-tos'ter-ohn) The steroid male sex hormone of vertebrates which is secreted by the testes.

tetanus (tet'an-us) Sustained, steady maximal contraction of a muscle, without distinct twitching, resulting from a rapid succession of nerve impulses.

tetrad A group of four homologous chromatids formed during the first meiotic prophase.

tetraploid (tet'rah-ploid") An individual or cell having twice the normal number of chromosomes.

tetrapods (tet'rah-podz) Four-limbed vertebrates: the amphibians, reptiles, birds, and mammals.

thalamus (thal'uh-mus) The part of the brain that serves as a main relay center transmitting information between the spinal cord and the cerebrum.

thallus (thal'us) The body of a fungus or nonvascular plant without roots, stems, or leaves; for example, a liverwort thallus or a lichen thallus.

therapsids (ther-ap'sids) A group of mammal-like reptiles of the Permian period; gave rise to the mammals.

thermodynamics (thurm"oh-dy-nam'iks) Principles governing heat or energy transfer.

thigmotropism (thig"moh-troh'pizm) Plant growth in

response to contact with a solid object, such as plant tendrils.

thoracic (thor-as'ik)   Pertaining to the chest.

thorax   1. The upper body of vertebrates. 2. The second major division of the arthropod body.

thorn   A stem that is modified for protection; compare with spine.

threshold   The value at which a stimulus just produces a sensation, is just appreciable, or comes just within the limits of perception.

thrombin (throm'bin)   The enzyme derived from prothrombin that converts fibrinogen to fibrin; participates in blood clotting.

thrombus (throm'bus)   A blood clot formed within a blood vessel or within the heart.

thylakoids (thy'lah-koid)   Stacks of flat membraneous sacs inside the chloroplast where light energy is converted into ATP and NADPH used in carbohydrate synthesis.

thymine (thy'meen)   See nucleic acid.

thymus gland (thy'mus)   An endocrine gland that functions as part of the lymphatic system; important in the development of the immune response mechanism.

thyroid gland   An endocrine gland that lies anterior to the trachea and releases hormones that regulate the rate of metabolism.

thyroxine (thy-rok'sin)   The principal hormone of the thyroid gland.

tight junctions   Specialized structures that form between some animal cells, producing a tight seal that prevents materials from passing through the spaces between the cells.

tissue   A group of closely associated, similar cells that work together to carry out specific functions.

T lymphocyte   A type of white blood cell responsible for cell-mediated immunity. Also called T cell.

tonsils   Aggregates of lymph nodules in the throat region. The tonsils are located strategically to deal with pathogens that enter through the mouth or nose.

tonus (toh'nus)   The continuous partial contraction of muscle.

totipotency (toh-tip'oh-tunce-ee)   Ability of a cell (or nucleus) to provide information for the development of an entire organism.

trachea (tray'kee-uh)   (1) Principal thoracic air duct of terrestrial vertebrates; (2) one of the microscopic air ducts ramifying throughout the body of most terrestrial arthropods and some terrestrial mollusks.

tracheids (tray'kee-idz)   A type of water-conducting cell in the xylem of plants.

tracheophyte (tray'kee-oh-fite")   A vascular plant, having xylem and phloem.

transcription   The synthesis of RNA from a DNA template.

transducers   Devices receiving energy from one system in one form and emitting energy of a different form; e.g., converting radiant energy to chemical energy. Sense organs may be considered transducers.

transduction   The transfer of a genetic fragment from one cell to another; e.g., from one bacterium to another by a virus.

transfer RNA (tRNA)   A form of RNA composed of about 70 nucleotides which serve as adaptor molecules in the synthesis of proteins. An amino acid is bound to a specific kind of tRNA and then arranged in order by the complementary pairing of the nucleotide triplet (codon) in template or mRNA and the triplet anticodon of tRNA.

transformation   (1) The incorporation of genetic material by a cell that causes a change in its phenotype; (2) the conversion of a normal cell to a malignant cell.

translation   Conversion of information provided by mRNA into a specific sequence of amino acids in the production of a polypeptide chain; the information in the mRNA is translated into a certain kind of protein.

translocation   (1) The movement of materials (water, dissolved minerals, dissolved food) in the vascular tissues of a plant; (2) chromosome abnormality in which part of one chromosome has become attached to another.

transmission, neural   Conduction of a neural impulse along a neuron, or from one neuron to another.

transpiration   Evaporation of water from the leaves of a plant; aids in drawing water up the stem.

transport vesicles   Small cytoplasmic vesicles that move substances from one membrane system to another.

transposon (tranz-poze'on)   A DNA segment that is capable of moving from one chromosome to another, or to different sites within the same chromosome.

triacylglycerol (try-as"-il-glis'-er-ol)   The most common form of fat consisting of three fatty acid chains chemically linked with a glycerol. Also called triglyceride.

trichocyst (trik'oh-sist)   A cellular organelle found in certain ciliated protozoa such as *Paramecium* that can discharge a filament that may aid in trapping and holding prey.

trichome (trik'ome)   A hair or other appendage growing out from the epidermis of plants.

trilobite (try'loh-bite)   Marine arthropods of the Paleozoic era characterized by two dorsal longitudinal furrows that separated the body into three lobes.

triplet   A sequence of three nucleotides that serve as the basic unit of genetic information, usually signifying the identity and position of an amino acid unit in a protein.

triplet code   The sequences of three nucleotides that compose the codons, the units of genetic information in DNA that specify the order of amino acids in a peptide chain.

triploid   An individual or cell having three sets of chromosomes. Many angiosperms have a triploid endosperm in their seeds.

trisomy (try'sohm-ee)   Condition in which a chromosome is present in triplicate instead of the normal pair.

trochophore (troh'koh-for)   A larval form found in larva of mollusks and many polychaetes.

troop   The social unit of many primate species, consisting of several males, three or more females, and their offspring.

trophic level   The distance of an organism in a food chain from the primary producers of a community.

trophoblast (troh'foh-blast)   The outer layer of a late blastocyst which, in placental mammals, gives rise to the chorion and placenta.

tropic hormone A hormone that helps regulate another endocrine gland; e.g., thyroid-stimulating hormone, released by the pituitary gland, regulates the thyroid gland. Sometimes spelled trophic.

tropism (troh'pizm) A growth response in plants that is elicited by an external stimulus.

tropomyosin (troh-poh-my'oh-sin) A muscle protein involved in the control of contraction.

tubers Thickened underground stems that are adapted for food storage; found in plants such as the white potato.

tubulin (toob'yoo-lin) The protein dimers from which microtubules are constructed by cells.

tumor Mass of tissue that is growing in an uncontrolled manner; a neoplasm.

tundra A treeless plain between the taiga in the south and the polar ice cap in the north; characterized by low temperatures, a short growing season, and ground that is frozen most of the year.

turgor pressure (tur'gor) Hydrostatic pressure that develops within a walled cell, such as a plant cell, when the osmotic pressure of the cell's contents is greater than the osmotic pressure of the surrounding fluid.

ubiquinone (yoo-bik'kwin-ohn) Coenzyme Q, a component of the electron transport system which can take up and release electrons.

ultimate factors The long-term evolutionary causes of behavior.

undulipods (und"yoo-lih'podz) Collective term proposed for the cilia and flagella of eukaryotes.

ungulates (ung'yoo-lates) Four-legged mammals in which the digits may be more or less fused, and their ends protected with a horny coating, or hoof.

uniformitarianism The theory that geological changes have been caused by natural processes such as vulcanism, erosion, and glaciation which have occurred in the same way and at the same rate in the geological past as they do today.

upper epidermis The outermost, protective layer covering the top of the leaf blade.

uracil (yur'ah-sil) See nucleic acid.

urea (yur-ee'ah) The principal nitrogenous excretory product of mammals; one of the water-soluble end-products of protein metabolism.

ureter (yoo-ree'tur) One of the paired tubular structures that conducts urine from the kidney to the bladder.

urethra (yoo-ree'thruh) The tube which conducts urine from the bladder to the outside of the body.

uric acid (yoor'ik) The principal nitrogenous excretory product of insects, birds, and reptiles; a relatively insoluble end product of protein metabolism; also occurs in mammals as an end product of purine metabolism.

uterine tubes (yoo'tur-in) Paired tubes attached to each end of the uterus that receive the ovum after it has been ovulated into the peritoneal cavity; also called fallopian tubes or oviducts.

uterus (yoo'tur-us) The womb; the hollow, muscular organ of the female reproductive tract in which the fetus undergoes development.

utricle (you'trih-kl) The larger of the two divisions of the membranous labyrinth of the inner ear; contains the receptors for static equilibrium.

vaccine (vak-seen') The commercially produced antigen of a particular disease, strong enough to stimulate the body to make antibodies, but not sufficiently strong to cause the disease's harmful effects.

vacuole (vak'yoo-ole) A cavity enclosed by a membrane and found within the cytoplasm; may function in storage, digestion, or in water elimination.

vagina The elastic, muscular tube extending from the cervix to its orifice, that receives the penis during sexual intercourse, and serves as the birth canal.

valence The number of electrons that an atom can donate, accept, or share in the formation of chemical bonds.

Van der Waals interactions A weak attractive force between the atoms of nonpolar molecules in which fluctuations in charge interact.

vascular Pertaining to, consisting of, or provided with vessels.

vascular cambium A lateral meristem in plants that produces secondary xylem (wood) and secondary phloem (inner bark).

vas deferens (vas def'er-enz) Paired duct of the male reproductive system which conveys sperm from the testis to the ejaculatory duct.

vasoconstriction (vas-oh-kon-strik'shun) The narrowing of blood vessels; refers especially to the narrowing of the arteriolar lumen.

vasodilation (vas-oh-dy-lay'shun) The widening of blood vessels, especially the functional increase in the lumen of arterioles.

vector (1) Nucleic acid molecule such as a plasmid that transfers genetic information; (2) agent that transfers a parasite or a virus from one organism to another.

vegetal pole The yolky pole of a vertebrate or echinoderm egg.

vein A blood vessel that carries blood from the tissues toward the heart.

ventral Referring to the belly aspect of an animal's body.

ventricle A cavity in an organ, such as one of the several cavities of the brain, or one of the chambers of the heart that receives blood from the atria.

venule (ven'yool) A small vein.

vernalization Promotion of flowering in certain plants by exposing them to a cold period.

vertebrates Chordates that possess a bony vertebral column; fish, amphibians, reptiles, birds, and mammals.

vesicle (ves'ih-kl) Any small sac, especially a small spherical membrane-bounded compartment within the cytoplasm.

vessel element A type of water-conducting cell in the xylem of plants.

vestigial (ves-tij'ee-ul) Rudimentary; an evolutionary remnant of a formerly functional structure.

villus pl. villi A minute elongated projection from the surface of a membrane, e.g., villi of the mucosa of the small intestine.

viroids (vy′roids) Tiny, naked viruses consisting only of nucleic acid.

virus A tiny pathogen composed of a core of nucleic acid usually encased in protein, and capable of infecting living cells. A virus is characterized by total dependence upon a living host.

viscera (vis′ur-uh) Sing. **viscus** The internal body organs, especially those located in the abdominal or thoracic cavities.

visceral muscle (vis′ur-ul) Smooth or involuntary muscle.

vital capacity The total amount of air displaced when one breathes in as deeply as possible and then breathes out as completely as possible.

vitamin An organic substance necessary in small amounts for the normal metabolic functioning of a given organism.

vitreous body Mass of clear, jelly material that occupies the largest part of the eye, the posterior part of the eyeball between the lens and the retina.

viviparous (vih-vip′er-us) Bearing living young that develop within the body of the mother.

volitional behaviors Behaviors that depend on conscious choice or decision.

voltage-sensitive channel A sodium gate that opens in response to changes in voltage during cellular depolarization.

vulva The liplike external genital structures in the female.

water potential A measure of the free energy of water in different situations. The water potential of pure water is zero. The presence of dissolved materials in water lowers its water potential.

water vascular system Unique hydraulic system of echinoderms; functions in locomotion and feeding.

whorl An arrangement of several identical plant parts, such as petals, around a common center.

wild-type The phenotypically normal (naturally occurring) form of a gene or an organism.

woody A plant with secondary tissues, wood and bark.

xanthophyll (zan′thoh-fil) A yellow to brown pigment of plants.

X-linked gene Gene carried on an x-chromosome.

X organ Organ present in crustaceans that produces hormones which regulate molting, metabolism, reproduction, the distribution of pigment in compound eyes, and the control of pigmentation of the body.

xylem (zy′lem) Vascular tissue that conducts water and dissolved minerals in certain plants.

yolk sac One of the extraembryonic membranes; a pouchlike outgrowth of the digestive tract of certain vertebrate embryos that grows around the yolk, digests it, and makes it available to the rest of the organism.

zona pellucida (pel-loo′sih-duh) The thick, transparent membrane surrounding the cell membrane of the ovum.

zooplankton (zoh″oh-plank′tun) The nonphotosynthetic organisms present in plankton.

zoospore (zoh′oh-spore) A flagellated motile spore produced asexually.

zooxanthellae (zoh″oh-zan-thel′ee) An endosymbiotic dinoflagellate found in certain marine invertebrates.

zygote (zy′gote) The diploid (2n) cell that results from the union of two haploid gametes; a fertilized egg.

PHOTO CREDITS

Contents Overview

Page xiv: (top to bottom): Don Fawcett/Photo Researchers, Inc.; Dwight R. Kuhn; David Cavagnaro; M.L. Dembinsky, Jr. Page xv (top to bottom): David Cavagnaro; David Cavagnaro; Chip Clark; David Muench.

Table of Contents

Part I Openers:

Banded garden spider, David Cavagnaro; Yucca plant, Earth Scenes © 1988 Fred Whitehead; Transmission electron micrograph of bat cell, Don Fawcett/Science Source/Photo Researchers, Inc.; Human cheek cells, Eric Gravé/Science Source/Photo Researchers, Inc.; Adult mosquito resting on water surface, Dwight R. Kuhn; tropical fish, Guido Alberto Rossi/The Image Bank.

Chapter 1

Opener: Earth from space, NASA; 1–1a: Computer Graphics Group, Lawrence Livermore Laboratory/BPS; 1–1b: Courtesy of Dr. Daniel Williams and the Lilly Electron Microscope Laboratory; 1–1c, 1–9, 1–18a,d, 1–19: Charles Seaborn; 1–1d: Frans Lanting; 1–2a, 1–7: E.R. Degginger; 1–2b: Jeffrey L. Rotman; 1–3b: Phillip A. Harrington/Peter Arnold, Inc.; 1–4b: BPS (23-815); 1–5: Gary Milburn, Tom Stack & Associates; 1–6a: Biophoto Associates; 1–6b, 1–15a: Manfred Kage/Peter Arnold, Inc.; 1–8b, 1–11: CNRI/Science Photo Library, Photo Researchers, Inc.; 1–8c: Courtesy of Busch Gardens; 1–10: NASA; 1–14a: Carolina Biological Supply Company; 1–14b: Biophoto Associates/Science Source/ Photo Researchers, Inc.; 1–15a: Peter Arnold, Inc.; 1–15b, 1–17a: Carolina Biological Supply Company; 1–16a: Dwight R. Kuhn; 1–16b: Dr. Wilfred Little; 1–17b: Ed Reschke/Peter Arnold, Inc.; 1–17c: BPS (23-559); 1–17d: Werner H. Muller/Peter Arnold. Inc.; 1–18b,c,e: James H. Carmichael; 1–18f: Courtesy of Busch Gardens; Charles Seaborn 1–20: E.R. Degginger.

Chapter 2

Opener: Silicon atoms viewed through scanning tunnelling microscope, IBM Research/Visuals Unlimited; 2–1a: A. Jenik/The Image Bank; 2–1b: James H. Carmichael, Jr./The Image Bank; 2–4: Nikolaas J. van der Merive, *American Scientist* 70:596–606, 1982; 2–9: From Desaki, J., *Biomedical Research Supplement*, 139–143, 1981; 2–14: NASA; 2–15: Animals Animals © 1988, G.J. Bernard; 2–17a,b: Redrawn with permission of Irving Geis; 2–17c: Barbara O'Donnell/BPS [44–020]; 2–19: Townsend P. Dickinson/Photo Researchers, Inc.; 2–20: Charles D. Winters.

Chapter 3

Opener: Computer representation of a protein, Evans and Sutherland; 3–1a: General Electric; 3–1b: E.R. Degginger; 3–4: From Kotz, J., and K. Purcell, Chemistry & Chemical Reactivity; 3–9c: Ed Reschke; 3–10: Omikron/Photo Researchers, Inc.; 3–11b: James L. Castner; 3–12c: Charles D. Winters; 3–14a: After Lehninger; 3–16b: Howard Sochurek/Medichrome, The Stock Shop, Inc.; 3–21: After Bettelheim and March, Introduction to General, Organic, & Biochemistry; 3–23a,b: After Arms and Camp, Biology: A Journey Into Life; 3–23c: R.K. Burnard, Ohio State University/BPS; 3–24b: McKay, D.B., Weber, I.T., and Steitz, T.A.: *J. Biol. Chem.* 257:9518–9524, 1982; 3–27: Courtesy of Computer Graphics Laboratory, University of California, San Francisco; 3–29: From Bettelheim and March.

Chapter 4

Opener: Phil Gates/BPS [6–212]; 4–2a: Carolina Biological Supply Company; 4–2b,c,d,e: BPS [2–191, 2–192, 2–193, 2–194]; 4–4: Courtesy of Dr. F. George Zaki; Squibb Institute for Medical Research; 4–5: David Scharf; 4–10: Dr. A. Ryter; 4–13a: Dr. Susumu Ito, Harvard Medical School; 4–14a: Kenneth Miller, Brown University; 4–14c: Biophoto Associates; 4–16: Dr. Susanne M. Gollin and Wayne Wray, Kleberg Cytogenetics Laboratory, Baylor College of Medicine, and Biology Department, The Johns Hopkins School of Medicine; 4–17a: Dr. E. Anderson; 4–20e, 4–21a: Don Fawcett/Photo Researchers, Inc.; 4–21b: Dr. Paul Gallup; 4–22: Walker England/Photo Researchers, Inc.; 4–23: BPS [507–032#2]; 4–24b: Keith Porter; 4–25a: L.K. Shumway/Photo Researchers, Inc.; 4–26: Dr. Mary Osborn, Max-Planck-Institute for Biophysical Chemistry; 4–27a: B.F. King, School of Medicine, University of California/BPS; 4–28b: Omikron, Photo Researchers, Inc.; 4–28c: Dr. M.C. Holley; 4–29a–c: Courtesy of Dr. Ulf Nordin, *Acta Otolaryngol.* 94, 1982; 4–29d: Courtesy of Drs. James A. Papp and Joseph T. Martin, *Am. J. Anat.* 169, 1984; 4–30: M. Schliwa/Visuals Unlimited; 4–31a: After Keith Porter; 4–31b: M. Schliwa/ Visuals Unlimited; 4–32a: Omikron/Photo Researchers, Inc.; 4–32b: Biophoto Associates; Focus Box: Photo by Charles Seaborn.

Chapter 5

Opener: Desmosomes. R.M. Tektoff, CNRI/Science Photo Library, Photo Researchers, Inc.; 5–1a: Omikron/Photo Researchers, Inc.; 5–7b: Don Fawcett/Photo Researchers, Inc.; 5–13: Dr. R.F. Baker, University of Southern California Medical School; 5–14d: M.E. Bayer, Institute for Cancer Research/BPS; 5–16a and 5–17: After Alberts; 5–18: J.F. Gennaro/Photo Researchers, Inc.; 5–19d: J.G. Hadley, Battelle/BPS; 5–21a: From M.M. Perry and A.B. Gilbert, *J. Cell Sci.* 39:257–272, 1979; 5–21b: After Alberts; 5–22a: Anderson, E., *J. Morphol.* 150:135–166, 1976; 5–23a: G.E. Palade; 5–23b: Friend: *J. Cell Biol.* 53:758, 1972; 5–25a: Alberfini, D.F., and Anderson, E., *J. Cell Biol.* 63:234–250, 1974; 5–25c: Anderson, E., *J. Morphol.* 156:339–366, 1978; 5–26a: E.H. Newcomb, University of Wisconsin, Madison/BPS [507–008].

Part II Openers:

Sun and grass, Dwight R. Kuhn; Brown pelican with fish, Steven C. Kaufman/Peter Arnold Inc.; Snowgeese landing, Frans Lanting; Prayer plant, James L. Castner; Goldenrod crab spider with captured honeybee, Gregory K. Scott; Chloroplasts, energy centers of plant cells, Hugh Spencer/Science Source/Photo Researchers, Inc.

Chapter 6

Opener: Dwight R. Kuhn; 6–1: Animals Animals © 1988 Stouffer Productions; 6–4b: Charles Steele; 6–5: Richard Coomber/Seaphot Ltd.; 6–6: Charles D. Winters; 6–9b: Computer Graphics Laboratory, University of California, San Francisco; 6–11: Peter David/Seaphot, Ltd.; 6–13: Thomas Eisner, Daniel Aneshansley; 6–16: Dr. Arthur Lesk, Laboratory for Molecular Biology/Science Photo Library/Photo Researchers, Inc.; 6–22: Courtesy of Drs. Victor Lorian and Barbara Atkinson, with permission of *The American Journal of Clinical Pathology.*

Chapter 7

Opener: Don Fawcett/Science Source/Photo Researchers, Inc.; 7–1: Fred Bavendam/Peter Arnold, Inc.; 7–9: Dr. Jeremy Burgess/Science Photo Library/Photo Researchers, Inc.; 7–17: H. Fernandez-Moran; 7–20: M.P. Kahl/ VIREO.

Chapter 8

Opener: Elodea photosynthesizing. E.R. Degginger; 8–1a: Dave Lyons/Planet Earth Pictures; 8–1b: Sinclair Stammers/Science Photo Library/Photo Researchers, Inc.; 8–2: Biophoto Associates; 8–3b: L.K. Shumway/Photo Researchers, Inc.; 8–5: E.R. Degginger; 8–11a: Brian Porter/Tom Stack & Associates; 8–12a: E.R. Degginger: 8–12b: Ron Goulet/M.L. Dembinsky, Jr., Photography Associates; 8–19a,b: J. Robert Waaland, University of Washington/BPS.

Part III Openers

Pea seeds with genetic variation, Dwight R. Kuhn; Radial loop model of human chromosome, Joel Katz; Meiosis in human sperm cells, Manfred Kage/Peter Arnold, Inc.; Group courtship in Galapagos flamingos,

David Cavagnaro; Mitosis in whitefish blastula, Robert Knauft/Biology Media/Photo Researchers, Inc.; Genetic variation in Petunias, James L. Castner.

Chapter 9

Opener: Dividing lymphocyte. CNRI/Science Photo Library, Photo Researchers, Inc.; 9–3: Jonathan G. Izant; 9–4: Carolina Biological Supply Company; 9–5: Andrew S. Bajer, University of Oregon; 9–6: Courtesy of E.J. DuPraw; 9–8 and 9–9: Jonathan G. Izant; 9–10a,b: BPS [524–001, 507–011]; 9–11: Phillip A. Harrington/ Peter Arnold, Inc.; 9–12: C. Allan Morgan/Peter Arnold, Inc.; 9–14: A.H. Sparrow and R.F. Smith; 9–15: D. VonWettstein; 9–16: Dennis Gould; 9–17: Courtesy of J. Kezer.

Chapter 10

Opener: Mammal chromosomes. David M. Phillips/ Visuals Unlimited; 10–1: V. Orel, Mendelianum of the Moravian Museum; 10–4: Tom McHugh/Photo Researchers; 10–17: Courtesy of Dr. Ursula Mittwoch, Galton Laboratory, University College, London; 10–18: Browning/Photo Researchers, Inc.; 10–23: Connecticut Agricultural Experiment Station.

Chapter 11

Opener: Richard Hutchings/Photo Researchers, Inc.; 11–3: Montgomery County Assn. for Retarded Citizens, Norristown, Pa., Medslide, Inc.; 11–4: Medslide Incorporated; 11–7: Dr. Norman Gabelman; 11–8: David M. Phillips/Visuals Unlimited; 11–9: Cystic Fibrosis Foundation; 11–11: CNRI/Science Photo Library/ Photo Researchers, Inc.

Chapter 12

Opener: Transmission electron micrograph of *E. coli* bacterium. Dr. Gopal Murti/Science Photo Library/Photo Researchers, Inc.; 12–7: Courtesy of Biophysics Research Unit, Medical Research Council, King's College, London; 12–15: Courtesy of H.J. Kriegstein and D.S. Hogness; 12–16: Courtesy of D.E. Olins and A.L. Olins; 12–18: Courtesy of U.K. Laemmli, from *Cell* 12:817, 1988; 12–19: From Quigley, G.J.: Left-handed right-handed helical DNA. *Science*, January 1981, cover. © 1981 by the American Association for the Advancement of Science; 12–20: Dr. Kenneth Breslauer/Rutgers University

Chapter 13

Opener: K.G. Murti/Visuals Unlimited. O.L. Miller, Jr., et al, from *Science* 169:392, 1970.

Chapter 14

Opener: Polytene chromosomes of *Drosophila melanogaster*. BPS [31–794]; 14–4,6,8: Courtesy of Dr. Richard Ebright, Rutgers University.

Chapter 15

Opener: DNA strands synthesized from animal DNA. Phillip A. Harrington/Peter Arnold, Inc.; 15–1a: Hank Morgan/Photo Researchers, Inc.; 15–1b: Applied Biostem/ Peter Arnold, Inc.; 15–4b: Stanley N. Cohen, Science

Source/Photo Researchers; 15–9b: Michael Gabridge/ Visuals Unlimited; 15–12: Courtesy of Dr. Daniel C. Williams and the Lilly Microscope Laboratory; 15–11e: Dr. Barton Slatko/New England Biolabs; 15–13: R.L. Brinster; 15–14: E.J. Cable/Tom Stack & Associates.

Chapter 16
Opener: Computer-enhanced photo of protein gradient in *Drosophila* egg. W. Driever, from W. Driever and C. Nüsslein-Volhard, *Cell 54*, 1988; 16–2: Patrick O'Farrell, from Darnell, Lodish and Bathmore, *Molecular Cell Biology*, Fig. 12–1, Scientific American Books; 16–9: W.J. Gehring, from A. Fjose, W.J. McGinnis and W.J. Gehring, 1985, *Nature* 313:284; 6–10: W.J. Gehring, from *Science* 236 (1987):1245–1252; 16–11: Dr. Thomas Kauffman; 16–13: Dr. John Sulston, Medical Research Council, England; 16–14: E. Schierenberg, from G. von Ehrenstein and E. Schierenberg, 1980, in Nematodes as Biological Models, vol. 1, B. Zuckerman, ed. (New York: Academic Press); 16–19: R.L. Brinster, University of Pennsylvania School of Veterinary Medicine, from R.L. Brinster and R.D. Palmiter, "Introduction of Genes into the Germ Line of Animals," in *Harvey Lectures, Series 80*, 1986, pp. 1–38 (Alan R. Liss, Inc.); 16–20a: Carolina Biological Supply Company; 16–20b: Carolina Biological Supply Company; 16–21a: Dr. Elliot Meyerowitz, California Institute of Technology; 16–23a: Ed Reschke; page 409: Lennart Nilsson © Boehringer Ingelheim International, GmbH.

Part IV Openers:
Mandrill, Gary Milburn/Tom Stack & Associates; Wildebeest, Frans Lanting; Charles Darwin, John D. Cunningham/Visuals Unlimited; Painted lady butterfly, Skip Moody/Marvin L. Dembinsky, Jr., Photography Associates; Fossilized starfish, from North Museum, Franklin and Marshall College, photo by Grant Heilman; Galapagos tortoise, Photo Researchers, Inc.

Chapter 17
Darwin Lake, Galapagos Islands, C.J. Collins/Photo Researchers, Inc.; 17–1: Christine Case/Visuals Unlimited; 17–2: Frans Lanting; 17–3: © Larry Burrows Collection; 17–4: Christopher Ralling; 17–6a: George H. Harrison from Grant Heilman; 17–6b: Frans Lanting; 17–6c E.R. Degginger; 17–6d and e: Frans Lanting; 17–6f: C.P. Hickman; 17–6g: M. Cavagnaro; 17–6h: William E. Ferguson; 17–7a: Jeanne White/Photo Researchers, Inc.; 17–7b,c: Miguel Castro/Photo Researchers, Inc.; 17–8a: Kate & Bill Deits/Photo Researchers, Inc.; 17–8b: Jerry Cooke/Photo Researchers, Inc.; 17–8c: John D. Cunningham/Visuals Unlimited; 17–8d: Townsend P. Dickinson/Photo Researchers, Inc.; 17–8e: Earl Roberge/ Photo Researchers, Inc.; 17–8f: Mary Bloom/Peter Arnold, Inc; 17–9a: John Colwell, from Grant Heilman; 17–9b: G.R. Roberts; 17–9c: Larry Lefever, from Grant Heilman; 17–9d,e, and f: G.R. Roberts; 17–10a: Arthur M. Siegelman; 17–10b: © David Muench 1988; 17–10c: John Cancalosi/Tom Stack & Associates; 17–12: John D. Cunningham/Visuals Unlimited; 17–13: No. V/C 993 (C.R. Knight painting), Courtesy Department of Library Services, American Museum of Natural History;

17–15a: Frans Lanting; 17–15b: Biophoto Associates; 17–17: Robert Pugh; 17–19a: Steve McCutcheon/Visuals Unlimited; 17–19b: Frans Lanting; 17–20a: Animals Animals © 1988, Michael Dick; 17–20b: Animals Animals © 1988, Fran Allan; 17–20c: Frans Lanting; 17–20d: Animals Animals © 1988, M.J. Coe, Oxford Scientific Films; 17–20e: Frans Lanting.

Chapter 18
Opener: Colony of California Sea Lions, Año Nuevo Island, Frans Lanting; 18–7: Dwight R. Kuhn; 18–9: Connecticut Agricultural Experiment Station; 18–10: NASCO; 18–12a: James L. Castner; 18–13a: E.D. Brodi, Jr., Adelphi University/BPS; 18–13b(left): L.E. Gilbert, University of Texas at Austin/BPS; 18–13b(right): P.J. Bryant, University of California–Irvine/BPS; 18–17a,b: Michael Tweedie/Photo Researchers, Inc.; 18–18: BPS (#9–127); page 447: Courtesy of Tina Waisman.

Chapter 19
Opener: Timber Wolf, Carl R. Sams, II/Marvin L. Dembinsky, Jr., Photography Associates; 19–1a: Animals Animals © 1988, Anthony Bannister; 19–1b: E.R. Degginger; 19–2a: Dwight R. Kuhn; 19–2b: J.R. Woodward/VIREO; 19–2c: Tom J. Ulrich/Visuals Unlimited; 19–3: H.C. Bennett-Clark; 19–4: Grant Heilman, Grant Heilman Photography; 19–6a: M.J. Rauzon/VIREO; 19–6b: R. Villani/VIREO; 19–8: Runk/ Schoenberger, from Grant Heilman; 19–11a: Dwight R. Kuhn; page 466: A.J. Copley/Visuals Unlimited.

Chapter 20
Opener: Primeval scenic, Frans Lanting; 20–2: Dr. Stanley L. Miller, University of California, San Diego; 20–3: Dr. Sidney Fox; 20–5a: BPS (30–028#3); 20–5b: A. Knoll, et al., *Science* 147:563, 1965; 20–6: William E. Ferguson; 20–9: Paul Chesley/Photographers Aspen: 20–13: Jeremy Pickett-Heaps, in Alberts et al., *Molecular Biology of the Cell*, 1983, Garland; 20–14: Dr. Houston Swift, University of Chicago; 20–16a: BPS (30–024); 20–16b: William E. Ferguson; 20–20: No. CK–49T (painting by Charles R. Knight), Field Museum of Natural History, Chicago; 20–29: John D. Cunningham/Visuals Unlimited; 20–30: Lawrence Berkeley Laboratory, University of California; 20–31: No. V/C 1326 (painting by Charles R. Knight), Courtesy Department of Library Services, American Museum of Natural History; 20–32: No. CK–21T (painting by Charles R. Knight), Field Museum of Natural History, Chicago; 20–35: No. 1016 (painting by Charles R. Knight), Courtesy of Department of Library Services, American Museum of Natural History; 20–36: Jane Burton/Bruce Coleman, Inc.

Chapter 21
Opener: Skulls of human evolution, E.R. Degginger; 21–1: Warren Garst/Tom Stack & Associates; 21–3a: Frans Lanting; 21–3b: Gary Milburn/Tom Stack & Associates; 21–4a: Animals Animals © 1988, Ted Levin; 21–4b: Animals Animals © 1988, Zig Leszczynski; 21–5a: E.R. Degginger; 21–5b: Tom McHugh/Photo Researchers, Inc.; 21–5c: John D. Cunningham/Visuals Unlimited; 21–5d: Animals Animals © 1988, Zig Leszczynski; 21–9:

Institute of Human Origins; 21–10: Peter Jones; 21–13a: photograph of Wenner-Ren Foundatikon replica by David G. Gant; 21–15: John D. Cunningham/Visuals Unlimited; 21–16a: Frans Lanting; 21–16b,c, and d: E.R. Degginger.

Part V Openers:

Horsetail strobilus, David Cavagnaro; Nudibranch on Tealia, Richard Herrmann; Lubber grasshopper, James L. Castner; Chip Clark; Chip Clark; Vorticella, T.E. Adams/Visuals Unlimited.

Chapter 22

Opener: Sea anemone, Guam. Animals Animals © 1988 Tim Rock; 22–1: T.E. Adams; 22–8a: E.R. Degginger; 22–8b: G.R. Roberts; 22–9: Dr. Ralph Somes, University of Connecticut.

Chapter 23

Opener: Viral synthesis in a cell. Tektoff-RM, CNRI/Science Photo Library, Photo Researchers, Inc.; 23–1: CNRI/Science Photo Library, Photo Researchers, Inc.; 23–2a,b: R. Williams; 23–2c: E.S. Boatman; 23–2d: Dr. Lyle C. Dearden; 23–3: From Kleinschmidt, A.K., Lang, D., Jacherts, D., and Zahn, R.K., *Biochim. Biophys. Acta.* 61:859, 1962; 23–4b: Courtesy of Dr. Lee D. Simon, Institute for Cancer Research, Philadelphia; 23–5: Photo by Lennart Nilsson, © Boehringer Ingelheim GmbH; 23–9: C.L. Case/Visuals Unlimited; 23–10: Courtesy of T. Koller and J.M. Sogo, Swiss Federal Institute of Technology, Zurich; 23–11: Lowell Georgia/Science Source/Photo Researchers, Inc.

Chapter 24

Opener: False-color transmission electron micrograph of the bacteria *Neisseria meningitis*. Tektoff-Merieux, CNRI/Science Photo Library/Photo Researchers, Inc.; 24–1: Courtesy of Monsanto Company; 24–2: Courtesy of J.W. Schopf; 24–4a: T.E. Adams; 24–4b: Carolina Biological Supply Company; 24–5a: Courtesy of G.B. Chapman; 24–5c: D.L. Balkwill, S.A. Nierzwicki-Bauer, and S.E. Stevens, Jr. Reproduced from *The Journal of Cell Biology,* 1983, vol. 97, pp. 713–722 by copyright permission of the Rockefeller University Press; 24–6: J. Robert Waaland/BPS [#1–467]; 24–7: Biophoto Associates; 24–9: Courtesy of Dr. Victory Lorian, with permission of *The American Journal of Clinical Pathology;* 24–10: Courtesy of V. Chambers; 24–11: CNRI/Science Photo Library/Photo Researchers, Inc.; 24–12: Courtesy of Hoskins, J.D., Henk, W.G., and Abdelbaki, Y.Z., *American Journal of Veterinary Research* 43:1715–1720, 1982; 24–13: Courtesy of C. Brinton, Jr., and J. Carnahan; 24–14: From Santo, L., Hohl, H., and Frank, H., *Journal of Bacteriology* 99:824, 1969; 24–15: David M. Phillips/Visuals Unlimited; 24–16: From Grilione, P.L., and Pangborn, J., *Journal of Bacteriology* 124:1558, 1975; 24–17: BPS [510–085#2]; 24–18: BPS [Gabridge: *M. pneumoniae*]; 24–19: BPS [CDC 94039]; 24–20: Courtesy of Dr. S.W. Watson; UN24–A: BPS [24–827]; UN24–B: BPS [24–828]; UN24–C: BPS [24–829]; page 555: Michael G. Gabridge/Visuals Unlimited.

Chapter 25

Opener: Diatoms. Jan Hinsch/Science Photo Library/Photo Researchers, Inc.; 25–1: J. Robert Waaland, University of Washington/BPS; 25–3: Eric V. Gravé/Photo Researchers, Inc.; 25–4: W.L. Dentler, University of Kansas/BPS; 25–5: From Clarkson, Allen B., Jr., and Frederick H. Brohn, *Science 194:*204, 1976. © 1976 by AAAS; 25–7: Michael Abbey, Photo Researchers, Inc.; 25–9: Eric V. Gravé/Photo Researchers, Inc.; 25–10 and 25–11: Manfred Kage/Peter Arnold, Inc.; 25–12b: Courtesy of Dr. Eugene Small; 25–13: Courtesy of Dr. H. Plattner; 25–14: Biophoto Associates; 25–15: From Luporini, P., and Dallai, R.: *Developmental Biology* 77:167–177, 1980. © 1980 by Academic Press, Inc.; 25–16a: Carolina Biological Supply Company; 25–16b: Eric V. Gravé/Photo Researchers, Inc.; 25–17: Paul W. Johnson, University of Rhode Island BPS; 25–19: John D. Dodge; 25–20: Paul W. Johnson, University of Rhode Island/BPS; 25–21: Kevin Schafer/Tom Stack & Associates; 25–22: Manfred Kage/Peter Arnold, Inc.; 25–23a,b: From Hufford, T.L., and Collins, G.B., *Journal of Psychology* 8:192–195, 1972; 25–23c,d: Jonathan G. Izant; 25–24b: T.E. Adams; 25–25a: J. Robert Waaland, University of Washington/BPS; 25–25b: James Bell/Photo Researchers, Inc.; 25–25c: Doug Wechsler; 25–25d: Ed Reschke; 25–25e: Biophoto Associates; 25–27: Carolina Biological Supply Company; 25–28: From Gantt, E., and Conti, S.F., reproduced from *The Journal of Cell Biology,* by copyright permission of The Rockefeller University; 25–29 and 25–30: Biophoto Associates; 25–31: Peter Scoones/Seaphot Ltd.; 25–32: J. Robert Waaland, University of Washington/BPS; 25–33: Richard Herrmann; 25–34: Patrick W. Grace/Photo Researchers, Inc.; 25–35: E.R. Degginger; 25–36: Courtesy of Dr. J. Gregg. *The Fungi.* New York, Academic Press; 25–37: Carolina Biological Supply Company.

Chapter 26

Opener: Scarlet waxy-cap mushrooms. E.R. Degginger; 26–1: L.E. Gilbert, University of Texas at Austin/BPS; 26–2: James L. Castner; 26–3b: Dr. Jeremy Burgess/Science Photo Library/Photo Researchers, Inc.; 26–5: Dwight R. Kuhn; 26–7: Carolina Biological Supply Company; 26–9a,b: E.R. Degginger; 26–10c,e: Carolina Biological Supply Company; 26–10f: From Martinez, A.T., Calvo, M.A., and Ramirez, C.: Scanning electron microscopy of *Penicillium conidia.* Antonie van Leeuwenhoek 48:245–255, 1982; 26–12a,b: E.R. Degginger; 26–12c: James L. Castner; 26–12d: Frank E. Toman/Taurus Photos, Inc.; 26–13: E.R. Degginger; 26–15a: Eric V. Gravé/Photo Researchers, Inc.; 26–15b: Photo Researchers, Inc.; 26–16a: From Ahmadjian, V., and Jacobs, J.B.: *Nature* 289:169–172, 1981; 26–16b: Carolina Biological Supply Company; 26–16c: Courtesy of Leo Frandzel; 26–16d: E.R. Degginger; 26–17a: John D. Cunningham/Visuals Unlimited; 26–17b: R. Ronacordi/Visuals Unlimited; 26–18: Dr. Jeremy Burgess/Science Photo Library/Photo Researchers, Inc.; 26–19a: Virginia P. Weinland/Photo Researchers, Inc.; 26–19b,c: Leo Frandzel; 26–20: Courtesy of Fermenta Products, Inc.; 26–21: Carolina Biological Supply Company; 26–22c: Dr. Jeremy Burgess/Science Photo Library/Photo Researchers, Inc.; 26–23:

Carolina Biological Supply Company; 26–24b: Courtesy of Dr. Wilfred D. Little.

Chapter 27

Opener: G.R. Roberts; 27–1: Courtesy of Dr. Linda Graham, University of Wisconsin, Madison; 27–4 and 27–5: David Cavagnaro; 27–7: G.R. Roberts; 27–8: BPS [#6–049]; 27–11a: W.H. Hodge/Peter Arnold, Inc.; 27–15a: Carolina Biological Supply Company; 27–15b: J.N.A. Lott, McMaster University/BPS; 27–15c: W.H. Hodge/Peter Arnold, Inc.; 27–15d: B.J. Miller, Fairfax, VA/BPS; 27–17a: Biophoto Associates; 27–17b: David M. Phillips/Visuals Unlimited; 27–18: Biophoto Associates; 27–19a: J. Robert Waaland, University of Washington/BPS; 27–19b: Charles Seaborn; 27–20a: Dwight R. Kuhn; 27–20b: Ed Reschke; 27–22: Carolina Biological Supply Company; page 630: No. Geo. 75400c, Field Museum of Natural History, Chicago.

Chapter 28

Opener: Willow catkins. David Cavagnaro; 28–1: Renee Purse, Photo Researchers, Inc.; 28–2: Gary R. Bonner; 28–3: P. William Davis; 28–4: R. Knauft/Biology Media/Photo Researchers, Inc.; 28–6: G.R. Roberts; 28–7: Manfred Kage/Peter Arnold, Inc.; 28–8: E.R. Degginger; 28–9: P. William Davis; 28–10: W.H. Hodge/Peter Arnold, Inc.; 28–11a: E.R. Degginger; 28–11b and 28–12: W.H. Hodge/Peter Arnold, Inc.; 28–13: John D. Cunningham/Visuals Unlimited; 28–14: No. Bot. 83024c, Courtesy, Field Museum of Natural History; 28–15a: P. William Davis; 28–15b: E.R. Degginger; 28–15c: E.R. Degginger; 28–15d,f: Leo Frandzel; 28–15e: BPS [24–1477]; 28–15g: R. Humbert/BPS; 28–15h: Don and Esther Phillips/Tom Stack & Associates; 28–18: Biophoto Associates, N.H.P.A.; 28–20: Thomas Eisner; 28–21: E.R. Degginger; 28–22: Steve Maslowski/Photo Researchers, Inc.; 28–23: Merlin Tuttle/Photo Researchers, Inc.; 28–24a: David Cavagnaro; 28–24b: Dwight Kuhn; 28–25: Biophoto Associates, N.H.P.A.; page 649 (top): Carolina Biological Supply Company; (bottom): Biophoto Associates.

Chapter 29

Opener: Sea anemone. E.R. Degginger; 29–1a: Robert and Linda Mitchell; 29–1b: Herman Eisenbeiss/Photo Researchers, Inc.; 29–1c: Robert Noonan/Photo Researchers, Inc.; 29–4,5,9,15,16: After Barnes.; 29–8a: Charles Seaborn; 29–8b: H.W. Pratt/BPS [33–151]; 29–10c: James Solliday/BPS [2–007]; 29–11a: Carolina Biological Supply Company; 29–11b,d: Charles Seaborn; 29–11c: Animals Animals © 1988 Oxford Scientific Films, Peter Parks; 29–11e: Doug Wechsler; 29–13a: Tom Branch/Photo Researchers, Inc.; 29–14a: Courtesy of G.B. Chapman, Cornell University Medical College. From Lenhoff, H.M., and Loomis, W.F. (eds.): *The Biology of Hydra.* Coral Gables, University of Miami Press, 1961; 29–17b: Roy Lewis III; 29–18a: Carolina Biological Supply Company; 29–18c: T.E. Adams; 29–19a: Carolina Biological Supply Company; 29–21: Cath Ellis, Dept. of Zoology, University/Science Photo Library/Photo Researchers, Inc.; 29–24: T.E. Adams; 29–26 and 29–27a:

Carolina Biological Supply Company; 29–28b: T.E. Adams; 29–28c: James Bell/Photo Researchers, Inc.

Chapter 30

Opener: Cone-headed grasshopper. Chip Clark; 30–1: Kenneth Lucas, Steinhart Aquarium/BPS; 30–2a,b: Robin Lewis; 30–2c: BPS [47–007]; 30–5: Charles Seaborn; 30–6a: James H. Carmichael, Coastal Creations; 30–6c: E.R. Degginger; 30–9: Jane Burton, Bruce Coleman, Inc.; 30–12a: Charles Seaborn; 30–12b: James H. Carmichael, Coastal Creations; 30–14: R.K. Burnard, Ohio State University/BPS; 30–15a: T.E. Adams; 30–15b: St. Bartholomew's Hospital/Science Photo Library/Photo Researchers, Inc.; 30–16: Animals Animals © 1988, Raymond A. Mendez; 30–17a: Robin Lewis; 30–17b: James H. Carmichael; 30–17c: Peter J. Bryant, University of California, Irvine/BPS; 30–19a: David P. Maitland, Seaphot; 30–19b: Ed Reschke; 30–19c: Dwight R. Kuhn; 30–19d: BPS [2–117]; 30–20: Skip Moody/M.L. Dembinsky, Jr., Photography Associates; 30–21a: Frans Lanting; 30–21b,c,d: Charles Seaborn; 30–23b: Chris Simon; 30–23c: Stephen Dalton/Photo Researchers, Inc.; 30–25c: T.E. Adams; 30–26a: E.R. Degginger; 30–26b–e: Leo Frandzel; 30–27: James L. Castner; 30–28a: Carolina Biological Supply; 30–28b: E.R. Degginger; page 708: (silverfish) Harry Rogers, Photo Researchers, Inc.; page 709: (aphid) Peter J. Bryant, University of California, Irvine/BPS; (Io caterpillar) Luci Giglio.

Chapter 31

Opener: Starfish. Jeffrey L. Rotman; 31–1a: BPS [33–089]; 31–1b: Brian Parker/Tom Stack & Associates; 31–1c: Phil Degginger; 31–2b, 31–6, 31–11d, 31–17, 31–18, 31–20, 31–26a: Charles Seaborn; 31–3: Susan Blanchet/M.L. Dembinsky, Jr., Photography Associates; 31–4: Fred Bavendam/Peter Arnold, Inc.; 31–5b: Richard Chesher/Seaphot, Ltd.; 31–7: Jeff Rotman/Peter Arnold, Inc.; 31–8, 31–20, 31–21: Peter Scoones/Seaphot, Ltd.; 31–10, 31–25a, 31–33a,c: Robin Lewis; 31–11c: Richard A. Cloney, University of Washington; 31–12: Robin Lewis; 31–13: After Romer; 31–15a: Tom Stack/Tom Stack & Associates; 31–15b: Courtesy of Dr. Kiyoko Uehara; 31–16, 31–24, 31–31: After Romer and Parsons; 31–25b: Carolina Biological Supply Company; 31–26b: L. Stone/The Image Bank; 31–27: Ed Reschke; 31–29a: D. Klees/Worldview/Planet Earth Pictures; 31–29b: Courtesy of Busch Gardens; 31–29c: Jonathan Scott/Seaphot, Ltd.; 31–29d: Christopher Crowley/Tom Stack & Associates; 31–30: From a painting by Rudolph Freund, Courtesy of the Carnegie Museum of Natural History; 31–32: Tom McHugh/Photo Researchers, Inc.; 31–33b: Photo by Robert Anderson, reprinted with permission of Hubbard Scientific Company; Focus on Placental Mammals: a: Merlin Tuttle/Photo Researchers, Inc.; b,e,f: E.R. Degginger; c,d: Charles Seaborn; Focus on Hemichordates: C.R. Wyttenbach, University of Kansas/BPS.

Part VI Openers:

Uluhe, David Cavagnaro; Sodom's apple, seeds blown by wind, Hans Christian Heap/Seaphot Ltd.; Desert wildflowers and saguaro cactus, Organ Pipe National

Monument, Marty Cordano; Sunflower seeds, G.R. Roberts; Hummingbird flower, Peru, James L. Castner; Gazania, David Cavagnaro.

Chapter 32

Opener: Scaliform performation plates in red alder, J.D. Litvay/Visuals Unlimited; 32–4, 32–7, 32–16: Courtesy of Triarch; 32–8: From Troughton, J., and Donaldson, L.A., *Probing Plant Structure.* New York, McGraw-Hill Book Co., 1972; 32–10: Phil Gates, University of Durham/BPS; 32–11, 32–12: Ed Reschke; 32–13: Biophoto Associates/ Photo Researchers, Inc.; 32–14: Courtesy of John Dwyer; 32–15: Biophoto Associates; 32–17a: BPS [523–011#2]; 32–17b: J. Robert Waaland, University of Washington/BPS; 32–18: J. Robert Waaland, University of Washington/BPS; 32–19: James Bell/Photo Researchers, Inc.; 32–28: Carolina Biological Supply Company; 32–29: From Constabel, F., et al., *C.R. Acad. Sci.* (Paris) 285:319–322, 1977.

Chapter 33

Opener: Lupine, Sierra Nevada, David Cavagnaro; 33–2: Geoff du Fue/Seaphot/Planet Earth Pictures; 33–3: Biophoto Associates; 33–4: Carolina Biological Supply Company; 33–5: Ed Reschke; 33–6: Robert and Linda Mitchell; 33–7: James L. Castner; 33–8: Dwight R. Kuhn; 33–9: Courtesy Dr. Klaus Raschke, from Humble and Raschke, *Plant Physiol.* 48:447–453, 1971. Michigan State University–Atomic Energy Commission Plant Research Laboratory, Michigan State University; 33–10: P. Dayanandan/Photo Researchers, Inc.; 33–11: Jack M. Bostrack; 33–13: Biophoto Associates; 33–14: John D. Lutz/M.L. Dembinsky, Jr., Photography Associates; 33–15: Dwight R. Kuhn; 33–16a: Skip Moody/M.L. Dembinsky, Jr., Photography Associates; 33–16b: Carolina Biological Supply Company; 33–17a: Howard A. Miller, Sr./Photo Researchers, Inc.; 33–17b: Nuridsany et Perennou/Photo Researchers, Inc.; 33–18: David A. Steingraeber.

Chapter 34

Opener: Polarized light micrograph of a corn stem. J. Bostrack/Visuals Unlimited; 34–1: Frans Lanting; 34–2, 34–3b: Ed Reschke; 34–3a, 34–7: Carolina Biological Supply Company; 34–4: E.R. Degginger; 34–8: Courtesy of Triarch; 34–9: Doug Wechsler; 34–10: Terry Ashley/ Tom Stack & Associates; 34–16a: Dwight R. Kuhn; 34–16b: M.H. Zimmerman, *Science* 133:73–79 (Fig. 4), January 13, 1961, Copyright 1988 by the American Association for the Advancement of Science; Focus Box: E.R. Degginger.

Chapter 35

Opener: Radish seedlings, Doug Wechsler; 35–1: James L. Castner; 35–2: Biophoto Associates; 35–3: W. Harlow/Photo Researchers, Inc.; 35–5, 35–11: Ed Reschke; 35–6: Robert and Linda Mitchell; 35–9a: Carolina Biological Supply Company; 35–16: Dr. J. Menge, University of California, Riverside; 35–17: W.R. Robbins, Rutgers University.

Chapter 36

Opener: Lupine seeds dispensed by exploding pods, David Cavagnaro; 36–1, 36–8, 36–9: E.R. Degginger; 36–3a: M.L. Dembinsky, Jr.; 36–3b: USDA Technical Bulletin 1286; 36–7, 36–12a, 36–19a,b,c,e: James L. Castner; 36–11: Doug Wechsler; 36–12b: David Cavagnaro; 36–12c: G.R. Roberts; 36–14a, 36–16, 36–19d: Dwight R. Kuhn; 36–15: Phil Degginger; 36–18: BPS [MAY, tumbleweeds blown against fence]; 36–21: USDA; 36–23: Borthwick, H.A., et al., *Proceedings of the National Academy of Science* USA 38:662–666, 1952.

Chapter 37

Opener: Test-tube schefflera, William D. Adams; 37–1, 37–2: E.R. Degginger; 37–3: BPS [9–227]; 37–4a: Richard F. Trump/Photo Researchers, Inc.; 37–5: Biophoto Associates; 37–13: Runk/Schoenberger, from Grant Heilman; 37–14: Dr. Clive Spray, UCLA; 37–15: Robert E. Lyons/Visuals Unlimited; 37–16a,b: Abbot Laboratories; 37–20: Dave Millert/Tom Stack & Associates.

Part VII Openers:

Chip Clark; Squat lobster, Seaphot, Ltd.; Giraffes sparring, Frans Lanting; Salamander egg development, Dwight R. Kuhn; Fruit fly laying eggs, Gregory K. Scott; Short-eared owl, Dwight R. Kuhn.

Chapter 38

Opener: Bengal tiger, Animals Animals © 1988 Anup Shah; 38–1a: R.K. Burnard/BPS; 38–1b: B.J. Miller/BPS; 38–3: CNRI/Science Photo Library/Photo Researchers, Inc.; 38–4: From Warwick, R., and P.L Williams, *Gray's Anatomy*, 36th ed. Edinburg, Churchill Livingstone, 1980; 38–8c: Al Lammé/Phototake, New York City; 38–9: Ed Reschke; page 858, 859 (top) and 862 (bottom): Ed Reschke.

Chapter 39

Opener: Hunting cheetah, G. Ziesler/Peter Arnold, Inc.; 39–1: Y. Momatiuk/Photo Researchers, Inc.; 39–3a: Photo Researchers, Inc.; 39–3b: Courtesy of Dr. Karen A. Holbrook; 39–4a: Carl R. Sams, II/M.L. Deminsky, Jr.; 39–4b: E.R. Degginger; 39–5: Erik L. Simm/Image Bank; 39–9: Chris Simon; 39–10a: Brian Parker/Tom Stack & Associates; 39–10b: Robert and Linda Mitchell; 39–11: Carolina Biological Supply Company; 39–14b: Ed Reschke; 39–14*c: Courtesy of Dr. Lyle C. Dearden; 39–15b: From Desaki, J., *Biomedical Research* (Supplement), 139–143, 1981; 39–15*c,d: Taken by Yashuo Uehara, Junzo Desaki, and Takashi Fukiwara of the Ehime University School of Medicine in Japan, from Desaki, J., and Uehara, Y., *Journal of Neurocytology* 10:101–110, 1981; Focus Box: From Heinrich, Bumblebee Economics. Harvard University Press, 1978. Electron micrograph by Mary Ashton. Courtesy of Dr. Heinrich; used by permission.

Chapter 40

Opener: Red slug feeding. Hans Pfletschinger/Peter Arnold, Inc.; 40–1a: Darwin Dale/Photo Researchers,

Inc.; b,c,d: Courtesy of Busch Gardens; 40–3a: Charles Seaborn; 40–3b: Courtesy of Mical Solomon and Trudi Segal; 40–3c: P.J. Bryant, UC-Irvine/BPS; 40–4a: E.R. Degginger; 40–4b: Jane Burton/Bruce Coleman, Inc.; 40–16: From Kessel, R.G., and Kardon, R.H.: *Tissues and Organs: A Text-Atlas of Scanning Electron Microscopy.* San Francisco, W.H. Freeman and Co., 1979; 40–17: Courtesy of J.D. Hoskins, W.G. Henk, and Y.Z. Abdelbaki, from the *American Journal of Veterinary Research*, Vol. 43, No. 10.

Chapter 41
Opener: "89" butterfly taking moisture, Brazil. BPS [3–005]; 41–1: CNRI/Science Photo Library/Photo Researchers, Inc.; 41–4: Food and Agriculture Organization of the United Nations; 41–7: Courtesy of E.R. Squibb and Sons; 41–9: Biophoto Associates

Chapter 42
Opener: False-color scanning electron micrograph of human red blood cells. Dr. Tony Brain/Science Photo Library/Photo Researchers, Inc.; 42–5a: Biophoto Associates; 42–5b, 42–7c: Ed Reschke; 42–7a: Manfred Kage/Peter Arnold, Inc.; 42–7b: Biophoto Associates/Science Source/Photo Researchers, Inc.; 42–9: Courtesy of Johan G. Hanstede and *The Anatomical Record*; 42–17a: Al Lammé/Phototake, New York City; 42–31: From E.K. Markell and M. Voge; page 125(*a*): Markin M. Rotker/Phototake–NYC; (*b*) and (*c*): M. Cubberly/Phototake–NYC.

Chapter 43
Opener: False-color scanning electron micrograph of mosquito stinging skin. Lennart Nilsson from p. 161 of *The Incredible Machine*, National Geographic Society; 43–3a: Lennart Nilsson/© Boehringer Ingelheim International GmbH, from *The Incredible Machine*, p. 170; 43–3b: Lennart Nilsson/©Boehringer Ingelheim International GmbH, from *The Incredible Machine*, p. 171; 43–3c, 43–6, and 43–12a: Nilsson/Boehringer; 43–5: Howard Sochurek/Medichrome, The Stock Shop, Inc.; 43–8: From W. Bloom and D. Fawcett; 43–15: Lennart Nilsson/© Boehringer Ingelheim International GmbH, from *The Incredible Machine*, p. 211; page 144: Lennart Nilsson.

Chapter 44
Opener: Wapiti with condensing breath. Stan Oslinski/M.L. Dembinsky, Jr., Photography Associates; 44–2: Chuck Davis; 44–3: Courtesy of Dr. James L. Nation and *Stain Technology*, vol. 58, 1983; 44–4: Animals Animals © 1988 G.I. Bernard; 44–5a: Dwight R. Kuhn; 44–7b: From Kessel, R.G., and Kardon, R.H.: *Tissues and Organs, A Text-Atlas of Scanning Electron Microscopy.* San Francisco, W.H. Freeman Co., 1979; 44–8a, 44–9a: Ed Reschke; 44–8b: From Port, C., and Corvin, I.: cover photo. *Science*, 77 (No. 4054, September 22, 1972. © 1972 by The American Association for the Advancement of Science; 44–13: Charles Seaborn; 44–14: © Bill Paulson, 1984 Telephoto; 44–15: Alfred Pasieka/Taurus Photos, Inc.; 44–16: Martin M. Rotker/Taurus Photos, Inc.; page 168: Chuck Davis.

Chapter 45
Opener: Glomerulus. Lennart Nilsson © Boehringer Ingelheim International GmbH, from *The Incredible Machine*, p. 139, National Geographic Society; 45–13: CNRI/Science Photo Library/Photo Researchers, Inc.; 45–15: From W. Bloom and D.W. Fawcett; page 196: Werner H. Muller/Peter Arnold, Inc.

Chapter 46
Opener: Neurons. Lennart Nilsson © Boehringer Ingelheim International GmbH, from *The Incredible Machine*, p. 265, National Geographic Society; 46–2b,c, and d: Courtesy of Dr. B.H. Gähwiler, *Experientia*, 40:1984; 46–5: CNRI/Science Photo Library/Photo Researchers, Inc.; 46–6: Photo Researchers, Inc; 46–7b: Courtesy of Dr. Lyle C. Dearden; 46–13a: Courtesy of Dr. John Heuser.

Chapter 47
Opener: Swallow drinking. Animals Animals © 1988 Stephen Dalton; 47–8: Dissection by Dr. M.C.E. Hutchinson, Department of Anatomy, Guy's Hospital Medical School, London, England. From Williams and Warwick (eds.): *Gray's Anatomy.*

Chapter 48
Opener: Bull frog. Stephen G. Maka; 48–1: Dr. Merlin Tuttle/Milwaukee Public Museum/Photo Researchers; 48–5a: Ed Reschke; 48–6: Courtesy of J. Derrenbacher; 48–9: G.R. Roberts; 48–19: Gregory G. Dimijian, M.D./Photo Researchers, Inc.; 48–20c: Terry Ashley/Tom Stack & Associates; 48–21a: Tom Stack/Tom Stack & Associates; 48–21c: Christian Petron/Seaphot, Ltd.; 48–21e: Peri Coelho/Planet Earth Pictures; 48–22b: David Scharf; 48–22c: Seaphot Ltd.; 48–23b: Manfred Kage/Peter Arnold, Inc.; 48–23c: Lennart Nilsson, © Boehringer Ingelheim International GmbH, from *The Incredible Machine*, p. 279, National Geographic Society; 48–23d: Courtesy of Francisco M. De Monasterio, S.J. Schein, and E.P. McCrane; 48–24a,b: T. Eisner 48–26a: Stephen G. Maka; 48–26b: Lennart Nillson; 48–26c: Manfred Kage/Peter Arnold; 48–28b: David Scharf; 48–28c: John Lythgoe/Seaphot, Ltd.; 48–30: Dr. Thomas Eisner.

Chapter 49
Opener: Anemone Shrimp. E.R. Degginger; 49–11b: Connie Toops; 49–18: From C.D. Turner and J.T. Bagnara; 49–19: Bettina Cirone/Photo Researchers, Inc.; 49–20: Courtesy of Dr. Gordon Williams; page 304: John Paul Kay/Peter Arnold, Inc.

Chapter 50
Opener: Human sperm and egg. Lennart Nilsson, from *The Incredible Machine*, p. 22; National Geographic Society. 50–1a: Richard Campbell/BPS [28–014]; 50–2a: Hans Pfletschinger/Peter Arnold, Inc.; 50–2b: Paul Kuhn/Tom Stack & Associates; 50–2c: BPS [9–036]; 50–4a: BPS [7–085]; 50–4b: Geoff du Feu/Seaphot, Ltd.; 50–4d: BPS [17–173]; 50–4c: William H. Amos/Bruce Coleman, Inc.; 50–7b: Kessel, R.G., and R.H. Kardon. *Tissues and Organs: A Text-Atlas of Scanning Electron Microscopy.* San

Index*

---

*Bold face page numbers indicate pages on which the index term is defined; "il" following the page number indicates an illustration, "t" a table, "f" a focus, and "n" a footnote.